Flora Capensis

*Being a Systematic Description
of the Plants of the Cape Colony,
Caffraria & Port Natal,
and Neighbouring Territories*

VOLUME 4: PART 1
VACCINIACEAE TO GENTIANEAE

WILLIAM H. HARVEY *ET AL.*

CAMBRIDGE
UNIVERSITY PRESS

CAMBRIDGE
UNIVERSITY PRESS

University Printing House, Cambridge, CB2 8BS, United Kingdom

Cambridge University Press is part of the University of Cambridge.

It furthers the University's mission by disseminating knowledge in the pursuit of
education, learning and research at the highest international levels of excellence.

www.cambridge.org
Information on this title: www.cambridge.org/9781108068093

© in this compilation Cambridge University Press 2014

This edition first published 1909
This digitally printed version 2014

ISBN 978-1-108-06809-3 Paperback

CAMBRIDGE LIBRARY COLLECTION

Books of enduring scholarly value

Botany and Horticulture

Until the nineteenth century, the investigation of natural phenomena, plants and animals was considered either the preserve of elite scholars or a pastime for the leisured upper classes. As increasing academic rigour and systematisation was brought to the study of 'natural history', its subdisciplines were adopted into university curricula, and learned societies (such as the Royal Horticultural Society, founded in 1804) were established to support research in these areas. A related development was strong enthusiasm for exotic garden plants, which resulted in plant collecting expeditions to every corner of the globe, some-times with tragic consequences. This series includes accounts of some of those expeditions, detailed reference works on the flora of different regions, and practical advice for amateur and professional gardeners.

Flora Capensis

This seminal publication began life as a collaborative effort between the Irish botanist William Henry Harvey (1811–66) and his German counterpart Otto Wilhelm Sonder (1812–81). Relying on many contributors of specimens and descriptions from colonial South Africa – and building on the foundations laid by Carl Peter Thunberg, whose *Flora Capensis* (1823) is also reissued in this series – they published the first three volumes between 1860 and 1865. These were reprinted unchanged in 1894, and from 1896 the project was supervised by William Thiselton-Dyer (1843–1928), director of the Royal Botanic Gardens at Kew. A final supplement appeared in 1933. Reissued now in ten parts, this significant reference work catalogues more than 11,500 species of plant found in South Africa. Volume 4 appeared in two parts, the first comprising sections published between 1905 and 1909, covering Vacciniaceae to Gentianeae.

Cambridge University Press has long been a pioneer in the reissuing of
out-of-print titles from its own backlist, producing digital reprints of books
that are still sought after by scholars and students but could not be reprinted
economically using traditional technology. The Cambridge Library Collection
extends this activity to a wider range of books which are still of importance to
researchers and professionals, either for the source material they contain, or as
landmarks in the history of their academic discipline.

Drawing from the world-renowned collections in the Cambridge University
Library and other partner libraries, and guided by the advice of experts
in each subject area, Cambridge University Press is using state-of-the-art
scanning machines in its own Printing House to capture the content of each
book selected for inclusion. The files are processed to give a consistently clear,
crisp image, and the books finished to the high quality standard for which the
Press is recognised around the world. The latest print-on-demand technology
ensures that the books will remain available indefinitely, and that orders for
single or multiple copies can quickly be supplied.

The Cambridge Library Collection brings back to life books of enduring
scholarly value (including out-of-copyright works originally issued by other
publishers) across a wide range of disciplines in the humanities and social
sciences and in science and technology.

DATES OF PUBLICATION OF THE SEVERAL PARTS
OF THIS VOLUME.

Part I., pp. 1–192, was published *May*, 1905.

Part II., pp. 193–336, was published *August*, 1905.

Part III., pp. 337–480, was published *October*, 1906.

Part IV., pp. 481–672, was published *November*, 1907.

Part V., pp. 673–864, was published *March*, 1908.

Part VI., pp. 865–end, was published *February*, 1909.

FLORA CAPENSIS.

VOL. IV. SECT. 1.

FLORA CAPENSIS:

BEING A

Systematic Description of the Plants

OF THE

CAPE COLONY, CAFFRARIA, & PORT NATAL

(AND NEIGHBOURING TERRITORIES)

BY

VARIOUS BOTANISTS.

EDITED BY

SIR WILLIAM T. THISELTON-DYER, K.C.M.G.,

C.I.E., LL.D., Sc.D., F.R.S.

HONORARY STUDENT OF CHRIST CHURCH, OXFORD,

LATE DIRECTOR, ROYAL BOTANIC GARDENS, KEW.

*Published under the authority of the Governments of the
Cape of Good Hope, Natal and the Transvaal.*

VOLUME IV. SECTION 1.

VACCINIACEÆ TO GENTIANEÆ.

LONDON

LOVELL REEVE & CO., LTD.

6, HENRIETTA STREET, COVENT GARDEN,

Publishers to the Home, Colonial and Indian Governments.

1909.

LONDON:
PRINTED BY WILLIAM CLOWES AND SONS, LIMITED,
DUKE STREET, STAMFORD STREET, S.E., AND GREAT WINDMILL STREET, W.

PREFACE.

As pointed out in the Preface to Section 2, the preparation of the preceding one, which is now completed, was delayed by unavoidable difficulties. It comprises two of the most important orders in the South African flora, the *Ericaceæ* and *Asclepiadeæ*, in both of which the specific characters require minute and therefore often lengthy description. The bulk of this Section has grown accordingly, and it is actually equivalent to that of two ordinary volumes.

Dr. HARRY BOLUS, who has given such munificent encouragement to Botany in South Africa, kindly undertook the elaboration of *Ericaceæ*, which for many years he had assiduously studied in the field. Professor FRANCIS GUTHRIE, LL.B., B.A., who had long collaborated with Dr. BOLUS, unhappily died on October 19, 1899. The elaboration of the intricate genus *Erica* is the result of their joint labours. The loss of his fellow-worker and his own indifferent health compelled Dr. BOLUS to abandon reluctantly proceeding with the remaining genera. These were undertaken by Mr. N. E. BROWN, A.L.S., who, in working them out, found it necessary to establish some new ones, and in other respects to depart from the published key to the order. He has accordingly prepared a new one which will be found in the " Additions and Corrections " at page 1123 of this Section.

Mr. N. E. BROWN has also worked out the *Asclepiadeæ* with immense pains. He has had the advantage of following Dr. R. SCHLECHTER, an acute botanist who has few rivals as a collector. Both have been disposed perhaps to cut their species rather fine. But South African botanists in the future will find less difficulty in uniting species which cannot be sustained than in separating those which have been injudiciously united. Mr. BROWN has himself been occupied with the study of the fascinating group of *Stapelieæ*, both under cultivation and in the herbarium, for the past forty years, and it may be hoped therefore that his conclusions will have reached some finality. He has been led to the important result that many supposed native species have arisen from cross-fertilisaticn in European gardens. It is, however, a singular circumstance that some of the species which were the earliest described, and which cannot be accounted for in ·this way, have never been met with again by subsequent collectors. It is to be feared that one of the most striking features of the South African flora is doomed to gradual and irremediable extinction.

Mr. W. P. HIERN, F.R.S., who is the acknowledged authority on the order, has undertaken the *Ebenaceæ*. In his monograph published in 1873 he had already taken up two species of *Royena* left by the late Professor HARVEY in manuscript.

For some of the smaller orders it has also been possible to make use of Harvey's work. It is somewhat remarkable that two species of *Jasminum* of which he left descriptions have remained since 1865 unanticipated in publication by any other describer.

Dr. STAPF, F.R.S., has elaborated the *Apocynaceæ*.

Lieut.-Colonel PRAIN, C.I.E., F.R.S., the Director of the Royal Botanic Gardens, has undertaken *Loganiaceæ* with Major H. A. CUMMINS, C.M.G., and *Gentianeæ* with Mr. A. W. HILL, the Assistant-Director.

For the limits of the regions under which the localities are cited in which the species have been found to occur, reference may be made to the Preface to Vol. VI.

I continue to be indebted for invaluable aid to Mr. C. H. WRIGHT, A.L.S., now Assistant Keeper, and to Mr. N. E. BROWN, A.L.S., Assistant in the Herbarium of the Royal Botanic Gardens, the former in reading the proofs, and the latter in working out the geographical distribution.

Besides the maps already cited in the Prefaces to Volumes VI. and VII., the following have also been used :—

Map of the Colony of the Cape of Good Hope and neighbouring territories. Compiled from the best available information. By JOHN TEMPLER HORNE, Surveyor-General. 1895.

Stanford's new map of the Orange Free State and the southern part of the South African Republic, &c. 1899.

Carte du Théâtre de la Guerre Sud-Africaine. Par le Colonel Camille Favre. 1902.

To many of the South African correspondents of Kew enumerated in previously published volumes I have again to tender my acknowledgments for the contribution of specimens in aid of the work to the Herbarium of the Royal Botanic Gardens.

I must further record my obligations to some new contributors, and to those whose kind assistance in various ways has been of the greatest value in the preparation of this section of Volume IV.

HARRY BOLUS, Esq., D.Sc., F.L.S., has continued his generous gifts of specimens, besides lending others from his herbarium of *Asclepiadeæ* and *Gentianeæ*, including many types of Dr. Schlechter's species.

Dr. JOHN BRIQUET, Director of the Botanic Garden, Geneva. Loan of specimens from the Delessert Herbarium.

PAUL CONRATH, Esq. Plants from the Transvaal.

JOSEPH BURTT DAVY, F.L.S., Government Agrostologist and Botanist, Transvaal. Plants from the Transvaal.

Dr. CASIMIR DE CANDOLLE, Geneva. Photograph and loan of specimen from the Candollean herbarium.

Geheimrath Dr. A. ENGLER, Director of the Botanic Garden and Museum, Berlin. Loan of *Gentianeæ*.

Dr. H. O. JUEL, Director of the Botanic Garden, Upsala. Loan of portions of Thunberg's herbarium.

Prof. PAUL HENRI LECOMTE, Jardin des Plantes, Paris. Photographs of Lamarck's types of *Chironia*.

Miss R. LEENDERTZ. Plants from the Transvaal.

Dr. H. W. C. LENZ, Director of the Museum, Lubeck. Loan of E. Meyer's types of *Asclepiadeæ* and *Gentianeæ*.

Dr. C. A. M. LINDMAN. Loan of specimen of *Chironia* from the herbarium of P. J. Bergius, Stockholm.

Dr. JOHN P. LOTSY, Director of the Royal Herbarium, Leiden. Loan of specimens of *Gentianeæ*.

Dr. RUDOLPH MARLOTH, Capetown. *Stapelieæ*, living and in fluid.

Miss A. Pegler.   Plants from the Transkei.

Dr. G. Albert Peter, Director of the Botanic Garden, Gottingen.   Loan of specimens of *Gentianeæ*.

N. S. Pillans, Capetown.   Large collection of *Stapelieæ*, living and in fluid, including many new species, and the loan of his large and valuable herbarium of the group.   The generous aid of this enthusiastic collector and the free use of his copious notes have been of the greatest value in the difficult task of working out the species.

Humphrey John Sankey, Esq.   Plants from the vicinity of Harrismith, Orange River Colony.

Dr. Hans Schinz, Director of the Botanic Garden and Museum, Zürich.   Numerous specimens collected by Dr. R. Schlechter ; loan of *Gentianeæ*.

Dr. Selmar Schönland, Curator of the Albany Museum, Grahamstown.   Small collection of *Asclepiadeæ* and loan of *Gentianeæ*.

Prof. Albert Charles Seward, F.R.S.   Loan of *Gentianeæ* from the University Museum, Cambridge.

Miss Ethel West.   Plants from Port Elizabeth.

John Medley Wood, A.L.S., Director of the Botanic Gardens, Durban.   Numerous specimens of *Asclepiadeæ* and loan of *Gentianeæ*.

Prof. E. Perceval Wright, Sec. R.I.A., Keeper of the Herbarium, University of Dublin.   Loan of portions of Harvey's herbarium.

Dr. Alexander Zahlbruckner, Keeper of the Botanical Collections of the Naturhistorische Hofmuseum, Vienna.   Loan of specimens of *Asclepiadeæ* and *Gentianeæ*.

On this occasion it is appropriate to pay a brief tribute to the memory of Sir Henry Barkly, G.C.M.G., K.C.B., F.R.S., who died on October 21, 1898.   It was at his instance that during his last period of official life as Governor of Cape Colony (1873–7) the preparation of the *Flora Capensis* was resumed, and it was due to his support that the approval and aid of the legislature of Cape Colony and Natal were secured. He was deeply interested in the *Stapelieæ*, of which at the time the study had been comparatively neglected.   He collected and cultivated as many as he could, and also sent a large living collection to Kew. Lady Barkly and Miss E. B. Barkly made water-colour drawings of them as they flowered. Of these copies were sent to Kew, together with specimens in alcohol, accompanied by copious descriptive notes. An account of this material was published in Hooker's *Icones Plantarum*, tt. 1901–25; and coloured figures of four of the species that flowered at Kew appeared in the *Botanical Magazine*. The whole of Sir Henry Barkly's material was a contribution to the study and elaboration of the group only second perhaps in value to that of Mr. Pillans.

I may be permitted more personally to express my indebtedness to Lieut.-Colonel PRAIN for kind and unfailing assistance in many ways, without which the task of editing a work of this kind at a distance from the resources of Kew would be one of peculiar difficulty.

The critics of niceties of nomenclature, which often seem to obscure the interest of larger problems, will probably notice that the Kew tradition has been adhered to, and the supposed right of priority of the original specific epithet has been disregarded where an existing name is available which has correctly placed a species in the genus to which its affinity is most obvious. The principle was laid down by Sir Joseph Hooker in 1872 in *The Flora of British India*. Its justification is based on technical grounds equally with those of common sense. It may be convenient to briefly state them :—

i. The so-called binominal nomenclature which we employ was devised by Linnæus, and, as with everything he did, on a logical and definite basis. Nothing but confusion can arise by departure from this. To the specific epithet, apart from its proper function, Linnæus attached no importance at all. He saw that the scientific problem was to get the species into its right genus. "Nomen specificum sine generico est quasi pistillum sine campana." The specific name taken alone is the clapper without the bell. A Linnean name, then, though it consists of two parts, must be treated as a whole. "Nomen omne plantarum constabit nomine generico et specifico." And the same principle obviously applies to all names constructed in accordance with Linnean rules. The supposed appeal to justice begins by repudiating the authority of the lawgiver. Alphonse de Candolle appreciated the true position when he said : " The real merit of Linnæus has been to combine, for all plants, the generic name with the specific epithet."

ii. But the claim for justice works the greatest injustice, and it is not even tempered with mercy. Any careless or incompetent botanist can tack on a blundering name to an undescribed plant, and his blunder with his name attached is to be handed down to posterity for all time. As Linnæus saw, the real scientific feat is to discover its true affinity, not to give it a haphazard label. And the author who does this successfully is the one whose insight deserves commemoration. It is impossible not to agree with Sir Joseph Hooker when he says : " I regard the naturalist who puts a described plant into its proper position in regard to its allies as rendering a greater service to science than its describer when he either puts it into a wrong place or throws it into any of those chaotic heaps, miscalled genera, with which systematic works still abound."

iii. Every revision of the contents of an order involves a reconsideration of the mutual affinities of its contents, and this usually involves some transposition of species from one genus to another, or the creation of new genera. It may be hoped that the process is generally judiciously accomplished. But in any case it yields a crop of synonyms. This is inevitable, and these in a work like the present have to be examined and quoted. The labour involved will be evident from many of its pages. There is said to be a species of *Fimbristylis* with 135 synonyms. Taxonomic science must in the end be crushed by its own literary top-hamper. The only remedy eventually will be to draw a line behind which synonymy will be ignored. But we need not add to the burden by the creation of a new specific name when one which is valid and available already exists in the genus. The appeal to justice lays itself open in such cases to the suspicion of being little more than a cloak for the vanity of the author.

I have equally resisted the wholesale manufacture of new specific names by the revival of obsolete or forgotten genera without any obvious necessity. Nomenclature is a mere means to an end; it is ignominious to become its slave, and we alienate the sympathy of the public, which we wish to secure, by changes of familiar names which, to its eyes, must seem simply wanton.

The expenses of preparation and publication of the present section of Volume IV. have been aided by grants from the Governments of Cape Colony and Natal. The Government of the Transvaal has not hitherto been associated with the work. It is a gratifying evidence of the appreciation of its usefulness in South Africa that that Government has now spontaneously also made a grant in aid of its publication.

W. T. T.-D.

Witcombe, 1909.

SEQUENCE OF ORDERS CONTAINED IN VOL. IV. SECT. 1, WITH BRIEF CHARACTERS.

Sub-class III. GAMOPETALÆ. Ord. LXXVII.—XC.

Series II. HETEROMERÆ. Ord. LXXVIII.—LXXXIV. Ovary usually superior. Carpels more than two.

COHORT iv. ERICALES. *Flowers* regular, hermaphrodite. *Stamens* twice as many as the corolla-lobes, or equal in number to and alternate with them. *Ovary* 2–∞ -celled. *Seeds* small or minute.

LXXVIII. VACCINIACEÆ (page 1). *Ovary* inferior. *Fruit* fleshy. (*Shrubs, undershrubs or small trees. Leaves alternate, exstipulate. Flowers racemose or fascicled, rarely solitary.*)

LXXIX. ERICACEÆ (page 2). *Ovary* superior. *Fruit* capsular, rarely fleshy. (*Shrubs, undershrubs or trees, often evergreen. Leaves alternate, opposite or verticillate, exstipulate. Inflorescence various.*)

COHORT v. PRIMULALES. *Flowers* hermaphrodite, or by abortion polygamous. *Stamens* as many as, and opposite to the corolla-lobes. *Ovary* polycarpellary, 1-celled; placenta 1, basal.

LXXX. PLUMBAGINEÆ (page 418). *Ovary* 1-ovuled; styles or style-branches 5. *Endosperm* floury. (*Herbs, usually stemless, rarely shrubs. Leaves rosulate or alternate. Inflorescence spicate, racemose, cymose or paniculate.*)

LXXXI. PRIMULACEÆ (page 426). *Ovary* 2–∞ -ovuled; style simple. *Fruit* capsular. *Endosperm* fleshy or horny. (*Herbs. Leaves radical, opposite or alternate. Flowers solitary and axillary, or in umbels, racemes or panicles.*)

LXXXII. MYRSINEÆ (page 431). *Ovary* 2–∞ -ovuled; style simple. *Fruit* indehiscent. *Endosperm* fleshy or horny. (*Shrubs or trees. Leaves usually alternate. Inflorescence various; flowers usually small.*)

COHORT vi. EBENALES. *Flowers* regular, hermaphrodite or unisexual. *Stamens* as many as the corolla-lobes and opposite to them, or numerous. *Ovary* 2–∞ -celled.

LXXIII. SAPOTACEÆ (page 436). *Flowers* hermaphrodite. *Stamens* epipetalous. *Ovules* solitary, ascending. (*Trees or shrubs with milky sap. Leaves alternate, entire, extipulate. Flowers axillary.*)

LXXXIV. EBENACEÆ (page 444). *Flowers* diœcious rarely hermaphrodite. *Stamens* hypogynous or on the very base of the corolla-tube. *Ovules* geminate or solitary, pendulous. (*Trees or shrubs; heartwood hard; sap not milky. Leaves alternate, entire, usually coriaceous. Flowers cymose, racemose or solitary.*)

Series III. BIOARPELLATÆ. Ord. LXXXV.—XC. Ovary usually superior. Carpels 2, rarely 1 or 3.

COHORT vii. GENTIANALES. *Corolla* regular. *Stamens* as many as the corolla-lobes and alternate with them, or if fewer alternate with the carpels. *Leaves* usually opposite.

LXXXV. OLEACEÆ (page 478). *Corolla-lobes* 4 to many. *Stamens* 2. *Ovary* 2-celled; style simple; stigma terminal; ovules usually 2 in each cell. (*Trees or shrubs. Leaves opposite, exstipulate, simple or compound. Inflorescence cymose, paniculate or fascicled.*)

LXXXVI. SALVADORACEÆ (page 488). *Corolla-lobes* 4. *Stamens* 4. *Ovary* 1-2-celled; ovules 1-2 in each cell. (*Trees or shrubs, sometimes spiny. Leaves opposite, entire; stipules rudimentary. Panicles trichotomous, axillary.*)

LXXXVII. APOCYNACEÆ (page 490). *Corolla-lobes* 5. *Stamens* 5; pollen-grains free, rarely cohering. *Carpels* usually 2, free or united below; style entire or divided at the base; ovules few or many. (*Trees, erect or climbing shrubs, very rarely herbs. Leaves opposite or whorled, entire, stipulate or not. Flowers cymose, small, or large and showy.*)

LXXXVIII. ASCLEPIADEÆ (page 518). *Corolla-lobes* 5. *Stamens* 5; pollen-grains usually united into waxy masses. *Carpels* usually 2, free b low; styles united and dilated above; ovules usually many. (*Herbs or shrubs, erect, prostrate or twining; sap milky or watery. Leaves opposite or whorled, sometimes absent, exstipulate. Flowers usually few cymosely arranged.*)

LXXXIX. LOGANIACEÆ (page 1036). *Corolla-lobes* 5-16. *Stamens* 5-16. *Carpels* 2, united; style simple; ovules many. (*Shrubs or trees. Leaves opposite, rarely whorled or fascicled; stipules usually reduced to an interpetiolar line. Flowers cymose, rarely racemose.*)

XC. GENTIANEÆ (page 1056). *Corolla-lobes* 4-6. *Stamens* 4-6. *Carpels* 2, united; style simple; ovary 1-2-celled; ovules numerous. (*Annual or perennial glabrous herbs, rarely shrubs. Leaves opposite, rarely alternate. Flowers cymose, paniculate, fasciculate or solitary terminal or axillary.*)

FLORA CAPENSIS.

ORDER LXXVIII. **VACCINIACEÆ.**

(By N. E. Brown.)

Flowers hermaphrodite, regular. *Calyx-tube* adnate to the ovary; limb 5-lobed or 5-toothed, persistent or deciduous. *Corolla* gamopetalous, tubular, campanulate, globose or urceolate, 4–7- (usually 5-) lobed or toothed, deciduous; lobes imbricate or valvate. *Stamens* 4–14, usually twice as many as the corolla-lobes, epigynous or inserted at the base of the corolla; filaments free or connate; anthers dorsifixed, 2-celled; cells usually produced into tubes at the apex, opening by terminal pores or elongated slits. *Ovary* inferior, 2–10-celled, often crowned with an epigynous disk; style slender or filiform; stigma simple, subcapitate or minutely 4–5-lobed; ovules usually numerous, rarely 2 to few in each cell, axile on tumid or bilobed placentas, anatropous. *Fruit* a many- (rarely few-) seeded berry. *Seeds* small or minute, usually compressed, testa coriaceous; albumen fleshy; embryo minute, axile.

Shrubs, undershrubs, or small trees, sometimes epiphytic. Leaves alternate; stipules 0. Inflorescence various, usually bracteate.

DISTRIB. Genera about 32; species about 350, chiefly natives of the Tropics on mountains, with a few in the temperate regions; only 3 or 4 occur in Africa.

I. **VACCINIUM**, Linn.

Calyx superior, 4–5-lobed. *Corolla* urceolate, campanulate or subtubular, 4–5-lobed or -toothed. *Stamens* 8–10, free; anthers produced into 2 tubes opening by pores or slits at the apex, sometimes spurred on the back. *Ovary* inferior, turbinate, hemispherical or globose, 4–5-celled or falsely 8–10-celled by spurious partitions from the placenta; style slender; stigma simple or subcapitate; ovules numerous or few in each cell. *Fruit* a berry.

Shrubs or undershrubs, rarely trees; leaves evergreen or deciduous, entire or serrulate; flowers in fascicles or racemes, rarely solitary, bracteate.

Species over 100, chiefly natives of the North Temperate zone and mountains of the Tropics, 3–4 in Africa.

1. V. Exul (Bolus in Hook. Ic. Pl. t. 1941); a shrub 2–5 ft. high; branchlets terete, becoming angular-striate when dried, pubescent, leafy; leaves rather rigidly coriaceous; petiole 1–1½ lin. long, pubescent; blade ¾–1¾ in. long, ⅓–¾ in. broad, ovate-lanceolate or lanceolate, acute or shortly acuminate, slightly thickened and recurved along the serrulate margins, more or less pubescent on the basal part of the midrib, otherwise glabrous on both sides, reticulately veined, with the veins more conspicuous beneath; racemes axillary, $\frac{1}{2}$–$\frac{3}{4}$ in. long, 5–8-flowered; rhachis angular, glabrous; flowers nodding; bracts 1–2 lin. long, $\frac{1}{2}$–1 lin. broad, lanceolate to elliptic-ovate, acute, boat-shaped, minutely ciliate; pedicels 2–3 lin. long, jointed under the ovary, glabrous; calyx glabrous; ovary-part $\frac{3}{4}$ lin. long, broadly obconical; lobes $\frac{3}{4}$ lin. long, $\frac{3}{4}$–1 lin. broad, deltoid-ovate, acuminate, erect; corolla urceolate, 4–5-lobed, glabrous, white; tube 2 lin. long, $1\frac{1}{2}$–$1\frac{3}{4}$ lin. in diam.; lobes $\frac{3}{4}$ lin. long, $\frac{1}{2}$ lin. broad, ovate-oblong, obtuse, recurved; stamens 8–10, epigynous, about $\frac{1}{4}$ lin. longer than the corolla-tube; filaments $1\frac{1}{2}$ lin. long, flattened, hairy; anthers $1\frac{2}{3}$ lin. long, minutely scabrous, produced into 2 tubes opening by oblique pores at the apex; ovary 4–5-celled; style filiform, 2 lin. long; stigma simple; fruit not seen.

KALAHARI REGION: Transvaal; amongst rocks on the summit of the Devils Kantoor, near Barberton, 5500–5600 ft., *Galpin*, 659! *Bolus*, 7616!

V. africanum (Britt. in Trans. Linn. Soc. ser. 2, Bot. iv. 23, t. 4, fig. 3–5) from Nyasaland, is a very close ally of this, differing in its glabrous stems, longer pointed leaves (which are much more conspicuously reticulate with more prominent veins on both sides) and longer racemes.

ORDER LXXIX. **ERICACEÆ.**

(By H. BOLUS, F. GUTHRIE, and N. E. BROWN.)

Flowers hermaphrodite, regular. *Calyx* free, 4–5-toothed, -lobed or -partite; teeth, lobes or segments equal or unequal, imbricate, valvate or open in bud, persistent. *Corolla* hypogynous or in *Lagenocarpus* arising from the middle of the ovary, gamopetalous and 2–5- (rarely 6–10-) lobed or of 3–7 free petals, very variable in form, imbricate, contorted or rarely valvate in bud, persistent or deciduous. *Stamens* 3–10, rarely 12 or more, hypogynous except in *Lagenocarpus*, sometimes adnate to the base of the corolla; filaments free or rarely connate; anthers 2-celled, basifixed or dorsifixed, free or connate; cells often free down to the middle or beyond, with or without a crest, spur or arista at the base or on the back at or towards the apex, sometimes produced at the apex into tubes, opening by pores or longitudinal slits. *Disk* none or annular or tumid, crenate or lobed. *Ovary* superior or in *Lagenocarpus* half-inferior to the corolla, 1–12-celled; style filiform or cylindric; stigma simple, capitate or peltate, entire, lobed or dentate. *Ovules* 1 to many in each cell, axile, pendulous or rarely erect, anatropous. *Fruit* a loculicidal or septicidal capsule, rarely baccate or drupaceous, 1- to many-seeded. *Seeds* usually minute, angled or compressed; testa sometimes lax and produced at the ends; albumen fleshy, embryo minute, axile.

Shrubs, undershrubs or trees, often evergreen; leaves alternate, opposite or verticillate; stipules 0; inflorescence various; pedicels often bracteate.

DISTRIB. Genera about 56; species about 1200, distributed in all parts of the world.

** Corolla and stamens hypogynous; ovary superior.*

† Ovary 2–4- (rarely 8-) celled.
‡ Ovules 2 or more in each cell.
§ Stamens normally 8, rarely 6–7.

I. **Erica.**—*Bracts* 3 (very rarely 1 or 0). *Calyx* equally 4-partite (rarely 4-lobed), very rarely longer than the variously shaped 4-lobed or -toothed corolla. *Ovary* 4- (rarely 8-) celled.

II. **Philippia.**—*Bracts* 0. *Calyx* unequally 4-lobed or -partite. *Corolla* very small, 4-lobed or -toothed. *Ovary* 4-celled.

§§ Stamens normally 4, occasionally 5–6.

III. **Ericinella.**—*Bracts* 0. *Calyx* unequally 3-4-partite or -lobed. *Corolla* very small, 3-4-lobed or -toothed. *Ovary* 3-4-celled.

IV. **Blæria.**—*Bracts* 3. *Calyx* equally 4-lobed or -partite. *Corolla* small, tubular or tubular-campanulate, 4-lobed, often 4-angled. *Ovary* 4-celled.

‡‡ Ovule solitary in each cell.
§ Stamens 6–8.

VII. **Eremia.**—*Bracts* 3. *Calyx* equally 4-partite; segments broad, ciliate or hairy, but not woolly. *Corolla* urceolate-campanulate or campanulate, 1½–2 times as long as broad. *Stamens* 8; anthers included or not exceeding the corolla-lobes. *Ovary* 2-4-celled. *Leaves* spreading, not woolly.

VIII. **Hexastemon.**—*Bracts* 3. *Calyx* equally and deeply 4-lobed; lobes rather narrow, white-woolly. *Corolla* narrowly ovoid-tubular, 3 times as long as broad. *Stamens* 6; anthers much exserted. *Ovary* 2-celled. *Leaves* densely imbricate, white-woolly.

XVII. **Salaxis.**—*Bracts* 0. *Calyx* unequally or subequally 4-toothed, -lobed or -partite. *Corolla* minute, obconic or subglobose. *Anthers* included or partly exserted. *Stigma* large, crater-like or peltate.

§§ Stamens 3–4.
‖ Corolla 4- (very rarely 3-) toothed or lobed.
¶ Bracts 0. Calyx unequally 4-partite, glabrous; tube 0.

V. **Coilostigma.**—*Corolla* small, ovoid, oblong-ovoid or tubular-oblong, often contracted at the apex. *Ovary* 2-celled.

¶¶ Bracts 3 (2 of them extremely minute in one species of *Thoracosperma*). Calyx equally and deeply 4-lobed or 4-partite; tube short or 0, not thick nor fleshy nor sharply 4-angled.

(*a*) Corolla more or less constricted below the limb or at the middle or inflated at the base. Anthers included or not exceeding the corolla-lobes.

IX. **Grisebachia.**—*Calyx-segments* broad (not linear), hairy or ciliate. *Corolla* campanulate or globose-campanulate above the constriction. *Ovary* 2-celled.

(*b*) Corolla not constricted at any part, sometimes narrowed below. Anthers entirely or partially exserted beyond the corolla-lobes (except in one species of *Thoracosperma*).

VI. **Thoracosperma.**—*Calyx* minute, inconspicuous, glabrous or ciliate or puberulous with very minute hairs. *Corolla* ovoid, suburceolate, tubular-oblong or campanulate, often contracted at the apex. *Ovary* 2-4-celled.

X. **Acrostemon.**—*Calyx* conspicuous, clothed with long hairs; segments linear or narrowly deltoid. *Corolla* elongated, tubular or clavate-tubular, often contracted at the apex. *Ovary* 2-celled.

¶¶¶ Bracts 1–3. Calyx tubular, campanulate or obconic, equally and shortly 4-toothed or lobed to the middle; tube usually rather thick or somewhat fleshy and often sharply 4-angled.

XI. **Simocheilus.**—*Corolla* much elongated and tubular or tubular-clavate or 4-angled and about twice as long as the calyx. *Ovary* 2-4-celled.

‖‖ Corolla equally but shortly 2-lobed, tubular or tubular-funnel-shaped. Anthers exserted. Ovary 2-celled.

XIV. **Aniserica.**—*Bract* 1, at the very base of the pedicel, or 0. *Calyx* campanulate, 4-angled, equally 4-toothed.

XV. **Sympieza.**—*Bracts* 0-3. *Calyx* dorsally much flattened, 2-angled, or if 3-angled, with the 3rd angle next the axis, equally 2–3-lobed.

†† Ovary 1-celled; ovule solitary.
a. Stamens 8.

XVI. **Lepterica.**—*Bracts* 0. *Calyx* subunequally 4-toothed. *Corolla* minute, obconic. *Anthers* included. *Stigma* very large, crater-like.

β. Stamens 3–5, very rarely 2.

XII. **Syndesmanthus.**—*Bracts* usually 0, rarely 1–3. *Calyx* obconic or turbinate, equally 3–4-toothed, usually hairy; tube distinctly 3-8-angled, thin or coriaceous, not fleshy. *Corolla* tubular or clavate. *Stigma* simple, or peltate with the centre produced.

XIII. **Anomalanthus.**—*Bracts* 3. *Calyx* ovoid or campanulate, thick and fleshy, scarcely or not at all angular, equally 4-toothed, glabrous. *Corolla* campanulate, tubular or narrowly funnel-shaped. *Stigma* simple.

XVIII. **Scyphogyne.**—*Bracts* usually 0, rarely 1. *Calyx* obconic or campanulate, thin or coriaceous, not fleshy, more or less angular, unequally or equally 3–4-toothed or -lobed. *Corolla* minute, globose, obconic, campanulate or urceolate. *Stigma* very large, crater-like or peltate.

** *Corolla arising from the middle of the ovary, which is half inferior to it, but free from the calyx.*

XIX. **Lagenocarpus.**—*Bracts* 0. *Calyx* campanulate, 4-toothed. *Corolla* minute, campanulate or subglobose. *Anthers* sessile, included. *Ovary* 2-3-celled.

I. ERICA, Linn.

(By F. Guthrie and H. Bolus.) *

Calyx free, mostly 4-partite, rarely 4-fid or 4-dentate. *Corolla* hypogynous, deciduous, or rarely persistent, tubular, ampullaceous, urceolate, globose, ovoid, campanulate, cyathiform, obconic, or funnel-shaped, 4-lobed, less commonly 4-fid, rarely sub-4-partite. *Stamens* hypogynous, normally 8, very rarely fewer or more, mostly arising from the base of a free, more or less elevated disk; filaments free; anthers terminal or lateral, dehiscing by lateral pores or slits, muticous, crested or aristate at or near the insertion of the filament or more rarely distant, the appendages often adnate to it, or sometimes

---

* During the progress of this work, I have had to lament the death of my friend and colleague, Prof. Guthrie; but not before he had contributed a large and important share towards its completion. It is a pleasing duty to acknowledge the valuable help rendered to me in the latter portion of the work by my relative, Miss L. Kensit, B.A.—*H.B.*

entirely free from the anther-cells. *Disk* more or less prominent, lobed or crenulate, rarely obsolete. *Ovary* sessile or stipitate, mostly 4- (very rarely 8-celled); cells 2-∞-ovuled. *Style* filiform; stigma simple, capitate, peltate or cyathiform, very rarely 4-fid. *Capsule* globose, conical or cylindrical, loculicidally 4-valved, valves separating from the axis, mostly many-seeded. *Seeds* minute, ellipsoidal, more rarely lenticular, or much compressed and margined or more rarely winged.

Perennial shrubs, from a few inches to 10 ft., or rarely more; leaves 3–6-nate, in whorls, less commonly scattered or opposite, most usually rigid and narrow, linear, trigonous, margins revolute and connate with the underside, leaving only a channel more or less wide and deep between them, or less commonly, flatter, broader and " open-backed," *i.e.* the margins revolute or reflexed, but leaving the underside visible, very rarely nearly flat; inflorescence mostly normally terminal, or often axillary (the flowers clustered in the axils, at the ends of partially or entirely arrested lateral branchlets), very rarely truly indefinite and racemose; flowers solitary, more commonly 2–4-nate, umbellate or capitate; pedicels 1-flowered; bracts 3, rarely fewer, very rarely wanting. The flowers are fertilized in some cases by the wind, in others by insects, and many of the longer and tubular (Subgenus I. SYRINGODEA) by birds (*cf. G. F. Scott-Elliot, in Ann. Bot.* iv. 269, 270).

The *Ericaceæ* are chiefly xerophilous mostly on rocky mountain-sides, lower hills, or sandy plains, very rarely on wet or marshy ground. They inhabit for the most part the littoral strip, some 50 or 60 miles in breadth, from the Olifants River on the west to the Van Stadensberg Range on the east, diminishing rapidly in number beyond these limits. Their greatest concentration may be on the Cape Peninsula, where 92 species have been recorded in an area of 198 square miles; but the home of the more beautiful, and now rarer, species is in the Caledon Division. Many species have a very small distribution-area.

DISTRIB. Species over 500, of which 469 are endemic in South Africa, 6 or 8 in Tropical Africa, and the rest dispersed from the Atlantic Isles through Europe and North Africa to the Orient.

This genus is remarkable for an unusual degree of variability in the form of almost all its organs. It is therefore one difficult of definition as to its species, and of arrangement into satisfactory natural groups. Many of the species are obviously allied to others in very different sections; and in most of the sections and subgenera it is necessary to note exceptions to the general technical characters. Many authors have treated of them with great divergency of views; and the earlier botanists unduly multiplied the species as they arrived from the Cape. At the end of the 18th, and the early part of the 19th centuries, the heaths became fashionable in European gardens, were much hybridized and copiously figured. This has added to the difficulty of definition, and still more to the confusion of the synonomy. In respect of the latter, we have had largely to rely upon others, and can but hope at most to have cleared up some few of the obscurities by which the genus has been surrounded, leaving many in which, owing to absence of types and imperfect descriptions, it may continue to be involved.

The following terms have been used in this genus :—the flowers are said to be *calycine* (*Aiton*) or *corolline* acording as the calyx or corolla predominates in the general appearance of the flower, which depends upon either the position or relative size of those organs. The relative height of the calyx and corolla is taken from the flowers when viewed horizontally, and, on account of the usual spreading of the sepals, appears at first sight at variance with the measurements of those organs when separated and flattened out. The shape of the anther-cell is described from its profile, unless otherwise stated. Anthers projecting forwards at the base are termed *prognathous*; these occur chiefly in § 8, *Euryloma.* Anthers neither distinctly included nor exserted, but plainly visible at the level of the mouth of the corolla-tube, are said to be *manifest.*

SUBGENERA.

(Exceptions disregarded and shown under the Sections.)

Flowers mostly corolline :
    Corolla tubular, over 4 lin. long, mostly from 6–12
      lin. long, limb not stellato-patent ; flowers
      always corolline ...   .   I. SYRINGODEA.
    Corolla various, not tubular, if exceeding 4 lin. in
      length then urceolate, or inflated, or with a
      stellato-patent limb ; flowers corolline, very
      rarely subcalycine :
        Corolla-segments stellato-patent ; tube mostly
          elongate, or ampullaceous, or urceolate   ...   II. STELLANTHE.
        Corolla-segments not, or very rarely, stellato-
          patent, mostly less than (rarely over) 4 lin.
          long   ...   .   III. EUERICA.
Flowers mostly calycine :
    Corolla mostly urceolate or subcyathiform, not
      usually widening to the mouth   ...   .   IV. CHLAMYDANTHE.
    Corolla mostly obconic or campanulate, usually
      widening to the mouth, more rarely suburceolate
      or broad-cyathiform   ...   .   V. PLATYSTOMA.

SYNOPSIS OF SECTIONS.

SUBGENUS I. SYRINGODEA.

§ 1. **Gigandra.** *Inflorescence* terminal or axillary. *Corolla* tubular, or tubular-inflated, rarely elongate-ovoid, glabrous, dry. *Anthers* far-exserted, terminal, muticous.   .   .   .   . Species 1–5.

§ 2. **Didymanthera.** *Inflorescence* terminal or axillary. *Corolla* tubular, glabrous, dry. *Anthers* exserted, subterminal or sublateral, sometimes (or always ?) articulated to the filament at their base   .   .   . Species 6–11.

§ 3. **Pleurocallis.** *Inflorescence* mostly axillary, more rarely also terminal on the same plant. *Corolla* tubular, mostly over 6 lin. long, rarely shorter. *Anthers* lateral, rarely terminal .   .   .   .   .   . Species 12–35.

§ 4. **Evanthe.** *Inflorescence* terminal, very rarely umbellate never sub-capitate. *Corolla* mostly tubular, rarely obconic or subcampanulate. *Anthers* mostly included, rarely exserted, lateral, free. *Ovary* usually glabrous.
Species 36–70.

§ 5. **Dasyanthes.** *Inflorescence* terminal, umbellate or subcapitate. *Corolla* tubular or subovoid-tubular, mostly roughly hispid, never glabrous. *Anthers* lateral, free. *Ovary* more or less hairy .   .   .   . Species 71–77.

§ 6. **Chona.** *Inflorescence* terminal, umbellate. *Corolla* clavate-tubular. *Anthers* exserted, lateral, cohering round the style in a conical truncated tube.
Species 78.

§ 7. **Bactridium.** *Inflorescence* terminal, umbellate, or pseudo-axillary and subverticillate. *Corolla* tubular or inflated-tubular, glabrous, viscid. *Anthers* included, lateral. *Ovary* long-stipitate .   .   .   .   . Species 79–80.

SUBGENUS II. STELLANTHE.

§ 8. **Euryloma.** *Inflorescence* terminal, mostly umbellate. *Corolla* mostly ampullaceous, or suburceolate, rarely tubular, contracted at the throat ; limb often large. *Anthers* included, lateral, usually more or less prognathous and bilobed at the base. *Ovary* stipitate or elongate .   .   . Species 81–95.

§ 9. **Ceramus.** *Inflorescence* terminal ; flowers umbellate, or 3–4-nate. *Corolla* urceolate, glabrous, dry. *Anthers* included, lateral, rarely prognathous. *Ovary* stipitate, or at least narrowed to the base.   .   .   .   . Species 96–100.

§ 10. **Callista.** *Inflorescence* terminal ; flowers usually 4-nate, rarely solitary. *Corolla* narrow-tubular or suburceolate, rarely subampullaceous. *Anthers* included, lateral, not prognathous at the base, usually very small. *Ovary* sessile.
Species 101–111.

§ 11. **Platyspora.** *Inflorescence* axillary, pseudo-racemose, pseudo-spicate or capitate; flowers corolline or rarely subcalycine. *Sepals* mostly in opposite pairs. *Corolla* various, 3–8 lin. long (mostly resembling that of §§ *Euryloma, Callista,* or *Lamprotis*). *Anthers* included, lateral. *Ovules* much compressed or lenticular, sometimes margined or winged   .   .   .   Species 112–117.

§ 12. **Myra.** *Inflorescence* mostly axillary, in two species strictly racemose, or sometimes terminal. *Corolla* subtubular, urceolate, or rarely subampullaceous, mostly pubescent. *Anthers* usually included, lateral, appendiculate; mostly glandular-viscid subshrubs   .   .   .   .   Species 118–122.

Subgenus III. **EUERICA.**

§ 13. **Ephebus.** *Inflorescence* terminal; flowers 3–4-nate or solitary, rarely clustered or umbellate, mostly corolline, rarely subcalycine. *Corolla* variously shaped, rarely over 3 lin. long, most usually more or less hairy, occasionally glabrous, dry, rarely viscid. *Anthers* mostly included, rarely exserted or sub-exserted, lateral.   .   .   .   .   Species 123–175.

§ 14. **Ceramia.** *Inflorescence* terminal or axillary, sometimes both on the same plant; flowers mostly 3–4-nate, or umbellate, rarely subcapitate or solitary, corolline rarely subcalycine. *Corolla* variously shaped, rarely over 2½ lin. long, mostly glabrous or less commonly pubescent. *Anthers* lateral, very rarely subterminal. Mostly diffuse and slender plants, less commonly erect and slender, very rarely stout and rigid   .   .   .   .   Species 176–207.

§ 15. **Desmia.** *Inflorescence* terminal; flowers 3-nate, umbellate or capitate, corolline or subcorolline. *Corolla* urceolate, glabrous, 1½–2½ lin. long. *Anthers* subexserted, terminal, muticous   .   .   .   .   .   Species 208–210.

§ 16. **Gypsocallis.** *Inflorescence* axillary, rarely also terminal; flowers corolline or rarely subcalycine. *Corolla* urceolate, ovoid, or subcampanulate, glabrous, dry; limb short, 1–3 lin. long. *Anthers* mostly exserted or sub-exserted, rarely subincluded, usually lateral, rarely subterminal. Species 211–218.

§ 17. **Pyronium.** *Inflorescence* terminal; flowers 3-nate, rarely umbellate or clustered, corolline, rarely subcorolline. *Corolla* mostly cyathiform, or sub-campanulate, suburceolate, or ovoid, ¾–2½ lin. long. Resembling § *Gypsocallis,* differing almost solely by its terminal flowers   .   .   .   Species 219–232.

§ 18. **Orophanes.** *Inflorescence* terminal; flowers 4-nate or sometimes um-bellate or irregularly clustered, very rarely in a congested pseudo-spike; flowers corolline. *Corolla* various, mostly urceolate, cyathiform or campanulate, usually 1½–2½ (rarely 4–5) lin. long. *Anthers* mostly included, rarely subexserted, lateral, always appendiculate. *Leaves* almost always 4-nate . . Species 233–259.

§ 19. **Leptodendron.** *Inflorescence* terminal; flowers 3-nate or solitary, corolline, very rarely subcorolline. *Corolla* various in shape, glabrous, dry, 1½–3½ lin. long. *Anthers* included, lateral. *Leaves* almost always 3-nate—these and its 3-nate flowers forming the chief distinction from the preceding section   .   .   .   .   Species 260–268.

§ 20. **Pachysa.** *Inflorescence* terminal, occasionally also pseudo-axillary; flowers mostly 3–4-nate or umbellate, rarely subcapitate, corolline, or very rarely subcalycine. *Corolla* more or less viscid, ovoid, urceolate, or very rarely obconic, 1–10 (usually 2–4) lin. long. *Anthers* mostly included, very rarely exserted, lateral, appendiculate or very rarely muticous. *Ovary* sessile, rarely sub-stipitate   .   .   .   .   Species 269–292.

§ 21. **Hermes.** *Inflorescence* mostly axillary, rarely both terminal and axillary, usually pseudo-spicate or pseudo-racemose towards the ends of the branches; flowers corolline, rarely subcalycine. *Corolla* various, mostly campanulate, obconic or subtubular, 1½–4 lin. long. *Anthers* more usually included, lateral or subterminal   .   .   .   .   Species 293–307.

§ 22. **Chlorocodon.** *Inflorescence* axillary; flowers 1–3-nate on the uppermost branches, occasionally a few also terminal, corolline, very small. *Corolla* cyathi-form, campanulate or subobconic, ¼–1 lin. long. *Anthers* mostly lateral, rarely subterminal. *Stigma* capitate, peltate or cyathiform.   .   .   Species 308–314.

§ 23. **Arsace.** *Inflorescence* terminal; flowers 3-nate, corolline. *Corolla* urceolate, cyathiform or campanulate, $\frac{1}{2}$-$1\frac{1}{2}$ lin. long. *Anthers* lateral or sub-lateral. *Stigma* as in the preceding section . . . Species 315–330.

§ 24. **Pseuderemia.** *Inflorescence* terminal, capitate; flowers mostly corolline. *Corolla* urceolate or campanulate, 1–5 lin. long. *Anthers* mostly included, lateral, appendiculate. . . . . . . Species 331–339.

§ 25. **Polydesmia.** *Inflorescence* terminal, capitate or subumbellate; flowers mostly corolline, rarely subcalycine. *Corolla* suburceolate or ovoid, $1\frac{3}{4}$-$2\frac{1}{2}$ lin. long. *Anthers* exserted or subexserted, terminal or subterminal. Species 340–343.

SUBGENUS IV. CHLAMYDANTHE.

§ 26. **Chromostegia.** *Inflorescence* terminal, capitate; heads 4-fld., mostly involucrated by the more or less enlarged and discoloured floral leaves; flowers calycine. *Bracts* closely approximate, broader than the narrow sepals. *Corolla* various, puberulous, 1–2 lin. loug. Procumbent or suberect; leaves 4-nate, strongly ciliate . . . . Species 344–346.

§ 27. **Oxyloma.** *Inflorescence* terminal; flowers subsessile, 3-nate, or capitate in many-fld. heads, calycine. *Bracts* approximate, and (with the sepals) closely adpressed to the corolla, coloured. *Corolla* short-tubular or subinflated, 2–5 lin. long; segments suberect, long, acute or acuminate. *Anthers* included, lateral, muticous . . . . . . Species 347–349.

§ 28. **Eriodesmia.** *Inflorescence* terminal; flowers 1–6-nate, subcapitate or umbellate, mostly calycine, rarely subcorolline. *Sepals* densely villous or woolly. *Corolla* urceolate or campanulate, hairy, $1\frac{1}{2}$-$2\frac{1}{4}$ lin. long. *Anthers* terminal. Plants more or less hairy in all parts . . . . Species 350–353.

§ 29. **Amphodea.** *Inflorescence* terminal, capitate; heads 3-fld.; flowers calycine. *Bracts* approximate, like the sepals scarious and coloured. *Corolla* narrow-ovoid or suburceolate, glabrous, dry, $1\frac{3}{4}$-$2\frac{1}{2}$ lin. long; segments erect, long, acute. *Anthers* terminal or subterminal, decurrent-aristate. *Leaves* 3-nate . . . . Species 354–356.

§ 30. **Geissostegia.** *Inflorescence* terminal; flowers mostly 3-nate, calycine, small. *Bracts* mostly approximate and imbricate, more rarely subremote. *Sepals* usually about as long as the corolla, rarely much shorter. *Corolla* various, mostly urceolate or cyathiform, 1–$2\frac{1}{2}$ lin. long. *Anthers* exserted, mostly terminal or subterminal, muticous. *Leaves* 3-nate . . . Species 357–371.

§ 31. **Elytrostegia.** *Inflorescence* terminal; flowers mostly 3-nate or clustered, calycine. *Bracts* closely approximate, sepal-like, paleaceous or cartilaginous. *Corolla* tubular-cyathiform or suburceolate, glabrous, dry, 1–$2\frac{1}{4}$ lin. long. *Anthers* lateral or subterminal. *Stigma* from subsimple to peltate. *Leaves* 3-nate . . . . Species 372–376.

§ 32. **Apœcus.** *Inflorescence* terminal; flowers mostly 4-nate or subumbellate, calycine. *Sepals* almost entirely concealing the corolla. *Corolla* cyathiform or subobconic, glabrous, $1\frac{1}{2}$-$2\frac{1}{4}$ lin. long. *Anthers* included or subincluded, lateral, appendiculate. *Leaves* 4-nate . . . . . Species 377–378.

§ 33. **Lamprotis.** *Inflorescence* terminal, rarely also pseudo-axillary; flowers 3-nate or clustered, capitate or spicate, calycine. *Corolla* mostly suburceolate or conical, more rarely subtubular or cyathiform, glabrous, $1\frac{1}{2}$–10 lin. long; segments spreading horizontally, recurved or erect. *Anthers* included, lateral, mostly appendiculate, rarely muticous. *Leaves* 3-nate or opposite, mostly adpressed and erect. All parts of the plants usually glabrous, smooth and glossy . . . . Species 379–398.

§ 34. **Eurystegia.** *Inflorescence* terminal, rarely also axillary; flowers usually 3-nate, rarely 4-nate or umbellate, calycine. *Sepals* large and prominent. *Corolla* various (in two species 4-partite nearly to the base), mostly $3\frac{1}{2}$–11 lin. long, rarely 2 lin. only. *Anthers* included, lateral or sublateral. *Leaves* 3-nate, mostly spreading, often long, rigid, and coarse. Species 399–408.

§ 35. **Adelopetalum.** *Inflorescence* terminal; flowers solitary or in pairs on short branchlets, usually in long dense leafy pseudo-racemes, more rarely somewhat lax on longer branchlets, calycine. *Sepals* in opposite pairs, about five times longer than the subovoid 4-fid corolla, and completely concealing it.   Species 409.

§ 36. **Trigemma.** *Inflorescence* terminal, occasionally also axillary; flowers 3-4-nate, rarely subumbellate, never capitate, calycine or rarely subcorolline. *Sepals* mostly large, prominent coloured. *Corolla* various, mostly glabrous, 1½–3½ lin. long; limb mostly somewhat spreading or recurved, not stellate-spreading. *Anthers* included, very rarely exserted. *Leaves* 3-4-nate, mostly spreading, seldom or never closely adpressed   .   .   .   Species 410–431.

SUBGENUS V. **PLATYSTOMA.**

§ 37. **Polycodon.** *Inflorescence* terminal; flowers 3-nate, mostly corolline or subcalycine, rarely calycine. *Corolla* obconic or broad-cyathiform, glabrous, dry, ¾–1¾ lin. long; segments equalling, or longer than, the tube, rarely a little shorter. *Anthers* lateral. *Leaves* 3-nate .   .   .   .   Species 432–439.

§ 38. **Eurystoma.** *Inflorescence* terminal; flowers usually 3-nate, mostly calycine or subcalycine, more rarely subcorolline or corolline. *Corolla* obconic, cyathiform or subcampanulate, wide-mouthed, very rarely subglobose-urceolate, glabrous, dry, 1–2½ lin. long. *Anthers* included or rarely subexserted, mostly manifest, not produced beyond the pore. *Leaves* 3-nate, very rarely opposite.
Species 440–450.

§ 39. **Melastemon.** *Inflorescence* terminal or sometimes also pseudo-axillary; flowers mostly 3-nate, more rarely umbellate or subcapitate, mostly corolline, rarely calycine or subcalycine, 1–4 lin. long, usually glabrous. *Corolla* mostly obconic or broad-cyathiform, wide-mouthed, rarely cyathiform. *Anthers* included, lateral, usually more or less produced at the apex beyond the pore.
Species 451-460.

§ 40. **Gamochlamys.** *Inflorescence* terminal, or sometimes also pseudo-axillary; flowers mostly 3-nate, rarely solitary, corolline or more rarely calycine. *Calyx* more or less gamosepalous. *Corolla* wide-mouthed, glabrous, dry, 1½–3 lin. long. *Anthers* mostly included, rarely subexserted, lateral, muticous, in one species produced beyond the pore. *Leaves* 3-nate .   .   .   Species 461–466.

§ 41. **Cyatholoma.** *Inflorescence* terminal, or sometimes pseudo-axillary; flowers 3-nate, subcalycine or subcorolline. *Corolla* subglobose below with large cup-shaped limb, or ovoid-urceolate, 2–5 lin. long. *Anthers* included, lateral, in two species subprognathous at the base. *Ovary* sessile or stipitate. *Leaves* 3-nate .   .   .   .   Species 467–469.

I. SYRINGODEA.

§ 1. GIGANDRA. Inflorescence terminal, or by arrest of the lateral branchlets sometimes pseudo-axillary; flowers 3-nate or solitary, at length cernuous. Bracts approximate, imbricate, or remote and small, sepal-like. Sepals scarious, rigid, imbricate at the base. Corolla mostly tubular or tubular-inflated, rarely elongate-ovoid, glabrous, dry; limb erect. Anthers long-exserted, terminal, linear, bifid, muticous. Ovary sessile, glabrous. Leaves 3-nate. Rigid, usually stout shrubs, 1–2 ft. high.

EXCEPTIONS : corolla elongate-ovoid in *E. scariosa;* less than 4½ lin. long in *E. Petiveri, var.* γ.

Bracts approximate, imbricate, usually large and prominent :
    Leaves 2–3½ lin. long, glabrous or nearly so, rigid   ...   (1) **Petiveri.**
    Leaves 4–5 lin. long, pubescent and pilose ...   .   (2) **vestiflua.**
Bracts remote, usually small :
    Sepals broad-ovate or oblong, flat, not keeled   ...   ...   (3) **lineata.**
    Sepals narrow-lanceolate to ovate, concave, keeled :
        Corolla 6 lin. long or more   ...   .   (5) **Plukeneti.**
        Corolla less than 5 lin. long   ...   .   (4) **scariosa.**

§ 2. DIDYMANTHERA. Inflorescence terminal or axillary; flowers 3-nate. Bracts approximate, sepal-like, adpressed, rigid, scarious, coloured. Corolla tubular, glabrous, dry. Anthers exserted, terminal or sublateral, often (or always ?) articulated to the filament; cells mostly divaricate. Ovary sessile. Erect or procumbent shrubs; leaves 3–4-nate or scattered.

Inflorescence terminal :
  Erect shrubs :
    Corolla 5 lin. long, crimson ... ... (6) **monadelphia.**
    Corolla 10–13 lin. long, green ... ... (9) **viridiflora.**
  Procumbent shrubs :
    Corolla wide, almost equally so from base to
      apex; filaments ½ lin. wide ... ... (7) **Banksia.**
    Corolla narrower, especially at the base, widen-
      ing upwards; filaments ¼ lin. wide or less ... (8) **primulina.**
Inflorescence axillary :
  Leaves 4-nate; anthers muticous, 1½ lin. long ... (10) **sphenanthera.**
  Leaves 6-nate or scattered; anthers denticulate,
    ½–⅔ lin. long ... ... (11) **cerviciflora.**

§ 3. PLEUROCALLIS. Inflorescence axillary, rarely both axillary and terminal; flowers mostly crowded below the ends of the branches, sometimes lax. Bracts mostly small, rarely somewhat large. Corolla usually tubular, from 6–12 lin. long, rarely shorter, glabrous or hairy, dry or viscid. Anthers usually included, rarely subexserted or exserted, lateral, rarely terminal, appendiculate or muticous. Leaves 6-nate, scattered or 4-nate. Stout shrubs, often of virgate habit, 1–3 ft. high.

Mr. Scott-Elliot (*Ann. Bot.* iv. 270) observes that "probably the whole *Pleurocallis* section is ornithophilous."

EXCEPTIONS : inflorescence terminal and axillary in *E. annectens ;* bracts rather large in *E. gilva,* and *E. sessiliflora ;* corolla short, 6 lin. or less in *E. filipendula var., coccinea var., conica, regia var., casta, nematophylla,* and *filamentosa ;* anthers subexserted in *E. longisepala, purpurea,* sometimes so in *E. coccinea, pinea,* and *vestita,* and exserted in *E. exsurgens, grandiflora,* sometimes so in *E. vestita.*

KEY TO SUBSECTIONS.

Anthers aristate; ovary glabrous ... ... (1) MAMMOSÆ.
Anthers muticous (in 32, sometimes minutely decurrent-
  denticulate) :
  Ovary glabrous, or at most puberulous :
    Ovary sessile, broadish at the base, mostly elon-
      gate ... ... (2) FLAMMULÆ.
    Ovary substipitate, or narrowed to the base,
      short ... ... (3) PINEÆ.
  Ovary villous, rarely puberulous :
    Sepals broadish, ovate-lanceolate to ovate ... (4) REGIÆ.
    Sepals narrow, linear to lanceolate ... ... (5) LONGIFOLIÆ.

† 1. Mammosæ.

Ovary sessile :
  Corolla more than 5 lin. long :
    Corolla more or less distinctly 4-foveolate at
      the base :
      Bracts about 1 lin. long; sepals 1½–2 lin.
        long ... ... (12) **mammosa.**
      Bracts 2–2½ lin. long; sepals about 3 lin.
        long ... ... (15) **gilva.**
    Corolla terete at the base or nearly so :
      Flowers in a very dense close spike ... ... (16) **sessiliflora.**
      Flowers not in a very dense spike :
        Corolla inflated below, tapering to a
          narrow mouth, red ... ... (13) **broadleyana.**

Corolla somewhat asymmetrically in-
flated medially, white :
　　Leaves 1½–3 lin. long, obtuse,
　　　spreading　　...　　...　　...　(14) **bowieana.**
　　Leaves 4–5 lin. long, acute, sub-
　　　erect　　...　　...　　...　　...　(15) **gilva.**
　　Corolla 3–4 lin. long, red or yellow ...　　...　　...　(17) **filipendula, var.**
Ovary stipitate, corolla of various length, white　　...　(17) **filipendula.**

† 2. *Flammulæ.*

Anthers subterminal; cells free; filaments forked at
　the apex :
　　Corolla-segments semiovate, bluntly acute ...　　...　(18) **grandiflora.**
　　Corolla-segments semilanceolate, very acute　　...　(19) **exsurgens.**
Anthers sublateral or lateral; cells not free to the
　base :
　　Corolla extremely viscid; sepals thick, leathery　...　(21) **Hibbertia.**
　　Corolla slightly viscid, or dry :
　　　Corolla tubular, 9–13 lin. long :
　　　　Sepals lanceolate, glabrous; ovary gla-
　　　　　brous　　...　　...　　...　　...　　...　(22) **purpurea.**
　　　　Sepals oblanceolate, villous; ovary his-
　　　　　pidulous ...　　...　　...　　...　　...　(23) **coccinea.**
　　　Corolla tubular, 5–8 lin. long :
　　　　Sepals linear-lanceolate, about equal to
　　　　　the corolla-tube　...　　...　　...　　...　(20) **longisepala.**
　　　　Sepals obcuneate, cuspidate, villous, much
　　　　　shorter than the corolla-tube　...　　...　(23) **coccinea, var. β.**
　　　Corolla obconic or subcampanulate, 3½–6 lin.
　　　　long　　...　　...　　...　　...　　...　　...　(24) **conica.**

† 3. *Pineæ.*

Ovary puberulous; sepals 2 lin. long　　...　　...　　...　(26) **hesseana.**
Ovary glabrous; sepals 2½–3½ lin. long :
　　Anthers dorsifixed at the base, ⅔–1 lin. long;
　　　flowers yellow or white　　...　　...　　...　　...　(25) **pinea.**
　　Anthers dorsifixed above the base, ½ lin. long;
　　　flowers red　...　　...　　...　　...　　...　　...　(27) **annectens.**

† 4. *Regiæ.*

Anthers prognathous; corolla 4–9 lin. long, colour
　various, seldom or never entirely white　　...　　...　(28) **regia.**
Anthers not, or scarcely, prognathous :
　　Corolla attenuate at the base, mostly white or rosy,
　　　inflated at the middle, mouth contracted, 4–8 lin.
　　　long ...　　...　　...　　...　　...　　...　　...　(29) **casta.**
　　Corolla-tube equal, bent, mouth not contracted,
　　　12 lin. long ...　　...　　...　　...　　...　　...　(30) **Mariæ.**

† 5. *Longifoliæ.*

Leaves ⅛–¼ lin. broad; petiole 1–2 lin. long by ¹⁄₁₂–⅛ lin.
　broad :
　　Corolla clavate-tubular, 8–12 lin. long, mouth
　　　more or less widened　　...　　...　　...　　...　(31) **vestita.**
　　Corolla clavate-tubular, 5 lin. long, mouth more
　　　or less contracted ...　　...　　...　　...　　...　(32) **nematophylla.**
　　Corolla obconic-tubular or subconic, 4–6½ lin. long,
　　　mouth much widened　　...　　...　　...　　...　(33) **filamentosa.**
Leaves distinctly broader; petioles shorter and broader :
　　Sepals linear; bracts linear and approximate　...　(34) **longifolia.**

Sepals lanceolate :
   Corolla trumpet-shaped ; mouth not contracted ;
      bracts approximate, long ... ... (34) **longifolia,** var. ζ.
   Corolla inflated ; mouth contracted ; bracts
      subremote, small ... ... (35) **onosmæflora.**

§ 4. EVANTHE. Inflorescence terminal ; flowers solitary or 2–4-nate, rarely clustered. Corolla mostly tubular, very rarely obconic or subcampanulate, at least $4\frac{1}{2}$ lin. long, usually longer. Anthers mostly included or more rarely sub-exserted or exserted, lateral, free, muticous or decurrent-denticulate or aristate, very rarely subcristate. Ovary sessile, glabrous or more rarely pubescent, 4- or more rarely 8-celled. Leaves 3–4-nate, rarely 4–6-nate. Usually erect, stout, woody shrubs, from 1–5 ft. high ; more rarely procumbent or slender.

Mr. G. F. Scott-Elliot believes all the large-flowered *Evanthes* to be almost certainly ornithophilous (*Ann. Bot.* iv. 270).

EXCEPTIONS : *E. verticillata,* var. β, has sometimes corollas less than $4\frac{1}{2}$ lin. long. Umbellate inflorescence may sometimes occur in *E. verticillata.*

Many of the species of this section are very variable and difficult of definition. Some, with normally 4-nate leaves, occasionally have 3-nate leaves ; more often the inflorescence varies according to the degree of luxuriance of growth, and the size of the corolla is frequently affected by the same cause. In using the key, therefore, it will often be found necessary to try under more headings than one.

A. LONGIFLORÆ. Corolla more than 6 lin. long.
1. *Ovary more or less pubescent.*

Anthers muticous :
   Sepals 3–4 lin. long ; filaments dilated at the base ... (58) **versicolor.**
   Sepals under 1 lin. long ; filaments not dilated at
      the base ... ... (40) **xanthina.**
Anthers appendiculate :
   Sepals broad-ovate ; leaves 3–4-nate, subglabrous... (65) **mertensiana.**
   Sepals ovate-lanceolate ; leaves 3-nate, cano-pu-
      bescent ... ... (41) **Maximiliani.**
   Sepals lanceolate-linear ; leaves 3-nate, shortly
      pilose... ... ... (67) **wendlandiana.**

2. *Ovary glabrous.*

* Flowers solitary (in 44 ζ unknown) :
   Anthers aristate, awns free :
      Leaves 3-nate :
         Awns shorter than the anther-cell :
            Leaves linear to lanceolate, $1\frac{1}{2}$–$2\frac{1}{2}$ lin.
             long ; sepals lanceolate or oblong,
             3–$3\frac{1}{2}$ lin. long ... ... (49) **densifolia.**
            Leaves linear-subfiliform ; sepals sub-
             linear, 2 lin. long ... ... (50) **xerophila.**
         Awns usually much longer than the anther-
           cell ... ... (51) **speciosa.**
      Leaves 4-nate ; awns as long as or exceeding
         the anther-cell :
         Sepals 2 lin. long ; leaves 2–$2\frac{1}{2}$ lin. long ... (37) **sacciflora.**
         Sepals 4–6 lin. long ; leaves 4–6 lin. long (36) **abietina.**
      Leaves 4-nate ; awns shorter than the anther-
         cell ... ... (45) **curviflora,** var γ.
   Anthers muticous or minutely decurrent-awned or
      toothed :
      Sepals $\frac{1}{2}$ lin. long, ovate or subcuneate ... (46) **sulcata.**
      Sepals under 1 lin. long, or sometimes a little
         longer ... ... (40) **xanthina.**

Sepals 1½–4 lin. long :
  Anthers distinctly curved forwards at the
    base ... ... (48) **conspicua.**
  Anthers not curved forwards at the base :
    Corolla-tube subequal in diam. :
      Sepals under 1 lin. long ; bracts
        minute ; corolla-tube yellow ... (40) **xanthina.**
      Sepals 3–4 lin. long ; bracts large ;
        corolla-tube red .. ... (57) **MacOwanii.**
    Corolla globosely swollen below the
      throat, glabrous, white to rosy ... (63) **colorans.**
    Corolla somewhat inflated at the base
      and apex, yellow ... ... (43) **dubia.**
    Corolla gradually widened to the
      mouth, rarely subequal :
      Sepals and tube of yellow corolla
        glabrous ; filaments attached at
        a right-angle to connective ... (44) **bibax.**
      Sepals, corolla, and filaments not so :
        Ovary 8-celled ... ... (62) **perspicua.**
        Ovary 4-celled ; sepals very
          variable in shape and size ... (45) **curviflora.**
          Corolla-limb 2–3½ lin. long (45) **curviflora, var. β.**
          Corolla-limb mostly less
           than 2 lin. long :
            Branches stout, erect ;
            leaves densely im-
            bricate :
              Anthers affixed
              about the mid-
              dle of the cell,
              appendiculate . (45) **curviflora, var. γ.**
              Anthers affixed
              below the middle
              of the cell, mu-
              ticous :
                Corolla about
                19 lin. long,
                nearly glab-
                rous, red ... (45) **curviflora, var. ζ.**
                Corolla under
                14 lin. long,
                densely villous,
                yellow ... (45) **curviflora, var. δ.**
            Branches diffuse and
             slender ; leaves sub-
             distant ... ... (45) **curviflora, var. ε.**
** Flowers 2-3-4 nate (in 64, sometimes umbellate, in
48 and 62 usually, in others, occasionally, solitary) :
  † Anthers muticous :
    Anthers affixed near the middle of the cell by
      the thick connective ; ovary 8-celled ... (64) **verticillata.**
    Anthers affixed near the base of the cell, and
      as long as the filament ; corolla hairy within (70) **brachialis.**
    Anthers, ovary and corolla not as in the two
      foregoing :
      ‡ Corolla dry (or in 64, viscidulous with
        glandular hairs) :
        Ovary 4-celled :
          Anthers straight or nearly so :
            Sepals 1–2 lin. long :

    Leaves 4-nate; anthers ½ lin.
      long :
        Corolla pubescent, yellow ;
          limb connivent...   ... (40) **xanthina.**
        Corolla glabrous, blood-red;
          limb erect or spreading . (68) **hæmatosiphon.**
    Leaves 3-nate ; corolla 11–14
      lin. long ; anthers 1–1½ lin.
      long ...   . (58) **versicolor.**
    Leaves 3-nate ; corolla 8–10 lin.
      long ; anthers ¾–1 lin. long . (66) **cruenta, var. β.**
    Sepals exceeding 2 lin. long :
      Leaves 3-nate ; flowers 3–6-
      nate, 10 lin. long   ...   ... (42) **pallens.**
      Leaves 3–4-nate; flowers 1–2-
      nate, 8 lin. long   ...   ... (43) **dubia.**
      Leaves 4-nate :
        Leaves glabrous; corolla
          about 19 lin. long   ... (45) **curviflora, var. ζ.**
        Leaves plumose-ciliate, or
          serrulate ; corolla 12 lin.
          long   ...   . (56) **serratifolia.**
    Anthers curved forwards at the base:
      Flowers usually 4-nate, gland-
      hairy ; anthers 1 lin. long .. (61) **glandulosa.**
      Flowers usually solitary; not
      gland-hairy ; anthers 1¼–1½
      lin. long   ...   . (48) **conspicua.**
    Ovary 8-celled ; corolla usually villous
      or pubescent   ...   . (62) **perspicua.**
  ‡‡ Corolla viscid :
    Sepals 2 lin. long :
      Anthers 1¼–1½ lin. long; leaves
      squarrose to recurved ...   .. (59) **berzelioides.**
      Anthers 2 or more lin. long ;
      leaves suberect, imbricate   ... (60) **diaphana.**
    Sepals 3–5 lin. long :
      Leaves 3-nate   ...   . (58) **versicolor.**
      Leaves 4-nate   ...   . (54) **dichrus.**
 †† Anthers appendiculate :
  Anthers ½ lin. long or less :
    Appendages shortly subulate, scarcely
      reaching to the base of the cell ...   ... (47) **macropus.**
    Appendages decurrent-denticulate on the
      filament ...   . (63) **colorans.**
    Appendages of free awns, as long as or
      longer than the cell   ...   . (38) **foliacea.**
  Anthers ¾–1 lin. long :
    Awns of anther about as long as or longer
      than the cell :
      Procumbent, dwarf; ovary sub-
      stipitate or constricted   ...   ... (39) **nana.**
      Erect, tall ; ovary sessile, broad-based (38) **foliacea.**
    Awns much shorter :
      Base of the anthers hollowed at the
      approximation of the cells :
        Leaves 3-nate; awns of anther
        ⅓–⅔ of length of cell ...   ... (53) **discolor.**
        Leaves 4-nate; awns minute,
        scarcely projecting beyond the
        cells   ...   . (54) **dichrus.**

Base of the anthers not, or scarcely,
hollowed :
　Anthers oblong, affixed to the
　　filament above their base;
　　bracts mostly small :
　　　Leaves ciliate with plumose
　　　　hairs; flowers yellow　...　(56) **serratifolia.**
　　　Leaves naked; flowers blood-
　　　　red...　　...　　...　　..., (66) **cruenta.**
　Anthers linear, affixed near their
　　base; bracts largish　...　...　(55) **unicolor.**
Anthers 1¼–2 lin. long :
　Leaves 4–6-nate; corolla contracted at
　　the throat...　...　...　...　...　(69) **chloroloma.**
　Leaves 3-nate :
　　Awns of anthers nearly as long as the
　　　cell ; cells blunt at the base　...　(51) **speciosa.**
　　Awns of anther shorter; cells mostly
　　　with a sharp projecting basal
　　　point :
　　　Sepals tomentose...　　...　　...　(52) **hebecalyx.**
　　　Sepals glabrous ...　　...　　...　(60) **diaphana.**

　　　B.　BREVIFLORÆ.　Corolla less than 6 lin. long.

Leaves 3-nate　　...　　...　　...　　...　　...　　...　(66) **cruenta,**vars. γ, δ.
Leaves 4-nate :
　Ovary 8-celled; connective of anther projecting
　　dorsally　　...　　...　　...　　...　　...　　...　(64) **verticillata,**var. β.
　Ovary 4-celled ; connective of anther not pro-
　　jecting　　...　　...　　...　　...　　...　　...　(61) **glandulosa,** var. β.

§ 5. DASYANTHES.　Inflorescence terminal; flowers mostly umbellate, or subcapitate.　Bracts approximate or subremote.　Corolla tubular, rarely subovoid-tubular, more or less hairy, sometimes setose-hispid, never quite glabrous,
mostly dry, rarely subviscidulous.　Anthers included or exserted, lateral, free,
muticous or aristate.　Ovary sessile, usually villous, or rarely puberulous, never
glabrous, 4-celled.　Leaves usually 4-nate, more rarely 3–6–∞-nate.　Rigid
woody shrubs, 1–4 ft. high.

Ovary villous :
　Sepals, bracts and leaves pectinate-ciliate :
　　Sepals and bracts oblong-linear　　...　　...　(71) **strigilifolia.**
　　Sepals and bracts unguiculate　　...　　...　(72) **pectinifolia.**
　Sepals and bracts not pectinate-ciliate :
　　Sepals and bracts thickly setose-hispid with
　　　long pallid hairs　　...　　...　　...　　...　(73) **Sparrmanni.**
　　Sepals and bracts not thickly setose-hispid :
　　　Anthers aristate :
　　　　Corolla 5–6½ lin. long　　...　　...　(76) **Oatesii.**
　　　　Corolla 6–16 lin. long (mostly
　　　　　10–12) ; anthers often muticous. (77) **cerinthoides.**
　　　Anthers cristate-aristate; awns subulate,
　　　　curved and lacerate　　...　　...　　...　(75) **splendens.**
　　　Anthers muticous or very minutely
　　　　aristate...　　...　　...　　...　　...　(77) **cerinthoides.**
Ovary puberulous ; anthers shortly decurrent-aristate (74) **doliiformis.**

§ 6. CHONA.

　Only species...　　...　　...　　...　　...　　...　(78) **embothriifolia.**

§ 7. BACTRIDIUM.　Inflorescence terminal and umbellate, or pseudo-axillary
and subverticillate by innovation of the branches.　Corolla tubular or inflated-

tubular, viscid. Anthers included, lateral. Ovary narrow, elongate, long-stipitate, glabrous.

The species are probably fertilized by birds. (*Cf. Scott-Elliot, Annals of Botany,* iv. 270).

Tall and virgate; leaves linear, 6–8 lin. long ... ... (79) **fascicularis.**
Shorter, more branched; leaves oblong, 2 lin. long ... (80) **Massoni.**

II. STELLANTHE.

§ 8. EURYLOMA. Inflorescence terminal; flowers generally umbellate. Corolla mostly ampullaceous, sometimes suburceolate, rarely tubular, contracted at the throat, usually viscid; limb spreading, often large. Anthers included, lateral, either subcrescent-shaped, prognathous and distinctly bilobed at the base, or less commonly nearly straight, equal and scarcely lobed, muticous or appendiculate. Ovary more or less distinctly stipitate, or elongate, usually glabrous. Leaves 3–4-nate, adpressed, squarrose or recurved. Usually glabrous shrubs.

1. *Anthers distinctly prognathous, also crescent-shaped and bilobed at the base.*

Corolla mostly ampullaceous, or at least tubular-inflated
  below :
    Anthers aristate; ovary short-stipitate; corolla 9–20
      lin. long ... ... (87) **Junonia.**
    Anthers muticous :
      Leaves 3-nate :
        Corolla ampullaceous, 12–15 lin. long :
          Corolla tapering into a long thin
            neck; leaves 3–4 lin. long ... ... (85) **lagenæformis.**
          Corolla with a shorter and thicker
            neck; leaves 4–6 lin. long ... ... (88) **shannonea.**
        Corolla subampullaceous or suburceolate,
          4-10 lin. long ... ... (90) **irbyana.**
      Leaves 4-nate :
        Corolla-tube 4 lin. long; leaves nearly erect (81) **Gysbertii.**
        Corolla-tube much longer; leaves spreading
          or recurved :
          Sepals pubescent; corolla-segments very
            acute ... ... (84) **retorta.**
          Sepals glabrous; corolla-segments obtuse
           or retuse :
             Corolla-segments wider than long,
              retuse ... ... (83) **aristata.**
            Corolla-segments longer than wide,
               obtuse ... ... (89) **ampullacea.**
Corolla-tube cylindrical, not inflated; limb very large ... (86) **jasminiflora.**

2. *Anthers not distinctly prognathous, straight or nearly so, obscurely bilobed at the base.*

Anthers appendiculate :
    Corolla 5–6 lin. long; anthers broad-cuneate; awns
      nearly as long as the cell ... ... (92) **pectinata.**
    Corolla 3–4 lin. long; anthers oblong or semiovate;
      awns much shorter ... ... (93) **trichroma.**
    Corolla 2–2¼ lin. long; awns of anther about equalling
      the cell ... ... (95) **rhodopis.**
Anthers muticous :
    Leaves 3-nate, minutely gland-ciliolate or serrulate :
      Leaves 2–3 lin. long; corolla 4–8½ lin. long ... (91) **curvifolia.**
      Leaves ¾–1 lin. long; corolla about 2 lin. long ... (94) **tubercularis.**
    Leaves 4-nate or sub-4-nate, ciliate with long hairs ... (82) **squarrosa.**

§ 9. **CERAMUS.** Inflorescence terminal; flowers umbellate or 3–4-nate, rarely pseudo-corymbose. Bracts usually remote, or subremote and small, rarely sub-approximate and rather large. Sepals narrow, small, rarely equalling the corolla-tube. Corolla ovoid-urceolate, or obovoid-urceolate, rarely urceolate, glabrous, dry; throat mostly much constricted. Anthers included, lateral, mostly equal at the base, rarely prognathous or slightly so, usually appendiculate, more rarely muticous. Ovary stipitate or at least narrowed at the base, glabrous. Leaves 3-nate, 4-nate or scattered.

EXCEPTIONS: anthers prognathous, or slightly so, in *E. incarnata.*

Leaves 3-nate ...　　...　　...　　...　　...　　... (96) **incarnata.**
Leaves 4-nate :
  Leaves 5–10 lin. long :
    Virgate; leaves **very** slender; corolla
    3–4 lin. long　　...　　...　　...　　... (99) **inflata.**
    Much branched; corolla 5–10 lin. long ... (100) **ventricosa.**
  Leaves 1–4 lin. long :
    Anthers very minutely cristate; leaves
    2–3 lin. long　　...　　...　　...　　... (97) **Savilea.**
    Anthers muticous; leaves 3–4 lin. long ... (98) **præcox.**

§ 10. **CALLISTA** (*Stellanthe*, Salisb.). Inflorescence terminal; flowers usually 4-nate, rarely solitary. Bracts approximate. Sepals linear, lanceolate-oblong, ovate, obovate or spathulate, mostly ciliate, denticulate or lacerate, scarious. Corolla narrow-tubular, urceolate, or narrow-ovoid-urceolate, from 2–10 lin. long, mostly glabrous and dry, the spreading limb often covered on the upper surface by a pallid thickish mealy film. Anthers included, lateral, basi-fixed or subdorsifixed, muticous or very rarely minutely aristulate, usually $\frac{1}{4}$–$\frac{1}{2}$ lin. long. Ovary sessile, rarely elongate, glabrous. Leaves 4-nate, mostly linear, rárely broader, usually glabrous. Erect, rigid shrubs with glabrous rarely pubescent branches.

The section is a natural one; but the last four or five species are difficult of definition and limitation.

EXCEPTIONS: flowers usually solitary in *E. Lawsonia;* ovary elongate, some-what contracted at the base in *E. cylindrica.*

Corolla tubular, not, or scarcely inflated; sepals linear
  or lanceolate-linear :
    Flowers solitary, or generally so ...　　...　　... (101) **Lawsonia.**
    Flowers 4-nate, or sub-4-nate :
      Sepals less than half as long as the corolla-
        tube :
        Corolla-tube 8-10 lin. long, about $\frac{1}{2}$ lin.
          wide, slightly inflated at the throat ... (102) **pavettæflora.**
        Corolla-tube 6–7 lin. long, about $\frac{3}{4}$ lin.
          wide, slightly contracted at the throat (103) **cylindrica.**
      Sepals from $\frac{1}{2}$–$\frac{5}{6}$ the length of the corolla-
        tube :
        Corolla-tube not inflated, 4 lin. by 1 lin. (104) **fastigiata.**
        Corolla-tube slightly inflated, 5–6 lin.
          by 1–1$\frac{1}{2}$ lin. :
          Corolla-segments ovate-oblong, 2
           lin. long　　...　　...　　...　　... (104) **fastigiata,** var. γ.
          Corolla-segments ovate-lanceolate,
           3 lin. long ...　　...　　...　　... (104) **fastigiata,** var. β.
  Corolla-tube scarcely inflated; sepals lanceolate or
    ovate-lanceolate :
    Corolla-tube 2–2$\frac{1}{2}$ lin. long　　...　　...　　... (105) **transparens.**
    Corolla-tube 3$\frac{1}{2}$–7 lin. long :
      Sepals 5–5$\frac{1}{2}$ lin., subequal to corolla-tube;
        corolla-segments 3 lin. by 2$\frac{1}{2}$ lin. ...　　... (106) **Vallis-Gratiæ.**

Sepals about ½ the length of the corolla-tube
   or less :
      Corolla-tube 4–5 lin. by 1½ lin. ; sepals
         about 2 lin. long   ...   . (109) **daphniflora,** var. ζ.
      Corolla-tube 7 lin. by 1½ lin. ; sepals
         about 3 lin. long   ...   . (107) **prænitens.**
Corolla-tube usually distinctly inflated below :
   Sepals not distinctly widest above the middle :
      Sepals subulate-linear, 6 lin. long   ...   ... (104) **fastigiata,** var. γ.
      Sepals broader, tapering from middle to
         apex ...   . (109) **daphniflora,** and
      Sepals oblong, elliptical or ovate-lanceolate,   vars. β, γ, and ζ.
        not tapering from middle to apex   ... (109) **daphniflora,** vars.
   Sepals, and generally the bracts also, distinctly   β, γ, δ, ε.
     widest above the middle, and more or less
     lacerate-edged :
      Sepals only slightly lacerate, or denticulate ;
      leaves usually spreading-incurved :
         Sepals with a long foliaceous cusp;
          corolla 5–6 lin. long   ...   . (110) **pellucida.**
         Sepals not so, or with a very short cusp;
          corolla 3–4 lin. long   ...   . (108) **Walkeria.**
      Sepals and bracts usually deeply pectinate-
      lacerate or fimbriate ; leaves usually erect
      or ascending ...   . (111) **denticulata.**

§ 11. PLATYSPORA. Inflorescence axillary or subaxillary ; flowers sometimes
arranged in long pseudo-racemes, or dense pseudo-spikes or heads, corolline or
rarely subcalycine. Bracts and sepals scarious or glumaceous, the former always
smaller, the sepals more or less distinctly imbricate and in opposite pairs, upper
pair somewhat narrower, all white or pallid. Corolla suburceolate, subsalver-
shaped, or ovoid, contracted at the throat, from 3–8½ lin. long ; limb mostly
spreading or recurved, sometimes large (with a general resemblance to that of
§§ *Callista* and *Lamprotis*). Filaments capillary, rarely rather broad. Anthers
included, lateral, shortly aristate or muticous. Style included ; stigma simple or
capitellate ; ovary mostly glabrous, rarely puberulous ; ovules flat or more or
less broadly-winged, or with a membranous margin, or lenticular and not
marginate. Leaves 3- or 4-nate, always narrow, acute or acuminate, glabrous.
Branches mostly virgate.

EXCEPTION : the flowers are subcalycine in *E. albens*.

Allied on various sides to the §§ *Euryloma*, *Callista*, and *Lamprotis*. From
the two former it is distinguished generally by the flat or compressed, mostly margined or winged, ovules and seeds. The character is not, however, always
capable of verification, and sometimes not well-marked. Similar seeds have been
found in other species, as in *E. dianthifolia*, *E. chlorosepala* (and perhaps in
others) which, owing to the predominance of other characters, have been placed
elsewhere. From § *Lamprotis* this section is distinguished by its usually
corolline flowers.

Sepals ovate or lanceolate :
   Pedicels about 5 lin. long ; inflorescence sub-
     corymbose   ...   . (112) **astroites.**
   Pedicels about 2 lin. long ; inflorescence pseudo-
     spicate or racemose :
      Flowers in a long lax pseudo-raceme, often
       secund, white   ...   . (113) **albens.**
      Flowers in a short crowded spike or head, red (114) **georgica.**
Sepals narrow-lanceolate :
   Leaves 4-nate ; anthers muticous ; seeds scarcely
     margined ...   . (115) **macilenta.**

Leaves 3- (or occasionally 4-) nate :
    Corolla-tube 4 lin. by 1 lin., 4-gonous; anthers
      muticous; seeds not (or narrowly) margined ... (116) **tetragona.**
    Corolla-tube 3 lin. by 1½ lin., cylindrical; anthers
      minutely aristate; seeds broadly winged   ... (117) **heliophila.**

§ 12. MYRA. Inflorescence axillary, in at least two species strictly racemose, or sometimes terminal and umbellate. Bracts small and inconspicuous, remote or approximate. Sepals small, glabrous or gland-hispid. Corolla subtubular, urceolate, rarely subampullaceous, mostly pubescent. Anthers generally included, lateral, appendiculate. Ovary sessile or shortly stipitate, glabrous or rarely thinly puberulous; limb mostly flattish and stellato-patent. Leaves 3–4-nate below, scattered on the upper branches, irregular and often distant. Well characterized by the branches, leaves, pedicels, and sometimes the bracts and sepals, being more or less densely hispid and viscid with gland-tipped hairs.

EXCEPTION : corolla-segments spreading, but scarcely stellato-patent in *E. armata.*

Sepals glabrous; anthers crested :
    Corolla-tube about 4 lin. long; anthers cuneate-
      oblong ...   . (121) **glutinosa.**
    Corolla-tube 7–12 lin. long; anthers narrow-cuneate.. (120) **rufescens.**
Sepals gland-hispid; anthers crested :
    Corolla-tube 6–8 lin. long; pedicels 4–6 lin. long   ... (119) **irrorata.**
    Corolla-tube 3–4 lin. long; pedicels 1½–2 lin. long ... (118) **glandulifera.**
Sepals gland-hispid; anthers aristate   ...   . (122) **armata.**

III. EUERICA, BENTH.

§ 13. EPHEBUS (*Lasianthæ*, Bartl.). Inflorescence mostly terminal, rarely terminal and axillary; flowers usually 3–4-nate, sometimes solitary, rarely clustered or umbellate, corolline rarely subcalycine. Sepals usually small and inconspicuous. Corolla variously shaped, rarely over 3 lin. long, most usually with indumentum, dry or rarely viscid. Anthers usually included, rarely exserted or subexserted, lateral, appendiculate or muticous. Leaves 3–4-nate. Generally erect and rigid shrubs.

EXCEPTIONS : inflorescence terminal and lateral in *E. catervæflora, Atherstonei, aggregata, globosa,* and sometimes so in *E. podophylla;* clustered or subcapitate in *E. catervæflora, distorta, turgida, nidularia, aggregata;* flowers solitary, or occasionally so, in *E. podophylla, propendens, pyramidalis, chrysocodon, trichophora* and *cyrillæflora;* subcalycine in *E. podophylla* and *E. Lerouxiæ;* corolla glabrous in vars. of *E. Alopecurus, pyramidalis, catervæflora, intervallaris, oresigena,* or becomes glabrous in *E. Atherstonei;* and exceeds 3 lin. in length in *E. propendens, pyramidalis, chrysocodon, oresigena, Flanagani,* and *coffra;* anthers exserted in *E. turgida* and subexserted in *E. algida, Atherstonei, perlata,* and *globosa;* ovary sometimes 8-celled in *E. propendens.*

The shape of the corolla is necessarily taken in most cases from dried or boiled specimens. In either state it often appears wider at the mouth than in the living plant, and, notwithstanding care, may be sometimes thus described.

I. SULCATÆ. Leaves sulcate, or only occasionally subopen-backed.

1. *Leaves 3-nate.*

Inflorescence a dense narrow pseudo-spike, 1–3 in. long ... (133) **Alopecurus.**
Inflorescence not a dense narrow pseudo-spike :
    Corolla-mouth more or less widened at maturity :
      Ovary glabrous :
        Leaves subpungent; corolla dark-purple, viscid-
          pubescent   ...   . (136) **sicæfolia.**
        Leaves not pungent; corolla white or pallid,
          very minutely puberulous   ...   . (134) **æmula.**

Ovary pubescent :
    Corolla densely tomentose ...   . (124) **Peziza.**
    Corolla minutely puberulous   ...   ... (160) **trichadenia.**
Corolla scarcely widened or contracted at the mouth
  or throat :
    Ovary never glabrous :
      Bracts remote and small :
        Corolla densely woolly and shaggy ... (125) **ovina.**
        Corolla hirsute, pubescent or puberu-
          lous :
          Anthers cristate or subcristate :
            Corolla strongly 4-nerved,
              pulverulent   ...   ... (138) **Atherstonei.**
            Corolla not nerved :
              Anthers broad-cristate ;
                petioles 1 lin. or more
                long ...   . (137) **podophylla.**
              Anthers narrow-cristate ;
                petioles ½ lin. long or
                less ...   . (129) **albescens.**
          Anthers aristate or subaristate :
            Sepals and young leaves
              copiously long-setose ; co-
              rolla 3 lin. long ...   ... (130) **oxyandra.**
            Sepals and leaves not long-
              setose ; corolla 1½ lin. long :
              Anthers over ⅓ lin. long (129) **albescens.**
              Anthers ⅕ lin. long   ... (147) **parviflora,** var. ζ.
      Bracts approximate, 1 lin. or more long :
        Corolla 2–2½ lin. long ; sepals 1½ lin.
          long ; anthers nearly ½ lin. long ... (131) **dysantha.**
        Corolla 1 lin. long ; sepals ¾ lin. long ;
          anthers ⅕ lin. long ...   . (161) **eriocodon.**
    Ovary glabrous ; corolla 3–4 lin. long ...   ... (154) **modesta.**
    Ovary sometimes glabrous ; corolla 1–1½ lin. long (147) **parviflora,** var.
Corolla contracted at the throat :
    Bracts approximate :
      Whole plant roughly hispid; sepals ½ lin. long (128) **Constantia.**
      Plant not roughly hispid ; sepals 1½–2 lin.
        long :
        Straggling plant ; leaves distant ;
          sepals linear ...   . (135) **auriculata.**
        Erect plant ; leaves imbricate ; sepals
          ovate ...   . (154) **modesta.**
    Bracts remote and small :
      Corolla-segments recurved, tips dark-
        coloured   ...   . (127) **pubigera.**
      Corolla-segments not recurved :
        Sepals nearly as long as the corolla ... (132) **dilatata.**
        Sepals much shorter than the corolla :
          Corolla globose-urceolate ; mouth
            much contracted   ...   ... (123) **nivalis.**
          Corolla urceolate or ovoid-urceo-
            late ; mouth not much con-
            tracted :
            Sepals 1–1¼ lin. long, ⅓ to
              nearly ½ as high as corolla (170) **pubescens.**
            Sepals ½ lin. long, ¼ as high
              as corolla ...   . (126) **tomentosa.**
          Corolla oval or cyathiform, densely
            woolly and shaggy   ...   ... (125) **ovina.**

2. Leaves 4-nate.

* Corolla-mouth widened; ovary glabrous :
 Anthers muticous :
  Corolla 1½ lin. long; sepals ½ lin. long   ... (143) **setulosa.**
  Corolla 3–5 lin. long; sepals 1–1½ lin. long :
   Bracts approximate; sepals obovate, gla-
    brous, scarious ...   . (139) **propendens.**
   Bracts remote; sepals pubescent :
    Sepals linear or acuminate from an
     ovate coloured base :
      Leaves 2½–3½ lin. long; corolla
       glabrous or puberulous   ... (140) **pyramidalis.**
      Leaves 1–1½ lin. long; corolla
       softly pilose   ...   . (142) **trichophora.**
     Sepals ovate or ovate-lanceolate;
      corolla funnel-shaped   ...   ... (141) **chrysocodon.**
 Anthers appendiculate :
  Anthers oblong, subcristate, ⅓ lin. long, pore
   about ½ as long as the cell   ...   . (145) **distorta.**
  Anthers oblong, aristate, ⅕ lin. long, pore about
   ½ as long as the cell ...   . (144) **pusilla.**
  Anthers broad-oblong, aristate, 1/10 lin. long,
   pore about ¾ as long as the cell   ...   ... (144) **pusilla,** var. β.
** Corolla-mouth scarcely widened or contracted :
 † Ovary glabrous; bracts remote and small :
  Corolla tetragonous or subtetragonous :
   Flowers densely clustered; corolla 2½ lin.
    long   ...   . (146) **catervæflora.**
   Flowers not densely clustered; corolla
    1½–2 lin. long   ...   . (148) **intervallaris.**
  Corolla not tetragonous :
   Anthers appendiculate :
    Corolla semiglobose-urceolate or broad-
     cyathiform, 1–1¼ lin. long ...   ... (145) **distorta.**
    Corolla ovoid-urceolate, or cyathiform
     or subtubular, 1½–2 lin. long :
      Corolla mostly ovoid-urceolate;
       hairs usually very rough;
       anthers ovate-cuneate   ... (156) **hirtiflora.**
      Corolla mostly narrow-cyathiform
       or subtubular; hairs less rough
       than in the above; anthers
       oblong ...   . (147) **parviflora.**
   Anthers muticous :
    Corolla subtubular or tubular-cam-
     panulate, 3 lin. long   ...   ... (149) **cyrillæflora.**
    Corolla suburceolate, 2 lin. long   ... (147) **parviflora,** var. γ.
 †† Ovary with indumentum :
  Bracts subapproximate, one longer and folia-
   ceous :
    Anthers ovate-oblong, mostly with a sharp
     anterior basal point   ...   . (151) **turgida.**
    Anthers obovate-oblong, without a sharp
     anterior basal point   ...   . (150) **algida.**
  Bracts remote; anthers longitudinally semi-
   ovate   ...   . (152) **nidularia.**
*** Corolla-mouth more or less contracted; corolla
 mostly urceolate :
  Bracts approximate or subapproximate :
   Corolla 3½–4 lin. long; very stout shrub   ... (155) **Flanagani.**

Corolla 2 lin. long; flowers few; branches with
long internodes ... ... (135) **auriculata.**
Corolla 1½ lin. long; flowers numerous; inter-
nodes short ... ... (151) **turgida.**
Bracts remote and small:
Corolla-segments acuminate; leaves ovoid,
thick, short ... ... (159) **oophylla.**
Corolla-segments not acuminate; leaves not
ovoid, thick and short:
Sepals 2 lin. or more long, gland-ciliate,
or soft subplumose-ciliate ... ... (153) **oresigena.**
Sepals 1 lin. or less long, not gland-ciliate:
Anthers with a sharp anterior basal
point ... ... (151) **turgida.**
Anthers without a sharp anterior basal
point:
Corolla-limb spreading, nearly ½
as long as the tube; stigma subpeltate ... ... (157) **mollis.**
Corolla-limb erect or rarely
spreading, much less than ½ the
length of the tube; stigma
capitate:
Sepals subequal with the
corolla ... ... (132) **dilatata.**
Sepals half as long as the
corolla or less:
Anthers ovate-cuneate;
awns long-ciliate:
Corolla puberulous,
bright crimson ... (158) **ribisaria.**
Corolla roughly tu-
berculate-hirsute,
pale purple or rosy (156) **hirtiflora.**
Anthers oblong; awns
rough-edged or short-
ciliolate ... ... (147) **parviflora.**

II. CERAMIOIDEÆ.   Leaves more or less open-backed and spreading.

1. *Leaves 3-nate.*

Bracts approximate:
Leaves ovate; sepals foliaceous, green, short-ciliate (167) **straminea.**
Leaves lanceolate; sepals coloured, at least in the
lower part:
Corolla-mouth wide; anthers subincluded,
manifest:
Sepals 1 lin. long ... ... (166) **setosa.**
Sepals 1½ lin. long ... ... (165) **fausta.**
Corolla moderately contracted or nearly equal
at the throat:
Sepals 1¾–2 lin. long, reaching to ½ the height
of the corolla; filaments broad, ciliate (168) **Lerouxiæ.**
Sepals 1–1¼ lin. long, reaching to ¼-⅓ the
height of the corolla; filaments narrow,
naked ... ... (170) **pubescens.**
Bracts remote, small:
Leaves open-backed, narrow, whether linear, oblong
or lanceolate:
Petioles 1 lin. long or more, ciliate (leaves
mostly sulcate only) ... ... (137) **podophylla.**

Petioles ½ lin. long or less :
    Anthers muticous ...   . (161) **eriocodon.**
    Anthers appendiculate :
        Corolla-mouth distinctly contracted :
            Corolla 3–3½ lin. long, long per-
                persistent :
                Anthers distinctly aristate ;
                    leaves up to 6 lin. long ... (169) **caffra.**
                Anthers obscurely aristate ;
                    leaves much shorter   ... (169) **caffra, var. β.**
            Corolla 2 lin. long ; leaves 2–3 lin.
                long   ...   . (172) **aggregata.**
        Corolla-mouth scarcely widened or
            contracted :
            Sepals broad-lanceolate, 1 lin.
                long ; corolla ovoid-urceolate .. (170) **pubescens.**
            Sepals linear, 1½ lin. long ; corolla
                broad-urceolate   ...   ... (171) **hirta.**
        Corolla-mouth widened ; leaves fim-
            briate-ciliate   ...   . (160) **trichadenia.**
Leaves open-backed, ovate or at least broad :
    Corolla-limb subequal to the tube ; mouth
      widened ...   . (162) **patens.**
    Corolla-limb much shorter than the tube ;
      mouth not widened :
        Corolla about 3 lin. long ; leaves broad,
          ovate to elliptical   ...   . (174) **marifolia.**
        Corolla 2–2½ lin. long ; leaves linear to
          narrow-ovate :
            Dwarf, diffuse ; pubescence silvery ... (175) **argyræa.**
            Taller, erect ; pubescence not silvery :
                Flowers crowded at the ends of
                  the branches ; pubescence fine
                  and soft, not glandular   ... (172) **aggregata.**
                Flowers not crowded ; pubescence
                  usually more or less coarsely
                  glandular-viscid :
                    Anthers oblong, 3 times as
                    long as broad   ...   ... (171) **hirta.**
                  Anthers linear, 4–5 times as
                    long as broad   ...   ... (173) **globosa.**

    2. *Leaves 4-nate (linear or linear-lanceolate).*

Anthers cristate, included but often manifest   ...   ... (164) **barbata.**
Anthers minutely aristate, subexserted ...   . (163) **perlata.**

§ 14. CERAMIA. Inflorescence variable, sometimes so on the same plant,
terminal or axillary ; flowers commonly 3–4-nate, umbellate or rarely subcapitate or solitary, corolline, rarely subcalycine. Sepals usually small and
inconspicuous. Corolla variously shaped, rarely exceeding 2½ lin. in length,
mostly glabrous or less commonly pubescent, or puberulous, dry or viscid.
Anthers exserted or included, lateral or rarely terminal or subterminal, appendiculate or muticous. Leaves 3-nate or less commonly 4-nate. Diffuse or often
weakly-growing shrubs, less commonly erect but then generally slender, very
rarely stout and rigid.

Notwithstanding the inconstancy of many of the sectional characters, this
section is a fairly natural one, and will give little trouble. Species not readily
found here may be sought for in § *Ephebus.*

EXCEPTIONS : flowers subcalycine in *E. cordata* and *E. Lehmannii ;* subcorolline in *E. oreophila ;* the corolla exceeds (at least occasionally) 2½ lin. in

length in *E. flexicaulis* and *E. macrophylla* ; stamens irregular in number (4 or 8) in *E. filiformis* ; anthers terminal or subterminal in *E. Lehmannii, E. Marlothii*, and rarely so in *E. filiformis* ; sublateral in several species ; habit erect, stout, and rigid in *E. strigosa*.

1. *Diffuse shrubs.*

*Anthers included, seldom even manifest :
  Leaves so rolled back at the sides as to appear inflated
    or bladdery   ...   .   (182) **physophylla.**
  Leaves broad, more or less flat, open-backed, some-
    times concave below :
     † Anthers aristate :
      Anthers ⅓ lin. long or less :
        Corolla-mouth widened :
          Corolla obconic-cyathiform   ... (183) **oxycoccifolia.**
          Corolla funnel-shaped :
            Bracts remote ...   . (187)leptoclada,var.β.
            Bracts approximate   ...   ... (188) **trichoclada.**
        Corolla-mouth contracted or not widened :
          Anthers obovate, ⅕ lin. long ; sepals
           ½ lin. long   ...   . (176) **cymosa.**
          Anthers short-cuneate, ¼ lin. long ;
           sepals 1 lin. long   ...   ... (177) **oreophila.**
      Anthers over ⅓ lin. long :
        Flowers terminal or mostly so, 3-nate ;
         bracts approximate ..   .. (179) **heterophylla.**
        Flowers axillary or mostly so ; bracts
         usually remote :
          Leaves 1½–2 lin. long ; filaments
           broader at the base ; flowers pale
           purple   ...   . (178) **planifolia.**
          Leaves ¾–1¼ lin. long ; filaments
           nearly equal ; flowers yellowish .. (180) **cryptanthera.**
     †† Anthers muticous :
      Leaves ovate to subrotund ; pedicels 2–3½ lin.
        long ...   ..–   ...   . (183) **oxycoccifolia.**
      Leaves linear to lanceolate ; pedicels 1½ lin.
        long ...   . (187) **leptoclada.**
      Leaves oblong ; pedicels 6–9 lin. long   ... (181) **tenuipes.**
  Leaves narrow, oblong or linear, sulcate, sometimes
    subopen-backed :
     Leaves generally 4-nate, small and very slender,
      about 1 lin. long ...   . (185) **myriocodon.**
     Leaves 3-nate, linear-oblong, 3–4 lin. long   ... (184) **tenuicaulis.**
     Leaves 3-nate, linear or lanceolate, 1½ lin. long ... (187) **leptoclada.**
** Anthers subincluded or subexserted, generally mani-
  fest :
    Corolla funnel-shaped ; mouth widened   ...   ... (188) **trichoclada.**
    Corolla urceolate ; mouth contracted ...   . (202) **latifolia.**
*** Anthers exserted :
   † Leaves ovate or lanceolate, open-backed :
     Bracts approximate ; leaves ½–¾ lin. long ;
      anthers minutely aristate   ...   . (186) **brachycentra.**
     Bracts remote or lax ; leaves more than ¾ lin.
      long :
       Anthers muticous :
        Corolla-mouth much contracted ; flowers
         mostly solitary :
          Corolla much collapsed in the upper
           part during flowering ; style stout,
           uncinate-decurved   ...   ... (189) **Marlothii.**

Corolla moderately collapsed during
flowering; style slender, straight (190) **oligantha.**
Corolla-mouth slightly contracted;
flowers terminal, 3-nate　　...　　...　(191) **debilis.**
Anthers minutely aristate; corolla-mouth
not contracted; filaments ovate at the
base ...　　...　　...　　...　　...　　...　(192) **thimifolia.**

†† Leaves linear or setaceous, sulcate (not open-
backed) :
　Leaves 3-nate :
　　Bracts approximate :
　　　Anthers muticous; stamens variable
　　　　in number (4–8)　　...　　...　　...　(193) **filiformis.**
　　　Anthers aristate; stamens regular　...　(194) **Tysoni.**
　　Bracts remote, minute　　...　　...　...　(199) **confusa.**
　Leaves 4-nate, setaceous, about $\frac{1}{100}$ in. wide　...　(195) **aspalathoides.**

2. Erect-branched, or scarcely diffuse shrubs.

* Corolla contracted at the mouth :
　† Anthers muticous:
　　Bracts remote :
　　　Ovary variably hairy :
　　　　Corolla less than 2 lin. long :
　　　　　Pedicels $\frac{3}{4}$–$1\frac{1}{2}$ lin. long :
　　　　　　Inflorescence densely capitate;
　　　　　　　sepals much shorter than
　　　　　　　the corolla ...　　...　　...　(207) **ocellata.**
　　　　　　Inflorescence subcapitate, or
　　　　　　　3-nate; sepals nearly equal
　　　　　　　to the corolla　　...　　...　(203) **cordata.**
　　　　　Pedicels 3–4 lin. long　　...　　...　(202) **latifolia.**
　　　　Corolla 2 lin. long or more :
　　　　　Anthers included; leaves 4-nate,
　　　　　　broad-linear　　...　　...　　...　(198) **flexicaulis.**
　　　　　Anthers exserted or subexserted :
　　　　　　Pedicels $\frac{1}{2}$–1 lin. long; leaves
　　　　　　　ciliate, hairy above ...　　...　(206) **macrophylla.**
　　　　　　Pedicels 2–4 lin. long; leaves
　　　　　　　ciliate, glabrous above　...　(204) **hirsuta.**
　　　Ovary glabrous　　...　　...　　...　　...　(198) **flexicaulis.**
　　Bracts all approximate; anthers far-exserted,
　　　subterminal　　...　　...　　...　　...　　...　(205) **Lehmannii.**
　　Bracts 2 approximate, 1 remote; leaves 4-nate,
　　　flowers umbellate　　...　　...　　...　　...　(198) **flexicaulis.**
　　Bracts subremote, very small; leaves 3-nate;
　　　flowers capitate ...　　...　　...　　...　　...　(207) **ocellata.**
　†† Anthers aristate :
　　Sepals foliaceous, $\frac{1}{2}$ lin. long, broad-ovate　...　(199) **confusa.**
　　Sepals scarious, 1–$1\frac{1}{4}$ lin. long, ovate-lanceolate ...　(200) **grata.**
** Corolla not contracted at the mouth :
　Anthers muticous :
　　Filaments spoon-bowl-shaped at the base　...　(201) **flacca.**
　　Filaments equal or nearly so :
　　　Bracts remote, minute　　...　　...　　...　(196) **Mundii.**
　　　Bracts approximate, rather large, scarious ...　(205) **Lehmannii.**
　Anthers aristate ; leaves 4-nate, narrow　...　...　(197) **strigosa.**
*** Corolla slightly widened at the mouth, subtubular-
campanulate ...　　...　　...　　...　　...　　...　...　(201) **flacca.**

§ 15. Desmia (*Confertæ*, Klotzsch).　Inflorescence terminal; flowers 3-nate,
umbellate or capitate, corolline or subcorolline.　Bracts approximate, or the lower

sometimes remote or lax, sepal-like, coloured. Sepals ovate or lanceolate, shorter than the corolla, subscarious or cartilaginous, coloured. Corolla urceolate, glabrous, sometimes viscid, $1\frac{1}{2}$–$2\frac{1}{2}$ lin. long; segments short. Anthers subexserted, terminal, muticous. Style exserted; stigma capitate. Ovary mostly glabrous or subglabrous. Leaves mostly 3- (rarely 4-) nate, mostly long and thin. Straggling or erect shrubs. Allied to § *Ceramia* and § *Pachysa*; but distinct from both by its terminal anthers.

Leaves 4-nate; heads densely many-flowered ... ... (208) **conferta.**
Leaves 3-nate; flowers umbellate or clustered, but
  scarcely dense or many :
    Leaves 5–11 lin. long, acute or acuminate ... ... (209) **polifolia.**
    Leaves 3–5 lin. long, obtuse ... ... (210) **obtusata.**

§ 16. **GYPSOCALLIS.** Inflorescence axillary towards the ends of the branches, rarely also terminal; flowers corolline or rarely subcalycine. Bracts small and (except in one species) remote. Sepals usually small and inconspicuous. Corolla urceolate, ovoid or subcampanulate, 1–3 lin. long, dry, glabrous; limb short or very short. Anthers exserted or subexserted, rarely subincluded, mostly lateral, more rarely subterminal, muticous, or (in two species only) aristulate; pore generally small proportionately to the cell. Generally erect and rigid shrubs, with 3–4-nate, erect, sulcate (or rarely subopen-backed) leaves.

EXCEPTIONS : leaves sometimes subopen-backed in *E. racemosa* and *E. dumosa* ; flowers subcalycine in *E. fucata*, and sometimes so in *E. scytophylla* ; anthers sometimes scarcely exserted in *E. fucata* and in *E. scytophylla*.

Leaves 4-nate ; anthers muticous :
  Ovary pubescent; anthers subexserted; leaves
    hairy ... ... (211) **racemosa.**
  Ovary glabrous; anthers quite exserted; leaves
    subglabrous ... ... (212) **aghillana.**
Leaves 3-nate :
  Bracts (or 2 at least) quite approximate; anthers
    aristate ... ... (213) **petræa.**
  Bracts remote; anthers muticous :
    Ovary pubescent, at least at the top; leaves
      sometimes subopen-backed ... ... (214) **dumosa.**
    Ovary glabrous; leaf-back always closed :
      Sepals subequal to the corolla :
        Pedicels 3–$5\frac{1}{2}$ lin. long ; anthers
          muticous ... ... (215) **fucata.**
        Pedicels 2–3 lin. long; anthers denticulate at the base ... ... (216) **scytophylla.**
      Sepals shorter than the corolla :
        Leaves 1–$1\frac{1}{2}$ lin. long; anthers pale-
          brown ... ... (217) **capillaris.**
        Leaves 2–4 lin. long; anthers dark-
          coloured :
            Anthers well-exserted, muticous (218) **nudiflora.**
            Anthers subexserted, squarrose-
              denticulate at the base ... (216) **scytophylla.**

§ 17. **PYRONIUM.** Inflorescence terminal; flowers corolline, rarely subcorolline, 3-nate, rarely umbellate or clustered. Bracts always small and inconspicuous. Sepals usually small. Corolla various, mostly cyathiform, sometimes subcampanulate, suburceolate or ovoid, mostly glabrous, from $\frac{3}{4}$–$2\frac{1}{2}$ lin. long. Anthers exserted, subexserted or just manifest, mostly lateral, sometimes subterminal, rarely terminal, muticous or appendiculate. Leaves 3- (rarely 3–4-) nate, linear-trigonous, sulcate, very rarely subopen-backed. Small rigid shrubs.

EXCEPTIONS: leaves sometimes 4-nate in *E. nutans* and *E. drakensbergensis;* sometimes open-backed in *E. bicolor;* flowers sometimes subcorolline in *E. paniculata.*

<pre>
Corolla mostly 1¾–2½ lin. long :
    Anthers muticous, well exserted   ...   ... (219) nutans.
    Anthers muticous, subexserted ...   . (220) deliciosa.
    Anthers appendiculate, subexserted :
        Appendages of anther minute, toothed or
            single   ...   . (220) deliciosa.
        Appendages of anther long-decurrent, with
            free points   ...   . (230) unilateralis.
        Appendages of anther aristate, entirely
            free   ...   . (221) drakensbergensis.
Corolla not exceeding 1½ lin. long, mostly less :
    Anthers appendiculate :
        Appendages decurrent, longer than the
            cells   ...   . (230) unilateralis.
        Appendages decurrent, much shorter than
            the cells ...   . (229) parvula.
        Appendages not decurrent, aristate or
            toothed :
            Sepals scarious or subscarious :
                Corolla much widened from the
                    base to the mouth   ...   ... (224) paniculata.
                Corolla not (or scarcely) widened
                    to the mouth :
                    Corolla tubular-campanulate,
                        narrow   ...   . (227) opulenta.
                    Corolla subcyathiform, broad-
                        ish :
                        Ovary cano-pubescent... (222) decipiens.
                        Ovary glabrous   ... (221) drakensbergensis.
            Sepals foliaceous or subfoliaceous :
                Corolla 1¼–2 lin. long   ...   ... (221) drakensbergensis.
                Corolla 1–1¼ lin. long ; anthers
                    broad, very obtuse   ...   ... (225) bicolor.
                Corolla ¾ lin. long ; anthers
                    narrow, tapering to the apex ... (228) harveiana.
    Anthers muticous :
        Sepals scarious or subscarious :
            Corolla obconic-cyathiform ...   ... (224) paniculata.
            Corolla narrow-cyathiform to subovoid
                or ovoid-urceolate :
                Pedicels ½–1 lin. long ; corolla
                    glabrous   ...   . (222) decipiens.
                Pedicels 1–2 lin. long ; corolla
                    very minutely hispidulous   ... (223) demissa.
        Sepals foliaceous or subfoliaceous :
            Sepals coloured ; flowers numerous,
                bright red, prominent   ...   ... (226) diotæflora.
            Sepals green ; flowers few, dull yellow
                or red, inconspicuous   ...   ... (231) brachysepala.
Imperfectly known species ; anthers muticous   ... (232) kraussiana.
</pre>

§ 18. OROPHANES. Inflorescence terminal ; flowers 4-nate, sometimes um-bellate or irregularly clustered, rarely in a congested pseudo-spike, corolline. Bracts generally remote, always small and inconspicuous. Sepals small, rarely as long as the corolla-tube. Corolla various, mostly urceolate, cyathiform or campanulate, mouth contracted or widened, glabrous, dry, mostly 1½–2½ (rarely 4–5) lin. long. Anthers mostly included, rarely subexserted, lateral, dorsifixed just above the base, or very rarely at or above the middle of the cell, usually small, often minute, always appendiculate. Ovary mostly glabrous and sessile, rarely shortly stipitate. Leaves mostly 4-nate, rarely scattered. § *Leptodendron* should also be consulted in cases of difficulty.

EXCEPTIONS : inflorescence in *E. aspalathifolia* forming a dense pseudo-spike; umbellate or clustered in *E. pilulifera*; corolla obconic in *E. adæquata*; anthers subexserted or sometimes manifest in *E. scabriuscula, E. gibbosa, E. adæquata,* and *E. lateralis*; inserted at or near the middle of the cell in *E. bergiana*; leaves scattered or partially so in *E. pilulifera* and in *E. subulata*.

1. *Leaves scattered, or partially so.*

Inflorescence umbellate, or clustered with some axillary
  flowers below the umbel ...   . (233) **pilulifera.**
Inflorescence 4-nate   ...   . (234) **subulata.**

2. *Leaves scabro- or tuberculate-hispid.*

Sepals ascending :
    Pedicels 2 lin. long or more; sepals oblong or
      lanceolate, 1 lin. long   ...   . (235) **scabriuscula.**
    Pedicels 1 lin. long; sepals ovate or subovate,
      $\frac{1}{4}$ lin. long ...   . (236) **gibbosa.**
Sepals reflexed   ...   . (237) **bergiana.**

3. *Inflorescence usually, at least partially, a congested pseudo-spike.*

Leaves **very** narrow, almost hair-like; flowers some-
  times 3-nate   ...   . (238) **aspalathifolia.**

4. *Sepals reflexed.*

Only species ...   .   ,.. (237) **bergiana.**

5. *Anthers subexserted.*

Corolla obconic; mouth widened ...   . (239) **adæquata.**
Corolla suburceolate; mouth contracted or not widened (246) **lateralis.**

6. *Leaves, inflorescence, sepals and anthers not as in the foregoing.*

Anther-appendages more or less adnate to the upper
  part of the filament, or to the produced connective,
  and standing out at a right angle, or nearly so to the
  cell :
    Corolla-mouth contracted :
      Pedicels 6–7 lin. long; ovary stipitate   ... (240) **rubens.**
      Pedicels 2–4 lin. long; ovary sessile ...   ... (241) **læta.**
    Corolla-mouth widened, or not contracted; sepals
      as long as the corolla-tube :
      Anthers crested ...   . (242) **turbiniflora.**
      Anthers long-decurrent-aristate   ...   ... (242) **turbiniflora,var.β.**
Anther-appendages not as above described :
    *Anthers subequilateral-triangular, with rounded
      angles, $\frac{1}{5}$–$\frac{1}{4}$ lin. long (in 242 sometimes $\frac{1}{2}$ lin.
      long) :
      Interstices of the corolla-limb rounded, con-
        spicuous in bud   ...   . (243) **viridipurpurea.**
      Interstices of the corolla-limb acute :
        Anthers crested; crests fringed or lacerate :
          Corolla 2–3 lin. long, urceolate;
          limb less than $\frac{1}{3}$ as long as the tube (248) **tenella.**
          Corolla 1$\frac{1}{2}$ lin. long, subglobose-urceo-
          late; limb more than $\frac{1}{2}$ as long as
          the tube   ...   . (249) **chionophila.**

Anthers aristate; awns not fringed ;
corolla ¾–1¼ lin. long :

    Corolla-mouth distinctly widened ;
leaves ¾–1⅓ lin. long ... ... (250) **quadrangularis.**

    Corolla-mouth scarcely, or very little
widened ; leaves ½–1 lin. long ... (251) **cyathiformis.**

****Anthers suboblong, or cuneate or ovate, always
longer than wide :

    Leaves aristate :

        Pedicels 1½ lin. long ; bracts approxi-
mate ; anthers oblong, aristate ... (234) **subulata.**

        Pedicels 2–3 lin. long ; bracts subremote ;
anthers cuneate, cristate ... ... (252) **hæmastoma.**

    Leaves not aristate :

        Corolla 3–4 lin. long ; anthers papillose-
scabrid :

            Flowers red or white ; anthers some-
what square ... ... (244) **sitiens.**

            Flowers yellow ; anthers longitudinally
semiovate ... ... (245) **Blandfordia.**

        Corolla 2–3 lin. long :

            Anthers crested ; leaves stout ; flowers
often subumbellate :

                Leaves mostly keeled and only
faintly sulcate ... ... (216) **lateralis.**

                Leaves scarcely keeled, deeply
sulcate ... ... (247) **verecunda.**

            Anthers aristate ; corolla 2–2½ lin.
long ; leaves slender ... ... (253) **velitaris.**

        Corolla 1–2 lin. long :

            Anthers about ⅙ lin. long ; leaves
glabrous :

                Anthers ovate ; corolla-mouth
slightly contracted ... ... (254) **gracilis.**

                Anthers elliptical or oblong ;
corolla-mouth widened ... (255) **leucantha.**

            Anthers ¼–⅓ lin. long ; leaves glabrous
or pubescent :

                Anthers oblong, length about
twice the width in the middle .. (256) **subdivaricata.**

                Anthers oblong, length about 4
times the width in the middle .. (259) **trichophylla.**

            Anthers about ½ lin. long ; leaves
glabrous :

                Corolla subtubular-oblong, not or
scarcely widened at the mouth,
white ... ... (257) **margaritacea.**

                Corolla campanulate ; mouth
widened... ... ... (258) **curvirostris.**

§ 19. LEPTODENDRON. Inflorescence terminal ; flowers 3-nate or solitary, corolline, very rarely subcorolline. Bracts generally remote, small. Sepals small, very rarely equal to the corolla-tube. Corolla various in shape, glabrous, dry, 1½–3½ lin. long. Anthers included, lateral, appendiculate or muticous. Leaves almost always 3-nate, occasionally also 4-nate, in one species opposite, in another sometimes so. A very artificial section, which has been separated from § *Orophanes* chiefly by its usually 3-nate flowers and leaves.

EXCEPTIONS : leaves, in *E. virginalis,* so far as known, always opposite ; in *E. rupicola,* sometimes 4-nate ; in *E. campanulata,* sometimes 4-nate, or opposite ; flowers in *E. polycoma,* subcorolline.

Leaves generally 3-nate, rarely 4-nate or opposite :
   Sepals silky pilose; anthers crested; flowers
     subcorolline ... ... (267) **polycoma.**
   Sepals glabrous or merely puberulous :
     Anthers distinctly crested; ovary glabrous :
       Corolla urceolate; throat contracted;
        leaves glabrous ... ... (264) **Vanheurckii.**
       Corolla campanulate - cyathiform ;
        throat not contracted; leaves
        tomentose-downy ... ... (266) **lachnæoides.**
     Anthers narrow-crested or subaristate :
       Corolla subtubular ... ... (262) **mira.**
       Corolla broad-cyathiform ... ... (263) **micrandra.**
     Anthers muticous :
       Corolla contracted towards the apex.. (260) **rupicola.**
       Corolla not contracted towards the
        apex :
         Corolla-segments $\frac{1}{2}$–$\frac{3}{4}$ as long as
          the tube ... ... (261) **condensata,** var. β.
         Corolla-segments less than $\frac{1}{2}$ as
          long as the tube :
           Filaments about 2 lin. long,
            sigmoid at the apex ... (260) **rupicola.**
           Filaments about $\frac{1}{2}$ lin. long,
            nearly straight :
             Corolla yellow; sepals
              scarious, $1\frac{1}{2}$–$1\frac{3}{4}$ lin.
              long ... ... (265) **campanulata.**
             Corolla red; sepals
               thick, subcoriaceous,
               $\frac{3}{4}$–1 lin. long ... (261) **condensata.**
Leaves regularly opposite; anthers with small
   crests... ... ... (268) **virginalis.**

§ 20. PACHYSA (*Physoideæ*, Klotzsch). Inflorescence terminal, occasionally also pseudo-lateral; flowers mostly 3–4-nate or umbellate, rarely subcapitate, corolline, or rarely with a subcalycine appearance. Sepals $\frac{1}{2}$ as long as the corolla or less, usually thick and viscid. Corolla more or less viscid, usually ovoid, urceolate or rarely obconic, generally thick in texture, 1–10 lin. long, mostly from 2–4 lin. long. Anthers usually included, rarely exserted, lateral, dorsifixed at or shortly above the base, appendiculate or rarely muticous. Ovary sessile, or rarely substipitate. Leaves 3–4-nate, rarely scattered.

EXCEPTIONS : leaves scattered in *E. obliqua*; inflorescence subcapitate and flowers subcalycine in *E. Fairii*; corolla obconic, with a very long limb, in *E. macroloma*; anthers exserted in *E. Alexandri*, muticous in *E. odorata*; ovary stipitate in *E. obliqua* and substipitate in *E. macroloma*.

1. *Flowers terminal, rarely also sublateral (in 286); corolla mostly under 3 lin. long (in 281 var. β, 4 lin.), white, red or reddish.*

* Leaves 3-nate :
   Anthers crested :
     Ovary hirsute; anther-crests narrow, de-
      current, bearded ... ... (281) **glomiflora.**
     Ovary glabrous, or nearly so :
       Flowers umbellate; corolla subglobose (269) **ramentacea.**
       Flowers 3-nate; corolla ovoid-urceolate (271) **macra.**
   Anthers aristate, or sometimes narrow-crested :
     Pedicels $3\frac{1}{2}$–6 lin. long :
       Awns of anther decurrent along the
        dilated connective; ovary glabrous ... (273) **spectabilis.**
       Awns of anther free; ovary hispidulous (277) **nubigena.**

Pedicels 2 lin. or less ; ovary densely hirsute :
 Corolla-tube globose, not tapering to a
  neck ... ... (278) **formosa.**
 Corolla-tube ovoid, more or less tapering to
  a neck :
   Anthers $2\frac{1}{2}$ times longer than the width
    in the middle, less than $\frac{1}{2}$ lin. long ;
    corolla 2–$2\frac{1}{2}$ lin. long ... ... (280) **tragulifera.**
   Anthers 3–$3\frac{1}{2}$ times longer than the
    width in the middle, $\frac{5}{8}$–$\frac{7}{8}$ lin. long ;
    corolla $2\frac{1}{2}$–4 lin. long ... ... (281) **glomiflora.**
 Anthers muticous, oblong ; pedicels short ; ovary
  glabrous ... ... (279) **umbelliflora.**
** Leaves 4-nate :
 Sepals densely floccose-woolly ; stamens included ... (274) **flocciflora.**
 Sepals hispid on the keel only ; stamens exserted ... (275) **Alexandri.**
 Sepals glabrous, sometimes ciliate :
  Corolla subglobose ; anthers densely and coarsely
   hairy ... ... (269) **ramentacea.**
  Corolla not subglobose :
   Anthers free aristate :
    Pedicels 1 lin. long ; bracts subapproxi-
     mate ... ... (276) **frigida.**
    Pedicels 5–6 lin. long ; bracts remote ... (277) **nubigena.**
   Anthers crested :
    Corolla elliptic-urceolate ; ovary gla-
     brous ... ... (270) **mucosa.**
    Corolla suburceolate-cyathiform ; ovary
     villous ... ... (272) **Schlechteri.**

2. *Flowers mostly terminal, or rarely a few sublateral at the ends of the
branches ; corolla medium-sized to large, white, green, or orange. Usually
stout shrubs.*

Sepals linear ; flowers subcapitate ; pedicels under 1 lin.
 long ... ... (284) **Fairii.**
Sepals not linear ; pedicels $1\frac{1}{2}$ lin. or longer :
 Anthers decurrent-aristate ; corolla narrow-ovoid ... (285) **oblongiflora.**
 Anthers crested ; corolla globose-urceolate or ovoid-
  conical :
  Pedicels mostly less than 3 lin. long ; compact
   small shrub ; flowers white ... ... (282) **physodes.**
  Pedicels mostly longer than 3 lin. :
   Leaves 4-nate ; tall straggling shrub ; flower
    green ... ... (283) **Urna-viridis.**
   Leaves 3-nate :
    Corolla red or white, 3–4 lin. long ... (287) **ardens.**
    Corolla orange, green-tipped, $4\frac{1}{2}$–$6\frac{1}{2}$ lin.
     long ... ... (288) **blenna.**
    Corolla orange, green-tipped, 9–10 lin.
     long ... ... (288) **blenna,** var. β.
 Anthers muticous ; stamens very short ... ... (286) **odorata.**

3. *Flowers terminal, long-pedicelled, umbellate or subracemose at the ends
of the branches. Bracts small and remote. Corolla 2–4 lin. long, red,
rosy or white. Mostly slender shrubs.*

Ovary sessile, oblate :
 Habit decumbent ; corolla urceolate-globose, contracted at the mouth ... ... (289) **ixanthera.**
 Habit suberect ; corolla urceolate, wide at the
  mouth ... ... (290) **carduifolia.**

Ovary stipitate, or at least narrow at the base, prolate:
    Corolla ovoid-urceolate, limb much shorter than the
        tube ... ... (291) **obliqua.**
    Corolla obconic, limb equal to or longer than the
        tube ... ... (292) **macroloma.**

§ 21. HERMES. Inflorescence axillary, rarely both terminal and axillary,
usually pseudo-spicate or pseudo-racemose towards the ends of the branches;
flowers corolline, rarely subcalycine. Bracts and sepals various, not prominent.
Corolla various, mostly campanulate, obconic or subtubular, usually glabrous,
often viscid, from 1½–4 lin. long. Anthers generally included, lateral or subterminal, muticous or appendiculate. Leaves 3-, 4-, 6-nate or scattered. Erect
or rarely prostrate shrubs.

EXCEPTIONS: inflorescence both terminal and axillary in *E. hæmantha;* sometimes subcapitate in *E. amœna;* flowers sometimes subcorolline in *E. deflexa;*
and subcalycine in *E. pulvinata* and *E. collina.*

Flowers subsessile in a congested pseudo-spike; pedicels
    ¼–1 lin. long:
        Leaves 6-nate; anthers oblong, narrow-crested ... (293) **empetrifolia.**
        Leaves 6-nate; anthers elliptical, very broad-
          crested ... ... (294) **pyxidiflora.**
        Leaves 4-nate; anthers setaceo-aristate ... ... (296) **Dodii.**
Flowers pseudo-racemose; pedicels over 1 lin. long:
    Ovary villous:
        Leaves filiform, tremulous, 5–7 lin. long, ¼ lin.
          broad .... (33) **filamentosa.**
        Leaves shorter, broader and rigid:
          Leaves, pedicels, bracts and sepals hirsute
            with long hairs; anthers long-crested-
            aristate ... ... (295) **amœna.**
          Leaves, &c., glabrous or pubescent; an-
            thers muticous or decurrent-denticulate:
            Sepals and corolla more or less pubes-
              cent or scaberulous; corolla nearly
              equal to the mouth ... ... (302) **viscaria.**
            Sepals and corolla glabrous; corolla
              usually widened to the mouth ... (303) **axilliflora.**
    Ovary glabrous or glabrescent, at most puberulous:
        Anthers muticous, or sometimes decurrent-
          denticulate:
          Corolla subobconic or subfunnel-shaped,
            3½–6 lin. long, red ... ... (24) **conica.**
          Corolla subtubular to campanulate, 2–4 lin.
            long, yellow ... ... (301) **parilis.**
        Anthers aristate, or broadish-aristate; awns
          free or partly decurrent:
          Corolla obconic, widened to the mouth ... (299) **longiaristata.**
          Corolla ovoid, or tubular-ovoid, contracted
            to the mouth ... ... (17) **filipendula.**
          Corolla various, equal or scarcely widened
            or contracted:
            Leaves squarrose or spreading, in-
             curved above the middle; flowers
            subcalycine:
             Corolla 1½–1¾ lin. long; leaves
              1½–2 lin. long ... ... (306) **collina.**
             Corolla 2–2½ lin. long; leaves
              3–5 lin. long ... ... (305) **pulvinata.**

Leaves not incurved; flowers corolline:
　Leaves 3–6 lin. long; branches
　　and racemes mostly very long ... (297) **regerminans.**
　Leaves 1–2½ lin. long, flat or
　　flattish:
　　　Corolla campanulate-tubular,
　　　　2–2½ lin. long, yellow;
　　　　sepals pallid　　... , 　... (300) **flavicoma.**
　　　Corolla urceolate-campanu-
　　　　late, 1½–2 lin. long, red;
　　　　sepals dark red　...　　... (298) **pulchella.**
　Leaves 1–1½ lin. long, thick, sub-
　　terete or semiterete; corolla
　　cyathiform-obconic,　1½　lin.
　　long　　...　　...　　...　　... (307) **deflexa.**
Anthers crested, or narrow-crested:
　Inflorescence lax; corolla 3½–4 lin. long,
　　crimson　...　　...　　...　　...　　... (304) **hæmantha.**
　Inflorescence dense, pseudo-racemose;
　　corolla 2 lin. long　　...　　...　　... (297) **regerminans.**

§ 22. CHLOROCODON. Inflorescence axillary; flowers 1–3-nate on the uppermost branches, occasionally a few also terminal, corolline, very small. Bracts
remote, small, sometimes deficient. Sepals shorter than the corolla. Corolla
cyathiform, campanulate or subobconic, from ½–1 lin. long. Anthers included or
often subincluded and manifest, mostly lateral (in one species subterminal),
muticous, or (in one species) aristate. Stigma capitate, peltate, or subcyathiform.
Leaves mostly 3-nate, occasionally also 4-nate on the same plant, and (in one
species) constantly 4-nate. Mostly small shrubs, but in one species reaching
5–7 ft. high.

EXCEPTIONS: leaves occasionally 4-nate in *E. coarctata* and constantly so in
E. alticola; inflorescence sometimes terminal as well as axillary in *E. Woodii*;
anthers aristate in *E. Woodii*, subterminal in *E. mæsta*.

Leaves 3-nate (occasionally in *E. coarctata* 3- and 4-
　nate):
　Anthers aristate　　...　　...　　...　　...　　... (308) **Woodii.**
　Anthers muticous:
　　Anthers subterminal; style shorter than the
　　　ovary ...　　...　　...　　...　　...　　... (309) **mæsta.**
　　Anthers lateral or sublateral; style longer
　　　than the ovary:
　　　Corolla cyathiform or campanulate, not
　　　　much narrowed at the base; sepals about
　　　　½ lin. long:
　　　　Corolla-segments shorter or a little
　　　　　longer than the tube:
　　　　　Leaves linear, 2½–4 lin. long;
　　　　　　pedicels ½–1½ lin. long　　... (310) **coarctata.**
　　　　　Leaves oblong, 1½–2 lin. long;
　　　　　　pedicels 1½–2 lin. long　　... (311) **curtophylla.**
　　　　Corolla-segments 3–4 times longer
　　　　　than the tube　　...　　...　　... (312) **Priori.**
　　　Corolla obconic, narrowed at the base;
　　　　sepals ¼ lin. long ...　　...　　... (313) **leptostachya.**
Leaves 4-nate:
　Pseudo-spikes short, dense at the ends of the
　　branches; style included ...　　...　　... (314) **alticola.**
　Pseudo-spikes long, lax below the ends of the
　　branches; style exserted ...　　...　　... (310) **coarctata.**

§ 23. **ARSACE.** Inflorescence terminal, occasionally by arrest of branchlets appearing axillary; flowers 3-nate, corolline, small. Bracts usually remote, rarely subapproximate, small. Sepals inconspicuous, small. Corolla urceolate, cyathiform or campanulate, mostly glabrous, $\frac{1}{2}$–$1\frac{1}{2}$ lin. long. Anthers mostly included, occasionally subexserted, lateral or rarely sublateral, appendiculate or muticous. Style exserted or more rarely included. Stigma capitate, peltate or cyathiform. Ovary glabrous or hispidulous. Leaves 3-nate; tall, or more usually, small shrubs, erect, very rarely diffuse.

EXCEPTIONS: anthers sublateral in *E. arachnoidea*; corolla sometimes puberulous in *E. arachnoidea*, *E. leucopelta* var. γ, *E. copiosa*, and *E. setacea*.

I. PELTATÆ.—Stigma peltate or cyathiform.

Corolla urceolate :
   Bracts 3 :
      Leaves open-backed :
         Leaves densely white tomentose below   ... (316) **arachnoidea.**
         Leaves glabrous   ...   . (315) **hispidula**, var. β.
         Leaves sulcate, usually linear ...   . (315) **hispidula.**
      Bract 1, cartilaginous, semi-amplexicaul ...   ... (317) **inops.**
   Corolla usually cyathiform, or obconic or campanulate :
      Anthers aristate; bracts subapproximate ...   ... (320) **salax.**
      Anthers muticous; bracts remote :
         Leaves glabrous; bracts subequal, quite basal (319) **maritima.**
         Leaves pubescent; bracts unequal, 1 much
           larger, 2 minute or wanting ...   . (318) **leucopelta.**

II. CAPITATÆ.—Stigma capitate or clavate-capitate.

Style included; plant entirely glabrous :
   Corolla cyathiform; anthers aristate :
      Pedicels $\frac{1}{2}$–1 lin. long; filaments usually ex-
        panded at the base; anthers $\frac{1}{7}$–$\frac{1}{5}$ lin. long ... (324) **tenuis.**
      Pedicels 1–2$\frac{1}{2}$ lin. long; filaments usually equal
        throughout; anthers $\frac{1}{4}$–$\frac{2}{5}$ lin. long ...   ... (322) **leptopus.**
   Corolla globose-urceolate; anthers crested ...   ... (321) **carinata.**
Style exserted; plant more or less hairy :
   Branches floccose with minute plumose hairs :
      Leaves roughly setose-hispid ...   . (330) **setacea.**
      Leaves glabrous, or rarely shortly hispid :
         Anthers $\frac{3}{8}$ lin. long; appendages very
           short, spreading; stigma large ... (325) **crateriformis.**
         Anthers $\frac{1}{5}$ lin. long; appendages long or
           short pendulous; stigma small ...   ... (326) **subverticillaris.**
   Branches with simple hairs :
      Anthers muticous :
         Sepals cartilaginous, pallid, smooth, gla-
           brous; anthers $\frac{1}{2}$ lin. long ...   ... (329) **microcodon.**
         Sepals mostly foliaceous; anthers $\frac{1}{4}$–$\frac{1}{3}$ lin.
           long   ...   . (327) **copiosa.**
      Anthers aristate or crested aristate :
         Whole plant densely covered with small
           flowers; leaves $\frac{1}{2}$–$\frac{3}{4}$ lin. long ; anthers
         $\frac{1}{5}$ lin. long   ...   . (328) **onusta.**
         Habit not so; leaves and anthers usually
           longer :
              Branches mostly glabrous; pedicels
                1–2$\frac{1}{2}$ lin. long   ...   . (322) **leptopus.**
              Branches hirsute; pedicels about 1 lin.
               long ...   . (327) **copiosa.**
      Anthers unknown; plant slender with very
        small flowers; pedicels longer than the
        corolla; style long-exserted ...   ... (323) **minutissima.**

§ 24. PSEUDEREMIA. Inflorescence terminal, capitate; heads few or many-flowered; flowers mostly corolline. Bracts mostly closely approximate, like the sepals narrow, and most usually ciliate with rather long plumose hairs. Corolla urceolate or campanulate, glabrous or pubescent, mostly 1–4 lin. long, in one species sometimes attaining to 5 lin. Anthers mostly included or rarely sub-exserted lateral, appendiculate. Habit various. Leaves 3-nate or 4-nate, some-times irregular or scattered. Pubescence variable, often characterized by plumose hairs. All the species (except one) are denizens of the higher mountains.

The section is allied to § *Chromostegia*, from which it is artificially divided by a different aspect and the absence of involucrating leaves below the heads.

EXCEPTIONS: flowers subcorolline or subcalycine in *E. Cooperi*; anthers sub-exserted in *E. Greyii*.

Leaves 3-nate (in *E. Maderi* often 4-nate or scattered) :
    Heads 10–23-flowered; corolla 2½–5 lin. long ... (332) **Maderi.**
    Heads usually fewer than 10-flowered; corolla
      2 lin. long or less :
        Anthers long-subulate-aristate; style in-
          cluded ... ... (337) **oxysepala.**
        Anthers very minutely toothed; style ex-
          serted ... ... (338) **Greyii.**
Leaves 4-nate (in *E. Maderi* sometimes 3-nate or
  scattered) :
    Corolla pubescent or puberulous :
        Not much branched, mostly weak and
          straggling ... ... (334) **Solandra.**
        Much branched, erect :
          Stout and rigid; corolla narrow-urceolate,
            2–2½ lin. long ... ... (335) **Cooperi.**
          Slender, subflexuous; corolla globose-
            urceolate, 1½ lin. long ... ... (336) **Baurii.**
    Corolla glabrous :
        Sepals filiform at the base, wider upwards ... (339) **clavisepala.**
        Sepals lanceolate or linear, not wider up-
          wards :
            Ovary villous or hispid ... ... (334) **Solandra.**
            Ovary glabrous :
              Branches more or less flexuous :
                Branches numerous, slender;
                  heads 4–8-flowered ... ... (331) **cernua.**
                Branches few, stoutish; heads
                  10–23-flowered, 7–10 lin. in
                  diam. ... ... (332) **Maderi.**
              Branches mostly straight, slender;
                heads 6–28-flowered, 4–6½ lin. in
                diam. ... ... (333) **sphærocephala.**

§ 25. POLYDESMIA (*Stellares*, Klotzsch). Inflorescence terminal, capitate or subumbellate; flowers mostly corolline, rarely subcalycine. Bracts sepal-like, mostly approximate, rarely remote and minute. Sepals usually narrow, dry or rarely viscidulous. Corolla suburceolate, or ovoid, glabrous or nearly so, dry or viscidulous, 1¾–2¼ lin. long. Anthers exserted or subexserted, terminal or sub-terminal, muticous, toothed or decurrent-aristate. Style exserted. Ovary glabrous or hairy. Leaves 3–6-nate or scattered. Small shrubs.

One species (*E. ustulescens*) has the aspect of § *Pyronium*, having scarcely either capitate or umbellate flowers, but is retained here by reason of the similarity of its floral characters to those of the species of this section.

Ovary glabrous :
 Heads densely many-flowered, hemispherical; leaves
  incurved ... ... (340) **incurva.**
 Heads laxly 5–9-flowered or subumbellate; leaves
  straight ... ... (341) **ustulescens.**
Ovary hairy :
 Leaves 4–5-nate, oblong or subterete; sepals ⅔ the
  length of the corolla ... ... (342) **stylaris.**
 Leaves 3-nate, lanceolate, flat; sepals ¼–⅓ the length
  of the corolla ... ... (343) **turmalis.**

IV. CHLAMYDANTHE.

§ 26. CHROMOSTEGIA (*Involucratæ*, Klotzsch). Inflorescence terminal, capitate; heads 4-flowered, usually involucrated by the more or less enlarged and discoloured floral leaves; flowers calycine. Bracts closely approximate, broader than the narrow sepals, and ciliate with long hairs. Corolla of various shapes, 1–2 lin. long, puberulous. Anthers exserted or subexserted, terminal or lateral, appendiculate. Leaves 4-nate, always strongly ciliate. Procumbent or suberect shrubs, generally roughly hairy. All the species belong to mountainous regions.

 Floral leaves only slightly, or not, different from the
  cauline; anthers lateral, aristulate ... ... (344) **eriophoros.**
 Floral leaves more or less enlarged, and discoloured :
  Leaves 1½ lin. long; anthers lateral, crested ... (345) **senilis.**
  Leaves 2–3 lin. long; anthers terminal, crested ... (346) **involucráta.**

§ 27. OXYLOMA (*Acutifissæ*, Klotzsch). Inflorescence terminal; flowers subsessile, 3-nate or capitate in many-flowered heads, calycine. Bracts approximate and with the sepals imbricate and closely adpressed to the corolla, coloured. Corolla short-tubular, slightly inflated, 4-fid, 2–5 lin. long; segments suberect, acute or acuminate, long. Anthers included, lateral or sublateral, dorsifixed at the base, muticous. Style exserted; stigma simple. Ovary glabrous or minutely hirtellous. Leaves 3–4-nate or scattered.

 This has some resemblances to § *Polydesmia.*

 Leaves scattered (4–6 lin. long); bracts and sepals
  nearly equalling or exceeding the corolla ... ... (347) **recurvata.**
 Leaves 3-nate; bracts and sepals shorter than corolla;
  style dilated and truncate at the apex ... ... (348) **genistæfolia.**
 Leaves 4-nate; style fusiform at the apex ... ... (349) **cumuliflora.**

§ 28. ERIODESMIA (*Capitatæ*, Klotzsch). Inflorescence terminal; flowers 1–6-nate, subcapitate or umbellate, mostly calycine or subcorolline. Bracts approximate or remote, small. Sepals densely villous or woolly. Corolla mostly urceolate or campanulate, puberulous or pilose, 1½–2¼ lin. long. Anthers mostly exserted, or included but manifest, terminal. Style exserted. Ovary more or less villous. Leaves mostly 3- (or rarely 4-) nate. Small shrubs, more or less hairy in all parts.

 EXCEPTION : flowers sometimes subcorolline in *E. villosa.*
  Leaves usually 4-nate, open-backed; flowers laxly
   capitate... ... ... (350) **lanata.**
  Leaves 3-nate, sulcate; flowers 3-nate or solitary :
   Anthers exserted or subexserted; flowers ovoid
    or oblong :
    Sepals lanceolate; hairs coarse, not very
     dense or very long ... ... (351) **villosa.**
    Sepals ovate or oblong; hairs fine, very dense
     and long ... ... (352) **bruniades.**
   Anthers included, just manifest; flowers globular (353) **capitata.**

§ 29. AMPHODEA (*Spumosæ*, Klotzsch). Inflorescence terminal, capitate; heads 3 flowered; flowers calycine. Bracts approximate, like the sepals adpressed, scarious, coloured, very concave, glabrous, dry, 1¾–2½ lin. long; segments mostly long, acute, erect. Anthers exserted, terminal or subterminal, decurrent-aristate. Style exserted; stigma simple. Ovary glabrous. Leaves 3-nate.

Bracts broad-ovate :
　Corolla-segments ovate, acute; flowers dull whitish (354) **sexfaria.**
　Corolla-segments lanceolate, acute; flowers reddish (355) **spumosa.**
Bracts obovate-oblong; corolla-segments lanceolate,
　acuminate...　　...　　...　　...　　...　　...　　... (356) **amphigena.**

§ 30. GEISSOSTEGIA. Inflorescence terminal; flowers most usually 3-nate, rarely (perhaps in one species) subumbellate, calycine, small. Bracts approximate, more rarely remote. Sepals prominent, usually enwrapping the corolla, and as long as or a little exceeding it, rarely shorter or much shorter, coloured. Corolla various, most commonly urceolate or cyathiform, 1–2½ lin. long. Anthers exserted, terminal or subterminal, rarely sublateral, always muticous. Style exserted. Stigma capitate or subclavate, small. Ovary most usually glabrous, rarely scantily pubescent. Mostly small shrubs, 1½ ft. high or less; leaves 3-nate.

Corolla about twice as long as the sepals, or longer :
　Corolla obconic-tubular, longer than wide　　... (357) **desmantha.**
　Corolla subglobose-urceolate, wider than long　... (358) **physantha.**
Corolla less than twice as long as the sepals :
　Corolla depressed below the apex, and contracted
　　round the stamens :
　　Sepals 1½ lin. long; leaves mostly 3–4 lin. long (368) **placentæflora.**
　　Sepals ½–¾ lin. long; leaves 1–1½ lin. long ... (367) **Guthriei.**
　Corolla not depressed below the apex nor con-
　　tracted round the stamens :
　　Anthers (viewed laterally) obtuse or widely
　　　rounded :
　　　Leaves (at least the older) recurved;
　　　　anthers broad-elliptical ...　　...　　... (364) **azaleæfolia.**
　　　Leaves and anthers not so :
　　　　Sepals about equalling the corolla, a
　　　　　little longer or shorter :
　　　　　Leaves subadpressed, erect, in-
　　　　　　curved ...　　...　　...　　... (360) **involvens.**
　　　　　Leaves spreading, straight or
　　　　　　nearly so :
　　　　　　Sepals transversely wrinkled;
　　　　　　　leaves rugulose or pitted ... (365) **sonderiana.**
　　　　　　Sepals smooth; leaves not
　　　　　　　rugulose or pitted　　... (369) **imbricata.**
　　　　Sepals much shorter than the corolla　(358) **physantha.**
　　Anthers (viewed laterally) acute or acuminate,
　　　rarely subobtuse :
　　　* Anther-cells narrow, 3½ times or more
　　　　longer than wide :
　　　　Corolla a little wider than its length;
　　　　　sepals with a thickened marginal
　　　　　line on either side, converging to
　　　　　the apex, instead of the usual
　　　　　median keel ...　　...　　...　　... (366) **crassisepala.**
　　　　Corolla longer than its width; keel on
　　　　　the sepals median :
　　　　　Leaves lanceolate or linear-lanceo-
　　　　　　late; sepals keel-tipped　　... (370) **triceps.**

Leaves narrow-linear or tri-
gonous :
  Corolla-segments broad-ob-
  long, rounded or sub-
  truncate :
    Anthers setaceo-acumi-
    nate ... ... (371) calyculata.
    Anthers subobtuse ... (371) calyculata,
  Corolla-segments not broad-   var. β.
  oblong ... ... (369) imbricata.

** Anther-cells broad or broadish, 3 times
or less longer than their width :
Style hooked or at least deflexed :
Sepals and bracts usually lax;
flowers somewhat few... ... (359) adunca.
Sepals and bracts usually adpressed; flowers somewhat
numerous ... ... (369) imbricata.
Style usually straight or nearly so :
Leaves recurved, spreading or
squarrose :
Sepals and corolla minutely
puberulous, or rough ... (363) pogonanthera.
Sepals and corolla glabrous,
smooth, shining ... ... (369) imbricata.
Leaves incurved or straight,
erect or spreading :
Sepals broad-ovate or elliptical ; leaves mostly spreading ... ... (361) chartacea.
Sepals obovate-oblong ; leaves
mostly incurved ... ... (362) suffulta.

§ 31. ELYTROSTEGIA. Inflorescence terminal ; flowers mostly 3-nate, or clustered, calycine or subcorolline. Bracts closely approximate, sepal-like, paleaceous or cartilaginous, mostly small. Sepals like the bracts, but larger, from half the corolla to equal to it in length. Corolla subtubular, cyathiform or suburceolate, glabrous, dry. Anthers lateral to subterminal, exserted or included, muticous or appendiculate. Style exserted. Stigma various, from subsimple to capitate, cyathiform or peltate. Ovary glabrous. Leaves 3-nate.

EXCEPTIONS : Bracts wanting in *E. accommodata, var. γ, ebracteata;* corolla may be viscidulous in *E. accommodata, var. β, subviscidula.*

Anthers included, or just manifest, aristate ... ... (376) diosmæfolia.
Anthers subexserted, or subincluded, muticous or very
minutely decurrent-toothed :
  Branches with plumose hairs ... ... (375) lepidota.
  Branches with simple hairs ... ... (373) accommodata.
Anthers fully exserted at maturity :
  Anthers muticous ... ... (372) lasciva.
  Anthers decurrent-aristate ... ... (374) glumæflora.

§ 32. APŒCUS. Inflorescence terminal, occasionally pseudo-lateral ; flowers usually 4-nate, sometimes subumbellate, calycine. Bracts remote, rarely sub-approximate, small. Sepals imbricate, adpressed to the corolla, nearly as long as and almost entirely concealing it, cartilaginous or subglumaceous. Corolla sub-obconic or cyathiform, mostly somewhat widened to the mouth, glabrous, dry or subviscidulous, 1½–2¼ lin. long or less ; segments longish, sometimes nearly equalling the tube. Anthers included or subincluded, lateral, appendiculate. Style exserted or perhaps occasionally included. Ovary glabrous ; ovules compressed ; mature seeds unknown.　Leaves 4-nate, small.　Small woody shrubs of various habit.

> Flowers 2-2½ lin. long; anther-crests ample, suborbicular, incised　　…　　…　　…　　…　　… (377) **Brownleeæ.**
>
> Flowers 1-1½ lin. long ; anthers subcrested or aristate, appendages narrow and small, or very minute　　… (378) **Caffrorum.**

§ 33. LAMPROTIS.　Inflorescence terminal, rarely also pseudo-lateral ; flowers usually 3-nate, often clustered, sometimes pseudo-capitate or pseudo-spicate, calycine.　Two upper bracts approximate or remote, the lower basal, wanting or represented by a floral leaf not adherent to the pedicels.　Sepals prominent, scarious, glabrous, smooth, coloured, from half the corolla to a little exceeding it in length.　Corolla mostly suburceolate or conical, more rarely subtubular or cyathiform, glabrous, usually dry, from 1½-10 lin. long ; limb spreading horizontally, or nearly erect or recurved, frequently discoloured or brownish.　Anthers included, lateral, mostly appendiculate, rarely muticous.　Style mostly included ; stigma capitate, rarely 4-fid.　Ovary glabrous.　Leaves 3-nate or opposite, generally adpressed or erect.　All parts of the plant most usually glabrous, smooth and glossy.—Closely allied to the § *Trigemma*, and being separated from it chiefly by secondary characters, is often difficult of distinction.　The "stellate-patent" character of the corolla-limb, attributed by former authors, is not constant, or at least is often obscure, disappearing by the connivance of the segments immediately after maturity, or after gathering, and therefore generally lost in dried specimens, and, besides, occurs more or less distinctly in species of other sections, outside of Bentham's subgenus *Stellanthe.*

EXCEPTIONS : leaves spreading or even squarrose in *E. taxifolia;* leaves, bracts and sepals, ciliate in *E. melanacme;* inflorescence pseudo-axillary (apparently always) in *E. chlorosepala,* sometimes so in *E. taxifolia, E. lævigata,* and *E. nodiflora;* flowers somewhat viscidulous in *E. nigrimontana* and *E. melanacme;* corolla viscidulous in *E. chlamydiflora;* style exserted in *E. nigrimontana.*　One or two other species have a slight pubescence on the younger branches.

A species with the habit of this section (405, *E. lanipes*) will be found in § *Eurystegia.*

> 1. *Leaves opposite (sometimes also 3-nate in 382, 394, and 395).*

Limb of the corolla concolorous ; leaves constantly opposite :
> Pedicels pubescent with minutely plumose hairs　… (379) **dianthifolia.**
> Pedicels without plumose hairs :
> > Corolla-tube inflated ; corolla 5-6 lin. long, red　　…　　…　　…　　…　　…　　… (380) **borboniæfolia.**
> > Corolla-tube conical, slightly inflated, slender ; corolla 3-3½ lin. long, white or pallid　　… (381) **lutea.**
> > Corolla ovoid-urceolate or subconical, 2½-3 lin. long, rosy　　…　　…　　…　　…　　… (382) **tenuifolia.**

Limb of the corolla discoloured or brownish ; leaves 3-nate or opposite :
> Corolla urceolate-tubular or narrow-urceolate, scarcely inflated below or contracted upwards, 1½-5 lin. long　　…　　…　　…　　…　　… (394) **corifolia.**
> Corolla urceolate-subconical, very wide below and much contracted upwards, 1¼-2 lin. long　　… (395) **nodiflora.**

> 2. *Leaves 3-nate (sometimes also opposite in 382, 394, and 395).*

Inflorescence pseudo-axillary ; flowers solitary below the ends of excurrent branches ; habit prostrate or decumbent (flowers golden-yellow)　　…　　…　　… (392) **chlorosepala.**

Inflorescence mostly terminal, only occasionally a few
  flowers pseudo-axillary; habit erect :
    Stigma 4-fid, cruciform   ...   . (389) **gnaphaloides.**
    Stigma capitate, rarely subsimple :
      Corolla from 3–10 lin. long :
        Anthers muticous :
          Corolla 9–10 lin. long; leaves 4–4½
            lin. long; sepals ovate-acuminate .. (383) **Alfredii.**
          Corolla 4–5 lin. long; leaves 1½–2½
            lin. long; sepals oblanceolate   ... (385) **steinbergiana.**
        Anthers minutely toothed at the base;
          sepals unguiculate-spathulate ...   ... (384) **bracteolaris.**
        Anthers broad-aristate, awns short and
          subulate; leaves 4–6 lin. long ...   ... (386) **taxifolia.**
        Anthers crested :
          Leaves ovate or broad-lanceolate :
            Sepals oblanceolate, 3 lin. long;
              anther-crests ⅓ as long as the cell (387) **pycnantha.**
            Sepals broad-obovate, 1½ lin.
              long; anther-crests ¾ as long
              as the cell   ...   . (391) **caledonica.**
          Leaves linear or linear-lanceolate   ... (394) **corifolia.**
      Corolla under 3 lin. (commonly 1½–2 lin.) long :
      Sepals as long as the corolla, or nearly so :
        Pedicels mostly 3–4 lin. long; sepals
          not or scarcely spreading :
            Sepals straight, naked, dry, apex
              concolorous   ...   . (396) **palliiflora.**
            Sepals incurved at the apex,
              ciliate, subviscid, apex darkly
              discoloured   ...   . (398) **melanacme.**
        Pedicels mostly 1½–2 (rarely 3) lin.
          long; sepals spreading :
            Flowers dry; style included   ... (394) **corifolia.**
            Flowers viscidulous; style ex-
              serted   ...   . (397) **nigrimontana.**
      Sepals mostly ½, or at most reaching ⅔,
        the length of the corolla :
        Corolla-limb concolorous or nearly so,
          not brownish :
            Anthers minutely appendiculate .. (393) **lævigata.**
            Anthers crested :
               Anthers 5–6 times longer
                than their greatest width;
                crests narrow and small... (388) **chlamydiflora.**
              Anthers relatively shorter;
                crests broader :
                Corolla subconical, much
                  widened at the base,
                  falling down in a fold
                  at maturity ...   ... (395) **nodiflora,** var. β.
                Corolla suburceolate or
                  subtubular, neither
                  much wider at the
                  base nor contracted
                  at the throat   ... (390) **articularis.**
        Corolla-limb discoloured or brownish :
          Corolla subconical, much wider
            at the base   ...   . (395) **nodiflora.**
        Corolla suburceolate or subtubu-
          lar, not much wider at the base (390) **articularis.**

§ 34. EURYSTEGIA (incl. *Eurylepsis,* Don). Inflorescence terminal, rarely also axillary ; flowers usually 3- (rarely 4-) nate, or umbellate, calycine. Bracts approximate or remote, mostly scarious, coloured, sometimes large. Sepals large and prominent, scarious, cartilaginous or subpetaloid, from half to nearly as long as the corolla. Corolla various, but never tubular (in 2 species quadripartite nearly to the base), pubescent or glabrous, dry, mostly from $3\frac{1}{2}$–11 lin. long, rarely 2 lin. only ; limb erect or slightly recurved. Anthers included, lateral or sublateral, appendiculate or rarely sometimes muticous. Style included, rarely subexserted. Ovary glabrous. Mostly strong-growing, large-flowered shrubs. Leaves 3- (or rarely 4-) nate, mostly spreading, often long, rigid and coarse. This section is scarcely separable from § *Trigemma,* except by its usually larger flowers.

EXCEPTIONS : leaves adpressed and the whole plant with the aspect of § *Lamprotis,* in *E. lanipes;* inflorescence often axillary in *E. Grisbrookii;* flowers 4-nate or umbellate in *E. papyracea,* only 2 lin. long in *E. holosericea,* var. *β ;* anthers muticous in *E. Halicacaba.*

<pre>
Leaves 4-nate; flowers 4-nate, clustered or umbel-
    late ...   . (406) papyracea.
Leaves 3-nate; flowers 3–2-nate or solitary :
    Corolla deeply parted nearly to the base :
        Corolla 7–9 lin. long ; bracts approximate (399) lanuginosa.
        Corolla 4–6 lin. long ; bracts remote   ... (400) Bodkinii.
    Corolla not deeply parted :
        Corolla 8–11 lin. long:
            Bracts approximate; flowers yellow
                or white :
                Corolla divided ½ its entire
                    length ; leaves 4–7 lin. long ... (401) Halicacaba.
                Corolla divided ⅙–⅕ its entire
                    length ; leaves 2–3 lin. long .. (402) monsoniana.
            Bracts remote, flowers rosy ...   ... (403) nobilis.
        Corolla 3½–6 lin. long ; sepals 2½–5 lin.
            long :
            Leaves closely adpressed ; bracts
                approximate   ...   . (405) lanipes.
            Leaves erect (but scarcely adpressed)
                or spreading ; bracts remote :
                Entirely glabrous ; leaves glau-
                    cous   ...   . (404) glauca.
                More or less pubescent; leaves
                    not glaucous :
                    Sepals puberulous; corolla
                        campanulate-cyathiform .. (407) holosericea.
                    Sepals glabrous ; corolla
                        conical-ovoid or globose-
                        urceolate ...   . (408) Grisbrookii.
        Corolla 2 lin. long ; sepals 1½ lin. long ... (407) holosericea, var. β.
</pre>

§ 35. ADELOPETALUM.

<pre>
Only species   ...   . (409) Nabea.
</pre>

§ 36. TRIGEMMA (*Calycinæ,* Klotzsch). Inflorescence terminal, or occasionally, by arrest of lateral branchlets, appearing axillary ; flowers 3-nate or less commonly 4-nate, rarely subumbellate, never capitate, calycine, rarely subcorolline. Bracts approximate or remote, usually small. Sepals mostly large, prominent and adpressed, seldom imbricate, never entirely concealing the corolla, scarious, membranous or rarely cartilaginous, glabrous, generally dry, rarely viscidulous, coloured. Corolla urceolate, cyathiform or campanulate, almost always glabrous, occasionally viscidulous, $1\frac{1}{2}$–$3\frac{1}{2}$ lin. long ; limb generally more or less spreading, or recurved, not horizontally stellate. Anthers included, or rarely (in one species only) exserted, mostly lateral, more rarely subterminal or terminal, the cells sometimes stalked upon the low or decurrent connective, variously appendiculate, more rarely muticous. Style mostly included, more rarely exserted. Ovary generally glabrous, more rarely lanate or villous ; ovules sometimes flat or compressed. Leaves mostly 3-nate, less commonly 4-nate, more or less spreading, or more rarely erect-incurved, seldom or never closely adpressed to the branches, glabrous or rarely pubescent. Tall or small shrubs of various habit.

EXCEPTIONS : anthers exserted in *E. gracilipes ;* subexserted in *E. petiolaris ;* terminal or subterminal in *E. petiolaris ;* subterminal in *E. Lycopodiastrum, E. brevifolia,* and *E. acuta.*

A. *Leaves 4-nate.*

Bracts approximate, large, equalling or longer than the
  sepals   ...   . (414) **gigantea.**
Bracts remote or subremote, smaller or shorter than the
  sepals :
    Style included or subincluded :
      Flowers 1¾ lin. long, subcorolline ; ovary
        densely villous   ...   . (431) **lasiocarpa.**
      Flowers 2–2½ lin. long, calycine or subcaly-
        cine ; ovary glabrous :
        Flowers terminal ; bracts 2–3 lin. long ... (415) **baccans.**
        Flowers pseudo-axillary and terminal ;
          bracts ¾ lin. long   ...   . (416) **irregularis.**
    Style at full maturity exserted or subexserted :
      Sepals ovate, obovate or oblanceolate, nearly
        covering the corolla :
        Stigma a short filiform point, much narower
          than the truncate apex of the style ;
          anther-crests broad   ...   . (417) **affinis.**
        Stigma subsimple or clavate ; anthers de-
          current-aristate ...   . (418) **propinqua.**
      Sepals lanceolate or lanceolate-linear, not
        nearly covering the corolla :
        Leaves oblong or lanceolate, 3–4 lin. long,
          1 lin. wide   ...   . (421) **depressa.**
        Leaves linear, 1½–2 lin. long, under ½ lin.
          wide   ...   . (431) **lasiocarpa.**

B. *Leaves 3-nate.*

I. Anthers included, or subincluded.

1. *Anthers muticous.*

Petioles at least ½ as long as the blade ...   . (422) **petiolaris.**
Petioles less than ½ as long as the blade :
  Corolla 3 lin. long, urceolate ; sepals 1¼ lin. long,
    small   ...   . (412) **crassifolia.**
  Corolla 3½ lin. long, campanulate ; sepals 2½ lin.
    long ...   . (411) **inclusa.**

2. *Anthers appendiculate.*

Pubescence of the branches floccose with minutely
  plumose or barbellate hairs   ...   . (410) **plumigera.**
Pubescence of branches (when present) not floccose :
  Corolla 3 lin. long, ovoid-inflated ; anthers crested (429) **pumila.**
  Corolla less than 3 lin. long :
    Ovary densely woolly or villous :
      Bracts approximate ; sepals broad-ovate,
        obtuse, pectinate-ciliate ; anthers aris-
        tate   ...   . (427) **fimbriata.**

<pre>
        Bracts remote; sepals lanceolate-acumi-
            nate, naked; anthers crested ...   ..._ (419) leucodesmia.
    Ovary glabrous :
        Anthers crested; crests broad-lanceolate
            to suborbicular :
                Leaves, bracts and sepals setaceo-
                    acuminate   ...   . (425) acuta.
                Leaves, bracts and sepals not so :
                    Anther-cells more or less dis-
                        tinctly stalked upon the basal
                        connective   ...   . (420) triflora.
                    Anther-cells not stalked :
                        Sepals with a deep depres-
                            sion at the base; pedicels
                            2½–3 lin. long;   corolla
                            1¾–2 lin. long   ...   ... (423) selaginifolia.
                        Sepals not foveolate; pedi-
                            cels 1½–2 lin. long; corolla
                            2–2½ lin. long :
                                Inflorescence terminal;
                                    bracts   approximate,
                                    2–2½ lin. long   ... (413) tegulæfolia.
                                Inflorescence   terminal
                                    and   pseudo-axillary;
                                    bracts remote, ¾ lin.
                                    long ...   . (416) irregularis.
        Anthers narrow-crested or broad-aristate :
            Anther-cells more or less distinctly
                stalked on the basal connective :
                    Leaves spreading, 3–5 lin. long;
                        corolla 2½ lin. long   ...   ... (420) triflora.
                    Leaves erect, closely imbricate,
                        1½–2 lin. long; corolla 1½ lin.
                        long   ...   . (428) Lycopodiastrum.
            Anther-cells not stalked on the con-
                nective :
                    Style exserted; sepals 2–2½ lin.
                        long   ...   . (421) depressa.
                    Style included; sepals 1¼–1½ lin.
                        long   ...   . (430) Thodei.
        Anthers decurrent-aristate;   cells   very
            obtuse, stalked   ...   . (426) brevifolia.

                    II. Anthers exserted.
    Flowers pseudo-axillary ...   . (424) gracilipes.
</pre>

V. PLATYSTOMA.

§ 37. POLYCODON. Inflorescence terminal; flowers 3-nate, corolline or sub-calycine, rarely calycine. Bracts small, approximate or remote. Sepals lanceo-late to orbicular, mostly wide, glabrous, prominent, coloured. Corolla obconic or broad-cyathiform, glabrous, dry, small, ¾–1¾ lin. long; segments equalling or somewhat longer than the tube, rarely a little shorter. Anthers included or exserted, lateral, more usually muticous, or (in two species) aristate or occa-sionally so. Style mostly exserted. Ovary glabrous or hairy. Leaves 3-nate.

Small shrubs with the aspect of §§ *Pyronium* and *Arsace*; distinguishable from the former by its more wide-mouthed corolla; from the latter by its broader, more prominent, and always coloured sepals.

EXCEPTIONS: flowers sometimes calycine in *E. leucanthera*, subcalycine in a few other species.

Style included, rarely as long as the corolla; flowers
   yellow ... ... (432) **leucanthera.**
Style exserted; flowers red or reddish :
   Leaves squarrose-recurved, ciliate (at least the
      younger) ... ... (433) **stenantha.**
   Leaves not so, not distinctly ciliate :
      Ovary glabrous or subglabrous :
         Anthers aristate; pore ⅔ of cell ... ... (434) **consobrina.**
         Anthers muticous; pore ⅞ of cell   ... (439) **macrotrema.**
      Ovary pubescent or bispidulous :
         Anthers muticous :
            Hairs on the pedicels simple :
               Stigma capitellate ... ... (435) **nemorosa,**
               Stigma   narrow-obconic-cyathi-
                  form; anther-pore ¼ of cell ... (439) **macrotrema.**
               Stigma very broad cyathiform-
                  peltate; anther-pore ⅔ of cell (438) **peltata.**
            Hairs on the pedicels minutely sub-
               plumose :
               Anthers oblong, obtuse; flowers
                  ¾–1 lin. long ... ... (436) **floribunda.**
               Anthers longitudinally semiovate,
                  acute or subacute; flowers
                  1½ lin. long ... ... (437) **rhodantha.**
         Anthers aristulate; stigma capitellate ... (435) **nemorosa.**

§ 38. EURYSTOMA. Inflorescence terminal; flowers usually 3-nate, mostly
calycine or subcalycine, more rarely corolline or subcorolline. Bracts and sepals
rarely as long as the corolla, mostly ½–¾ its length, cartilaginous or scarious, rigid,
glabrous, dry, coloured. Corolla from obconic to cyathiform or subcampanulate,
mostly wide-mouthed, rarely subglobose-urceolate, glabrous, dry, 1–2½ lin.
long; segments erect or spreading, shorter than the tube, or equalling or rarely
exceeding it in length. Anthers included, rarely subexserted, almost always
manifest, lateral, subterminal or rarely terminal, appendiculate or muticous.
Style usually exserted. Ovary glabrous or hairy. Leaves 3-nate or rarely opposite.

EXCEPTIONS: leaves opposite in *E. brevicaulis,* sometimes opposite in *E.
calycina, var. γ,* villous-pubescent in *E. comata;* flowers corolline or subcorolline in *E. argentea, E. media,* and *E. brevicaulis;* corolla sometimes
cyathiform or subglobose-urceolate in *E. lucida,* and subcyathiform in *E. media.*

1. *Leaves 3-nate.*

* Ovary glabrous :
   Branches floccose with subplumose or compound
      hairs :
         Corolla 2 lin. long; segments as long as the
            tube, white ... ... (448) **floccosa.**
         Corolla 1¼–1½ lin. long; segments ½–¾ as
            long as the tube, red ... ... (449) **lucida.**
   Branches when hairy having simple hairs :
      Leaves 4–10 lin. long; pedicels 3–5 lin. long (450) **mucronata.**
      Leaves and pedicels shorter :
         Sepals as long as the corolla ... ... (443) **nivea.**
         Sepals shorter than the corolla :
            Leaves very close-ranked, obovate,
               thick, ¾–1 lin. long ... ... (444) **lachneæfolia.**
            Leaves not so and also longer :
               Flowers corolline or subcorol-
                  line :
                  Sepals   ovate;   corolla
                     cyathiform;   segments
                     ¼ as long as the tube ... (440) **media.**

Sepals lanceolate; corolla obconic; segments nearly as long as the tube ... ... (441) **argentea.**
Flowers calycine or subcalycine :
   Anthers lateral :
      Anther-crests subulate, entire; sepals $\frac{1}{2}$ as long as the corolla, or less ... (440) **media.**
      Anther-crests broad, deeply incised; sepals $\frac{2}{3}-\frac{3}{4}$ as long as the corolla ... (449) **lucida.**
   Anthers generally terminal or subterminal, or very rarely lateral ... (445) **calycina.**

** Ovary more or less hairy :
   Anthers muticous; leaves villous-pubescent, tipped with a tuft of hairs ... ... (446) **comata.**
   Anthers appendiculate; leaves glabrous :
      Leaves erect, adpressed; corolla $1\frac{1}{2}$ lin. long ... ... (441) **argentea.**
      Leaves spreading or squarrose; corolla about 1 lin. long ... ... (447) **saxicola.**

2. *Leaves opposite.*

Leaves $\frac{3}{4}$–1 lin. long; pedicels glabrous; ovary entirely shortly-villous ... ... (442) **brevicaulis.**
Leaves mostly $1\frac{1}{2}$–2 lin. long; pedicels floccose or pubescent; ovary hirtellous on the apex or subglabrous ... ... (445) **calycina,** var. γ.

§ 39. MELASTEMON (*Lophandra*, Don; *Cornutarum*, Klotzsch). Inflorescence terminal, or by arrest of the lateral branchlets sometimes pseudo-axillary; flowers mostly 3-nate, more rarely umbellate or subcapitate, mostly corolline, rarely calycine or subcalycine, 1–4 lin. long. Bracts usually mediocre or inconspicuous. Sepals mediocre, scarious or cartilaginous, coloured, mostly glabrous, rarely hirsute. Corolla mostly obconic or broad-cyathiform, wide-mouthed, rarely cyathiform and subequal at the mouth, glabrous, dry, from 1–4 lin. long; segments of various length. Anthers mostly included, very rarely subexserted, lateral, usually more or less produced at the apex (i.e. more than in other sections) beyond the pore, acute or acuminate, usually roughly papillose or scaberulous, or coarsely shaggy, mostly muticous, very rarely aristulate. Style exserted. Stigma generally small. Ovary glabrous, or more rarely minutely puberulous. Leaves mostly 3- more rarely 4-nate or scattered, glabrous, or in one species hispid.

EXCEPTIONS : sepals in *E. trachysantha* densely hirsute; corolla in *E. trachysantha*, cyathiform and subequal at the mouth; anthers sometimes subexserted in *E. lavandulæfolia*; anther-cells in *E. nervata* and in *E. trachysantha* are acute, but very little produced beyond the pore; anthers are sometimes very minutely aristulate in *E. cubica.*

NOTE : *E. melanthera*, a species with anthers produced beyond the pore, as in this section, will be found in § *Gamochlamys* (462).

Flowers terminal or axillary, mostly more or less clustered, umbellate or subcapitate; leaves 3-4-nate or scattered :
   Leaves 1–2 lin. long, mostly uniformly erect-incurved; corolla $1\frac{1}{4}$–$1\frac{3}{4}$ lin. long ... ... (451) **seriphiifolia.**

Leaves 2–5 lin. long, mostly spreading or also
    incurved; corolla 1½–3¼ lin. long   ...   ... (452) cubica.
Flowers terminal and 3-nate, or axillary towards the
    ends of the branches, pseudo-umbellate, lax, not
    clustered; habit laxly diffuse   ...   . (453) tetrathecoides.
Flowers terminal, 3-nate or solitary; leaves 3-nate :
    Ovary more or less hairy :
        Leaves linear, sulcate :
            Branches and pedicels floccose with
                minute subplumose hairs   ...   ... (458) Gillii.
            Branches and pedicels pubescent with
                simple hairs   ..   .. (457) lavandulæfolia.
        Leaves lanceolate, more or less open-
            backed :
            Sepals glabrous; pedicels 2–2½ lin. long (459) nervata.
            Sepals densely hirsute; pedicels 4–6 lin.
                long   ...   . (460) trachysantha.
    Ovary glabrous :
        Leaves notably few, spreading, mostly
            shorter than the internodes, 3–4½ lin.
            long ...   . (453) tetrathecoides.
        Leaves not so and shorter :
            Leaves adpressed, ¾–1 lin. long; flower-
                ing branches mostly long and fili-
                form   ...   . (454) humifusa.
            Leaves not usually adpressed, erect or
                spreading, 1–3 lin. long; branchlets
                not long and thin :
                Pedicels pubescent with simple
                    hairs :
                    Corolla 2½–4 lin. long   ... (456) moschata.
                    Corolla 1–2¼ lin. long   ... (455) cristæflora.
                Pedicels floccose with minute plu-
                  mose hairs   ...   . (452) cubica.

§ 40. GAMOCHLAMYS. Inflorescence terminal, or rarely (by arrest of lateral
branchlets) pseudo-axillary; flowers mostly 3-nate, rarely solitary, corolline or
more rarely calycine. Bracts usually small and inconspicuous. Calyx more or
less gamosepalous, scarious or cartilaginous, glabrous or hairy; segments from
⅓–¾ the entire length. Corolla obconic, broad-cyathiform, cyathiform or subfunnel-shaped, glabrous, dry, from 1½–3 lin. long. Anthers most usually included, rarely subexserted, lateral, muticous; in one species distinctly produced at the apex beyond the pore (as in § *Melastemon*). Style exserted or
included. Ovary most usually hairy. Leaves 3-nate. Shrubs with the general
aspect of §§ *Polycodon* and *Melastemon*, except the first species, which is more
woolly than is usual in these sections.

EXCEPTIONS : anther-cell produced considerably beyond the pore in *E. melanthera* (as in § *Melastemon*); anthers subexserted in *E. canaliculata*.

Calyx with short white closely-matted tomentum   ... (461) Passerinæ.
Calyx not so :
    Stigma peltate or cyathiform   ...   . (466) natalitia.
    Stigma simple, obconic or capitellate :
        Calyx-segments about equal to the tube,
            or somewhat shorter or longer, mostly
            more or less hairy :
            Pedicels and calyx puberulous or gla-
                brous; anther produced beyond the
                pore   ...   . (462) melanthera.
            Pedicels and calyx villous; anther not,
                or slightly, produced beyond the pore (463) Chamissonis.

Calyx-segments about 2-4 times longer than
the tube, glabrous :
 Calyx-segments convex, strongly keeled
  from the base upwards　..　...　(464) **longipes.**
 Calyx-segments concave or flattish, not
  strongly keeled, margins recurved　...　(465) **canaliculata.**

§ 41. CYATHOLOMA (*Coloratæ*, Klotzsch). Inflorescence terminal; flowers
3-nate, sometimes (by arrest of the branchlets) pseudo-axillary, subcalycine or
subcorolline. Sepals lax, bright-coloured and prominent, more than ½ as long
as the corolla, or more adpressed and somewhat smaller. Corolla with a subglobose, constricted tube and larger, ascending, cup-like limb, or (in one species)
ovoid-urceolate with shorter limb, 2-5 lin. long. Anthers included, lateral, in
two species subprognathous at the base, appendiculate or muticous. Ovary
sessile or stipitate, glabrous. Leaves 3-nate.

A not very natural section even as proposed by Bentham on the two original
species, *E. Thunbergii* and *E. Corydalis*, which are utterly unlike in habit and
aspect, and have only the single character of a similar corolla in common. To
these we have added a third which is only related to the former in appearance,
sepals and bracts but differs from both in several important characters. We
have associated these rather than make three sections for the three species.

Corolla globose at the base with a large dilated cup-
 shaped limb :
  Sepals lax, dry, bright yellow　...　...　...　(467) **Thunbergii.**
  Sepals adpressed, viscidulous, horn-colour　...　(468) **Corydalis.**
 Corolla ovoid-urceolate with a small limb　...　...　(469) **flavisepala.**

Section I. GIGANDRA. (Sp. 1-5.)

1. E. Petiveri (Linn. Diss. Erica, n. 50 ; Mant. alt. 235) ; erect,
2-3 ft. high; branches stout, with many short mostly curved
branchlets ; leaves 3-nate, spreading or recurved, more rarely erect
and straight, linear, subglabrous, 2-4 lin. long ; flowers 1-3-nate,
mostly cernuous ; pedicels $2-2\frac{1}{2}$ lin. long ; bracts approximate,
ovate, $\frac{1}{2}-\frac{2}{3}$ the length of the sepals ; sepals ovate, oblong or narrow-
lanceolate, acute, scarious, concave, glabrous, $2\frac{1}{2}-3\frac{1}{2}$ lin. long, $\frac{1}{3}-\frac{2}{3}$ as
long as the corolla ; corolla narrow ovoid to tubular, glabrous,
greenish-yellow, orange or reddish, $3-8\frac{1}{2}$ lin. long ; limb erect,
2-3 lin. long. *E. coccinea, Linn. Sp. Pl. ed.* 1, i. 355, *not of Diss.
E. Sebana, Donn, Hort. Cantab.* 45, *ex Willd. Sp. Pl.* ii. 395 ;
Bauer, Exot. Pl. t. 10 ; *Wendl. Eric. Ic. fasc.* 10, 5 ; *Lodd. Bot.
Cab. t.* 23 ; *Benth. in DC. Prodr.* vii. 621. *E. Sebana, vars. aurantia,
lutea, viridis and spicata, Andr. Heathery, tt.* 83, 84, 85, 190, *and
Col. Heaths, tt.* 55, 56, 57, 129. *E. Sebana, var. nana, Andr.
Heathery, t.* 189. *E. Sebana, var. lutea, Lodd. Bot. Cab. t.* 266.
E. socciflora, Salisb. in Trans. Linn. Soc. vi. 347. *E. demissa,
Sinclair, Hort. Eric. Wob.* 8, *not of Klotzsch. E. baculiflora,
Salisb. in Trans. Linn. Soc.* vi. 346. *E. cothurnalis, Salisb. Prodr.*
297, *and in Trans. Linn. Soc.* vi. 347.—*E. africana angustifolia,*
etc., Seba, Thes. i. 32, *t.* 21, *fig.* 4.

VAR. β, **pubescens** (Bolus) ; bracts and sepals almost without keel, and cano-
tomentose.

VAR. γ, **intermedia** (Bolus) ; flowers frequently solitary, small, 3-4 lin. long.
E. intermedia, Klotzsch ex Benth. in DC. Prodr. vii. 621.

VAR. δ, **Willldenovii** (Bolus); leaves usually shorter and straighter; flowers mostly solitary, more rarely 3-nate; sepals oval or oblong, without keel or nearly so. *E. Petiverii, var. β, Thunb. Diss. Erica,* 21. *E. Petiverii, Willd. Sp. Pl.* ii. 394; *vars. fusca and lutea, Wendl. l.c. fasc.* 9, 35. *E. petiveriana, and var. aurantia, Andr. Heathery, tt.* 78, 79, *and Col. Heaths, tt.* 44, 117; *Wendl. l.c. fasc.* 14, 23. *E. Petiveri, Lodd. Bot. Cab. t.* 1426; *Benth. l.c.* 621. *E. follicularis, Salisb., l.c.* 348; *Rach in Linnæa,* xxvi. 772.

VAR. ε, **melastoma** (Benth. in DC. Prodr. vii. 622); habit of the preceding; bracts and sepals lanceolate, acuminate, the latter more than half the length of the corolla, coloured; limb of the corolla oblong, 3 lin. long. purple. *E. melastoma, Andr. Heathery, t.* 30, *and Col. Heaths, t.* 37; *Wendl. Eric. Ic. fasc.* 17, 67, *t.* 25; *Lodd. l.c. t.* 333.

SOUTH AFRICA: without locality, *Drège,* 7695! 7696! Var. δ: *Thunberg.* And *cultivated specimens* of the type and var. δ!

COAST REGION: on mountains from 500–5000 ft.: Tulbagh Div.; near Tulbagh, *Schlechter,* 7473! Cape Div.; Table Mountain and other places, *Thunberg, Burchell,* 612! 776! *Ecklon,* 99! *Drège! Bolus,* 3770! 4568! Stellenbosch Div.; Hottentots Holland, *MacOwan, Herb. Norm. Aust.-Afr.,* 5! *Zeyher,* 3185β! Caledon Div.; near Genadendal and Baviaans Kloof, *Burchell,* 8627! 7650! 7803! *Schlechter,* 10308! Klein River Kloof, *Zeyher,* 3185a! Swellendam Div.; near Swellendam, *Burchell,* 7302! Riversdale Div.; summit of Kampsche Berg, *Burchell,* 7125! George Div.; between Touw River and Kaymans River, *Burchell,* 5768! Var. β: Stellenbosch Div.; Lowrys Pass, *Burchell,* 8246! Hottentots Holland, *Niven,* 129! Caledon Div.; Houw Hoek, *Burchell,* 8130! Div.? *Bolus,* 8036! Var. γ: Swellendam Div., *Niven,* 127! George Div.; Outeniqualand, *Niven,* 128! Div.? *Guthrie,* 3015! Var. δ: Cape Div.; mountains near Cape Town, *Burchell,* 553! *Cooper,* 2681! *Wilms,* 3449! *Bolus,* 4476! Stellenbosch Div.; Hottentots Holland, *Mund,* 1! Bredasdorp Div.; Elim, *Schlechter,* 7741! Swellendam Div.; peak near Swellendam, *Burchell,* 7424! Var. ε: Caledon Div.; Zwart Berg, near Caledon, *Mund,* 2! *Pappe!* near Hemel en Aarde, *Zeyher,* 3186! Bredasdorp Div.; Elim, *Schlechter,* 7631!

CENTRAL REGION: Var. ε: Ceres Div.; Cold Bokkeveld, *Schlechter,* 8899.

2. E. vestiflua (Salisb. in Trans. Linn. Soc. vi. 346); erect, 1½ ft. high; branches ascending, pubescent, covered with shorter branchlets and leaf-tufts; leaves 3-nate, spreading-incurved, linear, slender, margins inflexed, pilose, 4–5 lin. long; petiole longish, pallid, flat, flowers solitary (or 3-nate ex *Andrews*), cernuous; pedicels about 1 lin.; bracts approximate, broad ovate, rigid, paleaceous, glabrous, 2 lin. long; sepals like the bracts, subobtuse, apiculate or minutely keel-tipped, 4 lin. long; corolla tubular, curved, about 8 lin. long, *Benth. in DC. Prodr.* vii. 622. *E. Petiveri, var. hirsuta, Andr. Heathery, t.* 80, *and Col. Heaths, t.* 118. *E. picta, Sinclair, Hort. Eric. Wob.* 19.

SOUTH AFRICA: without locality; growing on the mountains, *Niven,* 6! *Herb. Salisbury!* and *cultivated specimens!* All in Herb. Kew.

In floral structure this much resembles *E. Petiveri, var. Willdenovii,* but the flowers are usually smaller and the habit and leaves quite different.

3. E. lineata (Benth. in DC. Prodr. vii. 622); erect, glabrous, 1–1½ ft. high; branches subvirgate; leaves 3-nately fasciculate, erect-incurved, densely imbricate, linear-subulate, slender, 6–8 lin. long; inflorescence pseudo-axillary, on very short branchlets; pedicels 4–7–10 lin. long ; bracts remote, adpressed, slender, scarious ; sepals ovate or oblong, subobtuse, flat, not keeled, faintly nerved, keel-tipped, rigid, scarious, $3\frac{1}{2}$–5 lin. long; corolla tubular, slightly inflated at the base, glabrous, dry, $5\frac{1}{2}$–6 lin. long ; limb short, erect.

SOUTH AFRICA : without locality, *Mund !*
COAST REGION : Bredasdorp Div. ; near Elim, 300 ft., *Schlechter,* 7712 ! hills near Ratel River, 50 ft., *Schlechter,* 9711 !

4. E. scariosa (Thunb. Fl. Cap. ed. Schultes, 350) ; erect, glabrous, 1–$1\frac{1}{2}$ ft. high ; leaves erect-spreading, crowded, mostly on very short branchlets ; inflorescence strictly terminal, flowers solitary, sometimes by the suppression of the branchlets pseudo-lateral ; pedicels 3–4 lin. long ; bracts remote, minute ; sepals ovate to ovate-lanceolate, acute or acuminate, strongly keeled, scarious, $1\frac{1}{2}$–2 lin. long ; corolla ovoid to tubular-inflated, white or rosy, $3\frac{1}{4}$–$4\frac{1}{2}$ lin. long, 2–$2\frac{1}{2}$ lin. wide. *E. Plukenetii β, Thunb. Diss. Erica, t. 2. E. Plukenetii, var. inflata, Wendl. Eric. Ic. fasc. 22, 147, t. 55 ? E. Plukenetii, conferta, Wendl. l.c. fasc. 24, 183, t. 69 ?. E. fusiformis, vars. γ and δ, Salisb. in Trans. Linn. Soc. vi. 347. E. Plukenetii, var. eckloniana, Klotzsch, and var. dregeana, Klotzsch ex Rach in Linnæa,* xxvi. 772. *E. Petiverii var. β, Salisb. in Thunb. Diss. Erica, ed. altera, 20, fig. 5, not of Thunb. E. penicillata, Benth. in DC. Prodr.* vii. 622, *not of Andr.*

SOUTH AFRICA : without locality, *Thunberg ! Niven ! Zeyher,* 1090! *Drège,* 7694 ! *Bolus,* 2964 !
COAST REGION : Clanwilliam Div. ; Cederberg Range, *Leipoldt,* 357 ! and without precise locality, *Zeyher ! Mader !* Piquetberg Div. ; near Twenty-four Rivers, *Zeyher,* 1090 ! Tulbagh Div. ; Ceres Road, *Schlechter,* 9084 ! Worcester Div. ; Breede River Valley, *Bolus,* 5114 ! Paarl Div. ; French Hoek Mountains, 1000 ft., *MacOwan, Herb. Norm. Aust-Afr.,* 937 ! Cape Div. ; Muizen Berg, *Harvey !*
CENTRAL REGION : Ceres Div. ; near the Wagenbooms River, 5500 ft., *Schlechter,* 10151 !

We admit this species with doubt. Its principal distinction from *E. Plukeneti* lies in the shorter and proportionately broader corolla, and there does appear to be a break between the shortest-flowered forms of *E. Plukeneti,* and the longest-flowered of this. It is curious, however, that *Andrews' Heathery,* t. 135, which Bentham cites, is clearly *E. Plukeneti,* with corollas of $6\frac{1}{2}$–$7\frac{1}{2}$ lin. long. The specimens Bentham had in describing were those collected by Drège, Niven, and Zeyher (1098), which show corollas shorter than any others, viz. 3–5 lin. long, and sometimes somewhat globose, nearly 3 lin. in diameter. These are shorter and broader than any others we have seen. We are obliged to cite Wendland's figures with doubt : they are from cultivated specimens, and probably more luxuriant than any wild ones ; in fact, we have not seen any figure fairly representing the wild plant, which has a distinct facies, but little like that of *E. Plukeneti,* and on this account we allow it, with some reluctance, to stand.

[This is the plant figured by Salisbury as " *E. Petiverii,* var. *β,*" and the type specimen so named from his herbarium, now at Kew, is a branchlet broken from (as is evidenced by the ends fitting together) the type specimen of *E. scariosa* in Thunberg's herbarium, on the sheet of which Thunberg has written " *E. Plukenetii β,*" and has figured the plant under that name. Salisbury evidently made a mistake in copying and publishing the name as " *E. Petiverii,* var. *β,*" since the plant so named in Thunberg's herbarium is *E. Petiveri,* var. *Willdenovii,* Bolus.—*N. E. Brown.*]

5. E. Plukeneti (Linn. Sp. Pl. ed. 1, 356); erect, glabrous, 1–2 ft. high; leaves spreading, usually densely imbricate, linear, subtrigonous, mostly 6–8 rarely 11 lin. long; inflorescence generally pseudo-axillary, sometimes obviously terminal; pedicels 4–6 lin. long; sepals ovate-lanceolate to narrow-lanceolate, acuminate, keeled, very concave, 2–5 lin. long; corolla tubular, more or less inflated towards the base, 5½–9 lin. long, from twice to four times as long as the sepals, red-orange or whitish. *E. Plukenetii, var. pinifolia, Wendl. Beobacht.* 45, *var. pinea, Wendl. Eric. Ic. fasc.* 1, 9; *vars. interrupta and inflata, Wendl. Beobacht.* 46 *and Eric. Ic. fasc.* 2, 21, *and fasc.* 22, 147, *t.* 55. *E. Plukenetia, Andr. Heathery, t.* 186; *vars. nana and albens, Andr. l.c. t.* 139, 187, *and Col. Heaths, t.* 47, 256. *E. plukenetiana, Bauer, Exot. Pl. t.* 9; *Lodd. Bot. Cab. t.* 1274. *E. Plukenetii, Berg. Descr. Pl. Cap.* 91; *Benth. in DC. Prodr.* vii. 622. *E. penicillata, Andr. Heathery, t.* 135, *and Col. Heaths, t.* 116, *not of Benth.; Lodd. l.c. t.* 1918. *E. fusiformis, Salisb. Prodr.* 297, *and in Trans. Linn. Soc.* vi. 345. *E. adenostoma, Kuntze in Linnæa,* xx. 38.

Var. β, **bicarinata** (Bolus); sepals ⅝ as long as the corolla, with a deep median channel down the keel. *E. revolvens, Bartl. in Linnæa,* vii. 631.

Var. γ, **brevifolia** (Bolus); leaves 2½–3½ lin. long, crowded, subsquarrose; sepals short, about ¼ of the 8–9 line long corolla; lateral branchlets very short.

South Africa: without locality, *Drège,* 7692! 7693a! *Zeyher,* 1089! and *cultivated specimens!* Var. β: *Thom,* 611! *Herb. Salisbury!*

Coast Region, common, ascending from 20–5500 ft. : Clanwilliam Div.; Koude Berg, *Schlechter,* 8743! Tulbagh Div.; Witsen Berg, *Burchell,* 8689! Mitchells Pass, *Schlechter,* 8942! Paarl Div.; Great Britain Park, *Wilms,* 3444! Cape Div.; Table Mountain and other localities, *Thunberg, Ecklon,* 102! 275! *MacOwan, Herb. Norm. Aust.-Afr.,* 6! *Burchell,* 554! 281! 8410! 8574! *Bolus,* 2116! 2963! Stellenbosch Div.; Hottentots Holland, *Mund,* 4! Stellenbosch, *Niven,* 138! Caledon Div. ; *Burchell,* 7954! *Schlechter,* 5416! 7345! 10399! *Zeyher,* 3184β! *Bolus,* 5115! Swellendam Div. ; peak near Swellendam, *Burchell,* 7385! Var. β: Bredasdorp Div.; Elim, 2500 ft., *Schlechter,* 7743! Caledon Div.; Baviaans Kloof, *Burchell,* 7767! *Niven,* 231! Swellendam Div. ; near Swellendam, *Mund,* 3! Div.? *Guthrie,* 4575! Var. γ: Piquetberg Div.; Piquetberg Range, *Schlechter,* 5208!

Chiefly variable in the size of its flowers, and the relative proportions of the sepals and corolla. Mr. G. F. Scott-Elliot observed that this species was abundantly visited on the hills near Cape Town by a "sugar-bird" (*Nectarinia chalybea*). He says, "The distance from the extremities [of the anthers] to the base of the flower is 16 lines, which is exactly the length of the beak of *N. chalybea*. . . . . The bird always seizes the branch below the flowers and exhausts one branch before going to another. It is an important article of diet to the birds, as it blooms practically all the year." (*Ann. Bot.* iv. 269–270).

Section II. **DIDYMANTHERA.** (Sp. 6–11.)

6. E. monadelphia (Andr. Col. Heaths, t. 38); erect, 12–18 in. high; branches subvirgate, or spreading, stout; leaves 3-nate, sub-erect and imbricate, or squarrose, linear-trigonous, blunt, 1½–2½ lin. long; inflorescence mostly 1–2-flowered; flowers erect or cernuous; pedicels 1½ lin. long; bracts broad-ovate, obtuse, sulcate-keeled, thick, rigid, coloured, about half as long as the sepals; sepals like the bracts, but narrower, about $2\frac{1}{2}$ lin. long; corolla-limb short, erect, entirely crimson, 2–$2\frac{1}{2}$ times the length of the sepals; filaments $\frac{1}{2}$ lin. wide, slightly broader than the $2\frac{1}{4}$ lin. long linear anther, slightly curved at the articulation with it. *Andr. Heathery, t.* 129; *Benth. in DC. Prodr.* vii. 622; *Willd. Sp. Pl.* ii. 396. *E. furfurosa, Salisb. in Trans. Linn. Soc.* vi. 348.

COAST REGION : Cape Div. ; Simons Bay, *MacGillivray,* 441! Caledon Div. ; Klein Houw Hoek, *Niven,* 130! Klein River Mountains, *Zeyher,* 3187! Hermanus, *Guthrie,* 4124! Bredasdorp Div. ; Elands Kloof, 1200 ft., *Schlechter,* 9750! near the mouth of Ratel River, *Bolus,* 8509! Riversdale Div. ; above the waterfall at Garcias Pass, *Burchell,* 7002! also *cultivated specimens !*

Bentham remarks that the younger filaments are submonadelphous, whence the name. We have not observed this; perhaps it is less frequent in wild specimens.

7. E. Banksia (Andr. Col. Heaths, t. 5, and Heathery, t. 105); procumbent; branches stout, flexuous, rigid, brittle, cano-pubescent, 6–9 in. long; leaves erect-spreading, imbricate, broad-linear, acute, thick, pallid, somewhat glossy, 2–4 lin. long; flowers mostly cernuous; pedicels 1 lin. long or less; bracts ovate, acute, keeled, rigid, thick, concave, 3–4 lin. long; sepals like the bracts but more scarious and more acuminate (sometimes " very unequal," *Bentham*), 3–6 lin. long; corolla 7–10 lin. long; tube rosy, white or yellowish; segments erect or spreading, acute, sometimes discoloured; filaments $\frac{1}{2}$ lin. wide; anthers somewhat spathulate, cells at length very divaricate; ovary villous with white hairs, sometimes glabrescent. *E. Banksii, Willd. Sp. Plant.* ii. 395; *Bauer, Exot. Pl. t.* 29; *Wendl. Eric. Ic. fasc.* 14, 21; *Benth. in DC. Prodr.* vii. 623. *E. aristata, Salisb. Prodr.* 292. *E. fragilis, Salisb. in Trans. Linn. Soc.* vi. 346. *E. pilifera, Thunb. Fl. Cap. ed. Schultes,* 350; *Rach in Linnæa,* xxvi. 772.

VAR. β, purpurea (Andr. Heathery, t. 106); corolla-limb bright purple; pore of the anther shorter; ovary glabrous. *Andr. Col. Heaths, t.* 151. *E. monadelpha, Curt. Bot. Mag. t.* 1370, *not of Andr. E. banksiana, Paxt. Mag.* vii. 243, *with plate.*

SOUTH AFRICA : without locality, *Thunberg, Mund!* and *cultivated specimens* of the type and var. β !

COAST REGION : Stellenbosch Div.; Hottentots Holland Mountains, *Zeyher,* 3189! *Niven,* 131! *Masson* (ex *Salisbury*). Caledon Div.; mountains near Lowrys Pass, 1800 ft., *Guthrie,* 2025! Houw Hoek, 3600 ft., *Schlechter,* 7572! *Bolus,* 5457! Var. β : Caledon Div. ; on Babylons Tower, *Guthrie,* 4092!

8. E. primulina (Bolus); procumbent, much branched, under 1 ft. high; branches spreading, flexuous, rigid, brittle, cano-puberulous, glabrescent, 3–6 in. long; leaves 3-nate or somewhat irregular, crowded, spreading or suberect, incurved, oblong, subobtuse, thick, flattish above, rounded and deeply sulcate below, pale grey-pubescent, 3–4 lin. long; flowers 2–3-nate, subcernuous, pedicels copiously woolly, $1\frac{1}{2}$–2 lin. long; bracts approximate, imbricating the sepals, unequal, ovate, acute, concave, subscarious, ciliate, 2–$2\frac{1}{2}$ lin. long; sepals erect, adpressed, in opposite pairs, the lower imbricating the upper, oblong or narrow-ovate, acute, concave, keel-tipped, subscarious, ciliate, pallid, $2\frac{1}{2}$–3 lin. long; corolla trumpet-shaped, from a narrow base widening upwards, straight or curved, nerved, glabrous, dry, pale-primrose, about 9–10 lin. long; segments erect, ovate or oblong, obtuse, about $\frac{1}{4}$ the length of the tube; filaments linear, about $\frac{1}{4}$ lin. wide, slightly widening and becoming reddish-brown at the apex, with a small tooth-like projection on either side below the anther; anthers subexserted (in our specimens which may probably not be quite mature), oblong, subobtuse, cells becoming somewhat divaricate, about $1\frac{1}{2}$ lin. long, pore $\frac{2}{3}$ as long as the cell; style subexserted or manifest; stigma turbinate; ovary subobconic, glabrous.

CoAST REGION: Ladismith Div.; on rocky mountain slopes of the Klein Zwartebergen, near Seven Weeks Poort, 3250 ft., *Marloth*, 2937!

This has the habit of *E. Banksia*, but more nearly the floral structure of *E. viridiflora*, yet is well distinguished from either. It is one of the few species which straggle far inland, the only known station being about 70 miles from the nearest coast. It bears a remarkable superficial resemblance to *E. Maximiliani*, which grows not far from the present, but still further (83 miles) from the sea. The habit of the two is different, but a small frustule of either may be easily mistaken for the other. The anthers afford a ready means of distinction.

9. E. viridiflora (Andr. Heathery, t. 299); 3–4 ft. high; branches ascending, stout; leaves erect-spreading, imbricate, linear, acute, glabrous, $2\frac{1}{2}$–4 lin. long; flowers cernuous; pedicels $1\frac{1}{2}$ lin. long; bracts lanceolate, acute, scarious, 3–4 lin. long; sepals like the bracts, but acuminate, ciliate, 4–6 lin. long; corolla-tube equal or widening upwards, glabrous, dry, green, 11–13 lin. long; limb erect or spreading; segments oblong, $2\frac{1}{2}$ lin. long; filaments narrow, $\frac{1}{5}$ lin. wide; anthers sublateral, spreading, linear, with a short spreading appendage at the base, or more rarely muticous; ovary glabrous. *Andr. Col. Heaths, t.* 287; *Lodd. Bot. Cab. t.* 917; *Benth. in DC. Prodr.* vii. 623. *E. clavata, Andr. Heathery, t.* 208, *and Col. Heaths, t.* 159.

CoAST REGION: George Div.; mountains near George, 1600 ft., *Guthrie*, 3657! *Bolus*, 8672! *Pappe!* Montagu Pass, 2500 ft., *Schlechter*, 5801!

E. clavata, Audr., differs chiefly from having been described and figured from a plant not fully developed, except as to its muticous anthers, in respect of which it is probably variable.

10. E. sphenanthera (Tausch in Flora, 1839, 626); leaves 4-nate, erect, linear, acute, glabrous, $\frac{1}{2}$ in. long; inflorescence axillary, below the ends of the branches, secund; flowers cernuous; pedicels 1 lin. long; bracts approximate, ovate, acuminate, scarious, nerved, rigid, glabrous, $2\frac{1}{2}$ lin. long; sepals ovate, cuspidate, keel-tipped, scarious, rigid, glabrous, coloured, 4 lin. long; corolla tubular, glabrous, dry, $10\frac{1}{2}$ lin. long, including the erect obtuse $2\frac{1}{2}$ lin. long limb-segments; anthers exserted, subterminal, linear, apex obtusely rounded, muticous, $1\frac{1}{2}$–2 lin. long; ovary glabrous.

SoUTH AFRICA? Described from a specimen without indication of origin (possibly cultivated and a hybrid), preserved in the herbarium of the University of Prague, which we have examined.

11. E. cerviciflora (Salisb. in Trans. Linn. Soc. vi. 362); branches somewhat slender; leaves 4-nate (?), sub-6-nate or scattered, erect-spreading, linear, acute, 1-nerved above, sulcate below, subglabrous, about ½ in. long; pedicels puberulous, about 2 lin. long; bracts subapproximate, lanceolate, acute; sepals lanceolate, acuminate, glandular-ciliolate, 2–2½ lin. long; corolla tubular, slightly inflated in the middle, narrowed at the throat, glabrous or minutely puberulous, about ½ in. long; segments slightly spreading, obtusely rounded; anthers exserted (*Salisbury*), (included according to *Bentham*, but the filaments in the type are certainly longer than the corolla-tube), sublateral, cells bipartite, at length somewhat spreading above, muticous or decurrent-denticulate; filament just below the anther dilated into a small point on either side; ovary glabrous. *Benth. in DC. Prodr.* vii. 664. *E. inaperta, Hort. ex G. Don in Sweet, Hort. Brit. ed. 3, 429.*

COAST REGION : Stellenbosch Div.; Hottentots Holland, *Mulder !*

A singular species only known to us from the type in Salisbury's herbarium at Kew. In a note on the ticket the author observes : "I do not know any to which it is allied; the peduncles are certainly axillary in the upper leaves, not alternate with them. In appearance it does not differ much from *E. melliflua* [*E. cruenta, Soland*]. The anthers resemble those of *E. grandiflora*, muticous, but with a rudimentary spur on the filament, almost as in *E. hottoniæflora* [*E. cubica, L.*]." The shape of the anther is, however, remarkably like that of *E. viridiflora*, and in spite of its lateral inflorescence it is, perhaps, better placed here.

Section III. PLEUROCALLIS. (Sp. 12–35.)

12. E. mammosa (Linn. Mant. alt. 234); erect, 1–4 ft. high; branches generally few, ascending, sometimes virgate, leafy above, naked below; leaves 4-nate or scattered, erect-spreading, linear-lanceolate, subacute, 3–5 lin. long; flowers in a loose or congested raceme; pedicels 1–2 lin. long; bracts remote, small; sepals ovate, acute, keeled, scarious, 1½–2 lin. long; corolla tubular, slightly inflated, glabrous, dry, 4-foveolate at the base, 6–12 (mostly 8–10) lin. long, purple-red, scarlet or white; limb short, erect, concolorous; filaments capillary; anthers included, lateral, oblong or cuneate, mostly with a sharp basal point in front, about 1 lin. long, aristate, awns of variable length; ovary glabrous. *Andr. Heathery,* tt. 124, 125 (*var. minor*), *and Col. Heaths, t.* 33, *and* 184, *not of Thunb.; Lodd. Bot. Cab. t.* 125, 951 (*var. pallida*). *Benth. in DC. Prodr.* vii. 624. *E. abietina, Berg. Pl. Cap.* 105; *Thunb. Diss. Erica,* 42, *not of Linn. E. laxa, Lam. Encycl.* i. 480, *not of others. E. verticillata, Andr. Heathery, t.* 48, *and Col. Heaths, t.* 66; *Lodd. Bot. Cab. t.* 145, *not of Berg. E. quadrifossa, Salisb. Prodr.* 293. *E. coralloides, Hort. ex Tausch in Flora,* 1834, 600. *E. speciosa, Schneev. Ic. Pl. Rar. t.* 3.

SOUTH AFRICA : without locality, *Thunberg,* and *cultivated specimens !*
COAST REGION, common from near sea level to 4600 ft.: Piquetberg Div.; *Zeyher,* 1087! 1088! Paarl Div.; Drakensteen Mountains, *MacOwan, Herb. Norm. Aust.-Afr.,* 7 ! 7b ! Paarl Mountain, *Drège!* Cape Div.; on hills and flats,

Burchell, 551 ! 704 ! 8540 ! *Bolus,* 2951 ! 3354 ! Stellenbosch Div.; near Jonkers Hoek, *Burchell,* 8309 ! Caledon Div.; Baviaans Kloof, &c., *Burchell,* 7786 ! 7582 ! Bot River, *Zeyher,* 1075 !

CENTRAL REGION : Ceres Div.; Cold Bokkeveld, *Schlechter,* 10115 !

13. E. broadleyana (Andr. Heathery, t. 206, not of Benth.); erect, 1½ ft. high, apparently quite glabrous; branches numerous, ascending; leaves 3-nate (in the description, or 4-nate in the figure), fascicled, erect-spreading, linear, acute, incurved, glabrous, 4–5 lin. long; flowers solitary, pendulous, forming a loose pseudo-raceme in the middle of the branches (or perhaps sometimes at the end of short branchlets?); pedicels coloured, curved, 3–5 lin. long; bracts remote, nearly basal, small, coloured; corolla inflated-tubular, much contracted at the mouth, red-purple, 7–8½ lin. long; segments small, erect, yellow; anthers included, lateral, oblong, muticous; filament with a double sigmoid curve above the middle; ovary cylindrical, sessile, glabrous. *Andr. Col. Heaths, t.* 154.

SOUTH AFRICA : locality and collector unknown.

The type of this species is Andrews' figure above cited, which exhibits a very distinct plant we have nowhere met with. Bentham mistakenly identified with this a specimen of Thom's in herb. Kew, which on dissection proved to be, without doubt, a form of *E. filipendula, Benth.* The author noted its want of agreement with Andrews' figure and framed his description partly upon the specimen and partly upon the figure, placing it in § *Bactridium.* We have confined ourselves to the figure and description. It is impossible to see in Andrews' figure whether the inflorescence is terminal or lateral. But the ovary is so very different from that of the other species of § *Bactridium* that it seems more properly a member of this section.

14. E. bowieana (Lodd. Bot. Cab. t. 842); branches virgate, 10–12 in. long, with dense pseudo-racemes of flowers below the ends of the branches; leaves 4-nate, squarrose to recurved, crowded, oblong or lanceolate, acute, sulcate, thick, glabrous, 2–2½ lin. long; pedicels glabrous, about 3 lin. long; bracts remote, small, ovate-rhomboidal or suborbicular, subacute, keel-tipped, scarious, 2–2½ lin. long; corolla tubular asymmetrically inflated, mouth contracted, oblique, white, glabrous, dry, 8–10 lin. long; limb erect, short; anthers included, oblong, sharply prognathous, about 1 lin. long, aristate; awns about as long as the cells; ovary very shortly stipitate, glabrous. *Benth. in DC. Prodr.* vii. 624. *E. Bowia, Sinclair, Hort. Eric. Wob.* 4. *E. Bauera, Andr. Heathery, t.* 252, *Col. Heaths, t.* 221. *E. Bowerii, Donovan in Nat. Repos.* iii. (1824) *t.* 83.

SOUTH AFRICA : without locality, *Herb. Banks* in British Museum. Also *cultivated specimens !*

COAST REGION : Riversdale Div.; Aasvogel Berg, *Galpin,* 3564 ! mountains near Riversdale, *Hewitt in Herb. Bolus,* 3067 !

15. E. gilva (Wendl. Beobacht. 48); erect, glabrous, 1½–3 ft. high; branches stout, ascending, leafy; leaves 4-nate, erect-spreading to squarrose, imbricate, often flexuous, lanceolate-linear, acute, 3–5 lin. long; inflorescence lateral, mostly solitary, though somewhat crowded, rarely subterminal; pedicels decurved, 2–5½ lin. long; bracts subremote, ovate, scarious, 2–2½ lin. long; sepals rhomboid-oval, acute, keeled, scarious, 3 lin. long, greenish; corolla tubular, slightly asymmetrically inflated, mouth somewhat contracted, more or less deeply 4-foveolate at the base, glabrous, dry, 6–12 lin. long, white or greenish-white; limb a somewhat darker green; anthers nearly as in *E. bowieana,* but somewhat wider; ovary sessile, glabrous. *Wendl. Eric. Ic. fasc.* 13, 13; *var. angustata, Rach in Linnæa,* xxvi. 773. *E. alveiflora, Salisb. in Trans. Linn. Soc.* vi. 366. *E. gelida, Andr. Heathery, t.* 24, *and Col. Heaths, t.* 96.

Coast Region: Cape Div.; on the Devils Peak and Table Mountain, 2000–3000 ft., *Niven,* 204! *Bolus,* 4567! *Guthrie,* 1520! *Treleaven in MacOwan, Herb. Aust.-Afr.,* 1761! *Wilms,* 3435! *Herb. Salisbury!* and *cultivated specimens!*

16. E. sessiliflora (Linn. f. Suppl. 222); erect, 1–1½ ft. high; branches ascending, densely leafy; leaves 6-nate or scattered, erect-spreading or incurved, imbricate, linear or subulate, acute or acuminate, often mucronate, glabrous or pubescent, ciliate or naked, 2–7 lin. long; inflorescence generally a dense congested spike, thickening and persisting long after the fall of the corolla; pedicels always very short and stout; bracts and sepals closely adpressed, usually similar in shape and texture, from rhomboid, orbicular-spathulate, or broad-obovate to unguiculate-spathulate (the lamina small, ovate, acute), oblanceolate, and narrow-oblanceolate, ciliate or naked, rarely lacerate, scarious, nerved, pale coloured or at length deep red, from 1½–3 lin. long; corolla tubular trumpet-shaped, mostly narrow at the base and widening upwards, rarely more equal, glabrous, dry, green to greenish-white or yellowish-green, 8–15 (usually 10–12) lin. long; limb small, erect; anthers usually included, sometimes just visible, sub-cuneate-oblong, prognathous at the base, with a more or less sharp deflexed point, ⅔–1 lin. long, aristate; awns variable, usually shorter than the cell; ovary sessile, glabrous. *Benth. in DC. Prodr.* vii. 625. *E. cephalotes, Willd. ex Steud. Nomen. ed.* 2, i. 570, *not of Thunb. E. spicata, Thunb. Diss.* 43, *t.* 4; *Andr. Heathery, t.* 45, *and Col. Heaths, t.* 61; *Wendl. Eric. Ic. fasc.* 2, 27; *Lodd. Bot. Cab. t.* 1203. *E. favosa, Salisb. Prodr.* 298, *and in Trans. Linn. Soc.* vi. 365.

Var. β, clavæflora (Bolus); sepals broadly rhomboid, orbicular-spathulate or obovate, margin naked, 3–3½ lin. long, 2½–3¾ lin. wide. *E. sessiliflora, Andr. t.* 86, *and Col. Heaths, t.* 130. *E. clavæflora, Salisb. in Trans. Linn. Soc.* vi. 365.

Var. γ, oblanceolata (Bolus); sepals oblanceolate, ciliate or naked, about 2½ lin. long.

Var. δ, sceptriformis (Bolus); sepals broadly spathulate, margins lacerate. *E. sceptriformis, Salisb.,* and *E. enneaphylla, Roxb. ex Salisb. in Trans. Linn. Soc.* vi. 365. *E. spicata, Thunb.,* var. *ciliata, Benth. in DC. Prodr.* vii. 625.

Coast Region: Paarl Div.; French Hoek, *Niven,* 207! Cape Div.; on mountain sides, *Thunberg, Bolus,* 4486! and *in Herb. Norm. Aust.-Afr.,* 41!

Stellenbosch Div.; Lowrys Pass, *Mund,* 8! Caledon Div.; on mountains, *Burchell,* 7719! 8104! *Zeyher,* 3176! *Drège,* 7707a! *Bolus,* 5343! *Schlechter,* 7583! George Div.; Keurbooms River, *Burchell,* 5130! Knysna Div.; various localities, *Burchell,* 5197! 5354! *Bolus,* 2386! Humansdorp Div.; near Kromme River, *Drège,* 7707b! and *cultivated specimens!* Var. β: Caledon Div.? top of hills (Houw Hoek?), *Niven,* 17! *Herb. Salisbury!* Var. γ: Div.? *Guthrie,* 3795! Var. δ: Tulbagh Div.; Witsen Berg, *Burchell,* 8652! Roode Zand, *Niven,* 208! *Herb. Salisbury* (at Kew)!

We adopt the older name of Linnæus the younger, and have little hesitation in grouping the several forms above described. The type and var. γ almost pass into each other by gradations; var. β is more distinct, but it differs chiefly in its broader sepals. It appears to have been described by Salisbury, and figured by Andrews, from cultivated plants; it is scarcely possible in a group where the sepals vary so greatly to rely upon them for specific differences. We have dissected eight gradations of form as indicated in the text above, without finding any other uniformly correlated differences of any weight. The mere greater or less width of the corolla is of even less value.

17. E. filipendula (Benth. in DC. Prodr. vii. 663); glabrous, 1–2 ft. high; branches erect, subvirgate, bearing in their upper portion a somewhat lax pseudo-raceme of flowers; leaves 4-nate, erect-spreading, not densely crowded, linear, acute, somewhat round-backed, 3–5 lin. long; pedicels 3–4 lin. long; bracts remote, nearly basal, small; sepals lanceolate or linear-lanceolate, keeled, acute, scarious, 1½–2 lin. long, whitish; corolla ovoid to ovoid-urceolate, canary-yellow, 4 lin. long, mouth somewhat contracted, texture papery, glabrous, dry, finally brownish; limb short, erect; anthers included, oblong, bilobed at the base, prognathous, and sometimes sharp-pointed, about $\frac{7}{8}$ lin. long, aristate; awns nearly as long as the cells; ovary on a short stipe.

VAR. β, **major** (Bolus); sepals broadly ovate; corolla tubular-urceolate to inflated-tubular, 5–9 lin. long; ovary somewhat long-stipitate. *E. broadleyana, Benth. in DC. Prodr.* vii. 637, *not of Andr.*

VAR. γ, **minor** (Bolus); like the preceding but the corolla smaller, hardly 3 lin. long, white, becoming pink near the apex; anthers slightly or scarcely prognathous at the base; ovary substipitate or contracted at the base.

SOUTH AFRICA: without locality, *Bowie!* Var. β: *Thom,* 1094! and *cultivated specimen!*

. COAST REGION: Bredasdorp Div.; hills near Elim and Ratel River, 300–400 ft., *Guthrie,* 3785! *Bolus,* 8450! Var. β: Bredasdorp Div.; fairly abundant on the downs between Elim and Ratel River, 300–600 ft., *Guthrie,* 3786! *Bolus,* 8452! *Schlechter,* 7618! 7726! and in *MacOwan, Herb. Aust.-Afr.,* 1920! Var. γ: Bredasdorp Div.; between Elim and Ratel River, 200–300 ft., *Guthrie,* 3784! *Bolus,* 8451! *Schlechter,* 10472; here also apparently belongs *Zeyher,* 1090, of which the locality is unknown to us.

These three forms, unlike as they appear at first sight, can hardly be separated specifically. Beyond the size and colour of the corolla there is scarcely anything to distinguish them excepting a small difference in the length of the stipes of the ovary, a very variable character, where it exists, throughout the genus. They afford a fresh example of the cases, which a larger supply of material has brought to light, of the connection and gradation between supposed species with long- and others with short-flowered corollas which have hitherto been placed in different sections. In this instance the affinity is shown with § *Hermes,* and other cases occur in *E. casta* and *E. coccinea,* in this section; besides several in § *Evanthe.* There is a considerable resemblance between our var. β and *E. gilva,* and *E. mammosa,* and the almost identical shape of anthers in all three, as well as in the type and var. γ, leaves little room for doubt as to the true position of this species.

18. **E. grandiflora** (Linn. f. Suppl. 223); erect, robust, 3–5 ft. high; branches ascending, sometimes virgate; leaves 6-nate or scattered, erect-incurved, spreading or squarrose, linear, acuminate, round-backed, glabrous, 8–10 lin. ·long; inflorescence generally crowded towards the ends of the branches; pedicels about 2 lin. long; bracts subremote or subapproximate, linear, about $1\frac{1}{2}$ lin. long; sepals lanceolate, acuminate, scarious, glabrous, 1–3 lin. long; corolla attenuate below, glabrous, viscidulous or dry, 12–14 lin. long, orange-red; limb erect; segments semi-ovate, bluntly acute; stamens far exserted; anthers subterminal or sublateral, obovate-oblong, the long wide pore (from $\frac{3}{4}$ to nearly the full length of the cell) giving an ear-like appearance; cells often bipartite from below the base, muticous, 1–$1\frac{1}{3}$ lin. long; ovary most usually minutely puberulous with reversed hairs, more rarely glabrous. *Bauer, Exot. Pl. t. 8; Bot. Mag. t. 189; Andr. Heathery, t. 117, and Col. Heaths, t. 26; Lodd. Bot. Cab. t. 498; var. longifolia, Wendl. Eric. Ic. fasc. 6, 5; var. monstrosa, Rach in Linnæa, xxvi. 774. E. strictifolia, Klotzsch in Linnæa, ix. 646. E. strictiflora, Steud. Nomencl. ed. 2, i. 580. E. spectabilis, Waitz. Beschr. Heid. 220.*

SOUTH AFRICA: without locality, *Thunberg.* Also *cultivated specimens!*
COAST REGION: Ceres Div.; near the Breede River, *Masson* (ex *Salisbury*), Mitchells Pass, 1200–1800 ft., *Bolus,* 4400! *Guthrie,* 2285! *Schlechter,* 9958! Worcester Div.; on Matroos Berg, 5200 ft., *Marloth,* 2211!

19. **E. exsurgens** (Andr. Col. Heaths, t. 22); characters of *E. grandiflora,* except: size somewhat smaller, habit sometimes more diffuse; sepals longer (5 lin.) and more acuminate; corolla dry, more uniformly widening upwards, with larger lanceolate more acute and more spreading segments; stamens less exserted. *Andr. Heathery, t. 20; Lodd. Bot. Cab. t. 835; Benth. in DC. Prodr. vii. 627; var. minor, Andr. l.c. t. 21, and var. longifolia, Andr. l.c. t. 215. E. grandiflora, var. brevifolia, Wendl. Eric. Ic. fasc. 7, 7. E. pharetræformis, Salisb. Prodr. 295, and in Trans. Linn. Soc. vi. 361. E. acutiflora, Tausch in Flora, 1839, 626.*

VAR. β. **diffusa** (Bolus); branches more diffuse and straggling; corolla viscid.
COAST REGION, ascending from the Flats to 3000 ft.: Tulbagh Div.; Tulbagh, *Pappe!* Worcester Div.; near Bains Kloof, *Bolus,* 4054! *Schlechter,* 9197! Dutoits Kloof, *Bolus,* 5295! Paarl Div.; French Hoek, *Niven,* 195! *Miss Fair in Herb. Bolus,* 6321! Stellenbosch Div., *Burchell,* 8340! 8370! *Drège! Bolus and MacOwan Herb. Norm. Aust.-Afr.,* 10! 11! *Schlechter,* 7808! Caledon Div.; near Caledon, *MacOwan, Herb. Aust.-Afr.,* 1717! Also *cultivated specimens!* Var. β: Riversdale Div.; Platte Kloof, Tygerberg, without collector's name, in Cape Govt. Herb.
Very near the preceding, of which it ought perhaps to be regarded as a variety.

20. **E. longisepala** (Guthrie & Bolus); erect, virgately branched with largish glossy leaves and verticillate flowers with corolla

scarcely longer than the green sepals; branches reversedly puberulous, 10–12 in. (or more) long; leaves 6-nate or sub-6-nate, erect-incurved, densely imbricate on the upper part of the branches, broadly linear, subacute, glabrous, 5 lin. long; inflorescence somewhat crowded below the ends of the branches; pedicels stout, puberulous, about 3 lin. long; bracts subremote, one infra-median, two supra-median, foliaceous, 3½ lin. long; sepals foliaceous, linear-lanceolate, subacute, keeled, concave, rigid, glabrous, 4½ lin. long; corolla tubular-inflated, viscidulous-puberulous, about 5 lin. long; limb short, generally erect; anthers subexserted, subterminal, obliquely tapering to the filament, semi-ovate, scarcely ear-like, about 1 lin. long, minutely decurrent-denticulate for about ¾ lin. below the cells; ovary cylindrical, lobed, puberulous; style 4-gonous, puberulous; stigma capitate.

COAST REGION: Clanwilliam Div.; without precise locality, *Mader!* in the Cape Govt. Herb.

A very distinct species connecting the sections *Pleurocallis* and *Hermes*; it forms a link between *E. grandiflora* of the former, and *E. viscaria, var. β, decora* and *E. parilis* of the latter. It is described from a single, but excellent, specimen; though there is some ground for suspecting that it may not be quite fully developed, in which case the corolla might become somewhat longer, and the stamens more fully exserted.

21. E. Hibbertia (Andr. Heathery, t. 118); stout, erect, 1–1½ ft. high; branches ascending, puberulous; leaves 6-nate, spreading-incurved, imbricate, linear, blunt, glabrous or minutely puberulous, glandular-ciliate, 3–4 lin. long; inflorescence verticillate towards the ends of the branches, sometimes apparently terminal; pedicels viscid-puberulous, about 5 lin. long; bracts, two approximate, one remote (in our specimen and in Andrews' figure, but, according to Bentham, all remote), linear, about 3 lin. long; sepals oblong or lanceolate, subobtuse, concave, thick, rigid, foliaceous, glabrous, viscid, pallid, 3 lin. long, about 1 lin. wide; corolla clavate-tubular, slightly curved, glabrous, extremely viscid, 12–15 lin. long by 2 lin. wide, red with a green erect limb; anthers subincluded, lateral, oblong, muticous, about 1 lin. long; ovary very shortly substipitate, with prominent glands alternating with the bases of the filaments, glabrous. *Andr. Col. Heaths, t.* 172. *E. hibbertiana, Ait. Hort. Kew. ed.* 2. ii. 378; *Bot. Mag. t.* 1758; *Benth. in DC. Prodr.* vii. 628.

COAST REGION: Caledon Div.; high rocky places near the Zondereinde River, *Niven,* 171! in clayey soil, near the eastern end of French Hoek Pass, 1300 ft., *Bolus,* 5168! Also *cultivated specimens!*

This species shows an affinity with those of the section *Bactridium.*

22. E. purpurea (Andr. Col. Heaths, t. 50); erect, virgately branched, 2–3 ft. high; leaves sub-6-nate, erect-incurved or erect-spreading to squarrose, linear, acute, glabrous, 4–7 lin. long; flowers verticillate towards the end of the branches; pedicels about 1 lin. long ; bracts subapproximate to subremote, linear, acute, scarious, 1½ lin. long; sepals lanceolate or linear-lanceolate, acuminate, foliaceous above, submembranous below, glabrous, about 3 lin. long ; corolla clavate-tubular, curved, glabrous, dry or viscidulous, red or red-purple, 10–12 lin. long ; anthers subexserted, lateral, oblong, muticous, pallid, pore more than half the length of the cell, about 1 lin. long; ovary cylindrical, longer than wide, often or always shortly stipitate, glabrescent. *Andr. Heathery, t.* 81 ; *Wendl. Eric. Ic. fasc.* 15, 39, *t.* 15 ; *Benth. in DC. Prodr.* vii. 627.   *E. phylicæfolia, Salisb. in Trans. Linn. Soc.* vi. 364.   *E. formosa, Andr. Heathery, t.* 114, *and Col. Heaths, t.* 94, *not of Thunb.*   *E. Salisburia, Andr. Heathery, t.* 288, *and Col. Heaths, t.* 271.   *E. pinea, var. purpurea, Lodd. Bot. Cab. t.* 1259, *not of Thunb., nor of Wendl.*   *E. acutifolia, Bartl. in Linnæa,* vii. 651 ?   *E. rigidiuscula, Wendl. ex Klotsch in Linnæa,* ix. 647.   *E. coccinea, a, Thunb. Herb. ex Rach in Linnæa,* xxvi. 774.   *E. superba, Sweet, Hort. Brit. ed.* 2, 340.

COAST REGION, frequent, ascending to 3500 ft. : Cape Div.; region of Cape Town, &c., *Thunberg, Burchell,* 455 ! *Niven,* 179 ! *Drège! Bolus,* 4475 ! 4516 ! *Wilms,* 3439 ! *MacOwan, Herb. Norm. Aust.-Afr.,* 192 ! Stellenbosch Div.; Hottentots Holland, *Mulder!* Lowrys Pass, *Galpin,* 3526 ! Also *cultivated specimens!*

CENTRAL REGION : Ceres Div.; Cold Bokkeveld, *Schlechter,* 8878 !

Allied to *E. coccinea* and *E. annectens,* from either of which it may be distinguished by the sepals. The best figure of the plant, as known to us in its wild state, is Andr. Heathery, t. 114.

Mr. G. F. Scott-Elliot observes of this species : " I have often seen the flowers visited by *Nectarinea chalybea* [a species of ' sugar-bird '] at Wynberg Butts and Muizenberg. Owing to the upward curvature the bird has to seize the branch above the flowers and suck them head downwards. This is an advantage to the flower, as self-fertilization is quite impossible, while in *E. Plukeneti* it must occasionally happen." (*Ann. Bot.* iv. 270.)

23. E. coccinea (Berg. Descr. Pl. Cap. 92) ; erect, 2–3 ft. high ; branches ascending, thickly pubescent ; leaves 6-nate, erect-spreading or squarrose-upcurved, crowded, linear, acute, stout and rigid, glabrous, 5–6 lin. long ; inflorescence verticillate, crowded at the ends of the branches ; pedicels 2–3 lin. long ; bracts from approximate to remote, oblanceolate, acuminate, nerved, viscid, pubescent, ciliolate, 2–3 lin. long ; sepals like the bracts, villous, glandmargined, very viscid, 4–5 lin. long; corolla tubular, widening upwards, slightly curved, pubescent, or sometimes glabrous, viscidulous, bright red, 10–13 lin. long ; limb erect, minutely erosulate ; anthers included, sometimes becoming subexserted, sublateral or lateral, oblong, somewhat narrowly ear-shaped with a wide pore, muticous, a little more than 1 lin. long ; style thinly puberulous ; ovary very shortly stipitate, minutely and reversedly puberulous. *Bauer, Exot. Pl. t.* 25 ; *Andr. Heathery, t.* 57, *and Col. Heaths, t.* 13 ; *Wendl. Eric. Ic. fasc.* 3, 9 ; *Lodd. Bot. Cab. t.* 1375 ; *Benth. in DC. Prodr.* vii. 627, *and others, not of Linn.*   *E. frondosa, Salisb.*

Prodr. 296, *and in Trans. Linn. Soc.* vi. 364. *E. pulviniformis,*
Salisb. in Trans. Linn. Soc. vi. 364.

VAR. β, echiiflora (Benth. in DC. Prodr. vii. 627); sepals obcuneate, cuspidate,
villous; corolla 5-8 lin. long; anthers subterminal, smaller, about ¾ lin. long,
scarcely ear-shaped, dark coloured. *E. cephalotes, Thunb. Diss. Erica,* 21. *E.*
echiiflora, Lodd. Bot. Cab. t. 364. *E. echiflora, var. purpurea, Andr. Heathery,*
t. 260, *and Col. Heaths,* t. 164. *E. glandulosa, Wendl. Eric. Ic. fasc.* 13, 5, *not*
of Thunb. *E. ostrina, Lodd. l.c.* t. 1218. *E. puberula, Klotzsch in Linnæa,* ix.
629; *Benth. in DC. Prodr.* vii. 625. *E. coccinea, Thunb. Diss. Erica,* 23, *ex*
Rach in Linnæa, xxvi. 773.

SOUTH AFRICA : without locality, *Forster !* *Berg !* and *cultivated specimens !*
Var. β : *Thunberg,* and *cultivated specimens !*

COAST REGION, frequent on mountains from 800-3000 ft. : Cape Div. ;
around Cape Town, *Thunberg, Burchell,* 99 ! 582 ! 8429 ! 8508 ! *Drège ! Ecklon,*
277 ! *Bolus,* 2966 ! 3366 ! *Wilms,* 3438a ! *Wolley Dod,* 178 ! 2124 ! Swellendam
Div. ; near the River Zondereinde, *Zeyher,* 1084 ! Var. β : Cape Div. ; Devils
Peak, 2000-3300 ft., *Bolus,* 3772 ! and *Herb. Norm. Aust. Afr.,* 189 ! *Wilms,*
3437 ! Stellenbosch Div. ; Hottentots Holland (ex *Salisbury*), *Mulder !*

Allied to *E. purpurea* and *E. annectens,* but generally distinguishable from
either by its sepals, and the indumentum of its ovary. Of *E. puberula,* Klotzsch,
we have seen specimens in Herb. Kew of Mund and Maire so named by Bentham,
which are probably authentic. These, as also Klotzsch's description, agree very
well with var. *echiiflora.*

24. E. conica (Lodd. Bot. Cab. t. 1179) ; erect, 1-1½ ft. high;
branches subvirgate ; leaves 4-6-nate, mostly erect-incurved, more
rarely spreading, linear, acute, glabrous, 4-6 lin. long ; inflorescence
crowded towards the ends of the branches ; flowers usually erect or
ascending ; pedicels less than 1 lin. long ; bracts subapproximate or
remote, linear, 1½ lin. long ; sepals ovate-lanceolate, acuminate or
cuspidate, keeled, keel broadly cuneate at base tapering upwards,
viscidulous, green, 2-3 lin. long ; corolla subobconic, sometimes
subcampanulate, narrow-tubular in the lower half, then somewhat
suddenly and obliquely widened upwards, oblique at the apex,
viscidulous, red, ¼-½ in. long ; limb erect ; anthers included,
subterminal to sublateral, semiobovate, muticous, dark-coloured,
about ¾ lin. long ; style glabrous, included ; ovary turbinate, sessile
or substipitate, puberulous, glabrescent. *Benth. in DC. Prodr.* vii.
664. *E. coccinea, Berg., var. breviflora, Rach in Linnæa,* xxvi.
774.

COAST REGION : Cape Div. ; mountains near Cape Town, 1200 ft., *Bolus,*
3715 ! 7949 ! above Tokay Plantation, *Wolley Dod,* 1280 ! Also *cultivated*
specimen !

A species precisely upon the meeting-point of the present section and § *Hermes.*
We place it here because of its close affinity with *E. coccinea,* var. *echiiflora,*
from which there is little to separate it, except the habit. Loddiges' figure is
good. It has a strong external resemblance to *E. axilliflora, Bartl.*

25. E. pinea (Thunb. Diss. Erica, 23, not of Andr. nor Wendl.) ;
erect, 1 ft. or more high ; branches stout, rigid, often flexuous ;
leaves sub-6-nate, or scattered, densely imbricate, erect-incurved,
lanceolate-linear or linear, acute, round-backed, glabrous, 6-8 lin.
long ; flowers crowded at the ends of the branches ; pedicels 1½ lin.

long; bracts, 2 approximate, linear, $2\frac{1}{2}$ lin. long, the third medial and smaller; sepals linear from an ovate base with subscarious margins, acute, foliaceous, glabrous, $2\frac{1}{2}$–$3\frac{1}{2}$ lin. long; corolla clavate-tubular, slightly curved, glabrous, dry, yellow, or white, 10–12 lin. long; anthers subexserted or subincluded, lateral, oblong, basifixed, muticous, from $\frac{2}{3}$–1 lin. long, pore about $\frac{1}{3}$ the length of the cell; ovary turbinate, shortly stipitate, glabrous. *E. aurea, and var., Andr. Heathery, t.* 153, 204, *and Col. Heaths, t.* 76, 149; *Benth. in DC. Prodr.* vii. 628.

VAR. β, **argentiflora** (Bolus); sepals somewhat shorter, $2\frac{1}{2}$ lin. long, the base proportionately broader, the linear foliaceous part shorter; corolla white; ovary sometimes puberulous. *E. argentiflora, Andr. Heathery, t.* 202, *E. purpurea, var. argentiflora, Benth. in DC. Prodr.* vii. 628.

VAR. γ, **viscosissima** (Bolus); sepals shorter, shortly acuminate; corolla more viscid. *E. aurea, Andr., var. viscosissima, Benth. in DC. Prod.* vii. 628.

SOUTH AFRICA: without locality, *Thunberg, Drège*, 7705b! and *cultivated specimens!*

COAST REGION, on mountains 1000–2000 ft.: Tulbagh Div.; vicinity of Tulbagh, *Niven*, 184! *Scott Elliot*, 20! Worcester Div.; Dutoits Kloof, *Drège!* Bains Kloof, *Schlechter*, 9197! 10254! Paarl Div.; Drakensteen Range, *Mac-Owan, Herb. Norm. Aust.-Afr.*, 9! Caledon Div.; Baviaans Kloof, *Burchell*, 7798! Var. β: Paarl Div.; French Hoek, *Miss Fair in Herb. Bolus*, 6320! *Niven*, 185! *Schlechter*, 10275! Var. γ: Paarl Div.; French Hoek, *Masson*, 120!

We owe to Rach's paper on the examination of Thunberg's collection of heaths (*Linnæa*, xxvi. 774) the identification of Andrews' *E. aurea* with this species. That *E. argentiflora* belongs here is shown by its anther with small pore, as well as by its sepals. The species may readily be known from *E. purpurea* by the anther pore alone; from *E. annectens*, by the shape of the anther, besides other characters.

26. E. hesseana (Wendl. ex Klotsch in Linnæa, ix. 634); erect, $1\frac{1}{2}$ ft. high; branches verticillate, densely covered with leaf-scars; leaves squarrose, 4–5-(rarely 6-)nate, elliptic-linear, subobtuse, glabrous, $1\frac{3}{4}$ lin. long; inflorescence axillary, verticillate, crowded at the tops of the branches; pedicels pubescent, 1 lin. long; bracts approximate, linear; sepals adpressed, lanceolate, obtuse, glabrous, thick, subfoliaceous, 2 lin. long; corolla clavate, glabrous, purple, 9 lin. long; limb short, erect, crenulate; anthers exserted, lateral, elongate, cells parted to the base, muticous; stigma simple, obtuse, ovary turbinate, puberulous.

SOUTH AFRICA! without locality, *Hesse.*

We only know this plant from Klotzsch's description which we have condensed and cited above.

27. E. annectens (Guthrie & Bolus); erect, 1 ft. high; branches spreading, diffuse, but stout and rigid; leaves 4–6-nate, spreading-recurved, subulate, subobtuse, glabrous, 3–5 lin. long; inflorescence both terminal and lateral, flowers 4-nate or verticillate; pedicels striate, glabrous, 1 lin.; bracts strictly remote, but approximate in aspect, subulate, foliaceous, about 3 lin. long; sepals narrow linear from a short ovate scarious base, foliaceous, keeled, $3\frac{1}{2}$ lin. long; corolla tubular, subequal, curved, glabrous, viscidulous in the lower part, red, 10–12 lin. long; limb erect, $1\frac{1}{2}$ lin. long; anthers sub-included, lateral, dorsifixed, semiovate, dark coloured, muticous, about $\frac{1}{2}$ lin. long; style glabrous; ovary broadly turbinate, wider than long, shortly stipitate, glabrous.

COAST REGION: Cape Div.; mountains near Kalk Bay, *Guthrie*, 1002!

This species, from its inflorescence, might be placed in either of the sections *Pleurocallis* or *Evanthe*; in its leaves and general aspect it seems better included here. It is allied to *E. purpurea* and *E. coccinea*, but the sepals and anthers are quite different from either; also to *E. pinea*, from which it may be known by its more spreading and blunter leaves, and smaller, differently set and shaped anthers; lastly, to *E. hesseana*, from which it is to be distinguished by its longer sepals, shorter anthers, &c.

28. E. regia (Bartl. in Linnæa, vii. 630); erect, 2–3 ft. high; branches straggling, somewhat slender, except on the smaller generally denuded of leaves; leaves 6-nate, erect or slightly spreading, imbricate, linear, blunt or subacute, round-backed, glabrous or minutely puberulous, 3–6 lin. long; inflorescence verticillate, mostly towards the ends of the branches, somewhat lax; flowers spreading or pendulous; pedicels $1\frac{1}{2}$–$2\frac{1}{2}$ lin. long; bracts from subapproximate to remote, linear, small; sepals lanceolate to ovate, acuminate, keeled, thick, rigid, glossy, $1\frac{1}{2}$–$2\frac{1}{2}$ lin. long; corolla broadly tubular, subequal, contracted at the mouth, more or less viscid, crimson, about 7–9 lin. long by 2 lin. wide; limb small, spreading or recurved, crispulate; anthers included, lateral, basifixed, oblong, somewhat prognathous, convex at the base, muticous, pallid, $\frac{3}{4}$–1 lin. long; ovary turbinate, villous with copious erect grey hairs. *Benth. in DC. Prodr.* vii. 626.

VAR. β, **variegata** (Bolus); corolla white or pink from the base shading to purple, red and occasionally green towards the summit.

VAR. γ, **williana** (Bolus); pedicels somewhat longer, $2\frac{1}{2}$ lin. long; inflorescence looser; corolla shorter, $\frac{1}{4}$–$\frac{1}{2}$ in. long, of various colours (mostly crimson) but not variegated; anthers somewhat shorter and narrower.

COAST REGION: Bredasdorp Div.; near Zoetendals Vlei, *Miss Joubert*; hills near Elim, 150–500 ft., *Guthrie*, 3796! *Schlechter*, 7667! *Will in MacOwan, Herb. Aust-Afr.*, 1728! var. β: Bredasdorp Div.; hills near Elim, 300–400 ft., *Pappe*, 60! *Bolus*, 6754! *Guthrie*, 2362! *Will in MacOwan, Herb. Aust.-Afr.*, 1718! *Schlechter*, 7621! var. γ: same locality, *Guthrie*, 3788! *Bolus*, 8448! *Schlechter*, 7680! *Will in MacOwan, Herb. Aust.-Afr.*, 1719!

The var. β, locally known as "the Elim heath" is of great beauty, but does not appear to have been ever cultivated in England.

29. E. casta (Guthrie & Bolus); erect, 1–$1\frac{1}{2}$ ft. high; branches virgate, slender; leaves 6-nate, erect, imbricate, linear, subacute, keeled, glabrous, $\frac{1}{4}$–$\frac{1}{2}$ in. long; inflorescence verticillate towards the ends of the branches; pedicels viscid, puberulous, about 1 lin. long; bracts approximate, linear, less than 1 lin. long; sepals lanceolate, subacuminate, sulcate, keel-tipped, glabrous, viscid, pallid or reddish, 2–3 lin. long ; corolla tubular, unequally inflated in the upper part, mouth widened or contracted, attenuated at the base, glabrous, more or less viscid, white or pale rosy, 5–7 lin. long by 1–1½ lin. broad ; limb-segments less than 1 lin. long, suberect, not imbricate at the base ; anthers included, lateral, basifixed, oblong, not prognathous, muticous, dark-coloured, ½–⅔ lin. long ; ovary villous with erect white hairs.

VAR. β, **breviflora** (Guthrie & Bolus) ; sepals ovate, acuminate ; corolla cyathiform or obconic, about ¼ in. long.

COAST REGION : Bredasdorp Div. ; maritime downs and hills near Elim, about 300 ft., *Guthrie,* 3719 ! *Bolus,* 6752 ! 6762 ! 8446 ! Var. β : Bredasdorp Div. ; hills near Elim, 300 ft., *Guthrie,* 3790 ! *Bolus,* 8449 ! 8460 !

Closely allied to *E. regia,* the chief difference, besides that of colour, being in the usually coarser and stouter leaves, corolla more attenuated to the base, the limb less spreading, and the absence of a projecting base to the anther in this species. The two species are each abundant, grow very near each other, and though the look of extreme forms is different, we propose the species with some doubt. The var. *breviflora,* had it stood alone, would naturally have been placed in the § *Hermes.*

30. E. Mariæ (Guthrie & Bolus) ; apparently a rather large shrub, with rigid subvirgate branches ; leaves 6-nate, erect, mostly imbricate, broadish linear, blunt, sulcate, glabrous, about ¼ in. long ; pedicels puberulous, 2–2½ lin. long ; bracts 2 subapproximate, one remote, linear, 1 lin. long ; sepals ovate, acuminate, thick, glabrous, red, about 2½ lin. long ; corolla tubular, subequal, slightly curved, mouth not contracted, glabrous, dry, red, 10–12 lin. long ; limb short, erect or somewhat spreading ; anthers included, lateral, basifixed, oblong, not prognathous at base ; cells somewhat ear-like, pale brown, about 1 lin. long ; ovary turbinate, densely villous except in the lower part with long pale silky hairs.

COAST REGION : Riversdale Div. ; at Milkwood Fontein, 600 ft., *Galpin,* 3565 !

This species, with *E. regia* and *E. casta,* forms a fairly well-defined group of the section. It connects with the next group through *E. onosmæflora.* From *E. regia* it is most readily known by its longer, more erect corolla, not contracted at the mouth, with less spreading limb, and anthers not prognathous at the base. We name the species in honour of Mrs. E. E. Galpin of Queenstown.

31. E. vestita (Thunb. Diss. Eric. 22) ; erect, virgately branched, 1–2½ ft. high ; leaves 6-nate, crowded, erect to spreading, tremulous, slender, linear-subtrigonous, acute, glabrous, ½–1⅓ in. long, scarcely ¼ lin. wide ; petioles pallid, capillary, 1½–2 lin. long, $\frac{1}{12}$–$\frac{1}{10}$ lin. wide ; pedicels 1½–4 lin. long ; bracts from approximate to remote, mostly small ; sepals linear from an ovate base, acuminate, foliaceous, 4–5 lin. long ; corolla clavate-tubular, mouth more or less widened, straight or curved, more or less pubescent, dry, white, yellow, rosy or crimson, ⅔–1 in. long ; limb erect or slightly spreading ; anthers included, subexserted, or rarely exserted, dorsifixed at or shortly above the base, oblong, more or less curved, $\frac{3}{4}$–1 lin. long, muticous or decurrent-denticulate on the filament very near the apex; pore about $\frac{1}{2}$ the length of the cell; ovary turbinate, deeply lobed, villous on the upper half, usually glabrous below. *Bauer, Exot. Pl. t.* 26; *Benth. in DC. Prodr.* vii. 626. *E. vestita, vars. incarnata, rosea, alba, purpurea, carnea, and lutea, Andr. Heathery, tt.* 97, 98, 147, 198, 246, 247, *and Col. Heaths, tt.* 68, 69, 138, 139, 215. *E. vestita, var. purpurea, Wendl. Eric. Ic. fasc.* 10, 7; *Lodd. Bot. Cab. t.* 217; *var. alba, Wendl. l.c. fasc.* 12, 3, *and Lodd. l.c. t.* 243; *var. coccinea, Wendl. l.c. fasc.* 11, 5; *var. formosa, Wendl. l.c. fasc.* 23, 167, *t.* 63, *and vars. carnea and blanda, Lodd. l.c. tt.* 1696, 1716. *E. pinifolia, Salisb. in Trans. Linn. Soc.* vi. 362. *E. eckloniana, Tausch in Flora,* 1839, 625.

VAR. β, **fulgida** (Andr. Col. Heaths, ii. t. 137, and Heathery, ii. t. 96); anthers exserted; ovary entirely silky or puberulous. *E. fulgida, Sinclair, Hort. Eric. Wob.* 11; *Benth. in DC. Prodr.* vii. 626; *Lodd. Bot. Cab. t.* 1633. *E. mera and E. speciosissima, Klotzsch in Linnæa,* ix. 644, 645. *E. vestita, var. coccinea, Andr. Heathery, t.* 199, *and Col. Heaths, t.* 70.

SOUTH AFRICA: without locality, *Thunberg, Masson,* 122! *Zeyher,* 3169! *Niven,* 182! and *cultivated specimens!* Var. β: *Mund & Maire, Lichtenstein* (ex *Klotzsch*), and *cultivated specimens!*

COAST REGION, on mountains 600–5000 ft.: Tulbagh Div.; Witsen Berg, *Burchell,* 8729! Caledon Div.; Donker Hoek Mountain, *Burchell,* 7949! near Caledon, *Mund,* 7! Zoetemelks Valley, *Niven,* 181! and various localities, *Bolus,* 3192! 9226! *Guthrie,* 2500! 4576! Swellendam Div.; near Swellendam, *Pappe!* Bredasdorp Div.; Elim, 800 ft., *Schlechter,* 7634!

A distinct and well-marked species; yet often confused with *E. longifolia.* From this it may be most readily distinguished by its more slender leaves, and their longer and more slender petioles; they are also tremulous in the wind, while in *E. longifolia* they are rigid. The two next species are closely allied by the same peculiar foliage, their sepals, anthers and ovary. But they differ by their corolla. We have specimens connecting, by subexserted anthers, var. β with the type.

32. E. nematophylla (Guthrie & Bolus); branches slender, glabrous, the younger deeply channelled between the long prominent leaf-cushions; leaves 6-nate or somewhat scattered, erect or spreading, with a long hair-like pallid petiole, slender-linear, subtrigonous, keeled, acute, glabrous, $\frac{1}{2}$–$\frac{3}{4}$ in. long; flowers more or less crowded towards the ends of the branches; pedicels slender, puberulous, $\frac{1}{4}$–$\frac{1}{3}$ in. long; bracts remote, narrow-linear, $1\frac{1}{2}$ lin. long; sepals linear from an ovate base, long-acuminate, minutely gland-ciliate, deep-red, $1\frac{1}{2}$–$2\frac{1}{2}$ lin. long; corolla clavate-tubular, slightly inflated above the middle, thence gradually contracted towards the apex, glabrous or minutely velvety or puberulous, dry, 5 lin. long; limb short, rounded, erect; filaments 3 lin. long; anthers included, dorsifixed near the base, oblong, obtuse, about $\frac{4}{5}$ lin. long, muticous or sometimes minutely decurrent-denticulate shortly below the anther; pore $\frac{2}{5}$ the length of the cell; style exserted, stout; stigma rather large, capitate; ovary exactly that of preceding and next species.

COAST REGION: Riversdale Div., 1000 ft.; slopes of the Langeberg Range near Riversdale, *Schlechter,* 1728! road-side, Garcias Pass, *Galpin,* 3643!

This is in many respects intermediate between the preceding and the next species. The leaves, sepals, anthers and ovary are closely similar. The shape of the corolla is, however, different, being so far as our specimens go, always narrowed (instead of widened) towards the apex.

33. E. filamentosa (Andr. Heathery, t. 22); slender, erect, 1–1½ ft. high; branches subvirgate, puberulous, channelled between the long prominent leaf-cushions; leaves 6-nate or irregular and somewhat scattered, erect-incurved, crowded, tremulous, slender-linear, acute, round-backed or keeled, sulcate, glabrous, 5–7 lin. long, about ¼ lin. wide; petioles pallid, about 1 lin. long, $\frac{1}{12}$ lin. wide; inflorescence axillary, crowded towards the ends of the branches; pedicels slender, 4–6 lin. long; bracts remote, small; sepals linear from a short ovate or lanceolate base, acuminate, coloured, 2 lin. long; corolla obconic- or subobconic-tubular, much wider at the mouth than at the base, glabrous or minutely puberulous, dry, rosy, 4–4½ lin. long; anthers included, dorsifixed just above the base, oblong, obtuse, $\frac{2}{3}$–$\frac{4}{5}$ lin. long, muticous; pore a little under or over ½ the length of the cell; style at length exserted; stigma capitellate; ovary exactly that of the two preceding species. *Andr. Col. Heaths, t.* 91; *Bot. Reg. t.* 6; *Lodd. Bot. Cab. t.* 395; *Benth. in DC. Prodr.* vii. 664.

VAR. β, **longiflora** (Bolus); branches, leaves and pedicels stouter than in the type; corolla obconic-tubular, 6–6½ lin. long.

COAST REGION: Swellendam Div.; mountains near Swellendam, 400–1500 ft., *Niven*, 183! *Mund*, 12! *MacOwan, Herb. Aust.-Afr.*, 1494! also *cultivated specimens!* Var. β: Swellendam Div.; mountain ridges along the lower part of the River Zondereinde, *Zeyher*, 3171! Caledon Div.: without collector's name or number, Cape Govt. Herb.!

Closely allied to the two preceding species; and not improbably, when more ample material of the two latter is available, it will be thought better to unite all three as forms of one variable species. Here as elsewhere in this section we have a distinct link with the § *Hermes*, in which indeed the present species was included by Bentham.

34. E. longifolia (Ait. in Bauer, Exot. Pl. t. 4); erect, 1–3 ft. high; branches virgate or spreading; leaves 6-nate, mostly erect and imbricate, sometimes spreading, rarely squarrose, linear, acute, mostly glabrous, rigid, 4–10 lin. long, from ⅓ to over ½ lin. wide; pedicels 1–3 lin. long; bracts linear, approximate or subapproximate, from ½ to ¾ the length of the sepals; sepals linear or linear-subulate, foliaceous, dry or viscid, glabrous or pubescent, 3–5 lin. long; corolla tubular or clavate-tubular, asymmetrically inflated, more rarely subequal, contracted or widened at the mouth, pubescent or villous, usually more or less viscid, variously coloured, 6–11 lin. long; anthers included, lateral, oblong, blunt, rarely semiovate and acute, muticous, ½–1 lin. long; ovary ovoid or turbinate, villous with long hairs, mostly glabrous towards the base. *Donn, Hort. Cantab. ed.* i. 42; *Benth. in DC. Prodr.* vii. 625. *E. pulchella, Thunb. Diss. Erica*, 22, *not of others, fide Rach. E. pityophylla, Spreng. Syst.* ii. 181. *E. Leea, Andr. ex Willd. Sp. Pl.* ii. 400; *also Heathery, t.* 74 *and Col. Heaths, t.* 31. *E. leeana,* [*Dryand. in*] *Ait. Hort. Kew,* *ed.* 2, ii. 376; *Bauer, l.c. t.* 24; *Benth. l.c.* 625; *var. longifolia, Rach in Linnæa,* xxvi. 773.   *E. vestita, Drège ex Benth. l.c.* 625.   *E. costæflora,* and *E. argutifolia, Salisb. in Trans. Linn. Soc.* vi. 363, 364.   *E. candida, Bartl. in Linnæa,* vii. 628.   *E. glaucescens, Bartl. l.c.* 651?   *E. piniformis, Wendl. ex Klotzsch in Linnæa,* ix. 626.

VAR. β, **contracta** (Bolus); corolla asymmetrically inflated, mouth contracted, white, red, orange or purple.   *E. pinea, Wendl. Eric. Ic. fasc.* 1, 11, *and fasc.* 13, 9, *not of Thunb., nor Andr.*   *E. pinifolia, Andr. Heathery, t.* 184, *and Col. Heaths, t.* 199 *and E. pinifolia, var. coccinea, Heathery, t.* 137, *and Col. Heaths, t.* 119.   *E. leeana, Lodd. Bot. Cab. t.* 298.   *E. glutinosa, var. minor, Andr. Heathery, t.* 67, *and Col. Heaths, t.* 171.   *E. pura, Sinclair, Hort. Eric. Wob.* 20; *Benth. in DC. Prodr.* vii. 626 (*probably an error for E. pinifolia, Andr. Heathery,* iv. *t.* 184).

VAR. γ, **amplicata** (Bolus); corolla clavate-tubular, curved, not inflated, gradually widened to the mouth.   *E. vestita, var. coccinea, Bot. Mag. t.* 402, *and Lodd. Bot. Cab. t.* 55, *not of Thunb.*   *E. pinea, Andr. Heathery, t.* 35, *and Col. Heaths, t.* 46, *not of Thunb.*   *E. pinifolia, var. discolor, Andr. Heathery, t.* 138, *and Col. Heaths, t.* 200.

VAR. δ, **viridis** (Bolus); corolla tubular, asymmetrically inflated, sometimes wider than in other forms, contracted at the mouth, hispid, and covered with more or less prominent rough tubercles, viscid, green or greenish, or sometimes purple.   *E. viridis, Andr. Heathery, t.* 148; *Col. Heaths, t.* 140.

VAR. ε, **squarrosa** (Bolus); leaves narrower than usual, crowded, squarrose; corolla clavate-tubular, curved, mouth widened, white or red with a white tip.

VAR. ζ, **maritima** (Bolus); leaves glabrous, 5–7 lin. long; including the 1–1¼ lin. long petiole; sepals lanceolate, acuminate, subscarious, with a linear foliaceous cusp of variable length, 3–4 lin. long; corolla clavate-tubular, velvety, pale purple, 9–11 lin. long; anthers minutely decurrent-denticulate on the filament.

SOUTH AFRICA: without locality, *Thunberg, Mund! Drège!* and *cultivated specimens!* Var. γ: *Niven,* 40! Var. ε: *Herb. Lehmann!*

COAST REGION, mountain sides to 5000 ft. : Stellenbosch Div.; Lowrys Pass, *Burchell,* 8203! 8251! *Bolus, Herb. Norm. Aust.-Afr.,* 8! Hottentots Holland, *Zeyher,* 3172! Caledon Div.; common, *Burchell,* 7752! 7866! 8082! *Zeyher,* 3177! *Niven,* 177! 178! 180! Var. β: Div. and locality? *Bolus,* 6952! *Guthrie,* 2024! 2293! 3794! 4094! 4969! Var. γ: Caledon Div.; near Genadendal, *Schlechter,* 9824! *Bolus, Herb. Norm. Aust.-Afr.,* 349! Div. and locality? *Bolus,* 5398! *Guthrie,* 4972! Var. δ: Caledon Div.; Houw Hoek Mountains, *Burchell,* 8039! *Schlechter,* 7342! locality? *Guthrie,* 4970! Var. ε: Div. and locality? *Bolus,* 6474! *Schlechter,* 4789! Var. ζ: Bredasdorp Div.; hills near Cape Agulhas, *Schlechter,* 10556!

We have united with this *E. Leea,* Andr., which consists of nothing but short-leaved forms; and even most of the varieties we have tried to distinguish run into each other.   The last, however, is more distinct, and might by some be held as a good species.   It has the sepals of *E. vestita,* but not its leaves; the corolla and leaves of the present species, but broader sepals, and differs from *E. onosmæflora* chiefly in its somewhat differently-shaped, not inflated, and not viscid corolla. Yet it appears to be too near to all of these to make it worthy of distinction as a species.

35. **E. onosmæflora** (*Salisb. in Trans. Linn. Soc.* vi. 363); erect, 1–2 ft. high; branches somewhat virgate; leaves 6-nate, erect-incurved to spreading, linear, subacute, round-backed, glabrous, 5–7 lin. long by ⅓–½ lin. broad; flowers towards the ends of the branches, mostly ascending; pedicels 1–2 lin. long; bracts sub-approximate, small; sepals lanceolate with a long linear acuminate tip, thick, glabrous, viscid and at length glossy and wrinkled, pallid,

3–5 lin. long; corolla tubular, more or less asymmetrically inflated and slightly contracted at the mouth, rarely equal, viscid, glabrous or puberulous, pale purple or ochraceous yellow, 8–11 lin. long; anthers included, lateral, oblong, muticous, about $\frac{3}{4}$ lin. long; ovary villous in the upper portion, glabrous below. *E. glutinosa, Andr. Heathery, t.* 66, *and Col. Heaths, t.* 25, *not of Thunb. E. viscida, Wendl. Eric. Ic. fasc.* 13, 7. *E. pulchella, Willd. ex Steud. Nomencl. ed.* 2, i. 578. *E. viscosa, Steud. Nomencl. ed.* i. 311 ?.

SOUTH AFRICA: without locality, *Herb. Salisbury!* also *cultivated specimens!*

COAST REGION, on mountains, 400–4500 ft.: Worcester Div.; Dutoits Peak *Marloth,* 2415! Paarl Div.; French Hoek, *Alexander!* Cape Div.; Camp ground, *Grey!* Stellenbosch Div.; Lowrys Pass, *Guthrie,* 2295! Caledon Div.; common, *Burchell,* 7579! 7799! 7950! 7995! *Zeyher,* 3178! *Bolus,* 5395! Bredasdorp Div.; near Elim, *Bolus,* 6751! 8447!

CENTRAL REGION: Ceres Div.; Cold Bokkeveld, *Schlechter,* 8902!

Section VI. EVANTHE. (Sp. 36–70.)

36. E. abietina (Linn. Sp. Pl. ed. i. 355); erect, glabrous, 2 ft. or more high; branches virgate, with a dense pseudo-raceme of flowers, several inches long, not reaching to the apex; leaves 4-nate, densely imbricate erect-incurved, linear, acute, 4–6 lin. long; flowers solitary, on short crowded branchlets; pedicels $\frac{1}{2}$ lin. long or less; bracts approximate, linear, foliaceous, 2–2$\frac{1}{2}$ lin. long; sepals linear from a broad ovate scarious base, acuminate, foliaceous, 3$\frac{1}{2}$–6 lin. long; corolla tubular, contracted towards the apex, mostly straight or nearly so, glabrous, dry, yellow with an orange limb, 7–9 lin. long (or according to *Salisbury* 6–8 lin. and to *Bentham* 9–12 lin.); limb short, spreading, at length revolute; anthers included, oblong, papillose, $\frac{1}{2}$–$\frac{3}{4}$ lin. long, aristate; awns slightly longer than the cells; ovary glabrous. *Benth. in DC. Prodr.* vii. 630. *E. Patersonia, Andr. Col. Heaths, t.* 43, *and Heathery, t.* 181; *Wendl. Eric. Ic. fasc.* 1, 15. *E. Patersonia, var. major, Andr. Col. Heaths, t.* 195 *and Heathery, t.* 228. *E. Patersoni, Lodd. Bot. Cab. t.* 1727. *E. spissifolia, Salisb. Prodr.* 293, *and in Trans. Linn. Soc.* vi. 355.

SOUTH AFRICA: without locality, *Thunberg!*

COAST REGION: Cape Div.; hills near Smitswinkle Bay, *Bolus,* 7201! and in *Herb. Norm. Aust.-Afr.,* 12! Caledon Div.; Klein River Mountains, 1000–3000 ft., *Zeyher,* 3179! mountains near Vogel Gat, 3200 ft., *Schlechter,* 9551! also *cultivated specimens!*

This is one of the few winter-flowering heaths.

37. E. sacciflora (Salisb. in Trans. Linn. Soc. vi. 355); erect, 2–3 ft. high; branches mostly virgate, furrowed by cushions below the leaves, the older scarred, bearing many flowers in a pseudo-raceme on short crowded branchlets below the apex; leaves 4-nate, erect-incurved, imbricate, linear, glabrous, 2–2$\frac{1}{2}$ lin. long; flowers solitary; pedicels 1 lin. long; bracts subremote, small, scarious; sepals broadly ovate, acute, keeled, scarious, glabrous, coloured, 2 lin. long; corolla tubular, contracted at the apex, straight or

slightly curved, glabrous, dry, yellow, darkened at the throat, 8–10
lin. long; limb short, subcordate-reniform, spreading; anthers included, oblong, papillose, ¾ lin. long, with spreading awns as long as
the cells; ovary glabrous. *Benth. in DC. Prodr.* vii. 630. *E.
pauciflora, Steud. Nomencl. ed.* 2, ii. 656 *in syn.*

CoAST REGION: Paarl Div.; in a valley near Kehrwieder, French Hoek,
Le Roux in Herb. Huguenot Seminary, 302! and in *Herb. Bolus,* 5949! mountains
of French Hoek, 1600–1700 ft., *Bolus,* 6333! 6900! *Mac Owan, Herb. Norm. Aust.-
Afr.,* 958! Stellenbosch Div.; Hottentots Holland, *Masson* (ex *Salisbury*).

38. E. foliacea (Andr. Heathery, t. 263); erect, 2–3 ft. high;
branches stout, spreading or virgate, glabrous or puberulous (the
plant entirely glabrous elsewhere); leaves 4-nate, somewhat spreading, curved, linear, rather stout, 3–5 lin. long; flowers 4–5-nate,
on distant spreading lateral branchlets ½–1 in. long, upon pedicels
½ lin. long; bracts approximate, narrow-lanceolate, acute, rigid,
keeled, margins scarious, 2–3 lin. long; sepals like the bracts but
broader and a little longer; corolla tubular, mostly somewhat inflated, straight or slightly curved, distinctly contracted at the throat,
dry, stout in texture, clear sulphur-yellow, 7–8 lin. long; limb
subspreading, semi-ovate, short; awns included, oblong, scaberulous, ¾ lin. long, aristate; awns 1½–2 times the length of the cells;
ovary glabrous, broad-based, sessile. *Andr. Col. Heaths, t.* 235;
Benth. in DC. Prodr. vii. 630.

VAR. β, **fulgens** (Klotzsch in Linnæa, xii. 507); more virgate; flowering
branchlets more numerous, shorter, mostly densely crowded and forming a
pseudo-spike of flowers several inches long; leaves shorter and straighter,
2½–3½ lin. long, closely imbricate; sepals somewhat smaller in proportion to the
corolla; corolla orange or reddish orange; anthers smaller, about ½ lin. long.
E. fulgens, Klotzsch, l.c.
CoAST REGION: Caledon Div.; without precise locality, *Bolus,* 6870! Var. β:
Caledon Div.; Klein Houw Hoek, *Niven,* 173! near the mouth of Klein River,
Masson, 114! mountains near Palmiet River and near Lowrys Pass, *Ecklon &
Zeyher, Guthrie,* 3551! 3766!
CENTRAL REGION: Ceres Div.; Cold Bokkeveld, in moist places, 5000 ft.,
Schlechter, 8925!
Taking Andrews' figure as the type, the specimens first above cited are the
only ones we have seen which entirely agree with it. Klotzsch's var. has a very
different appearance, usually more resembling *E. abietina,* yet quite distinct from
that by its 4-nate flowers, besides other characters; but the floral differences of
var. *fulgens* from the type are not considerable. Bentham appears to have
described from the variety, which seems to be much more frequent than
Andrews' type. In its flowers it also resembles *E. nana,* but the corolla is
somewhat smaller, the ovary quite sessile and broad-based, and the habit very
different.

39. E. nana (Salisb. in Trans. Linn. Soc. vi. 355); procumbent,
robust, a span high or less; branches divaricate, flexuous, rigid,
glabrous; leaves 4-nate, erect-incurved or spreading, imbricate,
linear, glabrous, 2–4 lin. long; flowers 3–4-nate; pedicels 1 lin.
long; bracts approximate, lanceolate, 3–4 lin. long; sepals ovate,
acuminate, keeled, 3–3½ lin. long, margins wide, scarious, sublacerate; corolla subclavate-tubular, wide, thick in texture, slightly
constricted at the wide throat, glabrous, dry, yellow, 10–11 lin.

long; limb short, spreading or reflexed; anthers included, oblong, curved, $\frac{3}{4}$–1 lin. long, with setiform awns somewhat shorter than, or equalling the cells; ovary substipitate, glabrous. *Wendl. Eric. Ic. fasc.* 25, 9, *t.* 4; *Benth. in DC. Prodr.* vii. 631. *E. depressa, Andr. Heathery, t.* 17, *not of Linn.*

SOUTH AFRICA: without locality, *Roxburgh,* and *Paterson* (ex *Bentham*). Also *cultivated specimens!*

COAST REGION: Stellenbosch Div.; Hottentots Holland, procumbent on the rocks, *Niven,* 172! *Masson!* Caledon Div.; mountains near Lowrys Pass, *native collector in Herb. Guthrie,* 3554!

A well-marked species allied to *E. foliacea* and *E. sacciflora.* It seems to be now rare, and a small twig, with a few flowers, brought in by a native collector in 1895, appears to be the only recorded gathering during a period of nearly a century.

40. E. xanthina (Guthrie & Bolus); branches 12–14 in. long, subvirgate, stout, densely covered with short pubescent branchlets; leaves 4-nate, mostly erect-incurved, imbricate, sometimes spreading, linear, subobtuse, convex and slightly sulcate below, flat above, roughly puberulous, ciliolate, 2–3 lin. long including the longish pallid slender petiole; flowers 1–2-nate at the ends of the numerous branchlets; pedicels less than 1 lin. long, pubescent; bracts remote, almost basal, minute; sepals ovate to subrotund, cuspidate-acuminate, keeled, pubescent, ciliate, subscarious, coloured, less than 1 lin. long; corolla tubular, slightly narrowed in the middle, and again towards the apex, pubescent, dry, yellow or orange; segments of the limb ovate, subacute, subconnivent (in age?), apparently discoloured, the whole 7–7$\frac{1}{2}$ lin. long; filaments tapering upwards, sharply bent near the anther, glabrous; anther included, dorsifixed just above the base, oblong or subsemiovate, muticous, $\frac{1}{2}$ lin. long; style subexserted; stigma capitellate; ovary obovoid or subspherical, minutely and thinly puberulous, probably becoming glabrous.

COAST REGION: Caledon Div.; Genadendal, *Alexander,* 5! in Herb. Kew.

41. E. Maximiliani (Guthrie & Bolus in Engl. Jahrb. xxvii. 173); erect, 2–4 ft. high; branches rigid, spreading, curved, naked and dark-coloured below; branches cano-tomentose, densely leafy; leaves 4-nate or occasionally scattered, erect-spreading and slightly incurved, imbricate, linear-subterete, obtuse, sulcate, cano-pulverulent, the younger villous, 4–6 lin. long; inflorescence 1–4-nate, on short branchlets; pedicels cano-tomentose, 3–4 lin. long; bracts subremote, small, upper one lanceolate, obtuse, lower two smaller, nearly basal, pubescent; sepals ovate to ovate-lanceolate, bluntish, imbricate at base, keeled, pubescent, 2$\frac{1}{2}$–3 lin. long; corolla tubular, widening upwards, straight or curved, glabrous, dry, pale sulphur-yellow, 14–16 lin. long; limb short and broad, at length spreading; anthers at length subexserted, linear, slender, brown, smooth, 1$\frac{3}{4}$–2 lin. long, aristate, awns inserted with the filament about $\frac{1}{4}$ of the length of the cells above their base, and reaching about the same length below it; ovary densely villous, except towards its base, with somewhat long erect hairs.

COAST REGION: Clanwilliam Div.; Koude Berg, near Wupperthal, 3700 ft., *Schlechter*, 8739! Pakhuis Pass on the Cederberg Range, at the highest part of the public road, about 2500 ft., *Bolus*, 8683!

CENTRAL REGION: Worcester Div.; rocky places on the Witteberg Range, near Matjes Fontein, 3750 ft., *Marloth*, 2950!

42. E. pallens (Andr. Col. Heaths, t. 194); erect, 6 in. high; branches numerous, spreading; leaves 3-nate, spreading or recurved, subulate, downy, whitish-green, 2–3 lin. long; inflorescence umbellate (3–6, Andrews' fig. shows as many as 9 flowers), subsessile; flowers spreading horizontally; bracts subapproximate, sepaloid, largish; sepals lanceolate, subacuminate, apparently glabrous, about 3 lin. long; corolla clavate-tubular, slightly constricted at the throat, straightish, glabrous, pale sulphur-yellow, 10 lin. long; limb short, spreading; anthers included, muticous; ovary glabrous. *Andr. Heathery, t.* 182; *Benth. in DC. Prodr.* vii. 635. *E. Dickinsoni, Lodd. Bot. Cab. t.* 1793?.

SOUTH AFRICA: Only known from Andrews' figures and description.

Evidently a very distinct species, and may perhaps be allied to *E. Maximiliani.*

43. E. dubia (Rach in Linnæa, xxvi. 776); erect, glabrous, with subvirgate branches, 2–3 ft. high; leaves 3–4-nate, erect-spreading, linear, 3–4 lin. long; flowers solitary or geminate, spreading; bracts remote, foliaceous, small; sepals ovate-lanceolate, acuminate, keel-tipped, ciliate-serrulate, foliaceous, $2\frac{1}{2}$–3 lin. long; corolla tubular, straight, wider at the base and apex, the middle somewhat contracted, clear yellow, 8 lin. long; limb short, erect; anthers subovate, muticous, included; ovary glabrous. *E. cylindrica, Wendl. Eric. Ic. fasc.* 11, 7, *not of Thunb. nor Andr.*

SOUTH AFRICA: Known to us only from cultivated specimens and Wendland's excellent figure and description.

Allied to *E. bibax*, yet apparently distinct by its more foliaceous sepals and the different manner in which the anthers are affixed to the filament.

44. E. bibax (Salisb. in Trans. Linn. Soc. vi. 358); erect, virgate, 1–2 ft. high; branches and limb of the corolla puberulous, glabrous elsewhere; leaves 4-nate, linear-trigonous, erect-spreading, 2–4 lin. long; flowers solitary, spreading; bracts approximate, like the sepals but smaller; sepals lanceolate, acute, keel-tipped, scarious, shining, coloured, $3\frac{1}{2}$–$4\frac{1}{2}$ lin. long; corolla clavate-tubular, curved or nearly straight, puberulous or velvety, dry, yellow with a paler limb, 7–10 lin. long; filaments sharply bent at a right angle at the junction with the thickened dorsally-projecting connective; anthers included, oblong, $\frac{3}{4}$ lin. long, attached to the filament near the base of the cells, muticous; ovary glabrous. *E. flammea, Andr. Heathery, t.* 23, *and Col. Heaths, t.* 92; *Benth. in DC. Prodr.* vii. 633.

SOUTH AFRICA: without locality, *Thunberg.*

COAST REGION: Stellenbosch Div.; Lowrys Pass, 1300 ft., *Schlechter*, 7813! *Grisbrook in Herb. Bolus*, 6473! Caledon Div.; mountains near Grietjes Gat, *Zeyher!* on the Zwart Berg near Caledon, *MacOwan, Herb. Aust.-Afr.*, 1721!

Very near to the more glabrous forms of *E. curviflora,* of which it might, perhaps, be regarded as a variety. The angular attachment of the anther to the filament is, however, more marked than in any specimens of that which we have seen.

45. E. curviflora (Linn. Sp. Pl. ed. 1, 354); $1\frac{1}{2}$–5 ft. high; branches virgate, stout, more rarely spreading or slender; leaves 4-nate, erect, incurved or spreading, imbricate, linear or lanceolate-linear, acute, subtrigonous, keeled, or somewhat open-backed, ciliate or glabrous or pubescent, $1\frac{1}{2}$–$3\frac{1}{2}$ lin. long; flowers solitary, usually subsessile and ascending, sometimes at length decurved; bracts small, approximate, or subremote; sepals variable, linear and foliaceous from a broad scarious base, or lanceolate or ovate, scarious with a broad foliaceous keel, or entirely scarious and coloured, glabrous or pubescent, $1\frac{1}{2}$–$4\frac{1}{2}$ lin. long; corolla clavate-tubular or trumpet-shaped, curved or more rarely straight, mostly pubescent, or villous, more rarely glabrous, dry, red, orange or yellow, 10–19 (commonly 11-13) lin. long; limb $1\frac{1}{2}$–$2\frac{1}{2}$ lin. long; filaments slender, glabrous or pilose; anthers included or subexserted, oblong, straight or slightly curved forwards at the base, inserted on the filament above their base, about $\frac{3}{4}$ to a little over 1 lin. long, muticous or minutely decurrent-denticulate along the filament; ovary glabrous, sessile or substipitate, sometimes excavate at the apex, with callous bosses rising above the base of the style. *Thunb. Diss. Erica,* 24; *Wendl. Eric. Ic. fasc.* 3, 7; *Benth. in DC. Prodr.* vii. 633; *Andr. Heathery, tt.* 16, 211 (*var.* **rubra**), *and Col. Heaths, tt.* 19, 161; *Lodd. Bot. Cab. t.* 1663. *E. tubiflora, Linn. Sp. Pl. ed.* 2, 505; *Thunb. Diss. Erica,* 25. *E. cuspidigera, Salisb. in Trans. Linn. Soc.* vi. 358. *E. procera, Wendl. Eric. Ic. fasc.* 17, 71, *t.* 27. *E. simpliciflora, Willd. Sp. Pl.* ii. 402, *and Wendl. Eric. Ic. fasc.* 17, 69, *t.* 26. *E. simplicifolia, Steud. Nomencl. ed.* i. 309. *E. fastuosa, Salisb. in Trans. Linn. Soc.* vi. 359. *E. buccinæformis, Salisb. Prodr.* 294, *and in Trans. Linn. Soc.* vi. 359; *Benth. in DC. Prodr.* vii. 633. *E. ignescens, Wendl. Eric. Ic. fasc.* 8, 3, *and var.* atropurpurea, *fasc.* 20, 117, *t.* 44; *Andr. Heathery, t.* 27, *and Col. Heaths, t.* 103. *E. laniflora, F. W. Schmidt, Neue und selt. Pfl.* 44, *not of Wendl. E. tubulosa, Sm. ex Steud. Nomencl. ed.* 1, 310.—*Erica, Seb. Mus.* 2, *t.* 19, *fig.* 5, *and* 1, *t.* 20, *fig.* 4.

VAR. β, **Burchellii** (Bolus); tall, stout, virgate; leaves often more or less open-backed; sepals $1\frac{1}{2}$–$4\frac{1}{2}$ lin. long; corolla hairy or rarely glabrous; limb oblong, 2–$3\frac{1}{2}$ lin. long, the sinus acute, or obtuse and somewhat open; ovary mostly callous-bossed at the apex. *E. tubiflora, Willd. Sp. Pl.* ii. 403, *and others, not of Linn.; Bauer, Exot. Pl. t.* 28; *Andr. Heathery, t.* 46, *and Col. Heaths, t.* 64; *Wendl. Eric. Ic. fasc.* 4, 7; *Benth. in DC. Prodr.* vii. 634. *E. coccinea, of Linn. Herb. not of Sp. Pl. ed.* 1, *according to Benth. E. Burchellii, Benth. l.c.* 632. *E. sordida, Drège ex Benth. l.c.* 634.

VAR. γ, **versatilis** (Bolus); bracts scarious, coloured; sepals ovate-lanceolate, acute, keeled, scarious, coloured, pubescent, 3 lin. long; filaments inserted higher on the anther than usual, about the middle of the cell, with decurrent awns slightly free at the apex, but scarcely reaching to the base of the cells.

VAR. δ: **sulfurea** (Bolus); very villous in all parts; leaves subdistant;

branches spreading; sepals ovate-lanceolate, 2½–3 lin. long, with a channelled keel; corolla yellow. *E. sulfurea, Andr.* Heathery, t. 241, *and Col.* Heaths, t. 278; *Bot. Mag.* t. 1984; *Lodd. Bot. Cab.* t. 1762; *Benth. in DC. Prodr.* vii. 634. *E. stagnalis, Salisb. in Trans. Linn. Soc.* vi. 359? *E. tubiflora, Roxb. ex Salisb. l.c.*

VAR. ε, **diffusa** (Bolus); more slender than any other form; branches diffuse; leaves spreading and subdistant, 6 lin. long, glabrous; sepals lanceolate-ovate to ovate, acute, scarious, ciliate, coloured; corolla glabrous except for the ciliate and hair-tufted limb, clear rosy red.

SOUTH AFRICA: without locality, *Herb. Salisbury!* Var. β, *Sieber,* 76! Var. δ, *Mund, Herb. Salisbury!* Var. ζ, *Mund & Maire!* and *cultivated specimens* of the type and vars. β and δ!

COAST REGION: in Clanwilliam, Tulbagh and Piquetberg Div., and in all the coast Divisions from Cape to Albany Div., at from 300–2500 ft., *Thunberg, Burchell,* 546! 3719! 5747! 7046! 7777! *Niven,* 193! 194! 229! *Zeyher,* 3160! *Drège,* 7700! 7701! 7702! 7710! *Galpin,* 3585! *MacOwan,* 81! and in *Herb. Norm. Aust.-Afr.,* 751! *Bolus,* 1577! 3993! 4168! 5851! 8668! *Mulder! Schlechter,* 1918! 2417! 10379! Var. β: Cape Div.; near Cape Town, *Niven,* 191! *Mund,* 11! *Bolus,* 3716! *MacOwan, Herb. Norm. Aust.-Afr.,* 13! George Div.; Cradock Berg, *Burchell,* 6023! near Touw River, *Burchell,* 5738! Var. γ: Worcester Div.; Goudini, *Schlechter,* 9948! Caledon Div.; mountains near the Zondereinde River, 4500 ft., *Bolus in Herb. Guthrie,* 4595! Var. δ: Paarl Div.; mountains around French Hoek, *MacOwan, Herb. Norm. Aust.-Afr.,* 957! *Schlechter,* 9264! Caledon Div.; Klein Houw Hoek, *Niven,* 10! Var. ε: Caledon Div.; near Genadendal, 3000 ft., *Galpin,* 3573! mountains near Greytown, *Herb. Bolus,* 6750!

A very variable species of wide distribution. Earlier authors described and figured the forms as they successively appeared, as so many distinct species. With a very large amount of material before us we have with difficulty distinguished 3 or 4 varieties, and even some of these run into each other.

46. E. sulcata (Benth. in DC. Prodr. vii. 632); erect; leaves 4-nate, incurved, narrow or broadish linear, blunt, deeply sulcate, sometimes slightly open-backed, the upper copiously and softly long ciliate, 1½–2½ lin. long; flowers solitary, subsessile; bracts sub-remote, minute, sepal-like: sepals ovate, or subcuneate and retuse, apiculate, scarious, about ½ lin. long; corolla clavate-tubular, incurved, tomentose-pubescent, dry, 10–12 lin. long; limb 1¼ lin. long, spreading, somewhat fleshy; filaments widening towards the anther; anthers subexserted, oblong, mid-dorsifixed, muticous, ¾–1 lin. long; ovary glabrous.

COAST REGION: George Div.; Devils Kop, near George, *Niven,* 190!
Closely allied to *E. curviflora,* var. *Burchellii,* and might perhaps be regarded as another variety of that with minute sepals. Bentham also says that "it has the habit of *E. tubiflora, Willd.*" [*E. curviflora, L.*]

47. E. macropus (Guthrie & Bolus); erect, rigid, much-branched; branches pallid, pubescent; leaves 4-nate, incurvo-patent, linear, subtrigonous, glabrous, 2 lin. long; flowers sub-4-nate, numerous; pedicels 4 lin. long, slender; bracts linear, remote; sepals lanceolate, acute, foliaceous, scarious-edged, concave, glabrous, 2½ lin. long; corolla clavate-tubular, incurved, glabrous, dry, red, 9 lin. long; limb short, erect: anthers included, oblong-cuneate, less than ½ lin. long, with short and broad subulate appendages reaching to the base of the cells, or slightly longer.

COAST REGION: Clanwilliam Div.; near Clanwilliam, *Mader*, 2180! in the Cape Gov. Herb., not of Herb. MacOwan.

Well distinguished by its long pedicels, and broad, short anther-appendages.

48. E. conspicua (Soland. in Ait. Hort. Kew. ed. 1, ii. 22); 2-3 ft. high; branches subvirgate, variably pubescent, rarely glabrous, usually covered with many short branchlets, the uppermost only bearing flowers; leaves 4-nate, slightly spreading, imbricate, linear to sublanceolate, pubescent, villous or rarely glabrous, 1-2 lin. long; flowers solitary, more rarely 2-4-nate; pedicels 1-2 lin. long; bracts subremote, sepal-like, oblong, very obtuse, or linear, 2-$2\frac{1}{2}$ lin. long; sepals from ovate subacute to lanceolate acuminate, subscarious, with a thick keel or boss at the apex, pilose or glabrous, 3-$4\frac{1}{2}$ lin. long; corolla clavate-tubular, incurved, variably hairy or more rarely glabrous, dry, 9-20 (mostly 15-18) lin. long, rosy, dull red, or yellow; limb oblong, spreading or recurved, 2-$2\frac{1}{2}$ lin. long; filaments somewhat broad at the base, occasionally thinly pilose; anthers included or subexserted, oblong, affixed shortly below the middle of the cell, distinctly curved forwards in front, $1\frac{1}{4}$-$1\frac{1}{2}$ lin. long, muticous or sometimes minutely decurrent-denticulate; ovary glabrous. *Bauer, Exot. Pl. t.* 12; *Wendl. Eric. Ic. fasc.* 4, 9; *Andr. Heathery, t.* 12, *and Col. Heaths, t.* 14; *Salisb. in Trans. Linn. Soc.* vi. 359. *E. inconspicua, Thunb. Prodr.* 71 (*sphalm.*). *E. longiflora, Salisb. l.c.* 359; *Andr. Heathery, t.* 222, *and Col. Heaths, t.* 183; *Lodd. Bot. Cab. t.* 983. *E. elata, Andr. Heathery, t.* 112, *and Col. Heaths, t.* 87; *Lodd. l.c. t.* 1788. *E. splendens, Wendl. Eric. Ic. fasc.* 8, 5, *not of Andr. E. laniflora, Wendl. l.c. fasc.* 2, 23. *E. laniflora, var. glabra, Wendl. l.c. fasc.* 19, 105, *t.* 40. *E. lanata, Wendl. l.c. fasc.* 5, 5. *E. sordida, Andr. Heathery, t.* 191, *and Col. Heaths, t.* 60; *Lodd. l.c. t.* 1973. *E. floccosa and E. verticillaris, Salisb. l.c.* 360, *not of Bartl.*

COAST REGION, on mountains, 700–1000 ft.: Worcester Div.; near Worcester, *Rehmann*, 2529! in the Goudini, *Bolus*, 5171! *MacOwan, Herb. Norm. Aust.-Afr.*, 231! Paarl Div.; near Wellington, in *Cape Gov. Herb.*; French Hoek, *MacOwan, Herb. Norm. Aust.-Afr.*, 936! *Bolus*, 5169! *Schlechter*, 9320! Caledon Div.; Zwart Berg, *Masson*, 51! near the Zondereinde River, *Schlechter*, 9888! Also *cultivated specimens!*

A distinct species well characterized by its small subremote bracts and its anthers. We are unable, however, to find any serviceable definitions for its varieties. Var. *glabra* (Benth. in DC. Prodr. vii. 633) appears only to be known from garden specimens.

49. E. densifolia (Willd. Sp. Pl. ii. 359); erect, 2-3 ft. high; branches stout, erect, pubescent; leaves 3-nate, mostly densely imbricate on short branchlets, incurved or spreading, lanceolate to linear, the younger pubescent becoming glabrous and shining, $1\frac{1}{2}$-$2\frac{1}{2}$ lin. long; flowers mostly solitary at the ends of short branchlets forming a pseudo-raceme; pedicels $\frac{1}{2}$ lin. long or less; bracts approximate, $\frac{1}{2}$ lin. long; sepals lanceolate to oblong, acute, thick and rigid, keeled or thickened at the margins and sulcate, pubescent or glabrous, sometimes forming a tetragonous calyx, 3-$3\frac{1}{2}$ lin. long;

corolla clavate-tubular, curved, more or less hairy, viscidulous chiefly
in the lower half, tube red, the throat and limb greenish-yellow,
12–15 lin. long; anthers subexserted, linear-lanceolate, tapering to
each extremity, aristate; awns straightly pendulous, $\frac{1}{2}$–$\frac{3}{4}$ the length
of the cell; style usually decurved at the apex; ovary glabrous. *E.
decora, Salisb. in Trans. Linn. Soc.* vi. 354, *not of Andr. E. Uhria,
Andr. Heathery, t.* 149, *and Col. Heaths, t.* 141; *Wendl. Eric. Ic.
fasc.* 18, 91, *t.* 35; *Benth. in DC. Prodr.* vii. 628, *and var. pilosa,
Andr. Heathery, t.* 150, *and Col. Heaths, t.* 142. *E. ewerana
[Dryand. in] Ait. Hort. Kew, ed.* 2, ii. 368; *Lodd. Bot. Cab. t.*
1303, *and var. pilosa, t.* 1992.

CoAST REGION: George Div.; Outeniqualand, *Niven,* 202! Montagu Pass,
1400 ft., *Bolus,* 8669! *Schlechter,* 5837! Uniondale Div.; near Groot Fontein,
in Lange Kloof, 2000 ft., *Drège,* 7713! Knysna Div.; near Knysna, *Pappe!
Buchanan!* and in *Herb. Bolus,* 5834! Also *cultivated specimens!*

50. E. xerophila (Bolus); erect, reaching 6 ft. high; branches
ascending, rigid, the younger puberulous, glabrescent; leaves 3-nate,
crowded, erect or spreading, incurved, narrow-linear or subfiliform,
subobtuse, deeply sulcate, minutely hispidulous and tuberculate,
shining, 3–4 lin. long; petioles pallid; flowers solitary, few; pedicels puberulous, 1 lin. long; bracts subremote, linear, incurved,
foliaceous, $\frac{3}{4}$–1 lin. long; sepals linear from a short broader base,
leaf-like, incurved, sulcate, sparsely ciliate, 2 lin. long; corolla
trumpet-shaped, or nearly straight, narrow at the base, thinly puberulous, dry, thin in texture, crimson, 10–11 lin. long; segments erect
or spreading, subdeltoid, concolorous, under 1 lin. long; anthers
affixed to the filament about $\frac{1}{5}$ of the length of the cells above their
base, included, narrow-oblong, slightly bilobed but not at all pro-
jecting at the base, $\frac{4}{5}$ lin. long, aristate; pore about $\frac{2}{3}$ the length of
the cell; awns setiform, $\frac{2}{3}$ the length of the cell; style exserted,
thickened and channelled at the base; stigma capitellate; ovary
cylindrical, slightly contracted at the base, pallid, glabrous.

CoAST REGION: Ladismith Div.; slopes of the Klein Zwartberg Range,
near Ladismith, about 2000 ft., *Marloth,* 2935!

From *E. cruenta* it differs by its much more slender rough leaves, solitary
flowers and narrower sepals. The species is one of the few stragglers far
inland, the station above recorded being about 63 miles from the nearest sea-
coast.

51. E. speciosa (*Andr. Heathery, t.* 192); erect, 2–4 ft. high;
branches virgate or spreading; leaves 3-nate, suberect and imbricate
or spreading, linear, narrow-lanceolate, or oblong, sulcate, glabrous or
viscid-puberulous, ciliate or naked, 2–4$\frac{1}{2}$ lin. long; flowers 3-nate,
occasionally solitary; bracts usually approximate, rarely remote,
like the sepals, but shorter; sepals narrow- to broad-lanceolate,
scarious below with a leaf-like keeled tip, glabrous or pubescent,
ciliate or naked, usually coloured, 2$\frac{1}{2}$–4$\frac{1}{2}$ lin. long; corolla clavate-
tubular, curved, dry or viscid, glabrous or rarely sparsely pilose,
10–15 lin. long, crimson, limb usually greenish-yellow; filaments
slender, linear; anthers included or subexserted, oblong, subobtuse

at the base, $1\frac{1}{4}$–2 lin. long, aristate ; awns straight, from $\frac{2}{3}$ to nearly the length of the cell ; ovary glabrous. *Andr. Col. Heaths, t.* 133. *E. hirta, Andr. Heathery, t.* 165, *and Col. Heaths, t.* 101 ; *Lodd. Bot. Cab. t.* 1116, *not of Thunb. E. polytricha, Sweet, Hort. Brit. ed.* ii. 338, *ex Ind. Kew.*

COAST REGION, on mountains from 500–4500 ft. : Oudtshoorn Div. ; Zwartberg Pass, *Marloth,* 2423 ! George Div. ; Cradock Berg, *Burchell,* 5926 ! *Galpin,* 3578! near Touw River, *Burchell,* 5731 ! Knysna Div. ; near the Keurbooms River, *Burchell,* 5160 ! 5169 ! near Knysna, *Newdigate in MacOwan, Herb. Aust.-Afr.,* 1922 ! The Glebe, *Galpin,* 3577 ! Uniondale Div. ; near Vlugt, *Bolus,* 2382 ! 2383 ! 2385 ! Humansdorp Div. ; Lottering Bush, *Galpin,* 3587 ! Uitenhage Div. ; Witte Klip, *Bolus !* Also *cultivated specimens !*

52. E. hebecalyx (Benth. in DC. Prodr. vii. 630) ; erect, virgately branched ; leaves 3-nate, erect, subincurved, imbricate, linear or linear-lanceolate, sulcate, puberulous, 2–3 lin. long ; flowers usually 3-nate, occasionally solitary by abortion ; pedicels 3 lin. long ; bracts approximate, ovate, acute, keel-tipped, with scarious margins, like the sepals very concave and covered with a soft greyish tomentum, ciliate, about $\frac{1}{2}$ as long as the sepals ; sepals oblong or obovate, obtuse, keel-tipped, apiculate, scarious, $2\frac{1}{2}$–3 lin. long ; corolla subclavate-tubular, incurved, glabrous, viscid, 12–15 lin. long ; limb nearly erect ; filaments capillary, nearly equally linear ; anthers included or subexserted, oblong, truncate at the base and somewhat sharply pointed in front, $1\frac{1}{2}$–$1\frac{3}{4}$ lin. long, aristate ; awns capillary and somewhat turned backwards ; ovary glabrous. *E. speciosa, Lodd. Bot. Cab. t.* 575 (*good*), *not of Andr. E. speciosa, var. tomentosa, Klotzsch in Linnæa,* ix. 656, *fide Benth., not of Andr.*

SOUTH AFRICA : without locality, *Mund & Maire !* Also *cultivated specimens !*

COAST REGION : George Div. ; Montagu Pass, near George, 1000–2500 ft., *Alexander,* 22 ! *Schlechter,* 2268 ! *Marloth,* 2407 ! *Tyson,* 3164 !

The broad softly tomentose sepals give this a distinct appearance ; the awns of the anther are usually shorter than in *E. speciosa,* and are not straight as in that species but spread backwards. It seems to be very local.

53. E. discolor (Andr. Col. Heaths, t. 20) ; erect, 2–3 ft. high ; branches stout, pubescent ; leaves 3-nate, suberect and imbricate, or somewhat spreading, linear or narrow-lanceolate, glabrous or the younger viscid-puberulous, often glossy, $1\frac{1}{2}$–$3\frac{1}{2}$ lin. long ; pedicels 1–$1\frac{1}{2}$ lin. long ; bracts approximate, like the sepals but about $\frac{2}{3}$ their length ; sepals ovate-lanceolate, acute or acuminate, keeled-tipped, scarious, coloured, glabrous, ciliolate or naked, 2–$2\frac{1}{2}$ lin. long ; corolla clavate-tubular or trumpet-shaped, glabrous, dry, red, 9–12 lin. long ; limb erect or slightly spreading, red or greenish ; filaments equal at the base or but slightly dilated ; anthers included or just manifest, linear to oblong, more or less hollowed at the base, about $\frac{3}{4}$ lin. long, aristate ; awns reaching $\frac{2}{3}$ the length of the cell below it, or much shorter ; style usually curved at the apex ; ovary glabrous. *Andr. Heathery, t.* 160 ; *Wendl. Eric. Ic. fasc.* 5, 9 ; *Lodd. Bot. Cab. t.* 1453. *E. cupressiformis, Salisb. in Trans. Linn. Soc.*

vi. 354. *E. densiflora, Drège, Zwei Pflanzengeogr. Documente,* 105.

VAR. β, **puberula** (Benth. in DC. Prodr. vii. 630); leaves and sepals puberulous.

COAST REGION: Caledon Div.; Palmiet River, *Niven,* 103! Houw Hoek, *Zeyher,* 3159; *Bolus,* 5346! and in *Herb. Norm. Aust.-Afr.,* 193! Bredasdorp Div.; near Elim, *Bolus,* 6761! George Div.; Outeniqualand, *Niven,* 197; west end of Lange Valley, *Burchell,* 5686; Knysna Div.; hills near Melville, *Burchell,* 5460! Humansdorp Div.; near Kromme River, *Burchell,* 4846, 4875, 4877. Port Elizabeth Div.; near Port Elizabeth, *Holland!* Also *cultivated specimens!* Var. β: Mossel Bay Div.; Attaquas Kloof, *Mund & Maire!*

Very near to *E. unicolor*; also allied to *E. versicolor,* from which it may be most readily distinguished by its aristate anthers not sharp pointed at the base in front, and by its filaments nearly equal at the base; its flowering season also appears to be almost constantly some months earlier.

54. E. dichrus (Spreng. Syst. Veg. ii. 179); erect, 3–5 ft. high; branches subvirgate, pubescent; leaves 4-nate, erect and imbricate, or somewhat spreading, linear, sulcate, glabrous, the younger ciliate, 3–5 lin. long; flowers 4-nate, mostly erect; bracts approximate, like the sepals, but about $\frac{2}{3}$ of their length; sepals linear-lanceolate and acuminate, or ovate-lanceolate or oblong and subobtuse, sub-scarious below with a foliaceous keeled apex, pilose or glabrous, ciliate, $2\frac{1}{2}$–$4\frac{1}{2}$ lin. long; corolla tubular or clavate-tubular, glabrous, viscid, red below, yellow above, 10–12 lin. long; filaments very slightly wider at the base; anthers included or manifest, broad-linear, truncate and somewhat hollowed at the base, but scarcely pointed in front, about 1 lin. long, minutely aristate or muticous; awns scarcely reaching beyond the base of the cells; style often decurved at the apex; ovary glabrous. *E. bicolor, Andr. Heathery, t.* 54, *and Col. Heaths, t.* 79; *Lodd. Bot. Cab. t.* 1001, *not of Thunb. E. dichromata, Lodd. l.c. t.* 1813. *E. quadriflora, Klotzsch in Linnæa,* ix. 665. *E. refulgens, Andr. Heathery, t.* 284?.

COAST REGION: Mossel Bay Div.; Hagel Kraal, near Attaquas Kloof, *Niven,* 200! Attaquas Kloof, *Drège,* 7716! near Mossel Bay, 800 ft., *Galpin,* 3579! George Div.; Outeniqualand, *Niven,* 201! Montagu Pass, 1000 ft., *Bolus,* 8667! Also *cultivated specimens!*

55. E unicolor (Wendl. Eric. Ic. fasc. 25, 7, t. 3); erect, 3–4 ft. high, with ascending branches; leaves 4-nate ("3–4-nate," *Bentham*), laxly spreading, linear, hispid with long hairs, 3–4 lin. long; flowers 4-nate (or "sub-3-nate," *Bentham*), sessile or pedicels very short; bracts approximate, linear, foliaceous, hispid, $2\frac{1}{2}$ lin. long; sepals like the bracts, $2\frac{1}{2}$–3 lin. long (or in Andrews' fig. 5 lin. long); corolla tubular, straight, glabrous, viscid, green, 8–11 lin. long; anthers narrow-oblong, 1 lin. long, included, aristate; awns capillary, $\frac{1}{6}$–$\frac{2}{3}$ the length of the cell; ovary glabrous. *E. virescens, Link, Enum. Hort. Berol.* i. 359, *not of Thunb. E. viridescens, Lodd. Bot. Cab. t.* 233. *E. hirta, var. viridiflora, Andr. Heathery, t.* 166, *and Col. Heaths, t.* 173.

SOUTH AFRICA: without locality, *Mund & Maire!* Also *cultivated specimens!*

COAST REGION: Mossel Bay Div.; Attaquas Kloof, *Masson*, 104!

We retain this species, which seems too closely allied to *E. dichrus*, with much doubt. Excepting in its more spreading, more hairy leaves, and straighter green corolla, there is little difference, and both come from the same locality. The material, however, at disposal, is very poor.

56. E. serratifolia (Andr. Col. Heaths, t. 58); erect, 1 ft. or more high; branches numerous; leaves 4-nate, spreading or re-curved, linear, ciliate with plumose hairs, ("ciliate-serrate," *Bentham*) 2–3 lin. long; flowers 4-nate [2–3-nate (*Andrews*) or subsolitary (*Bentham*)]; bracts approximate or subremote, foliaceous, 2 lin. long; sepals lanceolate, acuminate, mucronate, foliaceous, gland-ciliate, $2\frac{1}{2}$–3 lin. long; corolla tubular, nearly straight, red below, yellow above, 9–12 lin. long; limb short, spreading; filaments capillary; anthers oblong, not pointed at the base, $\frac{3}{4}$–1 lin. long, muticous or shortly aristate (*Salisbury*); ovary glabrous. *Andr. Heathery, t.* 44; *Lodd. Bot. Cab. t.* 1703. *E. plumosa, Wendl. Eric. Ic. fasc.* 12, 5, *not of Andr. E. cylindriflora, Salisb. in Trans. Linn. Soc.* vi. 356.

SOUTH AFRICA: *Cultivated specimens!*

Unknown to us in the wild state, unless it be a cultivated form of the next species. Andrews shows the leaves as 4-nate; Wendland as 3-nate; both describe the anthers as muticous. Salisbury finds them always calcarate. In one cultivated specimen examined we found the anthers minutely 2-toothed at the base.

57. E. MacOwanii (Cufino in Bull. Soc. Bot. Ital., 1903, 290); erect, probably 2–3 ft. high; branches stout, sometimes virgate, densely leafy, pubescent, with many lateral flowering branchlets from $\frac{1}{3}$–$\frac{2}{3}$ in. long; leaves 4-nate, imbricate, erect-incurved, narrow-lanceolate, acute, keeled, sulcate, glabrous, the younger ciliolate with simple hairs, 3–$4\frac{1}{2}$ lin. long; flowers solitary, erect or spreading; pedicels pubescent, under 1 lin. long; bracts adpressed, narrow-lanceolate, acute, scarious, translucent, pale green, ciliate, about $\frac{2}{3}$ the length of the sepals; sepals like the bracts, but larger and narrow ovate, keeled with a dark green band, 4 lin. long; corolla 10–12 lin. long, tubular, usually straight, rarely slightly curved; tube nearly equal, dilated at the throat, the lower portion thinly pubescent and dark red, the upper pilose, passing into orange at the throat and limb; throat oblique; segments spreading, semiorbicular, $1\frac{1}{2}$–2 lin. long; anthers included or manifest, rarely subexserted, inserted on the filament just above the base of the cell, broad oblong, subobtuse, rounded at the base, scaberulous, muticous, dark-coloured, about $\frac{7}{8}$ lin. long; pore $\frac{1}{3}$–$\frac{2}{5}$ of the length of the cell; style shortly exserted; stigma capitellate; ovary subglobose, glabrous.

COAST REGION: Caledon Div.; Zwart Berg, *MacOwan!* Locality and collector unknown, *Herb. Bolus*, 6899!

This appears to be closely allied to the preceding species, differing by the much shorter and simple hairs on the leaf-margin, by its broader sepals, and solitary flowers. From *E. perspicua*, to which it is compared by the author, it differs by its 4-celled ovary, and by the colour of the flower.

58. E. versicolor (Wendl. Eric. Ic. fasc. 11, 3); erect, 2–4 ft. high; branches mostly flexuous, sometimes straight; leaves 3-nate, erect and imbricate, or spreading, linear to narrow-elliptic, acute, glabrous or the younger puberulous, ciliolate or naked, 2–$3\frac{1}{2}$ lin. long; flowers 3-nate, erect or spreading; bracts approximate, like the sepals but about $\frac{2}{3}$ of the length; sepals usually lanceolate, acuminate, more rarely oblong and subobtuse, keel-tipped, scarious, glabrous, coloured, 3–4 lin. long; corolla tubular and equal, or widening to the apex, or inflated below, glabrous, dry or viscidulous, costate or smooth, mostly red below and greenish-yellow near the tip, 11–14 lin. long; limb short, usually more or less erose or crispulate; filaments usually more or less dilated and ovate at the base, or at least wider and gradually contracted and tapering upwards; anthers included or subexserted, linear, usually with a sharp projecting point in front at the base, muticous, 1–$1\frac{1}{2}$ lin. long; ovary mostly glabrous, rarely minutely hirtulous. *Andr. Heathery, tt. 47, 249 (var. longiflora), and Col. Heaths, t. 67; Lodd. Bot. Cab. tt. 208, 1316 (var. major); Benth. in DC. Prodr. vii. 631. E. versicolor, var. ciliata, Wendl. Eric. Ic. fasc. 20, 115, t. 43. E. costata, Andr. Heathery, t. 13, and Col. Heaths, t. 16. E. ovata, Wendl. ex Steud. Nom. ed. 2, i. 577.*

VAR. β, **monticola** (Bolus); anthers truncate or rounded at the base in front; leaves more closely and regularly adpressed than in the usual form.

SOUTH AFRICA: without locality, *Thunberg, Herb. Salisbury!* and *cultivated specimens!*

COAST REGION: Swellendam Div.; near and around Swellendam, *Masson,* 105! *Niven,* 198! *Zeyher,* 3157! 3158! near Kinko River, *Zeyher,* 3156! Riversdale Div.; Garcias Pass, 1000 ft., *Galpin,* 3581! near Zoetemelks River, *Burchell,* 6571! near Tygerfontein, *Galpin,* 3580! *Schlechter,* 1981! Kampsche Berg, *Burchell,* 7087! Var. β: Worcester Div.; Matroos Berg, 3500 ft., *Bolus in Herb. Guthrie,* 3951! Swellendam Div., *Shand in Herb. Bolus!*

A very variable species; yet the lanceolate, scarious, coloured sepals seem constant, and serve chiefly to distinguish it from its more immediate allies. The spoon-shaped base of the filaments is also decisive where it exists, but is not a constant character, or at least is not distinctly visible in some specimens.

59. E. berzelioides (Guthrie & Bolus); branches stout, ascending, puberulous; leaves 3-nate, spreading, squarrose, the lower reflexed, oblong, blunt, glabrous, smooth and glossy, sulcate, the younger gland-ciliate, $1\frac{1}{2}$–2 lin. long; flowers 3-nate, pedicels, bracts, sepals and corolla very viscid; pedicels $1\frac{1}{2}$–2 lin. long; bracts approximate, like the sepals but with a larger foliaceous point, $1\frac{1}{2}$ lin. long; sepals lanceolate, acuminate, scarious, about 2 lin. long; corolla trumpet-shaped, glabrous, viscid, red below, paler above, 1 in. long; limb erect; segments 1 lin. long, rounded; filaments slender, scarcely widened at the base; anthers subexserted, linear-oblong, rounded at the base in front, muticous, $1\frac{1}{4}$ lin. long; ovary glabrous.

COAST REGION: Bredasdorp Div.; Mierkraal, 200 ft., *Schlechter,* 10530!

60. E. diaphana (Spreng. Syst. Veg. ii. 178); erect, 2–5 ft. high; branches spreading, rigid, puberulous; leaves 3-nate, erect and imbricate, or somewhat spreading, linear to elliptic-oblong, sulcate, thick, smooth, 2–3 lin. long; flowers 3-nate, at length nodding; pedicels $2\frac{1}{2}$ lin. long; bracts approximate, foliaceous, ovate, very thick, viscid, $\frac{2}{3}$ the length of the sepals; sepals like the bracts, 2 lin. long; corolla subclavate-tubular, glabrous, thick, very viscid, 10–12 lin. long; tube pink; limb suberect, green; stamens included; filaments slightly wider at the base; anthers narrow-oblong, with a projecting point at the base, 2 lin. long, muticous or minutely aristulate; ovary glabrous. *E. transparens, Lodd. Bot. Cab. t. 177; Andr. Heathery, t. 296, and Col. Heaths, t. 283, not of Berg.*

CoAST REGION: Mossel Bay Div.; near Mossel Bay, *Alexander*, 14! Humansdorp Div.; by streams near the Gamtoos River, *Niven*, 199! near the Kromme River, *Drège*, 7714! rocky places near Storms River, *Schlechter*, 5993! Kadies Berg, *Galpin*, 3584! Witte Els Bosch, *Galpin*, 3589! Also a *cultivated specimen!*

61. E. glandulosa (Thunb. Diss. Erica, 25); erect, 2 ft. or more high, all parts more or less covered with gland-tipped hairs; leaves 4-nate, spreading, linear, obtuse, sulcate or sometimes somewhat open-backed, 2–5 lin. long; flowers usually 4-nate (sometimes 2-5-nate); pedicels 2–3 lin. long; bracts approximate, foliaceous, linear or broad linear; sepals broad linear, acute, 3–$4\frac{1}{2}$ lin. long; corolla clavate-tubular, mostly pubescent, rarely glabrous, dry or glandular-viscid, 9–13 lin. long; filaments slender, not, or only very slightly, dilated at the base; anthers included or subexserted, oblong, curved forwards at the base, muticous, about 1 lin. long; ovary glabrous. *E. pellucida, Andr. Heathery, tt. 183, 277 (var. rubra), and Col. Heaths, t. 197; Lodd. Bot. Cab. t. 276; Klotzsch in Linnæa, ix. 675. E. exudans, Andr. Heathery, t. 216. E. exsudans, Lodd. l.c. t. 287. E. droseræfolia, Tausch in Flora, 1834, 600?*

VAR. β, **breviflora** (Bolus); bracts remote; sepals 2 lin. long; corolla tubular-obconic, suboblique, less than 5 lin. long; anthers decurrent-denticulate, about $\frac{3}{4}$ lin. long.

SOUTH AFRICA: without locality, *Thunberg*. Also *cultivated specimens!*

CoAST REGION, frequent between 150–800 ft.: Mossel Bay Div.; Attaquas Kloof, *Masson*, 109! George Div.; between George and Malgat River, *Burchell*, 6097! *Bolus*, 5520! 8671! Outeniqualand, *Niven*, 189! Humansdorp Div.; near the Gamtoos River, *Niven*, 186! by the Kromme River, *Masson*, 108! Uitenhage Div.; near Van Stadens Hoogte, *MacOwan*, 1124! Port Elizabeth Div., *Burchell*, 4576! *Pappe!* Var. β: Humansdorp Div.; between Gamtoos River and Leuwenbosch River, *Burchell*, 4803!

62. E. perspicua (Wendl. Eric. Ic. fasc. 1, 7); erect, 2 ft. high or more; branches virgate or sometimes spreading; leaves 3- or 4-nate, erect or spreading, narrow-linear, subtrigonous, pubescent or glabrous, lanceolate and roughly hispid, ciliate or naked, $1\frac{1}{2}$–2 lin. long; flowers solitary on numerous short lateral branchlets, or sometimes (according to *Bentham*) 1-3-nate; pedicels $\frac{1}{2}$–$\frac{3}{4}$ lin. long;

bracts small, subapproximate, scarious, keel-tipped; sepals very variable, linear from a broad subrotund scarious lacerate base, more or less elongate, keeled, or ovate, or oblong-lanceolate, acute, pilose or glabrous, $1\frac{1}{4}$–$2\frac{1}{2}$ lin. long; corolla subclavate-tubular, more or less pubescent or villous, dry, at length curved, pale rosy or purplish, 8–12 lin. long; limb erect or spreading; filaments dilated and bent near the anther; anthers included, broadly oblong or cuneate, muticous; ovary 8-celled (or, according to *Salisbury*) 6–8-celled, glabrous. *Andr. Heathery, t.* 230, *and Col. Heaths, t.* 255; *Benth. in DC. Prodr.* vii. 634. *E. transparens, Thunb. Prodr.* 71. *E. lituiflora, Salisb. in Trans. Linn. Soc.* vi. 356. *E. Linnæa, Andr. Heathery, t.* 75?, *and Col. Heaths, t.* 106? *E. linnæana, Lodd. Bot. Cab. t.* 102?

VAR. *β*, **latifolia** (Benth. in DC. Prodr. vii. 634); leaves linear-lanceolate, broader, more rigid, incurved, hispid; sepals lanceolate, slightly dilated at the base; corolla 7–8 lin. long.

VAR. *γ*, **lanceolata** (Bolus); leaves lanceolate, broader than in var. *β*, incurved; sepals narrow-lanceolate, not dilated at the base, the lower half scarious, foliaceous above; corolla 9–10 lin. long.

SOUTH AFRICA: without locality, *Mund & Maire!* and *cultivated specimens!*

COAST REGION, from near sea-level to 3000 ft.: Caledon Div.; near the mouth of Klein River, *Masson,* 50! Klein River Kloof, *Zeyher,* 3165! near Hawston, *Schlechter,* 9477! between Houw Hoek and Bot River, *Galpin,* 3576! Houw Hoek, *Schlechter,* 9426! near Babylon's Tower, *Templeman in Herb. MacOwan,* 2747! Var. *β*, Caledon Div.; near the mouth of the Klein River, *Masson,* 112! Var. *γ*, Caledon Div.; Zwart Berg, *Miss Borcherds in Herb. Bolus,* 6286!

Judging from numerous flowers examined, this species seems to differ from its allies by its constantly 8-celled ovary. The alternate dissepiments are sometimes not quite complete to the central column; but are usually so, and always nearly so. In other respects the species resembles some forms of *E. curviflora,* but the anther is slightly different.

63. E. colorans (Andr. Heathery, t. 209); 1–2 ft. high; branches virgate, hirsute; leaves 4-nate, erect-spreading or incurved, linear-subulate, subtrigonous, keeled, glabrous, ciliate, $1\frac{1}{2}$–2 lin. long; flowers 1–4-nate, on short branchlets densely crowded into a pseudo-raceme below· the ends of the branches; pedicels $\frac{1}{2}$–$\frac{3}{4}$ lin. long; bracts subremote, small and narrow; sepals lanceolate, acuminate, scarious, keel-tipped, ciliate, about $2\frac{1}{2}$ lin. long; corolla tubular, with a slight globose swelling below the apex (not always visible in dried specimens), mostly straight, glabrous, dry, white, rosy towards the apex, somewhat transparent, 7–8 lin. long; limb short, acute, subincurved, 1–$1\frac{1}{2}$ lin. long; anthers included, subsemiovate, truncate at the broad base, black, less than $\frac{1}{2}$ lin. long, decurrent-denticulate along the filament; ovary glabrous. *Andr. Col. Heaths, t.* 223; *Bot. Reg. t.* 601; *Lodd. Bot. Cab. t.* 224; *Benth. in DC. Prodr.* vii. 634.

COAST REGION: Bredasdorp Div.; near streams on the downs and mountains near Elim, 100–1600 ft., *Bolus,* 6760! *Schlechter,* 7693! and in *MacOwan, Herb. Aust.-Afr.,* 1921! Also *cultivated specimens!*

A well-marked species, most nearly allied to *E. perspicua* (with which it appears sometimes to hybridize naturally), and also to *E. xanthina*. The globose swelling of the corolla when in the living state, or when (as occasionally happens) visible in the dried state, and the appendiculate anthers, will always serve to distinguish it. Andrews' figure cited is excellent; the others fairly good.

64. E. verticillata (Berg. Descr. Pl. Cap. 99, not of Andr.); erect, 4–5 ft. high; branches spreading or erect; leaves 4–5–6-nate, densely imbricate, erect or spreading, linear, convex and sulcate below, glabrous, the younger ciliate, 2–3 lin. long; flowers mostly 4-nate in wild specimens (in cultivated plants umbellately 3–10-fld., *Andrews*), usually crowded on short branchlets in dense oblong pseudo-racemes below the ends of the branches; flowers erect or spreading; pedicels 1 lin. long; bracts approximate, linear, foliaceous; sepals linear from an ovate scarious base, foliaceous, ciliate, 2–2½ lin. long; corolla tubular, mostly straight, pubescent, dry, rosy, 7–10 lin. long; limb erect or spreading; anthers included, oblong, curved; filament inserted about the middle of the cell to the thick dorsally-projecting connective, about ½ lin. long, muticous; ovary completely 8-celled; capsule at length splitting into 8 valves, crowned at the apex by a callous cup-like process ("concave-truncate" (*Bentham*), glabrous. *E. concinna*, [*Soland. in*] *Ait. Hort. Kew.* ed. 1, ii. 23; *Andr. Heathery, t.* 58, *and Col. Heaths, t.* 82; *Wendl. Eric. Ic. fasc.* 9, 9; *Benth. in DC. Prodr.* vii. 636. *E. paludosa, Salisb. Prodr.* 293, *and in Trans. Linn. Soc.* vi. 356. *E. abietina, Andr. ex G. Don, Gen. Syst.* iii. 823, *not of Linn. nor Berg.*

? VAR. β, **Roxburghii** (Bolus); leaves 4-nate; corolla tubular-oblong or tubular-obconic, 4–5 lin. long; ovary callous-crowned, sub-8-celled, glabrous. *E. Roxburghii, Benth. in DC. Prodr.* vii. 682.

SOUTH AFRICA: without locality, *Thunberg*.

COAST REGION: Cape Div.; on the Flats in the vicinity of Cape Town, Wynberg and Rondebosch, *Burchell*, 700! 765! *Niven*, 206! *Zeyher*, 1085! *Bolus*, 2965! 3755! and in *Herb. Norm. Aust.-Afr.*, 14!: *Schlechter*, 7534! Also *cultivated specimens!* Var. β: Stellenbosch Div.; in marshy places, *Roxburgh! Niven!*

Readily recognized by its peculiar anther, and callous-crested 8-celled ovary. (There is, however, a less marked form of these callous protuberances in *E. curviflora*, var. *Burchellii*, which has a 4-celled ovary.) Respecting the var β, Bentham says that on the authority of the collectors it was found growing with *E. verticillata* and *E. pyramidalis*, and was thought by them to be a hybrid between those species. He adds that it is intermediate between them, and seemed to have perfect seeds. It appears to us better to regard it as a short-flowered form of *E. verticillata* (not a few instances of this variation occurring in other sections); and its sepals, anthers and ovary agree far better with this species than with *E. pyramidalis*. The anther, especially, of *E. Roxburghii*, agrees almost exactly with the peculiar anther of this species.

65. E. mertensiana (Wendl. ex Klotzsch in Linnæa, ix. 659); erect, 2 ft. high; branches puberulous, reddish; leaves 3–4-nate, erect-spreading, linear, flat above, convex and sulcate below, subglabrous, 3–4 lin. long; flowers 3-nate, erect; pedicels 2½–3 lin. long; bracts subremote, linear, acute, pubescent; sepals broad ovate, acuminate, puberulous, keel-tipped, purplish-green, 2 lin. long;

corolla clavate, straight, glabrous, blood-red, faintly ribbed, 9 lin. long; limb short, erect, crenulate, concolorous; anthers exserted, aristate, awns short; ovary silky. *Benth. in DC. Prodr.* vii. 635.

SOUTH AFRICA: cultivated specimen in *Herb. Wendland.*

Only known to us from the above description. We have not seen any specimen which agrees with it.

66. E. cruenta (Soland. in Ait. Hort. Kew. ed. 1, ii. 16); erect, 2-3 ft. high; leaves 3-nate (3-4-nate, *Bentham*), spreading or incurved, imbricate, linear, glabrous or the younger pubescent, $2\frac{1}{2}$-4 lin. long; pedicels 3 lin. long; bracts mostly small and remote, rarely approximate; sepals ovate, acute, or lanceolate, acuminate, about 2 lin. long; corolla clavate-tubular or trumpet-shaped, glabrous, dry, 10-12 lin. long, blood-red; filaments slender, wider at the base, usually affixed at an angle to the anther and shortly below its middle; anther included or subexserted, oblong, about 1 lin. long, aristate; awns sometimes short and partially subdecurrent along the filament, or longer and hanging below the cell; more rarely muticous; ovary glabrous. *Bauer, Exot. Pl. t.* 13; *Wendl. Eric. Ic. fasc.* 4, 11; *Andr. Heathery, t.* 110, *and Col. Heaths, t.* 17; *Lodd. Bot. Cab. t.* 1656; *Benth. in DC. Prodr.* vii. 629. *E. melliflua, Salisb. Prodr.* 293, *and in Trans. Linn. Soc.* vi. 354. *E. coccinea, Drège; Zwei Pflanzengeogr. Documente,* 182.

VAR. β, **mutica** (Bolus); leaves broader; pedicels shorter, less than 1 lin. long; corolla 8-10 lin. long; anthers muticous.

VAR. γ, **buccinula** (Bolus); pedicels $1\frac{1}{2}$ lin. long; sepals 2 lin. long; corolla tubular-funnel-shaped, curved, oblique at the mouth, paler purplish-red, 5-8 lin. long.

VAR. δ, **campanulata** (Bolus); flowers 2-3-nate; corolla wide campanulate or obconic, purple or pink, $3-3\frac{1}{2}$ lin. long by $3-3\frac{1}{2}$ lin. wide at the mouth (when flattened); anthers semi-obovate, muticous.

SOUTH AFRICA: without locality, *Thunberg, Drège! Herb. Salisbury!* and *cultivated specimens!*

COAST REGION: Caledon Div.; from Palmiet River to Zoetemelks Valley, 700-2500 ft., *Burchell,* 7580! 7928! *MacOwan & Bolus, Herb. Norm. Aust.-Afr.,* 188! *Schlechter,* 5573! 5599; 7783! 10449! Swellendam Div.; near Swellendam, *Masson,* 118! near Zuurbraak, *Galpin,* 3575! Var. β: Caledon Div.; Houw Hoek, 1000 ft., *Guthrie,* 4590! Var. γ: Caledon Div.; Palmiet River, 800 ft., *Guthrie,* 2297! 3553! 4163! Var. δ: Bredasdorp Div.; near Elim, 300 ft., *Bolus,* 6738! Caledon Div.; Diep Gat, 300 ft., *Galpin,* 3650!

The species is variable as to the length of sepals and corolla, and of anther-appendages; in var. β, which differs little, the anthers are muticous. In var. γ the shorter and paler corolla gives so different an appearance that we should have hesitated to place it here but for the presence of one specimen which showed corollas on the same branch ranging from 6-8 lin. in length. Otherwise the structure of the flowers is similar. This var. is intermediate in character between the normal form and var. δ, and connects two forms which appear widely different, though we cannot doubt their close relationship.

67. E. wendlandiana (Klotzsch in Linnæa, ix. 652); erect, 2 ft. high; branches crowded, elongate, flexuous, whitish puberulous; leaves 3-nate, erect-spreading, linear, shortly pilose,

2–3 lin. long; inflorescence sub-3-nate; flowers erect; pedicels 2 lin. long; bracts remote, linear, puberulous; sepals leaf-like, lanceolate-linear, shortly pilose, 1½ lin. long; corolla clavate, puberulous, subcostate, subdilated at the apex, brick-red, 8–9 lin. long; limb short, recurved, concolorous; filaments (and style) pilose; anthers far-exserted, aristate; awns straight; ovary " glabrous, capitate, the capitulum pilose " (*Klotzsch*). (*Bentham* says " ovary capitate-dilated at the apex, minutely puberulous "). *Benth. in DC. Prodr.* vii. 629.

COAST REGION: Tulbagh Div.; near Tulbagh, *Mund & Maire!* in Herb. Kew.

We have only seen one poor specimen, which could not be examined, and have been obliged to confine ourselves to copying the author's original description. Bentham saw a specimen in the Berlin Herbarium. It appears to be allied to *E. cruenta*.

68. E. hæmatosiphon (Guthrie & Bolus); erect, somewhat straggling, 1–1½ ft. high; branches spreading, pubescent; leaves 4-nate, erect, incurved, imbricate, linear or linear-lanceolate, glabrous, 1–2½ lin. long; flowers 3–4-nate; pedicels slender, coloured, 2–2½ lin. long; bracts remote, small; sepals broad ovate or ovate-lanceolate, subfoliaceous with scarious margins, leathery, keeled, glabrous or pubescent, 1–1½ lin. long; corolla clavate-tubular, slightly and obliquely curved or nearly straight, glabrous, dry, blood-red, ½–1 in. long; limb short, rounded, erect; filaments capillary; anthers included, ovate to subcuneate, about ½ lin. long, muticous; style included, nearly straight; ovary substipitate, glabrous.

COAST REGION, from 1700–5500 ft.: Clanwilliam Div.; without precise locality, *Leipoldt*, 622! Ceres Div.; near Ceres, *Guthrie*, 3182! Gydouw Mountain, *Schlechter*, 10045! Worcester Div.; Matroos Berg, *Bolus in Herb. Guthrie*, 3946!

A well-marked species, somewhat resembling *E. chloroloma* and *E. cruenta*, but easily distinguished from either. Schlechter's specimens from a higher altitude, but scarcely more than 10 or 12 miles distant, have shorter leaves, corollas and anthers than Guthrie's 3182, but otherwise agree. Guthrie's 3946 might be regarded as a distinct variety with laxer, weaker habit, pedicels up to 4 lin. long, broader sepals, and corolla 11 lin. long.

69. E. chloroloma (Lindl. Bot. Reg. 1838, t. 17); erect, 2 ft. high; branches ascending, stout, pubescent, furrowed with ridges below the somewhat prominent leaf-cushions; leaves 4–6-nate, erect-spreading, imbricate, linear, minutely scabrid-pubescent, 3–4½ lin. long; flowers mostly 3-nate, occasionally 4-nate, or solitary; pedicels 1½–2 lin. long; bracts subremote, minute; sepals linear from an ovate scarious-edged ciliate base, acute, keeled, 1½ lin. long; corolla clavate-tubular, very slightly constricted at the throat, little curved, glabrous, dry, red, 7–8 lin. long; limb very short, green; anthers included, linear, 1¼ lin. long, aristate; awns less than half the length of the cell; style at length exserted; ovary glabrous. *Benth. in DC. Prodr.* vii. 629. *E. dregeana, Klotzsch in Linnæa,* xii. 501.

COAST REGION: Uitenhage Div. ; without precise locality, *Zeyher!* Port Elizabeth Div.; hills near Port Elizabeth, *Zeyher! Pappe! Mrs. Holland,* 23 ! *Miss West!* and in *MacOwan, Herb. Aust.-Afr.,* 1919 !

70. E. brachialis (Salisb. in Trans. Linn. Soc. vi. 367) ; erect, 1½–2½ ft. high ; branches numerous, erect or divaricate, very stout and rigid, densely leafy, tomentose, becoming glabrous ; leaves scattered or 4–6-nate, erect-incurved, densely imbricate, narrow-oblong, subobtuse, convex and sulcate below, minutely serrulate, glabrous or minutely puberulous, 2½ lin. long ; inflorescence umbellate ; umbels 1–5- (mostly 3–4-) flowered ; flowers erect or spreading ; pedicels 3–4 lin. long, pubescent ; bracts subapproximate, leaf-like ; sepals lanceolate, subobtuse, sulcate-keeled in the upper half, scarious below, pubescent, 3½ lin. long ; corolla tubular, slightly narrowed in the middle, cano-pubescent externally and within, dry, thick and rigid, pale ochre-yellow, 7–8 lin. long ; limb short, slightly spreading, less than 1 lin. long ; filaments short, contracted from a wide base, thickened and bent into a subsigmoid flexure below the anther ; anthers included, dorsifixed, linear, smooth, muticous, 2 lin. long, equalling in length the filaments ; style slender, shortly exserted ; ovary 4-celled, glabrous. *Benth. in DC. Prodr.* vii. 635.

COAST REGION : Cape Div. ; Camps Bay, *Alexander,* 4 ! on low granite hills near the western coast, between Table Bay and Hout Bay, *Marloth in Herb. Bolus,* 4922! *Bolus Herb. Norm. Aust.-Afr.,* 347 ! Chapmans Peak, *Wolley Dol,* 949! Stellenbosch Div.; Hottentots Holland, *Masson in Herb. Salisbury!*

A singular species, unlike any other known to us. Bentham places it in this section with doubt, and suggests *Dasyanthes,* in which, however, we cannot concur. It seems to be confined to the Cape Peninsula. No recent collector has found it elsewhere, and Salisbury's citation of "Hottentots Holland" is probably quite erroneous in this as in many other instances.

Section V. **DASYANTHES.** (Sp. 71–77.)

71. E. strigilifolia (Salisb. in Trans. Linn. Soc. vi. 367) ; erect, 1–1½ ft. high ; branches slender ; leaves 4-nate, erect or slightly spreading, closely imbricate, oblong or cuneate-oblong, acute, roughly hispid and pectinate-ciliate with rigid pallid hairs, thick and rigid, 1½–2 lin. long ; flowers 4-nate, cernuous ; pedicels stout, under 1 lin. long ; bracts approximate, like the leaves and as long ; sepals like the bracts, linear-lanceolate, acuminate, pectinate-ciliate, 2–2½ lin. long ; corolla tubular, slightly inflated in the middle, contracted at the throat, puberulous all over and in the upper portion also clothed with longer shaggy or barbellate hairs, white, 7–9 lin. long ; limb short, somewhat spreading ; anthers included, oblong, ⅞ lin. long, muticous ; style included (or possibly at length exserted ?) ; ovary obconic, villous with long silky white hairs. *E. elongata, Lodd. Bot. Cab. t.* 738 ; *Benth. in DC. Prodr.* vii. 637. *E. venusta, Klotzsch in Linnæa,* ix. 695, *not of Salisb. nor of Sinclair. E. cerinthoides, Berg. Descr. Pl. Cap.* 104, *not of Linn. E. cerinthoides, var. γ, Thunb. Diss. Erica,* 26, *and Fl. Cap. ed. Schultes,* 354. *E. strigiliflora, Steud. Nomencl. ed.* 2, i. 580.

VAR. β, **rosea** (Bolus) ; sepals oblong-lanceolate, subobtuse, softly puberulous, rosy ; corolla rosy ; anthers short-aristate, awns ¼–⅓ lin. long.

SOUTH AFRICA: without locality, *Thunberg, Drège*, 7730! 7731! *Mund &*
Maire!

COAST REGION: Oudtshoorn Div.; Zwartberg Pass, 4200 ft., *Marloth*, 2406!
Kolbe! Var. β: same locality, 4900 ft., *Marloth*, 2404!

Much resembling the following species, but distinguishable by its sepals. The
variety described may be connected by intermediate forms, but though we have
several specimens we have seen none.

72. E. pectinifolia (Salisb. in Trans. Linn. Soc. vi. 367); erect,
$1\frac{1}{2}$–$2\frac{1}{2}$ ft. high; branches slender; leaves 4-nate, erect, imbricate or
sometimes shorter than the internodes, variable in size, from linear to
lanceolate and ovate, acute, more or less open-backed, thick and
rigid, roughly and thinly hispid, and pectinate-ciliate with long
pallid cartilaginous hairs, the floral leaves wider and passing gradually
into the form of the bracts, $1\frac{1}{2}$–3 lin. long; flowers mostly 4-nate,
cernuous or spreading; pedicels $\frac{1}{2}$ lin. long; bracts approximate,
unguiculate, broad ovate or subtriangular, margins scarious, rigidly
pectinate-ciliate, with a thick central nerve, about 3 lin. long;
sepals like the bracts but the claw longer, and the blade somewhat
smaller, about $3\frac{1}{2}$ lin. long; corolla tubular, slightly inflated about
the middle, constricted below the short spreading limb, pubescent,
the upper part with longer barbellate or shaggy hairs, white or rosy,
7–9 lin. long; anthers included, oblong, $\frac{7}{8}$ lin. long, shortly aristate;
style exserted; ovary obconical, villous with long white silky hairs.
E. erubescens, Andr. Heathery, t. 113; *Benth. in DC. Prodr.* vii.
637, *not of Lodd. E. cerinthoides, var. β, Thunb. Diss. Erica,* 26,
and Fl. Cap. ed. Schultes, 354.

COAST REGION, on mountains at 400–600 ft.: Uniondale Div.; Lange Kloof
Mountains, *Thunberg, Bolus*, 2391! Humansdorp Div.; Kromme River, *Niven*,
167! Kruisfontein, *Galpin*, 3594! Clarkson, 400 ft., *Schlechter*, 6006! Zitzikamma, *Pappe!* Uitenhage Div., near the Lead Mines, and between there
and Van Stadens River, *Burchell*, 4511! 4580! 4621! Van Stadens Mountains,
MacOwan, 1126! 2059! *Bolus*, 1578! Witte Klip, *Bolus*, 9132!

There has often been confusion of this species with the preceding. Yet
according to Salisbury's own description the peculiar sepals and bracts of this
species separate the two quite effectually, and we have seen no intermediate or
doubtful forms. The resemblance is considerable at first sight, but is only
superficial. This species, so far as we know, is much commoner and more widely
distributed than the preceding.

73. E. Sparrmanni (Linn. f. in Vet. Acad. Handl. Stockh. 1778,
21, t. 2); erect, 1–$1\frac{1}{2}$ ft. high, with the habit of the preceding;
leaves 4-nate, suberect, imbricate, linear-lanceolate, subobtuse, deeply
sulcate or somewhat open-backed, rigidly hispid and pectinate-
ciliate, 2–3 lin. long; flowers 4-nate on short branchlets, cernuous;
pedicels about 1 lin. long; bracts approximate, lanceolate, hispid and
ciliate like the leaves, paler and more naked below, about 2 lin. long;
sepals like the bracts but linear-lanceolate, 3 lin. long; corolla
tubular, inflated about the middle, densely hispid with long coarse
setose yellow hairs, yellow or greenish-yellow, $\frac{1}{2}$–$\frac{3}{4}$ in. long;
segments erect, deltoid, subacute; anthers included, oblong, 1 lin.
long, muticous; style exserted; ovary obconic, villous with white

silky hairs. *Linn. Amœn. Acad. ed. Schreb.* x. *App.* 123, *t.* 6; *Thunb. Diss. Erica*, 26; *Benth. in DC. Prodr.* vii. 636. *E. hystriciflora, Salisb. in Trans. Linn. Soc.* vi. 367. *E. aspera, Andr. Heathery, t.* 104, *and Col. Heaths, t.* 148. *E. armata, Spreng. Syst. Veg.* ii. 184? *not of Klotzsch.*

COAST REGION: Uniondale Div.; mountains of Long Kloof *Thunberg, Drège.* Humansdorp Div.; mountains near Storms River, *Herb. MacOwan,* 3370! near Kromme River, *Drège, Masson,* 101! *Niven,* 166! near Clarkson, 250–800 ft., *Schlechter,* 6010! *Galpin,* 3593! Also *cultivated specimens!*

A distinct species; recognized by the inflated, almost narrow-ovoid, corolla, and its peculiar long bristly yellow hairs.

74. E. doliiformis (Salisb. in Trans. Linn. Soc. vi. 368); under 1 ft. high; branches erect or spreading; leaves sub-6-nate, crowded, spreading and upcurved, or subsquarrose, linear, glabrous, glandular-ciliate, 5–6 lin. long; flowers umbellate; pedicels 5–7 lin. long, glandular-pubescent; bracts approximate, linear, foliaceous, $2\frac{1}{2}$–3 lin. long; sepals linear-lanceolate, acuminate, glandular, slightly exceed-ing the bracts; corolla tubular-inflated, narrowed to base and apex, glandular-pubescent, 5–7 lin. long, rosy; anthers subincluded, oblong, very shortly decurrent-aristate, nearly 1 lin. long; style shortly exserted; ovary turbinate, thinly puberulous. *E. blanda, Andr. Heathery, t.* 107, *and Col. Heaths, t.* 152, *not of Salisb.; Lodd. Bot. Cab. t.* 13; *Benth. in DC. Prodr.* vii. 636. *E. mam-mosa, Thunb. Diss. Erica,* 42, *not of Linn. E. metulæflora, Andr. Heathery, t.* 224, *and Col. Heaths, t.* 185; *Bot. Mag. t.* 612; *and var. discolor, Andr. Heathery,* 269.

SOUTH AFRICA: without locality, *Thunberg.*
COAST REGION: Worcester Div.; Dutoits Kloof, 2000–3000 ft., *Drège,* 1146! 7736! Paarl Div.; French Hoek, *Niven,* 149! Also *cultivated specimens!*

Apparently a rare species, not found by any recent collector. There are good specimens in the Cape Govt. Herb. but without any indication of origin. It seems to be usually distinguishable by its long slender glandular-hairy pedicels and puberulous ovaries.

75. E. splendens (Andr. Heathery, t. 240, not of Wendl.); erect, 1–2 ft. high; branches spreading, flexuous, rigid, pubescent or pilose; leaves 4-nate, densely crowded, erect or spreading, imbricate, linear, oblong or lanceolate, subobtuse, pubescent, ashy-grey, hispid-ciliate, $1\frac{1}{2}$–2 lin. long; flowers 4-nate or umbellate, cernuous, spreading or erect; pedicels pubescent and glandular-pilose, 4–8 lin. long; bracts from approximate to subremote, linear or oblong, obtuse, pubescent and glandular-pilose, $1\frac{1}{2}$–$2\frac{1}{2}$ lin. long; sepals like the bracts, but linear-lanceolate, 2–3 lin. long; corolla wide-tubular-inflated, sometimes asymmetrically so, straight or curved, pubescent, dry, blood-red or pink, $\frac{1}{2}$–1 in. long; anthers subexserted, linear, about 1 lin. long, cristato-aristate; awns curved-subulate, lacerate on the outer edge, $\frac{1}{5}$–$\frac{1}{3}$ lin. long; filaments capillary; style exserted; ovary oval, villous. *Andr. Col. Heaths, t.* 275. *E. tumida, Ker-Gawl. in Bot. Reg. t.* 65; *Benth. in DC. Prodr.* vii. 636.

VAR. *β*, **minor** (Bolus); smaller in all parts; corolla somewhat less inflated, narrower, shorter and scarcely curved, pink, 5-6 lin. long.

COAST REGION : Tulbagh Div.; slopes of Winterhoek Mountains, 5000 ft., *Bodkin in Herb. Bolus,* 5905 ! *Bolus in Herb. Guthrie,* 4174 ! Var. *β*: Clanwilliam Div. ; Suecuwkop, Cederberg Range, *Leipoldt,* 623 !

CENTRAL REGION: Ceres Div.; Table Mountain in the Cold Bokkeveld, 5500 ft. *Schlechter,* 10102 !

The var. *β* has some curious points of resemblance to *E. armata,* Klotzsch in the section *Myra.* The species appears to be somewhat rare.

76. E. Oatesii (Rolfe in Oates, Matabeleland, ed. 2, 402, t. 11) ;

erect, 1-4 ft. high ; branches spreading, pubescent and hispid ; leaves 3-4-nate (sometimes on the same branch), spreading and recurved or suberect, narrow-linear, subacute, puberulous or roughly hispidulous, naked or ciliate with long gland-tipped hairs, margins reflexed, the broader forms often open-backed, 3-4½ lin. long ; flowers in more or less close, few-flowered umbels ; pedicels pubescent, 3 lin. long ; bracts, two approximate, the lower remote, or all subremote, oblong or linear-pubescent, gland-ciliate, 2-3 lin. long ; sepals lanceolate, acute or acuminate, somewhat scarious below and tapering to a linear foliaceous point, pubescent, gland-ciliate, 2½-3 lin. long ; corolla ovoid-tubular (or "ellipsoid-urceolate," *Rolfe*), glabrous or puberulous, scarlet, 5-6½ lin. long ; limb short, suberect ; anthers included, dorsifixed shortly above the base, semiovate, $\frac{3}{4}$-$\frac{7}{8}$ lin. long, shortly awned ; style included, rarely subexserted ; ovary villous.

VAR. *β*, **latifolia** (Bolus); leaves somewhat shorter and broader than in the type, sublanceolate, more or less open-backed, suberect, pubescence more copious and more grey ; corolla usually somewhat smaller, 5-5½ lin. long, scarlet.

EASTERN REGION : Natal; between Pietermaritzburg and Crocodile River, *Oates!* Amajuba Hill, *Todd in Herb. Wood,* 1638 ! *Mrs. Steinbank in Herb. Wood,* 3640 ! Buffalo River, *McKen,* 3 ! Var. *β* : Natal; Amawahqua Mountain, 6000-7000 ft., *Wood,* 4622 ! Weenen County, *Evans,* 399 ! Cathkin Peak, 8000-10000 ft., *Thode,* 62a ! Bushmans River, 6000-7000 ft., *Thode,* 62b !

KALAHARI REGION : Orange River Colony; Allanvale Farm, near the Drakensberg, 5500 ft., *Crisbrook in Herb. Guthrie,* 3005 !

This species almost unites the two sections *Dasyanthes* and *Ephebus,* and is placed here on account of its affinity to *E. splendens,* which, like it, is also a plant of the higher mountains, though the stations are widely separated. The variety is not very distinct, and Evans' 399 connects the two. It may be known from all forms of *E. cerinthoides* by its usually shorter, broader and less hairy corolla, and the distinct awns of its anther.

[The rosy-purple colouring of the original plate was adapted from the dried specimens. Collectors state that the flowers are scarlet.—*N. E. Brown.*]

77. E. cerinthoides (Linn. Sp. Pl. ed. 2, 505) ; erect, 2-3 ft.

high ; branches ascending, often virgate and simple ; leaves 4-5-6-nate, from erect and closely imbricate to spreading, squarrose or recurved, linear or linear-lanceolate, blunt, deeply sulcate to subopen-backed, variably pubescent and mostly glandular, hispid and ciliate, more rarely glabrous and naked, 3-8 lin. long by $\frac{1}{3}$-$\frac{1}{2}$ lin. wide ; flowers umbellate ; pedicels glabrous or pubescent, 1-6 lin. long ; bracts approximate, or one subremote, linear-lanceolate, oblong or lanceolate, acute or subobtuse, gland-hispid and ciliate, about 2 lin. long; sepals like the bracts but larger, from 2–3½ lin. long; corolla tubular, more or less inflated, slightly constricted at the mouth, more or less pubescent, usually both shortly so and also pilose with longer (sometimes glandular) hairs, mostly crimson, more rarely rosy, ½–1⅓ in. long; limb spreading or erect; anthers included, dorsifixed just above the base, oblong, pallid, about 1 lin. long, muticous or with very short or rudimentary minute horns; style included or exserted; ovary broadish turbinate, villous. *Bauer, Exot. Pl. t.* 30; *Andr. Heathery, t.* 207, *and Col. Heaths, t.* 11; *Wendl. Eric. Ic. fasc.* 7, 9; *Bot. Mag. t.* 220; *Lodd. Bot. Cab. t.* 1679; *Herb. Amat. t.* 108; *Thunb. Diss. Erica,* 25, *and Fl. Cap. ed. Schult.* 354; *Benth. in DC. Prodr.* vii 636. *E. pulviniformis, Salisb. Prodr.* 295. *E. crinifolia, Salisb. in Trans. Linn. Soc.* vi. 367. *E. verecunda, Lodd. Bot. Cab. t.* 1827? *not of Salisb. E. Meuroni, Benth. in DC. Prodr.* vii. 636? *E. metulæflora, Klotzsch in Linnæa,* xii. 502, *not of Andr. nor others.*

VAR. β, **barbertona** (Bolus); corolla broadly tubular, slightly inflated in the middle and constricted at the throat, viscose-pubescent, 6 lin. long. *E. barbertona, Galpin in Kew Bulletin,* 1895, 148.

COAST REGION, frequent on plains and mountains, ascending to 6000 ft.: Worcester Div., *Cooper,* 1656! 1746! *Drège!* Malmesbury Div.; Groene Kloof, *Drège;* Cape Div.; around Cape Town, *Thunberg, Burchell,* 53! 395! 583! 8469! 8536! *Niven,* 168! *Drège! Marloth in MacOwan Herb. Aust.-Afr.,* 1489! *Ecklon,* 100! *Wilms,* 3443! Caledon Div., *Burchell,* 7609! 7648! 7695! 7827! 8181! Riversdale Div., *Burchell,* 6768! George Div.; near Kaymans River, *Burchell,* 5794! Knysna Div., *Pappe!* Uitenhage Div., *Cooper,* 1486! *Bolus,* 1575! Port Elizabeth Div., *Burchell,* 4474! Albany Div., *MacOwan! Scott-Elliot,* 749!

KALAHARI REGION: Basutoland, *Cooper,* 758! Transvaal; Waterfall Creek, *McLea in Herb. Bolus,* 3036! Macmac Falls and head of Sabie River, *Mudd!* Lydenburg, *Wilms,* 913! Var. β : Transvaal! near Barberton, 4500 ft., *Galpin,* 598!

EASTERN REGION: Tembuland; Bazeia Mountain, *Baur,* 539! Pondoland, *Bachmann,* 1003! Natal; Inanda, *Wood,* 507! near Murchison, *Wood,* 3042! 3113! Var. β : Swaziland; Havelock Concession, *Saltmarshe in Herb. Galpin,* 1046!

An extremely variable species, chiefly as to pubescence, leaves, and size of corolla. With a very large suite of specimens before us we cannot define any distinct varieties, except that above mentioned; this is probably a stunted form caused by burning by grass-fires of which the specimens show evidence; the corolla is shorter and relatively broader, but there is little else to distinguish it, and other specimens show corollas as short but relatively narrower. *E. Meuroni* (judging from Bentham's description) can only be a rather more glabrous form of this species. He himself appeared to be doubtful of it and the garden specimens marked by him, which we have examined, show scarcely any differences. The species is one of the most widely spread in the genus.

Section VI. CHONA. (Sp. 78.)

78. E. embothriifolia (Salisb. in Trans. Linn. Soc. vi. 379); erect, 9–18 in. high; branches slender, ascending or spreading, glabrous, puberulous or viscid with gland-tipped hairs; leaves 3-nate, often subdistant, or shorter than the internodes, erect-spread-

ing to squarrose-recurved, linear, acute or obtuse, slender, subterete, sulcate, ciliate and tipped with long glandular hairs, 4–7 lin.
long; flowers umbellate; pedicels slender, striate, viscid, 5–8 lin.
long; bracts remote, lanceolate, 1 lin. long; sepals linear to lanceolate, subacute, tipped with a glandular seta, margins involute, gland-
ciliate, viscid, polished, 2–3 lin. long; corolla tubular below, inflated above, constricted at the throat, puberulous, rather viscid,
red, about ½ in. long; segments short, broad, recurved-spreading;
anthers exserted, cohering into a conical tube round the style,
dorsifixed, linear, slender, 1½–3 lin. long, aristate; awns very
slender, of variable length; style exserted beyond the anthers;
ovary cylindrical, glabrous. *E. Nivenia, Andr. Heathery, t. 76, and
Col. Heaths, t. 112. E. Niveni, Benth. in DC. Prodr. vii. 635.*

Var. β, **longiflora** (Bolus); corolla-tube longer and more slender in the lower
portion, 10–12 lin. long; anthers 3 lin. long. *E. Nivenia, var. longiflora, Andr.
Heathery, t. 227, and Col. Heaths, t. 189.*

Var. γ, **subæqualis** (Bolus); corolla with a less inflated subequal tube,
slightly widening from the base upwards, 7–8 lin. long, 2 lin. wide; awns of
the anther very short; ovary pilose near the top.

South Africa: without locality, *Herb. Salisbury!* Var. β: *Cape Gov.
Herb.!*

Coast Region: Caledon Div.; mountain tops near Genadendal and Baviaans
Kloof, *Niven*, 154 partly! *Masson*, 39! *Burchell*, 7749! *Bolus*, 5399! and in
Herb. Norm. Aust.-Afr., 343! *Schlechter*, 10304! *Galpin*, 3592! near Appels
Kraal, *Zeyher*, 3173! Var. β: mountains of Zoetemelks Valley, *Masson*, 38!
Niven, 154 partly! *Grisbrook in Herb. Guthrie*, 3300!

Section VII. BACTRIDIUM. (Sp. 79–80.)

79. E. fascicularis (Linn. f. Suppl. 219); 2–6 ft. high;
branches long, virgate, slender, terminating in a conspicuous rosette
of large pinkish green-tipped flowers; leaves 6-nate or scattered,
erect-spreading; petioles capillary, 1½ lin. long; blade linear, subterete, blunt, glabrous, the whole ½–⅔ in. long; flowers 10–25 in
the umbel, sometimes pseudo-lateral owing to new growth; pedicels
3–4 lin. long; bracts foliaceous, linear, remote or subremote;
sepals narrow-lanceolate, acute, foliaceous, glabrous, rather viscid,
about 3 lin. long; corolla tubular, equal, viscid, glossy, 1–1⅓ in.
long, rosy below, paler upwards; segments subdeltoid, incurved,
greenish; anthers oblong; awns straight, shorter than the cells.
*Bauer, Exot. Pl. t. 6; Wendl. Eric. Ic. fasc. 14, 29; Benth. in DC.
Prodr. vii. 637. E. octophylla, Thunb. Diss. Erica, 44, t. 3. E.
coronata, Andr. Heathery, t. 109, and Col. Heaths, t. 15. E.
radiiflora, Salisb. in Trans. Linn. Soc. vi. 360.*

Var. β, **imperialis** (Bolus); corolla inflated about the middle; segments a
little longer and more acute. *E. imperialis, Andr. Col. Heaths, t. 239, and
Heathery, t. 266.*

South Africa: without locality, *Herb. Salisbury!*

Coast Region, ascending to 3400 ft.: Stellenbosch Div.; in and near
Lowrys Pass, *Burchell*, 8219! 8272! *MacOwan & Bolus, Herb. Norm. Aust.-
Afr.*, 15! Hottentots Holland, *Thunberg*. Caledon Div.; Baviaans Kloof,
Burchell, 7670! near Hemel-en-Aarde, *Zeyher*, 3195! Houw Hoek, *Scott-
Elliot!* Genadendal Mountains, *Galpin*, 3596! Bredasdorp Div.; near Hagel

Kraal, *Mund!* near Elim, *Bolus*, 6755! near the mouth of the Ratel River, *Bolus!* near Brand Fontein, *Schlechter*, 10582! Var. β, Caledon Div.; mountains near Appels Kraal, *Zeyher*, 3195! and without precise locality, *Herb. Bolus*, 6381!

Bentham regarded *E. imperialis*, Andr., as a garden hybrid between this and some other species. It has been found at least twice since growing wild, and exhibited at the Caledon Flower Show in 1897, and is probably an ordinary form of the present species. Zeyher's specimens of 3195 are from two stations cited above, and those in Herb. Berlin are certainly of the var. β; of the other we are now somewhat uncertain.

80. E. Massoni (Linn. f. Suppl. 221); erect, 1–1½ ft. high; branches spreading; leaves 4–6-nate, closely crowded and densely imbricate, erect or subspreading, oblong, obtuse, flat, long-ciliate, 2 lin. long; umbels 5–10-flowered; pedicels lanate, 4 lin. long; bracts subremote, long-haired, 2 lin. long; sepals lanceolate or oblong, blunt, sulcate, glossy, very viscid, more or less long-hairy, 2–2½ lin. long; corolla tubular, asymmetrically inflated, subglobosely swollen below the mouth, extremely viscid, red or orange with a green limb, about 1 in. long; stamens subincluded; filaments capillary, thickened and curved at the insertion of the oblong dorsifixed muticous anther. *Thunb. Diss. Erica*, 27, *t. 3; Bauer, Exot. Pl. t. 20; Wendl. Eric. Ic. fasc. 26, 6; Bot. Mag. t. 356; Lodd. Bot. Cab. t. 1069; Benth. in DC. Prodr. vii. 638. E. Massonia, Andr. Heathery, t. 128, and Col. Heaths, t. 36. E. lycopodiifolia, Salisb. Prodr. 294, and in Trans. Linn. Soc. vi. 361.*

VAR. β, **minor** (Benth. in DC. Prodr. vii. 638); corolla shorter and less inflated; bracts and sepals linear and less ciliate. *E. gemmifera, Lodd. Bot. Cab. t. 457; Bot. Mag. t. 2266.*

SOUTH AFRICA: without locality, *Herb. Salisbury!* Var. β: *cultivated specimen!*

COAST REGION, on mountains, 1000–3500 ft.: Stellenbosch Div.; Hottentots Holland, *Thunberg! Niven*, 170! *Zeyher*, 3180! *Bolus*, 4171! *and in Herb. Norm. Aust.-Afr.*, 16! Caledon Div.; near Palmiet River, *Burchell*, 8173! Klein Houw Hoek, *Niven*, 169! Houw Hoek Mountains, *Burchell*, 8155! *Guthrie*, 2298!

Very distinct from the preceding by its habit and its hairy, usually closepressed leaves; the anther is also different. The var. β is unknown to us except from the figures quoted. As the large-flowered form is related to *E. squarrosa*, so is this variety to our *E. Gysbertii*, of the same section.

Section VIII. **EURYLOMA.** (Sp. 81–95.)

81. E. Gysbertii (Guthrie & Bolus); erect; branches slender, subvirgate; leaves 4-nate, erect or slightly spreading, imbricate, obtuse, thick, glabrous, sparsely ciliate with thickish white caducous hairs, and tipped with a long brown bristle, including the longish petiole 1½–2½ lin. long; umbels 3–4-flowered; pedicels slender, dark, 2 lin. long; bracts subapproximate, linear, ciliate and bristletipped with hairs much longer than themselves; sepals linear, ciliate and tipped like the bracts, 1½ lin. long, hairs sometimes 3½ lin. long; corolla tubular, only slightly inflated, suddenly constricted at the throat, viscidulous; tube 4 lin. long, rosy; throat purple; limb paler; segments ovate, acuminate, 1 lin. long; filaments capillary, dilated towards and near the anther; anthers semiovate, subacute, curved or subcrescent-shaped, prognathous, bilobed at the base, pallid, muticous, little exceeding $\frac{1}{2}$ lin. long; ovary elongate, shortly stipitate.

CoAST REGION : Stellenbosch Div.; on the western foot of the Hottentots Holland Mountains, 200 ft., *Guthrie*, 3654!

Allied to *E. squarrosa* but the corolla has a different shape, the anthers distinctly prognathous, and the leaves not squarrose. Also allied to *E. Massoni*, var. *minor*, judging by the figures of the latter, but seems distinct.

82. E. squarrosa (Salisb. in Trans. Linn. Soc. vi. 380); leaves sub-4-nate, squarrose or spreading, more or less crowded, oblong, obtuse, thickish, somewhat thinly ciliate with long spreading rusty hairs, 2–3 lin. long; umbels many-flowered; pedicels slender, viscid, 4–5 lin. long; bracts remote, linear, ciliate like the leaves, $1\frac{1}{2}$ lin. long; sepals lanceolate, acute, barbed with long hairs at the apex, keeled, $1\frac{1}{2}$ lin. long; corolla narrow-ovoid-urceolate, viscid, glabrous, 5–6 lin. long; tube rosy; throat purple; segments subacute, apparently white; filaments capillary, dilated near the anthers; anther dorsifixed, oblong, slightly bent, scarcely prognathous or bilobed at the base, $\frac{2}{3}$ lin. long, muticous; ovary distinctly stipitate, glabrous. *E. ferruginea, Andr. Heathery, t.* 162, *and Col. Heaths, t.* 168.

CoAST REGION: Paarl Div.; French Hoek Kloof, *Niven*, 151! Caledon Div.; mountains near Zondereinde River, *Masson*, 40! Also *cultivated specimens!*

A distinct and apparently rare species, no specimens having been collected, so far as we know, during the last 100 years. It is closely allied to *E. Gysbertii*, and has a curious resemblance in its leaves, anthers and ovary to *E. Massoni*.

83. E. aristata (Andr. Heathery, t. 152); erect, glabrous; leaves 4-nate, squarrose-recurved, narrow-oblong, acute, glabrous, rigidly ciliate, $2\frac{1}{2}$–3 lin. long; floral leaves more erect, longer and bract-like; umbels 4-flowered; pedicels viscid, 3 lin. long; bracts remote or subremote, linear-lanceolate, rigid, coriaceous, viscid, 3–4 lin. long; sepals linear-lanceolate, acuminate, coriaceous, viscid, 3–5 lin. long; corolla tubular-inflated, only slightly narrowed above the middle, a little constricted at the throat, viscid, rosy, with 8 darker red veins; tube 1 in. long; throat dark purple; limb white; segments spreading, oblong, retuse, wider than long, $1\frac{1}{2}$ lin. long; filaments slender, much dilated and darker coloured at the apex near the anther, with a distinct dark central nerve; anthers subcrescent-shaped, deeply bilobed, and subprognathous at the base, over $1\frac{1}{2}$ lin. long, muticous; style equalling the stamens.; stigma capitellate, very small; ovary elongate, not or scarcely stipitate, 3 lin. long. *Andr. Col. Heaths, t.* 147 ; *Bot. Mag. t.* 1249 ; *Lodd. Bot. Cab. t.* 73 (*the two last excellent*) ; *Benth. in DC. Prodr.* vii. 643.

CoAST REGION : Riversdale Div.; amongst shrubs, Platte Kloof, in the Cape

Gov. Herb.! Caledon Div.? locality unknown, exhibited at the annual show of wild-flowers at Caledon, Sept., 1895, *Herb. Guthrie*, 3770!

We have seen no authentic specimens of this species, nor had Bentham, but the excellent ones from Caledon (certainly wild) quoted above and from which we describe, agree so exactly with the figures of Loddiges and the Bot. Mag. that we have no doubt as to their identity. It is the more interesting as leading to the supposition that it is a genuine species and not, as Bentham thought probable, a hybrid. Andrews' t. 152 is not quite so representative of the wild plant as known to us, as are the others. His t. 203, cited by Bentham, appears to belong to *E. retorta, Linn. f.*, and the two species have been confused in herbaria.

84. E. retorta (Montin, in Kongl. Vet. Acad. Handl., 1774, 297, t. 7); erect, subglabrous, 12–15 in. high; branches many, straggling, closely leafy; leaves 4-nate, mostly revolute or squarrose-recurved, crowded, ovate-lanceolate or oblong, acute, tipped with a long bristle, ciliate, thick, rigid, about 2 lin. long; umbels 4–8-flowered; pedicels slender, glandular-viscid, 3 lin. long; bracts remote, oblanceolate, acute, with a terminal bristle of their own length, scarious, softly pubescent, ciliate, 2 lin. long; sepals lanceolate, otherwise like the bracts but more foliaceous, 2–2½ lin. long; corolla ampullaceous, attenuate above but variable and the neck sometimes almost absent, throat contracted, viscid, of almost hyaline texture; tube ½–1 in. long, pale rosy, throat and limb darker coloured; segments ovate, acute, 1–1½ or even 2 lin. long; filaments slender, only slightly widened towards the anther; anther subcrescent-shaped, acute, bilobed and slightly prognathous at the base, $\frac{3}{4}$–$\frac{7}{8}$ lin. long, more or less irregularly woolly with short white hairs, pallid; ovary elongate, on a thickish stipe of nearly its own diameter. *Linn. f. Suppl.* 220 ; *Andr. Heathery, t.* 144, *and Col. Heaths, t.* 54 ; *Bot. Mag. t.* 362 ; *Wendl. Eric. Ic.* 15, 45, *t.* 18 ; *Lodd. Bot. Cab. t.* 804 ; *Benth. in DC. Prodr.* vii. 644. *E. aristata, var. minor, Andr. Heathery, t.* 203, *and Col. Heaths, t.* 219, *not of Linn. f. E. gorteriæfolia, Salisb. in Trans. Linn. Soc.* vi. 381. *E. eximia, Lodd. l.c. t.* 1105.

SOUTH AFRICA: without locality, *Mund!* *Herb. Salisbury!* and *cultivated specimens!*

COAST REGION: Stellenbosch Div.: Hottentots Holland, *Thunberg.* Caledon Div.; Klein Houw Hoek, *Niven*, 152! *Zeyher*, 3196! southern slopes of Great Houw Hoek Mountains, 2800 ft., *Bolus*, 6954! 6955!

In habit and external appearance this much resembles *E. aristata*; but the leaves and corolla are somewhat differently shaped, and the corolla-segments afford an immediate means of distinction. It is apparently rare, or at least very local; and only once have we seen some dozens of plants together. Specimens with large and small corollas of either extreme in size, were found growing intermixed.

85. E. lagenæformis (Salisb. in Trans. Linn. Soc. vi. 382); erect, glabrous; branches flexuous; leaves 3-nate, erect-recurved, oblong-linear, obtuse, 3–4 lin. long; bracts subremote; sepals broad-lanceolate, bluntish, coloured, about 2½ lin. long; corolla elongate-ampullaceous, or inflated at the base, gradually narrowing to the apex into an elongate neck, viscidulous; tube rosy with darker red veins, 12–14 lin. long ; segments ovate, subobtuse, spreading, 4–5 lin. long; anthers narrow-subcrescent-shaped, somewhat prognathous, bilobed at the base, muticous; ovary subsessile (in the figures cited), stipitate (*Bentham*). *E. jasminiflora, Andr. Heathery, t.* 26, *and Col. Heaths, t.* 29, *not of Salisb.; Wendl. Eric. Ic. fasc.* 27, 12; *Benth. in DC. Prodr.* vii. 644. *E. pulcherrima, G. Don in Sweet, Hort. Brit. ed.* 3, 434.

CoAST REGION : Riversdale Div. ; Platte Kloof, *Masson in Herb. Salisbury ! Roxburgh. Also cultivated specimens !*

86. E. jasminiflora (Salisb. Prodr. 293); erect, glabrous, 2 ft. or more high; branches few, subvirgate or flexuous, slender, closely leafy ; leaves 3-nate, adpressed, imbricate or scarcely longer than the internodes, linear, obtuse, mucronulate, thick, gland-ciliate, 3–5 lin. long ; flowers 2–3-nate, erect ; pedicels slender, glandular, 4 lin. long ; bracts remote, linear, erect; sepals linear-oblong, acute, concave, sulcate, viscid, gland-ciliate, $\frac{1}{2}$ in. long ; corolla-tube cylindrical, nearly equal, 15–16 lin. long, $1\frac{1}{2}$ lin. wide, slightly enlarged at the throat, white or pale rose with deeper red veins ; segments spreading, ovate, obtuse, white or striped, about 5 lin. long by 3 lin. wide ; filaments slender, dilated at the anther ; anthers subcrescent-shaped, acuminate, acutely bilobed and prognathous at the base, pallid, muticous, about $1\frac{1}{2}$ lin. long ; ovary cylindrical, subsessile. *Salisb. in Trans. Linn. Soc.* vi. 382. *E. Aitonia, Masson in Bot. Mag. t.* 429 ; *Andr. Heathery, t.* 102, *and Col. Heaths, t.* 1 ; *var. β recta, Klotzsch in Linnæa,* x. 348. *E. aitoniana, Lodd. Bot. Cab. t.* 144 ; *Benth. in DC. Prodr.* vii. 645. *E. Aitoni, Willd. Sp. Pl.* ii. 398.

CoAST REGION: Caledon Div. ; on the Zwart Berg, near the warm baths, *Paterson, Masson ! Niven ! Also cultivated specimens !*

A very distinct and handsome species; now, apparently, so rare, that we have heard of no collector since Niven's time, about the early part of the 19th century. Bentham placed it in a special section, *Platyloma,* but the grounds for separation from § *Euryloma* appear insufficient.

87. E. Junonia (Bolus in Journ. Bot. 1894, 234) ; erect, glabrous, $1\frac{1}{2}$–2 ft. high; branches spreading and ascending; leaves 3-nate, erect-spreading, imbricate, linear, 2–$3\frac{1}{2}$ lin. long; umbels copious, 3–6-flowered ; pedicels 4–7 lin. long ; bracts remote, linear, $2\frac{1}{2}$–$3\frac{1}{2}$ lin. long ; sepals linear-lanceolate, acuminate, red with a green nerve, 3–$5\frac{1}{2}$ lin. long ; corolla subampullaceous, much inflated below, with a very long much attenuate neck, dilated in the throat, dry or slightly viscid, bright red, ribbed with 8 darker red veins ; tube 9–20 lin. long ; segments ovate, acute or acuminate, stellato-patent, 2–5 lin. long ; filaments very narrow at the base, widened upwards but not dilated towards the anther; anthers oblong, nearly straight, bilobed, not prognathous at the base, pallid, about 1 lin. long or less, aristate ; awns very short, not reaching to the base of the anther ; ovary on a broadish, longer or shorter stipe, glabrous.

VAR. β, **minor** (Bolus); flower smaller in all parts; corolla-tube 8–10 lin. long, with a smaller and shorter neck.

COAST REGION: Ceres Div.; rocky ridges of the Skurfde Berg, near Ceres, 5000 ft., *Bodkin, Bolus in Herb. Norm. Aust.-Afr.*, 1309! Var. β: Worcester Div.; on Matroos Berg, 6000 ft., *Marloth*, 2208!

CENTRAL REGION: Var. β: Ceres Div.; Table Mountain, in the Cold Bokkeveld, 6200 ft., *Schlechter*, 10100!

88. **E. shannonea** (Andr. Heathery, t. 239); erect, glabrous; branches somewhat flexuous; leaves 3-nate, erect-spreading, curved, linear-trigonous, acute, somewhat concave on the upper surface, ciliate or naked, 4–9 lin. long (in our wild specimens not exceeding ½ in.); pedicels slender, 7–9 lin. long; bracts remote, linear; sepals linear-lanceolate, acute, keeled, thick, red, 5–7 lin. long; corolla subampullaceous or oblong-tubular, inflated, tapering upwards (but not into a long thin neck), contracted at the throat, viscidulous, white with a rosy tinge (not apparently red-veined); tube 12–14 lin. long; segments spreading, ovate, subacute, white, 2–3 lin. long; filaments slender, dilated at the anther, crumpled and bent; anthers subcrescent-shaped, acute, prognathous and bilobed at the base, pallid, nearly 1½ lin. long, muticous; ovary subturbinate, enlarging gradually above the short stipe. *Andr. Col. Heaths, t.* 273. *E. Shannoni, Lodd. Bot. Cab. t.* 168. *E. shannoneana, Spreng. Syst.* ii. 185. *E. Shannoniæ, Sweet, Hort. Brit. ed.* i. 261. *E. shannoniana, Benth. in DC. Prodr.* vii. 644. *E. obesa, Tausch in Flora,* 1839, 628, *not of Salisb. E. dianthiflora, Tausch in Flora,* 1834, 594. *E. muscicapa, Tausch, l.c.* 595.

COAST REGION: Caledon Div., rather rare; near Klein River, *Masson,* 42! mountains near Hartebeest River, *King in Herb. Bolus!* Bredasdorp Div.; Elands Kloof, 1200 ft., *Schlechter,* 9753! Also *cultivated specimens!*

We describe chiefly from King & Schlechter's excellent specimens which agree well with Masson's, and with Andrews' figure. The species is very near to *E. lagenæformis,* differing chiefly in its smaller corolla-limb and narrower leaves, but we scarcely know the last-named well enough to warrant us in uniting them here.

89. **E. ampullacea** (Curt. Bot. Mag. t. 303); erect, glabrous, 1–1½ ft. high; branches of straggling habit, rigid; leaves 4-nate, adpressed at the base, erect-recurved to subsquarrose, oval to lanceolate, blunt, thick, ciliate, 2–4 lin. long, the floral more or less (sometimes much) dilated; umbels 3–4-flowered; pedicels stout, 3–4 lin. long; bracts approximate, oblanceolate or lanceolate, ciliate, reddish, 3–4 lin. long; sepals oblong, obtuse, ciliate, crimson, 3–4 lin. long; corolla ampullaceous; tube ovoid or subglobose below, more or less gradually narrowed to the neck, then slightly dilated at the throat, viscidulous, 9–12 lin. long, pale rose with darker red veins; segments spreading, broad ovate, very obtuse or subreniform, crenulate, white (or spotted or edged with red), 2–3 lin. long; filaments widened at the anther; anthers subcrescent-shaped, acute, slender, prognathous and bilobed at the base, pallid, 1–1¼ lin. long, muticous; ovary subsessile or on a short broad stipe. *Andr. Heathery, t.* 103, *and Col. Heaths, t.* 3; *Wendl. Eric. Ic. fasc.* 23, 169, *t.* 64; *Lodd. Bot Cab. t.* 508; *Tratt. Archiv. t.* 285; *Benth.*

in DC. Prodr. vii. 644. *E. ampullæformis, Salisb. in Trans. Linn. Soc.* vi. 381. *E. andrewsiana, Tausch,* and *E. ampullacea, Tratt. ex Tausch in Flora,* 1834, 594 ?

Var. β, obbata (Bolus); floral leaves usually more dilated than in the assumed type; corolla with a somewhat thicker neck and throat; limb-segments sometimes shorter, subreniform (and variously spotted or margined ?). *E. obbata, Andr. Heathery, t.* 32, and *Col. Heaths, t.* 113; *Benth. in DC. Prodr.* vii. 644, and var. *umbellata, Andr. ll. cc. tt.* 132, 190. *E. capax, Salisb. in Trans. Linn. Soc.* vi. 381. *E. prægnans, Soland. ex Salisb. l.c.*

Coast Region: Caledon Div., somewhat rare; dry rocky places on the Zwart Berg, *Niven,* 153! *Bolus,* 7405! *Bodkin, in Herb. Bolus,* 6953! Genadendal Mountain, 2500 ft., *Bodkin in Herb. Guthrie,* 3610! Var. β: Stellenbosch Div.; Hottentots Holland, *Masson in Herb. Salisbury!* Caledon Div.; Klein River Kloof, *Zeyher,* 3198! Zwart Berg, *Zeyher,* 3197! Bredasdorp Div.; frequent on hills near Elim, *Bolus,* 6757! and in *Herb. Aust.-Afr* , 1628! Also *cultivated specimens* of type and var.!

The var. *obbata* is scarcely separable by any constant characters, and is noted chiefly for convenience of reference to the old figures. It is merely a maritime form, and specimens occur between Caledon and the sea which are intermediate in one or the other character. We have not seen any wild plants coloured as in Andr. l.c. tt. 32 and 132, and have little doubt that these are merely horticultural variations.

90. E. irbyana (Andr. Heathery, t. 219); erect, glabrous, 1–1½ ft. high; branches straighter, more slender and longer than in *E. shannonea ;* leaves 3-nate, spreading or erect or subadpressed, linear, subtrigonous, acute, mucronate, 2–4 lin. long (or in *Andrews'* figs. 6–9 lin.); umbels 3–8-flowered; pedicels 5–7 lin. long; bracts remote, slender; sepals lanceolate, acuminate, transversely wrinkled, viscid, deep red, 2½–4 lin. long; corolla somewhat variable, suburceolate, not much inflated, or subampullaceous with a thin neck, viscid, pale flesh colour, 4–7 (or, in cultivated specimens, 10) lin. long; segments ovate, acute or obtuse, 1½–3 lin. long; filaments and anthers as in *E. shannonea,* the latter about 1 lin. long; ovary elongate, substipitate. *Andr. Col. Heaths, t.* 176; *Lodd. Bot. Cab. t.* 816 (*probably a garden hybrid with a more slender corolla*).

South Africa: without locality, *Masson,* 43! and *cultivated specimens!*

Coast Region, on mountains 500–3500 ft.: Caledon Div.; Babylons Tower, *Guthrie,* 4093! Vogel Gat, near Klein River mouth, *Schlechter,* 9552! *Zeyher!* Bredasdorp Div.; near Elim, *Bolus,* 6753! near Koude River, *Schlechter,* 9620! 9729! (the latter a short-flowered form with corollas 4–5 lin. long)!

Very like *E. shannonea* in miniature, but the habit and set of the leaves is different, and the corolla with its relatively wider neck tends more towards an urceolate than an ampullaceous shape.

91. E. curvifolia (Salisb. in Trans. Linn. Soc. vi. 380); erect, ¾–1½ ft. high; branches few, slender, ascending; leaves 3-nate, curved and spreading, a little longer or a little shorter than the internodes, linear, blunt, glabrous, gland-ciliolate, thick, 2–3 lin. long; umbels 3–5-flowered; pedicels slender, 2½–3 lin. long; bracts subremote, slender, small ; sepals linear or linear-lanceolate, sulcate, viscid, 2–2½ lin. long; corolla tubular-inflated below, attenuate above, or tubular-urceolate, glabrous, viscid, tube rosy, throat purple,

4 lin. long; segments spreading, white, under 1 lin. long; anthers
oblong, pallid, dorsifixed, straight or slightly curved at the base, but
not crescent-shaped nor prognathous, slightly bilobed at the base,
about $\frac{1}{2}$ lin. long, muticous; style at length shortly exserted;
ovary stipitate, glabrous ("minutely hairy," *Salisbury*); cells
several-ovuled. *Benth. in DC. Prodr.* vii. 642. *E. comptoniana,*
Andr. Heathery, t. 255?, *and Col. Heaths, t.* 224 ?. *E. terminalis,*
Klotzsch in Linnæa, xii. 509. *E. angusticollis, Bartl. in Linnæa,* vii.
638.

Var. β, **Zeyheri** (Bolus); flowers larger, 7–8 lin. long. *E. Zeyheri, Spreng.*
Tent. Suppl. Syst. Veg. 12; *Benth. l.c.* 643.

South Africa : without locality, *Roxburgh! Thunberg.*
Coast Region: Stellenbosch Div.; Hottentots Holland, *Mulder!* Caledon
Div.; Baviaans Kloof, *Niven,* 148! *Bodkin in Herb. Norm. Aust.-Afr.,* 346!
Genadendal Mountain, *Bolus,* 5401! Swellendam Div.; Tyger Hoek, *Masson,*
33! Var. β : Caledon Div.; tops of mountains near Genadendal, *Burchell,*
7750! *Zeyher,* 437 (ex *Sprengel*)! *Bolus,* 5400! and in *Herb. Norm. Aust.-Afr.,*
345! *Schlechter,* 10322!

Chiefly variable in the size of the flowers. Salisbury appears to have founded
the species on the smaller size (corolla 4 lin. long) and Sprengel, his *E. Zeyheri,*
on the larger (7–8½ lin.). *E. comptoniana, Andr. Heathery, t.* 255, figured from
a cultivated specimen, only differs slightly in the corolla.

92. E. pectinata (Klotzsch in Linnæa, xii. 509, not of Bartl.);
erect, 1½–2 ft. high; branches slender, ascending, subvirgate;
leaves 3-nate, about as long as the internodes, erect, adpressed, tips
slightly recurved, linear or narrow-elliptical, blunt, glabrous, pectinate-ciliate, thick, about 2 lin. long; umbels 4–6-flowered; pedicels
glandular-hispid, 2–3 lin. long; bracts remote, sublinear, small,
viscid, gland-ciliate; sepals linear or linear-spathulate, glabrous,
foliaceous, dark purple, 1–1½ lin. long; corolla elongate-elliptic or
tubular-urceolate, mouth contracted, viscid, tuberculate-hispid, red,
5–6 lin. long; segments short, suberect, scarcely stellate-patent;
filaments capillary; anthers oblong or cuneate, prognathous and
bilobed at the base, about $\frac{2}{5}$ lin. long, decurrent-aristate; awns
about $\frac{3}{4}$ the length of the cell; ovary cylindrical, glabrous, on a
long slender stipe from 1–2 times its own length; cells 4-ovuled.
Benth. in DC. Prodr. vii. 642.

Coast Region : Stellenbosch Div.; Lowrys Pass, *Schlechter,* 7241! Caledon
Div.; mountains near Grietjes Gat, between Palmiet River and Steenbrass River,
Zeyher, 3203! Klein River Mountains, 2000 ft., *Bodkin in Herb. Guthrie,*
4110!

93. E. trichroma (Benth. in DC. Prodr. vii. 642); erect, about
1 ft. high; branches slender, irregularly spreading, 6–8 in. high;
leaves 3-nate, erect and adpressed, or spreading, somewhat imbricate,
or only as long as the internodes, linear, acute or obtuse, glabrous,
ciliate, about 1 lin. long; bracts subremote, gland-ciliate, small;
sepals linear, obtuse, margins incurved and gland-ciliate, foliaceous,
subviscid, red, 1–1¼ lin. long; corolla tubular-ovoid, inflated below the
middle, attenuate above, very minutely tuberculate-hispid, viscidulous,

rosy and purple, 3–4 lin. long; segments erect, scarcely (in the dried state) stellate-spreading; filaments, for the greater part of their length, adherent to the corolla-tube; anthers dorsifixed, semiovate, straight, not prognathous and scarcely bilobed at the base, pallid, $\frac{1}{4}$–$\frac{1}{3}$ lin. long, subdecurrent-aristate, or free aristate; awns much shorter than the cell, spreading; style included; ovary on a stipe of variable length, glabrous, cells 2–4-ovuled. *E. tricolor, Niven ex Benth. in DC. Prodr.* vii. 642.

VAR. β, **imbricata** (Bolus); leaves more constantly imbricate and adpressed, not spreading at the apex, and wider; corolla somewhat more inflated.

SOUTH AFRICA: without locality, *Masson!*
COAST REGION: Paarl Div.; French Hoek, 2000 ft., *Niven*, 147! *Schlechter*, 10278! Var. β: Tulbagh Div.; New Kloof, 2500 ft., *Schlechter*, 7530!

Bentham describes the cells of the ovary as 2-ovuled. This is not constant. An examination of Niven's type 147, showed 12 ovules in the 4 cells, of Masson's 17 ovules; of Schlechter's 10278, 17 ovules, and in his 7530, 6 and 7. In Schlechter, 10278, the corolla frequently has one or more short shred-like blunt processes, which have not been seen upon the other specimens. This is possibly insect-work.

94. E. tubercularis (Salisb. in Trans. Linn. Soc. vi. 330); apparently a dwarf shrub, glabrous; branches slender, subflexuous, spreading; leaves 3-nate, imbricate, adpressed, linear-oblong, acute, glabrous, thick, concave above, cartilagineo-serrulate, $\frac{3}{4}$–1 lin. long; flowers 2–3-nate, or more rarely subumbellate; pedicels $\frac{1}{2}$ lin. long, downy; bracts 3, minute; sepals lanceolate-linear, about $\frac{1}{2}$ lin. long; corolla ovoid, subacute at the discoloured apex, more or less covered with minute wart-like tubercles, rosy, about 2 lin. long; segments erect and connivent (in the dried state), very short; anthers dorsi-fixed, ovate, muticous, about $\frac{1}{5}$ lin. long; ovary on a slender stipe exceeding it in length; ovules 2 (or sometimes by abortion 1) in each cell. *E. notabilis, Wendl. in Spreng. Syst.* ii. 184. *Eremia tubercularis, Benth. in DC. Prodr.* vii. 700.

SOUTH AFRICA: without locality, *Roxburgh!*
COAST REGION: Stellenbosch Div., *Zeyher!*

Like the preceding and the next species, this appears to be irregular in the number of the ovules in each cell. Bentham apparently found flowers with one only, and placed it in *Eremia*. But besides the fact of there being undoubtedly two ovules present, it has very little resemblance to the other plants of that genus, while it is very similar to the preceding species, than which it is generally smaller in all parts.

95. E. rhodopis (Bolus); much branched, 4–8 in. high; branches ascending, flexuous, slender, glabrescent; leaves 3-nate, adpressed, broadly linear, glabrous, somewhat shorter than the internodes, $1\frac{1}{2}$ lin. long; flowers usually terminal, or sometimes by the abortion of lateral branchlets pseudo-lateral, 2–3-nate; pedicels very slender, $\frac{3}{4}$ lin. long; bracts remote, minute; sepals ovate-lanceolate, obtuse, glabrous, about $\frac{1}{4}$ the length of the corolla; corolla ovoid, glabrous, dry, rosy, 2–2$\frac{1}{4}$ lin. long; segments deltoid, connivent in the dried state, about $\frac{1}{4}$ the length of the tube; anthers ovate-oblong, about

$\frac{1}{5}$ lin. long, aristate; awns spreading, a little longer than the cells; style included; stigma capitate; ovary on a thick stipe about as long as itself, cells 2- (or sometimes 1-?) ovuled. *Eremia rhodopis, Bolus in Journ. Bot.* 1894, 239.

COAST REGION: Caledon Div.; Honw Hoek, *Pappe,* 36! *Guthrie,* 2210! *Schlechter,* 7548! near Bot River, *Schlechter,* 9440! Babylons Tower Mountain, *Zeyher,* 3230!

This has the aspect of the § *Orophanes,* from which it is separated by its distinctly stipitate ovary and some other characters.

Section IX. CERAMUS. (Sp. 96–100.)

96. E. incarnata (Thunb. Diss. Erica, 50, not of Andr., nor of Benth.); erect, a foot or more high; branches virgate, subflexuous, thinly villous; leaves 3-nate, adpressed or slightly spreading, narrow-ovate, obtuse, somewhat concave on the upper surface, glabrous, $1\frac{1}{4}$–$1\frac{1}{2}$ lin. long; flowers usually 3-nate, more rarely in 5–6-flowered umbels, erect or suberect; pedicels slender, glabrous, about 3 lin. long; bracts leaf-like, narrow-lanceolate, acute, the two upper approximate, adpressed, the lowest at or near the base, larger and more spreading; sepals narrow-lanceolate, acute, glabrous, margins naked, $1\frac{1}{2}$ lin. long; corolla inflated-urceolate, not tapering but suddenly contracted at the neck, or narrow-ovoid, 3–$4\frac{1}{2}$ lin. long; limb spreading, small; segments rounded; anthers oblong, somewhat incurved, slightly prognathous, a little over $\frac{1}{2}$ lin. long; pore about $\frac{1}{3}$ of the cell; cristate at the base; crests lanceolate, acuminate, dentate on the outer margin, about $\frac{1}{2}$ the length of the cell; ovary upon a stipe of about its own length, or in some specimens longer, and about $\frac{1}{3}$ of its own diameter. *E. amœna, Salisb. Trans. Linn. Soc.* vi. 329, *not of Wendl.*

SOUTH AFRICA: without locality, *Thunberg!*
COAST REGION: Clanwilliam Div.; on the Cederberg Range near Sneeuw Kop and Wupperthal, 3500–4500 ft., *Bodkin in Herb. Bolus,* 8630! *Leipoldt,* 621!

We have compared this with Thunberg's type, which Bentham does not appear to have seen. The plants so named by him, or distributed by others under this name, are forms of *E. Savilea,* from which this differs by the position of its bracts, by the shape of its sepals, anthers and their appendages, and by the longer and narrower stipe of the ovary. The pedicels, bracts and sepals, and in some specimens the flower-buds, are uniformly deep red.

97. E. Savilea (Andr. Heathery, t. 238); about 1 ft. high, much branched; branches pubescent; leaves 4-nate (or sometimes 3-nate, *Bentham*), linear, subtrigonous, acute, glabrous, ciliate, especially the younger, with long soft hairs, 1–3 lin. long; umbels 3–4-flowered; pedicels slender, puberulous, 3–4 lin. long; bracts remote, linear, small; sepals lanceolate, acuminate, keeled, scarious-edged; corolla urceolate or ovoid-urceolate, throat much contracted, glabrous, dry, 4–5 lin. long; limb broad, under 1 lin. long; anthers broad-cuneate, purple, $\frac{1}{3}$–$\frac{1}{2}$ lin. long, very minutely cristate or aristate, the appendages not reaching below the base of the cells;

pore about $\frac{1}{3}$ the length of the cell; ovary subsessile or broadly short-stipitate. *E. Savileia, Andr. Col. Heaths. t.* 205. *E. Savilliæ, Lodd. Bot. Cab. t.* 96. *E. savileana, Sweet, Hort. Brit. ed.* i. 261; *Benth. in DC. Prodr.* vii. 641. *E. Behen, E. Meyer ex Benth. l.c.* 642 *partly. E. delecta, Tausch in Flora,* 1839, 633. *E. tristis, Klotzsch in Linnæa,* xii. 519?

VAR. β, **grandiflora** (Bolus); corolla 6–7 lin. long.

COAST REGION : Tulbagh Div.: Mosterts Hoek Mountain, *Bolus.* 6479! Ceres Div.; near Ceres, *Miss Liesching in Herb. Guthrie,* 3367! Paarl Div.; near Wellington, *MacOwan, Herb. Norm. Aust.-Afr.,* 17! Worcester Div. ; Dutoits Kloof, *Drège ;* Matroos Berg, *Cook in Herb. Bolus,* 6887! Swellendam Div.; near Swellendam, *Borcherds in Herb. Bolus,* 6266! Also *cultivated specimens!* Var. β: Paarl Div.; around French Hoek, *MacOwan,* 2920! Worcester Div.; Matroos Berg, *Bolus in Herb. Guthrie,* 3947!

98. E. præcox (Klotsch in Linnæa, xii. 517, not of Lodd.) ; erect, 6–8 in. high ; branches ascending, mostly curved ; branchlets pubescent, and, as are the leaves, pedicels, bracts and sepals ciliate or more or less pilose with soft white hairs, sometimes nearly glabrous, or becoming so ; leaves 4-nate, erect-spreading or slightly incurved, linear, subacute, 3–4 lin. long; flowers 3-nate, sometimes by abortion of the lateral branchlets appearing axillary and clustered; pedicels slender, 3–4 lin. long; bracts remote, adpressed, small; sepals lanceolate, acuminate, keeled, scarious-edged, $2\frac{1}{2}$ lin. long; corolla ovoid-urceolate ; tube pale or bright red, about 4 lin. long ; throat purple ; limb red ; segments ovate, acute, $\frac{3}{4}$–1 lin. long ; anthers oblong or subcuneate, $\frac{1}{3}$–$\frac{1}{2}$ lin. long, purple, muticous; pore about $\frac{1}{2}$ the length of the cell ; ovary stipitate. *Benth. in DC. Prodr.* vii. 641. *E. Behen, E. Meyer ex Benth. l.c.* 642 *partly.*

COAST REGION : Worcester Div.; Dutoits Kloof, 3000–4000 ft., *Drège !* Paarl Div.; French Hoek, *Le Roux, in Herb. Bolus,* 5950! Caledon Div.; Houw Hoek, *Kennedy,* 28! near Palmiet River, *Schlechter,* 7455!

Chiefly distinguished from the preceding by its longer leaves, narrower and muticous anthers with larger pore, and more evidently stipitate ovary; the pubescence on the younger parts is also generally more copious. Our material, though good, is not copious enough to afford good evidence that these small differences are constant.

99. E. inflata (Thunb. Diss. Erica, 41, t. 2) ; erect, 1–3 ft. high ; branches slender, erect, often virgate, leafy, terminating in dense clusters of flowers ; leaves 4-nate or scattered, suberect or incurved, not crowded, often subsecund, very slender, narrowly linear, acute, mucronate or aristate, glabrous, flat above, convex and faintly sulcate below, petioles long and very slender, the whole 5–10 lin. long ; flowers in dense many-flowered corymbose umbels ; pedicels slender, pubescent, red, 6–7 lin. long ; bracts subremote, slender, red, aristate ; sepals lanceolate or subulate, acuminate, glabrous or pubescent, crimson or pallid, aristate, 1–$1\frac{1}{2}$ lin. long ; corolla ovoid-urceolate; throat contracted, dry ; tube pale rose, 3–4 lin. long, throat and limb darker red; segments about $\frac{1}{2}$ lin. long; anthers semiovate or broad, subcuneate, purple, about $\frac{1}{2}$ lin. long, aristate;

pore $\frac{1}{3}$—$\frac{2}{5}$ the length of the cell; awns inserted above the middle of the cell and equalling or exceeding it in length, capillary, much involute; style subexserted; ovary on a distinct, broad, shortish stipe. *Benth. in DC. Prodr.* vii. 641. *E. amabilis, Salisb. in Trans. Linn. Soc.* vi. 385. *E. sainsburyana, Andr. Heathery. t.* 287, *and Col. Heaths, t.* 270. *E. carniula, Lodd. Bot. Cab. t.* 926. *E. carinula, Steud. Nomencl. ed.* 2, i. 570. *E. ollula, Andr. Heathery, t.* 275, *and Col. Heaths, t.* 251; *Lodd. l.c. t.* 1646.

SOUTH AFRICA: without locality, *Thunberg*. Also *cultivated specimens!*

COAST REGION, ascending from 1000–2500 ft. : Clanwilliam Div.; Cederberg Range, *Drège*, 7737! *Leipoldt*, 62! *Mader*, 68! Piquetberg Div.; foot of Twenty-four Rivers Mountains, *Zeyher*, 1091! Tulbagh and Ceres Div.; Mitchells Pass, *Bolus*, 6923! and in *Herb. Norm. Aust.-Afr.*, 385! *Schlechter*, 9967! *Guthrie*, 2178!

Distinct in the section by its very long leaves, but with a curious resemblance to *E. obliqua, Thunb.*, in the § *Pachysa* (which also has a stipitate ovary) and might be placed with it, but that the sepals and anthers are so different; the pedicels and leaves also are longer. The leaves are very like those of *E. vestita, Thunb.*, but far fewer. We have not seen any wild specimens with flowers either so long, or so short, as those represented in the figures of Andrews and Loddiges.

100. E. ventricosa (Thunb. Diss. Erica, 27, t. 1); erect, 2–6 ft. high; branches stout, rigid; leaves 4-nate, spreading or squarrose, sometimes undulate, crowded, linear-subulate, acuminate, margins white, pilose, rarely subglabrous, 6–8 lin. long; flowers in dense umbels at the ends of the branchlets, sometimes forming close pyramidal masses; pedicels pubescent, 3–4 lin. long; bracts remote, small, linear, ciliate; sepals linear-lanceolate, acuminate, scarious-edged, glabrous, 3 lin. long; corolla ovoid-urceolate, attenuate upwards and constricted at the throat, glabrous, dry, white, rosy, or red; tube 6–8 lin. long; segments ovate, acute, sometimes mealy, $1\frac{1}{2}$ lin. long; anthers cuneate-oblong, purple, about $\frac{2}{3}$ lin. long, minutely crested; crests not reaching to the base of the cell; style included, stigma subsimple; ovary turbinate, stipitate. *Andr. Heathery, t.* 197, *and Col. Heaths, t.* 65; *Bot. Mag. t.* 350; *Wendl. Eric. Ic. fasc.* 3, 11; *Lodd. Bot. Cab. t.* 431; *Herb. Amat. t.* 62; *Benth. in DC. Prodr.* vii. 642. *E. prægnans, Andr. Heathery, t.* 231, *and Col. Heaths, t.* 202; *Lodd. l.c. t.* 945. *E. venusta, Salisb. Prodr.* 297, *and in Trans. Linn. Soc.* vi. 385, *not of Sinclair, nor Klotzsch. E. densa, Andr. Heathery, t.* 212, *and Col. Heaths, t.* 163. *E. glabra, Link, Enum. Hort. Berol.* i. 362. *E. translucens, Wendl. ex Spreng. Syst.* ii. 186, *in syn.?*

SOUTH AFRICA: without locality, *Thunberg, Herb. Salisbury!* and *cultivated specimens!*

COAST REGION, ascending from 1000–5200 ft.: Worcester Div.; Dutoits Peak, *Marloth*, 2416! Paarl Div.; mountains about French Hoek, *Schlechter*, 9236! *MacOwan, Herb. Norm. Aust.-Afr.*, 939! Stellenbosch Div.; hills about False Bay, *Musson and Niven* (ex *Bentham*). Caledon Div.; near Amandel River, *Bolus*, 5172!

Chiefly variable in the size of its flowers and the degree of copiousness or density of the inflorescence. But we cannot discover any constant characters on

which to form varieties. The smaller flowered forms verge towards *E. Walkeria*
in § *Callista.* Andrews' t. 231 looks distinct, but is possibly merely a garden
hybrid; this species having been much in cultivation. From the great number,
and the porcelain-like texture of its delicately coloured flowers, it is one of the
most beautiful of heaths, and a fine pyramidal bush of 6 ft. high was once seen
by the writer, covered from base to apex with thousands of flowers. *E. ventricosa,* var. *bothwelliana,* Carr. in Rev. Hort. 1882, 363, probably belongs
here.

Section X. CALLISTA. (Sp. 101–111.)

101. E. Lawsonia (Andr. Heathery, t. 267); erect, 1–2 ft. high;
branches slender, subvirgate; leaves erect, acute, keeled, puberulous, ciliate, $1\frac{1}{2}$–2 lin. long; flowers mostly solitary on short lateral
branchlets, forming a pseudo-raceme (or rarely 4-nate, ex *Bot. Mag.*',
subsessile; bracts approximate, leaf-like, scarious-edged, 1–$2\frac{1}{2}$ lin.
long; sepals lanceolate-linear, acuminate, scarious, about 3 lin. long;
corolla-tube cylindrical (or when old, somewhat inflated at the base
round the swollen capsule), mouth not contracted, dry, puberulous or
glabrous, red, 5–7 lin. long, $\frac{2}{3}$ lin. broad; segments ovate-lanceolate,
acute, about 2 lin. long; anthers oval, $\frac{1}{4}$–$\frac{1}{3}$ lin. long, muticous.
Andr. Col. Heaths, t. 242. E. lawsoniana, Benth. in DC. Prodr.
vii. 646. *E. Lawsoni, Sims, Bot. Mag. t.* 1720; *Lodd. Bot. Cab.*
t. 488. *E. infundibuliformis, Bartl. in Linnæa,* vii. 638, *not of
Andr., and E. leptocarpha, Spreng. f. Tent. Suppl. Syst. Veg.* 13,
fide Klotzsch in Linnæa, xii. 519.

COAST REGION : Caledon Div.; Baviaans Kloof, *Masson,* 30 ! near Caledon,
Ecklon & Zeyher! mountains near Genadendal, 3200 ft., *Bolus,* 5407! and in
Herb. Norm. Aust. Afr., 344! *Bodkin in Herb. Guthrie,* 3612 ! Zwart Berg, near
Caledon, 2000 ft., *Schlechter,* 10363!

102. E. pavettæflora (Salisb. in Trans. Linn. Soc. vi. 382);
erect, 1 ft. or more high; branches spreading, slender, glabrous,
8–9 in. long; internodes on the ultimate branchlets becoming much
elongated; leaves erect, linear, acute, 2–3 lin. long; flowers sessile;
bracts approximate, linear, 2–3 lin. long; sepals linear to linear-
lanceolate, acuminate, keeled, rigid, ciliate, 3-4 lin. long; corolla-
tube narrow-cylindrical, slightly and gradually widened at the throat,
glabrous, dry, red, 8–10 lin. long, $\frac{1}{2}$–$\frac{3}{4}$ lin. broad; segments ovate,
subacute, about 2 lin. long, 1 lin. broad; anthers semiovate, $\frac{1}{3}$ lin.
long, muticous or (according to *Salisbury* and *Andrews*) minutely
aristulate; pore $\frac{1}{3}$ the length of the cell; style subincluded. *Benth.
in DC. Prodr.* vii. 645. *E. infundibuliformis, Andr. Heathery, t.*
218, *and Col. Heaths, t.* 240; *Lodd. Bot. Cab. t.* 589. *E. hypocra-
teriformis, Tausch, Flora,* 1837, 492.

SOUTH AFRICA : without locality, *Roxburgh! Herb. Salisbury !* and *cultivated
specimens!*
COAST REGION : Worcester Div. ; Dutoits Peak, 4200 ft., *Marloth,* 2414!
Caledon Div.; Klein River, *Masson!* Houw Hoek, 2500 ft., *Niven,* 81!
Schlechter, 7422!

The specimens last cited have a more slender corolla-tube than others, and
also longer than those described by Bentham. There is, however, no doubt of
their identity.

103. E. cylindrica (Thunb. Diss. Erica, 24, not of Wendl. nor Andr.); erect, 1–2 ft. high, glabrous in all parts; branches subvirgate with elongating internodes; leaves erect, subacute, 2–$2\frac{1}{2}$ lin. long; pedicels slender, over $\frac{1}{2}$ lin. long; bracts approximate, linear, 1–$1\frac{1}{2}$ lin. long; sepals linear-lanceolate, acuminate, keeled, scarious, ciliolate, $2\frac{1}{2}$ lin. long; corolla-tube cylindrical, mouth slightly contracted, glabrous, dry, yellow or white, 5–6 lin. long, about $\frac{3}{4}$ lin. broad; segments ovate, subacute, $\frac{3}{4}$ lin. long; anthers narrow-oblong, slightly under $\frac{1}{2}$ lin. long, muticous; pore $\frac{3}{4}$ the length of the cell; style included; ovary elongate, somewhat contracted at the base. *E. tenuiflora, Andr. Heathery, t.* 146, *and Col. Heaths, t.* 210; *Lodd. Bot. Cab. t.* 1717. *E. tenuiflora, var. alba, Andr. Heathery, t.* 194, *and Col. Heaths, t.* 211. *E. cliffordiana, Lodd. l.c. t.* 34. *E. fistulæflora, Salisb. in Trans. Linn. Soc.* vi. 383. *E. stenantha, Sweet, Hort. Brit. ed.* 2, 340.

SOUTH AFRICA: without locality, *Thunberg in Herb. Salisbury!* Also *cultivated specimens!*

COAST REGION: Tulbagh Div.; near Tulbagh Waterfall, 1500 ft., *Niven,* 82! *Masson! Bolus,* 5461! and in *Herb. Norm. Aust.-Afr.,* 1306!

Allied to and resembling *E. pavettæflora,* but quite distinct by its more equal corolla-tube, smaller limb, and different anthers. It appears to be very local. Rach, who examined Thunberg's herbarium, *in Linnæa,* xxvi. 777, identified this with *E. tenuiflora, Andr.,* and the descriptions agree well.

104. E. fastigiata (Linn. Mant. 66, not of Andr.); erect, 1–$1\frac{1}{4}$ ft. high; branches flexuous or subvirgate; internodes elongating on the upper branches; leaves erect or spreading, crowded below, incurved, linear, acute, glabrous, about 3 lin. long; flowers subsessile; bracts approximate, lanceolate-linear or subulate-linear, acute, keeled, scarious, 3 lin. long; sepals like the bracts, but somewhat longer, 3–6 lin. long; corolla-tube cylindrical, sometimes slightly inflated below, sometimes at the throat, glabrous, red or rosy, about 5 lin. long; limb subovoid, mealy, white, dark-centred, $1\frac{1}{2}$ lin. long, $1\frac{1}{2}$–$2\frac{1}{2}$ lin. wide; anthers oblong or suboval, pallid, $\frac{5}{12}$ lin. long, muticous; pore $\frac{1}{4}$–$\frac{1}{3}$ the length of the cell; style included. *Wendl. Eric. Ic. fasc.* 19, 103, *t.* 39?; *Bot. Mag. t.* 2084; *Benth. in DC. Prodr.* vii. 646; *var. ciliata, Rach in Linnæa,* xxvi. 778? *E. fasciformis, Salisb. in Trans. Linn. Soc.* vi. 382. *E. mundula, Andr. Heathery, t.* 273, *and Col. Heaths, t.* 249; *Lodd. Bot. Cab. t.* 114. *E. humeana, Lodd. l.c. t.* 389.

VAR. β, **coventryana** (Bolus); leaves straighter, more acute, 5 lin. long; bracts and sepals ciliate, 4–6 lin. long; corolla-tube slightly inflated below, tetragonous, about 6 lin. long, throat contracted, limb without any dark centre, segments ovate, acute, 3 lin. long by 2 lin. broad, red below, white and somewhat mealy above; anthers nearly $\frac{1}{2}$ lin. long, pore $\frac{1}{2}$ the cell. *E. Coventrya, Andr. Heathery, t.* 210, *and Col. Heaths, t.* 226. *E. coventryana, Lodd. Bot. Cab. t.* 423.

VAR. γ, **immaculata** (Bolus); branches less virgate, more spreading; leaves stouter, longer, more distinctly sulcate below and more spreading, 4–5 lin. long; sepals mostly longer, equalling and often exceeding the corolla-tube, 5–6 lin. long; corolla-tube more inflated, subviscid; segments oblong or ovate, bluntish, white without a dark centre; anthers oblong or elliptical, sometimes very minutely aristate, $\frac{1}{2}$–$\frac{7}{12}$ lin. long; ovary subelongate.

SOUTH AFRICA: without locality, specimen from *Herb. Linnæus in Herb. Salisbury! Thunberg,* sheet β. Also *cultivated specimens* of type and var. β!

COAST REGION, ascending from 1500–3500 ft. : Ceres Div.; near Ceres, *Bolus,* 9811! Worcester Div.; near Worcester; *Bolus in Herb. Guthrie,* 3887! Paarl Div.; French Hoek, *Thunberg, Guthrie,* 3238! Cape Div.; Table Mountain, *Schlechter,* 933! Stellenbosch Div.; Hottentots Holland, *Mund,* 16! *Thunberg.* Lowrys Pass, *Guthrie,* 3875! Caledon Div.; Houw Hoek, *Bolus,* 5455! *Schlechter,* 5385! *Guthrie,* 3874! Riversdale Div.; Platte Kloof, *Masson* (ex *Salisbury*), *Thunberg.* Var β: Caledon Div.; Klein River Mountains, *Zeyher,* 3199! mountains near Vogel Gat, *Schlechter,* 9549! and near Hermanus, *Bodkin in Herb. Bolus,* 6484! Var. γ: Paarl Div.: mountains around French Hoek, *Bolus,* 9183! *Bolus in Herb. Guthrie,* 4974!

This species has been much confused in herbaria. In Salisbury's herb. at Kew is a branchlet marked "from Linnæus's type." We have dissected a flower and compared it with the specimens first cited above as the typical form. Our var. β was regarded by Bentham as a hybrid from *E. pavettæflora.* It does not seem to be a hybrid, but a maritime form; very local, its stations being within five, or possibly three, miles from each other. Var. γ, in aspect, resembles *E. hyacinthoides,* but the leaves are larger and the sepals very different. Wendland's fig. in Eric. Ic. fasc. 19, can only be cited with doubt. Its sepals are broader and shorter than in the wild specimens before us.

105. E. transparens (Berg. Descr. Pl. Cap. 108, not of Andr. nor Lodd.); erect, about 1 ft. high; branches many, dense, slender, fastigiate, villous or finely tomentose, becoming glabrous; leaves suberect to spreading, crowded, linear, keeled, ciliate or naked, about $1\frac{1}{2}$ lin. long; flowers very numerous and densely clustered, sub-sessile; bracts, 2 approximate, the lowest remote, linear, ciliate, under $\frac{1}{2}$ lin. long; sepals lanceolate to ovate, acute, scarious, softly ciliate, usually red, more rarely pale rosy, 1–$1\frac{1}{2}$ lin. long; corolla tubular, slightly inflated, glabrous, white, pale rosy to deep red, 2–$2\frac{1}{2}$ lin. long; segments ovate, acute, 1–$1\frac{1}{2}$ lin. long; anthers oblong or subovate, about $\frac{1}{5}$ lin. long, shortly aristate near the base; style included. *E. comosa, Linn. Mant. alt.* 234; *Bauer, Exot. Pl. t.* 18; *Andr. Heathery, t.* 10, *and Col. Heaths, t.* 80; *Benth. in DC. Prodr.* vii. 650; *var. rubra, Andr. ll. cc. tt.* 11, 81. *E. galiiflora, Salisb. in Trans. Linn. Soc.* vi. 383.

SOUTH AFRICA: without locality, *Bergius in Herb. Salisbury!* Also *cultivated specimens!*

COAST REGION, mountains at 2000–4000 ft. : Worcester Div.; Dutoits Kloof, *Drège! Bolus,* 5174! Paarl Div.; near French Hoek Pass, *Mann in Herb. Bolus,* 6391! *Schlechter,* 9290! Cape Div.; Table Mountain, *Thunberg! Burchell,* 624! *Wolley Dod,* 796! Caledon Div.; near Genadendal, *Burchell,* 7727! *Bolus, Herb. Norm. Aust.-Afr.,* 356! Zwart Berg, *Schlechter,* 9781! Swellendam Div.; Tradouw Mountains, *Bowie!* Riversdale Div.; Kampsche Berg, *Burchell,* 7082! Platte Kloof, *Thunberg.*

106. E. Vallis-Gratiæ (Guthrie & Bolus); erect, branching, probably 1–$1\frac{1}{2}$ ft. high; leaves suberect, incurved, imbricate, linear-trigonous, acute, ciliate, 4–5 lin. long, $\frac{1}{3}$ lin. broad; flowers 4-nate, generally densely clustered into heads; pedicels about 1 lin. long; bracts linear-lanceolate, acuminate, scarious, ciliate, 4–5 lin.

long; sepals ovate-lanceolate or oblanceolate, somewhat suddenly narrowed to an acuminate point, keeled above and pubescent on the keel, scarious, ciliate, 5–5½ lin. long, 1⅔ lin. broad, somewhat coloured; corolla-tube subcylindrical, slightly larger at the base and more or less narrowed to the throat, glabrous, red-purple, 5½–7½ lin. long, 2–2½ lin. broad; segments broad-ovate, acute, 2½–3 lin. long, more than 2 lin. broad, white, sometimes with red bands below; anthers included, oblong, dark, ⅔–¾ lin. long, muticous; pore about ⅖ the length of the cell; style shortly exserted; stigma small, capitellate. *E. ventricosa, Thunb.,* var. *grandiflora, Benth. in DC. Prodr.* vii. 642.

CoAST REGION: Caledon Div.; on the summit and upper part of the great mountain of Baviaans Kloof near Genadendal, 4700–4800 ft., *Burchell,* 7724! *Schlechter,* 9833! *Galpin,* 3599!

A fine and well-marked species, perhaps nearest to *E. Walkeria,* but larger in all parts, and especially distinguishable by the straighter, more erect, longer leaves. It is also near to *E. fastigiata,* var. *coventryana,* but the sepals and the anthers are very different.

107. E. prænitens (Tausch in Flora, 1834, 596); erect, 1–2 ft. high; branches stout, densely flowered; leaves erect to patent, subflexuous, linear-trigonous, subacute, glabrous, over 2 lin. long; flowers subsessile; bracts approximate, unilateral, linear-subulate, glabrous, 2½–3 lin. long; sepals narrow-lanceolate, long-acuminate, keeled, scarious, very shortly lacerate, 3 lin. long; corolla-tube cylindrical, very slightly inflated, glabrous, dry, 7 lin. long, about 1½ lin. broad; segments ovate, scarcely mealy, 1½ lin. long; anthers oblong, dark, ½ lin. long, muticous; pore ⅓ the length of the cell; style at length exserted. *E. walkeriana,* var. δ? *grandiflora, Benth. in DC. Prodr.* vii. 647.

CoAST REGION: Swellendam Div.; Tyger Hoek, *Masson,* 26!

This can hardly be a variety of *E. Walkeria.* It is indeed nearer to *E. fastigiata,* but seems to be sufficiently distinct. Bentham, to whom Tausch's species was, apparently, unknown, quoted it with doubt, and says " an species propria?" In general appearance it has some resemblance to *E. denticulata,* var. *grandiflora,* but is immediately distinguishable by the different sepals, besides the less inflated corolla.

108. E. Walkeria (Andr. Heathery, t. 50); erect, 1–1½ ft. high; branches mostly slender, densely leafy and flowered; leaves usually spreading and incurved, linear, subacute, keeled, about 2 lin. long; flowers subsessile; bracts approximate, subequal, lanceolate, scarious; sepals oblong, lanceolate, or sometimes suboblanceolate, acute, scarious, minutely ciliate-lacerate, thickly keeled at the apex, 2–3 lin. long, bracts and sepals often uncinately-incurved; corolla-tube urceolate, rosy or red, 3–4 lin. long; segments ovate or suborbicular, rarely mealy above, about 1 lin. long; anthers oblong, about ⅖ lin. long, muticous; pore ½ the length of the cell. *Andr. Col. Heaths,* t. 72. *E. walkeriana, Benth. in DC. Prodr.* vii. 647. *E. Walkeria,* var. *rubra, Andr. Heathery,* t. 100, and *Col. Heaths,* t. 144. *E. fastigiata,*

Andr. Heathery, t. 62, and Col. Heaths, t. 90; Lodd. Bot. Cab. t.
207, not of Linn. E. Walkeri, Lodd. l.c. t. 256. E. juliana,
Lodd. l.c. t. 799. E. pulchra, Salisb. in Trans. Linn. Soc. vi. 384.

VAR. *β*, **præstans** (Bolus); corolla-segments larger, about half the length
of the tube. *E. præstans, Andr. Heathery, t. 232, and Col. Heaths, t. 257;*
Benth. in DC. Prodr. vii. 647.

SOUTH AFRICA: without locality, *Drège! Herb. Salisbury!* Also *cultivated*
specimens!

COAST REGION: Ceres Div.; near Ceres, *Bolus in Herb. Guthrie,* 2366!
Paarl Div.; on the Drakensteen Mountains near Wellington, 1000 ft., *MacOwan,*
Herb. Norm. Aust.-Afr., 18! Swellendam Div.; without precise locality, *Herb.*
Huguenot Seminary, 84! Var. *β*: Tulbagh Div.; mountains near Tul-
bagh, *Ecklon & Zeyher* and Worcester Div.; Dutoits Kloof, *Drège* (ex
Bentham).

We have not seen specimens of var. *β*.

109. E. daphniflora (Salisb. in Trans. Linn. Soc. vi. 384, in an
extended sense); erect, 1–2 ft. high; leaves erect or spreading,
straightish (seldom or never uniformly spreading-incurved), linear,
2–3 lin. long; flowers in scattered clusters, sometimes congested
into large dense masses towards the ends of the branches; pedicels
1–2½ lin. long; bracts approximate, or subremote and lax; sepals
very variable, from lanceolate-linear, lanceolate, narrow-ovate or
subelliptical, never widest above the middle, long-acuminate to
acute, always scarious, shortly ciliate, denticulate or naked, equalling
or half as long as the corolla-tube; corolla urceolate, narrow-ovoid-
urceolate or subconical, red or rosy; tube 3–7 lin. long; segments
obtuse or acute, variable in size, concolorous or white; anthers
oblong, about ½ lin. long, or rarely longer, minutely aristate or
muticous. *Benth. in DC. Prodr. vii.* 646. *E. daphneflora, Wendl.*
Eric. Ic. fasc. 26, 3. E. daphnæflora, Lodd. Bot. Cab. t. 543.
E. bartlingiana, and E. incerta, Klotzsch in Linnæa, xii. 521. *E.*
indigesta, Klotzsch in Linnæa, xii. 523.

VAR. *β*, **pedicellata** (Bolus); branches virgate, slender, the floriferous often or
always with long internodes, the barren closely leafy; leaves about 2 lin. long;
pedicels 2 lin. long or less; sepals oblong, acute or acuminate, denticulate,
2½ lin. long; corolla narrow-ovoid-urceolate; tube 6–7 lin. long, white or pale
rose; segments ovate, acute or acuminate, 2 lin. long; anthers (sometimes at
least) very minutely aristate. *E. pedicellata, Klotzsch in Linnæa, xii.* 525;
Benth. in DC. Prodr. vii. 646. *E. distans, Benth. l.c.* 646.

VAR. *γ*, **Muscari** (Bolus); branches usually virgate and slender; leaves erect,
mostly straightish and slender; inflorescence often lax, flowering branchlets
clustered, clusters more or less remote; pedicels 1–1½ lin. long; sepals narrow-
to broad-lanceolate, usually somewhat less than half the corolla-tube; corolla
commonly narrow-ovoid-urceolate, usually ochreous or yellow, more rarely white
(or rosy?); tube 3–4 lin. long. *E. Muscari, Andr. Col. Heaths, t.* 40, *and*
Heathery, t. 130; *Wendl. Eric. Ic. fasc.* 18, 85, t 32; *Benth. in DC. Prodr.*
vii. 648. *E. fragrans, Salisb. in Trans. Linn. Soc. vi.* 383; *not of Andr.*
E. Bonplandia, Lodd. Bot. Cab. t. 345. *E. bonplandiana, Bot. Mag. t.* 2126?
E. scoliostoma, Klotzsch in Linnæa, x. 322, *fide Benth. E. moschata, Lodd. l.c.*
t. 614? *E. nidiflora, Salisb. l.c., and E. denticulata, Roxb. ex Salisb. l.c. (fide*
Index Kew.)

VAR. *δ*, **retusa** (Bolus); habit of var. *γ*; leaves mostly shorter, 1½–2 lin.
long; flowers subsessile; bracts and sepals broader, broad-lanceolate or oblong,

acute, subdenticulate or slightly and shortly lacerate; corolla mostly shorter and broader, 3–4 lin. long, ovoid or broad-ovoid-urceolate, often but not always, much contracted at the throat, bright yellow. *E. retusa, Tausch in Flora,* 1834, 598.

This variety connects this species with *E. denticulata,* which it approaches almost as nearly as it does this species.

VAR. ε, **latisepala** (Bolus); branches somewhat flexuous; leaves broader and shorter than in other vars., oblong to elliptical, obtuse, 1–1½ lin. long; inflorescence somewhat lax; pedicels ½ lin. long; sepals (in the fully developed flowers) somewhat spreading at the apex, broad-lanceolate or elliptical, acute, thickly-keeled, orange below, red towards the apex, about 2 lin. long, 1 lin. wide; corolla urceolate, red or rosy; tube 4 lin. long; segments broad-ovate, very obtuse, short, red.

VAR. ζ, **Leipoldtii** (Bolus); habit of the type, but sepals somewhat broader, ovate-lanceolate, 2 lin. long; corolla urceolate, red or rosy; tube 4 lin. long; segments suboblong, tapering to the apex, not imbricating at the base but with somewhat open sinuses.

SOUTH AFRICA: without locality, *Zeyher,* 1094! *Herb. Salisbury!* and *cultivated specimens* of vars. β and γ! Var. γ, *Herb. Salisbury!*

COAST REGION, on mountains at 1500–6000 ft.: Clanwilliam Div.; between Ezelsbank and Dwars River, *Drège,* 7739! near Oliphants River, *Zeyher,* 1095! Tulbagh Div.; near Tulbagh waterfall, *Niven,* 77! Witsen Berg, *Burchell,* 8668! Ceres Div.; near Ceres, *Bolus,* 5499! and *Herb. Norm. Aust. Afr.,* 601! Swellendam Div.; near Swellendam, *Burchell,* 7299! 7390! *Niven,* 75! *Mund,* 14! near the Zondereinde River, *Zeyher,* 3201! Zuurbraak Mountain, *Galpin,* 3600! Var. γ: Tulbagh and Ceres Div.; near Tulbagh, *Guthrie,* 2079! Mitchells Pass, *Bolus,* 5291! *Schlechter,* 8934! 9953! near Ceres, *Bolus,* 7447! 8482! *Tyson in Herb. Norm. Aust.-Afr.,* 996! Paarl Div.; near Wellington, *MacOwan, Herb. Norm. Aust.-Afr.,* 19! Cape Div.? Zout River, *Masson* (ex *Salisbury*). Swellendam Div.; Grootvaders Bosch, *Mund!* Var. ε, Worcester Div.; Matroos Berg, *Bolus in Herb. Guthrie,* 4417! Var. ζ, Clanwilliam Div.; Kerskop Flats on the Cederberg Range, *Leipoldt,* 132! Pakhuis Berg, *Schlechter,* 10813.

CENTRAL REGION: Var. β: Ceres Div.; Cold Bokkeveld, *Masson!* near Wagenbooms River, *Schlechter,* 10158! Var. δ: Ceres Div.; Skurfdeberg Range, near Gydouw, *Martin in Herb. Bolus,* 7344! *Bodkin in Herb. Bolus,* 7553! Gydouw Mountain, *Schlechter,* 10044! Var. ζ: Ceres Div.; Gydouw, *Schlechter,* 10004! Wagenbooms River, *Schlechter,* 10168!

We have endeavoured to characterize the varieties named above, but fear that some of them are mere forms, which seem to run into each other.

110. E. pellucida (Soland. ex Salisb. in Trans. Linn. Soc. vi. 384, not of Andr.); erect, with numerous, spreading branches, 6–12 in. high; leaves close-set, spreading to squarrose and most usually incurved, linear, subacute, 2–3 lin. long; flowers 4-nate, often in dense masses at the ends of short branchlets; bracts approximate, linear-lanceolate, keel-tipped, lacerate, somewhat shorter than the sepals; sepals obovate or obovate-cuneate, with a long sublinear foliaceous cusp, scarious, lacerate or fimbriate-lacerate, about 3 lin. long; corolla-tube subinflated, bright rose, 5–6 lin. long; segments ovate, subobtuse, about 2 lin. long; anthers nearly square, a little over ½ lin. long; pore ⅓ the length of the cell; style at length exserted. *E. Parmentierii, Lodd. Bot. Cab. t.* 197. *E. venusta, Sinclair in Hort. Eric. Wob.* 27? *not of Salisb.; Benth. in DC. Prodr.* vii. 647.

Coast Region: Stellenbosch Div.; Hottentots Holland, *Masson in Herb. Salisbury!* Caledon Div.; summit and upper part of the mountains of Baviaans Kloof near Genadendal, 3300-3500 ft., *Burchell*, 7706! *Bolus*, 5402! and in *Herb. Norm. Aust.-Afr.*, 350! *Bodkin in Herb. Guthrie*, 3611! *Schlechter*, 9811! Zoetemelks Valley, *Burchell*, 7591!

Near to *E. denticulata*, but sepals less decidedly pectinate-lacerate, always (as to our specimens) with a long cuspidate point (not present in that species), with a longer, brighter-coloured corolla, and a smaller anther-pore. Specimens in herbaria have been named *E. hyacinthoides*, Andr., but that, according to Andrews' figure, has very different sepals. Bentham regarded the latter as a garden-hybrid, and we have seen no wild specimens like it. With Bentham's short description our specimens agree, though he does not mention the cuspidate sepals. This plant is at least an entity, with fairly constant characters, and must be distinguished.

111. E. denticulata (Linn. Mant. Alt. 229, in a wider sense); erect, branches spreading or virgate, glabrous, striate; leaves suberect or spreading, linear, acute, trigonous or round-backed, 3-3½ lin. long; flowers 4-nate, subsessile; bracts approximate; sepals and bracts variable in outline, from sublinear to obovate or spathulate, almost invariably broadest above the middle, sometimes unguiculate, scarious, more or less deeply pectinate-lacerate or fimbriate-lacerate, 2-3 lin. long, not cuspidate; corolla urceolate to tubular, dry, glabrous, white or rosy; tube 3-4 lin. long, 1¾ lin. wide; limb variable, sometimes half as long as the tube, mostly mealy above; segments 1 lin. long or less, suborbicular; pore of the anther almost as long as the cell. *Wendl. Eric. Ic. fasc.* 25, 5, *t.* 2; *Lodd. Bot. Cab. t.* 1090; *Benth. in DC. Prodr.* vii. 647. *E. dentata, Thunb. Diss. Erica*, 28; *Lam. Encycl.* i. 485? *E. pavettæflora, Salisb. Prodr.* 297, *not elsewhere. E. denticularis, Salisb. in Trans. Linn. Soc.* vi. 384. *E. venusta, Hort. ex Benth. l.c.* 647.

Var. β, **longiflora** (Bolus); subvirgate; leaves about 2 lin. long; bracts (shorter than in the type) and sepals obovate-cuneate, not unguiculate, keel not prolonged into a linear cusp; corolla-tube 6-7 lin. long; segments broad-ovate and very obtuse, or narrower, longer and subacute, 1½-1¾ lin. long; anther-pore nearly ⅔ the length of the cell. *E. dentata, Wendl. Eric. Ic. fasc.* 19, 101, *t.* 38, *not of Thunb.*

Var. γ, **grandiloba** (Bolus); leaves erect, incurved, usually very caducous (the branches being for the greater part naked), narrow-lanceolate or linear, 3 lin. long; sepals subobovate or oblanceolate, slightly lacerate; corolla-tube 3½-4½ lin. long, somewhat inflated; segments broad-ovate, obtuse, large, half as long as the tube or more, 2-2½ lin. long; anthers less than ½ lin. long; pore ⅓ the length of the cell.

South Africa: without locality, *Bergius* and *Thunberg* (both in Herb. Salisbury)!

Coast Region, on mountains at 800-2400 ft.: Paarl Div.; French Hoek, *Schlechter*, 9231! *Bolus*, 6990! Caledon Div.; tops of Baviaans Kloof Mountains, *Burchell*, 7702! Zwart Berg, near Caledon, *Zeyher*, 3200! *Bolus*, 6756! between Villiersdorp and French Hoek, *Bolus*, 5173! near Lowrys Pass. *Guthrie*, 2023! Var. β: Caledon Div.; Genadendal Mountain, *Bodkin in Herb. Bolus*, 6485! Zwart Berg, near Caledon, *Guthrie*, 2501! *Bolus*, 6956! Var. γ: Tulbagh Div.; above Tulbagh Waterfall, *Bolus*, 5460! and in *Herb. Norm. Aust.-Afr.*, 1307! mountains of Tulbagh Kloof, *Guthrie*, 2075!

E. maculosa (Tausch in Flora, 1839, 629), is probably a garden-hybrid between *E. denticulata* and *E. Walkeria*. It has the habit and spreading leaves of the latter, while the sepals (oblong-linear, slightly wider above the middle and lacerate) and the corolla approach our var. γ of the former.

Section XI. **PLATYSPORA.** (Sp. 112–117.)

112. E. astroites (Guthrie & Bolus) ; entirely glabrous ; branches pallid, longitudinally wrinkled (in the specimen before us 8–9 in. long) ; leaves 4-nate, spreading or reflexed, irregularly curved and bent round the branches, linear, acute, sulcate, glaucous ; younger petioles ciliate, 6–7 lin. long, much exceeding the internodes ; inflorescence axillary, flowers spreading at the ends of the branches, corolline ; pedicels slender, 4–5½ lin. long; bracts remote, lanceolate-linear, acuminate, 1–2 lin. long ; sepals broad-lanceolate, acuminate, scarious, pallid, with a darker keel on the upper half, 3 lin. long ; corolla somewhat salver-shaped ; tube narrow-ovoid, dry, rosy, 4½ lin. long ; segments large, stellate-spreading, elliptic-oblong, obtuse or subacute (like those of § *Euryloma*), paler than the tube, 3½–4 lin. long, 1½ lin. wide ; filaments capillary, 4½ lin. long ; anthers broad-oblong, very obtuse, smooth, a little over ½ lin. long, minutely aristulate ; pore ⅜ the length of the cell ; awns situated above the middle of the cell, linear or tooth-like ; style included, slender, thickened at the apex ; stigma capitellate ; ovary glabrous ; ovules obovate, with a white papery wing or margin.

VAR. β, **minor** (Guthrie & Bolus) ; smaller in all parts ; leaves 3- or 4-nate, 4–5 lin. long ; pedicels 1½ lin. long ; corolla-tube 4 lin. long ; segments more acute, 1½ lin. long ; anthers just as in the type, but ⅔ of the size. *E. albens, var. longiflora, Benth. in DC. Prodr.* vii. 649.

COAST REGION : Oudtshoorn Div. ; Meirings Poort, on the right side of the road to Oudtshoorn, near the Waterfall, *Stoney !* (in the Cape Gov. Herb.) Var. β : George Div.; on mountains at Barbiers Kraal, *Niven,* 136 ! mountains near George, *Alexander,* 18, and without collector's name or number in the Cape Gov. Herb. !

The larger-flowered form is very distinct in this section ; there is, however, a certain resemblance in the flower to *E. Vallis-Gratiæ* and *E. Alfredii,* but the sepals and bracts in each of these are much wider and longer, besides other differences. Our var. β most resembles *E. heliophila,* but has a larger corolla and the anther is somewhat different in shape, and has the peculiarly high and small appendages of the type (which seem to indicate a close relationship), while that of the species last named is muticous. Possibly, however, that also may prove to be another form of the present species.

113. E. albens (Linn. Mant. Alt. 231) ; erect, glabrous, 1–1½ ft. high ; branches virgate, somewhat slender, the younger subangular, but not deeply channelled by the leaf-cushions, closely leafy above, naked below ; leaves 3-nate, nearly erect, imbricate, internodes elongating, linear-trigonous, acute, sulcate, finely ciliolate, 5–7 lin. long ; inflorescence axillary, in sub-3-nate tufts, forming a lax, mostly subsecund, false raceme, 2–4 in. long ; flowers subcalycine, from erect to nodding ; pedicels about 2 lin. long ; bracts remote, lanceolate, scarious, small ; sepals obovate or ovate, acute, distinctly imbricate, concave, keeled, scarious, pallid, about 2 lin. long, reaching from ½–¾ the length of the corolla-tube ; corolla ovoid-urceolate, much narrowed to the mouth, white or pink-tinged, about 3 lin. long; segments spreading (at least slightly), connivent in age, oblong and obtuse, or lanceolate and acute, about $\frac{1}{5}$ of the tube in length; filaments capillary; anthers longitudinally broad-semiovate, obtuse, smooth, brown, $\frac{3}{8}$ lin. long, minutely aristate; pore about $\frac{1}{3}$ the length of the cell; awns subulate, spreading, not reaching below the base of the cell; style slender, clavate at the apex; stigma minute, capitellate or subconical; ovules flat, winged. *Wendl. Eric. Ic. fasc.* 6, 3; *Andr. Heathery, t.* 2, *and Col. Heaths, t.* 2; *Bot. Mag. t.* 440; *Lodd. Bot. Cab. t.* 95. *E. viminalis, Salisb. Prodr.* 298, *and in Trans. Linn. Soc.* vi. 387.

COAST REGION: Tulbagh Div.; Tulbagh Waterfall, *Niven,* 133! Swellendam Div.; Riet Kuil, *Niven,* 134! mountains near Swellendam, *Borcherds in Mac-Owan & Bolus, Herb. Norm.,* 1310! *Burchell,* 7345! *Zeyher,* Tradouw Pass, near Zuurbraak, *Galpin,* 3604! Riversdale Div.; summit of Kampsche Berg, *Burchell,* 7121! George Div.; Cradock Berg, near George, *Burchell,* 5912! 6015!

Usually recognizable in the section by its exceptionally elongate inflorescence, and more calycine flowers, the sepals being longer in proportion to the corolla than in the other species.

114. E. georgica (Guthrie & Bolus); 1 ft. or more high, entirely glabrous; branches virgate below, sometimes spreading near the apex, the younger deeply channelled by the decurrent leaf-cushions; leaves 3-nate or sometimes scattered, erect to spreading, linear, acuminate, incurved, concave above, sulcate below, somewhat glaucous, 4-6 lin. long; inflorescence axillary; flowers corolline, verticillate in close ovoid or oblong heads or false spikes, $\frac{1}{2}$-2 in. long, $\frac{7}{8}$ in. in diam., spreading in every direction; pedicels slender, $1\frac{1}{2}$ lin. long; bracts remote, small, pallid; sepals broad-lanceolate, acuminate, concave, keeled towards the apex, cartilaginous, pallid, only slightly imbricate at the base, about 2 lin. long; corolla-tube ovoid, 3-4 lin. long, throat contracted, rosy; segments stellate-spreading or reflexed, narrow-ovate, acute, somewhat paler than the tube, $1\frac{1}{2}$ by 1 lin.; anthers exactly elliptical, broadly rounded at either extremity, about $\frac{1}{3}$ lin. long, muticous; pore $\frac{1}{3}$ the length of the cell; stigma capitate, just manifest in the throat; seeds with a wide suborbicular membranous margin.

COAST REGION: George Div.; mountain slopes, Montagu Pass, 4500 ft., *Schlechter,* 5852!

115. E. macilenta (Guthrie & Bolus); erect, 1-1½ ft. high; branches virgate, slender, few, sometimes quite simple, not prominently channelled between the leaf-cushions, naked below, leafy above, white-pubescent, glabrescent; leaves 4-nate, the lower erect and more distant, sometimes only slightly exceeding the internodes, the upper more spreading and more closely imbricate, linear, acute, sulcate, ciliate, 3-4½ lin. long; inflorescence axillary (or sometimes terminal, on very short arrested branchlets?) forming a dense ovoid pseudo-raceme, about 9 lin. long and wide, situated shortly beneath the ends of excurrent branches; flowers corolline, spreading horizontally; pedicels slender, $\frac{1}{2}$–1 lin. long; bracts remote, linear, sepal-like, about 1 lin. long; sepals linear from a short ovate scarious ciliate base, leaf-like, keeled, acuminate, about 2$\frac{1}{2}$ lin. long, reaching to $\frac{3}{4}$ the length of the corolla-tube; corolla tubular-inflated below; tube about 3 lin. long, red or rosy; segments spreading, ovate, acute, from a narrow minutely unguiculate imbricating base, about 1$\frac{1}{4}$ lin. long; anthers as in *E. georgica,* but slightly smaller; stigma minutely capitellate; ovules membranous-margined.

CoAST REGION: Swellendam Div.; moist places on the Lange Bergen, near Zuurbraak, 2500 ft., *Schlechter,* 2043!

The corolla is almost exactly that of *E. steinbergiana,* but the leaves, inflorescence and sepals differ, the latter being, in this, narrower and more foliaceous.

116. E. tetragona (Linn. f. Suppl. 223); erect, glabrous; branches virgate, slender, somewhat channelled by the decurrent leaf-cushions, 6 in. or more long; leaves erect or spreading, imbricate, slender, linear, acute, keeled, 4–6 lin. long; inflorescence axillary, in a somewhat lax pseudo-raceme, 1–2 in. long, situated shortly beneath the ends of excurrent branches; flowers corolline, spreading in every direction; pedicels 1 lin. long; bracts approximate, linear, subfoliaceous, 1–2 lin. long; sepals lanceolate or linear-lanceolate, acuminate, keeled, the lower part scarious, foliaceous upwards, 2$\frac{1}{2}$ lin. long; corolla narrow-ovoid or tubular-inflated, more or less contracted at the throat, tetragonous, pale yellow, 3–4 lin. long; segments spreading-recurved, ovate, obtuse or subacute, $\frac{3}{4}$–1 lin. long; anthers oblong, obtuse at either extremity, pallid, $\frac{1}{3}$ lin. long, muticous; pore $\frac{1}{3}$ the length of the cell; stigma capitellate, sometimes minute; ovary substipitate, or at least contracted at the base; seeds oblong, with a narrow wing. *Thunb. Diss. Erica,* 14, *t.* 4; *Andr. Heathery, t.* 95, *and Col. Heaths, t.* 212; *Wendl. Eric. Ic. fasc.* 23, 163, *t.* 61; *Lodd. Bot. Cab. t.* 1239; *Benth. in DC. Prodr.* vii. 649. *E. pugionifolia, Salisb. in Trans. Linn. Soc.* vi. 387.

SOUTH AFRICA: without locality, *Thunberg, Masson! Herb. Lamarck!* and *culivated specimens!*

CoAST REGION: Swellendam Div.; Tradouw, *Mund & Maire!* Riet Kuil, *Niven,* 135! Humansdorp Div.; mountains near Kromme River, *Drège!*

117. E. heliophila (Guthrie & Bolus); branches slender, virgate, white-pubescent, not prominently channelled between the leaf-cushions, 9 in. or more long; leaves 3-nate, the upper crowded, imbricate, erect, the lower more distant and subspreading, linear-trigonous, acuminate, ciliate on the blade and on the petiole, about 3 lin. long; inflorescence axillary, laxly pseudo-racemose below the ends of excurrent branches, flowers corolline; pedicels pubescent, 1$\frac{1}{2}$ lin. long; bracts remote, linear, 1 lin. long or less; sepals somewhat spreading, lanceolate, acuminate, keeled, concave, scarious, ciliate, about 2 lin. long; corolla ovoid-urceolate; tube not angled, contracted at the throat, about 3 lin. long, $1\frac{1}{2}$ lin. wide ; segments spreading-recurved, ovate, acute, about $\frac{3}{4}$ lin. long ; anthers oblong or subcuneate-oblong, truncate at the base, very obtuse at the apex, $\frac{1}{4}$ lin. long, minutely aristulate ; pore about $\frac{1}{3}$ the length of the cell ; awns spreading, not reaching below the base of cell ; ovary sessile, seeds elliptical, with a rather wide margin or wing.

COAST REGION : Swellendam Div. ; marshy places on the Lange Bergen, near Zuurbraak, 2000 ft., *Schlechter*, 2112 !

This has the aspect of *E. tetragona*, but differs by its floral characters. It also resembles somewhat *E. astroites*, var. *minor*.

Section XII. **MYRA.**   (Sp. 118–122.)

118. E. glandulifera (Klotzsch in Linnæa, x. 333) ; erect, 1–$1\frac{1}{2}$ ft. high ; branches slender, rusty brown, together with the leaves, pedicels, bracts and sepals more or less densely covered with gland-tipped hairs ; lower leaves 3-nate, upper scattered, erect or spreading, imbricate, linear-subulate, obtuse, sulcate, 2–3 lin. long ; flowers in the axils of the upper leaves on one or more branchlets, forming a racemose panicle 4–5 in. long, or often simply racemose ; pedicels $1\frac{1}{2}$–2 lin. long ; bracts, two sometimes approximate, sometimes remote, the third basal, oblong, obtuse, foliaceous ; sepals oblong or oblanceolate, subobtuse, sulcate, $1\frac{1}{2}$–2 lin. long ; corolla tubular, more or less (but never much) inflated below the middle, somewhat contracted at the throat, viscid, puberulous ; tube 3–4 lin. long ; limb short, erect-spreading ; anthers included, cuneate-linear, acute, $\frac{3}{4}$ lin. long ; crests ovate, acute, about $\frac{1}{5}$ the length of the cell ; ovary sessile, not contracted above the disk, glabrous.

COAST REGION : Ceres Div. ; Witsen Berg or Skurfde Berg, *Mund & Maire* ; *Zeyher*, 1092 !

CENTRAL REGION : Ceres Div. ; Cold Bokkeveld, near Wagen Drift, 5000 ft., *Schlechter*, 10069 !

This and the next species are interesting as exhibiting a truly racemose inflorescence, and constituting, in that respect, the only exceptions we have observed in the genus.

119. E. irrorata (Guthrie & Bolus) ; erect, 1–$1\frac{1}{2}$ ft. high ; branches decumbent from the base, ascending, subvirgate, rusty brown, covered (as are the leaves, pedicels, bracts and sepals) with gland-tipped hairs ; leaves 3-nate or scattered and crowded on the lower parts of the branches, solitary on the upper parts, spreading, linear, bluntish, sulcate, puberulous as well as gland-hispid, 3–4 lin. long, the axils gemmiferous ; inflorescence axillary ; flowers solitary, with lengthening internodes, forming a long loose raceme, 4–7 in. long ; pedicels slender, 4–6 lin. long ; bracts two only, remote, small, about $1\frac{1}{2}$ lin. long ; sepals oblong or lanceolate, subacute, 2–$2\frac{1}{2}$ lin. long ; corolla tubular, wider at the base, gradually but slightly attenuated upwards, again widened below the throat and constricted above it, pubescent, dry ; tube rosy, from 6–8 lin. long, by 2–$2\frac{1}{2}$ lin. wide in the widest, and $1\frac{1}{2}$ lin. wide in the narrowest part ; throat purple ;

segments cordate-reniform, spreading, twice as wide as long, white or pallid; anthers included, cuneate, acute, 1 lin. long; crests crimson, ovate, acuminate, lacerate, about ½ the length of the cell; style shortly exserted; ovary subglobular or turbinate, substipitate, or at least constricted above the disk, glabrous.

COAST REGION: Tulbagh Div.; Lower Winterhoek Mountain, 4000 ft., *Bodkin in Herb. Bolus,* 5906! Great Winterhoek Mountain, 3500 ft., *Bolus in Herb. Guthrie,* 4172!

120. E. rufescens (Klotzsch in Linnæa, x. 332); erect, about 1 ft. high; branches virgate, rusty brown, more or less thickly beset (as are the leaves and pedicels) with gland-tipped hairs; leaves alternate, erect-spreading, densely crowded below, more distant above, linear, sulcate, 3–6 lin. long; flowers "panicled" (*Klotzsch*) or in 4–6-flowered umbels; pedicels slender, flexuous, 5–7 lin. long; bracts remote, erect, small; sepals ovate or oblong, acute, concave, keeled, glabrous, viscid, green or reddish, 1½–2 lin. long; corolla tubular-inflated below, or subampullaceous, either attenuated to a short narrow neck, thinly pubescent with fine soft hairs, rosy below, darker at the throat; tube 7–12 lin. long; segments rounded, wider than long, white or pallid, about 1 lin. long; anthers included, narrow-cuneate, subacute, pale brown, membranous, nearly 1 lin. long, crested; crests about ⅓ the length of the cell, darker-coloured; pore ⅔ the length of the cell; style shortly exserted; ovary cylindrical, elongate, glabrous, shortly stipitate. *Benth. in DC. Prodr.* vii. 640.

VAR. β, **minor** (Bolus); umbels simple, 4–6-flowered; corolla 7–8 lin. long, suddenly contracted to the throat.

SOUTH AFRICA: without locality, *Zeyher!*

COAST REGION: Var. β: Caledon Div.; mountains near the River Zonder-einde, 5400 ft., *Bolus,* 6480!

We have seen Klotzsch's type in Herb. Berlin, and it is certainly conspecific with the plants cited under var. β. The difference in size of the inflorescence and flowers might easily be accounted for by a greater luxuriance of growth; yet, as the material is scanty, it is better to distinguish them. A far greater difficulty is the question whether either should be separated from *E. glutinosa.* On this point there are grave doubts, the solution of which, since our material is scanty, may be left to future workers.

121. E. glutinosa (Berg. Descr. Pl. Cap. 98, not of Andr.); erect, generally under 1 ft. high; branches ascending, together with the leaves and pedicels more or less densely covered with viscid hairs; leaves sub-4-nate below, irregularly scattered on the upper parts, recurvo-patent, somewhat crowded, linear, obtuse, sulcate or some-what open-backed, puberulous and gland-ciliate, 2–3 lin. long; umbels many-flowered, subracemoso-corymbose at the ends of the branches; pedicels slender, 3–6 lin. long; bracts remote, small; sepals lanceolate, thick, glabrous, glossy, red, about 1 lin. long; corolla urceolate, sparsely pubescent, viscid; tube rosy, 4–5 lin. long; throat contracted, purple; segments very short, rounded, stellately spreading; anthers included, cuneate-oblong, slender, pale brown, membranous, a little over ½ lin. long, narrow-cristate, crests less than ½ the length of the cell; ovary glabrous, substipitate. *Bauer, Exot. Pl. t.* 17; *Wendl. Eric. Ic. fasc.* 22, 157, *t.* 60; *Benth. in DC. Prodr.* vii. 641.   *E. droseroides, Lam. Encycl.* i. 489; *Andr. Heathery, t.* 18, *and Col. Heaths, t.* 21; *Lodd. Bot. Cab. t.* 1685. *Andromeda droseroides, Linn. Mant.* 239.

VAR. β, **parviflora** (Benth. in DC. Prodr. vii. 641); flowers under 3 lin. long; anthers cuneate, crests longer than the cells.   *E. droseroides, var. minor, Andr. Heathery, t.* 259, *and Col. Heaths, t.* 229.

COAST REGION, between 1000 and 4500 ft.:   Worcester Div.; Dutoits Kloof, *Drège!* Cape Div.; mountains around Cape Town, *Thunberg, Burchell,* 629! *Drège! Guthrie,* 1006! *Bolus,* 4170! 8030! *Ecklon,* 280! *Wolley Dod,* 862! Paarl Div.; French Hoek, *Schlechter,* 10277! Stellenbosch Div.; Lowrys Pass, *Burchell,* 8198! 8271! *Schlechter,* 7234! Caledon Div.; Houw Hoek, *Niven,* 159! *Bolus!* near Palmiet River, *Bolus,* 4170! Genadendal Mountain, *Bolus!* Also *cultivated specimens!* Var. β: *Herb. Salisbury!*

CENTRAL REGION: Ceres Div.; near Schoongezigt, in the Cold Bokkeveld, *Schlechter,* 10191 !

122. E. armata (Klotzsch ex Benth. in DC. Prodr. vii. 672, not of Spreng.); erect, the whole, except the essential organs, more or less densely hispid and viscidulous, with stiff rigid gland-bearing hairs; leaves 4-nate, spreading, imbricate, subulate, acute, flat above, convex and sulcate beneath, puberulous as well as gland-hairy, $2\frac{1}{2}$–$3\frac{1}{2}$ lin. long; flowers 4–7 at or near the ends of the branches, suberect; pedicels $1\frac{1}{2}$ lin. long; bracts remote, small, foliaceous; sepals narrow-ovate, acute, puberulous, $1\frac{1}{2}$–2 lin. long; corolla suburceolate or tubular-inflated, viscose-puberulous, red, the whole $3\frac{1}{2}$ lin. long; limb very small, somewhat spreading but scarcely stellate; anthers subincluded, oblong, slightly curved backwards at the base, about $\frac{3}{4}$ lin. long, aristate; awns $\frac{1}{3}$–$\frac{1}{2}$ the length of the cell; ovary subturbinate, thinly puberulous.

VAR. β, **breviaristata** (Bolus); uppermost leaves subscattered, or arranged in unilateral pairs, not opposite nor 4-nate; inflorescence somewhat more elongate; pedicels longer; bract sometimes 1, or wholly absent; corolla finely velvety, or subglabrous; limb suberect; awns of the anther very minute, scarcely reaching to the base of the cell.

COAST REGION, on mountains between 2000 and 5000 ft.: Worcester Div.; Dutoits Kloof, *Drège,* 1148! Paarl Div.; French Hoek, *Niven,* 158! Var. β: Caledon Div.; near Genadendal, *Bolus,* 5413! and in *Herb. Norm. Aust.-Afr.,* 606! *Schlechter,* 9821! *Masson,* 37 !

This species was placed by Bentham in the § *Ephebus,* to which, without doubt, it is allied, and to which it affords a transition; but it seems better included here.   It has also affinities with *E. tumida, Ker.*

Section XIII. **EPHEBUS.** (Sp. 123-175.)

123. E. nivalis (Andr. Heathery, t. 274); erect, branched, 1 ft. or more high; leaves 3-nate, from subpatent to squarrose, linear, obtuse, glabrous, straight, 2–3 lin. long; flowers 3-nate, subcernuous; pedicels about 2 lin. long; bracts remote, small; sepals lanceolate, acute, about 1 lin. long; corolla globose-urceolate, mouth much contracted, covered with soft hairs, white, $2\frac{1}{2}$ lin. long; limb very small; segments erect or subconnivent; anthers included, oblong-cuneate, aristate; style included; stigma capitellate. *Andr. Col. Heaths, t.* 250; *Benth. in DC. Prodr.* vii. 674.

South Africa : Locality unknown.

The specimens we have seen under this name belong to *E. Peziza, Lodd.* We have seen no authentic specimen of this species, but accept Andrews' t. 274, with its accompanying description as the type. The shape of the corolla is there so different from that of *E. Peziza,* that it is quite impossible to regard the two as one species.

124. E. Peziza (Lodd. Bot. Cab. t. 265); erect, densely flori-ferous, $1-1\frac{1}{2}$ ft. high; branches ascending or nearly erect, tomentose; leaves 3-nate, erect or subspreading, linear, obtuse, sulcate, glabrous, mostly shining, $1\frac{1}{2}-2$ lin. long; flowers sub-3-nate; pedicels tomentose, $1\frac{1}{2}$ lin. long; bracts remote, linear, small; sepals ovate-lanceolate, acute, sulcate and keel-tipped, pubescent, $\frac{3}{5}$ lin. long; corolla cyathiform to obconic, not contracted but often widened at the mouth, densely long-tomentose, white, mostly about $2\frac{1}{2}$ lin. long; limb from nearly half, to as long as the tube; segments rounded, erect or slightly reflexed at the tips only; filaments bent at the apex; anthers included, ovate, dark brown, less than $\frac{1}{3}$ lin. long, aristate; awns affixed near the middle of the cells, subulate with roughened margins, nearly as long as the cell; style included; ovary minutely tomentose.

Coast Region, on mountains between 800–1000 ft. : Caledon Div.; near the Zondereinde River, *Schlechter,* 5637 ! near Caledon, *MacOwan, Herb. Aust.-Afr.,* 1492! Robertson Div.; near Montagu, *Tyson,* 3033! *Bolus,* 3992! 6387! 6720 ! Swellendam Div.; near Hessaquas Kloof, *Zeyher,* 3242 !

This species was included by Bentham (*DC. Prodr.* vii. 674) as a synonym of *E. nivalis,* Andr. We have numerous specimens, agreeing well with Loddiges' figure, but not, as to the shape of the corolla, with Andrews' fig. of *E. nivalis* (t. 274). So great a difference in the shape of the corolla we have not seen in any species; and we feel bound to maintain both as distinct, although Andrews' species appears to be known only from his figure.

125. E. ovina (Klotzsch ex Benth. in DC. Prodr. vii. 674); stout, erect, $1-1\frac{1}{2}$ ft. high; branches straight or flexuous, densely grey-woolly, becoming naked below; leaves 3-nate, spreading, linear or subfiliform, sulcate, obtuse, very slender, puberulous, glabrescent, $2-2\frac{1}{2}$ lin. long; flowers 3-nate; pedicels pubescent, $1-1\frac{1}{2}$ lin. long; bracts remote, small; sepals ovate or ovate-lanceolate, acute, keel-tipped, pubescent, $\frac{3}{4}-1$ lin. long; corolla ovoid or ovoid-oblong, mouth slightly contracted, densely but not very closely lanate with somewhat long twisted hairs, white, 3 lin. long; limb erect, short, about $\frac{1}{4}$ as long as the tube; anthers included, sub-cuneate-oblong, dark-brown, less than $\frac{1}{3}$ lin. long, aristate; awns narrow, about $\frac{3}{4}$ the length of the cell; style included, compressed; stigma capitate; ovary turbinate, somewhat truncate, hispidulous.

Var. β, purpurea (Bolus); habit more branched; corolla cyathiform, equal or perhaps slightly widened to the mouth, pale purple, $2-2\frac{1}{4}$ lin. long.

COAST REGION : Caledon Div.; mountain ridges along the lower part of Zondereinde River, *Zeyher*, 3210! Var. β : Caledon Div.; near Zoetmelks Vlei, *Grisbrook in Herb. Guthrie*, 3284!

126. E. tomentosa (Salisb. in Trans. Linn. Soc. vi. 327) ; erect, about 1 ft. high ; branches numerous, paniculate, tomentose-pubescent, densely many-flowered ; leaves 3-nate, suberect, linear, obtuse, sulcate, puberulous, glabrescent, 1–1½ lin. long ; flowers sub-3-nate ; pedicels pubescent, 1–1½ lin. long; bracts small, remote ; sepals narrow-lanceolate to ovate, pubescent, coloured, about ½ lin. long ; corolla urceolate or narrow-urceolate, mouth slightly contracted, densely and shortly pubescent, almost velvety, dull red to lilac, 1½–2 lin. long; limb very short ; segments spreading ; filaments capillary, bent at the apex ; anthers included, dorsifixed near the base, broad oblong, very obtuse, about ⅓ lin. long, aristate ; awns short, subulate, sometimes recurved upwards, the apices touching the back of the anther ; style subexserted, stigma capitate ; ovary from pubescent to subglabrous. *Benth. in DC. Prodr.* vii. 673. *E. velutina, Bartl. in Linnæa*, vii. 645.

COAST REGION, between 1000–2000 ft. : Stellenbosch Div.; Hottentots Holland, *Mulder*, ex *Salisbury!* Caledon Div.; mountains of Baviaans Kloof and near Genadendal, *Burchell*, 7611! *Drège*, 7770a! *Galpin*, 3659! mountains near Zoetmelks Vlei, *Grisbrook in Herb. Guthrie*, 3295! at the foot of Babylons Tower, *Ecklon* (ex *Bartling*). Also cultivated specimens, *Herb. Salisbury!*

127. E. pubigera (Salisb. in Trans. Linn. Soc. vi. 372) ; erect, 6–10 in. high ; branches slender, erect, tomentose ; leaves 3-nate (or 4-nate, *Bentham*) ; erect-spreading, or squarrose below, narrow semi-terete, sulcate, obtuse, glabrous, under 2 lin. long; flowers 3-nate ; pedicels bent, 1 lin. long ; bracts 3, remote, median, one foliaceous, the others shorter, coloured ; sepals linear or linear-lanceolate, acute, sulcate, glabrous or pubescent, foliaceous, 1 lin. long ; corolla urceolate, mouth contracted, densely hispid, subviscid, white, about 2 lin. long ; segments ovate, recurved, dark purple, short ; filaments capillary, curved at the apex ; anthers included, dorsifixed near the base, broad-oblong, very obtuse, about ¼ lin. long, aristate ; pore half the length of the cell ; awns subulate, about ¾ the length of the cell ; style included ; stigma capitellate ; ovary pilose-pubescent. *Benth. in DC. Prodr.* vii. 673. *E. præusta, Bartl. in Linnœa*, vii. 645.

COAST REGION : Swellendam Div.; Voormans Bosch, *Ludwig & Beil* (ex *Bartling*); on the Lange Bergen, near Swellendam, *Niven*, 224! *Borcherds in Herb. Bolus*, 6326! *Schlechter*, 5678! *Galpin*, 3664 ! *Ecklon & Zeyher.*

128. E. Constantia (Nois. ex Benth. in DC. Prodr. vii. 672) ; branches erect, rigid, the younger tomentose ; leaves 3-nate, erect-spreading, incurved, oblong, sulcate, obtuse, naked above, below furnished with 2 rows of rigid tuberculate-setose hairs, 2–2½ lin. long; flowers subumbellate ; pedicels stout, pubescent, about 1 lin. long ; bracts approximate, one large almost exactly like the leaves, the two smaller like the sepals ; sepals small, narrow-ovate, acute, pubescent, ciliate, $\frac{1}{2}$ lin. long; corolla urceolate, mouth slightly contracted, coarsely and rigidly pubescent, of thick and tough texture, 2–$2\frac{1}{2}$ lin. long; limb small; segments spreading or recurved, rounded; filaments narrow, nearly equal, slightly bent; anthers included or subincluded, laterally basifixed, oblong, subobtuse, slightly curved, muticous; pore about $\frac{1}{2}$ the length of the cell, $\frac{1}{2}$ lin. long; style exserted; stigma capitate; ovary pubescent.

CoAST REGION : Worcester Div.; on the Matroos Berg, 5200 ft., *Marloth*, 1960!

[This number was erroneously cited as *E. oresigena, Bolus,* in Journ. Bot. 1894, 238, 239.]

The specimens of Dr. Marloth agree excellently with those of the type in Herb. Berlin marked "ex horto Kennedy, 1816," which we have seen and dissected. Bentham calls the bracts remote, but the type shows them clearly approximate. This rare species is well marked by the one large bract, the minute sepals, and thick rough corolla. Its rediscovery is very interesting. It has a general resemblance to *E. trichadenia.*

129. E. albescens (Klotzsch ex Benth. in DC. Prodr. vii. 673); prostrate, much branched, more or less tomentose with shorter and longer hairs intermixed; leaves 3-nate (or 3-4-nate, *Bentham*), from subpatent to squarrose, rather close-set, incurved or straight, linear, obtuse, sulcate, thick, puberulous or subglabrous, ciliate or naked, pale green, 1–2 lin. long; flowers fasciculate-racemose, 3-nate; pedicels puberulous, 1–1$\frac{1}{2}$ lin. long; bracts remote, two small, one larger and foliaceous; sepals from an ovate base, sulcate-keeled, and foliaceous in the upper part, thick, ciliate, subglabrous, 1–1$\frac{1}{4}$ lin. long; corolla suburceolate, mouth not (or very slightly) contracted, puberulous, white, 1$\frac{1}{2}$ lin. long; limb erect, $\frac{1}{2}$ to $\frac{1}{3}$ the length of the tube, sinuses sometimes rounded; filaments slender, straight; anthers subincluded, oblong, obtuse or subacute, a little over $\frac{1}{2}$ lin. long, cristate; crests with a lacerate margin or subaristate and closely ciliolate, about $\frac{1}{2}$ the length of the cell; style exserted; stigma capitate; **ovary sparingly villous.**

VAR. β, **erecta** (Bolus); branches erect, virgate. 1 ft. or more long; leaves sometimes lanceolate and somewhat open-backed, tuberculate-hispid and rough : sepals linear to linear-lanceolate; anthers obtuse, subaristate; awns closely ciliolate.

CoAST REGION, from 1700-2500 ft. : Swellendam Div.; Tradouw or Kanna-land, *Mund & Maire* in Herb. Berlin! Var. β : Piquetberg Div.; Piquetberg Mountain, *Bodkin in Herb. Bolus,* 6336! *Guthrie,* 2657! Tulbagh Div.; mountains near Saron, *Schlechter,* 10654!

We have seen some leaves and dissected a flower from a specimen in Herb. Berlin, which is probably that which Bentham used for his description. The plants we have placed under var. β agree in the most important characters with this type, and we think are too near to make it advisable to separate them specifically. The greatest difference is in the habit. Kannaland is a dry tract behind the Lange Bergen, and we may conjecture that the specimens found since Mund's time, further to the west, are more luxuriant forms due to a moister and less severe climate.

130. E. oxyandra (Guthrie & Bolus) ; erect, 1 ft. or more high; branches thinly tomentose, also with sparsely scattered long hairs;

leaves 3-nate, spreading-incurved, linear, obtuse, deeply sulcate, or the lower somewhat open-backed, setaceous-pilose, with long spreading tubercle-based hairs, glabrescent but the tubercles persistent, 2–3 lin. long; flowers sub-3-nate; pedicels pubescent, about $2\frac{1}{2}$ lin. long; bracts remote, subbasal, minute; sepals like the younger leaves, setose-pilose and ciliate with mostly very long hairs, $1\frac{1}{2}$ lin. long; corolla suburceolate-cyathiform, with only a slight contraction at the throat, sub-4-gonous at the base, setaceous-pilose, white, $2\frac{1}{2}$–3 lin. long; segments erect or slightly spreading, broadly rounded; filaments bent at the apex; anthers included, cuneate-oblong, acute, brown, about $\frac{3}{4}$ lin. long, cristate; crests short, ovate, lacerate; style exserted; ovary thinly hispidulous.

CoAST REGION : Swellendam Div. ; slopes of the Lange Bergen near Swellendam, about 3000 ft., *Bolus*, 8088 !

The long hairy indumentum gives some resemblance to *E. fausta* and *E. setosa* ; but it may be known from both by its less expanded leaves, and especially by its sharply cuneate acute anthers.

131. E. dysantha (Benth. in DC. Prodr. vii. 674); branches straggling, slender, 12 in. or more long, red, shortly pubescent and also pilose with longer hairs; leaves 3-nate, squarrose, linear, sulcate below, hispid, ciliate, over 1 lin. long; flowers 3-nate; pedicels less than 1 lin. long; bracts approximate, one foliaceous, the others small, linear, pilose; sepals lanceolate, acuminate, pilose, $1\frac{1}{2}$ lin. long; corolla urceolate, mouth only slightly contracted, pilose with long stiffish hairs, nearly $2\frac{1}{2}$ lin. long; limb erect, $\frac{1}{4}$ the length of the tube; segments rounded; filaments widened below the anthers, straight; anthers just included, oblong, very obtuse, less than $\frac{1}{2}$ lin. long, with two very short squarrose tooth-like appendages at the base; style exserted; stigma capitate; ovary villous with long hairs.

CoAST REGION : Riversdale Div.; on the slopes and summit of Kampsche Berg, *Burchell*, 7093! 7119!

A well-marked species which seems to have evaded observation before Burchell's time (1814) and since. Except for its almost shaggy corolla it might go into § *Ceramia*.

132. E. dilatata (Wendl. fil. ex Benth. in DC. Prodr. vii. 656); erect, 1 ft. or more high; branches spreading, flexuous; leaves 3-nate (3-4-nate, *Bentham*), erect or spreading, linear, acute, sulcate, puberulous, ciliate, 2–3 lin. long; flowers 3-nate; pedicels much longer than the flowers, pubescent; bracts remote, small; sepals lanceolate, acuminate, ciliate, coloured, 1–$1\frac{1}{2}$ lin. long; corolla urceolate, mouth slightly contracted, silky pubescent, $2\frac{1}{2}$ lin. long; limb short, erect; filaments slender, bent below the anther; anthers included, shortly aristate; style included; stigma capitate; ovary puberulous. *E. hirta, Wendl. Eric. Ic. fasc.* 27, 16? *not of Thunb.*

SoUTH AFRICA : without locality or collector's name in the Berlin Herbarium.

Bentham's description is based on what is probably a garden plant in Herb.
Berlin. There is also a specimen in Herb. Kew, labelled "Koenig," from
Salisbury's Herbarium with the name " *E. dilatata*, Wendl.," added by Bentham.
These may or may not be the same as the figure of *E. hirta* in Wendland's *Eric.
Ic. fasc.* 27, quoted by Bentham. This last is a good figure, and much resembles
E. affinis and *E. propinqua*, near which Bentham doubtfully placed this species.
But the sepals of this species are not like those of § *Trigemma*, nor does the
presence of an indumentum on the corolla agree. Bentham suspected this
might be a hybrid. We have endeavoured to draw up a description embracing
both Wendland's and Bentham's, but have relied chiefly on the excellent figure
of the former.

133. E. Alopecurus (Harv. Thes. Cap. i. 31, t. 48); 1 ft. or more
high; branches virgate, imbricate throughout their whole length with
small and short or minute erect branchlets, pubescent and also setose-
hirsute; leaves 3-nate, incurved-ascending, linear, obtuse, sulcate,
midrib visible beneath, coarsely pubescent or hirsute, 2–3 lin. long;
flowers 3-nate, crowded on very short branchlets towards the ends
of the branches into a cylindrical false spike from 1–3 in. long,
3–5 lin. wide; pedicels under $\frac{1}{2}$ lin. long; bracts remote, small or
sometimes larger and foliaceous; sepals linear-subulate, pubescent
or setose, setose-tipped, ciliate, about 1 lin. long; corolla broad-
ovoid or suburceolate-ovoid, mouth much contracted, pubescent,
about $1\frac{1}{2}$ lin. long, or rarely longer; limb very short; segments
spreading; filaments capillary; anthers included, broad ovate, pale
brown, less than $\frac{1}{5}$ lin. long, aristate; awns rough, nearly as
long as the cell; style included, curved; stigma capitate; ovary
villous.

VAR. β, **glabriflora** (Bolus); corolla glabrous.

COAST REGION: Cathcart Div.; Bontebok Flats, *Kennedy*, 106! King
Williamstown Div.; Perie, *Sim*, 101! Stutterheim Div.; Dohne Mountain,
3000 ft., *Sim*, 172! Kaffraria, *Brownlee!*
CENTRAL REGION: Somerset Div.; summits of Bruintjes Hoogte and Bosch
Berg, 4500 ft., *MacOwan*, 1550! and in *Herb. Norm.* 34!
KALAHARI REGION: Basutoland, *Cooper*, 756! Transvaal; Gemsbok Spruit,
Nelson, 389! Komati, *Mrs. Stainbank in Herb. Wood*, 3652! Var. β : Trans-
vaal; Witwatersrand, *Mrs. H. Hutton*, 875!
EASTERN REGION, from 7000–8500 ft.: Griqualand East; Luhana Pass,
Drakensberg Range, *Galpin*, 2321! summit of Ingeli Mountain, *Tyson in Mac-
Owan & Bolus, Herb. Norm.*, 457! Natal; Rovelo Hills, *Sutherland!* Mooi
River, *Gerrard*, 731; Kar Kloof, *Rehmann*, 7365! near Van Reenen, *Schlechter-*
6992! Var. β : Pondoland; Murchisons Plain, in Fakus Territory, *Suther-
land!*

This is very distinct by its peculiar inflorescence, which resembles that of
Stœbe cinerea, Thunb., and by its very small broad anther.

134. E. æmula (Guthrie & Bolus); erect, $1\frac{1}{2}$–2 ft. high; branches
flexuous or nearly straight, pubescent with pallid spreading hairs;
leaves 3-nate, spreading, not crowded, linear, subacute, rounded at
the back and channelled, shining, glabrous or pilose with spreading
hairs, $2\frac{1}{2}$–3 lin. long; flowers umbellate or 3-nate; pedicels puberulous, about 1 lin. long; bracts remote or subremote, small; sepals
lanceolate, acute, varying to linear-subulate from a broader

base, $\frac{3}{4}$–1 lin. long; corolla obconic, with a wide mouth, minutely puberulous or pubescent, white or pale ochraceous, 2 lin. long; segments erect, rounded, $\frac{1}{2}$ the length of the tube; anthers included, cuneate-oblong, pale brown, $\frac{1}{2}$ lin. long, aristate; awns about $\frac{2}{3}$ the length of the cell, rough; style included; stigma capitate; ovary glabrous.

VAR. β, **pubescens** (Guthrie & Bolus); all parts, including the corolla, more copiously pubescent.

COAST REGION: Worcester Div.; Bains Kloof, *Miss Cummings in Herb. Huguenot Seminary*, 171! Stellenbosch Div.; Fish Hoek, Gordons Bay, 100 ft., *Guthrie*, 3108! Var. β: Clanwilliam Div.; in wet places, on Sneeuwkop, Cederberg Range, *Leipoldt*, 613!

In general appearance and floral structure, this much resembles *E. subdivaricata*, Berg. But, besides its 3-nate and more laxly set leaves, it has larger flowers, somewhat different sepals, an obconic corolla, and never (so far as our specimens go) quite glabrous.

135. E. auriculata (Guthrie & Bolus); erect, 6–9 in. high; branches ascending, somewhat straggling, sparingly leafy, with long internodes, puberulous, greyish; leaves 3-nate (rarely 4-nate?) erect or subpatent, mostly distant, narrow-oblong, obtuse, sulcate, thick, closely hispidulous, the younger shortly gland-ciliate, about 3 lin. long; flowers 3-nate?; pedicels pubescent, about $1\frac{1}{2}$ lin. long; bracts approximate, unequal, linear, foliaceous, glandular-pubescent, the longest about $1\frac{1}{2}$ lin.; sepals like the longer bract, $1\frac{1}{2}$–2 lin. long; corolla urceolate, mouth contracted, pubescent, apparently white or whitish, 2 lin. long; limb erect, $\frac{1}{4}$ the length of the tube or less; filaments gradually much expanded and wider at the apex than the anther, furnished at the apex with two erect free adpressed ear-like lobes, slightly coloured, thickened and fringed with short hairs on the outer margin, and covering the base of the anther at its back; anthers included, lateral, inserted on the filament above the base of the cells, oblong, subacute, nearly $\frac{1}{20}$ in. long, strictly muticous, but apparently cristate by the filamentary processes above described; pore $\frac{2}{3}$ of the cell; ovary softly woolly.

CENTRAL REGION: Ceres Div.; on the Skurfde Bergen, near Klein Valley, 5800 ft., *Schlechter*, 10207!

A very distinct species, with a filamentary appendage different from any other known to us. This appears like an anther-crest, arising in the more usual way from the cells, but on dissection it is plainly found to come away with the filament. Our material is poor in respect of flowers; but the species is in every way well marked.

136. E. sicæfolia (Salisb. in Trans. Linn. Soc. vi. 326); suberect, straggling, slender, with scanty foliage, 4–8 in. high; branches ascending, glabrescent; leaves 3-nate, spreading-erect, linear, acuminate, setose-mucronate, mostly curved, sulcate, glabrous, shining, $2\frac{1}{2}$–5 lin. long; flowers 3-nate; pedicels viscid, puberulous, mostly curved, about 2 lin. long; bracts subapproximate or remote, small; sepals ovate-lanceolate, obtuse, viscid-pubescent, dark-coloured,

$\frac{1}{2}-\frac{2}{3}$ lin. long ; corolla broad-cyathiform, mouth wide, pubescent, sub-viscidulous, about 2 lin. long, 2 lin. broad, dark purple; limb short, rounded ; segments ciliolate, about $\frac{1}{3}$ the length of the tube ; anthers included, oblong, about $\frac{1}{2}$ lin. long, aristate ; awns subulate from $\frac{1}{2}-\frac{2}{3}$ the length of the cell ; style exserted ; ovary glabrous. *Benth. in DC. Prodr.* vii. 672. *E. pygmæa, Andr. Heathery, t.* 279, *and Col. Heaths, t.* 259. *E. sanguinolenta, Lodd. Bot. Cab. t.* 468; *Bot. Mag. t.* 2263.

Coast Region : Caledon Div.; mountain tops of Baviaans Kloof, near Genadendal, *Masson,* 9 ! *Burchell,* 7746! *Bolus,* 5404! and in *Herb. Norm. Aust.-Afr.,* 353 ! Also *cultivated specimens !*

Andrews' figure does not well represent the wild plant; the others are better.

137. E. podophylla (Benth. in DC. Prodr. vii. 681) ; "a small alpine plant about 6 in. high" (*Niven*) ; branches rigid, pubescent; leaves 3-nate, crowded on short branchlets as if fascicled, spreading, long-petioled; petioles ciliate, 1 lin. or more long ; blade broadly linear, thick, rigid, hispid, shining, obtuse, callous and recurved at the apex, the whole $2\frac{1}{2}-3\frac{1}{2}$ lin. long ; flowers terminal (or "irregu-larly axillary and terminal, subracemose," *Bentham*), subcalycine, solitary, on short branchlets ; pedicels pilose with spreading or reversed hairs, $1\frac{1}{2}-2$ lin. long; bracts remote, adpressed, linear, ciliate, scarious ; sepals ovate, acuminate, keel-tipped, thick and leathery, glabrous, smooth, shining, coloured, ciliate, 1 lin. long ; corolla rather broad urceolate-campanulate, mouth only slightly con-tracted, 4-gonous, texture thick, densely puberulous, about 2 lin. long ; limb subpatent, more than $\frac{1}{2}$ the length of the tube ; segments ovate, rounded; filaments bent at a right angle at their apex ; anthers included, suboblong, obtuse, $\frac{2}{5}$ lin. long, cristate ; pore $\frac{2}{3}$ the length of the cell ; crests ovate in outline deeply incised at the apex, with linear lobes, in length $\frac{3}{5}$ of the cell ; ovary hispidulous.

Coast Region : Swellendam Div.; Grootvaders Bosch, *Masson!* moist rocky places at Riet Kuil and near Swellendam, *Niven,* 30a! 297 ! summit of the great mountain near Swellendam, *Burchell,* 7328 !

A very distinct and well-marked species. Bentham placed it in § *Leptoden-dron ;* but it seems to fit better here. Like *E. setosa* and *E. fausta*, the species has escaped all recent collectors.

138. E. Atherstonei (Diels MSS.) ; a small, probably procumbent, creeping, branched shrub, a few inches high ; branches flexuose, rather stout, the younger (as also the pedicels and margins of the sepals) densely covered with short white floccose hairs ; leaves 3-nate, erect-spreading, crowded, imbricate, linear-lanceolate, concave, with a prominent mid-nerve above, convex and sulcate below, ciliate on the margin and on the dorsal reflexed edge, becoming glabrous and glossy, $1\frac{1}{2}-2$ lin. long ; flowers both terminal and axillary, solitary at or near the ends of the branches ; pedicels decurved, $2\frac{1}{2}-4$ lin. long ; bracts remote, thick, very small ; sepals narrow-lanceolate to oblong, subacute, upper margins recurved and thickened, foliaceous, subcoriaceous, ciliate, $1\frac{1}{2}$ lin. long, or more than half as long as the corolla ; corolla tubular-campanulate to obconic, the middle of each petal strongly nerved, and hence appearing sub-tetragonous, minutely pulverulent, becoming glabrous, $1\frac{3}{4}$–$2\frac{1}{4}$ lin. long, 1–$1\frac{1}{4}$ lin. wide at the apex; limb erect; segments oblong, blunt, about $\frac{2}{5}$ the length of the tube, at length splitting nearly to the base; anthers subexserted, narrow-oblong, nearly straight, scaberulous, nearly $\frac{3}{4}$ lin. long, cristate at the base ; pore nearly $\frac{1}{2}$ the length of the cell ; crests lanceolate, acuminate, dentate, curved, about $\frac{1}{3}$ the length of the cells ; style exserted, rather stout, decurved at the apex ; stigma capitate ; ovary pubescent.

KALAHARI REGION : Transvaal; near Lydenburg, *Atherstone!* Devils Knuckles, near Spitzkop, *Wilms,* 904 !

Allied to *E. podophylla,* which is somewhat remarkable from the fact that their native localities are so far apart. The strongly-nerved corolla of the present species would alone serve to distinguish it, and is very unusual in the genus. This species affords a transition to the § *Gypsocallis,* and might almost equally well have been placed there on account of its partially axillary in-florescence, but the anthers are not fully exserted, and it has hardly the appearance of that section.

139. E. propendens (Andr. Col. Heaths, t. 122) ; erect, 12–18 in. high ; branches nearly straight, the younger pubescent; leaves 4-nate, erect-spreading, imbricate, linear-trigonous, ciliate, glabrous or hairy, 1–$1\frac{1}{2}$ lin. long; flowers 1–4-nate ; pedicels $\frac{1}{2}$–$\frac{3}{4}$ lin. long ; bracts approximate, scarious, coloured ; sepals obovate, scarious, keel-tipped, sometimes with a long excurrent point, lacerate, coloured, $1\frac{1}{2}$ lin. long; corolla broad-campanulate, pubescent, red, 4–5 lin. long ; segments large, erect or spreading, rounded, about 1 lin. long ; filaments slightly thickened below the anther; anthers included, dorsifixed near the base, oblong, a little over $\frac{1}{2}$ lin. long, dark coloured, muticous ; style straight, included ; stigma capitellate ; ovary 4–8-celled, glabrous, rough. *Andr. Heathery, t.* 141 ; *Bot. Mag. t.* 2140; *Lodd. Bot. Cab. t.* 63; *Benth. in DC. Prodr.* vii. 682.

COAST REGION : Caledon Div. ; Houw Hoek, 800 ft., *Grisbrook in Herb. Guthrie,* 2363 ! near the mouth of the Palmiet River, *Bolus,* 9935 ! and without precise locality, *Bolus,* 5926 ! 6311 ! *Guthrie,* 4579 ! *Niven !* Also *cultivated specimens !*

This much resembles *E. chrysocodon* in habit and general appearance ; but is perhaps nearer to *E. perspicua.* Like the latter it has often an 8-celled ovary, though some specimens are 4-celled. One specimen before us with the corolla 5 lin. long may be a hybrid between *E. perspicua* and some other species ; but the calyx is very different.

140. E. pyramidalis (Soland. in Ait. Hort. Kew. ed. 1, iii. 491) ; erect, 1–2 ft. high ; branches mostly virgate ; leaves 4-nate, erect, subincurved, linear-trigonous, slender, glabrous or puberulous, $2\frac{1}{2}$–$3\frac{1}{2}$ lin. long ; flowers 4-nate, abundant, on short branchlets ;

pedicels 1 lin. long; bracts minute, remote; sepals linear-acuminate from an ovate scarious base, foliaceous, $1\frac{1}{4}$–$1\frac{1}{2}$ lin. long; corolla campanulate-obconic, glabrous or minutely puberulous, rosy-red, $3\frac{1}{2}$–5 lin. long; segments ovate, slightly spreading, $1\frac{1}{2}$ lin. long; anthers included, dorsifixed considerably above the base, oblong, about $\frac{3}{4}$ lin. long, muticous; ovary glabrous. *Bauer, Exot. Pl. t.* 27 ; *Andr. Heathery, t.* 142, *and Col. Heaths, t.* 51 ; *Wendl. Eric. Ic. fasc.* 5, 3 ; *Lodd. Bot. Cab. t.* 319 ; *Benth. in DC. Prodr.* vii. 682. *E. transparens, Thunb. Prodr.* 71, *not of Berg. E. obpyramidalis, Salisb. Prodr.* 298, *and in Trans. Linn. Soc.* vi. 356. *E. campanulata, F. W. Schmidt, Neue und Selt. Pfl.* 47.

VAR. β, **vernalis** (Benth. in DC. Prodr. vii. 682); flowers much smaller in all parts, sepals 1 lin. long, corolla subcampanulate-cyathiform, $2\frac{1}{2}$–3 lin. long, the limb shorter and more spreading, anthers about $\frac{1}{2}$ lin. long. *E. vernalis, Lodd. Bot. Cab. t.* 1608. *E. pyramidiformis, Wendl. Eric. Ic. fasc.* 24, 181, *t.* 68.

SOUTH AFRICA: without locality, *Sieber*, 169! *Herb. Salisbury!* and *cultivated specimens! Var. β : cultivated specimens!*

COAST REGION: Cape Div.; Zwart River near Rondebosch, *Masson! Niven*, 100! between Wynberg and Muizenberg, *Werner!* (in Cape Herb.). Cape Flats, *Bolus*, 2953! 8019! Paarl Div.; Drakensteen Mountains, *Thunberg.* Stellenbosch Div.; Hottentots Holland, *Thunberg.*

There can be no question of the identity of Wendland's *E. pyramidiformis*, and of Loddiges' *E. vernalis;* Bentham held the latter as a variety of this species, and maintains the former as distinct. We have seen a specimen marked *E. pyramidiformis* in the Berlin Herbarium, but it has clearly nothing to do with this species, and is probably *E. pusilla, Salisb.*

141. E. chrysocodon (Guthrie & Bolus); erect, 1–$1\frac{1}{2}$ ft. high; branches rather slender, virgate or spreading, the upper hirsute, the lower glabrescent ; leaves 4-nate, erect, imbricate, linear or subulate, acute, keeled, hairy on the keel and ciliate, $1\frac{1}{2}$–2 lin. long; flowers on short branchlets, mostly solitary, or in pairs, usually forming a long dense cylindrical pseudo-raceme from 3–6 in. long at the ends of the branches; pedicels $\frac{1}{2}$–$\frac{3}{5}$ lin. long; bracts remote, subbasal, small; sepals ovate-lanceolate, acuminate, keel-tipped, subfoliaceous, thin, hirsute and ciliate, $1\frac{1}{2}$ lin. long; corolla funnel-shaped, lower half of the tube distinctly contracted, upper half widened to the mouth, variable between obconic and subcampanulate, thinly pilose with longish hairs, bright yellow, 3–$3\frac{1}{2}$ lin. long; limb ample; segments rounded or occasionally acute, mostly spreading at maturity, occasionally recurved, less than $\frac{1}{2}$ as long as the tube; anthers manifest, just reaching beyond the corolla-tube, dorsifixed a little above the base, cuneate-oblong, subacute, minutely scaberulous, about $\frac{2}{3}$ lin. long, muticous; pore narrow, $\frac{1}{3}$–$\frac{1}{2}$ the length of the cell; style shortly exserted, slender; stigma small, capitellate; ovary spherical, small, glabrous.

COAST REGION: Paarl Div.; mountains around French Hoek, *Bolus in MacOwan Herb. Aust.-Afr.*, 1629! Caledon Div.; in wet places at the foot of the French Hoek Pass on the eastern side, 1200 ft., *Fair in Herb. Bolus*, 6334!

142. E. trichophora (Benth. in DC. Prodr. vii. 673); erect, 12–14 in. high or more; branches virgate, pubescent, with longer setose hairs intermixed; leaves 4-nate, suberect or spreading, linear, obtuse, slender, keeled, hirsute, 1–1½ lin. long; flowers usually solitary in the specimens seen, but occasionally 2-nate and probably normally 4-nate; pedicels rather stout, ¼ lin. long; bracts subbasal, one long linear, two very short; sepals linear, acuminate, keeled, foliaceous from an ovate scarious denticulate-edged coloured base, pubescent, 1–1½ lin. long; corolla tubular-obconic or suburceolate-cyathiform, widened to the mouth, more or less copiously clothed with coarse white hairs, about 3 lin. long; limb erect; segments rounded, about ⅕ the length of the tube; anthers included, dorsifixed shortly above the base, cuneate-oblong, barely ½ lin. long, muticous; pore about ½ the length of the cell; style included, slender; stigma very small, subsimple; ovary oblate-spheroidal, glabrous. *E. pilifera, Klotzsch in Linnæa*, xii. 505, *not of Thunb.*

COAST REGION: Caledon Div.; Klein River Mountains, 1000–3000 ft., *Ecklon & Zeyher* (in Herb. Berlin). Swellendam Div.; mountains near Puspas Vley, 1000–4000 ft., *Ecklon & Zeyher* (in Herb. Trin. Coll., Dublin).

Somewhat intermediate between *E. chrysocodon* and *E. setulosa,* differing from the former by its sepals and corolla, and from the latter by its leaves and corolla. It seems possible that all three may be states of one variable species; but the evidence is scarcely strong enough at present.

143. E. setulosa (Benth. in DC. Prodr. vii. 682); dwarf; branches straggling, setaceous-pilose; leaves 4-nate, narrow-oblong to oblong, obtuse, minutely puberulous, and sparsely setaceous-ciliate with long pallid hairs, 1¼ lin. long including the pubescent petiole; flowers 4-nate; pedicels puberulous, under 1 lin. long; bracts approximate, small; sepals ovate-lanceolate, acuminate, glandular-ciliate, pubescent, foliaceous, coloured, ½ lin. long; corolla obconic, widened to the mouth, puberulous or somewhat tomentose, about 1½ lin. long; segments erect, subobtuse, ⅓–½ the length of the tube; filaments slender, nearly straight; anthers included, dorsifixed near the base, oblong, obtuse, about ⅜ lin. long, muticous; pore about ½ the length of the cell; style shortly exserted; stigma capitellate; ovary subspherical, glabrous.

SOUTH AFRICA: without locality, *Niven!* in Herb. Kew.

We have dissected and examined flowers from the type, which consists of a branch 6 in. in length.

144. E. pusilla (Salisb. in Trans. Linn. Soc. vi. 374); erect, 1–2 ft. high; branches spreading or subvirgate, puberulous, and also pilose (intermixed with longer hairs); leaves 4-nate, from suberect to spreading, linear, slender, faintly sulcate, roughly hirsute, 1½–2 lin. long; flowers 4-nate; pedicels hirsute, ½–1 lin. long; bracts remote, small, sometimes one, or two, wanting; sepals from ovate to linear, hirsute, ½–1 lin. long; corolla subcampanulate, cyathiform or obconic, widened to the mouth, hirsute, pale rosy, about 1 lin. long; limb ample, slightly spreading or erect, from equal to the tube to about ½ its length; filaments capillary, bent below the anther; anthers included, dorsifixed above the base, oblong, about ⅕ lin. long, about 2½ times as long as broad, aristate; awns slender, rough-edged, straight or divaricate, nearly as long as or a little longer than the cells; style included or rarely just visible, slender; stigma capitellate; ovary glabrous. *Benth. in DC. Prodr.* vii. 675. *E. villosiuscula, Lodd. Bot. Cab. t.* 1844. *E. canescens, Wendl. Eric. Ic. fasc.* 23, 171, *t.* 65?

VAR. β, **micranthera** (Bolus); anther-cells somewhat divaricate, broad-oblong, narrower towards the apex but very obtuse, truncate at the base, $\frac{1}{10}$ lin. long, about twice longer than broad; pore ¾ the length of the cell; awns spreading outwards and backwards, longer than the cell; in all other respects like the type, or the corolla a little narrower than is commonly the case in that.

COAST REGION, ascending from 20 to 5000 ft. : Worcester Div.; Dutoits Kloof, *Bolus in Herb. Guthrie,* 4669! Cape Div.; mountains and flats around Cape Town, *Burke! Bolus,* 4749! 4011! *Guthrie,* 127! *Pappe! Ecklon,* 93! Stellenbosch Div.; Bottelary Hill, *Burchell,* 8364! south-west cascade, *Niven,* 2! Caledon Div.; near Caledon, *Bolus,* 9153! Klein River Mountains, *Zeyher,* 3226! Bredasdorp Div.; near Ratel River, *Schlechter,* 9713! Var. β, Caledon Div.; near Papies Vley, *Schlechter,* 10705!

145. E. distorta (Bartl. in Linnæa, vii. 644); mostly dwarf, diffuse, under 1 ft. high; branches numerous, ascending or divaricate, hirsute; leaves 4-nate, erect-incurved or spreading, linear, sulcate, thick, hirsute, 1½–2½ lin. long; flowers clustered, subcapitate; pedicels ½–1 lin. long or less; bracts remote, small; sepals linear to cuneate, acute, foliaceous, hirsute, under ½ lin. long; corolla urceo-late-broad-cyathiform, pubescent, sometimes prominently veined, not widened to the mouth in the living state (when dried and com-pressed appearing widened), white turning brown, pale rose or crimson, 1–1¼ lin. long, and about as broad; limb usually spreading at maturity, broad and rounded, rarely subacute; anthers included, dorsifixed above the base, oblong, subobtuse, about ⅓ lin. long, sub-cristate; pore ⅗ the length of the cell; appendages from a broader ciliolate base narrowed to subulate-acuminate, about ½ the length of the cell; style exserted; stigma capitate; ovary glabrous. *E. murina, Klotzsch ex Benth. in DC. Prodr.* vii. 679. *E. setacea, E. Meyer ex Benth. l.c.* 679. *E. persoluta, var. hispidula, Benth. l.c.* 679, *partly.*

COAST REGION, at 1500–2500 ft. : Worcester Div.; Dutoits Kloof, *Drège.* Cape Div.; Table Mountain, *Ecklon! Bolus,* 4924! Constantia Berg, *Bodkin in Herb. Bolus,* 8061! *Marloth!* Paarl Div.; Drakensteen Mountains near Bains Kloof, *Bolus,* 4052! French Hoek, *Guthrie,* 3739!

The shape of the corolla is nearly that of *E. pusilla;* but the habit (usually more divaricately-spreading), inflorescence, size of the anthers, and the shape of their appendages, are different. The latter closely resemble those of *E. pubescens* and *E. hirtiflora.* The species does not seem to occur frequently.

146. E. catervæflora (Salisb. in Trans. Linn. Soc. vi. 372); arborescent, 8–12 ft. high; branches hirsute, the older glabrescent;

leaves 4-nate, suberect or spreading-incurved, stout, semiterete, sulcate, hirsute, glabrescent or glabrous, $2\frac{1}{2}$–$3\frac{1}{2}$ lin. long ; flowers terminal and lateral, congested in clusters at the ends of the branches ; pedicels usually hirsute, 1–2 lin. long ; bracts remote and small ; sepals ovate-lanceolate, acute, keel-tipped, ciliate, hirsute, under 1 lin. long ; corolla oblong-tubular, or clavate-tubular, mouth not (or very slightly) contracted, subtetragonous, ribbed, roughly hirsute with tuberculate hairs like those of *E. hirtiflora*, white or pale rose, $2\frac{1}{2}$ lin. long ; segments generally spreading, short, rounded ; filaments capillary, bent below the anther ; anthers included, dorsifixed shortly above the base, broadly oblong or subelliptical, very obtuse, $\frac{1}{4}$ lin. long, subcristate ; pore very wide, more than $\frac{1}{2}$ the length of the cell ; appendages from a broader denticulate base suddenly narrowed to subulate-acuminate, about $\frac{2}{3}$ the length of the cell ; style included, slender ; stigma small, capitellate ; ovary glabrous. *Benth. in DC. Prodr.* vii. 673. *E. catervæfolia, Pers. Syn.* i. 421. *E. pubescens, var. villosa, Thunb. Diss. Erica,* 39, *t.* 5 ; *Wendl. Eric. Ic. fasc.* 5, 11. *E. compacta, Bartl. in Linnæa,* vii. 644.

VAR. β, **glabrata** (Benth. l.c. 673) ; all parts, including the corolla, glabrous. *E. corymbosa, Bartl. in Linnæa,* vii. 648. *E. sieberiana, Klotzsch ex Benth. l.c.* 673. *E. corymbosa, Tausch in Flora,* 1834, 618.

SOUTH AFRICA : without locality, *Masson ! Hesse ! Mund ! Lehmann !* Var. β : *Sieber,* 167 !

COAST REGION : Cape Div. ; Clanwilliam Div., *Leipoldt !* hills below Table Mountain, *Thunberg,* by the Hout Bay stream, 2000 ft., *Marloth in MacOwan & Bolus, Herb. Norm.,* 348 ! Swellendam Div. ; locality and collector unknown, *Cape Herb. !*

A very distinct species and generally recognizable at once by its clustered flowers and rather large, coarse leaves. The anther-crests are almost exactly those of *E. pubescens, L.,* and *E. distorta, Bartl.* It appears to be somewhat rare.

147. E. parviflora (Linn. Sp. Pl. ed. 2, 506) ; erect, 1–3 ft. high ; branches ascending or slender and diffuse, usually with long and short hairs intermixed ; leaves 4-nate, spreading to squarrose, linear, sulcate, closely or sparsely hairy, 1–$2\frac{1}{2}$ lin. long ; flowers 3-nate ; pedicels $\frac{1}{2}$–$1\frac{1}{2}$ lin. long ; bracts remote and small ; sepals linear, or linear-lanceolate, rarely ovate-lanceolate, subfoliaceous, coloured, $\frac{1}{2}$–1 lin. long ; corolla urceolate, mouth slightly contracted, finely puberulous to coarsely hirsute, red or pink, 1–2 lin. long ; segments rounded, generally spreading, about $\frac{1}{4}$ the length of the tube ; filaments slender ; anthers included, dorsifixed at or near the base, oblong, obtuse, aristate, rarely muticous ; pore about $\frac{1}{2}$ the length of the cell ; cell from $\frac{1}{5}$ to over $\frac{1}{4}$ lin. long ; awns linear, rough-edged, from $\frac{2}{3}$ the length of the cell to about as long as it, style included to subexserted ; stigma capitate ; ovary glabrous. *Benth. in DC. Prodr.* vii. 675. *E. pubescens, Bot. Mag. t.* 480, *not of Linn. ; Wendl. Eric. Ic. fasc.* 6, 11. *E. pubescens, var. minima, Andr. Heathery, t.* 39, *and Col. Heaths, t.* 124. *E. tubiuscula,*

Sinclair, Hort. Eric. Wob. 26 ; *Lodd. Bot. Cab. t.* 1157. *E. tardiflora, Salisb. in Trans. Linn. Soc.* vi. 373.

VAR. β, **exigua** (Bolus) ; corolla cyathiform or cyathiform-obconic, mostly narrower and shorter than in the other varieties ; mouth scarcely contracted ; anthers aristate ; ovary glabrous. *E. exigua, Salisb. l.c.* 373.

VAR. γ, **inermis** (Bolus); characters as in β, but anthers muticous. *E. inermis, Klotzsch in Linnæa,* xii. 505.

VAR. δ, **puberula** (Bolus); characters of the type, but differing in having long internodes and tufted foliage, pubescence rough and hispid. *E. tardiflora, Salisb. Herb., scarcely of Trans. Linn. Soc.* vi. **373.** *E. puberula, Bartl. in Linnæa,* vii. **641,** *not of Klotzsch. E. grisea, Klotzsch ex Benth. in DC. Prodr.* vii. 674.

VAR. ε, **hispida** (Bolus) ; characters of the type, but corolla broader and ovary hispid.

VAR. ζ, **ternifólia** (Bolus) ; leaves 3-nate ; corolla narrow-ovoid-urceolate or narrow-ovoid, puberulous (or perhaps sometimes cyathiform-conical and villous) ; anther suboblong, broader than usual in other varieties, less than ¼ lin. long. *E. angustata, Bartl. in Linnæa,* vii. 651 ?

SOUTH AFRICA : without locality, the type and vars. β and δ, *Drège! Herb. Salisbury!* and *cultivated specimens!* Var. ε : *Bolus,* 4174 !

COAST REGION : Worcester Div. ; near Worcester, *Rehmann,* 2527! Cape Div.; hills and flats around Cape Town, *Thunberg, Burchell,* 70! 100! 312! 817! 8440! *Niven,* 28! *Zeyher,* 1099! *Bolus,* 3707! *Sieber,* 75! 198! Caledon Div.; Houw Hoek, *Zeyher,* 3206! near the Palmiet River, *Bolus,* 4174! Zwart Berg, *Bolus,* 6748! Swellendam Div.; near Swellendam, *Mund,* 51! near the Zondereinde River, *Zeyher,* 3225! Var. β : Paarl Div. ; near Paarl, *Burchell,* 951! Caledon Div.; near Villiersdorp, *Bolus,* 5179 ! Bredasdorp Div. ; between Elim and Napier, 8458! *Schlechter,* 7692 ! Var. γ : Caledon Div. ; near Grietjes Gat, *Ecklon & Zeyher* (ex *Klotzsch*). Var. δ : Worcester Div. ; Breede River Valley, near Bains Kloof, *Bolus,* 2958! *Guthrie,* 2076 ! Cape Div.; Cape Flats, *Ecklon* (ex *Bartling*). Var. ε : Stellenbosch Div. ; Lowrys Pass, *Schlechter,* 7809 ! Var. ζ: Paarl Div. ; near Bains Kloof, *Bolus,* 4050 ! Tulbagh or Ceres Div. ; Mitchells Pass, *Schlechter,* 8954 !

Somewhat resembling *E. pusilla* but with longer and narrower corollas ; also in some of its forms approaching *E. intervallaris.*

148. E. intervallaris (Salisb. in Trans. Linn. Soc. vi. 375) ; erect, 1–1½ ft. high ; branches many, virgate, slender, pubescent or pilose ; leaves 4-nate, incurved-erect to squarrose, narrow-linear, sulcate, pubescent, glabrescent, 1–1½ lin. long ; flowers 4-nate, mostly densely crowded on short branchlets along the branches, forming a pseudo-raceme ; pedicels very short ; bracts remote, small ; sepals linear-lanceolate or lanceolate, rarely broad-ovate, acuminate, glabrous, ciliate, or sometimes naked, usually smooth and glossy, coloured, 1–1¼ lin. long ; corolla narrow-cyathiform, or suburceolate-cyathi-form, mouth scarcely contracted, tetragonous, glabrous or more rarely with a fine and scanty pubescence, bright rosy-red, 1½–2 lin. long ; segments erect or spreading, rounded, ¼–½ the length of the tube ; anthers included, dorsifixed above the base, narrow-oblong, obtuse, about ⅓ lin. long, aristate ; awns slender, rough-edged, ½ the length of the cell ; style included ; stigma capitate ; ovary glabrous or puberulous. *E. alopecuroides, Wendl. Eric. Ic. fasc.* 20, 119, *t.* 45, *not of Lodd.*

VAR. β, **grandiflora** (Bolus); corolla suburceolate-oblong, about 2 lin. long, over 1 lin. in diam. ; anthers nearly ½ lin. long ; style 4-angled.

Coast Region, at 1000–4800 ft.: Cape Div.; near Wynberg, *Niven*, 27!
Stellenbosch Div.; mountains of Lowrys Pass, *Burchell*, 8211! *Niven*, 26!
MacOwan, Herb. Aust.-Afr., 1773! Caledon Div.; Houw Hoek Mountains,
Burchell, 8141! *Schlechter*, 7780! Hottentots Holland Mountains, eastern side,
Guthrie, 2048! *Bolus*, 5549! Zoetemelks Vlei, *Grisbrook!* near Appels Kraal,
Zeyher, 3205β! Genadendal Mountain, *Galpin*, 3704! Bredasdorp Div.; near
Elim, *Guthrie!* Var. β: Caledon Div.; near the Palmiet River, *Guthrie*,
3548!

This species is so near to some forms of *E. parviflora* that we admit it with
some doubt. Yet even as a variety of that species it would be well marked by
its usually larger and more membranous sepals, its much less hairy leaves, sepals,
and (usually quite glabrous) corolla, its pseudo-racemose inflorescence, and its
more virgate graceful habit. It is usually found, and often in abundance,
growing along the margins of streams, or watercourses which are streams in
winter, in which respect also it differs from the preceding. Our var. β looks
different in habit from the ordinary form, and seems to connect with *E.
cyrillæflora*. But the anther is exactly that of this species, and as we have
only a single specimen we do not venture to make a species of it.

149. E. cyrillæflora (Salisb. in Trans. Linn. Soc. vi. 357); dwarf,
diffuse, under 1 ft. high; branches very slender, numerous, pubescent, red when young; leaves 4-nate, spreading-incurved, slender,
linear-subterete, puberulous or glabrescent, $2\frac{1}{2}$–3 lin. long; flowers
1–3–4-nate, on short branchlets; pedicels red with white pubescence,
very short; bracts remote, minute; sepals linear from an ovate
ciliate membranous base, acuminate, foliaceous in the upper part,
1 lin. long; corolla subtubular or tubular-campanulate, mouth not
(or slightly) contracted, pubescent, red, 3 lin. long, about 1 lin.
wide; segments small, spreading or erect, rounded; filaments long,
flexuous; anthers just included, dorsifixed above the base, oblong,
obtuse, $\frac{2}{5}$ lin. long, muticous; style included; stigma very small,
capitellate; ovary glabrous. *E. palustris, Andr. Heathery, t. 77,
and Col. Heaths, t. 114; Benth. in DC. Prodr. vii. 682; Lodd.
Bot. Cab. t. 4.*

Coast Region: Cape Div.; marshy plain on Simons Bay Mountains, *Niven*,
28! 99! Ruined Cottage Marsh, *Grey!* Stellenbosch Div.; Hottentots Holland,
Mulder (ex *Salisbury*)! Also *cultivated specimens!*

A very distinct species approaching nearest to the preceding, but of very
different habit. It is probably now very rare: no records of any more recent
collectors are known to us.

150. E. algida (Bolus in Journ. Bot. 1894, 238); erect, 1 ft.
or more high; branches erect, rigid, pubescent, sometimes (as also
the pedicels and the ciliate hairs of the leaves) with compound
subplumose or floccose hairs; leaves 4-nate, erect, incurved, imbricate, linear to lanceolate or oblong, sulcate, the lower sometimes
slightly open-backed, puberulous and ciliate with branched or subplumose hairs, the older glabrescent, $1\frac{1}{2}$–2 lin. long; flowers sub-
4-nate; pedicels hairy, 1–$1\frac{1}{2}$ lin. long; bracts approximate, lanceolate, villous, viscid, nearly as large as the sepals; sepals ovate, obtuse,
subfoliaceous, viscid, villous, glandular-ciliate, nearly 1 lin. long;
corolla broadly oval or oval-cyathiform, mouth not much contracted,

coarsely pubescent, pale red or dull purple, nearly 2 lin. long; segments suberect, broad, rounded, about ½ lin. long; filaments slender, straight; anthers subincluded (just visible above the tube), dorsifixed close to the base, obovate-oblong, obtuse, scaberulous, dark-coloured, slightly over ½ lin. long, aristate; awns subulate, rough, less than ½ the length of the cells; style subexserted; stigma capitate; ovary densely woolly with white hairs.

KALAHARI REGION: Orange River Colony; on the higher slopes of Mont-aux-Sources, 8000–9000 ft., *Flanagan*, 2030! *Thode*, 33! Nelsons Kop, *Cooper*, 854! Basutoland; Machacha, *Bryce.*

EASTERN REGION, at 7000–10000 ft.: Griqualand East; summit of Mount Currie, *Tyson*, 1193! 1769! Natal(?); near the source of the Umtjesi River, *Thode*, 63!

151. **E. turgida** (Salisb. in Trans. Linn. Soc. vi. 345, not of Link); erect, 12–18 in. high; branches many, spreading, hirsute, reddish, somewhat slender; leaves 4-nate, incurved-erect, equalling or often shorter than the internodes, linear, blunt, sulcate, hirsute, 1–2 lin. long; flowers 4-nate, subcapitate; pedicels ½ lin. long; bracts subapproximate to remote, one long and leaf-like on a slender petiole, long hirsute; sepals linear, long ciliate, ⅔ lin. long; corolla urceolate to cyathiform, mouth contracted or only slightly so, hispidulous, red, 1½ lin. long; segments erect or subpatent, rounded, ¼ the length of the tube; filaments capillary, straight; anthers exserted, dorsifixed above the base, ovate-oblong, with a sharp anterior point at the base, over ⅓ lin. long, cristate-aristate; awns from a broader denticulate base narrowing to subulate, acuminate, rough-edged, ¾ the length of the cell; style exserted, slender; stigma large, capitate; ovary hispidulous. *Benth. in DC. Prodr.* vii. 675. *E. mellifera, Link, Enum. Hort. Berol.* i. 370. *E. fusco-rubens, Roxb. ex Salisb. in Trans. Linn. Soc.* vi. 345.

SOUTH AFRICA: without locality, *Roxburgh!* *Harvey*, 174! 187! *Herb. Salisbury!* and *cultivated specimens!*
COAST REGION: Cape Div.; Cape Flats, near Wynberg, 60 ft., *Niven*, 47! *Bolus*, 7100! 7100b! and in *Herb. Norm. Aust.-Afr.*, 1302! (erroneously marked "*E. fimbriata, Andr. ?*").

Allied by its exserted anthers to § *Pyronium*, and by its capitate inflorescence to § *Pseuderemia*; but best placed, as by Bentham, in this section.

152. **E. nidularia** (Lodd. Bot. Cab. t. 764); branches slender, puberulous, reddish; leaves 4-nate, spreading or incurved, linear-oblong, obtuse or subacute, widely sulcate, or occasionally open-backed, pubescent or (according to *Bentham*) subglabrous and somewhat hoary, 2 lin. long; flowers in clusters of 6–8; pedicels slender, 1–1½ lin. long; bracts remote, linear, rather long; sepals linear-subulate or lanceolate, acute or obtuse, puberulous, 1–1¼ lin. long; corolla urceolate-campanulate or cyathiform, mouth little or not at all constricted, finely puberulous, white, 1½–2 lin. long; limb erect or slightly spreading; anthers included, dorsifixed shortly above the base, cells deeply parted, longitudinally semiovate, the dorsal edge

straight, smooth, somewhat membranous and pallid, nearly $\frac{2}{5}$ lin. long, aristate; pore about $\frac{1}{3}$ the length of cell; awns somewhat subulate, rough, about $\frac{1}{2}$ the length of the cell; style included; stigma capitate; ovary (according to *Bentham*) puberulous. *Benth. in DC. Prodr.* vii. 678.   *E. nidicularia, G. Don, Gen. Syst.* iii. 799. *E. nudiflora* and *E. marioides, Hort. ex Benth. l.c.   E. multumbraculata, Tausch in Flora,* 1839, 635.

SOUTH AFRICA: ex *Loddiges; cultivated specimens!*

This species has been founded upon garden specimens, from which alone it is known to us.   It may be allied to *E. turgida* or *E. distorta*, but it seems so distinct that we are unable to find a close connection with any species.   Bentham thought it allied to *E. persoluta;* and also suggested that it might be a garden hybrid.   Our description is drawn partly from Bentham's, partly from a garden specimen named by him, and partly from Loddiges' figure; and even these do not well agree.   We have also seen specimens at the British Museum marked "Hort. Reg. Bot. Berol."   The type of *E. multumbraculata, Tausch,* is a garden specimen in Herb. Royal Bot. Inst. Prague, kindly sent us for examination, and which we find almost identical.

153. E. oresigena (Bolus in Journ. Bot. 1894, 238); erect, rigid, branched, more or less hoary-pubescent in most parts; branches stout; leaves 4-nate, from suberect to subpatent, imbricate, narrow-obovate-oblong, very obtuse, sulcate, thick, cano-puberulous, glandular-ciliate with short rigid hairs, $2\frac{1}{2}$–3 lin. long; flowers 4-nate; pedicels glandular-pilose, 3–$4\frac{1}{2}$ lin. long; bracts remote, unilateral, oblong, cano-pubescent, rigidly glandular-ciliate, or more rarely ciliate with plumose hairs, $1\frac{1}{2}$–2 lin. long; sepals like the bracts but more lanceolate and longer, 2–$2\frac{1}{2}$ lin. long; corolla urceolate, mouth not very much contracted, minutely puberulous, reddish, $2\frac{1}{2}$–4 lin. long; segments erect, broad and rounded, about $\frac{1}{4}$ the length of the tube; filaments tapering upwards, slightly bent at the apex; anthers included, lateral, dorsifixed near the base, cuneate-oblong, obtuse, over $\frac{1}{3}$ lin. long, cristate; pore $\frac{1}{3}$ the length of the cell; crests broad and lacerate at the base, with a terminal subulate lobe, the whole about $\frac{3}{4}$ the length of the cell, nearly white; style included; stigma capitate or clavate; ovary thinly hispid on the top, or glabrous.

VAR. β, **intermedia** (Bolus); leaves and sepals as in the type; corolla ovoid-urceolate, glabrous or nearly so; mouth very much contracted; segments rounded, spreading, short.

VAR. γ, **mollipila** (Bolus); leaves and sepals ciliate with soft subplumose hairs; corolla urceolate, glabrous, or nearly so; mouth not much contracted; segments erect, about $\frac{1}{4}$ the length of the tube, subacute; or, corolla pubescent, with short and spreading segments as in var. β.

COAST REGION, at 5500–6000 ft.: Tulbagh Div.; amongst rocks on Winterhoek Mountain, *Marloth,* 1645! Tulbagh or Ceres Div.; on Mosterts Hoek Mountain, *Bolus in Herb. Guthrie,* 4259! Var. γ, Worcester Div.; on the Matroos Berg, *Bolus!* Clanwilliam Div.; Sneeuw Kop, Cederberg Range, near Wupperthal, *Bodkin in Herb. Bolus,* 6492!

CENTRAL REGION: Var. β: Ceres Div.; Table Mountain, in the Cold Bokkeveld, 6200 ft., *Schlechter,* 10086!

A somewhat variable species, yet well-marked in its chief characters. It is quite confined to the higher mountains.

154. E. modesta (Salisb. in Trans. Linn. Soc. vi. 352); erect, dwarf; stem stout and woody, probably under 1 ft. high; branches numerous, flexuous, sometimes intricate, covered with persistent leaf-cushions, the younger (also the pedicels) floccose with small subplumose or minutely barbellate hairs, younger leaves, bracts and sepals similarly ciliate; leaves 3-nate (also 4-nate, according to *Bentham*), erect-spreading, closely imbricate, sexfarious, oblong, sub-obtuse, thick, round-backed, sulcate, the younger ciliate and incano-pulverulent, glabrescent, $1–1\frac{1}{2}$ lin. long; flowers 3-nate; pedicels floccose, $1\frac{1}{2}–2$ lin. long; bracts closely approximate, ovate, keel-tipped, cartilaginous, pubescent, ciliate, very concave, pale rosy, about 1 lin. long; sepals like the bracts, but somewhat larger, $1\frac{1}{2}–2$ lin. long, about $\frac{1}{2}$ the length of the corolla; corolla ovoid-urceolate, tapering somewhat gradually to the contracted throat, or tubular-urceolate, and then only very slightly contracted at the throat, pubescent, pale rosy, 3–4 lin. long; segments spreading, rounded, about $1\frac{1}{4}$ lin. long; filaments slender, sigmoid below the anther; anthers included, lateral, dorsifixed well above the base, cuneate-oblong, subobtuse, dark-coloured, scaberulous, a little over $\frac{1}{2}$ lin. long, aristate; pore $\frac{2}{5}$ the length of the cell; awns narrow-subulate, ciliolate, about $\frac{1}{2}$ the cells in length; style included, or at length manifest; stigma capitate; ovary very pale, glabrous. *Benth. in DC. Prodr.* vii. 655. *Raspalia angulata, E. Meyer in Drège, Zwei Pflanzengeogr. Documente,* 215; *and see Harv. & Sond. Fl. Cap.* ii. 320.

COAST REGION: Caledon Div.; tops of the mountains of Baviaans Kloof, near Genadendal, 2200–5000 ft., *Niven,* 53! 212e! *Masson,* 14! *Burchell,* 7711! *Drège, Bolus,* 5425! *Bodkin in Herb. Guthrie,* 3620! and in *Herb. Bolus,* 6957! *Schlechter,* 9859!

Bentham quotes *E. leucophylla, Klotzsch,* with a sign of doubt, as a synonym for this species. We have not seen it (nor had Bentham), but from the description it seems to be nearer to *E. oresigena,* and may possibly be our var. β of that species, yet there are differences which prevent identification.

155. E. Flanagani (Bolus in Journ. Bot. 1894, 238); erect, stout, rigid, branched, 2–3 ft. high; stem 3 lin. in diam.; branches subflexuous, pubescent, densely leafy; leaves 4-nate, erect, crowded, linear, subobtuse, sulcate, glabrous, 4–5 lin. long; flowers 4-nate, subcernuous; pedicels thick, 2 lin. long; bracts approximate, broad-linear, thinly pubescent, coriaceous, thick, pallid, $3–3\frac{1}{2}$ lin. long; sepals like the bracts, 3 lin. long; corolla urceolate or ovoid-urceo-late, mouth contracted, densely white-pubescent, white, about 4 lin. long; segments erect, rounded with imbricate margin, scarcely $\frac{1}{5}$ the length of the tube; filaments rather broad, bent at the apex; anthers included, affixed above the base, narrow-oblong, slightly curved, pale brown, over 1 lin. long, cristate-aristate; awns from an oblong denticulate base, suddenly contracted to subulate-acuminate, $\frac{2}{3}$ or $\frac{3}{4}$

the length of the cell ; style stout, included ; stigma simple ; ovary deeply lobed, pubescent.

KALAHARI REGION : Orange River Colony ; on the landward slopes of the Mont-aux-Sources, above Elands River Valley, 8000 ft., *Flanagan*, 2031 !

A very distinct species, somewhat resembling *E. triflora* in external appearance, and unlike any other in this section.

156. E. hirtiflora (Curt. Bot. Mag. t. 481) ; erect, 1–2 ft. high ; branches stout, hirsute ; leaves 4-nate, incurvo-patent to squarrose, linear, obtuse, sulcate, hirsute and rough with minute whitish tubercles at the base of the hairs, 1–2 lin. long ; petiole somewhat long ; flowers 4-nate ; pedicels hirsute, $1\frac{1}{2}$–$2\frac{1}{2}$ lin. long ; bracts remote, small ; sepals lanceolate, hirsute, foliaceous, $\frac{1}{2}$–1 lin. long ; corolla ovoid-urceolate, or more rarely broad-urceolate, throat contracted, roughly tuberculate-hirsute (as the leaves), pale purple, usually darker at the apex, $1\frac{1}{2}$–2 lin. long ; segments spreading, about $\frac{1}{4}$ or more rarely $\frac{1}{2}$ the length of the tube ; filaments slender ; anthers included, dorsifixed, ovate-cuneate, subacute or obtuse, sparsely hispidulous below, $\frac{1}{5}$ lin. long, aristate or subcristate ; pore narrow-elliptical, less than $\frac{1}{2}$ the length of the cell ; awns broad-linear, pale brown, ciliate with long thickish hairs, as long as or longer than the cell ; style included, slender ; stigma capitate ; ovary glabrous. *Benth. in DC. Prodr.* vii. 674. *E. pubescens, Andr. Heathery, t.* 37, *and Col. Heaths t.* 48 ; *Lodd. Bot. Cab. t.* 167, *not of Linn. E. pubescens, var. pilosa, Wendl. Eric. Ic. fasc.* 15, 41, *t.* 16. *E. pubescens, var. hispida, Thunb. Diss. Erica,* 39, *and of his Herb. (according to Rach). E. mitræformis, Salisb. in Trans. Linn. Soc.* vi. 372. *E. tardiflora, Salisb. Herb., but scarcely of his descr. l.c.* 373 (*according to Benth.*). *E. puberula, Bartl. in Linnæa,* vii. 644, *not of Klotzsch. E. grisea, Klotzsch, ex Benth. l.c.* 674. *E. minima, Pritz, Ic. Bot. Ind.* i. 417. *E. ovata, Lodd. Bot. Cab. t.* 417.

COAST REGION : Malmesbury Div. ; Saldanha Bay, *Niven,* 24 ! Cape Div. ; mountains near Cape Town, *Thunberg, Ecklon,* 291 ! *Burchell,* 627 ! 8407 ! *Bolus,* 3296 ! 3718 ! and in *Herb. Norm. Aust-Afr.,* 33 ! *Guthrie,* 1450 ! 1177 ! *Wolley Dod,* 588 ! 824 ! Simons Bay, *Milne,* 110 ! *MacGillivray,* 447 ! Caledon Div. ; Nieuw Kloof, Houw Hoek Mountains, *Burchell,* 8142 ! Swellendam Div. ; near the Zondereinde River, *Zeyher,* 1098 !

Nearest to *E. mollis.* This is the only heath known to us which, by its sub-social growth, gives, in some favourable seasons, a colouring to the eastern slopes of Table Mountain which may be seen at a considerable distance, like that of the similar *E. cinerea, L.,* on the Scottish mountains. Bentham makes two varieties, dependent on the size of the flower ; but we find intermediate sizes. The synonymy, as Bentham observes, is very confused. We have done our best to unravel it, but have been obliged, in great part, to follow him.

157. E. mollis (Andr. Heathery, t. 272) ; erect, 1–$1\frac{1}{2}$ ft. high, all parts more or less hirsute with rather long hairs ; leaves 4-nate, from suberect to squarrose, linear, subterete, obtuse, obscurely sulcate, somewhat rough from the tubercles at the base of the hairs, about $2\frac{1}{2}$ lin. long ; flowers 4-nate, clustered ; pedicels scarcely 1 lin. long ;

bracts remote, small; sepals lanceolate, hirsute, long-ciliate, coloured, under 1 lin. long; corolla globose-urceolate, contracted at the throat, coarsely hirsute, dull red, $1\frac{1}{2}$ lin. long; segments spreading, rounded, less than $\frac{1}{2}$ the length of the tube; filaments slightly bent at the apex; anthers included, laterally basifixed, trigonous, sparsely hispidulous, about $\frac{1}{6}$ lin. long, aristate; pore orbicular, about $\frac{1}{2}$ the length of the cell; awns straight, linear, closely serrulate, as long as the cell; style mostly included, rarely subexserted; stigma large, subpeltate; ovary glabrous. *Andr. Col. Heaths*, t. 247; *Benth. in DC. Prodr.* vii. 674. *E. ramosissima, Wendl. Eric. Ic. fasc.* 18, 93, t. 36. *E. modesta, Sinclair, Hort. Eric. Wob.* 15, *not of Salisb. E. albiflora, Klotzsch in Herb. Berlin. E. pubescens, Sieber ex Benth. l.c., not of Linn.*

SOUTH AFRICA: without locality, *Thunberg, Sieber*, 171! and *cultivated specimens!*

COAST REGION: Cape Div.; Devils Peak, *Niven*, 30! Table Mountain, 3000 ft., *Bolus*, 4618! Constantia Mountain, *Wolley Dod*, 3427!

A species very easily recognized by its stigma being unusually large for the section, and its rather wide flowers, which, in drying, commonly lie flattened from above, so that the mouth is exhibited open.

158. E. ribisaria (Guthrie & Bolus); erect, about $1\frac{1}{2}$ ft. high; branches many, weak-growing and slender, sparsely hirsute, glabrescent, reddish; leaves 4-nate, spreading, narrow-linear, subterete, blunt, sulcate, hispidulous, 1–2 lin. long; internodes usually becoming much longer than the leaves; flowers 4-nate; pedicels very slender, $2–2\frac{1}{2}$ lin. long; bracts remote, infra-median, very small; sepals linear from a broader base, foliaceous at least in the upper portion, pubescent, $\frac{1}{2}–\frac{3}{4}$ lin. long; corolla ovoid to globose-urceolate, contracted at the mouth, thinly puberulous, dry, $1\frac{1}{2}–2$ lin. long, bright crimson; segments erect, connivent, broad and rounded, very short, darker than the tube; filaments slender; anthers included, dorsifixed, ovate-cuneate, very obtuse, about as broad at the base as their length, sparsely hispidulous, about $\frac{1}{5}$ lin. long, aristate; pore wide, about $\frac{1}{2}$ the length of the cell; awns linear, straight, somewhat long ciliate, about as long as the cells (the anther much resembling that of *E. mollis*, but slightly larger and with longer hairs on the awns); style included, straight; stigma small, capitellate; ovary hispidulous.

COAST REGION: Caledon Div.; Houw Hoek Mountains, *Zeyher*, 3227! in Herb. Berlin, *Guthrie*, 2283! *Schlechter*, 7360!

This species occurs in a well-searched locality, and has probably only escaped examination on account of the resemblance of its flowers to those of *E. ramentacea*, from which in some other respects (especially its habit) it is very different. It appears to be very local.

159. E. oophylla (Benth. in DC. Prodr. vii. 672); 3–5 in. high; branches puberulous; leaves 4-nate, spreading, ovoid or subterete, very thick, obtuse, obscurely sulcate, densely hispidulous, scarcely 1 lin. long ; flowers 3-nate ? or umbellate ? ; pedicels puberulous, $1\frac{1}{2}$–2 lin. long ; bracts remote, linear, about 1 lin. long ; sepals lanceolate-oblong, obtuse, puberulous, viscid, coloured, 1 lin. long ; corolla urceolate ("ovate," *Bentham*), throat somewhat contracted, pubescent, viscid, about $1\frac{1}{4}$ lin. long ; segments erect or perhaps spreading, shortly acuminate, darker coloured, about $\frac{1}{3}$ the length of the tube ; filaments rather broad, bent at the apex ; anthers included, dorsifixed, oblong, obtuse, over $\frac{1}{3}$ lin. long, crested ; pore less than $\frac{1}{2}$ the length of the cell ; crests ovate, ciliolate, about $\frac{1}{2}$ the length of the cell ; style included ; stigma subsimple ; ovary short, turbinate, pilose chiefly towards the summit.

COAST REGION : Swellendam Div. ; on the slopes and summit of the great mountain peak near Swellendam, *Burchell*, 7326 ! 7347 ! (not 1347 as quoted by *Bentham*) 7403 !

It seems strange that this species, so well-marked by its leaves and peculiar corolla-limb, should have escaped all collectors but Burchell. But the locality has not been well explored.

160. E. trichadenia (Bolus in Journ. Linn. Soc. xxiv. 183) ; erect, much-branched, rough, 2–3 ft. high ; branches flexuous or divaricate, pubescent and glandular-setose ; leaves 3-nate, spreading or squarrose, linear or linear-lanceolate, acute, sulcate to open-backed, puberulous and ciliate with rigid gland-tipped hairs, $2\frac{1}{2}$–$3\frac{1}{2}$ lin. long ; flowers umbellate, 3–7-flowered ; pedicels glandular-pubescent, 2–$2\frac{1}{2}$ lin. long ; bracts remote, mostly very small, sometimes one larger and foliaceous, or solitary ; sepals linear-oblong, acute, pubescent and gland-ciliate, 1 lin. long ; corolla narrow-campanulate, mouth slightly widened, sparsely and very minutely hispidulous, pale flesh-coloured, $1\frac{1}{2}$–$2\frac{1}{2}$ lin. long ; segments rounded, about $\frac{1}{3}$–$\frac{1}{2}$ the length of the tube ; anthers included, subterminal, oblong, tapering much to the filament at the base, $\frac{1}{2}$ lin. long, subulate-aristate ; awns smooth ; style exserted, rather stout, curved over to an angle of 90°, red ; ovary pubescent.

COAST REGION : Ceres Div. ; mountain slopes above Mitchells Pass, near Ceres, 2000 ft., *Bolus*, 5297 !

161. E. eriocodon (Bolus in Journ. Linn. Soc. xxiv. 186) ; erect, somewhat robust, about 1 ft. high ; branches virgate, pubescent and also pilose with longer and sometimes glandular hairs ; leaves 3-nate, spreading, crowded, linear to lanceolate, from sulcate to open-backed, pubescent and sparsely glandular-setose, ciliate, grey, $1\frac{1}{2}$–2 lin. long ; flowers 3-nate ; pedicels less than 1 lin. long ; bracts subapproximate, minute ; sepals ovate or ovate-lanceolate, pubescent and glandular-ciliate, foliaceous, about $\frac{3}{4}$ lin. long ; corolla broad-cyathiform, mouth neither contracted nor much widened, pubescent, somewhat viscidulous, 1 lin. long and a little wider ; segments broad, nearly flat, erect or at length incurved, about $\frac{1}{2}$ as long as the tube ; filaments capillary ; anthers included, obovate (in front view nearly orbicular), about $\frac{1}{5}$ lin. long, muticous ; style shortly exserted ; stigma capitate ; ovary hirsute. *E. eriopodon, Ind. Kew. Suppl. i.* 156.

COAST REGION : Tulbagh Div.; on the slopes of the Winterhoek Mountain, 1500 ft., *Bolus,* 5190!

This has somewhat the habit and appearance of *E. bicolor,* but the flowers are usually fewer and less prominent. It is very distinct in this section by its anthers : none of those which are muticous being so small and few even of those which are appendiculate.

162. E. patens (Andr. Heathery, t. 133); erect, "2 ft. high";

branches pubescent and pilose ; leaves 3-nate, spreading, elliptic-oblong, acute, open-backed, puberulous, sparingly gland-ciliate, $1\frac{1}{2}$ lin. long; flowers 3-nate ; pedicels pubescent, $1\frac{1}{4}$ lin. long ; bracts remote, small; sepals lanceolate-ovate, acute, foliaceous, pubescent, gland-ciliate, less than 1 lin. long ; corolla broadly obconic, mouth widened, puberulous, rosy red, 2 lin. long ; segments subacute or rounded and obtuse, erect or spreading, about equal to the tube ; anthers included, semi-ovate, sublateral, lobed above the base, dark-coloured, scabrid, under $\frac{1}{2}$ lin. long, aristate ; awns subdecurrent, subulate, ciliolate, short ; ovary densely villous with long hairs. *Andr. Col. Heaths, t.* 115; *Wendl. Eric. Ic. fasc.* 26, 7 ; *Lodd. Bot. Cab. t.* 1228 ; *Benth. in DC. Prodr.* vii. 686. *E. expansa, Hort. ex Benth. l.c.*

SOUTH AFRICA : without locality, *Niven* (ex *Loddiges*), *cultivated specimens!*

This species is only known to us from the descriptions, figures, and several garden specimens in Herb. Kew. It is very distinct and peculiar, having the corolla-form of § *Eurystoma,* but in every other respect the characters of this section. The anther is unusual in shape, tapering very much to the base, but scarcely subterminal.

163. E. perlata (Sinclair, Hort. Eric. Wob. 18); erect, 1–2 ft.

high ; branches ascending, often straggling, tomentose and glandular-pilose ; leaves 4-nate, squarrose, narrow-oblong, or the shorter subovate, open-backed, obtuse, gland-pilose, 2–3 lin. long; flowers umbellate ; umbels 4–10-flowered, often clustered ; pedicels 2–3 lin. long ; bracts remote ; sepals lanceolate-oblong, pubescent and gland-ciliate with long hairs, about 1 lin. long, pedicels, bracts and sepals all glandular-pilose ; corolla oblong or cyathiform-campanulate, mouth not contracted, pubescent, yellowish, about $1\frac{1}{2}$ lin. long; segments spreading, more than $\frac{1}{2}$ as long as the tube ; anthers subexserted, oblong, narrowed to the base, minutely aristulate or sometimes muticous (?) ; awns not reaching beyond the base of the cell, or sometimes with a minute tooth on either side of the filament near its apex; style exserted ; stigma clavate-capitate ; ovary lanate. *Benth. in DC. Prodr.* vii. 670. *E. barbata, var. minor, Andr. Heathery, t.* 5, *and Col. Heaths, t.* 78. *E. barbata, Drège in Linnæa,* xx. 187. *E. pura, Lodd. Bot. Cab. t.* 72? *E. procumbens, Lodd. l.c. t.* 1993? *E. ephemera, Tausch, in Flora,* 1839, 635.

COAST REGION: Caledon Div.; near the Zondereinde River, *Zeyher,* 3243! *Schlechter,* 5642! mountains near Genadendal, 1500 ft., *Drège, Galpin,* 3663!

E. pura, Lodd., and *E. procumbens,* Lodd., were described and figured from garden specimens, and may have been hybrids. No wild specimens have been seen. They are clearly closely related here; and the garden specimens preserved show this even more fully than the plates.

164. E. barbata (Benth. in DC. Prodr. vii. 671); characters of *E. perlata,* but anthers crested, crests curved and about $\frac{1}{3}$ the length of the cell; whole plant somewhat more robust, and larger in all its parts; indumentum longer, coarser and more copious; stamens more included, though sometimes visible; the lower bract sometimes wanting. *E. barbata, var. major, Andr. Heathery, t. 4, and Col. Heaths,* 77; *Lodd. Bot. Cab. t.* 124. *E. pannosa, Salisb. in Trans. Linn. Soc.* vi. 339.

SOUTH AFRICA: without locality, *Masson! Niven! Bowie.* Also *cultivated specimens!*

COAST REGION: Caledon Div.; mountains near Genadendal, *Burchell,* 7775! *Bolus,* 5433! *Schlechter,* 7752! near Zoetemelks Vlei, *Grisbrook in Herb. Guthrie,* 3282! 3283!

The only absolute character distinguishing this from the preceding is the anther-crests, which can hardly be regarded as of specific value. The plants have, however, a somewhat different aspect, and we follow Bentham, though with doubt, in keeping them separate.

165. E. fausta (Salisb. in Trans. Linn. Soc. vi. 326); branchlets rather straight, finely pubescent intermixed with long coarser white hairs, the pubescence said by *Bentham* to be subviscid; leaves 3-nate, spreading, ovate-elliptical to narrow-lanceolate, acute, open-backed, puberulous, setose-ciliate with long hairs, 2–3 lin. long; flowers 3-nate; pedicels $1\frac{1}{2}$ lin. long; bracts approximate, small and slender; sepals subulate, acute, densely setose-hispid, coloured on the lower part, foliaceous above, $1\frac{1}{2}$ lin. long; corolla urceolate-campanulate, puberulous, mouth wide, scarcely contracted at the throat, about 2 lin. long; segments somewhat spreading, about $\frac{1}{2}$ as long as the tube; anthers included, longer than the corolla-tube and manifest, dorsifixed well above the base, oblong, obtuse, about $\frac{2}{3}$ lin. long, cristate; crests sublanceolate, acuminate, distantly toothed on the outer margin, incurved about $\frac{2}{3}$ the length of the cell; style subexserted; stigma capitellate; ovary densely hispid.

SOUTH AFRICA: without locality, *Masson! Niven! Roxburgh! Herb. Salisbury!*

COAST REGION: Stellenbosch Div.; Hottentots Holland, *Mulder* (ex *Salisbury*).

Said by Bentham to resemble in habit *E. marifolia,* but to be more rigid and more branched. It is closely allied to the three next; but our material, with the single exception of *E. Lerouxiæ,* is too scanty to enable us to decide satisfactorily whether or not they should all be regarded as varieties of this, or be maintained as distinct.

166. E. setosa (Bartl. in Linnæa, vii. 646) ; a somewhat straggling small shrub ; branches pubescent and setose with long white hairs ; leaves 3-nate, spreading, broad-lanceolate or elliptic-oblong, acute, very open-backed, the margin only recurved, puberulous, paler beneath, margin ciliate with distant rather long white hairs, and apex similarly tipped, about 2 lin. long, less than 1 lin. wide ; flowers 4-nate or umbellate ; pedicels over 1 lin. long ; bracts approximate or one subremote, linear, ciliate ; sepals subulate, acute, adpressed, flat, not keeled, ciliate with long coarse hairs, 1 lin. long ; corolla urceolate-campanulate, mouth wide, scarcely contracted at the throat, puberulous, $1\frac{3}{4}$ lin. long ; segments wide, rounded and frequently reflexed, $\frac{1}{2}$ as long as the tube ; filaments slightly curved ; anthers included, manifest, oblong, widened and rounded at the apex, over $\frac{1}{2}$ lin. long, crested ; pore wide, over $\frac{1}{2}$ the length of the cell ; crests lanceolate, acuminate, toothed on the outer margin, not spreading, about $\frac{3}{5}$ the length of the cell ; style exserted ; stigma capitellate ; ovary pubescent. *Benth. in DC. Prodr.* vii. 671.

COAST REGION : Stellenbosch Div. ; near Somerset West, *Ecklon & Zeyher,* 167 ! in Herb. Berlin ; Hottentots Holland, *Ecklon* (ex *Bartling*).

Closely allied to the preceding, differing chiefly by its more slender habit, umbellate or 4-nate flowers, much smaller sepals, and less hairy ovary. The habit is that of § *Ceramia ;* the corolla somewhat that of § *Eurystoma.* Ecklon & Zeyher's specimens are mixed (at least as to one sheet, which seems to be E. *barbata*), but we take that numbered 167 to be the type.

167. E. straminea (Wendl. Eric. Ic. fasc. 27, 19) ; erect, 1–2 ft. high ; branches brown, silky ; leaves 3-nate, spreading, ovate or elliptical, flat, the margin only revolute, silky, ciliate, dark green above, pale green below, 2–3 lin. long ; flowers 3–4-nate ; pedicels $2\frac{1}{2}$–3 lin. long ; bracts subremote (about median), like the sepals but narrower and smaller ; sepals ovoid, almost exactly like the leaves, but smaller, about 2 lin. long ; corolla between globose-urceolate and urceolate, throat moderately contracted, pubescent with rather long fine silky hairs, straw-coloured, about $2\frac{1}{2}$ lin. long ; segments slightly spreading, rounded, less than $\frac{1}{3}$ the length of the tube ; anthers included, dorsifixed at or near the base, oblong-cuneate, cells deeply parted, scaberulous, about $\frac{1}{2}$ lin. long, subcristate ; crests narrow, ciliate, about $\frac{1}{3}$ the length of the cell ; ovary silky.

SOUTH AFRICA : without locality, ex *Wendland.*

Not having seen any specimen, we have drawn up the description from the excellent one of Wendland, modified or supplemented by his good figure. The species has been entirely overlooked by all other writers so far as we know.

168. E. Lerouxiæ (Bolus in Journ. Linn. Soc. xxiv. 182) ; erect, stout, branched, 2–3 ft. high ; branches puberulous and sparsely pilose ; leaves 3-nate, spreading, linear-lanceolate, acute, open-backed, puberulous above, velvety or nearly glabrous, and often paler, beneath, long ciliate or naked, 3–4 lin. long ; flowers 3-nate, subcalycine ; pedicels pubescent, 2 lin. long ; bracts subapproximate ; sepals lanceolate, acute, pubescent, coloured, 2 lin. long ; corolla broadish-ovoid-suburceolate, mouth moderately contracted, pubescent, white, about $2\frac{1}{2}$ lin. long ; segments suberect, obtuse, about $\frac{3}{7}$ the length of the tube ; filaments broad-sigmoid below the anther, sparsely pilose and ciliate ; anthers included, oblong, not tapering to the base ; cells bipartite, dark-coloured, $\frac{3}{4}$ lin. long ; crests spreading, ovate in outline, acute, deeply incised, about $\frac{1}{2}$ the length of the cell ; style exserted ; stigma bluntly conical, rather large ; ovary silky-pubescent.

COAST REGION: Paarl Div. ; mountains near Beukenhout Kloof, French Hoek, 1000–1200 ft., *Miss E. Le Roux* in Herb. Huguenot Seminary, 303 ! *Bolus*, 6986 ! and in *Herb. Norm. Aust.-Afr.*, 605 !

The species differs from the three preceding by its more robust habit, larger, coloured, and finely pubescent sepals, ciliate filaments, and somewhat larger anthers which do not at all taper towards the filament. The large size of the sepals, although they are coloured like the corolla, give the flower a somewhat more calycine aspect than is usual in the section.

169. E. caffra (Linn. Sp. Pl. ed. 1, 353, not of Salisb. nor of others) ; erect, 4–12 ft. high, greyish in all parts ; branches virgate or spreading, pubescent ; leaves 3-nate, suberect to squarrose, usually spreading, linear to linear-lanceolate, acute, open-backed, greyish-pubescent, occasionally subglabrous, commonly 4–6 (sometimes 7) lin. long ; flowers 3-nate or umbellate ; pedicels tomentose, about 2 lin. long ; bracts remote, minute, adpressed ; sepals lanceolate, acuminate, margin reflexed at the apex, pubescent, 1–$2\frac{1}{2}$ lin. long ; corolla long-persistent, ovoid-tubular-urceolate, throat contracted, laxly villous, or rarely finely pubescent, most usually white or yellowish-white, $2\frac{1}{2}$–$3\frac{1}{2}$ lin. long ; segments very small, spreading ; filaments broader at the base, tapering upwards ; anthers included, oblong, somewhat wider at the base, pallid, under $\frac{1}{2}$ lin. long, aristate ; awns subulate, about $\frac{1}{2}$ as long as the cell ; style just exserted ; ovary shortly villous or strigose. *E. urceolaris, Berg. Descr. Pl. Cap.* 107 ; *Bauer, Exot. Pl. t.* 16 ; *Salisb. Prodr.* 292 ; *Wendl. Eric. Ic. fasc.* 9, 11 ; *Lodd. Bot. Cab. t.* 1894 ; *Benth. in DC. Prodr.* vii. 671. *E. pentaphylla, Linn. Sp. Pl. ed.* 2, 506. *E. lamellaris, Salisb. in Trans. Linn. Soc.* vi. 327.

VAR. β, **auricularis** (Bolus) ; leaves shorter ; corolla smaller, more glabrous ; awns of the anther minute. *E. auricularis, Salisb. in Trans. Linn. Soc.* vi. 327. *E. urceolaris, var. auricularis, Benth. in DC. Prodr.* vii. 671.

COAST REGION, on mountains, usually near streams, 1000–6000 ft. : Clanwilliam Div., *Leipoldt*, 215 ! Tulbagh Div. ; near streams below Winterhoek, *Thunberg* ; Mitchells Pass, *Bolus*, 2614 ! Paarl Div. ; Paarl Mountain, *Drège*, 7727a ! Cape Div. ; Table Mountain, *Marloth in Herb. Norm. Aust.-Afr.*, 401 ! Caledon Div. ; Genadendal, *Drège !* Swellendam Div. ; near Puspas River, *Thunberg.* George Div. ; between Malgat River and Great Brak River, *Burchell*, 6140 ! Albany Div. ; near Grahamstown, *Burchell*, 3560 ! 3605 ! *Zeyher*, 3204 ! *MacOwan*, 802, 1230 ! Stutterheim Div. ; near Klaklazele River, *Cooper*, 72 ! Queenstown Div. ; Andries Berg, near Bailey, *Galpin*, 2165 !

CENTRAL REGION : Molteno Div. ; Broughton, *Flanagan*, 1611 ! Aliwal North

Div.; Kraai River, *Drège*. Var. *β*: Willowmore Div.; Aasvogel Berg, *Drège*, 7727d!

EASTERN REGION: Tembuland; near Gat Berg, *Baur*, 240! Pondoland; Egossa Forest, *Sim*, 2517! Natal; Umbilo Falls, *Wood*, 1723! 1961! Fields Hill, *Rehmann*, 8018!

Var. *β* scarcely seems worthy of note. The species is widely distributed, and one of the very few which straggle into the Central Region. It is noticeable by its persistent flowers.

170. E. pubescens (Linn. Sp. Pl. ed. 2, 506, not of others); erect, 1–1½ ft. high; branches clothed with a close greyish tomentum, and long scattered hairs; leaves 3-nate, upper suberect, lower spreading, linear, blunt, sulcate, or sometimes lanceolate and open-backed, pubescent, pale green, about 1½ lin. long; flowers 3-nate to sub-umbellate; pedicels pubescent, 3 lin. long; bracts subapproximate or remote, foliaceous, less than 1 lin. long; sepals lanceolate, subobtuse, margins recurved, foliaceous, densely velvety-puberulous, ciliate or naked, 1–1¼ lin. long, reaching from ¼ to nearly ½ the height of the corolla; corolla narrow- or broadish-ovoid-urceolate, not, or slightly, contracted at the throat, velvety-pubescent, white or pale rosy, mostly 2½ or rarely 2 lin. long; segments erect, obtuse, ¼–⅓ the length of the tube; filaments broadest at the base, bent at the apex; anthers included, oblong, obtuse, somewhat tapering to the base, the cells subdistant, about ¾ lin. long, subcristate; pore about ½ the length of the cell; appendages oblong and denticulate below, subulate-acuminate towards the apex, about ½ as long as the cells, or subulate-aristate throughout, ciliolate, and only ⅓ the length of the cells; style stout, just exserted; stigma capitate; ovary villous. *E. pubescens, var. pilosa, Thunb. Diss. Erica,* 39. *E. incana, Wendl. Eric. Ic. fasc.* 18, 89, *t.* 34 (*good!*). *E. pallida, Salisb. in Trans. Linn. Soc.* vi. 326; *Benth. in DC. Prodr.* vii. 673. *E. purialis, Hort. ex Sinclair, Hort. Eric. Wob.* 20; *Benth. l.c.*

SOUTH AFRICA: without locality, *Sieber*, 174! *Herb. Salisbury!*
COAST REGION, from 800–2800 ft.: Clanwilliam Div.; Middle Berg, Cederberg Range, *Leipoldt*, 742! Cape Div.; mountains around Cape Town, *Thunberg, Niven*, 20! 127! *Bolus*, 2959! and in *Herb. Norm. Aust.-Afr.*, 32! *Guthrie*, 1378! *Wolley Dod*, 2336! *Ecklon*, 290!

Allied to *E. tomentosa*, but easily distinguished by the size and relative proportions of the sepals and corolla. Also near to *E. Lerouxiæ*, differing, however, in several characters. Leipoldt's specimens agree in all substantial characters, but the flowers are a little smaller than the others, either from their not being fully developed, or from their station in a drier climate. They were distributed as *E. hirta, Thunb.*, which may possibly prove to be a form of the present species.

171. E. hirta (Thunb. Diss. Erica, 36, t. 2); branches viscous-pubescent and setose; leaves 3-nate, spreading, from linear (margins revolute and approximate) to lanceolate and open-backed, acute, sub-viscous-puberulous, the younger setose, 2–3 lin. long; flowers terminal, 3-nate; pedicels puberulous, 2½ lin. long (in *Salisbury's* specimen, probably from Roxburgh), to 5–7 lin. (in *Thunberg's* figure); bracts remote, small; sepals somewhat lax and spreading, foliaceous, linear or narrow-lanceolate, acuminate, pubescent and setulose, $1\frac{1}{2}$–2 lin. long; corolla broad-urceolate, not much contracted at the throat, pubescent, $2\frac{1}{2}$–3 lin. long; segments spreading, about $\frac{1}{4}$ the length of the tube; filaments narrow-linear; anthers included, lateral, oblong, tapering slightly to the apex, subobtuse (but in front view acuminate), smooth and submembranous, about $\frac{3}{4}$ lin. long, aristate; pore $\frac{1}{2}$ the length of the cell; awns ciliolate about $\frac{2}{5}$ the length of the cell; style included; stigma capitellate; ovary hispid. *Benth. in DC. Prodr.* vii. 672. *E. urceolaris, Salisb. in Trans. Linn. Soc.* vi. 326, *not of Berg. E. dura, Soland. ex Salisb. in Trans. Linn. Soc.* vi. 326.

South Africa: without locality, *Thunberg! Niven! Roxburgh!*

We have seen Thunberg's type, and the specimens of Niven and Roxburgh agree in floral characters. We suspect that the two neighbouring and more recent species (*E. globosa* and *E. aggregata*) are nothing more than forms of this. But all, with the exception of *E. globosa*, are poorly represented in the herbaria to which we have had access, and we are compelled therefore to follow the older authors.

172. E. aggregata (Wendl. Eric. Ic. fasc. 13, 11); erect, 2 ft. high; pubescence soft and short, not rough or glandular; branches spreading-erect, pubescent; leaves 3-nate (or " 3–4-nate," *Wendland*) like those of *E. urceolaris*, but greener and about 3 lin. long; flowers umbellate, or the terminal umbellate, the lower axillary, more or less densely clustered or crowded at the ends of the branches; pedicels 2 lin. long; bracts remote, foliaceous, at the base of the pedicel; sepals lanceolate, acute, foliaceous, pubescent and ciliate, less than $\frac{3}{4}$ lin. long; corolla urceolate, puberulous, rosy red, $1\frac{3}{4}$–2 lin. long; segments spreading, very small; filaments narrow, widened at the apex; anthers included, dorsifixed, oblong, narrowed to the apex, dark-coloured, about $\frac{2}{5}$ lin. long, minutely aristulate; style exserted; stigma capitate; ovary hirsute. *Lodd. Bot. Cab. t.* 1678; *Benth. in DC. Prodr.* vii. 671.

South Africa: according to *Wendland* and *Loddiges;* but only known from garden specimens and the figures above cited.

Bentham compares this with *E. urceolaris* (*E. caffra,* Linn.), and suggests that it may be a garden hybrid. This may be; but it seems distinct from the last named by its corolla, which is smaller and much broader in proportion to its length. It appears to be much more closely allied to *E. hirta* and *E. globosa.*

173. E. globosa (Andr. Heathery, t. 116); erect, strong-growing, 1–$1\frac{1}{2}$ ft. high; branches rather long, glandular-setose, or closely pubescent; leaves 3-nate, mostly squarrose or spreading, longer than the internodes, oblong or lanceolate and open-backed with reflexed margins, puberulous or glabrous, 3–$5\frac{1}{2}$ lin. long; flowers terminal, and axillary towards the ends of the branches; pedicels slender, puberulous or glandular-pubescent, $2\frac{1}{2}$–4 lin. long; bracts subapproximate to remote, rather long or short, slender; sepals narrow-lanceolate, pubescent, viscid, gland-ciliate, foliaceous, $\frac{3}{4}$–$1\frac{1}{2}$ lin. long; corolla globose-urceolate or ovoid-urceolate; mouth more or less contracted, viscid, puberulous, or nearly glabrous, 2–3 lin. long; segments erect, rounded, about $\frac{1}{4}$ the length of the tube; anthers subexserted, rarely exserted, lateral, dorsifixed near the base, linear, obtuse or subacute, nearly straight, nearly 1 lin. long, 4–5 times longer than their width in the middle, aristate; pore $\frac{1}{3}$–$\frac{1}{2}$ the length of the cell; awns rather rough, about $\frac{1}{4}$ the length of the cell; style exserted, slender; stigma capitellate; ovary varying from villous to glabrous. *Andr. Col. Heaths, t.* 170; *Benth. in DC. Prodr.* vii. 672. *E. perlata, Lee ex Benth. l.c.* 672, *not of Forbes.*

VAR. β, **subterminalis** (Bolus); leaves linear, sulcate, glabrous, margins revolute; bracts remote, nearly basal; sepals rather thick, $\frac{1}{2}$–$\frac{3}{4}$ lin. long; anther-pore $\frac{1}{5}$ the length of the cell; ovary glabrous. *E. subterminalis, Klotzsch ex Benth, l.c.* 670.

SOUTH AFRICA? Var. β, *in Herb. Berlin.*

COAST REGION: Stellenbosch Div.; Hottentots Holland Mountains, *Niven,* 157! *MacOwan & Bolus, Herb. Norm.,* 31! Lowrys Pass, *Schlechter,* 7800! *Bolus,* 6494! Knysna Div.? without precise locality, *Buchanan!*

A somewhat variable species, especially as to the size of the leaves and flowers. It has a strong resemblance to *E. hirta,* but seems to have a more spreading habit, broader leaves, and a usually larger and more globose corolla; also to *E. aggregata,* but has larger flowers and differently shaped anthers. The corolla in Andrews' figure is more globose and the mouth more contracted than any shown in our specimens, which is probably (partially at least) a result of compression in drying.

174. E. marifolia (Soland. in Ait. Hort. Kew. ed. 1, ii. 15); of straggling diffuse habit, under 1 ft. high; branches spreading and divaricate, puberulous and sparsely setose; leaves 3-nate, spreading, ovate or oblong, subacute, open-backed, pubescent above, paler and velvety or shortly hoary-tomentose beneath, $2\frac{1}{2}$–3 lin. long; pedicels $2\frac{1}{2}$–3 lin. long; bracts subremote, slender, spathulate-linear; sepals linear, or subspathulate-linear, apex slightly widened by a foliaceous tip, pubescent, $1\frac{1}{2}$ lin. long; corolla from narrow-ovoid-urceolate to urceolate, mouth contracted, pubescent, white, about 3 lin. long; segments erect, ciliolate, small; filaments tapering upwards; anthers included, narrow-oblong, pale brown, over $\frac{2}{3}$ lin. long, aristate; awns subulate about $\frac{1}{2}$ the length of the cell; ovary hispid. *Andr. Heathery, t.* 127, *and Col. Heaths, t.* 34; *Bauer, Exot. Pl. t.* 14; *Wendl. Eric. Ic. fasc.* 11, 9; *Benth. in DC. Prodr.* vii. 671. *E. helianthemifolia, Salisb. in Trans. Linn. Soc.* vi. 328.

SOUTH AFRICA: without locality, *Herb. Salisbury! Niven,* 16 (9)! and *cultivated specimens!*

COAST REGION: Cape Div.; at about 1400 ft.; rocky places and near the Waterfall on the Devils Peak, *Niven,* 38! *Mund,* 169! *Bolus,* 2960! *Guthrie,* 1491! *Wolley Dod,* 785! rocks above Tokay Plantation, *Wolley Dod,* 450!

This species appears to be confined to the Cape Peninsula. *E. marifolia* var. *longifolia, Wendl. Eric. Ic. fasc.* 26, 4, is almost certainly *E. triflora, L.*

175. E. argyræa (Guthrie & Bolus); habit straggling and diffuse, with a general silvery aspect, a few inches high; branches pubescent, a few inches long; leaves 3-nate, spreading, crowded, narrow-ovate to oblong, subobtuse, open-backed, copiously but shortly pilose, paler beneath, $1\frac{1}{2}$–$2\frac{1}{4}$ lin. long; flowers sub-3-nate; pedicels puberulous, $1\frac{1}{2}$–2 lin. long; bracts very remote, sometimes basal, very small; sepals lanceolate to ovate-lanceolate, glandular-pubescent and ciliate, viscidulous, about $\frac{3}{4}$ lin. long; corolla cyathiform to suburceolate-cyathiform, mouth not (or very slightly) contracted, puberulous, white to pale rosy, 2–$2\frac{1}{2}$ by $1\frac{1}{2}$ lin.; segments erect, rounded, small; filaments tapering upwards; anthers included, semiovate or sub-cuneate, dark-coloured, $\frac{1}{2}$ lin. long, narrow crested; crests cuspidate and only slightly denticulate on the margin; style well exserted, curved above; ovary villous.

Coast Region: Stellenbosch Div.; mountain slopes near Jonkers Hoek, about 1000 ft., *Bolus in Herb. Guthrie,* 4767!

Allied to *E. marifolia* and resembling it in miniature; but differing in the sepals, corolla, and the anthers, the latter being both shorter and broader; also to *E. hirta,* to the flowers of which (especially the short-sepaled form) it is very near; but the habit, pubescence and anthers are different. The pubescence appears to be more silvery than in either of the two species above named.

Section XIV. **CERAMIA.** (Sp. 176–207.)

176. E. cymosa (E. Meyer ex Benth. in DC. Prodr. vii. 670); diffuse, weak and straggling, a few inches high; branches slender, as thick as a stout hog's bristle, glabrous or thinly hairy; leaves 3-nate, distant, spreading, elliptical, obovate or suborbicular, very obtuse, mucronulate, gland-ciliate, membranous, veined, 1–3 lin. long; flowers both terminal and axillary; pedicels capillary, 5 lin. long; bracts remote, minute, inframedian; sepals broad-ovate, cuspidate, foliaceous, $\frac{1}{2}$ lin. long; corolla oblate-spheroidal or globular-cyathi-form, subglabrous, the mouth contracted, 1 lin. long, $1\frac{1}{2}$ lin. wide; segments broad, rounded, incurved; filaments tapering upwards from the base; anthers included, dorsifixed a little above the base, sub-obovate, very obtuse, about $\frac{1}{5}$ lin. long, aristate; pore about $\frac{2}{3}$ the length of the cell; awns somewhat thick and rough-edged, about $\frac{2}{3}$ the length of the cell; style included, short, stout; stigma capitate; ovary globose, glabrous.

Coast Region: Worcester Div.; Dutoits Kloof, 2000–3000 ft., *Drège,* 1185!

A well-defined species, allied to *E. oxycoccifolia.* Bentham remarks that he found no cymose inflorescence, nor have we.

177. E. oreophila (Guthrie & Bolus); diffuse, about 6 in. high; branches from a stout short stem, filiform, spreading, flexuous, puberulous, reddish, the older with small but prominent scars of leaf-cushions; leaves 3-nate, spreading or squarrose, oblong some-what widened upwards, or lanceolate, acute, the older open-backed with reflexed margin, the younger smaller, thicker, and only sulcate, gland-ciliate, 2–3 lin. long; flowers 3-nate, terminal (? sometimes lateral), subcorolline; pedicels slender, viscid, $2\frac{1}{2}$ lin. long; bracts remote, small, 2 inframedian, 1 subbasal; sepals ovate, subcordate, acute, glabrous, viscid, nerved, subscarious, about 1 lin. long, equal to the corolla-tube ; corolla oblate-urceolate or suburceolate-cyathiform, mouth somewhat contracted (in the dried state appearing scarcely so), glabrous, viscid, pale rose below, darker tinted above, $1\frac{1}{2}$ lin. long, 2 lin. broad; segments more than $\frac{1}{2}$ as long as the tube, deltoid, subacute; filaments tapering upwards, bent below the anther ; anthers included, lateral, short-cuneate, dorsifixed at or near the base, about $\frac{1}{4}$ lin. long, aristate; awns subulate, rough-edged, as long as the cell; style included ; stigma capitellate or subsimple; ovary turbinate, very pallid, hispidulous.

COAST REGION: Paarl Div.; mountains about French Hoek, 2400 ft. *Schlechter, 9261* !

178. E. planifolia (Linn. Sp. Pl. ed. 2, 508); diffuse, much branched, 6–12 in. high ; branches pubescent ; leaves 3-nate, spreading, ovate to lanceolate, acute, open-backed, margins revolute, midrib thickened upwards into a smooth prominent callosity, puberulous, ciliate, $1\frac{1}{2}$–2 lin. long; flowers axillary, solitary or usually so; pedicels slender, hispidulous, 2–$2\frac{1}{2}$ lin. long ; bracts 2, remote, foliaceous, shortly petiolate; sepals lanceolate, acute, foliaceous, gland-ciliate, margins revolute, about 1 lin. long; corolla cyathiform, mouth neither contracted nor widened, hispidulous, viscid, $1\frac{1}{2}$ lin. long; limb suberect, less than $\frac{1}{2}$ as long as the tube ; filaments from a short dilated base tapering upwards; anthers included, lateral, dorsifixed close to the base, oblong, $\frac{2}{5}$ lin. long, aristate; awns setiform, about $\frac{1}{2}$ as long as the cell ; style included ; stigma capitate; ovary hispidulous. *Berg. Descr. Pl. Cap.* 100 ; *Benth. in DC. Prodr.* vii. 669. *E. thymifolia, vars. α and β, Salisb. in Trans. Linn. Soc.* vi. 325, *acc. to Bentham. E. thymifolia, Andr. Heathery, t.* 195, *and Col. Heaths, t.* 136, *not of Wendl. E. distans, Spreng. fil., Fl. Tent. Suppl. Syst.* 13, *not of Benth.*

VAR. β, **calycina** (Bolus); sepals about as long as the corolla, $1\frac{1}{2}$ lin. or more long; corolla puberulous, about 1 lin. long, $1\frac{1}{2}$ lin. wide; filaments equal; awns of the anther about as long as the cell, minutely notched; ovary white hirsute.

SOUTH AFRICA: without locality, *Herb. Salisbury!* and *cultivated specimens !*

COAST REGION, at 2000–3000 ft. : Worcester Div.; Dutoits Kloof, *Drège! Bolus,* 5178! Paarl Div.; French Hoek, *Schlechter,* 9239 ! Cape Div. ; on mountains around Cape Town, *Thunberg, Roxburgh! Niven ! Burchell,* 602 ! 8435 ! *Bolus,* 3334! *Schlechter,* 223 ! *Guthrie,* 557 ! Stellenbosch Div. ; Hottentots Holland, *Zeyher,* 334 (ex *Sprengel*). Var. β : Caledon Div.; mountains near the Zondereinde River, *Zeyher,* 3215! (in Cape Govt. and Berlin Herbaria).

Most nearly allied to *E. cryptanthera,* and *E. thymifolia.* It is more robust than the former and distinguished from the latter by its shorter corollas and included anthers.

179. E. heterophylla (Guthrie & Bolus) ; diffuse, 6 in. or more high ; branches flexuous, straggling ; hirsute, laxly leafy ; leaves

3-nate, spreading or squarrose, as long as or often shorter than the
internodes, from lanceolate and acute to broadly elliptical and obtuse,
or suborbicular, thinly hispid and hispid-ciliate, or nearly glabrous,
1–2 lin. long; flowers 3-nate; pedicels slender, $1\frac{1}{2}$ lin. long; bracts
remote or subapproximate, setaceous to spathulate, glandular, hairy;
sepals linear to oblanceolate, acute or acuminate, long gland-ciliate,
$\frac{2}{3}$–1 lin. long; corolla urceolate-campanulate to subcyathiform, mouth
scarcely contracted or widened, puberulous, $1\frac{1}{4}$–$1\frac{3}{4}$ lin. long; limb
$\frac{1}{2}$–$\frac{2}{3}$ the length of the tube; segments spreading or recurved; filaments equal; anthers subexserted, nearly as long as, or slightly
exceeding the corolla-limb, lateral, dorsifixed near the base, oblong,
incurved, about $\frac{2}{3}$ lin. long, cristate-aristate; pore $\frac{3}{5}$ the length of
the cell; awns sublinear, curved outwards, and deeply toothed on
the outer side only, about $\frac{1}{2}$ the cell in length; style exserted;
stigma capitellate; ovary hirsute.

COAST REGION : Swellendam Div. ; mountains near Puspas Vlei Grootvaders-
bosch, &c., *Zeyher*, 165 ! in Cape Govt. Herb.

180. E. cryptanthera (Guthrie & Bolus); somewhat diffuse, drying
a pale yellowish hue, 4–6 in. high; branches very slender, sparsely
pubescent, the ultimate subcapillary; leaves 3-nate, spreading,
shorter than the internodes, lanceolate, acute, open-backed, strongly
revolute-edged, ciliate and awned at the apex, $\frac{3}{4}$–$1\frac{1}{4}$ lin. long; flowers
axillary, solitary or binate; pedicels slender, bent, $1\frac{1}{2}$–2 lin. long;
bracts 2, remote, basal, foliaceous, small; sepals lanceolate or oblong,
acute, $\frac{1}{2}$–$\frac{3}{4}$ lin. long; corolla broad-cyathiform, not widened at the
mouth, puberulous, ochraceous yellow, about $1\frac{1}{2}$ lin. long; segments
semiorbicular, erect, about $\frac{1}{2}$ as long as the tube; filaments slender,
equal; anthers included, oblong, dorsifixed a little above the base,
under $\frac{1}{3}$ lin. long, aristate; awns slender, upcurved, less than $\frac{1}{3}$ the
length of cell; style included; stigma subsimple upon the slightly
dilated and truncate apex of the style; ovary hispidulous.

COAST REGION: Paarl Div.; mountains about French Hoek, 1800 ft.,
Schlechter, 9357! Caledon Div.; mountains by the Zondereinde River, near
Zoetemelks Vlei, *Grisbrook* in Herb. Guthrie, 2297 !

In general appearance, and in the flowers, resembling *E. myriocodon*, but
distinct by its constantly 3-nate and broader leaves, and its larger oblong anthers.
Also very near to *E. planifolia*.

181. E. tenuipes (Guthrie & Bolus) ; laxly straggling, suberect,
or perhaps twining ; stouter branches up to $\frac{3}{4}$ lin. in diam., glandular-
pubescent, subviscid, very pallid ; leaves 3-nate, spreading, shorter
than the internodes, oblong or more rarely lanceolate, obtuse, thin
or submembranous, thinly puberulous on both sides, gland-ciliate,
margin only slightly recurved, midrib prominent but not large,
2–3 lin. long, up to about 1 lin. wide ; flowers terminal, 3-nate? or
umbellate? (most flowers dropped from the specimens) ; pedicels
very slender, pubescent, viscidulous, persistent and elongating, 6–9
lin. long; bracts remote, adpressed, minute ; sepals lanceolate,

scarious, coloured, viscid, $\frac{3}{4}$–1 lin. long; corolla cyathiform or short-and wide-suburceolate, mouth neither contracted nor widened, dia-phanous, viscid, glabrous, lilac, about $1\frac{1}{2}$ lin. long; segments recurved, subdeltoid, obtuse, about $\frac{1}{2}$ the length of the tube; filaments slender, equal, pallid; anthers included or just manifest, dorsifixed just above the base, oblong or dorsally curved and semiovate, $\frac{2}{5}$ lin. long, muticous; pore $\frac{1}{2}$ as long as the cell; style exserted, compressed; stigma capitellate; ovary pubescent, chiefly at the apex.

COAST REGION: Paarl Div.; mountains about French Hoek, 2500 ft., *Schlechter*, 10601!

Recognizable by its pallid branches, oblong leaves, long pedicels and muticous anthers.

182. E. physophylla (Benth. in DC. Prodr. vii. 682); procum-bent or prostrate, probably growing on rocks; branches sometimes cæspitose and intricately matted, sometimes more straggling, shortly gland-hispid, reddish; leaves 3-nate, spreading, crowded, ovate, appearing inflated or bladder-like on account of the recurved margin, glabrous, ciliate with many long spreading white hairs longer than the leaf itself, $1\frac{1}{2}$–2 lin. long; flowers terminal, 3-nate; pedicels about 1 lin. long; bracts approximate, one spathulate, about 1 lin. long, 2 smaller; sepals lanceolate, subscarious, viscid, sulcate-keeled, shortly gland-ciliate, coloured, about 1 lin. long; corolla broad-cyathiform, mouth widened, glabrous, scarcely $1\frac{1}{2}$ lin. long; limb erect, a little shorter than the tube; anthers included, lateral, dorsi-fixed close to the base, cuneate-ovate, pale brown, diaphanous, $\frac{1}{6}$ lin. long, aristate; pore $\frac{1}{3}$ the length of the cell; awns slender, upcurved, about $\frac{1}{2}$ the length of the cell; style included, short; stigma simple; ovary glabrous.

COAST REGION: Caledon Div.; tops of the mountains of Baviaans Kloof at Genadendal, *Burchell*, 7745! *Bolus*, 5414! and in *Herb. Norm. Aust.-Afr.*, 610! *Schlechter*, 9827!

The peculiar leaves, which are apparently inflated, distinguish this species from any other.

183. E. oxycoccifolia (Salisb. in Trans. Linn. Soc. vi. 324); dwarf, straggling, procumbent or decumbent in clefts of wet rocks; branches numerous, spreading, glandular-hispid, 8 or 10 in. long; leaves 3-nate, spreading, crowded, elliptical, orbicular or subovate, very obtuse, thin, flat, with reflexed margins, ciliate, otherwise nearly glabrous, 1–$1\frac{1}{2}$ lin. long; flowers usually axillary; pedicels slender, hispid, 2–$3\frac{1}{2}$ lin. long; bracts remote, minute; sepals ovate, ciliate, pubescent, $\frac{1}{2}$–$\frac{3}{4}$ lin. long; corolla obconic-cyathiform, mouth widened, puberulous, $1\frac{1}{2}$ lin. long; segments erect, broad, rounded, more than $\frac{1}{2}$ the length of the tube; anthers included, lateral, dorsifixed close to the base, very nearly semiorbicular, about $\frac{1}{4}$ lin. long, muticous (or minutely aristate, *Bentham*); style included; stigma capitellate; ovary glabrous.

COAST REGION: Cape Div.; Table Mountain, 3000 ft., *Ecklon & Zeyher;*
Niven, 119! *Milne,* 207! *Bolus,* 4541! *Wolley Dod,* 2120!

Closely allied to *E. physophylla,* of which it may be regarded as a Table
Mountain representative. This is one of those species which has only been
found on one mountain, and which it may be feared will probably soon, like
E. sexfaria and some others, become extinct. It is of very slow growth, bears
but few flowers, and is often subjected to fires.

184. E. tenuicaulis (Klotzsch ex Benth. in DC. Prodr. vii. 669);
diffuse, almost entirely glabrous; branches few, straggling, slender,
2–3 in. long; leaves 3-nate, spreading or deflexed, usually longer
than the internodes, broadly linear, subacute, slightly curved, flat
above, sulcate below, somewhat rigid and coriaceous, 3–4 lin. long;
flowers axillary, racemoso-umbellate, often reduced to 1 or 2 in each
axil; pedicels slender, 2–3 lin. long; bracts remote, minute;
sepals lanceolate, keeled, ciliolate, coloured, rigid, about $\frac{2}{3}$ lin.
long; corolla obconic-campanuloid, mouth widened, glabrous or
minutely puberulous, 1 lin. long; segments erect, rounded, from
about $\frac{2}{3}$ the length of the tube to a little longer than it; filaments
scarcely exceeding the anther; anthers included, lateral, dorsifixed
near the base, narrow-ovate, about $\frac{1}{3}$ lin. long, muticous; pore about
$\frac{2}{3}$ the length of the cell; style included; stigma capitate; ovary
sparsely hispid, glabrescent (or villous, *Bentham*).

COAST REGION : Caledon Div.; Baviaans Kloof, near Genadendal, *Niven,*
244! near the Zondereinde River and Eksteens, *Zeyher!* in Herb. Berlin.
Riversdale Div.; clefts of rocks on the Lange Bergen near Riversdale, 2000 ft.,
Schlechter, 1844!

This species has more external resemblance to those of the § *Desmia* than to
this section; but the floral structure is different. The leaves are peculiar and
unlike those of § *Ceramia* generally.

185. E. myriocodon (Guthrie & Bolus); diffuse, 6–10 in. high,
densely and intricately branched, with numerous flowers; branches
slender, puberulous, dark red; leaves usually 4- occasionally 3-nate,
spreading, incurved or recurved, linear, blunt, sulcate, hispid, 1–1¼
lin. long; flowers terminal, 4-nate to solitary; pedicels pubescent,
about 1 lin. long; bracts remote, minute; sepals lanceolate or linear,
coloured or greenish, under 1 lin. long; corolla campanulate-cyathiform, mouth scarcely widened or contracted, hirsute with short
squarrose hairs, about 1 lin. long; segments erect or spreading,
rounded, $\frac{1}{2}$–$\frac{2}{3}$ the length of the tube; filaments capillary; anthers
included, lateral, dorsifixed just above the base, ovate-cuneate, about
$\frac{1}{5}$ lin. long, aristate; pore about $\frac{1}{2}$ the length of the cell; awns
slender, rough-edged, more than $\frac{1}{2}$ the length of the cell; style
included; stigma capitellate; ovary hirsute.

COAST REGION: Paarl Div.; mountains about French Hoek, 2000 ft.,
Bolus, 8619! Caledon Div.; marshy places on the mountains near Appels
Kraal, *Zeyher,* 3225!

This is in many respects like *E. cryptanthera;* but differs by its usually
4-nate and sulcate leaves, terminal inflorescence and shorter but proportionately
broader anthers.

186. E. brachycentra (Benth. in DC. Prodr. vii. 688); diffuse, dwarf; branches intricate, rather rigid, and, together with the leaves, pedicels, bracts and sepals glandular-pilose; leaves 3-nate, recurved-squarrose, usually shorter than the internodes, oblong to ovate, acute, subopen-backed with broad thick margins, rigid, rather thick, gland-ciliate, glabrous on the upper, velvety on the lower surface, $\frac{1}{2}$–$\frac{3}{4}$ lin. long; flowers terminal, 3-nate; pedicels slender, about 1 lin. long; bracts closely approximate, elliptical, thick; sepals ovate, foliaceous, less than $\frac{1}{2}$ lin. long; corolla subcampanulate-cyathiform or broad sub-urceolate, mouth scarcely widened, glabrous, viscidulous, about 1 lin. long; segments very short, almost truncate; filaments slender, equal; anthers exserted, dorsifixed half-way between the base and the middle, very shortly ciliolate on the front margin, $\frac{2}{5}$ lin. long, aristate; pore about $\frac{1}{3}$ the length of the cell; awns very short scarcely reaching to the base of the cell; style long-exserted, slender; stigma capitellate; ovary glabrous.

COAST REGION: Mossel Bay Div.; Attaquas Kloof, *Masson*, 56! and *Niven*, 84! in Herb. Kew.

This obscure species, quite distinct from any other, does not fit very well into any of the sections. Bentham placed it, with doubt, in § *Polycodon*, and it might almost equally well go into § *Pyronium*. Its spreading open-backed leaves, its straggling habit, and general viscidity, induce us to place it here.

187. E. leptoclada (Van Heurck & Muell. Arg., Obs. Bot. et Descr. Pl. Nov. 34); diffuse, procumbent, 3–10 in. high; branches glandular-hispid, slender; leaves 3-nate (rarely some 4-nate on the same branch), spreading, linear, linear-lanceolate or broad-lanceolate, acute, sulcate, hispid and ciliate with glandular hairs, $1\frac{1}{2}$ lin. long, margins revolute; flowers 3-nate, axillary towards the ends of the branches; pedicels very slender, about as long as the leaves; bracts remote, basal, foliaceous, hispid; sepals lanceolate, acuminate, folia-ceous, hispid, gland-ciliate, distinctly cohering at the base, sometimes with obtuse sinuses, about $\frac{3}{4}$ lin. long; corolla broadly funnel-shaped, expanded from a point below the middle, minutely puberu-lous, $1\frac{1}{2}$–2 lin. long; segments erect or spreading, broad and rounded, from $\frac{1}{3}$ to $\frac{1}{2}$ the length of the rest of the corolla; anthers included, dorsifixed just above the base, semiovate, dark-coloured, $\frac{1}{3}$ lin. long, muticous; pore about $\frac{1}{3}$ the length of the cell; style slender, exserted, often bent below the stigma; stigma capitellate; ovary glabrous.

VAR. β, **aristata** (Bolus); leaves broader than in the type, commonly open-backed; sepals linear or subulate, aristate; corolla-limb mostly spreading or reflexed; anthers aristulate; awns $\frac{1}{4}$ the length of the cell or less.

COAST REGION: Piquetberg Div.; on the Oliphants River (Kardouw) Moun-tains, *Zeyher*, 1108! Var. β: Paarl Div.; mountains about French Hoek, 2700 ft., *Bolus*, 9996! *Guthrie*, 4701! Caledon Div.; mountains near Gena-dendal, 2000 ft., *Pappe! Galpin*, 3655!

This species, like the next, is well-marked by its very distinctly funnel-shaped corolla, with a rather large limb. It appears to be rare and our material is somewhat scanty; but we have availed ourselves of the full and careful description of the authors. Galpin's specimens are small plants of one year's growth and look more luxuriant than the others, but they differ little in floral structure. They exhibit a few whorls of 4-nate leaves above those which are 3-nate.

188. E. trichoclada (Guthrie & Bolus); diffuse, dwarf, the whole plant viscidulous; branches decumbent, divaricate, roughly hispid, rusty-brown; branchlets set at nearly a right angle from the stem; leaves 3-nate, spreading or squarrose, elliptical or lanceolate, acute, sparingly hispid above or glabrescent, glabrous below, margins recurved, open-backed, ciliate with gland-tipped hairs, 1–1¼ lin. long; flowers axillary, 1–2–3 in the whorls of the uppermost leaves; pedicels slender, hispid, 2 lin. long; bracts approximate, foliaceous; sepals lanceolate, acute, foliaceous, ciliate with long gland-tipped hairs, ¾ lin. long; corolla funnel-shaped, abruptly widening above the middle, glabrous, dry, "pink," 1½ lin. long; segments erect, broad-oblong, obtuse, from ½ the tube to equal to it in length; anthers subincluded (shorter than the whole corolla but manifest), lateral, dorsifixed near the base, oblong, pale brown, ⅓ lin. long, aristate; pore about ¼ the length of the cell, awns rough, about ⅓ the length of the cell; style slender, included, a little shorter than the stamens; stigma subsimple, very small; ovary subglobose, glabrous.

Eastern Region: Natal; in a ravine at Liddesdale, 4000 ft., *Wood*, 3933!

A species singularly interesting as being an outlier from the great central home of heaths; and as curiously resembling *E. leptoclada*, from the station of which it is separated by some 800 miles, while both appear to be rare. It is distinguishable from that species by its broader leaves, its approximate (not basal) bracts, and its apparently stouter and stronger habit; yet the shape of the corolla, and the shape and size of the anthers are almost exactly the same.

189. E. Marlothii (Bolus in Journ. Bot. 1894, 237); diffuse, under 1 ft. high; branches sometimes rather stout, divaricate, rigid, pilose; leaves 3-nate, spreading or squarrose, close-set, ovate, open-backed, margins reflexed, pubescent, sparsely ciliate, 1 lin. or less long; flowers terminal, solitary; pedicels about 1 lin. long; bracts remote, small; sepals broad-ovate, thick, viscid, villous, ¾ lin. long; corolla ovoid in bud, becoming oblate-urceolate-depressed; upper part of the tube falling in; segments about ¼ the length of the tube, connivent with upturned apices closely surrounding the exserted filaments and style, with 4 depressions at the base as in *E. baccans*, mouth much contracted; after the swelling of the ovary, the corolla assumes the ordinary urceolate shape, pubescent, 2 lin. long, about the same in width; filaments rather broad, equal, bent inwards over the ovary, about 2 lin. long, far exserted; anthers terminal or subterminal, oblong, or (from the pore being nearly the length of the cell) somewhat earshaped; cells deeply partite, about ⅔ lin. long, muticous; style far exserted, hooked and thinly pubescent near the apex ; stigma capitate ; ovary pubescent.

South Africa : without locality, *Marloth,* 2244 !

Coast Region : Worcester Div.; shady places between stones on the Matroos Berg, 5200 ft., *Marloth,* 1956!

A very distinct species, its nearest ally being apparently *E. oligantha.* The peculiar corolla in which the apex falls in occurs also in some species of §§ *Lamprotis* and *Geissostegia,* but is very rare in the genus.

190. E. oligantha (Guthrie & Bolus) ; diffuse, twining amongst low shrubs, under 1 ft. high ; branches slender, straggling, pubescent ; leaves usually 3-nate (occasionally 4-nate on the main stem), spreading, not crowded, ovate to oblong in apparent outline according to the degree to which the margins are reflexed or revolute, open-backed or sulcate, pubescent and ciliate with long white hairs, some as long as the leaves, 1–1½ lin. long ; flowers usually terminal and solitary, occasionally 2–3-nate, very rarely axillary ; pedicels ½–1 lin. long ; bracts remote, small ; sepals broad-ovate, viscid, pubescent, coloured, about ¾ lin. long, margins reflexed ; corolla oblate-spheroidal, pubescent, viscid, the summit slightly falling inwards, red, about ¾ lin. long and wide ; segments in full flower horizontally connivent, becoming more erect ; filaments straight, dilating upwards ; anthers exserted, sublateral, dorsifixed close to the base, oblong, but curved on the dorsal edge, minutely ciliolate, over ⅖ lin. long, muticous ; pore ⅔ the length of the cell ; style exserted, slender ; stigma subsimple ; ovary pallid, pubescent.

Coast Region: Bredasdorp Div.; mountain slopes near Elim Mission Station, 1400 ft., *Bodkin,* in Herb. Bolus, 6735 !

191. E. debilis (Guthrie & Bolus) ; apparently a low growing diffuse shrub, with slender, scarcely puberulous branches ; leaves 3-nate, spreading, broadly ovate, obtuse, glabrous above, pallid and concave below, 1 lin. long and wide ; flowers terminal, 3-nate ; " pedicels, bracts, sepals and the white corolla minutely glandular-puberulous " (*Bentham*) ; pedicels 2 lin. long ; bracts remote, small ; sepals 2 ovate, 2 narrower oblanceolate, subacute, membranous, coloured, 1¼ lin. long ; corolla subconical oblate-urceolate, mouth somewhat contracted, 1½ lin. long ; limb very short, slightly spreading ; filaments straight, equal, nearly as broad as the anther ; anthers exserted, lateral, dorsifixed close to the base, suboblong, subacute, about ⅖ lin. long, muticous ; pore as long as the cell ; style exserted, generally bent ; stigma capitellate ; ovary coarsely villous. *E. lycopodioides, Lodd. ex Benth. in DC. Prodr.* vii. 619, *not of Hornemann.   E. suaveolens, Lodd. Bot. Cab. t.* 24 ?

South Africa : *cultivated specimen !*

In Herb. Kew are two branches marked " 209-a Hort. Bot. Edinb. 18 June '38 " and below " Graham, 1839," which are those described by Bentham. From those and from the description, we have drawn up the above. Nothing more seems to be known of the species, which is chiefly distinguished by the shape of its corolla. We are compelled to give it a new name because it is impossible from Hornemann's very brief description to come to the conclusion that this is even probably the species so named by him at an earlier date (*Hortus Hafniensis*, i. (1813) 374). He describes the leaves of his plant as 4-nate, while on Graham's specimens they are clearly 3-nate; nor does the shape of the corolla seem to agree.

192. E. thimifolia (Wendl. Bot. Beobacht. 48) ; diffuse, 1 ft. or more high ; branches numerous, straggling, puberulous and also sparsely pilose, glabrescent, red ; leaves 3-nate, spreading, not crowded, linear-lanceolate to ovate, acute, open-backed, margins recurved or often almost flat, midrib prominent, thickening upwards as a smooth callosity, sparsely glandular-hairy and ciliate, 1–2 lin. long; flowers terminal, 3-nate, sometimes also axillary ; pedicels slender, pilose, about 2 lin. long ; bracts 2, remote, basal, linear (or sometimes a third, foliaceous?) ; sepals lanceolate, acute, foliaceous, margins revolute, long glandular-ciliate, about 1 lin. long; corolla urceolate-campanulate or tubular-campanulate, mouth scarcely widened, glabrous, $1\frac{1}{2}$ lin. long ; segments erect or slightly spreading, $\frac{1}{4}-\frac{1}{3}$ the length of the tube ; filaments much dilated at the base, tapering upwards ; anthers exserted, dorsifixed above the base, oblong, about $\frac{3}{5}$ lin. long, aristate ; pore $\frac{1}{3}$ the length of the cell; awns small, ciliolate, scarcely reaching to the base of the cell; style exserted, slender ; stigma capitellate ; ovary pubescent. *E. thymifolia, Salisb. in Trans. Linn. Soc.* vi. 325, *var.* γ *only*; *Benth. in DC. Prodr.* vii. 669. *E. planifolia, Wendl. Eric. Ic. fasc.* 16, 59, *t.* 23 ; *Andr. Heathery, t.* 185 ? *and Col. Heaths, t.* 201 ? *E. distans, Spreng. f. Tent. Suppl. Syst. Veg.* 13, *not of Benth.*

SOUTH AFRICA : without locality, *Sieber*, 146 ! 181 ! *Herb. Salisbury !* and *cultivated specimens!*

COAST REGION : Tulbagh Div. ; Witsenberg Range, *Pappe!* Cape Div. ; foot of Table Mountain, east side, 300 ft., *Bolus*, 4487! near Hout Bay Nek, 800 ft., *Guthrie*, 145! Orange Kloof, *Wolley Dod*, 2176 ! Constantia Berg, *Wolley Dod*, 461 !

Closely allied to *E. planifolia*, from which it differs chiefly in its longer and narrower corolla and its exserted and slightly different anthers. There has been some confusion between the two species, possibly due to hybridization under cultivation. The "*E. thymifolia*" of *Andr. Heathery*, t. 195, is, there is little doubt, *E. planifolia, L.*; while "*E. planifolia*" of *Andr. l.c. t.* 185, is either a form of *E. thimifolia, Wendl.*, or a hybrid between the two. We have seen no good figure of the present species as it occurs wild. The best is that of *Wendl. Eric. Ic. fasc.* 16, where, while the analytical figure of the corolla is not very incorrect, the shape of those on the branches, is very much so ; the corollas not being much contracted at the mouth as there represented. The species is not so common as *E. planifolia ;* we have seen six different gatherings, and amongst them none which are not clearly distinguishable by the characters above stated.

193. E. filiformis (Salisb. in Trans. Linn. Soc. vi. 345, not of Bartl.) ; procumbent, sometimes almost prostrate, much branched, spreading 6–8 in. from the stem in every direction ; branches puberulous, leafy ; leaves 3-nate, erect-spreading, usually shorter than the internodes, linear-subterete, sulcate, hirsute or glabrescent, mostly with a tuft of white hairs at the apex, $1\frac{1}{2}$–3 lin. long ; flowers

terminal, 3-nate or in clusters of 3–6 ; pedicels slender, about 1 lin.
long; bracts closely approximate, small ; sepals linear-subulate or
narrow-lanceolate, acute, foliaceous, hirsute, shortly gland-ciliate,
about 1 lin. long ; corolla suburceolate, tubular-urceolate or ovoid-
urceolate, more or less contracted at the throat, usually tetragonous,
viscid, glabrous, $1\frac{1}{4}$–$1\frac{3}{4}$ lin. long ; segments more or less spreading or
suberect, semiovate, rounded, from $\frac{1}{3}$–$\frac{1}{2}$ the length of the tube ;
stamens usually 8, but often 4–7 ; filaments narrow, a little dilated
and dark-coloured below the anther ; anthers exserted or subexserted,
lateral, sublateral or rarely subterminal, narrow-oblong, $\frac{1}{3}$–$\frac{2}{5}$ lin.
long, muticous ; pore about $\frac{3}{4}$ the length of the cell ; style slender,
exserted ; stigma small, subsimple ; ovary globose, glabrous. *Benth.
in DC. Prodr.* vii. 670.   *E. humilis, Benth. l.c.* 615, *not of Neck.
nor Salisb.   E. divergens, Wendl., E. flavida, Klotzsch, and E.
connivens, Klotzsch, ex Benth. l.c.* 670.

VAR. β, **maritima** (Bolus) ; leaves somewhat longer and more distant than
usual ; flowers somewhat smaller, $1\frac{1}{4}$–$1\frac{1}{2}$ lin. long ; stamens 4 ; anthers narrow-
elliptic, subcuneate at the base; cells approximate, pallid, $\frac{1}{3}$ lin. long ; pore $\frac{1}{3}$–$\frac{1}{2}$
the length of the cell.

VAR. γ, **longibracteata** (Bolus) ; bracts slender, but longer, one foliaceous and
reaching to the top of the corolla ; ovary pubescent.

SOUTH AFRICA : without locality, *Roxburgh! Herb. Salisbury! Drège!*
COAST REGION, from 800 to 2000 ft. : Caledon Div. ; Klein Houw Hoek,
Zeyher, 3217 ! Nieuw Berg, near Palmiet River, *Zeyher*, 3332 ! hills near
Grabouw, *Bolus*, 4177 ! 4178 ! *Guthrie*, 4169 ! Houw Hoek, *Bolus*, 7369 !
6958 ! *Schlechter*, 9425 ! Var. β : Bredasdorp Div.; hills near Cape Agulhas,
250 ft., *Schlechter*, 10559 ! Var. γ : Stellenbosch Div.; Lowrys Pass, 1500 ft.,
Schlechter, 7247 !

This is a curiously variable species in respect of the number of its stamens.
We have found them 4 to 8. Those specimens with 4 stamens (*Bolus*, 4178, and
Zeyher, 3332) are so in most, if not all, the flowers. They are then technically
Blæria and not *Erica*. Even if these stood alone it would seem a forcing of
nature to separate them from *E. filiformis*, with which they agree in all other
respects. But intermediate forms appear to indicate at once an unstable
condition, which induces us to abandon (for systematic purposes in this
particular case) a character based upon stability in the number of the stamens.

194. E. Tysoni (Bolus in Journ. Linn. Soc. xxiv. 181) ; diffuse,
decumbent, a few in. high ; branches numerous, straggling, slender
but woody and rigid, scabrid-hispid or glabrescent, 6–12 in. long ;
leaves 3-nate, erect-spreading, linear or narrow-lanceolate, sulcate,
viscidulous, scabrid-hispidulous, sometimes glabrescent, about 1 lin.
long ; flowers axillary, 3-nate ; pedicels $\frac{1}{2}$ lin. long ; bracts approximate, oblong or obovate, foliaceous ; sepals oblong or lanceolate,
obtuse, foliaceous, gland-ciliate, about 1 lin. long ; corolla tubular-
campanulate, slightly widened to the mouth, generally more or less
(sometimes strongly) tetragonous, viscidulous, purple (*Evans*), $1\frac{1}{4}$–2
lin. long ; segments broad, erect-spreading, from $\frac{2}{3}$–$\frac{3}{4}$ the length of
the tube ; anthers subexserted, lateral, dorsifixed shortly above the
base, oblong, scaberulous, ciliolate, about $\frac{1}{3}$ lin. long, aristate ; pore
$\frac{1}{3}$–$\frac{1}{2}$ the length of the cell ; awns ciliolate, curved, about $\frac{1}{3}$–$\frac{2}{3}$ the
length of the cell ; style exserted ; stigma capitellate ; ovary

glabrous. *E. satureioides, Sond. ex Bolus in Journ. Linn. Soc.*
xxiv. 182.

EASTERN REGION, in rocky places at 6000–7500 ft. : Griqualand East;
summit of Ingeli Mountain, *Tyson*, 1290! and in *Herb. Norm. Aust.-Afr.*, 469!
Pondoland ; Insizwa Mountains, *Schlechter*, 6493 ! Fakus Territory, *Sutherland !*
Natal; on the Drakensberg Range at Polela, *Evans*, 673 !

Allied to *E. filiformis*, from which it differs by its fewer flowers in the whorl,
and its appendiculate anthers. The corolla is singularly like it in shape and
texture, except that the limb in this is considerably longer.

195. E. aspalathoides (Guthrie & Bolus in Engl. Jahrb. xxvii.
173) ; diffuse, very slender, weakly-trailing, almost herbaceous in
appearance ; branches spreading, about $\frac{1}{4}$–$\frac{1}{3}$ lin. in diam. at their
thickest part, 4–8 in. long, closely leafy, thinly hairy, red-brown ;
leaves 4-nate, squarrose or recurved, linear-setaceous, 1–2 lin. long,
$\frac{1}{8}$–$\frac{1}{6}$ lin. broad, longer than the internodes, with 2–3 obsolete teeth
on either side, each tooth and the apex bearing a long fine white
gland-tipped hair ; flowers few ; pedicels short, about $\frac{1}{2}$–1 lin. long ;
bracts approximate, leaf-like and with similar white hairs, but pale
brown coloured, 1$\frac{1}{2}$ lin. long ; sepals leaf-like, coloured like the
bracts, and about as long, reaching to the top of the corolla ; corolla
cyathiform, slightly widened at the mouth, glabrous, dry (when
dried, pale brown) 1$\frac{1}{4}$ lin. long ; segments erect, rounded, $\frac{1}{3}$–$\frac{1}{2}$ the
length of the tube ; filaments capillary, dilated just below the
anther ; anthers semi-exserted, lateral, dorsifixed close to the base,
narrow-oblong, very pale brown, less than $\frac{2}{5}$ lin. long, aristate ; pore
about $\frac{1}{3}$ the length of the cell ; awns about $\frac{1}{3}$ the length of the cell ;
style exserted, slender, bent ; stigma small, capitellate ; ovary
puberulous.

COAST REGION : Clanwilliam Div. ; Cederberg Range, at Ezelsbank, near
Wupperthal, 5000 ft., *Schlechter*, 8812 !

Easily recognized by the extreme fineness of its leaves. This species must not
be confused with *E. aspalathifolia, Bolus.*

196. E. Mundii (Guthrie & Bolus) ; erect, rigid ; branches
puberulous and hispid ; leaves 3-nate, spreading, not crowded,
oblanceolate or oblong, acute, rarely subobtuse, open-backed with
reflexed margins, sometimes almost flat, ciliate on margins and on
midrib, occasionally hairy below, glabrous above, the upper sometimes oblong, sulcate, and apparently subterete, by the more strongly
revolute margins, or even all the leaves so, 1$\frac{1}{2}$–2 lin. long ; flowers
terminal, 3-nate, or umbellate and 4–6-flowered ; pedicels slender,
$\frac{3}{4}$–2 lin. long ; bracts remote, small, adpressed ; sepals linear-subulate
or lanceolate, acute, foliaceous, callous-pointed, coloured, viscid,
about 1 lin. long ; corolla cyathiform, mouth scarcely contracted or
widened, glabrous or minutely hairy, viscidulous, $\frac{3}{4}$ to a little
over 1 lin. long, and about the same in width ; segments erect,
subdeltoid, about $\frac{1}{3}$ the length of the tube ; filaments slender, equal ;
anthers included or subexserted, lateral, dorsifixed shortly above

the base, cuneate-oblong, $\frac{1}{2}$ lin. long or slightly less, muticous; pore $\frac{1}{2}$ the length of the cell; style shortly exserted, stout, curved; stigma capitate, somewhat large; ovary subturbinate, glabrous below, thinly and shortly hairy above, or entirely glabrous.

SOUTH AFRICA: without locality, *Kennedy in Herb. MacOwan*, 1714!

COAST REGION : Swellendam Div. ; mountains near Voormans Bosch, *Zeyher*, 3258! Voormans Bosch, *Mund*, 36! in Cape Govt. Herb.

The different aspect of the broad- or narrow-leaved forms is sometimes puzzling ; but the characters of the flowers are fairly constant. In general appearance the broad-leaved forms resemble *E. planifolia*, but are structurally different and easily distinguishable. The narrow-leaved forms are probably the result of a drier season.

197. E. strigosa (Soland. in Ait. Hort. Kew. ed. 1, ii. 17, not of Wendl.); erect, stout, 2–3 ft. high, by far the largest in this section ; branches thickish, erect or divaricate, puberulous and also pilose with glandular hairs ; leaves 4-nate, mostly squarrose or spreading, generally longer than the internodes, linear or linear-lanceolate, acuminate, puberulous above, velvety and paler beneath, narrowly open-backed with thick revolute margins, minutely puberulous, ciliate with setose gland-tipped hairs, 2–3 lin. long ; flowers axillary and terminal, appearing subracemose on the ends of the branches, sometimes lax and interrupted, sometimes congested into a short pseudo-spike ; pedicels from 1–2 lin. long ; bracts remote, minute, or some occasionally wanting ; sepals linear or linear-lanceolate, acuminate, foliaceous, tipped with a glandular hair, about 1 lin. long ; corolla short-cyathiform or suburceolate-cyathiform, scarcely con-tracted or widened to the mouth, glabrous, dry, about $1\frac{1}{2}$ lin. long, rather less in width ; segments short, erect ; anthers from sub-included to subexserted, lateral, dorsifixed near the base, about $\frac{1}{2}$ lin. long or less, aristate ; pore about $\frac{1}{2}$ the length of the cell ; awns straight, $\frac{1}{2}$ the length of the cell or less; style exserted, straight; stigma capitate ; ovary turbinate, glabrous. *Benth. in DC. Prodr.* vii. 678. *E. arborea, Thunb. Diss. Erica*, 40, *not of Linn. nor of others, fide Salisb. and Rach. E. axillaris, Salisb. in Trans. Linn. Soc.* vi. 325, *not of Thunb. E. præcox, Lodd. Bot. Cab. t.* 1413. *E. pilulifera, Wendl. Eric. Ic. fasc.* 19, 107, *t.* 41? *not of Linn. E. Chamætetralix, Tausch in Flora*, 1834, 616 ? *E. lasiophylla, Spreng. Syst. Veg.* ii. 195, *fide Benth. E. scabriuscula, Drège ex Benth. in DC. Prodr.* vii. 678.

SOUTH AFRICA : without locality, *Thunberg, Drège! Herb. Salisbury!*

COAST REGION, between 1000 and 2000 ft.: Cape Div.; Devils Peak, *Burchell*, 8465! *Guthrie*, 1167! *Wolley Dod*, 1740! Table Mountain, *Masson! Bolus*, 4752! *Kässner*, 162! Orange Kloof, *Wolley Dod*, 3416! Div.? Wilde River, *Niven!*

This species is well-marked in the section by its robust habit; and in this respect it is certainly exceptional. Bentham placed it in the § *Orophanes;* but by its usually axillary (as well as terminal) flowers, its spreading and open-backed leaves, it seems to us more conveniently arranged here.

198. E. flexicaulis (Dry. in Ait. Hort. Kew. ed. 2, ii. 395);

erect, about 1 ft. high ; stem and branches flexuous ; leaves 4-nate, close-set, spreading, squarrose or decurved, broad-linear, acute, open-backed with revolute margins, thinly gland-ciliate, $2\frac{1}{2}$–3 lin. long ; flowers terminal, umbellate, cernuous, 4–6 in the umbels ; pedicels 2–3 lin. long ; bracts, two approximate, third remote, coloured, gland-ciliate ; sepals lanceolate or ovate, coloured, gland-ciliate, about $1\frac{1}{2}$ lin. long, margins revolute ; corolla urceolate or ovoid-urceolate, mouth contracted, glabrous?, shining, $3\frac{1}{2}$–4 lin. long ; segments erect or slightly spreading, about $\frac{1}{3}$ as long as the tube ; anthers included, lateral, oblong ; pore about $\frac{1}{2}$ as long as the cell, muticous ; style included ; stigma small, subsimple ; ovary glabrous? *Benth. in DC. Prodr.* vii. 692. *E. glandulosa, Andr. Col. Heaths, t.* 97, *and Heathery, t.* 115, *not of Thunb. nor of Wendl.*

SOUTH AFRICA: without locality, *Niven, ex Dryander.*

The type of this species is Andrews' figure first above cited, which is quoted by Dryander. It is otherwise unknown to us, for we have been unable to recog-nize it amongst any collections we have seen. A small frustule in Herb. Brit. Mus., so marked, is certainly not this species, but possibly *E. strigosa,* which this certainly resembles, differing in its inflorescence and also in its muticous anthers. We place it here with some doubt.

199. E. confusa (Guthrie & Bolus) ; branches slender, flexuous, clothed, as are the leaves, pedicels, bracts and sepals, with longish gland-tipped hairs ; leaves 3-nate, spreading, linear or oblong, obtuse, sulcate, minutely puberulous and also long-ciliate, 1–$1\frac{1}{2}$ lin. long ; flowers axillary and terminal, 2–3-nate ; pedicels slender, $1\frac{1}{2}$–$2\frac{1}{2}$ lin. long ; bracts 3, remote, median, minute ; sepals broad ovate, with thick strongly revolute margins, dark-coloured when dry, about $\frac{1}{2}$ lin. long ; corolla ovoid-urceolate, at first not much contracted to the mouth, afterwards becoming more so, and also more inflated, glabrous, very viscid, $1\frac{1}{2}$–$1\frac{3}{4}$ lin. long ; segments slightly spreading, bluntly rounded, short ; filaments equal, slender ; anthers semi-exserted, lateral, dorsifixed just above the base, oblong, from about $\frac{2}{5}$ to nearly $\frac{3}{5}$ lin. long, aristulate ; pore from $\frac{2}{5}$ to $\frac{3}{5}$ the length of the cell ; awns short, sometimes extremely minute ; style exserted, slender ; stigma capitellate ; ovary glabrous. *E. filiformis, Drège in Linnæa,* xx. 187, *not of Salisb.*

COAST REGION : Caledon Div. ; Sweet Melks Valley, *Niven,* 42 ! on the mountains near Appels Kraal, by the Zondereinde River, *Zeyher,* 3212 ! Gena-dendal, *Schultz in Herb. Bolus,* 6493 !

200. E. grata (Guthrie & Bolus) ; erect or subdecumbent, stout, branched, with bright green leaves, and pseudo-racemose dense flowering branches, 1–$1\frac{1}{2}$ ft. or more high ; branches with a short tomentum, interspersed with longer hairs ; leaves 3-nate, from spreading to squarrose, close-set or with longish internodes, linear to narrow-lanceolate, sulcate or more commonly open-backed, roughly and thickly or thinly pilose with tubercle-based hairs, 2–3 lin. long ; flowers terminal, 3-nate, on short branchlets ; pedicels pubescent, sometimes glandular, $1\frac{1}{2}$–$2\frac{1}{2}$ lin. long; bracts remote, small, scarious, coloured; sepals ovate-lanceolate, keeled and keel-tipped, scarious, coloured, glabrous, shortly gland-ciliate, dry or viscid, 1–$1\frac{1}{4}$ lin. long, somewhat shorter than the corolla-tube; corolla broad-urceolate, mouth more or less contracted, glabrous, dry or viscid, $1\frac{1}{4}$–$1\frac{3}{4}$ lin. long; segments erect or somewhat spreading, $\frac{1}{5}$–$\frac{1}{4}$ as long as the tube; filaments equal, straight; anthers subexserted, dorsifixed above the base, obovate-oblong, very obtuse, ciliolate along the margins, cells deeply partite and at length subdistant, under $\frac{1}{2}$ lin. long, aristate; pore very wide, about $\frac{1}{2}$ the cell in length; awns subulate, incurved, rough-edged, about $\frac{1}{4}$ as long as the cell; style exserted, stout, dark red; stigma capitate; ovary lanate with long white hairs.

CoAST REGION : Riversdale Div.; Garcias Pass, near the toll-house, 1000 ft., *Galpin*, 3652!

In the general appearance of the flowers this species bears a great resemblance to *E. oreophila*. But the habit, and especially the structure and appearance of the anthers and ovary are very different. The size and shape of the corolla, and the texture of the sepals, separate it from *E. globosa*.

201. E. flacca (E. Meyer ex Benth. in DC. Prodr. vii. 670); erect, slenderly branched, but not diffuse or straggling; branches pubescent or glandular-hairy; leaves usually 3-nate, rarely also 4-nate on the same plant, spreading, not crowded, sometimes distant and much shorter than the long internodes, linear as if subterete and sulcate with strongly revolute margins, or in more luxuriant plants lanceolate or oblong, and open-backed with reflexed margins, acute, aristate, pubescent or pilose, usually gland-ciliate, $1\frac{1}{2}$–$2\frac{1}{2}$ (occasionally 4) lin. long; flowers axillary, mostly 3-nate in the whorls towards the ends of the branches; pedicels slender, 2–$2\frac{1}{2}$ lin. long; bracts sometimes, or always (?) one approximate, small, usually two basal, large, leaf-like; sepals mostly somewhat loose or spreading, lanceolate, acute, foliaceous, pubescent, gland-ciliate, under 1 lin. long, margins reflexed; corolla subtubular-campanulate, only slightly widened to the mouth, puberulous, $1\frac{1}{2}$–2 lin. long; segments slightly spreading, about $\frac{1}{3}$ as long as the tube; filaments from a short ovate or obovate base thence abruptly contracted upward; anthers exserted, or sometimes long exserted, lateral, dorsifixed shortly above the base, semiovate, dorsally curved, anteriorly straight, glabrous, mostly pale brown, about $\frac{2}{5}$ lin. long, muticous; pore less than $\frac{1}{2}$ as long as the cell; style exserted, slender; stigma capitellate, small; ovary glabrous or pubescent on the top. *E. thymoides, Klotzsch ex Benth. in DC. Prodr.* vii. 670.

CoAST REGION, between 1000 and 3000 ft.: Clanwilliam Div.; Cederberg Range, at Ezelsbank, *Drège !* Blue Berg, *Drège !* Krakadouw Pass, *Leipoldt*, 206! Pakhuis Pass, *Leipoldt*, 624! Tulbagh Div.; Old Kloof, Roode Zand (near Tulbagh), *Niven*, 43! Ceres Div.; Mitchells Pass, *Bolus*, 5286!

A very distinct species. The shape of the corolla is somewhat like that of *E. filiformis*, but other differences are great. The filaments with their wide spoon-bowl-shaped base are something like those of *E. thimifolia*.

[*E. ciliaris, Thunb. Diss. Erica,* 19, *partly, and Fl. Cap. ed. Schult.* 349, as to sheet *a* of Herb. Thunb., on which *E. planifolia,* var. *robusta, Rach in Linnæa* xxvi. 783, was founded, is (according to Thunberg's type specimen, which I have examined) identical with *E. flacca,* E. Meyer. The locality where it was collected is not mentioned by Thunberg.—*N. E. Brown.*]

202. E. latifolia (Andr. Heathery, t. 72); erect, 1 ft. high; branches flexuous, pubescent; leaves 3-nate, laxly spreading or squarrose, broad-ovate or lanceolate to oblong, acute, open-backed, margins revolute, densely pilose above, pale below, up to 6 lin. long; flowers axillary, umbellate in threes in the middle of the branches, cernuous; pedicels decurved, 3–4 lin. long; bracts remote, foliaceous, small; sepals broad-ovate, obtuse, leaf-like in shape and texture, 1–1½ lin. long; corolla urceolate-globose, much contracted to the mouth, glabrous, bright-red, 1½–1¾ lin. long and equally wide; limb very small, erect; filaments rather broad; anthers subexserted, lateral? narrow-oblong, muticous; style exserted; stigma small, subsimple; ovary villous with long straight hairs. *Andr. Col. Heaths, t.* 105; *Benth. in DC. Prodr.* vii. 669. *E. crassifolia, Klotzsch ex Benth. l.c.,* 669, *not of Andr.*

COAST REGION : Swellendam Div., *Ecklon & Zeyher,* ex *Bentham.*

This species is known to us only from Andrews' figure and brief description, from which we have drawn up the foregoing. Bentham also cites Niven's 43; we have seen and dissected the flowers of the specimen so marked by him and find it to be without doubt *E. flacca,* E. Meyer. Ecklon and Zeyher's specimen we have not seen. Bentham cites *E. suaveolens, Lodd. Bot. Cab. t.* 24, as a synonym; but this figure seems to us nearer to our *E. debilis.*

203. E. cordata (Andr. Heathery, t. 158); erect, 1¼ ft. or more high; branches more or less slender, sometimes spreading, but not diffuse, closely pubescent and also densely pilose with gland-tipped hairs; leaves 3-nate, spreading, somewhat crowded, from ovate to lanceolate, acute, open-backed, more or less strongly revolute at the margins, sometimes subcordate at the base, upper surface hispid with tubercle-based hairs, becoming scabrid, closely pale tomentose below, 1½–2½ lin. long, or, in luxuriant specimens, 4 lin. long; flowers terminal, 3-nate, more rarely clustered or capitate, 3–6-flowered, subcalycine; pedicels pilose, 1½–2 lin. long; bracts remote, small, scarious; sepals ovate, acute or acuminate, scarious, subviscid, coloured, ciliate, 1–1¼ lin. long; corolla broad-urceolate-campanulate or globose-urceolate, mouth not, or very slightly, contracted, subviscidulous, shining, pale red, 1¼–1½ lin. long; segments rounded, mostly spreading, about ⅓ as long as the tube; filaments rather broad, dilated just below the anther; anthers subexserted, lateral, oblong, tapering towards the apex, not quite ½ lin. long, muticous; pore over ½ as long as the cell; style exserted; stigma capitellate; ovary villous. *Benth. in DC. Prodr.* vii. 669; *Andr. Col. Heaths, t.* 160. *E. punctata, Bartl. in Linnæa,* vii. 646.

SOUTH AFRICA : without locality, *Masson! Drège,* 2315 !
COAST REGION, between 2000–3000 ft. : Swellendam Div.; mountains near

Swellendam, *Mund*, 47 ! *Shand in Herb. Bolus*, 6254 ! *Galpin*, 3658 ! Voormans Bosch, *Zeyher*, 3213 ! Riversdale Div.; mountains near Kaffirkuils River, *Niven*, 37 ! Garcias Pass, *Burchell*, 7032 ! Kampsche Berg, *Burchell*, 7077 ! mountains near Riversdale, *Schlechter*, 2196 ! Uitenhage Div.; Vanstadens Mountains, *Zeyher*, 787 !

204. E. hirsuta (Klotzsch ex Benth. in DC. Prodr. vii. 669, not of Thunb., nor Salisb., nor Lodd.); erect, 1 ft. or more high; branches ascending; branchlets spreading or sometimes divaricate, hirsute; leaves 4-nate, or sometimes 3-nate, spreading, not close-set, ovate and more or less flat with reflexed margins, or lanceolate with more strongly revolute margins, acute, open-backed, midrib prominent, pubescent below, glabrous above, setose-ciliate, 2-$2\frac{1}{2}$ lin. long, 1-$1\frac{1}{2}$ lin. wide; flowers terminal, sub-4-nate or umbellate (acc. to *Bentham* occasionally axillary); pedicels slender, gland-pilose, 2-4 lin. long; bracts remote, lax, small, variable in shape and position; sepals ovate-lanceolate, scarious, shining, viscid, coloured, more or less copiously setose-ciliate with longish subdistant hairs, rarely almost naked, otherwise glabrous, about 1 lin. long; corolla urceolate or ovoid-urceolate, mouth much contracted, viscid, 2-$2\frac{1}{2}$ lin. long; limb erect or slightly spreading, $\frac{1}{6}$-$\frac{1}{5}$ as long as the tube; filaments narrow, dilated just below the anther; anthers exserted or subexserted, lateral, dorsifixed just above the base, oblong, curved, subacute, about $\frac{4}{5}$ lin. long, minutely aristulate or perhaps sometimes (as described by *Bentham*) muticous; pore narrow, more than $\frac{1}{2}$ as long as the cell; awns scarcely reaching below the base of the cell; style slender, well exserted; stigma capitellate; ovary hispidulous.

SOUTH AFRICA : without locality, *Mund!*
COAST REGION : George Div.; on the mountain near George, *Alexander*, 12 ! Montagu Pass, 1200–1500 ft., *Young in Herb. Bolus*, 5524 ! *Schlechter*, 5786 !

205. E. Lehmannii (Klotzsch ex Benth. in DC. Prodr. vii. 618); erect, 2–3 ft. high; branches ascending, subvirgate, pilose; leaves 3-nate, suberect, imbricate, ovate to lanceolate, acute, open-backed, margins revolute, roughly ciliate with longish tubercle-based hairs, cano-puberulous below, glabrous above, $1\frac{1}{2}$-2 lin. long; flowers terminal, 3-nate, subcalycine; pedicels $\frac{1}{2}$ lin. long; bracts approximate, lanceolate, acuminate, scarious, ciliate, larger than usual in the section, $1\frac{1}{2}$-2 lin. long; sepals lanceolate, subulate-acuminate, scarious, ciliate, $1\frac{1}{2}$-2 lin. long; corolla suburceolate-campanulate, mouth scarcely or not at all contracted, glabrous, dry, about 2 lin. long; segments slightly spreading, about $\frac{1}{3}$ as long as the tube; filaments broader at the base, tapering upwards; anthers exserted, terminal or subterminal, narrow-oblong or subobovate-oblong, incurved, about $\frac{2}{3}$ lin. long, muticous; pore less than $\frac{1}{2}$ as long as the tube; style slender, exserted; stigma very small, subsimple or capitellate; ovary glabrous.

COAST REGION : George Div.; on the Post Berg (now Cradock Berg) near

George, *Burchell*, 5908 ! 5981 ! mountains near George, *Drège*, 7784 ! Montagu
Pass, 1200 ft., *Tyson*, 3163 !

Placed by Bentham in § *Geissostegia ;* but the bracts are not those of that
section, and sublateral anthers, approaching these, occur elsewhere in this
section, to which its open-backed leaves seem to indicate a natural alliance.

206. E. macrophylla (Klotzsch ex Benth. in DC. Prodr. vii.
669); procumbent or sometimes suberect ? ; branches stout, rigid,
villous; leaves 3-nate, spreading, crowded, ovate, acute, open-
backed, margins revolute, bullate-convex, and densely villous with
long hairs on the upper surface, closely pale tomentose below,
$2\frac{1}{2}$–$3\frac{1}{2}$ lin. long including the rather long petiole; flowers terminal,
clustered or sometimes (acc. to *Bentham*) also axillary; pedicels
stoutish, $\frac{1}{2}$–1 lin. long; bracts remote, 2 very small, 1 larger; sepals
oblong, in opposite pairs, 2 longer and 2 shorter, or ovate-lanceolate,
obtuse, viscid, coloured, about $\frac{1}{2}$ lin. long; corolla ovoid-urceolate,
mouth contracted, viscid, minutely puberulous or glabrous, $2\frac{1}{2}$–3 lin.
long; segments suberect, $\frac{1}{6}$–$\frac{1}{5}$ as long as the tube; anthers subexserted, sublateral, broad-linear, about $\frac{2}{3}$ lin. long, muticous; pore
about $\frac{1}{2}$ as long as the cell; style exserted, decurved; stigma capitate; ovary densely and closely white-woolly.

COAST REGION : Swellendam Div. ; summit of a mountain peak near Swellendam, *Burchell*, 7330! Grootvaders Bosch, *Masson !* Riversdale Div. ; mountains near Kafferkuils River, on moist rocks, *Niven*, 36 ! summit of Kampsche
Berg, *Burchell*, 7118! George Div.; Cradock Berg, *Burchell*, 5909 !

A distinct species, not found by any recent collector. We have chiefly
described from Burchell's 7330 !

207. E. ocellata (Guthrie & Bolus) ; erect ; branches of medium
thickness, not slender, nor diffuse, pilose ; leaves 3-nate, spreading,
the older distant, the younger imbricate, ovate or lanceolate, open-
backed, margins recurved or revolute, rough and ciliate with tubercle-
based hairs, upper surface somewhat concave, glabrous, shining,
closely tomentose and pale below, from 2–4 lin. long, 1–2 lin. wide;
flowers capitate ; heads 6–10-flowered ; pedicels stoutish, under
1 lin. long; bracts remote to subapproximate, small, glandular-
pilose ; sepals lanceolate or oblong, viscid, pale, very inconspicuous,
$\frac{1}{2}$–$\frac{2}{3}$ lin. long ; corolla urceolate, mouth contracted, glabrous, viscid,
$1\frac{3}{4}$ lin. long, $1\frac{1}{2}$ lin. wide, limb and generally also a small portion of
the upper part of the tube recurved or revolute ; segments about
$\frac{1}{6}$ as long as the tube, broad, shortly subacute ; filaments broadish,
slightly dilated below the anther; anthers subexserted, sublateral,
basifixed at the dorsal side, or lateral and dorsifixed near the base,
oblong, tapering somewhat to the apex, about $\frac{1}{2}$ lin. long, muticous;
pore over $\frac{1}{2}$ as long as the cell; style exserted, at length decurved ;
stigma clavate-capitate; ovary loosely white-woolly.

COAST REGION: Swellendam Div. ; Tradouw Pass, *Borcherds* in Herb. Bolus,
6496 ! Zuurbraak Mountain, 2500 ft., *Galpin*, 3657 !

The anthers do not seem quite constant, some being distinctly lateral, others
affixed at the very base of the cell, though dorsal.

Section XV. **DESMIA**. (Sp. 208–210.)

208. E. conferta (Andr. Col. Heaths, t. 83); erect, entirely glabrous, 1–2 ft. high; branches slender, 4-sulcate from the long prominent decurrent leaf-cushions; leaves 4-nate, spreading or reflexed, crowded or distant, linear, acuminate, sulcate, aristate, 6–11 lin. long; flowers capitate, corolline; heads 6–20-flowered; pedicels about 1 lin. long; bracts approximate, lanceolate, a little longer than the sepals; sepals lanceolate, acute, about $\frac{1}{2}$ lin. long, coloured; corolla subglobose-urceolate, constricted at the throat, white (*Andrews*), $1\frac{1}{2}$–$1\frac{3}{4}$ lin. long; segments short, broad, rounded, revolute, $\frac{1}{6}$–$\frac{1}{5}$ as long as the tube; anthers exserted or subexserted, narrow-elliptical, subobtuse, acute at the base, smooth, pale brown, about $\frac{2}{7}$ lin. long, muticous; pore $\frac{3}{5}$ as long as the cell; seeds foveolate, the pits oblong. *Heathery, t.* 59; *Lodd. Bot. Cab. t.* 1335; *Benth. in DC. Prodr.* vii. 615.

South Africa: without locality, *Masson!* and *cultivated specimens!*
Coast Region: Riversdale Div.; alpine shady places, near the Kafferkuils River, scarce, *Niven,* 104! and at Riet Kuil, *Niven,* 104!

209. E. polifolia (Salisb. ex Benth. in DC. Prodr. vii. 615); weakly diffuse and probably trailing amongst other shrubs or in long grass, or stouter and erect; branches up to 12 in. long, slender, pale, the younger furrowed by the decurrent leaf-cushions, distantly leafy with internodes $\frac{1}{2}$–1 in. long, ultimate flowering branches sometimes pedunculoid, with leaves reduced to small (3-nate) bracts; leaves 3-nate, spreading or recurved, linear, acute, aristate, flattish, sulcate, 6–11 lin. long, mostly $\frac{3}{4}$ lin. in width, but sometimes reaching $1\frac{1}{2}$ lin.; flowers subcorolline, in 3–6-flowered umbels; pedicels 1–2 lin. long; bracts all approximate, or 1 remote, basal, ovate, acuminate, about 2 lin. long, the lower very caducous; sepals ovate, cuspidate, acuminate, or aristate, often gland-ciliate, $1\frac{1}{2}$–2 lin. long; corolla broad-urceolate, slightly contracted at the throat, $2\frac{1}{4}$–$2\frac{1}{2}$ lin. long; segments broadly rounded, spreading or recurved, about $\frac{1}{2}$ lin. long; filaments narrow, or broader than the anther (in front view); anthers exserted, narrow-elliptical, darkbrown, scaberulous, about $\frac{3}{4}$ lin. long; pore $\frac{2}{3}$ as long as the cell. *Benth. in DC. Prodr.* vii. 615. *E. caduca, Thunb.? Prodr.* 71; *Flor. Cap. ed. Schult* 356; *Rach in Linnæa,* xxvi. 769. *E. cuspidata, Klotzsch ex Benth. l.c.* 615. *E. æqualis, Benth., l.c.* 615.

Var. β, **angustata** (Bolus); sepals lanceolate or linear-lanceolate, tapering gradually to the aristate apex.

South Africa: without locality, *Masson!*
Coast Region: Var. β, Swellendam Div.; trailing slender plant, watery places among long grass, alpine situations near Swellendam, *Niven,* 232! Voormans Bosch, *Zeyher,* 3247! Langeberg Range, near Zuurbraak, 3000 ft., *Schlechter,* 2046! 5673! *Galpin,* 3532!

Masson's specimens, being the earliest known, were doubtless Salisbury's type. The sepals in it are somewhat differently shaped from most of the others

we have examined, being more ovate and shorter; but we can find no other
differences, nor can we in any way separate *E. æqualis* by definite characters.
The habit of the plant varies according to the locality,—specimens from drier
open places being more erect, those from moist grassy spots more diffuse, and
there are intermediate forms, such as Ecklon & Zeyher's 3247 and Galpin's
3534, from mountains near Swellendam. Thunberg's *E. caduca*, above cited,
must remain, a very doubtful species : it was collected on Table Mountain,
near Cape Town. His specimen is without flowers and his description is quite
inadequate. It may be either this, or possibly *E. obtusata, Klotzsch.*

210. E. obtusata (Klotzsch ex Benth. in DC. Prodr. vii. 615);
erect; branches many, divaricately spreading, puberulous or glan-
dular-scaberulous; leaves 3-nate, spreading or squarrose, linear or
subterete, rarely flattish, obtuse, thick, sulcate, subviscidulous,
shining, minutely gland-scabrous, 3–5 lin. long, leaf-cushions short
and crescent-shaped, not decurrent; flowers subcorolline, 3–6 in
short umbels or subcapitate; pedicels 1–1½ lin. long; bracts subremote, small, scarious, like the pedicels, sepals and corolla viscidulous; sepals lanceolate, acute, subscarious, nerved, margins reflexed
and minutely gland-ciliolate, 1–1½ lin. long, reaching to the height
of the corolla-tube or less; corolla globose-urceolate, viscid, white,
about 1½ lin. long; segments spreading, about ½ the length of the
tube; filaments broad, curved; anthers exserted or subexserted,
broadly elliptical, very obtuse, light brown, smooth, about ⅜ lin. long;
pore very wide and large, occupying nearly the whole of the cell;
style shortly exserted; stigma capitate, large.

CoAST REGION, between 2000 and 5000 ft.: Cape Div.; Table Mountain,
Bergius, Harvey, 59! head of Waai Vley, 3000 ft., *Wolley Dod,* 3257! Caledon
Div.; mountains near Genadendal, *Drège, Galpin,* 3533! Houw Hoek,
Schlechter, 5467! Klein River, *Niven,* 233!

Niven marks his ticket " glutinous plant, 2–3 ft. high."

Section XVI. **GYPSOCALLIS.** (Sp. 211–218.)

211. E. racemosa (Thunb. Diss. Erica, 31, t. 5); erect, 6–12 in.
high; branches somewhat slender, pubescent, and hirsute with
longer often gland-tipped hairs; leaves 4-nate, the upper erect-
spreading, the lower often subsquarrose, crowded, linear, sulcate,
rarely narrow-lanceolate and subopen-backed, pubescent and ciliate
with gland-tipped hairs, 1–2 lin. long; flowers 3–4 in each whorl
of the leaves at the ends of the branches, generally forming a more
or less dense pseudo-raceme (not truly racemose); pedicels puberulous, 1½–2½ lin. long; bracts remote, small; corolla urceolate, mouth
somewhat contracted (when young subtubular or cyathiform, mouth
not or scarcely contracted), glabrous, 1½–1¾ lin. long; segments
⅛–⅙ as long as the tube; filaments very slender, equal; anthers
commonly subexserted, occasionally exserted, lateral, suboblong,
distinctly incurved at the back, scabrid, a little more than ⅖ lin.
long, muticous; pore about ⅓ as long as the cell; style exserted
beyond the anthers, straight; stigma capitellate; ovary hispid.

Benth. in DC. Prodr. vii. 668; *Wendl. Eric. Ic. fasc.* 10, 3. *E. flexilis, Salisb. Prodr.* 296, *and in Trans. Linn. Soc.* vi. 342. *E. hispida, Andr. Heathery, t.* 69, *and Col. Heaths, t.* 100, *not of Burm. f.?, nor of Thunb.; Lodd. Bot. Cab. t.* 1982.

South Africa: without locality, *Thunberg, Herb. Salisbury!* and *cultivated specimens!*

Coast Region, between 300 and 1500 ft.: Caledon Div.; mountains of Baviaans Kloof near Genadendal, *Burchell,* 7808! *Niven! Drège,* 7755! Zoetemelks Vlei, *Grisbrook in Herb. Guthrie,* 3542! near Villiersdorp, *Bolus,* 5177! near the Zondereinde River, *Schlechter,* 9890! Swellendam Div.; near Swellendam, *Mund,* 46! near Grootvaders Bosch, *Burchell,* 7215! mountains near Voormans Bosch, *Zeyher,* 3235! near Zuurbraak, *Galpin,* 3654! Langeberg Range, *Schlechter,* 5668! Riversdale Div.; Garcias Pass, *Burchell,* 7049! near Riversdale, *Schlechter,* 1777!

Bentham remarks that in cultivation the leaves become more open-backed, showing an affinity with § *Ceramia.* Following our predecessors we cite Thunberg as the author, though it is most probably the *E. hispida, Burm. f. Prodr. Fl. Cap.* 11 (1768). It is, however, unlikely that any certainty is attainable, and Burmann's brief description of seven words is almost as useless as a bare name.

212. E. aghillana (Guthrie & Bolus); erect, apparently 6–8 in. high; branches subglabrous, channelled, pale brown; leaves 4-nate, the upper erect and adpressed, the lower spreading or squarrose, linear, subobtuse, sulcate, glabrous, the younger ciliate, 2–3 lin. long; flowers 1–2 in each whorl of leaves towards the ends of the branches, forming a pseudo-raceme; pedicels about $1\frac{1}{2}$ lin. long; bracts remote, minute, hairy; sepals narrow-lanceolate, obtuse, pubescent, foliaceous, reddish, about $\frac{3}{4}$ lin. long; corolla broad-urceolate-campanulate, mouth slightly, or not at all, contracted, the younger tetragonous, the older becoming round, glabrous, red, $1\frac{1}{4}$–$1\frac{1}{2}$ lin. long; segments erect, obtuse, rounded, $\frac{1}{4}$–$\frac{1}{3}$ the length of the tube; anthers exserted, lateral, oblong, slightly incurved, scabrid, nearly $\frac{2}{3}$ lin. long; pore $\frac{1}{4}$ the length of the cell, muticous; style well exserted, stoutish; stigma capitate; ovary glabrous, lobed, pale.

Var. β, latifolia (Guthrie & Bolus); leaves oblong, obtuse, $1\frac{1}{2}$ lin. long, $\frac{2}{3}$ lin. wide, thick, closely adpressed to the branches (at least the upper ones); pedicels longer, 4–$5\frac{1}{2}$ lin. long.

South Africa: without locality, Var. β: *Mund!* in the Cape Govt. Herb.

Coast Region: Bredasdorp Div.; Rhenoster Kop, near Cape Agulhas, 400 ft., *Schlechter,* 10571!

213. E. petræa (Benth. in DC. Prodr. vii. 668); erect, 1 ft. or more high; branches ascending but somewhat spreading or straggling, stoutish, glabrous or slightly pubescent; leaves 3-nate, erect, imbricate, obovate-oblong, subacute, the upper face concave with a thick nerve, the lower sulcate, thick, glabrous, smooth and shining, margin cartilaginous, 2–$2\frac{1}{2}$ lin. long; flowers terminal and axillary towards the ends of the branches; pedicels slender, pubescent, $1\frac{1}{2}$ lin. long; bracts 2, closely approximate, small, foliaceous; sepals lanceolate, acute, keel-tipped, ciliolate, rigid, thick, ¾ lin. long; corolla ovoid, glabrous, 1¼–1½ lin. long; segments erect, short, rounded; anthers exserted, lateral, oblong, slightly incurved, bifid, minutely scaberulous, pale brown, nearly ⅗ lin. long, very shortly aristulate; pore over ⅓ as long as the cell; style straight, exserted beyond the anthers; stigma capitellate; ovary glabrous.

COAST REGION: George or Uniondale Div.; dry stony places, Kamanassie Mountains, *Masson*, 66!

214. E. dumosa (Andr. Heathery, t. 213, not of Salisb.); erect, 6–12 in. high; branches slender but rigid, mostly spreading and flexuous, the younger glandular-pubescent; leaves 3-nate, erect-spreading, imbricate, more rarely distant, linear, narrow-oblong or linear-lanceolate, generally tipped with a rigid white callosity, thick and deeply sulcate, with strong revolute margins, or more rarely subopen-backed, somewhat bullate, tomentose-pubescent, occasionally with gland-tipped hairs towards the apex, at length somewhat glabrescent, 1½–2½ lin. long; flowers 1–2 in each leaf-whorl towards the ends of the branches, mostly somewhat lax and distant; pedicels capillary or slender, ascending, straight or curved, persistent, elongating after flowering, 2½–6 (sometimes even 9) lin. long; bracts 3, basal, foliaceous; sepals ovate-lanceolate, acute or subobtuse, foliaceous, with revolute margins, dark-coloured, gland-pubescent, 1–1½ lin. long, generally about ½ the length of the corolla; corolla varying from ovoid to tubular-ovoid, with contracted mouth, glabrous, rosy to darker red, 3–3½ lin. long; filaments slightly dilated at the base and tapering upwards, ciliate or naked; anthers exserted, lateral, oblong or linear, nearly straight, 1–1⅛ lin. long, muticous; pore ⅕–¼ as long as the cell; style exserted; stigma capitellate; ovary glabrous. *Andr. Col. Heaths, t.* 230; *Benth. in DC. Prodr.* vii. 668. *E. longipedunculata, Lodd. Bot. Cab. t.* 103.

VAR. β, **intermedia** (Bolus); leaves with an apical callosity only; corolla urceolate-campanulate or campanulate-cyathiform, mouth slightly widened, or rarely a few flowers ovoid or tubular-ovoid, with contracted mouth, as in the type, glabrous, 1¾–2 lin. long; filaments wider at the base than in the type, sometimes distinctly ovate at the base and more abruptly narrowed upwards; anthers exserted or subincluded, but manifest, curved, smaller, ½–⅔ lin. long; ovary pubescent.

VAR. γ, **setifera** (Bolus); leaves tipped with an apical white callosity produced into a longish white bristle, tapering to a fine point; sepals bristle-tipped; pedicels somewhat shorter and capillary; corolla like the campanulate forms of var. β, but smaller, 1½ lin. long, pubescent or sometimes glabrous; filaments ovate-lanceolate at the base, ciliate; anthers exserted or subincluded, curved, about ⅔ lin. long; ovary pubescent or villous with longish hairs.

SOUTH AFRICA: without locality, *cultivated specimens!*

COAST REGION: Clanwilliam Div.; Ezelsbank, near Wupperthal, *Schlechter*, 8810! Ceres Div.; near Ceres, *Bolus*, 8480!

CENTRAL REGION, at 4000–5800 ft.: Ceres Div.; Cold Bokkeveld, *Schlechter*, 8920! Var. β: Ceres Div.; Gydouw Mountain, *Schlechter*, 10224! near Klein

Vlei, *Schlechter*, 10068 ! Var. γ : Ceres Div.; Cold Bokkeveld, near Sandfontein, *Schlechter*, 10148 !

A very variable yet well-marked species. To Bentham it was only known by Andrews' figure and by two cultivated specimens in Herb. Kew. With these, the specimens of *Bolus*, 8480 agree very well, due allowance being made for the effects of cultivation. The agreement of the other specimens is not obvious, but only revealed upon search. Regarding only extreme forms, the propriety of uniting them might be doubtful. One character is common to all and serves to distinguish the species from its allies, viz., the presence of a callous white point to the leaves, which is seldom wanting. The difference in the shape of the corolla is the real crux; and here the specimens of *Schlechter*, 10068, solved the difficulty, since they exhibit, in some cases, corollas like those of the type, and in others, of a form nearly like those of var. γ.

215. E. fucata (Klotzsch ex Benth. in DC. Prodr. vii. 667, not of Thunb.) ; erect or procumbent, glabrous in all parts, 6–12 in. high ; leaves 3-nate, gemmiferous in the axils, erect-incurved, linear, subobtuse, flat above, round-backed, faintly sulcate, 2–3 lin. long ; flowers subsolitary and somewhat distant or crowded towards the ends of the branches, subcalycine ; pedicels slender, erect-decurved, red, 3–5½ lin. long ; bracts remote, minute ; sepals lanceolate, acute or obtuse, concave, cartilaginous, margins hyaline, keel-tipped, red, 1–1½ lin. long ; corolla broad-cyathiform or subglobose-urceolate, 1–1¼ lin. long, and as wide, mouth slightly widened ; segments rounded, spreading or erect, ⅓ the length of the tube ; anthers exserted or subexserted, from sublateral to subterminal, oblong or semilanceolate, subacute, tapering towards the base, ½–⅗ lin. long, muticous ; pore nearly ½ the length of the cell ; style exserted, stoutish, decurved, red ; stigma subsimple or capitellate ; ovary hemispherical, truncate above, glabrous, dark red.

VAR. β, **cæspitosa** (Bolus) ; procumbent on rocks, subcæspitose ; corolla subtetragonous towards the base, 1½ lin. long ; segments a little smaller, ⅕–¼ the length of the tube.

COAST REGION : Caledon Div.; mountains of Klein River Kloof, *Zeyher*, 3341 ! Bredasdorp Div.; hills near Elim, 300 ft., *Bolus*, 6739 ! Var. β : Bredasdorp Div.; on rocks near Elim, 250 ft., *Bolus*, 8507 ! *Schlechter*, 9709 !

There has been some confusion as to the type of this species, and some specimens in herbaria thus marked do not well agree. Bentham unfortunately does not quote Zeyher's number. We have described from a specimen with fully developed flowers in the Kew Herb., marked as coming from the Berlin Herb. (but not bearing any number) and which no doubt, as it agrees well with his description, was what Bentham described. We have also seen a specimen in the Cape Govt. Herb., numbered 3341 and marked " stony places in Klein Rivers Kloof, Aug." with quite undeveloped flowers but which no doubt belongs to this. The species is allied to *E. scytophylla*, from which it is distinguished by its generally longer leaves and pedicels, larger flowers, and muticous anthers. Var. β has scarcely any differences from the type beyond its habit, and somewhat paler flowers.

216. E. scytophylla (Guthrie & Bolus) ; except the branches, entirely glabrous ; branches erect or spreading, few, subvirgate, rigid, very little divided and leafy above, soon naked below, downy, glabrescent, 9–10 in. long ; leaves 3-nate, erect, incurved, imbricate, the younger elliptical or oblong, subobtuse, callous-denticulate, the older and lower narrow-oblong, somewhat longer and naked, all sulcate, round-backed, thick, leathery, rigid, smooth, 2–3 lin. long ; flowers corolline, or subcalycine, solitary in the axils of the leaves, with 3 minute bract-like leaves above and at the base of each pedicel ; pedicels slender, coloured, downy, 2–3 lin. long ; bracts remote, minute, 2 infra-median, 1 nearly basal ; sepals adpressed, linear, thick, rigid, red, $1\frac{1}{4}$–$1\frac{1}{2}$ lin. long, reaching to about the top of the corolla-tube ; corolla urceolate-campanulate, scarcely contracted at the throat at full maturity, subtetragonous, thick, sub-fleshy, rosy, $1\frac{1}{2}$ lin. long ; segments spreading, ovate, from $\frac{1}{3}$ of the tube in length to nearly equal to it ; filaments lanceolate ; anthers subexserted or just manifest, subterminal, oblong, tapering to the dorsal margin ; cells bipartite, approximate, dark-coloured, about $\frac{1}{2}$ lin. long, denticulate at the base, teeth short, squarrose ; pore less than $\frac{1}{2}$ the length of the cell ; style exserted ; stigma small, capitellate ; ovary glabrous.

COAST REGION : Bredasdorp Div. ; hills near Mier Kraal, 300 ft., *Schlechter*, 10526 !

This bears some resemblance to *E. curtophylla*.

217. E. capillaris (Bartl. in Linnæa, vii. 647); erect, entirely glabrous, 6–12 in. high ; branches numerous, erect, corymbose ; leaves 3-nate, erect, imbricate, slender, linear, acute, straight, 1–$1\frac{1}{2}$ lin. long ; flowers axillary, pseudo-racemose towards the ends of the branches ; pedicels slender, about 1 lin. long ; bracts remote, small, sometimes one or more wanting ; sepals linear-lanceolate, sulcate, glabrous or ciliate, foliaceous, rigid, about $\frac{3}{4}$ lin. long; corolla tubular-campanulate or obconic-campanulate, commonly widened to the mouth, rarely equal, subtetragonous, $1\frac{1}{2}$ lin. long ; segments spreading, $\frac{1}{4}$–$\frac{1}{3}$ the length of the tube ; anthers exserted, lateral, linear or oblong, bipartite, slightly curved, pale brown, from under $\frac{1}{2}$ lin. to nearly $\frac{2}{3}$ lin. long, muticous ; pore about $\frac{1}{4}$ the length of the cell ; style exserted, slender ; stigma capitellate ; ovary glabrous.

VAR. β, **compacta** (Bolus); leaves oblong or elliptic, incurved, often shorter than the internodes ; sepals and corolla somewhat shorter and broader than in the type ; flowers more closely crowded at the ends of the branches; anthers less than $\frac{1}{2}$ lin. long, more curved.

VAR. γ, **poliotes** (Bolus); habit stronger and somewhat taller, 1–$1\frac{1}{2}$ ft. high, general colour ashy-grey ; branches pubescent ; leaves erect-incurved, internodes and petioles longer, oblong to elliptical, thick and somewhat concave above, convex below, the younger ciliate, 1–$1\frac{1}{2}$ lin. long; inflorescence mostly lateral but sometimes also terminal ; anthers curved as in var. β and about as long, but somewhat broader.

COAST REGION : Cape Div. ; plains near Wynberg, *Niven*, 10 ! Cape Flats, *Zeyher*, 1105 ! near Cape Town, *Bolus*, 4614 ! *Harvey*, 161 ! Var. β : Caledon Div. ; Klein River, 1000 ft., *Schlechter*, 7607 ! Bredasdorp Div. ; Elim, 300 ft., *Schlechter*, 7700 !

CENTRAL REGION : Var. γ : Ceres Div. ; Mountains near Klein Vlei, in the Cold Bokkeveld, 5000 ft., *Schlechter*, 10056 !

The type is a neat little shrub, very uniform in character. Until quite lately it had only been found on the Cape Peninsula. Recently Mr. Schlechter has discovered what seems another form further eastwards,—our var. β, which we cannot separate by any tangible characters, though it looks somewhat different on account of its shorter and more distant leaves, &c. It is intermediate between the typical form and our var. γ. But for it we should probably have made a species of Schlechter, 10056, the flowers on which are, however, mostly undeveloped; one or two show an almost exactly identical corolla, and only the anther is shorter and broader in shape.

218. E. nudiflora (Linn. Mant. Alt. 229); erect, 6–18 in. high; branches numerous, erect, subcorymbose or subvirgate, hirsute; leaves 3-nate, erect to wide-spreading, linear-subulate or linear-semiterete, subacute, flattish and glabrous above, round-backed, sulcate, generally more or less hispid below, ciliate, hairs tubercle-based, sometimes almost entirely glabrous, 2–4 lin. long; pedicels slender, puberulous, $1\frac{1}{2}$–$2\frac{1}{2}$ lin. long; bracts remote, small; sepals lanceolate, sometimes united for a short distance above the base, acute, ciliate, rigid, keeled, $\frac{1}{2}$–$\frac{3}{4}$ lin. long; corolla subcampanulate-cyathiform, mouth not contracted, sometimes varying to tetragonous-tubular, narrow-ovoid, or even ovoid with contracted mouth, glabrous, usually bright red, rarely pale rose, $1\frac{1}{2}$–2 lin. long; segments slightly spreading or erect, $\frac{1}{8}$–$\frac{1}{6}$ the length of the tube; filaments capillary, subequal; anthers wholly exserted, lateral, dorsifixed just above the base, oblong or slightly wider above, scaberulous, rounded, very obtuse, sometimes longitudinally semiobovate, tapering very much to the base and thus appearing subterminal, about $\frac{1}{2}$ lin. long, muticous; pore $\frac{1}{4}$–$\frac{1}{3}$ the length of the cell; style slender, exserted; stigma subsimple; ovary glabrous. *Benth. in DC. Prodr.* vii. 668. *E. floribunda, Wendl. Eric. Ic. fasc.* 14, 19, *not of Lodd. E. alopecuroides, Lodd. Bot. Cab. t.* 874, *not of Wendl. E. microstoma, Berg. ex G. Don, Gen. Syst.* iii. 801. *E. sertiflora, Salisb. in Trans. Linn. Soc.* vi. 342.

South Africa: without locality, *Thunberg; Herb. Salisbury!* and *cultivated specimens!*

Coast Region, on mountains from 400 to 2500 ft. : Clanwilliam Div. ; Cederberg Range, *Leipoldt*, 131 ! Tulbagh Div. ; Witsen Berg, *Burchell*, 8674 ! Roode Zand, *Niven*, 9 ; Mitchells Pass, *MacOwan & Bolus, Herb. Norm.*, 30 ! Paarl Div. ; Paarl Mountain, *Drège!* Drakensteen Mountains, *Bolus*, 4057 ! French Hoek, *Guthrie*, 3435 ! Cape Div. ; *Burchell*, 816 ! *Sieber*, 182 ! *Drège! Bolus*, 3713 ! *MacOwan*, 2300 ! *Guthrie*, 816 ! 817 ! *Wolley Dod*, 1009 ! Stellenbosch Div. ; Lowrys Pass, *Burchell*, 8209 ! *Schlechter*, 4811 ! Caledon Div. ; near Genadendal, *Burchell*, 7610 ! 7794 ! 8620 ! Donkerhoek Mountain, *Burchell*, 7998 ! Hemel-en-Aarde, *Schlechter*, 10375 !

Chiefly variable in the shape of the corolla; but generally recognizable by its numerous flowers and well-exserted stamens. The pedicels are always shorter and the corolla generally longer, both absolutely and relatively to its width, than in *E. fucata.*

Section XVII. **PYRONIUM.** (Sp. 219–232.)

219. E. nutans (Wendl. Eric. Ic. et Descr. fasc. 3, 5); erect, 3–4 ft. high; branches slender, pubescent; leaves 3-nate (or 3–4-nate), erect-spreading, linear-filiform, glandular-ciliolate, scabrid,

2–3 lin. long ; flowers terminal, sub-3-nate, cernuous ; bracts sub-remote, foliaceous, small ; sepals narrow-ovate, acuminate, keel-tipped, foliaceous, pubescent, slightly over $\frac{1}{2}$ lin. long ; corolla urceolate or ovoid-urceolate, more or less contracted at the mouth, viscid, white or rosy, $1\frac{1}{4}$–$2\frac{1}{2}$ lin. long ; segments nearly erect, $\frac{1}{7}$–$\frac{1}{5}$ of the length of the tube ; anthers well exserted, subterminal, obovate, or (according to a specimen in Herb. Berol.) oblong, very obtuse, scaberulous on the margins, $\frac{2}{5}$ lin. long, muticous ; pore about $\frac{1}{3}$ the length of the cell ; style slender, exserted ; stigma capitellate, small ; ovary pubescent, seated on a largish dark-coloured disk. *Benth. in DC. Prodr.* vii. 618. *E. pudibunda, Salisb. in Trans. Linn. Soc.* vi. 345. *E. padibunda, Pers. Syn.* i. 431.

SOUTH AFRICA : without locality, *cultivated specimen* from Herren-hausen !

Closely allied to *E. deliciosa,* from which it is chiefly distinguishable by its relatively smaller sepals and broader anthers ; and we are by no means sure that it should not be regarded as a variety of that species. Our description is drawn from Wendland's, as well as from his figure, and from the specimen we have seen.

220. E. deliciosa (Wendl. fil. ex Benth. in DC. Prodr. vii. 666) ; erect, 2–3 ft. high ; branches white-pubescent ; leaves 3-nate, erect-spreading, imbricate, slender, linear, blunt, sulcate, glabrescent, $1\frac{1}{2}$–2 lin. long ; flowers 3-nate, abundant ; pedicels $1\frac{1}{2}$–$2\frac{1}{2}$ lin. long ; bracts approximate or sometimes subremote, foliaceous ; sepals from broadish- to narrow-lanceolate, subacute, sometimes foliaceous, greenish and dry, sometimes subscarious, coloured and slightly viscid, $\frac{2}{3}$–$1\frac{1}{3}$ lin. long ; corolla urceolate, mouth very slightly con-tracted, or ovoid with a narrower mouth, glabrous, white or rosy, $1\frac{3}{4}$–$2\frac{1}{4}$ lin. long ; segments erect, very short, about $\frac{1}{5}$ the length of the tube ; anthers subexserted, subterminal, or sometimes strictly terminal, suboblong or oblong-obovate, dorsally curved, very obtuse, about $\frac{1}{2}$ lin. long, quite muticous, or very minutely toothed on the cells above the base, or with a single projecting tooth between the cells at their base in front ; pore $\frac{1}{3}$–$\frac{2}{5}$ the length of the cell ; style exserted ; stigma clavate-capitellate, small ; ovary hispid-puberu-lous. *E. umbellata, Bartl. in Linnæa,* vii. 647 ? *not of Linn.*

COAST REGION, between 500 and 1500 ft. : Riversdale Div. ; near Garcias Pass, *Burchell,* 7155 ! Langeberg Range, 1000 ft., *Schlechter,* 1729 ! George Div. ; Montagu Pass, 1500 ft., *Schlechter,* 5799 ! Oudtshoorn Div. ; Olifants River, *Gill !* Mossel Bay Div. ; Attaquas Kloof, *Niven,* 12 ! Humansdorp Div. ; Kromme River, *Masson,* 55 ! near Clarkson, 500 ft., *Galpin,* 3653 ! Uiten-hage Div. ; Van Stadens Berg, *MacOwan,* 1034 ! *Holland in Herb. Bolus,* 1209 ! *Bolus,* 1576 !

This species is singularly variable in respect of its anther. The white-flowered form from Van Stadens Berg has usually the curious single basal free tooth. The red-flowered forms from the stations further west are occasionally either quite muticous, or with the single basal tooth adnate to the filament, or with two very minute teeth upon the cell, distant from the filament and some-times hardly visible. They are constant in being very obtuse and scaberulous on the front margin.

221. E. drakensbergensis (Guthrie & Bolus); erect, usually 1–2 ft., sometimes 3 ft. high; branches erect or spreading, pallid, puberulous; leaves 3–4-nate, often varying on the same plant, mostly erect and imbricate, or sometimes shorter than the internodes, more rarely spreading, linear, semiterete, glabrous, pallid, 1–2 lin. long; flowers 3–4–6-nate; pedicels straight or decurved, puberulous, 1–2 lin. long; bracts linear, remote, small; sepals ovate to lanceolate, keel-tipped, acute or obtuse, ciliate, subscarious or subfoliaceous, mostly glabrous, sometimes pubescent, $\frac{1}{2}$–$\frac{3}{4}$ lin. long; corolla suburceolate-cyathiform, not (or very slightly) constricted at the throat, glabrous, white, 1$\frac{1}{4}$–2 lin. long, $\frac{3}{4}$–1 lin. wide; segments broad, erect, sometimes erosulate, about $\frac{1}{3}$ the length of the tube; filaments straight; anthers subexserted, rarely exserted, sublateral, dorsifixed very close to, but above, the base, obliquely oblong or narrow-elliptical, sparsely ciliate on the front margins, $\frac{2}{5}$–$\frac{1}{2}$ lin. long; aristate; pore about $\frac{1}{2}$ the length of the cell; awns curved, subulate, acuminate, ciliate, $\frac{1}{2}$–$\frac{2}{3}$ the length of the cell; style exserted, sometimes decurved; stigma peltate-capitate; ovary glabrous.

KALAHARI REGION: Transvaal; widely distributed in the eastern part of the colony, *Bolus,* 7677! *Roe in Herb. Bolus,* 2643! 3136! *Schlechter,* 4115! *Wilms,* 902! 905! 910! 911! 912! *Nelson,* 362! *Rehmann,* 6573! 6621! *Wood,* 1639! *Galpin,* 453! *Thode,* 66!

EASTERN REGION, from 2000 to 6000 ft.: Griqualand East; Pumagwan Mountain, 3000 ft., *Tyson,* 2976! Natal; near Richmond, *Wood,* 1857! Little Noods Berg, *Wood,* 4125! and without precise locality, *Gerrard,* 1777!

This species is perhaps best placed in this section (as first suggested to us by Mr. N. E. Brown of Kew), as an ally of both the preceding species and of *E. decipiens,* Spreng., while quite distinct from either. The number of leaves in the whorl is very variable, some plants exhibiting only 3-nate, others only 4-nate leaves, others with both forms.

222. E. decipiens (Spreng. fil. Tent. Suppl. Syst. Veg. 13); erect, 1–1$\frac{1}{2}$ ft. high; branches straggling or virgate, cano-pubescent; leaves 3-nate, erect to spreading, linear, blunt, sulcate, glabrous, glossy, 2–3 lin. long; flowers 3-nate; pedicels $\frac{1}{2}$–1 lin. long; sepals narrow-lanceolate to ovate-lanceolate, glabrous, subscarious, $\frac{2}{3}$–1 lin. long; corolla at first narrow-cyathiform, ultimately ovoid-suburceolate, mouth slightly contracted, glabrous, white (or rarely pale rosy) 1$\frac{1}{2}$–1$\frac{3}{4}$ lin. long; segments erect, $\frac{1}{7}$–$\frac{1}{5}$ the length of the tube; anthers exserted or subexserted, lateral, narrow-oblong, acute, $\frac{3}{4}$ lin. long, muticous, scaberulous, mostly muticous; pore about $\frac{1}{2}$ the length of the cell; style exserted, slender; stigma clavate-capitate, lobed, large; ovary cano-pubescent. *Benth. in DC. Prodr.* vii. 666.

VAR. β, **trivialis** (Bolus); corolla suburceolate-cyathiform or cyathiform-campanulate; anthers broad-oblong or elliptical, very obtuse, generally muticous, $\frac{1}{4}$–$\frac{1}{3}$ lin. long. *E. trivialis and E. atroviridis, Klotzsch ex Benth. in DC. Prodr.* vii. 666.

VAR. γ, **tetragona** (Bolus); corolla cyathiform, tetragonous, 1$\frac{1}{4}$ lin. long; anthers subexserted (? whether mature), oblong, obtuse, $\frac{2}{3}$ lin. long, aristate; awns about $\frac{1}{3}$ the length of the cell.

Coast Region : Uitenhage Div.; without exact locality, *Zeyher,* 75 (ex *Sprengel*), but in Herb. Berlin without a number, stated to be the type ! Var. β : Uitenhage Div. ; between Galgebosch and Melk River, *Burchell,* 4783, probably, and without precise locality, *Alexander,* 9 ? *Miss Kensit in Herb. Bolus,* 6467 ! Port Elizabeth Div. ; near Port Elizabeth, *Cooper,* 1480 ! *Bolus,* 1675 !

Central Region : Somerset Div. ; on the Bosch Berg, 3000 ft., *MacOwan,* 1231 ! Div. ? *Drège,* 7799 ! Var. γ : Humansdorp Div. ; Clarkson, 900 ft., *Galpin,* 3711 !

The type of this species (of which a portion has been kindly sent us by Prof. Engler, from Berlin), exhibits an anther so unusually different from the specimens hitherto associated with it, that we have hesitated whether to separate them or not. Besides, the leaves upon the specimen are clearly 3-nate, though Sprengel describes them as 4-nate. Finally, the specimen standing alone, and our material being a fragment merely, we have decided to leave them as they were, for the present. Further, our var. γ (Galpin, 3711), does not well agree with either the type or var. β, and may hereafter be separated.

The species is closely related to *E. paniculata,* but the inflorescence is usually less dense, the flowers larger, the anthers more generally muticous; also to *E. demissa.*

223. E. demissa (Klotzsch ex Benth. in DC. Prodr. vii. 666); erect, much branched, probably under 1 ft. high ; branches puberulous or shortly floccose, glabrescent ; leaves 3-nate, erect to spreading, or sometimes recurved, somewhat crowded, linear to oblong, subobtuse, sulcate, glabrous, glossy, ciliate, $1\frac{1}{2}$–$2\frac{1}{2}$ lin. long ; flowers terminal, or by partial abortion of floriferous branchlets sometimes apparently lateral ; pedicels 1–2 lin. long ; bracts two approximate, or nearly so, third remote and longer, or all subapproximate ; sepals lanceolate, acute, keeled, rigid and somewhat scarious, glabrous, about $\frac{3}{4}$ lin. long ; corolla narrow-cyathiform to subovoid, the mouth in the latter case contracted, minutely hispidulous, rarely (perhaps only in age) glabrous, 1–$1\frac{1}{2}$ lin. long; anthers exserted, lateral, oblong, subobtuse or acute, a little under or over $\frac{1}{2}$ lin. long, muticous ; pore less than $\frac{1}{2}$ the length of the cell; style exserted, stoutish ; stigma peltate-capitate ; ovary puberulous, rarely subglabrous. *E. globuliflora, Klotzsch ex Benth. in DC. Prodr.* vii. 667.

Coast Region : Humansdorp Div.; near the Gamtoos River, *Masson,* 58! by the Kromme River, *Burchell,* 4838! Uitenhage Div. ; Van Stadens Mountains, *Zeyher,* 780! 785! 3224! *Burchell,* 4709! Albany Div.; near Grahamstown, *Burchell,* 3576! *Guthrie,* 2370!

There are two forms, or perhaps three, of this species so far as we know it. The two first differ chiefly in the leaves, the one being broader and somewhat recurved, the other narrower and suberect. These are connected by Masson's specimens. The third form is from Grahamstown; the habit is somewhat different, the branches ternate and more slender, the flowers paler in colour, the anthers a little longer, and more acute. Guthrie's specimens gathered in the same month, 78 years after Burchell's, almost exactly agree with the latter. In floral characters very much like *E. decipiens,* but looks different; the branches more rigid ; leaves usually longer, stouter, and more glossy : pedicels twice as long.

224. E. paniculata (Linn. Sp. Pl. ed. 2, 508, not of Wendl. nor Thunb. nor Lodd.); erect, 1–$1\frac{1}{2}$ ft. high ; branches numerous, virgate or fastigiate-paniculate, densely long-pilose or shortly floccose, covered with abundant corolline or subcorolline flowers, in close subcylindrical panicles ; leaves 3-nate, erect, imbricate or at least close-set, linear or subulate, sulcate, glabrous, 1–2 lin. long ; flowers sub-3-nate on short branchlets ; pedicels about 1 lin. long ; bracts approximate or subremote, small ; sepals lanceolate, keeled, sub-scarious, glabrous or puberulous, about $\frac{2}{3}$ lin. long ; corolla cyathi-form or obconic-cyathiform, mouth more or less widened, straight, not curved upwards from the base, about $1\frac{1}{4}$ lin. long ; segments generally continuous, about $\frac{1}{2}$ the length of the tube or more, usually red, more rarely pallid or white ; anthers subexserted or rarely exserted, lateral, obliquely oblong or subcuneate, or semiovate, about $\frac{1}{3}$ lin. long, aristate ; pore about $\frac{1}{2}$ the length of the cell ; awns variable, sometimes very minute or even wanting, sometimes nearly $\frac{2}{3}$ the length of the cell ; style exserted ; stigma broad-capitate, truncate and often lobed above ; ovary mostly glabrous, very rarely (in only one specimen of many examined), slightly puberulous. *Benth. in DC. Prodr.* vii. 690. *E. milleflora, Berg. Descr. Pl. Cap.* 96. *E. sodalis, Klotzsch ex Benth. l.c.* 690.

CoAST REGION : Tulbagh Div.; near Tulbagh, *MacOwan*, 2442 ! Paarl Div.; near Wellington, *Herb. Huguenot Seminary*, 9 ! Cape Div.; frequent on the lower mountain slopes near Cape Town, from 200 to 700 ft., *Drège*, 7781 ! *Bolus*, 2946 ! 3691 ! *Guthrie*, 1174 ! 1219 ! *Wolley Dod*, 176 ! Stellenbosch Div., *Niven*, 1 ! Lowrys Pass, *Guthrie*, 2505 ! Hottentots Holland Mountains, *Mund*, 33 ! Caledon Div.; near Caledon, *Bolus*, 8498 !

225. E. bicolor (Thunb. Diss. Erica, 36) ; erect, 1–2 ft. high ; branches generally virgate, the upper part laden with numerous flowers, cano-puberulous ; leaves 3-nate, erect-spreading, crowded, mostly linear, sulcate, puberulous, 1–$1\frac{1}{2}$ lin. long, more rarely broad-linear, slightly open-backed and from $1\frac{1}{2}$–$2\frac{1}{2}$ lin. long ; flowers 3-nate ; pedicels 1–$1\frac{1}{2}$ lin. long ; bracts remote, small ; sepals broad-ovate, blunt, keeled, viscid-puberulous or shortly tomentose, folia-ceous, $\frac{1}{2}$–$\frac{3}{4}$ lin. long, reaching from $\frac{1}{3}$–$\frac{1}{2}$ the height of the corolla ; corolla cyathiform, broad-cyathiform or subcampanulate-cyathiform, mouth a little widened or sometimes not at all so, the outline always curved, not tapering direct from the base, glabrous, dull red, 1–$1\frac{1}{4}$ lin. long ; segments erect or slightly spreading, $\frac{1}{3}$–$\frac{2}{3}$ the length of the tube ; anthers usually subincluded and manifest, more rarely subexserted, lateral, curved-oblong, very obtuse, scaberulous, $\frac{2}{5}$ lin. long ; pore $\frac{2}{5}$ the length of cell ; awns curved, subulate, ciliolate, $\frac{1}{2}$–$\frac{3}{4}$ the length of the cell ; style slender, well-exserted ; stigma capitellate, small ; ovary pubescent. *Benth. in DC. Prodr.* vii. 688 ; *Wendl. Eric. Ic. fasc.* 21, 139, *t.* 53. *E. calathiflora, Salisb. in Trans. Linn. Soc.* vi. 328. *E. concava, Lodd. Bot. Cab. t.* 134 ? and *Bot. Mag. t.* 2149 ? *E. canaliculata, var. minor, Andr. Heathery, t.* 157, *and Col. Heaths, t.* 158.

SOUTH AFRICA : without locality, *Mund* ! and *cultivated specimens* !

COAST REGION, from 800 to 3000 ft. : Clanwilliam Div.; Krakadouw Pass, *Leipoldt*, 214 partly ! Tulbagh and Ceres Div.; on the Witsen Berg, *Pappe* !

mountains near Tulbagh, *Guthrie*, 2734! Mitchells Pass, *Bolus*, 5298!
Schlechter, 8948! 8962! Worcester Div.; Dutoits Kloof, *Drège!* Paarl Div.;
French Hoek Mountains, *Thunberg*, Paarl Mountain, *Drège.* Stellenbosch Div.;
Lowrys Pass, *Bolus*, 5558! Caledon Div.; Zwarte Berg, *Schlechter*, 5588!
Knysna Div.; between Knysna and Plettenberg Bay, *Cape Govt. Herb.*

Generally recognizable by its numerous small, well-rounded flowers, bowl-
shaped to the base, and its slender long-exserted style. The shape of the corolla
is somewhat variable, the limb sometimes not at all spreading, at others distinctly though shortly spreading. But we have not seen any wild specimens
resembling in this respect those named *E. concava*, cited above, with very wide-
spreading campanulate corollas and figured from garden specimens. We follow
Bentham in quoting them, though we are uncertain whether this character
be only an effect of cultivation or hybridization, or whether it represents a
distinct species. Schlechter's 8948 looks like a species of the § *Ceramia*, but a
careful examination shows it to belong here. This species also connects with the
§ *Eurystoma*, and does not entirely agree with this group, owing to its usually
half-included stamens. Andrews' fig. *l.c. t.* 157, represents the plant in its wild
state very well, except that we find the sepals more foliaceous, always greenish
and pubescent.

226. E. diotæflora (Salisb. in Trans. Linn. Soc. vi. 342); erect,
$1-1\frac{1}{2}$ ft. high; branches many, suberect, often flexuous, somewhat
slender, pubescent; leaves 3-nate, suberect to spreading, not crowded,
slender, linear, subacute, sulcate, glabrous, 2–3 lin. long; flowers
3-nate; pedicels slender, $1\frac{1}{2}$–2 lin. long; bracts remote, small;
sepals lanceolate, puberulous, subfoliaceous, coloured, about $\frac{1}{2}$ lin.
long; corolla from campanulate-cyathiform with the mouth slightly,
or not, contracted, to ovoid-suburceolate with a moderately contracted
mouth, glabrous, minutely pulverulent or impressed-punctate, dry,
rich ruby-red or purple, $1\frac{1}{4}$–$1\frac{1}{2}$ lin. long; segments $\frac{1}{5}$–$\frac{1}{4}$ the length
of the tube; anthers usually included, about equal to the corolla,
sometimes subexserted, sublateral, attached to the filaments dorsally
at the base, longish oblong, wider at the base, from 4 to 7 times as
long as the width in the middle, over $\frac{1}{2}$ lin. long, muticous; pore
$\frac{1}{7}$–$\frac{1}{5}$ the length of the cell; style exserted, decurved; stigma capitate; ovary hispidulous. *Benth. in DC. Prodr.* vii. 667. *E.
pistillaris, Soland. ex Salisb. l.c.* 342.

SOUTH AFRICA: without locality, *Masson* or *Niven*, 44!

COAST REGION: Caledon Div.; mountains about the Zondereinde River,
near Zoetemelks Vlei, *Grisbrook in Herb. Guthrie*, 3298 (or 3278)! and *MacOwan
Herb. Aust.-Afr.*, 1627!

It is closely allied to the §§ *Gypsocallis* and *Ceramia*, and might almost be
placed in either. Our specimens show but little variation.

227. E. opulenta (Wendl. ex Klotzsch in Linnæa, xii. 499?);
branches slender, densely clothed with submatted short grey plumose
hairs; leaves 3-nate, erect, imbricate, linear, subobtuse, sulcate,
glabrous, about 1 lin. long; flowers numerous, densely clustered;
pedicels slender, less than 1 lin. long; bracts, 2 subapproximate,
1 remote, small; sepals lanceolate, acute, keeled, scarious, glabrous,
coloured, about $\frac{1}{4}$ lin. long; corolla tubular-campanulate, slightly
widened at the mouth, glabrous, about 1 lin. long; segments about
$\frac{1}{4}$ the length of the tube; anthers exserted, lateral, oblong, obtuse,

slightly incurved, $\frac{3}{5}$ lin. long, aristate; pore varying from nearly orbicular $\frac{1}{5}-\frac{1}{4}$ the length of the cell, to elliptical $\frac{1}{4}-\frac{1}{3}$ the length of the cell; awns about $\frac{1}{3}$ the length of the cell, recurved; style exserted; stigma capitellate; ovary glabrous. *Benth. in DC. Prodr.* vii. 667. *Microtrema opulentum, Klotzsch, l.c.* 499.

SOUTH AFRICA : without locality, *Hesse.*

COAST REGION : George Div.; between Touw River and Kaymans River, *Burchell,* 5774! near George, *Alexander,* 1!

There is a little uncertainty about this species which was made the type of a new genus by Klotzsch, as having a 1-celled, 1-ovuled ovary, 6 stamens and anthers with a minute pore. Bentham noted on Hesse's ticket in the Berlin Herb. : "I have examined 2 flowers,—in one I found 7 stamina and in the other 8; in both 4 cells and 2 ovules in each." With it he identified Burchell's 5774. Possibly Hesse's plant was variable; we have been able only to see a single flower of it, in which the ovary was imperfect; it had 7 stamens and though the anther-pore was unusually small, it was closely connected as regards size with some of Burchell's specimens, which are also variable. The species is probably a true *Erica,* with somewhat the habit of *E. paniculata.*

228. E. harveiana (Guthrie & Bolus); apparently a small shrub; branches slender, densely covered with rather long, straight, squarrose simple hairs; leaves 3-nate, very crowded, suberect, imbricate, slender, linear, subobtuse, flat above, convex and sulcate below, glabrous, ciliate, shortly aristate-apiculate, 1 lin. long; flowers 3-nate; pedicels about $\frac{1}{2}$ lin. long; bracts approximate, foliaceous, minute; sepals foliaceous, lanceolate, united a little above the base, glabrous, about $\frac{1}{2}$ lin. long; corolla broadish cyathiform when young, mouth neither widened nor contracted, when older becoming somewhat broadly-ovoid with the mouth slightly contracted, glabrous, about $\frac{3}{4}$ lin. long; segments broadly deltoid, slightly spreading, $\frac{1}{3}-\frac{1}{2}$ the length of the tube; anthers exserted, lateral, longitudinally semilanceolate, subacute, slightly curved at the back, straightish in front, glabrous, smooth, $\frac{2}{5}$ lin. long, aristate; pore about $\frac{1}{3}$ as long as the cell; awns less than $\frac{1}{3}$ as long as the cell, incurved (? always); style exserted, slender, decurved; stigma capitate, lobed or punctate above, somewhat large; ovary white-pubescent, 4-celled, cells several-ovuled.

COAST REGION : Uitenhage Div.; on the Van Stadens Mountains, *Zeyher,* 790!

229. E. parvula (Guthrie & Bolus); dwarf, decumbent, 6–12 in. high; stem stout and woody, $2\frac{1}{2}-3\frac{1}{2}$ lin. in diam. at the base; branches many, much-spreading, glabrous, red when young; leaves 3-nate, erect, adpressed, about equalling the internodes, oblong or elliptical, blunt, sulcate, ciliate, glabrous, glossy, thick, $\frac{1}{2}-\frac{3}{4}$ lin. long; flowers 3-nate, not numerous; pedicels at length deflexed, slender, under 1 lin. long; bracts remote, small; sepals from ovate to lanceolate, connate at the base, foliaceous, keeled, about $\frac{1}{2}$ lin. long; corolla tubular-campanulate, distinctly widened at the mouth, $1\frac{1}{2}$ lin. long, glabrous, white; segments spreading, rounded, about $\frac{1}{5}$ the length of the tube; anthers subexserted (possibly at length exserted), subterminal, longitudinally semiobovate, nearly $\frac{1}{2}$ lin. long, dark-coloured, decurrent-aristate at the base; pore $\frac{1}{3}$ the length of the cell; awns rough, with only a short free tooth-like point; style exserted, slender, straight; stigma small, capitellate or subsimple; ovary glabrous.

COAST REGION: Stellenbosch Div.; on a rock near the mouth of the Steen-brass River, 20–30 ft. above the sea, *Guthrie*, 3710!

230. E. unilateralis (Klotzsch ex Benth. in DC. Prodr. vii. 667); erect, 6–12 in. high; branches puberulous, slender, straight or flexuous; leaves 3-nate, erect-spreading, linear, keeled, sulcate, glabrous, 2–2$\frac{1}{2}$ lin. long; flowers 3-nate (rarely 4-nate?); pedicels puberulous, about 2 lin. long; bracts approximate or sometimes subremote, subscarious, small; sepals obovate or ovate, acute, keeled, glabrous, scarious, $\frac{1}{2}$–$\frac{3}{4}$ lin. long; corolla variable, subtubular, tubular-campanulate and not contracted or widened at the mouth, suburceolate and slightly contracted in the throat, or subovoid and more contracted at the mouth, glabrous, dry, 1$\frac{1}{2}$–2 lin. long; segments erect, $\frac{1}{5}$–$\frac{1}{3}$ the length of the tube; anthers subexserted, subterminal or sometimes sublateral, linear-clavate, obtuse, about $\frac{1}{2}$ lin. long, decurrent-aristate; pore about $\frac{1}{2}$ the length of the cell; awns very narrow, dark-coloured, connate with the filament for a length somewhat greater than that of the cells and terminating in short spreading free points; style exserted, slender; stigma sub-simple; ovary puberulous.

COAST REGION: Uitenhage Div.; Van Stadens Mountains, *Zeyher*, 791! 3192! Albany Div.; on the rocks in Zwartwater Poort, north of the Zuurberg Range, *Burchell*, 3376! 3422!

There has been some confusion in herbaria between this species and *E. inconspicua, Bartl.,* of which we have not seen a type specimen, and place amongst the imperfectly known species.

231. E. brachysepala (Guthrie & Bolus); erect, of stunted appearance, 6–8 in. high; branchlets flexuous, rigid, roughly pubescent, or glabrescent; leaves 3-nate, erect, often adpressed, subimbricate or sometimes not exceeding the internodes, from linear to oblong or elliptical, blunt, sulcate, glabrous, distantly gland-ciliate, thick, $\frac{3}{4}$–1$\frac{1}{4}$ lin. long; flowers 3-nate, or in clusters of 5–6; pedicels viscous-pubescent, $\frac{1}{2}$–1 lin. long; bracts from subapproximate to subremote, small; sepals like the bracts and leaves, oblong, thick, obtuse, glandular, subviscid, $\frac{1}{2}$–$\frac{2}{3}$ lin. long; corolla cyathiform or broad-cyathiform-campanulate, mouth widened or not contracted, glabrous, subviscid, dull-yellow or rosy, 1–1$\frac{1}{4}$ lin. long; segments erect or slightly spreading, broadly rounded, about $\frac{1}{3}$ the length of the tube; anthers well-exserted, lateral, dorsifixed just above the base, oblong, obtuse, from a little over $\frac{1}{3}$ to nearly $\frac{1}{2}$ lin. long, muticous; pore $\frac{1}{3}$ the length of the cell, style exserted, slender, unusually straight (in this group); stigma subpeltate-capitate; ovary glabrous.

COAST REGION: Caledon Div.; Houw Hoek, 2000 ft., *Schlechter*, 7789! Hemel-en-Aarde, 1500 ft., *Schlechter*, 10377!

A distinct species, although it is difficult of location. In general appearance and in its small inconspicuous dull-coloured flowers, it resembles species of the § *Arsace,* but is separated by the exserted anthers. It approaches *E. demissa,* but is smaller and more slender in habit and the corolla and sepals are quite different. There appear to be two forms: one, with yellowish flowers, and paler and shorter anthers, is Schlechter's 10377; the other, with rosy flowers, darker and longer anthers, is his 7789.

232. E. kraussiana (Klotzsch in Walp. Rep. ii. 728); branches everywhere plumose-puberulous; leaves 3-nate, short, thick, glabrous, densely imbricate, evanescent, plumose-denticulate on the margin; bracts approximate, lanceolate; sepals lanceolate, lilac, glabrous, shortly ciliate; corolla oblong-globose, small, contracted at the throat; limb very short, straight; anthers exserted, muticous; style exserted; ovary glabrous.

COAST REGION: Caledon Div.; mountain-sides near Genadendal, 1000–2000 ft., *Krauss,* 957!

Section XVIII. OROPHANES. (Sp. 233–259.)

233. E. pilulifera (Linn. Sp. Pl. ed. 1, 355); erect, virgate, 1-1½ ft. high, simple and naked below, branched and leafy above, surmounted by umbels of red flowers; branches slender, channelled, glabrous; leaves 4-nate (or sometimes 3-nate?) or scattered, crowded, suberect, linear, subacute, sulcate, pubescent, glabrescent, ciliate on the edge and on the folds at the meeting of the margins, 2½–4 lin. long; flowers 4-nate or more commonly umbellate with a few axillary immediately below the umbel; pedicels glabrous, 3–3½ lin. long; bracts 2, remote, linear, scarious, over 1 lin. long (the third appears to be wanting); sepals lanceolate to ovate-lanceolate, acuminate, keeled, scarious, glabrous, red, 1¼–1¾ lin. long; corolla suburceo-late-cyathiform, very little constricted at the throat or widened at the mouth, about 2½ lin. long; segments erect or subspreading, ¼–⅓ the length of the tube; anthers included, oblong-cuneate, smooth, glabrous, less than ½ lin. long, aristate; pore about ½ the length of the cell; awns minutely ciliolate, about equal to the cell; style included; stigma capitate; ovary glabrous. *E. nudicaulis, Berg. Descr. Pl. Cap.* 113. *E. piluliformis, Salisb. in Trans. Linn. Soc.* vi. 378. *E. unica, Spreng. Syst. Veg.* ii. 188. *E. pedunculata, Andr. Heathery, t.* 229, *and Col. Heaths, t.* 252. *E. pilulæformis, Benth. in DC. Prodr.* vii. 680.

SOUTH AFRICA: without locality, *Roxburgh! Bowie! Miller! Herb. Salisbury!* and *cultivated specimens!*

COAST REGION: Cape Div.; moist or marshy places on the Lower Plateau of Table Mountain, about 2500 ft., *Niven,* 221! *Zeyher,* 5003! *Bolus,* 3712! *Wolley Dod,* 1372! mountains between Cape Town and False Bay, *Thunberg.*

234. E. subulata (Wendl. Eric. Ic. fasc. 20, 123, t. 47); erect, rigid, much branched, 1-2½ ft. high; branches puberulous, with long hairs sparsely intermixed; leaves erect or spreading, always rigidly curved, linear, acuminate, subpungent, sulcate, pallid, glabrous, 4–7 lin. long; flowers usually 4-nate, sometimes clustered on short branchlets in heads $\frac{1}{2}$ in. in diam.; pedicels about $1\frac{1}{2}$ lin. long; bracts approximate, linear, tapering into a long bristle, nearly $1\frac{1}{2}$ lin. long; sepals linear from a broad scarious somewhat toothed base, and tapering upwards into a longer and finer bristle than that of the bracts, $1\frac{3}{4}$–2 lin. long; corolla tubular-suburceolate, slightly asymmetrical, nearly equal at the mouth, more or less pale rosy, 2–$2\frac{1}{4}$ lin. long; limb short, erect or very little spreading; filaments bent below the anther; anthers included, oblong, obtuse, scabrid on the margins, 3–$3\frac{1}{2}$ times longer than the width at the middle, about $\frac{1}{2}$ lin. long, aristate; pore $\frac{2}{5}$ the length of the cell; awns setaceous, smooth, about $\frac{1}{3}$ the length of the cell; style exserted; stigma capitate; ovary elongate, glabrous. *Benth. in DC. Prodr.* vii. 675.

COAST REGION, at 1300–3250 ft.: Tulbagh Div.; on the Witsen Berg, near Tulbagh, *Burchell*, 8666! Ceres Div.; rocky places at the foot of the mountains near Ceres, *Bolus*, 6726! *Guthrie*, 3177! Worcester Div.; Dutoits Kloof, *Drège!* *Marloth*, 603! Paarl Div.; hills near the Berg River, *Drège.*

235. E. scabriuscula (Lodd. Bot. Cab. t. 517); erect, virgate, with many ascending, densely glandular-hispid branches, 2–3 ft. high; leaves spreading or squarrose, crowded, linear, margins revolute, just touching, or oblong or narrow-linear and more or less open-backed, obtuse, somewhat thickly glandular-hispid, dark green, $1\frac{1}{2}$–2 lin. long; flowers 4-nate, sometimes clustered; pedicels sparsely hairy, 2–$2\frac{1}{2}$ lin. long; bracts subremote to remote, small; sepals oblong or lanceolate, obtuse, tipped with a greenish-keel, subscarious, coloured, pubescent, about 1 lin. long; corolla ovoid-urceolate, throat more or less (but not much) constricted, pale rosy, about 2 lin. long; segments short, erect or very slightly spreading; filaments bent below the anther; anthers included, rarely just manifest, subcuneate or longitudinally subsemiovate, rounded or somewhat narrowed at the apex, cells minutely papillose, ciliate on the front margin, $\frac{2}{5}$ lin. long, aristate; pore $\frac{2}{5}$ the length of the cell; awns straightish, deflexed or spreading, ciliolate, about equal to the cell; style exserted; stigma capitate; ovary hispid, chiefly at or towards the apex. *Link, Enum. Hort. Berol.* i. 372? *not of Drège's Exsicc.; Benth. in DC. Prodr.* vii. 678.

SOUTH AFRICA: without locality, *Mund!* and *cultivated specimens!*
COAST REGION, between 200 and 1200 ft.: George Div.; between Touw River and Kaymans River, *Burchell*, 5773! Barbiers Kraal, *Niven*, 46! near George, *Burchell*, 5995! *Schlechter*, 2242! 5772! *Bolus*, 8670! Montagu Pass, *Young in Herb. Bolus*, 5525! Knysna Div.; near Knysna, *Tyson in MacOwan & Bolus, Herb. Norm. Aust.-Afr.*, 995! Humansdorp Div.; Storms River, *Schlechter*, 5960!

236. E. gibbosa (Klotzsch ex Benth. in DC. Prodr. vii. 678); erect, reaching 5 ft. high; branches puberulous and also beset with glandular-setulose squarrose hairs; leaves spreading to squarrose, linear-oblong, obtuse, thick, sulcate or subopen-backed, hispid with coarse shortish tubercle-based, frequently gland-tipped hairs, 1–1½ lin. long; flowers 4-nate; pedicels glabrous, 1 lin. long; bracts 2 approximate, 1 remote, small; sepals ovate or subovate, acute, keel-tipped, lacerate, subscarious, ½ lin. long; corolla subcyathiform (when young) to urceolate, mouth equal or slightly widened when young, contracted after flowering, 1–1¾ lin. long; segments erect or spreading, short; filaments much bent below the anthers; anthers included or just manifest, obtusely-cuneate, minutely papillose, ciliolate, a little over ⅓ lin. long, aristate; pore ½–⅗ as long as the cell; awns setaceous, ciliolate, about ½ the length of the cell; style exserted; stigma capitate; ovary hispidulous, chiefly at the apex.

Sᴏᴜᴛʜ Aғʀɪᴄᴀ: without locality, *Mund!* *Drège,* 7785a!

Cᴏᴀsᴛ Rᴇɢɪᴏɴ : Oudtshoorn Div.; between Oudtshoorn and Zwartberg Pass, *Kolbe in Herb. Bolus!* George Div. ; near the west end of Lange Vlei, *Burchell,* 5698! near George, 600 ft., *Schlechter,* 5861! Knysna Div. ; between Knysna River Ford and Goukamma River, *Burchell,* 5560! near Forest Hall, *Miss Newdigate,* 68! Humansdorp Div.; near Storms River, 600 ft., *Galpin,* 3666!

237. E. bergiana (Linn. Diss. de Erica, n. 6); erect, 1–3 ft. high; branches mostly virgate, hirsute ; leaves generally spreading, linear, obtuse, sulcate, or subopen-backed, rough with tubercle-based hairs, or sometimes smooth and subglabrous, ciliate, 1½–3 lin. long ; flowers 4-nate ; pedicels pubescent, red, 1½–3 lin. long ; bracts remote, small ; sepals from a broad ovate scarious lacerate ciliate base, tapering into a shorter or longer foliaceous green cusp, ½–1¼ lin. long ; corolla urceolate-globose, or in smaller specimens subovoid, red, 1¾–3½ lin. long ; segments erect or slightly spreading, ovate to semi-orbicular, imbricate at the base, ¼–⅓ as long as the tube ; anthers included, dorsifixed mostly rather high above the base, subovate, very obtusely rounded at the top, smoothish, ciliate or naked, ⅕ to nearly ½ lin. long, cristate or subaristate, the appendage broad-lanceolate, lacerate or subulate and fringed ; pore ⅖–½ as long as the cell; style included or slightly exserted ; stigma capitate ; ovary on a broad disk, glabrous. *Amœn. Acad.* viii. 55 ; *Mant. Alt.* 235 ; *Wendl. Eric. Ic. fasc.* 2, 29 ; *Lodd. Bot. Cab. t.* 939 ; *Benth. in DC. Prodr.* vii. 662 ; *var. glabra, Wendl. l.c. fasc.* 24, 189, *t.* 72. *E. lacunæflora, Salisb. Prodr.* 297, *and in Trans. Linn. Soc.* vi. 377. *E. quadriflora, Andr. Heathery, t.* 41, *and Col. Heaths, t.* 125. *E. florida, Thunb. Diss. Erica,* 40. *E. campylophylla, Spreng. Syst. Veg.* ii. 190 ? *E. turrigera, Salisb. in Trans. Linn. Soc.* vi. 377. *E. cupressina, Sinclair, Hort. Eric. Wob.* 7. *E. incurva, Andr. Col. Heaths,* iii. *t.* 175, *and Heathery, t.* 169. *E. quadrifolia, Pritz. Ic. Ind.* i. 421. *E. nitens, Lee ex Steud. Nomencl. ed.* 2, i. 576 ?

Sᴏᴜᴛʜ Aғʀɪᴄᴀ: without locality, *Thunberg, Herb. Salisbury!* and *cultivated specimens!*

Cᴏᴀsᴛ Rᴇɢɪᴏɴ, 1000–4000 ft.: Clanwilliam Div.; Ezelsbank, *Drège,* 7769!

Pakhuis Berg, *Schlechter*, 10815! Tulbagh Div.; Roode Zand, *Thunberg;*
Tulbagh Kloof, *Zeyher*, 1097! *Guthrie*, 2187! *Drège;* Tulbagh Waterfall,
Niven, 23! Mitchells Pass, *Bolus*, 5176! *Nelson*, 34! Worcester Div.; Dutoits
Peak, *Marloth*, 2413! Dutoits Kloof, *Drège.* Paarl Div.; French Hoek,
MacOwan Herb. Norm. Aust.-Afr., 935! Stellenbosch Div.; Jonkers Hoek,
Bolus in Herb. Guthrie, 4957! Caledon Div.; mountains near Zoetemelks
Vlei, *Grisbrook in Herb. Guthrie*, 3296! *Zeyher*, 3218! Zwart Berg,
Pappe!

This species is easily recognized from all others in this section by its reflexed
sepals. These specimens, which we have carefully examined, appear to show
a gradation of difference in almost every organ, which renders even a distinction
of varieties with constant characters, apparently impossible.

238. E. aspalathifolia (Bolus in Journ. Linn. Soc. xxiv. 182);
diffuse, sometimes weak-growing, with a few long subsimple erect or
subdecurrent branches, 12–18 in. long, and a few short branchlets
above, more or less setulose in all parts; leaves 3–4-nate or scattered,
spreading or squarrose, crowded, very narrow-linear, acute, flattish
and faintly sulcate below, ciliate with long white hairs, minutely
puberulous, 1–1½ lin. long; flowers 3–4-nate, crowded on very
short close branchlets and forming a dense pseudo-spike 6–8 in.
long, ½ in. in diam., sometimes (in poorly grown specimens) 3–4
flowers simply terminal; pedicels puberulous, ½–1 lin. long; bracts
subapproximate, foliaceous, 1 lin. long; sepals linear, acute, foliaceous, long-ciliate, ¾–1 lin. long; corolla subampullaceous-tubular
(somewhat inflated below) to tubular, sometimes gradually tapering
to the apex, sometimes nearly equal and little or not at all contracted
at the mouth, of thin texture, white (*Wood*), mostly 2¼–2½ lin.
long; segments erect, rounded, ⅕–¼ the length of the tube; filaments
only slightly curved below the anther; anthers included, broad-
cuneate-oblong, widely rounded at the apex, scaberulous, ¼ lin. long,
aristate; pore about ½ the length of the cell; awns capillary, ⅔–¾ the
length of the cell; style included, slender; stigma capitellate;
ovary hispidulous, chiefly on or towards the apex.

VAR. β, **Bachmannii** (Bolus); sepals lanceolate, ¾ lin. long; corolla tubular
or subclavate-tubular, 5 lin. long; anthers nearly ⅓ lin. long, inserted on the
filament higher above the base of the cell.

EASTERN REGION : Eastern Pondoland; Egossa Forest, *Sim*, 2454! Natal,
1000–4000 ft., Inanda, *Wood*, 693! 7519! damp stony places on Great and
Little Noods Berg, 3000–4000 ft., *Wood*, 911! Var. β: Pondoland; probably
near the mouth of St. Johns River, 800–1600 ft., *Bachmann*, 359! (Herb.
Berlin).

A very distinct species, with the aspect of plants of the § *Hermes*, from which
it is separated by its strictly terminal inflorescence. Wood's 693 appears to be
a short-stamened form of the species.

239. E. adæquata (Tausch in Flora, 1839, 634); branchlets
slender, subglabrous or minutely hispidulous; leaves spreading,
narrow-linear, subobtuse, faintly sulcate, glabrous, remotely and
shortly ciliolate, 1¾ lin. long; flowers terminal, ("and also lateral"
Tausch), sub-4-nate, cernuous; pedicels slender, straight, puberulous, over 1 lin. long; bracts narrow-linear, upper subremote, small,

lower median ; sepals lanceolate, acuminate, keeled, scarious-edged, and sparsely subdenticulate, 1 lin. long; corolla obconic ("campanulate," *Tausch*), widened at the mouth, $1\frac{1}{2}$ lin. long; segments erect, continuous, less than $\frac{1}{2}$ the length of tube ; filaments not very slender, straight ; anthers subexserted, cuneate-oblong, straight, glabrous, $\frac{2}{5}$ lin. long, cristate-aristate at the base ; pore $\frac{2}{5}$ the length of the cell ; awns subulate, curved, denticulate or slightly lobed, fringed, less than $\frac{1}{2}$ the length of the cell; style straight, tapering upwards, about equalling the anthers ; stigma simple ; ovary thinly hispidulous. *E. flexicaulis, Hort. ex Tausch, l.c., not of Dryander.*

South Africa : without locality, *cultivated specimens !*

We insert this species with some hesitation, but having seen and dissected Tausch's type (by the courtesy of the authorities of the Royal Bot. Inst. of the Univ. of Prague), it gives us the impression of being a good and distinct species. The anthers are only very little exserted, or for about half their length.

240. E. rubens (Thunb. Diss. Erica, 49, not of others); erect, almost entirely glabrous, about 1 ft. high ; branches numerous, slender, reddish, glabrescent ; leaves erect or suberect, imbricate below or on the upper branchlets, often shorter than the internodes, linear or narrow-oblong, acute, sulcate, the uppermost sometimes pubescent, about 2 lin. long; flowers terminal, 3-nate or umbellate, 3–6-flowered, at length usually cernuous ; pedicels slender, glabrous, crimson, 3–7 lin. long ; bracts remote, slender, linear, adpressed ; sepals lanceolate, sulcate, glossy, dark red, about 1 lin. long ; corolla urceolate-ovoid to urceolate-globose, mouth much contracted, bright red, $2\frac{1}{2}$–$3\frac{1}{2}$ lin. long; segments very short, suberect or connivent ; filaments bent ; anthers included, cuneate-oblong, thinly hairy chiefly at the base, about $\frac{1}{2}$ lin. long; pore $\frac{3}{4}$–$\frac{7}{8}$ the length of the cell ; connective produced backwards at a right angle near the base and carrying a broad pale lacerated papery crest, winged laterally, with a linear terminal lobe, the whole about as long as the cell ; style included ; stigma capitellate, very small ; ovary obovate, on a distinct stipe $\frac{1}{2}$ lin. or more long. *E. peduncularis, Salisb. in Trans. Linn. Soc.* vi. 329 ; *Benth. in DC. Prodr.* vii. 676.

South Africa : without locality, *Thunberg !*
Coast Region : Clanwilliam Div. ; on the Cederberg Range, *Shaw in Herb. Bolus,* 5665 ! Honig Valei, near Krakadouw, *Leipoldt,* 136 ! 208 ! *MacOwan, Herb. Aust.-Afr.,* 1917 !
Central Region : Ceres Div. ; mountains in the Cold Bokkeveld, *Masson !*

Of this species we have seen Thunberg's type and also poor specimens of Masson's in the British Museum. These were almost certainly Salisbury's types of his *E. peduncularis,* and the latter quotes Thunberg's *E. rubens* as a synonym. As to the specimens marked α and β in Thunberg's herbarium this is confirmed by Rach ; and the species is indeed so distinctly marked that the descriptions alone would have sufficed. Our other specimens agree well, and the only variable character is the length of the pedicel.

241. E. læta (Bartl. in Linnæa, vii. 648) ; a glabrous erect shrub, 6-12 in. or more high ; leaves mostly erect, imbricate, linear, acute, sulcate, keeled, 2–3½ lin. long; flowers 4-nate, or shortly umbellate; pedicels slender, 2–4 lin. long; bracts remote; sepals subulate, or linear-acuminate from a wide base, or lanceolate-acuminate, keel-tipped, 1–1½ lin. long; corolla urceolate to ovoid-urceolate, mouth more or less contracted, sometimes subtetragonous, usually bright red, 2–2½ (rarely 3) lin. long; limb erect, ¼–⅓ the length of the tube; anthers included, oblong, obtuse, naked on the margins, ⅖ lin. long; pore ½–⅞ the length of the cell; connective produced backwards at a right angle from above the base of the cells, bearing a broadish ovate or subfalcate-ovate, acute, serrulate, but not fimbriate, white, papery crest, as long as or longer than the cells; style included or shortly exserted; stigma capitate; ovary sessile, glabrous, smooth or scabrid.

VAR. β, **incisa** (Bolus); habit and leaves somewhat stouter, corolla somewhat larger than in the type; crests of the anthers more or less deeply inciso-lobulate. *E. rubens, Benth. in DC. Prodr.* vii. 676; *Andr. Heathery, t.* 43? *and Col. Heaths, t.* 127? *Wendl. Eric. Ic. fasc.* 15, 43, *t.* 17, *not of Thunb. nor of others.*

COAST REGION, between 200 and 3000 ft. : Stellenbosch Div.; between Gordons Bay and Hanglip, *Guthrie,* 3723! Caledon Div.; sandy places between Great Houw Hoek and Bot River, *Zeyher,* 3228! Klein River Mountains, *Zeyher,* 3232! Vogel Gat, *Schlechter,* 10413! Var. β, Cape Div.; Simons Bay, *Niven,* 16! Muizenberg Mountain, *Bolus,* 4474! near Simons Town, *MacOwan,* 2285! on the Steen Berg, *Fair!* Smitswinkel Vley, *Wolley Dod,* 792! Klaver Vley, *Wolley Dod,* 805!

Our variety β is based upon Niven's 16, which served as the plant for Bentham's description of *E. rubens, Andr.* (not of Thunb.). But there is very little difference between the two, as Bentham himself observed, and amongst numerous specimens we have sought in vain for any distinctive characters of specific value.

242. E. turbiniflora (Salisb. in Trans. Linn. Soc. vi. 377); erect, glabrous, about 1 ft. high; branches corymbose, channelled, reddish; leaves erect or suberect, imbricate, narrow-linear, acute, round-backed, faintly sulcate, 3–4 lin. long; flowers 4-nate, sometimes clustered; pedicels slender, 1–2 lin. long; bracts subapproximate to subremote, rather long, linear; sepals narrow-linear, sometimes with a short broad scarious-margined base, 1–2 lin. long, generally longer than the corolla-tube; corolla broad-cyathiform; tube turbinate, 4-angled (8-angled, *Salisbury*) or 8-nerved, mouth widened, red, about 1½ lin. long; segments erect, about ⅓ the length of the tube; anthers included, oblong, very obtuse, membranous, smooth, pallid, ¼ lin. long; pore about ⅖ the length of the cell; connective produced at the base backwards at nearly a right angle to the cell, and bearing falcate-lanceolate crests, which are free for the greater part of their length; style subincluded, usually just equal to the corolla; stigma capitate; ovary glabrous. *Benth. in DC. Prodr.* vii. 678.

VAR. β, **aristata** (Bolus); anthers aristate; awns nearly as long as the cells, decurrent along the connective for the greater part of their length.

SOUTH AFRICA : without locality, *Niven,* 77! *Roxburgh! Herb. Salisbury!*

COAST REGION : Var. β : Stellenbosch Div. ; by the foot-path from Gordons Bay towards Cape Hanglip, 200 ft., *Guthrie*, 3723b !

243. E. viridipurpurea (Linn. Sp. Pl. ed. 1, 353); erect, graceful, not much spreading, 1–2 ft. high ; branches subvirgate, pubescent ; leaves more or less erect or incurved, imbricate, linear, round-backed, ciliate or naked, glabrous, 1–2 lin. long; flowers 4-nate, often clustered ; pedicels about 1 lin. long ; bracts remote, very small ; sepals ovate-lanceolate, keel-tipped, ciliate or naked, usually rosy, ½–¾ lin. long; corolla obconic-campanulate, slightly widened at the mouth, red, 1–1¼ lin. long ; segments slightly spreading, nearly equal to the tube, with the interstices rounded at the base, open and obvious in bud ; anthers almost exactly those of *E. cyathiformis*, but slightly larger (.018 in. long); style included ; stigma peltate-capitate ; ovary glabrous. *E. mauritanica, Linn. Syst. ed.* 10, 1002, *Diss. Erica, n.* 9, *with fig. of fl., and Amœn. Acad.* viii. 59, *n.* 9.   *E. pelviformis, Salisb. Prodr.* 298, *and in Trans. Linn. Soc.* vi. 376.   *E. persoluta, Curt. Bot. Mag. t.* 342, *and Wendl. Eric. Ic. fasc.* 14, 27, *not of Linn.   E. regerminans, Andr. Heathery, t.* 235, *and Col. Heaths, t.* 265, *not of Linn.   E. leucantha, G. Don, Gen. Syst.* iii. 792.   *E. virescens, Thunb. Diss. Erica,* 37 ?   *E. multumbellifera, Hort. ex Benth. in DC. Prodr.* vii 679.

SOUTH AFRICA : without locality, *Drège*, 7760! *Herb. Salisbury !* and *cultivated specimens !*

COAST REGION, from 100–2800 ft. : Worcester Div., *Cooper*, 1596! Cape Div. ; common on the mountains and Flats around Cape Town, *Burchell*, 8564! *Bolus*, 1203 ! 2956! *Herb. Norm. Aust.-Afr.*, 758 ! *Wolley Dod*, 175 ! Caledon Div.; mountains of Klein River Kloof, *Zeyher*, 3260 !

244. E. sitiens (Klotzsch in Linnæa, xii. 505); erect, 1–2 ft. high ; branches stout, ascending, subvirgate, pubescent or puberulous ; leaves generally erect, straight and imbricate, sometimes flexuous or squarrose-patent, linear, acute, keeled or round-backed, ciliate or naked, glabrous, 1½–4 lin. long; flowers 4-nate, or occasionally 3-nate or solitary ; pedicels about 1 lin. long ; bracts remote or subapproximate, small ; sepals lanceolate, acute or acuminate, with a long tapering point, keeled, subscarious, glabrous, 1–2 lin. long ; corolla most commonly oblong-urceolate, asymmetrically inflated in the middle, constricted at the throat and distinctly oblique at the mouth, sometimes varying to shortly tubular with nearly equal sides and but little constricted or oblique, white to red, limb when red often white-edged, 3–4 lin. long ; segments short, usually spreading ; filaments bent below the anther ; anthers included, broadly cuneate but with a more or less rounded apex, in some examples nearly as broad as long, in others longer, rough with small papillæ, about ⅓ lin. long, cristate-aristate ; pore about ½ the length of the cell; appendages subulate, entire or occasionally lobed, closely ciliate, spreading, about ⅔–¾ the length of the cell; style included ; stigma capitate ; ovary generally turbinate, glabrous. *Benth. in DC. Prodr.* vii. 677.

COAST REGION, between 1000 and 3000 ft. : Stellenbosch Div.; Lowrys Pass, *Schlechter,* 7226! *Galpin,* 4940! Caledon Div.; mountains near Palmiet River, *Zeyher! Guthrie in MacOwan Herb. Aust.-Afr.,* 1722! *Schlechter,* 7453! *Bolus,* 4175! mountains near the mouth of Bot River, *Bolus,* 8490!

245. E. Blandfordia (Andr. Heathery, t. 154); branches stout, erect, puberulous, 10 in. or more long; leaves mostly squarrose or spreading, the uppermost only suberect, narrow-linear, acute, keeled, glabrous, about 2 lin. long; flowers 4-nate to solitary; pedicels glabrous, under 1 lin. long; bracts remote (subbasal), small; sepals ovate, shortly acuminate, strongly keeled the whole length, scarious, glabrous, $1\frac{1}{2}$ lin. long; corolla broad-urceolate, yellow, $3\frac{1}{2}$–4 lin. long; throat slightly constricted; tube symmetrically or sometimes somewhat asymmetrically inflated and the mouth then somewhat oblique; segments short, rounded, spreading; filaments slightly bent below the anther; anthers included, longitudinally semiovate, the posterior margin straightish, obtuse, rough with small papillæ and ciliolate on the margin, about $\frac{1}{2}$ lin. long, aristate; pore about $\frac{2}{5}$ the length of the cell; awns nearly setaceous, spreading, smooth or minutely ciliolate, about $\frac{1}{2}$ the length of the cell; style included; stigma capitate; ovary glabrous. *Andr. Col. Heaths, t.* 153. *E. blanfordiana, Sims in Bot. Mag. t.* 1793; *Lodd. Bot. Cab. t.* 115; *Benth. in DC. Prodr.* vii. 648.

COAST REGION : Worcester Div.; near Breede River, *Niven,* 268! Bains Kloof, 2000 ft., *Miss Cummings in Herb. Bolus,* 5854!

246. E. lateralis (Willd. Sp. Pl. ii. 380); erect, $1\frac{1}{2}$ ft. or more high; branches stout, pubescent; leaves from erect to spreading, usually incurved, linear, subacute, stout, keeled or obscurely sulcate, ciliolate or naked, glabrous or rarely slightly puberulous, glossy, 2–4 lin. long; flowers terminal, 4-nate, sometimes umbellate, sometimes pseudo-racemose; pedicels $1\frac{1}{2}$–4 lin. long; bracts remote, small; sepals lanceolate to ovate, subacute, glabrous and glossy, or puberulous, keeled, scarious-edged, ciliate, mostly deeply-coloured, 1–$1\frac{1}{2}$ lin. long; corolla urceolate, ovoid-urceolate or suburceolate-cyathiform, more or less (sometimes only very slightly) contracted at the throat, rosy, 2–3 lin. long; segments small, erect, or slightly spreading; anthers included, more rarely just manifest, oblong, subobtuse, dark-coloured, $\frac{1}{2}$–$\frac{2}{3}$ lin. long, crested shortly above the base; pore about $\frac{1}{2}$ the length of the cell; crests lanceolate or ovate, lacerate, lobed or serrulate, $\frac{1}{2}$–$\frac{3}{4}$ the length of the cells, rosy; style exserted; stigma capitate; ovary glabrous. *Benth. in DC. Prodr.* vii. 676; *Andr. Col. Heaths, t.* 30, *and Heathery, t.* 71. *E. guttæflora, Salisb. in Trans. Linn. Soc.* vi. 374. *E. incarnata, Andr. Heathery, t.* 168? *and Col. Heaths, t.* 28; *Lodd. Bot. Cab. t.* 1655? *not of Thunb. E. rubens, var. humilis, Wendl. Eric. Ic. fasc.* 3, 13, *not of Thunb. E. pedunculata, Wendl. l.c. fasc.* 23, 173, *t.* 66? *not of Andr. E. pendula, Lodd. l.c. t.* 902? *E. nutans, G. Don, Gen. Syst.* iii. 796. *E. arbuscula, Lodd. l.c.*

t. 843? *E. Fibula, Link, Enum. Hort. Berol.* i. 368 (*fide Benth.*).

SOUTH AFRICA: without locality, *Herb. Salisbury!* and *cultivated speci-mens!*

COAST REGION, between 1000 and 4000 ft. : Tulbagh Div. ; Mitchells Pass, *MacOwan, Herb. Norm. Aust.-Afr.,* 37! *Bolus,* 5354! Ceres Div. ; near Ceres, *Guthrie,* 2286b! Worcester Div. ; Dutoits Kloof, *Drège,* 7763! Paarl Div. ; French Hoek Mountains, *Schlechter,* 10270! Caledon Div. ; Palmiet River, *Guthrie,* 3872! Houw Hoek Mountains, *Zeyher,* 3207! mountains near Gena-dendal, *Burchell,* 7714! 7756! *Bolus,* 5423!

247. E. verecunda (Salisb. in Trans. Linn. Soc. vi. 379) ; erect, reaching 3–5 ft. high ; branches ascending, cano-puberulous, glabrescent; leaves erect, incurved-erect or spreading, imbricate, linear, rather blunt, sulcate, glabrous, 2–3 lin. long; flowers in many-flowered umbels, sometimes shortly pedicelled and appear-ing subcapitate or subspicate, usually cernuous at maturity; pedicels pubescent, crimson, 2–4 lin. long ; bracts small and quite remote, or two small and approximate, one median, long and foliaceous; sepals ovate, acute, sulcate, ciliate, about $\frac{1}{2}$ lin. long, or lanceolate, acuminate and about 1 lin. long ; corolla ovoid-urceolate, mouth contracted, $1\frac{3}{4}$–3 lin. long; segments erect or spreading, white or very pale rose ; anthers included, cuneate, subacute from $\frac{2}{5}$ to nearly $\frac{3}{5}$ lin. long, crested ; pore about $\frac{3}{4}$ the length of the cell ; crests ample, lobed and erosulate at the base, linear above, white, sub-diaphanous, about equalling the cell in length ; style slightly exserted or just manifest ; stigma capitate ; ovary glabrous. *Benth. in DC. Prodr.* vii. 676. *E. cernua, Andr. Heathery, t.* 9, *and Col. Heaths, t.* 12; *Lodd. Bot. Cab. t.* 822, *not of Linn. f.* ; *var. lanceolata, Wendl. Eric. Ic. fasc.* 8, 13. *E. ignorata, Klotzsch ex Benth. l.c.* 676.

SOUTH AFRICA: without locality, *Drège! Herb. Salisbury!* and *cultivated specimens!*

COAST REGION: Vanrhynsdorp Div. ; by rivulets on Wind Hoek Mountain, *Niven,* 21! 22! Clanwilliam Div., *Mader in Herb. MacOwan,* 2184! *Leipoldt,* 133! Cederberg Range at Pakhuis Pass, 2300 ft., *Schlechter,* 8607! *Bolus,* 8682! *Marloth,* 2639! 2691! Div. ? Pinaars Kloof, *Burke!*

CENTRAL REGION: Ceres Div.; Gydouw Mountain, 6000 ft., *Schlechter,* 10221!

The occasional closeness of the inflorescence is sometimes misleading. Schlechter's 10221 is a smaller-flowered form from a much higher elevation. The three figures cited are all good, but being from garden plants they are better grown than wild specimens.

248. E. tenella (Andr. Heathery, t. 94) ; erect, 1–1$\frac{1}{2}$ ft. or more high, generally more slender than any of the four preceding species; branches ascending, grey-pubescent ; leaves erect or spreading, linear, acute or subobtuse, keeled, glabrous or sparsely hairy and becoming so, mostly about 2 lin. long ; flowers crowded on short branchlets at the ends of the branches, mostly forming a pseudo-raceme ; pedicels puberulous, red, 1–2 lin. long ; bracts remote, small ; sepals mostly ovate-lanceolate, rarely subovate, acute or acuminate, keeled, glabrous, glossy, usually dark red, $\frac{3}{4}$–$1\frac{1}{4}$ lin. long; corolla urceolate or ovoid-urceolate, constricted at the throat, but always somewhat wide-mouthed, white or pale rose, 2–3 lin. long; segments mostly rather large and well-spreading, more rarely erect, sometimes darker-coloured than the tube, from $\frac{1}{5}$–$\frac{1}{4}$ the length of the tube; anthers included, somewhat variable in shape and length, from transversely semiovate-cuneate to oblong-cuneate, minutely ciliolate on the margin, the width at the base sometimes equal to the length, some-times a little narrower, from $\frac{1}{4}$–$\frac{2}{5}$ lin. long, crested; pore $\frac{1}{3}$–$\frac{2}{3}$ the length of the cell; crests always immediately deflexed close to the cells, subulate to lanceolate, or lanceolate-falcate, finely fimbriate, mostly entire or slightly lobed or lacerate, $\frac{1}{2}$–$\frac{2}{3}$ the length of the cells, rarely even a little longer than them; style mostly included, rarely subexserted; stigma capitate; ovary glabrous. *Andr. Col. Heaths, t.* 135; *Wendl. Eric. Ic. fasc.* 22, 155, *t.* 59; *Benth. in DC. Prodr.* vii. 677. *Lodd. Bot. Cab. t.* 375? *E. umbellifera, Wendl. Eric. Ic. fasc.* 25, 13, *t.* 6. *E. lactiflora, Lodd. l.c. t.* 991. *E. lactea, Lee ex Klotzsch, and E. kennedyana, Klotzsch in Linnæa,* xii. 506. *E. lasiandra, Klotzsch ex Benth. l.c.* 677. *E. Sprengelii, Sweet ex Steud. Nomencl. ed.* 2, i. 580.

[*E. pulchella, Salisb. in Trans. Linn. Soc.* vi. 379, according to the type specimen, belongs to this species.—*N. E. Brown.*]

VAR. β, **gracilior** (Bolus); branches leaves and pedicels more slender; leaves mostly spreading or squarrose and more distant; pedicels more spreading; bracts minute; anthers as in the type, but the crests often here and there lobed, as well as finely fringed.

SOUTH AFRICA: without locality, *cultivated specimens!*

COAST REGION, from 300 to 3000 ft. : Stellenbosch Div.; Hottentots Holland, *Mulder!* Lowrys Pass, *Burchell,* 8269! Caledon Div.; Houw Hoek Mountains, *Zeyher,* 3207! *Bolus,* 5358! *Guthrie,* 2286! Zondereinde River Mountains, *Guthrie,* 3280! Onrust River, *Schlechter,* 10384! mountains near Hermanus, *Guthrie,* 4653! Bredasdorp Div.; hills near Elim, *Bolus,* 6732! *Schlechter,* 7638! Var. β : Caledon Div.; mountains near the mouth of the Klein River, *Zeyher,* 3229!

Our variety β has a very different look, but the shape and size of the sepals, corolla and anthers are so similar that we cannot venture to separate it as a species. We have only Zeyher's specimen found " in shady places," to which unusual circumstance its vegetative differences are probably due.

249. E. chionophila (Guthrie & Bolus); erect, 6–7 in. or more in height; branches glabrous; leaves erect-incurved, d nsely imbri-cate, linear, subobtuse, sulcate, pubescent, ciliate, 2–$2\frac{1}{4}$ lin. long; flowers terminal, 3-nate, on short clustered branchlets; pedicels red, under 1 lin. long; bracts subapproximate, small, ciliate, red; sepals lanceolate, acute, subglabrous, softly ciliate, red, $\frac{3}{4}$ lin. long; corolla subglobose-urceolate, contracted at the throat, glabrous, deep red, $1\frac{1}{2}$ lin. long; segments rounded, erect, $\frac{2}{3}$–$\frac{7}{8}$ the length of the tube; filaments broad; anthers included, obliquely ovate, very obtuse, between $\frac{1}{5}$–$\frac{1}{4}$ lin. long, cristate; pore about $\frac{1}{2}$ the length of the cell; crests broad-subulate to cuneate, subentire or lobed, ciliate,

about $\frac{1}{2}$ the cell in length ; style included, short, thickened towards
the apex ; stigma clavate-capitate, large ; ovary glabrous.

COAST REGION : Worcester Div. ; on Sneeuwkop Mountain, near Bains Kloof,
Marloth in Herb. Bolus, 6497 !

A very well-marked species, if its characters are constant ; we have only one
small specimen from the single station cited. It verges towards the Subgenus
Platystoma, and is distinguished in the present section by the combination
of a corolla-limb rather large for this section, with a very small cristate
anther.

250. E. quadrangularis (Salisb. Prodr. 297, not of Andrews) ;
erect, 12–18 in. high ; branches mostly much spreading or divaricate, rarely subvirgate, puberulous ; leaves usually spreading, or
only the younger erect, linear, glabrous, the younger ciliate, $\frac{3}{4}$–$1\frac{1}{2}$ lin.
long ; flowers 4-nate ; pedicels slender, about 1 lin. long ; bracts
remote, minute ; sepals ovate or lanceolate-ovate, acute, subscarious,
ciliate or naked, keeled, glabrous, $\frac{1}{2}$–$\frac{3}{4}$ lin. long ; corolla generally
broad-cyathiform, more rarely subcampanulate-cyathiform, or the
tube sometimes obconic ("poculiform," *Salisbury*), mostly widened
at the mouth, sometimes tetragonous, 1–$1\frac{1}{4}$ lin. long, white to rosy ;
limb erect, variable in shape and length, $\frac{1}{2}$–$\frac{3}{4}$ the length of the tube,
the interstices of the limb-segments acute at the base ; anthers
included, from obliquely cuneate-oblong, $\frac{1}{6}$ lin. long, in *Salisbury's*
type to subtriangular with very rounded angles, $\frac{1}{7}$ lin. long and
proportionately broader (the front view of the anthers showing a
greater width than their length), in most other specimens quite
smooth and glabrous, aristate ; pore $\frac{3}{4}$–$\frac{7}{8}$ the length of the cell ; awns
narrow, rough-edged, but not fringed, about $\frac{1}{2}$ the length of the cell ;
style straight, included ; stigma capitellate, mostly small ; ovary
glabrous. *E. quadræflora, Salisb. in Trans. Linn. Soc.* vi. 375 ;
Benth. in DC. Prodr. vii. 679. *E. corusca, Lichtenst. ex Benth.
l.c.* 679.

SOUTH AFRICA : without locality, *Roxburgh! Herb. Salisbury!*
COAST REGION, between 150 and 1500 ft. : Clanwilliam Div. ; near Clan-
william, *Leipoldt*, 211 ! Worcester Div. ; Matroos Berg, *Marloth !* Paarl Div. !
French Hoek, *Bolus*, 6983 ! Stellenbosch Div. ; Lowrys Pass, *Schlechter*, 5353 ;
Caledon Div. ; near Zoetemelks Vlei, *Guthrie*, 3285 ! near Caledon, *Bolus*,
6749 ! *Schlechter*, 5604 ! Bredasdorp Div. ; near Elim, *Bolus*, 6734 ! Swellendam
Div. ; Zuurbraak, *Galpin*, 3673 ! Riversdale Div. ; near Riversdale, *Schlechter*,
1932 ! Knysna Div. ; Millwood, *Tyson !*

Salisbury says "the flowers are inconspicuous, but in the whole genus very
distinct," neither of which statements are intelligible to us, although we have
seen and dissected his type, from which our specimens differ only slightly in the
anthers. If we are correct in referring them here, of which we have little doubt,
the inflorescence is abundant and conspicuous, though the flowers themselves are
not large, and the corolla greatly resembles that of several other species. Salisbury doubtless described from a single specimen, and possibly a poor one. The
Index Kewensis, i. 870, puts this species down as a synonym of *E. denticulata*.

251. E. cyathiformis (Salisb. in Trans. Linn. Soc. vi. 376) ;
erect, much branched, 1–$1\frac{1}{2}$ ft. high ; branches puberulous, the

ultimate very slender; leaves erect or spreading, linear, obtuse, glabrous, slender, often curved, ½–1 lin. long; flowers 4-nate, sometimes by abortion of branchlets pseudo-axillary; pedicels slender, ¾–1 lin. long; bracts remote, very small; sepals ovate, obtuse, ciliate, shining, less than ½ lin. long; corolla subglobose-campanulate or broad-urceolate-campanulate, not (or hardly) constricted at the throat, somewhat widened at the mouth, ¾–1⅛ lin. long; segments about ⅔ as long as the tube; anthers included, subtriangular, with broadly rounded angles, subscabrid, ciliate at the base, ⅕ lin. long, aristate; pore about ⅔ the length of the cell; awns spreading, distant, naked on the margins, shorter than the cell; style included, short; stigma subpeltate-capitate; ovary glabrous or puberulous. *E. lœvis, Andr. Heathery, t. 221 and Col. Heaths, t. 182; Spreng. Pugill. i. 30? Lodd. Bot. Cab. t. 1393. E. caffra, Lodd. l.c. t. 196, not of Linn. nor of others. E. paniculata, var. alba, Wendl. Eric. Ic. fasc. 14, 25? E. paniculata, Wendl. ex Steud. Nom. ed. 2, i. 577. E. persoluta, var. major, Wendl. l.c. fasc. 24, 187, t. 71? E. persoluta, var. lævis, Benth. in DC. Prodr. vii. 679. E. stenophylla, Benth. l.c. 679. E. odorata, Spreng. Neue Entdeck. i. 271.*

SOUTH AFRICA : without locality, *Herb. Salisbury!* and *cultivated specimens!*

COAST REGION : Worcester Div.; banks of streams in Hex River Valley, 1600 ft., *Tyson,* 637! Caledon Div.; mountains near Grabouw, *Bolus,* 4176! and without exact locality, *Thom,* 978! Tulbagh or Worcester Div.; *Bolus,* 5191!

There has been some confusion about this species. Our specimens from several localities agree well with each other, with the specimen in the Berlin Herbarium marked *E. lœvis,* Spreng., and with Bentham's type of *E. stenophylla.* Salisbury himself saw the resemblance of his species to *E. viridipurpurea, L.,* though he observes : "Closely allied to [it], the anthers being almost exactly the same; yet on account of the difference in the interstices of the corolla-limb, I can scarcely regard it as a variety of that." To this we can only add that the leaves appear to be constantly smaller,—about half the size. It is much nearer to *E. viridipurpurea* than to *E. subdivaricata,* under which it was also placed by Bentham. It is also allied to *E. quadrangularis,* but the habit and general aspect are different, leaves constantly smaller, and the corolla less wide at the mouth.

252. E. hæmastoma (Wendl. Eric. Icon. fasc. 27, 18); branches few, erect, slender, glabrous; leaves erect to spreading, linear, sulcate, the younger pubescent and ciliate like the bracts and sepals with minutely plumose hairs, glabrescent, terminating in a sharp awn, 3–6 lin. long; flowers sub-4-nate; pedicels 2–3 lin. long; bracts subremote, two very small, the third longer, foliaceous; sepals lanceolate, acuminate, awned, coloured, 1–1½ lin. long; corolla ovoid to globose-urceolate, mouth much contracted, 2½–3 lin. long or more; segments erect, rounded; anthers included, subcuneate or longitudinally semiovate, dorsally curved, ⅖ lin. long, crested at the base; pore ¾ as long as the cell; crests more than ½ the length of the cell, lobed, acute, erosulate; style included or just manifest;

stigma capitate ; ovary glabrous.   *E. aristifolia, Niven ex Benth. in DC. Prodr.* vii. 676.

SOUTH AFRICA: without locality, *Drège*, 7766!
COAST REGION : Vanrhynsdorp Div. ; Wind Hoek Berg, *Niven*, 19!

253. E. velitaris (Salisb. in Trans. Linn. Soc. vi. 357) ; branches 13 in. long or more ; leaves erect or spreading, linear-trigonous, thin, acute, subglabrous, $2\frac{1}{2}$ lin. long ; pedicels $1-1\frac{1}{2}$ lin. long ; bracts subremote, the lower long and leaf-like ; sepals linear from an ovate subscarious base, acuminate, upper part foliaceous ; corolla subcyathiform, mouth about equal, "pubescent, $2\frac{1}{2}-3$ lin. long " (*Salisbury*), glabrescent, 2 lin. long ; segments about $\frac{1}{3}$ as long as the tube ; anthers included, oblong, smooth, glabrous, a little over $\frac{2}{5}$ lin. long, with slender rough awns about as long as the cell ; style included ; stigma capitate ; ovary glabrous.   *Benth. in DC. Prodr.* vii. 678.

VAR. β, **hemisphærica** (Bolus) ; branches and leaves somewhat stouter and more rigid ; upper bracts approximate, lower equal to the pedicel ; sepals narrow-lanceolate, acuminate, strongly keeled, about $1\frac{1}{4}$ lin. long ; corolla suburceolate, slightly contracted at the throat, glabrous, $2\frac{1}{2}$ lin. long ; anther-cells cuneate, obtuse, about $\frac{1}{4}$ lin. long.   *E. velitaris, var. β, Benth. in DC. Prodr.* vii. 678. *E. hemisphærica, Klotsch ex Benth. l. c.*

SOUTH AFRICA : without locality, *Roxburgh !*
COAST REGION: Var. β: Caledon Div. ; about Bot River, 500–2000 ft., *Zeyher!* in Herb. Trin. Coll. Dubl. ; near Palmiet River, *Zeyher,* in Herb. Berlin (ex *Bentham*).

A somewhat uncertain species.   We have not seen Salisbury's type, and his description is too brief to avail, but he cites in *Trans. Linn. Soc.* vi. 357, "*E. glabra*, Roxb. MS.''   If the corolla be usually pubescent, the species would be better in § *Ephebus.*   But, though we have examined a few flowers from Roxburgh's plant, we have only found a microscopic down on one corolla ; and the corolla of the var. β is perfectly glabrous.   We leave it in this section where Bentham placed it.

254. E. gracilis (Wendl. Bot. Beobacht. 47) ; a slender branched erect shrub, under 1 ft. high ; branches ascending or spreading, pubescent ; leaves erect or sometimes incurved or spreading, imbricate, linear, keeled or round-backed, slender, glabrous, 1–2 lin. long ; flowers 4-nate, on rather short branchlets ; pedicels $1-2\frac{1}{2}$ lin. long ; bracts remote, small ; sepals ovate-lanceolate to lanceolate, ciliolate or naked, about $\frac{1}{2}$ lin. long ; corolla urceolate, mouth usually slightly (rarely much) contracted, rosy, $1\frac{1}{4}-2$ (mostly $1\frac{1}{2}$) lin. long ; segments small, erect or spreading (probably always spreading at full maturity); anthers included, ovate, obtuse, glabrous, margins naked, little more than $\frac{1}{5}$ lin. long, broad-aristate or subcristate ; pore about $\frac{1}{2}$ as long as the cell ; awns deflexed close to the cells, broadish-subulate, entire or minutely lacerate, from $\frac{1}{2}$ as long to nearly as long as the cells ; style included ; stigma capitellate : ovary sessile on a broad disk, glabrous.   *Benth. in DC. Prodr.* vii. 677 ; *Willd. Sp. Pl.* ii. 365 ; *Salisb. in Trans. Linn. Soc.* vi. 375 ; *Andr. Heathery, t.* 68 ? *and Col. Heaths, t.* 98 ? ; *Wendl. Eric. Ic. fasc.* 8, 9.   *E. tenuis-*

sima, Wendl. l.c. fasc. 6, 9.   *E. tenera, Klotzsch ex Benth. l.c.* 677.
E. imbecilla, Sweet. Hort. Brit. ed. i. 258.   *E. neglecta, G. Don,
Gen. Syst.* iii. 792.

SOUTH AFRICA : without locality, *Herb. Salisbury!* and *cultivated specimens!*
COAST REGION : common from Swellendam Div. eastward to Humansdorp
Div., from 250–3000 ft., *Niven,* 14! *Alexander,* 19! *Zeyher,* 3255! *Bolus,* 2392!
5840! 9233! *Schlechter,* 2228! 2066! 5835! 5991! *Galpin,* 3674!

In the wild state this species is fairly constant in its characters and is
generally recognizable by its small sepals and anthers. In cultivation, however,
it seems to have varied, and this has given rise to some doubt and confusion as
to the figures. It is very uncertain whether Andrews' figure is the same as
Wendland's plant; we have seen no specimens with such globose corollas; nor
can we cite, as does Bentham, *Loddiges, Bot. Cab.* t. 244.

255. E. leucantha (Link, Enum. Hort. Berol. i. 371); erect,
somewhat slender, entirely glabrous, 12–18 in. high; branches
mostly virgate, often covered with a white papery semitransparent
and deciduous or separable epidermis; leaves 4-nate (we have seen
none 3-nate as mentioned by *Bentham*), mostly erect and imbricate,
rarely spreading, linear, blunt, faintly sulcate, $1\frac{1}{2}$–2 lin. long; flowers
4-nate, often clustered; pedicels about 1 lin. long; bracts remote,
small; sepals ovate to lanceolate, acuminate, keeled, very concave,
scarious-edged, 1 lin. long or less; corolla cyathiform, or sometimes
(probably in immature flowers) somewhat obconic, widened or not
contracted at the mouth, subtetragonous, white, about $1\frac{1}{2}$ lin. long;
segments erect or scarcely spreading, about $\frac{1}{2}$ as long as the tube;
anthers elliptical or oblong, smooth, glabrous, about $\frac{1}{5}$ lin. long,
aristate; pore about $\frac{2}{3}$ as long as the cell; awns subulate, entire or
nearly so, about $\frac{1}{2}$ as long as the cell; style included, slender; stigma
capitellate; ovary glabrous. *Klotzsch ex Benth. in DC. Prodr.* vii.
677. *E. leucanthera, Andr. Heathery,* t. 28, *and Col. Heaths,* t.
108, *not of Linn.* (*a very good fig. of the wild plant*). *E. lutea, var.
alba, Sinclair, Hort. Eric. Woburn.* 14.

SOUTH AFRICA : without locality, *cultivated specimens!*
COAST REGION : Paarl Div.; French Hoek, *Schlechter,* 9221! Caledon
Div.; mountains near Grietjes Gat, *Zeyher,* 3255! near Villiersdorp, *Bolus,*
5180! near Genadendal, 800–2200 ft., *Bolus,* 5430! 6959!

Bentham supposed this to be probably a garden hybrid; but from our
numerous wild specimens this seems improbable.

256. E. subdivaricata (Berg. Descr. Pl. Cap. 114); erect, much-
branched, 1–2 or even 3 ft. high; branches generally hirsute; leaves
spreading or erect, linear, subobtuse, round-backed, glabrous or
sparsely pubescent, about 2 lin. long; flowers 4-nate on short
branches, becoming clustered; pedicels glabrous, about 1 lin. long;
bracts remote, linear, ciliolate, small; sepals lanceolate, acute,
foliaceous or subscarious on the margins, keeled, ciliate or naked,
$\frac{1}{2}$–$\frac{7}{8}$ lin. long; corolla campanulate-cyathiform or subturbinate,
mouth widened, sometimes only slightly so, white, more rarely rosy,

1¼–2 lin. long; limb spreading or erect, about ⅔ as long as the tube, the interstices between the segments acute, closed; anthers included, longitudinally semi-ovate, cuneate or oblong-cuneate, smooth, glabrous, a little over ¼ lin. long, aristate; pore about ⅔ the length of the cell; awns curved, slightly rough on the inner edge, mostly somewhat shorter than the cell; style exserted; stigma capitate; ovary glabrous or thinly puberulous. *E. persoluta, Linn. Mant. Alt.* 230; *Benth. in DC. Prodr.* vii. 678. *E. prolifera, Salisb. in Trans. Linn. Soc.* vi. 376. *E. assurgens, Link, Enum. Hort. Berol.* i. 372. *E. congesta, Wendl. Eric. Ic. fasc.* 17, 75, *t.* 29, *not of Lodd. E. strigosa, Wendl. l.c. fasc.* 2, 25, *not of Soland. E. caffra, Andr. Heathery, t.* 7, *and Col. Heaths, t.* 7, *not of Linn. E. exserrens and E. pallidiflora, Klotzsch ex Benth l.c.* 679. *E. regerminans, Hort. ex Benth. l.c.*

COAST REGION: frequent near Cape Town, and extending eastward to Knysna Div.; ascending to 2500 ft., many collectors, *Thunberg, Niven,* 93! *Burchell,* 699! 702! 737! 766! 6542! 7048! 7581! 8376! 8588! *Zeyher,* 3245! *Sieber,* 174! *Drège,* 7761! *Bolus,* 2947! *Schlechter,* 2041! 7687! *Guthrie,* 44! 2281! *Wolley Dod,* 2269! Knysna Div.; *Buchanan!* And without locality, *cultivated specimens!*

Somewhat variable in habit and in the size of the flowers, and especially so as to indumentum. *E. persoluta, L.* var. *lævis,* Benth. in *DC. Prodr.* vii. 679, seems to be identical with *E. cyathiformis* and *E. persoluta,* var. γ *subcarnea, Benth. l.c.* may be a form of *E. curvirostris.* The species is closely allied to *E. viridipurpurea,* but is generally at once distinguishable by the acute interstices between the segments of the corolla-limb; the flowering season is also quite different.

257. E. margaritacea (Soland. in Ait. Hort. Kew. ed. 1, ii. 20); erect, glabrous, 1–1½ ft. high; branches slender, mostly straight and fastigiate; leaves usually erect and closely imbricate, linear, keeled, slender, 1–2 lin. long; flowers 4-nate or sometimes umbellately clustered; pedicels 1–2 lin. long; bracts remote, linear, about 1 lin. long; sepals linear-acuminate or lanceolate from an ovate denticulate-edged base, sometimes laxly set, ¾–1 lin. long; corolla cyathiform or suburceolate-campanulate, equal at the throat or nearly so, white, sometimes faint pink, about 2 lin. long; limb short, erect or very slightly spreading; anthers included, oblong, very obtuse, glabrous, smooth, nearly ½ lin. long; pore about ⅗ as long as the cell; crests oblong or ovate, acute, variously lacerate, papery, whitish, from ½–⅔ the length of the cell; style subexserted; stigma capitate; ovary glabrous. *Benth. in DC. Prodr.* vii. 677; *Andr. Heathery, t.* 126, *and Col. Heaths, t.* 35; *Wendl. Eric. Ic. fasc.* 8, 11 (*all good figures*). *E. obesa, Salisb. Prodr.* 294, *and in Trans. Linn. Soc.* vi. 375, *not of Tausch.*

COAST REGION: Cape Div.; Cape Flats, *Burchell,* 173! 412! 703! 821! *Ecklon,* 72! *Mund,* 37! *Bolus,* 2957! 3774! *Guthrie,* 380! *Wolley Dod,* 906! Stellenbosch Div.; Hottentots Holland, *Thunberg, Niven,* 13! Also without locality, *Herb. Salisbury!* and *cultivated specimens!*

258. E. curvirostris (Salisb. in Trans. Linn. Soc. vi. 375); erect, stout and rigid, 9–12 in. or more high; branches ascending, pubescent with reversed hairs; leaves suberect to spreading or squarrose, broadly linear, subobtuse, keeled, sulcate, glabrous, 2–3 lin. long; flowers 3-nate to umbellate; pedicels slender, $1\frac{1}{2}$–2 lin. long; bracts remote, adpressed, small; sepals ovate to lanceolate-oblong, acute or obtuse, subscarious, thick-keeled, glabrous, ciliolate or naked, often yellowish, $\frac{3}{4}$–$1\frac{1}{2}$ lin. long; corolla subcampanulate, widened above the hemispherical tube, $1\frac{1}{2}$–2 lin. long; segments slightly spreading, from $\frac{2}{3}$–$\frac{4}{5}$ the length of the tube; anthers included, cuneate and subacute, or oblong and subobtuse, ciliolate or naked, $\frac{2}{5}$–$\frac{1}{2}$ lin. long, cristate; pore $\frac{2}{5}$–$\frac{3}{5}$ the length of the cell; crests either spreading backwards at an angle or immediately deflexed, lanceolate, acuminate, lacerulate or fringed or pubescent, sometimes with one or more short lobes $\frac{1}{2}$ as long as the cells or somewhat longer; style subexserted often decurved, stoutish; stigma capitate; ovary glabrous. *Benth. in DC. Prodr.* vii. 677. *E. declinata, Lodd. Bot. Cab. t.* 1662. *E. decunata, Steud. Nom. ed.* i. 304. *E. thyrsoidea, Tausch in Flora,* 1834, 619.

South Africa: without locality, *Sieber,* 178! *Herb. Salisbury!* and *cultivated specimens!*

Coast Region: Cape Div.; Table Mountain, 2400 ft., *Bolus,* 4480! *Schoennberg in Herb. Galpin,* 4907! Stellenbosch Div.; Hottentots Holland, *Niven,* 144! Lowrys Pass, *Burchell,* 8243! Caledon Div.; Houw Hoek Mountains, *Burchell,* 8078, 8079; *Bolus,* 5355! near Lowrys Pass, 2000 ft., *Schlechter,* 4827!

The corolla is represented as white in Loddiges' fig. and was so in Bolus, 5355; but in other specimens it seems to have been red or rosy. This species flowers in the autumn and winter.

259. E. trichophylla (Benth. in DC. Prodr. vii. 679); erect, slender, 6–14 in. high; branches chiefly virgate, the ultimate very slender, hispidulous, rusty red; leaves from suberect and imbricate above to spreading below, the lowest subsquarrose, linear, semi-terete, very slender, pubescent, glabrescent, about 1 lin. long; flowers sub-4-nate; pedicels slender, about $\frac{1}{2}$–$\frac{3}{4}$ lin. long; bracts remote, linear, one sometimes exceeding the pedicel; sepals linear-lanceolate, long-acuminate, green-keeled, scarious-edged, ciliate, about 1 lin. long, or little shorter than the corolla; corolla broad-cyathiform, nearly equal at the mouth, somewhat hyaline, white, $1\frac{1}{4}$ lin. long; segments erect, from $\frac{1}{2}$ as long to as long as the tube; anthers included, very straight, narrow-oblong, pallid, smooth, glabrous, nearly $\frac{1}{3}$ lin. long, aristate; pore $\frac{2}{5}$ the length of the cell; awns entire, about $\frac{1}{3}$ the length of the cell; style included, slender; stigma capitellate, very small; ovary glabrous.

South Africa: without locality, *Masson* (in Herb. Kew)!
Coast Region: Caledon Div.; Zoetemelks Vlei, *Guthrie,* 3301! 3541!

Our specimens agree almost exactly with Masson's.

Section XIX. LEPTODENDRON. (Sp. 260–268.)

260. E. rupicola (Klotzsch in Linnæa, xii. 504); erect; branches stoutish, ascending, either few and subvirgate, or many, spreading and subfastigiate, glabrous or puberulous; leaves 3-nate, or occasionally 4-nate on the same plant, crowded, erect or spreading-incurved, narrow-linear, keeled, glabrous, $1\frac{1}{2}$–2 lin. long; flowers on short branchlets along the branches, with a subracemose appearance, corolline; pedicels $\frac{3}{4}$ lin. long; bracts remote, subbasal, minute; sepals broad-lanceolate, acuminate, glabrous, ciliate, coloured, $\frac{3}{4}$ lin. long; corolla ovoid or suburceolate-ovoid, mouth contracted, slightly oblique, "deep rosy red" (*Klotzsch*), about $2\frac{1}{2}$ lin. or rarely $1\frac{3}{4}$ lin. long; segments slightly spreading, or erect and at length connivent, about $\frac{1}{3}$ as long as the tube; filaments very slender, sigmoid near the apex and thickened above the curve, about 2 lin. long; anthers oblong, obtuse, scaberulous; cells partite to the base, $\frac{2}{5}$ lin. long, muticous; pore $\frac{2}{5}$ the length of the cell; style included, often just manifest; stigma capitate; ovary glabrous. *Benth. in DC. Prodr.* vii. 681.

COAST REGION: Swellendam Div.; mountains along the lower part of the Zondereinde River, near Appels Kraal, *Zeyher*, 3234! 209! (the latter a form with 4-nate leaves). Bedford Div.; near Bedford, *Weale*, 1! (in Herb. Albany Museum, Grahamstown, a specimen with smaller flowers $1\frac{3}{4}$ lin. long, not fully or poorly grown).

261. E. condensata (Benth. in DC. Prodr. vii. 681); branches stoutish, erect, puberulous, 10 in. or more long; leaves 3-nate, erect or subincurved-spreading, crowded, linear, subobtuse, sulcate, glabrous, about 2 lin. long; flowers sub-3-nate, corolline; pedicels about 1 lin. long; bracts approximate, all sepal-like, or the lower sometimes longer and equalling the sepals; sepals broad-ovate or ovate-rhomboid, acute, cuspidate-keeled, thick, subcoriaceous, somewhat viscid (*Bentham*), $\frac{3}{4}$–1 lin. long; corolla subcampanulate-cyathiform (or "ovate," *Bentham*), equal or slightly widened at the mouth, or (in *Burchell's* type specimens, apparently over mature) contracted towards the apex, 4-foveolate at the base, about 2 lin. long; segments variable in length, about $\frac{1}{4}$ the length of the tube; filaments nearly straight; anthers oblong, obtuse, scaberulous, muticous; cells deeply parted, about $\frac{1}{2}$ lin. long; pore about $\frac{1}{3}$ the length of the cell; style included; stigma capitate; ovary glabrous.

VAR. β, **quadrifida** (Bolus); corolla-limb more deeply cleft; segments from $\frac{1}{2}$–$\frac{3}{4}$ the length of the tube.

COAST REGION: Swellendam Div.; on a lofty craggy peak near Swellendam, *Burchell*, 7391! Riversdale Div.; summit of Kampsche Berg, *Burchell*, 7112! Var. β, Swellendam Div.; on the Langeberg Range, near Zuurbraak, 2500 ft., *Schlechter*, 2146!

262. E. mira (Klotzsch ex Benth. in DC. Prodr. vii. 680); apparently erect, 8 in. or more high; branches straggling, slender, reddish, white-puberulous; leaves 3-nate, suberect, mostly scanty with longish internodes, linear-oblong, blunt, sulcate, glabrous, pallid, under 2 lin. long; flowers 3-nate, scanty, corolline; pedicels puberulous, $1\frac{1}{2}$ lin. long; bracts remote, small; sepals subulate-

linear, acute, glabrous, about 1 lin. long; corolla subtubular, marked
with 8 longitudinal lines, a little over 2 lin. long; segments erect,
rounded, about $\frac{1}{5}$ the length of the tube; filaments long and slender,
$1\frac{2}{3}$ lin. long; anthers sublinear, a little widened towards the apex,
smooth, nearly $\frac{1}{2}$ lin. long, aristate; pore nearly equal in length to
the cell; awns strap-shaped, equal throughout, closely approximate
but free, the points shortly spreading, a little shorter than the cells;
style exserted; stigma clavellate, small; ovary globose, minutely
hispidulous or glabrous.

COAST REGION: Worcester Div.; on a mountain, between de Draai and
Driekoppen, *Drège,* 7724! in Herb. Berlin.

We have seen a good specimen of the type of this apparently rare and little-
known species. It is very distinct and might almost be recognized from
the anther alone, which is not quite like any other we remember to have
seen.

263. E. micrandra (Guthrie & Bolus); erect, entirely glabrous,
except for a slight down on the younger branches, about 1 ft. high;
branches mostly straggling or subdivaricate, more rarely ascending
or suberect, thinly and distinctly leafy, with naked internodes of
from $\frac{1}{2}$–1 in. long; leaves 3-nate, erect, incurved, never crowded,
nor (except a few of the youngest) imbricate, linear-semiterete,
obtuse or subacute, flat above, round-backed, faintly sulcate, $1\frac{1}{2}$–2
lin. long; flowers 3-nate, corolline; pedicels slender, about 1 lin.
long; bracts subapproximate, linear, about $\frac{1}{2}$ lin. long; sepals
adpressed, linear, acute, sulcate, foliaceous, sometimes coloured,
reddish, about $\frac{3}{4}$ lin. long, not reaching so high as the corolla-tube;
corolla broad-cyathiform, flesh-coloured, about $1\frac{1}{2}$ lin. long, $2\frac{1}{4}$ lin.
wide at the mouth when flattened down; segments continuous,
subsemiovate, broadly-rounded, about equal to the tube; filaments
capillary; anthers manifest, basifixed near the ventral margin,
semiorbicular, ventral margin straight; cells subdistant and divari-
cate, about $\frac{1}{100}$ in. long, aristate; pore $\frac{3}{4}$ the length of the cell;
connective at length bent downwards and projecting as an acute
tooth in front; awns decurrent for less than $\frac{1}{2}$ their length along
the filament, then free, spreading, subulate, acuminate, the whole
appendage nearly $2\frac{1}{2}$ times the length of the cells; style exserted,
slender; stigma capitellate; ovary glabrous.

COAST REGION: Ceres Div.; at the foot of the mountains near Ceres, 1700 ft.,
Bolus, 9198!

This very distinct species has the habit and general appearance of *E. quadrangularis,* with the corolla of the § *Eurystoma,* but very different sepals. Its
peculiar and very small anther alone distinguishes it at once from every other
species known to us, except *E. pusilla,* var. β, the anther of which is curiously
similar and is even slightly smaller; but in other respects the plant is very
different.

264. E. Vanheurckii (Muell. Arg. in Van Heurck, Obs. Bot. &
Plant. Nov., fasc. i. 36); erect, about 1 ft. high; branches slender,
subvirgate, glabrous; leaves 3-nate, erect or subspreading, mostly

shorter than the internodes on the slender flowering branches, or sometimes imbricate, linear-elliptic, subacute, sulcate, thickish, glabrous, 1–1½ lin. long; flowers 3-nate or umbellate (4–8-flowered), corolline; pedicels decurved, slender, glabrous, 1½–4 lin. long; bracts subapproximate to subremote, linear, small, coloured; sepals narrow-lanceolate, acute, glabrous, ciliate or naked, mostly deep red, ¾–1 lin. long; corolla urceolate or obovoid-urceolate, constricted at the throat, red, 1¼–2 lin. long ; segments spreading, rounded, $\frac{1}{6}$–$\frac{1}{4}$ the length of the tube; filaments slender, mostly ciliate, with short or long hairs, about 1 lin. long; anthers oblong or cuneate-oblong, obtuse, smooth, pale brown, bearded at the base or naked, $\frac{1}{4}$–$\frac{2}{5}$ lin. long, crested at the base; pore $\frac{1}{2}$–$\frac{2}{3}$ the length of the cell; crests ovate to lanceolate, acuminate, variously incised, from ½ as long to nearly as long as the cell; style mostly included, sometimes just manifest; stigma subsimple or clavellate, small; ovary subturbinate, constricted at the base, glabrous.

COAST REGION: Tulbagh Div.; flats between the Witsenberg and Skurfdeberg Ranges, *Zeyher*, 1103! 1103a!

CENTRAL REGION: Ceres Div.; Cold Bokkeveld, Skurfdeberg Range near Gydouw, 5000 ft., *Schlechter*, 10006! Gydouw Berg, 6000 ft., *Schlechter*, 10223! near Wagenbooms River, *Schlechter*, 10152!

This species has been confused with *E. leptopus, Benth.*, but the corolla is quite different in shape. The anther, as also the filament, is much more copiously bearded in Ecklon & Zeyher's specimens than in Schlechter's; but of these latter only his 10223 appears to have a few fully developed flowers, and those showing them clearly as described.

265. **E. campanulata** (Andr. Col. Heaths, t. 9) ; erect, slender, somewhat variable in habit and appearance, from 6 to probably 24 in. in height; branches numerous, erect or spreading, or when young and grown in damp places subsimple, virgate, and very slender with fewer and smaller flowers, glabrous, reddish ; leaves commonly 3-nate, rarely 4-nate or opposite on the same plant, erect or subspreading, narrow-linear, acute, keeled, glabrous, 1½–2½ lin. long; flowers generally solitary, corolline ; pedicels puberulous, 1–1½ lin. long ; bracts remote or subremote, rarely subapproximate; sepals from ovate and acute to lanceolate and long-acuminate, scarious, glabrous, margins mostly naked, rarely ciliate, coloured, 1½–1¾ lin. long; corolla very variable in size and shape, tubular-campanulate, cyathiform, broad-cyathiform or suburceolate, bright yellow, more rarely cream-colour, 2½–3½ lin. long ; segments erect, more rarely spreading, $\frac{1}{4}$–$\frac{1}{2}$ the length of the tube ; filaments nearly straight, about ½ lin. long ; anthers oblong-cuneate, obtuse, scaberulous, $\frac{1}{2}$–$\frac{2}{3}$ lin. long, most generally muticous (said to be "sometimes minutely toothed at base"); pore about ½ the length of the cell; style included, sometimes compressed ; stigma capitellate, small; ovary glabrous. *Andr. Heathery, t.* 55 ; *Wendl. Eric. Ic. fasc.* 13, 3 ; *Lodd. Bot. Cab. t.* 184; *Benth. in DC. Prodr.* vii. 681. *E. campanularis, Salisb. in Trans. Linn. Soc.* vi. 330. *E.*

tenuifolia, Hort. ex Salisb. l.c. 331.　*E. flavicans, Klotzsch ex Benth. l.c.* 681.

COAST REGION: Caledon Div.; mountains near Palmiet River, *Ecklon & Zeyher,* 95 ! in Herb. Berlin; Babylons Tower Mountain, 1200 ft., *Templeman in Herb. Norm. Aust.-Afr.,* 753 ! Zwart Berg, near Caledon, *MacOwan, Herb. Aust.-Afr.,* 2016! between French Hoek and Villiersdorp, *Bolus,* 6989 ! Div. ? *Bolus,* 6872 !　Also *cultivated specimens !*

This is a very variable species, but with numerous specimens before us we are unable to define even sufficiently well-marked or stable varieties.　It has been ordinarily recognized (from Audrews' figure) by its somewhat slender habit and leaves, and solitary yellow flowers.　It varies gradually up to stouter forms, with 4-nate leaves and clustered (3–4-nate) flowers, and these look very different from the others, but without any differences in the structure of the flowers. Extreme forms of these are represented in *Bolus,* 6872.

266. E. lachnæoides (G. Don, Gen. Syst. iii. 795) ; erect, 1 ft. or more high; branches numerous, ascending, with short floriferous branchlets ; leaves 3-nate, erect to spreading, crowded, imbricate, linear, obtuse, sulcate or subopen-backed, tomentose-downy and of "a mealy green," $1\frac{1}{2}$–$2\frac{1}{2}$ lin. long; flowers corolline, spreading or cernuous ; pedicels very short ; bracts adpressed, broad-lanceolate, subobtuse, sepaloid, ciliate, coloured, pink, about $\frac{1}{2}$ the length of the sepals ; sepals like the bracts, from $\frac{1}{4}$–$\frac{1}{3}$ the length of the corolla ; corolla suburceolate-cylindrical, not contracted at the throat, "purple-pulverulent," pink, 3–$3\frac{1}{2}$ lin. long, $1\frac{1}{4}$ lin. in diam.; seg- ments spreading, $\frac{1}{4}$–$\frac{1}{3}$ the length of the tube ; anthers oblong, obtuse, length not discernible from the figure, cristate ; pore about $\frac{2}{3}$ the length of the cell ; crests lanceolate, acute, serrulate, pallid, nearly as long as the cells; style included, short ; stigma capitellate ; ovary glabrous. *Benth. in DC. Prodr.* vii. 693.　*E. Lachnæa, var. purpurea, Col. Heaths, t.* 178, *and Andr. Heathery, t.* 170.

SOUTH AFRICA : without locality, ex *Andrews.*

This is only known from the figures and descriptions cited above.

267. E. polycoma (Benth. in DC. Prodr. vii. 655); dwarf, probably under 1 ft. high; whole plant silvery-grey ; branches erect, stoutish ; branchlets short and flexuous, these and the leaves, pedicels, bracts and sepals pilose with short and long simple, some- what silky, grey hairs ; leaves 3-nate, erect, closely imbricate, adpressed, lanceolate, subobtuse, pubescent, the hairs towards the apex long and tufted, 1–$1\frac{1}{2}$ lin. long; flowers 3-nate, subcorolline ; pedicels curved, $1\frac{1}{2}$–2 lin. long; bracts adpressed, ovate, concave, keeled, subscarious, long-ciliate, about 1 lin. long ; sepals like the bracts but broad-ovate, obtuse, apiculate, 1–$1\frac{1}{4}$ lin. long, about $\frac{1}{2}$ the length of the corolla ; corolla ovoid-cyathiform, or by the con- nivence of the segments in dried specimens appearing ovoid-inflated, glabrous, dry (or "subviscid," *Bentham*), rosy, about 2 lin. long ; segments erect, or connivent in age (and in dried specimens), semi- ovate, obtuse, longer than broad, equal or nearly so to the tube in length ; filaments slender; anthers subcuneate-oblong, slightly incurved, nearly $\frac{2}{5}$ lin. long, crested; pore $\frac{1}{2}$–$\frac{2}{3}$ the length of the cell; crests narrow-lanceolate, acute, pendulous, long-ciliate, brown, about $\frac{2}{3}$ the length of the cells; style included, 4-gonous below, swollen and terete above; stigma subsimple, small; ovary glabrous, or thinly hispid on the apex.

SOUTH AFRICA: without locality, *Masson! Niven!*
COAST REGION: Caledon Div.; rocky summit of the Genadendal Mountain, 4800 ft., *Bolus*, 5421! *Schlechter*, 9835!

This species does not come very happily into this or any section. Though distinct, it is allied to *E. lachnœoides* (*E. Lachnœa, var. purpurea, Andr.,* as figured in *Andr. Heathery,* t. 170). Scarcely any of the specimens, excepting Schlechter's, exhibit fully-developed flowers.

268. E. virginalis (Klotzsch ex Benth. in DC. Prodr. vii. 653); apparently a small slender plant, with filiform, white-tomentulose branches ("quite glabrous," *Bentham*); leaves opposite, erect, adpressed, oblong, obtuse, keeled, glabrous, glossy, under 1 lin. long; flowers 2-nate, corolline; pedicels decurved, nearly 1 lin. long; bracts subremote, small; sepals narrow-lanceolate, acute, adpressed, round-backed and sulcate, or "keeled," thickish, under 1 lin. long; corolla subtubular, 4-gonous, "red," 2 lin. long; limb erect or " stellate-spreading?" (*Bentham*); segments rounded, imbricate at the base, $\frac{1}{4}$–$\frac{1}{3}$ the length of the tube; filaments slender, tapering to the apex, about $1\frac{1}{3}$ lin. long; anthers narrow-oblong, obtuse, somewhat rough, under $\frac{2}{5}$ lin. long, crested; pore $\frac{3}{5}$ the length of the cell; crests ovate, bluntish, about $\frac{1}{2}$ the length of the cell; style apparently included; ovary sessile, glabrous.

COAST REGION: Tulbagh Div.; mountains near Tulbagh, *Ecklon & Zeyher!*

We have seen and examined a frustule with a flower from a specimen marked as above, which is very probably the type. We find, however, no trace of a stellate-patent corolla-limb, a character which is often deceptive and lost in dried specimens, and which Bentham mentions with a sign of doubt. In other respects it does not agree well with the § *Lamprotis* where Bentham provisionally placed it, the flowers being distinctly corolline; but it fits better and fairly well here. The species is a very distinct one.

Section XX. **PACHYSA.** (Sp. 269–292.)

269. E. ramentacea (Linn. Mant. 65); erect, 1–$1\frac{1}{2}$ ft. high; mostly much branched from shortly above the base; branches usually much spreading, sometimes erect, puberulous or glabrous; leaves 4- (rarely 3-) nate, erect or spreading, linear, very slender, glabrous, $1\frac{1}{2}$–3 lin. long; flowers terminal, umbellate, with a heavy, somewhat musk-like odour; pedicels erect or spreading, slender, red, $1\frac{1}{4}$–$1\frac{1}{2}$ lin. long; bracts 2 subapproximate, 1 median, linear, red; sepals lanceolate, keeled, glabrous, naked or ciliate, reddish, about $\frac{3}{4}$ lin. long; corolla mostly subglobose, occasionally ovoid, contracted at the throat, glabrous, slightly viscid, crimson, $1\frac{1}{2}$ to nearly 2 lin. long; limb short, erect, darker-coloured; anthers included, dorsifixed near

the middle of the cell, connivent round the stigma, from elliptical
to bluntly triangular (in front view nearly orbicular) densely covered
and the crests ciliate with longish shred-like hairs, about $\frac{1}{4}$ lin. long;
pore $\frac{7}{8}$ the length of the cell; crests sublanceolate, acute, pale brown,
about as long as the cell; style included; stigma capitate; ovary
glabrous (or pubescent, *Salisbury*), sessile on a prominent disk.
Linn. Syst. Nat. ed. 12, ii. 269; *Bauer, Exot. Pl. t.* 23; *Wendl.
Eric. Ic. fasc.* 1, 17; *Lodd. Bot. Cab. t.* 446; *Andr. Heathery, t.*
143, *and Col. Heaths, t.* 53; *Benth. in DC. Prodr.* vii. 658. *E.
granulata, Linn. Mant. Alt.* 234. *E. multumbellifera, Berg. Descr.
Pl. Cap.* 110. *E. pilulifera, Berg. l.c.* 111. *E. bullularis, Salisb.
Prodr.* 296, *and in Trans. Linn. Soc.* vi. 377. *E. bullaris, Steud.
Nomencl. ed.* 2, i. 569. *E. multiumbellata, Benth. in DC. Prodr.*
vii. 658. *E. pilularis, Benth. l.c. sub E. ramentacea.*

SOUTH AFRICA: without locality, *Thunberg*, Herb. *Salisbury!* and *cultivated specimens!*

COAST REGION: Paarl Div.; between Paarl and French Hoek, *Drège!*
Cape Div.; common on the Cape Flats and occasional ou the hills near Cape
Town, *Burchell*, 377! 820! *Ecklon*, 85! *Mund*, 40! *Sieber*, 170! *Zeyher*, 1104!
Guthrie, 191! *Wolley Dod*, 946! *Bolus*, 2962! Stellenbosch Div.; between
Lowrys Pass and Jonkers Hoek, *Burchell*, 8316! Caledon Div.; near Gena-
dendal, *Burchell*, 7776! Zwart Berg, near Caledon, 2300 ft., *Bolus*, 5429!
Swellendam Div.; mountain peak, near Swellendam, *Burchell*, 7389! Rivers-
dale Div.; Garcias Pass, *Galpin*, 3676!

This has the look of the § *Orophanes*, but the corolla is viscidulous. Specimens
from eastern stations and growing at a greater distance from the sea exhibit
fewer and somewhat smaller flowers and a laxer habit than those from the Cape
Peninsula.

270. E. mucosa (*Linn. Mant. Alt.* 232); erect, glabrous in all
parts, commonly from 1–2 (more rarely 3–4) ft. high; branches
erect, spreading or straggling, rigid; leaves 4-nate, erect or somewhat
spreading, imbricate or shorter than the internodes, linear to linear-
oblong, obtuse, keeled, viscid, ciliate or naked, thick, rigid, 1–3
lin. long; flowers terminal, umbellate; umbels 3–6-flowered; pedicels 2–3 lin. long; bracts approximate to remote, scarious, or
foliaceous in the upper part, viscid; sepals narrow-ovate or lanceolate, viscid, tipped with a thick green keel-like point, scarious,
coloured, 1–1½ lin. long; corolla urceolate, or subovoid-urceolate,
very viscid, 1½–2½ lin. long, dull red or purple; limb short, erect
or subpatent; anthers included, dorsifixed a little above the base,
cuneate (suboblong in front view) obtuse, glabrous, over $\frac{1}{4}$ lin. long
or somewhat less than 1½ times its width at the base; pore $\frac{3}{5}$ the
length of the cell; crests sublanceolate, coarsely serrate, long-
acuminate, as long as the cells; style included, short; ovary
glabrous. *Bauer, Exot. Pl. t.* 15; *Andr. Heathery, t.* 174, *and
Col. Heaths, t.* 39; *Lodd. Bot. Cab. t.* 35; *Wendl. Eric. Ic. fasc.*
4, 13; *Benth. in DC. Prodr.* vii. 658, *and var. brevifolia, Benth.
l.c.* 659. *E. pilularis, Lodd. Bot Cab. t.* 1563? *E. ferrea, Berg.
Descr. Pl. Cap.* 112. *E. mucosoides, Lodd. Bot. Cab. t.* 1202. *E.
pilulifera, Andr. Heathery, t.* 278? *fide Benth.*

VAR. β, **crenata** (Benth. l.c. 659); leaves elliptical, minutely glandular-serrulate. *E. crenata, E. Meyer ex Benth. l.c.* 659.

SOUTH AFRICA: without locality; *Herb. Salisbury!* and *cultivated specimens* of the type and var. β!

COAST REGION: Cape Div.; occasional on the Cape Flats and mountains up to 1100 ft., *Niven*, 41! *Burchell*, 698! 8352! 8353! *Mund*, 41! *Drège*, 7758! *Zeyher*, 1100! *Sieber*, 173! *Bolus*, 4519! *Schlechter*, 575! 7301! and in *Herb. Norm. Aust.-Afr.*, 1301! *Guthrie*, 292! *Wolley Dod*, 463! Var. β: Cape Div.; sandy plains near Cape Town, *Niven*, 40!

271. E. macra (Guthrie & Bolus); dwarf, erect, 6–8 in. high; branches ascending, dark-coloured, naked below, leafy above, the younger cano-puberulous; leaves 3-nate, suberect or spreading, elliptical, blunt, thick and rounded, sulcate, glabrous, 1–1½ lin. long; flowers terminal, 3-nate; pedicels slender, viscid, red, 2½–3 lin. long; bracts remote, 2 median, 1 basal, scarious, 1 lin. long; sepals lanceolate-oblong, acute, scarious, viscid, glabrous, 1½ lin. long; corolla narrow-ovoid-urceolate, glabrous, viscid, purple, 2½ lin. long; limb short, spreading; anthers included, dorsifixed shortly above the base, cuneate, subacute, ½ lin. long or about twice the width at the base, glabrous, smooth, subaristate or narrow-crested; pore about ½ the length of the cell; awns lanceolate at the base, tapering to a long fine point, with (sometimes at least) a single median saw-like tooth, the whole about as long as the cell; style included; stigma capitellate; ovary minutely puberulous or glabrous.

COAST REGION: Worcester Div.; on the Matroos Berg, 7000 ft., *Bolus in Herb. Guthrie*, 3948! rocky places on Dutoits Peak, 6200 ft., *Marloth*, 2418!

272. E. Schlechteri (Bolus in Journ. Bot. 1894, 235); erect, 2–3 ft. high; branches ascending, flexuous, puberulous, glabrescent; leaves 4-nate, erect or subspreading, imbricate, linear, bluntish, sulcate, glabrous, dry or the younger subviscid, 1½–2 lin. long; flowers terminal, 3–4-nate; pedicels viscid-pubescent, 1–2 lin. long; bracts approximate or remote, linear, viscid-pubescent, 1 lin. long; sepals lanceolate, glabrous, viscid, coloured, 1–1½ lin. long; corolla urceolate-campanulate, only slightly or not contracted at the throat, viscidulous, pale rosy, 2¼–3¾ lin. long; segments somewhat large and well-spreading, rounded, from ⅓–⅖ the length of the tube; anthers included, dorsifixed at the base, longitudinally subsemi-ovate, obtuse, glabrous, over ⅓ lin. long, crested; pore about ⅞ the length of the cell; crests lanceolate, acuminate, serrulate, nearly equal to the cell; style included; stigma capitate; ovary densely hirsute.

COAST REGION: Queenstown Div.; Hangklip Mountain, 6000–6300 ft., *Galpin*, 1611!

EASTERN REGION, from 2500 to 6000 ft.: Natal; by the Umzimkulu River at Handcocks Drift, *Tyson*, 3066! by the Mooi River, near Weston, *Wood*, 5476! 5326! *Rehmann*, 7357! *Schlechter*, 3341!

KALAHARI REGION: Orange River Colony; Mont-aux-Sources, 6000 ft., *Bolus!*

Mr. Schlechter describes it as overhanging the streams in Natal.

273. E. spectabilis (Klotzsch ex Benth. in DC. Prodr. vii. 659) ;
stout, erect ; branches straight or flexuous, cano-puberulous, glabrescent ; leaves 3-nate, usually erect-incurved, imbricate, oblong or
obovate-oblong, obtuse, sulcate, thick, glossy or minutely puberulous, gland-ciliolate, 1–2½ lin. long ; flowers terminal, 3-nate ;
pedicels spreading, gland-pubescent, 2½–3½ lin. long ; bracts subremote or remote, sepal-like, whitish ; sepals ovate or subobovate,
obtuse, keel-tipped, gland-ciliolate, glabrous, white, reaching to
about ½ the height of the corolla, 1–1¼ lin. long ; corolla globose-
urceolate, moderately contracted at the throat, glabrous, viscidulous
or nearly dry, 1½–2½ lin. long ; segments short, erect, rounded ;
anthers included, dorsifixed at the base, subtriangular-cuneate or
oblong, obtuse, ciliolate or naked, $\frac{2}{5}$ lin. long, decurrent-aristate ;
pore $\frac{7}{8}$ the length of the cell ; awns decurrent along the produced
widened deflexed dark-coloured connective, free for about ½ their
length, ciliate or scaberulous, nearly as long as or somewhat exceeding the cells ; style included ; stigma capitellate or subsimple ; ovary
glabrous, pallid on a dark disk.

COAST REGION : Bredasdorp Div.; on the chalk hills between Cape Agulhas
and Potte Berg, below 500 ft., *Drège !* Marcus Bay, *Fry in Herb. Galpin*, 4968 !
Riversdale Div. ; on a ridge overlooking Stil Bay at Milkwood Fontein, 600 ft.,
Galpin, 3625 !

Galpin's 4968 closely resembles the type which was distributed by Drège
as "*E. lachnæifolia*"; his 3625 has larger leaves, but otherwise differs
little.

274. E. flocciflora (Benth. in DC. Prodr. vii. 660) ; branches
long and somewhat virgate, the younger cano-pubescent ; leaves
4-nate, erect (at least the younger), oblong-linear, obtuse, keeled,
sulcate, the younger tomentose-lanate, the older glabrous and glossy,
about 2 lin. long ; flowers mostly terminal, 3-nate (or " occasionally
lateral ? " *Bentham*) ; pedicels tomentose, stout, 4 lin. long ; bracts
subapproximate, floccose-lanate ; sepals ovate-oblong, subobtuse,
keel-tipped, densely floccose-lanate, subscarious, 2 lin. long ; corolla
urceolate, mouth at maturity much contracted, tomentose, viscid,
thick, 2½–3 lin. long ; limb short, erect ; anthers " broadly-crested "
(*Bentham*) ; style exserted ; stigma clavate ; capsule nearly as large
as the corolla, with a conical top, smooth, glossy, glabrous.

COAST REGION : Uniondale Div.; rocky hill near Groot River in the Long
Kloof, near the village of Haarlem, *Burchell*, 4992 !

Very different from any other species in this section by the peculiar indumentum on its sepals. Unfortunately, Burchell's specimens are fructiferous,
and we have been unable to find an anther.

275. E. Alexandri (Guthrie & Bolus) ; subdecumbent, 10–13 in.
high ; branches numerous, ascending, weak and somewhat straggling,
leafy, dark red, pubescent and also hispid with long gland-tipped
hairs ; leaves 4-nate, spreading or squarrose, linear to broad-elliptical,
blunt, pubescent or glabrescent, ciliate with long glandular hairs,
about 1 lin. long ; flowers terminal, sub-4-nate or umbellate :

pedicels slender, red, thinly pilose, 1½ lin. long; bracts approximate, foliaceous, spreading, gland-hairy; sepals ovate-lanceolate, acute, scarious, keeled, glandular-pilose on the keel only, ciliolate with short sessile brown glands, 1 lin. long; corolla urceolate, well contracted at the mouth, viscid, glabrous, 2 lin. long, 1½ lin. wide; limb spreading, about ⅓ the length of the tube; filaments broad, subequal; anthers subexserted, dorsifixed considerably above the base, oblong, obtuse, smooth, glabrous, about ⅖ lin. long, crested; pore ½–⅔ the length of the cell; crests semiorbicular, lacerate, very little longer or wider than the base and sides of the cells; style slender, long exserted; stigma capitate; ovary hispid, chiefly on the apex.

COAST REGION : Paarl Div.; sandy flats below Paarl Mountain, *Alexander*, 17!

276. E. frigida (Bolus in Journ. Bot. 1894, 235); suberect, decumbent or sometimes at higher elevations densely matted and procumbent; branches straggling, slender, the younger thinly pilose, up to 12 in. long; leaves 4-nate, subrecurved, spreading or squarrose, elliptical or oblong, thick, sulcate, margins rough with tuberclebased glandular hairs, rarely subnaked and glossy, ¾–1½ lin. long; flowers terminal, sub-4-nate or umbellate (4–5-flowered); pedicels viscid, about 1 lin. long; bracts subapproximate, glandular-ciliate, small; sepals broad- to narrow-lanceolate, acute, scarious, coloured, viscid, keel-tipped and gland-ciliate near the apex, about ¾ lin. long; corolla urceolate or ovoid-urceolate, gradually contracted towards the mouth and well-constricted at the throat, viscid, glabrous or rarely very minutely puberulous, 1¼–1¾ lin. long, deep red to rosy and whitish; segments spreading, short, about ⅕ the length of the tube; anthers included, dorsifixed above the base, subcuneate, obtuse, or subnarrow-ovate, smooth, glabrous, over ¼ lin. long, aristate; pore nearly ½ the length of the cell; awns rough, spreading, about ⅔ the length of the cell; style included; stigma subsimple; ovary densely villous.

EASTERN REGION, between 6000 and 7000 ft. : Griqualand East; summit of Mount Currie, *Tyson*, 1255! 1771! Insiswa Mountain, *Schlechter*, 6451! Natal; summit of Amawahqua Mountain, *Evans*, 675! *Wood*, 4580!

KALAHARI REGION : Orange River Colony; summit of the Mont-aux-Sources, 9500 ft., *Flanagan*, 2029! *Thode*, 32! 65! *Bolus!*

277. E. nubigena (Bolus in Journ. Bot. 1894, 236); erect, 6–12 in. high; branches usually much-spreading and numerous, stout, glabrous, brittle; leaves mostly 3-nate, more rarely 3–4-nate on the same plant, suberect to squarrose, linear-oblong or narrow-elliptical, subobtuse, sulcate, glabrous, thick, viscid, glossy, sparsely glandciliolate or naked, the younger black-apiculate, rather long-petiolate, 1½–2½ lin. long; flowers terminal, 3–4-nate; pedicels slender, viscid, coloured, 4–6 lin. long; bracts remote, 2 median, 1 subbasal, viscid, adpressed; sepals lanceolate, acute, coriaceous, viscid, minutely keel-tipped, glabrous, gland-ciliolate, 1½–2 lin. long; corolla ovoid-urceolate, very viscid, glabrous, red or purple, 3–3½ lin. long ; limb short, spreading, about $\frac{1}{12}$ the length of the tube ; filaments rather narrow ; anthers included, dorsifixed near the base, oblong-cuneate, obtuse, smooth, glabrous, ⅔ lin. long, aristate ; pore nearly ½ the length of the cell ; awns subulate-acuminate from a broadish base, curved, about ½ the length of the cell ; style included ; stigma capitate ; ovary deeply 8-lobed, velvety, hispidulous at the apex.

COAST REGION : Tulbagh Div. ; rocks on the Great Winterhoek Berg, 6100 ft., *Marloth,* 1630! *Bolus in Herb. Guthrie,* 4171! Worcester Div. ; on the Matroos Berg, 5500 ft., *Marloth,* 2209!

CENTRAL REGION : Ceres Div. ; on the Tafelberg in the Cold Bokkeveld, 6200 ft., *Schlechter,* 10092!

278. E. formosa (Thunb. Diss. Erica, 49, t. 3) ; erect, 1–2 ft. high ; branches erect or spreading, hispid or pubescent, somewhat slender ; leaves 3-nate, the younger suberect, the older spreading or squarrose, elliptical or oblong, obtuse, thick, sulcate, glabrous, glossy, the younger ciliolate, 1–1½ lin. long ; flowers 3-nate ; pedicels slender, pubescent, 2 lin. long ; bracts approximate, like the sepals but smaller ; sepals spreading or reflexed, ovate or broad-lanceolate, acute, coriaceous, coloured, subviscid, glabrous, about 1 lin. long ; corolla globose-urceolate, distinctly or suddenly contracted at the throat, with 8 longitudinal channels, viscid, glabrous, white, about 2 lin. long ; segments spreading or recurved, about ⅛ the length of the tube ; filaments bent round the ovary and again contracted round the stigma ; anthers included, dorsifixed at the base upon the thickened projecting connective, oblong, smooth, ciliate at the base, nearly ⅓ lin. long, aristate or subcrested ; pore nearly as long as the cell ; awns broad linear, densely ciliate, partially decurrent along the connective which stands out at a wide angle from the cells, with long free spreading points, the whole about ⅔ of the cell in length ; style included, much thickened towards the truncate apex ; stigma subsimple ; ovary densely hirsute. *Benth. in DC. Prodr.* vii. 659. *E. grandinosa, Andr. Heathery, t.* 265, *and Col. Heaths, t.* 238 ; *Lodd. Bot. Cab. t.* 627. *E. quadrata, Lodd. l.c. t.* 1943.

SOUTH AFRICA : without locality, *Thunberg,* and *cultivated specimens!*

COAST REGION : Swellendam Div. ; Voormansbosch, *Ecklon & Zeyher,* 136! George Div. ; Outeniqualand, *Masson,* 82! Cradock Berg, *Mund!* Kaymaus Gat, *Drège,* 7756! Malgaten River, 400 ft., *Young in Herb. Bolus,* 5521! Knysna Div. ; near Knysna, *Burchell,* 5453! 5500! *Tyson in MacOwan & Bolus, Herb. Norm. Aust.-Afr.,* 994!

279. E. umbelliflora (Klotzsch ex Benth. in DC. Prodr. vii. 659) ; branches cano-pubescent with minutely plumose hairs ; leaves 3-nate, erect or spreading, oblong, obtuse, glabrous, glossy, thick, coriaceous, mostly 1–1½ (occasionally 2–3, *Bentham*) lin. long ; flowers in few-flowered umbels, more or less viscid ; pedicels floccose with minutely plumose hairs, 1–1½ lin. long ; bracts approximate, somewhat spreading, broad-ovate or suborbicular, shortly acute, keeled in the upper half, coriaceous, margins scarious, concave, horn-coloured, ½–¾ lin.

long; sepals like the bracts, but larger and more distinctly, but shortly, ciliolate, about 1 lin. long and nearly as wide, not reaching half the height of the corolla; corolla subglobose, subtetragonous, shortly tapering to the apex, scarcely constricted at the throat, about $1\frac{3}{4}$ lin. long; segments (in the dried state) connivent, incurved, broad, $\frac{1}{4}-\frac{1}{3}$ the length of the tube; anthers included, dorsifixed considerably above the base, oblong, obtuse, nearly $\frac{1}{2}$ lin. long, muticous; pore about $\frac{1}{3}$ the length of the cell; style well-exserted, slender; stigma capitellate; ovary depressed, much wider than its length, glabrous.

COAST REGION : Mossel Bay Div.; Attaquas Kloof, *Masson*, 61! Humansdorp Div. ; dry sandy eminences near Gamtoos River, *Masson*, 60!

CENTRAL REGION : Prince Albert Div. ; on the Great Zwartberg Range, near Vrolykheid, 4000–5000 ft., *Drège*, 7757 !

280. E. tragulifera (Salisb. in Trans. Linn. Soc. vi. 374); somewhat slender, with the habit, general appearance and the leaves of the preceding; flowers 3-nate; pedicels pubescent, 1–2 lin. long; bracts approximate, scarious, viscid; sepals spreading or erect, lanceolate, acute or acuminate, subscarious, viscid, glabrous, about 1 lin. long; corolla ovoid-urceolate, tapering gradually to the constricted throat, viscid, glabrous, sometimes with 8 longitudinal channels, white, 2–$2\frac{1}{2}$ lin. long; segments erect or slightly spreading, short; anthers included, dorsifixed at the base upon the thickened projecting connective, oblong, smooth, ciliate below, aristate; cells deeply partite and at length separating down to the connective at the base, nearly $\frac{1}{2}$ lin. long; pore $\frac{2}{3}-\frac{3}{4}$ the length of the cell; awns broad-linear, densely ciliate, partially decurrent along the projecting connective, spreading, as long as in the preceding species; style included, straight, equal to the apex; stigma small, subsimple or capitellate; ovary densely hirsute. *E. nitida, Andr. Heathery, t.* 131, *and Col. Heaths, t.* 188; *Lodd. Bot. Cab. t.* 1131; *Benth. in DC. Prodr.* vii. 659. *E. Gordonia, Forbes, Hort. Woburn.* 81.

SOUTH AFRICA : without locality, *cultivated specimens !*

COAST REGION : George Div. ; near George, *Alexander*, 13 ; Montagu Pass, 1200 ft., *Young in Herb. Bolus*, 5523! *Schlechter*, 5783! Uniondale Div.; Keurbooms River, in the Long Kloof, *Masson*, 81! Long Kloof, *Mund!*

CENTRAL REGION : Prince Albert Div.; near Klaarstrom, 3000–4000 ft., *Drège*, 7765.

We have carefully examined only the specimens of *Young* and *Schlechter*, and cannot be certain of the others, though they are probably correctly named.

281. E. glomiflora (Salisb. in Trans. Linn. Soc. vi. 330) ; erect, $1-1\frac{1}{2}$ ft. high ; branches suberect or spreading, floccose with very short subplumose hairs, beset with many short flowering branches ; leaves 3-nate, erect or spreading, crowded, subulate-linear, obtuse, glabrous and glossy, or puberulous, the younger cartilagineo-ciliolate, 1–3 lin. long ; flowers terminal, 3-nate ; pedicels slender, pubescent. $1\frac{1}{2}-3\frac{1}{2}$ lin. long; bracts approximate, coriaceous, coloured, rigid, sepal-like, sometimes equalling or even exceeding the sepals in length; sepals ovate or ovate-lanceolate, acuminate, keel-tipped, thick, coriaceous, adpressed or subsquarrose, glabrous, viscid, coloured, $\frac{3}{4}$–$1\frac{1}{2}$ lin. long; corolla ovoid-urceolate, sometimes slightly narrowed above, at others into a distinct and somewhat narrow neck, viscid, white or red, $2\frac{1}{2}$–3 lin. long; segments spreading or nearly erect; anthers included, dorsifixed near the base, narrow-oblong, subacute, 3–$3\frac{1}{2}$ times longer than the width at the middle, smooth, ciliate at the base, $\frac{5}{8}$–$\frac{7}{8}$ lin. long, aristate at the base; pore about $\frac{3}{4}$ the length of the cell; awns broad-linear, densely ciliate, partially decurrent along the filament or connective, which projects at an angle from the cell, thence free and bent downwards, in total length about equal to the cells; style included; stigma small, subclavate-capitellate; ovary densely hirsute. *E. reflexa, Link, Enum. Hort. Berol.* i. 371; *Andr. Heathery, t.* 283, *and Col. Heaths, t.* 263; *Lodd. Bot. Cab. t.* 1787; *Benth. in DC. Prodr.* vii. 659. *E. vesicaria, Soland. ex Salisb. in Trans. Linn. Soc.* vi. 330.

VAR. β, cantharæformis (Bolus); corolla 4 lin. long, with a longer neck than in the type. *E. cantharæformis, Lodd. Bot. Cab. t.* 1961.

SOUTH AFRICA: without locality, *cultivated specimens!* Var. β: *Masson* (in Herb. Trinity College, Dublin)!

COAST REGION: Mossel Bay Div.; Attaquas Kloof, *Masson,* 79! *Niven,* 48! George Div.; Devils Kop, near George, *Masson,* 89! Outeniqua Mountains, *Herb. Bolus,* 6305! Oudtshoorn Div.; Zwartberg Pass, 4200 ft., *Marloth,* 2410! *Kolbe! Atherstone,* 266! Uniondale Div.; near Haarlem, in Long Kloof, *Galpin,* 3642!

282. E. physodes (Linn. Syst. Nat. ed. x. 1002); erect, $1\frac{1}{2}$–$2\frac{1}{2}$ ft. high; branchlets and pedicels puberulous, otherwise glabrous; branches strong, rigid, erect, straightish, and, like the branchlets, densely leafy; leaves 4- (more rarely 3-) nate, erect-spreading, imbricate, linear, obtuse, more or less triquetrous, keeled, not sulcate, shining, about 3 lin. long; flowers terminal (sometimes sublateral, *Bentham*), 3–4-nate, mostly cernuous; pedicels $\frac{1}{2}$–3 lin. long; bracts remote or sometimes subapproximate, like the sepals but narrower and smaller; sepals ovate-lanceolate, obtuse, green-keel-tipped, thick, coriaceous, viscid, whitish, $1\frac{1}{2}$–2 lin. long; corolla globose-urceolate or suburceolate-ovoid, much constricted at the throat, very viscid, white, 3–4 lin. long; segments erect or slightly spreading, at length connivent and almost closing the mouth, $\frac{1}{5}$–$\frac{1}{7}$ the length of the tube; anthers included, dorsifixed at the base, oblong, subacute, glabrous, smooth, a little over $\frac{1}{2}$ lin. long; pore about $\frac{3}{5}$ the length of the cell; crests semiorbicular and crenulate at the base with a long terminal linear spreading lobe, glabrous, about $\frac{1}{2}$ the length of the cell; style included or slightly exserted, equal; stigma capitellate; ovary turbinate, on a very short contracted foot, glabrous. *Andr. Heathery, t.* 34, *and Col. Heaths, t.* 45; *Wendl. Eric. Ic. fasc.* 7, 13; *Bot. Mag. t.* 443; *Lodd. Bot. Cab. t.* 223; *Benth. in DC. Prodr.* vii. 659. *E. sequax, Salisb. Prodr.* 293, *and in Trans. Linn. Soc.* vi. 378.

SOUTH AFRICA : without locality, *cultivated specimens!*
COAST REGION : Cape Div.; Table Mountain, *Thunberg, Ecklon,* 12! Simons
Bay, *Alexander,* 6! *Bolus,* 3711! and in *Herb. Norm. Aust.-Afr.,* 49! *Guthrie,*
554! Noord Hoek Mountains, *Wolley Dod,* 1070!

283. E. Urna-viridis (Bolus in Journ. Linn. Soc. xxiv. 180);
erect, 3–5 ft. high; stem unbranched and leafless below for about
half its height; branches few and spreading, straggling, leafy,
puberulous; leaves and flowers as in the preceding species ; pedicels
viscidulous, 3–4 lin. long; bracts remote, small; sepals as in the
preceding; corolla ovoid or narrow-ovoid, subobtuse, viscid, bright
green, 4–6 lin. long; segments erect, connivent; anthers as in the
preceding species, but larger, about 1 lin. long; the rest as in the
preceding.

COAST REGION : Cape Div.; Muizenberg Mountain, *Zeyher,* 3194! 4997!
Bolus, 3355! *MacOwan & Bolus, Herb. Norm. Aust.-Afr.,* 42! *Guthrie,* 350!
Wolley Dod, 573! Cape Flats, *Burke!*

Dried specimens of this are not easily distinguished from the preceding. In
the living state it is at once known by its habit and green flowers. The pedicels
in this are longer; the corolla larger, with no tendency towards a globular
shape. It is noteworthy that while this species has a flowering season from
September to March, and a zone of growth in altitude of from 600 to 2300 ft.,
E. physodes flowers from June to Sept., and has a zone of from 2300–
2800 ft. So far as is known to us the two species do not grow on the same
mountain.

284. E. Fairii (Bolus in Journ. Bot. 1894, 236) ; erect, 1–1½ ft.
high, glabrous; branches ascending, sometimes spreading or straggling, stout, leafy, rigid leaves 4–6- (usually 5-) nate, spreading,
recurved, elliptical or oblong, obtuse, thick, rigid, sulcate, margins
cartilagineo-denticulate, 2–3 lin. long, the floral obovate and somewhat dilated, closely enveloping the pedicels and bracts, and together
with bracts, sepals and corolla, very viscid ; flowers terminal, 4-nate
or subcapitate, subcernuous, subcalycine ; pedicels ½–1 lin. long;
bracts closely approximate, oblanceolate, scarious, keeled and keel-
tipped, denticulate or sublacerate, 3 lin. long; sepals linear, acute,
margins inflexed and glandular, keel-tipped, subscarious, greenish,
about 3 lin. long; corolla tubular-urceolate, only slightly inflated
below and very little constricted at the throat, white, about 4 lin.
long; segments erect or slightly spreading, erosulate, about ⅐ the
length of the tube; filaments broad, tapering upwards ; anthers
included, dorsifixed at the base, oblong, obtuse, roughish, ¾ lin. long,
aristate ; pore about the length of the cell; awns free, spreading,
rough-edged, about ⅖ the length of the cell; style slender, at length
slightly exserted ; stigma capitate; ovary turbinate, pallid, glabrous.

COAST REGION : Cape Div.; rocky ridges on the mountain near Simons
Town, 800 ft., *Fair* in Herb. Bolus, 7191! and in *Herb. Norm. Aust.-Afr.,*
1308! *Wolley Dod,* 1118!

285. E. oblongiflora (Benth. in DC. Prodr. vii. 632); erect,
2–3 ft. high; branches spreading or straggling, stout, leafy and

pubescent above, naked and glabrescent below, dark-coloured and rough with leaf-scars; leaves mostly 4-nate, occasionally somewhat scattered, crowded, spreading, recurved, oblong or narrow-elliptical, obtuse, deeply and widely sulcate, glabrous, thick, shining, margins cartilagineo-denticulate, 1–2½ lin. long; flowers terminal, 4-nate; pedicels 2½–4 lin. long; bracts approximate, oblong, thick, scarious, about 1 lin. long; sepals lanceolate, acute, keeled, thick, scarious, coloured, about 2 lin. long; corolla narrow-ovoid or suburceolate-ovoid, viscid, glabrous, of thick consistence, greenish, with 16 faint longitudinal veins, 4–5½ lin. long; segments small, erect, semiorbicular, about $\frac{1}{10}$ the length of the tube; filaments broad at the base, tapering upwards with a sigmoid curve; anthers included, dorsifixed at the base, linear-cuneate, subacute, smooth, glabrous, $\frac{3}{4}$ lin. long, decurrent-aristate; pore about $\frac{2}{3}$ the length of the cell; awns adnate to the filament for about $\frac{1}{4}$ lin., with acute free tooth-like points; style shortly exserted; stigma clavate-capitate; ovary glabrous. *E. decurrens, Klotzsch ex Benth. in DC. Prodr.* vii. 632. *E. oblongifolia, Steud. Nom. ed.* 2, i. 577.

SOUTH AFRICA : without locality, *Bowie!*

COAST REGION : Bredasdorp Div. ; rocky hill near Bredasdorp and Mier Kraal, 250 ft., *Bolus,* 8453 ! *Guthrie,* 3781 ! *Schlechter,* 10504 !

286. E. odorata (Andr. Heathery, t. 177, and Col. Heaths, t. 191); erect, glabrous, ½–1 ft. high ; branches ascending, not much spreading ; leaves 3-nate (or also 4-nate,' *Bentham*), squarrose-recurved, crowded, linear to oblanceolate, obtuse and callous at the apex, thick, keeled and faintly sulcate, closely ciliate with stiff gland-tipped hairs, which sometimes extend also to the reduplications on either side of the keel, including the long pallid petiole (of about 1 lin.) 3–4 lin. long; flowers terminal and lateral, subracemose-umbellate, 3-4-∞ -flowered ; pedicels slender, varying much in length during flowering, 3–7 lin. (or 1 in. in cultivated specimens) ; bracts remote and distant from each other, linear-lanceolate ; sepals lanceolate, acute, subcoriaceous, green-keel-tipped, closely gland-ciliolate, whitish, 2½ lin. long; corolla suburceolate, scarcely constricted at the throat, subviscid, white, about 4 lin. long; segments erect or slightly spreading, broad and short (in cultivated specimens the throat is more constricted and the segments more spreading); filaments narrow, equal, about $\frac{1}{3}$ the length of the corolla; anthers included, dorsifixed above the base, oblong-ovate or narrow-elliptical, obtuse, smooth, about $\frac{3}{5}$ lin. long, muticous; pore about $\frac{2}{3}$ the length of the cell; style included; stigma capitellate; ovary glabrous. *Bot. Mag. t.* 1399; *Lodd. Bot. Cab. t.* 633; *Benth. in DC. Prodr.* vii. 660. *E. adenophylla, Bolus in Journ. Linn. Soc.* xxiv. 181. *E. spirans, Hoffmansegg, Verz. Pfl. Nachtr.* ii. 28.

SOUTH AFRICA : without locality, *Niven !* and *cultivated specimens !*

COAST REGION : Caledon Div.; mountains about Houw Hoek, 3000–4000 ft., *Bolus,* 5453 ! *Schlechter,* 5445 !

The rediscovery of the wild plant by two recent collectors enables us to describe it as it grows in its native home, and to perceive that such differences as exist between it and the figures and descriptions above referred to are almost entirely those of size and luxuriance due to the effects of cultivation.

287. E. ardens (Andr. Heathery, t. 51, and Col. Heaths, t. 75); erect, 1–1½ ft. high; branches stoutish, subvirgate, puberulous or glabrous; leaves generally 3- (occasionally 4-) nate, erect to spreading, linear or linear-lanceolate, sulcate, somewhat thick, glossy, rigid, the younger minutely cartilagineo-serrulate, 3–5 lin. long; flowers terminal, 3-nate, or sometimes solitary and by arrest of the lateral branchlets pseudo-lateral; pedicels 3–5 lin. long; bracts remote or subapproximate, like the sepals but smaller; sepals ovate or ovate-lanceolate, acute or obtuse, keel-tipped, thick, coriaceous, viscid, coloured, 1½–2 lin. long; corolla ovoid-urceolate to globose-urceolate, thick, very viscid, at length much inflated, coral-red or white, 3–4 lin. long; segments suberect, broad, from $\frac{1}{4}$–$\frac{1}{3}$ the length of the tube; filaments more or less dilated, sometimes much so, lanceolate, much bent below the anther; anthers included, dorsifixed above the base, oblong, scaberulous, obtuse, $\frac{3}{4}$ lin. long, crested; pore $\frac{3}{5}$–$\frac{2}{3}$ the length of the cell; crests lanceolate, deeply lacerate, more or less densely hairy, $\frac{1}{3}$–$\frac{3}{5}$ the length of the cell; style included, slender; stigma small, capitellate; ovary glabrous. *Bot. Reg. t.* 115; *Lodd. Bot. Cab. t.* 47; *Benth. in DC. Prodr.* vii. 660.

SOUTH AFRICA: without locality, *cultivated specimens!*
COAST REGION, from 1000–4000 ft.: Swellendam Div.; rocky places above Voormans Bosch, *Zeyher*, 3208; mountains near Swellendam, *Mund*, 78 or 98! *Bolus Herb. Norm. Aust.-Afr.*, 602! *Niven*, 50! between Sparrbosch and Tradouw, *Drège! Kennedy in Herb. Bolus*, 1200! Riversdale Div.; Garcias Pass, *Bain in Herb. Bolus*, 2961! *Galpin*, 3637! *Schlechter*, 1715!

288. E. blenna (Salisb. in Trans. Linn. Soc. vi. 379); erect, 1–1½ ft. high; branches stout, ascending, virgate or flexuous, puberulous or glabrous; leaves 3-nate, mostly erect and imbricate or subspreading, linear, subobtuse, flat above, keeled and sulcate beneath, glabrous, 4–5 lin. long; flowers terminal, 3-nate ("here and there sublateral," *Bentham*); pedicels about 4 lin. long; bracts remote, lanceolate, about 3 lin. long; sepals ovate, acuminate, keel-tipped, thickish, subscarious, viscid, coloured or greenish, about 2½ lin. long; corolla conical-ovoid or suburceolate-conical, much contracted to the mouth but only slightly constricted at the throat, very viscid, 4½–6½ lin. long, bright orange-red, the limb and some distance below it green; segments spreading or erect, about $\frac{1}{8}$ the length of the tube; filaments broad at the base tapering upwards, bent below the anther; anthers included, dorsifixed well above the base, cuneate, subacute, scaberulous, ciliolate, about 1 lin. long, crested; pore $\frac{3}{5}$–$\frac{2}{3}$ the length of the cell; crests quite free from the filament, subsemiorbicular in outline, deeply inciso-lacerate, about $\frac{1}{2}$ the length of the cell; style included, straight; stigma capitellate; ovary glabrous. *E. vernix and var. longiflora, Andr. Heathery, tt.* 248 *and* 250, *and Col. Heaths, tt.* 214 *and* 285; *Benth. in DC. Prodr.* vii. 660. *E. resinosa, Bot. Mag. t.* 1139; *Lodd. Bot. Cab. t.* 679.

VAR. β, **grandiflora** (Bolus); flowers usually solitary, rarely in pairs; corolla 8–10 lin. long; anthers 1½ lin. long.

SOUTH AFRICA: without locality, *Bain in Herb. Bolus*, 3193! and *cultivated specimens!*

COAST REGION: Swellendam Div.; Grootvaders Bosch, *Masson*, 47! near Tradouw Pass, *MacOwan & Bolus, Herb. Norm. Aust.-Afr.*, 603! Zuurbraak, 3000 ft., *Galpin*, 3639! Riet Kuil, *Niven*, 51! Var. β, Riversdale Div.; exact locality unknown, *Miss Borcherds in Herb. Bolus*, 6310! and in *Herb. Guthrie*, 2364! Bredasdorp Div.; *De Villiers* in Cape Govt. Herb.

Our var. β appears to be somewhat rare; there is a striking difference in size, but in structure it is too close to be specifically distinguished.

289. E. ixanthera (Benth. in DC. Prodr. vii. 660); decumbent, glabrous; branches slender, simple, curved, distantly leafy, 6 in. or more long; leaves 3-nate, spreading, curved, distant, but mostly longer than the internodes, slender, linear, obtuse, subcallous at the apex, flattish above, sulcate below, somewhat thin, shining, 3–5 lin. long; flowers terminal, umbellate, 5–8-flowered (or sometimes sub-racemose with the terminal branch continued, *Bentham*); pedicels slender, spreading, 3–6 lin. long; bracts remote, small; sepals lanceolate, subacute, keel-tipped, thick, rigid, viscid, about 1 lin. long; corolla " urceolate-globose " (*Bentham*) or suboblate-urceolate, throat slightly constricted, viscid, about 2 lin. long, 2½ lin. wide; segments rather large, ½ the length of the tube, suberect; filaments subequal, bent in a simple semicircular curve round and close to the ovary; anthers included, dorsifixed at the base upon a prolongation of the connective, spreading at nearly a right angle from the cell, oblong, pubescent, about ⅗ lin. long; pore about ⅔ the length of the cell; awns decurrent along the connective with free points (of about ½ lin. long), the whole about as long as, or sometimes much exceeding, the cells; style included; ovary oblate, much depressed, width twice the length, lobed, glabrous.

COAST REGION: Riversdale Div.; lower part of the Langeberg Range, at Garcias Pass, *Burchell*, 6942!

290. E. carduifolia (Salisb. in Trans. Linn. Soc. vi. 330); erect, 6–12 in. high; branches ascending or spreading, somewhat flexuous, roughly hispid; leaves 3-nate, mostly somewhat distant, either longer or shorter than the internodes, spreading, squarrose or recurved, linear-oblong, obtuse, convex above and thick, sulcate or open-backed, long-setose-ciliate, the younger with gland-tipped hairs, 2–4 lin. long; flowers terminal, umbellate, or (by the arrest of lateral branchlets) sometimes pseudo-racemose; pedicels slender, viscid, 4–5 lin. long; bracts remote, small, the lowest basal or nearly so; sepals linear-lanceolate to subovate, acute, thick, viscid, minutely gland-ciliolate, about 1½ lin. long; corolla suburceolate or urceolate-cyathiform, wide-mouthed, scarcely contracted at the throat, viscid, red, 3–3½ lin. long, 2–2¼ lin. wide; segments erect, about ¼ the length of the tube; filaments remarkably short, broad-lanceolate, scarcely over 1 lin. long; anthers included, dorsifixed above the base upon a prolongation of the connective bent downwards at an

angle of about 45° with the cell, oblong, smooth, glabrous, under $\frac{1}{2}$ lin.
long, aristate ; pore about $\frac{2}{3}$ the length of the cell ; awns decurrent
for a short distance along the connective, with longer free spreading
points, the whole about equal to, or shorter than, the cell; style
included ; stigma subsimple or clavate, small; ovary glabrous.
Benth. in DC. Prodr. vii. 660. *E. aprica, Klotzsch ex Benth. l.c.*
660.

CoAST REGION : Caledon Div. ; high on the Baviaans Kloof Mountains, near
Genadendal, *Niven,* 21 ! summit of Genadendal Mountain, *Galpin,* 3636 !
Zwart Berg, near Caledon, *Bodkin in Herb. Bolus,* 6731 ! *Galpin,* 3635 !
Swellendam Div. ; mountains near Appels Kraal, *Zeyher,* 3214 !

291. E. obliqua (Thunb. Diss. Erica, 44, t. 1) ; erect, entirely
glabrous, 1–1$\frac{1}{2}$ ft. high ; branches few, subsimple, virgate, densely
leafy ; leaves scattered, spreading, crowded, long-petiolate, narrow-
linear, obtuse, slender, sulcate, glabrous, 3–5 lin. long ; flowers
terminal and sublateral at the ends of the branches, umbellate ;
pedicels spreading, slender, viscid glandular, 4–8 lin. long ; bracts
remote, linear ; sepals ovate-oblong, obtuse, somewhat laxly set,
thick, concave, viscid, $\frac{1}{2}$–$\frac{3}{4}$ lin. long ; corolla ovoid-urceolate, constricted at the throat, thick, viscid, glabrous or often minutely
hispidulous on the upper part, 3$\frac{1}{2}$–4 lin. long ; segments erect or
subspreading, large, $\frac{1}{3}$–$\frac{1}{2}$ the length of the tube ; filaments dilated at
the apex ; anthers included, dorsifixed shortly above the base, linear-
cuneate or semiovate, acute, smooth, glabrous, $\frac{2}{3}$–1 lin. long, cristate ;
pore $\frac{1}{2}$–$\frac{2}{3}$ the length of the cell; crests semiorbicular, coarsely
dentate, about $\frac{1}{2}$ the length of the cell; style included ; stigma
capitellate ; ovary stipitate, glabrous. *Bauer, Exot. Pl. t.* 3 ;
Wendl. Eric. Ic. fasc. 17, 77, *t.* 30 ; *Andr. Heathery, t.* 33, *and
Col. Heaths, t.* 42 ; *Lodd. Bot. Cab. t.* 14 ; *Benth. in DC. Prodr.*
vii. 661.

SOUTH AFRICA: without locality, *Herb. Salisbury !*
COAST REGION : Cape Div. ; Muizenberg Mountain, 1400 ft., *Bolus,* 4518 !
Guthrie, 321 ! *MacOwan, Herb. Norm. Aust.-Afr.,* 25 ! Table Mountain, *Wolley
Dod,* 814 ! near Simons Town, *Wolley Dod,* 812 ! Stellenbosch Div. ; Hottentots
Holland Mountains, *Zeyher,* 3211 ! *Thunberg.* Caledon Div. ; between Palmiet
River and Lowrys Pass, *Burchell,* 8184 ! Houw Hoek Mountains, *Burchell,*
8032 ! 8101 ! *Galpin,* 3641 ! *Guthrie,* 2288 !

In structure, and especially by the stipitate ovary, this and the following species
are near to *E. inflata, Thunb.,* and form a link with the § *Ceramus.*

292. E. macroloma (Benth. in DC. Prodr. vii. 661) ; erect, 12 in.
or more high ; branches long, straightish, leafy, minutely gland-
pubescent ; leaves 3-nate, rather long-petiolate, imbricate, erect-
spreading, linear-oblong or linear-obcuneate, obtuse, viscidulous,
margins neatly, prominently and shortly gland-ciliolate, 2–3 lin.
long ; flowers as in the preceding species ; pedicels very slender,
viscid, 3–9 lin. long ; bracts remote, small, foliaceous ; sepals foliaceous, lanceolate or oblong, obtuse, adpressed to the corolla, viscid,
gland-ciliolate, 1$\frac{1}{2}$ lin. or more long ; corolla obconic, viscid, glabrous,

or sometimes scantily pilose on the tube, pale rose, 3–3½ lin. long (dried and pressed specimens 2½–3 lin. wide at the apex); segments widened but not curved, oblong or ovate, acute or obtuse, nearly 2 lin. long, or somewhat exceeding the tube, a gibbous thick callus at the base of each sinus; filaments broadish, equal, only a little longer than the anther; anthers included, dorsifixed just above the base, broad-linear, subincurved, smooth, glabrous, ¾–1 lin. long, cristate; pore about ½ the length of the cell; crests sublanceolate, lacerate, with a sharp terminal lobe, more than ½ the length of the cell; style manifest, about equalling the corolla; stigma capitate; ovary glabrous, somewhat constricted at the base, but scarcely stipitate.

COAST REGION: Clanwilliam Div.; near Clanwilliam, *Leipoldt in MacOwan, Herb. Aust.-Afr.*, 1631! Caledon Div.; Great Houw Hoek Mountains, *Masson,* 78! mountains near Lowrys Pass, *Grisbrook in Herb. Guthrie,* 3028! mountains near Palmiet River, *Guthrie in MacOwan, Herb. Aust-Afr.,* 1916!

Section XXI. HERMES. (Sp. 293–307.)

293. E. empetrifolia (Linn. Sp. Pl. ed. ii. 507); erect, 6–12 in. high; branches erect or spreading, rough with the scars of prominent leaf-cushions, puberulous or glabrous; leaves 6-nate, ascending, imbricate, the lower squarrose-incurved, linear, blunt, sulcate, thick, hispid, the younger ciliate, 2–4 lin. long; inflorescence a dense pseudo-spike towards the ends of the branches, mostly 1 in. or less (rarely 2 in.) long, ½ in. or less wide; pedicels glabrous, straight, 1 lin. or less long; bracts remote, linear, long-ciliate; sepals narrow-lanceolate, pilose near the apex, long-ciliate, 1¼ lin. long; corolla suburceolate, only slightly constricted at the throat with a somewhat tapering neck, sparsely pilose or glabrous, red, about 2½ lin. long; segments spreading, rounded, ⅓–½ the length of the tube; anthers lateral, dorsifixed near the base, oblong, obtuse, glabrous, about ⅓ lin. long, narrow-crested; pore ½ the length of the cell; crests small, lanceolate-acuminate, lacerate or lobed below, about ⅔ the length of the cell; style exserted; stigma capitate; ovary hispidulous. *Benth. in DC. Prodr.* vii. 663; *Bot. Mag. t.* 447; *Wendl. Eric. Ic. fasc.* 5, 13; *Lodd. Bot. Cab. t.* 1875. *E. malleolaris, Salisb. in Trans. Linn. Soc.* vi. 370. *E. empetrina, Linn. Syst. ed.* x. 1003. *E. mollearis, Pers. Syn.* i. 422.

SOUTH AFRICA: without locality, *Herb. Salisbury!*
COAST REGION: Cape Div.; mountains between Cape Town and False Bay, *Thunberg;* near Simons Bay, *Milne,* 121! Table Mountain, *Burchell,* 533! *Bolus,* 4613! *Wolley Dod,* 3335!

294. E. pyxidiflora (Salisb. in Trans. Linn. Soc. vi. 371); erect, 1–2 ft. high; branches virgate, glabrous, rough, scarred; leaves 6-nate, densely crowded, erect, imbricate, linear-oblong, blunt, sulcate, hispidulous, ciliate, about 2½–3 lin. long; inflorescence as in *E. empetrifolia* but the spikes longer, up to 3½ in.; pedicels ¼ lin.

long; bracts subremote, sepal-like, rather large; sepals narrow-lanceolate, pilose and ciliate, about 1½ lin. long; corolla subcampanuloid,
mouth widened, or the tube obconic, suddenly dilated just below the
limb into a wider bowl, glabrous, about 2 lin. long; limb more or less
spreading; segments very broad and rounded, nearly equal to the
tube; anthers dorsifixed above the base, broadly elliptical, with a
large broad cushion-like hispid base, about ⅖ lin. long, crested; pore
nearly equal to the cell exclusive of the cushion; crests unusually large
and broad, semi-orbicular, lacerate, resembling bat's wings, about as
long as the cells; style exserted; stigma capitate; ovary glabrous.
E. empetroides, Andr. Heathery, t. 19, *and Col. Heaths, t.* 88; *Benth.
in DC. Prodr.* vii. 663 ; *Lodd. Bot. Cab. t.* 1758. *E. empetrifolia,
var. glauca, Wendl. Eric. Ic. fasc.* 11, 11.

SOUTH AFRICA: without locality, *Herb. Salisbury!* and *cultivated specimens!*
COAST REGION, between 800 and 1500 ft. : Cape Div.; Simons Bay, *Wright!*
on the Muizen Berg, *MacOwan, Herb. Norm. Aust.-Afr.,* 27! *Guthrie,* 671!
near Simons Town, *Bolus,* 4865! sources of Slangkop River, *Wolley Dod,*
3256! Ruined Valley, *Grey.*

295. E. amœna (Wendl. Bot. Beobacht. 48, not of Salisb.);
erect, 12–18 in. high ; branches erect, rigid, subglabrous ; leaves
4–6-nate, subpatent or incurved, crowded with longish narrow pale
petioles, linear, blunt, sulcate, very hirsute, 2–3 lin. long; flowers
axillary towards the ends of the branches, often somewhat dense or
subcapitate ; pedicels curved, hirsute, red, 2½ lin. long; bracts
remote, hirsute, red; sepals lanceolate, acuminate, keeled, hirsute,
red, 1½ lin. long; corolla broad-campanulate, glabrous, red, 2–2½ lin.
long; limb oblong, broadly rounded, about equalling the tube;
filaments dilated at the apex; anthers included, dorsifixed above the
base, oblong, very obtuse, glabrous, dark-coloured, about ½ lin. long,
cristate-aristate ; pore broadly elliptical, about ⅝ the length of the
cell ; awns broad at the base and subdecurrent along the dilated
filament for ¼ of their length, thence subulate-acuminate, spreading,
the whole equal to the cell; style subincluded; stigma capitate;
ovary turbinate, villous with red-tinged hairs. *Wendl. Eric. Ic.
fasc.* 17, 73, *t.* 28. *E. nolæflora, Salisb. in Trans. Linn. Soc.* vi.
371. *E. plumosa, Andr. Heathery, t.* 36, *and Col. Heaths, t.* 120;
Lodd. Bot. Cab. t. 1702 ; *Benth. in DC. Prodr.* vii. 663, *not of
Thunb., nor of Wendl. E. scholliana, Lodd. l.c. t.* 538. *E. glomerata, Sinclair, Hort. Eric. Wob.* 11, *fide Ind. Kew.*

VAR. β, **pusilla** (Bolus); smaller in all parts, the virgate subsimple stems
almost filiform, 3–6 in. high, leaves and flowers few at the summit, the corolla
barely 2 lin. long.

SOUTH AFRICA: without locality, *Masson, Herb. Salisbury!* and *cultivated
specimens!*
COAST REGION, from 800–1400 ft. : Cape Div.; Simons Bay, *Wright!* by
streams on the Muizen Berg, 1400 ft., *Bolus,* 4605! and in *Herb. Norm. Aust.-
Afr.,* 191! Steen Berg, *Guthrie,* 1360! Silvermine River, *Wolley Dod,* 1926!
Red Hill, *Mrs. Jameson.* Var. β: Cape Div.; Cape Point, 800 ft., *Schlechter,* 7317!

Var. β may be merely a state of 1 or 2 years old, or grown in an exposed
position, or other more severe circumstances,

296. E. Dodii (Guthrie & Bolus); erect, 10–12 in. high, glabrous in all parts; branches numerous, spreading or straggling, striate with long prominent linear leaf-cushions, deeply channelled between; leaves 4-nate, spreading- or squarrose-incurved, linear, acute, sulcate, 4 lin. long; inflorescence as in *E. regerminans*, but the pseudo-spikes shorter (the longest about 1½ in.), narrower and denser, and the intermixed leaves longer and more prominent; pedicels decurved, ½–1¼ lin. long; bracts remote, small; sepals ovate-lanceolate, acute, keel-tipped, scarious, greenish with rosy tips, ¾–1 lin. long; corolla 1½–2 lin. long, campanulate-cyathiform, sometimes subobconic, about equal at the throat and mouth; tube pale rosy becoming darker upwards; limb bright red; segments slightly spreading or suberect, rounded, ¼–⅓ the length of the tube; anthers dorsifixed near the base, oblong, obtuse, glabrous, ¼ lin. long, aristate; pore about ½ the length of the cell; awns subsetiform, pallid, about equal to the cell; style included; stigma capitellate; ovary oblate-spheroidal, glabrous.

Coast Region: Cape Div.; on rocks at the head of Waai Vlei, on Table Mountain, 3000 ft., *Wolley Dod*, 3333! western ledges of Table Mountain, *Galpin*, 3647! Caledon Div.; summit of Genadendal Mountain, 5000 ft., *Galpin*, 3646!

Closely allied to the next; distinguishable by its spreading not virgate habit, its shorter and narrower pseudo-spikes, corolla not at all constricted at the throat, besides some minor characters.

297. E. regerminans (Linn. Mant. Alt. 232, not of Andr.); erect, glabrous, 1–2 ft. high; branches long, virgate, the younger striate with long prominent leaf-cushions; leaves mostly sub-6-nate, sometimes 3–4-nate or scattered, incurved-erect, linear, acuminate, slender, 3–6 lin. long; inflorescence a long and mostly dense pseudo-raceme or spike, 4–5 in. long and ½ in. or more in diam.; pedicels slender, 2–3 lin. long; bracts remote, small; sepals ovate to lanceolate, acuminate, keel-tipped, margins scarious, about ¾ lin. long; corolla from globose-urceolate to ovoid-urceolate, more or less constricted at the throat, from pale rosy to bright-red, about 2 lin. long; limb rounded, slightly spreading, about ⅙ the length of the tube; anthers narrow-subovate, subacute, smooth, about ¼ lin. long, crested-aristate; pore a little over ½ the length of the cell; awns subulate, irregularly lobulate, bearded, about ¾ the length of the cell; style included; stigma capitellate; ovary glabrous. *Benth. in DC. Prodr.* vii. 662. *E. uncifolia, Salisb. in Trans. Linn. Soc.* vi. 369. *E. smithiana, Lodd. Bot. Cab. t.* 1614. *E. racemifera, Andr. Heathery, t.* 188, *and Col. Heaths, t.* 204. *E. Lichtensteinii, Klotzsch in Linnæa,* xii. 504. *E. juncea, Bartl. in Linnæa,* vii. 648? (*from description*). *E. rosea, Lichtenst. ex Klotzsch, l.c.*

South Africa: without locality, *Thunberg, Herb. Salisbury!* and *cultivated specimens!*
Coast Region, between 1000 and 5000 ft.: Swellendam Div.; mountains near Swellendam, *Masson*, 18! *Burchell*, 7314! Grootvaders Bosch, *Zeyher*,

3233! on the Langeberg Range, *Schlechter*, 5663! Zuurbraak Mountain, *Galpin*, 3644! 3645! Riversdale Div.; on the Kampsche Berg, *Burchell*, 7090!

Generally distinguishable by the virgate habit, and the closely-set inflorescence. *E. regerminans*, var. *grandiflora*, *Benth. l.c.*, founded upon Masson's 18, is noted as having broader leaves and a larger corolla (2½–3 lin. long).

298. E. pulchella (Houttuyn, Handl. iv. 504, t. 23, fig. 1, not of Thunb.); erect, mostly 1 (rarely 1½–2) ft. high; branches many, erect, subvirgate or flexuous, pubescent; leaves 3-nate, mostly erect and adpressed, about equalling the internodes or subimbricate, broadish-linear or ovate-oblong, acute or obtuse, thick, nerved above, sulcate below, glabrous, mostly 1–2 lin. long, rarely longer; inflorescence closely pseudo-racemose at the ends of the branches; flowers 1–3-nate in the axils of the leaves; racemes mostly 1–1½ in. long, rarely longer; pedicels pubescent, about 2 lin. long; bracts remote, small; sepals oblong, lanceolate or linear-lanceolate, obtuse, keeled, very concave, scarious on the margin, ciliate, about 1 lin. long; corolla urceolate-campanulate, mouth slightly contracted, subtetragonous, glabrous, rosy or dark red, 1½–2 lin. long; limb slightly spreading, ¼–⅓ the length of the tube; anthers included, sublateral, dorsifixed at or just above the base, narrow-ovate or suboblong; cells partite to or below the base, about ¼ lin. long, decurrent-aristate; pore over ½ the length of the cell; awns nearly equal to the cell, the free portion of variable length; style included; stigma capitate; ovary glabrous or thinly and minutely puberulous. *Andr. Heathery, t. 40, and Col. Heaths, t. 49; Lodd. Bot. Cab. t. 307; Benth. in DC. Prodr. vii. 662. E. articularis, Thunb. Diss. Erica, 37, not of Linn., nor Curt. E. caduceifera, Salisb. in Trans. Linn. Soc. vi. 370. E. retroflexa, Wendl. Eric. Ic. fasc. 8, 7. E. phylicoides, Willd. Sp. Pl. ii. 361? E. furcæflora, Salisb. Prodr.* 294.

SOUTH AFRICA : without locality, *Thunberg*, *Herb. Salisbury!* and *cultivated specimens!*

COAST REGION: Cape Div.; on the flats and lower mountains near Cape Town, common, *Masson*, 94! *Burchell*, 826! *Ecklon*, 86! *Zeyher*, 1102! *Sieber*, 197! *Harvey*, 162! *Guthrie*, 379! *Wolley Dod*, 279! Stellenbosch Div.; between Stellenbosch and Cape Flats, *Burchell*, 8353/²! Hottentots Holland, *MacOwan*, *Herb. Norm. Aust.-Afr.*, 26! Caledon Div.; between Palmiet River and Lowrys Pass, *Burchell*, 8171! *Guthrie*, 2285! *Bolus*, 2124!

Thunberg appears to have been the earliest to note that in this genus the appendages (sometimes at least) belong to the filament. He observes (*Diss. Erica*, 37) : "The anthers in this species are singular, since they seem to be connate with the filament, so that the latter, rather than the anthers, should be called aristate." (Salisbury repeats the remark as if it were original.)

299. E. longiaristata (Benth. in DC. Prodr. vii. 663); characters of *E. pulchella*, except: habit usually more spreading, branches more slender; corolla from obconic to subcampanulate, mostly considerably widened to the mouth at full maturity, usually rosy, more rarely crimson; segments larger, oblong, from $\frac{1}{3}-\frac{1}{2}$ the length of the tube; anthers $\frac{3}{8}$ to nearly $\frac{1}{2}$ lin. long.

SOUTH AFRICA: without locality, *Mund!*
COAST REGION, on mountains from 200–1200 ft. : Caledon Div. ; near Hermanus, *Guthrie*, 4119 ! Bredasdorp Div.; near Elim, *Guthrie*, 3787! *Bolus*, 8454 ! *Schlechter*, 7619 ! near Napier, *Schlechter*, 9657!

This comes very near to the preceding, and might perhaps be regarded as a variety of it. The only distinction of importance is in the shape of the corolla, which seems, however, to be constant.

300. E. flavicoma (Bartl. in Linnæa, vii. 639); erect, 6–10 in. high ; branches virgate, slender, glabrescent ; leaves mostly 3-nate, but occasionally also 4-nate on the same plant, subimbricate, erect, adpressed, suboblong, width more than $\frac{1}{4}$ of the length, subacute, concave and midnerved above, round-backed and sulcate, the younger slightly pubescent, soon glabrous, about 2 lin. long ; flowers axillary at the ends of the branches, spreading in a somewhat close pseudo-raceme 1 in. or less long, $\frac{1}{2}$ in. wide ; pedicels $1\frac{1}{2}$ lin. long ; bracts remote, small ; sepals oblong-obovate, concave, keeled, subscarious, ciliate, $1\frac{1}{4}$ lin. long ; corolla campanulate-tubular, neither inflated below nor constricted at the throat, pale yellow, $2-2\frac{1}{2}$ lin. long ; segments somewhat spreading, $\frac{1}{5}-\frac{1}{4}$ the length of the tube ; anthers dorsifixed just above the base, oblong-recurved, obtuse and equal at base and apex, pallid, smooth, about $\frac{1}{4}$ lin. long, aristate ; pore $\frac{2}{5}$ the length of the cell ; awns squarrose on the upper part of the filament, short and ascending like cock's spurs ; style included, stout below, tapering upwards ; stigma simple ; ovary glabrous. *Benth. in DC. Prodr.* vii. 663.

COAST REGION : Bredasdorp Div. ; near Zoetendals Vlei, *Miss Joubert!* on hills near Elim, 300 ft., *Bolus*, 6729 !

We have not seen the original specimen (which was probably gathered more than 70 years ago), nor does the species seem to have been collected again until 1894. But our Elim specimens, found within a few miles of the first, agree so well with Bartling's short description, and the species is so distinct, that we have no hesitation in identifying them and describing them more fully. At first sight they seem as if they might be a small-flowered variety of *E. parilis, Salisb.*, but the structure of the flowers is quite different. It is not improbable that this may be either *E. fallax*, or *E. festa, Salisb.*, but, in the absence of specimens, the short descriptions of that author are seldom of use, and we are compelled to leave it under Bartling's later name.

301. E. parilis (Salisb. in Trans. Linn. Soc. vi. 371); erect, 1–2 ft. high ; branches stoutish, more or less virgate, pubescent ; leaves 3–4–6-nate or sometimes scattered, incurved, erect or spreading, linear, blunt, keeled or round-backed, sulcate, somewhat rigid, glabrous, pale green, $2\frac{1}{2}-4$ lin. long; flowers axillary, crowded towards the ends of the branches ; pedicels pubescent, 2–3 lin. long ; bracts more or less remote, one often much longer than the others, linear, foliaceous ; sepals linear, acuminate, foliaceous, $1\frac{1}{4}-2\frac{1}{2}$ lin. long, shorter than (or sometimes even a little exceeding) the corolla-tube ; corolla from subtubular to campanulate, mouth scarcely contracted or widened, viscidulous, glabrous or puberulous, yellow, 2–4 lin. long; segments rounded, slightly spreading, about ½ lin. long; anthers included, or sometimes subincluded and just manifest, subterminal, longitudinally semiovate, very acute or acuminate, or subobtuse; cells partite nearly to the base, about ⅝ lin. long, either muticous or decurrent-denticulate along the filament; pore about ⅔ the length of the cell; style included or subexserted; stigma capitate, largish; ovary puberulous. *Benth. in DC. Prodr.* vii. 664. *E. festa, Salisb. in Trans. Linn. Soc.* vi. 371, *fide Benth. E. flava, Andr. Heathery, t.* 64, *and Col. Heaths, t.* 93; *Lodd. Bot. Cab. t.* 882.

VAR. β, **parviflora** (Benth. in DC. Prodr. vii. 664); corolla campanulate, 2 lin. long; anthers subsemielliptical, very obtuse, about ⅜ lin. long; style generally shortly exserted. *E. fallax, Salisb. in Trans. Linn. Soc.* vi. 371, *fide Benth.*

COAST REGION, from 1500 to 6000 ft.: Clanwilliam Div.; Blaauw Berg, *Drège;* Sneeuw Kop, *Leipoldt,* 616! Tulbagh Div.; Witsen Berg, *Burchell,* 8669! Mitchells Pass, *MacOwan, Herb. Norm. Aust.-Afr.,* 29! Tulbagh Waterfall, *Niven,* 115! near Tulbagh, *Niven,* 137! *Guthrie,* 2365! Worcester or Paarl Div.; Dutoits Kloof, *Drège!* Bains Kloof, *Bolus,* 4053! Var. β: Worcester Div.; on the Matroos Berg, 6000 ft., *Bolus,* 9284!

CENTRAL REGION: Ceres Div.; Cold Bokkeveld, near Wagen Drift, *Schlechter,* 10071!

This species is somewhat variable in the length, though little in the width, of the corolla, and also in the shape of the anthers. It may generally be distinguished, when in the living state, from all others of this section, except *E. flavicoma,* by its yellow corolla. From the latter it may be known by its quite different leaves and sepals, as well as by its anthers; the corolla only, both in shape and colour, is very similar. We have not seen any authentic specimen of Bentham's var. *parviflora;* our specimens above cited agree with his short description.

302. E. viscaria (Linn. Mant. Alt. 231); erect, 8–18 in. high; branches straightly ascending, stout, pubescent; leaves 4–6-nate, erect-incurved, crowded, linear, sulcate to round-backed, glabrous, about 4 lin. long; inflorescence pseudo-spicate, 1–2 in. long, ½–¾ in. wide; pedicels puberulo-viscid, usually under 1 (rarely 1½) lin. long; upper bracts at least approximate or subapproximate, foliaceous, 1–2½ lin. long; sepals linear to broad-lanceolate from a broad base, sulcate-keeled, viscidulous, glabrous or puberulous, 1–3 lin. long; corolla campanulate or tubular-campanulate, often strongly nerved, more or less viscid, subglabrous, 2–2½ lin. long; limb erect or slightly spreading, about ⅓ the length of the tube; filaments narrow or rather broad, dilated below the anther; anthers narrow-oblong, cuneate-linear or subelliptical, obtuse or acute, subterminal to lateral, glabrous, ½ lin. long, muticous or sometimes minutely decurrent-denticulate; pore ½–⅔ the length of the cell; style included; stigma capitate; ovary turbinate, villous. *Andr. Heathery, t.* 49, *and Col. Heaths, t.* 71; *Wendl. Eric. Ic. fasc.* 12, 9; *Lodd. Bot. Cab. t.* 726; *Benth. in DC. Prodr.* vii. 664. *E. viscida, Salisb. in Trans. Linn. Soc.* vi. 372. *E. cubitalis, Linn. Amœn. Acad.* viii. 51?

VAR. β, **decora** (Bolus); leaves and sepals usually puberulous; inflorescence pseudo-racemose; racemes broader and more lax than in the type; pedicels 2 lin. or more long; corolla 3–4 lin. long; anthers somewhat longer than in the type. *E. decora, Andr. Heathery, t.* 159, *and Col. Heaths, t.* 162; *Lodd. Bot. Cab. t.* 1385; *Benth. in DC. Prodr.* vii. 664. *E. viscaria, Bauer, Exot. Pl. t.* 1. *E. secundiflora, Tausch in Flora,* 1834, 617.

VAR. γ, **hispida** (Bolus); characters of var. β, but the sepals and bracts usually longer, the former nearly as long as the corolla-tube; corolla 4 lin. long, densely hispid on the lower part; filaments broader; anthers cuneate-linear, subacuminate, about 1 lin. long. *E. pulchella, Thunb. Diss. Erica,* 22, *t.* 4, *not of Houtt. E. argutifolia, Salisb. in Trans. Linn. Soc.* vi. 364. *E. leeana, var. pulchella, Benth. in DC. Prodr.* vii. 626.

SOUTH AFRICA : without locality; Var. γ : *Thunberg !*
COAST REGION: Cape Div.; hills near Cape Town, *Thunberg !* Cape Flats, *Burchell,* 8454/²! *Zeyher,* 1107! *Sieber,* 72 partly ! *Bolus,* 3706! 4396! *Guthrie,* 128! *Wolley Dod,* 2820! Var. β : Cape Div.; on the Flats and mountains up to 1400 ft., *Burchell.* 767! 8558! *Sieber,* 72 *partly !* 90! 199! *MacGillivray,* 442! *MacOwan, Herb. Norm. Aust.-Afr.,* 28! *Bolus,* 4610! *Guthrie,* 458! 733! *Wolley Dod,* 1737! Var. γ : Stellenbosch Div.; mountains near Lowrys Pass, 900 ft., *Bolus,* 5548! *Galpin,* 3524!

A very variable plant. Forms have been found with 5-fid sepals and corollas ; others with double flowers, the inner corolline series being an 8-fid tube ; others exhibiting petalody in the stamens, and lastly others with phyllody in the ovaries. The anthers are also unusually variable.

303. E. axilliflora (Bartl. in Linnæa, vii. 640) ; erect, about 1 ft. high ; branches numerous, ascending, subvirgate, fastigiate, puberulous, glabrescent, mostly very floriferous ; leaves 4–5–6-nate or sometimes scattered, much crowded upwards, erect-incurved, linear, subobtuse, keeled or round-backed and sulcate, glabrous, 2–4 lin. long ; inflorescence pseudo-racemose, resembling that of *E. viscaria, var. decora;* pedicels solitary or binate, coloured, 2–2½ lin. long ; bracts 2 supra-median, 1 median, linear, small ; sepals broad-ovate to ovate-lanceolate, acuminate, keeled or keel-tipped, concave, subcoriaceous, glabrous, glossy, red, 1¼–1½ lin. long ; corolla campanulate or widely obconic, glabrous, more or less viscid, deep red, 2½–3½ lin. long ; segments erect or slightly spreading, broadly rounded, from ⅓ to nearly ½ the length of the tube ; anthers included, subterminal, basifixed at the side on the dilated and thickened apex of the filament, suboblong, subacute, the base ascending, ⅝–¾ lin. long, mostly muticous, sometimes minutely decurrent-denticulate, on the filament near the apex ; pore ½–⅔ the length of the cell ; style included or subexserted, straight ; stigma capitate ; ovary globose, villous, especially towards the summit, dark red. *Benth. in DC. Prodr.* vii. 664.

COAST REGION: Bredasdorp Div.; Zoetendals Vlei, *Ecklon, ex Bartling;* on hills near Elim Mission Station, 300–400 ft., *Bolus,* 8449! 6730! *Guthrie,* 3789!

We have seen no specimen of the type, nor any well authenticated. Our specimens agree with Bartling's meagre description ; and the station of Ecklon's plant is within a few miles of those of ours. The species is allied to *E. conica* and to *E. viscaria.* From the former it is distinct by its less robust habit, its leaves usually fewer in the whorl, and always shorter and less acuminate, by its shorter and more strictly obconic corolla, relatively broader at the base, longer pedicels and hirsute ovary. From the latter it is separated by its entirely different bracts and sepals, and its corolla more regularly widening from the base upwards, whereas in *E. viscaria* it is usually wider and more cup-shaped at the base.

304. E. hæmantha (Bolus in Journ. Linn. Soc. xxiv. 181); erect, 1–1½ ft. high ; branches ascending, puberulous ; leaves 4-nate, erect-incurved, imbricate, not crowded, linear, round-backed, faintly sulcate, glabrous, 2–3 lin. long ; inflorescence axillary and also terminal, flowers 4-nate; pedicels slender, puberulous, 2½–3 lin. long ; bracts remote, small ; sepals narrow-ovate, acute, very concave, subscarious, glabrous, keeled, about 1 lin. long; corolla suburceolate-tubular, mouth scarcely widened or contracted, glabrous, viscidulous? or dry, crimson, 3½–4 lin. long ; segments erect, about ½ lin. long ; anthers dorsifixed just above the base, oblong, subobtuse ; cells deeply partite, glabrous, about ⅖ lin. long, crested at the base; pore nearly ½ the length of the cell ; crests sublanceolate, acuminate, with one or two short side teeth, ½ as long as the cell ; style subexserted, tapering upwards ; stigma capitate ; ovary glabrous.

Coast Region: Ceres Div. ; rocky mountain slopes, near Ceres, 1700 ft., *Bolus*, 5344 !

This species is a somewhat anomalous member of this section, owing to its partially terminal inflorescence. But there are similar difficulties in regard to other sections, and we place it here chiefly on account of its apparent connection with *E. axilliflora* and *E. viscaria*.

305. E. pulvinata (Guthrie & Bolus) ; erect, ½–1 ft. high ; branches ascending, glabrous, roughly scarred and channelled by the cushions of old leaves ; leaves scattered, or sometimes 4–6-nate, crowded, spreading or squarrose or reflexed, flexuous, commonly incurved towards the apex, linear, callous at the apex subpungent-mucronate, round-backed, sulcate, sometimes white-tomentose along the channel, thick, rigid, glabrous, 3–5 lin. long ; inflorescence a short pseudo-raceme at or below the ends of the branches, sometimes crowded ; flowers subcalycine ; pedicels slender, 4–5 lin. long ; bracts small, remote, almost basal ; sepals lanceolate, acuminate, scarious, keeled, 2 lin. long ; corolla broadish-urceolate, throat not much constricted, but gradually narrowed to the mouth, 2–2½ lin. long ; segments slightly spreading, rounded, about ¼ the length of the tube ; anthers dorsifixed above the base, cuneate-ovate, subscaberulous, ⅖ lin. long, broad-aristate ; pore about ½ the length of the cell ; awns subulate-acuminate, irregularly rough-edged or smooth, about ⅔ the length of the cells ; style straight, shortly exserted ; stigma capitate, 4-lobed ; ovary glabrous.

VAR. β, **montana** (Guthrie & Bolus) ; leaves more ascending ; sepals shorter, 1–1¼ lin. long ; anthers oblong, ½ lin. long, much narrower than in the type.

Coast Region : Bredasdorp Div. ; hills at Riet Fontein Poort, near Elim, 200 ft., *Schlechter*, 9704 ! *Bolus*, 8512 ! Var. β : Riversdale Div.; summit of Kampsche Berg, *Burchell*, 7111 !

There is an unusual difference in the form of the anther in the two varieties, but both are so well marked in several other characteristics and agree so closely, that we cannot separate them specifically. The station of var. β was probably of not less altitude than, and may have exceeded, 3000 ft., and the distance from the sea about 30 miles ; that of the type is about 200 ft., and 5 or 6 miles from the sea.

306. E. collina (Guthrie & Bolus) ; erect, 1–2 ft. high ; branches slender, ascending, subflexuous, pubescent ; leaves 4-nate, erect to spreading, sometimes squarrose, incurved towards the apex, linear-oblong, acute, keeled, glabrous, concave above, the younger minutely ciliolate, $1\frac{1}{2}$–2 lin. long; inflorescence clustered at the ends of the branches and then subumbellate, or axillary below the end with the branch excurrent beyond, both forms on the same plant; flowers subcalycine ; pedicels very slender, pubescent, 3–4 lin. long ; bracts frequently (or always ?) only 2, remote, linear, scarious, adpressed ; sepals ovate-lanceolate or lanceolate, acute or acuminate, coriaceous, coloured, keeled, not imbricate at the base, glabrous, reaching mostly to a little below the corolla-tube, about $1\frac{1}{4}$ lin. long; corolla sub-ampullaceous-urceolate, with a longish neck, tetragonous, rosy red, $1\frac{1}{2}$–$1\frac{3}{4}$ lin. long; segments suberect, rounded, $\frac{1}{5}$–$\frac{1}{4}$ the length of the tube ; anthers oblong, obtuse, smooth, about $\frac{1}{3}$ lin. long ($\cdot$03 in.), aristate ; pore $\frac{3}{5}$ the length of the cell ; awns subulate, about $\frac{1}{3}$ the length of the cell, entire ; style included, very short ; stigma capitellate ; ovary subglobose, puberulous or glabrescent.

COAST REGION : Caledon Div. ; hills between Babylons Tower and Hermanus, about 500 ft., *Bolus*, 8491 ! rocky places at Hemel-en-Aarde, 1000 ft., *Schlechter*, 10381 ! (flowers undeveloped).

The rather large sepals give to the flowers a subcalycine aspect. They greatly resemble those of *E. selaginifolia, E. acuta*, and *E. brevifolia* in the § *Trigemma*, and the species is only separated from their neighbourhood by its axillary inflorescence. The whole aspect of the plant is, however, even more strikingly like *E. seriphiifolia* (§ *Melastemon*), but the flowers are very different.

307. E. deflexa (Sinclair, Hort. Eric. Wob. 8) ; branches slender, tomentose-puberulous ; leaves 3-nate, spreading to subsquarrose, linear, obtuse, thick, convex on the upper side with rounded back and margins, deeply sulcate, the younger ciliate, hirtulous or glabrous, shining, 1–$1\frac{1}{4}$ lin. long ; inflorescence axillary ; flowers solitary or scanty and subdistant towards the ends of the branches, corolline or sometimes subcorolline ; pedicels decurved, hirtulous, $1\frac{1}{2}$ lin. long ; bracts remote, small ; sepals narrow-lanceolate or oblong, subacute, foliaceous, margins revolute and thickened at the apex, $\frac{3}{4}$–1 lin. long ; corolla cyathiform-obconic, subtetragonous, glabrous, dry, " white," about $1\frac{1}{2}$ lin. long ; segments continuous, rather large, rounded, $\frac{1}{4}$–$\frac{1}{3}$ the length of the tube ; anthers included (or " subexserted," *Sinclair*), dorsifixed near the base, narrow-oblong, obtuse ; cells parted to the base, membranous, smooth, pallid, almost $\frac{2}{5}$ lin. long ($3\frac{1}{2}$ times longer than their width in the middle), aristate ; pore $\frac{2}{5}$ the length of the cell ; awns setiform, rough-edged, over $\frac{1}{2}$ the length of the cell; style exserted; stigma capitellate; ovary obovate, shortly and sparsely puberulous, or " glabrous." *Benth. in DC. Prodr.* vii. 680.

SOUTH AFRICA : without locality, cultivated specimen from *Loddiges !*

Little is known of this species, which has been described from Loddiges' garden specimens only, and the material is scanty. It seems well-marked. Sinclair describes the anthers as subexserted, but they are not so in the specimens we have seen, nor does Bentham mention the character.

Section XXII. **CHLOROCODON.** (Sp. 308–314.)

308. E. Woodii (Bolus in Journ. Bot. 1894, 237) ; habit various, branched from the base with diffuse spreading branches, 6–8 in. high, or with erect, virgate or subsimple branches, 12–18 in. high ; branches puberulous or hispidulous, sometimes both intermixed, the longer hairs generally more or less compound (subplumose or bifurcate) or gland-tipped ; leaves mostly 3-nate (rarely 4-nate on the same plant), spreading, narrow-lanceolate to ovate-lanceolate, obtuse, margins revolute or reflexed, closed or open-backed, hispid-ciliate, scabrid-hispidulous or glabrous and at length shining on the upper surface, 1–2 lin. long ; flowers mostly axillary, solitary, along the upper part of the branches, and often crowded, occasionally here and there with a few terminal flowers at the ends of the branches or on the short, produced, lateral branchlets ; pedicels curved, about 1 lin. long ; bracts remote, small, occasionally one wanting ; sepals linear, subspathulate-linear, subulate or narrow-lanceolate, hispid or gland-ciliate, foliaceous, about $\frac{1}{2}$ lin. long ; corolla broad-cyathiform or subcampanulate, equal or slightly wider at the mouth, glabrous, dry, white, mostly about $\frac{3}{4}$ lin. long ; segments rounded, erect or slightly spreading, $\frac{1}{2}$–$\frac{2}{3}$ the length of the tube ; anthers included or sub-included and manifest, dorsifixed just above the base, oblong, obtuse, smooth, $\frac{1}{4}$–$\frac{2}{5}$ lin. long, aristate ; pore about $\frac{3}{5}$ or $\frac{2}{3}$ the length of the cell ; awns variable in length from $\frac{1}{6}$–$\frac{3}{4}$ the length of the cell ; style exserted, longer than the ovary, somewhat rigid, sometimes compressed ; stigma capitate, large ; ovary sessile, or substipitate by contraction just above the disk, pubescent or glabrous.

COAST REGION : Clanwilliam Div.; on the Cederberg Range, 2250 ft., *Marloth*, 2685 ! Stutterheim Div.; summit of Dohne Mountain, *Bolus*, 8773 ! *Sim*, 2132 !

CENTRAL REGION : Graaff Reinet Div.; on Koudveld Mountain, *Bolus*, 2583 ! 5189 !

EASTERN REGION : Griqualand East ; Mount Currie. *Tyson*, 1253 ! Insiswa Mountain, *Schlechter*, 6492 ! Natal ; Ismout, *Wood*, 1839 ! Inanda, *Wood*, 873 ! Little Noods Berg, *Wood*, 4136 ! near Umkomaas River, *Wood*, 4611 ! Polela, *Evans*, 674 ! *Schlechter*, 6832 !

KALAHARI REGION : Orange River Colony ; Oliviers Hock and Mont-aux-Sources, *Thode*, 60 ! 61 ! and without precise locality, *Cooper*, 1043 ! Transvaal ; Saddleback Mountain, near Barberton, 5000 ft., *Galpin*, 817 ! Houtbosch, *Schlechter*, 4749 !

Closely allied, and often very similar in aspect to *E. hispidula* (§ *Arsace*). It differs chiefly by its constantly capitate stigma and its more usually axillary flowers, which are only occasionally terminal. In Wood's 873 the stamens are irregular in number, 4–7 and 8 on the same plant.

309. E. mæsta (Bolus in Journ. Bot. 1894, 239); erect, 5–7 ft. high ; branches subvirgate, ashy-grey pubescent, sometimes also pilose with longer white hairs, glabrescent ; leaves erect-spreading, subimbricate, linear, subacute, thick, sulcate, pubescent, ashy-grey, ciliate with a few longer tubercle-based hairs, the older glabrescent, 1–2 lin. long; flowers solitary or binate ; pedicels pubescent, under 1 lin. long ; bracts remote, small ; sepals ovate-lanceolate, pubescent, about $\frac{1}{2}$ the length of the corolla ; corolla cyathiform, mouth scarcely widened or contracted, glabrous, sordid greenish-yellow, $\frac{1}{2}$–$\frac{3}{4}$ lin. or rarely nearly 1 lin. long; segments connivent after maturity, from $\frac{1}{2}$ to as long as the tube ; filaments capillary ; anthers subincluded, manifest, subterminal, tapering to the base, narrow-subobovate, smooth, nearly $\frac{1}{3}$ lin. long, muticous ; pore less than $\frac{1}{2}$ the length of the cell; style elongating, at length exserted ; stigma large, cyathiform ; ovary globose, pallid, glabrous except for a few scattered hairs on the summit.

Coast Region, 300–6000 ft. : Humansdorp Div.; slopes near the river at Humansdorp, *Galpin*, 3708 ! Bedford Div. ; Kaga Berg, *Weale!* Queenstown Div. ; Hangklip Mountain, near Queenstown, *Galpin*, 1610 !

Central Region : Graaff Reinet Div.; Oude Berg, 5000 ft., *Bolus*, 628 ! Koudveld Berg, *MacLea !*

Kalahari Region : Basutoland, *Cooper*, 759 ! 760!

Much resembles in its flowers *E. leucopelta, Tausch* (§ *Arsace*) ; and detached flowers are with difficulty distinguishable. But the leaves and inflorescence are different, the latter being in this always axillary; the anthers also in *E. leucopelta* are lateral, but in this species nearly terminal.

310. E. coarctata (Wendl. Eric. Ic. fasc. 19, 99, t. 37) ; usually less than 1 ft. high ; branches many, slender, suberect, mostly sub-virgate, puberulous or glabrous; leaves mostly 3-nate, more rarely 4-nate (in some specimens 3-nate on the barren, 4-nate on the flowering branches), spreading or incurved, uniformly crowded or gemmiferous in clusters separated by distinct internodes, linear, blunt, trigonous or subterete, $2\frac{1}{2}$–4 lin. long ; flowers mostly in pairs, more or less crowded along a great part of the branches, and forming a narrow and sometimes dense pseudo-spike ; pedicels $\frac{1}{2}$–$1\frac{1}{2}$ lin. long ; bracts remote, basal, minute; sepals ovate or lanceolate, acute, keeled, glabrous, greenish, about $\frac{1}{2}$ lin. long, or about equalling the corolla-tube ; corolla globose-cyathiform to campanulate-cyathiform, the mouth probably nearly equal at maturity but becoming some-what contracted shortly after, glabrous, rosy or sordid yellow, about $\frac{3}{4}$ lin. long ; segments erect or slightly spreading, about equal to the tube ; anthers mostly just manifest, or a little longer or shorter than the corolla, subovate, smooth, brown, $\frac{2}{5}$ lin. long, muticous ; pore $\frac{1}{3}$ to nearly $\frac{1}{2}$ the length of the cell; style exserted, mostly at length decurved ; stigma cyathiform, peltate (or perhaps becoming

calyptriform ?); ovary glabrous. *Benth. in DC. Prodr.* vii. 692.
E. axillaris, Soland. in Herb. Banks., acc. to Benth. l.c. E. minutæ-flora, Andr. Heathery, t. 270, *and Col. Heaths, t.* 245. *E. brevipes, Bartl. in Linnæa,* vii. 643.

VAR. β, **longipes** (Benth. l.c.); "pedicels longer; anthers exserted; stigma subequal." *E. longipes, Bartl. in Linnæa,* vii. 643, *not of Klotzsch.*

COAST REGION, ascending to 1000 ft.: Tulbagh Div.; Saron, 900 ft., *Schlechter,* 7864! Malmesbury Div.; Groot Vallei, Riebeeks Kasteel, *Zeyher,* 1122! Cape Div.; Cape Flats (*Ecklon?*), in *Cape Govt. Herb.!* Flats near Wynberg, *Bodkin in Herb. Bolus,* 9282! Caledon Div.; Houw Hoek, and near Caledon, *Bolus!* Klein River Kloof, *Zeyher,* 3341! Bredasdorp Div.; near Elim, *Bolus,* 8459! Riversdale Div.; near Muis Kraal, *Galpin,* 3696! Var. β: Uitenhage Div.; Elands Kloof, *Ecklon & Zeyher,* 250! in Herb. Brit. Mus.

A somewhat variable species, of which we take as the type Wendland's excellent figure above-cited. With this Bentham identified, with some doubt, Andrews' *E. minutæflora,* but, as we think, after examination of a considerable number of specimens, quite rightly. The only difference of any importance is in the form of the stigma, represented by Wendland as calyptriform, a form we have nowhere seen in the genus. But in wild specimens the shape varies from cyathiform to peltate, which may be due to the spreading and deflection of the margin of the stigma consequent upon age. This view seems to receive confirmation from the fact that Klotzsch, who described the species from a living plant in the Berlin Bot. Gardens, uses for the stigma the term "umbraculiforme." It is easy to conjecture that a further modification might take place and result in a calyptriform stigma. Those who see only Wendland's and Andrews' figures might naturally suppose them to be distinct; but these are from cultivated specimens, which cannot be relied upon for habit or vegetative characters. Bentham's var. β, we quote on his authority, not having examined it.

311. **E. curtophylla** (Guthrie & Bolus): probably under 1 ft. high; branches numerous, incurved and leafy in the upper portion, puberulous; leaves 3-nate, erect, incurved, oblong, very obtuse, flat or concave above, rounded and sulcate beneath, thick, rigid, glabrous, smooth, pallid, $1\frac{1}{2}$–$2\frac{1}{2}$ lin. long; inflorescence a subdense pseudo-raceme towards the ends of the branches; flowers mostly binate, small; pedicels slender, $1\frac{1}{2}$–2 lin. long; bracts remote, small; sepals lanceolate, coloured, shorter than the corolla-tube, concave, sub-scarious, about $\frac{1}{2}$ lin. long; corolla from urceolate-cyathiform to campanulate, slightly contracted at the throat and expanded at the mouth, or subequal, glabrous, dry, minutely pitted, pale red, about 1 lin. long; segments about $\frac{1}{2}$ the length of the tube; anthers included, dorsifixed close to the base, longitudinally semiovate, smooth, $\frac{2}{5}$ lin. long, muticous; pore $\frac{1}{3}$ the length of the cell; style exserted; stigma rather large, subcapitate or sometimes subpeltate; ovary glabrous.

SOUTH AFRICA: without locality, specimen without collector's name, in Herb. Trinity College, Dublin!
COAST REGION: Riversdale Div.; ridge near Milkwoodfontein, near Stil Bay, *Galpin,* 3651!

312. **E. Priori** (Guthrie & Bolus); branches slender, scaberu-

lous, 6–8 in. or more long ; leaves 3-nate, spreading or subsquarrose,
straight, linear, blunt, flat above, deeply sulcate below, glabrous,
3–3½ lin. long ; flowers solitary, somewhat scanty and lax ; pedicels
rather slender, 2 lin. long or more ; bracts remote, minute, between
the middle and base ; sepals ovate, acute, keeled, scarious, concave,
glabrous, about ½ lin. long ; corolla broad-cyathiform or subobconic,
mouth slightly widened, glabrous, dry, about 1 lin. long ; segments
oblong, rounded, 3–4 times the length of the very short tube ;
anthers included to subincluded, manifest, lateral, dorsifixed near the
base, narrow-obovate, smooth, $\frac{2}{5}$ lin. long, muticous ; pore $\frac{2}{5}$ the
length of the cell ; style exserted, slender ; stigma broad-capitate to
subpeltate ; ovary glabrous.

Coast Region : George Div.; near George, *Alexander !* in Herb. Kew.

Of this we have only seen the single specimen above-cited. It resembles, and
is clearly allied to *E. fucata* (§ *Gypsocallis*) ; but is quite distinct by its leaves
not being fasciculate as in that, and more deeply sulcate ; by its different sepals ;
corolla-segments much longer ; and by its subincluded quite lateral and somewhat differently shaped anthers. We have named this after the late Dr. R. C.
Alexander, who subsequently assumed the name of Prior, and who, during a
long journey through the colony in 1847, made many interesting discoveries. He
died in 1903.

313. E. leptostachya (Guthrie & Bolus) ; erect ; branches 5–6 in.,
or more, long ; branchlets subsimple, erect, delicately slender, almost
capillary, striate, subglabrous ; leaves 3-nate, erect to spreading-
incurved, somewhat distant (except the uppermost), narrow-linear,
subacute, deeply sulcate, glabrous, about 2 lin. long ; flowers few,
solitary or binate ; pedicels about $\frac{2}{3}$ lin. long ; bracts remote, submedian, minute ; sepals equal or subunequal, lanceolate, subscarious,
about $\frac{1}{4}$ lin. long ; corolla obconic, mouth widened, $\frac{1}{2}$–$\frac{2}{3}$ lin. long ;
segments variable from a little shorter to a little longer than the
tube ; anthers subincluded to included, lateral, dorsifixed close to
the base, oblong, very obtuse, incurved, smooth, a little over $\frac{1}{4}$ lin.
long, muticous ; pore $\frac{2}{5}$ lin. long ; style shortly exserted ; stigma
large, peltate or subcyathiform ; ovary glabrous or thinly hirtulous
near the apex.

South Africa? : without note of origin ; specimen marked "*ex Herb.*
MacNab, 405a, 109," in Herb. Kew !

With the aspect of several species of *Salaxis* this seems a true *Erica*. Though
its origin is somewhat uncertain, it is most probably South African, and
appears to be a quite distinct species, which may yet be refound. The anthers
occasionally exhibit clearly four complete cells, and such are of a different
form.

314. E. alticola (Guthrie & Bolus) ; dwarf, under a span high ;
branches spreading, then ascending, flexuous, puberulous, with
prominent lunate leaf-cushions ; leaves 4-nate, spreading, straight or
subrecurved, not crowded, slender, linear, subobtuse, flat above,
rounded and sulcate beneath, glabrous, 3–4 lin. long ; inflorescence
a leafy, oblong, pseudo-spike at the ends of the branchlets, about
$\frac{3}{4}$ in. long by $\frac{1}{4}$ in. wide ; flowers solitary ; pedicels pubescent, about

$\frac{3}{4}$ lin. long; bracts variable, mostly (in our specimens) entirely absent, occasionally 2, subapproximate, foliaceous, sometimes one only, long and reaching nearly to the top of the corolla ; sepals sometimes somewhat unequal, slightly united at the base, lanceolate, acute, keeled, often reduced to three, about $\frac{1}{2}$ lin. long; corolla urceolate-cyathiform, scarcely constricted at the throat or widened at the mouth, red, about $\frac{3}{4}$ lin. long; segments rounded, suberect, about $\frac{1}{3}$ the length of the tube ; filaments slender, equal, a little shorter than the anther; anthers sublateral, obovate-oblong, obtuse, smooth, over $\frac{1}{4}$ lin. long, muticous; pore about $\frac{1}{2}$ the length of the cell ; ovary subglobose, glabrous.

EASTERN REGION : Transvaal; in rocky places on the Drakensberg Range, near the Devils Kantoor, about 5500 ft., *Bolus,* 7678 !

In floral structure this is near to *E. mœsta,* but it is separated by its much longer and 4-nate leaves, and different habit; also to *E. coarctata.*

Section XXIII. ARSACE. (Sp. 315–330.)

315. **E. hispidula** (Linn. Sp. Pl. ed. 2, 1672); strong-growing, much-branched, of various aspect, commonly 2–3 ft. high, less in dry places, or reaching 5–6 ft. in favourable situations; branches pubescent or hirsute, usually rigid ; leaves more or less spreading, from linear and sulcate, to ovate or oblong and then open-backed, often incurved, usually scabrid and hirsute, rarely glabrescent, 1–$2\frac{1}{2}$ lin. long ; flowers on short more or less distinguishable branchlets, not in dense crowded masses ; pedicels $\frac{1}{3}$–$\frac{1}{2}$ (very rarely 1) lin. long; bracts 3, variable, mostly subremote or remote, occasionally subbasal; sepals linear to lanceolate, foliaceous or cartilaginous, usually reaching as high as the corolla-tube, more rarely equalling the corolla, often viscid ; corolla mostly broad-urceolate, sometimes narrower, or occasionally globose-urceolate, at maturity almost always more or less contracted at the throat, generally pale rosy, rarely red, $\frac{3}{4}$–1 lin. long; segments more or less spreading, rounded, about $\frac{1}{4}$–$\frac{1}{3}$ the length of the tube ; filaments capillary, mostly about as long as the anthers; anthers included, mostly lateral, rarely sublateral, oblong and obtuse, or longitudinally semiovate, tapering at the base, subacute, smooth, pallid, submembranous, $\frac{1}{4}$–$\frac{1}{3}$ lin. long, muticous; pore about $\frac{1}{2}$ the length of the cell ; style shortly exserted ; stigma peltate or cyathiform; ovary more or less hispidulous, rarely glabrous. *Wendl. Eric. Ic. fasc.* 21, 133, *t.* 50; *Benth. in DC. Prodr.* vii. 691. *E. hispida, Thunb. Diss. Erica,* 19 ; *Wendl. Eric. Ic. fasc.* 27, 13. *E. virgata, Thunb. l.c.* 18; *var. hirta, Wendl. l.c. fasc.* 15, 35, *t.* 13. *E. absinthoides, Linn. Mant.* 66, *fide Benth. E. virgularis, Salisb. in Trans. Linn. Soc.* vi. 324. *E. Colleter, Spreng. Syst.* ii. 192. *E. approximata, Schlecht. ex Spreng. Syst.* ii. 196. *E. serrata, Thunb. Fl. Cap. ed. S·hult.* 346 ? *E. minuta, Klotzsch ex Benth. l.c.*

VAR. *β*, **serpyllifolia** (Benth. in DC. Prodr. vii. 691); leaves broadish open-backed with narrow reflexed margins, or sometimes nearly linear, closed and sulcate, glabrous and smooth above, pallid or puberulous below. *E. serpyllifolia, Andr. Heathery, t.* 289; *Lodd. Bot. Cab. t.* 744.

SOUTH AFRICA : without locality, *Thunberg* (*E. serrata* and *E. absinthoides*).

COAST REGION, from Paarl Div., eastward to George Div., generally on the mountains up to 3500 ft.; many collectors :—*Thunberg* (*E. virgata* and *E. hispida*), *Zeyher*, 3259! 3333! *Bolus, Herb. Norm. Aust.-Afr.*, 609 partly! *Bolus*, 4051! 4582! 6987! *Schlechter*, 2041 (or 2042, ticket uncertain)! 10382! *Galpin*, 3707! *Guthrie*, 2278! 4658! *Wolley Dod*, 178! Var. *β* : Cape Div.; near Cape Town, *Burchell*, 8404! Table Mountain, *Bolus*, 3705! 4479! 4756! and *Herb. Norm. Aust.-Afr.*, 609 partly! *Galpin*, 4808! *Guthrie*, 1010! *Wolley Dod*, 892 ! 1004 !

We have not seen any type of this species; and Linnæus' brief description gives no account of the shape of the corolla, nor of the stigma. Hence we have had to depend upon the figures and descriptions of post-Linnean writers. We quote only such specimens as have passed through our hands. Briefly stated, we have included here only such as have three usually remote bracts, small inconspicuous sepals, and small urceolate corollas, with a cyathiform or peltate stigma. Even as thus restricted the species remains a variable one, and our var. *β* is connected with what we can only term the commoner and more widely spread form, by intermediate specimens, most, if not all, of which have narrow leaves, closed and only sulcate below, though they appear to be invariably smooth, shining and usually quite glabrous.

316. E. arachnoidea (Klotzsch ex Benth. in DC. Prodr. vii. 691) ; stout, much branched; branches hirsute with longish, coarse, tawny, hairs; leaves spreading or squarrose, those upon the primary branches mostly gemmiferous with internodes longer than the leaves, those on the ultimate branchlets more crowded, broad-ovate, subacute, subcordate at the base, open-backed, margins reflexed, hirsute and rough with raised tubercles above, becoming glabrous and shining, closely-felted, white-tomentose below, $1\frac{1}{2}$–$2\frac{1}{2}$ lin. long ; inflorescence pseudo-spicate on short lateral branchlets, or in interrupted tufts, flowers solitary to 3-nate ; pedicels $\frac{1}{2}$ lin. long ; bracts 3, basal or remote, small; sepals oblong, tapering to the inflexed apex, subpubescent or glabrous, ciliate, greenish, $\frac{2}{3}$–$\frac{3}{4}$ lin. long, reaching mostly about $\frac{1}{2}$ the height of the corolla; corolla broad-urceolate, throat contracted, glabrous or hirtulous, viscid, $1\frac{1}{4}$ lin. long ; segments deltoid, acute, $\frac{1}{4}$–$\frac{1}{3}$ the length of the tube ; filaments rather broad, more or less bent below the anther, $\frac{2}{3}$–$\frac{7}{8}$ lin. long ; anthers lateral or sublateral, subovate or ovate-cuneate, obtuse, smooth, about $\frac{2}{5}$ lin. long, muticous ; pore $\frac{1}{2}$–$\frac{2}{3}$ the length of the cell ; style exserted, dilated at the apex, rather slender; stigma obconic, and truncate or slightly cyathiform at the apex, or subpeltate ; ovary densely woolly. *E. hispidula, var. crassifolia, Benth. in DC. Prodr.* vii. 691.

COAST REGION : George Div.; Cradock Berg, 3000 ft., *Galpin*, 3695! Devils Kop, near George, *Niven*, 35! Uitenhage Div.; Van Stadens Berg, *Zeyher*, 787! 3216 !

This is very near to *E. cordata* (§ *Ceramia*) both in appearance and in floral structure and might almost be regarded as a variety of that. In spite of the fact that Bentham quotes *E. punctata, Bartl.*, as a synonym of *E. cordata* and

E. arachnoidea as a var. of *E. hispidula*, L., we suspect that the two may be the same. This only affects the priority of name, and we have not been able to see Bartling's type. This is certainly Klotzsch's plant, and while we have some hesitation in separating it from *E. cordata*, we have none in distinguishing it from *E. hispidula*, L.

317. E. inops (Bolus in Journ. Linn. Soc. xxiv. 186); erect, much-branched, 2–3 ft. high; branches pubescent, sometimes floccose with minute plumose hairs; leaves spreading, linear to narrow-lanceolate, obtuse, deeply sulcate, with a raised nerve on the flat upper surface, glabrous and smooth or roughly hispid with tubercle-based hairs, ciliate or naked, $1\frac{1}{2}$–3 (mostly 2–$2\frac{1}{2}$) lin. long; flowers at the ends of short branchlets; pedicels $\frac{1}{2}$–$\frac{3}{4}$ lin. long; bract one only, approximate or subremote, sometimes subamplexicaul, sepal-like but smaller, pallid; sepals adpressed, narrow-ovate, acute, keel-tipped, cartilaginous, glabrous, mostly smooth and shining, whitish, from $\frac{3}{4}$–$\frac{7}{8}$ the length of the corolla; corolla ovoid-urceolate to globose-urceolate, whitish, about 1 lin. long; segments rounded or deltoid, spreading, about $\frac{1}{2}$ the length of the tube; filaments rather broad, shorter than the anther; anthers lateral, dorsifixed a little above the base, obliquely ovate, subacute, smooth, under $\frac{1}{3}$ lin. long; pore small, less than $\frac{1}{3}$ the length of the cell; style shortly exserted; stigma peltate; ovary hispidulous.

COAST REGION: Cape Div.; Table Mountain, 2500 ft., *Bolus*, 3719! Muizenberg Mountain, 600–1400 ft., *Bolus, Herb. Norm. Aust.-Afr.*, 50! *Guthrie*, 521! *Miss Mansergh!* Steen Berg, *Wolley Dod*, 1277!

318. E. leucopelta (Tausch in Flora, 1834, 616); erect; branches and leaves pubescent, or also hispid with scanty or dense, coarse, stiff, white, sometimes gland-tipped hairs; leaves erect or spreading, oblong-lanceolate or linear, sulcate or very rarely subopen-backed, 1–$2\frac{1}{2}$ lin. long; flowers mostly scanty, or at least not in dense masses; pedicels $\frac{1}{2}$–$\frac{3}{4}$ lin. long; bracts remote, mostly 3, one rather large and foliaceous, and two usually small or very minute, or perhaps sometimes wanting; sepals linear or oblong, foliaceous, pubescent, hispid-ciliate or naked, about $\frac{1}{3}$ lin. long; corolla broad-cyathiform or hemispherical, widened to the mouth, glabrous, smooth or minutely papillose, dull yellow, $\frac{1}{2}$–$\frac{3}{4}$ lin. long; segments erect, broadly rounded, equalling or slightly longer than the tube; filaments capillary, shorter than the anther; anthers subexserted or subincluded, elliptical or subovate, obtuse, smooth, pale brown, $\frac{1}{4}$–$\frac{2}{5}$ lin. long, muticous; pore about $\frac{1}{3}$ the length of the cell; style included or exserted; stigma cyathiform or peltate, large; ovary thinly hispidulous, glabrescent. *E. barbata, Hort. ex Tausch in Flora*, 1834, 616. *E. unibracteata, Klotzsch ex Benth. in DC. Prodr.* vii. 692. *E. hispidula, var. foliacea, Benth. l.c.* 691. *E. foliacea, Klotzsch, and E. galioides, Klotzsch ex Benth. l.c. partly, not of Lam.*

VAR. β, **pubescens** (Bolus); branches and leaves with a uniform pubescence, not hispid; corolla glabrous.

VAR. γ, **ephebioides** (Bolus) ; indumentum of branches and leaves as in either of the preceding ; corolla puberulous. *E. galioides, Klotzsch ex Benth. l.c.*, *partly.*

COAST REGION: Stockenstrom Div.; Kat Berg, *Ecklon & Zeyher !* (in Herb. Berlin; the type of *E. unibracteata, Kl.*) Var. β : King Williamstown Div.; summit of Perie Mountain, 3000 ft., *Flanagan*, 2163 ! Buffalo River Mountains, *Murray in Herb. MacOwan*, 1045 ! Var. γ : Caledon Div.; mountains near Hemel-en-Aarde, Aug., *Zeyher*, 3337 ! Albany Div.; mountains near Grahamstown, *Zeyher*, 882 !

EASTERN REGION: Var. β : Tembuland; Entwanazana, near Gat Berg, 4000 ft., *Baur*, 517 !

The type of Tausch (a cultivated specimen in the Herb. of the Royal Univ. of Prague) and of *E. unibracteata, Klotzsch*, agree well. Both have distinctly three bracts, though two are usually very minute. Our other specimens chiefly differ in the indumentum of the branches, leaves and corolla. The bracts are somewhat variable in size and position, but are of one general type. The flowers much resemble those of *E. mæsta* (§ *Chlorocodon*).

319. E. maritima (Guthrie & Bolus); stem stout, much and intricately branched ; branches divaricate, rigid, cinereo-puberulous, soon glabrescent ; leaves erect, subadpressed, only a little longer than the internodes, oblong, subacute, slightly curved from the middle outwards, sulcate, thickish, glabrous, $1-1\frac{1}{4}$ lin. long ; flowers in few-flowered clusters at the ends of lateral branchlets ; pedicels about $\frac{1}{2}$ lin. long ; bracts basal or nearly so, small, subscarious, shining ; sepals lanceolate, subacute, subscarious, concave, rigid, keel-tipped, glabrous, $\frac{1}{3}$ lin. long ; corolla broad-cyathiform, widened to the mouth, veined, glabrous, about $\frac{5}{8}$ lin. long ; segments erect, semi-ovate, narrowed to the apex but scarcely acute, about $\frac{2}{3}$ the length of the tube ; filaments linear, rather broad, veined, in length equalling the anther ; anthers subexserted, ovate, obtuse, smooth, pale brown, membranous, from $\frac{1}{4}-\frac{1}{3}$ lin. long, muticous ; pore broad-elliptical, large for the section, about $\frac{2}{3}$ the length of the cell; style exserted ; stigma peltate ; ovary minutely hispidulous.

COAST REGION : Bredasdorp Div.; hills near Zeekoe Vley, 100 ft. *Schlechter*, 10544!

This has a distinct appearance, but its somewhat stunted habit (our specimens are not above 6 in. high) may be due to bush-fires. The bracts, sepals, and anthers with their large pores and broad filaments are, however, different from any other of this section.

320. E. salax (Salisb. in Trans. Linn. Soc. vi. 336) ; branches numerous, fastigiate, pubescent ; leaves subspreading, lanceolate, acute, sulcate, thick, shining, minutely serrulate, glabrous, $1-1\frac{1}{4}$ lin. long ; flowers 3-nate ; pedicels pubescent, $\frac{1}{2}$ lin. or more long ; bracts subapproximate, small ; sepals ovate, acute, very concave, keeled, scarious, ciliolate, $\frac{2}{3}$ lin. long ; corolla broad-cyathiform, broader than its length, barely 1 lin. long ; segments broad-ovate, obtuse, erect, in length equalling or twice that of the tube ; filaments rather broad, about as long as the anther ; anthers narrow-ovate, subacute, minutely ciliolate on the front margin, about $\frac{2}{5}$ lin. long, aristate ; pore over $\frac{1}{2}$ the length of the cell; awns setiform, roughly ciliolate, about $\frac{1}{3}$ the length of the cell; style subincluded; stigma peltate, just manifest; ovary glabrous. *Benth. in DC. Prodr.* vii. 691. *E. ramosissima, Roxb. ex Salisb. l.c.*

SOUTH AFRICA: without locality, *Niven, 2!* *Roxburgh! Herb. Salisbury!* in Herb. Kew.

COAST REGION: Stellenbosch Div.; Hottentots Holland Mountains, *Masson.*

321. **E. carinata** (Klotzsch ex Benth. in DC. Prodr. vii. 680, not of Lodd.); dwarf, glabrous; branches ascending; leaves mostly erect, the lower spreading, slender, linear, subacute, keeled, 1–2 lin. long; flowers on short lateral branchlets, 3–4-nate or subumbellate, sometimes crowded and copious; pedicels slender, glabrous, 1½ lin. long; bracts remote, small; sepals elliptic, acute, keeled, scarious, very concave and rigid, 1· lin. long; corolla globose-urceolate, well constricted at the throat, including the substellate-spreading limb, 1½ lin. long; segments broad, ovate, subacute, about $\frac{1}{3}$ lin. long; filaments slender, $\frac{2}{3}$ lin. long; anthers obliquely ovate, subacute, scaberulous on the front margin, about $\frac{1}{5}$ lin. long, crested; crests affixed somewhat high and on the margin of the cells, subulate, curved, roughly ciliolate, about $\frac{2}{3}$ the length of the cells; style included, short; stigma clavate-capitate; ovary glabrous.

COAST REGION: Stellenbosch Div.; Hottentots Holland Mountains, *Zeyher,* 3231!

322. **E. leptopus** (Benth. in DC. Prodr. vii. 680); erect, 1 ft. high or less; branches ascending, slender, puberulous or nearly glabrous; leaves mostly erect, generally subdistant on the flowering branches, linear, sulcate, glabrous, 1–2 lin. long; umbels 4–6-flowered; pedicels at length decurved, slender, 1–2½ lin. long; bracts variable, small, either all subremote or subapproximate, or the lowest larger and median, the upper minute; sepals linear-subulate, acute, rather thick, $\frac{3}{4}$ lin. long, reaching about $\frac{1}{2}$ the length of the corolla; corolla broad-cyathiform, about 1 lin. long, when dry and compressed wider at the mouth than its length; segments continuous with and about $\frac{2}{3}$ the length of the tube; filaments very narrow, equal, nearly straight; anthers obliquely ovate, $\frac{1}{4}$ lin. long; awns subulate, coarsely serrulate, nearly $\frac{2}{3}$ the length of the cell; style included; stigma capitellate or subsimple; ovary glabrous. *E. patula, E. Meyer ex Benth. l.c.*

VAR. β, **piquetbergensis** (Bolus); sepals ovate-lanceolate, about $\frac{1}{2}$ lin. long; corolla broad-cyathiform; segments ovate, about equal to the tube; anthers longitudinally semilanceolate, $\frac{2}{3}$ lin. long; pore $\frac{1}{3}$ the length of the cell; awns setiform, minutely ciliolate, about $\frac{1}{3}$ the length of the cell; style exserted.

VAR. γ, **breviloba** (Bolus); sepals as in var. β but narrower; corolla very broad-cyathiform; segments broader and shorter than in the preceding, about $\frac{2}{3}$ or $\frac{3}{4}$ the length of the tube; anthers subtriangular with rounded angles, $\frac{1}{4}$ lin. long; awns about $\frac{1}{3}$ the length of the cell; style included.

SOUTH AFRICA: without locality, distributed as "*E. patula, E. M.,*" *Drège!*

Coast Region : Var. β : Piquetberg Div. ; summit of Piquet Berg, 2200 ft., *Bodkin in Herb. Bolus*, 6964 ! Var. γ : Clanwilliam Div. ; summit of Sneeuw Kop, 6300 ft., *Leipoldt*, 615 ! Worcester Div. ; on the Matroos Berg, 5700 ft., *Bolus*, 6365 !

We have had only a frustule of Drège's type specimen ; but good material of the others cited.

323. E. minutissima (Klotzsch ex Benth. in DC. Prodr, vii. 691) ; "dwarf, much branched"; branches up to 9 in. long, very slender, puberulous ; leaves erect-spreading, nearly straight, linear, obtuse, faintly sulcate, rather thick, glabrous, $\frac{3}{4}$–$1\frac{1}{4}$ lin. long ; pedicels slender, glabrous, $\frac{3}{4}$ lin. long ; bracts median, nearly verticillate, sub-scarious, coloured, small ; sepals ovate, acute, scarious except the green keel, ciliolate, glabrous, about $\frac{1}{3}$ lin. long, or a little over $\frac{1}{2}$ the length of the corolla ; corolla cyathiform or broad-cyathiform, more or less widened at the mouth, $\frac{1}{2}$–$\frac{2}{3}$ lin. long, pallid ; segments ovate, 2–3 times as long as the tube ; anthers "muticous ?" (*Bentham*—in our specimens of the type all appear to have fallen off) ; style exserted, dilated towards the apex ; stigma truncate and lobulate, rather large ; capsule globose, glabrous.

Coast Region : Stellenbosch Div. ; around Somerset West, *Ecklon & Zeyher* (locality no. 83) in Herb. Berlin.

324. E. tenuis (Salisb. in Trans. Linn. Soc. vi. 329) ; glabrous, erect or diffuse, from 1–5 ft. high ; leaves erect or spreading, slender, linear and acute or rather thick and obtuse, sulcate, 1–2 (rarely $2\frac{1}{2}$) lin. long ; flowers terminal and lateral ; pedicels decurved, glabrous, about 1 lin. long ; bracts remote, occasionally subapproximate, small ; sepals lanceolate, acute, $\frac{1}{2}$–$\frac{3}{4}$ lin. long ; corolla campanulate or cyathiform, mouth widened, white, about 1 lin. long ; segments slightly spreading, rounded and obtuse, or narrower and subacute, about equal to the tube ; filaments broader at the base or subequal, flexuous, about $\frac{1}{2}$ lin. long ; anthers broad-ovate or subtriangular, nearly as wide at the base as their length, subacute, pallid, mem-branous, $\frac{1}{7}$–$\frac{1}{6}$ lin. long, aristate ; pore $\frac{2}{5}$–$\frac{3}{5}$ the length of the cell ; awns setiform, rough, about equal to the cell ; style included, rarely just exserted ; stigma clavate-capitellate ; ovary glabrous. *Benth. in DC. Prodr.* vii. 680. *E. divaricata, Sinclair, Hort. Eric. Wob.* 8. *E. longifissa, Klotzsch, and E. capillaris, Drège ex Benth. l.c.* 680.

South Africa : without locality, *Herb. Salisbury !* and *cultivated speci-mens !*
Coast Region, from 500–3300 ft. : Clanwilliam Div. ; Ezelsbauk, *Drège*, Tulbagh Div. ; Winterhoek Mountain, *Bolus*, 5181 ! near Tulbagh Waterfall, *Bolus*, 5462 ! and *Herb. Norm. Aust.-Afr.*, 1303 ! *Schlechter*, 9004 ! Cape Div. ; Devils Peak, *Bolus*, 4873 ! Constautia Berg, *Wolley Dod*, 1959 ! Caledon Div. ; Genadendal, *Drège* ; Zwart Berg, *Bodkin in Herb. Bolus*, 9227 ! Rivers-dale Div. ; Garcias Pass, *Burchell*, 6945 ! *Galpin*, 3677 ! Humansdorp Div. ; Elands River, *Galpin*, 3713 ! ·
Central Region : Ceres Div. ; Skurfdeberg Range, 5600 ft., *Schlechter*, 10174 !

There are two chief forms of this species : one with diffuse habit and longer and more spreading leaves; the other with more erect and rigid habit, and leaves more erect and shorter.   The corolla-limb is also variable.

325. E. crateriformis (Guthrie & Bolus) ; erect, 14 in. or more high ; branchlets slender, floccose with minutely plumose hairs; leaves erect, adpressed, oblong, keeled, thickish, glabrous, about 1 lin. long ; flowers on short lateral branchlets ; pedicels $\frac{3}{4}$ lin. long ; bracts subremote, foliaceous, rather long ; sepals lanceolate, acute, keel-tipped, scarious-edged, ciliate, $\frac{1}{2}$ lin. long, not quite reaching to the height of the corolla-tube ; corolla broad-cyathiform, mouth widened, nearly 1 lin. long ; segments erect, subdeltoid, obtuse, $\frac{1}{2}$–$\frac{2}{3}$ the length of the tube ; filaments capillary, with a sharp semi-circular bend about the middle, about $1\frac{1}{2}$ times the length of the anther ; anthers subincluded, just manifest, broad-elliptic, very obtuse, smooth, about $\frac{1}{2}$ lin. long, aristulate ; pore about $\frac{1}{2}$ the length of the cell ; awns very minute, curved outwards, and not reaching to the base of the cell, caducous ; style exserted, slender ; stigma capitate, lobulate, rather large ; ovary globose, glabrous.

COAST REGION : Caledon Div. ; grassy hills at the foot of the Klein River Mountains, *Zeyher*, 3340!

Near to some forms of *E. copiosa.*

326. E. subverticillaris (Diels) ; erect, 10–12 in. high ; branches slender, leafy, thinly floccose with minute plumose hairs; leaves usually spreading or sometimes erect, slender, linear or narrow-lanceolate, semiterete, sulcate, glabrous, $\frac{3}{4}$–1 lin. long ; flowers clustered on short branchlets towards the ends of the main branches, in appearance more or less interruptedly subspicate ; pedicels 1–$1\frac{1}{4}$ lin. long ; bracts, one foliaceous, nearly 1 lin. long, two minute, adpressed or often wanting ; sepals adpressed, linear-oblong, tapering upwards, ciliate, $\frac{5}{8}$–$\frac{2}{3}$ lin. long, reaching to (or a little above) the base of the corolla-segments ; corolla broad-cyathiform, glabrous, papillose, about 1 lin. long, apparently rosy ; segments continuous with tube, ovate or subtriangular, acute, about equalling the tube or but little shorter or longer ; anthers included, ovate-elliptic or sometimes subtriangular with a broader base and rounded angles, $\frac{1}{5}$ lin. long, aristate ; pore about $\frac{1}{2}$ the length of the cell ; awns setiform, ciliate, about equalling the cell ; style exserted ; stigma capitellate ; ovary glabrous or very minutely puberulous.

VAR. β, **revoluta** (Bolus); older leaves on the main branches lanceolate, acute, more loosely revolute, shortly hispid, shining, 2 lin. or less long ; anthers tapering to the base ; awns not more than $\frac{1}{4}$ the length of the cell.

KALAHARI REGION : Transvaal ; mountains near Lydenburg, *Wilms*, 903 ! Sabie Valley, dry rocky hills below the Mauchs Berg, *Burtt Davy*, 487 ! Var. β : Spitz Kop, near Lydenburg, *Wilms*, 908 !

Closely allied both to *E. leptopus* and to *E. crateriformis.*   The var. β looks different, and possibly may be distinct ; but the floral characters are in close resemblance.

327. E. copiosa (Wendl. Eric. Ic. fasc. 25, 3, t. 1); erect, about 1 ft. high, somewhat slender; branches numerous, often diffuse, mostly hirsute with coarse spreading hairs, with many short lateral branchlets bearing abundant flowers ; leaves more or less spreading, linear to lanceolate, thick, sulcate, rarely narrow-ovate and open-backed, mostly hispidulous, sometimes glabrescent, mostly 1–1½ lin. long ; inflorescence strictly terminal, often pseudo-racemose by crowding on the short branchlets ; pedicels about 1 lin. long ; bracts variable, mostly remote and small, or minute, occasionally (as in the type specimen) the lowest one larger and foliaceous ; sepals lanceolate or oblong-lanceolate, usually less than ¾ lin. long, leaf-like or coloured ; corolla narrow-campanulate (in the type) to broad-campanulate, or cyathiform, from a little longer than its width to a little shorter, the mouth more or less widened, glabrous, or rarely puberulous, about 1 lin. long; segments mostly distinctly and sometimes widely spreading, from ½ the length of the tube to equal its length ; filaments slender, bent below the anther; anthers included, sometimes manifest, oblong, with more or less obliquity, subcuneate, or narrow-elliptic, usually ¼–⅓ (rarely ⅕) lin. long, muticous or minutely or long aristate ; pore less than ½ the length of the cell, style exserted ; stigma capitate; ovary usually hispidulous, or at least towards the summit. *E. incomta, Klotzsch ex Benth. in DC. Prodr.* vii. 690.

VAR. β, **linearisepala** (Bolus); leaves sometimes rather flat and open-backed ; sepals linear, mostly coloured with green tips, ½–¾ lin. long ; anthers sometimes subexserted, muticous or broad-cuneate and long aristate.

VAR. γ, **parvisepala** (Bolus); sepals ovate or broad-lanceolate, acute, ciliate, coloured, ¼–⅓ lin. long; corolla sometimes puberulous; anthers aristate ; awns ½–⅔ the length of the cell, setiform, about ¼ lin. long.

VAR. δ, **longicauda** (Bolus); leaves deeply sulcate or slightly open-backed, linear or subterete, obtuse ; sepals narrow-ovate, obtuse or subacute, ciliate, deep crimson or with a green foliaceous tip, about ½ lin. long ; anthers with rather broad subulate, serrulate or lobed awns, or subcrested nearly as long as the cells.

SOUTH AFRICA : without locality, *Lichtenstein !*
COAST REGION : Caledon Div.; Caledon, *Zeyher*, 7 ! in Herb. Trin. Coll., Dublin. Mossel Bay Div.; near Great Brak River, *Galpin*, 3701 ! George Div.; near George, *Alexander*, 7 ! Knysna Div.; Woodlands, *Galpin*, 3712 ! Uitenhage Div.; Elands River Mountains, *Ecklon & Zeyher !* in Herb. Berlin. Van Stadensberg Range, *West*, 4 ! Var. β : Swellendam Div., *Schlechter*, 2089 ! Uitenhage Div.; Zuurberg Range, *Bolus*, 9112 ! Var. γ : Stellenbosch Div., *Pappe !* Tulbagh Div.; Witsenberg Vlakte, *Pappe*, 39 ! George Div., *Schlechter*, 5813 ! Knysna Div. ; near Plettenberg Bay, *Burchell*, 5334 ! near Touw River, *Burchell*, 5722 ! Var. δ : Clanwilliam Div.; Cederberg Range, near Sneeuw Kop, 4500 ft., *Bodkin in Herb. Bolus*, 8678 !
EASTERN REGION : Griqualand East, *Tyson*, 1783 ! 2859 !

The type is Wendland's excellent figure and description ; and we have besides, by the courtesy of Prof. Dr. Engler, Director of the Royal Bot. Mus. of Berlin (whose generous assistance in this and many other instances we gratefully acknowledge) been favoured with the opportunity of seeing the type of Klotzsch's *E. incomta* and of dissecting the flower. This agrees very well with Wendland's. The stigma is clearly capitate, not peltate as described by Bentham. The varieties are fairly distinct as to the sepals, but are closely connected in

almost every other character.   Most of them are marked by a general roughness
to the touch.

328. E. onusta (Guthrie & Bolus) ; erect, about 1 ft. high ; stem
rather stout ; branches ascending, slender, subflexuous, white-pubescent ; leaves erect-spreading, narrow-lanceolate and sulcate, or the
older occasionally lanceolate and open-backed, rather thick, hirtulous,
glabrescent, drying pallid, mostly $\frac{1}{2}$–$\frac{3}{4}$ (a few older rarely 1) lin.
long ; flowers so densely covering the whole plant that but few
leaves are visible, 3-nate or solitary on very short branchlets ;
pedicels $\frac{3}{4}$ lin. long ; bracts remote, small, the lowest larger and
leaf-like ; sepals narrow-lanceolate, acute, sulcate, mostly coloured,
puberulous, $\frac{2}{5}$ lin. long ; corolla cyathiform, glabrous, red, mostly
$\frac{3}{5}$ (rarely $\frac{4}{5}$) lin. long ; segments erect, or only very slightly spreading, broad and rounded, about $\frac{2}{3}$ the length of the tube ; filaments
capillary, $1\frac{1}{2}$–2 times as long as the anthers ; anthers subexserted,
narrow-elliptic, obtuse at the base and apex, smooth, minutely
ciliolate on the anterior margin, a little over $\frac{1}{5}$ lin. long, aristate ;
pore about $\frac{2}{5}$ the length of the cell ; awns setiform, spreading, more
than $\frac{1}{3}$ the length of the cell ; style exserted, slender ; stigma
clavate-capitellate, small ; ovary glabrous.

CoAST REGION : Knysna Div. ; near Forest Hall, *Miss Newdigate*, 60 ! in
Cape Govt. Herb. and Herb. Bolus.

Near to *E. copiosa, Wendl., var. parvisepala,* but the whole appearance of the
good specimens before us is so diverse that we hesitate to regard it as a variety of
that species.

329. E. microcodon (Guthrie & Bolus) ; erect, probably 1–$1\frac{1}{2}$ ft.
high ; branches usually slender, ascending, hirsute ; leaves erect-
spreading, mostly narrow-lanceolate, acute, thick and sulcate, varying
more rarely to broad-lanceolate and open-backed, commonly glabrous,
smooth and shining, sometimes puberulous but glabrescent, 1–$1\frac{1}{2}$ lin.
long, more rarely 2 lin. long ; flowers mostly dense on short branchlets ; pedicels about 1 lin. long ; bracts remote, small, cartilaginous,
pallid ; sepals lanceolate, tapering much to the apex but scarcely
acute, glabrous and often shining, thick, cartilaginous, rosy, about
$\frac{3}{4}$ lin. long, reaching to $\frac{2}{3}$ the length of the corolla-tube or higher ;
corolla from broad-suburceolate to subcampanulate, not (or only
slightly) constricted at the throat, glabrous, $\frac{3}{4}$–$1\frac{1}{4}$ lin. long ; segments
rounded, more or less spreading, about $\frac{1}{2}$ the length of the tube ;
filaments slender, at full maturity $1\frac{1}{2}$–$1\frac{3}{4}$ times the length of the
anthers ; anthers subexserted, cuneate-oblong, acute or subacute,
smooth, about $\frac{1}{2}$ lin. long, muticous ; pore about $\frac{1}{2}$ the length of the
cell ; style exserted ; stigma small, clavate-capitellate ; ovary hispidulous, chiefly on the summit.

CoAST REGION, from 900–2000 ft. : Swellendam Div. ; woods, Voormans
Bosch (*Ecklon & Zeyher ?*) in Cape Govt. Herb. ! mountains near Swellendam,
Galpin, 3697 ! 3714 ! *Schlechter*, 5660 ! Riversdale Div. ; Garcias Pass, *Galpin*,
3700 !

330. E. setacea (Andr. Col. Heaths, t. 59) ; erect, $1-1\frac{1}{2}$ ft. high ; branches often flexuous, tomentose and also usually more or less floccose with minute plumose hairs ; leaves incurved-erect, oblong, blunt, thick, rigid, deeply sulcate or rarely subopen-backed, mostly roughly setose-hispid, and also pubescent or floccose-tomentose, $1-1\frac{1}{2}$ lin. long ; flowers sometimes lateral by arrest of the branchlets ; pedicels tomentose, 1–2 lin. long ; bracts remote, small ; sepals somewhat lax, slightly coherent at the base, foliaceous, oblong-lanceolate, acute or subobtuse, keeled, puberulous and setose-ciliate, $\frac{1}{2}-\frac{3}{4}$ lin. long ; corolla cyathiform or subcampanulate-cyathiform, mouth scarcely widened, usually glabrous, more rarely minutely puberulous or velutinous, whitish, pale yellow or pink, $\frac{3}{4}-1\frac{1}{2}$ lin. long ; segments erect, rounded, $\frac{1}{2}-\frac{2}{3}$ the length of the tube ; anthers included, often just manifest, oblong, obtuse, oblique at the base, smooth, pale brown, $\frac{1}{2}$ lin. long, aristate at the base ; pore about $\frac{1}{2}$ the length of the cell ; awns setiform, ciliate, mostly $\frac{1}{5}-\frac{1}{3}$ the length of the cell, sometimes more minute and caducous ; style shortly exserted ; stigma capitate, 4-lobed, rather large ; ovary hispidulous or glabrous. *Heathery, t.* 87 ; *Benth. in DC. Prodr.* vii. 690. *E. asperifolia, Salisb. in Trans. Linn. Soc.* vi. 324 (*who quotes Andr. name*). *E. holocalycina* and *E. cumulata, Klotzsch ex Benth. l.c.* 690.

SOUTH AFRICA : without locality, *Herb. Salisbury!*

COAST REGION : frequent on mountains from Paarl Div. eastwards to Oudts-hoorn Div., from 500–4500 ft., *Mund, Niven, Zeyher,* 3256 ! 3257 ! *Guthrie,* 2504 ! 2507 ! 3908 ! 4587 ! *Bolus,* 6388 ! 6498 ! 6988 ! *Herb. Huguenot Seminary,* 293 ! *Galpin,* 3702 ! 3703 ! *Marloth,* 2405 !

The difficulty felt by Bentham in upholding *E. variabilis* as a separate species, has increased with the acquisition of fresh material, and the species is there-fore abandoned. The larger flowered forms have, as a rule, the more copious indumentum, and these have been generally named *E. setacea;* the smaller flowered and less hairy forms going by the name of *E. variabilis.* But there is no constant character to separate them even as varieties.

Section XXIV. **PSEUDEREMIA.** (Sp. 331–339.)

331. E. cernua (Montin in Nov. Act. Reg. Soc. Upsal., ii. 292, t. 9, fig. 3, 1775) ; about 1 ft. high or less ; branches numerous, widely spreading, subcorymbose, the ultimate flexuous and cernuous, finely puberulous, glabrescent, with occasionally tufts of short subplumose hairs beneath the leaf-cushions ; leaves 4-nate, mostly erect, sometimes adpressed, straight or slightly curved, linear, acute or subacute, sulcate, glabrous, the younger ciliate with subplumose hairs, at length naked, 2–3 lin. long ; heads 4–8-flowered, cernuous ; pedicels puberulous, 1 lin. long ; bracts subulate to lanceolate, acuminate, subscarious, ciliate with long soft subplumose hairs, $2\frac{1}{2}-3$ lin. long ; sepals like the narrower bracts, or linear, $2\frac{1}{2}$ lin. long ; corolla broad-urceolate or ovoid, only slightly constricted at the throat, glabrous, rosy, $2\frac{1}{2}$ lin. long ; segments erect, or perhaps spreading in the living state, ovate, rounded, about $\frac{1}{4}$ the length of

the tube; anthers included, cuneate-oblong, obtuse, smooth, some-
what under $\frac{1}{2}$ lin. long, crested; pore $\frac{1}{2}-\frac{3}{5}$ the length of the cell;
crests lanceolate-acuminate, serrulate, equalling or longer than the
cells; style included; stigma capitellate, small, or subsimple; ovary
glabrous. *Linn. f. Suppl.* 222; *Benth. in DC. Prodr.* vii. 658, *not
of Andr. E. cernua, var. lanceolata, Wendl. Eric. Ic. fasc.* 8,
13.

Soᴜᴛʜ Aғʀɪᴄᴀ : without locality, *Masson !*
Cᴏᴀsᴛ Rᴇɢɪᴏɴ : Tulbagh Div.; Witsenberg Vlakte, *Ecklon & Zeyher.*
Clanwilliam Div., *Mader,* 211! Cederberg Range, 2500 ft., *Leipoldt,* 210!
between Sneeuw Kop and Wupperthal, 3000–4000 ft., *Bodkin in Herb. Bolus,*
8632 !
Cᴇɴᴛʀᴀʟ Rᴇɢɪᴏɴ : Ceres Div.; Cold Bokkeveld, *Thunberg !*

There is some doubt as to the identity of this species because Montin's figure
does not well represent Thunberg's and Masson's specimens so marked, and
indeed appears to us to resemble more the next species; and the description is
insufficient. But as the present has been known under Montin's name for so
many years, and was so known to Thunberg and Linnæus, we have judged it
better to adhere to it.

332. E. Maderi (Guthrie & Bolus); 6–12 in. high; branches
ascending, rather stout, pubescent; leaves 4-nate or scattered,
irregular, erect or subspreading, crowded, imbricate, linear or lanceolate-linear, acute, sulcate, pubescent or glabrous, ciliate with subplumose hairs (as are the bracts and sepals), or becoming naked,
2–3 lin. long; heads erect or slightly cernuous, globose or semi-
globose, 10–23-flowered, 7–10 lin. in diameter; pedicels $\frac{1}{2}$–1 lin.
long; lower bract long and sometimes foliaceous (chiefly so in the
outer flowers of the head), 2–3$\frac{1}{2}$ lin. long, two upper much smaller,
submembranous, pubescent; sepals from a broader ovate or lanceo-
late base, very acuminate, membranous, pubescent, coloured, $\frac{3}{4}$–1$\frac{1}{2}$
lin. long; corolla ovoid-urceolate, obliquely inflated, glabrous, white
to rosy, 2$\frac{1}{2}$–5 lin. long; segments subspreading, $\frac{1}{6}$–$\frac{1}{4}$ the length of
the tube; anthers included, cuneate to oblong-cuneate, obtuse,
smooth, $\frac{2}{5}$ to nearly $\frac{1}{2}$ lin. long, aristate; pore about $\frac{1}{2}$ the length of
the cell; awns setiform or subulate, $\frac{1}{2}$–$\frac{3}{4}$ the length of the cell; style
included; stigma capitellate; ovary glabrous.

Cᴏᴀsᴛ Rᴇɢɪᴏɴ, between 3000 and 6800 ft. : Clanwilliam Div., *Mader* in
Herb. MacOwan, 2185! Cederberg Range between Sneeuw Kop and Wupperthal,
3000–4000 ft., *Bodkin in Herb. Bolus,* 8631! *Leipoldt,* 619! Worcester Div.;
marshy places on the summit of the Matroos Berg, *Marloth,* 2272! *Bolus in
Herb. Guthrie,* 4421!

333. E. sphærocephala (Wendl. ex Benth. in DC. Prodr. vii.
658); erect, about 1 ft. high; branches ascending, slender, and
mostly straight, sometimes subvirgate, pubescent, occasionally with
soft plumose hairs under the leaf-cushions, leaves, bracts and sepals
often ciliate with the same; leaves 4-nate, erect or spreading,
straight or incurved, densely imbricate or as long as or shorter than
the internodes, linear to narrow-lanceolate, acute, sulcate, ciliate or
naked, pubescent or glabrous, 1$\frac{1}{2}$–2$\frac{1}{2}$ lin. long; heads subglobose,

few- to (in finer specimens) 28-flowered, 4–6½ lin. in diam.; pedicels ½–¾ lin. long; bracts, the lower larger and more or less leaf-like, the upper smaller and membranous; sepals narrow-lanceolate, acuminate, pubescent, ciliate and terminated by a long setiform awn, membranous, coloured, ½–¾ lin. long; corolla urceolate, those of the outer flowers obliquely inflated, glabrous, pale to deep rosy, 2¼–3 lin. long; segments slightly spreading, ⅙–⅕ the length of the tube; filaments capillary, sometimes flexuous; anthers included, cuneate-oblong or oblong and incurved, obtuse, smooth, glabrous, ⅖ lin. long, crested or crested-aristate; pore about ½ the length of the cell; crests narrow-lanceolate, acuminate, notched at the base, about ¾ the length of the cell; style included, or just manifest; stigma capitate; ovary glabrous.

COAST REGION: Ceres Div.; Skurfdeberg Range, *Zeyher*, 1096! near Ceres, *Guthrie*, 2177! *Bolus*, 7342! 7448! *Marloth*, 1687! Prince Alfred, *Schlechter*, 9981! Cape Div.; Constantia Berg, 2000 ft., *Schlechter*, 784!

334. E. Solandra (Andr. Col. Heaths, ii. 132); usually weak, straggling, diffuse; branches slender, pubescent or setulose-hispid, 6–9 in. long; leaves 4-nate, erect or spreading, usually incurved, linear, obtuse, sulcate, occasionally subopen-backed, glabrous on the upper surface, more or less densely setulose-hispid with rigid tubercle-based hairs below, 1–2 lin. long; heads subglobose, 3–4½ lin. in diam., several-flowered; pedicels 1 lin. long or less; bracts like the sepals but shorter; sepals linear, acuminate, beset with numerous long spreading setulose white hairs, foliaceous or sub-membranous, 1½–2 lin. long; corolla ovoid-urceolate to urceolate, not much constricted at the throat, puberulous, subscabrid or glabrous, 1½–2 lin. long; segments slightly spreading, ⅕–⅓ the length of the tube; filaments capillary, flexuous; anthers almost exactly those of *E. Maderi*, but a little smaller; style mostly included, rarely exserted; stigma capitellate; ovary villous or hispidulous. *Andr. Heathery, t.* 89. *E. solandriana, Benth. in DC. Prodr.* vii. 658.

SOUTH AFRICA: without locality, *cultivated specimens!*
COAST REGION: George Div.; Montagu Pass, 3500 ft., *Schlechter*, 5814! rocky elevated places at Barbiers Kraal, near Devils Kop, *Niven!*
EASTERN REGION: Natal; Van Reenens Pass, 7000 ft., *Schlechter*, 6938!

The occurrence of this species so far eastward as Natal is unexpected and interesting.

335. E. Cooperi (Bolus in Journ. Linn. Soc. xxiv. 179); stout, rigid, much branched, rough to the touch, 2–3 ft. high; branches spreading or subdivaricate, roughly hispid with spreading simple or more or less compound hairs; leaves 4-nate, mostly spreading, linear to linear-lanceolate, sulcate to open-backed, ciliate with rough simple or forked or subplumose hairs, the under surface paler or livid and subglabrous, 2–2½ lin. long; flowers subcorolline or sub-calycine; heads mostly 4-flowered, cernuous, semiglobose, about 4½ lin. in diam.; pedicels 1 lin. long; bracts linear-lanceolate, acute, coloured, densely ciliate with long soft plumose hairs, about equal in length to the sepals; sepals like the bracts, but narrow-linear or subulate, $2\frac{1}{2}$–3 lin. long; corolla narrow-urceolate, puberulous, white or pale rose, 2–$2\frac{1}{2}$ lin. long; segments spreading, short; filaments slender, $1\frac{1}{2}$–2 times longer than the anthers; anthers included, cuneate or cuneate-oblong, subacute, glabrous, from a little under to a little over $\frac{1}{2}$ lin. long; cells deeply parted, crested; pore about $\frac{3}{5}$ the length of the cell; crests varying from narrow-lanceolate, acuminate, serrulate or more usually suborbicular, incised, with one longer terminal subulate lobe, the whole $\frac{2}{3}$–$\frac{3}{4}$ the length of the cell; style included, short, straight; stigma capitellate, small; ovary turbinate, truncate, hirsute.

VAR. β, **Missionis** (Bolus); habit usually more erect; branches straighter; leaves more erect, often adpressed, usually linear, sulcate, 1–$1\frac{1}{2}$ lin. long; anthers as in the type or sometimes (as in Tyson's 1252) subtriangular, about $\frac{1}{8}$ lin. long; pore smaller; crest proportionately larger and a little longer than the cell. *E. Missionis, Bolus in Journ. Linn. Soc.* xxiv. 179.

KALAHARI REGION: Orange River Colony, *Cooper,* 2528! 3531!

EASTERN REGION, between 2000 and 7000 ft. : Natal; sources of the Umgeni and Umvoti Rivers, *Sutherland!* Mid Illovo, *Wood,* 1890! Noods Berg. *Wood,* 888! and without precise locality, *Buchanan,* 37! *Cooper,* 1101! Var. β: Tembuland; near Cala, *Galpin,* 2318! Griqualand East; near St. Augustine, *Baur,* 218! summit of Mount Currie, *Tyson,* 1252! between Elliot and Maclear, *Bolus,* 8730! *Flanagan,* 2872!

Allied to *E. Baurii,* but a stronger, more woody plant, with larger and differently shaped corolla. Our var. β differs chiefly in aspect, the floral characters being almost identical. The anther varies unusually, both in size and shape.

336. E. Baurii (Bolus in Journ. Linn. Soc. xxiv. 178); erect, 1 ft. or more high; branches ascending, slender, subflexuous, densely beset with rather rough subplumose setulose hairs, and numerous side branchlets bearing globose 4-flowered heads, 4–5 lin. in diameter; leaves 4-nate, erect or spreading, imbricate, linear-lanceolate, acute, more or less open-backed, roughly hispid-ciliate, 1–$1\frac{1}{2}$ lin. long; pedicels $\frac{1}{2}$ lin. long; bracts linear, subscarious, coloured, green-tipped, densely ciliate with long soft subplumose hairs, $1\frac{1}{2}$–2 lin. long; sepals like the bracts but narrower, $1\frac{1}{2}$ lin. long; corolla globose-urceolate, well constricted at the throat, pubescent, $1\frac{1}{2}$ lin. long and nearly as wide; segments rounded, short, more spreading than in any other of this section; filaments slender, flexuous, $1\frac{3}{4}$ times the length of the anther; anthers included, narrow-elliptic, obtuse at either extremity, dark-coloured, over $\frac{1}{4}$ lin. long, amply crested; pore about $\frac{1}{3}$ the length of the cell; crests as in the broader-crested forms of *E. Cooperi;* style included, short; stigma capitellate or subsimple; ovary depressed-globose, wider than its length, hispidulous, very small.

EASTERN REGION: Tembuland, Umtata Div.; margins of woods, near Bazeia, 3000 ft., *Baur,* 639!

337. E. oxysepala (Guthrie & Bolus); branches slender, ascending, glabrous, ashy-grey, 4–6 in. or more long; leaves 3-nate, spreading-recurved, not crowded, linear, setaceous-acuminate, convex and sulcate below, flat above, rigid, glabrous, smooth, $2\frac{1}{2}$–$3\frac{1}{2}$ lin. long; heads 5–10-flowered; pedicels $\frac{1}{2}$–1 lin. long; bracts linear, leaf-like, a little shorter than the sepals; sepals somewhat lax, subulate, from a broad short lacerate base, acuminate, the younger gland-ciliate, about $1\frac{1}{2}$ lin. long; corolla broad-urceolate or sub-campanulate, only slightly constricted at the throat, glabrous, sordid yellow, nearly 2 lin. long; segments slightly spreading, rounded, overlapping at the base to the right (viewed externally), about $\frac{2}{3}$ the length of the tube; filaments short, about equal to the ovary; anthers included, cuneate-oblong; cells deeply parted, smooth, $\frac{2}{5}$ lin. long, aristate; pore less than $\frac{1}{2}$ the length of the cell; awns subulate-acuminate from a broader base, about or nearly as long as the cell; style included or just manifest; stigma clavate-capitellate; ovary obovate, subconstricted at the base, glabrous.

VAR. β, **pubescens** (Guthrie & Bolus); the whole plant more softly downy; branches subvirgate, slender, pubescent; leaves pubescent; bracts and sepals densely ciliate with long soft plumose hairs.

COAST REGION : Tulbagh Div.; mountains of New Kloof, 3000 ft., *Schlechter*, 7495!

CENTRAL REGION : Var. β : Ceres Div.; Cold Bokkeveld, near Wagenbooms River, 6000 ft., *Schlechter*, 10157!

Our var. β is based upon a specimen with mostly undeveloped flowers; but we have little doubt of its relation to this species.

338. E. Greyii (Guthrie & Bolus); decumbent, wide-spreading; branches chiefly from the base, rather slender, minutely pubescent, 6–12 in. long; leaves 3-nate, few and subdistant, erect-spreading, linear-subterete, acute, sulcate, distantly and obsoletely ciliate, shining, glabrous, 3–$4\frac{1}{2}$ lin. long; heads subglobose, 5–6 lin. in diam., 6–8-flowered; pedicels 1 lin. long; bracts narrow-lanceolate, gland-hispid and ciliate with long white hairs, 1–$1\frac{1}{2}$ lin. long; sepals like the bracts, but shortly connate at the base, about 1 lin. long; corolla urceolate, not much constricted at the throat, puberulous, 2 lin. long; segments spreading, erosulate, $\frac{1}{6}$–$\frac{1}{5}$ the length of the tube; filaments dilated at the base and apex, narrowed in the middle, thinly pilose, about three times the length of the anther; anthers subexserted or included (but probably always exserted at maturity), oblong, incurved, obtuse at either extremity, glabrous, nearly $\frac{1}{3}$ lin. long, denticulate-aristate; pore $\frac{3}{5}$ the length of the cell; teeth minute on either side of the apex of the wide filament, about $\frac{1}{8}$ the length of the cell and not reaching to its base; style exserted, straight, slender; stigma capitellate; ovary subglobose, thinly hispidulous.

CENTRAL REGION : Ceres Div.; Cold Bokkeveld, *Grey*, 658! in Herb. Kew.

339. E. clavisepala (Guthrie & Bolus); erect, much branched, all parts more or less viscidulous, and except the corolla, more or less glandular-hairy, under 1 ft. high; leaves 4-nate, mostly equalling the internodes, erect, incurved or the lower spreading, linear to ovate, blunt, thick, roughly hispidulous, viscid, ciliate with long gland-tipped hairs, sulcate to slightly open-backed, 1–1½ lin. long; heads mostly 6-flowered, 3–3½ lin. long, often clustered; pedicels ½–¾ lin. long; bracts leaf-like, longer and wider than the sepals; sepals filiform from the base, gradually widened and thickened upwards, clavate obtuse at the apex, viscid and gland-hairy, 1 lin. long; corolla urceolate, only slightly constricted at the throat, viscidulous, glabrous, dull red, 1½ lin. long; segments suberect, rounded, about ¼ the length of the tube; filaments capillary, straight, 1 lin. long; anthers included, broad-oblong, very obtuse, truncate at the base, smooth, ¼ lin. long, crested; pore ⅔–¾ the length of the cell; crests ovate and incised at the base, with a terminal subulate lobe, the whole about as long as the cell; style included; stigma capitate; ovary subovoid on a wide disk, glabrous.

COAST REGION: Cape Div.; near the western shore of the Cape Peninsula, about due west from Simons Town, 50 ft., *Guthrie,* 1304!

Section XXV. **POLYDESMIA.** (Sp. 340–343.)

340. **E**. **incurva** (Wendl. Bot. Beobacht. 47, not of Thunb. nor of Andr.); erect, ½–1½ ft. high; branches ascending, straight or subflexuous, densely leafy, pallid, pubescent or glabrescent; leaves 3–4–6-nate or scattered, generally erect, imbricate and strongly incurved, sometimes spreading, linear, subobtuse, glabrous, ciliate, 2–2½ lin. long; heads subglobose, cernuous, 4–7 lin. in diam., densely many-flowered; flowers corolline; pedicels ½–1 lin. long; bracts approximate, lower spathulate, larger than the others and exceeding the sepals, sepaloid, pallid, 1½–2 lin. long; sepals narrow-linear or subspathulate, more or less hispid, ciliate, sometimes viscidulous, 1½ lin. or more long, as long as the corolla-tube or longer; corolla urceolate-cyathiform, glabrous, about 2 lin. long; segments broadly-rounded, suberect, about ⅓ the length of the tube; anthers subincluded or subexserted, spathulate-linear or oblong, obtuse, smooth, pale-brown, ⅝–⅞ lin. long, muticous or decurrent-denticulate; pore ½–⅔ the length of the cell; stigma simple or capitate; ovary glabrous. *Eric. Ic. fasc.* 22, 151, *t.* 57; *Willd. Sp. Plant.* ii. 407; *Waitz, Beschreib. Heid.* 242; *Benth. in DC. Prodr.* vii. 616. *E. hemisphærica, Soland. ex Salisb. in Trans. Linn. Soc.* vi. 341. *E. bruniæfolia, Salisb. l.c.* 341.

VAR. β, **solandroides** (Bolus); leaves rather straight; bracts and sepals all linear; anthers decurrent-aristate, with free points; stigma capitellate. *E. solandroides, Andr. Heathery, t.* 290, *and Col. Heaths, t.* 274.

VAR. γ, **stellata** (Bolus); leaves more pubescent, straight or incurved; bracts?; sepals linear-subulate; anthers decurrent-aristate, with short free points; stigma capitellate. *E. stellata, Lodd. Bot. Cab. t.* 893. *E. stellaris, Nois. ex G. Don, Gen. Syst.* iii. 827. *E. setifera, Klotzsch ex Benth. in DC. Prodr.* vii. 616.

VAR. δ, **barbigera** (Bolus); leaves hirsute on the under-side, long-ciliate;

lower bract larger, lanceolate or linear and only a little longer; sepals linear, slightly dilated at the apex; anthers muticous; stigma apparently simple (may be undeveloped in our specimens). *E. barbigera, Klotzsch ex Benth. in DC. Prodr.* vii. 616.

VAR. ε, **subglabra** (Bolus); entirely glabrous, except the branches; leaves, bracts and sepals naked; leaves drying pallid; lower bract large, lanceolate, concave, 3–3½ lin. long; sepals subulate-linear, acute or acuminate, shorter than the corolla-tube; corolla tubular-urceolate; anthers muticous, decurrent-denticulate, or decurrent-aristate, with minute free spreading points; stigma obconic, larger than in the other varieties.

SOUTH AFRICA: without locality, *Roxburgh!* Var. *β*: *cultivated specimens!*

COAST REGION: Stellenbosch Div.; Hottentots Holland (ex *Salisbury*), *Masson!* Var. *γ*: Clanwilliam Div.; Ezelsbank, Cederberg Range, 3000–4000 ft., *Drège*, 7728! Var. *δ*: Bredasdorp Div.; limestone hills between Potte Berg and Cape Agulhas, under 500 ft., *Drège*, 3553! Caledon Div.; near the Palmiet River at Grabouw, *Guthrie*, 3877! Houw Hoek Mountains, 2000 ft., *Schlechter*, 9428! Var. *ε*: Bredasdorp Div.; hills near Elim, 300 ft., *Bolus*, 6727!

Of the foregoing Bentham admitted four species, but observes of three, that they are probably varieties of one. Of *E. setifera* we have seen no specimens, none existing either in the Kew or Berlin Herb.; and, the figure of Loddiges having no analyses, there are no types excepting of *E. barbigera* and *E. bruniæfolia;* that of the latter being at Kew. The materials of all, except that of the few recent collectors cited, is scanty and poor. Our var. *subglabra* is the most distinct, but, in our view, does not merit specific rank.

341. E. ustulescens (Guthrie & Bolus); 6–12 in. high; branches ascending, virgulate or flexuous, closely leafy, puberulous, soon glabrescent; leaves scattered, suberect, imbricate, linear, obtuse, apex subinflexed, minutely and shortly viscous-pubescent, at length glabrous, gland-ciliolate, 1½–2¼ lin. long; flowers subcapitate-umbellate, 4–9, corolline; pedicels 1 lin. long; bracts approximate, rarely subremote, linear, one slightly larger, foliaceous; sepals linear, foliaceous; ½–¾ lin. long; corolla at first tubular, then urceolate, finally subovoid, viscid, white, ustulescent, 1¾–2 lin. long; limb erect, short, ⅑–⅐ the length of the tube; anthers exserted, clavate, subacute, smooth, pallid, about ¾ lin. long, muticous or decurrent-denticulate for a length about ⅙ that of the cell; pore ⅖–½ the length of the cell; teeth very small; stigma capitate, lobed; ovary glabrous.

COAST REGION: Caledon Div.; near the mouth of Bot River, *Zeyher*, 3191! hills near the Bot River Bridge, 400 ft., *Bolus*, 5456! *Guthrie*, 4111! and in *MacOwan, Herb. Aust.-Afr.*, 1734! hills near Houw Hoek, 1300 ft., *Bolus*, 8084!

342. E. stylaris (Spreng. Syst. ii. 198); dwarf; branches spreading, hirsute, 6 in. or more long; leaves 4- (rarely 5-) nate, much spreading, incurved, oblong or subterete, obtuse, thick, sulcate, hispid with tubercle-based hairs, viscidulous, scarcely 2 lin. long; flowers in hemispherical, densely many-flowered heads, about 6 lin. in diam., subcalycine; pedicels ½ lin. long; bracts approximate, lanceolate, gland-ciliate, larger and longer than the sepals, reaching with them to the top of the corolla-tube; sepals linear-lanceolate or broad-linear, acute, gland-ciliate, viscidulous, about 2 lin. long;

corolla campanulate-urceolate, wide-mouthed or ovoid-urceolate, well-contracted at the throat, glabrous or sparsely hispidulous, 2–2½ lin. long; anther-cells deeply partite and spreading, oblong, obtuse, brown, about ¾ lin. long, muticous; pore ½ the length of the cell; stigma capitate; ovary silky-villous. *Benth. in DC. Prodr.* vii. 616. *E. congesta, Lodd. Bot. Cab. t.* 1743, *not of Wendl.*

SOUTH AFRICA: without locality, cultivated specimen, *Herb. MacNab,* 72 !

The specimen quoted is in Herb. Kew. From this, from Bentham's description and from Loddiges' figure, we have compiled the above description. There is little doubt the species belongs to the Cape, and by its leaves and bracts is quite distinct.

343. E. turmalis (Salisb. in Trans. Linn. Soc. vi. 342); erect; branches villous, 12 in. or more long; leaves spreading, squarrose or decurved, ovate or lanceolate, acute, open-backed, margins thick, revolute, hispid on the upper surface with tubercle-based hairs, becoming glabrous and glossy, white-tomentose beneath, ciliate, 1½–2½ lin. long; heads suboval, 4 lin. in diam., 8-flowered, erect or cernuous; flowers corolline; pedicels about 1 lin. long; bracts median, minute, triangular, densely hispid; sepals linear-subulate, acute, pubescent, gland-ciliolate, under ¾ lin. long; corolla sub-urceolate or ovoid, not (or only slightly) contracted at the throat, subglabrous or minutely rough, about 2½ lin. long; segments sub-spreading, about ⅕ the length of the tube; anthers subexserted, oblong, obtuse or subacute, incurved, about ¾ lin. long, muticous; pore ½ the length of the cell; stigma capitellate; ovary densely woolly with rather long hairs. *Benth. in DC. Prodr.* vii. 616.

COAST REGION: Stellenbosch Div.; Hottentots Holland, *Mulder* (ex *Salisb.*). "Cape," *Mund !* in Herb. Kew.

Salisbury describes this species as having 4–5 stamens; but we have found 8 in Mund's specimens. The plant has much external resemblance to *Blæria ericoides.*

Section XXVI. CHROMOSTEGIA. (Sp. 344–346.)

344. E. eriophoros (Guthrie & Bolus); erect or possibly decumbent; branches stout, dark-coloured, naked and glabrescent towards the base, the ultimate flowering branchlets 1–3 in. long, densely leafy; leaves erect, incurved, closely imbricate, narrow-oblong, obtuse, pilose except on the upper surface and ciliate with long curled white hairs, the floral leaves very slightly enlarged, 1½ lin. long; heads 4-flowered, about 2½ lin. in diam.; flowers subsessile; bracts approximate, leaf-like, tomentose and ciliate, 1½–2 lin. long; sepals narrow-linear, slightly wider at the apex, glabrous on the lower, tomentose on the upper half, 1½ lin. long; corolla ovoid-cyathiform, minutely puberulous, 1 lin. long; segments rounded, erect, more than ½ the length of the tube; filaments rather wide, 1½ times longer than the anther; anthers well-exserted, lateral, linear, acute, widened towards the base; cells deeply parted and at length spreading and horn-like, smooth, light brown, $\frac{3}{5}$ lin. long, aristulate; pore $\frac{1}{5}$ the length of the cell; awns very small, not reaching below the base of the cell and about $\frac{1}{6}$–$\frac{1}{5}$ of its length; style exserted, longer than the stamens; stigma subsimple; ovary cylindrical, minutely pilose.

CENTRAL REGION : Ceres Div. ; Gydouw Mountain in the Cold Bokkeveld, 6000 ft., *Schlechter,* 10240 !

345. E. senilis (Klotzsch ex Benth. in DC. Prodr. vii. 617) ; procumbent or erect, 6–8 in. high ; branches some ascending, some spreading with many ascending branchlets, 2–3 in. long, pubescent ; leaves very uniformly erect-incurved, adpressed, scarcely longer than the internodes, oblong-lanceolate, deeply sulcate, flat and glabrous above, finely downy and hispid with tubercle-based hairs below, at length subglabrescent, 1–1$\frac{1}{2}$ lin. long, a few of the upper leaves under the heads enlarged, bract-like and discoloured ; heads hemispherical, cernuous, 4–4$\frac{1}{2}$ lin. wide at the top, about 4-flowered ; pedicels $\frac{3}{4}$ lin. long ; bracts lanceolate, like the sepals thickly beset in the upper part with bristly straight hairs, 2$\frac{1}{4}$ lin. long; sepals narrow-linear, 2$\frac{1}{2}$ lin. long ; corolla broad-ovoid or subglobose, puberulous, pale yellow, 1$\frac{1}{2}$ lin. long ; segments large, rounded, ciliolate, about equal to the tube ; filaments broad ; anthers subexserted, lateral, narrow-elliptic or longitudinally semiovate, smooth, membranous, about $\frac{3}{5}$ lin. long, crested ; pore $\frac{3}{4}$ the length of the cell; crests suborbicular, incised, scabrid, dark coloured, $\frac{1}{2}$ the length of the cell ; style far exserted, straight, tapering upwards, 3 lin. long ; stigma capitellate ; ovary pubescent.

COAST REGION : Clanwilliam Div.; Ezelsbank, on the Cederberg Range, 2600–4000 ft., *Drège,* 2966! *Shaw in Herb. Bolus,* 5661 ! *Marloth,* 2686! near Sneeuw Kop, *Bodkin in Herb. Bolus,* 8627 !

CENTRAL REGION : Ceres Div.; mountains near Tweefontein, in the Cold Bokkeveld, 5800 ft., *Schlechter,* 10129 !

Our specimens all agree well with Drège's type, and with each other, and in all we find the anthers distinctly lateral.

346. E. involucrata (Klotzsch ex Benth. in DC. Prodr. vii. 617) ; procumbent, or younger plants erect ; branches spreading, roughly hispid ; leaves, the younger erect, the older spreading, oldest often reflexed, incurved, broad-linear to lanceolate-oblong, obtuse, sulcate, rather flat, glabrous above, ciliate on the margins and on the edges of the fold with stiff bristly hairs, 2–3 lin. long, the floral longer, broader and discoloured, enwrapping the flower-heads ; heads cernuous, cyathiform, 4-flowered, 5–6 lin. in diam.; pedicels 1 lin. long ; bracts lanceolate, ciliate, coloured, 3 lin. long ; sepals obovate-linear, ciliate, coloured, 2 lin. long ; corolla urceolate-campanulate, puberulous, 1$\frac{1}{2}$–2 lin. long ; segments erect, rounded, about as long as the tube ; filaments linear, narrow, dilated and thickened under the anther ; anthers subexserted, terminal, longitudinally semiovate, tapering at the base into the thickened filament, scabrid, $\frac{3}{5}$ lin. long;

crested at the base; pore $\frac{1}{2}$–$\frac{2}{3}$ the length of the cell; crest entirely adnate to the filament below the base of the cells, narrow, incised, scabrid; style exserted; stigma capitellate; ovary minutely puberulous.

COAST REGION : Tulbagh Div. ; on the Winterhoek Mountain, 4000–6200 ft., *Ecklon & Zeyher* ; *Bolus*, 5107 ! *Marloth in Herb. Norm. Aust.-Afr.*, 342 ! Tulbagh Waterfall, 2000 ft., *Schlechter*, 9009 ! Mosterts Hoek Mountain, 5000 ft., *Bolus !*

Section XXVII. OXYLOMA. (Sp. 347–349.)

347. E. recurvata (Andr. Heathery, t. 282); erect, in cultivation from 1–2 ft. high; branchlets densely leafy, puberulous; leaves scattered, spreading or subsquarrose, linear, subobtuse, obscurely denticulate, keeled, glabrous or minutely puberulous, 4–6 lin. long; flowers capitate; heads many-flowered, cernuous, 6–9 lin. in diam.; pedicels less than 1 lin. long; bracts approximate, linear or linear-lanceolate, acute, keeled, gland-ciliolate, yellowish, about equalling the sepals, or one sometimes longer, 4–5 lin. long; sepals like the bracts, a little shorter or longer than the corolla; corolla tubular, inflated below, not (or very little) constricted at the throat, deeply 4-fid, thinly puberulous, yellowish, 4–5 lin. long; segments erect or slightly spreading, variable in length from $\frac{1}{2}$ the tube to $1\frac{1}{2}$ times its length, oblong, acute, apiculate, dark brownish-purple; filaments long, narrow; anthers included, but in our specimens, reaching to nearly $\frac{3}{4}$ the height of the corolla, oblong or lanceolate, acute, apiculate; pore nearly as long as the cell; cells about $\frac{4}{5}$ lin. long, muticous; style long exserted; stigma subsimple; ovary glabrous. *Andr. Col. Heaths, t.* 262 ; *Lodd. Bot. Cab. t.* 1093 ; *Bot. Mag. t.* 3427 ; *Benth. in DC. Prodr.* vii. 657.

SOUTH AFRICA : without locality, *cultivated specimens !*

348. E. genistæfolia (Salisb. in Trans. Linn. Soc. vi. 337); erect, 6–9 in. high; branches mostly slender, puberulous; leaves 3-nate, spreading, not crowded, linear, acute, rather flat, keeled, glabrous, 2–3 lin. long; flowers 3-nate, usually rather few; pedicels under $\frac{1}{2}$ lin. long; bracts approximate, glumaceous, small; sepals lanceolate, acute, concave, keeled, rigid, glumaceous, about $1\frac{1}{2}$ lin. long, reaching somewhat over $\frac{1}{2}$ the height of the corolla; corolla tubular or subinflated-tubular, dry, glabrous, 2–$2\frac{1}{2}$ lin. long; tube whitish or pale yellow; segments from an oblong ciliate base, dilated above to a cordate-lanceolate acuminate dark-brown tip, about $\frac{4}{5}$ the length of the tube; stamens included, nearly as long as the corolla; filaments capillary; anthers narrow-elliptic or longitudinally semiovate, acute, minutely hirtellous, pale brown, $\frac{1}{2}$ lin. long, muticous; pore about $\frac{2}{3}$ the length of the cell; style exserted, dilated at the apex; stigma simple; ovary glabrous. *Benth. in DC. Prodr.* vii. 657. *E. tetraloba, Roxb. ex Salisb. in Trans. Linn. Soc.* vi. 337.

SOUTH AFRICA: without locality, *Miller! Roxburgh* (ex *Salisbury*), *Mund &*
Maire, Herb. Salisbury!

COAST REGION: Cape Div.; Table Mountain, *Niven*, 121! Simons Berg,
Grey! Muizen Berg, 1600 ft., *Bolus*, 7029! Constantia Berg, *Wolley Dod*,
1960!

This plant seems to be confined to the Cape Peninsula, and is becoming scarce.
We have only once found it.

349. E. cumuliflora (Salisb. in Trans. Linn. Soc. vi. 336); erect,
1–1½ ft. high; branches somewhat flexuous, puberulous; leaves
4-nate, spreading or recurved-squarrose, not much crowded, mostly
oblong, more rarely linear, subacute, flat above, keeled, the younger
ciliolate, glabrous, 1½–2 lin. long; flowers capitate; heads densely
5–12-flowered; pedicels ½ lin. long; bracts approximate, ovate or
obovate, acute, cartilaginous or submembranous, glabrous, white,
about 1½ lin. long; sepals like the bracts but obovate-oblong, 1½–2
lin. long; corolla almost exactly as in *E. genistæfolia*, but the tube
slightly more inflated and somewhat more contracted at the throat;
anthers as in the last named, but a little narrower at the base, and
¼ larger; style exserted, cartilaginous, rigid, fusiform near the apex;
stigma simple, remarkably small, reduced to a mere point; ovary
glabrous or minutely hirtellous. *Benth. in DC. Prodr.* vii. 657.
E. horizontalis, Andr. Heathery, t. 70, *and Col. Heaths, t.* 102
(*with depauperated heads*). "*E. tricolor, Spreng. Syst. Veg.* ii. 193,
not of Niven" (*according to Bentham*). *E. sessiliflora, Wendl. Eric.*
Ic. fasc. 22, 149, *t.* 56, *not of Linn., nor of Andr.* (*a colour-variety,*
due probably to cultivation). *E. aggregata, Roxb. ex Salisb. in*
Trans. Linn. Soc. vi. 336, *not of Wendl.*

SOUTH AFRICA: without locality, *Roxburgh! Herb. Salisbury!* and *cultivated*
specimens!

COAST REGION: Paarl Div.; French Hoek Mountains, 2600 ft., *Schlechter*,
9252! Caledon Div.; Baviaans Kloof, Genadendal, *Niven*, 120b! mountains
near Appels Kraal, *Zeyher*, 3297! Zwart Berg, 2600 ft., *Bolus*, 5403! and in
Herb. Norm. Aust.-Afr., 351! mountains near the mouth of the Klein River,
Bodkin!

Closely allied to *E. genistæfolia;* may be readily distinguished by its 4-nate
leaves and its minute stigma; usually also by its more numerously-flowered
heads, but there are examples where the flowers are reduced to three in number.
Wendland's figure was cited by Bentham under *E. genistæfolia*, but its 4-nate
leaves and many-flowered heads point rather to the present species. It is
also very near to *E. amphigena*, and, through it, connects this section with
Amphodea.

Section XXVIII. **ERIODESMIA.** (Sp. 350–353.)

350. E. lanata (Andr. Heathery, t. 121); erect, 2–3 ft. high;
branches stout, pilose; leaves mostly 4-nate (or sometimes, ex
Andrews' fig., 3-nate), erect-spreading, lanceolate to oblong, acute,
open-backed, margins revolute, pilose above, canescent beneath,
2–3 lin. long, 1 lin. wide or less; flowers subcapitate, calycine; heads
laxly 4–6-flowered, subsessile; bracts approximate, oblanceolate,
densely villous, equalling or exceeding the sepals and corolla; sepals

spathulate, acute, margins revolute, densely villous in the upper part, whitish, nearly equal to the corolla, 2 lin. long; corolla cyathiform-campanulate (globose-urceolate, ex *Andrews'* fig.), hispidulous or puberulous, white, about 2 lin. long; segments rounded, spreading, $\frac{1}{3}$–$\frac{3}{8}$ the length of the tube; filaments nearly straight or strongly bent below the anther; anthers exserted; cells bipartite to the base, narrow-elliptical, acute, scaberulous, dark-brown, $\frac{7}{8}$ lin. long; stigma subsimple, small; ovary pubescent or pilose. *Andr. Col. Heaths, t.* 179; *Benth. in DC. Prodr.* vii. 617. *E. flaccida, Link, Enum. Hort. Berol.* i. 367.

SOUTH AFRICA : without locality, *cultivated specimens !*
COAST REGION, from 1000–2500 ft. : George Div.; Barbiers Kraal, *Niven,* 164! *Masson,* 75 ! Outeniqua Mountains, *Schlechter,* 2336 ! near George, *Bolus,* 8674! and *Guthrie,* 2456! Montagu Pass, *Schlechter,* 5798! Knysna Div. ; mountains near Millwood, *Tyson in Herb. Norm. Aust.-Afr.,* 972 ! and without precise locality, *Buchanan !* and in *Herb. Bolus,* 5832 !

351. E. villosa (Andr. Heathery, t. 200); 1–1½ ft. high; branches ascending or straggling, pubescent, somewhat slender; leaves 3-nate, erect or spreading, imbricate or lax, linear, obtuse, sulcate, pilose or pubescent, glabrescent, 1½–2 lin. long; flowers umbellate, subcalycine or sometimes (by the more spreading sepals) subcorolline or corolline on the same plant, 3–4-nate; umbels lax, spreading; pedicels pilose, 3–5 lin. long; bracts remote, spathulate-linear, pilose, 1 lin. or more long; sepals broad-lanceolate, subobtuse, pilose, the lower half pallid, the upper green, leaf-like, with revolute margins, about equal to the corolla-tube, or occasionally only reaching to about half its height, 1½–2 lin. long; corolla urceolate, subtetragonous, pilose, white, 2 lin. long; segments rounded, erect or spreading, about $\frac{1}{4}$ the length of the tube; anthers exserted; cells deeply parted, oblong and obtuse, or sublanceolate and subacute, scaberulous, dark-brown, about 1 lin. long; pore $\frac{3}{4}$–$\frac{7}{8}$ the length of the cell; style tetragonous; stigma simple, or (rarely ?) capitate; ovary pilose with long erect hairs. *Andr. Col. Heaths, t.* 216; *Wendl. Eric. Ic. fasc.* 16, 55; *Benth. in DC. Prodr.* vii. 617. *E. canescens, Dryand. in Ait. Hort. Kew. ed.* 2, ii. 407 ? *not of Wendl. E. eriocephala, Andr. Heathery, t.* 61 ? *and Col. Heaths, t.* 89 ? *not of Lam. ; Lodd. Bot. Cab. t.* 1270. *E. pilosa, Lodd. l.c. t.* 606.

SOUTH AFRICA : without locality, *cultivated specimens !*
COAST REGION : Caledon Div.; mountains near Klein River, *Zeyher,* 1114 partly ! near Vogel Gat, *Schlechter,* 9510! hills near Babylons Tower Mountain, *Bolus,* 8493 ! near Hermanus, *Bolus,* 9843! *Guthrie,* 4109!

A fairly distinct and rather local species. Bentham remarks that cultivated specimens are with difficulty distinguishable from *E. bruniades,* but that the wild specimens appear different. We have never found any difficulty whatever in separating the two species. We quote *E. eriocephala, Andr.,* with great doubt. It looks very different, and seems undeveloped; the author himself describes the stamens as " exserted," but figures them as included ! It may be a distinct species, with the aspect of some species in the § *Ephebus,* but can hardly belong there.

352. E. bruniades (Linn. Sp. Pl. ed. 1, 354); erect or diffuse, 1–1½ ft. high ; branches ascending, sometimes slender, flexuous and straggling, pubescent ; leaves 3-nate, erect and adpressed, or spreading, rarely squarrose, imbricate or rarely shorter than the internodes, linear, obtuse, sulcate, pilose, 1–2½ lin. long ; flowers 3-nate, calycine, spreading ; pedicels villous, 2–4 lin. long ; bracts remote, small, pilose ; sepals broad-oblanceolate or obovate-oblong, acute, very densely covered with fine spreading white or pinkish hairs, the whole entirely hiding the corolla except near the apex, 1¼–1½ lin. long, about equalling the corolla-tube ; corolla urceolate, pilose, white or pale pink, 1½–1¾ lin. long ; segments recurved, ¼–⅓ the length of the tube ; anthers exserted, almost exactly those of *E. lanata,* or sometimes more acuminate ; stigma subsimple ; ovary villous, chiefly on the summit. *Wendl. Eric. Ic. fasc.* 16, 53, *t.* 20 ; *Andr. Col. Heaths, t.* 6 ; *Lodd. Bot. Cab. t.* 1365. *E. bruinades, Andr. Heathery, t.* 6. *E. velleriflora and E. carbasina, Salisb. in Trans. Linn. Soc.* vi 333. *E. capitata, Thunb. Diss. Erica,* 17, *not of Linn., ex Salisb. and Rach. E. lasiocephala and E. eriantha, Klotzsch in Herb. Berol. ex Benth. in DC. Prodr.* vii. 617. *E. villosa, Wendl. Eric. Ic. fasc.* 16, 55, *t.* 21, *ex Ind. Kew.*

Plukenet seems to have first given this name (*Almag. Bot. Mant.* 69, t. 347, fig. 9, 1700), but Bentham refers the plant to *E. villosa.* As there may be some doubt about it, and as Plukenet was pre-Linnean, we have not cited the figure under either species. Plukenet says he named his plant after Alexander Brown, an Englishman, who first brought it from Africa. Aiton (*Hort. Kew. ed.* 2, ii. 365) calls it the " Brunia-like Heath."

353. E. capitata (Linn. Sp. Pl. ed. 1, 355); erect, 10–15 in. high ; branches ascending or slightly spreading, slender, woolly, glabrescent ; leaves 3-nate, erect or spreading, linear-oblong, obtuse, sulcate, rather thick, subglabrous above, densely woolly below, becoming glabrous and scaberulous, 1½–2 lin. long ; flowers 1–3-nate, calycine, spreading, globular, the whole of a greenish-yellow, 2½–3 lin. in diam. ; pedicels woolly, 2½–3½ lin. long ; bracts from subapproximate to subremote, oblong, woolly, small ; sepals ovate to suborbicular, densely woolly-villous with greenish-yellow hairs, about equalling the corolla-tube ; corolla globose-urceolate, pilose, creamy, 2–2¼ lin. long ; segments rounded, recurved, ⅕–¼ the length of the tube ; filaments broader than in the others of this section ; anthers subincluded but manifest above the corolla-tube, oblong and obtuse, or longitudinally semi-ovate and acute, dorsally bearded, ⅞–1 lin. long ; pore about ½ the length of the cell ; stigma capitellate ; ovary villous, chiefly on the summit. *Andr. Heathery, t. 56, and Col. Heaths, t. 10 ; Wendl. Eric. Ic. fasc. 3, 3 ; Lodd. Bot. Cab. t. 1519. E. byssina, Salisb. in Trans. Linn. Soc. vi. 333.*

SOUTH AFRICA : without locality, *Thunberg,* and *cultivated specimens!*
COAST REGION : Cape Div. ; sandy downs near Cape Town and elsewhere on the Cape Peninsula ; generally below 100 ft., *Ecklon,* 283 ! *Zeyher,* 1115 ! *Mac-Gillivray,* 440 ! *Wright! Bolus,* 3287 ! 8027 ! and in *Herb. Norm. Aust.-Afr.,* 2 ! *Guthrie,* 19 !

Section XXIX. AMPHODEA. (Sp. 354–356.)

354. E. sexfaria (Bauer, Exot. Pl. t. 11) ; erect, 8–15 in. high ; branches ascending, flexuous, rigid, brittle, the younger pubescent, the older covered with scars of leaf-cushions, blackish ; leaves spreading or suberect, sometimes densely imbricate and, owing to the shortness of the internodes, standing in six close ranks, broad linear-trigonous, thick, glabrous, $1\frac{1}{2}$ lin. long ; heads subglobose ; bracts broad-ovate ; sepals obovate, keeled, $1\frac{1}{2}$–2 lin. long, reaching nearly or quite to the top of the corolla ; corolla narrow-ovoid or suburceolate, not (or but slightly) contracted at the throat, white, 2 lin. long ; segments suberect, ovate, $\frac{2}{3}$–$\frac{7}{8}$ of the tube in length ; anthers narrow-elliptic, bifid to the middle, tapering acutely to either end, about $\frac{3}{4}$ lin. long, decurrent-aristate ; pore over $\frac{1}{2}$ the length of the cell, nearly black ; free portion of the awns very variable in length, always much shorter than the adnate portion, sometimes reduced to a mere point, and the anther then (except for the widened filament) appearing muticous. *Andr. Heathery, t. 88, and Col. Heaths, t. 131 ; Benth. in DC. Prodr. vii. 618. E. spumosa, Thunb. Diss. Erica, 17, not of Linn.*

SOUTH AFRICA : without locality, *cultivated specimens!*
COAST REGION : Clanwilliam Div. ; Wupperthal, *Wurmb,* ex *Drège.* Cape Div. ; top of Table Mountain, *Thunberg, Burchell,* 537 ! *Bowie ! Drège,* 7747 ! *McGillivray,* 445 ! *Cooper,* 2682 ! *Milne,* 122 ! *Bolus,* 2950 ! *Guthrie,* 1539 ! *Schlechter,* 548 ! Caledon Div. ; Houw Hoek Mountains, 3200 ft., *Bolus,* 5458 ! *Schlechter,* 5459 ! summit of Genadendal Mountain, *Galpin,* 3540 ! Swellendam Div. ; on mountains, *Bowie !* Riversdale Div. ; Garcias Pass, 2300 ft., *Galpin,* 3541 !

This species varies in the setting of the leaves, which are sometimes spreading and less closely-set, as in Bauer's fine figure of the type, sometimes densely imbricate, 3 times longer than the internodes, and more erect ; the latter form gave rise to the term "sexfarious," for while all the heaths with 3-nate leaves are strictly so, they do not equally distinctly show it. Galpin's 3540 and 3541 are in almost exact agreement with the type ; while plants from the rocky, barren summit of Table Mountain have leaves commonly more erect and close-set. It also varies in the length of the free portion of the awns of the anther, as may be seen by comparing Bauer's with Andrews' figures.

355. E. spumosa (Linn. Pl. Afr. Rar. 11 and Amœn. Acad. viii. 56, n. 35, not of Thunb., nor of Wendl.) ; erect, 1–$1\frac{1}{2}$ ft. high ; branches more spreading and slender than in the preceding, pubescent ; leaves erect-spreading, more rarely erect and adpressed, somewhat lax, rarely crowded, linear-trigonous, subobtuse, sulcate, thick, 1–1½ lin. long, the floral sometimes wider, coloured, sepaloid; heads subglobose; bracts and sepals as in *E. sexfaria*, but a little smaller, more wrinkled, and more rosy, and the latter sometimes only ½–⅔ the length of the corolla; corolla as in *E. sexfaria*, but about 1¾ lin. long; segments lanceolate, more acute; anthers broad-linear, subobtuse; cells deeply partite, black, about 1 lin. long, aristate; awns decurrent for a short distance, then free, about ⅓ the length of the cells. *Mant.* 375; *Benth. in DC. Prodr.* vii. 618; *Lodd. Bot. Cab. t.* 566. *E. scariosa, Berg. Descr. Pl. Cap.* 102; *Wendl. Eric. Ic. fasc.* 21, 137, *t.* 52. *E. turbinata, Andr. Heathery, t.* 297.

SOUTH AFRICA: without locality, *Herb. Salisbury!* and *cultivated specimens!*

COAST REGION, between 1700 and 4000 ft.: Paarl Div.; French Hoek Mountains, *Bolus*, 9812! Cape Div.; Table Mountain, *Bolus*, 4482! near Simons Town, *Bolus*, 8049! Stellenbosch Div.; Lowrys Pass, *Guthrie in Herb. Bolus*, 6010! Caledon Div.; Genadendal, *Drège;* Zwart Berg, near Caledon, *Zeyher*, 3296! and without exact locality, *Bolus*, 9813!

356. E. amphigena (Guthrie & Bolus); erect, about 1 ft. high; branches straight or spreading; branchlets mostly decurved at the apex, pubescent; leaves imbricate and erect, or sometimes more lax and spreading, linear, keeled, subobtuse, 2–3 lin. long, a few floral scarious and bract-like; flowers subcapitate or clustered, at length mostly cernuous; bracts oblong, keel-tipped, a little shorter than the sepals; sepals obovate-oblong, subobtuse, keeled or keel-tipped, about 2 lin. long, reaching to a little below the apex of the corolla-limb; corolla narrow-ovoid, white, 2½ lin. long; segments subulate or lanceolate, acute or acuminate, nearly equal in length to the tube; anthers subterminal or sublateral; cells deeply partite, narrow-oblong, obtusely rounded, about ¾ lin. long, decurrent aristate; awns decurrent for less than ½ their length, then divaricately spreading, the whole from ⅓–½ the length of the cells.

SOUTH AFRICA: without locality, *Stanger !* in Herb. Kew.

COAST REGION, between 1800 and 3500 ft.: Stellenbosch Div.; mountains near Lowrys Pass, *Guthrie*, 2047! Caledon Div.; Houw Hoek Mountains, *Schlechter*, 5449! *Galpin*, 3542! (a more spreading form).

This species connects this section with *Oxyloma*, approaching to *E. cumuliflora*, which it much resembles, but differing by its 3-nate flowers in fewer-flowered heads, and its appendiculate, less distinctly lateral, anthers.

Section XXX. **GEISSOSTEGIA.** (Sp. 357–371.)

357. E. desmantha (Benth. in DC. Prodr. vii. 620); branches spreading; leaves spreading to squarrose, long-petiolate, linear, thick, 2–3 lin. long; pedicels 2 lin. long; bracts approximate, scarious, small; sepals ovate, acute, rigid, scarious, 1½–1¾ lin. long; corolla obconic-tubular, subtetragonous, glabrous (viscid, acc. to *Bentham*), 2½ lin. long, or (owing to the spreading of the sepals at base) appearing twice as long as the sepals; segments slightly spreading; filaments slightly curved; anthers subterminal, oblong, obtuse, $\frac{5}{8}$–$\frac{3}{4}$ lin. long; pore more than $\frac{1}{2}$ the length of the cell; stigma clavate; ovary glabrous.

SOUTH AFRICA: without locality, *Masson !* in Herb. Kew.

We have only seen a branch 7 in. long, and have dissected a single flower. Quite distinct in the section by its long corolla, and the long ($\frac{3}{4}$-1 lin.) petioles of the leaves. The material is scanty, and the inflorescence not properly seen; it may be umbellate, 5–6-flowered.

358. E. physantha (Benth. in DC. Prodr. vii. 619); erect, 6 in. or more high; branches flexuous, subglabrous; leaves somewhat spreading, oblong, obtuse, thick, glabrous, $1\frac{1}{2}$–2 lin. long; pedicels under 1 lin. long; bracts approximate, sepal-like, but smaller; sepals broad-ovate, subobtuse, keel-tipped, scarious, ciliate, glabrous, $1\frac{1}{2}$ lin. long; corolla oblate, subglobose-urceolate, glabrous, wide-mouthed, $2\frac{1}{4}$ lin. long, or reaching nearly twice the height of the sepals; filaments with a sigmoid bend just below the anther; anthers sub-lateral, shortly oblong, obtuse or obliquely subacute, rounded at the base, minutely ciliolate, a little over $\frac{1}{2}$ lin. long, dark-coloured; pore about $\frac{3}{5}$ the length of the cell; stigma clavate-capitate; ovary globose, thinly hispidulous.

COAST REGION: Riversdale Div.; on hills near Zoetmelks River, *Burchell*, 6794!

359. E. adunca (Benth. in DC. Prodr. vii. 618); branches 8 in. or more long; leaves from erect to squarrose, linear, subulate or lanceolate-oblong, subtrigonous, the younger ciliolate, 1–$2\frac{1}{2}$ lin. long; pedicels $1\frac{1}{2}$ lin. long; bracts subremote, sepal-like but smaller; sepals somewhat lax and spreading, narrow-ovate, acute or acuminate, keel-tipped, subviscid, glabrous, white, $1\frac{1}{2}$ lin. long; corolla urceolate or globose-urceolate, longer or sometimes shorter than its width, glabrous, viscid (or viscid-puberulous, acc. to *Bentham*), $1\frac{1}{2}$–$1\frac{3}{4}$ lin. long, usually somewhat exceeding the sepals, occasionally shorter; limb spreading; filaments dilated just below the anther, slightly curved; anther subterminal, longitudinally semiovate, the dorsal margin curved, acute, ciliolate, more or less hispidulous, $\frac{2}{3}$–$\frac{3}{4}$ lin. long; pore $\frac{3}{5}$ the length of the cell; style generally deflexed or hooked; ovary on a conspicuous black disk, glabrous or hispidulous.

SOUTH AFRICA: without locality, *Niven*, 100!
COAST REGION: Swellendam Div.; mountains near Swellendam, *Bolus*, 7304! Riversdale Div.; Garcias Pass, 1000 ft., *Galpin*, 3546! 3547! *Schlechter*, 1731! George Div.; Montagu Pass, 2500 ft., *Schlechter*, 5802!

Variable as to the length and position of the leaves, length of corolla and hairiness of the anther; it may generally be recognized by its loose sepals, deflexed style (in well-matured flowers), and the prominent dark disk below the ovary.

360. E. involvens (Benth. in DC. Prodr. vii. 619); branches

4 in. or more long; leaves erect, incurved, imbricate, linear, obtuse, glabrous, $1\frac{1}{2}$–2 lin. long; pedicels $\frac{1}{2}$ lin. long; bracts approximate, but somewhat lax, sepal-like, but smaller; sepals broad-ovate, sub-obtuse, thick, rigid, cartilaginous, subviscid, slightly longer than the corolla; corolla broad-cyathiform, viscid, $1\frac{3}{4}$ lin. long; anthers terminal, ovate, very obtuse, tapering at the base, smooth, $\frac{3}{4}$–$\frac{7}{8}$ lin. long; pore broadly elliptical, $\frac{1}{2}$ the length of the cell; stigma sub-simple; ovary glabrous.

SOUTH AFRICA: without locality, *Niven!* in Herb. Kew.

Very near to *E. suffulta*, and we retain the species with some doubt; but the sepals are broader, especially at the base, anthers much more obtuse; and the flowers larger.

361. E. chartacea (Guthrie & Bolus); branches ascending, pubescent, 12 in. or more long; leaves erect-spreading, linear, obtuse, $1\frac{1}{2}$–3 lin. long; pedicels $2\frac{1}{2}$–3 lin. long; bracts approximate, some-what lax, sepal-like but smaller; sepals somewhat lax, broad-ovate, keel-tipped, puberulous, papery, ciliate-lacerate, $1\frac{1}{2}$ lin. long; corolla suburceolate or subcampanulate-cyathiform, glabrous or puberulous, $1\frac{1}{4}$–$1\frac{1}{2}$ lin. long, and about as wide; segments ciliate-lacerate; filaments slender below, dilated upwards; anthers subterminal, obliquely lanceolate, acute or acuminate, glabrous, or hispidulous near the apex, naked or ciliolate, $\frac{3}{4}$–$\frac{7}{8}$ lin. long; pore about $\frac{3}{4}$ the length of the cell; stigma clavate; ovary glabrous.

COAST REGION: Swellendam Div.; near Swellendam, *Borcherds in Herb. Bolus*, 5959! Tradouw Pass, 1200 ft., *Galpin*, 3555!

362. E. suffulta (Wendl. ex Benth. in DC. Prodr. vii. 619); erect, 1–$1\frac{1}{2}$ ft. high; branches and leaves somewhat hoary-puberulous, at length glabrescent, pallid; leaves erect and incurved, or spreading, linear to oblong, obtuse, thick, keeled, sulcate, $1\frac{1}{2}$–2 lin. long; pedicels 1–$1\frac{1}{2}$ lin. long; bracts approximate, somewhat lax, sepal-like; sepals obovate-oblong, keel-tipped, rather thick, ciliolate, about as long as the corolla; corolla suburceolate-cyathiform, scarcely con-tracted at the throat, minutely puberulous, white, 1–$1\frac{1}{4}$ lin. long, the width somewhat exceeding the length; anthers subterminal, obliquely lanceolate, acute, about $\frac{3}{4}$ lin. long, minutely ciliolate; pore $\frac{1}{2}$–$\frac{2}{3}$ the length of the cell; stigma clavate; ovary glabrous.

SOUTH AFRICA: without locality, *Masson!*
COAST REGION: Caledon Div.; Houw Hoek Mountains, 1500 ft., *Bolus*, 5450! *Scott-Elliot*, 1136!

363. E. pogonanthera (Bartl. in Linnæa, vii. 634); small shrub, 1–$1\frac{1}{2}$ ft. high; branches pubescent; leaves spreading or squarrose, mostly recurved, linear or oblong, obtuse, thick, glabrous, the younger ciliolate, 1–$1\frac{1}{2}$ lin. long; pedicels 1–$1\frac{1}{2}$ lin. long; bracts approximate, ovate, white; sepals ovate-orbicular, acute, narrowed at the base, white, $1\frac{1}{4}$ lin. long; corolla broad-urceolate, puberu-lous or glabrous, white, 1–$1\frac{1}{4}$ lin. long, a little wider than its length, equalling or a little shorter than the sepals ; anthers terminal, longitudinally semi-elliptical, acute, tapering at the base, dark brown, covered with shaggy purplish hairs, or sometimes glabrous and only shortly ciliate, $\frac{5}{8}-\frac{3}{4}$ lin. long ; pore about $\frac{2}{3}$ the length of the cell ; stigma capitate ; ovary glabrous. *Benth. in DC. Prodr.* vii. 619.

COAST REGION : Caledon Div. ; Zwart Berg, 2000 ft., *Ludwig & Beil, Niven!* *Bolus*, 6747 ! mountains near Appels Kraal, *Zeyher*, 3301 !

364. E. azaleæfolia (Salisb. in Trans. Linn. Soc. vi. 334) ; erect, 6–12 in. high ; branches diffusely spreading, pubescent ; leaves mostly squarrose-recurved or erect-spreading, ovate to oblong, subobtuse, thick, shining, the younger puberulous, 1–1$\frac{1}{2}$ lin. long ; pedicels $\frac{1}{2}$ lin. long ; bracts approximate, ovate-orbicular, subscarious, ciliate ; sepals oblong-spathulate, acute, ciliate, 2 lin. long, about equal to the corolla ; corolla suburceolate, puberulous, 1$\frac{1}{2}$–2 lin. long, its width less than its length, pink ; segments ciliate ; anthers subterminal ; cells deeply parted and subdivaricate, broad-elliptical, very obtuse, smooth, polished, dark-coloured, about $\frac{1}{2}$ lin. long ; pore unusually large occupying nearly the whole cell ; ovary glabrous. *Benth. in DC. Prodr.* vii. 619. *E. spumosa, Roxb. ex Salisb. l.c.* 334, *not of Linn. E. azaleæflora, Steud. Nom. ed.* 2, i. 618.

SOUTH AFRICA : without locality, *Herb. Salisbury !*
COAST REGION : Caledon Div.; tops of mountains near Genadendal, *Burchell*, 7689 ! Houw Hoek Mountains, 2000 ft., *Bolus*, 5459 ! *Schlechter*, 9396 ! Klein Houw Hoek, *Niven*, 120a ! Zwart Berg, 2500 ft., *Bolus, in Herb. Norm. Austr.- Afr.*, 352 !

365. E. sonderiana (Guthrie & Bolus) ; 1 ft. or more high ; branches ascending, pubescent ; leaves erect-spreading, imbricate, linear, subobtuse, thick, subrugulose, pallid, glaucous, viscidulous or dry, the younger gland-ciliate, 2–2$\frac{1}{2}$ lin. long ; pedicels pubescent, 1 lin. long ; bracts approximate, ovate, viscid ; sepals broad-ovate to nearly orbicular, obtuse, keel-tipped, cartilaginous, transversely wrinkled when dry, very thick, viscid, shining (in some specimens both bracts and sepals have an unusually long excurrent keel-point), about as long as the corolla ; corolla suburceolate, glabrous, viscid, 1$\frac{1}{2}$ lin. long, white, width about equal to the length ; segments ciliolate ; anthers sublateral, oblong-incurved, obtuse, smooth, about $\frac{1}{2}$ lin. long ; pore variable, $\frac{3}{5}-\frac{7}{8}$ the length of the cell ; style exserted ; stigma clavate ; ovary glabrous.

SOUTH AFRICA : without locality, *Niven*, b ! in Herb. Trin. Coll. Dublin.
COAST REGION : Ceres Div. ; near Ceres, *Bolus*, 9110 ! Worcester Div. ; Matroos Berg, 5200 ft., *Marloth*, 2205 ! *Bolus in Herb. Guthrie*, 3945 ! Swellendam Div. ; mountains near Swellendam, *Bolus*, 7506 ! (immature).

366. E. crassisepala (Benth. in DC. Prodr. vii. 619) ; branches ascending, pubescent, rigid, 6–8 in. high ; leaves 3-nate or " scattered " (acc. to *Bentham*), erect-spreading, imbricate, linear or somewhat semiterete, obtuse, margins somewhat thickened, the younger gland-ciliate, 1–2 lin. long; pedicels 1–1½ lin. long; bracts approximate, sepal-like but smaller; sepals ovate-lanceolate, obtusely acuminate, with a thickened marginal line on either side, converging upwards to the apex, instead of the usual median keel, cartilaginous, thick, viscid, very rigid, 1½ lin. long, not reaching to the top of the corolla; corolla cyathiform or hemispherical, minutely puberulous, "viscid" (acc. to *Bentham*), 1½ lin. long, somewhat greater in width; anthers terminal, very narrow-elliptical, acute, tapering at either end; cells deeply partite and somewhat distant, membranous, smooth, pale brown, about ⅞ lin. long; pore about ½ the length of the cell; ovary minutely puberulous.

Coast Region: Swellendam Div.; summit of the great mountain near Swellendam, *Burchell*, 7354 ! 7407!

Besides the numbers of Burchell quoted above, Bentham also cites 8665 and 8683. The first of these is without flowers; the specimens under the latter are a quite different species from those under 7354 and 7407. As these latter agree better with the description, we assume them to be the type of this species; and we have placed 8683 under *E. imbricata*. The specimens have a pale stripe on the underside of the leaves, on either side of the channel, which is peculiar and unusual; but we do not know whether it is constant, or visible in the fresh state.

367. E. Guthriei (Bolus in Journ. Bot. 1894, 234); erect, a foot high, glabrous; branches diffuse, ascending; leaves erect, adpressed and closely imbricate, or slightly spreading, linear to oblong, obtuse, the younger viscid, thick, shining, 1–1½ lin. long; pedicels ½ lin. long; bracts approximate, ovate, coloured, very small; sepals ovate or obovate, viscid, coloured, shorter than the corolla, ½–¾ lin. long; corolla oblate-globose, depressed above the middle, viscid, red, 1½ lin. long, width somewhat greater; segments erect, closely contracted round the stamens; anthers sublateral, oblong, obtuse, a little over ½ lin. long; pore large and wide, about ⅞ the length of the cell; ovary glabrous.

Var. β, **strictior** (Bolus); branches straighter; leaves broader, thicker, more obtuse and more densely imbricated.

Coast Region: Piquetberg Div.; summit of Piquet Berg, 1500 ft., *Guthrie*, 2659! Var. β: Clanwilliam Div.; Ezelsbank, near Wupperthal, 4000 ft., *Schlechter*, 8805!

The peculiar corolla approaches in shape that of *E. placentæflora*, and of *E. Marlothii*. The small adpressed leaves and the glossy aspect of the whole plant are good marks for its identification.

368. E. placentæflora (Salisb. in Trans. Linn. Soc. vi. 348); erect, 1–2 ft. high; branches mostly lax and spreading, pubescent; leaves erect or spreading, imbricate or shorter than the internodes, linear, acute or obtuse, glabrous, 1–6 (mostly 3–4) lin. long; pedicels often decurved, 2–3 lin. long; bracts subapproximate or the lowest subremote, somewhat lax; sepals broad-ovate or subobovate, keeled, glabrous, 1½ lin. long, a little longer than the corolla; corolla somewhat turnip-shaped or depressed-globose ; limb closely contracted round the stamens, or sometimes suburceolate and scarcely depressed, glabrous, 1 lin. long, about $1\frac{1}{2}$ lin. wide; anthers subterminal or sublateral, longitudinally semiovate or obliquely lanceolate, acute or subobtuse, smooth, ciliolate or naked, $\frac{3}{8}-\frac{1}{2}$ lin. long; pore about $\frac{2}{3}$ the length of the cell; style exserted, rather short and stout; stigma capitate; ovary glabrous. *Benth. in DC. Prodr.* vii. 620. *E. tiaræflora, Andr. Heathery, t.* 196, *and Col. Heaths, t.* 213. *E. leptophylla, Klotzsch ex Benth. l.c.*

SOUTH AFRICA : without locality, *Herb. Salisbury !*

COAST REGION : Tulbagh Div.; Tulbagh Waterfall, *Niven,* 124 ! Witsen Berg, *Zeyher,* 1112 ? *Bolus in Herb. Norm., Austr.-Afr.,* 4 ! Winterhoek Mountain, 2500 ft., *Bolus,* 5112 ! Cape Div. ; Table Mountain, 800 ft., *Schlechter,* 932 ! Caledon Div.; Houw Hoek, *Schlechter,* 5491 ! Onrust River, *Schlechter,* 9493 ! Hermanus, *Galpin,* 3549 ! Swellendam Div.; near Swellendam, *Mund,* 2 ! *Zeyher,* 3299 ! Zuurbraak, *Galpin,* 3554!

A variable species as to habit and corolla, and having forms which run into each other. Some specimens which we cite above have been referred to *E. imbricata,* with some forms of which there is considerable affinity and we find it difficult to distinguish them.

369. E. imbricata (Linn. Sp. Pl. ed. 2, 503) ; erect, 1–3 ft. high ; branches virgate or diffuse, variably pubescent, most usually with abundant flowers ; leaves erect or spreading, linear, $1\frac{1}{2}-2\frac{1}{2}$ lin. long ; pedicels $1\frac{1}{2}$ lin. long ; bracts approximate, more rarely subremote, sepal-like but smaller; sepals ovate, keel-tipped, subcartilaginous, rigid, glabrous, white, brown, or red, mostly about $1\frac{1}{4}$ lin. long, a little longer or shorter than the corolla ; corolla cyathiform, ovoid, urceolate or globose-urceolate, white, foxy or reddish, $1-1\frac{1}{4}$ lin. long ; anthers varying from terminal to nearly lateral, linear, oblong, lanceolate, semiovate or narrow-ovate, mostly acute or acuminate, more rarely obtuse, glabrous, smooth, light to dark-brown, mostly about $\frac{1}{2}$ (rarely $\frac{3}{4}$) lin. long; pore from $\frac{1}{2}-\frac{2}{3}$ the length of the cell; stigma clavate-capitellate ; ovary usually glabrous, rarely puberulous. *Benth. in DC. Prodr.* vii. 620 ; *Andr. Heathery, t.* 119, *and Col. Heaths, t.* 27 ; *Wendl. Eric. Ic. fasc.* 21, 135, *t.* 51 ; *Lodd. Bot. Cab. t.* 1243. *E. quinquangularis, Berg. Descr. Pl. Cap.* 117, *and E. laricifolia, Lam. Encycl.* i. 487, *acc. to Ind. Kew. E. flexuosa, Andr. Heathery, t.* 65, *and Col. Heaths, t.* 23 ; *Lodd. l.c. t.* 1495. *E. divaricata, Wendl. l.c. fasc.* 7, 5. *E. cæsia, Wendl. l.c. fasc.* 24, 179, *t.* 67. *E. pyramidalis, E. squamæflora, E. cæsia, E. flexuosa, E. stylosa, Salisb., E. imbricata, Roxb., and E. bracteata, Roxb. ex Salisb. in Trans. Linn. Soc.* vi. 349, 350. *E. ramulosa, E. densiflora, E. brunneo-alba, E. myriantha, E. porrigens, Bartl. in Linnæa,* vii. 632, 633. *E. Actea, Sinclair, Hort. Eric. Wob.* 1. *E. violacea, E. leptocephala, E. trifaria, E. sparsa, E. paleacea, and E. glaucifolia, Klotzsch ex Benth. in DC. Prodr.* vii. 620. *E. imbricata, var. elongata, Rach in Linnæa,* xxvi. 771.

A most variable species, difficult of distinction into varieties. We have merely thrown the very large suite of specimens before us into four groups, which, in the absence of constant floral characters, may frequently be recognized by their general appearance :—

A. The form represented by a named specimen in Linnæus' Herbarium ; by the plates of Wendland in *Eric. Ic. fasc.* 7 and 24, and by specimens named *E. flexuosa, E. ornata, Klotzsch ex Benth. in DC. Prodr.* vii. 620, *and E. densiflora, Bartl. in Linnæa,* vii. 632. The inflorescence is usually copious and dense ; the flowers have a foxy colour ; the anthers are pale brown, and the branches are commonly flexuous.

B. The form represented by the plates in *Andrews' Heathery, t.* 65, and *Wendland, Eric. Ic. fasc.* 21. Inflorescence less copious and less dense ; flowers white, more rarely rosy, contrasting with dark-brown anthers.

C. The form shown in *Andrews' Heathery, t.* 119. Flowers red or rosy ; anthers dark ; leaves often spreading-recurved.

D. Habit more robust and more virgate than in the other forms ; flowers white or rosy, more globose and less crowded ; anthers usually more lateral than in the preceding forms, rarely subterminal. This form approaches *E. placentæflora.*

SOUTH AFRICA : without locality ; A, *Herb. Linnæus ! Herb. Salisbury !* and *cultivated specimens !*

COAST REGION, common on plains and mountains : A—Cape Div. ; *Thunberg, Ecklon,* 287 ! Caledon Div., *Zeyher,* 3304 ! 3305 ! *Bolus,* 4173 ! 5113 ! Swellendam Div., *Masson,* 2 ! Div. ? *Bolus,* 1196 ! B—Stellenbosch Div. ; Lowrys Pass, *Schlechter,* 5388 ! Div. ? *Bolus,* 6740 ! 7471 ! C—Caledon Div., *Schlechter,* 9493 ! 10400 ! Bredasdorp Div., *Schlechter,* 10479 ! 10542 ! *Bolus,* 8445 ! Div. ? *Bolus,* 6741 ! D—Tulbagh Div. ; Witsen Berg, *Burchell,* 8683 ! Paarl Div., *Bolus,* 6969 ! Cape Div., *Burchell,* 24 ! 628 ! Caledon Div., *Bolus,* 5111 ! and in *Herb. Norm. Austr.-Afr.,* 607 ! *Zeyher,* 3298 ! Div. ? *Guthrie,* 2026 ! 3768 ! 4168 !

370. E. triceps (Link, Enum. Hort. Berol. i. 371) ; stem stout ; branches ascending, pubescent, 10 in. or more long ; leaves erect-spreading, imbricate, lanceolate to oblong, subtrigonous, the younger ciliate, 2–2½ lin. long ; pedicels 1–1½ lin. long ; bracts approximate, ovate, acute, cartilaginous, large ; sepals lanceolate to ovate, acute or acuminate, subscarious or cartilaginous, with a strong excurrent short keel-tip, glabrous, ciliate or naked, 2 lin. long, a little shorter or longer than the corolla ; corolla suburceolate, glabrous, viscid ? (acc. to *Bentham*), about 2 lin. long, longer than its width ; filaments rather broad, dilated and shortly bent just below the anther ; anthers terminal, longitudinally narrow-semi-elliptical, acute or acuminate ; cells deeply partite and usually at length spreading in the upper part, hispidulous, brownish, 1 lin. long or a little over ; pore about ⅔ the length of the cell ; ovary glabrous. *Benth. in DC. Prodr.* vii. 620 ; *Lodd. Bot. Cab. t.* 962. *E. spumosa, Wendl. Eric. Ic. fasc.* 16, 51, *t.* 19, *not of Linn.*

SOUTH AFRICA : without locality, *Masson !* and *cultivated specimens !*
COAST REGION : George Div. ; Montagu Pass, 4000 ft., *Schlechter,* 5834 ! near George, *Alexander !*

371. E. calyculata (Wendl. Eric. Ic. fasc. 4, 5) ; erect, 12–18 in. or more high ; branches ascending, pubescent ; leaves erect-spreading and straight or recurved, linear, acute, 1½–3 lin. long ; pedicels 1–3 lin. long ; bracts approximate ; sepal-like but smaller ; sepals oblong or ovate, acute or acuminate, keel-tipped, cartilaginous, glabrous, $1\frac{1}{2}$–$2\frac{1}{2}$ lin. long, a little longer or shorter than the corolla; corolla subcyathiform or ovoid, glabrous, $1\frac{1}{4}$–2 lin. long; segments broad-oblong, rounded or subtruncate, always more or less concave or with inflexed margins, naked or ciliate, $\frac{1}{4}$–$\frac{2}{5}$ the length of the tube; filaments usually dilated at the apex; anthers terminal, narrow-lanceolate, tapering much to the apex and setaceous-acuminate, glabrous and smooth, light brown, $1\frac{1}{2}$–2 lin. long; pore about $\frac{2}{3}$ the length of the cell; ovary glabrous. *E. penicilliformis, Salisb. Prodr.* 297. *E. penicilliflora, Salisb. in Trans. Linn. Soc.* vi. 348; *Benth. in DC. Prodr.* vii. 621. *E. calyculata, Wendl. Eric. Ic. fasc.* 4, 5. *E. rostella, Sinclair, Hort. Eric. Wob.* 21. *E. rastellum, Spreng. Syst. Veg.* ii. 196. *E. rostrata, Bartl. in Linnæa,* vii. 632.

Var. β, **chrysantha** (Bolus); leaves broader; anthers subobtuse or scarcely acute, not setaceous-acuminate, minutely scaberulous, $\frac{3}{4}$ lin. long; ovary sometimes minutely hirtellous. *E. chrysantha, Klotzsch ex Benth. in DC. Prodr.* vii. 620.

South Africa: without locality, *Herb. Salisbury!*
Coast Region, between 1200 and 4500 ft. : Caledon Div.; Genadendal Mountain, *Galpin*, 3553! Sweetmilks Valley, *Niven*, 122! Swellendam Div.; mountains near Swellendam, *Masson*, 48! *Mund*, 30! Zuurbraak Mountain, *Galpin*, 3552! *Schlechter*, 2151! George Div.; Montagu Pass, *Schlechter*, 5851! Uniondale Div.; mountains near Avontuur, *Bolus*, 2388! Var. β : Clanwilliam Div.; on the Cederberg Range, 2500 ft., *Bolus*, 9108! *Leipoldt*, 209! Caledon Div.; Zwart Berg, *Ecklon & Zeyher! Galpin*, 3545! near Hawston, 50 ft., *Schlechter*, 9470!

Section XXXI. **ELYTROSTEGIA.**   (Sp. 372–376.)

372. E. lasciva (Salisb. in Trans. Linn. Soc. vi. 349); erect, 2–3 ft. high; branches ascending, subvirgate, **tomentulose,** with numerous lateral, erect or decurved, short flowering branchlets of from $\frac{1}{2}$–1 in. long; leaves erect-spreading, imbricate, linear-lanceolate, trigonous, subobtuse, sulcate, glabrous, smooth, $1\frac{1}{2}$–$2\frac{1}{2}$ lin. long; flowers often clustered, calycine; pedicels 1 lin. long; bracts and sepals broad-ovate, keel-tipped, cartilaginous, very concave, pallid, completely concealing the corolla, the former $\frac{1}{2}$–$\frac{3}{4}$ lin., the latter about 1 lin. long; corolla subtubular, white, about 1 lin. long; segments erect, rounded, $\frac{1}{3}$–$\frac{1}{2}$ the length of the tube; filaments very broad; anthers exserted, sublateral, oblong, subacute, nearly smooth, brown, about $\frac{2}{3}$ lin. long, muticous; pore $\frac{3}{4}$ the length of the cell; stigma narrow-peltate, 4-dentate. *Benth. in DC. Prodr.* vii. 689. *E. pachycephala, Klotzsch ex Benth. l.c.* 689. *E. exserta, Hort. ex Benth. l.c. E. brachycrossa, Tausch in Flora,* 1839, 634.

Coast Region : Cape Div.; sandy downs on the Cape Flats, *Thunberg in Herb. Salisbury! Niven*, 123! *Mund*, 31! *Bolus*, 3292! and in *Herb. Norm. Austr.-Afr.*, 46! *Guthrie*, 1448! *Wolley Dod*, 986!

We have seen and examined Tausch's type of his *E. brachycrossa*, evidently a garden specimen, and find the identity complete.

373. E. accommodata (Klotzsch ex Benth. in DC. Prodr. vii. 620) ; erect, under a foot high ; branches ascending or somewhat spreading, puberulous or somewhat tomentulose ; leaves mostly erect and straight, or incurved, more rarely spreading, linear or linear-oblong, obtuse, sulcate, glabrous, ciliolate or naked, dry or viscidulous, sometimes wrinkled, 1–2 lin. long ; flowers scarcely or seldom clustered, calycine ; pedicels 1–1½ lin. long ; bracts and sepals as in *E. lasciva* or sometimes lanceolate and often wrinkled on the back ; corolla urceolate or subcyathiform, not much contracted at the throat, white, 1 lin. or a little less long ; segments erect, rounded, ½–⅔ the length of the tube ; filaments rather narrow ; anthers subexserted never entirely beyond the corolla, sublateral, otherwise as in *E. lasciva*, but a little smaller ; stigma simple at the end of the dilated truncate style.

VAR. *β*, **subviscidula** (Bolus) ; all parts viscidulous ; branches denser ; leaves more spreading, more distinctly wrinkled at the sides, glandular-ciliolate ; sepals lanceolate, ciliate, subviscidulous, 1–1¼ lin. long ; corolla suburceolate ; segments ovate, more tapering to the apex but scarcely acute and a little longer ; filaments broader ; anthers muticous, or more or less distinctly decurrent-toothed ; stigma subpeltate, 4-punctate, smaller than in the type.

VAR. *γ*, **ebracteata** (Bolus) ; leaves shorter, more adpressed ; bracts none ; sepals with a more prominent, foliaceous, keeled apex ; corolla slightly smaller ; stigma rather larger.

COAST REGION : Caledon Div. ; near the Bot River, *Bolus*, 5448 ! between Bot River and the Zwart Berg, *Zeyher*, 3306 ! Bredasdorp Div. ; hills near Elim, *Bolus*, 8444 ! *Schlechter*, 7630 ! Var. *β* : Caledon Div. ; mountains near Genadendal, 4500 ft., *Schlechter*, 10325 ! Var. *γ*, Caledon Div. ; Klein River, 1000 ft., *Schlechter*, 7606 !

374. E. glumæflora (Klotzsch ex Benth. in DC. Prodr. vii. 689) ; erect, about a foot high ; branches erect or spreading, with short ashy-grey tomentum ; leaves erect or spreading, linear-subulate, keeled or subtrigonous, sulcate, glabrous, 2–4 lin. long ; flowers on short lateral branchlets, often cernuous, not very dense, varying from calycine to subcorolline ; pedicels pubescent, ½–1½ lin. long ; bracts sepal-like, ½–⅔ the length of the sepals ; sepals from lanceolate to elliptical, oblong-obovate or broad-obovate, acuminate, acute or subobtuse, keeled or only keel-tipped, cartilaginous or glumaceous, in some specimens (*Burchell's*) drying dark, in others nearly white, 1½–2 lin. long, or from ½ the length of the corolla to equalling it ; corolla urceolate-cyathiform, or ovoid-cyathiform, apparently white, 1½ to nearly 2½ lin. long ; segments suberect, semi-orbicular or semi-ovate and tapering to the apex, obtuse or acute, ¼–½ the length of the tube ; anthers exserted, subterminal or sublateral ; cells bipartite sometimes divaricate, linear to oblong, acute or subobtuse, smooth, brown, about ½–¾ lin. long, decurrent-aristate ; pore ½ the length of the cell ; appendage hirsute, in length from ⅕–⅓ of the cell, free portion spreading, of variable length ; stigma either cyathiform with or without 4 short filiform processes (springing from the base of the cup and sometimes exserted beyond its margin), or simple at the clavate, truncate apex of the style. *E. leucosepala, Klotzsch,* and *E. setula, Klotzsch ? ex Benth. in DC. Prodr.* vii. 689.

SOUTH AFRICA: without locality, *Krebs,* 224! *Wallich ! Bowker in Herb. Huguenot Seminary,* 224 !

COAST REGION: Knysna Div.; near Knysna, *Burchell,* 5511! 5656! Uitenhage Div.; Van Stadens Berg, *Zeyher,* 3282! Port Elizabeth Div.; near Port Elizabeth, *Sim!* Albany Div.; near Riebeek, *Burchell,* 3466! near Grahamstown, *Burchell,* 3537! *MacOwan,* 300! *Galpin,* 3086! East London Div.; near Amalinda, *Miss Arnold in Herb. Flanagan,* 2352! Stutterheim Div.; Fort Cunynghame, 3000 ft., *Sim,* 1895 !

Burchell's specimens 5656 and 5511 and Flanagan's 2352, appear to be quite distinct; yet they are connected so closely by numerous intermediate forms that it is impossible to define satisfactory varieties.

375. E. lepidota (Rach in Linnæa, xxvi. 786); robust, erect, 1 ft. or more high; branches ascending, straight, densely clothed with short, thick, glandular, plumose, dirty-yellow hairs; leaves adpressed, imbricate, narrow-linear-trigonous, glabrous, smooth, shining, the younger ciliolate, sulcate-keeled, leaden-olive-green (in the dried state), 2 lin. long; pedicels whitish-silky, very short; bracts adpressed, oblong, glabrous, very obtuse, ciliate, ½ the length of the sepals; sepals ovate, obtuse, glabrous, membrane-ciliate, a little shorter than the corolla; corolla urceolate-cylindrical, scarcely 1 lin. long, ½ lin. wide; segments ovate, very obtuse, ½ the length of the corolla; anthers subincluded, oblong, very obtuse at either end, saccate at the base, hirtillous, muticous; pore ½ the length of the cell; style subexserted, thick; stigma cyathiform-peltate, quadrangular; ovary glabrous.

SOUTH AFRICA: without locality, *Thunberg!*

We have seen the type specimen in Thunberg's herbarium, named by Rach. The leaves are clearly not scurfy, but smooth. The plant is nearest to *E. accommodata.*

376. E. diosmæfolia (Salisb. in Trans. Linn. Soc. vi. 350); erect, 4-8 in. high; branches many, stout, rigid, corymbose, tomentulose; leaves erect-spreading, crowded, closely and neatly imbricate, broad-linear-trigonous, sulcate, rigid, glabrous, glossy, ciliate-serrulate, 2-3 lin. long; flowers usually scanty at the ends of the longer branches, subcorolline; pedicels stout, tomentulose, 1-1½ lin. long; bracts approximate, adpressed, narrow-ovate, acute, keel-tipped, ciliolate or naked, concave, paleaceous; sepals like the bracts, but longer, pallid, about 1¼ lin. long, reaching to a little lower than the top of the corolla-tube; corolla cyathiform, slightly widened upwards from the throat, white, 1½-1¾ lin. long; segments ovate, very little spreading, about as long as the tube; filaments capillary; anthers lateral, included, shorter than the corolla-segments, but just manifest (or acc. to *Bentham,* sometimes exserted); cells deeply partite, oblong, obtuse, slightly narrowed to the base, subscaberulous, dark red, ½ lin. long, subulate-aristate; pore about ⅓ the length of the cell; awns curved, acuminate, serrulate or ciliolate, about ½ the length of

the cell; stigma broad-capitate or peltate; ovary pallid. *E. serrulata,
Sinclair, Hort. Eric. Wob.* 23. *E. subserrata, Roxb. ex Salisb. in
Trans. Linn. Soc.* vi. 350.

SOUTH AFRICA: without locality, *Mund! Herb. Salisbury!*
COAST REGION: Cape Div.; Table Mountain, 3500 ft., *Bolus*, 2954! 3877!
Harvey! Guthrie, 1462!

This species forms a transition to the § *Eurystoma* to which it is closely
allied.

Section XXXII.   **APŒCUS.**   (Sp. 377–378.)

377. E. Brownleeæ (Bolus in Journ. Linn. Soc. xxiv. 185);
erect, robust, 1–4 ft. high; branches subvirgate, pubescent; leaves
4-nate, erect or subspreading, crowded, laxly imbricate, incurved,
subulate, acute, nearly flat, sulcate, glabrous, or the younger sometimes thinly pilose, $2\frac{1}{2}$–7 (mostly 4–6) lin. long; flowers 4-nate,
or umbellate, 5–7-flowered; pedicels pubescent, $1\frac{1}{2}$–2 lin. long;
bracts subapproximate, ovate, acute, scarious, $1\frac{1}{2}$ lin. long; sepals in
opposite pairs, imbricate and somewhat closely enwrapping the
corolla, obovate or suborbicular, keel-tipped, acute, thick and
wrinkled below, concave, scarious, subviscidulous, glabrous, about
$2\frac{1}{2}$ lin. long, reaching to the top of the corolla-segments or nearly
so; corolla cyathiform, widened to the mouth, or subobconic, white,
$1\frac{3}{4}$–$2\frac{1}{4}$ lin. long; limb slightly expanded, but not recurved; segments
ovate, 1–$1\frac{1}{4}$ lin. long, or about $\frac{3}{4}$–$\frac{4}{5}$ the length of the tube; filaments
slender, bent; anthers subincluded, sometimes manifest, oblong,
obtuse, scaberulous, brown, $\frac{1}{2}$–$\frac{4}{5}$ lin. long, crested; pore about $\frac{2}{3}$ the
length of the cell; crests suborbicular, incised, less than $\frac{1}{2}$ the
length of the cell; style decurved; stigma clavate-capitellate,
4-dentate; ovules compressed; seeds unknown.

COAST REGION: Albany Div.; mountains near Grahamstown, *MacOwan*,
1260! King Williams Town Div.; edge of the Perie Forest, 2500 ft., *Miss
Brownlee in Herb. Tyson*, 2858! Perie Mountain, *Tyson in Herb. Norm. Austr.-
Afr.*, 858! Stockenstrom Div.; southern slopes of Kat Berg, *Scully*, 142!
Stutterheim Div.; Dohne Mountain, 4000 ft., *Flanagan*, 2301! Kabousie River,
Murray, 24!

378. E. Caffrorum (Bolus in Journ. Linn. Soc. xxiv. 184); erect,
branched or subvirgate, of various aspects, from 1–3 ft. high;
branches puberulous or tomentulose; leaves 4-nate, mostly recurved-
spreading, linear, acute, sulcate, mostly ciliate, glabrous, glossy,
2–4 lin. long; flowers 4-nate, or subumbellate, or occasionally
pseudo-lateral and clustered at the ends of the branches; pedicels
puberulous, 1–$1\frac{1}{2}$ lin. long; bracts remote or rarely subapproximate,
about 1 lin. long; sepals lanceolate to ovate, acute, glumaceous,
rigid, glabrous, 1–$1\frac{1}{2}$ lin. long, reaching generally $\frac{3}{4}$ the height of
the corolla; corolla obconic or obconic-cyathiform, mostly somewhat
widened to the mouth, more rarely cyathiform, apparently dirty-
white, about $1\frac{1}{2}$ lin. long; limb spreading or suberect, subequal to

the tube in length ; anthers included, subovate or oblong, membranous, smooth, scabrid on the margins only, $\frac{1}{3}$–$\frac{3}{5}$ lin. long, very shortly aristate; awns more or less hairy ; pore about $\frac{1}{2}$ the length of the cell; style exserted (probably always so at full maturity, but often included, in our specimens) ; stigma clavate, capitellate or simple, sometimes distinctly 4-toothed or lobulate ; ovules compressed ; seeds unknown.

VAR. β, **glomerata** (Bolus); branches more spreading; flowers mostly clustered ; corolla broader, cyathiform ; limb shorter than the tube ; anthers as in the type.

VAR. γ, **luxurians** (Bolus); branches long, erect, subflexuous ; leaves more crowded, more erect and imbricate, more distinctly ciliate; pedicels woolly; bracts larger, sometimes subapproximate ; corolla more tubular; limb shorter than the tube; anthers minutely crested; crests lanceolate; stigma more distinctly 4-lobed.

VAR. δ, **aristulata** (Bolus); branches and leaves as in γ; pedicels pubescent ; corolla of the type but limb shorter ; anthers very minutely aristulate ; awns not reaching to the base of the cell.

COAST REGION : Queenstown Div., *Cooper*, 211 !

EASTERN REGION: Tembuland; Bazeia, *Baur*, 507 ! Pondoland ; Insiswa Mountain, 6800 ft., *Schlechter*, 6495 ! Var. β : Tembuland ; near the Gat Berg, *Bolus*, 8731 ! Griqualand East ; Mount Currie, 7500 ft., *Tyson*, 1194 ! Var. γ: Natal ; foot of the Drakensberg Range, *Wood*, 3519 ! Mahwaqua Peak, Polela, *Wood*, 4281 ! Var. δ : Drakensberg Range, *Bolus in Herb. Guthrie*, 4976 !

Section XXXIII. **LAMPROTIS.**   (Sp. 379–398.)

379. E. dianthifolia (Salisb. in Trans. Linn. Soc. vi. 338) ; erect, 1–1$\frac{1}{2}$ ft. high ; branches numerous, subvirgate, the younger pubescent with short subplumose or barbellate hairs, glabrescent ; leaves opposite, erect, linear, acuminate, keeled, glabrous, 4–7 lin. long ; inflorescence terminal; flowers solitary, or 2-nate, calycine ; pedicels tomentose, 2 lin. long ; bracts approximate, ovate-lanceolate, acuminate, concave, rigid, glumaceous, whitish, shorter than the sepals ; sepals like the bracts, but larger and somewhat spreading, the keel produced into a stout cuspidate point, the whole 2$\frac{1}{2}$–3 lin. long, reaching to below the apex of the corolla-segments ; corolla ovoid-urceolate, white, 3$\frac{1}{2}$–4 lin. long ; segments spreading, ovate-cordate, nearly as long as the tube ; anthers lateral, affixed well above the base of the cells, oblong, subobtuse, $\frac{3}{5}$ lin. long, crested ; pore under $\frac{1}{2}$ the length of the cell; crests ovate, acute, serrulate, nearly $\frac{1}{2}$ the length of the cell; ovary glabrous ; ovules subovate, membrane-margined. *Benth. in DC. Prodr.* vii. 649. *E. oppositifolia, var. major, Andr. Heathery, t.* 179. *E. biflora, Link, Enum. Hort. Berol.* i. 367 ; *Lodd. Bot. Cab. t.* 683.

SOUTH AFRICA: without locality, *Herb. Salisbury!* and *cultivated specimens!*

COAST REGION : Caledon Div.; Baviaans Kloof, near Genadendal, *Niven*, 19 ! 70a ! mountains near Hartebeest River, *Grisbrook!* Swellendam Div. ;

near Storms Vlei, *Zeyher*, 3265! mountains near Swellendam, *Shand in Herb.
Bolus*, 6253 ! *Schlechter*, 5682 ! Riversdale Div.; summit of Kampsche Berg,
Burchell, 7114 !

Allied to *E. Nabea*, of which it has somewhat the appearance in miniature;
but the sepals are much smaller, and the corolla larger and differently
shaped.

380. E. borboniæfolia (Salisb. in Trans. Linn. Soc. vi. 386); stout,
erect, entirely glabrous; branches numerous, somewhat crowded,
ascending, virgate or flexuous, and subdiffuse; leaves opposite, erect,
mostly adpressed, linear, acute, keeled, 2–3 lin. long, the floral often
subpetaloid; flowers mostly 2-nate, crowded or clustered, often in
dense masses; pedicels 2 lin. long; bracts 2, median, ovate or
spathulate (*Bentham*), acute, coloured, 3 lin. long; sepals like the
bracts, or subobovate, petaloid, 4–5 lin. long; corolla tubular-
inflated, slightly contracted at the throat, 5–6 lin. long by about
$1\frac{3}{4}$ lin. in the widest diam.; segments ovate, acuminate, $\frac{1}{3}$–$\frac{3}{8}$ the
length of the tube; anthers oblong, tapering above the middle,
truncate at the base, about $\frac{1}{2}$ lin. long; crested high above the base;
pore nearly $\frac{1}{3}$ the length of the cell; crests ovate-acuminate, more
than $\frac{1}{3}$ the length of the cell, reaching very little below its base;
style included; stigma capitellate; ovary cylindrical, elongate,
8-lobed. *E. togata, Sims, Bot. Mag. t.* 1626. *E. aperta, Spreng.
Syst. Veg.* ii. 200.

COAST REGION, from 1200 to 4700 ft.: Caledon Div.; Zwartberg Range,
near Caledon, *Pappe in Herb. Norm. Austr.-Afr.*, 20b! tops of the mountains of
Baviaans Kloof, near Genadendal, *Masson*, 24! *Burchell*, 7762! *Bolus, Herb.
Norm. Austr.-Afr.*, 354! *Schlechter,* 9872 ! *Roser !*

The relative length of the sepals and corolla is apparently variable; partly
due to different stages of maturity, and partly to the degree of spread of the
sepals.

381. E. lutea (Berg. Desc. Pl. Cap. 115); about 1–$1\frac{1}{2}$ ft. high,
entirely glabrous; branches numerous, ascending, subvirgate or
flexuous, slender; leaves opposite, imbricate, slightly spreading or
adpressed, linear, acute, keeled, 1–3 lin. long; flowers 2-3-4-nate,
often clustered; pedicels 1–$1\frac{1}{2}$ lin. long; bracts 2, submedian, more
rarely a third subbasal, 1 lin. or less long; sepals broad-oblanceolate
or more commonly broad-ovate or obovate, keeled-tipped, shortly
acuminate, $1\frac{1}{4}$–$1\frac{3}{4}$ lin. long, from $\frac{1}{3}$–$\frac{2}{3}$ the length of the corolla; corolla
conical-tubular, somewhat inflated at the base, tapering to a more or
less narrow mouth, white or yellow, $3\frac{1}{2}$–4 lin. long; segments
spreading, ovate, obtuse, $\frac{1}{3}$–$\frac{2}{5}$ the length of the tube; anthers oblong,
very obtuse, pallid, about $\frac{1}{4}$ lin. long, shortly crested; pore under
$\frac{1}{2}$ the length of the cell; crests affixed near the base of the corolla,
and (in the flowers examined) spreading somewhat horizontally,
lanceolate, subentire, curved, less than $\frac{1}{3}$ the length of the cell;
stigma capitellate; ovary cylindrical, lobed. *Linn. Mant. Alt.* 234;
Thunb. Fl. Cap. ed. Schult. 359; *Wendl. Eric. Ic. fasc.* 1, 13;
Andr. Heathery, t. 29, *and Col. Heaths, t.* 32; *Lodd. Bot. Cab.
t.* 64; *Benth. in DC. Prodr.* vii. 651. *E. imbellis, Salisb. Prodr.*

298, *and in Trans. Linn. Soc.* vi. 385. *E. oppositifolia, Andr.
Heathery. t.* 178, *and Col. Heaths, t.* 192 ; *var. alba, Lodd. Bot.
Cab. t.* 1343.

COAST REGION : Cape Div.; frequent on the mountains of the Cape Peninsula, *Thunberg, Niven,* 69 ! 70 ! *Sieber,* 195! *Zeyher,* 1093 ! *Bolus,* 3295! *Guthrie,* 483 ! *Wolley Dod,* 819 ! Stellenbosch Div.; mountains near Lowrys Pass, *Guthrie!* Paarl or Worcester Div.; Dutoits Kloof, *Drège!* Caledon Div.; Zwart Berg, *Mund,* 17 ! *Pappe !*

382. E. tenuifolia (Linn. Syst. Nat. ed. 10, 1002) ; 6–18 in. high, entirely glabrous ; leaves opposite (occasionally some 3-nate on the same plant), erect or adpressed, imbricate, linear-trigonous, acute, keeled, sulcate, the younger with a reddish tinge in the upper part, 1–2½ lin. long ; flowers 3-nate, mostly crowded ; pedicels 1–1½ lin. long ; bracts 2, remote, about 1 lin. long ; sepals spreading, obovate, very concave, acute, apiculate, keeled, 1½–2 lin. long, mostly reaching about ½ the length of corolla ; corolla ovoid-urceolate or subconical, inflated at the base, contracted at the throat, 2½–3 lin. long ; segments ovate, subacute, nearly ½ the length of the tube ; anthers oblong, obtuse, pallid, ⅖ lin. long, minutely toothed or crested ; pore ½ the length of the cell ; crests very variable in size, from 1/10 to ⅓ the length of the cell ; stigma small, subsimple ; ovary subtruncate, lobed. *Berg. Descr. Pl. Cap.* 116 ; *Linn. Sp. Pl. ed.* 2, 507 ; *Benth. in DC. Prodr.* vii. 651. *E. oppositifolia, var. rubra, Andr. Heathery, t.* 180 (*not characteristic*). *E. linifolia, Salisb. in Trans. Linn. Soc.* vi. 386.

COAST REGION : Cape Div. ; frequent on the mountains of the Cape Peninsula, up to the summit of Table Mountain, *Thunberg, Burchell,* 594 ! *Cooper,* 3047 ! *Bolus,* 3875 ! 4700 ! *Guthrie,* 338 ! 814 ! *Wolley Dod,* 2156 ! 3336 ! *Galpin,* 3619 ! 3620 ! Caledon Div. ; mountains near Appels Kraal, *Ecklon & Zeyher,* near Zoetmelks Vlei, *Grisbrook !* near Onrust Kiver, *Schlechter,* 9491 !

Closely allied to *E. lutea,* and perhaps to be regarded as a variety of it. It is also allied to *E. borboniæfolia,* and all three resemble each other in floral structure, but differ in aspect.

383. E. Alfredii (Guthrie & Bolus) ; erect, entirely glabrous ; branches virgate or incurved, the younger pallid, smooth, strongly channelled between the decurrent leaf-cushions ; leaves 3-nate, imbricate, erect, subadpressed, linear-lanceolate, acute, callous-tipped, margins inflexed with a narrow subscarious edge, glaucous, 4–5½ lin. long ; flowers umbellate ; umbels 6–9-flowered, erect-spreading ; pedicels 7–9 lin. long ; bracts, 2 subapproximate, 1½ lin. from apex of pedicel, the third quite basal, caducous or sometimes wanting, lanceolate, acuminate, subscarious, foxy-brown, 5–6 lin. long ; sepals like the bracts but ovate, acuminate, 6–7 lin. long, 3–4 lin. wide ; corolla tubular-conical or subampullaceous, rosy, 9–10 lin. long, about 3–4 lin. in diam. above the base, rosy ; segments probably stellato-patent, ovate, acuminate, about ⅕ the length of the tube ; filaments capillary, flexuous, 8 lin. long ; anthers broad-oblong, very obtuse, ⅔ lin. long, muticous ; pore ⅔ the length of the cell ; style included; stigma capitate; capsule globose, glabrous; seeds subglobose, ecorbiculate, muricate on the margins of the pits.

COAST REGION : Caledon Div.; mountains near the Zondereinde River, 5400 ft., *Bolus in Herb. Guthrie*, 4574!

384. E. bracteolaris (Lam. Encycl. i. 481); stout, 1½ ft. or more high, entirely glabrous; branches ascending, often virgate, fastigiate; leaves 3-nate, adpressed, imbricate, or the lower as long as the internodes, linear or narrow-lanceolate, acute, sulcate, 2–3 lin. long; floral leaves sometimes dilated or even subsepaloid; flowers 3-nate, pseudo-capitate by close packing in very short or arrested terminal branchlets; pedicels 1 lin. long; bracts, two approximate, oblanceolate, acuminate, the third basal, longer and broader; sepals spathulate, unguiculate, keel-tipped, subobtuse, apiculate, with a rather broad diaphanous margin, about 4 lin. long, scarcely reaching to the top of the corolla-tube; corolla subtubular, almost equal or a little inflated above the base, 3½–5 lin. long; segments ovate, acute, scarcely ⅓ the length of the tube; anthers broad-oblong, very obtuse, the length 1¾ times the width, muticous or very minutely subdenticulate at the base; pore about ½ the length of the cell; appendages not reaching nearly to the base of the cell; stigma capitate; ovary cylindrical, lobed. *E. glomerata, Andr. Heathery, t. 264, and Col. Heaths, t. 237, not of Salisb. nor of Sinclair; Benth. in DC. Prodr.* vii. 650. *E. rubella, Bot. Mag. t.* 2165; *Lodd. Bot. Cab. t.* 658. *E. calycinades, Sinclair, Hort. Eric. Wob.* 4.

COAST REGION : Swellendam Div.; Riet Kuil, *Niven*, 74! mountain peak near Swellendam, *Burchell*, 7300! 7333! *Borcherds in Herb. Bolus*, 5958! mountains near Zuurbraak, *Schlechter*, 2102! *Galpin*, 3608! Also *cultivated specimens!*

We have seen and dissected Lamarck's type in Paris Herbarium; it agrees well with Andrews' figures.

385. E. steinbergiana (Wendl. f. ex Klotzsch in Linnæa, xii. 531); 1–1½ ft. high, entirely glabrous; branches few, virgate; leaves 3-nate, adpressed, imbricate, or the lower sometimes scarcely longer than the internodes, lanceolate, 1½–2½ lin. long, acute or acuminate, thick, round-backed, sulcate; flowers 3-nate, subsessile, somewhat clustered; bracts 2, approximate, linear-oblanceolate, acute, 2–2½ lin. long; sepals oblanceolate or lanceolate, keel-tipped, sharply acuminate, the margins somewhat involute, 2½–3 lin. long, about as long as the corolla-tube; corolla subtubular, more or less inflated at the base, subtetragonous, 4–5 lin. long; segments ovate, acuminate, about ⅓ the length of the tube; anthers broad-oblong, very obtuse, pallid, smooth, a little over ¼ lin. long, muticous; pore about ½ the length of the cell; stigma capitellate; ovary short-cylindrical, truncate, prominently 8-lobed. *Benth. in DC. Prodr.* vii. 650.

VAR. β, **abbreviata** (Bolus); leaves broad-lanceolate, 1½ lin. long ; sepals broad-oblanceolate, mostly ½–⅔ the length of the corolla-tube.

SOUTH AFRICA : without locality, *Hesse.*

COAST REGION : Riversdale Div.; Kafferkuils River, *Niven*, 73! Langeberg

Range, 1000 ft., *Schlechter*, 1911! Var. *β* : Riversdale Div.; on the Kampsche Berg, *Burchell*, 7086!

386. E. taxifolia (Bauer, Exot. Pl. t. 19); erect, rigid, generally under a foot high; branches numerous, short, nearly straight, stout, pallid, the younger channelled by the prominent decurrent leaf-cushions, pubescent or glabrous; leaves 3-nate, erect, spreading or the older squarrose, crowded, imbricate, linear-trigonous, acute or acuminate, sulcate, glabrous, 4–6 lin. long; flowers umbellate, or subumbellately-clustered, with occasionally a few axillary; pedicels 3–4 lin. long, pubescent or glabrous; bracts remote, oblanceolate, coloured, 2–4 lin. long; sepals ovate or obovate, acute, keel-tipped, mucronate, coloured, $2\frac{1}{2}$–$4\frac{1}{2}$ lin. long; corolla ovoid-urceolate 3–$4\frac{1}{2}$ lin. long, $1\frac{1}{2}$–2 lin. wide; segments ovate, subacute, darker-coloured or brownish, a little less than $\frac{1}{2}$ the length of the tube; anthers narrow-elliptical, rounded at the base and apex, smooth, $\frac{1}{3}$–$\frac{2}{5}$ lin. long, subaristate; pore less than $\frac{1}{2}$ the length of the cell; awns subulate, about $\frac{1}{3}$ the length of the cell; style included; stigma capitate; ovary turbinate, or depressed-globose, glabrous. *Wendl. Bot. Beobacht.* 44, *and Eric. Ic. fasc.* 2, 19; *Salisb. in Trans. Linn. Soc.* vi. 387; *Andr. Heathery, t.* 93, *and Col. Heaths, t.* 63, *also var. major, l.c. t.* 243, *and l.c. t.* 279; *Benth. in DC. Prodr.* vii. 651. *E. corifolia, ζ, Herb. Thunb. ex Rach in Linnæa,* xxvi. 779. *E. turgida, Link, Enum. Hort. Berol.* i. 365, *not of Salisb. E. juniperifolia, Salisb. Prodr.* 296. *E. laxifolia, Steud. Nom. ed.* 2, ii. 7.

SOUTH AFRICA: without locality, *Niven*, 64! and *cultivated specimens* !
COAST REGION, from 1000 to 3000 ft.: Tulbagh Div.; Witsen Berg, *Burchell*, 8664! Worcester Div.; Dutoits Kloof, *Drège! Bolus*, 5175! Paarl Div.; near Wellington, *MacOwan, Herb. Norm. Austr.-Afr.,* 20! French Hoek, *Bolus*, 6966! *Schlechter*, 10266! Stellenbosch Div.; Gordons Bay, *Miss Guthrie!* Lowrys Pass, *Burchell*, 8233! Caledon Div., *Roser!* Knysna Div., *Buchanan in Herb. Bolus,* 5837!

Bentham, following Andrews, made a variety of the larger-flowered forms, but we find the size of the flowers variable by small gradations; as a rule, the specimens from the higher altitudes appear to have the larger flowers.

387. E. pycnantha (Benth. in DC. Prodr. vii. 653); entirely glabrous; floriferous branchlets densely fastigiate-corymbose, 1–2 in. long; leaves 3-nate, erect, adpressed, imbricate, ovate (*Bentham*) or lanceolate, very acute, sulcate, concave, the upper 2 lin. long; flowers 3-nate, pseudo-corymbose in our specimens, erect; pedicels 2–$2\frac{1}{2}$ lin. long; bracts narrow-lanceolate, remote, 2 lin. long; sepals oblong-lanceolate, acute, apiculate, keeled, margins hyaline, 3 lin. long; corolla tubular, slightly inflated, 4–$4\frac{1}{2}$ lin. long; segments cordate-ovate, acute, $\frac{1}{4}$ the length of the tube; anthers broad-oblong, very obtuse, $\frac{1}{4}$ lin. long, narrow-crested; pore $\frac{1}{2}$ the length of the cell; crests minute, $\frac{1}{3}$ the length of the cell, and not reaching below its base; style included; stigma capitate.

COAST REGION: Caledon Div.; near the Klein River, *Masson*, 21 !

388. E. chlamydiflora (Salisb. in Trans. Linn. Soc. vi. 338);
slender, erect, 1–1½ ft. or more high; branches subvirgate, pallid,
pubescent, glabrous or glabrescent; leaves 3-nate, erect, often
adpressed, linear-lanceolate or oblong, subobtuse, sulcate, glandular-
hirsute or glabrous, ciliate or at length naked, imbricate in the uppermost, or equalling or shorter than the internodes in the lower parts
of the branches, 2–2½ lin. long; flowers sub-3-nate, clustered or
more often lax; pedicels 2–3 lin. long; bracts, 2 upper subapproximate, lowest median, lanceolate, ciliate, 1½ lin. long; sepals
lanceolate, acute or acuminate, not or scarcely imbricating at the
base, spreading or sometimes somewhat recurved towards the apex,
keeled or the keel sometimes obsolete, mostly faintly wrinkled across
the keel near the base, 1¾–2 lin. long, pale purple or lilac, reaching
a little below or above the corolla; corolla urceolate-tubular, very
little inflated below or contracted at the throat, subtetragonous,
prominently 4-nerved, viscidulous, 1½–1¾ lin. long; segments
rounded, stellate-patent or subrecurved, concolorous, about ¼ the
length of the tube; anthers longitudinally semilanceolate, 5½ times
as long as their greatest width, tapering to the apex but scarcely
acute, smooth; cells deeply partite, between ⅖ and ½ lin. long,
appendiculate; pore more than ½ the length of the cell; appendages
lanceolate, acute, incurved, incised, ⅓ the length of the cell; style
included, or just manifest above the corolla-tube; stigma capitate;
ovary glabrous. *Benth. in DC. Prodr.* vii. 656. *E. viscaria,.Roxb.
ex Salisb. l.c.*

CoAST REGION : Cape Div.; eastern slopes of Table Mountain, above Constantia, in the Cape Govt. Herb., without collector's name, but with the locality
number 85 of *Ecklon & Zeyher* on the label! Stellenbosch Div.; Hottentots
Holland Mountains, *Masson in Herb. Salisbury!* Caledon Div. ; mountains near
the Zondereinde River, *Grisbrook in Herb. Guthrie,* 4975 !

We have seen and examined Salisbury's type. The species is allied to *E.
gnaphaloides.* The slenderness of the whole flower, the sepals more spreading
and less imbricate at the base than is usual, and their almost lilac colour,
serve generally to distinguish this species, which is by no means common.

389. E. gnaphaloides (Linn. Sp. Pl. ed. 1, 356) ; erect, 1–1½ ft.
high; .branches mostly erect and fastigiate, slender, puberulous,
glabrescent, the younger angular by prominent leaf-cushions; leaves
3-nate, erect and mostly adpressed, not crowded, equalling or longer
than the internodes, linear, blunt, sulcate, glabrous, 1–1½ lin. long;
flowers 3-nate, often clustered ; pedicels ¾–1½ lin. long; bracts, 2
subapproximate, 1 remote, narrow-lanceolate; sepals obovate, elliptical or lanceolate, acute or acuminate, keeled, concave, more or less
imbricate at the base, incurved but often somewhat spreading at the
apex, about 1½ lin. long, pallid or rosy; corolla somewhat variable,
campanulate-tubular to obconic, sometimes subtetragonous, 1¼–1½ lin.
long, rarely in fully-matured and well-grown plants (*Schlechter,*
5397) 1¾ or nearly 2 lin. long; segments at maturity recurved,
rounded, from ¼–½ the length of the tube, concolorous or nearly so ;

anthers oblong, obtuse, ascending at the base, $\frac{1}{4}$ lin. long, with the pore less than $\frac{1}{2}$ the length of the cell, or longitudinally semi-elliptical, tapering to base and apex, $\frac{2}{5}$ lin. long, with the pore $\frac{2}{3}$ the length of the cell; cells deeply parted, crested; crests subovate, acute in outline, incised, naked or ciliolate, about $\frac{1}{2}$ the length of the cell; style included, angular; stigma 4-fid, cruciform; ovary pallid, lobed, glabrous. *Benth. in DC. Prodr.* vii. 656. *E. gnaphalodes, Berg. Descr. Pl. Cap.* 119; *Wendl. Eric. Ic. fasc.* 19, 109, *t.* 42. *E. gnaphaliiflora, Salisb. in Trans. Linn. Soc.* vi. 337. *E. calycina, var. minor, Andr. Heathery, t.* 108, *and Col. Heaths, t.* 156. *E. paniculata, Lodd. Bot. Cab. t.* 1194? *not of Linn. E. tetrastigmata, Bolus in Journ. Linn. Soc.* xxiv. 178. *E. lilacina, Klotzsch ex Benth. l.c.* 657.

CoAST REGION, ascending to 2500 ft.: Paarl Div.; French Hoek, *Bolus,* 6984! Cape Div.; Cape Flats, near Rondebosch, &c., *Burchell,* 170! *Zeyher,* 1111! *Bolus,* 3747! *Guthrie,* 88! 381! Steen Berg, *Wolley Dod,* 3431! Muizen Berg, *Bolus,* 7300! Stellenbosch Div.; Lowrys Pass, *Schlechter,* 5397! Caledon Div.; Houw Hoek, *Bolus,* 5451! mountains near Hemel-en-Aarde, *Zeyher,* 3276! Zwart Berg, *Schlechter,* 5577! Bredasdorp Div.; near Potte Berg, *Mund,* 28! Riversdale Div.; Garcias Pass, *Burchell,* 6943!

This species has been much confused, and is often marked "*E. articularis*" in herbaria. The 4-fid cruciform stigma is generally a diagnostic character, and was noted by Linnæus. Drège's 7743, while it exhibits this character, has flowers much more like those of *E. chlamydiflora.* We regard it as a hybrid, and merely note it. We do not cite Thunberg's *E. gnaphalodes, Diss. Erica,* 45, although Rach's determination of this as *E. articularis, L.,* is probably wrong. Our plants agree well with Wendland's figure cited above. There are two forms, one represented by the last-named figure, slender and virgate; and a stouter and more branching one, with more numerous and crowded flowers, resembling some species of § *Trigemma.* But the two forms are by no means distinct, and are connected by intermediates.

390. E. articularis (Linn. Mant. 65); mostly erect, 1–1$\frac{1}{2}$ ft. high; branches usually virgate, not numerous, slender, glabrous, rarely puberulous; leaves 3-nate, erect, mostly incurved and adpressed, generally about equal to the internodes, linear or narrow-lanceolate, keeled or sulcate, mostly 1$\frac{1}{2}$–3 lin. long; flowers 3-nate, more or less clustered, often so crowded on short branchlets as to form a dense long pseudo-spike; pedicels slender, 1–1$\frac{1}{2}$ lin. long; bracts 2, subremote or subapproximate, linear-subulate to oblanceolate, third subbasal, longer or wanting; sepals obovate, keel-tipped, acute or obtuse, concave, spreading at full maturity, generally reaching from $\frac{1}{2}$–$\frac{3}{5}$ the length of the corolla (including the limb), $\frac{3}{4}$–1$\frac{1}{2}$ lin. long; corolla suburceolate, only slightly contracted at the throat, sometimes subtetragonous, 1$\frac{1}{2}$–2 lin. long; segments erect or spreading, subovate, deep flesh-colour (*Andrews*), concolorous or dark-coloured, $\frac{1}{3}$–$\frac{1}{2}$ the tube in length; anthers sublateral, almost basifixed; cells deeply parted, semi-ovate or elliptical, $\frac{1}{5}$–$\frac{2}{5}$ lin. long, crested; pore about $\frac{1}{2}$ the length of the cell; crests subovate in outline, more or less deeply incised; style included; stigma capitate; ovary lobed. *Benth. in DC. Prodr.* vii. 653; *E. teretiuscula,*

Wendl. Eric. Ic. fasc. 21, 141, *t.* 54. *E. flagelliformis, Andr.
Col. Heaths, t.* 234. *E. flagellata, Andr. Heathery, t.* 262.
E. flagellaris, Link, Enum. Hort. Berol. i. 365. *E. corifolia,
Salisb. in Trans. Linn. Soc.* vi. 386. *E. struthiolæfolia and E.
filiformis? Bartl. in Linnæa,* vii. 636 (*acc. to Benth.*). *E. rubicunda, Klotzsch in Linnæa,* xii. 540. *E. strutiæfolia, Steud.
Nomencl. ed.* 2, i. 580.

VAR. β, **meyeriana** (Bolus); dwarf, 4–8 in. high; branches numerous, short,
flexuous and intertwined; leaves ovate, oblong, or narrow-oblong, ¾–1 lin. long;
flowers laxer and fewer than in the type; corolla-limb dark-coloured and
brownish; anthers smaller than in the type, the crests less incised. *E. meyeriana,
Klotzsch in Linnæa,* xii. 541. *E. pallescens, Klotzsch, l.c.* 541. *E. gnaphaloides, E. Meyer ex Klotzsch, l.c.* 541.

VAR. γ, **implexa** (Bolus); dwarf, 6 in. high; branches many, slender,
flexuous or intertwined; leaves small, subovate, acute, thick, keeled or sulcate,
¾–1 lin. long; corolla from ovoid to subtubular, thin and hyaline in texture,
subtetragonous, white, 1½ lin. long; limb spreading, concolorous.

SOUTH AFRICA: without locality; Var. γ (*Niven ?*), 65!
COAST REGION: frequent from Clanwilliam Div. to Caledon Div., *Thunberg,
Burchell,* 8610! *Ecklon & Zeyher! Drège! MacOwan & Bolus, Herb. Norm.
Austr.-Afr.,* 22! 355! *Bolus,* 4055! 4748! 5285! 5422! *Guthrie,* 1007! 1504!
2188! 3178! *Marloth,* 1682! *Leipoldt,* 611! Var. β: much less frequent;
Worcester Div.; Dutoits Kloof, distributed as "*E. gnaphalodes, L.,*" *Drège!*
Paarl Div.; French Hoek, 1000–1500 ft., *Guthrie,* 3166! Caledon Div.; near
the Baths at Caledon, *Guthrie,* 2503! Riversdale Div.; Kampsche Berg,
1000 ft., *Galpin,* 3607! Var. γ: Worcester Div.; on the Matroosberg, 5700 ft.,
Bolus, 6897! (*Schlechter,* 10073, may belong here, but the flowers are too young
for identification).

In Linnæus' herb. is a specimen marked *E. articularis,* which agrees with his
description, so far as that goes; also with the figure of Wendland's *E. teretiuscula,* and with numerous specimens marked by Bentham and others with
the latter name. Our var. β is hardly more than a colour variety. Galpin's
3607 and Guthrie's 3166 connect the two varieties as to leaf-form. Var. γ
may be a more stunted, subalpine form of var. β (which also was found at a
higher elevation); had it stood alone, it would doubtless have been deemed of
specific rank. It has the aspect of *E. humifusa, Salisb.,* otherwise a very
different plant.

391. E. caledonica (Spreng. fil. Tent. Suppl. Syst. Veg. 13);
apparently of dwarf habit, entirely glabrous; branches ascending,
scarred with the short prominent leaf-cushions, the ultimate
fastigiate and crowded; leaves 3-nate, erect, sexfariously, densely
and uniformly imbricate, ovate, acute, concave, thick, sulcate, 2½ lin.
long by 1 lin. or less broad; flowers 3-nate, somewhat clustered;
pedicels 3–4 lin. long, exceeding the sepals; bracts remote, the two
upper linear-spathulate, acute, the lower absent or replaced by a
bract-like floral leaf at the base; sepals broadly obovate, keel-tipped
and apiculate, very concave and rigid, scarious-edged, about 1½ lin.
long, or ½ the length of the corolla-tube; corolla urceolate, 3½–4
lin. long; segments stellate-patent, ovate, subacute, about ¼ the
length of the tube; anthers broad-oblong, suddenly tapering to a
subacute apex, ⅖ lin. long, crested; pore about ½ the length of the
cell; crests affixed about the middle of the cell, not spreading, narrow-

lanceolate, incised, in length $\frac{3}{4}$ of the cell; ovary columnar, glabrous. *Benth. in DC. Prodr.* vii. 653. *E. marginata, Bartl. in Linnæa,* vii. 637. *E. cochleariformis, Wendl. ex Klotsch in Linnæa,* xii. 542.

SOUTH AFRICA: without locality (*Niven ?*), 66! *Ecklon & Zeyher,* ex *Bartling.*

COAST REGION: Caledon Div.; tops of the mountains of Baviaans Kloof, near Genadendal, *Masson,* 22! *Burchell,* 7682!

392. E. chlorosepala (Benth. in DC. Prodr. vii. 649); prostrate (*Burchell*), or decumbent (*Masson & Schlechter*); branches straggling, slender, rigid, pallid or yellowish, minutely downy, channelled between the leaf-cushions, and the older covered with their persistent scars, 4 in. or more long; leaves 3-nate, erect, imbricate, elliptic-oblong, subacute, rather flat, but rather thick and coriaceous, often shrinking and wrinkled in drying, sulcate, glabrous, the stout pallid petiole ciliate, $2\frac{1}{2}$–3 lin. long; flowers subaxillary, solitary, the pedicels subtended at the base by several pallid bract-like aborted leaves, forming a short pseudo-raceme below the ends of excurrent branches, calycine; pedicels curved, tomentulose, $1\frac{1}{2}$–2 lin. long; bracts subapproximate, like the sepals but smaller, $1\frac{1}{2}$ lin. long; sepals oblong-lanceolate, acute, keeled in the upper half, cartilaginous, rigid, concave, pale-margined, white or yellowish (or acc. to *Bentham,* subfoliaceous, greenish), nearly $2\frac{1}{2}$ lin. long, reaching to a little below the top of the corolla-tube; corolla ovoid-urceolate, well-contracted at the throat, "golden-yellow" (*Schlechter*), $2\frac{1}{2}$ lin. long; segments spreading-recurved, ovate-orbicular, under 1 lin. long, the whole $3\frac{1}{2}$ lin. long; anthers bluntly subtriangular or obliquely subovate, obtuse, membranous, pallid; cells deeply partite, nearly $\frac{1}{3}$ lin. long, aristate; pore $\frac{3}{8}$ the length of the cell; awns affixed a little above the base of the cell, straightly dependent, setiform, nearly as long as the cells; style stout, clavate, 4-gonous; stigma simple or conical; ovary finely puberulous, dark-coloured; ovules lenticular, or at least compressed, immarginate.

COAST REGION: Swellendam Div.; Langeberg Range, near Zuurbraak, 3500 ft., *Schlechter,* 2044! Riversdale Div.; summit of Kampsche Berg, *Burchell,* 7098! Div.? "Boschjesmans Pad," *Masson,* 12!

A very distinct species, somewhat anomalous by its inflorescence in this or any of our sections. It was placed by Bentham in § *Platyspora,* but is unlike that by its calycine flowers. Bentham had not seen mature seeds, nor have we.

393. E. lævigata (Bartl. in Linnæa, vii. 638); entirely glabrous, or the pedicels minutely hairy; branches ascending, slender, sometimes very numerous and filiform; leaves 3-nate, erect, sub-adpressed, imbricate, or only a little longer than the internodes, linear or narrow-lanceolate, acute, mucronate, very smooth, often with a narrow pellucid margin and apex, $1\frac{1}{2}$–$2\frac{1}{2}$ lin. long; flowers

3-nate, or by arrest of branchlets often axillary, somewhat more lax and the flowers less crowded than in the allied species; pedicels slender, minutely hairy, 2–3 lin. long; bracts 3, remote, the 2 upper small, the lower basal, larger, or absent; sepals ovate, acuminate, acute, keel-tipped, concave, somewhat spreading, mostly about $1\frac{1}{4}$ lin. long, reaching to about $\frac{1}{2}$ the length of corolla-tube, pale or darker rosy-red, concolorous; corolla elongate-ovoid-urceolate, the neck rather long-tapering, only slightly contracted at the throat, rosy or darker red, 2–$2\frac{1}{4}$ lin. long; segments ovate, obtuse or subacute, spreading or early connivent, $\frac{1}{3}$–$\frac{1}{2}$ the length of the tube, concolorous; anthers oblong, obtuse, pallid, $\frac{1}{4}$–$\frac{1}{3}$ lin. long, minutely aristate; pore over $\frac{1}{2}$ the length of the cell; awns free, tooth-like; style included; stigma capitate. *Benth. in DC. Prodr.* vii. 650. *E. oppositifolia, var. rubra, Lodd. Bot. Cab. t.* 1060?

VAR. β, **elongata** (Bolus); sepals obovate; corolla longer than in the type; tube less inflated or nearly equal, reaching nearly $2\frac{1}{2}$ lin. in length; segments rounded and very obtuse anthers smaller, $\frac{1}{4}$ lin. long, more ovate, with minute free awns.

VAR. γ, **decurrens** (Bolus); anthers minutely decurrent-denticulate along the filament.

SOUTH AFRICA: without locality, *Mund!*

COAST REGION: Caledon Div.; Klein River, *Masson*, 35 ! Genadendal Mountain, 2500 ft., *Bodkin in Herb. Bolus*, 6898 ! near Caledon Baths, 800 ft., *Guthrie*, 2502B ! and without exact locality, *Ecklon & Zeyher*, ex *Bartling*. Var. β: Knysna Div.; without locality, *Buchanan in Herb. Bolus*, 5836 ! Var. γ: Caledon Div.; Zwart Berg, near Sandfontein, 2000 ft., *Schlechter*, 10346 !

Closely allied to *E. articularis, L.,* differing chiefly by its laxer and more frequently axillary inflorescence, its relatively shorter sepals, and the smaller and differently-shaped anther-appendages.

394. E. corifolia (Linn. Sp. Pl. ed. 1, 355, in an extended sense); usually 1–$1\frac{1}{2}$ ft. high; branches erect or spreading, mostly glabrous, rarely puberulous; leaves mostly 3-nate, in one form (*E. patula, Klotzsch*) often opposite, erect or adpressed, often imbricate, linear to lanceolate, rarely narrow-ovate, acute, keeled or sulcate, concave, the younger ciliate or naked, $1\frac{1}{2}$–$4\frac{1}{2}$ lin. long, the floral sometimes larger, scarious or sepaloid; flowers in terminal clusters, rarely lax; pedicels $1\frac{1}{2}$–2 (rarely 3) lin. long; bracts, two median or nearly so, third basal or wanting, spathulate or oblanceolate, mostly naked, rarely softly ciliate, $1\frac{1}{2}$–4 lin. long; sepals widespreading, imbricate below, a little longer or a little shorter than the corolla, rarely about $\frac{1}{2}$ its length, obovate or lanceolate, acute, keeled or keel-tipped, very smooth, sometimes shining bright rosy to white, $1\frac{1}{2}$–4 lin. long; corolla varying between tubular-urceolate and ovoid-urceolate, never much inflated at the base or contracted at the throat, bright red to white, $1\frac{1}{2}$–5 lin. long; segments mostly ovate, acute or acuminate, sometimes short and broader than long, discoloured or at length brownish, mostly from $\frac{1}{4}$–$\frac{1}{3}$ the length of the tube; anthers from cuneate to oblong, $\frac{1}{3}-\frac{2}{5}$ lin. long, crested; pore $\frac{1}{2}$ the length of the cell or more; crests lanceolate to ovate, incised or serrulate, about $\frac{1}{2}$ the length of the cell; style included; stigma capitate. *Berg. Pl. Cap.* 108; *Bauer, Exot. Pl. t.* 21; *Benth. in DC. Prodr.* vii. 652. *E. corifolia, Thunb. Diss. Erica,* 46, *partly; see Rach in Linnæa,* xxvi. 779. *E. corifolia, var. spicata, Wendl. Eric. Ic. fasc.* 24, 185, *t.* 70. *E. calycina, Andr. Heathery, t.* 8, and *Col. Heaths, t.* 8; *Wendl. l.c. fasc.* 10, 11; *Lodd. Bot. Cab. t.* 594. *E. calycina, var. melastoma, Andr. Heathery, t.* 254. *E. articularis, Curt. Bot. Mag. t.* 423. " *E. tunicata, Bartl. in Linnæa,* vii. 636, *and E. Alopecias, Tausch in Flora,* 1837, 498," *acc. to Benth. E. bracteata, Thunb. Diss. Erica,* 13. *E. obvallaris, Salisb. in Trans. Linn. Soc.* vi. 386, *and* " *E. hyssopifolia, Salisb. l.c.* 387, *and E. obcordata, Sinclair, Hort. Eric. Wob.* 17," *acc. to Benth. E. polygalæflora, Klotzsch in Linnæa,* xii. 537. *E. patula, Klotzsch, l.c.* 538. *E. calycanthoides, Klotzsch, l.c.* 539. *E. obtecta, Tausch in Flora,* 1839, 638. *E. togatoides, Forbes, Hort. Wob.* 86. *E. erectiuscula, Wendl. ex Klotzsch, l.c.,* 536. *E. pigra, Soland. ex Salisb. in Trans. Linn. Soc.* vi. 387.

COAST REGION: frequent on the plains and mountains from Tulbagh Div. to George Div., *Thunberg, Burchell,* 7661! 7938! *Ecklon & Zeyher,* 3267! 3268! 3271! *MacOwan & Bolus, Herb. Norm. Austr.-Afr.,* 21! 1304! *Guthrie,* 1305! 2282! 2287! 2502! 2794! 3613! 4591! *Bolus,* 1199! 4172! 5431! 6742! 6895! 7512! 8494! 9168! *Schlechter,* 2267! 5414! 7262! 9508! 10360! *Galpin,* 3615! 3617! 4941! *Wolley Dod,* 295! 894! 904!

A very variable species, which has been unduly divided by the earlier authors, and even Bentham observed under *E. corifolia:* " A variable species, nor is it separated by any certain characters from its neighbours, since it is connected by intermediate forms with the preceding species [*E. bracteata, Thunb.*] and with the four following [*E. polygalæflora, patula, nodiflora,* and *teretiuscula*]." We have examined a very large suite of specimens, and in the result find that it runs into three chief forms, which cannot even be separated into varieties, since they are all connected by intermediates. These are :

A. Corolla $3\frac{1}{2}$–5 lin. long; floral leaves often larger and sepaloid; sepals from a little shorter than, to a little exceeding the corolla. This form answers broadly to *E. bracteata,* Thunb.

B. Corolla mostly 2–3 lin. long; floral leaves like the cauline; sepals from $\frac{3}{4}$ of the corolla to equal to it in length. This seems to be *E. corifolia* as understood by Linnæus and the older authors, and as figured by Bauer, by Andrews, *l.c. t.* 8, and by Wendland, *l.c. fasc.* 24, *t.* 70.

C. Corolla $1\frac{1}{2}$–2 lin. long, its segments sometimes, but not always, obtuse and broader than long; sepals often white or pallid, in length as in B; leaves often all opposite, or 3-nate and opposite on the same plant. These answer mostly to *E. patula,* Klotzsch, and in part to *E. polygalæflora,* Klotzsch, besides others.

The nearest allies to the species, as we understand it, are *E. nodiflora* and *E. articularis.*

395. E. nodiflora (Klotzsch in Linnæa, xii. 539, not of Salisb.); erect, 6–12 in. high, entirely glabrous; branches numerous, ascending, slender, sometimes filiform; leaves 3-nate or rarely opposite, mostly about as long as the internodes, or occasionally imbricate, linear or narrow-oblong, acute, keeled, sulcate, $1-2\frac{1}{2}$ lin. long;

flowers 3-nate, or occasionally (by arrest of lateral branchlets) pseudo-axillary; pedicels 1–2 lin. long; bracts 2, rarely 3, remote, usually small and often very small; sepals spreading, obovate or suborbicular, obtuse, with a keel-like discoloured apiculate tip, 1–1¼ lin. long, reaching from ⅓–⅔ the height of the corolla; corolla conical-urceolate, or in over-mature specimens (as Zeyher's 3287) ovoid-urceolate, very broad at the base, and more or less (often much) contracted at the throat, of thin and hyaline texture, rosy or pale red, 1¼–1½ lin. long; segments in the living state spreading, not stellate-patent, at length connivent, ovate, acute, short, more or less discoloured, about ¼ the length of the tube ; anthers oblong, or oblong-cuneate, obtuse, from less than ¼ to nearly ⅓ lin. long, crested ; pore ½–⅔ the length of the cell; crests variable, ovate, acute, serrulate or incised, ⅓–½ the length of the cell; style included; stigma capitate ; ovary subturbinate. *Benth. in DC. Prodr.* vii. 652.

VAR. β, **delapsa** (Bolus); leaves somewhat longer, up to 2½ lin. long; bracts a little larger and more coloured; corolla 1½–2 lin. long, the upper part of the tube at length falling down, making a fold or wrinkle just above the middle.

COAST REGION : Caledon Div.; sandy places near Onrust River, under 500 ft., *Zeyher*, 3287! *Schlechter*, 10388! Bredasdorp Div.; hills near Elim, *Bolus*, 8457! Ratel River Valley, *Guthrie*, 3792! Swellendam Div.; Potte Berg, *Mund*, 25! Var. β : Worcester Div.; on the Matroos Berg, about 5000 ft., *Bolus in Herb. Guthrie*, 3952!

396. **E. palliiflora** (Salisb. in Trans. Linn. Soc. vi. 351); erect, 1–1½ ft. high; branches ascending or spreading, the upper angular; leaves 3-nate, erect or adpressed, imbricate or shorter than the internodes, linear or linear-lanceolate, keeled, 1½–3 lin. long; flowers 3-nate, mostly clustered; pedicels mostly glandular or puberulous, 1–2½ lin. long; bracts subremote, lax, linear to lanceolate, 1–2 lin. long; sepals erect or slightly incurved, not spreading, ovate, acuminate, the alternate pair a little narrower, keeled, in the dried state more or less transversely wrinkled across the keel, about 2 lin. long, usually about as long as the corolla; corolla campanulate-tubular, sometimes marked with 12 more or less prominent parallel veins, rosy or white, 2–2¼ lin. long; segments slightly spreading, rounded, concolorous, about ⅓ the length of the tube; anthers longitudinally subsemiovate, subacute; cells deeply parted, smooth, ½ lin. long, crested; pore ⅔ the length of the cell; crests sublanceolate, incised, ⅖ the length of the cell ; style included ; ovary ovoid, with a conical projection at the apex, depressed in the centre. *Benth. in DC. Prodr.* vii. 651. *E. pallida, Wendl. Eric. Ic. fasc.* 25, 11, *t.* 5, *not of Salisb. E. candida, Soland. ex Salisb. l.c.*

COAST REGION : Cape Div.; mountains near and south of Cape Town, at 300–1500 ft., *Masson*, 70! *Bolus*, 4654! 7002! 7918! *Guthrie*, 649! *Schlechter!* *Guthrie*, 646! *Wolley Dod*, 1874! 2877! Caledon Div.; mountains between Caledon and Elim, 400 ft., *Bolus*, 6743! Riversdale Div.; on the Langeberg Range, 1000 ft., *Schlechter*, 1771! George Div.; Montagu Pass, 4200 ft., *Schlechter*, 5847!

397. E. nigrimontana (Guthrie & Bolus); erect, 1 ft. or more high, glabrous except on the younger branches and the ovary; branches ascending, rather straight, subfastigiate, puberulous, glabrescent; leaves 3-nate, erect, adpressed, not crowded, about as long as the internodes, linear-trigonous, obtusely acute, sulcate, the younger minutely gland-ciliate, $1\frac{1}{2}$–2 lin. long; flowers 3-nate; pedicels subviscidulous, about $1\frac{1}{2}$ lin. long; bracts, 2 subapproximate, 1 more remote, ovate, cartilaginous, viscidulous, rosy, about 1 lin. long; sepals like the bracts but broad-ovate, keel-tipped, imbricate at the base, about 2 lin. long, reaching nearly to the top of the corolla; corolla urceolate-campanulate or globose-suburceolate, not much constricted at the throat, viscidulous, rosy, 2 lin. long; segments spreading or stellate-patent, or sometimes subrecurved, broadly rounded, concolorous, a little over $\frac{1}{2}$ the length of the tube; filaments dilated and bent just below the anther; anthers manifest, about as long as the corolla, dorsifixed at or near the base, oblong, obtuse, incurved, about $\frac{5}{8}$ lin. long, crested; pore $\frac{2}{3}$ the length of the cell; crests free, lanceolate in outline, deeply inciso-partite; lobes linear, the whole less than $\frac{1}{2}$ the length of the cell; style shortly exserted, thickened towards the apex; stigma clavate-capitellate; ovary glabrous; ovules much compressed.

COAST REGION: Caledon Div.: on the Zwart Berg, between 1000 and 3000 ft., *Zeyher*, 3278! *Bodkin in Herb. Bolus*, 6745!

Allied to and resembling *E. brevifolia*, but differs by (1) more adpressed and narrower leaves, (2) wider sepals, (3) wider corolla, (4) anthers and appendages, (5) exserted style. The flowers are somewhat like those of *E. selaginifolia*, but in this the pedicels are much shorter, the sepals much smaller and not foveolate at the base. It has also been confused in herbaria with *E. lucida*, which it approaches in structure; but the habit of this is different by its more fastigiate growth, the flowers larger and viscidulous.

398. E. melanacme (Guthrie & Bolus); erect, 1 ft. or more high; branches slender, the younger subtrigonous by the prominent decurrent leaf-cushions; leaves 3-nate, erect, somewhat adpressed, the uppermost imbricate, the lower equalling or shorter than the internodes, narrow-oblong or narrow-elliptic, blunt, sulcate, viscid-pubescent becoming glabrous, softly ciliate, about 2 lin. long; flowers 3-nate, somewhat clustered; pedicels thinly pubescent, viscidulous, $3\frac{1}{2}$–4 lin. long; bracts lax, 2 subapproximate, 1 inframedian, lanceolate, all or the lowest tipped with a green keel, ciliate, viscid, $1\frac{1}{2}$ lin. long; sepals somewhat unequal, the outer pair ovate, the inner lanceolate, all softly ciliate, viscid, tipped with a sulcate, green keel, more or less transversely wrinkled in the dried state, rosy below, deeper upwards, dark purple at the apex, about 2 lin. long; corolla suburceolate, very slightly contracted at the throat, prominently 12-veined, subviscid, 2 lin. long; segments almost invariably spreading or reflexed, rounded, erosulate, about $\frac{1}{6}$ the length of the tube; anthers narrow subelliptical, wider at the base, slightly recurved, about $\frac{2}{5}$ lin. long, crested; pore $\frac{1}{2}$ the length of the cell;

crests suborbicular, deeply incised, a little over ½ the cell in length; style included, short; ovary ovoid, with a conical apex, nearly as in *E. palliifolia*, very pallid, almost glabrous.

COAST REGION: Bredasdorp Div.; hills near Elim, 300 ft., *Schlechter*, 9637 ! *Bolus*, 8506!

Closely allied to, and greatly resembling *E. palliiflora*. It presents the following differences: the leaves are smaller and blunter, the younger pubescent, and also (as are the bracts and sepals) softly ciliate; bracts larger, tipped with a green keel and (together with the sepals and corolla) more or less viscidulous; sepals incurved, and more adpressed (so that the outline of the flower is more ovate) much more obtuse, their apex always dark-coloured; corolla more inflated below, limb smaller, more spreading and dark-coloured; anthers and crests somewhat different. This species is notable as affording the nearest transition to the § *Trigemma*.

Section XXXIV. EURYSTEGIA. (Sp. 399–408.)

399. E. lanuginosa (Andr. Heathery, t. 122); erect, about 1 ft. high; branches subflexuous, glabrous; leaves 3-nate, spreading, incurved, or at length recurved, linear, acute, the younger ciliate, 10–12 lin. long; flowers 3-nate or often solitary, cernuous; pedicels puberulous, 2–3 lin. long; bracts approximate, ovate, acute, scarious, keeled, minutely velvety as are the sepals and corolla, 3–4 lin. long; sepals imbricate, erect, ovate, acute, scarious, keeled near the apex, ciliate or naked, 6 lin. long; corolla ovoid-inflated at the base, 4-parted nearly to the base, dull red-brown, 7–9 lin. long; segments erect, ovate-acuminate, much tapering to the apex, acute, alternating at the base with short triangular, acute, reflexed teeth, situated in the sinuses, convex on the upper, concave on the lower surface; filaments much curved, about 4 lin. long; anthers included, dorsifixed above the attenuated base, but appearing as if somewhat terminal, broad-linear, subobtuse, smooth, $1\frac{1}{3}$–$1\frac{1}{2}$ lin. long, aristate; pore about ½ the length of the cell; awns setiform, decurrent for a short distance along the filament, thence free and spreading, nearly as long as the cells; style elongate, a little shorter than the corolla-segments; stigma subsimple, small; ovary glabrous. *Benth. in DC. Prodr.* vii. 623; *Andr. Col. Heaths, t.* 180. *E. fuscata, F. G. Dietr. ex Steud. Nom. ed.* I. 305, *acc. to Ind. Kew.*

COAST REGION: Caledon Div.; mountains near the mouth of the Klein River, *Masson*, 110 ! Klein River Mountains, *Zeyher*, 3240 !

The teeth between the sinuses of the corolla are very peculiar when full-grown, but are not easily distinguishable in young or badly-dried specimens; we have not seen a corolla in any other species similarly furnished, and the character is not well portrayed in Andrews' figure.

400. E. Bodkinii (Guthrie & Bolus); erect, 1–2 ft. high; branches ascending, the younger puberulous, glabrescent; leaves 3-nate, erect and imbricate or somewhat spreading-incurved, linear-subulate, acute, mucronate, sulcate, thinly pilose and softly ciliate, becoming glabrous and naked, 6–11 lin. long; flowers 1–3-nate, cernuous; pedicels cano-pubescent, reddish, 4–5 lin. long; bracts remote, linear to ovate, pubescent, subscarious, 1–2½ lin. long; sepals ovate, acutely keel-tipped, thick and rigid, concave, finely and closely tomentose, dull red, 3–5 lin. long, mostly from ⅔–¾ the length of the corolla; corolla from subcyathiform to ovate, 4-partite very nearly to the base; segments lanceolate, oblong-lanceolate, obtusely acute, suberect, at length connivent, externally puberulous, dull pink or nearly white, 4–6 (mostly about 5) lin. long; filaments strongly bent at the base, much bent and dilated at the apex, about 2½ lin. long; anthers lateral, dorsifixed well above the base, but (owing to the narrowness of the cells at the base) appearing sub-terminal; cells deeply partite and subdistant, oblong, about 2 lin. long, crested; pore about ½ the length of the cell; crests partially adnate to the filament above, then free and spreading, oblong, lacerate, subtruncate, ⅔–¾ the length of the cells; style included, or rather subexserted; stigma capitellate.

COAST REGION, at 400–1200 ft. : Bredasdorp Div.; mountains between Elim and Napier, *Bodkin in Herb. Bolus,* 8456! mountains between Fairfield and Elim, *Bolus,* 6892! near Napier, *Schlechter,* 9660! Caledon Div.; Shaws Mountain, near Caledon, *Bodkin!*

Habit and general appearance of *E. holosericea,* from which it is to be known by its larger flowers, deeply parted corolla, with more acute segments, three times larger and differently shaped anthers. In floral structure, however, it is allied to the preceding, distinguishable by its remote bracts and smaller corolla.

401. E. Halicacaba (Linn. Plant. Rar. Afr. 11); procumbent, or erect against the steep sides of rocks; branches divaricate, flexuous, stout, woody, rigid, brittle, puberulous, densely leafy or covered with the scars of old leaf-cushions, 6–12 in. long; leaves 3-nate, very closely set, spreading or squarrose-recurved, linear, round-backed, sulcate, rigid, glabrous, 4–7 lin. long; flowers mostly solitary, some-times 2–3-nate, at length pendulous; pedicels puberulous, 2–3 lin. long; bracts approximate, ovate, acuminate, keeled, scarious, about 3 lin. long; sepals like the bracts but much larger and more adpressed, 4–5 lin. long; corolla ovoid or ovoid-tubular, inflated, glabrous, pale yellow, 8–11 lin. long, by about 3 lin. in diam. at the widest part; segments erect, and at length connivent, long, tapering, obtuse, nearly ½ the length of the entire corolla; filaments rather broad, dilated and thickened below the anther, very little bent, 5–6 lin. long; anthers oblong, acute, affixed dorsally at the very base of the deeply parted divaricate cells, 1¼–1¾ lin. long, aristate or muticous; pore nearly ⅔ the length of the cell; awns spreading and upcurved, or decurved, from ⅛–⅓ of the cell in length; style straight, shorter than the corolla-segments, but mostly manifest; stigma subsimple, small; ovary globose, rough, glabrous. *Amœn. Acad.* vi. 88, *and. Mant. Alt.* 374; *Bauer, Exot. Pl. t.* 2; *Wendl. Eric. Ic. fasc.* 6, 7; *Andr. Heathery, t.* 164, *and Col. Heaths, t.* 99; *Benth.*

in DC. Prodr. vii. 623. *E. rupestris, Salisb. in Trans. Linn. Soc.* vi. 353, *not of Andr. E. grossa, Salisb. Prodr.* 292.

COAST REGION : Cape Div. ; in rocky places on the mountains of the Cape Peninsula from Table Mountain to Simons Town, from 1000 ft. upwards, now becoming scarce : *Masson, Thunberg, Burchell,* 625 ! *Zeyher,* 3241 ! *Harvey,* 175 ! *Bolus,* 4473 ! *Guthrie,* 672 ! *Wolley Dod,* 2037 ! Bentham mentions Stellenbosch Div. as a locality, but we have not been able to find any record of stations outside of the Cape Peninsula.

The anthers are called muticous by Bentham ; yet Linnaeus describes them as aristate, and they are so figured by Bauer and by Wendland ; Andrews shows them as muticous, and we have found both forms.

402. E. monsoniana (Linn. fil. Suppl. 223) ; 2–4 ft. high ; branches erect, rather straight, when young covered with a floccose pubescence of minutely plumose hairs ; leaves 3-nate, mostly erect and closely imbricate, sometimes subspreading, linear to oblong or narrow-elliptic, sulcate, shortly ciliate, glabrous, smooth, shining, the floral often dilated, with a scarious margin, 2–3 lin. long ; flowers 3-nate, on short branchlets for some distance along the branches, at length cernuous ; pedicels floccose, $1\frac{1}{2}$–2 lin. long ; bracts approximate, broad ovate, 4–5 lin. long ; sepals erect, oval-oblong, obtuse, glabrous, white, 6–7 lin. long ; corolla elongate-ovoid-urceolate, contracted at the throat, glabrous, white, 9–11 lin. long ; limb nearly erect, obtuse, about $\frac{1}{8}$ of the tube in length ; filaments rather broad, red-nerved, 4–5 lin. long ; anthers mostly included, linear-cuneate, subacute, smooth, shining, black, 2–$2\frac{1}{4}$ lin. long, crested ; pore $\frac{1}{3}$–$\frac{1}{2}$ the length of the cell ; crests pendulous, lanceolate, acute, serrulate, $\frac{1}{2}$ the length of the cells or less ; style included ; stigma simple ; ovary globular, glabrous, small. *Andr. Heathery, t.* 173, *and Col. Heaths, t.* 110 ; *Wendl. Eric. Ic. fasc.* 10, 9 ; *Benth. in DC. Prodr.* vii. 624. *E. Monsoniæ, Bauer, Exot. Pl. t.* 7. *E. variifolia, Salisb. Prodr.* 298, *and in Trans. Linn. Soc.* vi. 353.

VAR. *β*, **exserta** (Klotzsch in Linnæa, ix. 701) ; anthers exserted or sub-exserted, their apices spreading. *E. Monsoniæ, Bot. Mag. t.* 1915.

SOUTH AFRICA : without locality, *Herb. Salisbury !* and *cultivated specimens !* Var. *β* : *Mund & Maire.*

COAST REGION : Paarl Div. ; French Hoek, *Thunberg.* Stellenbosch Div. ; Hottentots Holland, *Thunberg ;* Lowrys Pass, 2000 ft., *Schlechter,* 4805 ! Caledon Div. ; mountains near Houw Hoek, *Guthrie,* 2040 ! near Bot River, *Guthrie,* 2290 ! Babylon's Tower Mountain, *Templeman in Herb. Norm. Austr.-Afr.,* 752 ! Worcester Div. ; near Darling Bridge, *Bain in Herb. Bolus,* 3165 ! Robertson Div. ; near Montagu, *Bain in Herb. Bolus,* 3162 ! Riversdale Div.; near Riversdale, *Herb. Bolus,* 3162 !

CENTRAL REGION : Ceres Div. ; Cold Bokkeveld, 5200 ft., *Schlechter,* 8924 !

403. E. nobilis (Guthrie & Bolus) ; erect, about 2 ft. high, entirely glabrous ; branches ascending or spreading, pallid, angular by the prominent leaf-cushions ; leaves spreading, recurved, linear, acute, sulcate, glaucous, 4–6 lin. long, the floral commonly larger and subpetaloid, pink ; flowers umbellate, at length cernuous, often

numerous; pedicels 6–8 lin. long ; bracts large, remote, lax, broad-
lanceolate or subovate, coloured, 3–6 lin. long; sepals ovate or
obovate, acute or obtusely apiculate, often with a broad white
margin, rosy, 5–6 lin. long; corolla elongate-ovoid, or subovoid-
urceolate, more rarely the tube somewhat conical above the base,
generally much narrowed at the throat, 8–10$\frac{1}{2}$ lin. long, rosy ; segments spreading, ovate, obtuse, about $\frac{1}{8}$–$\frac{1}{7}$ of the tube in length;
filaments narrow, not rigid, rising well above the ovary, 4–6 lin. long ;
anthers oblong-cuneate, acute, about 1 lin. long, crested; pore $\frac{1}{3}$–$\frac{1}{2}$
the length of the cell ; crests affixed well above the base of the cell,
straightly pendulous, lanceolate to oblong, free from the filament,
variously incised, as long or nearly as long as the cells; style
included, sometimes just manifest, somewhat 4-angled ; stigma
capitellate.

Coast Region : Clanwilliam Div.; Cederberg Range, *Leipoldt,* 216!
Marloth ! Herb. Bolus, 6325 ! *Mader,* 209!

Allied to *E. glauca,* and much resembling our var. β of that species.

404. E. glauca (Andr. Col. Heaths, t. 24) ; stout, erect, entirely
glabrous, 2–3 ft. high ; branches rigid, ascending ; leaves 3-nate,
from suberect to spreading, mostly curved, linear-semiterete, acute or
mucronate, sulcate, glaucous, the floral often larger and subpetaloid,
3–6 lin. long; flowers in 4–7-flowered umbels, at length cernuous;
pedicels mostly about 4 (rarely 10–12) lin. long ; bracts remote, the
lowest subbasal, laxly spreading, ovate ; sepals spreading, elliptic
or subovate, obtusely acute, keel-tipped, about 3 lin. long, or $\frac{2}{3}$ the
length of the corolla more or less ; corolla subglobose-urceolate, with
a rather long neck, not much contracted at the throat in the mature
state, but becoming more so as the capsule ripens, 4–6 lin. long, of
very thick texture and becoming indurated in age, variable in colour,
dull red and more or less livid purple about the throat and limb, or
pallid with green tips, or entirely green ; segments short, spreading,
about $\frac{1}{5}$–$\frac{1}{4}$ of the tube in length; filaments broad, rigid, strongly
bent below the anther, 1$\frac{1}{2}$–2$\frac{1}{4}$ lin. long, about as long as the capsule,
so that the anther-crests rest upon its apex ; anthers oblong-cuneate,
acute or subacute, 1–1$\frac{1}{3}$ lin. long, crested ; pore over $\frac{1}{2}$ the length of
the cell ; crests spreading, broad-ovate or semiorbicular, affixed at the
base of the cell and adnate to the upper part of the filament, or
partially free at the apex, finely crenulate or incised, from $\frac{1}{3}$–$\frac{2}{3}$ the
length of the cells ; capsule globular, glabrous. *Andr. Heathery,*
t. 25 ; *Bot. Mag. t.* 580 ; *Wendl. Eric. Ic. fasc.* 26, 9; *Benth.*
in DC. Prodr. vii. 654. *E. elegans, Bot. Mag. t.* 966, *not of*
Andr.

Var. β, **elegans** (Bolus); aspect of *E. nobilis ;* sepals nearly as long as the
corolla-tube, longer than in the typical form, somewhat less spreading, and mostly
of paler colour ; corolla like the type, but sometimes reaching to 5$\frac{1}{2}$ or 6 lin. in
length ; for the rest as in the type. *E. elegans, Andr. Heathery, t.* 111, *and*
Col. Heaths, t. 165 ; *Benth. in DC. Prodr.* vii. 654.

South Africa : without locality, *cultivated specimens* of the type and
var. β!

COAST REGION : Clanwilliam Div. ; Brakfontein, near Oliphants River (*Ecklon & Zeyher?* in Cape Govt. Herb.) ! Ceres Div. ; Skurfdeberg Range, *Masson,* 134! near Ceres, *MacOwan & Bolus, Herb. Norm. Austr.-Afr.,* 1079! *Bolus,* 7343 ! Piquetberg Div. ; top of the Piquetberg Range, *Niven,* 62! Paarl Div. ; Drakensteen Mountains, near Wellington, *MacOwan, Herb. Norm. Austr.-Afr.,* 23 ! Var. β : Paarl Div. ; on Sneuwkop Mountain, near Wellington, 4900 ft., *Miss Cummings in Herb. Bolus,* 5853 ! *Marloth,* 651 ! Drakensteen Mountains, *MacOwan,* 2416 !

CENTRAL REGION : Ceres Div. ; Cold Bokkeveld, *Schlechter,* 8923 !

Allied to *E. nobilis,* to which our var. β bears so great an external resemblance as often to have been confused with it. It may possibly be a natural hybrid between the two species.

405. E. lanipes (Guthrie & Bolus) ; the branch before us 7 in. long, virgate ; side branches few, erect, straight, swollen at the nodes, very pallid, mostly glabrous, or slightly pubescent near the summit of the flowering branches ; leaves 3-nate, erect, adpressed, slightly imbricate, linear-lanceolate, subobtuse, concave above, sulcate-keeled, glabrous, 3–3½ lin. long ; flowers 3-nate, cernuous ; pedicels decurved, stout, thickened upwards, densely covered with coarse woolly barbellate white hairs, 6–7 lin. long ; bracts approximate, large, ovate, scarious, coloured, concave, amplexicaul, 3–3½ lin. long ; sepals slightly spreading, ovate, acute, subimbricate at the base, cartilaginous, coloured, 4 lin. long ; corolla ovoid-suburceolate, not much contracted at the throat, glabrous, rosy or lilac, 4½–5 lin. long ; segments erect (in the dried specimens, probably somewhat spreading in the living plant), cordate-ovate, imbricate at the base, acute, $\frac{3}{8}$–$\frac{3}{7}$ of the tube in length ; filaments broad, rigid, much bent below the anther, about 2 lin. long, considerably longer than the ovary ; anthers narrow-oblong, subacute, about 1 lin. long, crested ; pore about ½ the length of the cell ; crests obovate, obtuse or truncate at their apex, serrulate or variously toothed, inserted above the base of the cells, free from the filament, nearly ½ the length of the cells ; style straight, angular ; stigma clavate-capitellate ; ovary smooth, pallid, with an apical conical projection.

COAST REGION : Caledon Div. ? Our only example was a living specimen from the Caledon " Wild-Flower Show," 1896, *Herb. Bolus,* 6894 !

A very distinct species, with the habit and foliage of § *Lamprotis,* but with floral characters so similar to those of *E. glauca,* though quite unlike in general appearance, that they must be placed in juxtaposition. The anther-crests are almost those of the latter, but they are placed well above the base of the cell, and free, as in *E. nobilis ;* while it is immediately known from both by its much smaller, erect and adpressed leaves. It is to be regretted that our material (though good) is so small ; but a fragment sufficient for identification has been sent to the Kew Herbarium.

406. E. papyracea (Guthrie & Bolus) ; erect, reaching 7–8 ft. high (*Schlechter*) ; branches stout, rather straight, puberulous ; leaves 4-nate, erect-incurved, crowded, densely imbricate, linear-subulate, acuminate, keeled, glabrous, 4–5 lin. long ; flowers 4-nate, sometimes clustered and umbellate, mostly at length cernuous ; pedicels puberulous or villous, 3–4 lin. long ; bracts subapproximate, spreading, lax, broad-ovate, acute or acuminate, keeled, papery, white, $2\frac{1}{2}$–$3\frac{1}{2}$ lin. long; sepals like the bracts, but nearly erect, imbricating for about $\frac{1}{2}$ their length more or less, much wrinkled in the lower part (in the dried state), 4–5 lin. long, mostly reaching $\frac{1}{2}$–1 lin. lower than the corolla or sometimes equalling it; corolla varying between campanulate, broad-cyathiform and subobconic, always widening to the mouth, glabrous, $4\frac{1}{2}$–5 lin. long, apparently varying from pale rose to nearly white; segments erect or very slightly spreading at maturity, shortly afterwards becoming connivent and giving the corolla a somewhat ovoid shape, semiovate, about $\frac{1}{2}$ the tube in length; filaments tapering upwards from a broader base, sigmoid, 1–$1\frac{1}{4}$ lin. long; anthers dorsifixed well above the base, broad-linear, much tapering to the acuminate point, and produced sometimes much beyond the pore, scabrid, about $1\frac{1}{2}$ lin. long, or sometimes a little more, aristate; pore about $\frac{1}{2}$ the length of the cell; awns upcurved, $\frac{1}{5}$–$\frac{1}{4}$ of the length of the cell; style slender; stigma simple; ovary glabrous.

COAST REGION: Caledon Div. ? from the Caledon " Wild-Flower Show," 1893, *Herb. Guthrie*, 2148! and 1897, *Herb. Bolus*, 6383! Riversdale Div.; Garcias Pass, 2000 ft., *Bain in Herb. Bolus*, 3163! *Schlechter*, 1917! *Galpin*, 3622!

Allied to *E. inclusa*, but differs a good deal in appearance by its stouter habit, much longer and acuminate leaves, larger flowers, differently coloured sepals, and aristate (but otherwise very similar) anthers. Has been confused with *E. holosericea* from which, however, it is even more distinct.

407. E. holosericea (Salisb. in Trans. Linn. Soc. vi. 352); erect, apparently 1–$1\frac{1}{2}$ ft. high; branches rather stout, puberulous, glabrescent; leaves 3-nate, suberect, linear-subulate, acute, mucronate, pilose-pubescent or glabrous, the younger ciliate, 5–8 lin. long; flowers 3-nate, or sometimes solitary, cernuous; pedicels puberulous, $2\frac{1}{2}$–4 lin. long; bracts remote, ovate-lanceolate, membranous, puberulous, about 1 lin. long; sepals ovate, acute, imbricate, puberulous, mostly somewhat spreading at full maturity and shorter than the corolla, rosy red, $2\frac{1}{2}$–$4\frac{1}{2}$ lin. long; corolla campanulate-cyathiform, sub-4-gonous, glabrous or less commonly puberulous, rosy, $3\frac{1}{2}$–5 lin. long; segments slightly spreading, at length connivent, ovate, $\frac{1}{2}$–$\frac{2}{3}$ of the tube in length; filaments about $1\frac{1}{2}$ lin. long; anthers oblong, obtuse, dorsifixed above the base, scabrid, about $\frac{2}{3}$ lin. long, crested; pore over $\frac{1}{2}$ the length of the cell; crests free, straight or spreading, oblong, toothed towards the apex, from $\frac{1}{2}$–$\frac{3}{4}$ the length of the cell; stigma small, clavate-capitellate; ovary on a prominent dark-coloured disk. *E. andromedæflora, Andr. Heathery, t.* 151, *and Col. Heaths, t.* 146; *Lodd. Bot. Cab. t.* 521; *Benth. in DC. Prodr.* vii. 654. *E. pomifera, Hort. ex Benth. l.c.*

The following appear to be mere colour varieties, or, as Bentham supposed, garden hybrids:—*E. triumphans, Lodd. Bot. Cab. t.* 257. *E. andromedæflora, var triumphans, Bot. Mag. t.* 2322. (So far as we know they have not been met with in a wild state).

Var. β, **parviflora** (Bolus); almost exactly as in the forms described and figured in the plates first above cited, except that all parts of the flower are smaller : sepals 1½ lin. long; corolla 2 lin. long.

South Africa : without locality, *Roxburgh!* and *cultivated specimens!*
Coast Region, between 1000 and 1500 ft. : Caledon Div. ; Klein River Mountains, *Masson, Niven! Galpin,* 3621 ! *Schlechter!* mountains near Caledon, *Bolus,* 9162 ! Var. β : Bredasdorp Div. ; mountains near Elim, *Bolus,* 6737 !

408. E. Grisbrookii (Guthrie & Bolus); 1 ft. or more high ; branches stout, ascending, pubescent, channelled by the prominent decurrent leaf-cushions ; leaves 3-nate, spreading or occasionally recurved, linear, acute, deeply sulcate, softly ciliate, becoming naked, 6–7 lin. long ; flowers solitary, terminal, or by arrest of the branchlets also axillary, at length subcernuous ; pedicels puberulous, 4–5 lin. long ; bracts very remote, below the middle or subbasal, lanceolate, acute, scarious, about 2 lin. long ; sepals erect, incurved, closely imbricate, cordate at the base and inserted upon the enlarged angular apex of the pedicel, ovate to suborbicular, shortly acute and apiculate, strongly nerved (nerves prominent on the inner surface), rigid, cartilaginous, glabrous, 4–5 lin. long, and reaching from $\frac{3}{4}-\frac{7}{8}$ the height of the corolla ; corolla conical-ovoid, or perhaps in the living state globose-urceolate, 4-gonous below, thick in texture, glabrous, 5–5½ lin. long, white ; segments oblong, obtuse, erect (in the dried specimens), about ½ the tube in length or somewhat less ; filaments rather broad, bent below the anther ; anthers cuneate or ovate-cuneate, subacute, black, $\frac{3}{4}-\frac{7}{8}$ lin. long, crested ; pore ⅓ the length of the cell; crests orbicular, more or less deeply incised, black, about ½ the length of the cells; style 4-gonous, included ; stigma subsimple ; ovary on a broad disk.

Coast Region : Caledon Div.; mountains near Zoetemelks Vlei, *Grisbrook in Herb. Guthrie,* 2744!

We have also seen specimens exhibited at the Caledon and other Wild-Flower Shows, without locality.
This resembles *E. Bodkinii* and *E. holosericea;* but differs in floral characters from both, and especially in the anthers and their appendages. In the coarse, strong habit of its growth it seems to be like *E. gigantea.* But our specimens of this are but small branches.

Section XXXV. ADELOPETALUM. (Sp. 409.)

409. E. Nabea (Guthrie & Bolus) ; erect, 3–5 ft. high ; branches virgate, the younger villous, glabrescent, the older with many short lateral, floriferous branchlets ; leaves 3-nate, erect-spreading, linear-subulate, acuminate, sulcate, ciliolate, 5–7 lin. long ; inflorescence terminal, flowers 1–2 on the short branchlets, mostly in long dense leafy pseudo-racemes, or more rarely somewhat lax and interrupted on longer branchlets, suberect to spreading, subsessile, calycine ; bracts closely approximate, ovate to lanceolate, like the sepals glumaceous, 3–4 lin. long; sepals in opposite pairs, the outer pair lanceolate, acute or acuminate, the inner pair more oblong and more obtuse, concave, 7–8 lin. long, much exceeding the corolla; corolla subovate, 4-fid (or at length, being ruptured by the swelling ovary, 4-partite), $1\frac{1}{3}$–$1\frac{1}{2}$ lin. long; segments erect, ovate, obtuse, about as long as the tube; stamens 5–6 lin. long, shorter than the sepals, much exceeding the corolla; anthers terminal, longitudinally semi-lanceolate, acute, smooth, submembranous; cells deeply partite, about $\frac{3}{4}$ lin. long; pore nearly as long as the cell; style subexserted, hooked; stigma capitate; ovary glabrous; seeds broadly margined. *Nabea montana, Lehm. Ind. Sem. Hort. Hamb.* 1831, 5, *name only; Klotzsch in Linnæa,* viii. 667. *Macnabia montana, Benth. in DC. Prodr.* vii. 612.

COAST REGION: George Div.; Montagu Pass, 3500 ft., *Schlechter,* 5820! Devils Kop, near George, *Mund,* ex *Klotzsch.* Uniondale Div.; mountains between Avontuur and Vlugt, *Bolus,* 2381! Uitenhage Div.; Van Stadens Berg, *Zeyher,* 3308!

There seems no sufficient reason for retaining the genus *Nabea.* It has no characters which are not found in a greater or lesser degree in several species of *Erica;* and the only one for which it is remarkable is the great length of the sepals compared with the corolla. But the range of variation in the relative size and length of these organs in the genus (as generally admitted) is very great, and it does not seem well to uphold the separation on that ground. The much compressed and wide-margined seeds connect the species with the § *Platyspora,* but from that it is separated both by its calycine flowers, and the very different shape of its corolla. In some respects it approaches to the § *Eurystegia.* The name *E. montana,* having been pre-occupied, it is necessary to give the plant a new specific name, and it seems better to revert to the generic name of Lehmann.

Section XXXVI. TRIGEMMA. (Sp. 410–431.)

410. E. plumigera (Bartl. in Linnæa, vii. 636); $1\frac{1}{2}$ ft. or more high; stem erect, not much branched; branches subflexuous, finely floccose with minutely plumose or barbellate hairs; lateral branchlets $\frac{1}{2}$–1 in. long, numerous, cernuous, floriferous; leaves 3-nate, erect or subspreading, imbricate, linear to oblong, subtrigonous, keeled, ciliolate or naked, 1–$1\frac{1}{2}$ lin. long; flowers 3-nate, numerous at the ends of short decurved branches; pedicels cernuous, floccose, about 1 lin. long; bracts 3, approximate, elliptic, apiculate, about 1–$1\frac{1}{2}$ lin. long; sepals elliptic, acute, keeled, subcomplicate, approximate or imbricate, $1\frac{3}{4}$–2 lin. long, at full maturity a little lower than the top of the corolla; corolla urceolate or conical-urceolate, somewhat contracted at the throat, reddish, $1\frac{1}{2}$–2 lin. long; segments variable in length, spreading-recurved, but not stellate-patent; filaments rather broad, tapering to the apex; anthers included, lateral, sub-cuneate, $\frac{2}{5}$ lin. long, the length $2\frac{1}{4}$ times the greatest width, crested; pore about $\frac{1}{2}$ the length of the cell; crests affixed rather high above the base of the cell and half its length, ovate or suborbicular, serrulate with a terminal longer subulate lobe; style subincluded, sometimes just manifest above the corolla-tube; stigma clavate;

ovary globose, minutely hairy or scabrous; ovules pallid, lenticular or flat, not margined. *E. lamprotes, Klotzsch in Linnæa,* xii. 536; *Benth. in DC. Prodr.* vii. 651.

COAST REGION : Cape Div.; amongst bushes on the eastern side of the Devils Peak, between 1000 and 2000 ft., *Ecklon & Zeyher,* ex *Bartling.* Caledon Div.; mountains above Palmiet River, *Zeyher,* 3274! *Guthrie,* 4167! Caledon Wild-Flower Show, *Herb. Bolus,* 6835! (the only fully matured specimens we have seen).

We have not seen Bartling's type, but the chief of the peculiar characters of this species are so well described by him that we can have no doubt of the identity of the specimens cited with his type.

411. E. inclusa (Wendl. fil. ex Benth. in DC. Prodr. vii. 654); 6–10 in. high; branches few, slender, ascending or erect, sometimes subvirgate, puberulous; leaves 3-nate, crowded, suberect and imbricate, or spreading or squarrose, linear, blunt, glabrous, the younger minutely ciliolate, $1\frac{1}{2}$–2 lin. long; flowers sub-3-nate on short branchlets, by arrest of branchlets sometimes appearing axillary; pedicels puberulous, about 3 lin. long; bracts approximate, lax, ovate, membranous, $1\frac{1}{2}$–2 lin. long; sepals erect or subspreading, ovate, acute, keel-tipped, submembranous, about $2\frac{1}{2}$ lin. long; corolla at full maturity campanulate, afterwards becoming somewhat ovate, $3\frac{1}{2}$ lin. long, and nearly as wide when flattened; segments slightly spreading, semiorbicular, $\frac{2}{5}$ the tube in length, the whole flower a delicate pale rose, corolla darker than the sepals; filaments $\frac{1}{2}$–$\frac{3}{4}$ lin. long, straight and spreading at a low angle, dilated at the apex; anthers included, dorsifixed above the base, broad-linear, tapering to the subacute apex, slightly recurved, scabrous, dark-coloured, 1–$1\frac{1}{2}$ lin. long, muticous; pore $\frac{1}{3}$ the length of the cell; style straight, slender; stigma subsimple; ovary globose, pallid, glabrous.

SOUTH AFRICA : without locality, *Hesse,* in Herb. Berlin.
COAST REGION : Riversdale Div.; on the Kampsche Berg, *Burchell,* 7081! 7113! Mozambique Kop, 2500 ft., *Galpin,* 3623! Swellendam Div.; without exact locality, *Tyson !*

412. E. crassifolia (Andr. Heathery, t. 257); branches flexuous; leaves 3-nate, erect-spreading, oblong, obtuse, short, fleshy, glaucous, 2–$2\frac{1}{2}$ lin. long; flowers 3-nate, on short lateral branchlets; pedicels about $1\frac{1}{2}$ lin. long; bracts remote, lanceolate, acute, coloured, spreading, smaller than the sepals; sepals narrow-ovate, acute, incurved, with wide interspaces, rosy, about $1\frac{1}{4}$ lin. long; corolla broad-urceolate or globose-urceolate, somewhat contracted at the throat, but wide-mouthed, rosy, about 3 lin. long; segments spreading or recurved, rounded, about $\frac{1}{2}$ the length of the tube; filaments curved, rather broad, at least twice the length of the anther; anther included, oblong, obtuse, muticous; pore $\frac{1}{2}$ the length of the cell; style included, short; stigma capitellate; ovary (apparently) glabrous. *Andr. Col. Heaths, t.* 227; *Benth. in DC. Prodr.* vii. 685.

SOUTH AFRICA : Locality and collector unknown.

The species is only known to us from Andrews' figure cited above, and his short description, from both of which we have drawn the preceding. Bentham thought it "perhaps some hybrid"; but he wrongly identified with it specimens in Herb. Kew., which are really *E. Corydalis, Salisb.*, a quite different species.

413. E. tegulæfolia (Salisb. in Trans. Linn. Soc. vi. 351); stout, erect, $1\frac{1}{2}$–$2\frac{1}{2}$ ft. high; branches nearly straight, pubescent; leaves 3-nate, erect and closely imbricate, or spreading, mostly incurved, linear or oblong, to broad-elliptic, obtuse, ciliolate, sulcate, keeled, mostly glabrous (or "pubescent," *Salisbury*), $1\frac{1}{2}$–3 lin. long, the floral a little larger, coloured, sepaloid; flowers 3-nate (in our wild specimens, but described by *Andrews* from cultivated plants as "umbellate 3–6-nate"), sometimes abundant and clustered; pedicels pubescent, red, $1\frac{1}{2}$–2 lin. long; bracts approximate, ovate, acute, strongly keeled, incurved and deeply concave at the apex, scarious, rigid, coloured, 2–$2\frac{1}{2}$ lin. long; sepals like the bracts but larger, much wrinkled, $2\frac{1}{2}$ lin. long, reaching to $\frac{2}{3}$–$\frac{3}{4}$ of the height of the corolla; corolla urceolate or globose-urceolate, slightly contracted at the throat, 4-gonous at the base, glabrous, rosy-red, about $2\frac{1}{2}$ lin. long; segments slightly spreading, at length connivent, ovate, $\frac{2}{3}$–$\frac{7}{8}$ the length of the tube; filaments tapering upwards from a broader base, bent below the anther; anthers included, dorsifixed above the base, subcuneate, acute, about $\frac{5}{8}$ lin. long, crested; pore $\frac{3}{4}$–$\frac{7}{8}$ the length of the cell; crests broad-ovate or suborbicular, spreading, free, margin toothed, as long as or longer than the cells; style included, thickened towards the apex, 4-gonous; stigma simple; ovary glabrous. *E. squamosa, Andr.? Heathery, t.* 91, *and Col. Heaths, t.* 207; *Benth. in DC. Prodr.* vii. 655.

CoAST Region : without locality, *Masson*, in Herb. Brit. Mus. Paarl Div.; French Hoek, *Bolus*, 6891 ! Caledon Div.; Hottentots Holland Mountains, near Lowrys Pass, *Guthrie*, 3549! 3764 !

We have not been able to find Salisbury's type in his Herbarium, but have seen Masson's specimens, in the British Museum, which are probably the same which Salisbury quotes. These agree substantially with Andrews' figure of *E. squamosa*, cited by Bentham. Both, however, appear immature and poorly grown. But in floral characters they agree so nearly with the finer specimens quoted under Guthrie, 3549 and 3764, and Bolus, 6891, that we assume them to be the same species, and have chiefly described from these latter.

414. E. gigantea (Klotzsch ex Benth. in DC. Prodr. vii. 656); 4–7 ft. high; branches stout, rigid, erect, cano-puberulous, all except the lowest parts densely leafy; leaves 4-nate, erect-spreading, crowded, imbricate, oblong, obtuse, thick, rigid, sulcate, glabrous, gland-ciliolate, 2–4 lin. long; flowers 4-nate, on short branchlets; pedicels 3 lin. long; bracts approximate, scarious, acute, the two upper longer than, and exceeding in height the sepals; sepals erect-incurved, broad-ovate and obtuse to broad-lanceolate and acute, subscarious, 2–$2\frac{1}{2}$ lin. long, somewhat shorter than the corolla; corolla urceolate, throat slightly constricted, of thick texture, white, about $2\frac{1}{2}$ lin. long; segments slightly spreading, broad, rounded, yellow, a little less than $\frac{1}{2}$ the length of the tube ; filaments tapering from a broad base, flexuous ; anthers included, dorsifixed almost at the base, cuneate, brown, about $\frac{1}{2}$ lin. long, narrow-crested ; pore $\frac{3}{4}$ the length of the cell ; crests subulate, acute, serrulate, about $\frac{1}{2}$ the length of the cell ; style included ; stigma subsimple ; ovary glabrous.

Coast Region : Riversdale Div. ; alpine dry stony places near Platte Kloof, *Masson*, 123 ! Kannaland, *Ecklon & Zeyher*, ex *Bentham* ; summit of Muiskraal Ridge, near Garcias Pass, 1500 ft., *Galpin*, 3626 !

415. E. baccans (Linn. Mant. Alt. 233) ; erect, entirely glabrous, 2–5 ft. high ; branches numerous, ascending, subflexuous ; leaves 4-nate, erect-incurved, imbricate, narrow-linear, subobtuse, sulcate, entire or obscurely serrulate, $2\frac{1}{2}$–$3\frac{1}{2}$ lin. long ; flowers terminal, 4-nate ; pedicels 3 lin. long ; bracts remote, lanceolate, keeled, scarious, coloured, 2–3 lin. long ; sepals like the bracts but broad-lanceolate to obovate-oblong, acute, imbricate, keel-tipped, very concave, about $2\frac{1}{2}$ lin. long, mostly reaching to the length of the corolla-tube ; corolla globose-urceolate, distinctly constricted at the throat, bluntly 4-gonous below with deep intermediate depressions between the angles, about $2\frac{1}{2}$ lin. long ; segments short, broad, spreading, about $\frac{1}{4}$ the length of the tube ; filaments rather broad, bent below the anther ; anthers included, lateral, broad-oblong, very obtusely rounded ; cells completely divided to below the base, each on a branch of the decurrent connective, and at length becoming subdistant, pale brown, about $\frac{1}{2}$ lin. long, crested ; pore large and wide, $\frac{2}{3}$ the length of the cell ; crests obliquely lanceolate-acuminate, serrulate, pallid, somewhat longer than the cell ; style included, very short and stout ; stigma capitellate or subsimple ; ovary glabrous. *Bauer, Exot. Pl. t. 22 ; Wendl. Eric. Ic. fasc. 6, 13 ; Andr. Heathery, t. 53, and Col. Heaths, t. 4 ; Bot. Mag. t. 358 ; Benth. in DC. Prodr. vii. 656. E. baccæformis, Salisb. in Trans. Linn. Soc. vi. 352. E. moniliformis, Salisb. Prodr. 293, acc. to Index Kew.*

Coast Region : Clanwilliam Div. ; near Wupperthal, *Wurmb*, ex *Drège ;* Cape Div. ; frequent on the mountains and occasionally on the Flats, *Thunberg, Burchell*, 43 ! 376 ! 8425 ! *Ecklon*, 281 ; *Zeyher*, 1110 ! *Drège*, 7746 ! *Mc-Gillivray*, 444 ! *Bolus*, 2969 ! *Guthrie*, 146 ! *MacOwan, Herb. Norm. Austr.-Afr.*, 24 ! Worcester Div. ; *Cooper*, 1595 ! Also *cultivated specimens !*

A well-marked species, allied to *E. tegulæfolia*, differing in corolla and anthers. Salisbury says the leaves are sometimes 3-nate ; we have not seen such. It is also allied to *E. irregularis.*

416. E. irregularis (Benth. in DC. Prodr. vii. 680) ; erect, a foot or more high ; branches subvirgate, the younger, with the pedicels, shortly and softly pubescent, otherwise glabrous ; leaves 3-4-nate on the same branch, erect-incurved, linear, semiterete, tapering upwards, somewhat obtuse, sulcate, $2\frac{1}{2}$–3 lin. long ; flowers abundant, in a somewhat long irregular pseudo-raceme, on very short

suppressed branchlets, solitary in each axil, with one or two minute
bract-like leaves, at the base of the pedicel, or occasionally also
terminal; pedicels slender, downy, 2–2½ lin. long; bracts remote,
minute, scarious, erect, the upper pair about ¾ lin. long, the lowest
smaller; sepals subspreading, not imbricate at the base, lanceolate-
oblong, acute, scarious, brown-keeled, ciliolate, about 1½ lin. long,
reaching ⅗–⅔ the height of the corolla; corolla ovoid, not (or very
slightly) constricted at the throat, with a rather wide mouth, rosy,
about 2 lin. long; limb erect; segments broadly rounded, about
¼ the length of the tube; anthers lateral, oblong-cuneate, contracted
at the base; cells ½ lin. long; pore narrow, ¾ the length of the cell;
crests lanceolate-acuminate, denticulate on the outer margin, as
long as the cell; style included; stigma capitellate; ovary
glabrous.

SOUTH AFRICA: without locality, *Mund!* in Herb. Kew.

In the structure of the flowers this comes very near to *E. baccans;* but
the different inflorescence, the minute bracts, shorter sepals, &c., serve to
distinguish it.

417. E. affinis (Benth. in DC. Prodr. vii. 656); branches straggling, diffuse, flexuous, woody, but somewhat slender for their size,
white-pubescent, 6–8 in. long; leaves 4-nate, spreading, slender,
linear, subobtuse, sulcate, nearly flat, glabrous, 2½–3½ lin. long;
flowers mostly 4-nate, occasionally 3-nate, sometimes by arrest of the
branchlets axillary; pedicels suberect, pubescent, 4–5 lin. long;
bracts remote, lanceolate, acuminate, scarious, 1½–2 lin. long; sepals
erect, not imbricate at the base, obovate or oblanceolate, shortly
acute, keeled, scarious, 2–2½ lin. long, about equalling the tube of
the corolla; corolla broad-urceolate or subglobose-urceolate, throat
slightly constricted, several-nerved, glabrous, dry, red, about 2½ lin.
long; segments slightly spreading, broad, rounded, about ⅖ the
length of the tube; filaments broad below, tapering upwards, about
1 lin. long; anthers included, lateral, longitudinally semilanceolate
to semiovate, subacuminate; cells deeply parted, about ½ lin. long,
crested; pore ⅔ the length of the cell; crests broad-ovate or suborbicular, minutely serrulate, about ⅔ the length of the cell, brown;
style at length exserted, angular, swollen a little at the apex; stigma
a short filiform point, much narrower than the truncated end of the
style; ovary glabrous. *E. Zeyheri, Bartl. in Linnœa,* vii. 635, *not
of Spreng. fil.*

COAST REGION: Uitenhage Div.; stony bed of the Zwartkops River, *Zeyher,*
1072! 3279!

418. E. propinqua (Guthrie & Bolus); erect, 3–4 ft. high;
branches white-pubescent, glabrescent, angled and the older scarred
by prominent leaf-cushions; leaves 4-nate, either erect, adpressed
and about as long as the internodes, or somewhat spreading, oblong,
or narrow-lanceolate, acute, glabrous, 2–3 lin. long; flowers 3–4-

nate or umbellate ; pedicels white-tomentose, from 2–10 lin. long, apparently elongating during and after flowering ; bracts mostly remote, rarely subapproximate, lanceolate, acuminate, strongly keeled, scarious, very concave, incurved, 1–2 lin. long ; sepals like the bracts but broader, 2–2$\frac{1}{2}$ lin. long ; corolla suburceolate, only slightly contracted at the throat, glabrous, rosy-red, 2–2$\frac{1}{2}$ lin. long ; segments spreading, oblong, obtuse, with usually more or less wide sinuses which are sometimes obtuse or semicircular at the base, from $\frac{1}{2}$–$\frac{3}{4}$ the length of the tube ; anthers subincluded, or reaching above the corolla-tube, dorsifixed at or near the base ; cells deeply partite, oblong, obtuse, scaberulous, blackish, from $\frac{1}{2}$ to nearly $\frac{2}{3}$ lin. long, aristate ; pore $\frac{1}{2}$ the length of the cell ; awns decurrent for over $\frac{1}{2}$ their length, then free and spreading, subulate, acute, about $\frac{1}{2}$ as long as the cells ; style exserted ; stigma subclavate, small ; ovary glabrous.

COAST REGION: Bredasdorp Div. ; in rocky places near the mouth of Ratel River, *Bolus*, 6728 ! 8514 ! *Schlechter*, 9721 ! hills near the Koude River, near Elim, 700 ft., *Schlechter*, 9598 !

419. E. leucodesmia (Benth. in DC. Prodr. vii. 681) ; apparently a small straggling shrub, with hoary tomentulose branches ; leaves 3-nate, erect, incurved, often fasciculate, linear, obtuse, rather thick, sulcate, uniformly covered with a fine short hoary or grey tomentum, becoming at length glabrous, 2$\frac{1}{2}$–3 lin. long ; flowers 3-nate on short lateral branchlets, at length cernuous ; pedicels white-pubescent, 2–3 lin. long ; bracts remote, small ; sepals lanceolate, acuminate, keeled, rigid, tomentulose, about 1$\frac{1}{2}$ lin. long ; corolla suburceolate, only slightly contracted at the throat, glabrous ("white," *Marloth*) about 1$\frac{1}{2}$ lin. long ; segments rounded, very short, slightly spreading, from $\frac{1}{6}$–$\frac{1}{4}$ the length of the tube ; filaments slender, nearly twice as long as the anthers ; anthers included, lateral, cuneate-oblong, somewhat rough, under $\frac{1}{2}$ lin. long, crested at the base ; pore more than $\frac{1}{2}$ the length of the cell ; crests ovate-lanceolate, acuminate, distantly dentate, thin, hyaline, $\frac{2}{3}$ the length of the cells ; style included, short, dilated at the base ; stigma capitate ; ovary densely woolly.

SOUTH AFRICA : without locality, *Niven !* in Herb. Kew.
COAST REGION : Worcester Div. ; rocky places on the Matroos Berg, 5200 ft., *Marloth*, 2214 !

420. E. triflora (Linn. Sp. Pl. ed. 1, 354) ; erect, 3–10 ft. high ; stem sometimes 2–3 in. in diam. at the base ; branches ascending, pallid, the upper parts tomentose with dense long and short barbellate hairs ; leaves 3-nate, crowded, spreading, incurved, subulate-linear, acute or acuminate, somewhat trigonous, sulcate, glabrous, 3–5 lin. long ; flowers 3-nate ; pedicels pubescent, 1$\frac{1}{2}$–2$\frac{1}{2}$ lin. long ; bracts approximate, or the lowest subremote, lanceolate, scarious, keeled, 2–2$\frac{1}{2}$ lin. long ; sepals like the bracts but ovate, incurved, 2$\frac{1}{2}$–3 lin. long, about as long as the corolla ; corolla campanulate-

urceolate, slightly contracted at the throat, glabrous, white, about
$2\frac{1}{2}$ lin. long; segments spreading, $\frac{1}{3}$–$\frac{2}{3}$ the length of the tube;
filaments tapering upwards, bent above; anthers variable, included,
dorsifixed at or near the base; cells deeply partite and sometimes
spreading or incurved, oblong, tapering to the base or longitudinally
semiovate, obtuse or subacute, scaberulous, about $\frac{1}{2}$ lin. long, crested;
pore $\frac{2}{3}$ the length of the cell; crests from linear to broadly ovate,
denticulate, or more or less incised, equalling or little shorter than
the cells, brown; style short, included; stigma capitellate; ovary
glabrous. *Thunb. Diss. Erica,* 47, *t.* 5; *Wendl. Eric. Ic. fasc.* 12,
13; *Andr. Heathery, t.* 245, *and Col. Heaths, t.* 284; *Lodd. Bot.
Cab. t.* 1733; *Benth. in DC. Prodr.* vii. 656. *E. triflora, var.
aristata, Wendl. Beobacht.* 47, *and Eric. Ic. fasc.* 22, 153, *t.* 58. *E.
fugax and E. pyrolæflora, Salisb. in Trans. Linn. Soc.* vi. 351.
E. marifolia, var. longifolia, Wendl. Eric. Ic. fasc. 26, 4, *not of
Soland.*

VAR. β, **rosea** (Benth. in DC. Prodr. vii. 656); leaves 3–4-nate; pedicels
longer, up to 4 lin.; flowers rosy. *E. arbutiflora, Wendl. Eric. Ic. fasc.* 26,
10.

SOUTH AFRICA : without locality, *Drège! Herb. Salisbury!* and *cultivated
specimens!*
COAST REGION : Cape Div.; Table Mountain and Devils Mountain, up to
2500 ft., *Thunberg, Bolus,* 3704! 3782! and in *Herb. Norm. Austr.-Afr.,* 190!
Guthrie, 813! kloof beyond Kirstenbosch, *Wolley Dod,* 1684! 1685! Caledon
Div. ; Grabouw near Palmiet River, *Grisbrook!*

In admitting the var. β in this place, we have followed Bentham, with some
doubt; it is only known to us from Wendland's figure and description. The
very good figure seems to unite some of the characters of this species and of *E.
affinis* and *E. baccans,* differing from all, though scarcely sufficiently so for
specific distinction.

E. stenoma, Tausch (*Flora,* 1839, 633) appears to be this species or near to it,
but the type is too small for satisfactory examination.

421. E. depressa (Linn. Mant. Alt. 230, not of Andr.); a rigid
depressed glabrous shrub, generally less than a foot long; branches
stout, brittle, divaricate, usually flexuous; leaves mostly 3-nate,
occasionally also 4-nate on the same plant, crowded, spreading,
straight or curved, oblong or lanceolate, keeled and faintly sulcate,
coriaceous, rigid, pallid, smooth, 3–4 lin. long; flowers 3-nate,
generally fewer, when fully expanded subcorolline in aspect; pedicels
about 1 lin. long; sepals oblong or lanceolate, acuminate, keeled,
scarious, pallid, 2–$2\frac{1}{2}$ lin. long, mostly as long as the tube of the
corolla; corolla cyathiform or subcampanulate, mouth scarcely or
only very slightly widened, white, 2–$2\frac{1}{2}$ lin. long; limb subspreading
or erect, $\frac{1}{3}$–$\frac{1}{2}$ the length of the tube; filaments rather broad; anthers
included, lateral, cuneate-oblong, $\frac{1}{2}$ lin. long, narrow-cristate; pore
about $\frac{1}{3}$ the length of the cell; crests subulate-acuminate, with a few
teeth at the outer margin near the base, or subaristate; awns about
$\frac{1}{2}$ the length of the cell; style shortly exserted; stigma capitate;
ovary glabrous; capsule subglobose. *Thunb. Diss. Erica,* 33, *t.* 6;

Benth. in DC. Prodr. vii. 680.　*E. humilis, Salisb. in Trans. Linn. Soc.* vi. 329, *not of Benth.　E. rupestris, Andr. Heathery, t.* 145, *and Col. Heaths, t.* 128 ; *Lodd. Bot. Cab. t.* 855.　*E. semisulcata, Drège ex Benth. l.c.*

COAST REGION : Cape Div. ; in clefts of rocks on summit of Table Mountain and Devils Peak, *Drège, Harvey! Alexander,* 15 ! *Bolus,* 4484 ! *Guthrie,* 1165 ! *Wolley Dod,* 855 ! Muizenberg Mountain, *Bolus,* 4472 ! Also *Herb. Salisbury !* and *cultivated specimens !*

Andrews' *l.c. t.* 145, drawn from a cultivated plant, does not give a good idea of the wild plant, especially as to the corolla ; whereas Thunberg's *t.* 6, evidently drawn from a dried wild specimen, though small, is excellent.

422. E. petiolaris (Lam. Encycl. i. 487) ; $\frac{1}{2}$–$1\frac{1}{2}$ ft. high ; branches many, erect or divaricately-spreading, rigid, white-pubescent ; leaves 3-nate, from suberect to spreading or squarrose, linear, lanceolate or elliptic, acute, sulcate or more usually open-backed, with rounded or revolute margins, thick, rigid, white-tomentose beneath, glabrous and glossy above, from 5–8 lin. long, $\frac{1}{2}$–$1\frac{1}{2}$ lin. wide ; petiole long, pallid, adpressed, from $\frac{1}{2}$ the length of the blade, to (in the smaller leaves) nearly as long ; flowers 3-nate, often clustered in dense heads, subcalycine ; pedicels $1\frac{1}{2}$ lin. long ; bracts approximate, lanceolate, acute, cartilaginous, rigid, glabrous, white or pallid ; sepals like the bracts, but ovate, acute or acuminate, wrinkled or striate, reaching a little lower than the top of the corolla, $1\frac{1}{2}$–2 lin. long ; corolla cyathiform-urceolate, glabrous, white, 2–$2\frac{1}{2}$ lin. long ; segments deltoid and subobtuse, or broadly-rounded, recurved, $\frac{1}{4}$–$\frac{1}{3}$ the length of the tube ; filaments rather broad, often dilated, somewhat darkened and angled or bluntly toothed near the apex ; anthers subexserted, terminal or subterminal, longitudinally semioval, subacute or obtuse, smooth, about $\frac{3}{5}$ lin. long, muticous ; pore $\frac{2}{3}$ the length of the cell, style exserted ; stigma capitate, toothed ; ovary minutely hairy on the top. *Salisb. in Trans. Linn. Soc.* vi. 334.　*E. petiolata, Thunb. Diss. Erica,* 15, *t.* 6 ; *Wendl. Eric. Ic. fasc.* 27, 15 ; *Andr. Heathery, t.* 136, *and Col. Heaths, t.* 198 ; *Lodd. Bot. Cab. t.* 1150 ; *Benth. in DC. Prodr.* vii. 615.

SOUTH AFRICA : without locality, *Mund ! Drège ! Herb. Salisbury !* and *cultivated specimens !*
COAST REGION : Cape Div.; clefts of rocks on Table Mountain, 3500 ft., *Thunberg, Bolus,* 4483 ! *Harvey ! Wolley Dod,* 3482 ! Devils Peak, *Niven,* 214 ! Constantia Berg, *Bodkin ! Wolley Dod,* 1936 !

423. E. selaginifolia (Salisb. in Trans. Linn. Soc. vi. 338) ; erect, $1\frac{1}{2}$ ft. or more high ; branches incurved-erect, tomentose-puberulous ; leaves 3-nate, nearly erect, imbricate, linear-trigonous, blunt, glabrous, minutely gland-ciliolate, 2–4 lin. long ; flowers 3-nate ; pedicels puberulous, $2\frac{1}{2}$–3 lin. long ; bracts remote, ovate-lanceolate, acuminate, keel-tipped, cartilaginous, subviscidulous, coloured, about 1 lin. long ; sepals like the bracts, but ovate, less acuminate, not imbricate in the fully matured flower, with a peculiar depression in the centre at the base, nearly 1½ lin. long, reaching as high as ½–⅔ of the corolla-tube ; corolla globose-urceolate, narrowed not much constricted at the throat, 4-gonous towards the base, subviscidulous, rosy red, 1¾–2 lin. long ; limb slightly spreading, broad and rounded, about ½ the length of the tube ; filaments from a broader base tapering upwards, not much longer than the ovary ; anthers included, or scarcely exceeding the corolla-tube, dorsifixed near the base ; cells deeply parted, suboblong, obtuse or subacute, brown, under ½ lin. long, crested ; crests ovate-lanceolate, denticulate, pale brown, longer than the cells ; style subincluded ; stigma subsimple, small ; ovary glabrous. *Benth. in DC. Prodr.* vii. 657.

SOUTH AFRICA : without locality, *Masson!*

COAST REGION : Clanwilliam Div. ; on the Cederberg Range, 2600 ft., *Marloth*, 2683 ! Worcester Div. ; on the Matroos Berg, *Marloth in Herb. Bolus*, 6375 ! Riversdale Div. ; Muiskraal, near Garcias Pass, 1500 ft., *Galpin*, 3627 ! Uniondale Div. ; Long Kloof, *Paterson in Herb. Salisbury !*

424. E. gracilipes (Guthrie & Bolus) ; erect, 3 ft. or more high ; glabrous, except the puberulous pedicels ; branches stout, ascending, mostly marked with the scars of prominent leaf-cushions ; leaves 3-nate, erect-incurved, imbricate, not very crowded, linear, obtuse or subacute, round-backed, faintly sulcate, 3–4 lin. long ; flowers 1–3 pseudo-axillary, on short arrested branchlets, somewhat copious, calycine ; pedicels spreading, slender, curved, puberulous, red, 4–6 lin. long ; bracts remote, 2 submedian, 1 nearly basal, lanceolate, scarious, under 1 lin. long ; sepals adpressed, imbricate for about ½ their length, ovate, acute, shortly keel-tipped, cartilaginous, coloured, rosy, wrinkled when dried, reaching to the top of the corolla-tube ; corolla urceolate, slightly constricted at the throat, 4-gonous, rosy, 1½ lin. long ; segments spreading, about ⅓ the length of the tube ; anthers exserted, subterminal, longitudinally semi-lanceolate, the dorsal margin curved, acute, dark-coloured ; cell produced beyond the pore, ⅞ lin. long, very shortly aristulate at the base ; pore nearly ⅓ the length of the cell ; awns free, about ⅛ of the cell in length ; style slender, well-exserted, dilated near the apex ; stigma subsimple ; ovary glabrous.

COAST REGION : Bredasdorp Div. ; hills near Riet Fontein Poort, 200 ft., *Schlechter*, 9687 ! *Bolus*, 8511 !

The anthers are caducous, or most of them in our specimens have dropped from age ; in this state the flowers resemble those of *E. selaginifolia* or *E. affinis*. But the species is remarkable, and even anomalous here, by its exserted anthers and pseudo-axillary inflorescence. Yet its calycine flowers, with their large imbricate sepals, indicate a nearer approach to this section than to § *Gypsocallis*.

425. E. acuta (Andr. Heathery, t. 1) ; erect, probably 1–1½ ft. high ; branches numerous, ascending, pubescent ; leaves 3-nate, crowded, subrecurved-spreading, linear-subulate, acute, mucronate or setaceous-acuminate, keeled, sulcate, glabrous, 2–3 lin. long ; flowers

3-nate; pedicels thickened upwards, viscous-pubescent, 3 lin. long; bracts varying from subapproximate to subremote, lanceolate-acuminate, 1–1½ lin. long; sepals ovate or ovate-lanceolate, acuminate, with a strong wrinkled incurved keel throughout its entire length, produced into a rather long spreading setiform point, scarious, subviscid, margin entire or sublacerate, coloured, nearly 2 lin. long, about as long as (or a little exceeding) the corolla-tube; corolla urceolate, throat contracted (in age becoming, by the swelling of the ovary, widened, and the tube more conical, as in § *Eurystoma*), reddish, about 2 lin. long; segments suberect in dried specimens, probably spreading in life, semiovate, obtuse, ½–⅔ the length of the tube; filaments slender, tapering upwards; anthers included, subterminal, dorsifixed at or near the base; cells bipartite but not much spreading, subcuneate-oblong with an oblique ascending base, subacute, a little over ½ lin. long, crested; pore about ⅗ the length of the cell; base of the crests affixed considerably above the base of the cells, broad-lanceolate, acute, pendulous, toothed at the apex, a little over ½ the cell in length; style included; stigma capitellate, small; ovary glabrous. *Andr. Col. Heaths, t.* 73; *Benth. in DC. Prodr.* vii. 685. *E. scariosa, Lodd. Bot. Cab. t.* 477 ? *not of Berg. nor of Thunb.* " *E. crossota, Spreng. Syst. Veg.* ii. 201 " *ex Benth.*

SOUTH AFRICA : without locality, *Niven !* and *cultivated specimens!*

COAST REGION : Tulbagh or Worcester Div.? without precise locality, *Bolus,* 5184! Paarl Div. ; French Hoek Mountains, 4000 ft., *Schlechter,* 9272!

Allowing for differences of age the specimens cited all agree fairly well with Niven's and with Andrews' figure. Niven's, however, are too old, and led Bentham to place the species doubtfully in § *Eurystoma.* It varies chiefly in the length and acumination of the leaves and the length of the corolla-limb.

426. E. brevifolia (Soland. ex Salisb. in Trans. Linn. Soc. vi. 338); erect, 1–1½ ft. high; branches glabrous or occasionally puberulous; leaves 3-nate, mostly erect or subspreading, longer or sometimes shorter than the internodes, mostly ovate-oblong or oblong, blunt, sulcate, thick, glabrous, usually 1–1½ lin. long, rarely in "drawn-up" specimens linear, subspreading-recurved, 2 lin. long; flowers usually 3-nate, sometimes clustered on short approximate branchlets, sometimes 2-nate or solitary; pedicels puberulous, 1½–2 lin. long; bracts remote, small, about 1 lin. long; sepals ovate or obovate, acute, not keeled, but shortly keel-tipped, and like the bracts mostly cartilaginous, thick, viscid, shining, about 1½ lin. long, a little shorter than, to as long as the corolla; corolla sub-cyathiform-campanulate or urceolate-campanulate, throat scarcely contracted, viscidulous, pale pink to rosy red, 1¾–2 lin. long; segments slightly spreading, semiovate, about ½ the tube in length; filaments slender; anthers included, subterminal, ovate-oblong, very obtuse; cells bipartite, stalked upon the decurrent connective (as in *E. baccans* and *E. Lycopodiastrum*), a little over ¼ lin. long, aristate; pore nearly as long as the cell; awns decurrent along the filament for about $\frac{1}{2}$ their length, then free, spreading, subulate, about as long as the cell; style included, short; stigma capitellate; ovary glabrous. *Benth. in DC. Prodr.* vii. 657. *E. callosa, Wendl. Eric. Ic. fasc.* 18, 87, *t.* 33. *E. obtusa, Lodd. Bot. Cab. t.* 1027? *E. pachyphylla, Spreng. Syst. Veg.* ii. 199.

South Africa : without locality, *Drège,* 1155! and *cultivated specimens!*

Coast Region : Cape Div. ; frequent on the mountains of the Cape Peninsula, up to 2700 ft., *Bolus,* 3771! 3876! 4894! *Guthrie,* 691! *Wolley Dod,* 1714! Paarl Div. ; mountains at French Hoek, *Schlechter,* 9298! Stellenbosch Div.; Lowrys Pass, *Schlechter,* 5399! Caledon Div. ; mountains near Genadendal, 5000 ft., *Burchell,* 7684/2! *Galpin,* 3628! near Vogel Gat, 3000 ft., *Schlechter,* 9564! Swellendam Div. ; summit of a mountain peak, near Swellendam, *Burchell,* 7343! George Div.; mountains near George, *Alexander,* 23 !

427. E. fimbriata (Andr. Heathery, t. 63); erect or subdecumbent, 6–8 in. high ; branches numerous, spreading, flexuous, rigid, pubescent ; leaves 3-nate (or, also 4-nate, *Andrews*), erect-spreading, imbricate, elliptic or oblong, subobtuse, trigonous or round-backed, sulcate, thick, glabrous, gland-ciliolate, about 1 lin. long ; flowers 3-nate, calycine ; pedicels tomentose with subplumose or barbellate hairs, about 1 lin. long ; bracts approximate, ovate to lanceolate, cartilaginous, deeply and prominently pectinate-ciliate, whitish or tinged-lilac, about 1 lin. long; sepals like the bracts but broad-ovate, or nearly orbicular, keel-tipped, the closely-ranked cilia on the margin in well-grown plants barbellate and gland-tipped, concave, a little over 1 lin. long and broad ; corolla campanulate-cyathiform (in age becoming more globose) glabrous, dry, white or rosy, mostly $1\frac{1}{2}$ (rarely 2) lin. long ; segments slightly spreading, rounded, about $\frac{2}{3}$ of the tube in length ; filaments slender, sometimes dilated at the base, thickened and bent near the anther ; anthers included, lateral, dorsifixed at, or just above the base, oblong or subsemiovate, sometimes recurved, smooth, ciliate at the base, $\frac{1}{2}$ lin. long or a little more or less, awned ; pore $\frac{3}{4}$ the length of the cell ; awns spreading, subulate, ciliolate, about $\frac{1}{2}$ the cell in length ; style included, 4-gonous ; stigma simple (or sometimes capitellate, or " subpeltate," *Bentham*); ovary depressed, densely woolly with longish white hairs. *Andr. Col. Heaths, t.* 169; *Lodd. Bot. Cab. t.* 1047; *Benth. in DC. Prodr.* vii. 658.

South Africa : without locality, *Masson, Mund & Maire,* ex *Bentham,* and *cultivated specimens !*

Coast Region : George Div. ; Devils Kop, *Niven,* 115! Montagu Pass, 3500 ft., *Schlechter,* 5833 ! Oudtshoorn Div. ; Zwartberg Pass, near the Victoria Hotel, 3900 ft., *Marloth,* 2408 !

Andrews' figure is the type, but is a very poor representation of the plant as known from Schlechter and Marloth's excellent wild specimens. Loddiges' figure is much better. With these plates Bentham identified Niven's specimens, which we have examined, and also one from the Berlin Herbarium, probably Mund's ; these are both past maturity. The species is very distinct, yet somewhat anomalous wherever placed.   To its neighbours in this section it has a general resemblance, its peculiar sepals and woolly ovary being the chief exceptional characters. From the § *Pseuderemia* it differs somewhat in appearance, also by the inflorescence not being capitate, and its calycine, not corolline, flowers.   It is also connected with the § *Eurystoma.*

428. E. Lycopodiastrum (Lam. Ill. ii. 428, t. 287, fig. 4); "a dwarf plant" (*Bentham*); branches erect, rigidly flexuous, tomentose-puberulous, 10–15 in. long; leaves 3-nate, nearly erect, closely and very regularly imbricate, oblong, subacute, concave above, round-backed, sulcate, thick, glabrous, shining, $1\frac{1}{2}$–2 lin. long; flowers 3-nate, terminal on short lateral branchlets, or sometimes (by their partial arrest) sublateral; pedicels tomentose, decurved, $1\frac{1}{2}$–2 lin. long; bracts mostly approximate, more rarely submedian, ovate, acute, keeled, rigid, scarious, 1–$1\frac{1}{4}$ lin. long; sepals like the bracts, but slightly larger, $1\frac{1}{4}$ lin. long, $\frac{1}{2}$–$\frac{3}{4}$ the length of the corolla; corolla obconic-cyathiform, slightly widened to the mouth, glabrous, $1\frac{1}{2}$ lin. long (flesh-coloured, *Wendland*); segments slightly spreading, rounded, $\frac{1}{2}$–$\frac{2}{3}$ of the tube in length; filaments slender; anthers included, subterminal; cells bipartite, distant, stalked upon the slender connective, narrow-obovate, subobtuse, pale brown, membranous, about $\frac{1}{3}$ lin. long, broad-aristate; pore $\frac{1}{3}$–$\frac{1}{2}$ the length of the cell; awns affixed to the connective and free from the filament, pendulous, subulate, acuminate, toothed, a little over $\frac{1}{2}$ the cell in length; style included, short; stigma capitellate; ovary glabrous or with a few hairs at the summit; ovules ovoid (*Bentham* found them compressed, which our observations do not confirm). *E. fabrilis,* Salisb. in Trans. Linn. Soc. vi. 338; Benth. in DC. Prodr. vii. 655. *E. confertifolia, Wendl. Eric. Ic. fasc.* 20, 121, t. 46.   *E. montana, Sinclair, Hort. Eric. Wob.* 15.   *E. confertiflora, Steud. Nom. ed.* 2, i. 570.

SOUTH AFRICA: without locality, *Masson, Drège, Herb. Salisbury! Herb. Lamarck* (641)! and *cultivated specimens!*
COAST REGION: Paarl Div.; French Hoek, by the river, *Niven,* 55!

Connects §§ *Trigemma* and *Eurystoma,* yet has hardly the longer corolla-limb of the latter.   In appearance it has some resemblance to the two succeeding species, but is distinct by its peculiarly thin membranous subterminal anthers, and their narrow appendages.   Wendland describes the leaves (from a cultivated plant) as 4-nate, which we have not seen.   The plant appears to be now rare.

429. E. pumila (Andr. Heathery, t. 234); dwarf, probably only a few inches high; branches numerous, short; leaves 3-nate, erect or spreading, or the older squarrose, obovate, obtuse, thick, sulcate, subglaucous, pallid, glabrous (or in cultivated specimens linear and green, *Andrews*), 2–$2\frac{1}{2}$ lin. long; flowers 3-nate; pedicels short; bracts short, approximate, cartilaginous, white, glabrous; sepals ovate, obtuse, keel-tipped, cartilaginous, viscidulous, white (or in cultivated specimens acute, rosy), about $\frac{1}{2}$ the length of the corolla; corolla ovoid-inflated or subcylindric, glabrous, subviscid, tube rosy, the whole 3–3½ lin. long; segments erect or slightly spreading, about ¼ the length of the tube, green; anthers included, longitudinally semiovate, acute, crested; crests semiorbicular, denticulate on the outer margin, about ⅔ of the length of the cell, brown; style included; stigma subsimple or capitellate; ovary glabrous. *Andr. Col. Heaths, t.* 258; *Benth. in DC. Prodr.* vii. 655. *E. coniflora, Klotzsch in Herb. Berol. ex Benth. l.c.*

SOUTH AFRICA: without locality, *Masson*, 96! in Herb. Kew; also, collector and locality unknown, in Herb. Berlin!

Our material is very scanty; Masson's specimens have no flowers; we have only seen one flower and a few loose leaves in the Berlin Herbarium; and our description is therefore drawn from these, from Bentham's brief description and Andrews' figure. Bentham remarks that it seems to be allied to *E. modesta* and to § *Pachysa.*

430. E. Thodei (Guthrie & Bolus); dwarf (the single specimen 7 in. high); stout, erect; branches numerous, nearly straight, rigid, pubescent, glabrescent, leafy above, covered with scars of prominent leaf-cushions below; leaves 3-nate, crowded, subspreading but closely imbricate, oblong or narrow-elliptic, subobtuse, thick, keeled, not sulcate or the keel only faintly channelled, cartilaginous-ciliolate, the flat scales on the younger leaves minutely branched or lacerate, glabrous, shining, petiole more than ½ the length of the blade, the whole 1½–2 lin. long; flowers strictly terminal and 3-nate, but in young and undeveloped or arrested branchlets sometimes appearing almost axillary, long-persistent; pedicels puberulous, lengthening during or after flowering, attaining 1½ lin. long; bracts remote or subapproximate, erect, linear, scarious, 1 lin. or less long; sepals erect, adpressed, narrow-lanceolate, acute, concave, scarious, keeled, not imbricate at the base, glabrous, ciliolate as the leaves, 1¼–1½ lin. long, about ⅔ the corolla; corolla tubular-cyathiform, equal to the mouth, glabrous, dry, "white," 2 lin. long; segments broader than long, rounded, ⅕–¼ the length of the tube; filaments capillary, over 1 lin. long; anthers included, lateral, longitudinally semiovate, smooth, membranous, pale brown, about ½ lin. long, crested-aristate; pore under ½ the length of the cell; awns broadly subulate, acute, long-ciliate, nearly ⅔ the length of the cells; style included; stigma subsimple; ovary glabrous.

KALAHARI REGION: Orange River Colony; rocky places on the summit of the Mont-aux-Sources, near the sources of the Tugela River, 9000–10000 ft., *Thode*, 64! in Herb. Bolus and Kew, but may be distributed under other numbers in other herbaria.

In floral structure this is near the next species, but the sepals are relatively shorter, the indumentum on the leaves different, the ovary glabrous, and the habit much more erect and rigid. The specimen bears persistent old flowers; but we cannot be sure that they are those of the preceding year.

431. E. lasiocarpa (Guthrie & Bolus); branches flexuous, puberulous, pallid, 8–10 in. long; leaves 4-nate, subspreading, linear, acute, widely sulcate, puberulous and thinly hispid with short stiffish hairs, glabrescent, $1\frac{1}{2}$–2 lin. long; flowers 4-nate on short lateral branchlets, subcorolline; pedicels 1 lin. long; bracts subremote, linear, erect, subglumaceous, about 1 lin. long; sepals somewhat spreading, not imbricate unless slightly at the very base, and not concealing the corolla-tube, lanceolate-linear, tapering to a narrow keeled point, subglumaceous, ciliate, glabrous, $1\frac{1}{2}$–$1\frac{3}{4}$ lin. long, usually reaching as high as the top of the corolla-tube or higher; corolla narrow-cyathiform or suburceolate, the throat not (or scarcely) spreading, tetragonous, veined, glabrous, dry, $1\frac{3}{4}$ lin. long; segments erect (in the dried specimens), or subspreading, semiovate, over-lapping at the base, about $\frac{1}{3}$ the length of the tube; filaments capillary; anthers included, lateral, cuneate-oblong, obtusely rounded, smooth, membranous, brown, scaberulous on the margins, $\frac{2}{5}$ lin. long, aristate; pore $\frac{1}{2}$ the length of the cell; awns subulate, hairy, recurved at an angle of 45°–60° from the cells, about $\frac{1}{2}$ the length of the cells; style included or sometimes just equal to the corolla and manifest; stigma capitellate; ovary densely and shortly villous.

EASTERN REGION : Natal; rocky places near Van Reenen, 7000 ft., *Schlechter*, 6941 !

This has somewhat the habit and aspect of *E. caffrorum*. The flowers on our specimens are rather scanty.

Section XXXVII. **POLYCODON.** (Sp. 432–439.)

432. E. leucanthera (Linn. f. Suppl. 223); erect, 1–$1\frac{1}{2}$ ft. high; branches numerous, ascending, often dense, pubescent with short barbellate hairs; leaves mostly erect or subspreading, crowded, imbricate, linear-trigonous or linear-lanceolate, subacute or obtuse, sulcate-keeled, glabrous, 1–$1\frac{1}{2}$ lin. long; flowers calycine or sub-calycine on numerous short branchlets, often crowded in dense masses; pedicels less than 1 lin. long; bracts from subapproximate to subremote, sepal-like, small; sepals narrow-ovate, acute, keeled, pallid, cartilaginous, margin scarious, under 1 lin. long, or about $\frac{2}{3}$ the length of the corolla; corolla obconic, pale yellow, 1–$1\frac{3}{8}$ lin. long; segments spreading to the apex, straight, semiorbicular or broad semiovate, from about as long as to a little longer than the tube; stamens far-exserted and recurved or subincluded; filaments slender, long or short; anthers oblong, slightly wider at the base; cells deeply partite and spreading, or approximate, subacute or sub-obtuse, membranous, pale yellow, about $\frac{1}{2}$ lin. long, muticous; pore small, about $\frac{1}{4}$ the length of the cell; style included, or just manifest; stigma capitate; ovary minutely hairy, chiefly on the summit. *Benth. in DC. Prodr.* vii. 689. *E. spiræœflora,*

Salisb. in Trans. Linn. Soc. vi. 350.　*E. staminea, Andr.* Heathery,
t. 193, *and Col. Heaths, t.* 208.　*E. thalictriflora, Lodd. Bot. Cab.
t. 1294.

South Africa : without locality, *Thunberg, Herb. Salisbury !* and *cultivated
specimens !*

Coast Region, from 900–1200 ft. : Tulbagh Div. ; Roode Zand, *Masson,*
133 ! Winterhoek Mountain, *Niven,* 87 ! *Bolus,* 5188 ! Tulbagh Waterfall,
MacOwan, Herb. Norm. Austr.-Afr., 232 ! *Bolus,* 5370 ! *Schlechter,* 9008 !
Guthrie, 2072 ! Worcester Div., *Ecklon & Zeyher !* Paarl Div. ; French Hoek
Mountains, *Bolus, Herb. Norm. Austr.-Afr.,* 608 !

Central Region : Ceres Div. ; Cold Bokkeveld, near Elandsfontein, 4000 ft.,
Schlechter, 10027 !

There appear to be two forms of this species :—the one with slightly smaller
flowers, shorter corolla-limb, subincluded stamens, anther-cells smaller and
approximate and broader stigma ; the other with larger flowers, longer corolla-
limb, far-exserted stamens, anther-cells separate and spreading, stigma somewhat
smaller. The first may be the original form of *Linn. f.* ; the latter is certainly
E. staminea, Andr. These are either heterostyled forms, or the first are plants
not yet fully developed ; to decide, observation of the living plant is necessary.
In either case, we think they are one species. Bentham says this was Klotzsch's
conclusion (though we have not been able to trace the statement) ; but he
himself thought they were sufficiently distinct.

433. E. stenantha (Klotzsch ex Benth. in DC. Prodr. vii. 685);
small, branched, 8 in. or more high ; branches ascending, subflexuous, villous or pubescent ; leaves crowded, squarrose-recurved,
rigid, linear-lanceolate, subobtuse, flattish, sulcate, glabrous, ciliate,
glossy, about 1 lin. long ; flowers numerous, on short crowded
branchlets towards the ends of the branches, subcorolline ; pedicels
slender, $\frac{3}{4}$–$2\frac{1}{2}$ lin. long ; bracts remote, lanceolate, subscarious, ciliate,
small ; sepals ovate to subrhomboidal, acute, keeled, concave,
scarious, cilate, paler than the corolla, $\frac{1}{2}$–1 lin. long ; corolla broad-
obconic to obconic-campanuloid, rosy or bright red, 1–1$\frac{3}{4}$ lin. long ;
segments continuous with the tube or slightly recurved, ovate or
lanceolate, obtuse or subacute, keeled, from as long as, to nearly
twice as long as the tube ; anthers included, longitudinally semiovate, subacute, minutely hairy, $\frac{1}{2}$–$\frac{3}{5}$ lin. long, muticous ; pore
$\frac{1}{4}$ the length of the cell ; style exserted ; stigma subsimple or
minutely obconic or distinctly obconic-cyathiform ; ovary villous.

Coast Region : Caledon Div. ; Zwart Berg, near Caledon, *Zeyher !* Swellendam Div. ; mountains near Swellendam, 3000 ft., *Niven,* 107 ! *Masson,* 1 !
Ecklon & Zeyher, 221 ! *Burchell,* 7298 ! 7332 ! *Bolus,* 7505 ! Zuurbraak Mountain, 4600 ft., *Galpin,* 3694 !

All our specimens agree fairly well, except that in Galpin's 3694 the stigma is
much larger and distinctly obconic-cyathiform and the pedicels shorter ; but as
there is a tendency towards enlargement in some of the others, it does not seem
well to separate them. Bentham placed this species in the § *Melastemon,* but
the anther-cell is scarcely produced beyond the pore and the corolla seems more
to resemble that of the present section.

434. E. consobrina (Guthrie & Bolus); branches (in our only specimen) incurved, cano-puberulous with simple, intermixed with minute compound, hairs on the main branches, 10–12 in. long; leaves spreading, linear, acute, sulcate, subglabrous, the younger erect, imbricate, crowded, 1–1½ lin. long; flowers corolline; pedicels pubescent, 1 lin. long; bracts, 2 subapproximate, 1 median, linear, ⅜ lin. long; sepals broad-ovate, acute, keel-tipped, scarious, concave, coloured, ¾ lin. long, reaching to less than ½ the height of the corolla; corolla campanulate-cyathiform, rosy, 1⅜ lin. long; segments slightly spreading nearly ⅔ the length of the tube; filaments narrow-linear, equal; anthers subexserted, dorsifixed well above the base, obliquely broad-lanceolate, acute, scaberulous, blackish, ⅖ lin. long, aristate; pore ⅗ the length of the cell; awns straightly pendulous, subulate, ciliolate, ¼ the length of the cell; style very shortly exserted, scarcely exceeding the anthers; stigma subsimple, of 4 short erect tooth-like processes, blackish; ovary glabrous, pallid.

COAST REGION: Clanwilliam Div.; on the Cederberg Range near Krakadouw, 3000 ft., *Bodkin in Herb. Bolus*, 8679!

This resembles *E. rhodantha*, but the flowers are somewhat broader, stamens subexserted, anthers aristate, style shorter, stigma smaller and of different structure, and the ovary glabrous. It has also some similarity to *E. cristæflora*, var. *β*.

435. E. nemorosa (Klotzsch ex Benth. in DC. Prodr. vii. 688); erect, reaching 6 ft. high (*Bentham*); branches subvirgate or more rarely spreading, tomentose with simple soft hairs; leaves suberect, slender, linear, obtuse, sulcate, glabrous, scarcely 1 lin. long; flowers corolline; pedicels ½–¾ lin. long; bracts remote, linear, small; sepals from subovate and acute to oblong-lanceolate and sub-acuminate, keel-tipped, glabrous or puberulous, mostly strongly incurved towards the apex, more rarely straight, coloured, about ½ the height of the corolla, ½–⅔ lin. long; corolla varying from broad- to narrow-cyathiform, red or rosy, ¾–1¼ lin. long; segments erect, very variable, from ½ the tube to somewhat longer; anthers included or sometimes subexserted, suboblong, obtuse, ⅕–⅓ lin. long, muticous or minutely aristulate; pore ⅖ the length of the cell; style slender, exserted; stigma capitellate; ovary minutely hairy, or at least on the summit. *E. floribunda, var. micrantha, Benth. l.c.* 688, *not of Lodd. E. polycodon, Benth. l.c.* 688.

COAST REGION: Caledon Div.; near Caledon, *Bolus*, 8497! Humansdorp Div.; near the Gamtoos River, *Niven*, 95! near Clarkson, 900 ft., *Galpin*, 3711! Uitenhage Div.; stony channel of the Zwartkops River, *Ecklon & Zeyher*, 30! *Zeyher*, 3338! *Alexander*, 9? Van Stadens Berg, *Zeyher*, 784! Albany Div.; near Grahamstown, *Burchell*, 3549! *Atherstone*, 2! Bathurst Div.; Karega River, *Zeyher*, 3339! Stutterheim or Komgha Div.; banks of the Kabousie River, 3000 ft., *Murray*, 53!

This seems to be a variable species, chiefly in respect of the shape and size of the corolla and its limb; and is consequently somewhat difficult of distinction.

The type of Klotzsch was most probably Ecklon and Zeyher's 30, and a careful comparison of this with Niven's 95 (the type of *E. polycodon*, Benth.), appears to show that these can hardly be separated. The corolla is a little different, but intermediate forms are present. In Galpin's 3711 the anthers are subexserted, as in Alexander's 9, and these approach *E. decipiens* in structure and appearance; but in the latter the sepals are glabrous and more scarious and the corolla longer. All the forms (except Niven's 95) appear to be characterized by similarly incurved sepals. In general appearance it resembles *E. floribunda*, but may be distinguished by the simple hairs on its branches as well as by its capitellate, not obconic, stigma.

436. E. floribunda (Lodd. Bot. Cab. t. 176); erect, 2 ft. or more in height; branches many, ascending, covered with a short pubescence of floccose shortly plumose ashy-grey hairs; leaves suberect to subspreading, linear, obtuse, obscurely sulcate, glabrous, 1–$1\frac{1}{2}$ lin. long; flowers subcalycine, mostly very numerous and crowded in dense masses on short branchlets; pedicels $\frac{1}{2}$–1 lin. long; bracts remote or subapproximate, erect, ovate, small; sepals broad-ovate to orbicular, keel-tipped, subacute, concave, subcartilaginous, coloured (mostly paler than the corolla) $\frac{1}{3}$–$\frac{1}{2}$ lin. long; corolla cyathiform to broad-cyathiform, mouth widened, pale rosy, $\frac{3}{4}$–1 lin. long; segments broadly rounded, erect, about equal to the tube; filaments short, straight; anthers subincluded, generally manifest, oblong or ovate, obtuse, rough, nearly black, $\frac{1}{3}$–$\frac{2}{5}$ lin. long, muticous; pore less than $\frac{1}{2}$ the length of the cell, style slender, well-exserted; stigma obconic-cyathiform, or sometimes obconic (solid), small; ovary minutely hairy. *Benth. in DC. Prodr.* vii. 688. *E. sparsa, Lodd. Bot. Cab. t.* 1467. *E. macrostoma, Klotzsch ex Benth. l.c.* 688.

SOUTH AFRICA : without locality, *cultivated specimens!*
COAST REGION : frequent, from George Div. eastwards to Albany Div., *Niven,* 96! *Burchell,* 5523! *MacOwan,* 272! *Bolus,* 8293! 8673! *Schlechter,* 2244! 5886! The following, which we have not been able to examine, probably also belong here : *Niven,* 98; *Burchell,* 5157, 5371, 5455! Oudtshoorn Div. ; Zwartberg Pass, *Tyson!*

Loddiges' figure is the type of this species; but for critical purposes is all but useless. Bentham identified with it *E. galliiflora, Bartl.*, and we have a specimen so marked from the Cape Gov. Herb., most probably Zeyher's. As far as it goes we cannot confirm from it Bentham's identification. It has the stigma " capitato-4-lobo " as described by Bartling, and in this respect differs from all our other specimens. The species much resembles *E. paniculata* and is easily confused with it. But it may generally be distinguished by its broader sepals (more calycine flowers), its longer, more slender style, its pale obconic stigma, and minutely hairy ovary. Both have a like floccose indumentum. The stigma of this is somewhat variable; usually it is obconic, quite hollow or depressed within; sometimes it is solid and has a truncate top. But it is always small, as compared, for instance, with *E. peltata,* with which it has some affinity.

437. E. rhodantha (Guthrie & Bolus); erect; branches ascending, somewhat straight, with numerous short floriferous subdecurved branchlets of from 1–$1\frac{1}{2}$ in. long, pubescent with simple hairs; leaves erect or subspreading, crowded, imbricate, linear-trigonous, subobtuse, glabrous, $1\frac{1}{2}$–2 lin. long; flowers somewhat copious, subcalycine; pedicels softly floccose with minutely plumose hairs, $1\frac{1}{2}$ lin.

long ; bracts remote, laxly spreading, lanceolate, concave, scarious, keel-tipped, coloured, the lowest more erect, about $\frac{3}{4}$ lin. long ; sepals broad-ovate, 1 lin. long, otherwise like the bracts, reaching about $\frac{2}{3}$ the height of the corolla ; corolla cyathiform or campanulate-cyathiform, about $1\frac{1}{2}$ lin. long ; segments erect or slightly spreading, broadly rounded, about as long as the tube ; anthers included, dorsifixed well above the base, longitudinally semiovate, acute or subacute, ciliolate on the margins, dark-coloured, $\frac{1}{2}$ lin. long, muticous ; pore about $\frac{1}{2}$ the length of the cell ; style exserted ; stigma subcapitellate, 4-lobed, rather flat on the summit ; ovary oblate-spheroidal, puberulous.

COAST REGION : Riversdale Div. ; Garcias Pass, 1200 ft., *Galpin,* 3706 !

In floral structure allied closely to *E. floribunda,* yet with certain differences, and especially in its general aspect, which, after much consideration, have obliged us to distinguish it.

438. E. peltata (Andr. Heathery, t. 276) ; erect, 1–2 ft. high ; branches ascending, straight or flexuous, usually clothed with a pubescence of very short compound (rarely simple) hairs ; leaves erect, imbricate, not crowded, linear, round-backed, $1–1\frac{1}{2}$ lin. long ; flowers generally in interrupted numerous clusters at the ends of the branchlets, subcorolline ; pedicels puberulous, about $\frac{1}{2}$ lin. long ; bracts variable, sometimes 1 only, close to the calyx, or with 2 smaller and subremote, or, entirely wanting ; sepals from ovate-lanceolate to ovate, or suborbicular, acute or subobtuse and apiculate, concave, hyaline, pallid, ciliolate or naked, about as long as the corolla-tube ; corolla broad-cyathiform or subhemispherical, nearly 1 lin. long ; segments broadly rounded, erect, a little shorter than the tube ; anthers subexserted, from shortly oblong to subovate, subacute, scabrid, dull red, about $\frac{1}{2}$ lin. long, muticous ; pore about $\frac{2}{3}$ the length of the cell ; style exserted, straight, red ; stigma cyathiform, large, with 4 intramarginal short points, bright crimson, about $\frac{1}{2}$ lin. in diam. ; ovary small, minutely hairy. *Andr. Col. Heaths, t.* 254 ; *Benth. in DC. Prodr.* vii. 691. *E. Actæa, Link, Enum. Hort. Berol.* i. 371. *E. exserta, Sinclair, Hort. Eric. Wob.* 9.

COAST REGION: Swellendam Div. ; mountains near Swellendam, 1000 ft., *Niven,* 31 ! *Bolus,* 7302 ! and in *Herb. Norm. Austr.-Afr.,* 604 ! Garcias Pass, 1500 ft., *Schlechter,* 2194 ! Riversdale Div. ; rocky places, *Schlechter,* 2172 ! George Div. ; near George, *Schlechter,* 2232 ! near Great Brak River, *Guthrie,* 3555 ! Uniondale Div.; near the Keurbooms River, *Burchell,* 5132 !

The habit varies from that with more erect and straightish, to that with more slender, whip-like, flexuous branchlets ; but there are intermediate forms.

439. E. macrotrema (Guthrie & Bolus) ; erect, apparently 1 ft. or more high ; branches few, ascending, subflexuous ; branchlets many, the younger white-pubescent ; leaves slightly spreading, sub-incurved, oblong or narrow-elliptic, subobtuse, thick, round-backed, faintly sulcate, glabrous, cartilaginous-denticulate, $\frac{1}{2}–1\frac{1}{4}$ lin. long, about 2 to $2\frac{1}{2}$ times longer than the internodes; flowers not numerous, on short branchlets thinly scattered over the whole plant, subcalycine; pedicels curved, stout, red, shortly white-tomentose, 1 lin. long; bracts 3, approximate, oblong and lanceolate, ciliolate, shorter than the sepals; sepals ovate, acute, concave, glumaceous, keeled or keel-tipped, rosy suffused with bright red, white-margined and ciliolate with scale-like, forked or subplumose hairs, about $\frac{3}{4}$ lin. long; corolla hemispherical or subobconic-cyathiform, about 1 lin. long; segments equal to the tube or slightly longer, rounded, suberect; filaments rather broad, lanceolate; anthers subincluded, or at full maturity subexserted, longitudinally semiovate, obtuse, minutely scaberulous, a little over $\frac{1}{2}$ lin. long, muticous; pore large, $\frac{7}{8}$ the length of the cell; style well-exserted, stout; stigma obconic-cyathiform; ovary sometimes glabrous, sometimes pubescent.

COAST REGION : Clanwilliam Div. ; Kers Kop, Cederberg Range, near Wupperthal, 3000 ft., *Schlechter*, 8788 !

In floral structure very like the preceding; but the branches are closer, more rigid and erect, the leaves broader and mostly shorter, pedicels thicker, longer, more densely tomentose; bracts more constant; sepals more prominent, more ciliate; anthers larger with a much longer pore; stigma smaller. It has some resemblance to the § *Melastemon* and *E. lavandulæfolia.*

Section XXXVIII. **EURYSTOMA.** (Sp. 440–450.)

440. E. media (Klotzsch ex Benth. in DC. Prodr. vii. 682); branches cano-puberulous; leaves erect, adpressed, oblong, subobtuse, round-backed, thick, concave and nerved above, sulcate below, glabrous, about 1 lin. long; flowers subcorolline; pedicels rather stout, puberulous, 1 lin. long; bracts approximate, closely adpressed to the calyx, sepal-like, ovate, obtuse; sepals ovate, acute, keeled, concave, scarious or subcoriaceous, $\frac{3}{4}$ lin. long; corolla cyathiform, mouth somewhat widened, $1\frac{1}{2}$ lin. long; segments erect, deltoid, obtuse, about $\frac{1}{3}$ the length of the tube; filaments rather broad, subequal, about $\frac{2}{3}$ lin. long; anthers lateral, oblong, obtuse, about $\frac{3}{8}$ lin. long, appendiculate at the base; pore $\frac{1}{4}$–$\frac{1}{3}$ the length of the cell; appendages between crested and aristate, oblong, entire, short, blunt, brown, placed rather high on the cell, in length about $\frac{1}{3}$ of the cell; style shortly exserted; stigma small, capitellate; ovary glabrous.

COAST REGION : Caledon Div.; Cape Hangklip, *Mund & Maire !* in Herb. Berlin and Kew.

This species is closely allied to *E. argentea, Klotzsch,* and the latter may possibly be a form of this. The fragments we have before us are poor, and we therefore leave the species as Klotzsch and Bentham had it.

441. E. argentea (Klotzsch ex Benth. in DC. Prodr. vii. 686); erect, 1–$1\frac{1}{2}$ ft. high ; stem and branches flexuous ; branchlets slender, the ultimate and floriferous generally decurved ; leaves mostly erect

and adpressed or incurved, imbricate or often only as long as the
internodes, oblong or elliptic, acute, thick, flat above, convex and
glossy below, glabrous, generally $\frac{1}{2}$–1 (sometimes $1\frac{1}{2}$) lin. long;
flowering branchlets subdistant and somewhat scanty; pedicels under
$\frac{1}{2}$ lin. long; flowers corolline; bracts approximate, elliptic, small;
sepals lanceolate, acute, keeled, rigid, thick with hyaline margin,
about 1 lin. long; corolla obconic, about $1\frac{1}{2}$ lin. long; segments continuous with the tube, semiovate, subacute, somewhat shorter than
the tube; anthers included, manifest, lateral, narrow-oblong, $\frac{1}{2}$ lin.
long, incurved, obtuse, pale brown, shortly aristate; pore $\frac{1}{6}$–$\frac{1}{5}$ the
length of the cell; awns subulate, about $\frac{1}{6}$ the length of the cell;
style exserted, very slender; stigma subsimple; ovary pubescent or
glabrous.

VAR. β, **rigida** (Bolus); stouter and more rigid, less flexuous : branchlets not
usually decurved; anthers shorter and broader, oblong, $\frac{1}{3}$ lin. long.

COAST REGION: Piquetberg Div.; Oliphants River Mountains, near
Piqueniers Kloof, *Zeyher*, 1106! Tulbagh Div.; foot of Winterhoek Mountain,
Ecklon & Zeyher, 224! Ceres Div.; mountains near Ceres, 1800 ft., *Guthrie*,
2413! *Bolus*, 7472! Var. β : Ceres Div.; near Ceres, *Bolus*, 8481!

The whole plant has a somewhat shining silvery-grey appearance, yet not
quite glabrous. It is very near the preceding species.

442. E. brevicaulis (Guthrie & Bolus); apparently a dwarf
stunted shrub, 3 in. high; caudex stout, about $1\frac{1}{4}$ lin. in diam.,
divided immediately into several rather stout main branches and
very numerous ascending filiform branchlets, puberulous, glabrescent, marked with prominent scars of leaf-cushions; leaves opposite,
erect, adpressed, the lower about as long as the internodes, the
uppermost imbricate, not crowded, lanceolate, acute, thick, concave
with a middle nerve above, sulcate below, glabrous, $\frac{3}{4}$–1 lin. long;
flowers terminal, 3-nate? subcorolline; pedicels slender, mostly
bent, red, $1\frac{1}{2}$–$1\frac{3}{4}$ lin. long; bracts remote, 2 inframedian, 1 basal,
linear, 1 lin. long; sepals narrow-lanceolate, acute, keeled, concave,
glabrous, glossy, red, $1\frac{1}{4}$ lin. long, reaching a little beyond the top
of the corolla-tube; corolla broad-campanulate, or between that and
broad-cyathiform, glabrous, veined, apparently red or crimson,
nearly $1\frac{1}{2}$ lin. long; segments slightly spreading, ovate, a little more
than $\frac{1}{2}$ the length of the tube, colour darkened towards the apex;
filaments slightly dilated just below the anther; anthers included,
manifest, sublateral, dorsifixed at or near the base, oblong, subobtuse,
smooth, about $\frac{1}{2}$ lin. long, crested; pore about $\frac{1}{2}$ the length of the
cell; crests partially adherent at their base to the filament, then
spreading-incurved, narrow-lanceolate or subulate, acuminate, with
2 or more incised teeth near the base, the whole about $\frac{1}{3}$ the length
of cell; style shortly exserted, slender; stigma capitellate; ovary
subglobose, shortly villous.

COAST REGION : Worcester Div.; rocky places on the Matroos Berg, 5500 ft.,
Marloth, 2213!

U 2

Our only specimen is in poor condition and has dropped its flowers ; but the species is so distinct that we do not hesitate to describe it.   It has the corolla of this section, but the sepals are so narrow as to give the flowers a somewhat corolline appearance.   In this respect it is nearest to *E. argentea,* and in so far both recede from one of the chief characters of the section.

443. E. nivea (Sinclair, Hort. Eric. Wob. 16) ; erect, almost entirely glabrous, 1–2 ft. high ; branches ascending, rather straight or subvirgate, sometimes puberulous, glabrescent ; leaves erect, mostly subadpressed, not crowded, linear-subulate, acute, trigonous, deeply sulcate, 2–3 lin. long ; flowers somewhat densely crowded on short branchlets, calycine ; pedicels slender, subglabrous, 2–3 lin. long ; bracts laxly subapproximate, lanceolate, about 2 lin. long ; sepals imbricate, obovate or oblong, subacute, keeled on the upper half and keel-tipped, white, about 2 lin. long, about equalling the corolla ; corolla subobconic or between rotate and salver-shaped from a broad base ; tube subtetragonous, wider than its length ; segments broad, obtuse, spreading at an angle of about 45°, about $1\frac{1}{2}$ times the length of the tube, the whole about 2 lin. long, entirely white ; filaments from a broader base tapering upwards, about 1 lin. long ; anthers included, manifest, sublateral, obovate-oblong, obtuse, smooth ; cells bipartite, about $\frac{2}{3}$ lin. long, crested ; pore about $\frac{2}{3}$ the length of the cell ; crests about $\frac{1}{4}$ the length of the cell, spreading ; style exserted ; stigma capitellate ; ovary glabrous, pallid on a dark disk.   *Benth. in DC. Prodr.* vii. 687.

South Africa : without locality, *Mund & Maire,* ex *Bentham.*
Coast Region : Cape Div.; Table Mountain, 2500 ft., *Marloth,* 214! Simons Bay, *Wright!* hills west of Simons Town, *Wolley Dod,* 1452! Caledon Div.; mountains near Lowrys Pass, 2000 ft., *Guthrie,* 3536! moist places near Steenbrass River, *Schlechter,* 5380 ! and without precise locality, *Bolus,* 6384!

444. E. lachneæfolia (Salisb. in Trans. Linn. Soc. vi. 335) ; erect, probably under 1 ft. high ; branches ascending, leafy, pubescent, covered with the scars of small but prominent leaf-cushions; leaves 6-farious, closely crowded, imbricate, obovate or subelliptical, very obtuse, thick, somewhat concave above, rounded below, minutely puberulous, sulcate, gland-ciliolate, very shortly petiolate, $\frac{3}{4}$–1 lin. long ; flowers subcalycine ; pedicels stout, tomentose, under 1 lin. long ; bracts approximate or subapproximate, ovate, leathery, rigid, keeled, with green or coloured leaf-like sulcate thickened tips, shorter than the sepals ; sepals like the bracts but sometimes obovate, larger, ciliolate, nearly $1\frac{1}{2}$ lin. long ; corolla obconic-cyathiform, white (*Andrews*), about 2 lin. long; segments spreading, oblong, rounded, a little longer than the tube; anthers included, manifest, sublateral ; cells bipartite, oblong, tapering a little towards the subobtuse apex, papillose, thick, rigid, blackish, about $\frac{2}{5}$ lin. long, with a thick dorsal band produced downwards into two thick subulate awns decurrent along the filament for $\frac{1}{2}$–$\frac{2}{3}$ their length, the acuminate apices free and spreading (or entirely free according to *Andrews'* figure, and some imperfect specimens seen), the whole projecting below the cells for a distance shorter than the length of the latter; pore about $\frac{1}{2}$ the length of the cell; style exserted; stigma clavate-capitate, truncate; ovary glabrous. *E. lachnœa, Andr. Heathery,* t. 120, *and Col. Heaths, t.* 177 ; *Benth. in DC. Prodr.* vii. 685.

SOUTH AFRICA: without locality, *Masson, Mund!* and *cultivated speci-mens!*
COAST REGION : Caledon Div. ; Zwart Berg, *Bowie!* tops of the mountains of Baviaans Kloof, near Genadendal, *Burchell,* 7729.

Very distinct in the section by its peculiar leaves. Our description is from Bowie's and Mund's specimens in Herb. Kew., which agree well. Bentham quotes with a sign of doubt, *E. Pohlmanni, Lodd. Bot. Cab. t.* 1852; but this seems to us quite different, and is a plant unknown to us. The present species is apparently rare and has not been gathered by any recent collector.

445. E. calycina (Linn. Sp. Pl. ed. 2, 507, not of some others) ; generally stout, erect, rigid, $1\frac{1}{2}$–2 ft. high, with numerous, ascending, subvirgate, rarely white-pubescent branches ; leaves suberect and imbricate, or sometimes spreading, or squarrose-recurved, crowded, linear or narrow-lanceolate, subtrigonous, acute or subobtuse, sulcate, glabrous, ciliolate or naked, mostly $1\frac{1}{2}$–2 lin. long; flowers often crowded on short branchlets, more rarely scanty, calycine or subcaly-cine ; pedicels mostly densely white-tomentose, more or less with plumose hairs intermixed with simple (occurring sometimes on the branches), spreading, $1\frac{1}{2}$–$2\frac{1}{2}$ lin. long; bracts subremote (or the lowest remote) to subapproximate, lax or more rarely imbricating the sepals, ovate or oblong, coloured, scarious or cartilaginous, about 1 lin. long ; sepals ovate to orbicular or subrhomboidal, imbricate for a greater or less distance from the base upwards, scarious or carti-laginous, keeled or keel-tipped, white, ciliate or naked, 1–$1\frac{1}{2}$ lin. long, or mostly from $\frac{2}{3}$–$\frac{3}{4}$ the length of the corolla (rarely $\frac{1}{2}$ only) ; corolla obconic-cyathiform, white, $1\frac{1}{2}$–2 (rarely $2\frac{1}{2}$) lin. long ; segments in full flower spreading or recurved, but very soon becom-ing erect, oblong or ovate, obtuse or subacute, from equal to the tube to a little shorter than it ; anthers included, but manifest, terminal or subterminal; cells partite to the base, oblong or semi-lanceolate, acute or obtuse, scaberulous, $\frac{2}{3}$–1 lin. long, cristate ; pore $\frac{1}{2}$–$\frac{2}{3}$ the length of the cell; crests free, coarsely toothed or incised or crested-aristate ; awns free, short, subulate, ciliate or hairy ; style stout, straight, shortly exserted; stigma simple or clavate; ovary pallid, glabrous or hispidulous at the apex. *Thunb. Diss. Erica,* 47. *E. nigrita, Linn. Mant.* 65 ; *Andr. Heathery,* t. 31, *and Col. Heaths, t.* 41 ; *Wendl. Eric. Ic. fasc.* 12, 11 ; *Lodd. Bot. Cab. t.* 54 ; *Benth. in DC. Prodr.* vii. 687. *E. laricina, Berg. Desc. Pl. Cap.* 94, *not of Spreng. f. E. vespertina, Linn. f. Suppl.* 221 ; *Benth. l.c.* 686. *E. gnidiœfolia, Salisb. in Trans. Linn. Soc.* vi. 336. *E. lyrigera, Salisb.,* and *E. nigrita, Roxb. ex Salisb. l.c. E. volutœflora, Salisb. l.c.* 335. *E. munda, Salisb. l.c.* 337. *E. acutangula, Lodd. Bot. Cab. t.* 1868. *E. emarginata, Andr.*

Heathery, t. 214.　*E. marginata, Benth., and E. nitidula, Hort. ex Benth. l.c.* 687.　*E. Dickensonia, var. alba, Sinclair, Hort. Eric. Wob.* 8.　*E. cucullata, Tausch in Flora,* 1834, 615, acc. to *Ind. Kew. E. nigrescens, Steud. Nom. ed.* 2, ii. 7.　*E. nigricans, Lodd. ex G. Don in Loud. Hort. Brit.* 151, *fide Ind. Kew. E. divaricata, Lodd. ex Steud. Nom. ed.* 1, 305.

VAR. β, **periplocæflora** (Bolus); flowers reddish or dull red-purple, numerous; anthers terminal, decurrent-crested or aristate, the appendages usually (or always?) more or less adherent to the filament, only partially free, or more rarely muticous. *E. periplocæflora, Salisb. in Trans. Linn. Soc.* vi. 337.

VAR. γ, **fragrans** (Bolus); leaves opposite or 3-nate; flowers fewer and somewhat larger, $2\frac{1}{2}$ lin. long, blueish-purple, or livid: anthers subexserted, terminal, decurrent-aristate, with short free points. *E. fragrans, Andr. Col. Heaths, t.* 95, *and Heathery, t.* 163.

SOUTH AFRICA: without locality, *Herb. Salisbury!* and *cultivated specimens!*

COAST REGION : from Clanwilliam Div. on the north, throughout the Coast Region generally, eastward as far as Grahamstown, usually at an elevation of from 2000 to 5000 ft.; *Thunberg, Niven,* 114, 219*! Burchell,* 7685*! Drège,* 7778*b! Zeyher,* 341! 3253! 3254! 3280! 3286! *MacOwan,* 1260! *Bolus,* 1579! 3294! 3294β*!* 3771! 3991! 5186! 5466! 6337! 9368! *MacOwan, Herb. Norm. Austr.-Afr.,* 36! *Guthrie,* 648a! *Schlechter,* 1733! 5509! 8915! *Wolley Dod,* 1272! 3258! *Galpin,* 3688! *Leipoldt,* 550! 1046! *Marloth,* 2687! Var. β : *Niven,* 57! *MacOwan, Herb. Norm. Austr.-Afr.,* 938! *Bolus,* 5185! 5454! *Guthrie,* 3669! *Schlechter,* 5592! Var. γ : Swellendam Div. ; summit of Zuurbraak Mountain, 4900 ft., *Galpin,* 3693! (*Schlechter,* 5400! and *Zeyher,* 3295! probably belong to this species, but are immature).

The examination of a very large number of specimens has obliged us to unite the four species to which Bentham reduced the still larger number cited above. It is evident that he still suspected the identity of three of these; and our additional material has confirmed the suspicion.　Like all widely distributed species it varies somewhat in habit and in every organ, so that if every variation were regarded as distinctive, it would be necessary to make 8 or even 10 varieties or species.　Those forms which have been collected northwards towards Clanwilliam diverge from the others by a poorer growth and smaller leaves and flowers, the result probably of climatic influences ; but the floral structure differs little ; the anthers are more lateral, but even these pass by gradations into the usual or commoner, more southern, form.　The var. γ looks distinct, but the real differences are not great ; and it is connected with var. β by such a form as *Bolus,* 5454, which has almost exactly the same anther.　One of Bentham's distinctions of *E. fragrans* was the opposite leaves ; but *Galpin's* 3693, which is undoubtedly the same as *Andrews'* fig. t. 163, has the 3-nate leaves of the section.

446. E. comata (Guthrie & Bolus) ; branches densely and shortly tomentulose, at length glabrescent; the younger leaves, also the bracts and sepals ciliate and furnished with a tuft of rather long white hairs at the apex; leaves erect-spreading, imbricate, linear-subulate, sulcate, greyish-pubescent, ciliate, at length naked ; petiole red, 2-$2\frac{1}{2}$ lin. long; flowers on short branchlets, subcorolline; pedicels woolly, scarcely 1 lin. long; bracts subapproximate, incurved, usually lax and spreading, oblanceolate, acute, keeled, ciliate, reddish, $\frac{3}{4}$-$1\frac{1}{4}$ lin. long; sepals somewhat lax and incurved, suborbicular, very concave, keeled, the keel red, the rest hyaline, prominently and somewhat rigidly ciliate, about $\frac{3}{4}$ lin. long; corolla broad-cyathiform, apparently red, $1\frac{1}{2}$–$1\frac{3}{4}$ lin. long ; segments erect, broadly rounded, about as long as the tube or a little less ; filaments linear, equal, about as long as the anther ; anthers included, dorsi-fixed well above the base, broadly oblong, obtuse, scaberulous, ciliolate, dark-coloured, under $\frac{1}{2}$ lin. long, muticous ; pore about $\frac{1}{2}$ the length of the cell ; style exserted, very slender ; stigma simple ; ovary depressed-obconic, wider than its length, silky-villous.

COAST REGION : Swellendam Div. ; on the summit of the craggy peak near Swellendam, 4000–5000 ft., *Burchell,* 7329 !

A very distinct species, the flowers somewhat too corolline for this section, but difficult to place elsewhere. Its nearest ally seems to be *E. lachneæfolia, Salisb.,* which was found by Burchell on the same range of mountains, but further westwards, from which, so far as our material goes, it is easily distinguished at sight.

447. E. saxicola (Guthrie & Bolus) ; dwarf, 3–4 (rarely 6) in. high ; branches many, decumbent or ascending, diffuse, flexuous, often intricate, the ultimate slender, rigid, white-pubescent ; leaves spreading or squarrose, oblong-trigonous, subacute, mostly thick and fleshy, glabrous, mostly 1 (rarely $1\frac{1}{2}$) lin. long ; flowers moderately crowded, on short lateral branchlets, subcalycine or subcorolline ; pedicels about $\frac{1}{2}$ lin. long ; bracts approximate, $\frac{1}{2}$ lin. long ; sepals somewhat spreading, ovate, acute, concave, thick, rigid, keel-tipped, scarious, glabrous, ciliolate, less than 1 lin. long, reaching about as high as the corolla-tube ; corolla broad-cyathiform, white, mostly about 1 (rarely $1\frac{1}{5}$) lin. long, wider when expanded than its length ; segments ascending, rounded, somewhat shorter than the tube ; anthers included, lateral, oblong-cuneate, or narrow-elliptic, obtuse, smooth, brown, $\frac{1}{5}$–$\frac{1}{3}$ lin. long, crested or subulate-aristate ; pore $\frac{1}{3}$ the length of the cell ; crests or awns as long as the cells or a little shorter ; style subincluded or shortly exserted ; stigma capitellate ; ovary oblate-globose, minutely puberulous.

COAST REGION : Bredasdorp Div. ; on the ledges of steep rocks, Riet Fontein Poort, near Elim, 200 ft., *Bolus,* 8508 ! *Schlechter,* 9705 ! Caledon Div. ; hills near Papies Vlei, 300 ft., *Schlechter,* 10445 !

Very distinct in the section by its small leaves and flowers. The anther-cells are very similar in shape and size to those of *E. lachneæfolia* and *E. floccosa,* but there are considerable differences in other respects. Schlechter's specimens show not a few flowers with 6-lobed corollas, 11 or 12 stamens and a 6-celled ovary ; and it is his plants, also, which have the smaller elliptic aristate anthers, mentioned in the description ; these occur in the normal, as well as in the abnormal, flowers.

448. E. floccosa (Bartl. in Linnæa, vii. 640, not of Salisb.) ; erect, 1–$1\frac{1}{2}$ ft. high ; branches many, ascending, straight or flexuous, floccose-tomentose with dense short plumulose hairs ; leaves erect to subpatent, crowded, linear-trigonous, subobtuse, slender, glabrous, $1\frac{1}{2}$–2 lin. long; flowers abundant, crowded on numerous side-branchlets, subcorolline ; pedicels pubescent, about 1 lin. or less long; bracts approximate or subapproximate, ovate or lanceolate, scarious, $\frac{1}{2}-\frac{3}{4}$ lin. long; sepals ovate-orbicular, keeled and keel-tipped, scarious, whitish, about 1 lin. long, reaching about or nearly half-way up the corolla; corolla obconic-campanulate, white, 2 lin. long; segments widely-spreading or subrecurved, ovate, very obtuse, in length equal to the tube; filaments very slender; anthers included, submanifest, sublateral; cells bipartite, broadly oblong, incurved, with a thick dorsal ridge, scaberulous, dark brown or reddish, nearly $\frac{3}{8}$ lin. long, subulate-aristate; pore $\frac{1}{2}-\frac{2}{3}$ the length of the cell; awns hairy or ciliolate or naked, sometimes decurrent for a short distance along the filament at its apex, nearly as long as the cell; style exserted; stigma capitate; ovary glabrous. *Benth. in DC. Prodr.* vii. 686.

COAST REGION: Caledon Div.; on the Zwart Berg, near Caledon, 1300–3400 ft., *Mund*, 35! *Bolus*, 6746! 9167! *Schlechter*, 5542! mountains near the Zondereinde River, *Guthrie*, 3281! and without precise locality, *Ecklon & Zeyher!* Robertson Div.; mountains near Montagu Bath (warm springs), *Bolus*, 8294!

A fairly distinct species. From others which resemble it externally it may be known by its anthers; these are somewhat like those of *E. lachneæfolia*, Salisb., but the awns are not or only slightly adherent to the filament, and the whole aspect of the plant is otherwise very different.

449. E. lucida (Salisb. in Trans. Linn. Soc. vi. 337); erect, stout, 1–1$\frac{1}{2}$ ft. or more high; branches ascending, mostly flexuous, sometimes subvirgate, usually floccose-tomentose, as are the pedicels, with short greyish plumulose hairs, more rarely puberulous; leaves erect, often adpressed, imbricate, rarely somewhat spreading, linear-trigonous, subacute, sulcate, glossy, 1–3 (mostly 2–2$\frac{1}{2}$) lin. long; flowers numerous, mostly crowded at the ends of short lateral branchlets, calycine, glossy; pedicels tomentose or rarely subglabrous, 1–1$\frac{1}{2}$ lin. long; bracts approximate to subremote, ovate, coloured, shorter than the sepals; sepals imbricate below and somewhat spreading at the apex, ovate or obovate, acute or obtuse and cuspidate, keeled, concave, thick, rigid, ciliate or rarely naked, glossy, coloured, 1–1$\frac{1}{4}$ lin. long, reaching $\frac{2}{3}-\frac{3}{4}$ the length of the corolla; corolla campanulate, cyathiform or subglobose-urceolate, only in the last-named form, very slightly contracted at the throat, 1$\frac{1}{4}$–1$\frac{1}{2}$ lin. long, red or rosy; segments spreading, sometimes horizontally, ovate, $\frac{1}{2}-\frac{3}{4}$ the length of the tube; anthers included, lateral, oblong, obtuse; cells bipartite, $\frac{3}{8}-\frac{1}{2}$ lin. long, crested; pore about $\frac{1}{2}$ the length of the cell; crests suborbicular in outline, deeply incised, about $\frac{2}{3}$ the length of the cell; style shortly exserted; stigma capitellate; ovary glabrous. *Andr. Heathery, t.* 172, *and Col. Heaths, t.* 109; *Benth. in DC. Prodr.* vii. 687. *E. rigidifolia, Wendl. Eric. Ic. fasc.* 20, 125, *t.* 48. *E. nitens, Bartl. in Linnæa,* vii. 652.

VAR. β, **pauciflora** (Bolus); branches more slender; leaves fewer, more adpressed and smaller; flowers fewer, less glossy, and a little smaller (1 lin.

long); anthers either as in the type or crested-aristate; awns subulate, acuminate, slightly incised.

VAR. γ, **laxa** (Bolus); as in var. β, but the anthers cristate; crests narrow-ovate, crenulate, about ⅖ the length of the cell. *E. laxa, Andr. Heathery, t.* 73, *and Andr. Col. Heaths, t.* 181; *Benth. in DC. Prodr.* vii. 687. *E. rigescens, Bartl. in Linnæa,* vii. 635, *fide Benth. E. rigens, Benth. l.c.*

SOUTH AFRICA: without locality, *Niven, Herb. Salisbury!* and *cultivated specimens!* Var. γ : *Niven!* and *cultivated specimen!*

COAST REGION: Clanwilliam Div.; Cederberg Range, 2600 ft., *Marloth,* 2690! *Leipoldt!* Tulbagh Div.; Winterhoek Mountain, *Bolus,* 6351! mountains near Saron, *Schlechter,* 10679! Cape Div.; Table Mountain, *Pappe!* Devils Peak, 1100–1500 ft., *Guthrie,* 1217! *Bolus,* 4470! 4590! Paarl Div.; Sneeuw Kop, *Marloth!* French Hoek, *Bolus,* 6985! Caledon Div.; Zwart Berg, *Zeyher,* 3278! Var. β : Tulbagh Div.; Mitchells Pass, 1000 ft., *Bolus,* 5187! mountains near Tulbagh Waterfall, 800 ft., *Bolus,* 5464!

The larger-flowered form of this has some resemblance to *E. calycina,* but the flowers are smaller. Our var. β is more lax in habit and is fewer-flowered, but otherwise seems to agree with the type. Niven's specimen in Herb. Kew., marked *E. laxa* by Bentham, is the only one we have seen in which the anther-crest agrees with Andrews' figure, and we have distinguished it accordingly. Taken as a whole the species differs from all the rest in the section by its corolla, which is not so uniformly spreading upwards, but has a tendency to become cyathiform or even globose-urceolate with a slight contraction at the throat; and in so far it approaches to § *Trigemma.*

450. E. mucronata (Andr. Col. Heaths, t. 186, and Heathery, t. 225); erect, 1½ ft. high ; branches erect or spreading, downy; leaves spreading, incurved, narrow-lanceolate, setaceous-acuminate, mucronate, keeled, glabrous, 4–10 lin. long ; flowers 3-nate (or 3–6-nate, *Andrews*) on short arrested branchlets, subcorolline ; pedicels downy, decurved, 3–4 lin. long ; bracts, 2 approximate imbricating the sepals, 1 remote, lanceolate, setaceous-acuminate, scarious, keeled ; sepals like the bracts, but larger, including the bristle-like point, 2½ lin. long ; corolla obconic-campanulate, rosy, 2½ lin. long (in *MacNab's* 233, a cult. sp. ?); segments slightly spreading, ovate, subacute, about ½ the length of the tube; filaments rather broad at the base, tapering upwards, a little longer than the anther ; anthers included, lateral, narrow-oblong, subacute, minutely scaberulous, dark-coloured, 1 lin. long, 4½ times longer than its greatest width, muticous; pore small, about 4½ times shorter than the cell and near its apex ; style slender, well-exserted ; stigma small, obconic ; ovary glabrous. *Benth. in DC. Prodr.* vii. 683. *E. eriopus, Benth. l.c.* 650.

SOUTH AFRICA: without locality, *Masson!* and *cultivated specimens!*

COAST REGION: Riversdale Div.; mountains near the Kaffirkuils River, *Niven,* 103 ! summit of Kampsche Berg, *Burchell,* 7117 !

This species was placed in § *Melastemon* by Bentham and next to *E. tetrathecoides,* which, in the external appearance of its flowers, it somewhat resembles. But it has not the anther of that section; and also differs in leaves, bracts and sepals. It also resembles in some respects *E. acuta,* but has larger flowers and a different corolla and anthers. Burchell's specimens are not fully developed, and on this Bentham established *E. eriopus* and placed it in § *Platyspora.* But Niven's specimens, 103, afford mature seeds which are ellipsoidal and not at all compressed as in that section.

Section XXXIX. **MELASTEMON.** (Sp. 451–460.)

451. E. seriphiifolia (Salisb. Prodr. 297); erect, 1–1½ ft. high; branches straight and subvirgate, or spreading, subcorymbose or fastigiate; leaves mostly 4-nate or sometimes scattered, erect, imbricate, usually strongly incurved, more rarely straight, linear, acute, glabrous, mostly 1–1½ (rarely 2) lin. long; flowers mostly axillary, generally some also terminal, in dense clusters as if capitate at the ends of the branches, corolline; pedicels slender, floccose (as in *E. cubica*), 2–4 lin. long; bracts remote, linear-spathulate, about 1 lin. long; sepals ovate or lanceolate and acute, or sometimes obovate and retuse with a produced keel-point, often deeply concave, rigid, subscarious, coloured, ½–¾ lin. long, reaching $\frac{2}{5}$–½ the height of the corolla; corolla obconic or sometimes somewhat funnel-shaped, bright red or crimson, 1¼–1¾ lin. long; segments continuous, ovate, rounded or subacute, from ½ the length of the tube to a little longer than it; anthers included, longitudinally semiovate, acute, the cell produced beyond the pore for ½–⅔ the length of the latter, papillose, hispidulous or nearly glabrous, dark-coloured, about ½ lin. long; pore from ½–⅔ the length of the cell; style well-exserted, very slender; stigma subsimple; ovary globular, glabrous, small. *Salisb. in Trans. Linn. Soc.* vi. 331; *Benth. in DC. Prodr.* vii. 683. *E. cubica, var. minor, Andr. Heathery, t. 15; and Col. Heaths, t. 84. E. cubica, Thunb. Diss. Erica, 31, acc. to Salisb. and also Rach. E. bella, Spreng. Syst. Veg.* ii. 197, *acc. to Benth. E. incurva, Thunb. Prodr.* 188, *and Fl. Cap. ed. Schult.* 359? *not of Wendl. nor of Andr. E. inflexa, Pers. Syn.* i. 428?

South Africa: without locality, *Thunberg, Drège,* 7772! 7783! and *cultivated specimens!*

Coast Region: Riversdale Div.; near Garcias Pass, *Burchell,* 7150! George Div.; Outeniqualand, *Niven,* 105! Montagu Pass, 2000 ft., *Schlechter,* 2270! *Guthrie in Herb. Bolus,* 8675! Knysna Div.; near Plettenberg Bay, *Bowie! Burchell,* 5378! mountains near Millwood, 2000 ft., *Tyson in MacOwan, Herb. Austr.-Afr.,* 1496! Uniondale Div.; Long Kloof, *Ecklon & Zeyher,* 211! Uitenhage Div.; Van Stadens Berg, *Burchell,* 4692!

452. E. cubica (Linn. Diss. Erica, 45); erect, 1–1½ ft. or more high; branches straight and subvirgate or spreading, subcorymbose and fastigiate, glabrous, pallid or dark, with more or less prominent, sometimes decurrent, leaf-cushions; leaves 4-nate or scattered (or sometimes apparently 3-nate), very variable in size and setting, from 2–5 lin. long, the shorter erect, crowded, imbricate, strongly incurved, the longer either straighter, erect and imbricate, or incurved and squarrose, all linear, acute, sulcate, glabrous; flowers terminal and axillary, densely umbellate at the ends of the branches, corolline; pedicels slender, floccose with minute distantly-branched or subplumose hairs, 2–9 lin. long; bracts 2 only (? always), remote, linear-spathulate, foliaceous, about 1¼–2½ lin. long; sepals narrow-lanceolate, lanceolate, ovate, obovate or suborbicular, acute, acuminate, keeled or sometimes retuse with an excurrent keel-tip, occasionally by projection of the keel-angles, appearing tetragonous, the broader flattish, the smaller more concave, margins entire or lacerate, scarious, glabrous and glossy, pallid or deep coloured, 1–2 lin. long, reaching from about $\frac{2}{5}$–$\frac{3}{4}$ the height of the corolla; corolla campanuloid or obconic or subfunnel-shaped, pale rosy or deep red, $1\frac{1}{2}$–$3\frac{1}{4}$ lin. long; segments continuous, ovate or semiorbicular, mostly equal in length to the tube, but occasionally shorter or longer; filaments rather broad, tapering upwards, rigid, nerved; anthers included, oblong to semiovate, acute or acuminate, papillose or minutely hairy, $\frac{1}{2}$–1 lin. long, muticous, or sometimes with minute pallid squarrose awns at the apex of the filament; pore about $\frac{2}{5}$ the length of the cell, situate in the middle; style exserted, slender; stigma subsimple, small; ovary glabrous. *Linn. Amœn. Acad.* viii. 56, *and Mant. Alt.* 233; *Wendl. Eric. Ic. fasc.* 11, 13; *Andr. Heathery, t.* 14, *and Col. Heaths, t.* 18; *Lodd. Bot. Cab. t.* 972; *Benth. in DC. Prodr.* vii. 683. *E. hottoniæflora, Salisb. in Trans. Linn. Soc.* vi. 331.

VAR. β, **coronifera** (Bolus); branches pallid, leaf-cushions decurrent, prominent; leaves squarrose or decurved, incurved, 3–5 lin. long; pedicels subsetaceous, 7–9 lin. long; bracts subapproximate or remote; sepals suborbicular and scarious at the base, with a long foliaceous point, the whole $2\frac{1}{2}$ lin. long, reaching from $\frac{3}{4}$ to the height of the corolla; anthers oblong, acuminate, somewhat longer and proportionately narrower and rougher than in the type, pale brown, nearly 1 lin. long. *E. coronifera, Benth. in DC. Prodr.* vii. 683.

VAR. γ, **natalensis** (Bolus); branches pallid, flexuous, slender, leaf-cushions decurrent, prominent; leaves more slender than in β, erect or spreading, but not squarrose, at least as to the uppermost, 3–5 lin. long; pedicels about 7 lin. long; bracts remote; sepals orbicular or lanceolate; other characters as in var. β.

COAST REGION: Swellendam Div.; mountains near Swellendam, *Masson,* 27! *Burchell,* 7392! *Zeyher,* 3246! *Schlechter,* 5677! *Borcherds in Herb. Norm. Austr.-Afr.,* 1305! Zuurbraak Mountain, *Galpin,* 3681! Riversdale Div.; Platte Kloof, *Masson,* 125! Garcias Pass, *Galpin,* 3679! George Div.; Montagu Pass, 2000 ft., *Young in Herb. Bolus,* 5522! Knysna Div.; Millwood, *Tyson,* 3165! Uniondale Div.; mountains near Avontuur, *Bolus,* 2393! Humansdorp Div.; Clarkson, 800 ft., *Galpin,* 3678! Uitenhage Div.; Van Stadens Berg, *Miss West,* 5! Var. β: George Div.; between Touw River and Kaymans River, *Burchell,* 5779!

EASTERN REGION: Var. γ: Natal; Fields Hill, Umbilo River, *Sutherland,* 922! north of Umzimkaba River, *Sutherland!* Inanda, *Wood,* 970! and without precise locality, *Mrs. K. Saunders!*

453. E. tetrathecoides (Benth. in DC. Prodr. vii. 683); diffuse, entirely glabrous, probably under a foot high; branches, pedicels and sepals smooth and somewhat glossy; branches slender, flexuous, straggling, distantly leafy above, naked below; leaves notably scanty and mostly shorter than the internodes, spreading, incurved, linear-subterete, acute, 3–$4\frac{1}{2}$ lin. long; flowers terminal, or a few also axillary at the ends of the branches, spreading or cernuous, corolline; pedicels very slender, 3–$4\frac{1}{2}$ lin. long; bracts remote, very small, 2 above the middle, 1 basal; sepals subspreading, oblong, scarious, with a thick keel far-excurrent at the apex, the whole 2 lin. long; corolla obconic, $2\frac{1}{2}$–3 lin. long, mouth much widened;

segments continuous, broadly rounded, equal to the tube or somewhat longer ; filaments rather broad, with a strong median nerve ;
anthers included, narrow-lanceolate, acuminate ; cell produced beyond
the pore for about $\frac{3}{4}$ the length of the latter, papillose, minutely
ciliolate on the margins, fox-coloured, $\frac{1}{2}$–1 lin. long, muticous ; pore
less than $\frac{1}{2}$ the length of the cell ; style exserted, very slender ;
stigma subsimple ; ovary glabrous.

COAST REGION : Riversdale Div. ; Garcias Pass, *Burchell,* 7027 !

454. E. humifusa (Hibbert ex Salisb. in Trans. Linn. Soc. vi.
332) ; erect or procumbent, 6–10 in. or more high, almost entirely
glabrous ; branches flexuous, subglabrous, with whitish scars of leaf-
cushions ; leaves 3-nate or opposite, erect, adpressed, somewhat
imbricate or about as long as the internodes, not crowded, oblong or
elliptic, obtuse, round-backed, thickish, sulcate, smooth, $\frac{3}{4}$–1 lin.
long ; flowers subcorolline ; pedicels slender, about 1 lin. long ;
bracts subapproximate, imbricating the sepals but shorter, ovate or
lanceolate, keeled, scarious, whitish, $\frac{1}{2}$–$\frac{3}{4}$ lin. long ; sepals like the
bracts, adpressed, very concave, $\frac{3}{4}$ lin. long or a little more, not
reaching to the top of the corolla-tube ; corolla funnel-shaped,
veined, red or rosy, $1\frac{1}{4}$–$1\frac{3}{4}$ lin. long ; tube narrow-obconic ; segments
more spreading, semiovate, very obtuse, from equal in length to the
tube to twice as long ; filaments slender, tapering to the apex ;
anthers included, manifest, $\frac{2}{5}$–$\frac{1}{2}$ lin. long or more, the shorter subovate, acute, the longer lanceolate or semilanceolate, acuminate ;
cell produced beyond the pore for from $\frac{3}{4}$–1 of the length of the
latter, muticous ; pore about $\frac{1}{2}$ the length of the cell, scabrid,
foxy-brown ; style exserted ; stigma capitellate ; ovary glabrous.

SOUTH AFRICA : without locality, *Herb. Salisbury !* procumbent on the rocks
in shady places among the mountains, *Niven,* 37 !
COAST REGION : Clanwilliam Div. ; on the Cederberg Range, 2500 ft.,
Marloth, 2682 ! Worcester Div. ; Matroos Berg, 5500 ft., *Marloth,* 2245 !
Stellenbosch Div. ; Hottentots Holland, *Mulder* ex *Salisbury.* Somerset Div. ;
rocky places, summit of Bruintjeshoogte Mountain, 4500 ft., *MacOwan,* 1648 !

We have seen and examined specimens of Salisbury's type, and of all the
others cited, and find them to agree fairly well. Marloth's specimens and
MacOwan's have 3-nate leaves. In the others they are opposite so far as we
have seen ; but Bentham must have seen some also 3-nate. By its thin whip-
like branches it looks distinct from any other species in the section.

455. E. cristæflora (Salisb. in Trans. Linn. Soc. vi. 332) ; erect,
1–2 ft. high ; branches ascending, stout, puberulous ; leaves erect-
spreading, imbricate, rarely subsquarrose, linear, subobtuse, glabrous,
rarely puberulous or canescent, glabrescent, 1–3 lin. long ; flowers
scanty or more or less densely clustered, corolline ; pedicels puberulous, 1–$2\frac{1}{2}$ lin. long ; bracts laxly subapproximate, sometimes subremote, ovate, acute, keeled, scarious, coloured, rosy or pallid, $\frac{1}{2}$–$1\frac{1}{4}$
lin. long ; sepals like the bracts but larger and broadly elliptic or
suborbicular, obtuse or subacute, $\frac{3}{4}$–$1\frac{1}{4}$ lin. long, mostly reaching to
about the top of the corolla-tube ; corolla very broadly obconic,

much widened to the mouth, very variable in size, 1–2¼ lin. long ; segments continuous with the tube, semiovate, varying from a little longer than the tube to a little less than ½ its length ; anthers included, subovate, acute or acuminate, more or less villous, or papillose and glabrous, ⅖–1 lin. long ; pore ¼–½ the length of the cell, which is produced above the pore for from ½–1½ times the length of the latter ; style shortly exserted ; stigma obconic, very small, rarely subpeltate ; ovary glabrous. *E. melanthera, Thunb. Diss. Erica*, 16, *not of Linn., teste Salisb. & Rach.*

Var. β, **blanda** (Bolus) ; habit mostly, but not always, more diffuse ; branches more flexuous ; leaves somewhat shorter and more spreading ; flowers smaller in all parts, 1–1½ lin. long ; stigma subpeltate, with 4 tooth-like interior processes. *E. blanda, Salisb. in Trans. Linn. Soc.* vi. 331. *E. cornuta, Roxb. ex Salisb. l.c.; Benth. in DC. Prodr.* vii. 684. *E. suavis, Bartl. in Linnæa,* vii. 641, *fide Benth.*

SOUTH AFRICA : without locality, *Herb. Salisbury !* Var. β : *Herb. Salisbury !*

COAST REGION : Tulbagh Div.; Winterhoek Berg, *Bolus,* 5182 ! Tulbagh Waterfall, *Bolus,* 5183 ! near Saron, *Schlechter,* 10666 ! Mitchells Pass, *MacOwan, Herb. Norm. Austr.-Afr.,* 35 ! *Bolus,* 5284 ! 9810 ! *Schlechter,* 8936 ! 9973 ! *Guthrie,* 625b ! Paarl Div.; French Hoek, *Niven,* 100 ! Bains Kloof, *Bolus,* 4056 ! Cape Div.; near Cape Town, *Thunberg,* Table Mountain, 2000 ft., *Schlechter,* 14 ! Var. β : Clanwilliam Div.; Cederberg Range, on Sneeuw Kop, 4500 ft., *Bodkin in Herb. Bolus,* 8679b ! *Marloth,* 2688 ! *Leipoldt,* 214 ! 744 ! Tulbagh Div.; near Saron, *Schlechter,* 10692 ! Cape Div.; Devils Peak and Table Mountain, 1200–2000 ft., *Bolus,* 3352 ! 4751 ! *MacOwan, Herb. Norm. Austr.-Afr.,* 754 ! *Wolley Dod,* 1681 !

CENTRAL REGION : Ceres Div.; Cold Bokkeveld, 5000 ft., *Schlechter,* 9996 !

This species varies considerably in aspect, and extreme forms with pallid often subsquarrose, sometimes even silky leaves (as in Schlechter's 9973), seem distinct, and worthy of separation as a variety, were they not connected by intermediate forms with the type. The same remarks apply in part to our var. β, which has little to distinguish it beyond the smaller size of the flowers. The subpeltate stigma even, relied upon by Bentham as a distinguishing character for *E. cornuta,* is exhibited in Schlechter's 9996 and in Bolus 5182, which in every other respect have the habit, leaves and flowers of the typical *cristæflora,* group.

456. E. moschata (Andr. Heathery, t. 226, not of Lodd) ; erect, 1 ft. or more high ; branches few, puberulous, with lateral floriferous branchlets 1–1½ in. long ; leaves erect or somewhat spreading, straight or incurved, imbricate but scarcely crowded, lanceolate-linear, subobtuse, pallid, glabrous, the younger ciliolate, 2–3 lin. long ; flowers mostly cernuous, not clustered, corolline ; pedicels pubescent, 1½–2½ lin. long ; bracts subremote or subapproximate, lax, ovate or cordate, acute, keeled, scarious, coloured, about 1¼ lin. long ; sepals, 2 orbicular, 2 narrower and ovate, obtuse, apiculate, otherwise like the bracts and about as long, rarely reaching to the top of the corolla-tube ; corolla obconic or subfunnel-shaped, widely spreading to the mouth, rosy, 2½–4 lin. long ; segments continuous, ovate, rounded, variable in length, from less than ½ the tube to a little longer than it ; anthers included, the lower portion suborbicular, produced above into a contracted acuminate apex for from 1–1½ times the length of the pore, coarsely shaggy and ciliate, foxy-brown, $\frac{2}{3}$–$\frac{3}{4}$ lin. long; pore about $\frac{1}{3}$ the length of the cell; style exserted, very slender; stigma obconic, very small; ovary glabrous. *Andr. Col. Heaths, t.* 248; *Benth. in DC. Prodr.* vii. 683. *E. florida, Lodd. Bot. Cab. t.* 234, *not of Thunb. E. anthina, Spreng. Syst. Veg.* ii. 196, *acc. to Benth. & Rach.*

SOUTH AFRICA: without locality, *Miller!* Herb. *Salisbury!* and *cultivated specimens!*

COAST REGION : Tulbagh Div.; Witsen Berg, *Ecklon & Zeyher!* Winterhoek Berg, 4000 ft., *Bolus,* 6350! Worcester Div.; Brand Vlei, *Masson,* 36! Dutoits Peak, 4800 ft., *Marloth,* 2412 !

Closely allied, perhaps too closely, to *E. cristæflora.* It differs by the larger flowers, larger, more acuminate and more shaggy anthers, and also somewhat in habit; the flowers being fewer, less clustered, more cernuous, the branches fewer, the leaves longer and paler. The flowers are fragrant as noted by Andrews, and also by Marloth, who likens their odour to that of *Disa graminifolia, Ker-Gawl.*

457. E. lavandulæfolia (Salisb. in Trans. Linn. Soc. vi. 332); subarboreous, 3–8 ft. high; branches subvirgate, pubescent, with many floriferous branchlets, forming towards the top of the branches a somewhat thyrsoid inflorescence; leaves erect, mostly incurved, linear, acute, round-backed, sulcate, glabrous, 2–3$\frac{1}{2}$ lin. long; flowers subcorolline; pedicels decurved, slender, white-pubescent, 1–1$\frac{1}{2}$ lin. long; bracts approximate, imbricating the sepals, or rarely the lowest remote, lanceolate, acute, keeled, ciliate, scarious, glabrous, whitish or pale rose, $\frac{1}{2}$ lin. long; sepals like the bracts but ovate, very concave, about 1$\frac{1}{4}$ lin. long, reaching to about $\frac{1}{2}$ the height of the corolla ; corolla nearly broad-obconic, or very slightly funnel-shaped, rosy, 1$\frac{3}{4}$–2 lin. long; segments continuous with the widening tube, ovate or oblong, about $\frac{1}{2}$ the length of the tube; filaments lanceolate at the base tapering upwards; anthers included or sometimes subexserted, longitudinally semiovate, the dorsal margin curved, the inner nearly straight, acute, the cell extending beyond the pore for about $\frac{1}{2}$ the length of the latter, rough but not very hairy, about $\frac{3}{4}$ lin. long, muticous; pore narrow, about $\frac{1}{2}$ the length of the cell; style slender, well-exserted; stigma capitellate, small; ovary small, minutely puberulous. *Benth. in DC. Prodr.* vii. 684. *E. monticola, Klotzsch in Herb. Berol. ex Benth. l.c.*

SOUTH AFRICA : without locality, *Masson !*

COAST REGION : Tulbagh Div.; near Tulbagh (Roode Zand), *Niven,* 102 ! 800 ft., *Guthrie,* 3113! *Marloth,* 2911 ! Stellenbosch Div.; Hottentots Holland, **ex** *Salisbury.*

458. E. Gillii (Benth. in DC. Prodr. vii. 684) ; "habit of *E. moschata*" (*Bentham*); branches floccose with minute compound, distantly subplumose greyish hairs; leaves subrecurved, spreading, imbricate and crowded (on the ultimate branchlets), linear-trigonous, obtuse, sulcate, glabrous, pallid, 1$\frac{1}{2}$ lin. (or probably more) long; pedicels floccose, 2 lin. long; flowers calycine; bracts subremote, the uppermost one only imbricating the sepals, lax and somewhat spreading, ovate, acute, complicate, keeled, membranous, pallid, about $1\frac{1}{2}$ lin. long; sepals like the bracts but larger and flatter, broad-ovate or suborbicular, imbricate above the base, $1\frac{5}{8}$ lin. long, by $1\frac{1}{2}$ lin. wide, reaching a little beyond the top of the corolla-tube; corolla broad-cyathiform, 2 lin. long; segments erect, broadly rounded, about $\frac{2}{3}$ the length of the tube; anthers included, sub-cuneate, acute, scaberulous or roughly papillose, nearly black, $\frac{5}{8}$ lin. long; pore about $\frac{1}{2}$ the length of the cell, placed in the middle of the cell; style slender, exserted; stigma minute, obconic or subsimple; ovary puberulous.

COAST REGION: Mossel Bay Div.; Attaquas Kloof, *Gill!* in Herb. Kew.

459. E. nervata (Guthrie & Bolus); branches hirsute (1 lin. in diam. in our small specimen), the lower parts covered with the persistent short stumps of old flowering branchlets, the upper densely leafy; leaves erect, rigid, strongly incurved, crowded, imbricate, axils gemmiferous, broad-lanceolate, acute, from deeply sulcate to open-backed, the revolute margins flat and wide, long-ciliate, the younger hispid, glabrescent, pallid, $1\frac{1}{2}$–2 lin. long; flowers terminal on very short arrested branchlets and thus appearing axillary, solitary, subcorolline; pedicels woolly with longish white hairs, 2–$2\frac{1}{2}$ lin. long; bracts remote, basal, minute; sepals incurved, broad-ovate or suborbicular, shortly acute, very concave, cartilaginous, rigid, with about 9 distinct parallel nerves, ciliate, pale rosy, about 1 lin. long or a little more, not reaching to the top of the corolla-tube; corolla broad-cyathiform, red, $2\frac{1}{2}$ lin. long; segments continuous with the tube, suberect, broad-ovate, about equal to it in length; filaments rigid, shorter than the anther; anthers included, longitudinally semilanceolate, acute, the cell produced beyond the pore for about $\frac{2}{3}$ the length of the latter, scaberulous, dark-coloured, $\frac{7}{8}$ lin. long, muticous; pore about $\frac{1}{4}$ the length of the cell; style included; stigma subsimple; ovary densely woolly.

COAST REGION: Oudtshoorn Div.; grassy places, Zwarteberg Pass, 4900 ft., *Marloth,* 2409!

460. E. trachysantha (Bolus in Journ. Linn. Soc. xxiv. 184); erect, rather slender, not much branched, under 1 ft. high; branches (except in the lowest parts), under surface of the leaves, pedicels and especially the sepals, densely hirsute with shaggy dirty-white hairs; leaves crowded, fasciculate in tufts, shorter or a little longer than the internodes, erect-incurved, ovate-lanceolate, acute, open-backed, with reflexed wide flattish margins, subglabrous on the upper surface, 2–$2\frac{1}{3}$ lin. long; flowers mostly solitary on short branchlets, spreading, erect or cernuous, calycine; pedicels 4–$6\frac{1}{2}$ lin. long; bracts remote, basal, very small; sepals slightly united at the base, ovate, acute, $1\frac{1}{2}$–2 lin. long; corolla cyathiform, nearly equal at the mouth, 2–$2\frac{1}{2}$ lin. long; segments equal to or a little longer than the tube; anthers included, subulate or sublanceolate, acuminate, recurved, 1 lin. long; pore $\frac{1}{4}$–$\frac{1}{3}$ the length of the cell ; style just exserted ; stigma small, capitellate or subsimple ; ovary subconic, hispidulous.

COAST REGION : Uniondale Div. ; Kouga Mountains, between Uniondale and Avontuur, *Bolus*, 2387 !

This is most nearly allied to *E. Passerinæ* (§ *Gamochlamys*), and is a connecting link between the present and that section. It may be readily known from that species by its different indumentum, larger and flatter leaves, longer pedicels, and more deeply divided calyx. It is also related to the preceding.

Section XL. GAMOCHLAMYS. (Sp. 461–466.)

461. E. Passerinæ (Montin in Act. Nov. Upsal. ii. 289, t. 9, fig. 1) ; erect, about 2 ft. high, tomentose in the upper part, glabrescent below ; branches, leaves, pedicels, bracts, sepals and ovary, but especially the sepals, uniformly covered with a short closely-matted white tomentum ; leaves ovate to elliptic, obtuse, thick, deeply sulcate, or sometimes subopen-backed, at length glabrescent, 1–1$\frac{1}{2}$ lin. long ; flowers on very short lateral branchlets, solitary or in pairs, erect or pendulous, from subcorolline to subcalycine ; pedicels 2 lin. long ; bracts remote, subbasal, small ; sepals more or less united at the base, ovate, about $\frac{1}{2}$ the length of the corolla, 1$\frac{1}{2}$ lin. long ; corolla cyathiform, mouth scarcely widened or contracted ("red," *Thunberg*), 3 lin. long ; segments erect, rounded, $\frac{1}{4}$–$\frac{1}{3}$ the length of the tube ; filaments short and rather broad, less than $\frac{1}{2}$ the length of the anther ; anthers subulate, acute, recurved, about $\frac{5}{6}$ lin. long ; pore $\frac{1}{4}$–$\frac{1}{3}$ the length of the cell ; style included ; stigma capitellate, very small ; ovary conical, densely tomentose. *Linn. f. Suppl.* 221 ; *Thunb. Diss. Erica*, 18 ; *Benth. in DC. Prodr.* vii. 681. *E. passerinœfolia, Salisb. in Trans. Linn. Soc.* vi. 332.

SOUTH AFRICA : without locality, *Thunberg*, *Paterson! Miss Cummings in Herb. Bolus*, 9286 !

COAST REGION : Uniondale Div. ; Long Kloof, *Masson*, 63 !

Closely allied to the two preceding species (*E. nervata* and *E. trachysantha*) all being distinct in the genus. It is also noteworthy that all inhabit the same general region, so far as their habitat is known to us.

462. E. melanthera (Linn. Diss. Erica, n. 37 ; Mant. Alt. 232, not of Lodd.) ; erect, about 1–2 ft. (acc. to *Niven*, 5–6 ft.) high ; branches ascending or widely spreading, pubescent ; leaves mostly spreading, more rarely squarrose, from linear to oval, blunt, thick, sulcate, minutely tuberculate-hispid or glabrous, at length usually glossy, 1–2 lin. long ; flowers generally abundant, corolline ; pedicels puberulous, 1–2 lin. long ; bracts from subremote to subapproximate, linear, small, sometimes larger and somewhat imbricating the calyx ; calyx obconic, 4-cleft, puberulous or glabrous, keeled throughout or only keel-tipped, coloured, about 1 lin. long ; segments reniform or obcuneate, apiculate, as long as the tube or somewhat longer or shorter; corolla obconic, broad-cyathiform or funnel-shaped, pale or bright red, $1\frac{3}{4}$–2 lin. long; segments continuous or. spreading, broadly rounded, about equal to the tube or from somewhat longer to shorter; filaments rather broad; anthers obliquely oblong or longitudinally semiovate, produced above into a contracted point about or nearly equal to· the pore, scabrid, nearly black, about $\frac{1}{2}$ lin. long; pore $\frac{1}{2}$–$\frac{5}{8}$ the length of the cell; style exserted; stigma obconic or subsimple, minute; ovary silky-villous, chiefly on the summit. *Benth. in DC. Prodr.* vii. 684. *E. lysimachiæflora, Salisb. in Trans. Linn. Soc.* vi. 332. *E. mundtiana, Klotzsch ex Benth. l.c.* 684. *E. varia, Lodd. Bot. Cab. t.* 1325; *Benth. l.c.* 685. *E. caroliniana, Hort. ex Benth. l.c. E. muricata, Wendl. fil. ex Benth. l.c. E. leiophylla, Benth. l.c.* 684. *E. jubata, Lodd. ex Spreng. Syst.* ii. 198, *acc. to Ind. Kew.*

SOUTH AFRICA: without locality, *Drège*, 7773! *Paterson! Herb. Salisbury!* and *cultivated specimens!*

COAST REGION: Swellendam Div.; near Swellendam, *Niven*, 101! *Mund*, 20! Voormans Bosch, *Zeyher*, 3249! Langeberg Range, *Schlechter*, 5659! Riversdale Div.; near Riversdale, *Hewitt in Herb. Bolus*, 3687! between Garcias Pass and Krombecks River, *Burchell*, 7189! Garcias Pass and Muis Kraal, 1000–1500 ft., *Burchell*, 7039! *Galpin*, 3683! 3684! George Div.; Devils Kop, *Niven!* Montagu Pass, 4000 ft., *Marloth*, 2403! Uniondale and Humansdorp Div.; Long Kloof and Kromme River, *Masson*, 62! Kouga Mountains, near Avontuur, *Bolus*, 2389! near Haarlem, *Galpin*, 3685! Knysna Div.; Paarde Berg, *Burchell*, 5191! Keurbooms River, *Burchell*, 5134! Uitenhage Div.; Van Stadens Berg, *Zeyher*, 792! 3250!

We have examined Niven's specimen of *E. varia*, so named in Bentham's hand, but can find no sufficient characters to separate it, and Loddiges' figure (upon which the species appears to have been founded) affords no additional evidence. Bentham observes that the leaves are not glossy, but in Niven's specimen some, at least, are· so. The type of *E. leiophylla, Benth.*, collected by Paterson appears to be nothing but a starved state of this, varying in no essential character.

463. E. Chamissonis (Klotzsch ex Benth. in DC. Prodr. vii. 685); erect, 1–2 ft. or more high; branches stout, ascending, pubescent; leaves crowded, suberect or often spreading-incurved, linear, sulcate or often more or less open-backed, usually hispid and rough with longish white tubercle-based hairs, the tubercles persistent after the hairs have dropped, $1\frac{1}{2}$–$2\frac{1}{2}$ lin. long; flowers copious, along the secondary and tertiary branchlets, corolline; pedicels densely villous, 2–$2\frac{1}{2}$ lin. long; bracts remote, villous, minute; calyx cyathiform, densely villous with white hairs, $\frac{3}{4}$–$1\frac{1}{4}$ lin. long; segments deltoid or semiovate, acute, variable from about equal to the tube to $\frac{1}{2}$ its length, reaching $\frac{1}{3}$–$\frac{1}{2}$ the height of the corolla; corolla broad-cyathiform, $1\frac{3}{4}$–$2\frac{1}{4}$ lin. long; segments erect, suborbicular, strongly nerved near the apex, as long as or sometimes a little shorter than the tube; anthers dorsifixed well above the base, oblong or suboblanceolate, obtuse, not or only very slightly, produced beyond the pore, rough and ciliolate on the margins or subglabrous, $\frac{1}{2}$–$\frac{3}{4}$ lin. long; style exserted, slender, dilated and obconic at the apex; stigma simple;

ovary turbinate, silky-villous. *E. polyantha, Klotzsch ex Benth. in DC. Prodr.* vii. 688.

SOUTH AFRICA : without locality, *Chamisso, Wallich!*
COAST REGION : Uitenhage Div. ; without exact locality, *Zeyher;* Van Stadens Berg (*E. polyantha, Kl.*) *Zeyher,* 218 ! 786 ! Port Elizabeth Div.; *Kemsley,* 357 ! *Miss West,* 3 ! Albany Div.; near Grahamstown, *Burchell,* 3557 ! *MacOwan,* 33 ! *Tyson in MacOwan & Bolus, Herb. Norm. Austr.-Afr.,* 983 ! *Guthrie,* 2369 !

We have examined the types of *E. polyantha,* and do not doubt that they are merely poorly-grown plants of this species, of which they have all the characters, but of a reduced size, and all parts (especially the calyx) less hairy.

464. E. longipes (Klotzsch ex Benth. in DC. Prodr. vii. 684, not of Bartl.) ; branches erect, rigid, rather rough, pubescent with simple hairs, 10 in. or more long ; leaves erect-incurved, imbricate, linear, deeply sulcate, hispidulous or glabrous, $\frac{3}{4}$–$1\frac{1}{4}$ lin. long ; flowers on short lateral branchlets, somewhat copious, corolline ; pedicels puberulous or densely woolly, $1\frac{1}{2}$–$2\frac{1}{2}$ lin. long ; bracts remote, very small ; calyx obconic, thick, rigid, scarious, glabrous, concave, strongly keeled, the keel prominent from the very base and the whole thus somewhat truncate at the base, 1 lin. long, or slightly less ; segments ovate, erect, acute, ciliolate or naked, 3–4 times the length of the tube ; corolla between funnel-shaped and campanulate, rosy, about 2 lin. long ; segments slightly spreading, rounded, about as long as the tube, or sometimes longer ; anthers dorsifixed above the base, oblong, obtuse, minutely scaberulous, reddish-brown, $\frac{1}{2}$–$\frac{3}{5}$ lin. long ; pore $\frac{1}{3}$–$\frac{2}{5}$ the length of the cell ; style exserted, slender ; stigma subsimple or capitellate ; ovary turbinate, silky-villous on the conical summit.

SOUTH AFRICA: without locality, *Masson ! Mund!*
COAST REGION : Knysna Div. ; near Forest Hall, *Miss Newdigate,* 63 ! Uitenhage Div. ; Grass Ridge, *Ecklon & Zeyher.*

We have not seen Ecklon & Zeyher's type ; but have examined Masson's and Mund's specimens, both named by Bentham, and Miss Newdigate's, all of which agree well.

465. E. canaliculata (Andr. Heathery, t. 156, and Col. Heaths, t. 157) ; erect, reaching to 6 ft. high (*Galpin*) ; branches ascending, greyish-puberulous, in some specimens floccose with minute compound hairs, with many subverticillate spreading copiously floriferous branchlets ; leaves erect-spreading, very straight, linear, deeply sulcate or more or less open-backed, showing a paler tomentulose under-surface, scabrid-puberulous above, sometimes becoming glabrous and smooth, 2–3 lin. long ; flowers apparently normally 3-nate, but often by arrest of the lateral branchlets so crowded as to appear umbellate with 4–6 flowers, subcorolline ; pedicels slender, puberulous, 2–$2\frac{1}{4}$ lin. long ; bracts remote, minute ; calyx glabrous, roughly papillose, pallid outside, red within, the central line closely adpressed to the corolla, deeply 4-fid ; segments ovate, acute, 2–4 times the length of the tube, ciliate at the base, sides more or less strongly reflexed and the whole then tetragonous, $\frac{3}{4}$–1 lin. long, reaching generally about $\frac{1}{2}$ the height of the corolla; corolla broadish-cyathiform or subcampanulate, tetragonous, rosy, $1\frac{1}{2}$–$1\frac{3}{4}$ lin. long; segments not, or only slightly, spreading, broad, about as long as the tube; anthers subexserted or only manifest, dorsifixed well above the base, oblong, obtuse, $\frac{2}{5}$–$\frac{1}{2}$ lin. long; style exserted, slender; stigma simple; ovary velvety or glabrous. *Benth. in DC. Prodr.* vii. 688. *E. melanthera, Lodd. t.* 867 ? *not of Linn. nor of Thunb.*

SOUTH AFRICA : without locality, *cultivated specimens!*

COAST REGION: George Div.; woods near George, 900 ft., *Schlechter,* 2329! Barbiers Kraal, near Devils Kop, *Niven,* 97! Humansdorp Div. ; Witte Els Bosch, Zitzikamma, 500 ft., *Galpin,* 3698!

We have taken as the type Andrews' figure and Niven's 97, named by Bentham, and have described from these and the two more recent gatherings cited above, which agree very closely. The species is a very distinct one by its peculiar calyx. Galpin's and Niven's specimens both exhibit compound hairs.

466. E. natalitia (Bolus in Journ. Linn. Soc. xxiv. 187); erect; branches numerous, ascending, cano-puberulous; leaves erect, or the older spreading, linear to oblong, acute, deeply sulcate or open-backed, mostly glabrous, sometimes hispidulous, 1–$2\frac{1}{2}$ lin. long; flowers 3-nate, sometimes umbellately clustered, subcalycine; bracts remote, basal, small; calyx more or less deeply 4-fid, about $\frac{3}{8}$ lin. long; segments deltoid or ovate, subacute, keeled or keel-tipped, about equal to the tube or only a little shorter or longer; corolla cyathiform or subobconic, pink, about $\frac{1}{2}$ lin. long; segments divergent, somewhat longer than the tube; anthers oblong, obtuse, smooth, about $\frac{1}{4}$ lin. long; pore less than $\frac{1}{2}$ the length of the cell; style shortly exserted, sometimes dilated at the apex; stigma peltate or cyathiform; ovary glabrous.

EASTERN REGION : Natal; Indwedwe, 2000 ft., *Wood,* 990! near Emberton, 1000 ft., *Schlechter,* 3230! on mountains occasionally snowed, 4000–5000 ft., *Sutherland!* Zululand; N'Kandhla, *Wood,* 7301!

Distinct in the section by its small flowers, and proportionately large peltate or cyathiform stigma. It connects with the § *Arsace,* differing by its calyx and its subcalycine flowers.

Section XLI. **CYATHOLOMA.** (Sp. 467–469.)

467. E. Thunbergii (Montin in Act. Nov. Upsal. ii. 292, t. 9, fig. 2); erect, 6–12 in. high, entirely glabrous or the branches sometimes downy; leaves erect, imbricate, linear, sulcate, pallid or subglaucous, 2–3 lin. long; flowers few or copious on short branchlets towards the ends of the branches, subcalycine; pedicels slender, spreading or decurved, 3–6 lin. long; bracts remote, lanceolate, acute or acuminate, keel-tipped, subscarious, bright canary-yellow, 2–3 lin. long; sepals like the bracts but oblong or ovate, large and conspicuous, more acuminate, $2\frac{1}{2}$–$3\frac{1}{2}$ lin. long; corolla subfunnel-shaped, the lower part somewhat globose, constricted above, yellowish, $1\frac{1}{2}$–2 lin. long, the upper part broad-cyathiform, red, 3 lin. long, the whole $4\frac{1}{2}$–5 lin. long; segments erect, broad-ovate, divided from a longer or shorter distance above the constriction, about 2–$2\frac{1}{2}$ lin. long; filaments slender, dilated at the apex; anthers dorsifixed above the base, oblong, tapering to the apex, obtuse or subacute, curved forwards (subprognathous) at the base, papillose, brown, nearly 1 lin. long, muticous; pore $\frac{1}{3}$–$\frac{2}{5}$ the length of the cell; style subincluded; stigma simple; ovary turbinate, glabrous, shortly stipitate. *Linn. f. Suppl.* 220; *Bot. Mag. t.* 1214; *Lodd. Bot. Cab. t.* 277; *Benth. in DC. Prodr.* vii. 648. *E. Thunbergia, Andr. Heathery, t.* 244, *and Col. Heaths, t.* 282. *E. medioliflora, Salisb. in Trans. Linn. Soc.* vi. 331.

SOUTH AFRICA: without locality, *Herb. Salisbury!* and *cultivated specimens!*

COAST REGION: Clanwilliam Div.; Ezelsbank, on the Cederberg Range, 3000 ft., *Drège!* Sneeuwkop, 5800 ft., *Leipoldt,* 617! *Thode,* 70! *Bodkin in Herb. Bolus,* 6882!

CENTRAL REGION: Ceres Div.; Cold Bokkeveld, *Thunberg,* chiefly on the farms Waarde Drift and Rietfontein, 24–42 miles north of Ceres, 3000–3500 ft., *Carson in Herb. MacOwan,* 2778! *MacOwan, Herb. Norm. Austr.-Afr.,* 959!

A very distinct species unlike any other in floral character, or in aspect, with the single exception of *E. flavisepala.* Bentham in DC. Prodr. vii. 649, quotes *E. celsiana, Lodd. Bot. Cab. t.* 1777, as a variety of this, with a larger rosy corolla-tube and darker limb, and with narrower, paler bracts and sepals. He thought it perhaps a garden hybrid, and we have seen no specimens like it. It is more unlike *E. flavisepala* in both aspects and structure.

468. E. Corydalis (Salisb. in Trans. Linn. Soc. vi. 334); erect, 8–10 in. high; branches ascending, subvirgate or sometimes spreading, puberulous; leaves recurved or squarrose, crowded or somewhat lax, lanceolate to oblong, acute, flat or subconcave above, sulcate below, thick, glabrous, glossy, $1\frac{1}{2}$–$2\frac{1}{2}$ lin. long; flowers somewhat scanty along the branches, either (in well-grown specimens) on short branchlets bearing only a few bract-like leaves, or (by arrest of the branchlets) pseudo-lateral, corolline; pedicels rather stout, puberulous, subviscid, 1–2 lin. long; bracts approximate, ovate, viscid, $\frac{1}{2}$–$\frac{3}{4}$ lin. long; sepals broad-ovate or obovate, acute, keel-tipped, leathery, viscid, coloured, 1–$1\frac{1}{4}$ lin. long; corolla subrotate; tube short hemispherical, about 1 lin. long; segments suddenly and widely spreading, very broad and obtuse, about $1\frac{1}{4}$–$1\frac{1}{2}$ lin. long, the whole dry, white, 2–$2\frac{1}{2}$ lin. long, when flattened out; anthers manifest, very broad-oblong or subquadrate, with a dorsal and apical ridge or entire crest, membranous at the apex, sometimes expanded into a wing-like process on either side, thickened downwards into awns or muticous; cells separate, papillose-scabrid, $\frac{1}{2}$ lin. long, aristate or subcristate; pore less than $\frac{1}{2}$ the length of the cell; awns or crests (when present) subulate, thick, entire or sometimes broader and dentate, acute, about $\frac{1}{4}$ the length of the cell; style exserted slender; stigma capitellate; ovary glabrous. *Benth. in DC. Prodr.* vii. 649. *E. complanata, Nois. ex Spreng. Syst. Veg.* ii. 196 ? *E. crassifolia, Benth. l.c.* 685, *not of Andr.*

SOUTH AFRICA: without locality, *Niven!* and *cultivated specimens!*

COAST REGION : Caledon Div., 3000–4000 ft.; southern slopes of the Houw Hoek Mountains, *Bolus,* 5452! *Schlechter,* 5453! mountains near Vogel Gat, Klein River, *Schlechter,* 9548!

It is probable that Salisbury's type is Niven's specimen in Herb. Kew. This we have dissected, as also all the specimens cited above, besides one in Herb. MacNab at Kew, and one in the Berlin Herbarium, both marked *E. crassifolia, Andr.,* and are satisfied that all are identical. Bentham describes this species under the name of *E. Corydalis,* and also under *E. crassifolia,* with which it has a superficial resemblance: an easy mistake in so large and intricate a genus, from which we ourselves can hardly expect to have escaped. The species is very well marked, and quite different, upon dissection, from the last-named; the hemispherical "tube" is differentiated from the limb by an interior thin circular ridge and longitudinal nerves, and the anther is very different. We cite *E. complanata, Nois.,* upon Bentham's authority, and can only say that the description agrees, as far as it goes. The species connects this section both with § *Pachysa* and § *Trigemma.*

469. E. flavisepala (Guthrie & Bolus); erect, slender, a foot or more high; branches subfiliform, glabrous or minutely puberulous at the apex; leaves 3-nate, erect, imbricate, linear or linear-lanceolate, acute, the younger mucronate and ciliate, the floral somewhat dilated, 2–3 lin. long; flowers 3-nate, clustered at the ends of short branchlets, subcorolline; pedicels glabrous, red, 2–3 lin. long; bracts remote, lanceolate, setaceous-acuminate, concave, scarious, ciliate, the hairs often barbellate or forked, pale yellow, about 3 lin. long; sepals like the bracts or sometimes a little broader and shorter; corolla ovoid-urceolate, but little contracted at the throat, dry, glabrous, red, 4–4½ lin. long; segments slightly spreading or erect, somewhat concave, ovate, acute, subcordate, about ⅓ of the tube in length; filaments capillary; anthers subcuneate-linear, or narrow-oblong; cells deeply parted, very slightly bilobed but scarcely prognathous at the base, about ¾ lin. long, aristate; pore scarcely ½ the length of the cell; awns inserted well above the base of the cell, subulate, projecting backwards, about ½ the length of the cell; style straight, red; stigma capitellate, small; ovary subturbinate, shortly stipitate, glabrous.

CENTRAL REGION : Ceres Div.; sent with *E. Thunbergii* and said to grow with it on the Cold Bokkeveld, *Herb. Bolus,* 6893! and *Cape Govt. Herb.!*

This species, of which we have seen only one (but a good) specimen, is a very well-marked one, but difficult to place satisfactorily in any of the sections. In general aspect it is strikingly similar to *E. Thunbergii;* but has not the peculiar globose corolla-tube of either of the other species of this section. From § *Ceramus* it recedes by its large loose bracts and sepals, by its corolla but little constricted at the throat, and in aspect is unlike any other species described there. From § *Trigemma* it differs by its lax and bright coloured bracts and sepals and its shortly stipitate ovary.

Imperfectly known species.

E. abrotanoides, Burm. f. Prodr. Cap. 11.

E. adenophora, Spreng. Syst. ii. 188.

E. Aphanes, Spreng. Syst. ii. 196.

E. appressa, Spreng. Syst. iv. Cur. Post. 146.

E. bicalyculata, Moench, Meth. Suppl. 18.

E. boucheana, Regel in Gartenfl. 1852, 73.

E. calyciflora, Tausch in Flora, 1834, 617.

E. candida, Spreng. Pugill. i. 30.

E. carneola, Sinclair, Hort. Eric. Wob. 5 ; G. Don, Gen. Syst. iii. 798.

E. cinerascens, Willd. Enum. Hort. Berol. Suppl. 21.

E. cistifolia, Link, Enum. Hort. Berol. i. 369.

E. comosa, Wendl. Eric. Ic. fasc. 12, 7.

E. conifera, Tausch in Flora, 1839, 632.   Said to be allied to *E. denticulata, L.*

E. decolorans, Willd. Enum. Hort. Berol. Suppl. 21 ; name only. Hab. ?

E. dicranifolia, Tausch in Flora, 1839, 634.   We have seen the type of this, which is not sufficient for treatment.

E. Edelinia, Bonpl. Descr. Pl. Rar. Malm. 43, t. 16.

E. epiptera, Willd. Enum. Hort. Berol. Suppl. 21.   Name only.

E. erubescens, Lodd. Bot. Cab. t. 1826.

E. exposita, Lodd. Bot. Cab. t. 1521.

E. exquisita, Carr. in Rev. Hort. 1882, 362.

E. faireana, Carr. in Rev. Hort. 1882, 363.

E. Fergusoni, Gentil. & Carr. in Rev. Hort. 1882, 407.

E. filifolia, Regel, Ind. Sem. Hort. Petrop. 1856, 29.   Hab. ?

E. finitima, Lodd. ex G. Don in Loud. Hort. Brit. 153.

E. flocciflora, Tausch in Flora, 1839, 629.   Hab.?   Near *E. daphniflora.*

E. galiiflora, Bartl. in Linnæa, vii. 643.

E. globosa, Burm. f. Prodr. Cap. 11.

E. globosa, Willd. Sp. Pl. ii. 408.

E. gracilis, Lodd. Bot. Cab. t. 244, not of Wendl. nor of Salisb.

E. hirtifolia, Hornem. Hort. Hafn. i. 370.

E. **hispida,** Burm. f. Prodr. Cap. 11.

E. **innocens,** Hoffmgg. Verz. Pfl. Nachtr. ii. 111.

E. **insulsa,** Sinclair, Hort. Eric. Wob. 12.

E. **jasminiflora,** var. **minor,** Andr. Heathery, t. 220.

E. **jasminoides,** Carr. in Rev. Hort. 1882, 362.   Hab. ?

E. **lanceolaris,** Steud. Nom. ed. i. 306.

E. **lanceolata,** Pers. Syn. i. 424.   Hab. ?

E. **laricea,** Burm. f. Prodr. Cap. 11.

E. **laricina,** Spreng. f. Tent. Suppl. Syst. 12.

E. **laxa,** Thunb. Prodr. 189.

E. **leucophylla,** Klotzsch in Linnæa, xii. 507.

E. **litoralis,** Regel, Cat. Pl. Hort. Aksakov. 55 (name only).

E. **longipedunculata,** Wender. ex Steud. Nom. ed. ii. i. 575.   *E. longipedicellata,* Hoffmgg. Verz. Pfl. 59.   Hab. ?

E. **lychnidea,** Wendl. ex Steud. Nom. ed. i. 307 (name only).

E. **lycopodioides,** Hornem. Hort. Hafn. i. 374.

E. **microcalyx,** Regel, Ind. Sem. Hort. Petrop. 1856, 29.

E. **mollissima,** Lodd. ex G. Don in Loud. Hort. Brit. 150.

E. **mutica,** Tausch in Flora, 1834, 597.

E. **nitens,** Sinclair, Hort. Eric. Wob. 16.   Hab. ?

E. **ochroleuca,** Wendl. f. ex Benth. in DC. Prodr. vii. 687.

E. **octagona,** Lodd. ex G. Don in Loud. Hort. Brit. 147.

E. **Pabsti,** Regel, Gartenfl. 1858, 50.

E. **pallens,** Spreng. Pugill. i. 30, not of Andr.

E. **Paxtoni,** Gentilh. & Carr. in Rev. Hort. 1882, 407.   Hab. ?

E. **pellucida,** var. **rubra,** Andr. Heathery, t. 277.

E. **pellucidoides,** Sinclair, Hort. Eric. Wob. 18.

E. **polytricha,** Sweet, Hort. Brit. ed. ii. 338.

E. **protrudens,** Link, Enum. Hort. Berol. i. 372.

E. **pulverulenta,** Sinclair, Hort. Eric. Wob. 20.

E. **quadriflora,** Willd. Sp. Pl. ii. 379.

E. **Rachii,** Regel, Ind. Sem. Hort. Petrop. 1856, 30.   Hab. ?

E. **retusa,** Tausch in Flora, 1834, 598.

E. **Rinzii,** Regel, Ind. Sem. Hort. Petrop. 1857, 48.   Hab. ?

E. **rollisonia,** Sinclair, Hort. Eric. Wob. 21.

E. **russelliana,** Lodd. Bot. Cab. t. 1013.

E. **saturejæfolia,** Tausch in Flora, 1839, 636.

E. **scabra**, F. W. Schmidt, Neue u. Selt. Pfl. 42.

E. **scabriuscula**, Link, Enum. Hort. Berol. i. 372.

E. **sphærantha**, Spreng. Syst. iv. Cur. Post. 146.

E. **spiralis**, Lodd. ex G. Don in Loud. Hort. Brit. 153 (name only).

E. **splendida**, Mackay ex G. Don in Loud. Hort. Brit. 146 (name only).

E. **struthiolæflora**, Lodd. ex G. Don in Loud. Hort. Brit. 148.

E. **sulcata**, Hornem. Hort. Hafn. i. 371.

E. **superba**, Hoffmgg. Verz. Pfl. Nachtr. iii. 35.

E. **syndriana**, Hort. ex Gentilh. & Carr. in Rev. Hortic. 1882, 306.   Hab.?

E. **tenera**, Steud. Nom. ed. 2, i. 580.   *E. sphærantha,* Link ex Steud. Nomencl. ed. 2, 580.   Hab.?

E. **tenuis**, Moench, Meth. Suppl. 17.   Hab.?

E. **teucriifolia**, Spreng. Pugill. i. 31.

E. **torosa**, Moench, Meth. Suppl. 18.   Hab.?

E. **tortuosa**, Lodd. ex G. Don in Loud. Hort. Brit. 153 (name only).

E. **triphylla**, Link, Enum. Hort. Berol, i. 360.   Probably belonging to § *Ephebus.*   Described from a cultivated specimen.

E. **uniflora**, Burm. f. Prodr. Cap. 11.

E. **ursina**, Lee ex Sweet, Hort. Brit. ed. i. 263 (name only).

E. **venosa**, Gentilh. & Carr. in Rev. Hortic. 1882, 406.   Hab.?

E. **ventrosa**, Sweet, Hort. Brit. ed. ii. 341.   **Hab.?**

E. **verniciflua**, Salisb. in Trans. Linn. Soc. vi. 319, 335.   *E. glutinosa,* Roxb. ex Salisb. l.c.

E. **Vernoni**, Gentilh & Carr. in Rev. Hort. 1882, 408.   Hab.?

E. **vesicularis**, Salisb. in Trans. Linn. Soc. vi. 319, 335; Benth. in DC. Prodr. vii. 661.   *E. conacea,* Hort. ex Salisb. l.c. 335.

CoAST REGION : Stellenbosch Div.; mountains near Stellenbosch, *Mulder,* ex *Salisbury.*

This is not in Herb. Salisbury at Kew, and we have been able to find no further clue to it.

E. **vestitoides**, Regel, Ind. Sem. Hort. Petrop. 1856, 30.   Hab.?

E. **virgata**, Wendl. ex Spreng. Syst. ii. 197.

Supposed Hybrids.

E. ACUMINATA, Andr. Heathery, t. 101, and Col. Heaths, t. 145 ; Lodd. Bot. Cab. t. 216.

E. ACUMINATA, var. ANGUSTIFLORA, Andr. Heathery, t. 251.

E. ACUMINATA, var. LONGIFLORA, Rolliss. in Gard. Chron. 1843, 461.

E. ADJUVANS, Klotzsch in Linnæa, xii. 502.

E. AMBIGUA, Wendl. Eric. Ic. fasc. 16, 61, t. 24.

E. AMPULLACEA, vars. RUBRA and VITTATA, Rolliss. in Gard. Chron. 1843, 461.

E. AMPULLACEOIDES, Rolliss. l.c.

E. ANDREWSII, Klotzsch in Linnæa, x. 314.   *E. andrewsiana*, Rolliss. l.c.

E. ARCHERIA, Andr. Heathery, t. 3.

E. ARCHERIANA, Lodd. Bot. Cab. t. 1466.

E. ARISTATA, var. MINOR, Rolliss. in Gard. Chron. 1843, 461.

E. ARISTELLA, Sinclair, Hort. Eric. Wob. 2.   *E. equestris*, Klotzsch in Linnæa, x. 350.   *E. eximia*, Lodd. Bot. Cab. t. 1105.

E. BANDONIA, Andr. Heathery, t. 205, and Col. Heaths, t. 220.

E. BATEMANIA, Rolliss. in Gard. Chron. 1843, 461.

E. BEAUMONTIA, Andr. Heathery, t. 253, and Rolliss. in Gard. Chron. 1843, 461.

E. BEAUMONTIANA, Rolliss. ex Lodd. Bot. Cab. t. 1686.

E. BLANDA, Rolliss. in Gard. Chron. 1843, 461.

E. BUCCINIFLORA, Bot. Mag. t. 2465; Lodd. Bot. Cab. t. 1127.

E. BURNETTII, Hort. ex Planch. in Fl. Serres, viii. 261, t. 845.   *E. Hartnello-hiemalis*, Planch, l.c.

E. CALOSTOMA, Lodd. Bot. Cab. t. 1759.

E. CARINATA, Lodd. l.c. t. 1071.

E. CAVENDISHIANA, Paxt. Mag. xiii. 3.   (*E. Cavendishii* on the plate.)

E. CELSIANA, Lodd. Bot. Cab. t. 1777.

E. CLOWESIANA, Rolliss. in Gard. Chron. 1843, 461.

E. CRINITA, Lodd. Bot. Cab. t. 1432.

E. CRUCIFORMIS, Andr. Heathery, t. 258, and Col. Heaths, t. 228.   *E. cuneiformis*, Benth. in DC. Prodr. vii. 648.

E. CULCITÆFLORA, Salisb. in Trans. Linn. Soc. vi. 357.

E. CURVIFLORA, var. RUBRA, Rolliss. in Gard. Chron. 1843, 461.

E. CYLINDRICA, Andr. Heathery, t. 60; Lodd. Bot. Cab. t. 1734; Rolliss. in Gard. Chron. 1843, 461, not of others.

E. DAPHNOIDES, Lodd. Bot. Cab. t. 154.

E. DENSA, Rolliss. in Gard. Chron. 1843, 461.

E. DEPRESSA, var. RUBRA, Rolliss. l.c.

E. DOUGLASIÆ, T. Moore in Moore & Ayres, Bot. Mag. iii. 9, t. 1, fig. 1.

E. DUCALIS, Klotzsch in Linnæa, x. 347.

E. DUNBARIANA, Rolliss. in Gard. Chron. 1843, 461.

E. ECHIIFLORA, var. CARNEA, Rolliss. l.c.

E. EFFUSA, Nichols. Dict. Gard. i. 522.

E. EPISTOMIA, Lodd. Bot. Cab. t. 1186.

E. EXCELSA, Tausch in Flora, 1834, 596.

E. EXSURGENS, var. COCCINEA, Rolliss. in Gard. Chron. 1843, 461.

E. FAVOIDES and vars. ELEGANS and PURPUREA, Rolliss. l.c.

E. FORBESIANA, Klotzsch in Linnæa, x. 349.

E. FORMOSA, Rolliss. in Gard. Chron. 1843, 461.

E. GRANDIFLORA, var. HUMILIS, Rolliss. l.c.

E. HARTNELLII, Rolliss. l.c.

E. HIRSUTA, Lodd. Bot. Cab. t. 754.

E. HYACINTHOIDES, Andr. Heathery, t. 167, and Col. Heaths, t. 174.

E. HYBRIDA, Rolliss. in Gard. Chron. 1843, 461.

E. IMPULSA, Rolliss. l.c.

E. INFLATA and var. RUBRA, Rolliss. l.c.

E. INGRAMI, Hort. ex Morr. Belg. Hortic. vii. 322, t. 52, fig. 1.

E. INTERTEXTA, Lodd. Bot. Cab. t. 1034.

E. JACKSONII, Paxt. Mag. viii. 149.

E. JASMINIFLORA, vars. NANA and RUBRA, Rolliss. in Gard. Chron. 1843, 461.

E. LAMBERTIA, Andr. Heathery, t. 171, and Col. Heaths, t. 104. *E. lambertiana,* Lodd. Bot. Cab. t. 3.

E. LAWRENCEANA, Rolliss. in Gard. Chron. 1843, 461.

E. LAXIFLORA, Buck in DC. Prodr. Index, iii. 167. *E. præstans,* var. *laxiflora,* Benth. in DC. Prodr. vii. 647.

E. LEUCOSTOMA, Tausch in Flora, 1834, 596.

E. LINNÆA, var. SUPERBA, Andr. Heathery, t. 268, and Col. Heaths, t. 243.

E. LINNÆANA, var. CURVIFLORA, Rolliss. in Gard. Chron. 1843, 461.

E. LINNÆANA, var. SUPERBA, Lodd. Bot. Cab. t. 1778 ; Rolliss. l.c.

E. LINNÆOIDES, Andr. Heathery, t. 123.

E. MACNABIANA, Paxt. Mag. vii. 125.

E. MAGNIFICA, Andr. Heathery, t. 223, and Col. Heaths, t. 244.

E. MARNOCKIANA, T. Moore in Moore & Ayres, Mag. Bot. iii. 9, t. 1, fig. 2.

E. METULÆFLORA, and var. BICOLOR, Rolliss. in Gard. Chron. 1843, 461.

E. MIRABILIS, Andr. Heathery, t. 271, and Col. Heaths, t. 246.

E. MOOREANA, Lem. Jard. Fleur. iii. t. 259, fig. 2.

E. MULTUMBELLIFERA, Tausch in Flora, 1839, 628.

E. MURRAYANA, Thunb. ex Paxt. Mag. xi. 77.

E. MUTABILIS, Andr. Heathery, t. 176, and Col. Heaths, t. 187 ; Lodd. Bot. Cab. t. 46; Bot. Mag. t. 2348 ; Rolliss. in Gard. Chron. 1843, 461, not of Salisb.

E. OBLONGA, Sinclair, Hort. Eric. Wob. 17.

E. OSTRINA, Rolliss. in Gard. Chron. 1843, 461.

E. PALLIDA, Lodd. Bot. Cab. t. 1355.

E. PATERSONIA, var. COCCINEA, Andr. Heathery, t. 134, and Col. Heaths, t. 196.

E. PATERSONIADES, Sinclair, Hort. Eric. Wob. 18.

E. PERSPICUA, Sinclair, l.c. ; Rolliss. in Gard. Chron. 1843, 461.

E. PERSPICUA, var. MAJOR, Klotzsch in Linnæa, ix. 674.

E. PERSPICUOIDES, Sinclair, Hort. Eric. Wob. 18.

E. PINEA, and var. PURPUREA, Rolliss. in Gard. Chron. 1843, 461.

E. PINGUIS, Klotzsch in Linnæa, x. 351. *E. clowieana,* Hort. ex Benth. in DC. Prodr. vii. 643.

E. PINIFOLIA, vars. COCCINEA, DISCOLOR and ELEGANS, Rolliss. in Gard. Chron. 1843, 461.

E. PRINCEPS, Andr. Heathery, t. 140, and Col. Heaths, t. 121 ; Lodd. Bot. Cab. t. 647.

E. PRINCEPS, var. CARNEA, Rolliss. in Gard. Chron. 1843, 461.

E. PRIMULOIDES, Andr. Heathery, t. 233, and Col. Heaths, t. 203 ; Bot. Mag. t. 1548 ; Lodd. Bot. Cab. t. 715. *E. dilecta,* Hort. ex Klotzsch in Linnæa, xii. 521.

E. PSEUDOVESTITA, Benth. in Maund, Bot. iii. 104.

E. PULCHERRIMA, Rolliss. in Gard. Chron. 1843, 461.

E. PUNICEA, Rolliss. l.c.

E. QUADRANGULARIS, Andr. Heathery, t. 280, and Col. Heaths, t. 260. *E. erosa,* Lodd. Bot. Cab. t. 133.

E. RADIATA, Andr. l.c. t. 42, l.c. t. 52. *E. calamiformis,* Salisb. in Trans. Linn. Soc. vi. 362.

E. RADIATA, var. DISCOLOR, Andr. l.c. t. 281, l.c. t. 261.

E. REFULGENS, Andr. l.c. t. 284, l.c. t. 264.

E. RETORTA, var. MAJOR, Rolliss. in Gard. Chron. 1843, 461.

E. RIGIDA, Lodd. Bot. Cab. t. 1286.

E. ROLLISSONI, Rolliss. in Gard. Chron. 1843, 461.

E. ROSEA, Andr. Heathery, t. 82; Lodd. Bot. Cab. t. 782.

E. RUBERCALYX, Andr. l.c. t. 285, and Col. Heaths, t. 266.

E. RUBIDA, Lodd. Bot. Cab. t. 1166.

E. RUBROCALLA, Rolliss. in Gard. Chron. 1843, 461.

E. RUBROCALYX, Gentilh. & Carr. in Rev. Hort. 1882, 306.

E. RUBROSEPALA, Sweet, Hort. Brit. ed. 2, 339.

E. RUGOSA, Andr. Heathery, t. 236, and Col. Heaths, t. 267.

E. RUSSELLIANA, Rolliss. in Gard. Chron. 1843, 461.

E. SANGUINEA, Lodd. Bot. Cab. t. 86.

E. SIMULATA, T. Moore in Moore & Ayres, Mag. Bot. iii. 9, t. 1, fig. 3.

E. SPENCERIANA, Planch. in Fl. Serres, t. 2323.

E. SPRENGELLII, Endl. in Harting. Parad. Vindob. t. 67, fig. 2; Rolliss. in Gard. Chron. 1843, 461.

E. SPURIA, Andr. Heathery, t. 90, and Col. Heaths, t. 62; Rolliss. l.c.

E. STELLIFERA, Andr. l.c. 291, l.c. 276; Lodd. Bot. Cab. t. 1622. *E. bibracteata*, Klotzsch in Linnæa, xii. 516.

E. SUAVEOLENS, Andr. l.c. t. 292, l.c. t. 277; Rolliss. in Gard. Chron. 1843, 461.

E. SWAINSONIA, Andr. l.c. t. 242, l.c. t. 209. *E. swainsoniana*, Rolliss. l.c.

E. TEMPLEA, Lee ex Andr. l.c. t. 293, l.c. t. 280.

E. TENUIFLORA, var. CARNEA, Andr. l.c. t. 294, l.c. t. 281.

E. THOMSONII, Lem. Jard. Fleur. iii. t. 259, fig. 1.

E. TORTULIFLORA, Rolliss. in Gard. Chron. 1843, 461.

E. TRANSLUCENS, Andr. Heathery, t. 295.

E. TRANSPARENS, Andr. Col. Heaths, t. 283.

E. TRICOLOR, and vars. ELEGANS, IMPRESSA, MAJOR and SUPERBA, Rolliss. in Gard. Chron. 1843, 461.

E. TROSSULA, Lodd. Bot. Cab. tt. 668, and var. RUBRA, t. 1742. *E. trassula*, Klotzsch in Linnæa, xii. 515.

E. TUBIFLORA, Rolliss. in Gard. Chron. 1843, 461.

E. TUBULOSA, Wendl. Eric. Ic. fasc. 16, 57, t. 22.

E. TURGIOLA, Rolliss. in Gard. Chron. 1843, 461.

E. UNDULATA, Andr. Heathery, t. 300, and Col. Heaths, t. 288; Lodd. Bot. Cab. t. 1792.

E. VENTRICOSA, vars. ALBA, CARNEA, NANA, PURPUREA, and STELLIFERA, Rolliss. in Gard. Chron. 1843, 461.

E. VERNIX, var. RUBRA, Rolliss. l.c.

E. VESTITA, vars. BLANDA, ELEGANS, FULGIDA, INCARNATA and ROSEA, Rolliss. l.c.

E. WEBBIANA, Rolliss. l.c.

E. WESTPHALINGIA, Hort. ex Benth. in DC. Prodr. vii. 639.

E. WILLMOREI, Knowles & Westcott, Fl. Cab. ii. (1838) 115. *E. vilmoreana, E. villmoriniana, E. willmoreana* and *E. willmoriana*, Hort. ex Carr. in **Rev. Hort.** 1892, 335.

II. **PHILIPPIA**, Klotzsch.

By N. E. BROWN, A.L.S.

Pedicels ebracteate. *Calyx* unequally 4-lobed or 4-partite, one segment distinctly larger than the others and outside them. *Corolla* very small, 4-lobed. *Stamens* 8; filaments free or connate; anthers bifid or bipartite, without dorsal or basal spurs, opening by oblique pores. *Ovary* 4-celled; style exserted, persistent; stigma large, peltate or saucer-shaped; ovules 2 or more in each cell. *Capsule* loculicidally 4-valved; seeds 1 or more in each cell.

Shrubs or undershrubs with the habit of *Erica*; leaves shortly petiolate, grooved down the very convex back; flowers very small, in small terminal clusters.

DISTRIB. Species between 30 and 40, several of them in Tropical Africa, more numerous in the Mascarene Islands.

Leaves hispid with long gland-tipped hairs ...   ... (1) **leeana.**
Leaves not hispid with gland-tipped hairs:
    Branchlets with short gland-tipped hairs   ... (2) **Evansii.**

Branchlets with minute white tomentum, without gland-tipped hairs :
    Calyx-lobes unequal; staminal-filaments all
      connate   ...   . (3) **Chamissonis**.
    Calyx-lobes subequal : staminal-filaments
      free or some connate for $\frac{1}{2}$ their length ... (4) **tristis**.

1. **P. leeana** (Klotzsch in Linnæa, xii. 213); branchlets ascending
or spreading, usually curved, minutely or conspicuously pubescent
with spreading glandular hairs; leaves 3-nate, spreading, $\frac{2}{3}$–1 lin.
long, ovate, oblong-ovate or ovate-lanceolate, obtuse or acute, somewhat turgid, hispid on the back with rather long gland-tipped hairs,
puberulous on the upper surface; flowers 1–6 together at the tips of
the branchlets; pedicels $\frac{1}{2}$–1 lin. long, varying from glandular-
pubescent to nearly glabrous; calyx-lobes thin, with the keel or tip
thick and coriaceous, very broadly ovate or subquadrate-oblong,
acute or obtuse, variably connate or nearly free, ciliate with long
gland-tipped hairs, otherwise glabrous, the larger $\frac{1}{2}$–$\frac{3}{4}$ lin. long,
$\frac{1}{3}$–$\frac{1}{2}$ lin. broad; corolla $\frac{2}{5}$–$\frac{3}{4}$ lin. long, $\frac{3}{4}$ lin. in diam., campanulate or
broadly cup-shaped, 4-lobed to about $\frac{1}{3}$ of the way down, glabrous;
lobes very obtuse, very minutely denticulate; stamens not exserted;
filaments broad, entirely connate; anthers $\frac{1}{3}$–$\frac{2}{5}$ lin. long, oblong,
bifid to nearly $\frac{1}{2}$-way, connate; ovary subglobose, 4-grooved, glabrous;
style very much exserted, 1–1$\frac{1}{4}$ lin. long, glabrous or pubescent;
stigma circular, $\frac{2}{5}$ lin. in diam., flat or nearly so, with 4 minute
central tubercles; seeds often solitary, rather large. *Benth. in DC.
Prodr.* vii. 695. *Erica absinthoides, E. Meyer ex Klotzsch in
Linnæa*, xii. 213.

COAST REGION : Stellenbosch Div.; Lowrys Pass, 1000–2000 ft., *Drège!*
Hottentots Holland Mountains, *Bolus in Herb. Norm. Austr.-Afr.*, 44! Caledon
Div.; by the Palmiet River, *Schlechter*, 5430! *Ecklon & Zeyher* (ex *Klotzsch*);
Houw Hoek Mountains, *Burchell*, 8143! *Schlechter*, 9395! Zwart Berg, near
Caledon Baths, 1000 ft., *Zeyher*, 3329! *Galpin*, 3715! *MacOwan & Bolus,
Herb. Norm. Austr.-Afr.*, 755! *Schlechter*, 9772! 10361! *Guthrie*, 2503! hills
near Caledon, *Bolus*, 9907! *Ecklon & Zeyher* (ex *Klotzsch*); near Genadendal,
Ecklon & Zeyher (ex *Klotzsch*).

This has been distributed in MacOwan & Bolus, Herb. Norm. as *Scyphogyne
divaricata*, Benth.

2. **P. Evansii** (N. E. Br.); branchlets spreading when young,
becoming ascending, shortly glandular-hairy; leaves 3-nate, ascending or spreading, $\frac{3}{4}$–1 lin. long, linear-oblong, very minutely gland-
denticulate on the margins, otherwise glabrous; flowers 1–3 to-
gether, terminal; pedicels $\frac{2}{3}$–$\frac{3}{4}$ lin. long, glabrous; calyx-lobes
coriaceous, broadly ovate, subacute, grooved down the back, minutely
gland-ciliate, the larger $\frac{1}{3}$–$\frac{1}{2}$ lin. long, $\frac{1}{3}$ lin. broad, nearly twice as
long as the rest; corolla globose-campanulate, about $\frac{1}{2}$ lin. long and
$\frac{2}{3}$ lin. in diam., 4-lobed to nearly $\frac{1}{2}$-way down, glabrous, pale greenish;
lobes broader than long, very obtuse; stamens not exserted; fila-
ments $\frac{1}{8}$ lin. long, free; anthers $\frac{2}{5}$ lin. long, bifid to $\frac{1}{2}$-way down,
free; ovary globose, 4-grooved, glabrous; style about equalling the
corolla; stigma peltate, orbicular, $\frac{1}{2}$ lin. in diam.

EASTERN REGION : Natal; near Ulundi, 5000–6000 ft., *Evans,* 62! in Herb. Bolus.

3. P. Chamissonis (Klotzsch in Linnæa, ix. 356); a large shrub (tree, ex *Niven*); branchlets densely crowded, erect or ascending, densely white-puberulous with minute deflexed hairs; leaves 3-nate, crowded, imbricately adpressed, $\frac{3}{4}$–$1\frac{1}{3}$ lin. long, linear, obtuse, flat, with a raised midrib above, glabrous and smooth or minutely scabrid, often with wrinkled sides in the dried state, subentire or very minutely scabrid on the margins; flowers in terminal clusters of 3–7; pedicels $\frac{1}{2}$ lin. long, glabrous; calyx-lobes coriaceous, the larger about $\frac{1}{2}$–$\frac{3}{4}$ lin. long, $\frac{1}{2}$ lin. broad, very broadly ovate, obtusely pointed, slightly grooved down the back towards apex, glabrous or minutely scabrid; corolla campanulate-globose, glabrous; tube $\frac{2}{5}$ lin. long; lobes $\frac{1}{4}$ lin. long, rounded or broadly ovate, obtuse; stamens as long as the corolla or slightly exserted; filaments connate; anthers $\frac{1}{3}$–$\frac{2}{5}$ lin. long, bifid to $\frac{1}{2}$-way down, connate below; ovary globose, glabrous; style shortly exserted, $\frac{1}{4}$–$\frac{1}{3}$ lin. long; stigma $\frac{1}{3}$ lin. in diam., crater-like, with or without 4 small papillæ or larger radiating processes at the bottom of the cup. *Klotzsch in Linnæa,* xii. 213; *Rach in Linnæa,* xxvi. 788; *Benth. in DC. Prodr.* vii. 695. *Erica virgata,* var. δ, *Thunb. Diss. Erica,* 19. *E. absinth-oides, Thunb. Fl. Cap. ed. Schult.* 349. *E. tristis, Bartl. in Linnæa,* vii. 643; *Benth. in DC. Prodr.* vii. 691. *E. cupressifolia, Wendl. ex Klotzsch, l.c.* 213.

SOUTH AFRICA: without locality, *Thunberg! Chamisso* (ex *Klotzsch*), *Mace! Roxburgh! Ecklon & Zeyher! Burchell,* seed 1808! and *cultivated specimen!*

COAST REGION : Cape Div.; Cape Flats, *Guthrie,* 1164! Table Mountain, *Niven,* 218! *Harvey!* Camps Bay, *Burchell,* 848! Fish Hoek, *Niven,* 90! Muizen Berg, 400 ft., *Bolus,* 4477! Caledon Div.; Hot springs near Caledon, *Ecklon!*

4. P. tristis (Bolus in Journ. Linn. Soc. xxiv. 187); branchlets crowded, subparallel, covered with a minute dense white tomentum, becoming brown and glabrous; leaves 3-nate, adpressed, $\frac{1}{2}$–$1\frac{1}{4}$ lin. long, linear, obtuse, glabrous, very minutely denticulate; flowers 2–3 in a cluster, lateral and terminal; pedicels $\frac{1}{3}$–$\frac{1}{2}$ lin. long, glabrous; calyx-lobes subequal, subcoriaceous, $\frac{1}{5}$–$\frac{1}{4}$ lin. long, $\frac{1}{5}$ lin. broad, ovate, obtuse, glabrous, minutely ciliate; corolla about $\frac{1}{2}$ lin. long and the same in diam., subglobose-campanulate; lobes about $\frac{1}{4}$ lin. long and broad, broadly ovate, obtuse; stamens about as long as the corolla; filaments nearly $\frac{1}{4}$ lin. long, adnate to the bottom of the corolla and some of them sometimes connate for half their length, free above; anthers $\frac{1}{4}$ lin. long, bifid to nearly half-way down, cohering at the middle, becoming free when in fruit; ovary globose, 4-grooved, glabrous; style $\frac{1}{4}$ lin. long, shortly exserted; stigma $\frac{1}{8}$–$\frac{1}{5}$ lin. in diam., funnel-shaped, with incurved margin or crater-like; seeds usually solitary in each cell.

CENTRAL REGION : Graaff Reinet Div.; Koudveld Mountains, 4500 ft., *Bolus,* 2594!

III. **ERICINELLA**, Klotzsch.

Pedicels ebracteate. *Calyx* unequally 3–4-partite or -lobed, one sepal distinctly larger than the rest. *Corolla* very small, 3–4-lobed. *Stamens* 4–6; filaments free; anthers free or connate, with or without basal spurs. *Ovary* 3–4-celled; style persistent; stigma peltate. *Ovules* several in each cell. *Capsule* 3–4-celled, 3–4-valved.

Shrubs or undershrubs with the habit of *Erica;* leaves grooved down the convex back; flowers very small, in small terminal clusters.

DISTRIB. Species 5 or 6, the others in Tropical Africa and Madagascar.

Corolla campanulate, with erect lobes; style much
    exserted ... ... (1) **multiflora.**
Corolla obconic-clavate, with incurved lobes.; style
    not exserted ... ... (2) **passerinoides.**

1. **E. multiflora** (Klotzsch in Linnæa, xii. 223); very densely branched ; branchlets puberulous, becoming glabrous; leaves 3-nate, adpressed, imbricate, including the petiole $\frac{3}{4}$–$1\frac{1}{4}$ lin. long, $\frac{1}{5}$ lin. broad, linear, obtuse, glabrous, ciliate when young; flowers in terminal clusters of 3–9 ; pedicels $\frac{1}{3}$–$\frac{1}{2}$ lin. long, minutely puberulous; sepals 4, unequal, minutely ciliate, the larger $\frac{1}{3}$–$\frac{3}{5}$ lin. long, linear or linear-lanceolate, more or less leaf-like, at least at the apex, the others about half as long and more narrowed to the obtuse apex ; corolla campanulate, glabrous; tube $\frac{1}{2}$ lin. long; lobes erect, about $\frac{1}{4}$ lin. long, rounded; stamens 4, equalling the corolla; filaments $\frac{2}{5}$ lin. long, free, filiform; anthers nearly $\frac{1}{3}$ lin. long and as much in breadth at the top, cuneate-subquadrate, bifid to half-way down, very obtuse, with awn-like spurs at the base, slightly and very minutely scabrid; ovary subglobose, adpressed, pubescent; style very much exserted, 1 lin. long; stigma broadly obconic, triangular, $\frac{1}{6}$ lin. in diam.; ovules about 8 in each cell. *Benth. in DC. Prodr.* vii. 697.

COAST REGION : Queenstown Div.; Winterberg, *Ecklon & Zeyher!*

2. **E. passerinoides** (Bolus in Journ. Linn. Soc. xviii. 393); a densely branched erect shrub, about 2 ft. high ; branchlets at first minutely tomentose, soon becoming glabrous; leaves 3-nate, adpressed, imbricate, including the petiole $\frac{3}{4}$–1 lin. long, $\frac{1}{3}$ lin. broad, oblong or oblong-ovate, obtuse, flat on the face, very minutely ciliate; flowers in small terminal clusters of 3–9, drooping; pedicels $\frac{3}{4}$–1 lin. long, curved, puberulous; sepals 4, unequal or rarely subequal, narrowly oblong, obtuse, minutely ciliolate, the larger $\frac{2}{3}$ to nearly 1 lin. long, usually leaf-like at the upper half ; corolla obconic-clavate, 4-lobed, glabrous, pink (*Bolus*); tube $\frac{3}{4}$ lin. long; lobes incurved, $\frac{1}{4}$ lin. long, rounded; stamens 4, included ; filaments $\frac{1}{3}$ lin. long, free, filiform; anthers nearly $\frac{1}{3}$ lin. long and broad at the top, broadly cuneate, bifid to half-way down, very obtuse,

with awn-like spurs at the base, very minutely scabrid; ovary subglobose, slightly 4-grooved, puberulous; style just reaching the tips of the corolla-lobes, about $\frac{1}{3}$ lin. long, glabrous; stigma peltate, $\frac{1}{4}$ lin. in diam.; ovules 8–9 in each cell.

CENTRAL REGION: Graaff Reinet Div.; Koudeveld Mountains, 5000 ft., *Bolus*, 2582!

IV. BLÆRIA, Linn.

Pedicels 3- (rarely 2-) bracteate. *Calyx* equally 4-lobed or 4-partite. *Corolla* small, tubular or campanulate, shortly 4-lobed, often 4-angled. *Stamens* 4–6; filaments free, glabrous; anthers usually much exserted, bipartite, with or without basal spurs, opening by oblique pores or short slits. *Ovary* 4-celled; cells 2–5-ovuled; style filiform, persistent, long and much exserted, except in one species, glabrous; stigma simple or slightly enlarged, peltate in one species. *Capsule* loculicidally 4-valved; seeds 1 to few in each cell.

Shrubs or shrublets, with the habit and foliage of *Erica*; leaves usually grooved down the back, rarely open-backed; flowers in 2- to many-flowered umbels, terminal and often head-like or terminating very short axillary branchlets, which are racemosely arranged along the branches, rarely axillary and whorled.

DISTRIB. Species 14, endemic.

I. Anthers spurred at the base.

 * Anthers with almost parallel sides down to the
  spurs :
   Leaves glabrous, but sometimes ciliolate :
    Leaves open-backed, $\frac{1}{2}$–1 lin. broad; corolla
    3 lin. long ... ... (1) **grandis.**
    Leaves convex and grooved down the back,
    $\frac{1}{4}$–$\frac{1}{2}$ lin. broad; corolla under 2 lin.
    long :
     Anthers $\frac{2}{3}$ lin. long; ovary puberulous (2) **fuscescens.**
     Anthers $\frac{1}{2}$ lin. long; ovary glabrous ... (7) **purpurea.**
   Leaves pubescent or puberulous, at least when
   young :.
    Leaves 3-nate; ovary puberulous at the
    top :
     Sepals and young leaves puberulous,
     without long hairs ... ... (3) **fastigiata.**
     Sepals and young leaves puberulous,
     beset with long white hairs ... ... (4) **coccinea.**
    Leaves 4-nate; ovary glabrous ... ... (14) **kraussiana.**
 ** Anthers distinctly narrowed at the base above the
  spurs :
   Corolla yellow, drying blackish-brown ... ... (6) **flava.**
   Corolla rosy-purple ... ... (8) **dumosa.**
II. Anthers without spurs at the base (see also *B. flava*,
   in which they are very minute and
   may be overlooked).

   * Bracts near or below the middle of the pedicel :
      Branchlets minutely greyish-tomentose ; ovary
        puberulous at the top ... ... (5) **pusilla.**
      Branchlets and ovary glabrous :
        Branchlets nearly straight, not interwoven ;
          anthers exserted, $\frac{1}{3}$ lin. long ... ... (9) **campanulata.**
        Branches flexuose, interwoven ; anthers not :
          exserted, $\frac{1}{8}$ lin. long ... ... (10) **flexuosa.**
  ** Bracts close under and adpressed to the calyx ;
    branchlets and leaves pubescent :
      Calyx ciliate with long hairs, besides glands,
        or with a row of hairs behind the glands ;
        corolla-tube 1–1$\frac{1}{2}$ lin. long :
          Flower-clusters about 2 lin. in diam. ... (11) **affinis.**
          Flower-clusters 3–4 lin. in diam. ... (12) **ericoides.**
      Calyx gland-ciliate, without hairs ; corolla-tube
        2 lin. long ... ... (13) **revoluta.**

1. **B. grandis** (N. E. Br.) ; laxly branched ; branches 3–10 in.
long, 4-angled, glabrous ; leaves 4-nate, varying from loosely imbri-
cate to spreading, 3–4$\frac{1}{2}$ lin. long, $\frac{1}{2}$–1 lin. broad, linear to lanceolate-
oblong, acute, open-backed and very concave (boat-like) beneath,
glabrous, shining green above, opaque and slightly reddish beneath,
with a minutely whitish-papillate surface ; petiole $\frac{1}{2}$ lin. long ;
flowers axillary, whorled, forming oblong lax clusters at or
towards the ends of the branches, which often grow out beyond the
flowers ; pedicels 3$\frac{1}{2}$–4 lin. long, 3-bracteate near and below the
middle, apparently angular, minutely puberulous ; bracts $\frac{3}{4}$–1 lin.
long, lanceolate, acute, convolute, glabrous ; sepals free,· 2 lin.
long, $\frac{2}{3}$–$\frac{3}{4}$ lin. broad, lanceolate, acute, acutely keeled, glabrous ;
corolla glabrous ; tube 2$\frac{1}{2}$–2$\frac{3}{4}$ lin. long, obscurely 4-angled ; lobes
erect, about $\frac{1}{2}$ lin. long, deltoid-ovate, acute ; stamens 4 ; filaments
flat ; anthers partly exserted, rather more than 1 lin. long, linear,
with almost parallel sides down to the base, glabrous, with short
puberulous basal spurs ; ovary oblong-obovoid, 4-angled, puberulous
on the apical part ; ovules 4–5 in each cell ; style 3 lin. long ; stigma
simple.

Coast Region : Uitenhage Div. ; Van Stadens Mountains, *Zeyher*, 718!
West in Herb. MacOwan, 3111 ! 3119 !

2. **B. fuscescens** (Klotzsch in Linnæa, viii. 657) ; an erect shrub,
1–3 ft. high ; branches erect, puberulous or minutely subtomentose ;
leaves 3-nate, $\frac{3}{4}$–2 lin. long, $\frac{1}{4}$–$\frac{1}{2}$ lin. broad, linear, acute or subacute,
erect or spreading, with adpressed petioles up to $\frac{1}{3}$ lin. long,
glabrous ; umbels 2–4-flowered, terminating very short branchlets,
which are racemosely arranged along the branches ; pedicels 1$\frac{1}{2}$ lin.
long, puberulous ; bracts $\frac{1}{3}$–$\frac{1}{2}$ lin. long, oblong-lanceolate, acute,
glabrous, minutely ciliate, the upper distant from the calyx but
seated above the middle of the pedicel ; sepals free, $\frac{3}{4}$–1$\frac{1}{3}$ lin. long,
coriaceous, lanceolate, acute, keeled down the back, minutely ciliate ;
corolla white, glabrous ; tube 1–1$\frac{1}{3}$ lin. long, tubular or tubular-
campanulate, not angular ; lobes $\frac{1}{3}$ lin. long, broadly ovate-oblong,

very obtuse, spreading; stamens 4, exserted; anthers $\frac{2}{3}$ lin. long, oblong, with nearly parallel sides down to the spurs, black or dark brown, smooth, with scabrid basal spurs, $\frac{1}{8}-\frac{1}{5}$ lin. long; ovary oblong, 4-angled, puberulous, with 2–3 ovules in each cell; style exserted beyond the anthers, $1\frac{1}{2}$–2 lin. long; stigma simple. *Benth. in DC. Prodr.* vii. 697. *Erica sagittata, Klotzsch ex Benth. in DC. Prodr.* vii. 681. *Blairia fuscescens, Dietr. Syn. Pl.* i. 443.

SOUTH AFRICA: without precise locality, *Reeves! Miller! Mund & Maire!*

COAST REGION: George Div.; on Cradock Berg (Post Berg), near George, *Burchell,* 5910! *Prior! Galpin,* 3717! Kuysna Div.; Zitzikamma, *Pappe!* Uitenhage Div.; Vanstaden Berg, *Drège,* ex *Bentham.*

The type of *Erica sagittata,* Klotzsch, in the Berlin Herbarium has been examined by Dr. H. Bolus, who informs me that there is only a single imperfect flower upon the specimen, but that there can be no doubt of its identity with *Blæria fuscescens,* Klotzsch.

3. B. fastigiata (Benth. in DC. Prodr. vii. 697); an erect shrub, about 2 ft. high, densely much branched; branchlets short, crowded along the primary and secondary branches in a dense spike-like manner, pubescent; leaves 3-nate, crowded, $\frac{1}{2}$–1 lin. long, linear, obtuse, finely puberulous, incurved-spreading, with short adpressed petioles; umbels 2–6-flowered, terminating the short lateral branchlets, and forming paniculate or racemiform masses; pedicels $\frac{3}{4}$–1 lin. long, puberulous; bracts seated below the middle of the pedicel, upper pair very minute; sepals free, $\frac{1}{3}-\frac{2}{5}$ lin. long, linear, obtuse, puberulous; corolla $1\frac{1}{4}$–$1\frac{1}{2}$ lin. long, tubular, 4-angled, puberulous; lobes very short, rounded, erect; stamens 4, exserted; anthers $\frac{1}{2}$ lin. long, linear-oblong, sides parallel down to the spurs, minutely scabrid, with slightly diverging basal spurs nearly half as long as the cells; ovary narrowly obovoid, 4-angled, puberulous at the top; ovules about 2 in each cell; style $1\frac{3}{4}$–2 lin. long, exserted.

COAST REGION: Swellendam Div.; summit of a mountain peak near Swellendam, *Burchell,* 7331! mountains near Swellendam, 2000 ft., *MacOwan,* 1671!

4. B. coccinea (Klotzsch in Linnæa, viii. 657); a shrub, 3–4 ft. high (*Masson*); branchlets erect, crowded, at first pilose, becoming glabrous; leaves 3-nate, $1\frac{1}{2}$–$1\frac{2}{3}$ lin. long, linear or linear-lanceolate, obtuse, margins and tips thinly beset with long white hairs, at least when young, minutely greyish-puberulous beneath; umbels 2–6-flowered, terminal on very short racemosely arranged branchlets; pedicels $\frac{2}{3}$–1 lin. long, puberulous, bracteate at or below the middle; bracts minute, linear-subulate, fringed or tipped with a few long hairs; sepals free, $\frac{1}{2}-\frac{2}{3}$ lin. long, linear or lanceolate, coriaceous, minutely puberulous, fringed with long white hairs; corolla $1\frac{1}{2}$ lin. long, $\frac{3}{4}$ lin. in diam., tubular, 4-angled, minutely puberulous, red; lobes erect, $\frac{1}{4}$ lin. long, rounded; stamens 4, shortly exserted ; anthers $\frac{1}{2}$ lin. long, linear-oblong, with nearly parallel sides down to the spurs, minutely scabrid; spurs about $\frac{1}{5}$ as long as the cells; ovary obovoid, 4-angled, puberulous; cells few-ovuled; style $1\frac{2}{3}$–$2\frac{1}{4}$ lin. long, much exserted ; stigma simple, slightly dilated. *Benth. in DC. Prodr.* vii. 697. *Blairia coccinea, Dietr. Syn. Pl.* i. 443.

SOUTH AFRICA : without precise locality, *Masson*, 4 ! *Mund & Maire* !
COAST REGION : Swellendam Div.; on the Lange Bergen, 2000 ft., *Schlechter*, 2055 ! mountains near Swellendam, *Marloth*, 3527 !

5. B. pusilla (Klotzsch in Linnæa, viii. 659, not of Linn.) ; a bush 6–12 in. high, densely much branched ; branchlets minutely greyish-tomentose ; leaves 3-nate, erect, straight or slightly incurved, $\frac{1}{2}$–1 lin. long, with short adpressed petioles, linear, subacute, keeled down the upper surface, ciliate, otherwise glabrous ; umbels 2–3-flowered, terminating extremely short ($\frac{1}{2}$–1 lin. long) axillary branchlets, arranged in interrupted spike-like racemes ; pedicels $\frac{1}{3}$–$\frac{1}{2}$ lin. long, bracteate below the middle, glabrous ; bracts 3, sometimes verticillate, $\frac{1}{6}$–$\frac{1}{4}$ lin. long, adpressed to the pedicel, linear, obtuse, ciliate, submembranous ; calyx $\frac{1}{2}$ lin. long, 4- (rarely 5-) lobed to $\frac{2}{3}$ of the way down, glabrous, ciliate on the ovate-lanceolate acute lobes ; corolla $\frac{3}{4}$ lin. long, campanulate, 4- (rarely 5-) angled, shortly 4- (rarely 5-) lobed, glabrous, purple ; lobes about as long as broad, erect, obtusely rounded, often very minutely denticulate ; stamens 4 or rarely 5, exserted ; anthers $\frac{2}{5}$ lin. long, obtusely rounded at the base, without spurs ; ovary broadly obovoid, 4-angled, minutely pubescent at the top ; ovules several in each cell ; style 1–$1\frac{1}{5}$ lin. long ; stigma peltate or funnel-shaped. *Benth. in DC. Prodr.* vii. 698, *not of Linn. nor Thunb.*

SOUTH AFRICA : without locality, *Lichtenstein !*
COAST REGION : Caledon Div.; Klein River Mountains, *Zeyher*, 3331! Bredasdorp Div. ; hills near Elim, 300–500 ft., *Bolus*, 8461 ! 8462! *Guthrie*, 3783 !

6. B. flava (Bolus in Journ. Bot. 1894, 239) ; a shrublet, 6–9 in. high, much branched, glabrous in all parts ; leaves 3-nate, more or less imbricate, very shortly petiolate, $\frac{3}{4}$–$1\frac{1}{4}$ lin. long, $\frac{1}{3}$–$\frac{1}{2}$ lin. broad, oblong, subacute ; umbels 3–6-flowered ; pedicels $\frac{3}{4}$–1 lin. long, with 2–3 linear bracts about $\frac{1}{2}$ lin. long below the middle ; calyx $\frac{3}{4}$ lin. long, rigidly coriaceous, campanulate, 4-lobed to half-way down ; lobes oblong, subacute, keeled and slightly grooved down the back ; corolla $1\frac{1}{3}$ lin. long, campanulate, 4-angled, pale yellow, drying dark brown ; lobes broadly rounded, broader than long, erect ; stamens 4 ; anthers about half exserted, $\frac{2}{5}$ lin. long, distinctly narrowed at the base above the very short spurs, minutely scabrid ; ovary obovoid, 4-angled, crowned by the abruptly dilated cushion-like base of the style, which forms a sort of cap in fruit, glabrous ; ovules 2, pendulous in each cell ; style much exserted, $1\frac{1}{2}$ lin. long, very slightly thickened at the apex ; stigma simple.

COAST REGION: Caledon Div.; on the Zwart Berg, near Caledon, 2500 ft., *Bolus*, 5417! and in *Herb. Norm. Austr.-Afr.*, 611! by the Steenbrass River, *Niven*, 6! 7!

7. B. purpurea (Linn. f. Suppl. 122); dwarf, densely much branched; branches somewhat trigonous, glabrous; leaves 3-nate, adpressed or imbricate, $\frac{1}{2}$–$1\frac{1}{4}$ lin. long, $\frac{1}{4}$–$\frac{1}{3}$ lin. broad, oblong or linear-oblong, obtuse, very shortly petiolate, minutely ciliolate or slightly scabrid on the margins, otherwise glabrous; umbels terminal, 3–9-flowered; pedicels 1–$1\frac{1}{4}$ lin. long, glabrous, bracteate near the middle; bracts $\frac{1}{4}$–$\frac{1}{2}$ lin. long, linear or the lower enlarged upwards, glabrous, minutely ciliate; calyx about $\frac{1}{2}$ lin. long, 4-lobed to below the middle, rigid, glabrous, minutely ciliate and sometimes with glands on the margins of the oblong, lanceolate or deltoid, acute or obtuse lobes; corolla $1\frac{2}{3}$ (very rarely $1\frac{3}{4}$–2) lin. long, $\frac{3}{4}$–1 lin. in diam., tubular, 4-angled, very shortly 4-lobed, glabrous, bright rosy purple; lobes erect, rounded, broader than long; stamens 4; anthers partly exserted, $\frac{1}{2}$ lin. long, bipartite nearly to the base, oblong, with subparallel sides, not constricted above the short subparallel or slightly divergent spurs, scabrid, dark brown or blackish, deciduous; ovary 4-angled, glabrous; ovules 3–4 in each cell; style $1\frac{1}{3}$–2 lin. long, dilated at the base and articulated to the ovary, much exserted, slightly thickened at the apex; stigma simple. *Murray, Syst. Veg. ed.* 14, 154; *Thunb. Diss. Blæria,* 8; *Willd. Sp. Pl.* i. 630; *Ait. Hort. Kew. ed.* 2, i. 249; *Roem. & Schult. Syst. Veg.* iii. 169, *and Mant.* 107; *Bartl. in Linnæa,* vii. 650; *Klotzsch in Linnæa,* viii. 658, *and* xii. 221; *G. Don, Gen. Syst.* iii. 804; *Benth. in DC. Prodr.* vii. 698 (*excluding synonyms B. dumosa, Wendl., and Erica dumosa, Salisb.*); *Rach in Linnæa,* xxvi. 788. *B. jucunda, Reichb. ex Benth. in DC. Prodr.* vii. 698. *B. equisetifolia, G. Don, Gen. Syst.* iii. 805. *Blairia purpurea, Dietr. Syn. Pl.* i. 443. *Erica purpurea, Thunb. Prodr.* 71, *and Fl. Cap. ed. Schultes,* 356. *E. equisetifolia, Salisb. in Trans. Linn. Soc.* vi 342.

SOUTH AFRICA: without locality, *Sieber,* 153! 165! *Herb. Salisbury!*
COAST REGION: Cape Div.; Table Mountain, *Thunberg! Guthrie,* 1166! *Burchell,* 552! mountains near Muizenberg, *Burke!* Stellenbosch Div.; Lowrys Bass, *Burchell,* 8261! Caledon Div.; Hottentots Holland Mountains, near Steenbrass River, *Bolus,* 5420! and in *Herb. Norm.· Austr.-Afr.,* 43! Houw Hoek, *Schlechter,* 9420! *Galpin,* 3718! mountains near Genadendal, *Schlechter,* 9834! *Drège.*

8. B. dumosa (Wendl. Collect. ii. 3, t. 38); in habit, foliage and inflorescence exactly as in *B. purpurea,* but differing as follows:— corolla $1\frac{1}{4}$–$1\frac{1}{2}$ lin. long, tubular, 4-angled, rosy-purple; anthers divided for $\frac{2}{3}$ of their length, distinctly narrowed or constricted at the base above the short divergent spurs and about half as broad at that part as at the apex. *Roem. & Schult. Syst. Veg.* iii. 170; *G. Don, Gen. Syst.* iii. 805.

VAR. β, **breviflora** (N. E. Br.); corolla 1–$1\frac{1}{4}$ lin. long, campanulate, 4-angled, purple; otherwise as in the type. *B. glabella, Drège ex Benth. in DC. Prodr.* vii. 698, *under B. campanulata.*

South Africa : without locality, the type and var. β, *Drège!*

Coast Region : Paarl Div.; French Hoek, *Schlechter*, 10265! Stellenbosch Div.; Lowrys Pass, *Burchell*, 8252! Caledon Div. ; tops of the mountains near Genadendal, *Burchell*, 7697 ! *Bolus*, 5416 ! and in *Herb. Norm. Austr.-Afr.*, 612 ! on the Zwart Berg, near Sandfontein, *Schlechter*, 10339 ! Bredasdorp Div. ; near Elim, *Schlechter*, 9639 ! Var. β : Caledon Div.; mountains near Genadendal, *Bolus*, 5419 ! and in *Herb. Norm. Austr.-Afr.*, 613 ! *Guthrie*, 3140 !

I retain this as distinct from *B. purpurea* (which it closely resembles) on account of its different anthers. Var. β is intermediate between typical *B. dumosa* and *B. campanulata* in structure, but differs from the latter in the colour of its flowers and fewer stamens.

9. B. campanulata (Benth. in DC. Prodr. vii. 698) ; plant 6–10 in. high, with erect slender rather straight glabrous branchlets; leaves 3-nate, adpressed, very shortly petiolate, $\frac{1}{2}$–$1\frac{1}{4}$ lin. long, $\frac{1}{4}$–$\frac{1}{3}$ lin. broad, oblong or linear-oblong, subacute, glabrous with minutely scabrid margins ; umbels or superposed whorls terminal, 3–6-flowered ; pedicels $\frac{3}{4}$–1 lin. long, with 3 linear-subulate ciliolate bracts near the middle ; calyx $\frac{2}{3}$ lin. long, equally 4-lobed to below the middle, glabrous ; lobes lanceolate or linear-oblong, acute, rigid, minutely gland-ciliate ; corolla 1–$1\frac{1}{4}$ lin. long, $\frac{3}{4}$–1 lin. in diam., campanulate, 4-angled, glabrous, white ; lobes erect, $\frac{1}{4}$ lin. long, broadly rounded ; stamens usually 6, sometimes 5 ; anthers exserted, $\frac{1}{2}$ lin. long, divided nearly to the base, distinctly constricted at the base of the cells into a subquadrate truncate spurless base, scabrid, dark brown ; ovary elliptic-obovoid, glabrous, 4-angled ; cells 2–3-ovuled ; style much exserted, $1\frac{1}{2}$–$1\frac{3}{4}$ lin. long, slightly dilated at the simple stigma.

Coast Region: Caledon Div.; tops of the mountains of Baviaaus Kloof, near Caledon, *Burchell*, 7693! 7773!

I do not find the corolla-lobes fimbriate as described by Bentham. The specimens collected by Drège which were referred to this species by Bentham, differ in having purple flowers, fewer stamens and spurred anthers. I place them with others like them as a var. of *B. dumosa*, Wendl.

10. B. flexuosa (Benth. in DC. Prodr. vii. 698) ; dwarf ; branches very slender, flexuose and interwoven, glabrous ; leaves opposite or 3-nate on the same plant, adpressed, very shortly and stoutly petiolate, $\frac{3}{4}$–$1\frac{1}{3}$ lin. long, linear, subacute, glabrous, minutely scabrid on the narrowly cartilaginous margins ; flowers terminal, solitary or 2–6 in an umbel; pedicels $\frac{3}{4}$–1 lin. long; bracts remote from the calyx, linear, glabrous, very minutely gland-ciliate ; calyx lobed to below the middle ; lobes $\frac{1}{2}$ lin. long, linear-lanceolate, acute, coriaceous, keeled, glabrous ; corolla $1\frac{1}{4}$–$1\frac{1}{3}$ lin. long, $\frac{3}{4}$ lin. in diam., campanulate, distinctly 4-angled, glabrous ; lobes erect, $\frac{1}{3}$ lin. long, broadly rounded ; stamens 4–5 ; anthers not exserted, $\frac{1}{3}$ lin. long, bipartite, indistinctly scabrid, spurless but with very minutely projecting angles at the base ; ovary 4-angled, glabrous ; cells about 4-ovuled ; style equalling or shortly exserted from the corolla, $\frac{3}{4}$ lin. long ; stigma simple.

Coast Region: Caledon Div.; near the Steenbrass River, *Niven*, 6 !

11. B. affinis (N. E. Br.) ; branchlets slender, compactly sub-erect, pubescent, greyish ; leaves 4-nate, and including the petiole $\frac{1}{2}$–$1\frac{1}{4}$ lin. long, $\frac{1}{4}$–$\frac{1}{3}$ lin. broad, imbricate or slightly spreading, ovate to linear, acute or obtuse, pubescent with rather long spreading hairs and glandular ; petiole half as long as the blade ; umbels terminal, head-like, 6–9-flowered, about 2 lin. in diam.; pedicels $\frac{1}{4}$–$\frac{1}{3}$ lin. long, villous-pubescent ; bracts 3, adpressed to the calyx, equal or unequal, $\frac{2}{5}$–$\frac{2}{3}$ lin. long, linear, obtuse, ciliate with glands and long hairs ; sepals free, $\frac{1}{3}$ lin. long, $\frac{1}{8}$ lin. broad, linear or linear-lanceolate, subobtuse, ciliate like the bracts ; corolla about 1 lin. long, $\frac{2}{3}$ lin. in diam., campanulate, not very distinctly 4-angled in the dried state, glabrous, apparently pink ; lobes scarcely $\frac{1}{4}$ lin. long, broadly rounded, erect or slightly recurved at the tips ; stamens 4 ; anthers exserted, $\frac{1}{3}$ lin. long, oblong, rounded in to the filament at the base, spurless ; ovary subglobose, 4-angled, glabrous; ovules 2 in each cell ; style much exserted, $1\frac{3}{4}$–2 lin. long ; stigma simple.

Coast Region : Caledon Div.; mountains near Vogel Gat, near the mouth of the Klein River, 1500 ft., *Schlechter*, 10418!

This differs from *B. ericoides* not only in its smaller flower-heads, but the branchlets are more erect and, including the spread of the leaves, are only $\frac{3}{4}$–1 lin. in diam., whilst in *B. ericoides* they are usually $1\frac{1}{2}$–3 lin. in diam., and are generally divergent, sometimes very widely so.

12. B. ericoides (Linn. Sp. Pl. ed. 1, 112); branchlets more or less diverging, at least at their base, villous or villous-pubescent ; leaves 4-nate, imbricate or spreading, and including the petiole $\frac{3}{4}$–$2\frac{3}{4}$ lin. long, linear or oblong-linear, obtuse or subacute, pubescent with rather long spreading hairs often mingled with minute glands ; umbels head-like, terminal, 6–12-flowered, dense, usually 3–4 lin. in diam. ; pedicels $\frac{1}{4}$–$\frac{1}{2}$ lin. long, pubescent ; bracts 3, usually unequal, adpressed to the calyx, $\frac{1}{2}$–1 lin. long, linear, obtuse, villous-ciliate ; sepals free, $\frac{1}{2}$–$\frac{2}{3}$ lin. long, linear-lanceolate, subacute, villous and minutely glandular ; corolla $1\frac{1}{4}$–$1\frac{2}{3}$ lin. long, $\frac{3}{4}$ lin. in diam., tubular-campanulate, 4-angled at the basal part, glabrous, purple ; lobes $\frac{1}{4}$–$\frac{1}{3}$ lin. long, broadly rounded, erect or slightly spreading, with slightly recurved margins ; stamens 4 ; anthers exserted, $\frac{1}{3}$–$\frac{3}{4}$ lin. long, oblong, divided almost to the base, spurless, glabrous or very minutely scabrid ; ovary 4-angled, glabrous ; cells 2–4-ovuled ; style much exserted, $1\frac{1}{2}$–2 lin. long, slightly enlarged at the simple stigma. *Thunb. Diss. Blæria,* 7 ; *Wendl. Collect.* i. 73, t. 25 ; *Willd. Sp. Pl.* i. 629 ; *Ait. Hort. Kew. ed.* 1, i. 149 ; *Roem. & Schult. Syst. Veg.* iii. 168, *and Mant.* 106 ; *Bartl. in Linnæa,* vii. 649 ; *Klotzsch in Linnæa,* viii. 663, *and* xii. 222 ; *Rach in Linnæa,* xxvi. 788 ; *G. Don, Gen. Syst.* iii. 804 ; *Benth. in DC. Prodr.* vii. 698. *B. fasciculata, Sieb., & B. scabra, Drège ex Benth. l.c. B. rubra, Hort. ex Steud. Nomencl. Bot. ed.* 2, i. 208. *Blairia ericoides, Dietr. Syn. Pl.* i. 444. *Erica Blæria, Thunb. Prodr.* 72, *and Fl. Cap. ed. Schultes,* 358. *E. dumosa, Salisb. Prodr.* 296, *and in Trans. Linn. Soc.* vi. 341, *not of Andr. E. orbicularis, Lodd. Bot. Cab. t.* 153.

13. **B. revoluta** (Bartl. in Linnæa, vii. 650); plant 6–12 in. high, viscid and pilose on the branches, leaves and bracts; leaves 4-nate or irregular, varying from ascending or imbricate to recurving, $1\frac{1}{2}$–$2\frac{3}{4}$ lin. long, $\frac{1}{4}$–$\frac{1}{3}$ lin. broad, linear, obtuse; flowers 6–12 in dense terminal subglobose clusters; pedicels $\frac{1}{4}$–$\frac{2}{3}$ lin. long; bracts 3, adpressed to the calyx, unequal, linear or linear-lanceolate, obtuse, narrowed at the base into rather long curved petioles, glandular and villous with white hairs which are usually longer than those on the other parts; middle bract larger than the rest, $1\frac{1}{2}$–2 lin. long; sepals free, $\frac{1}{3}$–1 lin. long, linear or tapering from the base, acute, ciliate with very viscid glands, but without hairs; corolla rosy-purple; tube 2 lin. long, 4-angled, glabrous; lobes $\frac{1}{3}$ lin. long, rounded, recurved or very spreading; stamens 4; anthers exserted, about 1–$1\frac{1}{4}$ lin. long, oblong-linear, spurless at the very shortly subcuneate base; ovary ellipsoid, obtuse, obtusely 4-angled, glabrous; ovules 3–4 in each cell; style exserted, $2\frac{1}{4}$–$2\frac{1}{2}$ lin. long, dilated at the base; stigma simple. *Klotzsch in Linnæa*, viii. 663, *and* xii. 222; *Benth. in DC. Prodr.* vii. 698. *B. barbigera, G. Don, Gen. Syst.* iii. 805; *Klotzsch in Linnæa*, xii. 246, *and Erica barbigera, Salisb. in Trans. Linn. Soc.* vi. 341, *ex Benth. l.c. Blairia revoluta, Dietr. Syn. Pl.* i. 444.

14. **B. kraussiana** (Klotzsch ex Walpers, Rep. Bot. ii. 728); plant dwarf, somewhat thinly pilose on the branches, leaves and calyx with fine short spreading hairs, those on the calyx much longer; leaves 4-nate, $\frac{3}{4}$–$1\frac{1}{4}$ lin. long with the petiole, incurved-erect, linear or linear-oblong, subobtuse, minutely rugose in the dried state; umbels globose, head-like, nodding, about 4 lin. in diam.; pedicels $\frac{1}{2}$ lin. long, forming nearly a right angle with the calyx, pilose; bracts 3, close to the calyx and all on one side of it, $\frac{1}{2}$–$\frac{3}{4}$ lin. long, narrowly linear, obtuse, glandular and pilose; sepals free, $\frac{3}{4}$ lin. long, $\frac{1}{5}$ lin. broad at the base, lanceolate, acute, glandular and pilose; corolla tubular, apparently pink, glabrous; tube $1\frac{1}{3}$ lin. long, 4-angled; lobes nearly $\frac{1}{2}$ lin. long and about as broad, spreading, with slightly recurved tips, broadly ovate, obtuse; stamens 4, exserted; anthers nearly 1 lin. long, linear-oblong, with subparallel sides down to the parallel basal spurs, minutely punctate; spurs about $\frac{1}{4}$ as long as the anther; ovary globose-quadrangular,

glabrous, with about 4 pendulous ovules in each cell; style $1\frac{2}{3}$–$1\frac{3}{4}$ lin. long, exserted, equalling the stamens; stigma simple. *Klotzsch in Flora*, 1844, 824; *Krauss, Beitr. Fl. Cap. und Natal.* 117.

COAST REGION: Caledon Div.; at the foot of Babylons Tower, near Hemel-en-Aarde, 600-800 ft., *Krauss,* 973 !

According to the type specimens in Thunberg's Herbarium, *Blæria caduca* (Thunb. Diss. Blæria, 10) is *Erica caduca,* Thunb., whilst *B. nudiflora* (Thunb. l.c. 6), and *B. pusilla* (Thunb. l.c. 9, as to sheet α of his Herbarium), both belong to *Erica nudiflora,* Linn. Sheets β and γ of *B. pusilla* may be a variety of *Erica nudiflora,* but they differ in being more hairy and have smaller flowers than is usual in that species. The specimens on all 3 sheets of *E. pusilla* have 8 stamens, not 4 as stated in *Thunberg, Fl. Cap. ed. Schultes,* 348.

V. **COILOSTIGMA**, Klotzsch.

Bracts none. *Calyx* unequally 4-partite, 1 segment much larger than the rest. *Corolla* ovoid or cylindric, 4-toothed. *Stamens* 4; filaments linear, glabrous, free; anthers free, divided nearly to the base, with parallel contiguous cells. *Ovary* 2–4-celled, with 1 pendulous ovule in each cell; style filiform, exserted, terminal, sometimes becoming lateral in fruit by the abortion of a cell; stigma rather large, peltate or crater-like.

Heath-like shrublets; leaves grooved down the convex back; flowers in small clusters.

DISTRIB. Species 4, endemic.

Corolla puberulous ... ... (1) **tenuifolium.**
Corolla glabrous:
　Branchlets glabrous; calyx $\frac{1}{3}$ as long as the
　　corolla ... ... (2) **glabrum.**
　Branchlets puberulous; calyx $\frac{1}{2}$–$\frac{3}{4}$ as long as the
　　corolla:
　　Branchlets slender, subflexuose ... ... (3) **zeyherianum.**
　　Branchlets rather stout, straight ... ... (4) **dregeanum.**

1. **C. tenuifolium** (Klotzsch in Linnæa, xii. 234); branches sub-flexuose, erect, puberulous, greyish; leaves 3-nate, $\frac{3}{4}$–2 lin. long with the petiole, erect, linear, subobtuse, glabrous, margins obscurely denticulate or slightly scabrid; flowers axillary and terminal, usually forming small clusters; pedicels not more than $\frac{1}{8}$ lin. long; calyx-segments ciliate, the larger $\frac{1}{3}$–$\frac{3}{4}$ lin. long, varying from lanceolate, acute and submembranous, to linear, obtuse and thick and leaf-like, the rest $\frac{1}{4}$ lin. long, lanceolate, subacute; corolla $\frac{3}{4}$ lin. long, ovoid, with 4 short incurved subacute teeth, puberulous outside; anthers exserted, oblong, with parallel sides and a short broad notch at the apex, subtruncate and spurless at the base; ovary 2-celled and compressed, broader than long or 3-celled and trigonous, glabrous; style exserted, glabrous; stigma peltate, 4-angled, with 2 central points. *Benth. in DC. Prodr.* vii. 708.

COAST REGION: Uitenhage Div.; between Vanstadens Mountains and Kra-kakamma, *Ecklon & Zeyher* (ex *Klotzsch*), and without precise locality, *Zeyher,*

719! Port Elizabeth Div.; near Emerald Hill, *Bodkin in Herb. Bolus,* 6693! Alexandria Div.; Oliphants Hoek, *Ecklon & Zeyher* (ex *Klotzsch*). Albany Div.; Slay Kraal, *Zeyher!*

2. C. glabrum (Benth. in DC. Prodr. vii. 708); branchlets very erect, straight, glabrous, whitish; leaves 3-nate, adpressed, $\frac{3}{4}$–$1\frac{1}{3}$ lin. long with the petiole, linear, obtuse, flat above, glabrous; flowers in axillary clusters of 2–4, subsessile on minute bracteolate peduncle-like branchlets which are much shorter than the leaves; calyx unequally 4-partite; lobes oblong or linear-oblong, obtuse, glabrous, very minutely ciliate, the longest about $\frac{1}{3}$ lin. long; corolla nearly 1 lin. long, $\frac{2}{5}$ lin. in diam., cylindric, 4-lobed, glabrous; lobes erect, broader than long, rounded; anthers exserted, $\frac{1}{2}$–$\frac{3}{5}$ lin. long, linear-oblong, notched at the apex, obtuse and spurless at the base; ovary subglobose, obtusely 4-angled, 4-celled, glabrous; style exserted, glabrous; stigma crater-like.

Coast Region: Riversdale Div.; between Little Vet River and Garcias Pass, *Burchell,* 6875!

3. C. zeyherianum (Klotzsch in Linnæa, xii. 234); branchlets slender, subflexuose, erect, puberulous, greyish; leaves 3-nate, adpressed or imbricate, $\frac{3}{4}$–$1\frac{1}{2}$ lin. long with the short petiole, linear, subtrigonous, subacute, glabrous; flowers axillary and terminal, usually forming small clusters; pedicels $\frac{1}{8}$ lin. long; one sepal $\frac{1}{2}$–$\frac{3}{4}$ lin. long, linear or spathulate and leaf-like, from a thin dilated base, the others $\frac{1}{4}$ lin. long, linear, obtuse, glabrous, minutely ciliate; corolla nearly 1 lin. long, ovoid, with 4 short rounded somewhat incurved teeth, glabrous; anthers exserted, oblong, with parallel sides and a short broad apical notch, spurless at the base; ovary compressed, broader than long, somewhat transversely rhomboid, glabrous; style exserted, glabrous; stigma peltate, 4-angled. *Benth. in DC. Prodr.* vii. 708.

Coast Region: Uitenhage Div.; Vanstadens Mountains, *Ecklon & Zeyher! Zeyher!*

4. C. dregeanum (Klotzsch in Linnæa, xii. 235); about a foot high; branchlets rather stout, straight, puberulous; leaves 3-nate, linear, acute, rather thick, glabrous; flowers axillary and terminal, subsessile; sepals unequal, the larger sometimes leaf-like and about equalling the corolla, the others lanceolate, acute, glabrous; corolla ovoid, glabrous; anthers exserted, spurless. *Benth in DC. Prodr.* vii. 708.

South Africa: without locality, *Drège,* 7753.

According to Bentham this only differs from *C. zeyherianum* in its somewhat larger leaves and more rigid habit. I have not seen it.

VI. **THORACOSPERMA**, Klotzsch.

Bracts 1–3, adpressed to the calyx. *Calyx* minute, equally 4-lobed or 4-partite; tube very short or none, rather thin, not angular.

Corolla very small, ovoid, ovoid-oblong, or subcampanulate, with 4 short connivent or erect lobes. *Stamens* 4; filaments linear or linear-filiform, free; anthers more or less exserted (except in *T. puberulum*), bipartite, with or without basal spurs; cells opening by short slits. *Ovary* 2–4-celled, by abortion 1–3-celled in fruit; cells 1-ovuled; style exserted; stigma simple or slightly obconically thickened or subpeltate, always small. *Fruit* globose, crustaceous, usually 1–2-seeded, perhaps indehiscent, but not seen mature.

Small shrubs or shrublets with the habit of *Erica;* leaves flat or convex above, convex and grooved on the back; flowers small, in small axillary clusters along the branches, on minute branchlets shorter than the leaves subtending them, or terminal on short lateral branchlets.

DISTRIB. Species 6, all endemic.

<pre>
Corolla puberulous or pubescent   ...   .   (6) puberulum.
Corolla glabrous:
    Leaves 4-nate, incurved-erect, thinly and minutely
        scabrid-pubescent   ...   .   (4) interruptum.
    Leaves 3-nate, glabrous:
        Leaves convex on both sides, ½–1 lin. long,
            spreading or incurved-erect; corolla not
            quite 1 lin. long   ...   .   (5) nanum.
        Leaves flat above, convex and grooved down
            the back, imbricate or adpressed; corolla
            1¼–1½ lin. long:
            Leaves narrowly-linear, 2–4½ lin. long;
                anther-spurs at the base of the cells,
                free from the filament   ...   ...   (1) paniculatum.
            Leaves linear-lanceolate or oblong-
                linear, 1–2¼ lin. long; anther-spurs
                spreading from the filament below the
                cells:
                Calyx lobed ⅔–¾ of the way down;
                    anthers cuneate-obovate, with
                    horizontal or upcurved spurs,
                    sometimes obsolete   ...   ...   (2) Marlothii.
            Calyx lobed nearly to the base;
                anthers with nearly parallel sides
                and descending-divergent spurs ..   (3) Galpini.
</pre>

1. **T. paniculatum** (Klotzsch in Linnæa, ix. 350, not of xii. 229); a small shrublet, 1–2 ft. high; branchlets minutely whitish-tomentose; leaves 3-nate, adpressed or erect, 2–4½ lin. long with the petiole, linear, obtuse, flat above, convex on the back, glabrous, not ciliate; flowers 1–3 together on minute peduncle-like axillary branchlets 0–⅓ lin. long or terminal, forming small elongated clusters at the ends of the branches; pedicels minute, ⅛ lin. long; bracts 3, adpressed to the calyx, equal or unequal, linear, obtuse, ciliate with exceedingly minute glands; calyx nearly ½ lin. long, 4-lobed to ⅔ of the way down, glabrous; lobes ovate or ovate-lanceolate, obtusely pointed, very minutely ciliate like the bracts; corolla 1¼ lin. long, ovoid-oblong, slightly narrowed at the mouth, glabrous, pinkish-white (*Thunberg*); lobes very small, quadrate, truncate, erect; anthers partly (or wholly?) exserted, rather more than ½ lin. long, linear-oblong, spurred at the base, scabrid; spurs free, awn-

like, ¼ as long as the anther; ovary ellipsoid, glabrous; style 1½ lin. long, glabrous; stigma obconically thickened, truncate. *Rach in Linnæa,* xxvi. 790. *Erica paniculata, Thunb. Prodr.* 72, *and Fl. Cap. ed. Schultes,* 360. *Blæria paniculata, Thunb. Diss. Blæria,* 10. *Simocheilus quadrifidus, Benth. in DC. Prodr.* vii. 703.

SOUTH AFRICA: without locality, *Thunberg!*
COAST REGION: George Div. ; Brak River Heights, between Great Brak River and Malgaten River, *Burchell,* 6126!

This is the type of the genus and must be very rare, as I have only seen the above two specimens.

2. **T. Marlothii** (N. E. Br.); about 1 ft. high; branches erect, minutely puberulous on the younger parts, brown or greyish; leaves 3-nate, imbricate or adpressed, 1–1½ lin. long with the petiole, linear-lanceolate or oblong-linear, obtuse, flat above, convex on the back, glabrous, not ciliate, flowers 1–6 together on minute axillary branchlets and terminal, forming short unilateral clusters at the ends of the branches; pedicels ¼–½ lin. long, minutely puberulous; bracts 3, unequal, shorter than and adpressed to the calyx or 1 lower than the rest, linear, obtuse; calyx ⅓–½ lin. long, 4-lobed to ⅔ or ¾ of the way down; lobes ovate, acute, glabrous, ciliate, pink; corolla 1¼–1½ lin. long, oblong or ovoid-oblong, glabrous, pink or red; lobes broader than long, rounded or broadly deltoid, obtuse, erect; anthers ultimately exserted much beyond the corolla, ⅓–⅖ lin. long, cuneate-oblong, scabrid, usually with minute awn-like spurs arising from the filament a short but variable distance below the anther-cells, horizontally spreading or slightly upcurved, sometimes nearly or quite obsolete; ovary subglobose, more or less 4-angled, glabrous; style 2½ lin. long, glabrous; stigma simple, not thickened.

CENTRAL REGION, between 4500 and 4900 ft. : Worcester Div.; Witteberg Range, near Matjesfontein, *Marloth,* 2955! 2956! Prince Albert Div.; Zwartberg Range, near Sevenweeks Poort, *Marloth,* 2976!

3. **T. Galpini** (N. E. Br.); a small shrublet; branchlets erect, minutely puberulous, brown or partly greyish; leaves 3-nate, adpressed, 1½–2¼ lin. long with the petiole, linear-lanceolate, obtuse, flat above, convex on the back, glabrous not ciliate; flowers 1–3 together on minute axillary branchlets ¼–⅓ lin. long, also terminal, forming short unilateral clusters at the ends of the branchlets; pedicels ¼–⅓ lin. long, minutely puberulous; bracts 3, adpressed to the calyx, equal or unequal, minutely ciliate; calyx ½ lin. long, 4-lobed almost to the base; lobes ovate or ovate-lanceolate, acute or subacuminate, glabrous, distinctly ciliate; corolla 1¼ lin. long, ovoid-oblong, slightly narrowed at the mouth, glabrous, white or pale pink (*Galpin*); lobes very small, deltoid-ovate, obtuse; anthers half-exserted, not quite ½ lin. long, linear-oblong, spurred at the base, scabrid; spurs awn-like, adnate to the filament for about ⅔ of their length then diverging (not horizontally spreading); ovary ellipsoid or

globose, very minutely puberulous ; style 1½ lin. long, glabrous ;
stigma simple, not thickened.

COAST REGION : Riversdale Div.; Garcias Pass, 200 ft., *Galpin*, 3732 !

4. T. interruptum (N. E. Br.) ; a shrublet, 15 in. or more high ;
branchlets pubescent, light fulvous-brown ; leaves 4-nate, incurved-
erect, in distant whorls, 1–2¼ lin. long with the petiole, linear, subacute, convex on both sides, scabrid-pubescent on the back ; flowers
in terminal clusters of 2–8 on short lateral branchlets ; pedicels
¼–⅓ lin. long, bearing 1 large leaf-like bract ½–⅔ lin. long at its apex ;
calyx ⅓ lin. long, equally 4-lobed nearly to the base ; lobes erect,
ovate, acute, grooved down the back, ciliate ; corolla nearly 1¼ lin.
long, ovoid, glabrous ; lobes about ¼ as long as the tube, obtuse,
connivent-erect ; stamens exserted ; filaments 1–1¼ lin. long, glabrous ;
anthers ¾ lin. long, linear-oblong, with parallel sides ; spurs very
short, basal, directed inwards, smooth ; ovary compressed-ellipsoid,
2-celled, glabrous ; style 2 lin. long, exserted for half its length,
filiform, glabrous ; stigma simple.

SOUTH AFRICA : without locality, *Bowie! Ward!* both in Herb. Kew.

5. T. nanum (N. E. Br.) ; a dwarf shrublet, about 6–8 in. high ;
branchlets very minutely greyish-tomentose ; leaves 3-nate, ½–1 lin.
long with the petiole, imbricate or rather spreading, linear, subacute, slightly incurved, convex above, glabrous, not ciliate ; flowers
1–3 together, terminal on the short lateral branchlets ; pedicels
about ¼ lin. long ; bracts 3, adpressed to the calyx, ¼–⅓ lin. long,
linear or slightly spathulate, thickened and grooved down the back
at the apex, obtuse, glabrous, minutely ciliate ; calyx ½ lin. long,
equally 4-lobed to ¾ of the way down ; lobes erect, linear, obtuse,
grooved down the back, glabrous, minutely ciliate ; corolla not quite
1 lin. long, subcampanulate or ovoid-campanulate, glabrous ; lobes
erect, broader than long, obtuse ; stamens much exserted ; filaments
1¼–1½ lin. long, glabrous ; anthers ⅓ lin. long, cuneate-obovate,
divided nearly to the spurless base, very minutely scabrid, falling
away with about ⅓ of the filament attached to them ; ovary obovoid,
glabrous, 2-celled ; style 1½–1¾ lin. long, glabrous ; stigma simple,
scarcely thicker than the style.

COAST REGION : Uitenhage Div.; Van Stadens River Mountains, *Bolus*,
1580!

6. T. puberulum (N. E. Br.) ; branchlets erect, crowded, densely
puberulous or minutely tomentose, greyish ; leaves 3-nate, adpressed
or subimbricate, rarely slightly spreading, ½–1 lin. long with the
petiole, linear, obtuse, convex on both sides, glabrous or very
minutely subscabrid-puberulous ; flowers in axillary clusters of 3–5
on minute branchlets shorter than the leaves, crowded together along
the short erect crowded lateral branchlets, in a dense paniculate

inflorescence on each main branch; pedicels very minute; bracts 3 or 1, the 2 lateral bracts being always very minute, sometimes absent, middle bract minute, $\frac{1}{8}$ lin. long, puberulous; calyx equally 4-partite; lobes $\frac{1}{4}$ in. long, lanceolate, obtuse or acute, puberulous; corolla $\frac{3}{5}$–$\frac{2}{3}$ lin. long, ovoid, with connivent or erect deltoid-ovoid obtuse lobes, varying from exceedingly minutely puberulous to distinctly pubescent with spreading hairs, purple; stamens included; filaments $\frac{1}{3}$ lin. long, doubled upon themselves, glabrous; anthers nearly $\frac{1}{2}$ lin. long, oblong, with acute tips to the cells, membranous, smooth, pale brown; ovary tetragonous-ellipsoid or -subglobose, obtuse, 4-celled, or by abortion in fruit 1–3-celled, minutely and sparsely puberulous at the top; style $\frac{2}{3}$ lin. long, much exserted, glabrous; stigma subpeltate. *Thamnium puberulum, Klotzsch in Linnœa,* xii. 223. *Coilostigma puberulum, Benth. in DC. Prodr.* vii. 708.

SOUTH AFRICA: without locality, *Thom!*

COAST REGION: Caledon Div.; mountains near the Zondereinde River and at Knoblauch (Knofflochs Kraal) on the Bot River, *Ecklon & Zeyher!* mountains between Caledon and Elim, 700 ft., *Bolus,* 6763! Bredasdorp Div.; hills near Elim, 300 ft., *Bolus,* 6764!

This differs from the other species of *Thoracosperma* in its puberulous corolla, and the very different texture of its included anthers, but these characters scarcely warrant its generic separation. The pubescence on the corollas of Dr. Thom's specimens is very different, and might almost be described as hairy, but I find no structural distinction.

VII. EREMIA, D. Don.

Bracts 3. *Calyx* 4-partite, rather large proportionately to the corolla, campanulate; segments equal or in one species with 1 rather broader than the rest, ciliate or hairy but not woolly. *Corolla* urceolate, campanulate or cup-shaped, shortly 4-lobed, not more than twice as long as broad. *Stamens* 8, included, equalling or very slightly exceeding the corolla-lobes; filaments and the bipartite anthers free. *Ovary* seated on a short or thin disk, 2–4-celled; ovules solitary in each cell, pendulous; style usually shortly exserted; stigma simple or minutely 4-lobed. *Fruit* not seen.

Small heath-like shrubs or shrublets; leaves grooved down the convex back, spreading, not woolly; flowers small in terminal clusters on short lateral branchlets.

DISTRIB. Species 4, endemic.

<pre>
Leaves 3-nate:
    Calyx shorter than the 1¼–1¾ lin. long corolla:
        Leaves 1–3 lin. long, hairy or subechinate
            on the back; corolla urceolate   ...   ...   (1) totta.
        Leaves ⅓–⅔ lin. long, at first ciliate, but
            without hairs on the back; corolla cam-
            panulate   ...   .   (2) recurvata
    Calyx subequalling the minute corolla   ...   ...   (3) parviflora.
Leaves 4-nate, ½–1 lin. long; corolla cup-shaped,
    ⅔ lin. long   ...   .   (4) brevifolia.
</pre>

1. E. totta (D. Don in Edinb. New Phil. Journ. xvii. 156, and
Gen. Syst. iii. 828); a bush, 2–4 ft. high; branches stout, hairy
with rather long spreading hairs or shortly subhispid; leaves 3-nate,
spreading or reflexed, 1–3 lin. long with the petiole, usually curved,
angular, or ridged and glabrous on the upper side, varying from
with long white spreading hairs to subechinate on the back; flowers
in compound clusters at the ends of short lateral branches, usually
composed of 2 to several clusters of 2–3 flowers on minute axillary
branchlets; pedicels $\frac{1}{4}$–$\frac{1}{3}$ lin. long; bracts 3, adpressed to the
calyx, $\frac{2}{3}$–1 lin. long and broad, elliptic-ovate, subacute, glabrous or
puberulous, serrulate-ciliate, white; calyx equally 4-partite; seg-
ments 1–1$\frac{1}{2}$ lin. long, $\frac{2}{3}$–1 lin. broad, elliptic-oblong, obtuse, white,
glabrous or puberulous, serrulate-ciliate, the cilia often gland-tipped
and minutely hispid; corolla 1$\frac{1}{4}$–1$\frac{3}{4}$ lin. long, with the greater part
included, urceolate, 4-angled, narrowed from the much inflated base,
but scarcely or not at all constricted below the erect rounded lobes,
glabrous or occasionally very minutely puberulous, white; stamens
about as long as the corolla-tube; filaments filiform, glabrous;
anthers $\frac{1}{3}$ lin. long, oblong, with contiguous parallel cells, spurless,
smooth; ovary rather broadly and obtusely conical, 4-angled, 4-celled,
thinly or densely covered with long and very fine woolly hairs at the
top; style included or shortly exserted, filiform, glabrous; stigma
simple, scarcely or but slightly enlarged. *Klotzsch in Linnæa*, xii.
218; *Benth. in DC. Prodr.* vii. 699. *Erica totta, Thunb. Diss.
Erica*, 18, *Prodr. Pl. Cap.* 70, *and Fl. Cap. ed. Schultes*, 348. *E.
pectinata, Bartl. in Linnæa*, vii. 647. *Euremia totta and E. bartlingiana, Rach in Linnæa*, xxvi. 789.

Var. β, **bartlingiana** (N. E. Br.); ovary glabrous, otherwise as in the type
and varying in the same way. *E. bartlingiana, Klotzsch in Linnæa*, xii. 218;
Benth. in DC. Prodr. vii. 699. *Erica ferox, Salisb. in Trans. Linn. Soc.* vi.
324. *E. totta, Bartl. in Linnæa*, vii. 647.

SOUTH AFRICA : without locality, *Thunberg! Roxburgh! Niven!* Var. β,
Zeyher, 1116!
COAST REGION, between 500 and 3600 ft.; Clanwilliam Div.; Koude Berg,
Schlechter, 8750! Tulbagh Div.; near Tulbagh, *Ecklon & Zeyher! Guthrie*,
2078! near Saron, *Schlechter*, 10687! Worcester Div.; Hex River Valley,
Tyson, 700! and without precise locality, *Zeyher!* Malmesbury Div.; Riebeeks
Castle, *Niven*, 83! Var. β : Tulbagh Div.; near Tulbagh, *Pappe!* Worcester
Div.; Dutoits Kloof, *Drège!* Paarl Div.; Paarl Mountain, *Bolus*, 2948! French
Hoek, *MacOwan, Herb. Norm. Austr.-Afr.*, 756!

Rach, at the place quoted, refers one of the specimens on sheet *a* and the
single specimen on sheet β of Thunberg's Herbarium to *E. bartlingiana*. I have
examined these specimens and find that the ovary of the specimen on sheet β has
a few long woolly hairs upon it and is not quite glabrous, whilst both specimens
on sheet *a* have densely woolly ovaries. I cannot distinguish them varietally, as
other specimens show various degrees of woolliness; nor can I find any such
difference as is described by Klotzsch and Bentham in the corollas of the woolly
and glabrous forms.

2. E. recurvata (Klotzsch in Linnæa, xii. 498); a shrublet, $\frac{1}{3}$–1
ft. high, somewhat loosely branched and apparently rather straggling

in habit; branchlets rather rigid, ascending or spreading, very minutely puberulous and sometimes thinly beset with rather long spreading white hairs, becoming glabrous, usually greyish; leaves 3-nate, spreading-recurved, $\frac{1}{3}$–$\frac{2}{3}$ lin. long, $\frac{1}{4}$–$\frac{1}{3}$ lin. broad, ovate, apiculate, thick and rigid, rather more convex above than beneath, at first ciliate with 3–4 long white stiff caducous hairs on each side, shining, but at first most minutely puberulous above, becoming glabrous, flowers 1–3 in terminal clusters; pedicels $\frac{1}{4}$–$\frac{2}{3}$ lin. long, puberulous with or without an intermingling of long white hairs; bracts 3, close to the calyx, usually more or less spreading, $\frac{1}{2}$ lin. long, leaf-like, but often thinner, more acute, and more beset with long white hairs; calyx 4-partite; segments lanceolate or ovate acute, $\frac{2}{3}$–$\frac{3}{4}$ lin. long, $\frac{1}{3}$–$\frac{1}{2}$ lin. broad, thinly covered with long white hairs; corolla $1\frac{1}{2}$ lin. long, campanulate, 4-angled, glabrous; lobes erect, $\frac{1}{3}$ lin. long, broadly rounded; stamens included; filaments linear-lanceolate, tapering into the filiform apex, glabrous; anthers $\frac{1}{3}$ lin. long, narrowed upwards, with contiguous cells, bigibbous in front at the base and dorsally spurred, smooth; spurs $\frac{2}{3}$ as long as the cells, narrowly lanceolate, acuminate, ciliate, inserted on the back of the cells at about $\frac{1}{4}$ above their base; ovary ellipsoid or globose-obovoid, obtuse, 2-celled, thinly pubescent at the top; style 1 lin. long, ultimately shortly exserted, filiform, glabrous, deciduous; stigma simple, not thickened. *Benth. in DC. Prodr.* vii. **700.**

Coast Region: Clanwilliam Div.; Blue Mountain, and at Ezelsbank on the Cederberg Range, *Drège*, 2965!

Central Region: Ceres Div.; at Klyn Vley in the Cold Bokkeveld, 5500 ft., *Schlechter*, 10053!

3. E. parviflora (Klotzsch in Linnæa, xii. 498); leaves 3-nate, very spreading, trigonous, subobtuse, evanescently scabrid on the margin; flowers in terminal clusters of 3, subsessile; bracts minute, ovate, acute, close to the calyx; calyx-lobes subequalling the corolla, oval, carinate-apiculate, glabrous, pectinate on the margins, white; corolla minute, tetragonous, glabrous; anthers included, aristate; ovary obtusely conical, hairy at the apex, 2-celled; style exserted. *Benth. in DC. Prodr.* vii. 700. *Erica shalliana, Hort. ex Klotzsch l.c.* 498.

Coast Region: Robertson or Swellendam Div.; mountains near Kogmans Kloof, *Ecklon & Zeyher*, and hills between Kogmans Kloof and Puspas Valley, *Ecklon & Zeyher, ex Klotzsch.*

I have not seen this species.

4. E. brevifolia (Benth. in DC. Prodr. vii. 700); a shrub, 3–4 ft. high; branches erect, puberulous, more or less intermingled with long spreading often gland-tipped hairs or subhispid, greyish; leaves 4-nate, incurved-spreading, $\frac{1}{2}$–1 lin. long, oblong-ovate, subacute, thick, flat and glabrous above, openly grooved on the very convex minutely hairy back; flowers in terminal clusters of 4 or fewer on

short lateral branchlets, subsessile or on very minute pedicels ; bracts
3, adpressed to and equalling the flower in length, the outer or
middle bract $\frac{3}{4}$ lin. long, $\frac{2}{3}$ lin. broad, broadly ovate, acute, partly
covering the rather smaller laterally excised side bracts, glabrous,
minutely gland-ciliate ; calyx 4-partite ; segments about $\frac{2}{3}$ lin. long,
one obovate-spathulate, rather broader than the other linear three,
puberulous, minutely gland-ciliate ; corolla $\frac{2}{3}$ lin. long, cup-shaped,
with 4 rounded lobes nearly twice as broad as long, glabrous ; stamens
as long as or very slightly exceeding the corolla ; filaments $\frac{1}{4}$ lin. long ;
anthers $\frac{1}{2}$ lin. long, oblong, with contiguous parallel cells, bearded at
the base and along the inner margins, opening to $\frac{2}{3}$ of the way down
by very large pores ; ovary subglobose, 2-celled, minutely puberulous
on the top ; style filiform, very slightly exceeding the anthers,
glabrous ; stigma minutely 4-lobed.

COAST REGION : Mossel Bay Div. ; Attaquas Kloof, *Masson,* 57 ! (*Niven ?*)
85 !

VIIA. **PLATYCALYX**, N. E. Br.

Bracts 3. *Calyx* nearly flat, nearly square in outline, 4-lobed.
Corolla subglobose or globose-ovoid, much contracted at the mouth,
4-toothed. *Stamens* normally 6, occasionally 5 or 7, exserted ;
filaments and anthers free. *Ovary* superior, seated on a thin disk,
2-celled, with 1 pendulous ovule in each cell ; style exserted ; stigma
simple.

A small Erica-like shrublet ; flowers small, 1–3 together, terminal, red.
DISTRIB. Monotypic, endemic.

1. **P. pumila** (N. E. Br.) ; a very small shrublet, 3–5 in. high,
loosely branched ; branches slender, clothed with very minute
greyish tomentum ; leaves minute, 3-nate, adpressed, $\frac{1}{2}$–1 lin. long
with the petiole, linear-oblong, obtuse, very thick, subquadrangular,
flattened above and on the grooved back, and slightly at the sides,
glabrous, not ciliate ; flowers in terminal clusters of 2–3 ; pedicels
$\frac{2}{3}$ lin. long, glabrous ; bracts 3, at or above the middle of the pedicel,
minute, linear, glabrous ; calyx $\frac{1}{2}$ lin. square ; lobes or angles
apiculate, much broader than long, glabrous, not ciliate, red ; corolla
$1\frac{1}{2}$–$1\frac{3}{4}$ lin. long, inflated globose-ovoid, contracted at the very shortly
4-lobed mouth, glabrous, red ; lobes scarcely $\frac{1}{4}$ lin. long, very broadly
rounded, connivent ; stamens more or less exserted ; filaments
$1\frac{1}{2}$–2 lin. long, filiform, glabrous ; anthers scarcely $\frac{1}{4}$ lin. long and
nearly as broad, shortly oblong, bipartite, with contiguous parallel
cells, spurless, basifixed, minutely scabrid, opening by minute
pores ; ovary compressed-globose, very obtuse, minutely white-
woolly, 2-celled, with 1 ovule in each cell ; style $1\frac{3}{4}$ lin. long,

slightly exceeding the anthers, filiform, glabrous; stigma simple, very slightly thickened.

COAST REGION: Riversdale Div.; near Riversdale, *Rust,* 543 !

In general appearance this peculiar plant might be likened to *Erica tubercularis,* Salisb., but in structural characters it is distinct from *Erica* and every other genus as at present established, and to no genus except *Erica* does it outwardly bear any general resemblance. The name is in allusion to the nearly flat calyx.

VIII. **HEXASTEMON**, Klotzsch.

Bracts 3. *Calyx* 4-partite; segments erect, white-woolly. *Corolla* elongated-ovoid or inflated at the base and tubular above, 4-toothed. *Stamens* 6, much exserted; filaments and anthers free. *Ovary* 2-celled, with 1 pendulous ovule in each cell; style much exserted; stigma minutely capitate or 2-lobed. *Fruit* not seen.

A small shrub or shrublet, densely covered with white woolly hairs on the leaves and flowers, which are in small terminal clusters.

DISTRIB. Monotypic, endemic.

1. **H. lanatus** (Klotzsch in Linnæa, xii. 220); a white-woolly shrub or shrublet; branches moderately stout, more or less divergent, at first covered with a minute tomentum, naked below, densely leafy on the upper part; leaves 3-nate, densely imbricate, $\frac{3}{4}$–$1\frac{1}{4}$ lin. long, $\frac{1}{4}$–$\frac{1}{3}$ lin. broad, linear-oblong, obtuse, slightly incurved, flat and glabrous above, convex, grooved and white-woolly or glabrous on the back, densely white-woolly-ciliate; flowers 1–6 together in terminal clusters; pedicels $\frac{1}{4}$–$\frac{1}{2}$ lin. long; bracts 3, adpressed to the calyx, equal or unequal, all 1 lin. long, linear or linear-lanceolate, or the middle one larger and oblanceolate-spathulate, obtuse or acute, densely fringed with long white woolly hairs, otherwise glabrous; calyx-segments erect and agglutinated to the corolla, about $1\frac{1}{4}$ lin. long, $\frac{1}{2}$–$\frac{2}{3}$ lin. broad, lanceolate, ovate or oblong-ovate, obtuse or acute, densely white-woolly on the margins and sometimes on the back, very viscid and broadly bordered with a series of papillæ on the inner face; corolla $1\frac{2}{3}$–$1\frac{3}{4}$ lin. long, acutely 4-angled, elongate-ovoid or obovoid or pear-shaped inflated at the base and tubular above, shortly 4-toothed, glabrous; stamens 6, much exserted, free; filaments $1\frac{3}{4}$–2 lin. long, linear, glabrous; anthers $\frac{1}{2}$–$\frac{3}{4}$ lin. long, linear, bipartite, spurless, dorsifixed close to the base, smooth, opening by longitudinal slits; ovary compressed-ellipsoid, obtuse, glabrous, 2-celled; style $2\frac{1}{2}$–$4\frac{1}{2}$ lin. long, exserted much beyond the anthers, filiform, glabrous; stigma minutely capitate and entire or 2-lobed and somewhat crutch-like. *Eremia lanata, Benth. in DC. Prodr.* vii. 700. *Erica xeranthemifolia, Salisb. in Trans. Linn. Soc.* vi. 339. *Blæria xeranthimifolia, G. Don, Gen. Syst.* iii. 805; *Klotzsch in Linnæa,* xii. 246.

SOUTH AFRICA: without locality, *Herb. Salisbury! Ecklon & Zeyher! Guthrie,* 4540 !

COAST REGION: Caledon Div.; Babylons Tower, *Zeyher*, 3294! Shaws Mountain, near Caledon, 1200–1300 ft., *Bolus*, 9145! *Galpin*, 3719! and without precise locality, *Zeyher!*

Salisbury has described the flowers of this plant as having only 4 stamens, but in his type specimen at Kew they have six.

IX. **GRISEBACHIA**, Klotzsch.

Pedicels 3-bracteate. *Calyx* 4-lobed or 4-partite; tube usually 4-angled; lobes or segments sometimes unequal in breadth, long-ciliate. *Corolla* shortly 4-lobed, often slightly oblique; tube usually constricted at or above the middle, with the part above the constriction cup-shaped or campanulate and the part below it sharply 4-angled, in a few species without any constriction and funnel-shaped, tubular or tubular-obconic. *Stamens* 4, included or rarely slightly exceeding the corolla-lobes; filaments shortly and broadly dilated or somewhat crutch-like at their attachment to the back of the anthers, often hairy; anthers nearly or quite as broad as long, with their cells entirely free from each other, dorsifixed near the base or tapering into a slender attachment to the filament, contiguous or separated, with or without basal awn-like spurs. *Ovary* seated on a disk, usually compressed and 2-celled, rarely trigonous and 3-celled; style filiform, exserted; stigma minute, simple or capitate; ovules solitary in each cell, pendulous. *Fruit* subglobose, obovoid or ellipsoid, 2–3-celled; pericarp somewhat fleshy, indehiscent?

Shrubs or shrublets with the habit of *Erica;* leaves grooved down the convex back; flowers small, pedicellate or subsessile, arranged either in 6–15-flowered globose heads, which are terminal or lateral and subsessile or on short branchlets, or only 3–4 together on minute axillary branchlets spicately arranged along the branches.

DISTRIB. Species 21, endemic.

Leaves 4-nate:
    Calyx divided nearly or quite to the base   ...   (1) **ciliaris.**
    Calyx lobed to half-way down, with a 4-angled
      tube   ...   .   (2) **hispida.**
Leaves 3-nate:
  * Corolla-tube distinctly constricted at or above
    the middle, puberulous outside and at the
    constriction within:
     † Calyx divided to the base:
      § Anthers without spurs:
        Sepals about $1\frac{1}{4}$ lin. long, ciliate
          and hairy on the back with
          hispid hairs $\frac{1}{3}$–$\frac{1}{2}$ lin. long   ...   (4) **Bolusii.**
        Sepals ciliate with simple or but
          slightly hispid hairs:
          Young leaves apiculate with
            a gland-tipped hair; se-
            pals $\frac{3}{4}$ lin. long, ciliate
            with gland-tipped hairs
            $\frac{1}{3}$ lin. long   ...   ...   (5) **apiculata.**
          Young leaves not apiculate:
            Young leaves puberulous,
             or not ciliate with
             sessile glands:
              Sepals 2 lin. long,

    with cilia $\frac{3}{4}$–1 lin.
    long ... ... (3) **involuta.**
   Sepals 1 lin. long,
    with cilia $\frac{3}{4}$–1$\frac{2}{3}$
    lin. long ... (6) **velleriflora.**
   Sepals 1 lin. long,
    with cilia $\frac{1}{3}$–$\frac{1}{2}$ lin.
    long ... ... (7) **dregeana.**
   Young leaves glabrous,
    ciliate with 4–6 sessile
    glands on each margin;
    sepals and their cilia
    each about $\frac{3}{4}$ lin. long (8) **zeyheriana.**
 §§ Anthers with awn-like spurs near
  the base :
   Anthers white (the only species
    in which they are so) ... ... (12) **alba.**
   Anthers dark-coloured :
    Sepals 1–1$\frac{1}{4}$ lin. long, rather
     rigidly ciliate with stout
     simple or slightly scabrous
     hairs ... ... (9) **rigida.**
    Sepals $\frac{1}{2}$–$\frac{3}{4}$ lin. long, ciliate
     with hispid hairs :
     Leaves glabrous ; sepals
      $\frac{1}{3}$–$\frac{1}{2}$ lin. broad ... (10) **Niveni.**
     Leaves puberulous or
      minutely tomentose
      when young; sepals
      not $\frac{1}{4}$ lin. broad ... (11) **incaua.**
 †† Calyx lobed $\frac{1}{3}$–$\frac{2}{3}$ of the way down ; tube
  4-angled :
   Anthers without basal spurs :
    Calyx-lobes ciliate and clothed on
     the back with long simple gland-
     tipped hairs; ovary slightly
     pubescent ... ... (13) **hirta.**
    Calyx-lobes ciliate and sometimes
     clothed on the back with long
     hispid hairs ; ovary glabrous ... (14) **plumosa.**
    Calyx-lobes ciliate with simple
     hairs as long as the lobes are
     broad ; ovary glabrous .. (15) **pilifolia.**
    Calyx-lobes ciliate with stout
     stiff simple or slightly hispid
     hairs not more than $\frac{1}{4}$ as long
     as the lobes are broad ; ovary
     slightly pubescent ... ... (16) **solivaga.**
   Anthers spurred near the base ... (14) **plumosa,** var. β
 ** Corolla-tube very obscurely or not at all
 constricted at any part; anthers with awn-
 like spurs near the base :
  Calyx divided to the base ; corolla-tube
   puberulous outside : glabrous within ... (17) **Thunbergii.**
  Calyx lobed $\frac{1}{2}$–$\frac{3}{4}$ of the way down :
   † Corolla glabrous outside and within :
    Flowers in globose head-like
     clusters of 6–15 on very short
     axillary branchlets or subses-
     sile at the nodes, not arranged
     in a spike-like manner :

Clusters 1½–2 lin. in diam.;
cilia of the calyx hispid
to the apex, not gland-
tipped ... ... (18) **nodiflora**.
Clusters not more than 1½ lin.
in diam.; cilia of the calyx
gland-tipped, hispid on the
lower part ... ... (19) **minutiflora**.
Flowers 1–4 at the ends of extremely short axillary branchlets, which are arranged in
a spicate manner along the
branches; leaves spreading, recurved ... ... (20) **eremioides**.
†† Corolla-tube puberulous outside,
glabrous within:
Leaves very spreading, recurved;
calyx-lobes ovate, puberulous
on the back ... ... (20) **eremioides**, var. γ.
Leaves ascending or spreading,
straight; calyx-lobes linear-
oblong, glabrous on the back .. (21) **similis**.

1. **G. ciliaris** (Klotzsch in Linnæa, xii. 225); branchlets pubescent, flexuose, the ultimate filiform; leaves 4-nate, adpressed, ½ lin.
long, elliptic, thinly pubescent, shining; flowers in terminal nodding
heads; bracts subremote from the calyx, linear, villous; calyx
4-partite; segments lanceolate, pubescent, ciliate with thick plumose
white hairs; corolla campanulate, 4-lobed, pink; anthers subexserted,
spurred; style much exserted. *Blæria ciliaris, Ait. Hort. Kew.
ed. 2, i. 249; Wendl. Collect. ii. 35, t. 49; Roem. & Schultes,
Syst. Veg. iii. 170, not of Mantissa 108; Klotzsch in Linnæa, viii.
658; G. Don, Gen. Syst. iii. 805, not of Linn. fil. Blairia ciliaris,
Dietr. Syn. Pl. i. 443.*

SOUTH AFRICA : without locality, *Herb. Willdenow* ex *Klotzsch*.
COAST REGION : Vanrhynsdorp Div. ; Gift Berg, 1500–2500 ft., *Drège* ex
Klotzsch.

I have not seen this species. Klotzsch also includes *Erica plumosa*, Thunb.,
in the synonymy, but Rach in *Linnæa*, xxvi. 789, separates that under the name
of *G. Thunbergii*, and it is possible that the plant of Aiton may not belong here.

2. **G. hispida** (Klotzsch in Linnæa, xii. 226); branchlets puberulous, sometimes intermingled with longer hispid hairs; leaves
4-nate, usually slightly spreading, 1¼–2 lin. long, ⅓–½ lin. broad,
linear-oblong, subacute, at first minutely puberulous all over, ciliate
with long hispid hairs and with 2 or more rows of deciduous
similar hairs on the back, finally nearly glabrous and minutely
echinate-tuberculate; flowers in dense globose terminal heads 5–5½
lin. in diam. ; pedicels ½ lin. long, bracteate at the middle ;
bracts applied to the calyx, very unequal, the middle one about
1½ lin. long, lanceolate to broadly-ovate, with a thickened leaf-
like tip, the lateral pair shorter and linear or linear-lanceolate, all
minutely puberulous and ciliate with long hispid hairs; calyx
1½ lin. long, obconic-campanulate, 4-angled, lobed to half-way down ;

lobes $\frac{1}{2}$–1 lin. broad, ovate to orbicular, densely ciliate with long hispid white hairs and with similar hairs on the back; corolla 2 lin. long, constricted above the middle, minutely puberulous outside on the sharply 4-angled part of the tube, pubescent within at the constriction; lobes $\frac{1}{2}$–$\frac{2}{3}$ lin. long, broadly ovate, obtuse, concave, incurved, glabrous; stamens rather shorter than the corolla; filaments ciliate with very fine hairs; anthers $\frac{2}{3}$ lin. long and nearly as broad, scabrous, without spurs; ovary compressed, ellipsoid, very obtuse, glabrous; style exserted, glabrous; stigma capitate. *Benth. in DC. Prodr.* vii. 701; *Rach in Linnæa,* xxvi. 790. *Blæria ptilota, E. Meyer ex Benth. l.c.*

SOUTH AFRICA : without locality, *Thunberg! Masson!*

COAST REGION: Clanwilliam Div.; near Zwartbast Kraal below 1000 ft., *Drège!* near the Oliphants River and Brakfontein, *Ecklon & Zeyher,* ex *Klotzsch.* Tulbagh Div.; Witsen Berg and near Vogel Valley, *Ecklon & Zeyher,* ex *Klotzsch.*

3. G. involuta (Klotzsch in Linnæa, xii. 227); branches and branchlets erect, puberulous; leaves 3-nate, adpressed or imbricate, straight or somewhat recurved, $1\frac{1}{4}$–$2\frac{1}{4}$ lin. long, oblong-linear, subacute, glabrous or at first puberulous all over and ciliate with long slightly hispid or simple white hairs, but no long hairs on the back, soon becoming glabrous, with minutely serrulate margins; heads globose, terminal, 5–8 lin. in diam., 6–12-flowered; pedicels $\frac{3}{4}$–1 lin. long, bracteate at the middle; bracts adpressed to the calyx, unequal, the larger $1\frac{1}{2}$–2 lin. long, all ovate-lanceolate or the lateral linear-oblong, glabrous, ciliate with very long simple or very slightly hispid hairs; calyx divided almost to the base; segments 2 lin. long, 1 lin. broad, lanceolate-oblong, acute, densely ciliate and more or less hairy on the back with simple or slightly hispid hairs, $\frac{3}{4}$–1 lin. long; corolla-tube $2\frac{1}{2}$ lin. long, constricted near the top, sharply angled, very slightly broadened at the basal part, minutely puberulous outside, pubescent on the upper half within; lobes $\frac{2}{3}$ lin. long and broad, ovate, obtuse, concave, incurved-erect, glabrous; stamens not exceeding the corolla-lobes; filaments hairy; anthers $\frac{2}{3}$–$\frac{3}{4}$ lin. long, broadly cuneate-oblong, spurless, almost hispid-scabrous; ovary obtuse, glabrous; style 3 lin. long, exserted, glabrous; stigma small, capitate. *Benth. in DC. Prodr.* vii. 701; *Drège, Zwei Pfl. Docum.* 72. *G. involucrata, Klotzsch ex Drège, l.c.* 188.

SOUTH AFRICA: without locality, *Drège* (distributed as *Blairia*) 7801!

COAST REGION: Clanwilliam Div.; Cederberg Range, near Krakadouw, *Bodkin in Herb. Bolus,* 8680! Bosch Kloof, *Drège,* ex *Klotzsch.*

Drège 7801 is the plant described by Bentham, and is possibly the same as the specimen from Bosch Kloof, although Klotzsch describes the calyx of that as being glabrous.

4. G. Bolusii (N. E. Br.); branchlets white-puberulous or minutely subtomentose; leaves 3-nate, $\frac{3}{4}$–$1\frac{3}{4}$ lin. long with the petiole, adpressed-imbricate, linear-oblong, obtuse, at first puberulous, minutely woolly on the margins and ciliate with a few short branched or

plumose-hispid hairs, becoming glabrous on the back and minutely
serrulate; heads globose, terminal, 4–5 lin. in diam., 6–8-flowered;
pedicels ½ lin. long, bracteate at the middle and base; bracts 1–1⅓ lin.
long, the larger one ¾–1 lin. broad, broadly ovate or rhomboid, acute,
the lateral oblong, all green-tipped, glabrous, ciliate with short hispid
hairs; calyx 4-partite; segments often somewhat unequal, about
1¼ lin. long, ½–¾ lin. broad, lanceolate, elliptic-ovate or subrhomboid-
ovate, acute, densely ciliate and hairy on the back with hispid hairs
⅓–½ lin. long; corolla-tube 1½ lin. long, constricted near the top,
oblong and sharply 4-angled below the constriction, pubescent outside and at the constriction within; lobes ½ lin. long, ovate, obtuse-
concave, connivent-incurved, glabrous; anthers ½ lin. long, cuneate,
subquadrate, scabrous, spurless; ovary obtuse, glabrous; style 2⅓ lin.
long; stigma capitate.

CoAST REGION: Clanwilliam Div.; Cederberg Range, near Pakhuis, *Bolus*,
8681!

5. **G. apiculata** (N. E. Br.); branchlets puberulous; leaves 3-
nate, as long as or shorter than the internodes, adpressed, with
slightly spreading tips, ¾–1 lin. long, linear, acute, puberulous, at
first usually with 2–4 minute gland-tipped hairs on each margin and
one at the apex, the latter remaining as an apiculus; heads subglobose, terminal, ¼–⅓ in. in diam., 3–9-flowered; pedicels ¾–1 lin.
long, bracteate at or below the middle; bracts ascending-spreading,
not reaching to the calyx, ½–¾ lin. long, linear, acute, ciliate with
simple hairs; calyx 4-partite; segments ¾ lin. long, ¼ lin. broad,
oblong-linear, subacute, ciliate and hairy on the back with minutely
gland-tipped simple hairs ⅓ lin. long; corolla-tube about ¾ lin. long,
constricted near the top, subglobose and sharply 4-angled below the
constriction, puberulous outside and at the constriction within; lobes
½ lin. long, ovate, obtuse, concave, erect, glabrous; anthers ½ lin.
long, scabrous, spurless; ovary glabrous, slightly emarginate at the
apex; style much exserted, 1½ lin. long; stigma capitate.

CoAST REGION: Piquetberg Div.; mountains near Piquiniers Kloof, 1200 ft.,
Schlechter, 4969!

6. **G. velleriflora** (Klotzsch in Linnæa, xii. 227); young branchlets puberulous; leaves 3-nate, adpressed-imbricate, ⅓–1 lin. long
with the petiole, linear-oblong, obtuse or subacute, at first more or
less puberulous and minutely ciliate with very fine woolly hairs,
becoming glabrous; heads loosely globose, terminal, about ⅓ in.
in diam., 4–6-flowered; pedicels ¾–1 lin. long, bracteate near
the middle; bracts ⅔–1 lin. long, with the tips of the upper
pair applied to the calyx and the lower one scarcely reaching it,
lanceolate or oblong-lanceolate, acute, glabrous, ciliate with very
long simple hairs; calyx 4-partite; segments about 1 lin. long, ⅓–½
lin. broad, ovate or lanceolate, acute, hairy on the back and ciliate
with simple or slightly scabrous hairs ¾–1⅔ lin. long; corolla-tube
1½–1¾ lin. long, constricted at the middle, subglobose or oblong-

ovoid and sharply 4-angled below the constriction, cup-shaped above
it, more or less puberulous outside and at the constriction within;
lobes $\frac{1}{3}$–$\frac{1}{2}$ lin. long, broadly ovate or rounded, concave, erect, glabrous;
anthers $\frac{1}{2}$ lin. long, cuneate-subquadrate, minutely scabrous, densely
ciliate or somewhat hispid around the inner margins, spurless; ovary
obtuse, glabrous; style $2\frac{1}{2}$ lin. long, much exserted, glabrous; stigma
thickened, scarcely capitate. *Benth. in DC. Prodr.* vii. 701. *Erica
ciliiflora, Salisb. in Trans. Linn. Soc.* vi. 339, *ex Benth. l.c.
Blæria ciliiflora, G. Don, Gen. Syst.* iii. 805; *Klotzsch in Linnæa,*
xii. 246.

SOUTH AFRICA: without locality (distributed as *Erica lanata*, Wendl.),
Drège!
COAST REGION: Clanwilliam Div.; between Twenty-four Rivers and Oliphants
River (between Lange Valley and Oliphants River, ex *Drège*), *Drège* ex *Klotzsch;*
near Clanwilliam, *Mader*, 181! *Leipoldt*, 213! Stellenbosch Div.; Hottentots
Holland, *Masson* ex *Salisbury*, but locality doubtful.

Probably *Schlechter*, 5127 from Alexanders Hoek in Clanwilliam Div. also
belongs here, but the flowers are not fully developed.

7. G. dregeana (Benth. in DC. Prodr. vii. 701); branchlets
whitish-pubescent; leaves 3-nate, adpressed or imbricate, $\frac{3}{4}$–$1\frac{1}{3}$ lin.
long, linear-oblong, obtuse or subacute, whitish-pubescent or subtomentose with minute subwoolly hairs, becoming glabrous on
the back; heads terminal, about 3–6-flowered; pedicels bracteate
above and below the middle, pubescent; bracts spreading or
the upper applied to the calyx, $\frac{2}{3}$–$\frac{3}{4}$ lin. long, linear-lanceolate,
obtuse, ciliate with hispid hairs; calyx 4-partite; segments 1 lin.
long, $\frac{1}{2}$ lin. broad, oblong-lanceolate, acute, hairy on the back and
ciliate with simple or slightly hispid hairs, $\frac{1}{3}$–$\frac{1}{2}$ lin. long; corolla-
tube $1\frac{1}{4}$ lin. long, constricted above the middle, ellipsoid or subglobose and sharply 4-angled below the constriction, pubescent outside and at the constriction within; lobes $\frac{1}{3}$ lin. long, ovate, obtuse,
concave, incurved-connivent, glabrous; anthers $\frac{1}{2}$ lin. long, cuneate-
subquadrate, scabrous, spurless; ovary obtuse, glabrous; style $1\frac{2}{3}$ lin.
long; stigma small, capitate.

SOUTH AFRICA: without locality, *Drège*, 7803! distributed as a "*Blairia.*"

8. G. zeyheriana (Klotzsch in Linnæa, xii. 227); branchlets
glabrous or glandular-puberulous; leaves 3-nate, adpressed-imbricate, $\frac{2}{3}$–1 lin. long, linear-oblong, obtuse, with 4–6 sessile glands
on each margin, otherwise glabrous, but apparently somewhat
viscid; heads globose, terminal, $\frac{1}{4}$–$\frac{1}{3}$ in. in diam., 4–6-flowered;
pedicels $\frac{1}{2}$–$\frac{3}{4}$ lin. long, bracteate at or below the middle; bracts
erect, with their tips applied to the calyx, $\frac{2}{3}$–$\frac{3}{4}$ lin. long, subequal
and linear-oblong or lanceolate or the middle one broader and
ovate, thickened and leafy at the tips, glabrous, ciliate with
long simple or slightly hispid hairs; calyx 4-partite; segments
$\frac{3}{4}$ lin. long, $\frac{1}{3}$ lin. broad, lanceolate, acute, ciliate with simple
or slightly hispid hairs $\frac{3}{4}$ lin. long, and with a few similar

hairs on the back at the base; corolla-tube $1\frac{1}{4}$ lin. long, constricted near the top, ovoid and sharply 4-angled below the constriction, pubescent below the constriction outside and at the constriction within; lobes $\frac{1}{2}$ lin. long, ovate, obtuse, concave, erect, glabrous; anthers $\frac{1}{3}$ lin. long, cuneate-subquadrate, scabrous, spurless; ovary subtruncate or subemarginate at the apex, glabrous; style 2 lin. long, glabrous; stigma small, capitate.

COAST REGION: Clanwilliam Div.; by the Oliphants River and near Brakfontein, *Eklon & Zeyher,* 269! in Berlin Herb.

The description given by Klotzsch is erroneous as to the back of the leaves being scabrid and the corolla glabrous.

9. G. rigida (N. E. Br.); branchlets puberulous, more or less intermingled with short hispid hairs; leaves 3-nate, adpressed, $\frac{3}{4}$–$1\frac{1}{4}$ lin. long, ovate or oblong, obtuse, very deeply grooved on the back, slightly viscid, at first puberulous and ciliate with minute gland-tipped hairs, becoming nearly glabrous on the back and minutely serrulate; heads globose, terminal, 3–4 lin. in diam., 3–6-flowered; pedicels $\frac{1}{2}$ lin. long, bracteate at the middle; bracts applied to the calyx, all about $\frac{3}{4}$ lin. long, oblong, obtuse, stiffly ciliate with simple or minutely gland-tipped hairs; calyx 4-partite, rather rigid; segments 1–$1\frac{1}{4}$ lin. long, $\frac{1}{2}$–$\frac{2}{3}$ lin. broad, elliptic or elliptic-oblong, obtuse, rather rigidly ciliate with stout simple or slightly scabrous often flexuous hairs $\frac{1}{3}$–$\frac{1}{2}$ as broad as the segment, and with or without some similar hairs on the back; corolla-tube 1 lin. long, constricted near the top, subglobose or ellipsoid and 4-angled below the constriction, puberulous outside and at the constriction within; lobes $\frac{1}{2}$ lin. long, ovate, obtuse, concave, incurved-connivent, glabrous; anthers nearly $\frac{1}{2}$ lin. long, with parallel cells, spurred on the back $\frac{1}{3}$ above the base, scabrous; spurs $\frac{1}{3}$ as long as the cells, awn-like, spreading; ovary obtuse, thinly puberulous at the top; style deciduous, not seen.

COAST REGION: Worcester Div.; near Brand Vley, 1000 ft., *Schlechter,* 9926! Caledon Div.; mountains between French Hoek and Villiersdorp, 2000 ft., *Bolus,* 5193!

10. G. Niveni (N. E. Br.); branchlets minutely tomentose or greyish-puberulous; leaves 3-nate, $\frac{1}{3}$–$\frac{3}{4}$ lin. long with the petiole, oblong or elliptic-oblong, obtuse or acute, thick, flat on the upper side and on the grooved back, glabrous, minutely dentate-ciliate; flowers in small globose terminal heads of 4–8; pedicels $\frac{1}{3}$–$\frac{1}{2}$ lin. long, bracteate at the middle; bracts applied to the calyx, about $\frac{2}{3}$ lin. long, linear, oblong or oblong-lanceolate, obtuse or acute, ciliate with hispid hairs; sepals often unequal in breadth, $\frac{3}{4}$ lin. long, $\frac{1}{3}$–$\frac{1}{2}$ lin. broad, linear, oblong or oblong-lanceolate, obtuse or acute, hairy on the back and densely ciliate with hispid hairs, which are shorter than the breadth of the sepal; corolla $1\frac{1}{2}$–$1\frac{3}{4}$ lin. long, constricted at the middle, pubescent with spreading hairs up to the base of the lobes outside and at the middle part within; tube inflated, sharply 4-angled; lobes about ½ lin. long, ovate-oblong, obtuse, erect or slightly incurved; stamens not quite as long as the corolla; filaments linear, hairy; anthers ⅓ lin. long, with subparallel or slightly diverging oblong cells, spurred on the back ⅓ above the base; spurs awn-like, curved; ovary compressed, very obtuse, glabrous or pubescent on the top; style exserted, about 1½ lin. long; stigma simple, very slightly thickened. *G. ciliaris, Benth. in DC. Prodr.* vii. 701, *not of Klotzsch.*

Coast Region : Stellenbosch Div.; Hottentots Holland, *Niven,* 128! Caledon or Swellendam Div.; near the Zondereinde River, *Gill!* near Swellendam, 600–800 ft., *Mund,* 3!

This is the plant described as *G. ciliaris* by Bentham, but according to description and locality it cannot be the same as *G. ciliaris,* Klotzsch.

11. G. incana (Klotzsch in Linnæa, xii. 225); branches densely whitish-puberulous or minutely tomentose; leaves 3-nate, ¾–1 lin. long with the petiole, ⅓ lin. broad, oblong, obtuse or subacute, flat above, grooved on the convex back, whitish-puberulous or minutely tomentose; flowers in small globose terminal heads of 3–6; pedicels ¼ lin. long; bracts adpressed to the calyx, unequal, the largest ½–⅔ lin. long, linear-oblong, obtuse, the others linear, whitish-puberulous, often ciliate with hispid or plumose hairs; sepals ½–⅔ lin. long, less than ¼ lin. broad, linear or narrowly oblong, acute, whitish-puberulous, ciliate with hispid hairs, which are sometimes gland-tipped, pink (*Burchell*); corolla 1¼–1⅓ lin. long, constricted at the middle, pink (*Burchell*); tube ellipsoid, puberulous outside and on the upper part within; lobes ⅓–⅖ lin. long, oblong-ovate, obtuse, concave, slightly spreading, glabrous; stamens equalling or very slightly exceeding the corolla; filaments linear, hairy; anthers ⅓ lin. long, with oblong nearly parallel cells, which apparently become inverted and divergent with age, spurred on the back just above the base, scabrous; spurs awn-like, sometimes ½ as long as the cells and curved, sometimes almost obsolete; ovary compressed, very obtuse, pubescent; style 1½ lin. long, much exserted; stigma simple, slightly thickened. *Benth. in DC. Prodr.* vii. 701. *Blæria incana, Bartl. in Linnæa,* vii. 650; *Klotzsch in Linnæa,* viii. 660. *Blairia incana, Dietr. Syn. Pl.* i. 443.

South Africa : without locality, *Mund & Maire !*
Coast Region : Worcester Div.; Dutoits Kloof, *Drège,* 6869 ! Cape Div.; Cape Flats, near Tyger Berg, *Ecklon,* ex *Bartling.* Stellenbosch Div.; between Cape Flats and Stellenbosch, *Burchell,* 8344 ! Hottentots Holland, *Ecklon,* ex *Bartling.* Swellendam Div.; mountains near Swellendam, *Mund !*

Drège, 6869 was distributed under the generic name of *Raspalia.*

12. G. alba (N. E. Br.); branchlets whitish-pubescent; leaves 3-nate, adpressed, scarcely or not at all imbricate, ¾–1 lin. long, ovate, elliptic or elliptic-oblong, obtuse, pubescent, at first minutely serrulate-ciliate on the margins; heads small, subglobose, terminal, about 2 lin. in diam., 3–6-flowered, apparently nodding, white; pedicels ¼ lin. long, bracteate near the top; bracts spreading or applied to the calyx, ⅔–¾ lin. long, linear, slightly broadest at the obtuse apex, ciliate with minutely gland-tipped simple or finely hispid hairs; calyx 4-partite, white; segments ¾–1 lin. long, ¼ lin. broad, linear, obtuse, ciliate and hairy on the back with minutely gland-tipped simple or slightly hispid hairs about ¼ lin. long; corolla white; tube 1¼ lin. long, constricted near the top, obconic-oblong and 4-angled below the constriction, very minutely puberulous outside and at the constriction within; lobes ⅓ lin. long, ovate, obtuse, concave, erect, glabrous; stamens shorter than the corolla; filaments glabrous or nearly so; anthers ¼ lin. long, cuneate-subquadrate, with minute awn-like spurs on the back above the base, white; ovary obtuse, with a few very minute hairs at the top; style 1¼–1¾ lin. long, exserted; stigma capitate.

SOUTH AFRICA: without locality, *Grey !*

13. G. hirta (Klotzsch in Linnæa, xii. 226); branchlets rather crowded, puberulous with an intermingling of short gland-tipped subhispid hairs; leaves 3-nate, imbricate or spreading, probably clammy, 1–1½ lin. long with the petiole, oblong, obtuse, minutely puberulous on both sides, ciliate with gland-tipped more or less hispid hairs and with 2 rows of similar hairs on the back, deciduous, leaving minute teeth and tuberculate points; heads terminal, 6–8-flowered, often drooping; pedicels ½–⅔ lin. long, bracteate at or above the middle, glandular-pubescent; bracts unequal, applied to the calyx, ciliate and hairy on the back with long gland-tipped hairs, the larger ¾–1 lin. long, lanceolate, subacute; calyx about ¾ lin. long, campanulate, 4-angled, lobed to ⅔ of the way down, ciliate and hairy outside with long simple gland-tipped hairs; lobes ⅓ lin. broad at the base, elongate-deltoid or deltoid-ovate, acute; corolla 1¼–1½ lin. long; tube constricted near the top, sharply 4-angled and puberulous outside below the constriction, pubescent at the constriction within; lobes ½ lin. long, broadly ovate, obtuse, apparently spreading; stamens 1¼ lin. long; filaments hairy; anthers ⅓ lin. long, scabrous, without spurs; ovary compressed, subacute, slightly pubescent at the apex; style exserted, 1½–1¾ lin. long, glabrous; stigma capitate. *Benth. in DC. Prodr.* vii. 701.

COAST REGION: Malmesbury Div.; sand flats near Groene Kloof (Mamre), *Drège,* 7795!

I have not seen an authentic specimen, but describe from specimens distributed by Drège as " *Blairia,* 7795 " named *G. hirta,* Klotzsch, by Bentham, which I assume to be correct, and quote the locality as given by Klotzsch, since Drège's number is not quoted in his Zwei Pflauzengeogr. Documente.

14. G. plumosa (Klotzsch in Linnæa, xii. 226); branchlets puberulous or minutely tomentose and beset with short hispid hairs; leaves 3-nate, imbricate, ¾–1½ lin. long, linear-oblong, lanceolate or ovate, obtuse, at first puberulous all over and ciliate with long hispid

hairs, soon becoming glabrous and smooth on the back and serrulate on the margins; heads globose, terminal, 3–5 lin. in
diam., usually 6–12-flowered; pedicels $\frac{1}{4}$–$\frac{1}{3}$ lin. long; bracts adpressed to the calyx, the larger $\frac{3}{4}$–$1\frac{1}{3}$ lin. long, linear-oblong, lanceolate or oblanceolate, acute, ciliate with long hispid hairs; calyx
about 1 lin. long, campanulate, acutely 4-angled, lobed to the middle,
rather rigid; lobes $\frac{1}{3}$–$\frac{2}{3}$ lin. broad, ovate or broadly deltoid, acute,
ciliate with long hispid hairs, which are occasionally gland-tipped,
and usually with some similar hairs on the back; corolla $1\frac{1}{2}$ lin.
long; tube constricted above the middle, sharply 4-angled and somewhat obovoid below the constriction, puberulous outside, pubescent
within at the constriction; lobes $\frac{1}{2}$ lin. long, broadly ovate, obtuse,
concave, incurved-erect, glabrous; stamens not exceeding the corolla-
lobes; filaments hairy; anthers $\frac{1}{2}$ lin. long, scabrous, without spurs;
ovary compressed-oblong, obtuse, glabrous; style exserted, $1\frac{3}{4}$–$2\frac{1}{4}$ lin.
long; stigma small, thickened or capitate. *Benth. in DC. Prodr.* vii.
701. *Erica nodiflora, Salisb. in Trans. Linn. Soc.* vi. 340, *ex
Benth.* E. *capitata, Salisb. Prodr.* 293. *Blæria nodiflora, G.
Don, Gen. Syst.* iii. 805; *Klotzsch in Linnæa*, xii. 246.

VAR. β, **serrulata** (N. E. Br.); anthers with curved awn-like spurs on the
back, just below the middle, otherwise as in the type. *G. serrulata, Benth. in
DC. Prodr.* vii. 701.

VAR. γ, **scabra** (N. E. Br.); young leaves beset with long slightly hispid
hairs on the back, which fall away so that the leaf becomes scabrous with
age.

SOUTH AFRICA: without locality; *Paterson!* Var. β: *Drège*, 7802! Var. γ:
Thom!

COAST REGION: Clanwilliam Div.; Oliphants River Mountains, 2000 ft.,
Schlechter, 5099! Malmesbury Div.; Saurevelder, near Hopefield, *Bachmann!*
Paarl Div.; between Mosselbanks River and Berg River, *Burchell*, 981! Cape
Div.; Cape Flats, near Doornhoogte, *Ecklon & Zeyher*, and *Drège*, ex *Klotzsch*.
Var. β: Malmesbury Div.; near Groene Kloof (Mamre), *Bolus*, 4242!

15. G. pilifolia (N. E. Br.); branchlets pubescent; leaves 3-nate,
adpressed-imbricate, 1–$1\frac{1}{2}$ lin. long with the petiole, linear-oblong or
oblong-lanceolate, or the smaller sometimes ovate, obtuse, rather thin
and flatter than in other species, most minutely puberulous all over
and thinly beset on the back and margins with long simple or slightly
hispid hairs; heads globose, terminal, $\frac{1}{4}$–$\frac{1}{3}$ in. in diam., 9-flowered
in the specimen seen; pedicels $\frac{1}{3}$ lin. long, bracteate at the middle;
bracts unequal, adpressed to the calyx, the larger $\frac{3}{4}$ lin. long, ovate-
lanceolate, acute, the lateral smaller, lanceolate, semitransparent,
with green tips, glabrous, ciliate with long hairs which are simple or
branched at their tips; calyx campanulate, nearly 1 lin. long, 4-lobed
to the middle, rather thin and semitransparent; tube 4-angled,
glabrous; lobes $\frac{1}{3}$–$\frac{1}{2}$ lin. broad, ovate, acute, ciliate with simple hairs
$\frac{1}{2}$ lin. long, and usually with a line of short hairs down the back;
corolla-tube 1 lin. long, constricted near the top, quadrangular-
obovoid below the constriction, puberulous outside and at the constriction within; lobes $\frac{1}{2}$ lin. long, ovate, obtuse, concave, erect,

glabrous; anthers $\frac{1}{2}$ lin. long, cuneate-subquadrate, scabrous, spurless; ovary obtuse, glabrous; style exserted, about 2 lin. long, glabrous; stigma small, subcapitate.

COAST REGION: Clanwilliam Div.; near Clanwilliam, *Leipoldt*, 46! in Herb. Bolus.

16. **G. solivaga** (N. E. Br.); branchlets minutely subtomentose, pallid; leaves 3-nate, adpressed or imbricate, with slightly recurving tips, $\frac{3}{4}$–$1\frac{1}{2}$ lin. long with the petiole, ovate to oblong, acute or subobtuse, at first puberulous with minute curved white hairs and serrulate or ciliate with 4–5 short very hispid hairs on each side, becoming glabrous on the back and minutely serrulate; heads subglobose, terminal, $2\frac{1}{2}$–3 lin. in diam., 6–8-flowered; pedicels $\frac{1}{3}$ lin. long, bracteate at or below the middle; bracts adpressed to the calyx, one of them $\frac{2}{3}$–$\frac{3}{4}$ lin. long, ovate or elliptic-ovate, subacute, with the apical part leaf-like, the others very much smaller, linear, all very minutely ciliate with hispid or branching hairs; calyx $\frac{3}{4}$–1 lin. long, campanulate, 4-angled, lobed to less than half-way down, rather rigid, glabrous, ciliate on the lobes with stout stiff simple or very slightly hispid hairs less than $\frac{1}{4}$ lin. long; lobes $\frac{1}{3}$ lin. long, broadly deltoid, acute and usually green at the apex; corolla-tube $1\frac{1}{2}$ lin. long, constricted near the top, cuneately oblong, sometimes oblique, 4-angled, puberulous outside and at the constriction within; lobes $\frac{1}{2}$ lin. long, deltoid-ovate, obtuse, concave, erect (or incurved-connivent?), glabrous; anthers $\frac{1}{3}$ lin. long, broadly obtriangular, with a broad apical notch, spurless, minutely scabrous; ovary obtuse, slightly pubescent at the top, 2–3-celled; style exserted, 2 lin. long; stigma subcapitate.

COAST REGION: Clanwilliam Div.; Zekoe Vlei, 400 ft., *Schlechter*, 8480!

17. **G. Thunbergii** (Rach in Linnæa, xxvi. 789); branchlets moderately stout, minutely greyish-tomentose; leaves 3-nate, adpressed or imbricate, $\frac{1}{2}$–1 lin. long, oblong, subacute, at first minutely white-tomentose, becoming glabrous on the back; flowers in terminal heads of 3–8 on short lateral branchlets; heads often somewhat crowded; pedicels $\frac{1}{2}$–$\frac{3}{4}$ lin. long, bracteate at the middle; bracts slightly spreading or more or less applied to the calyx, $\frac{1}{3}$–$\frac{2}{3}$ lin. long, linear or the middle one oblong, minutely woolly-ciliate; sepals $\frac{3}{4}$–1 lin. long, $\frac{1}{4}$–$\frac{1}{3}$ lin. broad, linear or linear-lanceolate, acute, white (*Thunberg*), ciliate with stout simple or slightly hispid hairs longer than the sepal is broad, some directed outwards and crossing at the back of the sepal, which has often a few similar hairs; corolla $1\frac{1}{4}$–$1\frac{1}{2}$ lin. long, very slightly constricted at the middle, puberulous outside on the sharply 4-angled part of the tube, glabrous within, white (*Thunberg*); lobes $\frac{1}{4}$ lin. long, broader than long, rounded, very minutely crenulate, concave, incurved-erect, glabrous; stamens about as long as the corolla; filaments hairy; anthers about $\frac{1}{4}$ lin. long, with parallel cells, spurred at or slightly above the

base, scabrous; spurs awn-like, curved, nearly $\frac{1}{2}$ as long as the cells; ovary compressed-ellipsoid, pubescent at the obtuse top; style exserted, glabrous; stigma simple, slightly thickened. *Erica plumosa, Thunb. Prodr.* 73, *and Fl. Cap. ed. Schultes,* 364 ; *Salisb. in Trans. Linn. Soc.* vi. 339. *Blæria plumosa, Thunb. Diss. Blæria,* 9. *B. ciliaris, Linn. f. Suppl.* 122 ; *Willd. Sp. Pl.* i. 631 ; *Roem. and Schult. Syst. Veg.* iii., *Mant.* 108, *not of* iii. 170.

COAST REGION: Clanwilliam Div.; Middel Berg; Cederberg Range, *Leipoldt,* 741!

CENTRAL REGION: Calvinia Div.; Bokkeland, *Thunberg!* Bokkeveld Mountains, 3000 ft., *Leipoldt,* 740! mountains near Willems River, 2300 ft., *Schlechter,* 10974!

Thunberg describes the leaves of *Erica plumosa* as 4-nate, but those on his type specimen are 3-nate. I have examined the type specimen of *Blæria ciliaris* in the Linnean Herbarium and find it to be identical with the type of *G. Thunbergii.* As Rach states that this is not the same as *G. ciliaris,* Kl., I retain them as distinct species.

18. G. nodiflora (N. E. Br.); apparently about 4–8 in. high; branchlets pubescent; leaves 3-nate, adpressed, usually shorter than the internodes, 1–1$\frac{1}{2}$ lin. long with the petiole, linear or linear-oblong, subobtuse, puberulous, becoming glabrous; flowers subsessile, in dense globose heads, sessile at the nodes or terminating very short lateral branchlets; heads 1$\frac{1}{2}$–2 lin. in diam., 6–12-flowered, often 2 at a node; bracts adpressed to the calyx at the base, with spreading tips $\frac{1}{2}$ lin. long, linear or linear-lanceolate, subacute, ciliate with minute branched hairs; calyx-tube $\frac{1}{3}$ lin. long, obconic, not angular, puberulous outside; lobes nearly $\frac{1}{2}$ lin. long, spreading, oblong, acute, puberulous on the back, with 3–5 stout hispid cilia on each margin ; corolla equalling or exceeding the calyx variable in the same cluster, funnel-shaped or obconic-tubular, not constricted at any part, glabrous outside and within; tube $\frac{2}{3}$ to nearly 1 lin. long; lobes $\frac{1}{3}$ lin. long, rounded or shortly oblong, obtuse, erect; stamens sometimes slightly exceeding the corolla; filaments glabrous; anthers $\frac{1}{4}$–$\frac{1}{3}$ lin. long, oblong, with parallel sides, spurred on the back near the base, minutely scabrous ; spurs awn-like, $\frac{1}{3}$–$\frac{1}{2}$ as long as the cells; ovary obtuse, puberulous at the top, 2–3-celled ; style 1–1$\frac{1}{4}$ lin. long, glabrous; stigma minute, very slightly thickened.

CENTRAL REGION : Ceres Div.; Schoongezicht, in the Cold Bokkeveld, 4500 ft., *Schlechter,* 10188 !

19. G. minutiflora (N. E. Br.); branchlets densely white-puberulous; leaves 3-nate, adpressed, equalling or shorter than the internodes, $\frac{3}{4}$–1 lin. long, linear-oblong, subobtuse, at first slightly puberulous in the groove down the back, soon becoming glabrous, minutely glandular on the margins, apparently viscid; heads globose, lateral on very short axillary branchlets and terminal, about 1$\frac{1}{2}$ lin. in diam., 6–15-flowered; pedicels $\frac{1}{8}$–$\frac{1}{4}$ lin. long, bracteate at the apex close to the calyx; bracts applied to the calyx, $\frac{1}{3}$ lin. long, linear, subacute, ciliate with minute gland-tipped

and simple hairs; calyx $\frac{1}{2}$ lin. long, obconic or funnel-shaped,
lobed to the middle, pubescent outside; lobes very slightly
spreading, oblong or ovate-oblong, ciliate with gland-tipped hairs
hispid on their basal half; corolla funnel-shaped, not constricted,
glabrous outside and within; tube $\frac{1}{2}$ lin. long; lobes $\frac{1}{4}$ lin. long,
ovate, obtuse, concave, erect; stamens about as long as the corolla;
filaments glabrous, doubled upon themselves above the middle;
anthers $\frac{1}{4}$ lin. long, oblong, very minutely scabrous on the outer
margins which are spurred at about $\frac{1}{3}$ above the base; spurs awn-
like, $\frac{1}{3}$ as long as the cells, spreading and more or less upcurved;
ovary obtuse, puberulous at the top; style $\frac{1}{2}$ lin. long, not exserted,
much curved (after fertilization ?); stigma simple.

CENTRAL REGION: Ceres Div.; near Klein Vley, in the Cold Bokkeveld,
4000 ft., *Schlechter*, 10064!

20. **G. eremioides** (MacOwan in Journ. Linn. Soc. xxv. 392);
branches erect, straight, pubescent or puberulous, beset with very
numerous short branchlets; leaves 3-nate, crowded, very spreading,
recurved or rarely straight, $\frac{1}{3}$–$1\frac{1}{4}$ lin. long with the petiole, linear or
rarely ovate, trigonous, subacute, glabrous or at first minutely puberu-
lous, with or without a few gland-tipped hairs on the margins and
back; flowers 1–4 at the ends of extremely short axillary branchlets,
which are arranged in a spike-like manner along the main and lateral
branches; pedicels less than $\frac{1}{4}$ lin. long; bracts 3, subequal, adpressed
to the calyx, $\frac{1}{4}$–$\frac{1}{3}$ lin. long, linear, acute, ciliate; calyx $\frac{1}{3}$–$\frac{1}{2}$ lin. long,
campanulate, lobed to or beyond the middle; lobes subquadrate and
cuspidate-apiculate, glabrous, ciliate with short gland-tipped hairs or
at the lower part with fine simple hairs; corolla $\frac{2}{3}$–1 lin. long, funnel-
shaped, not constricted at any part, glabrous outside and within;
lobes broader than long, obtuse, concave, erect or incurved; stamens
4; filaments glabrous; anthers $\frac{1}{4}$–$\frac{1}{3}$ lin. long, with parallel but
rather widely separated cells, spurred above the base, minutely
scabrous; spurs awn-like, $\frac{1}{3}$–$\frac{1}{2}$ as long as the cells or occasionally
rudimentary or absent; ovary more or less pointed and puberulous
at the top; style exserted, $\frac{1}{2}$–$\frac{2}{3}$ lin. long, glabrous; stigma minute,
capitate. *Eremia parviflora, Klotzsch in Linnæa,* xii. 498; *Benth.
in DC. Prodr.* vii. 700; *supra, p.* 334. *Erica shalliana, Hort. Berol.
ex Klotzsch l.c.* 498.

VAR. β, **eglandula** (N. E. Br.); leaves turgidly trigonous, recurved-spreading;
calyx-lobes subquadrate, subtruncate with a thickened apiculus, closely ciliate
with short simple fine hairs; corolla glabrous outside; ovary densely white-
pubescent on the upper half.

VAR. γ, **pubicalyx** (N. E. Br.); leaves turgidly trigonous, recurved-spreading;
calyx-lobes ovate, acute or acuminate, puberulous, ciliate with gland-tipped
hairs; corolla puberulous outside; ovary thinly puberulous.

COAST REGION: Tulbagh Div.; Witsen Berg, *Zeyher*, 1117! *Pappe!* near
Tulbagh Waterfall, *MacOwan*, 2685! and in *Herb. Norm. Austr.-Afr.*, 564!
Swellendam Div.; hills between Puspas Valley and Kokmans (Kogmans) Kloof,
Ecklon! Var. β: Clanwilliam Div.; Cederberg Range at Ezels Kop, 4000 ft.,
Schlechter, 8818! near Clanwilliam, *Leipoldt*, 135! Var. γ: Tulbagh Div.;
mountains of Tulbagh Kloof, 1400 ft., *Bolus*, 5304!

(also distributed as 8917)! Tafel Berg, *Schlechter*, 10091!

I have now (by the courtesy of the Berlin authorities) examined the type of
Eremia parviflora, Klotzsch, and find it to be this plant. The description of
Klotzsch is very erroneous, as the calyx is not subequal to the corolla, but con-
siderably shorter than it, and the stamens are 4, not 8 as Klotzsch states. It is
identical with Zeyher, 1117, except that the leaves are straight, like those of
Schlechter, 10091.

21. G. similis (N. E. Br.); branchlets puberulous; leaves 3-nate,
ascending or spreading, straight, not in the least recurved, $\frac{2}{3}$–$1\frac{1}{2}$ lin.
long with the petiole, slender, linear, trigonous, acute, glabrous,
minutely serrulate, or with 2 or more very short stiff or gland-tipped
hairs on each margin; inflorescence and pedicels exactly like that of
G. eremioides; bracts equal, $\frac{1}{4}$ lin. long, lanceolate, acute, ciliate
with short gland-tipped and simple hairs; calyx lobed to $\frac{3}{4}$ of the
way down; lobes $\frac{1}{2}$ lin. long, linear-oblong or narrowly oblong-
lanceolate, acute, glabrous, distantly ciliate with 4–5 short gland-
tipped hairs on each side; corolla subtubular, slightly narrowed
but scarcely constricted under the limb, obtusely 4-angled; tube
about $\frac{2}{3}$ lin. long, puberulous outside; lobes $\frac{1}{5}$ lin. long, ovate, obtuse,
concave, erect or slightly incurved, glabrous; stamens subequalling
the corolla; filaments glabrous; anthers $\frac{1}{5}$ lin. long, membranous,
with parallel noncontiguous cells, with awn-like spurs above the
base; ovary rather broader than long, very obtuse, glabrous; style
exserted, $\frac{3}{4}$ lin. long, glabrous; stigma minute, capitate.

VAR. β, **grata** (N. E. Br.); leaves often finely and minutely ciliate on the
basal part; calyx-lobes $\frac{1}{4}$–$\frac{1}{3}$ lin. long, oblong-linear and as well as the linear
bracts rather densely ciliate with short simple hairs; corolla narrowly tubular-
funnel-shaped, puberulous on the tube outside; ovary thinly and minutely
pubescent at the top.

COAST REGION: Var. β, Clanwilliam Div.; Cederberg Range, near Sneeuw
Kop and Wupperthal, about 4000 ft., *Bodkin in Herb. Bolus*, 8628!
CENTRAL REGION: Ceres Div.; Cold Bokkeveld, 5000 ft., *Schlechter*, 8896!

X. ACROSTEMON, Klotzsch.

Pedicels 3-bracteate. *Calyx* equally 4-lobed nearly or quite to
the base, villous with long simple hairs or beset with gland-tipped
hairs. *Corolla* hypogynous, tubular or ovoid-tubular and 4-angled
or cylindric, 4-toothed or -lobed, often contracted at the apex by the
lobes being connivent. *Stamens* 4, hypogynous; filaments linear;
anthers basifixed, wholly or partly exserted from the corolla,
deciduous, linear or oblong, 2–5 times as long as broad, some-
times spurred at the base, with the cells connate at the base for $\frac{1}{4}$–$\frac{1}{3}$
of their length, never free, parallel. *Ovary* 2–4-celled (1-celled in
A. eriocephalus); style exserted, filiform; stigma simple or capitate;
ovule solitary in each cell, pendulous. *Fruit* with a thin pericarp,
loculicidal.

Shrublets with the habit of *Erica*; leaves grooved down the convex back,
mostly more or less pubescent; flowers shortly pedicellate, in terminal heads,
with the calyx and bracts clothed with long white hairs.

Species 9, endemic.

This genus has been combined with *Grisebachia*, but is readily distinguished by its anthers alone, which are exserted, basifixed, deciduous, much longer than broad, and their cells are connate below; whilst in *Grisebachia* the anthers rarely exceed the corolla-lobes, are scarcely longer than broad, and their cells are quite disconnected from each other and dorsifixed near the base or taper into a slender attachment to the filament.

Anthers with short subulate spurs at the base; leaves mostly 4-nate, incurved ... ... (1) **concinnus.**

Anthers without spurs at the base:

Leaves 4-nate, not glabrous on the back when young; corolla glabrous:

Bracts 3; calyx lobed to $\frac{3}{4}$ or to the base; anthers very much longer than broad:

Middle bract leaf-like, 2–8 times as long as the small or minute lateral pair:

Leaves straight; sepals $\frac{1}{5}$ lin. broad, linear-lanceolate ... (2) **hirsutus.**

Leaves incurved; sepals $\frac{1}{3}$ lin. broad, ovate or ovate-lanceolate . (3) **incurvus**

Middle bract not twice as long as the lateral pair, all similar; leaves straight ... ... (4) **glandulosus.**

Bract solitary; calyx lobed to about $\frac{2}{3}$; anthers as long to twice as long as broad... (5) **Schlechteri.**

Leaves 3-nate:

Leaves glabrous and smooth on the back; corolla glabrous:

Young leaves ciliate with long white hairs; calyx-segments oblong-lanceolate, obtuse ... ... (6) **incanus.**

Young leaves edged with minute sessile glands and sometimes tipped with a few hairs; calyx-segments ovate, acute ... ... (7) **equisetoides.**

Leaves not glabrous on the back, at least when young; corolla puberulous outside:

Hairs on the leaves and calyx gland-tipped; staminal-filaments glabrous ... (8) **viscidus.**

Hairs on the leaves and calyx not gland - tipped; staminal - filaments hairy ... ... (9) **eriocephalus.**

1. A. concinnus (N. E. Br.); branchlets pubescent or finely pilose with more or less deflexed hairs; leaves 4-nate, or on lateral flowering branchlets sometimes 3-nate, usually shorter than the internodes, erect, incurved, 1–1$\frac{2}{3}$ lin. long with the petiole, linear, obtuse, tuberculate-rugose on the back, thinly pilose with long fine white hairs intermingled with subsessile glands when young; heads globose, 6–18-flowered; pedicels $\frac{1}{2}$–1 lin. long, bracteate at the apex, pilose; bracts loosely embracing the calyx, $\frac{2}{3}$–1 lin. long, the lateral pair nearly as long as the middle one, linear, slightly enlarging upwards, pilose with long white hairs, intermingled with subsessile glands, clammy; calyx divided almost to the base; segments $\frac{3}{4}$ lin. long, scarcely $\frac{1}{5}$ lin. broad at the base, thence tapering to the acute apex, ciliate with long white hairs and subsessile glands, with a few hairs on the back, clammy ; corolla 1½ lin. long, tubular, 4-angled, glabrous, purple; lobes erect, ¼–⅓ lin. long, broadly ovate, obtuse ; filaments of the stamens 1 lin. long, glabrous; anthers half-exserted, 1 lin. long, linear, spurred at the base, smooth, minutely ciliate on the inner margins of the cells ; spurs subulate, about ⅙ as long as the cells, parallel with or slightly incurved towards the filament ; ovary oblong, 4-angled and 4-celled, glabrous ; style much exserted, 2 lin. long, glabrous ; stigma simple.

CoAST REGION : Caledon Div.; on the Zwart Berg, near Caledon, *Bodkin in Herb. Bolus,* 9228 !

2. A. hirsutus (Klotzsch in Linnæa, xii. 228, excl. synonym) ; branchlets puberulous or pubescent; leaves 4-nate, ascending, laxly imbricate or shorter than the internodes, straight, 1–3½ lin. long with the petiole, linear, obtuse, thinly pilose or pubescent with soft hairs, becoming scabrous; heads globose, terminal and on short lateral branchlets, 4–16-flowered ; pedicels ¼ lin. long, bracteate at the apex ; bracts adpressed to the calyx, very unequal, the middle one leaf-like, ⅔–1 lin. long, 2–4 times as long as the lateral pair, all villous with long white hairs; calyx ¾–1 lin. long, divided nearly to the base, villous with long white hairs ; lobes ½–¾ lin. long, ⅕ lin. broad at the base, narrowly linear-lanceolate, acute ; corolla 1½–1¾ lin. long, tubular, 4-angled, glabrous ; lobes minute, deltoid, obtuse, connivent-erect; stamens much exserted ; filaments 1½–1¾ lin. long, glabrous; anthers 1 lin. long, linear, spurless, smooth ; ovary oblong, obtuse, 2-celled ; style ¼ in. long ; stigma simple. *Benth. in DC. Prodr.* vii. 702.

CoAST REGION : Cape Div. ; mountains near Simonstown, 800 ft., *Schlechter,* 1074 ! in Herb. Bolus. Caledon Div. ; Houw Hoek Mountains, 1200-2000 ft., *Burchell,* 8048 ! *Zeyher,* 3316 ! *Scott-Elliot,* 1139 ! *Schlechter,* 7408 ! *Bolus in Herb. Norm. Austr-Afr.* 195 ! *Galpin,* 3720 ! near the Palmiet River and near Grietjes Gat, *Ecklon & Zeyher,* ex *Klotzsch.* Bredasdorp Div. ; near Elim, *Schlechter,* 7715 !

3. A. incurvus (Benth. in DC. Prodr. vii. 702); branchlets pubescent or densely puberulous intermingled with longer hairs ; leaves 4-nate, closely placed or loosely imbricate, incurved-erect, 1⅓–2⅓ lin. long, linear, subobtuse, villous on the back with long hairs which often fall away, rugose, glabrous on the upper side; heads globose, terminating short lateral branches, 6–9-flowered ; pedicels ½–¾ lin. long, bracteate at the apex, pilose ; bracts adpressed to the calyx, very unequal, the middle one 1–1½ lin. long, equalling or exceeding the calyx, leaf-like, incurved, the lateral pair ⅛–⅓ lin. long, linear, all villous ; calyx lobed to more than ¾ of the way down; lobes ½ lin. long, ⅓ lin. broad, ovate or ovate-lanceolate, acute, villous with long simple hairs on the back ; corolla 1⅓–1¾ lin. long, tubular, 4-angled, faintly narrowed at the middle, glabrous; lobes very small, rounded or deltoid-ovate, obtuse, connivent-erect ; filaments 1⅔–2 lin. long, glabrous; anthers much exserted, ½–¾ lin. long, linear or oblong-linear, spurless, smooth, or minutely scabrous on the margins; ovary subquadrate, obtuse, 3–4-angled and -celled, glabrous; style $2\frac{1}{4}$ lin. long, glabrous; stigma simple. *Erica hirsuta, Thunb. Prodr.* 72, *and Fl. Cap. ed. Schultes,* 358, *partly* (*sheet β of Thunberg's Herbarium*); *Salisbury in Trans. Linn. Soc.* vi. 339. *Blæria hirsuta, Thunb. Diss. Blæria,* 8. *B. Thunbergii, G. Don, Gen. Syst.* iii. 805. *Comacephalus incurvus, Klotzsch in Linnæa,* xii. 224; *Rach in Linnæa,* xxvi. 790.

SOUTH AFRICA : without locality, *Thunberg! Roxburgh! Sieber,* 152!

COAST REGION : Tulbagh Div.; mountains near Saron, *Schlechter,* 10680! near Tulbagh Waterfall, *Bolus,* 7583! Winterhoek Mountain, Witsenberg Range and Vogel Valley, *Ecklon & Zeyher,* ex *Klotzsch.*

The descriptions of Thunberg are doubtless chiefly, and that of Salisbury probably exclusively, based upon the specimen on sheet β of *Erica hirsuta* in Thunberg's Herbarium. Salisbury has evidently dissected it and has written upon that sheet, but not on sheet *a,* which contains a very poor specimen (the type of *A. glandulosus*), with the flowers mostly destroyed. A specimen in the British Museum, collected by Masson and named *Erica hirsuta* by Salisbury, is *Grisebachia hispida,* Klotzsch.

4. **A. glandulosus** (Rach in Linnæa, xxvi. 790); apparently laxly branched; branchlets minutely tomentose; leaves 4-nate, adpressed or imbricate. $\frac{3}{4}$–$1\frac{1}{2}$ lin. long with the petiole, straight, linear-oblong, obtuse, flat with a median ridge above, very convex on the back (not compressed as originally described), shortly pubescent on both surfaces, ciliate with long simple hairs, and perhaps sometimes with similar long hairs on the back; heads several-flowered; pedicels $\frac{1}{3}$–$\frac{1}{2}$ lin. long, bracteate at the apex; bracts adpressed to the calyx, the middle one about $\frac{3}{4}$ lin. long, not much longer than the lateral pair, all linear, pilose and ciliate with long simple hairs, clammy; calyx $\frac{3}{4}$ lin. long, divided to the base; segments about $\frac{1}{3}$ lin. broad, lanceolate, obtuse, pilose and ciliate with long simple hairs, clammy; corolla $1\frac{3}{4}$ lin. long, tubular, 4-angled, glabrous; lobes $\frac{1}{5}$ lin. long, rounded, erect; stamens 4, much exserted; filaments nearly 2 lin. long, glabrous; anthers $\frac{3}{4}$ lin. long, linear, spurless; ovary ovoid, subacute, 2-celled, glabrous; style imperfect, glabrous. *Erica hirsuta, Thunb. Prodr.* 72, *and Fl. Cap. ed. Schultes,* 358 *partly, and Thunb. Herb. sheet a.*

SOUTH AFRICA : without locality, *Thunberg !*

Rach has described the hairs on the branches, bracts and calyx as glandular, but I do not find them so on the type specimen, they are all simple, but very clammy. The leaves are also distinctly 4-nate, not 3-nate as Rach describes them.

5. **A. Schlechteri** (N. E. Br.); about 6 in. high; branchlets erect or divergent and straggling, softly pubescent with spreading hairs; leaves 4-nate, suberect to spreading, 1–2 lin. long with the petiole, imbricate or shorter than the internodes, linear, obtuse, straight, about equally convex on both sides, minutely rugulose or tuberculate and pubescent with short spreading hairs, apparently viscid; flowers in globose terminal heads about 3 lin. in diam.,

shortly pedicellate ; bract solitary at the top of the pedicel, adpressed to
and about as long as the calyx, leaf-like, pubescent ; calyx cup-shaped,
lobed to $\frac{2}{3}$ (rarely only to half) of the way down, pubescent all over ;
lobes erect, oblong or deltoid-oblong, obtuse, ciliate with sessile
glands and having a whitish pulverulent patch on the back between
the slightly raised margins ; corolla 1–1$\frac{1}{4}$ lin. long, tubular, slightly
narrowing downwards, 4-angled, slightly oblique, glabrous ; lobes
erect, broader than long, rounded ; anthers shortly exserted, basifixed, about as long as broad, oblong or cuneate-subquadrate, bifid
to half-way, spurless, smooth ; ovary 2–3-celled, glabrous ; style
1 lin. long, stigma simple.

Coast Region : Bredasdorp Div. ; Rhenoster Kop, 50 ft., *Schlechter*, 10576!
Cape Agulhas, 250 ft., *Schlechter*, 10559!

The Cape Agulhas plant is much more slender and straggling than the other,
and has a different appearance, but I do not find any other distinction.

6. **A. incanus** (Klotzsch in Linnæa, xii. 228) ; branchlets rather
stout, whitish-tomentose ; leaves 3-nate, closely imbricate, about
1 lin. long with the petiole, $\frac{1}{3}$–$\frac{1}{2}$ lin. broad, oblong to elliptic-
ovate, obtuse, glabrous, at first ciliate with deciduous long
white hairs; heads globose, 6–12-flowered; pedicels $\frac{1}{2}$ lin. long,
bracteate at the apex ; bracts adpressed to the calyx, unequal,
oblong-linear, obtuse, $\frac{1}{2}$–$\frac{3}{4}$ lin. long, the lateral pair $\frac{3}{4}$ as long as the
middle one, all villous with long white hairs ; calyx $\frac{3}{4}$ lin. long, lobed
to $\frac{3}{4}$ of the way down, villous with long white hairs; lobes $\frac{1}{4}$ lin.
broad, oblong-lanceolate, obtuse ; corolla 2 lin. long, often slightly
curved, tubular, 4-angled, glabrous ; lobes very small, rounded,
connivent-erect ; filaments about as long as the corolla, glabrous;
anthers wholly exserted, 1 lin. long, linear, spurless ; ovary orbicular-
oblong, obtusely pointed, glabrous, 2-celled; style 2$\frac{1}{2}$ lin. long,
much exserted, glabrous ; stigma simple. *Benth. in DC. Prodr.*
vii. 702.

South Africa : without locality (distributed as *Blairia articulata*, L.),
Drège!
Coast Region : Tulbagh Div. ; near Tulbagh Waterfall, *Ecklon & Zeyher*, ex
Klotzsch.

7. **A. equisetoides** (Klotzsch in Linnæa, xii. 228) ; branchlets
puberulous ; leaves 3-nate, adpressed-imbricate, $\frac{3}{4}$–1$\frac{1}{2}$ lin. long with
the petiole, $\frac{1}{3}$–$\frac{1}{2}$ lin. broad, ovate-oblong to linear-oblong, subobtuse,
glabrous, ciliate with minute sessile glands and the youngest some-
times tipped with a few hairs ; heads globose, terminal, 6–9-flowered ;
pedicels $\frac{1}{3}$–$\frac{3}{4}$ lin. long, puberulous, bracteate at the apex ; bracts
adpressed to the calyx, unequal, the lateral pair $\frac{1}{2}$–$\frac{3}{4}$ as long as the
$\frac{3}{4}$–1 lin. long middle one, linear-oblong, subobtuse, ciliate with subsessile glands and long white hairs and more or less hairy on the
back; calyx nearly 1 lin. long, divided nearly to the somewhat
flattened base ; segments ovate or deltoid-ovate, acute, ciliate with
subsessile glands and long hairs, hairy on the back ; corolla 1$\frac{3}{4}$–2$\frac{1}{3}$

lin. long, straight or slightly curved, tubular, 4-angled, contracted at the very shortly 4-toothed mouth, and faintly narrowed at the middle, glabrous ; teeth broader than long, obtuse, connivent-incurved ; filaments $1\frac{1}{2}$–2 lin. long, linear, glabrous ; anthers wholly exserted, 1 lin. long, linear, without spurs ; ovary obtuse, glabrous, 2-celled ; style $2\frac{1}{2}$–3 lin. long, glabrous ; stigma simple. *Benth. in DC. Prodr.* vii. 702.

COAST REGION : Tulbagh Div. ; near Tulbagh Waterfall, *Zeyher ! Bolus,* 5194 ! *Ecklon & Zeyher, ex Klotzsch.*

8. A. viscidus (N. E. Br.) ; branchlets pubescent with spreading minutely gland-tipped hairs ; leaves 3-nate, ascending or somewhat spreading, often shorter than the internodes, 1–2 lin. long with the petiole, linear, obtuse, ciliate and tipped with gland-tipped hairs, with 2 rows of similar hairs on the back and sometimes minutely puberulous besides when young, viscid ; flowers 3–6 (or more ?) together, terminal and on short lateral shoots ; pedicels $\frac{1}{3}$ lin. long, bracteate at the apex, puberulous ; bracts unequal, the lateral pair $\frac{1}{2}$–$\frac{2}{3}$ as long as the $\frac{1}{2}$–1 lin. long middle one, linear, obtuse or acute, ciliate and beset with gland-tipped hairs ; calyx divided to the base ; segments $\frac{3}{4}$–1 lin. long, linear or linear-lanceolate, acute, ciliate and beset on the back with 2 rows of moderately stout minutely gland-tipped hairs ; corolla $1\frac{1}{3}$ lin. long, narrowly ovoid-tubular, sharply 4-angled, minutely puberulous outside, dull red (*Guthrie*) ; lobes minute, rounded or obtuse, erect ; filaments 1–$1\frac{1}{4}$ lin. long, glabrous ; anthers $\frac{1}{3}$ lin. long, partly exserted beyond the corolla-lobes, oblong, spurless, smooth ; ovary oblong, obtuse or slightly emarginate, glabrous, 2-celled ; style $1\frac{1}{4}$ lin. long, glabrous ; stigma capitate.

COAST REGION : Ceres Div. ; Ceres Flats, 1500 ft., *Guthrie,* 2181 !
CENTRAL REGION : Ceres Div. ; Skurfdeberg Range near Elandsfontein, 4500 ft., *Schlechter,* 10013 !

9. A. eriocephalus (N. E. Br.) ; plant 4–6 in. high, with spreading pubescent or minutely tomentose branchlets ; leaves 3-nate, spreading or the upper more or less adpressed, $\frac{3}{4}$–$2\frac{1}{2}$ lin. long with the petiole, linear to subovate, puberulous and beset (sometimes densely) with long fine hairs, becoming glabrous ; heads terminal, globose, about $\frac{1}{3}$ in. in diam., 3–12-flowered, very hairy ; pedicels $\frac{2}{3}$–$\frac{3}{4}$ lin. long, bracteate at the apex ; bracts unequal, small or minute, the lateral pair $\frac{1}{3}$–$\frac{1}{2}$ as long as the $\frac{1}{4}$–$\frac{3}{4}$ lin. long middle one, linear, densely ciliate with long simple hairs ; calyx-segments 1–$1\frac{1}{4}$ lin. long, $\frac{1}{4}$–$\frac{1}{3}$ lin. broad, linear or linear-lanceolate, acute, densely covered on the back with long white spreading hairs, and beset on the inner face with much shorter gland-tipped hairs ; corolla shorter than, equalling or rarely slightly exceeding the calyx, 1–$1\frac{1}{4}$ lin. long, cylindric, 4-toothed, but appearing subtruncate at the top, puberulous outside ; teeth very small, rounded ; stamens shortly or much exserted, according to age ; mature filaments about 1 lin. long, hairy ; anthers $\frac{1}{2}$–$\frac{2}{3}$ lin. long, linear, minutely scabrous on the margins, spurless; ovary ellipsoid, acute, 1- (very rarely 2-) celled, glabrous; style much exserted, $1\frac{1}{2}$ lin. long, glabrous, stigma simple. *Finckea bruniades, and F. eriocephala, Klotzsch in Linnæa,* xii. 238. *Grisebachia eriocephala and G. bruniades, Benth. in DC. Prodr.* vii. 702.

SOUTH AFRICA : without locality, *Ecklon & Zeyher !*

COAST REGION : Stellenbosch Div. ; Hottentots Holland Mountains, near Lowrys Pass, *Bolus,* 5555 ! 9928 ! Caledon Div. ; mountains between French Hoek and Villiersdorp, *Bolus,* 5108 ! mountains near Genadendal, *Drège,* 7804 ! Houw Hoek Mountains, *Galpin,* 3535 ! *Schlechter,* 5439 !

This plant differs from all other species of *Acrostemon* in its almost constantly 1-celled ovary, although flowers with a 2-celled ovary certainly occur, and Klotzsch has characterized the genus *Finckea* as being 2-celled, but it must be of rare occurrence. As the ovary has a somewhat oblique appearance as if 1 cell had aborted, I regard it as a degenerate species of *Acrostemon,* from which it differs in no other character than the ovary. The type of *Finckea bruniades* has its anthers faded to a paler brown and more spreading leaves than in *F. eriocephala,* but there is no specific distinction, and, excepting the paler anthers, is identical with *Galpin,* 3535.

Xa. THAMNUS, Klotzsch.

Bracts 3, adpressed to the calyx. *Calyx* campanulate, deeply 4-toothed. *Corolla* hypogynous, suburceolate-ovoid or -obovoid, minutely 4-lobed at the contracted mouth. *Stamens* 4, hypogynous, free, exserted; filaments linear ; anthers basifixed, divided to $\frac{3}{4}$ of the way down, cells parallel. *Ovary* seated on a disk, at first with 4 compressed angles, becoming globose-obovoid and obscurely angular in fruit, the grooves between the angles bulging out and disappearing, 1-celled, with 4 ovules suspended from the apex of a free central 4-angled placenta; style exserted, filiform ; stigma simple. *Capsule* thin, separating into 4 valves. *Seeds* 4 or fewer by abortion, pendulous from the apex of the placenta.

A small shrub with the habit of *Erica;* leaves grooved down the convex back; flowers small, 1–7 together, axillary and terminal on short lateral branchlets.

DISTRIB. Species 1, endemic.

This genus has been combined with *Simocheilus* by Bentham and others following him, but no author appears to have correctly understood the structure of the ovary, which (together with that of *Lagenocarpus*) differs from that of all other genera of South African *Ericaceæ* in having a free central placenta, unattached to the side wall of the ovary. I have examined the flowers of numerous specimens and always found the ovary to be as above described ; when young it is sharply 4-angled, with the sides doubled in towards the placenta, but I cannot find any trace of dissepiments connecting them with it. After fertilization, the sides bulge out and the angles almost disappear.

1. **T. multiflorus** (Klotzsch in Linnæa, xii. 235) ; about 2 ft. high (*Burchell*); branchlets erect, minutely subtomentose-puberulous; leaves 3-nate, erect, imbricate, 1–3 lin. long with the petiole, linear, acute, glabrous; flowers subsessile or distinctly pedicellate ; bracts

subequal and about as long as the calyx-tube or the middle one longer, linear, subacute, minutely ciliate ; calyx $\frac{1}{2}-\frac{2}{3}$ lin. long, campanulate, deltoidly toothed to about the middle, obscurely 8-ribbed, very minutely puberulous, very adhesive to the corolla ; corolla $1-1\frac{1}{4}$ lin. long, suburceolate-ovoid or -obovoid, contracted at the mouth, glabrous ; lobes broader than long, rounded, erect ; anthers exserted, basifixed, $\frac{1}{2}-\frac{3}{4}$ lin. long, linear, spurred at the base ; spurs awn-like, pendant, sometimes nearly obsolete ; ovary not longer than broad, with 4 compressed angles, becoming subglobose in fruit, glabrous. *Simocheilus obovatus, Benth. in DC. Prodr.* vii. 703. *Thoracosperma paniculatum, Klotzsch in Linnœa,* xii. 229, *ex Bentham, l.c., and O. Kuntze, Rev. Gen. Pl.* ii. 390, *but not the original plant of Klotzsch in Linnœa,* ix. 350.

COAST REGION : Humansdorp Div. ; hill near the Kromme River, 500 ft., *Galpin,* 3709 ! Clarkson, *Galpin,* 3710 ! *Schlechter,* 6016 ! Uitenhage Div. ; Vanstadens Mountains, *Burchell,* 4707 ! *Zeyher,* 789 ! 3322 !

XI. **SIMOCHEILUS**, Klotzsch.

Bracts 0, 1 or 3, adpressed to and coming away with the calyx. *Calyx* oblong, obconic, campanulate or tubular, sometimes becoming ovoid in fruit, coriaceous or thick and fleshy or thin, 4-toothed ; tube 4-angled or 8-ribbed ; teeth usually much shorter than the tube, sometimes equalling it. *Corolla* small, hypogynous, much longer than the calyx, tubular, tubular-campanulate or funnel-shaped, often curved, minutely 4-lobed, usually more or less 4-angled. *Stamens* 4, hypogynous, always exserted when mature ; filaments free, linear or filiform, glabrous ; anthers free, basifixed or dorsifixed, divided nearly to the base, with parallel contiguous cells, with or without spurs on the back or at the base ; cells opening by oblique pores. *Ovary* seated on a disk, 2-celled, often becoming 1-celled in fruit ; style exserted, filiform, glabrous ; stigma minute, simple, slightly thickened or capitate. *Ovule* solitary in each cell, pendulous. *Fruit* usually very small, elongated or ellipsoid, often 1-seeded ; pericarp usually very thin or in a few species hardened and crustaceous.

Small shrubs or shrublets resembling *Syndesmanthus ;* leaves often with a ridge down the middle of the flattened upper side, convex and grooved down the back ; flowers solitary or 2 to several in a cluster or head, axillary and terminal.

DISTRIB. Species 21, endemic.

In the measurements of the leaves the length of the petiole is always included. It should be understood that in this and in all other genera of South African *Ericaceæ,* there is no real morphological distinction between the solitary bract and a floral leaf, but when it comes away with the calyx it has been customary to call it a bract, and when persistent on the axis, a floral-leaf. When 3 bracts are present, the middle one probably represents the very reduced leaf from whose axil the flower arises, and the lateral pair two bracteoles developed on the sub-obsolete pedicel, which are sometimes present and sometimes absent in the same species as in *S. glabellus* and *S. depressus.* The remarkable change in the

develcopment of the calyx as the fruit matures in the last 4 species (and perhaps
with them *S. quadrisulcus* should also be associated), would almost seem to
warraut the retention of the genus *Pachycalyx*, Klotzscb, for them.

I. Calyx very distinctly 4- (rarely 3-) angled, with
   or without a more or less distinct midrib on the flat
   or grooved faces between the angles :
    * Leaves 3-nate ; bracts 3, coming away with the
      calyx ; calyx-teeth opposite the angles of the
      tube ; ovary minutely puberulous at the top,
      2-celled :
        Calyx-teeth as long as the tube ; stigma
          minute, peltate, shortly crater-like   ...   (1) **carneus.**
        Calyx-teeth $\frac{1}{3}$–$\frac{1}{2}$ as long as the tube ; stigma
          simple or obconically thickened :
          Leaves glabrous or rarely thinly pubes-
            cent on the back ; corolla contracted
            at the basal third :
              Young leaves ciliate and together
                with the bracts and calyx-teeth
                usually tipped with a few hairs ...   (3) **barbiger.**
              Young leaves not ciliate, and as
                well as the bracts and calyx-teeth
                without a tuft of hairs at their
                tips   ...   .   (2) **multiflorus.**
           Leaves finely and shortly pubescent or
            puberulous all over ; corolla slightly
            narrowed just at the base   ...   ...   (4) **pubescens.**
    * Leaves 4-nate ; calyx-teeth alternating with
      the angles of the tube ; ovary glabrous :
      † Stigma simple or very slightly thickened :
        Bracts 3, coming away with the calyx ;
          calyx-tube glabrous :
          Ovary 2-celled, not 4-angled ; calyx-
            teeth not ciliate ...   .   (5) **dispar.**
          Ovary 4-celled, with 4 much com-
            pressed angles ; calyx-teeth ciliate   (6) **piquetbergensis.**
        Bract solitary (in 7, *S. glabellus*, 8,
          S. depressus, and 10, *S. submuticus*,
          sometimes accompanied by 2 very
          small lateral bracts,) and coming
          away with the calyx or 0 except the
          floral leaves, which remain on the
          axis ; ovary 2- rarely 3-celled :
          ‡ Calyx-tube glabrous or only pubes-
            cent down the middle of the
            grooves or on the ribs between
            the angles, often ciliate on the
            teeth :
              § Leaves linear or linear-lan-
                ceolate ; calyx-oblong or
                slightly obconic - oblong,
                scarcely or but slightly nar-
                rowed downwards :
                  Leaves widely spreading ;
                    corolla $1\frac{1}{4}$ lin. long,
                    distinctly but not much
                    narrowed within the $\frac{1}{2}$–$\frac{2}{3}$
                    lin. long calyx ...   ...   (12) **patulus.**
                  Leaves spreading to erectly
                    imbricate ; corolla $1\frac{1}{4}$–2

lin. long, much nar-
rowed within the $\frac{3}{4}$ lin.
long calyx, with lobes
$\frac{1}{4}-\frac{1}{3}$ lin. broad ...　...　(7) **glabellus.**
Leaves erectly imbricate :
 Corolla　$1\frac{1}{2}-1\frac{3}{4}$　lin.
 long, scarcely or but
 slightly　narrowed
 within　the　calyx,
 with　lobes　about
 $\frac{1}{6}$ lin. broad :
  Calyx　$\frac{3}{4}-1$　lin.
  long ;　corolla
  very　slender,
  $\frac{1}{4}$ lin. in diam.　(9) **hirsutus.**
  Calyx　$\frac{2}{3}-\frac{3}{4}$　lin.
  long ;　corolla
  $\frac{1}{3}-\frac{1}{2}$　lin.　in
  diam....　...　(10) **submuticus.**
 Corolla $1-1\frac{1}{4}$ lin. long,
 cuneately tapering
 within the $\frac{1}{2}$ lin.
 long　calyx,　with
 lobes $\frac{1}{4}$ lin. broad .. (13) **globiferus.**
§§ Leaves usually ovate or lan-
ceolate, occasionally linear ;
calyx distinctly obconic or
obconic-pear-shaped,　much
narrowed downwards :
 Calyx-tube with very acute
 angles　and　pubescent
 along the grooves be-
 tween them　...　...　(14) **acutangulus.**
 Calyx-tube　with　obtuse
 angles and usually gla-
 brous in the grooves be-
 tween them　...　...　(15) **consors.**
‡‡ Calyx-tube hairy or pubescent
all　over,　oblong,　oblong-campanulate or ellipsoid-oblong :
 Corolla $1\frac{1}{3}-2$ lin. long, the
 exserted part from shorter
 than to $1\frac{1}{4}$ times as long as
 the calyx　...　...　...　(8) **depressus.**
 Corolla $2-2\frac{1}{2}$ lin. long, the
 exserted part twice as long
 as the calyx ...　...　...　(11) **oblongus.**
†† Stigma small but distinctly capitate ;
calyx　acutely 4-angled, as　long as or
longer　than　the　exserted　part　of　the
corolla　•　...　...　...　...　...　(16) **subrigidus.**
II. Calyx-tube with a distinct groove descending
from the sinuses between the teeth, very distinct
and cut-like in the much enlarged fruiting state ;
anther-spurs basal　...　...　...　...　...　(17) **quadrisulcus.**
III. Calyx-tube subcircular in transverse section,
obscurely or distinctly 8-ribbed, fleshy, becoming
much enlarged and ovoid or globose-ovoid with
thickened ribs in fruit :
 Exserted part of the corolla much longer than
 the calyx ...　...　...　...　...　...　(18) **klotzschianus.**

Exserted part of the corolla shorter than or subequalling (rarely slightly longer than) the calyx:
    Exserted part of the corolla conical-ovoid,
      subtruncately contracted at its base by
      the connivent teeth of the fruiting calyx;
      stigma distinctly capitate ...   . (20) **albirameus**.
    Exserted part of the corolla not conical-
      ovoid nor subtruncately contracted by
      the teeth of the fruiting calyx; stigma
      minute, simple, slightly thickened or sub-
      capitate:
        Flowering calyx-tube distinctly longer
          than broad; ribs on the fruiting-calyx
          not very stout nor tubercle-like;
          anthers spurred   ...   . (19) **bicolor**.
        Flowering calyx-tube scarcely longer
          than broad; ribs on the fruiting-calyx
          very stout somewhat tubercle-like;
          anthers spurless   ...   . (21) **glaber**.

1. **S. carneus** (Klotzsch in Linnæa, xii. 236); more than 1 ft.
high, branches and the 1–2 in. long lateral branchlets erect, probably
compact, minutely subtomentose-puberulous, greyish; leaves 3-nate,
erect, imbricate or shorter than the internodes, $\frac{2}{3}$–$1\frac{1}{2}$ lin. long, linear
to oblong-lanceolate, subacute, puberulous all over, with a tuft of
much longer hairs at the tips when young; flowers subsessile, 2–6
together, terminal on short branchlets; bracts 3, subequal or unequal, the middle one about as long as or exceeding the calyx-tube,
linear, subacute, ciliate with hairs and subsessile glands and tipped
with a tuft of hairs; calyx $\frac{1}{2}$ lin. long, thinly coriaceous, tubular-
campanulate; tube sharply 4-angled, glabrous or very slightly pubescent; teeth opposite the angles, as long as the tube, lanceolate, acute,
ciliate with hairs and glands and tipped with a tuft of hairs; corolla
about 1 lin. long, campanulate-funnel-shaped, contracted near the
base, straight, glabrous (pink ex *Klotzsch*); lobes about twice as
broad as long, broadly rounded, incurved in the dried flowers but
perhaps erect when living; anthers exserted, basifixed, $\frac{1}{2}$ lin. long,
cuneate from a slight divergence of the cells, spurred at the base;
spurs nearly half as long as the cells, slighty flattened, linear-lanceolate or sublanceolate, very acute; ovary compressed, obovoid,
puberulous, 2-celled; style twice as long as the corolla; stigma
minute, peltate, very shortly crater-like. *Benth. in DC. Prodr.* vii.
703. *Blæria carnea, Klotzsch in Linnæa*, viii. 661. *Blairia carnea,
Dietr. Synop. Pl.* i. 443. *Thoracosperma carnea, O. Kuntze, Rev.
Gen. Pl.* ii. 390.

Coast Region: Knysna Div.; Doukamma, *Mund!*
Described from the type in the Berlin Herbarium.

2. **S. multiflorus** (Klotzsch in Linnæa, xii. 236); 2–3 ft. high,
main branches stout, erect, often elongated, densely branched;
branchlets tomentose, greyish; leaves 3-nate, erect or imbricate,
1–2 lin. long, linear or oblong-linear, subacute, glabrous; flowers
in small axillary and terminal clusters crowded along the ends of
the lateral branchlets; bracts 3, minute, $\frac{1}{4}$–$\frac{1}{3}$ as long as the calyx-

tube, linear, obtuse, minutely ciliate; calyx $\frac{2}{3}$ lin. long, thinly coriaceous, obconic-tubular; tube sharply 4-angled, glabrous; teeth opposite the angles, $\frac{1}{3}$ as long as the tube, ovate, acute, minutely ciliate; corolla $1\frac{1}{2}$ lin. long, tubular, contracted at the basal third, slightly curved, almost gibbous on one side just above the calyx, glabrous (whitish, *Burchell*); lobes as long as broad, rounded, erect; anthers exserted, basifixed, nearly $\frac{1}{2}$ lin. long, with slightly divergent cells or somewhat narrowed to the spurred base; spurs awn-like; ovary very minutely puberulous, 2-celled; stigma simple, or slightly obconically thickened. *Benth. in DC. Prodr.* vii. 704. *Blæria multiflora, Klotzsch in Linnæa,* viii. 661. *B. aggregata, Wendl., and B. sessiliflora, Wendl. ex Steud. Nom. ed.* 2, i. 208. *Blairia multiflora, Dietr. Synop. Pl.* i. 443. *Thoracosperma multiflora, O. Kuntze, Rev. Gen. Pl.* ii. 390.

Var. β, **Atherstonei** (N. E. Br.); probably 2–3 ft. or more high; main stems or branches erect, stout, densely covered with erect branches and short branchlets, forming long oblong or oblong-linear masses; calyx $\frac{2}{3}$–$\frac{3}{4}$ lin. long, tubular, obtusely 4-angled, densely pubescent all over; stigma subcapitate; otherwise exactly as in the type.

South Africa: without locality, *Mund & Maire! Miller!*

Coast Region: George Div.; near George, *Prior!* near Grootfontein in Long Kloof, *Drège,* 7793! Montagu Pass, 2000 ft., *Schlechter,* 2259! Knysna Div.; near the Keurbooms River, *Burchell* 5164! on the Parde Berg, *Burchell,* 5188! and without precise locality, *Buchanan!* Var. β: Caledon Div.; "Caledon, R.W.R.," in the Cape Herb.! George Div.; without precise locality, *Atherstone!*

3. S. barbiger (Klotzsch in Linnæa, xii. 237); compactly much-branched; branchlets minutely tomentose-puberulous; leaves 3-nate, imbricate to shorter than the internodes, 1–$1\frac{1}{2}$ lin. long, glabrous or rarely thinly pubescent on the back, often tipped with a tuft of hairs and more or less ciliate when young; flowers in small clusters, axillary or terminal on short lateral branchlets, more or less crowded along the main branchlets; bracts 3, the middle one usually about $\frac{1}{2}$ as long as the calyx-tube, tipped with a tuft of hairs; calyx $\frac{1}{2}$–$\frac{2}{3}$ lin. long, coriaceous, tubular-oblong; tube 4-angled, glabrous or minutely pubescent between the angles; teeth opposite the angles, $\frac{1}{3}$–$\frac{1}{2}$ as long as the tube, ovate, acute, minutely ciliate, with or without a tuft of hairs at the apex; corolla about $1\frac{1}{2}$ lin. long, tubular, contracted at the basal third, straight or nearly so, the exserted part longer than the calyx; lobes as long as or rather longer than broad, ovate or ovate-oblong, rounded at the apex, erect; anthers exserted, basifixed, $\frac{1}{2}$ lin. long, with slightly diverging cells or somewhat narrowed to the spurred base; spurs awn-like; ovary very minutely puberulous, 2-celled; stigma obconically thickened. *Benth. in DC. Prodr.* vii. 704. *Thoracosperma barbigera, O. Kuntze, Rev. Gen. Pl.* ii. 390.

Coast Region: Uitenhage Div.; Vanstadens Mountains, *Burchell,* 4684! *Drège,* 7798! *Zeyher,* 716! 3323! Witte Klip, *MacOwan,* 1125! *Bolus,* 9131!

This may possibly be only a local variety of *S. multiflorus,* but besides other slight differences, appears to be dwarfer, with less numerous flowers, and the angles of the calyx are thicker and less acute.

4. S. pubescens (Klotzsch in Linnæa, xii. 236); more than 1 ft. high, branches and branchlets erect, tomentose-pubescent, greyish; leaves 3-nate, erect, imbricate or shorter than the internodes, $\frac{3}{4}$–$1\frac{2}{3}$ lin. long, linear, subacute, finely and shortly pubescent or puberulous all over, usually with the apical hairs longer than the rest, minutely glandular on the margins when young; flowers very shortly pedicellate or subsessile, in small axillary and terminal clusters, which are more or less crowded along the short lateral branchlets or terminal on very short axillary branchlets scattered along the main branches; bracts 3, unequal, $\frac{1}{6}$–$\frac{2}{5}$ lin. long, linear, obtuse, puberulous and ciliate with longer hairs; calyx $\frac{2}{3}$ lin. long, tubular-oblong, sharply 4-angled, thinly coriaceous, hairy, or the basal third of the tube glabrous; teeth opposite the angles, erect, $\frac{1}{3}$ as long as the tube or less, ovate, subacute or obtuse, ciliate with rather long hairs; corolla about $1\frac{1}{4}$ lin. long, tubular, slightly narrowed just at the base, slightly curved, glabrous, the exserted part about as long as the calyx; lobes erect or slightly incurved, rounded, about as long as broad; anthers exserted, basifixed, $\frac{1}{2}$ lin. long, linear or the cells slightly diverging with short pendant awn-like spurs at the base; ovary puberulous at the top, 2-celled; stigma obconically thickened. *Benth. in DC. Prodr.* vii. 704. *Thoracosperma pubescens, O. Kuntze, Rev. Gen. Pl.* ii. 390.

COAST REGION : Uitenhage Div. ; near the Elands River, *Ecklon & Zeyher !*
Described from the type in the Berlin Herbarium.

5. S. dispar (N. E. Br.); branchlets puberulous; leaves 4-nate, 1–2 lin. long, incurved-erect, about equalling the internodes or laxly imbricate, ovate to linear, acute or subobtuse, entire or more rarely with a very few minute marginal teeth, glabrous; flowers subsessile, 3–8 in a cluster, terminal; bracts 3, unequal, linear, acute, glabrous, not ciliate, the middle one at least $\frac{1}{2}$ as long as the calyx-tube; calyx nearly 1 lin. long, obconic, somewhat oblique, rigidly coriaceous; tube sharply 4-angled, glabrous; teeth connivent, alternating with the angles, scarcely $\frac{1}{2}$ as long as the tube, broadly deltoid, acute, not ciliate; corolla $1\frac{3}{4}$ lin. long, tubular, scarcely or but slightly narrowing downwards, glabrous, the exserted part about as long as the calyx; lobes erect, as long as broad, deltoid-ovate, emarginate or obtusely bifid; anthers exserted, basifixed, oblong, spurless; ovary glabrous, 2-celled; stigma simple.

COAST REGION : Riversdale Div. ; near Riversdale, *Rust,* 557 !

6. S. piquetbergensis (N. E. Brown); a small rigid shrublet, $\frac{1}{2}$–1 ft. high; branches subtomentose-puberulous, greyish, becoming brown and glabrous; leaves 4-nate, erectly imbricate, 2–$2\frac{1}{2}$ lin. long, linear, acute, minutely tuberculate-scabrid, pilose with white hairs when young, becoming nearly or quite glabrous; flowers 6 to many in terminal heads 3–5 lin. in diam., on short lateral branchlets; bracts 3, the middle one leaf-like nearly or quite as long as the calyx-tube, the others not half as long, more or less ciliate or with hairs

at the tips and usually with sessile glands on the edges ; calyx rather
more than 1 lin. long, 4-angled, rigidly coriaceous or somewhat
fleshy ; tube with rather thick obtuse angles, glabrous; teeth alternating with the angles, $\frac{1}{2}$ as long as the tube, erect, broadly deltoid,
acute, ciliate with hairs and sessile glands; corolla $2\frac{1}{3}$–$2\frac{1}{2}$ lin. long,
tubular, straight or nearly so, purple, the exserted part a little longer
than the calyx; lobes about as long as broad, rounded; anthers
exserted, basifixed, $\frac{2}{3}$ lin. long, oblong, spurless ; ovary with 4 compressed angles, 4-celled, with 1 ovule in each cell, glabrous; stigma
simple.

CoAST REGION : Piquetberg Div. ; Piquet Berg, 2200 ft., *Guthrie,* 2661 !

7. S. glabellus (Benth. in DC. Prodr. vii. 704) ; branchlets
varying from pubescent with rather long white hairs to puberulous ;
leaves 4-nate, usually spreading or occasionally imbricate, 1–3 lin.
long, straight or slightly recurving, linear or linear-lanceolate, obtuse,
often paler in colour and somewhat callus-like at the tips, glabrous
and smooth or minutely scabrid-tuberculate, or thinly pilose, all
forms sometimes present on the same specimen, minutely denticulate, sometimes ciliate or with a few hairs at the tips or with sessile
glands on the margins ; flowers subsessile, few or many in terminal
heads 3–4 lin. in diam. on short lateral branchlets of longer branchlets, often more or less crowded together; bracts solitary or sometimes accompanied by 2 minute lateral bracts, from half as long as
to equalling the calyx, glabrous, sometimes slightly ciliate ; calyx
about $\frac{3}{4}$ lin. long, slightly obconic-oblong, coriaceous ; tube 4-angled,
with slight ribs alternating with the rounded angles, glabrous ; teeth
alternating with the angles, nearly half the length of tube, deltoid,
acute, erect, with or without cilia ; corolla $1\frac{1}{4}$–2 lin. long, tubular,
much narrowed within the calyx, glabrous, the exserted part distinctly longer than the calyx in the typical form, in others about
equalling or shorter than it ; lobes $\frac{1}{4}$–$\frac{1}{3}$ lin. broad, erect, rounded ;
anthers exserted, basifixed, $\frac{2}{3}$–$\frac{3}{4}$ lin. long, spurred or rarely spurless
at the base ; spurs awn-like ; ovary glabrous ; stigma simple. *Erica
glabella, Thunb. Prodr.* 73, *and Fl. Cap. ed. Schultes,* 364. *E.
fasciculata, Thunb. Prodr.* 71, *and Fl. Cap. ed. Schultes,* 357.
E. scabra, Thunb. Prodr. 72, *and Fl. Cap. ed. Schultes,* 357.
E. embolifera, Salisb. in Trans. Linn. Soc. vi. 340. *E. exilis,
Salisb. in Trans. Linn. Soc.* vi. 340. *E. africana pumila, &c.,
Seba, Thes.* i. 30, *t.* 20, *fig.* 2. *Blæria pusilla, Linn. Mant.* i. 39 ;
Roem. & Schultes, Syst. Veg. iii. 169 ; *Dietr. Synop. Pl.* i. 443.
B. purpurea, Berg. Pl. Cap. 34. *B. glabella, Willd. Sp. Pl.* i. 631 ;
Thunb. Diss. Blæria, 8 ; *Roem. & Schultes, Syst. Veg.* iii. 170 ;
G. Don, Gen. Syst. iii. 805. *B. glabella, var. thunbergiana, Klotzsch
in Linnæa,* viii. 662. *B. fasciculata, Willd. Sp. Pl.* i. 629 ; *Thunb.
Diss. Blæria,* 7 ; *Roem. & Schultes, Syst. Veg.* iii. 169 ; *G. Don, Gen.
Syst.* iii. 804. *B. scabra, Willd. Sp. Pl.* i. 629 ; *Thunb. Diss. Blæria,*
7 ; *Roem. & Schultes, Syst. Veg.* iii. 169 ; *G. Don, Gen. Syst.* iii. 804,
not of Wendl. B. capitata, Thunb. Herb. ex Rach in Linnæa, xxvi.

791. *Blairia glabella, Dietr. Syn. Pl.* i. 443. *Octogonia glabella,
var. thunbergiana, Klotzsch in Linnæa,* xii. 233 ; *Rach in Linnæa,*
xxvi. 791. *O. glabella, var. mutica, Rach, l.c. Thoracosperma glabella,
T. fasciculata, and T. scabra, O. Kuntze, Rev. Gen. Pl.* ii. 390, 391.

SOUTH AFRICA: without locality, *Thunberg! Bergius! Herb. Salisbury!
Sieber,* 153! 155! 178! *Miller! Mund!*
COAST REGION : Cape Div.; Table Mountain and other mountains near Cape
Town, *Thunberg! Burchell,* 8405! *Ecklon,* 105! *Milne,* 108! *MacGillivray,*
457! *Prior! Bolus,* 2970! 3714! 4909! 4944! 4994! 7905! 8048! *Wolley
Dod,* 1289! 2636! *Guthrie,* 1460! *Wilms,* 3401! 3417!

Very variable in foliage and in the relative length of the exserted part of the
corolla to the calyx. Thunberg's type of *E. glabella* consists of 5 small branchlets,
on 2 of which the exserted part of the corolla is very distinctly longer than the calyx,
on the other 3 it is about as long or very slightly exceeds the length of the calyx.
E. fasciculata is described by Thunberg as having spurless anthers, but his type
specimen has the anthers very distinctly spurred, and is identical with a specimen
collected by Bergius and distributed from the Berlin Herbarium as *Blæria
glabella,* var. *thunbergiana,* Klotzsch. The name *Blæria capitata* in Thunberg's
Herbarium, quoted by Rach, is in Salisbury's handwriting. *E. scabra,* Thunb.,
seems somewhat intermediate between *S. glabellus* and *S. depressus,* being like
the latter species in habit and foliage, but the calyx is that of *S. glabellus,* and
glabrous with the exception of a few hairs on the ribs in the grooves between the
angles; the anthers are quite spurless. I have not seen any other specimen
exactly like it ; Sieber's 178 is the nearest approach, but it has spurred anthers,
which, however, is a very variable character.

8. S. depressus (Benth. in DC. Prodr. vii. 704) ; like *S. glabellus*
in all characters except :—leaves not paler in colour nor at all callus-
like at the tips, always minutely scabrous-tuberculate and often
hispid-pubescent all over, sometimes sprinkled with gland-tipped
hairs ; calyx $\frac{2}{3}$ to nearly 1 lin. long, oblong or ellipsoid-oblong,
more or less hairy all over the tube and ciliate with rather long hairs
on the teeth ; corolla $1\frac{1}{3}$–2 lin. long, the exserted part varying from
shorter than to $1\frac{1}{4}$ times as long as the calyx ; anthers often spurless. *Blæria depressa, Lichtenstein ex Roem. & Schultes, Syst. Veg.*
iii. 168. *B. scabra, Wendl. Collect.* i. 85, *t.* 31, *not of other authors.*
*B. fasciculata, Drège ex Benth. l.c., not Willd. B. glabella, var.
bartlingiana, Klotzsch in Linnæa,* viii. 662. *Erica exprompta,
Spreng. Syst. Veg.* ii. 195, *ex Klotzsch, & Benth. Gypsocallis exprompta, Don, Gen. Syst.* iii. 804. *Octogonia glabella, var. bartlingiana, Klotzsch in Linnæa,* xii. 234. *Thoracosperma depressa,
O. Kuntze, Rev. Gen. Pl.* ii. 390.

VAR. β, **patens** (N. E. Br.); calyx about $1\frac{1}{4}$ lin. long, its teeth more than half
as long as the hairy tube, ciliate with long hairs; corolla 2 lin. long, the
exserted part scarcely as long as the calyx; anthers spurless. *S. patens, Benth.
in DC. Prodr.* vii. 704. *Thoracosperma patens, O. Kuntze, Rev. Gen. Pl.* ii. 390.

SOUTH AFRICA : without locality, *Drège! Mund & Maire! Forbes!* Var. β :
Sieber, 156!
COAST REGION : Cape Div.; near Cape Town, *Burchell,* 915! *Prior!* neck
between Shoesters and Patrys Vley, *Wolley Dod,* 3040! Twelve Apostles, *Wolley
Dod,* 1199! hills near Simons Town, *Wright! MacOwan & Bolus, Herb. Norm.
Austr.-Afr.,* 40! *Bolus,* 4993! *Guthrie,* 1461! Caledon Div. ; Houw Hoek,
Scott-Elliot, 1194!

This plant varies in a similar manner to *S. glabellus* and may be only a form of that species with a constantly pubescent calyx-tube.

9. S. hirsutus (Benth. in DC. Prodr. vii. 704); branchlets glabrous or pubescent; leaves 4-nate, erectly imbricate, $1-2\frac{1}{2}$ lin. long, linear, subacute, minutely scabrid-tuberculate, more or less pubescent or subpilose when young; flowers in small terminal clusters, shortly pedicellate or subsessile; bract 1, leaf-like, $\frac{1}{2}$ as long to as long as the calyx-tube, ciliate; calyx $\frac{3}{4}$ to nearly 1 lin. long, rigidly and thickly coriaceous or subfleshy, oblong; tube obtusely 4-angled with 4 ribs alternating with the angles, glabrous; teeth alternating with the angles, erectly connivent, deltoid, acute, ciliate; corolla $1\frac{3}{4}$ lin. long, tubular, very slender, $\frac{1}{4}$ lin. square at the apex and of nearly equal diameter throughout, slightly curved, glabrous; lobes very small, erect, deltoid-ovate, obtuse; anthers exserted, basifixed, linear-oblong, spurless; ovary glabrous; stigma simple. *Octogonia hirta, Klotzsch in Linnæa,* xii. 233. *Thoracosperma hirsuta, and T. hirta, O. Kuntze, Rev. Gen. Pl.* ii. 390, 391.

COAST REGION : Caledon Div. ; mountain ridges between Babylons Tower and Caledon, *Ecklon & Zeyher, Zeyher!*

This chiefly differs from *S. glabellus* in its very slender corolla, which is scarcely enlarged in its upper part and only half as much in diameter there as it is in *S. glabellus.*

10. S. submuticus (Benth. in DC. Prodr. vii. 704); branchlets minutely tomentose to somewhat thinly pilose-pubescent; leaves 4-nate, erect or ascending and more or less imbricate, 1-2 lin. long, linear or linear-lanceolate, subobtuse, glabrous, sometimes with a few sessile glands on the margins, rarely with a very few hairs; flowers subsessile, 3 to many in terminal heads on the lateral branchlets; heads usually 3-4 lin. in diam.; bract solitary or accompanied by 2 very small lateral bracts at its base, leaf-like, half as long as to as long as the calyx-tube, glabrous, sometimes ciliate with minute hairs or sessile glands; calyx $\frac{2}{3}-\frac{3}{4}$ lin. long, tubular-oblong or ovoid-oblong, thickly coriaceous or . subfleshy; tube 4-angled, glabrous; angles obtuse, alternating with 4 small ribs; teeth alternating with the angles, deltoid, acute, about half as long as the tube, erect or sub-connivent, not ciliate; corolla $1\frac{1}{2}-1\frac{3}{4}$ lin. long, tubular, $\frac{1}{3}-\frac{1}{2}$ lin. in diam. at the apex, slightly narrowing from apex to base, curved, glabrous, the exserted part about twice as long as the calyx; lobes about as long as broad, rounded, subtruncate or emarginate, erect; anthers exserted, basifixed, $\frac{1}{2}$ lin. long, linear-oblong, spurless and broadly cuneate or subtruncate and minutely angular at the base, often in the same flower, rarely with short spurs; ovary glabrous; stigma simple. *Thoracosperma submutica, O. Kuntze, Rev. Gen. Pl.* ii. 390.

COAST REGION : Caledon Div.; Onrust River, 20 ft., *Schlechter,* 10397! near Swellendam, 500-800 ft., *Mund! Fry in Herb. Galpin,* 5008!

This differs from *S. hirsutus* by its glabrous leaves, smaller calyx and stouter corolla. The Onrust River plant has its branchlets thinly pubescent instead of tomentose and its calyx rather smaller than in the type, but I believe it to be this species.

11. S. oblongus (Benth. in DC. Prodr. vii. 705); branchlets puberulous or pubescent; leaves 4-nate, erect or imbricate, 1–2½ lin. long, linear or linear-lanceolate, obtuse, pubescent, at least when young; flowers usually numerous in terminal clusters, subsessile or very shortly pedicellate; bract solitary, leaf-like, as long as or longer than the calyx-tube, hairy; calyx ¾ lin. long, coriaceous, oblong-campanulate; tube 4-angled, with flat sides, no alternating ribs, hairy all over; teeth alternating with the angles, erect, not connivent, half as long as the tube, deltoid, acute, ciliate; corolla 2–2½ lin. long, narrowly tubular, slightly narrowing downwards, slightly curved, the exserted part at least twice as long as the calyx, glabrous, its lobes shorter than broad, rounded, erect; anthers exserted, basifixed, ⅔ lin. long, linear, spurless; ovary occasionally 3-celled, glabrous; stigma simple. *Syndesmanthus capitellatus, Klotzsch ex Benth. in DC. Prodr.* vii. 705. *Thoracosperma oblonga, O. Kuntze, Rev. Gen. Pl.* ii. 390.

Coast Region: Clanwilliam Div.; sandy hills between Berg Valley and Lange Valley, *Drège,* 7796! *Zeyher,* 1119!

12. S. patulus (N. E. Br.); branchlets puberulous, intermingled with a longer white pubescence; leaves 4-nate, very spreading, 1½–2 lin. long, linear, obtuse, rugulose or subtuberculate and thinly pilose-pubescent on the back; flowers subsessile, 3–12 in a head; heads 2–2½ lin. in diam., on very short axillary branchlets; bract solitary often equalling the calyx-tube, linear, obtuse, pilose-pubescent; calyx ½–⅔ lin. long, oblong or oblong-campanulate, coriaceous or subfleshy; tube 4-angled, with 4 slight ribs alternating with the angles, having a few hairs on the ribs, otherwise glabrous; teeth alternating with the angles, deltoid, acute, about half as long as the tube, ciliate with rather long hairs; corolla about 1¼ lin. long, tubular, narrowing downwards from the middle, curved, glabrous; lobes broadly deltoid, subobtuse, erect; anthers much exserted, basifixed, ⅓ lin. long, oblong, minutely angular at the truncate spurless base; ovary glabrous; style 2 lin. long; stigma simple.

Coast Region: Caledon Div.; Klein River, 1000 ft., *Schlechter,* 7605! hills near Papies Valley, 500 ft., *Schlechter,* 10434!

13. S. globiferus (N. E. Br.); less than 1 ft. high, not densely branched; branches and branchlets ascending or erect, pubescent or puberulous; leaves 4-nate, erectly imbricate, 1–2 lin. long, linear, subacute, glabrous or with a few hairs and cilia at the tips; flowers subsessile or very shortly pedicellate, 12–20 or more in small dense globose terminal heads 2–2½ lin. in diam., on short lateral branchlets, somewhat racemosely arranged along the main branches; bract solitary, ½–¾ as long as the calyx, lanceolate or rhomboid-lanceolate, acute, tipped with a few hairs and minutely ciliate; calyx ½ lin. long, oblong; tube 4-angled, with 4 slight ribs alternating with the angles, thin, but slightly fleshy, glabrous; teeth alternating with the angles, ½ as long as the tube, erect, deltoid, acute, ciliate with sessile glands and tipped with a few minute hairs; corolla 1–1¼ lin. long, tubular, cuneately tapering to the base at the lower third, glabrous, the exserted part as long as or slightly longer than the calyx ; lobes ¼ lin. broad, erect, rather broader than long, truncate or emarginate ; anthers exserted, basifixed, ⅓ lin. long, oblong, very shortly spurred at the base ; ovary glabrous ; stigma simple.

Coast Region : Swellendam Div. ; mountains along the lower part of the Zondereinde River, *Zeyher*, 3313 !

14. S. acutangulus (N. E. Br.) ; apparently of somewhat straggling growth, irregularly branched ; branchlets shortly and somewhat harshly pubescent ; leaves 4-nate, 1–2 lin. long, longer or shorter than the internodes, erect or spreading, incurved, linear-lanceolate or ovate-lanceolate, obtuse or subacute, minutely tuberculate and thinly covered with short spreading hairs on the back ; flowers minutely pedicellate, in terminal and axillary more or less clustered heads 3–5 lin. in diam. ; bracts 0, except the floral leaves ; calyx ¾ lin. long, stoutly obconic, coriaceous, not rigid ; tube 4-angled, with 4 slight pubescent ribs alternating with the thin acute glabrous angles ; teeth alternating with the angles, erect, about ½ as long as the tube, broadly deltoid, acute, ciliate; corolla about 2 lin. long, tubular, narrowed downwards, slightly curved, glabrous, the exserted part nearly twice as long as the calyx ; lobes short, erect, rounded ; anthers exserted, basifixed, ⅔ lin. long, linear, spurless at the truncate base ; ovary glabrous ; stigma simple.

Coast Region : Bredasdorp Div. ; hills near Mier Kraal, 200 ft., *Schlechter*, 10523 !

15. S. consors (N. E. Br.) ; compactly branched ; branches erect, rather stout, minutely subtomentose ; leaves 4-nate, imbricate, incurved, 1–2 lin. long, ovate, acute, thinly villous-pubescent on the back, with sessile glands on the margins when young, becoming glabrous, minutely rugulose ; flowers sessile, in terminal heads 2–2½ lin. in diam., on short lateral branchlets ; bracts 0 except the floral leaves ; calyx 1 lin. long, stiffly coriaceous, somewhat pear-shaped-obconic ; tube 4- or occasionally 3-angled, with a deeply impressed groove between the angles, glabrous; teeth alternating with the angles, half as long as the tube, erect, like an equilateral triangle, ciliate ; corolla 2 lin. long, tubular, much narrowed below the middle, 3–4-lobed, glabrous, but possibly with a velvet-like sheen ; lobes shorter than broad, rounded, erect ; stamens 3–4 ; anthers exserted, basifixed, ¾ lin. long, linear, spurless at the shortly cuneate base ; ovary glabrous ; stigma slightly thickened.

Coast Region : Caledon Div.; Shaws Mountain, 1300 ft., *Galpin*, 3724 !

The surface of the corolla differs from that of the other species and appears as if it may have something of the velvet-like nature of a *Pelargonium* petal when alive. It may possibly be slightly viscid.

16. S. subrigidus (N. E. Br.) ; main branches erect, irregularly and rather laxly branching at distant points ; branchlets usually 2–4 in a whorl, puberulous or pubescent ; leaves 4-nate, 1½–2 lin.

long, erect, shorter than the internodes or imbricate, linear, subacute, slightly incurved, subtuberculate-rugose and thinly covered
with spreading hairs on the back; flowers subsessile, in terminal
and axillary heads 2–3 lin. in diam.; bract 0, except the floral leaf;
calyx rather more than 1 lin. long, rigidly coriaceous, but not thick,
slightly obconic-tubular; tube sharply 4-angled, with slight ribs
alternating with the angles, glabrous; teeth alternating with the
angles, erect, broadly deltoid, acute, about $\frac{1}{4}$ as long as the tube,
ciliate; corolla $1\frac{2}{3}$ lin. long, tubular, gradually tapering downwards
from the apex, slightly curved, glabrous; lobes erect, not longer
than broad, rounded; anthers exserted, basifixed, $\frac{2}{3}$ lin. long, linear,
spurless; ovary glabrous; stigma capitate.

COAST REGION: Bredasdorp Div.; by the Koude River near Elim, 700 ft.,
Schlechter, 9583!

17. **S. quadrisulcus** (N. E. Br.); branches stout; branchlets
puberulous or also with a minute close greyish tomentum; leaves
3-nate, often with leafy tufts in their axils, imbricate or shorter than
the internodes, and erect or spreading, 1–2 lin. long, linear, acute or
subacute, glabrous or minutely puberulous; flowers subsessile, 1–4
together, axillary or terminal on very short axillary branchlets;
bracts 3, subequal, about $\frac{1}{3}$ lin. long, linear, subacute or obtuse,
minutely ciliate; calyx $\frac{3}{4}$ lin. long, becoming rather larger in fruit,
fleshy, ovoid, toothed to half-way down, with 4 deep cut-like
grooves alternating with the teeth, and the 4 faces opposite the teeth
flattened or slightly grooved in the upper part and rounded below,
glabrous; teeth erect, becoming connivent in fruit, ovate or deltoid,
acute, very minutely ciliate or without cilia in fruit; corolla $1\frac{1}{4}$ lin.
long, tubular, very slightly narrowed below the middle, scarcely or
but slightly curved, glabrous; lobes as long as broad, erect, ovate,
rounded; anthers basifixed, about $\frac{1}{2}$ lin. long, linear-oblong, with
minute or awn-like spurs at the truncate base; ovary glabrous;
stigma simple.

CENTRAL REGION: Prince Albert Div.; mountains near Seven Weeks Poort,
5000 ft., *Marloth*, 2977! Div.? Zwart Berg, *Atherstone*, 267!

18. **S. klotzschianus** (Benth. in DC. Prodr. vii. 703); rigid,
stoutly branched; branchlets minutely tomentose, whitish or pale
grey; leaves 3- (rarely 4-) nate, erect or slightly spreading, varying
from densely crowded to about equalling the internodes, mostly
incurved, $\frac{2}{3}$–$1\frac{1}{2}$ (rarely 2) lin. long, ovate-lanceolate to linear, serrulate, in the type specimen densely pubescent and mucronate with a
very short gland-tipped bristle when young, in others puberulous
and simply mucronate or acute or subobtuse, becoming glabrous with
age; flowers sessile, usually 3–7 together in small axillary clusters,
which are racemosely arranged and often crowded along the upper
part of the branchlets, which sometimes grow out beyond them;
bracts 3, subequal, $\frac{1}{3}$–$\frac{1}{2}$ lin. long, ovate or oblong, obtuse, shortly
ciliate; calyx $\frac{2}{3}$ lin. long, toothed to the middle or beyond, puberulous; tube hemispheric-campanulate, somewhat obscurely 8-ribbed

at the flowering stage, enlarging and becoming globose and distinctly 8-ribbed in fruit; teeth erect or very slightly spreading, deltoid-ovate or deltoid-lanceolate, acute or acuminate, thickened at the margins, shortly ciliate; corolla $1\frac{2}{3}$ lin. long, tubular, contracted at the base within the calyx-tube, slightly curved, glabrous, the exserted part much longer than the calyx; lobes short, obtuse, erect; anthers exserted, dorsifixed just above the base, $\frac{1}{2}$ lin. long, oblong, not narrowed at the base, with short dorsal awn-like spurs just above the base, sometimes nearly obsolete; ovary seated on a very thick disk, compressed, with a few very minute hairs at the obtuse apex; style 2–3 lin. long; stigma thickened or subcapitate. *Plagiostemon puberulus, Klotzsch in Linnæa,* xii. 232. *P. pubescens, Klotzsch et Benth. in DC. Prodr.* vii. 703.

VAR. β, **glabrifolius** (N. E. Br.); leaves adpressed-imbricate to very spreading, glabrous, but often minutely ciliate when young; calyx glabrous or puberulous; anthers often with rather longer spurs than in the type, sometimes $\frac{1}{3}$–$\frac{1}{2}$ as long as the cells, which occasionally diverge at the tips; style about 2 lin. long: otherwise as in the type.

COAST REGION : Clanwilliam Div.; near the Oliphants River or near Brakfontein, *Ecklon & Zeyher!* mountains near Modderfontein, 1200 ft., *Schlechter,* 4974! Lange Kloof, 500 ft., *Schlechter,* 8042 (has also been distributed by error as 8043)! Cederberg Range, near Krakadouw, 3000 ft., *Bodkin in Herb. Bolus,* 8677! Var. β : Vanrhynsdorp Div.; Knagas (Konaquas) Berg, *Zeyher,* 1120! Clanwilliam Div.; Blauw Berg, 1500 ft., *Schlechter,* 8464 (in *Herb. Bolus*)! Cederberg Range, *Shaw in Herb. Bolus,* 5667! Piquetberg Div.; Piquetberg Range, *Schlechter,* 7908!

CENTRAL REGION: Calvinia Div.; Oorlogs Kloof, 2200 ft., *Schlechter,* 10961! Ceres Div.; Cold Bokkeveld, 4500 ft., *Schlechter,* 8864!

The type specimen and Schlechter, 4974 are pubescent with much longer hairs than on any other specimens I have seen. The specimen distributed to Herb. Bolus as Schlechter, 8464 belongs to this plant and is identical with Schlechter, 8864 at Kew, but Schlechter, 8464 distributed to Kew is *Scyphogyne rigidula,* var. *breviciliata.*

19. S. bicolor (Benth. in DC. Prodr. vii. 703); probably dwarf, rigidly branched; branchlets puberulous, whitish or pale greyish; leaves 3-nate, erect or ascending, imbricate or shorter than the internodes, $\frac{2}{3}$–$1\frac{1}{2}$ lin. long, linear, acute or obtuse, with or without a minute simple or gland-tipped apiculus, glabrous and smooth or slightly scabrid-tuberculate on the back, or puberulous with or without an admixture of a few minute gland-tipped hairs on the back and margins; flowers sessile, 1–7 together in lateral or terminal clusters; bracts 3, unequal or subequal, $\frac{1}{3}$–$\frac{1}{2}$ lin. long, linear to narrowly ovate, obtuse or subacute, minutely ciliate; calyx nearly or quite 1 lin. long, subtubular or campanulate-tubular, at first of about equal diameter throughout, becoming conical-ovoid and contracted at the mouth in fruit, fleshy, rigidly so in fruit, 8-ribbed, glabrous or minutely puberulous; teeth $\frac{1}{2}$–$\frac{3}{4}$ as long as the tube, erect, somewhat elongated-deltoid, acute, very minutely and indistinctly ciliate with hairs or with sessile glands; corolla $1\frac{1}{2}$–$1\frac{3}{4}$ lin. long, tubular, much narrowed downwards within the calyx,

glabrous; lobes scarcely as long as broad, rounded, apparently
incurved or connivent; anthers exserted, dorsifixed near the base,
linear-oblong, dorsally spurred at the insertion of the filament; spurs
awn-like, $\frac{1}{6}$–$\frac{1}{2}$ as long as the cells; ovary subglabrous or **very** minutely
puberulous at the top; stigma minute, thickened or subcapitate;
fruit ellipsoid, crustaceous. *S. ecklonianus, Benth., and S. hispidus,
Benth. in DC. Prodr.* vii. 705. *Blæria bicolor, Klotzsch in Linnæa,*
viii. 660 (*by error* 606); *Dietr. Synop. Pl.* i. 443. *Plagiostemon
bicolor, Klotzsch in Linnæa,* xii. 232. *Pachycalyx pubescens,
Klotzsch, l.c.* 230, *and P. hispidus, Klotzsch, l.c.* 231.

CoAST REGION : Tulbagh Div.; Tulbagh Waterfall (ex *Bentham*), *Mund &
Maire!* near Tulbagh, *Ecklon & Zeyher!* New Kloof, *Burchell,* 1000! *Drège,*
7789! near Saron, 3000 ft., *Schlechter,* 10682! Cape Div.; mountain near
Simons Town, 800 ft., *Guthrie,* 1402!

The calyx appears to be red and the corolla white or pink.

At Kew, the type of *S. hispidus* (Drège, 7789) consists of 1 glabrous and
1 puberulous-leaved specimen. The glabrous specimen I cannot distinguish from
S. bicolor, whilst the puberulous specimen cannot, I think, be maintained as
distinct from *S. ecklonianus.* I therefore consider them all to be slight forms of
one species, which is evidently variable in its pubescence, for some specimens of
Schlechter's 10682 are pubescent, whilst others are glabrous, and they unquestionably belong to one species. Bentham describes the anthers of *S.
ecklonianus* and *S. hispidus* as spurless, whilst Klotzsch does not mention them,
except in the generic character of *Pachycalyx,* where this character is probably
taken from *S. glaber,* as there are no flowers with anthers on the type of
S. ecklonianus or on Drège's specimen of *S. hispidus* at Kew, all are in the
fruiting stage; in Burchell's 1000, however, which Bentham quotes under
S. hispidus, the anthers are shortly but distinctly spurred.

20. S. albirameus (N. E. Br.); apparently laxly branched;
branches and branchlets minutely tomentose or densely woolly-
tomentose with longer hairs, becoming subglabrous, white; leaves
usually 3- (rarely 4-) nate, erect or ascending, imbricate or shorter
than the internodes, 1–2 lin. long, linear, acute, serrulate on the
acute edges, softly pubescent on the back, slightly puberulous on the
upperside, often with small leafy branchlets or leaf-tufts in their
axils; flowers 1–3 together terminating the very short axillary
branchlets, subsessile; bracts 3, the middle one longest, $\frac{1}{3}$ lin. long,
lanceolate or ovate-lanceolate, the lateral linear-oblong, all subacute
or obtuse, ciliate; calyx only seen in a fruiting stage, rigidly coriaceous or fleshy, nearly 1 lin. long, conical-ovoid, 8-ribbed, glabrous;
teeth erectly connivent, half or rather more than half as long as the
tube, elongate-deltoid, acute, ciliate; corolla $1\frac{3}{4}$ lin. long, the part
exserted from the calyx inflated oblong or conical-ovoid, abruptly
contracted into a narrow tube within the calyx, oblique or slightly
curved, glabrous; lobes connivent, as long as broad, obtuse; anthers
not seen; young fruit ellipsoid, minutely puberulous at the top,
crustaceous, 2-celled; stigma capitate.

CoAST REGION : Clanwilliam Div.; by the Oliphants River or near Brakfontein, *Ecklon & Zeyher,* 282!

21. **S. glaber** (Benth. in DC. Prodr. vii. 705); branchlets puberulous, greyish; leaves 3-nate, erect or imbricate, $\frac{3}{4}$–$1\frac{1}{2}$ lin. long, oblong-linear, glabrous, minutely serrulate on the acute edges or the serratures tipped with a gland; flowers solitary, axillary and terminal (racemose ex *Thunberg*), subsessile; bracts 3, subequal, $\frac{1}{4}$–$\frac{1}{3}$ lin. long, linear, obtuse, ciliate; calyx $\frac{3}{4}$ lin. long, fleshy or rigidly coriaceous, glabrous; tube scarcely longer than broad, campanulate, distinctly but not stoutly 8-ribbed; teeth erect, more than half as long as the tube, deltoid-ovate, acute, very minutely ciliate; in fruit the calyx becomes ovoid-conical, not more than 1 lin. long, the tube thickening very much and the ribs forming prominent elongated tubercles, those opposite the teeth very broadly deltoid-ovate in outline, those alternating with the teeth much narrower, tapering downwards; corolla $1\frac{1}{2}$–$1\frac{3}{4}$ lin. long, tubular, scarcely narrowed at the base, nearly or quite straight, glabrous (white ex *Thunberg*), the exserted part shorter than or not longer than the calyx; lobes erect, subquadrate, more or less emarginate; anthers half-exserted, dorsifixed near the base, $\frac{1}{2}$ lin. long, oblong, spurless; ovary minutely puberulous at the top; stigma simple. *Erica glabra, Thunb. Prodr.* 69, *and Fl. Cap. ed. Schultes,* 346. *Blæria glabra, Thunb. Diss. Blæria,* 10. *Pachycalyx glaber, Klotzsch in Linnæa,* xii. 231; *Rach in Linnæa,* xxvi. 791.

SOUTH AFRICA: without locality, *Thunberg! Masson!*
COAST REGION: Vanrhynsdorp Div.; Gift Berg, *Drège,* ex *Klotzsch.*
CENTRAL REGION: Calvinia Div.; between Grasberg River and Watervals River, *Drège,* 7790!

Imperfectly known species.

22. **Pachycalyx inæqualis** (Klotzsch in Linnæa, xii. 231); this is referred to *Simocheilus glaber* by Bentham and by Rach. I have not seen it, but the statement that the flowers are in heads, whilst in all the specimens of *S. glaber* that I have seen they are solitary, and the different locality, seem to indicate that it may be distinct. The following is a translation of the original description :—
Leaves 3-nate, adpressed, linear, acute, glabrous, evanescently serrate-hispid; flowers in heads at the tips of the branchlets; bracts 3, unequal, ovate-lanceolate, glabrous, puberulous-ciliate on the margin, the 2 lateral smaller; calyx glabrous; corolla tubular, glabrous, yellowish. It was collected by Ecklon & Zeyher near Tulbagh Waterfall.

XII. **SYNDESMANTHUS**, Klotzsch.

Bracts 0, 1 or 3 adpressed to and coming away with the calyx. *Calyx* obconic, obconic-oblong, campanulate, tubular-oblong, pear-shaped or dorsally flattened, 3–4-toothed; tube with or without 3–4 acute or obtuse angles, thin or coriaceous, not fleshy. *Corolla*

B b 2

small, hypogynous, longer than the calyx, tubular and tapering from
the apex or middle to the base, funnel-shaped, or more or less
inflated or cup- or broadly funnel-shaped above and abruptly contracted into a very slender tube below, 3–4-lobed, often 3–4- (rarely
8-) angled. *Stamens* 3–4, hypogynous, always more or less exserted
when mature; filaments free, linear or filiform, glabrous; anthers
free, basifixed or dorsifixed just above the base, linear, oblong or
more or less cuneate, bipartite, with parallel or slightly divergent
cells, spurless; cells opening by oblique pores. Ovary seated on a
disk, 1-celled; style exserted, filiform, glabrous; stigma minute,
simple or capitate or thickened, rarely disk-like and produced into a
short terete or clavate point near the centre or towards the margin.
Ovule solitary, pendulous. *Fruit* (not seen fully ripe) elongated,
narrow, apparently indehiscent; pericarp thin.

Small shrubs or shrublets resembling *Erica*; leaves grooved down the convex
back; flower-heads terminal or axillary, few- or many-flowered, erect or nodding;
flowers subsessile or very shortly pedicellate.

DISTRIB. Species 19, endemic.

In the following descriptions the measurements of the leaves always include
the petiole. Some of the species with a 3-angled calyx might be supposed to
belong to the genus *Sympieza*, especially where, as in *S. breviflorus, S. sympi-
zoides* and *S. pumilus*, the calyx is dorsally flattened and the angle next the
axis but little developed, the 3-lobed corolla and 1-celled ovary, however, readily
distinguish them.

I. Flowers covered with very viscid matter and aggluti-
    nated together when dried; stigma capitate :
        Flower-heads 3–3½ lin. in diam.; corolla ovoid or
           ellipsoid, contracted at the mouth   ...   ...   (1) **viscosus**.
        Flower-heads 1–1½ lin. in diam.; corolla obconic-
           campanulate or tubular-funnel-shaped, not con-
           tracted at the mouth   ...   .   (2) **Schlechteri**.
II. Flowers not agglutinated together :
    * Bracts 3, coming away with the pedicel, except
      in 3. *S. ciliatus* :
        Calyx pear-shaped, pallid and nerved at the
          lower half :
             Calyx-tube with 4 small puberulous
               tubercles at the middle of the flattened
               faces, otherwise glabrous, 4-nerved   ...   (3) **ciliatus**.
             Calyx pubescent all over; tube 8-nerved .   (4) **squarrosus**.
             Calyx villous all over with long hairs;
               tube 4-nerved   ...   .   (5) **paucifolius**.
             Calyx obconic-campanulate, obscurely 4-
               angled, nerveless; teeth villous with
               long hairs ...   .   (6) **Niveni**.
    ** Bracts 0 or solitary at the base of the very
      minute pedicel and rarely coming away with the
      calyx; calyx sharply 3–4-angled :
        § Angles of the usually villous calyx opposite
          the teeth :
             Stigma peltate, disk-like, excentrically
               produced into a short point   ...   ...   (7) **Erinus**.
             Stigma simple :
               Corolla about ¾ lin. long, funnel-
                 shaped above, abruptly narrowed

into a slender tube below (or in
var. *β*, tapering from apex to base) (8) **elimensis.**
Corolla 1¼-2 lin. long :
 Corolla oblong-tubular to somewhat funnel-shaped above,
  rapidly narrowed into a slender
  tube below the middle ... (10) **articulatus.**
 Corolla gradually tapering from
  apex to base :
   Branchlets not crowded ;
    leaves thinly villous to
    nearly glabrous, rugulose (9) **scaber.**
   Branchlets rather crowded ;
    leaves glabrous or the
    younger tipped with 1–5
    hairs ... ... (11) **similis.**
§§ Angles of the calyx alternating with the
 teeth :
 Stigma minute, capitate or distinctly
  thickened ; calyx usually 3-angled :
  Corolla scarcely 1 lin. long, the exserted part shorter than or equal-
   ling the calyx ... ... (12) **breviflorus.**
  Corolla 1½–1¾ lin. long, the exserted
   part longer than the calyx ... (13) **venustus.**
 Stigma simple, not or scarcely thicker
  than the style ; calyx acutely 3–4-
  angled :
  Corolla 1¾–2¼ lin. long, the exserted
   part 2–3 times as long as the calyx,
   tubular, gradually tapering from or
   below the middle into the slender
   basal part, 3-lobed :
   Plant 8 in. or more high,
    copiously much branched ... (14) **Zeyheri.**
   Plant 2½–5 in. high, annual-
    like, with few (5–25) ultimate
    branchlets ... ... (15) **pumilus.**
  Corolla 1–1½ lin. long, the exserted
   part from shorter than to 1½ times
   as long as the calyx, funnel- or cup-
   shaped, abruptly narrowed at or
   above the middle into a very slender
   tube :
   Leaves 4-nate, or in *S. gracilis*
    sometimes also 3-nate, glabrous
    to pubescent ; calyx usually
    ciliate or villous on the angles
    of the tube as well as on the
    lobes :
    Leaves linear or linear-lan-
     ceolate :
     Calyx usually 3-angled,
      the dorsal face ⅓–½
      lin. broad, marked
      with a whitish stripe (16) **globiceps.**
     Calyx 4-angled, ¼ lin.
      in diam., without a
      white stripe on the
      faces ... ... (17) **gracilis.**

Leaves ovate to lanceolate;
　calyx usually 3-angled,
　$\frac{1}{2}$–$\frac{2}{3}$ lin. broad, without a
　white stripe on the faces . (18) **sympiezoides**.
Leaves 3-nate, small, thick,
　ovate, glabrous; calyx 3-4-
　angled, angles glabrous, teeth
　ciliate ...　　...　　...　　... (19) **pulchellus**.

1. **S. viscosus** (N. E. Br.); "about a span high; branchlets pubescent; leaves 3-nate, erectly spreading, $\frac{3}{4}$ lin. long, narrowly ovate, very obtuse, with subcartilaginous ciliate margins, glabrous, shining" ex *Bolus*, but the few loose leaves seen were $\frac{3}{4}$–1 lin. long with the petiole, rather broad, oblong or ovate, obtuse, with minute glands on the obtuse margins when young, becoming entire; flowers covered with glutinous matter on the calyx and corolla and in the dried state agglutinated together in terminal heads 3–3$\frac{1}{2}$ lin. in diam.; bracts unequal, rigidly coriaceous, the middle one outside, very large, 1 lin. long, $\frac{3}{4}$ lin. broad, broadly ovate, acute, the lateral smaller, oblong-lanceolate, obtuse; calyx $\frac{1}{2}$ lin. long, concealed by the bracts, campanulate, 4-toothed to about half-way down, glabrous, thinly coriaceous; teeth erect, deltoid, acute, one smaller than the rest, not ciliate; corolla 1$\frac{3}{4}$ lin. long, ovoid or ellipsoid, contracted at the mouth, glabrous; lobes 4, shorter than broad, suberect or subincurved, rounded; anthers much exserted, basifixed, $\frac{3}{4}$ lin. long, linear, recurving from the cuneate spurless base; ovary pubescent, 1-celled, with a very thick placenta down one side; style curved or tortuous; stigma minutely capitate. *Simocheilus viscosus, Bolus in Journ. Bot.* 1894, 240.

COAST REGION: Riversdale Div.; Drooge Vlakte, near Riversdale, 400 ft., *Schlechter*, 2142!

Of this distinct species I have only seen a few detached leaves and flowers in the Herbarium of Dr. Bolus. Dried flowers cannot be examined until they have been soaked in ether or some other solvent of the glutinous matter upon them.

2. **S. Schlechteri** (N. E. Br.); about 4–6 in. high, intricately branched; branchlets rather slender, tortuous, at first minutely puberulous, soon becoming glabrous; leaves 3-nate, spreading or deflexed, $\frac{1}{2}$–1 lin. long, mostly shorter than the internodes, very thick, subovoid or ellipsoid, obtuse or subacute, glabrous, ciliate on the petiole; flowers pedicellate, 1–3 agglutinated together in terminal heads 1–1$\frac{1}{2}$ lin. in diam.; pedicels $\frac{1}{4}$–$\frac{1}{3}$ lin. long; bracts unequal, adpressed to and as long as the calyx, $\frac{1}{4}$–$\frac{1}{2}$ lin. long, ovate, acute, glabrous, shortly ciliate, the middle one much larger than the others; calyx about $\frac{2}{5}$ lin. long, cup-shaped, 4-toothed to half-way down or rather more, glabrous, very glutinous, not ciliate on the ovate acute teeth; corolla rather more than 1 lin. long, obconic-campanulate or tubular-funnel-shaped, very glutinous, white; lobes 4, erect, deltoid-ovate, acute; anthers exserted, basifixed, $\frac{1}{2}$ lin. long,

oblong, cuneate at the spurless base ; ovary glabrous ; stigma rather large, capitate.

COAST REGION : Bredasdorp Div.; Vogel Vley, 150 ft., *Schlechter*, 10481 !

The glutinous matter upon the dried flowers of this species must be removed by soaking in ether before they can be dissected. I have also seen a specimen bearing the same number localized as having been collected at Zeekoe Vley, but this is an error for Vogel Vley.

3. S. ciliatus (Benth. in DC. Prodr. vii. 707) ; 6–8 in. high, much branched ; branchlets slender, puberulous, soon becoming glabrous, brown ; leaves 3-nate, spreading or somewhat deflexed, mostly shorter than the internodes, $\frac{1}{2}$–$1\frac{1}{4}$ lin. long, very thick and subterete or turgid, obtuse, glabrous, or when very young slightly puberulous, hairy on the petiole, flowers shortly pedicellate, in axillary and terminal clusters of 3–6, crowded together at the ends of the branchlets ; bracts subequal, $\frac{1}{4}$ lin. long, linear, densely ciliate, very rarely coming away with the flower, but remaining upon the persistent base of the pedicel ; calyx about $\frac{3}{4}$ lin. long, clavately pear-shaped, coriaceous, pallid and 4-nerved below ; tube 4-angled below, having 4 small puberulous tubercles on the flattened faces at about the middle, otherwise glabrous, upper part twice as broad, with 4 angles at the sinuses between the teeth alternating with those on the lower part of the tube ; teeth connivent, scarcely half as long as the tube, broadly triangular, acute, somewhat rigid, grooved down the back, glabrous, very shortly ciliate ; corolla $1\frac{1}{2}$ lin. long, narrowly suburceolate above, tapering into a slender tube below, glabrous, the exserted part about as long as the calyx, white ex *Klotzsch ;* lobes 4, about as long as broad, erect or slightly connivent, deltoid-ovate, subacute ; anthers much exserted, basifixed, oblong-linear, spurless ; ovary glabrous ; stigma simple. *Macrolinum ciliatum, Klotzsch in Linnæa*, xii. 243.

COAST REGION : Caledon Div.; Babylons Tower Mountain and between it and Caledon, *Ecklon & Zeyher, Zeyher !* hills near Bot River, 700 ft., *Bolus*, 5449 !

4. S. squarrosus (Benth. in DC. Prodr. vii. 707) ; a small and apparently somewhat flat-topped shrublet 5–10 in. high ; branchlets densely puberulous or subtomentose, whitish or pale grey ; leaves 3-nate, varying from very spreading to erect, mostly shorter than the internodes, or the terminal imbricate, $\frac{2}{3}$–$1\frac{1}{4}$ lin. long, linear, obtuse, rather turgid, straight or recurved, glabrous, or more or less adpressed-pilose with long hairs when young ; petiole ciliate with long hairs ; flowers subsessile or very shortly pedicellate, 2–9 in axillary and terminal clusters racemosely crowded along the branch- lets ; bracts unequal, minute, the middle one rarely more than $\frac{1}{5}$ lin. long, linear or narrowly deltoid, ciliate with long hairs ; calyx rather more than $\frac{3}{4}$ lin. long, pear-shaped, pubescent all over with very fine short ascending or subadpressed hairs ; tube obconic, pallid, with 8 darker nerves ; teeth as long as or rather longer than the

tube, rather large, connivent, oblong or ovate-oblong, obtuse, rather
thick and rigid, probably green and leafy, grooved down the back,
ciliate with hairs like those of the pubescence; corolla $1\frac{1}{4}$–$1\frac{1}{2}$ lin.
long, inflated ovoid or subglobose and faintly 8-angled at the upper
half, rapidly narrowed into a very slender tube at the lower half,
glabrous, with a minutely papillate surface, pink (*Grisbrook*);
lobes 4, more or less connivent, broadly ovate, obtuse or rounded,
the sinuses between them forming minute spreading teeth which are
very conspicuous on the buds; anthers exserted, subbasifixed, linear,
spurless, the very tips of the cells with a slight divergent curvature;
ovary surrounded by a rather thick disk, very small, glabrous;
stigma simple. *Macrolinum paucifolium, Klotzsch in Linnæa,* xii.
242 (*at least as to Ecklon & Zeyher's specimens distributed from
Berlin under this name*), excl. synonym *Blæria paucifolia, Klotzsch
in Linnæa,* viii. 664 ?

SOUTH AFRICA: without locality, *Thom ! Ecklon & Zeyher !*
COAST REGION : Caledon Div.; on the Zwart Berg and region of the Baths
near Caledon, *Zeyher,* 3326! *Guthrie,* 2509! Zoetemelks Valley, *Grisbrook !*
and in *Herb. Bolus,* 6313 ! near Caledon, *Bolus,* 8499 !

Possibly this plant may have been confused with *S. paucifolius,* Benth., by
Klotzsch.

5. **S. paucifolius** (Benth. in DC. Prodr. vii. 707); apparently
4–6 in. high, somewhat laxly branched, sometimes spreading over
the ground; ultimate branchlets $\frac{1}{2}$–2 in. long, slender, tortuous,
puberulous, with a longer pubescence sometimes intermingled, soon
becoming glabrous; leaves 3-nate, spreading, 1–$1\frac{1}{2}$ lin. long, mostly
$\frac{1}{3}$–$\frac{1}{2}$ the length of the long internodes, linear-subterete, obtuse or
subacute, at first thinly pilose, soon becoming glabrous; petiole
ciliate; flowers shortly pedicellate, about 6–9 together in terminal
globose heads 1–6 lin. in diam.; bracts unequal, adpressed to the
calyx, $\frac{1}{4}$–$\frac{1}{2}$ lin. long, linear, acute, densely ciliate with long white
hairs; calyx 1 lin. long, somewhat pear-shaped, villous with white
hairs, those on the tube not half as long as those on the teeth; tube
obconic, 4-angled, 4-nerved, pallid, semitransparent in water; teeth
opposite the angles (nerves), longer than or equalling the tube,
incurved-erect or connivent, thick and rigid, ovate, ovate-oblong or
lanceolate-oblong, acute or apiculate, ciliate with minute sessile
glands under the hairs; corolla $1\frac{1}{4}$–$1\frac{3}{4}$ lin. long, clavate, inflated and
somewhat obovoid or urceolate-obovoid and very distinctly 4-angled
at the exserted part (which equals or is shorter than the calyx, but
much exceeds the hairs on its teeth), rapidly tapering into a slender
tube at the lower half, glabrous, pink; lobes 4, connivent, often
with erect or recurved tips, broadly ovate, obtuse or acute; anthers
exserted, dorsifixed just above the base or basifixed, $\frac{1}{2}$ lin. long,
linear-oblong, spurless, minutely ciliate-scabrid; ovary with a thick
disk at its base, glabrous; stigma simple. *Blæria paucifolia,
Wendl. Collect.* ii. 17, *t.* 43; *Roem. & Schultes, Syst. Veg.* iii. 170
as *B. pauciflora, corrected in Mantissa,* 107 ; *Bartl. in Linnæa,* vii.

649; *Dietr. Synop. Pl.* i. 444; *G. Don, Gen. Syst.* iii. 804 (*pauciflora*);
Klotzsch in Linnæa, viii. 664.   *B. hirsuta, Lichtenstein ex Roem. &*
Schultes, Syst. Veg. iii. 170; *G. Don, Gen. Syst.* iii. 804.   *Blairia*
paucifolia, Spreng. Syst. i. 432.   *Excl. Erica hirsuta from all the*
synonymy.   Erica flosculosa, Salisb. in Trans. Linn. Soc. vi. 340, *ex*
Bentham.   Blæria flosculosa, G. Don, Gen. Syst. iii. 805; *Klotzsch*
in Linnæa, xii. 246.

COAST REGION : Caledon Div.; at the foot of the mountains near Hermanus,
200 ft., *Bolus*, 9847! hills near Houw Hoek, 1300–2500 ft., *Bolus*, 6960!
Schlechter, 5518!

I have not seen an authentic specimen of this plant. Bolus, 9847, and
Schlechter, 5518, however, agree well with Wendland's figure except as to the
calyx, which, as described by Bentham, has a very small tube in proportion to
the lobes, whilst Wendland figures the lobes smaller than the tube, but as they
are much concealed by the hairs, this may have been an error on the part of the
artist.   Bolus, 6960, has flowers agreeing with the smaller dimensions given,
but otherwise does not differ from the type.   Possibly this species may have been
confused with *S. squarrosus* by Klotzsch.

6. S. Niveni (N. E. Br.); 8 in. or more high; branches some-
what divergent-erect, flexuose, rather stout, $\frac{1}{2}$–$\frac{2}{3}$ lin. thick, bearing
at intervals groups of 1–3 branchlets 2–6 lin. long scattered along
them; leaves 3-nate, spreading to erect, closely placed on the branchlets, distant on the branches, $\frac{2}{3}$–1 lin. long, linear-oblong to oblong-
ovate, obtuse, thick, ciliate and pilose on the back with a few long
hairs when young, soon becoming glabrous and shining; petiole
ciliate; flowers subsessile, about 3–6 together, in small terminal
heads on the short lateral branchlets; bracts unequal, densely ciliate
with long and rather stiff hairs, the middle one about $\frac{1}{3}$ lin. long;
calyx $\frac{2}{3}$ lin. long; tube obconic-campanulate, with a tuft of adpressed
stiff hairs around its base, otherwise glabrous and smooth, at least
on the lower part, slightly or very obscurely 4-angled, coriaceous,
opaque and quite nerveless; teeth opposite the angles, erectly-
connivent, nearly as long as the tube, deltoid or deltoid-ovate, acute,
thickly villous with ascending-spreading hairs longer than the teeth
themselves, and ciliate with minute sessile glands under the hairs;
corolla 1 lin. long, clavate, the exserted part shorter than the calyx
and not exceeding the cilia on its teeth, campanulately or broadly
funnel-shaped, abruptly narrowed into the slender tubular included
part, glabrous; lobes or teeth 4, short, broadly triangular, acute,
erect; anthers exserted, dorsifixed to the broad filaments just above
the base, $\frac{2}{5}$ lin. long, oblong, with the cells slightly and somewhat
orbicularly dilated at the apex, spurless; ovary glabrous; stigma
simple.

SOUTH AFRICA : without locality, *Niven*, 95!

7. S. Erinus (N. E. Br.); apparently about 10–12 in. high, much
branched; branchlets slender, the ultimate scarcely more than $\frac{1}{8}$ lin.
thick, divergent-ascending, with a mingled pubescence of long and

short hairs, brown; leaves 3-nate, ascending, equalling or shorter than the internodes, $\frac{3}{4}$–$1\frac{1}{2}$ lin. long, $\frac{1}{4}$–$\frac{1}{3}$ lin. broad, linear-oblong, obtuse, somewhat flattened, thinly pilose with 2 or more irregular rows of long white spreading hairs on the back; flowers pedicellate, few or many in subglobose clusters 2–$2\frac{1}{2}$ lin. in diam., terminal or on short axillary branchlets; bracts solitary at the base of the pedicels, linear-subspathulate, gland-ciliate and tipped with 2–3 long hairs, usually not coming away with the flowers; calyx $\frac{2}{3}$ lin. long, obconic, sharply 4-angled, thin, 4-nerved, covered on the tube with very long very spreading hairs and ciliate on the teeth with similar hairs and minute sessile glands; teeth opposite the angles, erect, $\frac{1}{3}$ as long as the tube or rather more, deltoid-ovate, subacute or obtuse; corolla nearly 1 lin. long, narrow, gradually tapering from apex to base, 4-angled, glabrous, apparently purple; lobes 4, rather broader than long, erect, rounded; anthers far exserted, basifixed or dorsifixed close to the base, $\frac{1}{2}$–$\frac{2}{3}$ lin. long, linear-oblong, slightly narrowed to the base, spurless; ovary glabrous; stigma peltate, disk-like, produced into a short terete or clavate point usually towards the margin, rarely quite central. *Codonostigma Erinus, Klotzsch ex Benth. in DC. Prodr.* vii. 709.

VAR. **validus** (N. E. Br.); ultimate branchlets much stouter than in the type, $\frac{1}{5}$–$\frac{1}{4}$ lin. thick, more or less villous-pubescent; leaves 3–4-nate, mostly shorter than the internodes, 1–$2\frac{1}{2}$ lin. long, $\frac{1}{3}$–$\frac{1}{2}$ lin. broad, ovate to linear, varying from smooth and glabrous to tuberculate and more or less thinly pilose on the back or at least ciliate with long spreading hairs; flower-heads globose, $1\frac{1}{2}$–2 lin. in diam., more compact than in the type; flowers subsessile; calyx-tube varying from ciliate on the angles only to densely villous all over; anthers $\frac{1}{3}$–$\frac{1}{2}$ lin. long; style rather stouter than in the type.

COAST REGION: Caledon Div.; Klein River Mountains, *Ecklon & Zeyher!* mountains between Caledon and Elim, *Bolus,* 6765! Var. β: Caledon Div.; Papies Vlei, 700 ft., *Schlechter,* 10439! Bredasdorp Div.; mountains near Koude River, 500 ft., *Schlechter,* 9728!

By the courtesy of the Berlin authorities I have been enabled to examine Klotzsch's type of *Codonostigma Erinus,* and find it identical with the other specimen quoted. Var. β, although scarcely differing in its flowers from the type, is so very distinct in appearance as to be readily distinguishable at a glance, and, when both can be compared alive, may prove to be a distinct species.

8. S. elimensis (N. E. Br.); 4–6 in. high, very much branched; branches and branchlets straggling, flexuose, puberulous, intermingled with a longer pubescence, becoming glabrous; leaves 4-nate, ascending or spreading, equalling or $\frac{1}{4}$–$\frac{1}{2}$ the length of the internodes, $\frac{2}{3}$–$1\frac{1}{2}$ lin. long, linear-oblong, obtuse, thinly covered with long spreading hairs on the minutely tuberculate back, or merely ciliate or quite glabrous and smooth on the back; flowers shortly pedicellate, 9–12 in terminal globose heads about 2 lin. in diam.; bract solitary, sometimes persistent on the axis, sometimes on the pedicel and coming away with the flower; calyx about $\frac{1}{2}$ lin. long, tubular-oblong or obconic, sharply 4-angled, thin; tube varying from glabrous to thickly beset on the angles with long spreading rather stiff hairs; teeth opposite the angles, half as long as the tube, erect, ovate, obtuse, always ciliate with long stiff hairs; corolla about ¾ lin. long, upper part more or less funnel-shaped, abruptly narrowed into a slender tube below, glabrous; lobes 4, broader than long, rounded, erect; anthers exserted, basifixed, ⅓ lin. long, oblong, spurless; ovary glabrous; stigma simple.

VAR. β, **incertus** (N. E. Br.); leaves rather narrower, linear, subacute, rugulose and sometimes with a scanty minute pubescence on the back; calyx obconic, with finer, rather shorter and less spreading hairs than in the type; corolla sometimes gradually tapering from top to bottom, sometimes as in the type on the same specimen; anthers dorsifixed just above the base.

COAST REGION: Bredasdorp Div.; hills near Elim, 200–900 ft., *Bolus*, 8517! *Schlechter*, 7641! Var. β: Paarl Div.; mountains near French Hoek, *Schlechter*, 10281!

9. **S. scaber** (Klotzsch in Linnæa, xii. 241, excl. all synonyms); 4–12 in. high, laxly branched or straggling; branches varying from widely divergent to ascending, puberulous, somewhat villous-pubescent or subtomentose, often crooked; leaves usually 4- (occasionally 3-) nate, 1 to nearly 2 lin. long, very spreading or erect, linear, obtuse, varying from thinly villous to nearly glabrous, more or less rugulose; flowers in globose terminal heads about 3 lin. in diam., shortly pedicellate or subsessile; bracts 0, except the floral-leaves which sometimes come away attached to the base of the pedicel; calyx ⅔–¾ lin. long, slightly obconic-oblong, sharply 4-angled, thin, villous on the tube and ciliate on the teeth with long white spreading hairs; teeth opposite the angles, half as long as the tube, erect, deltoid-ovate to lanceolate-ovate, obtuse, usually with a small whitish minutely puberulous patch on the back; corolla 1¼–2 lin. long, tubular, gradually tapering from apex to base, curved, glabrous; the exserted part from slightly longer than to nearly twice as long as the calyx; lobes 4, erect, ovate, rounded; anthers exserted, basifixed, ⅓–½ lin. long, oblong-linear, spurless; ovary glabrous; stigma simple. *Benth. in DC. Prodr.* vii. 706, *excl. all synonyms.*

COAST REGION: Swellendam Div.; mountains near Swellendam, 1000–3000 ft., *Mund*, 53! Riversdale Div.; Langeberg Range, near Riversdale, *Schlechter*, 2192! George Div.; near the Great Brak River, 100 ft., *Guthrie*, 4348!

A specimen collected between Donker Hoek and Houw Hoek Mountains in Caledon Div. (Burchell, 8106), has the exserted part of the corolla much shorter than the calyx and more funnel-shaped than in the other specimens, but does not otherwise differ, and is badly infested by insect galls, which may have arrested development.

10. **S. articulatus** (Klotzsch in Linnæa, xii. 241); 6–10 in. high, compactly branched; branchlets puberulous or pubescent; leaves 4-nate, 1–1½ lin. long, straight or incurved, erect, imbricate or shorter than the internodes, linear or lanceolate, subacute, more or less rugulose and puberulous, pubescent or minutely scabrid on the back; flowers numerous, subsessile or very shortly pedicellate, in

axillary and terminal heads about $\frac{1}{4}$ in. in diam.; bracts 0, except
the floral leaves; calyx $\frac{2}{3}$–$\frac{3}{4}$ lin. long, obconic-tubular, sharply
4-angled, thin, hairy on the tube and ciliate on the teeth with long
white spreading hairs; teeth opposite the angles, erect, $\frac{1}{3}$–$\frac{1}{2}$ as long
as the tube, deltoid-ovate or **oblong-ovate**, obtuse or subacute;
corolla $1\frac{1}{4}$–2 lin. long, oblong-tubular to somewhat funnel-shaped in
the upper part, rapidly narrowed into a slender tube below the
middle, curved or almost straight, glabrous, the exserted part varying
from shorter to much longer than the calyx; lobes 4, erect, broadly
ovate, rounded; anthers exserted, $\frac{1}{3}$–$\frac{1}{2}$ lin. long, basifixed, linear-
oblong or sometimes by divergence of the cells narrowed to the base,
spurless; ovary glabrous; stigma simple. *Benth. in DC. Prodr.*
vii. 706 (*incl. var. hirtus*). *S. glaucus, Klotzsch, l.c.* 242; *Benth.
l.c.* 707. *Blæria articulata, Linn. Mant.* ii. 198; *Lam. Encycl.* i.
429, *Suppl.* i. 640, *and Illust.* i. 315, *t.* 78; *Wendl. Collect.* ii. 19,
t. 44; *Roem. & Schultes, Syst. Veg.* iii. 169; *Dietr. Synop. Pl.* i.
444; *Klotzsch in Linnæa,* viii. 666; *G. Don, Gen. Syst.* iii. 804,
*excl. the synonym Erica articulata, Thunb. from all. B. eriantha,
Willd. ex Steud. Nom. ed.* 2, i. 208. *Erica eriocephala, Lam.
Encycl.* i. 489, *and Suppl.* i. 640, *under Blæria. Erica paleacea,
Salisb. in Trans. Linn. Soc.* vi. 341.

VAR. β, fasciculata (N. E. Br.); leaves puberulous or glabrous and more or
less shining on the rugulose back; calyx-tube glabrous or with here and there
a hair on the angles; teeth deltoid, subacute, ciliate with long hairs and having
a small minutely puberulous patch on the back; otherwise as in the type.
S. *fasciculatus, Klotzsch in Linnæa,* xii. 240; *Benth. in DC. Prodr.* vii. 707.

SOUTH AFRICA : without locality; *Herb. Linnæus! Herb. Salisbury! Drège!
Mund!* Var. β : *Mund & Maire!*
COAST REGION : Cape Div.; Cape Flats, *Burchell,* 705 partly! 828! 8380!
Prior! Bolus, 2949! *Wolley Dod,* 903! Table Mountain, *Prior!* Red Hill, near
Simonstown, *Mrs. Jameson!* Stellenbosch Div.; Hottentots Holland Mountains,
MacOwan & Bolus, Herb. Norm. Austr.-Afr., 38! Var. β : Worcester Div.;
Dutoits Kloof, 3000–4000 ft., *Drège,* 7788! Cape Div.; Cape Flats, *Burchell,*
705 partly! *Burke! Wolley Dod,* 905! Lions Rump, *Schlechter,* 47!

Var. *hirtus,* Benth. l.c. founded upon *Erica paleacea,* Salisb., has rather more
hairy leaves than usual, but is otherwise indistinguishable from the typical form.
The type specimen of S. *fasciculatus* (which by the courtesy of the Berlin
authorities I have been able to examine) is labelled "7788 Drège," although
the number is not quoted by Klotzsch or Bentham.

11. S. similis (N. E. Br.); about 4–6 in. high; branchlets erect,
rather crowded, puberulous; leaves 4-nate, erectly imbricate, 1–$1\frac{1}{2}$
lin. long, linear, subacute, quite glabrous or the younger tipped with
1–5 erect white hairs; flowers in terminal heads about 3 lin. in
diam., subsessile, bractless except for the floral leaves; calyx $\frac{2}{3}$ lin.
long, obconic-oblong, sharply 4-angled, villous on the tube and
ciliate on the teeth with long white spreading hairs; teeth opposite
the angles, erect, half as long as the tube, deltoid-oblong, obtuse;
corolla $1\frac{2}{3}$–$1\frac{3}{4}$ lin. long, slender, 4-lobed, gradually tapering downwards from the $\frac{1}{2}$–$\frac{3}{4}$ lin. broad mouth to the base, more or less
curved, glabrous, the exserted part $1\frac{1}{4}$ to nearly twice as long as the

calyx ; anthers exserted, basifixed, linear-oblong, spurless ; ovary glabrous ; stigma simple.

COAST REGION : Swellendam Div. ; mountains near Swellendam, 800 ft. *Bolus,* 8098!

This much resembles *S. articulatus,* Linn., in general appearance, but the glabrous or hair-tipped leaves and more slender and more tapering corolla readily distinguish it.

12. S. breviflorus (N. E. Br.) ; apparently about 6 in. high, loosely branched ; branchlets more or less flexuose, puberulous ; leaves 4-nate, imbricate or as long as the internodes, erect, about $\frac{3}{4}$ lin. long, ovate or elliptic-ovate, acute, glabrous or with a few scattered hairs on the back and often ciliate with hairs or minute sessile glands when young, those under the flower-heads often tipped with a few hairs ; flower-heads many, $1\frac{1}{2}$–2 lin. in diam., shortly ovoid or ellipsoid ; bracts 0 except the floral leaves ; calyx of the lowest flowers $\frac{2}{3}$ lin. long, $\frac{1}{2}$ lin. broad, 3-angledand -lobed or of the upper flowers often 4-angled and lobed, dorsally flattened and cuneate-obovate, occasionally with an angle or keel down the dorsal face ; teeth alternating with the angles, the dorsal-tooth broad, obtuse or rounded or occasionally bifid, the others more deltoid, subacute, all ciliate, as are sometimes the angles ; corolla scarcely 1 lin. long, funnel-shaped or gradually tapering to the base, 3–4-angled and lobed, the exserted part shorter than or equalling the calyx ; lobes rounded ; anthers exserted, $\frac{1}{2}$ lin. long, cuneate-oblong, spurless ; ovary glabrous ; stigma capitate.

COAST REGION : Bredasdorp Div ; hills near Elim, 200 ft., *Schlechter,* 9727 !

13. S. venustus (N. E. Br.) ; about 6 in. high ; branchlets compact, erect, glabrous or nearly so, whitish, becoming brown ; leaves 3-nate, $\frac{2}{3}$–1 lin. long, imbricate, linear-oblong to elliptic, flat on the upper side, obtuse, entire or very minutely denticulate, glabrous ; flowers 6–15 in a cluster, terminal, sessile in the axils of the leaves of the head, otherwise bractless ; calyx $\frac{3}{4}$ lin. long, slightly obconic-tubular, thin, semitransparent, obtusely 3-angled and 3-toothed, with slight ribs alternating with the angles, glabrous ; teeth alternating with the angles, half as long as the tube, oblong or ovate-oblong, obtuse, ciliate, with some longer hairs at the tips ; corolla $1\frac{1}{2}$–$1\frac{3}{4}$ lin. long ; slender, tubular, slightly tapering downwards, much narrowed and very slender within the calyx, 3-lobed, glabrous, the exserted part longer than the calyx ; lobes erect, broader than long, rounded ; stamens 3 ; anthers exserted, basifixed, $\frac{1}{2}$ lin. long, narrowing to the base, spurless ; ovary glabrous ; stigma dilated or subcapitate.

COAST REGION : Caledon or Bredasdorp Div.; hills between Caledon and Elim, 400 ft., *Bolus,* 8466 !

14. S. Zeyheri (Bolus) ; about 8–12 in. high, compactly much branched ; branchlets erect, finely pubescent or puberulous ; leaves

4-5-nate, usually less than 1 lin. long, incurved-erect, shorter than or slightly exceeding the internodes, often giving a somewhat beaded appearance to the branches, ovate, acute, varying from smooth and glabrous to minutely scabrid and shortly pubescent on the back, usually ciliate ; flowers in terminal heads 3-4 lin. in diam., subsessile ; bracts 0, except the floral leaves ; calyx $\frac{2}{3}$ lin. long, cuneate, acutely 3-angled, 3-toothed, somewhat rigidly coriaceous, but not thick, reddish-brown with a whitish stripe between the angles ; tube glabrous or with a few hairs in the grooves between the angles ; teeth half as long as the tube, alternating with the angles, erect, broadly deltoid, acute, ciliate to entirely without cilia ; corolla 3-lobed, $1\frac{3}{4}$–$2\frac{1}{4}$ lin. long, tubular, gradually tapering from or below the middle to the slender base, the exserted part more than twice as long as the calyx, glabrous, pink (*Bolus*) ; lobes much broader than long, rounded, erect ; stamens 3 ; anthers much exserted, basifixed, $\frac{1}{2}$–$\frac{2}{3}$ lin. long, cuneate, bipartite, spurless ; ovary glabrous ; stigma simple.

CoAST REGION : Caledon Div.; mountains of Klein River Kloof, *Zeyher*, 3315! hills between Caledon and Elim, 600 ft., *Bolus*, 8464! Bredasdorp Div.; hills near Bredasdorp, *Bolus*, 8465!

15. S. pumilus (N. E. Br.) ; plant (excluding the root) $2\frac{1}{2}$–5 in. high, divided into 2-6 main branches at or near the base ; branches erect or ascending, simple or with 1-4 branchlets, and including the leaves, which teretely cover them to their base, $\frac{2}{3}$–1 lin. in diam.; leaves 3-nate, erectly imbricate, $\frac{2}{3}$–1 lin. long, lanceolate, acute, glabrous or the upper minutely puberulous and usually finely ciliate, those at the centre of the flower-spikes minutely glandular at the tips ; flower-spikes 5-20 to a plant, erect, ovoid or oblong; bracts 0, except the floral leaves ; calyx $\frac{1}{2}$–$\frac{2}{3}$ lin. long, $\frac{1}{4}$–$\frac{1}{3}$ lin. broad, 3-toothed ; tube cuneate, dorsally flattened, acutely 2-angled, with 1 keel on the dorsal side and 3 subequal keels (one of them being the 3rd angle) on the side next the axis, glabrous ; teeth alternating with the angles, deltoid, acute, densely ciliate with long and somewhat woolly hairs; corolla $1\frac{3}{4}$–$2\frac{1}{4}$ lin. long, the exserted part 2-3 times as long as the calyx, tubular, gradually tapering from below the middle to the very slender base, 3-lobed, red ; lobes about as long as broad, very obtuse ; anthers exserted, basifixed, $\frac{1}{2}$ lin. long, with a row of hairs down the back of each cell, spurless ; stigma simple.

CoAST REGION : Bredasdorp Div.; Elim, *Schlechter*, 7651 !

A very distinct species, having the appearance of an annual, but woody.

16. S. globiceps (N. E. Br.) ; much branched ; branchlets somewhat puberulous, perhaps slightly viscid ; leaves 4-nate, erect or spreading, imbricate or shorter than the internodes, $\frac{2}{3}$–$1\frac{1}{4}$ lin. long, linear-lanceolate, acute, glabrous or puberulous, sometimes tipped with a few hairs when young ; flower-heads rather copious, 2-3 lin.

hypogynous, tubular, tubular-campanulate, obconic-campanulate or the upper part inflated-oblong or subglobose and much contracted below, 4-lobed. *Stamens* 4, hypogynous, much exserted; filaments and anthers free. *Ovary* narrowly ovoid, 1-celled, with 1 pendulous ovule; style much exserted; stigma simple, thickened or capitate. *Fruit* small, apparently indehiscent; pericarp thin.

Small shrubs resembling *Erica*; leaves often with small leafy tufts in their axils forming distant whorls on the elongated branchlets; flowers solitary or 2 to several in small clusters, axillary or at the ends of short axillary branchlets; bracts coming away with the calyx; fruit probably falling off enclosed in the fleshy calyx.

Distrib. Species 10, endemic.

This genus is readily distinguished from *Syndesmanthus* by its relatively small ($\frac{1}{2}$–$\frac{2}{3}$ lin. long) fleshy calyx becoming much enlarged and thickened in fruit, as well as by the usually very different and somewhat peculiar habit. From *Simocheilus* it is distinguished by its 1-celled ovary, and, with the exception of those species of that genus which form the genus *Pachycalyx* of Klotzsch, also by habit. The corolla, although glabrous, apparently often has a velvet-like surface something like that of a *Pelargonium* petal. The length of the petiole is always included in the measurements given for the leaves.

<pre>
 * Flowering-calyx campanulate or cup-shaped, either
     without angles or ribs or obscurely or obtusely 4-angled:
     † Leaves minutely puberulous to finely pubescent,
         at least when young:
         Calyx-tube puberulous or pubescent:
           Corolla 1¼–1½ lin. long:
             Corolla tubular, narrowed at the
                 base; stigma minutely capitate ...   (1) Marlothii.
             Corolla globose or globose-ovoid, con-
                 tracted into a very narrow tube
                 below; stigma simple   ...   ...   (4) Galpini.
           Corolla nearly 1 lin. long, obconic-cam-
                 panulate; stigma capitate   ...   ...   (6) parviflorus.
         Calyx-tube glabrous; corolla 1 lin. long,
             tubular-campanulate; stigma simple   ...   (7) puberulus.
     †† Leaves glabrous (or on the back only in A.
         collinus), rarely with gland-tipped hairs on the
         margins; calyx-tube glabrous:
           Corolla 1⅓–1½ lin. long:
             Corolla straight, inflated-oblong, much
                 contracted at the basal third; anthers
                 spurless...   ...   .   (2) collinus.
             Corolla curved, tubular, subglobose-
                 clavate at the apex; anthers very
                 shortly spurred   ...   .   (3) curviflorus.
           Corolla ¾–1 lin. long:
             Dried corolla with a dark band around
                 the middle; anther-spurs arising from
                 the filaments below the cells ...   ...   (5) discolor.
             Dried corolla without a dark band around
                 the middle; anther-spurs at the base
                 of the cells, sometimes subobsolete:
                   Fruiting-calyx cylindric or cylin-
                       dric-ovoid ...   .   (8) scoparius.
                   Fruiting-calyx obovoid or subglobose   (9) turbinatus.
 ** Flowering-calyx tubular-campanulate or tubular-
     oblong, sharply 4-angled   ...   .   (10) anguliger.
</pre>

1. A. Marlothii (N. E. Br.); about a foot high, rather rigidly branched, only leafy on the short young branchlets, which are puberulous, greyish ; leaves 3- or occasionally 4-nate, suberect, more or less imbricate, $\frac{2}{3}$–$1\frac{1}{2}$ lin. long, linear, obtuse, minutely puberulous, becoming glabrous; flowers in small axillary clusters congested towards the ends of the short branchlets, subsessile ; bracts subequal, $\frac{1}{4}$ lin. long, puberulous ; calyx $\frac{1}{2}$ lin. long, cup-shaped, with short triangular acute teeth, not at all ribbed, puberulous; corolla $1\frac{1}{2}$ lin. long, tubular, narrowed at the base, glabrous ; lobes short, erect, obtuse ; staminal filaments $1\frac{1}{2}$ lin. long, glabrous ; anthers exserted, basifixed, nearly $\frac{1}{2}$ lin. long, slightly cuneate-oblong, with 2 minute teeth or angles at the base ; ovary puberulous all over; style about $1\frac{3}{4}$ lin. long ; stigma minute, capitate.

CENTRAL REGION: Worcester Div.; hills near Touws River, 2500 ft., *Marloth,* 2996 !

2. A. collinus (N. E. Br.); about 1 ft. high, main branches very stout, branchlets crowded, erect, puberulous, whitish ; leaves 3-nate, ascending, subimbricate, $\frac{1}{2}$–1 lin. long, linear-subterete with acute margins, obtuse, slightly incurved, minutely puberulous on the upper side, glabrous on the back ; flowers 3 in a cluster, axillary or on very short axillary branchlets, subsessile or minutely pedicellate ; bracts subequal and about half as long as the calyx-tube or one longer than the rest, puberulous ; calyx $\frac{1}{2}$ lin. long, campanulate, not angular nor ribbed, shortly 4-toothed, fleshy, glabrous, minutely ciliate on the teeth ; corolla $1\frac{1}{3}$–$1\frac{1}{2}$ lin. long, straight, somewhat inflated-oblong, much contracted at the basal third, glabrous ; lobes broader than long, emarginate or subtruncate at the apex, incurved ; anthers much exserted, $\frac{1}{3}$ lin. long, cuneately narrowed to the spurless base ; ovary pubescent ; stigma slightly thickened.

CENTRAL REGION : Worcester Div.; Touws River, 2500 ft., *Marloth,* 2995 !

3. A. curviflorus (N. E. Br.) ; branchlets puberulous, greyish ; leaves 3-nate, 1–3 lin. long, linear, semiterete, subacute, glabrous; whorls distant on the main branches, crowded on the short or tuft-like axillary branchlets ; flowers 3–15 in small clusters on the axillary branchlets, subsessile ; bracts very unequal, the larger often as long as the calyx-tube, leaf-like, glabrous, ciliate, at least at the base ; calyx nearly $\frac{1}{2}$ lin. long, campanulate, 4–5-toothed, glabrous, not angular, fleshy ; teeth $\frac{1}{2}$ as long as the tube, deltoid, acute, minutely ciliate ; corolla $1\frac{1}{3}$ lin. long, much curved, tubular, sub-globose-clavate at the apex, glabrous, rather thick, apparently purple ; teeth as broad as long, obtuse, erect ; anthers exserted, basifixed, $\frac{1}{3}$ lin. long, cuneately narrowed at the very shortly spurred base ; spurs awn-like, partly adnate to the filament, pendent ; ovary puberulous ; style much exserted ; stigma thickened or subcapitate.

COAST REGION : Caledon Div.; Houw Hoek, *Zeyher,* 3174 !

4. A. Galpini (N. E. Br.); branchlets softly pubescent, brown; leaves 3-nate, $\frac{3}{4}$–2 lin. long, linear, subterete, subobtuse, puberulous with short spreading hairs, intermingled with a few gland-tipped hairs on the margins and petiole; whorls distant on the main branches, usually with crowded tufts in their axils; flowers in clusters of 3 on the very short tuft-like axillary branchlets, subsessile; bracts $\frac{1}{4}$–$\frac{1}{3}$ lin. long, linear, obtuse, puberulous; calyx $\frac{1}{2}$ lin. long, campanulate, obtusely 4-angled, minutely adpressed-puberulous; teeth opposite the angles, very short, broadly deltoid, obtuse, ciliate with sessile glands; corolla $1\frac{1}{4}$ lin. long, glabrous, apparently purple, upper half inflated globose-ovoid or globose, obtusely 4-angled, nearly closed at the mouth by the connivent broadly deltoid-ovate lobes, basal half abruptly contracted into a very narrow tube; staminal filaments $1\frac{1}{2}$ lin. long; anthers exserted, basifixed, nearly $\frac{1}{2}$ lin. long, cuneately narrowed to the spurred base; spurs awn-like or slightly flattened, pendent; ovary glabrous; style $2\frac{1}{4}$ lin. long; stigma simple.

COAST REGION: Riversdale Div.; north spur of Mozambique Kop, near Garcias Pass, 2000 ft., *Galpin*, 3730!

5. A. discolor (Klotzsch in Linnæa, xii. 239); branches and branchlets erect, more or less flexuose, minutely puberulous and greyish when young, becoming glabrous and dark brown, soon becoming leafless at the lower part; leaves 3-nate, $\frac{2}{3}$–$1\frac{1}{2}$ lin. long, linear, acute, glabrous, incurved, spreading or imbricate, longer or shorter than the internodes, often with small tufts in their axils; flowers solitary or perhaps 2–3 together, axillary and terminal, very shortly pedicellate; bracts subequal, linear, obtuse, adpressed to and about half as long as the calyx, glabrous, minutely ciliate; calyx nearly $\frac{1}{2}$ lin. long, campanulate, not angular, becoming ellipsoid and about 1 lin. long in fruit, fleshy, glabrous; teeth half as long as the tube, like an equilateral triangle in outline, acute, thick, erect, ciliate with minute sessile glands, no hairs; corolla about 1 lin. long, somewhat clavate-tubular, the upper half oblong-ovoid, tapering below, slightly curved, the exserted part much longer than the calyx, except in fruit, glabrous, subviolaceous-rosy with a whitish limb, ex *Klotzsch*, but in the dried state with a broad dark band around the middle · on a paler ground; anthers exserted, basifixed, $\frac{1}{3}$ lin. long, slightly narrowed to the base, with awn-like curved spurs from the filaments a little below the cells; ovary puberulous; stigma simple. *Codonanthemum discolor, Benth. in DC. Prodr.* vii. 708.

COAST REGION: Riversdale Div.; not far from the Gauritz River in Kannaland, *Ecklon & Zeyher*, 292!

Described from the type in the Berlin Herbarium. It resembles *A. scoparius* in general appearance, but differs in its broader gland-ciliate calyx-teeth and the brown-banded corolla.

6. A. parviflorus (N. E. Br.); apparently over a foot high, rather laxly branched; branchlets puberulous, whitish or pale grey;

leaves 3-nate, $\frac{3}{4}$–$1\frac{1}{2}$ lin. long, linear, obtuse, puberulous, erect or ascending, varying from slightly imbricate to shorter than the internodes; flowers 3–6 (or 9?), sessile or subsessile in small head-like axillary and terminal clusters; bracts subequal or 1 longer or shorter than the others and often leaf-like, nearly or quite as long as the calyx-tube, linear, obtuse, often slightly thickened at the apex, pubescent; calyx $\frac{1}{2}$–$\frac{2}{3}$ lin. long, campanulate, slightly fleshy, not angular nor ribbed, densely white-pubescent, especially on the tube; teeth about $\frac{3}{4}$ as long as the tube, erect, deltoid, acute; corolla nearly 1 lin. long, obconic-campanulate, glabrous; lobes erect, much broader than long, rounded; anthers exserted, $\frac{1}{2}$ lin. long, dorsifixed just above the base, oblong, spurless; ovary somewhat turbinate or obconic with a short conical top, densely pubescent; stigma small, capitate. *Blæria parviflora, Klotzsch in Linnæa*, viii. 665. *Blairia parviflora, Dietr. Synop. Pl.* i. 444. *Codonanthemum parviflorum, Klotzsch in Linnæa*, xii. 240; *Benth. in DC. Prodr.* vii. 708.

Coast Region : Swellendam Div.; Tradouw, *Mund!*

Described from the type in the Berlin Herbarium.

7. A. puberulus (N. E. Br.); branchlets velvety-pubescent, pale brown; leaves 3-nate, 1–2 lin. long, linear, subterete, obtuse, finely pubescent with spreading hairs; whorls on the main branches distant, usually with crowded tufts in their axils; flowers in small clusters on the tuft-like axillary branchlets, subsessile; bracts unequal, linear, puberulous, the larger $\frac{1}{3}$ lin. long; calyx $\frac{1}{2}$ lin. long, probably enlarging in fruit, campanulate, shortly 4-toothed, minutely ciliate on the teeth, otherwise quite glabrous (not pubescent as described), smooth, not angular nor ribbed; corolla 1 lin. long, tubular-campanulate, glabrous; lobes very short and broad, obtuse, erect; staminal filaments about $1\frac{1}{3}$ lin. long; anthers exserted, basifixed, $\frac{1}{3}$ lin. long, cuneately narrowed to the spurred base; spurs pendent, sublanceolate, usually with 1–2 obtuse lobules or crenations on the inner margin; ovary minutely pubescent; style $1\frac{1}{2}$–$1\frac{3}{4}$ lin. long; stigma simple. *Blæria puberula, Klotzsch in Linnæa*, viii. 661. *Blairia puberula, Dietr. Synop. Pl.* i. 443. *Codonanthemum puberulum, Klotzsch in Linnæa*, xii. 240.

South Africa : without locality, *Mund & Maire!*

8. A. scoparius (Klotzsch in Linnæa, xii. 239); branchlets puberulous, greyish; leaves 3-nate, distant on the main stems, often with small dense tufts in their axils, imbricate at the tips of the shoots, linear, glabrous; flowers in small axillary clusters or at the ends of the very short axillary branchlets, subsessile or very shortly pedicellate; bracts 3, $\frac{1}{4}$–$\frac{1}{2}$ lin. long, linear, minutely ciliate; calyx $\frac{1}{3}$–$\frac{1}{2}$ lin. long, campanulate, becoming $\frac{3}{4}$–1 lin. long, cylindric or cylindric-ovoid and fleshy in fruit, shortly 4-toothed, glabrous, smooth, not ribbed or angled; teeth erect, deltoid, minutely ciliate; corolla $\frac{3}{4}$–1 lin. long, somewhat funnel-shaped or tubular-campanu-

late, straight or scarcely curved, somewhat 4-angled, glabrous, purple ;
lobes erect, short, obtuse ; staminal filaments $1\frac{1}{4}$–$1\frac{1}{2}$ lin. long ;
anthers exserted, basifixed, $\frac{1}{3}$ to nearly $\frac{1}{2}$ lin. long, cuneately
narrowed to the shortly spurred base ; spurs awn-like, pendent,
sometimes reduced to mere points ; ovary puberulous ; style $1\frac{1}{4}$–$1\frac{1}{2}$
lin. long; stigma simple. *Eremia parviflora, Drège in Linnæa*,
xx. 190, *not of Klotzsch. Codonanthemum tenue, Benth. in DC.
Prodr.* vii. 708.

SOUTH AFRICA : without locality, *Mund !*
COAST REGION : Tulbagh Div. ; Witsenberg, *Burchell*, 8725 ! Stellenbosch
Div. ; Lowrys Pass, 400 ft., *Guthrie*, 2280 ! Caledon Div. ; Donker Hoek Moun-
tain, *Burchell*, 7992 ! mountains near Genadendal, *Burchell*, 7898 ! *Drège*,
7791 ! Knoffloks Kraal, *Ecklon & Zeyher* (ex *Klotzsch*) ! Houw Hoek Moun-
tains, *Zeyher*, 5324 ! *Bolus, Herb. Norm. Austr.-Afr.*, 194 ! *Schlechter*, 7562 !
Zwart Berg, near Caledon, *Zeyher*, 3347 ! Bredasdorp Div. ; hills near Mier
Kraal, *Schlechter*, 10534 ! Swellendam Div. ; Tradouw Pass, *Schlechter*, 2077 !

9. **A. turbinatus** (N. E. Br.) ; branches spreading and ascending,
whitish, puberulous ; leaves 3-nate, $\frac{1}{3}$–$1\frac{1}{4}$ lin. long, linear-subterete,
with acute margins, obtuse, glabrous ; whorls usually distant, with
tuft-like branchlets in their axils ; flowers 1–3 on the axillary tuft-
like branchlets, subsessile ; bracts unequal, minute or with thick
clavate leaf-like tips ; calyx $\frac{1}{2}$ lin. long, campanulate, becoming
obovoid or globose-obovoid and $\frac{2}{3}$ lin. in diam. in fruit, fleshy,
shortly 4-toothed, glabrous, not angular nor ribbed, teeth erect in
flower, abruptly inflexed in fruit, with a few minute sessile glands
on the margins, but no hairs ; corolla nearly 1 lin. long, somewhat
obconic-campanulate, slightly oblique, glabrous ; lobes broader than
long, rounded, erect ; anthers exserted, basifixed, cuneately narrowed
to the spurred base ; spurs awn-like, very short or subobsolete,
pendent, straight or incurved to the filament ; ovary long, narrow,
puberulous ; stigma simple.

COAST REGION : Bredasdorp Div. ; hills near Elim, 400 ft., *Bolus*, 8463 !

10. **A. anguliger** (N. E. Br.) ; branches slender, spreading and
ascending, thinly covered with very minute gland-tipped hairs or
subscaberulous ; leaves 3-nate, $\frac{1}{2}$–$1\frac{1}{3}$ lin. long, linear-subterete,
apiculate-acute or gland-tipped, glabrous, with 1–3 minute stalked
glands on each margin ; whorls distant, with short leafy or tuft-
like branchlets in their axils ; flowers 1–3 together, axillary and
terminal on the short axillary branchlets, minutely pedicellate ;
bracts subequal, scarcely half as long as the calyx-tube, linear, acute,
or gland-tipped, minutely ciliate ; calyx at first $\frac{1}{2}$–$\frac{2}{3}$ lin. long, tubular-
campanulate, or tubular-oblong, sharply 4-angled, glabrous, fleshy,
soon enlarging ; angles opposite the teeth ; teeth half as long as the
flowering tube, deltoid, acute, erect, minutely gland-ciliate ; corolla
$\frac{3}{4}$ lin. long, tubular, glabrous, the tube becoming included in fruit,
rather thick and almost fleshy at the upper part ; lobes longer than
broad, oblong or narrowly deltoid-oblong, obtuse or subacute, erect ;

anthers exserted, basifixed, cuneately narrowed to the spurred base ; spurs awn-like, pendent ; ovary long and narrow, minutely puberulous ; stigma simple.

Coast Region : Riversdale Div. ; Garcias Pass, 1000 ft., *Galpin*, 3731 !

XIIIa. **EREMIOPSIS**, N. E. Br.

Bracts 3. *Calyx* very deeply and equally 4-lobed. *Corolla* small, hypogynous, campanulate, 4-lobed, longer than the calyx. *Stamens* 8, hypogynous, included ; filaments and anthers free ; anthers bipartite, the cells distant, separated from each other by the dilated, somewhat crutch-like apex of the filaments. *Ovary* 1-celled, with one pendulous ovule ; style included, abruptly curved down upon the side of the ovary at its base, then erect and recurved at the apex ; stigma simple. *Fruit* (not seen fully ripe) very small, subglobose, apparently indehiscent ; pericarp thin.

An erect branching shrublet, much resembling an *Eremia* or *Grisebachia eremioides* in general appearance ; leaves small, spreading ; flowers small, in small terminal clusters of 2–6.

Distrib. Species 1, endemic.

1. **E. curvistyla** (N. E. Br.) ; a small shrublet, apparently of lax straggling habit ; branchlets irregular, slender, puberulous ; leaves 3-nate, spreading-recurved, $\frac{2}{3}$–1 lin. long with the petiole, $\frac{1}{4}$–$\frac{1}{3}$ lin. broad, linear-oblong or oblong-lanceolate, acute, mucronate, nearly flat above, convex and grooved beneath, glabrous, ciliate with 2–3 long gland-tipped hairs on each side, not shining in the dried state ; flowers in terminal clusters of 2–6 on short lateral branchlets, subsessile or on very minute pedicels ; bracts 3, subequal, about $\frac{1}{4}$ lin. long, ovate, acute, glabrous, ciliate ; calyx $\frac{1}{2}$ lin. long, equally 4-lobed to more than $\frac{3}{4}$ of the way down ; lobes $\frac{1}{3}$ lin. broad, elliptic, carinate-apiculate, glabrous, ciliate, with a pale obscure keel and a dark ∧-shaped mark at the apex ; corolla campanulate, slightly 4-angled, 4-lobed, glabrous ; tube $\frac{1}{2}$ lin. long ; lobes $\frac{1}{4}$ lin. long, and about as broad, rounded, erect, entire ; stamens 8, included ; filaments nearly $\frac{2}{3}$ lin. long, filiform, dilated and somewhat crutch-like at the apex, glabrous ; anthers $\frac{1}{5}$ lin. long, bipartite, with distant parallel cells, dorsally spurred at the base ; spurs very slender, awnlike, rather more than half as long as the cells ; ovary globose-ovoid, 1-celled, with 1 ovule, glabrous ; style $\frac{3}{4}$–1 lin. long, filiform, glabrous, included from being curved down against the side of the ovary at the base, then erect and incurved at the apex ; stigma simple, not thickened.

South Africa : without locality, *Herb. Salisbury ! Ward !*
Coast Region : Worcester Div., *Niven !*

This is the plant of Niven's quoted by Bentham in DC. Prodr. vii. 700, under *Eremia parviflora*, Kl., but it is totally different from that plant in its structure ; the style is very remarkable.

XIV. **ANISERICA**, N. E. Br.

Bracts 0, except the minute floral leaves. *Calyx* tubular-campanulate or campanulate, coriaceous, equally 4-toothed or 4-lobed. *Corolla* hypogynous, much longer than the calyx, tubular, 2-lobed. *Stamens* 4, hypogynous, exserted at maturity; filaments free, filiform, glabrous; anthers free, basifixed, divided almost to the base, spurless, opening by short oblique pores. *Ovary* seated on a small disk, 2-celled; style exserted, filiform, glabrous; stigma minute, simple. *Ovules* solitary in each cell, pendulous. *Fruit* not seen.

A small shrub or shrublet resembling an *Erica ;* leaves grooved down the convex back ; flowers small, numerous in terminal clusters, mostly on short lateral branchlets which are often crowded at the ends of longer branchlets.

DISTRIB. Species 1, endemic.

1. **A. gracilis** (N. E. Br.) ; much branched ; branchlets erect, densely and minutely pubescent as well as pilose ; leaves 3-nate, imbricate or about as long as the internodes, ascending or erect, $\frac{2}{3}$–$2\frac{1}{2}$ lin. long with the petiole, linear, oblong-linear or rarely somewhat ovate, obtuse or subacute, glabrous, ciliate (at least when young) with a few long spreading hairs and also on the petiole ; flowers in dense head-like globose or ovoid terminal clusters 2–4 lin. in diam. ; clusters compound, leafy, the flowers solitary or 2–3 together on very short peduncles in the axils of the leaves, very shortly pedicellate ; bracts 0 ; calyx $\frac{1}{2}$–$\frac{2}{3}$ lin. long, tubular-campanulate or campanulate, coriaceous, nerveless, glabrous, apparently red ; tube variable in length ; teeth $\frac{1}{3}$–$\frac{1}{2}$ as long as the tube, erect, ovate, obtuse or subacute, not ciliate ; corolla 1–$1\frac{1}{2}$ lin. long, tubular or oblong-tubular, narrowed at the base, glabrous, the 2 lobes connivent and almost closed together and rounded on the back, or gaping and straight at the back, red (*Galpin*) ; anthers basifixed, less than $\frac{1}{2}$ lin. long, linear-oblong, more or less cuneate at the spurless base ; ovary compressed-oblong, glabrous ; stigma simple. *Blæria gracilis, Bartl. in Linnæa,* vii. 650. *Sympieza Kunthii, Klotzsch in Linnæa,* viii. 656 and xii. 230 ; *Benth. in DC. Prodr.* vii. 705.

VAR. β, **hispida** (N. E. Br.) ; leaves (at least when young) thinly beset with long spreading hairs on the back and sometimes on the upper side, often tuberculate ; corolla 1–$1\frac{1}{3}$ lin. long, white (*Burchell*), pale pink (*Galpin*) ; otherwise as in the type. *Sympieza Kunthii, Klotzsch, vars. hispida and brachyphylla, Benth. in DC. Prodr.* vii. 705. *Blæria depressa, Drège, and B. ericoides, Drège ex Benth. l.c.*

SOUTH AFRICA : without locality, *Zeyher !* Var. β, *Roxburgh ! Drège !*

COAST REGION : Caledon Div. ; Genadendal Mountain, 3200–5000 ft., *Galpin,* 3725 ! 3726 ! Bredasdorp Div. ; near Potteberg, *Mund,* 49 ! Swellendam Div. ; mountains near Voormansbosch, *Ludwig & Beil !* mountains near the Zondereinde River, *Zeyher,* 3318 ! Var. β : Tulbagh Div. ; mountains near Saron, 2400 ft., *Schlechter,* 10688 ! Stellenbosch Div. ; Stellenbosch, *Prior !* Lowrys Pass, *Schlechter,* 1162 ! Caledon Div. ; Houw Hoek, *Schlechter,* 5462 ! mountains near Caledon, 4000–5000 ft., *Drège !* Swellendam Div. ; mountain peak near Swellendam, *Burchell,* 7301 ! Riversdale Div. ; Garcias Pass, *Burchell,*

7037! Mozambique Kop, *Galpin*, 3723! summit of Kampsche Berg, *Burchell*, 7130! George Div.; Cradock Berg, near George, *Burchell*, 5911!

The corolla often appears to be pink when in bud, but usually dries white. In some specimens, as in the type, the corolla-lobes seem always to be more or less closed together, in others as wide apart as the diameter of the tube; but whether this is a condition of age or a varietal character must be determined from the living plants. The name *gracilis* is very inappropriate to this plant, but as it is the oldest, I retain it. *Burchell*, 7301, on which Bentham founded his var. *brachyphylla*, is only a stunted form of var. *hispida.*

XV. **SYMPIEZA**, Lichtenstein.

Bracts (besides the floral-leaves) 0 or rarely 2. *Calyx* of the lowest or of all the flowers dorsally flattened, 2-edged and 2-lobed, of the central or upper flowers sometimes 3–4-angled, 3–4-lobed. *Corolla* hypogynous, tubular to funnel-shaped (dorsally compressed?), 2-lobed; lobes rounded, gaping. *Stamens* 4, hypogynous; filaments free, filiform, glabrous; anthers free, more or less exserted, basifixed, bipartite, with parallel cells, spurless, often scabrid, opening by oblique pores. *Ovary* seated on a disk, dorsally compressed, oblong, obtuse or emarginate at the apex, 2-celled; style slender, filiform, exserted, glabrous; stigma simple or slightly thickened or minutely capitate; ovule solitary in each cell, pendulous. *Fruit* not seen fully ripe, dorsally much flattened, apparently usually 2-celled, with a very thin pericarp.

Small shrubs or shrublets resembling *Erica;* leaves grooved down the convex back; flower-heads terminal, subglobose, erect or nodding; flowers subsessile, solitary in the axils of the bracts (floral-leaves), the lower of which are leaf-like.

DISTRIB. Species 8, endemic.

In the following descriptions the measurements of the leaves always include the petiole, which is sometimes half as long as the blade; and the descriptions of the calyx only apply to the dorsally flattened 2-edged and 2-lobed calyces characteristic of the genus; the 3–4-angled and lobed calyces often found in the upper or central part of the head are not taken into consideration. All the species of this genus require to be carefully studied from life in their natural habitats. Scarcely any two specimens I have examined have flowers that are identical, and although it is not difficult to sort them into the species here retained, it is doubtful whether *S. articulata, S. capitellata,* and *S. tenuiflora* should be considered as distinct varieties of one species or be further subdivided. Probably the species hybridize freely and thus produce a considerable range of variation.

Corolla white, the exserted part not tapering down-
    wards ... ... (1) **eckloniana.**
Corolla red or pink, never white when dried :
    Corolla not more than 1 lin. long, funnel-shaped,
      the exserted part shorter than the calyx ... (2) **breviflora.**
    Corolla 1¼–2½ lin. long, the exserted part at least
      equalling and usually 1¼–3 times as long as the
      calyx :
        * Calyx-lobes distinctly ciliate as seen under

a lens, and often conspicuously so to the
naked eye :
   Leaves glabrous within the cavity in
   transverse section; corolla $1\frac{1}{4}$–$1\frac{3}{4}$ lin.
   long, abruptly or rapidly narrowed
   at the middle into a slender tube
   below ... ... (3) **vestita.**
   Leaves usually densely lined with hairs
   within the cavity in transverse section :
      Corolla scarcely narrower within
      the calyx than at the apex, $1\frac{1}{4}$–$1\frac{2}{3}$
      lin. long ... ... (4) **pallescens.**
      Corolla distinctly much narrower
      within the calyx than at the
      apex :
         Corolla $1\frac{1}{3}$–$1\frac{2}{3}$ lin. long, the
         exserted part not more than
         $1\frac{1}{4}$ times as long as the calyx
         and not tapering downwards (5) **brachyphylla.**
         Corolla $1\frac{3}{4}$–$2\frac{1}{2}$ lin. long, the
         exserted part nearly or quite
         twice as long as the calyx
         and often tapering down-
         wards ... ... (6) **articulata.**
  ** Calyx-lobes as seen under a lens very
  thinly or inconspicuously ciliate with
  minute hairs or sessile glands or both
  or without cilia (see also 5, *S. brachy-*
  phylla) :
      Corolla $1\frac{1}{2}$ to nearly 2 lin. long, the
      exserted part twice as long as the
      calyx or less ... ... (7) **capitellata.**
      Corolla about $2\frac{1}{3}$ lin. long, slender,
      tubular, the exserted part about 3
      times as long as the calyx ... ... (8) **tenuiflora.**

1. S. eckloniana (Klotzsch in Linnæa, xii. 229); densely much
branched; branchlets very copious and slender, villous-pubescent or
puberulous; leaves 3-nate, imbricate to shorter than the internodes,
ascending or erect, $\frac{1}{2}$–$1\frac{1}{2}$ lin. long, linear or linear-lanceolate, acute,
glabrous, sometimes minutely ciliate when young; flower-heads
very copious, $1\frac{1}{2}$–2 lin. in diam.; calyx $\frac{1}{2}$ lin. long, $\frac{1}{3}$ lin. broad;
tube subrectangular or subquadrate, thin, glabrous, apparently light-
reddish; lobes broadly deltoid, subacute or rounded, shortly ciliate;
corolla $1\frac{1}{3}$–$1\frac{1}{2}$ lin. long, tubular or oblong-tubular, narrowed at the
base, not broader across the short lobes than elsewhere, white;
anthers $\frac{1}{3}$ lin. long, slightly cuneate-oblong; stigma simple. *Benth.
in DC. Prodr.* vii. 705.

CoAST REGION : Caledon Div. ; Klein River Mountains, *Ecklon!* mountains near Hermanus, 300–500 ft., *Galpin*, 3727! *Guthrie*, 4115! *Bolus*,
9848!

2. S. breviflora (N. E. Br.); apparently about 6 in. high, much
branched; branchlets tortuous, pubescent or puberulous; leaves
3-nate, as long as or slightly exceeding the internodes, erect, 1–2

lin. long, lanceolate, acute, glabrous or slightly puberulous on the back, ciliate on the acute edges, the cilia often intermingled with minutely stalked glands; flower-heads numerous, subglobose, 2–$2\frac{1}{2}$ lin. in diam.; calyx of the lower flowers $\frac{2}{3}$–$\frac{3}{4}$ lin. long, $\frac{1}{2}$–$\frac{2}{3}$ lin. broad, obovate, ciliate on the obtuse or rounded lobes, otherwise glabrous; corolla 1 lin. long, $\frac{2}{3}$–1 lin. broad across the 2 widely gaping lobes, compressed funnel-shaped, gradually narrowing to the base, red or pink; anthers partly exserted, $\frac{1}{2}$ lin. long, oblong; stigma simple.

COAST REGION: Cape Div.; Simons Berg, *Wolley Dod*, 315!

3. **S. vestita** (N. E. Br.); apparently about 6–8 in. high, rather loosely branched; branchlets divergent or ascending, usually much recurved at the tips, puberulous, pale brown; leaves 3-nate, imbricate or shorter than the internodes, erect or ascending, $\frac{2}{3}$–$1\frac{1}{2}$ lin. long, ovate or lanceolate, acute, glabrous, ciliate with hairs and minute glands on the acute cartilaginous margins, usually with or sometimes without a tuft of 3–6 longer and stouter hairs at the apex of the younger leaves, the internal cavity in transverse section rather large, without a lining of hairs; flower-heads numerous, very nodding, subglobose, about $\frac{1}{4}$ in. in diam.; calyx of the lower or of all the flowers $\frac{3}{4}$ lin. long, $\frac{2}{3}$ to nearly 1 lin. broad, orbicular, somewhat rhomboid-orbicular or broadly obovate, concave on the side next the axis, ciliate with rather long hairs on the broadly rounded lobes, otherwise glabrous or obscurely and minutely puberulous; corolla $1\frac{1}{4}$–$1\frac{3}{4}$ lin. long, $\frac{2}{3}$–1 lin. broad across the very gaping lobes, narrowly to broadly funnel-shaped at the upper half, rapidly or suddenly narrowed into a slender tube at the lower half, red or pink; lobes broadly rounded; anthers $\frac{1}{2}$ lin. long, glabrous or minutely scabrid along the back of the cells; stigma simple.

COAST REGION: Bredasdorp Div.; mountains between Fairfield and Elim, *Bolus*, 8515! hills near Elim, *Bolus*, 8516! Napier, *Schlechter*, 9654!

In all the allied species with which this can be confused, the internal cavity of the leaves as seen in transverse section is smaller than in *S. ves'ita* and more or less densely lined with very minute hairs.

4. **S. pallescens** (N. E. Br.); apparently about 6 in. high, rather loosely branched; branches and branchlets somewhat tortuose, puberulous to minutely subtomentose; leaves 3-nate, imbricate, $\frac{3}{4}$–$1\frac{1}{4}$ lin. long, ovate, acute, minutely puberulous or glabrous on the back, minutely ciliate; flower-heads numerous, subglobose, 2–$2\frac{1}{2}$ lin. in diam.; calyx of the lower flowers $\frac{2}{3}$ lin. long, usually rather more than $\frac{1}{2}$ lin. broad; tube cuneately oblong or subquadrate, glabrous, sometimes minutely ciliate on the edges; lobes very obtuse or almost rounded, densely ciliate with rather woolly hairs and puberulous on the back; corolla $1\frac{1}{4}$–$1\frac{2}{3}$ lin. long, tubular, of nearly equal diameter ($\frac{1}{4}$–$\frac{1}{3}$ lin.) from apex to base, viewed dorsally, apparently pale pink; anthers $\frac{1}{3}$ lin. long; stigma simple.

Coast Region: Caledon Div.; Houw Hoek, 1500 ft., *Schlechter*, 7328!

This is distinguished from those specimens of *S. articulata* which have a some-what woolly-ciliate calyx by the more pallid colour of the flowers when dried, the tendency of the leaves to be puberulous and particularly by the corolla being nearly as broad around the ovary as it is across the lobes.

5. S. brachyphylla (Benth. in DC. Prodr. vii. 706); much branched; branchlets minutely puberulous; leaves 3-nate, imbricate, straight or incurved, $\frac{2}{3}$–$1\frac{1}{4}$ lin. long, usually ovate, sometimes linear-oblong, glabrous, sometimes with a minute denticulation on the acute edges, usually smooth and often shining; flower-heads 2–$2\frac{1}{2}$ lin. in diam.; calyx of the lower flowers $\frac{2}{3}$–$\frac{3}{4}$ lin. long, $\frac{1}{2}$–$\frac{2}{3}$ lin. broad, coriaceous; tube subquadrate, glabrous; lobes broadly deltoid, obtuse or subacute, usually distinctly ciliate, rarely nearly without cilia; corolla $1\frac{1}{3}$–$1\frac{2}{3}$ lin. long, the exserted part not more than $1\frac{1}{4}$ times as long as the calyx, $\frac{1}{2}$ to nearly 1 lin. broad at the tips of the lobes when flattened, and of nearly equal diameter nearly or quite down to the calyx, then narrowed to the base within the calyx; anthers $\frac{1}{2}$ lin. long, linear-oblong; stigma very slightly thickened. *Rach in Linnæa*, xxvi. 791. *Erica labialis, Salisb. in Trans. Linn. Soc.* vi. 340. *E. capitella, Thunb. Herb. ex Rach in Linnæa, l.c.*

South Africa: without locality, *Thunberg! Masson! Forbes! Mund!*
Coast Region: Cape Div.; mountains near Muizenberg, *Burke!* Smit-winkle Vley, *Wolley Dod*, 3039! mountain near Simonstown, *Bolus*, 7005! *MacOwan & Bolus, Herb. Norm. Austr.-Afr.*, 39!

Thunberg's specimen, named *E. capitella* in his herbarium, is identical with Masson's specimens on which Salisbury founded his *Erica labialis!*

6. S. articulata (N. E. Br); $\frac{1}{2}$–1 ft. high, much branched; branchlets erect or ascending, puberulous or minutely tomentose; leaves 3-nate, imbricate or as long as the internodes, erect or ascending, 1–2 lin. long, linear to ovate, acute, often incurved, glabrous or those under the flower-heads sometimes very minutely puberulous, sometimes minutely ciliate with hairs and sessile glands when young; flower-heads copious, subglobose, $\frac{1}{4}$–$\frac{1}{3}$ in. in diam.; calyx of the lower flowers about $\frac{3}{4}$ lin. long, $\frac{1}{2}$–$\frac{3}{4}$ lin. broad, elliptic, sub-orbicular or rhomboid-obovate; tube glabrous to pubescent; lobes broadly deltoid and subacute or broadly rounded, usually con-spicuously and sometimes very densely ciliate with white hairs; corolla $1\frac{3}{4}$–$2\frac{1}{2}$ lin. long, tubular, the exserted part nearly, or quite twice as long as the calyx, subcylindric or gradually tapering down-wards from the apex, where it is usually $\frac{1}{3}$–$\frac{1}{2}$ (rarely $\frac{2}{3}$) lin. in diam.; anthers $\frac{1}{2}$–$\frac{2}{3}$ lin. long; stigma simple. *S. capitellata, Rach in Linnæa*, xxvi. 721, *not of Lichtenstein. Erica articulata, Thunb. Prodr.* 71, *and Fl. Cap. ed. Schultes*, 357, *not of Linn. Blæria articulata, Willd. Sp. Pl.* i. 629; *Thunb. Diss. Blæria*, 7. *B. bracteata, Wendl. Collect.* ii. 3, *t.* 37.

Var. β, **hians** (N. E. Br.); corolla more funnel-shaped at the upper part, $\frac{3}{4}$–1 lin. in diam. at the tips of the lobes.

6991! *Schlechter*, 9246! Stellenbosch Div.; Hottentots Holland, *Prior!*
Lowrys Pass, *Schlechter*, 7818! Caledon Div.; Donker Hoek Mountain, *Burchell*, 7977! between Villiersdorp and French Hoek, *Bolus*, 5195! Houw Hoek,
Bolus, 9929! Zwart Berg, near Caledon, *Bodkin in Herb. Bolus*, 6961! *Bolus*,
5424! 9930! mountains near Genadendal, *Burchell*, 7801/2! by the Onrust
River, *Schlechter*, 10386! Swellendam Div.; mountains near Swellendam,
Mund, 48! Var. β: Caledon Div.; mountains near Houw Hoek, *Bolus*, 6962!
and in Herb. Guthrie, 3626!

In Burchell, 7801/2, Mund, 48, Bolus, 5195, 5424, 9930, and Schlechter,
7818, the cilia on the calyx-lobes are much shorter and less evident than in the
other specimens, but are much more conspicuous than in any specimen of *S.
capitellata*, and appear to be somewhat intermediate. Schlechter, 9246, as sent
to Kew, consists of 3 forms of this plant, one of which has puberulous leaves.
Sieber (?) 154, is somewhat intermediate between this species and *S. brachyphylla*.

7. S. capitellata (Lichtenstein ex Roem. & Schultes, Syst. Veg.
iii. 171); apparently 1–2 ft. high, densely branched; branchlets
puberulous or minutely tomentose, with or without a mingling of
a longer pubescence; leaves 3-nate, imbricate or as long as the
internodes, erect or ascending, $\frac{2}{3}$–2 lin. long, linear or linear-lanceolate, acute, straight, more rarely lanceolate or subovate and incurved,
glabrous, often with minute sessile glands on the margins when
young; flower-heads copious, subglobose or obovoid, about $\frac{1}{4}$ in. in
diam.; calyx of the lower flowers $\frac{1}{2}$–$\frac{2}{3}$ lin. long and broad, elliptic,
obovate or rhomboid-obovate, thinly coriaceous, glabrous; lobes as
long as the tube in the type, in other forms about $\frac{1}{2}$ as long, rounded
or broadly deltoid and subacute or obtuse, thinly ciliate with minute
hairs often mingled with minute sessile glands or without cilia;
corolla $1\frac{1}{2}$–2 lin. long, tubular, much contracted within the calyx,
the exserted part about twice as long as the calyx or less, sometimes
gradually narrowing downwards but often scarcely broader across the
lobes than elsewhere, purple or red; anthers $\frac{1}{2}$ lin. long, linear,
oblong; stigma simple. *Kunth in Königl. Acad. Wissensch. Berlin*,
1831, 211, *and in Linnæa*, viii. *Liter.-Ber.* 47; *Spreng. Syst. Veg.*
i. 432; *Bartling in Linnæa*, vii. 651; *Klotzsch in Linnæa*, viii.
655, *and* xii. 229; *Benth. in DC. Prodr.* vii. 706, *excl. synonyms
from all.*

VAR. β, **crassistigma** (N. E. Br.); apparently about 5–6 in. high; leaves
sometimes with a few hairs at the tips when young; flower-heads 2–3 lin. in
diam.; bracts 0 or 2, linear-filiform; calyx about $\frac{3}{4}$ lin. long and $\frac{2}{3}$ lin. broad;
lobes about $\frac{1}{3}$ as long as the tube, ciliate with minute sessile glands, without
hairs; stigma thickened or subcapitate, but minute.

VAR. γ, **angustata** (N. E. Br.); flower-heads about 2 lin. in diam.; calyx
about $\frac{3}{4}$ lin. long and broad, obtriangular; lobes about $\frac{1}{3}$ as long as the tube,
ciliate with very minute hairs and sessile glands; corolla funnel-shaped, rapidly
or abruptly narrowed at about the middle into a slender tube, which is partly
exserted beyond the calyx; stigma simple.

COAST REGION: Cape Div.; Cape Flats, *Guthrie*, 1012! Stellenbosch Div.;
Hottentots Holland Mountains, *Zeyher*, 3310! Lowrys Pass, 1000 ft., *Guthrie*,
2002! *Bolus*, 5557! Zwart Berg, near Caledon, *Guthrie*, 2511! mountains near
Zoetemelks Valley, *Lichtenstein!* Var. β: Caledon Div.; Summit Ridge, Zand-

fontein, 1500 ft., *Galpin*, 3721 ! Var. γ: Caledon Div.; Zwart Berg, near Caledon, 2500 ft., *Bodkin in Herb. Bolus*, 6767 !

8. S. tenuiflora (Benth. in DC. Prodr. vii. 706) ; much, but not densely branched ; branchlets tomentose-puberulous ; leaves 3-nate, imbricate or erect, $\frac{3}{4}$–$1\frac{1}{2}$ lin. long, linear to ovate, acute, incurved, glabrous, sometimes when young with some very minute glands on the edges ; flower-heads rather copious, about $\frac{1}{4}$ in. in diam. ; calyx of the lower flowers $\frac{3}{4}$ lin. long, $\frac{2}{5}$–$\frac{1}{2}$ lin. broad, coriaceous ; tube cuneate-oblong, glabrous ; lobes half as long as the tube, or rather more, broadly deltoid, acute, very minutely and inconspicuously ciliate ; corolla about $2\frac{1}{2}$ lin. long, red ; tube cylindric, $\frac{1}{3}$ lin. in diam., the exserted part nearly 3 times as long as the calyx ; lobes gaping, $\frac{1}{2}$–$\frac{2}{3}$ lin. across their tips ; anthers partly exserted, $\frac{2}{3}$ lin. long, linear ; stigma simple.

SOUTH AFRICA : without locality, *Forbes ! Mund !*

This may be only a long-flowered form of *S. capitellata,* but Burchell's specimen, referred to this species by Bentham, belongs to *S. articulata,* and has a long-ciliate differently shaped calyx.

XVI. **LEPTERICA**, N. E. Br.

Bracts 0. *Calyx* more or less unequally 4-lobed. *Corolla* very small, hypogynous, obconic, 4-lobed. *Stamens* 8, hypogynous, included ; filaments connate at the base ; anthers connate. *Ovary* 1-celled, with 1 pendulous ovule ; style stout, soon enlarging and forming a hollow conical top to the young fruit ; stigma large, soon appearing sessile, peltate. *Fruit* ovoid, apparently indehiscent ; pericarp thin, with resin or latex canals in its substance.

A shrub with very copious slender subparallel branchlets ; leaves small, adpressed ; flowers minute, 1–3 together, axillary or terminal.

DISTRIB. Species 1, endemic.

1. L. tenuis (N. E. Br.) ; 3 ft. high, densely much branched ; branchlets very numerous, long and very slender, straight or slightly curved, subparallel, very minutely puberulous or almost glabrous ; leaves 3-nate, usually shorter than the internodes, erect and adpressed or sometimes imbricate, $\frac{2}{3}$–$1\frac{1}{3}$ lin. long, narrowly linear, acute, straight or very slightly spreading at the tips, which are sometimes paler in colour, as if slightly indurated, glabrous ; flowers 1–3 together, axillary or on exceedingly short axillary branchlets and terminal, subsessile ; bracts 0, except the floral leaves ; calyx $\frac{1}{2}$–$\frac{2}{3}$ lin. long, obconic, 4-angled, unequally 4-toothed, glabrous ; teeth erect, broadly ovate, acute, keeled, very minutely subciliate, one of them much larger than the rest, the smaller more than half as long as the tube ; corolla obconic, glabrous ; lobes broader than long, rounded, abruptly inflexed over the margin of the stigma, rigid ; stamens 8, included ; filaments connate into a tube at the base, free above ;

anthers connate, acutely bifid and minutely ciliate at the apex; ovary at first small, shortly obconic, 8-ribbed, truncately contracted into a stout style longer than itself, glabrous, becoming much enlarged and rhomboid-obovate, from the style increasing and forming a broadly conical hollow top to the obconic lower part; stigma large, peltate with upturned margins, slightly 4-angled, becoming convex and smooth at the central part, with a narrow incurved rim, and, upon the enlargement of the ovary, appearing to be subsessile; fruit ellipsoid, with a rather thin scarcely crustaceous pericarp. *Lagenocarpus tenuis, Benth. in DC. Prodr.* vii. 710.

South Africa: without locality, *Mund!*
Coast Region: Riversdale Div.; summit of the Kampsche Berg, *Burchell,* 7126! Garcias Pass, *Burchell,* 7034!

The corolla is probably pink, as Burchell states on his label that the flowers are " roseo-herbacei."

XVIᴀ. **COCCOSPERMA**, Klotzsch.

Bracts 0. *Calyx* small, unequally 4-lobed, one lobe usually longer than the rest, often nearly free. *Corolla* small, campanulate, globose-campanulate or somewhat obovoid, longer than the calyx, shortly 4-lobed; lobes about half as long as the tube, broadly deltoid, obtuse or subtruncate, erect or slightly incurved. *Stamens* 4–6, not exceeding the corolla, except when pushed out by the fruit; filaments at first connate up to and with the base of the anthers, becoming more or less separated by the enlargement of the ovary, and then as broad as the anthers, which are connate at the basal half, opening by lateral pores. *Ovary* 1-celled, with 2 large collateral ovules on one side of the cavity, or 2–3-celled with 2 collateral ovules in each cell on the axile placenta; style short; stigma large, slightly exserted, flat or shallowly funnel-shaped at the bottom, with a narrow or deep and cup-like erect margin. *Fruit* 2-seeded; pericarp thin, closely investing the seeds, subcoriaceous or perhaps slightly fleshy in the living state. *Seeds* flat on one side by mutual pressure, very convex on the other; testa thick, hard, crustaceous, with rather large cells.

Erect much branched shrubs or shrublets, much resembling *Salaxis* in general appearance; leaves 3-nate, linear, grooved on the convex back; flowers axillary and terminal, sessile or very shortly pedicellate, on an extremely short minutely bracteolate branchlet, which, together with the flowers, is very much shorter than the leaf from whose axil it arises. Existing descriptions of the ovary of this genus are inaccurate. The collateral ovules always have their contiguous sides closely applied to each other and often cling together so completely that they appear like a single ovule, but may be readily separated with needles in water. The seeds of *Coccosperma* are unlike those of any other South African genus of *Ericaceæ.*

Distrib. Species 4, endemic.

Ovary 1-celled, with 2 collateral ovules on the side
  of the cell :
    Stamens 6–7 ; ovary glabrous ; fruit minutely
      rugulose all over ...   .   (1) **forbesianum.**
    Stamens 4–6 ; ovary minutely puberulous at the
      top ; fruit with elevated longitudinal lines on
      the upper part   ...   .   (2) **areolatum.**
Ovary 2–3-celled, with 2 collateral ovules in each cell
  on an axile placenta ; stamens 8 :
    Branchlets at first sprinkled with minute shortly
      stalked glands ; flowers 3–6 together in ter-
      minal heads   ...   .   (3) **subcapitatum.**
    Branchlets at first inconspicuously and very
      minutely puberulous or almost glabrous ;
      flowers 1–3 together, mostly axillary, some
      terminal ...   .   (4) **rugosum.**

1. **C. forbesianum** (Klotzsch in Linnæa, xii. 215); branchlets
very minutely and inconspicuously puberulous, greyish ; leaves
erectly imbricate, $1\frac{1}{4}$–$2\frac{1}{2}$ lin. long, linear, obtuse, glabrous ; flowers
axillary and terminal, 3–6 together, mostly aggregated at the ends of
the branchlets, very shortly pedicellate ; calyx (excluding the long
lobe) about $\frac{1}{2}$ lin. long, shorter than the corolla-tube, glabrous ; lobes
lanceolate or oblong, minutely ciliate ; corolla $\frac{2}{3}$–$\frac{3}{4}$ lin. long, campanulate or globose-campanulate, glabrous ; stamens 6–7 ; filaments
connate up to and with the anthers nearly or quite as long as the
$\frac{1}{4}$–$\frac{1}{3}$ lin. long subquadrate connate anthers, usually breaking off in
irregular lengths in fruit ; ovary glabrous, 1-celled ; style very
distinct but less than $\frac{1}{8}$ lin. long ; stigma peltate, flat or intruded,
with a narrow erect cup-like margin above and a slightly prominent
ring around the underside, glabrous ; fruit globose, irregularly transversely rugulose, glabrous ; seeds 2. *Salaxis hexandra, Klotzsch in
Linnæa,* ix. 352 ; *Benth. in DC. Prodr.* vii. 711 ; *Dietr. Synop.
Pl.* ii. 1261.

Sᴏᴜᴛʜ Aғʀɪᴄᴀ : without locality, *Forbes ! Wright,* 169!

2. **C. areolatum** (N. E. Br.) ; branchlets erect or ascending,
flexuose, puberulous, pale greyish ; leaves adpressed or erectly imbricate, 1–2 lin. long, linear, acute, glabrous ; flowers axillary, 1–2-
together, sessile ; calyx (excluding the long lobe) $\frac{1}{3}$–$\frac{1}{2}$ lin. long,
shorter than the corolla-tube, broad and shallow, glabrous ; lobes
broadly ovate or deltoid-ovate, acute, minutely ciliate ; corolla about
$\frac{3}{4}$ lin. long, cup-shaped or campanulate, glabrous ; stamens 4–6 ;
filaments about as long as or shorter than the oblong or subquadrate
$\frac{1}{3}$ lin. long anthers, and connate up to and with them ; anthers
connate ; ovary subglobose, usually minutely puberulous at the
top, abruptly contracted into the short glabrous style, 1 celled ;
stigma just exserted, cup-shaped, flat or intruded at the bottom,
with erect sides ; fruit ovoid, ellipsoid or more or less compressed,
obtuse, the upper half more or less evidently divided into areas by
several longitudinal elevated lines, which are usually minutely

puberulous; the lower part nearly smooth or slightly rugulose; seeds 2.

Coast Region: Clanwilliam Div.; Koude Berg, 4000 ft., *Schlechter*, 8757! 8772! Tulbagh Div.; on the Witsen Berg, near Tulbagh, *Burchell*, 8710!

3. **C. subcapitatum** (N. E. Br.); compactly branched; branchlets at first sprinkled with minute very shortly stalked glands, soon becoming glabrous and pale greyish; leaves 3-nate, erectly imbricate or rarely shorter than the internodes, $1-1\frac{3}{4}$ lin. long, linear, acute, glabrous, ciliate with sessile glands; flowers 3–6 together in small terminal heads, subsessile; calyx broad and shallow, glabrous, about $\frac{1}{2}$ lin. long to the tips of the short lobes, which are shorter than the corolla-tube and united for nearly half their length, the long lobe free to the base or united as high up as the others, all edged with minute hairs and sessile glands; corolla about $\frac{2}{3}$ lin. long and as much or more in breadth, cup-shaped, glabrous; stamens 8; filaments from rather more than half to nearly as long as the anthers, connate up to and with the base of them; anthers $\frac{1}{3}$ lin. long, broadly-oblong to subquadrate; ovary compressed, broader than long, glabrous, whitish, with a minutely rugulose surface, 2-celled; stigma flattened funnel-shaped, with a narrow erect or inflexed rim above and a slightly prominent margin beneath, glabrous, abruptly tapering to a very short style which is sometimes almost wanting.

Coast Region: Cape Div.; on the Steenberg Plateau, near Muizen Berg, *Wolley Dod*, 2723!

4. **C. rugosum** (Klotzsch in Linnæa, xii. 215); apparently 1 ft. or more high, rather thinly branched; branchlets, erect or ascending, slender, more or less flexuose, at first very minutely and inconspicuously puberulous or nearly or quite glabrous, whitish or pale grey; leaves 3-nate, adpressed and shorter than the internodes or erectly imbricate, $1\frac{1}{4}-2\frac{1}{2}$ lin. long, narrowly linear, acute, glabrous; flowers 1–3 together, axillary and terminal, very shortly pedicellate or subsessile; calyx (excluding the larger lobe) about $\frac{1}{2}$ lin. long, shorter than the corolla-tube, broad and shallow, unequally 4-lobed, glabrous; lobes broadly ovate, subobtuse or the larger one acute or acuminate, very minutely subdenticulate-ciliate or entire; corolla about $\frac{1}{2}$ lin. long, campanulate, glabrous; stamens 8; filaments connate up to and with the base of the anthers, becoming more or less separated by the enlargement of the ovary, $\frac{1}{3}-\frac{2}{3}$ as long as the $\frac{1}{5}-\frac{2}{5}$ lin. long anthers; ovary with several longitudinal ridges on the upper part, very obtuse at the apex, glabrous, 2–3-celled; style very short; stigma broadly funnel-shaped with erectly inflexed sides to the cup; fruit (not seen ripe) subglobose, wrinkled on the lower part. *Salaxis rugosa, Benth. in DC. Prodr.* vii. 711; *Dietr. Synop. Pl.* ii. 1261.

South Africa: without locality, *Zeyher!*

COAST REGION: Bredasdorp Div.; Mier Kraal, 150 ft., *Schlechter,* 10517 !

Zeyher's specimen at Kew (named by Klotzsch) is not localized, but on the label Klotzsch has written the locality-number 84 of Zeyher, which according to Linnæa, xx. 258, refers to the locality "mountains near Cape Town." But I suspect some error, as no recent collector has found it there, and Schlechter's specimens, which seem to be identical with Zeyher's, are from quite a different locality.

XVII. SALAXIS, Salisb.

Bracts 0. *Calyx* more or less unequally 4-lobed, the larger lobe variably connate with the others or free nearly or quite to the base. *Corolla* small, hypogynous, campanulate or cup-shaped, 4-lobed, persistent; lobes about half as long as the tube, incurved, broadly deltoid or deltoid-ovate, very obtuse to slightly emarginate at the subtruncate apex. Hypogynous *disk* none. *Stamens* usually 8, rarely 6, hypogynous, included or in fruit shortly pushed out by the enlarged ovary; filaments variably connate, sometimes up to, but not with the base of the anthers, becoming more or less separated with the enlargement of the ovary and then (except in *S. pumila*) always narrower at the apex than the base of the anthers, glabrous; anthers connate to the middle or beyond, sometimes becoming free in fruit, oblong or cuneately subquadrate, notched at the apex, opening by lateral pores, spurless. *Ovary* superior, generally angular at the middle viewed sideways, and the upper part often 3–4-angled, 2–4-celled; style distinct to almost none; stigma large, funnel-shaped, with the sides of the upper part of the cup erect or inflexed. *Ovule* solitary in each cell, pendulous. *Fruit* apparently indehiscent, compressed to subglobose, often more or less distinctly 3–4-angled on the upper half, 2–4-celled; pericarp not very thick, with a crustaceous or hardened endocarp. *Seeds* solitary in each cell, with a hard or somewhat bony testa.

Small Heath-like shrubs or shrublets; leaves 3-nate, grooved down the convex back, usually thinly ciliate with very minute hairs or points; flowers very small and inconspicuous, axillary and terminal, apparently greenish.

DISTRIB. Species 7, endemic.

The species of *Salaxis* so closely resemble others belonging to the genera *Philippia, Scyphogyne, Coccosperma* and *Lagenocarpus,* that they can only be distinguished by a careful examination of the ovary.

Calyx and corolla puberulous; stamens 8   ...   ...   **(1) puberula.**
Calyx and corolla glabrous:
   Plant (excluding the root) 3–5 in. high; bark of
     each internode quickly breaking up into 3 small
     oblong or rectangular flakes; stamens 6   ...   **(2) pumila.**
   Plant rarely less than 9 in. high; bark of each
     internode not breaking up into 3 flakes:
      Calyx (excluding the larger lobe) ¾ lin. long,
       longer than broad, about equalling the tube
       of the 1 lin. long corolla; stamens 8,
       rarely 6   ...   .   **(3) major.**
      Calyx (excluding the larger lobe) ⅓–½ lin.

long, broader than long; corolla $\frac{1}{2}$–$\frac{2}{3}$ lin.
long; stamens 8 or occasionally 6 in *S.
axillaris*:
    Upper part of the ovary and underside of
      the stigma quite glabrous :
        Ovary with several longitudinal ridges
          on the upper part, which are but
          faint on the compressed or obscurely
          3-4-angled fruit ... ... (4) **axillaris.**
        Ovary acutely and fruit distinctly
          3-4-angled on the upper part ... (5) **octandra.**
    Upper part of the ovary very minutely and
      not very distinctly scabrid-puberulous ;
      stigma glabrous ... ... (6) **Sieberi.**
    Upper part of the ovary distinctly puberu-
      lous; stigma glabrous or in var. β,
      puberulous on the underside ... ... (7) **flexuosa.**

1. S. puberula (Klotzsch in Linnæa, xii. 212); apparently about
1 ft. high or less, compactly branched ; branchlets erect, flexuose,
puberulous; leaves 3-nate, usually erect or imbricate, spreading
where there are flowers or young shoots in their axils, $1\frac{1}{4}$–$2\frac{1}{4}$ lin.
long, linear, acute, glabrous; flowers 1–3 together, axillary and
terminal, subsessile ; calyx (excluding the larger lobe) $\frac{1}{3}$–$\frac{1}{2}$ lin. long,
shorter than the corolla-tube, lobed to half-way down, puberulous;
corolla $\frac{1}{2}$–$\frac{3}{4}$ lin. long, cup-shaped, puberulous, green; stamens 8 ;
filaments $\frac{1}{2}$–$\frac{2}{3}$ as long as the anthers, sometimes broken off shorter in
fruit, at first more or less connate to half-way up or beyond, becom-
ing separated in fruit and narrower at the apex than the base of the
$\frac{1}{4}$–$\frac{2}{5}$ lin. long ovate-oblong or oblong anthers ; ovary distinctly
puberulous, 2-celled ; style very short ; stigma with a funnel-shaped
base and erect sides to the cup, $\frac{1}{3}$ as deep as its breadth, puberulous
on the underside ; fruit somewhat compressed or subglobose, puberulous on the upper half, slightly reticulately wrinkled below. *Benth.
in DC. Prodr.* vii. 711 ; *Dietr. Synop. Pl.* ii. 1261.

CoAST REGION : Caledon Div.; mountains near the Palmiet River or Steenbrass River, *Ecklon & Zeyher!* Hermanus, 300 ft , *Guthrie*, 4139 !

2. S. pumila (N. E. Br.); plant (excluding the root) 3–5 in.
high and 2–3 in. in diam.; branchlets short, spreading at the base,
then curved upwards, rather stout for the size of the plant, glabrous,
greyish, becoming rough from the bark breaking up into small
rectangular flakes ; leaves 3-nate, erectly imbricate, $\frac{3}{4}$–$1\frac{1}{2}$ lin. long,
oblong or linear-oblong, subobtuse, glabrous; flowers 1–3 together
in small terminal heads, subsessile ; calyx about $\frac{1}{2}$ lin. long, broadly
and shallowly obconic, as long as the corolla-tube, lobed to half-way
down, glabrous; lobes ovate, obtuse · or the larger one acute, not
ciliate ; corolla rather more than $\frac{1}{2}$ lin. long, $\frac{3}{4}$–1 lin. in diam.,
broadly cup-shaped, glabrous; stamens 6 ; filaments very broad,
connate below or almost up to the anthers and at least half as long
as them ; anthers about $\frac{1}{3}$ lin. long and broad, cuneately subquadrate ;
ovary broader than long, angular at the middle and with 2–4 rather

sharply angular ridges on the upper half, glabrous, 2-celled, with very thick walls; style very short; stigma nearly flat or very shallowly crater-like, with a narrow inflexed margin, glabrous; fruit subglobose, irregularly wrinkled all over except at the apex.

COAST REGION : Bredasdorp Div.; Zeekoe Vley, 100 ft., *Schlechter*, 10540!

3. S. major (N. E. Br.); 1 ft. or more high, densely or compactly branched; branchlets flexuose, erect, minutely puberulous, becoming glabrous, grey; leaves 3-nate, $1\frac{1}{2}$–$2\frac{1}{2}$ lin. long, erectly imbricate to shorter than the internodes, linear, acute; flowers 1–3 together, axillary and terminal, subsessile or very minutely pedicellate; calyx (excluding the larger lobe) about $\frac{3}{4}$ lin. long, obconic, longer than broad, usually as long as the corolla-tube, lobed to half-way down or less, or the larger one more deeply glabrous; lobes linear-oblong, obtuse or acute, ciliate with minute sessile glands; corolla $\frac{3}{4}$–1 lin. long, campanulate, glabrous; stamens 8, rarely 6; filaments quite half as long as the anthers, at maturity connate to about the middle; anthers $\frac{1}{2}$ lin. long, oblong; ovary minutely and rather thickly puberulous, 2-celled; style $\frac{1}{6}$–$\frac{1}{4}$ lin. long; stigma funnel-shaped or cup-shaped, about as broad as deep, slightly and minutely puberulous or glabrous on the under side.

COAST REGION : Cape Div.; hills about Simons Bay, *Prior*, 3!

4. S. axillaris (Salisb. ex G. Don, Gen. Syst. iii. 828); much branched; branchlets minutely tomentose, erect, often flexuose, greyish or whitish; leaves erectly imbricate or shorter than the internodes, $1\frac{1}{4}$–$2\frac{1}{4}$ lin. long, linear, acute to subobtuse; flowers 1–3 together, axillary and terminal, subsessile; calyx (excluding the larger lobe, which is sometimes free to the base) $\frac{1}{3}$–$\frac{1}{2}$ lin. long, broader than long, shorter than the corolla-tube, lobed to the middle or beyond, glabrous; lobes broadly ovate, acute or subobtuse or the larger one acuminate, minutely ciliate; corolla $\frac{1}{2}$–$\frac{2}{3}$ lin. long, campanulate, with incurved lobes, glabrous; stamens 6–8; filaments at least half as long as the anthers, connate at the base or nearly to the middle when young, becoming free with the enlargement of the ovary; anthers $\frac{1}{4}$–$\frac{1}{3}$ lin. long, becoming quite free with age; ovary with several longitudinal ridges on the upper part, quite glabrous, 2-4-celled; style usually distinct and sometimes up to $\frac{1}{6}$ lin. long; stigma more or less funnel-shaped, with a narrow rim to the cup, glabrous; fruit compressed or subglobose, obscurely angular on the upper part, but often with several faint longitudinal ridges, prominently reticulated on the lower part, 2-4-celled. *Klotzsch in Linnæa*, xii. 211; *Benth. in DC. Prodr.* vii. 711; *Dietr. Synop. Pl.* ii. 1261; *Rach in Linnæa*, xxvi. 792. *Erica axillaris, Thunb. Prodr. Pl. Cap.* 69; *Diss. Erica*, 16; *and Fl. Cap. ed. Schultes*, 345.

SOUTH AFRICA : without locality, *Thunberg! Harvey! Ecklon & Zeyher!*
COAST REGION : Stellenbosch Div.; between Stellenbosch and Cape Flats, *Burchell*, 8354! Knysna Div.; hills near Knysna, *Burchell*, 5456!

Burchell notes on his label that the Knysna plant grows to 4 ft. high. The longitudinal ridges on the ovary, generally 8 or more in number, are very characteristic of this species, but they are only faintly marked on the fruit.

5. S. octandra (Klotzsch in Linnæa, ix. 353); apparently about 1 ft. high; branchlets slender, erect, subparallel to very flexuose, distinctly or very minutely puberulous, sometimes becoming glabrous, pale brown or greyish; leaves 3-nate, varying from adpressed and shorter than the internodes to erectly imbricate, $\frac{3}{4}$–$3\frac{1}{2}$ lin. long, narrowly linear, acute, glabrous; flowers axillary and terminal, 1–3 (rarely 4–5) together, subsessile; calyx (excluding the long lobe) $\frac{1}{3}$–$\frac{1}{2}$ lin. long, shorter than the corolla-tube, unequally lobed to the middle or the long lobe nearly free, glabrous; lobes oblong or ovate, acute or subobtuse or the long one acuminate, all minutely ciliate; corolla about $\frac{2}{3}$ lin. long, campanulate or subglobose-campanulate, glabrous; stamens 8; filaments about $\frac{1}{2}$ as long as the anthers, rarely less, connate for $\frac{1}{3}$–$\frac{3}{4}$ of their length and sometimes up to (but not with) the base of the anthers, becoming more or less free in fruit and narrower at their apex than the base of the $\frac{1}{3}$ to nearly $\frac{1}{2}$ lin. long oblong anthers; ovary acutely 2–4-angled on the upper half, 2–4-celled, glabrous; style very short to almost none; stigma deeply funnel-shaped with erect or slightly inflexed sides to the cup, glabrous; fruit distinctly 3–4- (occasionally 2-) angled on the smooth hardened upper half, slightly reticulate-rugulose below. *Klotzsch in Linnæa,* xii. 212; *Benth. in DC. Prodr.* vii. 711; *Dietr. Synop. Pl.* ii. 1261.

VAR. β, **artemisioides** (N. E. Br.); flowers very shortly pedicellate; calyx (excluding the long lobe) nearly or quite as long as the corolla tube; corolla not quite $\frac{1}{2}$ lin. long, obconic-cup-shaped; staminal filaments at first connate almost up to the anthers; anthers $\frac{1}{4}$ lin. long; otherwise as in the type. *S. artemisioides, Klotzsch in Linnæa,* xii. 212; *Benth. in DC. Prodr.* vii. 711; *Dietr. Synop. Pl.* ii. 1261. *Erica artemisioides, E. Mey. ex Klotzsch l.c.*

SOUTH AFRICA : without locality, *Mund & Maire,* ex *Klotzsch.*

COAST REGION : Worcester Div.; mountains of Bains Kloof, 4000 ft., *Schlechter,* 9127! Caledon Div.; mountain near Greitjes Gat, 2000–4000 ft., *Ecklon & Zeyher!* Bredasdorp Div.; sand-dunes near the mouth of Ratel River, 50 ft., *Bolus,* 6770! Var. β : Worcester Div.; "Dietris Kloof" an error for Dutoits Kloof, *Drège,* 1168!

Drège has evidently mixed up three plants under the name of *Erica artemisioides,* for the two specimens at Kew, distributed under that name, are quite distinct from the type of *Salaxis artemisioides* at Berlin, and appear to belong respectively to *S. Sieberi* and *S. axillaris.* Bentham in DC. Prodr. vii. 711 referred these two specimens to *S. hexandra (Coccosperma forbesianum),* from which, however, they are entirely different.

6. S. Sieberi (Benth. in DC. Prodr. vii. 711); 1 ft. or more high, much branched; branchlets divergently ascending, minutely subtomentose; leaves 3-nate, erectly imbricate to adpressed and shorter than the internodes, 1–2$\frac{1}{4}$ lin. long, linear, acute, glabrous; flowers axillary and terminal, 1–5 together, subsessile; calyx (excluding the larger lobe which is variably connate with the others or free almost to the base) $\frac{1}{3}$–$\frac{1}{2}$ lin. long, shorter than the corolla-tube, glabrous; lobes ovate or deltoid-ovate, acute or subobtuse or the larger one acuminate, minutely ciliate; corolla $\frac{1}{2}$–$\frac{3}{4}$ lin. long, campanulate or cup-shaped, with incurved lobes, glabrous; stamens 8; filaments of the long-styled form less than half as long as the $\frac{1}{2}$ lin. long anthers, and of the short-styled form half as long as the $\frac{1}{3}$ lin. long anthers, varying from shortly connate at their base to almost up to (but not with) the base of the anthers, becoming more or less separated in fruit; anthers oblong, usually very minutely scabrid on the apical notch, not glandular as originally described; ovary very minutely scabrid-puberulous on the ridges of the upper part, 2–3-celled; style very short or distinct and up to $\frac{1}{6}$ lin. long, stigma funnel-shaped, with the erect sides of the cup, about $\frac{1}{3}$ of its diam., glabrous or nearly so; fruit usually distinctly trigonous, with the 3 angles or ridges on the upper part subacute, slightly but not prominently reticulate on the lower part. *Dietr. Synop. Pl.* ii. 1261. *Erica calyciflora, Tausch in Flora,* xvii. ii. 617.

SOUTH AFRICA: without locality, *Sieber*, 176! *Harvey*, 155!

COAST REGION: Cape Div.; Cape Flats, 0–100 ft., *Bolus*, 3313! *Guthrie*, 1379! Wynberg, 80 ft., *Schlechter*, 7536!

7. **S. flexuosa** (Klotzsch in Linnæa, xii. 213); apparently laxly branched; branchlets divaricate at the base, then ascending, flexuose, rather distant, very minutely puberulous, very pale greyish-brown; leaves 3-nate, erect, shorter than or slightly exceeding the internodes and then slightly imbricate, $1\frac{1}{2}$–2 lin. long, linear, acute, glabrous; flowers 1–3 together, axillary and terminal, sessile or subsessile; calyx (excluding the larger lobe) scarcely $\frac{1}{2}$ lin. long, broadly and shallowly cup-shaped, shorter than the corolla-tube, usually lobed to about half-way down, glabrous; lobes very broadly deltoid-ovate, mostly much broader than long, subacute or the larger one shortly acuminate, minutely ciliate; corolla nearly $\frac{2}{3}$ lin. long, cup-shaped, with incurved lobes, glabrous; lobes scarcely half as long as the tube, broadly deltoid, truncate or very obtuse; stamens 8; filaments at first connate almost or quite up to the anthers, afterwards becoming more or less separated by the enlargement of the ovary, about $\frac{1}{4}$ as long as the $\frac{1}{2}$ lin. long anthers; ovary 2–3-celled, very minutely puberulous on the upper half; style very short, scarcely any; stigma large, deeply funnel-shaped, with a narrow inflexed margin, glabrous; fruit somewhat compressed or subglobose-trigonous, smooth, minutely puberulous on the upper half. *Benth. in DC. Prodr.* vii. 711; *Dietr. Synop. Pl.* ii. 1261.

VAR. β, **cognata** (N. E. Br.); compactly or rather densely branched; branchlets erect, often flexuose, minutely tomentose, puberulous or subglabrous, pale brown or greyish; leaves usually erectly imbricate, sometimes crowded, 1–$2\frac{1}{2}$ lin. long; calyx $\frac{1}{2}$ lin. long; stamens 8; filaments $\frac{1}{4}$–$\frac{2}{3}$ as long as the anthers, variably connate, becoming more or less separated by the enlargement of the ovary; anthers $\frac{1}{3}$–$\frac{1}{2}$ lin. long; ovary conspicuously but minutely puberulous on the upper part; style almost none or short and distinct; stigma distinctly puberulous on the under side, with the inflexed sides of the cup $\frac{1}{4}$–$\frac{1}{3}$ of its breadth; otherwise as in the type.

COAST REGION : Tulbagh Div. ; collected either in New Kloof, near Tulbagh, on Winterhoek Mountain, on the Witsenberg Range or near Vogel Valley, *Ecklon & Zeyher; and Ecklon & Zeyher*, 308 partly ! Var. β : Cape Div. ; foot of Muizen Berg, near Fish Hoek, 100 ft., *Bolus*, 4478 ! mountains near Simons Town, 1700 ft., *Bolus*, 4684 ! Caledon Div. ; Houw Hoek, 1500 ft., *Schlechter*, 7327 ! Hermanus, *Guthrie*, 3420 ! Bredasdorp Div. ; near Elim, 150–800 ft., *Bolus*, 6771 ! *Schlechter*, 7643 !

The description of *S. flexuosa* is made from the type in the Berlin Herbarium, which may be only a weak laxly branched specimen of the form I have considered as a variety of it ; if so, the latter is evidently the more typical state of the plant, although in general appearance very different. A specimen from the Cape Flats (Bolus, 4010) seems to belong to var. β, but the style is long and the stigma is glabrous on the underside.

XVIII. **SCYPHOGYNE**, Brongn.

Flowers ebracteate. *Calyx* obconic or campanulate, often angular, unequally or rarely equally 3–4-toothed or 3–4-lobed, not fleshy. *Corolla* minute, hypogynous, globose, obconic, campanulate or urceolate, 3–4-lobed or toothed. *Stamens* 3–4, rarely 5, hypogynous; filaments free or connate ; anthers included or partly exserted, notched at the apex, free or connate. *Ovary* superior, 1-celled, with a solitary pendulous ovule ; style equalling the anthers or exserted beyond them ; stigma large peltate or crater-like.

Shrubs or shrublets, with the habit and foliage of *Erica;* flowers minute, axillary and terminal, 1–3 together, or several in small globose heads, sessile or minutely pedicellate, the corolla always less than 1 lin. long.

DISTRIB. Species 17, endemic.

I. Stamens 4, or varying in *S. glandulifera* and *S. remota* up to 6, and in *S. micrantha* up to 8 ; calyx and corolla 4-lobed :
    * Filaments of the stamens slender, free at maturity ; anthers free or connate ; stigma just exceeding the anthers :
        Back of the leaves beset with short gland-tipped hairs ...   . (2) **Schlechteri.**
        Back of the leaves without gland-tipped hairs :
            Flowers axillary and terminal, more or less mixed with the leaves, not in distinct globose heads :
                Corolla usually much longer than the calyx ; leaves linear :
                    Corolla hyaline, drying white and usually very conspicuous; stamens 4   ...   ... (1) **inconspicua**
                    Corolla not hyaline nor drying white, usually inconspicuous ; stamens 4–8...   ...   ... (13) **micrantha.**
            Corolla scarcely longer than the calyx, drying brown, not conspicuous (occasionally also in *S. inconspicua*) ; leaves very small, ovate, oblong or lanceolate :

Young leaves glabrous on
the back, minutely ciliate;
flower to apex of stigma
$\frac{2}{3}-\frac{3}{4}$ lin. long ... ... (8) **eglandulosa.**
Young leaves puberulous on
the back: flower to apex of
stigma scarcely $\frac{1}{2}$ lin. long ... (9) **fasciculata.**
Flowers in small terminal globose heads,
distinct from the leaves, drying
brown or blackish:
Leaves mostly adpressed, denticu-
late, without gland-tipped hairs . (11) **capitata.**
Leaves spreading, ciliate with 3–5
gland-tipped hairs on each side... (12) **viscida.**
** Filaments of the stamens more or less connate
at maturity:
Filaments connate for nearly half their
length; leaves not ciliate ... ... (10) **remota.**
Filaments connate up to the anthers:
Stigma at maturity just exceeding the
anthers; style $\frac{1}{6}-\frac{1}{4}$ lin. long, not or
scarcely exserted:
Leaves ciliate with 4–7 simple or
branched hairs on each margin;
calyx $\frac{1}{2}$ lin. long... ... ... (3) **rigidula.**
Leaves serrate or ciliate with 2–4
gland-tipped hairs on each margin; calyx $\frac{1}{3}$ lin. long:
Leaves flat above, very convex
on the back, glabrous;
corolla glabrous or minutely
puberulous on the lobes ... (4) **divaricata.**
Leaves nearly equally convex
on both sides, puberulous;
corolla very pubescent on
the upper half ... ... (5) **biconvexa.**
Stigma and part of the $\frac{1}{2}-\frac{7}{8}$ lin. long
style exserted much beyond the
anthers:
Calyx puberulous, ciliate on all the
lobes with long gland-tipped
hairs; style puberulous ... ... (6) **glandulifera.**
Calyx glabrous, 1 or 2 lobes ending
in a long gland-tipped hair;
style glabrous ... ... (7) **longistyla,**
II. Stamens 3; filaments free:
Calyx and corolla 3-lobed; flowers sessile or
subsessile, $\frac{1}{2}-\frac{2}{3}$ lin. long ... ... (14) **trimera.**
Calyx and corolla 4-lobed:
Flowers with distinct pedicels $\frac{1}{8}$ lin. long;
corolla urceolate:
Corolla $\frac{2}{3}-\frac{3}{4}$ lin. long, globose-inflated at
the middle, contracted below the lobes (15) **urceolata.**
Corolla $\frac{1}{2}$ lin. long, ovoid-inflated above,
not contracted below the lobes ... (16) **Burchellii.**
Flowers sessile or subsessile, $\frac{1}{3}$ lin. long;
corolla globose ... ... (17) **puberula.**

1. S. inconspicua (Brongn. in Duperrey, Voy. Coquille, Atlas, t.
54); densely much branched; branches clothed with a soft spreading

viscid pubescence or sometimes puberulous; leaves 3-nate, erect or spreading, $\frac{1}{2}$–$1\frac{1}{2}$ lin. long including the petiole, $\frac{1}{5}$–$\frac{1}{3}$ lin. broad, linear, often with a sessile gland at the obtuse apex, glabrous or with a slight pubescence on the upper side, often viscid, with or without a minute ciliation or a few long more or less gland-tipped hairs on the margins; flowers axillary or in clusters of 3–6 at the tips of exceedingly short (rarely elongated) axillary branchlets, sub-sessile or minutely pedicellate; calyx $\frac{1}{4}$–$\frac{2}{5}$ lin. long, obconic, un-equally or subequally 4-toothed, 4-angled, pubescent; teeth about $\frac{1}{2}$ as long as the tube, erect, ovate, obtuse or acute, sometimes tipped with a sessile gland, ciliate, one usually larger or smaller than the rest; corolla $\frac{2}{5}$–$\frac{3}{5}$ lin. long, obconic, hyaline, quite glabrous; lobes subquadrate or rounded, subtruncate, minutely erose or denticu-late, but not ciliate, white; stamens 4, free at maturity, all connate in young flowers, very deciduous; filamemts $\frac{1}{8}$ lin. long, glabrous; anthers half-exserted, $\frac{1}{2}$ lin. long, $\frac{1}{4}$ lin. broad at the broadly notched apex, cuneate, rarely with, usually without, minute subsessile glands along the margin of the notch; ovary sessile, ovoid, pubescent round the top; style $\frac{1}{4}$ lin. long, glabrous, terminated by a large broadly funnel-shaped or saucer-shaped stigma, equalling or just exceeding the anthers, $\frac{1}{3}$–$\frac{2}{5}$ lin. in diam., deciduous. *Benth. in DC. Prodr.* vii. 709. *S. muscosa, Steud. Nom. ed.* 2, i. 568 *under Erica albens, Thunb. Erica albens, Thunb. Prodr. Cap.* 70, *not of Linn. E. albida, Thunb. Fl. Cap. ed. Schultes,* 347 *partly (sheet a, Herb. Thunb. !). Omphalocaryon muscosum, Klotzsch in Linnæa,* xii. 243; *Rach in Linnæa,* xxvi. 792. *Blæria muscosa, Ait. Hort. Kew. ed.* 1. i. 150, *and ed.* 2, i. 249; *Willd. Sp. Pl.* i. 630; *Roem. & Schultes, Syst. Veg.* iii. 169; *Dietr. Synop. Pl.* i. 444; *Klotzsch in Linnæa,* viii. 665. *B. mucosa, G. Don, Gen. Syst.* iii. 804. *B. albida, Thunb. Diss. Blæria,* 9. *B. pusilla, Wendl. Collect.* iii. 13, *t.* 79.

VAR. β, **glabriflora** (N. E. Br.); calyx glabrous, minutely ciliate on the lobes; corolla-lobes erose-denticulate or subentire, not ciliate; otherwise as in the type. *Omphalocaryon muscosum,* var. *glabrum,* Klotzsch in Linnæa, xii. 244.

VAR. γ, **ciliata** (N. E. Br.); calyx glabrous or more or less pubescent on the angles, distinctly ciliate on the lobes, corolla-lobes ciliate; otherwise as in the type. *Erica albida, Thunb. Fl. Cap. ed. Schultes,* 347 *partly* (sheet β Herb. Thunberg !)

VAR. δ, **pubescens** (N. E. Br.); leaves glabrous or puberulous all over; calyx distinctly pubescent all over, not sharply angulate, distinctly ciliate on the lobes: corolla-lobes ciliate; otherwise as in the type.

VAR. ε, **vestita** (N. E. Br.); calyx and the upper part of the corolla outside densely pubescent; corolla-lobes ciliate; otherwise as in the type.

SOUTH AFRICA : without locality; Var. γ: *Sieber,* 179 ! *Zeyher,* 1121 !

COAST REGION, from 500 to 5500 ft. : Cape Div. ; on mountains, *Thunberg ! Burchell,* 617 ! *Alexander,* 23 ! *Bolus,* 3070 ! 4685 ! *Wolley Dod,* 893 ! 1008 ! 3216 ! Caledon Div. ; Zwart Berg at Sandfontein, *Schlechter,* 10359 ! mountains near Palmiet River, *Guthrie,* 4577 ! Bredasdorp Div. ; Elim, *Schlechter,* 7663 partly ! Swellendam Div. ; Zuurbraak Mountain, *Galpin,* 3728 ! Var. β : Cape Div. ; Muizenberg Mountain, *Bolus,* 4424 ! George Div. ; by the Maal-

gaten River at Wolve Drift, *Burchell*, 6113! Var. γ : Cape Div. ; Table Moun-
tain, *Thunberg*! *Drège*! Cape Flats, *Burchell*, 162 ! 192! *Alexander*, 2! *Bolus*,
Herb. Norm. Austr.-Afr., 47 ! Caledon Div. ; near Caledon, *Pappe!* Volgelgat,
Schlechter, 9545 ! near Steenbrass River, *Schlechter*, 5391 ! Var. δ : Clanwilliam Div. ; Ezelsbauk, *Schlechter*, 8828 ! Cape Div. ; Kasteel Mountain,
Ecklon, 296! Caledon Div. ; near Steenbrass River, *Guthrie*, 2725 ! Bredasdorp
Div. ; Elim, *Schlechter*, 7663 partly ! Riversdale Div. ; lower part of the mountains at Garcias Pass, *Burchell*, 6928! Var. ε : Ceres Div.; near Ceres, *Bolus*,
8483 !

Central Region, between 4500 and 5000 ft. ; Var. δ : Ceres Div. ; Cold
Bokkeveld, *Schlechter*, 8927 ! 10059 !

This appears to be a very variable plant, all the organs differing in relative
size ; in Schlechter's 5391 and 9545 the corolla scarcely exceeds the calyx, in
others it is twice as long, and there are all intermediate stages ; the corolla,
apart from the cilia, is glabrous in all the specimens quoted except var. *vestita*,
which is densely pubescent outside, and some specimens of Zeyher, 1121, which
have a few hairs on the back of the lobes. A specimen named "*muscosa*, Soland,"
from Salisbury's Herbarium is identical with the type of the species, but one
from the Herbarium of Bishop Goodenough labelled "*Blæria muscosa*, Kew
Gardens," and may possibly be from Aiton's type, has a glabrous calyx and a
ciliate corolla, and belongs to var. *ciliata*.

2. S. Schlechteri (N. E. Br.) ; branchlets glandular-puberulous,
brownish ; leaves 3-nate, ½–¾ lin. long including the petiole, ¼ lin.
broad, oblong-linear, obtuse, flattish and glabrous above, very convex, grooved, and sprinkled with very short gland-tipped hairs on
the back ; flowers solitary or clustered on the exceedingly short
axillary branchlets, sessile ; calyx about ¼ lin. long, obconic, unequally 4-toothed, glabrous or very slightly pubescent, ciliate with
simple hairs and the larger tooth often glandular ; corolla about
⅓ lin. long, obconic or funnel-shaped, membranous, with 4 short
broad subtruncate or very obtuse lobes, glabrous, ciliate ; stamens 4,
free ; anthers partly exserted, broadly obtriangular, deciduous ;
ovary angular and with a ring of minute hairs around the middle ;
style short, deciduous ; stigma large, peltate scarcely exceeding the
anthers.

Central Region : Ceres Div. ; near Sand River in the Cold Bokkeveld,
Schlechter, 10111!

3. S. rigidula (N. E. Br.) ; branchlets clothed with a minute
white tomentum intermingled with longer hairs ; leaves 3-nate,
½–1 lin. long including the petiole, ⅓–⅖ lin. broad, ovate, subacute,
flat above, very convex and grooved on the back, very minutely
puberulous, serrate-ciliate with 4–5 long simple or branched hairs on
each side and often with 2 rows on the back, with 1 at the apex,
which is often gland-tipped ; flowers 1–3 together, terminal and
axillary, sessile ; calyx ½ lin. long, obconic-campanulate, unequally
4-lobed, glabrous, ciliate with branched hairs ; corolla ⅔ lin. long,
obconic, 4-angled, with 4 broad apiculate lobes, glabrous ; stamens
4, slightly exceeding the corolla, connate ; ovary rhomboid-ovate,
thinly puberulous on the upper part and on the ¼ lin. long style

and underside of the large crater-like stigma, which scarcely exceeds the anthers.

VAR. β, **breviciliata** (N. E. Br.); leaves $\frac{1}{2}$–$\frac{3}{4}$ lin. long, broadly ovate, less evidently puberulous than in the type, serrate-ciliate with 5–7 very short branched hairs on each margin, otherwise as in the type.

COAST REGION: Var. β: Clanwilliam Div.; Blue Berg, *Schlechter*, 8464 (in Herb. Kew)! Tulbagh Div.; Mitchells Pass, 1200 ft., *Schlechter*, 8954 (Herb. Bolus)!

CENTRAL REGION: Ceres Div.; on the Skurfdeberg Range near Eland-fontein, in the Cold Bokkeveld, 5300 ft., *Schlechter*, 10028!

There is evidently some error with regard to Schlechter's numbers 8464 and 8954, as the plant distributed to Dr. Bolus as 8464 is *Simocheilus klotzschianus*, var. *glabrifolius*, and identical with 8864 distributed to Kew, whilst the 8464 distributed to Kew is *Scyphogyne rigidula*, var. β, and identical with 8954 in Herb. Bolus, whereas the 8954 distributed to Kew is *Erica parviflora*, L.

4. S. divaricata (Benth. in DC. Prodr. vii. 710); branchlets glandular-puberulous or minutely tomentose, mingled with longer simple and gland-tipped hairs, greyish; leaves 3-nate, $\frac{1}{3}$–1 lin. long, $\frac{1}{6}$–$\frac{1}{3}$ lin. broad, ovate or linear, flat above, grooved down the very convex back, glabrous, sometimes ciliate with 2–3 long gland-tipped hairs on each side and one at the apex, sometimes merely serrate, with or without a few similar hairs or tubercles on the back; flowers 1–3 together, sessile, terminal or axillary, or on exceedingly short axillary branchlets racemosely scattered along the branches; calyx $\frac{1}{3}$ lin. long, obconic or broadly campanulate, unequally 4-lobed, glabrous or puberulous, ciliate, sometimes rather densely, with simple and branched hairs and 1 or more of the lobes often with 1 or more long gland-tipped hairs; corolla $\frac{1}{3}$–$\frac{1}{2}$ lin. long, campanulate, with 4 concave, deltoid, acute or rounded lobes, which are sometimes very minutely denticulate and glabrous or minutely puberulous on the back; stamens 4, equalling or slightly exceeding the corolla; filaments and the quadrate anthers connate; ovary puberulous around the base of the $\frac{1}{6}$ lin. long style; stigma large, peltate, just exceeding the anthers. *Blepharophyllum divaricatum, Klotzsch in Linnœa,* xii. 216.

COAST REGION: Worcester Div.; Bains Kloof, 3200 ft., *Schlechter*, 9181! Dutoits Kloof, *Drège*, 7750! Stellenbosch Div.; mountains near Lowrys Pass, *Bolus*, 5556! and without precise locality, *Zeyher!* Caledon Div.; Houw Hoek Mountains, *Ecklon & Zeyher!*

5. S. biconvexa (N. E. Br.); branchlets irregular, somewhat flexuose, minutely tomentose, mingled with longer simple or glandular hairs, greyish; leaves 3-nate, very spreading and more or less recurved, $\frac{1}{2}$–1 lin. long, $\frac{1}{4}$–$\frac{1}{3}$ lin. broad, linear-oblong, obtuse, about equally convex above and beneath, grooved on the back, pubescent or puberulous, with 2–3 long gland-tipped hairs on each margin and 1 at the apex; flowers usually 2–3 terminating the short lateral branchlets, sometimes also axillary, sessile; calyx broadly cup-shaped, unequally 4-lobed, $\frac{1}{3}$ lin. long, villous-pubescent, often with a few gland-tipped hairs on the ciliate margins of the lobes; corolla rather more than $\frac{1}{2}$ lin. long, globose-campanulate, with 4 short deltoid acute teeth, thickly pubescent to $\frac{1}{2}$-way down; stamens 4, connate, shortly exceeding the corolla; style $\frac{1}{6}$ lin. long, deciduous; stigma large, crater-like, with slightly incurved margins, exserted just beyond the anthers.

COAST REGION : Paarl Div.; French Hoek, *Schlechter*, 9244!

This closely resembles *S. divaricata*, Benth., but differs in its pubescent leaves, which are distinctly recurving and have a convex upper surface, the corolla also is slightly larger and pubescent with very much longer and different hairs.

6. S. glandulifera (N. E. Br.) ; a small very dwarf plant, 4–6 in. high ; branches erect, flexuose, clothed with a fine pubescence intermingled with longer gland-tipped hairs ; leaves 3-nate, ascending or spreading, $\frac{3}{4}$–$1\frac{1}{4}$ lin. long with the petiole, $\frac{1}{4}$–$\frac{1}{3}$ lin. broad, oblong-linear, obtuse, turgidly biconvex, puberulous, ciliate with about 3 long gland-tipped hairs on each side, with 1 at the apex and a few on the back ; flowers in small heads or clusters of 3–6 terminating exceedingly short lateral branchlets of the main branches, sessile ; calyx $\frac{1}{2}$ lin. long, broadly obconic or funnel-shaped, subequalling the corolla, somewhat unequally 4-lobed, puberulous ; lobes ovate, acute, all ciliate with long gland-tipped hairs ; corolla $\frac{1}{2}$–$\frac{2}{3}$ lin. long, funnel-shaped-campanulate, very shortly 4- (occasionally 5-) lobed, puberulous on the upper part; lobes broadly deltoid, erect, twice as broad as long, subacute, very minutely denticulate ; stamens 4 (occasionally 5), at maturity slightly exceeding the corolla; filaments connate into a membranous tube which extends half-way up the sides of the anthers and connects them together; anthers $\frac{2}{5}$ lin. long, subquadrate, with a triangular notch at the minutely scabrid apex; ovary slightly obovoid, minutely puberulous at the top, rather abruptly contracted into the slender puberulous $\frac{1}{2}$ lin. long style; stigma exserted $\frac{1}{3}$–$\frac{1}{2}$ lin. beyond the anthers, large, broadly crater-like.

CENTRAL REGION : Ceres Div. ; Cold Bokkeveld, 5000 ft., *Schlechter*, 8897!

7. S. longistyla (N. E. Br.) : branchlets slender, flexuose, puberulous with or without the admixture of longer gland-tipped hairs ; leaves 3-nate in more or less distant whorls, sometimes alternate, spreading or ascending, $\frac{1}{3}$–$\frac{3}{4}$ lin. long with the petiole, $\frac{1}{6}$–$\frac{1}{4}$ lin. broad, linear or ovate-oblong, convex on both sides, sometimes very turgid and almost subcylindric, with a long gland-tipped hair at the apex and 1–3 on each margin, often also minutely ciliate, otherwise glabrous, or slightly puberulous on the upper side; flowers 1–3, terminal, sessile or with a minute pedicel $\frac{1}{8}$–$\frac{1}{5}$ lin. long; calyx $\frac{1}{2}$ lin. long, subequally 4-lobed to the middle, obconic-campanulate, glabrous, ciliate, 1 or 2 lobes slightly longer than the others, ovate, acute, tipped with a long hair, the rest subquadrate, very obtuse or subtruncate, apiculate; corolla slightly exceeding the calyx, $\frac{3}{5}$ lin. long, obconic-campanulate, 4-lobed, glabrous, not ciliate ; lobes short, rounded or deltoid, obtuse, very minutely denticulate; stamens 4, about as long as the corolla; filaments and basal part of the anthers connate, $\frac{1}{4}$–$\frac{1}{3}$ lin. long; ovary globose, minutely puberulous at the base of the style or glabrous; style $\frac{1}{2}$ to nearly 1 lin. long, much exserted, glabrous; stigma large, peltate, flat.

COAST REGION: Caledon Div.; Houw Hoek, 1500 ft., *Schlechter*, 7556! Bredasdorp Div.; Elands Kloof, 1000 ft., *Schlechter*, 9748!

8. **S. eglandulosa** (Benth. in DC. Prodr. vii. 710); erect; branchlets slender, erect, flexuose, crowded, white-puberulous or glabrous; leaves 3-nate, adpressed, $\frac{1}{2}$–$\frac{3}{4}$ lin. long with the petiole, $\frac{1}{4}$–$\frac{1}{3}$ lin. broad, ovate, subacute, flat and usually puberulous (at least when young) on the upper side, very convex and glabrous on the back, at first minutely ciliate; flowers 1–3-nate, terminal or axillary, sessile; calyx about $\frac{1}{2}$ lin. long, nearly or quite as long as the corolla, obconic, sharply 4-angled, 4-lobed, glabrous or slightly puberulous on the angles; lobes unequal, 1 larger than the rest, about $\frac{1}{2}$ as long as the tube, deltoid, acute, ciliate; corolla $\frac{1}{2}$ lin. long, obconic, 4-lobed, glabrous, drying brown, ciliate on the deltoid or broadly ovate erect lobes, which are $\frac{1}{3}$–$\frac{1}{2}$ as long as the tube; stamens 4, equalling or just exceeding the corolla; filaments free, slender; anthers free at maturity, $\frac{1}{3}$ lin. long, cuneate-oblong, broadly notched at the apex; ovary narrowly ovoid, glabrous, tapering into the glabrous style; stigma just exserted, large, shallowly concave with an incurved toothed margin. *Omphalocaryon glandulosum, Klotzsch in Linnæa*, xii. 244. *O. eglandulosum, Klotzsch ex Benth. in DC. Prodr. l.c. and on label in Herb. Kew.*

COAST REGION: Caledon Div.; on the Zwart Berg, near the Hot Baths, *Zeyher! Ecklon & Zeyher ex Klotzsch.*

9. **S. fasciculata** (Benth. in DC. Prodr. vii. 709); branches erect, rather straight, crowded, minutely puberulous; leaves 3-nate, adpressed, $\frac{1}{2}$ to nearly 1 lin. long, $\frac{1}{5}$–$\frac{1}{3}$ lin. broad, lanceolate or ovate-oblong, acute, minutely ciliate and at first puberulous, becoming glabrous; flowers axillary or congested into very small leafy heads, sessile; calyx $\frac{1}{3}$ lin. long, sharply 4-angled, unequally 4-lobed $\frac{1}{3}$–$\frac{1}{2}$-way down, glabrous, ciliate, one lobe broader and stouter than the rest; corolla $\frac{1}{3}$ lin. long, scarcely exceeding the calyx, obconic, 4-lobed, glabrous, ciliate, very thin, drying brown; lobes about $\frac{1}{2}$ as long as the tube, deltoid-ovate, obtuse; stamens 4, free, very slightly exceeding the corolla; anthers $\frac{1}{5}$–$\frac{1}{4}$ lin. long, cuneate-oblong, broadly notched at the apex, dark brown or blackish; ovary narrowly ovoid, glabrous, tapering into the glabrous style; stigma just exceeding the anthers, large, shallowly concave with the margin incurved and more or less toothed.

COAST REGION: Caledon Div.; summit of the great mountain of Baviaans Kloof, near Caledon, *Burchell*, 7716!

10. S. remota (N. E. Br.); apparently about 1 ft. high, rather laxly branched; branches erect, with rather distant whorls of ascending-spreading branchlets, puberulous, pale whitish-brown or ash-coloured: leaves 3-nate, $\frac{1}{3}$–$\frac{1}{2}$ the length of the internodes on the main branches, about equalling them on the branchlets, mostly adpressed, 1–2 lin. long, linear, acute, slender, glabrous; flowers axillary and terminal, usually 1–3 (rarely 4–6) together, pedicellate; pedicels $\frac{1}{8}$–$\frac{1}{6}$ lin. long, glabrous; calyx $\frac{1}{4}$ lin. long, truncate at the base, unequally 4-toothed to half-way down, glabrous, green; teeth deltoid or the longer one somewhat linear, acute; corolla $\frac{1}{2}$ lin. long, campanulate, glabrous, thin, greenish; lobes about $\frac{1}{3}$ as long as the tube, deltoid-ovate, very obtuse or rounded, incurved; stamens 4–6; filaments $\frac{2}{3}$ as long as the anthers, broad, connate for nearly half their length, torn asunder by the enlargement of the ovary; anthers included, $\frac{1}{4}$ lin. long, oblong, bifid, connate for half their length; ovary ovoid, conically tapering into the short style, with several minutely puberulous lines, which are often invisible when wetted; stigma not exserted beyond the corolla, large, concave-peltate.

COAST REGION: Caledon Div.; near the River Zondereinde, 7000 ft., *Schlechter*, 9897!

11. S. capitata (Benth. in DC. Prodr. vii. 710); a bush 1–2 ft. high; branches puberulous or pubescent; leaves 3-nate, adpressed or ascending, $\frac{2}{3}$–1 lin. long with the petiole, $\frac{1}{4}$–$\frac{1}{3}$ lin. broad, linear to narrowly ovate-oblong, obtuse or with a sessile truncate gland at the apex, denticulate on the margins, glabrous, perhaps viscid; flowers sessile, 2–20 in small terminal head-like clusters 1–2 lin. in diam., greenish-yellow; calyx $\frac{2}{3}$–$\frac{3}{4}$ lin. long and as much in diam., coriaceous, 4-lobed, broadly pear-shaped, with the tips of the lobes incurved, rather densely puberulous all over outside; lobes broader than long, the outer one much broader than the rest, all subtruncate, with or without an apiculus or the larger with a short thick point, ciliate; corolla $\frac{1}{2}$–$\frac{2}{3}$ lin. long, $\frac{1}{3}$ lin. in diam., pear-shaped, glabrous, 4-lobed, about as long as the calyx, with the very short rounded suberect lobes exserted when fully developed, not hyaline, drying brown; stamens 4, included, adhesive to the corolla; filaments free, linear; anthers at first connate, apparently at length free, $\frac{1}{3}$ lin. long, subrectangular, broadly notched at the apex; ovary ovoid, narrowed into the style, which is minutely puberulous at the base; stigma large, peltate, with an erect margin, just exserted beyond the corolla. *Omphalocaryon capitatum, Klotzsch in Linnæa,* xii. 244.

VAR. β, **brevifolia** (N. E. Br.); calyx glabrous or slightly puberulous on the basal part only; lobes minutely denticulate or entire, not ciliate or only at the sides; otherwise as in the type. *S. brevifolia, Benth. in DC. Prodr.* vii. 710.

COAST REGION: Caledon Div.; mountains near Caledon and Genadendal and between Hottentots Holland and Caledon, *Ecklon & Zeyher!* mountains near Genadendal, 1000 ft., *Bolus, Herb. Norm. Austr.-Afr.,* 614! Var. β: Caledon Div.; mountains near Genadendal, 1000–2000 ft., *Burchell*, 7621! *Bolus*, 5418! 6963! *Schlechter*, 7748! 9799!

The corolla is much more exserted in Schlechter 7748 than in any other specimen examined; in some it is no longer than the calyx, but in these it may be immature.

12. S. viscida (N. E. Br.); an erect shrub; ultimate branchlets spreading, $\frac{1}{2}$–$1\frac{1}{2}$ in. long, glandular-pubescent; bark reddish-brown; leaves 3-nate, spreading, 1–$1\frac{1}{2}$ lin. long with the petiole, $\frac{1}{3}$–$\frac{1}{2}$ lin. broad, linear or ovate-oblong, with a sessile concave-truncate gland at the apex, glabrous, very viscid, ciliate with 3–5 long gland-tipped hairs on each side; flowers in very short terminal oblong or sub-globose heads or spikes, sessile; calyx obconic or somewhat top-shaped, unequally 4-lobed, about as long as the corolla-tube with the longer lobe exceeding it, $\frac{2}{3}$ lin. long at the shorter part, puberulous; lobes subtruncate, minutely apiculate, ciliolate, the larger often much produced and leaf-like; corolla $\frac{3}{4}$ lin. long, somewhat pear-shaped, often more or less oblique, 4-lobed, glabrous; lobes equila-terally triangular, acute, erect; stamens 4, as long as the corolla-tube; filaments free, slender; anthers $\frac{1}{8}$–$\frac{2}{5}$ lin. long, subrectangular, or quadrate, broadly notched at the apex, free or connate at the lower part, sometimes adhering to the corolla; ovary ovoid, minutely puberulous on the upper part, tapering into the very stout conical puberulous style; stigma just exserted, large, peltate, entire.

COAST REGION: Robertson Div.; rocky hills near Montagu Bath, 800 ft., *Bolus*, 6721!

13. S. micrantha (N. E. Br.); probably $1\frac{1}{2}$ ft. or more high, with stout main branches, densely much branched above; branchlets slender, erect, often flexuose, very minutely subtomentose to sub-glabrous, greyish, becoming brown; leaves 3-nate, erect or ascending-spreading, usually longer than the internodes, $1\frac{1}{3}$–$2\frac{1}{2}$ lin. long, linear, acute or subacute, slender, glabrous; flowers 1–3 together, axillary or on very minute axillary branchlets, shorter than the petioles of the leaves subtending them, subsessile; calyx usually about $\frac{1}{3}$ lin. long, rather broad and shallow, truncate at the base, unequally or subequally 4-toothed to the middle, usually glabrous or sometimes puberulous on the same specimen; teeth ovate, acute, or the larger one sometimes with a linear point, very minutely ciliate; corolla much exceeding the calyx, $\frac{1}{2}$ lin. long, subovoid or subglobose-campanulate, 4-lobed, thin, usually glabrous, but sometimes puberu-lous on the same specimen; lobes $\frac{1}{3}$–$\frac{1}{2}$ as long as the tube, incurved, deltoid-ovate, obtuse or rounded; stamens 4–8, included; filaments free or very shortly and irregularly connate at the base; anthers connate, about $\frac{1}{4}$ lin. long, subquadrate or oblong, notched and very minutely scabrid or subhispid at the apex, not subglandular as originally described; ovary ovoid with a minute tuft of hairs at the apex, 1-celled; style slender, arising by the side of the tuft and becoming more and more oblique or sublateral as the ovary enlarges and becomes more globose, when it bends over to the apex of the ovary; stigma very large in proportion to the ovary, peltate, crater-like with the margin incurved, puberulous on the underside, not exserted from the corolla; fruit subglobose, coriaceous or perhaps crustaceous when ripe, 1-seeded. *Salaxis micrantha, Benth. in DC. Prodr.* vii. 711; *Dietr. Synop. Pl.* ii. 1261. *S. axillaris, b, Drège, Zwei Pflanz. Docum.* 217, *not of Salisbury.*

SOUTH AFRICA : without locality, *Roxburgh !*
COAST REGION : Worcester Div.; Dutoits Kloof, 3000–4000 ft., *Drège !*

14. S. trimera (N. E. Br.); a densely much-branched shrublet 1–1½ ft. high; branchlets puberulous; leaves 3-nate, erect or ascending, ½–1 lin. long with the petiole, linear, subacute or obtuse, glabrous or minutely puberulous, often at first tipped with a few fine hairs; flowers axillary or terminating exceedingly short axillary branchlets, 1–3 together, sessile or subsessile; calyx obconic, 3-angled and unequally 3-lobed, rarely imperfectly 4-lobed, glabrous or puberulous on the angles, ciliate on the lobes; tube $\frac{1}{8}$–$\frac{1}{4}$ lin. long, transparent between the angles, lobes $\frac{1}{8}$–$\frac{1}{4}$ lin. long, linear, obtuse; corolla $\frac{1}{2}$–$\frac{2}{3}$ lin. long, urceolate, 3-lobed, ovoid-inflated, broadest ($\frac{1}{4}$ lin. in diam.) slightly above the middle, contracted at the base, but scarcely so or not at all below the lobes, varying from nearly or quite glabrous to puberulous nearly all over; lobes $\frac{1}{5}$–$\frac{1}{4}$ lin. long, oblong or ovate-oblong, very obtuse, erect or slightly connivent; stamens 3, included; filaments free, slender; anthers connate, less than $\frac{1}{4}$ lin. long, subquadrate, shortly and obtusely bifid; ovary narrowly ovoid, tapering into the style which is puberulous at the apex; stigma just exserted from the corolla, broadly funnel-shaped.

SOUTH AFRICA : without locality, *Sieber,* 175 !
COAST REGION : Cape Div.; eastern slopes of Devils Mountain, 1500 ft., *Bolus,* 4496 !

The variation in the size of the corolla is probably due to different stages of maturity of the flower.

15. S. urceolata (Benth. in DC. Prodr. vii. 709); like *S. trimera* in appearance and foliage, but the pubescence on the branchlets is longer; leaves glabrous; flowers 1–3 terminating short or elongated axillary branchlets, with distinct pedicels $\frac{1}{8}$–$\frac{1}{6}$ lin. long; calyx obconic, unequally 4-lobed, 4-angled, glabrous, ciliate on the lobes; tube $\frac{1}{5}$ lin. long, subtransparent between the angles; lobes $\frac{1}{3}$–$\frac{1}{2}$ lin. long, linear, acute; corolla $\frac{2}{3}$–$\frac{3}{4}$ lin. long, urceolate, 4-lobed, distinctly contracted below the lobes and at the base, broadest ($\frac{1}{3}$–$\frac{2}{5}$ lin. in diam.) at the globose-inflated middle, puberulous on the upper half; lobes erect, $\frac{1}{8}$–$\frac{1}{6}$ lin. long, ovate, subacute; stamens 3, included; filaments free, slender; anthers connate at the base, $\frac{1}{3}$ lin. long, subquadrate; ovary narrowly ovoid, narrowed into the style, which is puberulous at the apex; stigma just exserted from the corolla, broadly funnel-shaped. *Tristemon urceolatus, Klotzsch in Linnæa,* xii. 245.

COAST REGION : Cape Div.; "mountains near Cape Town" (ex *Klotzsch*), *Bergius!*

I find the flowers to be distinctly although minutely pedicellate in Bergius' specimen (which I take to be the type), not sessile as described by Klotzsch.

16. S. Burchellii (N. E. Br.) ; an erect shrub about 3 ft. high ; in its branchlets, leaves, arrangement of the flowers and pubescence exactly as in *S. trimera;* flowers with distinct pedicels $\frac{1}{8}$ lin. long ; calyx obconic, unequally 4-lobed, 4-angled, glabrous, with ciliate lobes; tube $\frac{1}{8}$ lin. long or less ; lobes $\frac{1}{4}-\frac{1}{3}$ lin. long, linear, subacute or obtuse ; corolla $\frac{1}{2}$ lin. long, urceolate, 4-lobed, ovoid-inflated and broadest ($\frac{1}{4}$ lin. in diam.) above the middle, tapering or slightly contracted at the base, not contracted under the slightly connivent lobes, puberulous nearly all over ; lobes about $\frac{1}{8}$ lin. long and as much or more in breadth, deltoid-ovate, obtuse, ciliolate ; stamens 3, included ; filaments free, slender ; anthers free or scarcely connate, not quite $\frac{1}{4}$ lin. long, cuneately subquadrate, shortly and obtusely bifid at the apex ; ovary narrowly ovoid, glabrous, tapering into the style, which is glabrous or with only 2 or 3 minute hairs at the apex ; stigma just exserted from the corolla, broadly funnel-shaped.

COAST REGION : Worcester Div.; Matroos Berg, 7000 ft., *Bolus in·Herb. Guthrie,* 3942! Caledon Div.; tops of the mountains of Baviaans Kloof near Caledon, *Burchell,* 7747! *Schlechter,* 9874!

17. S. puberula (Benth. in DC. Prodr. vii. 709); branches erect, crowded, finely puberulous ; leaves 3-nate, $\frac{1}{2}-1$ lin. long with their petioles, $\frac{1}{8}-\frac{1}{4}$ lin. broad, erect, linear, obtuse or subacute, puberulous; flowers axillary or terminating exceedingly short axillary branchlets, sessile ; calyx $\frac{1}{4}$ lin. long, unequally divided into 2 parts, 1 lobe being separated nearly to the base and widely spreading from the other 3, which are connate nearly to the middle into a trifid lobe, with incurved tips, all ciliate with long hairs on the margins, otherwise glabrous ; corolla $\frac{1}{3}$ lin. long and as much or rather more in diam., globose, 4-lobed, puberulous, not hyaline ; lobes sometimes obscure, at others broadly ovate, obtuse, incurved, ciliate ; stamens 3, about equalling the corolla, **free** ; anthers about $\frac{1}{4}$ lin. long and broad, subquadrate ; ovary ovoid, glabrous, narrowed into the style ; stigma just exserted beyond the corolla, cup-shaped, toothed, minutely puberulous around its base. *Tristemon puberulus, Klotzsch in Linnæa,* xii. 245.

COAST REGION : Caledon Div.; on the Zwart Berg, near Caledon, *Ecklon!*

XIX. **LAGENOCARPUS**, Klotzsch.

Bracts 0. *Calyx* campanulate or obconic-campanulate, more or less unequally 4- (rarely 5-) toothed, obscurely angular. *Corolla* very small, slightly exceeding the calyx or shorter than its longest tooth, arising from the middle or lower part of the ovary, not hypogynous, shortly tubular-campanulate, with 4 (rarely 5) incurved tooth-like rounded to subquadrate lobes $\frac{1}{4}$–$\frac{1}{3}$ as long as the tube, separated by rounded interspaces. *Stamens* usually 8, occasionally 7, quite included, arising from the ovary at the insertion of the corolla; filaments $\frac{1}{4}$–$\frac{1}{3}$ as long as the anthers, connate throughout their length into a tube, often becoming more or less torn asunder by the enlargement of the ovary, more or less adhesive to the corolla; anthers oblong, bifid, connate for half their length, opening laterally at the apical half. *Ovary* on a short stalk within and quite free from the calyx, but half inferior to the corolla, 1-celled (not 2–3-celled as stated in the key on p. 4); style short; stigma large peltate, just exserted from the corolla. *Ovules* 4–5, pendulous from near the apex of a central placenta, which is quite free from the side wall of the ovary and only attached at the apex and base, dilated and thin at the basal part, filiform above. *Fruit* not seen, but evidently often 1-seeded.

Small shrubs or shrublets resembling a *Salaxis* in appearance; leaves grooved down the back; flowers 1–3 together, terminal or on minute axillary branchlets, which are shorter than the leaves and covered with minute bracts or very small leaves.

DISTRIB. Species 2, endemic.

The ovary of this genus is, so far as I am aware, unique in the Vegetable Kingdom, and has never previously been correctly described. It is so conspicuously half-inferior to the corolla that it seems remarkable that so unique a character should have escaped the notice of all previous authors. The placentation and number of ovules are rather difficult to determine, and cannot be correctly understood from transverse sections. In young ovaries, before the stigma has emerged from the corolla, by carefully cutting or tearing away the side wall of the ovary all round, a central slender placenta attached at its base and apex, but quite unattached to the sides of the ovary, can be clearly seen, bearing 4–5 ovules suspended from near its apex; after the stigma has become exserted, and before the formation of the embryo, 1–3 of the ovules become aborted and sometimes apparently disappear, and in transverse sections of the ovary made at this stage of development, the walls of the closely placed and perhaps slightly adhesive empty ovules often give the false appearance of the partitions of a 2-3-celled ovary, an error of observation, made after a hasty examination, which caused me to so describe it on page 4. The central (not axile) placenta is remarkable, and among South African *Ericaceæ* only occurs in this genus and in *Thamnus*; no other author appears to have noticed it.

Calyx and corolla glabrous   ...   .   (1) **imbricatus.**
Calyx and corolla puberulous   ...   .   (2) **ciliatus.**

1. **L. imbricatus** (Klotzsch in Linnæa, xii. 214); 1–1$\frac{1}{2}$ ft. high, much branched; branchlets minutely tomentose-puberulous; leaves 3-nate, erectly imbricate, ascending or more rarely spreading, longer than the internodes, 1–2 lin. long with the petiole, oblong or oblong-linear, subacute, very minutely scabrid or erose-denticulate on the margins and sometimes edged with minute sessile glands when young; flowers axillary and terminal, 1–3 together or sometimes several clustered at the ends of the branchlets; calyx $\frac{3}{4}$–1 lin. long, coriaceous, glabrous, green; teeth usually about half as long as the tube, very broadly deltoid-ovate, acute or obtuse, with or without

minute sessile glands on their edges ; corolla $\frac{2}{3}$–$\frac{3}{4}$ lin. long from its insertion on the ovary, glabrous, greenish ; lobes $\frac{1}{3}$ as long as the tube ; ovary shortly but distinctly stalked, somewhat globose or ellipsoid, abruptly narrowed into a stout puberulous style ; stigma puberulous on the underside. *Benth. in DC. Prodr.* vii. 710 ; *Bach in Linnæa,* xxvi. 792. *Erica serrata, Thunb. Prodr.* 69, *and Fl. Cap. ed. Schultes,* 346. *Blæria serrata, Thunb. Diss. Blæria,* 6. *Salaxis imbricatus, Drude in Engl. & Prantl, Pflanzenfam.* iv. i. 64, *fig. A—D inaccurate.*

SOUTH AFRICA : without locality, *Thunberg !*
COAST REGION : Paarl Div. ; mountains around French Hoek, 1900–4000 ft., *Bolus,* 7000 ! *Schlechter,* 9280 ! Stellenbosch Div. ; Gordons Bay, 100 ft., *Guthrie !* Caledon Div. ; Houw Hoek Mountains, *Ecklon & Zeyher, Schlechter,* 7329 ! *Bolus,* 5411 ! mountains near Genadendal, *Burchell,* 7683 ! *Drège ! Bolus,* 7406 ! Zwart Berg, near Caledon, *Ecklon & Zeyher, Pappe !* Babylons Tower, *Ecklon & Zeyher !* Shaws Mountain, *Galpin,* 3729 ! Bredasdorp Div. ; Elim, 600 ft., *Schlechter,* 7644 !

2. L. ciliatus (N. E. Br.) ; habit, branchlets, foliage and arrangement of the flowers exactly as in *L. imbricatus,* but perhaps more dwarf ; calyx about 1 lin. long, rather rigidly coriaceous, puberulous all over the outside ; teeth unequal, broadly deltoid-ovate, obtuse or acute, $\frac{1}{3}$–$\frac{1}{2}$ as long as the tube ; corolla $\frac{1}{2}$–$\frac{2}{3}$ lin. long from its insertion on the ovary, minutely puberulous ; lobes about $\frac{1}{4}$ as long as the tube, rounded ; ovary varying from slightly rhomboid-ellipsoid to subglobose, abruptly contracted into the stalk below and into the rather stout style above, slightly angularly ribbed and puberulous on the upper half and style ; stigma puberulous on the underside. *Salaxis ciliata, Benth. in DC. Prodr.* vii. 711. *S. brevifolia, Dietr. Synop. Pl.* ii. 1261.

COAST REGION : Caledon Div. ; mountains near Genadendal, *Burchell,* 7616 ! 7681 ! 7829 ! *Bolus,* 5432 !

ORDER LXXX. PLUMBAGINEÆ.

(By C. H. WRIGHT.)

Flowers regular, hermaphrodite. *Calyx* 5- 10- or 15-ribbed ; tube cylindrical, rarely campanulate ; limb usually 5-toothed or -lobed, membranous and hyaline between the lobes, introrsely or extrorsely rolled in bud, patent during flowering. *Petals* 5, connate at the very base only or into a tube shorter or longer than the calyx, imbricate. *Stamens* 5, hypogynous or adnate to the petals ; filaments filiform ; anthers ovate or oblong, dorsifixed ; cells 2, parallel, dehiscence longitudinal. *Ovary* sessile or stipitate, 1-celled, often 5-angled above ; ovule 1, pendulous from an erect basal funicle ; styles 5, distinct or more or less connate ; stigmas capitate or linear. *Fruit* usually enclosed by the calyx, dehiscing by an operculum or circumscissile near the base, or 5-valved, sometimes indehiscent. *Seed* filling the cell; testa membranous; endosperm floury, copious, scanty or 0; embryo straight, terete or slightly compressed.

Herbs, acaulescent or with short stem and leaves in a rosette, sometimes suffrutescent, more rarely herbs or shrubs with elongated branched stem and alternate leaves, which leave annular scars on falling off; scape or peduncle terminal, dichotomously branched, more rarely simple; flowers sessile or on a short thick pedicel, solitary or few in spikelets, often secund, rarely capitate, rose, violet, blue or yellow, rarely white; bracts usually rigid, scarious at the edge, one subtending each spikelet; bracteoles 1 or more to each flower.

DISTRIB. Species about 200, in maritime and desert places, especially in the Mediterranean region; a few cosmopolitan.

I. **Statice.**—Stemless herbs or dwarf shrubs. *Calyx-limb* patent, scarious, coloured. *Petals* almost free.

II. **Plumbago.**—Rambling shrubs. *Calyx* not scarious, glandular, shortly 5-toothed. *Corolla* salver-shaped.

III. **Vogelia.**—Erect undershrubs. *Calyx* papery, deeply 5-lobed, bearing 5 vertical wings, eglandular. *Corolla* funnel-shaped.

I. **STATICE**, Willd.

Calyx usually funnel-shaped; tube usually equally 10-ribbed at the base, ribs uniting in pairs above, rarely 10- or 5-ribbed throughout; limb scarious, plicate, patent, 5-toothed. *Petals* connate at the very base or free. *Stamens* adnate to the base of the petals. *Styles* 5, distinct, rarely shortly united. *Fruit* included in the calyx, indehiscent or variously dehiscing. *Seed* filling the cell; albumen scanty, or copious on one side of or all round the embryo.

Perennial, rarely annual, stemless herbs, or small shrubs branched from the base, glabrous, pulverulent or lepidote; leaves radical or cauline and alternate, flat; inflorescence racemose, cymose or paniculate; peduncles terete or winged; bracts scale-like, small, enwrapping one or more flowers.

DISTRIB. Species about 120, chiefly in the coast and desert regions of the northern hemisphere, some in Australia.

Calyx-limb at least 5 lin. in diam. :
　Panicles corymbose :
　　　Leaves and scape smooth or nearly so ...　　...　(1) **purpurata.**
　　　Leaves and scape papillose　　...　　...　　...　(2) **rosea.**
　　Panicles spicate　　...　　...　　...　　...　　...　(3) **amœna.**
Calyx-limb less than 5 lin. in diam. :
　Plants erect :
　　Panicles dense :
　　　Leaves linear-subulate　　...　　...　　...　(4) **kraussiana.**
　　　Leaves linear　　...　　...　　...　　...　(5) **linifolia.**
　　Leaves obovate or oblanceolate :
　　　　Spikelets 1-flowered　...　　...　　...　(6) **scabra.**
　　　　Spikelets 2-flowered :
　　　　　Panicle densely racemosely
　　　　　　branched　　...　　...　　...　(7) **equisetina.**
　　　　　Panicle less dense, corymbose :
　　　　　　Calyx 2 lin. long　...　　...　(8) **dregeana.**
　　　　　　Calyx 3½ lin. long ...　　...　(9) **avenacea.**
　　Panicles very diffuse　　...　　...　　...　...　(10) **anthericoides.**
　Plants decumbent　　...　　...　　...　　...　...　(11) **decumbens.**

1. S. purpurata (Linn. Mant. 59); stem woody, terete, clothed with the persistent leaf-bases; leaves obovate-lanceolate, obtuse, mucronate, smooth, coriaceous, obscurely 3-nerved, 1 in. long, 6 lin. wide; petiole nearly 2 in. long, sheathing at the base; panicle much branched, nearly 1 ft. high, smooth or obscurely scabrid; spikelets 1-flowered; bracts 3 lin. long, lanceolate; bracteoles ovate, the outer 1, the inner 3 lin. long, coriaceous with a broad white scarious border; calyx-tube pilose below; limb 6 lin. in diam., 5-ribbed; lobes 10, rounded, the alternate shorter; stamens and pistil as in *S. rosea,* Sm. *Boiss. in DC. Prodr.* xii. 666.

VAR. β, **longifolia** (Boiss. in DC. Prodr. xii. 667); leaves oblanceolate or nearly linear, up to 10 in. long and 2 lin. wide, usually much smaller. *S. longifolia, Thunb. Prodr.* 54, *Fl. Cap. ed.* 2, ii. 241, *ed. Schult.* 276; *Drège, Zwei Pflanzengeogr. Documente,* 110. *S. scabra, Drège, l.c.* 70.

SOUTH AFRICA: without locality, *Wallich!* Var. β, *Villette!*
COAST REGION: Clanwilliam Div.; between Kromme River and Berg Vallei, below 1000 ft., and between Pikiniers Kloof and Marcus Kraal, 1000–1500 ft., *Drège!* Worcester Div.; near Worcester, *Cooper,* 1618!
CENTRAL REGION: Var. β, Calvinia Div.; Uien Vallei, 2000–2500 ft., *Drège!*

2. S. rosea (Sm. in Rees, Cyclop. xxxiv.); whole plant usually under 1 ft. high; stem short, woody, covered by the leaf-bases; leaves obovate, up to 2 in. long and nearly 1 in. wide, tapering into a petiole 1 in. long, which expands into a long sheathing base, papillose on both surfaces; panicle corymbose, many-flowered, rhachis and branches tuberculate; spikelets 1-flowered; bracts short, lanceolate; bracteoles ovate, enwrapping the calyx-tube, 4 lin. long, with broad white scarious margins; calyx-tube cylindric, 4 lin. long, pilose in the lower half, strongly 5-ribbed; limb 5 lin. in diam.; lobes 10, rounded, crenate, five 1 lin. long, five $\frac{1}{2}$ lin. long; petals oblanceolate, obtuse, 6 lin. long, $1\frac{1}{2}$ lin. wide, united at the very base; filaments 3 lin. long, inserted between the anther-cells; anthers oblong, nearly 1 lin. long, verrucose, cells distinct in the lower half, mucronate at the base; ovary oblong, deeply 5-lobed; styles distinct, longer than the petals. *Boiss. in DC. Prodr.* xii. 667. *S. peregrina, Berg. Descr. Pl. Cap.* 80, *excl. syn. S. purpurata, Willd. Sp. Pl.* i. 1528; *Thunb. Fl. Cap. ed. Schult.* 277, *not of Linn. S. rytidophylla, Hook. in Bot. Mag. t.* 4055. *S. dicksoniana, Hort. ex Hook. l.c.*

SOUTH AFRICA: without locality, *Burchell,* 754! *Wallich!*
COAST REGION: Piquetberg Div.; St. Helena Bay, *Mader!* Clanwilliam Div.; between Lange Vallei and Heerenlogement, below 500 ft., *Drège!* Zuurfontein, 150 ft., *Schlechter,* 8528! Cape Div.; near Blue Berg, *MacOwan, & Bolus, Herb. Norm. Austr.-Afr.,* 235! Div.? Roeberg (? Koe Berg), *Bergius ex Boissier.*

3. S. amœna (C. H. Wright); branches few from the apex of the rootstock; leaves oblanceolate, up to $1\frac{1}{2}$ in. long and 2 lin. wide, more or less rugose when dry, mucronate; scape about 6 in. high, branched low down, tuberculate; spikelets distichous, 1-flowered, spicately spaced out on the branches; outer bracteoles 1 lin. long, suborbicular, acute, hyaline-margined; inner bracteole elliptic, $3\frac{1}{2}$ lin. long, enwrapping the flower, hyaline-margined in the upper half; calyx tubular-campanulate, 5 lin. long, 5 lin. in diam., rose-colour; tube hirsute outside below; lobes 5, very short, rounded, the alternate larger; anthers oblong; cells united only in the upper half.

CENTRAL REGION: Worcester Div.; stony places near Touws River Railway Station, 2500 ft., *Bolus in Herb. Norm. Austr.-Afr.*, 1080!

4. S. kraussiana (Buching. ex Krauss in Flora, 1845, 73); stem short; branches tufted, woody; leaves congested, rigid, linear-subulate, almost pungent, flat above, convex beneath, 6 lin. long; scape about 1 in. long, paniculately branched; spikelets 2-flowered; outer bract very short, inner 2 lin. long, both coriaceous with scarious margins; calyx narrowly funnel-shaped, 3 lin. long, 5-ribbed; lobes short, triangular; petals oblanceolate, obtuse, 4 lin. long; anthers oval; ovary obovoid; style filiform. *Boiss. in DC. Prodr.* xii. 657.

COAST REGION: Bredasdorp Div.; Elim, 150 ft., *Schlechter*, 9664! Zoetendals Vallei, *Krauss*.

5. S. linifolia (Linn. f. Suppl. 187); stem very short, woody, usually branched; leaves linear, up to $1\frac{1}{4}$ in. long and $\frac{1}{2}$ lin. wide, scabrid; scape about 6 in. high, much branched above, scabrid; bracts short, scarious-edged; spikelets 1–2-flowered, sessile or shortly pedicelled; outer bracteole 2 lin. long, coriaceous with a hyaline border, two inner slightly shorter, scarious; calyx $2\frac{1}{2}$ lin. long, 5-ribbed, pubescent outside in the lower half; lobes short, triangular; petals $3\frac{1}{2}$ lin. long, oblanceolate, obtuse; filaments filiform; anthers $\frac{1}{2}$ lin. long; ovary oblong, 5-angled; styles filiform. *Thunb. Prodr.* 54, *and Fl. Cap. ed. Schult.* 277; *Boiss. in DC. Prodr.* xii. 657; *var. areticæfolia, Boiss. l.c.; var. collina, Eckl. & Zeyh. ex Presl, Bot. Bemerk.* 105, *and in Linnœa*, xx. 203. *S. areticæfolia, Fries ex Boiss. l.c. S. sp., Drège, Zwei Pflanzengeogr. Documente*, 56, 129.

VAR. β, **brachyphylla** (Boiss. in DC. Prodr. xii. 657); leaves very short, narrowly subspathulate, obtuse; scape short, densely branched. *S. linifolia*? *Drège, Zwei Pflanzengeogr. Documente*, 129.

VAR. γ, **maritima** (Eckl. & Zeyh. ex Boiss. l.c.); branchlets incurved; spikelets smaller, usually 1-flowered, more distant; leaves flatter, wider.

VAR. δ, **robusta** (C. H. Wright); stem bearing many short stout branches; leaves linear, usually obtuse, 9 lin. long, 1 lin. wide.

COAST REGION: Caledon Div.; *Verreaux*; Bredasdorp Div.; Cape Agulhas, 100 ft., *Schlechter*, 10568! Riversdale Div.; hills near Zoetemelks River, *Burchell*, 6787! Knysna Div.; sand-hills at Plettenberg Bay, *Burchell*, 5315! Uitenhage Div.; Zwartkops River, under 100 ft., *Drège*, 8017! *Burchell*, 4426! and without precise locality, *Zeyher!* Var. β: Port Elizabeth Div.; on sand-hills near Port Elizabeth, under 100 ft., *Drège!* Albany Div.; Brak Kloof, near Grahamstown, *MacOwan*, 444!

CENTRAL REGION: Ceres Div.; Verkeerde Vley, *Rehmann*, 2841! Richmond Div.; vicinity of Styl Kloof near Richmond, 4000-5000 ft., *Drège!* Colesberg Div.; Colesberg, *Shaw!*

WESTERN REGION: Little Namaqualand; Steinkopf, *Schlechter!*

KALAHARI REGION: Var. δ: Griqualand West; Lower Campbell, *Burchell*, 1818!

Boissier does not cite any locality or collector for var. *maritima.*

6. S. scabra (Thunb. Prodr. 54); rootstock woody; leaves radical, obovate-oblong, obtuse, attenuate below into the petiole, up to $3\frac{1}{2}$ in. long and $\frac{1}{3}$ in. wide; scapes several springing from the rootstock, 6–12 in. high, fastigiately branched above, densely scabrid as are also the branches; bracts 1 lin. long or less, triangular; spikelets 1-flowered; bracteoles $2\frac{1}{2}$ lin. long, coriaceous with a scarious border; calyx 3 lin. long, including the $\frac{1}{2}$ lin. long oval subacute lobes; petals oblanceolate, obtuse, slightly longer than the calyx; anthers short, oval; ovary 5-ribbed; styles filiform. *Fl. Cap. ed. Schult.* 276; *Krauss in Flora*, 1845, 72; *Boiss. in DC. Prodr.* xii. 659.

SOUTH AFRICA: without locality, *Grey! Thom*, 86! *Villette! Webb!*

COAST REGION: Cape Div.; Camps Bay, *Burchell*, 849! Cape Peninsula, Millers Point, *Jameson!* Fish-hook Bay, *Jameson!* Simons Bay, *Milne*, 202! Riversdale Div.; Great Valsch River, *Burchell*, 6534! Mossel Bay Div.; between the landing-place at Mossel Bay and Cape St. Blaize, *Burchell*, 6278! Knysna Div.; Plettenberg Bay, on sand-hills, *Burchell*, 5314! near Knysna, *Burchell*, 5387! Uitenhage Div.; Sundays River, *Gill!* locality doubtful, *Zeyher*, 656! Port Elizabeth Div.; on hills near Port Elizabeth and on dunes near Cape Recife, up to 500 ft., *Zeyher*, 1430! Bathurst Div.; sea-coast, among sand-hills at Tharfield and near Kowie, *Atherstone*, 59! Kowie, *MacOwan*, 385! between Port Alfred and Kaffir Drift, *Burchell*, 3858! Peddie Div.; Fredericksburg, on the Gualana River, *Gill!* Div.? sand-flats, *Baur*, 176!

WESTERN REGION: Little Namaqualand; Steinkopf, *Max Schlechter*, 37! and without precise locality, *Scully*, 225!

7. S. equisetina (Boiss. in DC. Prodr. xii. 658); rootstock short, woody; leaves obovate or oblanceolate, obtuse or acute, sharply mucronate, up to $1\frac{1}{2}$ in. long and $\frac{1}{2}$ in. wide, smooth or nearly so; scapes many from each rootstock, $\frac{1}{3}$–1 ft. high, much branched nearly from the base, forming a subcylindrical panicle: branches about 1 in. long, smooth or slightly rough; spikelets 2-flowered; outer bract broadly ovate or suborbicular, shortly cuspidate, $\frac{3}{4}$ lin. long; inner bract elliptic, navicular, hyaline-margined, about 2 lin. long; calyx tubular, shortly 5-lobed; petals oblanceolate. *S. linifolia, Drège, Zwei Pflanzengeogr. Documente*, 110, *not of Thunb.*

VAR. β, **depauperata** (Boiss. l.c.); scape and its branches tuberculate; inflorescence more lax than in the type; flowers sometimes solitary.

SOUTH AFRICA: without locality, *Wallich!* var. β: *Villette! Harvey*, 492!

COAST REGION: Clanwilliam Div.; between Kromme River and Berg Vallei, under 1000 ft., *Drège!* Var. β: Cape Div.; Cape Flats, *Burke!* about the Windmills at Salt River, *Burchell*, 512!

A plant about 6 in. high collected by Major Wolley Dod (678) at Vyges Kraal in Cape Div. probably belongs to this species.

8. S. dregeana (Presl, Bot. Bemerk. 105); rootstock woody; leaves linear-spathulate, 1½ in. long, up to 2 lin. wide, obtuse, mucronate; scape 4–10 in. high, flexuous, rough, as are also the panicle-branches and bracts; panicle much-branched, branches secund; spikelets 2-flowered; bracts short, sheathing, acute; outer bracteoles 2, subequal, inner elliptic, much longer than the outer; calyx 2 lin. long, 5-ribbed; tube hairy outside, reddish; limb scarcely as long as the tube, obtusely and shortly 5-lobed, white. *S. tetragona, Drège, Zwei Pflanzengeogr. Documente, 94, not of Thunb.*

WESTERN REGION : Little Namaqualand ; Haazenkraals River, 2000–2500 ft., *Drège!*

9. S. avenacea (C. H. Wright); rootstock woody; leaves oblanceolate, 1½ in. long, ¼–¾ in. wide, rounded and shortly mucronate at the apex, smooth, margins slightly revolute; scapes several from one rootstock, flexuous, slightly tuberculate, 1–2 ft. high; branches more or less erect; spikelets congested near the ends of the branches, 2-flowered; bracts shortly triangular, acuminate; outer bracteoles 1 lin. long, broadly triangular, acute, hyaline-margined, puberulous on the back; inner bracteole broadly elliptic, acute, 2½ lin. long, enwrapping the flower, hyaline-margined; calyx tubular-campanulate, 3½ lin. long, subulate in bud; lobes 5, ½ lin. long, rounded, entire; petals much longer than the calyx, linear-lanceolate, violet ?

SOUTH AFRICA : without locality, *Grey !*
COAST REGION : Bredasdorp Div.; sandy places near the mouth of the Ratel River, 150 ft., *Bolus,* 8576!

This resembles *S. scabra,* Thunb., but is of more lax habit.

10. S. anthericoides (Schlechter in Engl. Jahrb. xxiv. 450); a stemless herb; leaves in one or more radical rosettes, obovate- or elliptic-spathulate, obtuse, scabrid, coriaceous, 9 lin. long, 3 lin. wide; scape erect, about a foot high, verrucose, much (but laxly) branched above; spikelets 2-flowered, distant; bracts broadly ovate or suborbicular, obtuse, margin wide and hyaline; calyx obconic, 2–3 lin. long, 5-ribbed, pilose at the base; lobes 5, triangular, acute, long mucronate, alternating with 5 very short obtuse lobes; petals nearly 4 lin. long, narrow, obtuse or shortly emarginate; filaments glabrous, filiform from a dilated base; anthers oblong; styles filiform, free to the base, glabrous; ovary 5-angled, obovoid, glabrous.

COAST REGION : Bredasdorp Div.; heathy places near Elim, *Schlechter,* 7709! *Bolus,* 8575 !

11. S. decumbens (Boiss. in DC. Prodr. xii. 659); herbaceous, with small tufts of short hairs on the scapes and branches; radical leaves narrowly lanceolate-spathulate, obtuse, tapering into the petiole; branches decumbent or prostrate, filiform, long, simple or sparingly branched; panicles erect from the upper side of the branches, bearing 2–6 spikelets; spikelets 2–3-flowered, short, not crowded; lower bract very short, acute, not scarious at the margin, upper ovate-rotundate, convex, rather acute and hyaline at the apex; calyx quite glabrous; tube longer than the bracts; limb expanded, very shortly lobed.

South Africa : without locality, *Drège,* 9374 ex *Boissier.*
Coast Region : Malmesbury Div.; Saldanha Bay, *Grey !*

II. **PLUMBAGO**, Linn.

Calyx tubular, glandular, hyaline between the 5 ribs, persistent; lobes 5, short, erect. *Corolla* salver-shaped; tube slender; lobes 5, patent, equal or nearly so. *Stamens* free from the corolla; filaments filiform from a dilated base; anthers oblong-linear. *Ovary* tapering upwards; style terminal, filiform, divided in the upper part into 5 branches stigmatic along their inner surfaces. *Capsule* membranous, circumscissile near the base, the deciduous part often splitting into 5 valves from below. *Endosperm* scanty.

Perennial (rarely annual) herbs or shrubs, sometimes climbing; leaves alternate, auricled at the base, or the petiole amplexicaul; flowers in spikes at the ends of the branches, blue, rose, violet or white.

Distrib. Species about 10 in the warmer regions of both hemispheres.

Corolla-tube three times as long as the calyx; axis of
    spike puberulous ... ... (1) **capensis.**
Corolla-tube twice as long as the calyx; axis of spike
    glandular ... ... (2) **zeylanica.**

1. **P. capensis** (Thunb. Prodr. 33); stem woody, subscandent, striate; leaves entire, oblong or oblong-spathulate, obtuse or subacute, shortly mucronate, tapering downwards into a very short petiole, about 2 in. long and $\frac{3}{4}$ in. wide, glabrous; spikes about 2 in. long; rhachis puberulous; bracteoles oblong, acute, half as long as the calyx; calyx 5 lin. long, pubescent, also bearing in the upper part glands on stalks nearly 1 lin. long; corolla pale blue; tube $1\frac{1}{4}$ in. long; limb 1 in. in diam.; lobes obovate, obtuse; anthers blue; capsule oblong-clavate, rounded above, tapering and pentagonal below. *Fl. Cap. ed. Schult.* 166; *Bot. Mag. t.* 2110; *Bot. Reg. t.* 417; *Drège, Zwei Pflanzengeogr. Documente,* 133, 136; *Boiss. in DC. Prodr.* xii. 693. *P. auriculata, Poir. Encycl.* ii. 270. *P. grandiflora, Ten. Cat. Hort. Neap.* 1845, 91. *Plumbagidium auriculatum, Spach, Hist. Veg. Phan.* x. 339, *t.* 149.

South Africa : without locality, *Villette ! Pappe !*
Coast Region : Uitenhage Div.; Uitenhage, *Burchell,* 4406 ! *Zeyher !* Addo, *Burke !* Coega River, *Pappe !* Enon, under 500 ft. *Drège !* Zuurberg Range, 2000–3000 ft., *Drège !* and without precise locality, *Zeyher,* 568 ! Albany Div.; Howisons Port, near Grahamstown, *MacOwan, Herb. Norm. Austr.-Afr.,* 589 ! British Kaffraria, *Cooper,* 3046 !

CENTRAL REGION: Somerset Div.; on the Bosch Berg, *Burchell*, 3182!
3192!

EASTERN REGION: Natal; Inanda, *Wood*, 333! and without precise locality,
Gerrard, 1962! *Sutherland!*

2. P. zeylanica (Linn. Sp. Pl. ed. i. 151); stem woody, subscandent, striate, much branched; leaves ovate or oblong, acute,
shortly and abruptly contracted into the petiole, about $1\frac{1}{2}$ in. long
and 1 in. wide, glabrous; petiole $\frac{1}{2}$ in. long, amplexicaul and slightly
auricled at the base; rhachis of spike glandular; bracteoles oblong,
acuminate, $\frac{1}{3}$ the length of the calyx; calyx $\frac{1}{2}$ in. long, bearing long-
stalked glands throughout its length, otherwise glabrous; corolla
white; tube 1 in. long; limb $\frac{1}{2}$ in. in diam.; lobes obovate, obtuse
or retuse, mucronate; capsule elongate-oblong, tapering and 5-sulcate
upwards. *Bot. Reg.* 1846, *t.* 23; *Boiss. in DC. Prodr.* xii. 692;
Wight, Ill. t. 179; *Benth. Fl. Austr.* iv. 267; *C. B. Cl. in Hook.
f. Fl. Brit. Ind.* iii. 480. *P. flaccida, Moench, Meth.* 429. *P. sar-
mentosa, Lam. Ill.* i. 470, *partly.* *P. capensis, var. alba, Williams,
Catal.* 1886, 26. *Thela alba, Lour. Fl. Cochinch.* 119.

KALAHARI REGION: Transvaal; Berea, Barberton, 2918 ft., *Thorncroft in
Herb. Wood*, 3774! Kaap Valley, Barberton, 2000 ft., *Galpin*, 1298! 1349!

EASTERN REGION: Natal; Weenen County, *Wood*, 3555! Delagoa Bay,
Monteiro, 41!

Also in Tropical Africa, India, Malay Archipelago, Pacific Islands and
Australia.

Imperfectly known Species.

3. P. tristis (Ait. Hort. Kew. ed. 2, i. 324); stem woody, slender,
flexuous, much and intricately branched, shortly glandular-scabrid;
leaves obcordate-cuneate, retuse, mucronate, tapering into the petiole,
minutely scabrid, glandular-punctate, $\frac{1}{3}$ in. long including the
petiole, 2 lin. wide at the apex; flowers few, remote, in short
terminal spikes; bracteoles pubescent, the lowest oblong, $\frac{1}{2}$ the
length of the calyx, the lateral filiform, very small; calyx 5 lin.
long, shortly and adpressedly viscid-pubescent, also with very long
reddish spreading setæ on the ribs; corolla-tube more than twice
as long as the calyx; limb unknown; capsule oblong, acutely 5-
angled, puberulous on the angles above. *Boiss. in DC. Prodr.* xii.
693. *P. vogeliæfolia, Eckl. & Zeyh. ex Boiss. l.c.* 694.

SOUTH AFRICA: without locality, *Masson!*

Said to have been introduced into cultivation in 1792, but is now lost.

III. **VOGELIA**, Lam.

Calyx 5-partite almost to the base, strongly 5-nerved, appearing
winged from the out-turned margins of the segments, strongly
transversely wrinkled, persistent. *Corolla* funnel-shaped; lobes 5.
Stamens hypogynous. *Ovary* more or less 5-angled; style terminal,

with 5 arms stigmatose on the inner side. *Capsule* circumscissile at the base, and splitting upwards into 5 valves.

More or less scaly shrubs or undershrubs; leaves alternate, entire; flowers in dense terminal spikes; bracts and bracteoles small.

DISTRIB. A second species in India and a third in Socotra.

1. **V. africana** (Lam. Ill. ii. 148, t. 149); stems slender, woody, much branched, 2 ft. or more high, sulcate, glaucous; leaves alternate, often distant and few, fleshy, broadly cuneate, rounded, emarginate or cordate at the apex, sometimes mucronate, more or less coated with calcareous granules, up to 9 lin. long and 6 lin. wide, tapering into the 4 lin. long petiole; flowers patent; calyx-segments ovate, abruptly acuminate, 4 lin. long, 2 lin. wide; corolla-tube subcylindrical, 4 lin. long; lobes cuneate or obcordate, cuspidate in the sinus, 2 lin. long, $1\frac{1}{2}$ lin. broad; filaments very slender; anthers $\frac{1}{2}$ lin. long, sagittate; ovary oblong, acuminate, 1 lin. long; style $3\frac{1}{2}$ lin. long including the 1 lin. long arms, spreadingly hairy; capsule longer than the persistent calyx. *Boiss. in DC. Prodr.* xii. 696; *Drège, Zwei Pfl. Docum.* 90; *Harv. Thes. Cap.* ii. 62, *t.* 198; *Hiern in Cat. Afr. Pl. Welw.* i. 635. *V. sp., Welw. in Journ. Linn. Soc.* v. 184. *Dyerophytum africanum, O. Kuntze, Rev. Gen. Pl.* ii. 394.

WESTERN REGION: Little Namaqualand; between Buffels (Koussies) River and Silverfontein, 2000 ft., *Drège!* near Spektakel, 800 ft., *Bolus,* 635! and without precise locality, *Wyley!*

CENTRAL REGION: Calvinia Div.; between Lospers Flats and Springbok Kuil River, *Zeyher,* 1431!

Also in Angola and Hereroland.

ORDER LXXXI. **PRIMULACEÆ.**

(By W. H. HARVEY, with additions by C. H. WRIGHT.)

Flowers hermaphrodite, usually regular. *Calyx* inferior or half-superior, 5-fid or 5-partite. *Corolla* rotate, campanulate or salver-shaped, 5-lobed. *Stamens* 5, inserted in the corolla-tube and opposite the lobes, often alternating with staminodes; filaments filiform or subulate, short; anthers 2-celled, dehiscing longitudinally. *Ovary* superior or half-inferior, 1-celled, with a free-basal placenta; style terminal; stigma undivided; ovules numerous, semi-anatropous. *Capsule* circumscissile or bursting by valves. *Seeds* impressed in the placenta; embryo straight, terete; endosperm fleshy or horny.

Herbs usually perennial, rarely suffrutescent; leaves radical, opposite or alternate; flowers solitary and axillary, or in umbels, racemes or panicles.

DISTRIB. Genera 22; species about 300, in temperate regions of the Northern Hemisphere, rare in the Southern or in the tropics.

I. **Lysimachia.**—*Ovary* superior.　*Stamens* glabrous, without staminodes. *Capsule* 5-10-valved.

II. **Anagallis.**—*Ovary* superior.　*Stamens* hairy, without staminodes.　*Capsule* circumscissile.

III. **Samolus.**—*Ovary* half-inferior.　*Stamens* alternating with staminodes. *Capsule* 5-valved.

I. **LYSIMACHIA**, Linn.

Calyx 5-partite.　*Corolla* deeply 5-partite, subrotate or campanulate, longer than the calyx; tube very short.　*Stamens* 5, on the base of the corolla; filaments sometimes connate at the base; sometimes with as many intermediate sterile filaments.　*Capsule* globose, opening by 5-10 valves at the apex, many-seeded, free.　*DC. Prodr.* viii. 60.

A large European, Asiatic and American genus of herbaceous, mostly perennial, plants; leaves alternate, opposite or whorled, entire; flowers either axillary, racemose, spiked or panicled, yellow or purple. Name derived from λυσις (loosening), μαχη (strife), referring to the supposed property of soothing pain.

Stamens exserted　　…　　…　　…　　…　　…　　…　(1) **nutans.**
Stamens included :
　　Corolla 4 lin. long　　…　　…　　…　　…　　…　(2) **Woodii.**
　　Corolla 2 lin. long　　…　　…　　…　　…　　…　(3) **parviflora.**

1. **L. nutans** (Nees, Del. Sem. Hort. Bonn. 1831); stem erect, subsimple; leaves opposite, alternate or in threes, lanceolate, acute or acuminate, narrowed at the base, pale beneath, with subrevolute roughish margins, glabrous; flowers in a dense, afterwards lengthening, terminal raceme; bracteoles linear-subulate, shorter than the pedicels; calyx much shorter than the corolla, with broadly linear, obtuse lobes; corolla tubular-bell-shaped; lobes eroso-denticulate at the apex; stamens exserted; style subulate. *Duby in DC. Prodr.* viii. 61; *Pax & Knuth in Engl. Pflanzenr. Primul.* 294.　*L. atropurpurea, Hook. Exot. Fl. t.* 180, *not of Linn. Lubinia atropurpurea, Link & Otto, Ic. Pl. Sel. t.* 27; *Sweet, Brit. Fl. Gard. ser.* 2, i. *t.* 34.　*Coxia, Endl. Gen.* ii. 733.

COAST REGION : Uitenhage Div.; between Hoffmanns Kloof and Drie Fontein, 1000–2000 ft., *Drège!* Zuurberg Range, near Bontjes River, 2000 ft., *Drège!* Albany Div.; banks of streams near Grahamstown, *Burke! MacOwan,* 137! *Glass in MacOwan, Herb. Austr.-Afr.,* 1497! near Assegai Bush, *Burchell,* 4158! and without precise locality, *Zeyher !*

CENTRAL REGION : Tarka Div.; bank of the Tarka River, *Cooper,* 336 !

Stem 1–2 ft. high, leafy throughout; leaves 2–2½ in. long, ½–¾ in. broad; raceme oblong, somewhat nodding; flowers dark purple, handsome.

2. **L. Woodii** (Schlechter ex Pax & Knuth in Engl. Pflanzenr. Primul. 292.) ; erect, quite glabrous, branched above; leaves lanceolate or ovate-lanceolate, acuminate, alternate or opposite, about 3 in. long, nearly 1 in. wide, sessile and sheathing at the base; racemes up to 8 in. long, dense above; pedicels erecto-patent, nearly 2 lin. long; calyx puberulous outside, divided nearly to the base, ¼ in. long;

corolla slightly longer than the calyx, dirty white ; lobes oblong, obtuse, from as long as the tube to twice its length ; stamens shorter than the corolla and adnate to the base of its lobes ; capsule globose, about 2 lin. in diam., smooth, beaked by the persistent style.

EASTERN REGION : Natal ; in a damp valley near Van Reenens Pass, 5000–6000 ft., *Wood,* 4522 !

3. L. parviflora (Baker in . Journ. Linn. Soc. xx. 196) ; a herb, 2–4 ft. high, branched above, entirely glabrous ; stem terete ; leaves usually alternate, sessile, broadly lanceolate, tapering to the base ; racemes up to 6 in. long, at first dense at the top ; bracteoles linear, nearly as long as the flowers ; pedicels 1 lin. long in flower, 2 lin. long in fruit ; calyx $1\frac{1}{2}$ lin. long ; lobes oblong, $\frac{1}{3}$ lin. wide, with brown oblique glands in the upper part ; corolla campanulate, 2 lin. long ; lobes obovate or nearly spathulate, longer than the tube ; stamens shorter than the corolla-lobes ; anthers cordate, shortly mucronate ; ovary globose ; capsule about 2 lin. in diam.; style about $\frac{1}{2}$ lin. long in fruit. *Pax & Knuth in Engl. Pflanzenr. Primul.* 291.

KALAHARI REGION : Transvaal ; Macmac, *Mudd !*

Also in Madagascar.

II. ANAGALLIS, Linn.

Calyx 5-partite. *Corolla* rotate or campanulate, deciduous, longer than the calyx, very deeply 5-partite ; tube scarcely any ; lobes oblong or lanceolate. *Stamens* 5, at the base of the corolla ; filaments bearded. *Capsule* globose, membranous, circumscissile. *Seeds* numerous, sunk in the central placenta. *DC. Prodr.* viii. 69.

Small herbaceous or rarely suffruticose plants of the Eastern Hemisphere ; leaves opposite or alternate, entire. Peduncles axillary, 1-flowered. Name, from αναγελαω, to laugh ; application obscure.

Leaves opposite :
    Annual ; leaves sessile ; flowers red or blue ... (1) **arvensis.**
    Perennial ; leaves petiolate ; flowers white ... (2) **Huttoni.**
Leaves usually alternate, ovate or oblong, acute and
    mucronate ... ... (3) **pumila, var.**

1. A. arvensis (Linn. Sp. Pl. ed. i. 148) ; annual ; stems diffuse, branched ; branches elongate, decumbent, quadrangular, with narrow wings ; leaves opposite or ternate, ovate, sessile, subacute, spreading ; peduncles longer than the leaves ; calyx-lobes lanceolate, acuminate, keeled, nearly as long as the corolla ; corolla crimson to pale rosy ; lobes twice as long as the stamens, obovate, obtuse, denticulate, usually glandular-ciliate at the apex. *Curt. Fl. Lond. t.* 12 ; *Plenck, Ic. t.* 82 ; *Duby in DC. Prodr.* viii. 69 ; *Benth. Fl. Austral.* iv. 270 ; *Oliv. Fl. Trop. Afr.* iii. 490 ; *Hook. f. Fl. Brit. Ind.* iii. 506 ; *Pax & Knuth in Engl. Pflanzenr. Primul.* 322. *A. pulchella, Salisb.*

Prodr. 120. *A. phœnicea, Scop. Fl. Carn. ed.* 2, i. 139. *A. arabica, Duby in DC. Prodr.* viii. 69. *A. sp., Drège, Zwei Pfl. Documente,* 87, 105.

VAR. β, cœrulea (Gren. & Godr. Fl. France, ii. 467); corolla blue; lobes slightly ciliate at the apex, but not glandular. *Pax & Knuth, l.c.* 323. *A. cœrulea, Schreb. Spicil. Fl. Lips.* 5. *A. capensis, E. Meyer in Drège, l.c.* 98, *not of Linn. A. indica, Sweet, Fl. Gard. t.* 132.

SOUTH AFRICA: without locality, var. β: *Pappe! Miller! Thom! Harvey,* 224! 432!

COAST REGION: Cape Div.; Table Mountain and Devils Mountain, *Drège,* 8015a! Green Point, *MacGillivray,* 575! Paarl Div.; Simons Bay, *Wright!* near Simons Town, *Milne,* 155! Paarl Mountain, *Drège,* 8015b! Riversdale Div.; near Great Valsch River, *Burchell,* 6552! road-side, Garcias Pass, 1000 ft., *Galpin,* 4315! Var. β: Cape Div.; Devils Peak, *Wilms,* 3538! Paarl Div.; between Paarl and Lady Gray Railway Bridge, *Drège!*

KALAHARI REGION: Transvaal; near Lydenburg, *Wilms,* 1244!

EASTERN REGION: Var. β: Griqualand East; Shawbury, *Baur,* 199!

Cosmopolitan. Not having seen specimens from the following localities, we are unable to decide to which form they belong:—Cape Div.; sand-flats between Tygerberg and Blueberg, *Drège;* between Rondebosch and Hout Bay, *Drège;* Worcester Div.; Hex River Kloof, 1000–2000 ft., *Drège,* 8015c.

2. A. Huttoni (Harv. in Proc. Dubl. Univ. Zool. & Bot. Assoc. i. 141); perennial; stems decumbent or creeping (rarely "erect," *Gerrard*), simple or branched, 4-angled; leaves roundish-ovate or subrotund, petioled, spreading, opposite or in threes; peduncle longer than the leaf; calyx-lobes lanceolate, acuminate, keeled, shorter than the corolla; corolla rotate, its lobes subacute or obtuse; filaments broadly subulate, pilose. *Thes. Cap.* i. 3, *t.* 4.

VAR. β, nummularia (Harv.); stem suberect; leaves subsessile, subcoriaceous, broader and blunter; corolla-lobes very obtuse.

COAST REGION: Uniondale Div.; Long Kloof, at Aapies River, *Burchell,* 4949! Albany Div.; Howisons Poort, *Hutton!* near Grahamstown, *MacOwan,* 1027! and *Herb. Austr.-Afr.,* 1632!

KALAHARI REGION: Transvaal; Lydenburg district, near O'Neill's Farm, *Wilms,* 1243!

EASTERN REGION: Griqualand East; banks of streams on Mount Currie, 6500 ft., *Tyson,* 1359! Natal; near Maritzburg, 2500 ft., *Stainbank in Herb. Wood,* 3570! Var. β: Transkei; Kreilis Country, *Bowker!* Zululand, *Gerrard & M'Ken,* 2099!

Stems 1–2 ft. long, creeping in damp places, where water lodges, or near springs, nearly always in blossom. Leaf-pairs subdistant; leaves 4–5 lines long and as broad or broader, membranous and veiny. Flowers white. Var. β, which at first I was disposed to regard as a species, is a stronger growing plant, more erect, with more opaque and tougher, broader and less distinctly petioled leaves, and broader and blunter corolla-lobes.—*W. H. H.*

3. A. pumila, var. **natalensis** (Knuth ex Pax & Knuth in Engl. Pflanzenr. Primul. 332); a branched herb; stem decumbent and rooting at the base; branches slender, about 6 in. long; leaves alternate or the lower opposite, sessile or nearly so, ovate, shortly cuspidate; pedicels patent, longer than the leaves, filiform; calyx 5–6-partite;

lobes linear-lanceolate, acuminate, about 1 lin. long ; corolla nearly twice as long as the calyx, deeply lobed, white ; lobes lanceolate, obtuse ; stamens about $1\frac{1}{2}$ lin. long ; style 1 lin. long, filiform.

EASTERN REGION : Natal ; Inanda, *Wood*, 1609 ! Palmiet River, Umgena, *Gerrard*, 741 ! near the sea, Umtwalumi, *Gerrard & McKen*, and without precise locality, *Sanderson*, 685 !

III. **SAMOLUS**, Linn.

Calyx half-superior, 5-fid. *Corolla* perigynous, salver- or bell-shaped, deciduous ; limb 5-partite, with 5 narrow scales (abortive stamens ?) rising from the mouth of the tube and alternating with the corolla-lobes and stamens. *Stamens* 5, on the corolla-tube ; anthers basifixed. *Ovary* half-inferior, many-seeded ; style filiform. *Capsule* opening above by 5 valves. *DC. Prodr.* viii. 72.

Herbs of both hemispheres, growing in marshy places, generally near the sea. Radical leaves petioled, cauline alternate, diminishing upwards or reduced to scales. Flowers racemose or panicled. Name, derivation obscure.

Softly herbaceous ; stems subsimple, leafy ; pedicels naked
 at the base, with a bracteole in the middle ... ... (1) **Valerandi.**
Rigid ; stem much-branched ; branches erect, rod-like,
 rough ; upper leaves subulate ; pedicels without a
 medial bracteole ... ... (2) **porosus.**

1. **S. Valerandi** (Linn. Sp. Pl. ed. i. 171) ; radical leaves numerous, oval or obovate, obtuse, tapering at the base into long petioles ; stem erect, simple or paniculately branched, terete, fistular ; cauline leaves diminishing upwards, obovate, subsessile ; flowers racemose or panicled ; pedicels (often knee-bent) bearing a small narrow lanceolate bracteole in the middle ; calyx half as long as the corolla, its lobes triangular, subacute ; stamens included ; scales subulate, as long as the stamens ; valves of capsule not reflexed. *Thunb. Fl. Cap. ed. Schult.* 176 ; *Duby in DC. Prodr.* viii. 73 ; *Pax & Knuth in Engl. Pflanzenr. Primul.* 337.

SOUTH AFRICA : without locality, *Wallich !*
COAST REGION : Cape Div. ; Table Mountain, *Ecklon*, 25 ! Cape Flats near Doornhoogte, *Wolley Dod*, 679 ! King Williamstown Div. ; near Kei Road, 2000 ft., *Schlechter*, 6131 ! Uitenhage Div. ; near the Lead-mine, *Burchell*, 4489 ! Port Elizabeth Div. ; sand-hills near Port Elizabeth, *Drège !*
CENTRAL REGION : Somerset Div. ; Somerset East, *Miss Bowker !* Albert Div., *Cooper*, 1779 !
KALAHARI REGION : Griqualand West ; Lower Campbell, *Burchell*, 1794 ! Griqua Town, *Burchell*, 1873 ! Orange River Colony, *Cooper*, 2822 ! Bechuanaland ; near the sources of the Kuruman River, *Burchell*, 2511 !
EASTERN REGION : Natal ; Inanda, *Wood*, 314 ! 465 ! Mount Edgecombe, *Wood*, 1100 ! and without precise locality, *Gerrard*, 164 !

Found in most parts of the world in wet spongy ground. Stems 1–2 ft. high the upper half occupied by the long simple or branched raceme. Leaves bright green and glossy, soft to the touch, the radical ones, including the petiole, 2–4 in. long, $\frac{3}{4}$–$1\frac{1}{4}$ in. wide.

2. S. porosus (Thunb. Fl. Cap. ed. 2, ii. 32); stems, from a horizontal rhizome, erect, much-branched, rigid; branches very erect, rough with little raised dots; radical leaves few, obovate-spathulate, obtuse or subtruncate, tapering into a winged petiole; rameal leaves small, diminishing upwards, passing from linear to subulate, sessile; flowers racemose; pedicels from the axils of depauperated subulate leaves, ebracteolate; calyx half as long as the corolla, its lobes lanceolate, acuminate. *Pax & Knuth in Engl. Pflanzenr. Primul.* 342. *S. africanus, Burm. f. Prodr. Cap.* 5? *S. Valerandi, var. africanus, Linn. Sp. Pl. ed.* 1, 172. *S. campanuloides, R. Br. ex Duby in DC. Prodr.* viii. 73. *Campanula porosa, Thunb. Prodr.* 39.

SOUTH AFRICA: without locality, *Pappe! Thom*, 816! *Harvey*, 217! *Armstrong!*

COAST REGION: Vanrhynsdorp Div.; Ebenezar, *Drège!* Cape Div.; Lion Mountain, *Ecklon*, 26! Chapman Bay, *Wolley Dod*, 1668! Raapenberg Vley, *Wolley Dod*, 322! Riversdale Div.; by the Zoetemelks River, *Burchell*, 6610! Mossel Bay Div.; between Duyker River and Gauritz River, *Burchell*, 6392! George Div.; in a wood by the west bank of Kaymans River, *Burchell*, 5808! Uitenhage Div.; near Enon, *Baur! Zeyher*, 665! Port Elizabeth Div.; Port Elizabeth, *Burchell*, 4373! *Drège!* Bathurst Div.; at Kaffir Drift Military Post, *Burchell*, 3762! near Barville Park, *Burchell*, 4141! British Kaffraria, *Cooper*, 63!

CENTRAL REGION: Somerset Div.; Somerset East, *Bowker!*

EASTERN REGION: Pondoland; near the mouth of Umtentu River, *Drège!* Natal; Durban, *Wood*, 1940! *and in MacOwan, Herb. Austr.-Afr.*, 1498! *Plant*, 33!

Stems 1–2 ft. high or more, rigid and broom-like, naked except for the scattered subulate leaf-scales. Racemes ending the cord-like branches, 4–6 in. long, many-flowered. Corolla white, 3–4 lines long.

The reference usually given for *S. campanuloides* is " *R. Br. Prodr.* 429 (1810)", but such a name does not occur amongst the species enumerated in that place.—*C. H. W.*

Order LXXXII. **MYRSINEÆ.**

(By W. H. HARVEY, with additions by C. H. WRIGHT.)

Flowers regular, hermaphrodite or diœcious. *Calyx* inferior or in *Mæsa* slightly adnate at its base to the ovary, 4–6-fid or -partite; lobes usually ciliate, valvate, contorted or imbricate, usually persistent. *Corolla* usually rotate, sometimes with 4–6 free petals, at others 4–6-lobed; lobes contorted or imbricate, rarely valvate. *Stamens* as many as the corolla-lobes and opposite them; filaments usually short, adnate to the disk or corolla-tube; anthers dorsifixed near the base, dehiscing longitudinally, rarely by pores, sometimes divided into superposed cells; staminodes 0 or one opposite each sinus of the corolla, rarely also opposite its lobes. *Ovary* superior or in *Mæsa* slightly adnate to the calyx, 1-celled; placenta free-central, often globose; style short or long; stigma simple, rarely obscurely lobed; ovules few or many. *Fruit* usually small; epicarp fleshy; endocarp crustaceous. *Seed* usually solitary, globose; testa thin; albumen fleshy or horny, uniform or ruminate; embryo cylindrical, more or less curved.

Trees, shrubs or undershrubs, usually glabrous; leaves alternate, very rarely opposite, sometimes nearly verticillate, entire or toothed, often bearing translucent round or elongated resinous glands, exstipulate; inflorescence cymose, umbellate, fascicled, racemose or panicled; flowers usually small, often glandular, usually white or rosy.

DISTRIB. About 600 species, chiefly in the tropics of both hemispheres, some extending into the south and a few into the north extra-tropical regions.

I. **Mæsa.**—*Ovary* inferior or half-inferior. *Seeds* many. *Petals* united.
II. **Embelia.**—*Ovary* superior. *Seeds* solitary. *Petals* free.
III. **Myrsine.**—*Ovary* superior. *Seeds* solitary. *Petals* united.

I. **MÆSA**, Forsk.

Calyx bibracteolate at the base, its tube adherent to the ovary, limb 5-lobed, the lobes quincuncial in æstivation, 2 outer overlapping 3 inner. *Corolla* 5-fid, subcampanulate, the lobes obtuse with inflexed apices. *Stamens* 5, included, free. *Ovary* inferior or half-inferior, with a basilar placenta; style short; stigma capitate; ovules numerous. *Berry* covered by the calyx, ovoid. *DC. Prodr.* viii. 77.

African or Asiatic trees and shrubs; leaves alternate, penninerved and netted-veined, sometimes pellucid dotted; racemes or panicles axillary or terminal, many-flowered; flowers small, white. Name, *maas*, the Arabic name of one of the species.

Young branches tomentose; leaves elliptic, acute or acuminate, coarsely toothed; panicles much-branched   (1) **rufescens.**
Young branches puberulous; leaves cuneate-obovate, crenato-dentate beyond the middle; racemes simple or subsimple   ...   .   (2) **alnifolia.**
Young branches glabrous; leaves elliptic or oblong elliptic, irregularly serrate; panicles branched   ...   (3) **angolensis.**

1. **M. rufescens** (A.DC. in DC. Prodr. viii. 81); branches entirely or only at the apex rusty-tomentose; leaves elliptic, acute or shortly acuminate, coarsely serrate rarely entire, up to 3 in. long and $1\frac{1}{2}$ in. wide, glabrous or pubescent beneath; petiole up to 1 in. long; panicle many-flowered, shorter than the leaves, rhachis and branches covered with patent hairs; pedicels less than a line long; sepals $\frac{1}{2}$ lin. long, triangular, acute, ciliate; corolla about 1 lin. long, ciliate; stamens slightly exserted; style about as long as the ovary; stigma obtuse. *Mez in Engl. Pflanzenr. Myrsinaceœ,* 25. *M. palustris, Hochst. in Flora,* 1844, 825. *Bæobotrys rufescens, E. Meyer ex A.DC. l.c. Choristylis rhamnoides, Harv. in Harv. & Sond. Fl. Cap.* ii. 308 *partly.*

KALAHARI REGION: Transvaal; Houtbosch, *Rehmann,* 6012!

2. M. alnifolia (Harv. Thes. ii. 20, t. 129); twigs and petioles
rufo-puberulous; leaves cuneate-obovate, subtruncate, rarely acute,
penninerved, beyond the middle crenato-dentate, glabrous, very
minutely puberulous beneath; racemes axillary, puberulous, simple
or slightly branched; flowers polygamo-diœcious; bracts much
shorter than the pedicels, deciduous; calyx-lobes ovate, downy;
corolla of the male twice as long as the calyx. *Mez in Engl.
Pflanzenr. Myrsinaceæ*, 23, *fig.* 1.

COAST REGION: Stockenstrom Div.; in woods on the Elandsberg Moun-
tains, 4000 ft., *Scully in MacOwan & Bolus, Herb. Norm. Austr.-Afr.*, 757!
Komgha Div.; Gonubie River, 2000 ft., *Schlechter*, 6151! British Kaffraria,
Cooper, 291! 423!
EASTERN REGION: Transkei Div.; Krcilis Country, *Bowker*, 233! Natal;
near Durban, *Sanderson*, 463! Umcomaas, *Gerrard*, 1401! Umzinyati, 1200 ft.,
Wood, 1400! and without precise locality, *Sanderson*, 314!

A large shrub or small tree; leaves on ½–1 in. long petioles, 1–1½ in. long,
1 in. wide, mostly of the shape represented in the figure above quoted, but
sometimes they are more extensively serrated and even acute, the form changing
to obovate-lanceolate.

3. M. angolensis (Gilg in Notizbl. Königl. Bot. Gart. Berlin, i.
72); branches glabrous; leaves elliptic or oblong-elliptic, shortly
acuminate, acute at the base, about $3\frac{1}{2}$ in. long, $1\frac{1}{2}$ in. wide,
irregularly serrate; petiole about 1 in. long; flowers in axillary
panicles up to 3 in. long; pedicels very short; bracteoles subulate;
calyx divided nearly to the base; segments ovate, subacute; corolla
white; lobes ovate, acute; stamens slightly exserted; ovary glabrous; style short; stigma subdiscoid. *Hiern in Cat. Afr. Pl.
Welw.* i. 637; *Mez in Engl. Pflanzenr. Myrsinaceæ*, 26.

EASTERN REGION: Natal, *Cooper*, 1209!
Also in Tropical Africa.

II. **EMBELIA**, Burm.

Calyx 5-partite, 5-fid or 5-toothed; lobes twisted in bud. *Petals*
5, spreading or reflexed, imbricate. *Stamens* 5; filaments attached
to the base of the petal, filiform above; anthers ovoid, much shorter
than the filament. *Ovary* ovoid or depressed, often minute; style
short; stigma capitellate or simple; ovules 4–1, on a minute basal
placenta. *Drupe* globose. *DC. Prodr.* viii. 83.

Shrubs or small trees, Asiatic and African; leaves alternate, mostly entire;
flowers small, in racemes or panicles. Name, from the Cingalese name of one of
the species.

1. E. Kraussii (Harv. Thes. Cap. ii. 17, t. 127); glabrous; leaves
obovate-oblong, acute, quite entire, with reflexed edges, cuneate at
the base, shortly petiolate; racemes lateral, shorter than the leaf;

calyx 5-toothed, its teeth broadly deltoid, acute; petals oblong, spreading, twice as long as the calyx. *E. ruminata, Mez in Engl. Pflanzenr. Myrsin.* 331. *Myrsine? ruminata, E. Meyer in Drège, Zwei Pfl. Documente,* 155, 159; *DC. Prodr.* viii. 104. *Celastrus oleoides, Lam. Ill.* ii. 93; *Hochst. in Flora,* 1844, 304.

COAST REGION: Komgha Div.; near Komgha, 2000 ft., *Flanagan,* 284!

EASTERN REGION: Pondoland; between Umtentu River and Umzimkulu River, *Drège.* Natal; near Durban, *Drège! Krauss,* 407! *Wood,* 1350! Mount Edgecombe, *Wood,* 1149! and without precise locality, *Gerrard,* 292! *Wood! Cooper!*

Leaves 1-1½ in. long, thin and membranous. Racemes from old leaf-scars, ⅓-¾ in. long, 8-12-flowered. Ovary half sunk in a fleshy disk. Fruiting specimens of *Myrsine? ruminata,* E. Meyer, agree in foliage with our *E. Kraussii,* and come from Natal; the account of the inflorescence given by A. Decandolle also agrees with our plant, and to my eye the two are identical.—*W.H.H.*

III. **MYRSINE**, Linn.

Flowers polygamo-diœcious. *Calyx* 4–5-partite. *Corolla* 4–5-partite, usually imbricate. *Stamens* free, inserted on the base of the corolla; anthers longer than the filaments. *Ovary* globose; style cylindrical, short, caducous; stigma capitate. *Placenta* spherical; ovules 4–5 round the apex of the placenta, peltate. *Drupe* globose, with a crustaceous stone, 1-seeded. *DC. Prodr.* viii. 92.

Trees or shrubs of warm climates. Leaves alternate, rarely toothed. Flowers small, in axillary tufts; calyx-lobes very small. Name, μυρσινη, myrrh, which is not a product of these plants.

M.? racemosa, Steud. Nomencl. ed. 2, ii. 176 is *Ocotea bullata,* E. Meyer (*Laurineæ*).

Leaves serrulate beyond the middle, elliptic. (A bush) ... ... (1) **africana.**
Leaves quite entire. (Shrubs or trees):
    Leaves veinless beneath, narrow; drupes ellipsoid, acute ... ... (2) **gilliana.**
    Leaves more or less veined beneath; drupes globose:
        Leaves 3–5 in. long, 1-1¼ in. wide, tapering at the base into longish petioles ... (3) **melanophlœos.**
        Leaves 1-1¼ in. long, ½ in. wide, acute at the base, on shortish petioles ... ... (4) **Gerrardii.**

1. **M. africana** (Linn. Sp. Pl. ed. i. 196); twigs slightly puberulous; leaves glabrous, coriaceous, acute or obtuse, elliptic, acute at the base, serrulate beyond the middle; tufts about 3-flowered; calyx- and corolla-lobes ovate, acute, spotted, the corolla longer than the calyx and shorter than the stamens. *Thunb. Fl. Cap. ed. Schultes,* 195; *Drège, Zwei Pfl. Documente,* 60, 115, 136; *DC. Prodr.* viii. 93; *C. B. Cl. in Hook. f. Fl. Brit. Ind.* iii. 511; *Mez in Engl. Pflanzenr. Myrsin.* 340. *M. retusa, Ait. Hort. Kew.* i. 271 (*with blunt leaves*); *Jacq. Hort. Schœnbr.* iv. t. 424. *M. africana, var. retusa, A.DC. in Trans. Linn. Soc.* xvii. 105. *M. dioica, Aschers.*

& Schweinf. in Verh. Berl. Ges. Erdk. xviii. (1891) 549. *Buxus dioica, Forsk. Fl. Ægypt.-Arab.* 159.

SOUTH AFRICA: without locality, *Pappe! Thom,* 296! 497!

COAST REGION: Cape Div.; Devils Peak, *Wolley Dod,* 2513! Kloof between the Lions Head and Table Mountain, *Burchell,* 246! Simons Bay, *Wright!* Caledon Div.; Bavaans Kloof, *Drège!* Swellendam Div.; Zuurbraak Mountain, 600 ft., *Galpin,* 4316! Knysna Div.; Korata (Karratera) River, *Drège!* Uitenhage Div.; Zuurberg Range, *Drège!* Albany Div.; mountains near Grahamstown, *Zeyher,* 215! *MacOwan,* 55! Queenstown Div.; Shiloh, *Baur,* 873! Cathcart Div.; between Windvogel Mountain and Zwart Kei River, *Drège!* and without precise locality, *Cooper,* 233! 234!

CENTRAL REGION: Somerset Div.; Bosch Berg, *MacOwan,* 55! Aberdeen Div.; Candeboo Mountains, *Drège!* Middelburg Div.; near Middelburg, *Burchell,* 2807!

KALAHARI REGION: Basuto Land, *Cooper,* 738! 2697! Traansvaal; Shiluvane, *Junod,* 1269! mountains north of Blauw Bank, *Nelson,* 265!

EASTERN REGION: Griqualand East; at the foot of Mount Currie, 5000 ft., *Tyson,* 1901! Natal; Tugela River, *Gerrard,* 1776! 1776 bis! near Van Reenen, *Schlechter,* 6999!

A bush, 2–4 ft. high. Very variable in the size and somewhat in the shape of the leaves. Commonly the leaves are about ½ in. long and 3 lines wide; sometimes half this size, and sometimes 1–1½ in. long and ¾ in. wide.

Also in Tropical Africa, and from Arabia to Central China.

2. **M. gilliana** (Sond. in Linnæa, xxiii. 76); glabrous; leaves narrow-oblong, tapering at the base into a short petiole, minutely recurved at the apex, subemarginate, coriaceous, veinless, dotted beneath, with subrecurved edges; tufts 3–5-flowered; pedicels angular, glabrous, longer than the petioles; calyx-teeth obtuse, ciliolate; corolla-lobes ovate, acute, twice as long as the calyx and longer than the anthers; drupe ellipsoid, acute. *Rapanea gilliana, Mez in Engl. Pflanzenr. Myrsin.* 376.

COAST REGION: Humansdorp Div.; Gamtoos River, *Gill!* Uitenhage Div.; sand-hills between the mouths of the Coega and Zwartkops Rivers, *Zeyher,* 554! 3371! Port Elizabeth Div.; downs near Cape Recife, *Zeyher,* 754! shore near Port Elizabeth, *Ecklon & Zeyher!* by the Baakens River, *Burchell,* 4347!

Well distinguished from *M. melanophlœos,* R. Br., by its smaller leaves (1½–2 in. long, 4–6 lines wide), the absence of obvious veins, the shorter petiole and the ovoid acute drupes.

3. **M. melanophlœos** (R. Br. Prodr. 533); leaves oblong or oblong-lanceolate, obtuse, coriaceous, quite entire, tapering at the base into a longish petiole, finely penninerved and obscurely veined beneath, with subrecurved edges; bracts ovate, subciliate; tufts 3–4-flowered; pedicels angular, glabrous, short; calyx-teeth acute; corolla twice as long as the calyx, as long as the anthers, its lobes lanceolate; drupe globose. *DC. Prodr.* viii. 97; *Drège, Zwei Pfl. Documente,* 85, 87, 135 (*melanophlæa*). *M. Samara, Roem. & Schult. Syst.* iv. 511. *M. venulosa, Spreng. Syst.* i. 663. *Sideroxylon melanophlœos, Linn. Mant.* 48, *excl. syn.; Jacq. Hort. Vind. t.* 71. *Rœmeria melanophlœa, Thunb. Fl. Cap. ed. Schult.* 194. *Rapanea melanophlœos, Mez in Engl. Pflanzenr. Myrsin.* 375.

SOUTH AFRICA : without locality, *Villette!*

COAST REGION: Paarl Div. ; Paarl Mountain, *Drège!* Caledon Div.; Klein River Mountains, *Zeyher*, 3370 ! Uitenhage Div.; between Hoffmanns Kloof and Driefontein, *Drège!* Albany Div.; between Riebeek East and Grahamstown, *Burchell*, 3464 ! Howisons Poort, *Cooper*, 2700! Kloof west of Grahamstown, *Burchell*, 3575 !

CENTRAL REGION : Somerset Div.; on Bosch Berg, *Burchell*, 3190 !

EASTERN REGION: Natal ; Inanda, *Wood*, 588! near Kettle Fontein, *Cooper*, 1221 ! Durban, *Gueinzius!* and without precise locality, *Cooper*, 2699 !

A large shrub. Leaves 3–5 in. long, 1–1¼ in. broad; the petiole ½–¾ in. long.

4. M. Gerrardii (Harv.) ; glabrous ; leaves rather small, oblong-lanceolate, subobtuse, acute at the base and shortly petiolate, with subrevolute margins, midribbed and faintly nerved beneath; flowers not seen ; pedicels about equalling the petiole ; drupe globose, with thin flesh, dry.

EASTERN REGION : Zululand ; Ingoma, *Gerrard*, 1157 !

"A tree 12–15 ft. high" (*Gerrard*). Leaves usually 1–1½ in. long, rarely 2 in., ½ in. wide, drying light green. Drupes as large as peppercorns.

ORDER LXXXIII. **SAPOTACEÆ.**

(By W. H. HARVEY, with additions by C. H. WRIGHT.)

Flowers hermaphrodite, regular. *Calyx* 4–8-partite. *Corolla* 4–8-lobed ; lobes imbricate, sometimes in two series. *Stamens* inserted on the corolla, as many as the corolla-lobes and opposite them, or more numerous and 2-seriate, sometimes with alternating staminodes; anthers usually extrorse, dehiscing longitudinally. *Ovary* many-celled; style conical or cylindrical ; stigma acute or capitellate ; ovules solitary, ascending. *Fruit* a 1- or many-celled berry. *Seeds* with a hard testa ; hilum often large ; albumen none or scanty ; cotyledons foliaceous.

Trees or shrubs, with milky juice; leaves alternate, entire, exstipulate; flowers axillary.

DISTRIB. Genera 18, species about 400, throughout the tropics and a little way outside.

I. **Chrysophyllum.**—*Calyx-lobes* 5 (rarely 6–7), not distinctly 2-seriate. *Stamens* as many as the corolla-lobes ; staminodes 0.

II. **Sideroxylon.**—*Calyx-lobes* 5 (rarely 6), not distinctly 2-seriate. *Stamens* as many as the corolla-lobes, alternating with as many staminodes.

III. **Mimusops.**—*Calyx-lobes* 6 or 8 in two series. *Stamens* 6 or 8, alternating with as many staminodes.

I. **CHRYSOPHYLLUM,** Linn.

Calyx 5- (rarely 6-) partite; lobes obtuse, pubescent, imbricate, 3 inside 2. *Corolla* tubular or campanulate-rotate, 5- (rarely 6–7-) lobed; lobes imbricate. *Stamens* as many as the corolla-lobes and opposite them, inserted on the tube ; sterile none ; anthers included,

sagittate anthers; staminodes long triangular, a little shorter than the stamens; ovary ovoid, shortly pilose; style slender, three times as long as the ovary.

EASTERN REGION: Natal; Inanda, *Wood*, 683!

4. M. caffra (E. Meyer ex A.DC. Prodr. viii. 203); twigs, petioles and under-surface of leaves clothed with minute adpressed pubescence; leaves obovate, cuneate at the base, coriaceous, with revolute edges; axils 2–3-flowered; pedicels recurved, rufo-tomentose, 2–3 times as long as the petiole; calyx-lobes lanceolate, acuminate, the 4 outer rufo-tomentose on the outside, the 4 inner narrower, with whitish tomentum; outer corolla-lobes equalling the calyx, inner rather longer. *Drège, Zwei Pfl. Documente,* 155; *Wood, Natal Plants,* i. 36, t. 43; *Engl. Sapot. Afr.* 72, t. 27, fig. B.

COAST REGION: Bathurst Div.; near Port Alfred, *Burchell,* 3805! at the mouth of the Great Fish River, *Burchell,* 3761! Kleinmund River, *Mac-Owan,* 393! King Williamstown Div.; Keiskamma, *Hutton!*
EASTERN REGION: Pondoland; between Umtentu River and Umzimkulu River, *Drège!* Natal; near Durban, *Krauss,* 76! *Peddie! Wood,* 1215! and without precise locality, *Gerrard,* 90! *Cooper,* 1243! Delagoa Bay, *Forbes!*

A large shrub or small tree. Leaves very leathery, 1–1½ in. long, ¾–1 in. wide, pretty constantly obovate and retuse, without obvious veins beneath. Pedicels nearly uncial, nodding. Drupe ovoid.

5. M. Zeyheri (Sonder in Linnæa, xxiii. 74); twigs and young leaves rufo-tomentulose; leaves on long petioles, oblong-lanceolate, obtusely acuminate, acute at the base, glabrous, coriaceous, margined, prominently midribbed and somewhat veined beneath; axils 3- or several-flowered; pedicels about equalling the petiole, recurved, as well as the calyx rufo-tomentose; calyx-lobes ovate, acuminate, the inner narrower and paler; corolla as long as the calyx; drupe ellipsoid; seed ovate, compressed, produced at the base. *Engl. Sapot. Afr.* 73.

KALAHARI REGION: Transvaal; Magalies Berg, *Zeyher,* 1130! *Burke,* 72!

A large shrub or tree. Leaves 3–4 in. long, exclusive of a subuncial petiole, 1–1½ in. broad. Drupe edible, sweetish, 1 in. long, quite smooth.

A variety occurs in Nyasaland.

6. M. marginata (N. E. Br. in Kew Bulletin, 1895, 108); a large shrub; branches brown or ashy-grey, more or less corrugated, glabrous; leaves elliptic-lanceolate or cuneate-oblanceolate, shortly and obtusely cuspidate or obtuse, acute at the base, fulvo-tomentose when young, glabrous when adult, 2–5 in. long, ¾–2¼ in. wide; petiole 2½–8 lin. long; flowers 6–16, umbellately arranged at the apex of the branches; pedicels rusty-tomentose, 1–1½ in. long; sepals 6–8, in two series, lanceolate, acuminate, 4–5 lin. long, 1½–2 lin. wide, the outer rusty-tomentose with grey margins, the inner grey-tomentose on both sides; petals 18–24 in three series, linear-oblong or lanceolate, acute, 4 lin. long, 1–1½ lin. wide, glabrous, yellow; stamens 6–8; anthers lanceolate, apiculate, much longer than the tomentose filaments; ovary globose-ovoid, densely hirsute; style elongate, glabrous; fruit rather large, ellipsoid, acute or acuminate. *Engl. Sapot. Afr.* 71.

COAST REGION : Komgha Div.; Komgha, *Flanagan,* 27!
EASTERN REGION : Natal; Inanda, *Wood,* 1661! near Umlaas, *Wood,* 5340! Tongaat, *Cooper,* 2479! Nonoti, *Gerrard,* 1186! and without precise locality, *McKen,* 6!

Native name, "Amapumbulo."

7. **M. obovata** (Sond. in Linnæa, xxiii. 75); twigs, petioles and both surfaces of the leaves quite glabrous; leaves obovate or obovate-oblong, obtuse or bluntly acuminate, subcoriaceous, with flat or scarcely subrecurved edges, beneath paler, prominently midribbed and obviously somewhat netted-veined; axils 1–2-flowered; pedicels ferruginous, erect, 4–5 times longer than the petiole; calyx-lobes lanceolate, acuminate, the outer rusty, the inner narrower, with pale tomentum; corolla as long as the calyx; drupe ovate or subglobose, apiculate; seed elliptic-oblong. *Engl. Sapot. Afr.* 72, *t.* 27, *fig. D. Imbricaria obovata, Nees ex Engl. l.c.*

VAR. β, **grandifolia** (Harv.); leaves longer and broader. *M. obovata, Harv. Thes. Cap.* i. 28, *t.* 44.

COAST REGION : Uitenhage Div.; Olifants Hoek, *Ecklon & Zeyher!* *Zeyher!* Albany Div. ; Featherstones Kloof near Grahamstown, *MacOwan,* 258!
KALAHARI REGION : Transvaal; Warm Bath, *Burtt-Davy,* 2625!
EASTERN REGION : Pondoland, *Beyrick,* 149. Natal; near Durban, *Gerrard & McKen,* 721! 869! *Sanderson,* 107! 678! and without precise locality, *Gerrard,* 96! *Williamson! Cooper,* 1242! *Gueinzius,* 583 and 101. Var. β : Natal ; near Durban, *Sanderson,* 117!

A shrub. Leaves very variable in shape and size, sometimes exactly obovate, 2½–3 in. long, 1½–2½ in. wide ; sometimes almost ovate-lanceolate, 1–1½ in. long, ½–¾ in. wide. Called "Masatola" by the Zulus. *Gerrard & McKen,* 2047, from Zululand (without flowers) is said to be "a very large tree, 40–50 ft. high, with *spherical* fruit"; it has the same foliage and native name as *M. obovata,* and I hesitate to separate it specifically, as the fruit in *M. obovata* varies in shape.—*W. H. H.*

8. **M. oleifolia** (N. E. Br. in Kew Bulletin, 1895, 109); a large shrub; branches ashy-grey, glabrous; leaves lanceolate, tapering to both ends, obtuse, acute at the base, coriaceous, glabrous, 1–2 in. long, 2½–4 lin. wide; petiole 2–3 lin. long; flowers axillary, solitary; pedicels about 3 times as long as the petioles, glabrous or at first minutely adpressed puberulous; sepals 8 in two series, the outer lanceolate, acute, densely brown pubescent outside, minutely puberulous within, the inner linear-lanceolate, acute, puberulous outside, glabrous within, ciliolate; corolla-lobes 24 in three series, those of the two outer series linear-lanceolate, acute, 2½–2¾ lin. long, ½ lin. wide, the innermost lanceolate, acuminate, 3–3½ lin. long, 1 lin. wide, glabrous; staminodes narrowly lanceolate-attenuate, adpressedly hirsute on the back and margins; stamens 8; filaments ¾ lin. long, subulate; anthers linear-oblong, apiculate, 2 lin. long ; ovary ovoid,

tapering into the adpressedly pubescent style. *Engl. Sapot. Afr.* 73, *t.* 34, *fig. B.*

EASTERN REGION : Natal ; Tugela, *Gerrard,* 1642 !

9. **M. dispar** (N. E. Br. in Kew Bulletin, 1895, 107) ; a small tree ; branches glabrous, ashy-grey ; leaves cuneate-oblanceolate, obtuse, $\frac{3}{4}$-2 in. long, $\frac{1}{4}$-$\frac{3}{4}$ in. wide, fulvo-tomentose when young, glabrous when adult, closely and finely reticulate when dry ; petiole 2-4 lin. long; flowers 12–16 umbellately arranged at the apex of the branches ; pedicels 6–8 lin. long ; calyx at first fulvo-tomentose, at length with ashy adpressed pubescence ; sepals 6–8 in two series, ovate, the outer acute, the inner obtuse ; petals 18–24 in three series, about 3 lin. long, $\frac{1}{2}$-$\frac{3}{4}$ lin. wide, subequal, linear-lanceolate, acute, glabrous, yellow ; stamens 6–8, shorter than the petals ; anthers lanceolate, acute, flexuous, almost versatile, much longer than the glabrous subulate filaments ; staminodes lanceolate, acuminate, channelled, glabrous with nearly woolly margins ; ovary globose, densely hirsute ; style long, glabrous ; fruit " yellow, well-flavoured " (*Wood*). *Engl. Sapot. Afr.* 71.

EASTERN REGION : Natal; near Mooi River in "Thorns," 3000–4000 ft., *Wood,* 4472 ! *Thresh in Natal Herb.* 5425 ! Upper Tugela River, *Gerrard,* 1482! and without precise locality, *Gerrard,* 1910!

Native name "Umpumbula."

10. **M. Schinzii** (Engl. Sapot. Afr. 70, t. 29, fig. A) ; a small tree ; branches with rough greyish bark, densely leafy at the apex ; leaves oblong-elliptic, obtuse at both ends, or sometimes very shortly acuminate, $2\frac{1}{2}$-$4\frac{1}{3}$ in. long, $1\frac{1}{4}$-2 in. wide, coriaceous, green on both surfaces ; petiole up to 5 lin. long ; lateral nerves many, prominent beneath ; reticulations dense, prominent; pedicels in axillary fascicles of 2–3, about 3 times as long as the buds, shortly and densely rusty-pilose ; sepals oblong-lanceolate ; outer corolla-lobes oblong, acute, denticulate in the upper third, inner lanceolate of equal length ; filaments shorter than the ovate-sagittate apiculate anthers ; staminodes long triangular, acute, a little shorter than the stamens, densely pilose outside ; ovary subovoid, 8–6-celled, tapering into a style $3\frac{1}{2}$ times its length. *M. natalensis, Schinz in Bull. Herb. Boiss.* iv. 441.

COAST REGION : Komgha Div. ; in woods near Komgha, *Schlechter,* 6220.

[The locality as originally published is quite erroneous.—*N. E. Brown.*]

11. **M. concolor** (Harv.) ; leaves petiolate, oblong or elliptic, emarginate, coriaceous, flat, green on both sides, on both sides somewhat netted-veined ; pedicels axillary, in 2–3's, rather longer than the petiole, glabrous ; calyx glabrous, 6-partite, the lobes ovate, subacute ; corolla 12- (or more) partite, the segments linear.

EASTERN REGION : Natal ; on the Tugela, *Gerrard & McKen,* 1662 !

"A low tree with rough bark" (*Gerrard*). Leaves $1\frac{1}{2}$-2 in. long, $\frac{3}{4}$ lin. wide, almost exactly oblong, obtuse at the base; veins and veinlets impressed on the upper surface, less obviously veined beneath. The flowers in my single specimen have been partially eaten by insects, but enough remains to establish the genus.— *W. H. H.*

ORDER LXXXIV. **EBENACEÆ.**

(By W. P. HIERN.)

Flowers usually diœcious, rarely polygamous, or in *Royena* normally hermaphrodite, regular, 3–8-merous. *Calyx* free, often more or less accrescent in fruit, not coralline, persistent. *Corolla* hypogynous, gamopetalous, coralline, deciduous; lobes entire, sinistrorsely contorted in bud as seen from above, usually spreading or reflexed in open flower. *Stamens* in the male and hermaphrodite flowers 3 or more, in the female flowers 0, or represented by usually few staminodes; filaments inserted at the base of the corolla-tube or on the receptacle usually in one or two rows, those in the same pair in different rows, rather short; anthers 2-celled, basifixed, not connate, mostly linear-lanceolate and dehiscing laterally, the connective shortly produced at the apex beyond the cells; pollen spherical or spheroidal, smooth, often marked with three furrows. *Disk* usually 0. *Ovary* free, sessile, entire, in the male flowers abortive, rudimentary or obsolete, in the female and hermaphrodite flowers 2–16-celled; ovules solitary or two together in the cells, pendulous, anatropous, twice as numerous as the styles or branches of the single style; stigmas small or somewhat dilated, emarginate. *Fruit* baccate, fleshy or coriaceous, usually few-celled; seeds usually few or solitary, pendulous, exarillate, albuminous, usually marked with two or three impressed veins or lines proceeding from the base to the apex; testa coriaceous, not rough, usually thin; albumen copious, cartilaginous, equable or sometimes ruminated; embryo dicotyledonous, axile or somewhat oblique, straight or somewhat curved, $\frac{1}{2}$–$\frac{3}{4}$ as long as the seed; radicle superior, cylindrical; cotyledons foliaceous, ovate or lanceolate, as long as or longer than the radicle, their medial plane perpendicular to that of the carpel.

Trees or shrubs, in a few species climbing, rarely spiny; wood of the middle of the trunk hard, in many species heavy, durable and often black; sap limpid, not turning milky nor coagulating; leaves normally alternate and entire, often distichous or with an angular divergence of two-fifths, simple, more or less coriaceous, usually opaque and unicostate at the base, pinniveined, in most species evergreen; hairs simple, usually 1-celled; stomata confined to the lower face of the leaves; stipules 0; flowers cymose, racemose or solitary, axillary or lateral, rather small or of moderate size, never blue; peduncles or pedicels usually bracteate.

DISTRIB. Genera 6; species about 350, widely distributed over the warmer regions of the world; more than 70 additional fossil species have been described.

I. **Royena.**—*Flowers* hermaphrodite or rarely subdiœcious, 4–8-merous, often solitary. *Calyx* normally accrescent. *Stamens* usually 10, in one row. *Phellogen* usually pericyclic.

II. **Euclea.**—*Flowers* normally diœcious, 4–7-merous, racemose or paniculate.

Calyx not accrescent. *Stamens* 8–30, usually not in one row. *Phellogen* usually pericyclic.
III. **Maba.**—*Flowers* diœcious, 3-merous. *Calyx* not accrescent. *Phellogen* subepidermal.
IV. **Diospyros.**—*Flowers* normally diœcious, 4- or 5-merous. *Calyx* more or less accrescent. *Phellogen* subepidermal.

Leucoxilon laurinum, E. Meyer ex Drège, Cat. Pl. Exsicc. Afr. Austr. 7, is *Ilex capensis, Sond. & Harv.*

I. ROYENA, Linn.

Flowers hermaphrodite or rarely subdiœcious, in most species 5-merous and solitary. *Calyx* campanulate or urceolate, pubescent, more or less accrescent in fruit. *Corolla* urceolate or campanulate or rarely oblong, toothed or lobed. *Stamens* 5–14, in most species 10, inserted in one row, usually two in front of each corolla-lobe, near the base of the corolla-tube; filaments short; anthers lanceolate-linear, usually hispid. *Ovary* conical, 2–10-celled, pubescent or hispid; ovules solitary in the cells; style 2–5-cleft at the apex. *Fruit* globose, ovoid, oblong or spheroidal, $\frac{1}{2}$–1 in. long, more or less covered or based with the enlarged persistent calyx, 1–5-celled, more or less fleshy when ripe, sometimes splitting from the apex with valves; seeds glabrous, with a thin testa.

Shrubs or small trees or sometimes rather large trees; leaves entire, normally alternate, very rarely opposite, more or less coriaceous; periderm of the stem arising from the pericycle.

It is one of the characteristic genera of the Karroo region. The Kaffir names, " Um-gambeza " and " Leeke " are quoted for several species of the genus by Thomas R. Sim, *Fl. Kaffr.* 54.

DISTRIB. In addition to the following a few species occur in Tropical Africa.

Royena foliis acute dentatis et veluti spinosis flore luteo, Houst. mss. ex Linn. Sp. Pl. ed. i. 628, is *Læselia ciliata, Linn.*

Leaves mostly alternate; stamens 6–14, usually 10; ovary
  4–10-celled :
    Flowers 5–8-merous, usually pentamerous; fruit not or
      rarely glandular :
      Calyx shortly lobed :
        Leaves lanceolate or elliptic-oblong; corolla-
          tube glabrous outside; stamens 6 or 7,
          glabrous ... ... **(1) Wilmsii.**
        Leaves oval or somewhat ovate; corolla-tube
          pubescent outside; stamens 10, hispid ... **(2) lucida.**
      Calyx divided half-way down or deeper :
        Leaves cordate or rounded or very obtuse at
          the base :
          Style bifid; leaves subsessile :
            Leaves oval or oblong, smooth, not
              strongly nerved at the base; flowers
              hermaphrodite, yellowish ... ... **(3) cordata.**
            Leaves ovate, scabrid, strongly 3–5-
              nerved at the base; flowers ap-
              parently diœcious, white ... ... **(4) scabrida.**
         Style 5–4-cleft; leaves distinctly petio-
          late :
           Flowers creamy white, $\frac{1}{2}$–$\frac{3}{4}$ in. long... **(5) Galpini.**

Flowers yellow, about $\frac{1}{3}$ in. long ... (6) **villosa.**
Leaves narrowed at the base, not cordate :
Peduncles short, not or scarcely longer
than the solitary flowers :
Leaves subsessile, $\frac{1}{4}$–1 in. long ;
anthers 10, hairy ... ... (7) **hirsuta.**
Leaves sessile, $1\frac{1}{2}$–2 in. long ; anthers
14, glabrous ... ... (8) **sessilifolia.**
Peduncles nearly as long as the leaves or
much longer than the flowers or 5–1-
flowered :
Flowers usually solitary :
Leaves not reticular-rugose :
Fruiting calyx-lobe $\frac{1}{5}$–$\frac{1}{2}$ in.
long :
Branches moderately
dense ; hairs pallid ;
fruit globular or ovoid :
Flowers usually
hermaphrodite;
leaves obovate or
oblanceolate ... (9) **pallens.**
Flowers subdioe-
cious ; leaves obo-
vate-oval ... (10) **ambigua.**
Branches very dense ;
hairs solitary ; fruit
ellipsoid or oblong ... (11) **nitens.**
Fruiting calyx-lobes $\frac{3}{4}$–1 in.
long ... ... (12) **Simii.**
Leaves reticular-rugose ... ... (13) **Guerkei.**
Flowers often cymose :
Leaves narrowly elliptic, $\frac{1}{2}$–$1\frac{3}{8}$ in.
long ... ... (14) **glabra.**
Leaves obovate, 2–$6\frac{1}{2}$ in. long ... (15) **parviflora.**
Flowers usually tetramerous ; fruit glandular-hispid ... (16) **glandulosa.**
Leaves opposite ; stamens 5 ; ovary 2-celled ... ... (17) **pentandra.**

1. **R. Wilmsii** (Gürke in Engl. Jahrb. xxvi. 60) ; a small shrub, bush or tree ; branches intricate, somewhat rugose or knotty, glabrate, dusky-ashy ; branchlets hirsute and puberulous-tomentellous with short dull tawny hairs, leafy ; leaves alternate, lanceolate or elliptic-oblong, obtusely narrowed or somewhat acuminate or apiculate at the apex, rounded or cordate at the base, firmly coriaceous, pale-green, subglaucescent, glabrate or ciliate, sometimes thinly pilose on the midrib and beneath, entire, minutely dotted, $\frac{2}{3}$–2 in. long, $\frac{1}{5}$–$\frac{3}{5}$ in. broad ; reticulation slender ; petiole $\frac{1}{12}$–$\frac{1}{6}$ in. long, tomentellous-pubescent or hirsute ; flowers hermaphrodite or subdioecious, axillary, solitary, $\frac{1}{2}$ in. long ; peduncle pubescent-tomentose, or in fruit glabrescent, $\frac{1}{5}$–$\frac{1}{4}$ in. long, bibracteate at or near the apex ; bracts opposite or alternate, ovate or lanceolate-ovate, sessile, apiculate, entire, cordate-auriculate at the base, pubescent-tomentellous and somewhat pulverulent on the back, glabrous within, ciliate, $\frac{1}{5}$–$\frac{1}{4}$ in. long, $\frac{1}{12}$–$\frac{1}{8}$ in. broad ; calyx shortly tubular-urceolate, obtuse at the base, shortly 5- or 6-lobed, pubescent-tomentose on both sides, sessile-glandular, $\frac{1}{8}$–$\frac{1}{2}$ in. long, accrescent and glabrate in

fruit; teeth erect, deltoid, acuminate; corolla tubular, $\frac{1}{6}-\frac{1}{4}$ in. long, somewhat longer than or shorter and narrower than, and enclosed in the calyx, 5- or 6-cleft; tube glabrous outside; lobes oval-oblong or lanceolate-oval, hairy on the upper part outside, ciliate, glabrous within, about as long as the tube; stamens or staminodes 6 or 7, inserted in one row near the bottom of the corolla-tube, erect, glabrous, $\frac{1}{24}$ in. long; anthers narrowly linear-lanceolate, acute, sessile; ovary ovoid-conical, hairy, glandular, 4-celled, $\frac{1}{8}$ in. long, together with the 4-cleft style $\frac{1}{5}$ in. long; ovules solitary; fruit enclosed in the enlarged calyx, $\frac{1}{3}-\frac{1}{2}$ in. in diam.; fruiting-calyx ovoid-globose, inflated, coriaceous, nearly or quite glabrous, somewhat plicate, very shortly lobed at the apex, 1 in. long, $\frac{3}{4}$ in. in diam. *R. lucida, S. Moore in Journ. Bot.* 1903, 403, *not of Linn.*

KALAHARI REGION: Transvaal; near Pretoria, *Wilms,* 923! hills near Greylingstad, *Rand,* 1325!

2. R. lucida (Linn. Sp. Pl. ed. i. 397); an evergreen shrub 4–12 ft. high or a tree; stem or trunk erect, terete, often somewhat torulose, 5–12 in. thick; bark nearly smooth, dusky grey or whitish; wood light, porous; branches numerous, spreading, flexuous, leafy; young parts, peduncles, bracts and calyx clothed with subferruginous pubescence; leaves alternate, oval or somewhat ovate, usually broadly pointed at the apex and obtuse or rounded at the base, coriaceous, glabrescent and shining above, beneath more or less hirsute especially along the midrib and margin or glabrate, entire, $\frac{1}{2}-2\frac{1}{4}$ in. long, $\frac{1}{3}-1\frac{1}{4}$ in. broad; lateral veins not conspicuous; petioles very short, $\frac{1}{12}-\frac{1}{6}$ in. long; mesophyll bifacial; peduncles $\frac{1}{4}-1$ in. long, axillary, on the branchlets, spreading or arching, bearing 1–3 small lanceolate or foliaceous bracts; flowers solitary, hermaphrodite, white or yellowish, $\frac{1}{4}-\frac{1}{3}$ in. long; calyx urceolate, $\frac{1}{5}-\frac{1}{4}$ in. long, 5-toothed, accrescent in fruit, teeth short and usually acute; corolla-tube urceolate, pubescent outside, shortly exceeding the calyx; corolla-limb reflexed, 5-partite, puberulous on both surfaces; segments rounded, $\frac{1}{5}-\frac{1}{4}$ in. long; stamens 10, inserted around the base of the corolla, equal, $\frac{1}{5}$ in. long; filaments glabrous, very short; anthers lanceolate-linear, hispid on the upper half of the back; ovary conical, included, pubescent, usually 4-celled, terminating in the cleft style, glabrous at the apex; stigmas puncti-form; fruit ovoid or subglobose, $\frac{1}{2}-1$ in. long, enclosed in the pubescent or subglabrate somewhat pentagonal inflated calyx, red or purple and fleshy when ripe, 2–4-celled; the flesh firm, whitish; seeds oblong, solitary, glabrous, shining, yellowish; testa thin; albumen cartilaginous, hard, white; embryo $\frac{1}{2}-\frac{2}{3}$ as long as the albumen, somewhat curved; cotyledons ovate, rather shorter than the radicle. *Gærtn. Fruct.* ii. 80, *t.* 94, *f.* 4; *Thunb. Prodr.* 80, *and Fl. Cap. ed. Schult.* 390; *Jacq. Fragm.* 3, *t.* 1, *f.* 6; *Poir. Encycl.* vi. 321; *Desf. in Ann. Mus.* vi. 450, *t.* 62, *f.* 3; *Lindl. Bot. Reg.* 1846, *t.* 40; *Alph. DC. Prodr.* viii. 211; *Pappe, Silva Cap.* 20; *Hiern in Trans. Cambr. Phil. Soc.* xii. 80; *Krauss in Flora,* 1844,

824; *Parmentier in Ann. Univ. Lyon,* vi. *fasc.* ii. 69; *Gürke in Engl. & Prantl, Pflanzenfam.* iv. i. 157, *fig.* 84, *A—E*; *Molisch in Sitzb. d. Mathem.-Naturw. Cl. Wien,* lxxx. *Abth.* i. 63, 66, 70; *Schnizlein, Iconographia,* t. 159, *f.* 1; *Drège in Linnæa,* xx. 191; *Melliss, St. Helena,* 294; *Drège, Zwei Pfl. Documente,* 89, 125, 139, 217.—*Pistachia africana, s. Staphylodendron Æthiopicum* Μονολασιοκαλληνομενοφυλλον *singulari hirsuto folio nitente, Plukenet, Almag.* 298, *Phytogr.* t. 63, *f.* 4, t. 317, *f.* 5. *Staphylodendron Africanum, folio singulari lucido, Herm. Parad. Bat.* 232, *with plate; J. Burm. Cat. Pl. Afr.* 22. *Staphylodendron Africanum sempervirens foliis splendentibus, J. Burm. l.c.* 33.

SOUTH AFRICA: without locality, *Oldenland! Thunberg! Roxburgh! Bowie! Mund! Harvey! Alexander! Herb. Linnæus!*

COAST REGION: Tulbagh Div.; Tulbagh Kloof, *Ecklon & Zeyher!* Cape Div.; Devils Mountain, *Drège, Ecklon,* 698! *Pappe! Wolley Dod,* 3463! near Cape Town, *Harvey!* Caledon Div.; mountains near Genadendal, *Krauss.* Swellendam Div.; near Grootvaders Bosch, 1000–4000 ft., *Zeyher,* 3352! Ruggens, near Zuurbraak, 600 ft., *Galpin,* 4320! Knysna Div.; near Kaatjes Kraal, *Burchell,* 5256! near Knysna, *Burchell,* 5415! near Bosch River, *Drège;* Albany Div.; top of Zwarthoogdens Mountain, near Grahamstown, *MacOwan,* 309! Fort Beaufort Div.; hills near Kat River, *Drège!* Stockenstrom Div.; near Philipton, *Ecklon & Zeyher!* Cathcart Div.; Cathcart, *Kuntze.*

CENTRAL REGION: Somerset Div.; at the foot of Bosch Berg, *MacOwan, Herb. Norm. Austr.-Afr.,* 233! upper part of Bruintjes Hoogte, *Burchell,* 3062/2!

KALAHARI REGION: Orange River Colony, *Cooper,* 1062! 2692! Basutoland, *Cooper,* 2157!

EASTERN REGION: Griqualand East; Vaal Bank, *Haygarth in Herb. Wood,* 4188! Natal; between Pietermaritzburg and Greytown, *Wilms,* 2227! Van Reenens Pass, *Rehmann,* 7242! and without precise locality, *Sutherland!*

It is the "Zwartbast" of the Cape colonists; the wood is hard and tough, of a yellow tint with brownish stripes when polished, and well adapted for furniture, tools, screws, &c. It is also called "Kraai-besjes" (*Burchell,* 5256). The Kaffirs call it "Omgugunga." It has also been called "African bladder-nut."

3. R. cordata (E. Meyer, ex Drège, Cat. Pl. Exsicc. Afr.-Austr. 7); an evergreen shrub, erect, rigid, densely branched, 2–6 ft. high; stem terete; branches erect-patent; branchlets leafy; young parts subferruginous-pubescent, soon glabrescent; leaves alternate, oval or oblong or sometimes rotund, rounded or obtusely pointed at the apex or apiculate, usually strongly cordate at the base, ascending and sometimes adpressed, firmly coriaceous, rigid, entire, glabrescent and glossy above, glabrate or pubescent beneath, subsessile, $\frac{1}{2}$–$2\frac{3}{8}$ in. long, $\frac{1}{3}$–$1\frac{1}{8}$ in. broad; the lateral veins inconspicuous or on the larger leaves clearly marked beneath; mesophyll bifacial; flowers hermaphrodite, axillary and quasi-terminal, sometimes forming bracteate quasi-racemes of 1–1$\frac{1}{4}$ in. long, yellow or white or pale-cream, $\frac{1}{3}$ in. long; peduncles arching, 1-flowered, hirsute or pubescent, bibracteate, $\frac{1}{6}$–1$\frac{1}{3}$ in. long; bracts ovate or oval-apiculate, more or less hirsute, ciliate, deciduous, alternate, $\frac{1}{8}$–$\frac{1}{4}$ in. long; flowering calyx 5-partite, shaggy on both surfaces, $\frac{1}{6}$–$\frac{1}{4}$ in. long, accrescent;

segments ovate or ovate-lanceolate and acute; corolla 5-cleft, about
$\frac{1}{3}$ in. long; tube $\frac{1}{12}$–$\frac{1}{6}$ in. long and glabrous; lobes ovate-oblong,
$\frac{1}{6}$–$\frac{1}{4}$ in. long, subglabrous or puberulous and often reflexed at the
tips; stamens 10, inserted at the base of the corolla, $\frac{1}{6}$ in. long;
filaments very short; anthers hairy; ovary ovoid-conical, hispid,
usually 4-celled; style bilobed or styles 2; ovules solitary; fruit
subglobose, $\frac{1}{2}$–$\frac{5}{8}$ in. in diam., enclosed in the enlarged calyx; fruiting
calyx dilated, deeply 5-cleft, 5-winged, $\frac{1}{2}$–1 in. long; segments
glabrate, deltoid-ovate, cordate or auriculate at the base, reddish on
the margin. *Drège, Zwei Pfl. Documente,* 56, 141, 217; *Alph.
DC. Prodr.* viii. 211; *Hiern in Trans. Cambr. Phil. Soc.* xii. 81;
Parmentier in Ann. Univ. Lyon, vi. *fasc.* ii. 70. *R. opaca, E.
Meyer in Drège, Zwei Pfl. Documente,* 135, 217; *Alph. DC., l.c.
R. supracordata, Burch. ex Hiern, l.c.*

SOUTH AFRICA : without locality, *Masson ! Bowie !*

COAST REGION : Uniondale Div.; Long Kloof, *Burchell,* 4907 ! Uitenhage
Div.; Zuurberg Range, between Enon and Driefontein, 2000–3000 ft., *Drège !*
Alexandria Div.; Olifants Hoek, *Zeyher !* Bathurst Div.; near the Kowie
River, *MacOwan,* 429 ! Glenfilling, *Drège !* Albany Div.; between the source
of Kasuga River and the Bushmans River, *Burchell,* 4166 ! 4186 ! Fish River
Heights, *Mrs. Hutton !* Fort Beaufort Div. ; near Fort Beaufort, 1000–2000 ft.,
Ecklon & Zeyher ! Queenstown Div. ; near Queenstown, *Galpin,* 1559 ! and
without precise locality, *Mrs. Barber,* 307 ! Cathcart Div. ; Windvogel Moun-
tain, *Baur,* 1120 ! East London Div.; near the mouth of the Keiskamma River,
Mrs. Hutton ! British Kaffraria, *Cooper,* 35 ! 186 ! 306 !

CENTRAL REGION : Graaff Reinet Div., 3000–4000 ft. ; Oude Berg, *Drège ;*
Cave Mountain near Graaff Reinet, *Bolus,* 527 !

KALAHARI REGION : Transvaal ; hills near Pretoria, 4100 ft., *McLea in
Herb. Bolus,* 3103 !

EASTERN REGION : Natal ; Tongaat, *Cooper,* 3485 ! Inanda, *Wood,* 1232 !
near Murchison, 2000 ft., *Wood,* 478 ! 3059 ! and without precise locality,
Gerrard & McKen, 12 ! 18 ! 99 ! *Gueinzius !* Zululand, 1000–2000 ft., *Wylie
in Herb. Wood,* 5686 !

Most of the specimens from the Eastern Region are rather more hairy than
those from the other regions.

4. **R. scabrida** (Harv. ex Hiern in Trans. Cambr. Phil. Soc.
xii. 82); a shrub, 8–15 ft. high; branches simple, pilose at the
extremities with pallid hairs; leaves alternate, entire, ovate, acute or
obtuse at the apex, cordate at the base, dotted and scabrid especially
beneath, silky when young, shining above, subsessile, strongly 3–5-
nerved at the base, $1\frac{1}{3}$–$2\frac{1}{3}$ in. long, $\frac{7}{8}$–$1\frac{3}{5}$ in. broad, with a narrowly
revolute margin; flowers apparently dioecious; the male flowers
nearly $\frac{1}{2}$ in. long, white, axillary; peduncles 1-flowered, much
shorter than the leaves, bracteate; bracts ovate, acuminate; calyx
5-partite, erect, pilose-setose, $\frac{3}{8}$ in. long; segments ovate, acuminate,
broad-based; corolla campanulate-urceolate, adpressedly pilose,
deeply 5-cleft; lobes $\frac{1}{3}$ in. long, ovate-oblong, acute; stamens 10, in
one row, equal, inserted at the base of the corolla-tube, $\frac{1}{6}$ in. long;
filaments very short, hairy at the apex; anthers linear, acute, hairy
towards the apex; ovary rudimentary, hairy; style bifid, glabrous
above, hairy below; female flowers and fruit not seen.

EASTERN REGION : Natal; on grassy plains near the Tugela River, *Gerrard & McKen*, 1609!

5. R. Galpini (Hiern); shrubby, tomentose, not very closely branched, leafy; leaves alternate, entire, oval, obtusely narrowed or nearly rounded at the apex, truncate, rounded or very obtuse at the base, thinly coriaceous, yellowish-green, strigosely pubescent and minutely glandular-pulverulent above, pallid-tawny and softly tomentose beneath, 2–5 in. long, 1–2½ in. broad; petiole ¼–⅝ in. long tomentose; peduncles 1-flowered, drooping, about ½–⅔ in. long, bracteate above the middle; bracts narrow; flowers axillary, solitary, white or creamy-white, ½–⅔ in. long; calyx shaggy-tomentose on both surfaces, ⅓–⅜ in. long, very deeply lobed; segments ovate-lanceolate, acute; corolla deeply lobed, partly pubescent outside, glabrous within; lobes oblong, obtuse, about ⅝ in. long and ⅙ in. broad, hairy at the apex and partly down the outer face, glabrous where covered by the adjoining petal in the bud, finally spreading or reflexed; stamens 10, in one row, equal, ⅙ in. long; filaments inserted at the base of the corolla, very short; anthers lanceolate, hispid along the back, glabrous on the face; ovary ovoid-conical, somewhat pyramidal, marked with 5 raised hairy vertical lines and intervening glabrous spaces, terminating in 5 glabrous styles, ¼ in. long altogether, 10-celled.

KALAHARI REGION : Transvaal; grassy plains and hill-sides near Barberton, 2800–4000 ft., alt., *Galpin*, 603 !
EASTERN REGION : Zululand; Umlalaas, below 1000 ft., alt., *Wylie in Herb. Wood*, 7900 ! 8630 !

6. R. villosa (Linn. Syst. Nat. ed. xii. ii. 302); a pubescent shrub, trailing; branches patent, 5–40 ft. long; branchlets leafy, patent; leaves alternate, obovate-oblong, rounded, emarginate or shortly pointed at the apex, more or less cordate at the base, thinly coriaceous, entire, glabrescent and dark green above, paler green and hairy beneath, sometimes with the veinlets pellucid or with minute pellucid points, often wrinkled with impressed venation in the dry state, narrowly revolute on the margin, 1–5 in. long, ½–2½ in. broad; petiole ⅕–¾ in. long, hairy; flowers hermaphrodite, yellow, fragrant, about ⅓ in. long, densely pubescent; peduncles axillary, 1-flowered and ⅓–½ in. long or 3-flowered longer and with pedicels of about ₁⁄₁₀ in., bracteate, hairy; bracts usually 2 and subopposite, smaller and narrower than the leaves, deciduous; calyx 5-partite, accrescent in fruit, ⅙ in. long in flower; segments lanceolate or ovate in flower and broadly ovate in fruit; corolla tomentose outside except near the base, glabrous within, deeply 5-lobed; lobes oblong, about ⅕ in. long and recurving; stamens 10; anthers densely shaggy; ovary 8- or 10-celled; style 5–4-lobed; stigmas punctiform; fruit globose-pentagonal, tomentose or hispid with yellowish hairs, ½–¾ in. long, sometimes dehiscent from the apex with 5 valves, surrounded with and not much exceeding the enlarged wrinkled calyx. *Poir. Encycl.* vi. 321 ; *Meerburg, Pl. Rar. t.* 43 ; *Thunb. Prodr.* 80, *and Fl. Cap. ed. Schultes,* 390 ; *Alph. DC. Prodr.* viii. 213 ; *Hiern in Trans. Cambr. Phil. Soc.* xii. 82 ; *Drège, Zwei Pfl. Documente,* 132, 139, 159, 217 ; *Krauss in Flora,* 1844, 824 ; *Parmentier in Ann. Univ. Lyon,* vi. *fasc.* ii. 70. *R. scabra, Burm. f. Fl. Cap. Prodr.* 13. *R. scandens, Burch. ex Hiern, l.c. R. longifolia, Cels, Cat.* 1817, 33, *name only, is probably this species.*

South Africa : without locality, *Masson ! Thunberg ! Alexander ! Miller ! Herb. Linneus !*

Coast Region : Uitenhage Div. ; near the lead-mine by Maitland River, *Burchell,* 4506 ! Addo, 1000–2000 ft., *Drège, Ecklon & Zeyher,* 464 ! Albany Div. ; in a ravine by the spring at Blaauw Krantz, *Burchell,* 3673 ! Kloof east of Woest hill, Grahamstown, *MacOwan,* 516 ! near the Bushmans River and Geelhoutboom, *Drège !* near Alicedale, 1500 ft., *MacOwan, Herb. Norm. Austr.-Afr.,* 234 ! Bathurst Div. ; near Port Alfred, *Burchell,* 3793 ! British Kaffraria, *Cooper,* 1 !

Kalahari Region : Orange River Colony, *Cooper,* 2006 ! Basutoland, *Cooper,* 2687 !

Eastern Region : Natal ; near Durban, *Drège! McKen,* 613 ! *Wylie in Herb. Wood,* 6262 ! near Murchison, 2000 ft., *Wood,* 3058 ! Inanda, *Wood,* 613 ! forests by the Umlaas River, *Krauss,* 226 ! 472 ! 482, and without precise locality, *Stuart ! Gerrard & McKen,* 614 ! 2013 ! *Gerrard,* 30 ! *Sanderson,* 150 ! 613 ! 715 ! *Cooper,* 2689 ! 2690 ! *Gueinzius !*

The Boer name is " Zwart-bast "; the wood is used for the bodies of waggons and for making yokes for draught oxen. (*Thunberg, Travels, English* 3rd *edit.,* ii. 111). A specimen is recorded from the neighbourhood of Brisbane, Queensland (*Journ. Bot.* 1875, 353).

7. R. hirsuta (Linn. Sp. Pl. ed. i. 397) ; a rigid shrub, closely branched, more or less downy-hoary or tomentose, $1\frac{1}{2}$–15 ft. high ; branches terete, spreading, dusky or ashy ; branchlets densely leafy, knotty ; leaves alternate, oblanceolate, obtuse, rounded, apiculate or subacute at the apex, wedge-shaped or somewhat narrowed at the base, coriaceous, rigid, entire, nearly flat or revolute along the margin, hairy and rugose with raised midrib and more or less conspicuous veins or pitted beneath, sometimes deciduous, subsessile or shortly petiolate, $\frac{1}{4}$–1 in. long, $\frac{1}{10}$–$\frac{1}{4}$ in. broad ; petiole very short or ranging up to $\frac{1}{4}$ in. long ; flowers usually hermaphrodite, occasionally diœcious, white, pink or scarlet, $\frac{1}{6}$–$\frac{1}{4}$ in. long ; peduncles axillary, 1-flowered, arching or deflexed, shorter than or equalling the flowers, $\frac{1}{10}$–$\frac{1}{4}$ in. long, usually bibracteate about or above the middle ; bracts narrow, often deciduous, about $\frac{1}{12}$ in. long ; flowering calyx subcampanulate, deeply 5-lobed, hairy on both sides, $\frac{1}{8}$–$\frac{1}{6}$ in. long ; lobes ovate and erect ; corolla urceolate, 5-cleft, grey-felted outside, puberulous within ; lobes ovate or lanceolate-oblong, rounded or obtuse at the apex, equalling the tube and reflexed ; stamens usually 10 ; filaments short, dilated ; anthers lanceolate, hairy ; ovary ovoid-conical, shaggy, 4–8-celled, in the female flowers with small glands at the base alternating with 6–9 staminodes ; styles usually 2, occasionally 3 or 4 ; stigmas glabrous, more or less dilated, emarginate ; fruit globose or rarely obovoid, $\frac{1}{3}$–$\frac{1}{2}$ in. long, more or less tomentose, red or pallid-tawny, often dehiscent from the apex with 2–5 valves, based with the accrescent calyx ; fruiting calyx deeply 5-lobed, pubescent, ¼ in. long or more ; lobes broad or oblong, erect or reflexed; seeds 4–8.   *Thunb. Prodr.* 80, *and Fl. Cap. ed. Schult.* 391 ; *Poir. Encycl.* vi. 321 ; *Lam. Encycl. t.* 370, *fig.* 2 ; *Drège, Zwei Pfl. Documente,* 47, 56, 62, 217 ; *Alph. DC. Prodr.* viii. 212 ; *Hiern in Trans. Cambr. Phil. Soc.* xii. 83 ; *Parmentier in Ann. Univ. Lyon,* vi. *fasc.* ii. 73, *not of Jacq., nor of Sieb., nor of Eckl. R. angustifolia, Willd. Sp. Pl.* ii. 633 ; *Poir. Encycl.* vi. 322 ; *Alph. DC., l.c.; Parmentier, l.c.* 72 ; *Willd. Herb. n.* 8367 ! ; *Drège, Zwei Pfl. Documente,* 67, 70, 217.   *Diospyros hirsuta, Desf. in Ann. Mus. Par.* vi. 449, *t.* 62, *f.* 2 ; *not of Linn. f.   D. pubescens, Pers. Syn. Pl.* ii. 625 ; *not of Pursh. R. microphylla, Burchell, Trav. S. Afr.* i. 348, *note ; Alph. DC., l.c.   R. rugosa, E. Meyer ex Drège, Cat. Pl. Exsicc. Afr.-Austr.* 7 ; *Drège, l.c.* 113, 217 ; *Alph. DC. l.c.; Krauss in Flora,* 1844, 824. *Cf. R. media, Cels, Cat.* 1817, 33, *name only.   Cf. R. cuneata, Poir. Encycl.* vi. 322 ; *Alph. DC. Prodr. l.c.* 215.   *Arbutus foliis lanceolatis integerrimis hirsutis, Linn. Hort. Cliff.* 163.   *Staphylodendron Africanum folio lanuginoso Rosmarini latiori, Boerh. Ind. alt.* ii. 235, *ex Mill. Gard. Dict. ed.* 7, *n.* 3.

VAR. β, rigida (Hiern in Journ. Bot. 1874, 239) ; habit very rigid ; branches patent, very numerous; leaves alternate, rather pale, ⅓–½ iu. long, 1/10–¼ iu. broad; flowers white; calyx-lobes linear-lanceolate ; corolla-lobes lanceolate-oblong, very feebly pubescent, reflexed from the middle ; ovary 6-celled, globose ; style 4-cleft ; fruit eaten but not very palatable.

SOUTH AFRICA : without locality, *Auge! Masson! Oldenburg! Thunberg! Verreaux!*

COAST REGION : Clanwilliam Div.; Clanwilliam, 300 ft., *Schlechter,* 8008 ! Olifants River and Brakfontein, *Ecklon & Zeyher ! Zeyher,* 3351 ! Malmesbury Div.; Groene Kloof (Mamre), and between Groene Kloof and Saldhana Bay, below 500 ft., *Drège!* Cape Div. ; Cape Flats, *Krauss,* 1719 ! *Herb. Linneus!* Caledon Div. ; by the Zondereinde River, *Burchell,* 7537 ! Swellendam Div. ; between Swellendam and the Breede River, *Burchell,* 7446 ! Hessaquas Kloof, *Burchell,* 7531 ! on plains and near rivers, *Bowie!* near the lower part of the Zondereinde River, *Zeyher,* 3350 ! Grootvaders Bosch and adjacent mountains, *Ecklon & Zeyher !* Unioudale Div.; Lange Kloof, *Burchell,* 4898 ! *Bowie !* Humansdorp Div.; near Humansdorp, *MacOwan,* 269 ! Albany Div. ; *Bowie!* Queenstown Div.; near Queenstown, *Cooper,* 212 ! Guildford, near Tylden, 3500 ft., *Galpin,* 1565 ! hills near the Zwart Kei River, *Barber,* 311 ! Winterberg Range, on the highest hills between Tarka and Kat Berg, *Ecklon & Zeyher !* Bowkers Park, 4750 ft., *Galpin,* 2570 !

CENTRAL REGION : Calvinia Div. ; Bokfontein, 2500 ft., *Drège!* Worcester Div. ; Baviaans Krantz, *Rehmann,* 2885 ! Beaufort West Div. ; Nieuweveld Mountains near Beaufort West, 3000–5000 ft., *Drège!* Somerset Div. ; Bruintjes Hoogte, 3500 ft., *MacOwan,* 1984 ! Graaff Reinet Div.; Oude Berg, 3000–4000 ft., *Drège;* mountains near Graaff Reinet, 3000–4000 ft., *Drege, Bolus,* 470 ! Aberdeen Div. ; Cambeboo, near Hamer Kuil, 3000 ft., *Drège !* Var. β : Graaff Reinet Div.; on hills near Graaff Reinet, 2500–3300 ft., *Bolus,* 616 !

WESTERN REGION : Little Namaqualand; between Pedros Kloof and Lily Fontein, 3000–4000 ft., *Drège!* near Ookiep, 3300 ft., *Bolus in Herb. Norm. Austr.-Afr.,* 636!

KALAHARI REGION : Griqualand West; Barkly West, 4000 ft., *Marloth,* 1010. Hay Div.; between Wittewater and Griqua Town, *Burchell,* 1696 ! Orange River Colony; Bloemfontein, *Kuntze ;* and without precise locality, *Cooper,* 2688 ! 844 ! Bechuanaland ; near the sources of the Kuruman River,

Burchell, 2502 ! Transvaal; Waterval River, *Wilms,* 919 ! hills near Aapies
River, *Rehmann,* 4334 ! Klipriver Berg, near Johannesburg, *Rand,* 884 !
EASTERN REGION: Natal; near Van Reenen, 5500 ft., *Schlechter,* 6956 !
Wood, 5658 ! near the Tugela River, 4000 ft., *Wood,* 3592 ! bank of the Mooi
River, 5000 ft., *Sutherland !* Washbank, Newcastle district, 3000–4000 ft., *Wood,*
7904 !

There is considerable variation, as was noted by Thunberg, both in the size of
the leaves and in the abundance of the tomentum; the form which has the
smallest leaves is *R. microphylla,* Burchell, 1696, 2502. *R. hirsuta, Herb.
Ecklon,* 698, is *R. lucida, Linn.*

The Hottentot name of var. *β* is " Grietie-Rom "; it flowers either in spring
(October) or, in autumn (April), according to the rains, the more usual season
being the spring (*Bolus, MS.*).

8. R. sessilifolia (Hiern in Trans. Cambr. Phil. Soc. xii. 84) ;
a shrub ; stem erect ; branches spreading at a wide angle, pubescent ;
periderm of the stem subepidermal ; leaves alternate, oblong-
obovate or obovate, sessile, rounded or retuse at the apex, narrowed
to the obtuse base, submembranous, pubescent beneath and when
young on both surfaces, $1\frac{1}{2}$–2 in. long, $\frac{1}{2}$–$1\frac{1}{8}$ in. broad ; lateral
veins not very conspicuous, impressed on the upper surface ;
flowers subdiœcious, pallid, fragrant, $\frac{1}{4}$ in. long ; peduncles
axillary, 1-flowered, shorter than the flowers, pubescent ; calyx
5-partite, $\frac{1}{6}$ in. long, pubescent outside; segments lanceolate, erect-
patent and 3-veined ; corolla urceolate, 5-lobed, pubescent outside,
glabrous within ; lobes recurved, obtuse and $\frac{1}{10}$ in. long; stamens
14, glabrous ; filaments short ; anthers dehiscing from the apex ;
pollen globular, smooth ; ovary rudimentary, rounded, pubescent.
Diospyros sessilifolia, Parmentier in Ann. Univ. Lyon, vi. *fasc.* ii. 73.

SOUTH AFRICA?: described from a plant of unknown origin, cultivated at
Kew, whence specimens were obtained in 1877 and 1880 !

A specimen in the Leiden herbarium with sessile leaves, which are coriaceous
and mostly pointed at the apex with the lateral veins raised in relief on both
surfaces, possibly belongs to this species; it was cultivated in the Leiden Garden
in 1785.

R. latifolia, Willd. Enum. Pl. Hort. Berol. Suppl. 23, *without description ;
Alph. DC. Prodr.* viii. 215; is unknown to me, but is perhaps best mentioned
here.

9. R. pallens (Thunb. Prodr. 80) ; a shrub or tree, usually not
exceeding 15 ft. high, sometimes a large tree ; bark reddish-brown
or turning black or sometimes grey, rough ; branches alternate, terete,
silky-pubescent with pallid hairs or glabrescent, more or less spreading, sometimes with a trailing or twining habit ; branchlets leafy,
densely hairy ; leaves oblanceolate, obtuse or rarely pointed at the
apex, more or less wedge-shaped at the base, silky especially beneath
or glabrescent, green below the hairs, rather thick and firm, subcoriaceous, alternate, opposite or subverticillate, evergreen or sometimes deciduous, narrowly revolute on the margin, entire, $\frac{1}{2}$–2 in.
long, $\frac{1}{6}$–$\frac{3}{4}$ in. broad ; petiole pubescent or puberulous, $\frac{1}{20}$–$\frac{1}{5}$ in.
long ; flowers whitish or yellow, hermaphrodite or subdiœcious,
$\frac{1}{4}$–$\frac{2}{5}$ in. long, pendulous ; peduncles axillary, arching, 1- (or

rarely 2-) flowered, bearing 2 or 3 bracts about or above the middle, longer than the flowers, pubescent, somewhat thickened at the apex, $\frac{2}{5}-\frac{4}{5}$ in. long; bracts lanceolate, alternate, deciduous, $\frac{1}{12}-\frac{1}{8}$ in. long; calyx deeply 5-lobed, rounded below, pubescent on both surfaces, $\frac{1}{6}$ in. long in flower, more or less accrescent, $\frac{1}{4}-\frac{1}{2}$ in. long in fruit; lobes ovate or lanceolate, acute, erect and adpressed to the corolla-tube, finally spreading or reflexed; corolla campanulate-rotate, deeply 5-lobed, rather thick and fleshy, somewhat hairy outside, glabrous within; tube very short; lobes ovate-oblong, revolute and $\frac{1}{4}$ in. long; stamens 10, equal, $\frac{1}{6}-\frac{1}{5}$ in. long; filaments very short, flat, glabrous; anthers linear-lanceolate, hairy on the back and at the apex, $\frac{1}{8}-\frac{1}{6}$ in. long; ovary ovoid-conical or subglobose, silky-pubescent, with the style about $\frac{1}{6}$ in. long, 6–10-celled; style deeply 3–5-cleft, terete; stigmas punctiform, glabrous; fruit subglobose or ovoid, pubescent or hispidulous or rarely glabrate, $\frac{1}{2}-1\frac{1}{4}$ in. in diam., sometimes bursting with valves from the apex. *Poir. Encycl.* vi. 322; *Thunb. Fl. Cap. ed. Schult.* 391; *Drège, Zwei Pfl. Documente,* 137, 217; *Alph. DC. Prodr.* viii. 213; *Hiern in Trans. Cambr. Phil. Soc.* xii. 85; *Parmentier in Ann. Univers. Lyon,* vi. *fasc.* ii. 71; *Hiern in Oliver, Fl. Trop. Afr.* iii. 510; *Hiern in Cat. Afr. Pl. Welw.* i. 647; *Melliss, St. Helena,* 294; *Schinz, Deutsch Südwest Afrika,* 66; *not of Herb. Willd. n.* 8363! *R. pallens, β. Dregei, Alph. DC. Prodr.* viii. 214. *R. hirsuta, Jacq. Collect.* v. 110, *t.* 13, *f.* 1; *not of Linn., nor of Sieber, nor of Ecklon. R. lycioides, Desf. Tabl. Ecole Bot.* 79; *Alph. DC. Prodr.* viii. 214. *Diospyros lycioides, Desf. in Ann. Mus. Paris,* vi. 448, *t.* 62, *f.* 1. *R. pubescens, Willd. Enum. Hort. Berol.* 457; *Drège, l.c.* 123, 142, 217; *Alph. DC. Prodr.* viii. 213; *Herb. Willd. n.* 8364!, *not of Edwards, Bot. Reg. t.* 500. *R. decidua, Burchell, Trav. S. Afr.* i. 317. *R. cuneata, Spreng. Syst. Veg.* ii. 360; *not of Poir. R. brachiata, E. Meyer ex Drège, Cat. Pl. Exsicc. Afr. Austr.* 7, *and Zwei Pfl. Documente,* 127, 217; *Alph. DC., l.c.* 213. *R. cuneifolia, E. Meyer, l.c., and ex Drège, Cat. Süd-Afr. Pfl.* 9, *and Zwei Pfl. Documente,* 58, 93, 132, 138, 217; *Alph. DC. l.c.* 214; *Krauss in Flora,* 1844, 824. *R. ramulosa, E. Meyer ex Alph. DC. Prodr.* viii. 212. *R. sericea, Bernh. ex Krauss in Flora,* 1844, 824; *Walp. Rep.* vi. 457. *R. oleifolia, Desf., and R. hispidula, Harv. ex Hiern in Trans. Cambr. Phil. Soc.* xii. 85. *Royena* (*sp.*), *Burchell, l.c.* i. 390, 381 (*fig.*).

SOUTH AFRICA : without locality, *Thunberg! Masson! Zeyher,* 3348! 3354! *Alexander! Ferraira,* 516.

COAST REGION: Worcester Div.; near Matjesfontein (Maggisfontein by error), *Rehmann,* 2915! 2917! Caledon Div.; Baviaans Kloof, *Burchell,* 7860? Robertson Div.; Kogmans Kloof, *Kuntze;* Riversdale Div.; by the Zoetemelks River, *Burchell,* 6813! Muis Kraal, *Galpin,* 4319! Mossel Bay Div.; Driefontein, *Drège;* dry channel of an arm of the Gauritz River, *Burchell,* 6490! Knysna Div.; Groene Valley, *Burchell,* 5632! near Knysna, *Bowie! Krauss,* 1721! Uitenhage Div.; various localities, *Burchell,* 4501! *Drège! Zeyher,* 164! 819! 1127! *Ecklon & Zeyher,* 14! *Bowie! Tredgold!* Port Elizabeth Div.; Krakkakamma, *Zeyher,* 544! Bathurst Div.; near Port Alfred, *Burchell,* 3789! Albany Div.; various localities, *Burchell,* 3396! 3472! 4184! *Williamson!*

Bolton! Hutton! Miss Bowker! Baur! MacOwan, Herb. Norm. Aust.-Afr., 759! Stockenstrom Div. ; Chumie (Tyumie) Berg, *Ecklon & Zeyher!* Queenstown Div., *Cooper*, 272! Cathcart Div. ; Toise River, *Kuntze*; Goshen, *Baur*, 925! King Williamstown Div. ; between Buffalo and Yellowwood Rivers, *Drège* ; East London Div. ; East London, *Kuntze.*

CENTRAL REGION : Somerset Div. ; by Little and Great Fish Rivers, 2000–3000 ft., *Drège, Burke!* Commadagga, *Burchell*, 3301! 3325! Bruyntjes-Hoogte, *Burchell*, 3080! between Vogel River and Loots Kloof, and by the Klein Fish River, *MacOwan*, 1646! Graaff Reinet Div. ; mountains south-west of Graaff Reinet, *Burchell*, 2930! near Graaff Reinet, 2500 ft., *Bolus*, 128! Cradock Div. ; near Cradock, *Kuntze, Cooper*, 2691! Beaufort West Div. ; Nieuwveld, *Drège*; Beaufort West, *Kuntze.* Victoria West Div. ; between Brak River and Uitvlugt, *Drège!* Aliwal North Div. ; Aliwal North, *Kuntze.* Hopetown Div. ; near Hopetown, 4500 ft., *Muskett in Herb. Bolus*, 128!

WESTERN REGION : Little Namaqualand; Orange River, near Verleptpram, below 500 ft., *Drège!* Streeuwberg, *Wyley*, 103! Great Namaqualand ; Tiras, *Schinz.*

KALAHARI REGION : Herbert Div. ; by the Vaal River, at Blauwbosch Drift, *Burchell*, 1750! Kimberley, *Marloth*, 816! Bechuanaland ; Maadje Mountain, *Burchell*, 2371! Orange River Colony ; Leeuw Spruit and Vredefort, *Barrett-Hamilton!* Bloemfontein, *Kuntze.* Transvaal ; various localities, *Galpin*, 1094! 6042! *Rand*, 708! *Rehmann*, 4090! 4091! *Wilms*, 921! *Holub! Burke!*

EASTERN REGION : Tembuland ; Klip Kraus near Bazeia, *Baur*, 260! Natal ; near Durban, *Krauss*, 423! Weenen County, *Sutherland!* Mawbys Boot, *Sanderson*, 318! Inauda, *Wood!* Illovo Forest, *Wood*, 6434! Tugela River, *Wood*, 3853! Umsondus River, *Rehmann*, 7641! and without precise locality, *Gerrard*, 129! 1607! 1610! 6151! *Peddie! Sanderson*, 140! 511! 423! 527! 717! *Cooper*, 1238!

This tree is the *Royena* mentioned by Burchell, *Trav. S. Afr.* i. 390, and represented in his engraving, p. 381, and by his n. 1750. It is said to be naturalized in the island of St. Helena, and to be called " Poison Peach " by the colonists. In parts of Cape Colony it is called " Monkey Plum," and in the Kalahari Region it is known as " Zwartbast " (black bark) ; it occurs also in South tropical Africa.

10. R. ambigua (Venten. Jard. Malm. n. 17) ; a shrub ; branches numerous, erect-patent or ascending ; branchlets more or less tomentose throughout, leafy ; leaves alternate, obovate-oval, rounded or apiculate at the apex, somewhat narrowed at the base, dull green, thinly coriaceous, sometimes minutely pellucid-punctate, softly pubescent at least on the principal veins beneath and on the margin, 1–2 in. long, $\frac{1}{2}$–$\frac{3}{4}$ in. broad ; petiole $\frac{1}{10}$–$\frac{1}{5}$ in. long, pubescent ; flowers subdiœcious ; the female ones drooping, orange-yellow or yellowish, somewhat fragrant, $\frac{1}{4}$–$\frac{1}{3}$ in. long ; peduncles axillary, 1-flowered, pubescent, arching, $\frac{3}{10}$–$\frac{3}{4}$ in. long, bearing 2 or 3 alternate linear silky bracts near or above the middle ; calyx deeply 5- or 6-cleft, hairy, $\frac{1}{8}$ in. long, accrescent in fruit ; lobes lanceolate or ovate, or in fruit ovate-oblong ; corolla broadly urceolate, 5–8-cleft, pubescent outside, containing honey ; lobes oblong, obtuse, reflexed and rather shorter than the tube ; staminodes 10–14, shorter than the corolla-tube, inserted at its base, converging towards the apex round the ovary ; filaments flat, glabrous below, hairy above ; anthers (sterile) brown ; ovary globose, pubescent, pale yellow, 5–7-furrowed, 10-celled ; styles 5–7, slender, united at the base, pale yellow, longer than the staminodes ; fruit globular, pubescent, bright pale brown, nearly $\frac{1}{2}$ in. in diam., sometimes dehiscent from the apex with 5 valves; fruiting calyx reflexed, 5-partite; segments oblong-lanceolate, $\frac{1}{2}$ in. long; seeds 3, oblong, pendulous, $\frac{1}{4}$ in. long. Male flowers not seen. *Poiret in Lam. Encycl. Méth. Suppl.* iv. 722; *Alph. DC. Prodr.* viii. 214; *Hiern in Trans. Cambr. Phil. Soc.* xii. 86. *R. polyandra, β. ambigua, Pers. Syn. Pl.* i. 486. *Diospyros ambigua, Venten. Jard. Malm. t.* 17. *Diplonema ambigua, G. Don, Gen. Syst.* iv. 42.

KALAHARI REGION: Transvaal; Magalies Berg, *Zeyher*, 1126! Wonderboom Poort, near Pretoria, *Rehmann*, 4527! and without precise locality, *Burke!*

Perhaps only a variety of the previous species.

11. R. nitens (Harv. ex Hiern in Trans. Cambr. Phil. Soc. xii. 87); a shrub or undershrub, $1\frac{1}{2}$–4 ft. high; branches ascending, dense, terete, leafy; young shoots silky-tomentose, shining with silvery-white hairs; bark dusky, somewhat glossy; leaves alternate, narrowly elliptic, narrowed more or less towards both ends, coriaceous, dark-green and glossy above, silky-tomentose beneath with persistent pale hairs, subsessile, entire, $\frac{1}{3}$–$1\frac{1}{8}$ in. long, $\frac{1}{10}$–$\frac{3}{8}$ in. broad; flowers hermaphrodite or subdiœcious, numerous, white, $\frac{1}{4}$–$\frac{1}{3}$ in. long; peduncles axillary, 1-flowered, pubescent, arching in fruit, about $\frac{1}{4}$–$\frac{1}{2}$ in. long, bibracteate above the middle; bracts linear, pubescent, deciduous; calyx very deeply 5-lobed, pubescent on both surfaces, pallid, about $\frac{1}{5}$ in. long in flower, accrescent, more or less spreading and about $\frac{1}{2}$ in. in diam. under the fruit; segments lanceolate, subacute, $\frac{1}{5}$–$\frac{2}{5}$ in. long; corolla deeply 5-lobed, somewhat pubescent outside, glabrous within; lobes ovate-oblong, obtuse, spreading or reflexed; stamens 10, erect, about $\frac{1}{8}$ in. long or shorter in quasi-female flowers; filaments very short, glabrous, flattened; anthers hispid; ovary hirsute; styles 3–5, glabrous; fruit spheroidal or oblong, puberulous with very short inconspicuous hairs, splitting from the somewhat pointed apex, $\frac{1}{2}$–1 in. long, $\frac{1}{4}$–$\frac{2}{5}$ in. thick, 1-celled, like a plum in colour, 1-seeded.

KALAHARI REGION: Transvaal; plains around Barberton, 2800 ft., *Galpin*, 568! Mount Sheba, near Barberton, 3900 ft., *Bolus*, 7842!
EASTERN REGION: Zululand; White Umvolosi River, *Gerrard*, 1158!

12. R. Simii (O. Kuntze, Rev. Gen. Pl. iii. ii. 196); a dense, leafy shrub; branches strigose-pilose when young, at length glabrate; branchlets patent, densely leafy; bark pallid; leaves alternate, crowded, oblanceolate, obtusely acuminate or very obtusely pointed or nearly rounded at the apex, wedge-shaped at the base, thinly coriaceous, dark green and glabrescent above, somewhat paler and strigose-pilose at least along the midrib beneath, revolute along the margin, entire, 1–$2\frac{1}{8}$ in. long, $\frac{1}{4}$–$\frac{3}{5}$ in. broad; petiole $\frac{1}{8}$–$\frac{1}{2}$ in. long, strigose-pilose or glabrate; fruiting peduncles axillary, solitary, more or less spreading or recurved, puberulous or glabrate, robust, thickened towards the apex, bracteate, dusky, $\frac{2}{3}$–1 in. long; bracts alternate, deciduous; fruit solitary, globose, shortly velvety and sparingly pilose with deciduous hairs, about $\frac{3}{5}$ in. in diam.; fruiting calyx accrescent, deeply 5-lobed; tube about $\frac{1}{2}$ in. in diam., puberulous outside; lobes ovate-lanceolate, thinly coriaceous, 9–13-nerved, obtuse, glabrate or puberulous, erect, $\frac{3}{4}$–1 in. long, $\frac{3}{8}$–$\frac{1}{2}$ in. broad; seeds $\frac{1}{4}$–$\frac{1}{3}$ in. long, $\frac{1}{4}$–$\frac{1}{5}$ in. broad, dusky.

COAST REGION : King Williamstown Div. ; Perie Bush, 2000 ft., *Kuntze!*

EASTERN REGION : Pondoland, *Bachmann*, 1016 ; Natal ; Tugela River, *Gerrard*, 1611 !

R. Simii should be compared with *R. pubescens, Edwards, Bot. Reg. t.* 500, *vix Willd.*; also with the following specimens in fruit :—

COAST REGION : Knysna Div. ; near Knysna, *Burchell*, 5490 ! near the Knysna River Ford, *Burchell*, 5529 ! Fort Beaufort Div., *Cooper*, 418 !

The " Kraai beesies " of the Kaffirs.

The *R. pubescens* mentioned by Burchell, *Trav. S. Afr.* i. 24, *note*, may be this species ; it is 745 ! of his Catalogue, and was obtained in the Government garden at Cape Town. Burchell in his notes described it as a tree, 10 ft. high, with greenish-yellowish flowers, and with the fruit as in *Bot. Reg. t.* 500.

13. R. Guerkei (O. Kuntze, Rev. Gen. Pl. iii. ii. 196) ; a shrub, 6–10 ft. high ; branches numerous, dusky, puberulous or glabrate, terete ; branchlets hirsute, alternate, more or less spreading, densely leafy ; leaves alternate, elliptic-oblanceolate, rounded, obtuse or apiculate at the apex, wedge-shaped at the base, sparingly hirsute and rugose with impressed reticulation above, hirsute especially on the raised reticulation beneath, thinly coriaceous, narrowly revolute along the margin, entire, 1–2$\frac{2}{5}$ in. long, $\frac{1}{3}$–$\frac{4}{5}$ in. broad; petiole hirsute, $\frac{1}{12}$–$\frac{1}{6}$ in. long ; fruiting peduncles axillary on the branchlets, solitary, shortly hirsute, more or less spreading or recurving, $\frac{1}{4}$–$\frac{1}{2}$ in. long, marked with the scars of 2 or 3 alternate fallen bracts ; fruiting calyx puberulous and sparingly hirsute, divided nearly to the base ; segments lanceolate, strongly 3-nerved, dusky, $\frac{3}{5}$ in. long, $\frac{1}{6}$ in. broad; fruit solitary, spheroidal, puberulous, rather more than $\frac{1}{2}$ in. long, scarcely $\frac{1}{2}$ in. in diam.; seeds 4, reddish-brown, minutely wrinkled, $\frac{1}{3}$–$\frac{3}{8}$ in. long, $\frac{1}{6}$ in. broad.

EASTERN REGION : Natal ; Charlestown, 6000 ft., *Kuntze!*

14. R. glabra (Linn. Sp. Pl. ed. i. 397); an evergreen shrub, 2–6 ft. high, very apt to send up suckers from the roots ; stem erect, ranging up to 6 in. thick, terete, much branched ; bark thin, glabrous, grey or purplish, smooth ; branches scattered, erect or erect-patent, virgate ; branchlets comparatively slender, densely leafy ; young parts pilose ; leaves alternate, narrowly elliptic, usually narrowed at both ends, entire, thinly coriaceous, rigid, flat, glossy above, at length glabrous, subsessile, myrtle-like, $\frac{1}{2}$–1$\frac{3}{8}$ in. long, $\frac{1}{7}$–$\frac{1}{3}$ in. broad ; flowers hermaphrodite or polygamous, nodding, whitish, numerous, $\frac{1}{5}-\frac{1}{4}$ in. long; peduncles axillary, slender, hairy, arching, 1–5-flowered, usually about as long as the leaves ; pedicels as long as or shorter than the common peduncle ; bracts lanceolate; calyx 5-partite, hairy, persistent, $\frac{1}{8}-\frac{1}{6}$ in. long, slightly accrescent in fruit ; segments lanceolate or lanceolate-subulate, acute, erect; corolla urceolate-campanulate, glabrous, 5-cleft nearly to the middle ; lobes ovate-oblong, obtuse, spreading or recurved ; tube pentagonal ; stamens usually 10, bearded, oblong, about $\frac{1}{13}$ in. long, not always fertile ; filaments very short ; glands about the base of the ovary very small ; ovary subglobose or ovoid-conical, nearly glabrous, 4-celled ; style bilobed, hairy below, together with the ovary about $\frac{1}{9}$ in. long; stigmas obtuse, emarginate ; fruit globose or oblong, thinly glandular-pubescent, purple or reddish, $\frac{1}{3}-\frac{2}{3}$ in. long ; fruiting calyx subglabrate, usually reflexed, $\frac{1}{6}-\frac{1}{3}$ in. long ; seed usually solitary, spheroidal, black. *Bery. Pl. Cap.* 144 ; *Thunb. Prodr.* 80, *and Fl. Cap. ed. Schult.* 390 ; *Poir. Encycl.* vi. 322 ; *Mill. Gard. Dict. ed. Martyn,* ii. *part* ii. *n.* 4 ; *Burchell, Trav. S. Afr.* i. 15, 19, 62, 124 ; *Alph. DC. Prodr.* viii. 214 ; *Hiern in Trans. Cambr. Phil. Soc.* xii. 88 ; *Parmentier in Ann. Univers. Lyon,* vi. *fasc.* ii. 70 ; *Drège, Zwei Pfl. Documente,* 70, 71, 87, 101, 217 ; *Krauss in Flora,* 1844, 824. *Vaccinium pensylvanica, Mill. Gard. Dict. ed.* viii. *n.* 3. *V. pensylvanicum, Mill. Gard. Dict. ed.* ix. *R. myrtifolia, Cels, Cat.* 1817, 33 ; *Steud. Nomencl. Bot. ed.* 1, 705, *and ed.* 2, ii. 475 ; *Alph. DC. l.c.* 215. *R. falcata, E. Meyer ex Drège, Zwei Pfl. Documente,* 115, 217 ; *Alph. DC., l.c.* 211. *R. hirsuta, Sieber, Fl. Cap. Exsicc. n.* 94 ; *not of Linn., nor of Jacq., nor of Ecklon. Vitis Idæa æthiopica Myrtinis foliis flosculis dependentibus, Plukn. Almag.* 391, *and Phytogr. t.* 321, *fig.* 4. *Vitis Idæa æthiopica Buxi minoris folio floribus albis, Commel. Hort. Amstelod.* i. 125, *t.* 65. *Vitis Idæa foliis angustissimis longis alternis, Linn. in Herb. Hort. Cliff.* ? *Buxus africana folio oblongiori non serrato, Linn. in Herb. Gronov. Vitis Idæa æthiopica seu africana Buxi minoris folio floribus albidis, J. Burm. Cat. alt. Pl. Afric.* 33.

SOUTH AFRICA : without locality, *Oldenburg,* 495 ! *Masson ! Forster ! Sieber,* 94 ! *Niven,* 48 ! *Herb. Linneus !*

COAST REGION : Vanrhynsdorp Div. ; Gift Berg, 1500–2500 ft., *Drège !* Clanwilliam Div. ; Clanwilliam, *Mader,* 74 ! Piquetberg Div. ; Verloren Valley, *Thunberg !* Tulbagh Div. ; near New Kloof, 1000 ft., *MacOwan, Herb. Norm. Aust.-Afr.,* 538 ! Worcester Div. ; Brand Valley, within the influence of the Hot-spring, *Burchell.* Paarl Div. ; Paarl Mountain, *Drège, Bolus,* 2973 ! flats near the Berg River, *Drège !* Paarl, *Elliot !* Cape Div. ; Table mountain and other places around Cape Town, *Burchell,* 2 ! 808 ! *Ecklon,* 699 ! *Zeyher ! MacGillivray,* 610 ! *Krauss ! Rehmann,* 778 ! *Wolley Dod,* 2008 ! Caledon Div. ; mountains near Genadendal, *Drège !* Robertson Div. ; Kogmans Kloof, *Kuntze ;* Swellendam Div. ; by the Buffeljagts River, *Burchell,* 7288 ! *Zeyher,* 3349b ; near Grootvaders Bosch, *Zeyher,* 3349a ! Swellendam, *Kuntze ;* Riversdale Div. ; various localities, *Burchell,* 6788 ! 7186 ! 7208 ! George Div. ; various localities, *Burchell,* 5093 ! 5784 ! *Bowie !* Montagu Pass, *Rehmann,* 157 ! 158 ! Knysna Div. ; between Plettenbergs Bay and Knysna, *Burchell,* 5367 !

CENTRAL REGION : Calvinia Div. ; Bokkeveld, near Groene River and Waterval River, 2500–3000 ft., *Drège !*

The wood is light and porous, and is used for fuel.—*Dr. Pappe.*

15. R. parviflora (Hiern in Trans. Cambr. Phil. Soc. xii. 88) ; a large shrub, climbing ; branches terete ; young parts and inflorescence softly, shortly and adpressedly pubescent ; leaves alternate, obovate, rounded or very shortly and abruptly narrowed to the emarginate apex, wedge-shaped at the base, membranous, or the smaller ones subcoriaceous, green, glabrous, and inconspicuously veined above, puberulous, somewhat paler and delicately veined beneath, 2–$6\frac{1}{2}$ in. long, 1–$3\frac{4}{5}$ in. broad ; petiole $\frac{1}{3}$–$\frac{2}{3}$ in. long ; cymes axillary on the young shoots, $\frac{1}{2}$–$\frac{3}{4}$ in. long, 3–5-flowered ; common peduncles $\frac{1}{3}$–$\frac{1}{2}$ in. long ; lateral pedicels $\frac{1}{6}$–$\frac{1}{4}$ in. long, with a narrow bract at the base about as long as themselves ; flowers hermaphrodite, small, creamy-white, articulated at the base to the pedicel, depressedly conical, and about $\frac{1}{7}$ in. long and broad in the bud ; calyx short, depressedly hemispherical, 5-fid, flat at the base, puberulous outside ; lobes deltoid ; corolla shortly pubescent outside except the overlapped sides of the lobes, glabrous within, deeply 5-lobed ; lobes rounded, and three times as long as the tube ; stamens 10, hairy, equal, inserted in one row at the base of the corolla ; ovary depressedly conical, 10-celled, shortly hairy ; cells 1-ovuled ; style shortly hairy, 5-lobed at the apex.

Eastern Region : Zululand ; at Incansla, *Gerrard*, 2015 !

16. R. glandulosa (Hiern in Trans. Cambr. Phil. Soc. xii. 89, t. 2) ; a large shrub, 8–10 ft. high ; branchlets patent or spreading ; the young shoots, peduncles and fruit glandular-hispid, with jointed hairs, subferruginous ; leaves alternate, oval or ovate, obtusely pointed at the apex, rounded or obtuse at the base, thinly coriaceous or firmly membranous, ciliate, somewhat pilose beneath, $\frac{1}{2}$–1 in. long, $\frac{1}{4}$–$\frac{1}{2}$ in. broad ; petiole hirsute, about $\frac{1}{16}$ in. long ; flowers hermaphrodite, axillary on the young shoots, about $\frac{1}{4}$ in. long, urceolate, articulated to the peduncle, yellow, usually tetramerous, rarely pentamerous ; peduncles spreading, 1-flowered, solitary, $\frac{2}{5}$ in. long ; calyx deeply lobed, pilose outside, pubescent within, accrescent in fruit ; lobes in flower lanceolate, acute, rather spreading, and about $\frac{1}{7}$ in. long ; corolla urceolate, deeply lobed, glabrous except on the minutely ciliate margin ; lobes rounded and recurved above ; stamens usually 8, inserted in one row at the base of the corolla, short, equal, pilose, two of them opposite each lobe of the corolla ; filaments short ; ovary 8-celled, hairy on most parts ; style hairy, 4-lobed and glabrous at the apex ; fruit spheroidal, scarcely $\frac{1}{2}$ in. long by $\frac{1}{3}$ in. thick ; fruiting calyx loosely enclosing the fruit or reflexed, deeply lobed, $\frac{4}{5}$ in. long ; lobes ovate-oblong, foliaceous, reddish, and with about 8 inconspicuous nerves.

Eastern Region : Natal ; by the Tugela River, *Gerrard*, 1608 !

17. R. pentandra (Gürke in Engl. Jahrb. xxvi. 61) ; a tree or shrub ; branches spiny, short, nodose ; bark grey-white ; leaves opposite, cuneate-obovate, obtuse or rounded at the apex, shortly petiolate, entire, tolerably firm, scarcely coriaceous, dull-green and finely pubescent above, canescent-velvety beneath, $1\frac{1}{5}$–2 in. long, $\frac{3}{5}$–$1\frac{2}{5}$ in. broad ; petiole $\frac{1}{5}$–$\frac{1}{3}$ in. long; cymes lax, 6–8-flowered ; bracts lanceolate-subulate, sessile, hairy, $\frac{1}{25}$–$\frac{1}{12}$ in. long; peduncles $\frac{1}{5}$–$\frac{1}{2}$ in. long, slender, pubescent; flowers hermaphrodite, pentamerous ; bracts lanceolate-subulate, sessile, hairy, $\frac{1}{25}$–$\frac{1}{12}$ in. long ; calyx cup-shaped, velvety outside, $\frac{1}{3}$ in. long, or rather more ; lobes short, spathulate, and $\frac{1}{12}$ in. broad ; corolla twice as long as the calyx, finely hairy outside, glabrous within ; lobes ovate, obtuse, and about $\frac{1}{3}$ in. broad ; tube about $\frac{1}{4}$ in. long, or rather more, pilose both inside and out ; stamens 5, inserted about the top of the corolla-tube, alternating with the lobes ; filaments $\frac{1}{25}$ in. long ; anthers linear, $\frac{1}{7}$ in. long, obtuse at the apex ; ovary conical, hairy, 2-celled ; style $\frac{1}{3}$ in. long, or rather more, bilobed at the apex ; lobes broadly oval and compressed ; ovules solitary in the cells. *Schinz & Junod in Mém. Herb. Boiss.* x. 55.

EASTERN REGION : Delagoa Bay, *Junod*, 412 !

Also in Tropical Africa.

II. **EUCLEA**, Murr.

Flowers diœcious, or rarely polygamous, 4–7-merous, arranged in axillary racemes or panicles. *Calyx* campanulate or cup-shaped, or small and shallow, usually 4- or 5-cleft, not accrescent in fruit. *Corolla* campanulate or hemispherical, or shortly oblong. *Stamens* in the male flowers 10–30, usually 12–20, either free, or in pairs or combined at the base of the filaments, in one or two rows inserted at the base of the corolla, or around the base of the ovary ; anthers more or less hairy or glabrous, lanceolate or oblong, 2-celled, dehiscing laterally ; filaments short, usually slender and glabrous ; ovary usually abortive or rudimentary ; styles 2 or 1. In the female flowers staminodes usually obsolete, sometimes 2–4 and glabrous ; anthers 0. *Ovary* ovoid or globular, hairy or glabrous, usually 4- (rarely 2-) celled ; styles 2, or 1 and bifid, or rarely 3 ; stigmas emarginate or bifid at the apex ; ovules solitary in the cells, pendulous. *Fruit* globular, or rarely ovoid-conical, usually 1-celled and 1-seeded ; pericarp fleshy ; seeds globular, usually marked outside with three longitudinal impressed lines ; albumen usually with an intrusion of the testa at the micropyle, distinctly ruminated in a few species ; embryo somewhat curved, tending to be incumbent ; radicle superior, about as long as the foliaceous cotyledons.

Shrubs or small trees with alternate or opposite or rarely verticillate leaves, entire (except sometimes in *E. ovata* and *E. coriacea*), evergreen; periderm of the stem usually arising from the pericycle.

DISTRIB. In addition to the following there are some Tropical African, and one Arabian, species. *Diospyros suberifolia, Decaisne*, a species supposed to be Chilian, is *Euclea suberifolia, Parment. in Ann. Univers. Lyon*, vi. *fasc.* ii. 62, 116, 142.

Euclea is one of the characteristic genera of the Karroo region.

Corolla shortly lobed :
　Calyx cleft about half-way down :
　　Leaves oval or elliptic or obovate :
　　　Stamens 20–30 ; male racemes $\frac{1}{2}$–$1\frac{1}{2}$ in.
　　　　long ; female racemes $\frac{1}{10}$–$\frac{1}{4}$ in. long ;
　　　　fruit globose　…　…　…　…　(1) **polyandra.**
　　　Stamens 16–18 ; male racemes short ; fe-
　　　　male racemes very short ; fruit ovoid,
　　　　somewhat conical at the apex …　…　(2) **tomentosa.**
　　Leaves ovate　…　…　…　…　…　(3) **coriacea.**
　　Leaves oblong-lanceolate　…　…　…　(4) **acutifolia.**
　　Leaves linear-lanceolate or -oblanceolate　…　(5) **lancea.**
　　Leaves linear :
　　　Calyx pubescent ; flowers usually penta-
　　　　merous, rarely hexamerous　…　…　(6) **Pseudebenus.**
　　　Calyx glabrous ; flowers tetramerous,
　　　　rarely pentamerous　…　…　…　(7) **linearis.**
　　Calyx shortly toothed ; leaves obovate or oblanceo-
　　　late …　…　…　…　…　…　…　(8) **Guerkei.**
Corolla cleft about half-way down or lower :
　Leaves rather large, usually exceeding an inch, or
　　if small not strongly wavy :
　　Branches more or less pubescent, or if glabrous
　　　not reddish :
　　　Ovary hairy :
　　　　Calyx cleft about half-way down or
　　　　　less, the tube in fruit not con-
　　　　　solidated ; leaves lanceolate, elliptic
　　　　　or ovate :
　　　　　Male inflorescence racemose, 3–9-
　　　　　　flowered :
　　　　　　Leaves quite entire, obtuse
　　　　　　　or subacute　…　…　(9) **lanceolata.**
　　　　　　Leaves often crenulate, acute
　　　　　　　or apiculate　…　…　(10) **ovata.**
　　　　　Male inflorescence many-flowered,
　　　　　　often compound :
　　　　　　Leaves opposite or sub-
　　　　　　　opposite, narrowly elliptic,
　　　　　　　glabrous or nearly so, some-
　　　　　　　what glaucous　…　…　(11) **Divinorum.**
　　　　　　Leaves alternate or rarely
　　　　　　　subopposite, oval, elliptic
　　　　　　　or oblong, more or less
　　　　　　　pubescent, not glaucous …　(12) **multiflora.**
　　　　Calyx deeply lobed, the tube in fruit
　　　　　consolidated and articulate to the
　　　　　pedicel ; leaves obovate-oblong　…　(13) **natalensis.**
　　　Ovary glabrous :
　　　　Leaves obovate-oblong or oblanceo-
　　　　　late, $1\frac{1}{2}$–4 in. long, $\frac{3}{8}$–$1\frac{1}{2}$ in. broad ;
　　　　　petioles $\frac{1}{10}$–$\frac{1}{4}$ in. long　…　…　(14) **macrophylla.**
　　　　Leaves linear- or oblanceolate-oblong,
　　　　　1–3 in. long, $\frac{1}{6}$–$\frac{1}{2}$ in. broad, sub-
　　　　　sessile　…　…　…　…　(15) **daphnoides.**
　　　Branches glabrous, often reddish　…　…　(16) **racemosa.**
　　Leaves small, usually about an inch long or less
　　　and strongly wavy …　…　…　…　…　(17) **undulata.**

1. E. polyandra (E. Meyer ex Drège, Cat. Pl. Exsicc. Afr. Austr.

7); a shrub, 3–7 ft. high, or on mountain sides 9–12 in. only, pubescent, and often ferruginous, but sometimes glabrescent; branches terete or subterete, alternate or subopposite, leafy, spreading at an angle of 40–60°; leaves alternate or subopposite, entire, oval, obtuse, and often subapiculate at the apex, somewhat wedge-shaped, rounded, or rarely cordate at the base, coriaceous, 1–3 in. long, $\frac{1}{2}$–$1\frac{1}{2}$ in. broad; petiole $\frac{1}{12}$–$\frac{1}{4}$ in. long; flowers diœcious, scented, white. Male cymes racemose, axillary, pubescent, 3–9-flowered, usually drooping, $\frac{1}{2}$–$1\frac{1}{2}$ in. long; pedicels $\frac{1}{10}$–$\frac{1}{2}$ in. long, the lower ones the longer; bracts lanceolate, deciduous; calyx hemispherical, deeply 5–7-cleft, glabrous within, $\frac{1}{20}$–$\frac{1}{10}$ in. long; lobes deltoid, rounded, ovate or lanceolate; corolla urceolate, $\frac{1}{6}$–$\frac{1}{5}$ in. long, shortly 5–7-toothed; stamens 20–30, more or less united at the base in pairs or otherwise, hairy; ovary more or less abortive; styles 2, slender. Female cymes 3–5-flowered, axillary, pubescent or tomentose, $\frac{1}{10}$–$\frac{1}{4}$ in. long, usually drooping; pedicels short; bracts deciduous; calyx shorter than the corolla, 5–7-fid; lobes ovate or deltoid; corolla ellipsoid, $\frac{1}{8}$ in. long, shortly 5–7-toothed; staminodes 0; ovary ovoid-conical, 4-celled, hairy, $\frac{1}{10}$ in. long; ovules solitary; styles 2, short, glabrous above, level with the corolla-mouth; stigmas emarginate; fruit usually solitary, occasionally 2 or 3 together, tomentose, usually ferruginous, globular, $\frac{1}{5}$–$\frac{1}{2}$ in. in diameter, edible, but not pleasant to the palate, 1-celled, 1-seeded; seed globular; albumen somewhat ruminated. *Alph. DC. Prodr.* viii. 216 *under E. dregeana; Hiern in Trans. Cambr. Phil. Soc.* xii. 92. *Parmentier in Ann. Univ. Lyon,* vi. *fasc.* ii. 79; *Drège, Zwei Pfl. Documente,* 78, 87, 184. *Royena polyandra, Linn. f. Suppl. Pl.* 240; *Poir. Encycl.* vi. 322; *Thunb. Prodr.* 80, *and Fl. Cap. ed. Schult.* 391; *non Willd. Herb. n.* 8366! *Diplonema elliptica, G. Don, Gen. Syst.* iv. 42. *Rymia polyandra, Endl. Cat. Hort. Acad. Vindob.* ii. 123. *E. elliptica and E. dregeana, Alph. DC. Prodr.* viii. 216. *E. ferruginea, Bernh. ex Krauss in Flora,* 1844, 825. *Brachycheila pubescens, Harv. ex Drège in Linnæa,* xx. 192. *Euclea,* 22, *Drège, l.c.*

South Africa : without locality, *Niven,* 47! *Alexander!* *Herb. Linneus!*

Coast Region: Clanwilliam Div. ; Jackhals Vlei, *Niven,* 53! Malmesbury Div.; near the Berg River, *Thunberg! Masson!* Tulbagh Div.; New Kloof, *Drège!* near Tulbagh Waterfall, *Ecklon & Zeyher!* Paarl Div. ; Paarl Mountain, *Drège;* Caledon Div.; Zwart Berg and near the Baths, *Zeyher,* 3362! near Hemel-en-Aarde, *Zeyher,* 3364! near Genadendal, 3000 ft., *Schlechter,* 9852! Hermanus, *Galpin,* 4321! Vogel Gat, 500 ft., *Schlechter,* 10402! Robertson Div.; mountains near Montagu Bath, 2300 ft., *Bolus,* 6722! Riversdale Div.; Garcias Pass, *Burchell,* 6941! mountains near Riversdale, *Schlechter,* 1975! Uniondale Div.; hill near Haarlem, *Burchell,* 4998! Humansdorp Div.; between the Gamtoos and Leeuwenbosch Rivers, *Burchell,* 4807! near Kromme River, *Burchell,* 4873! Uitenhage Div.; Van Stadens Mountains, *Zeyher,* 727! 3363! Winterhoek Mountains, *Krauss!* Witte Klip, *MacOwan,* 1929! Alexandria Div.; hills of Quaggas Flats, *Zeyher,* 727!

This shrub is called "Kersbosch" in Cape Colony, according to Thunberg.

2. E. tomentosa (E. Meyer ex Drège, Cat. Pl. Exsicc. Afr. Austr.

7) ; a shrub, about 4 ft. high or more ; bark smooth, dusky ; branches alternate, spreading, tomentose or glabrate ; branchlets crowded, ashy-tomentose or puberulous, erect-patent, rigid, sometimes slightly flexuous, leafy ; leaves alternate or subopposite, oval or elliptic or obovate, rounded or obtusely narrowed or apiculate at the apex, rounded or wedge-shaped at the base, erect-patent, coriaceous, rigid, more or less tomentellous at least beneath, entire, flat, but narrowly revolute along the margin, $\frac{3}{4}$–2$\frac{1}{2}$ in. long, $\frac{3}{8}$–1$\frac{1}{2}$ in. broad ; petiole $\frac{1}{20}$–$\frac{1}{4}$ in. long, tomentellous; flowers diœcious, axillary. *Male* racemes 5–8-flowered, very short, or much shorter than the leaves; flowers tetramerous or pentamerous, subsessile, or on short recurving pedicels, bracteolate ; bracteoles concave, pubescent on the back, glabrous on the face, often imbricate, deciduous, $\frac{1}{24}$–$\frac{1}{8}$ in. long, $\frac{1}{24}$–$\frac{1}{8}$ in. broad ; calyx hirsute-tomentellous outside, glabrous and shining within, deeply lobed, $\frac{1}{16}$ in. long ; lobes ovate or ovate-deltoid ; corolla urceolate, $\frac{1}{8}$–$\frac{1}{5}$ in. long, cleft less than half-way down, more or less hairy outside at least above, glabrous within ; lobes rounded ; stamens 16–18, free, or somewhat connate at the base, $\frac{1}{15}$–$\frac{1}{12}$ in. long ; filaments unequal, $\frac{1}{40}$–$\frac{1}{30}$ in. long, glabrous ; anthers very sparingly setulose, $\frac{1}{30}$–$\frac{1}{20}$ in. long, sometimes in pairs ; ovary rudimentary, small, ovoid-conical, clothed with whitish hairs ; styles 2, short, glabrous. *Female* racemes very dense and short, one- or few-flowered, much shorter than the leaves, pendulous ; flowers 5–7-merous, $\frac{3}{16}$–$\frac{1}{5}$ in. long, subsessile, or on pedicels of $\frac{1}{24}$–$\frac{1}{15}$ in. long ; bracteoles rounded or apiculate, short, broad, deciduous or imbricate ; calyx tomentellous, deeply lobed, $\frac{1}{8}$ in. long ; segments ovate-acute or deltoid ; corolla urceolate or campanulate, cleft less than half-way down, $\frac{1}{6}$–$\frac{1}{5}$ in. long ; tube tomentellous outside, glabrous within ; lobes rounded, tomentellous on both sides ; staminodes 0 ; ovary shortly ovoid-conical, tomentellous, $\frac{1}{10}$ in. long, 4-celled ; styles 2, short, thick, nearly or quite glabrous ; stigmas dilated, bifid ; ovules solitary in the cells ; fruit ovoid, somewhat conical at the apex, hoary-tomentose, 4-celled, $\frac{2}{5}$–$\frac{1}{2}$ in. long, $\frac{1}{4}$–$\frac{3}{10}$ in. thick. *Drège, Zwei Pfl. Documente*, 72, 95, 104, 110, 184 ; *Alph. DC. Prodr.* viii. 216 ; *Hiern in Trans. Cambr. Phil. Soc.* xii. 93 ; *Parmentier in Ann. Univ. Lyon*, vi. *fasc.* ii. 81. *E. kraussiana, Bernh. ex Krauss in Flora*, 1844, 824.

SOUTH AFRICA : without locality, *Thunberg ! Ecklon & Zeyher !*

COAST REGION : Clanwilliam Div. ; on sand-hills near the Olifants River, 1000–1500 ft., *Drège !* Blaauw Berg, 1000–2000 ft., *Drège.* Piquetberg Div. ; flats between Twenty-four Rivers and Pikeniers Kloof, below 500 ft., *Drège !* Malmesbury Div. ; Zwartland, *Masson !* hills near Moorreesburg, 600 ft., *Bolus*, 9972 ! Tulbagh Div. ; between Kasteels Kloof and New Kloof, *Burchell*, 987 ! Cape Div. ; places at the foot of Koe Berg and Tiger Berg, *Krauss*, 1788 !

WESTERN REGION : Little Namaqualand ; Modderfontein (Garies), 1500–2000 ft., *Drège, Whitehead !* Ezels Fontein, *Whitehead !* near Spektakel, 3500 ft., *Bolus*, 9508 !

Called by the Boers " Kersboschjes " and " Jakhalsbosch."

3. E. coriacea (*Alph. DC Prodr.* viii. 216); a shrub ; branches dense, strong, dusky-ashy ; branchlets slightly pubescent ; leaves alternate, strongly wavy, entire or obscurely crenulate, ovate, more or less acute or apiculate at the apex, broad and subcordate at the base, coriaceous, pubescent, or nearly glabrescent, 1–2 in. long, $\frac{1}{2}$–$1\frac{1}{2}$ in. broad; veins inconspicuous and pallid above, dully marked beneath; petiole ranging up to $\frac{1}{8}$ in. long; flowers dioecious; bracts ovate, small, deciduous. Male cymes 1–5-flowered, rather dense, much shorter than the leaves; pedicels about as long as the flowers, $\frac{1}{6}$–$\frac{1}{5}$ in. long; calyx deeply 5- or 6-fid; lobes ovate, acute; corolla urceolate, four times the length of the calyx, shortly 5- or 6-lobed at the apex; stamens 16–22, sometimes in pairs; anthers linear-lanceolate, silky at the back; ovary rudimentary. Female cymes 3–7-flowered, very short; pedicels ranging up to $\frac{1}{10}$ in. long; fruiting calyx 5- or 6-fid, nearly flat, stellate, $\frac{1}{5}$ in. in diam.; lobes ovate, or lanceolate and acute; fruit globose, $\frac{1}{3}$–$\frac{2}{5}$ in. in diam., subglabrate or minutely puberulous, 1-celled, 1-seeded; seed subglobose, about $\frac{1}{4}$ in. in diam., marked outside with impressed curved lines; albumen somewhat ruminated. *Hiern in Trans. Cambr. Phil. Soc.* xii. 94, 289. *Euclea n.* 9140, *E. Meyer ex Drège, Zwei Pfl. Documente,* 48, 184. *Royena n.* 9140, *Drège ex Alph. DC. Prodr.* viii. 216.

COAST REGION: Queenstown Div.; Table Mountain, 6000-7000 ft., *Drège,* 9140!

CENTRAL REGION: Graaff Reinet Div.; Oude Berg, near Graaff Reinet, 4500 ft., *Bolus,* 638!

4. E. acutifolia (E. Meyer ex Drège, Cat. Pl. Exsicc. Afr. Austr. 7); a shrub; branches lax, glabrous or glabrescent; leaves alternate or subopposite, entire, oblong-lanceolate, acute and apiculate at the apex, wedge-shaped at the base, erect or ascending, coriaceous, sometimes subfalcate, glabrous, subglaucescent beneath, $1\frac{1}{2}$–$2\frac{1}{2}$ in. long, about $\frac{1}{3}$–$\frac{1}{2}$ in. broad; nerves not very distinctly marked; petiole $\frac{1}{20}$–$\frac{1}{8}$ in. long; flowers dioecious. Male flowers unknown. Female cymes racemose, $\frac{1}{4}$ in. long, 3–7-flowered; pedicels shortly hairy, about $\frac{1}{8}$ in. long; flowers $\frac{1}{6}$ in. long, pubescent, 5–7-merous; calyx short, deeply cleft; corolla cylindrical-urceolate, shortly lobed at the apex; ovary densely pilose; styles 2, erect, glabrous; stigmas dilated; fruit globose, glabrescent, finely wrinkled, dusky, $\frac{1}{3}$–$\frac{1}{2}$ in. in diam.; fruiting calyx $\frac{1}{8}$ in. in diam., adpressed to the base of the fruit, glabrescent; seed about $\frac{1}{4}$ in. in diam., dusky, the surface unequally divided by three impressed lines; albumen slightly ruminated. *Drège, Zwei Pfl. Documente,* 104, 109, 184; *Alph. DC. Prodr.* viii. 217; *Hiern in Trans. Cambr. Phil. Soc.* xii. 94.

COAST REGION: Clanwilliam Div.; hills between Lange Valley and Olifants River, 1000–1500 ft., *Drège!* Piquetberg Div.; on the flats between Twenty-four River and Pikeniers Kloof, below 500 ft., *Drège!* Tulbagh Div.; Tulbagh (New) Kloof, &c., *Ecklon & Zeyher!* Humansdorp and Uitenhage Divs.; between Kromme River and Uitenhage, *Ecklon & Zeyher!*

5. E. lancea (Thunb. Prodr. 85); a glabrous or subglabrous shrub, erect, 3–8 ft. high; branches densely intricate, alternate, terete, erect-patent, leafy; leaves alternate, entire, subsessile, linear-lanceolate or -oblanceolate, unequal, lower obtuse, upper acute at the apex, attenuate at the base, rigidly coriaceous, somewhat glaucescent, $\frac{2}{3}$–2 in. long, about $\frac{1}{8}$–$\frac{1}{4}$ in. broad, sides often unequal, net-veins inconspicuous; flowers in short axillary cymes, nearly glabrous or puberulous, urceolate or ovoid or campanulate, $\frac{1}{8}$–$\frac{1}{7}$ in. long, $\frac{1}{12}$–$\frac{1}{9}$ in. broad, diœcious. Male cymes $\frac{1}{3}$–$\frac{3}{8}$ in. long, 3–5-flowered; pedicels $\frac{1}{8}$–$\frac{1}{5}$ in. long, minutely puberulous; bracts very small, ovate-deltoid, deciduous; calyx small, cupuliform, $\frac{1}{12}$–$\frac{1}{10}$ in. in diam., shortly 5- or 6-fid, glabrous within; lobes ovate-deltoid; corolla shortly 5- or 6-toothed at the apex, glabrous within; stamens 15–22, inserted at the bottom of the corolla, some united at the base in pairs one in front of the other, others free; styles 2, glabrous; ovary rudimentary. Female cymes very short; 1–3-flowered; pedicels very short; calyx shortly 5- or 6-cleft; corolla shortly 5- or 6-lobed; staminodes 0; ovary 3-celled, shortly pubescent; styles 3, short, glabrous; fruit subglobose, nearly glabrous, $\frac{1}{6}$–$\frac{1}{3}$ in. in diam.; seed solitary. *Thunb. Fl. Cap. ed. Schultes,* 401; *Alph. DC. Prodr.* viii. 219; *Hiern in Trans. Cambr. Phil. Soc.* xii. 95, 289. *E. rigida, E. Meyer ex Drège, Cat. Pl. Exsicc. Afr. Austr.* 7; *Alph. DC. Prodr.* viii. 217; *Hiern, l.c.* 289; *Drège, Zwei Pfl. Documente,* 67, 71, 95, 184.

SOUTH AFRICA: without locality, *Thunberg! Masson!*

COAST REGION: Vanrhynsdorp Div.; Gift Berg, 1500–2500 ft., *Drège.*

WESTERN REGION: Little Namaqualand; between Uitkomst and Geelbeks Kraal, 2000–3000 ft., *Drège;* mountains near Kaspars Kloof, Elleboog Fontein, and Geelbeks Kraal, 3000–4000 ft., *Drège!* between Pedros Kloof and Lily Fontein, 3000–4000 ft., *Drège!* Kamies Berg, by rivulets, *Niven,* 46!

6. E. Pseudebenus (E. Meyer ex Drège, Cat. Pl. Exsicc. Afr. Austr. 7); a shrub 4–8 ft. high, or a tree much branched throughout; trunk 10–12 in. in diam.; branchlets slender, pubescent or glabrescent, leafy; bark thin; periderm subepidermal; leaves alternate, entire, linear, narrowed at both ends, coriaceous, glabrous or puberulous, 1-nerved, moderately reticulate, 1–2$\frac{1}{4}$ in. long, $\frac{1}{10}$–$\frac{1}{4}$ in. broad; petiole $\frac{1}{20}$–$\frac{1}{7}$ in. long; cymes racemose, axillary, shorter than the leaves, pubescent; flowers diœcious. Male racemes $\frac{1}{4}$–$\frac{1}{2}$ in. long, 3–7-flowered; pedicels slender, $\frac{1}{19}$–$\frac{1}{5}$ in. long; flowers $\frac{1}{10}$–$\frac{1}{7}$ in. long, usually pentamerous, rarely hexamerous, calyx pubescent, about half the length of the corolla, cleft half-way down; lobes ovate-deltoid; corolla globose-urceolate, shortly lobed, rather fleshy, white, hoary-tomentose outside; stamens 12–20, inserted at the bottom of the corolla; anthers erect, hirsute or glabrous; filaments short; pistil obsolete. Female racemes very short, $\frac{1}{6}$–$\frac{1}{3}$ in. long, 1–3-flowered; pedicels $\frac{1}{16}$–$\frac{1}{7}$ in. long; flowers $\frac{1}{8}$–$\frac{1}{7}$ in. long, pentamerous; calyx shortly 5–6-lobed; corolla urceolate-subglobose, very densely hirsute outside, slightly hairy inside about the throat, otherwise glabrous, with very short, broad obtuse lobes; staminodes 0; ovary sessile, globose, densely shaggy, 4-celled; styles 2, glabrous; ovules solitary; fruit globose, glabrescent, $\frac{1}{5}$–$\frac{1}{4}$ in. in diam., black-bluish, drupaceous, a little juicy, edible, sweet and slightly astringent; seeds solitary, marked with three depressed lines. *Drège, Zwei Pfl. Documente*, 93, 184; *Alph. DC. Prodr.* viii. 217; *Hiern in Trans. Cambr. Phil. Soc.* xii. 95, 289; *Hiern in Oliver, Fl. Trop. Afr.* iii. 512; *Hiern, Cat. Afr. Pl. Welw.* i. 647. *E. angustifolia, Benth. in Hook. Niger Fl.* 441. *Diospyros Pseudebenus, E. Meyer sec. Parment. in Ann. Univ. Lyon*, vi. *fasc.* ii. 81.

Western Region: Great Namaqualand; Gubub, *Schinz.* Little Namaqualand; by the lower part of the Orange River, below 800 ft., *Drège! Atherstone, 2! Schenck*, and without precise locality, *Wyley! Whitehead!*

It occurs also in South-west Tropical Africa. The heart-wood is extremely hard and black. It is known by the names of Orange River ebony, sneezewood, and zwartebbenhout. This valuable tree, though it does not grow very high, attains a diameter of from 10 to 12 inches and more; it has a very thin bark, and its jet-black hard and durable wood is very beautiful; Dr. Livingstone sent large circular specimens from the interior to the South African Museum (*Pappe, Silva capensis, ed.* ii. 26).

7. E. linearis (Zeyh. in Linnæa xx. 192, without description); a quite glabrous and somewhat glaucous shrub, $2\frac{1}{2}$–3 ft. high, much branched; branchlets leafy, slender; leaves alternate, opposite or subopposite, entire, linear, acutely or obtusely narrowed at the apex, more or less wedge-shaped at the subsessile or sessile base, often somewhat falcate, coriaceous, minutely gland-dotted, 1–$2\frac{1}{2}$ in. long, $\frac{1}{10}$–$\frac{1}{5}$ in. broad; cymes racemose, 3–7-flowered; bracts linear, acute, small; flowers diœcious, tetramerous, or rarely pentamerous. Male racemes $\frac{1}{3}$–$\frac{3}{5}$ in. long, usually drooping; pedicels $\frac{1}{20}$–$\frac{1}{10}$ in. long; calyx short, about $\frac{1}{12}$ in. in diam., cleft about half-way down with broad lobes; corolla shortly tubular, shortly lobed, $\frac{1}{10}$ in. long; stamens usually 12–16, subglabrous; ovary rudimentary, slightly hairy; styles 1 or 2. Female racemes $\frac{1}{5}$–$\frac{3}{10}$ in. long; calyx campanulate, $\frac{1}{25}$–$\frac{1}{20}$ in. long; lobes rather shorter than the tube, from a broad base dentiform; corolla openly campanulate, shortly cleft, $\frac{1}{12}$ in. long; lobes spreading, oval or ovate; staminodes 0; ovary ellipsoidal, hairy, 4-celled; styles 2, thick, glabrous; fruit subglobose, $\frac{1}{5}$ in. in diam.; seed solitary, about $\frac{1}{6}$ in. in diam. *Hiern in Trans. Cambr. Phil. Soc.* xii. 96; *Parmentier in Ann. Univ. Lyon*, vi. *fasc.* ii. 77.

South Africa: without locality, *Masson!*
Coast Region: Vanrhynsdorp Div.; Windhoek, *Zeyher*, 1125! Clanwilliam Div.; Lange Kloof, 600–700 ft., *Schlechter*, 8044!

8. E. Guerkei (Hiern); shrubby, glabrate, except on the puberulous extremities and inflorescence; branchlets smooth, dusky, leafy; leaves alternate, or the upper subopposite, obovate or oblanceolate or sometimes subelliptic, rounded, or very obtuse at the apex, more or less wedge-shaped at the base, dark green, delicately veined and net-veined above, pale but marked with dark and delicate veins and reticulation beneath, entire, nearly flat, except the narrowly revolute margin, $\frac{1}{2}$–2 in. long, $\frac{1}{4}$–$\frac{5}{6}$ in. broad; petiole $\frac{1}{10}$–$\frac{1}{6}$ in. long; flowers diœcious. Female cymes axillary, $\frac{1}{4}$–$\frac{1}{2}$ in. long, puberulous, or nearly glabrous, 3–5-flowered; pedicels short, or very short; common peduncle $\frac{1}{6}$–$\frac{1}{4}$ in. long, arising from the upper axils, often at length recurved; flowers $\frac{1}{12}$–$\frac{1}{10}$ in. long, tetramerous or pentamerous; calyx subhemispherical, $\frac{1}{24}$–$\frac{1}{20}$ in. long, $\frac{1}{16}$ in. in diam., ciliolate, otherwise nearly glabrous, shortly toothed; corolla about $\frac{1}{12}$ in. long, tubular-urceolate, nearly glabrous, but with short scattered pallid hairs outside, glabrous within, shortly lobed; lobes rounded; staminodes 0; ovary subglobose, hairy; ovules 4; style thick, glabrous, as long as the ovary, 4-cleft, deciduous. Male flowers and fruit not seen.

KALAHARI REGION: Transvaal; near Lydenburg, Rustplaats Farm, *Wilms,* 916!

EASTERN REGION: Natal; between Pietermaritzburg and Greytown, *Wilms,* 1923! (doubtful).

9. **E. lanceolata** (E. Meyer ex Drège, Cat. Pl. Exsicc. Afr. Austr. 7); an evergreen shrub, 1 ft. high and upwards, or a tree reaching 20–25 ft., with a trunk 10–15 in. in diam.; branches rigid, opposite or alternate, or occasionally ternate or quaternate, sometimes virgate or tortuous, or rarely thorny, pubescent or glabrescent or lepidote, sometimes resinous, leafy; leaves opposite or alternate, lanceolate or ovate or narrowly oval, obtuse or subacute at the apex, obtuse or narrowed at the base, pubescent or glabrous and glossy, often lepidote when young, coriaceous, wavy or nearly flat, entire, sometimes glaucescent, usually paler beneath, 1–3 in. long, $\frac{1}{10}$–$1\frac{1}{10}$ in. broad, occasionally smaller, those near the base of the stem sometimes two or three times broader than the upper; petiole $\frac{1}{15}$–$\frac{1}{5}$ in. long; cymes axillary, racemose; bracts small or foliaceous; flowers dioecious, white, or whitish-reddish or yellowish. Male racemes $\frac{1}{4}$–1 in. long, 3–10-flowered; pedicels $\frac{1}{10}$–$\frac{3}{10}$ in. long; flowers tetramerous or occasionally pentamerous, $\frac{1}{8}$–$\frac{1}{6}$ in. long; calyx broadly campanulate, cleft about half-way down, $\frac{1}{25}$–$\frac{1}{15}$ in. long; lobes deltoid; corolla campanulate, deeply cleft, somewhat pubescent outside, $\frac{1}{10}$–$\frac{3}{20}$ in. long; lobes rather spreading, ovate or oval; stamens usually about 16, rarely 8–10, inserted mostly in pairs at or near the base of the corolla, $\frac{1}{12}$–$\frac{1}{9}$ in. long; filaments glabrous, $\frac{1}{48}$–$\frac{1}{16}$ in. long; anthers lanceolate, pilose towards the apex, turgid, $\frac{1}{16}$–$\frac{1}{20}$ in. long; ovary rudimentary, hirsute; styles 2, glabrous. Female racemes $\frac{1}{4}$–$\frac{1}{3}$ in. long, 3–7-flowered; pedicels $\frac{1}{20}$–$\frac{1}{15}$ in. long; flowers tetramerous or pentamerous, whitish-reddish or whitish, about $\frac{1}{6}$ in. long; pedicels $\frac{1}{20}$–$\frac{1}{15}$ in. long, the lateral opposite; bracteoles linear-lanceolate, shortly pubescent, about $\frac{1}{10}$ in. long; calyx cleft about half-way down, $\frac{1}{20}$ in. long, not accrescent, pubescent; lobes deltoid; corolla deeply cleft; staminodes 0; ovary subglobose, clothed with short erect hairs, 4-celled; styles 2, glabrous; fruit pisiform, pubescent or glabrate, reddish or dark-purple or yellowish, edible, 1-celled, $\frac{1}{5}$–$\frac{1}{4}$ in. in diam., or rather larger; seed solitary, globose; testa intruded some distance into the albumen, blackish. *Drège, Cat. Südafr. Pfl.* 9, *and Zwei Pfl. Documente,* 48, 124, 131, 134, 142, 145, 184; *Alph. DC. Prodr.* viii. 218; *Hiern in Trans. Cambr. Phil. Soc.* xii. 97,

and in Oliv. Fl. Trop. Afr. iii. 512, *and Cat. Afr. Pl. Welw.* i. 648;
Parmentier in Ann. Univ. Lyon, vi. *fasc.* ii. 77; *Krauss in Flora,*
1844, 824; *Sim, Sketch Fl. Kaffraria,* 54. *E. Desertorum, and
E. humilis, Eckl. & Zeyh. ex Drège in Linnæa,* xx. 192. *E. rufescens,
Drège, Zwei Pfl. Documente,* 62, *not* 57, *nor* 130. *E. ovata, var.
glabra, Alph. DC. Prodr.* viii. 218. *E. ochrocarpa E. Meyer ex
Drège, Zwei Pfl. Documente,* 70, 184; *Alph. DC. Prodr.* viii. 217.
E. lanceolata, β. glabrescens, Alph. DC. Prodr. viii. 218.

SOUTH AFRICA : without locality, *Masson! Mund! Bowie! Zeyher,* 3355!

COAST REGION, 200–4000 ft. : Knysna Div.; Ruigte Valley, *Drège,* near the
west end of Groene Valley, *Burchell,* 5648! Uniondale Div.; Long Kloof,
Burchell, 4938! *Ecklon & Zeyher!* Humansdorp Div.; near Kromme River,
Burchell, 4880! Uitenhage Div.; Addo, *Drège!* between Enon and the Zuurberg
Range, *Drège!* Van Stadens Mountains, *Zeyher,* 575! 690! *MacOwan,* 1931!
Zeyher, 3357! between the Coega and Zwartkops Rivers, *Zeyher,* 3359! hills
between the Zwartkops and Sunday Rivers, *Zeyher,* 1125! Albany Div.; Assegai
Bosch, *Baur,* 1094! Bothas Berg, *MacOwan,* 902! and without precise locality,
Williamson! Queenstown Div.; mountain at Bowkers Park, *Galpin,* 2567!
hills near Klipplaat River, *Drège.* King Williamstown Div.; between Buffalo
River and Yellowwood River, *Drège!* Perie Bush, *Hutchins in Herb. Scott-
Elliot,* 948!

CENTRAL REGION, 2000–5000 ft. : Calvinia Div.; between Grasberg River and
Waterval River, *Drège!* Somerset Div.; Bruintjes Hoogte, *MacOwan,* 1740!
near the Brak River, *Krauss,* 1796. Beaufort West Div.; Nieuwveld, near
Beaufort West, *Drège.*

KALAHARI REGION : Basutoland; *Cooper,* 3464! 3481! Transvaal; Magalies
Berg, *Burke,* 379! *Zeyher,* 1123! near Middelburg, *Wilms,* 218! Aapies Poort,
Pretoria, *Rehmann,* 4210! 4214! 4216! Bosch Veld, Menaars Farm, *Rehmann,*
4857! near Johannesburg, *Rand,* 1107!

EASTERN REGION : Transkei; near Gekau (Geua) River, *Drège!* Kreilis
Country, *Bowker,* 324! Natal; Inanda, *Wood,* 394! Umhlauga, *Wood,* 1416!
by the Tugela River, *Gerrard,* 1155! 1605! Mooi River, *Gerrard,* 1156! Nonoti
River, *Gerrard,* 33! and without precise locality, *Gerrard,* 528!

It occurs also in Tropical Africa. Sim, *Fl. Kaffr.* 54, marks this species as a
forest tree reserved by Government, and as a plant of known or reputed medicinal
value. He quotes, as Dutch and Kaffir names for it, " Bosch quarre," " um-
Gwali," " i-yezalokuxaxazisa." It is called " Omgwali " by the Kaffirs (Dr.
Pappe). The wood is dark in colour and heavy, and is used for yokes, triggers,
&c.

This is a very variable species, and in some cases is difficult to distinguish from
E. ovata, Burch.

10. E. ovata (Burchell, Trav. S. Afr. i. 387); a densely branched
and leafy shrub, with the habit of a myrtle, 3–7 ft. high, pubescent,
subferruginous, or sometimes glabrescent; foliage like that of box;
leaves opposite or alternate, elliptic or narrowly ovate, acute or
apiculate, and obtuse at the apex, mostly obtuse at the base,
coriaceous, rigid, rather thick, flat or wavy, minutely crenulate,
or quite entire, $\frac{1}{2}$–2 in. long, $\frac{1}{5}$–1 in. broad; petioles $\frac{1}{20}$–$\frac{1}{8}$ in.
long; cymes axillary, 3–7-flowered, racemose, $\frac{3}{10}$–$\frac{3}{5}$ in. long, at
length drooping; pedicels $\frac{1}{20}$–$\frac{1}{6}$ in. long; bracts lanceolate, small,
deciduous; flowers diœcious or subhermaphrodite, tetramerous, or
occasionally pentamerous, pubescent, $\frac{1}{10}$–$\frac{1}{6}$ in. long, the female flowers
nodding; calyx hemispherical, shortly cleft, $\frac{1}{20}$–$\frac{1}{10}$ in. long; lobes

deltoid; corolla campanulate, cleft half-way down or more, twice as long as the calyx, greenish or whitish-herbaceous; lobes broadly ovate or rounded and apiculate, recurved at the apex; stamens 16 or 20–22 in the male flowers, about 12 in the subhermaphrodite, none in the female; filaments short, slender, glabrous; anthers lanceolate, shortly hairy; ovary shortly conical or ovoid, hairy, 2–4-celled; styles 2, or rarely 3, bifid at the apex, glabrous; stigmas trifid, but little exserted; ovules usually 4; fruit globose, dusky or brown, about $\frac{1}{5}$ in. in diam., at first pubescent, at length glabrate, edible, the flesh somewhat astringent; seed solitary, comparatively large, sometimes with vestiges of 2 or 3 abortive ovules. *Alph. DC. Prodr.* viii. 218; *Hiern in Trans. Cambr. Phil. Soc.* xii. 98; *Parmentier in Ann. Univ. Lyon,* vi. *fasc.* ii. 78. *Celastrus crispus, Thunb. in Hoffm. Phytogr. Blätt.* (i.) 23, *and in Roemer, Archiv. Botanik,* iii. 429; *DC. Prodr.* ii. 5; *Thunb. Fl. Cap. ed. Schult.* 217. *E. rufescens, E. Meyer ex Drège, Cat. Pl. Exsicc. Afr.-Austr.* 7, *and ex Drège, Zwei Pflanzengeogr. Documente,* 57, 130, 184, *not* 62. *Royena rufescens, E. Meyer ex Drège, Zwei Pflanzengeogr. Documente,* 154, 217. *E. ovata, β. hispida, Alph. DC. Prodr.* viii. 218. *E. crispa, Gürke in Engl. & Prantl, Pflanzenfam.* iv. i. 158. *E. ovata, forma undulata, Marloth in Engl. Jahrb.* x. 243.

SOUTH AFRICA: without locality, *Thunberg.*

COAST REGION: Uitenhage Div.; near the mouth of the Zwartkops River, below 500 ft., *Drège.*

CENTRAL REGION: Richmond Div.; Winterveld, near Limoen Fontein and Table Mountain, 3000–4000 ft., *Drège!* Graaff Reinet Div.; mountains near Graaff Reinet, 3500–3700 ft., *Burchell,* 2920! *Bolus,* 572! and in *Herb. Norm. Austr.-Afr.,* 1312! Somerset Div.; upper part of Bruintjes Hoogte, *Burchell,* 3058–1! 3058–2! 3102!

KALAHARI REGION: Griqualand West, Hay Div.; between Griqua Town and Spuigslang Fontein, *Burchell,* 1706! Bechuanaland; near the sources of Kuruman River, and between them and Kosi Fontein, *Burchell,* 2542! 2487/2! 2487/7! Orange River Colony; between Kimberley and Boshof, 4200 ft., *Marloth,* 795. Bloemfontein, *Kuntze.* Transvaal; Houtbosch, *Rehmann,* 6053!

EASTERN REGION: Pondoland or Natal; between Umtentu River and Umzimkulu River, below 500 ft., *Drège!*

E. ovata, γ, glabra, Alph. DC., l.c., is better referred to *E. lanceolata, E. Meyer;* it was founded on *E. rufescens, Drège, Zwei Pflanzengeogr. Documente,* 62, *not* 57, 130.

The species varies considerably in the shape of its leaves and in the amount of its pubescence; according to Dr. Bolus, the plant becomes more glabrous as the season advances and the fruit ripens. This, as well as the other species of the genus which bear edible fruits, is called " Guarri " by the Hottentots.

11. E. Divinorum (Hiern in Trans. Cambr. Phil. Soc. xii. 99); a dense shrub, or small tree, nearly glabrous, somewhat glaucous; branches rather slender; branchlets leafy; leaves opposite or sub-opposite, narrowly elliptic, obtusely narrowed at the apex, wedge-shaped at the base, coriaceous, green-glaucescent above, paler or slightly reddish and minutely glandular-pulverulent beneath, wavy on the margin, entire, inconspicuously veined, $1\frac{1}{2}$–3 in. long, $\frac{2}{5}$–$\frac{4}{5}$ in. broad; petioles $\frac{1}{6}$–$\frac{1}{4}$ in. long; flowers diœcious, white. Male cymes axillary, puberulous, $\frac{1}{3}$–$\frac{2}{3}$ in. long, 7–16-flowered, sessile or sub-sessile, in somewhat compound or simple racemes; pedicels $\frac{1}{16}$–$\frac{1}{8}$ in. long, spreading, mostly opposite and decussate; bracts very short, caducous; flowers spheroidal in the bud, $\frac{1}{5}$ in. long, $\frac{1}{6}$ in. thick, hemispherical when expanded, 4–5-merous; calyx short, shortly lobed, $\frac{1}{16}$ in. in diam., minutely glandular outside; lobes depressedly ovate-triangular; corolla cleft more than half-way down, shortly and sparingly hairy outside, glabrous or subglabrous within; stamens 12–17, inserted at the base of the corolla singly, or some in pairs, $\frac{1}{16}$–$\frac{1}{10}$ in. long; anthers somewhat hairy; filaments short, or very short, glabrous; ovary rudimentary, consisting of a bunch of pallid hairs. Female flowers tetramerous, $\frac{1}{12}$ in. long; calyx hemispherical, $\frac{1}{20}$ in. in diam.; corolla campanulate, deeply lobed, slightly hairy outside; ovary hairy. *Hiern in Oliv. Fl. Trop. Afr.* iii. 513.

KALAHARI REGION: Transvaal; by the Crocodile River, near Lauws Creek, 1400 ft., *Bolus, Herb. Norm. Austr.-Afr.*, 1311!

EASTERN REGION: Delagoa Bay, *Forbes*, 56!

Also in Tropical Africa.

12. **E. multiflora** (Hiern in Trans. Cambr. Phil. Soc. xii. 100, t. 3); a shrub or small tree, with the habit of a laurel and the branches usually subferruginously pubescent, sometimes subglabrous, or even subglaucous, leafy, 2–15 ft. high; leaves alternate or rarely subopposite, oval, elliptic or oblong, usually rounded or obtusely narrowed at the apex, but sometimes apiculate, narrowed or rounded at the base, coriaceous, wavy or flat at the narrowly revolute margins, entire, evergreen, 1–4 ˙ in. long, $\frac{1}{3}$–$1\frac{3}{8}$ in. broad; venation not very conspicuous; petioles $\frac{1}{6}$–$\frac{1}{2}$ in. long; flowers usually diœcious, occasionally hermaphrodite, 4–6-merous, greenish or sulphur-yellow, $\frac{1}{6}$–$\frac{1}{4}$ in. long; cymes axillary, usually paniculate, 10–30-flowered, 1–$1\frac{2}{3}$ in. long, pubescent, subsessile; pedicels spreading or drooping, shorter than or as long as the flowers; bracts small, pointed, deciduous; calyx campanulate or hemispherical, $\frac{1}{12}$–$\frac{1}{10}$ in. long, pubescent, cleft about half-way down; lobes ovate or deltoid, obtuse; corolla about twice as long as the calyx, deeply lobed; lobes oval or oblong, glabrous, or with a few hairs; stamens erect, shorter than the corolla, four times as many as the corolla-lobes in the male or hermaphrodite flowers, none in the female; anthers pallid, dehiscing on each side from the apex, subglabrous or hairy above; filaments glabrous, short, and inserted in pairs at the base of the corolla or around the ovary; ovary in the male flowers abortive, in the female or hermaphrodite flowers globose, densely hairy, 2- or 4-celled; styles 2, short, rather thick, spreading, glabrous or with a few hairs; stigmas obtuse; ovules solitary; fruit at first usually ferruginously pubescent, subsequently dusky and glabrate,

globose, $\frac{1}{3}$ in. in diam., 1-celled, 1-seeded ; embryo curved and tending to be incumbent. *Hiern in Oliv. Fl. Trop. Afr.* iii. 513 ; *Hiern, Cat. Afr. Pl. Welw.* i. 649; *Parmentier in Ann. Univ. Lyon,* vi. *fasc.* ii. 78 ; *Gürke in Engl. & Prantl, Pflanzenfam.* iv. i. 159, *fig.* 85, *A—C. Diospyros sp., Salt, Trav. Abyss.* 14.

Compare with this species *Kiggelaria integrifolia, Jacq. Collect.* ii. 296, *and Ic. Pl. Rar.* ii. 19, *t.* 628, *not of Eckl. & Zeyh., nor of Drège. E. pubescens, Eckl. & Zeyh. ex Drège in Linnæa,* xx. 192 (*without description*).

SOUTH AFRICA : without locality, *Wallich! Verreaux! Harvey!*

COAST REGION : Clanwilliam Div. ; Modder Fontein, 1000 ft., *Schlechter,* 7971 ! Cederberg Range, 2000–5000 ft., *Ecklon & Zeyher!* Tulbagh Div. ; mountains near Tulbagh Waterfall, 1500 ft., *Bolus,* 5387 ! Humansdorp Div. ; between Twee Fontein and Essenbosch, *Burchell,* 4835 ! Uitenhage Div. ; by the Coega and Zwartkops Rivers, *Ecklon & Zeyher!* Addo hills, *Zeyher,* 767 ! 3361 ! Albany Div. ; Bushmans River, *Ecklon & Zeyher,* 778 ! near Grahamstown, *Burchell,* 3572 ! *MacOwan,* 244 ! Howisons Poort, *Hutton!* between Tea Fontein and Kurukuru River, *Burchell,* 3510 ! and without precise locality, *Miss Bowker!* Bathurst Div. ; near Port Alfred, *Burchell,* 3980 ! British Kaffraria, *Cooper,* 44 !

KALAHARI REGION : Basutoland ; (doubtful) *Cooper,* 3488 !

EASTERN REGION : Natal ; near Durban, *Gerrard,* 699 ! by the coast, *Mrs. K. Saunders !*

Also in Tropical Africa.

13. **E. natalensis** (Alph. DC. Prodr. viii. 218); a shrub, 2–10 ft. high or more ; branches glabrate, dusky ; branchlets puberulous at the apex, leafy ; leaves alternate, obovate-oblong, rounded or obtuse at the apex, wedge-shaped at the base, coriaceous, glabrescent, dark green and somewhat glossy above, paler or slightly reddish beneath, wavy or nearly plane on the narrowly revolute margin, not very conspicuously veined, 2–4 in. long, $\frac{1}{2}$–$1\frac{1}{4}$ in. broad ; petioles $\frac{1}{4}$–$\frac{1}{2}$ in. long ; cymes axillary, $\frac{1}{2}$–1 in. long, racemose or paniculate, 8- to many-flowered, shortly pubescent ; pedicels very short or about as long as the flowers, dilated upwards in fruit to the articulation with the consolidated tube of the calyx ; bracts small, deciduous ; flowers 4–5-merous, $\frac{1}{8}$–$\frac{1}{6}$ in. long, diœcious or hermaphrodite, yellow ; calyx deeply lobed, about or less than half the length of the corolla, shortly hairy in flower, glabrescent in fruit; lobes ovate or lanceolate, acute or pointed ; corolla deeply lobed, nearly or quite glabrous ; segments at length spreading in a stellate manner, ovate-oblong, obtuse ; stamens 16 ; anthers more or less hispid ; filaments glabrous, short or very short ; in the female and hermaphrodite flowers the ovary hirsute, subglobose, 4-celled ; styles 2, slender, glabrous ; fruit globose, dusky, subglabrous, $\frac{1}{4}$–$\frac{1}{3}$ in. in diam., 1-celled ; seed solitary, globose, black, marked outside with three longitudinal lines ; albumen somewhat ruminated. *Hiern in Trans. Cambr. Phil. Soc.* xii. 101. *E. macrophylla, Drège, Zwei Pfl. Documente,* 159, *not* 133 *nor* 146. *Royena macrophylla, E. Meyer ex Alph. DC. Prodr.* viii. 215, 218.

KALAHARI REGION : Transvaal ; hill-sides near Barberton, 2200 ft., *Galpin,* 484 !

EASTERN REGION: Natal; woods near Durban, *Drège! Wood,* 363! on the Bluff, Durban Bay, *Cooper,* 2794! Groen Berg, *Wood,* 953! The Creek, *Wood,* 1014! and without precise locality, *Peddie! Cooper,* 1253! 2695! *Gueinzius! Williamson! Gerrard,* 92! Delagoa Bay, *Junod,* 460. Between Lourenço Marques and Matolla, *Bolus,* 7845! 7846!

It is called "Nhlangulane" at Delagoa Bay (Schinz and Junod in *Mém. Herb. Boiss.* x. 55).

14. E. macrophylla (E. Meyer ex Drège, Cat. Pl. Exsicc. Afr. Austr. 7); a glabrous or subglabrous shrub or low tree of 15 ft.; branches nodose, rarely somewhat pubescent; branchlets leafy, often verticillate three together, bark grey, smooth; leaves opposite or verticillate three together, obovate-oblong or oblanceolate, rounded at the apex, wedge-shaped at the base, rigidly coriaceous, dark green and glossy or reddish-brown above, paler and dull beneath, wavy or flat along the margin, entire, $1\frac{1}{2}$–4 in. long, $\frac{3}{8}$–$1\frac{1}{2}$ in. broad; lateral veins delicate; petioles $\frac{1}{10}$–$\frac{1}{4}$ in. long; racemes axillary, 5–15-flowered, $\frac{1}{2}$–$1\frac{1}{2}$ in. long; pedicels $\frac{1}{25}$–$\frac{1}{6}$ in. long, often opposite; flowers diœcious, tetramerous. Male flowers about $\frac{1}{8}$ in. long; calyx cleft half-way down, $\frac{1}{12}$ in. in diam.; lobes deltoid and pointed; corolla urceolate or campanulate, cleft half-way down, sparingly pilose outside or subglabrous; lobes obtuse and mucronate; stamens 4–16; anthers subsessile, hairy towards the apex; ovary rudimentary. Female flowers $\frac{1}{12}$–$\frac{1}{8}$ in. long; calyx about $\frac{1}{24}$ in. long, cleft nearly half-way down; lobes shortly deltoid, pointed, subglabrous; corolla urceolate, cleft about half-way down, subglabrous; lobes rounded; staminodes 0; **ovary glabrous, usually 4-(rarely 6-) celled, often only 2-celled in the upper part, rounded or obtusely conical; cells 1-ovuled; styles 2 or rarely 3, as long as the ovary or shorter, bifid at the apex; stigmas obtuse; fruit slightly apiculate, about $\frac{1}{5}$ in. in diam., minutely glandular.** *Alph. DC. Prodr.* viii. 218; *Drège, Zwei Pflanzengeogr. Documente,* 133, 146, not 159; *Drège, Cat. Südafr. Pfl.* 9; *Hiern in Trans. Cambr. Phil. Soc.* xii. 102; *Parmentier in Ann. Univers. Lyon,* vi. *fasc.* ii. 75; *Krauss in Flora,* 1844, 824.

SOUTH AFRICA: without locality, *Masson !*

COAST REGION: Humansdorp Div.; Zitzikamma Forest, *Krauss,* 1765. Div.? *MacOwan,* 51? Uitenhage Div.; in woods at Enon, *Drège !* Albany Div.; Grahamstown, *Atherstone,* 461!

KALAHARI REGION: Transvaal; Kaap Valley, near Barberton, 2000 ft., *Galpin,* 1329!

EASTERN REGION: Transkei; between Gekau (Geua) River and Bashee River, 1000–2000 ft., *Drège!* Natal; Tugela River, *Gerrard,* 1604! (doubtful), near Durban, *McKen,* 673! Pondoland; valleys between Umtata River and St. Johns River, 1000–2000 ft., *Drège.*

The last station is implied by Drège, *Zwei Pflanzengeogr. Documente,* 184, *letter* c; but the name of the species does not occur among those mentioned under the station on the place referred to by Drège [p. 150].

According to a note of Atherstone the ovary has frequently only two cells at the upper part in consequence of the dissepiments being incomplete.

15. E. daphnoides (Hiern in Trans. Cambr. Phil. Soc. xii. 102);

a glabrous, usually shining shrub of 2–7 ft. or more, or a low tree, sometimes thorny; branches ashy; branchlets numerous, alternate, opposite or subverticillate, leafy; bark white; leaves alternate, opposite or verticillate three together, linear- or oblanceolate-oblong, rounded or obtuse at the apex, wedge-shaped at the subsessile thickly articulate base, often suberect, rigidly coriaceous, flat or somewhat wavy on the narrowly revolute margin, entire, 1–3 in. long, $\frac{1}{6}$–$\frac{1}{2}$ in. broad; racemes axillary, shorter than the leaves, $\frac{2}{5}$–$1\frac{1}{5}$ in. long, spreading or drooping, 9–21-flowered; flowers diœcious or sub-hermaphrodite, whitish or yellowish; pedicels $\frac{1}{20}$–$\frac{1}{4}$ in. long. Male flowers tetramerous, nearly glabrous, $\frac{1}{8}$ in. long; calyx cleft half-way down, short; corolla campanulate, cleft nearly half-way down; lobes rounded, erect; stamens 12–16, glabrous below; ovary rudimentary, glabrous. Female flowers 4–5-merous, numerous, glabrous, $\frac{1}{10}$ in. long; calyx $\frac{1}{30}$ in. long, cleft half-way down; corolla openly campanulate, $\frac{1}{15}$ in. long, cleft nearly half-way down; lobes obtuse; staminodes 0 or 4 or 8, glabrous, very short; ovary ovoid, glabrous, $\frac{1}{25}$ in. long, 4–6-celled; styles 2 or 3, $\frac{1}{25}$ in. long, thick; stigmas bilobed at the apex, exserted, reddish, truncate, furrowed along the inner side; ovules or seeds solitary; fruit globose, $\frac{1}{4}$ in. in diam., dusky, glabrous, 1-celled; seed solitary; fruiting calyx small or minute; albumen not ruminated, but the testa is introverted at the apex of the seed.

COAST REGION : Swellendam Div.; without precise locality, *Bowie!* Humansdorp Div.; mountains near Wagenbooms River, in Lange Kloof, *Burchell*, 4909! Uitenhage Div.; Uitenhage, Coega River and Winterhoek Mountains, *Ecklon & Zeyher*, 28! Port Elizabeth Div.; by the Baakens River, *Burchell*, 4356!

EASTERN REGION : Natal; "Thorns" near Mooi River, 3000–4000 ft., *Wood*, 4471! 5327! Upper Tugela River, *Gerrard*, 1506! 1606!

16. E. racemosa (Murr. Syst. Veg. ed. 13, 747); an evergreen densely branched glabrous and shining shrub, usually about 6–9 ft. high with the main stem 5–6 in. in diam., sometimes a small tree of 18 ft.; bark usually grey, smooth; branchlets often reddish, angular, ascending, leafy; leaves alternate, subopposite or opposite, obovate or oblong-obovate, rounded at the apex, wedge-shaped or obtusely narrowed at the base, rigidly coriaceous, green above, pale beneath, not wavy on the revolute margin, entire, $\frac{1}{2}$–$2\frac{1}{2}$ in. long, $\frac{1}{3}$–$1\frac{1}{4}$ in. broad; venation slightly prominent on both surfaces; petioles $\frac{1}{20}$–$\frac{3}{20}$ in. long; racemes $\frac{1}{2}$–$1\frac{1}{4}$ in. long, axillary, at length drooping, 4–13-flowered; flowers diœcious, white, drooping. Male flowers 4- (or rarely 5–6-) merous, campanulate, glabrous, $\frac{1}{10}$–$\frac{1}{7}$ in. long; pedicels $\frac{1}{10}$–$\frac{1}{4}$ in. long; calyx short, cleft half-way down; lobes deltoid; corolla deeply lobed; lobes oval, obtuse or subacute, spreading or erect; stamens 12–18, in two rows; filaments $\frac{1}{50}$–$\frac{1}{20}$ in. long; anthers lanceolate, thick, $\frac{1}{24}$–$\frac{1}{14}$ in. long, erect, with a few hairs or glabrous; pollen white; ovary rudimentary; styles 2, erect, terete, white. Female flowers tetramerous or rarely pentamerous, ovoid, rather smaller than the male ones; pedicels $\frac{1}{10}$–$\frac{1}{8}$ in. long; calyx hemispherical, cleft half-way down; lobes ovate, acute; corolla deeply lobed; lobes not reflexed; staminodes 2–4, glabrous; ovary usually hairy, 4-celled; styles 2; ovules solitary in the cells; fruit globular, glabrescent or glabrous, black, 1-celled, 1-seeded, $\frac{1}{5}$ in. in diam. *Jacq. Fragm. t.* 1, *f.* 5, *t.* 63, *f.* 3; *Thunb. Fl. Cap. ed. Schult.* 401; *Alph. DC. Prodr.* viii. 219; *Chamisso in Linnæa,* vi. 350, 351; *Hiern in Trans. Cambr. Phil. Soc.* xii. 104; *Parmentier in Ann. Univ. Lyon,* vi. *fasc.* ii. 75; *Drège, Zwei Pfl. Documente,* 112, 113, 184; *Krauss in Flora,* 1844, 824; *not of L'Hérit; Herb. Willd. n.* 18478; *Burchell, Trav. S. Afr.* i. 29 *note,* 62 *note.—Padus foliis subrotundis, fructu racemoso, Burm. Afr.* 238, *t.* 84, *f.* 1.

VAR. *β,* **Burchellii** (Hiern in Trans. Cambr. Phil. Soc. xii. 105); a tree, 18 ft. high; trunk erect; branches ascending, dense; bark entire, turning whitish; leaves oblong-obovate; staminodes 0–4, inserted on the base of the corolla or around the base of the ovary; ovary globose, glabrous; styles 2 or 3, short; young fruit purple.

SOUTH AFRICA : without locality, *Oldenburg! Nelson! Banks! Siekmann! Alexander! Boivin!*

COAST REGION : Piquetberg Div.; St. Helena Bay, *Hove!* Malmesbury Div.; Groene Kloof (Mamre), *Drège!* Cape Div. ; Camps Bay, *Burchell,* 397 ! *Harvey!* Table Mountain and other places around Cape Town, *Thunberg! Burchell,* 807 ! *Ecklon! Krauss, MacOwan, Herb. Austr.-Afr.,* 1499! Simons Bay, *Wright!* Simons Town, *Meiklejohn!* Durbanville, *Schlechter,* 7828 ! Stellenbosch Div.; near Lowrys Pass, *Burchell,* 8295 ! near Somerset West, *Drège!* Fish Hoek, *Ecklon.* Bredasdorp Div.; near Cape Agulhas, *Ecklon,* Mier Kraal, *Schlechter,* 10500 ! 10501 ! Swellendam Div.; between the Breede and Duivenhoek Rivers, *Ecklon,* and without precise locality, *Bowie!* Uitenhage Div.; near the mouth of the Coega River, *Zeyher,* 574 ! Grass Ridge, *Zeyher,* 3356 ! Bathurst Div.; near Port Alfred, *Burchell,* 3806! Albany Div. ; near Grahamstown, *MacOwan,* 1511 !

CENTRAL REGION : VAR. *β* : Somerset Div. ; Bosch Berg, *Burchell,* 3219!

WESTERN REGION : Little Namaqualand ; Hondeklip Bay, *Whitehead !*

The wood is hard and heavy, and is employed by wheelwrights and turners, and serves very well for wooden screws; but it is chiefly used as fuel (*Pappe, Silva Capensis,* 21).

E. racemosa, L'Hérit. Sert. Angl. 32, is perhaps *Kiggelaria dregeana, Turcz.,* var. *obtusa, Harv. Fl. Cap.* i. 71.

17. E. undulata (Thunb. Nova Gen. Pl. 85) ; a glabrous shrub of 1–10 ft. high or a tree of moderate size up to 30 ft. high, densely branched; branches erect or ascending, terete, alternate or opposite; branchlets more or less spreading, opposite or alternate, leafy, minutely glandular towards the apex; bark whitish-grey, somewhat rough; leaves opposite and alternate, evergreen, obovate or oblanceolate, rounded or obtusely narrowed at the apex, wedge-shaped at the base, coriaceous, strongly wavy along the margin or when narrow nearly flat, entire, green above, pallid beneath, glabrous, minutely gland-dotted, $\frac{1}{2}$–$1\frac{1}{2}$ in. long, $\frac{1}{5}$–$\frac{3}{5}$ in. broad, narrower in the variety; lateral veins not very conspicuous, numerous; petioles $\frac{1}{25}$–$\frac{1}{10}$ in. long; racemes axillary; bracts deciduous, occasionally large and foliaceous; flowers dioecious, tetramerous. Male flowers hemispherical, nearly glabrous, $\frac{1}{10}$ in. long; racemes lax, 5–7-flowered, $\frac{2}{5}$–$\frac{4}{5}$ in. long; pedicels slender, $\frac{1}{10}$–$\frac{1}{4}$ in. long; **calyx**

broadly cup-shaped, short, cleft half-way down ; lobes deltoid and pointed ; corolla cleft more than half-way down ; lobes oval ; stamens 10–15, mostly in pairs ; filaments slender, $\frac{1}{100}$–$\frac{1}{50}$ in. long ; anthers oblong or obovate-oblong, apiculate, $\frac{1}{20}$–$\frac{1}{16}$ in. long, with a few hairs towards the apex ; ovary rudimentary, hairy ; styles 2. Female flowers campanulate, $\frac{3}{40}$–$\frac{1}{10}$ in. long, nearly glabrous ; racemes 3–8-flowered, suberect in flower, drooping in fruit, $\frac{1}{4}$–$\frac{1}{2}$ in. long ; pedicels mostly opposite, more or less patent in open flower and fruit, $\frac{3}{40}$–$\frac{1}{10}$ in. long ; bracteoles narrow, small, deciduous ; calyx campanulate, cleft scarcely half-way down, $\frac{1}{30}$ in. long, not accrescent ; lobes deltoid ; corolla deeply cleft ; segments oblong, more or less recurved near the apex, pale chestnut colour ; staminodes 0 ; ovary ovoid, 2–4-celled, $\frac{1}{25}$ in. long, $\frac{1}{30}$ in. in diam., glabrous above, with some short whitish slender hairs around the base ; ovules 4, oblong ; styles 2, united at the base, $\frac{1}{20}$ in. long, glabrous ; stigmas 2, bifid at the apex ; fruit globose, purple or red, or at length black, glabrous, edible, $\frac{1}{6}$–$\frac{1}{5}$ in. in diam., 1- or 2-celled, at length 1-celled and 1-seeded ; seed $\frac{1}{6}$ in. in diam. ; albumen equable. *Thunb. Fl. Cap. ed. Schult.* 401 ; *Drège, Zwei Pflanzengeogr. Documente,* 64, 129, 132, 134, 144, 184 ; *Alph. DC. Prodr.* viii. 219 ; *Krauss in Flora,* 1844, 824 ; *Burke in Hook. Lond. Journ. Bot.* v. (1846) 20 ; *Hiern in Trans. Cambr. Phil. Soc.* xii. 105, *excl. syn. E. humilis* ; *Parmentier in Ann. Univers. Lyon,* vi. *fasc.* ii. 27–30, 80, 138 ; *Kew Bulletin,* 1887, *Sept.,* 11 ; *Herb. Willd. n.* 18479. *Euclea n.s., Burchell, Trav. S. Afr.* i. 465, *note.*

VAR. β, **myrtina** (Hiern, l.c. 106) ; a densely branched shrub, 1–5 ft. high, in habit like a myrtle, leafy, diœcious ; leaves oblanceolate, obtuse, opposite, nearly flat or less strongly wavy than in the type, $\frac{1}{2}$–1 in. long, $\frac{1}{10}$–$\frac{1}{5}$ in. broad, glabrous ; fruit pisiform, black, glabrous, $\frac{1}{5}$ in. in diam., edible, sweet but subastringent. *E. myrtina, Burchell, l.c.,* ii. 588, *note* ; *Alph. DC. Prodr.* viii. 217. *Euclea, Burchell, l.c.* ii. 264, *note.*

SOUTH AFRICA : without locality, *Zeyher,* 3358 ! *Thom,* 243 ! 386 !

COAST REGION : Caledon Div. ; Caledon, *Ecklon.* Swellendam Div. ; Swellendam, *Lichtenstein.* Riversdale Div. ; hills near Spiegel River, *Burchell,* 7198 ! near Riversdale, *Schlechter,* 1983 ! by the Kaffirkuils River, *Krauss,* 1758. Mossel Bay Div. ; near Mossel Bay, *Masson* ! Uitenhage Div. ; various localities, *Drège* ! *Zeyher,* 218 ! *Alexander Prior* ! *Tredgold,* 35 ! Albany Div. ; Fish River Heights, *Hutton* ! near Grahamstown, *MacOwan* ! Komgha Div. ; hills near the Kei River, *Drège* !

CENTRAL REGION : Calvinia Div. ; between Lospers Plants and Springbok Kuil River, 2000–3000 ft., *Zeyher,* 1124 ! Prince Albert Div. ; Great Zwartberg Range, near Klaarstroom, *Drège.* Somerset Div. ; Bosch Berg, *Burchell,* 3168 ! Graaff Reinet Div. ; between Kruid Fontein and Melk River, *Burchell,* 2943 ! near Graaff Reinet, 2900 ft., *Bolus,* 655 !

KALAHARI REGION : Griqualand West ; Lower Campbell, *Burchell,* 1792 ! Transvaal ; Magalies Berg, *Burke* ! flats west of Blaauw Berg and Hang Klip, *Baines* ! Var. β : Griqualand West ; Klip Fontein, *Burchell,* 2162 ! Bechuanaland ; Kosi Fontein, *Burchell,* 2573 !

The fruit of this, as also of other species of the genus, is called "guarri" ; bruised and fermented it yields a kind of vinegar ; see *Thunberg, Travels,* English edition, i. 203. According to Burchell, *loc. cit.* ii. 588–589, the variety produces one of the only two edible fruits found wild in Bechuanaland. The wood is brown, hard, close-grained, and fit for joiners' fancy work, veneering, &c. See *Pappe, Silva Capensis,* 21.

III. **MABA**, J. R. & G. Forst.

Flowers nearly always diœcious and usually trimerous, solitary or cymose. *Calyx* campanulate, oblong or cup-shaped, lobed or truncate. *Corolla* campanulate or tubular; lobes sinistrorsely contorted in bud as seen from above. *Stamens* in the male flowers 3 or more, usually glabrous, in the female flowers obsolete or represented by staminodes. *Ovary* in the male flowers abortive, in the female flowers 3- or 6-celled, 6-ovuled; styles 1–3 or in the male flowers obsolete. *Fruit* baccate, usually moderate in size.

Trees or shrubs, with alternate and quite entire leaves, and axillary or rarely lateral inflorescence.

Species about 70, widely distributed over the warmer parts of the world.

1. **M. natalensis** (Harv. Thes. Cap. ii. 7, t. 110); a shrub, about 10–20 ft. high, soon becoming glabrous; branches numerous, pallid, slender, spreading at a wide angle; branchlets divaricate, flexuous, puberulous, leafy; leaves oval or ovate, obtuse or mucronate at the apex, firmly membranous, flat except at the narrowly revolute margin, shining and deep green above, paler beneath, delicately veiny, $\frac{1}{2}$–$1\frac{1}{6}$ in. long, $\frac{1}{4}$–$\frac{1}{12}$ in. broad: petioles puberulous, $\frac{1}{20}$–$\frac{1}{10}$ in. long; flowers white, diœcious. Female flowers solitary, axillary, $\frac{1}{5}$ in. long; peduncle $\frac{1}{25}$–$\frac{1}{20}$ in. long; calyx $\frac{1}{10}$–$\frac{1}{8}$ in. long, cup-shaped, truncate, entire, deep green, glabrous, semi-ellipsoidal; corolla $\frac{2}{11}$ in. long, silvery-silky, 3-lobed; lobes $\frac{1}{8}$ in. long, diverging, oblong, acute, apiculate or rounded; stamens or staminodes 6–9, free from or inserted at the base of the corolla, uniseriate, $\frac{1}{10}$ in. long, glabrous; filaments short, subulate; anthers erect, without pollen; ovary conical, glabrous; style as long as the ovary, glabrous, trifid at the apex, 3-celled; cells 2-ovuled; fruit spheroidal, glabrous, pale-chestnut in colour, $\frac{1}{3}$–$\frac{3}{8}$ in. long, $\frac{1}{5}$–$\frac{1}{4}$ in. in diam., tipped with the persistent style, based with the cup of the non-accrescent calyx, 1-seeded. *Hiern in Trans. Cambr. Phil. Soc.* xii. 131. *Ebenus natalensis, O. Kuntze, Rev. Gen. Pl.* ii. 408.

COAST REGION: East London Div.; in a wooded gulley near the mouth of the Nahoon (Kahoon) River, 20 ft., *Galpin*, 5665!

EASTERN REGION: Natal; near Durban, *Gerrard & McKen*, 675! *Gerrard*, 110! *Cooper*, 2695! Inanda, *Wood*, 1414!

IV. **DIOSPYROS**, Linn.

Flowers usually diœcious, cymose or solitary, usually tetramerous or pentamerous. *Calyx* lobed or truncate, often accrescent in fruit. *Corolla*-lobes usually obtuse and more or less spreading, sinistrorsely contorted in bud as seen from above. *Stamens* in the male flowers 4 or more, usually about 16 and in two rows, in the female flowers obsolete or usually represented by 4–8 staminodes. *Ovary* in the male flowers abortive or rudimentary or obsolete, in the female flowers 4–16-celled, usually 8- or 10-celled and with solitary ovules; styles 1–4 or in the male flowers obsolete. *Fruit* usually globose, oblong or conical, often pulpy and edible; seeds 1–10, usually oblong with dusky more or less shining testa; albumen cartilaginous, white and equable or in some species ruminated.

Trees or shrubs with alternate, or in a few species opposite, entire leaves, and axillary or lateral inflorescence.

Species more than 200, widely distributed over the warmer regions of both the eastern and western hemispheres.

Male flowers cymose; stamens 10–16; leaves elliptic-
  oblong or obovate-oblong　　...　　...　　...　　...　　(1) **mespiliformis**.
Flowers solitary; stamens about 30; leaves obovate-
  rotund ...　　...　　...　　...　　...　　...　　...　　(2) **rotundifolia**.

1. **D. mespiliformis** (Hochst. in Plant. Schimp. Abyss. Exsicc. Sect. ii. nn. 655, 1243); a shrub or tree, ranging up to 50 ft. high with a roundish head; trunk nearly straight, ranging up to 1½ ft. in diam.; bark becoming rough and much cracked; wood compact, very hard, heavy, white but often black in the centre; branches terete, more or less patent, glabrate; branchlets alternate, subterete, erect-patent, pallid or ashy, glabrescent; young shoots and inflorescence ferruginous-tomentose; leaves alternate, elliptic-oblong or obovate-oblong, obtusely pointed at the apex, somewhat narrowed or nearly rounded and sometimes unequal at the base, thinly coriaceous, rather glossy especially above, glabrescent or thinly silky beneath, entire, often somewhat reddish, deciduous, or sometimes evergreen, 2–6 in. long, ⅔–2½ in. broad, midrib depressed, veins delicate; young leaves membranous, very soft, often dusky-red; petioles ⅙–⅜ in. long. Inflorescence axillary, short; flowers white or slightly greenish, diœcious. Male inflorescence few- or many-flowered, ferruginous-tomentose, ½–1 in. long; flowers about ⅓ in. long, usually tetramerous or pentamerous; calyx about ⅙ in. long, cleft about half-way down, campanulate or narrowly so, hairy on both sides; lobes ovate or lanceolate; corolla urceolate-oblong, shortly cleft, silky outside, glabrous within; lobes spreading, pointed; stamens 10–16, often in pairs, nearly glabrous but with a narrow band of pale hairs along the back of the anthers; filaments short, inserted at the base of the corolla; connective produced at the apex; pollen spheroidal, smooth; ovary rudimentary, hairy. Female flowers solitary or in very short 2- or 3-flowered cymes; peduncles ⅛–⅜ in. long; bracts narrow, caducous; calyx campanulate, deeply lobed, hairy on both sides; lobes ovate, acuminate, with undulated margins, 3–6, usually 4 or 5; corolla pubescent outside, glabrous within, rather small but shortly exceeding the calyx; lobes pointed; staminodes 6–8, glabrous, inserted in one row at the base of the corolla; ovary ovoid or conical, silky, 4- or 8-celled, terminating in 2 short hirsute bilobed styles; ovules solitary in the cells. Fruit subglobose or spheroidal, solitary, subsessile, 3–8-celled, rather hard, glossy, glabrate, often slightly wrinkled, green, orange-yellow, olive-

tawny, or black-purple, edible, sparingly pulpy within, $\frac{5}{8}$–1 in. in
diam.; pulp gummy, very viscid; seeds 1–8, solitary in the cells,
$\frac{1}{2}$–$\frac{2}{3}$ in. long; testa glossy outside, somewhat intruded into the
white cartilaginous albumen; embryo straight; cotyledons linear-
lanceolate, longer than the radicle; fruiting calyx somewhat or but
little accrescent, adpressed to the base of the fruit or spreading,
$\frac{2}{3}$–$\frac{3}{4}$ in. in diam., somewhat pubescent outside, with wavy or subauriculate lobes. *Alph. DC. Prodr.* viii. 672; *Schweinf. Beitr. Fl.
Aethiop.* 85, 273; *Hiern in Trans. Cambr. Phil. Soc.* xii. 165, *and
in Oliv. Fl. Trop. Afr.* iii. 518; *Schweinf. Piante Utili dell'Eritrea,*
41, 48; *Parmentier in Ann. Univ. Lyon,* vi. *fasc.* ii. 46, 50, 56, 66,
131, 148; *Hiern in Cat. Afr. Pl. Welw.* i. 651. *D. senegalensis,
Perrott. ex Alph. DC., l.c.,* 234, 672. *D. bicolor, Klotzsch in
Peters, Reise Mossamb. Bot.* i. 184.

KALAHARI REGION: Transvaal; Crocodile Poort, near Barberton, *Pauling
in Herb. Galpin,* 1354!

EASTERN REGION: Portuguese East Africa; between Lourenço Marques and
Komati River Drift, *Bolus,* 7847!

Also in Tropical Africa.

2. **D. rotundifolia** (Hiern in Trans. Cambr. Phil. Soc. xii. 181);
young parts puberulous; branches pallid-ashy, glabrescent; branchlets leafy; leaves alternate, obovate-rotund, rounded at both
ends, coriaceous, glabrous, $\frac{3}{4}$–$1\frac{1}{2}$ in. long, $\frac{9}{16}$–$1\frac{1}{4}$ in. broad, inconspicuously veined beneath; petioles $\frac{1}{10}$–$\frac{1}{6}$ in. long; peduncles
axillary, solitary, crowded, in the upper axils, 1-flowered, puberulous, recurved, $\frac{1}{10}$–$\frac{1}{6}$ in. long; bracts caducous; flowers
dioecious. Male flowers glabrous, about $\frac{3}{8}$ in. long; calyx hemi-
spherical-campanulate, about $\frac{1}{4}$ in. long; lobes 5, shallow, apiculate;
corolla 5-fid; lobes oval, spreading; stamens about 30, nearly equal,
glabrous; filaments short, straight, inserted in one row on the
glabrous receptacle; anthers about $\frac{1}{8}$ in. long; ovary 0. Female
flowers: calyx accrescent, undulate, exceeding the young fruit
and then $\frac{1}{2}$ in. long and broad, shortly 5-lobed at the apex but
appearing 5-fid and 5-winged from folding of the calyx-tube;
styles 5, connate at the base, $\frac{1}{10}$ in. long, glabrous; stigmas
bifid; young fruit depressedly globose, $\frac{1}{8}$ in. long, glabrous, 8-
celled; ripe fruit globose, umbilicate at the apex, shining, $\frac{3}{4}$–1 in.
in diam.; fruiting calyx patelliform, $\frac{3}{4}$–1 in. in diam., $\frac{1}{12}$–$\frac{1}{3}$ in. high,
with a raised edge at the top, plicate; seeds compressed, $\frac{1}{2}$ in. long,
about $\frac{1}{3}$ in. broad; albumen not ruminated. *Parmentier in Ann.
Univ. Lyon,* vi. *fasc.* ii. 131; *not of Lesquer. Cretac. Fl.* 89, *t.* 6,
fig. 6.

EASTERN REGION: Delagoa Bay, *Forbes,* 34! *Owen!*

ORDER LXXXV. **OLEACEÆ.**

(By W. H. HARVEY, with additions by C. H. WRIGHT.)

Flowers mostly bisexual. *Calyx* inferior, 4- to many-lobed.
Corolla regular, gamopetalous and 4- to many-lobed, rarely poly-

petalous or absent. *Stamens* 2, epipetalous. *Ovary* superior, 2-celled; style simple; stigma thickened, often shortly 2-fid; ovules 2 in each cell, rarely 1 or 4–8. *Fruit* a berry, drupe or capsule. *Seeds* usually exalbuminous; embryo straight; cotyledons flat, ovate or oblong.

Trees or shrubs, sometimes climbing; leaves opposite, simple or compound, exstipulate; inflorescence cymose, paniculate or fascicled.

DISTRIB. Genera about 20, species about 300, found throughout the hot and temperate regions of the world.

I. **Jasminum.**—Shrubs. *Corolla* salver-shaped. *Stamens* included. *Fruit* a twin berry.

II. **Schrebera.**—Shrubs. *Corolla* salver-shaped. *Stamens* exserted. *Fruit* a woody capsule splitting lengthwise.

III. **Menodora.** — Small undershrubs. *Corolla* funnel-shaped. *Fruit* of 2 globose membranous capsules splitting across.

I. JASMINUM, Linn.

Calyx bell-shaped, 5–8–10-lobed or toothed. *Corolla* salver-shaped; limb flat, 5–12-partite; lobes oblique, twisted in bud. *Stamens* 2, included. *Ovary* 2-lobed; style filiform, 2-lobed at the apex. *Berries* twin; cells 1-seeded. *DC. Prodr.* viii. 301.

Shrubs of the old world, mostly climbers. Leaves opposite or alternate, compound; the petiole either jointed above the base and bearing one terminal leaflet, or trifoliolate or imparipinnate. Cymes few- or many-flowered. Corolla white or yellow, often sweetly scented. Name of uncertain origin: said to be from ἴον, a violet, and ὀσμή, smell.

Leaves unifoliolate (or apparently simple):
  Leaves glabrous:
    Calyx-lobes slender, longer than the tube:
      Twigs glabrous; calyx-lobes 5–7, subulate; corolla-lobes 6–7 ... ... (1) **glaucum.**
      Twigs microscopically puberulous; calyx-lobes 8–12, setaceous; corolla-lobes 8–12 ... ... (2) **multipartitum.**
      Calyx-lobes 4–5, short and tooth-like ... ... (3) **Gerrardi.**
  Leaves (and twigs) pubescent:
    Calyx-lobes 5–6, broadly triangular, much shorter than the glabrous tube ... ... (4) **breviflorum.**
    Calyx-lobes 5–7, subulate, as long as the pubescent tube ... ... (5) **streptopus.**
    Calyx-lobes 9–11, much longer than the tube ... ... (6) **stenolobum.**
  Leaves trifoliolate:
    Leaflets ovate; corolla-tube 1½ in. long (glabrous or hairy) ... ... (7) **angulare.**
    Leaflets lanceolate; corolla-tube about 1 in. long ... (8) **tortuosum.**
    Leaflets ovate; corolla-tube ¾–1 in. long ... ... (9) **mauritianum.**

1. **J. glaucum** (Ait. Hort. Kew. ed. 1, i. 9); quite glabrous; branches terete, straight or scandent; leaves opposite or subopposite, variable in form, lanceolate, ovate-lanceolate, oblong or ovate, mucronate, faintly 3- or several-nerved at the base; peduncle terminal, 1–3-flowered; calyx-lobes 5–7, subulate, twice as long as the tube;

corolla-lobes 6–7, acute or acuminate. *DC. Prodr.* viii. 305.
Nyctanthes glauca, Linn. f. Suppl. 82.

Stem generally climbing. Leaves 1–1½ in. long, ½–1¼ in. wide. Corolla-tube 1–1¼ in. long; flowers white.

Var. β, lanceolatum (E. Meyer ex DC. Prodr. viii. 305); leaves lanceolate; petiole short, articulated at the apex. *J. glaucum, Vent. Hort. Cels. t.* 55. *J. ligustrifolium, Lam. Encycl.* iii. 218.

Var. γ, latifolium (E. Meyer ex DC. l.c.); leaves ovate and many-nerved at the base or ovate-lanceolate; petiole short, articulated near the middle.

Var. δ, parvifolium (E. Meyer ex DC. l.c); scrubby, densely much-branched; leaves very small, ¼–½ in. long, ovate or ovate-lanceolate.

Coast Region: Tulbagh Div.; near Tulbagh, *Zeyher,* 1149! British Kaffraria, *Cooper,* 378! Var. β: Vanrhynsdorp Div.; Ebenezer, *Drège!* Var. γ: Clanwilliam Div.; near Honig Vallei and on the Koude Berg, 3000–4000 ft., *Drège!* Piquetberg Div.; near Piqueniers Kloof, *Dickson in Herb. Bolus,* 5699! Var. δ: Albany Div.; between Blauwkrantz and Kowi Poort, *Burchell,* 3657! near Blauwkrantz Bridge, *Galpin,* 266!
Kalahari Region: Transvaal; near Lydenburg, *Wilms,* 925!
Eastern Region: Natal; near Durban, *Plant,* 46! *Krauss,* 458!

2. **J. multipartitum** (Hochst. in Flora, 1844, 825); glabrous, the young twigs microscopically puberulous; leaves ovate or ovate-lanceolate, acute or obtuse, 3-nerved at the base; peduncle terminal, 1–3-flowered; calyx-lobes 8–10, setaceous, nearly twice as long as the tube; corolla-lobes 8–12, lanceolate, acute; tube 1¼–1½ in. long. *Walp. Rep.* vi. 463.

Coast Region: Uitenhage Div.; Sandfontein, *Burke!* Albany Div.; Fish River Heights, *Hutton!*
Eastern Region: Pondoland, *Bachmann,* 1030! Natal; Inanda, *Wood,* 356! Mooi River Valley, 2000–3000 ft., *Sutherland! Sanderson,* 83! Tugela River, *Gerrard,* 1154! and without precise locality, *Gerrard,* 264! *Cooper,* 1258!

Very closely allied to *J. glaucum,* from which it differs in the scarcely perceptible indumentum of the twigs, and the more numerous calyx-teeth and corolla-lobes, both of which latter characters however vary.

3. **J. Gerrardi** (Harv.); glabrous; branches terete, straight; leaves (small) ovate-lanceolate, acute at each end, mucronate, faintly nerved beneath; peduncle terminal, slender, 1- (rarely 2–3-) flowered; calyx-lobes 4–5, short, tooth-like; corolla-tube about ½ in. long.

Eastern Region: Natal; Nonoti River, *Gerrard,* 1477! Camperdown, *Rehmann,* 7706!

A low much-branched erect shrub. Leaves ¾ in. long, 2–4 lin. wide. Flowers white; corolla-lobes 7–8.

4. **J. breviflorum** (Harv.); branches terete, scandent, pubescent, becoming naked; twigs, leaves and pedicels softly albo-pubescent; leaves very shortly petiolate, ovate, obtuse or acute; peduncle terminal, 3-flowered; calyx glabrous, its lobes 5–6, broadly triangular, much shorter than the tube; corolla-tube about ½ in. long; lobes oblong, subacute.

KALAHARI REGION: Transvaal; Magalies Berg, *Burke!* Crocodile River, *Zeyher,* 1131 !

With much of the aspect of *J. streptopus,* but with much more copious whitish pubescence; more ovate, blunter leaves, and different calyx and corolla.

5. J. streptopus (E. Meyer ex DC. Prodr. viii. 307); branches terete, scandent, glabrate, ash-colour; twigs, leaves and calyces shortly, but closely pubescent; leaves very shortly petiolate, oval-oblong or ovate, acute at each end, penninerved, mucronate, the petiole twisted or incurved; peduncle terminal, 3–5-flowered; calyx-lobes 5–7, broadly subulate, equalling the calyx-tube, enlarged in fruit; corolla-tube 1–1¼ in. long; lobes 6–7, lanceolate, acute.

EASTERN REGION: Natal; near Durban, *Gerrard & McKen,* 628 ! *Peddie !* Inanda, *Wood,* 1191 ! Umlaas, *Wood,* 1827 ! and without precise locality, *Drège ! Gerrard,* 718 !

Young branches pubescent, as the rest of the plant, older quite glabrous. Leaves 1–2½ in. long, thinly membranous, ½–1¼ in. wide, variable in shape. Berries twin, globose; the fruiting calyx-lobes broadly triangular.

6. J. stenolobum (Rolfe in Oates, Matabeleland, ed. 2, 403); young branches villous-pubescent, afterwards glabrous; leaves uni-foliolate, ovate or ovate-lanceolate, obtuse or sometimes acute, pubescent, ½–1½ in. long, shortly petioled; flowers terminal, generally solitary; calyx pubescent; lobes about 10, acicular, obtuse, 3–4 lin. long; corolla-tube slender, 9–10 lin. long; lobes 10–15, narrowly lanceolate-linear, acute, a little shorter than the tube. *Baker in Dyer, Fl. Trop. Afr.* iv. i. 4.

KALAHARI REGION: Transvaal; Berea Ridge, Barberton, *Galpin,*
EASTERN REGION: Natal; Tugela River, *Gerrard,* 1968 !

Also in Tropical Africa.

7. J. angulare (Vahl, Symb. iii. 1); glabrous or pubescent, scandent; twigs 4–6-angled; leaves 3-foliolate; leaflets ovate, mucronate, 3-nerved at the base; cymes terminal or in the upper axils, 3–7-flowered; calyx glabrous, 5–6-toothed; corolla-tube 1½ in. long. *DC. Prodr.* viii. 311.

VAR. β, **glabratum** (E. Meyer ex DC. Prodr. l.c.); leaves glabrous. *Bot. Mag. t.* 6865. *J. capense, Thunb. Prodr.* 2, *and Fl. Cap. ed.* i. 41.

COAST REGION: Uitenhage Div.; between Enon and the Zuurberg Range, *Drège!* Addo, *Drège!* banks of the Zwartkops River, *Zeyher,* 231 ! Uitenhage, *Burchell,* 4411 ! Sundays River, *Gill!* Port Elizabeth Div.; Algoa Bay, *Cooper,* 2701 ! Albany Div., *Cooper,* 2702 ! Var. β: Uitenhage Div.; Enon, *Drège!* British Kaffraria, *cultivated specimens!*
CENTRAL REGION: Somerset Div.; on Bosch Berg, near Somerset East, *Bur-chell,* 3164 ! 3193 ! mountains above the spring of Commadagga, *Burchell,* 3342 !
KALAHARI REGION: Orange River Colony, *Cooper,* 2704 ! Transvaal; Komati Poort, *Burtt Davy,* 360 !
EASTERN REGION: Griqualand East; Clydesdale, 2500 ft., *Tyson,* 2017 ! Natal; Olivers Hoek Pass, *Wood,* 3515 ! and without precise locality, *Gerrard,* 280 ! *Cooper,* 1166 !

8. J. tortuosum (Willd. Enum. i. 10); glabrous or puberulous, scandent; twigs angular; leaves 3-foliolate; leaflets lanceolate, mucronate, glabrous; peduncle terminal, 3–5-flowered; calyx 5–6-toothed; corolla-tube about 1 in. long. *DC. Prodr.* viii. 311. *J. flexile, Jacq. Hort. Schœnbr. t.* 490, *not of Vahl.*

COAST REGION: Humansdorp Div.; Kabeljouws River, *Bolus*, 1667! Bedford Div.; near Bedford, *Mrs. Hutton!* British Kaffraria, *Mrs. Hutton!*

CENTRAL REGION: Somerset Div.; on the Bosch Berg, 3000 ft., *MacOwan,* 1946!

KALAHARI REGION: Transvaal; near Lydenburg, *Wilms,* 924! Waterval Boven, *Rogers,* 2511!

Scarcely distinct from *J. angulare* by the narrower leaflets and shorter corolla-tube.

9. J. mauritianum (Bojer, Hort. Maurit. 204); branchlets densely pubescent; leaves opposite, 3-foliolate; leaflets ovate, acute, subcoriaceous, pubescent or subglabrous, the end one 2–3 in. long, the side one slightly smaller; petiole $\frac{1}{4}$–$\frac{3}{4}$ in. long; petiolules $\frac{1}{2}$–1 in. long; cymes terminal and axillary, forming an ample panicle at the end of the branchlets; pedicels very short, pubescent; calyx 1 lin. long; teeth minute; corolla white; tube $\frac{3}{4}$–1 in. long; lobes 6–8, oblong or linear-oblong, half as long as the tube; berry small, globose. *DC. Prodr.* viii. 310; *Hiern in Cat. Afr. Pl. Welw.* i. 655; *Baker in Dyer, Fl. Trop. Afr.* iv. i. 10. *J. auriculatum, DC. l.c.* 309, *partly; Baker, Fl. Maurit.* 220, *not of Vahl. J. tettense, Klotzsch in Peters, Reise Mossamb. Bot.* i. 284. *J. zanzibarense, Bojer ex Klotzsch, l.c.* 283. *J. gratissimum, Deflers, Voy. Yemen, Bot.* 162.

KALAHARI REGION: Transvaal; Crocodile River Drift, between Komati River Drift and Barberton, *Bolus,* 7848!

Also in Tropical Africa, Mauritius and the Seychelles.

II. **SCHREBERA**, Roxb.

Calyx bell-shaped, shortly 5–8-toothed or subtruncate, splitting in fruit. *Corolla* salver-shaped; tube terete, longer than the calyx; limb 6–8-fid, the lobes inside more or less densely clothed with short swollen hairs (velvety to the touch), oblong, obtuse, twisted in bud. *Stamens* 2, inserted near the summit of the tube; filaments very short; anthers exserted or half included. *Ovary* 2-celled; cells each with 4 pendulous ovules; style filiform, included; stigma thickened, bifid at the apex. *Capsule* obovoid, coriaceous, compressed, 2-celled, separating through the middle into 2 septiferous boat-like valves. *Seeds* 4 in each cell, pendulous, compressed, winged at the apex. *DC. Prodr.* viii. 674. *Nathusia, DC. l.c.* 281.

Trees or climbing shrubs of India and Africa. Leaves opposite, imparipinnate. Flowers in trichotomous terminal panicles.

I have carefully examined flowers of *S. swietenioides*, Roxb., the type of the genus, and I find them to agree in structure in every respect with Decandolle's account of *Nathusia*, above quoted. The purple hairs on the corolla are more copious and more diffused in the Indian species; the filaments shorter, and the anthers partly included in the tube; the ovules are 4 and pendulous, exactly as in *Nathusia*. Named in honour of *Von Schreber*, a celebrated botanist and editor of "Linnæus."—*W. H. H.*

> Panicle-branches glabrous or finely puberulous   ... (1) **Saundersiæ.**
> Panicle-branches tomentose   ...   . (2) **argyrotricha.**

1. S. Saundersiæ (Harv. Thes. ii. 40, t. 163); petiole with a narrow wing; leaflets 3–5, sessile, oblong-lanceolate or ovate, scarcely oblique at the base, obtusely acuminate; calyx glabrous, truncate or very minutely denticulate. *S. latialata, Gilg in Engl. Jahrb.* xxx. 73.

K ALAHARI R EGION : Transvaal; Houtbosch, *Rehmann*, 5950!

E ASTERN R EGION : Natal; Durban, *Gerrard*, 1153! Berea, near Durban, *Wood*, 5201! Inanda, *Wood*, 819! and without precise locality, *Cooper*, 3030!

A partly scandent large shrub, or small tree 25–30 ft. high. Leaflets 1½–2 in. long, ¾–1 in. wide, the terminal larger, all subcoriaceous, netted-veined beneath, Panicle many-flowered, trichotomously much-branched. Bracts 2 lin. long, deciduous. Corolla 4 times longer than the calyx, the lobes commonly 6, white, with dark purple velvety patches at the base forming a star in the centre of the flower. Stamens exserted.

I have hesitated whether or not to keep this distinct from the Abyssinian *S. alata* (*Nathusia alata*, Hochst.!), with which it agrees in many particulars. That species, however, is said to be a "tall tree." A fruiting specimen of it in the Dublin Herbarium is much more robust, with larger leaves than the Natal plant, but otherwise similar; I have not seen its flowers.—*W. H. H.*

2. S. argyrotricha (Gilg in Engl. Jahrb. xxx. 74); a divaricate shrub; leaves opposite, imparipinnate; rhachis winged; leaflets 5, oval or obovate-oval, rounded at the apex, cuneate at the base, 1¼ in. long, about ½ in. wide, the terminal rather larger, more or less coriaceous, puberulous above, densely and shortly pubescent beneath; nerves numerous and prominent beneath, reticulate; flowers in terminal many-flowered dense panicles at the ends of the branchlets; panicle-branches tomentose, dichasial; pedicels very short, tomentose; calyx campanulate, 1½ lin. long, irregularly lobed or emarginate, puberulous, at length glabrescent; corolla glabrous; tube narrowly cylindrical, 5 lin. long; lobes half as long as the tube; capsule small, woody, glabrous.

K ALAHARI R EGION : Transvaal; by the great waterfall, near Lydenburg, *Wilms*, 201 !

III. **MENODORA**, Humb. & Bonpl.

Calyx deeply 5- to many-lobed, persistent. *Corolla* funnel-shaped, the short tube often hairy inside; limb 5-partite, imbricate in bud. *Stamens* 2, on the tube. *Ovary* 2-celled; ovules 4 in each cell; style filiform; stigma capitate. *Capsule* membranous, didymous, splitting across, sometimes irregularly ; cells 4-seeded, or by abortion fewer. *DC. Prodr.* viii. 316. *Bolivaria, Cham. & Schlecht. in Linnæa,* i. 207, *t.* 4, *fig.* 1.

Undershrubs, chiefly of the American continent, where they are found from Mexico to Patagonia. Habit and foliage very variable.

Stems erect, nearly naked, rigid ; leaves minute,
   linear, distant ...   . (1) **juncea.**
Stems diffuse, closely leafy :
     Leaves bipinnately multifid   ...   . (2) **africana.**
     Leaves entire or tripartite ...   . (3) **heterophylla.**

1. M. juncea (Harv. Gen. S. Afr. Pl. ed. 2, 220) ; stems erect, rigid, terete, striate, glabrous, irregularly branched, nearly naked ; branches erect ; leaves minute, distant, close-pressed, shortly linear, acute ; corymbose cymes terminal, several-flowered, minutely puberulous ; calyces semi-5-fid, the lobes subulate, entire ; capsule irregularly torn across.

WESTERN REGION : Little Namaqualand ; Modderfontein, *Whitehead !* and without precise locality, *Zeyher !*

Stems pale green, 2–3 ft. high, naked, except for the minute and distant, scarcely obvious leaves. Inflorescence panicled, terminal, its branches minutely puberulous. Calyx-lobes 2½ lin. long, rather longer than the tube.

2. M. africana (Hook. Ic. Pl. t. 586) ; root simple ; stems diffuse, many from the crown, branched, closely leafy throughout ; leaves alternate, short, bipinnately multifid, glabrous ; lobes narrow-linear, nerved beneath, acute ; flowers terminal, solitary ; calyx-limb multipartite, with filiform or pinnatifid lobes.

KALAHARI REGION : Griqualand West ; Dutoits Pan, *Tuck !* Barkly, *Nelson,* 175 ! on Flats, *Mrs. Barber,* 2 ! Orange River Colony ; Great Vet River, *Burke,* 134 ! Boshof, *Mrs. Barber !* Bechuanaland ; near Kuruman, *Burchell,* 2430 ! near the sources of the Kuruman River, *Burchell,* 2460 ! 2478 ! 2485 ! 2533 ! Transvaal ; near Pretoria, *McLea in Herb. Bolus,* 3104 ! Derde Poort, *Leendertz,* 369 ! near Irene, *Burtt-Davy,* 2308 ! Magalies Berg, *Zeyher,* 1132 ! near Lydenburg, *Wilms,* 1068 ! Crocodile Valley, near Barberton, *Galpin,* 1071 ! near Potchefstroom, *Nelson,* 342 !

EASTERN REGION : Natal ; by the Tugela River, near Colenso, *Wood,* 758 ! 3550 ! Bushmans River, *Gerrard,* 631 ! Zululand, *Miss Owen !*

Root fusiform, deeply descending. Stems 4–12 in. long, diffuse or ascending, scaberulous. Leaves scarcely ½ in. long, multifid. Flowers yellow.

3. M. heterophylla (Moric. ex DC. Prodr. viii. 316) ; a small shrub, 4–5 in. high, branched from the base ; stems ribbed, sparingly papillose-scabrid ; leaves alternate, entire or tripartite, shortly and broadly petiolate or subsessile ; lobes lanceolate or linear-lanceolate, acute, the central longest and often trifid ; calyx 10-fid or 10-partite ; lobes linear, scaberulous ; corolla-lobes obovate-oblong or -elliptic, obtuse, sometimes shortly apiculate. *Oliv. in Hook. Ic. Pl.* xv. 47, *t.* 1459.

KALAHARI REGION : Transvaal ; Matebe Valley. *Holub !* Woodstock, near Rustenburg, *Miss Pegler,* 950 !

Also in North America.

IV. **OLEA**, Linn.

Calyx cupular or shortly campanulate, 4-toothed. *Corolla-tube* short; lobes 4, spreading, induplicate-valvate. *Stamens* 2, inserted in the corolla-tube; filaments short; anthers ovate, dorsifixed. *Ovary* 2-celled; style short; stigma capitate or shortly bifid; ovules 2 in each cell, pendulous from the inner wall. *Drupe* ovoid, oblong or globose; endocarp bony or crustaceous. *Seeds* usually solitary; cotyledons flat; albumen fleshy, sometimes ruminate.

Trees or shrubs; leaves opposite, usually quite entire; flowers hermaphrodite or polygamous, in trichotomous panicles.

Species about 40 in Central or Tropical Asia, the Mediterranean region, Tropical Africa, the Mascarene Islands and New Zealand.

This genus (which was omitted from the key) differs from the other South African genera in having the corolla-lobes induplicate-valvate in bud and the ovules pendulous. From *Schrebera* and *Menodora* it also differs in having a drupaceous fruit, and from *Jasminum* in having 4-merous flowers.

Inflorescence axillary :
　　Leaves elliptic to obovate, up to 4½ by 1¾ in. 　　... 　(1) **Pegleri.**
　　Leaves oblong-elliptic, up to 1½ by ¾ in. ... 　　... 　(2) **foveolata.**
　　Leaves linear-lanceolate, up to 3 by 1 in. ... 　　... 　(3) **verrucosa.**
Inflorescence terminal :
　　Leaves linear-oblong ... 　　... 　　... 　　... 　　... 　(4) **exasperata.**
　　Leaves from lanceolate to broadly elliptic :
　　　　Branches of the inflorescence comparatively
　　　　　　stout :
　　　　　　Leaves obtuse, shortly cuneate at the base 　(5) **capensis.**
　　　　　　Leaves acuminate at both ends 　　... 　　... 　(6) **laurifolia.**
　　　　Branches of the inflorescence very slender :
　　　　　　Leaves lanceolate, 2½ in. by 7 lin., acumi-
　　　　　　　　nate 　　... 　　... 　　... 　　... 　　... 　(7) **Mackenii.**
　　　　　　Leaves elliptic, 1 in. by 5 lin., shortly
　　　　　　　　cuspidate ... 　　... 　　... 　　... 　　... 　(8) **enervis.**

1. O. Pegleri (C. H. Wright); branches ashy-white; leaves elliptic to obovate, shortly and obtusely acuminate, cuneate at the base, 4½ in. long, 1¾ in. wide, quite entire with a thickened marginal nerve, glabrous; petiole ½ in. long; cymes trichotomous, 2 in. long, axillary near the apex of the branchlets; pedicels 1½ lin. long; bracteoles lanceolate, 1 lin. long, brownish; calyx 1½ lin. long, deeply 4-lobed, glabrous; lateral lobes ovate, obtuse, 1 lin. broad, other lobes ½ lin. wide, acute; corolla 2½ lin. long; lobes short, cucullate at the apex.

Eastern Region : Transkei; near Kentani, 100 ft., *Miss Pegler*, 819 !

2. O. foveolata (E. Meyer, Comm. Pl. Afr. Austr. 176); a shrub about 12 ft. high; branches ashy-grey; leaves oblong-elliptic, obtuse, cuneate at the base, 1½ in. long, 9 lin. wide, sometimes with small pits in the axils of the veins on the lower surface, coriaceous, entire, glabrous; flowers in short axillary cymes; calyx-lobes rotundate, puberulous; corolla 1½ lin. long; lobes ovate, obtuse, cucullate;

fruit oblong, smooth, 6 lin. long, 3 lin. diam. *Drège, Zwei Pfl. Documente,* 135; *DC. Prodr.* viii. 285.

SOUTH AFRICA: without locality, *Thom! Miller!*

COAST REGION: Knysna Div.; forest near the Knysna River ford, *Burchell,* 5539! near the Goukamma River, *Burchell,* 5573! hills near Plettenbergs Bay, *Burchell,* 5328! Uitenhage Div.; near Uitenhage, *Burchell,* 4242! between Hoffmanns Kloof and Drie Fontein, 1000–2000 ft., *Drège!* Port Elizabeth Div.; Krakakamma, *Zeyher,* 571! *Burchell,* 4515!

KALAHARI REGION: Transvaal; Masetana River, near Shilovane, *Junod,* 1266!

EASTERN REGION: Transkei; near Kentani, 50 ft., *Miss Pegler,* 826! Natal; Tugela, *Gerrard,* 1665! Groen Berg, *Wood,* 1290!

3. **O. verrucosa** (Link, Enum. Hort. Berol. i. 33); a tree; branchlets terete, or sometimes quadrangular towards the apex, greyish, verrucose; leaves linear-lanceolate, obtuse or shortly acuminate, acuminate at the base, quite entire, glabrous above, minutely yellow-lepidote beneath, 3 in. long, 4–6 lin. wide; petioles up to 4 lin. long; panicles axillary, up to about 2 in. long; calyx cupular, obscurely 4-lobed, glabrous; corolla-lobes oblong, obtuse, 2 lin. long, glabrous; fruit subglobose, shortly apiculate, 3 lin. in diam., scarcely fleshy. *Drège, Zwei Pfl. Documente,* 85, 87; *DC. Prodr.* viii. 285. *O. europæa, Thunb. Fl. Cap. ed. Schult.* 3, *not of Linn. O. similis, Burch. Trav.* i. 177, *and* ii. 264. *O. woodiana, Knobl. in Engl. Jahrb.* xvii. 532; *Wood, Natal Plants, t.* 237.

SOUTH AFRICA: without locality, *Foster! Labillardière! Thom,* 751!

COAST REGION: Cape Div.; Hout Bay, *Wolley Dod,* 1638! Devils Mountain, *Ecklon,* 41! *Wilms,* 3484! Paarl Div.; Paarl Mountain, 1000–2000 ft., *Drège!* Uitenhage Div.; Addo, *Zeyher!* banks of the Zwartkops River, *Zeyher,* 633! Albany Div.; *Atherstone,* 95! Fort Beaufort Div., *Cooper,* 479! Queenstown Div.; Bowkers Park, 4200 ft., *Galpin,* 2566! Cathcart Div.; Goshen, *Baur,* 839!

CENTRAL REGION: Somerset Div.; Kloof of Bosch Berg, 3500 ft., *MacOwan,* 1364! Graaff Reinet Div.; mountains south-west of Graaff Reinet, *Burchell,* 2919!

KALAHARI REGION: Griqualand West; Upper Campbell, *Burchell,* 1826! Diamond Fields, *Shaw!* Transvaal; Magalies Berg, *Burke!* and without precise locality, *Sanderson!*

EASTERN REGION: Natal; Berea, near Durban, *Wood,* (548) 3156! 7579, and without precise locality, *Gerrard,* 1152!

4. **O. exasperata** (Jacq. Hort. Schoenbr. iii. 1, t. 251); from a small shrub to a tree 30 ft. high; branches slender, rough with elevated lenticels; leaves linear-oblong, about 2 in. long and 3 lin. wide, acute and mucronate, glabrous, shining above, nerves inconspicuous; panicle terminal, small; bracteoles very small; calyx minutely 4-dentate, verrucose; corolla $1\frac{1}{2}$ lin. in diam., white; lobes oblong, acute; stigma conical, $\frac{1}{3}$ lin. in diam. *E. Meyer, Comm. Fl. Afr. Austr.* 175; *Drège, Zwei Pfl. Documente,* 86, 130; *DC. Prodr.* viii. 287; *Knobl. in Engl. Jahrb.* xvii. 533.

SOUTH AFRICA: without locality, *Herb. Forsyth!*

COAST REGION: Paarl Div.; Paarl Mountain, *Drège!* Mossel Bay Div.;

near the landing-place at Mossel Bay, *Burchell*, 6313! Knysna Div. ; sand hills, Plettenbergs Bay, *Burchell*, 5308! Uitenhage Div. ; Addo, *Zeyher!* hill near the mouth of the Zwartkops River, *Drège!* Bathurst Div.; near Port Alfred, *Burchell*, 3829 !

5. O. capensis (Linn. Sp. Pl. ed. 1, 8) ; a small shrub; stem slender, glabrous, dark brown, lenticellate ; leaves obovate or oval, obtuse, quite entire, coriaceous, glabrous, up to $2\frac{1}{2}$ in. long and $1\frac{1}{2}$ in. wide ; petiole 3 lin. long ; panicle terminal, trichotomous, many-flowered ; corolla white ; tube very short ; lobes ovate, obtuse, patent, $1\frac{1}{2}$ lin. long ; anthers cordate ; fruit globose, glabrous, 4 lin. in diam. *Berg. Descr. Pl. Cap.* 1 ; *Thunb. Prodr.* 2, and *Fl. Cap. ed. Schult.* 3 ; *Pappe, Silva Cap.* 24 ; *E. Meyer, Comm.* 176 ; *Bot. Reg. t.* 613 ; *DC. Prodr.* viii. 287 ; *Knobl. in Engl. Jahrb.* xvii. 533 ; var. *coriacea, Ait. Hort. Kew. ed.* 1, i. 13.

SOUTH AFRICA : without locality, *Sieber*, 220! *Wallich! Bergius* ex *Knoblauch.*

COAST REGION : Cape Div.; mountains near Cape Town, *Drège! Milne*, 204! *Wolley Dod*, 171! 3641! *Burchell*, 781! Red Hill, *Jameson!* Simons Bay, *MacGillivray*, 665! Camps Bay, *Zeyher!* Stellenbosch Div.; Lowrys Pass, *Burchell*, 8236! *Schlechter*, 7267! Caledon Div.; between Houw Hoek Mountains and Palmiet River, *Burchell*, 8161! Knysna Div.; near Knysna, *Burchell*, 5497 ex *Knoblauch* ; Albany Div. ; Grahamstown, *MacOwan!*

6. O. laurifolia (Lam. Ill. i. 29) ; a tree with hard wood ; branches tuberculate, greyish, terete ; leaves usually elliptic-oblong and acute, but varying much in outline, entire, sometimes slightly undulate, glabrous, paler beneath, about 3 in. long and $1\frac{1}{4}$ in. wide ; petiole 6 lin. long ; panicles terminal, many-flowered, not much longer than the leaves ; calyx campanulate, shortly and acutely 4-toothed ; corolla white, about 2 lin. in diam. ; lobes lanceolate, acute, patent ; fruit globose, 3 lin. in diam. *DC. Prodr.* viii. 287. *O. undulata, Jacq. Hort. Schoenbr.* i. 1, *t.* 2 ; *Bot. Mag. t.* 3089 ; *Lodd. Bot. Cab. t.* 379 ; var. *β, E. Meyer in Drège, Zwei Pfl. Documente*, 136, 205 ; var. *planifolia, E. Meyer, Comm.* 176. *O. capensis, var. β, undulata, Ait. Hort. Kew. ed.* 2, i. 21.—*Sideroxylum foliis oblongis, &c., J. Burm. Rar. Afr. Pl.* 233, *t.* 81, *fig.* 1.

VAR. *β*, **concolor** (Harv.); branches brownish, verrucose ; leaves obovate, tapering into a petiole 3 lin. long, quite entire, sometimes shortly cuspidate, of the same dull green on both surfaces, minutely glandular-punctate beneath. *O. concolor, E. Meyer in Drège, Zwei Pfl. Documente*, 77, *and Comm.* 176 ; *DC. Prodr.* viii. 286.

SOUTH AFRICA : without locality, *Mund ! Villette ! Thom*, 484! *Sieber*, 219! *Pappe!*

COAST REGION : Cape Div.; Cape Town, *Zeyher!* Swellendam Div.; Groot-vaders Bosch, *Burchell*, 7227! George Div.; in the forest near George, *Burchell*, 6077! lower part of the Cradock Berg, *Burchell*, 6013! Knysna Div. ; between Keurbooms River and Bitou River, *Burchell*, 5225! 5227 ! near Knysna, *Burchell*, 5497! Uitenhage Div. ; Zuurberg Range, *Drège!* Van Stadens River, *Bolus*, 1210! Port Elizabeth Div.; sand hills by the sea-shore, *Burchell*, 4298! Krakakamma, *Burchell*, 4516! Albany Div.; between the source of Kasuga River and Assegai Bush, *Burchell*, 4159! Bathurst Div. ; near Port Alfred, *Burchell*, 3815 ! Var. *β* : Tulbagh Div. ; between New Kloof and Elands Kloof,

1000–2000 ft., *Drège!* Piquetberg Div.; Olifants River Mountains, *Zeyher,*
1150!

EASTERN REGION : Natal; Inanda, *Wood,* 500!

The variety *concolor* is connected with the typical form by Burchell's 7227, in
which, however, the shining green upper surface of the leaf contrasts with the
dull lower surface.   Under cultivation the leaves become undulate at the margin.

7. O. Mackenii (Harv.); branches slender, greyish; leaves
lanceolate, acuminate at both ends, more rarely elliptical and obtuse
at the apex, $2\frac{1}{2}$ in. long, 7 lin. wide, margins revolute, lateral nerves
inconspicuous; petiole 3 lin. long; panicles at and near the apex of
the branches; bracteoles minute; calyx cupular, $\frac{3}{4}$ lin. in diam.,
glabrous; teeth 4, broadly triangular; corolla-lobes short, rounded;
fruit oblong, 7 lin. long, 2 lin. in diam.

EASTERN REGION : Natal; Tugela, *Gerrard,* 380! 1666!

8. O. enervis (Harv.); a low tree; branches terete, ashy-grey,
lenticellate; leaves elliptic, very shortly cuspidate, cuneate below,
up to 1 in. long and 5 lin. wide, glabrous on both surfaces, thickened
at the margin, nerves (except the midrib) obscure below; petiole
2 lin. long; panicles terminal, 1 in. long; calyx cupular; corolla
$1\frac{1}{2}$ lin. in diam.; lobes oblong, acute; fruit ovoid-oblong, 3 lin.
long.

EASTERN REGION : Natal; Buffalo River and Mooi River, *Gerrard,* 1151!

ORDER LXXXVI. **SALVADORACEÆ.**

(By C. H. WRIGHT.)

Flowers hermaphrodite or dioecious, regular. *Calyx* inferior,
campanulate or ovoid, 2–4-toothed. *Corolla* 4-lobed, or of 4 free
petals, imbricate or valvate in bud. *Stamens* 4, inserted in the
corolla-tube and alternate with the lobes, or hypogynous; filaments
free or connate at the base; anthers ovoid, dorsifixed, sometimes
apiculate, dehiscing longitudinally. *Ovary* superior, 1–2-celled;
style short; stigma entire or bifid; ovules 1–2 in each cell, erect,
anatropous. *Berry* fleshy, usually 1-seeded; endocarp membranous or chartaceous. *Seed* globose or compressed; testa thin or
cartilaginous; albumen none.

Shrubs or trees, unarmed or **spiny**; leaves opposite, entire; stipules rudimentary; panicles trichotomous, **axillary**, often reduced to sessile fascicles.

DISTRIB. Genera 3, species about 10, in tropical and subtropical Asia, Africa
and the Mascarene Islands.

I. **Salvadora.**—*Petals* united.   *Ovary* 1-celled, 1-ovuled.
II. **Azima.**—*Petals* free.   *Ovary* 2-celled; cells 2-ovuled.

I. **SALVADORA**, Garcin.

Calyx campanulate, 4-fid. *Corolla* campanulate; tube short; lobes wide, obtuse, imbricate. *Stamens* 4, fixed to the base or middle of the corolla-tube; filaments slightly flattened; anthers ovoid. *Disc* of 4 glands or scales alternate with the filaments or none. *Ovary* 1-celled; style very short; stigma wide, truncate or nearly peltate; ovule solitary, erect. *Drupe* globose; endocarp chartaceous. *Seed* erect, globose; testa thin.

Shrubs or trees; leaves opposite, quite entire, rather thick, usually pallid; flowers small, on the branchlets of axillary panicles, sessile or pedicellate.

1. **S. persica** (Garcin ex Linn. Sp. Pl. ed. i. 122); a large, much-branched shrub or small tree; branchlets drooping, glabrous, whitish; leaves from ovate to oblong, coriaceous, $1\frac{1}{2}$–3 in. long, $\frac{3}{4}$–$1\frac{1}{4}$ in. wide, obtuse, often mucronate, pale green; petiole $\frac{1}{2}$–1 in. long; flowers in terminal and axillary panicles, greenish-yellow; calyx cleft half-way down; lobes rounded; corolla $\frac{1}{8}$ in. long, deeply cleft; lobes oblong, obtuse, reflexed; stamens shorter than the corolla; drupe $\frac{1}{8}$ in. in diam., globose, smooth, red. *DC. Prodr.* xvii. 28; *Schweinf. Beitr. Fl. Aethiop.* 163; *Oliv. in Trans. Linn. Soc.* xxix. 106; *C. B. Cl. in Hook. f. Fl. Brit. Ind.* iii. 619; *Hiern in Cat. Afr. Pl. Welw.* i. 659; *Baker in Dyer, Fl. Trop. Afr.* iv. i. 23. *S. crassinervia, Hochst. in Schimp. Pl. Abyss. Exsicc.* 2218; *T. Anders. in Journ. Linn. Soc.* v. *Suppl.* i. 30. *S. paniculata, Hochst. in Schimp. l.c.* 2325.

EASTERN REGION : Delagoa Bay; near Shepherd's, about 18 miles from Lorenzo Marques, *Bolus,* 9701 !

Also in Tropical and North Africa, Arabia and India.

II. **AZIMA**, Lam.

Flowers diœcious. *Calyx* campanulate, shortly 4-fid or in the female flower irregularly 2-4-partite. *Petals* 4, distinct. *Stamens* 4, alternate with the petals; filaments slender; anthers ovoid. *Disk* none. *Ovary* 2-celled; style very short; stigma large, 2-fid; ovules 2 in each cell, erect. *Berry* globose, usually, by abortion, 1-seeded; endocarp thinly membranous. *Seed* somewhat compressed; testa coriaceous.

Glabrous shrubs, very much branched or sarmentose; spines axillary, solitary or in pairs; leaves opposite, quite entire; flowers small, axillary or along the branchlets of a short panicle.

1. **A. tetracantha** (Lam. Encycl. i. 343 and Ill. t. 807); a low much-branched spiny shrub; branchlets tetragonal, pubescent when young; spines 2-4 from a node, straight, very sharp, 1 in. or more long; leaves oblong or elliptic, acute, mucronate, pale green, acute at the base, 1–$1\frac{3}{4}$ in. long, $\frac{1}{2}$–$\frac{3}{4}$ in. wide; petiole very short; flowers

axillary, the male fascicled, the female solitary or in pairs; male flower: calyx ⅛ in. long, lobes 4, ovate, acute; petals linear-lanceolate, ciliolate, a little longer than the calyx; female flower: calyx usually 2-lobed, lobes broadly ovate, apiculate; petals as in the male; ovary glabrous; stigma nearly sessile; berry globose, ¼ in. in diam., whitish, edible. *Wight, Illustr.* ii. *t.* 152; *Hook. f. Fl. Brit. Ind.* iii. 620; *Hiern in Cat. Afr. Pl. Welw.* i. 659; *Baker in Dyer, Fl. Trop. Afr.* iv. i. 22. *Monetia barlerioides, L'Hérit. Stirp. Nov.* i. *t.* 1; *Harv. & Sond. Fl. Cap.* i. 474.

Var. β, **laxior** (C. H. Wright); flowers in trichotomous panicles rather shorter than the leaves.

South Africa: without locality, *Drège,* 6749a! 6749b!

Coast Region: Uitenhage Div.; in valleys, 1500 ft., *Bolus,* 1659! *Zeyher,* 477! Bathurst Div.; by the Kowie River at Port Alfred, *Galpin,* 2969! *Burchell,* 4017! Var. β: East London Div.; East London Park, *Wood in Herb. Galpin,* 3129!

Central Region: Somerset Div.; by the Great Fish River, *Burchell,* 3247!

Eastern Region: Natal, *Gerrard,* 1775! Delagoa Bay, *Forbes!*

Also in Tropical Africa, Madagascar and India.

Order LXXXVII. **APOCYNACEÆ**.

(By Otto Stapf.)

Flowers hermaphrodite, regular. *Calyx* inferior; sepals 5 (very rarely 4), free or slightly (rarely more) united, more or less imbricate, equal or more or less unequal, often with (usually scale-like) glands near the base inside. *Corolla* salver- or funnel-shaped, rarely campanulate, urceolate or subglobose, glabrous or more or less hairy within, sometimes with scales or callous protuberances or ridges in the tube or mouth; lobes usually convolute, overlapping and frequently also twisted to the right or to the left when viewed from the side, very rarely valvate. *Stamens* 5 (very rarely 4), inserted in the corolla-tube or mouth; filaments filiform or more often flattened and short or reduced to a callous swelling, often passing at the base into more or less decurrent ridges projecting into the tube (filamental ridges); anthers frequently conniving in a cone, either linear or oblong (rarely elliptic), shortly and obtusely 2-lobed at the base with the anther-cells polliniferous and dehiscing to the base, or sagittate with barren tails (very frequently formed by the continuation of the outer halves of the cells), leaving the front basal part of the connective (foot) free; foot of the connective smooth or with various shaped projections or regular groups of spreading hairs. *Pollen* nearly always spherical with 3 pores, loose or rarely more or less cohering. *Disc,* if present, annular or cupular, 5-lobed or consisting of 2–5 scales, sometimes more or less adnate to the ovary. *Ovary* superior, or slightly inferior, of 2 (very rarely 3–5) united or distinct carpels, if syncarpous, 1-celled with parietal or 2-celled with central placentas, if apocarpous with ventral placentas. *Style* 1, entire or divided at the base ; stigma various, with or without a usually bifid apiculus and frequently with a ring or other appendages, viscous on the surface or exuding much glutinous matter and agglutinated to the anthers or adnate to the projections of the foot of the connective. *Ovules* anatropous, usually pendulous, few or many in each carpel. *Fruit* entire, baccate, drupaceous, samaroid, or consisting of 2 (rarely 3–5) baccate or follicular mericarps, rarely breaking up into 2 or 4 valves. *Seeds* various, frequently compressed, very often with a tuft of hairs (coma) at one or both ends, or winged, rarely with a plumose apical or basal awn ; testa coriaceous, crustaceous or membranous. *Endosperm*, if present, cartilaginous or fleshy. *Embryo* straight ; cotyledons usually flat, rarely convolute or contortuplicate ; radicle superior.

Trees, erect or scandent shrubs or perennial (very rarely annual) herbs, more or less laticiferous ; leaves simple, generally opposite, sometimes whorled, rarely spirally arranged, entire, pinnati-nerved; stipules (if present) short, intrapetiolar and often joining around the stem in a transverse ridge, very rarely one on each side of the petiole, or represented by spines; inflorescences made up of (often much reduced) cymes, terminal or pseudolateral or truly axillary; cymes solitary or clustered or gathered in loose or congested, often 2–3-tomous, panicles, corymbs or pseudo-umbels; bracts usually small and deciduous; flowers small to large and then often very showy.

DISTRIB. Genera about 180, comprising over 1000 species, chiefly in the tropics of both hemispheres.

Tribe 1. PLUMERIOIDEÆ.—*Corolla* salver-shaped, rarely funnel-shaped; lobes overlapping to the left, rarely to the right. *Anthers* linear, oblong or elliptic, shortly and obtusely 2-lobed (rarely subsagittate) at the base; anthercells polliniferous and dehiscing to the base or nearly so, not diverging below. *Ovary* syncarpous, 2-celled, or apocarpous with 2 (rarely 3–5) free or partly connate carpels ; stigma various, usually distinctly apiculate, rarely hairy or with frill-like appendages, often exuding more or less glutinous matter and then sometimes sticking to the anthers in the dry state, otherwise free. *Fruit* baccate, drupaceous or dry and follicular. *Seeds* not comose, exarillate ; endosperm (if any) smooth, rarely grooved and ruminate; cotyledons flat.

* *Ovary syncarpous.*

† Ovary 1-celled; stigma glabrous.

I. **Landolphia.**—*Style* short, not or shortly exserted from the eglandular calyx, filiform or subcolumnar, glabrous. *Inflorescences* terminal or pseudo-axillary.

†† Ovary 2-celled; stigma tips hairy.

II. **Carissa.**—Armed shrubs with simple or forked spines, rarely almost spineless.
III. **Acokanthera.**—Unarmed shrubs.

** *Ovary apocarpous, rarely imperfectly syncarpous (species of* Rauwolfia).

IV. **Rauwolfia.**—*Mericarps* drupaceous ; carpels 2-ovuled.
V. **Gonioma.**—*Mericarps* follicular. *Seeds* numerous, flat, winged. Shrubs.
VI. **Lochnera.**—*Mericarps* follicular. *Seeds* numerous, terete, wingless. Herbs.

Tribe 2. TABERNÆMONTANOIDEÆ.—*Corolla* salver-shaped, rarely funnel-shaped or campanulate with a cylindric basal tube; lobes overlapping to the left, very rarely to the right. *Anthers* linear, oblong or sagittate; anther-cells not or very slightly diverging below, and polliniferous and dehiscing to the base or nearly so, or diverging below and passing into barren tails leaving the glabrous foot of the connective free. *Ovary* apocarpous, rarely syncarpous;

carpels 2; stigma various, often exuding more or less glutinous matter and sticking tightly to the anthers at least in the dry state, otherwise free. *Fruit* baccate or follicular (but more or less fleshy). *Seeds* not comose, arillate in follicular fruits; endosperm ventrally grooved and more or less ruminate; cotyledons flat.

VII. **Conopharyngia.**—*Sepals* free, not circumscissile at the base.

VIII. **Voacanga.**—*Sepals* united into a tubular or subcampanulate, 5-lobed calyx, circumscissile at the base.

Tribe 3. ECHITOIDEÆ.—*Corolla* various; lobes overlapping to the right, very rarely to the left, or induplicate-valvate or valvate. *Anthers* usually sagittate; anther-cells diverging below, the outer halves passing into barren tailed appendages; foot of the connective free, generally provided with projections and regularly arranged groups of spreading hairs. *Ovary* apocarpous; rarely syncarpous; stigma various, exuding a glutinous matter and tightly agglutinated or adnate to the foot of the connective, very rarely to the base of the filaments. *Fruit* dry, follicular. *Seeds* comose, very rarely not so; or with a basal or apical plumose awn; endosperm smooth, often scanty; cotyledons flat, semiterete, convolute or contortuplicate.

    * *Leaves opposite, rarely whorled; trees or shrubs with woody stems.*

IX. **Wrightia.**—*Corolla* with appendages in the throat. *Anther-cone* exserted from the corolla-tube. *Seeds* with a persistent basal coma; cotyledons convolute.

X. **Strophanthus.**—*Corolla* fairly large, with paired appendages between the usually long-tailed lobes. *Anther-cone* included in the corolla-tube. *Seeds* with an apical plumose awn and a deciduous basal coma; cotyledons planoconvex.

XI. **Oncinotis.**—*Corolla* small, with 5 ligulate scales in the throat. *Seeds* with an apical coma.

    ** *Leaves spirally arranged; stem succulent.*

XII. **Adenium.**—Unarmed.

XIII. **Pachypodium.**—Armed with spines at the base of the leaves.

I. **LANDOLPHIA**, Beauv.

Calyx $\frac{3}{4}$–$1\frac{1}{2}$ lin. long (or $2\frac{1}{2}$–3 lin. in § *Mesandrœcia*); sepals 5, free or connate below, usually more or less ovate, hairy or glabrous, eglandular. *Corolla* salver-shaped; tube more or less cylindric, $1\frac{1}{2}$–12 lin. long, slender or rather stout, widened and staminiferous near the base or mouth or between them without correlation to the length of the corolla, but usually just above the calyx (higher up by $\frac{1}{2}$–2 calyx-lengths in §§ *Vahea* and *Mesandrœcia*); mouth naked, sometimes much constricted by a callous ring; lobes 5, narrow or broad, overlapping to the left. *Stamens* included; anthers ovate to lanceolate, minutely 2-lobed at and dehiscing to the base. *Disc* 0. *Ovary* entire, hairy or glabrous, 1-celled; placentas 2, parietal; style filiform or subcolumnar, usually short to very short, or longer and very slender (in §§ *Vahea* and *Mesandrœcia*); stigma at a level with the base of the anthers, conical from a slightly thickened base, 2-lobed; ovules numerous, pluriseriate. *Fruit* a globose or pear-shaped berry, sometimes of a large size. *Seeds* few or many, embedded in a juicy pulp, ovoid or oblong, smooth; endosperm horny; cotyledons foliaceous, very thin; radicle short.

Hairy or glabrous shrubs, often of a large size, usually climbing by flagelli-form, hook-branched, terminal or pseudo-axillary tendrils (modified inflorescences) or by sensitive inflorescences, rarely dwarf shrubs or undershrubs with partly herbaceous branches. Leaves opposite, of varying size; secondary nerves usually distant, rarely very close; axillary stipules 0; axillary glands minute, obscure, or (in § *Ancylobothrys*) subulate to filiform. Flowers small to middle-sized, rarely 2 in. long in bud, pedicelled or sessile in few- or many-flowered corymbs at the ends of the branches (sometimes overtopped by barren shoots and then occasionally pseudo-axillary), or gathered in more or less elongate panicles at the ends of their distant branches which are (like the rhachis) sensitive, and act as hooks or tendrils.

DISTRIB. Species about 50, the others in tropical Africa and the Mascarene Islands. In South Africa only the sections *Ancylobothrys* and *Eulandolphia*.

§ 1. ANCYLOBOTHRYS. Flowers sessile in dense clusters on the ends of the branches of terminal, mostly elongate, more or less sensitive panicles; calyx up to 1½ lin. long, hairy all over. Corolla-tube 3–12 lin. long, slender, cylindric, slightly widened and staminiferous just above the calyx; lobes 3–11 lin. long, ciliate along the outer edge. Ovary hairy; style not or very slightly exceeding the calyx. Fruit globose or obovoid, with a leathery rind, having no sclerenchy-matic layer; endosperm smooth.

Leaves elliptic or elliptic-oblong, 2½–4½ in. by 1–2 in. .. (1) **petersiana**.
Leaves more or less oblong, ¾ 2½ in. by ⅓–1 in. :
    Leaves 1¾–2½ in. by ¾–1 in.; lateral nerves sub-
    oblique, 8–10 on each side; corolla in the mature
    bud up to 18 lin. long　　…　　…　　…　　… (2) **Monteiroi**.
    Leaves ¾–1¾ in. by ⅓–⅔ in.; lateral nerves spread-
    ing more or less at right angles, 12–14 on each
    side; corolla in the mature bud 11–12 lin. long… (3) **capensis**.

§ 2. EULANDOLPHIA. Flowers many or few in dense terminal corymbs, rarely in elongated panicles. Calyx up to 2 lin long, hairy or almost glabrous; sepals usually very broad. Corolla-tube 1½–3 lin. long, inflated and stamini-ferous between the middle and the mouth, rarely at the middle; lobes 1½–3 lin. long, not ciliate. Ovary hairy or glabrous; style very short, not or scarcely exceeding the calyx. Fruit globose to pear-shaped, with a hard rind, having a concentric sclerenchymatous layer.

Only species in South Africa　　…　　…　　…　　… (4) **Kirkii**.

1. L. petersiana (Dyer in Kew Report, 1880, 42); a scandent shrub with sensitive inflorescences acting as tendrils; young branches minutely rusty-pubescent to tomentose, soon glabrescent, reddish or finally greyish-brown with numerous small lenticels; leaves elliptic or elliptic-oblong, subacute or more commonly obtuse at both ends, 3½–4½ in. long, 1¼–2 in. broad, coriaceous, loosely pubescent on both sides when quite young, soon glabrous, usually blackish and some-what glossy above when dry, pallid beneath; midrib channelled above, raised below; secondary nerves oblique, distinctly curved, 6–8 (rarely 10) on each side, finely channelled above, distinctly raised below; tertiary nerves more or less irregular and like the fine network of veins usually distinct, brown, scarcely raised; marginal arches obscure; petiole about 3 lin. long; panicle short or elongate, peduncled, bearing clusters of many sessile flowers at the ends of short spreading or recurved branches; branches or rhachis often acting as tendrils, finely rusty-pubescent or glabrescent all over; peduncle 1–4 in. long, slender; rhachis 1–1½ in. long, lowest

branches from a few lines to 1 in. long; bracts ovate-lanceolate or
ovate-oblong, acute or subacute, rusty-pubescent or tomentose;
calyx about 1 in. long; sepals ovate-oblong, obtuse, more or less
laterally compressed in the upper half, rusty-pubescent or tomentose;
corolla white, sweet-scented, rather variable in size, 8–13 lin. long
(rarely longer) in bud; tube slender, cylindric, about 3–4½ lin. long,
slightly wider and staminiferous 1–1½ lin. above the base, minutely
and equally pubescent above the widening; lobes obliquely oblong
or linear-oblong, acute or obtuse, 6–8 lin. long, curled-ciliate;
anthers ovate-oblong, acute, not quite 1 lin. long; ovary globose,
very minutely rufo-tomentose; style and stigma 1 lin. long, the
latter cylindric from a thicker base, bifid; fruit globose, up to 2½ in.
in diam., finely velvety; pericarp leathery, up to 2 lin. thick, without a sclerenchymatous layer; seeds 4–9 lin. long. *Ficalho, Pl.
Uteis Afr. Portug.* 219; *K. Schum. in Engl. Jahrb.* xv. 406; *L.
Planchon, Prod. Apocyn.* 319; *Dewèvre, Caoutch. Afr. Monogr.
Landolph.* 27 *partly ; Jumelle, Pl. à Caoutch. et à Gutta,* 57 *partly ;
Morris in Journ. Soc. Arts,* xlvi. 775, 780; *Warb. in Tropenpfl.*
iii. (1899) 222, *and Kautschukpfl.* 118 *partly ; Sadebeck, Kulturg.
Deutsch. Kolon.* 272 *partly? Henriques, Kautschuk,* t. iv.;
Schinz in Mém. Herb. Boiss. x. 1900, 57; *Dyer in Hook. Ic. Pl. t.* 2756;
Stapf in Dyer, Fl. Trop. Afr. iv. i. 47. *L. petersiana, var. crassi-
folia, Dewèvre, l.c.* 29 *partly ; Engl. Glied. Veg. Fl. Usambara,* 26 ;
Mikosch in Wiesner, Rohstoffe, ed. 2, i. 362, *not of K. Schum.
L. petersiana, var. rotundifolia, Dewèvre, l.c.* 30. *L. scandens,
vars. petersiana, rotundifolia, and stuhlmanniana, Hallier f. Kautschu-
klianen in Jahrb. Hamburg. Wissensch. Anstalt.* xvii. (1890), 3,
Beih. 82, 83. *L. senensis, K. Schum. in Engl. Pfl. Ost-Afr. B.*
453. *Ancylobothrys petersiana, and var. forbesiana (partly), Pierre
in Bull. Soc. Linn. Paris* 1898, 91; *Schinz, l.c. A. rotundifolia,
Pierre, l.c.* 92. *Willughbeia petersiana and W. senensis, Klotzsch
in Peters, Reise Mossamb. Bot.* i. 281, 282.

EASTERN REGION: Delagoa Bay, *Forbes !*
Throughout East Africa as far north as Mombasa, and a variety to Jur.

2. **L. Monteiroi** (Dyer, MS.); a shrub rambling among other
shrubs, with sensitive inflorescences acting as tendrils; young
branches rusty-pubescent, soon glabrescent and dotted with brown
lenticels; leaves oblong to lanceolate-oblong, obtuse to subacute,
more or less rounded at the base, 1¾–2½ in. long, ¾–1 in. broad,
coriaceous, rusty-pubescent in bud, soon quite glabrous, dark green
above, paler below; midrib channelled above, raised below; lateral
nerves 8–10 on each side, somewhat oblique, very faintly raised on
both sides; reticulation delicate; petiole 2½–3 lin. long; panicle
peduncled, bearing 2–3 dense clusters of subsessile flowers at the
ends of short sometimes spreading or recurved branches, the
peduncles (1–2 in. long) or · branches acting as tendrils, rusty-
pubescent when quite young; calyx rusty-tomentose, 1½ lin. long;
sepals ovate-oblong, subacute; corolla white, sweet-scented, up to
18 lin. long in bud; tube 6–7 lin. long, slightly inflated just above

the calyx, slender, delicately pubescent; lobes linear-oblong, sub-acute, about 1 in. long, margin sparingly ciliate; stamens inserted $1\frac{1}{2}$ lin. above the base of the corolla-tube; ovary very delicately silky-tomentose; ovules about 30 on each placenta, in 6 rows; fruit "yellow, about the size of a small orange, pulp yellow" (*R. Monteiro*). *Ancylobothrys petersiana, var. forbesiana, Pierre in Bull. Soc. Linn. Paris,* 1898, 91 (*partly; without description*).

EASTERN REGION : Delagoa Bay, *Monteiro, 37*!

The fruit is, according to Mrs. Monteiro, edible, acid, and refreshing.

3. L. capensis (Oliv. in Hook. Ic. Pl. t. 1228); a climbing or rambling shrub with sensitive inflorescences acting as tendrils; all the young parts finely rusty-tomentose, ultimately glabrescent with the exception of the cymes, adult branches reddish or greyish with few brown lenticels; leaves oblong to elliptic-oblong, obtuse at both ends or subacute at the base, $\frac{3}{4}$–$1\frac{3}{4}$ in. long, 4–8 lin. wide, coriaceous, dark green above, paler below, midrib slightly channelled above, raised below, lateral nerves 12–14 on each side, subhorizontal, faint; petiole about $1\frac{1}{2}$ lin. long; panicle on long or short peduncles, frequently reduced to a single, very dense, many-flowered cyme, the peduncles or the spreading or recurved branches often acting as tendrils; flowers sessile or subsessile; calyx $1\frac{1}{2}$ to almost 2 lin. long, rusty-tomentose; sepals ovate-oblong, subobtuse; corolla white, sweet-scented; tube widened below the middle, finely pubescent without, 5 lin. long; lobes oblong to elliptic-oblong, obtuse, 6–7 lin. long, margins ciliolate; stamens inserted about $1\frac{1}{4}$ lin. above the base of the corolla; ovary finely reddish-tomentose; fruit yellowish, of the size of a small peach, few-seeded; seeds rather flat, orbicular, 5 lin. in diam. *Dewèvre, Caoutch. Afr. Monogr. Landolph.* 55 ; *Hallier f. Kautschuklianen in Jahrb. Hamburg. Wissensch. Anstalt.* xvii. (1899), 3, *Beih.* 85.

KALAHARI REGION : Griqualand West; Diamond Fields, *Tuck,* 4! Transvaal, 4000–6000 ft.; Magalies Berg, *Zeyher,* 1186 ! *Burke,* 405 ! hills near Pretoria, *Rehmann,* 4295! *McLea in Herb. Bolus,* 3098! *Bolus,* 9702 ! *Burtt Davy,* 2182 ! near Johannesburg, *Adlam, 7*! *Gilfillan in Herb. Galpin,* 1472!

4. L. Kirkii (Dyer), var. **delagoensis** (Dewèvre, Caoutch. Afric. Monogr. Landolph. 48) ; a scandent shrub; young branches fulvo-pubescent or finely tomentose, at length glabrescent, reddish-brown with small whitish lenticels; leaves lanceolate to oblong, usually gradually tapering into a short obtuse acumen, shortly acute or obtuse at the base, $\frac{3}{4}$–$1\frac{1}{2}$ in. long, 5–6 lin. broad, thinly coriaceous, very loosely pubescent on both sides (except the midrib which is generally densely pubescent to villous below), finally more or less glabrescent, chiefly above, glossy above; midrib shallowly channelled above, prominent below; secondary nerves 10–12 on each side, very slender, slightly oblique, like the delicate network of the veins slightly raised on both sides; marginal arches obscure; petiole slender, 2–3 lin. long; corymbs dense, subsessile, many-flowered, about $\frac{1}{2}$ in. across, fulvo-pubescent or finely tomentose all over; bracts minute, ovate; pedicels very short; calyx scarcely 1 lin. long; sepals ovate, acute or subacute, membranous except the acutely edged midrib, pubescent; corolla whitish; tube cylindric below to the middle, then much inflated and distinctly constricted close to the mouth, up to 2 lin. long, minutely pubescent without in the upper half; lobes linear-oblong, subacute, as long as the tube, finely pubescent without along the middle, mouth very narrow, very minutely pubescent; stamens in the upper third of the tube; anthers linear-oblong, acute; ovary ovoid, glabrous; style and stigma about 1 lin. long, the latter cylindric from a thicker base, bifid; fruit obovoid-globose, $1\frac{1}{4}$–3 in. in diam.; seeds angular, 6–8 lin. long, numerous. *Jumelle, Pl. à Caoutch. et à Gutta,* 320. *L. delagoensis, Pierre in Bull. Linn. Soc. Paris,* 1898, 15; *Warburg, Kautschukpfl.* 120 (*French transl.* 236).

EASTERN REGION : Delagoa Bay, *Monteiro,* 3! *Junod.*

This variety, like the typical form, yields an excellent rubber, and it is also known by the same name, viz. " pink rubber " (Pierre, l.c.). It differs from the type only in the much smaller size of the leaves.

II. CARISSA, Linn.

Calyx small, eglandular, very rarely multiglandular within; sepals 5, very rarely 4, free or nearly so, imbricate, acute or acuminate. *Corolla* salver-shaped; tube slightly widened below the mouth or near the middle; lobes usually overlapping to the right, rarely to the left. *Stamens* enclosed in the widened part of the corolla-tube; filaments short, slender; anthers oblong, acute; cells obtuse at the base, polliniferous and dehiscing to the base. *Disc* 0. *Ovary* entire, 2-celled; ovules 1–4 in each cell, from the middle of the septum, rarely more in 2–3 rows; style filiform; stigma at the level of the anthers, or rarely some way below them, oblong, papillose and viscous, with a 2-lobed hairy tip. *Fruit* baccate, globose to oblong. *Seeds* usually 1–4, rarely more, peltate, plano-convex; hilum central; endosperm horny; cotyledons ovate; radicle superior.

Much branched, straggling and usually very spinous shrubs or small trees, rarely climbing; spines opposite, simple, rarely forked, often very stout. Leaves coriaceous, very variable in the same individual; axillary stipules 0; axillary glands very minute and few, or 0. Inflorescence often umbelliform, or corymbiform, and much contracted, terminal or pseudo-axillary, rarely cymose, lax and few-flowered; flowers subsessile, white or tinged with pink. Berries often edible.

DISTRIB. Species about 19, extending into the tropics of the Old World and Australia.

§ 1. EU-CARISSA. Corolla-lobes overlapping to the right; ovules 1–4 in each cell; spines simple.

 Only South African species   ... ... (1) **edulis.**

§ 2. ARDUINA. Corolla-lobes overlapping to the left; ovules 1 to many in each cell; spines bifurcate.

Corolla-limb 2–3 in. across; segments 2–3 times longer
  than the tube; fruit ovoid, pointed, up to 2 in.
  long, many-seeded ... ... (2) **grandiflora.**
Corolla-limb much smaller than in the preceding
  species; segments slightly longer or shorter than
  the tube; fruit oblong, obtuse, not much over $\frac{1}{2}$ in.
  long, 1–2-seeded :
    Corolla-limb 1–1$\frac{1}{2}$ in. across; segments as long as
      or slightly longer than the tube; spines small,
      few ... ... (3) **Wyliei.**
    Corolla-limb $\frac{1}{3}$ in. across; segments much shorter
      than the tube; spines numerous, $\frac{1}{2}$–1$\frac{1}{2}$ in. long ... (4) **Arduina.**

1. **C. edulis** (Vahl), var. **tomentosa** (Stapf in Dyer, Fl. Trop.
Afr. iv. i. 90) ; a very much branched straggling or climbing shrub ;
branches and leaves tomentose, at least in the young state ; spines
simple, straight or recurved, 1–2 in. long, rarely almost suppressed ;
leaves ovate to ovate-elliptic or sublanceolate, rarely orbicular,
9–24 lin. long, 9–18 lin. broad, sometimes much smaller, rounded at
the base or subcuneate, acute and often mucronate, rarely obtuse,
coriaceous ; nerves 3–5, faint on both sides ; petiole 1–1$\frac{1}{2}$ lin. long ;
calyx 1$\frac{1}{2}$–2 lin. long ; sepals lanceolate, acuminate, ciliolate, glabrous
or puberulous ; corolla white or purple turning white, glabrous or
minutely hairy at the mouth and on the inner surface of the lobes,
6–9 (rarely 4–6) lin. long ; lobes ovate or oblong, acute, 1$\frac{1}{4}$–4 lin.
long ; berry globose, purple to black, $\frac{1}{4}$–$\frac{2}{5}$ in. in diam., edible ; seeds
2–4. *C. tomentosa, A. Rich. Tent. Fl. Abyss.* ii. 30 ; *Engl. Hochgebirgsfl. Trop. Afr.* 340. *C. pilosa, Schinz in Verhandl. Bot. Ver.
Brandenb.* xxx. 258. *Jasminonerium tomentosum, O. Kuntze, Rev.
Gen. Pl.* ii. 415.

Kalahari Region : Transvaal ; Houtbosch, *Rehmann*, 6455! 6456! near
Shiluvane, 2000 ft., *Junod*, 620!

The area of this variety extends westwards to Damaraland and northwards to
Eritrea. The typical form is common throughout Tropical Africa, and also
occurs in Southern Arabia.

2. **C. grandiflora** (A.DC. Prodr. viii. 335) ; a glabrous shrub, up
to 15 ft. high, of compact habit, with strong simply or twice
bifurcate spines ; spines up to 1$\frac{1}{2}$ in. long ; leaves broad-ovate,
rarely ovate-lanceolate, rounded (rarely acute) at the base, mucronate,
1–3 in. long, $\frac{3}{4}$–2 in. wide, coriaceous, dark green above, paler below ;
secondary nerves about 6 on each side, like the veins very obscure,
particularly below; petiole 1–1$\frac{1}{2}$ lin. long ; cymes terminal, few-
flowered or reduced to a single flower, sometimes in the fork of a
spine ; pedicels 1–1$\frac{1}{2}$ lin. long; calyx 1$\frac{1}{2}$–3 lin. long ; sepals ovate
to ovate-lanceolate, acuminate, mucronate, margins revolute ; corolla
white, fragrant, very variable in size ; tube 4–6 lin. long, hairy
within ; lobes oblong to elliptic, narrowed towards the base, $\frac{1}{2}$–1$\frac{1}{2}$ in.
long, overlapping to the left ; stamens inserted slightly above the
middle of the corolla-tube ; ovules numerous ; berry ovoid, pointed,
scarlet, up to 2 in. long, about 16-seeded ; seeds plano-convex, rather
flat, elliptic in outline, 2$\frac{1}{2}$ lin. long. *L. Planch. Prod. Apocyn.* 140;
Wood, Natal Pl. i. 14, *t.* 14 ; *Sim, For. Fl. Cape Col.* 269, *t.* clv.

fig. 1. *C. macrocarpa*, *A.DC. Prodr.* viii. 336. *Arduina macro-carpa*, *Eckl. in South Afr. Quart. Journ.* i. (1830), 372. *A. grandiflora*, *E. Meyer in Drège Zwei Pflanzengeogr. Doc.* 154; *K. Schum. in Engl. & Prantl, Pflanzenfam.* iv. ii. 126. *Jasmino-nerium grandiflorum*, *O. Kuntze, Rev. Gen. Pl.* ii. 415.

EASTERN REGION: Natal; woods near Durban, *Krauss*, 88! *Peddie! Nelson*, 44! *Grant! Wilms*, 2133! between Umtentu River and Umzimkulu River, *Drège!* and without precise locality, *Gerrard*, 755! *Cooper*, 1235!

The fruit is known as Natal Plum or Amatungulu, and used for making jam. The plant is a valuable hedge shrub. Medley Wood states that there are long and short-styled flowers, the long-styled ones being functionally female, the others male.

3. C. Wyliei (N. E. Br. in Kew Bulletin, 1906, 165); a glabrous shrub with slender green almost unarmed branches; spines very scanty and small; leaves ovate, rounded or shortly acute at the base, very acute or mucronate at the tip, 2–4½ in. long, 1–2 in. wide, thinly coriaceous, dull or almost glaucous-green above, paler below; secondary nerves about 7–9 on each side, like the veins slightly prominent; petiole 1–1½ lin. long; cymes terminal, 3–6-flowered, subsessile; pedicels very slender, 2–3 lin. long; calyx 1½ lin. long; sepals lanceolate, long-acuminate; corolla white or pinkish (?); tube 6 lin. long, slender, hairy within; lobes lanceolate-oblong, acute, 6–7 lin. long, 2 lin. wide; fruits oblong, subacute, up to 7 lin. long and 3 lin. in diam., red, 2-seeded; seeds plano-convex, very flat, up to 5 lin. long.

EASTERN REGION: Zululand; Ngoya, 1000–2000 ft., *Wylie in Herb. Wood*, 7898! 8631!

4. C. Arduina (Lam. Encycl. i. 555); a glabrous, rarely pubes-cent shrub up to 10 ft. high, with numerous green simply or repeatedly bifurcate spines ½–1½ in. long; leaves very variable, generally ovate, acute, mucronulate, obtuse or sometimes cordate, and as long as broad, sometimes oblong or nearly lanceolate, ¾–3 in. long, ½–2 in. broad, dark shining green above, paler below; secondary nerves about 6–8 on each side, like the veins faintly prominent or quite obscure, mainly below; petiole 1–3 lin. long; cymes sub-sessile or shortly peduncled, few- to many-flowered, glabrous or sub-glabrous; pedicels slender, 1–2½ lin. long; calyx ¾–1 lin. long; sepals lanceolate, acuminate, ciliolate, glabrous or very finely puberu-lous, pale pink; corolla white, fragrant; tube 3–4 lin. long, densely hairy at the throat; lobes overlapping to the left, ovate to lanceolate, 2–2½ lin. long; stamens inserted between the middle and the mouth; ovule 1 in each cell; berry scarlet, oblong, acute, 4–7½ lin. long, 1–2-seeded, edible. *DC. Prodr.* viii. 334; *L. Planch. Prod. Apocyn.* 141, 258; *Lewin in Engl. Jahrb.* xvii. *Beibl.* 41, 49, 50; *Schinz in Mém. Herb. Boiss.* x. 56; *Stapf in Dyer, Fl. Trop. Afr.* iv. i. 91; *Sim, For. Fl. Cape Col.* 269. *C. bispinosa, Desf. Tabl. Écol. Bot.* 78. *C. ferox, DC. l.c.* 335; *Sim, For. Fl. Cape Col.* 269. *C. acuminata. DC. l.c.* 335; *Wood, Natal Pl. t.* 203. *C. myrtoides, Desf. Cat. Hort. Paris, ed.* 3, 398; *DC. l.c.* 335. *C. erythrocarpa, DC. l.c. C. hæmatocarpa, DC. l.c.* 336. *C. oblongifolia, Hochst.*

in Flora, 1844, 827.   *C. bispinata* (*by error*), *Lewin in Virchow's Arch. Path. Anat. u. Physiol. Bd.* 134, 246.   *Arduina bispinosa, Linn. Mant.* i. 52 ; *Lodd. Bot. Cat. t.* 387 ; *K. Schum. in Engl. & Prantl, Pflanzenfam.* iv. ii. 126 ; *in Engl. Pfl. Ost-Afr. C.* 315. *A. erythrocarpa, Eckl. in South Afr. Quart. Journ.* i. (1830), 372. *A. hæmatocarpa, Eckl. l.c.   A. ferox, E. Meyer, Comm.* 191 ; *K. Schum. in Engl. & Prantl, l.c.   A. acuminata, E. Meyer, l.c.   K. Schum. in Engl. & Prantl, l.c.   Lycium cordatum, Mill. Gard. Dict. ed.* viii. (1768), *no.* 10. *Jasminonerium acuminatum, J. bispinosum, J. erythrocarpum, J. ferox, J. hæmatocarpum, and J. oblongifolium, O. Kuntze, Rev. Gen. Pl.* ii. 415.

COAST REGION : Swellendam Div. ! between Rietkuil and Hemel-en-Aarde, *Zeyher,* 3415! Bredasdorp Div. ; Elim, *Schlechter,* 9667 ! Mossel Bay Div. ; Mossel Bay, *Burchell,* 6305 ! George Div. ; between Malgaten River and Great Brak River, *Burchell,* 6141! Uitenhage Div.; Zuurberg Range, at Boutjes River, *Drège !* Zwartkops River, *Zeyher,* 3414a ! 3416 ! and without precise locality, *Zeyher,* 308 ! 428 ! Albany Div. ; between Assegai Bosch and the Bushmans River, *Burchell,* 4188 ! Grahamstown, *Miss Daly,* 61 ! *MacOwan,* 2813 ! *and in Herb. Norm. Aust.-Afr.* 760 ! Bathurst Div. ; Blaauw Krantz, *Burchell,* 3679 ! between Blaauw Krantz and Kasuga River, *Burchell,* 3898 ! British Kaffraria, *Cooper,* 47 !

CENTRAL REGION : Somerset Div. ; Bosch Berg, *Burchell,* 3122 ! 3223 ! 3246 ! *MacOwan,* 1768 ! near the Little Fish River and Great Fish River, *Drège !* between Little Fish River and Commadagga, *Burchell,* 3287! 3288! Commadagga, *Burchell,* 3315 ! 3317 !

KALAHARI REGION : Transvaal ; Klerksdorp, mountain summit, *Nelson,* 308 ! Houtbosch Berg, *Nelson,* 421 !

EASTERN REGION : Pondoland ; between St. Johns River and Umtsikaba River, below 1000 ft., *Drège !* Natal ; Durban, *Peddie !* Inanda, *Wood,* 300 ! Nottingham, *Buchanan,* 137 ! between Pietermaritzburg and Newcastle, *Wilms,* 2132! and without precise locality, *Gueinzius ! Cooper,* 1103 ! 1235 ! *Sanderson ! Gerrard,* 147 !

Extends to Rhodesia and British Central Africa.

III. **ACOKANTHERA**, G. Don.

(ACOCANTHERA, K. Schum.)

Calyx small, eglandular within ; sepals free or almost so, imbricate, acute or acuminate, more or less scarious. *Corolla* salvershaped ; tube slightly widened near the mouth ; lobes short, overlapping to the left. *Stamens* enclosed in the widened part of the tube ; anthers ovate-oblong, connective produced into a short minutely pilose point, shortly 2-lobed at and dehiscing to the base. *Disc* 0. *Ovary* entire, 2-celled ; style filiform ; stigma short, conic or cylindric, with a ring of papillæ at the base and a minutely 2-lobed hairy apiculus ; ovule 1 in each cell, pendulous, attached to the centre of the septum. *Fruit* a globose or ellipsoid berry. *Seeds* 2, or 1 (by abortion), peltate, sessile on the septum, plano-convex ; hilum oblong, rather large ; endosperm bony ; cotyledons broadly ovate or subcordate ; radicle superior.

Unarmed shrubs or small trees. Leaves thickly coriaceous ; axillary stipules 0 ; axillary glands 0 or very scanty and minute. Corymbs very shortly peduncled or sessile, axillary, often reduced to clusters ; flowers subsessile, white or tinged with pink, usually sweet-scented.

DISTRIB. Species 3, in Tropical Africa and South Africa, 1 extending to Arabia.

Leaves not more than twice as long as broad ; corolla-
tube 3½–5 lin. long ... ... (1) **venenata.**
Leaves about three times as long as broad ; corolla-tube
7–9 lin. long ... ... (2) **spectabilis.**

1. A. venenata (G. Don, Gen. Syst. iv. 485) ; a shrub or a gnarled tree, up to 14 ft. high, glabrous (except sometimes the inflorescence) ; young branches compressed or ancipitous, smooth ; leaves mostly ovate or elliptic, sometimes oblong, rarely lanceolate, acute and usually mucronulate, rarely obtuse, acute at the base, 1½–4 in. long, ¾–2 in. broad, pale or olive-green when dry, somewhat shining above or on both sides ; secondary nerves 6–10 on each side, often with similar interposed tertiary nerves, oblique, parallel, prominent on both sides ; veins distinct or obscure ; petiole stout, 1–2 lin. long ; clusters glabrous or puberulous, sessile or subsessile, usually many-flowered ; bracts ovate, brown or the upper pinkish ; calyx glabrous or puberulous, 1 lin. long ; sepals ovate to ovate-lanceolate, acute to subacuminate, distinctly to very obscurely ciliolate ; corolla white to pink, sweet-scented ; tube 3½–5 lin. long, puberulous or glabrous without, hairy within ; lobes broad-ovate, acute or shortly acuminate, somewhat over 1 lin. long ; anthers ½–¾ lin. long ; stigma short, obtuse, conic ; berry globose, 1 in. in diam., purplish-black ; seeds semi-globose or semi-ellipsoid, 4–6 lin. long. *Vatke ex Schweinf. in Engl. Jahrb.* xvii. *Beibl.* 41, 46 (*footnote*) ; *Holmes in Pharm. Journ. ser.* 3, xxiv. 42 ; *K. Schum. in Engl. & Prantl, Pflanzenfam.* iv. ii. 126 (*not of Schweinf. ex Lewin in Engl. Jahrb.* xvii. *Beibl.* 41, 46, and *Lewin, l.c.* 49–51, 47, *fig. A., nor L. Planchon, Prod. Apocyn.* 255, *nor Vogtherr in Köhler, Mediz. Pfl.* iii. *t.* 64) ; *Stapf in Dyer, Fl. Trop. Afr.* iv. i. 94 ; *Sim, For. Fl. Cape Col.* 270, *t.* cliv. *fig.* 1. *A. Lamarkii, G. Don, l.c. A. Schimperi, Schweinf. in Boll. Soc. Afr. Italia,* x. (1891) xi.-xii. 12 (*the Taita plant*) ; *Pax in Engl. Pfl. Ost.-Afr. B.* 519 (*the Taita plant*). *A. abyssinica, K. Schum. in Engl. Pfl. Ost.-Afr. A.* 48 (*partly ?*). *Cestrum venenatum, Thunb. Prodr.* 36 ; *Fl. Cap. ed. Schult.* 193. *C. oppositifolium, Lam. Ill.* ii. 5, *t.* 112, *fig.* 2 ; *Poir. Suppl.* ii. 182. *Toxicophlæa Thunbergii, Harv. in Hook. Lond. Journ. Bot.* i. 24, *and Thes. Cap.* 10, *t.* 16. *T. cestroides, A.DC. Prodr.* viii. 336. *Sideroxylum toxiferum, Thunb. Trav. ed.* 3, i. 156.

COAST REGION : Mossel Bay Div. ; near Mossel Bay, *Burchell,* 6228 ! 6300 ! George Div. ; near Kaymans River Gat, *Drège !* Knysna Div. ; near the Bitou River, *Burchell,* 5303 ! near the Goukamma River, *Burchell,* 5603 ! near Knysna River Ford, *Burchell,* 5541 ! Uitenhage Div. ; Zuur Berg, *Cooper,* 1549 ! and without precise locality, *Zeyher,* 1184 ! Albany Div. ; Grahamstown, *MacOwan,* 433 ! and without precise locality, *Cooper,* 1549 ! near the mouth of the Kowie River, *MacOwan, Herb. Norm. Austr.-Afr.* 814 ! Bathurst Div. ; at the mouth of the Great Fish River, *Burchell,* 3747 ! Queenstown· Div. ; Finchams Nek, 4000 ft., *Galpin,* 1888 ! Eastern districts, *Cooper,* 49 !

KALAHARI REGION : Transvaal ; near Pretoria, *McLea in Herb. Bolus,* 5700 !

EASTERN REGION : Natal ; Inanda, *Wood,* 982 ! and without precise locality, *Gerrard,* 139 !

Also in Tropical Africa.

According to Thunberg, the root is used by the Hottentots for poisoning arrows. The Dutch call it "Gift-boom."

2. A. spectabilis (Hook. f. in Bot. Mag. t. 6359); a tall shrub or small tree up to 15 ft. high, glabrous (except sometimes the inflorescences); young branches compressed; leaves elliptic or oblong-lanceolate, acute, rarely obtuse, generally mucronate, acute at the base, $2\frac{1}{2}$–5 in. long, 1–2 in. broad, very coriaceous, dark green, paler and sometimes purplish beneath; secondary nerves usually 7–10 on each side, sometimes with similar tertiary nerves between them, faint or like the reticulating veins slightly prominent on both sides; petiole stout, 2–4 lin. long; corymbs or clusters short, dense, many-flowered, subsessile, glabrous or puberulous; bracts ovate, caducous, ciliolate; calyx more or less pubescent, green or whitish, $1\frac{1}{2}$ lin. long; sepals ovate-lanceolate, ciliolate; corolla white, tinged with pink, fragrant; tube 7–9 lin. long, pubescent or almost glabrous without, hairy within; lobes ovate to oblong, acute, 2–3 lin. long; anthers $\frac{3}{4}$ lin. long; stigma short, cylindric, obtusely apiculate; berry ellipsoid, 1 lin. long or longer, purplish-black; seeds 1–2, semi-ellipsoid, 5–9 lin. long. *K. Schum. in Engl. & Prantl, Pflanzenfam.* iv. ii. 126; *Wood, Natal Pl.* 60, t. 74; *Stapf in Dyer, Fl. Trop. Afr.* iv. i. 95. *A. venenata, Schweinf. ex Lewin in Engl. Jahrb.* xvii. *Beibl.* 41, 46, *and Lewin, l.c.* 47-51; *L. Planchon, Prod. Apocyn.* 255; *Vogtherr in Köhler, Mediz. Pfl.* iii. (*text to t.* 64); *not of G. Don. A. venenata, var. spectabilis, Sim, For. Fl. Cape Col.* 270, *t.* cliv. *fig.* 2. *A. sp., Benth. et Hook. f. Gen. Pl.* ii. 696. *Carissa oblongifolia, Hochst. in Flora,* 1844, 827, *and ex Walp. Rep.* vi. 466. *Toxicophlæa spectabilis, Dyer ex Gard. Chron.* 1872, 363; *Flor. Mag. new ser. t.* 20; *Gard. Chron.* xv. (1894), 209, *fig.* 23; *Rev. Hort.* 1879, 270, *with plate;* 1888, 517 *with fig. T. Thunbergii, Sonder in Linnæa,* xxiii. 79; *Gartenflora,* 1878, *t.* 940; *Rev. Hort.* 1880, 370, *with plate; Ill. Hort.* xxxii. (1885) *t.* 553, *not of Harvey.*

Coast Region: Bathurst Div.; wooded sand-hills near the sea, *Bowker!* Bathurst Div.; mouth of the Great Fish River, *Burchell,* 3760! East London Div.; coast at East London, *Galpin,* 1850!

Eastern Region: Natal; near The Point, *Krauss,* 361! near Durban, *Wood,* 1017! *Wilms,* 2005! in woods along the coast, *Wood in MacOwan, Herb. Austr.-Afr.* 1501! and without precise locality, *Cooper,* 1247! 1262! 1263! *Gerrard,* 88!

Sim (l.c.) considers this as "the eastern coast form" of *A. venenata,* into "which it merges gradually . . . in accordance with surroundings."

O. Kuntze indicates this species from Hereroland (Jahrb. Berl. Bot. Gart. iv. 1886, 268). I suspect that this is due to a confusion with *A. venenata.* The plant contains a deadly poison.

IV. **RAUWOLFIA**, Linn.

Calyx small, eglandular within, more or less herbaceous; sepals 5, almost free and imbricate or united into a flat 5-toothed cup. *Corolla* salver-shaped; tube slightly widened below the mouth, very rarely just below the middle; mouth constricted, without appendages, usually villous; lobes 5, twisted and overlapping to the left. *Stamens* in the widened part of the tube; filaments short; anthers free from the stigma, ovate, usually rather obtuse, shortly and obtusely 2-lobed at the base; anther-cells polliniferous and dehiscing to the base. *Disc* annular or cup-shaped, entire or slightly lobed. *Carpels* 2, free or more or less coherent; style filiform or columnar; stigma capitate, shortly cylindric, minutely papillose and slightly viscous, with a basal deflexed rim or membrane and a usually very short slightly bilobed apiculus rising from a shallow depression; ovules 2 in each cell, collateral. *Mericarps* 2 (or often 1 by abortion), free or more or less united, drupaceous; pyrenes crustaceous, 1–2-seeded, more or less compressed. *Seeds* ovoid; endosperm fleshy; cotyledons flat; radicle straight or recurved.

Mostly glabrous trees or shrubs. Leaves opposite or verticillate, those of a whorl often very unequal; axillary stipules 0; axillary glands numerous, in a dense fringe or in clusters, frequently secreting resin. Inflorescences terminal or pseudo-axillary, peduncled, few- or many-flowered, often repeatedly 2-3-chotomous, compound, umbelliform or corymbose, rarely racemiform; flowers small.

DISTRIB. Over 50 species, in the tropics of both hemispheres, 2 of them also in South Africa.

Leaves narrow-lanceolate ($\frac{3}{4}$–$1\frac{1}{2}$ in. broad), long and
  gradually acuminate at both ends; petioles $\frac{1}{2}$–$1\frac{1}{2}$ in.
  long, slender ... ... (1) **caffra.**
Leaves oblanceolate ($1\frac{1}{4}$–$2\frac{1}{2}$ in. broad), acute or subacuminate, long attenuated at the base and more or
  less decurrent on the usually short and stout petiole.. (2) **natalensis.**

1. **R. caffra** (Sonder in Linnæa, xxiii. 77); a large tree, 50–60 ft. high, quite glabrous; young branches angular or almost terete, stout, blackish-brown when dry; leaves in whorls of 3–5, unequal, lanceolate, long acuminate, long cuneate at the base, 3–8 in. long, $\frac{3}{4}$–$1\frac{1}{2}$ in. broad, membranous, rather firm; secondary nerves 20–30 on each side, straight or curved, subhorizontal; veins obscure, or more or less distinct below; petioles up to $1\frac{1}{2}$ in. long; cymes dense, at the ends of the secondary or tertiary rays of large umbels; flowers shortly pedicelled or subsessile; peduncles 1–$2\frac{1}{2}$ in. long; primary rays $\frac{3}{4}$–$1\frac{1}{2}$ in. long, secondary rays $\frac{1}{2}$–1 in. long; calyx $\frac{1}{2}$–$\frac{2}{3}$ lin. long; lobes ovate, acute; corolla white; tube $1\frac{1}{2}$–2 lin. long, mouth very hairy; lobes ovate, very short, subacute; carpels connate at the base in the flower, more or less fused in the fruit or usually only one developing; style glabrous, $\frac{1}{2}$ lin. long; stigma truncate with a reflexed membrane; fruit a simple drupe (by abortion), obovoid or almost globose, 3 lin. long, or an obcordate twin drupe. *Stapf in Dyer, Fl. Trop. Afr.* iv. i. 110.

KALAHARI REGION: Transvaal; Magalies Berg, *Zeyher*, 1183! *Burke*, 113! Crocodile River, *Burke!* bank of a river at Avoca, near Barberton, *Galpin*, 1061!

Also (in a slightly different form) in the Katanga District of the Congo Free State.

2. **R. natalensis** (Sonder in Linnæa, xxiii. 78); a tree, 30–40 ft. high, quite glabrous; young branches terete, stout, blackish or brown

when dry; leaves in whorls 3–4, oblanceolate, acute or subacuminate, long attenuated towards the base and more or less decurrent on the petiole, 5–12 in. long, $1\frac{1}{4}$–$2\frac{1}{2}$ in. broad, firmly membranous, pale dull green; secondary nerves 18–30 on each side, slightly curved, subhorizontal; veins quite obscure or very faint, loosely anastomosing; petiole 2–12 lin. long, stout; cymes very dense, at the ends of the secondary rays of large umbels; peduncle 2–$3\frac{1}{2}$ in. long, stout; primary rays 1–2 in. long; secondary rays 3–6 lin. long; pedicels in flower up to $\frac{1}{2}$ lin. long, in fruit up to 1 lin. long; calyx $\frac{1}{2}$ lin. long; lobes broad, ovate, subacute; corolla-tube about 2 lin. long, densely villous at the mouth; lobes small, rounded; carpels connate at the base or half-way up in flower; fruit a more or less obovoid or subglobose drupe, 4 lin. long (semimature). *K. Schum. in Engl. & Prantl, Pflanzenfam.* iv. ii. 154; *Stapf in Dyer, Fl. Trop. Afr.* iv. i. 111; *Sim, For. Fl. Cape Col.* 270, *t.* clvi.

EASTERN REGION : Pondoland; through the Egossa Forest to Port S. John, ex *Sim.* Natal; by the Umzinyati River, *Wood,* 648! and without precise locality, *Bowker!* *Gerrard,* 1585!

Locally known as the Quinine-tree.   Also in Tropical Africa.

V. **GONIOMA**, E. Meyer.

Calyx small, eglandular within, more or less herbaceous; sepals 5, free, imbricate, obtuse. *Corolla* salver-shaped; tube cylindric, scarcely widened above the middle; mouth constricted, without appendages, glabrous; lobes 5, overlapping to the left, auricled at the base of the inner half. *Stamens* inserted at the middle of the corolla-tube; filaments filiform, short; anthers free from the stigma, ovate-lanceolate, acute, shortly and obtusely 2-lobed at the base; anther-cells polliniferous and dehiscing to the base. *Disc* 0. *Carpels* 2, free, glabrous; style filiform; stigma cylindric-oblong, exannular, apiculus subulate; ovules numerous in each cell, pluriseriate. *Mericarps* 2, follicular, coriaceous, oblong or linear-oblong, apiculate, subterete, straight or slightly curved. *Seeds* pluriseriate, imbricate, quite flat, broad-cuneate or rectangular, with a broad wing at each end; nucleus elliptic, oblique or subhorizontal in the middle of the seed; endosperm fleshy; cotyledons flat, ovate, slightly longer than the cylindric radicle.

A glabrous shrub with opposite or 3–4-nate, coriaceous, shining leaves; axillary stipules 0; axillary glands few, minute. Inflorescences terminal, densely corymbose; flowers small.

DISTRIB. Species 1, endemic.

1. **G. Kamassi** (E. Meyer, Comm. 189); leaves oblong to lanceolate, acute to subobtuse, attenuated into the short petiole, $1\frac{1}{2}$–$3\frac{1}{2}$ in. long, $\frac{1}{2}$–1 in. wide, lateral nerves usually quite obscure, 12–15 on each side, oblique; petiole 2–4 lin. long; corymbs subsessile, up to $1\frac{1}{2}$ in. in diam.; pedicels slender, $1\frac{1}{2}$–$2\frac{1}{2}$ lin. long; calyx 1 lin. long; corolla yellow, fragrant; tube 3–$3\frac{1}{2}$ lin. long; lobes orbicular, over 1 lin. in diam.; anthers $\frac{3}{4}$ lin. long; follicles 1–$2\frac{1}{2}$ lin. long, 4–5

in. wide ; seeds (including wings) 7–9 lin. long, 3–3½ lin. wide ; embryo 3 lin. long. *DC. Prodr.* viii. 388 ; *L. Planchon, Prod. Apocyn.* 185, 259 ; *K. Schum. in Engl. & Prantl, Pflanzenfam.* iv. ii. 137 ; *F. Müller, Select. Extratrop. Pl. ed.* ix. 241 ; *Sim, For. Fl. Cape Col.* 271, *t.* cx. *G. Kamassi, var. brachycarpum, E. Meyer, l.c.* 189. *Gonioma, Harv. Gen. South Afr. Pl. ed.* 2, 246. *Tabernæmontana Camassi, Eckl. in South Afr. Quart. Journ.* i. (1830), 371.

COAST REGION: George Div. ; near George, *Burchell,* 6044! 6071! Knysna Div. ; Bosch River, *Drège!* Uitenhage Div. ; Van.Stadens Berg, in forest, 1000 ft., *MacOwan,* 1055! *Zeyher,* 730! near Sand Fontein and Matjes Fontein, *Drège!* Galgebosch, *Drège!* in the forests of Krakakamma and Adow, *Zeyher,* 730! between Van Stadens River and Galgebosch, *Burchell,* 4676! Bathurst Div. ; between Blaauw Krantz and Kowie Port, *Burchell,* 3659! near Barville Park, *Burchell,* 4138! East London Div. ; near East London, *Wood in Herb. Galpin,* 3352! *Leighton in MacOwan & Bolus, Herb. Norm. Austr.-Afr.,* 236!

This yields the hard, yellow Kamassi wood.

VI. **LOCHNERA**, Reichb.

(VINCA, Benth. et Hook. f. Gen. Pl. ii. 703 partly.)

Calyx middle-sized, herbaceous, eglandular within ; sepals 5, subulate, scarcely imbricate. *Corolla* salver-shaped ; tube slender, cylindric, slightly widened below the constricted callous velvety mouth ; lobes overlapping to the left. *Stamens* in the widened part of the corolla-tube ; filaments very short ; anthers free from the stigma ; ovate-lanceolate, acute, shortly and obtusely 2-lobed at the base ; anther-cells polliniferous and dehiscing to the base. *Disc* replaced by two long linear glands alternating with the carpels. *Carpels* 2, free ; style filiform ; stigma slightly below the level of the anthers, depressed-capitate, viscous with a long hyaline reflexed frill at the base and a minute obtuse 2-lobed apiculus surrounded by a very short erect membranous rim ; ovules numerous, 2-seriate. *Mericarps* follicular, cylindric, slightly spreading. *Seeds* numerous, small ; testa rugose ; hilum lateral ; endosperm fleshy ; cotyledons oblong, flat, shorter than the thick radicle.

Annual or perennial herbs or small undershrubs ; leaves opposite ; axillary stipules 0 ; axillary glands numerous in a fringe, the outer long, filiform, the inner minute. Flowers axillary, solitary or paired, white or pink.

DISTRIB. Species 3, indigenous in tropical America, India and Madagascar. One species widely diffused as a weed throughout the tropical and subtropical regions of both hemispheres.

1. **L. rosea** (Reichb. Conspectus, 134) ; a small undershrub, up to 3 ft. high ; leaves obovate or oblong (rarely subacute) and apiculate, acute at the base, 1½–3 in. long, ¾–1⅔ in. broad, herbaceous, finely pubescent to tomentose ; petioles 1–4 lin. long ; pedicels up to 1 lin. long, pubescent ; corolla white or pink ; tube puberulous, 1 in. long ; lobes broad, obliquely obovate, apiculate, ¾ in. long ; follicles up to 1¼ in. long, spreadingly pubescent, striate. *Schnizl.*

Iconogr. t. 132, *figs.* 2–16 ; *K. Schum. in Engl. & Prantl, Pflan-zenfam.* iv. ii. 145, *fig.* 57, *A–D.,* and in *Engl. Pfl. Ost-Afr. C.* 316 ; *L. Planchon, Prod. Apocyn.* 231, 284 ; *Durand & Schinz, Études Fl. Congo,* i. 190 ; *De Wild. & Durand, Contrib. Fl. Congo, fasc.* ii. 39, and *Reliq. Dewevr.* 151 ; *Stapf in Dyer, Fl. Trop. Afr.* iv. i. 118. *Vinca rosea, Linn. Syst. Nat. ed.* x. 944 ; *Gærtn. De Fruct.* ii. 172, *t.* 117 ; *DC. Prodr.* viii. 382 ; *Muell. Arg. in Mart. Fl. Bras.* vi. i. 69, *t.* 25 ; *Hook. Niger Flora,* 450 ; *Grisebach, Fl. Brit. West Ind.* 410 ; *Cardoso jun. in Bolet. Soc. Brot.* xiii. 144 ; *Hiern in Cat. Afr. Pl. Welw.* i. 667. *Catharanthus roseus, G. Don, Gen. Syst.* iv. 95.—*Vinca fol. oblongo-ovatis, Mill. Ic. t.* 186.

EASTERN REGION : Natal ; near Durban, *Wilms,* 2131 ! and without precise locality, *Cooper,* 2748 ! 2749 ! *Grant !* Delagoa Bay, *Wilms,* 926 !

Probably a native of the West Indies, now widely naturalized in the tropics of both hemispheres, chiefly near the coast.

VII. **CONOPHARYNGIA**, G. Don.

(TABERNÆMONTANA, Benth. et Hook. f. Gen. Pl. ii. 706 partly.)

Calyx small (at least comparatively), subcoriaceous ; sepals united at the base only, imbricate, obtuse, each with several minute glands inside the base. *Corolla* salver-shaped, small to large, often very fleshy ; tube cylindric, spindle- or barrel-shaped, widest at or below the middle, sometimes twisted, naked at the mouth, usually more or less tomentose inside ; lobes overlapping to the left, inflexed and descending into the corolla-tube in bud. *Stamens* in the widened part of the corolla-tube ; anthers conniving into a cone, subsessile, included, rarely shortly exserted in species with a short corolla-tube, lanceolate, acute, sagittate ; tails solid, barren ; filaments reduced to a callous swelling ; filamental ridge usually distinct. *Disc* 0. *Carpels* 2, free ; style filiform to columnar, short (at least com-paratively) ; stigma cylindric, grooved, with an entire or lobed projecting rim or short frill at the base and a minute 2-lobed apiculus, more rarely (§ *Leptopharyngia*) elliptic or globose, delicately papillose, not grooved, with a usually toothed rim at the base and a conspicuous 2-fid papillose apiculus as long as or rather longer than the rest of the stigma. *Mericarps* baccate, usually more or less globose or ovoid, smooth, rarely keeled or warty and tardily dehiscent when drying up. *Seeds* numerous, embedded in a usually fleshy pulp, more or less ellipsoid, deeply grooved ventrally ; testa crustaceous ; endosperm fleshy, ruminate ; cotyledons ovate, longer or shorter than the radicle.

Trees, often tall, or shrubs. Leaves opposite, more or less coriaceous, some-times very large ; axillary stipules distinct, very obtuse, united into a very short tubular sheath, usually with very numerous resiniferous glands within. In-florescences terminal or pseudo-axillary, corymbose, rarely panicled or reduced to few-flowered cymes. Flowers large and showy to middle-sized, rarely small, usually white and fragrant.

DISTRIB. Species about 25, mostly in tropical Africa; a few in the Mascarene Islands.

§ 1. SARCOPHARYNGIA. Corolla-tube $\frac{1}{3}$–$3\frac{1}{2}$ in. long, fleshy; stigma more or less cylindric, grooved, with a projecting entire or lobed rim at the base, and a minute apiculus.

Corolla-tube $\frac{1}{3}$ in. long ... ... (1) **ventricosa.**

§ 2. LEPTOPHARYNGIA. Corolla-tube $\frac{1}{4}$–$\frac{1}{2}$ in. long, comparatively thin. Stigma ellipsoid or globose, minutely papillose, not grooved, with a usually toothed rim at the base and a conspicuous papillose 2-fid apiculus as long as or longer than the rest of the stigma.

Only species in South Africa... ... ... (2) **elegans.**

1. C. ventricosa (Stapf in Dyer, Fl. Trop. Afr. iv. i. 149); a tree, 20–25 ft. high, perfectly glabrous; branches terete, fistular, olive-green when dry; leaves oblong, obscurely acuminate, acute at the base, 4–7 in. long, $1\frac{1}{2}$–3 in. broad, subcoriaceous; secondary nerves 13–17 on each side, suboblique; petiole 3–7 lin. long; inflorescence corymbose, dense; peduncle rather stout, about 2 in. long; bracts ovate, acute, 2–3 lin. long; pedicels 1–2 lin. long; calyx 2–$2\frac{1}{2}$ lin. long; sepals rotundate, ovate-obtuse; corolla white; tube ellipsoid, constricted at both ends, 4–5 lin. long, velvety between the filamental ridges, otherwise glabrous; lobes ovate-oblong, slightly longer than the tube, wavy; stamens inserted just below the middle of the corolla-tube; anthers 3 lin. long, reaching slightly beyond the mouth; style $1\frac{1}{2}$ lin. long; fruit unknown. *Tabernæmontana ventricosa, Hochst. ex DC. Prodr.* viii. 366; *Krauss in Flora,* 1844, 828; *Sim, For. Fl. Cape Col.* 271, *t.* clv. *fig.* iii.

EASTERN REGION: Pondoland; in swamps, ex *Sim.* Natal; woods near the Umgeni River, *Krauss,* 146! Inanda, *Wood,* 787! and without precise locality, *Gerrard,* 1423!

2. C. elegans (Stapf in Dyer, Fl. Trop. Afr. iv. i. 149); a shrub, 8–10 ft. high; branches terete, rather stout, drying more or less black; leaves narrowly oblong to lanceolate, subacuminate or obtuse, acute at the base, 4–6 in. long, $1\frac{1}{3}$–2 in. broad, subcoriaceous, opaque when dry, much paler beneath; secondary nerves 16–22 on each side, almost horizontal and straight; inflorescence corymbose or paniculate, terminal, more or less overtopped by young shoots, many-flowered, loose; peduncle 1–2 in. long; bracts small, scarious, caducous; pedicels slender, up to 6 lin. long; calyx 1–$1\frac{1}{2}$ lin. long; sepals rotundate-ovate, obtuse, not ciliolate, with 1 bifid or 2–3 entire basal glands within; corolla yellowish-white; tube short, cylindric, 3–$3\frac{1}{2}$ lin. long, constricted and thin and glabrous within below the stamens, more fleshy and hairy within in the upper part; lobes oblong, $4\frac{1}{2}$ lin. long; stamens inserted 1 lin. above the base of the corolla-tube; anthers subsagittate, with short solid basal points, 1 lin. long; style $\frac{1}{2}$ in. long; stigma subulate, densely papillose, 2-fid, from a globose or ellipsoid viscid base, supported by a ring of 10 small spreading lobes; berries obliquely ovoid, with apiculate recurved tips and 1 dorsal and 2 lateral ridges, up to 2 in.

long, fleshy, covered with numerous suberous warts, at length dehiscing ventrally. *Tabernæmontana elegans, Stapf in Kew Bulletin,* 1894, 24 ; *K. Schum. in Engl. Pfl. Ost-Afr. C.* 316.

EASTERN REGION : Delagoa Bay, *Monteiro*, 55 ! and specimen cultivated in Durban Botanic Garden, *Wood*, 8598 !

Also in East Africa, as far as Mombasa.

VIII. **VOACANGA**, Thouars.

(TABERNÆMONTANA, Benth et Hook. f. Gen. Pl. ii. 706 partly. PIPTOLÆNA, Harv. in Hook. Journ. Bot. iv. (1842) 135, and in Hook. Lond. Journ. i. 25.)

Calyx tubular or subcampanulate, 5-lobed, early circumscissile at the base and deciduous, or more persistent, tardily circumscissile or splitting longitudinally, with a ring or zone of (often numerous) small glands at or above the base ; lobes obtuse, imbricate. *Corolla* salver-shaped ; tube staminiferous above the middle, constricted below the stamens and at the mouth, with callous thickenings round the often very narrow orifice, more or less twisted from left to right, very rarely straight, with prominent filamental ridges ; lobes broad, obtuse, as long as or longer than the tube and more or less spreading, rarely much shorter and tightly reflexed, tips not inflexed in bud. *Anthers* sessile, adnate by a broad base to the corolla-tube, deeply sagittate ; tips subulate, usually more or less exserted, or reaching close to the mouth ; basal tails horny, slender, solid. *Disc* usually annular, very fleshy, surrounding the base of the ovary, or cupular and concealing the ovary, very rarely reduced to an inconspicuous ring, more or less confluent with the base of the ovary. *Carpels* 2, semi-ovoid or semi-globose, free, very rarely connate to the middle ; style columnar, thickened upwards; stigma subcapitate, 5-grooved with a fleshy wavy ring or frill at the base ; ovules multiseriate, very numerous on bifid placentas. *Mericarps* baccate, globose or pear-shaped, more or less oblique, sometimes with short recurved beaks, sometimes very tardily dehiscing along the ventral suture ; pericarp thick and fleshy or thinner and coriaceous when dry. *Seeds* numerous, embedded in a pulpy mass, oblong-ellipsoid, deeply grooved ventrally; testa crustaceous, more or less grooved longitudinally, often coarsely honeycombed by transverse partitions across the grooves, more or less intruding into the endosperm ; endosperm fleshy ; cotyledons foliaceous, thin, ventrally concave, shorter than the radicle.

Shrubs or trees, dichotomously branched ; leaf-buds sometimes coated with resin. Leaves opposite, herbaceous to coriaceous ; axillary stipules distinct, like those of *Conopharyngia* or quite obscure or 0 ; leaf-bases united into a rim or very short sheath ; axillary glands small, numerous. Flowers large to rather small, in terminal, frequently paired, peduncled, racemiform, umbelliform or corymbose inflorescences, usually from the young branch-forks. Corollas white, yellow or greenish or the limb violet-brown.

DISTRIB. Species 12, in Tropical Africa, Natal, and the Mascarene Islands, and about 4 in the Malay Archipelago.

1. **V. Dregei** (E. Meyer, Comm. 189) ; a glabrous tree, 30–40 ft.

high; branches stout, almost spongy, pallid, or the youngest blackish when dry; leaves crowded towards the tips of the branches, oblong, obtuse, subcuneate towards the base, 4–6 in. long, 1–2 in. broad, subcoriaceous, dull when dry; secondary nerves subhorizontal or rather oblique, almost straight, slender; petiole 4–6 lin. long; inflorescences usually geminate from the branch-forks, shortly racemiform or umbelliform, few-flowered; peduncle stout, 2–3 in. long; rhachis stout, gradually lengthening up to 1½ in. as the lower flowers fall; bracts ovate, concave, up to 5 lin. long, caducous; pedicels stout, finally up to 6 lin. long; calyx wide-tubular, about 4½–6 lin. long, early circumscissile at the base, with very numerous glands within; lobes rotundate, 1½–2 lin. long; corolla waxy, yellowish white; tube subcylindric, slightly exserted from the calyx, more or less constricted above the middle, twisted above the constriction, glabrous; limb broadly ovoid in bud, 6 lin. long; lobes somewhat asymmetric, broadly obcordate, narrow at the base, 6–9 lin. long, 9–12 lin. broad, sinus shallow; anthers inserted close to the mouth of the corolla, exserted for half their length, 3 lin. long; disc cupular, entire or almost so, shorter than the ovary, persistent; style up to 7 lin. long; stigma shortly cylindric, grooved, with a frill at the base; berries of the size of a fist (*Sutherland*), with a thick rind; seeds numerous. *K. Schum. in Engl. & Prantl, Pflanzenfam.* iv. ii. 149; *Stapf in Dyer, Fl. Trop. Afr.* iv. i. 154. *Annularia natalensis, Hochst. in Flora,* 1841, 671. *Piptolæna Dregei, DC. Prodr.* viii. 358; *Harv. Gen. S. Afr. Pl. ed.* 2, 246. *Cyclostigma natalense, Hochst. in Flora,* 1844, 828.

EASTERN REGION: Natal; between Umtentu River and Umkomanzi River, *Drège!* forest near the Umlaas River, *Krauss,* 27! near Pinetown, *Rehmann,* 8027! *Wood,* 3841! and without precise locality, *Gueinzius! Gerrard,* 408! *Sutherland!*

V. Dregei and *V. obtusa,* K. Schum., differ so little that the latter will probably have to be reduced to *V. Dregei.*

IX. **WRIGHTIA**, R. Br.

Calyx ½–3 lin. long, 5-lobed or more often of 5 free or almost free, obtuse, imbricate sepals, with 5–10 intracalycular glands. *Corolla* salver-shaped, small to over 1 in. long; tube cylindric, short, very rarely long and slender, with variously divided or entire, connate or free appendages in the throat (very rarely without); lobes overlapping to the left, obtuse or subobtuse, not tailed. *Stamens* inserted in the mouth of the corolla; anthers conniving in a cone, exserted, lanceolate, acute or acuminate, sagittate; wings rather thin with thick incurved margins; tails slender, long, incurved; foot of connective rather flat, with a decurrent dense line of hairs in the centre; filaments short, stout, passing into long slender ridges on the corolla-tube. *Disc 0. Carpels* free; style filiform; stigma capitate with a small or obscure frill at the base and a minute cleft tip, agglutinated to the hair-tuft of the anthers. *Ovules* pluriseriate, numerous. *Follicles* cylindric or spindle-shaped, slender, subparallel, coriaceous. *Seeds* cylindric, glabrous with a basal tuft of hairs; endosperm fleshy, scanty; cotyledons convolute; radicle short.

Shrubs or small trees with slender branches, glabrous or pubescent. Leaves opposite, loosely nerved; axillary stipules 0. Flowers small to rather large and then very showy, in terminal or pseudo-axillary cymes.

DISTRIB. Species about 20, all except the one below, in the warmer parts of Asia and Northern Australia.

1. **W**. **natalensis** (Stapf in Kew Bulletin, 1907, 51); a shrub; branches rusty-pubescent when quite young, soon glabrous and ultimately covered with a fine grey bark, scantily lenticellate; leaves lanceolate, subacute, cuneate at the base, up to 3 in. long to 9 lin. broad (not quite mature), rusty-pubescent on the back towards the base, otherwise quite glabrous; secondary nerves about 9 on each side; petiole slender, 2–4 lin. long; panicle short, subcorymbose, not over 1 in. across, rusty-pubescent; bracts linear-oblong to linear, $1\frac{1}{2}$–2 lin. long; pedicels up to 4 lin. long; calyx pubescent at the base, otherwise almost glabrous, $2\frac{1}{2}$–3 lin. long; sepals oblong, obtuse, margins membranous, each with 1 fleshy scale within; corolla yellow, densely and minutely papillose without; tube 2 lin. long; lobes oblong, subobtuse, 4 lin. long with 5 epipetalous emarginate scales which are united for half their length into a short corona and bear each 2 filiform appendages shorter than the scales; anthers 2 lin. long, finely acuminate.

EASTERN REGION: Natal; Umzinyati Falls, 800 ft., *Haygarth in Herb. Wood*, 7861!

I have placed this plant in *Wrightia* rather than in *Pleioceras* on account of the general appearance, which is more that of a *Wrightia*, but it is quite possible that the fruit, when known, will prove it to be a *Pleioceras*.

X. **STROPHANTHUS**, DC.

Calyx middle-sized to large, herbaceous, rarely scarious, with few to many glands at the base within; sepals 5, imbricate, sometimes foliaceous. *Corolla* funnel-shaped or campanulate, with a short or long cylindric basal tube; mouth with paired appendages alternating with the lobes; lobes 5, acuminate or produced into very long filiform tails, rarely obtuse. *Stamens* inserted at the upper end of the cylindric portion of the corolla-tube; anthers conniving in a cone and projecting into the widened part of the tube, lanceolate, acuminate or sometimes produced into a long bristle, sagittate; wings long, firm, obtusely edged; tails short; foot of connective with a central tuft of closely packed hairs in the upper part and a more or less hairy longitudinal crest below it; filaments distinct but short, filiform, passing into a prominent, more or less hairy ridge decurrent on the corolla-tube. *Disc* 0. *Carpels* 2, free; style filiform; stigma capitate, 5-grooved, with a membranous reflexed frill at the base, and a minutely bifid apiculus; ovules numerous, pluriseriate. *Mericarps* follicular, oblong or spindle-shaped, divaricate. *Seeds* spindle-shaped, slightly compressed, with an apical plumose awn and a deciduous basal coma; endosperm scanty, fleshy; cotyledons oblong, plano-convex; radicle short.

Shrubs, often scandent, glabrous or more or less hairy, with persistent or deciduous foliage. Leaves opposite, rarely ternate; axillary stipules 0; axillary glands subulate or conical, 2–6, rarely more, at the base of each petiole. Inflorescences terminal, often on the ends of short branches, corymbose, many- or few-flowered or reduced to solitary flowers; flowers mostly showy.

DISTRIB. Species about 45, in Tropical and South Africa and Tropical Asia.

Leaves opposite, not coriaceous :
    Sepals subfoliaceous, lanceolate, 7–8 lin. long;
      corolla-tube 1 in. long; lobes inclusive of the
      tails 6–7 in. long ... ... (1) **grandiflorus**.
    Sepals subulate-acuminate, 2–4 lin. long; corolla-
      tube about 6 lin. long; lobes inclusive of the
      tails from less than 1 in. to 1½ in. long ... ... (2) **Gerrardii**.
Leaves in whorls of 3–4, coriaceous ... ... (3) **speciosus**.

1. S. grandiflorus (Stapf in Dyer, Fl. Trop. Afr. iv. i. 182, 608); a dense shrub, 5–6 ft. high, quite glabrous; branches long, slender, reddish-brown with numerous white lenticels; leaves ovate to elliptic-oblong, shortly acuminate, acute or subobtuse at the base, about 2 in. (in cultivated specimens 3 in.) long, ¾–1 in. broad, membranous; secondary nerves 6–9 on each side, slender; reticulation rather delicate; petiole 3–4 lin. long; cymes terminal on leafy (in the wild specimens short) branches, usually reduced to a single flower; bracts linear-lanceolate, about 4 lin. long; pedicels up to 4 lin. long; calyx 7–8 lin. long; sepals oblong to lanceolate-oblong, subacute, 2 lin. broad, erect; corolla wide, purplish without, milk-white or creamy within; infra-staminal part of the tube 2 lin. long, supra-staminal part 10 lin. long; lobes ovate, produced into filiform tails, about 6 in. long; throat-scales subulate from a triangular base, 4 lin. long; anthers included, glabrous, 3 lin. long, produced into a fine bristle. *Gilg in Engl. Jahrb.* xxxii. 161, *and in Engl. Monogr. Afr. Pflanzenfam.* vii. (1903), 28 (*the Delagoa Bay plant*). *S. petersianus, var. grandiflorus, N. E. Br. in Kew Bulletin,* 1892, 126 (*the Delagoa Bay plant*); *Hook. f. in Bot. Mag. t.* 7390.

EASTERN REGION : Delagoa Bay, *Monteiro!* and cultivated specimen, *Mrs. Monteiro!*

Concerning the identity of this plant and the East African *S. verrucosus,* Stapf, my note in Flora of Tropical Africa, iv. i. 607 may be consulted. No more material of the Delagoa Bay plant has been received since the publication of this note.

2. S. Gerrardii (Stapf in Kew Bulletin, 1907, 52); a glabrous shrub with slender branches, reddish and dotted all over with whitish lenticels; leaves narrowly ovate to ovate-lanceolate, acute or obscurely acuminate, rounded or subacute at the base, 1½–2½ in. long, ¾–1 in. broad, papery, rather thin; secondary nerves about 5 on each side, like the veins very faint or quite obscure; petiole 2 lin. long; cymes

reduced to a solitary flower, terminal on short lateral branches, sometimes overtopped by leafy branchlets produced with the leaves;
peduncle (if any) slender, 2–6 lin. long; bracts early deciduous;
pedicels very slender, up to 6 lin. long; calyx 4–5 lin. long; sepals
subulate from a broader base; corolla glabrous without, puberulous
within; infra-staminal part of the tube $1\frac{1}{2}$ lin. long, supra-staminal
part almost tubular-campanulate, $4\frac{1}{2}$ lin. long, up to 4 lin. wide in
the dried state; lobes attenuate from an ovate or lanceolate base
into linear tails, total length $1\frac{1}{2}$ in.; throat-scales linear-subulate,
1–$1\frac{1}{2}$ lin. long; anthers terminating in a fine point.

EASTERN REGION: Natal or Zululand; without precise locality, *Gerrard,*
1795!

Very similar to *S. petersianus,* Klotzsch, but differing in the narrower leaves,
more subulate sepals and the shape of the supra-staminal part of the corolla-tube,
which is more that of *S. Schuchardti,* Pax.

3. S. speciosus (Reber in Der Fortschritt, iii. (1887), 299);
a rambling glabrous shrub; branches trailing on other shrubs,
sometimes running high up, olive-green, densely lenticellate;
leaves in whorls of 3–4, rarely the uppermost opposite, oblong-
lanceolate to lanceolate, acute, rarely acutely acuminate, long
attenuate at the base, $1\frac{1}{2}$–$3\frac{1}{2}$ in. long, $\frac{1}{2}$–1 in. broad, coriaceous,
glossy above; secondary nerves about 10–16 on each side, very faint
or like the veins obscure; petiole 2–3 lin. long; cymes terminal or
pseudo-axillary, shortly peduncled, few- to 12-flowered, corymbiform; bracts lanceolate to ovate, acuminate, 3–$\frac{1}{2}$ lin. long, usually
deciduous; pedicels up to $\frac{1}{2}$ in. long; calyx 3–5 lin. long; sepals
lanceolate to linear, narrow, acuminate, sometimes recurved; corolla
cream-coloured to yellow or orange, spotted with red; infra-staminal
part of the tube 2 lin. long, supra-staminal part funnel-shaped-campanulate, 4 lin. long; lobes attenuate from a somewhat broader base
into linear spreading tails from less than 1–$1\frac{1}{2}$ in. long; throat-
scales linear-subulate, up to 1 lin. long; anthers covered with silky
hairs above the middle; follicles very slender, lanceolate, long
acuminate, about 6 in. long, $\frac{1}{2}$ in. in diam., terete, tips somewhat
thickened; seeds oblong-lanceolate, glabrous, 6 lin. long, 2 lin.
wide, pale reddish-brown, with a sessile plume, hairs $1\frac{1}{2}$ in. long on
a very fine awn up to $\frac{1}{2}$ in. long; cotyledons oblong, flat, thin.
Autran in Bull. Soc. Bot. Belg. xxix. ii. 45; *Pax in Engl. Jahrb.* xv.
376; *Franch. in Nouv. Arch. Mus. Paris,* 3e *sér.* v. (1893), 287;
K. Schum. in Engl. & Prantl, Pflanzenfam. iv. ii. 182; *Gilg in
Engl. Jahrb.* xxxii. 157, *and in Engl. Monogr. Afr. Pflanzenfam.*
vii. 34. *S. capensis, A.DC. Prodr.* viii. 419; *Hook. Bot. Mag. t.*
5713; *Sim, For. Fl. Cape Col.* 272. *Christya speciosa, Ward &
Harv. in Hook. Journ. Bot.* iv. (1842), 134, *t.* 21; *A.DC. l.c.* 416.

COAST REGION: Bedford Div.; in forest on the Kaga Berg, 5000 ft., *Mrs.
Hutton!* Stockenstrom Div.; in woods on the Kat Berg, 4000 ft., *Scully in
MacOwan & Bolus, Herb. Norm. Austr.-Afr.* 761! *MacOwan,* 2020! *Mrs. Barber,*
28! Stutterheim Div.; Kabousie Forest, *Murray in Herb. MacOwan,* 2023!   .
EASTERN REGION: Transkei; Kentani, 1000 ft., *Miss Pegler,* 915! Natal;

Blinkwater, near York, *Wood,* 4305! Zululand, Qudeni Forest, 6000 ft., *Wood,* 7895! Eshowe, *Hon. Mrs. Evelyn Cecil,* 275! and without precise locality, *Mrs. K. Saunders!*

Only one other species (*S. gratus,* Baill.) having glabrous seeds is known in the genus, whilst *S. speciosus* is unique in having a sessile plume.

XI. **ONCINOTIS**, Benth.

Calyx small, eglandular within, rarely with 5 minute glands alternating with the sepals; sepals imbricate, ovate, acute or obtuse. *Corolla* salver-shaped; tube short, widest at the middle, densely tomentose within except at the very base, with 5 ligulate scales in the mouth alternating with the lobes and projecting obliquely into the mouth; lobes overlapping to the right, spreading or reflexed. *Stamens* inserted somewhat above the corolla-base; filaments very short, stout, arching over the ovary, densely hairy on the inner side; anthers conniving in a cone, included, sublinear, sagittate; appendages as long as the polliniferous part; tails short, very obtuse, recurved; foot of the connective with a faint central ridge in the upper part and a cushion of short papillæ at the base. *Disc* cupular, 5-lobed or 5-partite. *Carpels* 2, free, shortly exserted and free from the disc; style very short, passing into the short spindle-shaped stigma; apiculus 2-lobed. *Mericarps* follicular, spindle-shaped, divaricate, seeds lanceolate with an apical cone; embryo unknown.

Glabrous or hairy scandent shrubs; leaves opposite; secondary nerves usually distant; axillary stipules and glands 0. Panicles axillary or axillary and terminal, consisting of opposite or subopposite, few to many-flowered contracted cymes; flowers inconspicuous.

DISTRIB. Species about 12, in Tropical Africa, Natal and Madagascar.

1. **O. inandensis** (J. M. Wood & Evans in Journ. Bot. 1899, 254); a shrub climbing over trees; branches terete, the youngest parts firmly and densely fulvous-tomentose, soon glabrescent, at length dull brown or reddish-brown with whitish lenticels; leaves obovate-oblong to oblanceolate, abruptly acuminate, acute at the base, 3–3½ in. long, 1–1½ in. broad, firmly membranous, rusty-tomentellous in the very young bud, soon glabrous; secondary nerves 3–4 on each side, very oblique; petiole 2–2½ lin. long; flowers in numerous axillary minutely fulvous-tomentellous racemes or panicles, about 1 in. long and borne on usually very short peduncles; bracts ovate-lanceolate, small, early deciduous; pedicels very short; calyx 1 lin. long, finely fulvous-tomentose; sepals ovate, subacute; corolla greenish, minutely tomentellous without, very slender, almost subulate in bud; tube almost 1½ lin. long; lobes linear, 1½ lin. long; follicles cylindric, divaricate, 6–7 in. long, 3 lin. in diam., delicately tomentellous, at length glabrescent; seeds oblong-linear, 6–7 lin. long; coma 1½ in. long. *O. natalensis, Stapf in Kew Bulletin,* 1907, 52.

EASTERN REGION: Natal; Inanda, in woods, *Wood,* 1009! 6159!

Very similar to the West African *O. gracilis,* K. Schum. & Pax, but differing in the delicate, adpressed tomentum, much more acute leaf-bases, fewer and more oblique side-nerves, shorter petioles and shorter racemes.

XII. **ADENIUM**, Roem. & Schult.

Calyx 3–4 lin. long, herbaceous, eglandular within; sepals 5, lanceolate or subulate. *Corolla* funnel-shaped or campanulate from a short cylindric base, with paired, small or obscure scales in the mouth; scales more or less confluent at the base and forming obtriangular pockets, alternating with the lobes; lobes 5, broad, twisted, overlapping to the right. *Stamens* inserted at the base of the widened portion of the corolla-tube; filaments very short; anthers conniving in a cone, projecting into the widened part of the corolla-tube, lanceolate, with long filiform terminal appendages, sagittate; basal appendages much longer than the polliniferous part, distinctly tailed; foot of the connective channelled and glabrous in the upper part, with a projection in the centre and a brush-like cushion below, decurrent on the filament and passing into the tomentose filamental ridges. *Disc* 0. *Carpels* 2, free; style filiform, short; stigma campanulate, capitate, with a basal rim and a minute bifid apiculus, agglutinated to the foot of the connective; ovules very numerous, pluriseriate. *Mericarps* follicular, divaricate or reflexed. *Seeds* linear-oblong, covered with reversed hairs, with a deciduous coma at either end; endosperm very thin; cotyledons short, convolute; radicle much longer than the cotyledons.

Succulent shrubs, often with swollen stems and fleshy branches. Leaves in spirals and terminal fascicles, rather fleshy; axillary stipules 0; axillary glands subulate, conspicuous, several in each leaf axil. Cymes few-flowered, terminal, subsessile; flowers pink or purple, showy.

DISTRIB. Species about 12, extending through Tropical Africa to Socotra and Arabia, some of them very closely allied.

The upper wide part of the corolla quite glabrous and
  dark purple within; apical anther-tails short, in-
  cluded   ...   .   (1) **swazicum.**
The upper wide part of the corolla-tube hairy within;
  apical anther-tails reaching the mouth or exserted:
    Flowers from a bunch of coetaneous, linear-oblong,
      tomentose leaves   ...   .   (2) **oleifolium.**
    Flowers produced before the obovate-cuneate glab-
      rous leaves ...   .   (3) **multiflorum.**

1. A. swazicum (Stapf in Kew Bulletin, 1907, 53); a shrub; stem and branches glabrous with the exception of the youngest parts, whitish when dry; leaves obovate-cuneate, rounded at the apex, gradually narrowed into the basal cuneate portion, 3–4 in. long, $\frac{3}{4}$–$1\frac{1}{4}$ in. broad, somewhat coriaceous, softly whitish-tomentose on both sides when young, ultimately more or less glabrescent, glaucous; secondary nerves oblique, faint or quite obscure, $1\frac{1}{2}$–$2\frac{1}{2}$ lin. apart; petiole up to $1\frac{1}{2}$ lin. long; cymes terminal or pseudo-axillary, greyish-tomentose; bracts lanceolate, $2\frac{1}{2}$ lin. long; pedicels up to 5 lin. long; calyx 4 lin. long; sepals lanceolate, acuminate, pubescent to subtomentose; corolla pink, purple in the throat; infra-staminal part of the tube 4 lin. long, with 5 densely tomentose lines below the stamens, glabrous at the base, supra-staminal part about 1 lin. long, wide-tubular, glabrous within; throat-scales obscure; lobes rotundate-obovate, almost 1 in. long; anthers about 3 lin. long, hairy on the back, apical tails 3–5 lin. long, obtuse, densely villous.

EASTERN REGION: Swaziland; without precise locality, *Mrs. Rathbone in Herb. Bolus,* 6203! *and in Herb. Wood,* 3511!

Similar to *A. bœhmianum,* Schinz, from Ambo and Hereroland, but differing in the much narrower glabrescent leaves and oblique, very indistinct lateral nerves.

2. A. oleifolium (Stapf in Kew Bulletin, 1907, 53); a softly-pubescent shrub; leaves linear to oblong-linear, obtuse, more or less narrowed at the base, sessile or almost so, 3–4 in. long, $\frac{1}{3}$–$\frac{1}{2}$ in. broad, rather thick, glaucous, softly pubescent to subtomentose all over; secondary nerves quite obscure; cymes terminal, few-flowered, subsessile, whitish tomentose; bracts linear, about 3 lin. long; pedicels very short; calyx 3–3$\frac{1}{2}$ lin. long, whitish tomentose; sepals lanceolate, subacute; corolla pink, pubescent without; infra-staminal part of the tube 4–4$\frac{1}{2}$ lin. long with 5 hairy lines descending to the middle, glabrous below, supra-staminal part wide funnel-shaped-campanulate, 8 lin. long, mealy-papillose within; throat-scales small forming an obversely triangular pocket; lobes broad-ovate, cuspidate-acuminate, 6–7 lin. long; anthers 3 lin. long, apical tails 6 lin. long, exserted, loosely hairy.

KALAHARI REGION: Transvaal; without precise locality, *Todd,* 23! Bechuanaland; Bakwena Territory, near the Sirorume River, 3500 ft., *Holub!*

A. oleifolium differs from *A. somalense,* Balf. f., which it resembles to some extent, in the pubescent, broader and obtuse leaves and the mealy indumentum of the wider upper part of the corolla-tube.

3. A. multiflorum (Klotzsch in Peters, Reise Mossamb. Bot. 279, t. xliv.); a shrub 4–8 ft. high, leafless when in flower; branches succulent, stout, glabrous except at the very tips when young; leaves obovate to oblong, subacute or obtuse, acute or subacuminate at the base, 2–3 in. long, 9–15 lin. broad, subcoriaceous; midrib rather thin; nerves faint, though usually distinct, very oblique; petiole 1–2 lin. long; cymes much contracted, 5–15-flowered, terminal, sessile, preceded by more or less numerous deciduous triangular indurated bud-scales, hairy; bracts lanceolate, 2–3 lin. long; pedicels up to 2 lin. long; calyx 3–4 lin. long, hairy; sepals lanceolate; corolla white or pink, conspicuously cuspidate in bud; infra-staminal part of the tube 3–4 lin. long, with 5 hairy lines descending almost to the middle, glabrous below, supra-staminal part funnel-shaped, villous within, 9–12 lin. long; throat-scales small, confluent and forming an obversely triangular pocket; lobes broad ovate, shortly acuminate and conspicuously cuspidate; anthers 3 lin. long, densely villous; apical tails exserted; follicles reflexed, spindle-shaped, finely tomentose, 7 lin. long; seeds 6 lin. long, coma 15 lin. long. *Stapf in Dyer, Fl. Trop. Afr.* iv. i. 229.

KALAHARI REGION: Transvaal; Zoutpansberg, *Gray in Transvaal Herb.,* 2999! Woodbush Mountains, *Barber,* 20!

Eastern Region: Zululand; Lebombo Country, *Saunders!* Lower Pongolo River, in Amatonga Country, *Sanderson!* Delagoa Bay, *Forbes!*

XIII. **PACHYPODIUM**, Lindl.

Calyx small, herbaceous, eglandular within; sepals 5, ovate or lanceolate. *Corolla* salver-shaped, with a cylindric tube widened below the middle, constricted at the base, or funnel-shaped to campanulate with a short cylindric basal tube, naked at the mouth; lobes 5, twisted, overlapping to the right. *Stamens* inserted above the constriction; filaments very short, flattened; anthers conniving in a cone, included, linear-lanceolate, acuminate or acute, sagittate; appendages longer or shorter than the polliniferous part; tails sometimes very short; foot of the connective channelled and glabrous above, with a tongue-shaped projection and sometimes a brush-like cushion below, decurrent on the filament and passing into the hairy filamental ridges. *Disc* cupular, slightly 5-lobed or replaced by 2-5 distinct glands. *Carpels* 2, free; style filiform, short; stigma subcylindric with an annular rim or membrane at the base, very obscurely and obtusely 2-lobed; ovules numerous, pluriseriate. *Mericarps* follicular, 2, erect or spreading, elongate, spindle-shaped. *Seeds* ovate to oblong, with an apical coma; endosperm scanty; cotyledons ovate-cordate, flat; radicle short.

Succulent shrubs, sometimes with a much swollen trunk. Leaves in spirals, subsessile, stipulate; stipules transformed into rigid spines or the leaves suppressed with the exception of a terminal rosette, and the spinous stipules crowded more or less irregularly on the swollen branches. Cymes terminal, few- or many-flowered, sessile or peduncled; flowers pink, white, or yellow.

Distrib. Species about 12, the others in Tropical Africa and Madagascar.

Stem columnar, not branched; leaves velvety on
   both sides; corolla-tube wide-tubular above the
   stamens ...   .   (1) **namaquanum.**
Stem a low, massive bole, throwing out comparatively
   slender branches frequently dividing again :
      Leaves obovate-elliptic, twice as long as broad;
        corolla-tube somewhat widened above the inser-
        tion of the stamens, then gradually attenuate
        towards the mouth   ...   .   (2) **Saundersii.**
      Leaves lanceolate to linear, 4-6 times as long as
        broad :
          Leaves mucronate-acute; corolla-tube wide-
            funnel-shaped above the stamens   ...   (3) **bispinosum.**
          Leaves obtuse; corolla-tube narrow-cylindric   (4) **succulentum.**

1. **P. namaquanum** (Welw. in Trans. Linn. Soc. xxvii. 45); stem erect, 5-6 ft. high, thick and fleshy, tapering upwards, tubercled throughout, each tubercle armed with a pair of long straight spreading spines.; leaves crowded at the summit, obovate-oblong to oblong, obtuse or acute, shortly attenuate at the base, wavy, 4-5 in. long, 2-2½ in. broad, densely velvety on both sides, yellowish in the dry state; secondary nerves very slender, oblique, 1½-2 lin.

distant; petiole indistinct; flowers in scanty cymes from the axils of the leaves, on short villous peduncles; bracts oblong, acute, 6 lin. long, villous, with membranous margins; pedicels very short; calyx 4–5 lin. long; sepals resembling the bracts in structure and shape; corolla reddish, tinged with yellow and green, tubular, slightly widened upwards, loosely pubescent without; tube about 1 in. long, silky just above the stamens and along five lines below them; lobes ovate-elliptic, obtuse, 3 lin. long; stamens inserted $3\frac{1}{2}$ lin. above the corolla base, 3 lin. long. *Adenium namaquanum, Wyley in Harv. Thes. Cap.* ii. 11, *t.* 117; *Paters. Trav.* 124 *with plate.*

WESTERN REGION: Great Namaqualand; by the " Lions River," a tributary of the Orange River, *Paterson!* Little Namaqualand; without precise locality, *Wyley!*

2. P. Saundersii (N. E. Br. in Kew Bulletin, 1892, 126); a shrub, up to 4–5 ft. high with a ball-shaped bole rising little above the ground and densely spinous branches, 3–4 lin. thick, succulent, glabrous, covered with a thin, papery, greyish bark; leaves sub-sessile, obovate to obovate-elliptic, acute, constricted towards the base, with spinulous margins, $1\frac{1}{2}$–3 in. long, $\frac{3}{4}$–$1\frac{1}{2}$ in. broad, thin when dried, glabrous with the exception of the sparingly hairy midrib; stipules spiny, strong, up to $1\frac{1}{2}$ in. long; cymes sessile, several-flowered, contracted; pedicels hardly any; calyx $2\frac{1}{2}$ lin. long; sepals ovate, acutely acuminate; corolla white, tinged with pink; tube $1\frac{1}{4}$–$1\frac{1}{2}$ in. long, narrow below the stamens (for 5–6 lin.), widened above them, then attenuate towards the mouth, hairy within; lobes obliquely obovoid, much narrowed at the base, almost 1 in. long and wide; follicles spindle-shaped, 5 in. long; seeds about 3 lin. long, ovate in outline, coma 1–$1\frac{1}{2}$ in. long.

EASTERN REGION: Zululand; South-eastern Lebombo Mountains, in very stony places, *Saunders!*

3. P. bispinosum (DC. Prodr. viii. 424); a shrub of moderate height, branched; branches 2–3 lin. thick, the youngest pubescent, soon glabrous, the old covered with smooth, papery bark; leaves scattered on long shoots and in sessile fascicles (short shoots), the latter from axils of the former, all lanceolate, acute, mucronulate, with margins recurved, up to 1 in. long to $2\frac{1}{2}$ lin. broad, coriaceous, minutely asperulous above, loosely hirsute below, particularly along the midrib; stipules spiny, those of the long shoots spreading, $\frac{1}{4}$–$\frac{3}{4}$ (rarely to 1) in. long, fine, springing from an almost square leaf-cushion, usually under 1 lin. high, those of the short shoots much shorter; cymes few- to 1-flowered, terminal on the long and short shoots and hence often apparently axillary, sessile; bracts small, lanceolate to subulate, deciduous; pedicels 1–2 lin. long, glabrous; calyx $1\frac{1}{2}$ lin. long, glabrous or scantily puberulous; sepals ovate, acute to acuminate; corolla, infra-staminal part cylindric, slender, 3 lin. long, gradually passing into the upper funnel-shaped portion,

the whole tube 6–9 lin. long, purple, hairy within below the stamens ;
limb 5–7 lin. across ; lobes white to purple (the outer half), broad
elliptic, rounded at the tips ; anthers almost 2 lin. long ; disc deeply
5-lobed, as high as the ovary ; follicles, unknown. *Echites bis-*
pinosa, Linn. f. Suppl. 167 ; *Thunb. Prodr.* 37, *in Nov. Act. Imp.*
Soc. Sc. Petersb. xiv. (1805), 505, *and Fl. Cap. ed. Schult.* 232 ;
Ait. Hort. Kew. ed. 2, ii. 69. *Belonites bispinosa, E. Meyer, Comm.*
188. *P. glabrum, G. Don, Gen. Syst.* iv. 77.

COAST REGION : Uitenhage Div. ; between Gamtoos River and Sundays
River, *Thunberg* ; between Uitenhage and Algoa Bay, *Burchell,* 4281 ! near
the Zwartkops River, *Zeyher,* 261 ! Albany Div. ; without precise locality,
Zeyher ! Bathurst Div. ; between Blaauw Krantz and Kowie River, *Burchell,*
3880 !
CENTRAL REGION : Graaff Reinet Div. ; dry rocky hills near Graaff Reinet,
2500–4000 ft., *Bolus,* 113 ! Somerset Div. ; between the Zuurberg Range and
Klein Bruintjes Hoogte, *Drège !*

4. **P. succulentum** (DC. Prodr. viii. 424) ; a shrub, 1–2 ft.
high with a tuberous base rising about ½ ft. above the ground, as
thick as a man's arm, brownish, producing several more or less
branched stems ; branches finally greyish-tomentellous when young,
later on covered with a thin smooth brownish bark ; leaves scattered
on long shoots and in sessile fascicles (short shoots), the latter from
the axils of the former, all linear or oblong-linear, obtuse or (rarely)
acute, with recurved margins, 1½ in. long, 2–3 lin. broad, subcoriaceous, green and pubescent above, pale and tomentose below ;
stipules 3 (1 intrapetiolar), spiny, the lateral 2 of the long shoots
spreading, ½–1 in. long, the intrapetiolar shorter and erect, all
three rising from a conspicuous decurrent leaf-cushion, those of the
short shoots shorter, rapidly decreasing towards the base of the
shoots ; cymes terminal, sessile, contracted, few-flowered, finely
tomentose ; bracts subulate or lanceolate, small, early deciduous ;
pedicels 4–6 lin. long ; calyx 3 lin. long, finely tomentellous ; sepals
narrow-lanceolate, acuminate ; corolla-tube cylindric, pubescent without, purple, slender, infra-staminal part 2–3 lin. long, with 5 hairy
ridges below the stamens, supra-staminal very slightly wider, 4-6
lin. long ; limb to more than 1 in. across ; lobes white and purple
(the outer half), oblong, contracted into a distinct claw at the base,
6–10 lin. long, obtuse ; anthers 2–2½ lin. long ; disc 5-partite,
shorter than the ovary ; follicles spindle-shaped, long acuminate,
over 2 in. long. *Echites succulenta, Thunb. Prodr.* 37, *in Nov. Act.*
Imp. Soc. Sc. Petersb. xiv. (1805) 505, *t. 9, fig.* 2, *and Fl. Cap. ed.*
Schult. 232 ; *Jacq. Fragm.* 74, *t.* 117. *Belonites succulenta, E.*
Meyer, Comm. 187. *P. tomentosum, G. Don, Gen. Syst.* iv. 78.
Barleria rigida, Spreng. ex Schlecht. in Linnæa, xiv. 304.

COAST REGION : Uitenhage Div. ; near Uitenhage, *Burchell,* 4410 ! and
without precise locality, *Zeyher,* 282 ! Albany Div. ; between Sidbury and the
Bushmans River, *Burchell,* 4181 ! Bothas Hill, near Grahamstown, *MacOwan,*
1151 ! and without precise locality, *Bowker !* Stockenstrom Div. ; Katberg,
Miss Sole, 430 ! Div.? Karroo between Gauritz River and Sundays River,
Thunberg.

Central Region : Beaufort West Div. ; Nieuweveld, between Rhinoster Kop and Ganzefontain, *Drège ;* between Beaufort West and Rhenoster Kop, *Drège.* Graaff Reinet Div.; near Graaff Reinet, *Bolus,* 115 ! Somerset Div.; between the Zuurberg Range and Klein Bruintjes Hoogte, *Drège !* Somerset East, *Miss Bowker !* Craddock Div. ; without precise locality, *Cooper,* 1288 !

Kalahari District : Griqualand West ; Asbestos Mountains, *Burchell,* 1662 ! near Griquatown, *Burchell,* 1918 ! between Spuigslang Fontein and the Vaal River, *Burchell,* 1719 !

Order LXXXVIII. **ASCLEPIADEÆ.**

(By N. E. Brown.)

Flowers regular, hermaphrodite. *Calyx* of 5 free sepals or rarely 5-lobed, persistent ; segments imbricate, usually with minute processes at their base within. *Corolla* hypogynous, gamopetalous, 5-lobed, vary variable in shape and size, the lobes imbricate, contorted or valvate in æstivation, often recurved or reflexed, sometimes with the sides folded backwards (replicate), in a few genera connate at the tips, rarely connate into a column at the middle then free and again connate at the tips, the sinuses between them sometimes produced into teeth, with 1–3 series of free or connate lobes, processes, keels, tubercles or flaps arising from the corolla or the staminal whorl, forming the *corona,* sometimes absent. *Stamens* 5, inserted at or near the base, rarely at the middle or mouth of the corolla-tube, alternating with the corolla-lobes ; filaments sometimes free, but usually connate at their base or throughout into a *staminal column,* the apex often united to the dilated part of the style ; anthers not connate or only by their appendages, free or united to the dilated part of style, 2-celled, opening by apical, longitudinal or transverse slits ; margins of the cells or their basal prolongations more or less horny and wing-like (*anther-wings*), usually projecting outwards ; adjacent wings of each pair of anthers nearly meeting, leaving very narrow fissures leading to the stigmatic cavities ; connectives often produced into membranous or rarely fleshy or inflated terminal appendages or apiculate or unappendaged ; appendages free or connate. *Pollen* granular or united into 1 or 2 waxy masses, attached in pairs or in fours, sometimes directly, but more usually by means of arm-like processes (*caudicles*) to each of the 5 small or minute, horny or rarely soft bodies (*pollen-carriers*) at each angle of the dilated part of the style ; when granular, each granule consists of 4 pollen-grains or of 3–5 grains in a row, and then held in horny pollen-carriers with a spoon-, trumpet- or trowel-shaped entire or bifid blade, tapering downwards into a short or long stalk attached to a soft adhesive gland. *Pistil* superior, formed of 2 one-celled carpels, free below, but their styles united above and dilated at the middle or apex into a pentagonal disk ; *style-apex* flat or depressed in the centre, with or without a central simple or bilobed apiculus, or convex, pyramidal or prolonged into a long beak of variable form, which is entire, bifid or dilated, rarely there arises from the disk 2, 5 or 7 style-like processes; stigmatic cavities below the angles of the style-apex, behind the fissures between the anther-wings. *Ovules* numerous or very rarely few or solitary, anatropous, pendulous, imbricate in several series on the projecting placenta. *Fruit* of 2 follicles or by abortion of 1, variable in form, smooth, echinate or winged, opening by the ventral suture and usually liberating the placenta. *Seeds* usually numerous, very rarely few or solitary, imbricate, flat or cochleate, usually with a broad or narrow margin, crowned with a tuft of long silky hairs at one end, or rarely densely fringed all round, very rarely without a tuft of hairs; testa rather thick, sub-crustaceous or sometimes thin; albumen usually thin or none, rarely thick; embryo straight or rarely slightly curved, usually nearly or quite filling the seed; cotyledons flat; radicle superior.

Erect, prostrate, twining or scrambling herbs or shrubs, with milky or watery juice; stems simple or branched, sometimes leafless or with very minute leaves, then often succulent, with terete or angular branches, often toothed or spiny at the angles; leaves opposite or whorled, rarely alternate, thin, coriaceous or fleshy; flowers very variable in size and form, solitary or few or many together in umbels, umbel-like cymes, fascicles or racemes, axillary, more or less lateral between the bases of the leaves, or terminal.

DISTRIB. An Order of over 1,800 species widely spread throughout the Tropical and Sub-tropical regions; a few in the Temperate regions.

The Order is well marked by the peculiar structure of its pollen apparatus, coronal-appendages and stigma, but in other characters it is allied to *Apocynaceæ*. In having the pollen-contents of each anther-cell united into a waxy mass attached by caudicles in pairs to the pollen-carriers, it is unique among Dicotyledons, and resembles the *Orchideæ* among the Monocotyledons. The 5 stigmas or stigmatic cavities correspond to only 2 carpels, are completely hidden from view behind the anther-wings, and can only be seen by careful dissection or by making transverse sections of the dilated part of the style; the only openings to the stigmas (except in the tribe *Periploceæ*) are the 5 narrow fissures formed by the contiguous anther-wings. No other Order has a similar structure, a detailed account of which, as well as the manner of fertilization, has been given by T. H. Corry (Trans. Linn. Soc. ser. 2, Bot. ii. 75, and 173, tt. 16, 24–26). All previous authors seem not to have understood this structure and have erroneously described the *style-apex* as the stigma. The pollen-carrier is also described as a gland, which it is not, in any sense of the word, but a hard, horny, elastic structure. The *Asclepiadeæ* are a very difficult group to study, and no unknown member of the Order can be generically determined until the character and position of the pollen as seen seated in place in the anther-cells, not as withdrawn, have been first ascertained. By the pollen the Order may readily be divided into distinct primary groups:—(1) Granular and loosely contained in the more or less spathulate entire or bifid pollen-carriers, but not attached to the latter (*Periploceæ*).—(2) United into very minute waxy masses, 4 or 2 of which are sessile upon a very minute quadrate pale-coloured pollen-carrier (*Secamoneæ*).—(3) United into waxy masses, which are opaque (or rarely with a pellucid area or linear space at one end or on the inner margin, and then distinctly pendulous), usually not very minute, and attached in pairs by caudicles to dark-coloured pollen-carriers (*Cynancheæ* with pendulous pollen-masses, and *Marsdenieæ* with erect or horizontal pollen-masses; *Tylophora* is intermediate between these two tribes, having minute pollen-masses that are sometimes pendulous, sometimes horizontal).—(4) United into waxy masses, which are pellucid along one margin or just beneath the apex and attached in pairs by caudicles to dark-coloured pollen-carriers, erect, ascending or horizontal, never pendulous (*Ceropegieæ* and *Stapelieæ*).

In the following descriptions, the dimensions of the dried flowers I have examined, unless otherwise stated, are always taken after having boiled the flowers in water, thus more nearly approaching the dimensions of living flowers. The measurement of the staminal column is taken from its base to the level attained by the anther-appendages, whether these latter are erect or inflexed, and in the latter case does not comprise the full length of the appendages. Owing to their fleshy nature, the corona-lobes, when subjected to pressure, often have their original form altered, if not destroyed, and many species cannot be identified from descriptions because sufficient care has not been taken to ascertain the true form of the coronal structure, and there is sometimes much discrepancy between descriptions of the same plant.

Series I. *Filaments* of the stamens free. *Pollen* of loose granules, each granule formed of 4 pollen-grains united in a tetrad or of 3–5 united in a row.

Tribe I. PERIPLOCEÆ.—*Anthers* produced at the apex into a fleshy apiculus or small fleshy or filiform or rarely membranous appendage, connivent over the style-apex and frequently connate at their tips. *Pollen-carriers* spathulate, trumpet-shaped or trowel-shaped, sometimes bipartite, horny, furnished with an adhesive gland at the base, not attached to the pollen-grains, but holding them loosely in the blade or concave part.

* *Corolla with a distinct campanulate or cylindric tube* $\frac{1}{2}$–3 *lin. long.*
　† Corona-lobes inserted on the corolla-tube distinctly above the base of the stamens.

I. **Cryptolepis.**—*Corona-lobes* inserted at about the middle of the corolla-tube. Erect or twining shrubs or shrublets.
II. **Stomatostemma.**—*Corona-lobes* clavate, inserted at the mouth of the corolla-tube. *Stamens* inserted near the base of the corolla-tube; anthers with a glabrous apiculus. A climbing or bushy shrub.
III. **Ectadium.**—*Corona-lobes* subulate, inserted at the mouth of the corolla-tube. *Stamens* inserted at the base of the corolla-tube; anthers with long filiform hairy appendages. Erect shrubs.

　†† Corona-lobes inserted on the corolla-tube with the stamens or occasionally upon their filaments.

IV. **Raphionacme.**—*Corona-lobes* and *stamens* inserted above the middle or at the mouth of the corolla-tube. Herbs with a tuberous rootstock and dwarf or twining stems.

** *Corolla rotate or subrotate, lobed nearly to the base, without a distinct campanulate tube. Erect or climbing shrubs.*

V. **Tacazzea.**—*Corona-lobes* filiform and simple or linear and divided above into 2 or 3 filiform segments, inserted at or near the base of the staminal filaments and often shortly adnate to them. *Stipules* represented by a line connecting the bases of the petioles.
VI. **Chlorocodon.**—*Corona-lobes* broadly obcordate or obreniform with or without an erect or incurved dorsal process. *Stipules* well developed between the leaf-bases, toothed or frill-like.

Series II. *Filaments* of the stamens, when present, connate around the ovary into a tube, which is sometimes very short and ring-like, the top of the tube or the anthers or both adnate to the dilated part of the style. *Pollen-contents* of each anther-cell united into 1 or 2 waxy masses. *Pollen-carriers* quadrate, turgid or rarely dorsally flattened, with a suture down the back, never spathulate or trumpet-shaped.

* *Pollen-masses* 2 *in each anther-cell, exceedingly minute and distinct or the 2 are combined into one marked with a longitudinal suture and sometimes separable, seated directly on the pollen-carriers or upon a broad flap-like caudicle on their upper part in fours or in pairs. Pollen-carriers very minute, subquadrate, rather soft, pale-coloured.*

Tribe II. SECAMONEÆ.—*Anthers* erect or ascending, with more or less fimbriated appendages. *Style-apex* often exserted beyond the anthers.

VII. **Secamone.**—Climbing shrubs. *Flowers* small, in cymes. *Corona* of 5 simple lobes more or less adnate to the staminal column.

** *Pollen-masses solitary in each anther-cell, attached in pairs to each of the pollen-carriers or to lateral expansions of them by long or short caudicles, or sessile in* Fockea. *Pollen-carriers hard, horny, sometimes pale, but usually black, brown or dark-coloured.*

Tribe III. CYNANCHEÆ.—*Anthers* erect, connivent-erect or rarely divergent-erect, tipped with entire or rarely toothed or fringed membranous appendages, which are often inflexed over the style-apex or upon its rim. *Pollen-masses* pendulous in the anther-cells, opaque, except in some species of *Schizoglossum* which have a small pellucid area just beneath the apex, and in *Cordylogyne* and *Periglossum* which are subtransparent at one end.

† Corona none or of 5 inconspicuous tubercles or scales on the corolla-tube alternating with tufts of hairs near its base, middle or top, none on the staminal column.

VIII. **Astephanus.**—*Corolla-lobes* erect or spreading, not partly closing the mouth of the tube and without a hump on the back. *Corona* none. Slender twiners.

IX. **Microloma.**—*Corolla-lobes* connivent over or spirally arranged around and partly closing the mouth of the tube, often with a hump on the back. *Corona-tubercles* often present. Slender twiners or much branched and often rigid shrublets.

†† Corona of 5 broad simple lobes seated at the sinuses of and adnate to the corolla-tube, none on the staminal column.

X. **Parapodium.**—*Corolla-tube* shortly hemispherical or globose-campanulate; lobes recurved or spreading at the tips. Dwarf erect tuberous-rooted herbs. *Stem* simple. *Flowers* of moderate size, in lateral umbels.

††† Corona distinct, arising from the staminal column or in the angle between it and the corolla or both, none on the corolla.

‡ Stems erect, decumbent or prostrate, never twining, herbaceous (rarely woody in *Cynanchum*). Rootstock often a tuber.

§ Corona of 5 distinct free lobes in one series, often with one or more appendages or keels on their inner face; outer corona none or reduced to very minute lobules, teeth or (in *Periglossum*) filiform processes alternating with the lobes.

XI. **Woodia.**—*Corona-lobes* not exceeding the incurved-ascending, suberect or spreading corolla-lobes, dorsally flattened, not very fleshy in dried flowers, deeply trifid or quadrate with square shoulders and an incurved mid-lobe, with the margins incurved at the base and ascending the column as narrow wings.

XII. **Xysmalobium.**—*Corona-lobes* shorter than the erect, spreading or reflexed corolla-lobes, very fleshy, often nearly or quite as thick as broad, or if dorsally flattened, then comparatively thick, solid, without a fissure or cavity on the inner side, variously shaped, keelless or with 1 median keel, very rarely with 2 keels or a tooth (but no other process) on the inner face, not acutely keeled down the back. (See also *Schizoglossum periglossoides*).

XIII. **Periglossum.**—*Corona-lobes* shorter than the suberect corolla-lobes, dorsally flattened, spathulate; blade oblong, subcordate-oblong or ovate-sagittate, with 2 keels capped by a transverse ridge or tubercle (rarely without keels) on the inner face. *Staminal column* subglobose or globose-obovoid, constricted under the anther-appendages. *Pollen-masses* subtranslucent at the free end and much shorter than their caudicles. Style not exserted beyond the anther-appendages.

XIV. **Cordylogyne.**—*Corona-lobes* hastately subspathulate, otherwise as in *Periglossum*. *Staminal column* cylindric, not constricted under the anther-appendages. *Pollen-masses* subtranslucent at the attached end and much longer than their caudicles. *Style* exserted much beyond the anther-appendages.

XV. **Krebsia.**—*Corona-lobes* exserted beyond the recurved tips of the erect corolla-lobes, narrowly-lanceolate below, tapering into a subulate point, incurved at the apex, acutely keeled down the back, with a triangular wing on each side of the keel at the base forming a rhomboid expansion, no appendage or keel on the inner face.

XVI. **Schizoglossum.**—*Corona-lobes* longer or shorter than the corolla-lobes; dorsally flattened, thin (very rarely somewhat thick and fleshy) in dried flowers, never complicate, but occasionally with slightly incurved margins, often produced into a short or long subulate or filiform point at the apex and having 2 keels (rarely without them) and frequently also 1 or 2 filiform or other appendages or 2 basal teeth on their inner face. *Pollen-masses* opaque or with a small translucent space just below the apex.

XVII. **Fanninia.**—*Corona-lobes* shorter than the suberect corolla-lobes, dorsally flattened, not thick, linear-oblong, with a distinct midrib and 2 erect basal lobules, but no keels. *Flowers* rather large and showy, in terminal umbels, white with a purple corona.

XVIII. **Asclepias.**—*Corona-lobes* laterally flattened or at least measuring as much from front to back as in breadth, with the sides folded together throughout or at the base or apex, and often produced into teeth directed to the centre of the flower, forming an open cavity, fissure or channel between them (cucullate or complicate), with or without a horn or other appendage within the cavity, sometimes with the apex of the lobe prolonged, but never dilated or petaloid.

XIX. **Pachycarpus.**—*Corona-lobes* at their basal part or throughout horizontally radiating or ascending from the stout staminal column, sometimes consisting of a pair of contiguous fleshy erect lobes (keels) with or without a short point beyond, or more usually long and dorsally flattened, at least beyond the keels, linear, linear-oblong or spathulate with a distinct claw and often petaloid blade, with or without 2 parallel keels or large fleshy or wing-like erect contiguous lobes or keels at their base and the upper part often curving over them.

    §§ Corona-lobes united into a tube (at least at the base), truncate, toothed or lobed at the top, with or without keels or other processes within.

XXII. **Cynanchum.**—Small *herbs* 2–8 in. high, branching at the base, or a *shrublet* with the rigid arching woody branches. *Flowers* small.

    §§§ Corona complex or of 3 series of erect lobes :—outer series of 5 simple lobes opposite the corolla-lobes and a middle series of 5 deeply trifid or auriculate lobes alternating with them, both arising in the angle between the corolla and staminal column, and distinct or more or less united at the base in apparently 1 series ; inner series of 5 simple lobes on the staminal column opposite the anthers.

XXV. **Eustegia.**—Dwarf *herbs* branching at the base with branches 2–6 in. long. *Leaves* linear-filiform to linear-hastate. *Carpels* with several ovules in each. *Follicles* not seen.

XXVI. **Emicocarpus.**—*Herb*, with prostrate branches 2–4 ft. long. *Leaves* palmately divided. *Carpels* with 1 ovule in each. *Follicles* small, obtriangular, with 3 spreading spines. *Seed* solitary, without a tuft of hairs at either end.

    ‡‡ Stems twining or rambling.
    § Corona in one series; no outer corona.

XX. **Glossostephanus.**—*Corona* of 5 free simple lobes, laterally compressed and dorsally grooved. *Style* produced beyond the anther-appendages into an acutely bifid beak.

XXI. **Pentarrhinum.**—*Corona* of 5 free obconic or trumpet- or slipper-shaped lobes, with infolded margins and truncate or rounded at the apex, with a horn directed over or towards the anthers. *Style* not exceeding the anther-appendages.

XXII. **Cynanchum.**—*Corona* annular, cup-shaped or tubular and truncate or toothed at the top, or divided nearly or quite to the base into 5 dorsally flattened entire or toothed lobes, with or without keels or other processes within the tube or on the inner face or at the base of the lobes. *Style* shorter than or exceeding the anther-appendages.

XXVII. **Tylophora.**—*Corona-lobes* usually of 5 tubercles adnate to or radiating from the staminal column at or above its base and usually not exceeding the filament part of it, rarely with free tips or of free dorsally flattened lobes and attaining to the level of the anther-tips. *Pollen-masses* pendulous or subhorizontal. *Style* rarely slightly exceeding the anther-appendages.

§§ Corona in 2 distinct series.

XXIII. **Sarcostemma.**—*Corolla* rotate or rotate-campanulate. *Outer corona* annular or cup-shaped, pentagonal, truncate or shortly lobed. *Inner corona* of 5 simple laterally compressed or keeled lobes, embraced at the base by the outer corona. *Stem* leafless, fleshy.

XXIV. **Pergularia.**—*Corolla* with a campanulate or cylindric tube and widely spreading lobes. *Outer corona* annular, shortly 5-lobed. *Inner corona* of 5 erect fleshy lobes, produced into a subulate incurved horn at the apex and with a spreading or deflexed spur at the base. *Stem* not fleshy. *Leaves* large, cordate.

Tribe IV. Marsdenieæ.—*Anthers* erect or incurved-ascending, with (or without in *Sphærocodon*) a membranous appendage at the apex. *Pollen-masses* distinctly erect in the anther-cells, opaque and usually not very minute, or in *Emplectanthus* minute with a pellucid margin on one side.

N.B.—*Tylophora* is intermediate between *Marsdenieæ* and *Cynancheæ* and is placed in the latter tribe, as in most of the South African species the pollen-masses are distinctly pendulous, in others, however, they are very minute, subglobose, attached at their middle to very slender caudicles and neither erect nor pendulous.

† Corona double, arising from the staminal column.

XXVIII. **Emplectanthus.** - *Outer corona-lobes* with pocket-like bases and minute spreading emarginate or bifid tips. *Inner corona-lobes* adpressed to the backs of the anthers and not exceeding them, dorsally connected to the basal margins of the outer corona-lobes. *Follicles* long and slender, linear-terete, probably constricted between the seeds. *Stem* twining.

†† Corona of 5 distinct lobes or tubercles in one series arising from the staminal column.

XXIX. **Sphærocodon.**—*Corona* of 5 small fleshy tubercles on the staminal column much above its base. *Stamens* erect or tortuous, not twining.

XXX. **Marsdenia.**—*Corona* of 5 erect lobes adnate to the staminal column with free margins and tips. *Follicles* stout, coriaceous, with 4 broad wings. *Stem* very long, twining.

XXXI. **Prageluria.**—*Corona* of 5 erect lobes, adnate to the staminal column below, free above, with a subulate process on the inner face. *Stem* twining.

††† Corona tubular, arising in the angle between the staminal column and the corolla, lobed or toothed at the top, with 1–2 superposed series of teeth or filiform processes and 5 pairs of wing-like keels within the tube.

XXXII. **Fockea.**—*Rootstock* a large tuber. *Stem* erect or twining. *Anther-appendages* very large, membranous, inflated.

†††† Corona of 5 narrow grooved lobes upon the corolla-tube and adnate up to its mouth, incurved at the tips, ciliate on the adnate part; none on the staminal column.

XXXIII. **Gymnema.**—Twining shrubs. *Umbels* often 2 from the same node and opposite, sublateral. *Flowers* small.

††††† Corona none.

XXXIV. **Rhyssolobium.**—A much-branched shrublet, with rigid woody whitish-puberulous branchlets. *Leaves* small and thick. *Flowers* very small, 1–3 together, subsessile at the nodes.

Tribe V. CEROPEGIEÆ.—*Anthers* erect, connivent, or incumbent on the top of the style, with or without an apiculus or terminal appendage, sometimes tipped with hairs. *Pollen-masses* erect, ascending or horizontal in the anther-cells, pellucid along the inner margin or at the apex (see also *Emplectanthus*). *Pollen-carriers* sometimes with a wing-like expansion on each side. *Stems* herbaceous or fleshy, erect, prostrate or twining, with well-developed leaves, rarely leafless and then without distinct angles.

    * Corona-lobes all in 1 series on the staminal column opposite (or on or below the back of) the anthers, none alternating with them.

    † Corona of 5 lobes, without a dorsal arm or an appendage on their inner face.

XXXV. **Orthanthera.**—*Corolla-tube* elongated, not bearded at the mouth, more or less inflated at the base. *Corona-lobes* Λ-shaped, adnate to the staminal column, with free reflexed wing-like margins. *Stems* prostrate with broad leaves, or erect and shrubby with narrow leaves.

XXXVI. **Sisyranthus.**—*Corolla-tube* short, campanulate or globose-campanulate, sometimes bearded at the mouth and on the lobes. *Corona-lobes* erect, partly adnate to the staminal column, ovate or oblong and entire, or broadly-cuneate or subrhomboid and 3-toothed, with the middle tooth sometimes bifid. *Roots* thick, fleshy, clustered. *Stems* erect, rather slender, simple or branched. *Leaves* long, linear. *Flowers* small, in pedunculate umbel-like clusters.

XXXVIII. **Macropetalum.**—*Corolla-tube* exceedingly short; lobes very long, reflexed straight back. *Corona-lobes* exserted, adnate to the staminal column at the base, free and linear-lanceolate above. *Rootstock* a tuber. *Stem* erect, slender. *Leaves* filiform.

    †† Corona of 5 broad bifid lobes with a long linear filiform appendage on their inner face, or the dorsal part of the lobe short and subulate, appearing to arise from the back of the long linear-spathulate inner part.

XLII. **Anisotoma.**—*Corolla-tube* short, not bearded. *Herb* branching at the base into many prostrate stems. *Leaves* cordate. *Flowers* small in sessile or pedunculate umbel-like clusters. (See also *Brachystelma*.)

    ** Corona-lobes in 2 or 3 series, or falsely in 1 series of 3-fid lobes, all on the staminal column; 1 series or its teeth alternating with the anthers, the other 1 or 2 series opposite to them. Inner corona-lobes shorter than, equalling or longer than the anthers.

    † Corolla-tube tubular, 2 to several times as long as its diameter at the middle, often inflated at the base; lobes free or united.

XXXIX. **Riocreuxia.**—*Outer corona* sometimes 1-seriate, of 5 subulate simple or bifid lobes radiating horizontally or of 5 minute ascending or spreading often pouch-like bifid lobes; sometimes 2-seriate, with 5 minute bifid or emarginate lobes alternating with and just beneath the radiating subulate lobes, forming an irregularly 15–20-toothed frill. *Inflorescence* a branching cyme or panicle with few or many clusters of flowers scattered along its branches, or the clusters racemosely arranged along a single peduncle or sessile at the nodes.

XL. **Ceropegia.**—*Outer corona* always 1-seriate, cup-shaped and entire or 5–10-toothed, or of 5 bifid lobes often pouch-like at the base or reduced to minute pouches, or the lobes divided to the base and the halves adnate to the adjacent sides of the inner corona-lobes so as to apparently form 1 series of 5 trifid lobes opposite the anthers, or the halves or teeth of 2 adjacent lobes connate and forming 5 lobes immediately behind the inner corona-lobes which are shorter than, equalling or longer than the anthers. *Peduncle* bearing 1 to many flowers in a cluster or umbel-like cyme or occasionally with pairs of flowers racemosely scattered along it, or the clusters sessile or the flowers solitary or in pairs at the nodes.

†† Corolla-tube not twice as long as its diameter at the middle, very rarely tubular and slightly inflated at the base, sometimes none.

XXXVII. **Tenaris.**—*Corolla-tube* very short; lobes free, not reflexed. *Outer corona* of 5 small lobes or minute pouches; inner corona of 5 subulate, linear or filiform lobes shorter to longer than the anthers. *Rootstock* a tuber. *Stem* erect, slender.

XLI. **Brachystelma.**—*Corolla-tube* campanulate or saucer-shaped or expanded nearly flat, very rarely shortly tubular; lobes free or united. *Corona* varying exactly as in *Ceropegia*. *Rootstock* a tuber or cluster of fleshy roots. *Stems* dwarf, erect or prostrate. *Flowers* small or of moderate size, few or many together in lateral or terminal clusters or umbel-like cymes, or solitary or in pairs at the nodes.

Tribe VI. STAPELIEÆ.—*Anthers* suberect or incumbent upon the style-apex, without appendages. *Pollen and pollen-carriers* as in *Ceropegieæ*. *Stems* thick and fleshy, 3- to many-angled, usually dwarf, erect or procumbent or diving underground, tuberculate-tessellate, or toothed along the angles, leafless or the teeth tipped with rudimentary or small subulate fleshy leaves, or the tubercles or teeth stout and conical or spine-like or ending in slender bristles. *Flowers* fleshy. *Corolla-lobes* valvate in bud. *Corona* arising from the staminal column, none on the corolla.

* Corolla-lobes connate at the tips.

XLIII. **Pectinaria.**—*Stems* decumbent or procumbent, often diving underground, 4-angled or tessellately tuberculate, with prominent buds in the axils of all the teeth or tubercles. *Flowers* small, developing between the tubercles or angles. *Corona* double, variable.

** Corolla-lobes not connate at the tips. Stems without prominent buds in the axils of all the teeth or tubercles.

† Stems covered with crowded pointless tubercles or with 6 to many angles formed of closely-placed tubercles, each tipped with a slender spine or 3 bristles. (See also *Huernia Pillansii*.)

‡ Corolla-tube very small or none. Outer corona of 5 spreading bifid or bipartite lobes, concave or pouch-like at the base, or the lobes united into a cup with emarginate or bifid lobes.

XLV. **Trichocaulon.**—*Tubercles* of the stem pointless or tipped with a simple slender spine. *Corolla* less than ¾ in. in diam., lobed to half-way, with a very short tube just enclosing the corona or the united part very shallowly saucer-shaped without a tube.

XLVI. **Hoodia.**—*Tubercles* of the stem tipped with a simple slender spine. *Corolla* more than 1 in. in diam., cup- or saucer-shaped or nearly flat, subentire or obsoletely lobed, with 5 slender subulate points on the margin; tube a very small depression just enclosing the corona.

‡‡ Corolla-tube 1½–3 in. long. Outer corona divided into 10 filaments terminating in knobs.

XLVII. **Tavaresia.**—Tubercles of the stem tipped with 3 bristles.

†† Stems obtusely or acutely 4–6- (or in *Huernia* rarely 8–24-) angled; teeth of the angles not tipped with slender spines or bristles (except in *Huernia Pillansii*), but sometimes stout and conical with hardened spine-like tips.

‡ Corolla with the angles between the lobes produced into distinct teeth.

XLVIII. Huernia.—*Corolla* wholly campanulate or with a campanulate or globose-campanulate tube and spreading lobes or saucer-shaped limb, the latter sometimes raised into a ring around the mouth of the tube. *Outer corona* 5-lobed or 10-toothed, adnate to the very base of the corolla, absent in one species.

‡‡ Corolla not produced into distinct teeth at the angles between the lobes.

§ Corolla-tube with another tube within nearly as long as itself.

L. Diplocyatha. *Corolla* large, with a campanulate tube and very spreading lobes, ciliate with vibratile clavate hairs. *Outer corona* of 5 spreading bifid lobes. *Inner corona* of 5 ovate acuminate lobes incumbent on the anthers and produced beyond them into short erect points.

§§ Corolla-tube none or, when present, without another tube inside it.

‖ Outer corona present, distinct.

XLIV. Caralluma.—*Corolla-tube* varying from almost none to campanulate or subglobose. *Outer corona* of 5 small lobes or pouches alternating with the anthers and more or less adnate at their base or sides to the inner corona-lobes, rarely quite free to the base, usually bifid, sometimes so deeply that the whole corona appears to consist of 5 trifid lobes, or the lobes are united into an entire or 5–20-toothed ring or cup and connected to the backs of the inner corona-lobes.

LI. Stapelia.—*Corolla* usually star-like with a flattened or saucer-shaped disk, with or without a depression or cavity containing the corona or with a raised ring around it, rarely with a broad cup-like or shortly campanulate tube; lobes flat or with revolute margins, never replicate. *Outer corona-lobes* always very distinct, free to their base.

LIII. Duvalia.—*Corolla* with the disk raised into a tube-like rim supporting the corona; lobes more or less folded lengthwise (replicate) and often into narrow vertical plates. *Outer corona* in one piece, disk-like, subcircular or obtusely 5–10-angled (in *D. angustiloba* reduced to a mere margin), resting on or rarely below the rim of the raised ring on the corolla-disk and closing the spurious tube formed by it. *Inner corona-lobes* rhomboid-ovoid.

‖‖ Outer corona none. (See also *Huernia simplex* and *Duvalia angustiloba*).

XLIX. Huerniopsis.—*Corolla-tube* campanulate, as long as the lobes. *Corona-lobes* stout, square in transverse section below the middle, produced above the anthers into subulate points, without a dorsal crest at their base.

LII. Piaranthus.—*Corolla-tube* much shorter than the lobes or none. *Corona-lobes* not square in transverse section, with (or in *P. grivanus* without) a thick dorsal transverse crest at the base.

I. CRYPTOLEPIS, R. Br.

Calyx 5-partite. *Corolla* with a campanulate tube; lobes more or less twisted and overlapping to the left in bud, the angles between them usually provided with a small pocket-like flap within. *Corona* of 5 fleshy lobes arising from about the middle of the corolla-tube at some distance above the insertion of the stamens, alternating with the corolla-lobes. *Stamens* arising from the lower part of the corolla-tube; filaments free; anthers more or less deltoid or triangular, with the connective produced into a fleshy apiculus, united to the dilated part of the style at the base, more or less connivent in a cone. *Pollen* granular; pollen-carrier more or less spathulate. *Style* not exceeding the anthers, shortly conical at the apex. *Follicles* linear-terete, subfusiform or ovate, smooth. *Seeds* crowned with a tuft of hairs.

Erect or twining shrubs; leaves opposite; flowers small or of moderate size, arranged in subaxillary or terminal cymes, or rarely subsolitary.

Distrib. A genus of several species, ranging through the warmer parts of the Old World as far north as Cashmere.

Corolla-lobes 3–5 lin. long, linear-lanceolate, acuminate :
    Leaves acute; pedicels ½–1 in. long ... ... (1) **capensis.**
    Leaves obtuse or retuse, apiculate; pedicels
      1–2 lin. long ... ... (2) **obtusa.**
Corolla-lobes under 2 lin. long, oblong or linear-
  oblong:
    Petioles ½–1 lin. long :
      Stems twining; corona-lobes oblong, obtuse (3) **delagoensis.**
      Stems erect, not twining; corona-lobes sub-
        terete or clavate ... ... (5) **oblongifolia.**
    Petioles 2–4 lin. long; stems scrambling or
      twining; corona-lobes clavate ... ... (4) **transvaalensis.**

1. C. capensis (Schlechter in Verhandl. Bot. Ver. Brandenb. xxxv. 47); stem twining, slender, woody, glabrous, minutely tuberculate; leaves glabrous, thin; petiole 2–6 lin. long; blade 1½–3½ in. long, ½–1¾ in. broad, lanceolate to elliptic, acute, acuminate or shortly cuspidate at the apex, acute at the base; cymes lateral at the nodes or subterminal, very lax, once or twice dichotomously or trichotomously forked, pedunculate, glabrous; branches ¾–3 in. long, racemosely few-flowered; peduncle ¾–2¼ in. long; bracts ½–1 lin. long, subulate, glabrous; pedicels ½–1 in. long, glabrous; sepals about 1½ lin. long, ⅔ lin. broad, oblong-lanceolate, acute, glabrous; corolla long-acuminate and twisted in bud, glabrous; tube 2–2¾ lin. long, campanulate; lobes spreading, twisted, 4–5 lin. long when unrolled, lanceolate, acuminate, or tapering from a 1¼–1½ lin. broad base to a linear apex; corona-lobes affixed at the middle of the corolla-tube, ⅓–½ lin. long, ovate or rhomboid-ovate, obtusely pointed, fleshy; stamens arising ¾ lin. above the base of the corolla; filaments very short; anthers nearly ½ lin. long, ovate, very acuminate, connivent over the very short conical apex of the style; follicles 4–8½ in. long, 2–2½ lin. thick, terete, acuminate, slightly nodose; seeds 5 lin. long, 1 lin. broad, linear-oblong, keeled and concave on one side, convex, with raised irregular lines on the other side, glabrous, dark brown. *Schlechter in Journ. Bot.* 1896, 315.

Eastern Region: Transkei; near Kentani, *Miss Alice Pegler,* 663! Natal; Nonoti River, *Gerrard,* 1319! Inanda, *Wood,* 761! 886! 1583! *McKen,* 6! and without precise locality, *McKen,* 21 ! *Sanderson!*

The flowers are stated by Mr. Wood on one label to be white, on the other yellow, so they probably vary.

2. C. obtusa (N. E. Br. in Kew Bulletin, 1895, 110); stem twining, glabrous; leaves thinly coriaceous, glabrous; petiole 2–5 lin. long; blade $\frac{3}{4}$–3 in. long, $4\frac{1}{2}$–16 lin. broad, oblong, subtruncately obtuse, retuse or emarginate, mucronate; cymes 1–2 in. long, pedunculate, dichotomous, laxly 6–10-flowered, glabrous in all parts, axillary, often from both axils, and often arranged in elongate leafless (leaves fallen away?) narrow panicles 3–12 in. long, at the ends of the lateral shoots; peduncles $\frac{1}{8}$–1 in. long; bracts $\frac{1}{2}$–1 lin. long, lanceolate, acute; pedicels 1–2 lin. long; sepals $\frac{3}{4}$ lin. long, ovate, subacute; buds acuminate, twisted; corolla-tube 1 lin. long; lobes 3 lin. long, linear-lanceolate, acuminate, glabrous; corona-lobes arising at the middle of the corolla-tube, $\frac{1}{2}$ lin. long, fleshy, lanceolate, acuminate; anthers acuminate, connivent; follicles 3–5 in. long, $2\frac{1}{2}$–3 lin. thick, fusiform-terete, acuminate, glabrous, smooth, reflexed when ripe; seeds 3 lin. long, $\frac{2}{3}$–1 lin. broad, oblong-lanceolate, flat with a central keel on one side, convex on the other, minutely tuberculate, blackish-brown. *N. E. Br. in Dyer, Fl. Trop. Afr.* iv. i. 246; *Schlechter in Journ Bot.* 1896, 315; *K. Schum. in Engl. Pfl. Ost-Afr. C.* 320.

Eastern Region: Delagoa Bay, *Speke,* 12! *Monteiro! Schlechter!*
Kalahari Region: Transvaal; Komati Poort, 600 ft., *Rogers,* 900!
Also in Tropical Africa.

3. C. delagoensis (Schlechter in Engl. Jahrb. xxxviii. 26); stem twining, branched, glabrous; leaves spreading; petiole $\frac{1}{2}$–1 lin. long; blade 6–10 lin. long, 2–$3\frac{1}{2}$ lin. broad, elliptic or lanceolate-elliptic, acute or mucronulate, glabrous on both sides, pallid beneath, papery in texture; cymes many times shorter than the leaves, 2–4-flowered, glabrous; pedicels $\frac{1}{2}$–$\frac{3}{4}$ lin. long, glabrous; sepals scarcely 1 lin. long, ovate, subobtuse, glabrous; corolla $1\frac{1}{2}$ lin. long, lobed to below the middle, campanulate, glabrous; lobes oblong, obtuse; corona-lobes inserted below the corolla-throat, oblong, obtuse, glabrous; filaments of the stamens very short; anthers hastate-lanceolate, subacute, glabrous, without an appendage; style-apex shortly conical.

Eastern Region: Delagoa Bay; near Lourenco Marquez, *Schlechter.*

4. C. transvaalensis (Schlechter in Journ. Bot. 1896, 315); a woody climber, glabrous in all parts; leaves spreading, subcoriaceous; petiole 2–4 lin. long; blade $\frac{3}{4}$–$1\frac{1}{4}$ in. long, $\frac{1}{2}$–1 in. broad, elliptic or elliptic-obovate, obtusely rounded and shortly cuspidate-apiculate at the apex, broadly cuneate at the base; apiculus more or less recurved; undersurface rather pale, densely reticulate; cymes axillary or sublateral, many-flowered, small, about $\frac{1}{2}$ in. long including the short peduncle, 6–8 lin. broad, trichotomously or dichotomously branched, glabrous, including the flowers; bracts opposite, spreading, somewhat membranous, $\frac{1}{2}$–1 lin. long, $\frac{1}{4}$–$\frac{1}{3}$ lin. broad, oblong, sub-obtuse, very minutely ciliate ; pedicels $\frac{3}{4}$–1 lin. long, glabrous ; sepals $\frac{1}{2}$ lin. long, $\frac{1}{3}$–$\frac{2}{5}$ lin. broad, ovate, obtuse, rather membranous, very minutely ciliate ; corolla small, glabrous, yellowish ; tube $\frac{3}{4}$ lin. long, campanulate ; lobes $1\frac{1}{4}$ lin. long, $\frac{1}{2}$ lin. broad, oblong, obliquely obtuse at the apex ; corona-lobes arising from the middle of the corolla-tube, $\frac{1}{4}$ lin. long, clavate, fleshy, connivent over the stamens ; filaments of the stamens very short ; anthers $\frac{1}{4}$ lin. long, deltoid, very acuminate, connivent over the very short conical apex of the style ; follicles widely diverging, $1\frac{3}{4}$–2 in. long, $3\frac{1}{2}$–$5\frac{1}{2}$ lin. thick at the swollen base, whence they gradually taper into a long slender acute beak, with a very small hook at the apex; seeds 3–4 lin. long, $1\frac{1}{4}$–$1\frac{1}{2}$ lin. broad, narrowly ovate to somewhat rhomboid-ovate, flattened with a central keel on one side, convex on the other, scabrid-tuberculate all over, brown. *Ectadiopsis cryptolepioides, Schlechter in Engl. Jahrb.* xx. *Beibl.* 51, 10.

KALAHARI REGION : Transvaal ; Hout Bosch, *Rehmann,* 5879! 5880! near Botsabelo, 5000 ft., *Schlechter,* 4082 ! Bush Veld, near Warm Bath, 3900 ft., *Bolus,* 12154! among shrubs on Elandspruit Mountains, 6000 ft., and Magaliesberg Range, near Apies River, 4900 ft., *Schlechter,* ex *Schlechter.* Wonderboom Poort, *Leendertz,* 519! *Rogers in Transvaal Herb.* 2502!

5. **C. oblongifolia** (Schlechter in Journ. Bot. 1896, 315) ; an erect branching shrub ; branches usually rather long, slender, reddish-brown, very minutely scabrous ; leaves ascending, glabrous ; petiole $\frac{1}{2}$–1 lin. long ; blade 1–2 in. long, $\frac{1}{4}$–$\frac{2}{3}$ in. broad, varying from narrowly lanceolate to oblong-elliptic, acute, or obtuse and apiculate, cuneate or rounded at the base, pallid beneath ; cymes subaxillary, subsessile, trichotomous, $\frac{1}{2}$–$\frac{3}{4}$ in. in diam. ; bracts $\frac{1}{2}$–1 lin. long, ovate, obtuse ; pedicels 1–$1\frac{1}{2}$ lin. long ; sepals $\frac{2}{3}$–$\frac{3}{4}$ lin. long, oblong, obtuse, minutely ciliate at the apex ; corolla quite glabrous, yellowish-green ; tube $\frac{2}{3}$–1 lin. long, campanulate ; lobes $1\frac{1}{3}$–$1\frac{3}{4}$ lin. long, $\frac{2}{3}$–$\frac{3}{4}$ lin. broad, oblong or lanceolate-oblong, obtuse ; corona-lobes inserted at the middle of the corolla-tube, $\frac{1}{4}$–$\frac{1}{3}$ lin. long, sub-terete or clavate, truncate, obtuse or acute, fleshy ; anthers deltoid, very acuminate ; follicles diverging at an angle of about 80°, 3–$3\frac{3}{4}$ in. long, about $\frac{1}{4}$ in. thick, narrowly fusiform, gradually tapering from about the middle to a subacute point, glabrous ; seeds about 4 lin. long, 1 lin. broad, narrowly oblong, convex on one side, concave, with a central ridge on the other, minutely scabrous. *N. E. Br. in Dyer, Fl. Trop. Afr.* iv. i. 249. *Ectadium oblongi-folium, Meisn. in Hook. Lond. Journ. Bot.* ii. 1843, 542 (*by error* 442); *Hochst. in Flora,* 1844, 827 ; *Walp. Rep.* iv. 481. *Secamone acutifolia, Sond. in Linnæa,* xxiii. 76 ; *Walp. Ann.* iii. 48. *Ecta-diopsis oblongifolia, Benth., and E. acutifolia, Benth. in Benth. & Hook. f. Gen. Pl.* ii. 741. *E. oblongifolia, Schlechter in Engl. Jahrb.* xviii., *Beibl.* 45, 14 ; xx. *Beibl.* 51, 10.

SOUTH AFRICA : without locality, *Zeyher,* 1188 !
KALAHARI REGION : Transvaal ; Magalies Berg, *Burke,* 322 ! in stony places

near the Mooi River and on the Magalies Berg, *Zeyher*, 1182! near Apies River, 6000 ft., *Schlechter*, 3590, ex *Schlechter* in stony places near Little Olifant River, 5000 ft., *Schlechter*, 3804! near Pretoria, *McLea in Herb. Bolus*, 5701! *Leendertz*, 145! 345! Apies Poort, *Rehmann*, 4159! hills near Barberton, 3000–3300 ft., *Galpin*, 383, ex *Schlechter*; stony places near Botsabelo, 5000 ft., *Schlechter*, 4096! near Johannesburg, *Conrath*, 1061! *Gilfillan in Herb. Galpin*, 6144! *Rand*, 959! Shiluvane, *Junod*, 864!

EASTERN REGION: Natal; at the borders of woods near the Umgeni River, *Krauss*, 132! Inanda, *Wood*, 446! Fields Hill, *Wood in Natal Herb.*, 9! Palmiet River, *Gerrard*, 10! near Durban, *Gerrard*, 595! and without precise locality, *Gerrard*, 132!

Also in Tropical Africa.

II. **STOMATOSTEMMA**, N. E. Br.

Calyx 5-partite. *Corolla-tube* broadly campanulate; lobes overlapping in bud. *Corona* of 5 fleshy clavate lobes inserted in the sinuses between the corolla-lobes. *Stamens* inserted near the base of the corolla-tube; filaments free; anthers united at their base to the dilated part of the style, connivent in a cone, triangular, with the connective produced into a fleshy apiculus, glabrous. *Pollen* granular. *Pollen-carriers* with the margins inrolled so as to nearly form an oblique-mouthed tube, broadly ovate when flattened out, grooved down the back. *Style* shortly conical at the apex, shorter than the anthers. *Follicles* and seeds not seen.

A climbing or bushy shrub, with milky juice; leaves opposite; flowers of moderate size, in few-flowered cymes, axillary (often from both axils) and terminal, sometimes forming an elongated terminal narrow panicle, sometimes somewhat corymbose.

DISTRIB. Species 1, also in Tropical Africa.

I have separated this plant from *Cryptolepis* chiefly on account of the position of the corona-lobes, which are much more distant from the stamens than they are in *Cryptolepis* and occupy the same position that the sinus-pockets do in that genus, so that they probably represent the same organs, whilst the corona-lobes which arise from the middle of the corolla-tube in *Cryptolepis* are not represented in *Stomatostemma*. The corolla also differs from that of *Cryptolepis* in being more inflated and more obtuse when in bud, and has a much broader tube and broader lobes than in any species of that genus. The name is formed from στομα (a mouth) and στεμμα (a crown), in allusion to the position of the corona at the mouth of the corolla-tube.

1. **S. Monteiroæ** (N. E. Br. in Dyer, Fl. Trop. Afr. iv. i. 253); a climber, glabrous in all parts; leaves spreading, moderately distant; petiole about 2 lin. long; blade $1\frac{1}{4}$–3 in. long, $\frac{1}{6}$–$\frac{3}{4}$ in. broad, linear, oblong, lanceolate or cuneate-obovate, acute or obtuse and apiculate at the apex, acute at the base; cymes terminal, sub-corymbose, or spaced out along the terminal part of the stem in a racemose manner, few-flowered, 1–$1\frac{1}{2}$ in. long, including the peduncle; pedicels $1\frac{1}{2}$–2 lin. long, with two minute bracts at about the middle; sepals 1–$1\frac{1}{2}$ lin. long, ovate or ovate-oblong, obtuse; buds ellipsoid; corolla 1–$1\frac{1}{4}$ in. in diam., campanulate, cream-coloured, densely dusted with purple-brown along the middle of the lobes and entirely

purple-brown in the tube, or pink spotted with brown (*Galpin*); tube 2–3 lin. long; lobes 5–6 lin. long, oblong, obtuse, margins revolute; corona-lobes inserted at the sinuses of the corolla, 1 lin. long, clavate, fleshy, dark purple-brown or blackish; anthers acuminate, connivent over the short conical apex of the style, whitish or yellowish. *Cryptolepis Monteiroæ, Oliv. in Hook. Ic. Pl.* xvi. *t.* 1591; *Schlechter in Engl. Jahrb.* xviii., *Beibl.* 45, 14, *and Journ. Bot.* 1896, 315.

KALAHARI REGION : Transvaal; Avoca, near Barberton, 1900 ft., climbing over trees 20 ft. high, *Galpin,* 1250! Bush Veld, near Warm Bath, 3500 ft., *Bolus,* 12156!

EASTERN REGION : Delagoa Bay, *Mrs. Monteiro!*

Also in Tropical Africa near Lake Ngami.

III. ECTADIUM, E. Meyer.

Calyx 5-partite; sepals erect. *Corolla* hypocrateriform; tube cylindric; lobes overlapping to the left and slightly twisted in bud. *Corona* of 5 subulate lobes at the mouth of the corolla-tube, alternating with the corolla-lobes. *Stamens* inserted at the bottom of the corolla-tube; filaments very short, free; anthers adnate at their base to the dilated part of the style, produced above into a long erect appendage, hairy; pollen granular; pollen-carriers spathulate-oblong, emarginate at the apex. *Style* very much shorter than the stamens; apex rather slender, bifid. *Follicles* widely divergent.

Shrubby plants, with erect, virgate stems, opposite leaves, and lateral, forked cymes.

DISTRIB. Two species, endemic.

Leaves 1–2 lin. broad, not pitted in the adult stage	...	(1) **virgatum.**
Leaves 5–7 lin. broad, pitted in the adult stage...	...	(2) **latifolium.**

1. **E. virgatum** (E. Meyer, Comm. Pl. Afr.-Austr. 188); stems erect, simple or sparingly branched, 1–1¼ lin. thick, minutely puberulous or glabrous; leaves ascending, 1½–3 in. long, 1–2 lin. broad, linear, acute, narrowed at the base into a petiole 1–2 lin. long, microscopically puberulous or glabrous; cymes 3–6 to a stem, bifurcate, lateral at the nodes, pedunculate, whitish-tomentose, many flowered; peduncles ⅓–1 in. long, ascending; bracts 1–2½ lin. long, subulate; sepals 2–2½ lin. long, ½–⅔ lin. broad, lanceolate, acute, tomentose, ciliolate; corolla hypocrateriform; tube 2–2¼ lin. long, about 1¼ lin. in diam., cylindric, minutely pubescent outside; lobes about 2½ lin. long, ⅔–1 lin. broad, obliquely oblong, spathulate-oblong or lanceolate, obtuse or subacute, usually somewhat twisted, glabrous on both sides; corona-lobes ¾–1 lin. long, erect, subulate, with a shortly decurrent keel at the base; stamens with filaments ¼ lin. long; anthers deltoid-sagittate at the base, produced into a filiform, densely hairy appendage about 1¼ lin. long, connivent-erect; apex of the style deeply bifid, ¼ lin. long; follicles widely

divergent, 2–2½ in. long, 3–4 lin. thick, terete, tapering to an obtuse·
point, smooth, minutely puberulous or glabrous ; seeds about 3–4
lin. long, 1–1¼ lin. broad, oblong-lanceolate, keeled down the concave
face, smooth on both sides, blackish-brown. *DC. Prodr.* viii. 500;
Harv. Gen. S. Afr. Pl. ed. 2, 230 ; *Schlechter in Engl. Jahrb.* xxi.,
Beibl. 54, 1, *and Journ. Bot.* 1896, 314.

WESTERN REGION : Little Namaqualand ; in muddy places on the banks of
the Orange River, below 300 ft., *Drège,* 3047 ! and without precise locality,
Wyley, 76 !

2. **E. latifolium** (N. E. Br.) ; a shrub 1½–3 ft. high ; leaves
2¼–3¼ in. long, 5–7 lin. broad, lanceolate, acute, tapering to the
base, very thick and coriaceous, tomentose on both sides when
young, becoming glabrous in the adult stage and densely reticulate-
punctate on both sides ; petioles 1½–2 lin. long ; cymes similar to
those of *E. virgatum,* but larger, white-tomentose ; bracts 2 lin.
long, subulate ; pedicels 1–1½ lin. long ; sepals 3–3½ lin. long, ¾–1
lin. broad, lanceolate, acuminate, tomentose ; corolla hypocrateriform ;
tube 2½–2¾ lin. long, cylindric, pubescent outside ; lobes 2½–3 lin.
long, 1–1⅙ lin. broad, oblong, obtuse, glabrous on both sides ;
corona-lobes ½–⅔ lin. long, erect, subulate from a deltoid-oblong
base, usually with an indication of a minute tooth on each side at
the termination of the dilated part ; anther-appendages 1½–1¾ lin.
long, connivent-erect, filiform, densely white-hairy ; apex of the
style deeply bifid, ¼ lin. long ; follicles about 2½ in. long ; seeds
4½ lin. long. *E. virgatum, E. Meyer, var. latifolium, Schinz in
Verhandl. Bot. Ver. Brandenb.* xxx. 261.

WESTERN REGION : Great Namaqualand ; near Angra Pequena, in loose sand,
Schenck, 11 ! 30, *Schinz.*

Although very similar to *E. virgatum,* E. Meyer, this plant essentially differs
by its larger flowers, and especially in its leaves, which are at least twice as
thick as those of *E. virgatum,* and, in the adult state, are densely covered with
minute irregular pits, which appear, when viewed under a lens, as if placed
between a very dense network of veins. I find no trace of such structure in the
leaves of *E. virgatum.*

IV. **RAPHIONACME**, Harv.

Calyx 5-partite. *Corolla* with a distinct campanulate tube, and
erect, spreading or reflexed lobes, often with 2 narrow ridges on
their basal half, overlapping to the left and more or less twisted in
bud. *Corona* of 5 free, entire, bifid or trifid lobes, inserted with
the stamens and occasionally upon their filaments above the middle
of the corolla-tube, or at its mouth, alternating with the corolla-
lobes. *Stamens* inserted above the middle or at the mouth of the
corolla-tube ; filaments free ; anthers adnate to the dilated part of
the style at their base, connivent in a cone, and connate at their
tips. *Pollen* granular ; pollen-carriers spathulate. *Style* shortly
conical at the apex, not exceeding the anthers. *Follicles* often

solitary by abortion, short or long, lanceolate, fusiform or linear-terete. *Seeds* crowned with a tuft of hairs.

Perennial herbs; rootstock a depressed or ovoid tuber, often having a long woody neck, or consisting of a cluster of long fusiform fleshy roots $\frac{1}{2}$–1 in. thick; juice milky; stem usually dwarf, simple or branched, sometimes twining; leaves opposite; flowers small or of moderate size, in few- or many-flowered cymes or clusters, subaxillary from one axil, lateral at the nodes or terminal or seated in the forks of the stem.

DISTRIB. Species about 25, all the others in Tropical Africa.

Closely related to *Tacazzea*, from which it chiefly differs in its usually distinct campanulate corolla-tube, by the cymes never arising from both axils, and by habit.

The cymes in *Raphionacme* are not truly axillary, although often apparently so; morphologically they terminate the axis bearing the pair of leaves, from one of whose axils they appear to arise; the bud in the axil of one of these leaves rapidly develops, and falsely appears to be a direct continuation of the axis below, forming a sympodial stem; the bud in the other axil is arrested, and thus the terminal cyme takes its place and appears to be axillary; in a few species, however, the buds in both axils grow out and form branches, the cymes are then placed in the forks of the stem, and their truly terminal nature is made evident.

<pre>
Stem distinctly twining, minutely puberulous:
    Corolla-lobes 5–6 lin. long, 2½–3 lin. broad   ...   (1) Monteiroæ.
    Corolla-lobes 2½–3½ lin. long, ¾–1½ lin. broad ...   (2) Flanagani. .
Stem procumbent, up to 1 ft. long, conspicuously
    pubescent; corolla-lobes about 1¼ lin. long   ...   (3) procumbens.
Stem and branches erect:
    Plant usually with a solitary stem, simple or with
      1 branch near or above the middle:
        Plant 1–2 ft. high, rarely less, with a
          minute inconspicuous pubescence on the
          stem and leaves   ..   ..   (4) elata.
        Plant 2–9 in. high, with a very conspicuous
          pubescence on the leaves ...   .   (5) Galpinii.
    Plant branching at the base into 2 or more
      stems or the stem branching from the base
      upwards:
        Flowers green or yellowish-green; corona-
          lobes deeply trifid:
            Sepals less than twice as long as broad,
              broadly ovate or oblong-ovate, obtuse
              or subacute   ...   .   (6) Zeyheri.
            Sepals generally 2 or more times as
              long as broad, narrow, very acute:
                Leaves glabrous above, puberulous
                  beneath; sepals shorter than the
                  corolla-tube or rarely equalling it   (7) Burkei.
                Leaves softly and conspicuously
                  pubescent on both sides; sepals
                  slightly longer than the corolla-
                  tube   ...   .   (8) velutina.
        Flowers purple; corona-lobes variable,
          entire or toothed at the apical part only ...   (9) divaricata.
                                                            :
</pre>

1. **R. Monteiroæ** (N. E. Br.); rootstock tuberous; stem slender, twining, minutely puberulous; leaves distant, spreading; petiole

1–2 lin. long; blade $\frac{2}{3}$–$1\frac{1}{2}$ in. ($1\frac{1}{2}$–$2\frac{1}{2}$ in. ex *Oliver*) long, $3\frac{1}{2}$–7 lin. broad, oblong-obovate, obtuse, shortly apiculate, thinly puberulous above, softly pubescent beneath; cymes from one axil, few-flowered; peduncle and branches of the cyme each about 2–3 lin. long, pubescent; bracts $\frac{1}{2}$–$\frac{2}{3}$ lin. long, ovate, acute, brown, pubescent; pedicels $1\frac{1}{2}$–2 lin. long, pubescent; sepals 1 lin. long, $\frac{2}{3}$–$\frac{3}{4}$ lin. broad, ovate, acute, pubescent; corolla about 1 in. in diam., minutely pubescent outside, glabrous within, green; tube about 3 lin. long and broad, campanulate, internally with 5 decurrent broadly 2-winged nectaries alternating with the stamens; lobes ascending, 5–6 lin. long, $2\frac{1}{2}$–3 lin. broad, ovate-oblong, obliquely subemarginate at the obtuse apex, strongly revolute along the margins; corona-lobes inserted at the sinuses of the corolla, very broad at the base, and laterally adnate to the base of the corolla-lobes, 3-fid; lateral teeth about $\frac{1}{2}$ lin. long, deltoid-oblong, obtuse, incurved, middle tooth 2 lin. long, subulate, incurved over the anthers; staminal filaments $\frac{2}{3}$ lin. long; anthers $1\frac{3}{4}$–2 lin. long, oblong, apiculate, connivent and connate at the tips; pollen-carriers $1\frac{1}{2}$ lin. long; blade broadly ovate or cordate, obtuse; stalk slender, dilated towards the base; gland large, rectangular; style broadly and obtusely conical at the apex, shorter than the anthers. *Chlorocyathus Monteiroæ, Oliv. in Hook. Ic. Pl.* xvi. *t.* 1557; *Schlechter in Journ. Bot.* 1896, 314.

EASTERN REGION: Delagoa Bay, cultivated specimen, *Mrs. Monteiro!*

Described from a plant sent to Kew in 1882 by Mrs. Monteiro, which flowered in July, 1886.

2. R. Flanagani (Schlechter in Engl. Jahrb. xviii., Beibl. 45, 2); stem twining, puberulous on the young parts; leaves spreading, petiolate; petiole $1\frac{1}{2}$–6 lin. long; blade 2–$3\frac{1}{2}$ in. long, $\frac{1}{2}$–$1\frac{1}{3}$ in. broad above the middle, elongate-oblong, oblong-oblanceolate or obovate, acute, or obtuse and apiculate or very shortly cuspidate, more or less cuneate at the base, flat or wavy along the margins, rather thinly puberulous and dark green above, velvety puberulous and pale brown or pale greyish beneath in the dried state; cymes subaxillary, subsessile or on peduncles 1–6 lin. long, $\frac{1}{2}$–1 in. in diam., many-flowered; bracts 1–2 lin. long, lanceolate, acute, pubescent; pedicels 1–2 lin. long; sepals $\frac{2}{3}$–1 lin. long, ovate, acute or subobtuse, pubescent; corolla velvety-pubescent outside; tube 1 lin. long, campanulate; lobes spreading, $2\frac{1}{2}$–$3\frac{1}{2}$ lin. long, $\frac{3}{4}$–$1\frac{1}{2}$ lin. broad, oblong or lanceolate, obtuse or subacute; corona-lobes tripartite, middle segment 2–$3\frac{1}{4}$ lin. long, tortuous, lateral segments $\frac{1}{4}$–1 lin. long, all subulate or filiform; anthers oblong, apiculate; follicles about 2 in. long, lanceolate, acuminate, very minutely puberulous. *Schlechter in Journ. Bot.* 1895, 302, *and* 1896, 315. *R. scandens, N. E. Br. in Kew Bulletin,* 1895, 111.

COAST REGION: Komgha Div.; among shrubs along the Kei River, 1800 ft., *Flanagan,* 118!

EASTERN REGION: Natal; Tugela, *Gerrard.* 1312! near the Umkomanzi River, 3000 ft., *Schlechter,* 6691!

I have examined a type specimen of *C. Flanagani*, but do not find the leaves "ovate- or lanceolate-elliptic" nor the corolla-lobes 2 lin. broad, as described by Schlechter.

3. R. procumbens (Schlechter in Engl. Jahrb. xx., Beibl. 51, 11) ; plant branching at the base ; branches ½–1 ft. long, procumbent, softly villous-pubescent ; leaves somewhat rigidly coriaceous when dried ; petiole 1–4 lin. long ; blade 1–2½ in. long, 3½–7 lin. broad, narrowly oblong, oblong-lanceolate or elliptic-lanceolate, acute or obtuse and apiculate, both sides covered with rather long soft spreading hairs, but more densely beneath ; cymes subglobose, about ½ in. in diam., rather densely many-flowered, apparently pendent (always ?), on recurved or spreading peduncles 2–5 lin. long, shortly villous-pubescent on all parts to the outside of the corolla ; bracts 1–1½ lin. long, subulate ; pedicels 1¼–1½ lin. long, curved ; sepals 1¼–1½ lin. long, sometimes united into a short tube at the base, ovate-lanceolate, acuminate ; corolla-tube ½ lin. long ; lobes about 1¼ lin. (2 lin. ex *Schlechter*) long, ⅗ lin. broad, lanceolate or oblong-lanceolate, obtuse, with slightly incurved margins, glabrous on the inner face ; corona-lobes arising at the mouth of the corolla-tube, trifid, with the middle tooth scarcely ½ lin long and 3–4 times as long as the lateral teeth, but shorter than the anthers, which are connivent into an acute cone ½ lin. long, with connate tips. *Schlechter in Journ. Bot.* **1896, 315.**

KALAHARI REGION : Transvaal ; stony places on the Elandspruit Mountains, 5300 ft., *Schlechter*, 3867 !

Described from part of the type in Herb. Bolus. The pubescence on the leaves and flowers is like that of *R. Galpinii*, but the hairs on the stem are much longer and the flowers smaller.

4. R. elata (N. E. Br.) ; stems 1–2 ft. high, usually simple, erect, sometimes with a slight indication of twining at the terminal part, minutely and softly puberulous ; leaves petiolate ; petiole 2–4 lin. long ; blade 1¾–3¼ in. long, ⅙–2¼ in. broad, linear-lanceolate, lanceolate, oblong, obovate-oblong or elliptic, acute or obtuse and apiculate, cuneate to rounded at the base, minutely puberulous on both sides, paler beneath ; cymes lateral at the nodes or subaxillary and terminal, subglobose, ¾–1 in. in diam., subsessile or on peduncles 1–6 lin. long, many-flowered ; bracts 3–4 lin. long, subulate, pubescent ; pedicels 1½–2 lin. long, softly pubescent ; sepals 2–2½ lin. long, subulate, acute ; corolla velvety pubescent outside ; tube about 1½ lin. long, campanulate ; lobes 2½–3 lin. long, 1¼ lin. broad, oblong or somewhat ovate-oblong, obtuse, spreading, green, purple at the base ; corona-lobes trifid, shortly transverse-rectangular at the green base ; middle segment 2–2½ lin. long, filiform, white ; lateral segments ⅔ lin. long, filiform or subulate ; anthers oblong, apiculate ; pollen-carriers 1 lin. long, with an ovate-lanceolate blade and slender stalk of about equal length ; follicles solitary (always ?), 4½–5 in. long, ¼ in. thick, terete-fusiform, tapering into a beak, smooth, puberulous. *R. Galpinii, Schlechter in Engl. Jahrb.* xx., *Beibl.* **51, 10,** *not of* xviii. *Beibl.* **45, 14.**

KALAHARI REGION : Transvaal ; hills near Rustenburg and between there and Pretoria, *Miss Alice Pegler,* 1054! near the Olifants River, 3000 ft., *Schlechter,* 3768! (a specimen from Lydenburg, *Wilms,* 955, may belong here).

EASTERN REGION : Pondoland ; near the mouth of St. John's River, *Bolus,* 8308! Griqualand East ; hills around Clydesdale, 4300 ft., *Tyson,* 1248 ! Natal ; Verulam and Inanda, *Wood,* 849 ! and without precise locality, *Gerrard & McKen,* 1301 ! Zululand ; Sebundini, *Haygarth in Herb. Wood,* 7567 ! Swaziland ; near Bremersdorp, 2200 ft., *Bolus,* 12153 ! on a ridge between Bremersdorp and Miles, *Burtt Davy,* 2940 !

Specimens collected near Fort Bowker in Transkei (*Bowker,* 377) are only 6-9 in. high, but must, I think, be referred to this species, which is readily distinguished from *R. Galpinii* by its much larger stature and very minute pubescence.

According to Miss Alice Pegler the flowers are cream-coloured.

5. **R. Galpinii** (Schlechter in Engl. Jahrb. xviii., Beibl. 45, 14, not elsewhere) ; tuber up to 6 in. in diam., depressed ; stems 2-9 in. high, puberulous, usually simple ; leaves 3-7 pairs to a stem, shortly petiolate, 1-3 in. long, 3-9 lin. broad, linear-lanceolate, narrowly oblong, oblanceolate-oblong or narrowly obovate, acute, or obtuse and apiculate, conspicuously and softly pubescent on both sides ; petioles $1\frac{1}{2}$-3 lin. long ; cymes very dense, subglobose, about $\frac{3}{4}$ in. in diam., subaxillary, or crowded together at the top of the stem, many-flowered, sessile or pedunculate ; peduncles 2-6 lin. long, pubescent ; bracts 2-3 lin. long, subulate, pubescent ; sepals $1\frac{3}{4}$-$2\frac{1}{4}$ lin. long, $\frac{1}{3}$-$\frac{1}{2}$ lin. broad, linear-subulate, pubescent ; corolla pubescent outside, bright green ; tube campanulate, 1-$1\frac{1}{3}$ lin. long ; lobes $1\frac{3}{4}$-$2\frac{2}{3}$ lin. long, $\frac{3}{4}$-$1\frac{1}{4}$ lin. broad, oblong or elongate-ovate, obtuse, ascending ; corona-lobes arising at the mouth of the corolla-tube, very shortly transverse-rectangular at the base, 3-fid, the middle segment or tooth $1\frac{1}{4}$-2 lin. long, filiform, the lateral varying from mere acute shoulder-teeth to filiform and $\frac{1}{3}$-$\frac{1}{2}$ as long as the middle segment, erect or more or less divergent, sometimes connate with those of the adjacent lobes and forming 5 short bifid lobules alternating with 5 long filiform teeth ; anthers ovate, acute. *R. macrorrhiza, Schlechter in Engl. Jahrb. xx., Beibl. 51, 10, and in Journ. Bot.* 1896, 315.

KALAHARI REGION : Transvaal, 4000-6000 ft. ; slopes of the Saddleback Range near Barberton, *Galpin,* 613 ! near and around Johannesburg, *Gilfillan in Herb. Galpin,* 6043 ! *Rand,* 711 ! 1123 ! near Pretoria, *Leendertz,* 296 !

EASTERN REGION : Natal ; Groen Berg, *Wood,* 1032 ! Inanda and near Verulam, *Wood,* 1060! Greenwich Farm, Riet Vlei, *Fry in Herb. Galpin,* 2750 ! near Clairmont, 100 ft., *Wood,* 4925 ! *Schlechter,* 3084 ! near Krantz Kloof, 1400 ft., *Schlechter,* 3213 ! and without precise locality, *Sanderson,* 177 !

I can find no specific distinction between *R. Galpinii* and *R. macrorrhiza,* the side-teeth of the corona-lobes being variable. But it would appear that at the time of publishing *R. macrorrhiza* Dr. Schlechter mistook *R. elata* for *R. Galpinii,* since he quotes his 3768 as being the latter species on the same page.

6. **R. Zeyheri** (Harv. in Hook. Lond. Journ. Bot. i. 1842, 23) ; plant 2-5 in. high, branching from the woody neck of the tuber, very

minutely puberulous on all parts to the outside of the corolla;
leaves 1–1¼ in. long, 1–4 lin. broad, linear or linear-lanceolate,
subacute, tapering at the base into a very short petiole; cymes
small, axillary, 5–10-flowered; peduncles ½–1 lin. long; bracts
½–⅔ lin. long, subulate; pedicels ½–¾ lin. long; sepals ⅔ lin. long,
½ lin. broad, ovate or oblong-ovate, obtuse or subacute, puberulous;
corolla from nearly glabrous to puberulous outside, apparently green;
tube ¾–1 lin. long, campanulate; lobes 1⅔ lin. long, ¾ lin. broad,
oblong, subobtuse; corona-lobes arising at the mouth of the tube,
trifid, shortly transverse rectangular at the base; middle segment
1½–2 lin. long, filiform, lateral segments ½ lin. long, subulate, or
reduced to minute obtuse teeth; anthers ovate, acuminate; pollen-
carriers spathulate, with an elliptic or suborbicular blade. *Schlechter
in Journ. Bot.* 1896, 315.

COAST REGION : Uitenhage Div. ; near the Zwartkops River, *Zeyher!* near
Uitenhage, *Prior!* Albany Div. ; by the Fish River, *Burke!*

7. R. Burkei (N. E. Br.); plant 3–7 in. high, branching from
the base; stems minutely pubescent; leaves spreading; petiole
1–1½ lin. long; blade ½–1¾ in. long, 1–6 lin. broad, linear or linear-
oblong, subacute, narrowed into the petiole at the base, longitudinally
folded in the dried state, glabrous above, minutely puberulous
beneath; cymes lateral or subaxillary, sessile, many-flowered, about
4–5 lin. in diam. ; bracts ½ lin. long, ovate, acute, pubescent;
pedicels ½–1 lin. long, pubescent; sepals ⅔–¾ lin. long, ⅓ lin. broad,
narrowly deltoid or lanceolate-attenuate, acute, pubescent; corolla
distinctly pubescent outside, green; tube 1–1¼ lin. long; lobes
1¼–1⅔ lin. long, oblong-ovate, obtuse, spreading; corona-lobes
arising at the mouth of the corolla-tube, tripartite; middle segment
equalling or longer than the lateral segments, ¼–1 lin. long, filiform;
lateral ¼–⅓ lin. long, subulate; anthers narrowly ovate, acuminate;
follicles solitary (only 1 seen), erect, about 2½ in. long and ¼ in.
thick, terete-fusiform, scarcely beaked, smooth, glabrous.

KALAHARI REGION : Bechuanaland ; near the sources of the Kuruman River,
Burchell, 2444 ! 2455/1 ! 2497 ! Transvaal; northern slopes of the Magalies
Berg, 6000–7000 ft., *Zeyher,* 1141 ! *Burke,* 64 ! Modderfontein, *Conrath,* 978 !
near Rustenburg, 4000 ft., *Miss Alice Pegler,* 982 !

This is closely allied to *R. Zeyheri* and has been so named by Dr. Schlechter,
but the very differently shaped sepals, shorter corona-lobes, longer (although
minute) pubescence and different geographical area well distinguish it. Burchell's specimens have narrower leaves and fewer and rather smaller flowers than
the typical Transvaal plant, but I can find no other difference.

8. R. velutina (Schlechter in Engl. Jahrb. xx., Beibl. 51, 12);
plant 3–5 in. high, with 1 or more erect branching stems from the
woody neck of the tuber, conspicuously velvety-pubescent on the
stems, both sides of the leaves, and on the inflorescence, including
the outside of the corolla; leaves spreading or ascending; petiole
1–3 lin. long; blade ¾–1¾ in. long, 2½–4 lin. broad, more or less

folded lengthwise, narrowly oblong-linear or linear, obtuse or acute, acutely tapering into the petiole at the base, grey-green (from the pubescence) beneath, darker above; cymes sublateral at the nodes and terminal, pedunculate, rather densely 10–25-flowered; peduncles 2–4 lin. long; bracts subulate, $1\frac{1}{2}$–2 lin. long; pedicels 1–$1\frac{3}{4}$ lin. long; sepals slightly exceeding the corolla-tube, 1–$1\frac{1}{4}$ lin. long, $\frac{1}{2}$ lin. broad, lanceolate, attenuate-acute; corolla-tube 1 lin. long, campanulate; lobes suberect, $1\frac{3}{4}$–2 lin. long, 1 lin. broad, oblong, obtuse, glabrous on the inner face, green, not twisted in bud; corona-lobes arising at the mouth of the corolla-tube, trifid, very shortly transverse-rectangular at the base; segments filiform, the middle one 1–$1\frac{1}{2}$ lin. long, the lateral $\frac{2}{3}$ lin. long, erectly spreading; anthers ovate, acuminate, somewhat wavy at the margins; pollen-carriers pale ochreous yellow, spathulate, with a lanceolate obtuse or slightly bifid blade, revolute at the margins, and a rather slender stalk $\frac{2}{3}$ as long as the blade. *Schlechter in Journ. Bot.* 1896, 315.

KALAHARI REGION; Transvaal; near Heidelberg, 4900 ft., *Schlechter,* 3509; near Irene, *Conrath,* 979!

I have not seen a type specimen of *R. velutina,* and the above description is made entirely from Conrath's 979, which is the only specimen I have seen that at all corresponds with Schlechter's description of *R. velutina* and has a really "velvety" pubescence. In the British Museum, however, specimens of *R. Burkei* are named by Schlechter himself as being "*R. velutina,*" whilst the same plant in Herb. Conrath is named by him "*R. Zeyheri,*" Conrath's 979 being named "*R. procumbens.*" Upon comparison with the type of *R. procumbens* (Schlechter, 3867) I find that to be a totally different plant, whilst the pubescence on *R. Burkei* is scarcely velvety and the leaves are glabrous above. I am, therefore, uncertain as to what plant *R. velutina* is, but if I am right in identifying Conrath's specimen with it, the much longer and more evident pubescence, which resembles that on *R. Galpinii* and is present on both sides of the leaves, readily distinguishes it from its allies.

9. R. divaricata (Harv. in Hook. Lond. Journ. Bot. i. 1842, 23); a dwarf herb $2\frac{1}{2}$–8 in. high, much branched, pubescent with short spreading hairs on the stems, under surface of the leaves, peduncle, pedicels, calyx and outside of the corolla; leaves spreading or ascending; petiole $\frac{1}{2}$–3 lin. long; blade $\frac{1}{2}$–$1\frac{1}{4}$ in. long, $\frac{1}{4}$–1 in. broad, elliptic, orbicular or broadly obovate, acute or obtuse and apiculate, rounded or cuneately narrowed at the base; cymes sublateral at the nodes, or central in the forks of the stem, 5- to many-flowered; subsessile or on peduncles up to 1 in. long; bracts $\frac{2}{3}$–$1\frac{1}{2}$ lin. long, subulate, acute; pedicels $\frac{3}{4}$–6 lin. long; sepals 1–2 lin. long, $\frac{1}{3}$–$\frac{1}{2}$ lin. broad, lanceolate-subulate, acute; corolla very variable in size, purple; tube 1–$1\frac{2}{3}$ lin. long, campanulate; lobes $1\frac{1}{2}$–4 lin. long, $\frac{3}{4}$–$1\frac{1}{2}$ lin. broad, oblong, obtuse, spreading; corona-lobes exceedingly variable in size and form, varying even in the same flower, connivent over the staminal column, white or purple-tinted, $\frac{2}{3}$–2 lin. long, $\frac{1}{3}$–1 lin. broad, oblong-lanceolate, oblong-obovate, oblong-subspathu-late, or lanceolate, gradually or abruptly contracted into a short or long, simple or bifid filiform point, or simply bifid to about $\frac{1}{3}$ the way down, entire, or the broad terminal part more or less denticulate on each side of the cusp, or distinctly trifid, or the whole apex divided into an irregular fringe of teeth and filiform processes; anthers oblong, or slightly broader at the base, acute; pollen-carriers about $\frac{1}{2}$ lin. long, constricted at the middle, the small elliptic blade being no broader than the ovate or oblong stalk; follicles solitary, 1–2½ in. long, 3½–5 lin. thick, narrowly lanceolate or ovate-lanceolate, tapering to an acute point, smooth, puberulous; seeds about $\frac{1}{3}$ in. long, oblong, obtuse, flattened with a very prominent keel on one face, glabrous. *Walp. Rep.* vi. 480; *Schlechter in Engl. Jahrb.* xviii., *Beibl.* 45, 2, *and in Journ. Bot.* 1896, 315; *Rand in Journ. Bot.* 1903, 198. *R. pubescens, Hochst. in Flora,* 1844, 827; *Walp. Rep.* vi. 480. *R. obovata, Turcz. in Bull. Soc. Mosc.* 1848, *pt.* 1, 250; *Walp. Ann.* iii. 45; *Schlechter in Engl. Jahrb.* xx., *Beibl.* 51, 12. *Apoxyanthera pubescens, Hochst. in Flora,* 1843, 78. *Brachystelma? hirsutum, E. Meyer, Comm. Pl. Afr. Austr.* 197; *Dietr. Synop. Pl.* ii. 888; *Decne in DC. Prodr.* viii. 647.

VAR. β, glabra (N. E. Br.); plant glabrous in all parts or the stem more or less puberulous. *R. purpurea, Harv. Thes. Cap.* i. 41, *t.* 66. *Mafekingia parquetiana, Baill. in Baill. Hist. des Pl.* x. 303.

SOUTH AFRICA: without locality, *Ecklon,* 64 (ex *Turczaninow*) *Pearson!* the type and var. β, *Zeyher,* 1140!

COAST REGION: Alexandria Div.; Quagga Flats, *Bowie!* Albany Div.; Broekhuisons Poort, and in sandy rocky places near Grahamstown, 2000 ft., *MacOwan,* 707! *Hutton! Miss Daly,* 57! Bathurst Div.; between Kasuga River and Port Alfred, *Burchell,* 3968! Komgha Div.; grassy hills near Komgha, 1900 ft., *Flanagan,* 394 ex *Schlechter;* Queenstown Div.; Queenstown Flats, *Mrs. Barber,* 89! near the Zwart Kei River, *Cooper,* 319! British Kaffraria, *Cooper,* 2708!

KALAHARI REGION: Orange River Colony; Harrismith, *Sankey,* 19! Bethlehem, *Richardson!* Transvaal; Magalies Berg, *Burke! Zeyher!* Pilgrims Rest, *Greenstock!* Matebe Valley, *Holub!* Christiana on the Vaal River, *Nelson,* 202! Zuikerbosch Rand (and also var. β), *Schlechter,* 3499! Var. β: Transvaal; various localities, *Burke! Zeyher! Wilms,* 962a! *Bolus,* 8309! *MacLea in Herb. Bolus,* 8310! *Conrath,* 977! *Schlechter,* 6385! *Olive Nation,* 232! 274! *Leendertz,* 406! *Burtt Davy,* 731! 752! 1060! 1229! 1529! 2314! *Rand,* 861! 1230! *Rogers,* 240! 300! 801!

EASTERN REGION: Transkei; near Kentani, *Miss Alice Pegler,* 876! Tembuland; near Bazeia, *Baur,* 382 *bis* partly! Griqualand East, near Kokstad, *Tyson!* Natal; Weenen County, 3000–5000 ft., *Sutherland!* hills near Pieter Maritzburg, *Krauss,* 106b ex *Hochstetter;* Inanda, *Wood,* 368! 527! Port Natal, *Miss Owen!* Dargle Farm, *Mrs. Fannin,* 38! Var. β: Transkei; Kreilis Country, *Bowker!* Tsomo, *Bowker,* 775! and without precise locality, *Hallack!* Tembuland; on hill-sides, &c., near Bazeia, *Baur,* 382! 382 *bis* partly! Griqualand East; mountain slopes around Kokstad, *Tyson,* 1851! Natal; various localities, 500–3000 ft., *Sutherland! Sanderson,* 84! *Gerrard,* 1808! *Wilms* (glabrous and pubescent), 2010!

This is the most widely distributed of the South African species, and is very variable. Sometimes it is nearly leafless at the time of flowering, and some specimens (*Cooper,* 319, and *Mrs. Barber,* 89) in this state are reduced, probably from starvation, to a small cushion-like mass of flowers about 1–1¼ high. The corona-lobes are exceedingly variable, the same flower often having 2 or more variations. E. Meyer has stated that the flowers of *Brachystelma hirsutum* are unknown, but I find some withered flowers on his type, which upon dissection prove it to belong to this species. Mr. Burtt Davy informs me that the tuber attains a diameter of 15 inches.

V. **TACAZZEA**, Decne.

Calyx 5-partite. *Corolla* 5-lobed almost to the base, rotate; lobes overlapping to the left and slightly twisted. *Corona-lobes* 5, arising from the corolla at or near the base of the staminal filaments and usually shortly adnate to them, alternating with the corolla-lobes, filiform and simple, or linear and divided above into 2 or 3 filiform segments. *Stamens* inserted at or near the base of the corolla; filaments free above and united at their base into a ring with 5 minute, subquadrate, emarginate or bifid, alternating lobules, or entirely free; anthers attached at their base to the dilated part of the style, connivent in a cone, cohering at their tips, glabrous. *Pollen* granular; pollen-carrier more or less spathulate. *Style* shortly conical at the apex, shorter than the anthers. *Follicles* divergent. *Seeds* crowned with a tuft of hairs.

Twining or erect shrubs, with milky juice; leaves opposite or whorled; stipules usually represented by a transverse line (often with fleshy glands along it) connecting the petioles; flowers small, in axillary, paniculate, or corymbose cymes, often from both axils.

Distrib. Species about 15, 2 in South Africa, the others in Tropical Africa.

The genus *Tacazzea* is very closely related to *Raphionacme* and difficult to distinguish technically; the chief points of difference are, that in *Tacazzea* the cymes are truly axillary and frequently produced from both axils, the united part of the corolla does not form a distinct campanulate tube, and most species have 5 minute lobules alternating with and united to the base of the stamens; the leaves also, of most but not all of the species, have small (fleshy ?) processes scattered along their midribs.

Leaves tomentose beneath; corolla-lobes 2½ lin. long　(1) **Kirkii.**
Leaves glabrous on both sides; corolla-lobes 4–5 lin.
　long　...　...　...　...　...　...　...　(2) **natalensis.**

1. T. Kirkii (N. E. Br. in Kew Bulletin, 1895, 248); stem twining, more or less tomentose; leaves distant, spreading; petiole ½–1½ in. long; blade 1¼–4 in. long, 1–2¼ in. broad, varying from oblong to elliptic, obtuse, apiculate, or rarely acute, cordate or obtusely rounded at the base, subcoriaceous, glabrous above, whitish or greyish-tomentose beneath; panicles 2–4 in. long, more or less pubescent or subtomentose, or the branches subglabrate; bracts 1–1½ lin. long, broadly ovate, acute, often with a few hairs down the back, ciliate; pedicels 1½–4 lin. long, glabrous or nearly so; sepals 1 lin. long, broadly ovate, obtuse or acute, more or less hairy on the back; corolla glabrous; lobes 2½ lin. long, oblong, obtuse; corona-lobes 4 lin. long, filiform, sometimes bifid at the apex, erect, tortuous in the upper part, adnate at the base to the lower half of the staminal filaments, which are united with the alternating, shortly bifid lobules into a ring; follicles 1½–2¾ in. long, 3½–4½ lin. thick at the base, gradually tapering to a point, slightly reflexed, tomentose or softly pubescent. *N. E. Br. in Dyer, Fl. Trop. Afr.* iv. i. 268. *T. Welwitschii, Schlechter in Journ. Bot.* 1896, 314, *not of Baill. Leptopætia, Harv. Gen. South Afr. Pl. ed.* 2, 231.

EASTERN REGION : Natal or Zululand; without precise locality, *Gerrard,*
1796 !

Also in Tropical Africa.

2. T. natalensis (N. E. Br.); rootstock tuberous, 2 ft. in diam.
(*Miss Pegler*), just appearing above the ground; stem moderately
stout, twining to great heights, glabrous; bark reddish-brown;
leaves petiolate, spreading, glabrous; petiole $\frac{1}{2}$–$\frac{3}{4}$ in. long; blade
$3\frac{1}{2}$–5 in. long, $1\frac{1}{4}$–$2\frac{1}{2}$ in. broad, oblong, with nearly parallel sides,
very abruptly cuspidate, the point 4–5 lin. long, acute, broadly
rounded and shortly cordate at the base, with small overlapping
lobes; midrib very stout, without glands along its upper side;
veins numerous, horizontally spreading; cymes lateral, 5–9-flowered,
1–$1\frac{1}{2}$ in. in diam., glabrous in all parts; peduncle $\frac{1}{4}$–$\frac{1}{2}$ in. long;
pedicels 2–4 lin. long; sepals 1 lin. long, ovate-lanceolate, acute;
corolla rotate, 5-lobed nearly to the base, $\frac{3}{4}$ in. in diam., light green;
lobes 4–5 lin. long, $1\frac{3}{4}$ lin. broad, very spreading, lanceolate, acute,
somewhat twisted and curved; corona-lobes inserted midway
between the base of the stamens and the sinuses between the corolla-
lobes, 4 lin. long, filiform, erect, flexuose; stamens connivent over
the style; filaments $\frac{1}{3}$ lin. long, filiform, free, not connected into a
ring at the base; anthers 1 lin. long, lanceolate, acute, subtruncate
or slightly emarginate at the base, glabrous; pollen-carriers oblong,
obtuse, channelled down the middle; apex of the style 5-angled,
obtusely conical, much shorter than the anthers; follicles about
$3\frac{1}{2}$ in. long, $\frac{1}{2}$ in. thick, widely diverging, fusiform-lanceolate,
tapering to a blunt point, glabrous, rigidly coriaceous or almost
woody. *Pentopetia natalensis, Schlechter in Journ. Bot.* 1894,
257, *and* 1896, 315.

EASTERN REGION: Transkei; near Kentani, 1000 ft., *Miss Alice Pegler,*
916! Natal; in a forest near the Umbogintwini River, Umlazi native location,
Wood, 3634! and without precise locality, *Gerrard,* 780!

This plant was raised at Kew from seeds taken from Gerrard's specimen, and
flowered in March, 1867. I cannot distinguish *Pentopetia* by any technical
characters from *Tacazzea.*

VI. **CHLOROCODON**, Hook. f.

Calyx 5-partite. *Corolla* subrotate, 5-lobed nearly to the base;
lobes overlapping to the left in bud. *Corona* of 5 lobes arising
from the base of the staminal filaments, free, very broadly obcordate
or obreniform, with or without an erect or incurved dorsal process.
Stamens arising from the base of the corolla; filaments very short
and broad; anthers large, triangular, adnate to the dilated part of
the style, connivent in a cone, connate at the tips. *Pollen* granular.
Style shortly conical at the apex, not exceeding the anthers.

Tall climbers; leaves large, opposite, cordate; stipules well developed, toothed
or frill-like; flowers of moderate size, in paniculate cymes.

DISTRIB. A genus of two species, both Tropical African, of which only one
occurs in South Africa.

1. C. Whyteii (Hook. f. in Bot. Mag. t. 5898); stem climbing,
minutely pubescent; leaves petiolate, stipulate; petiole 1–2¼ in.
long, puberulous; blade 4–7 in. long, 3–5 in. broad, ovate or elliptic,
shortly cuspidate, broadly rounded or cordate at the base, glabrous
or minutely pubescent on both sides, or softly pubescent beneath;
upper surface of midrib with a few deciduous membranous scales;
stipules forming a reflexed toothed frill connecting the bases of the
petioles; panicles 2½–6 in. long, minutely puberulous; bracts 2–3
lin. long, oblong-lanceolate, acute or acuminate; pedicels 6–9 lin.
long; sepals 2–2½ lin. long, 1–1½ lin. broad, ovate, acute, glabrous or
puberulous; corolla 5-lobed nearly to the base, subrotate; lobes
5–6 lin. long, 3 lin. broad, ovate or ovate-oblong, subobtuse, glabrous,
very minutely ciliate along one margin, purple, with the margins and
a short central stripe at the base green; corona-lobes very broad,
obcordate, white, fleshy, 1 lin. long, 1¾–2 lin. broad, with a subulate, spreading, purple dorsal process 2–2½ lin. long, acute or bifid
at the apex; follicles 3–4 in. long, 1½–1¾ in. thick, ovoid-lanceolate, obtuse, widely divergent. *N. E. Br. in Dyer Fl. Trop. Afr.* iv.
i. 255; *Engl. & Prantl, Pflanzenfam.* iv. ii. 217, *f. 64, O–Q; Gard.
Chron.* 1895, xviii. 234 *and* 243, *fig.* 48; *Wood & Evans, Natal
Pl.* i. 27, *t.* 31; *Schlechter in Journ. Bot.* 1896, 314; *Hiern, Cat.
Afr. Pl. Welw.* i. 680. *Periploca latifolia, K. Schum. in Engl.
Pflanzenw. Ost-Afr. C.* 321, *and in Engl. Jahrb.* xxiii. 232.

EASTERN REGION: Natal; Wentworth, near Durban, *Wood!* Karkloof
Forest, and Inanda, near Mr. Groom's Farm, ex *Wood*, and various cultivated
specimens! Zululand; Ungoya Forest, ex *Wood*.

Also in Tropical Africa.

This plant is known by the native name of "Mundi" or "Umondi" and is
used as a tonic. In the *Botanical Magazine* the flowers are represented as being
of a pale greenish-white, but in the type specimen from which that drawing
was made, and in every other that I have seen, the flowers are coloured as above
described. Messrs. Wood & Evans, however, describe a form with "dull
greenish-white flowers." In the *Gardeners' Chronicle* the dorsal processes of
the corona-lobes are inaccurately represented as incumbent on the backs of
the anthers, instead of spreading, which is their natural position in the open
flower.

VII. SECAMONE, R. Br.

Calyx 5-partite. *Corolla* small, rotate or campanulate, 5-lobed to
the middle or beyond; lobes variably overlapping in bud, with fleshy
submarginal ridges and often a central one on their basal half, which
are decurrent on the tube within. *Corona* of 5 small or minute
simple lobes arising from and more or less adnate to the staminal
column, variable, but often laterally compressed. *Stamens* arising
from the bottom of the corolla, united with the dilated part of the
style but scarcely connate with each other, or only at the very
base; anthers minute, erect or connivent around the dilated part of
the style, terminated by fimbriate membranous appendages, which
are sometimes connate. *Pollen-masses* 20 (10 in all other genera

except *Toxocarpus* and the *Periploceæ*), exceedingly minute, globose
or oblong, attached in fours to the rather soft pale-coloured pollen-
carriers; caudicles none or flap-like. *Style* usually produced beyond
the dilated part and often exserted beyond the anthers; apical part
terete or clavate, obtuse, broadly truncate, bilobulate or bifid.
Follicles acuminate, smooth. *Seeds* crowned with a tuft of hairs.

Climbing shrubs; leaves opposite, often pellucid-dotted, from the presence of
crystals of lime in some of their cells; flowers small, usually in 3- to many-
(rarely 2-) flowered cymes, very rarely solitary; cymes axillary and terminal.

DISTRIB. Species many, widely spread through the hotter regions' of the Old
World.

The pollen masses of this genus are exceedingly minute and their true
structure is somewhat difficult to determine, and so far as I am aware, there is
not a single published figure of them that is correct.

Leaves ¼–1 in. broad (see also 5, *S. delagoensis*), lanceolate, ovate-lanceolate, oblong-lanceolate, oblong
  or elliptic:
    Corolla-lobes glabrous on the inner face; style-
    apex exserted much beyond the anther-tips:
      Corolla-lobes 2½ lin. long; style-apex 2-
      lobed ... ...   (1) **Gerrardi.**
      Corolla lobes 1 lin. long; style-apex entire
      or minutely bifid ... ...   (2) **zambesiaca,** var.
    Corolla-lobes white-pubescent on the inner face;
    style-apex equalling or shortly exceeding the
    anther-tips, entire ... ...   (3) **Alpini.**
Leaves less than ¼ in. broad, linear or linear-lanceolate; corolla-lobes glabrous on the inner face:
    Corolla-lobes longer than the flattened or saucer-
    shaped united part; style-apex cushion-like,
    just exceeding the anther-tips ... ...   (4) **frutescens.**
    Corolla-lobes not longer than the distinctly
    campanulate tube: style-apex tapering, exserted
    much beyond the anther-tips ... ...   (5) **delagoensis.**

1. S. Gerrardi (Harv. ex Benth. & Hook. f. Gen. Pl. ii. 746,
name only); a woody climber, with a greyish bark; leaves spreading,
subcoriaceous, glabrous; petiole 1–3 lin. long; blade 1–2 in. long,
5–10 lin. broad, oblong or oblong-lanceolate, obtuse, apiculate, or
shortly cuspidate-apiculate, cuneate or rounded at the base; cymes
subaxillary, or terminating short axillary shoots, 3–6-flowered, lax,
glabrous; peduncle ½–1 in. long; bracts ½–¾ lin. long, ovate, obtuse
or subacute, minutely ciliate; pedicels 5–10 lin. long, rather slender;
sepals 1–1¼ lin. long, ¾–1¼ lin. broad, broadly ovate or orbicular,
obtuse, minutely ciliate; corolla about ½ in. in diam.; tube 1–1¼ lin.
long, 2 lin. in diam., broadly campanulate, with a looped ring of
deflexed hairs in the throat, otherwise glabrous; lobes spreading,
2½ lin. long, 1¼ lin. broad, oblong, obtuse, somewhat twisted,
glabrous on both sides; corona-lobes 1½ lin. long, adnate to the
staminal column for rather more than half their length, fleshy,
flattened on the inner face, very convex on the back, obtuse, shortly
exserted from the corolla-tube; staminal column 1½ lin. long; style
exserted 1–1¼ lin. beyond the crenulate ring formed by the united

anther-appendages, simply divided into 2 linear-lanceolate lobes, or dilated into a 2-lobed head at the apex, with the lobes flattened and elliptic-oblong or short, thick and conical, acute or obtuse. *Schlechter in Journ. Bot.* 1895, 353, *and* 1897, 290, *and in Engl. Jahrb.* xviii., *Beibl.* 45, 2.

Coast Region: Komgha Div. ; woods near the mouth of the Kei River and in woods near Komgha, 2000 ft., *Flanagan,* 376 (ex *Schlechter*); among shrubs near the Kei River, 2000 ft., *Schlechter,* 6249 !

Eastern Region : Tembuland ; near Clarkbury, *Hallack!* Natal; near Durban, climbing over the tops of the trees, *Gerrard & McKen,* 86 ! 513 ! *Gerrard,* 20 ! *Wood,* 4497 ! *and in Natal Herb.* 365 ! *McKen,* 710 ! Inanda, *Wood,* 175 ! 622 !

The anther-appendages in this species are produced downwards into large quadrate flaps that cover up the anther-cells and partly hide the pollen-carriers.

2. S. zambesiaca (Schlechter), var. **parvifolia** (N. E. Br.); shrubby or twining ; branchlets glabrous or nearly so ; leaves thin ; petiole 1–1½ lin. long ; blade ⅔–1½ in. long, ¼–½ in. broad, narrowly lanceolate to ovate-lanceolate, acute or acuminate, rounded or sub-cuneate at the base, glabrous ; cymes axillary and terminal, ¼–½ in. in diam., 3–5-flowered, with very short ascending branchlets; peduncles ½–3 lin. long, glabrous ; bracts ⅓–½ lin. long, ovate or ovate-lanceolate, acute or subobtuse, minutely ciliate ; pedicels 1–2 lin. long, glabrous ; sepals ½ lin. long, elliptic or broadly ovate, obtuse to subacute, minutely ciliate ; corolla about 2 lin. in diam., quite glabrous ; tube ¼–⅓ lin. long ; lobes 1 lin. long, oblong, obtuse ; corona-lobes minute, subulate, nearly reaching to the top of the staminal column ; style-apex exserted ¼–⅓ lin. beyond the anther-tips, cylindric, entire or minutely bifid, obtuse or acute ; follicles 2½–3 in. long, 1½–2 lin. thick, narrowly fusiform, tapering into a long slender beak, smooth, glabrous ; seeds ½ in. long, scarcely 1 lin. broad, linear-lanceolate, truncate at the coma-end, smooth, glabrous, dark brown.

Kalahari Region : Transvaal ; Strydpoort, Makapans Berg, *Rehmann,* 5410 ! Waterval River, near Lydenburg, *Wilms,* 930 !

Eastern Region : Delagoa Bay ; Lourenço Marques, 100 ft., *Schlechter,* 11669 !

This only differs from typical *S. zambesiaca* (Schlechter in Journ. Bot. 1895, 303, and N. E. Br. in Dyer, Fl. Trop. Afr. iv. i. 285) by its more glabrous stems, peduncles and pedicels, smaller leaves with shorter petioles, and shorter peduncles ; in all other characters the two are identical.

3. S. Alpini (Schultes, Syst. Veg. vi. 125, excl. syn. *Secamone, Alpin. Ægypt.* ed. 1640, 133 and 134 with fig.) ; a scrambling shrub ; stem woody, glabrous or with a rust-coloured deciduous pubescence on the very young parts ; leaves subcoriaceous ; petiole 1–4 lin. long ; blade ½–2½ in. long, ¼–1 in. broad, oblong, elliptic or lanceolate, usually obtuse, apiculate, occasionally acute, cuneately rounded or acute at the base, glabrous on both sides in the adult stage, rusty-puberulous when very young ; cymes paniculate, pyramidal or corymbose, terminal or axillary, rusty-puberulous or subglabrous on all parts except the corolla; peduncles $\frac{1}{4}$–$\frac{3}{4}$ in. long; bracts $\frac{1}{2}$ lin. long, ovate, acute; pedicels $1\frac{1}{2}$–$2\frac{1}{2}$ lin. long; sepals $\frac{1}{2}$ lin. long, ovate, obtuse or subacute; corolla $1\frac{3}{4}$–2 lin. in diam., glabrous outside, pubescent with white hairs inside; tube scarcely $\frac{1}{4}$ lin. long; lobes spreading, $\frac{2}{3}$–$\frac{3}{4}$ lin. long, $\frac{1}{2}$ lin. broad, oblong-ovate, subacute or minutely and obliquely emarginate at the apex; corona-lobes $\frac{1}{4}$–$\frac{1}{3}$ lin. long, subulate, erect or incurved over the tips of the anthers; apex of the style about equalling or slightly exceeding the anther-tips, stout, truncate; follicles widely divergent or slightly reflexed, 3–4 in. long, about $1\frac{1}{2}$ lin. thick, terete, tapering to a rather long point, glabrous; seeds $\frac{1}{2}$ in. long, $\frac{3}{4}$ lin. broad, linear-lanceolate, channelled down the face, very convex on the back, blackish-brown, glabrous, crowned with a tuft of long white hairs. *Spreng. Syst. Veg.* i. 837; *Dietr. Synop. Pl.* ii. 884; *N. E. Br. in Dyer, Fl. Trop. Afr.* iv. i. 279. *S. Thunbergii, E. Meyer, Comm. Pl. Afr. Austr.* 224; *Harv. Gen. S. Afr. Pl. ed.* 1, 221; *Decne in DC. Prodr.* viii. 501; *Schlechter in Engl. Jahrb.* xviii. *Beibl.* 45, 2 *and* 15, xx. *Beibl.* 51, 12, *and* xxi. *Beibl.* 54, 1; *Journ. Bot.* 1897, 290, *and in Ann. Nat. Hist. Hofmus. Wien,* xviii. 398; *K. Schum. in Engl. & Prantl, Pflanzenfam.* iv. ii. 263. *S. ægyptica, G. Don, Gen. Syst.* iv. 160, *not of Ait. Periploca Secamone, Linn. Mant.* ii. 216, *excluding both synonyms; Thunb. Prodr. Pl. Cap.* 47; *Fl. Cap. ed.* 2, ii. 153, *and ed. Schultes,* 233; *Poir. Encycl. Meth.* v. 189; *Willd. Sp. Pl.* i. 1249; *Ait. Hort. Kew. ed.* 1, i. 301.

SOUTH AFRICA : without locality, *Herb. Linneus! Masson!*

COAST REGION : Clanwilliam Div. ; Gift Berg, 1500–2000 ft., *Drège!* Paarl Div; Paarl Mountains, *Drège! Prior (Alexander)*! Tulbagh Div. ; Tulbagh Waterfall, *Ecklon & Zeyher!* Mitchells Pass, 800 ft., *Bolus,* 5203! Worcester Div.; Hex River Mountains, *Rehmann,* 2702! Cape Div.; Table Mountain, *Harvey!* near Cape Town? *Burchell,* 8406! George Div.; Outeniqua woods, *Thunberg!* Montagu Pass, *Rehmann,* 285! Knysna Div. ; in the forest at Knysna, *Burchell,* 5396! Plettenberg Bay, *Bowie!* near Loeri River, *Penther,* 1926 (ex *Schlechter*). Uitenhage Div.; Van Stadens Berg, *Zeyher,* 603! near Uitenhage, *Burchell,* 4237! *Prior!* Port Elizabeth Div. ; on sand-hill along the coast, *E. S. C. A. Herb.,* 197! Albany Div.; in woods near Grahamstown, 2000 ft., *MacOwan,* 293! *Zeyher,* 603! *Galpin,* 354 (ex *Schlechter*), *Schönland,* 648! Glenfilling, *Drège,* 3474! Komgha Div. ; in woods near Komgha, 2000 ft., *Flanagan,* 377 (ex *Schlechter*). Bathurst Div.; near Barville Park, *Burchell,* 4136! Stockenstrom Div. ; Wellsdale Mountain, *Scully,* 337 ! Queenstown Div.; Finchams Nek, 3900 ft., *Galpin,* 1891 !

CENTRAL REGION : Somerset Div. ; on the Bosch Berg, *Burchell,* 3131!

KALAHARI REGION : Transvaal; near Botsabelo, 4900 ft., *Schlechter* (ex *Schlechter*) ; Rietfontein, Zoutpansberg Range, *Leendertz,* 875!

EASTERN REGION : Transkei ; near Kentani, *Miss Alice Pegler,* 662 ! Natal; near Durban, *Wood,* 138! 1415! *Wilms,* 2230 ! *Peddie!* Shepstone, 100 ft., *Rogers,* 599 ! Inanda, *Wood,* 1415 !

Also in Tropical Africa.

The plant figured and described by Alpino has been erroneously referred to this species by Schultes, and has been misunderstood by all authors. Alpino's figure is undoubtedly a representation of *Leptadenia heterophylla,* Decne, a well-known inhabitant of Upper Egypt and the Nile Region, whilst the description of *S. Alpini* given by Schultes, so unmistakably refers to this South African

plant, that it evidently was made entirely from it. The var. *retusa* (E. Meyer,
l.c.), founded upon Drège, 3474, with obtuse or slightly retuse leaves, is a
common form not worth varietal distinction. The follicles distributed with Miss
Pegler's 662 are much longer (5–6 in. long) and more slender (1 lin. thick)
than in any other specimen seen, but I find no difference in the flowers.

4. S. frutescens (Decne in DC. Prodr. viii. 501); stems woody,
often twining; leaves glabrous, rather thin; petiole $\frac{1}{2}$–$1\frac{1}{2}$ lin. long;
blade $\frac{1}{2}$–$1\frac{3}{4}$ in. long, $\frac{2}{3}$–2 lin. broad, linear, acute or obtuse, rounded
or cuneate at the base; cymes numerous, small, 3–5 lin. in diam.,
5–12-flowered, subsessile, or on peduncles rarely more than 1 lin.
long, glabrous in all parts; bracts minute, ovate; pedicels 1–2 lin.
long, slender; sepals about $\frac{1}{2}$ lin. long and broad, suborbicular, very
obtuse, minutely ciliate; corolla rotate, about $1\frac{1}{2}$ lin. in diam., quite
glabrous; lobes longer than the united part, about $\frac{1}{2}$ lin. long, oblong,
obtuse; corona-lobes not quite half as long as the staminal column,
erect from its base, wing-like, subquadrate, truncate at the apex;
style-apex large, cushion-like, slightly exceeding the anthers;
follicles $1\frac{1}{4}$–$1\frac{3}{4}$ in. long, about $1\frac{1}{2}$ lin. thick, terete, tapering into a
long slender beak. *Schlechter in Engl. Jahrb.* xviii. *Beibl.* 45, 2,
xx. *Beibl.* 51, 12, xxi. *Beibl.* 54, 1, *and Journ. Bot.* 1897, 290.
Astephanus frutescens, E. Meyer, Comm. Pl. Afr. Austr. 223; *D.
Dietr. Syn. Pl.* ii. 909. *A. fruticosus, Steud. Nom. Bot. ed.* 2, i.
153.

CoAST REGION: Uitenhage Div.; by the Zwartkops River, *Zeyher*, 538!
3808! *Drège*, 2230! near Uitenhage, *Prior!* among shrubs near Van Stadens
River, *Bolus*, 2398! Bathurst Div.; Trapps Valley, *Miss Daly*, 647! Alexandria
Div.; on the Zuurberg Range, near the Bontjes River, 2000 ft., *Drège!* Albany
Div.; near Grahamstown, *Galpin*, 2924! Fort Beaufort Div.; margins of a
wood near the Kat River, 2500–3000 ft., *Drège!* and without precise locality,
Cooper, 535! East London Div.; near East London, 100 ft., *Wood in Herb.
Galpin*, 3144! 3370! Komgha Div.; in woods near Komgha, 1800 ft., *Flanagan*,
18 (ex *Schlechter*). British Kaffraria, *Cooper*, 131!
CENTRAL REGION: Somerset Div.; on the Bosch Berg, near Somerset East,
Burchell, 3119! southern side of the Bosch Berg, *Burchell*, 3169! near
Somerset East, *Bowker!*
KALAHARI REGION: Transvaal; mountains near Tsacoma, 3600 ft., *Schlechter*,
4540 (ex *Schlechter*).
EASTERN REGION: Transkei; Kentani District, *Miss Alice Pegler*, 845!
Natal; Inanda, *Wood*, 638! and without precise locality, *Gerrard*, 133!
Cooper, 1158! Delagoa Bay region; Macocololo, 100 ft., *Schlechter*, 12059!

5. S. delagoensis (Schlechter in Engl. Jahrb. xxxviii. 35, fig. 3);
stem climbing, much branched, glabrous; leaves thinly coriaceous,
glabrous; petiole $\frac{1}{2}$–$\frac{2}{3}$ lin. long; blade $\frac{2}{3}$–$1\frac{1}{2}$ in. long, $\frac{3}{4}$–$2\frac{1}{2}$ lin.
($1\frac{1}{2}$–4 lin. ex *Schlechter*) broad, linear, linear-lanceolate or linear-
oblong, obtuse or acute at the apex, cuneate or rounded at the base;
flowers usually in 2-flowered cymes, occasionally solitary, subaxillary or terminal; peduncle 0–2 lin. long, glabrous; bracts
minute; pedicels $1\frac{1}{3}$–$2\frac{1}{2}$ lin. long, glabrous; sepals about $\frac{1}{2}$ lin. long
and broad, suborbicular, very obtuse, glabrous; corolla campanulate,
quite glabrous, apparently white or yellowish; tube rather more

than $\frac{3}{4}$ lin. long, faintly narrowed upwards; lobes suberect, $\frac{2}{3}$–$\frac{3}{4}$ lin. long, oblong, obtuse, alternating with minute pocket-like flaps at the sinuses; corona-lobes fleshy, ovate, subacute, entirely adnate to the basal half of the $\frac{2}{3}$ lin. long staminal column; anther-appendages erect, connate into a ring around the base of the beak-like style-apex, which is exserted $\frac{1}{4}$ lin. beyond them.

EASTERN REGION: Delagoa Bay; Lourenço Marques, 150 ft., *Schlechter*, 11625 (not 1125 as quoted under the original description)!

VIII. **ASTEPHANUS**, R. Br.

Calyx 5-partite. *Corolla* campanulate or suburceolate, 5-lobed, with 5 hairy patches inside the tube; lobes erect or spreading, overlapping in bud. *Corona* none. *Staminal column* arising near the base of the corolla-tube; anthers erect, with erect membranous appendages. *Pollen-masses* solitary in each cell, minute, oblong, not larger than the pollen-carrier, to which they are attached by short caudicles. *Style* produced at the apex into a beak, much exceeding the anther-appendages. *Follicle* solitary (always?), terete-fusiform, tapering into a beak. *Seeds* crowned with a tuft of hairs.

Slender twiners; leaves small, opposite, more or less coriaceous; flowers small, in small, subaxillary umbels.

DISTRIB. A small genus; 3 species in South Africa, the rest in tropical and subtropical America.

Corolla-tube $1\frac{3}{4}$–$2\frac{1}{4}$ lin. long, longer than the lobes ... (1) **marginatus.**
Corolla-tube $1\frac{1}{4}$ lin. long, as long as the lobes   ... (2) **triflorus.**
Corolla-tube scarcely 1 lin. long, shorter than the
lobes   ...   . (3) **neglectus.**

1. **A. marginatus** (Decne in DC. Prodr. viii. 508); stem slender, twining, branched, glabrous except at the tips of the branches; leaves spreading, glabrous, coriaceous; petiole $1\frac{1}{2}$–$3\frac{1}{2}$ lin. long; blade $\frac{1}{2}$–$1\frac{1}{4}$ in. long, $1\frac{1}{2}$–$7\frac{1}{2}$ lin. broad, varying from linear-lanceolate to elliptic, acute or obtuse, apiculate, thickened or slightly revolute at the margins; umbels subaxillary, 2–9-flowered; peduncle $\frac{1}{4}$–$\frac{1}{2}$ in. long, slender, with a pubescent line down one side; bracts minute, subulate; pedicels $1\frac{1}{2}$–$2\frac{1}{2}$ lin. long, pubescent down the inner side; sepals erect, $1\frac{1}{2}$ lin. long, $\frac{1}{2}$ lin. broad, lanceolate, acute, glabrous; corolla tubular-campanulate, glabrous outside; tube $1\frac{3}{4}$–$2\frac{1}{4}$ lin. long, 5-angled, with 5 small hairy patches at the base inside; lobes 1–$1\frac{1}{4}$ lin. long, $\frac{3}{4}$ lin. broad, oblong, obtuse, slightly spreading, with a minutely velvety surface inside, but without a pubescence; staminal column 1–$1\frac{1}{4}$ lin. long; filament part rather shorter than the anthers, stout, cylindric; apex of the style rather stout, beak-like, produced $1\frac{1}{4}$–$1\frac{1}{2}$ lin. beyond the anther-tips and reaching to about the middle of the lobes; follicle solitary, 2–$2\frac{1}{2}$ in. long, 3–4 lin. thick, fusiform-lanceolate, tapering from about the middle

into a long beak, glabrous. *Harv. Thes. Cap.* i. 57, *t.* 91 ; *Schlechter in Journ. Bot.* 1896, 418. *A. Zeyheri, Turcz. in Bull. Soc. Mosc.* 1852, xxv. *pt.* 2, 314 ; *Schlechter, l.c.,* 418.

COAST REGION : Knysna Div. ; on sand-hills near the west end of Groene Vallei, *Burchell,* 5662 ! Uitenhage Div. ; near the mouth of the Coega River, *Zeyher,* 3406 ! Albany Div. ; without precise locality, *Atherstone,* 10 ! *Cooper,* 2721 ter. ! King Williamstown Div. ; Keiskamma, *Mrs. Hutton !* Bathurst Div. ; Kowie sand-hills, in the bush, *MacOwan,* 405 ! Port Alfred, *Mrs. White,* 74 !
CENTRAL REGION : Somerset Div. ; near Somerset East ? *Bowker !*

2. **A. triflorus** (R. Br. in Mem. Wern. Soc. i. 54) ; stem slender, twining, minutely pubescent on the young parts, becoming glabrous ; leaves spreading, subcoriaceous, not very thick ; petiole about 1½ lin. long; blade 4–8 lin. long, 1–2 lin. broad, linear-lanceolate or linear-oblong, acute, rounded and often broadened or obscurely subhastate at the base, more or less revolute along the margins ; umbels subaxillary, 2–5-flowered ; peduncle 1–6 lin. long, slender, minutely puberulous ; bracts minute, subulate ; pedicels 1½–2 lin. long, minutely puberulous ; sepals erect, about 1 lin. long, ½ lin. broad, lanceolate, acute, thinly puberulous ; corolla campanulate, glabrous on both sides, with the exception of 5 small hairy patches near the base of the tube inside; tube 1¼ lin. long, 5-angled ; lobes 1¼ lin. long, ⅔ lin. broad, oblong, slightly oblique at the obtuse apex, spreading ; staminal column ¾ lin. long, filament part very short ; apex of the style beak-like, produced about ½ lin. beyond the anther-tips. *Schultes, Syst. Veg.* vi. 122; *Spreng. Syst. Veg.* i. 855 ; *G. Don, Gen. Syst.* iv. 158 ; *Dietr. Synop. Pl.* ii. 909 ; *Decne in DC. Prodr.* viii. 508 ; *Schlechter in Journ. Bot.* 1896, 418. *A. pauciflorus, E. Meyer, Comm. Pl. Afr. Austr.* 224 ; *Dietr. Synop. Pl.* ii. 909 ; *Decne in DC. Prodr.* viii. 508; *Schlechter in Engl. Jahrb.* xxi. *Beibl.* 54, 2 ; *in Journ. Bot.* 1895, 353, *and* 1896, 418. *Apocynum triflorum, Linn. f. Suppl.* 169 ; *Thunb. Prodr. Fl. Cap.* i. 47, *in Nov. Act. Acad. Petrop.* xiv. (1805), 512, *and Fl. Cap. ed.* 2, ii. 161, *and ed. Schultes,* 237 ; *Lam. Encycl. Meth.* i. 215 ; *Willd. Sp. Pl.* i. 1261.

SOUTH AFRICA : without locality, *Oldenburgh,* 393 ! *Thunberg ! Zeyher,* 1181 !
COAST REGION : Clanwilliam Div. ; near Modderfontein, between Lange Vallei and Kook Fontein, 400 ft., *Drège,* 6389 ! by the Olifants River, *Schlechter, and Penther,* 826 (ex *Schlechter*). Stellenbosch and Paarl Div. ; between Stellenbosch and Paarl, *Prior !* and between Paarl Mountain and Paarde Berg, 500 ft., *Drège* (ex *E. Meyer*).
WESTERN REGION : Little Namaqualand ; Van Rhyns Dorp Div., on the Karee Bergen, 1500 ft., *Schlechter,* 8236 !

3. **A. neglectus** (Schlechter in Engl. Jahrb. xviii. Beibl. 45, 26) ; stem slender, twining, branching, minutely pubescent, becoming glabrous ; leaves thick, coriaceous, spreading ; petiole about 1 lin. long ; blade ¼–1 in. long, usually 1½–2 lin. broad, oblong, occasionally 3–6 lin. broad and elliptic, obtuse at both ends, apiculate, thickened along the margin beneath, glabrous ; umbels subaxillary, 3–7-flowered ; peduncle 1–2 lin. long, slender, slightly

pubescent; bracts minute, subulate; pedicels 1–2 lin. long, more or less pubescent; sepals 1–1¼ lin. long, ½ lin. broad, lanceolate or oblong-lanceolate, subacuminate, with a few scattered hairs on the back; corolla campanulate, white outside, rosy-purple within (*Burchell*), glabrous with the exception of the 5 hairy patches at the base of the tube inside; tube scarcely 1 lin. long; lobes spreading or somewhat recurved, 1¼ lin. long, ⅔ lin. broad, oblong-lanceolate, minutely emarginate at the apex; staminal column ⅔ lin. long, filament part very short; apex of the style beak-like, produced about ½ lin. beyond the anther-tips; follicle terete-fusiform, beaked, glabrous, not seen in a ripe condition. *Schlechter in Journ. Bot.* 1896, 418.

SOUTH AFRICA: without locality, *Nelson! Forbes*, 342! *Grey! Hooker*, 417!

COAST REGION: Paarl Div.; road to Paarde Berg, *Prior!* Cape Div.; Camps Bay, *Prior!* Muisen Berg, *Harvey!* climbing among shrubs by the sea-shore, near Simons Town, 20 ft., *Bolus*, 4934! and in *MacOwan & Bolus, Herb. Norm. Austr.-Afr.*, 1081! *Schlechter*, 12, 13, and 1080 ex *Schlechter*; at the top of Red Hill Road, *Wolley Dod*, 2650! Knysna Div.; on sand-hills at Plettenberg Bay, *Burchell*, 5307! Bredasdorp Div.; Mier Kraal, 200 ft., *Schlechter*, 10510!

IX. **MICROLOMA**, R. Br.

Calyx 5-partite. *Corolla* urceolate or tubular, 5-lobed; tube 5-angled, often more or less inflated at the base, furnished inside with 5 tufts of deflexed hairs opposite the lobes; lobes broad or narrow, spirally arranged and more or less closing the mouth of the tube. *Corona* 0 or of 5 small tubercles or fleshy scales at the middle or near the apex of the corolla-tube, alternating with the hairy tufts. *Staminal column* arising from the base of the corolla, more or less contracted under the anthers; anthers connivent, terminated by erect membranous appendages, which are more or less connate at the tips; anther-wings horny, produced at the middle or at the base into triangular or spur-like projections. *Pollen-masses* solitary in each cell, pendulous, long, linear, attached to the small, rather narrow pollen-carrier, by short caudicles. *Style* with a conical, minutely bifid apex, not produced beyond the anther-appendages. *Follicle* solitary by abortion, fusiform, beaked. *Seeds* crowned with a tuft of hairs.

Perennials with slender twining stems, or dwarf, much-branched shrublets; leaves small, opposite; flowers in small umbel-like cymes, subaxillary.

DISTRIB. Species 10, endemic.

  * Plants distinctly twining (see also *M. incanum*); branches nearly or quite glabrous, except in *M. sagittatum*:

    Corolla-lobes not laterally compressed, sub-orbicular, very obtuse; leaves linear:

      Hairy tufts inside the corolla placed just below the mouth of the tube ... ... (1) tenuifolium.

Hairy tufts inside the corolla placed at about
the middle of the tube ... ... (2) **namaquense.**
Corolla-lobes laterally compressed, narrowly lanceolate or deltoid-lanceolate, acute, more or less
gibbous behind at the base :
Sepals about as long as or longer than the
entire corolla :
Leaves obtusely subhastate at the subtruncate base ; corolla with broad
semicircular gibbosities upon the backs
of the lobes, very abruptly contracted
into their apiculus-like tips ... ... (3) **gibbosum.**
Leaves narrowed or rounded into the
petiole; corolla with slight gibbosities
at the very base of the lobes, gradually
tapering into their tips ... ... (4) **calycinum.**
Sepals shorter than the fully-developed
corolla ; leaves linear-sagittate or linear-
hastate, with rounded basal auricles :
Pubescence very minute or none; sepals
from ⅔ as long to as long as the
corolla-tube ... ... (5) **glabratum.**
Pubescence distinct ; sepals ½–⅔ as long
as the corolla-tube ... ... (6) **sagit atum.**
** Plants not twining (or but slightly so in *M.
incanum*); apparently much-branched shrubs or
shrublets; branchlets minutely puberulous or
densely and minutely white-tomentose :
Corolla-tube ¼ in. long :
Branchlets erect or ascending-spreading ;
corolla-lobes 1½ lin. long ... ... (7) **incanum.**
Branchlets recurved-divaricate ; corolla-
lobes ½ lin. long ... . (8) **longituba.**
Corolla-tube 1–1¾ lin. long :
Corolla shortly conical-acute at the apex,
with connivent-erect lobes ... ... (9) **Burchellii.**
Corolla truncate at the apex, with abruptly-
inflexed gibbous-based lobes ... ... (10) **Massonii.**

1. **M. tenuifolium** (K. Schum. in Engl. & Prantl, Pflanzenfam.
iv. ii. 222) ; rootstock a cluster of stout fleshy roots ; stem slender,
twining, branching, subherbaceous, ½–¾ lin. thick, glabrous ; leaves
spreading or reflexed, glabrous ; petiole about 1 lin. long ; blade
1–2¾ in. long, ½–⅔ lin. broad, linear, with strongly revolute margins,
rarely flat, and then about 1½ lin. broad, acute, narrowed into the
short petiole at the base, very rarely auricled ; cymes subaxillary,
3–7-flowered ; peduncle 2–3 lin. long, glabrous or slightly pubescent ;
bracts 1½–2 lin. long, lanceolate, acute, pubescent ; pedicels 2–3½
lin. long, thinly covered with a somewhat adpressed pubescence ;
sepals 2½ lin. long, ½–⅔ lin. broad, linear-lanceolate or oblong-lanceolate, acute, thinly pubescent or nearly glabrous ; corolla urceolate,
somewhat fleshy, glabrous outside, bright carmine-red ; tube 2½ lin.
long, inflated at the base, contracted at the mouth, 5-angled, with
5 contiguous elongated tufts of deflexed fulvous hairs just below the
mouth inside ; lobes 1 lin. long and broad, orbicular-cordate, very
obtuse, hairy at the base of the inner half, much overlapping and

forming a spiral rosette closing the mouth of the tube; corona-tubercles small, obtuse, ascending, placed at the middle of the corolla-tube, on a level with the base of the tufts of hairs; staminal column 2 lin. long, 5-spurred at the middle; filament part stout, conical; follicles solitary, $2\frac{1}{4}$–3 in. long, $3\frac{1}{2}$–4 lin. thick, fusiform, tapering into a rather long point, minutely bifid at the apex; seeds 2 lin. long, ovate, concave on one face, very convex on the other, tuberculate. *M. lineare, R. Br. in Mem. Wern. Soc.* i. 53; *Schultes, Syst. Veg.* vi. 121; *Spreng. Syst. Veg.* i. 855; *E. Meyer, Comm.* 222; *Harv. Gen. S. Afr. Pl. ed.* 1, 228; *Thes. Cap.* i. 58, *t.* 92; *G. Don, Gen. Syst.* iv. 158; *Dietr. Synop. Pl.* ii. 909; *Decne in DC. Prodr.* viii. 510; *Krauss in Flora,* 1844, 827. *M. linearis, O. Kuntze in Jahrb. Bot. Gart. Berl.* iv. 268. *M. tenuiflora, O. Kuntze in Jahrb. Bot. Gart. Berl.* iv. 268. *M. tenuifolia, Schlechter in Journ. Bot.* 1896, 418. *Periploca tenuifolia, and var. β, Linn. Sp. Pl. ed.* i. 212; *Thunb. in Nov. Act. Acad. Petrop.* xiv. (1805), 517. *Ceropegia tenuifolia, Linn. Mant.* ii. 215, 346; *Thunb. Prodr.* i. 37, *Fl. Cap. ed.* 2, ii. 147, *and ed. Schultes,* 231; *Pers. Syn.* i. 277; *Linn. Syst. Veg. ed.* 14, 255; *Lam. Encycl.* i. 687, *and Tabl. Encycl.* ii. 325, *t.* 179, *fig.* 2. *C. tenuiflora, Willd. Sp. Pl.* i. 1276. *C. sinuata, Poir. Suppl. Encycl. Meth.* ii. 178; *Schultes, Syst. Veg.* vi. 4; *G. Don, Gen. Syst.* iv. 112.—*Cynanchum radice glandulosa, &c., J. Burm. Rar. Afr. Pl. Dec.* 2, 36, *t.* 15. *Cynanchum linearibus foliis, &c., J. Burm. l.c.* 37, *t.* 16, *fig.* 1. *Apocynum frutescens, &c., Pluk. Almagest. Bot. Mant.* 17, *t.* 335, *fig.* 5.

South Africa: without locality, *Forster! Masson! Mund! Harvey,* 538! *Forbes,* 262!

Coast Region: Van Rhynsdorp Div.; Gift Berg, 1500–2500 ft., *Drège!* Clanwilliam Div.; Cederberg Range, near Ezels Bank, 2000–3000 ft., *Drège* (ex *E. Meyer*). Tulbagh Div.; near Roodezand Kloof, *Lichtenstein* (ex *Schultes*). Worcester Div.; without precise locality, *Cooper,* 1662! 1744! Paarl Div.; Paarl Mountains, 500–1000 ft., *Drège!* Cape Div.; on the Flats and mountains around and south of Cape Town, *Thunberg! Burchell,* 140! 811! *Pappe! Ecklon,* 525! (*Alexander*) *Prior! Wilms,* 3481; *Wolley Dod,* 3283! *Krauss* (ex *Krauss*). Stellenbosch Div.; Stellenbosch Flats, *Lloyd in Herb. Sanderson,* 970! Caledon Div.; Zwart Berg, near Caledon, *Galpin,* 4326! Swellendam Div.; between Riet Kuil and Hemel-en-Aarde, *Zeyher,* 3407! Tradouw Mountains, *Drège* (ex *E. Meyer*). Riversdale Div.; near the Zoete-melks River, *Burchell,* 6647! Uniondale Div.; Lange Kloof, *Burchell,* 5025! *Bunbury,* 166! Humansdorp Div.; near the Kromme River, *Burchell,* 4836! Uitenhage Div.; various localities, *Burchell,* 4641! *Zeyher! Cooper,* 1477! *MacOwan,* 1090! *Drège* (ex *E. Meyer*). Port Elizabeth Div.; about Baakens River, *Burchell,* 4359! *MacOwan,* 1090! *Lloyd,* 369!

I do not find a double series of hairy patches within the corolla-tube of this species as implied by Bolus in the Journal of the Linnean Society, xxv. 163, in a note under *M. namaquense,* nor do they occur in any other species that I have examined.

2. **M. namaquense** (Bolus in Journ. Linn. Soc. Bot. xxv. 163); stem twining, very slender, glabrous; leaves rather distant, $\frac{3}{4}$–$1\frac{1}{2}$ in. long, including the very short petiole, $\frac{1}{2}$–$\frac{2}{3}$ lin. broad, linear, acute,

minutely auriculate-hastate at the base, with strongly revolute
margins, coriaceous, glabrous; cymes subaxillary, 3–6-flowered,
very minutely puberulous; peduncle 2–3 lin. long; bracts 1 lin.
long, subulate; pedicels 3–4 lin. long; sepals spreading or erect,
2 lin. long, rather more than $\frac{1}{2}$ lin. broad, linear-lanceolate, acute,
very minutely puberulous; corolla urceolate, flame-coloured, glabrous
outside; tube 3 lin. long, 5-angled, inside with 5 tufts of deflexed
hairs at about the middle; lobes $1\frac{1}{4}$ lin. long, $1\frac{3}{4}$ lin. broad, sub-
orbicular-reniform, very obtuse, much overlapping and forming a
spiral rosette, nearly or quite closing the mouth of the tube, minutely
pubescent on the basal part inside; corona-tubercles erect, obtuse,
placed rather below the level of the middle of the hairy patches;
staminal column 2 lin. long, 5-spurred at the middle; filament
part stout, cylindric. *Schlechter in Journ. Bot.* 1896, 418.

WESTERN REGION: Little Namaqualand; near Ookiep?, *Morris in Herb.
Bolus,* 5703! Spektakel Mountain, 3600 ft., *Bolus in MacOwan & Bolus, Herb.
Norm. Austr.-Afr.,* 639!

3. M. gibbosum (N. E. Br.); stem twining, branched, $\frac{1}{2}$–$\frac{3}{4}$ lin.
thick, very minutely puberulous on the youngest parts, soon becom-
ing glabrous, as do the spreading subcoriaceous leaves; petiole
$\frac{1}{2}$–$1\frac{1}{2}$ lin. long; blade 5–10 lin. long, 1–$3\frac{1}{2}$ lin. broad at the base,
linear-sagittate or narrowly oblong-sagittate, acute, with rounded
auricles at the subtruncate base; cymes subaxillary, 4–6-flowered;
peduncles 1–2 lin. long, thinly and very minutely puberulous, as are
also the $1\frac{1}{2}$–2 lin. long, linear-lanceolate bracts and 2–4 lin. long
pedicels; sepals $3\frac{1}{2}$–$4\frac{1}{2}$ lin. long, as long as or longer than the
corolla, 1 lin. broad at the base, lanceolate, acute or acuminate,
glabrous, reddish; corolla tubular, 5-angled, glabrous outside;
tube $\frac{1}{4}$ in. long, with 5 elongated hairy patches at the base within;
lobes $\frac{2}{3}$–$\frac{3}{4}$ lin. long, connivent-erect, laterally compressed, dorsally
very abruptly and broadly gibbous on the basal part from the angles
of the tube being suddenly and semicircularly rounded into the
small apiculus-like apex of the lobe, which does not exceed the
gibbosity .by more than $\frac{1}{4}$ lin., puberulous and minutely ciliate on
the inner face; corona-tubercles small, ovate, acute, placed at the
level of the middle of the hairy patches; staminal column 2 lin.
long, shortly spurred at the middle.

CENTRAL REGION: Calvinia Div.; Nieuwoudtville, *Leipoldt in Herb. Bolus,*
8311!

This differs from all its nearest allies by the very abrupt manner in which the
angles of the corolla-tube are rounded into the minute tips of the lobes; in all
the others they more or less taper into the tips, which exceed the gibbosity by
$\frac{2}{3}$ lin. or more. The corolla may have been yellow or red with green tips to the
lobes.

4. M. calycinum (E. Meyer, Comm. 223); a woody, much-
branched twiner; branches $\frac{1}{3}$–$\frac{2}{3}$ lin. thick, glabrous; leaves reflexed,
subcoriaceous or fleshy? glabrous; petiole $\frac{1}{4}$–$\frac{3}{4}$ lin. long; blade

2½–7 lin. long, ½–1⅓ lin. broad, linear, acute, cuneate or rounded into the petiole at the base; cymes subaxillary, 3–9-flowered; peduncle ½–3 lin. long, glabrous; bracts 1½ lin. long, subulate; pedicels 1½–4 lin. long, glabrous; sepals as long as the corolla or longer, 5 lin. long, ¾ lin. broad, linear, acute, glabrous, erect or more or less spreading, red; corolla tubular, 5-angled, glabrous outside; tube 2½–3 lin. long, rosy, furnished inside with 5 tufts of deflexed hairs near the base; lobes spirally connivent-erect, 1½ lin. long, ½ lin. broad, narrowly lanceolate, laterally compressed with a slight gibbosity at their very base which gradually tapers to their acute tips, very minutely velvety inside, green, margined with red; corona-tubercles minute, subhemispherical, placed at the level of the top of the hairy patches; staminal column 1½–2 lin. long, shortly spurred a little below the middle. *Harv. Gen. S. Afr. Pl. ed.* 1, 228; *Dietr. Syn. Pl.* ii. 909; *Decne in DC. Prodr.* viii. 511; *Schlechter in Engl. Jahrb.* xxi. *Beibl.* 54, 2, *and Journ. Bot.* 1896, 418.

VAR. β, **flavescens** (E. Meyer, Comm. 223); sepals shorter than the corolla, 2½–3 lin. long, yellowish.

WESTERN REGION: Little Namaqualand; Kaus (Steinkof) Mountains, near Goedemans Kraal, 2500–3000 ft., *Drège,* 3052b! stony places near Klip Fontein, 2500 ft., *Bolus in MacOwan & Bolus, Herb. Norm. Austr.-Afr.,* 638! by the Buffel River, *Schlechter,* 11261! Great Namaqualand; near Aus, *Schinz,* 30! Var. β: Little Namaqualand; Kamies Berg near Geelbeks Kraal, 3000–4000 ft., *Drège,* 3052!

5. **M. glabratum** (E. Meyer, Comm. 222); stem branching, twining, woody; branches ½–1 lin. thick, at first very minutely puberulous, becoming glabrous; leaves subcoriaceous, very minutely puberulous; petiole ½–1 lin. long; blade 6–11 lin. long, 1¼–2½ lin. broad, linear-hastate, acute; cymes subaxillary, 3–5-flowered; peduncle ½–2½ lin. long, puberulous; bracts 1–1½ lin. long, subulate; pedicels 1½–3 lin. long; sepals 3–3¼ lin. long, ¾ lin. broad, narrowly lanceolate, acuminate, erect, thinly puberulous; corolla tubular, slightly enlarged below, 5-angled, glabrous outside, red; tube 3½ lin. long, inside with 5 tufts of yellowish, deflexed hairs below the middle; lobes spirally connivent-erect, 1 lin. long, ½ lin. broad at the base, deltoid-lanceolate, laterally compressed, with gibbosities at the base of their lobes which gradually taper to their acute tips, pubescent inside, green; corona-tubercles small, obtuse, placed· at the level of the middle of the tufts of hairs; staminal column 2 lin. long, 5-spurred at the middle; filament part subcylindric. *Harv. Gen. S. Afr. Pl. ed.* 1, 228; *Dietr. Syn. Pl.* ii. 909; *Decne in DC. Prodr.* viii. 511; *Schlechter in Engl. Jahrb.* xxi. *Beibl.* 54, 2, *and Journ. Bot.* 1896, 418.

CENTRAL REGION: Prince Albert Div.; near Klaarstroom on the Great Zwart Bergen, 2000–3000 ft., *Drège!* Kendo (probably Kandos Mountain), 2500–3500 ft., *Drège* (ex E. *Meyer*).

WESTERN REGION: Little Namaqualand; in stony places near Ookiep, *Bolus in MacOwan & Bolus, Herb. Norm. Austr.-Afr.,* 637!

6. **M. sagittatum** (R. Br. in Mem. Wern. Soc. i. 53); stem twining, $\frac{2}{3}-\frac{3}{4}$ lin. thick, pubescent; leaves spreading, thinly pubescent to almost tomentose; petiole $\frac{1}{2}-2$ lin. long; blade $\frac{1}{3}-1\frac{1}{4}$ in. long, $\frac{1}{2}-3$ lin. broad, linear or linear-oblong, shortly sagittate or hastate at the base, obtuse, more or less revolute at the margins; cymes subaxillary, 3–9-flowered; peduncle 2–4 lin. long, pubescent; bracts 1 lin. long, linear-subulate, pubescent; pedicels $1\frac{1}{2}-2$ lin. long, pubescent; sepals $1\frac{1}{2}-3$ lin. long, $\frac{1}{2}-\frac{3}{4}$ lin. broad, lanceolate or oblong-lanceolate, acute, erect, adpressed to the corolla, pubescent; corolla tubular, slightly inflated below, 5-angled, pink, glabrous or puberulous along the angles only outside; tube $3\frac{1}{2}-3\frac{3}{4}$ lin. long, furnished inside below the middle with 5 tufts of deflexed hairs; lobes spirally connivent-erect, 1 lin. long, about $\frac{1}{2}$ lin. broad at the base, deltoid-lanceolate, laterally compressed, with a gibbosity at their base which gradually tapers to their acute tips, pubescent inside, green; corona-tubercles small, obtuse, placed at the level of the middle of the hairy patches; staminal column about 2 lin. long, shortly 5-spurred at the middle; filament part stout, cylindric; follicles $2-2\frac{3}{4}$ in. long, 4–5 lin. thick, lanceolate, tapering into a long beak, glabrous; seeds $2\frac{1}{2}-3$ lin. long, ovate, flattish, minutely tuberculate, black. *Ait. Hort. Kew. ed.* 2, ii. 76; *Schultes, Syst. Veg.* vi. 120; *Spreng. Syst. Veg.* i. 855; *Harv. Gen. S. Afr. Pl. ed.* 1, 228; *Dietr. Syn. Pl.* ii. 909; *E. Meyer, Comm.* 222; *G. Don, Gen. Syst.* iv. 158; *Krauss in Flora,* 1844, 827; *DC. Prodr.* viii. 510; *Schlechter in Engl. Jahrb.* xxi. *Beibl.* 54, 2, *and in Journ. Bot.* 1896, 417. *M. hastatum and Periploca hastata, Decne in DC. Prodr.* viii. 499 ?. *Ceropegia sagittata, Linn. Mant.* ii. 215; *Syst. Veg. ed.* 14, 255; *Willd. Sp. Pl.* i. 1276; *Thunb. Prodr.* i. 37; *Fl. Cap. ed.* 2, ii. 148, *and ed. Schultes,* 231; *and in Nov. Act. Acad. Petrop.* xiv. 515; *Lam. Encycl.* i. 686, *and Tabl. Encycl.* ii. 325. *t.* 179, *f.* 1; *Jacq. Hort. Schoenbr.* i. 17, *t.* 38; *Ait. Hort. Kew. ed.* 1, i. 300; *Pers. Syn. Pl.* i. 277. *Eustegia hastata, Sieber. ex Decne in DC. Prodr.* viii. 511, *not of R. Br.*

Coast Region : Clanwilliam Div.; near Sand Berg in Lauge Valley, 500 ft., *Drège,* 3054! near Clanwilliam, *Mader,* 199! by the Olifants River, *Ecklon & Zeyher!* Piquetberg Div.; Pieterstontein, 500 ft., *Drège,* 3054b! Pickeniers Kloof, *Penther,* 827, and near Undersberg Valley, *Penther,* 822 (ex *Schlechter*). Malmesbury Div.; Groene Kloof, below 200 ft., *Drège!* Worcester Div.; Hex River, *Burke!* Cape Div.; various localities on the Flats and sand dunes, *Thunberg! Prior (Alexander)! Zeyher,* 1180! *Burchell,* 968! *Krauss* (ex *Krauss), Bolus,* 2874! *Pappe! MacOwan & Bolus, Herb. Norm. Austr.-Afr.,* 366! *Wilms,* 3480! Caledon Div.; without precise locality, *Thom,* 942! Riversdale Div.; between Great Vals River and Gauritz River, *Burchell,* 6513! Tyger Fontein, 800 ft., *Galpin,* 4325!

Western Region : Little Namaqualand; Kamies Bergen, near Geelbeks Kraal, 2500–3000 ft., *Drège* (ex *E. Meyer*). Kaus Mountains, between Karakuis and Goedemans Kraal (near Steinkof), 2000–3000 ft., *Drège* (ex *E. Meyer*).

I have seen a specimen of this plant distributed by Drège as *M. calycinum.*

7. M. incanum (Decne in DC. Prodr. viii. 511); stem shrubby, branching; branches erect or ascending-spreading, occasionally twining, $\frac{3}{4}$–$1\frac{1}{4}$ lin. thick, whitish from a very dense tomentum of minute hairs; leaves spreading, coriaceous, puberulous; petiole 1–2 lin. long; blade 6–10 lin. long, $\frac{1}{2}$–$1\frac{1}{2}$ lin. broad, linear, slightly hastate-auriculate at the base, obtuse or acute, revolute at the margins; cymes subaxillary, subsessile, 3–6-flowered; pedicels about $1\frac{1}{2}$ lin. long, whitish-tomentose; sepals 2–$2\frac{1}{2}$ lin. long, scarcely $\frac{1}{2}$ lin. broad, linear-lanceolate, acute, whitish-tomentose, erect, applied to the corolla, sometimes recurved in the upper part; corolla tubular, 5-angled, minutely puberulous outside; tube 3 lin. long, with 5 tufts of deflexed fulvous hairs near the base inside; lobes spirally connivent-erect, $1\frac{1}{2}$–$1\frac{3}{4}$ lin. long, $\frac{1}{3}$ lin. broad, linear-lanceolate, acute, channelled down the face, minutely ciliate on one margin, puberulous inside, green, margined with red; corona-tubercles none; staminal column $1\frac{3}{4}$ lin. long, 5-spurred just below the middle; follicle solitary (always ?), 2–$2\frac{1}{2}$ in. long, about $2\frac{1}{2}$ lin. thick, narrowly fusiform, tapering to a long acute point and to an acute base, glabrous; seeds $2\frac{1}{2}$–3 lin. long, $1\frac{1}{2}$ lin. broad, ovate, flattish or plano-convex, minutely tuberculate on both sides. *Schlechter in Journ. Bot.* 1896, 418. *M. sajittatum, R. Br., var. β, E. Meyer, Comm.* 222; *M. sagittatum, R. Br., var. incanum, E. Meyer ex Decne in DC. Prodr.* viii. 511, *under M. incanum; Schlechter in Engl. Jahrb.* xxi. *Beibl.* 54, 2.

CENTRAL REGION : Calvinia Div.; between Lospers Plaats and Springbok Kuil River, 2000–3000 ft., *Zeyher,* 1179 !

WESTERN REGION: Little Namaqualand; between Koussie (Buffels) River and Silver Fontein, near Ookiep, 2000–3000 ft., *Drège,* 3051 ! hills near I'us (Tc Ous Mountain ?), *Schlechter,* 11424 ! Great Namaqualand; Gubub, *Schinz,* 28 !

8. M. longituba (Schlechter in Bull. Herb. Boiss. iv. 445); an erect much-branched shrub; branchlets recurved-divaricate, minutely velvety; leaves distant, subsessile, 2–$3\frac{1}{2}$ lin. long, scarcely 1 lin. broad, lanceolate or lanceolate-elliptic, acute, minutely velvety on both sides; flowers in few-flowered subaxillary fascicles; pedicels short, velvety; sepals half as long as the corolla-tube, linear-lanceolate, velvety; corolla tubular; tube 3 lin. long, $\frac{2}{3}$ lin in diam., very thinly puberulous outside, furnished with a ring of deflexed hairs (? 5 patches) at the base inside; lobes $\frac{1}{2}$ lin. long, subtriangular, subacute, erect, incurved at the apex, thinly velvety-puberulous; apex of the style produced into a beak.

WESTERN REGION : Great Namaqualand; among rocks near Keetmanshoop, *Fleck,* 244a (ex *Schlechter*).

I have not seen this plant, which, from the description, appears to be very similar to *M. incanum,* Decne, but differing in its foliage.

9. M. Burchellii (N. E. Br.); a dwarf, much-branched woody shrub, $\frac{1}{2}$–1 ft. high; branches decumbent (*Burchell*), $\frac{2}{3}$–$\frac{3}{4}$ lin. thick, rigid, subspinescent, minutely puberulous in the young state, green;

juice not milky; leaves minute, with very short petioles, 1–2½ lin.
long, ½–1 lin. broad, ovate or elongated deltoid, acute, subcordate or
truncate at the base, puberulous, rather thick and fleshy, soon
deciduous; cymes sessile, umbellately 3–6-flowered ; bracts minute,
subulate ; pedicels ½–¾ lin. long, puberulous; sepals ¾ lin. long,
scarcely ⅓ lin. broad, lanceolate, acute, puberulous ; corolla tubular,
5-angled, scarcely constricted at the mouth, greenish, very minutely
puberulous outside ; tube 1–1¼ lin. long, thinly puberulous inside,
with 5 tufts of reflexed hairs near the base ; lobes spirally connivent-
erect into a short acute cone, each about ⅔ lin. long, ⅓ lin. broad,
compressed-complicate, ovate-lanceolate, acute, ciliate along one
margin, slightly gibbous-keeled on the back at the base ; corona-
tubercles none ; staminal column rather more than ½ lin. long ;
filament part short, stout, cylindric ; anthers obtusely gibbous at the
base ; anther-wings inconspicuous, triangular, but scarcely spur-like ;
follicles not seen.

KALAHARI REGION: Griqualand West; Hay Div.; at the Kloof Village in
the Asbestos Mountains, *Burchell,* 2063! Bechuanaland; in rocky and stony
places at Kosi Fontein *Burchell,* 2585!

10. **M. Massonii** (Schlechter in Journ. Bot. 1896, 418) ; a dwarf,
much-branched woody shrub, 6–10 in. high ; branchlets erect,
straight, rather rigid and subspinescent at the apex, minutely puberulous, at least on the young parts, green ; juice not milky ; leaves
minute, with very short petioles, opposite, and occasionally alternate
on the same plant, 1–2½ lin. long, ½–1¼ lin. broad, cordate or lanceolate, acute, with revolute margins, rather thick and fleshy, puberulous or subglabrous ; cymes sessile, fasciculately 3–6-flowered ; bracts
minute, subulate ; pedicels ½–1 lin. long, puberulous ; sepals about
1 lin. long, ⅓ lin. broad, linear-lanceolate, acute, minutely puberulous ;
corolla urceolate, truncate, glabrous or rarely puberulous outside ;
tube 1¼–1¾ lin. long, ¾ lin. in diam., puberulous in the upper half
inside, with 5 tufts of deflexed hairs just below the middle ; lobes
inflexed at a right angle and spirally arranged at the mouth of the
tube, nearly closing it, ⅓–½ lin. long, cucullate-ovate, acute, slightly
gibbous at the base, minutely ciliolate along one margin ; corona-
tubercles none ; staminal column ⅔–¾ lin. long ; filament part stout,
cylindric ; anthers obtusely gibbous at the base ; anther-wings small,
inconspicuous, triangular, but scarcely spur-like ; follicle solitary,
1¾–2 in. long, 2–2½ lin. thick, fusiform, beaked, glabrous ; seeds
2½ lin. long, 1⅔ lin. broad, flattish, minutely rugulose-tuberculate.
Astephanus Massoni, Schultes, Syst. Veg. vi. 124; *Dietr. Syn. Pl.*
ii. 909 ; *G. Don, Gen. Syst.* iv. 158. *Hæmax Massoni, E. Meyer,*
Comm. 223 ; *Harv. Gen. S. Afr. Pl. ed.* 1, 228 ; *Decne in DC.*
Prodr. viii. 509.

VAR. β, **Dregei** (N. E. Br.); plant 12–16 in. (or perhaps more) high, with
the main branches a foot or more high and less branched than in the type,
puberulous to the base. *Hæmax Dregei, E. Meyer, Comm.* 223 ; *Harv. Gen.*
S. Afr. Pl. ed. i. 228; *Decne in DC. Prodr.* viii. 509. *Astephanus Dregei,*
D. Dietr. Syn. Pl. ii. 909.

CENTRAL REGION : Calvinia Div. ; between Lospers Plaats and Springbok
Kuil River, 2000–3000 ft., *Zeyher*, 1177! Prince Albert Div. ; in the Karoo, by
the Gamka River, 2000–3000 ft., *Burke! Zeyher*, 1178 ! Beaufort West Div. ;
Karoo, near Beaufort West, *Henderson*, 3 ! Middelburg Div. ; on the Sneeuwberg Range, between Compass Berg and Rhenoster Berg, 4500–5000 ft., *Drège*,
3417 ! plains, Rosmead Junction, *Sim in Herb. Galpin*, 5630 ! Middelburg,
Sim, 2590 ! Aberdeen Div. ; Camdeboo Mountains, near Hamer Kuil, 3000–
3500 ft., *Drège* ex *E. Meyer ;* Graaff Reinet Div. ; hills near Graaff Reinet,
2600 ft., *Bolus,* 365 ! Colesberg Div.; near Colesberg, *Fraser !* Naauw Poort,
Burchell, 2780 !

WESTERN REGION : Var. *β* : Little Namaqualand ; between Pedros Kloof
and Silver Fontein, near Buffels (Kousie) River, 2000–2500 ft., *Drège*, 3050! and
without precise locality, *Scully*, 246! and in *MacOwan & Bolus, Herb. Norm.
Aust.-Afr.*, 1313 !

The corolla of this plant varies considerably in length, actually, and relatively
to the sepals. The var. *Dregei* from Namaqualand, distinguished by E. Meyer
as a species, is taller and less densely branched than the other specimens, and has
a different appearance, but I do not find any difference in the flowers. The
stems, however, are not more puberulous than some examples of the ordinary
M. Massoni, in which they are never quite glabrous as described by E. Meyer,
although the pubescence is often exceedingly minute. Decaisne has wrongly
quoted R. Br. as the authority for *Astephanus Massoni*.

X. **PARAPODIUM,** E. Meyer.

Calyx 5-partite. *Corolla* 5-lobed ; tube hemispherical or globose-
campanulate ; lobes recurved or spreading at the tips. *Corona-lobes*
5, adnate to the corolla-tube in their lower part, free at the tips
alternating with the corolla-lobes, quite free from the staminal
column. *Stamens* 5, united to the dilated part of the style in a
column ; anthers with membranous appendages at the apex. *Pollen-
masses* solitary in each cell, pendulous, attached by rather long
caudicles to the pollen-carriers. *Style* conical at the apex, sometimes produced beyond the anther-appendages. *Follicles* large,
coriaceous.

Erect herbs ; leaves opposite, coriaceous; umbels lateral at the nodes.

DISTRIB. Species 3, endemic.

This remarkable genus is stated not to have been seen by Decaisne or Harvey,
and the Kew specimens of it (on one of which Decaisne has written the name
Parapodium costatum) were overlooked by Bentham when he worked up the
order for the Genera Plantarum, whilst Schlechter in the Journal of Botany,
1896, 454, erroneously refers the plant on which E. Meyer founded the genus to
Asclepias orbicularis, Schlechter (i.e. *Xysmalobium orbiculare*, Dietr.), a plant
with which the generic description of E. Meyer does not at all agree. But
Schlechter seems to have entirely misunderstood the genus *Parapodium*, since in
Engler's Jahrb. xx. Beibl. 51, p. 41, he founds the supposed new genus *Rhombonema* upon the very plant which E. Meyer had long before described as
Parapodium costatum.

Style-apex exserted beyond the anther-appendages ;
　　leaves lanceolate or oblong, flat or very slightly
　　undulated ...　　...　　...　　...　　...　　...　　...　　(1) **costatum.**
Style-apex not exserted beyond the anther-appendages :
　　Leaves oblong or oblong-linear, flat, not crisped　...　　(2) **simile.**
　　Leaves linear-lanceolate or lanceolate, very much
　　　undulated and crisped　　...　　...　　...　　...　　(3) **crispum.**

1. P. costatum (E. Meyer, Comm. 222); stem 3–11 in. high, erect, simple, stout, bifariously puberulous, at least on the upper part; leaves in 6–8 pairs, coriaceous, very slightly fleshy, thick and rigid when dry ; petiole $\frac{1}{2}$–$1\frac{1}{2}$ in. long, spreading or upcurved, stout, flattened and puberulous above ; blade 2–$5\frac{1}{2}$ in. long, $\frac{1}{2}$–$1\frac{1}{4}$ in. broad, ascending, sometimes almost making a bend at its junction with the petiole, elongated-oblong or lanceolate, acute or obtuse and apiculate at the apex, rounded or broadly cuneate at the base, flat or very slightly undulated, but not at all crisped on the margins, glabrous or slightly puberulous on the very stout midrib beneath ; umbels 3–5, lateral at the nodes, pedunculate, 3–7-flowered; peduncles $2\frac{1}{2}$–8 lin. long, puberulous on one side; bracts $1\frac{1}{4}$–3 lin. long, linear-subulate or lanceolate, slightly pubescent; pedicels 3–5 lin. long, glabrous or puberulous ; sepals ascending and nearly equalling the corolla, $3\frac{1}{2}$–$4\frac{1}{2}$ lin. long, $\frac{3}{4}$–$1\frac{1}{3}$ lin. broad, lanceolate, acuminate, slightly pubescent or almost glabrous on the back ; corolla subglobose or globose-campanulate, about 5 lin. in diam., obtusely 5-angled, lobed to $\frac{2}{3}$ of the way down, glabrous, dull purple-brown or violaceous outside, dull green within ; tube $1\frac{1}{2}$ lin. long, subhemispherical; lobes 3–4 lin. long, $1\frac{1}{2}$–$1\frac{3}{4}$ lin. broad, oblong-lanceolate, subacute, erect, closely contiguous and concave at the lower part, recurved or revolute and convex at the tips ; corona-lobes fleshy, $1\frac{1}{4}$–$1\frac{1}{3}$ lin. long, $1\frac{1}{2}$–2 lin. broad, broadly cuneate, truncate or rounded at the apex, adnate to the corolla up to the sinuses, spreading, closely con- tiguous, forming a short pentagonal truncate cup, with 5 rather stout keels, formed by the inflexed contiguous margins of the lobes com- posing it, white ; in some flowers, between the subcontiguous margins of each pair of lobes, is an exceedingly narrow lobule channelled down its middle and also adnate to the corolla-tube ; staminal column 2–$2\frac{1}{4}$ lin. long, conical; anther-appendages 1 lin. long, ovate, oblong or more or less pandurate, acute or subobtuse, erect and applied to the sides of the stout conical style-apex ; anther-wings somewhat hatchet-shaped, about half as broad just below the apex as they are at the hooked angular base, rounded on their outer margin ; style-apex projecting $\frac{1}{3}$–1 lin. beyond the anther-appendages, shortly bifid ; follicles not seen. *Harv. Gen. S. Afr. Pl. ed.* 1, 227 ; *Decne in DC. Prodr.* viii. 512 ; *N. E. Br. in Hook. Ic. Pl.* xxviii., *in note under t.* 2744 ; *Rand in Journ. Bot.* 1903, 336. *Metastelma costatum, Dietr. Syn. Pl.* ii. 908. *Rhombonema luridum, Schlechter in Engl. Jahrb.* xx. *Beibl.* 51, 41, *and Journ. Bot.* 1896, 418.

CENTRAL REGION : Aliwal North Div. ; on the Wittebergen Range, *Drège!*

KALAHARI REGION : Orange River Colony ; Rhenoster River, *Burke!* Trans- vaal; near Lydenburg, *Wilms*, 952! Vereeniging, 4750 ft., *Gilfillan in Herb. Galpin*, 6149! foot of the Magaliesberg Range, *Schlechter*, 3610! Modder- fontein, *Conrath*, 980! Veld north of Johannesburg, *Rand*, 965! near Pieters- burg, *Bolus*, 11124! and without precise locality, *Zeyher*, 1163!

Partly described from fresh flowers in fluid. As stated by Schlechter in Engler's Jahrb. xxi. Beibl. 54, 3, the specimen in E. Meyer's Herbarium named *Parapodium costatum*, by E. Meyer himself, is a piece of *Xysmalobium orbiculare*. But it is evident that some error has been made in placing the label bearing that name (Drège, 3421) with that specimen, because E. Meyer's manuscript description of *Parapodium* placed with it, describes a totally different floral structure, and his dissected flower, which is also on the same sheet, agrees with his description and with the flowers of the plant I have above described as *P. costatum*. Specimens of it, collected by Drège, have been erroneously distributed as *Pachycarpus rigidus*, E. Meyer, so that great confusion seems to have been made in the distribution of these three species.

2. **P. simile** (N. E. Br.) ; stems apparently about 6–10 in. high, 2–3 lin. thick, bifariously puberulous, leafy to the base ; leaves spreading or ascending ; petiole ⅓–¾ in. long, channelled and puberulous above ; blade 1½–3⅓ in. long, 2½–10 lin. broad, oblong, lanceolate-oblong or oblong-linear, acute, rounded or cuneate at the base, coriaceous, flat, not crisped at the margins, glabrous on both sides ; umbels 4–6, lateral at the nodes, pedunculate, 8–9-flowered ; peduncles ¼–1 in. long, moderately stout, puberulous on one side ; bracts 2–3 lin. long, linear-subulate, finely pointed, glabrous, minutely and sparsely ciliate ; pedicels ⅓–½ in. long, puberulous on one side ; sepals ¼ in. long, ¾–1 lin. broad, lanceolate, acuminate, with a few hairs on the back or subglabrous, applied to the corolla, with spreading tips ; corolla subglobose-campanulate, quite glabrous, "creamy-brown" according to *Miss Pegler*, dark purple-brown with pale margins to the lobes in the dried state ; tube 1–1¼ in. long ; lobes about 3 lin. long, erect and contiguous at the very concave lower part, recurved at the flat tips, revolute along the margins ; corona-lobes arising just below the sinuses between the corolla-lobes and decurrent-adnate to the bottom of the corolla, ¾ lin. high, 1⅓ lin. broad, transversely oblong, subtruncately rounded at the top, which has an acute edge, fleshy, white, erect, contiguous at their sides, which are slightly reflexed, forming a pentagonal cup surrounding, but free from, the 2 lin. long staminal column ; anther-appendages ovate, with the margins recurved as if pinched just below the truncate apex, erect, applied to the sides of the style-apex ; anther-wings triangular in outline, acute at the apex and not ¼ as broad just below as at the acutely hooked-angular base, nearly straight on their outer margin ; style-apex shortly columnar or subconical, minutely bilobed, not exserted beyond the anther-appendages ; follicle solitary, erect, very large in proportion to the plant, 5 in. or more long, 1 in. thick, lanceolate, truncate and somewhat toothed at the apex, with about 5 coriaceous irregularly toothed wings, teeth ⅛–¼ in. long ; seeds (unripe) ovate, very concave on one side, convex on the other, rugose, dark brown, probably about 2 lin. long when ripe.

KALAHARI REGION : Transvaal ; near Rustenburg, 4000 ft., *Miss Pegler*, 1022 ! Orange River Colony ; near Besters Vallei, *Bolus*, 6348 !

Although similar to *P. costatum*, the short style-apex and rather smaller differently shaped anther-wings clearly point to a specific distinction. The follicles of *P. costatum* may also prove to be different from those of this plant.

3. **P. crispum** (N. E. Br. in Hook. Ic. Pl. t. 2744) ; plant about

6–8 in. high; stem simple or branching at the base only, rather stout, pubescent along two broad lines; leaves rather close together, incurved-ascending, $2\frac{1}{4}$–$3\frac{1}{4}$ in. long, $2\frac{1}{2}$–5 lin. broad, linear-lanceolate, acute, narrowed into the $1\frac{1}{2}$–3 lin. lo ıg petiole at the base, coriaceous, very much crisped or undulate along the margins, glabrous on both sides; umbels several, subaxillary, 3–4-flowered; peduncles 1–2 lin. long, moderately stout; bracts subulate, $1\frac{1}{2}$–2 lin. long; pedicels 3–4 lin. long; sepals 3–4 lin. long, $\frac{3}{4}$ lin. broad, gradually tapering from base to apex, glabrous; corolla 4–6 lin. in diam., 5-lobed to rather more than half-way down, glabrous on both sides; tube about $1\frac{1}{2}$ lin. long, globose-campanulate; lobes 2–$2\frac{1}{2}$ lin. long, $1\frac{1}{2}$–$1\frac{2}{3}$ lin. broad at the base, ovate-lanceolate, subacute, concave and somewhat incurved in their lower part, recurved at the apex; corona-lobes adnate to the corolla-tube for half their length, and in that part confluent, forming a sort of disk lining the base of the corolla-tube, apical part free, erect and rising slightly above the sinuses of the corolla, $\frac{1}{2}$–$\frac{2}{3}$ lin. long, $1\frac{1}{6}$–$1\frac{1}{2}$ lin. broad, transversely oblong, truncate, in dried flowers rather thin, flat or with the margins incurved; staminal column $1\frac{1}{2}$ lin. long, stout, conical; anther-appendages small, oblong, acute, loosely inflexed over and covering the small depressed-truncate style-apex; follicles not seen.

COAST REGION : Queenstown Div.; on grassy hills, near Shiloh, 4000 ft., *Drège !*

CENTRAL REGION : Somerset Div., foot of Bosch Berg, *MacOwan,* 1343! and without precise locality, *Cooper,* 531! Cradock Div.; near Cradock, *Cooper,* 1285! Graaff Reinet Div.; on the Sneeuwberg Range near Graaff Reinet, 4100 ft., *Bolus !*

This remarkable plant bears such a close resemblance to *Xysmalobium gomphocarpoides,* Dietr., and *Woodia mucronata,* N. E. Br., as to be readily mistaken for either of them. It was distributed by Drège as *Pachycarpus gomphocarpoides* under letter " *a.*"

XI. **WOODIA**, Schlechter.

Calyx 5-partite. *Corolla* 5-lobed nearly to the base; lobes incurved-ascending or spreading ?, concave or flattish, with revolute margins. *Corona-lobes* 5, arising from the base of the staminal column opposite the anthers, dorsally flattened, concave at the basal part, with the incurved margins ascending the staminal column as narrow wings, deeply trifid, or quadrate with square shoulders and an incurved middle tooth, with one stout keel on or decurrent from the inner face of the middle tooth. *Stamens* arising from the base of the corolla, their filaments connate in a tube around the ovary; anthers terminated by a membranous appendage. *Pollen-masses* solitary in each anther-cell, nearly as broad as long, subquadrate or rectangular, pendulous, attached in pairs to the pollen-carriers by rather stout caudicles. *Style* depressed-truncate or truncate at the apex, shorter than the anther-appendages. *Follicles* where known, covered with short spine-like processes. *Seeds* crowned with a tuft of hairs.

Perennial herbs, probably with a tuberous rootstock ; stem usually solitary, erect, simple or rarely with one branch ; leaves opposite ; umbels lateral at the nodes and terminal, sessile or pedunculate.

DISTRIB. Species 3, endemic. The Tropical African *Woodia trilobata,* Schlechter, does not belong to this genus, but is a *Xysmalobium.*

Corona-lobes distinctly 3-toothed or -lobed :
 Peduncles ½–1½ in. long, 2 to several times as long
  as the petioles ; sepals as long as or exceeding
  the corolla ... ... (1) **verrucosa.**
 Peduncles 0–⅔ in. long, rarely longer than the
  petioles ; sepals shorter than the corolla... ... (2) **mucronata.**
Corona-lobes subquadrate, with square shoulders and
 a thick incurved obtuse middle point ... ... (3) **singularis.**

1. **W. verruculosa** (Schlechter in Engl. Jahrb. xviii. Beibl. 45, 31) ; stem simple, 6–10 in. high, apparently slightly compressed, glabrous with the exception of a puberulous line down one side ; leaves in 4–7 pairs ; petiole 1–3½ lin. long ; blade 1⅓–2¼ in. long, ½–1¼ in. broad, ovate, ovate-oblong or oblong-lanceolate, acute, with a somewhat pungent apiculus, rounded to subcordate at the base, glabrous on both sides, slightly scabrid on the slightly thickened margin ; umbels 2–4, erect, lateral and terminal, racemose or occasionally subcorymbose, 4–5-flowered ; peduncles ½–1½ in. long, puberulous down one side ; bracts 2–3 lin. long, subulate, glabrous ; pedicels 7–10 lin. long, puberulous on one side ; sepals as long as or longer than the corolla, suberect, about ¼ in. long and ¾ lin. broad, tapering from the base to the very acute apex, more or less revolute at the margins, glabrous or very thinly sprinkled with minute hairs ; corolla-lobes erect with more or less incurved tips, 2½–2¾ lin. long, 1 lin. broad at the base, thence tapering to the acute apex, flattish, with revolute margins, glabrous on both sides, apparently purple-brown with green or pale margins on the back, perhaps greenish on the inner face ; corona-lobes arising at the base of the staminal column and their margins ascending it as very narrow wings, 1¼ lin. long, 3-lobed, erect ; middle lobe lanceolate-subulate with a very prominent stout keel on the inner face at the lower part, slightly incurved and reaching to the top of the anthers ; lateral lobes much shorter, linear, obtuse, very slightly falcate, quite erect on each side of the central keel, with their flat sides facing each other ; staminal column 1¼ lin. long ; anther-appendages broader than long, reniform, obtuse, very abruptly inflexed over the margin of the truncate style-apex ; follicle solitary, 3–3½ in. long, ¾ in. thick, stoutly fusiform, acute, beset with several series of spine-like processes, ¾–1½ lin. long, glabrous. *Schlechter in Journ. Bot.* 1896, 456.

EASTERN REGION : Griqualand East ; mountains around Clydesdale, 2500 ft., *Tyson,* 2173 ! Natal ; hills near Camperdown, 2000–3000 ft., *Wood,* 4079 ! 4966 !

2. **W. mucronata** (N. E. Br.) ; stems 1–2 to a plant, 6–10 in.

high, bifariously puberulous, leafy nearly to the base, with internodes ¼–1 in. long; leaves widely spreading or more or less upcurved,
coriaceous; petiole ⅓–½ in. long, channelled and puberulous above;
blade 1½–3 in. long, ¼–1¼ in. broad, varying from linear-lanceolate
to elliptic or elliptic-oblong, and from nearly or quite flat to very
much undulate and crisped, glabrous on both sides or slightly
puberulous on the midrib beneath, often scabrous on the margins;
umbels 3–8, pedunculate or sessile, lateral at the nodes, racemose,
4–7-flowered; peduncles 0–8 lin. long, puberulous; bracts 1½–3 lin.
long, linear-subulate, glabrous or nearly so; pedicels 4½–6 lin. long,
puberulous; sepals shorter than the corolla, 2–2½ lin. long, ⅔–1 lin.
broad, lanceolate, acuminate, slightly puberulous; corolla-lobes
incurved-ascending, 2½–3½ lin. long, 1¼–2 lin. broad, ovate or oblong-
ovate, acutely pointed owing to the revolute margins, concave,
glabrous on both sides; corona-lobes arising at the base of the
staminal column and shortly adnate to the very base of the corolla,
1–1½ lin. high, fleshy, trifid, concave and broadly cuneate at the
base, with the margins ascending the staminal column as narrow
wings, about 1 lin. broad across the ascending-spreading ⅓–½ lin. long
lateral deltoid teeth, with the middle tooth 1–1¼ lin. long, stout,
keeled down the inner face or compressed, resembling a bird's claw,
incurved-erect, obtuse, attaining to about the level of the style-apex;
staminal column about 1¼ lin. long, pentagonal at the top, suddenly
constricted under the small rectangular very prominent anther-
wings; anther-appendages subreniform, inflexed over the rim of the
depressed-truncate style-apex; caudicles broad and flat, attached to
the inner apical angle of the subquadrate pollen-masses; pollen-
carrier surrounded by a hyaline margin; young follicles divergent,
ovoid or ellipsoid, tapering to a stout stalk-like base and to an acute
apex, glabrous, beset with several rows of soft subulate processes
about 1 lin. long. *Asclepias mucronata, Thunb. Prodr.* 47, *in Nov.
Act. Acad. Petrop.* xiv. (1805) 508; *Fl. Cap. ed.* 2, ii. 155, *and ed.
Schultes,* 235; *Poir. Encycl. Suppl.* i. 479; *Willd. Sp. Pl.* i. 1263;
Pers. Syn. Pl. i. 275; *Schultes, Syst. Veg.* vi. 80; *G. Don, Gen.
Syst.* iv. 142. *Pachycarpus marginatus, E. Meyer, Comm.* 213
partly. Xysmalobium marginatum, Dietr. Syn. Pl. ii. 902. *Gom-
phocarpus marginatus, Decne in DC. Prodr.* viii. 560, *not of
Schlechter.*

VAR. β, **trifurcata** (N. E. Br.); like the type, but blade of leaf 1¾–4 in.
long; sepals 2⅓–4 lin. long, ¾–1¼ lin. broad; corolla-lobes 3½–5 lin. long,
1¾–2½ lin. broad; corona-lobes 1½–1¾ lin. high and about 1¼–1½ lin. broad
across the natural spread of the lateral teeth, which are ¾–1 lin. long, linear or
linear-oblong, suberect, with their margins ascending the staminal column in
broader wings than in the type, the middle tooth 1–1½ lin. long, more compressed and broader from the keel to the channelled back than in the type;
staminal column 1½–2 lin. long; follicles 3¼–4¾ in. long, about ¾ in. thick,
lanceolate-fusiform or subequally tapering at both ends, acute, with several
longitudinal rows of spine-like processes ½–1 lin. long, glabrous; seeds small,
2–2¼ lin. long, 1–1¼ lin. broad, concave, with very incurved margins on one side,
very convex on the other, minutely tuberculate on both sides, dark brown.
Woodia trifurcata, Schlechter in Journ. Bot. 1895, 337 *in note, and var.*

planifolia, Schlechter in Engl. Jahrb. **xx.** *Beibl.* 51, 39. *W. marginata, Schlechter in Engl. Jahrb.* **xxi.** *Beibl.* 54, 9; *Journ. Bot.* 1896, 456, *and Ann. Naturhist. Hofmus. Wien,* **xv.** 69. *Gomphocarpus trifurcatus, Schlechter in Engl. Jahrb.* xviii. *Beibl.* 45, 9. *Pachycarpus marginatus, E. Meyer, Comm.* 213, *partly. Xysmalobium linguæforme, Weale in Journ. Linn. Soc.* xiii. 50.

SOUTH AFRICA : without locality, *MacOwan !* Var. β, *Drège,* 4939 !

COAST REGION : Humansdorp Div. ; across the Gamtoos River and near Kromme River, *Thunberg !* Uitenhage Div. ; Addo, *Drège,* 2224! Albany Div. ; without precise locality, *Drège ! Hutton ! Cooper,* 2718 ! Queenstown Div. ; Haugklip Mountain, near Queenstown, *Galpin,* 1807 ! Var. β : Queenstown Div. ; near Queenstown, 3800 ft., *Bolus !* Stutterheim Div. ; Dohne Mountain, 3000 ft., *Bolus,* 10213 ! East London Div.; near East London, 100 ft., *Wood in Herb. Galpin,* 3385 ! Komgha Div. ; hills by the Kei River, near Komgha, 1800 ft., *Flanagan,* 399 ! *Krook,* 792 (ex *Schlechter*).

CENTRAL REGION : Somerset Div. ; near Somerset East, *Miss Bowker !*

KALAHARI REGION : Var. β : Transvaal ; Little Olifants River, 5000 ft., *Schlechter,* 3799 (ex *Schlechter*).

EASTERN REGION : Var. β : Transkei ; near Kentani, 1000 ft., *Miss Alice Pegler,* 572 ! Tembuland ; near Bazeia, 2000 ft., *Baur,* 569 ! between Cala and Elliot, *Bolus,* 10186 ! near Umtata, *Bolus,* 10187 !

Although Thunberg described this remarkable plant more than 100 years ago, no one appears to have identified it until now. Thunberg's type has flat leaves without undulations, but is otherwise identical with Miss Bowker's and Hutton's specimens. Var. *trifurcata* seems to be the eastern form of the plant, with larger flowers and much longer side teeth to the corona-lobes. Some specimens bear a remarkable resemblance to *Parapodium crispum* and *Xysmalobium gomphocarpoides*; all three may easily be mistaken for each other unless dissected.

3. **W. singularis** (N. E. Br.) ; stem solitary, about $1\frac{1}{2}$ ft. high, 2 lin. thick at the base, simple, slightly bifariously puberulous on the upper part, glabrous below ; leaves in about 16 pairs, ascending-spreading, thinly coriaceous, quite glabrous on both sides ; petiole $\frac{1}{3}$–$\frac{1}{2}$ in. long ; blade of the middle leaves 2–$2\frac{3}{4}$ in. long, $\frac{3}{4}$–$1\frac{1}{4}$ in. broad, the upper and lower smaller, oblong or sublanceolate-oblong, acute or obtuse and apiculate, rounded or cuneate at the base, flat or undulated at the margins ; umbels sessile, lateral at the nodes and terminal, 5–6-flowered ; pedicels 3–5 lin. long, minutely puberulous ; sepals $\frac{1}{4}$ in. long, $\frac{2}{3}$ lin. broad, linear-lanceolate, very acute, glabrous ; corolla-lobes incurved-ascending or spreading?, about $3\frac{1}{2}$ lin. long, $1\frac{2}{3}$–$1\frac{3}{4}$ lin. broad, oblong, shortly subcuspidate-acute, concave, with revolute margins, glabrous on both sides, greenish?, tinted with purple-brown on the back ; corona-lobes arising at the base of the staminal column, with their margins ascending it as narrow wings, spreading, transversely oblong or subquadrate with square shoulders and about 1 lin. long at the concave basal part, produced at the middle for about $\frac{3}{4}$ lin. into a stout fleshy linear-oblong obtuse point, incurved at the apex and keeled down the inner face, rising to about the base of the anthers ; staminal column $1\frac{1}{2}$ lin. high, abruptly much constricted under the very broad transverse anthers, whose broad triangular-ovate appendages are inflexed over the margin of the broad 5-angled style-apex.

EASTERN REGION : Swaziland ; ridge between Bremersdorp and Mac Nab's Store, 2500 ft., *Burtt Davy,* 2933 !

XII. **XYSMALOBIUM**, R. Br.

Calyx 5-partite. *Corolla* 5-lobed nearly to the base; lobes sub-erect, spreading or reflexed, flat or slightly concave, or the lower part very concave and the upper part recurved, overlapping or rarely subvalvate in bud. *Corona-lobes* 5, arising from the staminal column opposite the anthers, variable in shape, very thick and fleshy, as thick as broad or laterally compressed or dorsally flattened, always solid, keelless or with one longitudinal median keel on the inner face, never cucullate or complicate, nor with any filiform horn or long tongue-like process on the inner face. *Staminal column* arising from the base of the corolla. *Anthers* terminated by a membranous appendage. *Pollen-masses* solitary in each anther-cell, pendulous, attached in pairs to the pollen-carriers by elongated caudicles. *Style* usually shorter than the anther-tips, rarely exserted beyond them. *Follicles* variable in shape, smooth or more or less covered with soft bristle-like processes. *Seeds* crowned with a tuft of hairs.

Perennial herbs with milky juice and tuberous rootstock or a cluster of thick fleshy fusiform roots; stems erect or decumbent, simple or branched at the base, sometimes solitary; leaves opposite; umbels few or many and lateral at the nodes with 1 or 2 terminal, sessile or pedunculate, or solitary and terminal on a long peduncle.

Distrib. Species 38, half of them in Tropical Africa.

Xysmalobium as understood by recent authors is very unsatisfactorily defined. It was separated by Robert Brown from *Asclepias* to include those species in which the corona-lobes are "fleshy, subrotund, simple on the inner face," and have 5 minute teeth or lobules alternating with them. He referred only 2 species to the genus, viz. :—*Asclepias undulata*, Linn., and *A. grandiflora*, Linn. f., which according to modern views cannot both belong to the same genus. As the characters given by R. Brown do not at all agree with the structure of *A. grandiflora*, whilst they do roughly accord with that of *A. undulata*, I take this latter as the type of *Xysmalobium*. The minute alternating lobes are of no generic importance, as they are absent from some and present in other species of this and many other genera. The definition of the genus as given by Bentham and Hooker, and copied in Engler & Prantl, Pflanzenfamilien, does not distinguish it from *Schizoglossum*, nor apply to all the species. Dr. Schlechter has repeatedly asserted that *Xysmalobium* cannot be distinguished from *Asclepias*, but does not define either genus. In the structure of the corona-lobes, the two genera are entirely different, always appearing solid in *Xysmalobium*, and always either cucullate or with a fissure or space between the inflexed sides of them in *Asclepias*. From their fleshy nature it will probably be found that some of the descriptions of the corona-lobes do not accord with those organs in living flowers, as they alter in the process of drying.

<pre>
  * Leaves 11–40 times as long as broad, all linear or
      narrowly linear-lanceolate, ½–3 lin. broad, not
      crisped on the margins :
          Corolla reflexed, glabrous on the inner face ;
              corona-lobes shorter than the staminal column :
                  Anther-wings broadest and distinctly an-
                      gular at the middle...   ...   .   (1) involucratum.
                  Anther-wings broadest at the base, not at all
                      angular at the middle   ...   .   (2) Zeyheri.
          Corolla-lobes erect, with recurved minutely
              velvety tips; corona-lobes overtopping the
              staminal column ...   .   (3) carinatum.
</pre>

** Middle and upper leaves 4–14 times as long as
  broad, varying from linear to narrowly elongate-
  lanceolate or lanceolate:
    Leaves 1½–5½ in. long, ⅛–⅔ in. broad, crisped on
      the margins; corolla glabrous on the inner
      face:
        Corolla-lobes about 2½ lin. long; corona-
          lobes about as long as broad, apiculate:
            Corona-lobes much shorter than the
              staminal column, suborbicular, with
              a very prominent keel on the inner
              face; anther-wings ½ lin. long   ...   (4) **winterbergense.**
            Corona-lobes nearly or quite as long as
              the staminal-column, cuneate-obovate
              or subquadrate, with or without a
              slight keel on the inner face; anther-
              wings ¼ lin. long   ...   .   (5) **brownianum.**
        Corolla-lobes 3½ lin. long; corona-lobes
          twice as long as broad   ...   .   (6) **gomphocarpoides.**
    Leaves 3–8½ in. long, ⅔–2½ in. broad, with or
      without crisped or undulate margins; corolla-
      lobes erect or suberect, densely white-pubescent
      at the recurved tips on the inner face:
        Plant ½–1½ ft. high; corolla glabrous out-
          side, with the hairs at the tips of the
          inner face not more than $\frac{1}{12}$ lin. long   ...   (7) **stockenstromense.**
        Plant 3–5 ft. high; corolla pubescent or
          puberulous outside, with the hairs at the
          tips of the inner face ¼–½ lin. long:
            Corolla-lobes 3–3½ lin. long; corona-
              lobes closely adpressed to the backs of
              the anthers   ...   .   (8) **ensifolium.**
            Corolla-lobes 4½–6 lin. long; corona-
              lobes standing free from the anthers ..   (9) **undulatum.**
*** Leaves 1½–3 (rarely 3–4) times as long as broad,
  all ovate, ovate-lanceolate, lanceolate, oblong or
  elliptic, ½–4 in. long:
    Umbels all sessile, numerous, racemosely ar-
      ranged along the stem, leaves elliptic or oblong (10) **confusum.**
    Umbels pedunculate:
        Leaves usually oblong or elliptic, rarely
          lanceolate or ovate, glabrous; umbels
          numerous and racemosely arranged along
          the stout simple stem   ...   .   (11) **orbiculare,**
        Leaves ovate, lanceolate or oblong-lanceo-
          late, pubescent or scabrous on both sides;
          plant branching at the base:
            Umbels 2 in a terminal pair, or 2–6 and
              racemosely arranged along the stems,
              occasionally solitary on weak branches;
              peduncles ¼–1½ in. long:
                Corolla-lobes suberect; corona-
                  lobes not contiguous, about
                  twice as long as broad, semi-
                  terete or subtrigonous, slightly
                  incurved-erect:
                    Corolla-lobes with a longi-
                      tudinally divided gibbosity
                      on the inner face   ...   ...   (12) **acerateoides.**

Corolla-lobes without a gibbosity on the inner face :
   Corolla-lobes 1¾–2 lin.
    long, yellow ... ... (13) **Gerrardi.**
   Corolla-lobes 1–1⅓ lin.
    long, greenish - white,
    often purplish down the
    back ... ... (14) **parviflorum.**
Corolla-lobes very spreading or
reflexed ; corona-lobes subcontiguous, ovoid to subglobose, not
nearly twice as long as broad ... (15) **asperum.**
Umbel solitary, terminal ; peduncle 1½–4
in. long :
   Corolla yellow ; corona-lobes oblong, erect, distinctly exceeding
   the staminal column, with slightly
   incurved tips ... ... (16) **tysonianum.**
   Corolla brownish or greenish-brown ;
   corona-lobes linear-oblong to
   narrowly ovate, slightly divergent-spreading, not exceeding
   the staminal column nor incurved
   at the tips... ... ... (17) **prunelloides.**
   Corolla white or greenish-white,
   tipped with purple, at least on
   the backs of the lobes ; corona-
   lobes erect, orbicular in dorsal
   outline, rather shorter than the
   staminal column ... ... (18) **Baurii.**

1. **X. involucratum** (Decne in DC. Prodr. viii. 520); tuber
somewhat carrot-like ; stems 1–2 to a tuber, erect, 4–12 in. long,
leafy, glabrous with 2 pubescent lines, or pubescent all round ;
leaves numerous, opposite, 2–4½ in. long, ¾–3 lin. broad, linear,
acute, erect or ascending, revolute along the margins, varying from
glabrous to pubescent on one or both sides ; umbels 1–5 to a stem,
usually subcorymbose at its summit, occasionally spaced along the
upper part, pedunculate, hemispherical, 20–30-flowered, involucrate
with several linear-subulate spreading or reflexed bracts 2–2½ lin.
long ; peduncles ½–1½ in. long, pubescent ; pedicels 4–5 lin. long,
pubescent ; sepals 1 lin. long, nearly ½ lin. broad, lanceolate, acute,
pubescent, reflexed ; corolla entirely reflexed, slightly pubescent on
the back, glabrous within, yellowish-green ; lobes 1½–1¾ lin. long,
¾ lin. broad, oblong-lanceolate, subacute ; corona-lobes arising about
¼ lin. above the base of the staminal column, and only reaching to
about its middle, ¾ lin. long, ovoid, acute, entire or bifid, thick and
fleshy, erect, laterally sessile, with the apical part free and applied to
the basal part of the anthers ; staminal column 1½ lin. long ; filament
part short, cylindric; anther part somewhat barrel-shaped ; anthers
erect, rhomboid ; their appendages small, elliptic-ovate, acute,
inflexed over the apex of the style, and their wings very broadly
triangular, very prominent and minutely notched at the middle,
narrowing towards the base and apex ; style-apex small, pentagonal,

excavated; follicle solitary, erect, 3–3½ in. long, about 5 lin. thick, fusiform, tapering into a beak, smooth, thinly and minutely puberulous; seeds 2½ lin. long, 2 lin. broad, broadly ovate, flat with a broad, plicately wrinkled margin, and minutely papillate on the disk. *Lagarinthus involucratus, E. Meyer, Comm.* 203. *Gomphocarpus involucratus, D. Dietr. Syn. Pl.* ii. 900; *Schlechter in Engl. Jahrb.* xviii. *Beibl.* 45, 8, *and* 19, xx. *Beibl.* 51, 32. *Asclepias chloroglossa, Schlechter in Journ. Bot.* 1896, 454.

COAST REGION: Uniondale Div.; between Welgelegen and Onzer, 2000 ft., *Drège!* Albany Div.; Heights of Albany, *Bowie!* near Grahamstown, *Bolton! MacOwan,* 654! 1217! *Galpin,* 249 (ex *Schlechter*), and without precise locality, *Zeyher! Cooper,* 2736! *Bowker!* Bathurst Div.; between Riet Fontein and the source of the Kasuga River, *Burchell,* 4147! Komgha Div.; near Komgha, 1900 ft., *Flanagan,* 390 (ex *Schlechter*). Fort Beaufort Div.; on the Winter Bergen, *Mrs. Barber,* 83! Queenstown Div.; mountain sides near Queenstown, 4000 ft., *Galpin,* 1656! Stockenstrom Div.; on the Kat Berg, 4000–5000 ft., *Drège* (ex *E. Meyer*). British Kaffraria, *Cooper,* 472!

EASTERN REGION: Transkei; Krelis Country, *Bowker!* and without precise locality, *Mrs. Barber,* 36! *Hallack!* Kentani Div.; hill-sides, *Miss Pegler,* 118! Tembuland; Bazeia, *Baur,* 383! Griqualand East; Malowe Mountain, 4000 ft., *Tyson,* 3114! Mount Currie, 5000 ft., *Tyson in MacOwan & Bolus, Herb. Norm. Austr.-Afr.,* 546! Pondoland; near the mouth of St. Johns River, *Bolus,* 10191! Natal; Inanda, *Wood,* 364! and in Natal Herb. 429! Mooi River, 4000 ft., *Wood,* 4063! Shafton, Howick, *Mrs. Hutton,* 403! near the mouth of the Umzimkulu River, *Drège* (ex *E. Meyer*), Tugela, *Gerrard,* 1801! and without precise locality, *Sanderson,* 373! *Peddie!*

According to Mrs. Barber, the Kaffirs eat the tubers at times when food is scarce.

2. X. Zeyheri (N. E. Br.); plant about 5–6 in. high, with 1 simple erect stem to a tuber, pubescent all round; leaves numerous, opposite, 3–4½ in. long, ½–⅔ lin. broad, linear, acute, erect or ascending, revolute along the margins, glabrous above, more or less pubescent on the midrib beneath; umbels 3–4 to a stem, subaxillary and terminal, pedunculate, hemispherical, about 20-flowered, involucrate with several linear-subulate spreading or reflexed bracts about 2 lin. long; peduncles 6–7 lin. long, pubescent; pedicels 3–5 lin. long, pubescent; sepals 1½ lin. long, ½ lin. broad, lanceolate, acute, pubescent, reflexed; corolla entirely reflexed, pubescent outside, glabrous within; lobes 2 lin. long, rather more than 1 lin. broad, oblonglanceolate, acute; corona-lobes arising ⅔ lin. above the base of the staminal column and reaching to about ¾ of its length, ¾ lin. long, oblong-ovoid, acute, thick and fleshy, erect, laterally attached at the base to the staminal column by a short distinct stalk, and standing free from the anthers; staminal column 1⅔ lin. long; filament part tapering upwards, constricted under the anthers; anthers oblong with nearly parallel sides, slightly broadest at the base, erect, their appendages very broadly ovate, acute, inflexed over the apex of the style, and their wings broadest at the base, gradually narrowed upwards; style slightly excavated at the pentagonal, truncate apex; follicles and seeds not seen. *Lagarinthus involucratus, b, in Herb. Drège.*

In all characters, except those of the corona and staminal column, this is quite like *X. involucratum*, Decne, but in that species the filament part of the staminal column is shorter than in this and cylindric, not tapering upwards, the anthers are distinctly rhomboid, their wings being broadest and most projecting at the middle, at about the level of the apex of the corona-lobes, whilst in *X. Zeyheri* they are slightly broadest at the base nearly at the level of the base of the corona-lobes, and do not project at all at the middle; the corona-lobes are also sessile in *X. involucratum* and distinctly stalked in *X. Zeyheri*.

3. X. carinatum (N. E. Br.); plant 9–10 in. high, branching at the base only; stems erect from a slightly decumbent base, slender, glabrous with 2 pubescent lines, leafy; leaves numerous, erect, 2–$3\frac{1}{2}$ in. long, $\frac{1}{2}$–1 lin. broad, linear, acute, revolute along the margins, glabrous; umbels several to a stem, subaxillary, shortly pedunculate or subsessile, subglobose, 7–10-flowered; peduncles 1–$2\frac{1}{2}$ lin. long; bracts about $1\frac{1}{4}$ lin. long, subulate or filiform; pedicels 2–$2\frac{1}{2}$ lin. long, minutely puberulous with curved hairs; sepals 1–$1\frac{1}{4}$ lin. long, $\frac{1}{2}$ lin. broad, lanceolate, acute, minutely puberulous, spreading or perhaps slightly recurved; corolla about $2\frac{1}{2}$ lin. in diam., bell-shaped, 5-lobed almost to the base; lobes erect or but slightly spreading, 2 lin. long, 1 lin. broad, oblong-lanceolate, obtusely pointed, slightly concave at the base, recurved and slightly thickened at the apex, and somewhat reflexed along the margins, glabrous outside, very minutely velvety at the thickened apex within; corona-lobes arising at the very base of the staminal column and incurved over it, $1\frac{1}{2}$ lin. long, linear-falcate when viewed sideways, fleshy, flat on the face, very convex on the back; staminal column about 1 lin. long; filament part above the insertion of the corona-lobes about $\frac{1}{5}$ lin. long; anthers quadrate, erect, their appendages broadly ovate, acute, erect, slightly inflexed at the tips over the rim of the style-apex; their wings broadened into a triangular projection a little below the middle; pollen-masses obliquely deltoid, shaped something like a ham of bacon; style-apex pentagonal, excavated, not exceeding the anther-appendages; follicles not seen. *Krebsia carinata, Schlechter in Journ. Bot.* 1895, 269, *and Ann. Naturhist. Hofm. Wien*, xv. 67.

This plant is so exactly like *Krebsia stenoglossa* and *Schizoglossum periglossoides* in the dried state, that it is scarcely possible to distinguish them unless their flowers are dissected, when the great differences of their corona-lobes, staminal column and pollen-masses are convincing enough of their distinctness. Probably when alive they are more easily distinguished.

4. X. winterbergense (N. E. Br.); plant branching into several stems at the base; stems 4–5 in. long and bearing 4–7 pairs of leaves in the examples seen, bifariously pubescent; leaves ascending, coriaceous ; petiole 2–6 lin. long; blade $1\frac{1}{4}$–$2\frac{3}{4}$ in. long, 1–4 lin. broad, linear, linear-lanceolate or the lowest narrowly lanceolate, acute, tapering into the petiole at the base, undulate or crisped, on the very scabrid margins, slightly scabrid on the midrib, otherwise glabrous on both sides ; umbels about 4 to a stem, lateral and terminal, 10–12-flowered, sessile ; pedicels 3–5 lin. long, puberulous on one side ; sepals $1\frac{1}{2}$ lin. long, $\frac{1}{2}$ lin. broad, lanceolate, acute, pubescent; corolla-lobes spreading, $2\frac{1}{2}$ lin. long, 1–$1\frac{1}{4}$ lin. broad, oblong, subacute, with narrowly revolute margins and a few hairs on the back, glabrous on the inner face, apparently greenish-yellow or green ; corona-lobes arising at the base of the staminal column and much shorter than it, erect, about 1 lin. long and broad, fleshy, suborbicular or orbicular-rhomboid, obtusely apiculate, flat, with a very prominent keel decurrent from the apiculus on the inner face and somewhat angular at its middle, apparently dark brown ; staminal column $\frac{1}{2}$ lin. long, constricted at the middle, with the upper half broadly obconic and the lower subcylindric ; anther-appendages very short, transverse, rounded, inflexed over the rim of the truncate style-apex, anther-wings $\frac{1}{2}$ lin. long.

COAST REGION : Fort Beaufort Div. ; in a valley of the Winterberg Range, *Mrs. Barber*, 86 !

According to Mrs. Barber's note with the specimen the flowers are " dark brown and green." She states that " it is very uncommon; I have only seen this one plant." Allied to *X. brownianum*, but differs in its corona-lobes, anther-wings and locality.

5. X. brownianum (S. Moore in Journ. Bot. 1903, 309); stem $\frac{1}{4}$–1 ft. high, simple or with 1–2 basal branches, bifariously puberulous, leafy to the base ; leaves in 7–9 pairs, coriaceous, erectly upcurved or ascending-spreading; petiole $\frac{1}{4}$–$1\frac{1}{2}$ in. long, channelled and puberulous above, spreading; blade of the lower leaves $\frac{1}{2}$–$2\frac{3}{4}$ in. long, $\frac{1}{2}$–$1\frac{1}{2}$ in. broad, ovate, elliptic, oblong, or elongated deltoid-lanceolate, obtuse and mucronate or acute, rounded to truncate at the base, spreading or ascending, of the upper leaves $1\frac{1}{2}$–$3\frac{1}{2}$ in. long, $1\frac{1}{2}$–10 lin. broad, linear, linear-oblong, or narrowly lanceolate, acute, cuneate or rounded at the base, where they somewhat abruptly curve upwards into an erect position, crisped on the slightly scabrid margins, glabrous on both sides ; umbels sessile, lateral at the nodes, 7–12-flowered ; bracts 1–2 lin. long, linear or linear-lanceolate, acute ; pedicels 3–5 lin. long, puberulous, intermingled with longer hairs on one side ; sepals $1\frac{1}{2}$ lin. long, lanceolate, acuminate, with a few hairs on the back ; corolla-lobes campanulately spreading or ascending, about $2\frac{1}{2}$ lin. long, 1 lin. broad, oblong, slightly notched at the obtuse recurved point, recurved at the margins, concave on the inner face and slightly gibbous at the base, glabrous on both sides or with a few short hairs on the back ; corona-lobes arising at the base of the staminal column and nearly as long as it, erect, $1\frac{1}{4}$ lin. long, 1–$1\frac{1}{4}$ lin. broad at the top, broadly cuneate-obovate or subquadrate, sloping or truncately contracted into a short and more or less inflexed apiculus, flat, sometimes with an indication of a slight keel down the inner face, apparently ochreous or orange; staminal column $1\frac{1}{4}$–$1\frac{1}{2}$ lin. long, constricted below the middle, with the upper part broadly obconic and the basal part shortly conical, pentagonal at the top; anther-appendages much broader than long, subreniform, obtuse, inflexed over the rim of the depressed-truncate style-apex; anther-wings $\frac{1}{4}$ lin. long, very small and prominent, broadest and angular at the base; caudicles nearly as broad as the minute pollen-carrier is long, attached to the erect, obliquely tapering apex of the relatively very large oblong pollen-masses, which are $\frac{2}{5}$ lin. long and quite straight along the inner margin. *Rand in Journ. Bot.* 1903, 339. *Asclepias anisophylla, Conrath & Schlechter in Engl. Jahrb.* xxxviii. 31.

KALAHARI REGION : Transvaal; Magalies Berg, *Burke ! Zeyher,* 1165 partly ! Irene, *Conrath,* 984 ! open veld, south of Johannesburg, *Rand,* 1053 ! Witkleifontein, *Burtt Davy,* 3134 !

The corona-lobes are very similar in outline to those of *Schizoglossum eustegioides.*

6. **X. gomphocarpoides** (Dietr. Syn. Pl. ii. 902) ; stem 1–$1\frac{1}{2}$ ft. or more high, erect, simple, moderately stout, glabrous or slightly pubescent along two broad lines; internodes short; leaves 4–$5\frac{1}{2}$ in. long, 3–7 lin. broad, ascending, shortly petiolate, linear-lanceolate, gradually tapering to an acute point, cuneate or sub-cordate at the base, undulated, scabrid on the margins, glabrous; umbels several, racemose along the stem, subaxillary, 4–8-flowered; peduncles 1–4 lin. long, stout, puberulous along one side; pedicels 4–6 lin. long, puberulous along one side; sepals 2 lin. long, $\frac{2}{3}$ lin. broad, lanceolate, acute, thinly puberulous; corolla 5-lobed almost to the base, glabrous on both sides; lobes somewhat reflexed, $3\frac{1}{2}$ lin. long, $1\frac{1}{2}$ lin. broad, oblong, acute, greenish in the dried state; corona-lobes arising from the base of the staminal column and about equalling it in length, erect, $1\frac{1}{2}$–$1\frac{3}{4}$ lin. long, auriculately dilated and about 1 lin. broad at the base, contracted above, linear-oblong, obtuse, very fleshy, very convex on the back, keeled down the inner face; staminal column $1\frac{1}{2}$–2 lin. long; anthers quadrate; anther-appendages rounded, incumbent upon the depressed-truncate pentagonal style-apex; follicle solitary (always?), $2\frac{1}{4}$–3 in. long, $\frac{3}{4}$–1 in. thick, obliquely ovoid or ovoid-lanceolate, nearly straight along the side turned to the stem, very bulging on the other, laxly covered with soft bristle-like processes, glabrous; seeds $2\frac{3}{4}$–3 lin. long, ovate, concave on one face, convex on the other, rugulose-tuberculate, dark-brown. *Decne in DC. Prodr.* viii. 519. *Pachycarpus gomphocarpoides, E. Meyer, Comm.* 213. *Gomphocarpus longifolius, Schlechter in Engl. Jahrb.* xviii. *Beibl.* 45, 9, *in note, not of Spreng. Asclepias gomphocarpoides, Schlechter in Engl. Jahrb.* xxi. *Beibl.* 54, 7, *and Journ. Bot.* 1896, 456.

COAST REGION : Oudtshoorn Div.; near the Cango Caves, 2100 ft., *Bolus,* 12133 !

CENTRAL REGION: Graaff Reinet Div.; Sneeuwberg Range, 4100 ft., *Bolus*, 635! Victoria West Div.; Nieuw Veld, between Brak River and Uitvlugt, 3000–4000 ft., *Drège!* Wodehouse Div.; northern slopes of the Andries Berg, 5800 ft., *Galpin*, 2220! Philipstown Div.; at "Washbanks River," *Burchell*, 2740! Colesberg Div.; near Colesberg, *Shaw!*

Some other localities are mentioned by E. Meyer, but as this plant is mixed with *Parapodium crispum* in his herbarium, without properly localized labels, I have only quoted the locality of the specimens distributed by Drège under letter *c*, which certainly belong here. This plant is exceedingly like *Parapodium crispum* and *Woodia mucronata* in the dried state, and may easily be mistaken for either of them unless dissected.

7. X. stockenstromense (Scott-Elliot in Journ. Bot. 1890, 364); plant 7–15 in. high; stem erect, simple, stout, rather fleshy, puberulous along two broad lines; internodes short; leaves subsessile or petiolate, 3–8 in. long, $\frac{3}{4}$–$1\frac{3}{4}$ in. broad, lanceolate-attenuate, acute, more or less undulated or flat, coriaceous, glabrous, scabrid on the margins; umbels 2 to several, lateral and terminal, pedunculate, many-flowered, often crowded in a corymbose manner; peduncles $\frac{1}{2}$–$2\frac{1}{2}$ in. long, stout, erect, puberulous down one side; bracts $1\frac{1}{2}$–3 lin. long, subulate; pedicels $\frac{1}{2}$–$\frac{3}{4}$ in. long, puberulous; sepals 2–3 lin. long, 1 lin. broad, lanceolate, acute, glabrous; corolla 5-lobed almost to the base, green, tipped with brown; lobes erect, 3–4 lin. long, $1\frac{3}{4}$–2 lin. broad, oblong, acute, glabrous outside, concave and glabrous in the lower part within, recurved and densely white-puberulous at the tips, with hairs not more than $\frac{1}{12}$ lin. long; corona-lobes arising at the base of the staminal column, about half as long as and standing free from it, reaching to the base of the anthers, in fresh flowers about $1\frac{1}{2}$ lin. long, 2 lin. broad, in dried ones 1 lin. long, 1–$1\frac{1}{4}$ lin. broad, sessile, erect, contiguous, broadly deltoid-ovate, obtuse, fleshy, dorsally flattened, and in dried flowers with the margins somewhat reflexed, and indistinctly keeled on the back, having a very prominent subacute keel on the face, which is adnate to the base of the column, in fresh flowers the lobe is triangular in transverse section and half as thick as broad, glabrous; staminal column $2\frac{1}{4}$–$2\frac{1}{2}$ lin. long, much constricted under the quadrate anthers; anther-appendages rounded, incumbent on the truncate pentagonal style-apex; fruit not seen. *Gomphocarpus stockenstromensis, Schlechter in Engl. Jahrb. xx. Beibl. 51, 37. Asclepias stockenstromensis, Schlechter in Journ. Bot. 1896, 454.*

COAST REGION: Stockenstrom Div.; on the scrubby slopes of Lushington Mountain, near Stockenstrom, *Scully*, 169! Stutterheim Div.; Dohne Mountain, 4500 ft., *Bolus*, 10204!

KALAHARI REGION: Basutoland; grassy slopes of Mount-aux-Sources, 7800 ft., *Thode*, 36! Transvaal; near Belfast, 6500 ft., *Bolus*, 12120! near Carolina, 5600 ft., *Bolus!* Donker Hoek, *Schlechter*, 4134 (ex *Schlechter*).

EASTERN REGION: Griqualand East; near Kokstad, *Haygarth in Herb. Wood*, 4172! Natal; on the South Downs in Weenen County, 4000 ft., *Wood*, 4419!

8. X. ensifolium (Burch. ex Scott-Elliot in Journ. Bot. 1890, 364); plant about 5 ft. high; stem stout, pubescent with spreading hairs; leaves ascending, very shortly petiolate, 7–8 in. long, 1–1$\frac{3}{4}$ in. broad, elongated ovate-lanceolate or linear-lanceolate, subacute, rounded or shortly cordate at the base, thinly pubescent on both sides, margins not very scabrid and not undulated; umbels several, the upper rather crowded, many-flowered, globose, 1$\frac{1}{2}$ in. in diam.; peduncle $\frac{1}{2}$–1$\frac{1}{4}$ in. long, erect, stout, hairy-pubescent; bracts $\frac{1}{4}$ in. long, subulate, pubescent; pedicels 6–8 lin. long, subtomentose; sepals 3 lin. long, 1 lin. broad, lanceolate, acuminate, channelled at the tips, pubescent; corolla campanulate, 5-lobed nearly to the base, about 4$\frac{1}{2}$ lin. in diam.; lobes 3–3$\frac{1}{2}$ lin. long, 1$\frac{1}{2}$ lin. broad, pubescent on the back, glabrous on the concave lower part within, densely bearded with white hairs $\frac{1}{4}$–$\frac{1}{2}$ lin. long on the recurved tips; corona-lobes 1–1$\frac{1}{4}$ lin. long, 1$\frac{1}{3}$ lin. broad, arising a little above the base of the staminal column, erect, closely applied to the backs of the anthers and reaching to the base of the anther-appendages, sub-orbicular, flat, fleshy, obtusely gibbous-keeled on the back, and having a very slight linear keel on the face at the base; staminal column about 2 lin. long; anthers subquadrate, with the wings produced into deltoid truncate teeth; anther-appendages ovate, acute, inflexed-connivent over the short conical obtuse style-apex; pollen-masses scarcely $\frac{1}{2}$ lin. long, roundish-pear-shaped, attached to the black ovoid pollen-carrier by rather long crooked caudicles; follicles not seen. *X. undulatum, R. Br., var., Scott-Elliot in Journ. Bot.* 1890, 364.

KALAHARI REGION: Griqualand West, Herbert Div.; at upper Campbell, *Burchell,* 1834! Bechuanaland; near the sources of Kuruman River, *Burchell,* 2491!

In habit, foliage, and flowers, this is exceedingly like *X. undulatum,* R. Br.; the flowers, however, are considerably smaller, and the corona-lobes are quite different, the stout fleshy keel, present on the upper part of the face in *X. undulatum,* being entirely absent in *X. ensifolium,* and the lobes themselves are pressed close to the anthers; the pollen-masses are also different in the two species.

9. **X. undulatum** (R. Br. in Mem. Wern. Soc. i. 39); stems 3–5 ft. high, stout, erect, hairy-pubescent; leaves more or less ascending, subsessile or with very short stout petioles, 4–8$\frac{1}{2}$ in. long, 1$\frac{1}{4}$–3 in. broad, elongated, lanceolate or ovate-lanceolate, or sometimes ovate, obtuse with a short point, or subacute, rounded or more or less cordate at the base, pubescent on both sides, scabrid along the margins, which are often undulated; umbels several, the upper often crowded, many-flowered, globose, 1$\frac{3}{4}$–2$\frac{1}{2}$ in. in diam.; peduncles $\frac{1}{3}$–1$\frac{1}{4}$ in. long, erect, stout, hairy-pubescent; bracts about 3 lin. long, subulate, pubescent; pedicels $\frac{1}{2}$–$\frac{3}{4}$ in. long, hairy-pubescent; sepals 3–4 lin. long, 1$\frac{1}{4}$–1$\frac{3}{4}$ lin. broad, lanceolate, acute, pubescent; corolla globose-campanulate, 5-lobed nearly to the base, 6–7 lin. in diam.; lobes 4$\frac{1}{2}$–6 lin. long, 2$\frac{1}{2}$–3 lin. broad, oblong, acute, pubescent outside, glabrous in the concave lower part within, dull green (*Rand*), densely bearded with white hairs $\frac{1}{4}$–$\frac{1}{2}$ lin. long on the recurved tips; corona-lobes 1$\frac{1}{2}$ lin. long, 1$\frac{2}{3}$–2$\frac{1}{3}$ lin. broad,

scarcely half as long as the staminal column, except when pressed out of position by drying, shortly spreading from its base, then erect, broadly cuneate, obovate or transverse, subtruncate or broadly rounded at the apex, fleshy, flat on the back, thickened into a triangular mass or stout keel on the inner face, dull white, tinged with purple externally (*Rand*); staminal column 2½–3 lin. long; anthers subquadrate, with their wings produced into a truncate tooth at or near the base; anther-appendages elliptic or suborbicular, incumbent upon the stout convex or cushion-like style-apex; follicle usually solitary, 3–5 in. long, 1½–1¾ in. thick, obliquely ovoid, usually very obtuse, sometimes somewhat tapering to a stout obtuse point, pubescent and thickly covered with long filiform pubescent processes; seeds 2½ lin. long, 1¼ lin. broad, convex on one side, flattish with thick raised margins on the other, covered with raised irregular lines and points on both sides, dark brown. *Ait. Hort. Kew. ed.* 2, ii. 79; *Harv. Gen. S. Afr. Pl. ed.* 1, 226; *Dietr. Syn. Pl.* ii. 902; *Spreng. Syst. Veg.* i. 850; *Schultes, Syst. Veg.* vi. 89; *E. Meyer, Comm.* 215; *G. Don, Gen. Syst.* iv. 146; *Scott-Elliot in Journ. Bot.* 1890, 364; *Rand in Journ. Bot.* 1903, 338, *not of Decne. X. lapathifolium, Decne in DC. Prodr.* viii. 519; *K. Schum. in Engl. & Prantl, Pflanzenfam.* iv. ii. 232, *f.* 67, A–D, *very inaccurate. Asclepias undulata, Linn. Sp. Pl. ed.* 1, i. 214; *Mant.* ii. 346; *Ait. Hort. Kew. ed.* 1, i. 304; *Thunb. Prodr.* 47; *in Nov. Act. Acad. Petrop.* xiv. (1805), 508; *Fl. Cap. ed.* 2, 155, *and ed. Schultes,* 234; *Willd Sp. Pl.* i. 1263; *Lam. Encycl. Meth.* i. 280; *Pers. Syn. Pl.* i. 275; *Schlechter in Engl. Jahrb.* xxi. *Beibl.* 54, 9; *Journ. Bot.* 1896, 451, *and Ann Naturhist. Hofmus. Wien,* xv. 68, *not of Jacquin. A. ciliata, Murray, Syst. Veg. ed.* 15, 271. *Gomphocarpus arborescens, Sprenger in Bull. Soc. Tosc. Hort. ser.* 2, v. 70, *fig.* 7, *and Wien. Ill. Gart. Zeit.* 1889, 494, *fig.* 78, *not of R. Br. G. undulatus, Schlechter in Engl. Jahrb.* xviii. *Beibl.* 45, 10, *not of Turcz.—Apocynum africanum, lapathi folio, Commelin, Hort. Med. Amstelœdam.* 16, *t.* 16.

EASTERN REGION; Transkei; near Colossa, *Krook*, 810 (ex *Schlechter*), around Kentani, *Miss Pegler*, 691! Natal; near Estcourt, 3800 ft., *Wood*, 3474! near Bothas, 2000 ft., *Wood*, 4809! and without precise locality, *Cooper*, 2745! Swaziland, *Burtt Davy*, 2931! 2926! *Bolus*, 12121!

10. X. confusum (Scott-Elliot in Journ. Bot. 1890, 363); stem 2–3 ft. high, stout, $\frac{1}{3}$ in. thick in the specimens seen, pubescent along two broad lines; leaves not very spreading, coriaceous; petiole 1–2 lin. long, stout; blade 2–4 in. long, 1–2$\frac{1}{2}$ in. broad, elliptic, elliptic-oblong or oblong, obtuse, shortly and abruptly cuspidate-mucronate, cuneately or very obtusely rounded at the base, very scabrid along the margins, glabrous on both sides; umbels numerous, many-flowered, sessile, the upper more or less crowded; bracts 3–3$\frac{1}{2}$ lin. long, subulate; pedicels 6–8 lin. long, minutely pubescent; sepals 3–4 lin. long, $\frac{3}{4}$ lin. broad at the ovate-oblong base, above which the margins are incurved, forming a channelled subulate point, spreading, glabrous, sparingly ciliolate; corolla 5-lobed almost to the base, quite glabrous; lobes suberect, 4–5 lin. long, 2$\frac{1}{2}$–2$\frac{3}{4}$ lin. broad, oblong-ovate, obtuse, with a minute oblique apical notch; margins reflexed or revolute; corona-lobes arising at the base of the staminal column and about half as long as it, fleshy, scarcely 1 lin. long, about 1$\frac{1}{4}$ lin. broad, broadly cuneate, truncate, with a minute quadrate lobule at the base and a fleshy prominent keel at the top on the inner face, triangular when seen from above and as if formed from the inflexed adnate apex; staminal column 1$\frac{1}{2}$ lin. long, constricted under the anthers; anther-appendages subreniform, obtuse, inflexed upon the outer part of the broad style-apex, the centre of which is produced slightly beyond them in a very short stout pentagonal truncate column; follicles not seen. *Gomphocarpus rectinervis, Schlechter in Engl. Jahrb.* xx. *Beibl.* 51, 38. *Asclepias confusa, Schlechter in Journ. Bot.* 1896, 454, *and A. rectinervis, Schlechter, l.c.* 456.

EASTERN REGION: Natal; Highlands of Natal, *Gerrard*, 1282! Inanda, *Wood*, 1163! and without precise locality, *Mrs. K. Saunders!* Swaziland; near Embabaan (Mbabane), 4700 ft., *Bolus*, 12123!
KALAHARI REGION: Transvaal; Houtbosh Berg, 6400 ft., *Schlechter*, 4429.

11. X. orbiculare (D. Dietr. Syn. Pl. ii. 902); stem 1–3 ft. high, stout, 2–5 lin. thick, simple or rarely branching towards the base, puberulous along two rather broad lines, or rarely glabrous; leaves rather closely placed, much longer than the internodes, spreading, coriaceous, glabrous; petiole 1–3 lin. long, stout; blade 1$\frac{1}{2}$–3 in. long, $\frac{1}{2}$–2$\frac{1}{4}$ in. broad, oblong, broadly ovate, elliptic or lanceolate, obtuse and apiculate or acute, rounded or cuneate-acute (rarely cordate) at the base; margins very scabrid; umbels numerous, subaxillary, racemosely arranged along the upper part of the stem, many-flowered; peduncle $\frac{1}{3}$–1 in. long, stout, erect, pubescent along the inner side; pedicels $\frac{1}{3}$–$\frac{1}{2}$ in. long, puberulous along the inner side; sepals 1$\frac{3}{4}$–2 lin. long, $\frac{3}{4}$ lin. broad, lanceolate, acute, nearly glabrous; corolla 5-lobed almost to the base, reflexed, glabrous on both sides, chocolate-coloured with a green blotch at the base of each lobe ; lobes 3–3½ lin. long, 1½ lin. broad, oblong, acute ; corona-lobes arising at the base of the staminal column and slightly shorter than it, 1–1¼ lin. long, 1½ lin. broad, erect, fleshy, dorsally flattened, suborbicular, obtuse and slightly incurved at the apex, obtusely keeled down the face, glabrous, whitish ; staminal column 1½ in. long, stout ; anther-appendages very broad and short, rounded, inflexed over the raised rim of the truncate apex of the style ; pollen-masses subfalcate-oblong, obtusely pointed, scarcely ½ lin. long, attached to the black pollen-carrier by stout caudicles ; follicles 3½–4 in. long, about 1 in. thick, lanceolate-fusiform, obtusely pointed, sparingly echinate with stout erect acute processes about 1 lin. long, glabrous ; seeds (immature) oblong-ovate, minutely tuberculate on both sides. *Decne in DC. Prodr.* viii. 519. *X. padifolium, Scott-Elliot in Journ. Bot.* 1890, 363 ; *Benth. ex Schlechter in Engl. Jahrb.* xx. *Beibl.* 51, 35, *in note. Pachycarpus orbicularis, E. Meyer, Comm.* 212. *Gomphocarpus padifolius, Baker in Saunders, Ref. Bot. t.* 254 ; *Schlechter in Engl. Jahrb.* xviii. *Beibl.* 45, 9, *and* xx. *Beibl.* 51, 35, *in note. G. orbicularis, Schlechter in Engl. Jahrb.* xx. *Beibl.* 51, 34. *Asclepias orbicularis, Schlechter in Journ. Bot.* 1896, 454, *and Ann. Naturhist. Hofmus. Wien,* xv. 68.

CoAST REGION: East London Div.; near East London, *Wood in Herb. Galpin,* 3151! Komgha Div.; near Komgha, *Krook,* 820 (ex *Schlechter*).
EASTERN REGION : Transkei (Krielis Country), *Bowker,* 215! around Kentani, *Miss Pegler,* 653! 887! Tembuland ; on hills near Bazeia, 2000 ft., *Baur,* 819! Pondoland ; between St. Johns River and Waterfall Bluff, *Drège,* 4980! Natal ; near Verulam, *Wood,* 912! at Umhlongwe, *Wood,* 3013! Inanda, *Wood,* 1254! and without precise locality, *Gueinzius! Gerrard,* 1281! Zululand ; near Umgoya, *Wood,* 747! 3924! 5677!

12. X. acerateoides (N. E. Br.) ; stem 2–2½ ft. high (*Galpin*), only pieces 6–9 in. long seen, ⅛–¼ in. thick, subhispid-pubescent ; leaves spreading ; petiole 2–7 lin. long ; blade 1⅔–4 in. long, ⅓–2 in. broad, lanceolate, ovate or oblong-ovate, acute or the upper acuminate, rounded to subcordate at the base, flat or revolute at the scabrid-ciliate margins, roughly pubescent on both sides ; umbels varying from 2 in a terminal pair to 6, lateral and terminal, occasionally solitary, ⅞–1⅙ in. in diam., 12–30-flowered ; peduncles ⅓–1¼ in. long, scabrid-pubescent ; bracts 1½ lin. long, subulate ; pedicels 3–5 lin. long, pubescent ; sepals 1½–2 lin. long, ½ lin. broad, lanceolate, acuminate, pubescent ; corolla-lobes erect, 1½–2½ lin. long, ¾–1 lin. broad, oblong, obtuse or subacute, with a longitudinally divided convexity at the middle of the inner face and correspondingly concave on the back, glabrous on both sides, greenish-yellow ; corona-lobes arising at the base of the staminal column and shorter than or subequalling it, erect, slightly incurved, ¾–1 lin. long, fleshy, solid, in fresh flowers (seen preserved in fluid) ½ lin. broad and nearly as thick, subcylindric, scarcely flattened on the inner face, very obtuse, with a very faint indication of lateral compression at the apex, but in dried flowers often about ⅓ lin. broad, sometimes appearing to be linear-oblong, obtuse, convex on the back and flat on the inner face, at others laterally much compressed and as if pinched together at the truncate apex, which often angularly projects on the inner face, so that the lobe appears somewhat linear-subfalcate in side view, with the inner side flattened and the back convex ; staminal column about 1 lin. long ; anther-appendages ovate to suborbicular, obtuse or acute, inflexed upon and concealing the small truncate style-apex. *Gomphocarpus acerateoides, Schlechter in Engl. Jahrb.* xviii. *Beibl.* 45, 16, *and G. ovatus, Schlechter, l.c.* 20. *Asclepias acerateoides, Schlechter in Journ. Bot.* 1896, 454, *and A. scabridifolia, Schlechter, l.c.* 455.

KALAHARI REGION : Transvaal ; De Kaap Flats, near Barberton, 2600 ft., *Galpin,* 664 ! summit of Saddleback Mountain, near Barberton, 5000 ft., *Galpin,* 674 ! near the top of Mauch Berg, 6600 ft., *Atherstone !*

EASTERN REGION : Swaziland ; near Embabaan (Mbabane), 4900 ft., *Bolus,* 12142 ! *Burtt Davy,* 2768 !

Similar to *X. Gerrardi,* but apparently a much stouter plant, and easily recognized by the peculiar gibbous intrusion of the middle of the petals, a character not mentioned by Schlechter, whose types of *G. acerateoides* and *G. ovatus* I have examined, and find to be identical. The differences in the corona-lobes mentioned may be observable, but are due to the remarkable changes which take place in drying, and are seen to vanish if a sufficient number of flowers are examined, but neither in the fresh nor dried flowers do I find any trace of the keels described by Schlechter under *G. ovatus,* which were probably raised lines due to shrinkage.

13. X. Gerrardi (Scott-Elliot in Journ. Bot. 1890, 364) ; plant branching at the base ; branches simple or sparingly branched, erect or perhaps decumbent at the base, more or less pubescent ; leaves shortly petiolate, 1¼–2½ in. long, ⅓–1 in. broad, lanceolate, oblong-lanceolate or oblong-ovate, obtuse or acute, varying from acute to almost subcordate at the base, thinly and harshly pubescent on both sides, slightly scabrid at the margins ; umbels usually 2–4 to a branch, occasionally solitary on weak branches, lateral and terminal, pedunculate, 6–14-flowered ; peduncles ⅓–1 in. long, pubescent on one side ; bracts 1½–2 lin. long, filiform-subulate ; pedicels rather slender, 3–4 lin. long, pubescent ; sepals ascending, 1½ lin. long, ½ lin. broad, linear-lanceolate, acute, pubescent ; corolla-lobes suberect, 1¾–2 lin. long, about 1 lin. broad, ovate-oblong, obtuse, with recurved margins, without a gibbosity on the glabrous inner face, more or less pubescent on the back near the apex, yellow or yellowish ; corona-lobes arising close to or slightly above the base of the staminal column and very distinctly longer than it, erect, slightly incurved, not contiguous, ¾–1⅓ lin. long, ⅓ to nearly ½ lin. broad, fleshy, oblong-linear, very convex on the back and flattened on the inner face or perhaps subterete, not clavate as originally described, obtuse ; staminal column 1 lin. long ; anthers subquadrate, slightly broader at the base, with orbicular appendages inflexed over the truncate style-apex ; follicles not seen. *Gompho-*

carpus ochroleucus, Schlechter in Engl. Jahrb. xviii. *Beibl.* 45, 30. *Asclepias ochroleuca, Schlechter in Journ. Bot.* 1896, 455, *and Ann. Naturhist. Hofmus. Wien,* xv. 68.

EASTERN REGION : Griqualand East; Insizwa Mountains, *Krook,* 803, 814 (ex *Schlechter*).   Natal; Fields Hill, near Pinetown, and on a hill near Gillets, *Wood,* 3398 ! and in *Herb. Natal,* 682 ! hill near Krantz Kloof, *Wood,* 5002 ! between Pietermaritzburg and Greytown, *Wilms,* 2136 ! and without precise locality, *Gerrard,* 1289 ! *Sanderson,* 179 ! *Mrs. K. Saunders!*

Scott-Elliot quotes 3 specimens for *X. Gerrardi,* viz. :—Mrs. K. Saunders' and Gerrard, 1289 and 1951; of these, Mrs. K. Saunders' specimen and Gerrard 1289 agree with the original description as to the corona-lobes (the other characters in the description will equally answer for 3 other species), whilst Gerrard 1951 does not, and differs entirely in its conspicuously reflexed corolla as well as in the corona-lobes.   I therefore take the above two specimens as the type of *X. Gerrardi,* and refer Gerrard 1951 to *X. asperum.*   In the Journal of Botany, 1896, 454, Schlechter enumerates *X. Gerrardi, X. parviflorum,* and *Gomphocarpus parviflorus,* Schlechter (three entirely different plants), as synonyms under the new name of *Asclepias sulphurea,* without description or any specimen quoted; he seems to have entirely misunderstood both *X. Gerrardi* and *X. parviflorum,* and has redescribed both under new names.

14. X. parviflorum (Harv. ex Scott-Elliot in Journ. Bot. 1890, 363); plant 3–10 in. high, branching at the base ; branches erect or decumbent at the base, pubescent, chiefly along 2 broad lines ; leaves in 3–7 pairs to a branch, often distant, ascending; petiole 1–2 lin. long ; blade $\frac{2}{3}$–$1\frac{1}{2}$ in. long, 4–9 lin. broad, ovate or ovate-lanceolate, acute, rounded or subcordate at the base, thinly and harshly pubescent on both sides, scabrid on the slightly thickened or revolute margin ; umbels 1–4 to a stem, subaxillary and terminal, pedunculate, 10–16-flowered ; peduncles $\frac{1}{3}$–$1\frac{1}{4}$ in. long, shortly hairy along one side ; bracts $\frac{2}{3}$–$1\frac{1}{2}$ lin. long, subulate ; pedicels 2–3 lin. long, shortly hairy along one side, deflexed in fruit ; sepals ascending, $1\frac{1}{3}$–$1\frac{1}{2}$ lin. long, $\frac{1}{2}$ lin. broad, narrowly lanceolate, acute, shortly hairy on the back, usually reddish ; corolla not much longer than the calyx, greenish-white, sometimes purplish down the back ; lobes incurved-erect, 1–$1\frac{1}{3}$ lin. long, about $\frac{3}{4}$ lin. broad, ovate, obtuse, flattish, without a gibbosity on the glabrous inner face, sparsely hairy on the back ; corona-lobes arising at the base of the staminal column and usually slightly exceeding it, falcately incurved-erect, distant, $\frac{3}{4}$ lin. long, $\frac{1}{3}$–$\frac{1}{2}$ lin. broad, linear or linear-spathulate, obtuse, viewed dorsally, fleshy, apparently somewhat compressed-keeled on the back, or perhaps very convex, flat or slightly concave on the inner face, apparently yellow or ochreous ; staminal column $\frac{2}{3}$–$\frac{3}{4}$ lin. long ; anthers subquadrate, their appendages broadly ovate, inflexed over the truncate apex of the style ; follicle solitary, erect, about $2\frac{1}{4}$ in. long, 5 lin. thick, lanceolate, acuminate, pubescent ; seeds 2 lin. long, scarcely 1 lin. broad, very convex on one side, having a raised rim around a flattish disk on the other, rugulose with slender, irregular ridges, dull brown.   *Gomphocarpus* *pachyglossus, Schlechter in Engl. Jahrb.* xx. *Beibl.* 51, 35.   *Asclepias pachyglossa, Schlechter in Journ. Bot.* 1896, 455.

KALAHARI REGION : Basutoland, *Cooper*, 934! Orange River Colony, near Witzies Hoek, 6300 ft., *Bolus*, 8112 ! Harrismith, *Sankey*, 134! Transvaal; between Waterval River and Zuikerbosch Rand, *Schlechter*, 3493!

EASTERN REGION : Transkei; near Tsomo, *Mrs. Barber*, 864! Griqualand East; stony slopes around Kokstad, 5100 ft., *Tyson*, 1247! *Haygarth in Herb. Wood*, 4212! Natal; Dargle Farm, *Fannin*, 41! Greenwich Farm, Riet Vlei, *Fry in Herb. Galpin*, 2742 ! Weenen County, 3000–6000 ft., *Sutherland!* *Wood*, 4370! near Van Reenen, *Wood*, 6633! near Nottingham Road, *Wood*, 6805! between Greytown and Newcastle, *Wilms*, 2137! Klip River, 3500–4500 ft., *Sutherland!* and without precise locality, *Gerrard*, 1288!

Harvey's type of this species is *Fannin* 41, with which the other specimens quoted are identical, but *Bolus*, 5704, and *Burchell*, 4151, quoted by Scott-Elliot under this species, do not belong to it. The former is *X. asperum*, the latter is too immature for determination; it was collected between Rietfontein and the source of Kasuga River, in Bathurst Div., a region whence *X. parviflorum* is unknown.

15. X. asperum (N. E. Br.); plant probably branched at the base ; branches ½–1 ft. long, apparently more or less decumbent, sparingly subhispid, chiefly bifariously; leaves ascending-spreading; petiole 1–4 lin. long; blade ⅔–2¼ in. long, 4½–9 lin. broad, ovate, ovate-lanceolate or oblong-lanceolate, acute or acuminate, rounded or subcordate at the base, scabrid on both sides, narrowly revolute or slightly thickened at the scabrid margins ; umbels 2–6 to a branch or occasionally solitary, lateral at the nodes and terminal, pedunculate, ¾–1 in. in diam., 10–20-flowered ; peduncles ½–1½ in. long, scabrid-pubescent on one side ; bracts 1½–2 lin. long, subulate ; pedicels 3–4 lin. long, scabrid-pubescent on one side ; sepals very spreading or reflexed, 1–1⅓ lin. long, ⅓–⅔ lin. broad, lanceolate or linear-lanceolate, acute, thinly and harshly pubescent on the back ; corolla-lobes spreading or reflexed, 1½–1¾ lin. long, 1¼–1⅓ lin. broad, ovate, subobtuse, without a gibbosity on the inner face, glabrous on both sides or with a very few hairs on the back, green (*Rand*), often purple at the tips on the back ; corona-lobes arising close to the base of the staminal column and about equalling or shorter than it, erect or ascending-spreading, more or less contiguous, ¾–1¼ lin. long, about ¾ lin. broad, fleshy, ovoid, ellipsoid or subglobose, obtuse, nearly as thick as broad, perhaps slightly compressed dorsally, sometimes appearing slightly flattened on the back, at others on the inner face, not keeled, apparently orange, " dull yellow " (*Rand*); staminal column ¾–1 lin. long, not broadened at the base ; anther-appendages transverse, rounded, inflexed on the margin of the truncate style-apex ; follicles 2½–3 in. long, about ¾ in. thick, lanceolate, acuminate, with about 9 longitudinal rows of stiff spine-like processes ½–1 lin. long, glabrous. *Gomphocarpus parviflorus, Schlechter in Engl. Jahrb.* xx. *Beibl.* 51, 35, *excl. syn. Xysmalobium parviflorum, Harv. Asclepias sulphurea, S. Moore in Journ. Bot.* 1903, 312, *and Rand, l.c.* 339, *doubtfully of Schlechter, see note on* p. 577.

KALAHARI REGION : Transvaal; around Pretoria, 4100 ft., *McLea in Herb.
Bolus*, 5704! Modderfontein, *Conrath*, 990! near Vlakfontein Beacon, Carolina
District, 6000 ft., *Burtt Davy*, 2966! Nelspruit, *Rogers*, 549! near Johannes-
burg, *Rand*, 1046! 1124! near Brug Spruit, 4600 ft., *Schlechter*, 3760!

EASTERN REGION: Natal, *Gerrard*, 1951! Swaziland; near Embabaan
(Mbabane), 5000–5600 ft., *Bolus*, 12143.

The specimen in Herb. Bolus, mentioned by Dr. Schlechter under *Gomphocarpus parviflorus*, without quotation of number or locality, I learn from Dr.
Bolus is his No. 5704, from near Pretoria. Gerrard 1951 has oblong-lanceolate leaves, rounded at the base, but does not otherwise differ from the more
cordate-leaved typical form, and is connected with it by Burtt Davy's 2966 with
lanceolate, rounded-based leaves.

16. X. tysonianum (N. E. Br.); stems probably several from a
tuber, branching at the base only, 6–9 in. long, apparently decumbent,
more or less pubescent, with short spreading jointed hairs; leaves
5–8 pairs to a stem or branch, ¾–1¾ in. distant, spreading; petiole
1–2 lin. long; blade ¾–2 in. long, 4–8 lin. broad, ovate-lanceolate
or oblong-lanceolate, acute at the apex, broadly cuneate, rounded or
slightly subcordate at the base, flat, thinly pubescent on both sides
with spreading hairs; umbel solitary, terminal, pedunculate, hemispherical, 16–20-flowered; peduncle 1½–4 in. long, pubescent;
bracts 1½ lin. long, subulate, pubescent, very deciduous; pedicels
3 lin. long, pubescent with short spreading hairs; sepals 1¼–1½ lin.
long, about ⅓ lin. broad, lanceolate, acute, pubescent; corolla about
2 lin. in diam., pubescent outside, glabrous within, yellow or
greenish-yellow; lobes 1⅓–1½ lin. long, 1–1¼ lin. broad, ovate or
elliptic-ovate, subacute, erect, incurved at the apex over the corona,
the sides slightly folded back at the lower half; corona-lobes arising
about ⅓ lin. above the base of the staminal column and shortly
overtopping it, ¾–1 lin. long, ⅓–⅖ (or when dorsally pressed ⅔) lin.
broad, erect, not contiguous (unless dorsally pressed), thick and
fleshy, compressed-oblong (or elliptic from dorsal pressure), obtuse,
slightly incurved upon the anther-tips at the apex, flat on the inner
face, very convex on the back; staminal column 1 lin. long; anthers
very small, erect, quadrate, their appendages reniform, erect, applied
to the margin of the style-apex, and their wings small and not very
conspicuous; style-apex cushion-like, not exceeding the anther-
appendages; follicle solitary, erect upon a deflexed pedicel, lanceolate, gradually tapering into a beak, smooth, puberulous. *Gomphocarpus tysonianus, Schlechter in Journ. Bot.* 1895, 271 *in note,
name only. Asclepias tysoniana, Schlechter in Journ. Bot.* 1895,
358. *Pachyacris capensis, Schlechter, l.c.* 358 *in note.*

EASTERN REGION : Griqualand East; Mount Currie, 6500 ft., *Tyson*, 1353!
and in MacOwan & Bolus, Herb. Norm. Austr.-Afr., 1314! in stony places
around Fort Donald, 4500 ft., *Tyson*, 1748! Ingeli Mountain, 7000 ft., *Sutherland!* Vaal Bank, near Kokstad, *Haygarth in Herb. Wood*, 4184!

17. X. prunelloides (Turcz. in Bull. Soc. Mosc. 1848, xxi., pt. i.
254); plant 6–9 in. high, branching at the base only into numerous
stems, which are roughly pubescent along two broad lines; leaves

3–7 pairs to a stem, 1–2 in. distant, spreading; petiole $1\frac{1}{2}$–$2\frac{1}{2}$ lin. long; blade $\frac{1}{2}$–$1\frac{1}{4}$ in. long, 4–7 lin. broad, ovate, acute, broadly rounded, slightly subcordate or occasionally broadly cuneate at the base, flat, rather thinly covered with rather soft spreading hairs on both sides; umbels solitary on each branch, terminal, pedunculate, 14–30-flowered, hemispherical or subglobose, about 1 in. in diam.; peduncle usually 2–3 in. long, in weak specimens much shorter, erect, thinly covered with spreading hairs, and often with a minute pubescence besides; bracts about $1\frac{1}{2}$ lin. long, subulate, deciduous; pedicels 3–$4\frac{1}{2}$ lin. long, pubescent with spreading hairs; sepals 1–$1\frac{1}{4}$ lin. long, $\frac{1}{3}$–$\frac{1}{2}$ lin. broad, lanceolate, acute, pubescent; corolla about $2\frac{1}{2}$ lin. in diam., glabrous inside, thinly pubescent with rather long spreading hairs outside, brownish or greenish-brown; lobes $1\frac{1}{2}$–$1\frac{3}{4}$ lin. long, 1 lin. broad, ascending, elliptic, subobtuse; corona-lobes arising slightly above the base of the staminal column, and reaching to about the same level, slightly divergent-spreading, $\frac{2}{3}$–$\frac{3}{4}$ lin. long, $\frac{1}{4}$ to rather more than $\frac{1}{3}$ lin. broad, linear-oblong to narrowly ovate, obtuse, rather thick and fleshy, in the dried state they sometimes appear to be slightly keeled down the inner face, probably due to shrinkage in drying; staminal column $\frac{3}{4}$ lin. long; anthers quadrate, their appendages suborbicular, inflexed over and nearly concealing the truncate apex of the style; follicles not seen. *X. pedunculatum, Harv. Thes. Cap.* ii. 8, *t.* 112. *Gomphocarpus prunelloides, Schlechter in Engl. Jahrb.* xx. *Beibl.* 51, 35, *in note. G. harveyanus, Schlechter in Journ. Bot.* 1895, 270. *Asclepias harveyana, Schlechter in Journ. Bot.* 1896, 456.

Coast Region: Fort Beaufort Div.; common in the valleys of the Winter-berg Range, 5000–6000 ft., *Ecklon,* 41! *Mrs. Barber,* 82! Stockenstrom Div.; grassy slopes of the Kat Berg, 4000 ft., *Galpin,* 1731! ridges east of Great Kat Berg, *Scully,* 178!

18. X. Baurii (N. E. Br.); stems several, branching at the decumbent base only, 4–6 in. long, more or less pubescent with short spreading jointed hairs; leaves about 4 pairs to a stem, $\frac{3}{4}$–$1\frac{3}{4}$ in. distant, spreading; petiole 1–2 lin. long; blade $\frac{1}{2}$–1 in. long, 3–6 lin. broad, ovate or elongated-ovate, acute, subcordate at the base, subcoriaceous, flat, thinly covered on both sides with short spreading hairs and ciliate with the same; umbel solitary, terminal, pedunculate, hemispherical, about 20-flowered; peduncle $1\frac{1}{2}$–$2\frac{1}{2}$ in. long, shortly hairy with jointed hairs; bracts 1–$1\frac{1}{4}$ lin. long, subulate or linear, pubescent, persistent during the flowering period; pedicels 3 lin. long, pubescent with short spreading hairs; sepals 1 lin. long, $\frac{1}{2}$ lin. broad, lanceolate, acute, pubescent with spreading hairs on the back; corolla about $\frac{1}{4}$ in. in diam., rotate, thinly pubescent outside, glabrous within, white, tipped with purple at least on the outside; lobes very spreading, but not reflexed, $1\frac{1}{3}$–$1\frac{1}{2}$ lin. long, 1 lin. broad, ovate, minutely notched at the subobtuse apex; corona-lobes arising near the base of the staminal column and nearly reaching to the same level, erect, contiguous, $\frac{2}{3}$ lin. long and broad, fleshy, orbicular in outline, and probably subglobose in the living state, very obtuse; staminal column $\frac{3}{4}$ lin. long; anthers erect, quadrate, their appendages reniform, erect, applied to the margin of the very dilated style-apex, and their wings small and nearly concealed by the corona-lobes; style-apex dilated so as to somewhat overhang the anther-cells, truncate, cushion-like, slightly depressed in the centre.

EASTERN REGION: Tembuland; near Bazeia, 3500 ft., *Baur,* 730!

Very similar to *X. prunelloides,* Turcz., in general appearance, but the different colour of the rotate (not campanulate) corolla, and the very different corona-lobes, easily distinguish it.

Imperfectly known species.

19. X. ambiguum (N. E. Br.); "stem simple, firm, somewhat compressed above, pubescent; leaves oblong-lanceolate or linear-lanceolate, attenuate, undulate, coriaceous, glabrous, tapering into a very short petiole; peduncles none; pedicels short, pubescent, accompanied by setaceous bracteoles; corona-lobes rounded, shorter than the gynostegium; corolla-lobes glabrous; follicles of *Gomphocarpus fruticosus.*" *X. undulatum, Decne in DC. Prodr.* viii. 519, *not of R. Br.*

SOUTH AFRICA: without locality, *in Herb. Delessert,* ex *Decaisne.*

Unknown to me; the above is a translation of Decaisne's description.

XIII. **PERIGLOSSUM,** Decne.

Calyx 5-partite. *Corolla* 5-lobed nearly to the base; lobes narrowly overlapping in bud. *Corona* of 5 lobes arising at the base of the staminal column, alternating with the corolla-lobes, very shortly connate at the base, erect, spathulate-oblong, spathulate-sagittate or linear-oblong, with a transverse ridge or flap and often 2 keels on the inner face, with a long and distinct or minute and rudimentary slender filiform process alternating with them at the base. *Staminal column* arising from the base of the corolla, subglobose from being constricted at the apex and base; anthers with a broad thin and membranous connective, terminated by a membranous appendage, the cells firmer, extending quite to the base of the column, slightly inflated and much curved outwards, not horny on the margins, except at the apex, where they project as small triangular acute wings. *Pollen-masses* solitary in each cell, pendulous, subclavately subterete, semicircularly curved, opaque at the attached end, much flattened and semitransparent at the truncate free end, attached in pairs to a very minute black horny pollen-carrier by rather stout caudicles, which are twice as long as the pollen masses, but entirely contained in the anther-cells, doubly curved or somewhat S-like, enlarged and slightly excavated where attached to the pollen-masses. *Style* not exserted beyond and mostly covered by the anther-appendages.

Perennial erect herbs; rootstock a tuber; stem solitary, simple; leaves linear, erect or spreading; umbels 2–4, racemosely arranged along the upper part of the stem, pedunculate, globose, densely 6- to many-flowered; pedicels $\frac{1}{2}$–$\frac{3}{4}$ lin. long.

DISTRIB. Species 4, endemic, very similar to one another in general appearance.

Closely allied to *Cordylogyne* and united with that genus by Bentham, but the presence of filaments alternating with the corona-lobes and the remarkable differences in the anthers and pollen-masses (which 3 characters are not noted by any previous author), combined with the much longer style, have induced me to retain *Periglossum* as a distinct genus. The exsertion of the style is the only distinction other authors have noted, but as species with and without an exserted style occur in several genera, it would seem to be of no generic value if taken alone. The umbels of *Periglossum* are also usually more numerous, and their peduncles and pedicels very much shorter than in *Cordylogyne*. Dr. Schlechter also upholds the two genera, but gives no word of distinction.

Blade of the corona-lobes sagittate-ovate, as long as the stalk; alternating filaments filiform, $\frac{2}{3}$–$1\frac{3}{4}$ lin. long ... ... (1) **angustifolium.**
Blade of the corona-lobes oblong or lanceolate-oblong to elliptic-oblong, 2–2$\frac{1}{2}$ times as long as the stalk:
   Filaments alternating with the corona-lobes, filiform, about 1 lin. long ... ... (2) **mossambicense.**
   Filaments alternating with the corona-lobes, minute or rudimentary:
      Leaves suberect; corona-lobes with a stout longitudinal double keel on the upper half and 2 slender contiguous keels below (3) **kassnerianum.**
      Leaves very spreading; corona-lobes thickened at the apex, without any longitudinal keel ... ... (4) **McKenii.**

1. **P. angustifolium** (Decne in DC. Prodr. viii. 520); stem solitary, 1–2$\frac{1}{2}$ ft. high, simple, more or less unifariously or bifariously puberulous at the upper part, glabrous below; leaves erect to spreading, 1$\frac{3}{4}$–5 in. long, $\frac{1}{2}$–1$\frac{1}{2}$ lin. broad, linear, acute, with revolute margins, glabrous on both sides or the midrib puberulous beneath; umbels 2–5, racemosely arranged, pedunculate, dense, globose, about $\frac{3}{4}$ in. in diam., many-flowered; bracts 1$\frac{1}{2}$–2 lin. long, subulate, slightly puberulous; pedicels $\frac{1}{2}$–$\frac{3}{4}$ lin. long, puberulous; sepals 2 lin. long, $\frac{3}{4}$–1 lin. broad, lanceolate, acute, puberulous with somewhat tortuous hairs; corolla-lobes erect or but slightly spreading, about 3 lin. long, 1$\frac{1}{4}$–1$\frac{1}{3}$ lin. broad, linear-oblong, somewhat abruptly acute, glabrous on both sides; corona-lobes arising at the base of the staminal column and slightly overtopping it, 1$\frac{3}{4}$–2$\frac{1}{4}$ lin. long, erect, stalked, with the blade about as long as the linear stalk, 1–1$\frac{1}{2}$ lin. broad, broadly sagittate-ovate, obtuse, with 2 prominent contiguous keels on the inner face, which are capped at their apex by a transverse keel or flap across the thickened top of the lobes; alternating filaments at the base of the lobes $\frac{2}{3}$–1$\frac{3}{4}$ lin. long, filiform, often resting in the channel on the back of the anthers; staminal

column 1½ lin. long, subglobose ; anther-cells slightly inflated, much curved on the back ; anther-appendages rather narrow, oblong-lanceolate, subobtuse, horizontally inflexed on the small style-apex from the triangularly thickened truncate apex of the connective ; pollen-masses semicircularly curved, subterete at the attached opaque end, thin, flat, translucent and truncate at the free end, attached by very long stout doubly curved or S-like caudicles to a very minute black pollen-carrier. *Schlechter in Engl. Jahrb.* xviii. *Beibl.* 45, 10, *and Journ. Bot.* 1896, 450. *Cordylogyne globosa, Meisn. in Hook. Lond. Journ. Bot.* ii. 1843, 546 (*by error* 446); *Krauss in Flora,* 1844, 827, *not of E. Meyer.*

COAST REGION : Komgha Div.; near Komgha, *Flanagan,* 590, and in *MacOwan, Herb. Austr.-Afr.,* 1502 !
KALAHARI REGION : Transvaal; Magaliesberg Range, *Burke!* Johannesburg, *Mrs. Hutton,* 302 ! Modderfontein, *Conrath,* 981 !
EASTERN REGION : Transkei ; by the Illetooli stream, *Mrs. Barber,* 823 ! Tsomo, *Mrs. Barber* (or *Bowker*), 860 ! Tembuland ; Jackals Kop, *Bolus,* 10184 ! Engcobo Mountain, *Bolus,* 10185 ! Griqualand East ; around Kokstad, 5000 ft., *Tyson,* 1360 ! Natal; Lynedoch, 4000–5000 ft., *Wood,* 4547 ! near Charlestown, *Wood,* 5548! Congella, 40 ft., *Schlechter,* 3070 ! Hilton Road, *Schlechter,* 6768 ! Maritzburg, *Krauss,* 171 ! *Mrs. Hutton,* 307! and without precise locality, *Gerrard,* 1293 ! *Mrs. Fannin,* 89 !

2. **P. mossambicense** (Schlechter in Engl. Jahrb. xxxviii. 33, fig. 2) ; plant 1–2 ft. high, with 1–4 simple stems, unifariously puberulous on the upper part, with 5–7 internodes ; leaves erect or ascending-spreading, 2½–4½ in. long, ½–1 lin. broad, linear, acute, with revolute margins, glabrous ; umbels 1–3 to a stem, lateral at the nodes and terminal, pedunculate, globose, about 6–8-flowered ; peduncles ½–1 in. long, puberulous on one side ; pedicels scarcely ½ lin. long ; sepals 1½ lin. long, ⅔ lin. broad, lanceolate, acute, with a few scattered minute hairs on the back ; corolla-lobes 2½–3 lin. long, 1 lin. broad, suberect, narrowly oblong, subacute, glabrous on both sides ; corona-lobes erect, overtopping the staminal column, 2–2¼ lin. long, ¾ lin. broad, stalked, with a lanceolate-oblong obtuse blade about 2½ times as long as the linear stalk and abruptly rounded (or tapering, according to *Schlechter*) into it, with a small transverse semicircular tubercle at the middle on the inner face, decurrent as 2 slight keels, which are not discernible on some dried specimens ; alternating with the lobes at the base are 5 filiform segments about 1 lin. long ; staminal column 1¾ lin. long, much contracted at the base and under the anther-appendages, which are very large, deltoid or triangular in outline, but longer than broad, obtuse, keeled down the inner face, erectly connivent much above the small obtuse style-apex, which is shortly produced beyond the pollen-carriers ; pollen-masses flattened, curved, obtuse at both ends, semitransparent at the free end ; caudicles nearly twice as long as the pollen-masses, doubly curved near the middle, the upper part straight, the lower incurved ; pollen-carrier minute, black.

EASTERN REGION : Delagoa Bay, *Junod,* 189 !

Also in Tropical Africa.

The figures of the anther and pollinia, as given by Schlechter, are very inaccurate.

3. P. kassnerianum (Schlechter in Engl. Jahrb. xx. Beibl. 51, 40); plant ½–1 (⅔–1½ ft., *Schlechter*) ft. high; stem solitary, simple, slender, more or less puberulous on the upper part; leaves erect or ascending, 2¼–3½ in. long, ½ lin. broad, linear-filiform, with revolute margins, acute, glabrous; umbels 1–3, pedunculate, globose, densely many-flowered, ½–⅔ in. in diam.; peduncles ⅔–2¼ in. long; bracts 1½ lin. long, linear-subulate; pedicels very short, scarcely ½ lin. long, puberulous; sepals 1¾–2 lin. long, ⅔ lin. broad, lanceolate, acute, puberulous; corolla-lobes 3 lin. long, 1¼ lin. broad, oblong-lanceolate, acute, suberect, slightly concave at the basal part, slightly convex with revolute margins above, recurved at the apex, glabrous on both sides, apparently greenish, more or less tinted with purple-brown on the back; corona-lobes much overtopping the staminal column, erect and slightly incurved-connivent at the middle, 2–2¼ lin. long, 1 lin. broad, stalked, with the oblong or ovate-oblong obtuse blade about 2½ times as long as the linear stalk, cordate or cordate-sagittate at the base, concave on the back, thickened on the upper half of the inner face into a stout 2-ridged keel, which ends at the middle of the lobe in a small flap resting on the top of the basal keels; alternating with the lobes at their base are 5 minute almost rudimentary filaments; staminal column 1¼ lin. long, subglobose; anther-cells curved on the back, very firm, with a membranous connective between them, triangularly thickened at the truncate apex, from which, in young flowers, the broadly ovate obtuse anther-appendages are horizontally inflexed, overlap and form a pentagonal white canopy, entirely concealing the small knob-like style-apex and top of the column, but are soon eaten away by insects; pollen-masses flattened, crescent-like, obtuse at both ends, slightly translucent near the free end; caudicles about twice as long as the pollen-masses, slender, slightly curved at the upper half, shortly and abruptly very much curved at the lower half, attached to a minute black pollen-carrier. *Schlechter in Journ. Bot.* 1896, 450.

KALAHARI REGION: Transvaal; near Rustenberg, 4000 ft., *Miss Pegler*, 1010! 1022! near Klein Olifants River, 5000 ft., *Schlechter*, 4043! Modderfontein, *Conrath*, 982!

4. P. McKenii (Harv. Thes. Cap. ii. 7, t. 111); stem 2–2½ ft. high, unifariously (bifariously, *Harvey*) puberulous on the upper part, with internodes 1½–3½ in. long; leaves sessile, widely spreading, 3½–4½ in. long, ¾–1¼ lin. broad, linear, acute, with revolute margins, glabrous above, puberulous on the midrib beneath; umbels about 5, lateral at the nodes, pedunculate, globose, about ¾ in. in diam., many-flowered; peduncles ½–2 in. long, puberulous along one side; pedicels ⅓–½ lin. long, puberulous; sepals 1⅓–1¼ lin.

long, $\frac{2}{3}$–$\frac{3}{4}$ lin. broad, lanceolate, acute, thinly puberulous; corolla-lobes suberect, $2\frac{1}{3}$–3 lin. long, $\frac{3}{4}$–1 lin. broad, narrowly oblong or lanceolate-oblong, somewhat acute, glabrous on both sides; corona-lobes erect, overtopping the staminal column, $1\frac{3}{4}$–$2\frac{1}{4}$ lin. long, $\frac{3}{4}$ lin. broad, spathulate, with an oblong or elliptic-oblong obtuse blade about twice as long as the linear stalk and variably tapering or subtruncately contracted into it, or sometimes linear-oblong with scarcely any evident stalk, thickened around the apex and with a transverse tubercle or ridge at the middle on the inner face, but no keels; alternating filaments at the base of the lobes very minute or rudimentary; staminal column $1\frac{1}{2}$ lin. long, subglobose, contracted under the anther-appendages, which are horizontally inflexed over the very small knob-like style-apex; anther-cells curved; pollen-masses flattened, crescent-like, obtuse at both ends, slightly trans-lucent at the free end; caudicles about twice as long as the pollen-masses, sigmoid-curved at the middle, straight above and below. *Schlechter in Journ. Bot.* 1896, 450.

EASTERN REGION; Natal; near Durban, *McKen,* 664! *Gueinzius!*

The floral details as figured by Harvey are not very accurate, the anther and pollen-masses are quite wrong; the figure of the plant, however, is excellent. Possibly *Wood,* 7377, from near Durban, may belong here, but I have not examined it. By a misprint the name in the text is spelt *M'Kenii,* but not on the plate.

XIV. CORDYLOGYNE, E. Meyer.

Calyx 5-partite. *Corolla* 5-lobed nearly to the base; lobes narrowly overlapping in bud. *Corona* of 5 free lobes arising at the base of the staminal column, alternating with the corolla-lobes, erect, subspathulate-hastate or linear-oblong with a triangular tooth on each side at the middle, with two keels and a transverse ridge or flap on the inner face, without any processes alternating with them at the base. *Staminal column* arising from the base of the corolla, cylindric, not contracted at the apex or base; anthers with a broad thin membranous connective and slightly firmer cells, terminated by a membranous appendage; cells extending quite to the base of the corona-lobes, not inflated, straight, not curved outwards. *Pollen-masses* solitary in each cell, pendulous, flattened, linear-oblong, slightly curved, tapering to an acute point at the attached end, which is thinner than the rest of the mass and semitransparent, attached in pairs to the very minute black horny pollen-carriers by slender caudicles, which are shorter than the pollen-masses and angularly bent, with an angular projection at the middle. *Style* produced much beyond the erect anther appendages. *Follicles* narrowly fusiform. *Seeds* crowned by a tuft of hairs.

Perennial erect herb; rootstock tuberous; leaves linear, erect or ascending; umbels globose, terminal, long-peduncled, solitary or 2 (rarely 3-4) to a stem; pedicels $\frac{1}{8}$–$\frac{1}{4}$ in. long.

DISTRIB. Species 1, endemic.

1. **C. globosa** (E. Meyer, Comm. 218); plant ½–2½ ft. high, with 1 or more stems, simple or branched at the base, puberulous on one side of the upper part, with the lower internodes short and the middle and upper 1¼–3¼ in. long; leaves erect or ascending, 1–3½ in. long, ⅓–1½ lin. broad, linear, acute, with revolute margins, glabrous; umbel solitary and terminal or 2–4 to a stem and racemose, subglobose, 10–25-flowered; peduncles 1–10 in. long, erect, puberulous on one side or all round; pedicels 1½–3 lin. long, puberulous; sepals suberect, 1–1½ lin. long, ½ lin. broad, lanceolate or ovate, acute, puberulous; corolla-lobes 2–2½ lin. long, ¾–1 lin. broad, oblong, obtuse, erect, recurved or shortly revolute at the tips, glabrous on both sides; corona-lobes erect, 1 lin. long, about ⅓ lin. broad at the upper part, subspathulate-hastate or linear-oblong, with a triangular projection on each side at about the middle, very obtuse, flat on the back, with a transverse ridge-like or broadly deltoid flap above the middle of the inner face pressed down upon the tops of the closely contiguous and rather inconspicuous keels; staminal column 1–1¼ lin. long; anther-appendages ovate or oblong, very obtuse, concave, erect around the base of the stout clavate style-apex, which is produced ½–¾ lin. beyond them and the corona-lobes, often globose at the top; follicles solitary, erect, about 3 in. long, ¼ in. thick, narrowly fusiform, nearly equally tapering at both ends, puberulous; seeds about ¼ in. long and ⅛ in. broad, ovate, flattish, with broad margins, minutely papillate rugulose or sub-scabrid on the disk on both sides, light brown. *Harv. Gen. S. Afr. Pl. ed.* 1, 226; *Decne in DC. Prodr.* viii. 518; *Delessert, Ic. Pl.* v. 27, *t.* 64; *Schlechter in Engl. Jahrb.* xx. *Beibl.* 51, 41; xxi. *Beibl.* 54, 10; *and Journ. Bot.* 1896, 450; *K. Schum. in Engl. and Prantl, Pflanzenfam.* iv. ii. 234, *fig.* 68, *L.*

South Africa : without locality, *Zeyher,* 1170!

Coast Region : Queenstown Div.; between Table Mountain and the Zwart-kei River, *Drège,* 3431! damp places by streams and springs, *Mrs. Barber!*

Central Region : Graaff Reinet Div.; Sneeuwberg Range, 5000 ft., *Bolus,* 2634! Colesberg Div. ; Slengersfontein, *Drège!* near Colesberg, *Shaw!* Hanover Div.; near Hanover, *Sim,* 2862 !

Kalahari Region : Griqualand West ; Campbell, *Burchell,* 1809! 1836! Orange River Colony; Sand Drift, *Burke!* Transvaal ; Bergendal, 6000 ft., *Schlechter,* 4015!

Eastern Region : Natal; near Charlestown, 5000–6000 ft., *Wood,* 4797! 5142! 5548! Greenwich Farm, Riet Vley, *Fry in Herb. Galpin,* 2738 ! Mohlamba Range, *Sutherland !.* near Newcastle, *Wood,* 6227 !

According to Mr. Medley Wood the flowers are green and white.

XV. KREBSIA, Harv. (not of Eckl. & Zeyh.)

Calyx 5-partite. *Corolla* 5-lobed nearly to the base ; lobes suberect or campanulately spreading, concave at the lower part, recurved at the tips. *Corona-lobes* 5, arising at or near the base of the staminal column, exceeding the corolla, simple, erect, narrowly lanceolate, strongly keeled all down the back and more or less acutely triquetrous in transverse section, with a rhomboid wing-like dilation on the back at the base, tapering above into an acute incurved-hooked apex, without an appendage on the inner face. *Stamens* united into a cylindric column around and with the style; filamental part very short or almost none; anthers erect, tipped with a membranous appendage. *Pollen-masses*. solitary in each cell, pendulous, attached in pairs by slender caudicles to the pollen-carriers, which are seated much below the margin of the style-apex. *Style* not exceeding the anther-appendages, depressed-truncate or shallowly crater-like with a thickened margin at the apex. *Follicles* not seen.

Perennial herbs, probably with a tuberous rootstock; leaves linear; umbels lateral at the nodes, subglobose, very shortly pedunculate; flowers small.

DISTRIB. Species 2, endemic.

Harvey (*Gen. S. Afr. Pl. ed.* 2, 233) founded the genus *Krebsia* upon Mrs. F. W. Barber and H. Bowker's 293, and Gerrard's 1309, without giving a specific name, and I retain it solely on account of its coronal structure, for besides the dorsal keel and basal wings at the back of the corona-lobes, there is little to distinguish it from *Schizoglossum*. Schlechter has referred to this genus (entirely on account of its similarity in appearance and disregarding its coronal structure) another species, which has the structure of *Xysmalobium*, under which genus I have placed it.

Pedicels about 1 lin. long; corona-lobes 3–4 lin. long   (1) **corniculata.**
Pedicels 1½–3 lin. long; corona-lobes 2½ lin. long   ...   (2) **stenoglossa.**

1. K. corniculata (Schlechter in Engl. Jahrb. xx. Beibl. 51, 40); stems 4–8 in. high, simple or branched at the base, bifariously puberulous; leaves ascending or somewhat spreading, 2–3½ in. long, ½–1¾ lin. broad, linear, acute, with revolute margins; umbels pedunculate or subsessile, 3–8-flowered, globose; peduncle 1–4½ lin. long; bracts 1–3 lin. long, linear-subulate, glabrous; pedicels about 1 lin. long, rather stout, puberulous; sepals 2–2½ lin. long, ¾ lin. broad, lanceolate, acute, glabrous; corolla-lobes 3 lin. long, 1¼–1⅓ lin. broad, oblong, obtusely pointed, erect, concave at the basal part, convex and recurved-revolute at the tips, glabrous on both sides with a velvety sheen on the surface of the apical part, whitish; corona-lobes erect, exserted beyond the corolla, 3–4 lin. long, nearly 1 lin. broad at the base, lanceolate, gradually tapering to a subulate incurved-hooked apex, prominently and acutely keeled down the back, with a rhomboid expansion on the back at the base, flattish or perhaps slightly channelled on the inner face, apparently purple-brown with paler margins; staminal column about 1½ lin. long; anther-appendages ovate, acute, with their tips inflexed over the rim of the shallowly crater-like style-apex. *Schlechter in Engl. Jahrb. xxi. Beibl. 54, 3, and Journ. Bot.* 1896, 450. *Lagarinthus corniculatus, E. Meyer, Comm.* 208. *Gomphocarpus corniculatus, Dietr. Syn. Pl.* ii. 901; *Decne in DC. Prodr.* viii. 561.

COAST REGION: Fort Beaufort Div.; hills near the Kat River, *Drege* (ex *E. Meyer*); Queenstown Div.; Table Mountain, 5000 ft., *Drège*, 3423!

CENTRAL REGION : Aliwal North Div.; Elands Hoek, near Aliwal North, *Bolus,* 10570!

KALAHARI REGION: Transvaal; Makapans Poort, near Potgeiters Rust, 4250 ft., *Schlechter,* 4316 and near Barberton, *Thorncroft* (ex *Schlechter*).

EASTERN REGION : Natal; by the Buffalo River, *Gerrard,* 1309! 2162!

2. K. stenoglossa (Schlechter in Engl. Jahrb. xx. Beibl. 51, 40, in note); stems 6–10 in. high, simple or branched near the base, rather slender, bifariously puberulous except at the base; leaves erect or ascending, 2–3½ in. long, ½–1¼ lin. broad, linear, acute, with revolute margins, glabrous; umbels pedunculate or subsessile, 5–10-flowered, subglobose ; peduncles 1–5 lin. long; pedicels 1½–3 lin. long; sepals 1½ lin. long, ½–⅔ lin. broad, lanceolate, acute, slightly puberulous; corolla-lobes 2–2½ lin. long, about 1 lin. broad, erect, oblong-lanceolate, subacute, concave at the basal part, convex and recurved at the apex, which is minutely papillate on the inner face, elsewhere glabrous, yellowish-green (*Mrs. Barber*); corona-lobes erect, exserted beyond the corolla, about 2½ lin. long and ⅔–¾ lin. broad near the base, lanceolate, gradually tapering to a subulate incurved-hooked apex, prominently and acutely keeled down the back, with a rhomboid expansion on the back at the base, flattish or perhaps slightly concave-channelled on the inner face, apparently dark purple with whitish margins ; staminal column 1–1⅓ lin. long; anther-appendages broadly ovate, obtuse or acute, erect, with their tips incurved over the margin of the depressed style-apex. *Schlechter in Journ. Bot.* 1896, 450. *Gomphocarpus stenoglossus, Schlechter in Journ. Bot.* 1894, 257.

COAST REGION : Queenstown Div.; Lesseyton Nek, 3900 ft., *Schonberg in Herb. Galpin,* 2572 !

EASTERN REGION : Transkei; Kreilis County, *Mrs. Barber,* 293! *Bowker,* 293! Swaziland ; near Mafutane, 1500 ft., *Bolus,* 12115!

I follow Schlechter in retaining this as specifically distinct from *K. corniculata,* for although it can scarcely be distinguished by structural characters, yet its different appearance, much smaller flowers, longer pedicels and puberulous sepals, seem to point to some other difference than is noticeable in dried specimens. Harvey (by his quotation of *Mrs. Barber,* 293, and *Gerrard,* 1309) evidently considered both to belong to one species. In appearance *K. stenoglossa* is exactly like *Xysmalobium carinatum,* N. E. Br., and *Schizoglossum periglossoides,* and dried specimens can scarcely be distinguished from each other, except by dissecting the flowers.

XVI. SCHIZOGLOSSUM, E. Meyer.

Calyx 5-partite. *Corolla* deeply 5-lobed, often nearly to the base, rotate, reflexed, or the lobes campanulately spreading or erect, or rarely connate at the tips, overlapping in bud. *Corona* of 5 lobes arising from the staminal column opposite the anthers, erect, dorsally flattened, usually thin, at least in dried flowers, rarely very thick and fleshy, never complicate, but sometimes with slightly infolded margins, often produced into a subulate or filiform point at the apex, and furnished with 2 keels and often with 1 or 2 lobes, filiform points, teeth or other appendages on the inner face, rarely without keels or appendages. *Staminal column* arising from the base of the corolla, united above with the dilated top of the style; anthers erect, terminated by membranous appendages, which are inflexed over the top of the style or erect. *Style* truncate or depressed at the apex. *Follicle* usually solitary by abortion, narrowly fusiform, beaked, smooth or covered with subulate processes. *Seeds* crowned with a tuft of hairs.

Perennial herbs, with a tuberous rootstock and erect or rarely decumbent, usually slender stems; leaves opposite, alternate or whorled, linear to elliptic, sometimes cordate or hastate at the base; flowers often small, in pedunculate or sessile umbels, lateral at the nodes and terminal.

DISTRIB. Species about 120, the others in Tropical Africa.

This genus presents to the student unusual difficulties in the determination of the species, because the external appearance of the stem, leaves and flowers is often alike in a whole group of species, whilst the structure of their corona-lobes and often the staminal column is entirely different. On the other hand, their foliage is sometimes so variable that two individuals with flowers that are identical in structure are totally different in appearance. It is frequently scarcely possible to name a species by comparison, without dissecting the flowers. In dried specimens the coronal structure is often so much altered and more or less obliterated by pressure, that it is frequently necessary to examine several flowers before their true structure can be determined. In some cases this does not appear to have been done, hence the discrepancies in some existing descriptions. Sometimes it is nearly or quite impossible to decide from dried flowers if the corona-lobe is continuous with the point, or whether the latter is an appendage arising from the apex of the inner face; in a few species both conditions appear to occur. The length and entireness of the point or appendage of the corona-lobes are sometimes variable to a limited extent in the same species, as is also the pubescence on the inner face of the corolla-lobes. As pointed out in the *Flora of Tropical Africa* (iv. i. 353), it is not possible to draw a rigid technical distinction between *Schizoglossum* and *Xysmalobium*, and *S. crassipes* might with almost equal right be placed in the latter genus, its corona-lobes being very similar to those of *X. brownianum*. Some plants which might perhaps be sought for here should be looked for in *Krebsia*, which only differs in the peculiar character of its corona-lobes. The papillation or pubescence on the inner face of the corolla-lobes is often nearly or quite invisible when the flowers are wetted for dissection. I have been enabled to identify the majority of the species (of this and other genera) described by Dr. Schlechter, owing to the kindness of Dr. Harry Bolus of Cape Town, who has portions of the type specimens of most of them, presented to him by Dr. Schlechter, and has most liberally lent his collection of Asclepiads to Kew.

Umbel solitary, terminal; corona-lobes produced into
    a long subulate or filiform point behind a shorter
    subulate-filiform appendage on the inner face (see also
    S. pilosum) ... ... (59) capitatum.
Umbels 2 or more to a stem or branch, occasionally
    solitary and terminal in 42, *S. pilosum;* 43, *S. pulchellum;* 49, *S. Conrathii;* 60, *S. restioides,* and
    72, *S. Buchanani:*
    * Umbels all pedunculate or the uppermost occa-
      sionally sessile; corona-lobes never produced into
      a subulate point behind their appendage when the
      latter is present:

† Corona-lobes with an appendage (which is simple,
bifid or divided into a pair, sometimes on the
same specimen) arising from their inner face,
or if apparently apical, then the true apex of
the lobe at least forms a transverse rim or
margin behind the base of the appendage :

‡ Appendage at least half as long as the corona-
lobe and exceeding the latter (but often inflexed from it) by from half to nearly all its
length when placed in a line with it :

§ Leaves varying from linear-hastate to oblong
or orbicular-ovate, hastate, cordate or truncate at the base :

‖ Corolla-lobes replicate (folded back lengthwise) or with reflexed margins :

Corolla-lobes $2\frac{1}{2}$–$3\frac{1}{4}$ lin. long; umbels
compactly corymbose or racemosely
arranged :

Peduncles 5–15 lin. long; appendage of
corona-lobes $\frac{1}{2}$–1 lin. long; anther-
wings broadest at the base ... ... (1) cordifolium.

Peduncles 0–5 lin. long; appendage of
corona-lobes $1\frac{1}{8}$–$1\frac{1}{2}$ lin. long; anther-
wings broadest and angular near the
middle ... ... (2) ingomense.

Corolla-lobes less than 2 lin. long; umbels
racemosely very crowded ... ... (14) pachyglossum.

‖‖ Corolla-lobes concave or nearly flat, the
margins not reflexed :

Flowers 10–60 in an umbel; corona-lobes
with a dorsal channel and an apical
lobule overhanging it behind the base
of the acute or bifid appendage .. (3) atropurpureum.

Flowers 3–15 in an umbel; corona-lobes
without a dorsal channel or apical
lobule, but sometimes with recurved
margins :

Corolla-lobes 3–5 lin. long; corona-lobes
with an appendage $\frac{3}{4}$–$1\frac{1}{2}$ in. long, and
the keels on their inner face nearly
$\frac{1}{3}$ the breadth of the lobe distant from
the margin :

Corolla-lobes rounded or very obtuse
at the apex; appendage of corona-
lobes normally deeply bifid or
bipartite ... ... (4) hamatum.

Corolla-lobes acute or subobtuse; appendage of corona-lobes normally
shortly trifid ... ... (13) decipiens.

Corolla-lobes 2–3 lin. long: corona-lobes
with an appendage $\frac{1}{3}$–$\frac{2}{3}$ lin. long : ..

Corona-lobes with the base of the ap-
pendage $\frac{1}{4}$–$\frac{1}{3}$ as broad as the lobe,
and the keels on their inner face
about $\frac{1}{4}$ the breadth of the lobe
distant from the margins ... ... (5) tridentatum.

Corona-lobes with the base of the ap-
pendage $\frac{1}{2}$ or more than $\frac{1}{2}$ as broad
as the lobe, and the keels on their

          inner face much less than $\frac{1}{4}$ of the breadth of the lobe distant from the margins :

     Internodes $\frac{3}{4}$–$1\frac{1}{2}$ in. long; corona-lobes rounded at the apex   ...   (6) **virens**.

     Internodes mostly $\frac{1}{8}$–$\frac{1}{2}$ in. long; corona-lobes emarginate, bifid or subtruncate at the apex ...   ...   (7) **euphorbioides**.

§§ Leaves filiform or linear, $\frac{1}{4}$–2 lin. broad, not hastate at the base; corona-lobes with a subulate or deltoid-linear point or appendage at or just below the apex of the inner face, with a rim or margin behind its base :

Plant $\frac{1}{2}$–$1\frac{1}{4}$ ft. high; stem usually simple, or branching from injury; leaves on the flowering part shorter than or not twice as long as the internodes :

Peduncles $\frac{1}{2}$–$3\frac{1}{2}$ in. long; corolla-lobes $1\frac{1}{2}$–2 lin. long, minutely puberulous on the inner face or at its base ...   .   (21) **linifolium**.

Peduncles $\frac{1}{12}$–$\frac{3}{4}$ in. long; corolla-lobes 1–$1\frac{1}{4}$ lin. long, glabrous on the inner face :

    Corona-lobes subquadrate, truncate, forming a distinct transverse rim behind the $\frac{1}{8}$–$\frac{1}{3}$ lin. long apical appendage ... (66) **Bolusii**.

    Corona-lobes oblong-ovate or somewhat rhomboid-ovate, not forming a transverse rim behind the point or apical appendage   ...   .   (73) **parvulum**.

Plant branching at the base, 6–10 in. high, with the leaves on the flowering part several times as long as the internodes ... (24) **garcianum**.

‡‡ Appendage usually not half as long as the corona-lobe, or if half as long, then not exceeding the lobe by half its length when placed in a line with it, mostly shorter than, equalling or but shortly exceeding the lobe :

§ Corolla-lobes $2\frac{1}{2}$–$3\frac{1}{4}$ lin. long; leaves mostly $2\frac{1}{2}$–15 lin. broad at the middle :

Appendage of corona-lobes directed over the staminal column, or if ascending then not parallel with the body of the lobe, divided into 2 subulate segments, rarely linear and bifid :

    Corona-lobes cuneately subquadrate, broadest at the top...   ...   .   (1) **cordifolium,**
                             var. *β.*

    Corona-lobes elongate-ovate, broadest at the slightly cordate base   ...   .   (9) **quadridens**.

Appendage of corona-lobes erect, parallel with the body of the lobe, deeply bifid or divided into 2 broad segments :

    Peduncle shorter than the leaves ; segments of the appendage bifid or toothed at the apex   ...   .   (8) **nitidum**.

    Peduncle as long as the leaves ; segments of the appendage obtuse...   ...   ...   (15) **Wallacei**.

§§ Corolla-lobes $1\frac{1}{3}$–2 (rarely $2\frac{1}{4}$) lin. long; leaves $\frac{1}{4}$–1 (in 14, *S. pachyglossum*, and 16, *S. bidens* $\frac{1}{2}$–3) lin. broad at the middle :

Appendage of the corona-lobes directed over
the style-apex or horizontally inflexed :
  Appendage arising immediately below the
    apex of the lobe, minute, transversely
    oblong   ...   . (18) **atrorubens**.
  Appendage arising near or at the middle of
    the lobe, simple, subulate   ...   ... (21) **linifolium**,
                                                              var. β.
  Appendage of the corona-lobes erect, sub-
    parallel with the body of the lobe, usually
    flat, bifid to obtuse or truncate, rarely
    subulate and acute in *S. pachyglossum* :
    Stem with 10–40 leafy nodes below the
      rather crowded flowering part, $\frac{3}{4}$–$1\frac{1}{4}$
      lin. thick ; peduncles $\frac{1}{12}$–$\frac{1}{2}$ in. long ... (14) **pachyglossum**.
    Stem with 4–16 leafy nodes below the lax
      flowering part and, except in *S. bidens*,
      $\frac{1}{2}$–$\frac{3}{4}$ lin. thick ; peduncles $\frac{1}{3}$–$1\frac{1}{2}$ in.
      long :
      Plant 1–2 ft. high ; leaves $\frac{1}{3}$–$\frac{3}{4}$ lin. broad
        above the base ; umbels racemosely
        arranged ; corona-lobes bifid or emar-
        ginate at the apex :
        Leaves broadened to 1–$1\frac{1}{4}$ lin. and sub-
          cordate-hastate at the base ; pedi-
          cels 5–8 lin. long   ...   . (19) **Galpinii**.
        Leaves not broadened at the cuneately
          rounded base ; pedicels $1\frac{1}{2}$–4 lin.
          long ...   . (20) **diversum**.
      Plant usually much under 1 ft. high ;
        leaves $\frac{1}{4}$–2 lin. broad :
        Umbels usually somewhat corymbosely
          arranged ; corona-lobes narrowed to
          a bifid or truncate apex :
          Corolla-lobes glabrous on the inner
            face   ...   . (16) **bidens**.
          Corolla-lobes pubescent on the inner
            face   ...   . (17) **umbellatum**.
        Umbels racemosely arranged ; corona-
          lobes suborbicular, obtuse (see also
          25, *S. crassipes*)   ...   . (26) **orbiculare**.
††Corona-lobes without an appendage (besides keels
  or lateral or basal teeth) or with it only represented by an arched and slightly produced
  thickening at the apex of the inner face, the
  point, when present, being a direct (erect or
  inflexed) continuation of the lobe without a transverse rim or margin behind its base :
  Corona-lobes shorter than or about equalling the
    staminal column ; plants less than 1 ft. high :
    Corona-lobes oblong-ligulate, obtuse, with 2
      keels on the inner face   ...   . (27) **umbelluliferum**.
    Corona-lobes subquadrate, with an arched apical
      thickening on the inner face, sometimes
      forming a very obtuse short thick point ; no
      distinct keels ...   . (25) **crassipes**.
    Corona-lobes 3-toothed, middle tooth radiately
      spreading, about twice as long as the lateral
      teeth ...   . (22) **aschersonianum**,
                                                              var. β.
  Corona-lobes exceeding the staminal column, with
    the point inflexed upon or much overtopping it :

Corolla-lobes 2¾–4 lin. long; leaves 1–9 lin.
  broad :
Corolla white, yellow or greenish-yellow :
  Corona-lobes ovate-oblong, subtruncate to
    2–3-toothed at the apex, without keels . (10) elingue.
  Corona-lobes stalked and with a deltoid-
    hastate or rhomboid-lanceolate blade, or
    sessile and linear-lanceolate, 2-keeled on
    the inner face   ...   . (11) flavum.
  Corolla dark purple-brown; corona-lobes taper-
    ing from an ovate base to a subulate
    acute or minutely bifid erect point, 2-
    keeled on the inner face ...   . (12) stenoglossum.
Corolla-lobes 1–2½ lin. long; leaves ¼–2 lin.
  broad :
  Corona-lobes ovate or stalked and subcordate-
    ovate, tapering or abruptly narrowed into
    a subulate-filiform point :
  Stem solitary and usually simple; leaves
    shorter than or less than twice as long
    as the ¾–4½ in. long internodes :
    Leaves glabrous; corona-lobes with the
      point very abruptly inflexed on the
      staminal column and 2 basal teeth on
      the inner face ...   . (21) linifolium.
    Leaves scaberulous; corona-lobes with the
      point incurved much above the sta-
      minal column and without basal teeth
      on the inner face   ...   . (73) parvulum.
  Stem branched at the base; leaves 2–6 times
    as long as the ¼–½ in. long internodes ... (23) aciculare.
  Corona-lobes 3-toothed at the top; middle
    tooth usually much longer than the small
    obtuse side teeth   ...   . (22) aschersonianum.
** Umbels all sessile, forming fascicles of 2 to many
  pedicellate flowers, or sometimes 1 or more of the
  lowest are pedunculate in 39, *S. ciliatum;* 66,
  S. Bolusii; 73, *S. parvulum,* and 80, *S. virgatum*
  (see also 14, *S. pachyglossum*); usually several to
  a stem and lateral at the nodes, racemosely arranged, or occasionally 1–2 and terminal in 42, *S.*
  pilosum; 43, *S. pulchellum,* and 60, *S. restioides :*
  Corolla with a very distinct cup-shaped or cam-
    panulate tube ¾–1½ lin. long; lobes puberu-
    lous on the inner face, at least at the base :
    Corona-lobes   subquadrate   with   angular
      shoulders and a long subulate point behind
      the appendage   ...   . (47) tubulosum.
    Corona-lobes elliptic or ovate, obtuse, forming a
      mere rim behind the base of the appendage (48) Schlechteri.
  Corolla lobed nearly to the base; no very distinct
    cup-shaped tube, perhaps most evident and
    about ½ lin. long in 56, *S. biflorum :*
    Corolla-lobes 5–6 lin. long; corona-lobes cross-
      shaped, 2½–3 lin. broad across the extended
      arms ...   .   ·   ...   . (46) Davyi.
    Corolla-lobes 1–4 lin. long; corona-lobes never
      cross-shaped, ¼–1¾ lin. broad :

† Corona-lobes produced into a distinct free apical
    lobe, point or points immediately behind the
    appendage or appendages on the inner face :
   ‡ Free point or points of the corona-lobes
      behind the appendage or appendages
      subulate or filiform, very acute or acutely
      bifid, rarely linear and obtuse :
    § Corona-lobes subquadrate, distinctly 3-
       toothed at the top, with the shoulder-
       teeth minute, erect and usually sub-
       falcate (see also 49, *S. Conrathii* and
       50, *S. excisum*) :
      Middle tooth or point of the corona-lobes
        ¼–1 lin. long, shorter than or equalling
        the appendage :
        Pedicels 3½–5 lin. long ; appendage
          5–7 times as long as and recurved
          over and far behind the middle
          tooth of the lobe ...   . (38) **dissimile.**
        Pedicels 1½–2½ lin. long ; appendage
          as long as to twice as long as the
          middle tooth of the lobe, erect or
          inflexed   ...   . (41) **robustum,var.γ.**
      Middle tooth or point of the lobe 1¼–2½
        lin. long (length unknown in 34, *S.
        grandiflorum*), and longer than or
        rarely only equalling the appendage
        or appendages :
        Corona-lobes with tuberculate margins
          and 2 free or connate 2-toothed
          appendages forming a sort of pocket
          on the inner face and scarcely ex-
          ceeding the level of the shoulder-
          teeth ...   . (40) **striatum.**
        Corona-lobes not tuberculate on the
          margins (sometimes slightly tuber-
          culate or crested at the shoulders in
          39, *S. ciliatum*), with 1 entire or
          bifid appendage on the inner face,
          which always much overtops the
          level of the shoulder-teeth :
        Stem 1–2 ft. high, with the middle
          internodes usually 1 in. or more
          long ; pedicels ½–1½ lin. long   ... (56) **biflorum.**
        Stem 2–10 in. (or in 41, *S. robustum*
          ½–3 ft.) high, with internodes
          (except in 39, *S. ciliatum*) usually
          less than 1 in. long, and the
          leaves often rather crowded ;
          pedicels 2–5 lin. long :
         Corolla-lobes 1¾–2⅔ lin. long, very
          acute ; stem with 1–3 distant
          leafy nodes below the lowest
          flowering node ...   . (39) **ciliatum.**
        Corolla-lobes 2¼–3½ lin. long, acute
          or obtusely pointed :
         Stem 2–4½ in. high, with 3–5
          leafy nodes below the lowest
          flowering node ; leaves
          spreading   ...   . (33) **Macowani.**

Stem ½–3 ft. high, with 6–20
  leafy nodes below the lowest
  flowering node (unknown in
  S. grandiflorum) :
  Corolla glabrous on the inner
    fuce :
    Leaves ascending ...   ... (41) **robustum.**

    Leaves spreading ...   ... (34) **grandiflorum.**
  Corolla pubescent or puberu-
    lous on the inner face ... (41) **robustum, var. β.**

§§Corona-lobes not erectly 3-toothed at the
  top, transverse or quadrate with the
  shoulders rounded, rectangularly acute
  or rarely produced into short horizontally spreading teeth, or rhomboid
  and angular at or below the middle, or
  ovate and tapering into the point :
‖Stem 1–2½ ft. high, simple or rarely
  branching at the upper part; leaves
  linear, linear-filiform or linear-oblong,
  erect or ascending :
  Corona-lobes produced behind a long
    filiform appendage into 2 long fili-
    form points, which are free or
    united at the base ...   . (58) **araneiferum.**
  Corona-lobes produced behind the ap-
    pendage into 1 entire or shortly and
    acutely bifid point :
  Pedicels 2–5 lin. long; corolla-lobes
    2–3½ lin. long :
    Corolla-lobes glabrous on the inner
      face ...   . (41) **robustum.**
    Corolla-lobes puberulous or pubes-
      cent on the inner face :
      Umbels 6 or more to a stem;
        corona-lobes with horizontal
        teeth at the shoulders   ... (41) **robustum, var. β.**
      Umbels 1–3 to a stem; corona-
        lobes without teeth at the
        shoulders   ...   . (42) **pilosum.**
  Pedicels ⅓–1½ lin. long (length un-
    known in 57, *S. strictum*) :
    Corolla-lobes ¼ lin. long, glabrous
      on the back, minutely papillate-
      puberulous on the inner face,
      blackish-purple   ...   ... (80) **virgatum.**
    Corolla-lobes 1¼–2¾ lin. long :
      Corona-lobes with a filiform point
        appearing to come out of the
        back of the lobe below the
        base of the ¾–1 lin. long
        filiform appendage :
        Corolla-lobes 1¾–2 lin. long,
          puberulous on the lower
          part of the inner face ... (49) **Conrathii.**
      Corolla-lobes 2½–2¾ lin. long,
        glabrous on the inner face (50) **excisum.**
      Corona-lobes with a terminal point
        whose base is level with or
        above that of the appendage :

Q q 2

Corolla-lobes glabrous on the
   inner face ; corona-lobes
   angular below the middle,
   tapering into a recurved
   bifid or entire point much
   shorter than the appendage (60) **restioides.**
Corolla-lobes puberulous or
   pubescent on the inner
   face :
   Corona - lobes subquadrate
     with horizontal shoulder-
     teeth ... ... (56) **biflorum.**
   Corona-lobes ovate or oblong-
     ovate, tapering into a
     point longer than the
     appendage :
     Corolla-lobes glabrous on
       the back ; appendage
       of corona-lobes nearly
       as long as the point ... (57) **strictum.**
     Corolla-lobes pubescent on
       the back :
       Corolla-lobes 1⅔–2 lin.
         long ; corona-lobes
         1¾–2 lin. long   ... (56) **biflorum,** var. **γ**
       Corolla-lobes 1¼–1⅓ lin.
         long ; corona-lobes
         ⅖–1¼ lin. long   ... (68) **carinatum.**
‖ ‖ Stem or plant 1½–10 in. high ; corona-
   lobes produced behind the appendage
   into 1 entire point :
 ¶ Corolla-lobes glabrous or with a very
   minute whitish papillation (scarcely
   pubescence) on the inner face :
   Appendage of corona-lobes 2–3 times
     as long as the point behind it,
     incurved-hooked at the apex ;
     leaves 4–6 pairs to a stem or
     branch   ... ... (31) **uncinatum.**
   Appendage of corona-lobes shorter
     than (in *S. heterophyllum* some-
     times subequalling) the point
     behind it :
     Stem branching at the base with
       5–15 pairs or whorls of leaves
       to a branch ; corolla-lobes 1–2
       lin. long   ... ... (29) **heterophyllum.**
     Stem solitary unbranched (except
       from injury), erect :
       Leaves in 3–8 pairs or whorls
         to a stem ; appendage of
         corona-lobes inflexed over
         the staminal column :
         Leaves ¾–1½ lin. broad, linear
           or lanceolate ; corolla-
           lobes 1¾–2½ lin. long   ... (37) **pumilum.**
         Leaves ⅓–1 in. broad, oblong
           or elliptic ; corolla-lobes
           3–3½ lin. long   ... ... (32) **ovalifolium.**

Leaves in 8 to numerous whorls
and pairs to a stem :
  Appendage of corona-lobes as-
    cending and arched back-
    wards under the incurved
    tip of the filiform point
    behind it; corolla-lobes
    $2\frac{1}{4}$ lin. long ... ... (36) **contracurvum.**
  Appendage of corona-lobes
    inflexed over the staminal
    column :
    Corolla-lobes $2\frac{1}{4}$–$3\frac{1}{2}$ lin. long ;
      corona-lobes with dis-
      tinct horizontal teeth at
      the shoulders ... ... (41) **robustum.**
    Corolla-lobes $1\frac{3}{4}$–2 lin. long;
      corona-lobes angular but
      scarcely toothed at the
      shoulders ... ... (35) **Harveyi.**
¶¶Corolla-lobes distinctly puberulous
  or pubescent on the inner face :
  Stem 3–4 in. high, often with 1 (or
    more ?) basal branch; corolla-
    lobes $1\frac{3}{8}$–2 lin. long ... ... (30) **consimile.**
  Stem at least 6 in. high, simple :
    Pedicels 2–5 lin. long; corolla-lobes
      2–$3\frac{1}{2}$ lin. long :
      Leaves usually much longer
        (rarely shorter) than the
        internodes; umbels 5 to
        many to a stem, often
        crowded ... . (41) **robustum,** vars.
      Leaves shorter than or subequal-   β & γ.
        ling the internodes; umbels
        1–3 to a stem, not crowded (42) **pilosum.**
    Pedicels $\frac{1}{2}$–$1\frac{1}{2}$ lin. long; corolla-
      lobes 1–$1\frac{1}{3}$ lin. long :
      Appendage of the corona-lobes
        3-toothed ... ... (69) **tricuspidatum.**
      Appendage of the corona-lobes
        subulate and acute, or linear
        and obtuse to bifid ... ... (68) **carinatum.**
‡‡Free part of the corona-lobes behind the
  appendage usually broad, rounded,
  obtuse or obtusely bifid, never acute,
  subulate or filiform, sometimes very short
  or scarcely noticeable :
  Corolla-lobes 3–4 lin. long, puberulous on
    the inner face; corona-lobes with the
    appendage very much longer than the
    obtuse apex behind it ... ... (43) **pulchellum**
  Corolla-lobes $1\frac{1}{4}$–2 lin. long, glabrous or
    minutely puberulous on the inner face :
    Appendage of corona-lobes very much
      longer than the often minute obtuse
      or notched free part behind it :
      Stem with 10–16 leafy nodes below the
        lowest flowering node ... ... (64) **glabrescens.**
      Stem with 4–6 leafy nodes below the
        lowest flowering node :

Corolla blackish-purple; corona-lobes rhomboid, $\frac{2}{5}-\frac{1}{2}$ lin. long, angular on each side, with an appendage $\frac{1}{4}-\frac{1}{3}$ lin. long ... ... (72) **Buchanani.**

Corolla not blackish-purple, greenish?
  Corona-lobes oblong or rhomboid-oblong, $\frac{1}{2}-\frac{3}{5}$ lin. long, with 2 minute teeth on each side and an appendage $\frac{2}{3}-1$ lin. long ... (73) **parvulum** var. *β.*

  Corona-lobes subquadrate, $\frac{1}{4}-\frac{1}{3}$ lin. long, with an appendage about $\frac{1}{2}$ lin. long ... ... (74) **parcum.**

Appendage of the corona-lobes very much shorter than the free part of the lobe behind it:

  Appendage of corona-lobes subulate, with a tooth on each side of its base ... (62) **delagoense.**

  Appendage of corona-lobes transverse or subquadrate, truncate or rounded, without a tooth on each side of its base ... ... (63) **periglossoides.**

††Corona-lobes not produced into a distinct free point or lobe behind an appendage (besides keels) or other point:

‡ Corona-lobes with a distinct filiform subulate deltoid or linear entire or toothed appendage or point either arising at or just below the apex of the inner face, and the lobe forming a mere rim or thick truncate top behind it, or directly continuous with the body of the lobe:

Corolla-lobes $1\frac{3}{4}-2\frac{1}{2}$ lin. long:
  Body of corona-lobes transverse or broadly obcordate, thick at the subtruncate top; appendage $2\frac{1}{2}-3\frac{1}{2}$ lin. long, filiform-subulate, curved to the left ... (45) **verticillare.**

  Body of corona-lobes somewhat ovate or elongated deltoid, with 2 teeth on each side; appendage $1-1\frac{1}{4}$ lin. long, lacerately fringed ... ... (53) **filipes.**

  Body of corona-lobes oblong, obovate-oblong or subquadrate, obtuse; appendage $\frac{1}{3}-\frac{2}{3}$ lin. long, inflexed, 3-toothed at the apex or with teeth near the middle or base:

    Leaves $\frac{3}{4}-1\frac{1}{4}$ in. long, in 12–20 pairs below the lowest flowering node ... (54) **tridens.**

    Leaves $1\frac{1}{4}-3$ in. long, in 5–8 pairs below the lowest flowering node ... ... (83) **Woodii.**

  Body of corona-lobes subquadrate, truncate, with minute erect teeth at the shoulders; appendage $\frac{3}{4}$ lin. long, subulate, acute ... ... (55) **Flanagani.**

Corolla-lobes $\frac{3}{4}-1\frac{1}{2}$ lin. long:
§ Stem $1\frac{1}{2}-3\frac{1}{2}$ ft. high, simple or branching at the upper part (see also 64, *S. glabrescens*):
  Corolla-lobes glabrous (in 70, *S. commixtum*, sometimes minutely puberulous at the base) on the inner face:

Corona-lobe forming a very distinct
truncate transverse rim or top behind the base of the appendage :
Appendage nearly as broad as the
lobe, broadly deltoid-ovate,
acute ... ... (82) **Dregei.**
Appendage about ½ as broad as the
lobe, linear, oblong or narrowly
deltoid, acute or obtuse ... (66) **Bolusii.**
Corona-lobe continuous with the point
or not forming a distinct truncate rim behind it :
Pedicels 1½–2 lin. long :
Pedicels and upper part of the
stem tomentose; corona-
lobes 3-toothed at the top,
with the shoulder-teeth erect (85) **tomentosum.**
Pedicels and upper part of the
stem puberulous; corona-
lobes subquadrate with rectangular shoulders and a short
subulate inflexed point ... (67) **monticola.**
Pedicels 3–4 lin. long, puberulous ;
corona-lobes rhomboid or rhomboid ovate, angular on each
side at the middle, ... (70) **commixtum.**
Corolla-lobes puberulous on the inner
face, blackish-purple ; stem simple (81) **loreum.**
Corolla-lobes densely bordered with
white hairs and puberulous on the
disk of the inner face, not blackish-
purple ; stem branched ... ... (87) **altissimum.**
§§Stem 4–16 in. high :
Corolla-lobes puberulous on the inner
face; pedicels 2–3 lin. long ... (81) **loreum.**
Corolla-lobes glabrous on the inner face,
or in 65, *S. lamellatum*, with a
minute papillate pubescence at the
apex and base; pedicels 1–3 lin.
long :
Corona-lobes rhomboid, angular on
each side, with a subulate or
linear point shorter than or
scarcely longer than the body of
the lobes :
Corona-lobes with an inflexed wing-
like auricle on each side at the
angles ... ... (71) **parile.**
Corona-lobes without an auricle on
each side ... ... (72) **Buchanani.**
Corona-lobes oblong or rhomboid-
oblong with 2 minute teeth on
each side at the middle and an
inflexed point or appendage ⅔–1
lin. long ... ... (73) **parvulum, var. β.**
Corona-lobes subquadrate, with their
point or appendage about twice
as long as the body of the lobe ;
stem simple, with 3–6 leafy nodes
below the lowest flowering node :

Stem $\frac{1}{4}$-$\frac{1}{3}$ lin. thick at the base; point
of corona-lobe erect, toothed... (79) **exile.**

Stem $\frac{2}{5}$-$\frac{3}{5}$ lin. thick at the base;
point of corona-lobe incurved
upon the top of the staminal
column, not toothed ...   ... (74) **parcum.**

Corona-lobes subquadrate, usually with
minute teeth at the shoulders,
with their point or appendage
shorter than or subequalling the
body of the lobe :

Stem simple, with 7-16 leafy nodes
below the lowest flowering
node; umbels 2-7-flowered ... (78) **filifolium.**

Stem with 2-8 leafy nodes below
the lowest flowering node:

Stem laxly clothed with leaves,
those on the flowering part
shorter than or not much
longer than the internodes :

Stem $\frac{1}{2}$-$1\frac{1}{4}$ lin. thick and
branched at the base;
umbels 6-9-flowered   ... (75) **addoense.**

Stem $\frac{1}{3}$-$\frac{1}{2}$ lin. thick at the
base, simple; umbels 7-14-
flowered ...   . (76) **Burchellii.**

Stem rather thickly clothed with
leaves mostly 2-3 times as
long as the internodes;
umbels 9-16-flowered   ... (77) **Bowkeræ.**

Corona-lobes oblong or elliptic-oblong,
somewhat 3-toothed at the apex,
the middle tooth apiculus-like
or very shortly linear-deltoid ... (65) **lamellatum.**

‡‡ Corona-lobes transverse or subquadrate,
truncate, with a minute apiculus at the
apex of the inner face (see also 65, *S.
lamellatum* and 66, *S. Bolusii*); corolla-
lobes puberulous and with a tuft of long
hairs on the inner face   ...   ... (86) **interruptum.**

‡‡‡ Corona-lobes without a subulate, filiform
or linear point or apiculus or appendage
at their apex (except sometimes in 65,
S. lamellatum) :

Corolla-lobes 2-2$\frac{1}{2}$ lin. long :

Corona-lobes transverse or broadly subobcordate with the very short obtuse
apex inflexed upon the top of the
stout double keel on the inner face ... (44) **anomalum.**

Corona-lobes rhomboid-lanceolate and obtuse or subquadrate with square
shoulders and a broadly-deltoid obtuse
erect point, with small inflexed side
auricles and a slight median keel :

Leaves $\frac{2}{3}$-$1\frac{1}{4}$ in. long, $\frac{1}{2}$-$\frac{3}{4}$ lin. broad,
shorter than the internodes   ... (51) **glanduliferum.**

Leaves up to 3 in. long and $1\frac{1}{2}$ lin.
broad, longer than the internodes ... (52) **biauriculatum.**

Corolla-lobes 1–1½ lin. long:
Leaves opposite; corona-lobes ovate, rhomboid-ovate, oblong or elliptic-oblong,
obtuse, acute or (in 65, *S. lamellatum*)
subtruncate and apiculate or 3-toothed
at the top :
Stem ½–1½ ft. high, simple, except from
injury :
Corolla-lobes with rather long hairs
at the apex and margins of the
inner face; keels of corona-
lobes diverging downwards from
the apex of the lobe   ...   ...   (61) **unicum.**
Corolla-lobes without long hairs on
the inner face; keels of corona-
lobes subparallel or slightly diverging upwards, not extending
beyond the shoulders   ...   ...   (65) **lamellatum.**
Stem 2–2½ ft. high, branching above;
corolla-lobes puberulous on the
inner face, with a tuft of long hairs
just below the apex...   ...   ...   (84) **Barberæ.**
Leaves mostly whorled; stem 4–8 in.
high ; corona-lobes subquadrate-ovate,
shortly 2-toothed at the apex ..   ...   (28) **Pegleræ.**

1. S. cordifolium (E. Meyer, Comm. 219); stem ½–1½ ft. high,
usually solitary, simple or branching, ¾–1½ lin. thick, pubescent ;
internodes ¼–2⅓ in. long; leaves usually spreading; petiole ½–3 lin.
long; blade ½–2 in. long, 1/12–1¼ in. broad, usually oblong, deltoid-
ovate, or deltoid-oblong to roundish-ovate, rarely linear-hastate,
acute or obtuse, often apiculate, cordate, subtruncate or more or less
hastate at the base, with rounded auricles, flat or narrowly revolute
along the margins, more or less pubescent or scaberulous on both
sides or only on the veins beneath ; umbels usually 3–6 to a stem or
branch, pedunculate, the lower racemose, the upper subcorymbose,
4–12-flowered ; peduncles 5–15 lin. long, pubescent or subtomentose ;
bracts filiform, 1½–2½ lin. long ; pedicels 3–5 lin. long, pubescent or
subtomentose ; sepals 1–1½ lin. long, ½ lin. broad, lanceolate or
linear-oblong, acute, pubescent ; corolla in bud globose with an
apical depression ; lobes 2–3 lin. long, 1¼–1⅔ lin. broad, oblong,
subacute, more or less replicate, spreading, with incurved tips,
glabrous or with a few hairs near the apex on the back, more or less
minutely puberulous at the base on the inner face, greenish according to collectors, in the dried state olive-brown with darker veins,
rarely entirely green ; corona-lobes ascending-spreading, or often
appearing erect in dried flowers, ½–1 lin. long, ⅔–1¼ lin. broad,
broadly cuneate-obovate, cuneately subquadrate or suborbicular,
rounded, subtruncate or obtusely pointed at the entire or notched
apex, rarely more or less deeply 2-lobed, often with slightly projecting side angles above the middle or at the top, with a divided or
entire appendage decurrent in 2 contiguous keels, which are not
always obvious, on the inner face, white or whitish ; appendage

arising near the top of the lobe, $\frac{1}{2}$–1 lin. long, usually deeply bifid or
divided to the base into 2 straight subulate contiguous segments,
rarely entire, or with a third smaller segment behind, ascending to
or slightly exceeding the top of the $\frac{3}{4}$–$1\frac{1}{4}$ lin. long staminal column ;
anther-wings broadest at the subtruncate base ; anther-appendages
reniform, erect, reaching to or slightly inflexed over the margin of
the broad depressed-truncate style-apex ; caudicles attached near or
above the middle of the oblong pollen-masses ; follicles about $2\frac{1}{2}$ in.
long, $\frac{1}{2}$ in. thick, fusiform-lanceolate, tapering into a beak, pubescent
and beset with subulate processes 2–$2\frac{1}{2}$ lin. long. *S. atropurpureum,*
var. *lineatum, Schlechter in Engl. Jahrb.* xviii. *Beibl.* 45, 2. *S.
virens, Schlechter, l.c.* 6 *and* 16, *and Journ. Bot.* 1896, 419 (*not of
E. Meyer*). *S. Hollandiæ, Harv. ex Schlechter in Engl. Jahrb.*
xviii. *Beibl.* 45, 6 *in note. S. æmulum, Schlechter in Journ. Bot.*
1894, 258, *and* 1896, 420. *S. æmulatum, Ind. Kew. Suppl.* i. 384.
S. furcatum, E. Meyer ex Schlechter in Engl. Jahrb. xxi. *Beibl.* 54,
3 *in a note. Cynanchum cordifolium, Dietr. Synop. Pl.* ii. 905.

VAR. β, **centralis** (N. E. Br.) ; sepals 2 lin. long ; corolla-lobes $3\frac{1}{2}$ lin. long,
blackish-purple with pale stripes in the dried state ; corona-lobes $1\frac{1}{4}$ lin. long,
$1\frac{1}{2}$ lin. broad, obovate, subtruncate, irregularly crenulate, with the appendage
inserted at their middle and scarcely exceeding their apex.

COAST REGION : Tulbagh Div. ; Great Winterhoek Mountain, 3500 ft.,
Bolus in Herb. Guthrie, 4249 ! Worcester Div. ; Dutoits Kloof, *Drège ! Tyson,*
852 ! Stellenbosch Div. ; Jonkers Hoek, *Bolus in Herb. Guthrie,* 4879 ! George
Div. ; Montagu Pass, *Bolus,* 8684 ! Uniondale Div. ; mountains of Long Kloof
near the Wagenbooms River, *Burchell,* 4904 ! Uitenhage Div. ; Vanstadens
Berg, *Burchell,* 4736 ! Addo (ex *E. Meyer,* but the specimen is labelled Galge-
bosch) *Drège!* Port Elizabeth Div. ; near Port Elizabeth, *Mrs. Holland,* 37 !
Alexandria Div. ; Zuur Berg, *Cooper,* 2726 ! Albany Div. ; hills near Botram,
Drège! near Grahamstown, *MacOwan,* 295 ! 662 ! *Bolton, Burke ! Galpin,*
248, ex *Schlechter.* Komgha Div. ; near the Great Kei River, *Flanagan,* 386 !
near Komgha, *Flanagan,* 387 ! Queenstown Div. ; Lesseyton Nek, 4000 ft.,
Galpin, 1755 ! Eastern Frontier, *Hutton !*
EASTERN REGION : Transkei ; Kentani, *Miss Pegler,* 1144 ! Natal ; Umzin-
yate, *Wood,* 1040 ! Shafton Howick, *Mrs. Hutton,* 43 ! and without precise
locality, *Gerrard,* 2166 ! Zululand ; Ngoya, *Wood,* 8271 ! Var. β : Tembuland ;
Bazeia Mountain, 3000 ft., *Baur,* 548 !

I have seen and dissected the type specimens of all the synonyms quoted side
by side with E. Meyer's type of *S. cordifolium,* and find them identical with
that species in appearance and structure. Schlechter has likewise seen E.
Meyer's type, but in Engl. Jahrb. xxi. Beibl. 54, 3, he erroneously refers *S.
cordifolium* to *S. euphorbioides,* and *S. Hollandiæ,* Harv. (of which the type is
Mrs. Holland's 37) to *S. virens,* so that *S. cordifolium,* Schlechter in Journ.
Bot. 1896, 419, can scarcely be the plant of E. Meyer, and as there is neither
description nor collector's number quoted, I cannot identify it. The minute
pubescence at the base of the inner face of the corolla-lobes cannot be seen on
flowers that have been wetted for dissection until they are dry.

2. **S. ingomense** (N. E. Br.) ; stem $1\frac{1}{2}$–3 ft. high, simple, hairy
along 3 lines with white spreading hairs, with internodes $\frac{1}{2}$–1 in.
long at the middle and upper part, those at the base longer ; leaves
3 in a whorl, ascending or slightly spreading, shortly petiolate,
1–$1\frac{1}{3}$ in. long, 2–6 lin. broad at the base, oblong or linear-oblong,

acute, truncate, subhastate or subcordate at the base, pubescent along
the revolute margins and on the midrib beneath, otherwise glabrous;
umbels several to a stem, sometimes more than 1 from the same
node, racemose along the terminal part of the stem, pedunculate, or
the uppermost nearly or quite sessile, 3–6-flowered; peduncles
0–5 lin. long, pubescent or subtomentose; bracts $1\frac{1}{2}$–4 lin. long,
filiform, linear or spathulate, pubescent; pedicels 3–4 lin. long,
pubescent or subtomentose; sepals $1\frac{1}{2}$–2 lin. long, $\frac{3}{4}$ lin. broad,
lanceolate, acute, pubescent; buds globose, with an apical depres-
sion; corolla-lobes suberect or campanulately spreading, 3–$3\frac{1}{4}$ lin.
long, 2 lin. broad, ovate, acute, with reflexed (perhaps replicate) and
somewhat wavy margins, glabrous on both sides, apparently olive-
brown; corona-lobes obcordate, with 2 parallel keels, 2 short
transverse keels near the top and a long subterminal appendage on
their inner face; obcordate part 1 lin. long, nearly $1\frac{1}{4}$ lin. broad,
about half as long as the staminal column; appendage $1\frac{1}{3}$–$1\frac{1}{2}$ lin.
long, linear-cuneate, bifid at the apex, incurved over the $1\frac{2}{3}$–$1\frac{3}{4}$ lin.
long staminal column; anther-wings with a very prominent and
rather acute angle at the middle; anther-appendages very broadly
cordate-ovate, entirely inflexed over the top of the depressed-truncate
style-apex; pollen-masses dark orange, very much compressed or
somewhat wing-like on the outer side near the caudicle, and this
wing-like part is often more or less folded outwards.

EASTERN REGION: Eastern border of the Transvaal at Ingoma, *Gerrard*,
1302 partly!

Similar to, but quite distinct from *S. cordifolium*, yet difficult to characterize.
S. virens has also been distributed as No. 1302.

3. **S. atropurpureum** (E. Meyer, Comm. 219); root a tuber;
stem $1\frac{1}{2}$–$3\frac{1}{2}$ ft. high, $1\frac{1}{4}$–$2\frac{1}{4}$ lin. thick, simple, pubescent all round
or nearly so; internodes 1–$2\frac{1}{2}$ in. long; leaves very shortly petiolate,
1–$2\frac{1}{4}$ in. long, $\frac{1}{2}$–1 in. broad, somewhat spreading, ovate-oblong or
elongate deltoid-oblong, obtuse or acute, subhastate or subcordate
with rounded auricles at the base, very broadly and cuneately contracted into the 1–$1\frac{1}{2}$ lin. long petiole, thinly puberulous all over or
only near the margins on the upper side, pubescent on the veins
beneath; umbels 5–12 to a stem, pedunculate, racemose, simple and
10–20-flowered, or compound with 50–60 flowers; peduncles $\frac{1}{4}$–$1\frac{1}{4}$
in. long, pubescent or tomentose; pedicels 5–6 lin. long, rather
slender, tomentose; sepals $1\frac{2}{3}$–$1\frac{3}{4}$ lin. long, $\frac{2}{3}$ lin. broad, lanceolate,
acute, pubescent; corolla-lobes $2\frac{1}{2}$–3 lin. long, $1\frac{1}{2}$–$1\frac{2}{3}$ lin. broad,
oblong, obtusely rounded at the apex, suberect, concave, somewhat
incurved at the tips, glabrous on both sides, blackish-purple; corona
lobes with 2 very prominent parallel keels and an appendage on their
inner face, apparently white; body of the lobe $\frac{3}{4}$ lin. long, $\frac{1}{2}$ lin.
broad, oblong or subquadrate, slightly broadest at the top, with the
sides very abruptly reflexed from the keels, forming a broad channel
between them at the back, at the apex is a small oblong obtuse
lobule recurving over the top of this channel, and on either side of

the lobule the apex is truncate with a very distinct pocket-like depression; appendage exceeding the staminal column, $\frac{3}{4}$ lin. long, arising at the base of the recurved lobule, erect or incurved, linear or linear-subulate, bifid or entire at the apex; staminal column 1 lin. long; anther-wings broadly rounded, not at all angular; anther-appendages broadly ovate, obtuse, erect, with their tips reaching to or shortly inflexed on the broad truncate style-apex, which has a sinuous-crenate margin and sometimes (always?) 5 small central tubercles; caudicles lateral at the obliquely truncate apex of the pollen-masses; follicles solitary, about 2 in. long, $\frac{1}{2}$ in. thick, ovoid-lanceolate, tapering to an obtuse point, covered with filiform processes 1–2 lin. long, everywhere pubescent. *DC. Prodr.* viii. 553; *Schlechter in Engl. Jahrb.* xxi. *Beibl.* 54, 3. *Cynanchum atropur-pureum, Dietr. Synop. Pl.* ii. 906.

CENTRAL REGION: Aliwal North Div.; on the Witte Bergen, 5000–6000 ft., *Drège!*

KALAHARI REGION: Orange River Colony; Besters Vley, near Witzies Hoek, *Bolus*, 8111! Harrismith, *Sankey*, 192! and without precise locality, *Cooper*, 2728!

EASTERN REGION: Natal; slope of the Drakensberg Range at Olivers Hoek Pass, *Wood*, 3471! Buffalo River Valley, near Charlestown, 5000–6000 ft., *Wood*, 5368!

The tuber is sweet-tasted and is eaten by the Basutos, according to a note in Herb. Bolus.

4. S. hamatum (E. Meyer, Comm. 220); stem $\frac{2}{3}$–$2\frac{1}{3}$ ft. high, simple, 1–$1\frac{1}{2}$ lin. thick, more or less compressed, puberulous with curved hairs; internodes $\frac{1}{2}$–$3\frac{1}{2}$ in. long; leaves shortly petiolate, $\frac{3}{4}$–2 in. long, $\frac{1}{12}$–$1\frac{1}{4}$ in. broad, linear, linear-oblong or oblong to roundish-oblong, obtuse or emarginate, or (when linear) acute at the apex, often apiculate, truncate, cordate or hastate at the base, with obtuse or rounded auricles, subscabrous or thinly sprinkled with minute hairs all over or towards the revolute margins of the upper surface and on the veins beneath; petiole 1–$2\frac{1}{2}$ lin. long; umbels 3–7, racemose or the upper subcorymbose, 3–7-flowered; peduncles $\frac{1}{2}$–$1\frac{3}{4}$ in. long, pubescent or subtomentose on one side; bracts $1\frac{1}{2}$–$2\frac{1}{2}$ lin. long, filiform, puberulous; pedicels 3–5 lin. long, pubescent or subtomentose; sepals $1\frac{2}{3}$–2 lin. long, $\frac{1}{2}$ lin. broad, oblong-lanceolate or ovate-lanceolate, acute, puberulous; corolla-lobes campanulately spreading or suberect, slightly concave, 3–4 lin. long, $1\frac{3}{4}$–$2\frac{1}{4}$ lin. broad, oblong, obtuse or rounded at the apex, apparently greenish or yellowish with purple-brown or blackish-purple veins or spots, or occasionally without markings, often drying brown or blackish-purple, quite glabrous; corona-lobes ascending-spreading, $\frac{1}{2}$–1 lin. long, $\frac{1}{3}$–$\frac{2}{3}$ lin. broad, narrowly oblong to suborbicular, somewhat recurved at the margins, with an appendage near the minutely and obtusely bifid or rounded apex and 2 parallel keels distant from the margins on the inner face, apparently whitish; appendage $\frac{3}{4}$–1 lin. long, from half to nearly as broad as the lobe at its base, erect or inflexed-erect with recurving or hooked and sometimes slightly diverging tips, divided to the base into 2 subulate segments or very deeply bifid or occasionally entire, overtopping the 1⅓ lin. long staminal column, which is somewhat excavated under the anthers; anther-appendages broadly rounded or broadly ovate, with their tips inflexed over the rim of the broad, depressed-truncate style-apex; anther-wings obtusely rounded at the very prominent basal part; caudicles attached below the apex or near the middle of the pollen-masses. *Decne in DC. Prodr.* viii. 554; *Schlechter in Engl. Jahrb.* xxi. *Beibl.* 54, 3, *and Journ. Bot.* 1896, 419. *S. atropurpureum, Harv. Thes. Cap.* i. 27, *t.* 42, *and Schlechter in Engl. Jahrb.* xviii. *Beibl.* 45, 2, *and Journ. Bot.* 1896, 419, *not of E. Meyer, and var. lineatum, Schlechter in Ann. Naturhist. Hofmus. Wien,* xv. 67 *partly. S. hastatum (error for S. hamatum, E. Meyer), Schlechter in Engl. Jahrb.* xviii. *Beibl.* 45, 5 *in note. Cynanchum hamatum, Dietr. Synop. Pl.* ii. 906.

VAR. β, **pallidum** (N. E. Br.); corolla-lobes 3½–5 lin. long, 2¼–2⅓ lin. broad, apparently yellowish or greenish, without a trace of purple; corona-lobes 1 lin. long, with an appendage 1–1½ lin. long, exceeding the staminal column by nearly all its length, its tips more recurved and rather more diverging than in the type.

VAR. γ, **elegans** (N. E. Br.); corolla-lobes 4–4½ lin. long, 2½ lin. broad, apparently white or creamy with purple tips and a few purple dots and lines on the veins.

COAST REGION : Queenstown Div.; Andriesberg Range, near Bailey, 6000 ft., *Galpin,* 2266! Komgha Div.; hills near the Kei River, *Flanagan,* 373!
CENTRAL REGION : Somerset Div.; Bosch Berg, 4000 ft., *MacOwan,* 1637!
EASTERN REGION: Transkei, Kreilis Country, *Bowker,* 11! Kentani district, *Miss Pegler,* 133! Kaffirland, *Brownlee!* Tembuland! near Cala, *Bolus,* 10218! Griqualand East; Kwenkwe Mountain, 5800 ft., *Bolus,* 10219! Mount Currie, *Tyson,* 1441! near Kokstad, *Tyson,* 1685! Insiswa Mountains, *Schlechter,* 6438! *Krook,* 800! Natal; Dargle Farm, *Fannin,* 63! Var. β: mountains of Kaffraria, *Mrs. Barber,* 29! Var. γ, Eastern Frontier, *Mrs. Barber!*

Easily distinguished from *S. cordifolium,* by not having the margins of the corolla-lobes folded backwards. The type specimen in E. Meyer's (Drège's) Herbarium has the leaves more crowded and very much narrower than usual, being ¾–1 in. long and 1–1½ lin. broad across the very base, and are linear-hastate in form. In floral structure it is identical with the other specimens quoted, some of which have similar leaves at the top of the stem, whose unusual narrowness may be due to an insufficiency of water. Other species vary in a similar manner.

5. **S. tridentatum** (Schlechter in Engl. Jahrb. xviii. Beibl. 45, 5); stem solitary, ⅔–1⅓ ft. high, simple or rarely branched, ⅔–1½ lin. thick at the base, pubescent, terete or slightly compressed; internodes numerous, usually much shorter than the leaves, which are ½–1¼ in. long, 1–6 lin. broad above the broader base, varying from oblong-hastate to linear-hastate, obtuse or acute, apiculate, somewhat scabrous above and on the veins beneath; petioles 1–2 lin. long; umbels 5–12 to a stem, pedunculate, racemose, usually 4–6-flowered; peduncles 3–9 lin. long, pubescent or subtomentose with minute

curved hairs; bracts 1–2 lin. long, filiform or subulate; pedicels 2½–5 lin. long, pubescent; sepals 1¼–1⅔ lin. long, about ⅗ lin. broad, oblong or oblong-lanceolate, acute, pubescent; corolla-lobes 2¼–2¾ lin. long, 1½ lin. broad, erect, rather deeply concave, incurved at the tips, very broadly rounded or subtruncate at the apex, glabrous on both sides or slightly and very minutely puberulous on the inner face, apparently dark brownish-purple more or less suffused with greenish at the tips; corona-lobes whitish or with a purple central stripe, shorter than or equalling the staminal column, ½ lin. long and as much or rather more in breadth, subquadrate or suborbicular, with the margin recurved at the top or all round or bent back thus Λ, with an appendage and 2 keels distant from the margins on the inner face, from the apex of the keels a short subhorizontal or deflexed transverse keel on each side extends to the margin; appendage scarcely ½ lin. long, ¼–⅓ as broad as the lobe at its base, erect, just overtopping the style-apex, linear, channelled down the inner face, more or less minutely bifid (rarely 3-toothed) at the apex or rarely divided to the base into 2 or 3 segments, or entire; staminal column ¾ lin. long; anther-appendages oblong or ovate, obtuse, erect, with their tips very slightly incurved on the margin of the broad truncate 5-crenate style-apex; caudicles lateral, attached above the middle of the oblong pollen-masses; follicles about 2 in. long and 5–6 lin. thick, fusiform-lanceolate, tapering into an obtuse beak, finely pubescent with minute curved hairs and rather thinly covered with erect filiform processes 1½–1¾ lin. long. *Schlechter in Journ. Bot.* 1896, 420.

COAST REGION, between 2000 and 5000 ft. : Komgha Div.; among rocks near Komgha, *Flanagan*, 1040! Stockenstrom Div.; Kat Berg, *Hutton!* Queenstown Div.; on damp slopes near Queenstown, *Galpin*, 1964! Stutterheim Div.; Kabusie, *Murray*, 1637! Dohne Mountain, *Bolus*, 10220! 10221! British Kaffraria, *Cooper*, 468! *Mrs. Hutton!*

EASTERN REGION : Transkei; Kentani, *Miss Pegler*, 184! Kreilis Country, *Bowker*, 94! 296! near the Tsomo River, *Bowker*, 859! Tembuland; Bazeia Mountain, 2500 ft., *Baur*, 818! Engcobo Mountain, *Bolus!*

6. S. virens (E. Meyer, Comm. 219); stem 2–3 ft. high, simple or branching in the upper part, stout, 1½–2½ lin. thick at the basal part, pubescent with minute curved hairs; internodes numerous, ⅔–1½ in. long; leaves ascending or spreading, those at the middle and lower part of the stem 1½–2 in. long, ½–¾ in. broad, oblong, obtuse or subacute, truncate or truncately subhastate at the base, the upper and those on the branches much smaller and more hastate, minutely scabrous near the margins above and on the veins beneath, otherwise glabrous; petiole 1–3 lin. long; umbels 3–9 to a branch or stem, racemose, pedunculate, 3–15-flowered; peduncles ¼–1¼ in. long, subtomentose; bracts 1½–2 lin. long, subulate or fili-form; pedicels ⅓–½ in. long, subtomentose; sepals 1½–1¾ lin. long, ¾ lin. broad, lanceolate or oblong-lanceolate, acute, minutely pubes-cent; corolla-lobes ascending, 2½–3 lin. long, 1½–2 lin. broad, oblong,

very obtusely rounded at the apex, slightly concave, glabrous on
both sides, sometimes minutely ciliate at the base, greenish or
yellowish-green; corona-lobes much shorter than the staminal
column, $\frac{1}{2}$–$\frac{2}{3}$ lin. long, $\frac{1}{3}$–$\frac{1}{2}$ lin. broad, ovate-oblong or oblong,
obtusely rounded at the apex, with an appendage and 2 distant
parallel keels placed near the margins and having a short keel on
each side extending from the base of the appendage obliquely downwards and out to the margins on the inner face, so that viewed sideways the margins appear to make a fold at this point; appendage
$\frac{2}{3}$ lin. long, arising just below the apex of the lobe and usually more
than half as broad as it is at the base, linear, deeply bifid or trifid
or divided to the base into 2–3 contiguous subulate segments, erect,
reaching to or slightly overtopping the style-apex, slightly recurved-
hooked at the apex; staminal column $\frac{3}{4}$–1 lin. long; anther-appendages subreniform, very obtuse, erect beneath the rim of the broad
flat or slightly concave style-apex; caudicles attached to the side of
the oblong pollen-masses between the middle and apex. *Decne in
DC. Prodr.* viii. 554; *Schlechter in Engl. Jahrb.* xxi. *Beibl.* 54, 3,
excl. syn. S. oblongum, Schlechter in Journ. Bot. 1894, 260, *and*
1896, 420. *Cynanchum virens, Dietr. Synop. Pl.* ii. 906, *not of*
905.

Eastern Region: Natal; between Umtsikaba River and Durban, *Drège,*
4956! on a hill at Umhlongwe, *Wood,* 3012! *and in Natal Herb.* 489! and
without precise locality, *Gerrard,* 1302 partly!

The specimen in E. Meyer's Herbarium, which Schlechter in *Engl. Jahrb.* **xxi.**
Beïbl. 54, 3, states was mixed with *S. euphorbioides,* E. Meyer (Drège, 4960),
and has named *S. tridentatum,* Schlechter, is *E. virens,* E. Meyer, and is evidently
a portion of the same gathering as Drège, 4596; it is quite distinct in its coronal
structure from *S. tridentatum,* with the type of which (Flanagan, 1040) I have
compared it. In E. Meyer's Herbarium, Schlechter has labelled the type
specimens of *S. cordifolium,* E. Meyer, as *S. virens,* and to the type of *S. virens*
he has not added a label.

7. S. euphorbioides (E. Meyer, Comm. 219); plant $1\frac{1}{2}$–2 ft. or
more high, branching or occasionally simple, 1–$1\frac{1}{2}$ lin. thick,
bifariously puberulous; internodes very numerous, mostly $\frac{1}{6}$–$\frac{1}{2}$ in.
long, or some of the upper $\frac{3}{4}$–1 in. long; leaves often somewhat
crowded, sessile or with petioles up to 1 lin. long, very spreading or
ascending, $\frac{1}{2}$–1 in. long, $1\frac{1}{2}$–4 lin. broad, often broadest near the
apex, spathulate-oblong, oblong or linear-oblong, obtuse or rounded
and minutely apiculate at the apex, truncate or subhastate at the
base, scabrous-pubescent along the revolute margins, otherwise
glabrous; umbels pedunculate, subcorymbose or clustered at the
apex of the branches or the lower racemose, 4–8-flowered; pe-
duncles $\frac{1}{3}$–$1\frac{1}{4}$ in. long, pubescent or subtomentose on one side, as
are the unequal $1\frac{1}{2}$–6 lin. long pedicels; bracts $\frac{3}{4}$–$1\frac{1}{2}$ lin. long,
subulate; sepals $1\frac{1}{2}$ lin. long, $\frac{1}{2}$ lin. broad, oblong or oblong-lanceolate, acute or subobtuse, slightly puberulous; corolla-lobes ascend-
ing, 2–$2\frac{1}{2}$ lin. long, $1\frac{1}{3}$–$1\frac{1}{2}$ lin. broad, oblong or slightly obovate-

oblong, slightly emarginate at the very obtuse apex, concave, glabrous on both sides ; corona-lobes $\frac{1}{2}$ lin. long, $\frac{2}{5}$ lin. broad, much shorter than the staminal column, ovate or ovate-oblong, subtruncate, emarginate or shortly bifid at the apex, with a divided appendage and 2 parallel distant keels placed near the margins and having a short keel on each side extending from the base of the appendage obliquely downwards and to the margin on the inner face ; appendage $\frac{1}{3}$–$\frac{2}{3}$ lin. long, arising near the apex of and more than half as broad as the lobe, much exceeding it and reaching to or slightly exceeding the style-apex, divided to the base into 2 subulate contiguous or slightly divergent segments straight or slightly recurved at the tips ; staminal column 1–$1\frac{1}{4}$ lin. long ; anther-appendages broadly ovate, obtuse, applied to the underside of and not exceeding the rim of the concave or crater-like style-apex. *Decne in DC. Prodr.* viii. 554 ; *Schlechter in Engl. Jahrb.* xxi. *Beibl.* 54, 3, *excluding syn. Cynanchum euphorbioides, Dietr. Synop. Pl.* ii. 906.

EASTERN REGION : Natal ; by the sea-shore between Umtentu River and Umzimkulu River, *Drège ;* between Umzimkulu River and Umcomaas River, *Drège,* 4960 ! Umcomaas River, *McKen,* 4 ! and without precise locality, *Drège,* 4959 !

This is very like *S. virens,* E. Meyer, in floral structure, of which it may possibly prove to be a peculiar maritime form, but its general appearance is so different, that until both are better known, I deem it best to follow E. Meyer and Decaisne in considering them distinct species. It differs from *S. virens* in its apparently more shrubby habit and more crowded leaves, which are smaller, much more shortly petiolate, and often somewhat spathulate in form ; the flowers are also smaller, but appear to be of the same colour as those of *S. virens.* Drège's specimen 4959 is stated by Schlechter in *Engl. Jahrb.* xxi. *Beibl.* 54, 3, to be distinct from *S. euphorbioides,* but in external appearance and in floral structure I find it to be identical with that species. There is no locality on the label of 4959, but it may be the plant collected between Umtentu and Umzimkulu Rivers. All the specimens of *S. euphorbioides,* distributed by Drège which I have seen, other than those in E. Meyer's Herbarium, are exactly like the branchlets of Drège's 4959. The specimen from Dutoits Kloof, placed under this species by E. Meyer, and named *S. æmulum* by Schlechter, belongs to *S. cordifolium.*

8. S. nitidum (Schlechter in Engl. Jahrb. xx. Beibl. 51, 18) ; stem 3–10 in. high, usually simple, flattened, pubescent, with about 4–6 internodes, the middle 2 mostly 2–$4\frac{1}{2}$ in. long, the others shorter ; leaves ascending, shortly petiolate, $\frac{1}{2}$–$1\frac{1}{2}$ in. long, 1–5 lin. broad, linear to oblong, acute or obtuse, apiculate, varying from rounded to obtusely hastate at the base, thinly pubescent or subscabrous all over, or only along the revolute margins above, pubescent or hairy on the midrib beneath ; umbels 2–3, subcorymbose or the lower one distant from the rest, pedunculate, 3–7-flowered ; peduncles 2–13 lin. long, pubescent or hairy ; pedicels 2–3 lin. long, pubescent or hairy ; sepals $1\frac{1}{2}$ lin. long, $\frac{1}{2}$ lin. broad, lanceolate, acute, shortly hairy ; corolla-lobes 3 lin. long, $1\frac{1}{2}$ lin. broad, suberect or erectly spreading, with recurved margins and incurved tips, oblong and obtusely pointed when flattened out, glabrous on both sides, or with very few hairs on the back, white with pink lines (*Wood*), in the dried state sometimes with purple-brown veins and borders on the back; corona-lobes erect, white (*Wood*), equalling or shortly exceeding the style-apex, 1–1¼ lin. long, ½ to nearly 1 lin. broad, ovate, elliptic-oblong or linear-oblong, bifid or variously toothed or subtruncate at the apex, with the teeth sometimes (always?) recurved, rounded, subtruncate or subcordate at the base, with two appendages and 2 parallel keels on the inner face; appendages arising near the apex of the lobe and not or but slightly exceeding it, ⅓–⅖ lin. long, oblong or subquadrate, contiguous or overlapping, bifid or minutely or distinctly toothed at the apex, rarely simple ; staminal column about 1¼ lin. long, broadly obconic; anthers blackish or dark coloured in the dried state, their appendages subreniform, very obtuse, erect, reaching to the rim of the style-apex, white with a purple-brown centre; anther wings broadly rounded, and not very prominent at the basal part; pollen-masses with a pellucid space at one end; caudicles lateral; style-apex broad, shortly 5-lobed, with a raised rim and depressed centre, pink (*Wood*). *Schlechter in Journ. Bot.* 1896, 421 ; *Rand in Journ. Bot.* 1903, 200.

KALAHARI REGION : Transvaal; in a marsh near Heidelburg, *Schlechter*, 3519! Modderfontein, *Conrath*, 1014! around Johannesburg, *Rand*, 706! 863! near Middelburg, *Schlechter*, 3796!

EASTERN REGION: Natal; Dargle Farm, *Fannin*, 18! 19! Mooi River District, 4000–5000 ft., *Wood*, 5578! hill near Mooi River, *Wood*, 5378! Shafton, Howick, *Mrs. Hutton*, 40a! 410 !

9. **S. quadridens** (N. E. Br. in Kew Bulletin, 1895, 252) ; plant 3½–6 in. high ; stems softly pubescent with spreading hairs along two broad lines, with about 6–8 internodes 3–13 lin. long; leaves ascending or suberect, 1–2 in. long, 1½–7 lin. broad, lanceolate, linear-lanceolate or oblong-lanceolate, acute, or the lower subobtuse, rounded, subtruncate or shortly cuneate at the base, revolute along the margins, rather softly pubescent with spreading hairs above and on the veins beneath; petiole 1–2½ lin. long ; umbels 2–4 to a stem, pedunculate, subcorymbose, about 6–8-flowered, softly pubescent like the stem (but more densely) on the peduncles, pedicels, bracts and calyx; lowest peduncle ½–1⅓ in. long, the others shorter ; bracts 2–3 lin. long, filiform; pedicels 3–5 lin. long ; sepals 2–3 lin. long, ⅔–¾ lin. broad, lanceolate, acute ; corolla-lobes somewhat campanulately spreading, 3–3¼ lin. long, 1½ lin. broad, oblong, obtuse, thinly pubescent on the back, finely and more densely pubescent on the inner face, except along one margin, which is glabrous, white; corona-lobes shortly overtopping the style-apex, 1¼ lin. long, ¾ lin. broad at the base, ovate, subcordate at the base, narrowing upwards to a bifid, trifid, or rarely subentire apex, about ⅖ lin. broad, with 2 slight keels and a bipartite appendage on the inner face, white; appendage of 2 falcate-subulate segments about ½ lin. long, arising near the apex of the lobe and not exceeding it, directed over the tips of the anthers with erectly recurved tips;

staminal column nearly 1 lin. long; anther-appendages subreniform, obtuse, inflexed over the rim of the depressed-truncate style-apex; anther-wings broadly angular below the middle; caudicles attached to the middle of one margin of the pollen-masses.

EASTERN REGION : Griqualand East; The Plateau, *Haygarth in Herb. Wood,* 4189 !

10. S. elingue (N. E. Br. in Kew Bulletin, 1895, 149); stems $3\frac{1}{2}$–6 in. high, simple, softly pubescent, with spreading hairs along 2 broad lines; internodes $\frac{1}{4}$–1 in. long; leaves in 5–10 pairs, ascending, $\frac{3}{4}$–$1\frac{3}{4}$ in. long, 2–7 lin. broad, oblong, or linear-oblong, subacute, rounded, truncate, or obscurely subhastate at the base, pubescent above and on the veins beneath, narrowly revolute along the margins; petiole $\frac{1}{2}$–2 lin. long ; umbels 2–3 to a stem, pedunculate, corymbose, 3–7-flowered; peduncles 3–7 lin. long, pubescent along one side ; bracts 3–4 lin. long, filiform or subulate, pubescent; pedicels unequal, $1\frac{1}{3}$–$3\frac{1}{2}$ lin. long, pubescent ; sepals 3–$3\frac{1}{3}$ lin. long, $\frac{3}{4}$ lin. broad, lanceolate, acute, pubescent, with spreading hairs; corolla-lobes erect, about $3\frac{1}{2}$ lin. long, $1\frac{1}{2}$–$1\frac{2}{3}$ lin. broad, oblong, obtuse, thinly adpressed-pubescent on the apical part of the back, ciliate on both margins, and rather densely adpressed-puberulous on the inner face, white ; corona-lobes erect, with the apical part slightly recurved-spreading, much overtopping the style-apex, $1\frac{3}{4}$–2 lin. long, $\frac{7}{8}$ lin. broad at the rounded or subcordate base, narrowing to $\frac{1}{2}$ lin. broad at the subequally 2–3-toothed emarginate or subtruncate apex, without keels or appendage on the inner face, thin and flat, white; staminal column 1 lin. long ; anther-appendages roundish-ovate, obtuse, inflexed over the style-apex; projecting angle of anther-wings rounded ; pollen-caudicles terminal. *Schlechter in Journ. Bot.* 1896, 449.

NATAL: slopes of the Drakensberg Range, 6000–7000 ft., *Evans,* 358 !

11. S. flavum (Schlechter in Journ. Bot. 1895, 355); stems 5–10 in. high, simple, flattened, more or less bifariously pubescent; leaves in 3–6 very irregularly distant pairs, ascending or rather spreading, $\frac{3}{4}$–$1\frac{3}{4}$ in. long, $\frac{1}{8}$–$\frac{3}{4}$ in. broad, ovate, oblong or linear-oblong, obtuse or subacute, rounded, subtruncate or subhastate at base, pubescent on both sides; petioles 1–2 lin. long ; umbels 1–4 to a stem, all, or all but the lower one, corymbose, 3–7-flowered; peduncle $\frac{1}{4}$–$1\frac{3}{4}$ in. long, spreading-pubescent; bracts 1–2 lin. long, filiform, pubescent ; pedicels $1\frac{1}{2}$–4 lin. long, pubescent ; sepals $1\frac{1}{2}$–$1\frac{3}{4}$ lin. long, $\frac{1}{2}$ lin. broad, lanceolate, acute, pubescent, and ciliate with spreading hairs; corolla-lobes 3–4 lin. long, $1\frac{1}{2}$ lin. broad, oblong, obtuse, suberect or campanulately spreading, flat, or slightly concave (with incurved tips?), glabrous on both sides, with or without a minute ciliation on one margin, yellow or green ; corona-lobes arising $\frac{1}{4}$–$\frac{1}{2}$ lin. above the base of the staminal column and much overtopping it, with a linear stalk $\frac{1}{2}$–$\frac{3}{4}$ lin. long, and a deltoid-

hastate or rhomboid-lanceolate, acute, or obtuse blade, 1–1¼ lin.
long, ¾–1 lin. broad, flat, with 2 contiguous parallel keels on the
inner face, but no appendage, apparently yellowish or whitish;
staminal column 1⅓ lin. long, much more slender in the ⅞ lin.-
long filamentous than in the head-like antheriferous part; anthers
much broader than long, their appendages cuneate-oblong, obtuse,
inflexed over the truncate 5-radiate style-apex, their wings very
small, prominently angular at the base; pollen-masses pear-shaped,
pellucid within the outer margin or at the oblique apex, with
terminal caudicles.

VAR. β, **lineare** (N. E. Br.); corona-lobes 1½–1¾ lin. long, ¼–⅓ lin. broad,
linear or linear-lanceolate, acute, erect to beyond the middle, then spreading,
with the tips erect, but in most dried flowers apparently erect; otherwise as in
the type.

EASTERN REGION: Griqualand East, at 4000–5500 ft. : Malowe Mountain,
Tyson, 3112! Mount Currie, *Tyson in Herb. Norm. Austr.-Afr.* 1085! Natal;
hill near Nottingham Road, 4000–5000 ft., *Wood,* 5358! Var. β : Natal; Dargle
Farm, *Fannin,* 48! Drakensberg, *Bolus in Herb. Guthrie,* 4878!

12. S. stenoglossum (Schlechter in Engl. Jahrb. xviii. Beibl.
45, 28); stems 1–2 to a tuber, simple, or with 1 branch at the base,
6–18 in. high, with a spreading pubescence; internodes above the
middle 1¼–2 in. long; leaves in 5–6 pairs, rarely more, ascending, or
slightly spreading, 1¼–2¾ in. long, 1–9 lin. broad, linear to ovate-
or oblong-lanceolate, acute or subobtuse, with revolute margins,
pubescent or somewhat scabrous above, hairy on the midrib beneath;
petiole ½–2 lin. long; umbels 2–5 to a stem, all corymbose, or the
lower racemose, pedunculate, 3–10-flowered; peduncles ⅓–2¼ in.
long, more or less densely villous-pubescent or subtomentose; bracts
2–3 lin. long, villous; pedicels 1½–3½ lin. long, densely villous or
tomentose; sepals 1¼–2 lin. long, ⅓–½ lin. broad, lanceolate, acute,
villous-pubescent; corolla-lobes slightly spreading, 2¾–3½ lin. long,
1⅓–1½ lin. broad, oblong, obtuse or subacute, glabrous on both sides,
slightly ciliate at the base, dark purple-brown, with yellowish-
green lines between the veins, becoming blackish-purple when dried;
corona-lobes much exceeding the style-apex, erect, 1–1¾ lin. long,
⅓–½ lin. broad at the ovate-lanceolate base, thence tapering to a
subulate acute point, with 2 prominent keels having a rather deep
groove between them, but no appendage on the inner face, at the
very base on each side the margin is very abruptly folded and continued inwards and upwards, uniting with the staminal column at
the base of the anther-wings; staminal column about 1 lin. long;
anthers broader than long, their appendages large, broadly ovate,
acute, connivent in a broad cone over the top of the shortly and
obtusely conical style-apex, their wings small, not angular at the
middle, but of nearly equal breadth throughout; caudicles laterally
attached below the truncate or rounded pellucid apex of the dark
orange pollen-masses; follicles 1¾ in. long, 5–6 lin. thick, fusiform-
lanceolate, densely pubescent, and covered with filiform pubescent

processes usually 3–4 (or occasionally 1–2) lin. long. *Schlechter in Engl. Jahrb.* xx. *Beibl.* 51, 22, *and in Journ. Bot.* 1896, 420. *Mackenia sp., Harv. Gen. S. Afr. Pl. ed.* 2, 233.

KALAHARI REGION: Orange River Colony; Harrismith, *Sankey,* 191 ! Basutoland, *Cooper,* 2722 ! Transvaal, *Sanderson !*

EASTERN REGION: on mountains at 2500–6000 ft.: Transkei, *Hallack! Bowker !* Tembuland; Bazeia Mountain. *Baur,* 384! Griqualand East, Malowe Mountain, *Tyson,* 3115! near Kokstad, *Tyson,* 1687 ! and in *Herb. Norm. Austr.- Afr.,* 1083 ! Natal; Dargle Farm, *Fannin,* 36! Riet Vley, *Fry in Herb. Galpin,* 2749! near Bothas and on a hill near Gillets, *Wood,* 3397! Mooi River, *Wood,* 4037! Van Reenen, *Wood,* 4561 ! 4778 ! 5009 ! 5603 ! near Emberton, *Schlechter,* 3228! between Pietermaritzburg and Greytown, *Wilms,* 2144 ! Zululand; Sebundini, *Haygarth in Herb. Wood,* 7554!

In the original place of publication no specimen or locality is quoted for this species, and I have not seen the type, but I have examined several specimens which have been named *S. stenoglossum* by Dr. Schlechter himself, and I therefore accept that name for this plant, although I do not find the corona-lobes at all as stated in the original description, which is as follows :—" Corona-lobes small, transverse, provided on the inner face with an erect fleshy linear very acute ligule, with inflexed margins, adorned at the base on each side with a small scale or gland." The variety *longipes,* Schlechter, in *Engl. Jahrb.* xviii. *Beibl.* 45, 29, is merely a form with the lower peduncles nearly as long as the leaves.

13. S. decipiens (N. E. Br.) ; stem 1–1½ ft. high, simple, compressed, with acute edges, pubescent, with spreading hairs, lower internodes 3–3½ in. long, those above the middle much shorter; leaves ascending or spreading, 1½–2¼ in. long, 1–4 lin. broad, linear or oblong-linear, subobtuse, more or less hastate (or the lower and upper rounded) at the base, scabrous-pubescent with spreading hairs above and on the midrib beneath; umbels 3–4 to a stem, pedunculate, racemose, 3–4-flowered; peduncles ⅓–1¼ in. long, pubescent; bracts 2–2½ lin. long, filiform, pubescent; pedicels 2–3 lin. long, densely pubescent with spreading hairs; sepals 2–2½ lin. long, ⅔–¾ lin. broad, lanceolate, acuminate, pubescent; corolla-lobes ascending, or but slightly spreading, 4–5 lin. long, 1½ lin. broad, lanceolate-oblong, acute or subobtuse, but not rounded at the apex, slightly concave, glabrous on both sides, sparsely and minutely ciliate on both margins, marked with 9 blackish-purple veins on a greenish or whitish ground; corona-lobes much shorter than the staminal column, ¾ lin. long and broad, suborbicular or subquadrate-orbicular, with a long appendage decurrent as 2 keels distant from the margins on the inner face, white or yellowish, with a dark purple margin; appendage arising near the apex of the lobe, and nearly ½ as broad as the latter at its base, ¾–⅞ lin. long, suberect, overtopping the staminal column, linear, complicate, or deeply channelled down the inner face, trifid at the apex (or sometimes entire?), with the teeth recurved and the middle tooth longest; staminal column 1⅓ lin. long; anther-appendages nearly twice as long as broad, ovate, obtuse, erect in the lower half, then loosely inflexed clear above and concealing the truncate style-apex; caudicles lateral at the middle of the pollen-masses, the point of attachment being produced into a rather stout cone; young follicles ovate-lanceolate, obtusely beaked, white-tomentose, and covered with pubescent filiform processes about 2 lin. long.

EASTERN REGION: Natal; South Downs in Weenen County, 5000 ft., *Wood,* 4395!

This may easily be mistaken for *S. stenoglossum,* Schlechter, unless dissected.

14. S. pachyglossum (Schlechter in Journ. Bot. 1894, 354); stems 6–15 in. high, erect, usually simple, or in strong plants sparingly branched at about the middle, rather densely leafy, somewhat villous-tomentose; leaves very numerous, alternate, opposite or whorled, spreading or ascending; petiole 1–1½ lin. long; blade ¾–1¾ in. long, ¾–3 lin. broad, linear-hastate or linear-oblong or lanceolate, acute or obtuse, often apiculate, hastate or subcordate-hastate at the base, revolute along the margins, pubescent with short spreading hairs, somewhat villous along the midrib beneath; umbels numerous, all pedunculate, or the upper sessile, subaxillary, crowded along the upper part of the stem, 5–15-flowered; peduncles 0–7 lin. long, pubescent or subtomentose; bracts 1½–2 lin. long, subulate; pedicels 2–4 lin. long, pubescent; sepals 1½ lin. long, ½ lin. broad, lanceolate, acute, pubescent; corolla pentagonal in bud, quite glabrous; lobes 1½–1¾ lin. long, 1 lin. broad, oblong, obtuse, erect, incurved at the apex, somewhat reflexed at the sides; corona-lobes arising near the base of the staminal column and slightly overtopping it, ⅔–¾ lin. long, usually contracted at the middle into an ovate basal part ½ lin. broad, and an upper oblong or rectangular part ⅓ lin. broad, bifid, emarginate or entire at the apex, rarely ovate with a truncate apex, flat, with a smaller rectangular or oblong, bifid or entire appendage on the upper part of the inner face, rather shorter than the lobe; staminal column ¾ lin. long; filament part short, subcylindric; anthers subquadrate, erect; their appendages subreniform, **very** obtuse, slightly incurved over the rim of the style-apex at their tips; their wings obliquely triangular, broadest at the base, scarcely horny; pollen-carriers seated much below the rim of the dilated, pentagonal, concave style-apex; follicles solitary, erect, lanceolate, beaked, covered with ascending, bristle-like processes, pubescent. *Schlechter in Journ. Bot.* 1896, 421.

VAR. *β,* **productum** (N. E. Br.); corona-lobes with the tapering or linear-subulate appendage on their inner face exceeding the lobe and overtopping the staminal column.

VAR. *γ,* **abbreviatum** (N. E. Br.); corona-lobes with a subulate appendage on their face, exceeding the lobe, but shorter than or not exceeding the staminal column.

COAST REGION: King Williamstown Div.; near King Williamstown, Krook, 817 (ex *Schlechter*).

KALAHARI REGION: Basutoland; without precise locality, *Cooper,* 3147! Orange Free State; on a hill near Harrismith, 5000–6000 ft., *Wood,* 5383! *Sankey,* 178! Besters Vley, near Witzies Hoek, 6000 ft., *Bolus,* 8110! *Flanagan,* 1881! Var. *β*: Transvaal; Houtbosch, *Rehmann,* 5867!

EASTERN REGION: Griqualand East; at the foot of Ingeli Mountains, *Tyson,*

1648! Insizwa, *Krook,* 799! Umzimhlava River, *Schlechter,* 6543! *Krook,* 794a!
Natal; Krans Kop, *McKen,* 16! Seven Fontein, near Boston, 3000–4000 ft.,
Wood, 5760! near Enon, *Wood,* 1873! and without precise locality, *Buchanan,*
151! Var. γ: Natal; without precise locality, *Gerrard,* 1303!

This plant is evidently variable in the colour of its flowers, as, according to
Wood's 5369, they are greenish-brown, in 5383 yellow and white, and in 5760
green; the corona-lobes also vary. In the size, shape and structure of its flowers
it is almost identical with *S. bidens,* E. Meyer, but it is a stouter and taller
plant, with shorter peduncles and pedicels, and more numerous flowers in an
umbel, and is much more pubescent than *S. bidens.* Krook, 794a, has been
erroneously distributed with the number 7945 and named *S. carinatum.*

15. S. Wallacei (Schlechter in Journ. Bot. 1895, 268); plant
dwarf, slightly branching from the base; stems erect, densely leafy,
unifariously villous; leaves exceeding the internodes, ¾–1 in. long,
2–4 lin. broad, linear or linear-lanceolate, obtuse, subtruncate-
hastate at the base, with revolute margins, scabrid-pilose above,
glabrous beneath; umbels lateral at the nodes, few-flowered;
peduncle as long as the leaves, erect, pilose; pedicels unequal, 3–4
times as long as the flowers, pilose; sepals half as long as the corolla,
lanceolate, acute, pilose; corolla-lobes erectly-spreading, 3 lin. long,
1½ lin. broad, ovate-oblong, obtuse, with reflexed margins, glabrous
on both sides; corona-lobes ovate, 3-toothed at the apex, with an
appendage on the inner face; lateral teeth erect, much larger than
the minute inconspicuous middle tooth; appendage deeply bifid,
with erect obtuse segments not exceeding the lateral teeth of the
lobe; anther-appendages ovate, very obtuse, inflexed on the style-
apex; caudicles attached to the middle of the ovoid pollen-masses.
Schlechter in Journ. Bot. 1896, 449.

KALAHARI REGION: Orange River Colony; near Heilbron, *Wallace.*

16. S. bidens (E. Meyer, Comm. 220); tuber fusiform, producing
1–3 simple or branched stems 4–9 in. high, pubescent with rather
minute curved hairs; internodes 2–14 lin. long; leaves ascending,
alternate, opposite or 3–4 in a whorl, often on the same stem, mostly
1–2¼ in. long, ½–2 lin. broad, linear, with revolute margins, obtuse
or acute, cuneate to subhastate at the base, minutely scabrous-
pubescent above and on the midrib beneath; petiole 1–1½ lin. long;
umbels 3–6 to a stem, usually subcorymbose, pedunculate, 2–6
flowered; peduncles usually 6–19 lin. long or the uppermost less,
puberulous; bracts 1–1½ lin. long, subulate; pedicels unequal,
slender, 3–9 lin. long, puberulous; sepals 2 lin. long, ½ lin. broad at
the base, thence tapering to an acute point, pubescent; corolla-lobes
1½–2¼ lin. long, ⅞–1⅛ lin. broad, oblong, acute, suberect, replicate,
with the tips incurved and the apex itself shortly recurved, quite
glabrous or with a very few minute hairs on the back, apparently
brownish; corona-lobes shortly overtopping the style-apex, 1–1⅓ lin.
long, ½–⅔ lin. broad, ovate or ovate-oblong, often narrowed into a
short linear-oblong upper part, bifid or emarginate at the rather broad
apex, with an appendage and 2 parallel keels on the inner face;

appendage shorter than or rarely exceeding the lobe, erect, $\frac{1}{4}$–$\frac{1}{2}$ lin. long, subquadrate or rectangular, truncate, emarginate or bifid; staminal column $\frac{3}{4}$ lin. long; anther-appendages reniform, obtuse, applied to the underside of the broad and rather deeply funnel-shaped style-apex, which overtops them; pollen-caudicles very slender, attached near or above the middle of the $\frac{1}{4}$–$\frac{1}{3}$ lin.-long linear or linear-oblong pollen-masses, which are of equal breadth throughout, obtuse at both ends, paler at the apex, but with no distinct pellucid space; follicles solitary, about $1\frac{3}{4}$ in. long, $\frac{1}{3}$ in. thick, lanceolate-fusiform, tapering to a beak, glabrous, thinly covered with filiform processes $\frac{1}{2}$–1 lin. long. *DC. Prodr.* viii. 554; *Schlechter in Engl. Jahrb.* xxi. *Beibl.* 54, 3, *and Journ. Bot.* 1896, 419. *Cynanchum bidens, Dietr. Syn. Pl.* ii. 906. ·

COAST REGION: Stockenstrom Div.; Kat Berg, *Drège!* Queenstown Div.; between the Klipplaat and Zwartkei Rivers, *Drège,* 3424! Table Mountain, *Drège* (ex *E. Meyer*). Cathcart Div.; Blesbok Flats near Windvogel Berg, *Drège,* 3422! British Kaffraria, *Bowker* (and *Mrs. Barber*), 771!

CENTRAL REGION: Somerset Div.; near Somerset East, 2100 ft., *Bolus!* Graaff Reinet Div.; summit of Oude Berg and other mountains near Graaff Reinet, *Bolus,* 168! 12149!

KALAHARI REGION: Orange Free State, *Mrs. Barber,* 737!

17. S. umbellatum (Schlechter in Journ. Bot. 1894, 356); stem 10–12 in. high, slender, erect, velvety on the upper part, densely leafy; leaves longer than the internodes, 1–2 in. long, erect, narrowly linear, acute, with revolute margins, narrowed at the base into a short petiole, very shortly pubescent; umbels lateral at the nodes, several-flowered; peduncles $1\frac{1}{4}$–$1\frac{1}{3}$ in. long, tomentose; pedicels 4–5 lin. long, tomentose; sepals lanceolate, acute, densely pilose; corolla-lobes erectly spreading, $1\frac{1}{2}$ lin. long, 1 lin. broad, ovate, obtuse, glabrescent on the back, pilose on the inner face; corona-lobes nearly as long as the staminal column, slightly narrowed above the middle from an ovate base to a bilobed apex, with a slightly narrower appendage on the inner face, about reaching to the apex of the lobe, shortly notched at the apex; anther-appendages large, very obtuse, inflexed upon the style-apex. *Schlechter in Journ. Bot.* 1896, 421.

EASTERN REGION: Kaffraria, probably Transkei, *Mrs. Barber* (ex *Schlechter*).

This only seems to differ from *S. bidens* by the corolla-lobes being pubescent on the inner face, but Dr. Schlechter states that the flowers are also smaller.

18. S. atrorubens (Schlechter in Journ. Bot. 1894, 353); stem 10–15 in. high, simple, puberulous with curved hairs; leaves 1–$2\frac{1}{2}$ in. long, $\frac{1}{2}$–1 lin. broad, linear, acute, often slightly dilated at the base, but not distinctly hastate, somewhat scabrous, ascending, those below the middle of the stem numerous and rather crowded, varying from alternate to verticillate, those above the middle usually in pairs 1–$1\frac{3}{4}$ in. apart; petioles $\frac{1}{2}$–1 lin. long; umbels 6–8 to a stem, the terminal 3 or 4 corymbose, the others racemose, 3–10-flowered; peduncles 6–22 lin. long, puberulous; bracts $\frac{1}{2}$–$1\frac{3}{4}$ lin. long, subulate, puberulous; pedicels unequal, 2–4 lin. long, puberulous; sepals $1\frac{1}{4}$–$1\frac{1}{3}$ lin. long, $\frac{1}{2}$ lin. broad, lanceolate, acute, pubescent; corolla-lobes $1\frac{3}{4}$–2 lin. long, 1 lin. broad, suberect, ovate-oblong, with reflexed margins when alive, minutely notched at the subacute incurved apex, glabrous on both sides, blackish-purple outside, with a peculiar lurid greenish tint on the inner face in the dried state; corona-lobes slightly overtopping the style-apex, $\frac{7}{8}$–1 lin. long, $\frac{1}{2}$–$\frac{2}{3}$ lin. broad, ovate-oblong, emarginate or subtruncate at the rather broad apex, erect, with an appendage and 2 slight keels (often invisible in the dried state) on the inner face, apparently whitish; appendage subapical, $\frac{1}{8}$ lin. long, $\frac{1}{4}$ lin. broad, transversely rectangular, emarginate, abruptly directed over the top of the large discoid slightly depressed pentagonal style-apex; staminal column $\frac{7}{8}$ lin. long; anther-appendages subreniform, applied to the underside of the style-apex, and scarcely exceeding its margin; pollen-caudicles very slender, attached near the middle of the $\frac{1}{3}$ lin.-long linear straight pollen-masses, which are obtuse at both ends, and have no pellucid space. *Schlechter in Journ. Bot.* 1896, 420.

EASTERN REGION: Tembuland; Bazeia Mountain, 4000 ft., *Baur,* 767!

19. S. Galpinii (Schlechter in Engl. Jahrb. xviii. Beibl. 45, 15); stem 12–15 in. high, about $\frac{1}{2}$ lin. thick near the base, minutely pubescent; internodes many, $\frac{1}{2}$–$1\frac{1}{4}$ in. long; leaves erect or ascending, very shortly petiolate, $\frac{3}{4}$–$1\frac{1}{3}$ in. long, linear, with strongly revolute margins, about $\frac{1}{2}$ lin. broad, dilated and subcordate-hastate at the 1–$1\frac{1}{4}$ lin. broad base, obtuse, minutely scabrous-pubescent; umbels 4–6 to a stem, pedunculate, racemose, 3–12-flowered; peduncle 8–11 lin. long, slender, densely puberulous; bracts about 1 lin. long, filiform; pedicels 5–8 lin. long, slender, densely puberulous or minutely subtomentose; sepals 1 lin. long, $\frac{1}{3}$ lin. broad, lanceolate, acute, puberulous; corolla-lobes suberect, slightly incurved at the tips, $1\frac{1}{2}$–2 lin. long, 1 lin. broad, ovate-oblong, minutely notched at the subobtuse apex, more or less reflexed along the margins, thinly puberulous above the middle on the back, glabrous on the inner face; corona-lobes slightly overtopping the style-apex, about 1 lin. long, $\frac{3}{4}$ lin. broad, flat, broadly elliptic, or somewhat rhomboid-elliptic, bifid to $\frac{1}{3}$ of the way down from the pointed apex, with 2 rather indistinct parallel keels and an appendage on the inner face; appendage $\frac{1}{2}$ lin. long, arising at the base of the notch in the lobe, and slightly exceeding the latter, erect, oblong, widening towards the base, bifid (or entire *Schlechter*); staminal column almost 1 lin. long; anther-appendages subreniform, obtuse, erect, pressed against the side of the thick rim of the broad truncate 5-crenate style-apex, which projects over the pollen-carriers; anther-wings prominent, triangular in outline; caudicles attached near the middle of the inner margin of the narrowly oblong pollen-masses, not at the end. *Schlechter in Journ. Bot.* 1896, 420.

KALAHARI REGION: Transvaal; near Barberton, on a hill-side near Rimers-

Creek, 3200 ft., *Galpin*, 861! and on the summit of Saddleback Mountain, 5000 ft., *Galpin*, 1326!

The flowers are white according to Mr. Galpin, but when dried, the corolla appears to be green and the corona white.

20. S. diversum (N. E. Br.); stems 1–2 ft. high, $\frac{1}{2}$–$\frac{3}{4}$ lin. thick at the base, simple, pubescent; internodes $\frac{3}{4}$–1$\frac{3}{4}$ in. long; leaves opposite or 3–4 in a whorl, erect or ascending, 1–2 in. long, $\frac{1}{3}$–$\frac{3}{4}$ lin. broad, linear or linear-filiform, with revolute margins, acute, cuneately rounded into the very short petiole at the base, scaberulously puberulous; umbels often 2–3 at a node, pedunculate, racemosely arranged, 3–9-flowered; peduncle $\frac{1}{3}$–1 in. long, adpressed-puberulous; bracts $\frac{3}{4}$–1$\frac{2}{3}$ lin. long, filiform, deciduous; pedicels 1$\frac{1}{2}$–4 lin. long, adpressed-puberulous; sepals 1–1$\frac{1}{4}$ lin. long, $\frac{1}{4}$–$\frac{1}{3}$ lin. broad, lanceolate, acuminate, thinly pubescent; corolla-lobes campanulately spreading, with incurved tips, 1$\frac{1}{3}$–1$\frac{1}{2}$ lin. long, $\frac{1}{2}$–$\frac{7}{8}$ lin. broad, oblong, narrowed to a minutely notched point, with the margins slightly recurved, glabrous on both sides, green (*Miss Pegler*); corona-lobes white (*Miss Pegler*), equalling or slightly overtopping the staminal column, $\frac{1}{2}$–$\frac{3}{4}$ lin. long $\frac{1}{2}$–$\frac{2}{3}$ lin. broad, rhomboid-ovate, angular on each side slightly below the middle, much narrowed above, bifid or emarginate at the apex, where the sides are somewhat pinched together, just above the middle of the inner face arises a linear-oblong or ovate-oblong, erect appendage, $\frac{1}{2}$–$\frac{2}{5}$ lin. long, $\frac{1}{4}$–$\frac{1}{3}$ lin. broad, exceeding the lobe, bifid, bipartite or emarginate at the apex, decurrent in 2 parallel keels at the base; staminal column $\frac{2}{3}$ lin. long; anther-appendages reniform, or very broadly ovate, very obtuse, with their tips slightly inflexed over or on the rim of the broad depressed style-apex; caudicles attached at or near the middle of the straight oblong-linear or oblong-lanceolate pollen-masses, which have a pellucid area immediately below the obtuse apex; young follicles fusiform, tapering into a long obtuse beak, puberulous, and beset with rather stiff ascending bristles $\frac{3}{4}$–1$\frac{1}{3}$ lin. long.

COAST REGION: King Williamstown Div.; Keiskamma, *Mrs. Hutton!*

EASTERN REGION, at 1000–3000 ft.: Transkei; Kreilis Country, *Bowker*, 6! valleys near Kentani, *Miss Pegler*, 661! Pondoland; hills between Umtata and St. Johns, *Bolus*, 8701!

Very similar to *S. Galpinii*, Schlechter, but the leaves are not at all hastate, the flowers are smaller, the corona-lobes more angular at the sides, and the anther-wings smaller and much less prominent, besides a different distribution.

21. S. linifolium (Schlechter in Engl. Jahrb. xviii. Beibl. 45, 4); tuber elongate-oblong or radish-like; stem solitary and simple or 2-branched at the base, usually 6–15 in. (rarely 1$\frac{1}{2}$–2 ft.) high, $\frac{1}{2}$–$\frac{3}{4}$ lin. thick at the base, glabrous; middle and upper internodes mostly 1$\frac{1}{4}$–4$\frac{1}{2}$ in. long; leaves erect, 1$\frac{1}{4}$–3 in. long, $\frac{1}{4}$–1 lin. broad, filiform or linear, acute, quite glabrous; umbels 2–9 to a stem, pedunculate, racemose, 3–6-flowered; peduncles $\frac{1}{2}$–3$\frac{1}{2}$ in. long, rather slender, glabrous, or with a few minute curved hairs; bracts $\frac{1}{2}$–1$\frac{1}{3}$ lin. long, filiform, deciduous; pedicels 3–6 lin. long, minutely puberulous; sepals 1–1¼ lin. long, ½ lin. broad, lanceolate, acute, more or less pubescent and ciliate; corolla-lobes 1½–2 lin. long, 1–1¼ lin. broad, spreading, with recurved margins or replicate in some dried flowers, flat in others, elliptic-oblong, subacute, glabrous on the back, minutely puberulous (rarely pubescent) on the inner face or its base, greenish; corona-lobes arising ¼–¾ lin. above the base of the staminal column, about ⅔ lin. long, rather thick and fleshy, ovate or oblong-ovate, produced at the apex on the inner face into a subulate point ½ lin. long, horizontally inflexed over the style-apex, without keels (in the dried state) on the inner face, but with two very minute teeth at the base, sometimes obliterated by pressure; staminal column ¾–1⅓ lin. long; anther-appendages very small, about ⅛–⅙ lin. in diam., roundish or subreniform, very obtuse, reaching to, or very slightly inflexed upon the margin of the convex style-apex. *Schlechter in Journ. Bot.* 1896, 420. *Asclepias filiformis, Linn. f. Suppl.* 169; *Thunb. Prodr.* 47, *in Nov. Act. Acad. Sc. Petrop.* xiv. (1805) 507, *Fl. Cap. ed.* 2, ii. 154, *and ed. Schultes,* 234; *Willd. Sp. Pl.* i. 1272; *Pers. Syn. Pl.* i. 277; *Poir. Suppl. Encycl. Meth.* i. 479; *Schultes, Syst. Veg.* vi. 81; *Spreng. Syst.* i. 847; *G. Don, Gen. Syst.* iv. 142; *Dietr. Syn. Pl.* ii. 898. *Asclepias tenuis, and A. Pachystephana, Schlechter in Journ. Bot.* 1896, 454. *Lagarinthus tenuis, E. Meyer, Comm.* 203. *Gomphocarpus tenuis, Dietr. Syn. Pl.* ii. 900; *Decne in DC. Prodr.* viii. 561. *G. pachystephanus, Schlechter in Verhandl. Bot. Brandenb.* xxxv. 52.

Var. β, **centrirostratum** (N. E. Br.); corona-lobes oblong-ovate, obtuse, with a subulate point ⅛ lin. long from the middle of their inner face produced over the style-apex; otherwise as in the type.

COAST REGION : Riversdale Div.; hills near Zoetemelks River, *Burchell,* 6730! 6777! Humansdorp Div.; Kromme River, *Thunberg!* Uitenhage Div.; Addo, *Drège,* 4976! between Coega (Kuga) River and Sunday River, *Drège.* Albany Div.; hills near Grahamstown, *MacOwan,* 672! *Misses Daly & Sole,* 343! and without precise locality, *Hutton!* Stockenstrom Div.; Kat Berg, 2000 ft., *Hutton!* Queenstown Div.; near Queenstown, *Mrs. Barber,* 91! *Cooper,* 2448! *Galpin,* 1580! Cathcart Div.; between Shiloh and Windvogel Mountain, *Drège,* 3429! Tambukiland, *Zeyher!* Komgha Div.; hills near Keimouth, *Flanagan,* 379! King Williamstown Div.; near King Williamstown, *Sim,* 290! 1642! *Flanagan,* 2179! Var. β: Bathurst Div.; Trapps Valley, *Miss Daly,* 677! Albany Div.; near Grahamstown, *Bolton!* Stockenstrom Div.; Katberg, *Miss Sole,* 388!

CENTRAL REGION : Somerset Div.; foot of Bosch Berg, *MacOwan,* 672! Cradock Div.; near Cradock, *Cooper,* 1283! Aliwal North Div.; Elands Hoek, *Bolus,* 155 (10475)! Albert Div.; Burghersdorp, *Guthrie,* 4199!

KALAHARI REGION : Basutoland, *Cooper,* 2727!

EASTERN REGION : Transkei; between the Gekau (Gcua) River and Bashee River, *Drège!* near Butterworth, *Bolus,* 10190! Kreilis Country, *Bowker,* 297! Tembuland; Bazeia, *Baur,* 343! Griqualand East; around Clydesdale, 2500 ft., *Tyson,* 2167!

The specimens from near Cape Town, quoted by Thunberg, belong to *S. aschersonianum.*

22. **S. aschersonianum** (Schlechter in Verhandl. Bot. Ver. Brandenb. xxxv. 48); tuber producing 1 or more erect, simple,

or slightly branched stems, $\frac{1}{4}$-1 ft. high, often zigzag, with internodes $\frac{1}{4}$-1 in. long, puberulous unifariously, bifariously, or all round; leaves erect or ascending, rarely spreading, $\frac{3}{4}$-1$\frac{3}{4}$ in. long, $\frac{1}{2}$-1$\frac{1}{4}$ lin. broad, linear or filiform, glabrous, or thinly and minutely scaberulous, acute; umbels 2-9, pedunculate, 3-8-flowered; peduncles $\frac{1}{6}$-1 in. long, puberulous; bracts minute, subulate; pedicels 1-3$\frac{1}{2}$ lin. long, puberulous; sepals $\frac{1}{2}$-$\frac{2}{3}$ lin. long, ovate-lanceolate, acute, puberulous; corolla-lobes very spreading or recurved, 1 lin. long, $\frac{3}{5}$ lin. broad, oblong-ovate, obtusely pointed, glabrous on both sides, but with a minutely papillate inner surface, apparently olive-green or brownish; corona-lobes white, overtopping the staminal column, with the basal part $\frac{1}{4}$-$\frac{1}{3}$ lin. long, $\frac{1}{3}$-$\frac{2}{3}$ lin. broad, broadly obovate or cuneate-subquadrate 3-toothed at the top, more or less reflexed at the sides, with 2 keels (often obliterated in dried flowers) on the inner face, but no appendage; lateral teeth short, oblong, obtuse, sometimes slightly falcate; middle tooth $\frac{1}{4}$-$\frac{3}{4}$ lin. long, filiform, or subulate and acute or linear and obtuse to bifid, incurved over the $\frac{1}{2}$ lin. long staminal column, at least at the tips; anther-appendages suborbicular or broadly oblong, very obtuse, inflexed on the truncate style-apex; pollen-masses scarcely curved; follicles solitary, 2$\frac{1}{2}$-3 in. long, 2$\frac{1}{2}$ lin. thick, narrowly fusiform, long-beaked, smooth, glabrous. *Schlechter in Journ. Bot.* 1896, 419. *Lagarinthus tenellus, Turcz. in Bull. Soc. Nat. Mosc.* 1848, i. 256.

VAR. β, radiatum (N. E. Br.); middle tooth of the corona-lobes $\frac{1}{4}$ lin. long, oblong, notched at the apex, stellately spreading and scarcely reaching to the level of the style-apex; otherwise exactly as in the type.

VAR. γ, pygmæum (N. E. Br.); plant usually 2-3 (rarely 5-6) in. high, usually with several simple or branched stems; pollen-masses much curved; otherwise as in the type. *S. pygmæum, Schlechter in Journ. Bot.* 1894, 355, and 1896, 421.

✕ VAR. δ, longipes (N. E. Br.); stem solitary, simple, 9-10 in. high, with internodes 1-2$\frac{1}{2}$ in. long, puberulous nearly all round; peduncles 1-2$\frac{1}{4}$ in. long; middle tooth of the corona-lobes purple; otherwise as in the type.

COAST REGION : Paarl Div.; near Paarl, *Prior!* Cape Div.; Flats and hills near Cape Town, *Thunberg! Burchell,* 898! *Ecklon,* 20! *Harvey!* Flats at Sweet Valley, *Harvey.* 233! Claremont Flats, *Schlechter,* 300! near Rondebosch, *Bodkin!* near Sandown Road, *Wolley Dod,* 2437! between Kalk Bay and Fish Hoek, *Bolus,* 4986! Stellenbosch Div.; hills near Gordons Bay, *Bolus!* Var. β: Cape Div.; Simons Bay, *Wolley Dod,* 847! Var. γ: Albany Div.; near Grahamstown, *MacOwan,* 247! 906! in low valleys. *Mrs. Barber,* 223! British Kaffraria, *Mrs. Barber (Bowker),* 777! *Cooper,* 3148! King Williamstown Div.; near King Williamstown, *Flanagan,* 2171! Var. δ : Humansdorp Div.; in Long Kloof near Kromme River Heights, *Bolus,* 2400!

CENTRAL REGION : Var. γ: Somerset Div.; near Somerset East and on the plain at the foot of Bosch Berg, 2000-2800 ft., *MacOwan,* 1654!

Probably *Lagarinthus hispidus*, Turcz. in *Bull. Soc. Nat. Mosc.* 1848, i. 257 (collected by Ecklon (21) near Cape Town, according to the locality number 64, which Turczaninow quotes) should be referred to *S. aschersonianum.* From the description it only appears to differ from the usual form in having hairs on the leaves and sepals, and in that respect quite agrees with Prior's specimen from Paarl. *S. pygmæum* seems only to be an eastern form of *S. aschersonianum;* it is usually dwarfer, with more numerous stems or branches, but I have seen specimens with a solitary simple stem, and others of *S. aschersonianum* that are as dwarf and as much branched as in *S. pygmæum*; in floral structure they are alike, so that there seems no tangible distinction. Var. *longipes* has a different appearance and may prove distinct, but I have only seen a single specimen, and the floral structure is the same as in *S. aschersonianum*. Living plants may exhibit distinctions not observable in dried specimens.

Gomphocarpus hispidus, Turcz., quoted by Schlechter in *Engl. Jahrb.* xx. *Beibl.* 51, 32 in a note, I have been unable to find, and suspect it to be intended for *Lagarinthus hispidus*, Turcz. Dr. Schlechter, however, refers it to *Gomphocarpus involucratus* (*Xysmalobium involucratum*).

23. S. aciculare (N. E. Br. in Dyer, Fl. Trop. Afr. iv. i. 363); plant 2–10 in. high, with 2 or more stems, or branching at the base; stems or branches with internodes $\frac{1}{4}$–$\frac{1}{2}$ in. long, puberulous, with minute curved hairs; leaves very spreading, subsessile, or very shortly petiolate, 1–3 in. long, $\frac{1}{3}$–2 lin. broad, linear, acute, thinly puberulous above, glabrous, or nearly so beneath; umbels 5–7 to a stem, racemose, pedunculate, 6–10-flowered; peduncles 5–10 lin. long, puberulous on one side; bracts 1–1$\frac{1}{2}$ lin. long, lanceolate-subulate; pedicels 1–1$\frac{1}{2}$ lin. long, puberulous; sepals 1$\frac{1}{4}$–1$\frac{1}{2}$ lin. long, $\frac{1}{2}$–$\frac{2}{3}$ lin. broad, ovate-lanceolate, acute; corolla usually, but not always lobed nearly to the base; lobes erect, with recurving or revolute tips, 1$\frac{3}{4}$–2$\frac{1}{2}$ lin. long, scarcely 1 lin. broad, lanceolate, acute or subobtuse, concave in the lower part, somewhat thickened, and very minutely papillate-puberulous on the inner face of the recurved apical half, elsewhere quite glabrous; corona-lobes erect, 2–2$\frac{1}{4}$ lin. long, subulate from a spathulate-subcordate base, which is $\frac{1}{2}$ lin. broad, with the auricles inflexed and forming a minute tooth on each side of the inner face, which is transversely thickened into a ridge just above them; staminal column 1$\frac{1}{4}$–1$\frac{1}{3}$ lin. long, of which the anther-appendages occupy more than half, much exceeding the style, $\frac{1}{3}$ lin. broad, ovate, acuminate, or lanceolate, acute, connivent-erect, white; anther-wings very prominent, angular at the middle; caudicles terminal at the tapering apex of the pollen-masses; style-apex $\frac{1}{3}$ lin. long, conical, minutely 2-lobed. *Stenostelma capense, Schlechter in Engl. Jahrb.* xviii. *Beibl.* 45, 6; xx. *Beibl.* 51, 41, *and in Bull. Herb. Boiss.* vii. 40.

KALAHARI REGION : Griqualand West ; between the Vaal River and Lower Campbell, *Burchell*, 1791! near Kimberley, *Flanagan*, 1693! Warrenton, *Miss Adams*, 1! Orange River Colony ; Leeuw Spruit or Vredefort, *Barrett Hamilton!* Transvaal ; at the foot of the Magaliesberg Range, *Schlechter*, 3689!
　　Also in Tropical Africa.

Schlechter has placed this plant in a separate genus, on account of the conical bifid style-apex (not the stigma as described by Schlechter) ; but this organ is very variable in several genera, e.g. *Cynanchum, Secamone, Xysmalobium, Pachycarpus*, and to some extent in *Schizoglossum*, that of *S. aciculare* being the most extreme variation in that genus, but not sufficiently so as to warrant its generic separation in my opinion. In one specimen of Schlechter's 3689 the corolla-lobes are connate nearly to their middle into a globose-campanulate tube, in all others examined the corolla is lobed nearly to the base.

24. S. garcianum (Schlechter in Engl. Jahrb. xxxviii. 28, and 30, fig. 1, A–K); plant 6–10 in. high, branching from the base, branches pubescent along 2 broad lines, with internodes $\frac{1}{4}$–$\frac{1}{2}$ in. long; leaves numerous, opposite, widely spreading, $1\frac{1}{2}$–$2\frac{3}{4}$ in. long, $\frac{3}{4}$–2 lin. broad, linear, acute, with revolute margins, glabrous or very slightly puberulous with minute scattered hairs; umbels 2–3 to a branch, pedunculate, subcorymbose, 6–8-flowered; peduncles 2–6 lin. long, puberulous; bracts about $1\frac{3}{4}$ lin. long, puberulous; pedicels $1\frac{1}{2}$–2 lin. long, puberulous; sepals $1\frac{1}{2}$ lin. long, $\frac{1}{2}$ lin. broad, lanceolate, acute, thinly puberulous; corolla-lobes 2–$2\frac{1}{4}$ lin. long, $1\frac{1}{4}$ lin. broad, ascending, oblong-ovate, acute, glabrous on both sides, but with an exceedingly minute papillation on the inner face; corona-lobes 1 lin. long, $\frac{2}{3}$ lin. broad, ovate, acute, with an appendage and 2 stout, slightly diverging keels on the inner face; appendage $\frac{3}{4}$–1 lin. long, arising close to the apex of the lobe, subulate from a stout conical base decurrent as keels, directed over the top of the $1\frac{1}{2}$ lin.-long staminal column, recurved at the apex; anther-appendages subreniform-orbicular, erect, applied to the sides, and not quite reaching up to the margin of the depressed or crater-like style-apex.

KALAHARI REGION: Transvaal; Komati Poort, *Kirk*, 108! hills near Komati Poort, 900 ft., *Schlechter*, 11730, 11734!

25. S. crassipes (S. Moore in Journ. Bot. 1902, 383); tuber oblong, with an elongated neck branching into many stems at the ground-level; stems $1\frac{1}{2}$–15 in. long, apparently decumbent or procumbent, but according to *Guthrie* up to 15 in. high, bifariously puberulous; leaves spreading; petiole $\frac{1}{2}$–1 lin. long; blade $\frac{1}{6}$–$2\frac{1}{2}$ in. long, $\frac{2}{3}$–3 lin. broad above the basal lobes, linear-hastate or linear, acute, cuneately narrowed below the short spreading obtusely triangular basal lobes or angles into the petiole, often revolute or slightly thickened along the margins, glabrous; umbels 1–5 to a stem, lateral at the nodes and terminal, pedunculate, 3–6-flowered; peduncle varying from $\frac{1}{2}$ lin. to $1\frac{3}{4}$ in. long, minutely puberulous or nearly glabrous; bracts 1–$1\frac{1}{2}$ lin. long, subulate, deciduous, or occasionally large, leaf-like, and persistent; pedicels 2–4 lin. long, minutely puberulous; sepals $\frac{3}{4}$–1 lin. long, $\frac{1}{2}$ lin. broad, lanceolate, acute, puberulous or almost glabrous on the back; corolla lobed nearly to the base, quite glabrous or occasionally with here and there a hair outside, purple-brown outside, white within (ex *Guthrie*); lobes erect or erectly-spreading, $1\frac{1}{2}$–$1\frac{3}{4}$ lin. long, $\frac{2}{3}$–$\frac{7}{8}$ lin. broad, oblong-lanceolate, minutely notched at the subacute apex, with revolute margins and recurved tips; corona-lobes arising about $\frac{1}{4}$ lin. above the base of the staminal column, and attaining the same level, $\frac{2}{3}$–1 lin. long, $\frac{3}{5}$–$\frac{3}{4}$ lin. broad, rectangular or subquadrate, very shortly stalked at the base, thickened in a somewhat horse-shoe or $\wedge$-shaped manner, and produced into a very short obtuse point on the inner face at the otherwise rounded or subtruncate apex, the side angles of which are sometimes slightly horizontally produced, and sometimes with a minute tooth below them, flat, not keeled down the face or back, and without an appendage; staminal column $\frac{3}{4}$–1 lin. long; anther-appendages broadly ovate, obtuse, inflexed over the depressed-truncate style-apex; anther-wings triangular, broadest at about the middle; caudicles attached to the apex of the large flat, pear-shaped pollen-masses; follicles solitary, $1\frac{1}{2}$–$1\frac{3}{4}$ in. long, narrowly ovoid-fusiform, beaked, puberulous, and beset with stout bristle-like processes. *Lagarinthus eustegioides, E. Meyer, Comm.* 207. *Gomphocarpus eustegioides, Dietr. Syn.* ii. 901; *Decne in DC. Prodr.* viii. 559. *Asclepias eustegioides, Schlechter in Engl. Jahrb.* xxi. *Beibl.* 54, 6, *and in Journ. Bot.* 1896, 452.

CENTRAL REGION: Graaff Reinet Div.; Sneeuwberg Range, *Drège,* 3438! Colesberg Div.; near Colesberg, *Shaw!* Albert Div.; Burghersdorp, *Guthrie,* 48B! 4880!

KALAHARI REGION: Orange River Colony; Leeuw Spruit or Vredefort, *Barrett-Hamilton!*

The corona-lobes of this plant are very similar to those of *Xysmalobium brownianum. S. crassipes* was described from a very small plant with stems $1\frac{1}{4}$–2 in. long, probably stunted by drought, whilst Guthrie, 48B, which is very different in general appearance, is from a very luxuriant plant, stated on the label to be 10–15 in. high; but all the specimens quoted unquestionably belong to the same species. E. Meyer's description and drawing in his Herbarium of the corona-lobes of *Lagarinthus eustegioides* is quite wrong. I have dissected his type and find it identical with the other specimens quoted above.

26. S. orbiculare (Schlechter in Engl. Jahrb. xx. Beibl. 51, 19); plant 3–4 in. high, with several erect, simple stems, puberulous on the upper part, densely leafy; leaves 7–14 lin. long, 1 lin. broad, erect, exceeding the internodes, linear, acute, with revolute margins, narrowed into a short petiole at the base, glabrescent; umbels lateral at the nodes, pedunculate, about 6-flowered; peduncle erect, much shorter than the leaves, thinly puberulous; pedicels $\frac{1}{2}$–$\frac{1}{3}$ as long as the peduncle, thinly puberulous; sepals 1 lin. long, lanceolate, acute, shortly pilose; corolla-lobes erectly-spreading, 2 lin. long, 1 lin. broad, oblong, subobtuse, with reflexed margins, glabrous; corona-lobes shorter than the staminal-column, suborbicular, obtuse, fleshy with a short obtuse appendage at the apex of the inner face slightly exceeding the lobe; anther-appendages suborbicular, very obtuse, inflexed on the style-apex. *Schlechter in Journ. Bot.* 1896, 421.

EASTERN REGION: Natal; Wessels Neck, 4300 ft. rare, *Schlechter,* 3395.

Of this I have only seen a copy of Dr. Schlechter's drawing of the floral dissections of his type, from which I partly describe. It seems to be allied to *S. crassipes,* but that species has rectangular or subquadrate shortly stalked corona-lobes thickened at the apex on the inner face, but without a distinct appendage.

27. S. umbelluliferum (Schlechter in Engl. Jahrb. xx. Beibl. 51, 24); a dwarf branching plant 3–4 in. high; branches erectly spreading, puberulous; leaves erect, very shortly petiolate, $1\frac{1}{4}$–$2\frac{3}{4}$ in. long, narrowly linear, acute, with revolute margins, glabrate, paler beneath; umbels lateral at the nodes, much shorter than the leaves, few-flowered; peduncle about 5 lin. long, deflexed after flowering;

pedicels shorter than the peduncle, puberulous, pendulous after flowering; sepals half as long as the corolla, lanceolate, acute, slightly puberulous; corolla-lobes very spreading, about 1 lin. long, ovate-lanceolate, obtuse, with recurved margins, glabrous; corona-lobes not as long as the staminal column, erect, fleshy, oblong-ligulate, obtuse, with 2 keels on the inner face; anther-appendages ovate, obtuse, with the tips incurved over the style-apex. *Schlechter in Journ. Bot.* 1896, 421.

KALAHARI REGION: Transvaal; on the plain at the foot of the Magalies Berg, 4600 ft., *Schlechter,* 3687.

I doubt if this is really distinct from *S. orbiculare,* Schlechter; only one specimen of each supposed species was found, and Dr. Schlechter states that at first he considered them both to belong to one species. I have not seen either. Are they distinct from *S. crassipes?*

28. S. Pegleræ (N. E. Br.); tubers small, often 2 or 3 to a plant, 2–4 lin. long, $1\frac{1}{2}$–2 lin. thick, ovoid; stem solitary, erect, simple, or branched at the base, 4–8 in. high, pubescent; leaves $\frac{3}{4}$–$1\frac{1}{4}$ in. long, 1–$2\frac{1}{2}$ lin. broad, lanceolate, linear-oblanceolate or linear-subspathulate, the lower opposite, the rest verticillate, slightly and minutely pubescent along the revolute margins and on the midrib beneath, otherwise glabrous, or nearly so; fascicles or sessile umbels often 2 at a node, 3–5-flowered; pedicels 2–$3\frac{1}{2}$ lin. long, puberulous; sepals about 1 lin. long, $\frac{1}{3}$ lin. broad, lanceolate, acute; corolla rotate, or perhaps slightly reflexed; lobes $1\frac{1}{3}$ lin. long, $\frac{3}{4}$ lin. broad, oblong-ovate, subacute, quite glabrous on both sides; corona-lobes $\frac{1}{2}$ lin. long, and as much in breadth, erect, subquadrate-ovate, bifid at the apex, with the teeth usually minutely denticulate on the inner margin, and inflexed upon the top of the staminal column, 2-keeled on the inner face, without any appendage, and with a rather broad shallow groove down the back; staminal column rather more than $\frac{1}{2}$ lin. long; anther-appendages minute, cuneate-oblong, truncate, abruptly inflexed over the rim of the crater-like style-apex; young follicles solitary, much swollen at the middle, and equally tapering to a long obtuse beak and stalk-like base, minutely subtomentose.

EASTERN REGION: Transkei; Kreilis Country, *Bowker,* 16! Kentani Div.; near the coast, on grassy veld, 100 ft., rare, *Miss Pegler,* 1289!

29. S. heterophyllum (Schlechter in Engl. Jahrb. xxi. Beibl. 54, 4); plant 2–6 in. high, branching at the base, rarely unbranched; branches erect, pubescent; leaves $\frac{1}{3}$–1 in. long, $\frac{1}{2}$–3 lin. broad, ascending or spreading, the upper often whorled, linear or linear-oblong, or the lower ovate-lanceolate or ovate, acute at the apex, subtruncate, subhastate or rounded at the base, revolute at the margins, subscabrous-pubescent to nearly glabrous above, pubescent on the midrib beneath; umbels sessile at the nodes, 3–14-flowered; pedicels $1\frac{1}{2}$–5 lin. long, pubescent; corolla-lobes rotately spreading, 1–$1\frac{1}{3}$ lin. long, $\frac{2}{3}$–$\frac{3}{4}$ lin. broad, oblong or ovate-oblong, subacute, with recurved margins, with or without a few hairs on the back, glabrous, or with a minute whitish papillation (not a distinct pubescence) on the inner face, not ciliate, green, or greenish-brown; corona-lobes exceeding the staminal column, $\frac{1}{2}$–$\frac{3}{4}$ lin. long to the tips of their points, with an appendage and 2 keels on their inner face; basal part varying from subquadrate with rectangular or rounded shoulders and truncately contracted or rounded into a subulate point to broadly ovate and acute, the point is probably directed outwards, but in some dried specimens appears erect or incurved; appendage shorter than, or subequalling the point of the lobe, $\frac{1}{4}$–$\frac{2}{5}$ lin. long, subulate or deltoid, acute, directed over the $\frac{1}{2}$ lin. long staminal column; anther-appendages suborbicular; follicles and seeds probably as in var. β. *Schlechter in Journ. Bot.* 1896, 419, *not of elsewhere. S. villosum, Schlechter in Engl. Jahrb.* xviii. *Beibl.* 45, 29, *and Journ. Bot.* 1896, 420. *Aspidoglossum heterophyllum, E. Meyer, Comm.* 200; *Decne in DC. Prodr.* viii. 555.

VAR. β, **schinzianum** (N. E. Br.); plant usually more branched than in the type; leaves spreading; corolla-lobes 1$\frac{1}{2}$–1$\frac{2}{3}$ lin. long, about $\frac{3}{4}$ lin. broad, apparently dark brown or greenish-brown with a minute whitish papillation on the inner face; corona-lobes 1$\frac{1}{4}$ lin. long to the tips of their points, their appendage subulate to linear, acute, obtuse or bifid; follicle solitary, erect, 1$\frac{1}{4}$–2$\frac{1}{2}$ in. long, about $\frac{1}{4}$ in. thick, fusiform, tapering to an acute beak, thinly beset with ascending fine subulate processes about 1 lin. long and puberulous all over; seeds about 2$\frac{1}{4}$ lin. long, 1$\frac{1}{4}$ lin. broad, ovate, concave on one side, convex on the other, rugose with long and short irregular acute ridges, glabrous, dark brown, otherwise as in the type. *S. schinzianum, Schlechter in Verhandl. Bot. Ver. Brandenb.* xxxv. 51.

VAR. γ, **majus** (N. E. Br.); leaves sometimes deflexed, with or without undulate margins, minutely subscabrous to very pubescent; corolla-lobes 2 lin. long, 1 lin. broad, dark brown or greenish-brown, with a minute whitish papillation on the inner face; corona-lobes 1$\frac{1}{2}$–1$\frac{3}{4}$ lin. long to the tips of their points, with their appendage subulate and entire or linear and bifid or trifid at the apex; otherwise as in the type.

SOUTH AFRICA: without locality, *Mrs. Barber*, 112! Var. γ, *Prior!*
COAST REGION: Stellenbosch Div.; hills near Gordons Bay, *Bolus!* Caledon Div.; mountains near Genadendal, *Bolus*, 8479! Zwart Berg, near Caledon, 1000 ft., *Bolus*, 6689! George Div.; Outeniqua Mountains, near Roodemuur, 2000–2500 ft., *Drège!* near George, 600 ft., *Schlechter* 2387! Port Elizabeth Div.; near Port Elizabeth, *Mrs. Holland*, 38! Bathurst Div.; Round Hill, *Bolus*, 6695! Albany Div.; Brookhuizens Poort, 2500 ft., *MacOwan*, 1021! King Williamstown Div.; Keiskamma, *Mrs. Hutton!* Kaffirland, *Brownlee!* Var. β: Cape Div.; Devils Mountain, *Rehmann*, 1057! Orange Kloof, *Wolley Dod*, 2406! near Rondebosch, *Fair in Herb. Bolus!* Caledon Div.; Zwart Berg, near Caledon, 1200 ft., *Bolus*, 6689! 7409! Var. γ: Uitenhage Div.; Van Stadens Mountains, *Zeyher*, 347!

This appears to be a variable species; the typical *S. heterophyllum* is very different in point of size and appearance from var. *majus*, but is distinctly connected with that by weak specimens of var. *schinzianum*, and I can find no characters but the size of the flowers to distinguish the three; the form of the corona-lobes is the same in all; possibly the living plants may exhibit differences not evident in dried specimens. There is an earlier reference to *S. heterophyllum* (Schlechter in Verhandl. Bot. Ver. Brandenb. xxxv. 51 in note under *S. schinzianum*, and in Engl. Jahrb. xviii. Beibl. 45, 16) which cannot belong to this plant, since from Dr. Schlechter's note under *S. villosum* (Engl. Jahrb. xviii. Beibl. 45, 30) it is evident that the corolla-lobes of his *S. hetero-phyllum* are not glabrous on the inner face, as they are in typical *S. hetero-*

phyllum. Therefore I only quote here the latter reference which distinctly refers to the type of *Aspidoglossum heterophyllum* in E. Meyer's Herbarium, which I have dissected side by side with the type of *S. villosum,* Schlechter, and find to be identical with that.

30. S. consimile (N. E. Br.); plant 3–4 in. high; stem simple or with 1 basal branch, erect, pubescent, with 2–5 leafy nodes below the lowest flowering node; leaves ½–1 in. long, ½–3½ lin. broad, all linear or linear-lanceolate or the lower ovate, lanceolate or oblong, acute, rounded or cuneate at the base, those at the flowering nodes often whorled, minutely pubescent above, roughly and sparsely pubescent on the midrib beneath; umbels sessile at the nodes, 3–9-flowered; pedicels 2½–3 lin. long, pubescent; sepals 1⅓ lin. long, lanceolate, acuminate, pubescent; corolla-lobes 1⅔–2 lin. long, ¾–1 lin. broad, oblong, subacute, more or less reflexed, glabrous outside, rather densely pubescent all over on the inner face, not ciliate; corona-lobes twice as long as the staminal column, basal part ½ lin. long, ¾ lin. broad, suborbicular or transversely or broadly rhomboid-ovate, abruptly narrowed into an erectly spreading (or in some flowers apparently erect with incurved tips) filiform point ¾–1 lin. long, and with 2 subparallel keels and a filiform appendage ½–⅔ lin. long on the inner face arising at the base of the terminal point and directed over the ½ lin.-long staminal column; anther-appendages rounded or subreniform, erect; anther-wings remarkably and abruptly projecting at the base in small rounded or subangular lobules.

COAST REGION: Albany Div.; mountains of Broekhuisens Poort, 2200 ft., near Grahamstown, *MacOwan,* 660! Coldstream, near Grahamstown, *South,* 627! Bathurst Div.; hills near the Great Fish River, "Asclepiad, 50" *Zeyher!*

Similar to *S. heterophyllum,* differing in having the inner face of the corolla very pubescent and much longer corona-lobes. Possibly this may be the plant intended for *S. heterophyllum* by Schlechter in *Verhandl. Bot. Ver. Brandenb.* xxxv. 51, under *S. schinzianum,* and in Engl. Jahrb. xviii. Beibl. 45, 16, see note under *S. heterophyllum.*

31. S. uncinatum (N. E. Br.); plant 4–5 in. high, usually branching at the base; stems with 3–6 internodes, pubescent all round; leaves usually opposite, occasionally whorled at the flowering nodes, ascending or slightly spreading; petiole ½–1½ lin. long; blade ½–1⅓ in. long, ⅛–½ in. broad, linear, oblong or elliptic, acute or obtuse, rounded or subtruncate at the base, revolute along the margins, pubescent all over above and on the veins beneath; umbels sessile at the nodes, 2–8-flowered; bracts filiform; pedicels 2½–4 lin. long, pubescent; sepals 1–1⅓ lin. long, ½ lin. broad, lanceolate, acuminate, pubescent; corolla slightly pubescent outside, glabrous within; lobes 1½ lin. long, ¾ lin. broad, campanulately ascending, oblong, with abruptly reflexed acute tips and broadly recurved margins; corona-lobes much longer than the staminal column, 1 lin. long, ½ lin. broad, oblong or subrectangular, with angular shoulders, narrowed above into an erect deltoid acute point, and with a subulate appendage and 2 parallel keels on the inner face; appendage ¾–1 lin.

long, 2–3 times as long as the apical point, ascending, abruptly incurved or hooked at its apex; staminal column $\frac{3}{4}$ lin. long; anther-appendages $\frac{1}{3}$ lin. long, broadly ovate, obtuse, loosely incurved over the depressed-truncate top of the style.

COAST REGION : Stockenstrom Div.; Kat Berg, *Zeyher!* British Kaffraria, *Mrs. Hutton!*

32. S. ovalifolium (Schlechter in Engl. Jahrb. xviii. Beibl. 45, 5); stems 5–8 in. high, simple, $\frac{1}{2}$–$\frac{3}{4}$ lin. thick, shortly villous; leaves in 5–8 pairs, very shortly petiolate, very spreading, $\frac{3}{4}$–$1\frac{1}{3}$ in. long, $\frac{1}{3}$–1 in. broad, elliptic, ovate or oblong, acute or obtuse and apiculate, rounded at the base, with narrowly revolute margins, thinly covered with short spreading hairs on both sides; umbels 3–4, sessile at the nodes, 3–5-flowered; bracts 2 lin. long, filiform; pedicels 3–4 lin. long, shortly villous; sepals $2\frac{1}{2}$–3 lin. long, $\frac{3}{4}$ lin. broad, narrowly lanceolate, acuminate, shortly villous; corolla-lobes spreading, 3–$3\frac{1}{2}$ lin. long, $1\frac{3}{4}$ lin. broad, oblong-ovate, subacute, pubescent on the back, glabrous on the inner face; corona-lobes with a broadly cuneate basal part 1–$1\frac{1}{4}$ lin. long, $1\frac{1}{3}$ lin. broad, having acutely angular shoulders, and truncately contracted into an erect or perhaps slightly recurving filiform point $1\frac{1}{2}$–2 lin. long, with a rather stout tooth-like appendage, $\frac{1}{2}$ lin. long on the inner face, horizontally extended over the top of the 1–$1\frac{1}{4}$ lin.-long staminal column and decurrent below as a keel; anther appendages orbicular, inflexed on the broad truncate style-apex. *Schlechter in Journ. Bot.* 1896, 420.

COAST REGION : Komgha Div.; grassy slope near Komgha, 2000 ft., *Flanagan*, 1307 !

In one flower of this plant I found the corona-lobes nearly all connate so as to form an imperfect cup.

33. S. Macowani (N. E. Br.); tuber elongated; stem solitary, simple, 2–$4\frac{1}{2}$ in. high, pubescent, with 3–5 leafy nodes below the lowest flowering node ; leaves spreading, those at the flowering nodes 3–5 in a whorl, $\frac{1}{2}$–$1\frac{1}{2}$ in. long, 1–3 lin. broad, linear or linear-lanceolate, acute or subobtuse, somewhat rounded at the base into a very short petiole, revolute along the margins, scabrid-pubescent above and on the midrib (or sparsely all over) beneath; umbels sessile at the nodes, 3–6-flowered; pedicels 3–$4\frac{1}{2}$ lin. long, pubescent; sepals $1\frac{1}{3}$–2 lin. long, $\frac{1}{3}$–$\frac{3}{4}$ lin. broad, reflexed with the corolla, but not permanently, lanceolate, acute, pubescent ; corolla-lobes reflexed, 3–$3\frac{1}{2}$ lin. long, $1\frac{1}{4}$ lin. broad, oblong-lanceolate, minutely notched at the subacute apex, recurved or somewhat revolute along the margins, glabrous on both sides or thinly pubescent on the back and slightly puberulous at the base on the inner face, purple (*Mrs. Barber*); corona-lobes 3 times as long as the staminal column; basal part $1\frac{1}{4}$–$1\frac{1}{3}$ lin. long, 1–$1\frac{1}{4}$ lin. broad, subquadrate or subrectangular, 3-toothed at the top, with 2 parallel keels and an appendage on the inner face; lateral teeth about $\frac{1}{6}$ lin. long, falcate, entire, thickened on the inner face into oblique keels which join the two from the appendage; middle tooth 1½–2 lin. long, erect, linear-subulate, shortly hooked or revolute at the apex; appendage 1½–2 lin. long, arising at the base of the middle tooth, and shorter than, or about equalling it, linear-subulate, simple, erect, straight, or recurved at the apex; basal alternating teeth, short and broad, erect, rounded or subtruncate; staminal column ¾–1 lin. long; anther-appendages broadly ovate, obtuse or acute, with their tips inflexed over the rim of the crater-like style-apex.

Var. β, **tugelense** (N. E. Br.); plant 3½ in. high, with only 3 leafy nodes below the lowest flowering node; leaves ½–⅔ in. long, 4–5 lin. broad, elliptic or elliptic-obovate, very obtuse, thinly puberulous on both sides; basal part of the corona-lobes 1¼ lin. long, ¾ lin. broad; otherwise as in the type.

Coast Region: Albany Div.; near Grahamstown, 2000 ft., *MacOwan,* 1042! *Bolton!*
Eastern Region: Transkei; Tsomo, rare (*Bowker*), *Mrs. Barber,* 789! Var. β: Natal; Tugela, *Gerrard,* 1807!

34. S. grandiflorum (Schlechter in Engl. Jahrb. xviii. Beibl. 45, 27); stem 6½–8 in. high, angulate, puberulous above, densely leafy; leaves spreading, 1–2¼ in. long, 2½–5 lin. broad, lanceolate, acute, glabrous, ciliate on the revolute margins; umbels sessile at the nodes, 2–5-flowered; pedicels 4 lin. long, pilose; sepals 1½ lin. long, lanceolate, acute, pilose; corolla-lobes reflexed-spreading, ovate, acute, with the margins reflexed in the upper part, glabrous; corona-lobes subquadrate, 3-toothed at the top, with an appendage above the middle on the inner face; lateral teeth short, linear, very acute, inflexed; middle tooth erect, much exceeding the staminal column, flexuose and inflexed above the style-apex at the tips; appendage ligulate, acute, shorter than the middle tooth of the lobe and inflexed over the style-apex; anther-appendages rounded, inflexed on the style-apex. *Schlechter in Engl. Jahrb. xx. Beibl. 51, 17, and Journ. Bot.* 1896, 420.

Coast Region: Bathurst Div.; grassy places near Port Alfred, *Schlechter,* 2747.
Eastern Region: Natal; Inchanga Hills, 3800 ft., *Schlechter,* 3246b.

Is this distinct from *S. robustum?* I have not seen it.

35. S. Harveyi (N. E. Br.); tuber elongated; stem solitary, 3–9 in. high, pubescent with 6–12 leaf-bearing nodes below the flowering part; leaves ½–¾ in. long, ½–1¼ lin. broad, all but the lower in whorls of 3–4, spreading or ascending, linear, acute, base rounded, pubescent on the revolute margins and on the midrib beneath, subsessile, or with petioles ½–1 lin. long; umbels sessile at the nodes, 3–5-flowered; pedicels 2–5 lin. long, pubescent; sepals 1–1¼ lin. long, ⅓–½ lin. broad, lanceolate or ovate-lanceolate, acute, pubescent; corolla-lobes spreading, 1¼–2 lin. long, 1 lin. broad, elliptic-oblong, minutely notched at the subobtuse apex, glabrous on the back, very minutely papillate-puberulous on the inner face; corona-lobes much overtopping the staminal column: basal part ¾ lin. long and broad, subquadrate, slightly broader at the top where it is truncately contracted into a subulate slightly incurved point $\frac{3}{4}$ lin. long, furnished on the inner face with 2 keels and a simple subulate or more or less deeply bifid appendage $\frac{1}{3}$ lin. long, directed over the $\frac{3}{4}$ lin.-long staminal column; anther-appendages suborbicular, or slightly transverse, erect, just reaching to the rim of the broad truncate apex of the style.

COAST REGION: Cape Div.; near Cape Town at Wynberg Hill, very rare, *Harvey! Bolus,* 4488! Camps Bay, *Harvey!*

36. S. contracurvum (N. E. Br.); stems erect, 5–6 in. high, 1 lin. thick, simple, shortly hairy; leaves verticillate at the middle and upper part of the stem, opposite below, erect or ascending, 1–1½ in. long, 1–3 lin. broad, linear or linear-lanceolate, acute, base subtruncate, margins revolute, subsessile, or with petioles up to 1 lin. long, coarsely pubescent above and on the midrib beneath with short spreading hairs; flowers in 1–2 pairs at each of the upper nodes, the pairs alternating with the whorled leaves; pedicels 3½–6 lin. long, coarsely pubescent or subglabrous; sepals reflexed, 1⅓ lin. long, $\frac{1}{2}$–$\frac{3}{4}$ lin. broad, ovate or ovate-lanceolate, acuminate, coarsely pubescent; corolla reflexed; lobes about 2¼ lin. long, 1¼–1⅓ lin. broad, elliptic-ovate or oblong-ovate, subacute, glabrous on both sides; corona-lobes much overtopping the staminal column; basal part $\frac{3}{4}$ lin. long and rather more in breadth, transversely oblong, truncately contracted into an erect or slightly incurved subulate point 1¼–1⅓ lin. long, and furnished on the inner face with 2 keels and a strongly recurving subulate appendage about 1½ lin. long, having a small auricle on each side of its base; staminal column $\frac{3}{4}$ lin. long; anther-appendages subreniform-ovate, obtuse, with their tips incurved over the rim of the slightly depressed style-apex.

EASTERN REGION: Natal; Greenwich Farm, Riet Vley, *Fry in Herb. Galpin,* 2747!

37. S. pumilum (Schlechter in Engl. Jahrb. xx. Beibl. 51, 21); stem solitary, 1½–6 in. high, simple, erect, shortly pubescent; leaves in 5–8 pairs or pairs and whorls, 3–5 of them below the lowest flowering node, erect or ascending, $\frac{1}{2}$–1½ in. long, $\frac{3}{4}$–1½ lin. broad, linear or linear-lanceolate, acute, with revolute margins, minutely subscabrous or thinly puberulous above, thinly pubescent on the midrib beneath; umbels 2–3, sessile, often crowded, 5–7-flowered; pedicels 2½–4 lin. long, pubescent; sepals 1 lin. long, lanceolate, acuminate, thinly pubescent; corolla-lobes very spreading or slightly reflexed, 1¾–2½ lin. long, 1–1¼ lin. broad, oblong-ovate or oblong, subacute, glabrous on both sides, or with a few hairs on the back, apparently dull or olive-green (green, *Sankey*), drying brown; corona-lobes greenish-white, with a darker dorsal stripe, with basal part cuneately subquadrate, ½ lin. long, $\frac{3}{4}$ lin. broad, with subrectangular shoulders, truncately contracted into a filiform point 1–1½ lin. long, which is probably spreading, but in some dried flowers appears erect, with a filiform-subulate appendage $\frac{3}{4}$ lin. long arising at the base of the point, and closely inflexed on the top of the 1 lin.-long staminal column, decurrent at the base into 2 keels; anther-appendages orbicular-ovate, very obtuse, inflexed on the top of the truncate 5-crenate style-apex. *Schlechter in Journ. Bot.* 1896, 421.

KALAHARI REGION: Orange River Colony; Bethlehem, *Richardson!* Harrismith, *Sankey*, 180! Transvaal; between Waterval River and Zuikerbosch Rand, 4600 ft., *Schlechter*, 3496!

38. S. dissimile (N. E. Br.); stems 5–8 in. high, simple or occasionally branched at the base, pubescent, very leafy; leaves ascending, opposite below, whorled or alternate above, shortly petiolate, $\frac{2}{3}$–1 in. long, $\frac{2}{3}$–$4\frac{1}{2}$ lin. broad, linear or narrowly oblong, acute, truncate or broadly rounded at the base, revolute along the margins, pubescent above and on the midrib beneath; umbels sessile at the nodes, 3–8-flowered; pedicels $3\frac{1}{2}$–5 lin. long, puberulous; sepals reflexed, $1\frac{1}{2}$–$1\frac{3}{4}$ lin. long, $\frac{1}{2}$ lin. broad, gradually tapering from the base to the very acute apex, pubescent; corolla-lobes reflexed, 2–3 lin. long, 1–$1\frac{1}{2}$ lin. broad, oblong or elliptic-oblong, acute, with more or less recurved margins, thinly pubescent on the back, glabrous on the inner face, apparently dark brown or dull greenish-brown; corona-lobes $\frac{3}{4}$–1 lin. long (including the teeth), 1 lin. broad, subquadrate, shortly 3-toothed at the top, with 2 central parallel keels, a small oblique auricle-like or crested keel from the base of each side tooth, and an appendage on the inner face, apparently whitish, with an elongated purple-brown median spot on the back; median tooth $\frac{1}{4}$–$\frac{1}{2}$ lin. long, broadly deltoid, acute; lateral teeth rather shorter, falcate, acute, often denticulate on the margin; appendage arising slightly above the middle of the lobe, much overtopping the staminal column, $1\frac{3}{4}$–$2\frac{1}{4}$ lin. long, subulate, erect at the base, then recurved over and far beyond the back of the lobe; staminal column $\frac{3}{4}$–1 lin. long; anther-appendages transverse, broadly rounded at the tips, which are inflexed over the concave style-apex. *S. fasciculare, Schlechter in Journ. Bot.* 1895, 354-5, *and* 1896, 420, *not elsewhere. Aspidoglossum fasciculare, Harv. Thes. Cap.* i. 57, *t.* 90, *not of E. Meyer.*

VAR. β, **pubiflorum** (N. E. Br.); corolla-lobes about $2\frac{1}{4}$ lin. long, $1\frac{1}{4}$ lin. broad, pubescent all over the inner face, thinly pubescent on the apical half at the back, otherwise as in the type.

COAST REGION: Stutterheim Div.; Fort Cunnynghame, 4000 ft., *Sim!* Var. β: Cathcart Div.; Blesbok Flats, near Windvogel Mountain, *Drège!*
EASTERN REGION: Transkei, *Mrs. Barber (Bowker)*, 93! Tembuland; Bazeia, 2500 ft., *Baur*, 380!

According to a note on the original drawing in Harvey's Herbarium, t. 90 in the *Thesaurus Capensis* was made from Mrs. Barber's 93, but the corona-lobes are not complicate as there represented, and the keel down their back I have only seen indicated in one specimen (that of Sim), where it may be due to the folding of the lobe in drying. In the other specimens I find no trace of it. Var. β is one of the two plants distributed by Drège under the name *Aspidoglossum fasciculare*, and was accepted as such by Harvey, but its floral structure is quite different from the type of that species.

39. S. ciliatum (Schlechter in Journ. Bot. 1895, 354); stem $4\frac{1}{2}$–8 in.
high, solitary, simple, hairy-pubescent, with 1–3 leaf-bearing nodes
below the flowering part; leaves in 5–7 pairs, or those at the flower-
ing nodes sometimes whorled, ascending, $\frac{2}{3}$–$1\frac{3}{4}$ in. long, 1–$4\frac{1}{2}$ lin.
broad, subsessile or shortly petiolate, linear to lanceolate-oblong or
oblanceolate-oblong, obtuse or acute, rounded at the base, pubescent
or minutely scabrous above, pubescent on the midrib and veins
beneath; umbels 3–6, lateral at the nodes, sessile, or the lower
occasionally pedunculate, 2–6-flowered; peduncles 0–$2\frac{1}{2}$ lin. long;
pedicels 3–4 lin. long, pilose or pubescent; sepals $1\frac{1}{2}$–$1\frac{3}{4}$ lin. long,
$\frac{1}{2}$ lin. broad at the base, thence gradually tapering to a very acute
apex, reflexed, pilose or pubescent; corolla-lobes reflexed or reflexed-
spreading, $1\frac{3}{4}$–$2\frac{2}{3}$ lin. long, $\frac{3}{4}$–1 lin. broad, narrowly or somewhat
attenuate lanceolate, very acute, pilose or slightly pubescent on the
back, puberulous all over or at the base and papillate or nearly
glabrous on the rest of the inner face, ciliate on one margin, greenish
or dark purple-brown; corona-lobes about 3 times as long as the
staminal column, whitish or pale yellowish, marked with purple-brown
on the points, or entirely dark purple-brown, the basal part $\frac{1}{2}$–$\frac{2}{3}$ lin.
long, $\frac{2}{3}$ lin. broad, quadrangular, 3-toothed at the top, with an appendage decurrent into 2 parallel keels on the inner face; lateral teeth
minute, falcate-deltoid, erect, minutely denticulate on the outer margin
or entire, with a short keel or crest (often minutely crenulated) extending obliquely inwards from their base to the principal keels on
the inner face, often very obscure in dried specimens; middle tooth
$1\frac{1}{2}$–2 lin. long, subulate, somewhat recurved spreading or suberect;
appendage arising at the base of the middle tooth, 1–$1\frac{2}{3}$ lin. long,
erect, subulate, entire or bifid at the apex; staminal column $\frac{2}{3}$–$\frac{3}{4}$ lin.
long; anther-appendages much longer than broad, ovate, obtuse or
acute, ascending-connivent over the style-apex, sometimes ciliate, or
with a minute tuft of hairs on each margin below the apex.
Schlechter in Engl. Jahrb. xxi. Beibl. 51, 22 *in note. Aspidoglossum fasciculare, E. Meyer, Comm.* 200; *Decne in DC.
Prodr.* viii. 555.

COAST REGION : King Williamstown Div.; Perie, *Sim*, 286! Cathcart Div.;
between Kat River and Klipplaats River, *Drège*, 3426!

EASTERN REGION : Transkei; Kreilis Country, *Bowker*, 300 partly! Willowvale Div.; Weza, *Miss Abernethy in Herb. Miss Pegler*, 1408! Natal; near
Howick, 3800 ft., *Wood*, 5357! Dargle Farm, *Fannin*, 40!

This very imperfectly known species differs from all its allies in having but 1–3
leaf-bearing nodes below the flowering part, in its very acute corolla-lobes and
small corona-lobes. In E. Meyer's Herbarium I find 3 very distinct species
mixed under the name of *Aspidoglossum fasciculare*, and 2 species have been
distributed by Drège under that name. I take as the type Drège, 3426,
because this is the only one of the three of which E. Meyer has made a drawing
of the corona-lobes, or appears to have dissected, or that completely agrees with
his description; the others I refer to *S. dissimile* and *S. anomalum*. A portion
of the type of *S. ciliatum*, sent by Mr. Wood to Kew, is identical with the type
of *Aspidoglossum fasciculare*, E. Meyer, and I find its sepals and corolla-lobes to
be as described above, not as originally described. *S. anomalum, S. dissimile*,
and *S. robustum* have been mistaken by Dr. Schlechter for *Aspidoglossum
fasciculare* at the places quoted under those species.

40. S. striatum (Schlechter in Journ. Bot. 1894, 356); tuber small, ovoid; stem solitary, about 1 ft. high, simple, somewhat coarsely pilose or villous; leaves in 7–9 pairs below the flowering part, sometimes whorled at the flowering nodes, subsessile, erect or ascending, $\frac{3}{4}$–$1\frac{1}{4}$ in. long, $\frac{3}{4}$–$1\frac{3}{4}$ lin. (3 lin. ex *Schlechter*) broad, linear or linear-lanceolate, acute at the apex, rounded or subauriculate at the base, revolute and pubescent along the margins; umbels 5–6, sessile, 3–5-flowered; pedicels 4–5 lin. long, puberulous (pilose ex *Schlechter*); sepals 2–$2\frac{1}{2}$ lin. ($1\frac{1}{2}$ lin. ex *Schlechter*) long, $\frac{1}{2}$–$\frac{2}{3}$ lin. broad, linear-lanceolate, acuminate, pubescent; corolla-lobes 2–$2\frac{3}{4}$ lin. long, 1–$1\frac{1}{3}$ lin. broad, spreading, ovate-oblong or ovate-lanceolate, subacute, recurved along the margins, thinly pubescent on the apical part at the back, glabrous on the inner face, green, striped with brown (*Wood*); corona-lobes about 4 times as long as the staminal column, white, edged with pink (*Wood*), with the basal part 1–$1\frac{1}{4}$ lin. long, 1 lin. broad, cuneately rectangular or subquadrate, 3-toothed at the top, thickened and tuberculate or crested on the margins and lateral teeth, with two appendages (sometimes united into one pocket-like appendage) decurrent into 2 keels on the inner face; shoulder-teeth $\frac{1}{3}$–$\frac{1}{2}$ lin. long, erect, obtuse; middle tooth $1\frac{1}{2}$–2 lin. long, subulate, erect, incurved at the apex, keeled or crested down the inner face; appendages $\frac{2}{3}$–$\frac{3}{4}$ lin. long, adnate for about $\frac{1}{3}$ lin. up the middle tooth, somewhat variable, sometimes free, with 1 or 2 unequal subulate teeth at the apex, the longer tooth $\frac{1}{3}$–$\frac{1}{2}$ lin. long, and the margins incurved so that the longer teeth are contiguous, or cross one another, sometimes with the margins connate, forming a little pocket with 1 or 2 contiguous erect teeth in front; staminal column very short, $\frac{2}{3}$ lin. long; anther-appendages rather large (minute ex *Schlechter*), roundish-ovate, inflexed over the top of the truncate style-apex. *Schlechter in Journ. Bot.* 1896, 421.

EASTERN REGION: Natal; Inanda, *Wood*, 863! 1210! and without precise locality, *Mrs. Saunders* (ex *Schlechter*).

I have not seen the type of this species, and although there are some small discrepancies in the above description and that of Dr. Schlechter, who makes no mention of the remarkable thickened cristate margins of the corona-lobes, yet I think there can be no doubt of the identity of Wood's plant with that of Mrs. Saunders.

41. S. robustum (Schlechter in Journ. Bot. 1895, 267); stem apparently solitary, usually $\frac{1}{2}$–1 ft., but sometimes $1\frac{1}{2}$–3 ft. high, 1–$1\frac{1}{3}$ lin. thick, rather coarsely pubescent, leafless on the lower part; leaves very numerous and usually rather crowded, but lax in tall specimens, mostly whorled at the flowering nodes, the others alternate, opposite or whorled, subsessile, ascending, $\frac{3}{4}$–$1\frac{1}{2}$ in. long, $\frac{2}{3}$–4 lin. broad, linear, oblong-linear or narrowly oblong-lanceolate, acute, rounded at the base, revolute at the margins, glabrous or slightly scabrous above, glabrous, or with a few hairs along the midrib and margins beneath; umbels several, sessile at the nodes, 3–6-flowered, rather crowded along the terminal part of the stem, and often shorter than the leaves, or forming a somewhat spike-like mass; pedicels

2–5 lin. long, pubescent to nearly glabrous; sepals 2–2½ lin. long, lanceolate, acuminate, pubescent to nearly glabrous; corolla-lobes 2¼–3½ lin. long, 1¼–1¾ lin. broad, reflexed-spreading, oblong or oblong-ovate, acute or subacute, thinly and minutely puberulous on the back, quite glabrous on the inner face, dark brownish-purple (violet ex *Sankey*); corona-lobes much overtopping the staminal column, their basal part subquadrate, 1–1½ lin. long, 1–1¾ lin. broad, narrowed to the base, 3-toothed at the top, with 2 subparallel keels and an appendage on the inner face, white, or with a dark purple line on the middle tooth and appendage; lateral teeth minute, erect, falcate, or occasionally horizontally spreading from the shoulders, entire, or minutely denticulate; middle tooth 1½–2½ lin. long, subulate or fili-form, erect or nearly so, appendage arising just below the base of the middle tooth, ¾–1 lin. long, subulate or filiform, entire, directed over the staminal column; alternating basal-teeth ¼ lin. long, deltoid or deltoid-oblong, acute or obtuse, inflexed over the opening to the stigmatic cavity; staminal column ¾–1⅓ lin. long; anther-appendages transversely elliptic or suborbicular-ovate, very obtuse, inflexed over the rim of the style-apex, which is crenate or divided into 5 small ovate lobes incurved around a central boss. *Schlechter in Journ. Bot.* 1896, 449.

VAR. β, **pubiflorum** (N. E. Br.); corolla-lobes puberulous or pubescent all over their inner face; middle tooth of corona-lobes 1½–2½ lin. long, sometimes hooked at the tips, longer than the 1–1½ lin.-long appendage, which is usually bifid, sometimes divided nearly to the base, sometimes entire.

VAR. γ, **inandense** (N. E. Br.); corolla-lobes pubescent on the inner face; middle tooth of the corona-lobes ⅓–1 lin. long, erect, subulate, shorter than or equalling the ⅔–1 lin.-long appendage, which is subulate and entire; follicles solitary, 2¾ in. long, ⅓ in. thick in the only example seen, fusiform, tapering into an obtuse beak and covered with several vertical series of ascending bristle-like processes ½–1 lin. long, minutely puberulous.

KALAHARI REGION : Orange River Colony ; Besters Vlei, near Witzies Hoek, 5500 ft., *Bolus*, 8108! near Harrismith, *Sankey*, 177! Transvaal; Donker Hoek, *Schlechter*, 3715! Var. β: Transvaal; Concession Creek, Barberton, *Galpin*, 787! near Johannesburg, *Gilfillan in Herb. Galpin*, 6229! 6230! mountains around Houtbosch, *Bolus*, 11123! near Lydenburg, *Wilms*, 959! Henops River, near Irene, *Conrath*, 985! Swaziland; between Bremersdorp and Mbabane (Embabaan), *Bolus*, 12145!

EASTERN REGION : Pondoland; between Umtata and Engcocos, 2500 ft., *Bolus*, 10222! Natal; Sevenfontein, near Boston, 3000–4000 ft., *Wylie in Herb. Wood*, 5369! Umzinto, *McKen*, 3! Howick, *Junod*, 194! near Ixopo, *Schlechter*, 6659, and Umgeni Valley, *Krook*, 824 (ex *Schlechter*); and without precise locality, *Mrs. K. Saunders!* Var. β: Natal; Krans Kop, *McKen*, 21! near Durban, *Guienzius!* Var. γ: Natal; Inanda, *Wood*, 316 partly! and without precise locality, *Gerrard*, 1316!

I believe the above to be *S. robustum*, Schlechter, but I have not seen a type specimen. It agrees with Dr. Schlechter's description, and some of the specimens quoted were so named by him, whilst others I have seen were named by him *S. verticillare* and *S. fasciculare*, and I suspect that this is the plant he has mistaken for *Aspidoglossum fasciculare* in *Engl. Jahrb.* xx. *Beibl.* 51, 22 in note, but not elsewhere.

42. **S. pilosum** (Schlechter in Engl. Jahrb. xx. Beibl. 51, 20); stem probably solitary, simple, 5–14 in. high, softly pubescent along

2 broad lines, with rather long adpressed and spreading hairs;
internodes ½–1¾ in. long; leaves erect or ascending, subsessile or
very shortly petiolate, ¾–1⅓ in. long, ½–2½ (2½–9½ ex *Schlechter*) lin.
broad, all linear, or the lower narrowly oblong, subacute (ovate,
acuminate or lanceolate, acute ex *Schlechter*), with revolute margins,
pubescent; umbels 2–3, sessile at the nodes, or solitary and terminal,
3–7-flowered, usually exceeding the leaves; pedicels 3–5 lin. long,
pubescent; sepals 2 lin. long, ⅔–¾ lin. long, lanceolate, acuminate,
pubescent, reflexed; corolla-lobes reflexed, 2–2½ lin. long, 1⅙–1⅓ lin.
broad, oblong, minutely notched at the subobtuse tips, pubescent on
the back, minutely puberulous on the inner face, except at the base,
brownish-green (*Wood*); corona-lobes greenish-white (*Wood*), with
a dark purple-brown stripe down the long point, the basal part ½–⅔
lin. long, 1 lin. broad, transversely oblong or subrectangular,
thickened at the top, with a depression on each side of the dilated
base of the point, which is 1½ lin. long, linear, spreading, with the
tip incurved, and has the margins of the depressions adnate to it
behind at the base in the form of 2 small rounded auricles, and an
appendage decurrent into 2 prominent keels arising at its base on the
inner face; appendage ½–¾ lin. long, ¼ lin. broad, linear-oblong,
bifid (or trifid ex *Schlechter's* drawing) at the apex, closely inflexed
over the staminal column; alternating basal teeth unusually large,
suborbicular, somewhat hood-like or concave, membranous, completely
covering the opening between the very protruding bases of the anther-
wings; staminal column ¾ lin. long; anther-appendages suborbicular,
inflexed over the style-apex, which is truncate, with a depressed
ring around a central boss. *Schlechter in Journ. Bot.* 1896, 421.

EASTERN REGION: Natal; near Emberton, 1800 ft., *Schlechter*, 3238!
Liddesdale, 4000–5000 ft., *Wood*, 4256 (not 4256 from Van Reenen)! Greenwich
Farm, Riet Vlei, on the Mooi River, *Fry in Herb. Galpin*, 2748! Shafton,
Howick, *Mrs. Hutton*, 404! Zululand, *Gerrard*, 1308!

I have not seen the type specimen of this species, but a copy of Dr. Schlechter's
drawing of the floral structure in Herb. Bolus, as well as his description, quite
agree with the specimens above quoted, except that Dr. Schlechter appears to
have figured and described a much flattened corona-lobe. The method of fertilization of this plant would appear to be of a very special and complicated nature,
as the unusually large basal teeth or lobules alternating with the corona-lobes so
completely cover and conceal the opening at the base of the anther-wings,
through which alone access to the stigma can be obtained, that it is difficult to
understand how the pollen-masses can be inserted in it by an insect, as they must
be for fertilization to take place.

43. S. pulchellum (Schlechter in Engl. Jahrb. xviii. Beibl. 45, 15);
tuber shortly carrot-shaped; stem solitary, ½–1¾ ft. high, simple,
pubescent; leaves mostly alternate, erect and more or less pressed
to the stem, subsessile, ¾–1½ in. long, ⅔–2 lin. broad, linear or linear-
lanceolate, acute to obtuse, rounded to subtruncate at the base, with
revolute margins, finely pubescent; umbels solitary and terminal or
1 terminal and 1–5 others lateral at the nodes, all sessile, 2–6-
flowered; pedicels 3–5 lin. long, pubescent; sepals 1½–2½ lin. long,
¾–1¼ lin. broad, ovate-lanceolate, acute, pubescent; corolla-lobes

3–4 lin. long, $1\frac{1}{2}$–$2\frac{1}{4}$ lin. broad, very spreading or reflexed, ovate or
oblong-ovate, subacute or obtusely pointed, with recurved margins,
puberulous, with minute black and longer white hairs on the back,
puberulous on the inner face, green (*Sankey*); corona-lobes erect,
contiguous, $1\frac{1}{4}$–$1\frac{3}{4}$ lin. long, and about as broad, ovate to subquadrate-
ovate, obtuse to slightly emarginate at the recurved or subrevolute
apex, obtusely keeled on the back, with an appendage arising from
the middle of the inner face at the top of a broad thick and fleshy
somewhat V-shaped keel, which is produced into a minute tooth
on each side of the base of the appendage, and connected to the
thickened margins by very short transverse ridges (these latter and
the keel are sometimes invisible in dried flowers); appendage
$2\frac{1}{2}$–$2\frac{3}{4}$ lin. long, stout, triquetrous-subulate, rather abruptly recurved
in a bold arch over the top of the lobe ; follicles lanceolate, tapering
into a long beak, pubescent and covered with ascending bristle-like
processes 1–$1\frac{1}{2}$ lin. long. *Schlechter in Engl. Jahrb.* xx. *Beibl.* 51,
21, *and in Journ. Bot.* 1896, 420 ; *Rand in Journ. Bot.* 1903, 336.

KALAHARI REGION : Orange River Colony ; Besters Vley, near Witzies Hoek,
Flanagan in Herb. Bolus, 8109 ! Harrismith, *Sankey*, 182 ! Basutoland, *Cooper,*
2729 ! Transvaal ; grassy slopes of Saddleback Mountain, near Barberton,
4500 ft., *Galpin*, 1089 ! near Botsabelo, *Schlechter*, 4097 ! High Veld, near
Belfast, 6500 ft., *Bolus*, 12146 ! valleys near Roodeport, *Rand*, 962 !

EASTERN REGION ; Natal ; Dargle Farm, *Fannin.* 16 ! Van Reenens Pass,
5500 ft., *Wood*, 4256 (not 4256 from Liddesdale) ! 5143 ! Shafton, Howick, *Mrs.
Hutton*, 406 !

The corolla is described by Schlechter as being " glabrous within," but in the
type and all other specimens examined, I find it very distinctly puberulous on
the inner face.

44. S. anomalum (N. E. Br.) ; stem solitary, 10–12 in. high,
simple, villous, very leafy ; leaves whorled at the flowering nodes,
opposite or alternate below, subsessile, erect or ascending, 1–$1\frac{1}{3}$ in.
long, $\frac{3}{4}$–4 lin. broad, linear to oblong, acute, subtruncate or rounded
at the base, usually revolute along the margins, villous or pubescent,
with rather long hairs ; umbels 4–6, sessile at the nodes, 4–6-
flowered ; pedicels 3–4 lin. long, villous or densely adpressed-pubescent with rather long hairs ; sepals 2 lin. long, $\frac{1}{2}$–$\frac{3}{4}$ lin. broad,
lanceolate, acuminate, villous ; corolla-lobes reflexed, 2–$2\frac{1}{2}$ lin. long,
$1\frac{1}{3}$–$1\frac{1}{2}$ lin. broad, ovate, minutely notched at the subacute apex,
thinly pubescent near the apex on the back, and very minutely
velvety or papillate-puberulous on the inner face ; corona-lobes about
equalling the staminal column, rather thick and fleshy, $\frac{3}{4}$–$\frac{7}{8}$ lin. long,
1–$1\frac{1}{6}$ lin. broad, broadly subobcordate, obovate or transversely
oblong, entire, flattish at the top from the rounded apex being incurved over the top of the very stout broad double keel on the inner
face ; staminal column $\frac{3}{4}$–$\frac{7}{8}$ lin. long ; anther-appendages suborbicular,
inflexed over the truncate style-apex. *S. fasciculare, Schlechter in
Engl. Jahrb.* xviii. *Beibl.* 45, 3, *not elsewhere.*

SOUTH AFRICA : without locality, *Drège* !

COAST REGION : Komgha Div. ; on grassy hills near the Kei River in the
vicinity of Komgha, *Flanagan*, 396 !

EASTERN REGION : Transkei ; Kreilis Country, *Bowker*, 210 ! 342 !

Although a specimen of this plant was mixed with *Aspidoglossum fasciculare* in E. Meyers' Herb., it cannot be easily confused with that species, as in its general appearance and flowers it is quite distinct, whilst the coronal structure is utterly different. The corona-lobes are much more fleshy than is usually the case in this genus.

45. S. verticillare (Schlechter in Engl. Jahrb. xx. Beibl. 51, 25) ; tubers small, ovoid ; stem solitary, simple, 4–14 in. high, pubescent ; leaves very numerous, usually crowded on the upper half of the stem, ascending, the lower opposite or alternate, those at the flowering nodes whorled, $\frac{1}{2}$–2 in. long, $\frac{1}{2}$–$\frac{3}{4}$ lin. broad, linear, subobtuse, with revolute margins, subsessile or very shortly petiolate, finely and rather softly adpressed-pubescent; umbels 3–6, sessile at the nodes, 2–8-flowered ; pedicels 3–6 lin. long, pubescent ; sepals $1\frac{3}{4}$–$2\frac{1}{4}$ lin. long, $\frac{3}{4}$ lin. broad, ovate-lanceolate, acuminate, pubescent ; corolla-lobes reflexed, $2\frac{1}{2}$ lin. long, $1\frac{1}{3}$ lin. broad, ovate, minutely notched at the subacute tips, pubescent on the back, glabrous, or very minutely puberulous on the inner face, apparently green or olive-green ; corona-lobes $\frac{3}{4}$ lin. long, 1–$1\frac{1}{3}$ lin. broad, transverse, subtruncate, thickened and somewhat excavated at the top ; appendage $2\frac{1}{2}$–$3\frac{1}{2}$ lin. long, arising at the top of the inner face, and decurrent as 2 slight keels down the thickened middle, filiform-subulate, with a minute tooth or shoulder on each side at the base, and acutely keeled down the back, the keel continuing across the depression at the top of the lobe ; as seen from above the appendages all curve to the left, forming a circle of loops around and above the $\frac{3}{4}$–1 lin.-long staminal column, not erect as described by *Schlechter;* anther-appendages subreniform, obtuse, inflexed upon the margin of the truncate style-apex. *Schlechter in Journ. Bot.* 1896, 449.

EASTERN REGION : Transkei ; by the Tsomo River, *Mrs. Barber*, 830 ! Tembuland ; between Cala and Elliot, 5200 ft., *Bolus*, 10217 ! Natal ; near Emberton, *Schlechter*, 3242 ! Inanda, *Wood*, 316 partly ! 863 ! 1434 ! Dargle Farm, *Fannin*, 6 ! Greenwich Farm, Riet Vlei, *Fry in Herb. Galpin*, 2746 ! and without precise locality, *Gerrard*, 1307 !

46. S. Davyi (N. E. Br.) ; stem solitary, 1–$1\frac{3}{4}$ ft. high, $\frac{1}{8}$ in. thick, densely white-tomentose ; leaves numerous, closely placed, mostly alternate, except at the flowering nodes, where they are verticillate, erect or ascending, very shortly petiolate, 1–$1\frac{3}{4}$ in. long, 1–4 lin. broad, linear to narrowly oblong-lanceolate, subacute, flat or with revolute margins, densely whitish-tomentose on both sides; umbels 2–5, sessile at the nodes, 2–4-flowered ; pedicels $\frac{1}{4}$–$\frac{1}{2}$ in. long, tomentose; sepals $4\frac{1}{2}$–5 lin. long, lanceolate, acute, tomentose ; corolla-lobes apparently campanulately spreading, 5–6 lin. long, 3–$3\frac{1}{2}$ lin. broad, ovate, slightly notched at the subacute apex, densely adpressed-pubescent or tomentose and green on the back, glabrous or at most with a very minute papillation and closely veined with dark purple or violet on a greenish ground on the inner face, with greenish edges ; corona-lobes arising at the base and somewhat under the staminal column, and including the appendages

attaining to about the same level, ascending-spreading, cross-shaped
in dorsal view, very obtuse, and sometimes with the margin slightly
notched at the apex, apparently white, with or without a dark purple
stripe down the back, 2–2½ lin. long, 1½ lin. broad across the main
body, 2½–3 lin. across the horizontal or upcurved arms, whose tips
touch those of the adjacent lobes, with a very prominent stout single
keel on the inner face united at its base to the staminal column
and ending at the level of the arms in a flat platform, rounded in
front, and bearing upon it a subulate or linear flattened appendage
or horn 1½–2 lin. long, directed outwards and slightly or abruptly
curved to one side like a sickle, or in a sort of loop, glabrous or
very minutely puberulous; staminal column about 2½ lin. long;
anther-appendages broadly rounded, inflexed on the margin of the
broad circular style-apex, which in a specimen in formalin was cushion-
like, but in dried flowers has a raised central boss from which radiate
5 ridges, or is irregularly crumpled; pollen-carriers large, very
broadly triangular at the upper half, suddenly contracted into
an oblong lower half; pollen-masses large, sausage-shaped, falcately
curved.

KALAHARI REGION : Transvaal; near Morgenzon, between Blesbok Spruit
and Ermelo, *Burtt Davy*, 964! High Veld, between Swazieland and Carolina,
5700 ft., *Bolus*, 12116!

This is the most distinct and largest-flowered of the South African species of
the genus. In a corona preserved in fluid in Herb. Bolus, the outspread
arms, touching at the tips, are strongly suggestive of 5 children with their
hands joined in a ring and dancing around a table.

47. S. tubulosum (Schlechter in Engl. Jahrb. xx. Beibl. 51, 23);
tuber small, ovoid; stem solitary, simple, 1–2¼ ft. high, puberulous,
with internodes ⅔–1 in. long; leaves erect or with spreading tips,
½–1¼ in. long, ⅓–⅔ lin. broad, linear, acute, with revolute margins,
glabrous or nearly so above, usually minutely puberulous on the
midrib beneath; umbels about 5, sessile at the nodes, 1–3-flowered;
pedicel 1½–2 lin. long, thinly adpressed-puberulous; sepals 1½ lin.
long, ¾ lin. broad, ovate-lanceolate, acute, pubescent ; corolla campanulate, green (*Wood*); tube 1¼–1½ lin. long, glabrous; lobes
erectly-spreading, 2 lin. long, 1–1¼ lin. broad at the base, ovate-
oblong or ovate-lanceolate, acute, with reflexed margins, thinly
pubescent on the back, puberulous at the base on the inner
face; corona-lobes twice as long as the staminal column, with the
basal part subquadrate, ¾ lin. long, about 1 lin. broad at the top,
slightly narrowed below, with an appendage and 2 parallel keels
on the inner face, abruptly narrowed or contracted at the top into
an erect subulate point 1 lin. long, with the lateral angles or
shoulders square, or very slightly projecting horizontally; appendage
arising at the base of the point, nearly 1 lin. long, linear-subulate,
very acute, directed over the staminal column, with upcurved tips;
alternating teeth very short, obtusely rounded ; staminal column
¾ lin. long; anther-appendages broadly ovate, subacute, with their

tips inflexed over the raised rim of the subtruncate or slightly depressed style-apex. *Schlechter in Journ. Bot.* 1896, 421.

EASTERN REGION; Natal; Attercliff, 800 ft., *Sanderson,* 449 partly ! Inanda, *Wood,* 740 ! near Newcastle, 4200 ft., *Schlechter,* 3410.

A copy of Dr. Schlechter's drawing of the analytical details of his type in the herbarium of Dr. Bolus, quite agrees with the specimens of Wood & Sanderson, except that he represents and describes the corolla-lobes as being puberulous to the tips on the inner face, which is not the case in the specimens I have seen.

48. **S. Schlechteri** (N. E. Br.); stem 15–20 in. high, simple, slender, with numerous internodes $\frac{3}{4}$–1 in. long, puberulous; leaves erect, $\frac{2}{3}$–1$\frac{1}{4}$ in. long, $\frac{1}{3}$–$\frac{1}{2}$ lin. broad, linear, acute, with thickened or revolute margins, glabrous, or with some very minute blackish hairs beneath; umbels 4–6, sessile at the nodes, 4–5-flowered; pedicels 1 lin. long, adpressed-puberulous with minute white and blackish hairs; sepals ascending, 1$\frac{1}{2}$ lin. long, $\frac{1}{2}$ lin. broad, lanceolate, tapering into a very acute point, sprinkled with very minute blackish hairs (not pilose, as originally described); corolla campanulate, glabrous outside, bearded with white hairs on the inner face of the lobes; tube almost 1 lin. long; lobes ascending or but slightly spreading, about 1$\frac{1}{4}$ lin. (2 lin. ex *Schlechter*) long, $\frac{3}{4}$ lin. broad, ovate, minutely notched at the subacute point; corona-lobes $\frac{3}{4}$ lin. long, $\frac{1}{2}$ lin. broad, elliptic or ovate, obtuse or obtusely pointed, forming a mere rim behind the base of the appendage, which arises near the apex of the lobe, is $\frac{1}{2}$ lin. long, subulate, inflexed over the top of the 1 lin.-long staminal column and decurrent at its base as 2 keels on the inner face of the lobe; anther-appendages longer than broad, ovate or oblong-ovate, erect, with only their tips inflexed on the truncate style-apex, which has a thick overhanging rim with the pollen-carriers seated a short distance below it; caudicles short, stout, attached at the side of the oblong pollen-masses at their apex. *S. barbatum, Schlechter in Engl. Jahrb.* xx. *Beibl.* 51, 14, *and Journ. Bot.* 1896, 421, *not of Britten and Rendle.*

KALAHARI REGION: Transvaal; Elands Spruit Mountains, 7000 ft. *Schlechter,* 3833 !

49. **S. Conrathii** (Schlechter in Engl. Jahrb. xxxviii. 27); stem apparently solitary, simple, 1–2 ft. high, slender, whitish-puberulous or minutely velvety-tomentose, with internodes 1$\frac{1}{4}$–3$\frac{1}{2}$ in. long; leaves suberect or ascending, subsessile, 1–2 in. long, $\frac{1}{2}$–$\frac{2}{3}$ lin. broad, linear, acute or obtuse, with revolute margins, puberulous or velvety; umbels sessile at the nodes, 3–8-(rarely 1–2-)flowered; pedicels 1–1$\frac{1}{2}$ lin. long, puberulous or minutely tomentose; sepals 1$\frac{1}{2}$ lin. long, $\frac{1}{2}$ lin. broad, lanceolate, acuminate, pubescent; corolla-lobes apparently campanulately spreading, perhaps subrotate, 1$\frac{3}{4}$–2 lin. long, $\frac{3}{4}$–1 lin. broad, oblong, obtuse, thinly pubescent on the back, puberulous on the inner face, except at the apex, slightly ciliolate on one margin; corona-lobes twice as long as the staminal column, about $\frac{1}{2}$ lin. long, $\frac{1}{2}$–$\frac{2}{3}$ lin. broad at the subquadrate basal part,

with the ¾–1 lin.-long filiform point apparently arising from near the middle of the back and erect or slightly spreading, sometimes incurved at the apex, whilst the ¾–1 lin.-long filiform erect appendage is dilated into a very broad base, with a square-shouldered tooth on each side, and falsely appears to be the apex of the lobe, recurved or slightly incurved at the tip, with 2 parallel keels and sometimes an oblique keel extending from them to the margin of the lobe on the inner face; staminal column ¾ lin. long; anther-appendages reniform or suborbicular, obtuse, their tips inflexed over the rim of the depressed-truncate style-apex; pollen-carriers long and narrow, linear.

KALAHARI REGION : Transvaal; Aapies Poort and elsewhere in the vicinity of Pretoria, *Rehmann*, 4154! *Conrath*, 989! *Miss Leendertz*, 278 ! and without precise locality, *McLea in Herb. Bolus*, 5705 !

This is closely allied to *S. excisum*, but has smaller flowers, puberulous inside, and comes from a different region; the colour of the flowers may also be different.

50. S. excisum (Schlechter in Journ. Bot. 1894, 259); stem simple, 15–18 in. high, with internodes ¾–2⅓ in. long, minutely velvety; leaves erect, subsessile, ⅔–2⅓ in. long, ½–⅔ lin. broad, subacute, with revolute margins, minutely puberulous; umbels 5–7, sessile at the nodes, 2–5-flowered; pedicels 1–1½ lin. long, pubescent; sepals 1⅓–1½ lin. long, ½ lin. broad, lanceolate, acuminate, pubescent; corolla-lobes 2½–2¾ lin. long, 1–1⅕ lin. broad, oblong, minutely notched at the obtuse apex, spreading, pubescent on the back, glabrous on the inner face, ciliate at the base on one margin, apparently brown; corona-lobes twice as long as the staminal column; basal part about ½ lin. long, ½ lin. broad, subquadrate, with the 1⅓–1½ lin.-long filiform point apparently arising from a notch near the middle of the back, and erect or slightly spreading, whilst the ¾–1 lin.-long filiform erect appendage is dilated to nearly as broad as the lobe at the base, with a square-shouldered tooth on each side, and falsely appearing to be the top of the lobe, on the inner face 2 prominent parallel keels are decurrent from the teeth, with an oblique keel extending from their apices to the middle of the lobe on each side; staminal column ¾–⅞ lin. long; anther-appendages suborbicular or broadly reniform-ovate, obtuse, inflexed over the truncate apex of the style; pollen-carriers long and narrow, linear; young follicles swollen at the base, tapering into a long beak, velvety-tomentose. *Schlechter in Journ. Bot.* 1896, 420.

EASTERN REGION : Transkei; near Tsomo or the Tsomo River, *Mrs. Barber!*

51. S. glanduliferum (Schlechter in Journ. Bot. 1894, 259); stem 1½–2 ft. high, simple, puberulous along 2 broad lines, with the internodes below the flowering part 1¼–2¾ in. long, the upper one longest; leaves erect, subsessile, ⅔–1¼ in. long, ½–¾ lin. broad, linear, acute, with thickened or revolute margins, glabrous; umbels 5–7, sessile, 4–8-flowered; pedicels 3–4 lin. long, adpressed-pubescent; sepals $\frac{3}{4}$ lin. long, $\frac{1}{2}$ lin. broad, ovate, acute, thinly pubescent; corolla-lobes widely spreading, 2 lin. long, $1\frac{1}{4}$ lin. broad, somewhat elliptic-oblong, subobtuse, with a few very minute hairs on the back, glabrous on the inner face, yellow; corona-lobes rather spreading, fleshy, about $\frac{3}{4}$ lin. long and the same in breadth, broadly cuneate-subquadrate, subtrilobed at the top, with the middle lobe broadly deltoid-ovate, obtuse, and the lateral lobes or angles very small, subtruncate; inner face much thickened below the apex with 2 small parallel central keels and an oblique inflexed auricle on each side at the lateral angles, no appendage; staminal column $\frac{1}{2}$ lin. long, $\frac{3}{4}$ lin. in diam.; anther-appendages nearly circular, inflexed on the truncate style-apex. *Schlechter in Engl. Jahrb.* xx. *Beibl.* 51, 22 *in note, and Journ. Bot.* 1896, 420.

EASTERN REGION: Natal; near Charlestown, 5000–6000 ft., *Wood*, 4804! Pondoland; near Fort William, 2500 ft., *Tyson*, 2827 ex *Schlechter.*

The specific name refers to the minute glands or processes within the calyx, but these are present in the majority of Asclepiads, and are of no specific value.

52. S. biauriculatum (Schlechter in Engl. Jahrb. xxxviii. 29); stem erect, slender, simple or slightly branched, subglabrous; leaves erect or erectly-spreading, up to 3 in. long and $1\frac{1}{2}$ lin. broad, longer than the internodes; umbels sessile, lateral at the nodes, 5–10-flowered; pedicels $2\frac{1}{2}$–$3\frac{1}{2}$ lin. long; sepals about 1 lin. long, lanceolate, acute, thinly puberulous, ciliate; corolla rotate, 2 lin. long, deeply 5-lobed; lobes oblong, obtuse, thinly pilose on the back, minutely ciliate on one margin, very minutely papillate-puberulous on the inner face especially towards the margins; corona-lobes as long as the staminal column, rhomboid-lanceolate, obtuse, with 2 small triangular auricle-like keels on the lower half of the inner face and a third and less prominent keel between them; anther-appendages suborbicular, very obtuse, incurved upon the depressed style-apex.

EASTERN REGION: Delagoa Bay; sandy places near Katembe, 50 ft., *Schlechter*, 11610.

Also near Beira in Tropical Africa.

53. S. filipes (Schlechter in Engl. Jahrb. xx. Beibl. 51, 16); stem 10–15 in. high, simple or slightly branched at the base, puberulous all round; leaves erect, $1\frac{1}{2}$–$2\frac{1}{2}$ in. long, $\frac{1}{3}$–1 lin. broad, linear, acute, with revolute margins, minutely puberulous, not pilose as originally described; umbels 4–5, sessile at the nodes, 3–4-flowered in the only specimen seen; pedicels 4–5 lin. long, puberulous; sepals $1\frac{1}{4}$–$1\frac{1}{3}$ lin. long, lanceolate, acute, minutely pubescent with curved hairs; corolla-lobes ascending-spreading, $2\frac{1}{2}$ lin. long, nearly 1 lin. broad, oblong, obtusely pointed, with reflexed margins, thinly pubescent on the back, glabrous on the inner face, greenish (*Schlechter*), drying brown; corona-lobes $\frac{3}{4}$ lin. long, $\frac{2}{3}$ lin. broad, somewhat ovate, obtuse, with 2 obtuse teeth on each side and a lacerate appendage decurrent in 2 keels on the inner face, the apex

merely forms a slight rim at the base of the appendage, which is
1–1¼ lin. long, complicate and deeply lacerate, slightly connivent
high over the ⅔ lin.-long staminal column; alternating basal-teeth
transverse, oblong, emarginate; anther-appendages oblong-spathulate,
obtuse, inflexed on and covering the truncate style-apex. *Schlechter
in Journ. Bot.* 1896, 421; *N. E. Br. in Dyer, Fl. Trop. Afr.* iv. i.
360.

KALAHARI REGION: Transvaal; stony places near Klipdam, 4600 ft.,
Schlechter, 4491!

I do not find the corolla-lobes "very shortly puberulous" on the inner face as
described by Schlechter. The complicate lacerate appendage of this species
distinguishes it from all others.

54. S. tridens (N. E. Br.) ; stem apparently solitary, 1–1½ ft.
high, simple, slender, with 12–20 leafy nodes below the lowest
flowering node, rather closely placed, minutely and more or less
bifariously puberulous; leaves opposite or 3 in a whorl, erect, ¾–1¼
in. long, ½–¾ lin. broad, linear, acute, with revolute margins, glabrous;
umbels sessile, 3–6-flowered; pedicels 3–4 lin. long, puberulous;
sepals 1½ lin. long, lanceolate, very acute, thinly puberulous or
almost glabrous; corolla-lobes apparently suberect, with a slight
twist at the apical half, 2 lin. long, nearly 1 lin. broad, oblong-
lanceolate, acute, with a few scattered hairs on the lower part of the
inner face, otherwise glabrous, brown with a green basal part;
corona-lobes slightly exceeding the staminal column, erect, contiguous
or slightly overlapping, 1 lin. long, ¾ lin. broad, oblong, very obtuse,
slightly notched on each side at the middle, with an appendage
decurrent in 2 keels and a minute marginal tooth on each side at the
notches on the inner face, apparently reddish or light purplish on
the upper part; appendage ⅔–¾ lin. long, very abruptly inflexed
over the style-apex from the very apex of the lobe, which forms a
mere marginal rim behind it, broadly linear, 3-toothed at the apex,
channelled down the inner face; staminal column nearly 1 lin. long;
anther-appendages ovate, subobtuse, with their tips inflexed on the
margin of the truncate style-apex.

KALAHARI REGION: Transvaal; Macamac Falls, 4300 ft., *Burtt Davy,* 1440
Gras Kop, near Pilgrims Rest, 5000 ft., *Burtt Davy,* 1461!

55. S. Flanagani (Schlechter in Engl. Jahrb. xviii. Beibl. 45, 3);
stem solitary, about 1 ft. high, simple, slender, puberulous, with
elongated internodes at the lower part and numerous shorter internodes and comparatively crowded leaves at the upper part; leaves
erect or ascending, 1–1¾ in. long, ⅓–¾ lin. broad, linear, acute,
narrowed at the base into a very short petiole, glabrous; umbels 5–6,
sessile at the nodes, 4–7-flowered; pedicels 1½–2 lin. long, pubescent; sepals 1 lin. long, ½ lin. broad, lanceolate, acute, sparsely
pubescent on the lower part; corolla-lobes ascending, 2–2½ lin. long,
¾–⅞ lin. broad, oblong-lanceolate, acute, with reflexed sides and
incurved tips, glabrous on both sides; corona-lobes ¾ lin. long, ½ lin.

broad, rectangular, truncate at the top, having a slight median notch and the shoulders slightly produced into minute teeth, with an appendage and 2 keels on the inner face; appendage arising below the apex of the lobe, $\frac{3}{4}$ lin. long, subulate, acute, directed over the staminal column, with upcurved tips; alternating basal teeth minute, truncate; staminal column $\frac{1}{2}$–$\frac{2}{3}$ lin. long; anther-appendages broadly rounded, inflexed on the truncate excavated style-apex; pollen-masses linear, slightly curved, with a pellucid area at the apex where they are laterally attached to the very short caudicles. *Schlechter in Journ. Bot.* 1896, 420.

COAST REGION : Komgha Div.; on a grassy sandy slope near the mouth of the Kei River, 100 ft., *Flanagan,* 1044!

The corona-lobes are described by Dr. Schlechter as " ovate-ligulate, obscurely 3-lobulate at the apex," but I do not find them so in the type. There is a very slight broad crenature on each side of the small notch, but the general effect is that they are truncate.

56. S. biflorum (Schlechter in Engl. Jahrb. xx. Beibl. 51, 25 in note); stem simple or rarely branched, 1–2 ft. high, with internodes $\frac{3}{4}$–2 in. long, minutely velvety-pubescent; leaves erect, subsessile, $\frac{3}{4}$–2$\frac{3}{4}$ in. long, $\frac{1}{2}$–1$\frac{1}{4}$ lin. broad, linear or linear-subfiliform, acute, with revolute margins, minutely pubescent or nearly glabrous; umbels 6–13, sessile at the nodes, usually 4–12- (rarely 2–3-) flowered; pedicels $\frac{1}{2}$–1$\frac{1}{2}$ lin. long, pubescent; sepals 1–1$\frac{1}{2}$ lin. long, $\frac{1}{3}$–$\frac{1}{2}$ lin. broad, ovate-lanceolate, acute, pubescent; corolla-tube $\frac{1}{2}$ lin. long, glabrous; lobes spreading or reflexed, 1$\frac{2}{3}$–2 lin. long, $\frac{3}{4}$–1 lin. broad, ovate-lanceolate, minutely notched at the subacute apex, thinly pubescent on the back, puberulous or pubescent on the inner face, apparently green or greenish-brown ; corona-lobes 3 times as long as the staminal column, white, purplish-tinted on the points ; basal part $\frac{1}{2}$–$\frac{3}{4}$ lin. long, $\frac{3}{4}$ lin. broad, subquadrate, 3-toothed at the top, with an appendage and 2 parallel keels on the inner face; lateral teeth minute, usually erect or incurved, rarely horizontally spreading; middle tooth 1$\frac{1}{4}$–2 lin. long, subulate, erectly spreading, with the upper half incurved; appendages connivent-erect, 1–1$\frac{1}{2}$ lin. long, subulate, sometimes bifid at the apex ; basal alternating teeth minute, obtuse; staminal column about 1 lin. long ; anther-appendages ovate or rounded, subacute or obtuse, inflexed over the marginal crenatures or tubercles surrounding the central boss of the style-apex ; young follicles narrowly fusiform, tapering into a beak, minutely tomentose. *Schlechter in Engl. Jahrb. xxi. Beibl.* 54, 4, *and in Journ. Bot.* 1896, 449. *S. guelense, S. Moore in Journ. Bot.* 1903, 310, *and Rand, l.c.* 336. *S. venustum, Schlechter in Engl. Jahrb.* xx. *Beibl.* 51, 24. *Aspidoglossum biflorum, E. Meyer, Comm.* 201 ; *Decne in DC. Prodr.* viii. 555.

VAR. β, concinnum (N. E. Br.); corolla and corona-lobes purple-brown or dark purple; otherwise as in the type. *S. venustum, var. concinnum, Schlechter in Engl. Jahrb.* xx. *Beibl.* 51, 25.

VAR. γ, **integrum** (N. E. Br.); corolla dark purple-brown; corona-lobes dark purple-brown or blackish-purple, 1¾–2 lin. long, ½ lin. broad, ovate or oblong-ovate at the lower half, not 3-toothed at the top, but gradually tapering into an erect subulate-filiform point incurved at the apex, without a trace of teeth at the sides; appendage ⅓–½ lin. long, much shorter than the point of the lobe, subulate, acute, or linear and bifid or subtruncate at the apex, ascending over the ¾ lin.-long staminal column; otherwise as in the type.

COAST REGION : Cathcart Div.; Windvogel Mountain, 4500 ft., *Drège,* 3427! Queenstown Div.; Intaba Magwela Mountain, near Queenstown, 3800 ft., *Galpin,* 1909! Var. γ: Bathurst Div.; Linch's Post, near the Kowie River, *Bowie!*

CENTRAL REGION: Aliwal North Div.; Elands Hoek, near Aliwal North, 4600 ft., *Bolus,* 10491 (*F. Bolus,* 238)!

KALAHARI REGION: Transvaal; near Klein Olifants River, *Schlechter,* 3794! near Johannesburg, 6000 ft., *Gilfillan in Herb. Galpin,* 6147! *Rand,* 1009! 1231! Magalies Berg, *Burke! Zeyher,* 1172! Ginsberg and near Johannesburg, *Miss Pegler,* 1053! Modderfontein, *Conrath,* 987! and without precise locality, *Burtt Davy,* 1162! Var. β: Orange River Colony; valleys and flats, *Mrs. Barber & Mrs. Bowker,* 733! Transvaal; Catos Ridge, *Schlechter,* 3262.

EASTERN REGION: Transkei; valleys near Tsomo, *Mrs. Barber,* 793! Natal; Shafton, Howick, *Mrs. Hutton,* 301!

S. gwelense, N. E. Br. in Dyer, *Fl. Trop. Afr.* iv. i. 360, must be referred to this species as a variety (var. *gwelense,* N. E. Br.), differing in having the basal part of the corona-lobes diamond-shaped and angular at the middle, not quadrangular with erect shoulder teeth; it may also occur in South Africa. Intermediate forms occur in which the shoulder-teeth spread horizontally.

There is no specimen in E. Meyer's Herbarium named *Aspidoglossum biflorum* by E. Meyer, but a specimen (Drège, 3427) named by him *Aspidoglossum virgatum,* accurately agrees with his description of *A. biflorum* and is identical with the plant distributed under that name by Drege, with the exception that there are (as he describes) only 2–3 flowers in a cluster on E. Meyer's specimen.

57. S. strictum (Schlechter in Engl. Jahrb. xx. Beibl. 51, 22); stem about 2 ft. high, slender, straight, glabrescent; leaves erect, very shortly petiolate, ¾–1¼ in. long, narrowly linear, with revolute margins, acute, glabrescent; umbels sessile at the nodes, few-flowered; pedicels short, unequal, pilose, deflexed after flowering; sepals 1 lin. long, lanceolate, acute, puberulous, ciliate; corolla-lobes 2 lin. long, 1 lin. broad, erectly spreading, ovate-oblong, obtusely pointed, white-margined, glabrous on the back, puberulous with white hairs on the inner face; corona-lobes ovate-oblong, tapering into a subulate point about as long as the oblong part and incurved over the staminal column, with an appendage decurrent in 2 keels on the inner face; appendage arising at the base of the point and nearly equalling it in length, inflexed on the top of the staminal column; anther-appendages ovate, very obtuse, inflexed on the style-apex. *Schlechter in Journ. Bot.* 1896, 421.

EASTERN REGION : Natal; near Ingagane, 4000 ft., *Schlechter,* 3405.

Of this plant I have only seen a copy in Herb. Bolus of Dr. Schlechter's drawing of the floral details.

58. S. araneiferum (Schlechter in Engl. Jahrb. xx. Beibl. 51, 13); stem 1–2⅓ ft. high, simple or occasionally branching in the upper part, slender, glabrous or very minutely unifariously puberulous on the upper part; leaves erect, $\frac{1}{2}$–1$\frac{1}{2}$ in. long, $\frac{1}{4}$–$\frac{1}{3}$ lin. broad, linear or linear-filiform, acute, with revolute margins, glabrous; umbels sessile, lateral at the nodes, 2–6-flowered; pedicels 1$\frac{1}{2}$–3 lin. long, adpressed-puberulous; sepals $\frac{3}{4}$–1 lin. long, $\frac{1}{2}$ lin. broad, ovate, acute, glabrous or thinly puberulous; corolla-lobes very spreading or perhaps reflexed, 1$\frac{1}{2}$–1$\frac{2}{3}$ lin. long, $\frac{2}{3}$–1 lin. broad, oblong-ovate, acute or obtusely pointed, glabrous on both sides, greenish; corona-lobes with the basal part closely contiguous, transversely rectangular, or subquadrate, $\frac{1}{3}$–$\frac{1}{2}$ lin. long, abruptly contracted into 2 filiform points 1–1$\frac{1}{2}$ lin. long, which are quite free or more or less united at their basal part, erect or directed slightly outwards below, then incurved high over the staminal column, with an erect filiform doubly bent appendage 1–1$\frac{1}{4}$ lin. long, on the inner face and 2 keels decurrent from its base; staminal column $\frac{2}{3}$ lin. long; anther-appendages orbicular-ovate, obtuse, with the tips inflexed over the rim of the depressed-truncate style-apex. *Schlechter in Journ. Bot.* 1896, 421. *S. polynema, Schlechter in Engl. Jahrb.* xxxviii. 30.

KALAHARI REGION : Transvaal; near Lydenburg, *Wilms*, 935! Orange River Colony; vicinity of Besters Vley, near Witzies Hoek (Mount aux Sources), *Bolus*, 8107!

EASTERN REGION: Natal; near Newcastle, 4100 ft., *Schlechter*, 3428. Portuguese East Africa; hill near Ressane Garcia, 1000 ft., *Schlechter*, 11907!

The corona of this is quite unlike any other known to me, and in fresh flowers forms a pretty miniature crown.

59. S. capitatum (Schlechter in Engl. Jahrb. xx. Beibl. 51, 15); stem up to 3 ft. or more high, subsimple, bifariously puberulous; leaves 1–1$\frac{3}{4}$ in. long, not half as long as the internodes, erect, narrowly linear, acute, shortly pilose (probably puberulous is intended); umbel (head ex *Schlechter*) solitary, terminal, many-flowered; pedicels short, unequal, puberulous; sepals half as long as the corolla, lanceolate, acute, pilose (puberulous?); corolla-lobes erectly spreading, 2$\frac{1}{2}$ lin. long, 1 lin. broad, ovate (oblong or lanceolate-oblong, according to a drawing), acute, glabrous on the back, bearded with white hairs on the inner face; corona-lobes subquadrate, 3-toothed at the top, with an appendage and 2 keels on the inner face; lateral teeth minute, spreading; middle tooth long, overtopping the staminal-column, subulate from a broad base, somewhat inflexed at the tip; appendage arising at the base of the middle tooth and shorter than it, subulate, incurved; anther-appendages oblong, obtuse, inflexed on the 5-lobed style-apex. *Schlechter in Journ. Bot.* 1896, 421.

KALAHARI REGION : Transvaal; on a hill near the Crocodile River, 4800 ft., *Schlechter*, 3905.

Of this plant I have only seen a tracing of Dr. Schlechter's original drawings of the dissections of a flower in Herb. Bolus, from which I have slightly modified the original descriptions of the corona-lobes.

T t 2

60. S. restioides (Schlechter in Verhandl. Bot. Ver. Brandenb.
xxxv. 50); tuber elongated ovoid; stem solitary, simple, 15–30 in.
high, glabrous or with a very minute adpressed pubescence on the
upper part, with internodes $\frac{3}{4}$–3 in. long; leaves erect, 5–11 lin.
long, $\frac{1}{4}$–$\frac{1}{2}$ lin. broad, linear, acute, with revolute margins, glabrous;
umbels 3–9, sessile at the nodes, 2–4-flowered, sometimes crowded
together into an oblong inflorescence $\frac{3}{4}$–1 in. long, sometimes distant;
pedicels $\frac{1}{3}$–$\frac{1}{2}$ lin. long, very minutely adpressed-puberulous; sepals
$1\frac{1}{4}$–$1\frac{1}{3}$ lin. long, $\frac{3}{4}$ lin. broad, ovate, acute or acuminate, rather thin,
very minutely puberulous with adpressed black hairs; corolla-lobes
apparently erectly spreading, $2\frac{1}{2}$ lin. long, $1\frac{1}{3}$ lin. broad, oblong-
ovate, obtuse, concave at the lower half, flat or with slightly recurved
margins above, glabrous on both sides or with a few minute blackish
adpressed hairs on the back, minutely ciliate on one margin at the
base; corona-lobes 1–$1\frac{2}{3}$ lin. long, $\frac{2}{3}$ lin. broad, rhomboid-ovate, with
a horizontally projecting angle on each side below the middle,
tapering above to a spreading or somewhat recurved (often apparently
erect in the dried state) bifid or entire point, with an appendage
and 2 keels on the inner face; appendage arising at about the middle
of the lobe and much longer than its point, 1–$1\frac{1}{2}$ lin. long, subulate,
linear or lanceolate, acute or bifid, inflexed and one crossing another
over the $1\frac{1}{2}$–$1\frac{2}{3}$ lin.-long staminal column; anther-appendages rather
large, broadly ovate, obtuse, with their tips inflexed over the style-
apex, which is rather small, crater-like, with the crenate margin
incurved; pollen-carriers partly concealed by the anther-appendages,
unusually large, $\frac{1}{3}$ lin. long, and larger than the pollen-masses.
Schlechter in Journ. Bot. 1896, 419. *S. pallidum, Schlechter in
Engl. Jahrb.* xx. *Beibl.* 51, 19, *and Journ. Bot.* 1896, 421. *S.
Randii, S. Moore in Journ. Bot.* 1903, 310, *and* 339.

Coast Region: Cape Div.; Camp-ground near Cape Town, *Schlechter,*
740!

Kalahari Region: Transvaal; near Schoen Strom, in the vicinity of Klerksdorp, *Burke,* 296! *Zeyher,* 1173! near Johannesburg, 6000 ft., *Gilfillan in
Herb. Galpin,* 6145! 6146! *Rand,* 1058! near Donker Hoek, *Schlechter,* 3708!
plain at the foot of the Magalies Berg, 4600 ft., *Schlechter,* 3681.

The Cape specimen (*Schlechter,* 740), on which the species was founded, is
doubtless a chance introduction from the Transvaal. I have examined flowers
from the type specimens of both supposed species and find them identical.

I do not find an "inflexed obtuse lobule on each margin" of the corona-lobes
as originally described. They are as described above in *Schlechter,* 740, and all
other specimens I have examined.

61. S. unicum (N. E. Br.); stem 15 in. high in the only
specimen seen, simple, slender, pubescent along two broad lines in
the upper part, glabrous below, with internodes $\frac{1}{2}$–$\frac{3}{4}$ in. long; leaves
erect, subsessile, $\frac{1}{2}$–$\frac{3}{4}$ in. long, $\frac{1}{2}$–$\frac{3}{4}$ lin. broad, linear, acute, with
revolute margins, glabrous on both sides; umbels about 6, sessile at
the nodes, 3–5-flowered; pedicels $1\frac{1}{2}$–$2\frac{1}{2}$ lin. long, adpressed-pubescent; sepals 1 lin. long, $\frac{1}{2}$ lin. broad, ovate-lanceolate, acute, slightly
puberulous; corolla-lobes $1\frac{1}{2}$ lin. long, $\frac{3}{4}$ lin. broad, oblong-ovate,

acute, apparently campanulately spreading, slightly puberulous on
the back, inner face with a tuft of rather long hairs at the apex, a
line of hairs along each margin and a few hairs down the centre, the
rest glabrous, one margin ciliolate; corona-lobes about $\frac{3}{4}$ lin. long,
$\frac{3}{5}$ lin. broad, broadly ovate, acute or subobtuse, rather fleshy, with
2 keels but no appendage on the inner face; the tips incurved over
the staminal column, apparently somewhat folded or channelled on
the back and thickened on the inner face; alternating teeth none or
obscure; staminal column $\frac{1}{2}$ lin. long; anther-appendages broadly
ovate, obtuse, inflexed over the truncate style-apex.

EASTERN REGION: Eastern border of the Transvaal at Ingoma, *Gerrard &
McKen,* 1317 !

62. **S. delagoense** (Schlechter in Bull. Herb. Boiss. iv. 446);
stem 15–24 in. high, slender, with 1–2 branches, puberulous all
round at the base and along one line alternating at the nodes in the
upper part; internodes 1–1$\frac{1}{2}$ in. long; leaves ascending (spreading,
ex *Schlechter*), subsessile, 1–2$\frac{1}{4}$ in. long, $\frac{1}{2}$–2$\frac{1}{2}$ lin. broad, linear to
narrowly lanceolate, acute, narrowed towards the base, revolute along
the margins, glabrous on both sides; umbels several, sessile at the
nodes, 1–5-flowered; pedicels 3–4 lin. long, puberulous; sepals 1 lin.
long, ovate-lanceolate, acute, puberulous; corolla-lobes campanulately
spreading, 2 lin. long, about 1 lin. broad, oblong-ovate, subacute,
thinly puberulous on the back, minutely papillate-puberulous on the
inner face (not glabrous as originally described), minutely ciliate on
one margin; corona-lobes about as long as the staminal column, 1 lin.
long, slightly spreading, connate at the base into a short cup,
rhomboid-ovate, obtuse, contracted below, with an erect subulate
tooth $\frac{1}{4}$ lin. long at the middle of the inner face and a small inflexed
obtuse tooth on each side of it below the middle, the inner margin of
which is decurrent as a keel down the contracted part of the lobe;
interspaces of the cup between the lobes broad, with 2 small
inflexed teeth or produced into a broad membranous emarginate or
shortly 2-toothed lobule of variable length and form; staminal
column 1 lin. long; anther appendages broadly rounded, with their
obtuse tips inflexed over the margin of the truncate style-apex.

EASTERN REGION: Delagoa Bay, *Junod,* 184! 484!

63. **S. periglossoides** (Schlechter in Engl. Jahrb. xx. Beibl. 51,
20); tuber elongated, radish-like, producing 1 or 2 simple stems
9–20 in. high, puberulous along 2 lines or glabrous below, with
internodes 1$\frac{1}{2}$–2$\frac{1}{4}$ in. long; leaves erect, 2$\frac{1}{3}$–4$\frac{1}{2}$ in. long, $\frac{1}{2}$–1$\frac{1}{2}$ lin.
broad, linear, acute, subsessile or shortly petiolate; umbels sessile,
subglobose, 10–12-flowered, lateral at the nodes; pedicels 1$\frac{1}{2}$–3 lin.
long, dusky-puberulous; sepals $\frac{3}{4}$–1$\frac{1}{4}$ lin. long, $\frac{1}{2}$ lin. broad, ovate or
ovate-lanceolate, acute, thinly puberulous; corolla-lobes suberect or
campanulately spreading, 2 lin. long, $\frac{3}{4}$–1 lin. broad, oblong or
rather narrowly oblong-ovate, concave at the lower $\frac{2}{3}$, narrowly
revolute along the margins, recurved at the subacute tips, thinly

puberulous on the back, minutely papillate-velvety on the margins and upper part of the inner face, glabrous in the concavity, green on the back, sometimes stained with purple-brown, whitish on the inner face; corona-lobes shorter than the staminal column, $\frac{2}{3}$–1 lin. long, $\frac{1}{3}$–$\frac{1}{2}$ lin. broad, erect, rather thick and fleshy, ovate, ovate-lanceolate or oblong-lanceolate, obtusely pointed, with a ∧-like thickening below the apex or slightly gibbous on the back, flat on the inner face, with a short fleshy erect truncate or broadly-rounded transverse appendage near the base and scarcely reaching to the middle; staminal column $1\frac{1}{4}$–$1\frac{1}{3}$ lin. long; anther-appendages rhomboid-ovate, acute, with their tips inflexed over the rim of the shallow crater-like style-apex; anther-wings angular at the middle; follicles (not quite mature in the specimens seen) solitary or in pairs, erect from the recurved pedicel, $2\frac{1}{2}$–3 in. long, $\frac{1}{4}$ in. thick, fusiform, tapering into a rather long beak, with about 6 longitudinal series of ascending subulate processes $\frac{3}{4}$–$1\frac{1}{2}$ lin. long and varying from fuscous-puberulous to nearly glabrous. *Schlechter in Journ. Bot.* 1896, 421.

KALAHARI REGION: Transvaal; Jeppestown Ridges, near Johannesburg, 6000 ft., *Gilfillan in Herb. Galpin,* 6231! Mundts Farm, near Pretoria, *Schlechter,* 4142! near Ermelo, south of Blesbok Spruit, *Burtt Davy,* 955! near Klein Olifants River, 5300 ft., *Schlechter,* 4027; Modderfontein, *Conrath,* 983! Beginsel Farm, near Standerton, *Burtt Davy,* 3156!

Dr. Schlechter has described the corona-lobes as "suborbicular, obtuse, furnished on the inner face at the base with a subquadrate fleshy scale, truncate at the apex and scarcely equalling the middle of the lobe," and "above the middle with a short ovate obtuse erect ligule a little overtopping the lobe." But in the specimens collected and named by himself, which I have examined, I find them as above described. This plant so closely resembles *Krebsia stenoglossa* and *Xysmalobium carinatum* in the dried state, as to be easily mistaken for either, until dissected.

64. **S. glabrescens** (Schlechter in Engl. Jahrb. xx. Beibl. 51, 17); stem solitary, $1\frac{1}{4}$–2 ft. high, slender, simple, with 15–18 internodes, varying from nearly glabrous to puberulous with very minute adpressed white or black hairs; leaves erect, $\frac{1}{2}$–$1\frac{1}{3}$ in. long, $\frac{1}{5}$–$\frac{2}{5}$ lin. broad, linear or filiform, acute, glabrous or the upper with a few minute blackish hairs; umbels several, sessile at the nodes, 3–8-flowered; pedicels $2\frac{1}{2}$–3 lin. long, very minutely adpressed-puberulous, often with blackish hairs; sepals $\frac{3}{4}$–$1\frac{1}{4}$ lin. long, $\frac{1}{2}$–$\frac{2}{3}$ lin. broad, ovate-lanceolate, acute to subulate-acuminate, puberulous like the pedicels; corolla-lobes rotately spreading, $1\frac{1}{4}$–$1\frac{1}{2}$ lin. long, about $\frac{3}{4}$ lin. broad, oblong or lanceolate-oblong, often minutely notched at the subacute apex, usually with a few very minute hairs on the back, very minutely and rather thinly puberulous on the basal half or all over the inner face or glabrous on both sides, minutely ciliate on one margin; corona-lobes arising a little above the base of the staminal column, $\frac{1}{2}$–$\frac{2}{3}$ lin. long and about as broad, broadly ovate or somewhat rhomboid-ovate or rhomboid-orbicular, notched and often recurved at the apex, rather abruptly contracted at the base, having a slight notch on each margin at about the middle, with 2 keels and an appendage on the inner face, a small tooth or transverse ridge usually

connects the marginal notches and keels; appendage arising near the top of the lobe, ½–1 lin. long, subulate, closely incumbent on the top of or loosely incurved over the staminal column and style-apex, often crossing at the tips, not erect, usually curved; anther appendages broadly ovate, obtuse or subacute, inflexed on the top of the truncate or umbonate style-apex; follicles 3–3¼ in. long, 2½ lin. thick at the lower part, narrowly fusiform, tapering into a long slender beak, puberulous. *Schlechter in Journ. Bot.* 1896, 421. *S. tenuissimum, Schlechter in Engl. Jahrb.* xx. *Beibl.* 51, 23, *and Journ. Bot.* 1896, 421.

VAR. β, **longirostre** (N. E. Br.); appendage of the corona-lobes erect or connivent at the base and then erect, or S-curved or somewhat tortuous at the tips, much exceeding the staminal column. *S. longirostre, Schlechter in Engl. Jahrb.* xx. *Beibl.* 51, 17, *and in Journ. Bot.* 1896, 421.

KALAHARI REGION : Transvaal; near Middelburg, 4900 ft., *Schlechter*, 4051; Jeppestown Ridges near Johannesburg, 6000 ft., *Gilfillan in Herb. Galpin*, 1480! 6227 ! Elandspruit Mountains, *Schlechter*, 3996 ! Var. β : near Botsabelo, *Schlechter*, 4074! Modderfontein, *Conrath*, 986 ! Meintjes Kop, near Pretoria, *Burtt Davy*, 4005 ! 4006 !

EASTERN REGION : Var. β : Natal; Shafton, Howick, *Mrs. Hutton*, 466!

The specimens above quoted have been described by Dr. Schlechter as 3 distinct species, differing in the pubescence on the inner face of the corona-lobes. *S. glabrescens* is described as having them puberulous all over, and is so represented in Dr. Schlechter's own drawings of his analyses of the flower, *S. tenuissimum* as puberulous on the lower half only, and *S. longirostre* as glabrous. This pubescence is always very minute, and is certainly very variable, as in different specimens of the same gathering made by Dr. Schlechter and others I find 2 or all 3 forms of it to exist. With the exception that the appendages of the corona-lobes or their tips are erect, I can find no specific distinction whatever between *S. longirostre* and the others, and the corona-lobes are not ovate-lanceolate nor lanceolate as originally described.

65. S. lamellatum (Schlechter in Verhandl. Bot. Ver. Brandenb. xxxv. 48) ; stem 7½–16 in. high, simple or slightly branched, minutely pubescent, with internodes ⅔–2 in. long; leaves erect or ascending, subsessile, ½–1¼ in. long, ¼–1 lin. broad, linear, acute, with revolute margins, nearly glabrous or very minutely scabrous; umbels 3–5, sessile at the nodes, 2–5-flowered ; pedicels ⅔–1½ lin. long, puberulous ; sepals ¾ lin. long, ½ lin. broad, ovate, acute ; corolla-lobes 1–1¼ lin. long, ⅔ lin. broad, oblong-ovate, subacute, campanulately spreading or suberect, sometimes apparently with incurved tips, glabrous or with very few hairs on the back, the inner face as seen under a strong lens is covered at the basal or apical half with exceedingly minute papilla-like hairs, but appears glabrous under a weak lens, apparently greenish, purple-tinted on the back, " brownish " (*Rand*); corona-lobes erect, overtopping the staminal column, about 1 lin. long, ⅔–¾ lin. broad, closely contiguous below, with slightly recurved margins, oblong or elliptic-oblong, usually narrowed into an acute, obtuse or minutely bifid point, but sometimes abruptly contracted into the point or 3-toothed at the top, the lateral teeth minute, obtuse, the middle one not more than ¼ lin. long, deltoid or narrow, with 2 thin wing-like keels on the inner face, slightly converging

towards the base and somewhat infolded, but no appendage ; staminal
column about ¾ lin. long ; anther appendages suborbicular, inflexed
over the rim of the depressed-truncate style-apex ; young follicles
swollen at the base, tapering into a long beak, minutely and densely
puberulous with white and dusky hairs. *Schlechter in Journ. Bot.*
1896, 419. *S. bilamellatum, Schlechter in Engl. Jahrb.* xx. *Beibl.*
51, 15, *incl. var. cordylogynoides, and Journ. Bot.* 1896, 421.
S. propinquum, S. Moore in Journ. Bot. 1903, 311, *and (Rand)* 338.

COAST REGION : Cape Div. ; on sandy reefs by the sea-shore near Muizenberg, *Schlechter,* 605 !
CENTRAL REGION : Wodehouse Div. ; Indwe, *Sim,* 2359 !
KALAHARI REGION: Transvaal ; Magaliesberg Range, *Burke* ! Bezuidenhout's
Valley, near Johannesburg, 5800 ft., *Gilfillan in Herb. Galpin,* 6044 ! hills
near Waterval River, 4600 ft., *Schlechter,* 3478 ! near Rustenburg, *Miss Pegler,*
25 ! Skinners Court, Pretoria, *Burtt Davy,* 1169 ! 1171 ! 1993 ! around Johannesburg, *Rand,* 860 ! 1125 ! Pilgrims Rest, *Greenstock !*
EASTERN REGION : Natal ; near Colenso, *Schlechter,* 3375.

Through the courtesy of Dr. Bolus I have been able to examine a flower from
the type of *S. lamellatum,* and find it to be identical with *S. bilamellatum,*
the differences mentioned by Dr. Schlechter being non-existent. As in the case of
S. restioides, Schlechter, and *S. pedunculatum,* Schlechter (=*Asclepias aurea,*
Schlechter), I believe it to have been introduced at the Cape locality among
ballast, &c. *S. bilamellatum,* var. *cordylogynoides* is merely a form with the
umbels crowded together into a head.

66. **S Bolusii** (Schlechter in Abhandl. Bot. Ver. Brandenb. xxxv.
48) ; tuber elongated ; stem usually solitary, 1-2⅓ ft. high, simple or
with 1 erect branch, slender, puberulous, with internodes ¾-3½ in.
long ; leaves erect, 1-2¾ in. long, ¼-½ lin. broad, linear-filiform, with
revolute margins, subobtuse or acute, glabrous ; umbels all sessile or
1 or more of the lower pedunculate, 4-9-flowered ; peduncles 0-9 lin.
long, puberulous ; bracts minute, deciduous ; pedicels 2-3 lin. long,
puberulous ; sepals ½-¾ lin. long, ¼-⅓ lin. broad, ovate-lanceolate,
acute, puberulous ; corolla-lobes spreading, 1-1¼ lin. long, ½-⅔ lin.
broad, oblong, minutely notched at the subobtuse incurved (always ?)
apex, sparsely pubescent on the back, glabrous on the inner face ;
corona-lobes ¼-½ lin. long and rather more in breadth, subquadrate,
truncate at the top forming a transverse rim behind the base of the
appendage, which is ⅛-⅓ lin. long, erect or ascending, linear-deltoid,
oblong or linear, acute, obtuse or emarginate, and decurrent in 2 keels
on the inner face, the keels often forming a slight angle or tooth at
the top, often obliterated in dried specimens ; staminal column ⅓-½
lin. long ; anther-appendages broadly ovate, obtuse or subacute,
inflexed upon the top of the truncate style-apex ; young follicles
fusiform, tapering into a long beak, minutely puberulous. *Schlechter
in Journ. Bot.* 1896, 419. *S. Guthriei, Schlechter, and S. lunatum,
Schlechter in Verhandl. Bot. Ver. Brandenb.* xxxv. 49, *and Journ.
Bot.* 1896, 419. *Cynanchum filiforme, Burch. Trav.* i. 37, *not of
Linn. f. Rhinolobium tenue, Arnott in Mag. Zool. and Bot.* ii. 421.
R. lineare, Decne in DC. Prodr. viii. 556, *under Lagarinthus
gracilis. Lagarinthus microdon, Turcz. in Bull. Soc. Nat. Mosc.*
1852, ii. 317.

South Africa: without locality, *Harvey,* 629 ex *Arnott.*

Coast Region: Cape Div. ; Wynberg, *Harvey !* Wynberg Hill, *Guthrie,* 316 ! between Newlands and Paradise, *Burchell,* 432 ! near Tokay, *Fair in Herb. Bolus !* Caledon Div. ; Baviaans Kloof, near Genadendal, 850 ft., *Bolus,* 5397 ! near Villiersdorp, 1500 ft., *Schlechter,* 9368 ! Swellendam Div. ; mountains along the lower part of the River Zondereinde, *Zeyher,* 3402 !

I have examined flowers from the types of all the names quoted, and can find no difference between them. Usually all the umbels are sessile, but occasionally 1 or more (as in the type specimen of *S. Bolusii*) of the lower umbels are pedunculate and the rest sessile. Some other species vary in the same way. I can find nothing of the nature of a "longitudinal linear gland" on the back of the corona-lobes of the type of *S. lunatum,* nor does such a structure occur in any species known to me. The specimen in Harvey's Herbarium named "*Rhinolobium tenue*" is that quoted from Wynberg, but bears no number ; it is identical with the type of *S. Guthriei.*

67. S. monticola (Schlechter in Engl. Jahrb. xxxviii. 27) ; stem 1½–2 ft. high, simple, puberulous on the upper part, otherwise glabrous, with internodes mostly 2–3 in. long ; leaves sessile, erect, 1–2 in. long, ⅓–½ lin. broad, linear-filiform, acute, with revolute margins, glabrous ; umbels 8–10, sessile at the nodes, few-flowered ; pedicels 1–1½ lin. long, puberulous ; sepals ¾ lin. long, ⅓ lin. broad, lanceolate, acute, puberulous ; corolla-lobes rather more than 1 lin. long, ½ lin. broad, lanceolate-oblong, subobtuse, slightly incurved at the tips, thinly adpressed-pubescent on the back, glabrous on the inner face ; corona-lobes subcontiguous, with their points incurved and crossing one another over the top of the staminal column, ¾ lin. long, ⅓ lin. broad ; basal half subquadrate, with rectangular shoulders, very abruptly contracted into a subulate point, inner face with 2 parallel keels, but no appendage ; staminal column ½ lin. long ; anther-appendages suborbicular, inflexed over the raised rim of the truncate style-apex ; young follicles fusiform, tapering into a long beak, puberulous.

Central Region: Ceres Div. ; on the Cold Bokkeveld at Klyn Vlei, 4000 ft., *Schlechter,* 10063 !

68. S. carinatum (Schlechter in Engl. Jahrb. xviii. Beibl. 45, 3) ; tuber elongated, radish-shaped ; stem solitary, simple, 7–18 in. high, pubescent all round, with internodes ¾–1¾ in. long ; leaves erect, subsessile or very shortly petiolate, ¾–2½ in. long, ⅓–2 lin. broad, linear, acute, with revolute margins, glabrous or thinly puberulous above and on the midrib beneath ; umbels 4–8, sessile at the nodes, 4–10-flowered, dense ; pedicels ½–1½ lin. long, adpressed puberulous ; sepals 1–1¼ lin. long, ½–⅗ lin. broad, ovate-lanceolate, acute, pubescent ; corolla-lobes rotately spreading, 1¼–1⅓ lin. long, ⅔ lin. broad, oblong or oblong-ovate, acute or obtuse, pubescent on the back, puberulous or shortly villous-pubescent on the inner face, apparently dark purple-brown ; corona-lobes very variable, equalling or overtopping the staminal column, erect, purple-brown, their basal part ⅖–½ lin. long and broad, subquadrate or rounded-oblong, with 2 keels and an appendage on the inner face, slightly keeled down the back, rather abruptly rounded or contracted into an acute deltoid subulate or filiform point $\frac{1}{5}$–$\frac{3}{4}$ lin. long, or simply acute or rarely obtuse without a point; appendage $\frac{1}{6}$–$\frac{1}{3}$ lin. long, varying from shorter than to much longer than the point of the lobe, subulate and acute or linear and obtuse or bifid, or reduced to a minute bifid process scarcely longer than broad, more or less incurved over the $\frac{1}{2}$–$\frac{3}{5}$ lin. long staminal column, or, when very short, erect; anther appendages orbicular, inflexed on the top of the truncate style-apex, which has a large slightly raised central boss with an impressed line across it; follicles $2\frac{1}{2}$ in. or more long, with a fusiform swelling below the middle, tapering into a very long beak, puberulous. *Schlechter in Engl. Jahrb.* xx. *Beibl.* 51, 16, *and Journ. Bot.* 1896, 420.

COAST REGION: Albany Div.; grassy slopes and Broekhuisens Poort near Grahamstown, 1800 ft., *MacOwan,* 639! Howisons Poort near Grahamstown, *Hutton!* in damp situations by New Year River, *Mrs. Barber,* 116! Komgha Div.; hill near Keimouth, *Flanagan,* 1043!

CENTRAL REGION: Somerset Div.: on the summit of Bosch Berg, rare, *MacOwan,* 639!

EASTERN REGION: Transkei; Kreilis Country, *Bowker!* Natal: Attercliff, 800 ft., *Sanderson,* 449 partly! Inanda, *Wood,* 1404! near Pinetown, *Schlechter,* 3165! Clairmont, *Wood,* 8268! *Schlechter,* 3085! Zululand, *Gerrard,* 1313!

The comparative length of the corona-lobes and staminal column, and the form of their appendages is very variable.

69. **S. tricuspidatum** (Schlechter in Journ. Bot. 1895, 267); stem erect, 6–8 in. high, simple, thinly velvety, glabrescent at the base, densely leafy; leaves erect, about $\frac{1}{2}$–1 in. long, narrowly linear, acute, with revolute margins, narrowed into a very short puberulous petiole at the base, glabrescent above, puberulous on the midrib beneath; umbels sessile, lateral at the nodes, subcapitate, several-flowered; pedicels $1\frac{1}{2}$ lin. long, velvety-puberulous; sepals half as long as the corolla, lanceolate, acute, puberulous; corolla-lobes 1 lin. long, $\frac{1}{2}$ lin. broad, ovate-oblong, obtuse, puberulous on both sides; corona-lobes erect, about twice as long as the staminal column, ovate-oblong, tapering into an erect linear obtuse point, with a ligulate tricuspidate appendage on the inner face, with its lateral teeth triangular, acute, and the middle tooth twice as large; anther-appendages transverse, very obtuse, inflexed on the style-apex. *Schlechter in Journ. Bot.* 1896, 449.

EASTERN REGION? Caffraria?, *Mrs. Barber,* in Grahamstown Herbarium.

Is this distinct from the variable *S. carinatum?*

70. **S. commixtum** (N. E. Br.); stem 1–1½ ft. high, simple, slender, bifariously puberulous; leaves erect, $\frac{3}{4}$–1$\frac{1}{3}$ in. long or the uppermost smaller, $\frac{1}{3}$–$\frac{1}{2}$ lin. broad, linear, with revolute margins, acute, glabrous; umbels 3–8, sessile at the nodes, 3–6-flowered; pedicels 3–4 lin. long, puberulous; sepals $\frac{3}{4}$–1$\frac{1}{4}$ lin. long, $\frac{1}{3}$ lin. broad, lanceolate, acute, pubescent; corolla-lobes apparently ascending, with abruptly incurved tips, 1$\frac{1}{3}$–1$\frac{1}{2}$ lin. long, nearly 1 lin. broad, oblong,

obtusely pointed, slightly concave at the lower part, with recurved margins at the tips, slightly pubescent on the back, glabrous or minutely ᵖuberulous at the base only on the inner face ; corona-lobes arising close to the base of the staminal column, the body of the lobe scarcely ½ lin. long, ⅖–½ lin. broad, somewhat rhomboid-ovate, with a distinct tooth or projection on each side at the middle, whence it deltoidly tapers into a filiform point or apical appendage ½ lin. long abruptly inflexed over the top of the staminal column, with 2 keels decurrent from the point on the inner face, besides a submarginal keel on each side descending from the lateral teeth ; staminal column ½ lin. long ; anther-appendages elliptic or suborbicular, inflexed upon and covering the truncate style-apex ; anther-wings very prominent, angular at their middle.

EASTERN REGION : Natal ; Shafton, Howick, *Mrs. Hutton,* 206 partly !

Mingled with this plant was one specimen of *S. Woodii,* Schlechter.

71. S. parile (N. E. Br.) ; stem 8–15 in. high, simple, pubescent, with 5–6 pairs of leaves below the lowest flowering node ; leaves erect, ½–1¼ in. long, ¼–¾ lin. broad, linear, acute, nearly or quite glabrous, all shorter than the internodes and those at the flowering nodes rather inconspicuous and scarcely exceeding the flowers ; umbels 3–5, sessile at the nodes, sometimes closely placed, 4–7-flowered ; pedicels 2–3 lin. long, rather densely villous-pubescent ; sepals ¾–1 lin. long, lanceolate, acute, pubescent ; corolla-lobes spreading, with incurved tips, 1½ lin. long, nearly 1 lin. broad, ovate or elliptic-ovate, subacute, thinly pubescent on the back, glabrous on the inner face, apparently striate with brownish on a paler ground ; corona-lobes ½ lin. long and ⅔ lin. broad, closely contiguous and very broadly or transversely rhomboid at the basal part, abruptly narrowed into a filiform-subulate point ½ lin. long inarching in a bold curve much above the style-apex and decurrent at its base to the middle of the inner face of the lobe in 2 stout contiguous keels, with a small inflexed wing-like auricle at each margin below the angular shoulders, apparently white ; staminal column rather short and comparatively broad, scarcely ½ lin. long.

EASTERN REGION : Natal ; Shafton, Howick, *Mrs. Hutton,* 40 ! 405 !

72. S. Buchanani (N. E. Br.) ; tuber about ½ in. long, ovoid ; stem solitary, 6–10 in. high, slender with 4–6 leafy nodes below the lowest flowering node, minutely pubescent ; internodes ¾–1½ in. long ; leaves erect, shorter than the internodes, subsessile, ½–1 in. long, ⅓–½ lin. broad, linear-filiform, acute, pubescent along the midrib beneath and sometimes with a few hairs scattered along the revolute margins, otherwise glabrous ; umbels 3–6, sessile at the nodes, 2–3- (or probably more) flowered ; pedicels 1½–3 lin. long, adpressed-puberulous ; sepals ¾–1 lin. long, ½ lin. broad, ovate-lanceolate, acute, adpressed-pubescent ; corolla-lobes spreading, 1⅓–1½ lin. long, ¾ lin. broad, oblong, minutely notched at the subacute apex, thinly adpressed-pubescent on the back, glabrous on the inner face, blackish-purple or dark purple-brown, paler at the basal part? one margin usually slightly ciliolate; corona-lobes with a rhomboid body $\frac{2}{5}$–$\frac{1}{2}$ lin. long, angular at the middle, having 2 keels (often obliterated in dried flowers) on the inner face and a subulate or linear point or appendage $\frac{1}{4}$–$\frac{1}{2}$ lin. long, inflexed over the staminal column, with or without a distinct rim or slightly produced obtuse apex of the lobe behind its base; staminal column $\frac{2}{3}$–$\frac{3}{4}$ lin. long; anther-appendages broadly ovate, erect, with their tips inflexed over the rim of the crater-like style-apex; follicle solitary, $1\frac{1}{3}$ in. long in the unripe examples seen, narrowly fusiform, tapering into a long beak, puberulous.

EASTERN REGION : Natal, *Buchanan!*

Very similar to *S. parcum,* N.E. Br., but the flowers are rather larger, entirely different in colour, and the corona-lobes are different in form.

73. S. parvulum (Schlechter in Journ. Bot. 1894, 354); stem 6–10 in. high, simple or (from injury) branched, $\frac{1}{2}$–1 lin. thick at the base, puberulous, with 4–6 leafy nodes below the lowest flowering node; leaves erect or ascending $\frac{3}{4}$–2 in. long or the lowest shorter, $\frac{1}{3}$–$1\frac{1}{2}$ lin. broad, linear-filiform or linear, obtuse or acute, puberulous or minutely scaberulous; umbels 3 or more, pedunculate, racemosely arranged, 3–6-flowered; peduncles 3–7 lin. long, puberulous or pubescent; bracts about 1 lin. long, filiform; pedicels 2–3 lin. long, pubescent; sepals $\frac{2}{3}$–$\frac{3}{4}$ lin. long, lanceolate, acute, pubescent; corolla-lobes slightly spreading, $1\frac{1}{3}$ lin. long, $\frac{2}{3}$ lin. broad, lanceolate-oblong, subobtuse, pubescent on the back, glabrous on the inner face, brownish (*Galpin*); corona-lobes about 1 lin. long, including the point, $\frac{2}{5}$ lin. broad at the oblong-ovate or somewhat rhomboid-ovate basal part, which has 2 minute teeth or a slight notch at the middle of each side, sometimes obliterated in drying, with 2 keels excurrent in the filiform point, which arises at the apex of the inner face or into which the lobe tapers and is abruptly inflexed over the $\frac{1}{2}$ lin.-long staminal column; anther-appendages broadly ovate or suborbicular, subacute or obtuse, closely inflexed on the depressed-truncate style-apex and concealing it; caudicles subapical or lateral near the apex of the rather long and slender pollen-masses. *Schlechter in Journ. Bot.* 1896, 421.

Var. β, **sessile** (N. E. Br.); umbels 3–6 to a stem or branch, all sessile or the lowest with a peduncle 1–2 lin. long; corolla-lobes $1\frac{1}{3}$–$1\frac{3}{4}$ lin. long, $\frac{2}{3}$–$\frac{3}{4}$ lin. broad; corona-lobes with the basal part $\frac{1}{2}$–$\frac{3}{5}$ lin. long and the inflexed point or appendage $\frac{2}{3}$–1 lin. long, usually arising just below the apex of the inner face of the lobe, which then forms a slight rim behind it; staminal column $\frac{2}{3}$–$\frac{3}{4}$ lin. long.

COAST REGION: Queenstown Div.; Finchams Nek, near Queenstown, 3800–4000 ft., *Galpin,* 1600! Var. β: Bathurst Div.; south of Blaauw Krantz, *Burchell,* 3887! Trapps Valley, *Miss Daly,* 628!

EASTERN REGION : Transkei; Kreilis Country, *Bowker,* 301! Tembuland; hills near Bazeia, 2000–2500 ft., *Baur,* 344!

74. S. parcum (N. E. Br.); stem 10–12 in. high, simple, with 3–5 leaf-bearing nodes below the flowering part; internodes 1–2¾ in. long, softly pubescent all round from base to apex; leaves erect, half as long as the internodes or less, subsessile, ½–1½ in. long, ⅓–1 lin. broad, linear, acute, with revolute margins, minutely scabrous or nearly glabrous; umbels 4–6, sessile at the nodes, 4–6-flowered; pedicels 1–2½ lin. long, adpressed-pubescent; sepals ⅔–¾ lin. long, ¼–⅓ lin. broad, ovate-lanceolate, acute, pubescent; corolla lobes spreading, 1¼–1½ lin. long, ½–¾ lin. broad, oblong, obtusely pointed, thinly pubescent on the back, glabrous on the inner face, apparently greenish, tinged with dull purple on the back; corona-lobes subquadrate, with an appendage having 2 minute teeth at its base and decurrent below as 2 keels on the inner face; subquadrate part ¼–⅓ lin. long, ¼–⅖ lin. broad, rounded or truncate at the top, appendage arising near the apex of the inner face, ⅖–½ lin. long, linear, obtuse or truncate, with the apex incurved over or upon the top of the ½ lin.-long staminal column; anther-appendages broadly ovate, subacute, inflexed on the subtruncate style-apex. *S. filifolium, Schlechter in Engl. Jahrb.* xx. *Beibl.* 51, 16, *not elsewhere.*

EASTERN REGION: Natal; Inanda, *Wood*, 287! near Pinetown, *Schlechter*, 3166! and without precise locality, *Mrs. K. Saunders!*

Similar in structure to *S. filifolium*, with which it was united by Dr. Schlechter, but the stems are rather stouter with a very different and more spreading pubescence and constantly fewer and very much longer internodes, most of them being 1½–2 in. long, giving the plant a different aspect, and in conjunction with the longer appendage to the corona-lobes and different geographic range are sufficient to discriminate it.

75. S. addoense (N. E. Br.); stem 10–15 in. high, branching at the base, ½–1 lin. thick; branches erect, subparallel, puberulous bifariously or nearly all round, with 2–4 leafy nodes below the lowest flowering node; internodes ¾–1½ in. long; leaves erect, ¾–2 in. long, ⅓–½ lin. broad, linear or linear-filiform, acute, with revolute margins, glabrous; umbels numerous, sessile at the nodes, 6–9-flowered; pedicels about 2 lin. long, with a short spreading pubescence; sepals ⅔–¾ lin. long, ¼–⅓ lin. broad, ovate-lanceolate, acute, pubescent; corolla-lobes ¾–1 lin. long, ½ lin. broad, apparently ascending-spreading, oblong or ovate-oblong, acute, with reflexed margins and incurved tips, thinly pubescent on the back, glabrous on the inner face; corona-lobes about ⅖ lin. long and broad, contiguous, very shortly connate at the base, quadrate, obtusely rounded at the top forming a slight rim (which is sometimes indistinguishable in dried specimens) behind the base of the almost terminal appendage, with a minute tooth near or at the shoulders, and 2 short parallel tooth-like keels on the inner face, often very obscure in dried specimens; appendage ¼ lin. long, deltoid-linear, acute or obtuse, inflexed on the backs of the anther-appendages and not exceeding them; staminal column ⅖ lin. long; anther-appendages suborbicular, inflexed on the top of the truncate style apex; follicles 2–2¼ in. long, ¼ in. thick, fusiform, tapering into a long beak, smooth, puberulous.

CoAST REGION: Uitenhage Div. ; Addo, *Bowie! Drège*, 2228!

This is the Addo plant enumerated by E. Meyer under *Lagarinthus gracilis* in *Comm. Pl. Afr. Austr.* 206, but is totally different from the plant he described under that name. It has the peculiarity of producing flowers nearly all along the branches.

76. S. Burchellii (N. E. Br.); stem 10–15 in. high, simple, $\frac{1}{3}$–$\frac{1}{2}$ lin. thick, bifariously puberulous, with about 4 pairs of leaves below the lowest flowering node; leaves erect or ascending, $1\frac{1}{4}$–2 in. long, $\frac{1}{3}$ lin. broad, shorter than or the lower exceeding the $\frac{3}{4}$–$2\frac{3}{4}$ in.-long internodes, linear-filiform, acute, with closely revolute margins, glabrous; umbels 5–7, sessile, lateral at the nodes, 7–14-flowered; pedicels 2–$2\frac{1}{2}$ lin. long, adpressed pubescent; sepals $\frac{3}{4}$ lin. long, lanceolate, very acute, adpressed-pubescent; corolla-lobes apparently ascending with incurved tips, 1–$1\frac{1}{4}$ lin. long, $\frac{1}{2}$ lin. broad, somewhat elliptic-oblong, acute or subacuminate, thinly adpressed-pubescent on the back, glabrous on the inner face, apparently dull greenish; corona-lobes subquadrate, contiguous, 3-toothed, with 2 minute teeth on the inner face, the quadrate part $\frac{1}{2}$ lin. long, $\frac{2}{3}$ lin. broad, perhaps forming a slight rim behind the middle tooth, which is less than $\frac{1}{4}$ lin. long and inflexed on the $\frac{1}{2}$ lin.-long staminal column; anther-appendages orbicular, inflexed on the top of and covering the style-apex.

CoAST REGION : Riversdale Div. ; between Vet River and Krombeks River, *Burchell*, 7181 !

Closely allied to *S. filifolium*, but less leafy, the stem and flowers dry brownish (not pale greenish), and the corona-lobes have the appendage arising at the very apex of the inner face, appearing continuous with the lobe or with a scarcely perceptible rim behind it. Burchell only collected 2 specimens.

77. S. Bowkeræ (N. E. Br.); stem solitary, simple or with 1 branch, very leafy, $4\frac{1}{2}$–9 in. high, $\frac{1}{2}$–$\frac{3}{4}$ lin. thick at the base, puberulous all round in all parts, with internodes $\frac{1}{4}$–1 in. long, and 6–8 pairs of leaves below the flowering part ; leaves ascending, sub-sessile, 1–2 in. long, $\frac{1}{3}$–2 lin. broad, 2–4 times as long as the inter-nodes, linear, acute, with revolute margins, puberulous on the midrib beneath, otherwise glabrous; umbels 4–10, sessile at the nodes, 9–16-flowered ; pedicels $1\frac{1}{2}$–$2\frac{1}{4}$ lin. long, adpressed pubescent ; sepals $\frac{2}{3}$–$\frac{3}{4}$ lin. long, $\frac{1}{3}$ lin. broad, ovate-lanceolate, acute, thinly adpressed-pubescent ; corolla-lobes apparently campanulately spreading, with incurved tips, 1 lin. long, $\frac{1}{2}$ lin. broad, oblong, subacute, glabrous on both sides or with a very few hairs on the back ; corona-lobes about $\frac{1}{3}$ lin. long and broad, subquadrate, obtuse or subtruncate, with a very minute tooth at the shoulders and 2 slight keels decurrent from the appendage on the inner face ; appendage inserted slightly below the apex of the lobe, which at least forms a transverse rim behind its base, $\frac{1}{6}$–$\frac{1}{5}$ lin. long, linear-deltoid, subacute, incumbent on the backs of the anthers or incurving over the $\frac{2}{5}$ lin.-long staminal column ; anther-appendages broadly cuneate-obovate, subtruncate or emarginate, inflexed on the top of the truncate style-apex.

COAST REGION : Albany Div. ; *Miss Bowker!* East London Div. ; on the river-bank near the Convict Station at East London, 25 ft., *Wood in Herb. Galpin*, 3125 !

78. S. filifolium (Schlechter in Engl. Jahrb. xviii. Beibl. 45, 4) ; stem apparently solitary, simple, 6–15 in. high, $\frac{1}{4}$–$\frac{1}{2}$ lin. thick, with 7–16 leafy nodes below the lowest flowering node, bifariously puberulous above, glabrous below ; leaves equalling to twice as long as the internodes, $\frac{1}{2}$–$1\frac{1}{4}$ in. long, $\frac{1}{4}$–$\frac{1}{2}$ lin. broad, linear or linear-filiform, with revolute margins, acute, glabrous or the uppermost with a minute scanty pubescence, erect or ascending ; umbels sessile at the nodes, 2–7-flowered ; pedicels 1–3 lin. long, puberulous ; sepals $\frac{3}{4}$–1 lin. long, ovate-lanceolate, acute, thinly adpressed pubescent ; corolla-lobes ascending-spreading, $\frac{3}{4}$–$1\frac{1}{4}$ lin. long, $\frac{1}{2}$–$\frac{2}{3}$ lin. broad, ovate-oblong, subacute, thinly adpressed-pubescent on the back, glabrous on the inner face ; corona-lobes about $\frac{1}{3}$ lin. long and broad, sub-quadrate, subtruncate or slightly rounded with a very minute tooth at each shoulder at the top and an appendage with 2 minute teeth at its base on the inner face ; appendage arising just below the apex of the lobe, which forms a rim behind it, $\frac{1}{5}$–$\frac{1}{3}$ lin. long, linear or deltoid-linear, pressed against the anthers and reaching to or directed over the top of the $\frac{1}{2}$ lin.-long staminal column ; anther-appendages suborbicular or broadly ovate, inflexed on the top of the depressed-truncate style-apex ; young follicles solitary, narrowly lanceolate-fusiform, tapering into a long acute beak, puberulous. *Schlechter in Journ. Bot.* 1896, 420, *not of Engl. Jahrb.* xx. *Beibl.* 51, 16.

COAST REGION : Komgha Div. ; hill near Keimouth, 200 ft,, *Flanagan*, 383 ! King Williamstown Div. ; on a flat at the foot of Perie Forest, *Flanagan*, 2169 ! Stockenstrom Div. ; Kat Berg, *Hutton* !

EASTERN REGION : Kentani Div. ; near Kentani, 1000 ft., *Miss Pegler*, 660 ! Willowvale Div. : Kreilis Country, *Bowker*, 213 !

One specimen of *Flanagan*, 2169, had corolla-lobes minutely and very sparsely puberulous within, and the corona-lobes larger than usual, $\frac{1}{2}$ in. long, $\frac{2}{3}$ lin. broad, subquadrate-rhomboid, emarginate at the obtuse apex, without teeth at the lateral angles or on the inner face, and the base of the appendage passing into 2 widely diverging keels, otherwise the specimen was identical with the usual form, with which it was found growing.

79. S. exile (Schlechter in Engl. Jahrb. xxi. Beibl. 54, 4) ; stem 10–13 in. high, very slender, $\frac{1}{3}$ lin. thick or less at the base, bifariously puberulous, with internodes $\frac{3}{4}$–2 in. long ; leaves erect, 8–10 lin. long, $\frac{1}{4}$–$\frac{1}{2}$ lin. broad, linear, acute, with revolute margins, glabrous ; umbels sessile at 4–7 of the upper nodes, 2–3-flowered ; pedicels $1\frac{1}{2}$–$1\frac{3}{4}$ lin. long, puberulous; sepals nearly 1 lin. long, $\frac{1}{3}$ lin. broad, lanceolate, acute, adpressed pubescent ; corolla-lobes suberect or campanulately spreading, rather more than 1 lin. long, $\frac{2}{5}$–$\frac{1}{2}$ lin. broad, lanceolate, acute, thinly adpressed-pubescent on the back, glabrous on the inner face ; corona-lobes with a subquadrate body $\frac{1}{4}$ lin. long and broad, having obtuse subtooth-like shoulders and produced at the apex of the 2-keeled inner face in front of a rim-like middle crenation into a slender erect linear slightly spathulate-linear or filiform point $\frac{1}{2}$ lin. long, usually having an inflexed tooth on one or both sides below the apex; staminal column $\frac{1}{3}$ lin. high; anther-appendages broadly ovate, obtuse, with the tips inflexed over the margin of the depressed style-apex. *Schlechter in Journ. Bot.* 1896, 450. *Lagarinthus gracilis, E. Meyer, Comm.* 206, *as to description and the specimens " d.";* Decne *in DC. Prodr.* viii: 556. *L. exilis,* Decne *l.c.* 556. *Gomphocarpus gracilis, Dietr. Syn. Pl.* ii. 901.

EASTERN REGION: Tembuland; between Bashee River and Umtata River, *Drège,* 4977 ! Natal; between Umzimkulu River and Umkomaas River, *Drège,* 4978!

The type of *Lagarinthus gracilis* in E. Meyer's Herbarium is *Drège,* 4978, and with it are placed 3 other specimens (2 of them belonging to different species), doubtless those quoted by E. Meyer under that species, although not so named by him, one (*S. addoense*) having no name, the other two, although bearing the same (unpublished) name, belong respectively to *S. exile* and *S. Dregei.* By the kindness of Mons. Casimir De Candolle a photograph has been sent to Kew of the type of *Lagarinthus exilis,* Decne, which demonstrates that it and *L. gracilis* are both founded upon the same plant, viz. *Drege,* 4978, distributed by Drège as "*Lagarinthus gracilis,* E. M., *d.*" A flower of *L. exilis* has also been compared with one from the type of *L. gracilis* by Mr. Buser and found to be identical, so that, although Decaisne has described the corona-lobes of the two as being very different, in all probability it was an imperfect flower of the specimen named *L. exilis* which he examined.

Lagarinthus gracilis, Meisn. in Hook. Lond. Journ. Bot. ii. 1843, 544 (by error 444), and Krauss in Flora, 1844, 826, collected by Krauss near the Knysna River in Knysna Div., is unknown to me. Meisner says of it—"Bad specimens, differing somewhat from Drège's plant in having the leaves longer than the internodia and glabrous."

80. **S. virgatum** (Schlechter in Engl. Jahrb. xviii. Beibl. 45, 6); stem $1\frac{1}{4}$–2 ft. high, usually simple, slender, puberulous, with numerous internodes $\frac{1}{4}$–1 in. long; leaves erect, $\frac{3}{4}$–$1\frac{2}{3}$ in. long, $\frac{1}{3}$–$1\frac{1}{4}$ lin. broad, linear, acute, quite glabrous or thinly and minutely puberulous on the revolute margins and midrib beneath; umbels 8 or more, sessile at the nodes, or the lower sometimes on peduncles $\frac{1}{2}$–$1\frac{1}{2}$ lin. long, 5–10-flowered; bracts $\frac{1}{2}$–1 lin. long, subulate; pedicels $\frac{3}{4}$–$1\frac{1}{4}$ lin. long, puberulous; sepals $\frac{2}{3}$–$\frac{3}{4}$ lin. long, $\frac{1}{3}$ lin. broad, ovate, acuminate, slightly puberulous: corolla very small, blackish-purple, glabrous outside, minutely papillate-puberulous on the lobes inside; lobes campanulately spreading, with the tips apparently incurved, $\frac{3}{4}$ lin. long, $\frac{1}{2}$ lin. broad, oblong-ovate, subacute; corona-lobes $\frac{1}{3}$–$\frac{2}{5}$ lin. long, dark-purple, subquadrate or somewhat elliptic, abruptly tapering into a short deltoid acute free point or mere apiculus behind and much shorter than the $\frac{1}{4}$ lin.-long deltoid-subulate appendage on the inner face, which is incurved over the $\frac{2}{5}$ lin.-long staminal column and decurrent at its base as 2 slight keels; anther-appendages orbicular, obtuse, inflexed over the margin of the depressed-truncate style-apex. *Schlechter in Engl. Jahrb.* xxi. *Beibl.* 54, 4, *and Journ. Bot.* 1896, 420. *Lagarinthus virgatus, E. Meyer, Comm.* 208; Decne *in DC. Prodr.* viii. 556. *Gomphocarpus virgatus, Dietr. Syn. Pl.* ii. 901.

COAST REGION: Komgha Div.; in a grassy valley near Komgha, 2000 ft., *Flanagan*, 1015! Ngqueleni Div.; in Mlengana Cutting, between St. John's and the Umgazi River, *Bolus*, 10199! between Umgazi River and Fakus Kraal, 1000–1500 ft. *Drège*, 4979!

Lagarinthus virgatus, var. *glabratus,* Meisn. in Hook. Lond. Journ. Bot. ii. 1843, 544 (by error 414); Krauss in Flora, 1844, 826, collected by Krauss, near Uitenhage, is unknown to me. There is no description beyond the following:—" Leaves glabrous, shorter than or scarcely exceeding the internodes."

81. S. loreum (S. Moore in Journ. Bot. 1903, 310); stem solitary, about 18 in. high, slender, about ¾ lin. thick at the base, bifariously puberulous or thinly sprinkled with very minute blackish or dusky hairs, with about 18 internodes 7–15 lin. long below the flowering part; leaves erect, 9–11 lin. long, ¼–½ lin. broad, linear or filiform, with revolute margins, acute, glabrous or the uppermost with a few very minute dusky adpressed hairs; umbels about 9, sessile, lateral at the nodes, 8–10-flowered; bracts 1–1¼ lin. long, setaceous, dusky; pedicels 2–3 lin. long, purple-brown, minutely puberulous with brown hairs; sepals ¾ lin. long, ½ lin. broad, ovate, acute, dark purple-brown, puberulous; corolla-lobes spreading, 1¼ lin. long, ⅞ lin. broad, ovate, minutely notched at the acute apex, puberulous on the inner face, with a few adpressed hairs down the middle of the back, blackish-purple; corona-lobes about 1½ lin. long, rather thick and fleshy, rhomboid, tapering above into a long filiform point much exceeding and loosely curving and crossing one another over the top of the staminal column, with 2 diverging keels but no appendage on the inner face, apparently white; staminal column about ½ lin. long; anther-appendages roundish-ovate obtuse, inflexed over the top of the truncate style-apex.

KALAHARI REGION: Transvaal; open veld to the northward of Johannesburg, only one specimen seen, *Rand*, 1122!

Allied to *S. virgatum,* of which it has the blackish-purple flowers, but differs in its longer pedicels and different corona-lobes.

82. S. Dregei (N. E. Brown); stem 3½ ft. high in the only specimen seen, 1 lin. thick at the base, simple, bifariously puberulous on the upper part, glabrous below, with the internodes below the flowering part 1¼–3¾ in. long; middle and lower leaves 1¼–2½ in. long, ⅓–½ lin. broad, the upper smaller, erect, linear, with revolute margins, acute, glabrous, umbels sessile at the nodes, 3–7-flowered; pedicels about 2 lin. long, pubescent; sepals ½–⅔ lin. long, ⅖ lin. broad, ovate, very acute, pubescent; corolla-lobes apparently somewhat spreading, with incurved tips, 1 lin. long, ½ lin. broad, oblong, minutely notched at the subacute apex, pubescent on the back, glabrous on the inner face; corona-lobes ½ lin. long (including the point-like appendage), ⅓ lin. broad at the base, shortly exceeding the staminal column, basal part subquadrate forming a slight but usually distinct rim or line at the base of the broadly deltoid-ovate acute or apiculate appendage, which is nearly as broad as the lobe, erect and continuous with it, apex recurved and nowhere inflexed, rounded or obscurely keeled on the back and decurrent as 2 keels on the inner face.

CoAST REGION : Paarl or Worcester Div. ; Dutoits Kloof, *Drège!*

The specimen upon which I found this very distinct species is the one in E. Meyer's Herbarium which he quotes from Dutoits Kloof under *Lagarinthus gracilis*, but it is certainly not the specimen from which he described that totally different species.   The corona-lobes are remarkable for the deceptive manner in which the very broad appendage appears to be the real point of the lobe.

83. **S. Woodii** (Schlechter in Engl. Jahrb. xx. Beibl. 51, 25 and 13 in note) ; stems $1\frac{1}{4}$–$2\frac{1}{2}$ ft. high, simple or branched above the middle, puberulous along 2 broad lines above, glabrous below, with 5–8 leafy distant nodes below the lowest flowering node ; leaves erect or ascending, subsessile, $1\frac{1}{4}$–3 in. long, $\frac{1}{4}$–$\frac{1}{2}$ lin. broad, linear or linear-filiform, acute, glabrous or the upper slightly puberulous ; umbels several, sessile at the nodes, 5–7-flowered ; pedicels 2–3 lin. long, pubescent ; sepals 1 lin. long, $\frac{1}{2}$ lin. broad, ovate, acute, pubescent ; corolla-lobes campanulately spreading, $1\frac{3}{4}$–2 lin. long, $\frac{3}{4}$–1 lin. broad, ovate-oblong or elliptic-oblong, with the margins recurved at the slightly incurved apex, pubescent on the back, ciliate, with a narrow border of hairs along one or both borders of the inner face ; corona-lobes $\frac{2}{3}$–$\frac{3}{4}$ lin. long, $\frac{1}{2}$–$\frac{3}{4}$ lin. broad, oblong, slightly obovate-oblong or subquadrate, obtusely rounded at the top, some-times forming a rim behind the appendage, which arises quite at the apex of the inner face, sometimes continuous with it ; appendage $\frac{1}{3}$–$\frac{1}{2}$ lin. long, inflexed over the staminal column, with the subulate apex incurved-hooked or erectly recurving, with broad infolded wing-like sides, at the base or up to the middle ending in prominent deltoid or rectangular teeth and decurrent as keels down the inner face of the lobe ; keels projecting in small teeth just below the middle ; staminal column $\frac{3}{4}$ lin. long ; anther-appendages subreniform, obtuse, inflexed over the margin of the depressed-truncate style-apex ; follicles 2–$2\frac{1}{4}$ in. long, 3–5 lin. thick at the basal part, fusiform, tapering into a long beak, pubescent. *Schlechter in Journ. Bot.* 1896, 449.

EASTERN REGION : Griqualand East ; near Clydesdale, *Tyson,* 2166 ! *and in Mac Owan and Bolus Herb. Norm. Austr.-Afr.* 1082 ! Natal ; near Howick, *Wood,* 3472 (not 3475 as originally quoted) ! 5382 ! *and in Herb. Natal,* 667 ! *Mrs. Hutton,* 206 partly ! near Mooi River, 6000 ft., *Schlechter,* 3339 ! near Emberton, *Schlechter,* 3231 (ex *Schlechter*), Inchanga, *Wood,* 7563 ! Attercliff, 800 ft., *Sanderson,* 508 !

I have examined the specimens quoted for this species by Dr. Schlechter, but do not find the corona-lobes to be as described by him.

84. **S. Barberæ** (Schlechter in Engl. Jahrb. xviii. Beibl. 45, 27) ; stems 2–$2\frac{1}{2}$ ft. high, erect, branching above, slender, $\frac{3}{4}$–1 lin. thick at the base, pubescent along one side on the flowering part, elsewhere

glabrous; leaves 1–1½ in. long on the middle part of the stem, gradually decreasing on the flowering part to ⅓ in. long, ¼ lin. broad, filiform, acute, erect, glabrous; umbels sessile at 3–9 of the upper nodes, 2–15-flowered; pedicels 1¼–1¾ lin. long, villous-pubescent; sepals ½ lin. long, ovate, acute, villous-pubescent; corolla-lobes spreading, 1 lin. long, ½ lin. broad, oblong-lanceolate, minutely notched at the subacute incurved apex, recurved along the margins, villous-pubescent on the back, puberulous on the inner face, with a tuft of long hairs just below the apex; corona-lobes arising above the base of the staminal column and incurved over it, scarcely ½ lin. long, subquadrate at the basal part, deltoid above, acute, with 2 minute teeth descending as 2 keels near the margins at about the middle on the inner face; appendage none; staminal column about ½ lin. long; anther-appendages semiorbicular, ¼ lin. broad, inflexed on the truncate apex of the style. *Schlechter in Journ. Bot.* 1896, 420.

EASTERN REGION: Transkei; near the Tsomo River, *Mrs. Barber*, 847! Tembuland; by the River Emgwali, near Engcobo, 2900 ft., *Bolus*, 10198!

85. **S. tomentosum** (Schlechter in Journ. Bot. 1894, 261, and 1896, 420); stem 2–2½ ft. high, simple or slightly branched and tomentose at the upper part, bifariously puberulous below, with internodes 1¾–4½ in. long; leaves erect, 1–2¾ in. long, ⅓–½ lin. broad, linear-filiform, acute, with revolute margins, glabrous or the uppermost slightly puberulous; umbels several, sessile at the nodes, 6–10-flowered; pedicels 1½–2 lin. long, shortly and densely villous-tomentose; sepals ¾–1 lin. long, lanceolate or ovate-lanceolate, acute, adpressed pubescent or tomentose; corolla-lobes spreading, with much incurved tips, 1–1¼ lin. long, ⅔ lin. broad, oblong or elliptic-oblong, minutely notched at the obtusely pointed tips, adpressed-pubescent on the back, glabrous on the inner face; corona-lobes (including the point) about equalling the staminal column, contiguous, very shortly connate at the base, basal part ¼ lin. long, ⅓–⅖ lin. broad, subquadrate, 3-toothed at the top, with 2 minute teeth decurrent as slight keels on the inner face (often evanescent in dried flowers), but no appendage on the inner face; shoulder-teeth minute, middle tooth ¼ lin. long, erect, narrowly deltoid-subulate, about ⅓ as broad as the lobe, very acute; staminal column ½–¾ lin. long; anther-appendages suborbicular, inflexed over the slightly depressed style-apex. *Cynanchum filiforme, Linn. f., Suppl.* 169; *Thunb. Prodr.* 46, *in Weber and Mohr, Archiv.* i. (1804) 28, *Fl. Cap. ed.* 2, ii. 157, *and ed. Schultes*, 235; *Willd. Sp. Pl.* i. 1253; *Pers. Syn. Pl.* i. 272; *Schultes, Syst. Veg.* vi. 107; *Spreng. Syst. Veg.* i. 853; *G. Don, Gen. Syst.* iv. 154; *Dietr. Syn. Pl.* ii. 906; *Decne in DC. Prodr.* viii. 549, *not of Jacquin. C. verticillare, Lam. Encycl.* ii. 236.

COAST REGION: Uniondale Div.; near Ongelegen in Long Kloof, *Bolus*, 2399! Humansdorp Div.; on hills below the mountains near Kromme River, *Thunberg!*

Thunberg describes the corona as monophyllous with 5 acute teeth. I find it as described above in his type specimen; but the corona-lobes being closely contiguous and very shortly connate at the base has caused him to mistake it for a monophyllous corona.

86. **S. interruptum** (Schlechter in Engl. Jahrb. xviii. Beibl. 45, 4 in note); stem 2–3 ft. high, usually branched in the upper part, with internodes $2\frac{1}{2}$–4 in. long, gradually reduced to $\frac{2}{3}$ in. long at the top, densely villous-pubescent or tomentose above, glabrous below, striate; leaves subsessile, erect, 1–$2\frac{1}{2}$ in. long, $\frac{1}{3}$ lin. broad, linear-filiform, with revolute margins, acute; umbels several to a stem or branch, sessile, 6–15-flowered; pedicels $1\frac{1}{2}$–3 lin. long, slender, very villous; sepals $\frac{2}{3}$–$\frac{3}{4}$ lin. long, about $\frac{1}{4}$ lin. broad, ovate-lanceolate, acute, villous; corolla-lobes $1\frac{1}{4}$–$1\frac{1}{2}$ lin. long, $\frac{1}{2}$ lin. broad, very spreading, narrowly oblong, subobtuse, with recurved revolute margins, villous on the back, puberulous and with a tuft of long white hairs at the tip on the inner face, green (*Haygarth*); corona-lobes shorter than or reaching to the top of the staminal column, about $\frac{1}{4}$ lin. long, $\frac{1}{3}$–$\frac{2}{5}$ lin. broad, transverse or subquadrate, obscurely toothed on each side near the middle, truncate at the top, with a minute obtuse tooth- or tubercle-like appendage at the apex on the inner face very slightly exceeding the lobe and 2 small transverse keels extending from the lateral teeth; staminal column about $\frac{2}{5}$ lin. long; anther-appendages suborbicular, closely inflexed over the truncate style-apex. *Schlechter in Engl. Jahrb.* xxi. *Beibl.* 54, 4, *and in Journ. Bot.* 1896, 450. *Lagarinthus interruptus, E. Meyer, Comm.* 208; *Meisner in Hook. Lond. Journ. Bot.* ii. 1843, 544 (*by error* 444); *Krauss in Flora,* 1844, 826; *Decne in DC. Prodr.* viii. 556. *Gomphocarpus interruptus, Dietr. Syn. Pl.* ii. 901.

CENTRAL REGION: Aliwal North Div.; Wittebergen Range, 5000 ft., *Drège,* 3428!

KALAHARI REGION: Transvaal; Ermelo, *Burtt Davy,* 994!

EASTERN REGION: Griqualand East; Vaal Bank near Kokstad, *Haygarth in Herb. Wood,* 4176! Natal; hills near Pietermaritzburg, *Krauss* (ex *Meisner*).

The plant collected by Krauss may be distinct. In the *Index Kewensis,* ii. 826, the name *Schizoglossum interruptum* is attributed to Bentham and Hooker f., *Genera Plantarum,* ii. 753, who merely state that *Lagarinthus interruptus,* E. Meyer, appears to be intermediate between *Xysmalobium* and *Schizoglossum.*

87. **S. altissimum** (Schlechter in Engl. Jahrb. xx. Beibl. 51, 13); tuber oblong or carrot-like, $\frac{1}{2}$–$\frac{3}{4}$ in. thick; stem solitary, $2\frac{1}{2}$–$3\frac{1}{2}$ ft. (up to 6 ft., *Schlechter*) high, branching at the upper half, pubescent on the flowering part, otherwise glabrous, with internodes 1–$2\frac{1}{2}$ in. long; branches erect; leaves erect, $\frac{2}{3}$–$2\frac{1}{4}$ in. long, $\frac{1}{4}$–$\frac{1}{2}$ lin. broad, linear-filiform, acute, with revolute margins, glabrous or the upper adpressed-pubescent; umbels numerous, sessile, 3–10- (or more) flowered; pedicels $1\frac{1}{2}$–2 lin. long, finely villous-pubescent; sepals $\frac{1}{2}$ lin. long, ovate, acute, pubescent; corolla-lobes spreading, $1\frac{1}{6}$–$1\frac{1}{4}$ lin. long, $\frac{1}{2}$–$\frac{2}{3}$ lin. broad, oblong, acute, with recurved margins, thinly

and finely pubescent on the back, densely bordered with white hairs (which are longest at the tips of the lobes) and minutely puberulous on the disk of the inner face; corona-lobes ovate or subquadrate, obtusely keeled down the back, with a filiform appendage about $\frac{1}{2}$ lin. long inflexed over the $\frac{1}{2}$ lin.-long staminal column from the very apex of their inner face and often appearing as a truly terminal point in dried flowers, decurrent as 2 closely placed keels to below the middle, abruptly diverging thence into horizontal sinuous keels to the margins; anther-appendages transverse, obtusely rounded, inflexed over the rim of the style-apex; follicles solitary, $1\frac{1}{2}$–$1\frac{3}{4}$ in. long, about $\frac{1}{4}$ in. thick, lanceolate-fusiform, tapering into a long beak, very softly and densely puberulous. *Schlechter in Journ. Bot.* 1896, 421. *S. lasiopetalum, Schlechter in Engl. Jahrb.* xxxviii. 29.

KALAHARI REGION: Transvaal; by streams near Lydenburg, 4800 ft., *Schlechter*, 3944! Jeppes Town Ridges near Johannesburg, 6000 ft., *Gilfillan, in Herb. Galpin,* 6228! near Modderfontein, *Conrath*, 988! Waterval Boven, *Burtt Davy,* 1411!

EASTERN REGION: Portuguese East Africa; Matolla, 20 ft., *Schlechter*, 11685!

I can find no difference whatever between *S. altissimum* and *S. lasiopetalum.*

Imperfectly known species.

88. **S. hirsutum** (Turcz. in Bull. Soc. Nat. Mosc. 1848, i. 256); stem simple or slightly branched, hairy; leaves subsessile or very shortly petiolate, oblong, obtuse or shortly acuminate, hairy on both sides, twice as long as the internodes, truncate or hastate at the base; umbels terminal or lateral at the upper nodes, about 6-flowered; corolla 5-partite, fleshy; corona-lobes concave, with an entire inflexed terminal point and a bipartite appendage on the inner face incumbent on the style-apex; pollen-masses oblong, laterally affixed; follicles solitary, covered with long soft bristles. *Schlechter in Journ. Bot.* 1896, 419.

SOUTH AFRICA: without locality, *Ecklon.*

This appears to be closely allied to *S. cordifolium*, E. Meyer, and may prove to be identical with that species.

89. **S. truncatum** (Schlechter in Engl. Jahrb. xviii. Beibl. 45, 28); plant branching from the base, 6–8 in. high; stems pilose, densely leafy; leaves erect, 1–$1\frac{5}{6}$ in. long, 1–$1\frac{1}{2}$ lin. broad, linear, acute, with revolute margins, minutely scabrous-pubescent; umbels pedunculate, lateral at the nodes (10–15-flowered according to *Schlechter*), 6-flowered with a puberulous peduncle 10 lin. long in the one examined; pedicels unequal, 3–5 lin. long, puberulous; bracts 1–2 lin. long, subulate, acute, ciliate with spreading hairs; sepals $1\frac{1}{4}$ lin. long, lanceolate, acute, shortly hairy; corolla-lobes erectly-spreading, $1\frac{3}{4}$ lin. long, $1\frac{1}{4}$ lin. broad, elliptic-ovate, subobtuse, with reflexed margins, incurved at the tips, glabrous on both sides,

apparently purplish; corona-lobes slightly exceeding · the staminal column, nearly 1 lin. long, $\frac{2}{3}$ lin. broad, ovate, obtusely bifid at the apex, with an oblong or subrectangular emarginate truncate or obtuse appendage $\frac{1}{4}$–$\frac{1}{3}$ lin. long decurrent into 2 keels on the inner face, and not exceeding the apex of the lobe; anther-appendages broadly rounded, suberect, applied to the underside of the slightly spreading margin of the depressed style-apex; caudicles attached just above the middle of the straight linear pollen-masses, which are obtuse at both ends. *Schlechter in Journ. Bot.* 1896, 420. *S. præmorsum, Schlechter in Engl. Jahrb.* xviii. *Beibl.* 45, 28, *in note.*

SOUTH AFRICA: without locality, *Mrs. Barber.*

Of this plant I have only seen one leaf and an umbel of flowers from the type, from an examination of which I suspect that it will prove to be only a slight form of *S. bidens.*

XVII. FANNINIA, Harv.

Calyx 5-partite; sepals narrow. *Corolla* 5-lobed nearly to the base; lobes overlapping in bud. *Corona* of 5 lobes arising from the staminal column, linear-oblong, flat, with a distinct midrib and bearing 2 erect lobules at their base. *Stamens* connate into a tube around the ovary, with the anthers united to the style; anthers erect, broader than long, tipped with a membranous fringed appendage. *Pollen-masses* solitary and pendulous in each anther-cell, attached to the pollen-carriers in pairs by short abruptly curved caudicles, which are broadly dilated at the basal half. *Style-apex* truncate or slightly depressed, not produced beyond the anther-appendages. *Fruit* not seen. *Panninia, Baill. Hist. Pl.* x. 258.

A tuberous-rooted herb, with opposite leaves and terminal umbels of showy white flowers with purple corona-lobes.

DISTRIB. Monotypic, endemic. This genus is very closely related to *Schizoglossum* and should perhaps be united with it, the flowers are exceedingly like those of *S. eximium, S. Grantii* and some other allied Tropical African species, but differ in having free basal lobules to the corona-lobes and no keels. The basal lobules are evidently the homologues of the minute free teeth at the base of the lobes in *Schizoglossum linifolium.*

1. **F. caloglossa** (Harv. Gen. S. Afr. Pl. ed. 2, 235); stem simple or with 1 branch at the base, 4–10 in. high, villous with jointed hairs, but not densely; leaves in 2–6 pairs, erect or slightly spreading; petiole 1–4 lin. long; blade 1–2$\frac{1}{3}$ in. long, $\frac{1}{6}$–1 in. broad, usually the lowest ovate, the middle pairs lanceolate or oblong-lanceolate and the uppermost linear, or occasionally all ovate or oblong-lanceolate, acute to obtuse, rounded or more rarely subcordate at the base, villous with jointed hairs on both sides, shortly ciliate; umbels solitary or a pair at the apex of the stem, pedunculate, 4–6-(rarely up to 10-) flowered; peduncles $\frac{3}{4}$–2$\frac{1}{3}$ in. long, villous; bracts 2–3 lin. long, filiform; pedicels 3$\frac{1}{2}$–7$\frac{1}{2}$ lin. long, villous; sepals

$2\frac{1}{2}$–$3\frac{1}{2}$ lin. long, $\frac{1}{2}$–$\frac{3}{4}$ lin. broad, linear or linear-lanceolate, acute, villous; corolla-lobes suberect or campanulately spreading, 5–$6\frac{1}{2}$ lin. long, 2–3 lin. broad, oblong-lanceolate, subacute, villous on the back, the inner face hairy on the borders and bearded with long hairs at the apex, otherwise glabrous, white, tinted with purplish on the back; corona-lobes arising about 1 lin. up the staminal column, suberect or very slightly spreading, $2\frac{1}{2}$–3 lin. long, 1 lin. broad, broadly linear, emarginate or notched at the apex, flat, with a well marked midrib, dark purple-brown, bearing at the base 2 erect linear acute whitish lobes 1 lin. long, $\frac{1}{3}$–$\frac{2}{5}$ lin. broad, with their tips incurved over the backs of the anther-appendages; staminal column 2 lin. long, contracted under the corona-lobes; anther-appendages much broader than long, rounded, fringed with long hairs at their tips, which are inflexed upon the truncate style-apex. *Schlechter in Journ. Bot.* 1894, 261, *and Engl. Jahrb.* xviii. *Beibl.* 45, 10 ; *K. Schum. in Engl. and Prantl, Pflanzenfam.* iv. ii. 233 *and* 234, *fig.* 68, E-F.

CoAST REGION : Cathcart Div.; near Thomas River, 2600 ft., *Flanagan,* 1685. Stutterheim Div.; Mount Dohne, 4000 ft., *Sim !*

EASTERN REGION : Transkei, *Hallack ! Mrs. Barber,* 30 ! Tembuland ; Bazeia, 2500 ft., *Baur,* 553 ! Griqualand East ; Mount Malowe, 4000 ft., *Tyson,* 2720 ! *and in MacOwan and Bolus, Herb. Norm. Austr.-Afr.,* 897 ; Natal, Dargle Farm, *Fannin,* 49 !

This is one of the most beautiful of South African Asclepiads and well worth cultivating.

XVIII. **ASCLEPIAS**, Linn.

Calyx 5-partite. *Corolla* 5-lobed, usually nearly to the base, often reflexed ; lobes narrowly overlapping in bud. *Corona* of 5 lobes arising at or above the base of the staminal column, erect or radiately spreading, cucullate or compressed-cucullate, at least at the basal part, or with a fissure down the inner face, with or without a horn, tooth, flap, keel or other process within the cavity. *Staminal column* arising from the base of the corolla, united above with the dilated top of the style ; anthers erect, terminated by membranous or rarely subpetaloid appendages, which are erect or inflexed upon or over the top of the style. *Pollen-masses* solitary and pendulous in each anther-cell, opaque, attached in pairs to the pollen-carriers by long or short caudicles of variable shape ; pollen-carriers seated upon or below the margin of the style-apex. *Style* truncate, depressed or rarely shortly 5-lobed at the apex. *Follicles* solitary or very rarely in pairs, globose to narrowly fusiform, usually beaked, smooth, winged or more or less covered with subulate processes or tubercles. *Seeds* crowned with a tuft of hairs.

Perennial herbs, often with a tuberous rootstock ; stems erect or decumbent, simple or branched ; leaves opposite or whorled, linear-filiform to elliptic, sometimes cordate or hastate at the base ; flowers usually of moderate size, in pedunculate or very rarely sessile umbels, lateral at the nodes or terminal or both.

DISTRIB. Species about 150, distributed throughout Africa and the warmer parts of North and South America, with 2 species in Arabia and the Orient, and 2 naturalised in many warm countries.

Until recently nearly all the South African species of this genus were placed under *Gomphocarpus,* but as stated in the *Flora of Tropical Africa,* iv. i. 314, this cannot be maintained as generically distinct from *Asclepias.* The only distinction between them is the presence in *Asclepias* or absence in *Gomphocarpus* of a horn, tooth or other process within the cavity of the corona-lobes. In some Tropical species, however, the process is present or absent in different flowers of the same species, sometimes even in different umbels on the same plant, so that both genera are represented in the same individual. This may possibly be the case with some of the South African species, but I have not yet noticed its occurrence. Dr. Schlechter has united these genera in the *Journal of Botany,* 1895, 334, and 1896, 451, but has also included with them *Pachycarpus* and *Xysmalobium,* without giving inclusive characters or a reason for doing so. The genera *Asclepias, Pachyglossum, Xysmalobium, Woodia, Schizoglossum, Fanninia* and *Krebsia* are all very closely allied, but as here limited are easily recognised by their coronal structure and usually also by habit; if the former character be ignored, then all must be merged into one unwieldy genus, as has been done by Baillon, which is not desirable. The absence of sharply defined characters for these genera has been productive of much confusion among recent writers, the same species having been described by the same author under two genera in more than one instance. See also *Flora of Tropical Africa,* iv. i. 299, 314, 353, 376.

Corolla not lobed beyond the middle, subglobose-campanu-
late; leaves linear   ..   ..   (1) **macra.**
Corolla lobed nearly to the base, often reflexed or the lobes
    spreading to suberect:
  * Corona-lobes with a compressed erect process or horn
      within the cavity:
      Leaves linear; corolla-lobes 4–5 lin. long ...   ...   (18) **navicularis,**
      Leaves filiform, linear or linear-lanceolate; corolla-   var. β.
        lobes 1½–3 lin. long.
        Stems numerous, 3–6 ft. high; umbels several,
          pedunculate, racemosely scattered along the stems   (25) **filiformis.**
        Stem solitary; umbels few and sessile in a
          terminal umbel-like cluster, with or without a
          distant sessile or rarely pedunculate umbel
          below them, or solitary and terminal.
        Stem more than 1½ ft. high; corolla not yellow,
          purplish on the back of the 3 lin.-long lobes;
          corona-lobes like a deep spoon-bowl, with a
          pair of nearly free linear teeth at the base ...   (22) **cognata.**
        Stem ½–1¼ ft. high; corolla yellow; teeth at
          the base of the corona-lobes not linear nor
          nearly free, or subobsolete.
        Corolla-lobes 1½–2⅓ lin. long; corona-lobes
          open, spoon-bowl-like, not compressed nor
          subquadrate, with short teeth at the base   (23) **flava.**
        Corolla-lobes 2⅓–3 lin. long; corona-lobes
          compressed-cucullate, subquadrate in side
          view, truncate at the top, scarcely produced
          into teeth   ...   .   (24) **schizoglossoides.**
      Leaves oblong, lanceolate or ovate-lanceolate;
        corolla-lobes 5½–6½ lin. long, bordered with white
        hairs ...   .   (35) **cultriformis.**
  **Corona-lobes with a median keel down the inner face
      of their cavity or upper part; leaves oblong,
      oblong-lanceolate or lanceolate.

Umbels 4–8–flowered ; corona-lobes twice as long as
the entire staminal column ... ... (39) **humilis.**
Umbels 8–15–flowered ; corona-lobes a little longer
than the stalk of the staminal column ... ... (40) **ulophylla.**
***Corona-lobes with a small transverse shelf-like flap
within the cavity near the top; leaves linear or
linear-lanceolate.
Leaves ½–1⅓ lin. broad, tapering into the petiole at
the base ; umbels not exceeding or shorter than
the leaves subtending them ... ... (18) **navicularis.**
Leaves 1½–4½ lin. broad, truncate or subhastate at
the base ; umbels much exceeding the leaves subtending them ... ... (14) **disparilis.**
****Corona-lobes without a keel, flap, horn or other
process (but sometimes puberulous) within the
cavity; umbels all distinctly pedunculate.
Leaves remarkably crisped-undulate, linear to lanceolate, truncate, rounded or cuneate at the base,
scabrous or pubescent ... ... (42) **crispa.**
Leaves not undulate nor crisped or but slightly so
(see 37, *A. meliodora,* 46, *A. Cooperi*).
† Leaves 9–10 times as long as broad, linear-
filiform, linear to lanceolate or linear-hastate :
‡ Umbels 2 or more to a stem or branch, rarely
solitary and then (except in 5, *A. præmorsa*)
lateral, with the stem bearing one or more
pairs of leaves beyond the base of the
peduncle :
§ Leaves tapering or rounded (never broadened,
hastate nor truncate) at the base :
‖ Plant 3–15 in. high ; corolla-lobes 1½–3½
lin. long :
Midrib on underside of dried leaves
very stout, ⅔–¾ lin. broad ; umbels
a terminal pair or 3 :
Upper internodes ½–¾ in. long; umbels not exceeding the subtending
leaves ... ... (2) **crassinervis.**
Upper internodes 2–4½ in. long; umbels exceeding the subtending
leaves ... ... (43) **Woodii.**
Midrib on underside of dried leaves
⅕–⅓ (or rarely ½) lin. broad :
Corona-lobes with the apical angles
of their inflexed sides produced
into slender subulate teeth 1 lin.
long, rising high over the style-
apex ... ... (8) **bicuspis.**
Corona-lobes with the angles at the
apex or middle of their inflexed
sides not produced into long
slender subulate teeth :
Corona-lobes measured along their
backs very much longer than
the staminal column, but on
account of their spread often
not overtopping it :
Leaves glabrous, linear-filiform
or linear ; peduncles 1¾–5¼ in.
long ... ... (21) **aurea.**

Leaves scabrous, scaberulous or
　　pubescent ; peduncles ⅓–2 in.
　　long ; tips of corona-lobes
　　radiate or in dried flowers
　　often erect :
　　　Plant branching into very
　　　　numerous stems at the
　　　　base; corolla-lobes glabrous
　　　　on both sides　...　...　(12) **meyeriana.**
　　　Plant branching into 2–8 stems
　　　　at the base ; corolla-lobes
　　　　pubescent on the back :
　　　　　Plant 3–4 in. high ; leaves
　　　　　　shortly pubescent ; co-
　　　　　　rona-lobes 2 lin. long ... (11) **velutina.**
　　　　　Plant 3–10 in. high ; leaves
　　　　　　more or less scabrous :
　　　　　　Corolla and corona-lobes
　　　　　　　2½–3 lin. long ...　...　(9) **stellifera.**
　　　　　　Corolla and corona-lobes
　　　　　　　1¾–2 lin. long ...　...　(10) **brevipes.**
　　Corona-lobes erect, shorter than to
　　　slightly exceeding the staminal
　　　column :
　　　　Cucullate part of corona-lobes
　　　　　about as long as broad　...　(3) **cucullata.**
　　　　Cucullate part of corona-lobes 2–
　　　　　3 times as long as broad　... (16) **brevicuspis.**
‖ ‖ Plant 1½–4 or sometimes up to 10 ft. high ;
　　leaves linear to lanceolate ½–10 lin.
　　broad :
　　Leaves in whorls of 3–4 or some of them
　　　(in 30, *A. crinita*) opposite ; umbels
　　　several, racemosely arranged :
　　　Leaves puberulous when young ;
　　　　umbels 4–6-flowered　...　... (30) **crinita.**
　　　Leaves entirely glabrous or minutely
　　　　ciliate when young; umbels 8–25-
　　　　flowered　...　...　...　... (31) **rivularis.**
　　Leaves all opposite :
　　　Umbels a pair or 3 (rarely solitary),
　　　　terminal, with a long peduncle-
　　　　like internode below them　... (43) **Woodii.**
　　　Umbels several or numerous race-
　　　　mosely arranged along the stems :
　　　　Young parts of stem, peduncles and
　　　　　pedicels densely white tomen-
　　　　　tose :
　　　　　Corona-lobes with backwardly
　　　　　　directed falcate teeth at the
　　　　　　inner apical angles ; follicles
　　　　　　ellipsoid, abruptly contracted
　　　　　　into a beak　...　...　... (26) **decipiens.**
　　　　　Corona-lobes D-shaped, without
　　　　　　teeth ; follicles ovoid-lanceo-
　　　　　　late, gradually tapering into
　　　　　　a beak　...　...　... (27) **Burchellii.**
　　　　Young parts of stem, peduncles and
　　　　　pedicels not densely white
　　　　　tomentose :

Corona-lobes with rather long
backwardly directed falcate
teeth at the inner apical
angles ; follicles ovoid or
lanceolate, tapering into a
beak or shortly acute ... (28) **fruticosa.**

Corona-lobes with erect obliquely
subtruncate or shortly falcate
teeth at the inner apical
angles; follicles subglobose,
obtuse or apiculate... ... (29) **physocarpa.**

§§Leaves truncate or hastately auriculate, but
not otherwise broadened at the base,
linear, subsessile, usually very long and
widely spreading or deflexed :

Anther-appendages longer than broad, subpetaloid, erect, much overtopping the
style-apex :

Corona-lobes produced at the dorsal apex
into a long linear-subulate point ... (4) **expansa.**

Corona-lobes truncate or obliquely
convex at the top, not produced at
the dorsal apex into a point ... (5) **præmorsa.**

Anther-appendages not longer than broad
nor subpetaloid, inflexed upon the
margin of the style-apex :

Corona-lobes compressed-cucullate, not
divided down the back nor with a
basal lobule ... ... (6) **patens.**

Corona-lobes dorsally divided to the base,
whence arises a fish-tail-like erect
lobule... ... ... (7) **peltigera.**

‡‡Umbel solitary and strictly terminal on each
stem or branch (see also 5, *A. præmorsa*,
very rarely in 16, *A. brevicuspis* with 1–2
lateral umbels below the terminal one);
plant branched at the base; leaves mostly
hastate, truncate or broadly cuneate and
often broadened at the base :

Corona-lobes measured along their backs
much longer than the staminal column :

Corona-lobes 6–7 times as long as broad
when viewed from the side, overtopping
the staminal column by half their length ;
corolla-lobes 5–7 lin. long ... ... (20) **eminens.**

Corona-lobes not 6–7 times as long as broad
across their side nor overtopping the
staminal column by half their length ;
corolla-lobes not more than 4½ lin. long :

Corona-lobes distinctly gibbous or curved
on the back with an erect subulate
point at their dorsal apex ¼-½ as
long as the rest of the lobe ... ... (19) **gibba.**

Corona-lobes not gibbous nor curved on
the back, with the terminal part
oblong or ovate, obtuse :

Leaves glabrous, smooth ... ... (15) **flexuosa.**

Leaves scabrous, at least on the mar-
gins and midrib ... ... (42) **crispa,** var. *β*.

Corona-lobes measured along their backs not
or but slightly longer than the staminal
column; corolla-lobes 2–4 lin. long :
Dorsal apex of the corona-lobes scarcely or
not at all exceeding the apical angles
of the inflexed sides :
Leaves linear, glabrous; umbels 3–11-
flowered ... ... (16) **brevicuspis.**
Leaves linear-lanceolate or lanceolate
mostly broadest at the base, scabrous
or harshly pubescent :
Umbel 1–1¼ in. in diam., 12–25-
flowered; peduncles 3½–7 in. long (44) **densiflora.**
Umbel 1½–2 in. in diam., 30–50-
flowered; peduncles 2–4 in. long (46) **Cooperi.**
Dorsal apex of the corona-lobes distinctly
produced above the apex of the
inflexed sides ... ... (17) **dissona.**
††Leaves from almost as broad as long to 6 or 7
times as long as broad, mostly broadest at the
base, lanceolate or rarely linear-lanceolate,
ovate, oblong, linear-oblong or elliptic (see
also occasional short-leaved specimens of 15,
A. flexuosa, 28, *A. fruticosa,* and 29, *A.
physocarpa*) :
‡Umbels a terminal pair or 3 by the forking of
1 peduncle, with a peduncle-like internode
2–4½ in. long below them ... ... (43) **Woodii.**
‡‡Umbels usually 2 to several to a stem or branch
mostly lateral at the nodes, or if solitary
then lateral below the apex of the stem :
Plant 1–3 ft. high, with stems 2–5 lin.
thick at the base :
Plant shrubbily branching, pubescent or
tomentose on the young branches and
back of the 2½–3½ lin.-long corolla-
lobes ... ... (32) **rotundifolia.**
Plant with a simple stem, glabrous on the
young parts and back of the 4½–6 lin.-
long corolla-lobes ... ... (33) **glaucophylla.**
Plant ¼–1¼ ft. high, usually branching at the
base, with stems rarely more than 1½
lin. thick at the base :
Corona-lobes in side view with the dorsal
apex below or not exceeding the level
of the teeth or apex of the inflexed
sides :
Peduncles 2–4¾ in. long, 2–4 times as
long as the leaves at their base ... (45) **fallax.**
Peduncles ½–2¾ in. long, shorter than to
1½ times as long as the leaves at
their base :
Pubescence on umbels of rather long
spreading white hairs; corona-
lobes in side view rounded or
scarcely angular at the dorsal
apex ... ... (36) **rara.**
Pubescence on umbels very short and
close, often **rust-coloured** or

tawny; corona-lobes distinctly angular in side view at the dorsal apex ... ... (34) **dregeana.**

Corona-lobes in side view with the dorsal apex rising much above or prolonged beyond the teeth or apex of the inflexed sides :

Leaf-blade 2½–10 lin. long, 1½–3½ lin. broad at the hastate or truncate base, glabrous above ... ... (13) **multicaulis.**

Leaf-blade ½–3¾ in. long, ¼–1 in. broad, more or less harshly pubescent or scahrous on both sides (sometimes glabrous above in 42, *A. crispa*, var. γ) :

Plant 2–5 in. high :

Peduncles as long as or longer than the leaves at their base ; corona-lobes 1 lin. long ... ... (38) **monticola.**

Peduncles much shorter than the leaves at their base ; corona-lobes 1½–2 lin. long ... ... (37) **meliodora.**

Plant ½–1¼ ft. high :

Peduncles 1–3 in. long, slightly shorter than to much longer than the leaves at their base ... (42) **crispa,**
vars. β and γ.

Peduncles ¼–¾ in. long, much shorter than the leaves at their base ... (41) **hastata.**

‡‡‡Umbel solitary and terminal on each stem or branch ; peduncles often long :

Corona-lobes subtruncate or rounded at the top and scarcely or not at all longer from the base to the dorsal apex than to the apex or teeth of the inflexed sides :

Corolla-lobes very distinctly puberulous all over the inner face ; umbel 1–1½ in. in diam. ... ... (48) **vicaria.**

Corolla-lobes glabrous on the inner face or very minutely puberulous at the base :

Umbels 1½–2 in. (rarely more) in diam. ; peduncle 1½–4 in. long ; coroua-lobes 1–1½ lin. long :

Leaves ovate to ovate-lanceolate, sub-cordate to broadly rounded at the base ; umbel 15–30-flowered (see also 50, *A. affinis*) ... ... (49) **albens.**

Leaves narrowly lanceolate, cuneate, subhastate or subtruncate at the base, often undulate ; umbel 30–50-flowered ... ... (46) **Cooperi.**

Umbels ¾–1¼ in. in diam., 12–25-flowered

Peduncles ½–2½ in. long ; corona-lobes not quite 1 lin. long ... ... (47) **adscendens.**

Peduncles 3–7 in. long ; corona-lobes 1¼–1⅓ lin. long :

Plant 1–1¾ ft. high ; corona-lobes entirely creamy-white or yellowish ... ... (44) **densiflora.**

Plant ½–1 ft. high ; corona-lobes dark violet-brown on the back (45) **fallax.**

Corona-lobes distinctly longer from the base
to the dorsal apex than to the teeth or
apex of the inflexed sides, and in dried
specimens often much overtopping the
staminal column ; corolla-lobes glabrous
on the inner face :
Leaves glabrous on the upper surface :
Leaves 2–10 lin. long, tapering to a very
acute point from a hastate or trun-
cate base   ...   .   (13) **multicaulis.**
Leaves $\frac{3}{4}$–2 in. long, linear-oblong,
oblong or oblong-lanceolate, obtuse
or subacute ...   .   (42) **crispa,** var. γ.
Leaves harshly or scabrous-pubescent or
scabrous above or on both sides :
Peduncles much shorter than the leaves
subtending them, $\frac{1}{4}$–$\frac{3}{4}$ in. long   ...   (41) **hastata.**
Peduncles as long as or longer than the
leaves subtending them :
Peduncles 1–3$\frac{3}{4}$ in. long ; corona-lobes
1–1$\frac{3}{4}$ lin. long   ...   .   (50) **affinis.**
Peduncles 3–6 in. long; corona-lobes
2$\frac{1}{4}$–2$\frac{1}{2}$ lin. long   ...   .   (51) **macropus.**

1. **A. macra** (Schlechter in Journ. Bot. 1896, 456) ; stem 6–12
in. high, simple, compressed, pubescent with spreading hairs, chiefly
bifariously, with internodes $\frac{1}{2}$–1$\frac{1}{2}$ in. long ; leaves erect or ascending,
apparently secund, 3–5$\frac{1}{2}$ in. long, 1–1$\frac{1}{2}$ lin. broad, linear with revolute
margins, acute, rounded or cuneate into a short broad petiole, more
or less scabrous-pubescent ; umbels 1–2 to a stem, pedunculate,
9–13-flowered ; peduncles $\frac{1}{2}$–1$\frac{1}{8}$ in. long, pubescent or scabrous-
pubescent ; bracts 3–4 lin. long, linear-subulate, acute, scabrous-
pubescent, deciduous ; pedicels 7–9 lin. long, scabrous-pubescent ;
sepals 3–3$\frac{1}{2}$ lin. long, 1 lin. broad, lanceolate, acuminate, pubescent ;
corolla subglobose-campanulate, $\frac{3}{4}$ in. in diam. across the lobes,
pubescent with spreading hairs and stained with purple-brown outside, glabrous and yellowish within ; tube 3–3$\frac{1}{2}$ lin. long, 5–6 lin. in
diam. at the mouth ; lobes 3 lin. long, 3$\frac{1}{3}$–3$\frac{1}{2}$ lin. broad, broadly
ovate, subobtuse, recurving at the tips, ciliolate along one margin ;
corona-lobes arising at the base of the staminal column and adnate to
it for $\frac{2}{3}$ of their length, compressed-cucullate, erect, 2$\frac{1}{2}$ lin. long,
1$\frac{1}{2}$ lin. broad across the side, subrectangular in general outline viewed
sideways, 3-lobed at the top, without a tooth or crest within, but the
inner surface minutely puberulous ; lateral lobes 1 lin. long, $\frac{3}{4}$ lin.
broad, falcate-oblong, obtuse, with their upper margin obliquely
spreading, dorsal lobe nearly 1 lin. long, oblong, obtusely bifid,
folded lengthwise, erect ; staminal column $\frac{1}{4}$ in. long ; anther-
appendages roundish-subquadrate, with their tips incurved over the
thickened crenulate rim of the depressed - truncate style-apex.
A. suaveolens, Schlechter in Engl. Jahrb. xxi. *Beibl.* 54, 9 *in note, not
of Leconte. Gomphocarpus suaveolens, Schlechter in Engl. Jahrb.*
xx. *Beibl.* 51, 38.

KALAHARI REGION: Transvaal; Elandspruit Mountains, 6800 ft., *Schlechter*, 4006; near the Olifants River, 4800 ft., *Schlechter*, 4109! Carolina district, near Bosses, *Burtt Davy*, 2956!

This remarkable species is so exceedingly like *Pachycarpus Gerrardi*, N. E. Br., as to be easily mistaken for that plant until the corona is examined.

2. **A. crassinervis** (N. E. Br.) ; plant 5–11 in. high ; stem solitary, simple, bifariously pubescent, leafless below the middle ; upper internodes $\frac{1}{2}$–$\frac{3}{4}$ in. long ; leaves 2–3$\frac{1}{2}$ in. long, 1$\frac{1}{2}$–3$\frac{1}{2}$ lin. broad, linear, acute, narrowed into a very short petiole at the base, more or less pubescent on both sides or nearly glabrous beneath, flat, with a very stout prominent midrib $\frac{2}{3}$–$\frac{3}{4}$ lin. broad beneath ; umbels usually a pair, occasionally 3, terminal, shorter than the leaves, corymbose, 4–6-flowered ; peduncles $\frac{1}{4}$–$\frac{3}{4}$ in. long, pubescent or puberulous down one side ; pedicels 3–7 lin. long, stout, puberulous on one side ; sepals 2–3 lin. long, 1 lin. broad, ovate-lanceolate, acuminate, pubescent ; corolla-lobes reflexed-spreading, 3–3$\frac{1}{2}$ lin. long, 1$\frac{3}{4}$–2$\frac{1}{3}$ lin. broad, oblong-ovate, subobtuse at the minutely notched apex, glabrous on both sides or with a few minute hairs at the tips on the back, apparently white or greenish-white, dark purple-brown on the back ; corona-lobes arising at the base of the staminal column and slightly exceeding it, erect, 1$\frac{3}{4}$–2$\frac{1}{2}$ lin. long, 1–1$\frac{1}{2}$ lin. broad across the side at the top, cucullate, rounded on the curved back, somewhat cornucopia-like in side view, with the dorsal apex produced into a very short obtuse point, the tips of the inflexed sides produced into teeth $\frac{1}{2}$ lin. long, resting on the anther-appendages, puberulous at the basal part within, but without a horn or other process, apparently dark purple-brown with the inflexed sides white ; staminal column 1$\frac{1}{2}$–2 lin. long ; anther appendages broadly ovate, acute, with the tips inflexed upon the style apex, which is depressed at the centre.

KALAHARI REGION : Transvaal ; Rehbokdraai, near Nelspruit, 3600 ft., *Burtt Davy*, 1629! between Carolina and Oshoek, *Burtt Davy*, 2988!

EASTERN REGION : Swaziland ; Embabaan (Mbabane), 4600 ft., *Burtt Davy*, 2775! 2330! between Bremersdorp and Mbabane, 2300–4000 ft., *Bolus*, 12135!

3. **A. cucullata** (Schlechter in Journ. Bot. 1896, 455); stems 2 or more to a plant, simple, 6–10 in. high, puberulous, with internodes $\frac{1}{2}$–1$\frac{1}{3}$ in. long ; leaves ascending or suberect, subsessile, 1$\frac{1}{2}$–4 in. long, $\frac{2}{3}$–2 lin. broad, linear, with revolute margins, acute, scaberulous or subglabrous ; umbels 1–4, lateral and terminal, 3–5-flowered ; peduncles $\frac{1}{2}$–2$\frac{3}{4}$ in. long, puberulous ; bracts 1–2 lin. long, subulate ; pedicels 4–8 lin. long, puberulous ; sepals about 1$\frac{1}{2}$ lin. long, lanceolate, acute, pubescent ; corolla-lobes apparently spreading, with upcurved tips, 3 lin. long, nearly 2 lin. broad, ovate, subacute, pubescent on the back, glabrous on the inner face ; corona-lobes arising from the base of and adnate high up the staminal column, shorter than to slightly overtopping it, 1$\frac{1}{2}$–2$\frac{1}{2}$ lin. long, 1–1$\frac{1}{4}$ lin. broad across the side, compressed-cucullate, with the sides bulging outwards and somewhat impressed (always?) below, semicircularly

rounded on the back, in side view subcircular and transversely notched at the top to half way down with the obtuse point incurved, or with a subcircular basal part dorsally produced above into a linear-oblong or ovate-oblong obtuse or slightly bifid point, erect or recurving at the tip, the part adnate to the column produced into short erect acute angles, densely puberulous inside, but without any appendage or process ; staminal-column 1¾–2 lin. long ; anther-appendages transverse, acute, inflexed over the rim of the depressed - truncate style - apex. *Gomphocarpus cucullatus, Schlechter in Engl. Jahrb.* xviii. *Beibl.* 45, 17, *and* xx. *Beibl.* 51, 28.

KALAHARI REGION: Orange River Colony; Harrismith, *Sankey*, 185! Transvaal; Saddleback Mountain near Barberton, 3500–4500 ft., *Galpin*, 1034! Frischgewacht in the Zoutpans Berg, *Miss Leendertz*, 825 !
EASTERN REGION : Transkei, *Hallack* ! Umtata Div. ; Bazeia, *Baur* ! Natal; near Van Reenen, 5000–6000 ft., *Wood*, 4820! 5384! 5667 ! between Greytown and Newcastle, *Wilms*, 2141! Pilgrims Rest, *Greenstock*! Greenwich Farm, Riet Vlei, *Fry in Herb. Galpin*, 2741 ! Shafton, Howick, *Mrs. Hutton*, 41! 347! 407! Swaziland; near Mbabane, 5000 ft., *Bolus*, 12131! *Miller*!

In the dried specimens the corolla appears to be whitish tinged with purple on the inside, purple on the back, and the corona whitish or dull yellow with a purple-brown stripe down the back, but Mr. Wood notes on one label "calyx brown, corolla and corona yellow " ; and on another label " corona pinky-white."

4. **A. expansa** (Schlechter in Engl. Jahrb. xxi. Beibl. 54, 7) ; stem solitary, simple, 10–24 in. high, puberulous on the flowering part, nearly or quite glabrous below, with internodes ½–3 in. long ; leaves varying from ascending to deflexed, 2½–5 in. long, ¾–1½ lin. broad, linear, acute, auriculate or subtruncate at the base, glabrous or scaberulous on the margins and midrib beneath ; umbels 2–5, racemose or occasionally subcorymbose, 4–8-flowered ; peduncles ¼–¾ in. long, puberulous ; bracts 1–1½ lin. long, subulate, very deciduous ; pedicels ½–¾ in. long, puberulous ; sepals 1⅓–1¾ lin. long, ¾ lin. broad, ovate or lanceolate, acute, thinly puberulous ; corolla-lobes spreading, 3½–5 lin. long, 2¼–2½ lin. broad, ovate or oblong-ovate, obtusely pointed, glabrous on both sides, green (*Bowker*) on the inner face, purplish on the back ; corona-lobes arising ½ lin. up the staminal column, apparently whitish or greenish-white, with the point and a dorsal stripe purple, 2–3 lin. long, 1–1⅓ lin. broad across the side of the complicate-cucullate part, the inflexed sides of which are produced at the apex into tapering or oblong and entire or unequally bilobulate acute or obtuse lobes ⅔–1½ lin. long, apparently more or less diverging and directed towards the centre, dorsal apex produced into a subulate spreading point 1–2 lin. long, no tooth or other process within ; anther-appendages subpetaloid, erect, 1½–1¾ lin. long, ⅔–1 lin. broad, oblong to subspathulate-elliptic, very obtuse, exceeding the style-apex by their whole length and slightly exceeding the cucullate part of the corona-lobes, " red brown " (*Wood*). *Schlechter in Journ. Bot.* 1896, 452.

Lagarinthus expansus, E. Meyer, Comm. 206 ; *Meisn. in Hook. Lond. Journ. Bot.* ii. 1843, 544 (*by error* 444), *and Krauss in Flora,* 1844, 826. *Gomphocarpus expansus, Dietr. Syn. Pl.* ii. 901 ; *Decne in DC. Prodr.* viii. 560 ; *Schlechter in Engl. Jahrb.* xviii. *Beibl.* 45, 7.

COAST REGION : Knysna Div. ; near the Knysna River, *Krauss* (ex *Meisner*). Uitenhage Div. : by the Zwartkops River, *Prior* ! Van Stadens Mountains, *Zeyher,* 591 ! *Bolus,* 1549 ! Albany Div. ; near Grahamstown, *Mrs. Hutton* ! *Mac Owan,* 675 ! *Galpin,* 3095 ! *Glass in MacOwan, Herb. Austr.-Afr.* 1504 ! *Bolus,* 9128 ! King Williamstown Div. : near Kachu (Yellowwood River) *Drège,* 4972 ! East London Div. ; near East London, *Wood in Herb. Galpin,* 3380 ! Komgha Div. ; near Komgha, *Flanagan,* 398 ! British Kaffraria, *Mrs. Hutton* !

EASTERN REGION : Transkei ; Fort Bowker, *Bowker,* 376 !

The petaloid anther-appendages of this species are very remarkable

5. **A. præmorsa** (Schlechter in Engl. Jahrb. xxi. Beibl. 54, 8) stem probably solitary, simple, $1\frac{1}{4}$–2 ft. high, slightly puberulous on the upper part, glabrous below, with internodes 2–4 in. long; leaves subsessile, spreading or upcurved, $2\frac{1}{2}$–5 in. long, $\frac{3}{4}$–1 lin. broad, linear, acute, truncate or auriculate at the base, with revolute margins, scaberulous ; umbels 3–5, lateral and terminal, racemose, or occasionally solitary and terminal in weak plants, 4–5-flowered ; peduncles $\frac{1}{3}$–1 in. long, puberulous ; bracts about $1\frac{1}{2}$ lin. long, subulate, very deciduous ; pedicels 6–7 lin. long, puberulous ; sepals about $1\frac{1}{2}$ lin. long, $\frac{2}{3}$ lin. broad, ovate-lanceolate, acute, pubescent ; corolla-lobes 2–$3\frac{1}{2}$ lin. long, $1\frac{1}{2}$–$2\frac{1}{4}$ lin. broad, elliptic-ovate, acute or subobtuse, glabrous on both sides, apparently white or greenish-white ; corona-lobes arising $\frac{1}{2}$ lin. up the staminal column, much overtopping the style-apex, erect, $1\frac{1}{4}$–$1\frac{1}{2}$ lin. long, 1–$1\frac{1}{4}$ lin. broad across the side, cucullate or complicate-cucullate, the top margins shortly produced into erect obtuse teeth at the inner apical margins, subtruncate and slightly recurved or concavely notched in front of the teeth and then convexly ascending to the dorsal obtuse apex, no tooth within ; anther-appendages subpetaloid, connivent-erect, overtopping the corona-lobes, $\frac{3}{4}$–$1\frac{1}{4}$ lin. long, $\frac{2}{3}$–$\frac{3}{4}$ lin. broad, oblong or elliptic-oblong, rounded at the apex, white. *Schlechter in Journ. Bot.* 1896, 453, *and Ann. Naturhist. Hofmus. Wien,* xv. 68. *Lagarinthus truncatus, E. Meyer, Comm.* 206. *Gomphocarpus truncatus, Dietr. Syn. Pl.* ii. 901 ; *Decne in DC. Prodr.* viii. 560.

COAST REGION : Komgha Div. ; near Komgha, *Krook,* 796 (ex *Schlechter*).

EASTERN REGION : Transkei ; between the Kei River and Gekau (Gcua) River, *Drège,* 4971 ! Pondoland ; between Umkwani River and Umsikaba River, 200 ft., *Tyson,* 2630 ! Natal ; Krantz Kloof and near Durban, *Wood,* 1162 ! hill near Murchison, *Wood,* 3040 !

Lagarinthus truncatus, Meisn. in Hook. Lond. Journ. Bot. ii. 1843, 544 (by error 444), and Krauss in Flora, 1844, 826, not of E. Meyer, collected by Krauss near Zitzikama in Humansdorp Div., is thus described :—" Perhaps a new species, differing from Drège's plant, which we have not seen, in having the leaves glabrous and the terminal umbella solitary, with an involucre of 4-5 linear leaflets of equal length as the pedicels, 4-5 lines ; but our specimens are insufficient." I have not seen Krauss' plant.

6. **A. patens** (N. E. Br.); stem solitary, erect, 1–1¾ ft. high, slender, puberulous on the upper part, glabrous below; leaves in 5–9 pairs, subsessile, widely spreading or slightly drooping, or the upper erect from a broadly curved base, 3–4¾ in. long, 1–4 lin. broad, linear, attenuate-acute, truncate or subhastate at the base, minutely scabrous above and on the midrib of the paler underside; umbels 3–5, lateral at the nodes and terminal, lower racemosely, upper corymbosely arranged, 4–6-flowered; peduncles usually 3–5 lin. long, adpressed-puberulous; pedicels ½–¾ in. long, slender, adpressed-puberulous; sepals about 1½ lin. long, lanceolate, acute, puberulous; corolla-lobes reflexed, 2½–2¾ lin. long, 1¾–2 lin. broad, elliptic or elliptic-ovate, subacute, glabrous on both sides, white on the inner face, dark purple-brown at the apical part on the back; corona-lobes arising ⅓–½ lin. up the staminal column, compressed-cucullate, 1–1¼ lin. long and 1½ lin. broad across the side at the top (including the teeth), somewhat quadrate in side view, with an outwardly descending truncate base, subtruncate at the top, with the dorsal apex slightly produced but not ascending to so high a level as the inner apical angles, which are produced over the style-apex into long falcate acute teeth, below which on each margin is a narrow wing ending in a minute erect tooth, the back and base apparently narrowly wing- or keel-margined on each side, no tooth or process within, white, or perhaps tinged with pink; staminal column about 1½ lin. long; anther-appendages small, about ⅓ lin. long and broad, suborbicular, inflexed upon the margin of the slightly depressed-truncate style-apex.

EASTERN REGION: Transkei; around Kentani, 1000 ft., *Miss Pegler*, 366! Kreilis Country, *Bowker*, 37! Pondoland; near Port St. John, 500 ft., *Galpin*, 3446!

This resembles *A. præmorsa*, but is at once distinguished by the non-petaloid inflexed anther-appendages.

7. **A. peltigera** (Schlechter in Engl. Jahrb. xxi. Beibl. 54, 8); stem solitary, simple, 1–3 ft. high, puberulous on the upper part, glabrous below, with internodes 1–3 in. long; leaves subsessile, mostly 3½–4½ in. long, ¾–1½ (rarely 2–3) lin. broad, linear, acute or obtuse at the apex, subtruncate or auriculate at the base, widely spreading or with the upper part curved upwards, sometimes deflexed, usually scabrous above and on the midrib beneath, rarely glabrous; umbels 2–5, lateral at the nodes and terminal, racemose, rarely solitary and terminal, 3–4-flowered; peduncles ½–1 in. long, puberulous; bracts none; pedicels ½–¾ in. long, puberulous; sepals 1½–1¾ lin. long, ⅔–¾ lin. broad, ovate-lanceolate, acute, pubescent; corolla-lobes 4–4½ lin. long, 2½–3 lin. broad, narrowly elliptic or elliptic-oblong, notched at the obtuse apex, very spreading or perhaps reflexed, pubescent on the back, glabrous on the inner face, apparently sometimes white, sometimes purple with white margins, purple on the back; corona-lobes arising at the base of the staminal

column shorter than and adnate to it, 2–2½ lin. high, oblong in side view, dorsally divided to the base, from which arises an erect lobe shaped like a fish-tail and reaching to about half way up, apparently white or yellowish, with the apex of the fish-tail-lobe purplish or orange-red tinted; staminal column 2¾–3 lin. long; anther-appendages half-orbicular, obtuse, inflexed over the margin of the slightly depressed-truncate style-apex; young follicles fusiform, minutely tomentose. *Schlechter in Journ. Bot.* 1896, 453, *and Ann. Naturhist. Hofmus. Wien,* xv. 68. *Lagarinthus peltigerus, E. Meyer, Comm.* 205. *Gomphocarpus peltigerus, Dietr. Syn. Pl.* ii. 901; *Decne in DC. Prodr.* viii. 561. *G. truncatus, Harv. Thes. Cap.* i. 42, *t.* 67; *Schlechter in Engl. Jahrb.* xviii. *Beibl.* 45, 10, *not of Dietr.*

COAST REGION: Uitenhage Div.; Vanstadens Berg, *Drège* (ex *E. Meyer*). Fort Beaufort Div.; near Fort Beaufort, *Cooper,* 2719 partly! Komgha Div.; near Komgha, *Flanagan,* 393! *Krook,* 787, 833 (ex *Schlechter*); British Kaffraria, *Cooper,* 470!

EASTERN REGION: Transkei; between Kei River and Bashee River, 1000–2000 ft., *Drège*! near Butterworth, *Mrs. Barber*! *Bowker,* 7! Kentani district, *Miss Pegler,* 365! Tembuland; Bazeia Mountain, 2500 ft., *Baur,* 385! near Engcobo, *Bolus,* 10200! Natal; between Umkomanzi River and Umlazi River, *Drège* (ex *E. Meyer*), Inanda, *Wood,* 362! and without precise locality, *Gueinzius*! *Gerrard,* 1984! *Sanderson,* 371!

Lagarinthus peltigerus, Meisn. in Hook. Lond. Journ. Bot. ii. 1843, 543 (by error 443), and Krauss in Flora, 1844, 826, collected by Krauss (105) around Durban Bay, is probably another species, as according to Meisner "Drège's plant" (i.e. the type of *L. peltigerus*) "has the flowers somewhat larger and the leaves shorter."

8. **A. bicuspis** (N. E. Br.); stem ½–1 ft. high, simple, somewhat harshly pubescent, more or less bifariously above; leaves apparently not more than 4–6 pairs to a stem, spreading, 1½–3 in. long, 1–2 lin. broad, linear, with revolute margins, subacute, subscabrous-pubescent above and with longer hairs on the midrib beneath; umbels 2–5, lateral and terminal, pedunculate, racemose or the upper sub-corymbose, 5–8-flowered; peduncles ⅓–1¼ in. long, pubescent on one side; pedicels 3½–5 lin. long, pubescent or puberulous on one side; sepals 1¼–1½ lin. long, ½ lin. broad, lanceolate, acute, pubescent; corolla-lobes spreading about 2½ lin. long, 1½ lin. broad, ovate, acute, perhaps with recurved margins, pubescent on the back, densely white-bearded or subwoolly along the borders and glabrous or most minutely puberulous on the disk of the inner face, not ciliate; corona-lobes arising about ⅓ lin. up the staminal column, suberect, complicate, ¾ lin. long and nearly as much across the side, subquadrate in side view, obliquely truncate at the top, with the inner apical angles produced into subulate teeth 1 lin. long, ascending high over the centre of the 1 lin.-long staminal column; no tooth within the cavity; anther-appendages exceedingly short, nearly 4 times as broad as long, transversely oblong-linear, erect, not at all inflexed over the margin of the truncate style-apex.

2 x 2

This species in coronal structure has a distinct alliance with *A. cultriformis,* Harv.

9. A. stellifera (Schlechter in Engl. Jahrb. xxi. Beibl. 54, 9); plant 3–10 in. high, branching at the base; stems or branches usually 2–6, rarely solitary, erect or spreading, often flexuose, rather minutely scabrous-puberulous all round; leaves subsessile or very shortly petiolate, 1½–4 in. long, ½–1 lin. broad, linear, acute, with revolute margins, scaberulous; umbels 1–4 lateral at the nodes below the uppermost pair or pairs of leaves, racemose or sub-corymbose, 4–7-flowered; peduncles ¼–2 in. long, puberulous; bracts none or filiform, 1–1½ lin. long; pedicels 3–8 lin. long, puberulous; sepals reflexed, about 2 lin. long, ½–¾ lin. broad, ovate-lanceolate, acuminate, pubescent; corolla-lobes reflexed, 2½–3 lin. long, 1⅓–1¾ lin. broad, ovate, minutely notched at the acute apex, pubescent on the back, glabrous on the inner face, apparently light purple; corona-lobes arising at the base of the staminal column, stellately wide-spreading, 2½–3½ lin. long, narrowly complicate-cucullate, acute, with the inflexed margins tapering into the point from the rectangularly truncate or slightly toothed base where pressed against the top of the staminal column, slightly concave or nearly straight along the top, minutely velvety-papillate within, but without a tooth or other process; staminal column 1 lin. long; anther-appendages transverse or orbicular, with their tips resting upon or inflexed over the margin of the crater-like style-apex; follicles 2½ in. or more long, 3–4 lin. thick, fusiform, acuminate or beaked, smooth, minutely puberulous. *Schlechter in Journ. Bot.* 1896, 453 ; *Rand in Journ. Bot.* 1903, 200. *A. simplex, Schlechter in Journ. Bot.* 1896, 455. *Gomphocarpus simplex, Schlechter in Engl. Jahrb. xviii. Beibl. 45, 21, and xx. Beibl. 51, 28, in note. Lagarin-thus revolutus, E. Meyer, Comm.* 205. *Gomphocarpus revolutus, Dietr. Syn. Pl. ii. 901 ; Decne in DC. Prodr. viii. 561 ; Schlechter in Engl. Jahrb. xviii. Beibl. 45, 9 ; xx. Beibl. 51, 36.*

COAST REGION, 3500–4000 ft. : Queenstown Div. ; hills near Shiloh, *Drège,* 3425 ! Finchams Nek, near Queenstown, *Galpin,* 1601 ! Tambukiland, *Zeyher* !
KALAHARI REGION, 4500–6000 ft. : Griqualand West ; Vaal River Flats, *Bowker (Mrs. Barber),* 706 ! Orange River Colony ; Bethlehem, *Richardson* ! Basutoland, *Cooper,* 2721 ! 2735 ! Transvaal ; near Johannesburg, *Rand,* 704 ! 705 ! 857 ! *Gilfillan,* 73 (in *Herb. Galpin,* 6047) ! Saddleback Mountain, near Barberton, 4000 ft., *Galpin,* 552 ! Zuikerbosch Rand, *Schlechter,* 3491 ! Heidel-berg, *Burtt Davy,* 3094 ! Springbok Flats, *Burtt Davy,* 2129 ! Bezuidenhout Valley, *Ommaney,* 143 ! Modderfontein, *Conrath,* 995 !
EASTERN REGION, 3500–4500 ft. : Transkei ; Tsomo, *Mrs. Barber,* 784 ! Griqualand East ; Mount Malowe, *Tyson,* 3116 ! Natal ; Klip Kiver, *Sutherland* ! Dargle Farm, *Fannin,* 17 !

The corona-lobes of *Gomphocarpus simplex* are described as " lanceolate-rhomboid, somewhat obtuse, verrucose on the thickened back, with erect margins produced into a short suberect tooth." I have examined a portion of the type and find them narrowly complicate-cucullate as described above, and the plant in every way identical with *A. stellifera.*

10. A. brevipes (Schlechter in Journ. Bot. 1896, 455); plant usually 3–6 or occasionally up to 10 in. high, branching at the base into 2–8 erect, decumbent or procumbent stems, or rarely with a solitary stem, often somewhat flexuous or even zigzag, minutely scaberulous; leaves 1½–4 in. long, ½–1 lin. broad, spreading or sub-erect, usually flexuous on dried specimens, linear, acute, with very revolute margins, coriaceous, minutely scabrous; umbels 2–4 to a branch, lateral at the nodes, and terminal or with a pair of leaves beyond the uppermost, subcorymbose or racemose, ⅔–¾ in. in diam., usually 4-flowered; peduncles ½–1¼ in. long, puberulous on one side; bracts about 1½ lin. long, subulate, caducous; pedicels 3½–6 lin. long, puberulous; sepals 1–1¼ lin. long, ½ lin. broad, lanceolate, acute, puberulous with curved hairs; corolla-lobes about 2 lin. long, 1¼ lin. broad, spreading, with incurved tips, ovate, acute, thinly puberulous or pubescent to nearly glabrous on the back, glabrous on the inner face, apparently greenish, tinted with purple-brown on the back; corona-lobes arising close to the base of the staminal column and much overtopping it, ascending to somewhat spreading, 1¾–2 lin. long, rounded in at the base, complicate-cucullate at the basal half, prolonged beyond into a linear or linear-oblong obtuse or subacute channelled point, the sides of the cucullate part making a right angle with the prolonged apex and produced into short ascending obtuse teeth resting against the backs of the anthers appear as if abruptly pinched together, leaving the dorsal body of the lobe standing out in a prominent ridge on each side (not visible in much-pressed specimens), their inner margin narrowly winged on the basal half, very minutely papillate within, but without a tooth or other process, apparently whitish with the point and dorsal part brown or dark-coloured; staminal column about ¾ lin. long; anther appendages transverse, obtusely rounded, inflexed just upon the margin of the depressed-truncate style-apex; follicles solitary, erect from a recurved pedicel, 2½–4 in. long, ⅓ in. thick, terete-fusiform, tapering into a long beak, smooth, very minutely puberulous; seeds about 2 lin. long, 1–1¼ lin. broad, ovate, slightly concave on one side, convex on the other, both sides covered with minute processes or short ridges, very minutely papillate-puberulous, pale brown. *Gomphocarpus brevipes, Schlechter in Engl. Jahrb.* xx. *Beibl.* 51, 28.

KALAHARI REGION: Transvaal; Magalies Berg, *Burke*! near Heidelberg, 4900 ft., *Schlechter*, 3516! Irene, *Burtt Davy*, 745! Springbok Flats, *Burtt Davy*, 2129! near Rustenburg, *Miss Pegler*, 2008! Pretoria Hills, *Miss Leendertz*, 295! Witbank, 5300 ft., *Rogers*! and without precise locality, *McLea in Herb. Bolus*, 5709!

In its stouter, less branched habit, this is much like *A. stellifera*, Schlechter, whilst the flowers are like those of *A. meyeriana*, Schlechter. It is probably a hybrid between these two species.

11. A. velutina (Schlechter in Journ. Bot. 1896, 455); 3–4 in. high, erect, branching at the base, shortly and somewhat harshly pubescent (not velvety) on the stems, leaves, peduncles, pedicels and sepals; leaves erect or ascending, 8–16 lin. long, $\frac{3}{4}$ lin. broad, linear, acute, with revolute margins; umbels lateral and terminal, 3–6-flowered; peduncle $\frac{1}{2}$ in. long; bracts subulate, 1 lin. long; pedicels 4–5 lin. long; sepals $1\frac{1}{2}$ lin. long, $\frac{1}{2}$ lin. broad, lanceolate, acute; corolla-lobes $2\frac{1}{2}$ lin. long, $1\frac{1}{3}$ lin. broad, in dried flowers suberect or campanulately spreading, with inflexed tips, perhaps spreading when alive, ovate, subacute, pubescent on the back, glabrous on the inner face, apparently purplish; corona-lobes arising at the base of the staminal column and in dried flowers overtopping it by about half their length, perhaps radiately spreading when alive, 2 lin. long, suberect, complicate-cucullate at the basal half, thence tapering to the obtuse point, the inflexed sides not or scarcely produced into teeth at the top angles, which are slightly below the level of the $1\frac{1}{4}$ lin.-long staminal column; anther-appendages subreniform, obtuse, inflexed over the margin of the depressed-truncate style-apex. *Gomphocarpus velutinus, Schlechter in Engl. Jahrb.* xviii. *Beibl.* 45, 22.

KALAHARI REGION: Transvaal; Saddleback Mountain, near Barberton, 3500–4000 ft., *Galpin*, 450 !

The above description is made from a portion of the type specimen presented to Kew by Mr. E. E. Galpin, but its dimensions do not accord with those given by Dr. Schlechter, who describes the peduncles as being 12–17 lin. long, the sepals as $2\frac{1}{2}$ lin. long, and the corolla-lobes 3 lin. long and 2 lin. broad. The flowers appear to become somewhat pulpy in boiling water. The pubescence on the example seen is by no means velvet-like, but short, spreading and inclined to harshness.

12. **A. meyeriana** (Schlechter in Engl. Jahrb. xxi. Beibl. 54, 8); plant 3–6 in. high, branched at the base into very numerous crowded erect scabrous-puberulous leafy stems; leaves 1–2 in. long, $\frac{1}{4}-\frac{2}{3}$ lin. broad, linear, acute, with revolute margins, scabrous; umbels 1–3 to a branch, all lateral below the uppermost leafy node, 3–4-flowered; peduncles $\frac{1}{3}-1\frac{1}{3}$ in. long, puberulous or scaberulous; pedicels 2–5 lin. long, puberulous or scaberulous; sepals reflexed, $1-1\frac{1}{2}$ lin. long, $\frac{1}{2}-\frac{2}{3}$ lin. broad, lanceolate, acute, pubescent; corolla-lobes reflexed or reflexed-spreading, $1\frac{1}{2}-2$ lin. long, $1-1\frac{1}{4}$ lin. broad, elliptic-ovate, subacute, glabrous on both sides; corona-lobes arising near the base of the staminal column and in dried flowers over-topping it by $\frac{1}{4}$ their length, but probably stellately radiating when alive, $1\frac{1}{2}-1\frac{3}{4}$ lin. long, narrowly complicate-cucullate at the basal part which is truncate at the top, making a right-angle with the prolonged linear concave obtuse or acute apex, apparently purple with whitish or yellowish sides; staminal column 1 lin. long; anther-appendages elliptic-ovate, with their obtuse tips inflexed over the margin of the small crater-like style-apex; follicles solitary, erect from a recurved pedicel, $1\frac{3}{4}-2\frac{1}{3}$ in. long, $\frac{1}{4}-\frac{1}{3}$ in. thick, fusiform, tapering into a beak, smooth, minutely puberulous; ripe seeds not seen. *Schlechter in Journ. Bot.* 1896, 453. *Lagarinthus revolutus,*

var. minor, E. Meyer, Comm. 205. *Gomphocarpus meyerianus, Schlechter in Engl. Jahrb.* xx. *Beibl.* 51, 33.

COAST REGION : Queenstown Div., 3000–4000 ft. ; near Shiloh, *Drège*, 3436 ! near Queenstown, *Cooper*, 268 ! *Mrs. Barber*, 90 ! *Galpin*, 1584 ! Eastern Frontier, *Mrs. Barber* !

CENTRAL REGION : Steynsburg Div. ; Zuurberg Range, *Burke*, 418 ! Albert Div. ; between the Zuurberg Range and Stormberg Spruit, *Zeyher*, 1156 ! Aliwal North Div. ; by the Orange River, *Burke* !

KALAHARI REGION : Griqualand West ; Warrenton, *Miss Adams*, 106 ! Transvaal ; Potchefstroom, *Burtt Davy*, 1818 ! 2168 !

EASTERN REGION : Natal ; plains by the Tugela River, near Colenso, 3000–4500 ft., *Wood*, 4108 ! *Schlechter*, 3378 ! *Krook*, 788 (ex *Schlechter*), near Ladysmith, *Schlechter* ! near Dundee, *Wood*, 6544 !

13. A. multicaulis (Schlechter in Engl. Jahrb. xxi. Beibl. 54, 8) ; plant 4–6 in. high, branching at the base ; stems often numerous, decumbent-spreading, bifariously pubescent, or sometimes thinly and rather harshly pubescent all round ; leaves secund-spreading ; petiole $\frac{1}{2}$–$\frac{3}{4}$ lin. long ; blade $2\frac{1}{2}$–10 lin. long, $1\frac{1}{2}$–$3\frac{1}{2}$ lin. broad at the truncate, subsagittate or subhastate base, thence tapering to a very acute apex, glabrous above, slightly scabrous on the margins and subhispid on the midrib beneath ; umbels solitary, rarely 2 to a stem, terminal, pedunculate, 6–9-flowered ; peduncle $\frac{1}{4}$–$1\frac{3}{4}$ in. long, pubescent, chiefly on one side ; bracts 1–2 lin. long, subulate, pubescent ; pedicels unequal, 2–6 lin. long, so that the umbel is flat-topped, pubescent or puberulous ; sepals 1–$1\frac{1}{2}$ lin. long, $\frac{1}{2}$–$\frac{3}{4}$ lin. broad, ovate-lanceolate, very acute, pubescent ; corolla-lobes reflexed, 2 lin. long, $1\frac{1}{4}$–$1\frac{1}{2}$ lin. broad, ovate, subacute, glabrous on both sides, but the inner face with a very minutely papillose surface, sometimes minutely ciliate on one margin ; corona-lobes arising at the base of the staminal column and shortly overtopping it, 1–$1\frac{1}{2}$ lin. long, erect, cucullate, but dorsally flattened, so that the inflexed sides are pressed against the inner face of the lobe and end in deltoid acute contiguous teeth, which incurve upon the back of the anther appendages, but do not exceed them, and are much shorter than the ovate obtuse apical part of the lobe ; staminal column $\frac{3}{4}$–1 lin. long ; anther-appendages subreniform or transversely elliptic, very obtuse, erect, not inflexed over the pentagonal style-apex ; follicles solitary, fusiform, tapering into a beak, smooth and glabrous, only an immature example seen. *Schlechter in Journ. Bot.* 1896, 453. *Lagarinthus multicaulis, E. Meyer, Comm.* 205. *Gomphocarpus multicaulis, Dietr. Syn. Pl.* ii. 901 ; *Decne in DC. Prodr.* viii. 559 ; *Schlechter in Engl. Jahrb.* xviii. *Beibl.* 45, 8 ; xx. *Beibl.* 51, 34.

COAST REGION : Queenstown Div. ; by the Klipplaat River, near Shiloh, *Drège*, 3437 !

CENTRAL REGION : Aliwal North Div. ; on the Witteberg Range, *Drège* (ex *E. Meyer*).

KALAHARI REGION : Basutoland, or Orange River Colony, *Cooper*, 935 ! Orange River Colony ; Bethlehem, *Richardson* ! Harrismith, *Sankey*, 183 ! Transvaal ; Witkleifontein, *Burtt Davy*, 3128 !

EASTERN REGION : Transkei ; Kreilis Country, *Bowker*, 213 ! Natal ; Greenwich Farm, Riet Vlei, *Fry in Herb. Galpin*, 2740 !

14. A. disparilis (N. E. Br.); plant 4–8 in. high, branching at the base; branches decumbent, somewhat harshly and thinly pubescent, chiefly along one side; leaves (including the 1–1½ lin. long petiole) ½–1½ in. long, 1½–4½ lin. broad, linear-attenuate, linear-lanceolate or lanceolate, acute, truncate or slightly hastate at the base, glabrous on both sides or with a few short stiff hairs on the midrib beneath; umbel solitary on each branch, terminal, erect, 4–8-flowered; peduncle 1–3 in. long; bracts 1½–3 lin. long, subulate or filiform; pedicels ½–¾ in. long, puberulous or subscabrous down one side; sepals 2 lin. long, ¾ lin. broad, lanceolate, acute, with a few spreading hairs on the back; corolla-lobes very spreading, 4 lin. long, 2 lin. broad, lanceolate or ovate-lanceolate, acute, glabrous on the back, with a very minute papillate-puberulous or velvet-like surface on the inner face, apparently pale purplish; corona-lobes arising close to the base of the staminal column and not overtopping it, 2–2¼ lin. long, nearly 1 lin. broad across the side, erect, complicate-cucullate, with infolded margins, narrowed into the column at the base, obliquely and sinuously subtruncate at the top, the dorsal subobtuse apex being very slightly produced, with a transverse shelf-like flap within the cavity above the middle; staminal column 2 lin. long.

EASTERN REGION : Griqualand East; meadows around Clydesdale, 2500 ft., *Tyson,* 2004 ! and in *MacOwan and Bolus, Herb. Norm. Austr.-Afr.,* 1319 !

Like *A. navicularis,* Schlechter, in floral structure, but the foliage and appearance of the plant is entirely different.

15. A. flexuosa (Schlechter in Journ. Bot. 1896, 453); plant ½–1 (rarely up to 2) ft. high, branching at the base; stems probably decumbent at the base, ascending or erect, more or less flattened, unifariously or bifariously puberulous; leaves very spreading or deflexed, very shortly petiolate, ½–2¼ in. long, ¾–2 lin. broad, linear, acute, slightly hastate or cordate at the base, somewhat scabrous-pubescent on the midrib beneath, otherwise glabrous, coriaceous; umbels solitary, very rarely 2, terminal, ¾–1⅓ in. in diam., 6–14-flowered; peduncle ½–1½ (rarely up to 3) in. long; bracts 1–3 lin. long, subulate; pedicels 3–7 lin. long, puberulous on one side; sepals ¾–2 lin. long, ½–⅔ lin. broad, lanceolate or ovate-lanceolate, acute, thinly pubescent; corolla-lobes reflexed or reflexed-spreading, 1¾–2½ lin. long, 1–1½ lin. broad, oblong-ovate to elliptic-ovate, subobtuse, glabrous on the back, minutely papillate-puberulous on the inner face; corona-lobes arising at the base of and nearly twice as long as the staminal column, erect or ascending, 1¼–2 lin. long; basal half cucullate, dorsally rounded into the column with ∧-shaped keels on the back, inflexed sides produced at the apex into short acute teeth resting on the anther-appendages, with a narrow wing descending obliquely from them nearly to the base, no tooth or process within; apical half lanceolate-oblong or elongate-deltoid-oblong, obtuse, concave, apparently marked with a dark mid-line; staminal column $\frac{3}{4}$–1$\frac{1}{4}$ lin. long; anther-appendages transversely oblong, emarginate or slightly bilobed, erect, not exceeding the margin of the truncate style-apex; follicles solitary, erect from a recurved pedicel, 2$\frac{3}{4}$–3$\frac{1}{2}$ in. long, $\frac{1}{4}$ in. thick, terete-fusiform, tapering into a beak, smooth, glabrous; seeds about 2$\frac{1}{2}$ lin. long, 1$\frac{1}{2}$ lin. broad, ovate, rather broadly margined, with minute tubercles and short linear ridges on both sides, brown. *Lagarinthus flexuosus, E. Meyer, Comm.* 207; *Meisn. in Hook. Lond. Journ. Bot.* ii. 1843, 544 (*by error* 444); *Krauss in Flora,* 1844, 826. *Gomphocarpus flexuosus, Dietr. Syn. Pl.* ii. 901; *Decne in DC. Prodr.* viii. 559. *G. fragrans, Schlechter in Engl. Jahrb.* xx. *Beibl.* 51, 30.

EASTERN REGION : Natal; between Umzimkulu River and Umkomaas River, *Drège,* 4965! near Durban, *Gerrard,* 514! *Sanderson,* 258! *Peddie!* near Umlaas River, *Krauss,* 343! Inanda, *Wood,* 280! near Howick, *Wood,* 5378! Clairmont, *Wood,* 8269! between Pietermaritzburg and Greytown, *Wilms,* 2139! hills near Pinetown, *Schlechter,* 3168, and without precise locality (in fruit only), *Gerrard,* 1292! Zululand; Ungoya, *Wylie in Herb. Wood,* 5680! and without precise locality, *Mrs. McKenzie!*

By different collectors the flowers are stated to be "pale purple and white or brown and white," "lilac and white or pinkish-white," "lilac-grey or brownish-grey," all mentioning 2 colours as if they were variable, and they are said to be fragrant. I have not seen the type of *Gomphocarpus fragrans,* but the specimens so named by Dr. Schlechter which I have examined, do not differ in any way from *Asclepias flexuosa.*

16. A. brevicuspis (Schlechter in Engl. Jahrb. xxi. Beibl. 54, 5); plant 4$\frac{1}{2}$–12 in. high, branching at the base; stems often decumbent at the base, unifariously pubescent; leaves 1$\frac{1}{4}$–3 in. long, $\frac{1}{2}$–2$\frac{1}{3}$ lin. broad, linear, acute, narrowed or rounded (rarely hastate) at the base into a short petiole, glabrous on both sides or scabrous on the midrib beneath, rarely scabrous on the margins; umbels solitary and terminal or very rarely 2–3 to a branch and racemose, 3–11-(usually 6–7-) flowered; peduncles $\frac{1}{2}$–2$\frac{1}{2}$ in. long, unifariously pubescent; pedicels 5–7 lin. long, unifariously puberulous; sepals 1$\frac{1}{2}$–2$\frac{1}{3}$ lin. long, $\frac{1}{2}$–$\frac{2}{3}$ lin. broad, lanceolate, acuminate, glabrous or pubescent and with or without short cilia; corolla-lobes very spreading, 3–4 lin. long, 1$\frac{1}{4}$–1$\frac{3}{4}$ lin. broad, oblong-lanceolate, acute, glabrous on the back, with a minutely papillate inner surface, sometimes minutely ciliate on one margin, lilac (*Wood*); corona-lobes arising at the base and reaching to the top of the staminal column, 1$\frac{1}{2}$–2 lin. long, linear-oblong viewed sideways and $\frac{2}{3}$–$\frac{3}{4}$ lin. broad across the top, nearly straight, slightly curved or slightly gibbous above the middle on the back, tapering into the column at the base, somewhat sinuate-truncate or with a concave notch at the top, with the inner apical angles rounded and the dorsal apex shortly acute, not produced into an erect subulate point, with an obliquely transverse or decurrent keel or narrow wing on each side; staminal column 1$\frac{1}{2}$–2 lin. long; anther-appendages oblong-ovate or suborbicular, obtuse, erect, with their tips inflexed over the crenate margin of the $\frac{3}{4}$–1 lin.-broad style-apex; pollen-carriers seated below the margin of the style-apex. *Schlechter in Journ. Bot.* 1896, 452. *Lagarinthus brevicuspis, E. Meyer, Comm.* 204. *Gomphocarpus brevicuspis, Dietr. Syn. Pl.* ii. 900; *Decne in DC. Prodr.* viii. 559.

COAST REGION: Albany Div.; plains of Albany, *Bowie*! King Williamstown Div.; Keiskamma, *Mrs. Hutton*! near King Williamstown, 1200 ft., *Sim*, 1294!

KALAHARI REGION: Basutoland, *Cooper*, 2721 bis! 2724! Transvaal: near Piet Retief, *Lady Barkly*!

EASTERN REGION: Pondoland, *Bachmann*, 1098! Natal; between Umzimkulu River and Umkomaas River, *Drège*, 4966! coast-land, *Wood*, 960! near Durban, *Peddie*! *Plant*, 54! *Wood*, 78! hills near Malvern, *Wood*, 4937! Umlazi River, *Sutherland*! Zululand, *Mrs. McKenzie*! and without precise locality, *Gerrard*, 322! *Sanderson*, 149! *Peddie*! *Mrs. K. Saunders*!

17. **A. dissona** (N. E. Br.); plant 8–9 in. high, branching at the base; stems unifariously puberulous; leaves very shortly petiolate, spreading or ascending, 1–2¼ in. long, 1–1½ lin. broad, linear, acute, subtruncate, very broadly and shortly cuneate or subhastate at the base, glabrous on both sides; umbel solitary and terminal on each branch, erect, 4–8-flowered; peduncle 1½–3¼ in. long; bracts few or none, 2–3 lin. long, subulate; pedicels ½–⅔ in. long; puberulous with curved hairs on one-side; sepals about 2 lin. long and ¾–1 lin. broad, lanceolate, acute, pubescent with short spreading jointed hairs on the back; corolla-lobes spreading or possibly reflexed, ¼ in. long, 1¾–2 lin. broad, ovate-lanceolate, acute, glabrous or with a few hairs on the back of the tips, very minutely papillate-puberulous on the inner face; corona-lobes arising at the base of the staminal column, erect, 2 lin. long, complicate-cucullate at the basal ⅔, with the apical third produced beyond into an erect narrowly deltoid channelled point, rising to about the level of the top of the 2 lin.-long staminal column, very broadly and abruptly rounded into the column at the base, without a tooth or other process within the cavity.

KALAHARI REGION: Transvaal; in damp meadows near the Crocodile River, 4800 ft., *Schlechter*, 3903!

Very similar to *A. navicularis*, Schlechter, and *A. brevicuspis*, Schlechter, in general appearance, but differs in its corona-lobes and more truncate-based leaves.

18. **A. navicularis** (Schlechter in Engl. Jahrb. xxi. Beibl. 54, 8); plant 6–10 in. high, branched at the base; branches ascending, more or less curved or decumbent at the base, unifariously pubescent; leaves ascending or somewhat spreading, 1¼–2¼ in. long, ½–1⅓ lin. broad, linear, acute, narrowed or rounded into a short petiole at the base, glabrous or sometimes shortly and sparsely subhispid on the midrib beneath; umbel terminal, solitary, 4–6-flowered; peduncle 2–10 lin. long, unifariously puberulous; bracts about 2 lin. long, subulate; pedicels 3–8 lin. long, unifariously puberulous; sepals 2–2½ lin. long, linear-lanceolate, very acute, subhispid on the back, ciliate at the base; corolla-lobes spreading

or ascending, 4–5 lin. long, 1½–2½ lin. broad, oblong-lanceolate or broadly ovate-lanceolate, subacute, reflexed along the margins, glabrous on the back, minutely papillate-puberulous on the inner face ; corona-lobes arising at the base of and equalling the staminal column, 1½–2 lin. long, erect, distant, alternating with a short flat broadly rounded lobule at their base, cucullate, oblong in outline, tapering into the column at the base, scarcely broader at the oblique apex than elsewhere, ⅔–1 lin. broad, viewed sideways, with the top margins slightly ascending to the dorsal apex, not produced into teeth at inner apical angles, which are much infolded forming a somewhat boat-shaped concavity on the inner face and when dorsally flattened in drying boat - shaped in outline, without oblique keels or wings on the sides, but with a transverse shelf-like flap or fold within a little below the apex ; staminal column about 1¾ lin. long ; anther appendages rather large, elliptic-ovate or transverse, with their obtuse tips inflexed over the margin of the style-apex, which is variable in size and concavity ; anther-wings rather long and broadly rounded. *Schlechter in Journ. Bot.* 1896, 453. *Lagarinthus navicularis, E. Meyer, Comm.* 204 ; *Meisn. in Hook. Lond. Journ. Bot.* ii. 1843, 543 (*by error* 443) ; *Krauss in Flora,* 1844, 826. *Gomphocarpus navicularis, Dietr. Syn. Pl.* ii. 901 ; *Decne in DC. Prodr.* viii. 559 ; *Schlechter in Engl. Jahrb.* xx. *Beibl.* 51, 34.

VAR. β, **compressidens** (N. E. Br.); corona-lobes shorter than the staminal column, with a broad laterally compressed obtusely rounded tooth within, instead of the transverse flap ; otherwise as in the type.

COAST REGION : Swellendam Div. ; near Swellendam, *Krauss* (ex *Meisner*). King Williamstown Div. ; between Chalumna River and Buffalo River, 1500 ft., *Drège*! Keiskamma, *Mrs. Hutton*! East London Div. ; grassy slopes near East London, 100 ft., *Wood in Herb. Galpin,* 1995 ! 3378 ! Var. β : Cathcart Div. ; between Windvogel Mountain and the Zwartkei River, *Drège*!
CENTRAL REGION : Var. β : Cradock Div., *Cooper,* 1284.

This much resembles *A. brevicuspis,* Schlechter, but it is at once distinguished by the corona-lobes.

19. **A. gibba** (Schlechter in Engl. Jahrb. xxi. Beibl. 54, 7) ; plant ¼–1 ft. high, branching at the base ; branches ascending or spreading, compressed, bifariously pubescent or subhispid ; leaves shortly petiolate, ¾–4 in. long, ½–2½ (rarely 4–6) lin. broad, linear, acute at the apex, hastate, truncate or cuneate at the base, flat or with revolute smooth (rarely scabrous) margins, glabrous on both sides or with a few hairs on the midrib beneath ; umbels solitary, terminal on the main branches, pedunculate, usually 3–6- (rarely 8–10-) flowered ; peduncles ⅓–3 in. long, pubescent ; bracts 1½–4 lin. long, subulate ; pedicels 4–10 lin. long, puberulous and often with some longer spreading hairs on one side ; sepals 2–3 lin. long, ⅖–¾ lin. broad, lanceolate or lanceolate-attenuate, acute, ciliate, more or less hairy ; corolla-lobes very spreading or somewhat reflexed, 3–4½ lin. long, 1⅔–2 lin. broad, lanceolate or ovate-lanceolate, acute, glabrous

on the back, with the inner face most minutely papillate, probably
having a velvety sheen, brownish-green (*Wood*); corona-lobes arising
at the base of the staminal column, erect, 2–3½ lin. long, complicate,
much compressed, with a semicircular hump on the back near the
middle, having a fold or keel extending obliquely downwards from
its base to the base of the inner margin, truncate at the top of
the infolded sides, which scarcely exceed the 1¾–2½ lin.-long staminal
column, dorsal apex produced into an erect straight subulate point
⅓–1 lin. long; anther-appendages ovate, subacute or obtuse, erect,
the tips inflexed over the margin of the excavated style-apex;
follicles about 3 in. long and ⅓ in. thick, fusiform, tapering into a
beak, smooth, very narrowly 6-winged on the beaked part, glabrous;
seeds 2 lin. long, deeply concave on one side, very convex on the
other, both sides sprinkled with minute tubercles or short linear
ridges, dark brown with a paler wrinkled margin and darker
tubercles. *Schlechter in Journ. Bot. 1896, 452, and Ann. Naturhist.
Hofmus. Wien*, xv. 67. *A. oxytropis, Schlechter in Journ. Bot.* 1896,
454. *Lagarinthus gibbus, E. Meyer, Comm.* 204. *Gomphocarpus
gibbus, Dietr. Syn. Pl.* ii. 900; *Decne in DC. Prodr.* viii. 559;
Schlechter in Engl. Jahrb. xviii. *Beibl.* 45, 8, *and* xx. *Beibl.* 51, 31.
G. oxytropis, Turcz. in Bull. Soc. Nat. Mosc. 1848, i. 259.

VAR. β, **media** (N. E. Br.); corona-lobes 3–4 lin. long, with the complicate part
distinctly (but shortly) overtopping the staminal column, curved but scarcely
gibbous on the back and without an oblique fold or keel upon the sides; otherwise
as in the type.

SOUTH AFRICA : without locality, *Herb. Thunberg*!
COAST REGION : Albany Div. ; near Grahamstown, *Bolus, Herb. Norm. Austr.-
Afr.*, 1318 ! Bedford Div. ; near Bedford, *Miss Nicol*, 101 ! Stockenstrom Div. ;
Kat Berg, *Hutton* ! *Galpin*, 1732 ! *Miss Sole*, 372 ! Queenstown Div. ; near Shiloh,
Ecklon, 28. Komgha Div. ; hills near Komgha, *Flanagan*, 397 ! British Kaffraria,
Cooper, 471 !
CENTRAL REGION : Somerset Div. ; on the Bosch Berg, 3000–4800 ft., *Mac
Owan*, 1992 ! Aliwal North Div. ; Witte Bergen, *Drège*, 4973 ! Molteno Div. ;
Broughton, near Molteno, *Flanagan*, 1615 !
KALAHARI REGION : Orange River Colony ; Leeuw Spruit and Vredefort,
Barrett Hamilton ! Basutoland, *Cooper*, 2725 ! Transvaal ; near Pretoria, *McLea
in Herb. Bolus*, 5707 ! near Ermelo, *Burtt Davy*, 956 ! 957 b ! near Standerton,
Burtt Davy, 895 ! 3084 ! Modderfontein, *Conrath*, 992 ! Var. β : Orange River
Colony ; Caledon River, *Burke* ! Besters Vlei near Witzies Hoek, *Bolus*, 8113 !
Transvaal ; Magalies Berg, *Burke* ! *Zeyher*, 1158 ! Yster Spruit, *Nelson*, 323 !
near Pretoria, *Rehmann*, 4155 ! *Burtt Davy*, 673 ! 795 ! 2131 ! *Miss Leendertz*,
328 ! near Nylstrom, *Burtt Davy*, 2045 ! Linokana and Matebe Valley, *Holub* !
Rustenburg, *Miss Pegler* !
EASTERN REGION, at 2000–4500 ft. : Transkei ; Kreilis Country, *Bowker*
(*Mrs. Barber*), 92 ! Tembuland ; Bazeia, *Baur*, 305 ! Pondoland ; between
Umtata River and St. John's River, *Drège*, 4967 ! Griqualand East ; near
Clydesdale, *Tyson*, 1656 ! Natal ; various localities, *Wood*, 4064 ! *Sanderson*, 164 !
Mrs. Clark ! *Fannin*, 46 ! *Wilms*, 2140 ! *Sutherland* ! *Krook*, 834 (ex *Schlechter*).
Swaziland ; between Carolina and Mbabane, 5300 ft., *Bolus*, 12151 ! Var. β :
Zululand, *Gerrard*, 1291 bis ! Swaziland : Bremersdorp, *Burtt Davy*, 3009 ! 3277 !

The variety β is probably a hybrid between *A. gibba*, Schlechter, and *A. eminens*,
Schlechter.

20. A. eminens (Schlechter in Journ. Bot. 1896, 453); plant 5–10 in. high, branching close to the ground; branches usually more or less decumbent, bifariously pubescent, with internodes ¼–1 in. long; leaves ascending, usually unilateral; petiole 1–3 lin. long; blade 1–4½ in. long, 1–2½ lin. broad above the basal auricles, linear, acute, shortly hastate, truncate or rounded at the base, subcoriaceous, thickened along the flat or slightly recurved margins, glabrous on both sides or thinly setulose-scabrous on the midrib beneath; umbels solitary on each branch, terminal, pedunculate, 3–6-flowered; peduncle ½–2¾ in. long, pubescent on one side; bracts 3–5 lin. long, ¼–⅓ lin. broad, linear, acute, glabrous or sparsely setulose-scabrous, deciduous; pedicels ⅓–⅔ in. long, pubescent on one side; sepals 2½–4 lin. long, ¾–1 lin. broad, lanceolate or linear-lanceolate, acute, glabrous or with a few hairs; corolla-lobes very spreading or reflexed, 4½–7 lin. long, 1⅔–3 lin. broad, lanceolate, acute or subobtusely pointed, glabrous on the back, minutely papillate on the inner face; corona-lobes arising ¾–1½ lin. up and decurrent to the base of the staminal column, and overtopping it by nearly or quite ½ of their length, 3–5 lin. long, ⅔–¾ lin. broad and of nearly equal breadth throughout across the side, compressed-complicate, with a small triangular cup-like mouth, erect, recurved-spreading from a little below the apex, and, viewed sideways, dorsally produced into a short deltoid point ½ lin. long; staminal column 2¾–3¼ lin. long; anther-appendages elliptic or suborbicular, obtuse, inflexed over the style-apex; anther-wings ½ lin. broad, obtusely rounded at the lower half; follicles 3½–4½ in. long, about ⅓ in. thick, fusiform, tapering into a long acute beak, smooth, glabrous. *Schlechter in Ann. Naturhist. Hofmus. Wien,* xv. 67; *N. E. Br. in Dyer, Fl. Trop. Afr.* iv. i. 351; *Rand in Journ. Bot.* 1903, 337. *Gomphocarpus eminens, Harv. Thes. Cap.* ii. 60, *t.* 195; *Schlechter in Engl. Jahrb.* xviii. *Beibl.* 45, 18, *and* xx. *Beibl.* 51, 29.

KALAHARI REGION: Orange River Colony; Sevenfontein, *Burke*! Transvaal; near Lydenburg, *Wilms,* 944! near Pretoria, *Burtt Davy,* 760! 826! 3060! *Miss Leendertz*! near Ermelo, *Burtt Davy,* 957! Modderfontein, *Conrath,* 993! plains around Barberton, 2800 ft., *Galpin,* 699! near Johannesburg, *Rand,* 1008! Pilgrims Rest, *Greenstock*!

EASTERN REGION: Natal; near Estcourt, 4000 ft., *Wood,* 3477! near Colenso, *Krook,* 823 (ex *Schlechter*). Zululand; on dry plains, *Gerrard,* 1291! Swaziland; Mbabane, *Miller*!

Also in Tropical Africa.

21. A. aurea (Schlechter in Journ. Bot. 1896, 455); stem 9–15 in. high, simple or occasionally branched at the base, rather slender, sometimes zigzag, puberulous along 2 lines; internodes ¾–2 in. long; leaves ascending-spreading, 1½–3¼ in. long, ½ lin. broad, linear-filiform, glabrous; umbels 3–6, long-pedunculate, racemose or the upper subcorymbose, 3–7-flowered; peduncles 1¾–5¼ in. long, slender, with a puberulous line on one side; bracts ½–1½ lin. long, subulate; pedicels 3–7 lin. long, slightly pubescent; sepals 1–1¼ lin. long, ⅓–⅖ lin. broad, lanceolate, acuminate, pubescent; corolla-lobes spreading, 1½–2 lin. long, 1–1¼ lin. broad, ovate, notched at the obtusely pointed apex, thinly pubescent and often more or less deeply stained with dull purple-brown on the back, glabrous and greenish-yellow or whitish on the inner face, very minutely ciliate on one margin; corona-lobes arising close to the base of the staminal column and twice as long as it, bright orange-yellow, compressed-cucullate, subtruncate, with the inner apical angles produced into short erect obtuse teeth and the dorsal apex into a long linear channelled erectly spreading or recurved point 1¼–1½ lin. long, the cucullate part ½–¾ lin. long; staminal column ¾–1 lin. long; anther-appendages short, broadly rounded, inflexed over the margin of the truncate style-apex; follicles solitary, or if in pairs sometimes abruptly incurved at the base so as to cross and then widely diverge, 2½–3 in. long, ⅛ in. thick, terete-fusiform, tapering into a long acute beak, smooth, glabrous. *N. E. Br. in Dyer, Fl. Trop. Afr.* iv. i. 345; *Rand in Journ. Bot.* 1903, 200. *Gomphocarpus aureus, Schlechter in Engl. Jahrb.* xviii. *Beibl.* 45, 17, *and* xx. *Beibl.* 51, 28. *Schizoglossum pedunculatum, Schlechter in Verhandl. Bot. Ver. Brandenb.* xxxv. 50, *and Journ. Bot.* 1896, 419.

Var. *β* : **vittata** (N. E. Br.) ; corona-lobes whitish or pale yellow, with a brown or dark-coloured dorsal stripe, the point produced ¾–1¼ lin. beyond the cucullate part. *Gomphocarpus schizoglossoides, Schlechter in Engl. Jahrb.* xviii. *Beibl.* 45, 21, *not elsewhere.*

Coast Region : Cape Div. ; between Claremont and Kenilworth Race Course, *Schlechter,* 351 !

Kalahari Region : Transvaal ; grassy plains around Barberton, *Galpin,* 580 ! Waterval Boven, *Rogers* ! Pilgrims Rest, *Greenstock* ! near Johannesburg, *Rand,* 703 ! 1042 ! *Gilfillan in Herb. Galpin,* 6046 ! Modderfontein, *Conrath,* 994 ! near Rustenberg, *Miss Pegler,* 954 ! Zoutpans Berg, at Fuschgewacht, *Miss Leendertz,* 830 ! Malelane, *Rogers,* 701 ! Matebe Valley, *Holub* ! Mooi River, *Burke* ! Magalies Berg, *Burke* ! Carolina district, *Burtt Davy,* 2985 ! Var. *β* : Basutoland, *Cooper,* 932 ! Orange River Colony ; Bethlehem, *Richardson* ! Transvaal ; Saddleback Mountain near Barberton, 3000–4000 ft., *Galpin,* 500 !

Eastern Region : Natal ; plains near Newcastle, *Schlechter,* 3409 ! Var. *β* : Natal ; Biggars Berg, 4000 ft., *Wood,* 4555 ! near Newcastle, *Wood,* 5892 ! Swaziland ; between Carolina and Mbabane (Embabaan), 5000 ft., *Bolus,* 12152 ! Embabaan, *Miller* !

Also in Tropical Africa.

The plant found by Dr. Schlechter between Claremont and Kenilworth was doubtless introduced among ballast or products brought from the Transvaal. I have examined a flower from the type of *Schizoglossum pedunculatum,* Schlechter, and find that the corona-lobes have been wrongly described, they (and other parts of the plant) are identical with those of *Asclepias aurea.* The var. *vittata* is distinct from var. *brevicuspis,* S. Moore in Journ. Bot. 1902, 255 (from Tropical Africa), which has the corona-lobes entirely yellow and does not differ from the type except that the tips of the lobes are shorter than usual, but in this character they vary considerably. According to Miss Pegler the tuber is eaten by the natives who call it " carrot."

22. **A. cognata** (N. E. Br.); stem of the only specimen seen $1\frac{3}{4}$ ft. high, slender, simple, puberulous bifariously or all round on the upper part; leaves in 6–7 pairs $2\frac{1}{4}$–4 in. apart, erect, sessile, 1–$2\frac{1}{4}$ in. long, $\frac{1}{2}$–$\frac{2}{3}$ lin. broad, linear, acute, with revolute margins, puberulous; umbels 4, lateral and terminal, the lowest shortly pedunculate, about 4-flowered; bracts $1\frac{1}{2}$ lin. long, subulate, puberulous; pedicels $\frac{1}{4}$–$\frac{1}{3}$ in. long, adpressed-puberulous; sepals 2 lin. long, $\frac{1}{2}$ lin. broad, lanceolate, acute or acuminate, pubescent; corolla-lobes apparently ascending-spreading, $\frac{1}{4}$ in. long, $1\frac{3}{4}$ lin. broad, narrowly elliptic, acute, glabrous and apparently dingy brownish-green or perhaps purplish-tinted on the inner face, with a line of minute hairs down the middle of the dull-purplish back; corona-lobes arising $\frac{1}{2}$ lin. up the staminal column, $1\frac{1}{2}$ lin. long, ascending-spreading apparently dull-purple or purplish-brown, narrowly wing-margined at the base, concave-cucullate, resembling a deep spoon-bowl, obtuse, the broadly-rounded margins inflexed and meeting at the base of the bowl and there produced into erect linear obtuse nearly free teeth $\frac{1}{3}$ lin. long, with a very compressed linear-oblong obtuse erect process arising from the bottom of the bowl in front of and attaining to the same level as the teeth; staminal column 1 lin. long; anther appendages roundish-subquadrate, subtruncate, inflexed on the truncate style-apex; very young follicles densely white-tomentose.

Eastern Region : Mount Ayliff Div. ; Insizwa Range, 6500 ft., *Schlechter*, 6496 !

The specimen from which I describe is in the Herbarium of Dr. Bolus and is named *A. schizoglossoides*, by Dr. Schlechter himself, but is not the type of that species (which see), and differs from it in being taller, with larger and differently coloured flowers, and in the corona-lobes not being truncate at the top.

23. **A. flava** (N. E. Br.); stems mostly solitary, simple, 7–15 in. high, slender, pubescent; leaves usually in 3–4 (rarely 5–6) distant pairs, erect, subsessile, $\frac{2}{3}$–2 in. long, $\frac{3}{4}$–$1\frac{1}{2}$ lin. broad, linear, acute, revolute at the margins, mostly puberulous above and on the midrib beneath, occasionally glabrous; umbel solitary and terminal or 2–3 sessile umbels crowded together at the apex of the stem, with or without a lateral sessile umbel 1–2 in. below, 3–16-flowered; bracts 1–2 lin. long, filiform; pedicels $2\frac{1}{2}$–6 lin. long, pubescent; sepals about $1\frac{1}{4}$ lin. long, $\frac{1}{2}$ lin. broad, lanceolate, acute, pubescent; corolla-lobes spreading, $1\frac{1}{2}$–$2\frac{1}{3}$ lin. long, 1–$1\frac{1}{4}$ lin. broad, elliptic or elliptic-oblong, acute, glabrous on both sides, yellow; corona-lobes arising $\frac{1}{4}$–$\frac{1}{2}$ lin. up the staminal column, $\frac{1}{2}$–$\frac{2}{3}$ lin. long, spreading, or perhaps ascending, thin, yellow, elliptic as seen from above, concave-cucullate, resembling a spoon-bowl, not compressed, rather broadly flattened with narrow keel-like margins on the back, very obtuse or subemarginate at the apex, with the sides inflexed at the base meeting and produced into erect linear obtuse teeth $\frac{1}{4}$ lin. long or less, with an erect much compressed linear-subclavate process arising from the bottom of the bowl and equalling the teeth;

staminal column $\frac{3}{4}$–1 lin. long ; anther-appendages transversely oblong, subtruncate, inflexed on the somewhat thick and puffy rim of the depressed-truncate style-apex ; very young follicles densely tomentose, brown-coloured at the apical part.

Coast Region : Albany Div. ; mountains near Grahamstown, 2000 ft., *Glass in MacOwan, Herb. Austr.-Afr.*, 1503, excluding "additional specimen " !

Eastern Region : Tembuland, 2500–4500 ft. ; Bazeia Mountain, *Baur*, 556 ! Engcobo Mountain, *Bolus*, 10216 ! Griqualand East, 4000–6000 ft.; Malowe Mountain, *Tyson*, 2723 ! and in *MacOwan and Bolus, Herb. Norm.*, 1086 ! Vaal Bank, *Haygarth in Herb. Wood*, 4230 ! Mount Kwenkwe, *Bolus*, 10215 ! Mount Currie, *Tyson*, 1686 ! Natal : Dargle Farm, *Fannin*, 13 ! 39 ! Liddesdale, *Wood*, 4249 ! near Lidgetten, *Wood*, 6255 ! Shafton, Howick, *Mrs. Hutton*, 408 ! and without precise locality, *Gerrard*, 1315 !

24. **A. schizoglossoides** (Schlechter in Engl. Jahrb. xviii. Beibl. 45, 32) ; stem solitary, simple, 6–12 in. high, slender, puberulous bifariously or all round ; leaves in 3–5 distant pairs, erect, $1\frac{1}{4}$–3 in. long, $\frac{1}{3}$–$1\frac{1}{2}$ lin. broad, linear, acute, often narrowed into a petiole at the base, glabrous, or puberulous above and on the midrib beneath ; umbel solitary and terminal, or 2–3 sessile umbels crowded together at the apex of the stem with or without 1–2 distant sessile lateral umbels below them ; bracts subulate, 1–2 lin. long ; pedicels 3–5 lin. long, pubescent or puberulous ; sepals $1\frac{1}{2}$ lin. long, $\frac{1}{2}$ lin. broad, lanceolate, acute, pubescent ; corolla-lobes spreading, $2\frac{1}{3}$–3 lin. long, $1\frac{1}{3}$–$1\frac{3}{4}$ lin. broad at the middle, elliptic, subacute, glabrous on both sides, yellow ; corona-lobes arising about $\frac{1}{3}$ lin. up the staminal column, erect, apparently deep orange-coloured, compressed-cucullate, subquadrate viewed sideways, obliquely and narrowly wing-margined at the base, truncate at the top, obtusely angular but not produced into distinct teeth at the apex of the sides, which do not incurve or meet at their edges, within and adnate to the apex of the lobe arises a much flattened deltoid-oblong very obtuse erect process, usually shortly exceeding the lobe ; staminal column 1–$1\frac{1}{4}$ lin. long ; anther-appendages transverse, subreniform, obtusely rounded, inflexed on the rather acutely edged rim of the depressed-truncate style-apex. *Schlechter in Journ. Bot.* 1896, 451. *Gomphocarpus diploglossus, Turcz. in Bull. Soc. Nat. Mosc.* 1848, i. 258.

Coast Region : Albany Div. ; Grahamstown hills, *MacOwan*, 850 ! Coldspring, *Glass*, 276 ! Fort Beaufort Div. ; peaks of the Winterberg Range, *Ecklon*, 23, *Mrs. Barber*, 84 ! Stutterheim Div. ; Dohne Mountain, 5000 ft., *Sim*, 1237 ! *Bolus*, 10214 ! Kaffraria ; mountain sides, *Mrs. Barber*, 35 !

Eastern Region : Natal ; Inanda, *Groom in Herb. Wood*, 1408 ! near Van Renen, *Wood*, 6576 ! and *MacOwan, Herb. Austr.-Afr.*, "1503, *additional specimen* " only !

Although originally described nearly 60 years ago, this species does not appear to be very common, and has remained unidentified with Turczaninow's plant until now.   I am indebted to Dr. Schlechter for flowers from his type of *A. schizoglossoides*, which agrees with all the specimens above quoted ; the locality and collector were omitted on publication, but Dr. Schlechter informs me that it was collected by "Mrs. Barber, probably in British Kaffraria," so it probably formed part of her gathering 35.

25. A. filiformis (Benth. et Hook. f. Gen. Pl. ii. 753 and 754 in notes under *Schizoglossum* and *Gomphocarpus*); stems numerous, 3–6 ft. high, slender, about ½ lin. thick at the flowering part, simple or rarely branched above, glabrous, covered with a thin white dry secretion; leaves 2–4 in. long, $\frac{1}{3}$–$\frac{3}{5}$ lin. broad, filiform or linear, acute, glabrous; umbels 4–8, pedunculate, racemose, 6–10-flowered; peduncles $\frac{1}{3}$–1 in. long; pedicels 4–6 lin. long, at first minutely white-tomentose, becoming glabrescent; sepals 1–1¼ lin. long, ovate or lanceolate, acute, white-tomentose; corolla-reflexed, sulphur-yellow; lobes 2–2$\frac{1}{3}$ lin. long, 1$\frac{1}{3}$ lin. broad, ovate, subacute or obtusely pointed, puberulous on the back; corona-lobes arising $\frac{1}{4}$–$\frac{1}{3}$ lin. up the staminal column and shortly exceeding it, 1½ lin. long, erect, complicate-cucullate, obliquely truncate, with the obtuse apex and the side apical angles slightly produced, and a compressed shortly exserted horn arising at or below the middle of the cavity within; staminal column 1¼–1½ lin. long; anther-appendages lanceolate, acute, inflexed on the depressed-truncate style-apex; follicles solitary, about 1½ in. long and ¼ in. thick, narrowly lanceolate, acuminate from the middle into a long acute beak, pulverulent or at length glabrate; seeds 2½–3 lin. long, ovate, plano-convex, smooth, brown, with a narrow darker brown border. *N. E. Br. in Dyer, Fl. Trop. Afr.* iv. i. 336; *Schinz in Verhandl. Bot. Ver. Brandenb.* xxx. 262, *under A. buchenaviana, Schinz; K. Schum. in Engl. and Prantl, Pflanzenfam.* iv. ii. 238. *A. flagellaris, Bolus ex Schlechter in Engl. Jahrb.* xviii. *Beibl.* 45, 32, *under A. schizoglossoides, and* xxi. *Beibl.* 54, 7, *and Journ. Bot.* 1896, 451. *Lagarinthus filiformis, E. Meyer, Comm.* 203. *Gomphocarpus filiformis, Dietr. Syn. Pl.* ii. 900; *Decne in DC. Prodr.* viii. 558.

Var. β, **buchenaviana** (N. E. Br. in Dyer, Fl. Trop. Afr. iv. i. 336); stems stouter, $\frac{3}{4}$ lin. thick at the flowering part, much branched above; otherwise as in the type, or with the horn of the corona-lobes arising above the middle of the cavity within. *A. buchenaviana, Schinz in Verhandl. Bot. Ver. Brandenb.* xxx. 261; *Engl. in Engl. Jahrb.* xix. 148; *K. Schum. in Engl. and Prantl, Pflanzenfam.* iv. ii. 238.

Central Region : Calvinia Div. ; between Lospers Plaats and Springbok Kuil River, *Zeyher*, 1167 ! Prince Albert Div. ; plain near Zwartbulletje, *Drège*, 892 ! near the Gamka River, *Burke*, 247 ! Boters Kraal, near Prince Albert, *Bolus*, 12137 ! Beaufort West Div. ; near Beaufort West, *Mrs. Barber*! near Riet Vley in the Gouph, *Bolus in MacOwan and Bolus, Herb. Norm.* 640 !

Western Region : Var. β : Great Namaqualand ; in the bed of a river, *Schinz*, 19 !

Also found in Tropical Africa.

This plant may prove to be of some economic value, since Mrs. Barber states as follows on a note with her specimens :—"This species ought to be useful as a fibre-plant, it is exceedingly tough, grows from 4–6 ft. high, and has upwards of two hundred, almost leafless slender stems on a single plant."

26. A. decipiens (N. E. Br.) ; a branching shrubby plant, 2–3 ft. high, with the habit of *A. fruticosa*, Linn. ; stems white-tomentose on the young parts ; leaves ascending, 3–5 in. long, ½–3 lin. broad, linear, tapering to an acute point, with revolute margins, narrowed at the base into a petiole 1–3 lin. long, puberulous with curved hairs above and on the midrib beneath ; umbels numerous, lateral at the nodes, racemose, 4–6-flowered ; peduncles ½–1 in. long, white-tomentose ; bracts ⅓–½ in. long, linear, tapering to an awn-like point, tomentose, very deciduous ; pedicels ¾–1⅙ in. long, white-tomentose ; sepals reflexed, 3 lin. long, ¾–1 lin. broad at the base, thence tapering to a very acute point, tomentose ; corolla-lobes reflexed ; 3–4 lin. long, 2–3 lin. broad, elliptic or elliptic-ovate, obtusely pointed, ciliate on one or both margins, apparently white or pale yellowish, tinted with purplish on the back ; corona-lobes arising ½–¾ lin. up the staminal column and reaching its top, 1½–1¾ lin. long, subquadrate and 1–1½ lin. broad when viewed side-ways, compressed-complicate, shallowly divided or grooved along the top and inner side, otherwise solid, without a tooth or process within, top margin slightly rounded, notched in front of a back-wardly directed falcate tooth at the inner apical angles, inner margins narrowly winged, the wing often ending in a small tooth at the apex ; staminal column 2–2½ lin. long ; anther-appendages suborbicular or elliptic, their obtuse tips inflexed on the rim of the truncate style-apex ; follicles ellipsoid or ellipsoid-ovoid at the lower part, abruptly contracted into a stout subulate beak, white-tomen-tose and covered all over with fine dark filiform processes 3–4 lin. long.

KALAHARI REGION : Transvaal ; Hennops Valley, near Irene, *Conrath,* 997 ! Wonderboom Poort, 4600 ft., *Miss Leendertz,* 607 ! near Johannesburg, *Rand,* 858 ! 1128 ! *Gilfillan in Herb. Galpin,* 6045 ! Pretoria, *Schlechter,* 3589 !

EASTERN REGION : Natal ; Mooi River, *Gerrard,* 1290 !

This closely resembles *A. Burchellii,* Schlechter, but differs from that species by the presence of the falcate recurved tooth at the top of the corona-lobes and by the follicles abruptly contracted into a beak, not gradually tapering to the apex, and, when they open, the basal part remains very much more concave and bag-like than in *A. Burchellii.*

27. **A. Burchellii** (Schlechter in Journ. Bot. 1895, 336 in note) ; stems 2–4 ft. high, branched ; branches white-tomentose ; leaves very numerous, erect or ascending, 1½–4 in. long, ¾–3 lin. broad, linear, acute, narrowed into the 1–2½ lin.-long petiole, revolute at the margins, at first softly puberulous, becoming more or less glabrous ; umbels several, racemosely arranged, white-tomentose, pedunculate, 3–7-flowered ; peduncles subequal, ½–1 in. long ; bracts 2–5 lin. long, linear or filiform, acute, deciduous ; pedicels ½–1 in. long, equal ; sepals 1⅔–2 lin. long, ½–¾ lin. broad, lanceolate, acute, reflexed ; corolla reflexed, tomentose on the back, glabrous on the inner face, white ; lobes 3½–4 lin. long, 2–2½ lin. broad, elliptic-ovate, obtusely pointed ; corona-lobes arising ½ lin. up the staminal column and reaching its top, 1½–1⅔ lin. long, broadly D-shaped and 1¼–1½ lin. broad viewed sideways, inflexed sides subrectangularly acute, but not

produced into distinct teeth, without a tooth or horn in the cavity ; follicles 2–2¾ in. long, about ¾ in. in diam., ovoid-lanceolate, tapering into a beak and covered with very numerous bristle-like processes, white-tomentose all over ; seeds about 2½ lin. long and 1 lin. broad, oblong-ovate, very convex on one side, concave or flat with a raised margin on the other, scrobiculate, dark brown, ochreous at the margin. *Schlechter in Engl. Jahrb.* xxi. *Beibl.* 54, 5, *and in Journ. Bot.* 1896, 452 ; *N. E. Br. in Dyer, Fl. Trop. Afr.* iv. i. 335. *Gomphocarpus tomentosus, Burch. Trav.* i. 543 ; *Schlechter in Engl. Jahrb.* xviii. *Beibl.* 45, 9, *and* xx. *Beibl.* 51, 38.　*G. fruticosus, var. tomentosus, K. Schum. in Engl. Pfl. Ost-Afr. C.* 322.　*G. lanatus, E. Meyer, Comm.* 202 ; *Dietr. Syn. Pl.* ii. 900 ; *Decne in DC. Prodr.* viii. 558.

CENTRAL REGION : Somerset Div. ; near Somerset East, *Atherstone,* 135 ! Graaff Reinet Div. ; hills near Graaff Reinet, *Bolus,* 378 ! and *in MacOwan and Bolus, Herb. Norm.* 1317 ! Aberdeen Div. ; hills in the Camdeboo, *Drège* ! at the foot of Camdeboo Mountain, *Drège,* Beaufort West Div. ; flats near Beaufort West, *Mrs. Barker* ! Hopetown Div., *Shaw* !

WESTERN REGION : Great Namaqualand, *Schinz,* 22 !

KALAHARI REGION : Griqualand West ; Asbestos Mountains, at the Kloof Village, *Burchell,* 2024 ! Transvaal ; Fourteen Streams, *Burtt Davy,* 1564 !

Also in Tropical Africa.

28. A. fruticosa (Linn. Sp. Pl. ed. i. 216) ; a shrub 3–10 ft. high ; branches erect, puberulous ; leaves more or less ascending ; petiole 2–4 lin. long ; blade 2–6 in. long, 2–9 lin. broad, linear or linear-lanceolate, acute or acuminate, mucronate, or rarely aristate, cuneate-acute at the base, with revolute margins, glabrous or minutely puberulous ; umbels pedunculate, lateral at the nodes, racemose, 6–10-flowered ; peduncles ¾–1½ in. long, pubescent or puberulous ; bracts 3–4 lin. long, linear-subulate, puberulous, deciduous ; pedicels ½–¾ in. long, puberulous ; sepals 1½–2 lin. long, lanceolate, acuminate, pubescent or puberulous ; corolla-lobes reflexed, 3½–4 lin. long, 2 lin. broad, ovate-oblong, obtuse, glabrous on both sides, usually but not always ciliate along one margin ; corona-lobes arising about ½ lin. above the base of the staminal column and reaching its summit, erect, compressed-cucullate, subrectangular in side view, with the apical angles of the sides produced into recurving narrowly falcate teeth, rising considerably above the rest of the lobe, their tips incurved towards each other ; margins of the sides narrowly winged outside ; no tooth or horn within ; follicles 2–3 in. long, ovoid, tapering into a beak, minutely tomentose and beset with bristles, which are nearly or quite glabrous. *Lam. Encycl.* i. 283 *and Ill.* ii. 323, *t.* 175, *fig.* 2 ; *Thunb. Prodr.* 47, *in Nov. Act. Acad. Petrop.* xiv. (1805) 507, *Fl. Cap. ed.* 2, ii. 154, *and ed. Schultes,* 234 ; *Ait. Hort. Kew. ed.* 1, i. 308 ; *Pers. Syn.* i. 276 ; *Willd. Sp. Pl.* 1271, *and Enum. Hort. Berol.* 279 ; *Schlechter in Engl. Jahrb.* xxi. *Beibl.* 54, 7 ; *Journ. Bot.* 1896, 451, *and Ann. Naturhist. Hofm. Wien,* xv. 67, *Rand in Journ. Bot.* 1903, 337 ; *N. E. Br. in Dyer, Fl. Trop. Afr.*

iv. i. 330. *A. glabra, Mill. Gard. Dict. ed.* 8, *no.* 12. *A. salicifolia, Salisb. Prodr.* 150. *A. crassifolia, Linn. ex Decne in DC. Prodr.* viii. 572. *Gomphocarpus fruticosus, R. Br. in Mem. Wern. Soc.* i. 38 ; *Ait. Hort. Kew. ed.* 2, ii. 80 ; *Schultes, Syst. Veg.* vi. 87 ; *Spreng. Syst. Veg.* i. 849 ; *G. Don, Gen. Syst.* iv. 143 ; *Decne in Ann. Sc. Nat. sér.* 2, ix. 324, *and in DC. Prodr.* viii. 557 ; *Reichb. Fl. Germ.* xvii. *t.* 1071 ; *Schlechter in Engl. Jahrb.* xviii. *Beibl.* 45, 7 ; xx. *Beibl.* 51, 31 ; *Rand in Journ. Bot.* 1903, 200. *G. frutescens* (*error for fruticosus*), *E. Meyer, Comm.* 202 ; *Dietr. Syn. Pl.* ii. 900 *partly* ; *Meisn. in Hook. Lond. Journ. Bot.* ii. 1843, 543 (*by error* 443).

South Africa : without locality ; *Herb. Linnæus*! *Herb. Miller*! *Thunberg*! *Masson*! *Niven*, 23 ! *Sieber*, 138 ! *Pappe*!

Coast Region : Cape Div. ; Simons Bay, *MacGillivray*, 590 ! Orange Kloof, *Wolley Dod*, 3286! Oudtshoorn Div. ; near the Olifants River, *Thom*, 185 ; Riversdale Div. ; near the Gauritz River, *Krauss*, 1258 (ex *Meisner*) ; near Riversdale, *Penther*, 832. George Div. ; near Silver River, *Penther*, 812. Knysna Div. ; near Knysna, *Penther*, 825 (all ex *Schlechter*). Queenstown Div. ; near Shiloh, 3500 ft., *Baur*, 10 ! 846 ! plains near Queenstown, *Galpin*, 1699 !

Central Region : Graaff Reinet Div. ; near Graaff Reinet, 2500 ft., *Bowie*! *Bolus*, 195 ! Colesberg Div. ; near Colesberg, *Knobel*, 7 !

Western Region : Little Namaqualand ; by the Orange River near Varleptpram, *Drège*, 3045b ! and without precise locality, *Wyley*!

Kalahari Region : Griqualand West ; Lower Campbell, *Burchell*, 1806 ! Orange River Colony ; Bethlehem, *Richardson*! Leeuw Spruit and Vredefort, *Barrett Hamilton*! Pargs, 4500 ft., *Rogers*, 802 ! Harrismith, *Sankey*, 241 ! Transvaal : near Pretoria, *Kirk*, 29 ! *Miss Leendertz*, 73 ! *Wilms*, 950 ! *Burtt Davy*, 2161 ! 2184 ! near Lydenburg, *Wilms*, 950 ! Platsand, *Rogers*, 629 ! Irene, *Burtt Davy*, 741 ! Fourteen Streams, *Burtt Davy*, 1565 ! Crocodile River, *Burtt Davy*, 235 ! near Johannesburg, *Rand*, 859 ! 1127 ! *Ommaney*! Modderfontein, *Conrath*, 996 ! Pilgrims Rest, *Greenstock* !

Eastern Region : Natal ; near Greytown, *Wilms*, 2142 ! near Ladysmith, *Wood*, 4755 ! and without precise locality, *Gueinzius* !

Also in North and Tropical Africa, the Mascarene Isles, Madeira, Canaries, Arabia and South Europe, perhaps naturalised in some of the localities. Specimens with ovoid or ellipsoid shortly and acutely pointed follicles are probably of hybrid origin between this species and *A. physocarpa.*

29. **A. physocarpa** (Schlechter in Engl. Jahrb. xxi. Beibl. 54, 8) ; plant up to 6 ft. high ; stems branched, whitish-pubescent or sub-tomentose on the young parts ; leaves ascending or spreading ; petiole 2–3 lin. long ; blade 2–4 in. long ; 3–10 lin. broad, linear-lanceolate or lanceolate, very acute, mucronulate, acute at the base, glabrous on both sides or with a very minute scanty pubescence above and often on the midrib beneath ; umbels lateral, racemose, 5–10-flowered ; peduncles $\frac{1}{2}$–1$\frac{3}{4}$ in. long, whitish-pubescent ; bracts 1$\frac{1}{2}$–2 lin. long, subulate or filiform, very deciduous ; pedicels $\frac{1}{2}$–1 in. long, puberulous ; sepals 1$\frac{1}{2}$–2 lin. long, lanceolate, acumi-nate, pubescent ; corolla-lobes reflexed, 3–4 lin. long, 2–2$\frac{1}{2}$ lin. broad, subobtuse, ciliate along one margin, otherwise glabrous, white ; corona-lobes arising $\frac{1}{2}$–$\frac{3}{4}$ up the staminal column and nearly or quite reaching to its top, 1$\frac{1}{2}$–1$\frac{3}{4}$ lin. long, complicate, subquadrate in the side view, with the apical angles of the inflexed sides shortly

produced above the rest of the lobe into a broad obliquely subtruncate or subfalcate lobule, no tooth within; staminal column
2-2½ lin. long; anther-appendages transverse, somewhat semicircular, partly inflexed over the margin of the depressed-truncate
style-apex; follicles solitary, inflated, subglobose or obliquely ovoid
1½-2½ in. in diam., puberulous and covered with long soft glabrous
bristles. *Schlechter in Journ. Bot.* 1896, 453; *Engl. Jahrb.* xxxix.
242, *and Ann. Naturhist. Hofmus. Wien,* xv. 68; *N. E. Br. in Dyer,
Fl. Trop. Afr.* iv. i. 328. *A. fruticosa, Mill. Dict. ed.* 8, *no.* 13,
not of Linn. Gomphocarpus physocarpus, E. Meyer, Comm. 202;
Dietr. Syn. Pl. ii. 900; *Decne in DC. Prodr.* viii. 558; *K. Schum. in
Engl. and Prantl, Pflanzenfam.* iv. ii. 236, *and in Engl. Pfl. Ost-Afr.
C.* 322; *Schlechter in Engl. Jahrb.* xviii. *Beibl.* 45, 9 *and* 20; *Neub.
Gart. Mag.* 1894, 73; *Wood, Natal Pl.* iii. 19, *t.* 217; *De Wild. Ic.
Hort. Then.* iii. *t.* 88. *G. fruticosus, Sims, Bot. Mag. t.* 1628, *not of
R. Br. G. frutescens* (*error for fruticosus*), *Dietr. Syn. Pl.* ii. 900
partly.

SOUTH AFRICA : without locality, *Forbes,* 43! *Herb. Miller*!
COAST REGION : Worcester Div. ; near Worcesler, *Cooper,* 1622! 2715! Uiten-
hage Div. ; by the Zwartkops River, *Niven,* 164! *Zeyher,* 524! Albany Div. ; near
Glenfilling, *Drège*! Kloofs near Grahamstown, *MacOwan,* 200! Bathurst Div. ;
Trapps Valley, *Miss Daly,* 689! Komgha Div. ; Kei River Valley, *Krook,* 813!
CENTRAL REGION : Somerset Div. ; Bosch Berg, *Mac Owan*!
EASTERN REGION : Transkei ; near Kentani, 1000 ft., *Miss Pegler,* 295!
Griqualand East ; around Clydesdale, *Tyson,* 2016! Pondoland, *Bachmann,* 1109!
1110! Natal ; near Durban, *Cooper,* 2716! *Wood,* 39! Umkomanzi River Valley,
Krook, 790 (ex *Schlechter*), Inanda, *Wood,* 33! Port Shepstone, *Rogers,* 533!
and without precise locality, *Gerrard,* 439! *Sutherland*! *Sanderson*! *Gueinzius*!
Delagoa Bay, *Langley*!

Also in Tropical Africa and the Cape Verd Islands.

The corona-lobes vary in the shape of the tooth or lobule at the apex of the
inflexed sides and sometimes approach *A. fruticosa* in this particular, but the
inflated follicles readily distinguish it.

30. **A. crinita** (N. E. Br. in Dyer, Fl. Trop. Afr. iv. i. 352);
stem erect, probably 2 ft. or more high, very much branched,
glabrous below, puberulous above, at length becoming hollow;
leaves opposite or 3-4 in a whorl, scarcely petiolate, 4-6 in. long,
about 3 lin. broad, the lower lanceolate, the upper linear, all
acuminate, revolute at the margins, puberulous when young, becoming glabrous; umbels lateral at the nodes, pedunculate, shorter
than the leaves, 4-6-flowered; peduncles rather stout; pedicels
about ¾ in. long; sepals lanceolate, acuminate, puberulous; corolla
scarcely exceeding the calyx (but represented as twice as long in
the figure), white; lobes very spreading-recurved, ovate-oblong,
obtuse; corona-lobes erect, crenate at the apex, at least in the
dried state; follicle (solitary?), ascending, ovoid-lanceolate, acuminate, covered with green bristles. *Gomphocarpus crinitus, Bertoloni
in Mem. Accad. Scienze Istit. Bologna,* 1851, iii. 253, *t.* 20, *fig.* 1;
K. Schum. in Engl. Pfl. Ost-Afr. C. 322.

EASTERN REGION: Portuguese East Africa; Inhambane, *Fornasini* (ex *Bertoloni*).

Probably also in Tropical Africa.

31. A. rivularis (Schlechter in Journ. Bot. 1896, 455); stems many from the same root and leafy to the base, growing to a height of 4 ft., simple, glabrous; leaves in whorls of 3–4 or the uppermost sometimes opposite, petiolate, $2\frac{1}{2}$–4 in. long, $1\frac{1}{2}$–$5\frac{1}{2}$ lin. broad, from almost linear to narrowly lanceolate, tapering to a very acute apex, cuneate-acute at the base, glabrous, sometimes minutely ciliate when young; umbels numerous, lateral at the nodes, racemose, pedunculate, 8–25-flowered; peduncles 1–$1\frac{3}{4}$ in. long, glabrous; bracts very caducous, only seen on very young umbels, numerous, $2\frac{1}{2}$–3 lin. long, linear-subulate, glabrous; pedicels $\frac{3}{4}$–$1\frac{1}{4}$ in. long, slender, glabrous; sepals $1\frac{1}{2}$–2 lin. long, $\frac{2}{3}$–$\frac{3}{4}$ lin. broad, ovate-lanceolate, acute or acuminate, glabrous; corolla-lobes reflexed, $2\frac{1}{2}$–$3\frac{1}{2}$ lin. long, $1\frac{1}{2}$–$1\frac{3}{4}$ lin. broad, lanceolate-oblong, obliquely notched at the obtuse apex, glabrous on both sides, often ciliate on one margin, white; corona-lobes arising $\frac{1}{3}$–$\frac{1}{2}$ lin. up the staminal column and shorter than it, erect, cucullate, about 1 lin. long from the truncate base to the obliquely truncate top, about $\frac{2}{3}$–$\frac{3}{4}$ lin. broad, the inflexed sides produced into erect acute points about $\frac{1}{2}$ lin. long, without a tooth or process within; staminal column 2–$2\frac{1}{4}$ lin. long; anther-appendages rather broader than long, subtruncately emarginate, rounded at the sides, slightly folded back longitudinally, loosely inflexed over the slightly depressed-truncate style-apex; follicles solitary, $1\frac{1}{4}$–$2\frac{1}{4}$ in. long, nearly 1 in. thick, ovoid, ellipsoid or elongate-obovoid, very obtuse, glabrous, apparently somewhat fleshy; seeds about $2\frac{1}{2}$ lin. long, $1\frac{3}{4}$ lin. broad, ovate, central part smooth and slightly convex on both sides, surrounded by a broad transversely plicate-wrinkled rim, glabrous, brown. *Schlechter in Ann. Naturhist. Hofmus. Wien,* xv. 68. *Gomphocarpus rivularis, Schlechter in Engl. Jahrb.* xx. *Beibl.* 51, 36.

KALAHARI REGION: Transvaal; Mooi River, *Burke*, 331! and probably from the same locality, *Zeyher*, 1166! Derde Poort, near Pretoria, *Rehmann,* 4782! in streams near Middelburg, *Schlechter*, 3789! Potchefstroom, *Burtt Davy*, 2174! Piet Retief, *Burtt Davy*, 1932! between Middelburg and Crocodile River, *Wilms*, 949! stream between Carolina and Wonderfontein, *Burtt Davy*, 2993! Potgeiters Rust, *Rogers*, 538!

EASTERN REGION, growing in water and very wet situations: Transkei; Bashee River, *Drège*, 4942! Tsomo River, *Bowker and Mrs. Barber*, 869! Illitooli River: *Bowker and Mrs. Barber*, 824! Tembuland; Bazeia, *Baur*, 487! Qumankwe River, near Engcobo, *Bolus*, 10202! Griqualand East; near Clydesdale, *Tyson.* 3119! banks of the Umzimhlava River, *Krook*, 795! *Schlechter*, 6541! near Kokstad, *Haygarth in Herb. Wood*, 4257! Pondoland, *Drège*, 4943!

32. A. rotundifolia (Mill. Dict. ed. 8, no. 15); a branching shrub about 2–3 ft. high; branches 2–$3\frac{1}{2}$ lin. thick, velvety-tomentose, with internodes $\frac{1}{4}$–$\frac{1}{2}$1 in. long; leaves spreading, subsessile or with petioles up to $1\frac{1}{2}$ lin. long; blade 1-2 in. long, $\frac{1}{2}$-$1\frac{1}{3}$ in. broad, varying from oblong-lanceolate to oblong, elliptic-oblong, or elliptic, acute or obtuse, apiculate, rounded or subcordate at the base, flat, coriaceous, rather rigid, sometimes pubescent, but more usually glabrous on both sides with the exception of the midrib beneath, whitish, especially beneath, with green veins (*Burchell*); umbels 1-7 to a branch, velvety-tomentose or villous, pedunculate, lateral at the nodes, racemose or subcorymbose at the apex, 12-30-flowered; peduncles $1\frac{1}{2}$-10 lin. long; bracts 3-4 lin. long, filiform or lanceolate, deciduous; pedicels $\frac{3}{4}$-1 in. long; sepals reflexed, 2-$2\frac{1}{2}$ lin. long, $\frac{3}{4}$-$1\frac{1}{4}$ lin. broad, ovate-lanceolate, acute, villous or tomentose; corolla-lobes reflexed, often with up-curved tips, $2\frac{1}{2}$-$3\frac{1}{2}$ lin. long, $1\frac{3}{4}$-2 lin. broad, ovate, obtusely pointed, villous or pubescent on the back, very minutely puberulous or glabrous on the inner face, white, often purplish on the back; corona-lobes arising at the base of the staminal column and as long, yellowish, $1\frac{1}{2}$-$1\frac{3}{4}$ lin. long, cucullate, oblique at the top, highest at the rounded apex, margins incurved and contiguous, produced into a short acute or obtuse tooth at the top, no tooth or process within, in the dried state a ridge or frill extends all round on the back below the middle; staminal column $1\frac{1}{2}$ lin. long; anther-appendages broadly ovate or suborbicular, obtuse, inflexed over the truncate 5-angled style-apex; anther-wings very prominent and angular at the base; follicles 2-$2\frac{3}{4}$ in. long, about $\frac{3}{4}$ in. thick, ovoid, tapering into an obtuse beak, pubescent and thinly beset with soft spines or processes 3-4 lin. long; seeds $2\frac{1}{2}$ lin. long, $1\frac{1}{3}$-$1\frac{1}{2}$ lin. broad, oblong-ovate, concave on one side, convex on the other, somewhat reticulately rugose. *A. cancellata, Burm. f. Prodr. Cap.* 7. *A. pubescens, Linn. Mant.* ii. 215, *as to description (excluding synonymy) and specimen in Herb. Linn.; Lam. Encycl.* i. 280; *Willd. Sp. Pl.* i. 1263; *Pers. Syn.* i. 275. *A. arborescens, Linn. Mant.* ii. 216; *Lam. Encycl.* i. 283; *Thunb. Prodr.* 47; *in Nov. Act. Acad. Petrop.* xiv. (1805), 509; *Fl. Cap. ed.* 2, ii. 156, *and ed. Schultes,* 235; *Ait. Hort. Kew. ed.* 1, i. 308; *Willd. Sp. Pl.* i. 1271, *and Enum. Hort. Berol.* 278; *Jacq. Hort. Schoenbr.* i. 25, *t.* 50; *Pers. Syn.* i. 276; *Schlechter in Engl. Jahrb.* xxi. *Beibl.* 54, 5, *and Journ. Bot.* 1896, 451. *A. arborea, Salisb. Prodr.* 150. *A. crassifolia, Hort. Paris. ex Decne in DC. Prodr.* viii. 572. *A. vestita, Hook. in Bot. Mag. t.* 4106, *not of Hook. and Arn. Gomphocarpus arborescens, B. Br. in Mem. Wern. Soc.* i. 38; *Ait. Hort. Kew, ed.* 2, ii. 79; *Schultes, Syst. Veg.* vi. 86; *Spreng. Syst. Veg.* i. 849; *Reichb. Mag. Bot. t.* 28; *E. Meyer, Comm.* 201; *G. Don, Gen. Syst.* iv. 143; *Dietr. Syn. Pl.* ii. 900; *Decne in DC. Prodr.* viii. 557; *Krauss in Flora,* 1844, 826.—*Apocynum frutescens, &c., Burm. Rar. Afr. Pl. Dec.* 2, 31, *t.* 13. *Apocynum radice longa, &c., Burm. l. c.* 32, *t.* 14, *fig.* 1.

SOUTH AFRICA: without locality; *Oldenburg! Nelson! Herb. Linnæus! Niven,* 24! *Forster!* and *cultivated specimen*!

COAST REGION: Cape Div.; hills around Cape Town, *Thunberg! Burchell,*

8454! *Burke*! *Prior*! *Rehmann*, 1581! Tyger Berg, *Krauss*. Riversdale Div. ; between Little Vet River and Garcias Pass, *Burchell*, 6882! Uniondale Div. ; Long Kloof, *Burchell*, 4965! Port Elizabeth Div. ; Baakens River, *Bowie*! Albany Div. ; Assegai Bosch, *Zeyher*, 896! near Brookhuizens Poort, 2100 ft., *MacOwan*, 320! on high hills, *Mrs. Barber*, 78!

CENTRAL REGION : Graaff Reinet Div. ; near Graaff Reinet, *Bolus*, 469!

WESTERN REGION : Little Namaqualand ; between Modderfontein and Uitkomst, 2000 ft., *Drège*! Modderfontein, *Whitehead*! Steinkopf, *Max Schlechter*! and without precise locality, *Scully*, 21!

The figure (*Plukenet, Phytog.* t. 139, fig. 1, and in *Morison, Hist.* iii. sect. 15, t. 3, fig. 35) quoted by Linnæus under *A. pubescens* certainly does not belong to the *A. pubescens* of his Herbarium and description, and appears to be a bad representation of *Xysmalobium undulatum*, R. Br. ; it cannot be *A. crispa* as stated by R. Brown in *Mem. Wern. Soc.* i. 38, but no specimen of the plant can be found in Plukenet's Herbarium at the British Museum.

33. A. glaucophylla (Schlechter in Journ. Bot. 1896, 455) ; a stout herb with 3–4 simple stems 1–2 ft. high (or more?), 3–5 lin. thick, glabrous in all parts except the corolla ; leaves sessile, 3–6 in. long, 1–2½ in. broad, elongate-ovate, acute, cordate at the base, with rounded stem-clasping lobes, flat, with smooth margins, glaucous, bluish-green ; umbels several, racemosely arranged, pedunculate, 7–15-flowered ; peduncles 1–1¾ in. long ; bracts 2½–4 lin. long, linear or filiform, acute ; pedicels 8–15 lin. long ; sepals 3 lin. long, 1–1½ lin. broad, oblong, subacute, ciliate, reflexed ; corolla reflexed, greenish, very minutely puberulous on the inner face, glabrous on the back ; lobes 4½–6 lin. long, 2¾–4 lin. broad, elliptic, somewhat obtusely pointed, minutely ciliate on one margin ; corona-lobes arising ½ lin. above the base of the staminal column and reaching its top, erect, 2–2½ lin. long, compressed-cucullate, subquadrate and 2–2½ lin. broad in side view, truncate at the top, with the apical angles of the sides produced into a short acute point over the top of the style-apex, no tooth within, yellowish-brown ; staminal column 2½ lin. long ; anther-appendages oblong or ovate, obtuse, inflexed over and concealing the rather small truncate style-apex ; anther-wings prominent and acutely angular at the base ; follicles 3½ in. or more long, about ¾ in. thick, fusiform, obtuse, with about 6 narrow entire or toothed wings, some of the upper teeth produced into compressed linear obtuse processes 2–4 lin. long. *S. Moore in Journ. Bot.* 1902, 255 ; *N. E. Br. in Dyer, Fl. Trop. Afr.* iv. i. 321. *Gomphocarpus glaucophyllus, Schlechter in Engl. Jahrb.* xviii. *Beibl.* 45, 19, *and* xx. *Beibl.* 51, 31.

KALAHARI REGION : Transvaal ; near Pretoria, *McLea in Herb. Bolus*, 3086! *Burtt Davy*, 2570! *Miss Leendertz*, 373! Flats of Kaap Valley, near Barberton, *Galpin*, 663! at the foot of the Magaliesberg Mountains near Rustenberg, *Miss Pegler*, 966!

EASTERN REGION : Swaziland ; near Kings Kraal, *Burtt Davy*, 2943!

Also in Tropical Africa.

34. A. dregeana (Schlechter in Journ. Bot. 1895, 337, in note, excl. synonym, and in Engl. Jahrb. xxi. Beibl. 54, 6) ; plant 3–15 in. high, or rarely taller, with 1–3 simple or branched stems, puberulous with usually rust-coloured or golden-brown minute curved hairs, or sometimes the leaves and corolla nearly or quite glabrous ; leaves in 4–8 pairs ; petiole $\frac{1}{2}$–1 lin. long ; blade $\frac{3}{4}$–2$\frac{1}{2}$ in. long, $\frac{1}{6}$–1 in. broad, linear-oblong, lanceolate, broadly oblong or ovate-oblong, acute or obtuse and mucronate, broadly rounded to slightly cordate at the base, often somewhat crisped at the scabrous margins ; umbels 2–6 to a branch, rarely solitary, lateral and terminal, racemose or sub-corymbose, 5–10-flowered ; peduncles $\frac{1}{2}$–2$\frac{3}{4}$ in. long ; bracts few, 1–3 lin. long, filiform or setaceous ; pedicels unequal, $\frac{1}{3}$–1 in. long ; sepals 1$\frac{3}{4}$–2$\frac{1}{2}$ lin. long, $\frac{1}{2}$–$\frac{2}{3}$ lin. broad at the base, tapering to a very acute point ; corolla-lobes rotately spreading or somewhat reflexed, 2$\frac{1}{2}$–3$\frac{1}{3}$ lin. long, 1$\frac{3}{4}$–2 lin. broad, elliptic or elliptic-oblong, minutely notched at the subacute point, glabrous on the inner face, brilliant green (*Wood*) ; corona-lobes arising $\frac{1}{2}$–$\frac{3}{4}$ lin. up the staminal column, complicate, subdeltoid or obliquely, or somewhat deltoidly subquadrangular viewed sideways, top margin sloping downwards and outwards, about 1 lin. high, 1$\frac{1}{4}$–1$\frac{1}{2}$ lin. broad, narrowly rim-margined at the truncate base, in dried flowers with the top margins horizontally spreading, forming a broad flat rim or margin, without a cavity, but fissured between the very acute or slightly hooked teeth into which the inner apical angles are produced ; staminal column 2–2$\frac{1}{2}$ lin. long ; anther-appendages small, suborbicular, obtuse, erect, inflexed at their tips on the rim of the truncate style-apex ; fruit about 3–3$\frac{1}{4}$ in. long, $\frac{1}{2}$–$\frac{2}{3}$ in. thick, lanceolate-fusiform, obtusely beaked, with 4 longitudinal serrate wings, rusty puberulous when young ; seeds about 2$\frac{1}{4}$ lin. long, 1$\frac{1}{4}$–1$\frac{1}{3}$ lin. broad, thick, with incurved margins, forming a deep groove on one side, very convex on the other, rugose all over with minute tubercles and thin ridges, dark brown. *Schlechter in Journ. Bot.* 1896, 452, *and Ann. Naturhist. Hofmus. Wien,* xv. 67. *Pachycarpus? viridiflorus, E. Meyer, Comm.* 214 ; *Meisn. in Hook. Lond. Journ. Bot.* ii. 1843, 545 (*by error* 445) ; *Krauss in Flora,* 1844, 827. *Xysmalobium viridiflorum, Dietr. Syn. Pl.* ii. 903. *Gomphocarpus viridiflorus, Decne in DC. Prodr.* viii. 561. *G. marginatus, Schlechter in Engl. Jahrb.* xviii. *Beibl.* 45, 8, *not of Decne.*

VAR. β, **Calceolus** (N. E. Br.) ; leaves $\frac{3}{4}$–3 in. long, $\frac{1}{6}$–1$\frac{1}{2}$ in. broad, ovate-lanceolate to elliptic-oblong, obtuse or acute ; corolla-lobes 3–4$\frac{1}{2}$ lin. long, 2–2$\frac{1}{3}$ lin. broad, green ; corona-lobes usually shaped as in the type, with a distinct fissure-like or open-mouthed cavity extending to half-way down, not solid to the top, usually with the top margins recurved forming a rim, but occasionally sub-quadrate in side view and truncate at the top, with the margins erect and slightly gaping, usually marked with violet on the basal rim ; follicles 4–4$\frac{1}{2}$ in. long, $\frac{3}{4}$ in. or more thick, lanceolate-fusiform, about equally tapering at both ends, with 4 serrate wings, puberulous. *A. Calceolus, S. Moore in Journ. Bot.* 1903, 312, 338. *Gomphocarpus marginatus, Schlechter in Engl. Jahrb.* **xx.** *Beibl.* 51, 33, *not of Decne.*

VAR. γ, **sordida** (N. E. Br.) ; plant as in var. β ; corolla-lobes 4 lin. long, 2$\frac{1}{2}$ lin. broad, closely veined with dark purple-brown on a green ground ; corona-lobes subquadrate in side view, top margins truncate, not recurved so as to

form a rim, not or but slightly produced into teeth at the inner top angles, with a
fissure-like cavity extending to the base on the inner side, marked with violet at
the top and on the basal rim.

SOUTH AFRICA : without locality, *Masson* ! *Thunberg* !
COAST REGION : Riversdale Div. ; near Zoetemelks River, *Burchell*, 6674 !
George Div. ; near George, *Prior* ! near Silver River, *Penther*, 828 (ex *Schlechter*).
Uitenhage Div. ; Zuurberg Range, *Drège* ! Bathurst Div. ; Kaffir Drift Military
Post, *Burchell*, 3771 ! Albany Div. ; near Grahamstown, *Bolton* ! *MacOwan*, 661 !
Galpin, 2903 ! *Misses Daly and Sole*, 325 ! *Schönland*, 343 ! and without precise
locality, *Hutton* ! *Mrs. Barber*, 115 ! *Bowker* ! East London Div. ; near East
London, *Wood in Herb. Galpin*, 3362 ! Komgha Div. ; near Komgha, 2000 ft.,
Flanagan, 372 !
CENTRAL REGION : Somerset Div. ; Somerset East, *Bowker* !
KALAHARI REGION : Var. β : Transvaal; open veld north of Johannesburg,
Rand, 966 ! Jeppestown Ridge, Johannesburg, *Gilfillan in Herb. Galpin*, 6148 ! near
Rustenburg, *Miss Pegler* ! Vlakfontein, near Carolina, *Burtt Davy*, 2969 ! Pilgrims
Rest, *Greenstock* ! near Pretoria, *Burtt Davy*, 3058 ! 3233 ! *Miss Leendertz*, 505 !
Schlechter, 3588 ; Modderfontein, *Conrath*, 998 !
EASTERN REGION : Transkei ; Kreilis Country, *Bowker*, 4 ! 294 ! Kentani, *Miss
Pegler*, 656 ! Natal ; at the foot of Table Mountain, *Krauss*, 470 (not dissected,
probably = var. β) ! Var. β : Griqualand East ; Mount Malowe, *Tyson*, 3113 !
Natal ; Dargle Farm, *Fannin*, 12 ! Inanda, *Wood*, 350 ! near Charlestown, *Wood*,
5643 ! near Krantz Kloof, *Schlechter*, 3195 ! and without precise locality, *Gerrard*,
1295 ! *Sanderson*, 241 ! 514 ! *Mrs. K. Saunders*, 8 ! Swaziland ; near Mbabane
(Embabaan) *Bolus*, 12119 ! Var. γ : Transkei ; Kentani, 1000 ft., *Miss Pegler*,
655 !

As stated by Dr. Schlechter in Engl. Jahrb. xxi. Beibl. 54, 6, the name
Gomphocarpus marginatus has been erroneously applied to this plant in many
Herbaria.   The flowers of the type and var. *sordida* look so very distinct in size
and appearance that they might perhaps be regarded as different species, but they
seem quite connected by the var. *Calceolus*, which sometimes (*Gerrard*, 1295,
Fannin, 12) also has subquadrate corona-lobes.

35. A. cultriformis (Harv. ex Schlechter in Engl. Jahrb. xviii.
Beibl. 45, 31) : stems 6–12 in. high, simple, hairy ; leaves ascending,
1–2 in. long with the short petiole, 3–7 lin. broad, oblong-linear,
lanceolate or ovate-lanceolate, acute, rounded to subcordate at the
base, thinly covered with long spreading hairs on both sides, ciliate
with shorter hairs ; umbel solitary and terminal or 2–3, the lower
lateral at the nodes, 3–10-flowered ; peduncle ½–1 in. long, hairy to
subglabrous ; bracts 3 lin. long, linear-subulate, ciliate or hairy ;
pedicels 4–9 lin. long, more or less hairy ; sepals 3–4 lin. long, 1–1½
lin. broad, lanceolate, acute, hairy ; corolla-lobes ascending-spreading,
perhaps slightly incurved, 5½–6½ lin. long, 3½–5 lin. broad, ovate,
acute, sometimes with recurved margins, more or less hairy on the
back, pubescent to densely white-hairy at the tips and along the
margins on the otherwise glabrous inner face, green, or sometimes
apparently dark dull-purple-brown ; corona-lobes arising ½ lin. up
the staminal column, erect, white, with a violet spot on the back
near the apex and more or less violet at the base, 1–2 lin. long and
about as broad across the side, compressed cucullate, subquadrate in
side view, truncate with a rim at the base, truncate to very
slightly concave and somewhat sloping outwards along the top,

angles of the sides produced into long subulate acute teeth, erect or directed over the backs of the anther-appendages, minutely puberulous within with a compressed deltoid acute or transverse truncate process not reaching to the top of the lobe; staminal column 2–2½ lin. long, with a truncate collar under each corona-lobe; anther-appendages broadly ovate, obtuse, inflexed on the truncate style-apex. *Schlechter in Engl. Jahrb.* xx. *Beibl.* 51, 26, *and in Journ. Bot.* 1896, 451; *Rand in Journ. Bot.* 1903, 340. *Gomphocarpus cultriformis, Harv. ex Schlechter in Engl. Jahrb.* xviii. *Beibl.* 45, 31, *and Journ. Bot.* 1896, 451.

KALAHARI REGION : Transvaal; Elandspruit Mountains, 7000 ft., *Schlechter*, 4005 ! Modderfontein, *Conrath*, 1002 ! High Veld near Belfast, 6500 ft., *Bolus* ! at the Beacon, Vlakfontein, *Burtt Davy*, 2970 ! near Johannesburg, *Rand*, 1126 !

EASTERN REGION : Transkei ; Tsomo, *Mrs. Barber*, 831 ! Natal ; Dargle Farm, *Fannin*, 85 ! Shafton, Howick, *Mrs. Hutton* ! Inanda, *Wood*, 405 ! and in *Herb. Natal*, 381 ! and without precise locality, *Sanderson*, 291 ! Zululand, *Gerrard*, 1296 ! Swaziland ; near Mbabane (Embabaan), 4700 ft., *Bolus*, 12118 ! *Burtt Davy*, 2758 !

36. **A. rara** (N. E. Br.) ; stem solitary, erect, 6–8 in. high, simple, rather densely and softly villous, at least above; leaves in 5–10 pairs; petiole ½–3 lin. long; blade ¾–2 in. long, ¼–1 in. broad, lowest usually ovate and obtuse, upper linear-lanceolate, narrowly oblong-lanceolate or elongate-ovate, acute, cordate or subcordate at the base, thinly pilose or subvillous on both sides or on the midrib only, ciliate ; umbels usually 3–4, lateral at the nodes and terminal, racemosely arranged, or sometimes 1 and lateral, with the stem prolonged beyond it, about 7-flowered; peduncles ½–1¾ in. long, villous ; pedicels ⅓–¾ in. long, villous ; sepals 1–1½ lin. long, ½–⅔ lin. broad, ovate-lanceolate, acute or acuminate, villous ; corolla-lobes reflexed-spreading, 2½–2¾ lin. long, 1½–2 lin. broad, ovate or elliptic-ovate, acute, thinly pilose on the back, glabrous on the inner face, brown or greenish-brown in dried flowers ; corona-lobes arising ¼–⅓ lin. up the staminal column, compressed cucullate, ¾–1 lin. long, somewhat D-shaped and ½–1 lin. broad in side view, the obliquely truncate top gradually rounded into the back or scarcely angular at the dorsal apex, inner top angles produced over the style-apex into rather long and very acute teeth, no tooth or process within the rather shallow fissure-like cavity ; staminal column 1–1¼ lin. long ; anther-appendages very small, transverse, rounded, erect, applied to the margin of the truncate style-apex, but not or very slightly inflexed over it.

COAST REGION : Albany Div. ; near Grahamstown, *Bolton* ! *MacOwan*, 713 ! *Miss Daly*, 735 !

37. **A. meliodora** (Schlechter in Journ. Bot. 1896, 455) ; plant with 1–2 stout stems 2–5 in. high, somewhat harshly subvillous-pubescent ; leaves in 5–7 pairs, spreading, petiolate, 1½–3¾ in. long,

¼–1 in. broad, lanceolate, acute, subtruncate, subcordate or sub-hastate at the base, more or less and somewhat harshly pubescent on both sides, shortly ciliate-scabrous on the margins ; umbels 2–3 to a stem, subcorymbose, 7–15-flowered ; peduncles ⅓–2 in. long, roughly pubescent ; bracts 2–3 lin. long, subulate or linear, pubescent ; pedicels 3½–7 lin. long, pubescent ; sepals about 2 lin. long, attenuate-lanceolate, acute, pubescent ; corolla-lobes reflexed, ¼ in. long, 1½ lin. broad, oblong, obtusely pointed, with a few hairs on the back or glabrous on both sides, apparently whitish, dull purple-tinted on the back ; corona-lobes arising at the base of the staminal column and much overtopping it, 1½–2 lin. long, erect, fleshy, subterete-oblong, very oblique at the top on the inner side, obtuse at the dorsal apex, scarcely cucullate, but with a narrow cavity or fissure down the inner face under the keel-like infolded edges, which end in a small erect tooth about 1 lin. below the apex, with a Λ-shaped rim on the back near the base ; staminal column 1½ lin. long ; anther-appendages ovate, acute or subobtuse, inflexed on the flattened cushion-like style-apex ; follicles 4–5 in. long, ½ in. thick, fusiform, tapering to an acute point, puberulous ; seeds 3½–4 lin. long, 2½–3 lin. broad, ovate, flat, broadly margined, scabrous on both sides, light brown. *Gomphocarpus meliodorus, Schlechter in Engl. Jahrb.* xx. *Beibl.* 51, 33.

Var. β, **brevicoronata** (N. E. Br.) ; corona-lobes scarcely overtopping the staminal column, subcucullate, with the inflexed sides as if pinched together and forming deltoid obtuse teeth at the top about ¼ lin. below the level of the apex ; otherwise as in the type.

Kalahari Region : Transvaal ; Magalies Berg, *Burke* ! *Zeyher,* 1165 partly ! near Pretoria, *Miss Leendertz,* 254 ! Pilgrims Rest, *Greenstock* ! *Nelson,* 72 ! Matebe Valley, *Holub* ! between Delagoa Bay and Barberton, *Bolus,* 7662 ! Sandloop, near Pietersburg, *Schlechter,* 4373 ; Ginsberg, East Rand, *Miss Pegler* ! Var. β : Transvaal ; near Rustenberg, *Miss Pegler in Herb. Bolus,* 10553 !

Xysmalobium brownianum, S. Moore, was also distributed by Zeyher as 1165.

38. **A. monticola** (N. E. Br.) ; plant dwarf, much branched at the base ; branches 2–5 in. long in the only example seen, probably more or less decumbent, rather slender, the stoutest not more than ¾ lin. thick, somewhat harshly pubescent ; leaves 4–6 pairs to a branch, shortly petiolate, ½–1⅓ in. long, 3–6 lin. broad, ovate or ovate-lanceolate, or the terminal pair lanceolate, acute, subcordate, subtruncate or broadly rounded at the base, thinly scabrous-pubescent on both sides ; umbels 2–3 to a stem, lateral and terminal, pedunculate, 6–12-flowered ; peduncles ⅔–1½ in. long, slender, puberulous on one side ; bracts not seen, very deciduous ; pedicels 5–6 lin. long, puberulous on one side ; sepals 1¼ lin. long, ½ lin. broad, lanceolate, acute, pubescent ; corolla-lobes reflexed, about 2½ lin. long, 1¼ lin. broad, elliptic-oblong, subacute, glabrous on the purple-brown back, minutely papillate-puberulous on the inner face, which appears to be greenish more or less tinged with purple ; corona-lobes arising ⅓ lin. up the staminal column and overtopping

it, erect, 1 lin. long, scarcely $\frac{1}{2}$ lin. broad across the side, cucullate, inflexed sides produced at the top into short falcate teeth resting on the backs of the anther-appendages, dorsal apex obliquely prolonged beyond them into an oblong obtuse slightly spreading concave tip, with a transverse curved wing-like ridge on each side at the middle of the lobe, apparently yellowish or whitish with their lower part and the base of the $1-1\frac{1}{4}$ lin.-long staminal column purple; anther-appendages about twice as broad as long, transversely oblong, emarginate, inflexed on the margin of the truncate style-apex.

COAST REGION : Queenstown Div. ; summit of the Andriesberg Range, 6800 ft., *Galpin*, 2262 !

39. **A. humilis** (Schlechter in Engl. Jahrb. xx. Beibl. 51, 26); plant 2–8 in. high with an elongated fleshy tuberous rootstock, producing 1–2 bifariously pubescent stems; leaves in 3–5 pairs, erect or ascending, mostly basal, $\frac{3}{4}$–3 in. long, 2–8 lin. broad, oblong or oblong-lanceolate, rarely linear or linear-lanceolate, obtuse or acute, rounded or narrowed at the base into a petiole 1–6 lin. long, pubescent on both sides and shortly ciliate ; umbels solitary and terminal or 2–4 and subcorymbose, pedunculate, 4–5- (6–8- *ex Schlechter*) flowered ; peduncles $\frac{1}{2}$–1 in. long, often with a long peduncle-like internode below them ; bracts 2 lin. long, linear-subulate, pubescent ; pedicels $\frac{1}{3}$–$\frac{3}{4}$ in. long, pubescent ; sepals 2–$3\frac{1}{2}$ lin. long, $\frac{2}{3}$–1 lin. broad, lanceolate, acuminate, pubescent ; corolla-lobes spreading, $3\frac{1}{2}$–5 lin. long, 2–$2\frac{1}{2}$ lin. broad, ovate or oblong-ovate, obtuse, thinly pubescent and dark purple on the back, glabrous and white on the inner face, more or less ciliate on one margin ; corona-lobes arising $\frac{1}{2}$ lin. up the staminal column and radiately spreading at about the level of its top, 3–$3\frac{3}{4}$ lin. long, complicate for half the length or beyond, and, viewed sideways, tapering from a 1 lin.-deep base to an acute apex, not produced into teeth at the rectangular basal angles of the inflexed sides, with a median keel for $\frac{3}{4}$ of its length between them, wing-like and as broad as the sides at the base, tapering and becoming undulated towards the apex, whitish ; staminal column $1\frac{1}{2}$ lin. long ; anther-appendages reniform, rounded, inflexed on the rim of the slightly depressed-truncate style-apex. *Schlechter in Engl. Jahrb. xxi. Beibl. 54, 8, and Journ. Bot. 1896, 452. Pachycarpus humilis, E. Meyer, Comm. 212. Xysmalobium humile, Dietr. Syn. Pl. ii. 902. Gomphocarpus humilis, Decne in DC. Prodr. viii. 561.*

CENTRAL REGION : Aliwal North Div. ; Witteberg Range, 6000–7000 ft., *Drège*, 3240 !

KALAHARI REGION : Orange River Colony ; Mont-aux-Sources, 8000 ft., *Flanagan in Herb. Bolus*, 8117 ! Caledon Range, *Thode*, 39 ! Basutoland, Machacha Mountain, *Bryce* !

The plant described as *A. humilis* by Schlechter in *Engl. Jahrb. xx. Beibl. 51, 26*, from the Elandspruit Mountains, Transvaal (*Schlechter*, 3880), may be this species, but according to description differs in having reflexed corolla-lobes only $1\frac{1}{2}$ lin. broad, erect corona-lobes more than twice overtopping the staminal column, and ovate anther-appendages. I have not seen it.

40. A. ulophylla (Schlechter in Engl. Jahrb. xxxviii. 32) ; stem 8 in. high, simple, erect, straight, bifariously puberulous, leafy ; leaves erectly spreading, 1¼–2⅓ in. long, 5–10 lin. broad below the middle, lanceolate-oblong, acuminate, at first minutely puberulous, becoming glabrous ; petiole 2½–5 lin. long ; umbels shortly pedunculate, not exceeding the leaves, 8–15-flowered ; pedicels 3½ lin. long, slender, puberulous, sepals 2½ lin. long, lanceolate, acute, sparsely pilose, not ciliate ; corolla-lobes 3 lin. long, oblong, somewhat obtuse, glabrous ; corona-lobes a little longer than the stipes of the staminal column, cuneate-obovate, 3-lobed at the apex, with the lateral lobes short and the middle one oblong, obtuse, incurved, fleshy, with a longitudinal median keel within ; staminal column conspicuously stipitate ; anther-appendages suborbicular, incurved.

KALAHARI REGION : Transvaal ; on a stony hill near Komati Poort, 1000 ft., *Schlechter*, 11788.

This seems to be very closely allied to *A. humilis*, Schlechter. I have not seen it.

41. A. hastata (Schlechter in Engl. Jahrb. xxi. Beibl. 54, 7) ; plant 9–15 in. high, branching ; branches somewhat compressed, scabrous-pubescent ; leaves spreading or ascending ; petiole 1–2½ lin. long ; blade ⅔–2½ in. long, 2½–8 lin. broad, oblong, deltoid-oblong or lanceolate, acute or obtuse, hastate or truncate at the base, narrowly or broadly revolute along the margins, scabrous above, subhispid on the midrib beneath ; umbel solitary or 2 terminating the main branches, rarely 3 and then racemose or subcorymbose, 8–20-flowered ; peduncles ¼–¾ in. long, more or less scaberulous ; bracts 1–2 lin. long, subulate, ciliate, deciduous ; pedicels 6–9 lin. long, scabrous-pubescent to nearly glabrous ; sepals reflexed, 1½–2 lin. long, ¾–1 lin. broad, ovate-lanceolate, acute, thinly pubescent ; corolla reflexed, quite glabrous ; lobes 2¼–3 lin. long, 1⅓–1¾ lin. broad, oblong, obtusely pointed, apparently purple, with the basal part whitish or greenish-white ; corona-lobes arising at the base of and much exceeding the staminal column, 1¾–2¼ lin. long, erect, cucullate, very obliquely produced from the middle into ovate obtuse slightly spreading tips, margins at the basal half infolded and meeting, produced at the apex into very short erect obtuse deltoid teeth, reaching to or slightly exceeding the middle of the lobe, on each side from a little below the teeth a narrow wing extends obliquely downwards and curved outwards towards the base, no tooth or other process within ; staminal column 1–1¼ lin. long ; anther-appendages transverse, much broader than long, emarginate or broadly rounded at the apex, applied to but scarcely inflexed over the margin of the slightly depressed-truncate style-apex ; follicles solitary, only seen in the young state, fusiform, acute, beset with spreading or curved minutely scabrous subulate processes 1–1½ lin. long. *Schlechter in Journ. Bot.* 1896, 452. *A. Flanaganii, Schlechter in Journ. Bot.* 1896, 454. *Gomphocarpus hastatus, E. Meyer, Comm.* 201 ; *Dietr. Syn.*

Pl. ii. 900; *Decne in DC. Prodr.* viii. 560. *G. asclepiaceus,
Schlechter in Engl. Jahrb.* xviii. *Beibl.* 45, 7. *G. geminatus,
Schlechter in Engl. Jahrb.* xviii. *Beibl.* 45, 8, *and* xx. *Beibl.* 51, 33
in note.

COAST REGION : King Williamstown Div. ; between Chalumna River and Buffalo
River, *Drège*, 4955 ! near King Williamstown, *Sim*, 275 ! 1565 ! Komgha Div. ;
near the Kei River, *Flanagan*, 391 ! *Krook*, 815 ! near Komgha, *Flanagan*,
1041 ! British Kaffraria, *Mrs. Hutton* !
EASTERN REGION : Transkei ; Kreilis Country, *Bowker*, 95 ! 295 !

I have examined type specimens of *Gomphocarpus asclepiaceus* and *G. geminatus*,
Schlechter (Flanagan, 1041 and 391) and find no distinction between them, except
that *G. asclepiaceus* is a luxuriant specimen with larger leaves than usual, the
flowers of both are identical with E. Meyer's type of *G. hastatus.*

42. **A. crispa** (Berg. Descr. Pl. Cap. 75); plant $\frac{1}{2}$–$1\frac{1}{4}$ ft. high,
branching at the base; branches erect or decumbent at the base,
villous or pubescent with short spreading hairs, leafy; leaves
spreading, shortly petiolate, $\frac{3}{4}$–$3\frac{1}{2}$ in. long, 1–9 lin. broad at the
obtuse subtruncate or subcordate-hastate base, thence gradually
tapering to the acute apex, with crisped or wavy margins, rarely
oblong-lanceolate, obtuse, with an apiculus and quite flat, rather
harshly pubescent or more or less scabrous on both sides, or at least
on the margins and midrib beneath ; umbels solitary and terminal
or 2–4 and racemosely arranged, 1–2 in. in diam., pubescent or
villous, pedunculate, 7–40-flowered ; peduncles $\frac{1}{2}$–3 in. long ; bracts
$1\frac{1}{2}$–3 lin. long, subulate or lanceolate ; pedicels 5–9 lin. long ; sepals
reflexed, $1\frac{1}{4}$–$2\frac{1}{4}$ lin. long, $\frac{1}{2}$–$\frac{3}{4}$ lin. broad, lanceolate, acute, villous-
pubescent and ciliate ; corolla-lobes reflexed, $2\frac{1}{2}$–$3\frac{1}{2}$ lin. long, $1\frac{1}{4}$–2
lin. broad, ovate or elliptic, acute or subobtuse, glabrous on both
sides or pubescent on the back, dull greenish or purplish-tinted on
the inner face, brownish-purple on the back ; corona-lobes arising
$\frac{1}{3}$–$\frac{1}{2}$ lin. above the base of the staminal column and much overtopping it, very variable in size, $1\frac{1}{2}$–3 lin. long, with the basal part
cucullate, $\frac{3}{4}$–$1\frac{1}{4}$ lin. long, and the inflexed sides produced into falcate
acute teeth projected over or against the top of the staminal column ;
upper part produced into an ovate or lanceolate acute or obtuse ascending-spreading blade $1\frac{1}{4}$–2 lin. long, flat at the apex, concave below, or
concave throughout, greenish or greenish-buff, with a dark purple
stripe down the middle and on the back ; staminal column 1–$1\frac{1}{2}$ lin.
long ; anther-appendages broader than long, usually bilobed, sometimes only emarginate, somewhat longitudinally folded or channelled
down the back, erect, with their tips just turned over the margin of
the truncate style-apex ; follicles 3–$4\frac{1}{2}$ in. long, fusiform, tapering
into a long beak, puberulous or glabrous, smooth or with several
series of short processes or tubercles along them ; seeds $\frac{1}{4}$ in. long,
$\frac{1}{8}$ in. broad, concave on one face convex on the other, rugulose-
scabrous. *Linn. Mant.* ii. 215 ; *Murr. Syst. Veg. ed.* 13, 213 ; *Linn. f.
Suppl.* 170 ; *Lam. Encycl.* i. 280 ; *Thunb. Prodr.* 47 ; *in Nov. Act.*

Acad. Petrop. xiv. (1805), 507; *Fl. Cap. ed.* 2, ii. 154, *and ed. Schultes,* 234; *Ait. Hort. Kew. ed.* 1, i. 305; *Willd. Sp. Pl.* i. 1263; *Pers. Syn.* i. 275; *Schlechter in Journ. Bot.* 1896, 451; *Engl. Jahrb.* xxi. *Beibl.* 54, 6; *and in Ann. Naturhist. Hofmus. Wien,* xv. 67. *A. sinuosa, Burm. fil. Prodr. Cap.* 7. *A. undulata, Murr. Syst. Veg. ed.* 13, 214, *not of Linn. A. repanda, Steud. Nom. Bot. ed.* 1, 77, *and Forsk. ex Steud. l.c. ed.* 2, i. 146. *Gomphocarpus crispus, R. Br. in Mem. Wern. Soc.* i. 38; *Ait. Hort. Kew. ed.* 2, ii. 79; *Schultes, Syst. Veg.* vi. 88; *Spreng. Syst. Veg.* i. 849; *G. Don, Gen. Syst.* iv. 144; *Dietr. Syn. Pl.* ii. 900; *Decne in DC. Prodr.* viii. 560. *G. hastatus, var. angustifolius, Meisn. in Hook. Lond. Journ. Bot.* ii. 1843, 543 *(by error* 443); *Krauss in Flora,* 1844, 826. *Pachycarpus crispus, E. Meyer, Comm.* 214; *Meisn. in Hook. Lond. Journ. Bot.* ii. 1843, 545 *(by error* 445); *Krauss in Flora,* 1844, 826. *Xysmalobium crispum, Dietr. Syn. Pl.* ii. 902.—*Apocynum erectum Africanum subhirsutum, &c., Commelin, Hort. Med. Amstel. Pl. Rar.* 17, f. 17.

Var. β, **pseudocrispa** (N. E. Br.); leaves 1–2½ in. long, 2–3½ lin. broad, linear or linear-lanceolate, acute at the apex, rounded, truncate or subhastate at the base, flat, usually revolute at the margins, but not undulate or crisped; umbels usually solitary and terminal on the branches, rarely 2 and 1 of them lateral; corolla-lobes 1¾–3 lin. long; corona-lobes ¾–1½ lin. long, with the cucullate and apical parts subequal. *Asclepias pseudocrispa, Schlechter in Journ. Bot.* 1895, 357.

Var. γ, **plana** (N. E. Br.); leaves ¾–2 in. long, 3½–8 lin. broad, linear-oblong, oblong, oblong-lanceolate or ovate-oblong, acute or obtuse, rounded to subtruncate at the base, flat, not undulate or crisped, somewhat hispid on the midrib beneath and scabrous on the margins, otherwise glabrous; umbels solitary and terminal or lateral or 2–3 lateral and terminal; corolla and corona as in var. β. *Gomphocarpus crispus, Schlechter in Engl. Jahrb.* xviii. *Beibl.* 45, 7, *not of R. Br.*

SOUTH AFRICA: without locality; *Herb. Linnæus*! *Oldenburg*! *Nelson*! *Masson*! *Zeyher,* 1157! *Harvey,* 559! 569!

COAST REGION: Vanrhynsdorp Div.; Gift Berg, *Drège*! Clanwilliam Div.; Bosch Kloof, *Drège*! Paarl Div.; Klein Drakenstein Mountains, *Drège*! near Paarl, *Schlechter,* 9212! Cape Div.; hills and flats near Cape Town, *Thunberg*! *Burchell,* 775! *Drège*! *Zeyher,* 4884! *Pappe*! *Bolus,* 3312! *Wolley Dod,* 349! 420! near Tyger Berg, *Prior*! Simons Bay, *Prior*! Stellenbosch Div.; near Eerste River, *Krauss,* 1259 (ex *Meisner*). Swellendam Div., *Krauss* (ex *Meisner*). Riversdale Div.; between Little Vet River and Garcias Pass, *Burchell,* 6901! Knysna Div.; near Knysna River, *Krauss* (ex *Meisner*). Humansdorp Div.; near Kromme River, *Bowie*! Uitenhage Div.; near Sunday River, *Bowie*! Uitenhage, *Burchell,* 4412! *Cooper,* 1488! *Schlechter,* 2520! Addo, *Drège*! near the Zwartkops River, *Zeyher,* 445! *Drège*! *Penther,* 789, 831 (ex *Schlechter*). Var. β: Bathurst Div.; near the mouth of Kowie River, *MacOwan,* 721! *Hutton*! Round Hill, *Bolus,* 6696! Trapps Valley, *Miss Daly,* 584! King Williamstown Div.; Keiskamma, *Mrs. Hutton*! East London Div.; near East London, *Wood in Herb. Galpin,* 3382! 3383! British Kaffraria, *Mrs. Hutton*! Var. γ: Komgha Div.; near Keimouth, *Flanagan,* 103! British Kaffraria, *Mrs. Hutton*!

CENTRAL REGION: Calvinia Div.; between Grasberg River and Waterval River, *Drège*!

EASTERN REGION: Transkei; Kreilis Country, *Bowker,* 12! 15!

43. A. Woodii (Schlechter in Journ. Bot. 1896, 456); stem simple, 1¼–2 ft. high, 1–1½ lin. thick at the base, often compressed, scabrous or scabrous-pubescent, with internodes 2–4½ in. long;

leaves 3–6 pairs to a stem, ascending or suberect, subsessile or very shortly petiolate, $2\frac{1}{4}$–$4\frac{1}{4}$ in. long, $2\frac{1}{2}$–8 lin. broad, broadly linear to lanceolate, acute or subacute, rounded or cuneate at the base, scabrous on both sides and at the narrowly revolute margins; umbel solitary and terminal or the stem forking once or twice dichotomously at the top and bearing 2–4 umbels, corymbosely arranged, 4–12-(mostly 7–8-) flowered; peduncles $\frac{1}{4}$–$2\frac{1}{4}$ or when solitary up to 7 in. long, scabrous-pubescent; bracts 2–3 lin. long, linear-subulate, acute, thinly pubescent; pedicels 3–5 lin. long, scabrous-pubescent or subhispid; sepals $2\frac{1}{4}$–$2\frac{3}{4}$ lin. long, 1–$1\frac{1}{4}$ lin. broad, lanceolate, acute, subhispid-pubescent; corolla-lobes reflexed-spreading, $2\frac{3}{4}$–$3\frac{1}{2}$ lin. long, $1\frac{3}{4}$–$2\frac{1}{4}$ lin. broad, ovate, subacutely pointed, glabrous on both sides, apparently whitish, tinted with purple on the back; corona-lobes arising close to the base of the staminal column and equalling or slightly exceeding it, $1\frac{1}{2}$–$1\frac{3}{4}$ lin. long, erect, about 1 lin. broad across the side, compressed-cucullate, truncately rounded at the top, rounded on the back, with the inflexed sides very slightly produced into obtuse teeth at the apex, no tooth or appendage within, yellow, with the basal part chocolate; staminal column $1\frac{1}{2}$ lin. long; anther-wings rectangularly angular at the middle; anther-appendages ovate to suborbicular, obtuse, with whitish granules embedded in them, producing a somewhat frosted appearance, closely inflexed over the concave style-apex. *Gomphocarpus Woodii, Schlechter in Journ. of Bot.* 1894, 258.

EASTERN REGION : Natal; Dargle Farm, *Fannin,* 34! near Howick, 3500–3700 ft., *Wood,* 4258! 5121! Greenwich Farm, Riet Vlei, *Fry in Herb. Galpin,* 2743! near Pietermaritzburg, *Wilms,* 2138!

The granules in the very thin membrane of the anther-appendages are remarkable, and I do not recollect having seen them in any other species; they appear to be crystals of oxalate of lime.

44. A. densiflora (N. E. Br. in Dyer, Fl. Trop. Afr. iv. i. 320); plant few-branched at the base, 1–$1\frac{3}{4}$ ft. high; stems erect or decumbent at the base, harshly or scabrous-pubescent; leaves 5–9 pairs to a stem, ascending or spreading; petiole 1–2 lin. long; blade 1–$2\frac{1}{2}$ in. long, 2–10 lin. broad at the base, the lower ovate or oblong-ovate, the middle elongated and gradually tapering from base to apex, the upper linear-lanceolate, all acute, cordate, subcordate, subtruncate or rounded at the base, narrowly revolute and undulate at the margins, harshly or scabrous-pubescent on both sides; umbel solitary, terminal, 1–$1\frac{1}{4}$ in. in diam., 15–25-flowered; peduncles on the main stems $3\frac{1}{2}$–7 in. long, more or less scabrous on one side; bracts $1\frac{1}{2}$–2 lin. long, subulate; pedicels 4–6 lin. long, glabrous or with a few stiff spreading hairs; sepals $1\frac{1}{4}$–$1\frac{1}{2}$ lin. long, lanceolate or ovate-lanceolate, acute, hairy-pubescent; corolla-lobes reflexed, $2\frac{1}{2}$ to nearly 3 lin. long, $1\frac{1}{3}$–$1\frac{3}{4}$ lin. broad, ovate or oblong-ovate, subacute, glabrous or with a few hairs on the back, glabrous on the inner face or most minutely puberulous at the base, white, tinted with dull purple on the back; corona-lobes arising near the base of the staminal column and equalling or slightly exceeding it, fleshy, about 1¼ lin. long, complicate-cucullate, with a narrow cavity, somewhat cuneate-oblong and 1 lin. broad at the top in side view, obliquely rounded in to the narrowed and almost claw-like base, rounded at the top, with the inner margin of the apical angles produced into very short deltoid obtuse teeth, below which, at about the middle, is a transverse upcurved ridge or keel on each side, no tooth or process within, entirely yellowish (or creamy-white?); staminal column 1–1¼ lin. long; anther-appendages somewhat rhomboid-orbicular, obtuse, inflexed upon and concealing the truncate 5-angled style-apex, which has 5 radiating depressed lines; anther-wings very broad and prominently angular just below the middle; follicles solitary, about 2½ in. long, ½ in. thick, narrowly lanceolate, tapering into an obtuse beak, echinate in several longitudinal series with rather stout conical obtuse or acute processes ½ lin. long, puberulous; seeds 1¾ lin. long, ¾–1 lin. broad, narrowly ovate or oblong-ovate, thick, deeply channelled on one side very convex on the other, with very prominent thin crowded ridges at the incurved margins and minutely tuberculate on the convex side, brown. *Rand in Journ. Bot.* 1903, 338.

Kalahari Region : Transvaal ; Magalies Berg, *Burke*! near Pretoria, *McLea in Herb. Bolus,* 5706! near Nylstroom, *Burtt Davy,* 2018! Warm Baths, *Burtt Davy,* 2221! Houtbosch (Woodbush), *Rehmann,* 5873! near Barberton, *Thorncroft,* 211 (*Herb. Wood,* 4287)!

Eastern Region : Swaziland ; near Mafutane, 1500 ft., *Bolus,* 12141 ! near Bremersdorp, 2500 ft., *Burtt Davy,* 3056 !

Also in Tropical Africa.

Very similar to *A. fallax,* Schlechter, but rather larger and stouter, and easily distinguished by the colour of the corona-lobes. The type of this species is Mrs. Evelyn Cecil's specimen from Rhodesia, with which the above quoted exactly agree, but in the original description the words "cuneate" and "subhastate" as applied to the leaves, and "50-flowered" and "1¼ in. in diam." as applied to the umbels, and "8 lin." long pedicels, refer to a specimen of *A. Cooperi,* N. E. Br., mixed with the above, from which I took those (but no other) characters without dissecting it and did not then take heed of its distinctness.

45. A. fallax (Schlechter in Journ. Bot. 1896, 455); plant few-branched at the base, ½–1 ft. high ; stems erect, harshly or scabrous-pubescent, leaves 5–7 pairs to a stem, ascending-spreading ; petiole 1–2 lin. long ; blade ¾–1¾ in. long, ⅓–⅔ in. broad at the truncate to broadly cuneate hastate base, thence tapering to the acute apex, harshly or scabrous-pubescent on both sides ; umbel solitary and terminal or 2, the lower lateral, both attaining to nearly the same level, ¾–1⅙ in. in diam., 12–15-flowered ; peduncle 3–4¾ in. long, scabrous-pubescent ; bracts 1–1½ lin. long, subulate ; pedicels 3–5 lin. long, harshly pilose ; sepals about 1½ lin. long, lanceolate, acute, hairy-pubescent ; corolla-lobes reflexed, with upcurved tips, 2–2⅓ lin. long, 1⅓–1½ lin. broad, ovate, subacute, glabrous on both sides or

with a few hairs on the back, white, tinted with purple on the back ;
corona-lobes arising near the base of the staminal column and
equalling or slightly overtopping it, fleshy, $1\frac{1}{4}$–$1\frac{1}{3}$ lin. long and about
1 lin. broad at the top in side view, complicate-cucullate, with a
narrow cavity, obliquely rounded at the back into the narrowed and
almost stalk-like base, truncate or slightly rounded at the top, with
the inner margin of the apical angles subrectangular or slightly
produced into very short deltoid teeth, no tooth within, greenish-
yellow at the top, dark violet-brown on the back and lower part ;
staminal column 1 lin. long ; anther-appendages suborbicular,
slightly emarginate, inflexed on the truncate obtusely pentagonal
style-apex, which is marked with 5 radiating slightly depressed
lines ; anther-wings rather broad, prominently angular below their
middle ; follicles (very immature) solitary, lanceolate-fusiform,
tapering to a beak, echinate with 6 (or more ?) longitudinal series of
small spine-like acute processes $\frac{1}{3}$ lin. long, puberulous. *Gomphocarpus fallax, Schlechter in Engl. Jahrb.* xx. *Beibl.* 51, 29.

KALAHARI REGION : Transvaal ; hills near Pretoria, *Schlechter*, 3604 !
Miss Leendertz, 294 ! Aapies Poort, *Rehmann*, 4157 ! Jeppestown Ridge, near
Johannesburg, *Gilfillan in Herb. Galpin*, 6048 ! hills near Rustenburg, 4000 ft.,
Miss Pegler, 980 !

In the original description the corolla-lobes are stated to be erectly-spreading
with reflexed tips, the anther-wings narrow and not at all enlarged at the base,
and in a note the style-apex is described as like a 5-rayed star with elongated
point-like bosses upon it. I have examined flowers from a part of the type, given
by Dr. Schlechter to Dr. Bolus, and do not find any of these characters agree with
that specimen ; the organs are as described above. The species is readily distinguished from those most closely allied to it by the peculiar coloration of the
corona-lobes.

46. A. Cooperi (N. E. Br.) ; plant $\frac{1}{2}$–1 ft. high, probably branching
at the base ; stems often decumbent below, bifariously subscabrous
to subhispid ; leaves 5–7 pairs to a stem, ascending or somewhat
spreading ; petiole 2–6 lin. long ; blade $1\frac{1}{2}$–3 in. long, $\frac{1}{4}$–$\frac{1}{2}$ in. broad,
lanceolate, cuneate or cuneately subhastate to subtruncate at the base,
thence tapering to the acute apex, undulate at the scabrous margins,
harshly pubescent with rather coarse hairs on both sides ; umbel
solitary, terminal, $1\frac{1}{2}$–2 in. in diam., semiglobose, 30–50-flowered ;
peduncle 2–4 in. long, subscabrous to thinly subhispid ; bracts about
2 lin. long, subulate, pubescent ; pedicels 6–8 lin. long, thinly and
somewhat harshly to densely and softly villous ; sepals about $1\frac{1}{2}$ lin.
long, lanceolate, acute, thinly hairy ; corolla-lobes reflexed, with
upcurved tips, $2\frac{1}{2}$–3 lin. long, $1\frac{1}{2}$ to nearly 2 lin. broad, elliptic-
oblong, subacute, most minutely puberulous at the base of the inner
face, otherwise glabrous or with a few hairs near the apex on the
back, apparently greenish-yellow, tipped with purple-brown on the
back ; corona-lobes arising about $\frac{1}{2}$ lin. up the staminal column,
erect, fleshy, complicate-cucullate, $1\frac{1}{4}$–$1\frac{1}{2}$ lin. long, about 1 lin. broad
across the side of the obliquely truncate top, which is slightly

highest at the dorsal apex, rounded in to the narrowed base at the back, in side-view somewhat oblong with a very short obliquely claw-like base, the inner margins produced at the top into short horizontal deltoid acute teeth, entirely deep-yellow; staminal column 1–1⅓ lin. long; anther-appendages subreniform, obtuse, inflexed upon and concealing the truncate style-apex; anther-wings remarkably prominent, angular at the middle.

COAST REGION: Fort Beaufort Div. ; Kat River Valley, *Cooper*, 473 ! tops of high hills above Water Kloof and at Kaal Neck near the Winter Berg, *Mrs. Barber*, 80 ! British Kaffraria, *Cooper*, 159 ! Stockenstrom Div. ; Lushington Mountain, *Scully*, 121 !

EASTERN REGION : Transkei ; Kreilis Country, *Bowker* (*Mrs. Barber*), 96 !

According to Mrs. Barber this is "a beautiful yellow-flowered species." It is allied to *A densiflora*, N. E. Br., but the different appearance, longer petioles, cuneate-based leaves and larger umbels of more numerous flowers, readily distinguish it.

47. A. adscendens (Schlechter in Journ. Bot. 1896, 455) ; plant 4–8 in. high, branching at the base into several erect or spreading-ascending stems, often decumbent at the base, compressed, subhispid ; leaves shortly petiolate, ½–1¾ in. long, 2–7 lin. broad, linear-lanceolate to ovate or ovate-lanceolate, acute, broadly rounded to cordate at the base, scabrous-pubescent, drying pale-green ; umbels solitary, terminal, semiglobose, 1–1¼ in. in diam., 12–20-flowered ; peduncle ½–2½ in. long, subhispid ; bracts several, 1½–3½ lin. long, linear, acute, pubescent ; pedicels 4–7 lin. long, hispid ; sepals 1½ lin. long, ⅔ to nearly 1 lin. broad, lanceolate or ovate-lanceolate, acute, subhispid ; corolla-lobes reflexed or reflexed-spreading, with up-curved tips, 2–2½ lin. long, 1½–1¾ lin. broad, elliptic-ovate, subacute, glabrous on both sides or with few hairs on the back, corona-lobes arising about ⅓ lin. up the staminal column and attaining to the same level, erect, not spreading at the tips, complicate-cucullate, subquadrate in side view, not quite 1 lin. long and about as broad at the truncate or truncately rounded top, rectangular or acute but scarcely produced into teeth at the apex of the inflexed sides, abruptly or subtruncately rounded into the column at the base, which has acute rim-like margins, with a narrow keel or wing near the inner margins curving outwards near the base, without a tooth or other process within the cavity; staminal column about 1¼ lin. long ; anther-appendages reniform or transverse, very obtusely rounded, inflexed upon the margin of the truncate pentagonal style-apex ; follicles mostly solitary, 2–2¼ in. long, about ½ in. thick, covered with stout spreading or recurving subulate processes 2–2½ lin. long, glabrous or thinly pubescent. *Gomphocarpus adscendens, Schlechter in Engl. Jahrb.* xviii. *Beibl.* 45, 16, *and* xx. *Beibl.* 51, 27.

KALAHARI REGION : Basutoland, *Cooper*, 2720 ! Transvaal ; around Barberton, 2500–3500 ft., *Galpin*, 596 ! *Bolus*, 7664 ! near Rustenberg, *Miss Pegler*, 978 ! near Heidelberg, 5200 ft., *Schlechter*, 3525 ! near Pretoria, *Burtt Davy*, 796 !

1992! 1998! 2235! 3068! 3075! Springbok Flats, *Burtt Davy,* 2507! Pinedene, near Irene, *Burtt Davy,* 2319! Matebes Valley, *Holub,* 1845! Modderfontein, *Conrath,* 991! and without precise locality, *McLea in Herb. Bolus,* 5708!

EASTERN REGION: Griqualand East; Mount Currie, *Tyson,* 3117! Swaziland; near Bremersdorp, 2100 ft., *Bolus,* 12138! *Burtt Davy,* 3026.

48. A. vicaria (N. E. Br.); plant probably branching at the base; stems decumbent at the base or for a considerable part of their length, $\frac{2}{3}$–1 ft. long, harshly pubescent; leaves in 5–7 pairs to a stem; petiole 2–3 lin. long; blade 1–1$\frac{3}{4}$ in. long, 5–11 lin. broad, of the lower leaves ovate, with a cordate base, of the upper lanceolate, with a rounded base, all acute, rather harshly pubescent on both sides; umbel solitary, terminal, half-globose, 1–1$\frac{1}{2}$ in. in diam., 25–40-flowered; peduncle 2$\frac{1}{2}$–3 in. long, harshly pubescent; bracts about 2 lin. long, lanceolate-subulate, pubescent; pedicels 4–5 lin. long, thinly covered with spreading rather stiff hairs; sepals 1$\frac{1}{2}$ lin. long, $\frac{3}{4}$ lin. broad, ovate-lanceolate, acute, pubescent with rather long hairs; corolla-lobes reflexed, 2–2$\frac{1}{3}$ lin. long, 1$\frac{1}{3}$–1$\frac{1}{2}$ lin. broad, ovate, subacute, pubescent on the back, puberulous on the inner face; corona-lobes arising a little above the base of the staminal column, erect, about $\frac{1}{2}$–$\frac{2}{3}$ lin. long, $\frac{2}{3}$–$\frac{3}{4}$ lin. broad at the top in side view, complicate-cucullate, truncate with a narrow rim at the base, obliquely truncate with recurved-spreading margins at the top, with the apical angles of the inflexed sides produced into short acute teeth resting upon the anther-appendages, and with an oblique wing near the inner margins at about the middle; staminal column 1$\frac{1}{4}$ lin. long; anther-appendages transversely oblong or reniform, rounded or slightly emarginate, inflexed upon (but not covering) the truncate style-apex.

EASTERN REGION: Pondoland (Mount Ayliff Div.); near Fort Donald, 4500 ft., *Tyson,* 1749!

49. A. albens (Schlechter in Engl. Jahrb. xxi. Beibl. 54, 5); plant 10–20 in. high, branching at the base; stems 1–3 lin. thick, often compressed, scabrous-pubescent or subhispid; leaves in 6–10 pairs; petiole $\frac{1}{2}$–3 lin. long; blade $\frac{3}{4}$–2 in. long, $\frac{1}{2}$–1 in. broad, ovate to ovate-lanceolate, acute, rarely obtuse, broadly rounded or subcordate at the base, rather coarsely and harshly or scabrous-pubescent on both sides, scabrous on the margins; umbels solitary, terminal, usually 1$\frac{1}{2}$–2 (rarely 3) in. in diam, 15–30-flowered; peduncle 1$\frac{1}{2}$–4 in. long, puberulous on one side and thinly subhispid; bracts 2–4 lin. long, linear-subulate or linear-lanceolate, acute; pedicels $\frac{1}{2}$–1 (rarely to 1$\frac{1}{2}$) in. long, varying from nearly glabrous to thinly hispid; sepals 1$\frac{1}{2}$–2 lin. long, $\frac{2}{3}$–1 lin. broad, ovate or lanceolate, acute, roughly pubescent; corolla-lobes reflexed, 3–3$\frac{1}{2}$ lin. long, 1$\frac{3}{4}$–2 lin. broad, ovate, acute, glabrous on both sides or thinly pubescent on the back, white or occasionally light purple; corona-lobes arising $\frac{1}{4}$–$\frac{1}{2}$ lin. up the staminal column and not exceeding it, erect, 1 lin. long, subquadrate, and 1 lin. broad at the top in side-view, scarcely longer than broad, cucullate, without a tooth or other process in the cavity, truncate and marginate at the base, produced into acute angles or short teeth at the apex of the inflexed sides, with the top margin subtruncate or sloping from the teeth outwards and downwards to the dorsal apex ; near the inner margins is a distinct wing, which curves outwards to the middle of the side at the base ; staminal column 1½ lin. long ; anther-appendages orbicular, obtuse or emarginate, more or less replicate, inflexed over and covering the truncate style-apex ; anther-wings broadest at the angular base. *Schlechter in Journ. Bot.* 1896, 452, *and Ann. Naturhist. Hofmus. Wien,* xviii. 398 ; *Rand in Journ. Bot.* 1903, 340. *Pachycarpus albens, E. Meyer, Comm.* 214 ; *Meisn. in Hook. Lond. Journ. Bot.* ii. 1843, 545 (*by error* 445) ; *Krauss in Fl.* 1844, 826. *Xysmalobium albens, Dietr. Syn. Pl.* ii. 902. *Gomphocarpus albens, Decne in DC. Prodr.* viii. 559 ; *Schlechter in Engl. Jahrb.* xviii. *Beibl.* 45, 7 and 17.

SOUTH AFRICA : without locality, *Herb. Thunberg*!

COAST REGION : Alexandria Div. ; Quagga Flats, *Bowie* ! Albany Div.; near Geelhoutboom,*Drège,* 3414 ! near Bothas Hill, *MacOwan,*227 ! near Grahamstown, *MacOwan* ! and without precise locality, *Hutton* ! East London Div. ; near East London, *Siefert in Herb. Conrath,* 991a ! King Williamstown Div. ; Keiskamma, *Mrs. Hutton* ! British Kaffraria, *Mrs. Hutton* !

EASTERN REGION : Transkei ; Kreilis Country, *Bowker,* 13 ! Griqualand East ; hills near Newmarket, *Krook,* 2407 (ex *Schlechter*). Pondoland ; near Umtsikaba River, *Drège* ! Natal ; near Umlaas River, *Krauss,* 84 ! near Durban, *McKen,* 826 ! *Wood in Herb. Norm. Austr.-Afr.,* 1029 ! Inanda, *Wood,* 101 ! Port Shepstone, *Rogers,* 523 ! and without precise locality, *Peddie* ! *Gueinzius* ! *Gerrard,* 564 !

The fruit described by E. Meyer belongs to a specimen of some species of *Pachycarpus,* which is mixed in his herbarium with those of this plant, and has the remains of several umbels lateral at the nodes, 2 of them with narrowly winged follicles as described. Another fruiting specimen in E. Meyer's herbarium, which may be that of the true *A. albens,* has a solitary terminal umbel, bearing a lanceolate acute follicle 3¼ in. long and ¾ in. thick, with one side straight, the other curved, smooth and glabrous,' not at all winged. Fruiting specimen distributed by Drège as *Pachycarpus albens,* under letter " C," from the Witteberg Range in Aliwal North, is like *A. albens* in habit, but the follicle is quite different from the others, being 2 in. long, ⅓ in. thick, fusiform-lanceolate and thickly covered with subulate processes 2-3 lin. long, and being from a region where *A. albens* is unknown is not likely to belong to that species.

50. A. affinis (Schlechter in Journ. Bot. 1896, 455, not of De Wild.) ; plant branched at the base ; branches more or less decumbent, 8-15 in. long, stout, subhispid or harshly pubescent ; leaves 1-3 in. long, including the short petiole, ½-1½ in. broad, ovate to ovate-lanceolate, acute, cordate or truncately rounded at the base, scabrous, subhispid or harshly pubescent on both sides ; umbel terminal, solitary, rarely 2, hemispheric, 1½-2½ in. in diam., 16-40-flowered ; peduncle 1-3¾ in. long ; bracts 2-3 lin. long, linear-lanceolate, acute ; pedicels ½-1 in. long and like the peduncle and bracts varying from very thinly subhispid to densely pilose-hispid, the hairs differing in harshness ; sepals 2-2½ lin. long, ⅔-1 lin. broad, lanceolate, acute, thinly to densely pilose-hispid ; corolla-lobes reflexed or reflexed-spreading, upcurved at the tips, $2\frac{1}{2}$–$3\frac{1}{2}$ in. long, $1\frac{2}{3}$–$2\frac{1}{4}$ lin. broad, ovate, subacute, thinly pubescent on the back, glabrous on the inner face ; corona-lobes erect, arising $\frac{1}{3}$–$\frac{1}{2}$ lin. up the staminal column and usually slightly exceeding it, 1–$1\frac{3}{4}$ lin. long, oblong and distinctly longer than broad in side view, cucullate, without a tooth or other process in the cavity, truncate, with a wing-like rim at the base, with the inner top angles of the sides produced into short and obtuse or longer falcate and acute teeth, from which the top margin slopes upwards and outwards in a distinct curve to the dorsal apex, which is often slightly recurved, with a narrow wing near the inner margins, which curves backwards towards the base, but is not easily seen in some dried flowers ; staminal column 1–$1\frac{1}{2}$ lin. long ; anther-appendages as broad or broader than long, subtruncate or emarginate, somewhat reflexed at the sides, inflexed upon the margin of the truncate style-apex. *A. macropus, Schlechter in Ann. Naturhist. Hofmus. Wien,* xv. 68 *not elsewhere. Gomphocarpus affinis, Schlechter in Engl. Jahrb.* xx. *Beibl.* 51, 27.

KALAHARI REGION : Orange River Colony, *Cooper,* 985 ! Transvaal ; Crocodile River, *Burke* ! by the Bushmans Spruit, west of Lake Chrissie, *Chamberlaine* ! near Wilge River, 4600 ft., *Schlechter,* 3751 ! Ginsberg, *Miss Pegler,* 1051 ! near Lydenburg, *Wilms,* 956 ! Modderfontein, *Conrath,* 991b ! Witkleifontein, *Burtt Davy,* 3126 ! White River Settlement, *Burtt Davy,* 1499 ! and without precise locality, *Sanderson* !

EASTERN REGION : Griqualand East ; at the foot of Mount Malowe, 3000 ft., *Tyson,* 3118 ! Natal ; Richmond, *Krook,* 816 ! Shafton, Howick, *Mrs. Hutton,* 402 ; Umkomanzi River, *Schlechter,* 6682 ! Swaziland, between Carolina and Mbabane, 5000–5600 ft., *Bolus,* 12139 ! near Kings Kraal, *Burtt Davy,* 2921 !

According to Miss Pegler the flowers are " yellow and lilac," whilst Dr. Schlechter describes the corolla as greenish-white. In dried specimens the corolla appears to be whitish, stained with purple on the back, and the corona brown or yellow-brown, often dark purple at the base. *A. affinis* is very closely allied to and may be only a variety of *A. albens,* but the usually more densely flowered umbels, and particularly the oblong shape and usually upward slope of the top margin of the corona-lobes seem constant characters whereby to distinguish them ; in *Burtt Davy,* 2921, however, which otherwise agrees, the corona-lobes are almost truncate at the top.

51. A. macropus (Schlechter in Journ. Bot. 1896, 456); plant branching at the base ; stems decumbent, 6–16 in. long, rather stout, unifariously puberulous and thinly hispidulous ; leaves spreading ; petiole 2–5 lin. long ; blade 1–$2\frac{1}{4}$ in. long, 10–15 lin. broad, ovate, ovate-oblong or elliptic-oblong, acute, rounded or cordate at the base, slightly scabrous or sparsely subhispid on both sides, scabrous on the margins ; umbels solitary, terminal, globose, $1\frac{3}{4}$–$2\frac{1}{4}$ in. in diam., 12–30-flowered ; peduncle 3–6 in. long, stout, pubescent along one side or subhispid all round ; bracts $\frac{1}{3}$ in. long, lanceolate or linear-lanceolate, acute, subhispid ; pedicels stout, 5–9 lin. long, subhispid ; sepals $2\frac{1}{2}$ lin. long, lanceolate, acute, more or less keeled and the sides infolded at the tips, subhispid ; corolla-lobes reflexed with upcurved tips, about $3\frac{1}{2}$ lin. long, 2 lin. broad, oblong-ovate, minutely notched at the obtusely pointed apex, glabrous on both sides, apparently yellowish-green, suffused or veined with purple-brown at the tips, often drying brown; corona-lobes arising ½–1 lin. up the staminal column, apparently yellow, with the back and basal part brownish-purple, erect, $2\frac{1}{4}$–$2\frac{1}{2}$ lin. long, 1–$1\frac{1}{4}$ lin. broad across the side of the complicate-cucullate basal half, which is very oblique with narrow wing-like margins at the base, has recurved margins at the top and is horizontally produced into subulate teeth directed over the staminal column, upper half dorsally produced into an oblong-lanceolate or ovate obtuse erect blade, no tooth or process within; staminal column about $1\frac{3}{4}$ lin. long; anther-appendages suborbicular, inflexed over and covering the truncate style-apex; anther-wings angular and very projecting at the base. *Gomphocarpus macropus, Schlechter in Journ. Bot.* 1894, 353.

KALAHARI REGION: Transvaal; Belfast, *Burtt Davy,* 1284! 1315! Ermelo, *Burtt Davy,* 984! High Veld, near Wonderfontein, 6000 ft., *Bolus,* 12140! Caledonia, Lake Chrissie, *Hamilton in Herb. Burtt Davy,* 1174!

EASTERN REGION: Transkei; Kreilis Country; *Bowker*! Tembuland; Engcobo Mountain, 3800 ft., *Bolus,* 10211! near Xalanga, 4500 ft., *Bolus,* 10212! Natal, 4000–5000 ft.; hill-sides at Lynedoch, *Wood,* 4544! near Mooi River, *Wood,* 5374! Dargle Farm, *Fannin,* 90!

Imperfectly known species.

52. **A. concinna** (Schlechter in Journ. Bot. 1896, 456); stem 12–14 in. high, simple, slender, sparsely leafy, pilose; leaves erect, subequalling or a little longer than the internodes, 2–$3\frac{1}{4}$ in. long, 1–$1\frac{1}{2}$ lin. broad, linear, acute, narrowed into a very short petiole at the base, with revolute margins, pilose on both sides; umbels lateral at the nodes or falsely terminal, pedunculate, 6–8-flowered; peduncle about 7 lin. long, erect, slender, pilose; pedicels about half as long as the peduncle, filiform, pilose; sepals half as long as the corolla, lanceolate, acute, pilose; corolla-lobes spreading, ¼ in. long, scarcely 2 lin. broad, ovate, somewhat obtuse, silky-pilose, with silky-villous margins, white; corona-lobes erect, slightly exceeding the staminal column, cucullate, obtuse, obliquely truncate at the apex, with a small porrect tooth on the margin on each side above the middle, inflexed upon the stigma (style-apex), greenish-brown; anther-appendages suborbicular, rounded, inflexed upon the stigma (style-apex). *Gomphocarpus concinnus, Schlechter in Journ. Bot.* 1895, 270.

EASTERN REGION: Griqualand East; near the Tina River, 4000 ft., *Schlechter,* 6418. Natal; without precise locality, *Wood.*

Only known to me by the above description.

53. **A. depressa** (Schlechter in Journ. Bot. 1896, 455); plant dwarf, branching at the base; branches decumbent, 4–8 in. long,

densely leafy, glabrescent; leaves spreading or adpressed to the
ground, very shortly petiolate, 5–10 lin. long, 3–5 lin. broad above
the base, hastate-ovate or hastate-lanceolate, shortly acute, glabrous;
umbel terminal, many-flowered; peduncle much longer than the
leaves, glabrescent; pedicels 5–7½ lin. long, filiform, glabrescent or
thinly pilose; sepals scarcely 2 lin. long, lanceolate, acute, puberulous; corolla-lobes spreading, twice as long as the calyx, ovate,
obtuse, dirty white and greenish; corona-lobes about twice as long
as the staminal column, very shortly clawed at the base, ovate-
lanceolate in outline, obtuse, concave, each margin produced below
the middle into a short ascending tooth inflexed over the stigma
(style-apex), brownish; anther-appendages broadly ovate, obtuse,
inflexed upon the stigma (style-apex); follicles 3 in. long, 3½ lin.
thick, fusiform, smooth, glabrous. *Gomphocarpus depressus,
Schlechter in Engl. Jahrb.* xx. *Beibl.* 51, 29.

KALAHARI REGION : Transvaal; Elandspruit Mountains, 7000 ft., *Schlechter*,
3835.

From the description this does not appear to be distinct from *A. multicaulis,*
Schlechter, but as Dr. Schlechter is well acquainted with that species I hesitate to
unite them.

54. A. multiflora (N. E. Br.); stem branching from the base;
branches terete or compressed, erect, pubescent; leaves ovate-
oblong, obtuse, subhastate or truncate at the base, with revolute
margins, pubescent on both sides; peduncles lateral at the nodes at
the apex of the branches, several-flowered, and together with the
pedicels whitish-puberulous; corona-lobes cuneate at the back,
3-lobed, with the inner (middle?) lobe acuminate, inflexed, arched,
lateral lobes obsolete, equalling the staminal column. *Gomphocarpus
multiflorus, Decne in DC. Prodr.* viii. 560.

SOUTH AFRICA : without locality, in the *Paris Herbarium.*

Stated to have the habit of *Schizoglossum bidens,* E. Meyer, but with lateral
(not terminal) peduncles.

55. A. sabulosa (Schlechter in Journ. Bot. 1896, 454); stems
erect or subdecumbent, 9–12 in. high, simple; leaves somewhat
crowded, erectly spreading, 1¼–3½ in. long, linear-filiform, acute,
with revolute margins, glabrous; umbels with long peduncles,
scarcely overtopping the leaves, about 4-flowered; pedicels much
longer than the calyx; sepals 1 lin. long, ovate-lanceolate, acute,
pilose; corolla-lobes 2½ lin. long, spreading, ovate, obtuse, concave,
pilose on the back, glabrous on the inner face; corona-lobes
cucullate, shortly 3-toothed at the apex, with the lateral teeth
smaller; anthers oblong, with obtuse inflexed membranous tips;
stigma (style-apex) depressed. *Gomphocarpus arenarius, Schlechter
in Abhandl. Bot. Ver. Brandenb.* xxxv. 52.

COAST REGION : Tulbagh Div.; at the foot of Mosterts Berg, 800 ft.,
Schlechter, 533.

56. A. Schlechteri (N. E. Br.) ; stem about 2 ft. high, 1 lin. thick at the base, simple, glabrous ; leaves sessile, 2–4¾ in. long, 1 lin. broad, linear, scabrous on the revolute margins, otherwise glabrous ; umbels lateral at the nodes, pedunculate, several-flowered ; peduncles 1–1¹⁄₆ in. long ; pedicels 4½ lin. long ; sepals 2 lin. long, subulate, puberulous ; corolla-lobes about ½ in. long, oblong, obtuse, fimbriolate on the upper part, clear green ; corona-lobes much overtopping the style-apex, 3½ lin. long, subulate, with the apex inflexed, dark purple red, with a pair of acute white wings on the inner face ; staminal column 1¼–1½ lin. long. *Gomphocarpus Schlechteri, K. Schum. in. Engl. Jahrb.* xxxiii. 325.

EASTERN REGION : Pondoland ; between Roskowe and Canham, *Bachmann,* 1083.

This appears (from description) to be very similar to *A. expansa,* Schlechter, and may prove to be that species.

57. A. villosa (Mill. Dict. ed. 8, no. 14) ; stem fruticose ; leaves lanceolate, acute, covered on both sides with short hairs ; umbel solitary, erect (probably terminal), small, loose ; flowers white.

SOUTH AFRICA, ex *Miller.*

A specimen from Miller's Herbarium named "*A. villosa*" in the British Museum is *A. physocarpa,* Schlechter, and does not agree with Miller's description of *A. villosa.*

Asclepias nivea, Burm. f. Prodr. Cap. 7, is the same as *A. nivea,* Linn. Sp. Pl. ed. i. 215, and is not a native of South Africa but of the West Indies.

Lagarinthus tenuis, Meisn. in Hook. London Journ. Bot. ii. 1843, 543 (by error 443), and Krauss in Flora, 1844, 826, not of E. Meyer, is only noted as follows :—"Our specimens differ from Drège's only in having many-flowered umbels, and the corolla turned down." I have not seen it. Possibly it may be the same as *A. filiformis,* Benth. ; but is stated to have been collected by Krauss near Swellendam, where *A. filiformis* is not known to occur.

GOMPHOCARPUS UNDULATUS (Turcz. in Bull. Soc. Nat. Mosc. 1848, i. 259, not of Schlechter) ; stem dwarf, firm ; lower leaves obovate, obtuse, mucronulate, tapering into the petiole, upper leaves oblong, tapering at both ends, acute, wavy on the margin ; umbels lateral and terminal, pedunculate, 6–7-flowered ; corolla connivent ; corona-lobes equalling the staminal column, connate at the base, concave-carinate, winged-dentate on the margin, with the teeth directed inwards, terminated by a short somewhat obtuse apiculus.

COAST REGION : Alexandria Div. ; Oliphants Hoek forest, below 300 ft., *Ecklon,* 36.

I think it probable that this plant will prove to be *Woodia mucronata,* N. E. Br., and therefore refrain from giving it a new name under *Asclepias.*

XIX. PACHYCARPUS, E. Meyer.

Calyx 5-partite. *Corolla* 5-lobed, subglobose, cup-shaped, rotate or reflexed ; lobes overlapping in bud. *Corona* of 5 lobes arising at or near the base of the staminal column, usually long and flattened dorsally beneath or at least beyond the keels, never complicate-cucullate, very spreading, or spreading at the base with erect or incurved tips, rarely wholly erect and then with reflexed sides and a single keel down the face, usually with a pair of contiguous often fleshy keels or a longitudinally fissured fleshy hump at the base and prolonged beyond into a linear, lanceolate, dilated or 3-lobed blade, occasionally reduced to the keels or hump only, rarely keeled to the apex or without keels. *Staminal column* arising from the bottom of the corolla, often with the filament part undeveloped, very broadly conical from the anther-wings being very broad and projecting at the base, or pentagonally cylindric with them less developed. *Anthers* with terminal membranous appendages. *Pollen-masses* solitary in each anther-cell, pendulous, attached in pairs to the pollen-carriers by well developed caudicles. *Style* truncate, excavated, 5-lobed or produced into a beak or column at the apex. *Follicles* solitary (in all the specimens seen), coriaceous, stoutly fusiform or ovoid-fusiform, very obtuse, winged, at least on the upper part, sometimes toothed or echinate along the wings. *Seeds* crowned with a tuft of hairs.

Erect perennial herbs ; rootstock probably tuberous or of thick fleshy roots ; leaves opposite ; flowers in pairs or in pedunculate 2- to several-flowered umbels, lateral at the upper nodes and terminal, large or of moderate size.

DISTRIB. Species 27, one extending into and another endemic to Tropical Africa.

Although *Pachycarpus* was established by E. Meyer as long ago as 1837, it has not been recognised as a distinct genus by other authors, probably because E. Meyer based the generic characters chiefly upon the fruit, and included some species that belong to other genera. Different authors have united it variously with *Gomphocarpus, Asclepias* or *Xysmalobium,* but (as it appears to me) without sufficient investigation, since the very different coronal structure, different character of the whole flower and fruit, united to the different habit of almost all the species, are, I think, sufficient to warrant its retention as a distinct genus, which can moreover be very easily recognised by the general appearance alone. *P. Gerrardi,* however, is so closely simulated by *Asclepias macra,* that dried specimens cannot be distinguished without dissection, when the former is seen to have the spreading dorsally flattened (but keeled) corona-lobes of *Pachycarpus,* and the latter the characteristic erect laterally compressed complicate-cucullate corona-lobes of *Asclepias.* For convenience I have used the term keels throughout for all the keel modifications or processes upon the base of the corona-lobes.

  * Leaf-blade $\frac{1}{3}$–$2\frac{1}{2}$ in. broad, $1\frac{1}{2}$–9 (rarely 10) times as long
      as broad, elliptic, oblong, lanceolate, linear-lanceolate
      or linear-oblong :
    Corolla lobed to the middle or slightly beyond, large,
      globose, yellow or greenish, with or without purple-
      brown spots ...　　...　　..　　...　　...　　... (21) **grandiflorus.**
    Corolla lobed at least to $\frac{2}{3}$ of the way down, often nearly
      to the base :
      Corona-lobes erect, oblong, with reflexed sides form-
        ing a single keel down the inner face and a
        linear tooth on each side at the base spreading
        out under the anther-wings　　...　　...　　... (1) **Galpinii.**
      Corona-lobes abruptly incurved from a spreading
        base, with erect tips, linear or slightly dilated at
        the 1–2 lin. broad tips, without keels ...　　　... (14) **dealbatus.**

Corona-lobes erect from a short spreading base, with
    very slight or inconspicuous keels :
  Terminal part of corona-lobes dilated into an
      orbicular-rhomboid or broadly cuneate entire
      or 3-lobed blade, dark chocolate   ...   ...   (2) **rigidus**.
  Terminal part of corona-lobes dilated into a
      lanceolate to subhastate-lanceolate blade, light
      brown when dried   ...   . (10) **macrochilus**.
Corona-lobes horizontally spreading, about twice as
    long as deep, narrowly lanceolate, deeply com-
    plicate-boat-shaped from the keels extending to
    the apex   ..   .. (20) **coronarius**.
Corona-lobes consisting of a pair of fleshy keels or a
    deeply fissured hump standing upon a recurved
    spreading base, not prolonged into a blade
    beyond them or only into a small or subulate
    point ; corolla-lobes reflexed or revolute :
  Sepals $3\frac{1}{2}$–$4\frac{1}{2}$ lin. long, about 2 lin. broad, lanceolate
      or ovate-lanceolate ; corolla-lobes about 7 lin.
      long   ...   .   (6) **inconstans**.
  Sepals 6–7 lin. long, 3–$3\frac{1}{2}$ lin. broad, ovate ;
      corolla-lobes 9–11 lin. long   ...   .   (7) **validus**.
Corona-lobes with a pair of very prominent or large
    contiguous keels or a deeply fissured hump at
    the base and prolonged beyond them into an
    erect spreading or much incurved linear, lanceo-
    late or variably dilated blade :
  Corolla reflexed nearly straight back from the
      base ; terminal part of the corona-lobes dilated
      into a thin lanceolate to suborbicular blade
      2–4 lin. broad :
    Keels or basal hump on the corona-lobes higher
        than broad viewed sideways   ...   ...   (4) **reflectens**.
    Keels or basal hump on the corona-lobes broader
        than high viewed sideways...   ...   ...   (5) **appendiculatus**.
  Corolla reflexed-spreading (not straight back) from
      the base ; terminal part of the corona-lobes
      linear or linear-spathulate, $\frac{3}{4}$–1 lin. broad   ...   (3) **scaber**.
  Corolla not wholly reflexed ; lobes campanulately
      spreading   to   slightly   incurved,   or   in
      P. natalensis with the basal half reflexed-
      spreading and the apical half curved upwards,
      sometimes with the tips spreading to revolute :
    Corona-lobes linear, replicate and not dilated
        at the incurved tips, with the rather dwarf
        keels extending from their base to beyond
        the middle and then excurrent at the
        margins   ..   .. (19) **McKenii**.
    Corona-lobes with the keels or deeply fissured
        hump rarely extending to the middle
        (usually to $\frac{1}{4}$–$\frac{1}{3}$) of the lobe :
      Keels on the corona-lobes subfalcate or
          obliquely oblong with their tips directed
          backwards :
        Blade of the corona-lobes 3-lobed, 3–$5\frac{1}{2}$ lin.
            broad across the side lobes when spread
            out   ...   . (11) **vexillaris**.

Blade of the corona-lobes linear or linear-
    lanceolate, truncate, 1 lin. broad   ... (12) **stenoglossus.**

Keels or fissured hump on the corona-lobes
    with the tips erect or directed inwards ;
    blade of the lobe not or (in *P. concolor*)
    rarely slightly and obtusely 3-lobed :
Flowers in pairs or the lower in 2-flowered
    pedunculate umbels, lateral at the
    nodes   ...   . (15) **concolor.**

Flowers usually 3–4 together, lateral at
    the nodes ; peduncle none ; corona-
    lobes dilated above the middle into an
    ovate blade :
Corona-lobes scarcely dilated at the base,
    with the narrow keels not con-
    spicuously folded ; anther-wings
    1 lin. long   ...   . (8) **natalensis.**

Corona-lobes dilated into distinct deltoid
    auricles at the base, with the narrow
    keels conspicuously folded longi-
    tudinally ; anther-wings $1\frac{3}{4}$ lin. long (9) **plicatus.**

Flowers in pedunculate usually 3–6-flowered
    umbels :
Corona-lobes dilated and concave or hood-
    like at the tips, white or pale yellow
    with a violet-purple stripe ...   ... (13) **schinzianus.**

Corona-lobes not dilated into concave or
    hood-like tips :
Corona-lobes dark purple-brown :
    Leaves scabrous with short points (no
    hairs) on both sides ; keels or
    hump on the corona-lobes large (16) **transvaalensis.**

    Leaves scabrous-pubescent with hairs
    on both sides ; keels or hump on
    the corona-lobes rather small ... (17) **insignis.**

Corona-lobes whitish or yellowish   ... (18) **decorus.**

** Leaf-blade $\frac{1}{8}$–$\frac{1}{3}$ in. broad, 12–24 times as long as broad,
  linear or narrowly linear-lanceolate ; corolla subglobose,
  pubescent outside :
Style-apex produced into a beak much beyond the
    anther-appendages, with 5 small connivent lobes
    at the apex   ...   . (25) **rostratus.**

Style-apex shortly exceeding the anther-appendages,
    very large, divided into 5 lobes $1\frac{1}{3}$–$1\frac{1}{2}$ lin. long,
    stellately radiating ...   . (26) **stelliceps.**

Style-apex shorter than and more or less covered by the
    anther-appendages, truncate or excavated, crenate
    or very shortly 5-lobed :
Corolla lobed to $\frac{2}{3}$ of the way down ; anther-appen-
    dages 3–$3\frac{1}{2}$ lin. long ; corona-lobes incurved-
    erect beyond the middle...   ...   . (22) **linearis.**

Corolla not lobed beyond the middle ; anther-appen-
    dages $\frac{1}{2}$–2 lin. long ; corona-lobes horizontally
    spreading, not incurved :
Umbels 3–7-flowered ; anther-appendages $1\frac{1}{2}$–2 lin.
    long :

Corolla 1¼–2 in. in diam., with lobes 7–13 lin.
long; corona-lobes 3½–7½ lin. long ... (23) **campanulatus.**
Corolla ¾–1 in. in diam., with lobes 4–5 lin.
long; corona-lobes 1½–3 lin. long ... ... (23) **campanulatus,**
var. β.
Umbels 8–13-flowered; corolla 7–8 lin. in diam.,
with lobes 3–3½ lin. long; corona-lobes about
2 lin. long; anther-appendages ½ lin. long ... (24) **Gerrardi.**

1. **P. Galpinii** (N. E. Br. in Dyer, Fl. Trop. Afr. iv. i. 377);
stem 10–12 in. high, simple, stout, glabrous or nearly so; leaves
erectly spreading, 1⅓–2½ in. long, 10–16 lin. broad, oblong or elliptic-
oblong, obtuse, or rounded at both ends, apiculate, slightly scabrous
on both sides, more scabrous at the margins; petiole about ¼ in.
long; umbels lateral at the nodes, 3–5-flowered; peduncle 1¾–2 in.
long; puberulous along one side; pedicels about 3 lin. long,
puberulous on one side; sepals 3½ lin. long, 2–3 lin. broad, ovate,
acute, pubescent; corolla broadly and shortly cup-shaped, about
7–8 lin. in diam., the united part about 2 lin. long; lobes 4 lin.
long, 3½ lin. broad, ovate, concave at the lower part, recurved or
revolute at the obtuse or emarginate tips, glabrous on the inner
face, minutely and thinly pubescent on the back, apparently dull
purple-brown at the basal part, yellowish-green at the apical part;
corona-lobes erect, much overtopping the style-apex, dark purple-
brown, 2½–3 lin. long, 1¼–1⅔ lin. broad, oblong, with reflexed sides,
forming an acute channel down the back, notched at the obtuse
apex and prominently keeled down the inner face, with a small
linear tooth ½–¾ lin. long on each side at the base, spreading under
the anther-wings close to the corolla; staminal column about 2 lin.
long; anther-appendages large, cordate-ovate, obtuse, inflexed upon
and nearly covering the depressed-truncate style-apex. *Gomphocarpus Galpinii, Schlechter in Engl. Jahrb.* xviii. *Beibl.* 45, 18.
Asclepias Galpinii, Schlechter in Journ. Bot. 1896, 455.

KALAHARI REGION: Transvaal; Saddleback Mountain, near Barberton, 3000–
4500 ft., *Galpin,* 692a!

EASTERN REGION: Swaziland; High Veld, near Mbabane, 4700 ft., *Bolus,*
12129! 12136!

2. **P. rigidus** (E. Meyer, Comm. 211); stem ½–2 ft. high, simple,
stout, nearly glabrous or somewhat scabrous; leaves opposite or 3
in a whorl, ascending or somewhat spreading; petiole 2–4 lin. long;
blade 1⅔–3½ in. long, ½–1½ in. broad, lanceolate, ovate-lanceolate or
ovate, acute or attenuately acute, wavy on the margins, rigidly
scabrous on the edges and midrib beneath, otherwise glabrous;
umbels lateral at the nodes, sessile or the lowest pedunculate,
5–6-flowered; pedicels about 5 lin. long, somewhat roughly
pubescent; sepals about equalling the corolla, 4–5 lin. long, 1¾–
2 lin. broad, lanceolate or ovate-lanceolate, acute or acuminate,
pubescent; corolla subglobose, lobed nearly to the base; lobes 4½–5

lin. long, 2½–3 lin. broad, erect, with spreading tips, ovate or elliptic-ovate, acute from the margins being revolute, concave at the lower part, glabrous on both sides, apparently pale-greenish, stained with purple-brown on the back ; corona-lobes about 3½–4 lin. long, erect, slightly curved, linear at the basal part, without keels or crests on the inner face, dilated in the upper half into a flat erect blade, 2–2½ lin. long, 2½–3 lin. broad, orbicular-rhomboid, entire or slightly trilobulate, usually very obtuse, rarely acute, cuneate at the base, dark chocolate-coloured ; staminal column 2–2½ lin. long ; anther-appendages broadly ovate, acute, closely inflexed over the crenate rim of the crater-like style-apex ; anther-wings slightly more than 1 lin. long, broadest at the base, with straight margins ; follicles solitary, in the immature specimen seen, 3 in. long, 1 in. thick, narrowly ellipsoid-fusiform, nearly equally narrowed at each end, subacute, broadly 6-winged, glabrous. *Xysmalobium rigidum, Dietr. Syn. Pl.* ii. 902. *Gomphocarpus rigidus, Decne in DC. Prodr.* viii. 563. *Asclepias rigida, Schlechter in Engl. Jahrb.* xxi. *Beibl.* 54, 9, *and in Journ Bot.* 1895, 357 *and* 1896, 453.

Var. *β*, **tridens** (E. Meyer, l.c.) ; blade of the corona-lobes very broadly cuneate, 3-lobed at the top ; lateral lobes horizontal or very slightly ascending, obtuse or rounded ; middle lobes ½–1½ lin. long, transverse to linear, obtuse, incurved or erect ; otherwise as in the type. *Gomphocarpus rigidus, var. tridens, Decne in DC. Prodr.* viii. 563.

COAST REGION : Var. *β* : Queenstown Div. ; Lesseyton Nek, 3800–4000 ft., *Galpin,* 2277 !
CENTRAL REGION : Tarkastad Div. ; Wildschuts Berg, *Drège* ! Aliwal North Div. ; Elands Hoek, near Aliwal North, *Bolus,* 10496 ! 10555 ! Var. *β* : Aliwal North Div. ; at the foot of the Witte Bergen, *Drège,* 6393 !
KALAHARI REGION : Orange River Colony ; near Bethlehem, *Bolus,* 8116 ! Basutoland ; Matelas Peak, *Thode,* 34 ! Transvaal ; Mooi River, *Burke* ! between New Denmark and Morgauson, *Burtt Davy,* 1004 ! Ermelo Road, *Burtt Davy,* 958 (including var. *β*) ! Var. *β* : Orange River Colony ; Bethlehem, *Richardson !* Transvaal ; Witkleifontein, *Burtt Davy,* 3122 !

Drège's localities are copied from the original labels in the Herbarium of E. Meyer, who has transposed them in his *Commentariorum,* p. 211.

3. **P. scaber** (N. E. Br. in Dyer, Fl. Trop. Afr. iv. i. 377) ; stem 1–2 ft. high, 2–4 lin. thick, glabrous or slightly scabrous ; leaves shortly petiolate, 2–3½ in. long, ¾–2¼ in. broad, oblong, elliptic-oblong or elliptic-ovate, obtuse, apiculate, rounded or very broadly cuneate at the base, glabrous above, more or less scabrous beneath, very scabrous at the margins ; umbels 1–4, subcorymbose at the top of the stem, 5–8-flowered ; peduncles ¼–1¼ in. long, more or less scabrous or almost glabrous ; bracts about ¼ in. long, linear or linear-subulate, acute ; pedicels ⅔–1 in. long, scabrous ; sepals about ⅓ in. long, 2–2½ lin. broad, ovate, acute, scabrous ; corolla-lobes reflexed-spreading, about 5 lin. long, 3 lin. broad, ovate, minutely notched at the obtuse apex, glabrous on both sides, revolute along the margins, white (*Wood, and Tyson*), primrose-yellow (*Bolus*) ; corona-lobes arising at the base of the staminal column, 4–5½ lin. long, white or

yellow, dilated into 2 deltoid rounded auricles or transversely rhomboid at the horizontally spreading base, which is concave beneath, then abruptly contracted into a linear or linear-spathulate blade $\frac{2}{3}$–1 lin. broad and subacute, obtuse, emarginate or more or less 3-lobed or 3-toothed at the apex, channelled down the back, convex or keeled down the inner face and abruptly incurved over the style-apex, with 2 erect contiguous keels 1–1$\frac{1}{2}$ lin. high and broad on the basal part, straight on the inner, rounded on the dorsal margin, obtuse; staminal column 2 lin. long; anther-appendages orbicular-reniform, very obtuse, inflexed on the rim of the depressed style-apex; follicles solitary, about 3$\frac{1}{2}$ in. long, 1$\frac{3}{4}$ in. thick, ovoid, obtuse, more or less winged at the upper part, coriaceous, glabrous. *Gomphocarpus scaber, Harv. Thes. Cap.* ii. 58, *t.* 192; *Schlechter in Engl. Jahrb.* xx. *Beibl.* 51, 37. *Asclepias scabra, Schlechter in Journ. Bot.* 1896, 454, *and Ann. Naturhist. Hofmus. Wien,* xv. 68.

KALAHARI REGION: Transvaal; Endholixana, near Piet Retief, *Lady Barkly,* 2! Elandspruit Mountains, 6200 ft, *Schlechter,* 3877.

EASTERN REGION: Griqualand East; around Clydesdale, *Tyson,* 2165! Natal; Inanda, *Wood,* 178! 468! Greenwich Farm, Riet Vlei, *Fry in Herb. Galpin,* 2745! Umkomanzi Valley, *Krook,* 821 (ex *Schlechter*). Zululand, *Gerrard,* 1285! Swaziland; Middel Veld. between Mbabane and Bremersdorp, 2600 ft., *Bolus,* 12124! High Veld, between Mbabane and Oshoek, 5000 ft., *Bolus,* 12125! between Bremersdorp and MacNabs store, *Burtt Davy,* 2932!

4. **P. reflectens** (E. Meyer, Comm. 210); stem 1–1$\frac{1}{2}$ ft. high, simple, stout, bifariously puberulous and more or less scabrous; leaves ascending or spreading; petiole 1–4 lin. long; blade 1$\frac{1}{2}$–3 in. long, $\frac{1}{2}$–1$\frac{1}{4}$ in. broad, lanceolate, ovate or oblong, or the uppermost linear or linear-lanceolate, scabrous on both sides and at the margins or glabrous above; umbels lateral at the nodes and terminal, sessile or the lowest pedunculate, 2–3-flowered; peduncle 0–$\frac{1}{2}$ in. long, pubescent and scabrous; bracts 3–4 lin. long, linear-subulate, deciduous; pedicels $\frac{1}{2}$–1 in. long, harshly pubescent; sepals 4–5 lin. long, 2 lin. broad, lanceolate, acute, roughly pubescent; corolla-lobes always reflexed nearly straight-back, 5–8 lin. long, rather narrowly lanceolate-attenuate from the very revolute margins, but oblong and 3$\frac{1}{2}$–4 lin. broad when flattened out, acute, slightly scabrous-pubescent on the back or glabrous on both sides, apparently pale-greenish or greenish-white, usually spotted with purple; corona-lobes 3$\frac{1}{2}$–5 lin. long; basal part horizontally spreading from the base of the staminal column, claw-like, dilated at the very base, bearing 2 erect closely contiguous oblong or subfalcate-oblong lobes always higher than broad, 1–2 lin. high, $\frac{2}{3}$–1 lin. broad; terminal part dilated into an orbicular-ovate or transversely elliptic erect blade 2$\frac{1}{2}$–3$\frac{1}{2}$ lin. long, 3–4 lin. broad, obtuse, or retuse with a short inflexed linear apical point, broadly rounded or subtruncate at the base; staminal column about 3 lin. long; anther-appendages rather large, ovate, obtuse, with the sides

folded back, loosely incurved over (in fresh flowers probably inflexed upon) and nearly concealing the style apex ; anther-wings $1\frac{2}{3}$–2 lin. long, much produced at the base, concave on the margin. *P. reflexus, Steud. Nom. Bot. ed.* 2, ii. 245. *Gomphocarpus reflectens, Decne in DC. Prodr.* viii. 563 ; *Schlechter in Engl. Jahrb.* xviii. *Beibl.* 45, 9. *Xysmalobium reflectens, Dietr. Syn. Pl.* ii. 902. *Asclepias reflectens, Schlechter in Engl. Jahrb.* xxi. *Beibl.* 54, 9 ; *Journ. Bot.* 1896, 453, *and Ann. Naturhist. Hofmus. Wien,* xv. 68.

COAST REGION : Stockenstrom Div. ; Kat Berg, 3500 ft., *MacOwan,* 860 ! *Galpin,* 1712 ! Queenstown Div. ; damp places by the Zwartkei River and on the Winter Berg, *Mrs. Barber,* 594 ! Komgha Div. ; near Komgha, *Flanaghan,* 16 ! *Krook,* 797, 803 (ex *Schlechter*).

EASTERN REGION : Transkei ; near Gekau (Geua) River (ex *Drège,* but according to *E. Meyer* between Fish River and Kap River in Bathurst Div.) *Drège,* 4934 ! Kreilis Country, *Bowker* ! near Colossa, *Krook,* 809 (ex *Schlechter*).

Very similar to *P. appendiculatus* and perhaps only a variety of that species, the flowers however are rather smaller and have a somewhat narrower appearance, whilst the fleshy keels on the base of the corona-lobes, viewed sideways, are always higher than broad. The fruit in E. Meyer's Herbarium, described as pyramidal, is less than 1 in. long and far too young to determine its mature shape. Drège's 4934 is E. Meyer's type, but another specimen, collected by Drège between Zandplaat and Komgha, and named "*reflectens?*" by E. Meyer, is *P. appendiculatus.*

5. **P. appendiculatus** (E. Meyer, Comm. 210) ; stem $\frac{3}{4}$–$1\frac{1}{2}$ ft. high, simple, stout, varying from nearly glabrous to bifariously puberulous and more or less scabrous ; leaves spreading or ascending ; petiole 1–4 lin. long ; blade $1\frac{1}{2}$–$3\frac{1}{2}$ in. long, $\frac{1}{3}$–$1\frac{3}{4}$ in. broad, varying from narrowly oblong or oblong-lanceolate to elliptic, acute, or obtuse and shortly apiculate, rounded or cuneate at the base, more or less scabrous beneath or on both sides, very scabrous on the margins ; umbels lateral at the nodes and terminal, sessile or the lowest pedunculate, 2–3-flowered ; peduncle 0–$\frac{3}{4}$ in. long, pubescent or scabrous ; bracts 2–4 lin. long, linear-subulate, very deciduous ; pedicels $\frac{1}{3}$–$1\frac{1}{4}$ in. long, scabrous or scabrous-pubescent on one side ; sepals $4\frac{1}{2}$–$5\frac{1}{2}$ lin. long, 2–$2\frac{1}{2}$ lin. broad, ovate-lanceolate or lanceolate, acute or acuminate, scabrous-pubescent ; corolla-lobes varying from recurved-spreading to reflexed straight back, 7–9 lin. long, 4–7 lin. broad when flattened out, oblong-lanceolate, elliptic-oblong or ovate, acute, usually nearly straight with very revolute margins, but sometimes rolled back from the tips, glabrous on both sides, apparently greenish white or light green on the inner face, often spotted with dark purple-brown on the back and sometimes on the inner face also ; corona-lobes 6–8 lin. long, spreading from the base of the staminal column at the elliptic or subquadrately rhomboid basal part, which is about 2 lin. broad and in fresh flowers seems formed of a fleshy ellipsoid mass with a cut-like longitudinal fissure, but in dried flowers appears to bear 2 erect parallel closely-contiguous wing-like keels always as broad or broader than high, $1\frac{1}{3}$–2 lin. high, $1\frac{1}{3}$–3 lin. broad, beyond which the lobe narrows into an

upcurved claw, then expands into a lanceolate, ovate, elliptic or suborbicular blade 3–7 lin. long, 2–4 lin. broad, acute, acuminate or more or less retuse and cuspidate with a linear point at the apex, cuneate or subtruncate at the base, entire or crenulate, abruptly incurved or arched over the basal keels and with the sides more or less folded back; staminal column $3\frac{1}{2}$–$4\frac{1}{2}$ lin. long; anther-appendages rather large, ovate, obtuse, inflexed upon or in dried flowers loosely arching over the top of the style-apex; anther-wings $2\frac{1}{2}$ lin. long, much produced at the base, concave along the margin; follicles solitary, subglobose or ellipsoid, very obtuse, narrowly winged, glabrous, very thick and spongy, $1\frac{3}{4}$ in. long, $1\frac{1}{2}$ in. thick in the immature specimen seen in E. Meyer's Herbarium. *Gomphocarpus appendiculatus, Decne in DC. Prodr.* viii. 562; *Schlechter in Engl. Jahrb.* xx. *Beibl.* 51, 39. *G. macroglossus, Turcz. in Bull. Soc. Nat. Mosc.* 1848, i. 259. *Xysmalobium appendiculatum, Dietr. Syn. Pl.* ii. 902. *Asclepias appendiculata, Schlechter in Engl. Jahrb.* xxi. *Beibl.* 54, 5; *and Journ. Bot.* 1896, 452.

COAST REGION: Bedford Div.; near Bedford, *Mrs. Hutton*! East London Div.; near East London, 100 ft., *Wood in Herb. Galpin*, 3386! Komgha Div.; near Komgha, 2000 ft., *Flanagan*, 1239! British Kaffraria, *Mrs. Hutton*!

EASTERN REGION: Tembuland; near Morley, *Drège*, 4933! Pondoland; near Isinuka, at the mouth of St. Johns River, *Bolus*, 10207! Natal; Krantz Kloof, 1400 ft., *Schlechter*, 3216. Zululand, *Gerrard*, 1286! Swaziland; near Bremersdorp, *Burtt Davy*, 2936! 3065! between Bremersdorp and Mbabane (Embabaan), 2500 ft., *Bolus*, 12122! near Mafutane, 1500 ft., *Bolus*, 12126!

The Bremersdorp specimen (*Burtt Davy*, 3065) may possibly be a hybrid between *P. appendiculatus* and *P. validus*, the foliage and corolla much resembling the latter, whilst the corona is like that of *P. appendiculatus*.

6. P. inconstans (N. E. Br.); stem simple, 1–2 ft. high, stout, more or less scabrous along 2 broad lines, sometimes nearly smooth; leaves spreading; petiole 2–4 lin. long; blade of the middle and lower leaves $1\frac{1}{2}$–$3\frac{1}{4}$ in. long, $\frac{2}{3}$–$1\frac{1}{2}$ in. broad, oblong or lanceolate, acute, or obtuse and apiculate, cuneate or rounded at the base, scabrous on the margins and thinly so on both sides or nearly or quite smooth above; umbels sessile or the lower pedunculate, lateral at the nodes, 5–6- (rarely 7–10-) flowered; peduncle 0–10 lin. long, scabrous on one side; bracts $2\frac{1}{2}$–4 lin. long, very deciduous, linear, acute; pedicels $\frac{2}{3}$–1 in. long, more or less scabrous; sepals $3\frac{1}{2}$–$4\frac{1}{2}$ lin. long, $1\frac{3}{4}$–2 lin. broad, ovate-lanceolate, somewhat acuminate, pubescent, ciliate; corolla-lobes about 7 lin. long, $3\frac{1}{2}$–$4\frac{1}{2}$ lin. broad, reflexed, oblong or ovate-oblong, obtusely pointed, revolute along the margins, twisted, glabrous on both sides or thinly pubescent on the back, often slightly rugulose (perhaps from shrinkage in drying) on the inner face; corona-lobes very variable, seated at the base of the staminal column, $1\frac{1}{4}$–$3\frac{1}{2}$ lin. long, usually consisting of 2 contiguous erect fleshy keels, varying from less than half as high to as high as the anthers, in side view broader than high or higher than broad, very obtuse or obliquely acute, arising from an ovate or lanceolate recurved-spreading base, which is more or less concave beneath and sometimes almost pointless, sometimes with a short recurving point, sometimes with a slender linear upcurved point 2–3 lin. long, with a small tooth on each side near the middle, and sometimes the keels are absent and the lobes are small, spreading, scarcely projecting beyond the anther-wings, ovate or oblong-ovate, recurved at the shortly pointed apex, fleshy, with a cut-like groove along the upper face and apparently somewhat rib-striate ; staminal column $2\frac{1}{2}$ lin. long ; anther-appendages ovate, acute, loosely inflexed over and concealing the style-apex, which sometimes (always ?) has a central cavity ; follicles not seen. *P. concolor, Meisn. in Hook. Lond. Journ. Bot.* ii. 1843, 544 (*by error* 444) ; *Krauss in Flora,* 1844, 826, *not of E. Meyer.*

COAST REGION : Komgha Div. ; hill near Keimouth, *Flanagan,* 757 !

EASTERN REGION : Transkei ; Kreilis Country, *Bowker* ! Pondoland ; hills near Port St. John, 300–1000 ft., *Flanagan* ! and in *Herb. Bolus,* 10205 ! *Galpin,* 3195 ! 3196 ! 3445 ! Natal ; margins of woods near Durban Bay, *Krauss,* 83 ! Inanda, *Wood,* 1075 !

The variability of the corona-lobes of this plant would seem to indicate that it may be a hybrid between *P. reflectens* and one or more other allied species. It is closely related to *P. validus,* but is not so tall, and has much smaller flowers and narrower leaves.

7. **P. validus** (N. E. Br. in Dyer, Fl. Trop. Afr. iv. i. 377) ; stem $2\frac{1}{2}$–$3\frac{1}{2}$ ft. high, stout, simple, finely scabrous, leafy to the base ; leaves numerous, spreading ; petiole $\frac{1}{8}$–$\frac{1}{2}$ in. long ; blade $2\frac{1}{2}$–$4\frac{1}{2}$ in. long, 1–$2\frac{1}{2}$ in. broad, oblong, very obtuse or subtruncately rounded at the apex, apiculate, rounded in and cuneately tapering into the petiole at the base, scabrous on both sides and on the margins ; umbels lateral at the nodes, sessile or the lower pedunculate, usually 4-flowered ; peduncles 0–$\frac{3}{4}$ in. long ; pedicels 1–$1\frac{1}{2}$ in. long, finely scabrous ; sepals about 7 lin. long, $3\frac{1}{2}$ lin. broad, ovate, acuminate, finely scabrous ; corolla-lobes sometimes reflexed and obliquely revolute or twisted, sometimes apparently spreading at the basal part then shortly upcurved, with the apical half closely and obliquely revolute, forming a somewhat bolster-like rim to a shallow cup, when unrolled 9–11 lin. long and 5–6 lin. broad, ovate or lanceolate, acute or obtusely pointed, subglabrous or inconspicuously scaberulous on the back, rugulose with a very minute velvet-like surface on the inner face, but glabrous to the eye, apparently greenish-white, sometimes tinted, with purple ; corona-lobes consisting of 2 contiguous erect wing-like (probably fleshy) keels on the upper face of a lanceolate acute lobe $2\frac{1}{2}$ lin. long, $1\frac{1}{2}$ lin. broad, spreading from the base of the staminal column and recurved at the apex ; keels about $\frac{1}{4}$ in. long, $1\frac{1}{2}$–2 lin. high, with the outer margins curving from the apex of the straight vertical puberulous margins next the anthers outwards and downwards to the recurved point, probably white ; staminal column $\frac{1}{4}$–$\frac{1}{3}$ in. long ; anther-appendages ovate, subobtuse, half-inflexed over the margin of the flat pentagonal style-apex. *Gomphocarpus validus, Schlechter in Engl. Jahrb.* xviii. *Beibl.* 45, 20. *Asclepias valida, Schlechter in Journ. Bot.* 1896, 455.

KALAHARI REGION : Transvaal ; grassy slopes at the base of mountains extending from Barberton to Avoca, 2000–2800 ft., *Galpin,* 707 !   White River Settlement, *Burtt Davy,* 1500 !

8. **P. natalensis** (N. E. Br.) ; plant $1\frac{1}{4}$–$1\frac{1}{2}$ ft. high, in stem, habit and general character of the leaves and inflorescence like *P. appendiculatus,* with the following differences ; leaves usually broader, $1\frac{3}{4}$–$2\frac{3}{4}$ in. long, 1–2 in. broad, oblong or elliptic, very obtuse or slightly retuse, apiculate, obtuse or very broadly cuneate at the base ; umbels sessile, usually 3–4-flowered ; pedicels $\frac{1}{3}$–$\frac{1}{2}$ in. long ; sepals 6–7 lin. long, 2 lin. broad, lanceolate, acuminate, rather coarsely pubescent, ciliate ; corolla-lobes apparently reflexed-spreading at the basal part with the terminal half upcurved, 7–8 lin. long, 5–6 lin. broad, ovate or elliptic-ovate, obtusely pointed, slightly concave or nearly flat, thinly pubescent on the back, glabrous on the inner face, green, spotted with purple-brown (*Wood*) ; corona-lobes 4–5 lin. long, with the basal part spreading, claw-like, scarcely or very slightly dilated and 1–$1\frac{1}{3}$ lin. broad at the very base, bearing 2 erect parallel subcontiguous oblong wing-like (or perhaps fleshy when alive) keels $1\frac{1}{4}$ lin. high and $\frac{2}{3}$ lin. broad, obliquely acute to truncate at their tips, with the inner and dorsal margins nearly parallel, and the inner margins with a narrow wing on the outer side extending to the broadest part of the claw, which at a short distance beyond the keels abruptly expands into an ovate or sub-cordate-ovate acute blade about $\frac{1}{4}$ in. long and broad, flat, inflexed over the claw and keels, apparently purple-brown with whitish keels ; staminal column about $2\frac{1}{2}$ lin. long ; anther-appendages large, $1\frac{1}{2}$ lin. long, ovate-acute, with the margins much folded back, loosely incurved over and much above the style-apex ; anther-wings 1 lin. long, with concave outer margins ; follicles solitary, horizontally spreading, $3\frac{1}{4}$ in. long, $1\frac{1}{6}$ in. in diam. in the immature specimen seen, stoutly fusiform-ellipsoid, obtuse with 4 entire wings $1\frac{1}{2}$–2 lin. broad, glabrous.

EASTERN REGION : Natal ; Inanda, *Wood,* 470 ! 1420 ! Intshanga, *Wood,* 6641 ! Greenwich Farm, Riet Vlei, *Fry in Herb. Galpin,* 2744 ! Dargle Farm, *Fannin,* 35 !

Probably *Wilms* 2135, collected near Pietermaritzburg, belongs to this species, but I have not dissected it.   Dr. Schlechter in various Herbaria has included this plant under *P. appendiculatus,* from which it is entirely different in floral structure.

9. **P. plicatus** (N. E. Br.) ; stem 8–15 in. high, simple or sparingly branched at the base, scabrous or subscaberulous along 2 broad lines ; leaves ascending or ascending-spreading ; petiole 2–6 lin. long ; blade $1\frac{1}{2}$–3 in. long, $\frac{2}{3}$–$1\frac{1}{2}$ in. broad, elongated oblong or oblong-lanceolate to elliptic-oblong or elliptic-ovate, shortly acute or obtuse or broadly rounded and apiculate at the apex, rounded to somewhat broadly cuneate at the base, scabrous to thinly scabrous-pubescent on both sides, densely so on the margins and midrib

beneath; flowers in 2–4 clusters of 2–4, lateral at the uppermost
nodes; pedicels 4–7 lin. long, scabrous; sepals 5–6½ lin. long, 2–2¾
lin. broad, lanceolate or ovate-lanceolate, acute or acuminate,
pubescent; corolla lobed nearly to the base; lobes campanulately
spreading, concave, with recurved or spreading tips, 7–8 lin. long,
4–5 lin. broad, ovate or lanceolate-ovate, acute, thinly pubescent on
the back, glabrous on the inner face, yellow (or in dried flowers
apparently whitish) spotted on the back with dark violet; corona-
lobes 5–6 lin. long, with the basal half claw-like, abruptly dilated
into ascending-spreading deltoid obtuse auricles at the base and
there 1¾–2 lin. broad, bearing 2 erect thin longitudinally folded
sublinear keels 1½–1¾ lin. high and ¾–1 lin. broad at the base
narrowing upwards to the truncate or emarginate apex, with the
halves (from the fold) directed outwards contiguous at the top,
gaping below, and the other halves placed nearly at right-angles
to them and extending to the tips of the auricles, apparently
yellowish; beyond the keels the margins of the claw incurve,
forming a channel or sometimes almost a tube, and the terminal
part of the lobe is dilated into an ovate acute concave blade 2½–3
lin. long, 2½ lin. broad, apparently incurved-erect, purple-brown;
staminal column about 2 lin. long; anther-appendages about 1 lin.
long, broadly ovate, obtuse or subacute, slightly folded longitudinally,
rather closely incurved over and concealing the style-apex; anther-
wings 1¾ lin. long, straight or slightly convex on their outer margin,
very prominent at the base.

Eastern Region: Griqualand East; near Kokstad, *Tyson* (a specimen distributed to the British Museum with *P. dealbatus* in *MacOwan and Bolus Herb.
Norm. Austr.-Afr.* 1315)! Natal; valley of the Buffalo River, near Charlestown,
5000–6000 ft., *Wood,* 4801! hill-side at Rock Fountain, Ixopo Div., *Mrs. Clarke*!

This resembles *P. natalensis* in appearance, but the corona-lobes and anthers are
quite different. Mr. Wood states on his label that the flowers are "yellow and
brown," the latter colour probably refers to the corona.

10. **P. macrochilus** (N. E. Br. in Dyer, Fl. Trop. Afr. iv. i. 376);
stem apparently about 10–12 in. high, simple, stout, slightly
scaberulous along 2 broad lines; leaves more or less spreading;
petiole 2–3 lin. long; blade 1½–2¾ in. long, about 1 in. broad near
the base, ovate-lanceolate, acute, broadly cuneate or rounded at the
base, often undulate, slightly scabrous on both sides or glabrous
above, scabrous on the margins; umbels subsessile or pedunculate,
lateral at the nodes, 3–6-flowered; peduncles 1–5 lin. long; pedicels
⅓–½ in. long, scabrous-pubescent; sepals about ½ in. long, 2–2½ lin.
broad, lanceolate, acuminate, thinly scabrous-pubescent; corolla-
lobes ⅔–¾ in. long, 4½–5 lin. broad, ascending or perhaps incurved-
erect, oblong-ovate, recurved at the obtuse tips, narrowly revolute
or recurved at the margins, glabrous on both sides or thinly scabrous-
pubescent on the back, apparently pale greenish-yellow more or less
spotted and tinted with purple; corona-lobes overtopping the
staminal column by about half their length, spreading out from its

base then upcurved and erect, 6–8 lin. long, linear-lanceolate or the basal part linear and claw-like, dilating above the bend into a flat lanceolate to subhastate-lanceolate blade, obtuse, cuneate at its base, at the very base of the claw and not rising higher than the base of the anthers are 2 slight contiguous fleshy keels; staminal column $2\frac{1}{2}$–3 lin. long; anther-appendages ovate, acute, loosely inflexed over the concave or crater-like style-apex. *Asclepias macrochila, Schlechter in Journ. Bot.* 1895, 355.

KALAHARI REGION : Orange River Colony; Bethlehem, *Richardson* ! Harrismith, *Sankey,* 181 ! Besters Vley, near Witzies Hoek, 6200 ft., *Bolus,* 8115 ! *Flanagan,* 2067 ! Mont-aux-Sources, 7800 ft., *Thode,* 35 !

11. **P. vexillaris** (E. Meyer, Comm. 212); stem $\frac{1}{2}$–1 ft. high, stout, simple or with 1 basal branch, hairy-pubescent; leaves rather numerous; petiole 1–4 lin. long; blade $1\frac{1}{2}$–3 in. long, $\frac{1}{4}$–1 in. broad, linear-lanceolate to oblong, acute to subobtuse, cuneate at the base, rather coarsely and thinly hairy-pubescent on both sides, undulate, with scabrous-ciliate margins; umbels lateral, sessile or pedunculate, 2–3-flowered; peduncles 0–$\frac{1}{2}$ in. long; bract solitary, 4–7 lin. long, linear or linear-subulate; pedicels $\frac{1}{3}$–$\frac{1}{2}$ in. long, hairy; sepals 5–10 lin. long, $1\frac{3}{4}$–$2\frac{3}{4}$ lin. broad, lanceolate, acuminate, hairy; lobes of the globose-campanulate corolla 7–8 lin. long, 4–$5\frac{1}{2}$ lin. broad, ovate, subacute, with narrowly revolute margins, glabrous on both sides, apparently whitish or pale greenish, stained with dull purple on the back; corona-lobes $\frac{1}{2}$ in. long, much overtopping the staminal column, spreading at the basal linear claw-like part, which bears 2 erect wing-like falcate white keels 2–$3\frac{1}{2}$ lin. long, $1\frac{1}{4}$–2 lin. broad, with the acute tips directed outwards, beyond them the lobe becomes erect and dilates into a cuneate 3-lobed concave dark purple blade 3–$5\frac{1}{2}$ lin. broad when flattened out, with the lateral teeth horizontally spreading or inwardly directed, varying from shortly and broadly deltoid to linear-deltoid, obtuse, middle tooth narrowly deltoid or linear-oblong, obtuse to bifid at the apex, erect or with the tip recurved; staminal column $2\frac{1}{3}$ lin. long; anther-appendages ovate, obtuse or subacute, with the tips inflexed over the emarginate angles of the pentagonal or shortly 5-lobed style-apex; follicles solitary, erect, $3\frac{1}{2}$ in. long, 1 in. thick, lanceolate-fusiform, tapering to a stout obtuse point, with 6 broad coarsely toothed or subentire wings, puberulous. *P. vexillatus, Steud. Nom. Bot. ed.* 2, ii. 245. *Xysmalobium vexillare, Dietr. Syn. Pl.* ii. 902. *Asclepias vexillata, Schlechter in Engl. Jahrb.* xxi. *Beibl.* 54, 9.

SOUTH AFRICA : without locality, Zeyher, 1162 !
COAST REGION : Queenstown Div.; Andries Berg, 5600 ft., *Galpin,* 2267 ! British Kaffraria, *Cooper,* 297 !
CENTRAL REGION : Somerset Div.; Bosch Berg, 3000 ft., *MacOwan,* 1687 ! Beaufort West Div.; Rhenoster Kop, *Burke* ! Graaff Reinet Div.; mountains near Graaff Reinet, *Bolus,* 229 ! 12132 ! Aliwal North Div.; Witteberg Range, *Drège,* 3418 ! Albert Div.; near Burghersdorp, *Guthrie in Herb. Bolus,* 10556 ! 10557 !

The specimens quoted by E. Meyer under letters "*b*" and "*c*" from George and Uitenhage Divisions respectively, probably belong to *P. dealbatus*, I have not seen them, but one specimen at Kew under letter "*a*" belongs to that species.

12. P. stenoglossus (N. E. Br.); plant 6–8 in. high; stems decumbent at the base then erect, harshly or somewhat scabrous pubescent with spreading hairs; leaves ascending; petiole $1\frac{1}{2}$–4 lin. long; blade $1\frac{3}{4}$–$2\frac{1}{2}$ in. long, or the lower shorter, $\frac{1}{4}$–$\frac{2}{3}$ in. broad, linear to elongate-lanceolate, with the basal ones ovate, acute or subobtuse, harshly pubescent with short spreading hairs on both sides; umbels 2–4 to a stem, lateral at the nodes and terminal, pedunculate, 3-flowered; peduncles $\frac{1}{4}$–1 in. long, and together with the $\frac{1}{2}$ in.-long pedicels harshly pubescent like the stem; sepals about $\frac{1}{2}$ in. long, $1\frac{1}{4}$–$1\frac{1}{2}$ lin. broad, enlarging in fruit, lanceolate, acuminate, spreading-pubescent; corolla cup-shaped, lobed to $\frac{3}{4}$ of the way down or more; lobes about 5 lin. long, $2\frac{1}{2}$–3 lin. broad, ovate, subacute, apparently with spreading or recurved tips, slightly pubescent on the back, glabrous on the inner face; corona-lobes 5 lin. long, 1 lin. broad at the broadest part, erect or suberect from a shortly spreading base, spathulately linear-oblong or linear-lanceolate at the upper part, truncate or very obtuse, with a pair of rather thin keels on the basal part, which are obliquely oblong, somewhat folded upon themselves, with their obtusely rounded tips directed outwards and the upper parts contiguous, as if pinched together, forming a cavity beneath with the opening outwards; staminal column 2 lin. long; anther-appendages rather small, ovate, obtuse, spreading from under with their tips curving over the margin of the style-apex, which is large, shortly 5-lobed, with a pit-like depression at the centre and the thick margin overhanging the top of the column and concealing the pollen-carriers as seen from above; young follicle ovoid, winged at the upper part, thinly puberulous. *Pachycarpus vexillaris, var. stenoglossus, E. Meyer, Comm.* 212.

CENTRAL REGION : Aliwal North Div. ; on the Wittebergen Range, 6000–7000 ft., *Drège*, 3419 !

The corona-lobes of this plant are entirely different from those of *P. vexillaris,* of which it was considered to be a variety by E. Meyer, from whose type I describe.

13. P. schinzianus (N. E. Br. in Dyer, Fl. Trop. Afr. iv. i. 376); $\frac{3}{4}$–$1\frac{1}{4}$ ft. high; stems often solitary, simple or with 1 branch at the base, stout, more or less scabrous; leaves shortly petiolate, 2–$4\frac{1}{2}$ in. long, $\frac{1}{3}$–$1\frac{1}{2}$ in. broad, linear-lanceolate, lanceolate or elongate-ovate, acute to subobtuse, broadly cuneate to subtruncate at the base, rigidly coriaceous, glabrous above or more or less scabrous on both sides, very scabrous and often wavy on the margins; umbels 2–4 and subcorymbose or solitary and terminal, pedunculate, 2–6-flowered; peduncles $\frac{1}{2}$–2 in. long, scabrous-pubescent; bracts 3–5 lin. long, subulate; pedicels $\frac{1}{2}$–1 in. long; sepals $4\frac{1}{2}$–7 lin. long,

2–3 lin. broad, ovate-lanceolate, acute, and together with the pedicels and bracts scabrous-pubescent; corolla-lobes suberect, forming a cup, recurved at the tips, ½–¾ in. long, 3½–5 lin. broad, ovate or oblong-ovate, acute, glabrous on both sides or more or less scabrous-pubescent on the back at the tips, white, purple-tinted at the tips on the back; corona-lobes arising at the base of the staminal column, ⅓–½ in. long, erect or slightly incurving upwards, white or pale lemon-yellow with a violet-purple stripe down the inner face; upper half elliptic-lanceolate or dilated into a rather large concave hood, obtuse or acute at the incurved apex, basal half linear-oblong, bearing 2 contiguous fleshy deltoid erect acute teeth not reaching to the top of the 2½–3 lin.-long staminal column; anther-appendages broadly ovate, acute or obtuse, inflexed on the top of the style-apex, which is truncate with a central star-like depression; follicles solitary, 3–4½ in. long, 1 in. thick, lanceolate or fusiform, obtusely-pointed, 6-winged, glabrous : seeds 2½ lin. long, 1–1¼ lin. broad, narrowly ovate or ovate-oblong, thick, with the margins much incurved, forming a rather deep channel on one side, very convex on the other, covered on both sides with minute raised tubercles and lines, dark brown. *Gomphocarpus schinzianus, Schlechter in Engl. Jahrb. xx. Beibl. 51, 37. Asclepias schinziana, Schlechter in Journ. Bot. 1896, 455; Rand in Journ. Bot. 1903, 200.*

KALAHARI REGION : Orange River Colony; Parys, *Rogers*, 800! Transvaal; Magaliesberg Range, *Burke*, 99! 359! *Zeyher*, 1161! various localities near Pretoria, *McLea in Herb. Bolus*, 3085! *Bolus*, 12131! *Rehmann*, 4405! *Burtt Davy*, 683! 816! 2137! 3059! *Miss Pegler*, 1020! *Miss Leendertz*, 325! 465! Pinedene, *Burtt Davy*, 2303! Irene, *Burtt Davy*, 751! 3242! Springbok Flats, *Burtt Davy*, 2128! near Heidelberg, *Schlechter*, 3528! Rustenberg, *Miss Pegler*, 1020! Yster Spruit, *Nelson*, 325! Linokana, *Holub*! Matebe Valley, *Holub*! High Veld, *Adlam*, 5! Modderfontein, *Conrath*, 1001! Johannesburg, *Ommanney*, 69! *Rand*, 862! 963! 964! 1010! 1229! Pilgrims Rest, *Greenstock*!

Burtt Davy 3059, from Bremersdorp, belongs to *P. concolor.*

14. P. dealbatus (E. Meyer, Comm. 211); stem ½–1½ ft. high, simple, stout, varying from nearly glabrous to densely hairy pubescent or sometimes scabrous-pubescent, at least along 2 broad lines; leaves ascending or spreading; petiole 2–5 lin. long; blade 1½–3¾ in. long, ⅓–1¾ in. broad, linear-lanceolate to broadly oblong or elliptic, acute, obtuse or retuse and apiculate, broadly rounded to cuneate at the base, scabrous on both sides and on the margins or glabrous above; umbels lateral at the nodes, sessile or the lowest shortly pedunculate, 3–6-flowered; pedicels 4–10 lin. long, subglabrous or scaberulous to densely subtomentose; sepals ¾ as long as the corolla, 5–7 lin. long, 2–2½ lin. broad, lanceolate, acuminate, from thinly and minutely pubescent to somewhat scabrous; corolla globose-campanulate, 7–9 lin. in diam., quite glabrous, pale green, spotted with purple-brown (*Wood*); lobes erect, with imbricating margins and recurved tips, 5–8 lin. long, 3½–4½ lin. broad, ovate, acute, very concave at the lower part; corona-lobes 4–5 lin. long, spreading from the base of the staminal column then abruptly incurved with erect tips, equalling or shortly exceeding the style-apex, linear or slightly dilated at the apex, or the apical part oblong or elliptic or sometimes constricted or with an obscure or distinct obtuse lobule on each side and 1–2 lin. broad, truncate, emarginate or rounded, flat from base to apex, without keels or other appendages, yellow with purple tips (*Wood*); staminal column 2–2½ lin. long ; anther-appendages large, broadly ovate, acute or subobtuse, inflexed over and covering the style-apex, which has 5 infolded contiguous lobes almost covering the cavity beneath them ; follicles solitary, erect, about 4 in. long and 1¼ in. thick, stoutly lanceolate-fusiform, obtusely pointed, 6-winged, glabrous, wings broad, shortly produced into obtuse teeth at the apex ; seeds about 2½ lin. long, concave on one side, very convex on the other, subreticulately rugose, brown. *P. ligulatus, E. Meyer, Comm.* 211 ; *Meisner in Hook. Lond. Journ. Bot.* 1843, 545 (*by error* 445) ; *Krauss in Flora,* 1844, 826. *Xysmalobium dealbatum and X. ligulatum, Dietr. Syn. Pl.* ii. 902. *Gomphocarpus dealbatus, Decne in DC. Prodr.* viii. 563 (*excl. syn. Pachycarpus vexillaris, E. Meyer*) ; *Schlechter in Engl. Jahrb.* xviii. *Beibl.* 45, 7 *and* 18, *and* (*by error G. albatus*) xx. *Beibl.* 51, 37 *in note.   G. alatus, Schlechter in Verhandl. Bot. Ver. Brandenb.* xxxv. 51.   *Asclepias dealbata, Schlechter in Engl. Jahrb.* xxi. *Beibl.* 54, 6 ; *Journ. Bot.* 1896, 452, *aud Ann. Naturhist. Hofmus. Wien,* xv. 67.

COAST REGION : Swellendam Div. ; near Zuurbraak, 800 ft., *Schlechter,* 2217. George Div. ; near the Keurbooms River, *Burchell,* 5127 ! Humansdorp Div. ; near the Kromme River, *Krauss* (ex *Meisner*).   Uitenhage Div. ; various localties, *Drège* ! *Bolus,* 9143 ! Port Elizabeth Div. ; near Port Elizabeth, *Burchell,* 4300 ! 4355 ! *Cooper,* 2717 ! Alexandria Div. ; Zwart Hoogte, *Zeyher,* 652 ! Bathurst Div. ; Trapps Valley, *Miss Daly,* 564 ! Albany Div. ; near Grahamstown, *MacOwan,* 614 ! *Haagner in Herb. Conrath,* 1000 ! Queenstown Div. ; near Queenstown, 4400 ft., *Galpin,* 2014 ! King Williamstown Div. ; near the Buffalo River and elsewhere, *Drège* ! Komgha Div. ; hills near the Kei River, *Flanagan,* 388 ! *Bolus,* 10203 ! near Komgha, *Flanagan in MacOwan, Herb. Austr.-Afr.,* 1505 ! *Krook,* 819 (ex *Schlechter*).

CENTRAL REGION : Cradock Div. ; near Cradock, *Cooper,* 1287 !

KALAHARI REGION : Basutoland, *Cooper,* 931 !

EASTERN REGION : Transkei ; Kreilis Country, *Bowker* ! near Butterworth, *Schlechter,* 6252 ! Tembuland ; near the Qumancu River, *Bolus* ! Pondoland ; near St. Andrews, *Tyson,* 3153 ! Griqualand East ; mountains around Kokstad, *Tyson,* 1682 ! 2704 ! and in *MacOwan and Bolus, Herb. Norm. Austr.-Afr.* 1315 ! *Haygarth in Herb. Wood,* 4173 ! *Krook,* 805, and foot of Insiswa Mountains, *Krook,* 808 (ex *Schlechter*). Natal ; near Hoffenthal, *Wood,* 3473 ! bank of the Tugela River, *Wood,* 3475 ! Biggars Berg, *Wood,* 4247 !

In Thunberg's Herbarium a specimen of this plant is mixed with *P. grandiflorus.* The typical and more western form has lanceolate acute leaves, whilst the eastern form usually has broad oblong or elliptic obtuse or retuse apiculate leaves, but does not otherwise differ.   *Krook's* 808 is also quoted by Dr. Schlechter for *Asclepias ochroleuca* (*Xysmalobium Gerrardi*).

15. P. concolor (E. Meyer, Comm. 210); stem 1–1½ ft. high, simple, stout, more or less scabrous ; leaves spreading ; petiole 1–4 lin. long ; blade 1¾–5 in. long, ½–1¾ in. broad, oblong, lanceolate,

narrowly oblong-lanceolate, elongated ovate or rarely elliptic-oblong,
acute or obtuse, rounded, broadly cuneate or subtruncate at the
base, flat or wavy, nearly glabrous above or more or less scabrous
on both sides, very scabrous on the margins; flowers in pairs or on
2-flowered peduncles, lateral at the nodes and terminal; peduncle
0–3 lin. long; pedicels 4–7 lin. long, roughly pubescent or scabrous;
sepals 4–7 lin. long, 2–4½ lin. broad, ovate or ovate-lanceolate, acute,
hairy-pubescent; corolla-lobes campanulately spreading or erect
with recurved tips, 7–8 lin. long, 4–5 lin. broad, elliptic-ovate or
elliptic-oblong, acute, glabrous on both sides or slightly pubescent
on the back, usually dull violet or dark purple-brown, varying to
(apparently) greenish or yellowish-green with purple-tinted tips;
corona-lobes variable, 4–5 lin. long, 2–3 lin. broad at the rhomboid
or ovate basal half, which is spreading and bears on its upper side
2 erect thick and fleshy or (in dried flowers) thin and wing-like
contiguous keels 1–2½ lin. high, 1–2 lin. broad, usually subfalcate-
deltoid viewed side-ways, and in fresh flowers as if pinched together
at the top, rounded on the dorsal margins and in dried flowers
narrowly winged (from shrinkage) on the basal margins; terminal
half of the lobe linear, linear-lanceolate, oblong, subovate or more
or less obtusely trilobulate, obtuse, variable in length, erect or
incurving over the top of the keels, green according to a drawing at
Kew, but in dried specimens apparently either entirely purple-
brown or yellow, or yellowish with the keels purple-brown and
with or without purple-brown margins; staminal column 2–2½ lin.
long; anther-appendages orbicular or broadly ovate, acute, inflexed
over and covering the truncate style-apex; follicles solitary, the
immature specimens seen 2½ in. long, ¾ in. thick, ovoid-lanceolate or
stoutly fusiform, very obtuse, about 4–6-winged, not distinctly
puberulous but thinly covered with exceedingly minute point-like
hairs quite invisible to the naked eye. *Gomphocarpus concolor,
Decne in DC. Prodr.* viii. 563; *Schlechter in Engl. Jahrb.* xviii.
Beibl. 45, 7.   *G. geminiflorus, Schlechter in Engl. Jahrb.* xx. *Beibl.*
51, 31.   *Asclepias concolor, Schlechter in Engl. Jahrb.* xx. *Beibl.* 54, 6,
and Journ. Bot. 1896, 452.   *A. geminiflora, Schlechter in Journ. Bot.*
1896, 455.   *Xysmalobium concolor, Dietr. Syn. Pl.* ii. 902.

COAST REGION : Stutterheim Div. ; near the Kabousie River, *Cooper,* 467 !
King Williamstown Div. ; between Chalumna River and Kachu (Yellowwood)
River, *Drège*! East London Div. ; near East London, *Wood in Herb. Galpin,*
3371 !

KALAHARI REGION : Orange River Colony ; Sand River, *Burke*! Bechuanaland :
near Kuruman, *Burchell,* 2489 ! Transvaal ; Springbok Flats, *Burtt Davy,* 1202 !
2145 ! Wonderboom Poort, *Miss Leendertz,* 417 ! *Miss Tennant*! near Aapies
River, *Schlechter,* 3669 ; Rustenburg, *Miss Nation,* 2 ! South African Gold Fields,
Baines! near Warm Bath, 3600 ft., *Bolus,* 12130 ! *Burtt Davy,* 2598 ! 2634 !
Daspoort Rand, *Burtt Davy,* 3190 !

EASTERN REGION : Transkei ; Kreilis Coontry, *Bowker,* 293 ! Kentani Div..
Miss Pegler, 644 ! Tembuland ; Bazeia Mountain, 2500–3000 ft., *Baur,* 581 !
Griqualand East ; near Clydesdale, *Tyson,* 2705 ! and in *MacOwan and Bolus,
Herb. Norm. Austr.-Afr.* 1316 ! Natal ; Shafton, Howick, *Mrs. Hutton,* 513 !
Rock Fountain, *Mrs. Clarke*! and without precise locality, *Gerrard,* 1282 bis !

Swaziland ; near Mafutane, 1500 ft., *Bolus,* 12127 ! near Bremersdorp, *Bolus,* 12128 ! *Burtt Davy,* 3059 ! 3060 !

Partly described from flowers preserved in fluid.   A few specimens above quoted have the terminal part of the corona-lobes slightly 3-lobed, these are probably of hybrid origin.   The number 293 was also given by Bowker to *Krebsia stenoglossa,* and 3059, 3060, by Burtt Davy also respectively to *P. schinzianus,* and *Asclepias eminens.*

16. **P. transvaalensis** (N. E. Br. in Dyer, Fl. Trop. Afr. iv. i. 376) ; stem 1 ft. or more high, simple, stout, more or less scabrous ; leaves in 7–8 pairs on the specimen seen, sessile or subsessile, mostly $2\frac{1}{2}$–3 in. long, $\frac{1}{2}$–1 in. broad at the base, the lower pairs smaller, narrowly or broadly oblong-lanceolate, acute, subtruncate or subcordate at the base, very scabrous on both sides and at the edges ; umbels 5 on the specimen seen, lateral and terminal, pedunculate, 4–5-flowered ; peduncles $\frac{1}{4}$–1 in. long, scabrous ; pedicels $\frac{1}{4}$–$\frac{1}{3}$ in. long, stout, scabrous ; sepals $4\frac{1}{2}$–$5\frac{1}{2}$ lin. long, $2\frac{1}{2}$ lin. broad, lanceolate or ovate-lanceolate, acuminate, scabrous ; corolla-lobes apparently slightly ascending-spreading, 7–8 lin. long, 5–$5\frac{1}{2}$ lin. broad at the middle, elliptic-ovate, subacute, pubescent (subscabrous?) with minute adpressed scattered hairs on the back, glabrous on the inner face, purple-brown ; corona-lobes 4–$4\frac{1}{2}$ lin. long, somewhat rigidly fleshy, dark purple-brown, rhomboid-ovate, 2 lin. broad and horizontally spreading at the basal part, thickened on the upper side into a large hump-like mass longitudinally divided by a deep fissure, truncate at the base with an excavation on each side of the fissure, and rising to about half the height of the $2\frac{1}{2}$ lin.-long staminal column, beyond the hump the lobe contracts into the narrowly oblong or tongue-shaped obtuse upper half $1\frac{1}{4}$–$1\frac{1}{2}$ lin. broad, which incurves over the basal part ; anther-appendages transverse, rounded, inflexed over the margin of the truncate style-apex. *Gomphocarpus transvaalensis, Schlechter in Engl. Jahrb.* xviii. *Beibl.* 45, 22, *and* 19 *in note. Asclepias transvaalensis, Schlechter in Journ. Bot.* 1896, 455.

KALAHARI REGION : Transvaal ; Saddleback Mountain, near Barberton, 3000–4500 ft., *Galpin,* 692b (592b ex *Schlechter*) !

17. **P. insignis** (N. E. Br.) ; stem 9–16 in. high, erect, simple, adpressed-scaberulous along 2 broad lines, or according to *Schlechter* " pilose-scabrous " ; leaves ascending, shortly petiolate, 2–$3\frac{1}{4}$ in. long, $\frac{1}{2}$–$\frac{3}{4}$ in. broad at the base, ovate, ovate-lanceolate, narrowly oblong or oblong-lanceolate, obtuse or subacute, subtruncate or broadly cuneate at the base, flat or slightly undulate at the margins, with a spreading pubescence (pilose-scabrous, *Schlechter*) on both sides, somewhat scabrous on the midrib beneath and scabrous-pubescent on the margins ; umbels lateral at the nodes and terminal, peduncu-late 3–4-flowered ; peduncles $\frac{2}{3}$–$1\frac{1}{2}$ in. long, scabrous-pubescent ; pedicels 4–5 lin. long, scabrous-pubescent ; sepals $3\frac{1}{2}$–4 lin. long, about 2 lin. broad, ovate, acute or acuminate, spreading-pubescent ;

corolla-lobes apparently spreading so as to form a broad cup, about 6 lin. long, $3\frac{1}{2}$–4 lin. broad, ovate, acute, glabrous on both sides, apparently whitish or sulphur-yellow, varying according to *Schlechter* to golden-yellow, purple-tinted on the back at the tips : corona-lobes 4–$4\frac{1}{2}$ lin. long, entirely dark purple-brown, with the basal half very spreading, oblong, $1\frac{1}{4}$–$1\frac{1}{3}$ lin. broad, bearing a pair of erect deltoid keels, contiguous and very fleshy at their base, thinner and rather widely gaping above, rising to about half-way up the anthers ; upper half of the lobes abruptly inflexed over the basal keels and staminal column, with their tips meeting at the centre, slightly constricted just above the bend, then dilated into an ovate obtuse blade $1\frac{1}{3}$–$1\frac{1}{2}$ lin. broad ; staminal column about 2 lin. long ; anther-appendages broadly ovate, acute, inflexed over the margin of the somewhat crater-like style-apex. *Gomphocarpus insignis, Schlechter in Engl. Jahrb. xx. Beibl. 51, 32. Asclepias insignis, Schlechter in Journ. Bot. 1896, 455.*

KALAHARI REGION : Transvaal ; Elandspruit Mountains, 7600 ft., *Schlechter*, 3847 ; near Lydenburg, *Wilms*, 954 !

This is doubtfully distinct from *P. transvaalensis*, and seems only to differ in its more pubescent leaves and the much smaller humps or keels on the base of the corona-lobes.   I have seen only one gathering of each.

18. **P. decorus** (N. E. Br.) ; stem scabrous to nearly glabrous, only the upper part seen ; leaves ascending or somewhat spreading, very shortly petiolate, $2\frac{1}{4}$–$3\frac{1}{2}$ in. long, $\frac{1}{2}$ to nearly 1 in. broad, lanceolate, narrowly oblong-lanceolate or linear-oblong, acute or subobtuse, cuneate to subtruncate at the base, thinly and rigidly scabrous on both sides, scabrous on the margins ; umbels lateral at the nodes and terminal, pedunculate, usually 3- (sometimes 4-) flowered ; peduncles $\frac{1}{4}$–2 in. long, scaberulous on one side ; bracts $\frac{1}{3}$ in. long, linear or linear-subulate, acute, scabrous-ciliate ; pedicels $\frac{3}{4}$–$1\frac{1}{3}$ in. long, scaberulous or scabrous ; sepals $\frac{1}{2}$ in. long, 2–$2\frac{1}{2}$ lin. broad, ovate or lanceolate, acuminate, pubescent ; corolla broadly cup-like, lobed nearly to the base, apparently yellow or creamy-white ; lobes about 7 lin. long, 4–$5\frac{1}{2}$ lin. broad, elliptic-ovate or oblong-ovate, acute, glabrous on both sides or slightly adpressed pubescent at the tips on the back ; corona-lobes apparently entirely yellow or white, about 5 lin. long, spreading or ascending-spreading, perhaps somewhat incurved at the apex, slightly dilated and 2–$2\frac{1}{4}$ lin. broad at the base, which bears a pair of fleshy contiguous erect deltoid very obtuse keels, rising to about $\frac{3}{4}$ of the height of the staminal column, faintly constricted just beyond the keels, with the terminal part $1\frac{1}{2}$–$1\frac{3}{4}$ lin. broad, varying from oblong-lanceolate to subhastate-lanceolate, obtuse ; staminal column 2 lin. long ; anthers with a large wedge-shaped blackish-violet spot on the upper part, and suborbicular or broadly rounded obtuse appendages inflexed over the truncate or slightly depressed style-apex.

KALAHARI REGION : Transvaal ; Endholeyana, near Piet Retief, *Lady Barkly* !
EASTERN REGION : Natal ; without precise locality, *Gerrard*, 1287 !

19. P. McKenii (N. E. Br. in Dyer, Fl. Trop. Afr. iv. i. 377);
stem 2 ft. or more high, stout, bifariously subhispid ; leaves spreading,
shortly petiolate, 1½–2 in. long, ¾–1½ in. broad, elliptic, obtuse,
apiculate, cuneate or rounded at the base, coriaceous, thinly
sprinkled with outstanding hairs on both sides, scabrous and ciliate
on the margins, the cilia disappearing with age ; umbels lateral at
the nodes, shortly pedunculate, 2–4-flowered ; peduncles 1½–2 lin.
long, roughly hairy ; pedicels 3–4 lin. long, roughly hairy ; sepals
5–6 lin. long, 2–2¼ lin. broad, lanceolate, acute, hairy and ciliate
with rather long and somewhat stiff hairs ; corolla 1 in. or more in
diam., thinly pubescent outside, glabrous within, apparently whitish,
spotted with purple ; tube about ¼ in. long, very broadly campanulate or cup-shaped ; lobes apparently ascending-spreading, about
5 lin. long, 4½ lin. broad, ovate, notched at the obtusely pointed
apex ; corona-lobes arising 1½ lin. up the staminal column, 7 lin.
long, 1–1¼ lin. broad, apparently purple-brown, horizontally spreading
at the base, then upcurved, with the upper half horizontally incurved
over the basal part, linear, acute, replicate at the upper half, bearing
2 keels extending from the base to beyond the middle and then
excurrent at the margins, rectangular and not reaching to the tops
of the anthers at the base, closely contiguous for ⅔ of their length,
then diverging ; staminal column ¼ in. long, anther appendages
ovate, acute, inflexed over the depressed-truncate style-apex ; anther-
wings abruptly widely divergent at the very base, but scarcely or
not at all broader or more prominent than at the apex. *Gompho-*
carpus M‘Kenii (*misprint for McKenii*), *Harv. Thes. Cap.* ii. 60,
(*McKenii*) *t.* 194.   *G. Mackenii, Pritz. Ind. Ic. Bot.* ii. 135.   *Asclepias*
Mackenii, Schlechter in Journ. Bot. 1896, 454.

Eastern Region : Zululand, *Gerrard*, 1284 !

Xysmalobium orbiculare was also distributed as *Gerrard*, 1284.

20. P. coronarius (E. Meyer, Comm. 209) ; plant 2–2½ ft. high,
exactly like *P. grandiflorus*, E. Meyer, in appearance ; leaves shortly
petiolate, spreading, 1½–3 in. long, up to 1 in. broad, oblong or
oblong-lanceolate, acute, somewhat scabrous or nearly glabrous on
both sides, undulate at the scabrous margins ; flowers in pairs at the
nodes ; pedicels 1–1¼ in. long, more or less pubescent ; sepals ¾–1 in.
long, 5–6 lin. broad, elliptic, shortly acuminate or very acute,
thinly pubescent ; corolla about 1½ in. in diam., inflated-globose, very
slightly pubescent outside or glabrous on both sides, apparently
whitish-green or yellowish-green, sometimes spotted with dark
purple-brown ; united part 4 lin. long ; lobes 1¼ in. long, 1 in. or
more broad, elliptic-ovate, acute, very concave, with overlapping
margins and recurved tips ; corona-lobes horizontally spreading, ¾ in.
long, lanceolate, deeply complicate boat-shaped, obtuse, somewhat
deltoid-ovate viewed sideways, narrowly wing-margined along the
linear-oblong under side, not produced at the apex beyond the broad
erect sides, which rise to the level of the style-apex and are 4½–5

lin. high at the broadly rounded (not angular) base, connate and perhaps slightly cucullate at the apex; staminal column 4 lin. long; anther-appendages suborbicular, with their rounded tips incurved over the margin of the truncate style-apex. *Gomphocarpus coronarius, Decne in DC. Prodr.* viii. 562. *Xysmalobium coronarium, Dietr. Syn. Pl.* ii. 902.

EASTERN REGION : Transkei ; near the Bashee River at Fort Bowker, *Bowker,* 379 ! Tembuland ; near the Umtata River, *Drège,* 4930 !

Although very like *P. grandiflorus,* E. Meyer, the very large sepals, much more deeply lobed corolla, and different corona-lobes readily distinguish it.

21. P. grandiflorus (E. Meyer, Comm. 209) ; plant 8–20 in. high ; stem usually simple, 1½–3 lin. thick below, pubescent, with internodes ½–1 in. long ; leaves spreading or somewhat deflexed ; petiole 1½–6 lin. long ; blade 1¾–4½ in. long, 4–14 lin. broad, oblong or lanceolate, acute, obtuse, or rounded and apiculate at the apex, rounded or broadly cuneate at the base, usually undulate at the margins, somewhat roughly pubescent or subscabrous on both sides or glabrous above ; umbels 2–4-flowered, sessile at the nodes or the lowest shortly pedunculate ; pedicels ½–1 in. long, pubescent ; sepals 6–9 lin. long, 3–3½ lin. broad, lanceolate, very acute, pubescent ; corolla about 1½ in. in diam., inflated-globose, thinly pubescent outside or glabrous on both sides, yellow or yellowish-green, spotted with dark purple-brown ; united part 6–7 lin. long ; lobes 8–9 lin. long, 9–11 lin. broad, broadly ovate, acute, incurved, with overlapping margins and recurved tips ; corona-lobes very spreading, with much incurved tips, ¾ in. long, linear or linear-lanceolate, sometimes somewhat spathulate at the tips, bearing upon their basal half 2 long contiguous keels, which at the angular base do not nearly rise to the level of the top of the anther-wings, then gradually tapering into the lobe, apparently yellow, often dotted with purple-brown ; staminal column 3½–4 lin. long, broadly conical ; anther-appendages suborbicular or elliptic-ovate, obtuse, inflexed over the rim of the truncate 5-crenate excavated style-apex ; pollen-carriers seated just below the rim of the style-apex ; follicles solitary, about 3½ in. long, ¾–1 in. thick, stoutly fusiform-lanceolate, equally tapering to base and apex, obtuse, with 6 entire wings 1–1½ lin. broad, thinly and minutely puberulous. *Meisn. in Hook. Lond. Journ. Bot.* ii. 1843, 544 (*by error* 444) ; *Krauss in Flora,* 1844, 826. *Asclepias grandiflora, Linn. f. Suppl.* 170 ; *Lam. Encycl.* i. 284 ; *Thunb. Prodr.* 47 ; *in Nov. Act. Acad. Petrop.* xiv. (1805), 509 ; *Fl. Cap. ed.* 2, ii. 156, *and ed. Schultes,* 235 ; *Willd. Sp. Pl.* i. 1264 ; *Pers. Syn.* i. 275 ; *Schlechter in Journ. Bot.* 1896, 451 (*excluding synonyms Pachycarpus coronarius and Gomphocarpus coronarius*) ; *Engl. Jahrb.* xxi. *Beibl.* 54, 7 : *and Ann. Naturhist. Hofmus. Wien,* xv. 67. *Xysmalobium grandiflorum, R. Br. in Mem. Wern. Soc.* i. 39 ; *Schultes, Syst. Veg.* vi. 90 ; *Spreng. Syst. Veg.* i. 850 ; *G. Don, Gen. Syst.* iv. 146 ; *Dietr. Syn. Pl.* ii. 902 ; *Decne in DC. Prodr.* viii. 519. *Gomphocarpus grandiflorus. Decne, l.c.* 562 ; *Schlechter in Engl. Jahrb.* xviii. *Beibl.* 45, 8.

Var. β, **elatocarinatus** (N. E. Br.) ; corolla yellow spotted with purple-brown ; keels of the corona-lobes rising to the level of the tops of the anther-wings at the base ; otherwise as in the type.

Var. γ, **chrysanthus** (N. E. Br.); corolla entirely yellow ; keels of the corona-lobes rising to the level of the tops of the anther-wings or slightly beyond ; otherwise as in the type. *Asclepias grandiflora, var. chrysantha, Schlechter in Engl. Jahrb.* xxi. *Beibl.* 54, 7, *and Ann. Naturhist. Hofmus. Wien,* xv. 68.

Var. δ, **tomentosus** (N. E. Br.); stem, leaves, peduncles, pedicels and sepals all thickly villous-tomentose, but slightly harsh to the touch ; corolla and corona-lobes as in var. β. *Gomphocarpus grandiflorus, var. tomentosus, Schlechter in Engl. Jahrb.* xviii. *Beibl.* 45, 19.

Coast Region : Humansdorp Div. ; near the Kromme River, *Thunberg* ! Uitenhage Div. ; region around Uitenhage, *Burchell,* 4421 ! 4440 ! 4449 ! *Zeyher* ! Addo, *Drège,* 2222 ! Van Stadens Berg, *Zeyher,* 630 ! Port Elizabeth Div. ; near Port Elizabeth, *Prior* ! *Miss West,* 35 ! Albany Div. ; Albany plains, *Bowie* ! near Grahamstown, *MacOwan,* 180 ! *Misses Daly and Sole,* 367 ! Bathurst Div. ; Linch's Post, near the Kowie River, *Bowie* ! Komgha Div. ; hills near Keimouth, *Krook,* 818 (ex *Schlechter*). Var. β : King Williamstown Div. ; Keiskamma, *Mrs. Hutton* ! Var. γ : East London Div. ; near East London, *Wood in Herb. Galpin,* 3387 ! Komgha Div. ; near Komgha and near Keimouth, *Flanagan,* 375 ! *Krook,* 802 (ex *Schlechter*).

Kalahari Region : Var. δ : Transvaal ; Carolina district, *Burtt Davy,* 2953 ! Umlomati Valley, near Barberton, 4000 ft., *Galpin,* 913 (ex *Schlechter*).

Eastern Region : Griqualand East ; hills near Umzimkulu River, *Krook,* 791 (ex *Schlechter*). Var. β : Tembuland ; Bazeia, *Baur,* 602 ! Natal ; hills above Byrne, *Wood,* 3169 ! hill near Lynedoch, *Wood in Herb. Natal,* 962 ! Peak of Byrne, *Wood in Herb. Natal,* 589 ! near Durban Bay, *Krauss,* 1260 (ex *Meisner*). Var. γ : Transkei ; Kreilis Country, *Bowker* !

Wood's 589 is more hairy than usual and approaches var. *tomentosus.*

22. P. linearis (N. E. Br.); stem 1–1¼ ft. high, simple, scabrous ; leaves erect, shortly petiolate, 3–5 in. long, 1–2½ lin. broad, linear, acute, with revolute margins, scabrous above and on the midrib beneath ; umbel solitary, terminal, nodding, 3–5-flowered ; peduncle ¾–1½ in. long, recurved at the apex, subscabrous ; bracts not seen ; pedicels ⅔–1 in. long, subscabrous ; sepals 3½–4½ lin. long, 1–1¼ lin. broad, lanceolate, acuminate, pubescent ; corolla probably subglobose and about 9–10 lin. in diam., but when dried often appearing broadly cup-shaped and 1¼–1½ in. in diam., lobed to ⅔–¾ of the way down, coarsely pubescent outside, glabrous within, "yellowish-green, brownish outside at the base" (*Baur*), "purple and buff" (*Bowker*); tube ¼ in. long; lobes 8–9 lin. long, 5–6 lin. broad, elliptic-ovate or ovate-oblong, obtuse ; corona-lobes spreading at the basal part then much incurved or incurved-erect, 5–7 lin. long, linear, often lanceolately dilated below the acute apex, with 2 keels extending to beyond the middle, rising at the base to the level of the top of the anther-wings (1½–2 lin. high) and there wing-like and toothed on the inner margin ; staminal column 4½–5 lin. long, of which length the anther-appendages take up more than half, being 3–3½ lin. long, 2 lin. broad, ovate-lanceolate or subhastate-lanceolate when flattened out, subacute, replicate, much exceeding and loosely curved over the large excavated style-apex, which has a very thick crenate rim. *Lagarinthus linearis, E. Meyer, Comm.* 207.

Gomphocarpus linearis, Dietr. Syn. Pl. ii. 901 ; *Decne in DC. Prodr.*
viii. 562, *not of Schlechter.   G. asper, Decne in DC. Prodr.* viii. 561.
Asclepias tenuiflora, Schlechter in Engl. Jahrb. xxi. *Beibl.* 54, 9, *partly.*
A. tenuifolia, Schlechter ex Ind. Kew. Suppl. ii. 19, *partly.   A. linearis,*
Schlechter in Journ Bot. 1896, 453, *partly.*

SOUTH AFRICA : without locality, *Drège,* 4969 !
EASTERN REGION : Transkei ; near Tsomo on the Tsomo River, *Mrs. Barber,*
826 ! *Bowker,* 343 ! 359 ! Tembuland ; hill-sides here and there near Bazeia and
Tabase, 2000–3000 ft., *Baur,* 381 !

The type specimen (*Drège,* 4969) in E. Meyer's Herbarium has no locality upon
the label ; Drège and E. Meyer, however, record it from between Zaandplaat and
Komgha, in Komgha Div., and between the Bashee River and Morley in
Tembuland.   Dr. Schlechter has confused this very distinct species with
P. campanulatus and a small-flowered variety of that plant, from which and from
all other species it is at once distinguished by the very different corona-lobes and
remarkable anther-appendages.

23. **P. campanulatus** (N. E. Br.) ; " root tuberous " (*Sanderson*) ;
stem ⅔–2 ft. high, simple, thinly and somewhat harshly pubescent
or scaberulous; leaves suberect, 3–6 in. long, 1⅓–4½ lin. broad,
linear, acute, tapering or rounded into the very short petiole at the
base, with revolute margins, harshly pubescent or more or less
scabrous above and on the margins and midrib beneath ; umbel
solitary, terminal, nodding, 3–5-flowered ; peduncle ¾–1¼ in. long,
pubescent or subscabrous-pubescent, as are also the pedicels, bracts
and calyx ; pedicels ¾–1½ in. long ; bracts 6–8 lin. long, 1 lin. broad,
linear or linear-lanceolate, acute, deciduous ; sepals 5–9 lin. long,
2–3 lin. broad, lanceolate, acute ; corolla globose, not lobed beyond
the middle, 1¼–2 in. in diam., somewhat coarsely pubescent
and usually purplish outside in dried flowers, but according to a
drawing at Kew dull green, glabrous and apparently whitish or
greenish-white inside, " corolla yellow " and " yellow-green " (*Wood*) ;
united part ¾–1 in. long ; lobes 7–13 lin. long and as much in
breadth, roundish-ovate, subacute, incurved, with overlapping
margins, perhaps sometimes recurved or spreading at the tips ;
corona-lobes arising 1–1½ lin. up the staminal column, horizontally
spreading, 3½–7½ lin. long, linear or linear-lanceolate, obtuse, with
a pair of erect obliquely deltoid wing-like keels at their base 1½–2
lin. high, reaching to the base of the anther-appendages, dark
purple-brown, with the keels paler ; staminal column 3–5 lin. high ;
anther-appendages 1½–2 lin. long, oblong, obtuse, with reflexed
sides, erect, with their tips exceeding and inflexed over the margin of
the crater-like 5-crenate style-apex. *Gomphocarpus campanulatus,*
Harv. Thes. Cap. i. 61, *t.* 97.   *Asclepias linearis, Schlechter in Journ.*
Bot. 1896, 453, *partly.   A tenuiflora, Schlechter in Engl. Jahrb.* xxi.
Beibl. 54, 9 *partly.   A. tenuifolia, Schlechter ex Ind. Kew. Suppl.* ii.
19, *partly.*

VAR. β, **Sutherlandi** (N. E. Br.) ; leaves 1–3½ lin. broad : umbels 3–7-flowered ;
peduncle ¾–2½ in. long ; pedicels ⅔–1¼ in. long ; corolla ¾–1¼ in. in diam., with

lobes 4–6 lin. long, 4–9 lin. broad ; corona-lobes 1½–3 lin. long, linear, with the keels extending to about the middle or beyond. *Gomphocarpus linearis, Schlechter in Engl. Jahrb.* xviii. *Beibl.* 45, 20, *and* xx. *Beibl.* 51, 32, *not of Decne. Asclepias linearis, Schlechter in Journ. Bot.* 1896, 453, *partly, and Ann. Naturhist. Hofmus. Wien,* xv. 68.

Coast Region : Var. β : King Williamstown Div. ; near Keiroad Station, *Krook,* 801 (ex *Schlechter*).

Kalahari Region : Var. β : Orange River Colony, *Cooper,* 2732 ! 2734 ! Basutoland, *Cooper,* 936 ! 2731 ! Transvaal ; Macmac, *Mudd* ! Houtbosch, *Rehmann,* 5871 ! Saddleback Mountain, near Barberton, *Galpin,* 1366, and Elands River Mountain, *Schlechter,* 3999 (ex *Schlechter*) ; Belfast, *Burtt Davy,* 1285 ! near Vlakfontein Beacon, *Burtt Davy,* 2967 ! near Spitz Kop, *Wilms,* 942 ! Graskop, near Pilgrims, *Burtt Davy,* 1463 !

Eastern Region : Griqualand East ; Enshlenzi Mountain, near Fort William, 2500 ft., *Tyson,* 3132 ! Natal ; near Durban, *Sanderson* ! Dargle Farm, *Mrs. Fannin,* 5 ! Inanda, *Wood,* 79 ! 1326 ! hills near Charlestown, *Wood,* 5151 mixed with var. β ! and without precise locality, *Gerrard,* 2161 ! 2164 ! *Sutherland* ! Var. β : Tembuland ; Engcobo Mountain, 4500 ft., *Bolus,* 10552 ! Griqualand East ; Mount Frere, *Schlechter,* 6412 ! *Krook,* 798 ! Natal ; Mohlamba Range, 5000–6000 ft., *Sutherland* ! Biggars Berg, *Gerrard,* 1298 !

The variety *Sutherlandi* is the plant that has hitherto been mistaken for *Gomphocarpus linearis* of Dietrich, and of Decaisne, from which it is readily distinguished by the very different corona-lobes and anther-appendages. It only appears to differ from typical *P. campanulatus* in its much smaller corolla and shorter corona-lobes, for although the specimens with the smallest flowers (*Sutherland, Cooper,* 936, 2374, &c.) seem quite distinct in appearance, other specimens (*Schlechter,* 6412, *Burtt Davy,* 1285, 2967, &c.) with larger flowers seem to connect it with typical *P. campanulatus.* In the process of drying, the globose form of the corolla of this and allied species is usually altered so that it appears to be broadly cup-shaped.

24. **P. Gerrardi** (N. E. Br.) ; stem usually simple, 9–15 in. high, rather roughly pubescent ; leaves erect, 2½–5 in. long, ¾–2½ lin. broad, linear, acute, scabrous above and on the midrib beneath ; umbels usually 2 or 3, racemose, or in weak plants solitary, 8–13-flowered ; peduncles ¾–1 in. long, often slightly recurved at the top, somewhat roughly pubescent ; bracts 2–3 lin. long, linear-subulate, very deciduous ; pedicels 8–9 lin. long, slightly scaberulous ; sepals 3–3½ lin. long, ¾–1 lin. broad, attenuate-lanceolate, very acute, pubescent ; corolla globose or globose-campanulate, campanulate in the dried state, lobed to about half-way down, 7–8 lin. in diam. ; apparently whitish, suffused with purplish-brown outside and there pubescent with white and brown hairs, glabrous within ; tube 3 lin. long ; lobes 3–3½ lin. long, 3½–4 lin. broad, broadly ovate, obtuse, probably incurved ; corona-lobes 1¾–2 lin. long, ¾ lin. broad, very spreading, oblong-linear, flat, rounded at the apex, purple-brown, with 2 paler broadly deltoid keels ½ lin. high at the base ; staminal column 1¼ lin. long ; anther-appendages ½ lin. long, oblong-linear, obtuse, connivent over and concealing the rather small truncate shortly 5-lobed style-apex ; young carpels covered with brown hairs. *Gomphocarpus Gerrardi, Harv. Thes. Cap.* ii. 59, *t.* 193. *Schizoglossum Gerardi, Benth. and Hook. f., ex Ind. Kew.* ii. 826. *Asclepias Gerrardii, Schlechter in Journ. Bot.* 1896, 453.

In Harvey's figure of this species the corolla-lobes are represented much more spreading and the tube less globose than they should be.

25. P. rostratus (N. E. Br.); habit, stem, leaves, pubescence, 8–12-flowered umbels, peduncles, pedicels and sepals all exactly as in *P. Gerrardi*; corolla apparently globose-campanulate, lobed to less than half-way down, about 8 lin. in diam., pubescent and purplish-brown outside, "greenish-white to light brown" (*Haygarth*); tube 4 lin. long; lobes $2\frac{1}{2}$–$2\frac{3}{4}$ lin. long, $3\frac{1}{2}$ lin. broad, deltoid-ovate, obtuse; corona-lobes arising at the base of the staminal column, $2\frac{1}{2}$ lin. long, $1\frac{1}{3}$ lin. broad, very spreading, flat, obovate-oblong, obtusely pointed, with 2 erect broadly deltoid wing-like keels $\frac{3}{4}$–1 lin. high at the base, dark purple-brown; staminal column $1\frac{1}{2}$ lin. long to the tips of the anther-appendages, which are $\frac{2}{3}$ lin. long, oblong-linear, subobtuse, erect and pressed against the sides of the columnar apical part of the style, which is produced about 1 lin. beyond them into a sort of beak divided at the apex into 5 erect contiguous lobes $\frac{1}{3}$–$\frac{1}{2}$ lin. long; young carpels covered with brown hairs.

The beaked style-apex at once distinguishes this from *P. Gerrardi,* and the flowers seem to be less globose and are larger.

26. P. stelliceps (N. E. Br.); stem solitary, $\frac{1}{2}$–1 ft. high, simple rather slender, pubescent; leaves erect or ascending, shortly petiolate, 2–$3\frac{1}{2}$ in. long, $1\frac{1}{2}$–$2\frac{1}{2}$ lin. broad, linear, acute, with revolute margins, somewhat scabrous-pubescent above and on the midrib beneath; umbels 1–2, terminal and lateral, pedunculate, 2–$2\frac{1}{4}$ in. in diam., 5–6-flowered; pedicels 5–7 lin. long, pubescent: sepals $2\frac{3}{4}$–4 lin. long, $1\frac{1}{2}$ lin. broad, lanceolate, acute, pubescent; corolla about $\frac{2}{3}$ in. in diam., subglobose or globose-campanulate, lobed to about half-way down or less, glabrous and apparently whitish inside, pubescent and dull violaceous-brown on the tube outside; tube 5–6 lin. long; lobes recurved, $2\frac{1}{2}$–3 lin. long, 4–5 lin. broad, very broadly ovate, obtuse; corona-lobes complicate, with the basal part $1\frac{1}{2}$ lin. long, $\frac{3}{4}$ lin. broad, linear, horizontally radiating from the base of the staminal column, slightly upcurved and bifid at the apex, from which the thin sides slope upwards ($1\frac{1}{2}$ lin. high) to the level of the base of the anther-appendages and are adnate to the column up to the base of the anthers, above which they form small free deltoid obtuse erect lobules, with a narrow double wing on the outer side, very minutely papillate-puberulous, on their inner surface, but there is no horn or other process within; staminal column $2\frac{1}{2}$ lin. long, not dilated at the base; anther-appendages elliptic-oblong, rounded at the apex, erect, shorter than the style-apex and adpressed to it between its lobes; anther-wings small, about $\frac{2}{3}$ lin. long, extending down to about the middle of the column ; style-apex convex, with a small crater-like pit at the top, and 5 stellately radiating compressed lobes $1\frac{1}{3}$–$1\frac{1}{2}$ lin. long, some what falcately oblong in side view, with a tubercle or gibbosity on each side at the base, very obtuse and upcurved at the tips, which spread out about as far as the corona-lobes, apparently whitish.

EASTERN REGION : Swaziland ; on the High Veld, near Embabaan (Mbabane), 5500 ft., *Bolus*, 12117 !

Very similar to *P. Gerrardi* and *P. linearis*, but at once distinguished by its very remarkable star-shaped style-apex and different corona-lobes.

Imperfectly known species.

27. P. asperifolius (Meisn. in Hook. Lond. Journ. Bot. ii. 1843, 544, by error 444) ; stem simple, glabrous ; leaves opposite, lanceolate-oblong, acute, tapering into a short petiole, flat, hispidulous-scabrous on both sides, with a thick midrib beneath ; pedicels 5–8 in an umbel-like cluster, axillary, 6–8 lin. long, half as long as the leaves and together with the calyx puberulous ; corolla spreading, as large as in *P. ligulatus* and *P. concolor*, with oblong acute segments, pale greenish, concolorous ; corona-lobes erect, terminated by an ovate acute pointless membranous erect blade, inflexed at the apex, with a tooth on each side at the base. *Krauss in Flora*, 1844, 826. *Gomphocarpus asperifolius, Walp. Rep. Bot.* vi. 486 ; *Ind. Kew.* i. 1050.

EASTERN REGION : Natal ; margins of woods around Durban Bay, *Krauss.*

Described as resembling *P. appendiculatus*, but differing by the leaves tapering below, the spreading corolla and form of the corona-lobes. I have not seen this, but think it may have been founded upon a specimen of *P. appendiculatus* in which the flowers were not fully expanded.

28. Gomphocarpus macroglossus (Turcz. in Bull. Soc. Nat. Mosc. 1848, i. 259) ; stem robust, erect, simple, hispid ; leaves oblong, somewhat acute, tapering into a short petiole, scabrous on both sides, with revolute margins ; peduncles scarcely any ; pedicels mostly 3 together, the lower deflexed ; corolla-lobes at length spreading, without spots ; corona-lobes with a long linear flattish base, keeled at the middle, passing into an ovate obtuse blade at the apex, twice as high as the staminal column.

COAST REGION : Bathurst Div. ; hills near the Great Fish River, between Kaffirs Drift and Governors Kop, 500–2000 ft., *Ecklon*, 34 (ex *Turczaninow*).

Stated to be allied to *P. concolor*, with flowers of the same size, but differing from that by its elongated pedicels and the corona-lobes narrower at the base, not dilated. From the description and locality this would appear to be referable to either *P. reflectens* or *P. appendiculatus*, and is possibly a specimen with the flowers not fully expanded.

3 B 2

XX. GLOSSOSTEPHANUS, E. Meyer.

Calyx 5-partite ; segments narrowly lanceolate. *Corolla* small, very deeply 5-lobed ; lobes suberect, linear. *Corona* of 5 simple lobes arising at the base of the staminal column, erect, laterally compressed and dorsally grooved. *Stamens* 5, united into a column around the ovary and adnate to the dilated part of the style ; anthers terminated by a minute membranous appendage. *Pollen-masses* pendulous, attached in pairs to the erect narrow pollen-carriers, which are dilated and somewhat crutch-like at the apex. *Style* produced much beyond the anther-appendages into a long beak-like point acutely bifid at the apex.

A perennial herb, with a slender twining stem ; leaves linear ; flowers in pedunculate cymes or racemes, lateral at the nodes.

DISTRIB. Monotypic, endemic.

1. **G. linearis** (E. Meyer, Comm. 218) ; a slender twining plant, glabrous in all parts except the corolla ; leaves spreading ; petiole $1\frac{1}{2}$–4 lin. long ; blade 1–$3\frac{1}{4}$ in. long, $\frac{1}{3}$–3 lin. broad, linear or linear-lanceolate, acute or subobtuse ; flowers in racemes or simple and subumbellate or compound cymes, lateral at the nodes and including the slender peduncles $\frac{1}{2}$–$2\frac{1}{2}$ in. long ; pedicels 2–4 lin. long, filiform ; sepals $\frac{3}{4}$–1 lin. long, linear, acute ; corolla-tube $\frac{1}{3}$–$\frac{2}{3}$ lin. long, pubescent within with fine deflexed hairs ; lobes suberect, $1\frac{3}{4}$–$2\frac{3}{4}$ lin. long, $\frac{1}{2}$–$\frac{3}{4}$ lin. broad at the base, linear, acute, very minutely papillate-scaberulous on the inner face, greenish-yellow, at length becoming reddish or purplish (*Burchell*) ; corona-lobes arising at the base of the staminal column and adpressed to it, united at the very base by a pocket-like fold under the anther-wings, erect, $\frac{1}{2}$–$\frac{3}{4}$ lin. long, fleshy, laterally flattened, linear (viewed sideways), acute, grooved down the narrow back, with an oblique keel or ridge on each side near the base, white ; staminal column $\frac{1}{2}$–$\frac{2}{3}$ lin. long ; anther-appendages minute, obtuse, erect, adpressed to the base of the tapering (angular?) style-apex, which is exserted 1–$1\frac{1}{2}$ lin. beyond them and minutely 2-lobed at the acute apex. *Harv. Gen. S. Afr. Pl. ed.* i. 226 ; *Meisn. in Hook. Lond. Journ. Bot.* ii. 1843, 546 (*by error 446*) ; *Krauss in Flora*, 1844, 827 ; *Decne in DC. Prodr.* viii. 521 ; *Schlechter in Engl. Jahrb.* xxi. *Beibl.* 54, 10, *and Journ. Bot.* 1896, 457. *Apocynum lineare, Linn. f. Suppl.* 169 ; *Thunb. Prodr.* 47 ; *in Nov. Act. Acad. Petrop.* xiv. 511 ; *Fl. Cap. ed.* 2, ii. 160, *and ed. Schultes,* 237 ; *Lam. Encycl.* i. 215 ; *Willd. Sp. Pl.* i. 1262 ; *Pers. Syn. Pl.* i. 275. *A. lanceolatum, Thunb. Prodr.* 47, *in Nov. Act. Acad. Petrop.* xiv. 512 ; *Flor. Cap. ed.* 2, ii. 162, *and ed. Schultes,* 237 ; *Willd. Sp. Pl.* i. 1261 ; *Pers. Syn. Pl.* i. 275. *Astephanus linearis and A. lanceolatus, R. Br. in Mem. Wern. Soc.* i. 54 ; *Schultes, Syst. Veg.* vi. 123 ; *Spreng. Syst. Veg.* i. 855 ; *G. Don, Gen. Syst.* iv. 158 ; *Dietr. Syn. Pl.* ii. 909 ; *Decne in*

DC. *Prodr.* viii. 508 ; (*A. lanceolatus*), *Schlechter in Journ. Bot.*
1896, 418.   *Oncinema Roxburghii, Arn.*, and *Periploca capensis, Roxb.
ex Arn. in Edinb. New Phil. Journ.* xvii. 261 ; *Decne in DC. Prodr.*
viii. 526.

SOUTH AFRICA : without locality ; *Masson* !
COAST REGION : Malmesbury Div. ; Zwartland, *Thunberg* ! Worcester Div. ;
Dutoits Kloof, *Drège*, 1873 ! mountains near Worcester, *Rehmann*, 2497 ! Paarl
Div. ; by the Berg River, *Pappe* ! Caledon Div., by the Zondereinde River,
Drège, 2229 ! Houw Hoek, *Schlechter*, 7375 ! Swellendam Div. ; Grootvaders
Bosch, *Thunberg* ! Riversdale Div. ; by the Great Vals River, *Burchell*, 6556 ! by
the waterfall, Garcias Pass, *Burchell*, 6983 ! George Div. ; mountains near George,
600 ft., *Bolus*, 8685 ! Uniondale Div. ; near the Keurbooms River, *Burchell*, 5114 !
Humansdorp Div. ; near Witte Els River, *Galpin*, 4328 ! Uitenhage Div. ; Van
Stadens River, *MacOwan*, 1930 !

Apocynum lanceolatum, Thunb., is merely the broader leaved form of the plant
and equivalent to *MacOwan*, 1930, *Schlechter*, 7375, &c.

XXI. **PENTARRHINUM**, E. Meyer.

Calyx 5-partite.   *Corolla* very deeply 5-lobed, rotate or reflexed ;
lobes overlapping to the left in bud.   *Corona* of 5 lobes arising from
the base of the staminal column and alternating with the corolla-
lobes, obconic or trumpet or slipper-shaped, with infolded margins,
forming a channel or a narrow funnel-shaped cavity within, truncate
or rounded at the apex, with a horn directed forwards over or
towards the anthers.   *Stamens* arising from the base of the corolla,
united into a tube around the ovary and style.   *Anthers* with a
terminal membranous appendage inflexed over the apex of the style.
Pollen-masses pendulous, solitary in each anther-cell, attached in
pairs to the pollen-carriers by slender caudicles.   *Follicles* lanceolate, more or less beaked and more or less echinate.   *Seeds* crowned
by a tuft of hairs.

Perennial twining herbs, with slender stems ; leaves opposite, petiolate, cordate ;
flowers small, arranged in pedunculate umbel-like corymbs, lateral at the nodes.

DISTRIB. Species 3, of which one extends into Tropical Africa and another is
endemic there.

1. **P. insipidum** (E. Meyer, Comm. 200) ; stem twining, usually
puberulous, sometimes glabrous ; leaves spreading ; petiole $\frac{1}{2}$–2 in.
long ; blade $\frac{3}{4}$–3 in. long, $\frac{1}{2}$–2 in. broad, cordate-ovate, acute or
acuminate, with broadly rounded basal lobes, glabrous on both
sides ; peduncles lateral at the nodes, 1–3 in. long, usually
puberulous along one side, occasionally glabrous ; corymbs 1–1$\frac{1}{4}$ in.
in diam., the axis elongating into a raceme as the flowers develop ;
bracts $\frac{1}{2}$–1 lin. long, subulate, deciduous ; pedicels 4–10 lin. long,
puberulous ; sepals 1–1$\frac{3}{4}$ lin. long, lanceolate, acute, more or less
puberulous, ciliate ; corolla reflexed ; lobes about 2 lin. long, 1–1$\frac{1}{2}$
lin. broad, oblong, obtuse, glabrous on both sides, minutely ciliate,

green or with the apical half more or less suffused with purple-brown; corona-lobes about 1 lin. long, equalling the staminal column, obconic, thick and fleshy, narrowly channelled down the face, truncate, with a slightly projecting rim at the top, from whence a stout central subulate tooth is directed forwards over the tips of the anthers, cream-colour or yellow; follicles solitary, $1\frac{1}{2}$–$2\frac{3}{4}$ in. long, about $\frac{3}{4}$ in. thick, lanceolate, acute or acuminate into a beak, more or less tuberculate-echinate, puberulous or glabrous: seeds 3–$3\frac{1}{2}$ lin. long, 2 lin. broad, ovate, plano-convex, with a thin wing-like margin, toothed at the broadest end, rugose or sub-reticulate on both sides with short irregular linear ridges, brown, glabrous. *Decne in DC. Prodr.* viii. 553; *Harvey, Gen. S. Afr. Pl. ed.* i. 224, *and Thes. Cap.* i. 7, *t.* 11; *Schlechter in Engl. Jahrb.* xviii. *Beibl.* 45, 23; xx. *Beibl.* 51, 43; *and* xxi. *Beibl.* 54, 10; *Journ. Bot.* 1896, 456; *K. Schum. in Engl. and Prantl, Pflanzenfam.* iv. ii. 244, *and in Engl. Jahrb.* xxx. 385; *N. E. Br. in Dyer, Fl. Trop. Afr.* iv. i. 378.—*Cynanchum foliis cordato-sagittatis, &c., Burm. Rar. Afr. Pl. Dec.* 2, 38, *t.* 16, *fig.* 2.

COAST REGION: Uitenhage Div.; near Enon, *Drège*, 2220! *Baur*, 1025! *Alexander* [*Prior*]! Klein Winterhoek, 800 ft., *Drège*, 3432!

CENTRAL REGION: Jansenville; by the Sundays River near Blue Krantz, *Drège*! Somerset Div.; near Somerset East, *Atherstone*, 155! Graaff Reinet Div.; near Graaff Reinet, *Bolus*, 186! along the Sundays River, *Bowie!* *Burchell*, 2878!

WESTERN REGION: Great Namaqualand; Tiras, *Schinz*, 25!

KALAHARI REGION: Griqualand West; Vetberg Hills, *Bowker*, 26! Orange River Colony; Olifantsfontein, *Rehmann*, 3489! and without precise locality, *Cooper*, 2714! Transvaal; at the warm baths on McCord's farm, *Burtt Davy*, 1201! near Pretoria, *Conrath*, 1003! Rustenberg, *Miss Nation*, 127! Rooiplaat, Pienaars River, *Miss Leendertz*, 758!

EASTERN REGION: Transkei; by the Bashee River, *Bowker*, 487! near Kentani, *Miss Pegler*, 288! Griqualand East; near Clydesdale, 2500 ft., *Tyson*, 2573! and in *MacOwan and Bolus Herb. Norm. Austr.-Afr.*, 1320! Natal; near Durban, *McKen*, 825! *Wood*, 266! 10167! Hilton, 3700 ft., *Dimock Brown*, 230! Tintern, 5000 ft., *Evans*! and without precise locality, *Gueinzius*! *Gerrard*, 319! 825! 1311! *Wood*, 1207! 10166!

Also in Tropical Africa.

Imperfectly known species.

2. **P. coriaceum** (Schlechter in Journ. Bot. 1894, 357); stem twining, with somewhat woody branches, thinly velvety towards the apex, rather densely leafy; leaves with a short velvety petiole, coriaceous, $1\frac{1}{4}$–$2\frac{1}{3}$ in. long, $\frac{3}{4}$–2 in. broad below the middle, ovate-cordate, acute or acuminate, glabrescent, with subrevolute margins; flowers smaller than in *P. insipidum*, in many-flowered cymes, not exceeding the leaves, lateral at the nodes; peduncle and pedicels unequal in length, thinly pilose; sepals scarcely 1 lin. long, lanceo-late, acute, puberulous; corolla-lobes spreading, 2 lin. long, 1 lin. broad at the middle; corona-lobes thinner and much more flattened out than in *P. insipidum*, subquadrate, emarginate at the apex, with a flattened appendage just below the top on the inner face, inflexed,

dilated and obtuse at the apex, decurrent as a keel below; anther-appendages suborbicular or oblong, very obtuse, inflexed upon the style-apex. *Schlechter in Journ. Bot.* 1896, 456.

EASTERN REGION : Natal, *Gerrard and McKen.*

From the description it appears somewhat doubtful if this plant belongs to the genus *Pentarrhinum.* According to Dr. Schlechter, it is only known to him "from a small scrap collected by Gerrard and McKen in Natal," but he does not state in what Herbarium it exists. A very full set of Gerrard and McKen's plants, including some uniques, are at Kew, but neither there nor in Harvey's Herbarium at Dublin, nor in the British Museum have I seen any specimen that will correspond with the description, and, according to Mr. Medley Wood, no specimen of it can be found in the Natal Herbarium.

3. **P. tylophoroides** (K. Schum. in Engl. and Prantl, Pflanzenfam. iv. ii. 244); inflorescence paniculate; leaves roundish.

SOUTH AFRICA, *Burchell* (ex *K. Schumann*).

The above is all the description given of this plant, which probably belongs to some other genus, but as no number is quoted, I cannot trace it among Burchell's plants.

XXII. CYNANCHUM, Linn.

Calyx 5-partite. *Corolla* very deeply 5-lobed, rotate or rotate-campanulate; lobes overlapping and straight, or more or less twisted in bud. *Corona* arising from the staminal column, near or at its base, either annular, cup-shaped or tubular and entire, toothed or lobed at the top, or divided nearly or quite to the base into 5 entire or toothed lobes, with or without a tooth, lobe, thickening or keels within the tube in front of each of the principal teeth or lobes, or on the inner face or at the base of the lobes when the corona is divided. *Staminal column* arising at or near the base of the corolla, constricted under the anthers into a short or long stipe (filament part) within the corona, or the anthers nearly or quite sessile without a stipe, tipped with membranous or slightly fleshy appendages, inflexed or connivent over the apex of the style or erect around it. *Pollen-masses* pendulous, solitary in each anther-cell, affixed in pairs by short or long caudicles to the pollen-carriers. *Style* shorter or longer than the anther-appendages, truncate, conical or beaked at the apical part. *Follicles* sometimes winged or keeled, smooth or setose. *Seeds* crowned with a tuft of hairs.

Stem twining, erect or decumbent, leafy, rarely leafless and succulent; leaves opposite; flowers rather small, fasciculate or in pedunculate simple umbel-like or compound and corymbose cymes, rarely in racemes, subaxillary or lateral at the nodes.

DISTRIB. Species more than 100, cosmopolitan.

* Plant with well developed leaves ; stems not fleshy :
  Plant shrubby, woody, not twining :
    Branches recurving, white - tomentose ; leaves
      puberulous ... ... (1) **Meyeri.**

    Branches spreading (glabrous ?) ; leaves glabrous,
      glaucous ... ... (2) **mucronatum.**

  Plant not shrubby, herbaceous :
    Plant less than 1 ft. high ; stems erect or procum-
      bent, not twining :
      Leaves linear ; corona subequally 10-lobed to half-
        way down ... ... (3) **orangeanum.**

      Leaves ovate to suborbicular ; corona shortly
        5-toothed ... ... (4) **Zeyheri.**

    Plant with twining stems usually more than 1 ft.
      long :
      † Leaf-blade $\frac{1}{2}$–2 in. long, cuneate, rounded, sub-
        truncate or shallowly cordate at the base,
        usually obtuse and apiculate at the apex, or if
        acute, then not cordate at the base :
      Staminal column stipitate within the corona :
        Corolla-lobes 1–$2\frac{1}{3}$ lin. long, glabrous on the
          inner face ; corona $\frac{2}{3}$–2 lin. long, tubular,
          5-toothed :
          Corolla-lobes 1–$1\frac{1}{4}$ lin. broad, dull greenish
            (or olive-brown when dried) ; teeth of
            the corona $\frac{1}{5}$–$\frac{1}{4}$ as long as the tube ... (5) **natalitium.**

          Corolla-lobes $\frac{3}{4}$ lin. broad, purple-brown ;
            teeth of the corona at least $\frac{1}{3}$ as long as
            the tube... ... ... (6) **intermedium.**

        Corolla-lobes 3–4 lin. long, glabrous on the
          inner face ; corona $2\frac{1}{2}$–$4\frac{1}{2}$ lin. long,
          tubular, 5–10-toothed ... ... (7) **africanum.**
      Staminal column not stipitate :
        Corolla - lobes $1\frac{1}{4}$ – 2 lin. long, minutely
          puberulous on the inner face ; corona
          acutely and deeply 5-lobed ... ... (8) **obtusifolium.**

        Corolla-lobes 1–$1\frac{1}{2}$ lin. long, glabrous on the
          inner face ; corona truncate or slightly
          5-crenate at the top ... ... (9) **capense.**

      †† Leaf-blade 1–$3\frac{1}{4}$ in. long, deeply cordate to sub-
        truncate at the base, acute or attenuate-acute
        at the apex ; staminal column not stipitate :
        Corolla-lobes 2–3 lin. long, twisted, puberulous
          on the inner face ; corona $1\frac{2}{3}$–2 lin. long... (10) **virens.**

        Corolla-lobes $\frac{2}{3}$ to nearly 1 lin. long, glabrous
          on the inner face ; corona $\frac{1}{2}$–$\frac{2}{3}$ lin. long ... (11) **schistoglossum.**

** Plant leafless, with fleshy twining stems ... ... (12) **sarcostemma-**
                                                     toides.

1. **C. Meyeri** (Schlechter in Engl. Jahrb. xx. Beibl. 51, 2) ; a shrub
or shrublet ; branches woody, strongly recurving, 1–$1\frac{1}{2}$ lin. thick,
minutely white-tomentose ; leaves rigid, probably fleshy ; petiole
1–2 lin. long ; blade $\frac{1}{4}$–$\frac{1}{2}$ in. long, $2\frac{1}{2}$–5 lin. broad, orbicular or ovate,
obtuse, apiculate, minutely puberulous on both sides ; flowers in
small fascicles, axillary or terminating short axillary branchlets ;

pedicels $\frac{1}{3}$–$\frac{1}{2}$ lin. long, puberulous; sepals $\frac{1}{2}$ lin. long, $\frac{1}{4}$–$\frac{1}{3}$ lin. broad, ovate, obtuse or subacute, puberulous; corolla lobed to $\frac{2}{3}$ of the way down, glabrous outside and within; lobes erect or campanulately spreading, $\frac{3}{4}$ lin. long, $\frac{1}{2}$ lin. broad, oblong, obtuse; corona $\frac{1}{2}$ lin. long, tubular-ovoid, 5-lobed, with subtruncate interspaces, no teeth or appendages within; lobes oblong, obtuse, revolute at the tips; staminal column $\frac{1}{3}$ lin. long, not stipitate; anther-appendages lanceolate, acute, connivent above the very obtuse style-apex. *Schlechter in Engl. Jahrb.* xxi. *Beibl.* 54, 11, *and Journ. Bot.* 1896, 456. *Sarcostemma ovatum, E. Meyer, Comm.* 221; *Harv. Gen. S. Afr. Pl. ed.* i. 224; *Dietr. Syn. Pl.* ii. 907. *Cynoctonum Meyeri, Decne in DC. Prodr.* viii. 531. *Vincetoxicum Meyeri, Benth. & Hook f. ex Ind. Kew.* ii. 1204; *Kuntze, Rev. Gen. Pl.* ii. 424.

Western Region: Little Namaqualand; hills near Arris, towards the mouth of the Orange River, *Drège*, 3048 !

2. **C. mucronatum** (N. E. Br.); " stem shrubby, terete, with spreading branches; leaves opposite, petiolate, subcordate-oblong, mucronate, glaucous, entire, 1-nerved, veiny, glabrous, about the size of a finger-nail; peduncles axillary, umbellate, about 6-flowered, shorter than the petioles; outer corolla brownish-red, rotate; inner corolla (corona) whitish, urceolate, 5-fid, the teeth alternating with the pollen-masses; pollen-masses pendulous, yellowish-white, oblong, united to a fulvous nodule concealed under folds of the column; genital column with acute erect membranes (anther-appendages) at the apex." *Metaplexis mucronata, Spreng. Neue Entdeck. Pfl.* i. 269, *and Syst. Veg.* i. 854; *Schultes, Syst. Veg.* vi. 112; *G. Don, Gen. Syst.* iv. 155; *Harv. Gen. South Afr. Pl. ed.* i. 227; *Dietr. Syn. Pl.* ii. 908, *excluding synonym Asclepias mucronata, Thunb., from all.*

South Africa: without locality, ex *Sprengel.*

This plant must be closely allied to *C. Meyeri*, and may perhaps be that species, but differs according to the description in having spreading (not recurved) branches, and veiny glabrous glaucous leaves.

3. **C. orangeanum** (N. E. Br.); rootstock woody; plant herbaceous, much branched at the base, $2\frac{1}{2}$–5 (6–8 ex *Schlechter*) in. high; branches erect or ascending, varying from glabrous to minutely puberulous all over or at the tips only, leafy; leaves $\frac{3}{4}$–$1\frac{3}{4}$ in. long, $\frac{1}{3}$–$\frac{1}{2}$ lin. broad, linear or linear-filiform, acute or obtuse, with thickened or perhaps revolute margins, glabrous; umbels sessile or pedunculate, lateral at the nodes, 2–4-flowered; peduncle 0–$2\frac{1}{2}$ lin. long; bracts minute; pedicels 1–3 lin. long, glabrous or minutely puberulous; sepals $\frac{2}{3}$–$\frac{3}{4}$ lin. long, ovate, acute, puberulous; corolla-lobes $1\frac{1}{2}$–$1\frac{3}{4}$ lin. long, $\frac{3}{4}$–1 lin. broad, suberect, ovate-lanceolate, obtusely pointed and often minutely notched at the apex, glabrous on both sides or with a few minute hairs on the back, apparently greenish on the inner face; corona arising at the base of the staminal column, $1\frac{1}{4}$–$1\frac{1}{3}$ lin. long, tubular-campanulate, subequally 10-lobed to nearly or quite half-way down, white; tube plicate, so

that the linear-terete lobes alternating with the corolla-lobes are folded inwards, solid at the apical half, with the margins incurved and forming a channel on the lower half (convex-cucullate, ex *Schlechter*), then decurrent as infolded-contiguous keels on the upper part of the tube; the lobes opposite the corolla-lobes with a broad deltoid base, narrowed into a short linear obtuse point, slightly convex on the inner face below, but without keels or other processes; staminal column $\frac{2}{3}$–$\frac{3}{4}$ lin. long, with the greenish anthers forming a short cone, having 5 small excavations beneath them; anther-appendages suborbicular, erect, adpressed to the sides of the stout conical obtuse greenish style-apex, which is shortly exserted beyond them, but is shorter than the corona; follicles $1\frac{1}{2}$–$2\frac{1}{4}$ in. long, lanceolate-fusiform, tapering into a long beak, smooth, minutely puberulous. *Flanagania orangeana, Schlechter in Engl. Jahrb.* xviii. *Beibl.* 45, 10, *and Journ. Bot.* 1896, 457.

CENTRAL REGION : Colesberg Div. ; near Colesberg, *Shaw,* 58 ! 59 !

KALAHARI REGION: Griqualand West; near Warrenton. *Miss Adams,* 129 ! 205 ! Dutoits Pan, *Tuck*! Orange River Colony; by the Orange River, near Bethulie, *Flanagan,* 1502 ! Bechuanaland; north of the Mashowing River, *Burchell,* 2330 !

I find no character to generically distinguish this plant from *Cynanchum.* Although having a wide range, it would appear to be a rare plant, as only four collectors seem to have met with it since its discovery by Burchell in 1812.

4. C. Zeyheri (Schlechter in Engl. Jahrb. xx. Beibl. 51, 3) ; rootstock woody, deeply descending, dividing at ground-level into several or many herbaceous branching stems 3–12 in. long, more or less procumbent or when short sometimes erect, slender, glabrous to puberulous; leaves small, thick, probably slightly fleshy when alive; petiole 1–$4\frac{1}{2}$ lin. long, channelled and puberulous above : blade $2\frac{1}{2}$-11 lin. long, 2–6 lin. broad, ovate, oblong-ovate or suborbicular, obtuse or acute, apiculate, glabrous ; umbels lateral at the nodes, pedunculate, 2–5-flowered ; peduncles 1–4 lin. long, glabrous or with a few very minute hairs ; bracts very minute ; pedicels 2–4 lin. long, glabrous; sepals about $\frac{2}{3}$ lin. long, ovate, acute, glabrous ; corolla-lobes spreading, or ascending-spreading, usually more or less twisted, or at least with revolute margins, $1\frac{1}{4}$–$1\frac{3}{4}$ lin. long, $\frac{1}{2}$ lin. broad, linear-oblong, subobtuse, glabrous on both sides, dark brown (*Harvey*) ; corona arising at the base of the staminal column, $\frac{3}{4}$–1 lin. long, tubular-campanulate or cup-shaped, shortly 5-toothed at the top, not subentire as originally described, without keels or other appendages within, white ; staminal column $\frac{3}{4}$–1 lin. long, contracted below the anthers into a rather long stipe : anther-appendages ovate, acute, connivent over and concealing the umbonate style-apex ; follicles solitary, $1\frac{3}{4}$–$2\frac{1}{2}$ in. long and 5–7 lin. thick, lanceolate-fusiform, tapering into a beak, with a keel on each side of the opening suture, smooth, glabrous ; seeds 3–$3\frac{1}{2}$ lin. long, 2 lin. broad, ovate, nearly flat on one side, very slightly convex on the other, finely rugulose-tuberculate all over, light brown. *Schlechter in Journ. Bot.* 1896, 457.

COAST REGION: Cape Div. ; Lion Mountain, *Ecklon and Zeyher*, 78 ! *Faure* ! *Schlechter*, 11504 ! Greenpoint, *Ecklon and Zeyher* ! *Pappe*, 67 ! near the Lighthouse at Cape Town, *Prior* ! by the Lion Battery, *Wolley Dod*, 1149 ! shore north of Camps Bay, *Wolley Dod*, 3445 !

The corona is described by Dr. Schlechter as subentire with lamella or keels on the inner face.   I do not find it so in any of the specimens above quoted.

5. **C. natalitium** (Schlechter in Engl. Jahrb. xviii. Beibl. 45, 32) ; stem twining, ½–1 lin. thick, puberulous on the young parts, becoming glabrous ; leaves thinly to thickly coriaceous when dried, evidently sometimes fleshy ; petiole ⅙–½ in. long ; blade ½–1¾ in. long, ⅓–1 in. broad, oblong, elliptic or suborbicular, very obtuse or broadly rounded at the apex, apiculate, rounded or somewhat cuneate at the base, thinly puberulous with minute curved hairs on both sides or glabrous ; umbels lateral at the nodes, pedunculate, 6–16-flowered ; peduncles ⅙–½ in. long, puberulous on one side ; bracts minute, deltoid ; pedicels 2–4 lin. long, glabrous or slightly puberulous ; sepals ½–¾ lin. long, ovate, acute, glabrous or puberulous ; corolla-lobes widely spreading, about 2 lin. long and 1–1¼ lin. broad, oblong, obtuse, glabrous on both sides, dull green, often drying brown ; corona tubular, usually slightly narrowed at the shortly 5-toothed mouth, white ; tube 1⅓–1½ lin. long ; teeth ¼–⅓ lin. long, truncate, the spaces between them broad, concave, with recurved margins ; staminal column 1 lin. long, contracted below the depressed subglobose antheriferous part into a fluted stipe ; anther-appendages lanceolate, acute, closely incumbent on the obtuse style-apex ; follicles 2–2½ in. long, ½–¾ in. thick, lanceolate-fusiform, beaked or acuminate, acute or obtuse, with a narrow wing on each side and a keel down the back, glabrous, puberulous when young. *Schlechter in Engl. Jahrb. xx. Beibl.* 51, 6 ; *Journ. Bot.* 1896, 457, *and Ann. Naturhist. Hofmus. Wien*, xv. 69 ; *Wood, Natal Pl.* iv. *t.* 301. *Cynoctonum capense, E. Meyer, Comm.* 216, *as to* " *d.*"

SOUTH AFRICA: without locality, *Drège*, 4953 !
COAST REGION: Bathurst Div. ; near Port Alfred, *Burchell*, 3819 ! *Penther*, 793 ! near the mouth of the Fish River, *Burchell*, 3753 ! Riet River, *Mrs. White*, 49 ! Albany Div. ; Sister Rocks near Fairfield, *Hutton* ! King Williamstown Div. ; Keiskamma, *Mrs. Hutton* ! East London Div. ; near East London, *Galpin*, 1831 !
EASTERN REGION: Transkei ; near the coast, *Miss Pegler*, 76 ! *Bowker* ! Pondoland ; between Umtentu River and Umzimkulu River, *Drège* ! Natal: near Durban, *Sanderson*, 547 ! *Schlechter*, 3082 ! Wentworth Bluff, *Sanderson*, 436 partly ! Bluff, *Wood*, 5387.

6. **C. intermedium** (N. E. Br.) ; plant about 1¼–2 ft. high ; stems twining, slender, ⅓–½ lin. thick, glabrous ; leaves thinly coriaceous, apparently not fleshy ; petiole 1½–2 lin. long, slender ; blade ½–¾ in. long, 4–5 lin. broad, oblong, obtuse or subacute, apiculate, broadly rounded at the base, glabrous on both sides or with a few minute hairs on the midrib beneath ; umbels lateral at the nodes, pedun-

culate, 2-5-flowered; peduncles $\frac{1}{4}$–$\frac{1}{2}$ in. long, glabrous; bracts minute, lanceolate, acute, glabrous; sepals $\frac{2}{3}$–$\frac{3}{4}$ lin. long, $\frac{1}{2}$ lin. broad, ovate, acute, glabrous; corolla-lobes 2–$2\frac{1}{3}$ lin. long, $\frac{3}{4}$ lin. broad, oblong-lanceolate, acute, glabrous on both sides, purple-brown; corona tubular, 5-toothed; tube $1\frac{1}{4}$–$1\frac{1}{3}$ lin. long; teeth $\frac{1}{2}$–$\frac{3}{4}$ lin. long, linear, obtuse, erect, the spaces between them rounded with recurved margins; staminal column 1–$1\frac{1}{3}$ lin. long, contracted under the conical antheriferous part into a slightly fluted stipe; anther-appendages ovate-lanceolate, acute, connivent over the obtusely conical style-apex; follicles solitary, about 2 in. long, lanceolate-fusiform, long-beaked, apparently narrowly 2-winged, glabrous.

COAST REGION: Port Elizabeth Div.; near Port Elizabeth, 250 ft., *West in MacOwan, Herb. Austr.-Afr.*, 1924!

Closely allied to *C. natalitium*, but evidently a smaller and much more slender plant, with less fleshy leaves. It is possibly a hybrid between *C. africanum* and *C. natalitium*, but does not quite agree with either species.

7. **C. africanum** (R. Br. Prodr. 463); plant branching from the base into several twining stems, thinly or densely covered with a soft spreading pubescence; leaves thinly coriaceous or rather thick and probably somewhat fleshy, more or less pubescent on both sides or glabrous; petiole 1–3 lin. long; blade $\frac{1}{3}$–1 in. long, $\frac{1}{4}$–$\frac{3}{4}$ in. broad, suborbicular, ovate, elliptic-oblong or oblong, acute or obtuse, apiculate, rounded to subcordate at the base; umbels lateral at the nodes, 5–7-flowered; peduncles $\frac{1}{4}$–$1\frac{1}{4}$ in. long, pubescent to nearly glabrous; bracts about 1 lin. long, ovate or deltoid-ovate, acute; pedicels 1–5 lin. long, pubescent; sepals 1–$1\frac{1}{4}$ lin. long, $\frac{2}{3}$–$\frac{3}{4}$ lin. broad, ovate, acute, pubescent to nearly glabrous; corolla-tube $\frac{1}{2}$–$\frac{3}{4}$ lin. long; lobes 3–4 lin. long, 1 lin. broad, oblong-linear, obtuse, recurving or twisted and suberect, glabrous on both sides or slightly pubescent on the back, dark purple-brown; corona $2\frac{1}{2}$–$4\frac{1}{2}$ lin. long, tubular, usually irregularly about 10-toothed, but sometimes with 5 entire or bifid teeth $\frac{1}{2}$–1 lin. long at the top, glabrous, white; staminal column $2\frac{1}{4}$–$3\frac{1}{2}$ lin. long, terete, suddenly contracted under the anthers into a slender stipe $1\frac{2}{3}$–$2\frac{3}{4}$ lin. long; anther-appendages lanceolate, acute, connivent over the shortly conical minutely 2-lobed style-apex; follicles about $2\frac{1}{2}$ in. long and 4–$4\frac{1}{2}$ lin. thick, narrowly lanceolate, tapering into a long slender acute beak, puberulous; seeds $2\frac{1}{4}$–$2\frac{3}{4}$ lin. long, ovate, concave on one side, convex on the other, minutely tuberculate-rugose, light brown. *Hoffmansegg, Verzeichn.* 54; *Schlechter in Engl. Jahrb.* xx. *Beibl.* 51, 5, *and* xxi. *Beibl.* 54, 11; *Journ. Bot.* 1896, 457, *and Ann. Naturhist. Hoffmus. Wien,* xviii. 398. *C. pilosum, R. Br. in Mem. Wern. Soc.* 1, 46; *Ait. Hort. Kew. ed.* 2, ii. 77; *Bot. Reg. t.* 111; *Schultes, Syst. Veg.* vi. 100; *Spreng. Syst.* i. 851; *G. Don, Gen. Syst.* iv. 153; *Dietr. Syn. Pl.* ii. 904. *Cynoctonum crassifolium, E. Meyer, var. pilosa, Decne in DC. Prodr.* viii. 530. *Cynoctonum pilosum, E. Meyer, Comm.* 216; *Meisn. in Hook. Lond. Journ. Bot.* ii. 1843, 545 (*by*

error 445); *Krauss in Flora*, 1844, 827. *Vincetoxicum africanum,
Kuntze, Rev. Gen. Pl.* ii. 422 ; *Bolus ex Schlechter in Engl. Jahrb.* xx.
Beibl. 51, 5, *and Journ. Bot.* 1896, 457, *in syn.   V. pilosum, Nichols.
Dict. Gard.* iv. 160.   *Periploca africana, Linn. Sp. Pl. ed.* i. 211 ;
Ait. Hort. Kew. ed. 1, i. 301 ; *Thunb. Prodr.* 47 ; *in Nov. Act. Acad.
Petrop.* xiv. 518 ; *Fl. Cap. ed.* 2, 152, *and ed. Schultes,* 233 ;
Willd. Sp. Pl. i. 1251 ; *Pers. Syn. Pl.* i. 272 ; *Poir. Encycl.* v. 190.
P. pallida, Salisb. Prodr. 148.—*Apocynum scandens africanum, &c.,
Commelin, Hort. Med. Amst. Pl. Rar.* 18, *t.* 18.

VAR. β, **crassifolium** (N. E. Br.) ; stem, leaves and inflorescence quite glabrous.
C. crassifolium, R. Br. in Mem. Wern. Soc. i. 46 ; *Schultes, Syst. Veg.* vi. 101 ;
Spreng. Syst. i. 851 ; *G. Don, Gen. Syst.* iv. 153 ; *Dietr. Syn. Pl.* ii. 904. *C.
rotundifolium, Thunb. ? ex Decne in DC. Prodr.* viii. 552.   *Cynoctonum crassifolium, E. Meyer, Comm.* 216 *partly* ; *Decne in DC. Prodr.* viii. 530, *excl.
synonyms ; Meisn. in Hook. Lond. Journ. Bot.* ii. 1843, 545 (*by error* 445). *C.
crassiflorum, Krauss in Flora*, 1844, 827.

SOUTH AFRICA : without locality. *Herb. Linnæus* ! *Kiggelaer* ! *Nelson* ! *Harvey,*
413 ! *Pappe* ! *Herb. Miller* ! Var. β : *Harvey*, 248 !

COAST REGION : Clanwilliam Div. ; near Modderfontein, *Penther*, 2411 (ex
Schlechter), by the Oliphants River, *Schlechter*, 4988 ! Malmesbury Div. ; Saldanha
Bay, *Drège*, 230b ! Tulbagh Div. ; New Kloof, near Tulbagh, *Burchell*, 1018 !
Tulbagh Waterfall, *Ecklon and Zeyher* ! Mosterts Berg, *MacOwan, Herb. Austr.-
Afr.* 1634 ! Paarl Div. ; Paarl, *Prior* (*Alexander*)! Cape Div. ; Cape Flats and
other parts of the Cape Peninsula, *Thunberg* ! *Burchell*, 370 ! *Zeyher*, 1176 !
Harvey, 236 ! *Bolus*, 4019 ! *and in Herb. Norm. Austr.-Afr.* 1087 ! *Galpin*, 4331 !
Wolley Dod, 1615 ! 2995 ! 3607 ! *Wilms*, 3479 ! *Wright*, 592 ! Knysna Div. ; by the
Gouwkamma River, *Krauss*, 1256 (ex *Meisner*).   Var. β : Clanwilliam Div. ;
Lamberts Bay, *Schlechter*, 8547 ! Piquetberg Div. ; St. Helena Bay, *Hove* ! Cape
Div. ; Paarden Island, *Drège*, 230 ! *Wolley Dod*, 3151 ! Humansdorp Div. ;
Zitzikamma, *Krauss*, 1257 (ex *Meisner*).

Var. β is only a glabrous maritime form.

8. **C. obtusifolium** (Linn. f. Suppl. 169) ; stems twining,
glabrous ; leaves spreading, thinly coriaceous, glabrous ; petiole
2–7 lin. long ; blade $\frac{2}{3}$–1$\frac{1}{2}$ in. long, $\frac{1}{3}$–1$\frac{1}{3}$ in. broad, oblong to broadly
elliptic or almost suborbicular, obtuse and apiculate at the apex,
rounded, subtruncate or cordate at the base, glabrous on both
sides ; flowers in small shortly pedunculate fascicles or umbel-like
cymes, the axis of which gradually elongates to 2–6 lin. long,
occasionally branches and is spirally marked with the scars of fallen
flowers ; peduncles $\frac{1}{2}$–5 lin. long, glabrous ; bracts minute ; pedicels
1–1$\frac{1}{2}$ lin. long, glabrous ; sepals $\frac{2}{3}$ to nearly 1 lin. long, $\frac{1}{2}$–$\frac{3}{4}$ lin.
broad, ovate, acute or subobtusely pointed, glabrous ; corolla-lobes
apparently very spreading, 1$\frac{1}{4}$–2 lin. long, $\frac{2}{3}$–1 lin. broad, ovate-
oblong, or oblong, subobtuse, glabrous on the back, minutely
puberulous on the inner face ; corona $\frac{2}{3}$–1$\frac{1}{4}$ lin. long, 5-lobed to $\frac{1}{2}$ or
$\frac{2}{3}$ of the way down, with a small two-toothed rounded or lanceolate
lobule alternating with the lobes, and the basal part shortly cup-
shaped : lobes ovate, acuminate, more or less denticulate on the
margins, incurved over the staminal column or rarely erect, with
2 keels on the inner face and rarely with a minute appendage at

the apex in front of the point; staminal column $\frac{1}{2}$–1 lin. long, not contracted into a stipe under the anthers; anther-appendages broadly ovate or suborbicular, obtuse, inflexed on the top of the style-apex, which has the centre slightly exserted as a minute bifid tubercle, from which 5 ridges radiate to the angles; follicles solitary, 2–2$\frac{1}{4}$ in. long, $\frac{1}{2}$–$\frac{2}{3}$ in. thick, lanceolate, obtusely beaked, leathery or perhaps somewhat fleshy, with a keel down each side of the ventral face, and one on the back, smooth, glabrous; seeds 2$\frac{1}{2}$ lin. long, 1$\frac{1}{2}$ lin. broad, ovate, flattened on one side, convex on the other very narrowly margined, nearly smooth, glabrous, very dark brown. *Lam. Encycl.* ii. 236; *Thunb. Prodr.* 46; *in Weber and Mohr, Archiv.* i. (1804), 30, *t.* 1; *Fl. Cap. ed.* 2, ii. 159, *and ed. Schultes,* 236; *Willd. Sp. Pl.* i. 1253; *Pers. Syn. Pl.* i. 272; *Schlechter in Engl. Jahrb.* xviii. *Beibl.* 45, 10; xx. *Beibl.* 51, 4; xxi. *Beibl.* 54, 11, *and Journ. Bot.* 1896, 457. *C. capense, R. Br. in Mem. Wern. Soc.* i. 46; *and perhaps also of Spreng. Syst. Veg.* i. 852; *Dietr. Syn. Pl.* ii. 905, *and G. Don, Gen. Syst.* iv. 153, *not of Linn. f. or Thunb. Periploca africana* β, *Linn. Sp. Pl. ed.* i. 211; *Willd. Sp. Pl.* i. 1251; *Poir. Encycl.* v. 190. *Cynoctonum Brownii, Meisn. in Hook. Lond. Journ. Bot.* ii. 1843, 546 (*by error* 446), *in note. C. dregeanum, Decne in DC. Prodr.* viii. 531. *C. crassifolium, E. Meyer, Comm.* 216, *partly, not of R. Br. Vincetoxicum dregeanum, Kuntze, Rev. Gen. Pl.* ii. 424.—*Cynanchum foliis planis, &c., Burm. Rar. Afr. Pl. Dec.* 2, 34, *t.* 14, *fig.* 2.

VAR. β, **pilosum** (Schlechter in Engl. Jahrb. xviii. Beibl. 45, 10); stem, leaves and inflorescence, excepting the corolla, thinly to densely pubescent or sub-tomentose, or the leaves nearly glabrous; corona $\frac{3}{4}$–1$\frac{2}{3}$ lin. long, with erect lobes; staminal column $\frac{2}{3}$–1 lin. long; follicles puberulous or pubescent; seeds 3–3$\frac{1}{2}$ lin. long, 2 lin. broad, rather light brown; otherwise as in the type.

SOUTH AFRICA: without locality, *Herb. Linn. fil! Thunberg! Masson! Nelson! Oldenberg! Forster! Herb. Miller!*
COAST REGION: Cape Div.; Camps Bay, *Burchell,* 357! *Harvey!* Kalk Bay, *Wolley Dod,* 2146! *Bolus,* 4914! near Simonstown, *Schlechter,* 320! Paarden Island, *Drège,* near Cape Town, *Rehmann,* 1212! George Div.; near George, *Prior!* Pacaltsdorp, *Schlechter,* 2457! Knysna Div.; near Knysna, *Burchell,* 5395! Uitenhage Div.; near Uitenhage, *Prior!* by the Zwartkops River, *Zeyher,* 582! Port Elizabeth Div.; near Port Elizabeth, *Bolus,* 2233! Albany Div.; Howisons Poort, near Grahamstown, *Hutton!* Bathurst Div.; coast, *Hutton!* Var. β: Cape Div.; Bushy Dell, Simonstown, *Wolley Dod,* 848! Knysna Div.; near Knysna, *Burchell,* 5480! Uitenhage Div.; near the Leadmine, *Burchell,* 4505! Komgha Div.; near the mouth of the Kei River, *Flanagan,* 382!
EASTERN REGION: Var. β: Transkei; near Kentani, *Miss Pegler,* 1294! Natal; Berea at Durban, *Cooper,* 1265! *McKen,* 7! near Durban, *Gerrard,* 516! *Grant! Wilms,* 2143! *Wood,* 1662! 3910! 5422 (also distributed as 5322)! Inanda, *Wood!* Sea-coast, *Gerrard,* 517! and without precise locality, *Gerrard,* 618! 712! 713!

This species appears to be dimorphic, as both in the type and in var. *pilosum* I find two kinds of flowers; one in which the corolla-lobes are about 1$\frac{1}{4}$–1$\frac{1}{3}$ lin. long and somewhat ovate in shape, combined with a staminal column $\frac{1}{2}$–$\frac{2}{3}$ lin. long, with very short anther-wings; the other with oblong corolla-lobes about 2 lin. long, combined with a staminal column 1 lin. long, having very much longer anther-wings. These two forms of flower do not correspond to different geographical areas, but are intermingled throughout the range of the species. Probably *Vincetoxicum obtusifolium*, var. *crassifolium*, Kuntze, Rev. Gen. Pl. iii. 200, from East London, should be referred to this species, as from the locality it can scarcely be the same as *Cynoctonum crassifolium*, E. Meyer, which Kuntze quotes as a synonym.

9. **C. capense** (Thunb. Prodr. 47, not of Linn. f., nor R. Br.); stem twining, becoming woody at the base, glabrous; leaves herbaceous or thinly coriaceous, glabrous; petiole 2–9 lin. long; blade ½–2 in. long, ⅓–1 in. broad, oblong to broadly elliptic-oblong, acute to very broadly rounded or emarginate and apiculate, rounded or slightly subcordate at the base; cymes simple and umbel-like or branched, lateral at the nodes, usually 2–14- or occasionally up to 30-flowered, glabrous in all parts; peduncle ⅛–¾ in. long; pedicels 2½–4½ lin. long; sepals ½–¾ lin. long, broadly ovate to oblong, obtuse, glabrous; corolla lobed nearly to the base, dark-coloured; lobes 1–1½ lin. long, ⅔ lin. broad, oblong, obtuse, spreading, half twisted, glabrous on both sides; corona 1¼–1½ lin. long, cup-shaped, truncate or slightly 5-crenate at the top, white; staminal column ½–⅔ lin. long, not stipitate; anther-appendages lanceolate, acute, connivent above the obtuse top of the style-apex; follicles solitary, 1¾–2 in. long, about ¼ in. thick near the base, thence tapering into a long acute beak, smooth, glabrous; seeds (immature) ¼ in. long, elongate-ovate, slightly rugose, blackish-brown. *Thunb. in Weber and Mohr, Archiv. i. (1804), 51; Fl. Cap. ed. 2, ii. 159; and ed. Schultes, 236; Schultes, Syst. Veg. vi. 102, partly, as to Thunberg's plant only; Schlechter in Engl. Jahrb. xviii. Beibl. 45, 32 in note; xx. Beibl. 51, 6; and xxi. Beibl. 54, 10; Journ. Bot. 1896, 457, and Ann. Naturhist. Hofmus. Wien, xv. 69. Cynoctonum capense, E. Meyer, Comm. 216, partly; Decne in DC. Prodr. viii. 530; Meisn. in Hook. Lond. Journ. Bot. ii. 1843, 545 (by error 445); Krauss in Flora, 1844, 827. Bunburia elliptica, Harv. Gen. S. Afr. Pl. ed. i. 417. Vincetoxicum capense, Kuntze, Rev. Gen. Pl. ii. 424; Schlechter in Engl. Jahrb. xx. Beibl. 51, 6.*

SOUTH AFRICA : without locality, *Thunberg*!

COAST REGION : George Div. ; near George, *Prior*! near Kaimans River, *Schlechter*, 2382! Knysna Div. ; near Knysna, *Burchell*, 5482! *Bowie*! Homtini Pass, *Galpin*, 4330! Humansdorp Div. ; Zitzikamma, *Krauss*, 1255 (ex *Meisner*) Uitenhage Div. ; by the Koega River and between the Zuurberg Range and Sunday River, *Drège*! near Uitenhage, *Haagner in Herb. Conrath*, 1004! *Krook*, 830 (ex *Schlechter*); Sam Tees Vlatke, *Drège*! Albany Div. ; near Grahamstown, *Zeyher*, 889! *Bunbury*, 163! *Atherstone*, 8! 462! *Cooper*, 1515! *MacOwan*, 1011! Komgha Div. ; near Komgha, *Flanagan*, 254 (ex *Schlechter*).

CENTRAL REGION : Somerset Div. ; near Somerset East, *Bowker*, 111! Graaff Reinet Div. ; near Graaff Reinet, *Burchell*, 2116! 2879! *Bolus*, 64! *Rattray*, 83! between Graaff Reinet and Melk River, *Burchell*, 2948! 2953/²!

KALAHARI REGION : Transvaal ; Rietfontein, Zoutpansberg Range, *Miss Leendertz*, 901!

EASTERN REGION : Natal ; between Pinetown and Umbilo, *Rehmann*, 8061! near Durban, *Rehmann*, 8797! *Wood*, 611 (not 611 from Inanda)! 4882 (ex *Schlechter*), and in *MacOwan and Bolus, Herb. Norm. Austr.-Afr.*, 1321! and without precise locality, *Guienzius*! *Cooper*, 1282! *Gerrard*, 313! 515!

The Natal specimens seem mostly to have a slightly 5-crenulate corona, which is not the case in those from the other regions. Specimens of *C. schistoglossum* have also been distributed by Mr. Wood under the number 611.

Until now all authors have erroneously identified *Cynanchum capense*, Linn. f., with this plant, no one having taken the trouble to examine his type The distinguishing characters in Linnæus' description of *C. capense* are: — "Leaves obsoletely cordate-ovate, mucronate, smooth, the younger ovate, the adult emarginate with a mucro" and "Pedicels capillary, longer than the peduncle." In the Linnean Herbarium are 2 sheets of good specimens under *Cynanchum* bearing the name "*capense*" upon them in Linnean handwriting. The first of these contains a specimen of *Pentatropis microphylla*, Wight and Arn., an Indian plant, collected by König, as quoted by Linnæus fil.; the second contains a specimen of the South African *Cynanchum obtusifolium*, Linn. f., probably the one from Sparrmann referred to by Linnæus fil. With the latter, the characters above noted, especially as to the pedicels, do not accord, whilst they exactly agree with the specimen of *Pentatropis microphylla*, from which latter it is clearly evident Linnæus fil. made his description of *C. capense*, therefore the name *C. capense*, Linn. f., must henceforth be quoted as a synonym of *Pentatropis microphylla*, Wight and Arn.

I have also examined the types of *Cynanchum capense* of Robert Brown and of Thunberg. The former I find to be *C. obtusifolium*, Linn. f., of which I have also seen the type. Thunberg's type of *C. capense* is, however, that which I have above described, and his name as the authority for this species must now replace that of Linn. f. There is no specimen of *C. capense*, Thunb., in the Linnean Herbarium under *Cynanchum* or the allied genera.

10. **C. virens** (Dietr. Syn. Pl. ii. 905, not of 906); stem twining, glabrous or slightly puberulous; leaves herbaceous; petiole $\frac{1}{3}$–2 in. long; blade 1–3$\frac{1}{4}$ in. long, $\frac{1}{2}$–1$\frac{1}{2}$ in. broad across the cordate base, thence tapering to an acute apex, glabrous or thinly sprinkled with minute curved hairs on one or both sides: cymes umbel-like, axillary, 3–15-flowered; peduncles 1–6 lin. long; pedicels 1$\frac{1}{2}$–2 lin. long, glabrous or with a few minute curved hairs; sepals 1–1$\frac{1}{4}$ lin. long, $\frac{1}{3}$ lin. broad, lanceolate, acuminate, glabrous; corolla-tube $\frac{1}{2}$ lin long; lobes 2–3 lin. long, $\frac{3}{4}$–1 lin. broad at the ovate-lanceolate base, tapering into a linear twisted point, glabrous on the back, puberulous on the inner face; corona very deeply 5-lobed; tube about $\frac{1}{3}$ lin. long, shortly cupular; lobes erect, 1$\frac{1}{2}$–1$\frac{3}{4}$ lin. long, $\frac{1}{2}$ lin. broad at the base, thence tapering to a very acute point, with a small oblong or rounded appendage near their base on the inner face, which varies from being entirely adnate to nearly free; staminal column not quite 1 lin. long, not stipitate; anther-appendages ovate, inflexed upon the short obtusely conical style-apex; follicles stout, solitary or in pairs, 2–2$\frac{1}{2}$ in. long, $\frac{2}{3}$–$\frac{3}{4}$ in. thick, lanceolate or ovate lanceolate, tapering into a beak, smooth, glabrous. *Steud. Nom. Bot. ed. 2*, i. 462; *Decne in DC. Prodr.* viii. 552; *Schlechter in Engl. Jahrb.* xx. *Beibl.* 51, 2 and 42; xxi. *Beibl.* 54, 11; *and Journ Bot.* 1896, 456. *Cynoctonum virens, E. Meyer, Comm.* 216; *Decne in DC. Prodr.* viii. 552. *Endotropis Meyeri, Decne, l.c.* 546. *Vincetoxicum virens, Kuntze, Rev. Gen. Pl.* ii. 424.

Coast Region: Queenstown Div.; between Klipplaat River and Zwartkei River, *Drège,* 3433!

CENTRAL REGION : Aliwal North Div. ; by the Orange River near Aliwal North, *Drège* ! *Burke*, 386 ! Albert Div. ; between the Zuurberg Range and Stormberg Spruit, *Zeyher*, 1155 ! Stormberg Spruit, *Burke*, 453 ! Hopetown Div. ; near Hopetown, *Muskett in Herb. Bolus*, 2566 !

KALAHARI REGION : Orange River Colony, *Hutton* ! *Mrs. Barber* ! Transvaal ; near Pretoria, *Schlechter*, 4149 ! *Conrath*, 1057 ! Irene, *Miss Leendertz*, 673 !

11. C. schistoglossum (Schlechter in Journ. Bot. 1895, 271) ; stem twining, minutely pubescent, becoming glabrous ; leaves thin, spreading; petiole $\frac{1}{4}$–$1\frac{1}{2}$ in. long, slender, puberulous along the channel on the upper side ; blade 1–$3\frac{1}{4}$ in. long, $\frac{1}{3}$–$1\frac{2}{5}$ in. broad, elongate-oblong or oblong-ovate, acute or acuminate, varying from subtruncate to deeply cordate at the base, with rounded basal lobes, glabrous or with a sparse pubescence on both sides ; cymes lateral at the nodes, umbel-like or the axis elongating with age into a short raceme, pedunculate or subsessile, 5- to many-flowered ; peduncles 1–8 lin. long, puberulous ; pedicels 1–4 lin. long, puberulous ; sepals $\frac{1}{2}$–$\frac{3}{4}$ lin. long, ovate, acute or acuminate, puberulous ; corolla $1\frac{1}{2}$–2 lin. in diam. ; lobes ascending-spreading, $\frac{2}{3}$ to nearly 1 lin. long, ovate or oblong, subacute, glabrous, green (*Schlechter*) ; corona arising from the base of the staminal column and equalling or shortly exceeding it, $\frac{1}{2}$–$\frac{2}{3}$ lin. long, cup-shaped, white, very variable in toothing and lobing, usually with 5 entire or bifid teeth from a broad base, with or without 5 minute teeth alternating with them, but sometimes with 5 pairs of filiform teeth, sometimes divided to the base into 5 cuneately subquadrate lobes alternating with the anthers, produced at the apical angles into teeth and entire or denticulate between them ; usually there are 2 slight keels within opposite each of the principal teeth, or when divided into quadrate lobes there is a keel near each margin, but in some flowers the keels seem absent or are obliterated in the process of drying ; staminal column not exceeding the corona, not stipitate ; anther-appendages suborbicular, inflexed over the style-apex, which has a minute central projection. *Schlechter in Journ. Bot. 1896, 457 ; N. E. Br. in Dyer, Fl. Trop. Afr. iv. i. 395 ; Hiern in Cat. Afr. Pl. Welw. i. 688. C. vagum, N. E. Br., and C. brevidens and var. zambesiacum, N. E. Br. in Kew Bulletin*, 1895, 257. *C. minutiflorum, K. Schum. in Engl. and Prantl, Pflanzenfam. iv. ii. 252, and in Bull. Soc. Bot. Belg.* xxxvii. 123.

EASTERN REGION : Natal ; near Phoenix, *Schlechter*, 7090 ! 7804 ! Inanda, *Wood*, 611 (not 611 from near Durban) ! near Umhlanga River, *Wood*, 5664 ! and without precise locality, *Gerrard*, 1306 !

Also in Tropical Africa.

As I have stated in the *Flora of Tropical Africa*, this is one of the most variable Asclepiads (as to its coronal structure) that I have examined. But the few Natal specimens I have seen usually have the corona as detailed in the first part of the above description, yet as I have found this form of corona and one or more variations of it upon the same individual among the Tropical African specimens, I have included the chief variations in the above description, as they are not unlikely to occur upon South African examples.

12. C. sarcostemmatoides (K. Schum. in Engl. Pfl. Ost-Afr. C, 323); stem very similar to that of *Sarcostemma viminale*, but more slender, $\frac{3}{4}$–$1\frac{1}{2}$ lin. thick, twining, succulent, leafless or with minute sessile ovate acute rudimentary leaves $\frac{1}{4}$–$\frac{3}{4}$ lin. long, glabrous; flowers few or several, umbellately fascicled on short lateral tubercles, which with age grow to $\frac{1}{8}$ in. long and are marked with the contiguous spirals of scars of fallen flowers; pedicels 1–2 lin. long, glabrous; sepals $\frac{1}{4}$–$\frac{1}{3}$ lin. long, ovate, acute, glabrous; corolla rotate or perhaps slightly recurved, about $2\frac{1}{2}$ lin. in diam.; lobes 1 lin. long, oblong-ovate, acute, glabrous on both sides, green; corona cup-shaped, 5-toothed, white, arising from the base of the staminal column and as long as it, $\frac{2}{3}$–$\frac{3}{4}$ lin. long, including the shortly ovate or oblong teeth, which are inflexed over the backs of the anthers, with 5 pairs of keels decurrent from their margins within the tube; below the keels the corona is attached to the staminal column by 5 short septa; anther-appendages broad and rounded, inflexed in 5 marginal depressions of the pentagonal style-apex, which has a small truncate central boss; follicles solitary (always?), 2 in. long, 3–$3\frac{1}{2}$ lin. thick, terete, acuminate, smooth, glabrous; seeds flattened, $\frac{1}{8}$ in. long, ovate, softly pubescent on both sides. *N. E. Br. in Dyer, Fl. Trop. Afr.* iv. i. 399. *C. sarcostemmoides, K. Schum. in Engl. and Prantl, Pflanzenfam.* iv. ii. 252. *Sarcostemma aphyllum, R. Br. in Mem. Wern. Soc.* i. 51, *partly; E. Meyer, Comm.* 221; *Harv. Gen. S. Afr. Pl. ed.* 1, 224; *Schultes, Syst. Veg.* vi. 116; *D. Dietr. Syn. Pl.* ii. 907; *Decne in DC. Prodr.* viii. 538; *K. Schum. in Engl. and Prantl, Pflanzenfam.* iv. ii. 256; *Schlechter in Engl. Jahrb.* xviii. *Beibl.* 45, 10; xxi. *Beibl.* 54, 11 *and in Journ. Bot.* 1896, 457. *Asclepias aphylla, Thunb. Prodr. Pl. Cap.* 47; *in Nov. Act. Acad. Petrop.* xiv. (1805), 506; *Fl. Cap. ed.* 2, ii. 153, *and ed. Schultes,* 234, *partly, as confirmed by one of the three specimens in his herbarium; Willd. Sp. Pl.* i. 1262. *Sarcostemma tetrapterum, Turcz.,* (*and Monostemma tetrapterum, Turcz., and M. aphyllum, Turcz. ex Index Kewensis,* ii. 260) *in Bull. Soc. Mosc.* 1848, i. 255. *S. Thunbergii, G. Don, Gen. Syst.* iv. 156. *Sarcocyphula Gerrardi, Harv. Thes. Cap.* ii. 58, *t.* 191, *and Gen. S. Afr. Pl. ed.* 2, 237.

Coast Region: Uitenhage Div.; between Sundays River and Addo, *Drège*! hills between the Zwartkops River and Sundays River, *Zeyher*, 671! near Uitenhage, *Ecklon*, 56 (ex *Turczaninow*). Bathurst Div.? Glenfilling, *Drège*! Albany Div.; hill near the Botanic Garden, Grahamstown, *Tidmarsh*! and without precise locality, *Bowker*! King Williamstown Div.; near King Williamstown, *Weale*!

Eastern Region: Natal; Tugela, *Gerrard*, 1321! Mooi River Thorns, *Wood*, 4339!

Also in Tropical Africa.

Thunberg gives the localities "Karoo across Hartequas (Attaquas) Kloof" in Oudtshoorn Div., and "near the Hex River" in Worcester Div. for his *Asclepias aphylla*, but I do not know which locality belongs to the present plant and which to *Sarcostemma viminale*, both being named *A. aphylla* in his herbarium.

Imperfectly known species.

13. **C. hastatum** (Pers. Syn. Pl. i. 273); leaves subhastate-lanceolate, with rounded auricles; flowers fasciculate, subsessile. *C. lanceolatum, Poir. Encycl. Suppl.* ii. 430 ; *Schultes, Syst. Veg.* vi. 111.

AFRICA.

This may not be a South African plant. It is impossible to determine it from the description.

XXIII. **SARCOSTEMMA**, R. Br.

Calyx 5-partite. *Corolla* 5-lobed nearly to the base, rotate or rotate-campanulate; lobes overlapping in bud. *Corona* double, arising from the filament part of the staminal column; outer corona annular or cup-shaped, pentagonal, truncate or shortly lobed; inner corona of 5 erect fleshy compressed or keeled lobes, embraced at the base by the outer corona. *Staminal column* arising from the base of the corolla; anthers with terminal membranous appendages inflexed upon the apex of the style or ascending and surrounding it. *Pollen-masses* pendulous, solitary in each anther-cell, attached in pairs by short slender caudicles to the pollen-carriers. *Style-apex* convex, shortly conical or shortly produced and often slightly bifid.

Leafless fleshy shrubs ; stems branching, terete, trailing or twining ; flowers in sessile umbels, terminal or lateral at the nodes.

DISTRIB. Species few, in the dry parts of the tropical and subtropical regions of the Old World.

1. **S. viminale** (R. Br. in Mem. Wern. Soc. i. 51); stems trailing or twining, or perhaps forming a bush, succulent, glabrous, becoming woody, and sometimes with a very thick corrugated corky bark when old ; flowering branches 1–3 lin. thick, leafless or with minute ovate rudimentary leaves $\frac{1}{2}$–1 lin. long ; umbels lateral and terminal, sessile, many-flowered ; pedicels $2\frac{1}{2}$–5 lin. long, minutely puberulous or subglabrous ; sepals $\frac{1}{2}$–$\frac{2}{3}$ lin. long, ovate to suborbicular, acute or obtuse ; corolla-lobes very spreading, 2–3 lin. long, 1–1$\frac{1}{2}$ lin. broad, ovate-oblong, obtuse or acute, with reflexed margins, greenish-white or sulphur-coloured, glabrous on both sides ; outer corona arising at or near the base of the staminal column, annular, pentagonal, $\frac{1}{3}$–$\frac{2}{3}$ lin. deep, truncate, enclosing the bases of the inner corona-lobes, which are $\frac{3}{4}$–1 lin. long, adnate to the column at the lower part, free above, ovate, acute, compressed-keeled at the base, rounded from apex to base on the back, with the slightly incurved tips applied to the backs of the anthers, white ; staminal column 1$\frac{1}{4}$–1$\frac{1}{3}$ lin. long ; anther-appendages ovate, acute, adpressed to the sides of the conical style-apex, which is sometimes produced much beyond them in a

subulate bifid point; follicles $2\frac{1}{3}$–$4\frac{1}{2}$ in. long, 2–4 lin. thick, linear-fusiform to subterete, acute or acuminate, smooth, glabrous; seeds $\frac{1}{4}$ in. long, $\frac{1}{8}$ in. broad, ovate, plano-convex, narrowly margined, slightly denticulate or entire at the broad end, minutely tufted-puberulous on the convex side, nearly glabrous on the flat side, minutely ciliate, dull brown. *Ait. Hort. Kew, ed. 2, ii. 76; Haw. Syn. Pl. Succ. 13; Schultes, Syst. Veg. vi. 113; Spreng. Syst. Veg. i. 853; E. Meyer, Comm. 220; Harv. Gen. S. Afr. Pl. ed. 1, 224; G. Don, Gen. Syst. iv. 156; Dietr. Syn. Pl. ii. 907; Decne in DC. Prodr. viii. 538; A. Rich. Tent. Fl. Abyss. ii. 34; Engl. Hochgebirgsfl. Trop. Afr. 342; Engl. and Prantl, Pflanzenfam. iv. ii. 251, fig. 73, N-O, and 256; Martelli, Fl. Bogos, 54; Hiern in Cat. Afr. Pl. Welw. i. 689; Penzig in Atti Congr. Bot. Internaz. 1892, 349; Schlechter in Engl. Jahrb. xviii. Beibl. 45, 10, and xxi. Beibl. 54, 11; Journ. Bot. 1896, 458, and Ann. Naturhist. Hofmus. Wien, xv. 69; S. Moore in Journ. Bot. 1902, 256; N. E. Br. in Dyer, Fl. Trop. Afr. iv. i. 384. S. aphyllum, R. Br. in Mem. Wern. Soc. i. 51, partly; Hochst. ex Decne in DC. Prodr. viii. 538. Euphorbia viminalis, Linn. Sp. Pl. ed. 1, 452. Cynanchum viminale, Linn. Mant. ii. 392; Lam. Encycl. ii. 233; Ait. Hort. Kew. ed. 1, i. 301; Willd. Sp. Pl. i. 1252, and Enum. Pl. Hort. Berol. 276. Cynanchum aphyllum, Linn. Syst. Nat. ed. 12, iii. 235. Apocynum viminale, Bassi in Comm. Bonon. ex Schultes, Syst. Veg. vi. 113. Asclepias nuda, Schumach. and Thonn. Beskr. Guin. Pl. 155. Asclepias aphylla, Thunb. Prodr. 47, and Fl. Cap. ed. Schultes, 234 partly, as to two of the three specimens in his Herbarium.*

SOUTH AFRICA : without locality, *Herb. Linnæus! Thunberg!*

COAST REGION : Mossel Bay Div.; between Dwyker River and Gauritz River, *Burchell,* 6384! Honig Klip, *Drège!* hills near Mossel Bay. *Penther,* 829, and Knysna Div.; near the Keurboom River, *Penther,* 811 (ex *Schlechter*). Uitenhage Div.; by the Zwartkops River, *Zeyher,* 670! near Uitenhage, *Prior.* Addo, *Drège!* Bathurst Div.; near Port Alfred, *Burchell,* 4000! Albany Div.; near Grahamstown, *Read in Herb. MacOwan!* and without precise locality, *Cooper,* 1531! *Atherstone,* 4! 97! Queenstown Div.; near Shiloh, 3500–4000 ft., *Baur,* 1171!

CENTRAL REGION: Prince Albert Div.; by the Gamka River, *Burke!* Graaff Reinet Div.; between Kruidfontein and Melk River, *Burchell,* 2949! near Graaff Reinet, *Bolus,* 420!

WESTERN REGION : Vanrhynsdorp Div.; north of Holl River, *Zeyher,* 1152!

KALAHARI REGION : Griqualand West; Lower Campbell, *Burchell,* 1823! Transvaal; Komati Poort, *Kirk,* 97! Potgeiters Rust, *Burtt Davy,* 2242! and without precise locality, *Sanderson!*

EASTERN REGION : Natal; near Durban, *Wood,* 6416! Umzinyati Falls, *Wood,* 1302! Mooi River Thorns, 2000–3000 ft., *Wood,* 4338! and without precise locality *Gerrard,* 447! Delagoa Bay, *Monteiro,* 23!

Also in Tropical Africa.

Dr. Schlechter (in *Engl. Jahrb.* xxi. *Beibl.* 54, 11) refers *S. tetrapterum,* Turcz., to this plant. I have not seen Turczaninow's type, but according to his description it should rather belong to *Cynanchum sarcostemmatoides,* where I have referred it in this work. See note under that species concerning Thunberg's localities.

XXIV. PERGULARIA, Linn. (not of other Authors).

(Dœmia, R. Br.)

Calyx 5-partite. *Corolla-tube* campanulate or cylindric; lobes 5, widely spreading, overlapping in bud. *Corona* double; outer corona arising at the base of the staminal column, membranous, annular, shortly 5-lobed; inner corona of 5 erect fleshy lobes adnate to the staminal column up to the anthers, free above and produced into subulate horns incurved over the staminal column, and at the base produced into spreading or deflexed spurs. *Staminal column* arising at the mouth of the corolla-tube, entirely exserted; anthers erect, terminated by a membranous appendage, inflexed over the apex of the style. *Pollen-masses* solitary in each anther-cell, pendulous, flattened, attached in pairs to the pollen-carriers by their tapering ends, without caudicles. *Follicles* lanceolate, smooth or echinate. *Seeds* crowned with a tuft of hair.

Perennial twining plants; leaves opposite, cordate; flowers of moderate size, in long-peduncled corymbs or racemes, sublateral at the nodes.

DISTRIB. Species 4, all occurring in Tropical Africa, 2 of them extending through Arabia and Syria into India, and 1 into Madagascar.

This genus has been misunderstood by all authors, including myself. When describing the Tropical African Asclepiads I did not examine Linnæus' definition of the genus *Pergularia* as given in his *Mantissa,* i. 8, but now that I have done so, I find that he has there so accurately and unmistakably described the floral structure of the plants upon which Robert Brown afterwards established the genus *Dœmia* (but with such a very erroneous description that it would be utterly impossible to correctly refer any plant to it, except for the synonymy given), that there can be no doubt whatever as to the identity of *Dœmia* with *Pergularia,* since the structure is peculiar and not to be confused with that of any other genus, and the characters of *Pergularia* as given by Linnæus do not at all accord with those of the plants hitherto supposed to belong to *Pergularia.* I have therefore re-established *Pergularia* in accordance with the definition of it given by Linnæus. For a more complete account see the Kew Bulletin, 1907, 323. For the genus hitherto known as *Pergularia* I have there proposed the anagrammatic name *Prageluria,* but it has since been discovered that the name *Telosma* (Coville in *Contrib. United States Nat. Herb.* ix. 385, published in 1905) has the prior claim.

Stem glabrous or very minutely puberulous ...   ... (1) **gariepensis.**
Stem with a very distinct spreading pubescence or
    somewhat hispid   ...   . (2) **extensa.**

1. **P. gariepensis** (N. E. Br.); stem twining, very finely puberulous or glabrous; leaves probably somewhat fleshy; petiole $\frac{1}{3}$–1 in. long, puberulous; blade $\frac{3}{4}$–1$\frac{3}{4}$ in. long, $\frac{1}{2}$–1$\frac{1}{4}$ in. broad, broadly ovate, acuminate, deeply cordate at the base, with a broad obtuse sinus more or less enclosed by the incurved rounded basal-lobes, green and glabrous on both sides or with a few very minute hairs on the veins beneath; racemes (including the peduncles) 3–6 in. long, glabrous or nearly so on all parts except the corolla; pedicels 6–10 lin. long; sepals erect, 1$\frac{1}{3}$–1$\frac{2}{3}$ lin. long, $\frac{2}{3}$ lin. broad, lanceolate, acute; corolla glabrous outside; tube 1$\frac{1}{2}$–2$\frac{1}{4}$ lin. long, campanulate;

lobes 3–4 lin. long, 2–2¼ lin. broad, ovate-oblong, obtuse, broadly bordered with white woolly hairs on the inner face, otherwise glabrous; outer corona at the base of the staminal column, submembranous, annular at the base, 5-lobed; lobes quadrate, truncate; inner corona-lobes arising ½–⅔ lin. above the base of the staminal column, 2 lin. long, fleshy, gradually tapering from the base to a subulate point, much exceeding and incurved over the staminal column, their basal spurs ½–¾ lin. long, horizontally spreading; staminal column 2 lin. long; follicles solitary (always?), about 2 in. long, lanceolate, tapering into a short beak, covered with soft bristles, otherwise glabrous; seeds 2½–3 lin. long, 2 lin. broad, ovate, plano-convex, minutely tomentose on both sides. *Dæmia garipensis, E. Meyer, Comm.* 220; *Decne in DC. Prodr.* viii. 544; *N. E. Br. in Dyer, Fl. Trop. Afr.* iv. i., 387. *D. gariepensis, Harv. Gen. S. Afr. Pl. ed.* 1, 224; *Engl. in Engl. Jahrb.* xix. 148. *D. extensa, Schlechter in Journ. Bot.* 1896, 458, *partly. Dimia caripensis, Dietr. Syn. Pl.* ii. 907.

WESTERN REGION : Great Namaqualand ; in a dry river-bed at Cannas, *Schinz,* 21 ! 23 ! Little Namaqualand ; by the Orange River at Verleptpram, *Drège* !

Also in Tropical Africa.

2. **P. extensa** (N. E. Br.); stem twining, and together with the petioles, peduncles and pedicels pubescent, setose-pubescent or hispid; leaves herbaceous; petiole ¾–3½ in. long; blade 1–6 in. long, ¾–5 in. broad, cordate-orbicular or cordate-ovate, cuspidate-acuminate, with a minute and rather thin pubescence on both sides or glabrous with hairs on the veins beneath; basal lobes incurved, semiorbicular, with a broad truncate-based sinus between them; peduncles sublateral, longer than the leaves, with the flowers developing in a corymbose manner at the apex, gradually elongating into a raceme 2–16 in. long, including the peduncle; bracts ½–1 lin. long, linear or subulate; pedicels 1–1¼ in. long, rather slender; sepals 1–2 lin. long, ovate or lanceolate, acute, glabrous or pubescent; corolla-tube 1½–2 lin. long, campanulate; lobes very spreading, 3–4 lin. long, 1½–2 lin. broad, ovate-oblong, acute, bearded along the margins; outer corona-lobes ½ lin. long, subquadrate or oblong, obtuse, truncate or denticulate; inner corona-lobes 2½–4 lin. long, fleshy, white, lanceolate, attenuate into subulate entire or bifid points, much exceeding and incurved over the staminal column, with an acute spur about 1 lin. long arising below the middle (⅔–¾ lin. above the base) of the staminal column; follicles 2–3 in. long, ½ in. thick, narrowly lanceolate, attenuate into a long beak, varying from densely echinate to nearly or quite free from tubercles or processes, pubescent; seeds ⅓ in. long, ovate, plano-convex, marginate, dentate or crenulate at the broad end, pubescent on both sides. *Dæmia extensa, R. Br. in Mem. Wern. Soc.* i. 50; *Ait. Hort. Kew, ed.* 2, ii. 76; *Schultes, Syst. Veg.* vi. 113; *G. Don, Gen. Syst.* iv. 156; *Decne in Ann. Sc. Nat. sér.* 2, ix. 336, *and in DC. Prodr.* viii. 544; *A. Rich. Tent. Fl. Abyss.* ii. 35; *Engl. Hochgebirgsfl. Trop. Afr.* 343;

Martelli, Fl. Bogos, 54 ; *Hiern in Cat. Afr. Pl. Welw.* i. 690 ; *Penzig in Atti del Congr. Internaz.* 1892, 348 ; *Schlechter in Engl. Jahrb.* xxi. *Beibl.* 54, 11 ; *Journ. Bot.* 1896, 458 (*excl. syn. D. garipensis, E. Meyer*) ; *De Wild. and Durand, Pl. Thonn.* 33, *and Reliq. Dewevr.* 159 ; *N. E. Br. in Dyer, Fl. Trop. Afr.* iv. i. 387. *D. æthiopica, Decne, and D. angolensis, Decne, in DC. Prodr.* viii. 544 ; *Hook. Niger Fl.* 454. *D. bicolor, Sweet, Hort. Brit.* ed. 1, 280. *D. scandens, G. Don, ex Loud. Hort. Brit.* 94. *D. guineensis, G. Don, Gen. Syst.* iv. 156. *D. barbata, Schlechter in Engl. Jahrb.* xx. *Beibl.* 51, 43, *not of Klotzsch. D. cordifolia, K. Schum. in Engl. Pfl. Ost-Afr. C.* 324, *in Engl. and Prantl, Pflanzenfam.* iv. ii. 257, *fig.* 74, *and* 258, *and in Ann. Istit. Bot. Roma,* vii. 40. *Dimia extensa, Spreng. Syst. Veg.* i. 853 ; *Dietr. Syn. Pl.* ii. 906. *Cynanchum extensum, Jacq. Miscell.* ii. 353, *and Ic. Rar.* i. *t.* 54 ; *Ait. Hort. Kew, ed.* 1, i. 303 ; *Salisb. Prodr.* 149 ; *Willd. Sp. Pl.* i. 1257. *C. cordifolium, Retz. Obs. Bot.* ii. 15. *C. bicolor, Andr. Bot. Rep.* ix. *t.* 562. *C. echinatum, Thunb. Obs. in Cynanch.* 8. *C. pendulum, Poir. Encycl. Suppl.* ii. 429 ; *Schultes, Syst. Veg.* vi. 108. *Asclepias scandens, Beauv. Fl. Oware et Benin,* i. 93, *t.* 56. *A. convolvulacea, Willd. Sp. Pl.* i. 1269 ; *Schum. and Thonn. Beskr. Guin. Pl.* 152. *A. muricata, Schum. and Thonn. Beskr. Guin. Pl.* 153. *Raphistemma ciliatum, Hook. f. Bot. Mag. t.* 5704.

KALAHARI REGION : Bechuanaland ; Banquaketse Territory, near Moshoneng, *Holub* ! Transvaal ; near Klippan, *Rehmann,* 5297 ! near Ramakopa, *Schlechter,* 4507 !

EASTERN REGION : Natal ; near the Tugela River, *Gerrard,* 1802 ! Delagoa Bay, *Monteiro,* 19 ! *Schlechter,* 11959 !

Also in Tropical Africa, Madagascar, and extending through Arabia into India. In the Tropical African forms the flowers vary from white or whitish-green, with or without a red blotch at the base of the lobes, to pale yellowish-green or lurid green with a red blotch at the base of the lobes. The follicles are exceedingly variable in their amount of echination.

XXV. EUSTEGIA, R. Br.

Calyx 5-partite ; sepals ovate or lanceolate. *Corolla* very deeply 5-lobed ; lobes spreading, overlapping in bud. *Corona* of 3 series of erect lobes arising from the base of the staminal column ; outer series of 5 entire lobes opposite the corolla-lobes ; middle series of 5 trifid or subsimple lobes alternating with the corolla-lobes ; inner series of 5 entire lobes also alternating with the corolla-lobes, pressed against the backs of the anthers. *Stamens* connate around the ovary, with the anthers adnate to the style-disk ; anthers erect, with short membranous appendages. *Pollen-masses* solitary and pendulous in each anther-cell, laterally attached at their attenuated tips to the strongly hooked ends of the long caudicles, which hang from the base of the very small pollen-carriers. *Style* produced much beyond the anther-appendages, tapering or stout. *Ovules* numerous in each carpel. *Follicles* not seen.

Small tuberous-rooted perennials, branching at ground-level into many decumbent or ascending stems ; leaves small, opposite, linear-filiform to linear-hastate ; flowers small, in pedunculate umbels, lateral at the nodes.

DISTRIB. Species 5, endemic.

The 3-seriate corona readily distinguishes this genus from all except *Emicocarpus*. The species are difficult to discriminate and require to be studied in the living state to determine their limits of variation.

Middle series of corona-lobes divided into 3 distinct
    segments :
  Leaves 2–5 lin. long, linear-hastate, acute ; lateral seg-
      ments of the middle corona-lobes with an (often
      inflexed) auricle near the base :
    Middle corona-lobes divided $\frac{1}{2}$–$\frac{3}{4}$ of the way down
        into segments $\frac{1}{3}$–$\frac{1}{2}$ lin. long, distinctly shorter
        than the style ... ... (1) **fraterna.**
    Middle corona-lobes divided almost to the base and
        rising as high as the top of the style ... ... (2) **minuta.**
  Leaves or most of them $\frac{1}{2}$–$1\frac{1}{2}$ in. long, filiform or
      filiform-hastate, rarely linear-hastate :
    Corolla-lobes $1\frac{1}{2}$–$1\frac{3}{4}$ lin. long ; middle series of
        corona-lobes without distinct auricles to their
        lateral segments ... ... (3) **filiformis.**
    Corolla-lobes $3\frac{1}{2}$ lin. long ; middle series of corona-
        lobes with rounded auricles to their lateral
        segments ... ... (4) **macropetala.**
Middle series of corona-lobes with inflexed triangular
    auricles at their base, but scarcely 3-lobed ; leaves
    filiform, rarely filiform-hastate ... ... (5) **plicata.**

1. **E. fraterna** (N. E. Br.) ; plant in habit, size and form of leaves, arrangement and size of flowers exactly as in *E. minuta*, differing as follows :—corona-lobes of the outer and middle series about $\frac{3}{4}$ lin. long, shorter than the style; outer series entire, $\frac{1}{3}$ lin. broad, linear-oblong, rounded at the apex, with a large gibbosity on the inner face near the top ; middle series divided $\frac{1}{2}$–$\frac{3}{4}$ of the way down into 3 linear segments about $\frac{1}{8}$–$\frac{1}{6}$ lin. broad, with the middle one much longer than the lateral and having a large gibbosity on its inner face at the base, and the base of the lateral segments or margin of the united part more or less dilated into inflexed auricles ; inner series entire, $\frac{1}{3}$ lin. long, linear or linear-oblong, obtuse, scarcely exceeding the anther-appendages ; style exserted $\frac{1}{2}$ lin. beyond the anther-appendages, $\frac{1}{3}$ lin. thick and oblong or sometimes nearly as thick as long, very obtuse. *E. hastata, Spreng. Neue Entdeck.* i. 268, *t.* 1, *fig.* 5–10, *and Syst. Veg.* i. 854, *not of R. Br.*

VAR. β : **pubescens** (N. E. Br.) ; rather densely pubescent on the stems, leaves, peduncles and pedicels ; outer corona-lobes spathulate, suborbicular and twice as broad at the apex as at the base, otherwise as in the type.

SOUTH AFRICA : without locality, *Drège*, 6391 !
COAST REGION : Cape Div. ; near Green Point, *Zeyher*, 4697 ! *Prior (Alexander)* ! near Cape Town, *Harvey* ! Lion Mountain, *Harvey* ! Var. β : Tulbagh Div. ; hills near Piquetberg Road (Gouda), 400 ft., *Schlechter*, 10710 !

This may be only a variety of *E. minuta,* but it conspicuously differs by the middle series of corona-lobes being shorter than the style and very much less deeply divided. The figure of *E. hastata* in Engl. and Prantl, *Pflanzenfam.* iv. ii. 234, fig. 68 N–P, may possibly be intended for this plant; but the lateral segments of the middle series of corona-lobes are not represented as being auriculate at the base.

2. E. minuta (R. Br. in Mem. Wern. Soc. i. 52); branches decumbent 2–3 in. long, unifariously puberulous; leaves 2–5 lin. long, those at the base sometimes ovate or lanceolate, the others usually linear-hastate with a short acute spreading tooth on each side at the base, or without them and linear, acute, cuneately narrowed into a very short petiole, glabrous, often irregularly ciliate with a few hairs; umbels pedunculate, 3–6-flowered; peduncles 2–8 lin. long; bracts $\frac{1}{2}$–$\frac{3}{4}$ lin. long, lanceolate-subulate, acute; pedicels 2–3 lin. long, puberulous on one side; sepals about $\frac{3}{4}$ lin. long, $\frac{1}{2}$ lin. broad, ovate, obtusely pointed or subacute, glabrous; corolla-lobes about $1\frac{1}{2}$ lin. long and $\frac{3}{4}$ lin. broad, oblong, obtuse, apparently slightly twisted, glabrous on both sides; corona-lobes of the outer and middle series $\frac{3}{4}$–1 lin. long, rising to about the level of the top of the style; outer series entire, $\frac{1}{4}$ lin. broad, linear, obtuse, with a gibbosity or thickening on the inner face at the top; middle series divided nearly to the base into 3 linear segments about $\frac{1}{6}$ lin. broad, with the middle one obtuse, having a gibbosity on the inner face at the middle, and the lateral acute or obtuse, with a small but distinct inflexed auricle or tooth on the outer margin near the base; inner series about $\frac{1}{2}$ lin. long, $\frac{1}{5}$ lin. broad, oblong-linear, very obtuse or rounded at the apex, scarcely exceeding the anther-appendages; style exserted $\frac{1}{3}$ lin. beyond the anther-appendages, about $\frac{1}{4}$ lin. thick, oblong, obtuse, apparently pentagonal. *E. hastata, R. Br. in Mem. Wern. Soc.* i. 52; *Schultes, Syst. Veg.* vi. 119; *G. Don, Gen. Syst.* iv. 158; *Dietr. Syn. Pl.* ii. 907; *Decne in DC. Prodr.* viii. 545, *excl. reference to Sprengel. Apocynum minutum, Linn. f. Suppl.* 169; *Murray, Syst. Veg. ed.* 14, 258; *Lam. Encycl.* i. 215. *A. hastatum, Thunb. Prodr.* 47; *in Nov. Act. Acad. Petrop.* xiv. (1805), 514, *t.* 9, *fig. b; Fl. Cap. ed.* 2, ii. 164, *and ed. Schultes* 238; *Willd. Sp. Pl.* i. 1259; *Pers. Syn. Pl.* i. 274.

South Africa: without locality, *Montin in Herb. Linnæus*!
Coast Region: Cape Div.; below Lion Mountain, *Thunberg*!

3. E. filiformis (Schultes, Syst. Veg. vi. 120); branches usually decumbent, 2–6 in. long, puberulous on one side; leaves $\frac{1}{2}$–$1\frac{1}{2}$ in. long, filiform, linear-filiform, filiform-hastate or rarely linear-hastate, glabrous or with a thin minute pubescence or a few small scattered hairs; umbels 4–8-flowered; peduncles $\frac{1}{4}$–$1\frac{3}{4}$ in. long, puberulous on one side; pedicels $1\frac{1}{2}$–2 lin. long, puberulous; sepals $\frac{3}{4}$–1 lin. long, about $\frac{1}{2}$ lin. broad, ovate or ovate-lanceolate, acute or obtuse, glabrous; corolla-lobes spreading, $1\frac{1}{2}$–$1\frac{3}{4}$ lin. long, $\frac{2}{3}$–$\frac{3}{4}$ lin. broad, oblong, obtuse, with more or less revolute margins, glabrous on both sides, apparently olive-green; corona-lobes apparently white; lobes of the outer series 1 lin. long, simple, narrowly oblong-linear, often bifid or emarginate at the apex, gibbous or thickened on the inner face; lobes of the middle series divided nearly to the base into 3 segments $\frac{3}{4}$–$1\frac{1}{4}$ lin. long, rising to the level of the apex of the style or exceeding it, the middle segment linear, with a compressed gibbosity or thickened keel at the middle, sometimes nearly obliterated in dried specimens, the lateral usually linear, without auriculate inflexed sides at the base; lobes of the inner series $\frac{2}{3}$–$\frac{3}{4}$ lin. long, entire, narrowly oblong-linear or linear-lanceolate, subacute or obtuse, with the tips pressed against the basal part of the style-apex, which is either stout and oblong or ovoid-oblong, or more slender, much elongated or narrowly conical, and apparently angular, produced about $\frac{3}{4}$ lin. beyond the very short erect rounded anther-appendages; staminal column $\frac{1}{6}$ lin. long. *Spreng. Syst. Veg.* i. 854; *G. Don, Gen. Syst.* iv. 158. *E. filiformis, E. humilis and E. lonchitis, E. Meyer, Comm.* 221; *Dietr. Syn. Pl.* ii. 907; *Decne in DC. Prodr.* viii. 545. *E. lonchitis, Schlechter in Ann. Naturhist. Hofmus. Wien,* xv. 67. *Apocynum filiforme, Linn. f. Suppl.* 169; *Murray, Syst. Veg.* ed. 14, 258; *Willd. Sp. Pl.* i. 1259; *Thunb. Prodr.* 47; *in Nov. Act. Acad. Petrop.* xiv. (1805), 510; *Fl. Cap. ed.* 2, ii. 160, *and ed. Schultes,* 237; *Lam. Encycl.* i. 215.

Coast Region : Clanwilliam Div. ; between Pakhuis and Clanwilliam, 200 ft., *Leipoldt,* 243 ! and in *MacOwan, Herb. Austr.-Afr.* 1923 ! Cederberg Range, 3000 ft., *Bodkin in Herb. Bolus,* 8312 ! near Olifants River, *Penther,* 2292 (ex *Schlechter*). Piquetberg Div. ; between Piquetberg Mountain and Kruis, *Drège,* 6392 ! Malmesbury Div. ; Zwartland, *Thunberg* ! near Hopefield, *Bachmann* ! *Schlechter,* 5306 ! *Penther,* 1689 and Krantzfontein, *Penther,* 1690 (ex *Schlechter*). Tulbagh or Ceres Div. ; Tulbagh Kloof, *Bolus,* 8313 ! near Ceres, *Bolus,* 7554 ! Cape Div. ; near Cape Town, *Harvey* ! *Bolus,* 4593 ! *Rogers* ! near Camps Bay, *Pappe* ! beyond Maitland Station, *Wolley Dod,* 3064 ! Paarl Div. ; between Paarl and Simons Berg, *Drège,* 6390 !

Western Region : Little Namaqualand ; Kamiesberg Range, between Leliefontein and Krakkeel Kraal, *Drège,* 3044 ! and at Riet Kloof, 2500 ft., *Schlechter,* 11186 !

The apical part of the style in this species seems to vary very much in stoutness, and E. Meyer founded 3 species upon the character of the style, but the distinction seems unsupported by other differences ; in Thunberg's type of *Apocynum filiforme* it is very stout.

4. **E. macropetala** (Schlechter in Journ. Bot. 1895, 358) ; plant $2\frac{3}{4}$–4 in. high, branching from the base ; branches erect or suberect, glabrescent, densely leafy ; leaves erect, $\frac{2}{3}$–$1\frac{1}{4}$ in. long, filiform, acute, with the margins usually revolute, glabrous ; umbels pedunculate, about 4-flowered ; peduncles short, bifariously puberulous ; pedicels 4 lin. long, filiform, glabrescent ; sepals 1 lin. long, ovate, acute, very thinly puberulous ; corolla-lobes erectly spreading, $3\frac{1}{2}$ lin. long, scarcely 2 lin. broad at the middle, ovate-oblong, obtuse, glabrous ; outer series of corona-lobes linear, obtuse, with a thick longitudinal keel on the inner face ; middle series a little longer

than the outer, deeply 3-partite, with the lateral segments linear, subfalcate, obtuse, enlarged into a rounded lobe on the outer margin below the middle, and with the middle segment suberect, linear, subobtuse; inner series linear-lanceolate, attenuate at the apex, subincurved, slightly exceediug the anthers.

CoAST REGION : Piquetberg Div. ; sandy places at the foot of Piquetberg Mountain, 1000 ft., *Schlechter,* 5213.

5. E. plicata (Schinz in Bull. Herb. Boiss. ii. 218); branches decumbent, 2–3½ in. long, more or less pubescent or puberulous; leaves ¼–⅔ in. long, filiform or linear-filiform, rarely hastate-filiform, acute, glabrous or pubescent; umbels pedunculate, 2–5-flowered; peduncles 2–5 lin. long, puberulous or pubescent on one side; bracts minute; pedicels 2–3 lin. long, puberulous; sepals about ¾ lin. long, ½ lin. broad, ovate, acute, glabrous; corolla-lobes 1½ lin. long, ⅔ lin. broad, very spreading, oblong, obtuse, slightly twisted, glabrous on both sides, apparently dull green; outer series of corona-lobes ½ lin. long, entire, oblong-linear, obtuse or subtruncate, purple, with white margins; middle series ¾ lin. long, oblong or linear, obtuse, with inflexed triangular sides at the base, but scarcely 3-lobed; inner series ⅓ lin. long, oblong-linear, concave, obtuse; style-apex very stout, oblong to obovoid-clavate, obtuse, probably 5-angled, produced ½ lin. beyond the anther-appendages.

CoAST REGION : Piquetberg Div.; hills near Piquiniers Kloof, *Schlechter,* 10755! Malmesbury Div. ; Hopefield, *Bachmann,* 1995 !

XXVI. **EMICOCARPUS,** K. Schum. and Schlechter.

Calyx 5-partite. *Corolla* 5-lobed nearly to the base ; lobes narrowly overlapping in bud. *Corona* very complicated ; outer arising in the angle between the corolla and the staminal column, consisting of 20 segments, all very shortly connected at the base or almost free and apparently belonging to 1 series, but probably composed of 2 whorls, the outer consisting of 5 simple segments opposite the corolla-lobes, and the inner of 5 tripartite segments alternating with them, with their lateral lobes infolded so as to stand in front of the simple outer lobe, as if they were appendages arising from the base of it ; inner of 5 simple lobes arising a short distance up the staminal column, near or at the base of the anthers and applied to their backs. *Stamens* arising at the base of the corolla, united into a tube around the ovary, with the anthers adnate at their base to the dilated part of the style, and tipped with a very minute scarcely visible appendage. *Pollen-masses* solitary in each anther-cell, pendulous, very minute, attached in pairs to the pollen-carriers by long abruptly bent caudicles. *Style* produced beyond the dilated part into a beak. *Carpels* with a single ovule in each, seated in a cavity at the top of a thickened placenta. *Follicles* small, obtriangular, with 3 spreading spines at the top, probably indehiscent. *Seed* solitary, curved, subterete, without a tuft of hairs at either end.

A procumbent herb; leaves palmately divided; flowers small, in small pedunculate umbels lateral at the nodes.

DISTRIB. Species 1, endemic.

This very remarkable Asclepiad, by its palmately lobed leaves, 1-ovuled carpels, 3-horned and 1-seeded follicles, and the curved seed without hairs, is quite unique in the Order. In the structure of its corona it is closely allied to *Eustegia*. K. Schumann and Schlechter describe and figure the carpels as having 2 ovules in each, but I have only been able to find one.

1. **E. fissifolius** (K. Schum. and Schlechter in Engl. Jahrb. xxix. Beibl. 66, 21, 22, with fig.); plant branching at the base; stems prostrate, 2–4 ft. long, rather slender, unifariously puberulous; leaves petiolate, palmately 5–7-lobed, cuneately narrowed into the $\frac{1}{4}$–$\frac{1}{3}$ in. long petiole, glabrous, with the exception of a few minute hairs on the midrib beneath and sometimes about the basal part; middle lobe $\frac{3}{4}$–$1\frac{1}{2}$ in. long, linear, acute; lateral lobes very much shorter, linear or attenuate from the base to an acute point, all spreading and the upper pair falcately incurved; umbels lateral at the nodes and terminal, pedunculate, 5–8-flowered; peduncle 5–10 lin. long, unifariously puberulous; bracts minute; pedicels $1\frac{1}{2}$–2 lin. long, unifariously puberulous; sepals $\frac{3}{4}$ lin. long, lanceolate or ovate-oblong, obtuse (not acuminate as originally described), glabrous; corolla-lobes widely spreading, 1–$1\frac{1}{4}$ lin. long, $\frac{1}{2}$ lin. broad; oblong or oblong-lanceolate, obtusely pointed, glabrous on both sides; outer corona-segments erect, the 5 alternating with the corolla-lobes $\frac{2}{5}$ lin. long, linear or subspathulate, obtuse; the 5 opposite the corolla-lobes $\frac{1}{2}$ lin. long, $\frac{1}{3}$ lin. broad, oblong, obtuse, with the infolded side lobes standing in front of them $\frac{3}{4}$ lin. long, linear, obtuse; inner corona-lobes erect, $\frac{1}{4}$–$\frac{1}{3}$ lin. long, oblong, obtuse or notched at the apex; staminal column $\frac{2}{5}$ lin. long; style produced into a beak $\frac{2}{3}$–$\frac{3}{4}$ lin. beyond the erect anther-tips; follicles usually solitary, 3–$4\frac{1}{2}$ lin. long, obtriangular, with 3 spreading spines at the subtruncate top, glabrous. *Lobostephanus palmatus*, N. E. Br. in Hook. Ic. Pl. xxvii. *t.* 2692.

EASTERN REGION : Delagoa Bay, *Junod*, 502! near Lourenço Marques, *Schlechter*, 11535!

XXVII. **TYLOPHORA**, R. Br.

Calyx 5-partite. *Corolla* lobed very deeply or nearly to the base, rotate or campanulate-rotate; lobes overlapping and sometimes twisted in bud. *Corona* usually of 5 tubercles adnate to or radiating from the staminal column at or above its base and usually not exceeding the filament part of it, rarely with free tips or of entirely free flat lobes and attaining to the level of the anther-tips. *Staminal column* arising from the base of the corolla ; anthers erect, tipped with a small membranous appendage, which is sometimes reduced to a mere hyaline margin, their fertile part not rising above the margin of the style-apex, usually opening by crescent-shaped transverse slits. *Pollen-masses* very minute, suberect, horizontal or pendulous, attached in pairs to the minute pollen-carriers by very slender caudicles affixed at their middle or near or at one end. *Style-apex* rarely slightly exceeding the anthers, pentagonal or 5-lobed, depressed, flattened and often with a small central boss, or convex. *Follicles* usually narrowly lanceolate or lanceolate-fusiform, smooth, sometimes winged. *Seeds* crowned with a tuft of hairs.

Twining or rarely erect perennials ; leaves opposite, petiolate ; inflorescence lateral at the nodes or axillary, sometimes consisting of a single pedunculate or sessile umbel-like cyme or flower-cluster, sometimes of 2 to several flower-clusters scattered along a single-jointed axis or along the branches of a dichotomously branched cyme or panicle.

DISTRIB. Species many, widely distributed in the tropical and subtropical regions of the Old World.

This genus is usually easy to recognise by the peculiar character of its inflorescence and the corona-lobes usually reduced to mere tubercles. Since the publication of the Tropical African Asclepiads I have examined some species in which the pollen-masses are distinctly pendulous ; as is also stated by Dr. Schlechter, and I now doubt if any of them have really erect pollen-masses ; in some they appear to be ascending, in others neither erect nor pendulous, in others still, they are distinctly pendulous, so that the genus is distinctly intermediate in character between the *Cynancheæ* and *Marsdenieæ*, and I now think that *Tylophoropsis* must be reduced to *Tylophora.* The pollen-masses are always very minute, and their position in the anther-cell is sometimes very difficult to determine. The caudicles are usually very slender, and at times I have failed to find any trace of them.

An erect, somewhat rigid branching shrublet with small
    leaves ¼–½ in. long, including the petiole ...   ... (1) **Fleckii.**
Stems twining ; leaf-blade ⅔–4 in. long :
    Corolla pilose on the inner surface ; leaves truncately
        rounded at the base...   ...   . (2) **inhambanensis.**
    Corolla puberulous on the inner surface ; leaves mostly
        cordate at the base ...   . (3) **anomala.**
    Corolla glabrous, or at most with a microscopic
        puberulence on the inner surface :
        Peduncle dichotomously or paniculately much
            branched, each branch bearing 2–4 small umbel-
            like cymes ...   . (4) **Flanagani.**
        Peduncle bearing 2–4 (or occasionally only 1) umbel-
            like cymes scattered along it ; bracts ⅓–1¼ lin.
            long, minute or inconspicuous, not forming an
            involucre :
        Corolla purple or rosy :
            Leaf-blade 1–3½ in. long, glabrous on both
                sides :
                Pedicels 4–7 lin. long ; corona-lobes very
                    prominent at the base   ...   . (5) **umbellata.**
                Pedicels 2½–4 lin. long ; corona-lobes slightly
                    prominent at the base   ...   . (6) **badia.**

Leaf-blade ⅔–1⅓ in. long, puberulous beneath ... (7) **simiana.**
Corolla green or yellowish-green ; leaf-blade 1½–4
    in. long, glabrous or pubescent on one or both
    sides   ...   . (8) **syringæfolia.**
Peduncle bearing 1 terminal umbel only ; bracts
    1–4 lin. long, very conspicuous, forming a kind
    of involucre   ...   . (9) **lycioides.**

1. **T. Fleckii** (N. E. Br.) ; a somewhat rigid erect branching shrublet; branches at first very thinly puberulous, becoming glabrous ; leaves (including the short petiole) 3½–6 lin. long, 2–3½ lin. broad at the middle, oblong-elliptic or oblong, obtuse, tapering at the base ; cymes lateral at the nodes, few-flowered, scarcely exceeding the leaves ; peduncles and pedicels about equal in length, very thinly puberulous ; sepals 1 lin. long, oblong, obtuse, very thinly puberulous ; corolla subrotate, with oblong obtuse glabrous lobes, about twice as long as the sepals ; corona-lobes fleshy, ovate, subacute, entirely adnate to the staminal column, not extending above the base of the anther-cells ; anther-appendages subreniform, incurved upon the style-apex, which is 5-angular, produced at the angles, conical at the apex. *Tylophoropsis Fleckii, Schlechter in Bull. Herb. Boiss.* vii. 39.

WESTERN REGION : Great Namaqualand ; Gansberg, *Fleck,* 431 (ex *Schlechter*).

Gansberg is just within the Tropical region, but as the plant will probably be found to occur south of the Tropic of Capricorn, it is here included.

2. **T. inhambanenis** (Schlechter in Engl. Jahrb. xxxviii. 52) ; stem herbaceous, twining ; branches filiform, flexuose, puberulous, laxly leafy ; leaves spreading ; petiole 5–10 lin. long, pilose ; blade 2⅓–3¼ in. long, 1–1¾ in. broad below the middle, ovate, acuminate, truncately rounded at the base, glabrous or subglabrous, thin in texture ; cymes subsessile or with a peduncle as long as the petiole ; pedicels up to 10 lin. long, filiform, glabrous ; sepals ¾ lin. long, ovate-lanceolate, acuminate, sparsely pilose ; corolla subrotate, deeply 5-lobed, 2 lin. long ; lobes oblong, subacute, glabrous on the back, pilose on the inner face ; corona-lobes fleshy, ovate-oblong, somewhat obtuse, adnate to the staminal column at the lower half, incurved at the apex, shortly exceeding the base of the anthers, which are tipped with a rounded obtuse appendage ; style-apex depressed.

EASTERN REGION : Portuguese East Africa ; near Machisugu, 100 ft., *Schlechter,* 12116 !

3. **T. anomala** (N. E. Br.) ; stem twining, slender, " not milky " (*Gerrard*), with a minute unifarious pubescence ; leaves herbaceous ; petiole ¼–½ in. long ; blade 1¼–3 in. long, ¾–2¼ in. broad, ovate or ovate oblong, acute to very obtuse or emarginate at the apiculate apex, usually cordate, occasionally rounded at the base, glabrous on both sides ; peduncles lateral at the nodes, ¼–¾ in. long, glabrous,

usually bearing 1, more rarely 2, umbel-like cymes of about 5–7 flowers developing in succession; pedicels 3–4 lin. long, slender, glabrous; sepals about $\frac{3}{4}$ lin. long, ovate or lanceolate, acute, glabrous; corolla about 2 lin. in diam., with the united saucer-shaped part about $\frac{2}{3}$ lin. long and the ovate acute (ascending-spreading?) lobes $\frac{3}{4}$ lin. long, $\frac{2}{3}$ lin. broad, glabrous outside, puberulous within, dull white (*Gerrard*); corona-lobes arising at the base of and equalling the staminal column, $\frac{2}{5}$ lin. long and nearly as broad, erect, broadly ovate or ovate-oblong, notched at the apex, nearly flat, with a stout keel (square in transverse section) down the middle of the inner face, by which they are attached to the $\frac{2}{5}$ lin.-long staminal column below the anthers, whose minute appendages are transverse and inflexed upon the margin of the depressed style-apex; pollen-masses distinctly pendulous, attached to the narrow elongated pollen-carrier by very slender caudicles; follicles solitary (only one immature example seen), about $2\frac{1}{2}$ in. long, $4\frac{1}{2}$ lin. thick, narrowly lanceolate, tapering at the upper third into an acute beak, smooth, glabrous; seeds (not ripe) about $3\frac{1}{2}$ lin. long, $1\frac{1}{2}$ lin. broad, plano-convex, ovate, wing-margined, smooth on both sides, dark brown.

EASTERN REGION : Natal ; "Buck-bush, near Durban, ex Herb. *M. J. McKen,* no. 4," ! and "Buck-bush, Umgeni," *Gerrard,* 1320 !

Gerrard's specimens are distributed without localities, those given above are copied from his original labels in the Dublin Herbarium.

4. **T. Flanagani** (Schlechter in Engl. Jahrb. xviii. Beibl. 45, 11) ; stem twining, glabrous ; leaves very thinly coriaceous ; petiole $\frac{1}{2}$–$1\frac{1}{3}$ in. long ; blade $1\frac{1}{2}$–$3\frac{1}{2}$ in. long, $\frac{7}{8}$–$2\frac{1}{2}$ in. broad, from ovate to elliptic or elliptic-oblong, very acute to very obtuse or rounded and apiculate at the apex, rounded to subtruncate at the base, glabrous ; flowers in lax branching panicles or forked cymes, lateral at the nodes, and (including the peduncles) 2–5 in. long, with a spread of 2–6 in., each branch bearing 2–4 small umbel-like cymes of 2–6 flowers ; bracts inconspicuous, subulate ; pedicels $\frac{1}{3}$ in. long, very slender, glabrous ; sepals about $\frac{3}{4}$ lin. long, lanceolate, acute, glabrous or with minute scattered adpressed hairs ; corolla rotate, about $4\frac{1}{2}$ lin. in diam., dark purple or dark crimson, glabrous ; lobes 2 lin. long, $\frac{3}{4}$ lin. broad, lanceolate-attenuate, twisted at the truncate apex, about 4 times as long as the united part ; corona-lobes arising at the base of the staminal column, equalling or slightly exceeding it and adnate to it at the basal part, erect, $\frac{1}{3}$–$\frac{1}{2}$ lin. long, slightly gibbous at the base, tapering into the subulate free incurved tips, dark purple-brown ; staminal column $\frac{1}{2}$ lin. long ; pollen-masses pendulous ; style-apex raised into a small central boss. *Schlechter in Journ. Bot.* 1898, 486.

COAST REGION : Komgha Div. ; in woods near Komgha, *Flanagan,* 378 !

EASTERN REGION : Tembuland ; summit of Engcobo Mountain, 4500 ft., *Bolus,* 10193 ! Natal ; at Westville, in marshy woods near Durban, 1000 ft., *Sanderson,* 2006 ! Zululand ; Ongoa, *Gerrard,* 2169 !

5. T. umbellata (Schlechter in Engl. Jahrb. xviii. Beibl. 45, 11);
stem twining, thinly puberulous or glabrous ; leaves thinly coriaceous ;
petiole $\frac{1}{3}$–1 in. long ; blade $1\frac{1}{4}$–$3\frac{1}{2}$ in. long, $\frac{2}{3}$–$1\frac{1}{2}$ in. broad, lanceolate,
ovate-lanceolate, broadly ovate or elliptic-ovate, acute or acuminate
to obtuse and apiculate, cuneate to broadly rounded at the base,
glabrous on both sides ; peduncles lateral, sometimes 2 at a node,
1–2 in. long, slender, usually bearing 2–3 umbels $\frac{1}{4}$–$\frac{3}{4}$ in. apart, or
occasionally with 1 umbel, puberulous on one side or glabrous ;
umbels 5–11-flowered ; bracts minute ; pedicels 4–7 lin. long, slender,
glabrous ; sepals $\frac{1}{2}$–$\frac{3}{4}$ lin. long, $\frac{1}{3}$ lin. broad, lanceolate-oblong,
obtuse or acute, more or less ciliate, otherwise glabrous ; corolla
rotate, dark dull purple, glabrous outside, with a thin microscopic
powder-like puberulence on the inner face, invisible when wetted ;
lobes 2 lin. long, 1 lin. or rather more in breadth, oblong or oblong-
lanceolate, obtuse ; corona-lobes entirely adnate to the upper half of
the filament part of the staminal column, wedge-shaped, in side
view sloping outwards from the acute apex to the very prominent
truncate base, where the margin on each side is prominent and rim-
like, the upper part being more compressed ; staminal column about
$\frac{3}{4}$ lin. long and the same in diam. at the top ; anther-appendages
reniform, abruptly inflexed upon the top of the truncate style-apex,
pollen-masses distinctly pendulous. *Schlechter in Journ. Bot.* 1898,
486.

Coast Region : Uitenhage Div. ; near Uitenhage, *Burchell*, 4263 ! Bedford
Div., near Bedford, *Mrs. Hutton* ! King Williamstown Div. ; along the Yellowwood
River near King Williamstown, *Flanagan*, 2191 ! Komgha Div. ; near Komgha,
1800 ft., *Flanagan*, 1702 !

In the original description the flowers are stated to be in simple umbels, but on
the specimen I have seen of the type (*Flanagan*, 1702) most of the peduncles
bear 2 umbels, as do all the other specimens quoted. This may be only a variety
of *T. badia*, but the flowers appear to be of a darker purple, the corona-lobes
are much more prominent at the base, and the staminal column of greater
diameter.

6. T. badia (Schlechter in Engl. Jahrb. xxi. Beibl. 54, 12); stems
twining, glabrous ; leaves apparently thinly coriaceous ; petiole $\frac{1}{4}$–$\frac{1}{2}$
in. long, channelled down the face, glabrous or with a few minute
hairs on the margins of the channel ; blade 1–$2\frac{1}{2}$ in. long, $\frac{1}{3}$–1 in.
broad, lanceolate, acute, cuneate or slightly rounded at the base,
glabrous ; peduncles or flowering axes $\frac{3}{4}$–1 in. long, very slender,
bearing 2–3 distant fascicles of 3–5 flowers, glabrous ; bracts minute ;
pedicels $2\frac{1}{2}$–4 lin. long, filiform, glabrous ; sepals $\frac{2}{3}$ lin. long, lanceolate or oblong-lanceolate, acute, glabrous ; corolla rotate, rosy
(*Flanagan*) ; lobes $1\frac{1}{2}$ lin. long, $\frac{2}{3}$ to nearly 1 lin. broad, lanceolate
or oblong-lanceolate, obtuse, slightly twisted at the apex, glabrous
on the back, with a microscopic puberulence on the inner face,
invisible when wetted ; corona-tubercles entirely adnate from close
to the base to the apex of the filament part of the staminal column
and not exceeding it, wedge-shaped, broadest at the somewhat
prominent subtruncate or broadly rounded base ; staminal column

⅔ lin. long and the same in diam. at the top; anther-appendages reniform or transversely oblong, abruptly inflexed on the truncate style-apex; pollen-masses distinctly pendulous. *Schlechter in Journ. Bot.* 1898, 486. *Astephanus badius, E. Meyer, Comm.* 224; *Dietr. Syn. Pl.* ii. 909; *Decne in DC. Prodr.* viii. 508.

VAR. β, **latifolia** (N. E. Br.); leaves with a petiole ½–¾ in. long, and a blade 2–3½ in. long, 1¼–2½ in. broad, elliptic, shortly and rather abruptly acute; flowers purplish (*Gerrard*); otherwise as in the type.

COAST REGION: Uitenhage Div.; without precise locality, *Brehm*! Komgha Div.; banks of the Kei River, *Drège*, 4954! valley between Impetu and Keimouth, 1000 ft., *Flanagan*, 1046!
EASTERN REGION: Var. β: Zululand, *Gerrard*, 2168!

7. T. simiana (Schlechter in Engl. Jahrb. xviii. Beibl. 45, 33); stem twining, branching, velvety-pubescent; leaves herbaceous; petiole 2–4 lin. long, pubescent like the stem; blade ⅔–1⅓ in. long, 3–8 lin. broad, ovate or ovate-lanceolate, obtuse or acute, broadly cuneate at the base, nearly glabrous above, velvety-puberulous beneath; peduncle ½–⅔ in. long, lateral at the nodes, bearing 1 or 2 umbel-like 3–7-flowered cymes, slightly puberulous; bracts minute; pedicels 2–2½ lin. long, glabrous; sepals about ¾ lin. long, ⅓ lin. broad, ovate, obtuse, thinly puberulous and ciliate; corolla lobed nearly to the base, dull purple; lobes very spreading, 2½ lin. long, 1¼ lin. broad, ovate-oblong, very obtuse, glabrous on both sides; corona-tubercles ⅓–⅖ lin. long, ovate in dorsal view, subacute, constricted on the inner face into a sort of keel by which they are adnate to their tips to (but do not exceed) the filament part of the ¾ lin.-long staminal column; anthers transversely oblong, with short transverse sub-truncate appendages inflexed upon the margin of the style-apex, which has a small truncate central boss; pollen-masses distinctly pendulous. *Schlechter in Journ. Bot.* 1898, 486.

COAST REGION: King Williamstown Div.; Mount Coke, 1500 ft., *Sim*, 1305!

I have examined a type specimen of this plant, named by Dr. Schlechter, but doubt if it is more than a mountain form of *T. syringæfolia*, the flowers seem identical, except in colour, besides which the only difference appears to be that the leaves are smaller and much more obtuse; in pubescence and other characters there is no distinction between it and some specimens of *T. syringæfolia*.

8. T. syringæfolia (E. Meyer, Comm. 198); stem herbaceous, twining, usually pubescent or tomentose, occasionally glabrous; leaves thinly coriaceous; petiole ⅓–1 in. long; blade 1½–4 in. long, ¾–2 in. broad, ovate, acute or acuminate, broadly rounded, sub-truncate or subcordate at the base, glabrous or thinly pubescent above, glabrous to densely pubescent beneath; peduncles lateral, ¾–2½ in. long, pubescent or glabrous, usually bearing 2 (rarely 3–4) umbel-like 5–8-flowered cymes ¼ to 1 in. apart; bracts minute; pedicels ¼–⅓ in. long, glabrous or slightly puberulous; sepals ⅔–¾ lin. long, ovate or oblong, obtuse, ciliate; corolla lobed nearly to the base, dull greenish (*Meyer*); lobes spreading, 2–$2\frac{1}{2}$ lin. long and $1\frac{1}{4}$ lin. broad, oblong, obtuse, glabrous on both sides; corona-tubercles $\frac{1}{3}$–$\frac{2}{5}$ lin. long, ovate in dorsal view, obtuse, constricted into a sort of keel on the inner side, by which they are adnate to their tips to (but do not exceed) the filament part of the staminal column, this keel-structure is often obscured or destroyed in dried specimens; staminal column $\frac{3}{4}$ lin. long; anthers transversely oblong, with short transverse rounded appendages inflexed upon the margin of the style-apex, which has a small truncate central boss; pollen-masses pendulous; follicles widely diverging or solitary, $1\frac{3}{4}$–$2\frac{3}{4}$ in. long, $\frac{1}{2}$ in. thick, lanceolate, acute, transversely deeply corrugated, glabrous; seeds $3\frac{1}{2}$ lin. long, 2 lin. broad, ovate, plano-convex, narrowly wing-margined, sometimes subdenticulate at the broad end, smooth, glabrous, brown. *Harv. Gen. S. Afr. Pl. ed.* 1, 223; *Dietr. Syn. Pl.* ii. 894; *Decne in DC. Prodr.* viii. 611; *Schlechter in Engl. Jahrb.* xxi. *Beibl.* 54, 12. *T. syringifolia, Schlechter in Engl. Jahrb.* xviii. *Beibl.* 45, 11, *and Journ. Bot.* 1898, 486. *Apocynum cordatum, Thunb. Prodr. Pl. Cap.* 47; *in Nov. Act. Acad. Petrop.* xiv. (1805) 513; *Fl. Cap. ed.* 2, ii. 163, *and ed. Schultes*, 238; *Willd. Sp. Pl.* i. 1261; *Pers. Syn. Pl.* i. 275. *Astephanus cordatus, R. Br. in Mem. Wern. Soc.* i. 54; *Schultes, Syst. Veg.* vi. 123; *Spreng. Syst. Veg.* i. 855; *G. Don, Gen. Syst.* iv. 158. *Dietr. Synop. Pl.* ii. 909; *Decne in DC. Prodr.* viii. 508; *Schlechter in Journ. Bot.* 1896, 418. *Vincetoxicum syringifolium, Kuntze, Rev. Gen. Pl.* ii. 425.

COAST REGION: George Div.; near George, *Burchell*, 6070! Kaimans Gat, *Prior*! Uitenhage Div.; Galgebosch, *Drège*, 3415! Bathurst Div.; mouth of Kleinemund River, 200 ft., *MacOwan*, 998! East London Div.; near East London, *Wood in Herb. Galpin*, 3134! 3369! Komgha Div.; near Komgha, 2000 ft., *Schlechter*, 6154! British Kaffraria, *Cooper*, 46!

CENTRAL REGION: Somerset Div.; Somerset East, *Bowker*! Cradock Div.; near the Tarka River, *Cooper*, 405.

EASTERN REGION: Transkei; Fort Bowker, *Bowker*, 608 (or 602?)! near Kentani, 1000 ft., *Miss Alice Pegler*, 1275!

Thunberg's type specimen of *Apocynum cordatum* is identical with *MacOwan's* 998 from Kleinemund River.

9. T. lycioides (Decne in DC. Prodr. viii. 608); stem twining, thinly adpressed-puberulous or glabrous; leaves herbaceous, glabrous; petiole $\frac{1}{4}$–1 in. long, rather slender; blade $\frac{2}{3}$–$2\frac{2}{3}$ in. long, $\frac{1}{6}$–$1\frac{1}{2}$ in. broad, varying from narrowly lanceolate to ovate-lanceolate or ovate, acute or acuminate, broadly rounded at the base; cymes lateral at the nodes, simple, umbel-like, glabrous, with a few minute hairs about the axils of the bracts; peduncles 1–8 lin. long, slender; bracts many, leafy, 1–4 lin. long, lanceolate, acute or obtuse; pedicels $\frac{1}{2}$–$\frac{3}{4}$ in. long, filiform; sepals $\frac{2}{3}$–$\frac{3}{4}$ lin. long, ovate or ovate-lanceolate, acute, glabrous; corolla lobed nearly to the base; lobes 2–3 lin. long, $\frac{3}{4}$–1 lin. broad at the base, linear-lanceolate, obtuse, appearing glabrous in dried flowers, but really with a very minute puberulence on the inner face; corona-tubercles entirely adnate,

¼ lin. long, subquadrate, with a dorsal transverse rim at the basal part, acuminate or deltoid above ; style-apex slightly or very distinctly exceeding the anthers, very obtuse or produced into a short acute bifid cone ; follicles about 2 in. long, ¼–⅓ in. thick, attenuate from a lanceolate base into a long slender acute beak, smooth, glabrous ; seeds about 3½ lin. long and 2 lin. broad, ovate, plano-convex, wing-margined, very minutely asperate-tuberculate on both sides, with a keel down the flat face, brown. *Schlechter in Engl. Jahrb.* xviii. *Beibl.* 45, 11 ; xxi. *Beibl.* 54, 12, *and Journ. Bot.* 1898, 486. *Cynoctonum lycioides, E. Meyer, Comm.* 217. *C. lycioides var. majus, Meisn. in Hook. Lond. Journ. Bot.* ii. 1843, 546 (*by error* 446) ; *Krauss in Flora,* 1844, 827. *Cynanchum lycioides, Dietr. Syn. Pl.* ii. 906 ; *Steud. Nom. Bot. ed.* 2, i. 462. *Vincetoxicum lyciodes, Kuntze, Rev. Gen. Pl.* ii. 424.

COAST REGION : Uitenhage Div. ; near Enon, *Drège,* 2235 ! *Prior* ! King Williamstown Div. ; near the Keiskamma River, *Cooper,* 340 ! East London Div. ; East London Park, *Galpin,* 3182 ! Komgha Div. ; near Keimouth, *Flanagan,* 441 !

EASTERN REGION : Pondoland ; by the River at Port St. John, *Galpin,* 3433 ! Natal ; by the Tugela River, *Gerrard and McKen,* 1800 ! *Rehmann,* 7166 ! near Durban, *Wood,* 7517 ! by the coast, *Wood,* 1207 ! and without precise locality, *Sanderson,* 445 ! 708 !

The pubescence on the corolla-lobes can rarely be seen on dried specimens, except by turning back the lobe in water and viewing the edge against the light.

XXVIII. EMPLECTANTHUS, N. E. Br.

Calyx 5-partite. *Corolla* deeply 5-lobed, the united part forming a broad shallow cupular tube ; lobes valvate in bud. *Corona* double, arising from the staminal column ; outer 5-lobed ; lobes small, pouch-like, spreading, alternating with the anthers ; inner of 5 simple lobes, more or less adnate to the stamens, with free tips, and dorsally connected with the base of the outer corona-lobes. *Stamens* united into a tube around the ovary and adnate to the enlarged part of the style ; anthers erect or incurved-ascending, applied to the sides of or sub-incumbent upon the small truncate style-apex, without a membranous appendage. *Pollen-masses* minute, subquadrate with a pellucid margin on one side, erect, solitary in each anther-cell, attached in pairs to the exceedingly minute pollen-carriers by extremely short caudicles. *Follicles* only seen in an immature state, long and slender, linear-terete, perhaps slightly constricted between the seeds when ripe.

Herbaceous twiners with the habit and general appearance of *Tylophora* ; leaves opposite, long-petioled, cordate ; peduncles lateral at the nodes, bearing 1 or more fascicles of flowers.

DISTRIB. Species 2, endemic.

This curious genus is exceedingly like *Tylophora* in general appearance and the species would probably be referred to it unless examined, but the corona is quite different from that of any *Tylophora*, and much resembles that of some species of *Brachystelma* and *Caralluma* ; the pollen-masses are also pellucid-margined as in those genera, whilst the follicles resemble those of the genus *Riocreuxia*, and it may possibly be of hybrid origin between that genus and *Tylophora*. The name is derived from ἔμπλεκτος, intermixed or perplexed, and ἄνθος, a flower, in allusion to the combination of the characters of different genera upon one plant.

Corolla-lobes ¼ in. long; inner corona-lobes produced
much above the anthers ... ... (1) **Gerrardi.**

Corolla-lobes ⅛ in. long ; inner corona-lobes not or very
slightly exceeding the anthers ... ... (2) **cordatus.**

1. **E. Gerrardi** (N. E. Br.) ; stem twining, unifariously puberulous ; leaves herbaceous, only 2 seen; petiole 1½ in. long ; blade 2¾ in. long, 1¾ in. broad, cordate, acuminate, with a deep broad-based sinus and rounded basal lobes, thinly and minutely adpressed-pubescent above, glabrous beneath ; peduncles lateral at the nodes, very short, 2–4 lin. long, 4- to very many- flowered, puberulous on one side ; pedicels 7–9 lin. long, very slender, glabrous ; sepals about 1⅓ lin. long, lanceolate, attenuate, glabrous ; corolla-tube shallowly cupular, about 1 lin. deep and 2 lin. in diam., faintly contracted at the top, glabrous inside and outside ; lobes apparently ascending-spreading, ¼ in. long, 1½ lin. broad at the base, lanceolate, acute, glabrous on both sides, apparently dull purple ; outer corona-lobes fleshy, ¼ lin. long, spreading, subquadrate-ovate, deeply concave, with a small bifid reflexed apex, in front of which, on the inner face, is a small channelled tubercle-like keel, dark purple-brown ; inner corona-lobes ½ lin. long, produced much beyond the anthers, linear-spathulate, or linear-lanceolate, obtuse, connivent-erect, sometimes only adnate up to the base of the anthers, at others up to their tips, dorsally connected at the base to the basal margins of the outer corona-lobes, dark purple-brown ; staminal column ½ lin. long ; anthers overtopping the small truncate style-apex.

EASTERN REGION : Zululand ; Qudeni, *Gerrard*, 2167 !

2. **E. cordatus** (N.E. Br.) ; stem twining, slender, unifariously puberulous ; leaves herbaceous ; petiole 1–3 in. long ; blade 1½–3¾ in. long, ⅞–2 in. broad, ovate, acuminate, deeply cordate, with a broad-based open sinus and rounded lobes at the base, sprinkled with very minute hairs above, glabrous and paler beneath ; peduncles ¾–2 in. long, lateral at the nodes, glabrous, or puberulous on one side, bearing 1 or 2 distant fascicles of 4–6 flowers ; pedicels very slender, ⅚–1 in. (or in fruit 1¼ in.) long, glabrous ; sepals ¾ lin. long, lanceolate, acuminate, glabrous ; corolla-tube cupular, about ¾ lin. deep and 1¾ lin. in diam., slightly contracted at the mouth, where it is thinly fringed with long very fine hairs, which are crumpled up in dried flowers and invisible when wetted ; lobes about 1½ lin. long, 1 lin. broad at the base, somewhat deltoid-ovate, acute,

glabrous on both sides; outer corona-lobes about $\frac{1}{6}$ lin. long, sub-quadrate, deeply concave, emarginate or bifid, fleshy, dark purple-brown; inner corona-lobes about $\frac{1}{4}$ lin. long, oblong, obtuse, adnate at the basal half to the stamens and dorsally to the basal margins of the outer corona-lobes, free above, incurved-ascending and closely applied to the backs of the anthers, but not or very slightly produced beyond them, purple-brown; young follicles linear-terete, hooked at the apex, glabrous, in the examples seen $2\frac{3}{4}$ in. long, $\frac{3}{4}$ lin. thick.

EASTERN REGION: Natal; Tugela, *Gerrard,* 1803!

XXIX. SPHÆROCODON, Benth.

Calyx 5-partite. *Corolla* with a campanulate tube and spreading lobes, or rotate; lobes overlapping and slightly twisted in bud. *Corona* of 5 small fleshy tubercles arising from the staminal column some distance above its base. *Staminal column* arising from the base of the corolla; anthers erect, obtuse, without appendages. *Pollen-masses* erect, solitary in each anther-cell, not very minute, united in pairs by very slender caudicles to the minute and rather thin pollen-carrier. *Style* shorter than the anthers, truncate, and shortly 5-rayed at the apex. *Fruit* unknown.

Perennial herbs, with a woody or tuberous rootstock; stems erect, often flexuose, but not twining; leaves opposite; flowers in pedunculate sublateral umbel-like cymes.

DISTRIB. Species 2, both in Tropical Africa, of which only 1 extends into South Africa.

1. S. obtusifolium (Benth. in Hook. Ic. Pl. xii. 78, t. 1190); stems one to several to a root, erect, 1–3 ft. high, simple or slightly branching at the base only, softly and shortly tomentose; leaves herbaceous; petiole 1–3 lin. long; blade $1\frac{1}{4}$–3 in. long, $\frac{1}{3}$–$1\frac{3}{4}$ in. broad, oblong to elliptic, subacute to very obtuse and apiculate, cuneate to rounded at the base, subglabrous or thinly puberulous above, softly and very shortly pubescent beneath; cymes umbel-like, 6–13-flowered; peduncles $\frac{1}{4}$–$\frac{3}{4}$ in. long, pubescent or tomentose; pedicels $\frac{1}{4}$–$\frac{1}{3}$ in. long, pubescent or tomentose; sepals 1–2 lin. long, linear-lanceolate, pubescent; corolla when dried blackish-purple; tube 2–$2\frac{1}{2}$ lin. long and about 3 lin. in diam., broadly campanulate; lobes spreading, $1\frac{3}{4}$–2 lin. long and as much in breadth at the base, ovate, obtuse; glabrous outside, loosely clothed within the tube and on the inner face of the lobes with very fine jointed (cobwebby?) hairs, not visible when wetted; corona-tubercles arising about $\frac{1}{2}$ lin. up the staminal column, radiating, compressed, obtuse. *Hiern in Cat. Afr. Pl. Welw.* i. 692; *N.E. Br. in Dyer, Fl. Trop. Afr.* iv. i. 412; *Bot. Mag. t.* 7925. *S. obtusifolia, K. Schum. in Engl. Pfl. Ost-Afr. C.*

326, *and in Engl. and Prantl, Pflanzenfam.* iv. ii. 283, *fig.* 85, J–L, and 285. *S. natalense, Benth in Hook. Ic. Pl.* xii. 79. *S. natalensis, K. Schum. in Engl. and Prantl, Pflanzenfam.* iv. ii. 285. *S. caffrum, Schlechter in Journ. Bot.* 1895, 339. *Tylophora caffra, Meisn. in Hook. London Journ. Bot.* ii. 1843, 542 (*by error* 442); *Decne in DC. Prodr.* viii. 612. *Vincetoxicum caffrum, Kuntze, Rev. Gen. Pl.* ii. 424. *Gongronema Welwitschii, K. Schum. in Engl. Jahrb.* xvii. 145.

KALAHARI REGION : Transvaal ; near Barberton, *Thorncroft* ! near Rustenburg, 4000–4500 ft., *Miss Pegler,* 1069 ! *Miss Nation,* 277 !

EASTERN REGION : Natal ; near Durban, *Krauss,* 85 ! and without precise locality, *Gerrard,* 1797 ! Swaziland ; between Mbabane (Embabaan) and Bremersdorp, 2500 ft., *Bolus,* 12144 !

Also in Tropical Africa.

XXX. MARSDENIA, R. Br.

Calyx 5-partite. *Corolla-tube* campanulate; lobes 5, erect, spreading or rotate, overlapping and straight, or slightly twisted in bud. *Corona* of 5 fleshy lobes arising from and adnate to the staminal column, with free tips and often with free margins, sometimes with tubercle-like projections at the base, which are sometimes confluent, producing more or less the appearance of an outer corona ; tips erect or somewhat connivent, applied to the backs of the anthers. *Staminal column* arising from or near the base of the corolla ; anthers erect, with the cells applied to the sides of the conical apex or more or less concealed under the margin of the dilated part of the style, terminated by membranous appendages, which are free or connate and more or less incumbent on the top of the style-apex or applied to the sides of its conical tip or beak. *Pollen-masses* erect, solitary in each anther-cell, attached in pairs to the pollen-carriers by short or elongated, moderately stout caudicles. *Style* depressed, convex, conical or produced into a long beak at the apex. *Follicles* with a thick or coriaceous pericarp, smooth, often winged. *Seeds* crowned with a tuft of hairs.

Climbing or erect perennials ; leaves opposite ; flowers small or of moderate size, arranged in umbel-like cymes or in small sessile umbels or clusters scattered along the branches of the cymes or panicles, which are lateral at the nodes or axillary.

DISTRIB. A large genus, widely distributed throughout the tropical and subtropical regions, only one species in South Africa.

1. **M. floribunda** (N. E. Br. in Dyer, Fl. Trop. Afr. iv. i. 422 in a note); stem climbing, woody, branched, rusty-puberulous at the young tips ; leaves herbaceous ; petiole $\frac{1}{4}$–$\frac{1}{2}$ in. long ; blade $\frac{2}{3}$–$2\frac{1}{2}$ in. long, $\frac{1}{2}$–$1\frac{1}{2}$ in. broad, ovate or ovate-lanceolate, obtuse, acute or acuminate, glabrous on both sides ; cymes umbel-like, densely many-flowered, $\frac{1}{2}$–1 in. in diam., lateral at the nodes ; peduncles

¼–⅔ in. long, adpressed-puberulous ; bracts 1½–4 lin. long, ½–1 lin. broad, lanceolate, obtuse, very deciduous ; pedicels 1¼–3 lin. long, adpressed-puberulous ; sepals ¾–1¾ lin. long, ½–1¼ lin. broad, elliptic, ovate, oblong or lanceolate, obtuse or acute, glabrous ; corolla quite glabrous, white ; tube ⅔–1¼ lin. long, campanulate ; lobes horizontally spreading, 1–1⅓ lin. long, ¾–1 lin. broad, oblong, very obtuse ; corona-lobes very small, inconspicuous, ⅓ lin. long, not exceeding the anther-cells, ovate, obtuse, erect and closely applied to the backs of the anthers, only adnate to the staminal column by the slightly impressed centre of the basal part ; staminal column ⅔–1 lin. long ; anther-appendages rather large, ovate or ovate-lanceolate, acute, connivent-erect and applied to the sides of and about equalling the stout conical acute style-apex ; follicles widely divergent or sometimes solitary, 2½–3 in. long, lanceolate, gradually tapering to an acute point, with 4 broad toothed wings and including the wings ¾–1 in. in diam. at the base, smooth, glabrous ; seeds ½ in. long, ¼ in. broad, ovate, flattened, narrowly margined, smooth, glabrous, pale brown, with the margin ochreous-brown. *Dregea floribunda, E. Meyer, Comm.* 199 ; *Decne in DC. Prodr.* viii. 618 ; *Harv. Gen. S. Afr. Pl. ed.* 2, 239 ; *Schlechter in Engl. Jahrb.* xviii. *Beibl.* 45, 12 *and* 23 ; xxi. *Beibl.* 54, 12 ; *Journ. Bot.* 1898, 486, *and Ann. Naturhist. Hofmus. Wien,* xviii. 398 ; *Wood and Evans, Natal Pl.* i. 69, *t.* 86, *as to description and figure of fruit only. Pterophora Dregea, Harv. Gen. S. Afr. Pl. ed.* i. 223.

COAST REGION: Uitenhage Div. ; Zuurberg Range, *Drège* ! *Cooper,* 2712 ! near Enon, *Drège* ! *Prior* ! between Port Elizabeth and Grahamstown, *Krook,* 2413 (ex *Schlechter*), Albany Div. ; in woods near Grahamstown, 2000 ft., *McOwan,* 697 ! near the Fish River, 2000 ft., *Schlechter,* 6108 ! Fort Beaufort Div. ; woods by the Kat and Blinkwater Rivers, *Mrs. Barber,* 79 ! Stockenstrom Div. ; "Ceded Territory," *Zeyher* ! Komgha Div. ; near the Kei River, 2000 ft., *Schlechter,* 6239 ! *Krook,* 2412 (ex *Schlechter*), British Kaffraria, *Cooper,* 104 !

CENTRAL REGION : Cradock Div. ; near the Tarka River, *Cooper,* 356 !

EASTERN REGION: Transkei ; Fort Bowker, *Bowker,* 607 ! near Kentani, *Miss Pegler,* 1195 ! Natal ; Coast, *Wood,* 1119 ! Palmiet, near Durban, 200 ft., *Wood,* 7384 ! and without precise locality, *Gerrard,* 12 ! 131 !

As stated in the Flora of Tropical Africa, I can find no character whereby to distinguish *Dregea* from *Marsdenia.* According to Mrs. Barber, *M. floribunda* climbs to the top of high forest trees and has greenish-yellow flowers ; Gerrard and Wood state on their labels that the flowers are white, and according to MacOwan they are "fetid, like mice." The figure in Wood and Evans, Natal Plants, represents a flowering branch and analytical details of *Gymnema sylvestre* and the fruit of *Marsdenia floribunda.*

XXXI. TELOSMA, Coville.

(PRAGELURIA, N. E. Br.)

Calyx 5-partite. *Corolla-tube* elongated, inflated at the base ; lobes 5, horizontally spreading, overlapping and straight or twisted in bud. *Corona* of 5 erect lobes arising from the base of the staminal column and adnate to it at their lower part, free above, with a narrow subulate process on their inner face. *Staminal column* arising from the base of the corolla; anthers oblong, erect, with long erect membranous appendages, connivent over the style-apex. *Pollen-masses* erect, solitary in each anther-cell, attached in pairs to the pollen-carriers by very short caudicles. *Style* with a stout penta-gonal-ovoid apical part, not exceeding the anther-appendages. *Follicles* smooth. *Seeds* crowned with a tuft of hairs.

Stem twining; leaves opposite; flowers of moderate size, numerous, in pedun-culate or subsessile umbel-like cymes; subaxillary or lateral at the nodes. The name is derived from τηλε, far, and οσμη, odour, in allusion to the great distance at which the aromatic odour of one of the species is perceptible; Mr. Coville stating that he "could always tell at a distance of two blocks (over 200 ft.) whether or not there was a bouquet of mil-leguas (the native name of *Telosma odoratissma*, Coville) in my house."

The name *Prageluria*, N. E. Br., entered in the key to the genera on p. 523 and provisionally indicated in the *Kew Bulletin*, 1907, 325, which I intended to apply to this genus, hitherto (as detailed on p. 757) mistaken for *Pergularia* (Linn.), was given in ignorance that Mr. Coville had in 1905 (*Contributions from the United States Herbarium*, ix. 384) already applied to it the name *Telosma*. But as that name merely appears in an alphabetical list, and the only indication that it is a new generic name is mixed up with the remarks upon the species, it had been entirely overlooked.

DISTRIB. Species several, mostly Indian and Malayan; the following is the only African species.

1. **T. africana** (N. E. Br.); stem twining, glabrous; leaves herbaceous, thin; petiole $\frac{1}{3}$–3 in. long; blade $1\frac{3}{4}$–$5\frac{1}{2}$ in. long, 1–$3\frac{1}{2}$ in. broad, oblong-ovate to broadly ovate, shortly cuspidate into an acute or obtuse point, rounded, subtruncate, cordate or more rarely cuneate-acute at the base, glabrous on both sides, or sparsely and minutely puberulous above; umbels lateral at the nodes, rather densely several- or many-flowered, subglobose; peduncles 0–5 lin. long, minutely adpressed-puberulous; pedicels 2–3 lin. long, glabrous or minutely puberulous; sepals $1\frac{1}{4}$–2 lin. long, $\frac{2}{3}$–1 lin. broad, lanceolate or ovate, glabrous, minutely ciliate; corolla green (*Wood*), yellow (*Barter*); tube usually about $\frac{1}{4}$ (more rarely $\frac{1}{3}$) in. long, globose-inflated and apparently somewhat plicate, constricted at the mouth or upper part, glabrous outside, densely hairy at the throat and upper part, with 5 lines of hairs alternating with the corona-lobes on the lower part; lobes $3\frac{1}{2}$–6 lin. long, $\frac{3}{4}$–$1\frac{1}{4}$ lin. broad at the base, linear-attenuate, obtuse, horizontally spreading, re-curved or revolute, often twisted, densely bearded at the base and more shortly along the rest of the inner face; corona-lobes erect, 1–$1\frac{3}{4}$ lin. long, elliptic, elliptic-lanceolate or more or less obovate, rounded, subtruncate or subacute at the slightly recurved apex, with a linear or attenuate appendage about 1 lin. long arising at about the middle of their inner face and connivent-erect over the $1\frac{1}{2}$ lin.-long staminal column; anther-appendages $\frac{3}{4}$–1 lin. long, ovate-lanceolate, acute or subobtuse, connivent over the ovoid obtuse style-apex; follicles not seen. *Pergularia africana, N. E. Br. in Kew*

Bulletin, 1895, 259, *and in Dyer, Fl. Trop. Afr.* iv. i. 426 ; *Schlechter in Journ. Bot.* 1898, 486.   *P. sanguinolenta, Britten in Trans. Linn. Soc. ser.* 2, *Bot.* iv. 29, *and K. Schum. in Engl. Pfl. Ost-Afr. C.* 326, *not of Lindley.*

Eastern Region : Natal ; Berea, near Durban, *McKen,* 2 ! 1996 ! *Wood,* 5147 ! 6591 ! Tugela Valley, *Gerrard,* 1804 ! in bush adjoining the Botanic Gardens, Durban, *Wood,* 3395 !

Also in Tropical Africa.

XXXII. FOCKEA, Endl.

Calyx 5-partite. *Corolla-tube* very short ; lobes oblong, linear-lanceolate or linear, valvate or overlapping and usually twisted in bud. *Corona* arising near the mouth or towards the base of the tube at the insertion of the staminal column and longer than the former, tubular, toothed at the top, and with 1–2 superposed series of 5 teeth or filiform processes placed between or decurrent as 5 pairs of wings within the tube. *Staminal column* inserted near the mouth or towards the base of the corolla-tube ; anthers erect, triangular, their appendages very large, erect, membranous, inflated. *Pollen-masses* erect, solitary in each anther-cell, very flat and thin, sessile, in pairs at the apex of the minute pollen-carriers. *Style* produced into a short thick point to about the level of the top of the anther-cells. *Follicles* not echinate. *Seeds* crowned with a tuft of hairs.

Perennials with large tuberous rootstock and twining or erect stems; leaves opposite ; flowers in small subaxillary clusters or cymes.

Distrib. Species 11, the others in Tropical Africa.

Leaves linear or linear-lanceolate, 4–20 times as long as
   broad :
  Stems or branches 2–6 in. long ; leaves truncate at the
    base, hooked at the apex, with remarkably sinuate-
    undulate margins ... ... (1) **undulata.**
  Stems or branches 6–18 in. long ; leaves narrowed or
    rounded at the base, not hooked at the apex nor
    undulate at the margins :
    Corolla-lobes 4 lin. long ; corona about 2½ times as
     long as the corolla-tube ... ... (2) **tugelensis.**
    Corolla-lobes 6–8 lin. long ; corona about 4 times as
     long as the corolla-tube ... ... (3) **angustifolia.**
Leaves oblanceolate, oblong-lanceolate, oblong, ovate or
   elliptic, 2–4 times as long as broad :
  Leaves glabrous on both sides ... ... (4) **glabra.**
  Leaves shortly pubescent on both sides ... ... (5) **capensis.**

1. **F. undulata** (N. E. Br. in Kew Bulletin, 1895, 260) ; tuber large, with an elongated woody neck, producing erect pale brown puberulous branches 2–6 in. long ; leaves sessile, ½–1¼ in. long, ⅔–1⅓ lin. broad, linear, incurved-hooked at the subacute apex, trun-

cate or subcordate at the base, remarkably undulate-sinuate along
the revolute margins, minutely puberulous above, glabrous except
on the midrib beneath ; cymes umbel-like, sessile at the nodes, 3–6-
flowered, minutely adpressed-whitish-puberulous on all parts to the
outside of the corolla ; pedicels $\frac{1}{2}$–1 lin. long ; sepals 1 lin. long,
lanceolate, acute ; corolla-tube 1 lin. long ; lobes erect, 2–2$\frac{1}{4}$ lin.
long, $\frac{3}{4}$ lin. broad, linear-lanceolate or oblong-linear, acute or obtuse,
glabrous on the inner face, green ; corona white, with a tube 1$\frac{1}{4}$ lin.
long, divided at the top into 5 trifid lobes alternating with 3 teeth
or with 5 smaller trifid lobes, the middle tooth of the 5 principal
lobes 1$\frac{1}{4}$–1$\frac{1}{2}$ lin. long, filiform, tortuous, all the other teeth $\frac{1}{3}$–$\frac{2}{3}$ lin.
long ; within the tube the 5 longest teeth are decurrent as stout
keels bearing above the middle of the tube an erect filiform process
arising to the level of the shorter teeth, and the teeth on each side
are decurrent as narrow wings ; anther-appendages nearly equalling
the corona-tube. *Brachystelma sinuatum, E. Meyer, Comm.* 196 ;
Dietr. Syn. Pl. ii. 887.

CENTRAL REGION: Beaufort West Div. (not Transvaal, as originally stated);
Rhenoster Kop, *Burke* !

The type specimen of *Brachystelma sinuatum* in E. Meyer's Herbarium, although
flowerless, is identical with *F. undulata*, and is labelled as having been collected
"between Kat and Zwart River" in Stockenstrom or Queenstown Div., whilst
E. Meyer, and also Drège, state that it was collected on hills near Dweka (Dwyka
River) in Prince Albert Div., and near Brak Vallei in Richmond Div.

2. **F. tugelensis** (N. E. Br.) ; in habit, stems, inflorescence and
pubescence not distinguishable from *F. angustifolia*, but differing as
follows :—leaves 1–2 in. long, 2–4 lin. broad at the base, narrowing
upwards, linear-lanceolate, acute or subobtuse, rounded at the base ;
sepals about 1 lin. long, narrowly lanceolate, acute ; corolla-tube
$\frac{2}{3}$ lin. long ; lobes $\frac{1}{3}$ in. long, $\frac{2}{3}$ lin. broad, linear, with recurved
margins, twisted, puberulous on the back and with a microscopic
puberulence on the inner surface, apparently greenish ; corona
about 2$\frac{1}{2}$ times as long as the corolla-tube, white, about 1$\frac{1}{2}$ lin. long,
tubular for rather more than half its length, usually with 20 or
sometimes 15 teeth at the top, 5 of which are about 3 times as long
as the others, filiform, decurrent as stout keels within the tube, and
bearing at about the middle of the tube 5 erect linear obtuse teeth or
processes, which rise to the level of the short teeth, 10 of which are
decurrent as narrow wings or keels within the tube ; anther-
appendages reaching to about $\frac{3}{4}$ of the way up the corona-tube.

EASTERN REGION: Natal ; Tugela, *Gerrard*, 1310 !

3. **F. angustifolia** (K. Schum. in Engl. Jahrb. xvii. 146) ; rootstock a large tuber ; stems several, clustered, usually simple, erect,
tortuous or sometimes twining, $\frac{1}{2}$–1$\frac{1}{2}$ ft. long, puberulous with
minute curved hairs ; leaves subsessile or very shortly petiolate, $\frac{2}{3}$–
4$\frac{1}{2}$ in. long, 1–3 lin. broad, linear, acute, narrowed or rounded into

the petiole at the base, quite glabrous or thinly puberulous, with minute curved hairs above and on the midrib beneath; flowers 2–6 together in sessile clusters at the nodes; pedicels $\frac{1}{4}$–$\frac{2}{3}$ lin. long, whitish-puberulous; sepals $\frac{2}{3}$–$1\frac{1}{2}$ lin. long, subulate to narrowly lanceolate, acute, whitish-puberulous; corolla-tube $\frac{1}{2}$–$\frac{3}{4}$ lin. long; lobes very spreading, $\frac{1}{2}$–$\frac{2}{3}$ in. long, $\frac{2}{3}$–$\frac{3}{4}$ lin. broad at the base, linear or linear-filiform from being revolute at the margins and somewhat twisted, whitish-puberulous or minutely tomentose on the back, glabrous on the inner face, apparently greenish; corona about 4 times as long as the corolla-tube, white, $2\frac{1}{2}$–3 lin. long, tubular for half its length, variably divided above, sometimes into 5 long trifid segments alternating with 5 short teeth, sometimes into many irregular teeth about $\frac{1}{2}$ lin. long, with 5 long simple linear-filiform or trifid processes arising within the tube at or near its top, and in front of these or of the trifid segments arise below the middle or near the base of the tube 5 other filiform processes, usually hooked or revolute at the apex and varying from half as long to as long as the tube; lateral teeth of the trifid segments or processes one quarter as long as the linear-filiform middle tooth; anther-appendages reaching to about the middle of the corona-tube. *K. Schum. in Engl. and Prantl, Pflanzenfam.* iv. ii, 296; *Schlechter in Journ. Bot.* 1898, 487. *Brachystelma circinatum, Marloth in Engl. Jahrb.* x. 244, *not of E. Meyer.*

SOUTH AFRICA; without locality, *Zeyher*, 1138!

KALAHARI REGION: Griqualand West; Groot Boetsap, *Marloth*, 1008. Orange River Colony; Rhenoster Kop, *Burke*, 510! *Zeyher*, 510! between Rhenoster River and Vaal River, *Zeyher*, 1135! Transvaal; hills near Potgeiters Rust, *Bolus*, 11014!

A specimen collected by *Burchell* (2465) near the sources of the Kuruman River in Bechuanaland, may belong to this species, but the flowers are only in very young bud. On a note with the specimen, Burchell states that it is called "*Gamroon* or *Gamrun*" by the Hottentots, and "has a root as large and of the shape of a large turnip, eatable, white, soft, spongy, sweet and watery (as a water-melon), and a fortunate resource for a thirsty traveller. The herb is milky, but the root is not so. It is usually eaten raw."

4. **F. glabra** (Decne in DC. Prodr. viii. 545); rootstock partly above ground, very large, up to 2 ft. in diam. (*Burchell*), more or less tuberculate, light brown; stems up to 2 ft. or more long, twining, procumbent or straggling, often branched, minutely puberulous on the younger parts; leaves often with leaf-tufts or very short leafy branchlets in their axils, usually spreading, sometimes deflexed; petiole $\frac{1}{2}$–2 lin. long; blade $\frac{1}{2}$–$1\frac{1}{2}$ in. long, $\frac{1}{6}$–$\frac{1}{2}$ in. broad, elliptic, lanceolate, oblong, oblanceolate or more or less obovate, acute, or obtuse or rounded and shortly apiculate at the apex, cuneate or rounded at the base, often wavy at the margins, glabrous on both sides, subcoriaceous; cymes subsessile or on puberulous peduncles up to $\frac{1}{8}$ in. long, 2–6-flowered, lateral at the nodes; bracts minute; pedicels $1\frac{1}{4}$–2 lin. long, minutely puberulous, as are the $\frac{3}{4}$–1 lin.-long lanceolate acute sepals; corolla puberulous

outside and within, dull light yellow; tube 1–1¼ lin. long, campanulate; lobes recurved-spreading, about ⅓ in. long, ¾ lin. broad at the base, linear-attenuate, obliquely subtruncate at the apex, with recurved margins and more or less twisted; corona arising near the base of the corolla, 2½–3 lin. long, white, the lower half tubular and longer than the corolla-tube, divided above into 5 trifid segments alternating with 5 minute entire or bifid lobules or 5 pairs of minute teeth, with the middle tooth of the trifid segments filiform or linear-subulate, shorter than or about equalling the tube and 3–4 times as long as the lateral teeth, which are decurrent as keels within the tube; between the keels at the middle of the tube arise 5 filiform teeth or processes connivent-erect over the white anther-appendages, which reach to the top of the corona-tube. *F. edulis, K. Schum. in Engl. Jahrb.* xvii. 146, *in note, and in Engl. and Prantl, Pflanzenfam.* iv. ii. 296; *Schlechter in Journ. Bot.* 1898, 487, *partly. Pergularia edulis, Thunb. Prodr.* 38; *in Nov. Act. Acad. Petrop.* xiv. (1805) 519; *Fl. Cap. ed.* 2, ii. 151 (*edalis*), *and ed. Schultes,* 233; *Willd. Sp. Pl.* i. 1247; *Pers. Syn. Pl.* i. 271; *Schultes, Syst. Veg.* vi. 56; *G. Don, Gen. Syst.* iv. 133. *Chymocormus edulis, Harv. in Hook. Lond. Journ. Bot.* i. 1842, 24.

South Africa: without locality, *Zeyher,* 239 ex *Decaisne,* and *cultivated specimens*!
Coast Region: Mossel Bay Div.; Gouds (Gauritz) River, *herb. Thunberg*! Uitenhage Div.; near Uitenhage, *Burchell,* 4443! 4457! near the Zwartkops Salt-pan and River, *Thunberg*! *Zeyher,* 965! *Ecklon and Zeyher,* 10!

Partly described from a living plant cultivated at Kew. The large tuberous rootstock is eaten by the natives, and, according to Zeyher, has a milky, somewhat sweetish juice.

5. **F. capensis** (Endl. Iconogr. Gen. Pl. t. 91, and Nov. Stirp. Dec. 3, 17); rootstock very large; stems 1–2 ft. long, tortuous or twining, puberulous or very shortly and softly pubescent; leaves very shortly petiolate, deflexed, ⅓–⅞ in. long, ¼–½ in. broad, elliptic-oblong, obtuse, apiculate, rounded at the base, slightly folded lengthwise, undulate, shortly pubescent on both sides; flowers 3–5 together, lateral at the nodes; pedicels 1–2 lin. long, minutely tomentose, as are the ¾–1 lin.-long lanceolate acute sepals; corolla puberulous outside and on the inner face; tube about 1⅓ lin. (scarcely 2 lin., *Endlicher*) long, campanulate; lobes very spreading, 3–3½ lin. long, linear-lanceolate, more or less twisted, truncate at the apex, with revolute margins, green; corona arising near the base of the corolla-tube, white, about 2 lin. long, the lower half tubular and about as long as the corolla-tube, divided above into 5 trifid segments alternating with 5 small recurving entire or bifid teeth, middle tooth of the trifid segments filiform, as long as the tube and 3–4 times as long as the lateral teeth, which are decurrent within the tube as 5 pairs of wing-like keels, between which, at the middle of the tube, arise 5 other filiform teeth in one series opposite the middle teeth and reaching to the level of the lateral teeth of

the trifid segments; anther-appendages as long as or slightly exceeding the coronal tube. *Decne in DC. Prodr.* viii. 545; *Wittmack in Gartenfl.* xlix. 344. *F. crispa, K. Schum. in Engl. and Prantl, Pflanzenfam.* iv. ii. 296. *F. edulis, Schlechter in Journ. Bot.* 1898, 487, *partly. Cynanchum crispum, Jacq. Fragm.* 31, *t.* 34, *fig.* 5.

SOUTH AFRICA: without locality, *cultivated specimen*!

I am indebted to the courtesy of Dr. A. Zahlbruckner, Head of the Botanic Department of the Vienna Hofmuseum, for the opportunity of examining the type specimen of this very rare plant, of which only a single individual is known. This was introduced from South Africa in the latter part of the 18th century and has been in cultivation in the Imperial Garden at Schönbrunn for over 100 years and was exhibited alive at the International Botanic Congress at Vienna in 1905. Dr. Zahlbruckner states that it has never produced seeds, and all efforts to propagate it have hitherto failed; no botanical collector seems to have refound it, so that the locality whence it came is unknown. Possibly, however, it may be only an individual variation of *F. glabra*, since with the exception of the relative proportions in size of the coronal parts, I do not find any great difference in floral structure; the appearance and pubescence of the plant are, however, decidedly different, and there does not seem to be the same tendency to produce short leafy branches in the axils of its leaves as there is in *F. glabra*. I have therefore maintained them as distinct species, leaving future discoveries to prove or disprove the correctness of this view.

Imperfectly known species.

6. **F. sessiliflora** (Schlechter in Engl. Jahrb. xx. Beibl. 51, 44); stem woody, glabrous; leaves erect or erectly spreading, $\frac{3}{4}$–$1\frac{2}{3}$ in. long, 4–7 lin. broad at the middle, lanceolate or lanceolate-oblong, acute, tapering into a very short petiole at the base, often revolute or undulate on the margins, glabrous; flowers few, sessile, clustered in the axils of the leaves (? lateral at the nodes); sepals deltoid-lanceolate, much shorter than the corolla, thinly puberulous, with mealy cushions ("*farinoso-pulvinatis*"); corolla-lobes erectly spreading, flexuose, scarcely $3\frac{1}{2}$ lin. long, linear-ligulate, somewhat obtuse, thinly puberulous on both sides; corona tubular, with 5 short trifid lobes at the top, having the lateral teeth a little shorter than the middle one; within the tube arise 10 erect filiform processes in front of each other, in two series; the upper series much exceeding the tube, with their margins decurrent as 5 pairs of parallel keels; the lower series half as long, scarcely equalling the tube. *N. E. Br. in Dyer, Fl. Trop. Afr.* iv. i. 429.

KALAHARI REGION: Transvaal; near Klipdam, 4500 ft., *Schlechter*, 4493.

I have not seen this species. In the British Museum a specimen of *F. tugelensis* has been named *F. sessiliflora* by Dr. Schlechter, whilst in the *Journal of Botany*, 1898, 487, he states that it is the same as *F. undulata*, but neither of these totally different species corresponds to his description of *F. sessiliflora*.

7. **F. Comaru** (N. E. Br.); tuber very large and fleshy, elongated, 3–6 in. thick, with a woody neck or permanent stem 2–3 lin. thick, from which the branches are produced; branches

$\frac{2}{3}$–1 lin. thick, apparently not more than 4–6 in. long, but broken in the type specimen, very minutely and densely puberulous with curved hairs, brown; leaves (broken) subsessile, $\frac{1}{2}$–$\frac{2}{3}$ lin. broad, linear, with revolute margins, minutely puberulous. *Brachystelma? Comaru, E. Meyer, Comm.* 195; *Dietr. Syn. Pl.* ii. 887; *Decne in DC. Prodr.* viii. 647.

CENTRAL REGION; Richmond Div.; vicinity of Styl Kloof, west of Richmond, *Drège*!

There are no flowers upon Drège's specimen, but its characteristics are so exactly those of a *Fockea* that I have no hesitation in placing it in this genus, and it may prove to be the same as *F. angustifolia,* which it very closely resembles, but the pubescence is much denser and the hairs composing it only about half as long as those on *F. angustifolia.* Drège's type, except in its more minute pubescence, is very similar to Burchell's 2465 noted under *F. angustifolia,* and the native name "*Comaru*" is not very unlike the native name "*Gamrun*" of that species.

XXXIII. **GYMNEMA**, R. Br.

Calyx 5-partite. *Corolla* 5-lobed to the middle or beyond; tube campanulate; lobes ascending or spreading, overlapping in bud. *Corona* arising from and adnate to the corolla-tube, either of 5 fleshy lobes adnate up to the mouth of the corolla-tube, with more or less incurved tips, or of 5 pairs of fleshy ridges on the lower part of the tube, alternating with the corolla-lobes, densely ciliate (always?). *Staminal column* arising from the base of the corolla; anthers short, erect, with short membranous appendages. *Pollen-masses* erect, solitary in each anther-cell. *Style* often produced beyond the anthers. *Follicles* smooth. *Seeds* crowned with a tuft of hairs.

Twining shrubs; leaves opposite; umbels lateral or subaxillary, often 2 from the same node and opposite; flowers small.

DISTRIB. Species several, in the tropical and subtropical regions of the Old World; only one in South Africa.

1. **G. sylvestre** (R. Br. in Mem. Wern. Soc. i. 33); stem climbing, woody, $\frac{2}{3}$–$1\frac{1}{2}$ lin. thick, shortly tomentose or densely pubescent, as are also the petioles, peduncles, bracts, pedicels and calyx; leaves herbaceous; petiole 3–10 lin. long; blade $\frac{3}{4}$–3 in. long, $\frac{1}{3}$–$1\frac{1}{2}$ in. broad, ovate, elliptic, ovate-oblong or ovate-lanceolate, acute or shortly and rather abruptly acuminate, rounded, or shortly and broadly cuneate at the base, glabrous or (usually thinly) pubescent above, varying through all stages from glabrous to softly and densely pubescent beneath; umbels subaxillary, usually opposite, about $\frac{1}{2}$ in. in diam., many-flowered; peduncles 1–5 lin. long; bracts minute; pedicels 1–3 lin. long; sepals 1 lin. long, elliptic-oblong, very obtuse, ciliate; corolla 2 lin. in diam., glabrous; tube campanulate, $\frac{3}{4}$–1 lin. long; lobes spreading-recurved, about as long as the tube, oblong-ovate or lanceolate-oblong, obtuse, fleshy, glabrous

on both sides, yellow or yellowish-white ; corona-lobes adnate up to
the mouth of the corolla-tube, fleshy, linear-cuneate, slightly
channelled down the middle, with the obtuse or acute apex incurved,
densely ciliate on each side of the adnate part, the hairs not
noticeable when wet; staminal column $\frac{2}{3}$ lin. long, columnar, with
the very short transverse rounded anther-tips erect and applied to
the base of the stout conical or ovoid style-apex, which protrudes
about $\frac{1}{3}$ lin. beyond them and is exserted from the corolla-tube,
entire or slightly bifid at the tip ; follicles 2–3½ in. long, $\frac{1}{4}$–$\frac{1}{3}$ in.
thick, narrowly lanceolate, attenuate into a beak, glabrous; seeds
3½–5 lin. long, ovate, plano-convex, with a rather broad marginal
wing, glabrous, brown. *Dietr. Syn. Pl.* ii. 895 ; *DC. Prodr.* viii.
621 ; *A. Rich. Tent. Fl. Abyss.* ii. 43 ; *Hook. f. Fl. Brit. Ind.* iv.
29 ; *Wight, Ic. Pl.* ii. i. 3, *t.* 349 ; *Martelli, Fl. Bogos*, 55 ; *K. Schum.
in Engl. Pfl. Ost-Afr. C.* 325, *and in Engl. Jahrb.* xxx. 385 ; *Schlechter
in Engl. Jahrb.* xviii. *Beibl.* 45, 23, *and Journ. Bot.* 1898, 486.
N. E. Br. in Dyer, Fl. Trop. Afr. iv. i. 413.　*G. rufescens, Decne,
and G. subvolubile, Decne in Ann. Sc. Nat.* 2 *sér.* ix. 277, *t.* 11, *fig. A,
and in DC. Prodr.* viii. 621.　*G. subvolubile, Hook. Niger Fl.* 455 ;
De Wild. and Durand, Reliq. Dewevr. 159.　*G. humile, Decne in DC.
Prodr.* viii. 621 ; *A. Rich. Tent. Fl. Abyss.* ii. 42.　*G. M'Kenii,
Harv. Gen. S. Afr. Pl. ed.* 2, 239.　*G. fruticulosum, Hochst. in Flora*,
1844, 101.　*G. geminatum, Hiern in Cat. Afr. Pl. Welw.* i. 691, *not
of R. Br.　Periploca sylvestris, Retz. Obs.* ii. 15 ; *Willd. Phytogr.* 7,
t. 5, *fig.* 3, *and Sp. Pl.* i. 1252.　*Cynanchum subvolubile, Schumach.
and Thonn. Beskr. Guin. Pl.* 150.　*C. senegalense, Sieber ex Decne in
DC. Prodr.* viii. 621.　*Asclepias geminata, Roxb. Hort. Beng.* 20,
and Fl. Ind. ii. 45 (1832).　*Dregea floribunda, Wood and Evans,
Natal Pl.* i. 69, *t.* 86, *excluding description and figure of the fruit.*

KALAHARI REGION : Transvaal; Avoca, near Barberton, 1900 ft., *Galpin,*
1239 ! South African Gold Fields, *Baines* !

EASTERN REGION : Natal ; near the Umgeni River, *Gerrard,* 1314 (mixed with
Marsdenia floribunda)! in a damp Ravine near Sydenham, *McKen,* 3 ! Umzinyati
Falls, *Wood,* 1249 ! Palmiet, 150 ft., *Wood,* 7559 ! Delagoa Bay, *Forbes* !

Also in Tropical Africa, Madagascar, and the drier parts of India ; the
Australian plant united with it by Bentham is quite distinct.

The fresh leaves of this plant, when chewed, have the property of destroying
the taste of sweetness, as was first observed by M. P. Edgeworth, see *Proc. Linn.
Soc.* i. 353 ; D. Hooper in *Nature*, xxxv. 565, and L. E. Shore in *Journ. of
Physiology*, xiii. 191.

XXXIV. RHYSSOLOBIUM, E. Meyer.

Calyx 5-partite ; sepals lanceolate, acute, thick. *Corolla* small,
somewhat fleshy, campanulate, with 5 oblong slightly spreading
lobes, densely bearded within. *Corona* none. *Stamens* united into
a short column, inserted near the base of the corolla-tube ; the
filament part with prominent horny margins on the upper part ;

anthers erect, terminated by a membranous appendage. *Pollen-masses* erect, attached in pairs by short caudicles to the elongated erect pollen-carriers. *Follicles* small, evidently thick and fleshy, only the dried flattened-out remains seen, which are covered with very crowded and very thin longitudinal lamella-like ridges outside and with fewer and much coarser ridges inside, both of which may be due to shrinkage. *Seeds* not seen.

A small, much-branched shrublet with woody whitish-puberulous branches. Leaves opposite or at the base of the branchlets subfasciculate, small and thick, subterete from the revolute margins ; flowers 1–3 together, subsessile at the nodes, very small.

DISTRIB. Species 1, endemic.

1. **R. dumosum** (E. Meyer, Comm. 217) ; a small, much-branched shrublet less than a foot high ; branchlets woody, rigid, whitish-puberulous ; leaves opposite or subfasciculate, subsessile, $1\frac{1}{2}$–3 lin. long, $\frac{1}{2}$–$\frac{3}{4}$ lin. broad, linear, obtuse, very thick, from the very revolute margins, puberulous ; flowers 1–3 together, subsessile at the nodes ; sepals about 1 lin. long, $\frac{1}{2}$ lin. broad, lanceolate or ovate-lanceolate, acute, thick, keeled on the back, puberulous ; corolla campanulate, glabrous outside and on the lower part of the tube within, the throat of the tube and inner face of the lobes, except at the tips, densely bearded with long hairs ; tube 1 lin. long ; lobes 1–$1\frac{1}{4}$ lin. long, $\frac{2}{3}$ lin. broad, oblong, obtuse, slightly spreading ; staminal column $\frac{3}{4}$ lin. long, cylindric ; anther-appendages oblong, subacute, connivent-erect over the short obtusely conical style-apex ; follicles 1–$1\frac{1}{4}$ in. long, probably lanceolate - fusiform, thick and fleshy, glabrous, the dried remains densely covered with very thin longitudinal ridges not thicker than paper. *Harv. Gen. S. Afr. Pl. ed.* i. 227, *and ed.* 2, 239 ; *Decne in DC. Prodr.* viii. 626 ; *Schlechter in Journ. Bot.* 1896, 458. *Astephanus dumosus, Dietr. Syn. Pl.* ii. 909.

WESTERN REGION : Little Namaqualand ; on dry hills near Arris, towards the mouth of the Orange River, below 200 ft., *Drège,* 3049 !

XXXV. ORTHANTHERA, Wight.

Calyx 5-partite. *Corolla-tube* elongated, slightly or globosely inflated at the base, often with the inflation extending between the sepals into 5 blunt angles ; lobes 5, erect or spreading, valvate in bud, sometimes with minute pocket-like thickenings at the sinuses between them. *Corona* of 5 erect Λ-shaped lobes arising at or near the base of the staminal column and adnate to it between the pairs of anther-wings, nearly or quite as long as the latter, with free out-standing or reflexed wing-like (when dried) margins. *Staminal column* arising from the bottom of the corolla-tube and enclosed in the inflated part ; anthers connivent-erect, linear or oblong at the fertile part, with a subulate or narrow appendage or apiculus; anther-wings below the cells decurrent on the filament part, sub-parallel. *Pollen-masses* erect, with a pellucid area at the apex, solitary in each anther-cell, attached in pairs to the pollen-carriers by short slender caudicles at their base. *Style* equalling or shorter than the anthers, with a slender conical or needle-like terminal part. *Follicles* terete and acute or fusiform and acuminate, smooth. *Seeds* crowned with a tuft of hairs.

Herbaceous with prostrate leafy stems, or shrubby with erect parallel branches, leafless or with linear leaves; leaves opposite; flowers small or of moderate size, in pedunculate or subsessile umbels or clusters lateral at the nodes.

DISTRIB.—Species 3, one in India, and the 2 following also in Tropical Africa.

Stems prostrate, pubescent or scabrous; corolla-tube
$\frac{1}{2}$–$\frac{3}{4}$ in. long　...　...　...　...　...　... (1) **jasminiflora.**

Stems erect, glabrous, white-glaucous; corolla-tube
rather less than $\frac{1}{4}$ in. long　...　...　...　... (2) **albida.**

1. **O. jasminiflora** (N. E. Br. ex Schinz in Verhandl. Bot. Ver. Brandenb. xxx. 265); stems prostrate, $1\frac{1}{2}$–15 ft. long, branching, pubescent or somewhat scabrous; leaves spreading, scabrous; petiole 1–5 lin. long; blade $\frac{2}{3}$–$2\frac{1}{2}$ in. long, $\frac{1}{8}$–1 in. broad, linear, lanceolate, oblong, ovate to nearly circular, acute or obtuse and apiculate at the apex, rounded, truncate, cordate or subhastate at the base, rather thick and rigid, often wavy or crisped on the margins; umbels pedunculate, 2–13-flowered; peduncle $\frac{1}{3}$–$1\frac{1}{2}$ in. long, scabrous or pubescent; bracts 1–$2\frac{1}{2}$ lin. long, $\frac{1}{4}$–$\frac{2}{3}$ lin. broad, subulate or ovate-lanceolate, acute; pedicels 2–7 lin. long, scabrous or pubescent; sepals $1\frac{1}{2}$–3 lin. long, $\frac{2}{3}$–1 lin. broad, lanceolate, acute or acuminate, pubescent; corolla pubescent all over the outside and within the upper part of the tube, cream-coloured; tube usually $\frac{1}{2}$–$\frac{3}{4}$ in. long, rarely less, globose-pentagonal (i.e. with 5 inflations between the sepals) at the base, cylindric and 1 lin. in diam. above; lobes usually $4\frac{1}{2}$–6 lin. long, rarely less, $1\frac{1}{4}$ lin. broad, more or less spreading, linear-lanceolate, acute, recurved along the margins, glabrous on the inner face; corona-lobes $\frac{1}{2}$–$\frac{2}{3}$ lin. long, broadly Λ-shaped, adnate to the staminal column, with broad free reflexed margins; staminal column $1\frac{1}{2}$ lin. long; anthers connivent-erect over and about twice as long as the conical acute style-apex, oblong, with a subulate or deltoid-lanceolate glabrous appendage $\frac{1}{2}$ as long as the cells; follicles 4 in. long, $\frac{1}{2}$ in. thick, fusiform, tapering into a long beak. *K. Schum. in Engl. and Prantl, Pflanzenfam.* iv. ii. 266 *and* 264, *fig.* 77, *D–E, inaccurate as to anther-tips*; *N. E. Br. in Dyer, Fl. Trop. Afr.* iv. i. 434. *Barrowia jasminiflora, Decne in DC. Prodr.* viii. 630; *Delessert, Ic.* v. 37, *t.* 88 (*inaccurate as to fig.* 1); *Harv. Gen. S. Afr. Pl. ed.* 2, 240; *Marloth in Engl. Jahrb.* x. 244; *Schlechter in Journ. Bot.* 1897, 291. *Pergularia? jasminiflora, Burch. ex Decne in DC. Prodr.* viii. 630.

Central Region: Hopetown Div. ; near Hopetown, *Shaw* ! *Muskett,* 120 ! and *in Herb. Bolus,* 2565 !

Kalahari Region: Griqualand West; near Kimberley, *Marloth,* 756 ! on plains near the Vaal River, *Bowker,* 20 ! Bechuanaland ; at Patani, near the Kuruman River, *Burchell,* 2416 ! near Kuruman, *Burchell,* 2422 ! 2427 ! near the sources of the Kuruman River, *Burchell,* 2505 ! Transvaal ; Magalies Berg, 6000–7000 ft., *Zeyher,* 1154 ! *Burke,* 63 ! Crocodile River, *Burke,* 129 ! Maquasi Mountain, *Nelson,* 47 !

Also in Tropical Africa.

The flowers are said to be strongly scented. A form with a corolla-tube 3½–4 lin. long and lobes 2½–3 lin. long, from Tropical Africa, has been described as *O. browniana,* Schinz in *Verhandl. Bot. Ver. Brandenb.* xxx. 264 ; this may also be found to occur in South Africa. I believe it to be a somewhat starved condition of the species.

2. **O. albida** (Schinz in Verhandl. Bot. Ver. Brandenb. xxx. 265) ; a much-branched bush ; branches erect, crowded, slender, white-glaucous ; leaves distant, erect or ascending, ¾–1½ in. long, ½–1 lin. broad, linear, pungent-acute, channelled down the face, subsessile, glabrous ; umbels 4–6-flowered, subsessile or on peduncles ½–1 lin. long, white-tomentose, except on the corolla ; bracts ½–1 lin. long ; pedicels ½ lin. long or less ; sepals 1 lin. long, lanceolate, acute ; corolla whitish-puberulous outside, glabrous within ; tube about 2½ lin. long, slightly inflated and about 1 lin. in diam. at the base, 5-grooved ; lobes 1¼–1½ lin. long, erect, linear, subacute, with 5 minute pocket-like thickenings at the sinuses between them ; corona-lobes minute, Λ-shaped, arising a little above the base of the 1 lin.-long staminal column and adnate to it between and scarcely as long as the pairs of anther-wings, with free reflexed margins ; anthers connivent-erect around and about equalling the slender terete style-apex, linear, with a short acute appendage, fringed or tipped with a tuft of long hairs. *Engl. in Engl. Jahrb.* xix. 148 ; *K. Schum. in Engl. and Prantl, Pflanzenfam.* iv. ii. 266 ; *N. E. Br. in Dyer, Fl. Trop. Afr.* iv. i. 434. *O. stricta, Hiern in Cat. Afr. Pl. Welw.* i. 694.

Western Region: Great Namaqualand ; Bysondermaid, *Schinz,* 26 ! Karakoes, *Schinz* !

Also in S. W. Tropical Africa.

XXXVI. SISYRANTHUS, E. Meyer.

Calyx 5-partite. *Corolla* 5-lobed to half-way ; tube campanulate or globose-campanulate, sometimes quite glabrous, but more often bearded with hairs at the mouth or with 5 small tufts of hair in the throat ; lobes ascending or spreading, glabrous or with hairs on the inner face. *Corona* of 5 lobes arising from the staminal column and partly adnate to it, with their sides and upper part free, ovate, deltoid-ovate or oblong, entire and obtuse or acute, or rhomboid or broadly cuneate and 3-toothed, with the lateral teeth very small and the middle tooth large, acute (rarely bifid), or prolonged into a subulate or filiform point, erect or the points incurved or connivent. *Stamens* arising from the base of the corolla, their filaments connate into a tube around the ovary and adnate to the top of the dilated part of the style; anthers free, erect, without an appendage, but often ciliate with long hairs at the apex. *Pollen-masses* solitary and erect in each cell, pellucid across the top, attached in pairs to the pollen-carriers by short caudicles. *Style* not exceeding or shorter than the anthers, depressed or concave at the apex. *Follicles* rather slender, narrowly fusiform, tapering into a beak, smooth. *Seeds* (not seen ripe), crowned by a tuft of hairs.

Erect herbs; rootstook a cluster of long thick fleshy roots; stem rather slender, usually simple; leaves long, linear, in distant pairs, those of the lowest 2–4 nodes very much reduced or rudimentary and closely adpressed to the stem; flowers in umbels, lateral and pedunculate or terminal and subsessile.

DISTRIB. Species 11, endemic. With the exception of *S. trichostomus, S. compactus,* and *S. anceps,* which are the most distinct in habit, the species of this genus are very much alike and cannot be determined without dissection.

Corolla-lobes with hairs all over their inner face or at
    least extending to their base near the margins, and
    the throat and mouth of the tube bearded with
    hairs :
  Umbels racemosely scattered along the stem, somewhat
    distant; corona-lobes exceeding and connivent or
    incurved over the tops and hairs of the anthers :
  Corona-lobes reaching to or shortly exserted from the
    mouth of the corolla-tube ... ... (1) **Saundersiæ.**

  Corona-lobes not nearly reaching to the mouth of the
    corolla-tube :
    Entire corolla 2½ lin. long; anthers broader than
      long, transversely elliptic-oblong, ciliate with
      very few hairs at the apex ... ... (3) **virgatus.**

    Entire corolla not more than 2 lin. long; anthers
      longer than broad, oblong, densely ciliate
      with rather long hairs at the apex ... ... (4) **barbatus.**

  Umbels clustered or crowded into a head at the top of
    the stem or the lowest distant; corona-lobes shorter
    than the anthers or scarcely exceeding the hairs at
    their tips; entire corolla 1¾–2 lin. long ... ... (5) **compactus.**

Corolla-lobes hairy at the tips or apical half, glabrous
    on the lower half of the inner face; mouth and throat
    of the tube bearded with hairs :
  Stem 2-edged; umbels crowded or 1 distant; corona-
    lobes with long filiform points much exceeding the
    anthers and reaching to or exserted from the
    mouth of the corolla-tube ... ... (2) **anceps.**

  Stem terete; umbels all distant; corona-lobes with a
    short inflexed point not exceeding the anther-tips
    and not or scarcely reaching to the mouth of the
    corolla-tube ... ... (6) **Fanninii.**

Corolla-lobes quite glabrous on their inner face:
 Corolla-tube densely bearded with hairs at the mouth
  and throat:
   Umbel solitary, rarely 2, terminal, densely 12–30-
    flowered; corona-lobes shorter than the staminal
    column ... ... **(7) trichostomus.**

   Umbels 2–3, racemosely arranged, 6–15-flowered;
    corona-lobes longer than and incurved or con-
    nivent over the staminal column ... ... **(8) Huttonæ.**

 Corolla-tube with only 5 small tufts of deflexed hairs
  at the throat; corona-lobes shorter than or not
  exceeding the anthers; umbels racemose ... ... **(9) imberbis.**

 Corolla-tube entirely destitute of hairs within:
  Leaves linear, $\frac{2}{3}$–$1\frac{1}{2}$ lin. broad; corona-lobes very
   broadly ovate and entire or 3-toothed, or sub-
   quadrate and shortly acuminate... ... ... **(10) macer.**

  Leaves filiform, rarely linear or more than $\frac{1}{2}$ lin.
   broad; corona-lobes oblong, obtuse or with an
   inflexed subacute apex channelled down the
   back, entire or rarely with minute teeth near
   the base ... ... **(11) Randii.**

1. **S. Saundersiæ** (N. E. Br.); stem $1\frac{1}{2}$–$2\frac{1}{2}$ ft. high, subterete, sometimes slightly 2-edged below each node or at the base, glabrous; developed leaves erect, 1–2 in. long, $\frac{1}{4}$–$\frac{1}{2}$ lin. broad, linear, acute, minutely ciliate-denticulate on the margins, glabrous; umbels 4–6 to a stem, erect, racemosely arranged, all rather distant, 3–5-flowered; peduncles 2–7 lin. long, glabrous; pedicels $\frac{1}{4}$–$\frac{1}{2}$ in. long, glabrous; sepals 1–$1\frac{1}{4}$ lin. long, lanceolate, acuminate, glabrous or with a few hairs on the back; corolla-tube $1\frac{1}{4}$–$1\frac{1}{2}$ lin. long, campanulate, faintly constricted at the mouth, densely bearded at the mouth with converging hairs and with 5 tufts of deflexed hairs at the throat within, otherwise glabrous; lobes suberect, 1–$1\frac{1}{4}$ lin. long, about $\frac{3}{4}$ lin. broad, oblong-ovate, acute, glabrous on the back, thinly or densely bearded with hairs at the apex and along within the margins to the base, and thinly sprinkled with scattered hairs or subglabrous on the middle part; corona-lobes $1\frac{1}{4}$ lin. long, reaching to or shortly exserted from the mouth of the corolla-tube, broadly or transversely rhomboid at the $\frac{1}{2}$–$\frac{2}{3}$ lin.-broad basal part, 3-toothed, lateral teeth minute, middle tooth broad-based, prolonged into a slender filiform point much longer than the basal part, connivent-erect high above the $\frac{1}{2}$ lin.-long staminal column; anthers sub-quadrate-ovate, tipped with a few hairs.

Eastern Region: Natal; Inanda, 1200 ft., *Wood*, 265, partly! and without precise locality, *Mrs. K. Saunders*!

Closely allied to *S. anceps* and has been distributed under the same number, but besides differing in the subterete stem, more numerous and distant umbels, and more hairy corolla-lobes, the lateral teeth of the corona-lobes are nearer to the top of the broad part of the lobe and also relatively to the top of the staminal column than they are in *S. anceps*.

2. **S. anceps** (Schlechter in Engl. Jahrb. xx. Beibl. 51, 45) ; stem solitary or 2 to a plant, simple, $1\frac{1}{4}$–$1\frac{3}{4}$ ft. high, compressed and distinctly 2-edged, glabrous ; leaves erect or ascending, 1–$3\frac{1}{2}$ in. long, $\frac{1}{4}$–$\frac{3}{4}$ lin. broad, linear or linear-filiform, acute, minutely denticulate and ciliate on the margins, glabrous ; umbels 2–3, all crowded at the top of the stem or the lower one distant, 4–9-flowered ; peduncle 0–2 lin. long, glabrous ; pedicels $\frac{1}{3}$–$\frac{1}{2}$ in. long, glabrous ; sepals 1–$1\frac{1}{4}$ lin. long, lanceolate, acute or acuminate, sparsely pubescent ; corolla-tube $1\frac{1}{4}$–$1\frac{1}{2}$ lin. long, campanulate, faintly contracted at the mouth, glabrous outside, densely bearded at the mouth and in the throat, with the hairs in the throat deflexed ; lobes $1\frac{1}{2}$ lin. long, suberect, oblong or oblong-ovate, acute, bearded with hairs on the apical half of the inner face, glabrous elsewhere ; corona-lobes reaching to or shortly exserted from the mouth of the corolla-tube, 1–$1\frac{1}{4}$ lin. long, basal half $\frac{2}{3}$–$\frac{3}{4}$ lin. broad, broadly rhomboid-ovate or obovate, 3-toothed, with the lateral teeth minute and the middle tooth filiform, erect, much exceeding the $\frac{2}{3}$–$\frac{3}{4}$ lin.-long staminal column and connivent above it ; anthers with 3 or 4 hairs at the apex. *K. Schum. in Engl. and Prantl, Pflanzenfam.* iv. ii. 264, *fig.* 77, A–C, *and* 265 ; *Schlechter in Journ. Bot.* 1897, 291.

EASTERN REGION : Natal ; Inanda, *Wood,* 265 partly ! near Gilletts, *Wood,* 6588 ! near Camperdown, 3000 ft., *Schlechter,* 3278 (ex *Schlechter*).

S. Saundersiæ, N. E. Br., has also been distributed by Mr. Wood as 265.

3. **S. virgatus** (E. Meyer, Comm. 197) ; stem of the type $2\frac{1}{2}$ ft. high, 1 lin. thick at the base, simple, glabrous ; leaves erect, sessile, $2\frac{1}{2}$–$3\frac{1}{2}$ in. long, $\frac{1}{2}$ lin. broad, linear, acute, with minutely toothed and slightly ciliate margins ; umbels lateral at the nodes and terminal, pedunculate, 2–8-flowered, sometimes slightly compound ; peduncles $\frac{1}{4}$– 1 in. long, glabrous ; bracts $\frac{3}{4}$–$1\frac{1}{2}$ lin. long, subulate, ciliate ; pedicels unequal, $4\frac{1}{2}$–9 lin. long, thinly puberulous along one side ; sepals about $1\frac{1}{4}$ lin. long, $\frac{2}{3}$ lin. broad, ovate, acuminate, with the margins inrolled at the apical part, glabrous ; corolla-tube $1\frac{1}{3}$ lin. long, globose-campanulate, glabrous outside, densely bearded with a ring of spreading hairs at the mouth and with 5 triangular patches of deflexed hairs on the upper part within, glabrous below ; lobes ascending-spreading, $1\frac{1}{4}$ lin. long, $\frac{2}{3}$–$\frac{3}{4}$ lin. broad, oblong, subacute, glabrous on the back, densely bearded with long hairs all over the inner face, those at the apex forming a distinct tuft ; corona-lobes reaching to about the middle of the corolla-tube, about $\frac{3}{4}$ lin. long, $\frac{2}{3}$ lin. broad, broadly cuneate-subquadrate, 3-toothed at the top, lateral teeth minute, middle tooth subulate, closely inflexed over the tops of the anthers ; staminal column $\frac{2}{3}$ lin. long ; anthers transversely elliptic or elliptic-oblong, thinly ciliate at the apex with very few hairs. *Decne in DC. Prodr.* viii. 649 ; *Schlechter in Engl. Jahrb.* xxi. *Beibl.* 54, 12, *not elsewhere* ; *K. Schum. in Engl. and Prantl, Pflanzenfam.* iv. ii. 265. *Brachystelma? virgatum, Dietr. Syn. Pl.* ii. 888.

EASTERN REGION : Pondoland; between St. John's River and the great waterfall, 500–1000 ft., *Drège*, 4975!

The above description is made from the type specimen in E. Meyer's Herbarium, which is the only one I have seen of this species. I do not know what plant is intended for *S. virgatus* by Schlechter in *Journ. Bot.* 1897, 291, where he records it as having a wide distribution, the name may possibly partly refer to *S. compactus*, N. E. Br.

4. S. barbatus (N. E. Br.); stem 16–20 in. high, usually simple, glabrous; leaves $2\frac{1}{4}$–4 in. long, $\frac{1}{2}$–1 lin. broad, linear, acute, glabrous, thinly ciliate but not toothed on the margins; umbels 2–4 to a stem, laxly racemose, pedunculate, 9–12-flowered, often subcompound; peduncles $\frac{1}{3}$–$1\frac{1}{4}$ in. long, glabrous; bracts $\frac{2}{3}$–1 lin. long, subulate, ciliate; pedicels 2–3 lin. long, pubescent on one side; sepals 1–$1\frac{1}{4}$ lin. long, $\frac{1}{3}$ lin. broad, lanceolate, acute or acuminate, thinly pubescent, ciliate; corolla-tube 1 lin. long, $1\frac{1}{3}$ lin. in diam., globose-campanulate, constricted under the limb, glabrous outside, densely bearded with deflexed hairs on the upper part within, glabrous below; lobes 1 lin. long, $\frac{2}{3}$ lin. broad, oblong, subacute, glabrous on the back, densely covered with long soft white hairs all over the inner face, except along the margins; corona-lobes not reaching to the mouth of the corolla-tube, $\frac{2}{3}$ lin. long, broadly cuneate, 3-toothed, lateral teeth very short, middle tooth subulate, shortly exceeding and incurved over the style-apex and hairs of the anthers; staminal column $\frac{2}{3}$ lin. long; anthers oblong, longer than broad, densely ciliate at the apex with long white hairs. *S. macer, Schlechter in Journ. Bot.* 1897, 291, *partly. Lagarinthus barbatus, Turcz. in Bull. Soc. Nat. Mosc.* 1848, i. 257.

COAST REGION : Bathurst Div. ; near or east of the Kowie River, *Ecklon* or *Zeyher*! Fort Beaufort or Queenstown Div. ; Winterberg Range, *Ecklon*, 24 (ex *Turczaninow*), Stutterheim Div. ; Mount Dohne, 4000–4500 ft., *Bolus*, 10189!
EASTERN REGION : Transkei ; Tsomo, *Mrs. Barber*, 863! Tembuland ; Engcobo Mountain, 4500 ft., *Bolus*!

This differs from *S. virgatus* by its much smaller flowers and oblong (not transverse) and more densely ciliate anthers. *S. imberbis*, var. *barbatus*, Schlechter in *Engl. Jahrb.* xx. *Beibl.* 51, 46, I am unable to identify, as no specimen is quoted, but from his description it is evident that Dr. Schlechter is mistaken in supposing it to be *Lagarinthus barbatus*, Turcz.

5. S. compactus (N. E. Br.); stems 6–14 in. high, simple, glabrous, marked with a transverse stipulary line at the nodes; leaves $2\frac{1}{4}$–4 in. long, $\frac{1}{3}$–$1\frac{1}{4}$ lin. broad, linear, acute, glabrous, minutely and somewhat obscurely toothed, with or without a slight ciliation; umbels 2–5 or 1 terminal compound umbel, 5–12- (or when compound very many-) flowered; peduncles 0–1 in. long, glabrous; bracts $\frac{3}{4}$–$1\frac{1}{4}$ lin. long, linear-subulate, ciliate; pedicels $1\frac{1}{2}$–4 lin. long, glabrous or pubescent along one side; sepals about 1 lin. long, $\frac{1}{3}$ lin. broad, ovate-lanceolate, acute, varying from thinly pubescent and ciliate to almost glabrous; corolla-tube 1 lin. long, $1\frac{1}{4}$–$1\frac{1}{2}$ lin. in diam., globose-campanulate, contracted at the mouth, glabrous outside, densely bearded with deflexed hairs on the upper part within, glabrous below ; lobes $\frac{3}{4}$–1 lin. long, $\frac{1}{2}$–$\frac{3}{4}$ lin. broad, ascending-spreading, ovate-oblong, acute, glabrous on the back, densely bearded with long hairs all over the inner face, darker coloured than the tube, with pale margins ; corona-lobes reaching to about the middle of the corolla-tube, $\frac{1}{2}$–$\frac{2}{3}$ lin. long, $\frac{1}{2}$–$\frac{2}{3}$ lin. broad, 3-toothed or broadly rhomboid with the side angles somewhat tooth-like, middle tooth deltoid, acute or shortly subulate-acuminate, erect, shorter than the anthers or scarcely exceeding the hairs at their tips, not incurved over them ; staminal column $\frac{2}{3}$ lin. long ; anthers oblong, densely ciliate with rather long white hairs at the tips. *S. virgatus, Harv. Thes. Cap.* ii. 10, *t.* 115 (*as to the plant figured*) ; *Schlechter in Engl. Jahrb.* xviii. *Beibl.* 45, 12, *not of E. Meyer or elsewhere.*

COAST REGION : Bathurst Div. ; between Kasuga River and Kowie River, *Bowie*! near Bathurst, *Hutton*! Trapps Valley, *Miss Daly*, 701! near the mouth of the Great Fish River, 100 ft., *MacOwan*, 1226! East London Div. ; coast at Panmure, *Mrs. Hutton*! near East London, *Wood in Herb. Galpin*, 3124! Komgha Div. ; hills near Komgha, *Flanagan*, 395 !

EASTERN REGION: Transkei ; Kreilis Country, *Bowker*! near Kentani, *Miss Pegler*, 658 !

This has been identified with *S. virgatus*, E. Meyer, by Harvey and Schlechter, but conspicuously differs in being only about half as tall, its inflorescence more compact, pedicels shorter, flowers smaller (they are represented too large in Harvey's figure), and the anthers are oblong (not transverse) and ciliate with much longer and more numerous hairs, which are erect and collectively form a little brush, usually projecting much above the corona-lobes.

6. **S. Fanninii** (N. E. Br.) ; stem 2 ft. or more high, simple, subterete, not 2-edged, slightly pubescent about the nodes and along one side at the upper part, otherwise glabrous ; leaves erect or ascending, $2\frac{1}{2}$–4 in. long, $\frac{1}{3}$–$\frac{1}{2}$ lin. broad, linear or linear-filiform, acute, minutely ciliate along the usually infolded margins ; umbels about 4, racemosely arranged, 4–9-flowered ; peduncles $\frac{1}{3}$–$1\frac{1}{4}$ in. long, thinly and minutely pubescent, as are the 4–7 lin.-long pedicels ; bracts 1–$2\frac{1}{2}$ lin. long, linear or subulate, ciliate ; sepals about $1\frac{1}{4}$ lin. long, ovate-lanceolate, acuminate, minutely pubescent, ciliate ; corolla-tube 1 lin. long, rather broadly cup-shaped, not contracted at the mouth, glabrous outside, bearded at the mouth with a ring of hairs, which descend very slightly into the tube as 5 broadly triangular patches ; lobes ascending, 1 lin. long and broad, ovate, subobtuse, with a thin tuft of a few hairs at the apex of the inner face, otherwise glabrous ; corona-lobes not or scarcely reaching to the mouth of the corolla-tube, $\frac{2}{3}$–$\frac{3}{4}$ lin. long, erect, deltoid-ovate or subhastate-ovate, with a small spreading tooth or angle near the base, acute, with a short point inflexed from the apex upon the anther-tips and not exceeding them ; staminal column $\frac{2}{3}$ lin. long ; anthers with or without 1 or 2 hairs at the apex.

EASTERN REGION : Natal ; Dargle Farm, *Miss Fannin*, 54 !

Allied to *S. virgatus*, differing in its shorter flowers and open cup-shaped corolla-tube.

7. **S. trichostomus** (K. Schum. in Engl. and Prantl, Pflanzenfam.
iv. ii. 265); stem 1⅙–2 ft. high, simple, thinly pubescent at the
top, elsewhere glabrous; developed leaves in 2–4 pairs, spreading
or reflexed, but ascending in some dried specimens, 2–6½ in. long,
½–1¼ lin. broad, linear, acute, narrowed at the base, glabrous,
minutely denticulate and ciliate on the infolded margins; umbel
solitary, rarely 2, terminal, subglobose, densely 12–30-flowered;
pedicels ⅓–½ in. long, shortly pilose; sepals 1¼–2 lin. long, ¼–⅔ lin.
broad, linear-subulate, lanceolate or ovate, acuminate; pubescent
or shortly pilose; corolla-tube ¾–1¼ lin. long, globose-campanulate,
slightly constricted and densely bearded at the mouth, with
the upper hairs ascending or converging and the lower deflexed,
apparently yellow; lobes 1–2 lin. long, ¾–1¼ lin. broad, spreading,
but often appearing erect in dried flowers, oblong or ovate-oblong,
obtuse, glabrous on both sides; corona-lobes shorter than the
staminal column, ⅓–½ lin. long, ½–⅔ lin. broad at the top, broadly
cuneate, 3-toothed at the top, with the middle tooth short, very
broad and bifid or deeply divided, lateral teeth minute; anthers
erect around and equalling the truncate style-apex, hairy at their
tips. *Schlechter in Journ. Bot.* 1897, 291. *S. virgatus, var.
trichostomus, Harv. Thes. Cap.* ii. 10. *S. rotatus, Schlechter in Engl.
Jahrb.* xx. *Beibl.* 51, 46. *S. expansum, Schlechter in Journ. Bot.*
1897, 291.

EASTERN REGION: Griqualand East; mountains around Clydesdale, 3000 ft.,
Tyson, 2146! and in *Bolus and MacOwan, Herb. Norm. Austr.-Afr.,* 1084!
Natal; Dargle Farm, *Mrs. Fannin,* 51! hill near Camperdown, 2500 ft., *Wood,*
5715! Greenwich Farm, Riet Vley, *Fry in Herb. Galpin,* 2736! between
Greytown and Newcastle, *Wilms,* 2145! near Pietermaritzburg, *Wilms,* 2146!
Rock Fountain, *Mrs. Clarke!* and without precise locality, *Gerrard,* 2165!
Sanderson, 136!

8. **S. Huttonæ** (S. Moore in Journ. Bot. 1907, 154); stem 1–1⅓
ft. high, simple, glabrous or with a few hairs about the uppermost
nodes; leaves ascending or slightly spreading, 2–3¼ in. long, ⅓–½
lin. broad, linear-filiform, glabrous, minutely and adpressedly ciliate-
denticulate; umbels 2–3, lateral and terminal, apparently nodding,
6–15-flowered; peduncles ½–1¼ in. long; pedicels 5–9 lin. long,
thinly puberulous along one side; bracts 1¼–2½ lin. long, subulate,
minutely ciliate; sepals 1½–1¾ lin. long, lanceolate, subulate-
acuminate from the margins being infolded, with a very few hairs
on the back or margins; corolla-tube 1–1¼ lin. long, subglobose-
campanulate, densely bearded at the mouth and throat, with the
upper hairs converging to the centre and the lower deflexed, otherwise glabrous; lobes 1–1½ lin. long, ¾–1 lin. broad, ovate or oblong-
ovate, subacute, quite glabrous on both sides; corona-lobes ¾–1⅓ lin.
long, triangular-ovate, with a minute tooth on each side near the
base, acuminate at the apex into subulate or filiform points
connivent above the anthers, with erect or recurving tips.
Schizoglossum Huttonæ, S. Moore in Journ. Bot. 1902, 383, *and*
1907, 154.

EASTERN REGION : Natal ; Shafton, Howick, *Mrs. Hutton*, 407 ! Greenwich Farm, Riet Vlei, *Fry in Herb. Galpin*, 2737 !

Asclepias cucullata, Schlechter, has also been distributed by Mrs. Hutton as 407. The Greenwich Farm plant has flowers that are larger in all their parts than the type, the points of the corona-lobes being very much longer, but in every other character the two are identical, and it is probably only a luxuriant specimen. Misled by its external resemblance to *S. Randii*, and without having dissected it, this specimen was unfortunately originally quoted by Moore under that species in *Journ. Bot.* 1903, 200, but the corolla and corona are quite different.

9. **S. imberbis** (Harv. Thes. Cap. ii. 11, t. 116); stems usually 1¼–2 (rarely up to 3¼) ft. high, ⅓–1 lin. thick at the base, usually simple, glabrous; leaves ascending or spreading, 1¼–5 in. long, ¼–1 lin. broad, linear, acute, apparently folded lengthwise, glabrous, minutely and sparsely ciliate and denticulate on the margins; umbels 2–6 to a stem, racemose, pedunculate, 2–7-flowered; peduncles ¼–1 in. long, rarely longer, slender, glabrous; bracts ½–2 lin. long, subulate or lanceolate, acuminate, ciliate; pedicels 2–4½ lin. long, glabrous; sepals 1–1⅔ lin. long, ½–⅔ lin. broad, ovate-lanceolate, convolute-acuminate at the apical half, glabrous, not ciliate; corolla glabrous outside, not bearded at the mouth, but furnished with 5 tufts of deflexed hairs at the top of the tube just below the lobes inside, otherwise glabrous; tube ¾–1 lin. long, 1½–1¾ lin. in diam., broadly campanulate or cup-shaped, slightly contracted at the mouth, 5-concave inside; lobes 1¼–1⅔ lin. long, ⅔–1 lin. broad, ascending-spreading, oblong or ovate-oblong, acute or subobtuse; corona-lobes erect, not exceeding or shorter than the anthers, ½–⅔ lin. long, broadly ovate or deltoid-ovate, obtuse or acute, with slightly reflexed sides and rounded basal angles; staminal column ¾–1 lin. long; anthers oblong, obtuse or with an obtuse apiculus, not ciliate; follicles 2¾–3½ in. long, 2 lin. thick, terete-fusiform, tapering into a long acute beak, smooth, glabrous. *K. Schum. in Engl. and Prantl, Pflanzenfam.* iv. ii. 265. *S. macer, Schlechter in Journ. Bot.* 1897, 291, *partly*.

COAST REGION : Albany Div. ; near Grahamstown, *MacOwan*, 664 ! *Misses Daly and Sole*, 542 ! *Hutton* ! Howisons Poort, *MacOwan* !

KALAHARI REGION : Basutoland, *Cooper*, 2730 ! Orange River Colony, *Cooper*, 959 !

EASTERN REGION : Transkei ; Kreilis Country, *Bowker*, 299 ! near Kentani, *Miss Pegler*, 657 ! Natal ; Inanda, *Wood*, 376 ! 1192 ! Clairmont, near Durban, *Wood*, 3907 ! and in *Herb. Natal*, 714 ! and without precise locality, *Gueinzius* !

Wood 714 is also an orchid.

10. **S. macer** (Schlechter in Journ. Bot. 1897, 291, as to E. Meyer's plant only) ; exactly like *S. imberbis* in habit, stem, umbels, and measurements of corolla ; leaves 4–5 in. long, ⅔–1½ lin. broad ; corolla entirely glabrous within and without ; corona-lobes erect, shorter than the stamens, ½ lin. long and as much in breadth, usually very broadly ovate and entire or somewhat angular at the sides or 3-toothed, with the side teeth very small and the middle

one broadly deltoid obtuse, or rarely subquadrate and shortly
acuminate; staminal column ¾ lin. long; anthers erect, exceeding
the style-apex, narrowly oblong, minutely apiculate, not ciliate.
S. imberbis, Schlechter in Engl. Jahrb. xviii. *Beibl.* 45, 12, *and* xxi.
Beibl. 54, 12, *with a false reference to E. Meyer, not of Harvey.*
Lagarinthus macer, E. Meyer, Comm. 206.   *Gomphocarpus macer,*
Dietr. Syn. Pl. ii. 901.   *Periglossum macrum, Decne in DC. Prodr.*
viii. 520.

COAST REGION: Komgha Div.; hills near Komgha, *Flanagan*, 374! Kaffraria,
Drège, 4974!

This may be only a variety or hybrid form of *S. imberbis*, but the absence of the
tufts of hair in the corolla-tube, which evidently play some important part in the
economy of this genus, seem to point to a specific distinction.   The corona-lobes
in different flowers of the single type specimen vary considerably.   It seems to be
rare, as I have only seen the two specimens above quoted.

11. S. Randii (S. Moore in Journ. Bot. 1903, 200, including var.
abbreviatus); stems one to several from a root, 1½–2½ ft. high,
simple or sparingly branched above, glabrous; developed leaves
in 5–8 pairs, 3–5½ in. long, decreasing upwards, erect or ascending,
¼–½ lin. broad, filiform or very narrowly linear, acute, channelled
down the face or longitudinally folded, minutely and distantly
ciliate, otherwise glabrous; umbels several, lateral at the nodes,
3–7-flowered, glabrous; peduncles ⅓–2 in. long; pedicels 3–5 lin.
long; sepals 1¼–2 lin. long, ovate or elliptic with a subulate point;
corolla quite glabrous outside and within, cream-coloured (*Miss
Pegler*), greenish-brown when dried; tube about 1 lin. long and
2 lin. in diam., depressed-globose, contracted at the mouth; lobes
erect or slightly spreading, 1–1¼ lin. long and broad, ovate or
oblong-ovate, obtuse; corona-lobes shorter than to about as long as
the staminal column, erect, ⅔ lin. long, scarcely ½ lin. broad unless
flattened out, oblong, entire or rarely with a minute tooth on each
side at the base, channelled down the back, obtuse or inflexed and
subacute at the apex; staminal column ⅔ lin. long; anthers conni-
vent-incumbent over the style-apex, oblong, usually ciliate with
2 or 3 minute bristle-like hairs at the tips.   *S. imberbis, Schlechter
in Engl. Jahrb.* xx. *Beibl.* 51, 45, *not of Harv.*

KALAHARI REGION: Transvaal; near a marsh, Ginsberg, East Rand, near
Johannesberg, 6000 ft., *Miss Pegler*, 1052! marsh near Elsburg, 5400 ft.,
Schlechter, 3543! open veldt, near Johannesberg, *Rand*, 856! Modderfontein,
Conrath, 1005! second water between Nelspruit and Sibthorps, *Burtt Davy*, 1611!
between Standerton and Pretoria, *Wilms*, 924! near Pretoria, *Schlechter*, 3692!

EASTERN REGION: Swaziland; near Mbabane (Embabaan), 4400 ft., *Bolus*,
12255!

This species is very similar to *S. macer*, but has narrower leaves and different
corona-lobes, the localities of the two are also widely separated.   The variety
abbreviatus is merely a rather smaller individual than the type but otherwise does
not differ.

XXXVII. TENARIS, E. Meyer.

Calyx 5-partite. *Corolla-tube* very short; lobes 5, linear, linear-spathulate or filiform, spreading or ascending, valvate or replicate-valvate in bud. *Corona* small, double, arising from the staminal column above its base; outer corona of 5 small concave lobes or minute pouches, more or less spreading, alternating with the anthers; inner corona of 5 linear, subulate or filiform lobes, incumbent on the backs of the anthers and shorter than, equalling or exceeding the latter. *Staminal column* arising from the base of the corolla, very small, included in the corolla-tube; anthers inflexed upon the style-apex, without an appendage or rarely with a very short one. *Pollen masses* minute, subascending or sub-horizontal, solitary in each anther-cell, with a pellucid linear area close to the apex, attached in pairs to the very minute pollen-carriers by very short slender caudicles at their middle or near their base. *Style-apex* usually shortly conical and acute, rarely comparatively large and obtuse. *Follicles* erect, somewhat diverging, slender, smooth. *Seeds* crowned with a tuft of hairs.

Perennial herbs with tuberous rootstock; stems slender, erect; leaves opposite, linear or filiform; inflorescence leafless and terminal, racemose or paniculate, or the flowers 1–3 together in sessile or pedunculate fascicles or in long slender racemes, lateral at the nodes of the terminal and leaf-bearing part of the stem.

DISTRIB. Species 7, the others in Tropical Africa.

Corolla-lobes linear-spathulate, $\frac{3}{4}$–$1\frac{1}{3}$ lin. broad at the
    apex, pink :
  Flowering part of the stem leafless and bractless or
    with some bracts $\frac{3}{4}$–1 lin. long; sepals lanceolate... (1) **rubella.**

  Flowering part of the stem with subulate or filiform
    bracts 1–6 lin. long; sepals filiform-subulate from
    a broader base   ...   . (2) **simulans.**
Corolla-lobes filiform or linear-filiform, $\frac{1}{4}$ lin. broad,
    purple-brown on the inner face; flowering part of
    stem with leaves :
  Flowers in fascicles of 3–5 at the nodes; inner corona-
    lobes shorter than or subequalling the anthers ... (3) **chlorantha.**

  Flowers in slender 2–6-noded racemes; inner corona-
    lobes much longer than the anthers   ...   ... (4) **filifolia.**

1. **T. rubella** (E. Meyer, Comm. 198); tuber depressed, 1–$1\frac{1}{2}$ in. in diam.; stem $\frac{1}{2}$–$1\frac{3}{4}$ ft. high, solitary, slender, simple or with 1 branch, leafless on the flowering part, glabrous; leaves in 3–6 pairs, ascending or erect, 1–$2\frac{1}{2}$ in. long, $\frac{1}{2}$–1 lin. broad, linear, acute, glabrous; flowers in 1 or 2 sessile or pedunculate pairs or sometimes a fascicle-like whorl of them at the upper nodes, forming a terminal leafless and often bractless raceme or narrow raceme-like panicle, glabrous in all parts, with the internode next below it peduncle-like and $1\frac{1}{2}$–5 in. long; peduncles or branches of the panicle 0–$\frac{3}{4}$ in.

long, 2-flowered; bracts and bracteoles (when present) $\frac{3}{4}$–1 lin. long; pedicels 2–4$\frac{1}{2}$ lin. long, slender; sepals $\frac{3}{4}$–1 lin. long, lanceolate, acute; corolla-tube $\frac{2}{3}$–$\frac{3}{4}$ lin. long, cup-like; lobes erectly-spreading, 3$\frac{1}{2}$–5 lin. long, 1–1$\frac{1}{3}$ lin. broad at the tips, linear-spathulate, very broadly rounded or subtruncate and subapiculate at the apex, apparently somewhat replicate, densely covered with minute papillæ on the lower half of the inner face, elsewhere smooth and glabrous, pink, with the papillæ darker; outer corona-lobes arising $\frac{1}{2}$ lin. up the staminal column, $\frac{1}{4}$ lin. long, subquadrate or subquadrate-ovate, notched at the apex, glabrous; inner corona-lobes $\frac{1}{5}$ lin. long, narrowly linear or linear-subulate, obtuse, incumbent upon and shorter than the backs of the anthers; young follicles (mature not seen) linear-terete, glabrous. *Decne in DC. Prodr.* viii. 606; *Harv. Gen. S. Afr. Pl. ed.* 1, 223, *and Thes. Cap.* i. 28, *t.* 43; *Schlechter in Engl. Jahrb.* xviii. *Beibl.* 45, 12; xxi. *Beibl.* 54, 12, *and Journ. Bot.* 1897, 291 (*excl. the Transvaal plant*).

COAST REGION : Uitenhage Div. ; Addo, *Drège*, 2227 ! Bathurst Div. ; near the Karega River, *Zeyher* ! Trapps Valley, *Miss Daly*, 693 ! Albany Div. ; near Grahamstown, *Reade*, 93 ! *Hutton* ! *MacOwan* ! *Misses Daly and Sole*, 438 ! *Bolus, Herb. Norm. Austr.-Afr.* 1322 ! common on sandy hills and flats, *Mrs. Barber*, 99 ! *Miss Bowker* ! *Prior (Alexander)* !

2. **T. simulans** (N. E. Br.); stems solitary, 1–1$\frac{1}{2}$ ft. high, simple, glabrous; leaves on the flowering part reduced to subulate or filiform bracts 1–6 lin. long, those below $\frac{3}{4}$–2$\frac{1}{2}$ in. long, $\frac{1}{4}$–$\frac{1}{3}$ lin. broad, linear or linear-filiform, acute, erect or with spreading tips, glabrous; inflorescence terminal, with the internode below it $\frac{3}{4}$–1$\frac{1}{2}$ in. long, glabrous in all parts, composed of 2–3 pedunculate umbel-like fascicles of 3–7 flowers, opening successively; peduncles $\frac{1}{4}$–$\frac{3}{4}$ in. long; bracteoles 1–1$\frac{1}{4}$ lin. long, subulate; pedicels 2–2$\frac{1}{2}$ lin. long; sepals 1$\frac{1}{4}$–1$\frac{1}{2}$ lin. long, filiform-subulate from a narrowly oblong base; corolla-tube 1–1$\frac{1}{4}$ lin. long, cup-like; lobes erectly-spreading, $\frac{1}{3}$ in. long, nearly 1 lin. broad near the apex, linear-spathulate, rounded and subapiculate at the apex, apparently somewhat replicate, thinly sprinkled with minute papillæ on the lower half of the inner face, elsewhere smooth and glabrous, apparently pale pink with darker veins ; outer corona-lobes arising $\frac{1}{2}$ lin. up the staminal column, nearly $\frac{1}{3}$ lin. long, $\frac{2}{5}$ lin. broad, subquadrate, emarginate to deeply bifid ; inner corona-lobes $\frac{1}{4}$–$\frac{1}{3}$ lin. long, linear-subulate, acute or subobtuse, closely incumbent on the backs of the anthers and not produced beyond them. *T. rubella, Schlechter in Engl. Jahrb.* xx. *Beibl.* 51, 45, *and Journ. Bot.* 1897, 291, *as to the Transvaal plant only.*

KALAHARI REGION : Transvaal ; Elands Spruit Mountains, 6000 ft., *Schlechter*, 3858 ! near Botsabelo, 5000 ft., *Schlechter*, 4071.

Similar to *T. rubella*, E. Meyer, but besides the widely different locality, differs conspicuously in the less naked appearance of its inflorescence and longer and much narrower sepals, the leaves are also narrower and the venation of the narrower less papillate petals rather different.

3. **T. chlorantha** (Schlechter in Engl. Jahrb. xx. Beibl. 51, 44) ;
tuber and stem as in *L. rubella*, but usually 1–1½ ft. high, sometimes
branched at the upper part and leafy from below the middle to the
apex ; leaves in 7–12 pairs, erect or ascending, the lower 2–3 in.
long, the upper shorter, ¼–⅔ lin. broad, linear-filiform or linear,
acute, glabrous ; flowers usually in fascicles of 3–7 at the 5–10
uppermost nodes ; peduncles 0–4 lin. long ; pedicels 3½–7 lin. long,
very slender, glabrous ; sepals ½–¾ lin. long, lanceolate, acute,
glabrous ; corolla glabrous ; tube ½–⅔ lin. long, campanulate ; lobes
2⅓–3¼ lin. long, ½ lin. broad at the base, ¼ lin. broad above, linear-
filiform, with recurved margins, acute, with a microscopic papillation
on the inner face, but not tuberculate ; outer corona-lobes ⅙ lin.
long, subquadrate, notched at the apex, glabrous ; inner corona-
lobes ⅙ lin. long, subulate, acute or obtuse, applied to the backs of
the anthers and shorter than or subequalling them. *Schlechter in
Journ. Bot.* 1897, 291. *Macropetalum Benthamii, K. Schum. in Engl.
and Prantl, Pflanzenfam.* iv. ii. 266.

Kalahari Region: Transvaal ; Magalies Berg, *Burke,* 326 ! *Zeyher,* 1175 !
ridge near Johannesberg, 6000 ft., *Gilfillan in Herb. Galpin,* 1458 ! *Rand,* 1121 !
near Pretoria, 4900 ft., *Schlechter,* 4152 ! near Klein Olifants River, *Schlechter,*
3812, and near Botsabelo, *Schlechter.*

Dr. Schlechter has stated that the flowers are green, but Dr. Rand in *Journ.
Bot.* 1903, 340, writing about the living plant, states that the petals are ''brown-
purple in colour,'' and all the dried specimens I have seen which distinctly show
any colour, have the inner surface of the corolla-lobes dark purple-brown.
Dr. Rand also states that the corona-lobes are ''continued up over stylar head,
swollen and glandular at the tips,'' but I do not find them so in his dried
specimen (1121), nor in any others that I have examined.

4. **T. filifolia** (N. E. Br.) tuber flattened, 1¼–1⅔ in. in diam. ;
stem solitary, 1½–2½ ft. high, slender, simple or sparingly branched
at the middle, ¾–1 lin. thick at the base, glabrous ; leaves erect or
ascending, in 10–20 or more pairs, from below the middle to the
apex of the stem, the lower 2¼–3¼ in. long, the others gradually
smaller, filiform, glabrous ; flowers in very slender glabrous racemes,
lateral at the nodes ; lower racemes 3–4 in. long, occasionally
branched, with 1–3 flowers at each of the 3–6 nodes, the upper much
reduced ; bracts very minute ; pedicels 2–3 (3–4, *Schlechter*) lin. long,
capillary, glabrous ; sepals ½–⅔ lin. long, lanceolate or ovate-lanceolate, acute, glabrous ; corolla quite glabrous, apparently greenish
outside and purple-brown on the inner face of the lobes ; tube about
⅔ lin. long, cup-shaped ; lobes ascending-spreading, 4–4½ lin. long,
filiform from a deltoid-ovate base ; outer corona of 5 very minute
prominent pouch-like lobes ; inner corona-lobes ¼ lin. long, filiform,
closely incumbent on the backs of the anthers and produced beyond
them into connivent-erect tips, dark-coloured ; staminal column ⅖ lin.
long ; follicles diverging, about 2½ in. long, 1¼ lin. thick, linear-
terete, beaked glabrous. *Macropetalum filifolium, Schlechter in
Engl. Jahrb.* xxxviii. 36, *fig.* 4.

KALAHARI REGION : Transvaal ; Komati Poort, *Schlechter,* 11733 !

In referring this plant to *Macropetalum* Dr. Schlechter appears to have mis-understood its structure, since he figures its corona as cup-shaped with 5 erect slightly incurving lobes, linear-subulate from a deltoid base, with the anthers adnate to them, erect and quite free from the style-apex, which is quite unlike any structure I have seen in any Asclepiad ; the pollen-masses are also inaccurately represented.   The corona is minute, and unless lightly pressed flowers are examined the outer corona-pouches are easily overlooked.   The whole coronal structure is different from that of *Macropetalum* and agrees with that of *Tenaris,* whilst specifically the plant is nearly allied to *T. chlorantha.*

XXXVIII. MACROPETALUM, Burchell.

Calyx 5-partite.   *Corolla-tube* exceedingly short ; lobes 5, linear-filiform, reflexed straight back from their base, valvate in bud. *Corona* small, of 5 simple lobes opposite the anthers, arising above the base of the staminal column and adnate to it up to the base of the anthers, free above.   *Staminal column* arising from the base of the corolla, small, and together with the corona, entirely exserted ; anthers erect, applied to the lower part of the sides of the stout style-apex, subquadrate or rather broader than long, tipped with a membranous appendage.   *Pollen-masses* minute, erect, solitary in each anther-cell, with a pellucid linear area close to the apex, attached in pairs to the very minute pollen-carriers by short slender caudicles at their middle.   *Style* truncate at the apex.   *Follicles* erect, slightly diverging, slender, smooth.   *Seeds* not seen.

Perennial herb with tuberous rootstock ; stems slender, erect, simple or branched above ; leaves opposite, filiform ; flowers small, with long corolla-lobes, in fascicles of 2–4 at the nodes.

DISTRIB. Species 1, endemic.

Allied to *Tenaris,* but differing in its reflexed corolla, simple and exserted corona, and in the very conspicuous anther-appendages.

1. M. Burchellii (Decne in DC. Prodr. viii. 627) ; tuber sub-globose, $1\frac{1}{2}$ in. or more in diam. ; stem 1–$1\frac{1}{2}$ ft. high, simple or sparingly branched at the middle, $\frac{1}{2}$–$\frac{3}{4}$ lin. thick, with some very minute adpressed scattered hairs near the nodes ; leaves 1–3 in. long, $\frac{1}{4}$–$\frac{1}{3}$ lin. thick, filiform, erect or ascending, glabrous ; flowers usually in pairs (more rarely 3–4) at the nodes ; pedicels $2\frac{1}{2}$–3 (in fruit 5–6) lin. long, very slender, with a very minute adpressed puberulence ; sepals $\frac{2}{5}$ lin. long, lanceolate, acute, erect at the basal and reflexed at the apical half, minutely puberulent ; corolla quite glabrous, apparently green or yellowish-green ; tube about $\frac{1}{3}$ lin. long ; lobes reflexed straight back from their base, 9–11 lin. long, $\frac{2}{5}$ lin. broad at the linear base, tapering into filiform tails slightly thickened at the apex ; corona-lobes arising near the base of the staminal column and much overtopping it, $\frac{3}{4}$–1 lin. long, wholly exserted from the corolla, erect, linear-lanceolate, acute, recurved or

revolute at the tips, slightly recurved at the margins, the adnate lower part prominent from the column, concave down the back or with the margins somewhat recurved, very truncate at the base ; staminal column about ½ lin. long ; anther-appendages deltoid or ovate-lanceolate, acute, erect, shortly overtopping the truncate style-apex; follicles erect and slightly diverging, about 3 in. long and 2 lin. thick, linear-fusiform, tapering into a slender beak, smooth, glabrous. *Harv. Gen. S. Afr. Pl. ed.* 2, 241 ; *Schlechter in Journ. Bot.* 1897, 291.

VAR. *β*, **grandiflora** (N. E. Br.) ; sepals ⅛ lin. long, ovate, acute ; corolla sparsely and minutely puberulent on the outside like the pedicels, with lobes 1½ in. long ; corona-lobes often with a minute tooth near the apex, apparently not recurved at the tips ; otherwise exactly as in the type.

KALAHARI REGION : Orange River Colony ; Sand River, *Burke* ! Bechuanaland ; near the sources of Kuruman River, *Burchell*, 2498 ! Var. *β* : Transvaal ; near Warm Baths, *Tandler in Transvaal Herb.* !

Also recorded from Tropical Africa (Rhodesia) by Dr. Rendle in *Journ. Bot.* 1905, 44, but I have not seen a specimen.

Burchell in a MS. note states that the Bechuanas call this plant "Klókwe," and adds, "These roots, being considered an article of food, were brought by a Bachapin, of whom I purchased them." According to Mr. Tandler the roots of var. *β* are "much eaten by the natives," who call the plant "Seruch." Burchell's specimens are nearly leafless at the flowering part, whilst the others quoted bear leaves at nearly all the flowering nodes.

XXXIX. RIOCREUXIA, Decne.

Calyx 5-partite. *Corolla* tubular, more or less inflated at the base, straight, 5-lobed ; lobes linear, erect or incurved-erect, connate at their tips, very slightly overlapping in bud, not strictly valvate. *Corona* 2–3- (very rarely 1-) seriate, arising from the staminal column ; outer corona-lobes 5, alternating with the anthers, either minute and bifid or deeply divided, or conspicuous and radiating, filiform, subulate, narrowly deltoid, ovate or subrectangular, free or connate at the base, with or without a series of 5 pairs of teeth alternating with them at their base ; inner corona-lobes 5, linear or subulate and usually equalling or exceeding the anthers and incurved over them, or shortly linear-oblong or minute and not reaching beyond the base of the anthers, rarely obsolete. *Staminal column* arising from or near the bottom of the corolla, very short and together with the corona included in the inflated part of the corolla ; anthers erect or ascending, oblong, tipped with a minute somewhat fleshy appendage or apiculus. *Pollen-masses* erect, pellucid just beneath the apex, solitary in each anther-cell, attached below their middle by exceedingly short caudicles in pairs to the pollen-carriers. *Style* not exceeding the anthers, truncate at the apex. *Follicles* linear-terete or linear-fusiform, acuminate, more or less beaded from constrictions between the seeds, smooth. *Seeds* crowned with a tuft of hairs.

Tuberous-rooted perennials with annual twining stems, often several feet in length; leaves opposite, herbaceous, cordate; flowers of moderate size, in laxly branching cymes, or in fascicles at the 2–3 nodes of simple peduncle-like flowering axes (which are bent at the nodes), with or without another fascicle at their base, or all compactly fasciculate at the nodes of the stem.

DISTRIB. Species 8, two of them in Tropical Africa.

This genus is closely related to *Ceropegia,* differing chiefly in the character of its inflorescence, for whilst some of the species have a different corona, in others it is scarcely distinct from that of some species of *Ceropegia.* Some of the species cannot be discriminated without dissecting the flowers.

Corolla marked with numerous longitudinal purple-brown veins; outer corona-lobes broadly ovate, connate into a sort of 5-lobed frill; inner corona-lobes obsolete ... ... (1) **picta.**

Corolla not marked with purple-brown veins :
*Outer corona-lobes conspicuous, radiating :
   Outer corona-lobes alternating with 5 pairs of minute teeth at their base; inner corona-lobes rather inconspicuous or minute :
      Peduncles cymosely or subpaniculately branching; branches elongating and bearing two flowers at each node ... ... (3) **Bolusii.**

      Peduncles simple, bent at the nodes, bearing 2–3 distant fascicles of 3 (rarely 2)–15 flowers (4) **torulosa.**

   Outer corona-lobes without any alternating teeth; peduncles simple, bent at the nodes, bearing 2–3 distant fascicles of 5–24 (on weak inflorescences fewer) flowers and usually another fascicle at their base :
      Inner corona-lobes not very conspicuous, only reaching to the base of the anthers ... ... (4) **torulosa,** var. δ.

      Inner corona-lobes conspicuous, linear or linear-subulate, nearly as long as the anthers ... (5) **Woodii.**

**Outer corona-lobes inconspicuous or minute, bifid or divided into 2 minute teeth turned face to face, without any alternating teeth at their base :
   Flowers 1–3 together at the nodes of the branches of laxly branched cymes 4–10 in. long ... ... (2) **polyantha.**

   Flowers 5 to many together in fascicles at the stem-nodes or on peduncles up to ½ in. long ... ... (6) **Flanagani.**

1. **R. picta** (Schlechter in Engl. Jahrb. xviii. Beibl. 45, 24); stem twining, usually glabrous, sometimes unifariously puberulous; leaves herbaceous or thinly subcoriaceous, glabrous on both sides or minutely scaberulous along the borders; petiole ¾–2 in. long; blade 1¼–7 in. long, ¾–5 in. broad, cordate-ovate, acuminate, peduncles or axis of the cymes 1–4 in. long, bearing 2–3 distant (or occasionally 1) umbel-like clusters of flowers, bent in a zigzag manner at the nodes, glabrous; pedicels ½–1¼ in. long, glabrous; sepals 1–1½ lin. long, subulate, glabrous; corolla glabrous, striped with dark brownish-red or purple-brown on a whitish or yellowish ground; tube 5–6 lin. long, somewhat globosely inflated at the base, the inflation sometimes disappearing in dried flowers; lobes connivent-

erect, connate at the tips, $3\frac{1}{2}$–4 lin. long, oblong-lanceolate or linear-lanceolate at the lower half, tapering into filiform tips ; outer corona-lobes connate into a sort of 5-lobed frill, with the lobes about $\frac{1}{3}$ lin. long, broadly ovate, concave or perhaps laterally somewhat compressed, acutely bifid or entire and acute ; inner corona-lobes none or represented by 5 ridges connecting the base of the anthers with the outer corona ; follicles $3\frac{1}{2}$–$4\frac{1}{2}$ in. long, $1\frac{1}{2}$–$2\frac{1}{2}$ lin. thick, terete-linear, shortly tapering at the base and at the apex into a long acute beak, with constrictions between the seeds 2–$2\frac{1}{2}$ lin. apart, smooth, glabrous. *Schlechter in Engl. Jahrb.* xx. *Beibl.* 51, 46, *and Journ. Bot.* 1897, 295.

KALAHARI REGION : Transvaal ; margins of Makwonga Forest, near Barberton, 3200 ft., *Galpin*, 908 ! in bush near Barberton, *Thorncroft*, 60 ! and *in Herb. Wood*, 4291 ! Houtbosch Berg, 6400 ft., *Schlechter*, 4423 ! Ingome Range, *Gerrard*, 1322 ! between Pilgrims Rest and Sabie, *Burtt Davy*, 1580 !

2. **R. polyantha** (Schlechter in Journ. Bot. 1895, 272) ; rootstock a cluster of numerous long narrow white tubers (*Mrs. Barber*) ; stem twining, unifariously puberulous ; leaves herbaceous, varying from nearly glabrous to thinly puberulous on both sides ; petiole 1–$2\frac{3}{4}$ in. long ; blade $1\frac{1}{4}$–6 in. long, $\frac{2}{3}$–5 in. broad, cordate-ovate, acute or acuminate ; cymes (including the $1\frac{1}{2}$–4 in.-long peduncle) 4–6 (or rarely up to 10) in. long, unifariously puberulous along the branches, otherwise glabrous in all parts, with 1–3 flowers at each node of the branches ; pedicels $\frac{2}{3}$–$1\frac{1}{2}$ in. long ; sepals 1–$1\frac{1}{2}$ lin. long, subulate ; corolla-tube 5–7 lin. long, cylindric, not or but slightly inflated at the base in dried flowers ; lobes connivent-erect, connate at the tips, 4–6 lin. long, linear-attenuate ; outer corona of 5 pairs of minute spreading teeth turned face to face or of 5 minute bifid concave lobes not more than $\frac{1}{8}$ lin. long, alternating with the inner corona-lobes, which are $\frac{1}{3}$–$\frac{1}{2}$ lin. long, filiform-subulate, incurved-erect, applied to the backs of the anthers and longer or occasionally slightly shorter than them. *Schlechter in Journ. Bot.* 1897, 295. *R. Burchellii, K. Schum. in Engl. and Prantl, Pflanzenfam.* iv. ii. 273. *R. torulosa, Schlechter in Engl. Jahrb.* xviii. *Beibl.* 45, 24.

SOUTH AFRICA : without locality, *Zeyher*, 1153 partly !

COAST REGION : King Williamstown Div. ; Gonubie Bridge, near Kei Road, 2000 ft., *Flanagan*, 2416 ! Albany Div. ; between Blaauw Krantz and Kowie Poort, *Burchell*, 2668 ! Queenstown Div. ; banks of the Zwartkei River, *Mrs. Barber*, 3 ! banks of Komani River, 3500 ft., *Galpin*, 2249 ! near Shiloh, *Baur*, 854 !

CENTRAL REGION : Aliwal North Div. ; Elands Hoek near Aliwal North, 4550 ft., *Bolus*, 301 ! (10501 ! 10522 !). Hopetown Div. ; by the Orange River, *Burchell*, 2652 !

KALAHARI REGION : Griqualand West ; by the Vaal River, *Bowker*, 17 ! Warrenton, *Miss Adams*, 18 ! Orange River Colony ; Rhenoster River, *Burke* ! Transvaal ; Crocodile River, *Burke*, 113 ! *Zeyher*, 113 ! Elands River, *Rehmann*, 4896 ! Umlomati Valley, *Galpin*, 1280 ! Modderfontein, *Conrath*, 1010 !

EASTERN REGION : Transkei ; near Tsomo, *Mrs. Barber*, 893 ! *Bowker* ! near the Bashee River, *Schlechter*, 6291 ! Griqualand East ; banks of the Umzimhlava River, near Kokstad, 4250 ft., *Tyson*, 1726 ! Natal, *Cooper*, 2718 partly !

The flowers are stated to be white by some collectors, by others yellow. *Woodia mucronata* was also distributed as 2718 by Cooper.

3. **R. Bolusii** (N. E. Br.); stem twining, pubescent all round; leaves herbaceous, thinly sprinkled with minute adpressed hairs above, puberulous or pubescent beneath; petiole $\frac{1}{2}$–$1\frac{1}{2}$ in. long; blade $1\frac{3}{4}$–$3\frac{3}{4}$ in. long, 1–$3\frac{1}{4}$ in. broad or perhaps larger, broadly cordate-ovate, acute or acuminate; cymes racemosely or dichotomously branched, and (including the $\frac{1}{2}$–$1\frac{1}{2}$ in.-long peduncle) 2–4 in. long, more or less puberulous on the branches and upper part of the $\frac{2}{3}$–1 in.-long pedicels; flowers in pairs racemosely scattered along the short or gradually elongating branches of the cyme; sepals $1\frac{1}{4}$ lin. long, subulate or lanceolate-subulate, puberulous; corolla-tube $\frac{1}{3}$–$\frac{1}{2}$ in. long, probably slightly ovoid-inflated at the base, but the inflation almost obliterated in dried specimens, glabrous, apparently pale yellowish; lobes connivent-erect and connate at the tips, 3–4 lin. long, attenuate-linear, glabrous, apparently slightly darker than the tube; outer corona-lobes about $\frac{1}{4}$ lin. long, narrowly deltoid or subulate-lanceolate, acute or bifid, concave-channelled down the face, with a minute tooth on each side at the base; inner corona-lobes oblong or linear-oblong, obtuse, adpressed to the backs of the anthers and about half as long as them.

EASTERN REGION: Tembuland; on the banks of the Umtata River near Umtata, 2200 ft., *Bolus*, 10194!

In its broad leaves and more cyme-like character of its inflorescence this species resembles *R. polyantha*, but the leaves are rather thinner and the corona entirely different, being more like that of *R. torulosa*.

4. **R. torulosa** (Decne in DC. Prodr. viii. 640); stem twining, pubescent nearly or quite all round; leaves herbaceous, somewhat adpressed-pubescent on both sides, often more densely so beneath; petiole $\frac{1}{2}$–$2\frac{1}{2}$ in. long; blade $1\frac{1}{4}$–5 in. long, $\frac{2}{3}$–$3\frac{3}{4}$ in. broad, elongated cordate-ovate or more rarely broadly cordate-ovate, acuminate; peduncles 1–3 at each stem-node, slender, $\frac{1}{3}$–$3\frac{1}{2}$ in. long, bent at the nodes, pubescent, bearing 2–3 distant umbel-like fascicles (rarely only 1 fascicle) of 2–15 flowers, with or without an additional fascicle sessile at the node; pedicels $\frac{1}{4}$–$1\frac{1}{3}$ in. long, glabrous or pubescent throughout, or minutely pubescent on the apical part and glabrous below; sepals about 1 lin. long, lanceolate, acuminate, pubescent; corolla glabrous, pale yellow, more or less tinged with red or purplish; tube $3\frac{1}{2}$–5 lin. long, globose or ovoid-inflated at the base; lobes connivent-erect, connate at the tips, $2\frac{1}{2}$–4 lin. long, linear-attenuate; outer corona of 5 subulate or filiform and entire or linear-subulate and bifid lobes, horizontally radiating high above the base of the staminal column, yellowish or purple, alternating with 5 pairs of minute or elongated teeth (sometimes nearly or quite obsolete) placed just below or forming one series with them; inner corona-lobes closely pressed against the staminal column, minute, reaching to about the base of the anthers, thin, oblong, obtuse; staminal column about 1 lin. long; follicles subparallel or curved, 3–7 in. long, $\frac{1}{8}$ in. thick, linear-terete, with constrictions

between the seeds 1–5 lin. long, tapering into a slender beak at the apex, smooth, glabrous or rarely pubescent; seeds $3\frac{1}{2}$–$4\frac{1}{2}$ lin. long, $\frac{3}{4}$ lin. broad, linear or linear-lanceolate, concave on one side, glabrous, dark brown. *Delessert, Ic. Pl.* v. 38, *t.* 91; *Saund. Ref. Bot.* iii. *t.* 157; *Schlechter in Engl. Jahrb.* xxi. *Beibl.* 54, 12; *Journ. Bot.* 1897, 295, *and in Ann. Naturhist. Hofmus. Wien,* xviii. 398, *not elsewhere. Ceropegia torulosa, E. Meyer, Comm.* 194; *Dietr. Syn. Pl.* ii. 891, *not of Haworth.*

VAR. β, **tomentosa** (N. E. Br.); leaves more or less tomentose or very densely pubescent beneath with longer hairs than in the type; outer corona-lobes subulate or filiform, rather long and the pairs of teeth alternating with them usually well developed.

VAR. γ, **longidens** (N. E. Br.); leaves very broadly cordate-ovate; outer corona-lobes filiform, very ($\frac{3}{4}$ lin.) long, with the alternating pairs of teeth nearly or quite obsolete; otherwise as in the type.

VAR. δ, **obsoleta** (N. E. Br.); leaves narrowly deltoid-lanceolate to ovate, acute, cordate with a broad shallow open sinus at the base, thinly and minutely puberulous above, nearly or quite glabrous beneath; outer corona-lobes narrowly deltoid-subulate, minutely notched at the apex, channelled down the face, without any trace of teeth alternating with them.

SOUTH AFRICA : without locality, *Zeyher,* 1153 partly !

COAST REGION : in George district, *Bowie* ! Knysna Div. ; near the Keurbooms River, *Burchell,* 5154 ! Humansdorp Div. ; Kromme River Valley, *Bolus,* 2401 ! Uitenhage or Alexandria Div. ; between Hoffmans Kloof and Driefontein, *Drège,* 2270 ! Albany Div., *Bowker* ! Fort Beaufort Div. ; *Cooper,* 1758 ! Stockenstrom Div. ; Kat Berg, 1000 ft., *Bowker* ! near Kachu (Yellowwood) River, *Drège,* 4944 ! Eastern Frontier, *MacOwan,* 434 ! Kaffraria ; Evelyn Valley, 4000 ft., *Sim,* 1645 ! and without precise locality, *Cooper,* 298 !

KALAHARI REGION : Transvaal, *Rehmann,* 5876 ! Var. β : Basutoland, *Cooper,* 2719 partly ! Orange River Colony ; Harrismith, *Sankey,* 316 ! Var. γ : Transvaal ; slopes of Marovunye near Shiluvane, *Junod,* 724 ! Var. δ ; Transvaal ; by the Crocodile River (probably near Pretoria), *Miss Leendertz,* 708 !

EASTERN REGION : Transkei ; around Kentani, 900 ft., *Miss Pegler,* 877 ! near Idutywa, *Schlechter,* 6269 ! and without precise locality, *Bowker,* 209 ! Tembuland ; near the Chwerka River, *Bolus,* 10195 ! Engcobo Mountain, *Bolus,* 10196 ! near Umtata, *Bolus,* 10197 ! Natal ; near Durban, *Drège* ! *Plant,* 73 ! *Wood in MacOwan, Herb. Austr.-Afr.,* 1732 ! Inanda, *McKen,* 3 ! *Wood,* 512 ! Nonoti River, *Gerrard,* 643 ! and without precise locality, *Gueinzius* ! *Gerrard,* 150 ! Var. β : Transkei ; Fort Bowker, *Bowker,* 572 ! Natal ; Greenwich Farm, Riet Vlei, *Galpin,* 5724 ! South Downs, 5000 ft., *Wood,* 4392 ! Lynedock, *Wood,* 4540 ! Weston, *Rehmann,* 7353 ! near Van Reenens Pass, *Wood,* 4539 !

The corona of this plant either varies considerably or more than one species is here included, but from dried material I am unable to discriminate more than the forms indicated. Dr. Bolus' specimens 10195, 10196, 10197, have a rather more inflated corolla, which dries of a more reddish tint than usual, but otherwise seem the same as the other specimens quoted. The genus requires to be studied from living specimens. *Asclepias peltigera,* Schlechter, has also been distributed by Cooper as no. 2719.

5. **R. Woodii** (N. E. Br.); stem twining, unifariously puberulous; leaves herbaceous, very thinly and shortly pubescent on both sides; petiole $\frac{1}{2}$–1 in. long; blade $1\frac{1}{2}$–$2\frac{1}{2}$ in. long, $\frac{1}{2}$–1 in. broad at the base, or perhaps larger, rather narrowly elongated deltoid-cordate or cordate-lanceolate, acute or acuminate; peduncle moderately stout,

1–2¾ in. long, bearing 2 distant umbels, with or without a third one at its base, puberulous along one side; umbels 8–24-flowered, somewhat flat-topped; pedicels 5–9 lin. long, glabrous, all more or less incurved; sepals 1–1½ lin. long, subulate or subulate-lanceolate, glabrous; corolla-tube 3½–4 lin. long, globose-inflated at the base, glabrous, apparently pale yellowish or whitish; lobes connivent-erect, 3½–4 lin. long, attenuate-linear, glabrous, drying dark brown; outer corona-lobes radiately spreading, ⅓ lin. long, ½ lin. broad at the base, narrowly deltoid or deltoid-ovate, obtuse, subtruncate or minutely notched at the apex, concave-channelled down the face, without any teeth alternating with them; inner corona-lobes linear or linear-subulate, acute or obtuse, nearly as long as the anthers and closely applied to their backs.

EASTERN REGION : Natal; Inanda, *Wood*, 338! and in *Herb. Natal*, 181!

A species of *Scilla* has also been distributed by Mr. Wood as no. 181.

6. R. Flanagani (Schlechter in Engl. Jahrb. xviii. Beibl. 45, 13); stem twining, bifariously puberulous; leaves thinly subcoriaceous, thinly sprinkled with minute hairs on both sides or nearly glabrous above; petiole ⅓–1½ in. long; blade 1–3½ in. long, ⅔–2¼ in. broad, cordate-ovate, acuminate; flower-fascicles umbel-like, ¾–1 in. in diam., compactly many-flowered, sessile or pedunculate; peduncle 0–⅔ in. long, bearing only 1 flower-fascicle; bracts subulate; pedicels 1½–2½ lin. long, glabrous; sepals 1¼–1½ lin. long, lanceolate, glabrous; corolla-tube 2½ lin. long, subcylindric, perhaps slightly contracted at the mouth, glabrous, apparently whitish or pale yellowish; lobes spreading at the base, then arched inwards and connate at the tips, 3 lin. long, linear-filiform, replicate? apparently orange or dull yellowish at the base and purple-brown above; outer corona-lobes minute, ⅙ lin. long, deeply divided into 2 deltoid teeth turned face to face, yellow; inner corona-lobes ½ lin. long, subulate or linear, incumbent on the backs of the anthers and much exceeding them, with their tips crossing one another or perhaps connivent, yellow. *Schlechter in Journ. Bot.* 1897, 295.

COAST REGION : Bathurst Div. ; Trapps Valley, *Miss Daly*, 629! Komgha Div. ; valleys near Komgha, 2000 ft., *Flanagan*, 381! East London Div. ; hillside beyond the lighthouse at East London, *Wood in Herb. Galpin*, 2821!

EASTERN REGION: Transkei ; valleys around Kentani, *Miss Pegler*, 214!

A specimen collected by Bowie "in moist thickets of George and the Knysna Forest" appears to belong to this species, but the outer corona-lobes are subquadrate and only bifid at the apex, not deeply divided.

XL. CEROPEGIA, Linn.

Calyx 5-partite. *Corolla* tubular, usually more or less inflated at the base, straight or curved, 5-lobed; lobes free or variably connate at their tips into a *canopy* or *cage*-like top, valvate in bud. *Corona* double, never 3-seriate; outer corona cup-shaped and entire or 5–10-toothed, or of 5 bifid lobes often pouch-like at the base or reduced

to minute pouches, or the lobes divided to the base and the halves
adnate to the adjacent sides of the inner corona-lobes, the whole
corona then apparently formed of 1 series of 5 trifid lobes opposite
the anthers, or the halves or teeth of 2 adjacent lobes connate and
forming 5 lobes immediately behind the inner corona-lobes, which
are shorter to longer than the anthers and incumbent upon them,
at least at the base, dorsally adnate at their base to the outer
corona. *Staminal column* arising at or close to the base of the corolla,
very short ; anthers ascending or incumbent on the top of the style,
oblong or subquadrate, without an appendage. *Pollen-masses* ascending or subhorizontal, solitary in each anther-cell, pellucid on the
inner margin, attached in pairs by very short caudicles to the pollen-
carriers or subsessile upon them. *Style* not exceeding the anthers,
truncate or shortly conical at the apex. *Follicles* lanceolate, very
narrowly fusiform or subterete, acutely acuminate or obtuse, smooth
or rugose. *Seeds* crowned with a tuft of hairs.

Perennial herbs ; rootstock a tuber or cluster of thick fleshy roots, or rarely
with ordinary stout root-fibres ; stems erect, twining or rambling among other
plants, prostrate or pendulous, herbaceous or fleshy ; leaves opposite or rarely
absent ; flowers usually of moderate size or large, of singular and varied forms,
solitary, in pairs, umbel-like cymes or rarely racemose, lateral at the nodes, rarely
terminal.

DISTRIB. Species over 160, the others in Tropical Africa, the Mascarene Isles,
Malay Archipelago and the hotter parts of Asia. In Tropical Africa the natives
eat the tubers of several species, but I do not know if they do so in South Africa.
The description of the basal inflation of the corolla-tube in most cases only applies
to dried flowers and may not at all agree with that part in the living plant, as it
alters very much and sometimes totally disappears (as in *C. ampliata*, E. Meyer) in
the process of drying. The hairs within the corolla-tube are often quite invisible
on dried flowers that have been wetted for examination, unless the open tube is
doubled back in water and the edge of the fold viewed against the light, or until
all moisture has evaporated from them and been replaced by air. This is also the
case with the pubescence on the corona.

```
  * Corolla-lobes free at the tips :
      Stem strictly erect, ½–1¾ ft. high, not fleshy :
        Leaves linear, ⅓–3½ lin. broad :
            Corolla-lobes  drooping  (erect  in  some  dried
                specimens),  linear-oblong,  or  narrowly  lan-
                ceolate-oblong, narrowed into a short stalk at
                the base :
              Corolla-tube ⅝–¾ in. long ; lobes greenish-yellow
                  without spots      ...      ...      ...      ... (1) Bowkeri.
              Corolla-tube 1–1½ in. long ; lobes with two rows
                  of glabrous blackish spots on a pubescent
                  green ground...      ...      ...      ...      ... (2) sororia.
            Corolla-lobes erect or somewhat spreading, repli-
                cate- or linear-filiform or filiform :
              Corolla-lobes ⅔–1¼ in.  long, puberulous  or
                  tomentose on the inner face, with a tuft of
                  red hairs near the base :
                Corolla smooth on the outside of the inflated
                    base ; lobes slightly thickened upwards    (3) tomentosa.
                Corolla papillate-scabrous on the outside of
                    the inflated base ; lobes not thickened
                    upwards   ...      ...      ...      ...      ... (4) scabriflora.
```

Corolla-lobes 2 in. long, filiform, thickened and
knob-like at the apex, glabrous except at
the base ... ... (5) **antennifera.**

Leaves elliptic-oblong or oblong, $\frac{3}{4}$–1 in. broad ;
corolla-lobes linear, replicate, pubescent on the
inner face ... ... (6) **Rudatisii.**

Stem not erect, thick and fleshy, tapering at the flowering tips, leafless ; corolla 2–3 in. long ... ... (7) **stapeliæformis.**

** Corolla-lobes free at the base, then connate into a
slender column, again becoming free and connate at
the tips, forming a small terminal cage-like body ... (8) **Haygarthii.**

*** Corolla-lobes united into an umbrella-like canopy
supported on 5 short stalks over the mouth of the
tube; stems twining or scrambling among other plants,
or in 9 *C. Rendallii*, sometimes short and erect :
Stem slender, $\frac{1}{2}$–$\frac{3}{4}$ lin. thick ; corolla less than 1 in.
long ... ... (9) **Rendallii.**

Stem rather fleshy, $\frac{1}{8}$ in. or more (when dried about
1 lin.) thick ; corolla 2 in. or more long :
Leaves $\frac{1}{2}$–2 in. long, $\frac{1}{3}$–1 in. broad ; corolla-tube
oblong - inflated at the base ; outer corona
truncate or nearly so :
Corolla-canopy with the centre depressed below the
level of the margin, light yellowish-green,
spotted with darker green ... ... (10) **Sandersoni.**

Corolla-canopy convex or broadly conical, with the
centre raised above the margin, mottled with
green and purple-brown on a whitish ground (11) **Monteiroæ.**

Leaves very small or rudimentary, $\frac{3}{4}$–3 lin. long ;
corolla-tube globose-inflated at the base ; outer
corona of 5 pairs of filiform hooked teeth ... (12) **fimbriata.**

**** Corolla-lobes connate at the tips only, often forming a cage-like top to the flower :
Stems erect, 2–5 in. high or under cultivation much
longer and requiring support :
Corolla-lobes 8–9 lin. long, as long as or longer than
the tube, ciliate at the base .. .. (33) **Barklyi.**

Corolla-lobes $3\frac{1}{2}$ lin. long, much shorter than the
tube, not ciliate ... ... (34) **Conrathii.**

Stems twining or straggling among other plants, rarely
straight and prostrate or pendulous :
Stem fleshy, leafless or with rudimentary leaves less
than $\frac{1}{4}$ in. long :
Corolla-tube 1–2 in. long, more than twice as long
as the lobes, glabrous outside ... ... (13) **ampliata.**

Corolla-tube scarcely $\frac{2}{3}$ in. long, not much longer
than the lobes, puberulous outside ... ... (14) **Zeyheri.**

Stem slender or fleshy, with well developed leaves
$\frac{1}{4}$–4 in. long, $\frac{1}{6}$–1$\frac{3}{4}$ in. broad :
† Corolla-tube puberulous outside, 9–10 lin. long :
Leaves very undulate or lobulate - undulate,
pubescent ; corolla-lobes 5–6 lin. long,
glabrous on the inner face ... ... (27) **undulata.**

Leaves not undulate ; corolla-lobes thinly
pubescent or tomentose-villous on the inner
face :
Leaves glabrous, minutely ciliate ; pedicels
1½–1¾ lin. long ; corolla-lobes 3½ lin.
long ... ... (28) **obscura.**

Leaves puberulous ; pedicels 3–4 lin. long ;
corolla-lobes 2½ lin. long ... ... (29) **pachystelma.**

†† Corolla-tube glabrous outside :
Corolla-lobes in living flowers spreading at the
base with the upper half abruptly or subhorizontally inflexed over the mouth of the
tube, in 30, *C. Meyeri*, sometimes connivent-erect in dried flowers :
Leaves herbaceous, cordate - ovate, entire,
toothed or lobed, pubescent ; corolla 1⅛–
1¾ in. long ; lobes linear... ... ... (30) **Meyeri.**

Leaves fleshy, linear-subterete or elliptic,
glabrous ; corolla ⅔–¾ in. long ; lobes
hair-like at the apical half ... ... (31) **multiflora.**

Corolla-lobes not spreading at the base, falcately
or incurved-connivent, connivent-erect and
straight or erect and curved so as to form
an ellipsoid cage-like top ;
‡ Corolla-lobes much shorter than the tube :
Corolla-tube 1–1⅓ in. long ; lobes ⅓–½ in.
long :
Leaves subsessile, linear or linear-lanceolate, very thick and fleshy, not undulate on the margins ... ... (15) **crassifolia.**

Leaves shortly petiolate, flat, lanceolate,
undulate-crisped at the margins ... (16) **crispata.**

Corolla-tube 1½–2 in. long ; lobes ¾–1⅛ in.
long ; stem prostrate, not twining ... (17) **radicans.**

Corolla-tube 5–9 lin. long :
§ Corolla-lobes 2½–4 lin. long :
Corolla-lobes lanceolate, pilose on the
outer and inner surfaces ; inner
corona-lobes scarcely exceeding the
outer corona ... . (18) **brachyceras.**

Corolla-lobes glabrous on the back,
often ciliate on the margins or
with hairs on the inner face or on
both ; inner corona-lobes much
exceeding the outer corona :
Corolla-lobes falcately or incurved-
connivent, tapering to the apex :
Leaves 1–3½ in. long, elliptic or
oblong-lanceolate to linear-
lanceolate ... ... (19) **setifera.**

Leaves ⅓–1 in. long, ovate or
lanceolate ... ... (20) **carnosa.**

Corolla-lobes connivent-erect, straight
or curved so as to form an
ellipsoid cage-like top :

Corolla-lobes distinctly enlarged
   at the tips, linear-spathulate,
   ciliate to the tips :
   Stem slender, very flexible, pen-
      dulous or prostrate, often
      bearing tubers ; leaves as
      broad as or broader than
      long, white-veined   ... (21) **Woodii.**
   Stem twining or straight and
      pendulous or prostrate, not
      bearing tubers ; leaves often
      longer than broad, not
      white-veined   ...   ... (22) **Caffrorum.**
Corolla-lobes not enlarged at the
   tips, narrowly linear :
   Corolla-lobes glabrous or with
      only a few hairs at the base
      of the keel on the inner face ;
      inner corona - lobes tri-
      quetrous-filiform ...   ... (23) **assimilis.**
   Corolla - lobes covered on the
      inner face and ciliate with
      purple hairs ; inner corona-
      lobes much compressed
      laterally, broadly falcate ... (24) **africana.**

§§ Corolla-lobes 2 lin. long or less, dilated
   at the apex ; stem twining. (If the
   lobes are connate at the tips in 36,
   C. leptocarpa, it should belong here,
   its corolla is 7 lin. long with thinly
   ciliate lobes) :

Leaves lanceolate or lanceolate-oblong,
   2½–4 lin. broad ; corolla-tube about
   ½ in. long ...   . (25) **euryacme.**
Leaves ovate or deltoid-ovate, 3–9 lin.
   broad ; corolla-tube ⅔–¾ in. long (26) **tenuis.**

‡‡ Corolla-lobes not much shorter to longer than
   the tube, very narrowly linear or linear-
   filiform :
Leaves ¼–¾ lin. broad, 10–30 times as long
   as broad ; corolla-lobes 7–10 lin. long (32) **infundibuli-**
                       formis.

Leaves 1–5 lin. broad, 3–9 times as long
   as broad ; corolla-lobes 8–9 lin. long... (33) **Barklyi.**

1. **C. Bowkeri** (Harv. Thes. Cap. i. 9, t. 14) ; rootstock a cluster
of stout fusiform fleshy roots (not a tuber as originally described) ;
stems ½–1 ft. (or more ?) high, erect, simple, glabrous ; leaves erect
or ascending, the lower 1½–2¾ in. long, ⅔–1½ lin. broad, the upper
smaller, linear, acute, glabrous ; flowers 1–5, solitary at the upper
nodes and terminal, distant ; pedicels ⅓–¾ in. long, glabrous ; sepals
2½–4 lin. long, subulate ; corolla-tube erect, ⅔–¾ in. long, globose-
inflated and ¼ in. in diam. at the base, contracted to ⅛ in. in diam.
above, and dilated to ⅓ in. across the short funnel-shaped mouth,
which has recurved bilobulate margins between the lobes, glabrous

outside and within, pale greenish, with the margin of the mouth, the veins in the throat and some markings at the top outside all dark violet-brown; lobes free, regularly or irregularly pendulous around the tube (not erect as originally figured), $\frac{3}{4}-\frac{7}{8}$ in. long, $1\frac{1}{2}-2$ lin. broad, linear-oblong, shortly acute, narrowed into a short stalk at the base, flat, with a row of large pits or depressions on each side of the midrib, light greenish yellow, glabrous on the back, shortly pubescent on the inner face and ciliate with yellow hairs; outer corona arising $\frac{1}{2}$ lin. up the staminal column, with 5 erect acutely bifid ovate lobes 1 lin. long, pouch-like at the base, rather densely villous on the inner face and ciliate with long fine hairs; inner corona-lobes about $1\frac{3}{4}$ lin. long, narrowly linear, obtuse, connivent-erect over and much overtopping the staminal column. *Harv. in Nat. Hist. Rev.* vi. 515, *t.* 27; *Bot. Mag. t.* 5407; *K. Schum. in Engl. and Prantl, Pflanzenfam.* iv. ii. 272, *fig.* 80, C; *Schlechter in Journ. Bot.* 1897, 294.

EASTERN REGION: Transkei; Kreilis Country, *Bowker*, 12! Caffraria, *Bowker*!

2. **C. sororia** (Harv. ex Hook. f. in Bot. Mag. t. 5578); rootstock a cluster of thick fleshy spindle-shaped roots; stem $1-1\frac{3}{4}$ ft. high, erect, simple or with 1 branch, glabrous; leaves $1\frac{1}{2}-5\frac{1}{2}$ in. long, $\frac{1}{2}-3\frac{1}{2}$ lin. broad, linear or narrowly linear-lanceolate, acute, slightly folded lengthwise, glabrous; flowers solitary at the upper nodes; pedicels usually 5–10 lin. long, or up to $2\frac{1}{4}$ lin. long on cultivated plants, glabrous; sepals 2–3 lin. long, subulate, glabrous; corolla-tube $1-1\frac{1}{2}$ in. long, $\frac{1}{4}-\frac{1}{3}$ in. in diam. at the elongated-ovoid inflated base, tapering into the cylindric upper part, which is about $\frac{1}{8}$ in. in diam., scarcely dilating at the $2-2\frac{1}{2}$ lin. wide mouth, glabrous outside and within, greenish-yellow at the base, densely dotted or suffused with violet on the other part or sometimes dotted all over; lobes free, pendulous on the living plant, but on dried specimens often erectly spreading, straight, $\frac{2}{3}-\frac{3}{4}$ in. long, $\frac{1}{8}$ in. broad, linear-oblong or narrowly lanceolate-oblong, acute, tapering into a short stalk at the base, somewhat replicate, blackish or purple-black, with a broad densely pubescent greyish-white line down the middle, giving off transverse bars on each side, according to the appearance of dried flowers and a note by Mr. Galpin, but according to the figure with 2 rows of black spots on a green ground, densely ciliate with purple or purple and yellowish hairs; outer corona overtopping the staminal column, shortly cupular at the base, with 5 pairs of erect narrowly deltoid-lanceolate teeth $\frac{2}{3}$ lin. long, villous-pubescent with rather long white hairs outside and on the inner face of the teeth, pale yellowish, with a few dark purple spots; inner corona-lobes about $1\frac{1}{4}$ lin. long, linear-lanceolate, obtuse, connivent-erect over the staminal column, glabrous, pale yellowish; follicles solitary (always ?), about 2 in. long and 2 lin. thick, narrowly fusiform, tapering into an acute beak, glabrous, smooth; seeds $1\frac{1}{2}$ lin. long, $\frac{3}{4}$ lin. broad, ovate, very pointed, concave on one side, very convex on the other,

rugose on both sides, dark brown. *K. Schum. in Engl. and Prantl, Pflanzenfam.* iv. ii. 272 ; *Schlechter in Journ. Bot.* 1897, 294.

COAST REGION: Queenstown Div. ; Lesseyton Nek, near Queenstown, *Galpin,* 1966 ! Bathurst or Albany Div. ; hills on the east side of Bushmans River, *Ecklon and Zeyher* !

EASTERN REGION: Transkei, *Mrs. Barber,* 15 ! *Bowker* ! and *cultivated specimen* !

The dense ciliation on the corolla-lobes is badly represented in the figure quoted.

3. C. tomentosa (Schlechter in Engl. Jahrb. xviii. Beibl. 45, 33) ; stem 1 ft. or more ? high, glabrous ; leaves ascending, $2\frac{1}{2}$–$3\frac{3}{4}$ in. long, $\frac{1}{2}$–$1\frac{1}{2}$ lin. broad, linear, acute, glabrous ; flowers solitary at the upper nodes, erect ; pedicels 3–5 lin. long, slender, glabrous ; sepals $\frac{1}{4}$–$\frac{1}{3}$ in. long, subulate, glabrous ; corolla-tube (dried) $\frac{2}{3}$–$\frac{3}{4}$ in. long, with an obovoid basal inflation $2\frac{1}{2}$ lin. in diam., abruptly contracted into the 1–$1\frac{1}{2}$ lin. in diam. cylindric part, and dilating to 3 lin. in diam. at the shortly funnel-shaped mouth, where the margin angularly projects between the lobes like 5 small pockets, glabrous and smooth outside and within, pale greenish or yellowish-green, with darker veins on the inflated part and dotted with purple-brown along the veins on the other part ; lobes free, erect, slightly diverging at the tips, 1–$1\frac{1}{4}$ in. long, linear from a deltoid base, replicate, at least in the lower half ; yellowish-green, densely tomentose with short curled or interwoven yellow hairs on the inner face, and ciliate for the space of about $\frac{3}{4}$ lin. with dense tufts of stout subclavate red hairs, about 2 lin. above the base ; outer corona $\frac{1}{2}$ in. long, and not exceeding the staminal column, very shortly cupular at the base, with 5 erect deeply bifid lobes or pairs of deltoid-lanceolate acute teeth $\frac{1}{3}$ lin. long, bearded at the tips on the inner face with long white hairs ; inner corona-lobes 1 lin. long, linear or subspathulate-linear, connivent-erect, dorsally connected with the outer corona at the base and forming a small tubercle between the outer corona-lobes, glabrous, apparently purple-brown on the basal half and where united to the outer corona, elsewhere the entire corona is white or yellowish. *Schlechter in Journ. Bot.* 1897, 294.

EASTERN REGION : Transkei ; near Butterworth, *Bowker* ! Bashee Ridge, near Fort Bowker, *Bowker,* 372 ! *Mrs. Barber,* 372 !

4. C. scabriflora (N. E. Br.) ; rootstock a cluster of thick fleshy fusiform roots 2–$3\frac{1}{2}$ in. long ; stem solitary, erect, simple or with 1 branch at the top, about 1 ft. high, scarcely 1 lin. thick, glabrous ; leaves in 6–7 pairs, suberect, 1–2 in. long, $\frac{1}{2}$–1 lin. broad, linear, acute, with recurved margins, glabrous ; flowers 1 or 2 to a stem, solitary, lateral at a subterminal node and terminal, erect ; pedicels $\frac{1}{4}$–$\frac{1}{3}$ in. long, minutely pubescent or subglabrous ; sepals $2\frac{1}{2}$–3 lin. long, lanceolate-subulate, glabrous ; corolla-tube $\frac{3}{4}$–1 in. long, globose-inflated and $\frac{1}{4}$–$\frac{1}{3}$ in. in diam. at the densely papillate-scabrous base, abruptly contracted into the $\frac{1}{8}$ in. in diam. smooth cylindric upper part, about 2–2$\frac{1}{2}$ lin. in diam. at the mouth, glabrous outside and within, but papillate within the inflated part, yellowish-green, apparently veined with purplish at the mouth; lobes free, spreading, $\frac{2}{3}$–1$\frac{1}{4}$ in. long, narrowly linear or replicate-filiform from a short deltoid 1–1$\frac{1}{4}$ lin.-broad base, pubescent with very fine purple hairs on the inner face and adorned with a tuft of stout clavate ruby-purple hairs near the base; outer corona arising about $\frac{1}{2}$ lin. up and overtopping the staminal column, 1 lin. long, shortly cupular at the base, with 5 broad deltoid acute (but at the same time bifid with short contiguous teeth), lobes $\frac{2}{3}$ lin. long, resembling in side view a funnel-shaped 5-lobed cup, ciliate and minutely pubescent inside, blackish-purple; inner corona-lobes $\frac{3}{4}$–1 lin. long, linear or slightly spathulate-linear, incumbent on the backs of the anthers at the base, then connivent-erect in a column, dorsally connected to the outer corona and thinly pubescent at the base, glabrous above, blackish-purple.

EASTERN REGION : Natal; near Verulam, 800 ft., *Wood,* 7908 !

5. **C. antennifera** (Schlechter in Engl. Jahrb. xx. Beibl. 51, 46) ; tuber depressed-globose, 1 in. or more in diam. ; stem solitary, erect, 5–6 in. (or more?) high, $\frac{2}{3}$ lin. thick, simple, glabrous; leaves in 5–6 pairs, erect, the lower very small, the upper $\frac{3}{4}$–1$\frac{1}{3}$ in. long, $\frac{1}{3}$–$\frac{1}{2}$ lin. broad, linear-filiform, acute, glabrous; flowers solitary, lateral at the 1–2 upper nodes, and terminal, erect, much over-topping the stem; pedicels 1–1$\frac{1}{4}$ in. long, glabrous; sepals $\frac{1}{4}$ in. long, subulate, very acute, glabrous; corolla-tube 1$\frac{1}{3}$ in. long, slightly inflated and about 2 lin. in diam. at the base, cylindric and 1 lin. in diam. above, becoming 1$\frac{1}{2}$ lin. in diam. at the mouth, glabrous and striped with purple-brown and white outside, pilose with fine purple hairs within; lobes free, 2–2$\frac{1}{4}$ in. long, erectly spreading, resembling the antennæ of a butterfly, $\frac{3}{4}$ lin. broad and bearded with purple hairs on the inner face of the linear-lanceolate base, filiform and glabrous above, with the tips dilated into a small knob, slightly notched at the apex and grooved on each side of a prominent keel on the inner face; " corona blackish-violet, outer lobes truncate at the apex with the margin on each side produced into a short acute tooth, ciliate; inner lobes erect, linear, acute, with the apex reflexed, much overtopping the outer lobes " (ex *Schlechter*). *Schlechter in Journ. Bot.* 1897, 295.

EASTERN REGION ; Natal; near Newcastle, 4100 ft., *Schlechter,* 3426 !

Of this I have only seen a part of the type in the Herbarium of Dr. Bolus, from which the corona had been removed for description.

6. **C. Rudatisii** (Schlechter in Engl. Jahrb. xl. 94) ; rootstock a tuber (*Wood*) ; stem solitary, simple, erect, 5–6 (about 20, *Schlechter*) in. high, 2 lin. thick at the base, glabrous (glaucescent, *Schlechter*) ;

leaves sessile or subpetiolate, $1\frac{1}{2}$ ($1\frac{3}{4}$–$2\frac{3}{4}$, *Schlechter*) in. long, $\frac{3}{4}$ (to rather more than 1, *Schlechter*) in. broad, oblong or elliptic-oblong, obtuse or subacute, apiculate, subcoriaceous or perhaps slightly fleshy, glabrous on both sides (glaucescent, *Schlechter*); flowers lateral at the nodes, apparently solitary, erect; pedicels $\frac{1}{2}$ ($1\frac{1}{2}$–$2\frac{1}{4}$, *Schlechter*) in. long, longer in fruit, glabrous; sepals 5–7 lin. long, $\frac{1}{2}$ lin. broad, linear, acute, erect, glabrous; corolla-tube 1–$1\frac{2}{3}$ in. long, ovoid-inflated, and probably about $3\frac{1}{2}$ ($2\frac{1}{2}$, *Schlechter*) lin. in diam. at the base, cylindric and $1\frac{1}{2}$ lin. in diam. above, gradually enlarging to $3\frac{1}{2}$ lin. across the mouth, glabrous outside and within, slightly subtuberculate along the veins in the inflated part, green, with the veins inside at the base dark purple-brown; lobes free, suberect (spreading, *Schlechter*), $\frac{7}{8}$–$1\frac{1}{3}$ in. long, about 1 lin. broad at the base, linear, acute, apparently replicate, glabrous on the back, pubescent with fine greenish-yellow hairs all over the inner face, and ciliate with long clavate yellow-based purple hairs for about $\frac{1}{4}$ in. at the base, a short row of the same hairs extending down the middle of the very base of the lobe, green, reticulately veined with dark purple-brown at the base and mouth of the tube; outer corona arising $\frac{1}{2}$ lin. up the staminal column and much over-topping it, shortly cupular at the base, with 5 erect deeply bifid lobes 1 lin. long, having parallel subulate teeth, dark purple-brown, ciliate with long fine white hairs and pubescent on the inner face; inner corona-lobes $1\frac{1}{2}$ lin. long, connivent-erect, linear, acute, exceeding the outer corona and dorsally connected to it at the base, glabrous, dark purple-brown; follicle solitary (always?), erect, 7 in. long, about $\frac{1}{4}$ in. thick, linear-terete, very slightly tapering to the obtuse apex, but distinctly so at the base; seeds $\frac{1}{2}$ in. long, 2 lin. broad, linear-oblong, broadly margined, flattened, smooth, glabrous, dark brown with paler borders.

EASTERN REGION: Natal; Hillcrest, 2000 ft., *Haygarth in Herb. Wood*, 9099! near Fairfield in Alexandra County, 2000 ft., *Rudatis*, 203 (ex *Schlechter*).

With the exception of the few words in brackets, the above description is entirely made from Mr. Haygarth's specimen, which so closely agrees with Dr. Schlechter's description, that I have little doubt that the few discrepancies are merely due to a difference of vigour in the two specimens.

7. **C. stapeliæformis** (Haw. in Phil. Mag. 1827, 121); stems and branches fleshy, up to $\frac{1}{2}$ in. thick, cylindric, tuberculate, jointed, decumbent or trailing, often tapering towards the flowering ends, which become rather slender and sometimes twine around other plants, glabrous, dull green, with darker mottling, often tinged with purplish; leaves rudimentary, seated on the tubercles, $\frac{1}{2}$–$1\frac{1}{2}$ lin. long, cordate-ovate or deltoid, acute, with minute globular stipules at their base, glabrous; cymes usually scattered along the tapering slender ends of the branches, pedunculate, successively developing 1 to several flowers; peduncles 1–9 lin. long, glabrous; pedicels 3–5 lin. long, glabrous; sepals about 2 lin. long, ovate-lanceolate, acute, glabrous; corolla-tube $1\frac{1}{4}$–$1\frac{3}{4}$ in. long, ovoid-inflated and $\frac{1}{4}$ in. in diam. at the base,

tapering into the cylindric ⅛ in. in diam. part above, and dilated
to ⅔–¾ in. across the funnel-shaped mouth, glabrous outside,
pubescent inside at the mouth and basal part with white curly
hairs, pale greenish-white, spotted with violet at the base and
upper part; lobes free, recurved-spreading, ¾–1½ in. long, linear
from a deltoid base, acute, replicate, pubescent on the lower part
of the inner surface, white, with the margins and apical part violet-
purple or dark purple-brown; outer corona about 1 lin. high and
slightly exceeding the staminal column, cup-like with 5 pairs of
short erect acute teeth, pubescent inside and ciliate all round the
margin with rather long hairs, alternately marked with broad
yellow and dark purple-brown stripes; inner corona-lobes 1½ lin.
(or more?) long, incumbent on the backs of the anthers at the base,
then connivent into an erect column with shortly recurved tips,
dorsally connected to the outer corona at the base, glabrous, yellow,
with a dark purple-brown stripe down the middle of the basal part.
Bot. Mag. t. 3567; *Maund, Bot.* iv. *t.* 154; *G. Don, Gen. Syst.* iv.
112; *Dietr. Syn. Pl.* ii. 890; *Decne in DC. Prodr.* viii. 643; *Fl. des
Serres,* ii. *June, t.* 4; *Rev. Hort.* 1869, 25, *fig.* 6, *and* 1901, 109, *fig.*
37; *Schlechter in Journ. Bot.* 1897, 294. *C. stapeliiformis, K. Schum.
in Engl. and Prantl, Pflanzenfam.* iv. ii. 270, *fig.* 80, D.

SOUTH AFRICA: without locality, *Thunberg*!
COAST REGION: Uitenhage Div.; among shrubs on the banks of the Zwartkops
River, *Zeyher,* 36! Albany Div.; by the Fish River, *Bowker*! Fort Beaufort
Div.; banks of the Koonap River, *Mrs. Barber,* 85! King Williamstown Div.;
near King Williamstown, *Barkly*!
CENTRAL REGION: Somerset Div.; Bruintjes Hoogte, *MacOwan*! Cradock
Div.; near Cradock, *Cooper,* 492! Graaff Reinet Div.; Magazine Hill, near Graaff
Reinet, 2700 ft., *Bolus,* 422!

Described from living plants.

8. **C. Haygarthii** (Schlechter in Engl. Jahrb. xxxviii. 46, fig. 7A);
stem climbing, fleshy, 1½–2 lin. thick, glabrous; leaves small, fleshy,
flat, ¼–⅔ in. long, ⅛–¼ in. broad (see note below), ovate or ovate-
lanceolate, acuminate; cymes 1–2-flowered, lateral at the nodes;
peduncles ¾–1½ in. long, glabrous; bracts 1–2 lin. long, subulate;
pedicels 5–7 lin. long, glabrous; sepals 5lin. (2½ lin., *Schlechter*)
long, subulate, glabrous; corolla abruptly bent at a right-angle near
the base; tube (following the bend) about 1⅓ in. long according to a
drawing, or about 1 in. long in dried flowers, globosely inflated at
the base, cylindric above, enlarging according to a drawing to about
¾ in. in diam. at the mouth, pinkish-white or greenish tinted, spotted
with violaceous, glabrous outside, pilose with very fine long hairs
within; lobes free at the base, abruptly inflexed over the mouth of
the tube and produced beneath into broad triangular partition-like
green plates or keels, meeting at the centre and connate into a
slender erect column 5–7 lin. long, then again becoming free and ex-
panding into elliptic-lanceolate replicate segments connate at the tips,
forming a small apical ellipsoid cage-like body 2½–3 lin. long, ciliate

on the margins, dull purple or purple-brown ; corona in the flowers
seen much eaten by insects, but apparently the outer corona is
cupular, with 5 acutely bifid lobes rising to the level of the top of
the staminal column, ciliate and hairy within with long fine hairs ;
inner corona-lobes 1 lin. long, linear or linear-spathulate, connivent-
erect over the staminal column, with very revolute tips.

COAST REGION : Stockenstrom Div. ; Maasdorp, in forest, *Scully*, 196 !
EASTERN REGION : Natal ; without precise locality, *McKen* ! *Sanderson* ! also
flower from a plant cultivated at Cape Town, *Barkly* !

In the Kew Herbarium is a copy of a drawing made by Mr. Sanderson (from
which drawing the figure published by Dr. Schlechter would seem originally to
have been also copied) in which the leaves are represented as subsessile, and as
described above, but in another drawing of this species at Kew they are represented as very distinctly petiolate. Dr. Schlechter described from the drawing
only, but the Kew Herbarium contains a few loose leaves sent with a flower by
McKen, which are as follows :—petiole 2-3 lin. long ; blade $\frac{3}{4}$-1$\frac{1}{4}$ in. long, $\frac{1}{3}$-$\frac{1}{2}$ in.
broad, ovate-lanceolate, acuminate to a very acute point, slightly subcordate at
the rounded base, glabrous on both sides ; the largest leaf with Scully's specimen is
about 1 in. long and $\frac{3}{4}$ in. broad, broadly ovate, and the peduncles are 2-4-flowered,
but it otherwise seems the same species.

9. C. Rendallii (N. E. Br. in Kew Bulletin, 1894, 100) ; stem
twining, or occasionally not more than 2-4 in. high and then erect,
slender, glabrous ; leaves somewhat fleshy, glabrous or with a few
scattered short hairs on both sides and with or without a short
ciliation ; petiole 1-2 lin. long ; blade $\frac{1}{2}$-1 in. long and $\frac{1}{4}$-$\frac{1}{2}$ in. broad
when ovate, 1-1$\frac{1}{2}$ in. long and 1$\frac{1}{2}$-3 lin. broad when linear or
linear-oblong, obtuse and apiculate or acute ; peduncles 1-3-flowered,
$\frac{1}{8}$-$\frac{3}{4}$ in. long, glabrous ; pedicels 1-2$\frac{1}{2}$ lin. long, glabrous ; sepals
$\frac{3}{4}$-1$\frac{1}{2}$ lin. long, subulate, glabrous ; corolla-tube $\frac{1}{2}$-$\frac{2}{3}$ in. long, curved,
globosely inflated and 1$\frac{1}{4}$-1$\frac{1}{2}$ lin. in diam. at the base, contracted to
$\frac{2}{3}$ lin. in diam. above and dilated to 2$\frac{1}{2}$-3 lin. in diam. at the
funnel-shaped mouth, glabrous outside, thinly pilose with long fine
hairs within, apparently dark purple at the base, but perhaps only
inside, whitish above ; lobes very slender and claw-like at the erect
basal half, inflexed, dilated and all united above into an umbrella-
like canopy $\frac{1}{3}$ in. in diam., raised 2$\frac{1}{2}$-3 lin. above the mouth of the
tube, with 10 rounded marginal lobes, which, together with the
claws, are shortly ciliate, but the hairs are easily detached, apparently dark purple or green ; outer corona $\frac{1}{2}$ lin. long, equalling the
staminal column, cup-like, apparently 5-angled, truncate or rising
into very short deltoid lobes at the angles, glabrous ; inner corona-
lobes $\frac{1}{2}$ lin. long, erect, much overtopping the staminal column,
laterally much flattened and about $\frac{1}{4}$ lin. broad, broadly recurved-
falcate, obtuse, dorsally connected at the base to the outer corona
and seeming to form part of it, apparently scarcely incumbent on
the backs of the anthers, glabrous. *C. Galpinii, Schlechter in Engl.
Jahrb.* xviii. *Beibl.* 45, 23 ; xx. *Beibl.* 51, 47, *and Journ. Bot.* 1897,
294.

KALAHARI REGION : Transvaal ; hills at Reimers Creek, near Barberton, *Galpin,* 1251, Barberton, *Miss Stainbank* ! near Lydenburg, *Schlechter,* 3936 ! and without precise locality, *Rendall* !

My original description of the corona of this plant is entirely wrong, owing to the specimen examined having been very much crushed. The colour of the corona is probably white, but becomes stained with the purple of the corolla when the flower is placed in boiling water for dissection. Dr. Schlechter describes the ovate leaves as being ¾–1⅓ in. long and ⅔–1⅛ in. broad, I have not seen them so large.

10. C. Sandersoni (Decne ex Hook. f. in Bot. Mag. t. 5792) ; " roots tuberous similar to those of a *Dahlia* " (*Gerrard*) ; stem twining, 1½–2 lin. thick, fleshy, glabrous, slightly rough to the touch ; leaves fleshy, glabrous ; petiole 1–3 lin. long, stout ; blade ⅔–1¾ in. long, ½–1 in. broad, ovate-lanceolate to broadly cordate-ovate, acute or shortly cuspidate-acute, light green ; cymes with 2–4 flowers, developed singly, glabrous ; peduncles 2–5 lin. long, 1½–2 lin. thick ; pedicels 3–5 lin. long, nearly or quite ⅛ in. thick, becoming stouter in fruit ; sepals 3–3½ lin. long, 1 lin. broad, narrowly oblong, acute, longitudinally folded, glabrous ; corolla-tube curved, 1½–2 in. long, with an oblong inflation ¼ in. in diam. at the base, narrowed above and enlarged to ⅔ or ¾ in. in diam. at the funnel-shaped mouth, glabrous with the exception of a few hairs at the very base inside ; striped with green and white on the upper part outside and within, light green on the inflation outside, dull greyish- or purplish-green within, with numerous ribs, which abruptly terminate at the base of the purple contracted part ; lobes united into a flattish 5-keeled umbrella-like canopy 1¼–1½ in. in diam., supported on 5 short claws, with 5 broad obtuse slightly bifid marginal much arched lobes, ciliate with vibratile white hairs, its centre distinctly depressed, with a 6-pointed tubercle above and a 5-ribbed projection beneath, yellowish-green, spotted with light green above and with brighter green underneath, with the ribbed projection beneath and some spots around it blackish-purple ; outer corona ½ lin. long, cup-shaped, not pentagonal, truncate, entire, whitish, with the margin and at its junction with the inner corona-lobes purple-brown, ciliate with white hairs ; inner corona-lobes 1⅓ lin. long, incumbent on the backs of the anthers, with erect filiform tips, recurved at the apex, dorsally connected to the outer corona at the base, glabrous, white ; follicles horizontally diverging, 3–5½ in. long, 3⅓–3¾ lin. thick, terete, tapering from about the middle to a slightly dilated umbonate apex about 2 lin. in diam., irregularly rugose and tuberculate, glabrous, green, stained with dull purplish. *Gard. Chron.* 1870, 173, *fig.* 29 ; *Baill. Dict. Bot. t.* 10 ; *Rev. Hort.* 1901, 111, *fig.* 39. *C. Sandersoniæ, Hook. in Bot. Mag. under t.* 6927. *C. Sandersonii, K. Schum. in. Engl. and Prantl, Pflanzenfam.* iv. ii. 273, *fig.* 80, B ; *Graebener in Monatssch. f. Kakt.* x. 71 *and* 73, *with fig.* ; *N. E. Br. in. Gard. Chron.* 1906, xl. 383, *fig.* 145 ; *C. fimbriata, Schlechter in Journ. Bot.* 1897, 294, *not of Engl. Jahrb.* xxi. *Beibl.* 54, 13, *nor of E. Meyer.*

EASTERN REGION: Natal; Tugela, *Gerrard*, 1798! *McKen*! and without precise locality, *Sanderson*! *Sutherland*! and *cultivated specimens*!

Described from living plants. In *Engl. Jahrb.* xxi. *Beibl.* 54, 13, and in *Journ. Bot.* 1897, 294 *C. Sandersoni* is referred by Dr. Schlechter to *C. fimbriata*, E. Meyer, but the two species are totally different in foliage and floral structures, and it is evident from the localities quoted that he intended the name to apply to *C. Sandersoni* only.

11. **C. Monteiroæ** (Hook. f. in Bot. Mag. t. 6927); stem twining, $1\frac{1}{4}$–$1\frac{1}{2}$ lin. thick, fleshy, glabrous, slightly rough to the touch; leaves distant, shortly petiolate, $\frac{1}{2}$–2 in. long, $4\frac{1}{2}$–10 lin. broad, ovate or ovate-lanceolate, acute, rounded or subcordate at the base, fleshy, glabrous; cymes shortly pedunculate or subsessile, 1–2-flowered; pedicels 3–$4\frac{1}{2}$ lin. long, 1 lin. thick, becoming 2 lin. thick when in fruit, fleshy, glabrous; sepals 3–4 lin. long, $\frac{1}{2}$–$\frac{3}{4}$ lin. broad, linear, acute, longitudinally folded, glabrous; corolla resembling that of *C. Sandersoni*; tube $1\frac{3}{4}$–2 in. long, slightly curved, with a basal oblong inflation slightly constricted at the middle and $\frac{1}{4}$ in. in diam., narrowed above, then dilated into a broad funnel-shaped mouth $\frac{3}{4}$–1 in. in diam., glabrous outside and within except in the inflation, which is rugosely ribbed throughout and pubescent with spreading hairs on the basal half, green at the base, striped with green and white on the upper part; lobes united into an umbrella-like convex canopy $1\frac{1}{3}$–$1\frac{1}{2}$ in. in diam., supported on 5 short claws, with 5 obtusely bifid somewhat flattened (not arched nor strongly keeled) marginal lobes, ciliate with long vibratile purple hairs, its centre raised above the margin, with a 6-pointed tubercle above and a 5-ribbed projection beneath, glabrous above and beneath, mottled with green and purplish-brown above and with bright green beneath on a whitish ground, the projecting centre being purple-black; outer corona $\frac{1}{2}$ lin. long, cup-shaped, pentagonal, with slightly projecting entire angles, yellow, marked with purple-brown where united to the inner corona-lobes and at the angles, ciliate with white hairs; inner corona-lobes 2 lin. long, incumbent on the backs of the anthers, with erect filiform tips, revolute at the apex, dorsally connected to the outer corona at the base, glabrous, yellowish; follicles widely diverging, $3\frac{1}{2}$–4 in. long, about 5 lin. thick, subterete, slightly tapering to a thick truncate apex, which is emarginate on the inner side, irregularly rugose and tuberculate, glabrous, green, stained and speckled with dull purplish. *Schlechter in Journ. Bot.* 1897, 294.

EASTERN REGION: Delagoa Bay, *Mrs. Monteiro, cultivated specimens*!

Described from a living plant. *C. Monteiroæ* may be distinguished from the closely allied *C. Sandersoni*, by the convex (not depressed-flattish and 5-keeled) top of the flower, and the flattish-roofed (not arched) openings between the claws. The figure in the *Botanical Magazine* was drawn from immature flowers, which become much larger than there represented.

12. **C. fimbriata** (E. Meyer, Comm. 194); stem twining, fleshy, glabrous; leaves minute or small, $\frac{3}{4}$–3 lin. long, $\frac{1}{2}$–$\frac{3}{4}$ lin. broad, lanceolate, acute, glabrous; flowers solitary at the nodes; peduncle

0 or exceedingly short ; pedicel about $\frac{1}{3}$ in. long, moderately stout, glabrous ; sepals 2–2$\frac{1}{2}$ lin. long, $\frac{2}{3}$ lin. broad, lanceolate, acuminate, glabrous ; corolla-tube 1$\frac{1}{2}$ in. or more long, apparently curved at about the middle, globose-inflated and $\frac{1}{3}$ in. in diam. at the base, cylindric and $\frac{1}{8}$ in. in diam. above, funnel-shaped in the upper part and about $\frac{1}{2}$ in. in diam. at the mouth, glabrous outside and within, but densely tuberculate on the inside of the inflated part ; lobes connate into a shortly conical umbrella-like canopy about $\frac{1}{3}$ in. long, supported on 5 short broad claws with their margins inrolled, scarcely keeled down the middle of the dilated part, margin of the canopy 10-crenate, fringed with long vibratile clavate purple hairs, otherwise glabrous ; outer corona-lobes 1–1$\frac{1}{4}$ lin. long, very deeply divided into 2 linear-filiform subparallel segments, incurved-hooked at the tips ; inner corona-lobes about 1$\frac{1}{3}$ lin. long, slightly over-topping the outer corona, linear-subulate, connivent-erect. *Dietr. Syn. Pl.* ii. 891 ; *Decne in DC. Prodr.* viii. 645 ; *Schlechter in Engl. Jahrb.* xxi. *Beibl.* 54, 13, *excl. synonym.*

Coast Region : Peddie Div. ; on dry stony hills near the Fish River not far from Trumpeters Drift, 600 ft., *Drège,* 4948 !

The colour of the flowers is not determinable from the only dried specimen (the type) seen. Dr. Schlechter has mistaken *C. Sandersoni,* Decne, for this species, but the two are totally different.

13. **C. ampliata** (E. Meyer, Comm. 194) ; stem succulent, twining or scrambling, leafless at the time of flowering, glabrous ; leaves only seen at the young tips of the stems, soon deciduous, minute, 1–1$\frac{1}{2}$ lin. long, lanceolate, acute, glabrous ; flowers 2–4 together at the nodes, successively developed ; pedicels 3–7 lin. long, glabrous ; sepals 1$\frac{2}{3}$–2$\frac{1}{4}$ lin. long, lanceolate, acuminate, glabrous ; corolla-tube in dried specimens 1–2 in. long, 4–6 lin. in diam., cylindric and slightly or not at all inflated at the base, but on the living plant, according to a drawing, 2 in. long, globosely and somewhat lobulate-inflated and about 1 in. in diam. at the base, cylindric and $\frac{1}{2}$ in. in diam. above, not dilated at the apex, pale green, with a narrow purple transverse band at the top of the inflation inside, glabrous outside, covered inside with long simple hairs, longer and more matted at the purple band and above than in the lower part ; lobes 4–6 lin. long, 2$\frac{1}{2}$–3 lin. broad at the base, lanceolate from a deltoid base, acute, erectly connivent and connate at the tips, replicate or with reflexed margins, glabrous on both sides and not ciliate, green, spotted with darker green, becoming olive-brown when dried, probably with a velvety sheen on the inner surface : outer corona cup-shaped, equally 10-toothed ; teeth $\frac{2}{5}$–$\frac{1}{2}$ lin. long, narrowly deltoid, acute, hairy on the inner surface ; inner corona-lobes 2–2$\frac{1}{2}$ lin. long, very slenderly filiform, connivent-erect, dorsally-connected by vertical plates to the outer corona at the base. *Dietr. Syn. Pl.* ii. 891 ; *Decne in DC. Prodr.* viii. 645 ; *Schlechter in Engl. Jahrb.* xviii. *Beibl.* 45, 12 ; xx. *Beibl.* 51, 49 ; xxi. *Beibl.* 54, 13, *and Journ. Bot.* 1897, 294.

E. Meyer describes the corolla-lobes as being "subciliate," I do not find them
so in his type specimen. The Natal specimens have much larger flowers than
those from the other localities, but appear to be otherwise the same. The basal
inflation of the corolla seems to disappear in the process of drying. Schlechter
records it from the Transvaal, but I have not seen a specimen from there.

14. C. Zeyheri (Schlechter in Engl. Jahrb. xxxviii. 48); stem
twining, branching, fleshy, glabrous, with the leaves reduced to
ovate acute spreading scales; cymes lateral at the nodes, pedunculate, 2–3-flowered; peduncle fleshy, $2\frac{1}{2}$–5 lin. long, glabrous;
pedicels 2–3 lin. long, glabrous; sepals about 2 lin. long, lanceolate,
acuminate, glabrous; corolla erect, about $1\frac{1}{8}$ in. long; tube subglobosely inflated and 3 lin. in diam. at the base, narrowed and
subcylindric above, puberulous outside, verrucose-spotted at the
base and glabrous within; lobes connate at the tips, about $\frac{1}{2}$ in.
long, narrowly linear, dilated at the apex, puberulous except at the
apex; outer corona-lobes connate into a cup at the base, deeply
bifid, with linear obtuse segments, glabrous; inner corona-lobes
shortly exceeding the outer corona, erect, linear, with spathulately
dilated obtuse tips, glabrous.

Dr. Schlechter has suppposed the Groote River to be synonymous with the
Vaal River, and has stated that the habitat is in the Transvaal, but this I believe
to be an error.

15. C. crassifolia (Schlechter in Journ. Bot. 1895, 273); stem
twining, apparently growing from a few inches up to 2 ft. high,
fleshy and as much as $\frac{1}{4}$ in. thick at the base, tapering upwards,
glabrous; lower leaves 2–4 in. long, the upper smaller, thick and
fleshy, subsessile, 2–$4\frac{1}{2}$ lin. broad, linear or linear-lanceolate, acute,
channelled down the face, glabrous; cymes lateral at the upper
nodes, pedunculate, about 3–5-flowered, glabrous in all parts
except inside the corolla; peduncle 2–4 lin. long, stout; pedicels
2–3 lin. long; sepals 2–$2\frac{1}{4}$ lin. long, lanceolate-attenuate; corolla
of dried flowers $1\frac{1}{3}$–$1\frac{7}{8}$ in. long; tube 1–$1\frac{1}{3}$ in. long, straight, with
an ellipsoid or subobovoid inflation $\frac{1}{4}$–$\frac{1}{3}$ in. in diam. at the base,
contracted to 2–$2\frac{1}{2}$ lin. in diam. above and gradually dilated into a
funnel-shaped mouth 4–5 lin. in diam., glabrous and whitish with
dark purple-brown spots and lines outside, inside pilose with long
fine curly hairs from the top of the inflated part to the mouth;
lobes 4–6 lin. long, connivent-erect, connate at the tips, narrowly
lanceolate, replicate and $1\frac{1}{2}$–2 lin. broad viewed sideways, glabrous
and whitish spotted with purple on the back, the inner face closely

reticulated with dark purple-brown on a cream-coloured or yellowish ground, or in dried flowers often appearing entirely purple-brown, sparsely covered with long purple hairs on the basal part and ciliate with them on the keel to the middle or beyond ; outer corona cup-shaped $\frac{3}{4}$–$1\frac{1}{4}$ lin. deep, with 5 large erect lobes, $\frac{1}{3}$–$\frac{1}{2}$ lin. long and $\frac{1}{2}$–1 lin. broad, truncate, very broadly rounded or notched at the top, glabrous ; inner corona-lobes $\frac{1}{3}$–$\frac{1}{2}$ lin. long, linear, connivent, shortly exceeding the anthers, but scarcely rising to the level of the outer corona, and dorsally connected at the base to the latter by vertical partitions, glabrous.

SOUTH AFRICA: without locality, specimen cultivated in Cape Town Botanic Garden !

COAST REGION: Uitenhage Div. ; in dry stony places, *Zeyher* ! "Korotra Hill, Uitenhage," *Prior* ! King Williamstown Div. ; near King Williamstown, 1500 ft., *Sim*, 312 ! Div. ? Chaka River, a tributary of the Fish River, *Mrs. Barber* !

EASTERN REGION : Natal ; Springvale, *Miss Button in Herb. Sanderson*, 2003 !

I describe the colour of the flowers from a drawing at Kew of the Natal specimen by Mr. Sanderson.

16. **C. crispata** (N. E. Br.) ; rootstock a cluster of fleshy spindle-shaped roots ; stem solitary, climbing, glabrous ; leaves (only one seen) fleshy, flat ; petiole 2 lin. (or more ?) long ; blade $1\frac{1}{3}$ in. long, $\frac{1}{2}$ in. broad or larger, lanceolate, acute, cuneate at the base, wavy at the scabrous-denticulate margins, glabrous on both sides ; cymes lateral at the nodes, 4–5-flowered ; peduncle $\frac{1}{8}$ in. long in the only example seen, glabrous ; bracts 1–2 lin. long, subulate ; pedicels about 2 lin. long, glabrous ; sepals about 2 lin. long, lanceolate-subulate, glabrous ; corolla-tube 1–$1\frac{1}{4}$ in. long, somewhat abruptly bent above the ellipsoid-inflated base (which in the larger of 2 flowers is 7 lin. long and 5 lin. broad), cylindric and $1\frac{1}{2}$–2 lin. in diam. above, enlarging to 4–5 lin. in diam. at the mouth, glabrous and whitish marked with elongated violet spots outside, inside covered with long white hairs èxcept in the inflated part ; lobes 5–6 lin. long, erect, connate at the tips, replicate, $1\frac{1}{4}$–$1\frac{3}{4}$ lin. broad across the side, oblong when flattened out, apparently blackish-purple or perhaps veined with that colour on a paler ground, fringed on the margins and keel and covered on the inner face of the basal half with long simple purple hairs ; outer corona cupular at the base, 5-lobed, glabrous ; lobes $\frac{1}{2}$ lin. long, oblong, deeply bifid, concave, rising to or above the level of the staminal column ; inner corona-lobes about $\frac{1}{2}$ lin. long, linear, acute, incumbent on the backs of the anthers and shortly produced beyond them, dorsally connected at the base with the outer corona, glabrous.

KALAHARI REGION : Orange River Colony or Griqualand West ; near the Vaal River, very rare, *Mrs. Barber*, 675 !

17. **C. radicans** (Schlechter in Engl. Jahrb. xviii. Beibl. 45, 12) ; stems prostrate, creeping, with fibrous roots at the nodes, fleshy, $\frac{1}{8}$ in. or more (about 1 lin. when dried) thick ; leaves thick, fleshy,

glabrous; petiole 2–4 lin. long; blade $\frac{2}{3}$–$1\frac{3}{4}$ in. long, $\frac{1}{2}$–$1\frac{1}{2}$ in. broad, ovate, elliptic or orbicular, acute or obtuse and apiculate, cuneate, rounded or subcordate at the base; peduncles obsolete or up to $\frac{1}{4}$ in. long, 1 lin. thick, 1–2-flowered; bracts minute; pedicels $\frac{1}{2}$–1 in. long, $\frac{3}{4}$ lin. thick, glabrous; sepals $\frac{1}{4}$–$\frac{1}{3}$ in. long, linear-subulate, glabrous; corolla-tube curved (often straight in dried specimens), $1\frac{1}{2}$–2 in. long, cylindrically or oblong-inflated for $\frac{1}{2}$–$\frac{2}{3}$ in. and $2\frac{1}{2}$–3 lin. in diameter at the base, 1–$1\frac{1}{2}$ lin. in diam. above and dilated to $\frac{1}{2}$ in. or more in diam. at the funnel-shaped mouth, outside glabrous, pale greenish below, whitish above, dotted or spotted with dark purple, inside pilose with white hairs at the base and upper part, glabrous elsewhere; lobes $\frac{3}{4}$–$1\frac{1}{8}$ in. long, connivent-erect, connate at the tips, linear from a 4–5 lin.-broad (when flattened out) deltoid base, closely replicate so as to form pocket-like openings between them, keeled down the inner face, purple-brown crossed by a broad transverse white band at the base, above passing into bright green or first into blackish-green and then to bright green, thinly hairy on the inner face below and ciliate to nearly half-way up with fine soft spreading hairs, glabrous above, but ciliate with long vibratile clavate purple hairs; outer corona $\frac{1}{4}$ lin. long, shorter than the staminal column, obtusely pentagonal or 5-lobed seen from above, pale yellowish-green; lobes pouch-like, shortly bifid, truncate or broadly rounded, glabrous; inner corona-lobes $1\frac{1}{3}$ lin. long, sub-linear or linear-spathulate, acute or obtuse, incumbent on the backs of the anthers at the base, then erect in a column, glabrous, pale yellowish-green. *Schlechter in Journ. Bot.* 1897, 294.

Coast Region : Komgha Div. ; prostrate under bush near Kei Bridge, 2000 ft., *Flanagan*, 384 ! and *cultivated specimen* !

Described from a living plant cultivated at Kew.

18. **C. brachyceras** (Schlechter in Engl. Jahrb. xxxviii. 45) ; stem twining, slightly fleshy, glabrous; leaves spreading or erectly-spreading, subsessile, slightly fleshy, 2–$2\frac{3}{4}$ in. long, $\frac{1}{2}$–1 in. broad, lanceolate-elliptic, lanceolate-oblong or sometimes ovate-lanceolate, acuminate, glabrous; cymes lateral at the nodes, pedunculate, few-flowered; peduncle about $2\frac{1}{2}$ lin. long, glabrous; pedicels about as long as the peduncle, glabrous; sepals scarcely more than $\frac{1}{8}$ in. long, lanceolate, acuminate, glabrous; corolla-tube about $\frac{2}{3}$ in. long, subglobosely inflated and about $\frac{1}{4}$ in. in diam. at the base, subcylindric above, glabrous outside, puberulous within; lobes $\frac{1}{3}$ in. long, connate at the tips, lanceolate, pilose on both sides, more or less undulate on the margins; outer corona shortly cupular, with 5 subquadrate lobes slightly bifid at the apex, glabrous; inner corona-lobes linear, acute, scarcely exceeding the outer corona.

Kalahari Region: Bechuanaland ; near Maritzani, *Duparquet* 432 (ex *Schlechter*).

19. C. setifera (Schlechter in Engl. Jahrb. xx. Beibl. 51, 48); stem several ft. long, twining, thinly and minutely pubescent along one side or glabrous; leaves herbaceous; petiole 3–10 lin. long, minutely pubescent in the channel only; blade 1–3¼ in. long, ½–1¾ in. broad, elliptic or elliptic-oblong, or the smaller sometimes lanceolate, acuminate, acute, or obtuse and apiculate, rounded or subcordate at the base, sometimes faintly sinuous at the margins, thinly puberulous (or glabrous, see note below) on both sides, very minutely adpressed-ciliate; peduncles lateral at the nodes, slender, ultimately 1–1¾ in. long, glabrous, racemosely 2–4-noded, with 2 flowers at a node, developing successively; pedicels ¼–⅓ in. long, glabrous; sepals 1¼–1½ lin. long, lanceolate, acuminate, very slightly pubescent; corolla-tube slightly curved, ⅔ in. long, slightly ovoid-inflated and 2–2½ lin. in diam. at the base, which gradually narrows into the ⅛ in. in diam. cylindric part above, enlarging to about ¼ in. in diam. at the mouth, glabrous outside, pilose within to the base with very fine hairs, invisible when wetted, apparently dark purple-brown or perhaps spotted with that colour at the base, shading into greenish above; lobes 2½–3 lin. long, connivent or incurved-connivent, connate at the tips, deltoid-lanceolate, deltoid-ovate, or oblong-ovate, tapering to the apex, replicate, ½–¾ lin. broad across the side, ciliate with long pale purple or white hairs along the fold or keel of the inner face, otherwise glabrous and apparently greenish or yellowish, with slightly darker edges; outer corona very shortly cupular at the base rising behind the inner corona-lobes into 5 cuneately subquadrate lobes ½ lin. long, with the shoulders produced into 2 widely spreading deltoid or deltoid-linear acute or obtuse teeth at the top, ciliate and hairy on the inner face with long white hairs; inner corona-lobes 1¼–1⅓ lin. long, triquetrous-subulate or compressed linear-falcate acute (or obtuse in *Schlechter,* 4543, see note below), connivent-erect, with falcately recurving or diverging tips, dorsally connected at the base with the outer corona, glabrous, dark purple-brown. *Schlechter in Journ. Bot.* 1897, 295; *N. E. Br. in Dyer, Fl. Trop. Afr.* iv. i. 457.

Var. β, **natalensis** (N. E. Br.); petiole 2–8 lin. long, puberulous all round; blade 1–3⅓ in. long, ¼–1¼ in. broad, linear-lanceolate, lanceolate or oblong-lanceolate, acute or obtusely pointed and apiculate, rounded or rarely subcordate at the base, puberulous on both sides; peduncles ultimately ½–2 in. long, racemosely 3–6-noded; pedicels ⅛–⅓ in. long; sepals about 1 lin. long; corolla "green with brown markings" (*Wood*); tube about ½ in. long; inner corona-lobes ¾–1 lin. long; follicles very widely divergent or slightly reflexed, 3¼–3¾ in. long and about ⅙ in. thick, linear-terete, tapering into a slender beak, smooth, glabrous, marked with short violaceous lines on a pale ground.

KALAHARI REGION: Transvaal; second water between Nelspruit and Sibthorps, *Burtt Davy,* 1619! Masetana Valley, near Shiluvane, *Junod,* 1021! near Barberton, *Thorncroft* (cultivated specimen in *Herb. Wood,* 10289)!
EASTERN REGION: Var. β: Natal; in thorny bush near the Tugela River, *Gerrard,* 1325! Umhloti River, *Wood,* 1318! near Durban, *Wood,* 8261!

This species is closely allied to *C. carnosa,* E. Meyer, but readily distinguished by its much larger leaves, it was originally collected by Dr. Schlechter in Tropical

Transvaal, and of the two specimens he quotes, I have only seen his 4543 from Valdezia, which, although named by himself, does not agree with his description, as the leaves are not "sparsely pilose" but glabrous on both sides, and on the lobes of the outer corona (which are described as "deeply bifid at the apex, with the lanceolate lobules divaricate-falcate setaceous-acuminate") I can find no trace of a setaceous point; they are exactly as described above for the Transvaal specimens, with which it is evidently conspecific, although differing from those quoted in its glabrous leaves and more compressed and therefore broader inner corona-lobes, all other characters being the same, and similar variations occur in other species. The variety *natalensis* may prove to be a distinct species, its narrower leaves giving it a different appearance, but a comparison of the living plants is needed to decide this point.

20. **C. carnosa** (E. Meyer, Comm. 193) : stem twining, glabrous; leaves thick and fleshy (thin, when dried), glabrous; petiole $\frac{1}{8}-\frac{1}{3}$ in. long; blade $\frac{1}{3}-1$ in. long, $\frac{1}{6}-1$ in. broad, lanceolate or ovate, rarely elliptic or rounded, usually acute or acuminate, or, in the rounded forms, obtuse or emarginate, rounded at the base; cymes pedunculate, lateral at the nodes, with 2–5 successively developed flowers; peduncle $\frac{1}{6}-\frac{2}{3}$ in. long, below the often elongated flowering part, glabrous; pedicels $\frac{1}{4}-\frac{1}{2}$ in. long, glabrous or with here and there a hair; sepals $1\frac{1}{2}-1\frac{3}{4}$ lin. long, $\frac{1}{3}-\frac{2}{5}$ lin. broad, subulate, glabrous or with a few scattered hairs; corolla-tube about 7 lin. long, curved above the (slightly in the dried state) inflated base, cylindric and $1\frac{1}{3}-1\frac{1}{2}$ lin. in diam. above, widening to about $3\frac{1}{2}$ lin. in diam. at the mouth, or in some dried specimens apparently only slightly broader at the mouth, glabrous outside, hairy at the base and upper part, but glabrous at the middle part within; lobes $2\frac{1}{2}-3$ lin. long, $1\frac{1}{2}$ lin. broad at the base, oblong-ovate, replicate, connivent or incurved-connivent, tapering to and connate at the tips, finely ciliate with white hairs (not easily seen in some dried specimens) on the upper part; outer corona of 10 teeth $\frac{1}{3}-\frac{1}{2}$ lin. long, adnate in pairs to the base of the inner corona-lobes, subhorizontally divergent and curved towards the teeth of other pairs so that the tips of 2 adjacent teeth nearly or quite meet, leaving a large opening below them, revealing the fissure between the anther-wings, hairy on the inner face; inner corona-lobes $1-1\frac{1}{4}$ lin. long, compressed or filiform-subulate, connivent at the base, then erect, with recurved slightly thickened tips, glabrous or more or less hairy at the base, apparently white. *Dietr. Syn. Pl.* ii. 891; *Decne in DC. Prodr.* viii. 645; *Schlechter in Engl. Jahrb.* xviii. *Beibl.* 45, 12 *and* 23; xxi. *Beibl.* 54, 13, *and Journ. Bot.* 1897, 294.

Coast Region: Uitenhage Div.; near the Zwartkops River, *Zeyher*, 654! near Uitenhage, *Schlechter*, 2494! Bathurst Div.; between Kowie river and Kap River, *Drège*, 4946! between Bushmans River and Karega River, *Zeyher*, 841! 3389! by the Kowie River, *Atherstone*! Albany Div.; near Blauw Krantz, *Mrs. Barber*, 13! Komgha Div.; near Komgha, 2000 ft., *Flanagan*, 1037! near Kei Bridge, 1800 ft., *Flanagan*, 1037!

Central Region: Somerset Div.; at the foot of Bosch Berg, 2000 ft., *MacOwan*, 924!

Eastern Region: Transkei; Fort Bowker, *Bowker*, 545! Natal: Inanda, *Wood*, 869! *Haygarth in Herb. Wood*, 7535! Zululand, *Wood*, 10288!

21. **C. Woodii** (Schlechter in Engl. Jahrb. xviii. Beibl. 45, 34);
tuber subglobose, producing many slender pendent, prostrate or
trailing glabrous stems 1–2 ft. long, which often bear small tubers
and root at the nodes ; leaves fleshy ; petiole $1\frac{1}{2}$–$4\frac{1}{2}$ lin. long ; blade
$\frac{1}{4}$–$\frac{2}{3}$ in. long and as much in breadth, broadly ovate to orbicular,
acute, or obtuse and apiculate, cordate at the base, glabrous on both
sides, dark green with whitish veins ; peduncles lateral at the
nodes, $2\frac{1}{2}$–7 lin. long, slender, glabrous, 1–4- (usually 2-) flowered ;
bracts minute ; pedicels $\frac{1}{8}$–$\frac{1}{3}$ in. long, slender, glabrous ; sepals
$\frac{3}{4}$–1 lin. long, narrowly lanceolate, acuminate ; corolla-tube very
slightly curved, 7–9 (or when dried 6–7) lin. long, globosely inflated
(or rarely the top of the inflation is pushed inwards or intruded),
and about $2\frac{1}{4}$ lin. in diam. at the base, cylindric and 1–$1\frac{1}{3}$ lin. in
diam. above, enlarging to 2 lin. in diam. at the mouth, glabrous
outside and nearly so within, pink or with darker lines on the basal
part ; lobes 3–$3\frac{1}{2}$ lin. long, connivent-erect, connate at the tips,
narrowly spathulate from a deltoid base, replicate, blackish-purple,
ciliate with long simple purple hairs and with similar hairs on the
basal part of the inner face ; outer corona of 5 small truncate
pocket-like lobes $\frac{1}{4}$ lin. long, alternating with the inner corona-lobes,
shorter than the staminal column, glabrous, white ; inner corona-
lobes 1–$1\frac{1}{4}$ lin. long, subspathulate-lanceolate or linear-lanceolate,
acute, connivent over the staminal column at the base, then erect,
with recurved tips, dorsally adnate at the base to the outer corona,
white, glabrous ; follicles diverging, 2–$2\frac{3}{4}$ in. long, $\frac{1}{8}$ in. thick, linear-
terete, with a slender beak, smooth, glabrous ; seeds $2\frac{1}{4}$–$2\frac{1}{2}$ lin. long,
narrowly oblong from being longitudinally folded, rather broadly
margined, smooth, glabrous, brown. *Schlechter in Journ. Bot.* 1897,
295 ; *N. E. Br. in Bot. Mag. t.* 7704 ; *Gartenfl.* 1901, *t.* 1486 ;
Wood and Evans, Natal Plants, iv. *t.* 357 ; *Roth in Monatssch.
Kakt.* xiii. 27.   *C. Woodi, Mast. in Gard. Chron.* 1897, xxii. 357–8,
fig. 104, *and* 1905, xxxvii. 244, *fig.* 103.

EASTERN REGION : Natal ; hanging from rocks on the Groen Berg, 2000 ft.,
Wood, 1317 ; Noods Berg, 2000–3000 ft., *Wood,* and *cultivated specimens* !

Described from living plants cultivated at Kew and Cambridge.   There are two
forms of this plant ; in one the corolla-lobes are erect and connivent at the connate
tips only ; in another form they close together near the base, separating above,
with connivent connate tips.

22. **C. Caffrorum** (Schlechter in Journ. Bot. 1894, 358) ; stem
twining, $\frac{1}{2}$–1 lin. thick, glabrous ; leaves spreading or deflexed,
apparently fleshy, glabrous ; petiole $\frac{2}{3}$–$1\frac{2}{3}$ lin. long ; blade $\frac{1}{3}$–$\frac{7}{8}$ in.
long, $\frac{3}{4}$–3 lin. broad, linear, lanceolate or ovate-lanceolate, acuminate
at the apex, cuneate, rounded or subcordate at the base ; peduncles
lateral at the nodes, $\frac{1}{4}$–$\frac{1}{2}$ in. long, 2–5-flowered, glabrous ; bracts
minute ; flowers developing successively ; pedicels $\frac{1}{8}$–$\frac{1}{4}$ in. long ;
sepals 1–$1\frac{1}{4}$ lin. long, lanceolate-subulate, with very acute recurved
tips, glabrous ; corolla-tube 7–8 lin. long, ovoid-inflated and (in
dried flowers) 1–$1\frac{1}{2}$ lin. in diam. at the base, cylindric above and

scarcely or but slightly widened at the mouth, glabrous and green with purple lines (*Wood*) outside, thinly covered with very fine long hairs within; lobes 3–3½ lin. long, straight, connivent-erect, connate at the enlarged tips, linear-spathulate, replicate, blackish-purple, ciliate throughout with short dark purple or violet hairs, otherwise glabrous; outer corona with pouch-like entire lobes alternating with the anthers, much shorter than the staminal column, but rising to its top behind the inner corona-lobes in deltoid-ovate obtuse lobes, glabrous; inner corona-lobes connivent-erect high above the staminal column or sometimes connate into a column, ¾–1 lin. long, spathulate-lanceolate, flat, with recurved or hooked tips, glabrous. *Schlechter in Journ. Bot.* 1897, 295.

VAR. β, **dubia** (N. E. Br.); stems 1–1⅓ lin. thick, occasionally twining, but usually weak, trailing or pendulous and straight, without the least signs of twining; leaves ⅓–⅞ in. long, ¼–⅔ in. broad, rather broadly triangular-ovate, very acute, slightly cordate at the base, slightly fleshy, undulate at the margins, deep green with impressed slightly darker veins; inner corona-lobes not recurved at the tips; otherwise as in the type.

EASTERN REGION: Natal; near Durban, *Wood*, 5376 (ex *Schlechter*), and without number, *Wood*! Var. β: Delagoa Bay, specimens cultivated at Kew, sent by *Mrs. Monteiro*!

The Delagoa Bay variety is very different from the type in habit, but does not appear to differ in its flowers.

23. **C. assimilis** (N. E. Br.); stems 1–2 ft. long, twining at the upper part, rather slender, simple or with 1 branch at or near the base, glabrous; leaves fleshy, flattish, shortly petiolate, ½–1⅓ in. long, 1–4½ lin. broad, linear to lanceolate, acute, glabrous; cymes lateral at the upper nodes, pedunculate, 2–4-flowered; peduncles ¼–¾ in. long, glabrous; bracts minute; pedicels 1½–3 lin. long, glabrous; sepals 1 lin. long, narrowly lanceolate, acute, glabrous; corolla-tube apparently nearly straight, ⅔ in. long, with an ellipsoid (or subglobose?) inflation ⅛ in. in diam. at the base, about ¾ lin. in diam. above, enlarged to ⅛ in. in diam. at the funnel-shaped mouth, apparently greenish, glabrous outside, thinly sprinkled with very fine hairs within except at the inflated base; lobes 3½–4 lin. long, narrowly linear from a small deltoid base, connivent-erect, connate at the tips, keeled down the inner face, with undulated margins, glabrous or perhaps with a few hairs at the base of the keel, apparently purplish-tinted; outer corona ⅓ lin. long, cup-shaped, with the margin of its pocket-like divisions truncate, glabrous; inner corona-lobes about 1 lin. long, triquetrous-filiform, acute, connivent over the style-apex at the basal half, then strongly recurved-spreading at the tips high above the outer corona, to which they are dorsally connected by their narrow partition-like keels, glabrous.

COAST REGION: Albany Div.; without precise locality, *Bowker*!

The hairs within the corolla-tube can only be observed in dried flowers under a strong lens, and are invisible when wet.

24. C. africana (R. Br. in Bot. Reg. t. 626); rootstock a tuber; stem ½–1 ft. long, twining, glabrous; leaves fleshy, glabrous; petiole 1–2 lin. long; blade ⅓–1 in. long, ⅛–½ in. broad, varying from ovate to linear-lanceolate, acute, mucronate or apiculate, rounded or broadly subcuneate at the base; cymes pedunculate, lateral at the nodes, 2–3-flowered; peduncles 1–3 lin. long, glabrous; pedicels 1–2 lin. long, glabrous; sepals ¾ lin. long, ¼ lin. broad, linear-lanceolate, acute, glabrous; corolla-tube 7–8 lin. long, straight, globosely inflated and about 2 lin. in diam. at the base, ½ lin. in diam. above, widening to 2½ lin. in diam. at the funnel-shaped mouth, glabrous outside and within, greenish, striate with violet-brown above; lobes 3–3½ (or according to the figure about 5) lin. long, straight, connivent-erect, connate at the tips, linear from a deltoid base, not enlarged at the apex, replicate, keeled down the inner face, glabrous on the back, dark violet-brown, ciliate on the margins and keel with dark purple hairs; outer corona cup-shaped, exceeding the staminal column, with 5 broad 3-crenate or broadly rounded erect lobes scarcely ¼ lin. long, alternating with the anthers; inner corona-lobes ¾ lin. long, erectly connivent over the staminal column, recurved at the apex, much compressed laterally, broadly falcate, very obtuse, ⅓ lin. broad, channelled on the inner and acute on the outer edge, dorsally connected at the base to the outer corona between its lobes. *Lodd. Bot. Cab. t.* 906 ; *Spreng. Syst. Veg.* i. 842 ; *G. Don, Gen. Syst.* iv. 110 ; *Dietr. Syn. Pl.* ii. 890 ; *Decne in DC. Prodr.* viii. 642 ; *Schlechter in Journ. Bot.* 1897, 294, *excl. syn. and localities.*

SOUTH AFRICA : without locality, *cultivated specimen* !

The very broad falcate inner corona-lobes and entire absence of hairs within the corolla-tube are distinctive characters of this species. According to Dr. Schlechter in *Engl. Jahrb.* xviii. *Beibl.* 45, 12, a specimen collected near Komgha, in Komgha Div. by Flanagan (714) belongs to this species, but I have not seen the specimen and doubt the identification ; it may be *C. linearis,* E. Meyer, as in *Engl. Jahrb.* xxi. *Beibl.* 54, 13, Schlechter refers that species to *C. africana.*

25. C. euryacme (Schlechter in Engl. Jahrb. xxxviii. 46); stem twining, filiform, glabrous; leaves spreading, somewhat fleshy, glabrous on both sides; petiole 2–3 lin. long; blade ¾–1 in. long, 2½–4 lin. broad at or below the middle, lanceolate or lanceolate-oblong, acute or acuminate; peduncles lateral at the nodes, about as long as the petioles, 1–2-flowered, glabrous; pedicels about 1½ lin. long, glabrous; sepals 1¼ lin. long, lanceolate, acuminate, glabrous; corolla-tube about ½ in. long, subglobosely inflated and 2 lin. in diam. at the base, narrower above and slightly enlarged at the mouth, glabrous outside, puberulous on the lower half within; lobes 2 lin. long, linear from the base, elliptically dilated at the apex, puberulous on the inner face; outer corona cupular, somewhat inconspicuously 5-lobed, glabrous; inner corona-lobes much exceeding the outer corona, linear-oblanceolate, obtuse, glabrous. *C. Woodii, Schlechter in Engl. Jahrb.* xx. *Beibl.* 51, 49, *not elsewhere.*

The author does not state if the corolla-lobes are free or connate, but as he at first named the plant *C. Woodii*, they are probably connate at the tips, as in that species.

26. C. tenuis (N. E. Br.); stems long, twining, very slender, $\frac{1}{3}-\frac{1}{2}$ lin. thick when dried, glabrous; leaves apparently somewhat fleshy, glabrous; petiole $1\frac{1}{2}-3$ lin. long; blade $\frac{1}{2}-1$ in. long, $\frac{1}{4}-\frac{2}{3}$ in. broad near the base, ovate or deltoid-ovate, acute or obtuse and apiculate; cymes pedunculate, subumbellately 3–6-flowered; flowers successively developed; peduncles $1\frac{1}{2}-4$ lin. long, glabrous; pedicels $1\frac{1}{2}-2$ lin. long, glabrous; sepals $1-1\frac{1}{4}$ lin. long, lanceolate, acute, glabrous; corolla-tube $\frac{2}{3}-\frac{3}{4}$ in. long, ovoid-inflated and probably about 2 lin. in diam. at the base, cylindric and $\frac{3}{4}$ lin. in diam. above, gradually widening to $1\frac{1}{2}$ lin. across the mouth, glabrous outside, thinly pilose with fine hairs within above the inflated part, which is smooth within, apparently whitish or pale greenish, passing into blackish-purple at the mouth; lobes 2 lin. long, erect, connate at the tips, linear-spathulate from a deltoid $\frac{3}{4}$ lin.-broad base, replicate, glabrous, ciliate with long purple hairs along the keel on the inner face formed by the folding, blackish-purple; outer corona scarcely $\frac{1}{2}$ lin. long, arising above the base of the staminal column, cupular, with entire subtruncate margins to the pocket-like spaces, glabrous, rising into 5 short deltoid obtuse lobules behind and adnate to the inner corona-lobes, which are 1 lin. long, dorsally flattened, lanceolate-spathulate, acute, connivent-erect, much overtopping the $\frac{2}{3}$ lin.-long staminal column, white.

27. C. undulata (N. E. Br.); tuber large, flattened; stem several feet long (*Gerrard*), slender, twining, pubescent; leaves apparently somewhat fleshy, only a few seen; petiole 2–3 lin. long, pubescent; blade $\frac{1}{3}-1\frac{1}{6}$ in. long, $\frac{1}{4}-\frac{3}{4}$ in. broad, or perhaps larger, ovate, very acute or obtuse and apiculate, cordate or rounded at the base, remarkably lobulate-undulate on the margins, pubescent on both sides; cymes pedunculate, umbel-like, 6–18-flowered, the flowers developing successively; peduncles $\frac{1}{3}-1$ in. long, pubescent; pedicels $1\frac{1}{2}-3$ lin. long, pubescent; sepals $1\frac{1}{3}-1\frac{1}{2}$ lin. long, subulate, pubescent; corolla-tube curved, about $\frac{3}{4}$ in. long, ovoid-inflated and about $1\frac{3}{4}$ lin. in diam. at the base, cylindric and $\frac{3}{4}$ lin. in diam. above, gradually widening to $1\frac{1}{2}-1\frac{3}{4}$ lin. in diam. at the mouth, puberulous outside, glabrous within; lobes erect, connate at the tips, 5–6 lin. long, $\frac{3}{4}$ lin. broad at the base, linear-spathulate, puberulous on the back, glabrous on the flat keelless inner face; outer corona not rising above the staminal column, cupular, probably obtusely pentagonal, rising into 5 minute erect teeth behind the inner corona-lobes, alternating with 5 truncate pocket-like spaces, glabrous, apparently white or yellowish ; inner corona-lobes ¾ lin. long, connivent-erect, spathulate-lanceolate, from a filiform base, recurved at the acute tips, glabrous, apparently white.

EASTERN REGION: Natal ; Tugela, *Gerrard,* 1799 !

According to Gerrard, the leaves and stem are eaten by the Kaffirs. The flowers may have been greenish.

28. **C. obscura** (N. E. Br.) ; stem twining, slender, thinly pubescent, chiefly near the nodes ; leaves (only a few from the upper part of the stem seen, probably larger below), fleshy ; petiole 1–1½ lin. long, puberulous ; blade ¼–½ in. long, 2–4 lin. broad, lanceolate, ovate-lanceolate or oblong, acute or obtuse and apiculate, minutely and sparsely ciliate, otherwise glabrous ; cymes lateral at the nodes, 3–5-flowered ; peduncles about ¼ in. long, thinly puberulous ; bracts minute ; pedicels 1½–1¾ lin. long, puberulous ; sepals ¾–1 lin. long, lanceolate, acuminate, puberulous ; corolla-tube slightly curved, ¾ in. long, with an ellipsoid inflation about 2 lin. in diam. at the base, cylindric and scarcely 1 lin. in diam. above, scarcely enlarging at the mouth, puberulous outside and sprinkled with some very fine curly hairs at the middle part only within, which are invisible when wetted, apparently pale greenish, spotted with violet on the basal part ; lobes erect, connate at the tips, 3½ lin. long, ⅓ lin. broad, linear, very slightly broadened near the apex, flat, puberulous on the back, pubescent on the inner face from the middle upwards with rather longer hairs, very shortly ciliate, purplish-brown, all the hairs colourless ; outer corona somewhat cup-shaped, with truncate pocket-like divisions, about as long as the staminal column, glabrous ; inner corona-lobes 1 lin. long, narrowly subspathulate-lanceolate, acuminate, connivent over the staminal column at the base, then erect, with recurved tips, dorsally connected at the base with the outer corona, glabrous.

EASTERN REGION: Delagoa Bay, *Mrs. Monteiro* !

29. **C. pachystelma** (Schlechter in Engl. Jahrb. xx. Beibl. 51, 47) ; stem twining, slightly branched, minutely pubescent ; leaves rather fleshy ; petiole 1½–3 lin. long ; blade ½–2 in. long, ¼–1 in. broad, ovate-oblong or elliptic-oblong, obtuse, abruptly apiculate, rounded at the base, puberulous ; cymes lateral at the nodes, 2–6 ?-flowered ; peduncle ½–6 lin. long, puberulous ; pedicels 3–4 lin. long, puberulous ; sepals ⅔ lin. long, deltoid-lanceolate, acuminate, puberulous ; corolla only seen in young bud, rather densely puberulous all over on the outside, but according to *Schlechter* has a tube 10 lin. long, subglobosely inflated and 2½ lin. in diam. at the base, cylindric above and slightly dilated at the mouth ; lobes 2½ lin. long, erect, connate at the tips, linear-oblong, obtuse, replicate, glabrous on the back (see above), tomentose-villous on the inner face ; outer corona-lobes connate in a short tube or ring, with free semiorbicular very

obtuse tips, thickened at the margins; inner corona-lobes erect, linear-oblong from a narrow base, dilated below the middle, obtuse, much longer than the outer corona. *Schlechter in Journ. Bot.* 1897, 295; *N. E. Br. in Dyer, Fl. Trop. Afr.* iv. i. 457.

KALAHARI REGION: Transvaal; Mailas Kop, 2500 ft., *Schlechter*, 4511! Makapans Poort, near Potgeiters Rust, 4300 ft., *Schlechter*, 4317.

The only specimen I have seen (*Schlechter*, 4511) has but one very immature flower upon it, which I was unable to dissect; the locality Mailas Kop is just within Tropical Transvaal, where the plant was also collected by Dr. Schlechter on the bank of the River Limvovo, near Valdezia.

30. **C. Meyeri** (Decne in DC. Prodr. viii. 645); rootstock a flattened tuber; stem herbaceous, up to 3 or 4 ft. long, twining, pubescent; leaves herbaceous; petiole $\frac{1}{3}$–$1\frac{1}{4}$ in. long, pubescent; blade $\frac{3}{4}$–$1\frac{3}{4}$ in. long, $\frac{1}{3}$–$1\frac{1}{6}$ in. broad, cordate-ovate or lanceolate-ovate, acute or somewhat acuminate, cordate or rounded at the base, entire, variably toothed, or somewhat lobed and the margin cut into numerous short linear-oblong lobules, more or less pubescent or rarely subglabrous on both sides; cymes sessile or subsessile at the nodes, 2–4-flowered; pedicels $\frac{1}{4}$–$\frac{2}{3}$ in. long, villous-pubescent; sepals $3\frac{1}{4}$–$4\frac{1}{2}$ lin. long, $\frac{1}{2}$ lin. broad at the base, subulate, pubescent; corolla-tube 1–$1\frac{1}{2}$ in. long, bottle-shaped, the basal $\frac{2}{3}$ inflated cylindric-oblong and $\frac{1}{4}$–$\frac{1}{3}$ in. in diam., narrowed above into a cylindric neck 1–$1\frac{1}{4}$ lin. in diam., abruptly dilated to $\frac{1}{4}$ in. in diam. at the mouth, glabrous outside and within, except at the slightly pubescent inner surface of the mouth, white or greenish-white on the basal part, striate or dotted with light or dark purple-brown or violet on the upper $\frac{2}{3}$, dark (purple?) at the base inside; lobes $4\frac{1}{2}$–5 lin. long, $1\frac{1}{2}$ lin. broad at the very base, linear, with reflexed margins, connate at the tips, in dried specimens usually appearing to be connivent-erect, but when alive, they are horizontally spreading at the base and horizontally incurved at the middle, almost black, with 2 central green longitudinal stripes on the inner face, glabrous, not ciliate; outer corona-lobes $\frac{1}{2}$ lin. long, deltoid, acute, ascending, glabrous, white; inner corona-lobes 1 lin. long, linear or slightly spathulate-linear, obtuse, connivent at the base, then slightly diverging and in a broad curve again connivent at the tips, blackish at the basal part, white above; follicles erect, subparallel, $3\frac{1}{2}$–4 in. long, $\frac{1}{8}$ in. thick, terete, tapering into a beak, smooth, glabrous. *Schlechter in Engl. Jahrb.* xviii. *Beibl.* 45, 12 *and* 24; xx. *Beibl.* 51, 47, *and* xxi. *Beibl.* 54, 13; *Journ. Bot.* 1897, 294; *N. E. Br. in Dyer, Fl. Trop. Afr.* iv. i. 454. *C. pubescens, E. Meyer, Comm.* 193; *Dietr. Syn. Pl.* ii. 891, *not of Wall.*

COAST REGION: Komgha Div.; near Komgha, *Flanagan*, 640!
KALAHARI REGION: Transvaal; hills at Riemers Creek, near Barberton, *Galpin*, 812; Elandspruit Mountains, 6000 ft., *Schlechter*, 3864; by the Litonandoa and Limvovo Rivers, 1800 ft., *Schlechter*, 4527, all ex *Schlechter*. Masetana Valley, near Shiluvane (with entire subglabrous and toothed pubescent leaves) *Junod*, 1020!

EASTERN REGION: Transkei; near Fort Bowker, *Bowker,* 483! 541! Butterworth, *Mrs. Barber (Bowker),* 942! 943! valleys around Kentani, *Miss Pegler,* 313! Tembuland; between the Bashee River and Morley, *Drège,* 4945! Natal; Umcomaas, *McKen,* 8! Inanda, *Wood,* 1307! Olivers Hoek Pass, *Wood,* 3476! and without precise locality, *Gerrard,* 430! *Mrs. K. Saunders*!

Also in Tropical Transvaal.

31. C. multiflora (Baker in Ref. Bot. i. t. 10); tuber 3–4 in. in diam., flattened; stem long, slender, twining, glabrous; leaves more or less reflexed, $\frac{3}{4}$–2 in. long, $\frac{1}{2}$–1 lin. broad and nearly as thick as broad, fleshy, subsessile, linear-subterete, channelled down the face, glabrous; cymes sessile or very shortly pedunculate, lateral at the nodes, 6–10 (or more?)-flowered, the flowers developing successively; peduncles 0–$\frac{1}{4}$ in. long, glabrous; pedicels $\frac{1}{4}$–$\frac{1}{3}$ in. long, glabrous; sepals 1 lin. long, subulate, glabrous; corolla-tube $\frac{1}{2}$–$\frac{2}{3}$ in. long, slightly curved or nearly straight, ovoid-inflated and $\frac{1}{8}$ in. in diam. at the base, narrowed to $\frac{2}{3}$ lin. in diam. and cylindric above, about 1 lin. in diam. at the mouth, glabrous outside, inside with deflexed white hairs in the slender part and some spreading fleshy subulate processes with purple tips in the inflated part; lobes with their basal part $1\frac{1}{2}$–$1\frac{3}{4}$ lin. long, linear-lanceolate or oblong-linear, ascending-spreading, abruptly tapering into very fine hair-like points $1\frac{1}{2}$ lin. long, which are horizontally inflexed and connate at their tips, green, covered on the inner face of the lower part with short adpressed retrorse white hairs; before expansion the hair-points form a fine needle-like point to the bud; outer corona $\frac{1}{3}$ lin. long, rising to the level of the top of the staminal column, cup-like, pentagonal, with very rounded angles, entire, truncate, glabrous, white; inner corona-lobes $\frac{1}{2}$ lin. long, linear or linear-spathulate, incumbent on the backs of the anthers at the base, then connivent-erect in a column, dorsally connected to the outer corona at the base, glabrous, white. *Schlechter in Journ. Bot.* 1897, 294. *Systrepha multiflora, Burchell ex Baker in Ref. Bot.* i. *under t.* 10.

VAR. β, **latifolia** (N. E. Br.); leaves with a petiole $\frac{1}{4}$–$\frac{1}{3}$ in. long and a blade $\frac{3}{4}$–1 in. long, $\frac{1}{2}$–$\frac{3}{4}$ in. broad, elliptic, obtuse, mucronulate, fleshy, flat, dark green above, paler beneath; midrib impressed above, prominent beneath.

SOUTH AFRICA: without locality, Var. β: cultivated at Kew!

CENTRAL REGION: Colesberg Div.; Colesberg, *Arnot* (ex *Baker*).

KALAHARI REGION: Bechuanaland; near the sources of the Kuruman River, *Burchell,* 2481!

The Colesberg locality appears to me doubtful, the plant was sent by Mr. Arnot (not Arnott as originally stated) from Colesberg to Kew, but I suspect it was brought from Griqualand West where Mr. Arnot was " Agent and Representative of the Griqua Chief and Government." There is a sketch at Kew of this plant, from the Transvaal.

32. C. infundibuliformis (E. Meyer, Comm. 194); rootstock a cluster of long fleshy roots $\frac{1}{3}$–$\frac{2}{3}$ in. thick; stem up to 2 ft. long, twining, glabrous; leaves $\frac{1}{2}$–$2\frac{1}{2}$ in. long, $\frac{1}{4}$–$\frac{3}{4}$ lin. broad, linear or linear-filiform, acute, glabrous, apparently fleshy; peduncle 3–5 lin.

long, 1–3-flowered, lateral at the nodes, glabrous; pedicels $1\frac{1}{2}$–3 lin.
long, glabrous; sepals $1\frac{1}{2}$–$2\frac{1}{2}$ lin. long, $\frac{1}{3}$ lin. broad at the base,
subulate or almost setaceous, glabrous; corolla-tube straight or but
slightly curved, $\frac{3}{4}$–1 in. long, globose-inflated at the base (*Burchell*),
but in dried specimens only very slightly inflated and about $2\frac{1}{2}$ lin.
in diam. at the base, cylindric and $1\frac{1}{2}$–$1\frac{3}{4}$ lin. in diam. above,
enlarging to $3\frac{1}{2}$ or 4 lin. in diam. at the mouth, glabrous outside and
within, minutely tuberculate on the veins inside the inflated part,
apparently white, thickly spotted with violet-purple above the
inflation, which is apparently striate with purple; lobes 7–10 lin.
long, erect, twisted together at the upper part, connate at the tips,
very narrowly linear from a deltoid base, replicate, with a rather
broad wing-like keel down the lower half of the inner face, glabrous,
not ciliate, dark purple-brown, the keel apparently white; outer
corona cup-shaped, with 5 erect linear lobes $\frac{2}{3}$–1 lin. long, divided to
below the middle into 2 linear obtuse parallel segments, hairy;
inner corona-lobes $\frac{3}{4}$ lin. long, exceeding the outer corona, linear,
connivent-erect over the staminal column, glabrous. *Dietr. Syn. Pl.*
ii. 891; *Decne in DC. Prodr.* viii. 645. *C. filiformis, Oliv. ex
Schlechter in Engl. Jahrb.* xxi. *Beibl.* 54, 13; *Schlechter in Journ.
Bot.* 1897, 295. *Systrepha filiformis, Burch. Trav.* i. 546. *S.
(Ceropegia ?) filiforme, Burch. Trav.* ii. 617.

CENTRAL REGION: Murraysburg Div.; near Murraysburg, 4100 ft., *Tyson*, 410!
Aberdeen Div.; Camdeboo, near Hamerkuil, 3000 ft., *Drège*, 5618! Cradock
Div.; near Cradock, *Cooper*, 2707!

KALAHARI REGION: Hay Div. in Griqualand West; on the plain at the foot of
the Asbestos Mountains, *Burchell*, 2092!

A specimen with immature flowers in the Albany Museum, Grahamstown,
collected by Mr. H. Hutton at Commadagga in Somerset Div., probably belongs
to this species.

33. **C. Barklyi** (Hook. f. in Bot. Mag. t. 6315, by error *Barkleyi*);
rootstock a tuber with 1 or more erect stems 2–6 in. long, simple,
sometimes slightly twining at the apex, under cultivation up to $1\frac{1}{2}$
ft. long and requiring support, or occasionally twining, pubescent
when dwarf, glabrous when elongated; leaves fleshy, very shortly
petiolate, $\frac{1}{2}$–$1\frac{1}{4}$ in. long, 1–5 lin. broad, linear to lanceolate, acute,
rounded or somewhat cuneate at the base, flat above, slightly convex
beneath, shortly and rather thinly pubescent to glabrous on both
sides, slightly and minutely ciliate, dark green with whitish veins;
peduncles lateral at the nodes, 2–9 lin. long, 2–3-flowered, glabrous;
bracts $\frac{1}{2}$–1 lin. long, subulate; pedicels $1\frac{1}{2}$–$2\frac{1}{2}$ lin. long, glabrous;
sepals $1\frac{1}{2}$ lin. long, subulate, glabrous; corolla-tube slightly curved
(straight in dried flowers), 7–8 lin. long, about 2 lin. in diam. at
the globose-inflated base, cylindric and about $\frac{3}{4}$ lin. in diam. above,
about $\frac{1}{4}$ in. in diam. at the funnel-shaped mouth, outside glabrous,
greenish-white at the base, pinkish above, passing into light green at
the mouth, inside with a few fine long hairs on the upper part; lobes
$\frac{2}{3}$–$\frac{3}{4}$ lin. long, very narrowly linear from a deltoid base, erect,
diverging above (in dried flowers subparallel), with incurved connate

tips, keeled on the inner face, which is veined with purple-brown on a greenish ground at the base and entirely purple-brown above, green on the back, ciliate on the margins for a short space just below the middle and on the basal part of the keel with fine simple purple hairs, otherwise glabrous; outer corona somewhat cup-like or its lobes pouch-like, subtruncate or slightly notched, glabrous; inner corona-lobes ½–⅔ lin. long, ⅓ lin. broad across the side, laterally much compressed, broadly falcate, obtuse, connivent over the style-apex at the base, then recurved in a semicircle above the outer corona and dorsally connected to it at the base, glabrous; follicles spreading, 1½ in. or more long, ⅛ in. thick, linear-terete, shortly acute; seeds 1½–1¾ lin. long, ⅔–¾ lin. broad, ovate, broadly margined, smooth, glabrous, brown. *Schlechter in Journ.* Bot. 1897, 294.

VAR. β, **tugelensis** (N. E. Br.); stem apparently long, very distinctly twining (a succulent climber, *Gerrard*), glabrous; leaves ¾–1¼ in. long, 2–5 lin. broad, lanceolate or ovate-lanceolate; peduncles 3–5-flowered; pedicels 2–3 lin. long; inner corona-lobes about ¾ lin. long and ⅛ lin. broad, erect, falcately recurved at the obtuse apex; follicles 2½–2¾ in. long, 2 lin. thick, terete-fusiform, tapering to a beak, smooth, glabrous; seeds 2½–3 lin. long, about 1 lin. broad, convex on one side, with incurved sides on the other, broadly margined, smooth, glabrous, light brown; otherwise as in the type.

EASTERN REGION: Transkei; near Old Morley, *Bowker*! and living plant sent to Kew by *Sir H. Barkly*! Var. β: Natal; thorny bush, Tugela River, *Gerrard*, 1323!

The follicles described, are from a plant cultivated by Mr. W. E. Ledger of Wimbledon, and appear to me not to have fully developed. The Tugela plant may prove to be a distinct species when better known; it looks very different from the wild specimens of *C. Barklyi*, but except in its very twining stems is scarcely distinguishable from that plant as it grows under cultivation, for no one would recognise that the figure in the *Botanical Magazine* (which is excellent of the culti-vated plant) belonged to the same species as the wild specimens, so very different are they in appearance, although in this case actually grown from the same tubers.

34. C. Conrathii (Schlechter in Engl. Jahrb. xxxviii. 45); whole plant including the tuber 3½–4½ in. high; tuber depressed-subglobose, about 2 in. in diam.; stems several or perhaps numerous, clustered, erect, simple or branched near the base, glabrous, about half their length buried in the ground in the specimen seen; leaves suberect or ascending, subsessile or very shortly petiolate, at the time of flowering ½–¾ in. long, 2–3 lin. broad, probably afterwards enlarging, lanceolate, very acute, cuneately narrowed at the base, wavy and crisped at the margins, sparsely ciliolate, glabrous on both sides, apparently somewhat fleshy; flowers in clusters of 4–6 at the nodes; pedicels erect, 4–7 lin. long, slender, glabrous; sepals nearly 1 lin. long, ⅖ lin. broad, ovate-lanceolate, acuminate, glabrous; corolla very slightly curved; tube ½ in. long, 1½–2 lin. in diam. at the ellipsoid-inflated base, cylindric above and about 1 lin. in diam., not enlarged at the mouth, glabrous outside, very sparsely hairy at the middle part and having a few very short cylindric processes on the inflated part within, apparently whitish or yellowish, dotted with dark violet-purple on the upper part, the spots becoming more

crowded below and perhaps confluent on the inflated part; lobes
about $3\frac{1}{2}$ lin. long, linear, scarcely dilated at the base, erect, slightly
curved, connate at the tips, glabrous, apparently purplish tinted
(flowers wine-red, *Conrath*); outer corona forming 5 small truncate
pockets less than $\frac{1}{4}$ lin. deep alternating with the anthers, glabrous;
inner corona-lobes $\frac{1}{2}$ lin. long, linear-spathulate, obtuse, connivent-
erect, dorsally adnate to the outer corona at the base, glabrous.

KALAHARI REGION: Transvaal; near Modderfontein, *Conrath*, 1008!

Kew is indebted to Mr. Paul Conrath for a portion of the type specimen of this
species, but I do not find the corolla-tube to be glabrous inside as described by
Dr. Schlechter, the tubercles are very conspicuous, but the hairs are not easily
discernible in dried specimens unless viewed against the light by doubling the
corolla back in water.

Imperfectly known species.

35. C.? aphylla (Haw. Syn. Pl. Succ. 13); "fleshy; branches
dichotomous, many-jointed, weak, terete; juice not very milky;
leaves stipule-like, about 1 lin. long, and at length almost deltoid,
adpressed, withering. *Schultes, Syst. Veg.* vi. 3; *G. Don, Gen. Syst.*
iv. 112.

SOUTH AFRICA? ex *Haworth.*

It is probable that this is not a South African plant. Link and Otto (*Ic. Pl.
Select.* 43, t. 18) figure *C. dichotoma*, Haw., from the Canary Isles, under the name
C. aphylla, but as Haworth describes both species on the same page and was a very
careful observer of plants, it is scarcely probable that they can be the same species,
and besides the juice of *C. dichotoma* is not at all milky.

36. C. leptocarpa (Schlechter in Engl. Jahrb. xxxviii. 47); stem
twining, filiform, glabrous; leaves spreading, scarcely fleshy,
glabrous; petiole $2\frac{1}{2}$–4 lin. long; blade 5–10 lin. long, 2–6 lin.
broad, ovate, lanceolate or lanceolate-elliptic, acuminate; peduncles
lateral at the nodes, a little longer than the petioles, few-flowered
at the apex, glabrous; pedicels $1\frac{1}{2}$ lin. long, glabrous; sepals $\frac{3}{4}$ lin.
long, narrowly lanceolate, acuminate, glabrous; corolla slender,
7 lin. long; tube ellipsoid-inflated at the base, narrowed above and
slightly dilating upwards to 1 lin. in diam. at the mouth, glabrous
outside, sparsely pilose within; lobes 2 lin. long, linear, obtuse,
thinly ciliate; outer corona broadly cup-shaped with semiorbicular
very obtuse lobes, glabrous; inner corona-lobes nearly three times
as long as the outer corona, falcate-erect, rather obtuse, slightly
dilated at the middle, glabrous; fruit not described.

EASTERN REGION: Delagoa Bay district; near Maramkene, 30 ft., *Schlechter*,
12077.

The author does not state if the corolla-lobes are free or connate at the tips,
but as they are described as obtuse they may be free.

37. C. linearis (E. Meyer, Comm. 194); stem twining, glabrous;
leaves fleshy, glabrous; petiole 1–2 lin. long; blade $\frac{3}{4}$–2 in. long,
$\frac{3}{4}$–3 lin. broad, linear, linear-ovate or narrowly ovate-lanceolate,

acute, obtuse or rounded at the base ; cymes pedunculate, lateral at the nodes, 2–3-flowered ; peduncle $\frac{1}{3}$–$\frac{1}{2}$ in. long, glabrous ; pedicels 1$\frac{1}{2}$–2$\frac{1}{2}$ lin. long, glabrous ; sepals 1$\frac{1}{4}$ lin. long, $\frac{2}{5}$ lin. broad, narrowly lanceolate, acuminate, or subulate, glabrous ; " corolla-tube slender, narrowed at the middle, striate with violaceous ; lobes of the limb very narrow, pilose, with reflexed sinuses." *Dietr. Syn. Pl.* ii. 891 ; *Decne in DC. Prodr.* viii. 644.

EASTERN REGION : Pondoland or Natal ; on sand-hills not far from the sea-shore between Umtentu River and Umzimkulu River, below 100 ft., *Drège,* 4947 !

I cannot identify this with any plant known to me, as the type specimen in E. Meyer's Herbarium is now destitute of corollas except in very young bud, but in general appearance it somewhat resembles *C. Caffrorum,* Schlechter, although the statement that the sinuses of the corolla are reflexed, indicates an entirely different species. E. Meyer suggests that it may be a variety of *C. africana,* R. Br., to which species Dr. Schlechter in *Engl. Jahrb.* xxi. *Beibl.* 54, 13, and *Journ. Bot.* 1897, 294 has referred it. This, however, is unlikely, as it comes from a region that was quite unexplored at the date (1822) when Robert Brown described *C. africana,* the plant is also different in appearance, and I find the sepals of *C. linearis* are much narrower, longer and more finely pointed than those of *C. africana,* whilst the sinuses of the corolla are not reflexed in the latter.

38. **C. ? torulosa** (Haw. Rev. Pl. Succ. 199), a much-branched succulent shrublet, with the appearance of a species of *Piper* ; branches about 9 in. long, erectly decumbent, effusely dichotomous, slender, terete ; leaves opposite, spreading ("expansa"), 1 in. long, shorter than the internodes, lanceolate-oblong, shaped like a pea-pod (probably depressed or compressed-terete in transverse section), copiously convex-subbullulate on both sides or torulose, dull green ; petiole 1 lin. long, filiform ; flowers unknown to *Haworth,* but according to *P. N. Don, Hort. Cantab. ed.* 13, 169, they are yellow, but are not described.

SOUTH AFRICA : without locality, ex *Haworth.*

I doubt if this is an Asclepiad, but cannot conjecture what it may be. I cannot discover that any author besides those quoted makes any mention of the plant, and it is omitted from the *Index Kewensis.*

It may be of interest to note here that in the Botanic Garden at Leiden a distinct hybrid (*C. hybrida,* N. E. Br.) has been raised from seed produced by *Ceropegia Sandersonii,* which had been fertilised by insect agency with pollen from *C. similis,* N. E. Br. The native country of the latter species is unknown. See the *Gardeners' Chronicle,* 1906, xl. 383, *figs.* 145–148.

XLI. BRACHYSTELMA, R. Br.

Calyx 5-partite. *Corolla* 5-lobed ; tube campanulate, cup-shaped or rarely shortly tubular, or the united part flat or saucer-shaped, or the whole corolla reflexed from the base ; lobes free and widely spreading, ascending or reflexed, or connate at the tips, forming a sort of cage, valvate or replicate-valvate in bud. *Corona* arising from the staminal column, double or falsely appearing 1-seriate ; outer corona cupular and 5–10-toothed or lobed or divided by 5 cut-

like fissures or rarely entire, or of 5 minute pouches or distinct entire
or bifid lobes, alternating with the inner corona-lobes or sometimes
divided to the base and the 2 adjacent halves of 2 lobes connate and
adnate to the sides or back of the base of the inner corona-lobes and
falsely appearing to form part of or to stand behind them ; inner
corona of 5 lobes incumbent upon the backs of the anthers and
shorter to longer than them, often dorsally connected at the base to
the outer corona or adnate to the sides of its lobes. *Staminal column*
arising from the base of the corolla ; anthers incumbent or inflexed
upon the apex of the style or suberect, oblong or subquadrate, without an appendage. *Pollen-masses* horizontal or ascending, solitary
in each anther-cell, pellucid on the inner margin or at their apex,
attached in pairs by very short caudicles to the pollen-carriers or
subsessile upon them. *Style* not exceeding the anthers, truncate or
convex at the apex. *Follicles* fusiform or linear-fusiform, smooth.
Seeds crowned with a tuft of hairs.

Perennial herbs, usually of dwarf habit ; rootstock a tuber or cluster of thick
fleshy roots ; stem solitary or two to several to a tuber, simple or branched, erect,
prostrate or rarely twining ; leaves opposite ; flowers small or of moderate size,
solitary or 2 to many together, lateral at the nodes or in terminal umbels or
umbel-like cymes.

DISTRIB. Species about 60, mostly South African, the others about equally
divided between Tropical Africa and India.

Brachystelma is closely allied to *Ceropegia,* differing chiefly in the corolla-tube
being very short or absent. As here understood it forms a compact and easily
recognised genus, but by Harvey and Schlechter no less than 6 other generic names
have been proposed for its various species. I find, however, no distinction of
structural importance by which they can be maintained. Four of them, viz. :
Decaceras, Harv., *Dichælia,* Harv., *Aulostephanus,* Schlechter, and *Lasiostelma,*
Benth. (= *Brachystelmaria,* Schlechter), are upheld by Dr. Schlechter. *Decaceras,*
however, merely differs in having no tube to the corolla, as is also the case in
B. caffrum, N. E. Br., and *B. pulchellum,* Schlechter, and a like difference is found
in several genera. *Dichælia* only differs in having the corolla-lobes connate at the
tips, and in *B. pygmæum* I find them free or connate on the same specimen !
B. Barberiæ, Harv., however, which has connate lobes, is placed by Dr. Schlechter
under *Brachystelma,* I can find no other difference, and *Ceropegia* and *Schizoglossum*
both contain species with free and connate lobes. *Aulostephanus* and *Lasiostelma* differ
in having a cluster of fleshy roots instead of a tuber, and more erect corolla-lobes, but
the same difference of habit is found in *Ceropegia, Asclepias* and *Xysmalobium,* and
the form and size of the corolla vary very much in many genera. Several large
genera, such as *Senecio, Pelargonium, Oxalis, Euphorbia,* etc., exhibit similar or
still more striking differences of habit and floral structure, so that there appears
nothing to warrant the retention of the above four as distinct from *Brachystelma.*
The tubers are eaten by the natives. Probably some of the following descriptions
of the corolla-tube, made from dried specimens, will not be found to agree with
that of the living plant, as it changes very much in drying ; see for example the
desciptions of *B. caudatum,* N. E. Br., and *B. Barberiæ,* Harv.

* Corolla-lobes free, see also 29, *B. pygmæum* :
  † Corolla with a distinct tube 1–8 lin. long :
    Corolla-tube about ½ in. long, campanulate or ovoid-
      campanulate, much longer than the lobes ;
      plants 1½–3 in. high :
      Corolla covered with long white hairs inside ; lobes
        ciliate   ...   . **(1) oianthum.**

Corolla covered within (at least below the middle)
with very short thick hairs; lobes not
ciliate   ...   .   (2) **campanulatum.**

Corolla-tube 1–3 (rarely 4) lin. long, campanulate:
Corolla-lobes 9–14 lin. long, pubescent on the inner
face, terminal half green   ...   .   (3) **crispum.**

Corolla-lobes 5–12 lin. long, glabrous or more rarely
puberulous on the inner face, terminal half
blackish or dark purple-brown   ...   ...   (4) **fœtidum.**
Corolla-lobes 2–3½ lin. long:
Corolla-lobes ciliate:
Corolla dark purple-brown, with yellow
transverse lines at the base of the lobes
and in the tube   ...   .   (5) **tuberosum.**

Corolla greenish or whitish   ...   .   (7) **meyerianum.**

Corolla-lobes not ciliate:
Corolla thinly pubescent on the inner surface
at the base of the lobes and throat of the
tube, otherwise glabrous; outer corona
10-toothed   ...   .   (8) **Thunbergii.**

Corolla quite glabrous on the inner surface:
Pedicels 1¼–2 in. long; corolla-lobes 2½ lin.
long, reflexed   ...   .   (10) **prælongum.**

Pedicels 2–2½ lin. long; corolla-lobes 3½ lin.
long, very spreading   ...   .   (6) **decipiens.**

Pedicels 1½ lin. long; corolla-lobes 3–3½ lin.
long, erect or erectly-spreading   ...   (25) **mafekingense.**

Corolla-lobes 1 lin. long, glabrous; tube 1½ lin.
long   ...   .   (9) **schonlandianum.**

†† Corolla without a distinct tube, the united part flattish
or saucer-shaped and except in 21, *B. Gerrardi,* and
22, *B. macropetalum* not more than ½ lin. deep:
‡ Corolla-lobes ⅔–2¼ lin. long:
Stems erect and 2–6 in. high or prostrate,
branching:
Pedicels ½–1½ in. long; flowers usually solitary,
yellow   ...   .   (14) **caffrum.**

Pedicels ⅛–½ in. long; flowers 2–7 together,
dark brown when dried, said to be yellow ...   (12) **Huttoni.**

Pedicels 1/12–¼ in. long (sometimes rather more
in 15, *B. pulchellum*) becoming longer in
fruit:
Corolla-lobes replicate, ciliate with minute
hairs on the basal part, white or pale
yellowish ...   .   (26) **ramosissimum.**

Corolla-lobes flat or with the margins incurved
near the middle or slightly recurved, not
replicate nor ciliate:
Corolla dark purple-brown:
Stems erect; corolla reflexed (sometimes
erect in dried flowers); lobes with
thickened green tips ...   .   (11) **Arnotii.**

Stems prostrate; corolla rotate; lobes
not thickened at the tips, with
yellow lines at the base   ...   ...   (15) **pulchellum.**

Corolla not dark purple-brown; stems
erect, 2–3 in. high:
Leaves lanceolate-elliptic; corolla-lobes
nearly 2 lin. long, yellowish ... (13) **flavidum.**

Leaves linear or linear-lanceolate; corolla-
lobes 1–1⅓ lin. long:
Leaves with a conspicuously prominent
midrib beneath; corolla apparently
pinkish or pale purplish, with
darker tips ... ... (16) **nanum.**

Leaves without a distinctly prominent
midrib beneath; corolla yellowish-
white, dotted with wine-red ... (17) **occidentale.**

Stem erect, ½ (rarely less) –2 ft. high, simple or
sparingly branched:
Pedicels 1–3 lin. long; corolla-lobes ciliate all
along or bordered at the tips with minute
thick weakly attached hairs:
Middle stem-leaves ⅓–1½ lin. broad, linear;
corolla-lobes ⅔–¾ lin. long ... ... (18) **schizoglossoides.**

Middle stem-leaves 1½–6 lin. broad, linear to
elliptic-oblong; corolla-lobes 1¼–2 lin.
long ... ... (20) **Sandersoni.**

Pedicels 4½–10 lin. long; middle leaves ⅔–1¾ in.
broad, roundish-ovate; corolla-lobes 1 lin.
long, not ciliate ... ... (19) **natalense.**

‡‡ Corolla-lobes 4–7 lin. long:
Corolla-lobes ciliate near the base with long clavate
purple hairs; leaves ½–1¾ in. long, ovate or
oblong to suborbicular:
Stem ¾–2¾ ft. high, with 10–25 pairs of
leaves ... ... (21) **Gerrardi.**

Stem 2–3 in. high, with 4–6 pairs of leaves ... (24) **comptum.**

Corolla-lobes ciliate on the inner face at the margin
to half way up with minute obovoid or clavate
hairs; leaves 1¼–3 in. long, linear ... ... (23) **longifolium.**

‡‡‡ Corolla-lobes 9–10 lin. long, puberulous on the inner
face, not ciliate with clavate hairs ... ... (22) **macropetalum.**

**Corolla-lobes linear or filiform, connate at the tips;
flower somewhat cage-like:
Stem and the elliptic leaves densely covered with long
soft spreading hairs; corolla-tube 2 lin. long, lobes
½–¾ in. long ... ... (40) **villosum.**

Stem and leaves varying from densely pubescent subtomentose or puberulous with short or minute
hairs to glabrous:
† Corolla with a distinct campanulate or cup-shaped
tube 1½–2½ lin. long (in some dried flowers of
41, *B. Barberiæ,* falsely appearing up to 4 lin.
long):
Leaves ¼–½ in. long, cuneate-ovate, undulate;
corolla-lobes 7 lin. long, sparsely ciliate ... (32) **undulatum.**

Leaves ¾–1½ in. long, linear or linear-lanceolate;
corolla-lobes 8–13 lin. long:
Corolla-lobes thickened at the apex, ciliate above
the middle; tube sparsely pilose ... ... (38) **elongatum.**

Corolla-lobes not thickened at the apex, with
minute clavate or subglobose hairs at their
base and throat of the tube within   ... (39) **distinctum.**

Leaves 1–4 in. long, cuneate-oblong to oblanceolate ; corolla-lobes ¾–1½ in. long, pubescent
to glabrous on the inner face ...   . (41) **Barberiæ.**

†† Corolla-tube almost none and the united part flattish
or saucer-shaped, or very shortly cup-shaped,
¼–¾ lin. long, or possibly rather more in 37,
B. cinereum :
Corolla-lobes in 31, *B. circinatum*, 4–5, in the
others 5–10 lin. long, glabrous (sometimes
subpuberulous in 35, *B. Galpinii*) on the inner
surface :
Leaves expanded, spathulately elliptic, oblong
or obovate, with very undulated margins ;
outer corona-lobes bifid to the middle   ... (33) **Bolusii.**

Leaves usually longitudinally folded, rarely flat,
not undulated on the margins, ¼–¾ in. long,
linear, linear - lanceolate, lanceolate or
elliptic :.
Leaves obscurely or thinly and very minutely
puberulous to glabrous ; inner corona-
lobes oblong - linear, obtuse, slightly
exceeding the anthers   ...   . (30) **commixtum.**

Leaves very distinctly (but minutely) pubescent
or puberulous on one or both sides :
Leaves glabrous on the upper or infolded
surface :
Flowers solitary (always ?), drooping ;
inner corona-lobes linear - oblong,
truncate or emarginate, not exceed-
ing the anthers   ...   . (34) **ovatum.**

Flowers 2 or more together at the nodes,
not drooping :
Inner corona-lobes twice as long as the
anthers, subulate, acute ...   ... (35) **Galpinii.**

Inner corona-lobes only half as long as
the anthers, linear, with a dorsal
thickening (or hump ?) at the
base   ...   . (31) **circinatum.**

Leaves puberulous or pubescent on both
sides ; inner corona-lobes much longer
than the anthers :
Inner corona-lobes rather shorter than
the outer corona   ...   . (36) **pallidum.**

Inner corona-lobes exceeding the outer
corona   ...   ...   ···   ... (37) **cinereum.**

Corolla-lobes 1¾–3 lin. long :
Leaves puberulous beneath or on outer side when
folded (on both sides, *Schlechter*) ; outer
corona-lobes divided to half way or beyond
into linear or linear-subulate segments :
Plant 2–5 in. high ; teeth of outer corona
glabrous   ...   . (27) **Zeyheri.**
Plant 6–8 in. high ; teeth of outer corona
minutely ciliate   ...   . (28) **filiforme.**

Leaves puberulous on the midrib beneath and
minutely ciliate, otherwise glabrous ; outer
corona-lobes minute or subquadrate, emarginate or shortly bifid, sometimes all
connate into a ring ... ... (29) **pygmæum.**

1. **B. oianthum** (Schlechter in Engl. Jahrb. xx. Beibl. 51, 53);
tuber flattened, about 2 in. in diam. ; stem 2–3 in. high, simple or
sparingly branched, comparatively stout, pubescent ; leaves 4–8
pairs to a stem or branch, ascending, $\frac{2}{3}$–2 in. long, 2–6 lin. broad,
linear-lanceolate or lanceolate, acute, tapering at the base into a
short petiole, longitudinally folded, minutely pubescent on both
sides ; flowers solitary at the nodes, comparatively large, nodding or
horizontal ; pedicels 1–2 lin. long, rather stout, minutely pubescent ;
sepals $1\frac{3}{4}$–2 lin. long, ovate-lanceolate, acute, thinly and minutely
pubescent ; corolla ovoid-campanulate, shortly 5-lobed, apparently
slightly fleshy, glabrous outside, dull yellow, spotted with dark
purple-brown, or (according to *Schlechter*) sometimes entirely dark
purple-brown ; tube (according to a drawing) about 6 lin. long and
7 lin. in diam. (smaller in dried flowers), clothed within with long
white hairs ; lobes free, spreading, about 2 lin. long and 3–3$\frac{1}{2}$ lin.
broad (in drawing), broadly deltoid-ovate, acute, ciliate with
long (vibratile ?) pale purple or rose-coloured hairs ; outer and inner
corona-lobes combined into one series, cupular, pentagonal, apparently
of 5 broad 3-lobed segments, with the lateral teeth $\frac{1}{3}$ lin. long,
deltoid, obtuse, contiguous in pairs with those of the adjacent
segments, having near one margin a small patch of minute deflexed
white hairs, otherwise glabrous, and the middle tooth (really the
inner corona-lobe) $\frac{2}{3}$ lin. long, shortly linear from a broadly deltoid
base, inflexed and closely incumbent upon the backs of the anthers
and not exceeding them, purple-brown. *Schlechter in Journ. Bot.*
1897, 292. *B. erianthum, Schlechter ex K. Schum. in Engl. and Prantl,
Pflanzenfam.* iv. ii. 268.

Kalahari Region : Orange River Colony ; near the Rhenoster River,
Sanderson, 7 ! Transvaal ; Modderfontein, *Conrath,* 1007 ! near Mooifontein,
5500 ft., *Schlechter,* 3557 and Elandspruit Mountains, 6000 ft., *Schlechter,* 3993
(ex *Schlechter*).

2. **B. campanulatum** (N. E. Br.) ; tuber flattened or turnip-shaped,
about 2 in. in diam., with one simple or sparingly branched stem
(together with the leaves) 1$\frac{1}{2}$–2 in. high ; branches about 1 in. long,
decumbent or spreading, puberulous ; leaves $\frac{3}{4}$–1$\frac{1}{2}$ in. long, 4–7 lin.
broad, elliptic, ovate or obovate, tapering or rounded at the base into a
petiole 2–3 lin. long, flattish or longitudinally folded and more or less
undulate, puberulous on both sides ; flowers solitary at the nodes,
drooping ; pedicels about 2 lin. long, puberulous ; sepals 2–2$\frac{1}{2}$ lin. long,
lanceolate, acuminate, puberulous ; corolla campanulate, shortly
5-lobed, densely subtomentose-puberulous and green outside, inside
covered with very short thick and probably stiff hairs, greenish-yellow,

spotted all over and longitudinally veined at the upper part with dark purple-brown; tube about $\frac{1}{2}$ in. long and nearly as much in diam.; lobes free, erect or very slightly spreading, 2–$2\frac{1}{2}$ lin. long and about 3 lin. broad at the base, deltoid-ovate, acute, not ciliate; outer and inner corona partially combined, dark purple-brown; outer lobes shortly bifid, pouch-like, with their sides united to the base of the shortly linear obtuse inner lobes, which are incumbent upon the backs of the anthers, with their tips just exceeding and incurved over them.

COAST REGION : Bathurst Div. ; in sandy soil at Linch's Post, near the Kowie River, *Bowie*!

Described partly from an excellent drawing at Kew of a plant introduced by Bowie in 1823 into Kew Gardens, which flowered there in July, 1824, and partly from an imperfect specimen in the British Museum, collected by Bowie. It must be a very rare plant, as although the region has been visited by several collectors, it has not been refound during the past 84 years. It is nearly allied to *B. oianthum,* Schlechter.

3. **B. crispum** (Grah. in Edinb. Phil. Journ. 1830, ii. 170) ; tuber flattened, 3–4 in. in diam. ; stems sometimes several, 4–6 in. high, simple or sparingly branched at the base, more or less pubescent ; leaves ascending-spreading, $\frac{1}{2}$–$1\frac{1}{2}$ in. long, $1\frac{1}{2}$–4 lin. broad, linear-lanceolate, oblong-lanceolate, lanceolate or obovate-spathulate, acute or obtuse, usually with wavy margins, shortly pubescent on both sides ; flowers 2–8 together, lateral at the nodes; pedicels $\frac{1}{4}$ to $1\frac{1}{4}$ in. long, pubescent ; sepals $1\frac{1}{2}$–2 lin. long, lanceolate-subulate, pubescent ; corolla pubescent and green outside, spotted with purple-brown on the tube and base of the lobes, the tube inside and dilated base of the lobes glabrous, yellow or whitish ? marked with dark purple-brown prominent or wart-like spots, narrow part of the lobes pubescent, green, without spots ; tube in living flowers 1–$1\frac{1}{4}$ lin. long and $1\frac{3}{4}$ lin. in diam. at the mouth, cup-like and but slightly exceeding the corona, but in dried flowers (owing to collapse from shrinkage of the horizontally spreading united part of the limb) appearing to be campanulate, $2\frac{1}{2}$–3 lin. long and much exceeding the corona, without a trace of the small actual tube, which when fresh occupies only the lower half of the apparent tube : lobes free, horizontally spreading or slightly recurving towards the tips, $\frac{3}{4}$–$1\frac{1}{4}$ (or from figures up to $1\frac{3}{4}$) in. long, broadly ovate at the base, tapering into long linear tails, with revolute or replicate margins ; outer corona somewhat cup-like, pentagonal, of 5 bifid pocket-like lobes alternating with the anthers, with deltoid acute teeth $\frac{1}{6}$ lin. long, purple-brown ; inner corona-lobes $\frac{1}{2}$–$\frac{2}{3}$ lin. long, linear, obtuse, inflexed upon the backs of the anthers and equalling or slightly exceeding them and incurved over their tips, dorsally connected to and appearing to be teeth inflexed from the outer corona, purple-brown with yellowish tips. *Bot. Mag.* t. 3016 ; *Decne in DC. Prodr.* viii. 647, *not of E. Meyer. B. caudatum, N. E. Br. in Journ. Linn. Soc.* xvii. 169 ; *K. Schum. in Engl. and Prantl,*

Pflanzenfam. iv. ii. 268; *Schlechter in Journ. Bot.* 1897, 292. *B. spatulatum, Lindl. Bot. Reg. t.* 1113; *G. Don, Gen. Syst.* iv. 125. *B. spathulatum, Decne in DC. Prodr.* viii. 646; *K. Schum. in Engl. and Prantl, Pflanzenfam.* iv. ii. 268. *Stapelia tuberosa, Meerburg, Pl. Rar. t.* 54, *fig.* 1. *S. caudata, Thunb. Prodr.* 46; *Fl. Cap. ed.* 2, ii. 171, *and ed. Schultes,* 241; *Willd. Sp. Pl.* i. 1286; *Pers. Syn. Pl.* i. 279; *Poir. Encycl. Meth.* vii. 384; *Schultes, Syst. Veg.* vi. 48; *G. Don, Gen. Syst.* iv. 117; *Decne in DC. Prodr.* viii. 663.

SOUTH AFRICA : without locality, *Thunberg*! *Harvey*, 515 ! *cultivated specimens*! COAST REGION : Malmesbury Div. ; sandy plain near Darling, 400 ft., *Bolus*, 11425 ! Cape Div. ? in the neighbourhood of Cape Town, locality unknown, *Pappe*!

Partly described from fresh flowers in formalin. The wart-like spots on the corolla are larger and less crowded in some specimens than in others. The figure in the *Botanical Register* was evidently drawn from a specimen on which the flowers had closed after expansion. Dried flowers are exceedingly deceptive as to the true form of the corolla.

4. **B. fœtidum** (Schlechter in Engl. Jahrb. xx. Beibl. 51, 52); tuber flattened, up to 4 or 5 in. in diam., producing 1 to several branching stems 3–6 in. high, variably pubescent with short spreading hairs; leaves spreading, ½–2 in. long, 1½–8 lin. broad, linear-lanceolate, lanceolate, somewhat elliptic or ovate, acute or obtuse, cuneately narrowed at the base into a short petiole, longitudinally folded, sometimes undulate at the margins, pubescent on both sides or glabrous above ; flowers in fascicles of 2–6 lateral at the nodes, very variable in size, according to the vigour of the plant ; pedicels ¼–⅓ in. long, pubescent ; sepals 1½–2½ lin. long, lanceolate, acute, pubescent ; corolla 1–2 in. in expanse, pubescent to subglabrous all over outside, inner surface of lobes and tube glabrous (rarely puberulous), dark purple-brown or nearly black on the lobes and marked with purple-brown wart-spots (not always visible in dried flowers) on a white or yellowish ground in the tube ; tube 3–4 lin. long, campanulate ; lobes free, spreading, ½–1 in. long, linear from a deltoid base, with revolute or reflexed margins, acute, ciliate with hairs like those on the back ; outer corona about equalling the staminal column, somewhat cup-like, 10-toothed or of 5 bifid lobes forming small pockets alternating with the anthers, apparently purple-brown ; teeth erect, ¼–⅓ lin. long, deltoid, acute or obtuse, usually pubescent on the inner face with minute white hairs (scarcely visible when wetted) or sometimes glabrous ; inner corona lobes up to ½ lin. long, linear, obtuse, incumbent on the backs of the anthers and not exceeding them, glabrous ; follicles in pairs or solitary, 2½–3 in. long, ¼ in. or more thick, terete-fusiform, tapering into an obtuse beak, pubescent. *Schlechter in Journ. Bot.* 1897, 292. *B. Rehmannii, Schlechter in Bull. Herb. Boiss.* iv. 449.

KALAHARI REGION: Orange River Colony ; near Harrismith, *Sanderson*, 6 ! 163 ! *Sankey*, 187 ! *Thode* ! Rhenoster River, *Sanderson* (drawing) ! Vals River, *Mrs. Barber*, 682 ! Besters Vallei near Witzies Hoek, *Miss Jacobsz* ! Parys, *Rogers*,

803 ! Basutoland, *Cooper*, 933 ! Transvaal ; Mooi Kiver, *Burke*, 292 ! near Elsburg, *Schlechter*, 3547 ! Potchefstroom, *Burtt Davy*, 1820 ! Houtbosch Berg, *Rehmann*, 5877 ! Modderfontein, *Conrath*, 1013 ! and without precise locality, *Nelson*, 13 !
EASTERN REGION : Zululand, *Thomas* !

A fruiting specimen from Cradock (*Cooper*, 2710) may also belong to this species.　According to Mr. Sanderson, the tuber "is sometimes eaten by the Hottentots and called Hottentot's bread, when eaten raw it tastes bitter."　The flowers are stated to be "abominably scented" or to emit "an awful stench." Dr. Schlechter founded *B. fœtidum* upon a small-flowered specimen with the inner surface of the corolla-lobes glabrous, and *B. Rehmannii* upon a large-flowered specimen with the inner surface of the corolla-lobes puberulous.　I have examined both and find no structural difference between the two extreme forms, nor does Dr. Schlechter mention any, and they are connected by specimens with flowers of intermediate sizes, which demonstrate that the size accords with the vigour of the plant, probably depending upon the amount of moisture.　Specimens with the inner surface of the corolla-lobes puberulous appear to be uncommon, the only examples I have seen are Rehmann's 5877 (the type of *B. Rehmannii*) and Rogers 803 ; I deem it to be a variable character, such as occurs in *B. Barberiæ*, Harv.

5. B. tuberosum (R. Br. in Bot. Mag. t. 2343 excl. syn.) ; tuber flattened, $1\frac{1}{2}$–2 in. in diam., producing 1 or more sparingly branched stems 3–4 in. high, or taller under cultivation, pubescent ; leaves spreading, $\frac{3}{4}$–$1\frac{1}{2}$ in. long, 2–6 lin. broad, mostly linear or linear-lanceolate, the lower sometimes larger and ovate or oblong, acute or subobtuse, tapering into a short petiole at the base, more or less longitudinally folded, puberulous on the under surface, glabrous above, minutely ciliate ; flowers 2–4 together at the nodes ; pedicels 2–3 lin. long, puberulous ; sepals about 1 lin. long, lanceolate, acuminate, puberulous ; corolla about $\frac{3}{4}$ in. in diam., glabrous and green speckled with purple-brown outside, inner face dark purple-brown on the lobes, yellow with transverse purple-brown lines in the campanulate tube, which appears to be 2–3 lin. long, and as much in diam. ; lobes free, $\frac{1}{4}$ in. long, very spreading or slightly reflexed, tapering from the base to the acute apex, with revolute margins, ciliate with rather long hairs at the base ; corona-lobes triangular, conniving at the points. *Bot. Reg. t.* 722 ; *Fl. des Serres*, iv. *t.* 340 ; *Geel, Sert. Bot.* ii. *t.* 7 ; *Spreng. Syst. Veg.* i. 842 ; *G. Don, Gen. Syst.* iv. 125, *fig.* 16 ; *Dietr. Syn. Pl.* ii. 887 ; *Decne in DC. Prodr.* viii. 646 *partly* ; *Schlechter in Journ. Bot.* 1897, 292.

SOUTH AFRICA : without locality, ex *Sims*.

This plant was probably introduced by Bowie, as there is a drawing of it at Kew, made in Sept. 1825 from a plant brought from the Cape by him in 1823.　No dried specimen of it seems to have been preserved.　I have, however, seen 3 specimens which much resemble the figure in the *Botanical Magazine*.　One of them distinctly differs in being quite destitute of cilia, this I have described as *B. decipiens*.　Another at the British Museum, labelled "*Huernia*, India" from Nuttall's Herbarium, agrees with the figure except as to the cilia, these appear to have fallen off, as here and there I notice a long deciduous and probably vibratile hair, otherwise the corolla is entirely glabrous ; the corona is blackish apparently of 5 three-toothed lobes opposite the anthers, with the lateral teeth (outer corona) very short, deltoid, puberulous on the inner face, and the middle tooth (inner corona-lobe) linear, obtuse, very abruptly inflexed upon the backs of the anthers and about as long as them. This may be the true plant, but all the figures appear to represent the cilia as simple fixed (not vibratile) hairs. The third specimen is from a plant cultivated at Kew in Sept. 1861, this differs as follows :—Leaves $\frac{1}{2}$–$\frac{3}{4}$ in. long, 1 lin. broad ; corolla as in the figure of *B. tuberosum* in size, form and colour, but covered with long white (or pale purple ?) hairs within the upper part of the tube, and probably ciliate with them at the base of the lobes, but of this latter point I am not certain ; the corona is pressed out of shape, but appears to have been very similar to that of the British Museum specimen above described. This plant is probably the *B. tuberosum* of K. Schum. in Engl. and Prantl, *Pflanzenfam.* iv. ii. 268, which is scarcely that of R. Br., as neither of the original drawings (preserved at Kew) of *B. tuberosum* nor their reproductions show any trace of hairs within the corolla-tube, but it cannot be definitely determined.

6. B. decipiens (N. E. Br.) ; stem solitary? simple or slightly branched, densely puberulous with spreading or deflexed hairs ; leaves spreading, about 1 in. long, 1–2 lin. broad, linear or linear-lanceolate, subacute, tapering at the base into a very short petiole, apparently longitudinally folded, somewhat harshly puberulous on the under surface, glabrous above, minutely ciliate ; flowers 2–4 together at the nodes ; pedicels 2–2$\frac{1}{2}$ lin. long, puberulous ; corolla glabrous outside and within, dark purple-brown on the lobes, with their base and inside of the tube yellow, marked with transverse purple-brown lines ; tube about 1$\frac{1}{2}$ lin. long, campanulate ; lobes free, very spreading, about 3$\frac{1}{2}$ lin. long, 1$\frac{1}{2}$ lin. broad at the base, linear-lanceolate, acute, not ciliate ; outer corona probably of 5 pocket-like bifid lobes, puberulous on the inner face, but in the dried flowers the outer and inner corona are apparently combined in 1 series of 5 dark purple-brown lobes, with their basal part $\frac{1}{4}$ lin. long and $\frac{2}{3}$ lin. broad, transversely rectangular with square obtuse shoulders, truncately contracted at the top into a linear obtuse middle tooth (really the inner corona-lobe) $\frac{1}{4}$ lin. long, incumbent upon the backs of the anthers and not exceeding them.

Coast Region : Albany Div. ; near Grahamstown, *Bolton* !

In general appearance this much resembles the figure of *B. tuberosum* in *Bot. Mag.* t. 2343, but the corolla is entirely destitute of cilia (which I have sought for in vain in unopened buds), the corolla-lobes appear to be more linear in form, and the corona-lobes are not triangular as described for *B. tuberosum.*

The Kew Herbarium contains a specimen collected at Grahamstown (*Glass*, 654), which seems closely related to *B. decipiens*, but has rather broader and more acute leaves, a rather larger corolla (which appears to be of a uniform colour), and the inner corona-lobes are subulate and much longer than the anthers. As there is but one flower upon the specimen I hesitate to describe it. There is also at Kew a drawing of a plant introduced by Bowie in 1823, which is likewise nearly related to *B. decipiens*, differing in its oblong leaves, $\frac{1}{2}$–$\frac{2}{3}$ in. broad, and the lobes of the drooping corolla 5 lin. long, suberect, not spreading and at least 3 times as long as the tube. I have not seen a specimen like it.

7. B. meyerianum (Schlechter in Engl. Jahrb. xxi. Beibl. 54, 14) ; plant dwarf, branching at the base, with several stems 1$\frac{1}{2}$–7 in. high, puberulous on the upper part ; leaves spreading ; petiole $\frac{1}{2}$–2 lin. long ; blade $\frac{1}{3}$–1 in. long, $\frac{2}{3}$–5 lin. broad, linear-lanceolate to elliptic, acute or obtuse, cuneate at the base, varying from rather harshly puberulous to glabrous ; flowers solitary or 2–4 together, subaxillary ; pedicels $1\frac{1}{2}$–$2\frac{1}{2}$ lin. long, minutely scabrous-pubescent ; sepals $1\frac{1}{4}$–$1\frac{1}{2}$ lin. long, $\frac{1}{2}$–$\frac{2}{3}$ lin. broad, lanceolate, acute, minutely pubescent to glabrous ; corolla-tube about 3 lin. long and $2\frac{1}{2}$ lin. in diam., campanulate, glabrous and "greyish-green" (*Schlechter*) outside, hairy within at the upper part, but when wetted the hairs become invisible ; lobes free, $2\frac{1}{2}$–3 lin. long, $1\frac{1}{4}$ lin. broad at the base, thence gradually tapering to an obtuse point, apparently with more or less reflexed margins, ciliate with long white hairs and sometimes hairy on the basal part, apparently greenish ; corona in dried flowers apparently of 5 transversely rectangular lobes $\frac{1}{4}$ lin. long, $\frac{3}{4}$ lin. broad (the outer corona, which is probably cupular in the living flowers), puberulous or pubescent on the inner face (always?), with a central linear obtuse point (the inner corona-lobe) $\frac{3}{4}$ lin. long, connivent over the $\frac{3}{4}$ lin.-long staminal column. *Schlechter in Journ. Bot.* 1897, 293. *B. tuberosum, E. Meyer, Comm.* 195, *as to description and the Katberg plant, the other quoted having no flowers, not of R. Br. ; Decne in DC. Prodr.* viii. 646 *partly ; B. caffrum, Schlechter in Engl. Jahrb.* xviii. *Beibl.* 45, 13, *not of N. E. Br.*

COAST REGION : Stockenstrom Div. ; Kat Berg, ex *E. Meyer*, "between Kat (River) and Klipplaat River" on the original label, *Drège*, 3439 ! King Williamstown Div. ; Mount Coke, 1500 ft., *Sim*, 1508 ! Div. ? ; Kaffrarian Mountains, *Mrs. Barber*, 37 !

The specimen collected by Drège between Kuga (Coega) River and Sunday River and quoted by E. Meyer for this species, is in fruit only and undeterminable, but appears to be quite distinct from *B. meyerianum*.

8. B. Thunbergii (N. E. Br.) ; plant 2–$3\frac{1}{2}$ in. high, branching at the base ; stems pubescent or puberulous ; leaves $\frac{1}{3}$–$1\frac{1}{4}$ in. long, 1–6 lin. broad, linear, lanceolate or elliptic, acute or obtuse, cuneate at the base, flat or undulated, ciliolate and thinly pubescent to nearly glabrous on both sides with minute spreading hairs, which are most abundant on the midrib beneath ; petiole $\frac{3}{4}$–$1\frac{1}{2}$ lin. long ; blade $\frac{1}{3}$–1 in. long, $\frac{1}{2}$–6 lin. broad ; flowers in fascicles of 2–8 or solitary at the nodes ; pedicels 1–4 lin. long, puberulous ; sepals 1–$1\frac{3}{4}$ lin. long, $\frac{1}{3}$ lin. broad, linear-lanceolate, acuminate, thinly puberulous or pubescent ; corolla-tube $1\frac{1}{4}$–2 lin. long, $1\frac{1}{2}$–$2\frac{1}{2}$ lin. in diam. at the mouth, campanulate, with a minute projecting tubercle or point at each sinus, glabrous outside, thinly pubescent on the upper part within, white ; lobes free, 2–3 lin. long, linear from the ovate-deltoid 1–$1\frac{1}{4}$ lin.-broad base, replicate, erect or very slightly spreading, straight, hooked at the apex, thinly pubescent at the very base within, otherwise glabrous and not ciliate, "greenish at the base, passing upwards to brownish or tawny yellow" (*Bolus*), drying olive-brown or blackish-brown ; outer corona cup-like, much shorter than the staminal column, 10-toothed ; teeth $\frac{1}{8}$ lin. long, deltoid, acute, erect, straight, glabrous, apparently white ; inner corona-lobes $\frac{1}{4}$ lin. long, linear, obtuse, closely incumbent on the backs of the anthers and shorter than or not exceeding them, dorsally connected with the outer corona, glabrous, apparently white; staminal column $\frac{2}{5}$ lin. long. *Cynanchum crispum, Thunb. Prodr.* 46; *Fl. Cap. ed.* 2, ii. 158, *and ed. Schultes,* 236, *and in Weber and Mohr, Archiv.* 1804, i. 29; *Willd. Sp. Pl.* i. 1253; *Pers. Syn. Pl.* i. 272; *Poir. Encycl. Suppl.* ii. 430; *Schultes, Syst. Veg.* vi. 106; *Spreng. Syst.* i. 852; *G. Don, Gen. Syst.* iv. 154; *Dietr. Syn. Pl.* ii. 904, *not of Jacquin.*

Coast Region: Malmesbury Div.; Zwartland, *Thunberg*! Caledon Div.; Genadendal, *Prior (Alexander)*! Riversdale Div.; Tygerfontein, 500 ft., *Galpin*, 4332! hills near Riversdale, 300 ft., *Bolus*, 11201!

Partly described from fresh flowers in fluid. This may be the same as *B. schonlandianum,* Schlechter. In many points the description agrees, but the corolla-lobes of that species are stated to be only 1 lin. long and glabrous, it may have been described from a starved specimen, but the description is insufficient to decide.

9. **B. schonlandianum** (Schlechter in Engl. Jahrb. xviii. Beibl. 45, 35); plant about $1\frac{3}{4}$ in. high, erect, branching; branches pilose; leaves $3\frac{1}{2}$–5 lin. long, 1–2 lin. broad at the middle, lanceolate, acute, tapering into the petiole at the base, subglabrous; flowers 2–4 together at the nodes; pedicels $1\frac{1}{2}$ lin. long, pilose; sepals $\frac{1}{4}$ lin. long, linear-lanceolate, acute, very shortly pilose; corolla-tube $1\frac{1}{2}$ lin. long, 1 lin. in diam. at the top, campanulate, glabrous; lobes free, 1 lin. long, lanceolate, tapering to an obtuse apex, glabrous; corona cupular, with the lobes 3-toothed at the apex; lateral teeth short, acute; middle tooth (really an inner corona-lobe) produced, incurved at the obtuse apex. *Schlechter in Journ. Bot.* 1897, 292.

Coast Region: Uitenhage Div.; dry stony places in a valley near Uitenhage, 520 ft., *Schlechter*, 2585.

10. **B. prælongum** (S. Moore in Journ. Bot. 1902, 384); plant (exclusive of the tuber) about 3 in. high, branching; branches shortly pubescent; leaves 5–6 lin. long, slender, subterete from the margins being infolded, acute, $\frac{1}{3}$ lin. in diam., glabrous on the inner (upper) surface, pubescent on the outer (under) surface; flowers in pairs or solitary, lateral at the nodes; pedicels $1\frac{1}{4}$–2 in. long, slender, shortly pubescent; sepals $1\frac{1}{2}$ lin. long, lanceolate, acuminate, pubescent; corolla-tube 1 lin. long, cup-shaped, apparently slightly 5-ribbed, with some very minute hairs along the ribs outside, otherwise glabrous, apparently pale greenish; lobes free, recurved or reflexed from their base, $2\frac{1}{2}$ lin. long, linear from a 1 lin.-broad base, replicate, glabrous on both sides, not ciliate, apparently blackish; outer corona-lobes arising about $\frac{2}{3}$ lin. up the staminal column and scarcely attaining to the level of its top, $\frac{1}{4}$ lin. long, divided into 2 widely diverging falcately deltoid acute teeth, minutely puberulous on the back and inner face as is also the

staminal column below them ; inner corona-lobes $\frac{1}{4}$ lin. long, linear, obtuse, closely incumbent upon the backs of the anthers and not exceeding them.

KALAHARI REGION : Orange River Colony ; near Vredefort Road, *Barrett Hamilton*!

Described from the type in the British Museum. In some particulars the above description does not agree with that originally published by Mr. Moore, but he has kindly re-examined the specimen with me and agrees that the above is correct. His original statement that both sides of the corolla-lobes are pilose, he believes to have been written in error for corona-lobes.

11. **B. Arnotii** (Baker in Ref. Bot. i. t. 9, by error *Arnottii*) ; tuber flattened about $2\frac{1}{2}$ in. in diam., producing 1 or more erect simple or branching stems about 3 (4–6, *Baker*) in. high, puberulous with minute deflexed hairs ; leaves (including the petiole) $\frac{1}{3}-\frac{1}{2}$ in. long, 1–2 lin. broad, lanceolate or ovate-lanceolate, acute, tapering into a distinct petiole at the base (not " nearly sessile " as originally described), longitudinally folded, with wavy margins, glabrous above, densely greyish-puberulous beneath ; flowers 2–4 together, lateral at the nodes, developing successively ; pedicels $2\frac{1}{2}-3$ lin. long, puberulous ; sepals $\frac{3}{4}$ lin. long, lanceolate, acute, puberulous ; corolla small, reflexed from the very base, but in flowers withered before being dried, falsely appearing to have erectly-spreading lobes and a saucer-shaped tube, puberulous on the back, glabrous on the inner face, dark-purple-brown, with a green tubercle-like thickening at the tips of the lobes, which are free, scarcely 1 lin. long, $\frac{1}{2}$ lin. broad, ovate or ovate-lanceolate, acute ; outer corona of 5 erect lobes alternating with the anthers, arising $\frac{1}{4}$ lin. up the staminal column and attaining to the same level, $\frac{1}{4}$ lin. long, linear, shortly bifid at the apex, with the segments diverging, greenish ; inner corona-lobes minute, reduced to transverse greenish cushion-like thickenings at the base of and about $\frac{1}{3}$ as long as the yellow anthers. *Decaceras Arnoldi, Schlechter in Engl. Jahrb.* xviii. *Beibl.* 45, 26 *in note. D. Arnottii, K. Schum. in Engl. and Prantl, Pflanzenfam.* iv. ii. 267, *fig.* 78 ; *Schlechter in Journ. Bot.* 1897, 292. *Anistoma Arnottii, Benth. and Hook. f. ex Ind. Kew.* i. 139.

CENTRAL REGION : Colesberg Div. ; near Colesberg (cultivated specimen) *Arnot* (not *Arnott*)!

12. **B. Huttoni** (N. E. Br.) ; rootstock not seen ; stem 2–3 in. (or more ?) high, simple or branched, zigzag on the flowering part, minutely puberulous ; leaves spreading and in the specimens seen more or less tortuous and often revolute at the tips, about $\frac{1}{2}$ in. long, $\frac{1}{3}-\frac{1}{2}$ lin. broad, linear or linear-filiform, acute, with revolute margins, glabrous or with a few very minute hairs, chiefly on the mid-rib and margins beneath ; flowers in fascicles of 2–7, lateral at the nodes and terminal ; pedicels $\frac{1}{3}-\frac{1}{2}$ in. long, minutely puberulous ; sepals $\frac{3}{4}-1$ lin. long, lanceolate, acute, minutely puberulous at the very

base, nearly or quite glabrous elsewhere ; corolla rotate, entirely glabrous and not ciliate, yellow according to *Hutton*, but the dried flowers appear to be dark purple-brown with the very small tube yellowish ; tube ½ lin. long, 1 lin. broad, cup-like ; lobes free, widely spreading, 1½ lin. long, not quite 1 lin. broad, linear or oblong-linear, shortly acute, with reflexed margins ; outer corona cup-shaped, with 5 pairs of erect linear obtuse teeth ⅓ lin. long, overtopping the staminal column, glabrous, apparently yellow, with darker (green ?) tips to the teeth ; inner corona-lobes scarcely ⅓ lin. long, linear, obtuse, closely incumbent on the backs of the anthers and not exceeding them, with a partition at their base dorsally connecting them with the outer corona, apparently yellow. *Decaceras Huttoni, Harv. Thes. Cap.* ii. 9, *t.* 114, *and Gen. South Afr. Pl. ed.* 2, 242 ; *K. Schum. in Engl. and Prantl, Pflanzenfam.* iv. ii. 267 ; *Schlechter in Journ. Bot.* 1897, 291.

Coast Region : Albany Div. ; Bothas Hill, near Grahamstown, *Hutton*, 3 !

The specimen figured by Harvey is much dwarfer than those at Kew. The "minutely bidentulate" interspaces between the pairs of corona-teeth I have observed in one flower, but think it due to the tissue having been torn.

13. **B. flavidum** (Schlechter in Engl. Jahrb. xl. 94) ; tuber depressed, up to 1¾ in. in diam. ; stems several (?), branching, about 2 in. high, slender, puberulous, laxly leafy ; leaves erectly spreading, not fully developed at the time of flowering, subsessile, up to 7½ lin. long, 1½–2 lin. broad, lanceolate-elliptic, acute, subglabrous ; flowers few together in fascicles at the nodes ; pedicels 2 lin. long, filiform, puberulous ; sepals scarcely 1¼ lin. long, lanceolate, acuminate, puberulous ; corolla 2½ lin. long, lobed to ⅔ of the way down, entirely glabrous, yellowish ; lobes free, spreading, long-acuminate from an ovate base ; corona-lobes connate high up, 3-lobed, with the lateral lobes minute, ciliate, and the middle lobe larger, ligulate, obtuse, incurved, glabrous.

Eastern Region : Natal ; Alexandra County, at Fairfield, 2200 ft., *Rudatis*, 68.

Said to be allied to *B. caffrum*, N. E. Br., and *B. pulchellum*, Schlechter, but distinguished by its more pointed corolla-lobes and different corona.

14. **B. caffrum** (N. E. Br. in Gard. Chron. 1894, xvi. 62) ; tuber depressed-globose, with a long neck, dividing into numerous procumbent branching minutely scaberulous stems 2–5 in. long ; leaves 2–8 lin. long, including the very short petiole, ¾–4½ lin. broad, mostly ovate or the upper lanceolate and some of the basal orbicular-ovate, acute, rather thick, glabrous, minutely and stiffly ciliate, with the midrib beneath scaberulous, more rarely with the undersurface slightly scaberulous ; flowers solitary or rarely in pairs at the nodes ; pedicels ½–1½ in. long, slender, erect, scaberulous ; sepals ¾–1 lin.

long, lanceolate, acute, sparsely scaberulous; corolla rotate, lobed to about ⅔ of the way down, ⅓–½ in. in diam., fleshy, bright clear yellow, without a distinct tube; lobes free, 1½–2¼ lin. long, 1–1¾ lin. broad, ovate (triangular in dried flowers), acute, flat, glabrous, with minutely ciliolate margins; outer corona-lobes very short, closely contiguous, forming a thick fleshy truncate ring not exceeding the staminal column, from which the linear obtuse inner corona-lobes are inflexed upon the backs of the anthers and do not exceed them, all yellow; staminal column ¾ lin. long; follicles erect, subparallel or slightly diverging, about 1¾ in. long, ⅛ in. thick, terete, tapering into a long beak, smooth, glabrous; seeds about ¼ in. long, 1 lin. broad, narrowly ovate or ovate-lanceolate, some-what concave on one side, smooth, glabrous, light brown, with an ochraceous margin; odour disagreeable. *Schlechter in Journ. Bot.* 1897, 292. *Tapeinostelma caffrum, Schlechter in Verhandl. Bot. Ver. Brand.* xxxv. 54; *K. Schum. in Engl. and Prantl, Pflanzenfam.* iv. ii. 267.

COAST REGION: King Williamstown Div.; abundant on rocky ground on a mountain at Perie, 4000–5000 ft., *Sim*, 315! and *cultivated specimen*!

Described from a living plant which flowered at Kew in July, 1894.

15. **B. pulchellum** (Schlechter in Engl. Jahrb. xx. Beibl. 51, 53); rootstock a small tuber, with 1 or more stems, divided at the base into 2 or more prostrate branches 3–9 in. long, which are simple or again branching, puberulous with very minute retrorse hairs to nearly glabrous; leaves spreading; petiole ½–2 lin. long; blade 1½–8 lin. long, ¾–7 lin. broad, mostly lanceolate or ovate-lanceolate, acute, tapering into the petiole at the base, under cultivation becoming broadly ovate or orbicular, acute or obtuse and apiculate, and rounded to subcordate at the base, thinly puberulous above or glabrous on both sides; flowers solitary or in pairs at the nodes, or occasionally in shortly pedunculate 2-flowered umbels; pedicels 1½–3½ lin. long, thinly puberulous; sepals 1 lin. long, lanceolate, acuminate, recurved at the tips, nearly glabrous; corolla rotate, about 3½ lin. in diam. when fresh, quite glabrous and not ciliate, dark purple-brown, with transverse dull yellowish lines on the basal part of the lobes, usually invisible on dried specimens; united part nearly flat; lobes free, 1¼–1⅓ lin. long, ¾ lin. broad, ovate, acute; outer corona of 5 small pouches alternating with and below the anthers, dark purple-brown; inner corona-lobes ¼ lin. long, linear, closely incumbent upon the backs of the anthers and not exceeding them, dark purple-brown; staminal column ½ lin. long. *Schlechter in Journ. Bot.* 1897, 293. *Micraster pulchellus, Harv. Gen. South Afr. Pl. ed.* 2, 242.

EASTERN REGION: Natal; on stony ground, near Bothas Hill, 2200 ft., *Wood*, 4536! fissures of rocks near Krantz Kloof, *Schlechter*, 3178 (ex *Schlechter*), and without precise locality, *McKen*, 2! *Sanderson*, 342! and *cultivated specimen*!

16. **B. nanum** (N. E. Br.); rootstock not seen; plant 2-3 ($2\frac{3}{4}$-$4\frac{3}{4}$, *Schlechter*) in. high, branching; branches covered with exceedingly minute recurved hairs, apparently reddish-brown; leaves spreading or ascending, $3\frac{1}{2}$-9 lin. long, $\frac{1}{3}$-1 (1-2, *Schlechter*) lin. broad, linear (linear-elliptic or linear-lanceolate, *Schlechter*), acute, tapering into the petiole, flat or slightly convex above, with revolute or thickened (sometimes slightly crisped, *Schlechter*) margins and a prominent midrib beneath, sometimes recurved or revolute at the tips, occasionally twisted, glabrous, with a sparse ciliation of very minute adpressed hairs on the margins; flowers in pairs (1-3, *Schlechter*) at the nodes, very small; pedicels $\frac{3}{4}$-$1\frac{1}{2}$ lin. long, with minute hairs, as on the stems; sepals about $\frac{1}{2}$ lin. long, lanceolate, acuminate, glabrous (puberulous, sparsely ciliate, *Schlechter*); corolla-tube about $\frac{1}{2}$ lin. long, cup-shaped; lobes free, campanulately spreading, about $1\frac{1}{3}$ lin. long, $\frac{2}{3}$-$\frac{3}{4}$ lin. broad at the base, thence gradually tapering to the acute apex or in some flowers ovate or ovate-lanceolate, acuminate, apparently thickened and incurved-hooked at the tips, below which the margins are inflexed, forming a slight constriction, glabrous on both sides, not ciliate, apparently pinkish or pale purplish with darker tips; outer corona-lobes arising $\frac{1}{4}$ lin. up the staminal column and much overtopping (or perhaps incurved over) it, $\frac{1}{3}$-$\frac{1}{2}$ lin. long, divided nearly to the base into 2 slender linear-subulate acute or obtuse segments, which are adnate in pairs at the base to the inner corona-lobes, puberulous on the basal part of the inner face with very minute flattened adpressed white hairs, not visible when wetted (glabrous, *Schlechter*); inner corona-lobes about $\frac{1}{4}$ lin. long, slenderly linear, incumbent upon the backs of the anthers and exceeding them, with erect or recurving tips (incurved at the apex, *Schlechter*), glabrous; staminal column $\frac{1}{2}$ lin. long; follicles divergent-erect, $1\frac{1}{2}$-2 in. long, $1\frac{1}{2}$ lin. thick, linear-terete, tapering at the base and into a beak at the apex, smooth, glabrous; seeds $\frac{1}{4}$ in. long, 1 lin. broad, linear-oblong with the margins much inflexed on one side, broadly margined, smooth, dark brown, with a tuft of rather short hairs at the apex. *Lasiostelma nanum, Schlechter in Engl. Jahrb.* xxxviii. 37.

KALAHARI REGION: Orange River Colony (not Transvaal, as originally stated); on stony ground near Rhenoster Kop, *Zeyher*, 509 (ex *Schlechter*), *Burke*, 509! and probably from the same locality, *Zeyher*, 1139!

In this case, as in that of *Fockea angustifolia*, K. Schum., Zeyher appears to have distributed the plant under the same number as Burke, as well as under a number of his own. I have not seen Zeyher's 509, from which Dr. Schlechter described, but so far as it goes, with the exception of the few discrepancies placed in brackets, his description quite agrees with Burke's plant, and I have no doubt of their identity. *B. nanum* is exceedingly like *B. occidentale* in appearance. It distinctly differs, however, in its leaves being slightly broader and having a conspicuous and prominent midrib beneath, whilst in *B. occidentale* there is no evident midrib, the outer corona-lobes are also puberulous within and the inner corona-lobes have erect tips. Living plants may exhibit other distinctive characters.

17. **B. occidentale** (Schlechter in Verhandl. Bot. Ver. Brandenb. xxxv. 53) ; tuber subglobose ; plant about 3 in. high, branching at the base ; stems erect, slender, covered with very minute often recurved hairs, slightly rough to the touch ; leaves ascending or spreading, 3–7 lin. long, about ⅓ lin. broad, linear, acute, tapering at the base, apparently somewhat fleshy and on the dried specimen often more or less twisted, flat on the upper side, slightly convex but without a distinct midrib on the back, with a very minute adpressed ciliation, otherwise glabrous ; flowers usually in pairs, sometimes solitary, lateral at the nodes or subaxillary ; pedicels 2–3 lin. long, slender, covered with exceedingly minute somewhat adpressed hairs, and having two small lanceolate acute bracts about ½ lin. long at their base ; sepals ½–⅔ lin. long, lanceolate, acuminate, glabrous ; corolla-tube scarcely ½ lin. long, cup-shaped ; lobes free, campanulately spreading or suberect, about 1 lin. long, ½ lin. broad at the base, oblong-lanceolate, acute, with the margins at the middle incurved and the apex much thickened and incurved-hooked, glabrous on both sides and not ciliate, " yellowish-white dotted with wine-red on the inner surface " (*Schlechter*) ; outer corona-lobes arising about ¼ lin. up the staminal column and much overtopping it, ⅖ lin. long, divided into 2 linear obtuse incurved segments, which are adnate in pairs at the base to the inner corona-lobes, glabrous ; inner corona-lobes ¼ lin. long, linear, obtuse, closely incumbent upon the backs of the anthers and scarcely exceeding them, glabrous. *Brachystelmaria occidentalis, Schlechter in Journ. Bot.* 1897, 293. *Lasiostelma occidentale, Schlechter in Engl. Jahrb.* xxxviii. 38 *in note.*

COAST REGION : Cape Div. ; stony places on a mountain near Smitswinkel Bay, 300 ft., very rare, only one plant seen, *Schlechter,* 666 !

Described from a branch and flowers from the type specimen in the Herbaria of Prof. Hans Schinz and Dr. Bolus, but I do not find either the stem, pedicels or calyx to be "pilose" as described by Dr. Schlechter, the hairs are exceedingly minute, and there are none on the sepals, except just at their very base.

18. **B. schizoglossoides** (N. E. Br.) ; roots fascicled, thick, fleshy, long and narrowly fusiform ; stem solitary, simple, 4–8 in. high, glabrous, naked at the basal part, leafy above ; leaves ascending, ⅓–1⅔ in. long, ⅓–1½ lin. broad, linear, acute, tapering at the base, glabrous ; flowers 1–3 together in fascicles or very shortly pedunculate umbels lateral at the nodes or rarely axillary ; peduncles 0–¾ lin. long, glabrous ; pedicels 1–2 lin. long, glabrous ; sepals ⅔–¾ lin. long, ovate-lanceolate, acuminate, recurved at the tips, glabrous ; corolla lobed to ⅔ of the way down, green (*Mrs. Barber*) ; tube much shorter than the staminal column, saucer-shaped ; lobes free, campanulately spreading, with incurved tips, ⅔–¾ lin. long and the same in breadth at the base, broadly ovate, shortly cuspidate-acute, bordered with minute very thick or subclavate weakly attached hairs at the tips on the inner face, otherwise glabrous ; outer corona-lobes arising ¼–⅓ lin. up the staminal column, opposite and adnate

to the base of the inner corona-lobes, $\frac{1}{2}$ lin. long, oblong, bifid to $\frac{1}{3}$ or half way down, with linear obtuse segments, very hairy; inner corona-lobes $\frac{1}{3}$–$\frac{1}{2}$ lin. long, linear, acute, incumbent upon the backs of the anthers at the base, with connivent-erect tips, glabrous; staminal column about $\frac{1}{2}$ lin. long. *Sisyranthus schizoglossoides, Schlechter in Journ. Bot.* 1894, 357, *and* 1897, 291.

COAST REGION: Albany Div.; near Grahamstown, *Glass* (ex *Schlechter*); Highlands above New Years River (Zwartwater Berg ?), *Mrs. Barber,* 117! Bathurst Div.; Round Hill, *Bolus,* 6694!

Described from part of the type.

19. **B. natalense** (N. E. Br.); roots thick, fleshy, spindle-shaped (*Schlechter*); stem solitary, simple or with 1 branch, $\frac{1}{2}$–2 ft. high, $\frac{3}{4}$–2 lin. thick at the base, with a spreading not very soft pubescence, leafy; leaves very shortly petiolate, spreading, 1–$2\frac{1}{4}$ in. long, $\frac{2}{3}$–$1\frac{3}{4}$ in. broad, or the upper smaller, ovate to orbicular-ovate, acute or obtuse and apiculate, slightly cordate to broadly rounded at the 5-nerved base, pilose on both sides, shortly and densely ciliate; flowers 2–4 together, lateral at the nodes, and more or less clustered at the apex of the stem by the crowding of the leaves, the internodes afterwards elongating; pedicels $4\frac{1}{2}$–10 lin. long, slender, often curved at the apex, villous-pubescent; sepals 1–$1\frac{1}{4}$ lin. long, lanceolate, acuminate, pubescent like the pedicels; corolla nodding, very small, about 2–$2\frac{1}{2}$ lin. in diam., lobed to $\frac{2}{3}$ of the way down, with the united part shallowly cup-shaped; lobes free, apparently campanulately spreading, 1 lin. long, $\frac{2}{3}$ lin. broad, ovate-lanceolate, tapering to an obtuse or minutely emarginate apex, glabrous on both sides or with a very few hairs on the back, apparently dull greenish; outer corona-lobes arising $\frac{1}{5}$ lin. up the staminal column, ascending, minute, about $\frac{1}{3}$ lin. long and broad, very broadly or subquadrately ovate, shortly bifid, with obtuse teeth; inner corona-lobes erect, much shorter than the anthers, slightly overtopping and nearly or about as large as the outer corona-lobes, of similar shape, but entire at the rounded apex, glabrous; staminal column $\frac{2}{5}$ lin. long; follicles solitary or in pairs, erect, about 6 in. long and $\frac{1}{4}$ in. thick, linear-terete, tapering to an acute apex, smooth, glabrous; seeds about 5 lin. long and 2 lin. broad, oblong, tapering at the apex, concave on one side, margined, glabrous, brown. *Aulostephanus natalensis, Schlechter in Bull. Herb. Boiss.* iv. 451.

EASTERN REGION: Natal; Inanda, 1700 ft., *Wood,* 410!

Dr. Schlechter also quotes Wood 176 for this plant, but that number, as sent to Kew, belongs to a species of *Senecio.* I find nothing in the structure of this plant to generically distinguish it from *Brachystelma.*

20. **B. Sandersoni** (N. E. Br.); rootstock a cluster of long fleshy terete or narrowly fusiform roots; stem erect, 8–18 in. high, 1–$1\frac{1}{2}$ lin. thick at the base, simple or branching above the middle,

rather leafy on the upper half, naked below, glabrous; leaves ascending, very shortly petiolate, subcoriaceous, $\frac{1}{2}$–$1\frac{1}{4}$ in. long, $\frac{1}{8}$–$\frac{1}{2}$ in. broad, linear, oblong, ovate-oblong or elliptic-oblong, acute or obtuse, rounded or subcuneate at the base, scabrous on the midrib beneath and usually minutely and stiffly ciliate or slightly scabrous on the margins, otherwise glabrous; flowers in axillary fascicles of 2–6 or occasionally solitary; pedicels 1–3 lin. long, apparently angular, glabrous; sepals 1 lin. long, lanceolate, acute, glabrous; corolla-tube very much shorter than the staminal column, $\frac{1}{3}$–$\frac{1}{2}$ lin. deep, saucer-shaped; lobes free, erect or but slightly spreading, $1\frac{1}{4}$–2 lin. long, 1 lin. broad at the base, ovate or oblong, shortly acute, with the sides folded back, forming a keel down the inner face, glabrous on both sides, ciliate with minute thick hairs, white, tipped with pink or light purple; outer corona divided into 10 erect teeth arising $\frac{1}{4}$ lin. up the staminal column and scarcely overtopping it, about $\frac{1}{3}$ lin. long, narrowly deltoid-lanceolate, or linear, acute, ciliate and usually covered on the back with minute thick or papilla-like hairs, and on the inner face with a finer and more minute pubescence, white; inner corona-lobes $\frac{1}{4}$–$\frac{1}{3}$ lin. long, oblong or oblong-linear, obtuse, very minutely puberulous at the base, glabrous above, incumbent upon the backs of the anthers and equalling or shortly exceeding them, white; staminal column $\frac{2}{3}$–$\frac{3}{4}$ lin. long. *Lasiostelma Sandersoni, Oliver in Hook. Ic. Pl.* xv. 39, *t.* 1449; *Schlechter in Journ. Bot.* 1898, 487, *and* 1899, 62. *L. Benthamii, K. Schum. in Engl. and Prantl, Pflanzenfam.* iv. ii. 297. *Dichælia natalensis, Schlechter in Engl. Jahrb.* xviii. *Beibl.* 45, 35. *Brachystelmaria natalensis, Schlechter in Engl. Jahrb.* xx. *Beibl.* 51, 50, *and Journ. Bot.* 1897, 293.

EASTERN REGION : Natal ; Tugela, *Gerrard,* 1805 ! Wentworth Bluff, *Sanderson,* 436 partly ! Verulam, *Wood,* 1161 ! Clairmont, near Durban, *Wood,* 3906 ! and in *Herb. Natal,* 713 ! and without precise locality, *Gueinzius* ! *Sanderson,* 915 !

Dr. Schlechter has described *Dichælia natalensis* as having the corolla "very shortly pilose outside" and the "corona narrowed at the base, then dilated into a cupuliform tube, with the 5 outer lobes dilated from a narrow base, deeply bifid, with erect linear somewhat acute glabrous partitions." I have examined a part of the type of *D. natalensis* (Wood, 3906) and find it identical with the other specimens quoted and as I have described above.

21. **B. Gerrardi** (Harv. Thes. Cap. ii. 61, t. 196); stem $\frac{3}{4}$–$2\frac{3}{4}$ ft. high, simple or branching in the upper part, 1–2 lin. thick at the base, with a short spreading pubescence, leafy; leaves numerous, shortly petiolate, more or less spreading, $\frac{3}{4}$–$1\frac{3}{4}$ in. long, $\frac{1}{3}$–$1\frac{1}{2}$ in. broad, ovate, ovate-oblong or the lower often suborbicular, obtuse or acute, or rounded and apiculate at the apex, rounded or subcordate at the base, glabrous or thinly pubescent above, pubescent beneath; flowers solitary at the nodes at the terminal part of the stem and branches; pedicels $\frac{1}{4}$–$\frac{3}{4}$ in. long, pubescent, with 1 or 2 subulate or filiform bracts 2–3 lin. long at their base; sepals 3–4 lin. long, filiform-subulate from a short broadly ovate or suborbicular

base, pubescent; corolla-tube about as long as the staminal column, apparently nearly flat at the base, broadly saucer-shaped and scarcely 1 lin. deep, glabrous, spotted with very dark green or blackish, or entirely blackish outside and within in dried flowers; lobes free, apparently incurved at the base then abruptly turned back and widely spreading, or in some dried flowers suberect, 4–7 lin. long, $1\frac{1}{2}$–2 lin. broad at the base and nearly as broad in the upper part, narrowly oblong, acute or abruptly acute at the apex, constricted about 1 lin. above the base, where they recurve, with the margins apparently slightly reflexed, glabrous and apparently whitish or pale greenish outside, very minutely puberulous (glabrous to the naked eye) and " bright metallic green " (*Wood*) on the inner face, ciliate at the base with long vibratile stout simple purple hairs; outer corona-lobes arising $\frac{1}{3}$–$\frac{1}{2}$ lin. up the staminal column, ascending-spreading, $\frac{1}{3}$–$\frac{2}{5}$ lin. long, linear-oblong, bifid at the apex, puberulous on both sides, with the hairs at the tips rather longer than the rest; inner corona-lobes 2–$2\frac{1}{2}$ lin. long, connivent-erect, linear, tapering to an obtuse apex, blackish, glabrous above, puberulous at the base, as is also the $\frac{3}{4}$ lin.-long staminal column below the corona. *Benth. and Hook. f. Gen. Pl.* ii. 781. *Dichælia Gerrardi, Harv. Gen. S. Afr. Pl. ed.* 2, 241. *Brachystelmaria Gerrardi, Schlechter in Engl. Jahrb.* xx. *Beibl.* 51, 50, *and Journ. Bot.* 1897, 293; *K. Schum. in Engl. and Prantl, Pflanzenfam.* iv. ii. 268. *Lasiostelma Gerrardi, Schlechter in Journ. Bot.* 1899, 62.

Eastern Region: Natal; Emyati, *Gerrard,* 1318! at or near Krans Kop, *McKen,* 18! Inanda, *Wood,* 455! 1607! and in *Herb. Natal,* 439!

According to Gerrard the flowers are "dark green, approaching black, velvety above, pale yellow below." The outer corona-lobes are not so deeply divided as represented by Harvey. *Gardenia Thunbergii* has also been distributed by Mr. Wood under no. 439.

22. **B. macropetalum** (N. E. Br.); rootstock not seen nor described; stem $\frac{3}{4}$–2 ft. high, 1–2 lin. thick at the base, erect, simple, softly pubescent, leafy; leaves ascending, shortly petiolate, $\frac{1}{2}$–$1\frac{1}{2}$ in. long, $\frac{1}{4}$–$1\frac{1}{4}$ in. broad, ovate, acute, rounded to subcordate at the base, pubescent on both sides, but more densely beneath; flowers solitary, lateral at the nodes; pedicels $\frac{2}{3}$–1 in. long, softly pubescent, bearing a petiolate lanceolate or linear-lanceolate acute leaf-like bract $\frac{1}{3}$–$\frac{1}{2}$ lin. long at their base; sepals ascending-spreading, $\frac{1}{2}$ in. long, $\frac{3}{4}$–1 lin. broad above the middle, linear-spathulate, acute, densely and softly pubescent; corolla-tube scarcely any, flattish or saucer-shaped, about $\frac{3}{4}$ lin. deep; lobes free, 9–10 lin. long, linear from a dilated $1\frac{1}{2}$–2 lin.-broad base, acute, erectly spreading, pubescent on the back, finely puberulous on the inner face, yellow (*Thorncroft*), greenish (*Schlechter*), brighter within, marked with reddish or brownish dots, but in the dried specimens apparently marked on the basal part only with dark green lines and spots or veining; outer corona-lobes arising $\frac{1}{2}$ lin. up the staminal column,

erect, rather more than 1 lin. long, divided to below the middle into
2 diverging linear-subulate segments, with the basal part linear-
oblong or rectangular, puberulous, greenish?; inner corona-lobes
much exceeding the outer corona, connivent-erect, 2–2½ lin. long,
spathulate-oblanceolate, convex on the back, flat or slightly concave
on the inner face, thickened at the apex, obtuse, puberulous, dark
violet-brown (*Schlechter*), blackish in dried specimens; follicles
solitary or in slightly diverging pairs, 5–5½ in. long, 2–2½ lin. thick,
linear-terete, acuminate, smooth, glabrous; seeds 4½–5 lin. long,
⅛ in. broad, oblong, narrowing to the apex, broadly margined,
concave on one side, smooth, glabrous, dark brown, crowned with a
tuft of hairs. *Brachystelmaria macropetala, Schlechter in Engl.
Jahrb.* xx. *Beibl.* 51, 51, *and Journ. Bot.* 1897, 293. *Lasiostelma
macropetalum, Schlechter in Journ. Bot.* 1899, 62.

KALAHARI REGION : Transvaal ; near Lydenberg, *Atherstone*! near Barberton,
2900 ft., *Thorncroft*, 267 (*Herb. Wood*, 4506)! Elandspruit Mountains, 5300 ft.,
Schlechter, 3869 !

Dr. Schlechter describes the outer corona-lobes as being glabrous, but I find
them to be puberulous in his type (3869) as in the other specimens quoted.

23. B. longifolium (N. E. Br.) ; rootstock not seen nor described ;
stem erect, 3½–5 in. high in the specimens seen (6–12 in., *Schlechter*),
branching above, glabrous or bifariously puberulous, naked below,
leafy above ; leaves more or less spreading, 1¼–3 in. long, ½–2 lin.
broad, linear, acute, narrowed or tapering into the petiole at the
base, scabrous on the midrib beneath and with very minute teeth
on the margins, otherwise glabrous ; flowers 1–2 together at the
nodes ; pedicels 2–4 lin. long, slender, glabrous ; sepals 1½–2 lin.
long, very spreading, subulate, glabrous ; corolla-tube almost none,
saucer-shaped, scarcely ½ lin. deep ; lobes erect, free, 5 lin. long,
about 1 lin. broad, linear-oblong, acute, replicate, ciliate with
minute thick obovoid or clavate purple hairs to half way up along
the inner face, otherwise glabrous on both sides, apparently greenish,
tinted with brown ; outer corona-lobes arising ¼ lin. up the staminal
column, with their subquadrate basal part adnate to it, erect, about
¾ lin. long, divided to the middle into two linear-attenuate segments,
which falcately curve inwards, with their points turned towards
each other by a slight twist, white, very minutely papillate-ciliate ;
inner corona-lobes 1⅓–1½ lin. long, linear, obtuse, connivent-erect,
much overtopping the outer corona, white, glabrous ; staminal
column about ¾ lin. long. *Brachystelmaria longifolia, Schlechter in
Engl. Jahrb.* xx. *Beibl.* 51, 50 ; *and Journ. Bot.* 1897, 293. *Lasiostelma longifolium, Schlechter in Journ. Bot.* 1899, 62.

KALAHARI REGION : Transvaal ; in stony places on the Elandspruit Mountains,
5300 ft., *Schlechter*, 3873 !

Described from a portion of the type in Herb. Bolus.

24. B. comptum (N. E. Br.); rootstock a cluster of long thick fleshy narrowly fusiform roots; stem in the only examples seen 2–3 in. high, simple or very sparingly branched, pubescent; leaves in 4–6 pairs, very shortly petiolate, spreading, 5–8 lin. long, 3–5 lin. broad, oblong, elliptic or orbicular-ovate, obtuse, rounded at the base, thinly pubescent on both sides and more densely on the midrib and margins with short spreading hairs; flowers solitary, lateral at the nodes; pedicels about $\frac{1}{3}$ in. long, pubescent; sepals $1\frac{1}{2}$ lin. long, lanceolate-subulate, pubescent; corolla-tube almost none, flattish or saucer-shaped, about $\frac{1}{2}$ lin. deep, glabrous; lobes free, erectly spreading, 4–$4\frac{1}{2}$ lin. long, 1 lin. broad at the base, linear-oblong, subacute, apparently with slightly incurved margins, glabrous on both sides, ciliate at about 1 lin. above their base with a tuft of long jointed subclavate purple hairs, which are inwardly directed; outer corona-lobes arising $\frac{2}{3}$ lin. up the staminal column and overtopping it by half their length, $\frac{2}{3}$–$\frac{3}{4}$ lin. long, erect, linear, truncate and very slightly dilated at the entire apex, minutely puberulous, ciliate with long hairs at the tips, apparently pale purplish; inner corona-lobes $\frac{1}{3}$ lin. long, linear, obtuse, closely incumbent upon the backs of the anthers and not or scarcely exceeding them, dark purple-brown; staminal column nearly 1 lin. long.

Coast Region: Uitenhage Div.; karoo-like places on hills by the Zwartkops River, *Zeyher*, 9!

The corolla-lobes of the only flower seen are dark-coloured and may have been either dark purple-brown or blackish-green.

25. B. mafekingense (N. E. Br.); tuber depressed, "about 3 in. in diam. and 2 in. thick"; stems "richly branched from near the base 4–7 cm. (i.e. $1\frac{2}{3}$–$2\frac{3}{4}$ in.) above the tuber" (*Schönland*), only 2 bits $\frac{1}{2}$ in. long seen, puberulous; leaves spreading, $\frac{1}{3}$–$\frac{2}{3}$ in. long, very shortly petiolate, $\frac{1}{8}$–$\frac{1}{4}$ in. broad, lanceolate, acute, cuneately rounded into the petiole at the base, apparently folded longitudinally, glabrous above, puberulous beneath; flowers (on the only example seen) in a terminal 10–12-flowered umbel, sessile between a pair of leaves, with a short axillary branch beside it; pedicels $\frac{1}{8}$ in. long, puberulous; sepals 1–$1\frac{1}{4}$ lin. long, narrowly lanceolate, acute, puberulous; corolla-tube 2–$2\frac{1}{2}$ lin. long, $1\frac{1}{2}$–$1\frac{3}{4}$ lin. in diam., obtusely pentagonal, slightly narrowing upwards, minutely puberulous on the upper part outside, glabrous within, apparently whitish; lobes 3–$3\frac{1}{2}$ lin. long, apparently free, erect, closely replicate and 1 lin. broad across the side, slightly obovate-oblong when flattened out, subacute, minutely puberulous on the back, glabrous and blackish on the inner face, not ciliate; outer corona arising at the base of and equalling the staminal column, consisting of 5 or rarely 6 bifid erect lobes $\frac{1}{2}$ lin. long at the back of and adnate to the inner corona-lobes, with the teeth narrow and not spreading, glabrous, apparently white; inner corona-lobes 1 lin. long, linear, obtuse, incumbent at the base on the backs of the anthers, then erect and contiguous in a column, with the tips free or connate, white.

KALAHARI REGION : Bechuanaland ; Mafeking, *Green in Herb. Schönland,* 1683 !

Described from a small branch preserved in formalin ; received from Dr. Schönland.

26. **B. ramosissimum** (N. E. Br.) ; roots 3–5 in. long, ¼ in. or more thick, fleshy, terete, fascicled ; plant up to about 6 in. high, usually much branched from the base and upwards ; branches erect, angular, puberulous thinly bifariously and on the stipulary line ; leaves very shortly petiolate or subsessile, ½–¾ in. long, 1–3 lin. broad, linear-oblong to lanceolate, acute, cuneate at the base, minutely ciliate and with a few minute curved hairs on the midrib beneath, otherwise glabrous ; flowers in fascicles of 2–4, lateral at the nodes ; pedicels 1½–2 lin. long ; sepals spreading, nearly 1 lin. long, ¼–⅓ lin. broad, lanceolate, acuminate, glabrous ; corolla about ¼ in. in diam., pale yellow (*Schlechter*), white (*Conrath*), the united part almost flat, not forming a tube, about 1⅓ lin. in diam. ; lobes free, suberect or slightly spreading, 1½–2 lin. long, 1 lin. broad at and replicate from the base, ovate-oblong when flattened out, acute and incurved at the apex, glabrous on both sides, ciliate near the base with short hairs directed inwards ; outer corona arising about ¼ lin. up the staminal column, comparatively rather large, about ⅓ lin. long, spreading or somewhat saucer-shaped, subequally and acutely 10-toothed, or with 5 broad emarginate or subtruncate lobes opposite the anthers, pubescent on the back and around the margin on the inner face, shortly ciliate ; inner corona-lobes scarcely ¼ lin. long, linear, obtuse or occasionally bifid, closely incumbent upon the backs of the anthers and shorter than them, not connected to the outer corona by any dorsal projection ; follicles erect, subparallel, 3–4½ in. long, ⅛ in. thick, terete, tapering into a slender beak, smooth, glabrous. *Brachystelmaria ramosissima, Schlechter in Engl. Jahrb. xx. Beibl. 51, 51, and Journ. Bot. 1897, 293. Lasiostelma ramosissimum, Schlechter in Journ. Bot. 1899, 62.*

KALAHARI REGION : Transvaal ; near Mooifontein, 5500 ft., *Schlechter,* 3554 ! near Modderfontein, *Conrath,* 1006 ! near the Vaal River Bridge at Vereeniging, 4750 ft., *Gilfillan in Herb. Galpin,* 6152 ! Walmaranstad, *Burtt Davy,* 1531 !

27. **B. Zeyheri** (N. E. Br.) ; plant 2–5 in. high, branching ; branches erect or spreading, puberulous ; leaves spreading, 2–6 lin. long, ¾–1 lin. broad, linear or linear-lanceolate, obtuse or subacute, longitudinally folded, glabrous above, puberulous beneath (" on both sides," *Schlechter*), narrowed into a very short petiole at the base ; flowers usually in pairs (1–3, *Schlechter*) at the nodes, small ; pedicels ½–1 lin. long, puberulous ; sepals nearly 1 lin. (½ lin., *Schlechter*) long, lanceolate, acute, puberulous ; corolla cage-like, lobed nearly to the base, with the lobes connate at the tips ; united part ¼ lin. long (as long as the sepals, *Schlechter*), saucer-shaped, glabrous

outside and within, green ; lobes 2–2¼ lin. long, linear or linear-filiform from a lanceolate or ovate-lanceolate base ½ lin. broad, perhaps replicate, thinly puberulous on the greenish back, glabrous or very minutely and sparsely papillate-puberulous and dull purplish or brownish-purple on the inner face ; outer corona-lobes arising ¼ lin. up the staminal column and equalling or slightly exceeding it, ⅓–½ lin. long, deeply divided (often nearly to the base) into 2 erect linear parallel or slightly diverging teeth, variable in different flowers, glabrous ; inner corona-lobes ⅓–½ lin. long, linear or linear-oblong, obtuse, closely incumbent on the backs of the anthers and equalling or exceeding them and curved downwards over their tips, glabrous ; follicles erect, slightly diverging, 2½ in. long, about 2½ lin. thick, tapering into a beak, smooth, glabrous ; seeds 3–3½ lin. long, 2 lin. broad, ovate or oblong-ovate, smooth and glabrous, dark brown with a broad pale brown margin. *Dichælia Zeyheri, Schlechter in Engl. Jahrb.* xxxviii. 43.

COAST REGION : Uitenhage Div. ; near Uitenhage, *Zeyher* ! Witte River Station (Enon), *Gill* ! stony places near the Zwartkops River, *Zeyher,* 3383 (ex *Schlechter*).

Zeyher's plant was quoted by Harvey under *B. filiforme,* Harv., but is not the plant he figured and described under that name. According to an excellent drawing of *B. Zeyheri* in the Kew Herbarium, made from a plant introduced by Bowie in 1823, it flowered at Kew in August, 1824.

28. **B. filiforme** (Harv. Thes. Cap. i. 58, t. 93) ; tuber flattened, 2–2¼ in. in diam., with 1 or more stems, branching at the base, erect, 6–8 in. high, minutely puberulous, with internodes up to 2 in. long ; leaves spreading, sometimes undeveloped at the time of flowering, 1½–6 lin. long, ½–1 lin. broad, linear or narrowly linear-lanceolate, acute or obtuse, tapering into a short petiole at the base, longitudinally folded, sometimes undulated at the margins, glabrous above, minutely puberulous beneath ; flowers 2–6 together at the nodes or forks of the branches ; pedicels ⅓–⅔ lin. long ; sepals ⅔ lin. long, lanceolate, acuminate, and like the pedicels puberulous ; corolla cage-like, lobed nearly to the base, with the lobes connate at the tips, " brownish-purple " (*Mrs. Barber*) ; united part flattish or saucer-shaped, scarcely ½ lin. long, shorter than the corona, glabrous outside and within ; lobes 2–2¼ lin. long, linear-filiform from a deltoid-ovate base ¾ lin. broad, with a few exceedingly minute papilla-like hairs on the back, otherwise glabrous ; outer corona-lobes ½ lin. long, apparently erect and equalling or slightly over-topping the staminal column, divided to the middle or beyond into 2 subparallel linear or subulate teeth, sparsely ciliate with very minute or papilla-like hairs ; inner corona-lobes ⅓ lin. long, linear or narrowly deltoid-linear, obtuse, closely incumbent on the backs of the anthers and exceeding them, with the tips incurved over their apex, very minutely papillate-puberulous at their base on the back. *Dichælia filiformis, Schlechter in Engl. Jahrb.* xviii. *Beibl.* 45, 36, *and Journ. Bot.* 1897, 293 ; *K. Schum. in Engl. and Prantl,*

Pflanzenfam. iv. ii. 269.   *D. macra, Schlechter in Engl. Jahrb.* xxxviii. 43.

CENTRAL REGION: Cradock Div. ; without precise locality, *Mrs. Barber,* 2 ! 88 ! Murraysburg Div. ; near Murraysburg, *Tyson* (ex *Schlechter*).

Harvey's figure and description of this plant is made entirely from Mrs. Barber's specimens. On her label of no. 2, she states that "it is eaten by the Dutch inhabitants and much esteemed as a preserve, under the name of 'kalkoujes.'" Zeyher's specimen, from near Uitenhage, which Harvey quotes under *B. filiforme,* belongs to *B. Zeyheri,* N. E. Br.   I have not seen a specimen of *Dichælia macra,* but from the description there seems nothing to distinguish it from *B. filiforme,* the only discrepancies I can find are that in *D. macra* the leaves are "1-1½ lin. broad, oblong-lanceolate, puberulous on both sides ; pedicels about 1 lin. long ; united part of corolla ¾ lin. long ; inner corona-lobes glabrous," but the minute puberulence at their base in *B. filiforme* is almost invisible when moistened for dissection, and with regard to the leaves, on account of the longitudinal folding of them, the real upper (infolded) surface may have been overlooked.

29. **B. pygmæum** (N. E. Br.) ; tuber turnip-shaped (*Bowker*) ; stems several, simple or slightly branched, erect, 1¼-3 in. high, puberulous ; leaves sometimes undeveloped at the time of flowering, spreading, 2-6 lin. long, ¾-2½ lin. broad, linear-spathulate, oblanceolate or oblong-lanceolate, acute or obtuse, cuneately narrowed into a short petiole at the base, glabrous on both sides except along the midrib beneath, minutely ciliate ; flowers 1-3 together, lateral at the nodes ; pedicels 1½-4 lin. long, puberulous ; sepals about ¾ lin. long, linear-lanceolate, acute, puberulous ; corolla usually cage-like, with the lobes connate at the tips, but sometimes with free and more or less spreading lobes, and both forms on the same specimen ; tube about ¾ lin. long, shallowly cup-shaped, glabrous outside, puberulous within (the hairs more or less invisible when wetted), apparently yellowish or greenish ; lobes 2-3 lin. long, linear from a deltoid base, when free varying from nearly straight to circinately inrolled at the tips, puberulous at the base on the inner face, otherwise entirely glabrous, apparently olive-green or purple-brown ; outer corona very variable, even in flowers upon the same branch, arising ¼-⅓ lin. up the ½ lin.-long staminal column, consisting of 5 minute entire pouches or 5 distinct lobes, which are sometimes very minute, at others more distinct and subquadrate, usually emarginate or shortly bifid, or the lobes are all connate into an entire narrow ring or cup entirely surrounding the column, sprinkled on the back and on the column immediately beneath them with minute but rather thick or papilla-like hairs and sometimes more or less ciliate with them ; inner corona-lobes ⅛-⅙ lin. long, linear-oblong, obtuse or notched at the apex, incumbent upon the backs of the anthers and not exceeding them, much larger than the outer corona-lobes, glabrous. *Dichælia pygmæa, Schlechter in Journ. Bot.* 1894, 262, *and* 1897, 293.

VAR. β, **breviflorum** (N. E. Br.) ; leaves undeveloped at the time of flowering ; tube and base of the 2-2½ lin.-long corolla-lobes glabrous on the inner face ; outer corona-lobes glabrous or with very few papilla-like hairs beneath, otherwise as in the type. *Dichælia breviflora, Schlechter in Engl. Jahrb.* **xx.** *Beibl.* 51, 49, *and Journ. Bot.* 1897, 294.

Coast Region : Bedford Div. ; near Bedford, *Hutton* ! British Kaffraria and Eastern Frontier, *Hutton* !

Kalahari Region : Var. *β* : Transvaal ; near the Leper Hospital on the way to Johannesburg, *Conrath,* 1009 ! near Mooifontein, near Heidelberg, *Schlechter,* 3568, Piet Retief, *Burtt Davy,* 1913 !

Eastern Region : Transkei ; near the Bashee River at Fort Bowker, *Bowker,* 593 ! and without precise locality, *Bowker,* 247 (*Mrs. Barber*) !

Dr. Schlechter has considered the Transvaal plant to be specifically distinct from *B. pygmæum,* but I find nothing to separate it except the absence of hairs on the inside of the corolla, and in some specimens of *B. pygmæum* they are rather scanty. The leaves of typical *B. pygmæum* are sometimes as undeveloped at the time of flowering as they are in var. *breviflorum.*

30. **B. commixtum** (N. E. Br.) ; plant 2–5 in. high, branching ; branches erect, thinly puberulous with exceedingly minute recurved hairs ; leaves ascending, $\frac{1}{3}$–1$\frac{1}{3}$ in. long, $\frac{1}{2}$–$\frac{2}{3}$ lin. broad, linear-subterete from the margins being infolded so as to meet, acute or subobtuse, sometimes twisted, obscurely or thinly and very minutely puberulous to glabrous ; flowers in pairs at the nodes ; pedicels $\frac{1}{2}$–$\frac{3}{4}$ lin. long, very minutely puberulous, as are the $\frac{3}{4}$ lin.-long ovate acuminate sepals ; corolla cage-like, lobed nearly to the base, with the lobes connate at the tips, thinly sprinkled with exceedingly minute papilla-like hairs outside and within or almost glabrous ; united part about $\frac{1}{3}$ lin. long, flattish ; lobes about 7 lin. long, filiform from a $\frac{2}{3}$ lin.-broad base, much contorted in the dried flowers, dark purple-brown ; outer corona-lobes shorter than the staminal column, about $\frac{1}{4}$ lin. long, divided nearly to the base into 2 linear-subulate slightly divergent teeth, glabrous ; inner corona-lobes $\frac{1}{3}$ lin. long, oblong-linear, obtuse, closely incumbent upon the backs of the anthers and slightly exceeding them, but not erect at the tips.

Central Region : Aliwal North Div. ; hills near Riet Valley, at the foot of the Witteberg Range, 5000 ft., *Drège,* 3440 partly !

This species is mixed with *B. circinatum* in E. Meyer's Herbarium and at Kew !

31. **B. circinatum** (E. Meyer, Comm. 196) ; plant 5–6 in. high, branching ; branches erect, densely puberulous with minute recurved hairs ; leaves ascending or suberect from a spreading base, 2$\frac{1}{2}$–6 lin. long, linear or linear-subterete from being longitudinally folded or having incurved margins, acute or subobtuse, $\frac{1}{3}$–$\frac{1}{2}$ lin. broad when folded, puberulous like the stems, with the upper (infolded) surface glabrous ; flowers mostly in pairs at the nodes ; pedicels 1 (in fruit 2–2$\frac{1}{2}$) lin. long ; sepals about $\frac{3}{4}$ lin. long, lanceolate, acuminate, and like the pedicels minutely puberulous ; corolla cage-like, lobed nearly to the base, with the lobes connate at the tips, minutely spreading-puberulous all over outside, glabrous within ; united part flattish or saucer-shaped, scarcely $\frac{1}{3}$ lin. deep ; lobes 4–5 lin. long, linear-filiform from a $\frac{2}{3}$ lin.-broad base ; outer

corona-lobes shorter than the staminal column, about ¼ lin. long, deeply divided into 2 subulate slightly divergent teeth, glabrous, apparently whitish ; inner corona-lobes ⅙–¼ lin. long, half as long as the anthers and resting in a concave channel on their backs, with a small linear or subulate acute or obtuse point, and a rather thick dark-coloured fleshy dorsal gibbosity at their base. *Dietr. Syn. Pl.* ii. 887, *excl. syn.* ; *Decne in DC. Prodr.* viii. 647 ; *K. Schum. in Engl. and Prantl, Pflanzenfam.* iv. ii. 268. *Dichælia circinata, Schlechter in Engl. Jahrb.* xxi. *Beibl.* 54, 13, *and Journ. Bot.* 1897, 294. *D. Galpinii, Schlechter in Engl. Jahrb.* xx. *Beibl.* 51, 49, *not elsewhere.* *D. brachylepis, Schlechter in Engl. Jahrb.* xxxviii. 42.

CENTRAL REGION : Aliwal North Div. ; hills near Riet Valley, at the foot of the Witteberg Range, 5000 ft., *Drège,* 3440 partly !

KALAHARI REGION : Transvaal ; stony places near Mooifontein, in the vicinity of Heidelberg, 5500 ft., *Schlechter,* 3556 !

Two distinct species mixed together have been distributed by Drège and are present in E. Meyer's Herbarium under the name *B. circinatum,* both are probably included in the original description, but I take as the type the larger and more puberulous plant, because it is the one to which E. Meyer's label is affixed and of which he has dissected a flower, although I can find nothing circinate about the latter. This plant I have dissected side by side with *Dichælia brachylepis,* Schlechter, and find them to be identical. The other (*B. commixtum,* N. E. Br.) appears to be a smaller plant and is more glabrous, with darker flowers and a different corona.

32. **B. undulatum** (N. E. Br.) ; stem erect, 8–10 in. high, branching, puberulous, remotely leafy ; leaves much shorter than the internodes, very shortly petiolate, 3–6 lin. long, 1½–3 lin. broad above the middle, cuneate-ovate, obtuse, with undulate margins ; flowers 2–4 together at the nodes ; pedicels scarcely as long as the calyx, hairy ; sepals 1 lin. long, linear-lanceolate, acute, pilose ; corolla cage-like, with the lobes connate at the tips ; tube 1½ lin. long, 1 lin. in diam. at the mouth, campanulate, glabrous ; lobes erect, 7 lin. long, linear-filiform, dilated at the apex, sparsely ciliate ; outer corona-lobes erect, deeply bifid, with linear subacute segments, glabrous ; inner corona-lobes slightly exceeding the outer corona, ovate at the base, narrowed to an obtuse apex, incumbent upon the backs of the anthers. *Dichælia undulata, Schlechter in Engl. Jahrb.* xviii. *Beibl.* 45, 36, *and Journ. Bot.* 1897, 293.

COAST REGION : Uitenhage Div. ; in fissures of rocks at Tzamas, near Uitenhage, 400 ft., *Schlechter,* 2709.

33. **B. Bolusii** (N. E. Br.) ; tuber flattened ; plant 4–6 in. (or under cultivation up to 1 ft.) high, branching at the base ; branches puberulous with minute decurved hairs ; leaves 3–6 lin. long, 1–2½ lin. broad, spathulately or cuneately elliptic, oblong or obovate, tapering into a short petiole at the base, obtuse or rounded at the apex, much undulated on the margins, glabrous above, minutely puberulous with curved hairs beneath ; flowers in pairs or occasion-

ally solitary at the nodes; pedicels 1–1½ lin. long, puberulous; sepals ⅔–¾ lin. long, narrowly lanceolate, puberulous; corolla cage-like, lobed nearly to the base, with the lobes connate at the tips; united part about ⅓ lin. long, flattish; lobes ½–⅔ in. long, narrowly linear, erect, sparsely covered with minute hairs on the back, glabrous on the inner face, whitish and dotted with pink on the basal part, light green above; outer corona-lobes arising ⅓ lin. up the staminal column, ½ lin. long, bifid to the middle, with slightly divergent linear-subulate segments, glabrous, whitish, with pale green tips; inner corona-lobes nearly ½ lin. long, linear, obtuse, closely incumbent on the backs of the anthers and slightly exceeding them, glabrous, light green.

34. **B. ovatum** (Oliver in Ref. Bot. iv. t. 226, by error *B. ovata*); tuber flattened, up to 5 in. in diam.; stems several, dichotomously or trichotomously branched, ⅔–1 ft. high under cultivation, puberulous with curved hairs; leaves 3–6 lin. long, 1½–3 lin. broad, elliptic or elliptic-lanceolate, subacute, narrowed into a short petiole at the base, glabrous above, puberulous beneath; flowers solitary (always ?) at the nodes, drooping; pedicels ½–1 lin. long, shortly pubescent; sepals ⅔–¾ lin. long, lanceolate, acute, shortly pubescent; corolla cage-like, subglobose or ellipsoid, subtruncate or depressed at the top, lobed nearly to the base, with the united part not more than ½ lin. long and shorter than the corona; lobes 5–6 lin. long, linear-filiform, much curved, connate at the tips, more or less replicate, with a very short spreading pubescence on the back, glabrous on the inner face, very pale yellowish or whitish on the lower half, light green above; outer corona-lobes about ⅖ lin. long, pouch-like at their base, divided to the middle into 2 falcate-subulate diverging teeth, glabrous, whitish; inner corona-lobes about ½ lin. long, linear-oblong, truncate or emarginate at the apex, closely incumbent on the backs of the anthers and not exceeding them. *Dichælia ovata, Schlechter in Engl. Jahrb.* xviii. *Beibl.* 45, 36, *and Journ. Bot.* 1897, 293 ; *K. Schum. in Engl. and Prantl, Pflanzenfam.* iv. ii. 269.

As Mr. Arnot was a Government agent for Griqualand West at the time he sent this plant to Kew, and obtained plants from there, it is just possible that this species is a native of that region rather than of Colesberg.

35. **B. Galpinii** (N. E. Br.); plant about 6 in. high, branching; branches minutely puberulous with recurved hairs; leaves ascending-spreading, ½–¾ in. long, 1–1½ lin. broad, linear-lanceolate, obtuse or subacute, cuneately tapering into a short petiole at the base, longitudinally folded, glabrous above, puberulous with minute

curved hairs beneath ; flowers 2 or more together at the nodes ;
pedicels ½–1 lin. long ; sepals about 1 lin. long, lanceolate, attenuate-
acuminate, puberulous with curved hairs ; corolla cage-like, lobed
nearly to the base, with the lobes connate at the tips ; tube ½–¾ lin.
long, cup-like, with a thin and exceedingly minute almost powder-
like puberulence or a few small hairs outside, glabrous within ;
lobes 7–10 lin. long, linear-filiform, with a microscopic puberulence
(scarcely pubescence) on both sides and ciliate with a few very
minute scattered curved hairs, or glabrous on the inner face and
thinly pubescent on the back, apparently dark purple-brown ; outer
corona of 5 pairs of erect linear-filiform teeth ¾ lin. long, thinly
ciliate with comparatively long hairs.; inner corona-lobes ½–⅔ lin.
long, linear-filiform, acute, about twice as long as the anthers and
incumbent upon them at the base, with the tips connivent-erect or
perhaps crossing each other. *Dichælia Galpinii, Schlechter in Engl.
Jahrb.* xviii. *Beibl.* 45, 25, *and Journ. Bot.* 1897, 293, *not elsewhere.*

KALAHARI REGION : Transvaal ; plains near Barberton, 2500–2800 ft., *Galpin,*
698 ! *Thorncroft,* 647 ! and in *Herb. Wood,* 10219 !

36. **B. pallidum** (N. E. Br.) ; tuber 2 in. or more in diam.,
flattened ; stem (often much) branched at the base or towards the
middle, 3–10 in. high, shortly velvety-pubescent ; leaves spreading,
¼–¾ in. long, ⅔–2½ lin. broad, linear-lanceolate, lanceolate or
narrowly elliptic, obtuse or subacute, cuneately narrowed into a
short petiole at the base, longitudinally folded, shortly pubescent or
velvety-pubescent on both sides, but more densely so beneath ;
flowers in fascicles of 2–8 at the nodes or occasionally solitary ;
pedicels ½–1½ lin. long ; sepals about 1 lin. long, subulate-lanceolate,
densely pubescent ; corolla cage-like, lobed nearly to the base, with the
lobes connate at the tips, pubescent like the leaves all over outside,
glabrous within, apparently yellowish or whitish at the lower half
and purplish-brown at the upper half ; united part flattish or
saucer-shaped, about ½ lin. long ; lobes 5½–8½ lin. long, linear-
filiform ; outer corona-lobes ½–⅞ lin. long, erect, divided from ⅔–⅚ of
their length into 2 parallel or slightly diverging linear obtuse or
acute segments, varying from minutely puberulous on the inner face
and ciliate with longer hairs to entirely glabrous, apparently whitish
or yellowish ; inner corona-lobes ⅓–⅔ lin. long, linear or linear-
spathulate, very obtuse, sometimes broader than the outer corona-
teeth, somewhat loosely incurved over the anthers and much longer
than them, but a little shorter than the outer corona-lobes, glabrous,
apparently whitish ; follicles 3–3½ in. long, ⅛–⅙ in. thick, terete,
tapering to the acute apex, smooth, glabrous. *Dichælia pallida,
Schlechter in Engl. Jahrb.* xx. *Beibl.* 51, 49, *and in Journ. Bot.* 1897,
294.   *D. microphylla, S. Moore in Journ. Bot.* 1903, 312.

KALAHARI REGION : Griqualand West ; Lower Campbell, *Burchell,* 1799 !
Bechuanaland ; near Kuruman, *Burchell,* 2499 ! near Lobatsi, *Marloth,* 3324 !
Transvaal ; Magalies Berg, *Burke*! *Zeyher,* 1136 ! Crocodile River, *Burke*!
Diamond Fields, *Tuck*! near Johannesburg, 6000 ft., *Gilfillan in Herb. Galpin,*

6049 ! 6050 ! *Rand*, 967 ! near Little Olifants River, 5200 ft., *Schlechter*, 3810 !
Modderfontein, *Conrath*, 1011 ! Pretoria Kopjes, *Miss Leendertz*, 658 !

I am very doubtful if this is more than a form of *B. Galpinii*, they are alike in
appearance and only appear to differ in the pubescence of the leaves and corolla,
and the form of the inner corona-lobes.   But *B. Galpinii* is a much rarer plant.

37. **B. cinereum** (N. E. Br.); plant about 6 in. high; branches
erect or erectly-spreading, minutely greyish-tomentose, leafy; leaves
ascending or spreading, $3\frac{1}{2}$–6 lin. long, 2–$2\frac{1}{2}$ lin. broad at the
middle, oblong or oblong-elliptic, acute or obtuse, narrowed into a
very short petiole at the base, minutely greyish-tomentose on both
sides; flowers in fascicles of ·2–5 at the nodes; pedicels about 1 lin.
long, minutely greyish-tomentose; sepals $1\frac{1}{4}$ lin. long, lanceolate,
acute, greyish-villous; corolla about 7 lin. long, cage-like, with the
lobes connate at the tips, puberulous outside, glabrous within; tube
about as long as the calyx, cup-shaped; lobes erect, linear from an
ovate base; outer corona-lobes connate into a cylindric annulus
(? arising above the base of the staminal column), deeply bifid, with
erect linear obtuse segments, sparsely ciliate; inner corona-lobes
slightly exceeding the outer corona, ligulate, obtuse, glabrous.
Dichælia cinerea, Schlechter in Engl. Jahrb. xxxviii. 42.

WESTERN REGION : Little Namaqualand ; on hills by the Orange River near
Ramonds Drift, 300 ft., *Schlechter*, 11505.

Stated to be nearly allied to *D. Galpinii*, Schlechter, and *D. pallida*, Schlechter,
but has a denser tomentum and the inner corona-lobes longer than the outer.

38. **B. elongatum** (N. E. Br.); plant 2–$3\frac{1}{2}$ in. high, branching,
velvety; branches erectly-spreading, densely leafy towards the apex;
leaves spreading, subfalcate, subsessile, $\frac{2}{3}$–$1\frac{1}{4}$ in. long, $\frac{1}{4}$–$\frac{1}{3}$ in. broad,
linear-lanceolate, subacute, narrowed at the base; flowers 2–3
together, lateral at the nodes; pedicels $2\frac{1}{2}$ lin. long; sepals 2 lin.
long, linear, acute, villous; corolla-tube $2\frac{1}{2}$ lin. long, 2 lin. broad at
the throat, subcylindric, slightly narrowed at the base, sparsely
pilose; lobes fleshy, connate at the tips, 9 lin. long, erect, linear-
filiform, thickened at the apex, ciliate above the middle; corona with
a cylindric tube; outer segments 5, short, bifid, glabrous; inner
segments slightly exceeding the outer, ovate, elongate-acuminate,
glabrous, adnate to the backs of the anthers, with free tips.
Dichælia elongata, Schlechter in Engl. Jahrb. xviii. *Beibl.* 45, 35,
and Journ. Bot. 1897, 293.

CENTRAL REGION : Somerset Div. ; Bosch Berg, 2500 ft., *Schlechter*, 2699.

This appears to be closely allied to *B. distinctum*, N. E. Br., and in many
particulars the description agrees with that species, but the corolla and corona
appear to be different.   I have not seen it.

39. **B. distinctum** (N. E. Br.); tuber flattened, up to about 3 in.
(or more?) in diam.; plant 2–6 in. high, branching at the base;

stems erect, pubescent; leaves 6–12 pairs to a stem, ascending-spreading, often falcately recurving, very shortly petiolate, $\frac{3}{4}$–$1\frac{1}{2}$ in. long, 1–3 lin. broad, linear or linear-lanceolate, obtuse or subacute, longitudinally folded, with a short spreading pubescence on the underside, glabrous above; flowers solitary or in fascicles of 2–6 at the nodes; pedicels 3–6 lin. long, pubescent; sepals 1–$1\frac{1}{2}$ lin. long, subulate or lanceolate-subulate, pubescent; corolla cage-like; tube 2 lin. long, much exceeding the corona, campanulate, apparently whitish, dotted (always?) with purple, glabrous or thinly sprinkled with short spreading hairs outside, glabrous on the basal half within and thickly covered above and for a short distance up the base of the lobes with inflated subglobose and clavate white (in dried flowers) hairs; lobes connate at the tips, 8–13 lin. long, narrowly linear, pubescent with short spreading hairs on the back, apparently dark-coloured; outer corona-lobes arising $\frac{1}{2}$–$\frac{3}{4}$ lin. above the base of the staminal column, scarcely $\frac{1}{4}$ lin. long, divided to the middle into 2 widely diverging falcately curved subulate teeth, very minutely and sparsely scabrous; inner corona-lobes $\frac{1}{4}$ lin. long, linear, sub-truncate, closely incumbent on the backs of the anthers and not or scarcely exceeding them; staminal column $\frac{3}{4}$–1 lin. long.

Coast Region : Albany Div. ; in rocky grassy places near Grahamstown, *MacOwan,* 1039 ! on mountains, *Mrs. Barber,* 242 ! Stockenstrom Div. ; Kat Berg, *Hutton* ! Eastern Frontier, *Hutton* !

Eastern Region : Transkei ; Fort Bowker, *Bowker,* 381 !

According to Mrs. Barber the flowers are "spotted with purple outside and a deep brown within." The long leaves of this species give it a very distinct appearance.

40. **B. villosum** (N. E. Br.); plant 10–12 in. high according to *Schlechter,* 3 in. high in the specimen seen, branching at the base into several stout stems densely clothed with long soft spreading hairs; leaves very shortly petiolate, 4–9 lin. long, $2\frac{1}{2}$–$3\frac{1}{2}$ lin. broad, elliptic or elliptic-oblong, obtuse or acute, glabrous above, villous beneath, especially along the midrib and margins, ciliate; flowers in fascicles of 2–6 at the nodes; pedicels 1–3 lin. long, thinly villous; sepals $1\frac{1}{2}$ lin. long, lanceolate-subulate, thinly villous; corolla cage-like, with the lobes connate at the tips; tube 2 lin. long, shortly tubular-campanulate, with the sinuses between the lobes produced into short recurved teeth, glabrous outside, thinly covered with spreading hairs within, apparently whitish; lobes $\frac{1}{2}$–$\frac{2}{3}$ ($\frac{7}{8}$, *Schlechter*) in. long, linear-filiform from a deltoid $\frac{3}{4}$–1 lin.-broad base, thinly ciliate and sprinkled on the back with short spreading hairs; outer corona-lobes arising $\frac{1}{3}$ lin. up the staminal column and shorter than it, about $\frac{1}{6}$ lin. long, erect, bifid, with the rather long teeth widely diverging and falcately recurved; inner corona-lobes about $\frac{1}{4}$ lin. long, linear, obtuse, closely incumbent upon the backs of the anthers and scarcely or not at all exceeding them. *Dichælia villosa, Schlechter in Engl. Jahrb.* xviii. *Beibl.* 45, 25, *and Journ. Bot.* 1897, 293.

KᴀLᴀHᴀR᳽ RᴇG᳽Oᴄ : Transvaal ; Mountain-tops at Upper Moodies, near Barberton, 4500–5000 ft., *Galpin*, 588 !

According to Mr. Galpin the flowers are sweetly scented.

41. B. Barberiæ (Harv. ex Hook. fil. in Bot. Mag. t. 5607); tuber large, flattened ; stems $2\frac{1}{2}$–$3\frac{1}{2}$ in. high, rather coarsely spreading-pubescent ; leaves 1–4 in. long, $\frac{1}{3}$–1 in. broad, cuneate-oblong to oblanceolate or obovate, acute to obtuse and apiculate, tapering below into the petiole, pubescent on both sides, but more coarsely so beneath ; flowers in opposite sessile umbels, forming a many-flowered ball 3–4 in. in diam. surrounding the stem ; pedicels $4\frac{1}{2}$–9 lin. long, spreading-pubescent ; sepals 3–$4\frac{1}{2}$ lin. long, $\frac{3}{4}$–1 lin. broad at the base, thence tapering to the acute apex, pubescent ; corolla 1–$1\frac{1}{2}$ in. long, cage-like, with the lobes connate at the tips, rich dark crimson-brown on the inner face, marked in the pale yellowish tube with irregular transverse purple-brown lines, outside pubescent, puberulous or glabrous and green ; tube (exclusive of the spreading united part of the limb) about 2 lin. long, subhemispheric, in dried flowers, from alteration in drying, appearing campanulate or cup-shaped and up to 4 lin. long, glabrous within ; limb around the mouth of the tube and base of the lobes covered with purple hairs ; lobes $\frac{3}{4}$–$1\frac{1}{2}$ in. long, $2\frac{1}{2}$–3 lin. broad at the shortly ovate spreading base, then contracted into linear erectly incurved tails with replicate margins, varying from pubescent to glabrous on the inner face ; outer corona cup-like, $\frac{1}{2}$ lin. high, divided to half way down by 5 cut-like fissures into 5 truncate or subtruncate segments, often slightly produced into obtuse teeth at the shoulders, adnate at the middle to the base of the inner corona-lobes, which are $\frac{1}{2}$ lin. long, linear, obtuse, inflexed on the backs of the anthers and not exceeding them, so that in dried flowers the corona often appears to consist of 5 transverse subtruncate lobes with a linear obtuse point abruptly inflexed from their middle. *K. Schum. in Engl. and Prantl, Pflanzenfam.* iv. ii. 268, *and* 269, *under Dichælia. B. Barberæ, Schlechter in Engl. Jahrb.* xviii. *Beibl.* 45, 25, *and Journ. Bot.* 1897, 292.

KᴀLᴀHᴀR᳽ RᴇG᳽Oᴄ : Transvaal ; grassy slopes near Barberton, 2900 ft., *Galpin*, 656 ! near Modderfontein, *Conrath*, 1012 ! near Pretoria, *Burtt Davy*, 2586 ! 3241 ! *Carnegie*, 1680 !

EᴀSTᴇRᴄ RᴇG᳽Oᴄ : Transkei ; on flats near Tsomo, *Mrs. Barber*, 806 ! Tembuland ; Bazeia, 2500 ft., *Baur*, 386 !

Described from a living plant, which flowered at Kew in Aug., 1905.

It also occurs in Rhodesia, in Tropical Africa.

Imperfectly known species.

42. B. micranthum (E. Meyer, Comm. 196) ; tuber flattened ; plant branching at the base into several stems, 2–3 in. high, $\frac{1}{2}$–$\frac{3}{4}$ lin. thick, puberulous ; leaves 5–6 lin. long, $\frac{2}{3}$–3 lin. broad, linear-lanceolate to elliptic, obtuse, tapering into a short petiole at the base, glabrous; flowers usually 2 (sometimes solitary) at the nodes; bracts $\frac{1}{2}$–$\frac{2}{3}$ lin. long, lanceolate, acute; pedicels 2–3 (in fruit becoming about 4) lin. long, puberulous; sepals about $\frac{3}{4}$ lin. long, $\frac{1}{4}$ lin. broad, lanceolate, acuminate, slightly puberulous; "corolla very minute, with ovate-lanceolate lobes a little longer than the calyx, spreading, corona projecting; flowers only 1 lin. long" (*E. Meyer*); follicles immature, $1\frac{1}{4}$–$2\frac{3}{4}$ in. long, $1\frac{1}{2}$–$1\frac{3}{4}$ lin. thick, linear-fusiform, tapering to an acute apex, glabrous. *Dietr. Syn. Pl.* ii. 888; *Decne in DC. Prodr.* viii. 647; *K. Schum. in Engl. and Prantl, Pflanzenfam.* iv. ii. 268. *Schlechter in Journ. Bot.* 1897, 293.

Coast Region: Queenstown Div.; near Shiloh, 4000–5000 ft., *Drège*! Table Mountain, *Drège* (ex *E. Meyer*).

The only flower of this species that I have seen is upon E. Meyer's type, but it is not describable, as the corona and all but the base of the corolla have been eaten by insects. The other specimens seen are in young fruit. Drège's original label with the type is localised "Klipprivier," intended probably for Klipplaat River, which is near Shiloh. I have not seen a specimen from the other locality.

43. B. crispum (E. Meyer, Comm. 196); tuber depressed; stem puberulous; leaves often reflexed, 3–4 lin. long, oblong, sinuate, irregularly undulate-crisped, puberulous; flowers scarcely 3 lin. long; corolla-tube campanulate; lobes elongated, very narrowly subspathulate; follicles less erect than in *B. meyerianum, Schlechter.* *Dietr. Syn. Pl.* ii. 888; *not of Graham.*

Central Region: Aberdeen Div.; hills near Hamerkuil, in the Camdeboo, 3000–3500 ft., *Drège*!

No specimen of this now exists in E. Meyer's Herbarium.

44. B.? macrorrhizum (E. Meyer, Comm. 197); tuber as large as a man's head, subglobose, slightly depressed; leaves 6–8 lin. long, shortly petiolate, reflexed, ovate, undulate-crisped, shining, canescent. *Dietr. Syn. Pl.* ii. 888; *Decne in DC. Prodr.* viii. 647.

Central Region: Graaff Reinet Div.; on the Oude Berg, 3500–4000 ft., *Drège.*

I found no specimen of this in E. Meyer's Herbarium when the latter was lent to Kew. Schlechter in *Engl. Jahrb.* xxi. *Beibl.* 54, 14, refers it to *Fockea glabra* (*F. edulis*, K. Schum.), but upon what grounds I do not know, for the depressed-globose tuber, reflexed canescent leaves and different geographic region, indicate a different plant.

XLII. ANISOTOMA.

Calyx 5-partite. *Corolla* deeply 5-lobed; tube campanulate; lobes linear or oblong-linear, valvate in bud. *Corona* arising near or at the base of the staminal column, really double, but apparently of 5 dorsally flattened entire, denticulate or bifid lobes opposite the anthers (really outer-corona-lobes), with a long appendage (really inner corona-lobes) on their inner face. *Staminal column* arising from the base of the corolla ; anthers suberect, oblong, without an appendage, but with two or more hairs at their tips. *Pollen-masses* erect, solitary in each anther-cell, with a small pellucid area immediately under the apex, attached in pairs to the pollen-carriers by very short slender caudicles. *Fruit* not seen.

Perennial herbs ; rootstock a cluster of thick fleshy roots ; stems several, prostrate ; leaves opposite, petiolate, cordate-ovate or cordate-orbicular, herbaceous ; flowers rather small, in pedunculate or sessile umbel-like cymes.

DISTRIB. Species 2, endemic.

The coronal structure of this genus is scarcely distinguishable from that of those species of *Brachystelma* (such as *B. schizoglossoides*, N. E. Br., *B. mafekingense*, N. E. Br., etc.), where the halves of the divided outer corona-lobes are connate with the halves of adjacent lobes immediately behind and partly adnate to the base of the inner corona-lobes, but the habit and general appearance is quite different from that of any species of *Brachystelma*.

Corona-lobes more or less rhomboid, bifid, with a
    glabrous filiform-subulate appendage...   ...   ... (1) **cordifolia.**

Corona-lobes oblong or lanceolate, entire or denticulate,
    with a puberulous linear-lanceolate acute or bifid
    appendage   ...   . (2) **pedunculata.**

1. **A. cordifolia** (Fenzl in Linnæa, xvii. 331) ; roots rather thick and numerous, clasping the rocks (*Mrs. Barber*) ; stems prostrate, branching at the base, slender, tortuous, unifariously puberulous and pilose-pubescent ; leaves spreading ; petiole $\frac{1}{4}$–$\frac{1}{2}$ in. long, more or less shortly pilose ; blade $\frac{1}{4}$–$1\frac{1}{4}$ in. long, $\frac{1}{4}$–1 in. broad, ovate to suborbicular, acute to rounded at the apex, cordate at the base, pubescent above, thinly and shortly pilose beneath ; cymes umbel-like, sessile or pedunculate, lateral at the nodes, 3–9-flowered ; peduncles 1–14 lin. long, pubescent or shortly·villous on one side ; bracts $\frac{1}{2}$–$1\frac{1}{4}$ lin. long, linear-subulate, acute, pubescent ; pedicels 2–10 lin. long, pubescent or puberulous along one side ; sepals $\frac{3}{4}$–1 lin. long, $\frac{1}{4}$–$\frac{1}{3}$ lin. broad, lanceolate, acute, puberulous ; corolla-tube $\frac{1}{2}$–$\frac{3}{4}$ lin. long, whitish ; lobes 1–2 lin. long, $\frac{1}{2}$ lin. broad, oblong-linear, subobtuse, glabrous or with a few hairs on the back, minutely puberulous on the inner face, whitish or yellowish at the base, dark brown or orange-brown above ; corona-lobes $\frac{1}{3}$–$\frac{1}{2}$ lin. long, erect, more or less rhomboid-ovate, bifid, minutely puberulous on the back, with a filiform-subulate glabrous appendage on the inner face $\frac{3}{4}$ lin. long ; appendages connivent-erect over the $\frac{1}{2}$ lin.-long staminal column ; anthers erect, linear-oblong, with 2 or 3 minute hairs at the apex. *A. mollis, Schlechter in Journ. Bot.* 1897, 292. *Cynoctonum molle, E. Meyer, Comm.* 216 ; *Decne in DC. Prodr.* viii. 530. *Cynanchum molle, Dietr. Syn. Pl.* ii. 905 ; *Steud. Nom. Bot. ed.* 2, i. 462. *Anisotomaria mollis, Presl, Bot. Bemerk.* 103. *Lophostephus mollis, Harv. Thes. Cap.* ii. 9, t. 113 ; *and Gen. S. Afr. Pl. ed.* 2, 242. *Anisotome mollis, Schlechter in Engl. Jahrb.* xviii. *Beibl.* 45, 26 ; *K. Schum. in Engl. and Prantl, Pflanzenfam.* iv. ii. 267.

COAST REGION : Uitenhage Div. ; Vanstadens Mountains, *Burchell*, 4724 ! *Drège* ! Zuurberg Range, near Doorn Nek, *Drège*, 2221 ! on high hills, *Mrs. Barber*, 433 ! Albany Div. ; by the Great Fish River, *Ecklon and Zeyher* ! near Grahamstown, *MacOwan*, 402 ! *Glass in MacOwan, Herb. Austr. Afr.* 1633 ! *Misses Daly and Sole*, 531 ! Stockenstrom Div. ; Kat Berg, *Drège*, 3435 ! Eastern Frontier, *Hutton* !

2. **A. pedunculata** (N. E. Br. in Kew Bulletin, 1895, 150) ; plant divided at the base into several prostrate branches $\frac{1}{2}$–2 feet or perhaps more long, simple or sparingly branched, spreading- or shortly pilose-pubescent on all green parts ; leaves unilaterally spreading ; petiole $\frac{1}{6}$–$\frac{3}{4}$ in. long; blade $\frac{1}{3}$–1$\frac{1}{2}$ in. long, $\frac{1}{4}$–1$\frac{1}{3}$ in. broad, cordate-ovate to cordate-orbicular, acute, obtuse or rounded at the apex ; cymes umbel-like, lateral at the nodes, pedunculate, 2–10-flowered, sometimes compound ; peduncles $\frac{1}{2}$–2 in. long ; pedicels $\frac{3}{4}$–1$\frac{1}{4}$ in. long ; sepals 1–1$\frac{1}{2}$ lin. long, lanceolate, acute ; corolla deeply lobed, thinly sprinkled with hairs outside, puberulous on the inner face, with the tube and base of the lobes cream-colour and the upper $\frac{2}{3}$ of the lobes brown (*Wylie*) ; tube $\frac{3}{4}$–1 lin. long, campanulate ; lobes spreading (ascending in dried flowers), 1$\frac{3}{4}$–2 lin. long, about $\frac{3}{4}$ lin. broad, oblong-linear, subobtuse, with recurved margins ; corona-lobes erect, with the body or basal part $\frac{1}{2}$–$\frac{2}{3}$ lin. long, lanceolate or oblong, acute, obtuse or denticulate at the apex, having a linear-lanceolate or linear-spathulate appendage $\frac{3}{4}$–1 lin. long on the inner face, acute or bifid at the apex, puberulous all over on the back ; connivent-erect high over the $\frac{1}{2}$ lin.-long staminal column, apparently whitish ; anthers tipped with 3–4 hairs. *A. peduncularis, Schlechter in Journ. Bot.* 1897, 292.

EASTERN REGION : Natal ; Drakensberg Range, Tyger-cave Valley, 6000–7000 ft., *Evans*, 379 ! Niginya, 5500 ft., *Wylie in Herb. Wood*, 10529 !

My original description of the corona-lobes of this plant is very incorrect, as in the flowers I then examined their true form had been obliterated by pressure, in better preserved flowers I find them to be as described above.

XLIII. PECTINARIA, Haworth (not of other authors).

Calyx 5-partite. *Corolla* small, bud-like, with a short cup-shaped, hemispheric or broad and shallow tube and 5 lobes connate at the apex, with narrow openings between them. *Corona* double ; outer cup-like and variably cut into 10 to numerous teeth or of 5 minute lobes ; inner of 5 lobes incumbent upon the backs of the anthers or erect and about equalling them or longer and connivent-erect over them. *Stamens* with the filaments connate into a tube around the ovary and adnate to the dilated top of the style ; anthers free, without appendages, incumbent upon the top of the style. *Pollen-masses* solitary in each anther-cell, minute, ellipsoid or subglobose, pellucid at the ends or sides turned towards each other, attached in pairs to the very minute brown pollen-carriers by very short caudicles. *Follicles* narrowly fusiform, acuminate, smooth, glabrous.

Succulent leafless herbs; stems tufted, usually procumbent and often with their tips or other portions buried in the ground, in one species sometimes (always ?) flowering underground, acutely, obtusely or obscurely 4–8-angled, acutely toothed or tessellately tuberculate along the angles, often with a distinct bud in the axil of each tooth or tubercle when the stems are fully developed; flowers small, solitary or in fascicles in the grooves or on the sides between the angles, small, bud-like, often with a "frosted" appearance on the inner surface and sometimes outside as well. The colour of the flower is taken from Mr. Pillans' notes, except in the case of *P. asperiflora.*

DISTRIB. Species 5, endemic.

Stems acutely 4-angled, with flat or slightly concave
    sides and distant acute deltoid teeth along the angles :
    Corolla broadly ovoid or subglobose, covered with fine
       hairs on the inner surface, blackish-purple or
       purple-brown   ...   . (1) **saxatilis.**
    Corolla broadly depressed, pear-shaped, papillate, but
       without hairs on the inner surface, light purple ... (2) **Pillansii.**
Stems very obtusely and somewhat obscurely 4-angled,
    tuberculate along the angles; corolla acuminate
    from a short ovoid base, glabrous and smooth, inside
    whitish, spotted with purple on the lobes, dark
    purple in the tube   ...   . (3) **arcuata.**
Stems cylindric, with 6–8 series of closely placed tubercles;
    corolla papillate outside and within :
    Corolla shortly and broadly conical from a subtruncate
       base, with simple papillæ within   ...   ... (4) **articulata,** var.
    Corolla pentagonally subglobose, with the papillæ on
       the inner surface covered with short spike-like
       processes   ...   . (5) **asperiflora.**

1. **P. saxatilis** (N. E. Br. in Gard. Chron. 1904, xxxv. 211); stems or branches prostrate, often growing downwards into the ground and then curving upwards or the young shoots commencing to develop underground, acutely 4-angled, $1\frac{1}{2}$–5 in. long, $\frac{1}{2}$–1 in. square, with flat or slightly concave sides and broadly deltoid acute teeth 1–$1\frac{3}{4}$ lin. long along the angles, having a small (and, with age, prominent) bud in each axil, glabrous, at first glaucous-green, becoming dull green, often tinged or marked with dull purplish, finally greyish; flowers in fascicles of 4–7 on the sides, near the base of the stems, developing successively, erect; pedicels $\frac{1}{2}$–1 lin. long, glabrous; sepals about 1 lin. long, lanceolate or ovate-subulate, with very acute recurved tips, glabrous; corolla bud-like, $4\frac{1}{2}$–6 lin. long, 4–5 lin. in diam., broadly ovoid or subglobose, acute or shortly acuminate or apiculate, entirely blackish-purple or very dark purple-brown, glabrous outside, covered with fine hairs and having a "frosted appearance" (*Pillans*) inside; tube 2–$2\frac{1}{2}$ lin. deep, broadly cup-shaped or hemispheric; lobes 3–5 lin. long, $2\frac{1}{2}$ lin. broad, deltoid to ovate-lanceolate, acute, cohering at the tips, with narrow fissures between them in the lower part, not recurved at the margins and not ciliate; outer corona-lobes at the base of the

staminal column, minute, not ¼ lin. long, spreading, broadly ovate or deltoid, subacute, glabrous, dark purple-brown ; inner corona-lobes 1⅓–1½ lin. long, much exceeding and erectly connivent (or sometimes connate at the tips) high above the staminal column, rather thick and fleshy, linear or subulate, obtuse or acute, with a short deltoid-subulate tooth or small thickened crest on the back below the middle, dark purple-brown ; staminal column ¾ lin. long ; anthers subquadrate, obtuse, without appendages, incumbent on the outer part of the truncate style-apex.

COAST REGION : Oudtshoorn Div. ; near Oudtshoorn, *Pillans,* 694 !

CENTRAL REGION : Laingsburg Div. ; on a hill north-west of Laingsburg at Zout Kloof, growing around and under stones, very scarce, *Pillans,* 115 ! along the road between Witte Poort and Laingsburg, ex *Pillans.*

Although there is a young bud in the axil of every tooth on the stems, yet according to Mr. Pillans only those on the underside develop into shoots.

2. **P. Pillansii** (N. E. Br.); stems up to 6 in. or more long, procumbent or partly subterranean, acutely 4-angled, ½–¾ in. square, with flat or slightly concave sides and acute deltoid-conical teeth 1½–2 lin. long, glabrous, dull green, more or less mottled or suffused with purple where exposed to the sun, with a slight glaucous bloom on the young parts, the teeth being at first of a paler green than the rest, becoming hardened and pale brown, having a bud in every axil ; flowers in clusters of three or more at the middle of the sides, often (always?) developed underground ; pedicels 1–1½ lin. long, glabrous ; sepals about 1 lin. long, ovate, acuminate, glabrous ; corolla 3½ lin. in diam., broadly depressed pear-shaped, light purple, with a frosted appearance all over the inside, from being thinly covered with watery or air-containing papillæ, almost smooth and glabrous outside ; tube about 1½ lin. long, cup-shaped ; lobes unusually thick and rigidly fleshy, 2–2¼ lin. long, 2–2½ lin. broad at the base, like an equilateral triangle in outline, acute, inflexed and connate (or occasionally, by accident? free) at the tips, forming a short broad dome-like top to the flower, their margins not recurved but separated so as to form very narrow fissures between them ; outer corona of 5 minute lobes about ¼ lin. long, broadly ovate, obtuse, spreading, apparently whitish ; inner corona-lobes 1⅓ lin. long, laterally much flattened, subfalcate, with a large rounded acute-edged hump on the back above the middle and there ½ lin. broad from front to back, erect, with the tips connivent high above the style-apex, dark purple.

CENTRAL REGION : Somerset Div. ; Glen Avon Estate, near Somerset East, growing flat on the ground under the dry leaves of an aloe, very rare, only one plant found, *Pillans,* 180 !

Mr. Pillans informs me that all the mature flowers he has seen have been developed underground, and that so far as he has observed, the young flower-buds which are formed above ground have not developed.

3. P. arcuata (N. E. Br.); stems 2–4 (or more) in. long, 3–5 lin. thick, arching-procumbent and then diving underground, sometimes rising again to the surface, but always finishing their growth underground, forming loops $\frac{1}{2}$–1 in. high, very obtusely and somewhat obscurely 4-angled, with slightly flattened sides, somewhat tessellately tuberculate along the angles, with the tubercles varying from scarcely prominent to $\frac{3}{4}$ lin. long, apiculate, glabrous, green, not glaucous, flowering along the sides or under the loops; flowers 1–3 in a fascicle; pedicels 1–1$\frac{1}{2}$ lin. long, glabrous, green; sepals 1 lin. long, ovate, subulate-acuminate, recurved at the tips, glabrous; corolla in general outline acuminate from a short ovoid base, glabrous and smooth outside and within; tube 1$\frac{1}{2}$ lin. long, cup-shaped or subhemispheric, straw-coloured or pale yellowish outside, dark purple within, the dark colour divided to half-way down into 5 ovate lobes or rounded crenations extending up the base of the lobes; lobes 3$\frac{1}{2}$ lin. long, 1–1$\frac{1}{4}$ lin. broad at the base, thence gradually tapering to the acute connate tips, not recurved at the margins, but separating so as to form narrow lanceolate openings between them, pinkish or pale purple-tinted on the back, creamy-white, spotted with purple at the tips on the inner face, not ciliate; outer and inner corona-lobes apparently in one series, bright yellow, with the 5 lobes alternating with the anthers (outer corona) minute, scarcely $\frac{1}{4}$ lin. long, deltoid, acute, ascending-spreading, the other 5 (inner corona) $\frac{1}{2}$ lin. long, more than $\frac{1}{4}$ lin. broad and almost as thick, erect, subclavate-oblong, very obtuse or subtruncate, scarcely exceeding the level of the anthers, with their rounded backs projecting like 5 stout ribs beyond the outer lobes.

COAST REGION : Bedford Div. ; found under a bush about 9 miles south-south-east of Bedford, very rare, *Pillans*, 182 !

The habit of this plant is extraordinary; several of the *Stapelieæ* produce underground shoots, but no other is known which seems to invariably finish the growth of its aerial shoots underground, as Mr. Pillans informs me (and as is the case in the cultivated plant seen) this one always does.

4. P. articulata (Haw. Suppl. Pl. Succ. 14); plant dwarf, tufted; branches procumbent or ascending, 1–2 in. long, $\frac{1}{2}$–$\frac{3}{4}$ in. thick, oblong, somewhat tessellately 6-angled or with 6 series of short stout conical tubercles, apiculate or acute, with a small bud in the axil of each, glabrous, ferruginous, with purplish branches; flowers solitary in the grooves between the angles at or towards the tips of the branches; pedicels apparently about 1 lin. long, thick, glabrous; sepals lanceolate, acute; corolla bud-like, about $\frac{1}{4}$ in. in diam., papillate and greenish outside, pearly-papillate and blackish-red within, glabrous, not ciliate; lobes deltoid, connate at the tips, with reflexed margins, leaving narrowly ovate acute openings between them ; outer corona pectinate, blackish-red, white below; inner corona-lobes blackish. *Sweet, Hort. Brit. ed.* i. 276 ; *G. Don, Gen. Syst.* iv. 122 ; *K. Schum. in Engl. and Prantl, Pflanzenfam.* iv. ii. 281. *Schlechter in Journ. Bot.* 1898, 486. *Stapelia articulata, Ait. Hort.*

Kew. ed. 1, i. 310, *and ed.* 2, ii. 90 ; *Masson, Stap.* 20, *t.* 30 ; *Hayne,*
Term. Bot. ed. 1799, *t.* 16, *fig.* 6 ; *Thunb. Trav. ed.* 3 (*English ed.*)
ii. 171 ; *Willd. Sp. Pl.* i. 1287 ; *Pers. Syn. Pl.* i. 279 ; *Poir. Encycl.*
vii. 384 ; *Haw. Syn. Pl. Succ.* 26 ; *R. Br. in Mem. Wern. Soc.* i. 24 ;
Schultes, Syst. Veg. vi. 26 ; *Spreng. Syst. Veg.* i. 841 ; *Loud. Encycl.*
Pl. 200, *fig.* 3299 ; *Dietr. Syn. Pl.* ii. 887 ; *Decne in DC. Prodr.*
viii. 663.

Var. β, **namaquensis** (N. E. Br.) ; branches $\frac{1}{3}$–$\frac{1}{2}$ in. thick, with the tubercles
$\frac{1}{2}$–1 lin. long, dull purple-brown ; flowers drooping (*Pillans*) ; pedicels 1$\frac{1}{2}$–2 lin.
long, stout, glabrous ; sepals $\frac{3}{4}$–1 lin. long, ovate or ovate-lanceolate, acute,
minutely papillate ; corolla bud-like, about 2 lin. long, 2$\frac{1}{2}$–3 lin. in diam.,
shortly and broadly conical, acute, subtruncate at the base, papillate and greenish
outside, covered with small (watery ?) tubercles on the (whitish ?) inner surface ;
tube about $\frac{1}{2}$ lin. deep, broad and shallow ; lobes 2 lin. long, 1$\frac{1}{4}$–1$\frac{1}{3}$ lin. broad at
the base, elongated deltoid ; outer corona cupular, scarcely $\frac{1}{2}$ lin. high, usually
more or less irregularly divided into numerous (30 or more) erect slender teeth,
glabrous, yellowish ; inner corona-lobes about $\frac{1}{4}$ lin. long, oblong, obtuse, closely
incumbent on the backs of the anthers and not exceeding them, dorsally
connected to the outer corona at the base, yellowish.

Central Region : Calvinia Div. ; near the edge of the Roggeveld Mountains,
Thunberg, Masson.

Western Region : Var. β : Little Namaqualand ; without precise locality
Templeman in Herb. Pillans, 22 !

According to Thunberg and Masson this plant is eaten by the Hottentots and
pickled like cucumbers by the colonists. There is no specimen of it in Thunberg's
Herbarium, nor in Masson's collection at the British Museum. Therefore until
the Roggeveld plant should be rediscovered and prove distinct, I place the
Namaqualand plant as a variety of it, since the colour of its corolla and corona
appear to be its chief distinction.

5. **P. asperiflora** (N. E. Br.) ; stems tufted, procumbent or erect,
$\frac{3}{4}$–3 in. long, $\frac{1}{2}$–$\frac{3}{4}$ in. thick, globose to cylindric, with 6–8 angles
formed of closely placed stout conical minutely apiculate tubercles
$\frac{3}{4}$–1 lin. long, glabrous, dark dull purple on the young parts ;
flowers solitary from the grooves between the angles on the young
growth ; pedicels 1$\frac{1}{2}$–4 lin. long, usually decurved or recurved at the
apex, slender, glabrous, very slightly and obscurely roughened, in
fruit ascending, $\frac{1}{2}$–$\frac{2}{3}$ in. long, 1$\frac{1}{2}$–2 lin. thick, purple-brown ; sepals
$\frac{3}{4}$–1 lin. long, ovate-lanceolate, acute, glabrous, obscurely papillate ;
corolla drooping, pentagonally subglobose, or with a campanulate
tube and a very shortly conical top, 2–2$\frac{1}{2}$ lin. in diam., abruptly
and shortly pointed in bud and when closed after opening, obtuse
when open, with the margins of the lobes slightly recurving, leaving
narrow lanceolate fissures between them, outside purple-brown,
minutely papillate ; inside white dotted with purple, densely covered
with large papillæ, which, when magnified, are seen to be covered
with spike-like processes ; tube 1$\frac{1}{2}$ lin. long, cup-shaped ; lobes
arching over the tube, connate at the tips, 1$\frac{1}{2}$–2 lin. long, 1$\frac{1}{2}$ lin.
broad, deltoid-ovate, acuminate ; outer corona about $\frac{1}{2}$ lin. long,
cupular, usually with 5 pairs of erect subulate teeth $\frac{1}{4}$ lin. long,
alternating with 5 broad transverse shorter lobules irregularly

5–8-denticulate at their subtruncate apices, but sometimes irregularly divided into many subulate teeth, dark purple-brown ; inner corona-lobes ¼ lin. long, deltoid, acute, closely incumbent upon the backs of the anthers and slightly exceeding them, broadly connected at the base to the outer corona, dark purple-brown ; follicles (immature) erect, slightly diverging, 3–3½ in. long, 2½ lin. thick, narrowly fusiform, tapering to an acute apex, smooth, glabrous, dull greyish-green, thickly marked with linear spots and lines of purple-brown.

CENTRAL REGION : Laingsburg Div. ; near Matjesfontein, *Pillans,* 70 !

Described from a living plant which flowered at Kew, Aug. 1907. The inner surface of the corolla is exceedingly beautiful when highly magnified, and unlike anything I have seen in any other plant.

XLIV. CARALLUMA, R. Br.

Calyx 5-partite. *Corolla* 5-lobed, with a small or very rarely large campanulate or saucer-shaped tube, or rarely without a distinct tube and the united part nearly flat and disk-like ; lobes free at the tips, very spreading or ascending, ovate to linear. *Corona* usually distinctly 2-seriate ; outer series of 5 distinct entire or bifid lobes or united into an entire or 5–15-toothed cup ; inner series of 5 filiform, subulate, linear, oblong, subquadrate or hemispheric lobes, shorter than to much longer than the anthers and often dorsally connected to the outer series ; or sometimes both series combined so as to falsely appear 1-seriate. *Stamens* with their filaments connate into a tube around the ovary and adnate to the dilated style-apex ; anthers free, ascending or inflexed upon the style-apex. *Pollen-masses* solitary in each anther-cell, ascending or subhorizontal, pellucid just under the apex or within the inner margin, attached in pairs by short slender caudicles to the pollen-carriers or to wing-like expansions of them.

Dwarf succulent leafless herbs, with stout 4–6- (very rarely 3-) angled stems and branches, toothed or toothless along the angles. Flowers variable, usually in fascicles, rarely solitary, near or at the base, apex, or along the grooves between the angles, in a few species racemose along the attenuated terminal part of the stems.

DISTRIB. Species about 80, the others distributed in Tropical and North Africa, South Europe, Socotra, and through Arabia to India.

Stems obscurely 6-angled or with obscurely 6-seriate
    tubercles (? tessellately 6-angled) ; pedicels 4–5 lin.
    long ; corolla densely pilose on the inner face,
    green ... ... (20) **chlorantha.**
Stems without distinct teeth, but sometimes somewhat
    crenate along their 4 very obtuse angles :
    Plant 2–2½ in. high ; pedicels 1–3 in. long, erect from
      near the base of the decumbent stems ... ... (23) **aperta.**
    Plant 9–12 in. high, bushily branched, flowering near
      or at the tips of the branches ; pedicels about 1
      (in fruit 3–4) lin. long ... ... (1) **ramosa.**

Stems with distinct conical or tubercle-like teeth along
   their angles :
Stems spirally 5–6-angled, with stout spine-like teeth
   $\frac{1}{4}$–$\frac{1}{2}$ in. long ; corolla-lobes 7–9 lin. long, blackish-
   purple or nearly black   ...   .   (2) **mammillaris.**
Stems 4- (rarely 5-) angled :
* Corolla-lobes $\frac{3}{4}$–1$\frac{1}{4}$ in. long ; plants 2–4 in. high :
   Pedicels less than $\frac{1}{4}$ in. long ; corolla dark purplish-
      olive-brown, very minutely ciliate at the base
      of the lobes   ...   . (17) **longicuspis.**
   Pedicels $\frac{1}{2}$–1 in. long ; corolla ciliate with vibratile
      clavate deciduous hairs :
      Corolla yellow ; inner corona-lobes with a short
         subulate tooth at their base   ...   ... (18) **lutea.**
      Corolla not yellow ; inner corona-lobes excavated
         at their base ...   . (19) **melanantha.**
** Corolla-lobes 3–7 lin. long :
   Pedicels 3–6 lin. long ; corolla-lobes deltoid-ovate,
      green, spotted with purple-brown (ciliate ?)   (5) **intermedia.**
   Pedicels $\frac{1}{2}$–2 lin. long :
   Teeth on the angles of the stem $\frac{1}{3}$–$\frac{1}{2}$ in. long :
      Corolla-lobes about 2 lin. broad, lanceolate,
         glabrous, with the apical half purple-
         brown and basal half greenish-yellow
         dotted with purple-brown   ...   ...   (3) **armata.**
      Corolla-lobes 3 lin. broad, oblong or oblong-
         ovate, thinly pubescent on the inner face,
         purplish-grey, spotted all over with
         purple-brown   ...   .   (4) **Pillansii.**
   Teeth on the angles of the stems $\frac{1}{8}$–$\frac{1}{4}$ in. long ;
      corolla-lobes linear, linear-lanceolate or
      attenuate, not ciliate :
      Corolla-lobes blackish-purple or purple-brown,
         quite glabrous :
         Corolla-lobes 3–4 lin. long ...   . (15) **linearis.**
         Corolla-lobes 5–7 lin. long ...   . (16) **arenicola.**
      Corolla-lobes creamy-white (or pale-pink ?),
         not spotted, pubescent at the very base   (14) **incarnata,** var.
*** Corolla-lobes 1$\frac{1}{4}$–2$\frac{1}{2}$ lin. long ; teeth on the angles
   of the stem 1–3 lin. long :
   Pedicels about 1 in. long ; corolla-lobes ovate, flat,
      pale yellow, not ciliate   ...   . (22) **longipes.**
   Pedicels 4–6 (in 11, *C. simulans*, 2–4) lin. long,
      longer in fruit :
      Plant 6–8 in. high ; pedicels deflexed ; corolla
         light greenish, spotted with purple-brown,
         lobes apparently glabrous on the inner
         surface, but ciliate ...   .   (7) **parviflora.**
      Plant 2–3 in. high ; pedicels erect :
      Corolla light green spotted with purple-brown,
         pubescent all over the inner surface with
         fine erect hairs   ...   . (21) **Marlothii.**
      Corolla pale yellowish, with purple-brown
         dots on the central part ; lobes with hairs
         at their tips   ...   . (10) **arida.**

Corolla pale yellow, without spots; lobes
with some rather long erect hairs at their
base and a few minute hairs at their tips (11) **simulans.**

Pedicels ½–2 lin. long, longer in fruit:
Corolla-lobes pubescent all over the inner face,
dark purple-brown ...   . (13) **pruinosa.**

Corolla-lobes glabrous on the inner face, except
sometimes at the very base:
Corolla-lobes ciliate with long soft purple
hairs:
Corolla-lobes yellowish-green with transverse purple-brown lines on the basal
half and entirely purple-brown beyond (8) **dependens.**

Corolla-lobes purple-brown on the basal
and green on the apical half, not
spotted ...   . (9) **inversa.**

Corolla-lobes not ciliate:
Corolla-lobes sublinear or linear-lanceolate,
pale pink (or whitish?), not spotted... (14) **incarnata.**

Corolla-lobes ovate or narrowly deltoid-
linear, pale yellowish or pale greenish-
yellow, not spotted   ...   . (12) **Hottentotorum**

Corolla-lobes deltoid-ovate, very acute, light
green or greenish-yellow, spotted with
purple-brown ...   . (6) **acutiloba.**

1. **C. ramosa** (N. E. Br. in Hook. Ic. Pl. t. 1904); plant about
1 ft. high, rather densely and bushily branched, glabrous, pale
greyish-green or somewhat glaucous, shaded with purple; branches
erect, ½–1 in. or more thick, 4-angled and 4-grooved; angles thick,
obtusely rounded, crenate, but scarcely toothed, except at the tips
of the young branches; flowers in small fascicles along the grooves
between the angles; pedicels about 1 lin. (in fruit becoming 3–4
lin.) long, glabrous; sepals 1–1¼ lin. long, lanceolate, acute, glabrous;
corolla small, glabrous outside; tube about ⅛ in. long, campanulate,
minutely pubescent around the mouth and apparently whitish within;
lobes 2½–3 lin. long, lanceolate, acute, replicate, with a rather acute
ridge down the face, glabrous and blackish-purple on the inner face,
not ciliate; outer corona-lobes minute, less than ¼ lin. long, broader
than long, bifid, pouch-like at the base; inner corona-lobes nearly
½ lin. long, linear or linear-subulate, closely incumbent upon the
backs of the anthers and slightly exceeding them, with a slight dorsal
transverse ridge at their base; follicles erect, subparallel or very
slightly diverging, 2–2¾ in. long, ⅓ in. thick, terete-fusiform, tapering
to an acute point, glabrous, pale green; seeds 2½ lin. long, 1 lin.
broad, irregularly and narrowly ovate-oblong, deeply concave from the
very thick margins being incurved on one side, very convex on the
other, not rugose, brown. *N. E. Br. in Gard. Chron.* 1892, xii. 370;
Schlechter in Journ. Bot. 1898, 477. *Stapelia ramosa, Masson, Stap.*
21, *t.* 32; *Willd. Sp. Pl.* i. 1288; *Pers. Syn. Pl.* i. 279; *Poir.*
Encycl. Meth. vii. 385; *Ait. Hort. Kew. ed.* 2, ii. 91; *Haw. Syn. Pl.*

Succ. 23 ; *Schultes, Syst. Veg.* vi. 22 ; *G. Don, Gen. Syst.* iv. 116 ; *Decne in DC. Prodr.* viii. 658 ; *Loud. Encycl. Pl.* 200, *fig.* 3292. *Piaranthus ramosus, Sweet, Hort. Brit. ed.* 2, 359 ; *Decne in DC. Prodr.* viii. 664.

COAST REGION: Riversdale or Ladismith Div. ; Karoo beyond Platte Kloof, around the Hot-springs, *Masson.*

CENTRAL REGION : Laingsburg Div. ; Zout Kloof and Karoo west of Laingsburg, *Pillans,* 131 ! between Ladismith and Laingsburg, *Pillans*! Prince Albert Div. ; near Grootfontein, *Dickson (Barkly,* 62)! *Pillans*! hill (Jacobs Kop ?) near Vlak Kraal (now Prince Albert Road Station), *Barkly,* 63 !

Described from living plants, and flowers preserved in fluid.

2. **C. mammillaris** (N. E. Br. in Hook. Ic. Pl. under t. 1902) ; plant $\frac{1}{2}$–$1\frac{1}{2}$ ft. high, branched in a bushy manner ; branches $\frac{3}{4}$–$1\frac{1}{2}$ in. thick, irregularly or spirally 5–6-angled, armed with stout conical hard-pointed acute spines $\frac{1}{4}$–$\frac{1}{2}$ in. long, glabrous, light green, faintly glaucous, with the apical half of the spines brown ; flowers in fascicles of 4–10 or more, along the grooves between the angles ; pedicels 1–$1\frac{1}{4}$ lin. long, glabrous ; sepals 1 lin. long, ovate, acuminate, glabrous ; corolla glabrous and pale green outside, with the tips, margins and a stripe down the middle of the lobes and some dots on the tube very dark purple-brown ; inside minutely papillate-setulose on the lobes and upper part of the tube, rich velvety black-purple on the lobes, pale yellowish dotted with blackish purple in the tube, which is 2–3 lin. long and as much in diam. ; lobes erectly spreading, 7–9 lin. long, about 2 lin. broad at the base, thence gradually tapering to a very acute apex, longitudinally folded, with revolute margins ; outer corona about as long as the staminal column, cupular, 10–15-toothed, glabrous, dark purple-brown ; inner corona-lobes $\frac{1}{2}$–$\frac{2}{3}$ lin. long, subulate or linear, acute to truncate, incumbent on the backs of the anthers and exceeding them, with more or less connivent-erect tips and a short dorsal projection at the base, adnate to the outer corona and forming 5 of its teeth, dark purple-brown. *N. E. Br. in Gard. Chron.* 1892, xii. 370 ; *Schlechter in Journ. Bot.* 1898, 477. *Stapelia mammillaris, Linn. Mant.* ii. 216 ; *Ait. Hort. Kew. ed.* 1, i. 310 ; *Thunb. Prodr.* 46 ; *Fl. Cap. ed.* 1, ii. 166, *and ed. Schultes,* 239 ; *Willd. Sp. Pl.* i. 1287 ; *Pers. Syn. Pl.* i. 279 ; *Poir. Encycl. Meth.* vii. 384, *and in Dict. Sc. Nat.* 1. 393 ; *Haw. Syn. Pl. Succ.* 25 ; *Schultes, Syst. Veg.* vi. 25 ; *Decne in DC. Prodr.* viii. 663. *S. pulla, Ait. Hort. Kew. ed.* 1, i. 310, *and ed.* 2, ii. 92 ; *Masson, Stap.* 21, *t.* 31 ; *Willd. Sp. Pl.* i. 1288 ; *Pers. Syn. Pl.* i. 279 ; *Poir. Encycl. .*vii. 385 ; *Alg. Teutsch. Gart. Mag.* 1818, 233, *t.* 26 ; *Bot. Mag. t.* 1648 ; *Spreng. Syst. Veg.* i. 841 ; *Dietr. Syn. Pl.* ii. 887. *Piaranthus pullus, R. Br. in Mem. Wern. Soc.* i. 23 ; *Haw. Syn. Pl. Succ.* 44 ; *G. Don, Gen. Syst.* iv. 113. *Decne in DC. Prodr.* viii. 650. *P. pulla, Schultes, Syst. Veg.* vi. 10. *P. mammillaris, G. Don, Gen. Syst.* iv. 114. *Pectinaria mammillaris, Sweet, Hort. Brit. ed.* 2, 357. *Boucerosia mammillaris, N. E. Br. in Journ.*

Linn. Soc. xvii. 165, *t.* 11, *figs.* 5–13.—*Stapelia aphylla ad nodos mammillaris, etc., Burm. Rar. Afr. Pl. dec.* ii. 27, *t.* 11.

South Africa : without locality, *Zeyher,* 1143 !

Coast Region: Clanwilliam Div. ; rocky hills near Olifants River, *Thunberg* ! near Clanwilliam, *Pillans* (ex *Pillans*). Worcester Div. ; Karoo, near Hex River, *Thunberg* ! lower slopes of hills east of Nuy, ex *Pillans,* hills north-west of Worcester, *Pillans,* 101 ! *and cultivated specimens* !

Central Region : Laingsburg Div. ; near Matjesfontein, ex *Pillans.* Prince Albert Div. ; near the Gamka River, *Burke,* 465 ! near Prince Albert, 2000 ft., *Bolus* (ex *Pillans*).

Western Div. : Little Namaqualand ; near Ookiep, 3000 ft., *Templeman in Herb. Pillans,* 1 ! Kamiesberg Range, *Barkly,* 30 !

Partly described from living plants. The corona is variable, especially in the length of the erect tips of the inner lobes. The odour of the flowers is extremely disagreeable.

3. C. armata (N. E. Br. in Hook. Ic. Pl. t. 1902); plant robust, branching ; branches 4-angled, $\frac{3}{4}$–1 in. square exclusive of the teeth, glabrous ; teeth $\frac{1}{3}$–$\frac{1}{2}$ in. long, stout, conical, with hard spine-like tips, horizontal or slightly recurved ; flowers in clusters of 8–10 (or more ?) along the grooves between the angles of the branches, several open together ; pedicels 2 lin. long, stout ; sepals $\frac{1}{8}$ in. long, lanceolate, acuminate ; corolla-tube about $\frac{1}{8}$ in. long, campanulate, greenish-yellow ; lobes $\frac{1}{3}$ in. long, about 2 lin. broad, somewhat spreading, lanceolate, acute, with the margins folded back, glabrous and not ciliate, greenish-white on the back, dark purple-brown or blackish-purple on the apical half of the inner face and greenish-yellow dotted with purple on the basal half ; outer-corona about equalling the staminal column, cup-shaped, sub-truncate, with 5 slight and broad erose crenations opposite and raised above the base of the inner corona-lobes and 5 pairs of minute teeth alternating with them, alternately blackish-purple and purple-brown in colour ; inner corona-lobes linear-oblong, notched at the apex, closely incumbent on the anthers and not produced beyond them. *Schlechter in Journ. Bot.* 1898, 477.

Western Region : Little Namaqualand ; at the foot of the Kamiesberg Range, *Barkly,* 47 !

4. C. Pillansii (N. E. Br.); plant $\frac{3}{4}$–1 ft. high, robust, with only 1 rooted main stem, bushily branched ; branches $\frac{3}{4}$–1 in. thick, exclusive of the teeth, 4-angled, glabrous, dull green mottled and spotted with purple-brown ; angles with stout much compressed deltoid-conical teeth $\frac{1}{3}$–$\frac{1}{2}$ in. long and (lengthwise of the stem) as much in breadth at the base, with short brown spine-like tips ; flowers several together in dense fascicles along the grooves of the stems, 2–3 of each fascicle open together ; pedicels $\frac{3}{4}$–1 lin. long, stout ; sepals $1\frac{1}{4}$ lin. long, $\frac{3}{4}$ lin. broad, ovate, acute or acuminate, glabrous ; corolla outside smooth, glabrous, whitish-green, dotted with purple-brown ; inside minutely tuberculate-rugulose, pubescent with short purple hairs ·fixed in every posture and spotted all over with " purple-brown on a purplish-grey ground" (*Pillans*); tube $2\frac{1}{2}$–3 lin. long, campanulate; lobes slightly spreading, 4–5 lin. long, $\frac{1}{4}$ in. broad, oblong or oblong-ovate, abruptly acute, with the sides folded back; outer corona-lobes minute, about $\frac{1}{4}$ lin. long and $\frac{1}{2}$ lin. broad, transversely rectangular, shortly and acutely bifid or emarginate, dark purple-brown, glabrous; inner corona-lobes $\frac{1}{2}$ lin. long, oblong, obtuse, erect, exceeding the erect anthers but not inflexed over them, with a small tubercle at the base where united to the outer corona, dark purple-brown; staminal column and entire corona very dwarf, scarcely 1 lin. high; style-apex not covered by the anthers.

COAST REGION: Robertson Div.; among bushes in Donker Kloof, near Montagu, *Pillans*, 678!

Described from living stems, and flowers preserved in fluid. Mr. Pillans states that this plant is attacked by the larva of some insect, which eats the centre of the stems and so destroys large plants.

5. **C. intermedia** (Schlechter in Journ. Bot. 1898, 478); stems or branches not seen, but according to a drawing, 5–6 in. long, 4-angled, about $\frac{1}{2}$ in. square, with stout spreading acute teeth about 2 lin. long, glabrous, flowering along the grooves between the angles towards the top; pedicels not seen, about $\frac{1}{2}$ in. long in the drawing; sepals about $1\frac{1}{2}$ lin. long, 1–$1\frac{1}{4}$ lin. broad, broadly ovate, acuminate, glabrous; corolla about 1 in. in diam., nearly flat, without a distinct tube, but with a depression in the disk, rugulose on the inner face, glabrous (ciliate?), according to the drawing dull green spotted with purple-brown; lobes 4–5 lin. long and $\frac{1}{4}$ in. broad at the base, ovate or deltoid-ovate, acute; outer corona-lobes $\frac{1}{2}$–$\frac{3}{4}$ lin. long, $\frac{2}{3}$–$\frac{3}{4}$ lin. broad, 3-toothed, with the middle tooth entire, crenulate or bifid, deltoid, much larger than the lateral teeth; sometimes all connate into a cup above the attachment to the inner corona-lobes, with 5 large deltoid acute teeth alternating with 5 pairs of minute teeth; inner corona lobes $\frac{1}{2}$ lin. long, lanceolate-subulate, closely incumbent on the backs of the anthers and not, or scarcely, produced beyond them, dorsally adnate at the base to the outer corona. *Stapelia intermedia*, *N. E. Br. in Hook. Ic. Pl. t.* 1910 *A.*

COAST REGION: Clanwilliam Div.; Olifants River, *Barkly* 8!

Of this plant I have only seen some dried flowers without pedicels, and a drawing made by Miss Barkly, from which latter I describe the stem and colour of the flowers. In my original description I stated that the lobes are "ciliate with vibratile, clavate, purple hairs," but I am now inclined to think that the very few cilia adhering to the margins of the lobes may have accidentally got there from the flowers of some other plant.

6. **C. acutiloba** (N. E. Br.); plant 6–8 in. high, bushily branched; branches $\frac{1}{2}$–$\frac{2}{3}$ in. thick, 4-angled, with stout conical acute teeth 2–$2\frac{1}{2}$ lin. long, hardened at the apex, glabrous; flowering cushions small, arranged along the grooves between the angles, 1–2 (or more?)-flowered; pedicels about $\frac{3}{4}$ lin. long, glabrous; sepals 1–$1\frac{1}{4}$ lin. long,

glabrous ; corolla small, entirely glabrous and not ciliate, light green (or greenish-yellow ?), spotted with dark purple-brown, more thickly at the tips of the lobes ; tube 1–1¼ lin. long, apparently broadly and shallowly cup-shaped ; lobes 2–2¼ lin. long, 1½ lin. broad at the base, deltoid-ovate, very acute or somewhat acuminate, probably spreading ; outer corona about equalling the staminal column, shortly cup-shaped, with 10 equidistant deltoid acute teeth ¼ lin. long, glabrous, dark purple-brown ; inner corona-lobes about ½ lin. long, linear or linear-subulate, acute, incumbent on the backs of the anthers and not produced beyond them, dorsally connected to the outer corona at the base, margined with (or entirely ?) purple-brown.

Western Region : Little Namaqualand, *Templeman in Herb. Pillans,* 8 !

7. **C. parviflora** (N. E. Br. in Gard. Chron. 1892, xii. 370) ; plant 6–8 in. high, with several erect stems from the base, which bear a few very spreading branches ; stems and branches 4-angled, 5–7 lin. square, with conical acute spreading or slightly recurved teeth 1½–2 lin. long at the angles, glabrous, dull green or purplish ; flowering-cushions arranged along the grooves between the angles on the upper part of the stems, producing 1–3 flowers (annually ?) ; pedicels 3–5 lin. long, curved downwards ; sepals about ¾ lin. long, ovate or ovate-lanceolate, acute ; corolla drooping about ⅓ in. in diam. ; tube or united part scarcely 1 lin. long, apparently saucer-shaped ; lobes somewhat spreading, ⅛ in. long, ¾ lin. broad, lanceolate, acute, flat, rugose, ciliate on the margins, yellowish-green spotted with purple-brown. *Schlechter in Journ. Bot.* 1898, 478 (*excl. syn. C. dependens, N. E. Br.*). *Stapelia parviflora, Masson, Stap.* 22, t. 35 ; *Willd. Sp. Pl.* i. 1283 ; *Pers. Syn. Pl.* i. 278 ; *Poir. Encycl.* vii. 381 ; *Ait. Hort. Kew. ed.* 2, ii. 89 ; *Haw. Syn. Pl. Succ.* 24 ; *Schultes, Syst. Veg.* vi. 24 ; *Spreng. Syst. Veg.* i. 841 ; *Dietr. Syn. Pl.* ii. 887. *Piaranthus parvi-florus, Sweet, Hort. Brit. ed.* i. 278 ; *G. Don, Gen. Syst.* iv. 113 ; *Decne in DC. Prodr.* viii. 650.

Western Region : Little Namaqualand, under shrubs, *Masson.*

Only known from Masson's figure and description.

8. **C. dependens** (N. E. Br. in Hook. Ic. Pl. t. 1903, B) ; plant 8–12 in. high, compactly branched ; branches erect, 4-angled, 5–9 lin. square excluding the teeth, glabrous, at first dull green, tinted with purple, becoming pale grey or pale purplish-grey with age ; angles rounded, with stout conical horizontal or slightly recurved acute teeth 1–2 lin. long, which become hardened at the point ; flowers successively developed from very small flowering points arranged along the grooves between the angles which produce flowers for two or more successive years, pendulous and closely applied to the stems ; pedicels 1–1½ lin. long, abruptly bent downwards from their base, becoming erect and 4–5 lin. long in fruit, glabrous ; sepals ¾ lin. long, lanceolate, acute, glabrous ;

corolla in bud oblong, obtuse, pendant and closely applied to the stem, when expanded rotate, without any tube, with 4 lobes reflexed and 1 (the lower) pressed against the stem, 2–$2\frac{1}{4}$ lin. long, 1 lin. broad, oblong-linear, subacute, convex, glabrous and green, tinted with purplish on the back; inner face smooth, light yellowish-green, marked with transverse purple-brown lines on the disk and basal half of the lobes, dark purple-brown on the apical half, ciliate with long soft twisted or curly purple hairs; outer corona-lobes erectly spreading, bifid to half-way down, basal part subquadrate, yellowish; teeth subulate, acute, arching-divaricate, purple-black; inner corona-lobes deltoid-subulate, acuminate, purple-black, closely incumbent on the backs of the anthers and not exceeding them; follicles diverging, $3\frac{1}{4}$–$5\frac{1}{3}$ in. long, about $\frac{1}{4}$ in. thick, terete, tapering to an acute point; seeds not seen. *K. Schum. in Engl. and Prantl, Pflanzenfam.* iv. ii. 278.

COAST REGION : Clanwilliam Div. ; Oliphants River, *Hesse*! near Clanwilliam, 250–300 ft., *Schlechter*, 8017 ! *Pocock in Herb. Pillans*, 68 ! on a farm 20 miles west of Clanwilliam, *Barkly*, 78 ! and *cultivated specimens*!

Described from living plants which flowered at Kew and with me in 1877 and 1906.

9. C. inversa (N. E. Br. in Gard. Chron. 1903, xxxiii. 354) ; an erect branching plant similar to *C. dependens*; branches 4-angled, with conical spreading teeth; flower-cushions arranged along the grooves between the angles; pedicels 1–2 lin. long, glabrous; sepals $\frac{2}{3}$–1 lin. long, lanceolate or deltoid-ovate, acute, glabrous; corolla-tube about $1\frac{1}{4}$ lin. long, campanulate or funnel-shaped, glabrous, white at the base, spotted with purple-brown above and entirely purple-brown at the mouth within; lobes about $2\frac{1}{2}$ lin. long, 1–$1\frac{1}{4}$ lin. broad, oblong or linear-oblong, shortly acute, dark purple-brown on the basal half, rather deep grass-green on the apical half, not spotted, glabrous on both sides, ciliate from base to apex with long jointed purple hairs; outer corona-lobes $\frac{1}{2}$–$\frac{2}{3}$ lin. long, slender, linear-filiform, entire, obtuse, glabrous, purple-brown; inner corona-lobes $\frac{1}{3}$–$\frac{1}{2}$ lin. long, ovate or ovate-oblong, obtuse, incumbent on the backs of the anthers and equalling or very slightly exceeding them, with a subquadrate or oblong projection about $\frac{1}{4}$ lin. long on the back at the base, which is obtuse, emarginate or minutely 3-toothed at its apex, purple-brown.

COAST REGION : Clanwilliam Div.; without precise locality, *Ayres in Herb. Pillans*, 92 !

10. C. arida (N. E. Br. in Gard. Chron. 1892, xii. 369) ; stems erect, somewhat crowded, 2–$2\frac{1}{2}$ in. high, 4-angled, about $\frac{1}{2}$–$\frac{2}{3}$ in. square, with conical acute spreading teeth 1–2 lin. long, glabrous, grey-green? flowering along the grooves between the angles at the upper part of the stems; flowers solitary, erect; pedicels 3–4 lin. long, elongating in fruit; sepals lanceolate, acute; corolla about

5 lin. in diam., apparently rotate, whitish-yellow, apparently dotted
with red (or purple-brown?) at the base of the lobes, which are
about 2 lin. long, ovate-lanceolate, acute, flat, pubescent at the tips.
Schlechter in Journ. Bot. 1898, 478. *Stapelia arida, Masson, Stap.* 21,
t. 33; *Willd. Sp. Pl.* i. 1283; *Pers. Syn. Pl.* i. 278; *Poir. Encycl.*
vii. 381; *Ait. Hort. Kew. ed.* 2, ii. 88; *Haw. Syn. Pl. Succ.* 24;
Schultes, Syst. Veg. vi. 23; *Spreng. Syst. Veg.* i. 838; *Dietr. Syn. Pl.*
ii. 885; *Loud. Encycl. Pl.* 200, *fig.* 3293. *Orbea?* arida, *Sweet, Hort.*
Brit. ed. 1, 277. *Obesia?* arida, *Sweet, Hort. Brit. ed.* 2, 358.
Piaranthus aridus, G. Don, Gen. Syst. iv. 114; *Decne in DC. Prodr.*
viii. 650.

CoAST REGION: Riversdale or Ladismith Div.; Kannaland, *Masson.*

Only known from Masson's figure and description.

11. C. simulans (N. E. Br.); stems erect, closely placed, branching
at the base, 4-angled, 2–3 in. high, 5–7 lin. square, with flattish
sides and acute angles with horizontally spreading conical acute
hard pointed teeth 1–1½ lin. long, glabrous, green or tinted with
purple-brown, slightly glaucous, bearing several flower-cushions
along the sides, apparently producing a succession of single flowers
during more than one year; pedicels 2–4 (in fruit 6–8) lin. long,
slender, erect, about ⅖ lin. thick, glabrous; sepals ¾ lin. long,
extending to the sinuses of the corolla, lanceolate or ovate-lanceolate,
acute, glabrous; corolla 3½ lin. in diam., deeply lobed, rotate,
without a distinct tube; lobes more or less recurving, 1½ lin. long,
1 lin. broad, ovate or ovate-lanceolate, acute, slightly thickened at
the apex, glabrous and smooth on the back, smooth on the inner
face, but thinly sprinkled with erect rather long hairs on the basal
half and with a few minute hairs at the tips, light yellow (*Marloth*);
outer corona-lobes erect, ½–¾ lin. long, linear, deeply bifid, with
slender diverging teeth; inner corona-lobes closely incumbent upon
the backs of the anthers and shorter than them, with the inflexed
part about ¼ lin. long, linear-oblong, obtuse, having a transverse
ridge on the back at the bend; follicles about 3½ in. long, erect,
subparallel, slender, about 1½–2 lin. thick, linear-terete, acute,
smooth, glabrous; seeds 2–2¼ lin. long, scarcely 1 lin. broad, ovate,
concave with incurved thickened margins on one side, convex on
the other, smooth, glabrous, brown.

CENTRAL REGION: Prince Albert Div.; near Prince Albert, *Marloth,* .4576!

This species is very closely allied to *C. arida*, N. E. Br., but differs from the
latter (according to Masson's figure and description of it) in having much smaller
flowers, without spots at the base of the lobes and with hairs at the base as well
as at the tips of the lobes.

12. C. Hottentotorum (N. E. Br. in Hook. Ic. Pl. under t. 1903);
plant 4–6 in. high, branching at the base; branches obtusely
4- (rarely 5-) angled, ½–1 in. square exclusive of the teeth, glabrous,
greyish-green or purplish-tinted; teeth 2–3 lin. long, stout, conical,

with hard spine-like tips, horizontal or slightly recurved; flowers in small clusters of 6–10 along the grooves of the stems, opening 1–2 at a time; pedicels less than 1 lin. long, becoming 4–6 lin. long in fruit, glabrous; sepals about $\frac{3}{4}$ lin. long, deltoid-ovate, acute, glabrous; corolla about $\frac{1}{4}$ in. in diam., glabrous with the exception of a few minute hairs at the mouth and throat of the tube, pale greenish-yellow; tube 1 lin. long, campanulate; lobes about $1\frac{1}{4}$ lin. long, ovate, acute, very spreading, faintly keeled on the inner face; outer corona-lobes very minute, pouch-like at the base, subquadrate, bifid, yellow; inner corona-lobes fleshy, roundish-ovate or sub-hemispheric, very obtuse and very convex on the back, becoming thin and nearly flat when dried, closely applied to the backs of the anthers and about half as long as them, yellow; follicles erect, subparallel, $2\frac{1}{2}$–3 in. long, about $2\frac{1}{2}$ lin. thick, narrowly fusiform, tapering at both ends, acute, greyish-green, streaked with purple-brown; seeds rather small, ovate, with a thickened margin. *N. E. Br. in Gard. Chron.* 1892, xii. 369; *Schlechter in Journ. Bot.* 1898, 477. *Quaqua Hottentotorum, N. E. Br. in Gard. Chron.* 1879, xii. 8 *and* 9, *fig.* 1.

VAR. β, **major** (N. E. Br.); corolla about 5 lin. in diam., straw-colour (*Pillans*); tube $1\frac{1}{4}$ lin. long; lobes 2 lin. long, 1 lin. broad at the base, thence gradually tapering to an acute apex; outer corona-lobes reduced to 5 pairs of minute teeth; inner corona-lobes stout, linear-oblong, obtuse, very convex on the back, nearly or quite as long as the anthers and closely incumbent on their backs; otherwise as in the type.

WESTERN REGION: Little Namaqualand; Klipfontein, *Barkly*, 50! 50 bis! and *cultivated specimens*! Ookiep, *Barkly*, 27! Var. β: Little Namaqualand; Mistkraal, near Kamaggas, *Rich in Herb. Pillans*, 10! 143! Oubiep, *Rich in Herb. Pillans*, 200!

Described from living plants and the variety from flowers preserved in fluid. According to Mr. Pillans this plant is known to the natives as "bitter gwagwa," and is not edible, being considered poisonous, with medicinal virtues.

13. **C. pruinosa** (N. E. Br. in Gard. Chron. 1892, xii. 370); plant 1–$1\frac{1}{2}$ ft. high, bushily branched, glabrous, greyish-green, tinted with purple; branches obtusely 4-angled, 5–7 lin. square, with very small teeth, about 1 lin. long, hard and brown at the tips, flowering along the grooves between the angles; flowers 1–3 together, successively developed; pedicels 1–$1\frac{3}{4}$ lin. long, glabrous; sepals about $\frac{3}{4}$ lin. long, ovate, acute, glabrous; corolla rotate, without a distinct tube, lobed to $\frac{3}{4}$ of the way down, 5–6 lin. in diam., glabrous and more or less mottled with purple-brown outside, pubescent and entirely dark purple-brown all over the inner face; united or disk part nearly flat; lobes about 2 lin. long and $1\frac{1}{3}$ lin. broad, deltoid-lanceolate, acute; outer corona shortly cupular and combined with the base of the inner corona-lobes, with 5 minute subrectangular bifid lobes about $\frac{1}{4}$ lin. long, apparently erect, glabrous, blackish; inner corona-lobes $\frac{1}{2}$ lin. long, linear-oblong, emarginate, bifid, irregularly toothed or subtruncate at the apex, closely incumbent upon

the backs of the anthers and slightly exceeding them, dorsally produced at the base into a short horizontally spreading truncate or rounded ridge or crest, blackish. *Schlechter in Journ. Bot.* 1898, 478. *Stapelia pruinosa, Masson, Stap.* 24, *t.* 41 ; *Willd. Sp. Pl.* i. 1287 ; *Ait. Hort. Kew. ed.* 2, ii. 91 ; *Pers. Syn. Pl.* i. 279 ; *Schultes, Syst. Veg.* vi. 35 ; *Spreng. Syst. Veg.* i. 840 ; *Dietr. Syn. Pl.* ii. 886 ; *Decne in DC. Prodr.* viii. 657 ; *Loud. Encycl. Pl.* 200, *fig.* 3317. *S. bruinosa, Poir. Encycl.* vii. 384. *Tromotriche pruinosa, Haw. Syn. Pl. Succ.* 37 ; *G. Don, Gen. Syst.* iv. 119.

WESTERN REGION : Little Namaqualand, *Masson, Pillans,* 21 !

Described from dried specimens.

14. **C. incarnata** (N. E. Br. in Gard. Chron. 1892, xii. 369) ; a bushy plant a foot or more high ; branches erect, $\frac{1}{2}$–$\frac{3}{4}$ in. thick, 4-angled, with stout conical spreading teeth $1\frac{1}{2}$–2 lin. long, hardened at the tips, glabrous, greyish-green (subglaucous ?) ; flowering-cushions small, arranged along the grooves between the angles, producing " usually solitary " (*Masson*), " 4-nate " (*Thunberg*) flowers, probably for 2 or more years in succession ; pedicels $\frac{1}{2}$–$\frac{3}{4}$ lin. long, glabrous ; sepals about $\frac{3}{4}$ lin. long, ovate-lanceolate, acuminate, glabrous ; corolla pale pink varying to white (*Masson*), with a few hairs around the mouth of the tube and at the very base of the lobes within, otherwise glabrous and not ciliate ; tube about 1 lin. long, campanulate ; lobes ascending-spreading, $1\frac{1}{2}$–$1\frac{3}{4}$ lin. long, $\frac{1}{2}$–$\frac{2}{3}$ lin. broad at the base, probably larger when alive, linear-lanceolate, subacute or obtuse, convex from the sides being reflexed or recurved ; outer corona-lobes about $\frac{1}{4}$ lin. long, erect or ascending-spreading, subquadrate, bifid, with a broad notch between the teeth, not exceeding the staminal column, "yellow" (*Thunberg*) ; inner corona-lobes $\frac{1}{3}$ lin. long, linear-subulate, acute, incumbent on the backs of the anthers and slightly exceeding them. *Schlechter in Journ. Bot.* 1898, 478. *Stapelia incarnata, Linn. f. Suppl.* 171 ; *Thunb. Prodr.* 46 ; *Fl. Cap. ed.* 2, ii. 167, *and ed. Schultes,* 240 ; *Masson, Stap.* 22, *t.* 34 ; *Willd. Sp. Pl.* i. 1289 ; *Pers. Syn. Pl.* i. 279 ; *Poir. Encycl. Meth.* vii. 386, *and in Dict. Sc. Nat.* 1. 392 ; *Ait. Hort. Kew. ed.* 2, ii. 92 ; *Haw. Syn. Pl. Succ.* 24 ; *Schultes, Syst. Veg.* vi. 23 ; *Spreng. Syst. Veg.* i. 840 ; *Loud. Encycl. Pl.* 200, *fig.* 3294 ; *Dietr. Syn. Pl.* ii. 886. *Podanthes incarnata, Sweet, Hort. Brit. ed.* 2, 358. *Piaranthus incarnatus, G. Don, Gen. Syst.* iv. 114 ; *Decne in DC. Prodr.* viii. 650. *Boucerosia incarnata, N. E. Br. in Journ. Linn. Soc.* xvii. 166, *t.* 11, *figs.* 14–17.—*Euphorbium erectum, quadrangulare, spinosum, &c., Burm. Rar. Afr. Pl. Dec.* i. 15, *t.* 7, *fig.* 1.

VAR. β, **alba** (N. E. Br.) ; plant branching near the base with branches 3–6 in. (or more ?) long, $\frac{1}{2}$–1 in. thick, erect, dark green with a greyish powdery covering ; flowers 3–5 together, developing in succession, pedicels and sepals about 1 lin. long ; corolla pink on the tube and lobes outside, uniformly creamy-white inside, covered on the very base of the lobes and upper part of the tube with small stiff pink hairs ; tube $1\frac{1}{2}$–2 lin. long ; lobes 3–5 lin. long, $\frac{3}{4}$–1 lin. broad at the base, linear, acute, with recurved margins ; outer and inner corona golden-yellow; in all other particulars exactly as in the type. *Piaranthus incarnatus, var. albus, G. Don, Gen. Syst.* iv. 114.

COAST REGION: near Verloren Valley in Piquetberg Div. and on mountains near Compagnies Post (East India Company's Fort?) at Saldahna Bay in Malmesbury Div., *Thunberg*! Var. β; Clanwilliam Div.; near Lamberts Bay, *Pillans* 86! Malmesbury Div.; among shrubs in sandy places near Hopefield, 200 ft., only one plant found, *Bolus*, 10729!

Possibly *Zeyher*, 1146, without locality, may belong to this species, but the specimens I have seen are flowerless. The variety *alba* is partly described from a fresh flower preserved in fluid and may not differ from the type in anything but colour, as the flowers of Thunberg's type may have originally been as large as those of the variety, but have become greatly shrunken from being less carefully pressed.

15. **C. linearis** (N. E. Br. in Hook. Ic. Pl. t. 1903, fig. A); stems obtusely 4-angled, with small teeth $\frac{1}{2}-\frac{3}{4}$ lin. long, hardened at the point, glabrous, flowering along the grooves between the angles; flowers probably fasciculate; pedicels $\frac{1}{2}-1$ lin. long, becoming about $\frac{1}{8}$ in. long in fruit, glabrous; sepals about $\frac{3}{4}$ lin. long, ovate, acute, glabrous; corolla-tube about $\frac{1}{8}$ in. long, campanulate, glabrous outside and within, apparently whitish; lobes ascending-spreading, $\frac{1}{4}$ in. (in dried flowers) long, $\frac{3}{4}$ lin. broad, linear, acute, replicate (longitudinally folded), apparently somewhat incurved at the apex, glabrous and apparently not ciliate, dark purple-brown; outer corona-lobes ascending, about $\frac{2}{5}$ lin. long and broad, subquadrate, bifid to half-way down, with or without a minute tooth in the notch and the teeth deltoid, obtuse, glabrous, apparently white or yellowish; inner corona-lobes $\frac{3}{4}$ lin. long, flat, linear, obtuse, incumbent on the backs of the anthers and connivent at the base, then erect with very revolute tips, purple or purple-brown; follicles about $1\frac{1}{2}$ in. long, narrowly fusiform, smooth, glabrous; seeds (perhaps immature) $\frac{1}{8}$ in. long, $\frac{1}{2}-\frac{2}{3}$ lin. broad, oblong, with very thick margins incurved on one side and a prominent central ridge on the other, glabrous, smooth, brown, tipped with a tuft of hairs scarcely $\frac{1}{2}$ in. long. *Schlechter in Journ. Bot.* 1898, 477.

CENTRAL REGION: Prince Albert Div.; Seven-weeks Poort, *Bain*, 8!

Of this species I have only seen a small piece of stem less than 1 in. long with follicles attached and a few loose flowers dried and in alcohol.

16. **C. arenicola** (N. E. Br.); plant about 3-4 in. high, with numerous erect 4-angled stems $\frac{3}{4}-1$ in. (or more?) square, glabrous, green, slightly glaucous, branching at the base; sides flat below, grooved above; teeth very spreading, $1\frac{1}{2}-2\frac{1}{2}$ lin. long, stout, conical, hardened at the acute tips; flowers in small clusters along the sides of the stems; pedicels $\frac{1}{2}-1\frac{1}{2}$ lin. (in fruit elongating to $\frac{1}{2}-1$ in.) long, stout, glabrous; sepals $1\frac{1}{4}-1\frac{1}{2}$ lin. long, $\frac{1}{2}$ lin. broad, ovate-lanceolate, acuminate, glabrous; corolla minutely papillate-tuberculate (or rarely with very minute hairs) at the base of the lobes on their inner surface and within the tube, elsewhere smooth and glabrous,

not ciliate, rich blackish-purple or "very deep maroon" (*Marloth*), with the base of the tube and a zone at its middle white or whitish; tube 2 lin. long, 2 lin. in diam. at the mouth inside, campanulate or cup-shaped; lobes ascending-spreading, 5–7 lin. long, 1¾–2 lin. broad at the base, thence tapering to the acute apex, with reflexed or replicate margins; outer corona somewhat saucer-shaped at the base, with 5 pairs of minute acute teeth about ¼ lin. long, with or without 5 other obtuse teeth (formed by the connection with the inner corona-lobes) alternating with them, yellowish, with purple tips to the teeth; inner corona-lobes ½ lin. long, oblong or linear-oblong, obtuse, subtruncate or emarginate at the apex, closely incumbent upon the backs of the anthers and shorter to longer than them, but not turned up at the tips, dorsally connected at the base to the very short cup of the outer corona, yellowish, more or less marked with purple-brown on the margins and dorsal projection; follicles very slightly diverging, 5½–6½ in. long, ¼ in. thick, linear-terete, tapering at the base and into an acute apex, smooth, glabrous.

CENTRAL REGION: Laingsburg Div.; near Matjesfontein, in soft sand accumulated under bushes, *Pillans*, 44! 62! Prince Albert Div.; near Prince Albert, *Marloth*, 4581!

Mr. Pillans notes that the plants sometimes spread to a diameter of 2 ft., having in some cases very robust stems.

17. **C. longicuspis** (N. E. Br.); "plant 2–3 in. high; stems erect, decumbent or descending underground, 1¼–6 in. long, ¼–¾ in. thick" (*Pillans*), 4-angled, with slightly concave or grooved sides, greyish-green (greyish-blue, *Pillans*) mottled with brown, angles rounded, with conical acute deciduous grey teeth 1–1½ lin. long; flowers in fascicles of 3–10, thickly distributed along the sides of the stems from the top to near the bottom, developing successively, erect; pedicels 2–2½ lin. long, glabrous; sepals 1¾ lin. long, lanceolate or ovate-lanceolate, acuminate, glabrous; corolla comparatively large, glabrous and greyish-green outside, inner surface densely and very minutely papillate, with a very minute ciliation at the base of the lobes, otherwise glabrous, "pale purple merging into purple in the tube" (*Pillans*), drying dark purplish-olive-brown; tube 2 lin. long and 4 lin. broad, shortly funnel-shaped; lobes erectly spreading, ¾–1 in. long, 2½–2¾ lin. broad, linear-lanceolate, acute, with recurved margins; outer corona dark purple-brown, shortly cupular at the base, 20-toothed, with the 5 pairs of teeth opposite the corolla-lobes ¾ lin. long, falcate-subulate, with the tips widely diverging, and the 5 alternating pairs (really short bifid projections from the base of the inner corona-lobes) very much shorter, straight and contained between the adjacent teeth, whose tips arch over them; inner corona-lobes 2½ lin. long, linear-lanceolate and closely incumbent at the base on the backs of the anthers, produced beyond them into long filiform tips connivent into an erect slender column, dark purple-brown.

Mr. Pillans informs me that he does not know if this species comes from the
Tropical or South-African part of the German colony, but believes that he also
has the same species in cultivation from Prieska.

18. C. lutea (N. E. Br. in Hook. Ic. Pl. t. 1901); stems crowded,
$1\frac{1}{2}$–4 in. long, sharply 4-angled, $\frac{1}{2}$–$\frac{3}{4}$ in. thick, excluding the teeth,
glabrous, green, mottled with dull purple; teeth $\frac{1}{4}$–$\frac{1}{3}$ in. long, stout,
acute, horizontal; flowers 3–26 in a cluster at the middle or lower
part of the young stems, mostly opening at the same time; pedicels
$\frac{1}{2}$–1 in. long, stout, glabrous; sepals $2\frac{1}{2}$–4 lin. long, lanceolate or
ovate-lanceolate, acuminate; corolla $1\frac{1}{2}$–$2\frac{1}{2}$ in. in diam., glabrous
outside, rugulose within, yellow; tube or united part 3–4 lin. long;
lobes spreading, $\frac{3}{4}$–$1\frac{1}{3}$ in. long, $\frac{1}{4}$–$\frac{1}{3}$ in. broad, narrowly lanceolate,
attenuate to an acute point, ciliate with vibratile clavate purple
hairs; outer corona cup-shaped with a recurved margin, consisting
of 5 subquadrate contiguous lobes $\frac{3}{4}$ lin. long, 1 lin. broad, minutely
toothed at the subtruncate apex and with several slight keels on
the inner face, yellow; inner corona-lobes 1 lin. long, filiform,
connivent-erect, dorsally connected with the sinuses of the outer
corona at their base, where there is a short erect subulate tooth,
yellow. *N. E. Br. in Gard. Chron.* 1892, xii. 370; *K. Schum. in
Engl. und Prantl, Pflanzenfam.* iv. ii. 278 *and* 276, *fig.* 83, *E–F;
Schlechter in Engl. Jahrb.* xx. *Beibl.* 51, 54, *and Journ. Bot.* 1898,
477..

Described from living plants, and flowers preserved in fluid.

In a letter from Mrs. Barber concerning this plant, she states:—"It is the
commonest of all the family up here (Kimberley) and occurs upon nearly every
grassy ridge upon the flats, and, although I have passed over acres of it, I have
never yet met with a seed-pod; the plant blossoms profusely in autumn." An
outline of the fruit, however, is represented on a drawing sent to Kew by
Mrs. Barber, from which the follicles appear to be about $3\frac{1}{2}$ in. long, fusiform and
moderately stout. The· odour of the flowers is stated to be very fetid, like that
of putrid fish.

19. C. melanantha (N. E. Br.); stems decumbent at the base,
2–$2\frac{3}{4}$ in. high, $\frac{3}{4}$–1 in. thick, 4-angled, with triangular acute spreading
teeth, glabrous; flowers in sessile umbels of 6–10 towards the tips
of the stems; pedicels about $\frac{1}{2}$ in. long, glabrous; sepals $3\frac{1}{2}$ lin.
long, lanceolate, acuminate, glabrous; corolla 2 in. in diam., rotate,
lobed to the middle, densely transversely rugose on the inner surface,
glabrous, ciliate with vibratile clavate hairs; lobes about $\frac{3}{4}$ in. long,
$\frac{1}{2}$ in. broad at the base, ovate-triangular, acuminate; outer corona-
lobes erectly spreading, connate at the lower half, semi-oblong

(transversely oblong may be intended) in outline above, irregularly
crenulate, slightly 4–6-grooved on the back; inner corona-lobes
ligulate, porrectly incurved above the anthers, intricately excavated
at the thickened base, which is dorsally adnate to the tube of the
outer corona. *Stapelia melanantha, Schlechter in Engl. Jahrb.*
xxxviii. 50.

KALAHARI REGION : Transvaal; stony Flats near Sandloop, 5000 ft., *Schlechter,*
4694 !

This is evidently closely allied to *C. lutea,* N. E. Br. The colour of the
flowers is not stated, but from the specific name it should be blackish or dark
purple-brown.

20. C. chlorantha (Schlechter in Engl. Jahrb. xviii. Beibl. 45, 37);
stems decumbent, about $1\frac{1}{2}$ in. long and 5 lin. thick, obscurely
6-angled or with obscurely 6-seriate tubercles; flowers solitary or
rarely in pairs; pedicels 4–5 lin. long, stout, glabrous; sepals 1 lin.
long, triangular-lanceolate, very acute, glabrous; corolla campanulate,
with a short tube, green; lobes erectly-spreading, $\frac{1}{4}$ in. long, linear-
lanceolate, flat, somewhat hooked at the obtuse apex, glabrous on
the back, densely pilose on the inner face; outer corona white,
shortly cup-shaped, 5-sulcate, with free subdivaricate lobes, shortly
notched at the truncate apex; inner corona-lobes shorter than the
staminal column, fleshy. *Schlechter in Journ. Bot. 1898, 478.*

COAST REGION: George Div.; in the Karoo near Klip Drift, 1800 ft.,
Schlechter, 2275.

21. C. Marlothii (N. E. Br. in Gard. Chron. 1903, xxxiv. 414);
plant about $3–3\frac{1}{2}$ in. high, branching at the base; branches rather
crowded, decumbent at the base, acutely 4-angled, $\frac{1}{2}–\frac{3}{4}$ in. square,
with nearly flat or slightly grooved faces, glabrous, entirely dull
violet or more or less mottled with that colour where exposed to the
sun; angles armed with conical acute teeth about 2–4 lin. apart and
1–2 lin. long, hardened and whitish at the tips; flowers in numerous
fascicles of 2–4, scattered along the middle of two or all four faces
of the branches, developing successively; pedicels $\frac{1}{3}–\frac{1}{2}$ in. long, erect,
slightly curved, slender, glabrous; sepals about $\frac{3}{4}$ lin. long, nearly
$\frac{1}{2}$ lin. broad at the base, thence tapering to an acute point, glabrous;
corolla rotate, with very revolute lobes, about $2\frac{1}{2}$ lin. (or $3\frac{1}{2}–4$ lin.,
with the lobes flattened out) in diam., light green, marked to the
tips of the lobes with small dark purple-brown lines and spots which
show through on the glabrous outside, thinly covered on the inner
face with erect stiffish but fine purple hairs; lobes rolled back to
the pedicel, $1\frac{1}{2}–2$ lin. long, 1 lin. broad, oblong-lanceolate, acute,
slightly thickened at the apex, where the hairs are slightly tufted;
outer corona-lobes about $\frac{2}{3}$ lin. long, $\frac{1}{5}$ lin. broad, erect, linear from
a broadened base, deeply bifid at the apex, with diverging and
recurving filiform points, glabrous, yellow; inner corona-lobes
somewhat hammer-shaped in side view, with the linear-lanceolate or

linear entire or subbifid tips horizontally incumbent on the backs of
the anthers and dorsally produced into a short stout obtuse arm,
which sometimes shows a tendency to become bifid, glabrous, yellow,
with faint brown markings; pollen-masses very conspicuous, globose,
dull yellowish.

CENTRAL REGION: Ceres Div. ; stony places near Zwartkops Drift in the
Bokkeveld Karoo, *Marloth,* 3307 ! between Ceres and Calvinia, *Pillans,* 66 !
Laingsburg Div. ; between Witte Poort and Laingsburg and near Matjesfontein,
ex *Pillans.*

Described from a living plant.

22. **C. longipes** (N. E. Br.); stems or branches decumbent or
procumbent, $\frac{3}{4}$–$1\frac{1}{4}$ in. long, 3–7 lin. thick, 5–6-angled, with small
tubercle-like conical teeth 1–1$\frac{1}{2}$ lin. long along the angles of the
young branches, becoming minute and distant with age (*Marloth*),
glabrous, dull-green, flowering between the angles at about the
middle; flowers 1–2 together, successively developed; pedicels
1–1$\frac{1}{4}$ in. long, erect, very slender, less than $\frac{1}{2}$ lin. thick, glabrous;
sepals about 1 lin. long, ovate, acute or acuminate, glabrous; corolla
about $\frac{1}{2}$ in. in diam., flat, without a distinct tube, but with a
depression in the disk containing the corona, smooth, quite glabrous
and not ciliate; lobes widely spreading, 2–2$\frac{1}{3}$ lin. long, 1$\frac{1}{2}$ lin. broad,
ovate, acute, flat, pale dull yellow with green tips (*Marloth*); corona
consisting of 5 pairs of minute slender parallel teeth, with a small
pocket-like cavity in front of them, alternating with 5 stout
transversely rectangular partition-like bodies (really the inner
corona-lobes), truncate and minutely denticulate on their dorsal
margin, truncate on their inner margin and not reaching half-way
up the backs of the anthers, orange.

CENTRAL REGION: Sutherland Div. ; on the Roggeveld, near Sutherland,
Marloth, 3799 !

23. **C. aperta** (N. E. Br. in Hook. Ic. Pl. t. 1905, fig. A); stems
erect, or decumbent at the base, 2–2$\frac{1}{2}$ in. long, very obtusely
4-angled, light glaucous green, scarcely toothed; flowers apparently
solitary near the base of the stems; pedicels 1–2 in. (3 in., *Masson*)
long, erect or ascending, glabrous; sepals 1–1$\frac{1}{2}$ lin. long, ovate,
acute, glabrous; corolla 1$\frac{1}{4}$–1$\frac{1}{2}$ in. in diam., quite glabrous; tube
about $\frac{1}{3}$ in. long and 5–6 lin. in diam., cup- or somewhat funnel-
shaped, densely papillate and usually purple-brown at the basal half
within, whitish or yellowish above, marked (sometimes labyrinthi-
cally) with small impressed irregular purple-brown lines and dots;
lobes very spreading, 5–7 lin. long, 3$\frac{1}{2}$–4$\frac{1}{2}$ lin. broad, ovate-oblong,
acute, with more or less reflexed margins, rugulose and with 3
impressed nerves on the inner face, apparently purplish-brown or
olive-brown; outer corona cupular, acutely pentagonal, about
equalling the staminal column, truncate and crenately denticulate
at the top with a short acute tooth at each angle apparently

whitish; inner corona-lobes $1\frac{1}{4}$–$1\frac{1}{2}$ lin. long, linear-filiform or very slightly clavate, closely incumbent on the anthers at the base then connivent-erect above them, dorsally connected with the outer corona, dark purple-brown. *Schlechter in Journ. Bot.* 1897, 478. *Stapelia aperta, Masson, Stap.* 23, *t.* 37; *Willd. Sp. Pl.* i. 1285; *Pers. Syn. Pl.* i. 279; *Poir. Encycl.* vii. 383; *Ait. Hort. Kew. ed.* 2, ii. 90; *Haw. Syn. Pl. Succ.* 23; *Schultes, Syst. Veg.* vi. 22; *Spreng. Syst. Veg.* i. 838; *G. Don, Gen. Syst.* iv. 116; *Dietr. Syn. Pl.* ii. 885; *Decne in DC. Prodr.* viii. 658. *Caruncularia aperta, Sweet, Hort. Brit. ed.* 2, 359. *Orbea?* *aperta, Sweet, Hort. Brit. ed.* i. 277.

WESTERN REGION: Little Namaqualand; near Kokfontein, south of Zwart Lintjes River, *Masson*, Kourkam, *Rich in Herb. Pillans*, 156! and without precise locality, *Barkly*, 19!

In its stems and very long erect pedicels this plant closely resembles *Stapelia pedunculata*, Masson, but the corona is quite different. Described from living stems and flowers preserved in fluid.

XLV. TRICHOCAULON, N. E. Br.

Calyx 5-partite. *Corolla* with a short tube or the united part flattish or saucer-shaped, rarely with a raised ring on the disk, 5-lobed; lobes valvate in bud. *Corona* double, or sometimes the outer and inner corona apparently belonging to one series of 5 shortly 3-toothed segments opposite the anthers; outer corona of 5 bipartite or bifid or emarginate lobes, with straight or divergent-arcuate segments; inner corona-lobes incumbent upon the backs of the anthers and sometimes produced beyond them, dorsally connected at the base with the outer corona and sometimes produced into a short tooth there. *Staminal column* arising from the bottom of the corolla, short; anthers inflexed upon the dilated part of the style, linear-oblong, without appendages. *Pollen-masses* ascending or subhorizontal, solitary in each anther-cell, pellucid along the inner margin, attached in pairs to the pollen-carriers by very short caudicles. *Style* not exceeding the anthers, truncate at the apex. *Follicles* fusiform, smooth. *Seeds* crowned with a tuft of hairs.

Succulent perennials, with thick cylindric leafless stems, having many vertical series of conical tubercles, tipped with a spine or a stout stiff bristle, or with crowded irregularly or spirally arranged short rounded pointless tubercles; flowers rather small, arising between the tubercles at or towards the top or all over the stems, subsolitary or 2 or more together, successively developed.

DISTRIB. Species 10, one or possibly two of them also occurring in Tropical Africa and another endemic there.

The various species of this genus are known by the name of *Guaap* and eaten by the Hottentot and Bushman races, and also used medicinally. I have also been informed by Sir Henry Barkly and Mr. Pillans that the stems, after having had their spine-tipped tubercles cut off, are sometimes preserved in sugar-syrup, and are very pleasant to eat.

* Stem-tubercles tipped with a stiff bristle or slender
    spine ; plants 5–24 in. high :
    Corolla with a very prominent raised ring on the
        disk, very dark purple-brown ...   .   (1) **annulatum.**
    Corolla without a (in *T. piliferum* with a very
        slightly) raised ring on the disk :
    † Corolla with a distinct cup-shaped or campanulate
        tube containing the corona :
        Stem-tubercles tipped with a rather stiff spine 3–5
            lin. long ; corolla yellow   ...   .   (4) **Alstoni.**
        Stem-tubercles tipped with spine-like bristles 2–3
            lin. long :
        Corolla dark purple-brown ; plants 5–8 in. high :
            Corolla-lobes 2–3 lin. long ; outer corona-lobes
                very spreading, with the divergent teeth
                arranged in pairs behind the inner corona-
                lobes   ...   .   (2) **piliferum.**
            Corolla-lobes 1½ lin. long ; outer corona of 5
                erect deeply bifid lobes or pairs of teeth
                alternating with the inner corona-lobes   (3) **rusticum.**
        Corolla yellow :
            Plant 5–13 in. high ; inner corona-lobes about
                half as long as the anthers   ...   ...   (5) **Pillansii.**
            Plant 1½–2 ft. high ; inner corona-lobes much
                longer than the anthers, with connivent-
                erect tips ...   .   (6) **grande.**
   †† Corolla without a distinct tube, the united part
        flattish or saucer-shaped ; plants 5–6 in. high :
        Corolla dull yellow, smooth or minutely papillate
            on the inner surface ; outer corona of 5 pairs
            of spreading falcately curved teeth behind the
            inner corona-lobes   ...   .   (7) **flavum.**
        Corolla purple-brown on the lobes, yellow around
            the corona, finely puberulous on the inner
            surface ; outer corona of 5 erect 3-toothed
            lobes behind the inner corona-lobes ...   ...   (8) **officinale.**
** Stem-tubercles short, broad, pointless or tipped with a
    minute point seated in a small depression or "stem
    smooth, only a little wrinkled" ; plants 2–4 in.
    high :
    Corolla without a distinct tube, the united part
        flattish or saucer-shaped, dark purple-brown ...   (9) **Marlothii.**
    Corolla with a distinct cup-shaped tube containing
        the corona :
        Corolla pale yellow spotted with crimson ; outer
            corona 10-toothed, teeth in pairs   ...   ...   (10) **cactiforme.**
        Corolla dark purple-brown ; outer corona 15-
            toothed, teeth in 5 groups of 3   ...   ...   (11) **simile.**

    1. **T. annulatum** (N. E. Br.) ; plant 6–18 in. high, branching
at the base ; stems light glaucous-green, 1¼–1¾ in. thick, cylindric,
with 23–30 vertical series of tubercles tipped with stiff light-brown
(darker when young) bristles 2½–3 lin. long ; flowers subsessile in
the grooves between the tubercles ; sepals 1–1¼ lin. long, subulate-
acuminate from a broadly ovate base, glabrous ; corolla about

10 lin. in diam., rotate, lobed to less than half-way down, with a
very prominent raised ring forming a cup enclosing the corona on
the disk, otherwise without a distinct tube, smooth and glabrous on
the back, densely covered with conical papillæ all over the very
dark purple-brown inner surface, except at the bottom of the cup
around the corona, most of them with a minute hair directed at a
right angle from their apex; lobes $2\frac{1}{2}$ lin. long, 4–$4\frac{1}{2}$ lin. broad,
very spreading, very broadly deltoid-ovate, shortly cuspidate-acute,
with recurved margins ; outer corona large, nearly 2 lin. high, rising
almost to the level of the rim of the cup, glabrous, very dark purple-
brown, cup-like at the base, with 5 rather broad lobes, divided
above their erect concave basal part into two sublanceolate
diverging and obliquely recurved-spreading teeth $\frac{2}{3}$ lin. long and
$\frac{1}{3}$ lin. broad ; inner corona-lobes about $\frac{1}{2}$ lin. long, linear or deltoid-
linear, obtuse, dark-purple brown, closely incumbent on the backs
of the anthers and equalling or slightly exceeding and incurved
over their tips.

2. **T. piliferum** (N. E. Br. in Journ. Linn. Soc. xvii. 164, t. 11,
fig. 1) ; plant 6–8 in. high, branching at the base ; stems cylindric
or slightly clavate-cylindric, $1\frac{1}{2}$–2 in. thick, with about 25 vertical
series of crowded tubercles, each ending in a stiff brown bristle
2–3 lin. long, dull-dark-green, somewhat glaucous ; flowers 1–3
together in the grooves between the tubercles ; pedicels about $\frac{3}{4}$ lin.
long, glabrous ; sepals 1–$1\frac{1}{2}$ lin. long, subulate-acuminate from an
ovate base, glabrous ; corolla in bud somewhat flattened at the top
and very abruptly cuspidate, when expanded $\frac{1}{2}$–$\frac{2}{3}$ in. in diam.,
glabrous and smooth on the back, densely papillate-tuberculate all
over the limb and lobes on the inner face, not ciliate, dark purple-
brown ; tube about 1 lin. long, cup-like, circular or subpentagonal
at the mouth ; limb spreading, lobed to $\frac{3}{4}$ of the way down, slightly
raised around the mouth of the tube ; lobes 2–3 lin. long and as
much in breadth, broadly deltoid-ovate, cuspidate-acuminate into a
short subulate point ; outer corona apparently consisting of 5
spreading deeply bifid (rarely entire) oblong, ovate or deltoid-ovate
acute lobes about $\frac{2}{3}$ lin. long, opposite the anthers, really formed by
the widely diverging adjacent segments of 2 lobes being connate
behind the inner corona-lobes for nearly half their length at the
base or rarely up to the apex, glabrous, dark purple-brown ; inner
corona-lobes $\frac{1}{4}$ lin. long, linear, obtuse, closely incumbent on the
backs of the anthers and shorter than them, dorsally connected at
the base to the outer corona, dark purple-brown. *Bot. Mag. t.* 6759 ;
K. Schum. in Engl. und Prantl, Pflanzenfam. iv. ii. 275 ; *Schlechter
in Journ. Bot.* 1898, 475. *Stapelia pilifera, Linn. f. Suppl.* 171 ;
Thunb. Prodr. 46 ; *Fl. Cap. ed.* 2, ii. 165, *and ed. Schultes,* 239 ;
Masson, Stap. 17, *t.* 23 ; *Willd. Sp. Pl.* i. 1286 ; *Ait. Hort. Kew.*

ed. 2, ii. 90 ; *Pers. Syn. Pl.* i. 279 ; *Poir. Encycl.* vii. 383, *and in Dict. Sc. Nat.* 1. 393 ; *Haw. Syn. Pl. Succ.* 24 ; *Schultes, Syst.* vi. 24 ; *Spreng. Syst.* i. 839 ; *G. Don, Gen. Syst.* iv. 116 ; *Dietr. Syn. Pl.* ii. 885 ; *Loud. Encycl. Pl.* 200, *fig.* 3296 ; *Decne in DC. Prodr.* viii. 655. *Piaranthus piliferus, Sweet, Hort. Brit. ed.* 2, 359.

SOUTH AFRICA : without locality, *Mund and Maire*, 288 ! and *cultivated specimen* !

COAST REGION : Oudtshoorn Div. ; on the Karoo the other side of Attaquas Kloof, *Thunberg* ! *Masson.*

CENTRAL REGION : Calvinia or Sunderland Div. ; Karoo below the Roggeveld, *Thunberg.*   Prince Albert Div. ; near Weltevrede, *Drège*, 5615 !

3. T. rusticum (N. E. Br.) ; plant about 5 in. high, similar to *T. piliferum ;* stems with 17–20 (or more ?) vertical series of conical tubercles, tipped with a slender dark-brown spine or stiff bristle about 2 lin. long, glaucous-green or purple-tinted ; pedicels about 1 lin. long, glabrous ; sepals $\frac{3}{4}$–1 lin. long, ovate, acuminate, glabrous ; corolla with a distinct campanulate tube about 1 lin. long, glabrous outside and within ; lobes spreading, about $\frac{1}{8}$ in. long and broad, deltoid-ovate, acute, glabrous on the back, with a microscopic pubescence on the inner face, not ciliate, dark purple-brown (*Marloth*) ; outer corona of 5 erect deeply bifid lobes or pairs of deltoid-ovate obtuse teeth about $\frac{1}{4}$ lin. long, alternating with the anthers and rising to about the level of the top of the staminal column ; inner corona-lobes about $\frac{1}{4}$ lin. long, linear-oblong, obtuse, closely incumbent on the backs of and about half as long as the anthers, with a broad dorsal connection with the outer corona at their base.

CENTRAL REGION : Kenhardt Div. ; Kenhardt, *Marloth*, 3764 !

Described from a living stem, and fresh flowers preserved in fluid. The pubescence on the inner surface of the corolla-lobes is exceedingly minute, and invisible when wetted.

4. T. Alstoni (N. E. Br. in Kew Bulletin, 1906, 166) ; plant about 6 in. high, branching at the base ; stems about $1\frac{1}{2}$–$1\frac{3}{4}$ in. thick, excluding the spines, many-angled, glabrous and apparently glaucous ; angles with closely placed tubercles tipped with a stiff spine 3–5 lin. long ; flowers in small fascicles between the angles of the stem ; bracts minute, subulate ; pedicels $1\frac{1}{2}$–2 lin. long, glabrous ; sepals erect, $1\frac{1}{2}$ lin. long, $\frac{2}{3}$ lin. broad, ovate, acuminate, glabrous ; corolla glabrous and smooth inside and out, yellow, acutely pointed in bud ; tube 2–$2\frac{1}{4}$ lin. long and about the same in diam. at the mouth, campanulately funnel-shaped ; lobes 2–$2\frac{1}{4}$ lin. long, $1\frac{1}{2}$–$1\frac{2}{3}$ lin. broad at the base, ovate, very acute or somewhat acuminate, ascending in the dried state, probably spreading when alive ; outer corona cup-shaped at the base, subequally 10-toothed or with 5 pairs of teeth behind the inner corona-lobes, glabrous ; teeth $\frac{1}{3}$–$\frac{2}{5}$ lin. long, deltoid or subulate, obtuse or acute, erect or incurved over the staminal column ; inner corona-lobes $\frac{1}{3}$ lin. long, oblong, obtuse, incumbent on the backs of the anthers and not produced beyond them, dorsally adnate at the base to the outer corona.

WESTERN REGION ; Little Namaqualand ; in stony fields near Namies, 3000 ft., *Alston in MacOwan, Herb. Austr.-Afr.*, 2017 ! and without precise locality, *Wyley*, 79 !

5. **T. Pillansii** (N. E. Br. in Gard. Chron. 1904, xxxv. 242) ; plant similar to *T. piliferum* ; stems 4–10 in. high, $1\frac{1}{4}$–$2\frac{1}{4}$ in. thick, with 25–30 series of tubercles ending in stiff pale brown (turning to grey) bristles $1\frac{1}{2}$–$2\frac{1}{4}$ lin. long, " dull green, hardly glaucous on the young growth, flowering nearly all over " (*Pillans*) ; flowers 1–3 in a fascicle ; pedicels $\frac{1}{2}$–$\frac{3}{4}$ lin. long, $\frac{3}{4}$ lin. thick, glabrous ; sepals 1–$1\frac{1}{4}$ lin. long, ovate, very acuminate, glabrous ; corolla $4\frac{1}{2}$ lin. in diam., bright canary-yellow, pinkish-green outside (*Pillans*), glabrous and smooth outside and within the tube, densely papillate and thinly and very minutely puberulous (always ?) on the inner face of the lobes, not ciliate ; tube distinctly cup-shaped, or campanulate, 1–$1\frac{1}{3}$ lin. long, $1\frac{1}{2}$ lin. in diam. at the mouth inside ; lobes spreading, $1\frac{1}{3}$–$1\frac{1}{2}$ lin. long, $1\frac{1}{2}$–$1\frac{3}{4}$ lin. broad, broadly deltoid-ovate, very acute ; outer corona shortly cupular at the base, equally 10-toothed, light yellow ; teeth $\frac{1}{4}$–$\frac{1}{3}$ lin. long, deltoid, acute or obtuse, ascending-spreading, quite straight and equidistant, not approaching each other in pairs ; inner corona-lobes $\frac{1}{4}$ lin. long, linear-oblong, obtuse, about half as long as the anthers and closely incumbent upon their backs, dorsally connected to the outer corona, light yellow ; follicles only developed upon the young growth (*Pillans*), erect, subparallel, $3\frac{1}{4}$ in. long and $3\frac{1}{2}$–4 lin. thick in the specimen seen, fusiform, tapering into a beak, incurved-hooked at the apex, smooth, glabrous ; seeds about 2 lin. long, scarcely 1 lin. broad, ovate or oblong-ovate, with thick margins, often incurved on one side, smooth, glabrous, light brown, with paler margins.

VAR. **major** (N. E. Br. in Gard. Chron. 1904, xxxv. 242) ; stems 9–13 in. high, $1\frac{1}{2}$–3 in. thick, with about 30 series of tubercles, tipped with brown bristles $\frac{1}{3}$–$\frac{1}{4}$ in. long (*Pillans*) ; corolla-tube $1\frac{1}{2}$ lin. long, about 2 lin. in diam. at the mouth inside ; lobes 2–$2\frac{1}{2}$ lin. long and about as broad ; teeth of the outer corona about $\frac{1}{2}$ lin. long, in slightly diverging pairs ; otherwise as in the type.

CENTRAL REGION ; Laingsburg Div. ; 6–8 miles west of Laingsburg, at Zout Kloof Farm, *Pillans*, 9 ! VAR. *β* : Laingsburg Div. ; western side of a hill southeast of Zout Kloof Farm, *Pillans*, 160 !

6. **T. grande** (N. E. Br.) ; plant $1\frac{1}{2}$–2 ft. high, branching at about 4 in. above the ground ; stems about 2 in. thick, with 30 or more vertical series of tubercles, each tipped with a very stiff pale brown (when young dull purplish) bristle or slender spine $2\frac{1}{2}$–3 lin. long, glabrous, glaucous-green ; flowers 1–3 together, developing successively ; pedicels $\frac{3}{4}$–1 lin. long, glabrous ; sepals $\frac{3}{4}$–1 lin. long, ovate-lanceolate, very acuminate, glabrous ; corolla 7–8 lin. in diam., glabrous and smooth outside and within the tube, densely papillate on the disk and lobes of the inner face, not ciliate, greenish-yellow ;

tube campanulate or cup-shaped, about 1½ lin. long and as much in diam. at the mouth, which is distinctly pentagonal and has a slight inwardly projecting rim; lobes spreading, with recurved tips, 2½–3 lin. long, 2½ lin. broad, very broadly ovate, shortly and somewhat abruptly acuminate; outer corona shortly cupular at the base, divided into 10 narrowly deltoid-subulate acute teeth ⅔ lin. long, which are straight, subequidistant, ascending-spreading and rising to the mouth of the corolla-tube, yellow; inner corona-lobes about ⅔ lin. long, subulate, incumbent upon the backs of the anthers at the base, produced beyond them for half their length in a connivent-erect column rising to the mouth of the corolla-tube.

CENTRAL REGION : Laingsburg Div. ; along the main road to Ladismith, about 12 miles south of Laingsburg, *Pillans,* 668 !

Partly described from fresh flowers preserved in fluid. This is the tallest species of the genus at present known.

7. **T. flavum** (N. E. Br. in Journ. Linn. Soc. xvii. 165, t. 11, fig. 2–4); plant about 6 in. high, with stems just like those of *S. piliferum* in size and other characters, but the slender spines terminating the tubercles are of a darker brown; pedicels ½–¾ lin. long, glabrous; sepals 1–1¼ lin. long, tapering from the ½ lin.-broad base to the very acute apex, glabrous; corolla about ½ in. in diam., lobed to half-way down, glabrous on both sides and not ciliate, smooth outside, minutely papillate or smooth on the inner face, dull yellow; tube none, the united part flat; lobes ascending-spreading, with recurved tips, about 2 lin. long and 2½ lin. broad at the base, broadly deltoid-ovate, rather abruptly or shortly subcuspidate-acuminate; outer corona-lobes 1 lin. long, very spreading or slightly recurving, shortly connate at the base, deeply divided into 2 linear diverging segments, which approach those of the adjacent lobes, forming pairs of teeth somewhat resembling the mandibles of a beetle, yellow, glabrous; inner corona-lobes scarcely ½ lin. long, linear, obtuse, closely incumbent upon the backs of the anthers and not exceeding them, dorsally adnate at the base to the outer corona, yellow, glabrous. *N. E. Br. in Hook. Ic. Pl. under t.* 1905; *K. Schum. in Engl. und Prantl, Pflanzenfam.* iv. ii. 275 ; *Schlechter in Journ. Bot.* 1898, 475.

CENTRAL REGION: Beaufort West Div. ; Rhenoster Kop, *Foster in Herb. Pillans,* 171 ! Prince Albert Div. ; near Prince Albert, *Marloth,* 4385 ! Div. ? Karoo, *Bain* !

Partly described from fresh flowers preserved in fluid ; those of the type are very minutely papillate on the inner surface, whilst in Mr. Pillans's specimens they are smooth.

8. **T. officinale** (N. E. Br. in Kew Bulletin, 1895, 264); stems about 5 in. high, like those of *T. piliferum,* but apparently with fewer vertical series of spine-tippped tubercles ; flowers 1–2 or perhaps more together between the tubercles on the sides of the stems ; pedicels ½–1 lin. long, glabrous; sepals 1–1⅔ lin. long, ovate

or ovate-lanceolate, acuminate, glabrous ; corolla 5–6 lin. in diam.,
rotate, without a distinct tube, glabrous and smooth outside, finely
to very minutely puberulous all over the inner surface, entirely dark
purple-brown, or with the part round the corona yellow or orange
nearly up to the sinuses (*Marloth*) ; united part flattish or saucer-
shaped ; lobes spreading or ascending-spreading, with recurved tips,
1¾–2 lin. long, 1½–2 lin. broad at the base, deltoid-ovate, acuminate,
outer corona arising near the base of the staminal column and attaining to about the same level, forming 5 very short entire or notched
pouches alternating with the anthers and rising into 5 erect subquadrate or subquadrate-ovate lobes behind the inner corona-lobes,
obtusely 3-toothed at the top, with the middle tooth inflexed upon
and adnate to the base of the inner corona-lobes, which are $\frac{1}{6}$–$\frac{1}{4}$ lin.
long, linear-oblong, obtuse, incumbent upon the backs of the
anthers and about half as long as them. *Schlechter in Journ. Bot.*
1898, 475 ; *N. E. Br. in Dyer, Fl. Trop. Afr.* iv. i. 489.

CENTRAL REGION : Prieska Div. ; near Prieska, *Marloth* !

KALAHARI REGION : Bechuanaland, ex *Holmes* ! Griqualand West ; Asbestos
Mountains, *Marloth*, 3773 !

I originally described this plant from some transverse slices of the stem with
flowers attached, which, dried and threaded on a string, were imported into
America from Bechuanaland as a remedy for piles, and were presented to the Kew
Herbarium by Mr. E. M. Holmes, of the Pharmaceutical Society, in 1889. Of the
plant from the Asbestos Mountains, fresh flowers preserved in formalin were sent
to Kew by Dr. Marloth, who informs me that the outer corona is of an orange
colour, and the corona of the type appears to have been of a similar colour, and
like that of Dr. Marloth's specimen is glabrous, whilst the Prieska specimen (also
received in formalin) has the outer corona minutely puberulous on the back,
and according to a coloured drawing of it is purplish.

9. T. **Marlothii** (N. E. Br.) ; plant 3–4 in. high, with 3–4
cylindric or cylindric-ovoid branches 1½–1¾ in. thick, narrowed
below, obtuse, covered with crowded obtuse tubercles 2–3 lin. in
diam., tipped with a minute point seated in a small depression,
glabrous, pale greyish ; flowers 2–5 together between the tubercles,
subsessile or with pedicels up to ½ lin. long, developing successively ;
sepals ½ lin long, ½–⅔ lin. broad, broadly ovate, acute, glabrous ;
corolla flat-topped and very shortly pointed in bud, when expanded
about 4 lin. in diam., lobed to half-way down, without a distinct
tube, the united part flattish or saucer-shaped, glabrous and smooth
outside and within, not ciliate ; lobes spreading, 1⅓ lin. long and
broad, nearly like an equilateral triangle in outline, acute, dark
purple-brown (*Marloth*) ; outer corona divided into 10 filiform-
subulate teeth ½ lin. long, arranged in pairs behind the inner
corona-lobes, very spreading, curving towards each other ; inner
corona-lobes ½ lin. long, linear, obtuse, closely incumbent on the
backs of the anthers and produced beyond them, with the tips
probably connivent-erect.

CENTRAL REGION : Kenhardt Div. ; near Kenhardt, *Marloth*, 3763 !

10. **T. cactiforme** (N. E. Br. in Hook. Ic. Pl. under t. 1905, by error *cactiformis*); plant 2–4 in. high, simple or branching at the base into 2–5 cylindric or clavate stems $1\frac{1}{2}$–2 in. thick, covered with crowded slightly prominent obtusely rounded pointless tubercles, glabrous, whitish-green, flowering at the very obtuse apex; pedicels $\frac{1}{2}$ lin. long, glabrous; sepals $\frac{3}{4}$ lin. long, ovate, acute, glabrous; corolla $\frac{1}{3}$ in. or more in diam., glabrous outside and within, not ciliate, papillate-rugulose on the inner surface, pale yellow spotted with blood-red; tube distinct, shallowly cup-shaped; lobes spreading, about $1\frac{1}{2}$ lin. long and as much or more in breadth at the base, very broadly ovate or deltoid-ovate, abruptly and shortly acuminate; outer corona-lobes ascending-spreading, $\frac{2}{5}$ lin. long, divided nearly to the base into 2 linear-falcate widely diverging segments which form mandible-like pairs of teeth with those of the adjacent lobes behind the inner corona-lobes, yellow, dotted with blood-red; inner corona-lobes $\frac{1}{4}$ lin. long, linear, obtuse, closely incumbent upon the backs of the anthers and not produced beyond them, dorsally connected at the base to the outer corona and there produced into a short suberect tooth; follicles diverging, slightly incurved, $1\frac{3}{4}$–2 in. long, 3–$3\frac{1}{2}$ lin. thick, terete-fusiform, acute, glabrous. *K. Schum. in Engl. und Prantl, Pflanzenfam.* iv. ii. 275; *Schlechter in Journ. Bot.* 1898, 475. *Stapelia clavata, Willd. Sp. Pl.* i. 1295; *Pers. Syn. Pl.* i. 280; *Poir. Encycl. Meth.* vii. 391; *Schultes, Syst. Veg.* vi. 49; *G. Don, Gen. Syst.* iv. 117. *S. cactiformis, Hook. Bot. Mag. t.* 4127; *Loud. Encycl. Pl.* 1330, *fig.* 18780; *Lemaire, Fl. des Serres,* i. 51, *t.* 20, *and Hort. Univ.* vi. 331, *with plate.*—*Stapelia, Paterson, Narrative of four journeys into the country of the Hottentots,* 60, *t.* 7.

Western Region: Little Namaqualand; between Koper Berg and Kookfontein, *Drège*, 6399! near Ookiep, ex *Pillans*, and without precise locality, *Barkly*, 37! Little Bushmanland (Gezelschaps Bank?), *Paterson*!

Also in Tropical German South-west Africa. This remarkable plant, so far as known to me, has only flowered in Europe once, in 1844, when it was figured in the *Botanical Magazine* from a plant sent by Zeyher from Little Namaqualand, all the other figures, except Paterson's, are copied from that one.

11. **T. simile** (N. E. Br.); stem not seen, but according to a sketch resembling that of *T. cactiforme*, oblong or globose-obovoid and up to $1\frac{3}{4}$ in. high and $1\frac{1}{2}$ in. thick, "quite smooth, only a little wrinkled" (*Marloth*), flowering at the top; pedicels $1\frac{1}{2}$ lin. long, glabrous; sepals $\frac{1}{2}$–$\frac{2}{3}$ lin. long and quite as broad, broadly ovate, acute, glabrous; corolla $4\frac{1}{2}$ lin. in diam., quite glabrous and not ciliate on the lobes, smooth outside, minutely papillate-rugulose on the inner face of the lobes, "deep purple-brown" (*Marloth*); tube $1\frac{1}{3}$ lin. long, $2\frac{2}{3}$ lin. in diam. outside, cup-shaped; lobes spreading, $1\frac{1}{2}$ lin. long and as much in breadth, triangular in outline, acute; outer corona shortly cupular at the base, acutely 15-toothed, with the teeth arranged in 5 groups of 3 behind the inner corona-lobes and the lateral teeth longer than and arching over the middle tooth,

suberect, rising to the level of the inner corona, $\frac{3}{4}$ lin. long; inner corona-lobes $\frac{1}{2}$ lin. long, flat, oblong-linear, obtuse, ascending-connivent, closely pressed to the backs of the anthers and shortly exceeding them, dorsally connected at the base to the cup of the outer corona.

COAST REGION: Van Rhynsdorp Div. ; near Van Rhynsdorp, *Marloth,* 4571 !

Of this I have only seen a single flower in fluid and a rough outline sketch of the plant.

XLVI. HOODIA, Sweet.

Calyx 5-partite. *Corolla* large, flat, concave or cup-shaped, with a very small proper tube, just enclosing the corona; limb obsoletely or but slightly 5-lobed; lobes abruptly terminated by a subulate point, valvate in bud. *Corona* double, arising from the staminal column; outer corona of 5 concave spreading lobes, emarginate or bifid at the apex; inner corona of 5 linear lobes incumbent upon the backs of the anthers, dorsally connected to the outer corona by 5 short partitions. *Staminal column* arising from near the base of the small corolla-tube; anthers incumbent on the top of the style, subquadrate, without an appendage at the apex. *Style* not produced beyond the anthers, truncate at the apex. *Follicles* elongated, terete-fusiform, solitary or in pairs, divaricate, smooth.

Stout succulent perennials, bushily branched, leafless; stems very thick, cylindric, with many tuberculate angles; tubercles spine- or bristle-toothed; flowers large and showy, in small clusters of 2–5 or sometimes solitary, arising from the grooves between the angles of the stem towards or at the top, developing successively.

DISTRIB. Species 10, the others in Tropical Africa.

Corolla hairy all over the inner surface, in dried flowers
    less than 2 in. in diam. ... ... (1) **Dregei.**

Corolla entirely without hairs :
  Corolla slightly rough (sprinkled with papillæ) on the
    centre of the inner surface :
    Corolla 3–4 in. in diam. :
      Plant 1–1½ ft. high, with light-brown spines;
        corolla very slightly concave or nearly flat,
        with or without revolute margins ... ... (6) **Gordoni.**

      Plant of unknown height, with white spines;
        corolla broadly cup-shaped ... ... (7) **albispina.**

    Corolla 2¼ in. in diam., apparently concave; plant
      5–6 in. high ... ... (4) **Pillansii.**

  Corolla smooth, without papillæ on the central area
    within :
    Corolla 3½–4 in. in diam., apparently slightly
      concave or flattish ... ... (5) **Burkei.**

    Corolla 2–3 in. in diam., distinctly cup-shaped :
      Outer corona-lobes deeply bifid, with deltoid
        acute teeth ... ... (2) **Barklyi.**
      Outer corona-lobes emarginate ... ... (3) **Bainii.**

1. **H. Dregei** (N. E. Br.); stems 1½ in. or more in diam., with 20–24 tuberculate ribs, glabrous; tubercles conical, tipped with a stiff bristle 2½–3 lin. long; pedicels short, becoming about 5 lin. long in fruit, glabrous; sepals 1½–2 lin. long, lanceolate-subulate, glabrous; corolla of the dried specimen 1¼–1⅓ in. in diam., probably cup- or saucer-shaped, slightly and broadly 5-lobed, glabrous outside, thickly covered all over inside with simple hairs ⅔–1 lin. long; lobes obtuse, tipped with a subulate point 1¼–1½ lin. long; outer corona cup-shaped, with 5 short transversely oblong emarginate or shortly and broadly bifid lobes, blackish-purple; inner corona-lobes ½ lin. long, linear, obtuse, incumbent on the backs of the anthers and not produced beyond them, dorsally adnate to the outer corona, blackish-purple.

South Africa : without locality, *Drège,* 5616 !

2. **H. Barklyi** (Dyer in Journ. Linn. Soc. xv. 252, t. 5, fig. 3); plant very much branched, probably not more than 1 ft. high; "main branches nearly 2 in. in diam., branchlets about 1 in. in diam., greyish-green" (*Barkly*), in the specimen seen with about 16 angles or series of tubercles, each tipped with a slender brown spine ¼–⅓ in. long; pedicels " not above ⅛ in. long, stout " (*Barkly*); " calyx very small, no bigger than the corona above it " (*Barkly*); corolla in bud 5-winged at the upper half, truncate with a short central point, when expanded about 2 in. in diam., cup-shaped, ¾ in. deep, truncate at the slightly 10-angled margin, with 5 slender points about 1 lin. long, glabrous and smooth outside and within, dull yellow; outer corona-lobes deeply bifid, forming 10 equidistant deltoid acute teeth, "nearly black" (*Barkly*); inner corona " a raised star of 5 strap-shaped rays of a light brown colour" (*Barkly*); lobes linear, subacute, broadened at the base and there dorsally connected with the inflexed sinuses of the outer corona. *N. E. Br. in Hook. Ic. Pl. under t.* 1905, *p.* 3 ; *K. Schum. in Engl. und Prantl, Pflanzenfam.* iv. ii. 275 ; *Schlechter in Journ. Bot.* 1898, 475.

Central Region : Worcester or Prince Albert Div. ; Karoo, *Lycett* (*Barkly,* 5) !

The habitat of this species is unknown. A living plant of it was brought from the Karoo and sent by Mr. Lycett of Worcester in 1873 to the Botanic Garden at Cape Town, where it flowered in 1874, and a drawing, branch and one flower were sent to Kew by Sir Henry Barkly, with the above information and a description. The calyx and corona belonging to the flower were unfortunately missing.

3. **H. Bainii** (Dyer in Bot. Mag. t. 6348); plant 6–8 in. high in the specimens seen (12–15 in., *Barkly*), bushily branched ; branches 1–1½ in. thick, with 12–15 tuberculate angles, glabrous, green, somewhat glaucous ; tubercles tipped with a slender pale brown spine 3½–5 lin. long ; flowers 1–2 together, glabrous in all parts ; pedicels ¼–½ in. long ; sepals 2–2½ lin. long, ovate-lanceolate, acuminate ; corolla in bud hemispheric at the basal part, 5-winged above, truncate, with a short central point, when expanded 2½–3 in. in diam.,

cup-shaped, about 1 in. deep, subtruncate at the margin with 5 subulate or awn-like points $1\frac{1}{2}$–3 lin. long, glabrous, smooth, not papillate on the central part, light yellow or pale buff, sometimes tinged with pinkish or very pale purple; tube obsolete, represented by a slight depression from which the blackish corona is exserted or its margin resting upon the rim, when dried contained in a very small cup; outer corona $1\frac{3}{4}$–2 lin. in diam., cupular, 5-lobed; lobes $\frac{1}{4}$–$\frac{1}{3}$ lin. long, nearly 1 lin. broad, transverse, emarginate; inner corona-lobes $\frac{2}{5}$ lin. long, oblong, obtuse, closely incumbent upon the backs of the anthers and not exceeding them, dorsally connected to the inflexed sinuses of the outer corona; follicles 4–5 in. long, 4–5 lin. thick, terete-fusiform, tapering to a beak, glabrous, smooth; seeds 3–$3\frac{1}{2}$ lin. long, $1\frac{1}{2}$ lin. broad, ovate, flat, with a slightly thickened margin, glabrous, smooth, light brown. *N. E. Br. in Hook. Ic. Pl. under t.* 1905, *p.* 3; *K. Schum. in Engl. und Prantl, Pflanzenfam.* iv. ii. 275; *Schlechter in Journ. Bot.* 1898, 475.

South Africa : without locality, *Drège,* 5617 !

Central Region : Prince Albert or Beaufort West Div. ; Dwyka River and Uitkyk, *Bain,* 11 ! and *cultivated specimens* ! near Grootfontein, *Pillans* ! Carnarvon Div. ; near Van Wyks Vlei, *Alston in Herb. Pillans,* 127 !

According to Mr. Bain this is called "Wolve n'Gaap" by the Hottentots. Both Sir Henry Barkly and Mr. Pillans state that this plant is 12–15 in. high, yet neither of the 4 living plants sent to England were more than half as tall, and Mr. Pillans speaks of it as "a stunted species." Have two closely allied forms been confused? Specimens collected by Orpen at St. Clair near Douglas in Herbert Div., and distributed by MacOwan under no. 3397, may possibly belong to this species, but living material is needed to properly identify the plant.

4. **H. Pillansii** (N. E. Br.); plant about 5–6 in. high, bushily branched; branches 1–$1\frac{1}{2}$ in. thick, with 15–18 tuberculate angles, whitish or glaucous-green; tubercles tipped with a slender brown spine 3–$4\frac{1}{2}$ lin. long; flowers 1–2 together; pedicels not seen; sepals 2–$2\frac{1}{2}$ lin. long, ovate, acuminate, glabrous; corolla about $2\frac{1}{2}$ in. in diam., apparently concave, with the margin obscurely pentagonal, with 5 awn-like points $1\frac{1}{2}$–2 lin. long, rough with small papillæ on the central part within, elsewhere smooth and glabrous, "salmon-coloured, with the centre a pretty peach-colour" (*Pillans*); tube in the dried flowers a small cup about 2 lin. in diam., containing the corona, which in living flowers probably rises to the level of the mouth or beyond; outer corona cupular, 5-lobed, with the lobes about $\frac{1}{4}$ lin. long and $\frac{2}{3}$ lin. broad, truncate, but appearing emarginate in dried flowers from the central part being slightly doubled inwards in the direction of their length, apparently dark purple-brown, as also the $\frac{2}{5}$ lin.-long oblong-linear obtuse inner corona-lobes, which shortly exceed the anthers and slightly cross each other at the tips.

Central Region : Prince Albert Div. ; Grootfontein, *Pillans,* 164 !

Described from living stems and dried flowers.

5. **H. Burkei** (N. E. Br.); plant more than 1 ft. high, resembling *H. Gordoni*; tubercles of the stems tipped with a slender brown spine $\frac{1}{4}$–$\frac{1}{3}$ in. long; pedicels 4–8 lin. long, glabrous; sepals $2\frac{1}{2}$–3 lin. long, ovate-lanceolate, acuminate, glabrous; corolla $3\frac{1}{2}$–4 in. in diam., slightly concave or flattish, subcircular, with the lobes very obscurely indicated by 5 very slight emarginations alternating with 5 awn-like points $\frac{1}{4}$ in. long, glabrous and smooth all over, not roughly papillate on the central area; corona equalling or perhaps slightly exserted from the small cup-like tube, apparently dark purple-brown; outer corona cupular, 5-lobed; lobes $\frac{1}{3}$ lin. long, $\frac{3}{4}$ lin. broad, transverse, very broadly notched to about half-way down; inner corona-lobes $\frac{1}{2}$–$\frac{2}{3}$ lin. long, oblong-linear, obtuse, closely incumbent upon the backs of the anthers, broader than and exceeding them and meeting or crossing at their tips, dorsally connected to the inflexed sinuses of the outer corona. *Stapelia Gordoni, Hook. in Lond. Journ. Bot. 1843, 164, and in Ic. Pl. under tt. 605–606, not of Masson. Scytanthus Gordoni, Hook. Ic. Pl. t. 625, and in Lond. Journ. Bot. 1846, 111.*

CENTRAL REGION: Beaufort West Div.; near the Gamka River, *Burke,* 464! Prince Albert Div.; Willow Fountain (Wilgefontein), *Burke,* 463! without locality (but probably collected at one of the above places when travelling with Burke), *Zeyher,* 1142! and *Zeyher,* 1144, in fruit only, is probably this species!

This species has hitherto been confused with *H. Gordoni,* but I find that it conspicuously differs from that species by the centre of the corolla being smooth, not roughly papillate as in *H. Gordoni,* the spines on the stems seem also to be more slender and of a different brown. Although Burke's localities are correctly cited in the *London Journal of Botany,* 1846, 111, yet it is erroneously stated in *Icones Plantarum,* under tt. 605–606, that Burke discovered this plant on the banks of the Orange River. But the point where Burke and Zeyher crossed the Orange River (according to Burke's diary) is over 450 miles from the localities where Burke collected it on the return journey from the Transvaal, nearly 600 miles from the home of the true *H. Gordoni,* and where no *Hoodias* have yet been discovered.

6. **H. Gordoni** (Sweet, Hort. Brit. ed. 2, 359); plant 1–$1\frac{1}{2}$ ft. high, with erect branches about 2 in. thick, glaucous-green, with the numerous angles beset with slender light brown spines 3–$4\frac{1}{2}$ lin. long; pedicels $\frac{1}{2}$–$\frac{3}{4}$ in. long, glabrous; sepals $2\frac{1}{2}$–3 lin. long, ovate-lanceolate, acuminate, glabrous; corolla in bud somewhat resembling a narrow pentagonal cone, with 5 very broad wings descending from the short central point to the base, truncate at the top and slightly hooked at the outer angles; when expanded 3–4 in. in diam., sub-circular with 5 very broad crenations, each very abruptly tipped with a slender arista-like point $2\frac{1}{2}$–3 lin. long, very slightly concave or nearly flat, with or without revolute margins, pale purple, radiately marked with pale greenish-yellow stripes along the veins, thickly sprinkled on the central part with minute dark red papillæ, elsewhere quite glabrous, but with a somewhat velvet-like appearance; tube very small, about $\frac{1}{4}$ in. in diam., just containing the corona, slightly raised around the mouth; outer corona-lobes

ascending-spreading, somewhat pouch-like at the base, about $\frac{1}{3}$ lin. long and $\frac{2}{3}$ lin. broad, transversely subrectangular, emarginate or shortly and obtusely bifid at the apex, with inflexed sinuses between them, purple-black, with a shining linear space down the centre of each; inner corona-lobes $\frac{1}{2}$–$\frac{2}{3}$ lin. long, linear, obtuse or truncate, incumbent upon the backs of the anthers and not exceeding them, dorsally connected to the outer corona at the inflexed sinuses between the lobes, purple-black; follicles $2\frac{1}{2}$–3 in. long, terete, fusiform, tapering to an acute beak, slightly hooked at the apex, smooth, glabrous; seeds not seen. *Decne in DC. Prodr.* viii. 665; *N. E. Br. in Gard. Chron.* 1875, iv. 452, *and in Hook. Ic. Pl. under t.* 1905, *p.* 3; *Dyer in Bot. Mag. t.* 6228 (*excl. analyses*), *and Journ. Linn. Soc.* xv. 252, *t.* 5, *fig.* 1; *Gard. and Forest,* x. 75; *K. Schum. in Monatsschr. Kakt.* iii. 57, *fig. excl. analyses; Schlechter in Journ. Bot.* 1898, 475. *H. Gordonii, K. Schum. in Engl und Prantl, Pflanzenfam.* iv. ii. 275, *fig.* 82, *excl. analyses. Stapelia Gordoni, Masson, Stap.* 24, *t.* 40; *Willd. Sp. Pl.* i. 1285; *Pers. Syn. Pl.* i. 279; *Poir. Encycl.* vii. 383; *Kerner, Hort. t.* 154; *Haw. Syn. Pl. Succ.* 25; *Schultes, Syst. Veg.* vi. 25. *S. Gordonii, Spreng. Syst. Veg.* i. 840; *Dietr. Syn. Pl.* ii. 886. *Gonostemon? Gordoni, Sweet, Hort. Brit. ed.* 1, 278. *Monothylaceum Gordoni, G. Don, Gen. Syst.* iv. 116.

WESTERN REGION : Little Namaqualand; near Henkries, 12 miles south of the Orange River, *Barkly*! also *cultivated specimen*! Great Namaqualand; towards the Orange River, *Gordon* (ex *Masson*).

Described from a living plant, sent by Sir Henry Barkly in 1874 from Little Namaqualand to Kew, where it flowered in August, 1875. The buds are wrongly represented in the *Botanical Magazine* from an immature example. Those on Masson's plate, although rough, are more nearly correct, for as they mature, the parts of the corolla corresponding to the midribs of the lobes are folded in close to the centre so that the bud appears to consist of 5 broad wings, truncate at the top.

The specimens collected by Burke, hitherto referred to this species, belong to *H. Burkei,* N. E. Br.

7. **H. albispina** (N. E. Br.); size of plant unknown, only one dried branch seen, $4\frac{1}{2}$ in. long, and $\frac{3}{4}$ in. thick, but evidently much shrunk in drying, with about 15 tuberculate angles; tubercles tipped with a white spine $2\frac{1}{2}$–3 lin. long; pedicels very short, 2 lin. long in the specimen seen; sepals 2–3 lin. long, attenuate from a 1 lin.-broad base, glabrous; corolla 3–$3\frac{1}{2}$ in. in diam., broadly cup-shaped, about 1 in. deep, with 5 very broad and slight crenations, each abruptly tipped with a subulate or awn-like point $1\frac{1}{2}$–2 lin. long, sprinkled and roughened on the central area within with small papillæ, otherwise glabrous and smooth; corona apparently slightly exserted from the very small tube, very dark purple-brown; outer corona cupular, with 5 transverse broadly emarginate or notched lobes $\frac{1}{3}$–$\frac{1}{2}$ lin. long and 1 lin. broad; inner corona-lobes scarcely $\frac{1}{2}$ lin. long, oblong, obtuse, not exceeding the anthers and not nearly meeting at the tips.

CENTRAL REGION : Carnarvon Div.; Van Wyks Vlei, *Alston in Herb. Pillans,* 18 !

XLVII. **TAVARESIA,** Welw.

Calyx 5-partite. *Corolla* large, tubular-funnel-shaped, 5-lobed; lobes valvate in bud, the sinuses between them forming acute projecting angles. *Corona* double, arising from the staminal column; outer corona shortly tubular at the base, then divided into 10 long filiform segments, each terminated by a knob, and usually with a minute tooth between the pairs of segments that alternate with the anthers; inner corona of 5 narrow lobes incumbent upon the backs of the anthers and dorsally connected with the basal part of the outer corona by short partitions. *Staminal column* arising from the base of the corolla; anthers ovate-oblong, without appendages, incumbent upon the top of the style. *Pollen-masses* solitary in each anther-cell, horizontal, rather large, compressed, pellucid-margined along one side, attached in pairs by short caudicles to subulate lateral processes from the pollen-carriers. *Style* not exceeding the anthers, flat and pentagonal at the apex. *Follicles* not seen, but produced in pairs (*Mrs. Barber*), somewhat diverging, narrowly lanceolate-fusiform, smooth. *Seeds* rather small, crowned with a tuft of hairs.

Leafless succulent perennials, branching at the base; stems erect or ascending, 6–12-angled; angles tuberculate-toothed, each tubercle furnished with 3 bristles; flowers 1–4 together at the base of the young stems, successively developed.

DISTRIB. Species 2, both occurring in Tropical Africa, one confined to it.

1. **T. Barklyi** (N. E. Br. in Dyer, Fl. Trop. Afr. iv. i. 494); stems 3–4 in. high, $\frac{2}{3}$–$\frac{3}{4}$ in. thick, cylindric, 10–12-angled, deep green or tinged with purplish; angles with very short closely set tubercles, each terminated by 3 white bristles; central bristle horizontal, the two lateral rather shorter, deflexed and diverging from each other at nearly a right angle; pedicels $\frac{1}{3}$–$\frac{1}{2}$ in. long, glabrous; sepals 3–3$\frac{1}{2}$ lin. long, lanceolate, acuminate, glabrous; corolla outside smooth, pale greenish or greenish white, spotted with purple-red; inside densely papillate, pale yellowish, covered with small (mostly linear in the tube) purple-red spots, except at the base, which is entirely purple-red; tube 1$\frac{1}{2}$–2 in. long, about 1 in. in diam. at the mouth; lobes spreading, $\frac{1}{2}$–$\frac{3}{4}$ in. long and about as broad, deltoid, acuminate; outer corona 4$\frac{1}{2}$–5 lin. long, shortly tubular at the base, divided into 10 filiform segments terminating in pendulous globose knobs, the filiform parts and knobs purple-brown, the lower part white, with the margins, some broad stripes alternating with each pair of filaments and a few linear spots all dark purple-brown; inner corona-lobes $\frac{3}{4}$ lin. long, linear, incumbent upon the backs of the anthers, which they slightly exceed, dorsally connected to the outer corona by short partitions, purple-brown. *Decabelone Barklyi, Dyer in Bot. Mag. t.* 6203, *and in Journ. Linn. Soc.* xv. 249–250, *t.* 5, *fig.* 4; *N. E. Br. in Hook. Ic. Pl. under t.*

1905, *p.* 3 ; *K. Schum. in Engl. und Prantl, Pflanzenfam.* iv. ii. 275 ; *Gard. Chron.* 1900, xxvii. 210, *fig.* 67 ; *Schlechter in Journ. Bot.* 1898, 476.

CENTRAL REGION: Richmond Div. ; Karoo, near Richmond Road Station, *Foster in Herb. Pillans,* 99 ! Colesberg Div. ; near Colesberg, *Shaw* ! *Barkly,* 26 ! Prieska Div. ; Karoo, by the Orange River, near Prieska, *Lichtenstein* !

WESTERN REGION : Little Namaqualand ; between Koper Berg and Kookfontein *Drège,* 6395 !

XLVIII. HUERNIA, R. Br.

Calyx 5-partite.   *Corolla* wholly campanulate or with a short tube more or less constricted at its mouth and a very abruptly or horizontally spreading limb, 5-lobed, with 5 small teeth alternating with the lobes, formed by the projecting sinuses, in one species with the teeth and lobes of nearly equal size, both small and tooth-like, smooth, papillate or beset with simple or clavate hairs or processes on the inner surface.   *Outer corona* spreading upon and more or less adnate to the bottom of the corolla-tube, disk-like and 5- or 10-crenate or toothed, or the disk with 5 distinct lobes, or the lobes free to the base, bifid, emarginate or rarely entire, entirely absent in one species. *Inner corona* arising from the upper part of the staminal column, of 5 simple lobes incumbent upon the backs of the anthers and equalling or exceeding them, subulate or clavate or thickened at the apex, often with a slight transverse dorsal ridge at their base, but no crest, wing or dorsal horn.   *Pollen-masses* solitary in each anther-cell, subhorizontal, pellucid just within the straight margin near the free end, attached in pairs by short triangular caudicles to lateral wing-like expansions of the pollen-carrier.

Perennial succulent dwarf plants ; stems leafless, angular, simply toothed or sometimes with bristle-pointed teeth along the angles, glabrous ; flowers solitary or in small clusters near the base or middle of the young stems, of moderate size.

DISTRIB. Species over 30, one in Arabia, 6 or 7 in Tropical Africa, the remainder in South Africa.

Outer corona none : ...   . (12) **simplex.**
Outer corona well developed :
   Inner corona-lobes wholly incumbent upon the anthers
     and not produced beyond their tips into erect
     points ; corolla-limb very abruptly spreading from
     the tube, saucer-shaped, cupular or subhorizontal :
   Corolla very distinctly raised into a broad smooth
     convex ring around the mouth of the tube.
     (See also 20, *H. venusta,* of which the corona is
     unknown) :
   Corolla-lobes marked with transverse purple-
     brown bars and distinctly pubescent or
     puberulous on the inner surface   ...   ... (23) **zebrina.**
   Corolla-lobes dotted with blood-red and minutely
     papillate on the inner surface, but without
     hairs   ...   . (17) **humilis.**

Corolla not or inconspicuously raised into a convex ring around the mouth of the tube, papillate-scabrous ... ... (16) **scabra,** var. δ.

Inner corona-lobes produced beyond the anther-tips into short or long points, connivent-erect or connivent with diverging or recurving tips:

Tips of the connivent-erect inner corona-lobes enlarged and somewhat resembling an inverted foot ; corolla-lobes recurved-spreading, covered with stout subulate acute processes ... ... (8) **Hystrix.**

Tips of the inner corona-lobes clavate (not foot-like) or linear and slightly thickened, obtuse, erect or recurving. (See also 24, *H. stapelioides*):

Stems with 20–24 vertical series of bristle-pointed tubercles ; corolla covered on the inner face with obtuse terete papillæ ¼–⅓ lin. long ... ... (6) **Pillansii.**

Stems acutely 4–5-angled:

Corolla wholly campanulate, with the tube gradually passing into the lobes or limb, covered with short stiff conical or somewhat flattened processes on the inner surface :

Corolla-tube of living flowers 4 lin. long ; lobes about 3 lin. long and broad ... (9) **loeseneriana.**

Corolla-tube of living flowers 9–10½ lin. long ; lobes about 5 lin. long and broad (10) **longituba.**

Corolla-limb saucer-shaped, lobed to half-way down, very abruptly or horizontally spreading from the globose-campanulate tube, densely covered with subulate processes around the mouth and in the throat of the tube ... ... (21) **Kirkii.**

Tips of the inner corona-lobes subulate, obtuse, acute or finely pointed, not at all thickened :

Stems (excluding the teeth) somewhat obtusely 8–9-angled ; corolla-lobes abruptly recurved-spreading from the tube, covered with hair-pointed papillæ ... ... (7) **distincta.**

Stems acutely 4–5- (rarely 6-) angled (unknown in 4, *H. decemdentata*):

* Corolla wholly campanulate, with the tube gradually passing into the limb or lobes, sometimes slightly constricted at the middle, with long clavate purple hairs on the inner surface :

Corolla subequally 10-toothed or the lobes not greatly larger than the alternating teeth ... ... (4) **decemdentata.**

Corolla-lobes several or many times larger than the alternating teeth :

Corolla spotted with purple outside and with blackish-purple on a whitish or sulphur-yellow ground within, with clavate hairs only at the middle, not nearly extending to the base of the lobes ... ... (1) **campanulata.**

Corolla not spotted outside, spotted with
crimson within :
  Clavate hairs very numerous or crowded,
    extending from below the middle of
    the tube to the base or middle of
    the lobes ; outer corona-lobes sub-
    truncate, emarginate or obtusely
    bifid, rarely 3-crenate :
    Corolla-lobes 6–8 lin. long ; ground-
      colour sulphur-yellow to pale
      buff... ... ... (2) **barbata.**

    Corolla-lobes 3–4 lin. long ; ground-
      colour ochreous or dull greenish-
      yellow ... ... (3) **clavigera.**

  Clavate hairs rather scattered and chiefly
    confined to the middle part of the
    corolla-tube, not nearly extending
    to the base of the lobes ; outer
    corona-lobes acutely bifid ... ... (2) **barbata,** var. γ.

** Corolla-limb or lobes very abruptly or horizontally spreading or recurving from the
  short campanulate or globose-campanulate
  tube :
† Corolla-disk raised into a convex ring often
  broad and very conspicuous around the
  mouth of the tube :
  Corolla with clavate purple hairs within
    and around the mouth of the tube,
    ring smooth or the spots on it slightly
    raised :
    Corolla with or without a minute bristle-
      like pubescence on the lobes ; spots
      large, irregular or confluent, pro-
      ducing a reticulated appearance ;
      outer corona black ... ... (22) **reticulata.**

    Corolla minutely papillate on the lobes ;
      spots small, rounded ; outer corona
      pale yellowish ? on the disk with
      dull red or crimson lobes ... ... (18) **ocellata.**

  Corolla with the ring or disk and throat of
    the tube covered with stiffly erect
    short thick clavate dark purple-brown
    hairs seated on conical papillæ ... (14) **præstans.**

  Corolla with some acute fleshy spine-like
    processes $\frac{1}{4}$–$\frac{1}{2}$ lin. long in the throat
    of the tube and just around its mouth
    on the inner margin of the smooth
    ring, but no hairs ... ... (19) **guttata.**

  Corolla very scabrous on the ring with
    acute conical papillæ, which at the
    mouth and in the throat of the tube
    are tipped with a short stiff acute hair (16) **scabra.**

  Corolla described as glabrous (without
    clavate hairs processes or conical tuber-
    cles ?) within the tube, but perhaps
    minutely papillate on the lobes ... (20) **venusta.**

†† Corolla-disk more or less convexly raised around the mouth of the tube, but not forming a very conspicuous ring, and as well as the throat of the tube scabrous with conical papillæ, those in the throat tipped with a minute acute hair... ... (16) **scabra**, vars.

††† Corolla-disk not convexly raised around the mouth of the tube, or only at the parts alternating with the lobes, not forming a perfectly continuous or conspicuous ring, sometimes narrow ; lobes widely spreading or recurved :

Corolla spotted or minutely dotted with crimson and with a crimson or purple area at the bottom of the tube :

Corolla papillate-scabrous from the throat of the tube to the middle or tips of the lobes, with or without a minute hair at the tips of the papillæ in the throat :

Outer corona-lobes purple - brown, united for half their length into a distinct disk which is often paler in colour ... ... (16) **scabra**, var.

Outer corona-lobes black, not united into a distinct disk ... ... (15) **brevirostris.**

Corolla minutely papillate on the lobes, smooth around the mouth of the tube, which is bearded with long stiff thick purple-brown hairs ... (5) **Piersii.**

Corolla smooth and without hairs on the inner surface, but the spots very slightly raised ... ... (13) **Thureti.**

Corolla pale greenish-yellow to deep primrose-yellow, without spots, smooth or papillate-scabrous on the disk and lobes :

Corolla $1\frac{1}{2}$–$1\frac{3}{4}$ in. in diam., faintly purple at the bottom of the tube ; papillæ in the throat of the tube hair-tipped ... ... (16) **scabra**, var. *β.*

Corolla $\frac{3}{4}$–$1\frac{1}{8}$ in. in diam., not purplish at the bottom of the tube ; no hair-tipped papillæ ... ... (11) **primulina.**

See also *Stapelia fasciculata*, Thunb., which may belong to this genus.

1. **H. campanulata** (R. Br. in Mem. Wern. Soc. i. 22) ; stems erect, $1\frac{1}{2}$–5 in. long, $\frac{1}{2}$–$\frac{3}{4}$ in. thick, acutely 4–5-angled, green or mottled with purplish ; angles compressed, with spreading deltoid acute teeth 1–2 lin. long ; flowers 2–3 together near the base of the young stems, developing in succession ; pedicels 3–6 lin. long, glabrous ; sepals narrowly lanceolate, acuminate, glabrous ; corolla somewhat pear-shaped and obtusely pentagonal in bud, with a short acute point and a small conical tooth at each angle ; when expanded, about $1\frac{1}{2}$ in. in diam., campanulate, with a slight

contraction at the middle, thence gradually passing into the broadly funnel-shaped limb and spreading lobes; outside smooth, pale greenish, spotted with purple; inside minutely papillate, bearded with stiff clavate purple hairs at the throat of the tube, whitish or pale sulphur-yellow, marked with raised blackish-purple or blackish-crimson spots, passing into transverse lines in the tube, which is entirely dark purple at the base; lobes 5–6 lin. long and about as broad at the base, deltoid, acuminate; outer corona-lobes much broader than long, subtruncate, emarginate, blackish, shining; inner corona-lobes subulate, acute, connivent over the style-apex, with recurved tips, slightly gibbous on the back at the base, dark purple. *Haw. Syn. Pl. Succ.* 28; *Allg. Teutsch. Gart. Mag.* 1818, 17, *t.* 2; *Schultes, Syst. Veg.* vi. 5; *Loud. Hort. Brit.* 97, *and Encycl. Pl.* 202; *G. Don, Gen. Syst.* iv. 112; *DC. Prodr.* viii. 651; *Schlechter in Journ. Bot.* 1898, 484. *Heurnia campanulata, Spreng. Syst. Veg.* i. 841; *Dietr. Syn. Pl.* ii. 887; *K. Schum. in Engl. und Prantl, Pflanzenfam.* iv. ii. 281. *Stapelia campanulata, Masson, Stap.* 11, *t.* 6; *Willd. Sp. Pl.* i. 1293; *Pers. Syn. Pl.* i. 280; *Bot. Mag. t.* 1227; *Jacq. Stap. t.* 1; *Ait. Hort. Kew. ed.* 2, ii. 95; *Poir. Encycl.* vii. 389, *and in Dict. Sc. Nat.* 1. 394.

VAR. *β*, **denticoronata** (N. E. Br.); outer corona-lobes as long as broad, deeply divided into two acute lobes, dark purple-brown, paler at the base.

SOUTH AFRICA: in dry regions, *Masson.*

CENTRAL REGION: Var. *β*: Prince Albert or Laingsburg Div.; near Laingsburg, *Pillans,* 157!

H. campanulata, Sprenger in *Monatsschr. Kakt.* iv. 37, which I have not seen, is evidently, from the description, not the same as *H. campanulata,* R. Br., and probably belongs to *H. barbata,* Haw., or its variety *tubata,* N. E. Br.

2. **H. barbata** (Haw. Syn. Pl. Succ. 31); stems erect, 1½–2½ in. long, ½–¾ in. thick, acutely 4–5-angled, glabrous, glaucous-green; angles with spreading deltoid acute teeth 1½ lin. long; flowers 2 or more, successively developed near the base of the young stems; pedicels ¼–⅓ in. long, rather slender, glabrous; sepals about ¼ in. long, lanceolate-subulate, glabrous; corolla acutely pointed and pentagonal in bud, when expanded 1¼–2 in. in diam., campanulate, with ascending-spreading or gradually recurving lobes, glabrous and without spots outside, clear or dirty sulphur-yellow to pale buff on the inside, marked on the lobes and upper part of the tube with blood-red dots, which pass into transverse interrupted lines on the lower part of the tube, covered from the middle of the tube to the base or to ⅔ of the way up the lobes with long stiff clavate purple hairs, which become minute upon the lobes, each seated on a purple-brown papilla, with the tips and margins of the lobes papillate; tube ⅔–1 in. long; lobes ½–⅔ in. long and nearly as broad, deltoid, acuminate; outer-corona 5-lobed, blackish or blackish-crimson; lobes subquadrate, emarginate or shortly and obtusely bifid; inner corona-lobes 1½–2 lin. long, subulate, slightly gibbous at the base,

connivent over the style-apex at the basal part, then divergent-ascending, with suberect tips, purple. *Schultes, Syst. Veg.* vi. 8; *Loud. Hort. Brit.* 97, *and Encycl. Pl.* 202; *G. Don, Gen. Syst.* iv. 113; *Decne in DC. Prodr.* viii. 651; *Schlechter in Journ. Bot.* 1898, 485.   *H. crispa, Haw. Syn. Pl. Succ.* 31; *Schultes, l.c.* 8; *G. Don, l.c.* 113; *Decne, l.c.* 651; *Schlechter in Journ. Bot.* 1898, 485.   *H. barbata, var. crispa, Loud. Encycl. Pl.* 202.   *Heurnia barbata, Spreng. Syst. Veg.* i. 841; *Dietr. Syn. Pl.* ii. 887 (*excl. syn. S. humilis, Masson*).   *Stapelia barbata, Masson, Stap.* 11, *t.* 7; *Willd. Sp. Pl.* i. 1293; *Pers. Syn. Pl.* i. 280; *Jacq. Stap. t.* 2; *Poir. Encycl.* vii. 389, *and in Dict. Sc. Nat.* 1. 395; *Ait. Hort. Kew. ed.* 2, ii. 95; *Lodd. Bot. Cab. t.* 225; *Bot. Mag. t.* 2401.

VAR. β, **tubata** (N. E. Br.); outer corona subequally 10-crenate or obscurely and very broadly 5-lobed, with the lobes rounded or emarginate, otherwise as in the type. *H. tubata, Haw. Syn. Pl. Succ.* 30, *and Suppl. Pl. Succ.* 10; *Schultes, Syst. Veg.* vi. 8; *G. Don, Gen. Syst.* iv. 113, *including var. duodecimfida, Loud. Encycl. Pl.* 202, *fig.* 3349; *Decne in DC. Prodr.* viii. 651; *Schlechter in Journ. Bot.* 1898, 485 (*excl. syn. S. crassa, Donn*).   *H. duodecimfida, Sweet, Hort. Brit. ed.* 2, 359.   *Heurnia tubata, Spreng. Syst.* i. 841 (*excl. syn. H. humilis*); *K. Schum. in Engl. und Prantl, Pflanzenfam.* iv. ii. 280.   *Stapelia tubata, Jacq. Stap. t.* 3; *Poir. Encycl. Suppl.* v. 233; *Willd. Enum. Pl. Hort. Berol.* 287; *Link, Enum. Pl. Hort. Berol.* i. 255.   *S. duodecimfida, Jacq. Stap. t.* 4.   *S. tubulosa, Hort. ex Steud. Nom. Bot. ed.* 2, ii. 632.

VAR. γ, **griquensis** (N. E. Br.); stems not seen; corolla-tube of dried flower about ⅓ in. long; lobes about 5 lin. long and 3½–4 lin. broad at the base, narrowly deltoid, very acuminate; clavate hairs much less numerous than in the type, thinly scattered about the middle part of the corolla and not nearly extending up to the base of the lobes; outer corona with 5 subquadrate lobes ⅔ lin. long and nearly as broad, acutely bifid; otherwise as in the type.

SOUTH AFRICA: without locality, *Masson,* and *cultivated specimens!* VAR. β, *cultivated specimens!*

CENTRAL REGION: Somerset Div.; near Sheldon, *Mrs. Hutton in Herb. Pillans,* 162! cultivated specimens, *MacOwan!* Colesberg Div.? cultivated specimens from a plant collected " in the Karoo towards Colesberg," *Shaw!*

KALAHARI REGION: Var. γ: Griqualand West; Diamond Fields, *Tuck in Herb. MacOwan,* 2245!

Described (except var. γ) from living plants. The ground colour of the corolla varies, as upon one plant sent to me by Dr. Shaw in 1874 I had open at the same time flowers with a clear sulphur-yellow ground and spreading limb and others with a dirty sulphur-yellow ground and recurved limb. Occasionally 6-lobed flowers are produced, and a plant of the variety *tubata,* with 6-lobed flowers was figured with a distinct name by Jacquin, but in my experience this is not a constant character, the plant always reverts to the normal 5-lobed flowers.

3. **H. clavigera** (Haw. Suppl. Pl. Succ. 10); plant about 2 (or under cultivation 2–4) in. high; stems erect or ascending, ½–1 in. thick, acutely 4–5-angled, glabrous, dull green, often blotched with purplish where exposed to the sun; angles with stout acute horizontal teeth, 1–3 lin. long; flowers 1–5 together at the base of the young stems, developing successively; pedicels 1½–7½ lin. long, 1 lin. thick, glabrous; sepals 2–3 lin. long, lanceolate or ovate-lanceo-

late, subulate-acuminate, glabrous ; corolla campanulate, a little
contracted at the middle so as to resemble a wide-mouthed bell,
$1\frac{1}{4}$–$1\frac{1}{3}$ in. in diam., with short slightly spreading or suberect lobes,
outside prominently nerved, glabrous, pale greenish or yellowish-
white; inside pale ochre-yellow or dull greenish-yellow, thickly
covered with very minute dark purple-brown papillæ and small
blood-red spots, which in the tube become confluent into irregular
blotches that nearly cover it or suddenly pass into narrow transverse lines or belts, bottom of the tube entirely blood-red, limb and
throat (but not the lobes) covered with stiff outstanding clavate
purple hairs ; tube $\frac{2}{3}$–1 in. long ; lobes 3–4 lin. long and 4–5 lin.
broad, deltoid, acute or subacuminate ; outer corona 5-lobed, black,
with the central part purple-brown velvety ; lobes about 1 lin.
long and slightly broader, subquadrate or transverse, bifid, with
obtuse teeth or rarely subequally 3-toothed, at each sinus between
the lobes is a minute tooth ; inner corona-lobes $1\frac{1}{2}$–2 lin. long,
subulate, with a dorsal ridge at their base, connivent over the
style-apex at the lower half, then diverging with erect tips,
dark purple-brown. *Schultes, Syst. Veg.* vi. 9 ; *G. Don, Gen. Syst.*
iv. 113 ; *Decne in DC. Prodr.* viii. 651 ; *Schlechter in Journ. Bot.*
1898, 485. *H. barbata, Fl. des Jard.* iii. 161, *t.* 11, *not of Masson.*
Heurnia clavigera, Spreng. Syst. Veg. i. 842 ; *Dietr. Syn. Pl.* ii. 887 ;
K. Schum. in Engl. und Prantl, Pflanzenfam. iv. ii. 281. *Stapelia*
clavigera, Jacq. Stap. t. 5 ; *Haw. Syn. Pl. Succ.* 26 ; *Hornem. Hort.*
Bot. Hafn. i. 252 ; *Poir. Encycl. Suppl.* v. 233 ; *Link, Enum. Pl.*
Hort. Berol. i. 255. *S. campanulata, Sims, Bot. Mag. t.* 1661, *not*
of Masson. S. clavata, Jacq. ex Decne in DC. Prodr. viii. 664.

VAR. β, **maritima** (N. E. Br.) ; stems $1\frac{1}{2}$–2 in. high, 4–6 lin. thick, acutely
5-angled, light green, somewhat glaucous ; outer corona ring-like, irregularly
and minutely toothed (not lobed) at the margin, black ; otherwise as in the
type.

COAST REGION : Var. β : Mossel Bay Div. ; near Mossel Bay, *Pillans*, 703 !

CENTRAL REGION : Laingsburg Div. ; near Matjesfontein and along the road
at Witte Poort between Laingsburg and Ladismith, *Pillans*, 59 ! 95 ! 600 ! *and*
cultivated specimens !

WESTERN REGION : Little Namaqualand, *Templeman in Herb. Pillans*, 95 !

Described from living plants ; the variety *maritima*, may prove to be distinct ;
I have not seen fresh flowers of it.

4. **H. decemdentata** (N. E. Br.) ; stems not seen ; pedicel 1 lin.
long in the only flower seen (? perfect), stout, glabrous ; sepals 2 lin.
long, 1–$1\frac{1}{4}$ lin. broad, broadly ovate, shortly subulate-acuminate,
glabrous ; corolla campanulate, about $\frac{3}{4}$ in. in diam., without any
constriction to the tube, subequally 10-toothed at the top,
apparently very minutely papillate-asperate outside, inside with the
basal part smooth and glabrous, above a belt about $\frac{1}{3}$ in. broad
densely covered with stout flattened clavate hairs $\frac{1}{2}$–1 lin. long, and
the apical fourth of the tube and the teeth densely and minutely

papillate; tube of the dried flower 10 lin. long, apparently purple-brown within, bordered with a paler colour (greenish-yellow?) at the mouth; larger teeth $1\frac{1}{2}$ lin. long and rather broader at the base, smaller teeth 1 lin. long and about $1\frac{1}{2}$ lin. broad at the base, all deltoid acute; outer corona 5-lobed, blackish; lobes $\frac{2}{3}$ lin. long, $1-1\frac{1}{4}$ lin. broad, transversely oblong, emarginate, crenately 3–4-toothed or acutely bifid at the apex, varying in the same flower; inner corona-lobes $1\frac{3}{4}$ lin. long, subulate, connivent-erect at the basal part, then diverging, with erect tips, dark purple-brown.

SOUTH AFRICA: without locality, cultivated, *Rabjohn*!

According to Mr. Rabjohn the stems are "short, stout, in dense clusters," and the "flowers dark brown, almost black."

5. **H. Piersii** (N. E. Br.); plant $1-1\frac{1}{2}$ in. high, with short erect crowded acutely 4-angled stems $\frac{3}{4}-1\frac{1}{4}$ in. long and about $\frac{1}{2}$ in. square, with acute spreading teeth about 1 lin. long, glabrous, green, mottled with dull purplish; flowers 2–3 together, near the base of the young stems; pedicels $\frac{1}{4}$ in. (or more?) long (only 1 detached flower seen), glabrous; sepals $\frac{1}{8}$ in. long, ovate-lanceolate, acuminate, glabrous; corolla $1-1\frac{1}{4}$ in. in diam.; tube campanulate, $\frac{1}{4}$ in. long, 5 lin. in diam. outside; lobes spreading very abruptly from the tube, $4-4\frac{1}{2}$ lin. long and broad, deltoid, very acutely (almost subulate-) acuminate, glabrous and smooth outside; inner surface very minutely papillate on the lobes only, smooth elsewhere, but thinly bearded with long stiff outstanding purple-brown hairs in the throat and around the mouth of the tube, deep ochreous-yellow, spotted and the minute papillæ tipped with purple or crimson, with the spots passing into transverse lines at the mouth of the tube, which is creamy-white with purple transverse lines and wholly purple around the corona; outer corona-lobes about $\frac{2}{3}$ lin. long and broad, shortly connate at the base, subquadrate and very deeply bifid, black; inner corona-lobes $\frac{1}{8}$ in. long, subulate, acute, with a transverse ridge at their base, connivent-erect much above the anthers, with diverging tips, purple.

CENTRAL REGION: Wodehouse Div.; Carnarvon Farm, near the Railway between Sterkstroom and Indwe, *Piers in Herb. Pillans*, 622!

Described from living stems and fresh flower preserved in fluid.

6. **H. Pillansii** (N. E. Br. in Gard. Chron. 1904, xxxv. 50, *Pillansi*, by error); plant densely tufted; stems $\frac{3}{4}-1\frac{1}{2}$ in. long, $\frac{1}{2}-\frac{2}{3}$ in. in diam., erect or ascending, subglobose when young, becoming ovoid or cylindric, densely covered with conical recurved bristle-pointed tubercles $1\frac{1}{2}-2$ lin. long, in 20–24 vertical or spiral series, glabrous, green, or dull purplish where exposed to the sun; flowers 1–3 together near the base of the young stems, successively developed; pedicels $1\frac{1}{2}-2$ lin. long, glabrous; sepals 2–3 lin. long, $\frac{2}{3}-1$ lin. broad, ovate-lanceolate at the base, attenuate into filiform recurving tips, glabrous; corolla $1-1\frac{1}{4}$ in. in diam.; tube campanulate, about $\frac{1}{3}$ in. long and

as much in diam. outside ; lobes abruptly and horizontally spreading from the tube, recurved, 5–6 lin. long, 3–3½ lin. broad at the base, whence they gradually taper to a fine acuminate point, glabrous outside ; inner surface covered on the lobes and in the throat of the tube with small fleshy terete obtuse erect processes ¼–⅓ lin. long, smooth in the lower part of the tube, pale yellow, becoming pinkish-cream in the tube, covered with small crimson spots and the tips of the processes also crimson ; outer corona 5-lobed, black or purple-black ; lobes about ¾ lin. long and broad, subquadrate or oblong, very shortly and obtusely bifid to acutely bifid to the middle, when the whole appears to consist of 5 pairs of small deltoid teeth ; inner corona-lobes about 1 lin. long, connivent-erect, with very slightly spreading tips, dorsally flattened, linear, slightly humped at the base, minutely papillate at the obtuse slightly thickened apex, dark purple-brown ; staminal column about 1¼ lin. long.

CENTRAL REGION : Laingsburg Div. ; stony places near Matjesfontein, *Pillans*, 23 ! *Marloth*, 3308 ! *and cultivated specimen*! Prince Albert Div. ; near Prince Albert, *Bolus* (ex *Pillans*).

This is one of the most distinct species of this genus, no other having similar stems, which resemble those of a *Trichocaulon* in miniature. The flowers are somewhat like those of *H. Hystrix*.

7. **H. distincta** (N. E. Br.) ; plant densely tufted ; stems erect, 2–3 in. long, ½–¾ in. thick, at first distinctly 8–9-angled, but with age the angles become very obtuse and scarcely prominent, formed of series of conical acute spreading teeth 1–2 lin. long, along the then nearly cylindric stems, glabrous, green, not glaucous ; flowers 1 (or probably more) at the base of the young stems ; pedicels ⅓–½ in. long, glabrous ; sepals 2–3 lin. long, lanceolate-subulate, glabrous ; corolla about 1¼ in. in diam. ; tube campanulate, ½ in. long and as much in diam. outside, ¼ in. in diam. at the mouth inside ; lobes horizontally spreading or recurved, 4½–5 lin. long and as broad at the base, much smaller in dried flowers, deltoid, very acute ; outer surface glabrous ; inside thickly covered from the middle of the tube to the tips of the lobes with small columnar papillæ terminated by a short stiff crimson hair-point, pale dull yellow, dotted with crimson, the spots fusing into short regular or irregular and labyrinthine lines on the smooth glabrous basal half of the tube ; outer corona-lobes ¾ lin. long, 1 lin. broad, subquadrate or transverse, subtruncate and entire to shortly bifid, with an acute notch between the rounded teeth, intense blackish-crimson ; inner corona-lobes 1½ lin. long, subulate, erect, connivent at the middle, with slightly spreading tips, dark purple-brown.

CENTRAL REGION : Laingsburg Div. ; near Matjesfontein, *Pillans*, 83 !

Described from a living plant. Mr. Pillans states, " Of this I have only been able to find one large clump. It seems to me to be a natural hybrid between *H. Pillansii* and *H. clavigera*, which bound it on both sides. Nevertheless it is a species well defined."

8. H. Hystrix (N. E. Br. in Gard. Chron. 1876, v. 795); stems usually 2–3 (rarely, under cultivation, up to 5) in. high, 4–6 lin. thick, 5-angled, glabrous, pale green, very slightly glaucous, sometimes tinted with dull purplish; teeth conical, acute, very spreading; flowers 2–5 together near the base of the stems, successively developed; pedicels $\frac{3}{4}$–$1\frac{1}{2}$ in. long, glabrous; sepals 4 lin. long, $\frac{3}{4}$ lin. broad, subulate from an ovate-lanceolate base, glabrous; corolla shortly pointed and with a small tooth at each angle in bud; when expanded 1–$1\frac{1}{2}$ in. in diam.; tube short, campanulate, about $\frac{1}{4}$ in. long and $\frac{1}{2}$ in. in diam. outside; lobes abruptly and horizontally spreading from the tube, with recurved tips, $\frac{1}{2}$–$\frac{2}{3}$ in. long and nearly as broad at the base, deltoid-ovate, acute; outside glabrous, but minutely papillate, pitted, prominently 3-nerved on the lobes; inner surface (except within the tube) covered with spine-like acute fleshy processes, ochreous-yellow, marked with crimson spots on the lobes and with numerous transverse crimson lines within the smooth and paler tube; outer corona-lobes half as long as broad, truncate, broadly emarginate or obscurely 3-toothed at the apex, black (apparently yellowish in a dried flower of *Wood* 10813); inner corona-lobes $1\frac{1}{2}$ lin. long, erect, much exceeding the staminal column, linear, with a transverse thickening or ridge on the back below and expanded at the apex into a horizontally spreading process somewhat resembling an inverted foot, yellow, dotted on the upper part with crimson. *Schlechter in Journ. Bot.* 1898, 485. *Heurnia Hystrix, K. Schum. in Engl. und Prantl, Pflanzenfam.* iv. ii. 280. *Stapelia Hystrix, Hook. f. Bot. Mag. t.* 5751.

Eastern Region : Natal; on mountains, *cultivated specimens*! *Wood*, 4118! *Pillans*, 25! near Weenen, 2000–3000 ft., *Haygarth in Herb. Wood*, 10813! Delagoa Bay; growing in the shade of bushes, *Mrs. Monteiro* !

Described from living cultivated plants.

9. H. loeseneriana (Schlechter in Engl. Jahrb. xx. Beibl. 51, 55); stems erect or ascending 1–$2\frac{1}{2}$ in. long, acutely 4-angled, $\frac{1}{2}$–$\frac{3}{4}$ in. square, green, purple-tinted at the tips; teeth deltoid, acute, 1–$1\frac{1}{2}$ lin. long; flowers 1–2 together, near the base of the young stems; pedicels 2–3 lin. (up to $\frac{1}{2}$ in., *Schlechter*) long; sepals $\frac{1}{4}$ in. ($\frac{1}{8}$ in., *Schlechter*) long, subulate from an ovate base; corolla campanulate, with the tube widening upwards and gradually passing into the spreading lobes, scarcely 1 in. in diam. (about $\frac{2}{3}$ in. long and nearly $\frac{1}{2}$ in. in diam., verrucose and brownish within, paler outside, *Schlechter*), very minutely asperate and brownish-purple outside, covered inside from the middle of the tube nearly to the tips of the lobes with short stiff compressed-conical obtuse processes, dull yellow, covered with very crowded narrow broken transverse lines and spots of dull brownish-crimson, and entirely of that colour at the base of the tube, but darker; tube $\frac{1}{3}$ in. long; lobes $\frac{1}{4}$ in. long and rather more in breadth at the base, broadly deltoid, very acute; outer corona-lobes about $\frac{3}{4}$ lin. long and broad, subquadrate, subtruncate at the apex, black, with a velvet-like surface; inner corona-lobes 1–1¼ lin. long, connivent-erect with diverging tips, much exceeding the anthers, linear-subulate, thickened and knob-like at the minutely scabrous apex, with a transverse ridge at the base, dull purplish-red, brownish at the tips. *Schlechter in Journ. Bot.* 1898, 485.

KALAHARI REGION: Transvaal; mountains around Pretoria, *Burtt Davy,* 2446 ! *Schlechter,* near the Olifants River, 5000 ft., *Schlechter,* 3774.

I have not seen an authentic specimen of this species and describe entirely from a living plant sent to Kew by Mr. Burtt Davy, which quite accords with Schlechter's description except as to the few characters inserted in brackets, and comes from the same general locality. In a note, however, Dr. Schlechter states that the corolla-tube of this species is longer than that of any other known to him and somewhat resembles that of a *Decabelone* (*Tavaresia*), although he describes the entire corolla as being only "about 1·5 cm. long," i.e. about ⅔ in.

10. **H. longituba** (N. E. Br.); tufted; stems erect, ¾–2 in. long, ⅓–¾ in. thick, excluding the teeth, sharply 4–5-angled, glabrous, light green, sometimes tinted with purple, scarcely glaucous; flowers 1–3 together at the lower part of the young stems, opening succes-sively; pedicels ½–¾ in. long, ¾ lin. thick, glabrous; sepals about ⅓ in. long, subulate from an ovate base, glabrous; corolla campanu-late, prominently 20-nerved on the glabrous smooth outside, inside (except at the base) covered with short stiff conical processes, decreasing in size towards the tips of the lobes, creamy yellow, spotted with purple, the spots becoming very crowded and darker purple at the middle part of the tube (*Pillans*), basal part smooth and whitish, with transverse dark purple (or blood-red?) lines, the processes tipped with dark purple; tube ¾–⅞ in. long, ¾ in. in diam. at the mouth; lobes about 5 lin. long and broad, erectly spreading in a continuous line with the tube, deltoid, very acute, the alternating teeth at the sinuses very prominent; outer corona 5-lobed, blackish-purple; lobes about ¾ lin. long, ovate, obtuse at the shortly bifid apex; inner corona-lobes 1½ lin. long, subterete, erect, connivent at the middle, with recurving slightly clavate obtuse tips, apparently blackish-purple or dark purple-brown.

KALAHARI REGION: Griqualand West; near Douglas, *Pillans,* 609 !

11. **H. primulina** (N. E. Br. in Hook. Ic. Pl. t. 1906); stems 1½–2½ in. high, ½–1 in. thick, excluding the teeth, acutely 4–5-angled, with more or less channelled sides, glabrous, glaucous-green, at least when young; angles with compressed deltoid teeth 1½–2½ lin. long, with dark-coloured very acute recurving tips; flowers in fascicles of 3–8 near the base of the young stems, 2–5 often being open at the same time; pedicels ½–1 in. long, glabrous; sepals 2–2½ lin. long, ¾ lin. broad, lanceolate-subulate, glabrous; corolla in bud obtusely to very acutely pointed, acutely 5-angled, when expanded ¾–1⅙ in. in diam.; tube globose-campanulate; limb horizontally

spreading abruptly from the tube, sometimes very slightly raised around its mouth, with recurving lobes 3–4½ lin. long, 3–5 lin. broad, deltoid, variably acute, smooth and glabrous outside and inside, entirely sulphur-yellow to golden-primrose, without markings or a purple area around the corona; tube 2½–3½ lin. long and as much in diam. outside, constricted at the mouth; outer corona-lobes ½–⅔ lin. long, subquadrate, bifid, crimson-black or purple-black; inner 1–1¼ lin. long, subulate, acute, with a slight transverse dorsal ridge at the base, connivent over the staminal column, with erect points, purple-brown. *Schlechter in Journ. Bot.* 1898, 485. *Heurnia primulina, K. Schum. in Engl. und Prantl, Pflanzenfam.* iv. ii. 280.

VAR. β, **rugosa** (N. E. Br.); disk of the corolla-limb covered with small tubercles, which gradually decrease in size towards the tips of the lobes, *H. flava, N. E. Br. in Kew Rep.* 1878, 7, *name only.*

COAST REGION : Albany Div. ; Brak Kloof, near Grahamstown, *Mrs. Gloheta,* 4 ! along the railway between Alicedale Junction and Grahamstown, *Pillans,* 12 ! Var. β: Albany Div. ; dry stony places near Hell Poort, Cawoods Hole and other places near Grahamstown, *MacOwan,* 910 ! *Pillans,* 43 ! and mixed with the type, *Barkly,* 13 ! Queenstown Div., *Barkly,* 13 *bis* !

The acumination of the buds and intensity of the colour of the corolla is variable. Mr. Pillans informs me that when alive "the stems of var. *rugosa* are distinguishable from the type by their larger size and darker colour."

12. **H. simplex** (N. E. Br.) ; stems erect, 1½–2 in. high, sharply 4–5-angled, glabrous, apparently glaucous-green ; teeth spreading, ¾–1 lin. long, compressed-deltoid ; flowers 3 or more together, successively produced near the base of the young stems ; pedicels about ½ in. long, apparently rather slender, glabrous ; sepals 1¼–1¾ lin. long, ovate-lanceolate, acuminate, glabrous ; corolla (of dried flowers) about 1 in. in diam., glabrous and smooth outside and within the tube, minutely papillate-puberulous on the inner surface of the lobes, "yellow, spotted with rosy in the centre" (*Miss Thomson*) ; tube about ⅓ in. long and ½ in. in diam., broadly campanulate ; lobes about ⅓ in. long and as much in breadth at the base, deltoid, very acute or acuminate, apparently spreading ; outer corona none ; inner corona-lobes arising at the middle of the staminal column, their basal half adnate to it as vertical semi-terete ridges between the pairs of anther-wings, and dilated into a transverse ridge or truncate rim at the top, upper half free, connivent over the anthers, ½–⅔ lin. long, ¼ lin. broad at the base, flat, tapering to the acute apex ; staminal column 1 lin. long, broadly conical at the basal half.

CENTRAL REGION : Victoria West Div. : near Gert Adriaans Kraal, *Miss Hester Thomson in Herb. Galpin,* 3056 !

The absence of an outer corona distinguishes this from all other species of *Huernia.* This character, taken alone, would place it in the genus *Huerniopsis,* but the inner corona and habit are so unlike that of the latter genus and so exactly as in *Huernia,* that I regard it as a member of that genus with the outer corona undeveloped.

13. **H. Thureti** (Cels in L'Horticult. Francais, 1866, 73, t. 3);
stems 1–2 in. high, 3–4 lin. thick, acutely 4–5-angled, glabrous,
glaucous-green; teeth 1–1½ lin. long, deltoid, acute; flowers 1–4
together, successively developed at the base of the stems; pedicels
6–8 lin. long, 1–1¼ lin. thick, glabrous; sepals 2½ lin. long,
lanceolate-subulate, glabrous; corolla 1 in. in diam.; tube subglobose-campanulate, about ⅓ in. long and 4½ lin. in diam. outside,
slightly constricted at the mouth, disk and lobes horizontally
spreading abruptly from the tube; lobes 4 lin. long, 4½ lin. broad,
deltoid, very acute; outside and within quite glabrous, without
processes or papillæ on the inner face, buff-coloured or yellow-ochre,
marked on the lobes with numerous small blood-red spots, those on
the disk transversely elongated, passing into transverse lines in
the tube, which is blood-red at the base; outer corona-lobes about
⅔ lin. long and 1 lin. broad, subquadrate, bifid, alternating with
5 minute tubercles, nearly black; inner corona-lobes about 1 lin.
long, subulate, connivent-erect, with the slightly spreading tips
produced ¾ lin. or more above the anthers, with a slight transverse
thickening at the base, purple-brown. *H. Thuretii, Schlechter in
Journ. Bot.* 1898, 485. *Stapelia Thureti, Croucher in The Garden,*
1877, xii. 524, *with fig. on* 525.

SOUTH AFRICA: without locality, *cultivated specimens*!

Described from living plants. A specimen (Watermeyer in Herb. Pillans,
1306) collected near Tsomo, in the Transkei, of which I have not seen flowers,
may belong to this species.

14. **H. præstans** (N. E. Br.); stems erect or ascending, 1–2 in.
long, about ½ in. thick including the teeth, acutely 4-angled,
glabrous, deep green (slate-coloured, *Pillans*) mottled with purple;
teeth 1½–2 lin. long, spreading, acute, slightly recurved at the
tips; flowers from near the base of the young stems; pedicels
short, glabrous; sepals 3–3½ lin. long, narrowly lanceolate-attenuate,
glabrous; corolla apparently somewhat turbinate and flat- or broad-
topped, with a short acute point, when expanded about 1½ in. in
diam.; tube campanulate about ¼ in. long and ½ in. in diam. outside;
limb abruptly and horizontally spreading from the tube, convex
and ring-like around its pentagonal mouth; lobes slightly ascending
and recurved, about ½ in. long and rather more in breadth, deltoid,
acute; outside glabrous and smooth, creamy, tinted with purplish
and marked with purple veins and spots; inside densely and very
minutely papillate on the lobes, with stiffly erect stout clavate dark
purple hairs ¼–½ lin. long seated on stout conical creamy papillæ
covering the disk or annulus and extending in 5 triangular areas
down to about the middle of the tube; lobes greenish-yellow,
marked with impressed purple dots and densely dusted with purple,
with the inside of the tube and the disk around its mouth creamy,
thickly marked with dull purple spots, much larger than those on
the lobes and more or less confluent into rather crowded irregular

lines in the tube, the base of which is entirely dark purple and shining; outer corona 5-lobed, with a small obtuse tooth alternating with the lobes, black; lobes ½ lin. long, 1 lin. broad, transversely oblong, emarginate; inner corona-lobes subulate, 1½ lin. long, somewhat connivent-erect at the lower half, with diverging tips, dark purple-brown, not nearly reaching to the mouth of the tube.

CENTRAL REGION : Laingsburg Div. ; near Witte Poort between Laingsburg and Ladismith, *Pillans,* 667 !

15. **H. brevirostris** (N. E. Br. in Gard. Chron. 1877, vii. 780, fig. 124); stems crowded, erect or ascending, 1½–2½ in. long, ½–1 in. thick, acutely 4–5-angled, glabrous, glaucous-green, becoming dull-green or purplish with age ; teeth of the angles 1–1½ lin. long, deltoid, acute ; flowers 3–6 in a fascicle, near the base of the young stems, successively developed ; pedicels ⅓–1 in. long, 1 lin. thick, often reddish-tinted, slightly glaucous ; sepals 1½–2½ lin. long, lanceolate-subulate, glabrous ; corolla shortly and acutely pointed in bud, when expanded 1–1¼ in. in diam. ; tube short, globose-campanulate, about ¼ in. long and 5 lin. in diam. outside, slightly constricted at (but not raised into a ring around) the mouth ; lobes abruptly horizontally spreading and often recurved, ⅓ in. long, ½ in. broad at the base, deltoid, very acute ; outside smooth, glabrous and often reddish-tinted or dotted with purple ; inner face rough with numerous papillæ on the lobes and in the throat of the tube, where they are generally tipped with (but sometimes entirely without) a minute hair, pale yellow on the lobes, pinkish-white in the tube, evenly dotted in variable density with crimson except at the bottom of the tube, which is entirely crimson and very shining, sometimes with crimson or purplish tips to the lobes ; outer corona black, subequally 10-toothed or with 5 deeply bifid lobes ⅔ lin. long, with deltoid-oblong obtuse teeth ; inner corona-lobes about 1 lin. long, subulate, with a transverse hump on the back at the base, incumbent upon the anthers and prolonged scarcely ½ lin. beyond them into connivent-erect tips, purple-brown or the upper part dull yellowish speckled with purple-brown. *N. E. Br. in Gard. Chron. 1908, xliv. 197–198, fig. 87 ; and Bot. Mag. t. 6379 ; Schlechter in Journ. Bot. 1898, 485. Heurnia brevirostris, K. Schum. in Engl. und Prantl, Pflanzenfam. iv. ii. 280.*

VAR. β, **intermedia** (N. E. Br.); corolla rugose with small conical papillæ within, except on the lower part of the tube, primrose-yellow with a minute purple dot at the tip of each papilla ; inside of the tube paler, with a light purple area at the base around the outer corona.

CENTRAL REGION : Graaff Reinet Div. ; dry rocky hills of Rynevelds Pass, near Graaff Reinet, 2700 ft., *Bolus,* 575 ! *Pillans,* 663 ! and *cultivated specimens* ! Var. β : Graaff Reinet Div. ; near Graaff Reinet, *Pillans,* 72 !

Described from living cultivated plants. The very short tips to the inner corona-lobes are a marked character in this species. Var. β seems intermediate between *H. brevirostris* and *H. primulina,* but is readily distinguished by the purple area at the base of the distinctly larger corolla-tube, which is absent in *H. primulina.*

16. H. scabra (N. E. Br.); stems compactly erect, 1–1¾ in. long, ½–¾ in. thick, sharply 4–5- (rarely 6-) angled, with compressed deltoid teeth 1½–2 lin. long, glabrous, dull green or greyish-green, sometimes mottled with dull purple; cymes with 3–5 successively developed flowers near the base of the stems; pedicels about 4 lin. long, 1 lin. thick, glabrous; sepals 2–2½ lin. long, lanceolate, acuminate, glabrous; corolla 1½–1⅔ in. in diam.; tube 4 lin. long and 5 lin. in diam. outside at the top, not wider at the mouth than below; limb spreading horizontally from the tube and raised around its mouth into a conspicuous broad convex ring, scabrous with rather stout conical acute papillæ, which, at the mouth and throat of the tube, are tipped with a short stiff acute dark purple hair; lobes somewhat spreading and recurving, about 5 lin. long and 7–8 lin. broad at the base, deltoid, very acute, with a minutely papillate inner surface; outside smooth, dotted with pale purple on a pale (flesh-coloured?) ground; inside with the ground colour varying from pale flesh-tinted to light canary-yellow, covered on the lobes and ring with round light purple (or crimson?) spots, which pass into transverse broken lines within the tube, which at the base, around the outer corona, is entirely purple; outer corona of 5 pairs of acute deltoid teeth or of 5 deeply and acutely bifid lobes ⅔–¾ lin. long, dark purple-brown, all united at their base into a circular disk of a pale pinkish tint, margined with purple-brown between the lobes; inner corona-lobes much overtopping the anthers, 1¾ lin. long, not reaching to the mouth of the tube, subulate, obtuse, with a transverse ridge at their base, connivent at the middle, slightly spreading at the upper half, pale purple.

Var. β, **immaculata** (N. E. Br.); corolla 1½–1¾ in. in diam., scarcely or inconspicuously raised into a convex ring around the mouth of the tube, primrose-yellow, without spots, tinged with very pale purple at the base of the tube around the corona, scabrous as in the type, with the papillæ at the mouth and in the throat of the tube tipped with short acute pinkish hairs; inner corona-lobes overtopping the anthers, about 1 lin. long, slightly gibbous behind; otherwise as in the type.

Var. γ, **pallida** (N. E. Br.); corolla about 1⅓ in. in diam., not or inconspicuously raised into a convex ring around the mouth of the tube, pale primrose-yellow with light purple markings, smaller and fainter than in the type, and the surface much less scabrous, with small slightly prominent papillæ, those in the throat being tipped with minute acute dark purple hairs; disk and lobes of outer corona dark purple-brown, with 5 slight ridges on the disk radiating from the base of the inner corona and ending in slight projections between the lobes; otherwise as in the type.

Var. δ, **ecornuta** (N. E. Br.); corolla 1⅓–1½ in. in diam., not or inconspicuously raised into a ring around the mouth of the tube, which widens upwards, pale canary-yellow with light purple spots, not dotted on the outside, papillate-scabrous, with the papillæ in the throat and at the mouth of the tube tipped with a short stiff clavate dark purple hair; disk and lobes of outer corona dark purple-brown; inner corona-lobes closely incumbent upon the anthers and not or scarcely exceeding them, without erect tips, purple; otherwise as in the type.

Var. ε, **longula** (N. E. Br.); pedicels 2–4 lin. long; corolla 1–1¼ in. in diam., slightly convex and somewhat ring-like around the mouth of the tube; lobes 4½–5 lin. long, 5–5½ lin. broad, yellow, with small crimson spots, papillate-scabrous, with the papillæ in the throat of the tube not hair-tipped; outer corona entirely blackish-crimson or blackish, with a narrow disk and 5 pairs of obtuse teeth scarcely $\frac{1}{2}$ lin. long; inner corona-lobes 1-1$\frac{1}{4}$ lin. long, rising nearly to the level of the mouth·of the tube (in all other forms not nearly attaining to that level); otherwise as in var. γ.

CENTRAL REGION: Victoria West Div.; in Biesjes Poort, *Pillans*, 632! Also growing in the same locality the following varieties:—Var. β, *Pillans*, 688! Var. γ, *Pillans*, 109! Var. δ, *Pillans*, 55! Var. ε: Beaufort West Div.; Rhenoster Kop, *Foster in Herb. Pillans*, 140!

This seems to be a very variable species, since besides the varieties I have described, there are others connecting them with one another and the type by gradations which it is scarcely possible to indicate clearly by words. It seems to merge into *H. brevirostris*, N. E. Br., through some of its varieties. I am informed by Mr. Pillans that all but one of these and probably other forms grow together in the same locality, and I believe there can be no reasonable doubt that they are hybrid forms resulting from cross-fertilization by insect agency, by which alone these plants are fertilized, see *Stapelia variegata*, Linn.

17. **H. humilis** (Haw. Syn. Pl. Succ. 30); plant 2-2$\frac{1}{2}$ in. high, compact; stems erect, 7-10 lin. thick, acutely 4-5-angled, glabrous, green, glaucous?; angles with acute spreading deltoid teeth 1-1$\frac{1}{2}$ lin. long; flowers solitary according to *Masson*, but in *Bain's* specimen 4-5 or more are successively produced on a very short peduncle near the base of the young stems; pedicels $\frac{1}{4}$-$\frac{1}{3}$ in. long, glabrous; sepals 2$\frac{1}{2}$-3 lin. long, lanceolate-subulate, glabrous; corolla in bud much flattened, pentagonal, very shortly abruptly and acutely pointed at the centre, with 5 small acute teeth at the angles and 5 very prominent ribs radiating from the point, glabrous, when expanded 1-1$\frac{1}{4}$ in. in diam., with a very shallow saucer-shaped tube and a very abruptly spreading limb, lobed to more than half-way and raised into a broad convex ring around the mouth of the tube, glabrous; lobes deltoid in *Masson's* figure, deltoid-ovate and minutely papillate in *Bain's* specimen, very acute, pale yellow, dotted with blood-red; ring smooth, blackish-purple, marked with undulated whitish spots; outer corona small, with very short transverse bifid lobes or subequally 10-toothed; inner corona-lobes about $\frac{1}{2}$ lin. long, ovate, acuminate, incumbent on the backs of the anthers and not produced beyond them. *Schultes, Syst. Veg.* vi. 7; *Loud. Hort. Brit.* 97, *and Encycl. Pl.* 202; *G. Don, Gen. Syst.* iv. 113; *Decne in DC. Prodr.* viii. 651; *N. E. Br. in Hook. Ic. Pl. t.* 1905, *fig.* B; *Schlechter in Journ. Bot.* 1898, 485, *partly. Heurnia humilis, K. Schum. in Engl. und Prantl, Pflanzenfam.* iv. ii. 280. *Stapelia humilis, Masson, Stap.* 10, *t.* 5; *Willd. Sp. Pl.* i. 1294; *Pers. Syn. Pl.* i. 280; *Poir. Encycl. Meth.* vii. 390; *Ait. Hort. Kew. ed.* 2, ii. 96.

SOUTH AFRICA: in dry regions, *Masson*.

CENTRAL REGION: Beaufort West Div.; Nieuwveld Mountains, *Bain*, X!

The details of the flowers, except as to colour, are described from Bain's specimen, which agrees so well with Masson's figure in general characters that I believe it to be this species, the colour of its flowers, however, is unknown to me, as the specimen was received preserved in alcohol.

H. humilis, Schlechter in *Engl. Jahrb.* xx. Beibl. 51, 54, is unknown to me, but it cannot be the same as *H. humilis*, Haw., the locality where it was collected (at the foot of the Houtboschberg Ranges, 5500 ft., in the Transvaal, *Schlechter*, 4761) is at least 600 miles distant from the home of the true *H. humilis*, and the colour of the flowers as described by Dr. Schlechter is totally different, varying between flesh-red and brownish, with large purple spots on the lobes and limb, the annulus varying between rose-red and dark purple-red.

18. H. ocellata (Schultes, Syst. Veg. vi. 9) ; stems acutely 4-angled, $1\frac{1}{2}$–2 in. long, $\frac{2}{3}$–$\frac{3}{4}$ in. square, glabrous ; angles with acute deltoid teeth 1–$1\frac{3}{4}$ lin. long ; flowers 2 (or more ?) together near the base of the young stems ; pedicels $\frac{1}{2}$ in. long, glabrous ; sepals 3 lin. long, $\frac{3}{4}$–1 lin. broad at the base, lanceolate-subulate, glabrous ; corolla about $1\frac{1}{2}$–$1\frac{3}{4}$ in. in diam. ; tube campanulate, about $\frac{1}{3}$ in. long and as much or more in diam. at the mouth ; limb saucer-shaped, lobed to half-way, very abruptly spreading from the tube and raised into a very conspicuous broad convex ring around its mouth, with ascending-spreading lobes, about $\frac{1}{3}$ in. long and more than $\frac{1}{2}$ in. broad at the base, broadly deltoid, acute ; outside glabrous and smooth ; inner face smooth on the raised ring, densely and minutely papillate on the lobes and rest of the limb, with some stiff clavate outstanding hairs about $\frac{1}{2}$ lin. long on the upper part of the tube, light yellow, marked all over with round crimson spots, those on the ring slightly raised ; outer corona of 5 oblong deeply bifid lobes nearly 1 lin. broad, with parallel obtuse teeth, pale yellowish on the disk with dull red or crimson lobes ; inner corona-lobes $1\frac{3}{4}$ lin. long, subulate, acute, slightly gibbous at the base, connivent-erect at the lower part, then diverging with erect tips, whitish, spotted with purple-brown. *G. Don, Gen. Syst.* iv. 113 ; *Decne in DC. Prodr.* viii. 651 ; *Schlechter in Journ Bot.* 1898, 485. *Heurnia ocellata, Spreng. Syst. Veg.* i. 841 (*excl. syns. S. guttata, Masson, and S. lentiginosa, Curt.*) ; *Dietr. Syn. Pl.* ii. 887 (*excl. syns.*) ; *K. Schum. in Engl. und Prantl, Pflanzenfam.* iv. ii. 281. *Stapelia ocellata, Jacq. Stap. t.* 6.

South Africa : without locality, *cultivated specimen* !

Chiefly described from a dried specimen, which completely agrees with Jacquin's figure.

19. H. guttata (R. Br. in Mem. Wern. Soc. i. 22) ; plant 2–3 in. high ; stems erect, $\frac{1}{2}$–$\frac{3}{4}$ in. thick, acutely 4–5-angled, glabrous, green, glaucous ? angles with acute deltoid slightly recurved teeth 1–$1\frac{1}{2}$ lin. long ; flowers 1–5 together near the base of the young stems, developing successively ; pedicels rather slender, $\frac{1}{2}$–$\frac{3}{4}$ in. long, glabrous ; sepals $2\frac{1}{2}$–3 lin. long, lanceolate-subulate, glabrous ; corolla flat-topped and 5-angled in bud, with 5 short (incurved ?) teeth at the angles and five prominent ribs radiating from the short central point, $1\frac{1}{4}$–$1\frac{1}{2}$ in. in diam. when expanded ; tube campanulate or subglobose-campanulate, slightly constricted at the mouth ; limb saucer-shaped, lobed to half-way, spreading very abruptly from the tube and raised into a very conspicuous broad convex ring around its mouth, with ascending-spreading lobes, 4–5 lin. long and broad, deltoid, acuminate ; outside smooth and glabrous; inner face with some scattered fleshy spine-like processes (not hairs) $\frac{1}{4}$–$\frac{1}{2}$ lin. long at the throat of the tube and around its mouth upon the otherwise smooth ring, minutely papillate on the rest of the limb and lobes, light yellow, dotted with crimson to the tips of the lobes ; outer corona divided into 5 pairs of narrow acute teeth nearly 1 lin. long, light pinkish or whitish with purple margins to the teeth and a blackish ring at the base around the staminal column, enlarging into 5 rounded spots opposite the pairs of teeth ; inner corona-lobes about $1\frac{1}{2}$ lin. long, subulate, acute, connivent at the lower part over the style-apex, then rather abruptly spreading, with up-curved tips, apparently yellow. *Bojer, Hort. Maur.* 213 ; *Haw. Syn. Pl. Succ.* 30 ; *Allg. Teutsch. Gart. Mag.* vii. t. 41 ; *Schultes, Syst. Veg.* vi. 7 ; *G. Don, Gen. Syst.* iv. 113 ; *DC. Prodr.* viii. 651 ; *Schlechter in Journ. Bot.* 1898, 484. *H. lentiginosa, Haw. Syn. Pl. Succ.* 29 ; *Schultes, Syst. Veg.* vi. 6 ; *G. Don, Gen. Syst.* iv. 113 ; *Decne in DC. Prodr.* viii. 651. *Stapelia guttata, Masson, Stap.* 10, *t.* 4 ; *Willd. Sp. Pl.* i. 1294 ; *Pers. Syn. Pl.* i. 280 ; *Poir. Encycl. Meth.* vii. 390, *and in Dict. Sc. Nat.* 1. 395 ; *Ait. Hort. Kew. ed.* 2, ii. 96 ; *Link, Enum. Pl. Hort. Berol.* i. 255. *S. venusta, Jacq. Stap. t.* 7, *and var. minor, t.* 64, *fig.* 4 ; *not of Masson. S. lentiginosa, Curt. in Bot. Mag. t.* 506 ; *Ait. Hort. Kew. ed.* 2, ii. 97 ; *Poir. Encycl. Suppl.* v. 234.

South Africa : without locality, *Masson, cultivated specimens* !

20. H. venusta (R. Br. in Mem. Wern. Soc. i. 22) ; stems erect or divaricate, 2–3 (or in cultivated plants sometimes up to 5) in. long, $\frac{2}{3}$–1 in. thick, acutely 4–5-angled, glabrous, green, probably somewhat glaucous ; angles compressed, with deltoid spreading teeth 1–$1\frac{1}{2}$ lin. long ; flowers 1–2 together near the base of the young stems, successively developed ; pedicels about $\frac{3}{4}$ in. long, glabrous ; sepals lanceolate, acute, glabrous ; corolla in bud somewhat turbinate, with a broad pentagonal flattened top, shortly pointed at the centre, 5-toothed at the angles, glabrous, when expanded about 2 in. in diam. ; tube very shortly campanulate, glabrous within, blackish-purple or dark blood-red at the bottom ; limb very abruptly spreading from the tube, saucer-shaped, raised into a conspicuous (smooth ?) broad ring around the mouth of the tube, with very acute deltoid lobes about 6–7 lin. long and broad, glabrous, but probably minutely papillate on the inner face, sulphur-yellow, covered to the tips of the lobes with small blood-red spots. *Haw. Syn. Pl. Succ.* 29 ; *Schultes, Syst. Veg.* vi. 6 ; *Loud. Encycl. Pl.* 202, *fig.* 3345 ; *G. Don, Gen. Syst.* iv. 112 ; *Decne in DC. Prodr.* viii. 651 ; *Schlechter in Journ. Bot.* 1898, 485. *Heurnia venusta, Spreng. Syst. Veg.* i. 841 ; *Dietr. Syn. Pl.* ii. 887 (*excl. ref. to Jacq.*). *Stapelia venusta, Masson, Stap.* 10, *t.* 3 ; *Willd. Sp. Pl.* i. 1294 ; *Pers. Syn. Pl.* i. 280 ;

Poir. Encycl. vii. 389, *and in Dict. Sc. Nat.* l. 395; *Ait. Hort. Kew.*
ed. 2, ii. 95.

REGION ? Karoo, *Masson.*

Only known to me from Masson's figure and description.

21. **H. Kirkii** (N. E. Br.); stems decumbent at the base, about
1–1½ in. long, ½–⅔ in. in diam. including the teeth, acutely 5-angled,
glabrous; teeth 1–1⅓ lin. long, stout, deltoid, acute, slightly
recurved; flowers 1–4 together at the base of the young stems,
developed successively; bracts 1½–2 lin. long, subulate or lanceolate-
subulate, glabrous; pedicels ½–⅔ in. long, glabrous; sepals 5–5½ lin.
long, very spreading, 1⅓ lin. broad at the ovate base, whence they
taper into long subulate points, glabrous; corolla 1¾–2 in. in diam.;
tube of a dried flower about ½ in. long, but according to *Mr. Kirk's*
drawing about ¾ in. long and 1 in. in diam. outside, subglobose-
campanulate, constricted at the mouth; limb saucer-shaped, lobed
to half-way, very abruptly spreading from the tube, with suberect
deltoid lobes, tapering into subulate points, according to the drawing
about ¾ in. long; outside glabrous; inside smooth and glabrous in
the lower part of the tube, densely covered with erect fleshy
subulate processes ½–¾ lin. long in the throat and around the mouth
of the tube, minutely papillate towards the margins, otherwise
glabrous, marked " with irregular red spots " (*Kirk*); outer corona
5-lobed; lobes subquadrate, obtusely 2-lobed at the apex; inner
corona-lobes 1½–1¾ lin. long, very much exceeding the anthers,
subterete, slightly clavate at the minutely papillate recurving apex,
with a transverse ridge at the base; staminal column 1 lin. long;
style-apex concave.

KALAHARI REGION : Transvaal ; Komati Poort, very rare, *Kirk*, 76 !

22. **H. reticulata** (Haw. Syn. Pl. Succ. 28); plant 3–4 in. high;
stems erect or ascending, ¾–1 in. thick including the teeth, acutely
5-angled, glabrous, green, mottled with purple; angles often spiral,
with large deltoid acute spreading teeth; flowers 1–4 together,
successively developed near the base of the young stems; pedicels
3–9 lin. long, ¾ lin. thick, glabrous; sepals 3–3½ lin. long, lanceolate,
acuminate, glabrous; corolla in bud somewhat turbinate, with a
very broad flat top, subpentagonal, shortly 5-toothed at the angles,
prominently nerved, whitish, speckled or tinged with purplish on
the glabrous outside, when expanded 1¼–1¾ in. in diam.; tube
campanulate, 4 lin. long, 5 lin. in diam.; limb saucer-shaped, lobed
to half-way down, very abruptly spreading from the tube and
raised around its mouth into a very conspicuous broad smooth and
shining ring; lobes 3–6 lin. long, broader than long at the base,
broadly deltoid, very acute; inside, the upper part of the dark
blood-red tube is thickly covered with stiff glassy clavate purple
hairs, those at the mouth converging and much larger than those

below, and the lobes are sometimes sprinkled with minute setules, elsewhere glabrous, pale yellow, thickly covered with rather large blood-red or deep crimson-brown spots, which are more or less confluent, producing a reticulated appearance, or sometimes the ring is entirely crimson; outer corona 10-toothed, purple-black, velvety; teeth deltoid, acute, upturned; inner corona-lobes reaching to about ⅔ of the length of the corolla-tube, subulate, acute, connivent over the style-apex at the lower part then widely diverging, with erect tips, red at the base, whitish, mottled with deep red above; odour strong, putrid. *Schultes, Syst. Veg.* vi. 5; *G. Don, Gen. Syst.* iv. 112; *Decne in DC. Prodr.* viii. 651; *Loud. Encycl. Pl.* 202, *fig.* 3343; *N. E. Br. in Hook. Ic. Pl. under t.* 1906. *H. barbata, Schlechter in Journ. Bot.* 1898, 485, *not of Haw. Heurnia reticulata, Spreng. Syst. Veg.* i. 841; *Dietr. Syn. Pl.* ii. 887; *K. Schum. in Engl. und Prantl, Pflanzenfam.* iv. ii. 280. *Stapelia reticulata, Masson, Stap.* 9, *t.* 2; *Willd. Sp. Pl.* i. 1294, *and Enum. Pl. Hort. Berol.* 287; *Pers. Syn. Pl.* i. 280; *Poir. Encycl.* vii. 390; *Jacq. Stap. t.* 8, *and var. deformis, Jacq. Stap. t.* 9; *Ait. Hort. Kew. ed.* 2, ii. 96; *Bot. Mag. t.* 1662. *S. crassa, Donn, Hort. Cantab. ed.* 3, 43, *ex Haw. Rev. Pl. Succ.* 199; *Decne in DC. Prodr.* viii. 663.

COAST REGION : Clanwilliam Div. ; among rocks near Oliphant's River, *Masson, cultivated specimens* ! on northern slopes of ridges near Clanwilliam, *Pillans*, 661 ! and without precise locality, *Barkly* !

Described from a living plant.

23. H. zebrina (N. E. Br.) ; stems 2–3 in. high, 4–5 lin. in diam. without the teeth, 5-angled? green; angles compressed, with stout conical acute spreading teeth 2–2½ lin. long; flowers 2 (or more?) together near the base of the young stems, successively developed; pedicels 6–7 lin. long, glabrous; sepals 3–4 lin. long, spreading, subulate-acuminate from an ovate-lanceolate base, glabrous; corolla 1½–1¾ in. in diam., puberulous on the lobes within, elsewhere glabrous; tube very shortly campanulate, constricted and about ¼ in. in diam. at the mouth; limb slightly saucer-shaped very abruptly spreading from the tube, raised around its mouth into a broad thick convex ring; lobes about 5 lin. long and ½ in. broad at the base, deltoid, very acute; inner surface sulphur-yellow or pale greenish-yellow, marked with transverse purple-brown broken bands which pass into spots upon the smooth shining annulus, and in the only dried flower seen are very evident on the back of the lobes; outer corona 5-lobed; lobes contiguous, ½–⅔ lin. long, 1⅓ lin. broad, transversely oblong, subtruncate with a notch at the middle; inner corona-lobes scarcely ½ lin. long, deltoid-lanceolate, acute, with a dorsal transverse ridge at the base, closely incumbent upon the backs of the anthers and not produced beyond them.

EASTERN REGION : "received from my daughter-in-law at Eshowe, Zululand, but I have as yet had no particulars as to where it was found," *Mrs. K. Saunders* !

A well-marked species, easily recognised by the zebra-like markings on the pubescent corolla-lobes.

Imperfectly known species.

24. H. stapelioides (Schlechter in Engl. Jahrb. xx. Beibl. 51, 55);
stems erect, about 1½ in. long and ½ in. thick, 4-angled, with
spreading tubercles, glabrous; flowers few, in extra-axillary fascicles; pedicels up to ½ in. long, glabrous; sepals 3½ lin. long,
linear, very acute, glabrous; corolla subrotate, scarcely 1¼ in. in
diam., glabrous outside, densely verrucose within; tube depressed;
lobes triangular, very acute; outer corona-lobes divaricate, subquadrate, obtusely emarginate at the apex; inner corona-lobes
filiform, thickened and verrucose at the reflexed apex, much overtopping the style-apex. *Schlechter in Journ. Bot.* 1898, 485.

KALAHARI REGION : Transvaal; near Nazareth, between Houtbosch (Woodbush) Mountains and Klipdam, 4500 ft., *Schlechter,* 4487.

In a note the flowers are said to be brownish, paler outside.

Stapelia fasciculata, Thunb., may belong to this genus, see *Stapelia.*

XLIX. HUERNIOPSIS, N. E. Br.

Calyx 5-partite. *Corolla* with a campanulate tube and 5 spreading
or recurved lobes, without small teeth at the sinal angles. *Corona*
of 5 simple lobes arising from the staminal column opposite the
anthers and adnate to it at their basal half, free above, no outer
corona. *Stamens* with their filaments connate into a tube around
the ovary and adnate at the top to the dilated part of the style;
anthers free, erect or connivent-erect around and exceeding the
small truncate style-apex. *Pollen-masses* solitary in each anther-
cell, erect or ascending, pellucid along the inner margin, attached in
pairs by very short broad cuneate caudicles to lateral processes of
the pollen-carrier.

A dwarf herb, with stout succulent 4–5-angled stems, flowering at the base,
middle or towards the top. Flowers of moderate size.

DISTRIB. Species 1, also in South-west Tropical Africa.

1. H. decipiens (N. E. Br. in Journ. Linn. Soc. Bot. xvii. 171,
t. 12, figs. 9–13); stems decumbent, more or less clavate, 1–3 in. long,
⅓–½ in. thick, obtusely 4-angled, with spreading teeth 1–1½ lin. long,
glabrous, dull green or purplish tinted; teeth with a minute tooth
on each side at the base of the ultimately marcescent tips; flowers
2–4 together at the middle or towards the top of the young stems,
opening successively; pedicels 1–3 lin. long, stout, glabrous; sepals
¼–⅓ in. long, lanceolate, acuminate, glabrous; corolla in bud glabrous,
ellipsoid, acuminate, 5-angled, when expanded about 1 in. in diam.,
glabrous outside and within, fringed at the base of the lobes with a
few vibratile clavate dark purple hairs, outside pale greyish-green
spotted and streaked with purplish-grey, inside brownish-red or rich

brownish-crimson, more or less mottled with yellow, the intensity of colour varying even in flowers of the same cluster; tube 4–5 lin. long, and as much in diam., campanulate; lobes 4–5 lin. long, 3–4 lin. broad at the base, deltoid-ovate, acuminate or very acute, recurved; corona-lobes about 2½ lin. long, very stout, nearly square in transverse section at the lower adnate part, with subulate acute connivent-erect tips much exceeding the anthers, purplish on the back and at the apex, yellowish on the slightly concave sides below the middle. *K. Schum. in Engl. und Prantl, Pflanzenfam.* iv. ii. 277; *Schlechter in Journ. Bot.* 1898, 476; *N. E. Br. in Dyer, Fl. Trop. Afr.* iv. i. 499.

SOUTH AFRICA : without locality, *MacOwan,* 2246 ! and *cultivated specimens* !

KALAHARI REGION : Griqualand East; near Griquatown, *Pillans,* 1313! near Douglas, *Pillans.* Bechuanaland; near Vryburg, *Miss Fry in Herb. Pillans,* 122 !

Also in Tropical German South-west Africa.

Described from living plants. Under cultivation the flowers emit their very nauseous odour most strongly during the evening, the next morning it has almost disappeared, so this plant is probably fertilised by a late-flying insect. The corona-lobes secrete a sweet fluid copiously upon their backs and sides.

L. DIPLOCYATHA, N. E. Br.

Calyx 5-partite. *Corolla-tube* campanulate, with another tube arising from its base within, reaching to the mouth and there thickened into a recurved rim; limb very spreading, 5-lobed, valvate in bud. *Corona* double, arising from the staminal column; outer of 5 broad bifid lobes connate at the base; inner of 5 ovate lobes incumbent on the backs of the anthers and dorsally adnate at the base to the outer corona. *Stamens* with their filaments connate into a tube around the ovary, and adnate above to the dilated top of the style; anthers subhorizontally inflexed upon the truncate style-apex. *Pollen-masses* solitary in each anther-cell, subhorizontal, with a pellucid submarginal line along the inner margin, attached in pairs by exceedingly short broad caudicles to (or subsessile upon) the rather large lateral wings of the pollen-carriers.

A dwarf succulent leafless perennial; stems decumbent, stout, 4-angled, acutely toothed; flowers large, pedicellate, arising from near the base of the young shoots.

DISTRIB. Species 1, endemic.

1. **D. ciliata** (N. E. Br. in Journ. Linn. Soc. xvii. 168, t. 12, figs. 1–3); stems decumbent and ascending, 1½–2½ in. long, ½–¾ in. thick, exclusive of the teeth, obtusely 4-angled, with stout conical acute teeth 2–3 lin. long, glabrous, green, mottled with purple; flowers subsolitary from near the base or middle of the stems; pedicels 6–8 lin. long, erect, glabrous; sepals about ¼ in. long, ovate or ovate-lanceolate, acute, glabrous; corolla about 3 in. in diam.,

smooth and glabrous outside, densely papillate-rugose on the inner-
face, according to *Thunberg* and *Masson*, greyish, with the tips of
the papillæ reddish, but according to *Masson's* figure, pale yellowish
with a greyish ring around the mouth of the tube, minutely dotted
with red ; tube campanulate, apparently slightly raised at its mouth
around the very thick recurved papillate-rugose rim of the inner
tube, which is densely covered with stiff purple hairs at the base
around and under the corona ; lobes about 1 in. long, $\frac{2}{3}-\frac{3}{4}$ in. broad,
spreading, ovate, acute, ciliate from base to apex with clavate
vibratile white hairs ; outer corona-lobes arising above the base of
the staminal column, connate at the base, somewhat spreading, with
the free part $\frac{2}{3}-\frac{3}{4}$ lin. long, 1 lin. broad, transverse or subquadrate,
very obtusely or subacutely bifid, glabrous, apparently yellowish
dotted with purple-brown ; inner corona-lobes incumbent on the
backs of the anthers, about $\frac{3}{4}$ lin. long, thick, ovate, acute, or
acuminate with the tips produced into a very short erect point,
apparently yellowish, dotted and marked with purple-brown.
Diplocyathus ciliatus, K. Schum. in Engl. und Prantl, Pflanzenfam.
iv. ii. 281 *and* 279, *fig.* 84, *G–H.* ; *Schlechter in Journ. Bot.* 1898,
485. *Stapelia ciliata, Thunb. Prodr.* 46 ; *Fl. Cap. ed.* 2, ii. 168,
and ed. Schultes, 240 ; *Masson, Stap.* 9, *t.* 1 ; *Willd. Sp. Pl.* i. 1277 ;
Pers. Syn. Pl. i. 277 ; *Poir. Encycl.* vii. 377 ; *Ait. Hort. Kew. ed.* 2,
ii. 85 ; *Schultes, Syst. Veg.* vi. 31 ; *Spreng. Syst. Veg.* i. 837 ; *Loud.
Encycl. Pl.* 200, *fig.* 3314 ; *Dietr. Syn. Pl.* ii. 884 ; *Decne in DC.
Prodr.* viii. 655. *Podanthes ciliata, Haw. Syn. Pl. Succ.* 34 ; *G. Don,
Gen. Syst.* iv. 118. *Tromotriche ciliata, Sweet, Hort. Brit. ed.* 2, 358.

CENTRAL REGION: Calvinia Div. ; near Hantam, *Meyer* ! (Herb. Berlin).
Ceres Div. ; Karoo between the Roggeveld Mountains and Paarde Berg, *Thun-
berg* ! Karoo between Bokkeveld and Paarde Berg (ex Banks' MSS. at British
Museum), *Masson.* Beaufort West Div. ; Nieuwveld, between Rhenoster Kop and
Ganzefontein, *Drège,* 6396 ! Prince Albert Div. ; Karoo near Prince Albert,
Marloth, 4570 !

Thunberg states that the flowers of this plant have scarcely any odour. The
Paarde Berg above mentioned, according to Thunberg's travels, is situated between
Ongeluks River and Doorn River.

LI. STAPELIA, Linn.

Calyx 5-partite. *Corolla* rotate or broadly cup-shaped or with a
short broad cup- or funnel-like tube or with a central cavity containing the corona, with or without a raised ring (*annulus*) on the
disk around the corona, 5-lobed ; lobes valvate in bud. *Corona*
double, arising from the staminal column ; outer corona of 5 simple
entire, bifid, 3-toothed or deeply 3-fid lobes alternating with the
anthers, quite free or rarely shortly connected at the base, very
rarely of 10–15 free segments or subequally 10-toothed ; inner
corona of 5 simple, bifid, 2-horned or dorsally winged or crested
lobes opposite and incumbent at their base or wholly upon the

anthers and longer or shorter than them. *Staminal column* arising from the bottom of the corolla ; anthers ascending or inflexed upon the top of the style, without a terminal appendage. *Pollen-masses* ascending or subhorizontal, pellucid along the upper part of the inner margin, solitary in each anther-cell, attached in pairs to the pollen-carriers by short stout caudicles ; pollen-carriers with a wing-like expansion on each side, blackish or dark brown. *Follicles* narrowly or stoutly fusiform, smooth. *Seeds* crowned with a tuft of hairs.

Dwarf succulent perennial herbs, branching at the base ; stems thick and fleshy, 4- (rarely, and then abnormally 5-6-) angled ; angles often compressed and usually toothed ; teeth spreading or ascending or tipped with minute rudimentary or rarely with distinct subulate leaves 2–5 lin. long ; flowers 1 to many together from near the base, middle or upper part of the stems or the clusters scattered along their sides, pedicellate, large or of moderate size, rarely small, usually with a carrion-like or disagreeable odour.

DISTRIB. Species about 56 in South Africa, with 3 or 4 in Tropical Africa, besides several hybrids that have originated in European gardens.

The genus *Stapelia* was established by Linnæus upon *S. variegata* and *S. hirsuta*, two forms so distinct in habit and details of structure that Haworth separated them (with their allies) into two genera, and established several others upon species that had been introduced by Masson at the end of the 18th century. But the characters of *Stapelia* and those upon which Haworth established his genera *Orbea, Tromotriche, Tridentea, Podanthes* and *Gonostemon* so completely pass into one another, that it becomes impossible to find any real distinguishing characters to separate them from *Stapelia* when the whole of the known species are reviewed. Thus understood, the genus is easily recognised and well distinguished from all the others, with the exception of *Caralluma*, which, although usually readily distinguished by habit alone, contains a few species which cannot ; *C. aperta* is a notable example. A large number of the Stapelias cultivated in European gardens (including several described by myself) supposed to have come from South Africa are really hybrids that have originated in Europe by the cross-fertilisation of different species or varieties. Since many of these have been described and figured as distinct species of South African origin, they are included in [ ], and their names in the key are not printed in thick type. Hitherto the toothing of the outer corona-lobes, the adhesion, freedom or length of the dorsal wing of the inner corona-lobes, and, to a certain extent, the markings upon the corolla have been considered as good specific characters. But I find from observations of living plants, that within certain limits these characters are all liable to vary, sometimes even upon the same plant Like other members of the tribe, some species of *Stapelia* are eaten by the natives, whilst others are poisonous and used for medicinal purposes only. The *Kew Bulletin*, 1903, 17, contains an interesting account by Mrs. M. E. Barber of these plants as they grow under natural conditions and of the manner in which they are gradually becoming exterminated in some parts of South Africa.

I. Corolla without a distinct raised ring (*annulus*) on the
　　disk, which sometimes forms a short broad tube or
　　has a depression or cavity containing the corona, but
　　is never raised around its mouth (dried flowers of
　　some species with an annulus often have the latter
　　nearly or quite obliterated, and then falsely appear
　　to belong here) :
　* Corolla with the very small cup-shaped united part
　　　and very base of the lobes covered with short erect
　　　clavate purple hairs ; lobes 8 lin. long, very rugose,
　　　dull yellow, not ciliate ; stems velvety　...　　... (39) **flavopurpurea.**

** Corolla with the cup-like united part and base of the
    lobes glabrous and the tips pubescent (unknown in
    36, *S. fissirostris*) on the rugose inner surface;
    inner corona-lobes erect, about ¾ lin. long, bifid
    or emarginate at the apex; stems velvety:
    Corolla entirely chocolate-red or with slender olive
        lines between the ridges ...   . (35) **rufa.**

    Corolla yellowish-green with purple-brown ridges ... (36) **fissirostris.**

*** Corolla glabrous and rugose or smooth on the inner
    surface of the disk and lobes (in 52, *S. Bayfieldii*,
    54, *S. engleriana*, 30, *S. incomparabilis* and some-
    times in 28, *S. conformis*, the depression around
    the corona or the cup-like cavity containing it is
    minutely pubescent):
    Pedicels 3–6 in. long; stems not or obscurely
        toothed, glabrous; corolla-lobes ciliate at the
        base with vibratile clavate hairs...   ...   ... (43) **pedunculata.**

    Pedicels ⅙–2 in. long:
    † Corolla with the lobes extended not more and
        usually less than 2 in. in diam. (except
        perhaps sometimes in 53, *S. bella*); lobes
        3–10 lin. long:

    ‡ Corolla-lobes ciliate with vibratile clavate hairs,
        easily detached and almost always absent
        from specimens that have been in fluid;
        stems glabrous, with spreading or ascending
        teeth, without erect rudimentary leaves:
    Corolla with a short distinct tube-like cavity
        enclosing the corona; lobes 7–9 lin. long,
        much recurved, smooth ...   . (55) **revoluta.**

    Corolla flattish, with the centre slightly
        depressed or very broadly and shallowly
        funnel-shaped, but without a distinct
        cavity enclosing the corona:
    Stems with ascending-spreading stout teeth
        4–6 lin. long; corolla-lobes 5–6 lin.
        long, very rugose and chocolate with
        a few yellow dots on the inner surface (46) **Woodii.**

    Stems with spreading teeth 1–1½ lin. long;
        corolla smooth or slightly rugose on
        the inner surface, spotted or dotted
        with purple-brown on a sulphur-white
        to pale greenish-yellow ground; lobes
        3–5 lin. long:
        Pedicels ½–1 in. long; outer corona-lobes
            tapering to an acute or minutely
            notched apex; inner corona-lobes
            erect, filiform, 3 times as long as
            the anthers ...   . (48) **jucunda.**
        Pedicels ¾–2 in. long; outer corona-lobes
            oblong, bifid, with diverging re-
            curved teeth; inner corona-lobes
            not erect and shorter to not much
            longer than the anthers   ...   ... (49) **parvipuncta.**

  ‡‡ Corolla-lobes 7–8 lin. long, nearly smooth deep
        purplish-red, ciliate with long flattened
        tapering hairs mingled with a few that are
        clavate; stems microscopically puberulous,
        glabrous to the eye, with spreading teeth... (53) bella.

‡‡‡ Corolla-lobes 5–10 lin. long, very distinctly ciliate
with simple hairs only, not vibratile ; teeth
of stems prominent or absent, but always
with erect rudimentary leaves ½–1 lin.
long :
Stems glabrous ; pedicels 8–10 lin. long ;
corolla-lobes smooth, pale flesh-colour to
dull purple ; inner corona-lobes not longer
than the anthers ... ... (24) **divaricata.**

Stems puberulous ; pedicels 2–4 lin. long ;
corolla-lobes rugose on the inner face :
Stem-angles broadly rounded, not toothed,
but with an impressed line at the base
of the rudimentary leaves ; corolla
pale to dark olive-green with brown
rugosities or entirely dull reddish-
brown ... ... (33) **olivacea.**

Stem-angles compressed, toothed ; corolla
purple-brown with transverse yellow
lines ... ... (34) **acuminata.**

‡‡‡‡ Corolla-lobes not ciliate or in 54, *S. engleriana*,
sometimes with a minute ciliation :
Inner corona - lobes incumbent upon the
anthers, with or without very short upcurved tips, scarcely or slightly humped
(not horned) at the base behind ; stems
glabrous ; pedicels spreading :
Stem-teeth ½–1 in. long, stoutly subulate,
ascending ; corolla spotted ... ... (45) **longidens.**

Stem-teeth small, with minute erect rudimentary leaves ; corolla purple, not
spotted ... ... (25) **stricta.**

Inner corona-lobes entirely wing-like, notched
at the apex ; corolla pale yellowish, with
transverse purple lines ... . (29) **Macowani.**

Inner corona-lobes 2-horned ; inner horn
slender, much longer than the anthers,
ascending-recurved, outer slender or
compressed ; corolla-lobes 4½–8 lin.
long :
Stems glabrous, with subulate spreading
leaves 2–5 lin. long at the tubercle-
like teeth ; pedicels erect ; corolla
greenish-yellow, deeply rugose ... (40) **virescens.**

Stems velvety - puberulous ; corolla not
deeply rugose :
Stems 5–6 lin. square, with erect rudimentary leaves at the small teeth ;
pedicels erect ; corolla-lobes 4½ lin.
long, 2 lin. broad, spreading ? ... (41) **surrecta.**

Stems 7–9 lin. square, with small spreading teeth ; pedicels not erect ;
corolla-lobes 5–8 lin. long, 5–6 lin.
broad, recurved upon the back of the
corolla ... ... (54) **engleriana**

†† Corolla with the lobes extended 2½–7 in. in
    diam. :
    Corolla-lobes 1–1¼ in. long ciliate with clavate
        vibratile hairs ; stems with spreading teeth
        or leaves, glabrous or with a scanty micro-
        scopic puberulence :
        Corolla with the united part broadly cup-like,
            entirely purple or with narrow transverse
            yellowish lines ; outer corona-lobes entire
            or bifid ... ... (30) incomparabilis.

        Corolla with a short distinct tube-like cavity
            enclosing the corona ; lobes very much
            recurved, smooth ... ... (55) **revoluta** and
                                      vars.

        Corolla with the united part flattish, usually
            slightly depressed at the centre, but not
            forming a distinct cavity or tube, blackish,
            or spotted or densely dotted with dark
            purple-brown on a pale greenish-yellow
            ground ... ... (31) **gemmiflora** and
                                        vars.

    Corolla-lobes ciliate with simple hairs, not
        vibratile ; stems velvety-puberulous :
    Corolla-lobes 1–1½ in. long, marked with
        transverse yellow lines :

        Stem-teeth tipped with an erect rudi-
            mentary leaf ; disk of corolla slightly
            and broadly cup-like ; lobes spreading
            or recurved ... ... (28) **conformis.**

        Stem-teeth spreading or ascending ; disk
            of corolla flattish, with a slightly
            depressed centre ; lobes recurved nearly
            or quite to the pedicel... ... ... (52) Bayfieldii.

    Corolla-lobes 1⅔–3½ in. long, entirely dark
        purple-brown; stem-teeth with erect
        rudimentary leaves ... ... (26) **Pillansii.**

    Corolla-lobes not ciliate :
        Stems (probably glabrous) with spreading
            conical acute teeth ; corolla-lobes
            ¾–1 in. long, yellow spotted with
            blood-red or purple-brown ... ... (62) **irrorata.**

        Stems distinctly velvety-puberulous or
            with a microscopic puberulence when
            viewed with a lens, their teeth with
            erect rudimentary leaves ½–2 lin.
            long:
        Inner corona-lobes with a free dorsal
            wing, or wing-like crest, ½–3 lin.
            long ; pedicels 6–12 lin. long :
            Corolla entirely blackish - purple,
                without transverse markings ;
                lobes 14–15 lin. long ; stems
                6–7 lin. square ... ... (32) **vetula.**

            Corolla with distinct transverse
                yellow or whitish lines ; stems
                9–12 lin. square :

Corolla-lobes 1½–2 in. long, marked
with transverse yellowish lines
to their tips ...　　..　　... (27) **glabriflora.**

Corolla-lobes 1–1¼ in. long, with
their apical third entirely dark
purple-brown　　...　　... (28) **conformis,** var. β.
Inner corona-lobes entirely wing-like
and merely notched at the apex ;
corolla pale yellowish, transversely
marked to the tips of the 9–11 lin.-
long lobes with slender purple or
purple-brown lines ; pedicels 2–5
lin. long　　...　　...　　...　　... (29) **Macowani.**

****Corolla with hairs on the inner surface of the disk or
disk and lobes besides the border of cilia, usually
transversely rugose ; stem-teeth unknown in 44,
S. furcata, spreading in 58, *S. maculosoides,* and
with erect or incurved rudimentary leaves ½–2 lin.
long in all the others :
Corolla 4½–16 lin. in diam. ; lobes not ciliate besides
the white hairs which cover the inner surface ;
stems velvety :
Stems flowering nearly from the base to the top ;
pedicels erect, 2–5 in. long ; corolla-lobes 2½
lin. long, rolled back to the pedicel, with
adpressed hairs...　　...　　...　　...　　... (42) **erectiflora.**

Stems flowering near the base ; pedicels not erect ;
corolla-lobes 5–7 lin. long, with rather stiff
erect hairs :
Pedicels 1½–3 in. long ; hairs on the corolla
clavate ; inner corona-lobes without a dorsal
wing ...　　..　　...　　...　　...　　... (37) **glanduliflora.**

Pedicels about ½ in. long ; hairs on the corolla
not clavate ; inner corona-lobes with a small
dorsal wing ...　　...　　...　　...　　... (38) **concinna.**

Corolla about 1½ in. in diam., lobes ciliate with long
vibratile clavate hairs, and covered on the inner
surface with short erect hairs resembling minute
spines ; inner corona-lobes incumbent on the
anthers, forked at their tips and behind　　... (44) **furcata.**

Corolla-lobes ciliate with long or short simple hairs
and those on the inner surface not spine-like ;
inner corona-lobes erect or recurving, very much
longer than the anthers ; stems normally flowering near the base or below the middle ; pedicels
not erect :
† Corolla with the lobes extended 8–14 in. in diam. ;
lobes 2⅔–5½ in. long, thinly covered with erect
purple hairs, light ochreous-yellow with transverse crimson lines ; stems velvety :
Corolla with a flattish disk, shallowly depressed
at the centre ...　　...　　...　　...　　... (14) **gigantea.**

Corolla with a distinct cup-like centre about
1⅙ in. deep　　...　　...　　...　　...　　.. (15) **nobilis.**

†† Corolla with the lobes extended 4½–6½ in. in
diam. ; lobes 1½–2¾ in. long ; stems velvety :

Corolla pubescent or hairy nearly or quite to
the tips of the lobes on the inner surface
besides the border of cilia :
Corolla-lobes 2–2¾ in. long, rather thickly
covered with hairs 1–3½ lin. long, all
directed towards the tips ; disk and
lower part of lobes with transverse
yellowish lines or in var. β entirely pale
greenish-yellow ... ... (6) **desmetiana.**

Corolla-lobes 1¾ in. long or less, thinly
sprinkled with erect hairs ¼–1½ lin. long,
and with or without transverse yellowish
lines ... ... (18) **Asterias** and
var.

Corolla with the hairs on the inner surface
(other than the border of cilia) not extending beyond ¾ of the length of the lobes and
sometimes not beyond their base :
Corolla without transverse yellow lines ; lobes
1½–2½ in. long ; stems ¾–1¼ in. square ;
pedicels ⅔–1 in. long :
Corolla dull purple or raw-beef colour,
somewhat thinly covered with erect
hairs on the disk and basal part of the
lobes ... ... (8) **ambigua.**

Corolla dark purple-brown, rather densely
covered with long fine purple hairs on
the disk and basal part of the lobes ... (10) **grandiflora.**

Corolla marked with transverse yellow or
whitish lines on the basal ½–⅔ of the
lobes :
Pedicels 1½–4 in. long :
Corolla-lobes very concave beneath and
humped above near the apex, caus-
ing the tips to reflex, their basal
half and the disk very densely
covered with a cushion of long fine
purple hairs ... ... (1) **pulvinata.**

Corolla-lobes not concave beneath nor
humped above near the apex, 1¼–2¼
in. long ; disk and base of the lobes
densely to rather thinly covered with
long fine hairs :
Inner corona-lobes with their inner
horns diverging to very widely
recurved-spreading, curved or
arched, not hooked at the apex ;
stems 4–10 lin. square ... ... (2) **hirsuta** and
vars.

Inner corona-lobes with their inner
horns erect or very slightly
diverging, distinctly hooked at
the apex :

Stems 4–6 lin. square ... ... (3) hamata.

Stems 6–10 lin. square ... ... (9) **sororia.**

Pedicels ½–1¼ (seldom 1½) in. long :

Outer corona-lobes minutely pubescent at
the apex and down the back ; corolla-
lobes 1¾ in. long ; disk thinly covered
with short erect purple hairs ... (13) **Plantii.**

Outer corona-lobes glabrous ; stems ¾–1¼
in. square (not seen in 5, *S. senilis*) ;
corolla-lobes 1½–2¾ in. long :
Inner corona-lobes yellow, shading into
purple-brown at the base ; hairs
on the disk and base of the lobes
erect ... ... (7) **flavirostris.**

Inner corona-lobes purple-brown, sometimes paler at the apex :
Inner horn of the inner corona-lobes
hooked at the apex ; hairs on
the disk and lobes of the corolla
erect, purple, scarcely silky ... (8) **ambigua**, var. *β.*

Inner horn of the inner corona-lobes
straight or curved, but not
hooked at the apex :
Hairs on the disk and lower ⅔ of
the lobes directed towards the
tips of the lobes, but not
adpressed, pale purple or
white, scarcely silky... ... (6) **desmetiana,**
var. *β.*

Hairs on the disk and base of the
lobes probably erect, pure
white, very fine and silky ;
pedicels ½ in. long (longer in
the 4 preceding) ... ... (5) **senilis.**

††† Corolla with the lobes extended 3–4½ in. in diam. ;
lobes 1⅛–1¾ in. long ; pedicels ⅓–1½ (or in 2,
S. hirsuta up to 4) in. long ; hairs on the
disk and base of lobes fine and soft or silky :
Corolla entirely yellow on the inner surface,
without markings ... ... (2) **hirsuta,** var.

Corolla entirely light purple to dark purple-
brown with or without darker tips to the
lobes or the disk or margins paler, without
markings :
Stems glabrous :
Corolla vinous-purple, somewhat ochreous
on the disk ; hairs on the disk and
base of the lobes long, dense and more
or less adpressed, light purple ... (20) **glabricaulis.**

Corolla apparently purple or purple-brown,
with 5 rows of hairs 1–1½ lin. long
radiating from the centre to the sinuses
and shortly pubescent around the corona
only ... ... (19) **Pegleræ.**

Stems velvety-puberulous or pubescent :
Corolla not or scarcely shining on the inner
surface ; disk and base of the lobes
densely covered with long hairs :

Corolla uniformly dark purple-brown ;
  inner horn of the inner corona-lobes
  not very stout, recurved, somewhat
  hook-like   ...   . (11) **fuscopurpurea.**

Corolla bright vinous purple with the
  tips of the lobes blackish or dark
  purple-brown ; inner horn of the
  inner corona-lobes very stout, nearly
  straight, acute   ...   . (12) **Arnoti.**

Corolla not shining, dull purple or raw-beef
  colour ; disk and base of lobes covered
  (but not densely) with short erect
  purple hairs :
Dorsal wing of the inner corona-lobes
  small, crest-like, $\frac{1}{2}$–1 lin. long   ... (23) **deflexa.**

Dorsal wing of the inner corona-lobes
  wing-like, 2–2$\frac{1}{2}$ lin. long   ...   ... (16) **Massoni,** var. *β.*

Corolla very shining, vinous - purple to
  purple-brown, sometimes with greenish
  or greenish-ochreous margins or tips ;
  the entire inner surface scantily
  covered with short erect purple hairs
  or the lobes more or less glabrous   ... (18) **Asterias,** vars. *β*
                                              and *γ.*

Corolla with transverse whitish or yellow lines
  on the lobes, thickly or thinly, but not
  densely, covered with short erect purple
  hairs on the disk and basal part of the lobes
  or in 18, *S. Asterias* nearly or quite to their
  tips, besides the cilia :
Corolla-lobes very shining on the inner surface,
  purple to purple-brown, marked often
  nearly to the scarcely darker tips with
  transverse yellowish-white lines ; inner
  horn of inner corona-lobes very obtuse
  viewed sideways ...   . (18) **Asterias.**

Corolla-lobes not shining on the inner surface,
  often darker at the tips :
Corolla-lobes raw-beef colour or dull purple,
  with a few yellowish-white transverse
  lines on the middle of the lobes :
Dorsal wing of the inner corona-lobes
  small, crest-like, $\frac{1}{2}$–1 lin. long   ... (23) **deflexa.**

Dorsal wing of the inner corona-lobes
  wing-like, 2–2$\frac{1}{2}$ lin. long   ...   ... (16) **Massoni.**

Corolla-lobes purple-brown with transverse
  yellow lines on the basal part of the
  lobes   ...   . (2) **hirsuta,** vars.

†††† Corolla with the lobes extended 2–3 in. in diam.,
  with lobes 1–1$\frac{1}{3}$ in. long :
Disk of the corolla shortly and usually thinly
  pilose or pubescent with erect hairs $\frac{1}{4}$–1 lin.
  long ; pedicels $\frac{3}{4}$–1$\frac{1}{2}$ in. long :
Stems velvety-puberulous, with erect rudimentary leaves at the teeth ; corolla
  entirely light to dark purple-red or with
  a few transverse whitish or yellowish
  lines   ...   . (23) **deflexa.**

Stems glabrous, with spreading teeth ; corolla
dark violet-purple with transverse yellow
markings .. .. (58) **maculosoides.**

Disk of the corolla with 5 rows of hairs 1–1½
lin. long radiating from the centre to the
sinuses and shortly pubescent around the
corona only ; lobes apparently without
yellow markings ; stems glabrous ... ... (19) **Pegleræ.**

Disk of the corolla thickly or densely covered
with fine long hairs :
　Stems glabrous or nearly so :
　　Pedicels 1–2 in. long; corolla-lobes 1¼–1⅓
　　in. long, reddish-purple, without mark-
　　ings ; disk somewhat ochreous with the
　　hairs more or less adpressed ... ... (20) **glabricaulis.**

　　Pedicels ¾–1 in. long; corolla-lobes 1–1⅛ in.
　　long, dull smoky brown, with or with-
　　out a few greenish or dirty yellowish
　　transverse lines ; disk not ochreous ... (21) **tsomoensis.**

　Stems velvety-puberulous ; corolla-lobes about
　1¼ in. long :
　　Stems 4–7 in. high ; pedicels ⅓–1 in. long :
　　　Corolla dark brownish-purple without
　　　markings ; dorsal wing of inner
　　　corona-lobes adnate to the inner
　　　horn ... ... (22) multiflora.

　　　Corolla dark purple-brown with a few
　　　transverse yellowish lines ; dorsal
　　　wing of inner corona-lobes free ... (17) virens.

　　Stems 1 ft. high ; pedicels more than 1 in.
　　long ; corolla purple, with yellowish
　　transverse lines ; inner horn of the
　　inner corona-lobes hooked at the apex (4) **dejecta.**

II. Corolla with a distinct raised ring (*annulus*) or cushion
on the disk around the corona or around the mouth
of the small cup-like cavity containing it, sometimes
obliterated in dried flowers ; stems with conical acute
spreading teeth 1–3 lin. long, without erect rudimentary leaves, and together with the pedicels and
sepals quite glabrous except in 61, *S. Barklyi* in
which they are glabrous to the eye, but very
minutely puberulous when viewed with a lens :
　Inner corona-lobes sometimes slightly humped at the
　base or connected to the outer corona, but not
　produced into a distinct dorsal horn (see also
　64, *S. pulchella* in which the outer horn is some-
　times not more than ¼ lin. long) :
　　Corolla-lobes 1–1¼ in. long, with or without a
　　minute ciliation ; annulus large, nearly circular
　　in transverse section, from its acute margin being
　　rolled back so as to nearly touch the disk ... (67) **namaquensis.**

　　Corolla-lobes 8–10 lin. long, not ciliate ; disk
　　shallowly basin-like ; annulus about ½ lin. high,
　　obtuse ... ... (63) **verrucosa.**

　　Corolla-lobes 4½–5 lin. long, not ciliate ; disk nearly
　　flat or very shallowly saucer-shaped ; annulus
　　about ½ lin. high, obtuse... ... ... (50) **fucosa.**

Corolla-lobes 5–7 lin. long, ciliate with rather long
vibratile clavate hairs ; annulus solid, broadly
cushion-like, sloping outwards, with a central
depression ... ... (47) **Cooperi.**
Corolla-lobes 2½ lin. long, not ciliate ; disk very small,
raised into a narrow obtuse annulus about ⅛ lin.
high ... ... (51) **miscella.**
Inner corona-lobes distinctly 2-horned, both horns
filiform or subulate, or in 61, *S. Barklyi* the outer
flattened and slightly wing-like, subequal or the
outer horn shorter than the inner, in 57, *S. trifida*
sometimes also with a tooth or short horn on each
side of the base of the inner horn :
Corolla 5–6 in. in diam. ; lobes 1¾–2 in. long, ciliate
with long simple hairs ; disk and the erect
obtuse annulus villous with long soft purple
hairs ... ... (61) **Barklyi.**
Corolla 1¼–4 in. in diam. :
Annulus ½–2 lin. high, erect, very obtuse or
rounded at the top, not cushion-like nor
with a spreading or recurved margin :
Corolla pale yellowish or light yellow with spots
and transverse lines on the basal part of
the lobes dark purple-brown to dull purple,
and their tips often entirely purple-brown,
or in 56, *S. mutabilis* var. *discolor* entirely
dark purple-brown with inconspicuous
transverse yellowish lines ; see also hybrids
74 to 96 :
Corolla with minute scattered pubescence on
and around the annulus ; lobes 1–1¼ in.
long, ciliate with simple hairs 1½ lin.
long ... . (59) maculosa.
Corolla glabrous on the inner surface :
Corolla-lobes 10–13 lin. long, ciliate with
a mixture of simple and slenderly
clavate hairs ½–⅔ lin. long ... ... (57) trifida.
Corolla-lobes ciliate with vibratile clavate
hairs ½–¾ lin. long :
Corolla-lobes 13–14 lin. long, nearly
twice as long as broad ... ... (60) tridentata.
Corolla-lobes 7–11 lin. long and nearly or
quite as broad ... ... (56) mutabilis.
Corolla pale yellow, sprinkled all over with dots
or small spots of purple-brown ; lobes 5–10
lin. long, not ciliate ; outer horn of inner
corona-lobes ¼–½ lin. long ... ... (64) **pulchella.**
Annulus very prominent, somewhat cup-like, with
an acute suberect (not at all or scarcely recurved) edge, together with the inner surface
of disk and lobes glabrous ; lobes ciliate with
clavate hairs ½ lin. or less long :
Corolla-lobes ⅔–¾ in. long, with the labyrinthic
yellow lines not very numerous ; teeth of
outer corona-lobes with parallel or slightly
diverging outer margins ... ... (65) cupularis.
Corolla-lobes 1 in. long, with very numerous
labyrinthic yellow lines ; teeth of outer
corona with their outer margins tapering to
the apex ... ... (66) angulata.

Annulus very prominent, with an acute edge to
the horizontally spreading or recurved margin,
which forms a broad rim to the cup ; inner
surface of corolla glabrous, rugose ; lobes $\frac{1}{2}$–1
in. long, with or without a minute ciliation of
thick or clavate hairs ; see also hybrid forms
74 to 96.
Outer horn of the inner corona-lobes not more
than half as long as the inner horn, usually
less ; corolla-lobes not ciliate ;
Corolla 1$\frac{1}{2}$ in. in diam., with lobes about 6 lin.
long, pale yellow, with rather small
scattered purple-brown spots ; pedicels
$\frac{3}{4}$–1 in. long　　…　　…　　…　　… (68) **lepida.**

Corolla 2$\frac{1}{4}$–2$\frac{1}{2}$ in. in diam., with lobes 8–10
lin. long, thickly covered with dark
purple-brown spots and irregular thick
lines on a pale greenish-yellow ground ;
pedicels 1$\frac{1}{4}$–1$\frac{1}{2}$ in. long …　　…　　… (69) **variegata,** var.
brevicornis.

Corolla 1$\frac{3}{4}$–2 in. in diam., blackish-purple or
very dark purple-brown, with a few
irregular yellow markings　　…　　… (69) **variegata,** var.
marmorata.

Outer horn of the inner corona-lobes from more
than half as long to quite as long as the
inner horn ; corolla 2–3$\frac{3}{4}$ in. in diam., lobes
with or without a minute ciliation…　　… (69) **variegata,** vars.
and hybrids,
see separate
key.

1. S. pulvinata (Masson, Stap. 13, t. 13) ; plant about 4 in. high,
branching at the base ; stems erect, decumbent at the base, 6–7 lin.
square, with compressed angles having erect rudimentary leaves
$\frac{1}{2}$–1 lin. long at the rather prominent teeth, softly pubescent, dull
green ; flowers solitary near the base of the stems ; pedicels 1$\frac{1}{2}$–2$\frac{3}{4}$
(in fruit about 3) in. long, 2$\frac{1}{2}$ (in fruit 4) lin. thick, velvety ; sepals
4–6 lin. long, 1$\frac{1}{4}$–1$\frac{1}{2}$ lin. broad, lanceolate, acute, velvety ; corolla in
bud flattened, with 5 depressions around the short conical point,
when expanded 4–4$\frac{1}{2}$ in. in diam., minutely puberulous on the back ;
disk on the inner face and basal part of the lobes very densely
covered with a thick cushion of fine soft purple hairs ; lobes 1$\frac{1}{2}$–1$\frac{2}{3}$
in. long, 1–1$\frac{1}{2}$ in. broad, roundish-oblong or ovate, usually rather
abruptly contracted into an obtuse point $\frac{1}{4}$–$\frac{1}{3}$ in. long, below which
they are remarkably concave on the back, so that the point is
abruptly recurved and then directed outwards, rugose and glabrous
on the inner face, but densely ciliate and bordered with long soft
purple hairs, dark purple-brown, marked with transverse yellow
lines on the basal two-thirds ; outer corona-lobes ascending-spread-
ing, 2$\frac{1}{2}$–3 lin. long, $\frac{3}{4}$–1 lin. broad, lanceolate, or narrowly oblong-
linear, acuminate or 3-toothed at the recurved apex, channelled
down the face, dark purple-brown ; inner corona-lobes ascending-
spreading, 5–5$\frac{1}{2}$ lin. long, fuscous or dark brownish-purple with
(always ?) dull ochreous tips, with the dorsal wing adnate for $\frac{3}{4}$ or

wholly to the inner horn, $1\frac{1}{2}$–2 lin. broad, subacute to truncate and usually denticulate at the top; inner horn triquetrous to broadly winged-triquetrous, acute, obtuse or toothed at the apex, sometimes but shortly exceeding the dorsal wing; follicles erect, 6–7 in. long, $\frac{1}{2}$–$\frac{2}{3}$ in. thick, tapering into a long beak and into a stalk-like base about $\frac{3}{4}$ in. long, pubescent. *Willd. Sp. Pl.* i. 1279; *Pers. Syn. Pl.* i. 278; *Poir. Encycl.* vii. 379, *and in Dict. Sc. Nat.* l. 390; *Bot. Mag. t.* 1240; *Ait. Hort. Kew. ed.* 2, ii. 86; *Haw. Syn. Pl. Succ.* 20; *Allg. Teutsch. Gart. Mag.* 1815, 110, *t.* 11 (*ex Schultes*); *Lodd. Bot. Cab. t.* 206; *Schultes, Syst. Veg.* vi. 19; *Spreng. Syst. Veg.* i. 837; *Geel, Sert. Bot.* ii.; *Reichenb. Fl. Exot.* v. 11, *t.* 303; *G. Don, Gen. Syst.* iv. 115; *Dietr. Syn. Pl.* ii. 884; *Decne in DC. Prodr.* viii. 654; *N. E. Br. in Hook. Ic. Pl. under t.* 1911; *Schlechter in Journ. Bot.* 1898, 483.

Western Region: Little Namaqualand; Kamiesberg Range, *Masson, Barkly,* 28! Springbokfontein, *Alston in Herb. Pillans,* 650! 651! 14 miles north-west of Concordia, *Alston in Herb. Pillans*! and without precise locality, *Scully*! *Herb. Pillans,* 2!

2. **S. hirsuta** (Linn. Sp. Pl. ed. 1, 217); stems erect, 4-angled, 5–8 or occasionally up to 12 in. high, $\frac{1}{2}$–$\frac{3}{4}$ in. square, with compressed angles and erect rudimentary leaves $\frac{1}{2}$–1 lin. long at the teeth, softly puberulous, green; flowers 1–3 together near the base of the young stems, successively developed; pedicels $1\frac{1}{2}$–4 in. long, 2–$2\frac{1}{2}$ lin. thick, velvety puberulous; sepals 4–6 lin. long, lanceolate or linear-lanceolate, acute, velvety-puberulous; corolla 4–5 in. in diam. with the lobes extended, rather minutely puberulous on the back, transversely rugose on the inner surface, with the flat disk and basal fourth part of the lobes densely covered with soft purple hairs 2–3 lin. long, somewhat intermingled or matted (woolly-looking in dried flowers), the remaining part of the lobes glabrous, but ciliate to their tips with rather coarser long simple pale purple or whitish hairs, half of them directed inwards, very dark purple-brown at the apical half and marked with transverse cream-coloured or yellowish and purple-brown lines on the basal half of the lobes, passing into cream-colour more or less tinted with purplish on the disk; lobes $1\frac{3}{4}$–$2\frac{1}{4}$ in. long, about 1 in. broad, lanceolate, acute, flat, recurved; outer corona-lobes ascending-spreading with recurved tips, about 3 lin. long, linear or narrowly oblong-linear, tapering or rounded at the apex into a subulate point, which sometimes has a minute tooth on each side at its base, channelled down the face, dark purple-brown, ochreous-tinted at the base; inner corona-lobes ascending, 4–$4\frac{1}{2}$ lin. long, blackish, with the dorsal wing adnate for the whole of its length to the inner horn and truncate and toothed at the top or adnate for $\frac{1}{2}$–$\frac{2}{3}$ of its length and produced beyond into a deltoid obtuse or acute point shorter than or equalling the inner horn and more or less toothed on the inner margin; inner horn with the free part triquetrous-subulate, acute, slightly curved, not hooked at the apex. *Linn. Syst. Nat. ed.* 12, ii. 194; *Burm. Fl. Cap. Prodr.* 7;

Miller, Dict. ed. 8, no. 2 ; Rottb. Bot. Udstrakte Nytte, 63, fig. 11, bad ; Jacq. Misc. i. 28, t. 3, Amer. Gewächse, t. 95, and Stap. t. 51, 52 ; Pl. Indig. et Exot. Ic. t. 111 ; Ait. Hort. Kew. ed. 1, i. 309, and ed. 2, ii. 85 ; Poir. Tabl. Encycl. ii. 325, t. 178 ; Encycl. vii. 377, and in Dict. Sc. Nat. 1. 389, Planches, iii. t. 60 ; Thunb. Prodr. 46, Fl. Cap. ed. 2, ii. 168, and ed. Schultes, 240, partly ; Willd. Sp. Pl. i. 1278, and Enum. Pl. Hort. Berol. 281 ; Thornton, Pict. Bot. Pl. t. 23 ; Pers. Syn. Pl. i. 278 ; R. Br. in Mem. Wern. Soc. i. 24 ; Haw. Syn. Pl. Succ. 19, and Suppl. 9 ; Schultes, Syst. Veg. vi. 17 ; Link, Enum. Pl. Hort. Berol. i. 254 ; Spreng. Syst. Veg. i. 839, excl. syn. ; Hook. Exot. Fl. iii. t. 230 ; DC. Pl. Grass. t. 160 (not seen) : G. Don, Gen. Syst. iv. 115 ; Dietr. Syn. Pl. ii. 885, excl. syn. ; Decne in DC. Prodr. viii. 653 ; N. E. Br. in Journ. Linn. Soc. Bot. xvii. 166, t. 11, fig. 18, 19 ; Schlechter in Journ. Bot. 1898, 481, partly this and partly the varieties following, excl. syn. S. sororia, Masson. S. hirsuta, Linn., var. atra, Lindl. Bot. Reg. t. 756 ; Loud. Encycl. Pl. 198, fig. 3280 β ; G. Don, Gen. Syst. iv. 115. S. lanifera, Haw. Suppl. Pl. Succ. 8 ; Decne in DC. Prodr. viii. 663 ; Schlechter in Journ. Bot. 1898, 481. S. lanigera, Loud. Hort. Brit. 96 ; G. Don, Gen. Syst. iv. 115. S. pulvinata, J. Donn, Hort. Cantab. ed. 4, 53 ; Gerard in Hortic. Univ. vii. 133, with fig., not of Masson. S. villosa, N. E. Br. in Hook. Ic. Pl. t. 1911.—Apocynum cauliculis longissimis, &c., Pluk. Almagest. Bot. 37. Asclepias aizoides cauliculis longissimis, &c., Moris. Hist. iii. 611. Crassa minor, Rupp. Fl. Jen. ed. 1718, 27. Asclepias africana aizoides longioribus foliis, &c., Tourn. Inst. Rei Herb. i. 94. Stapelia denticulis ramorum erectis, Linn. Hort. Cliff. 77 ; Royen, Fl. Leyd. Prodr. 409 ; Miller, Fig. Pl. ii. 172, t. 258. Stapelia tuberculis crassis, etc., Burm. Rar. Afr. Pl. 29, t. 12, fig. 1, bad.

Var. β, **affinis** (N. E. Br.) ; corolla usually 4–4½ (rarely 3–3½) in. in diam. ; inner corona-lobes with the dorsal wing free nearly or quite to the base, ascending-spreading, 3–4 lin. long, ¾–1 lin. broad at the base, linear-oblong or gradually tapering from the base to an acute or obtuse point, entire ; inner horn 4–5 lin. long, triquetrous-subulate, acute, recurved-ascending ; otherwise as in the type. *S. affinis, N. E. Br. in Hook. Ic. Pl. t. 1912. S. stellaris, Lodd. Bot. Cab. t. 1312, not of Haw. S. hirsuta, sheet b of Thunberg's Herbarium, N. E. Br. in Journ. Linn. Soc. xvii. 167, t. 11, fig. 22, 23.*

Var. γ, **lutea** (N. E. Br.) ; " stems rather pale green ; pedicel, calyx and back of the corolla greenish-primrose " (*Pillans*) ; corolla about 3½ in. in diam. with the 1¼–1½ in. long lobes extended, "uniformly bright primrose-yellow on the inner surface, without markings, hairs also yellow, like raw silk, turning greyish " (*Pillans*), outer and inner corona-lobes as in var. *affinis*, but slightly smaller, "bright canary-yellow " (*Pillans*).

Var. δ, **patula** (N. E. Br.) ; corolla smaller than in the type, 2½–3¾ (rarely 4) in. in diam. ; hairs on the disk rather shorter and less dense than in the type, this is especially observable in dried specimens ; lobes 1¼–1¾ in. long, usually lanceolate and nearly or quite twice as long as broad ; inner corona-lobes variable, 3–4 lin. long, with the dorsal wing either adnate from ⅓ to all of its length to the inner horn or free nearly or quite to the base, truncate and toothed at the top or deltoid, narrowly deltoid-lanceolate or oblong, entire or toothed ; follicles slightly diverging, about 5 in. long, 4–5 lin. thick, fusiform, shortly beaked, very minutely and rather thinly puberulous, marked with longitudinal purple-brown lines ; seeds not seen ; otherwise as in the type. *S. patula, Willd. Enum. Pl. Hort. Berol.* 281 ; *Poir. Encycl. Suppl.* v. 230 ; *Schultes, Syst. Veg.* vi. 15 ; *Link, Enum. Pl. Hort. Berol.* i. 254 ; *G. Don, Gen. Syst.* iv. 114 ; *Decne in DC. Prodr.* viii. 652 ; *N E. Br. in Hook. Ic. Pl. under t.* 1914 ; *Journ. Hort. ser.* 3, xxviii. 95, *fig.* 15. *S. hirsuta, Loisl. Deslongch. in Herb. Gen. L'Amat.* ii. *t.* 126 ; *Drapiez, Herb. L'Amat. de Fl.* v. *t.* 321.· *S. sororia and var., Jacq. Stap. t.* 56, 57, *and Dietr. Syn. Pl.* ii. 886 (*excl. syn. S. lucida, DC.*), *not of Masson nor of Willldenow. S. elongata, Sweet, Hort. Brit. ed.* 2, 357. *S. rufescens of gardens, not of Salm-Dyck, figured without a description in the thirteenth Report of Missouri Bot. Gard.* 1902, 23. *S. variegata, Gouas in Rev. Hort.* 1857, 42, *fig.* 18.—*Asclepias africana aizoides, &c., Commelin, Hort. Med. Amst. Pl. Rar.* 19, *t.* 19 ; *Bradley, Hist. Succ. Pl. Dec.* iii. 5, *t.* 23, *a poor figure.*

VAR. ε, **comata** (N. E. Br.) ; corolla and corona as in var. *patula,* but with the hairs on the disk of the corolla denser, more woolly and of a darker purple. *S. comata, Jacq. Stap. t.* 49 ; *Hornem. Hort. Hafn.* i. 248 ; *Schultes, Syst. Veg.* vi. 18 ; *G. Don, Gen. Syst.* iv. 115 ; *Decne in DC. Prodr.* viii. 653.

VAR. ζ, **grata** (N. E. Br.) ; stems up to 1 ft. high, "rather more robust and squarer than in *S. affinis* and *S. patula* " (*Pillans*) ; corolla $3\frac{3}{4}$–$4\frac{1}{2}$ in. in diam., with lobes $\frac{1}{5}$–$1\frac{1}{2}$ in. long, 1–$1\frac{1}{4}$ in. broad, ovate, acute, dark purple-brown at the apex, lighter with creamy-yellow transverse lines on the lower $\frac{2}{3}$ of the lobes ; disk creamy with the hairs bright purple and rather longer and more dense than in var. *patula,* with which it agrees in corona.

[VAR. η, **unguipetala** (N. E. Br.) ; corolla $3\frac{1}{2}$–$4\frac{1}{2}$ in. in diam. ; disk with the centre and 5 bands radiating to the sinuses pale greenish-ochre ; lobes with the margins of the apical half much revolute and the tips upcurved ; otherwise as in var. *patula. S. unguipetala, N. E. Br. in Gard. Chron.* 1877, vii. 334, *fig.* 54, *and* 1908, xliv. 169, *fig.* 69.]

VAR. θ, **depressa** (N. E. Br. in Gard. Chron. 1908, xliv. 169–170, fig. 70) ; corolla in bud subglobose, with 5 apical depressions ; inner corona-lobes widely spreading, with the inner horn suddenly recurved near its base and horizontally spreading almost to the sinuses of the corolla, recurved at the tip and about $\frac{1}{3}$ longer than the narrowly attenuate spreading wing-like outer horn, which is free to the very base ; plant and flowers otherwise as in var. *patula. S. depressa, Jacq. Stap. t.* 55 ; *Schultes, Syst. Veg.* vi. 33 ; *Spreng. Syst. Veg.* i. 839 ; *Kerner, Hort. Semp. t.* 828 ; *Dietr. Syn. Pl.* ii. 886 ; *Decne in DC. Prodr.* viii. 657. *S. sororia, Hook. f. Bot. Mag. t.* 5963, *not of Masson. S. patentirostris, N. E. Br., and S. Courcelli, Hort. ex N. E. Br. in Gard. Chron.* 1877, vii. 140, *fig.* 21 : *Schlechter in Journ. Bot.* 1898, 482. *S. patula, var. depressa, N. E. Br. in Hook. Ic. Pl. under t.* 1914. *Tridentea depressa, Schultes, Syst. Veg.* vi. 850 ; *G. Don, Gen. Syst.* iv. 118.

VAR. ι, **longirostris** (N. E. Br.) ; differs from var. *depressa* only in having the inner horn of the inner corona-lobes about twice as long as the outer horn and arching over and much beyond it or curved nearly in a semicircle so that its tip is placed close to the tip of the outer horn or curves under it. *S. patula, Willd., var. longirostris, N. E. Br. in Hook. Ic. t.* 1914.

SOUTH AFRICA : without locality ; *Herb. Thunberg,* partly ! *cultivated specimens* in the Herbaria of Kew and the British Museum ! Var. β : specimen cultivated in Cape Town Bot. Gard., *Barkly,* 16 ! and specimen cultivated in England ! Var. δ ; *Herb. Linnæus, Pillans,* 53 ! 649 ! and *cultivated specimens* ! Vars. ε, η, θ and ι, *cultivated specimens* !

COAST REGION : Caledon Div. ; near Genadendal, *Prior* ! Robertson Div. ; western aspect of Kogmans Kloof, *Pillans,* 648 ! Var. β : Worcester Div. ; Sandhills Farm, on the southern slopes of Hex River Mountains, *Ayres in Herb. Pillans,* 80 ! Robertson Div. ; western aspect of Kogmans Kloof (a small flowered form), *Pillans,* 620 ! Var. γ : Caledon Div. ; near the source of the Zondereinde River, *Piers in Herb. Pillans,* 618 ! Var. δ : Caledon Div. ; Vygeboom Farm, *Piers in Herb. Pillans,* 616 ! Robertson Div.; Karoo near Robertson, *Marloth,*

3789 ! Var. ε: Malmesbury Div. ; Paarde Berg, *Pillans*, 625 ! Robertson Div. ; western aspect of Kogmans Kloof, *Pillans*, 601 ! 633 ! Var. ζ: Robertson Div. ; hills near Robertson, *Pillans*, 603 ! Kruispad, *Pillans*, 677 ! Var. θ: Worcester Div. ; near Worcester, *Cooper*, 2703 bis !

CENTRAL REGION: Ceres Div. ; Mitchells Pass, near Ceres, all the following :— Var. δ: *Barkly*, 36 ! 54 partly ! *MacOwan*, 2244 ! 2255 ! *Pillans*, 36 ! and (a rather pale form), 630 ! also *cultivated specimens* ! Var. θ: *Pillans*, 89 ! 97 ! Var. ι: *Barkly*, 54 partly !

WESTERN REGION : Little Namaqualand ; *Barkly*, 28 bis !

S. hirsuta is the second species of the tribe *Stapeliæ* discovered in South Africa, apparently towards the end of the 17th century.   Like *S. variegata*, it varies considerably in its flowers, the principal varieties being enumerated, but other forms in cultivation raised from seeds produced by insect agency in Europe do not quite correspond to any of the above and are certainly hybrids between forms of *S. hirsuta* or between it and some other species.   Var. *unguipetala* I now believe to be one of these, as no wild example is known.   The specimen labelled *S. hirsuta* in the Linnean Herbarium consists of a single flower of the var. *patula*, from which the corona has been removed.   This, however, is not the form described by Linnæus in his *Hortus Cliffortianus* and preserved in the Hortus Cliffortianus Herbarium (now at the British Museum), and as that was evidently the form commonly cultivated (since all the other specimens I have seen bearing date between 1727 and 1800 belong to it), I have taken the *Hortus Cliffortianus* specimen as the type.   This form seems to have disappeared from cultivation, and I have never seen a living cultivated example of it.   Thunberg states that *S. hirsuta* grows "on mountains near Cape Town, near Paarl and elsewhere," but as I have shown in the *Journal of the Linnean Society*, xvii. 166, his Herbarium contains 3 unlocalised forms under that name, one of which is probably typical *S. hirsuta*, one the variety *affinis*, and the other I think is a distinct and at present otherwise unknown species.   So far as known to me no form of this species is now found on the mountains near Cape Town.   Commelin's figure (a good one for that period) quoted by Linnæus represents the form of var. *patula* in which the dorsal wing of the inner corona-lobes is entirely adnate to the inner horn, but this appears to be an unstable character.   The type specimen of *S. lanifera*, Haw. (a single flower), at Oxford, is very instructive as to variability of the inner corona lobes ; three have the dorsal wing adnate for all its length to the inner horn, one free for ¾ and the fifth quite free to the base, thus combining the characters of the type and var. *affinis*.   The var. *lutea* is a curious and distinct form, discovered in April, 1906, by Mr. C. Piers, who only found a single plant, growing amongst var. *affinis*, of which it is evidently a seedling form.   The variety *grata* may perhaps be only a slight form of typical *S. hirsuta*, but the corona more nearly resembles var. *patula*.   Of *S. villosa* I have only seen one flower preserved in alcohol, which I cannot now distinguish from typical *S. hirsuta*, and suspect that the locality "Little Namaqualand" may be an error.

[3. **S. hamata** (Jacq. Stap. t. 50) ; stems erect, 5–6 in. (up to 1 ft., *Jacquin*) high, 4–6 lin. square, with slightly compressed toothed angles when young, becoming flat-sided with age, velvety-puberulous; teeth with erect or incurved rudimentary leaves about 1 lin. long ; flowers 2 to several on a very short stout peduncle at the base of the stems, successively developed ; pedicels about 2 in. long, velvety ; sepals about 4 lin. long, lanceolate, acute, velvety ; corolla subglobose and subacute in bud, when expanded about 4–4½ in. in diam. with the recurved lobes extended, puberulous on the back, transversely rugose on the inner face, dark purple-brown (paler or yellowish-tinted at the centre), with numerous transverse yellow lines on the lower ⅔ of the lobes, covered on the flattish disk and very base of the lobes with rather short erect purple hairs, which are slightly longer and denser along 5 lines radiating from the centre to the sinuses between the lobes, the latter $1\frac{1}{2}$–$1\frac{3}{4}$ in. long, 9–11 lin. broad, lanceolate, acute, densely ciliate to their tips with long simple purple hairs, half of them directed inwards; outer corona-lobes ascending-spreading with recurved tips, 3 lin. long, linear or oblong-linear, acute to emarginate, dark purple-brown, reddish at the base; inner corona-lobes suberect, dark purple-brown with yellowish or paler tips to the inner horn; dorsal wing free to the base, 3 lin. long, 1–$1\frac{1}{2}$ lin. broad at the base, thence tapering to the acute apex, entire or with a tooth on the inner margin; inner horn parallel with the wing, 4–$4\frac{1}{2}$ lin. long, triquetrous-subulate, acute and recurved-hooked at the apex, sometimes (abnormally) with a small tooth above the middle. *Haw. Syn. Pl. Succ.* 19, *and Suppl.* 9; *Hornem. Hort. Bot. Hafn.* i. 249; *Kerner, Hort. Semp. t.* 550; *Poir. Encycl. Suppl.* v. 230; *Schultes, Syst. Veg.* vi. 17; *Link, Enum. Pl. Hort. Berol.* i. 254; *Lodd. Bot. Cab. t.* 242; *Spreng. Syst.* i. 839, *excl. syn.*; *G. Don, Gen. Syst.* iv. 115; *Dietr. Syn. Pl.* ii. 886, *excl. syn.*; *Decne in DC. Prodr.* viii. 653; *Loud. Encycl. Pl.* 198, *fig.* 3281; *Schlechter in Journ. Bot.* 1898, 481.

SOUTH AFRICA ? *cultivated specimen* !

This is only known to me from cultivated plants. Stated by Haworth to have been introduced in 1805, but may have been raised from seed produced in Europe by insect agency, and therefore a hybrid.]

4. S. dejecta (Salm-Dyck, Hort. Dyck, 372); stems about 1 ft. high, erect, thick, tomentose; angles with approximate erect teeth (rudimentary leaves); pedicels subsolitary near the base of the stems, longer than the corolla, tomentose; corolla 3 in. in diam., tomentose on the back, villous with purple hairs at the centre of the inner face; lobes variably spreading, deflexed or divaricate, lanceolate, acuminate, oblique at the apex, ciliate with reddish hairs along the revolute margins, otherwise glabrous on the inner face, shining, pale yellowish, with elevated transverse rosy lines; outer corona-lobes spreading, channelled, 3-toothed at the apex; inner corona-lobes blackish-crimson, with a broad toothed dorsal wing and a triquetrous hooked inner horn. *Decne in DC. Prodr.* viii. 659; *Schlechter in Journ. Bot.* 1898, 479.

ORIGIN : stated to have been raised from South African seeds in the garden of Prince Salm-Dyck.

5. S. senilis (N. E. Br.); stems 1 ft. high and much like those of *S. conformis* (*Pillans*); pedicels short, $\frac{1}{2}$ in. long, 2 lin. thick, velvety; sepals 4 lin. long, lanceolate, acute, velvety; corolla $4\frac{1}{2}$–5 in. in diam. with the lobes extended, velvety-puberulous on the back, transversely rugulose on the inner face, with the disk and lower $\frac{2}{3}$ of the lobes light purple with very numerous narrow crowded transverse light yellowish or straw-coloured lines not extending to the margins, and the apical third

dull purple or purple-brown with a small ochreous or greenish patch
at the apex; disk depressed at the centre and together with the
basal half or rather more of the lobes thickly covered with very fine
soft simple pure white hairs $\frac{1}{4}$ in. or more long; lobes recurving,
$1\frac{3}{4}$–2 in. long, 1–$1\frac{1}{8}$ in. broad, lanceolate, acute, ciliate with long
simple white hairs, much stouter and longer than those on their
face, part of them directed inwards; outer corona-lobes ascending-
spreading with recurving tips, about $\frac{1}{4}$ in. long, linear, acuminate,
concave down the face, dark purple-brown with paler (ochreous or
whitish?) margins; inner corona-lobes ascending-spreading, dark
purple-brown, perhaps slightly paler at the tips, with the dorsal
wing $2\frac{1}{2}$–$3\frac{1}{2}$ lin. long, adnate from $\frac{1}{4}$–$\frac{2}{3}$ of its length to the inner
horn, with the free part scarcely 1 lin. broad at the base, thence
tapering to an acute apex, entire or with a tooth on the inner
margin; inner horn 5–6 lin. long, triquetrous-subulate, acute, rather
straight and parallel with the dorsal margin of the wing below,
slightly recurved at the apex.

Coast Region : cultivated in Grahamstown Botanic Garden, said to have been
collected in Albany Div., *Herb. Pillans*, 198 !

Mr. Pillans writes: "In stem much like *S. conformis* and quite unlike
S. flavirostris or *S. desmetiana*. One of the most singular Stapelias I have seen
for colour. At a distance it looks as though it had been finely sprinkled with
flour." The hairs on the disk and face of the lobes are much finer and softer than
in *S. desmetiana*, var. *apicalis*, which it otherwise somewhat resembles when
dried.

6. **S. desmetiana** (N. E. Br. in Gard. Chron. 1889, vi. 684);
stems 6–12 in. high, $\frac{3}{4}$–$1\frac{1}{4}$ in. square, with the angles much compressed and the teeth with erect rudimentary leaves 1 lin. long,
green, conspicuously velvety-pubescent; flowers usually 3–10
successively developed from a short stout peduncle near the base of
the young stems; pedicels usually $\frac{3}{4}$–1 in. long, occasionally longer,
2 lin. thick, velvety; sepals 4–7 lin. long, lanceolate, acute, velvety;
corolla ovate, acuminate in bud, when expanded $4\frac{1}{2}$–$6\frac{1}{2}$ in. in diam.,
pubescent on the back, transversely rugose on the inner face, with
the disk and basal $\frac{1}{3}$–$\frac{2}{3}$ of the lobes varying from very pale purple to
dark purple-red, with numerous transverse pale yellowish (rarely
olive-coloured) lines, rather broadest on the lobes, apical part of the
lobes entirely dull purple to very dark purple, with the entire
surface nearly or quite to the tips of the lobes covered with straight
simple pale purple (or occasionally white) hairs 1–$3\frac{1}{2}$ lin. long, all
pointing towards the apex of the lobes, but are not adpressed; lobes
2–$2\frac{3}{4}$ in. long, 1–$1\frac{1}{4}$ in. broad, lanceolate, acuminate, flat, ciliate to
the tips with long simple white or pale purple hairs, half of them
directed inwards; outer corona-lobes ascending-spreading, slightly
recurved at the tips, 3–5 lin. long, linear-oblong to slightly ovate-
oblong or lanceolate, varying from acuminate or acute to very
obtuse, with a minute apiculus or slightly 3-toothed at the apex,
concave down the face, pale yellowish with a broad dull purple

stripe down the middle and some purple dots along the margins or purple with the apex or some mottlings there yellowish; inner corona-lobes connivent-erect with the tips slightly diverging, dark purple-brown, often paler (or perhaps yellowish) towards the tips, variable, sometimes with the dorsal wing adnate for $\frac{3}{4}$ or wholly to the inner horn, $1\frac{1}{3}$–2 lin. broad, truncate or oblique and denticulate or entire at the apex, sometimes free nearly or to the base or adnate for half its length, $\frac{3}{4}$–$1\frac{1}{2}$ lin. broad at the base, deltoid or deltoid-attenuate to oblong, acute or obtuse or toothed, inner horn 3–5 lin. long, varying from slightly longer to twice as long as the dorsal wing, triquetrous-subulate, acute, straight or slightly recurving. *N. E. Br. in Hook. Ic. Pl. t.* 1916.

VAR. β, **apicalis** (N. E. Br. in Gard. Chron. 1889, vi. 684); "stems very large" (*Pillans*); corolla-lobes with the apical $\frac{1}{4}$ to $\frac{1}{3}$ without hairs on their inner surface, sometimes marked with transverse yellow lines nearly to the tips; otherwise as in the type.

VAR. γ, **pallida** (N. E. Br.); "plant forming clumps sometimes yards across" (*Pillans*); stems erect, shortly decumbent at the base, very stout, 9–10 in. long, $1\frac{1}{4}$–$1\frac{1}{2}$ in. square; corolla $5\frac{1}{2}$–6 in. in diam., with the apex and margins of the lobes purple, the remainder of the lobes and whole of the disk pale greenish-yellow without markings of any kind; hairs pale greyish or whitish; outer corona-lobes light purplish on the sides with the acuminate tips and a line down the middle yellowish; inner corona-lobes dull purple or purple-brown at the base, yellowish above, with the deltoid acute dorsal wing nearly or quite free to the base; otherwise as in the type.

SOUTH AFRICA : without locality, *cultivated specimens*!
COAST REGION : Alexandria Div.; near Mimosa, *Mrs. Hollands in Herb. Pillans*, 608! Bedford Div.; Babians River, without indication of collector in *Albany Museum*, 2! *Herb. Pillans*, 188! Queenstown Div.; Shiloh or Oxkraal Mountains, 3500–4000 ft., *Baur*, 783!
CENTRAL REGION : Somerset Div.; near Klein Fish River, *MacOwan*, 1923 b! 2249! *Barkly*, 72! Glen Avon, near Somerset East, *Pillans*, 58! 77! 78! 134! 148! 169! 170! 177! 184! 187! 195! 626! Graaff Reinet Div.; Ryneveldts Pass and stony hills near Graaff Reinet, 2500–3200 ft., *Bolus*, 31! Var. β: Somerset Div.; Great Fish Riv., *Juby in Herb. MacOwan*! near Somerset East, *Pillans*, 174! 175! Aberdeen Div.; near Aberdeen, *Pillans*, 40! Var. γ: Willowmore Div.; between Willowmore and Kipplaat, *Pillans*, 155!

7. S. flavirostris (N. E. Br. in Gard. Chron. 1877, vii. 558, under *S. grandiflora*, var. *lineata*); stems erect, rather crowded, 4–7 in. high, $\frac{3}{4}$–$1\frac{1}{4}$ in. square, with compressed angles and concave sides, distinctly toothed, with erect rudimentary leaves 1–2 lin. long, green, with a distinctly visible dense velvety pubescence; flowers solitary or 3 together on a short stout peduncle, the two lateral (under cultivation) usually not developing, near the base of the young stems; pedicels $\frac{3}{4}$–$1\frac{1}{2}$ in. long, 2–$2\frac{1}{2}$ (in fruit 5–6) lin. thick, velvety; sepals 5–7 lin. long, lanceolate, acute, velvety; corolla in bud ovoid, obtuse and slightly depressed at the apex, when expanded $5\frac{1}{2}$–$6\frac{1}{2}$ in. in diam. with the lobes extended, velvety on the back; inner surface everywhere transversely rugose, with the apical $\frac{2}{3}$ of the lobes glabrous and the basal part and disk covered with erect soft simple purple hairs, dull purple-red on the disk, marked with transverse pale yellow and dull purple lines on the basal half and entirely very dark purple-brown on the apical half of the lobes, the dark colour extending towards the base in a broad triangle; lobes 2–2¾ in. long, 1–1¼ in. broad, lanceolate; acute, somewhat narrowed at the base, much recurved, ciliate to their tips with long simple hairs, half directed inwards and usually pale purple, the others white; outer corona-lobes ascending-spreading, recurved at the tips, 2½–3 lin. long, linear, linear-lanceolate or linear-oblong, acute, concave down the face, dull reddish-brown or dull purple at the base, ochreous at the apex; inner corona-lobes ascending-spreading, pale yellow or ochreous-yellow shading into dark purple-brown at the base, with the dorsal wing adnate usually from ⅔ to wholly (rarely for ½ or less) to the inner horn, very rarely free nearly or to the base, 2–4 lin. long, usually 1½–2 (rarely 1) lin. broad, entire, obtuse to truncate, rarely acute or toothed at the apex; inner horn 3–5 lin. long, triquetrous-subulate, acute or 3-toothed at the apex, varying from scarcely longer to twice as long as the wing, nearly straight or slightly recurved at the tips; follicles parallel, 6–7 in. long, ½ in. thick, terete, tapering to the base and to the incurved-hooked apex, velvety-pubescent. *N. E. Br. in Gard. Chron.* 1908, xliv. 187–188, *fig.* 80. *S. grandiflora, var. lineata, N. E. Br. in Gard. Chron.* 1877, vii. 558, *fig.* 85, *and in Hook. Ic. Pl. under t.* 1916.

SOUTH AFRICA: without locality, *cultivated specimens! Herb. Pillans,* 33 ! 34 ! *Herb. Albany Museum,* 11 !

COAST REGION : George or Oudtshoorn Div. ; between George and Oudtshoorn, *Pillans,* 664 ! Bedford Div. ; Babians River, *Albany Museum,* 14 ! Queenstown Div. ; near Queenstown, *Frost in Herb. Pillans,* 106 !

CENTRAL REGION: Somerset Div. ; near the Fish River, 2000 ft., *MacOwan,* 1197 ! *Barkly,* 21 ! near Pearston, *Pillans,* 39 ! Beaufort West Div. ; Rhenoster Kop, *Foster in Herb. Pillans,* 138 ! Victoria West, *Barkly,* 21 bis ! Richmond Div. ; near Richmoud Road Station, *Foster in Herb. Pillans,* 96 ! Middelburg Div. ; near Middelburg, *Chalwin in Herb. Pillans,* 135 ! near Conway, *Pillans,* 163 ! 635. Steynsburg Div. ; near Steynsburg, *Cumming in Herb. Pillans,* 680 ! Albert Div. ; New Hantam, *Drège,* 6397 ! Colesberg Div. ; near Colesberg, *Shaw! Barkly,* 21 ! and without precise locality, *Pillans,* 4 !

WESTERN REGION : Little Namaqualand, *Templeman in Herb. Pillans,* 11 !

KALAHARI REGION : Griqualand West ; near Griquatown, *Thompson in Herb. Pillans,* 123 ! *Arnot,* 7 ! and *cultivated specimens !* Basutoland ; Leribe, *Buchanan !*

Mr. Pillans states of his 664, "This is the most southern point from which I have had this species. The stems of this form are narrow and elongated, resembling those of *S. lucida,* so the growth is influenced by the heavy rainfall."

8. S. ambigua (Masson, Stap. 13, t. 12) ; plant 8–9 in. high, branching at the base ; stems ¾–1 in. square, softly pubescent ; angles compressed, with erect rudimentary leaves 1½–2 lin. long at the teeth ; flowers successively developed in a 3–5-flowered cyme near the base of the stems ; pedicels ¾–1 in. long, velvety ; sepals 4 lin. long, ¾–1 lin. broad, narrowly lanceolate, acute, velvety ; corolla in bud ovoid, acute, when expanded 4½–5 in. in diam. with the recurved lobes spread out, velvety on the back; inner surface transversely rugose, brownish-purple, with darker transverse lines (rugosities) on the lobes, somewhat thinly covered with erect purple hairs on the disk or disk and basal half of the lobes, glabrous on their apical half, ciliate to their tips with long simple light purple and white hairs, half of them directed inwards; lobes 1½–2 in. long, 9–10 lin. broad, lanceolate, acute; outer corona-lobes ascending-spreading, 4 lin. long, linear-lanceolate, acute and recurved at the apex, channelled down the face, dull red-brown, with ochreous apex and margins; inner corona-lobes diverging, dark purple-brown, with the dorsal wing nearly free or adnate for ⅓–½ of its length to the inner horn, ascending, deltoid or deltoid-oblong, more or less toothed on its inner margin; inner horn triquetrous-subulate, more or less recurved-hooked at the apex. *Willd. Sp. Pl.* i. 1279, *and Enum. Pl. Hort. Berol.* 282; *Pers. Syn. Pl.* i. 278; *Poir. Encycl.* vii. 379; *Ait. Hort. Kew. ed.* 2, ii. 86; *R. Br. in Mem. Wern. Soc.* i. 24; *Jacq. Stap. t.* 53; *Haw. Syn. Pl. Succ.* 17; *Kerner, Hort. Semp. t.* 372; *Schultes, Syst. Veg.* vi. 13; *Link, Enum. Pl. Hort. Berol.* i. 253; *G. Don, Gen. Syst.* iv. 114; *Decne in DC. Prodr.* viii. 652; *Loud. Encycl. Pl.* 198, *fig.* 3271; *N. E. Br. in Hook. Ic. Pl. under t.* 1916; *Schlechter in Journ. Bot.* 1898, 479.

Var. *β*, **fulva** (Sweet, Hort. Brit. ed. 2, 357); disk and basal half of the lobes pale yellowish-green with transverse purple-brown lines, apical half of the lobes entirely purple-brown. *S. ambigua, var., Jacq. Stap. t.* 54.

South Africa : without locality, *Masson,* and *cultivated specimens* (some dated 1812, 1813 and 1829 in Herb. Haworth and Kew are probably from plants introduced by Masson)! Var. *β* : *cultivated specimens* !
Central Region : Victoria West Div. ; near Victoria West, *Barkly,* 66 !

Specimens collected at the base of a mountain west of Queenstown (*Galpin,* 1935) may belong to this species. *S. ambigua* is easily distinguished from *S. grandiflora* by its smaller flowers, narrower corolla-lobes, much more thinly scattered and rather stiffer hairs on the disk and the recurved-hooked apex of the inner horn of the inner corona-lobes.

9. **S. sororia** (Masson, Stap. 23, t. 39); plant with a rather lax habit of growth, with stems 6–12 in. high, 6–10 lin. square, erect from a decumbent base, with compressed angles and erect rudimentary leaves 1–1½ lin. long at the teeth, probably velvety-puberulous; flowers 1–3 together near the base of the young stems; pedicels 1½–2½ in. long, pubescent; sepals lanceolate, acute, pubescent; corolla about 4½ in. in diam. with the recurved lobes extended, transversely rugose on the inner surface, purple-brown, marked from near the base to the tips of the lobes with transverse yellow lines, with soft purple hairs on the disk and base of the lobes and the remaining part of the inner surface glabrous, but ciliate to the tips of the lobes with long purplish simple hairs, half of them directed inwards; lobes 1¾ to nearly 2 in. long, about 1 in. broad, lanceolate, acute, much recurved, flattish, but slightly contracted and apparently with recurved margins at the base; outer corona-lobes ascending-spreading, with recurved tips, linear, acute or

obtuse, concave down the face, dark purple-brown; inner corona-lobes with the dorsal wing shortly adnate at its base to the inner horn, the free part oblong or deltoid-oblong, toothed at the apex, erectly spreading, dark purple-brown; inner horns connivent-erect with recurved-hooked tips, triquetrous-subulate, longer than the dorsal wings, yellowish, dotted with purple-brown. *Willd. Sp. Pl.* 1278, *and Enum. Pl. Hort. Berol.* 281; *Kerner, Hort. Semp. i.* 89, *and Ic. Pl. Sel. t.* 3; *Pers. Syn. Pl. i.* 278; *Poir. Encycl.* vii. 378; *R. Br. in Mem. Wern. Soc. i.* 24; *Hornem. Hort. Hafn. i.* 248; *Haw. Syn. Pl. Succ.* 17; *Schultes, Syst. Veg.* vi. 14; *Link, Enum. Pl. Hort. Berol. i.* 254; *G. Don, Gen. Syst.* iv. 114; *Decne in DC. Prodr.* viii. 652, *excl. reference to Loddiges and Jacquin partly. S. sororia, var. alia, Jacq. Stap. t.* 58. *S. uncinata, Jacq. Stap., Synop. Stap. and Disp. Tab.; Schlechter in Journ. Bot.* 1898, 483. *S. lunata, Sweet, Hort. Brit. ed.* 2, 357.

SOUTH AFRICA : without locality, *Masson.*

I have not seen this species and very little appears to be known concerning it. Masson states that it flowered in his garden at Cape Town in 1792 and at Kew in 1797, but it is not mentioned in Aiton's *Hortus Kewensis.* Haworth in 1812 merely copies Masson's diagnosis without any additional observation, so he probably never saw the plant, the specimen named *S. sororia* in his Herbarium is *S. Massoni* and is dated 1832. Jacquin devotes 3 plates to *S. sororia,* and under pl. 57 states that he has three plants under that name; those figured on pl. 56 and 57 are certainly not *S. sororia,* Masson, but represent two forms of *S. hirsuta,* var. *patula,* N. E. Br., whilst pl. 58 in its stout stems, lax habit and the corolla-lobes marked to their tips with transverse yellow lines agrees fairly well with the plant figured by Masson, who, however, has figured the more or less compressed stems with broadly rounded angles, but no reliance can be placed upon the accuracy of his figure. Willdenow knew the plant, and it is quite clear that his citation, " *Stapeliæ sororiæ varietas,* Jacq. Stap." must refer to pl. 58 of *Jacq. Stap.* and not to pl. 57, as Jacquin and others have supposed. The hairs upon the disk of the corolla are badly indicated by Jacquin by a few minute black lines, so that they are probably erect and not very evident unless viewed sideways, as is also the case in *S. ambigua* and some other species.

10. **S. grandiflora** (Masson, Stap. 13, t. 11); stems 6–12 in. high, 1–1¼ in. square, with very compressed angles and deeply trough-shaped sides, prominently toothed, with incurved-erect rudimentary leaves 1–2 lin. long, minutely velvety-pubescent; flowers 1–3 together near the base of the stems, successively developed; pedicels ⅔–1 in. long, stout; sepals 3½–5 lin. long, lanceolate, acute, and together with the pedicels velvety-pubescent; corolla in bud ovoid, acute, when expanded 5–6 in. in diam., flat or the lobes recurving and 1¾–2½ in. long, 1–1¼ in. broad, ovate-lanceolate, acute, not recurved or revolute at the margins, rather minutely velvety-pubescent on the back; inner face dark purple-brown, without any markings, darkest on the erect glabrous and transversely rugose apical half of the lobes, densely and softly villous with long erect purple hairs on the disk and basal half of the lobes and ciliate to their tips with long simple whitish or pale purple hairs, half of them directed inwards; outer corona-lobes ascending-spreading, 3–4

lin. long, linear-oblong, acuminate or rounded into a short subulate point at the apex, concave down the face, dark purple-brown, dull yellowish at the base; inner corona-lobes ascending-spreading, about 5 lin. long, very dark purple-brown, with the broad dorsal wing usually adnate for ¾ to the whole (more rarely for only half) of its length to and not much shorter than the slightly curved or straight (not hooked-tipped) triquetrous-subulate acute inner horn, irregularly toothed or entire at the top or free part. *Willd. Sp. Pl.* i. 1278, *and Enum. Pl. Hort. Berol.* 282; *Kerner, Hort. Semp. t.* 69; *Bot. Mag. t.* 585; *Pers. Syn. Pl.* i. 278; *Poir. Encycl.* vii. 378, *and in Dict. Sc. Nat.* l. 390; *Jacq. Stap. t.* 59, *and* 64, *fig.* 3; *Ait. Hort. Kew. ed.* 2, ii. 85; *Haw. Syn. Pl. Succ.* 16; *DC. Pl. Grass. t.* 172; *Schultes, Syst. Veg.* vi. 11; *Link, Enum. Pl. Hort. Berol.* i. 253; *Spreng. Syst. Veg.* i. 839, *excl. syns.*; *G. Don, Gen. Syst.* iv. 114; *Dietr. Syn. Pl.* ii. 886, *excl. syns.*; *Decne in DC. Prodr.* viii. 652; *Loud. Encycl. Pl.* 198, *fig.* 3269; *Groenland in Rev. Hort.* 1858, 152 *and* 154, *fig.* 37; *Schlechter in Journ. Bot.* 1898, 480, *excl. syns.*; *Berger in Monatsschr. Kakt.* xii. 126, *with fig.* S. spectabilis, *Haw. Syn. Pl. Succ.* 16; *Schultes, Syst. Veg.* vi. 13; *G. Don, Gen. Syst.* iv. 114; *Decne in DC. Prodr.* viii. 652; *Schlechter in Journ. Bot.* 1898, 483, *excl. syn.* S. desmetiana, *N. E. Br.* S. obscura, *N. E. Br. in. Gard. Chron.* 1877, vii. 558, *under* S. grandiflora, *var.* lineata, *N. E. Br.*

SOUTH AFRICA : without locality ; *Mund*! *Miller*! and *cultivated specimens* (one from a plant probably introduced by *Masson*)! *Herb. Haworth*!

COAST REGION : Uitenhage Div. ; near Sundays River, *Masson*, Barkly Bridge, *Rogers in Herb. Pillans*, 176! and without precise locality, *Herb. Pillans*, 684! Alexandria Div. ; Mimosa, *Mrs. Holland in Herb. Pillans*, 696!

CENTRAL REGION : Somerset Div. ; by the Great Fish River near Espags Drift, *MacOwan*, 1197! between Cookhouse and Somerset East, *Pillans*, 639! Karoo at Bruintjes Hoogte, *MacOwan*, 1923!

11. **S. fuscopurpurea** (N. E. Br. in Hook. Ic. Pl. t. 1913); stems probably 5–8 in. high, ¾–1 in. square, softly pubescent, with the angles apparently much compressed and erect rudimentary leaves ½–1 lin. long at the teeth; pedicels ¾–1 in. long, stout, pubescent ; sepals 3–4 lin. long, lanceolate, acute, pubescent ; corolla with the lobes spread out 3½–4 in. in diam., puberulous on the back, transversely slightly rugose on the inner face, uniformly dark purple-brown, without markings, densely covered with long soft purple hairs on the disk, ciliate with stiffer long purplish or whitish hairs to the tips of the otherwise glabrous lobes, which are reflexed, 1½ in. long, 10 lin. broad, ovate-lanceolate, acuminate, with revolute margins; outer corona-lobes ascending, 3 lin. long, ⅔–¾ lin. broad, linear, obtuse, with a minute apiculus at the recurved apex, concave down the face, with the sides somewhat inrolled as if pinched together at the middle, dark purple-brown; inner corona-lobes 4–5 lin. long, erect, blackish or dark purple-brown, with the dorsal wing entirely adnate to and about ⅓ shorter than the inner horn, broad and wing-like, truncate and denticulate at the top; inner

horn triquetrous, acute, recurved, hook-like. *Schlechter in Journ. Bot.* 1898, 480.

SOUTH AFRICA : without precise locality, *Barkly,* 55 !

Of this species I have only seen a single flower and a drawing sent to Kew by Sir Henry Barkly in 1875. The plant had been cultivated at the Cape Town Botanic Garden for a long time, but no note of its origin had been retained.

12. **S. Arnoti** (N. E. Br. in Hook. Ic. Pl. t. 1915) ; stems erect, 3–8 in. high, $\frac{1}{2}$–1 in. square, pubescent, green ; angles much compressed, with erect rudimentary leaves at the teeth ; flowers 2–3 or more together from near the base of the young stems, opening successively or in pairs ; pedicels about 1 in. long, stout, pubescent ; sepals about $\frac{1}{4}$ in. long, lanceolate, acute, pubescent ; corolla in bud very broadly ovate, obtuse, with a very short broadly obconical base, when expanded about 4 in. in diam., with a flattish disk and much recurved or revolute lanceolate acute lobes $1\frac{1}{3}$–$1\frac{1}{2}$ in. long, and 9–10 lin. broad ; disk and basal half of the lobes smooth, covered with pale purple (pink, *Barkly*) hairs, apical half of the lobes slightly rugose, glabrous, but ciliate like the basal part with long simple white hairs, half of them directed inwards, bright vinous-purple, without markings, with the tips of the lobes blackish ; outer corona-lobes about $\frac{1}{4}$ in. long, $\frac{1}{2}$–$\frac{2}{3}$ lin. broad, linear, acute or indistinctly 3-toothed, channelled down the face, recurving at the tips ; inner corona-lobes erectly spreading, about $3\frac{1}{2}$ lin. long, with the dorsal wing adnate for more than half its length to the inner horn, its free part broadly deltoid, acute or obtuse ; inner horn very stout, straight, triquetrous, acute.

KALAHARI REGION : Griqualand West, without precise locality, *Arnot* (*Barkly,* 70) !

According to Sir Henry Barkly the corona is " of a dark brown colour with yellow stripes in the centre," from which I suppose the outer corona-lobes to be yellow down their middle or at the basal part. I have not seen it alive.

13. **S. Plantii** (Hook. f. Bot. Mag. t. 5692) ; stems rather compact, 6–8 in. high, $\frac{3}{4}$–$\frac{7}{8}$ in. square when fully developed, with the angles much compressed, shortly toothed, pubescent, green ; rudimentary leaves about 1 lin. long, erect ; flowers 1–3 together at the base of the stems, developing successively ; pedicels 1–$1\frac{1}{4}$ in. long, 2 lin. thick, pubescent ; sepals about 5 lin. long, lanceolate-attenuate, pubescent ; corolla $4\frac{1}{2}$–5 in. in diam. with the lobes extended, minutely pubescent on the back ; inner surface transversely rugose, thinly covered with short erect purple hairs on the disk, glabrous on the lobes, which are about $1\frac{3}{4}$ in. long and 1 in. broad, lanceolate, acuminate, abruptly recurved at about $\frac{1}{3}$ above their base, thickly ciliate to their tips with long purple hairs, half of them directed inwards, dark purple-brown, more intense at the tips, with transverse irregular and often forked pale lemon-yellow lines (summits of the ridges) on the basal $\frac{2}{3}$ of the lobes ; disk

flattish with a slight central depression; outer corona-lobes ascending-spreading, straight, not recurved at the tips, about $2\frac{1}{2}$ lin. long, linear, deeply concave-channelled down the face, more or less 3-toothed at the apex, with the lateral teeth rounded, sometimes unequal and the middle tooth shortly subulate, blackish-purple, passing into dull orange at the base, minutely pubescent with blackish hairs on the middle tooth and down the middle of the back; inner corona-lobes blackish-purple, with the dorsal wing sometimes entirely adnate to and much shorter than the triquetrous-subulate acute erectly spreading and but slightly recurved inner horn, sometimes adnate for $\frac{1}{2}$–$\frac{1}{3}$ of its length and produced above into a broad oblong free wing, more or less denticulate at the oblique obtuse apex and not much shorter than the inner horn, glabrous. *Fl. des Serres,* xix. 137, *t.* 2012.   *S. Asterias, Journ. Hort. ser. 3,* xl. 53, *fig.* 10, *not of Masson.*

South Africa : without locality, *cultivated specimen* !

Described from a living plant sent by Mr. Tuck of Grahamstown to Kew in 1866, without indication of locality. It has not been rediscovered and no specimen was preserved, but there are excellent drawings of it at Kew.

14. S. gigantea (N. E. Br. in Gard. Chron. 1877, vii. 684, and 693, fig. 112); stems erect, branching and shortly decumbent at the base, 4–8 in. long, $\frac{3}{4}$–$1\frac{1}{4}$ in. square, pubescent, light dull green; angles much compressed, with erect rudimentary leaves 1–$1\frac{1}{2}$ lin. long on the small teeth; flowers 1–2 together, near the base or towards the middle of the stems; pedicels about 2 in. long, $2\frac{1}{2}$ lin. (in fruit about $\frac{1}{2}$ in.) thick, softly pubescent; sepals 4–5 lin. long, lanceolate, acuminate, pubescent; corolla very large, in bud pentagonally ovoid, acuminate, when expanded 11–16 in. in diam., with the united part disk-like, shallowly depressed at the centre; back pubescent; inner surface transversely rugulose and thinly covered all over with long fine erect pale purplish hairs and ciliate with similar but longer hairs, light ochreous-yellow, everywhere marked with transverse crimson lines; lobes 4–$6\frac{1}{4}$ in. long, 2–$2\frac{3}{4}$ in. broad, ovate-lanceolate, acuminate, very spreading or recurved, slightly convex, scarcely revolute at the margins; outer corona-lobes ascending-spreading, slightly recurved at the apex, $2\frac{1}{2}$–3 lin. long, $1\frac{1}{4}$ lin. broad, oblong, slightly concave down the face, 3-toothed at the apex, with the middle tooth subulate, acute and longer than the obtuse side teeth, glabrous, dark purple-brown; inner corona-lobes dark purple-brown, with the dorsal wing free to the base, ascending, 2–3 lin. long, 1–$1\frac{1}{2}$ lin. broad, oblong or subdeltoid-oblong, obtuse or acute, entire; inner horn 2–3 lin. long, erect, subulate, rather obtuse, nearly straight; follicles slightly diverging, 5–$6\frac{1}{2}$ in. long, $\frac{2}{3}$ in. thick, subterete, tapering to an obtuse slightly hooked apex, pubescent. *N. E. Br. in Gard. Chron.* 1888, iv. 728, *fig.* 101, *and* 1908, xliv. 187 *and* 182, *fig.* 77 ; *Bot. Mag. t.* 7068 ; *Journ. Hort.* 1890, xxi. 349 *and* 359, *fig.* 41 ; *Gard. and Forest,* viii. 515, *with fig.* ;

Cact. Journ. i. 23, *with fig.*; *N. E. Br. in Kew Bulletin*, 1899, 55, and in *Dyer, Fl. Trop. Afr.* iv. i. 501 ; *Schlechter in Journ. Bot.* 1898, 480.   *S. grandiflora, a plate without text in Rep. Missouri Bot. Gard.* 1902, *facing p.* 21.

EASTERN REGION : Zululand ; Umvelosi River, *Gerrard,* 717 ! 778 ! and without precise locality, *Plant* (cultivated specimens) ! and cultivated specimens, *Herb. MacOwan,* 2818 ! *Herb. Pillans,* 93 !

MacOwan, 2818, is from a plant cultivated in the Cape Town Botanic Garden said to have come from Namaqualand, but I suspect it is from the same plant as the specimens distributed in *MacOwan and Bolus Herb. Norm. Austr.-Afr.* 921, stated to have been introduced from Walfish Bay.   I now think it probable that both localities are erroneous and that all the plants at present in cultivation are descendants of the plant originally discovered by Plant in Zululand and at his death brought with his other effects by his native servants to Durban Botanic Garden.   Gerrard also found it in Zululand, and so far as I can learn no one else has yet found it growing wild in South Africa.

15. **S. nobilis** (N. E. Br. in Bot. Mag. t. 7771) ; stems erect, decumbent and branching at the base, 3–5 in. long, 5–10 lin. square, with concave sides, softly pubescent, green ; angles slightly compressed, with erect rudimentary leaves 1 lin. long at the teeth ; flowers 1–5 together, near the base or middle of the young branches, successively developed ; pedicels stout, 1 in. long, $2\frac{1}{4}$ lin. thick, velvety ; sepals $3\frac{1}{2}$–4 lin. long, $1\frac{1}{2}$–$1\frac{3}{4}$ lin. broad, ovate-lanceolate, acute, velvety ; corolla very large, in bud ovoid, acuminate, with 5 projecting teeth below the middle, when expanded 5-lobed to $\frac{2}{3}$ of the way down, the united part forming a distinct campanulate tube about $1\frac{1}{6}$ in. deep, $1\frac{1}{2}$ in. in diam.; lobes $2\frac{3}{4}$–4 in. long, $1\frac{1}{4}$–2 in. broad, ovate-lanceolate, acuminate, reflexed, puberulous and light reddish-purple on the back ; inner face transversely rugulose, light ochreous yellow, everywhere marked with irregular transverse crimson lines and somewhat thinly covered with very fine long erect purple hairs, ciliate on the lobes with rather stouter simple purple hairs ; outer corona-lobes ascending, 3 lin. long, $\frac{2}{3}$ lin. broad, linear, 3-toothed at the apex, concave down the face, glabrous, purple-brown ; inner corona-lobes dark purple-brown, with the dorsal wing free to the base, erect and parallel with the inner horn, $3\frac{1}{2}$–$4\frac{1}{2}$ lin. long, $\frac{3}{4}$–$1\frac{1}{2}$ lin. broad, narrowly oblong or oblong-lanceolate, acute or obtuse, entire or denticulate at the upper part ; inner horn 3–5 lin. long, subulate, triquetrous, erect, with slightly incurved tips, not contiguous at the middle part.

SOUTH AFRICA : without precise locality, *Pillans,* 6 ! and *specimens cultivated at Kew* ! and Durban, *Wood,* 10773 !

KALAHARI REGION : Transvaal ; rocky hill-sides of the Magaliesberg Range, *E. S. C. A. Herb.* 868 ! and possibly a specimen from Nylstroom River, *Nelson,* 114 belongs here.

EASTERN REGION : Natal ; probably from the Drakensberg Range, *Pillans,* 28 !

Closely allied to *S. gigantea,* N. E. Br., but the growth is more compact, the stems are not so stout and their angles very much less compressed than in that species ; the flower is very similar in colour, but smaller, and its distinctly campanulate tube at once distinguishes it, whilst the hairs which clothe the whole inner surface are much more evident than they are in *S. gigantea.* Described from a living plant. There is some uncertainty about this species being really a native of Natal.

[16. **S. Massoni** (Haw. Syn. Pl. Succ. 18); stems 6–8 in. high, 6–8 lin. square, with the angles much compressed, softly puberulous, green; teeth with incurved-erect rudimentary leaves 1–1½ lin. long; flowers 1–3 together near the base of the stems, successively developed; pedicels 1–1½ in. long, 1¼–1½ lin. thick, velvety-puberulous as are the 3½–4 lin.-long lanceolate acute sepals; corolla 4 in. in diam. with the lobes extended, rather minutely and not densely puberulous on the back; inner face transversely rugose, raw beef colour or dull purple, darker at the tips of the lobes, not shining, with transverse yellowish-white lines on the central part of the lobes, which are about 1½ in. long, ¾ in. broad, lanceolate, acute, flat, recurved or very spreading and the tips variably recurved, densely ciliate to the tips with long simple purple hairs, glabrous on the upper ¾, thickly covered at their base and on the disk with soft short erect purplish hairs; outer corona-lobes ascending-spreading, with recurved tips, about 2½ lin. long, linear, subacute, concave down the face, blackish-purple; inner corona-lobes blackish-purple, with the dorsal wing free nearly or quite to the base, rather spreading, 2–2½ lin. long, deltoid to linear-oblong, entire or more usually with 1 tooth on the inner margin; inner horn ascending-spreading and usually somewhat recurved, about 3 lin. long, triquetrous-subulate, acute. *Schultes, Syst. Veg.* vi. 16; *G. Don, Gen. Syst.* iv. 115; *Decne in DC. Prodr.* viii. 663; *N. E. Br. in Gard. Chron.* 1883, xx. 761; *Schlechter in Journ. Bot.* 1898, 481.

VAR. *β,* **livida** (N. E. Br.); corolla raw-beef colour to dull dark purple, without transverse whitish or yellowish lines; otherwise as in the type.

SOUTH AFRICA : *cultivated specimens* !

According to Haworth this plant was introduced before 1808. I suspect, however, that it did not come from South Africa, but probably originated in some European garden. It is very closely allied to *S. Asterias,* Masson, and may have been derived from that species, differing chiefly in the corolla not being shining and the inner horns of the inner corona-lobes more slender and much more acute than in *S. Asterias.* I describe the type and var. *β* from living plants. Although Haworth does not describe the flowers of *S. Massoni,* I believe the plant here described to be that species, as it is one that has been in cultivation for very many years under that name. No specimen named *S. Massoni* is contained in Haworth's Herbarium, but one named *S. sororia,* dated 1832, is identical with *S. Massoni.*]

[17. **S. virens** (Link, Enum. Pl. Hort. Berol. i. 258, by error *vlrens,* name only); stems 4–7 in. high, ½–⅔ in. square, with compressed angles and erect rudimentary leaves 1–1½ lin. long at the teeth, puberulous, rather light green; flowers 1–5 together at the lower part of the young stems, successively developed or sometimes 2 open at the same time; pedicels ½–1 in. long, velvety; sepals 3–3½ lin. long, lanceolate, acute, velvety; corolla about 3 in. in

diam., with recurved lobes, puberulous on the back, transversely
slightly rugose on the inner face, uniformly dark purple-brown or
with a few transverse whitish or very pale yellowish lines on the
lobes ; disk flattish, with the centre depressed, and together with
the basal third of the lobes thickly covered with erect soft simple
purple hairs about 2 lin. long ; lobes about $1\frac{1}{4}$ in. long, $\frac{1}{2}$–$\frac{2}{3}$ in. broad,
lanceolate-attenuate, very acute, glabrous on the upper $\frac{2}{3}$, thickly
ciliate to their tips with long simple pale purple hairs ; outer corona-
lobes ascending-spreading, recurved at the tips, $3\frac{1}{2}$ lin. long, linear-
oblong, acute or obtuse with a minute central point or very rarely
bifid at the apex, concave down the face, dark purple-brown ; inner
corona-lobes dark purple-brown, with the dorsal wing free to the
base, ascending-spreading, usually curved to one side, $1\frac{1}{2}$–$2\frac{1}{2}$ lin. long,
$\frac{2}{3}$ lin. broad, linear-oblong, obtuse or toothed at the apex ; inner
horns connivent at the base, then in a bold curve arching over and
sometimes beyond the dorsal wings, $4\frac{1}{2}$–5 lin. long, rather slender,
triquetrous-subulate, acute.

SOUTH AFRICA ? *cultivated specimens* !

I have been quite unable to find any published description of this plant,
and describe from living specimens which were in cultivation over 40 years ago
and probably long before under the name of *S. virens.*   I believe it to be a hybrid,
produced in some European garden.]

18. S. Asterias (Masson, Stap. 14, t. 14) ; stems 4–10 in. high,
7–10 lin. square, with compressed angles, nearly flat-sided with age,
not very prominently toothed, with erect rudimentary leaves 1–$1\frac{1}{2}$
lin. long, velvety-puberulous, green ; flowers 1–5 together near the
base of the stems, developing successively ; pedicels $\frac{2}{3}$–$1\frac{1}{4}$ in. long,
velvety ; sepals $3\frac{1}{2}$–$4\frac{1}{2}$ lin. long, lanceolate, acuminate, velvety ;
corolla in bud ovate, acuminate, when expanded 3–4 in. in diam.,
star-like with very spreading (or reflexed or somewhat contorted)
lobes, $1\frac{1}{4}$–$1\frac{3}{4}$ in. long, 7–9 lin. broad, back velvety-puberulous, inner
surface transversely rugulose on the lobes, dark purple-brown, with
narrow transverse yellow or whitish-yellow lines on the lobes, not
reaching the margins, very shining, rather thinly covered with short
erect purple hairs $\frac{1}{4}$–$1\frac{1}{2}$ lin. long on the disk, often more densely
and with rather longer hairs along 5 lines radiating to the sinuses ;
lobes more sparsely hairy to quite glabrous, lanceolate, acuminate,
convex from the revolute margins, or flattish at the basal part,
thickly ciliate to their tips with long simple pale purple hairs, half
of them directed inwards ; outer corona-lobes ascending-spreading
with recurved tips, 2–$2\frac{1}{2}$ lin. long, linear, slightly concave down
the face, very shortly 3-toothed or very obtuse and apiculate at the
apex, blackish or blackish-purple ; inner corona-lobes ascending-
spreading, $2\frac{1}{2}$–3 lin. long, dull dark purple-brown, paler towards
the tips with minute dotting on a lighter ground, dorsal wing free
or $\frac{1}{2}$-adnate to the inner horn, subparallel with and nearly or
quite as long, $\frac{2}{3}$–1 lin. broad, oblong or lanceolate-oblong, obtuse,

entire or occasionally toothed at the apex; inner horn slightly recurved-erect, triquetrous-subulate, very obtuse, viewed sideways. *Willd. Sp. Pl.* i. 1280, *and Enum. Pl. Hort. Berol.* 282; *Pers. Syn. Pl.* i. 278; *Poir. Encycl.* vii. 379, *and in Dict. Sc. Nat.* l. 390; *Ait. Hort. Kew. ed.* 2, ii. 86; *R. Br. in Mem. Wern. Soc.* i. 24; *Haw. Syn. Pl. Succ.* 18; *Jacq. Stap. t.* 47; *Bot. Mag. t.* 536; *Lodd. Bot. Cab. t.* 453; *DC. Pl. Grass. t.* 166; *Schultes, Syst. Veg.* vi. 16; *Link, Enum. Pl. Hort. Berol.* i. 254; *Spreng. Syst. Veg.* i. 839; *G. Don, Gen. Syst.* iv. 115; *Dietr. Syn. Pl.* ii. 886; *Decne in DC. Prodr.* viii. 653; *Schlechter in Journ. Bot.* 1898, 479. *S. stellaris, Haw. Syn. Pl. Succ.* 19, *and Suppl.* 9; *Link, Enum. Pl. Hort. Berol.* i. 254; *G. Don, Gen. Syst.* iv. 115; *Decne in DC. Prodr.* viii. 663; *Schlechter in Journ. Bot.* 1898, 483. *S. stellata, St. Lag. in Ann. Soc. Bot. Lyon,* vii. 135.

Var. β, **lucida** (N. E. Br.); corona entirely purplish-red or purple-brown without transverse markings on the lobes, sometimes dull greenish at the tips, very shining; outer corona-lobes dark purple-brown to nearly black, with or without greenish-ochre margins at the apical half; inner corona-lobes entirely very dark purple-brown or with paler tips as in the type, the dorsal wing free to the base or partly to almost entirely adnate to the inner horn, 1½–2½ lin. long, linear, linear-oblong, deltoid or deltoid-oblong, acute, obtuse or toothed at the apex and sometimes on the inner margin; inner horn 2½–3 lin. long, triquetrous-subulate, acute to very obtuse, viewed sideways, straight or slightly recurving; follicles parallel or slightly diverging, 5–7½ in. long, 5–6 lin. thick, terete tapering at the base and into a beak at the apex; seeds 3–3½ lin. long, 1⅓–1½ lin. broad, narrowly ovate, concave with thick incurved margins on one side, convex on the other, smooth, glabrous, light brown; otherwise as in the type. *S. lucida, DC. Cat. Pl. Hort. Monsp.* 148; *Poir. Encycl. Suppl.* v. 229; *Schultes, Syst. Veg.* vi. 15; *G. Don, Gen. Syst.* iv. 114; *Decne in DC. Prodr.* viii. 652; *N. E. Br. in Hook. Ic. Pl. t.* 1919; *Schlechter in Journ. Bot.* 1898, 481.

Var. γ, **gibba** (N. E. Br.); corolla with the lobes extended about 3 in. in diam., pale to dark vinous-purple, with the tips of the lobes, a border along their margins 1–1½ lin. broad and 5 stripes radiating to the sinuses all ochreous or dull greenish colour, without transverse yellow lines on the lobes; hairs on the disk "shorter and more purple" (*Pillans*) than in var. *lucida*; lobes about 1¼ in. long, ¾ in. broad, lanceolate, acuminate, with recurved but scarcely revolute margins, somewhat convexly humped just below the apex on the inner face, concave beneath; outer corona-lobes shortly bifid or 3-toothed at the apex, even in the same flower!; inner corona-lobes with the inner horn more or less recurved-spreading, acute or obtuse viewed sideways, sometimes slightly hooked at the apex; otherwise as in the type.

South Africa: Karoo, *Masson,* cultivated specimens dated 1811, 1812, 1813 and others much later, from plants probably introduced by Masson, in *Herb. Haworth*! and *Herb. Kew*! Var. β: *cultivated specimens*!

Coast Region: Var. β: Riversdale Div.; Muis Kraal, *Pillans,* 619! 666! Seven Weeks Poort, *Bain,* 9! and *cultivated specimens*! George Div.; Ezeljagts Poort, *Barkly,* 22! *MacOwan,* 2242! and *cultivated specimens*! Div.? Caledon Kloof, *Bain,* 5! Var. γ: Ladismith Div.; along the road between Ladismith and Laingsburg, *Pillans,* 607! 643!

Central Region: Var. β: Laingsburg Div.; Witte Poort, *Pillans,* 607!

The type and var. β both described from living plants. The type of *S. stellaris,* Haw., in Haworth's Herbarium at Oxford is identical with typical

S. *Asterias*, Masson. S. *stellaris*, Jacq. ex R. Br. in *Mem. Wern. Soc.* i. 24 (1811), and *Schultes, Syst. Veg.* vi. 17 (1820) name only, probably refers to the plant Haworth described, that being the only one R. Brown would be likely to be acquainted with. But the *S. stellaris, Jacq. Stap.* t. 48, which seems not to have been published until 1813, seems rather to belong to var. *lucida*, since Jacquin describes the corolla as dark purple, without mention of any transverse whitish or yellow lines upon it. The numerous transverse whitish lines on the lobes in Jacquin's figure, I recognise as the very shining transverse rugosities, which produce that effect on the living flower of var. *lucida*. Although the corolla-lobes of the type and var. *lucida* frequently have very revolute margins, as represented in Jacquin's figure of *S. Asterias*, they are sometimes much less so, or nearly flat at the base, as represented in Jacquin's figure of *S. stellaris* ; I have reen both forms upon the selfsame plant. The inner horns of the inner corona-lobes of var. *lucida* vary from very slender and twice as long as the dorsal wing and exactly as figured by Jacquin on his plate of *S. stellaris* to rather stout and scarcely longer than the dorsal wing, almost as in *S. Asterias*.

19. **S. Peglerae** (N. E. Br.) ; stems erect, 5–6 (or more ?) in. high, probably about $\frac{1}{2}$ in. square, with scarcely prominent teeth and rudimentary erect leaves about $\frac{1}{2}$ lin. long, quite glabrous ; flowers (unattached) 3–4 together, successively developed ; pedicels about $1\frac{1}{2}$ in. long, minutely and thinly puberulous ; sepals 3 lin. long, lanceolate, very acutely acuminate, glabrous or nearly so, with a few very minute hairs on the margins ; corolla $2\frac{1}{2}$ in. (or more ?) in diam., with lanceolate acute lobes 1 in. (or more ?) long, apparently transversely rugose and dark purple-brown on the inner face, glabrous on both sides, ciliate to the tips with long simple purple hairs, half of them directed inwards ; disk with 5 rows of long simple hairs radiating to the sinuses between the lobes, and thinly or shortly pubescent with erect soft hairs around the corona only ; outer corona-lobes $2\frac{1}{3}$–$2\frac{1}{2}$ lin. long, linear, acuminate, dark purple-brown ; inner corona-lobes with the outer horn or dorsal wing free to the base, ascending-spreading, about $2\frac{1}{3}$ lin. long, $\frac{1}{2}$ lin. broad at the base, tapering to the acute apex, entire, dark purple-brown ; inner horn recurved-spreading from close to the base over the dorsal wing, 3 lin. long, triquetrous-subulate, acute, dark purple-brown at the base, paler or becoming yellowish at the apex.

EASTERN REGION : Tembuland ; Mqánduli, in dry rocky ground, rare, 1000 ft., *Miss Pegler*, 760 !

20. **S. glabricaulis** (N. E. Br. in Hook. Ic. Pl. t. 1917) ; stems rather loosely branching, decumbent at the base, 4–8 in. high, $\frac{1}{2}$–$\frac{2}{3}$ in. square, with the angles much compressed, quite glabrous, green ; teeth with erect rudimentary leaves 1–$1\frac{1}{2}$ lin. long ; flowers 2–5 together, towards the base of the young stems, successively developed or sometimes 2 expanded at the same time ; pedicels 1–2 in. long, $1\frac{1}{2}$ lin. thick, glabrous or with very few minute scattered hairs only noticeable under a lens ; sepals 4–6 lin. long, lanceolate, acuminate, with a slight protuberance at the base, glabrous on the back, usually with a few hairs on the inner surface and sometimes on the margins ; corolla subglobose in bud, with 5 depressions just below the obtusely pointed apex, when expanded $2\frac{1}{2}$–$3\frac{1}{2}$ in. in diam.,

with stellately spreading (not recurved) lobes $1\frac{1}{4}$–$1\frac{1}{3}$ in. long, about $\frac{3}{4}$ in. broad, ovate-oblong, acute, with revolute margins, ciliate to their tips with long simple pale purple hairs, half of them directed inwards, glabrous on the back, transversely rugose and reddish-purple without markings on the inner face, becoming paler and somewhat ochreous at the centre, glabrous on the terminal half of the lobes, with their basal half and the disk densely covered with long fine silky purple or greyish-purple hairs, which are more or less adpressed and directed towards the tips of the lobes ; outer corona-lobes spreading, recurved at the tips, $2\frac{1}{2}$–3 lin. long, linear, variably obtuse, apiculate, erose, minutely 3-toothed or more rarely acuminate at the apex, purplish-brown with a yellow base and sometimes yellow-brown margins ; inner corona-lobes dark purple-brown, with the dorsal wing free to the base or adnate up to half its length to the inner horn, ascending, $2\frac{1}{2}$–3 lin. long, $\frac{2}{3}$–$\frac{3}{4}$ lin. broad at the base, gradually tapering to the acute apex, or oblong-linear and acute or slightly toothed at the apex ; inner horn $3\frac{1}{2}$–$4\frac{1}{2}$ lin. long, triquetrous-subulate, acute, erect below, with the apex recurved (often rather suddenly) over the dorsal wings ; follicles $4\frac{1}{2}$ in. long, 7–8 lin. thick, fusiform, acute (*Mrs. Barber*). *Schlechter in Journ. Bot.* 1898, 480, *excl. syn.*

Coast Region : Bathurst Div. ; near Port Alfred, *Mrs. Hutton* (cultivated specimens) *in Herb. Albany Museum*, 13 ! and *Herb. Pillans*, 653 ! Fort Beaufort Div. ; Blinkwater, *Barkly*, 52 ! and *cultivated specimens* !

21. **S. tsomoensis** (N. E. Br. in Gard. Chron. 1882, xviii. 168) ; stems 4–6 in. high, 5–8 lin. square, with compressed somewhat repand-dentate angles, opaque green, glabrous to the eye and touch, but very minutely puberulous on the angles viewed through a lens ; rudimentary leaves erect, 1 lin. long, ovate, acute, minutely puberulous ; flowers 4–9 together near the base of the young stems, successively developed ; pedicels $\frac{3}{4}$–1 in. long, velvety-puberulous ; sepals 3 lin. long, lanceolate, acuminate, velvety-puberulous, usually ciliate with much longer hairs ; corolla in bud globose, very obtusely pointed, when expanded and the lobes extended $2\frac{1}{2}$–3 in. in diam., minutely puberulous on the nerves of the glabrous light green back ; inner face with a few slightly raised transverse ridges on the apical half of the lobes, entirely dull smoky-purple and darker at the tips of the lobes, or with some of the ridges greenish or dirty-yellowish ; disk flattish and with the base of the lobes covered with long soft simple somewhat adpressed dark purple hairs ; lobes 1–$1\frac{1}{6}$ in. long, 7–8 lin. broad, ovate-lanceolate, acute, strongly recurved, not revolute at the margins, ciliate with long simple purple hairs, partly directed inwards ; outer corona-lobes about $2\frac{1}{2}$ lin. long, ascending-spreading with recurved tips, oblong-linear, obtuse and mucronate or acuminate, concave down the face, purplish-black or dark purple-brown ; inner corona-lobes purplish-black or dark purple-brown, divided into a free ascending-spreading deltoid or

linear or somewhat attenuate acute dorsal-wing 1–2 lin. long and
a subulate-filiform inner horn 3–4 lin. long, connivent at the base,
then recurving, often over and beyond the tips of the dorsal wings.
N. E. Br. in Hook. Ic. Pl. t. 1918.

EASTERN REGION : Transkei ; near Tsomo, *Mrs. Barber,* 8 ! *Bowker,* 11 ! *Barkly,*
32 ! 42 ! and *cultivated specimens* !

Described from living plants.

[**22. S. multiflora** (DC. Cat. Hort. Monsp. 149) ; stems 4–6 in.
high, ½–¾ in. square, with much-compressed angles and rudimentary
erect leaves ⅔–1 lin. long at the teeth, velvety-pubescent ; flowers
three to several together near the base of the stems, often 2 or
more open at the same time ; pedicels 4–8 lin. long, stout, velvety-
pubescent ; sepals about 3 lin. long, lanceolate, acute, velvety ;
corolla globose-ovoid, acute in bud, when expanded nearly flat,
with the lobes scarcely reflexed, about 3 in. in diam., velvety on
the back ; inner face slightly transversely rugose, dark brownish-
purple, without markings, thickly covered with fine soft simple
purple hairs on the disk and base of the lobes, which are about
1¼ in. long and ¾ in. broad, ovate-lanceolate, acute, with the
margins recurved at the apical half, glabrous, ciliate to the apex
with long pale purple or white hairs, half of them directed inwards ;
outer corona lobes ascending-spreading with recurved tips, about
3 lin. long, oblong-linear, minutely and subequally 3-toothed at the
very obtuse apex, concave down the face, blackish-purple or dark
purple-brown ; inner corona-lobes ascending-spreading, blackish-
purple or dark purple-brown, about 3 lin. long, with the dorsal
wing entirely adnate to the inner horn or produced at the apex
into a short free deltoid obtuse lobe, 1¼ lin. broad ; inner horn
stoutly triquetrous-subulate, acute, produced for about ⅓ of its
length beyond the adnate part of the dorsal wing and recurved
over it. *Poir. Encycl. Suppl.* v. 229 ; *Schultes, Syst. Veg.* vi. 18 ;
G. Don, Gen. Syst. iv. 115 ; *Decne in DC. Prodr.* viii. 653 ; *Schlechter
in Journ. Bot.* 1898, 482, *not of Rüst. S. comata, var. multiflora,
Loud. Encycl. Pl.* 198.

SOUTH AFRICA ? without locality, *cultivated specimen* !

Described partly from a specimen (probably from the type) cultivated in
Montpellier Garden in Nov. 1819, and partly from a photograph, natural size,
from a plant cultivated at Antibes between 1870 and 1880. I am not satisfied
that this is a native of South Africa, as no wild specimen is known to me, and am
inclined to believe it to be a hybrid.]

23. **S. deflexa** (Jacq. Stap. t. 20) ; plant laxly branching, with
erect stems, decumbent at the base, 4–7 in. long, 5–7 lin. square,
puberulous, green or tinged with purplish ; angles compressed, with
erect rudimentary leaves 1–1½ lin. long at the teeth ; flowers 3–6
together near the base of the stems, developing successively, as the
flowering proceeds a peduncle is formed, lasting and producing

flowers for 2 or 3 years in succession, sometimes growing to 1 in. or more in length; pedicels $\frac{3}{4}$–$1\frac{1}{2}$ in. long, $1\frac{1}{4}$–$1\frac{3}{4}$ lin. thick, puberulous; sepals $2\frac{1}{2}$–3 lin. long, lanceolate, acute, puberulous; corolla 2–$3\frac{1}{2}$ in. in diam. with the lobes extended, puberulous and pale green, often tinged with purplish on the back, slightly rugose on the inner face, pilose with short erect purple hairs on the disk, glabrous on the lobes, which are ciliate with long whitish or pale purplish simple hairs, either entirely of a light to dark livid purple-red or marked with a few transverse pale greenish-white or yellowish lines, tinged at the tips of the lobes and sometimes on the disk with greenish; lobes very strongly reflexed, 1–$1\frac{1}{2}$ in. long, $\frac{1}{2}$–$\frac{3}{4}$ in. broad, lanceolate or ovate-lanceolate, acuminate, with revolute margins, 5–9-nerved; outer corona-lobes $1\frac{3}{4}$–$2\frac{1}{3}$ lin. long, oblong-lanceolate or linear-oblong, acute or acuminate, concave-channelled down the face, with the apex recurved, dull purple-brown, ochreous at the base; inner corona-lobes $1\frac{1}{2}$–2 lin. long, subulate, recurving, with a short dorsal wing-like deltoid crest $\frac{1}{2}$–1 lin. long at the base, dull purple-brown; odour slight, but fetid. *Willd. Enum. Pl. Hort. Berol. Suppl.* 14; *Hornem. Hort. Bot. Hafn.* i. 249; *Bot. Mag. t.* 1890; *Lodd. Bot. Cab. t.* 135; *Schultes, Syst. Veg.* vi. 46; *Spreng. Syst.* i. 840; *Link, Enum. Pl. Hort. Berol.* i. 254; *Dietr. Syn. Pl.* ii. 886; *Decne in DC. Prodr.* viii. 654; *Schlechter in Journ. Bot.* 1898, 479. *S. deflexa, var. Brownii, Schinz in Bull. Herb. Boiss. sér.* 2, i. 879. *S. reflexa, Haw. Syn. Pl. Succ.* 18; *Schultes, Syst. Veg.* vi. 15; *G. Don, Gen. Syst.* iv. 114; *Decne in DC. Prodr.* viii. 652; *Schlechter in Journ. Bot.* 1898, 483. *S. brevirostris, Willd. Enum. Pl. Hort. Berol. Suppl.* 14; *Link, Enum. Pl. Hort. Berol.* i. 254; *G. Don, Gen. Syst.* iv. 117. *Duvalia? deflexa, G. Don, Gen. Syst.* iv. 121.

SOUTH AFRICA: without locality; *cultivated specimens! Herb. Haworth!*

Described from living plants. May be a native of Little Namaqualand, as I have seen a living specimen cultivated at Zurich (var. *Brownii*, Schinz), which was said to have been received from German South-West Africa. I have seen an authentic specimen of *S. brevirostris*, Willd., in the Berlin Herbarium, it is identical with *S. deflexa*, Jacq.

24. S. divaricata (Masson, Stap. 17, t. 22); plant very laxly branching at the base, with erect, straggling, spreading or diverging stems 3–4 in. long, $3\frac{1}{2}$–5 lin. square, glabrous; angles somewhat compressed, with minute erect rudimentary leaves on the small teeth; flowers 1–6 together, towards the base of the young stems; pedicels $\frac{1}{3}$–$\frac{3}{4}$ in. long, very minutely or papillate-puberulous; sepals $1\frac{1}{2}$–2 lin. long, lanceolate, acuminate, glabrous; corolla 1–2 in. in diam., according to the figures, smaller when dried, flattish, with recurved-spreading lobes, glabrous on the back and on the smooth shining inner surface, ciliate with simple white hairs except at the tips of the lobes, pale greenish on the back, varying from pale flesh-colour to dull purple with greenish tips on the inner face; lobes 6–10 lin. long, 4–$4\frac{1}{2}$ lin. broad, ovate-lanceolate, acuminate,

with revolute margins ; outer corona-lobes spreading, nearly 1 lin.
long, ¾ lin. broad, subquadrate, slightly broadest at the subtruncate
minutely 3-toothed or apiculate apex, reddish or perhaps brownish-
orange, with a yellow base ; inner corona-lobes about ¾ lin. long,
incumbent upon the backs of the anthers (and in dried flowers
shorter than them), with very short erect tips, ovate or narrowly
oblong-ovate, obtuse, with a very short dorsal projection or gibbosity
at the base, reddish or perhaps brownish-orange ; nearly odourless.
Willd. Sp. Pl. i. 1280, *and Enum. Pl. Hort. Berol.* 280 ; *Pers. Syn.
Pl.* i. 278 ; *Poir. Encycl.* vii. 380 ; *Ait. Hort. Kew. ed.* 2, ii. 87 ;
Jacq. Stap. t. 22 ; *Bot. Mag. t.* 1007 ; *Lodd. Bot. Cab.* x. *t.* 941 ;
Schultes, Syst. Veg. vi. 27 ; *Link, Enum. Pl. Hort. Berol.* i. 255 ;
Spreng. Syst. Veg. i. 841 ; *Dietr. Syn. Pl.* ii. 887 ; *Decne in DC.
Prodr.* viii. 655 ; *Loud. Encycl. Pl.* 200, *fig.* 3307 ; *Schlechter in
Journ. Bot.* 1898, 480. *S. pallida, Wendl. Coll. Pl.* ii. 39, *t.* 51 ;
Schultes, Syst. Veg. vi. 28 ; *Decne in DC. Prodr.* viii. 655 ; *Schlechter
in Journ. Bot.* 1898, 482. *S. pallens, Hort. ex Steud. Nom. Bot. ed.*
2, ii. 631. *Gonostemon divaricata, Haw. Syn. Pl. Succ.* 27, *and
Suppl. Pl. Succ.* 12 ; *G. Don, Gen. Syst.* iv. 117. *G. divaricatus and
G. pallidus, Loud. Hort. Brit.* 96. *G. divaricatum and G. pallidum,
Sweet, Hort. Brit. ed.* 1, 278. *G. pallida, G. Don, Gen. Syst.* iv.
117.

SOUTH AFRICA : without locality, *Masson, cultivated specimen* ! *Herb. Haworth* !

25. **S. stricta** (Sims, Bot. Mag. t. 2037) ; stems rather crowded,
erect, 3–5 in. high, slender, 3–3½ lin. square, with minute erect
rudimentary leaves at the small teeth, probably glabrous ; flowers 2
or more together at the base of the stems, developing successively ;
pedicels " shorter than the flower " ; corolla about 1⅔ in. in diam.,
flat, star-like, glabrous on the inner surface, faintly mottled with
dull purple on a slightly paler ground, with the centre of the disk
rather paler and the margins of the lobes dull greenish ; lobes about
7 lin. long and 4–4½ lin. broad, ovate, acute, flattish, not revolute
at the margins, not ciliate ; outer corona-lobes about 1 lin. long, and
⅔ lin. broad, oblong, apparently spreading, recurved at the obtuse or
emarginate tips, yellowish-brown, with the apex and a short median
stripe purple-brown ; inner corona-lobes incumbent upon the backs of
the anthers with very shortly upcurved tips, about ¾ lin. long, narrowly
ovate, acute, slightly gibbous at the base on the back, purple-brown.
Decne in DC. Prodr. viii. 655 ; *Schlechter in Journ. Bot.* 1898, 483.
Gonostemon stricta, Haw. Suppl. Pl. Succ. 12 ; *G. Don, Gen. Syst.* iv.
117. *G. strictum, Sweet, Hort. Brit. ed.* 1, 278. *G. strictus, Loud.
Hort. Brit.* 96.

SOUTH AFRICA : without locality, ex *Sims.*

I have seen no specimen of this species, and describe from Curtis' original
drawing at Kew. The colouring of the reproduction of this drawing in the
Botanical Magazine is too bright, and the centre of the disk is not white in the
original as represented in the reproduction, the mottling is very faint in the
original and is omitted in some and rendered too evident in others of the

published plate. No pubescence is represented on the stems in the drawing nor is it described, so they are probably glabrous as in the closely allied *S. divaricata,* Masson, from which it differs in its more compact erect habit, flatter and less acuminate corolla-lobes and the absence of ciliation.

26. **S. Pillansii** (N. E. Br. in Gard. Chron. 1904, xxxv. 242, fig. 100); stems rather crowded, 4–5½ in. (5–7 in., *Pillans*) high, 6–10 lin. (¾–1 in., *Pillans*) square, with concave sides and angles not much compressed, velvety pubescent, green; teeth not very prominent, with erect adpressed rudimentary leaves 1–1½ lin. long; flowers 2–5 together near the base of the stems and 2 or more sometimes open at the same time; pedicels ½–1½ in. long, 1½–2½ lin. thick, velvety; sepals about 4 lin. long, lanceolate, acute or acuminate, velvety; corolla in bud long-acuminate from a narrowly pentagonal-ovoid base, not twisted, when expanded 4–5½ in. in diam. with the lobes extended, velvety-puberulous on the back, smooth to slightly transversely rugose and glabrous on the inner face, purple-brown, without markings, shining, densely ciliate with soft light purple simple hairs, which are shorter and fewer and become absent towards the tips; disk flattish, depressed at the centre; lobes 1⅔–2⅓ lin. long, 7–9 lin. broad, lanceolate, attenuate into tail-like tips, variably spreading or recurved, not revolute at the margins; outer corona variable, blackish, shining, either of 10 free ascending lobes 2½ lin. long, 5 of them alternating with the inner corona-lobes, flattish, oblong or oblong-obovate, 1–1½ lin. broad at the top, which is abruptly acuminate or somewhat 3-toothed or abruptly rounded into a short central obtuse point, 5 others at the back of and adpressed to or united with the inner corona-lobes, about ¾ lin. broad, linear or oblong-linear, obtuse, truncate, notched or bifid at the apex; or the 5 lobes alternating with the inner corona-lobes narrower and oblong, and alternating with 5 pairs of filiform teeth; or of 5 very deeply 3-fid lobes with filiform lateral segments; or all the lobes connate at the basal part into a cup, divided above into 5 regularly 3-toothed or irregularly several-toothed lobes; inner corona-lobes about 3 lin. long, blackish, shining, with the dorsal-wing 1½ lin. broad, adnate for ½–⅔ of its length to the inner horn, its free part oblong or rounded, very obtuse, entire or obscurely crenulate; inner horn recurving, very stoutly triquetrous, acute to subobtuse, slightly longer than the dorsal wing; dried follicles subparallel, 6 in. long, puberulous; seeds 2¼ lin. long, 1¼ lin. broad, ovate, with thick incurved margins on one side, convex on the other, smooth, glabrous, light brown. *N. E. Br. in Gard. Chron.* 1908, xliv. 187, *fig.* 79.

VAR. β, **attenuata** (N. E. Br.); corolla in bud twisted at the apex, when expanded 6½–8 in. in diam., with the lobes 2¾–3½ in. long, with very long tapering points; otherwise as in the type.

CENTRAL REGION : Laingsburg Div. ; near Dwarsindeweg, *Marloth,* 3790 ! near Matjesfontein, *Pillans,* 38 ! Prince Albert Div. ; near Grootfontein, *Pillans,* 104 ! Var. β : Laingsburg Div.; by the road at Witte Poort, *Pillans,* 671 ! near Matjesfontein, *Marloth,* 4583 !

COAST REGION: Var. β: Ladismith Div. ; roadside between Muis Kraal and Ladismith, *Pillans,* 689 !

The flowers, according to Mr. Pillans, have a strong carrion-like odour ; he states that no flies ever lay their eggs on the flowers so far as he has observed.

[27. **S. glabriflora** (N. E. Br. in Gard. Chron. 1876, vi. 809, fig. 149) ; stems 4–9 in. long, erect, $\frac{3}{4}$–1 in. square, puberulous, green ; angles much compressed, with erect rudimentary leaves $\frac{2}{3}$–$1\frac{1}{2}$ lin. long at the teeth ; flowers 1–2 together at the middle or lower part of the young stems, opening successively ; pedicels $\frac{1}{2}$–$\frac{2}{3}$ in. long, $1\frac{1}{3}$–$1\frac{1}{2}$ lin. thick, velvety ; sepals $\frac{1}{4}$–$\frac{1}{3}$ in. long, lanceolate, acuminate, velvety ; corolla in bud ovate, acuminate, 5-angled, when expanded the lobes recurve and cross each other behind, but if laid flat the flower measures 3–$4\frac{1}{2}$ in. in diam., velvety on the back, transversely rugose and quite glabrous on the inner face, not ciliate, dull reddish-purple, darker at the tips of the lobes, the whole surface to the tips transversely marked with numerous very narrow linear irregular yellowish-white lines, those on the disk more slender and often dark-coloured instead of whitish ; lobes $1\frac{1}{2}$–2 in. long, $\frac{3}{4}$ in. broad, lanceolate, acuminate, much recurved, slightly revolute at the margins ; outer corona-lobes $\frac{1}{4}$ in. long, nearly 1 lin. broad, erectly spreading, recurved at the apex, linear, channelled down the face, obtuse or emarginate, with a very distinct apiculus, blackish-purple ; inner corona-lobes ascending-spreading, blackish-purple, with the dorsal wing $\frac{1}{4}$ in. long, erectly spreading, free to the base, oblong-linear or deltoid-lanceolate, obtuse or acute, entire or slightly toothed on the inner margin ; inner horn $\frac{1}{4}$–$\frac{1}{3}$ in. long, triquetrous-subulate, straight, with hooked or recurved tips. *N. E. Br. in Gard. Chron.* 1908, xliv. 187, *fig.* 78 ; *Schlechter in Journ. Bot.* 1898, 480.

ORIGIN UNKNOWN : *cultivated specimens* !

Described from living plants. In all probability a hybrid raised in Europe.]

28. **S. conformis** (N. E. Br.) ; stems erect, 5–11 in. long, $\frac{3}{4}$–$1\frac{1}{4}$ in. square, with much compressed toothed angles and rudimentary erect leaves 1–2 lin. long, velvety puberulous, green, sometimes tinted with purplish ; flowers 2 to several together, successively developed from a short stout peduncle or cyme near the base of the young stems ; pedicels $\frac{1}{2}$–1 in. long, stout, velvety ; sepals 2–4 lin. long, lanceolate or ovate-lanceolate, acute, velvety ; corolla 3–4 in. in diam., velvety on the back, transversely rugulose (except at the tips of the lobes) and glabrous or with a minute erect pubescence around the corona on the inner face, irregularly ciliate (sometimes very sparsely) on the very edge of the lobes with spreading white hairs 1–2 lin. long, mingled with others more minute ; disk and basal half of the lobes sulphur-yellow or dull yellowish-green or yellowish-white marked with numerous crowded narrow irregular transverse purple-brown lines, margins narrowly bordered with and apical half of lobes entirely dark purple-brown, usually with

a dull yellowish or greenish spot at apex; disk broadly and shallowly cup-like with a small star-like cavity at the centre; lobes very spreading and more or less recurved, 1–1½ in. long, 8–9 lin. broad, ovate-lanceolate, acute or somewhat acuminate; outer corona-lobes ascending-spreading, scarcely recurved at apex, 2½–3½ lin. long, ⅔–¾ lin. broad, linear, acuminate, acute or obtuse, with or without a minute apiculus deeply channelled down the face, brownish-ochreous or purple-brown with paler margins; inner corona-lobes dark olive-brown or purple-brown, with the dorsal wing free to the base, ascending-spreading, 1–2 lin. long, ¾–1 lin. broad at the base, deltoid to oblong-linear, acute or obtuse, entire or toothed on the inner margin; inner horns connivent at the base, then recurving, sometimes with the tips incurved, 1½–3 lin. long, triquetrous-subulate, acute.

VAR. β, **abrasa** (N. E. Br.); corolla-lobes without cilia and apparently more intensely coloured than in the type, marked for ⅔ of their length with yellowish lines and with a margin 1 lin. or more in breadth of dark purple-brown on their lower part, otherwise the same as the type.

COAST REGION: Albany Div.; near Grahamstown, *Cooper*, 1534! *Herb. Albany Museum*, 3! *Herb. Pillans*, 3! Curries Kloof, near Grahamstown, *MacOwan*! and *cultivated specimen*! Var. β: Bedford Div.; Patrys Hoogte, near the Great Fish River, *MacOwan*, 2247!

CENTRAL REGION: Somerset Div.; near Cookhouse Station, *Pillans*, 637!

Described from a living plant which flowered at Kew in Sept. 1898, and many dried flowers. Pillans, 637, is a form ciliate with very few long hairs and intermediate between the type and var. *abrasa*.

29. **S. Macowani** (N. E. Br. in Hook. Ic. Pl. t. 1920); stems erect, 6–12 in. high, 1 in. square, with very compressed angles, not very prominently toothed and the rudimentary leaves adpressed-erect, 1–1½ lin. long, velvety-puberulous; flowers 3–4 or more together, near the base of the young stems, successively developed; pedicels 2–5 lin. (in fruit up to 1¼ in.) long; sepals 2–3 lin. long, ovate-lanceolate, acute, and with the pedicels densely velvety-pubescent; corolla 2–2½ in. in diam., with the disk depressed into a very broad shallow somewhat funnel-shaped tube, having 5 slight grooves radiating from the centre to the sinuses between the flattish ovate acute very spreading or slightly recurved lobes, shortly pubescent on the back, but not so densely as the sepals, glabrous on the transversely rugulose inner surface, not ciliate or only by the short marginal pubescence of the back, pale "greenish-white" (*MacOwan*), or pale yellowish, marked to the tips of the lobes with slender crowded transverse wine-purple or purple-brown lines, and suffused with purplish in the cup of the disk; lobes 9–11 lin. long, 7–8 lin. broad; outer corona-lobes ascending-spreading, slightly recurved at the tips, with the apex itself upcurved, 3 lin. long, 1–1¼ lin. broad, slightly obovate-oblong, obtuse with a small apiculus or slightly 3-toothed, channelled down the face, purple-brown, with paler margins and yellowish at the base; inner corona-

lobes connivent-erect, 2–2¼ lin. long, 1½ lin. broad, entirely wing-like, broadly subobovate-oblong or subrectangular, entire, or slightly or distinctly notched at the more or less oblique apex, dark purple-brown ; follicles subparallel, about 5½ in. long and 10 lin. thick, fusiform, acute, puberulous. *Schlechter in Journ. Bot.* 1898, 481.

COAST REGION: Albany Div. ; Hell Poort and Bothas Berg, *MacOwan,* 909 ! *Barkly,* 49 !

CENTRAL REGION : Jansenville Div. ; Loots Kloof, *MacOwan,* 909 !

In my original description I stated that the pedicels were ½–¾ in. long, from a drawing, but none of the specimens have them more than 5 lin. long, mostly less.

[30. **S. incomparabilis** (N. E. Br. in Gard. Chron. 1901, xxx. 405) ; stems erect, 3–6 in. high, 4–6 lin. thick, 4-angled, with stout horizontal teeth 1–1½ lin. long, tipped with spreading or slightly ascending subulate rudimentary leaves 1–1¼ lin. long, which soon wither, glabrous to the eye, but very minutely and scantily puberulous under a strong lens ; flowers 3–4 together near the base of the stems, successively developed ; pedicels ½–1 lin. long, very minutely puberulous under a lens ; sepals 3–5 lin. long, 1 lin. broad, lanceolate, acuminate, glabrous, ciliate with white hairs ; corolla 3–3½ in. in diam., smooth and glabrous on the back, transversely rugose on the inner surface, entirely purple or purple-red (not purple-brown), sometimes with narrow irregular transverse yellowish or whitish lines on the edges of the rugosities, pubescent with short dark purple hairs in the cavity containing the corona, otherwise glabrous, ciliate to the tips of the lobes with long clavate vibratile dark purple (sometimes mingled with white) hairs ; united part broadly cup-shaped 1–1¼ in. in diam. and about ½ in. deep, suddenly contracted at the bottom into a very short subpentagonal cavity containing the corona ; lobes very spreading or reflexed, 1–1¼ in. long, 9–10 lin. broad, ovate, acute or acuminate ; outer corona-lobes ascending-spreading, 2 lin. long, ¾ lin. broad, linear, subtruncate or rounded with a slightly projecting tooth or emarginate or bifid at the apex, dull purple or purple-red with paler margins ; inner corona-lobes 2-horned, purple-brown, more or less speckled with a paler colour at the tips ; outer horns suberect, 1½ lin. long, laterally compressed, subulate, acute ; inner horns connivent-erect below, slightly diverging at the tips, 4 lin. long, filiform, slightly clavate and acute at the apex.

ORIGIN UNKNOWN : *cultivated specimens* !

Described from living plants first sent to me by Mr. Westcombe (an ardent cultivator of *Stapelieæ*) in 1875. It may be a native of South Africa, but I suspect that it is more probably a hybrid raised in Europe. The colour is brighter than and the flower totally different from that of any other species I have seen.]

31. **S. gemmiflora** (Masson, Stap. 14, t. 15) ; stems 3–4 in. high, obtusely 4-angled, ½–⅔ in. square, rather softly fleshy, glabrous, greyish- or subglaucous-green ; teeth ending in spreading subulate

acute leaves 2–4 lin. long ; flowers 1–4 together, successively
developed, near the base of the young stems ; pedicels erect or
spreading, 1¼–2 in. long, about 2 lin. thick, becoming 4–5 lin. thick
in fruit, glabrous ; sepals 2½–3 lin. long, ovate-lanceolate, acute or
acuminate, glabrous ; corolla in bud rhomboid-ovoid in side view,
pentagonal, with 5 short acute spreading or decurved points at the
angles, when expanded 2½–3½ in. in diam., flattish on the disk, with
spreading or recurving lobes, glabrous on both sides, smooth and
thickly covered with brownish-red spots (always ?) on a pale green
ground on the back, inner face densely and minutely rugose, intense
violet- or blackish-purple, minutely and indistinctly mottled with a
slightly lighter shade of the same colour and lighter along the
margins, with or without a few yellow spots at the base of the lobes
and on the disk ; lobes 1–1¼ in. long, 9–10 lin. broad, ovate or
oblong-ovate, acute, flat or slightly convex, beautifully ciliate from
base to apex with vibratile slightly clavate white or pale purple
hairs ; outer corona-lobes erectly spreading, 3½–4 lin. long, deeply
trifid, blackish-purple, with the tips of the teeth and margins of the
middle tooth below the middle ochreous ; middle tooth lanceolate or
linear-lanceolate, varying from entire and acute, obtuse or truncate
to bifid at the apex, channelled down the face and the sides, pinched
in at the base ; lateral teeth shorter, subulate or filiform ; inner
corona-lobes 2-horned, blackish-purple, with the tips of the inner
horn and a few dots at the base of both horns yellow ; outer horn
ascending, ¾–1 lin. long, compressed, linear- or narrowly deltoid-
subulate ; inner horns connivent-erect at the base, recurving above,
2–2¾ lin. long, filiform, acute or obtuse ; odour " horribly fetid "
(*MacOwan*), " musky, very agreeable " (*Pillans*) ; follicles erect,
subparallel, 4–5 in. long, ½–⅔ in. thick, fusiform, beaked, glabrous,
pale greenish, with numerous irregular blackish-green longitudinal
linear markings ; seeds ¼ in. long, ⅛ in. broad, elongate-ovate, flat,
with a wing-like margin, smooth, light-brown. *Willd. Sp. Pl.* i.
1280 ; *Pers. Syn. Pl.* i. 278 ; *Poir. Encycl.* vii. 379 ; *Jacq. Stap. t.*
24 ; *Ait. Hort. Kew. ed.* 2, ii. 87 ; *Curtis, Bot. Mag. t.* 1839 ;
Schultes, Syst. Veg. vi. 31 ; *Spreng. Syst. Veg.* i. 838 ; *Dietr. Syn. Pl.* ii.
885 ; *Decne in DC. Prodr.* viii. 656 ; *Loud. Encycl. Pl.* 200, *fig.* 3300 ;
N. E. Br. in Hook. Ic. Pl. under t. 1910, *p.* 3 ; *Schlechter in Journ.*
Bot. 1898, 480.   *S. stygia, Schultes, Syst. Veg.* vi. 32 (*by error* 562) ;
Decne in DC. Prodr. viii. 657 ; *Schlechter in Journ. Bot.* 1898, 483.
Tridentea gemmiflora, Haw. Syn. Pl. Succ. 34 ; *G. Don, Gen. Syst.* iv.
118.   *T. stygia, Haw. l.c.* 35 ; *G. Don, l.c.* 118.

VAR. β, **hircosa** (N. E. Br.) ; corolla-bud with spreading points at the sinal
angles ; inner face with some yellowish dots and irregular markings on the dark
purple-brown or blackish-purple ground colour or covered with large dark purple-
brown spots on a greenish-yellow ground ; corona in size, form and colour as in
the type ; odour "like that of a he-goat" (*Barkly*), "musky" (*Pillans*).
S. hircosa, Jacq. Stap. t. 25 ; *Willd. Enum. Pl. Hort. Berol.* 281 ; *Schultes,*
Syst. Veg. vi. 32 (*by error* 562) ; *Link, Enum. Pl. Hort. Berol.* i. 254 ; *Kerner,*
Hort. t. 330 ; *Spreng. Syst. Veg.* i. 839 ; *Dietr. Syn.* ii. 885 ; *Decne in DC.*
Prodr. viii. 656 ; *Schlechter in Journ. Bot.* 1898, 481 ; *N. E. Br. in Hook.*

Ic. Pl. under t. 1910, p. 2.   *S. hircola, Poir. Encycl. Suppl.* v. 234.   *S. moschata, J. Donn, Hort. Cantab.* ed. 3, 43 (*name only*) ; *Schultes, Syst. Veg.* vi. 32 (*by error* 562) ; *Lodd. Bot. Cab.* t. 1051.   *Tridentea hircosa, Schultes, l.c.* 850. *T. moschata, Haw. Syn. Pl. Succ.* 35, and *Suppl.* 10 ; *Schultes, l.c.* 850 ; *G. Don, Gen. Syst.* iv. 118.

VAR. γ, **densa** (N. E. Br.) ; corolla-bud with short upcurved points at the sinal angles ; inner face greenish-yellow densely speckled (except along the margins) with dark purple-brown ; outer and inner corona-lobes light ochreous-yellow, sparingly dotted with purple-brown and a line of it down the inner face of the bifid or entire and acute or obtuse middle tooth of the outer corona-lobes ; odour somewhat musk-like ; otherwise as in the type.   *S. hircosa, var. densa, N. E. Br. in Hook. Ic. Pl. under t.* 1910, p. 3.

SOUTH AFRICA : without locality, *Mund and Maire* ! *Barkly*, 48 ! *cultivated specimens* !   Var. *β* : *Barkly*, 79 ! and *cultivated specimen* !   Var. *γ* : *cultivated specimen* !

COAST REGION : Swellendam Div. ; near Swellendam, *Pillans*, 29 ! Var. *β* : Swellendam Div. ; near Swellendam, *Pillans*, 50 ! Ladismith Div. ; between Muis Kraal and Ladismith (with the type), ex *Pillans*.

CENTRAL REGION : Laingsburg Div. ; Witte Poort, *Pillans*, 624 ! 1214 ! Prince Albert Div. ; passes of the Zwartberg Range, *Bain*, 10 ! Graaff Reinet Div. ; hills near Graaff Reinet, 3000 ft., *Bolus*, 817 ! Zwartruggens, near the Sundays River, *MacOwan*, 2243 ! Steynsburg Div. ; near Steynsburg, ex *Pillans*.   Aliwal North Div. ; near Aliwal North, ex *Pillans*.   Albert Div. ; *Cooper*, 671 ! 3318 ! Var. *β* : Laingsburg Div. ; Witte Poort, ex *Pillans*.   Var. *γ* : Div. : near the Orange River and between Murraysburg and Richmond, *Barkly*, 10 ! *MacOwan*, 2263 !

The variety *densa* is described from a living plant sent to me by Sir Henry Barkly. I have not seen the type or var. *β* alive.

32. **S. vetula** (Masson, Stap. 15, t. 16) ; stems erect, obtusely 4-angled, 4–6 in. high, 6–7 lin. square, with a rather thin and exceedingly minute pubescence, green ; teeth tipped with incurved-erect rudimentary leaves ½–1 lin. long ; flowers 1–3 together near the base or the clusters scattered along the sides of the young stems, successively developed ; pedicels 6–10 lin. long, about 1 lin. thick glabrous ; sepals 2–3 lin. long, lanceolate, acute, glabrous ; corolla in bud conically pointed, when expanded 2½–2¾ in. in diam., flattish or with recurved lobes, glabrous on the smooth back, and on the transversely rugose inner surface, not ciliate, purplish-tinted on the outside, uniformly blackish-purple without markings on the inner face ; lobes 1⅙–1¼ in. long, 6–7½ lin. broad, ovate-lanceolate, acute or acuminate, with 3–5 longitudinal depressed veins on the inner face ; outer corona-lobes 2½–3 lin. long, ¾ lin. broad, linear-oblong, acute or obtuse and apiculate, dull reddish, with ochreous margins ; inner corona-lobes 2-horned, blackish, with the inner horns connivent at the base, then recurved in an arc, about 4 lin. long, subulate-filiform ; outer horn spreading, very small, ½–1 lin. long, crest- or wing-like, shortly deltoid, acute or obtuse.   *Willd. Sp. Pl.* i. 1291, *and Enum. Pl. Hort. Berol.* 280 ; *Pers. Syn. Pl.* i. 280 ; *Poir. Encycl.* vii. 387 ; *Jacq. Stap.* t. 27 ; *Ait. Hort. Kew.* ed. 2, ii. 93 ; *R. Br. in Mem. Wern. Soc.* i. 24 ; *Allg. Teutsch. Gart. Mag.* 1818, 17, t. 2 A, *ex Schultes* ; *Schultes, Syst. Veg.* vi. 32 (*by error* 562) ; *Lodd. Bot. Cab.* t. 428 ;

Link, Enum. Pl. Hort. Berol. i. 254; *Spreng. Syst. Veg.* i. 839;
Dietr. Syn. Pl. ii. 885; *Decne in DC. Prodr.* viii. 656; *Loud. Encycl.
Pl.* 200, *fig.* 3303; *Schlechter in Journ. Bot.* 1898, 484.   *S. juvencula,
Jacq. Stap. t.* 28; *Willd. Enum. Pl. Hort. Berol.* 280; *Poir. Encycl.
Suppl.* v. 231; *Schultes, Syst. Veg.* vi. 15; *Link, Enum. Pl. Hort.
Berol.* i. 254; *Spreng. Syst. Veg.* i. 839; *G. Don, Gen. Syst.* iv. 115;
Dietr. Syn. Pl. ii. 885; *Decne in DC. Prodr.* viii. 653.   *Tridentea?
vetula, Haw. Syn. Pl. Succ.* 35; *G. Don, Gen. Syst.* iv. 118.   *T.
juvencula, Sweet, Hort. Brit. ed.* 1, 277.

VAR. β, **Simsii** (N. E. Br.); outer horn of the inner corona-lobes half as long as
the inner horn, wing-like, irregularly deltoid-linear, slightly toothed on the inner
margin; otherwise as in the type.   *S. Simsii, Schultes, Syst. Veg.* vi. 33; *DC.
Prodr.* viii. 656; *Schlechter in Journ. Bot.* 1898, 483.   *S. vetula, Sims, Bot. Mag.
t.* 1234.   *Tridentea? Simsii, Haw. Syn. Pl. Succ.* 36; *G. Don, Gen. Syst.* iv. 118.

COAST REGION : Worcester Div.; mountains near Hex River, *Masson*! Var. β:
cultivated specimens!

Described from Masson's type, preserved in fluid at the British Museum, and
dried specimens.

33. **S. olivacea** (N. E. Br. in Gard. Chron. 1875, iii. 136 and 137,
fig. 24); plant rather compactly branched at the base; stems erect,
3–5 in. high, $\frac{3}{8}$–$\frac{1}{2}$ in. square, very obtusely 4-angled, with the faces
slightly grooved, not toothed on the angles, but with an impressed
transverse line at the base of the erect $\frac{1}{2}$–$\frac{3}{4}$ lin.-long rudimentary
leaves, minutely puberulous, very smooth-looking, greyish-green,
with the grooves and impressed lines darker, usually blotched
with purple where exposed to the sun; flowers 2–6 together at
the base of the young stems; pedicels 2–3 (in fruit 6–8) lin. long,
puberulous; sepals 2–2$\frac{1}{2}$ lin. long, lanceolate-subulate, puberulous;
corolla in bud ovoid, subobtuse, when expanded 1$\frac{1}{4}$–1$\frac{3}{4}$ in. in diam.,
with very spreading or slightly recurved lobes 5–7 lin. long, 4–5 lin.
broad, ovate, acute, puberulous and dull green on the back, inner
face deeply and closely rugose, glabrous, ciliate with white simple
hairs, varying from pale olive to dark olive-green with the rugosities
brown, or sometimes entirely dull brownish-red; outer corona-lobes
ascending-spreading, 1–1$\frac{1}{2}$ lin. long, narrowly deltoid-subulate, acute,
dark purple-brown, with a smooth polished central stripe from the
base to the middle; inner corona-lobes 2-horned, dark purple-brown;
inner horn 2–2$\frac{1}{2}$ lin. long, erect or very slightly recurving, slenderly
filiform; outer horn erectly recurved, $\frac{3}{4}$–1$\frac{1}{2}$ lin. long, falcately
subulate, laterally compressed; follicles 3$\frac{1}{2}$–4 in. long, about $\frac{1}{4}$ in.
thick, narrowly fusiform, acute, puberulous; seeds not seen; odour
strong, fetid. *Dyer in Bot. Mag. t.* 6212; *N. E. Br. in Hook. Ic.
Pl. under t.* 1920; *and Gard. Chron.* 1908, xliv. 196, *fig.* 86;
Schlechter in Journ. Bot. 1898, 482.   *S. eruciformis, Hort. ex N. E. Br.
in Gard. Chron.* 1875, iii. 136.

CENTRAL REGION : Beaufort West Div.; Rhenoster Kop, *Foster*! Richmond
Div.; near Richmond Road Station, *Foster in Herb. Pillans*, 98! Middelburg
Div.; two miles east of Conway Station, *Pillans*! Philipstown Div.; near
De Aar, living plant from *MacOwan*! Div.? common throughout the Karoo,
Barkly, 43! and *cultivated specimens*!

Described from living plants. In the *Gardeners' Chronicle*, 1875, iii. 206, it is stated by Mr. Croucher, and alluded to in *Bot. Mag.*, that another plant with black flowers, ciliate with black hairs, was also in cultivation under the name of *S. eruciformis*. I afterwards obtained flowers of Mr. Croucher's plant and found them to be identical with *S. olivacea*, and to have white cilia !

34. **S. acuminata** (Masson, Stap. 15, t. 17) ; stems erect, 3–6 in. high, $\frac{1}{2}$–$\frac{3}{4}$ in. square, puberulous, purplish-brown tinted ; angles compressed, toothed ; rudimentary leaves erect, $\frac{3}{4}$ lin. long ; flowers in fascicles of 2–5 at or above or occasionally below the middle of the stems, opening successively ; pedicels 2–4 lin. long, velvety-pubescent ; sepals 2–$2\frac{1}{2}$ lin. long, ovate-lanceolate, acute, velvety-pubescent ; corolla acuminate in bud, $1\frac{1}{6}$–$1\frac{3}{4}$ in. in diam. when expanded, rotate, softly puberulous on the back, glabrous and transversely rugose all over the inner surface, ciliate to the tips of the lobes, with long white simple hairs, dark purple-brown with transverse yellowish lines on the rugosities ; lobes 5–8 lin. long, 3–$3\frac{1}{2}$ lin. broad, ovate, acuminate into a short tail-like point ; outer corona-lobes $1\frac{1}{4}$–$1\frac{1}{2}$ lin. long, $\frac{1}{2}$–$\frac{3}{4}$ lin. broad, slightly ascending, linear-oblong, bifid or minutely 3-toothed at the apex, mottled with yellowish and purple-brown ; inner corona-lobes 2-horned, pale reddish, spotted with purple-brown ; outer horn $\frac{3}{4}$–1 lin. long, free to the base, spreading, laterally flattened, narrowly deltoid or deltoid-subulate, acute, more or less toothed on the upper margin ; inner horn $1\frac{1}{4}$–$1\frac{3}{4}$ lin. long, linear-filiform, dorsally flattened, connivent at the base, then erectly spreading or recurving in a bold arc above the outer horn. *Willd. Sp. Pl.* i. 1281 ; *Pers. Syn. Pl.* i. 278 ; *Poir. Encycl.* vii. 380 ; *Ait. Hort. Kew. ed.* 2, ii. 87 ; *Haw. Syn. Pl. Succ.* 23 ; *Schultes, Syst. Veg.* vi. 21 ; *Spreng. Syst. Veg.* i. 839 ; *G. Don, Gen. Syst.* iv. 116 ; *Dietr. Syn. Pl.* ii. 886 ; *Decne in DC. Prodr.* viii. 654 ; *Schlechter in Journ. Bot.* 1898, 479.

Var. *β*, **brevicuspis** (N. E. Br.) ; stems rather smaller than in the type (*Pillans*) ; corolla 9–10 lin. in diam. ; lobes 3–4 lin. long, $2\frac{1}{2}$ lin. broad, ovate, acute, not tapering into a point as in the type, and perhaps with brighter yellow lines, but with no other distinction.

WESTERN REGION : Little Namaqualand ; *Masson, Templeman in Herb. Pillans,* 5 ! 20 ! Var. *β* : Little Namaqualand ; *Templeman in Herb. Pillans,* 26 !

Masson described the stems as glabrous, but, although minute, the pubescence is easily discernible. Dietrich has erroneously described the outer corona-lobes as entire.

35. **S. rufa** (Masson, Stap. 16, t. 20) ; stems erect, not decumbent at the base, 4–9 in. high, $\frac{1}{2}$–$\frac{3}{4}$ in. square, with obtuse angles and nearly flat sides when mature, and the teeth with erect rudimentary leaves $\frac{3}{4}$–1 lin. long, softly and minutely puberulous, at first green, becoming olive-green or tinted with dull purple with age ; flowers 3–5 together at the base or near the middle of the young stems, successively developed ; pedicels $1\frac{1}{2}$–4 lin. long, stout, velvety ; sepals 2–$2\frac{1}{2}$ lin. long, tapering from the base to an acute apex,

velvety; corolla in bud ovate, acuminate, when expanded $1\frac{1}{4}$–2 in. in diam., velvety puberulous on the back, inner surface closely transversely rugose all over, with the shallowly and broadly funnel-like depressed disk and base of the lobes glabrous, their terminal part being pubescent or puberulous and minutely ciliate with purple hairs all along or sometimes with hairs $\frac{1}{2}$–1 lin. long, at the basal part, entirely dull dark red or chocolate-red, or with slender dull olive-green lines between the transverse ridges on the disk and basal half of the lobes, which are 7–8 lin. long, $3\frac{1}{2}$–4 lin. broad, attenuate-acuminate from about the middle to the rather slender tips from an ovate base, very spreading or recurving, with recurved margins; outer corona-lobes spreading, $\frac{1}{2}$–$\frac{3}{4}$ lin. long, $\frac{1}{2}$–$\frac{3}{4}$ lin. broad, subquadrate or transversely subrectangular, subtruncate, obscurely crenulate or with a very short obtuse-angled point at the apex, channelled down the face, very dark orange or reddish-orange (in dried flowers apparently yellow-ochre), microscopically ciliate (always?); inner corona-lobes erect, $\frac{3}{4}$ lin. long, narrowly oblong or subcuneate-oblong, viewed from the back, subtruncate or obtuse to shortly bifid at the apex, with a short compressed projection or gibbosity at the base on the back, minutely puberulous, nearly black, shining; anthers yellow. *Willd. Sp. Pl.* i. 1281; *Pers. Syn. Pl.* i. 278; *Poir. Encycl.* vii. 380; *Ait. Hort. Kew. ed.* 2, ii. 87; *Schultes, Syst. Veg.* vi. 18, *excl. reference to Haworth; Lodd. Bot. Cab. t.* 239; *G. Don, Gen. Syst.* iv. 115; *Decne in DC. Prodr.* viii. 653; *Loud. Encycl. Pl.* 198, *fig.* 3283; *N. E. Br. in Hook. Ic. Pl. t.* 1922; *Schlechter in Journ. Bot.* 1898, 483. *S. rufescens, Salm-Dyck, Hort. Dyck.* 373; *Decne in DC. Prodr.* viii. 654; *Schlechter in Journ. Bot.* 1898, 483.

VAR. β, **attenuata** (N. E. Br.); corolla-lobes 9–10 lin. long, 3 lin. broad at the deltoid or deltoid-ovate base, at about $\frac{1}{3}$ of the way up rapidly tapering into the narrow linear-attenuate tail-like tips, rather minutely ciliate; outer corona-lobes subrectangular, obtuse or with a broad obtuse-angled point, not ciliate; inner corona-lobes suberect, 1 lin. long, linear, very obtuse or slightly bifid, quite glabrous, blackish or very dark purple-brown; otherwise as in the type. *S. fissirostris, N. E. Br. in Hook. Ic. Pl. under t.* 1922, *not of Jacquin.*

COAST REGION: Riversdale Div.; beyond Platte Kloof, *Masson.* Var. β: Ladismith Div.; along the main road between Muis Kraal and Ladismith, *Pillans,* 685! Touws Berg (not Tomos Berg as originally printed by error) *Bain,* 3!

CENTRAL REGION: Laingsburg Div.; near Matjesfontein, ex *Pillans.* Prince Albert Div.; Karoo near Groot Fontein, *Barkly,* 65! *Pillans,* 17! Var. β: Laingsburg Div.; along the main road between Prins Poort and Laingsburg, *Pillans,* 685! Witte Poort, ex *Pillans.*

The type described from a living plant.

S. rufa, Haw. *Syn. Pl. Succ.* 20, not of Masson. The type sheet of this in Haworth's Herbarium contains 4 flowers; against one of them Haworth has written "'*S. rufa*' Donn" in ink and under that name in pencil the name "*hirsuta*;" under another flower is written "'*rufa*' Mr. Venes, 1812,'" and under all, the following:—"Pin dried circa 1812. 1814 sine nom. Mem. 1833, this is I now think not *S. rufa,* Masson, nor do I find it elsewhere described. H." Three of the flowers appear to belong to a form of *S. hirsuta,* var. *patula*, N. E. Br., and the fourth appears to me to be either a very shrunken flower of *S. ambigua* or some seedling variety of that species, the hairs on the disk, instead of being very fine and woolly as in the other 3 flowers, are of the same straight and rather stiff character as in *S. ambigua*, and the short pedicel and structure of the corona also agree with that species.

36. **S. fissirostris** (Jacq. Stap. t. 23); stems not seen, according to Jacquin growing to 2½ ft. high, but doubtless much elongated under conditions of cultivation and probably in nature about 6–8 in. high, laxly branching, 5–6 lin. square, with slightly compressed toothed angles, softly pubescent, whitish-green, teeth with erect rudimentary leaves 1–1½ lin. long; flowers in fascicles of 5–6 near the base of the young stems, opening successively; pedicels about ¼ in. long, velvety-pubescent, as are the 2 lin.-long lanceolate acuminate sepals; corolla in bud with a rather long conical-acuminate point, when expanded about 2 in. in diam., glabrous on the back; inner face rugose with transverse elevated ridges, yellowish-green with the ridges purplish-brown or fuscous; tube very short and broad; lobes widely spreading, about ¾ in. long and ⅓ in. broad at the base, long-attenuate from an ovate base, ciliate with short simple hairs, very slightly revolute at the margins near the base; outer corona-lobes spreading, less than 1 lin. long, oblong or ovate-oblong, subacute, dull orange; inner corona-lobes about as long as the outer, suberect, bifid to ⅓ of the way down, with divergent acute points and a short dorsal gibbosity or projection at the base, described as blood-red, but figured as dark purple-brown. *Haw. Suppl. Pl. Succ. 9; Schultes, Syst. Veg. vi. 20; Link, Enum. Hort. Berol. i. 254; Spreng. Syst. Veg. i. 841; Kerner, Hort. t. 434; G. Don, Gen. Syst. iv. 115; Dietr. Syn. Pl. ii. 887; Decne in DC. Prodr.* viii. 654; *Schlechter in Journ. Bot.* 1898, 480.

South Africa : without locality, *Schott* (ex *Jacquin*).

37. **S. glanduliflora** (Masson, Stap. 16, t. 19); stems erect, 4- (rarely 5–6-) angled, 3–6½ in. high, ½–¾ in. thick, softly pubescent; angles somewhat compressed; teeth rather prominent, tipped with an erect or incurved rudimentary leaf ½–1 lin. long; flowers 3–9 together, near the base of the young stems, successively developed; pedicels 1½–3 in. long, ¾ lin. thick, pubescent, reddish; sepals 1½–3 lin. long, lanceolate, acute, pubescent; corolla in bud ovoid, shortly pointed, when expanded 1–1½ in. in diam., flattish, minutely pubescent on the back, very slightly rugulose on the inner face and densely covered on the basal ⅓–½ of the lobes with long outstanding clavate stiff white hairs, and fringed all round with similar hairs, pale sulphur-yellow, marked all over with numerous dots and fine interrupted transverse lines of purplish-red; lobes 5–7 lin. long, 3½–4 lin. broad, ovate-lanceolate, acute or acuminate, very spreading, with the tips recurved; outer corona-lobes 1 lin. long, ½ lin. broad, lanceolate, acute, yellowish or purplish-orange at the base, purplish-brown or orange-brown at the apex; inner

corona-lobes 1–1½ lin. long, simple, subulate, without a dorsal gibbosity at the base, connivent-erect or crossing each other at the tips, dull yellowish, margined and dotted with red-brown; odour rather faint or scarcely any. *Smith, Exot. Bot.* ii. 23, *t.* 71; *Poir. Encycl.* vii. 382; *Ait. Hort. Kew. ed.* 2, ii. 89; *Jacq. Stap. t.* 21; *Haw. Syn. Pl. Succ.* 22; *Schultes, Syst. Veg.* vi. 21; *Spreng. Syst. Veg.* i. 840; *G. Don, Gen. Syst.* iv. 116; *Decne in DC. Prodr.* viii. 654; *Loud. Encycl. Pl.* 200, *fig.* 3287; *N. E. Br. in Hook. Ic. Pl. under t.* 1921; *Schlechter in Journ. Bot.* 1898, 480. *S. glandulifera, Willd. Sp. Pl.* i. 1284; *Pers. Syn. Pl.* i. 279; *Poir. Encycl.* vii. 382; *Haw. Syn. Pl. Succ.* 21; *R. Br. in Mem. Wern. Soc.* i. 24; *Schultes, Syst. Veg.* vi. 21; *G. Don, Gen. Syst.* iv. 116; *Dietr. Syn. Pl.* ii. 886. *S. hispidula, Hornem. Hort. Hafn.* i. 251; *Schultes, Syst. Veg.* vi. 21; *Spreng. Syst.* i. 840; *G. Don, Gen. Syst.* iv. 116; *Decne in DC. Prodr.* viii. 663.

VAR. β, **emarginata** (N. E. Br.); outer corona-lobes oblong or rectangular, emarginate or shortly bifid at the apex, otherwise as in the type.

COAST REGION: Clanwilliam Div.; deserts near North Olifants River, *Masson*; near Clanwilliam, *Pocock in Herb. Pillans*, 88! northern slopes of Olifants River Valley, *Pillans*! and *cultivated specimens*! Var. β: Clanwilliam Div.; near Clanwilliam, *cultivated specimen*!

Described from living plants. The flowers vary in the depth of tint of the ground colour, the distinctness and intensity of colour of the transverse markings, and the amount to which they are clothed with clavate hairs, which although white are somewhat translucent. Haworth has stated that the plant he had seen in cultivation and which he has carefully described as *S. glandulifera*, differs from Masson's *S. glanduliflora* by "its pubescent and otherwise different branches," and "entire stamina (inner corona-lobes), his having bifid ones." But he was misled by Masson's rather poor figure on which there is no indication (nor mention in the description) of pubescence, and the somewhat two-lobed spot behind the inner corona-lobes I believe to be intended for shading, as in a drawing of my own the shading has a somewhat similar appearance.

38. **S. concinna** (Masson, Stap. 15, t. 18); stems erect, 3–6 in. high, rather slender, 3–4 lin. square, grooved down the sides, with obtuse angles (probably slightly compressed) and erect rudimentary leaves ½–¾ lin. long at the teeth, velvety-puberulous, dull green, tinted with dull purple; flowers 2–3 together near the base of the young stems, successively developed; pedicels about ½ in. long, velvety puberulous, as are the 1½ lin.-long lanceolate acute sepals; corolla about 1¼ in. in diam., flattish, star-like, with very spreading lobes about 5–6 lin. long, 3½–4 lin. broad, ovate, acute, rather minutely velvety-puberulous on the back; inner surface transversely rugulose, dull purple (or purple-brown?), lobes with irregular transverse yellowish lines, tips yellowish with an indistinct reticulation of purplish, whole surface thickly covered with rather stiff erect simple white hairs, those at the margins longer than the others not forming spreading cilia; outer corona-lobes ascending-spreading, with recurved tips, 1½ lin. long, ½ lin. broad,

linear, very obtuse or subtruncate with a minute central tooth
or obscurely 3-toothed at the apex, very dark purple-brown or
blackish; inner corona-lobes very dark purple-brown, connivent
below, recurving above, $1\frac{1}{4}$–$1\frac{1}{2}$ lin. long, linear-subulate, dorsally
flattened (not triquetrous), with the dorsal wing at its base very
small, about $\frac{1}{4}$ lin. long, ascending-spreading, deltoid or deltoid-
oblong, obtuse. *Willd. Sp. Pl.* i. 1284, *and Enum. Pl. Hort. Berol.
Suppl.* 14; *Hayne, Term. Bot. ed.* 1799, *t.* 16, *fig.* 8; *Pers. Syn. Pl.*
i. 279; *Poir. Encycl.* vii. 382; *Ait. Hort. Kew. ed.* 2, ii. 89; *Haw.
Syn. Pl. Succ.* 21; *Schultes, Syst. Veg.* vi. 20; *Link, Enum. Pl.
Hort. Berol.* i. 254; *Spreng. Syst. Veg.* i. 839; *G. Don, Gen. Syst.*
iv. 116; *Dietr. Syn. Pl.* ii. 885; *Decne in DC. Prodr.* viii. 654;
Schlechter in Journ. Bot. 1898, 479.

VAR. *β*, **paniculata** (N. E. Br.); corolla entirely dull purple or purple-brown,
without transverse yellow lines on the lobes; otherwise as in the type. *S. paniculata, Willd. Enum. Pl. Hort. Berol. Suppl.* 13; *Jacq. Stap. t.* 26; *Schultes,
Syst. Veg.* vi. 34; *Link, Enum. Hort. Berol.* i. 255; *Spreng. Syst. Veg.* i. 839;
Dietr. Syn. Pl. ii. 885; *Decne in DC. Prodr.* viii. 657; *Schlechter in Journ. Bot.*
1898, 482. *Tridentea paniculata, Schultes, Syst. Veg.* vi. 850; *Sweet, Hort.
Brit. ed.* i. 277; *G. Don, Gen. Syst.* iv. 118.

SOUTH AFRICA: Karoo, *Masson*; *cultivated specimens* in *Herb. Haworth* (at
Oxford)! Var. *β*: *cultivated specimen* in *Herb. Berlin*!

Haworth's Herbarium contains 3 specimens, dated 1829, 1830 and 1832; on two
of the corolla-lobes of the specimen dated 1830 the transverse markings,
although faint, are very clearly visible under a lens, on the other flowers I cannot
trace them. The specimen dated 1832 is labelled "*S. paniculata,* Jacq. Ic. A. D.
Bevan 'letter J, *concinna.*'" The other two specimens were also from Mr. Bevan,
and all are probably from plants or possibly one plant long before introduced
by Masson. Jacquin received *S. paniculata* from Prince Salm-Dyck, who may
have obtained it from England. *S. concinna* has long ago disappeared from
cultivation and no other collector has refound it.

39. **S. flavopurpurea** (Marloth in Trans. S. Afr. Phil. Soc.
xviii. 48, t. 5, fig. 1); stems lax, decumbent at the base,
3–4 in. high, $\frac{1}{3}$–$\frac{1}{2}$ in. square, with flattish sides, glabrous to
the eye, but very minutely puberulous under a lens; rudimentary
leaves very minute, erect, about $\frac{1}{2}$ lin. long, deltoid, acute;
flowers 1–2 together at the lower part of the stems; pedicels $\frac{3}{4}$–$1\frac{1}{4}$
in. long, $\frac{3}{4}$ lin. thick, and together with the calyx minutely
puberulous like the stems; sepals $1\frac{1}{2}$–2 lin. long, lanceolate, acute;
corolla about $1\frac{1}{2}$ in. in diam., very deeply lobed, the united part
forming a very small funnel-shaped tube about 2 lin. deep and 3–$3\frac{1}{2}$
lin. broad, whitish, covered with short erect clavate purple hairs
on the upper part and around its mouth; lobes horizontally spread-
ing, $\frac{2}{3}$ in. long, about $2\frac{1}{2}$ lin. broad when flattened out, linear-
lanceolate, acute, margins much recurved, with an exceedingly
minute puberulence on the back, transversely rugose, with very
prominent ridges and glabrous on the inner surface, not ciliate,
dull yellow; outer corona-lobes nearly erect, $1\frac{1}{3}$–$1\frac{1}{2}$ lin. long,
oblong, with the sides much incurved and pressed against the base

of the inner corona-lobes forming deep nectar-cavities, subtruncate and somewhat toothed on the top margin on each side of the short acute apical spreading point, purple, yellowish at the base ; inner corona-lobes 2-horned, white, tinged with purple ; outer horn free to the base, about 2 lin. long, laterally flattened, linear-subulate, suberect ; inner horn about 3 lin. long, dorsally flattened, linear-filiform, connivent at the base, then recurving over the tip of the outer horn.

CENTRAL REGION : Laingsburg Div. ; Tanqua Karoo, *Marloth*, 4227 !

Described from a living stem and a flower preserved in fluid, aided by Dr. Marloth's description.

40. S. virescens (N. E. Br. in Hook. Ic. Pl. t. 1910 B) ; stems erect, $2\frac{1}{2}$–$3\frac{3}{4}$ in. high, 5–6 lin. thick, obtusely 4-angled, tuberculate-toothed, with the teeth tipped with subulate very spreading deciduous or withering leaves 2–5 lin. long, glabrous, greyish-green, mottled with dull purplish-brown, flowering below the middle; flowering axis or cyme ultimately elongating to $\frac{1}{2}$–$\frac{3}{4}$ in. long, erect, stout, racemosely 6–9-flowered ; pedicels erect, $1\frac{1}{4}$–2 in. long, glabrous; sepals $1\frac{1}{2}$–2 lin. long, lanceolate, acuminate, channelled down the inner face, glabrous, very sparsely ciliate with minute clavate hairs ; corolla 1–$1\frac{1}{6}$ in. (much less in dried specimens) in diam., very deeply 5-lobed, flat, without a distinct tube, smooth and whitish-green, tinted with purplish on the back, rugose with small densely crowded conical processes or stout papillæ and greenish-yellow on the inner face ; lobes about 5 lin. long and $3\frac{1}{2}$–4 lin. broad, ovate, acute, with reflexed margins, glabrous, not ciliate ; outer corona-lobes variable, $1\frac{1}{4}$–$1\frac{1}{2}$ lin. long, usually trifid with the middle lobe entire or variably 3–5-toothed and much longer than the lateral subulate lobes or teeth, or sometimes somewhat lanceolate, acute, with a minute tooth on each side below the middle, channelled down the face, yellow ; inner corona-lobes 2-horned, yellow ; outer horn $\frac{2}{3}$–1 lin. long, compressed, narrowly deltoid or subulate, ascending-spreading, straight or slightly recurved ; inner horns $1\frac{3}{4}$ lin. long, filiform, connivent at the base, then arching-recurved ; odour " disgusting " (*Barkly*).

CENTRAL REGION : Philipstown Div. ; near Philipstown, *Mrs. Cawood Giddy in Herb. Pillans*, 642 ! Britstown Div. ; near Britstown, *Waters* ! *Marloth*, 4375 ! Div. ? Karoo, on the road to the Diamond Fields, *Dickson* (*Barkly*, 35) ! and without precise locality, *Mrs. Barber* !

Described from a living plant and flowers preserved in fluid.

41. S. surrecta (N. E. Br.) ; stems (only 1 branch seen) $2\frac{1}{2}$–3 in. (or more ?) high, 5 lin. (or more ?) square, with slightly concave sides and the teeth on the rounded angles tipped with erect acute rudimentary leaves 1 lin. long, velvety-puberulous, green ; flowers 2–3 together, successively developed, along the sides near the top of the stems ; pedicels erect, $\frac{1}{2}$–$\frac{3}{4}$ in. long, rather slender, $\frac{1}{2}$–$\frac{3}{4}$ lin. thick, puberulous; sepals $1\frac{1}{2}$–$1\frac{3}{4}$ lin. long, lanceolate, acute; corolla about 1 in. in diam. with the lobes spread out, minutely puberulous on the outside, glabrous and slightly rugulose on the inner surface, not ciliate; tube distinct, about $2\frac{1}{2}$ lin. long, cup-shaped; lobes probably spreading, $4\frac{1}{2}$ lin. long in the only flower seen, 2 lin. broad at the base, thence tapering to a very acute point, rather prominently 3-nerved on the back; outer corona-lobes horizontally spreading, $\frac{1}{2}$ lin. long and broad, subquadrate, minutely 3-crenate at the truncate apex, concave down the face; inner corona-lobes 2-horned; outer horn ascending-spreading, 1 lin. long, wing-like, about $\frac{2}{5}$ lin. broad at the base, thence tapering to the obtuse apex, entire; inner horn $1\frac{1}{4}$ lin. long, subterete-subulate, ascending-recurved.

CENTRAL REGION : Laingsburg Div. ; Tanqua Karoo, *Marloth*, 3791 !

42. **S. erectiflora** (N. E. Br. in Gard. Chron. 1889, vi. 650); stems erect, decumbent at the base, 4–7 in. high, 4–5 lin. square, with concave-channelled sides and small teeth with erect rudimentary leaves about $\frac{1}{2}$ lin. long, puberulous, dull green, flowering all along their sides from base to apex, often with several cymes or fascicles of flowers on the same stem; cymes with a short peduncle 1–3 lin. long, 1–4- (rarely more) flowered; pedicels erect, 2–5 in. long, about 1 lin. thick at the base, slightly tapering upwards, puberulous, purplish; sepals $1\frac{1}{2}$ lin. long, lanceolate, acute, puberulous; corolla in bud, at first globose, ultimately broader than long, rather depressed, with a short blunt point, when expanded disk-like or somewhat resembling a Turk's cap, the lobes rolled back so closely as to entirely conceal the back of the flower and in this condition $4\frac{1}{2}$–5 lin. in diam., purplish-grey, from short dense adpressed pubescence of white hairs, on a smooth purple-brown ground; lobes about $2\frac{1}{2}$ lin. long, ovate, acute, without cilia other than the hairs on the inner surface; outer corona-lobes very spreading, with upcurved tips, $1\frac{3}{4}$ lin. long, $\frac{1}{2}$ lin. broad, linear, subobtuse, with 2 faint grooves down the face, brownish-red; inner corona-lobes 2 lin. long, subulate-filiform, quite simple, without a dorsal crest or horn, recurved-spreading, honey-yellow. *N. E. Br. in Hook. Ic. Pl. t.* 1921 ; *Schlechter in Journ. Bot.* 1898, 480.

COAST REGION : Clanwilliam Div.; Karoo, 6 miles beyond the Cederberg Mountains, *Bain (Barkly,* 80)! Clanwilliam, 500 ft., *Schlechter,* 8403! and *cultivated specimens* !

This is the most profuse-flowering species of the genus known to me, each flower remains open for from 8 to 14 days and is nearly or quite odourless.

43. **S. pedunculata** (Masson, Stap. 17, t. 21); stems erect or decumbent at the base, 2–5 in. (up to 8 in., *Barkly*) long, 4–8 lin. thick, somewhat obscurely 4-angled and very slightly or indistinctly toothed, smooth, glabrous, greyish-green, mottled or tinted with faint purplish-brown where exposed to the sun, slightly glaucous

when young; flowers 1–2 (and usually 2–5 buds which abort) together, near the base of the young stems, successively developed; pedicels erect, 3–6 in. long, much overtopping the stems, glabrous; sepals $2\frac{1}{2}$ lin. long, lanceolate-subulate, glabrous; corolla in bud ovoid, with 5 compressed angles; when expanded $1\frac{1}{2}$–$2\frac{1}{4}$ in. in diam., very deeply lobed, smooth on the back, faintly punctate-rugulose on the inner face, light olive-brown with a slight golden tinge, or olive-green or light greenish-yellow, the basal part of the lobes and the small cup pale greyish-white, dotted with reddish-brown, beautifully ciliate with vibratile clavate dark purple hairs for about $\frac{1}{4}$ in. along the margin near the base of the lobes, otherwise glabrous; tube scarcely any, the united part very shortly and broadly funnel-shaped or flattish; lobes horizontally spreading, with the margins rolled back so as to touch or overlap, forming a pointed cylinder, $\frac{7}{8}$–$1\frac{1}{4}$ in. long, 6–7 lin. broad when flattened out, lanceolate or elliptic-lanceolate, acute; outer corona-lobes spreading, comparatively very small, 1–$1\frac{1}{4}$ lin. long, linear, usually truncate or emarginate or minutely 3-toothed at the apex, more rarely acute, channelled down the face, purple-black; inner corona-lobes 2-horned, purple-black, shining; horns very similar, each terminating in a large knob covered with sharp angular projections, the inner erect or sub-connivent at the base, then recurved-spreading, about 2 lin. long, the outer rather shorter than and recurved-spreading under the inner horns; odour very nauseous, resembling stale dried salt-fish, but only present when the corolla is fully expanded. *Willd. Sp. Pl.* i. 1284; *Bot. Mag. t.* 793; *Pers. Syn. Pl.* i. 279; *Poir. Encycl.* vii. 382; *R. Br. in Mem. Wern. Soc.* i. 24; *Ait. Hort. Kew. ed.* 2, ii. 90; *Allg. Teusch. Gart. Mag.* 1811, 309, *t.* 30 (*ex Schultes*); *Haw. Syn. Pl. Succ.* 23; *Jacq. Stap. tt.* 60, 61, 62 *and* 63; *Schultes, Syst. Veg.* vi. 46; *Link, Enum. Pl. Hort. Berol.* i. 255; *Spreng. Syst. Veg.* i. 841; *Kerner, Hort. tt.* 501 *and* 730; *Geel, Sert. Bot.* ii.; *Reichenb. Fl. Exot.* v. 10, *t.* 302; *Dietr. Syn. Pl.* ii. 887; *Decne in DC. Prodr.* viii. 658; *Loud. Encycl. Pl.* 202, *fig.* 3339; *N. E. Br. in Hook. Ic. Pl. under t.* 1909; *Schlechter in Journ. Bot.* 1898, 482. *S. penduliflora, Steud. Nom. Bot. ed.* 2, ii. 631. *S. lævis, Decne in DC. Prodr.* viii. 658. *Caruncularia pedunculata, Haw. Syn. Pl. Succ.* 333; *G. Don, Gen. Syst.* iv. 122. *C. Simsii, Sweet, Hort. Brit. ed.* 2, 358. *C. Massoni, C. Jacquini and C. penduliflora, Sweet, l.c.* 359.

Western Region: Little Namaqualand; Spectakal, in the vicinity of the Kamiesberg Range, *Barkly,* 1! and *cultivated specimens!* Ookiep, *Morris* (*Barkly,* 75)! and *cultivated specimens!* also without precise locality, *Scully!*

The flowers of the remarkable species probably vary considerably, I have had several plants in cultivation, but no two had flowers quite alike. No specimen that I have seen had pendulous bright purple flowers like that figured in the *Botanical Magazine* at t. 793, and I am inclined to doubt the correctness of the colouring. The flowers remain open for about 3 days. From observations I made upon the growth of the very long pedicels I find that the lengthening is least at the early and later stages of development, when it varies from $\frac{1}{2}$–$1\frac{1}{2}$ lines in 12 hours, and greatest at the middle stage when it varies from 3–7 lines in 12 hours, it varies also from day to day with the temperature.

44. **S. furcata** (N. E. Br.); stems unknown; pedicels $\frac{3}{4}$–1 in. long, stout, glabrous; sepals 3–4 lin. long, ovate-lanceolate, acuminate, glabrous; corolla of dried flowers $1\frac{1}{3}$–$1\frac{1}{2}$ in. in diam., with a small cup-like cavity containing the corona and the united part apparently broadly and shortly funnel-shaped; lobes ovate, acute, 4–6 lin. long and 3–5 lin. broad, probably larger when alive, smooth and glabrous on the back, rugose and covered with rather stiff very acute and somewhat spine-like erect hairs $\frac{1}{3}$–$\frac{1}{2}$ lin. long on the inner face, ciliate on the lobes with long vibratile clavate purple hairs, apparently dark purple-brown, with the hairs probably purple; outer corona-lobes shortly connate at the base, spreading or ascending-spreading, about $1\frac{1}{4}$ lin. long and as much in breadth, 3-toothed, with the lateral teeth widely spreading and smaller than the middle tooth; inner corona-lobes $1\frac{1}{4}$–$1\frac{1}{2}$ lin. long, closely incumbent upon the backs of the anthers and shortly exceeding them, linear, deeply bifid at the apex, with one tooth directed forwards and the other to one side of it, dorsally connected with the outer corona at the base and there deeply notched or produced into 2 short teeth between the outer corona-lobes.

KALAHARI REGION: Transvaal, *Todd* !

This very remarkable species is evidently allied to *S. Woodii*, N. E. Br., but at the same time resembles *Caralluma intermedia*, N. E. Br., in its coronal structure. The short spine-like hairs on the corolla distinguish it from every known species.

45. **S. longidens** (N. E. Br. in Gard. Chron. 1895, xviii. 324, and 1898, xxiv. 7, fig. 3); stems erect or ascending from a decumbent base, $2\frac{1}{2}$–6 in. long, 4–5 lin. thick excluding the teeth, obtusely 4-angled, with long stout teeth $\frac{1}{2}$–1 in. long and about $\frac{1}{8}$ in. thick at the base, teretely tapering to a fine soft point, glabrous, green, mottled with purple; flowers 3 to several together at the base of the stems, successively developed; pedicels $\frac{3}{4}$ in. long, 1 lin. thick, becoming up to $1\frac{1}{2}$ in. long and $\frac{1}{4}$ in. thick in fruit, glabrous; sepals 4 lin. long, $1\frac{1}{4}$ lin. broad, lanceolate, acuminate; corolla $1\frac{1}{2}$–2 in. in diam., deeply lobed, quite glabrous and not ciliate on the lobes nor rugose on the pale greenish-yellow inner surface, spotted all over with dark purple-brown (cream-colour, spotted with crimson, *Mrs. Monteiro*), spots very small within the tube, increasing in size and more or less confluent towards the tips of the lobes; tube campanulate or cup-like, $2\frac{1}{2}$ lin. deep, with the bottom raised up to form a convex cushion supporting the corona; lobes 7–10 lin. long, $4\frac{1}{2}$–6 lin. broad, ovate-lanceolate, acute, flat; outer corona-lobes deflexed upon the basal cushion of the corolla, about $\frac{3}{4}$ lin. long and broad, quadrate, emarginate, purple-brown with yellowish margins; inner corona-lobes simple, closely incumbent upon the backs of the anthers, shortly exceeding them and meeting or crossing but not erect or very slightly so at the tips, lanceolate, acute, scarcely or but slightly gibbous behind, dull yellowish dusted with purple-brown; follicles

erect, slightly diverging, about 7 in. long, $\frac{1}{2}$ in. thick, terete-fusiform, gradually tapering into a beak, glabrous, pale brownish-white, streaked with purple; seeds 5 lin. long, 3 lin. broad, ovate or elliptic-ovate, flat, with broad wing-like margins, pale brown. *N. E. Br. in Gard. Chron.* 1908, xliv. 196, *fig.* 85.

EASTERN REGION : Delagoa Bay, *Mrs. Monteiro*! and *cultivated specimens*!

Described from a living plant. The stem-teeth are much longer in this species than in any other known to me, and give the plant a very distinct appearance. The follicle and seeds described were produced upon a cultivated plant, and may not accord with those of the wild plant, as they may possibly have been fertilised with the pollen of some other species.

46. **S. Woodii** (N. E. Br. in Gard. Chron. 1892, xi. 554); stems erect from a shortly decumbent base, $1\frac{1}{2}$–3 in. long, 3–5 lin. thick excluding the teeth, obtusely 4-angled, with ascending-spreading stout conical-subulate very acute teeth $\frac{1}{3}$–$\frac{1}{2}$ in. long, glabrous, green, striped and mottled with purple; flowers 3 or more together near the base of the young stems, successively developed; pedicels 1–$1\frac{1}{4}$ in. long, glabrous; sepals 2–$2\frac{1}{2}$ lin. long, ovate-lanceolate, acuminate, glabrous; corolla $1\frac{1}{2}$–$1\frac{2}{3}$ in. in diam., flattish, with a convex disk supporting the corona raised above the level of the horizontally spreading or slightly deflexed lobes, glabrous on both sides, very rugose with small transverse wrinkles on the inner surface, ciliate along the middle part only of the lobes with long vibratile clavate dark purple hairs, chocolate or dull purple-brown, with a few dull yellow dots on the lobes, which are 5–6 lin. long, 4–5 lin. broad, ovate, acute or shortly acuminate, convex, with recurving margins; outer corona-lobes horizontally spreading, about $\frac{3}{4}$ lin. long and nearly as broad, rectangular or subquadrate, slightly notched or subdenticulate at the apex, grooved down the face, dark purple-brown; inner corona-lobes simple, with the basal part horizontally incumbent on the backs of the anthers, flattish, ovate-oblong or elliptic-oblong, tapering into an erect subulate point about $\frac{1}{2}$ lin. long, entire or with a minute tooth on each side at the base of the point, and slightly produced behind into a minute ascending entire or bifid tubercle, purple-brown.

EASTERN REGION : Natal ; Noods Berg, *Wood*, 4119 !

Described from a living plant sent to Kew from Durban Botanic Garden by J. Medley Wood, which flowered in September, 1891.

47. **S. Cooperi** (N. E. Br.); stems erect or ascending, $1\frac{1}{2}$–2 in. high, 4–5 lin. thick excluding the teeth; angles obtuse, with spreading conical very acute teeth $2\frac{1}{2}$–3 lin. long, having a minute tooth on each side at their middle, glabrous, green or greyish-green with chocolate-coloured or dark green spots and lines; flowers gradually developed 1–3 at a time up to 10 or more from a sessile cyme or cluster at the base of the stems; pedicels 3–6 lin. long, glabrous ; sepals 2–$2\frac{1}{4}$ lin. long, ovate or ovate-lanceolate, acuminate, glabrous; corolla $1\frac{1}{3}$–$1\frac{1}{2}$ in. in diam., flat, with stellately spreading

or recurving lobes and an annulus on the disk, glabrous, smooth and " olive colour with dashes or longitudinal streaks of purple " (*Pillans*) on the back, rugose on the inner surface with minute crowded tubercles and short transverse ridges on the lobes and crowded minute tubercles on the annulus or lower part of it, ciliate on the basal half of the lobes with vibratile clavate purple hairs, otherwise glabrous, "light purple, with all the rugosities coloured yellow " (*Pillans*), "dull yellow, with fine purple-brown lines" and the annulus " purplish-pink " (*Marloth*); lobes 5–7 lin. long, 4½–5 lin. broad, ovate, acute, somewhat convex from the slightly recurring margins; annulus 4–4½ lin. in diam., pentagonal, solid, cushion-like, with sloping sides and a central depression containing the corona, outer corona resting upon the cup of the annulus, somewhat saucer-like and subequally 10-toothed or with 5 short bifid or emarginate lobes ½ lin. long and ¾ lin. broad, with obtuse or subacute deltoid teeth having a broadly triangular notch between them, "rich purple" (*Pillans*), "deep maroon" (*Marloth*), inner corona-lobes 1¼–1½ lin. long, with their basal part ascendingly incumbent on the backs of the anthers and dorsally connected to the outer corona by a stout basal projection, broadly ovate, oblong or ovate-oblong, entire or with 1–2 teeth on each side, apical part filiform or subulate, connivent-erect or with recurving and sometimes slightly clavate tips, pale yellow, minutely speckled with purple; follicles slightly diverging, about 3½ in. long and ⅓ in. thick, fusiform, acute, glabrous, smooth, densely streaked with dark purple-brown on a pale ground colour; seeds not seen.

COAST REGION : Van Rhynsdorp Div. ; near Van Rhynsdorp, *Marloth*, 4584 !

CENTRAL REGION : Cradock Div. ; near Cradock, *Cooper*, 3113 partly ! Middelburg Div. ; among bushes on level ground 2 miles east of Conway Station, *Pillans*, 181 ! Prieska Div. ; near Prieska, *Frank* (ex *Pillans*).

KALAHARI REGION : Griqualand West ; near Douglas, *Pillans*, 612 !

A specimen collected upon Umbumbula Mountain, near Queenstown, in Queenstown Div. (*Galpin*, 6942) probably belongs to this species, but the corona of the only flower seen is so much crushed that I cannot with certainty identify it. This remarkable plant does not quite agree with *Stapelia* or any other genus in structure, but I do not know where else to place it. The corona seems to be a combination of that of *Caralluma* and *Piaranthus*, whilst the corolla resembles somewhat that of *Stapelia Woodii* and its allies with an annulus added, but the difference is scarcely sufficient for generic distinction.

48. S. jucunda (N. E. Br.) ; stems compactly erect, 2–3 in. high, 4–7 lin. square, obtusely 4-angled, with spreading stoutly subulate acute deciduous teeth ½–¾ lin. long, leaving rather large and slightly prominent whitish tubercle-like scars, glabrous, green or greyish-green, spotted and mottled with purple, rather soft in substance ; flowers 1–3 together at the middle or upper part of the stems, successively developed ; pedicels ½–1 in. long, glabrous ; sepals 1–1¼ lin. long, ovate or ovate-lanceolate, acute or acuminate, glabrous ; corolla nearly flat, with the centre depressed, lobed to half-way down, about 1 in. in diam., glabrous on both surfaces, ciliate from near the base nearly to the tips of the lobes with long vibratile clavate purple hairs, smooth and light green on the back dotted and spotted with purple-brown; inner surface slightly rugulose, cream-colour, spotted all over and edged on the lobes with purple-brown, the spots around the corona not smaller than the rest; lobes 3–3½ lin. long and as much in breadth, broadly ovate or deltoid-ovate, acute; outer corona-lobes erectly spreading, 1½–1¾ lin. long, ½–¾ lin. broad at the base, thence gradually tapering to the minutely notched or acute apex, concave down the face, dark purple-brown; inner corona-lobes about 3 times as long as the anthers, 1½ lin. long, subulate-filiform, acute, connivent-erect, usually more or less spreading at the tips, and with a short dorsal gibbosity at their base, marked with cream-colour and purple and tipped with white.

VAR. β, **deficiens** (N. E. Br.); corolla greenish-primrose, with very numerous rather minute purple-brown dots, those around the corona rather smaller than the rest; disk very broadly and shallowly funnel-shaped; lobe ciliate from near the base to about the middle with subclavate purple hairs about ½ lin. long; outer corona-lobes ¾–1 lin. long, entire, acute; inner corona-lobes without a small dorsal gibbosity at their base, otherwise as in the type.

CENTRAL REGION: Var. β : Victoria West Div. ; Biesjes Poort, *Pillans,* 675 !
KALAHARI REGION: Griqualand West; near Douglas, *Pillans,* 644 !
Mrs. Barber !

49. S. parvipuncta (N. E. Br. in Hook. Ic. Pl. t. 1923); stems erect, somewhat crowded, 2–5 in. long, ½–⅔ in. square, obtusely 4-angled, grooved down the sides, very shortly toothed on the angles, the teeth tipped with rudimentary subulate leaves ½ lin. long, glabrous, dull green; flowers several together, successively developed on a stout gradually elongating peduncle near the base or middle of the young stems; pedicels ¾–2 in. long, glabrous, spreading or directed downwards; sepals 1–2½ lin. long, ovate or lanceolate, acute or acuminate, glabrous; corolla in bud flat, pentagonal, with very short points at the angles; when expanded 1–1¼ in. in diam., lobed to half-way down, flat, without a distinct tube, quite glabrous on both sides, but ciliate along ⅔ of the length of the lobes with long vibratile clavate purple hairs, smooth and pale green thickly spotted with purple-brown on the back; inner surface slightly rugulose, varying from very pale sulphur-white to pale greenish-yellow, everywhere covered with minute dust-like dots of purple-brown or those on the lobes much larger than those on the disk, sometimes the lobes are narrowly margined with purple-brown; lobes 3½–5 lin. long and as much in breadth, broadly ovate or deltoid-ovate, acute, flat, recurving; outer corona-lobes 1–1¼ lin. long, deeply bifid, with the rectangular basal part ascending-spreading and the subulate teeth recurving and diverging, dark purple-brown, shining; inner corona-lobes simple, ovate or ovate-lanceolate, acute or acuminate, horizontally incumbent on the backs of the anthers, in some flowers shorter than them, in others longer, with their tips crossing one another but not erect, not gibbous behind, purple-brown. *S. parvipunctata, K. Schum. in Engl. und Prantl, Pflanzen-fam.* iv. ii. 278 ; *Schlechter in Journ. Bot.* 1898, 482.

CENTRAL REGION : Laingsburg Div. ; Zout Kloof Farm, near Laingsburg, *Pillans,* 41 ! Matjesfontein, ex *Pillans.* Beaufort West or Frazerburg Div. ; Nieuwveld Mountains, *Bain* ! and *cultivated specimens* !

Described from a living plant, sent by Sir Henry Barkly in 1877.

50. S. fucosa (N. E. Br.) ; stems erect, $1\frac{1}{2}$–$2\frac{1}{2}$ in. long, $3\frac{1}{2}$–$4\frac{1}{2}$ lin. square, excluding the $\frac{3}{4}$–$1\frac{1}{2}$ lin.-long conical acute spreading teeth, very obtusely 4-angled, with flattish sides, glabrous, dull green, mottled with darker green or purple-brown, probably flowering near the base ; pedicels 5–6 lin. long, glabrous ; sepals $1\frac{3}{4}$–2 lin. long, ovate-lanceolate, acute or acuminate, glabrous ; corolla about $1\frac{1}{4}$ in. in diam., quite glabrous and not ciliate, smooth on the back, densely rugulose on the inner face ; disk nearly flat or very shallowly saucer-shaped, having a solid pentagonal annulus 4–$4\frac{1}{2}$ lin. in diam. and about $\frac{1}{2}$ lin. high around the corona ; lobes spreading, $4\frac{1}{2}$–5 lin. long and as much in breadth, like an equilateral triangle in form, acute ; pale greenish-yellow or sulphur-yellow, covered with numerous rounded blackish-purple spots about $\frac{1}{2}$ lin. in diam., which are densely crowded and confluent on the annulus so that it is almost entirely blackish-purple and narrowly margined with the same colour on the lobes ; outer corona-lobes very spreading, about $\frac{2}{3}$ lin. long and broad, subquadrate, bifid to half-way down, with parallel teeth, blackish, shining ; inner corona-lobes about $\frac{2}{3}$ lin. long, closely incumbent upon the backs of and not exceeding the anthers, ovate, acute or obtuse, not gibbous on the back at the shoulder, whitish or yellow on the disk, with the margins and base blackish.

EASTERN REGION : Pondoland ; upper western slope of the mountain (Tonti Mountain ?) half a mile south-east of Mount Ayliff settlement, only a few plants seen, *Pillans,* 173 !

Described from living stems and flowers preserved in fluid.

51. S. miscella (N. E. Br.) ; stems often several inches long, creeping underground and producing at intervals erect branching stems $1\frac{1}{2}$–$2\frac{1}{4}$ in. high, 2–4 lin. thick, obtusely 4-angled, with spreading acute teeth $\frac{1}{2}$–$\frac{3}{4}$ lin. long, glabrous, dull green suffused with purple ; flowers apparently solitary near the middle or base of the young branches ; pedicels 2–$2\frac{1}{2}$ lin. (lengthening in fruit to 6 lin.) long, glabrous ; sepals 1–$1\frac{1}{4}$ lin. long, $\frac{1}{2}$ lin. broad, ovate or ovate-lanceolate, acuminate, glabrous ; corolla $\frac{1}{2}$ in. in diam., rotate, without any tube, but with a raised cushion-like ring on the disk around the corona about $\frac{1}{3}$ lin. high, quite glabrous and not ciliate, dark purple-brown, somewhat lighter at the centre ; lobes spreading, $2\frac{1}{2}$ lin. long and nearly $1\frac{1}{2}$ lin. broad, ovate-lanceolate, acuminate or very acute, faintly rugulose on the inner face ; outer corona-lobes horizontally radiating, $\frac{1}{3}$ lin. long, $\frac{1}{2}$ lin. broad, transversely rectangular, slightly notched at the apex, very dark purple-brown or almost blackish; inner corona-lobes scarcely ½ lin. long, lanceolate, subobtuse, erect, with their tips slightly incurved upon the backs of the anthers but much shorter, coloured like the outer corona-lobes; follicles erect, slightly diverging, 1½ in. long, about 2 lin. thick, terete-fusiform, acutely beaked, smooth, glabrous, mottled with olive-green or purplish on a pale ground; seeds 2½ lin. long, 1¼ lin. broad, ovate, nearly flat, broadly margined, smooth, light ochreous-brown, with a tuft of hairs ⅓ in. long.

CENTRAL REGION : Jansenville Div. ; near Klipplaat, *Pillans,* 657 !

This plant is of very ambiguous affinity, and is almost certainly of hybrid origin between a *Duvalia* and some other genus, possibly a *Caralluma.* The creeping underground habit of its stems is like that of some species of *Duvalia* and *Pectinaria,* whilst the outer corona resembles that of *Stapelia verrucosa,* on which account and the presence of the ring on the disk of the corolla I place it under this genus.

[52. **S. Bayfieldii** (N. E. Br. in Gard. Chron. 1877, vii. 430–431, fig. 66) ; stems erect, branching at the base, 6–8 in. high, 7–9 lin. square, with concave sides and stout ascending or spreading conical teeth 1–1½ lin. long, puberulous ; flowers 3–5 together, successively developed from a 1½–2 lin.-long peduncle near the base of the young stems ; pedicels 8–10 lin. long, puberulous ; sepals 3½ lin. long, lanceolate, acute, puberulous ; corolla 2½–2¾ in. in diam. with the lobes spread out, puberulous and green tinged with purplish on the back, inner surface smooth or nearly so, with a few short scattered hairs in the slight depression around the corona, otherwise glabrous, ciliate with rather short simple pale purple hairs on the lobes, purple-red, darker towards the tips of the lobes, with the disk and basal half of the lobes marked with rather crowded wavy pale yellow transverse lines ; disk flattish, slightly depressed at the centre ; lobes 1–1¼ in. long, 9–10 lin. broad, ovate, acute or shortly acuminate, strongly recurved or revolute so as to nearly or quite touch the pedicel ; outer corona-lobes erectly spreading, with the apex slightly recurved, 2 lin. long, linear, abruptly or truncately contracted into a slender tooth at the apex, slightly concave-channelled at the upper part, dull brownish-red or dark purple-brown, mottled with dirty yellowish on the margins ; inner corona-lobes 2-horned, dark purple-brown, minutely mottled with dirty yellowish ; inner horn 2½–3 lin. long, subulate, erect, with the apex very slightly recurving ; outer horn about 1 lin. long, deltoid-subulate, free to the base. *N. E. Br. in Gard. Chron.* 1908, xliv. 169 & 168, *fig.* 68.

ORIGIN : From what I have learnt since publishing this species, I believe it to be of garden origin, raised from seed produced by *S. mutabilis,* doubtless from cross-fertilisation by insect agency with some other species. I have only seen cultivated specimens. The stems are very much like those of *S. mutabilis,* Jacq., but puberulous.]

[53. **S. bella** (A. Berger in Gard. Chron. 1902, xxxi. 137–138, fig. 40–41); stems erect, branching at or above the base, 5–7 in. high, $\frac{1}{2}$–$\frac{3}{4}$ in. square, with concave sides and slightly compressed angles, having small spreading acute teeth $\frac{1}{2}$–$\frac{3}{4}$ lin. long, with a minute tooth (rudimentary stipule) on each side, covered with exceedingly minute pubescence, dull green, not glaucous; flowers 3–4 together on a very short peduncle at the base of the young stems; pedicels $\frac{1}{4}$–$\frac{3}{4}$ in. long, 1 lin. thick; sepals 2–$2\frac{3}{4}$ lin. long, $\frac{3}{4}$–$1\frac{1}{4}$ lin. broad at the base, tapering to an acute point; corolla in bud much flattened, with a short apiculus and 5 ridges radiating from it to 5 short hooked teeth beneath the circumference, when expanded and the lobes extended $1\frac{3}{4}$–2 in. in diam., glabrous on both sides, with the inner face deep purplish-red, browner towards the tips of the lobes, paler at the centre and the small tube whitish, without markings, ciliate on the lobes with long flattened tapering purple hairs mingled with a few clavate, vibratile, easily detached; tube represented by a pentagonal depression or cavity about 1 lin. deep and 2 lin. in diam., from which the corona is exserted, with some short hairs within; disk slightly convex, nearly smooth; lobes recurved or revolute, flattish, 7–8 lin. long, 6–7 lin. broad, ovate, acute, slightly rugose on the inner face; outer corona-lobes $1\frac{2}{3}$ lin. long, $\frac{3}{4}$–1 lin. broad, oblong, with or without a tooth or angle on each side, thence narrowed to an obtuse or obscurely 3-toothed apex, slightly concave down the face, glabrous, blackish-brown; inner corona-lobes dull brownish-purple, with the dorsal wing free to the base, spreading, scarcely 1 lin. long, deltoid, obtuse or acute; inner horn 2–$2\frac{1}{2}$ lin. long, filiform, connivent with the others at the base, then arching-recurved over the outer horn. *N. E. Br. in Gard. Chron.* 1908, xliv. 169 & 168, *fig.* 66–67.

ORIGIN : a hybrid, raised in Europe, *cultivated specimen* !

Described from a living plant cultivated at La Mortola by the late Sir Thomas Hanbury, who received it under the name of *S. glauca*, Jacq. I believe it to have been raised from seed of that species, which had been cross-fertilised by insects, possibly with pollen of *S. deflexa*, Jacq.]

54. **S. engleriana** (Schlechter in Engl. Jahrb. xxxviii. 49, fig. 8); stems procumbent or decumbent with ascending tips, loosely or sparingly branched, $\frac{1}{3}$–1 ft. long, 7–9 lin. square, with flattish or concave sides, shortly toothed at the angles, minutely velvety-pubescent, greyish-green, flowering near the middle or towards the apex on one of the sides; flowers solitary, sometimes with 1–4 other buds at the base of the pedicel, which appear to abort; pedicels $\frac{1}{2}$–$\frac{3}{4}$ (in fruit up to $1\frac{1}{3}$) in. long, velvety puberulous, as are also the 2–3 lin.-long lanceolate or ovate-lanceolate acute sepals; corolla (from the reflexion of its lobes) somewhat Turk's-cap-like, circular in outline and 10–12 lin. in diam., with a cup-like cavity or tube, softly puberulous on the back, rugose and glabrous on the inner face of the lobes and disk, puberulous with short erect hairs at the

bottom of the tube, and with or without a minute ciliation on the
lobes, dark purple-brown with or without faint yellowish transverse
lines on the base of the lobes and rim of the tube, the bottom of the
tube whitish, spotted with purple-brown; tube 3–4 lin. long and
obtusely 5-angled outside and about 6 lin. in diam. at the acutely
5-angled mouth inside, broadly cup-like; lobes 5–8 lin. long, 5–6 lin.
broad, deltoid-ovate, acute, closely reflexed upon the back of the
tube; outer corona-lobes $1\frac{1}{2}$–2 lin. long, about $1\frac{1}{2}$ lin. broad,
rhomboid with side angles or subquadrate, variably bifid, with or
without a minute tooth in the notch or subtruncate and very
minutely 3-toothed at the apex, purple-brown, mottled with ochreous
or pale yellowish, or the central area and margins of that colour
dotted with purple-brown; inner corona-lobes 2-horned, yellowish,
dotted with purple-brown; outer horns erectly spreading, 2–$2\frac{1}{2}$ lin.
long, laterally compressed, linear, slightly clavate at the obtuse
apex; inner horns connivent-erect at the base, then arching back
nearly as far as the spread of the outer horns, $2\frac{3}{4}$–4 lin. long,
filiform, more or less clavate and obtuse or acute at the apex;
follicles $3\frac{1}{2}$–$4\frac{1}{4}$ in. long, 4–5 lin. thick, terete-fusiform, tapering
into a short slightly hooked beak, smooth, puberulous, streaked
with dark green or dull purple-brown on a pale-coloured ground;
seeds $2\frac{1}{2}$ lin. long, $\frac{3}{4}$ lin. broad, narrowly lanceolate from the
thickened margins being closely incurved on one face, smooth, light
brown. *Berger in Monatsschr. Kakt.* xvi. 176.

SOUTH AFRICA : without locality, *Burke* !
CENTRAL REGION : Laingsburg Div. ; Zout Kloof Farm, north-west of Laingsburg, *Pillans*, 60 ! two miles south-west of Laingsburg, *Pillans*, 679 ! Beaufort
West Div. ; Rhenoster Kop, *Foster in Herb. Pillans*, 145 ! Prince Albert Div. ;
near Prince Albert, *Marloth*, 4582 !

Described from living stems and flowers preserved in fluid. Dr. Schlechter
described and figured this species from a plant cultivated in the Botanic Garden
at Berlin, which he states was probably introduced by Dr. Stuhlmann from East
Tropical Africa, but this locality is almost certainly an error.

55. **S. revoluta** (Masson, Stap. 12, t. 10) ; stems $\frac{1}{2}$–$1\frac{1}{4}$ ft. high,
sometimes longer under cultivation, sparingly branched, acutely
4-angled, with slightly concave sides and acute spreading teeth
1–$1\frac{1}{2}$ lin. long on the angles, glabrous, glaucous-green ; flowers 1–3
together, from the sides of the upper part of the stems, successively
developed ; pedicels 3–5 lin. long, glabrous ; sepals $2\frac{1}{2}$–4 lin. long,
lanceolate, acuminate, glabrous ; corolla with the lobes very much
recurved (sometimes so as to touch the back of the flower) and
then $1\frac{1}{4}$–$1\frac{1}{2}$ in. in diam., glabrous and smooth on both surfaces,
pale purple, dull purple or purple-brown, paler on the disk, with
the central depression or a star-shaped central area pale greenish-
yellow or cream-colour, ciliate to the tips of the lobes with long
vibratile clavate purple hairs ; disk with a short tube-like depression
about $\frac{1}{3}$ in. in diam., enclosing the corona, not raised into a ring
around its mouth ; lobes 7–9 lin. long and as much in breadth,

ovate, acute or shortly acuminate ; outer corona-lobes $1\frac{1}{4}$–$1\frac{1}{2}$ lin.
long, 1–$1\frac{1}{4}$ lin. broad, subrectangular, slightly narrowing at the
3-toothed apex, purple-brown with a yellowish base ; inner corona-
lobes 2-horned, purple-brown at the base, dull yellowish dusted
with dull purple-brown on the upper part ; outer horns spreading,
$\frac{3}{4}$–$1\frac{1}{2}$ lin. long, laterally flattened, slenderly linear, obtuse ; inner
horns connivent-erect at the base, recurving above, 2–3 lin. long,
filiform, clavate and very minutely tuberculate at the apex ;
follicles subparallel, 5–6 in. long, about 5 lin. thick, fusiform,
tapering into an acute slightly hooked beak, smooth, glabrous,
glaucous ? ; seeds about $3\frac{1}{2}$ lin. long, 2 lin. broad, flattish, ovate,
with a thickened margin, smooth, brown. *Willd. Sp. Pl.* i. 1277 ;
Pers. Syn. Pl. i. 278 ; *Poir. Encycl.* vii. 378, *and in Dict. Sc. Nat.*
l. 389 ; *R. Br. in Mem. Wern. Soc.* i. 24 ; *Ait. Hort. Kew. ed.* 2, ii.
85 ; *Hornem. Hort. Hafn.* i. 247 ; *Jacq. Stap. t.* 45 ; *Schultes, Syst.
Veg.* vi. 34 ; *Link, Enum. Pl. Hort. Berol.* i. 256 ; *Spreng. Syst.
Veg.* i. 840 (*excl. syn.*) ; *Dietr. Syn. Pl.* ii. 886 ; *Decne in DC.
Prodr.* viii. 657. *N. E. Br. in Gard. Chron.* 1904, xxxvi. 206 ;
Schlechter in Journ. Bot. 1898, 483. *S. glauca, J. Donn, Hort.
Cantab. ed.* 3, 42 ; *Jacq. Stap. t.* 44 ; *Willd. Enum. Pl. Hort. Berol.*
279 ; *Poir. Encycl. Suppl.* v. 231 ; *Hornem. Hort. Hafn.* i. 247 ;
Kerner, Hort. Semp. t. 487 ; *Schultes, Syst. Veg.* vi. 34 ; *Link, Enum.
Pl. Hort. Berol.* i. 256 ; *Spreng. Syst. Veg.* i. 840 (*excl. syn.*) ; *Dietr.
Syn. Pl.* ii. 886 ; *Decne in DC. Prodr.* viii. 657. *S. protensa,*
Hornem. Hort. Hafn. Suppl. 30. *Tromotriche revoluta, Haw., and T.
glauca, Haw. Syn. Pl. Succ.* 36, 37, *and Suppl.* 11 ; *G. Don, Gen.
Syst.* iv. 119.

VAR. β, **tigridia** (N. E. Br.) ; corolla with a pale greenish-yellow central star
and the disk around it and the very base of the lobes marked with yellow dots
and short transverse lines, otherwise as in the type. *S. tigridia, Decne in DC.
Prodr.* viii. 657. *S. revoluta, Curtis, Bot. Mag. t.* 724.

VAR. γ, **fuscata** (N. E. Br.) ; corolla entirely reddish-brown on the inner face,
without markings ; lobes 1 in. long ; outer corona-lobes ascending with spreading
tips, oblong, twice as long as broad, 3-toothed at the apex, purple-brown ; horns
of the inner corona-lobes filiform, more slender and longer than in the type, pale
reddish-yellow dotted with purple-brown ; inner horn about 3 lin. long and twice
as long as the outer horn ; otherwise as in the type. *S. fuscata, Jacq. Stap. t.* 46 ;
Decne in DC. Prodr. viii. 657 ; *Schlechter in Journ. Bot.* 1898, 480. *Tromotriche
glauca, var.* β, *Haw. Syn. Pl. Succ.* 37.

COAST REGION : Clanwilliam Div. ; Karoo beyond North Olifants River, *Masson,*
south-western slopes of the ridge a mile south-east of Clanwilliam, *Pillans,* 158 !
and *cultivated specimens* !

Described from living plants. Haworth's type of *Tromotriche fuscata, Haw.
Suppl. Pl. Succ.* 10, *and G. Don, Gen. Syst.* iv. 119, is a colour-variety of
S. revoluta, as the inner corona-lobes have the stout horns of that form.

[56. **S. mutabilis** (Jacq. Stap. tt. 29, 30) ; stems 3–18 in. high,
$\frac{1}{2}$–$\frac{3}{4}$ in. square, acutely 4-angled, with slightly concave or flattish
sides and rather stout conical spreading teeth about 3 lin. long,
glabrous, green, sometimes tinged with purple, slightly glaucous ;
flowers 1–3 together, from near the base of the younger or higher

up on the older stems, successively developed; pedicels $\frac{1}{4}$–$1\frac{1}{6}$ in. long, glabrous; sepals $2\frac{1}{2}$–$4\frac{1}{2}$ lin. long, ovate-lanceolate, acuminate, glabrous; corolla with the lobes very much recurved and then about $1\frac{1}{2}$ in. in diam., glabrous on both surfaces, ciliate nearly or quite to the tips of the lobes with vibratile clavate purple hairs about $\frac{3}{4}$ lin. long, with the disk raised into an erect solid pentagonal annulus $1\frac{1}{2}$–2 lin. high around and forming a short tube containing the corona, but flat on the back beneath it; lobes 7–11 lin. long, 7–10 lin. broad, broadly ovate, acute or shortly acuminate; back smooth, pale green tinged with purplish on the nerves and tips of the lobes; inner face slightly rugose to nearly smooth, with the apical third of the lobes light or dark purple-brown and the remainder pale yellowish or greenish-yellow, covered with transversely elongated spots or thick or narrow (and then crowded) irregular transverse purple-brown lines, becoming fainter on the annulus; outer corona-lobes $1\frac{1}{4}$–$1\frac{1}{2}$ lin. long, $\frac{3}{4}$–1 lin. broad, rectangular or subquadrate, very variable in toothing, bifid, with parallel teeth, or with a small tooth in the notch and the outer edges of the lateral teeth tapering to the apex, or equally 3-toothed, or with 3 minute teeth projecting from the middle of the truncately contracted apex, pale yellowish, speckled with purple-brown and often with a broad central stripe of the same colour; inner corona-lobes 2-horned, pale yellowish, thickly dotted with purple-brown; outer horns suberect or ascending, 1–$1\frac{1}{2}$ lin. long, filiform, slightly clavate at the apex; inner horns 2 lin. long, filiform, conniventerect, recurving at the clavate tips; odour not very strong but disagreeable, sometimes odourless; follicles subparallel, 7–8 in. long, about 5 lin. thick, terete-fusiform, tapering into a long hooked beak, smooth, glabrous. *Haw. Rev. Pl. Succ.* 204; *Spreng. Syst. Veg.* i. 838 (*excl. syn.*); *Kerner, Hort. Semp. t.* 474; *Dietr. Syn. Pl.* ii. 884; *Decne in DC. Prodr.* viii. 661 (*excl. syn. S. rufa, Masson*); *Todaro, Hort. Panorm.* i. 47, *t.* 12, *fig.* 2; *Schlechter in Journ. Bot.* 1898, 482. *S. neglecta* and *S. Passerini, Todaro, S. fuscata, Hort. ex Todaro* (*not of Masson*) *and S. umbilicata, Thuret ex Todaro, l.c.* i. 47. *Tromotriche mutabilis* and var. *variabilis, Sweet, Hort Brit. ed.* 2, 358. *Orbea mutabilis, Sweet, Hort. Brit. ed.* 1, 276; *G. Don in Loud. Hort. Brit.* 96, *and Gen. Syst.* iv. 121.

VAR. β, **discolor** (N. E. Br.); corolla dark purple-brown, with inconspicuous yellowish transverse lines on some of the rugosities on the disk and base of the lobes; annulus yellowish with irregular purple lines; outer corona-lobes dark purple-brown, fuscous along the margins, with a median yellowish stripe. *S. discolor, Todaro, Hort. Panorm.* i. 49, *t.* 12, *fig.* 3; *Schlechter in Journ. Bot.* 1898, 479.

VAR. γ, **furva** (N. E. Br.); corolla marked as in the type, but the markings and apical half of the lobes of an exceedingly dark purple-brown, and the transverse yellowish lines very narrow; lobes about 9 lin. long and 8 lin. broad; annulus crenulately 5-lobulate; otherwise as in the type.

VAR. δ, **Nemesis** (Dammann ex Rüst in Monatsschr. Kakt. vi. 39); a hybrid form with a large pentagonal annulus 1 in. in diam and apparently with a recurving margin as in *S. variegata,* otherwise as in *S. mutabilis.*

Var. ε, **bicolor** (Dammann ex Rüst in Monatsschr. Kakt. vi. 37); a hybrid form with stems as in *S. variegata* and a corolla about 2¼ in. in diam.; lobes minutely ciliate, dark purple-brown, with about 6 longitudinal rows of small yellowish spots; annulus circular, brighter purple-brown with a few elongated yellow markings; outer corona-lobes oblong-linear, narrowed at the minutely bifid apex, pale yellow, with a purple-brown spot at the base and dots on the upper half; outer horn of inner corona-lobes suberect.

Var. ζ, **Circe** (Dammann ex Rüst in Monatsschr. Kakt. vi. 37); a hybrid form with stout stems and a corolla resembling those of *S. variegata*, having a solid and broadly 5 crenate annulus, not recurved at the margin, lobes not ciliate, pale yellow, thickly covered with short transverse purple-brown lines and small transverse spots passing into dots at the apex; annulus densely covered with dots or small rounded spots of purple-brown; outer corona-lobes linear-oblong, minutely 2–3-toothed at the apex, pale yellow, dotted with purple-brown at the apex and down the middle, with or without a spot at the base; outer horn of the inner corona-lobes erect.

Var. η, **Megara** (Dammann ex Rüst in Monatsschr. Kakt. vi. 39); like var. *Circe*, with the spots and transverse lines rather larger.

Var. θ, **Thetis** (Dammann ex Rüst in Monatsschr. Kakt vi. 37); a hybrid form with stems resembling those of *S. variegata*, and a corolla about 2¼ in. in diam.; lobes minutely ciliate, pale yellow, rather evenly spotted to the tips of the lobes with purple-brown, more darkly coloured on the disk around the annulus from the spots being confluent; annulus with broad crenulations, purple-brown with a pale yellowish interrupted ring on its erect rim; outer and inner corona as in *S. variegata*, var. *bufonia*.

Origin: raised from seeds in the Imperial garden at Vienna, numerous *cultivated specimens* including *Pillans*, 87 ! Vars. β and γ, *cultivated specimens* !

The type and vars. β and γ described from living plants; vars. δ, ε, ζ, η and θ described from Dr. Rüst's drawings. This is almost certainly a hybrid raised from seeds of *S. revoluta*, Masson, cross-fertilised by insect agency, since I had sent to me in 1877 two forms of it stated to have been raised from seeds of *S. glauca* (= *S. revoluta*). It has never been found wild in South Africa. There are several forms of it in cultivation, differing slightly in coloration and varying very much in the lobes of the outer corona, which in some forms rest upon the annulus and in others do not nearly touch it and are very variably toothed at the apex. The flowers remain expanded from 4–7 days. The varieties are doubtless only hybrid forms of it.]

[57. **S. trifida** (Tod. Hort. Panorm. i. 45, t. 12, fig. 1); habit and (glabrous) stems as in *S. variegata*; pedicels ¾–1 in. long; sepals 2–3 lin. long, ovate or ovate-lanceolate, acute, glabrous; corolla with the lobes extended about 3 in. in diam., flat, with recurving lobes and an annulus on the disk, smooth and glabrous on the back, inner surface slightly rugose, glabrous, pale yellow, thickly marked with purple-brown spots (less numerous on the annulus) and short thick irregular transverse lines, sometimes to the tips of the lobes or sometimes with that part entirely dark purple-brown; annulus about 10 lin. in diam., subcircular or broadly 5-crenate, 2–2½ lin. high, with the obtuse rim erect; lobes 10–13 lin. long, 9–11 lin. broad, ovate, acute or shortly acuminate, flattish, recurved from the middle, ciliate nearly or quite to the tips with thick or slightly clavate dark purple-brown hairs ½–⅔ lin. long; outer corona-lobes spreading, 2–2½ lin. long, 1 lin. broad, oblong or somewhat ovate-oblong, 2–3-toothed at the apex, with the lateral teeth parallel or

slightly diverging, longer than the middle tooth when the latter is
present, light yellow, dotted on the apical half with purple-brown
and with a central stripe of that colour on the basal half; inner
corona-lobes variable, unequally 2- or 4-horned, even in the same
flower, yellowish, dusted with purple-brown; outer horns ascending
or spreading, compressed-filiform, not clavate; inner horns about
$1\frac{3}{4}$ lin. long and scarcely longer than the outer, connivent-erect,
filiform, with slightly clavate recurving tips and with or without a
small tooth or short horn on each side at their base. *Schlechter in
Journ. Bot.* 1898, 483. *S. sanguinea, Pasq. Cat. Ort. Bot. Nap.* 1867,
99, *name only*.

ORIGIN: a hybrid raised in Europe; *cultivated specimens* (including
Pillans, 69)!

Described from living plants. Todaro states that he received this plant from
the Botanic Garden at Naples under the name of *Stapelia sanguinea*, so that it is
probably the plant intended by that name in Pasquale's Catalogue above quoted,
but the specimens I have seen cultivated as *S. sanguinea* in England have all
been forms of *S. variegata*.]

[58. **S. maculosoides** (N. E. Br. in Gard. Chron. 1901, xxx. 270);
stems erect, about 3 in. high, and 4–5 lin. thick, obtusely 4-angled,
with very slightly concave sides, glabrous; angles very shortly
toothed, with rudimentary deltoid acute ciliate leaves $\frac{1}{2}$–$\frac{2}{3}$ lin.
long at the teeth; pedicels $1\frac{1}{4}$–$1\frac{1}{2}$ in. long, $1\frac{1}{3}$ lin. thick, glabrous
to the eye, but with an exceedingly minute and very scattered
pubescence; sepals $2\frac{1}{2}$ lin. long, 1 lin. broad, ovate-lanceolate,
acuminate, glabrous; corolla about $2\frac{3}{4}$ in. in diam., glabrous
outside, 5-lobed to $\frac{2}{3}$ the way down, nearly flat; disk without
an annulus around the shallow central depression, dark violet-
purple, with concentric irregular paler lines and spots, thinly
covered with short purple hairs; lobes 12–13 lin. long, 8–9 lin.
broad, oblong-ovate, acute, slightly rugose, marked with dark violet-
purple transverse lines and spots on a light yellowish ground on the
central part and with dark violet-purple margins and tips, ciliate
with dark purple simple hairs $1\frac{1}{2}$–2 lin. long, otherwise glabrous;
outer corona-lobes $2\frac{1}{2}$ lin. long, $\frac{1}{2}$–$\frac{2}{3}$ lin. broad, linear, acute,
channelled down the face, blackish-purple or blackish with a faint
paler spot near the apex, glabrous; inner corona-lobes 2-horned,
blackish, faintly speckled with paler colour at the apex of the inner
horn, glabrous; outer horn 1–$1\frac{1}{8}$ lin. long, somewhat spreading,
laterally flattened, narrowly attenuate-deltoid, acute; inner horn
about 3 lin. long, subulate, strongly recurved just above the base,
neither thickened nor acute at the apex.

ORIGIN UNKNOWN : *cultivated specimen* !

Described from a living specimen received from Mr. Justus Corderoy in
Aug. 1901. The flower of this species much resembles that of *S. maculosa*, Jacq.,
but there is no trace of an annulus on the disk of the corolla and the corona and
stems are quite different. I believe it to be of hybrid origin raised from seed in
Europe.]

[59. **S. maculosa** (J. Donn, Hort. Cantab. ed. 3, 43); stems erect, 3–4 in. high, 5–7 lin. thick excluding the teeth, obtusely 4-angled, with spreading conical acute teeth 1–2 lin. long, glabrous, green; flowers solitary or 2–3 together near the base or towards the middle of the young stems, successively developed; pedicels 1¼–1¾ in. long, glabrous; sepals about ¼ in. long, ovate-lanceolate, acuminate, glabrous; corolla 3–4 in. in diam. with the lobes extended, having an annulus on the disk; glabrous and smooth on the back, with the nerves and some large more or less confluent blotches purple-red or suffused with that colour, slightly rugose on the inner face, with short or minute scattered pubescence on and around the annulus, otherwise glabrous, but ciliate to the tips of the lobes with simple purple hairs 1½ lin. long, pale greenish-yellow, tinted with purple-red on the annulus and disk, thickly covered with transverse spots or short (longer on and around the annulus) thick irregular lines of purple-brown, very narrowly margined with and the tips of the lobes almost entirely dark purple-brown; disk flat, with a solid pentagonal annulus 8–10 lin. in diam., about 1 lin. high, convex and slightly crenate-undulate at the top; lobes 1–1¼ in. long, ⅞–1 in. broad, ovate, shortly acuminate; outer corona-lobes about ¼ in. long, ¾–1 lin. broad, variable in different flowers on the same plant and often in the same flower, linear or linear-lanceolate, entire and acute, obtuse or sub-truncate, or bifid or minutely or shortly 3-toothed, pale greenish-yellow, with a large purple-brown spot at the base, a clear space above and on each side of it and the upper part variably speckled or dusted, or the whole dotted almost to the base with purple-brown; inner corona-lobes equally 2-horned, pale greenish-yellow, thickly dotted or dusted with dark purple-brown; inner horns 2–2¾ lin. long, connivent-erect, with recurved tips, filiform, more or less clavate at the apex; outer horns 2–2¾ lin. long, ascending-spreading, laterally slightly compressed, straight, linear-filiform, obtuse, occasionally bifid. *Jacq. Stap. t.* 31; *Willd. Enum. Pl. Hort. Berol.* 283; *Curtis, Bot. Mag. t.* 1833; *Schultes, Syst. Veg.* vi. 36; *Link, Enum. Pl. Hort. Berol.* i. 256; *Kerner, Hort. Semp. t.* 797; *Spreng. Syst. Veg.* i. 838; *Dietr. Syn. Pl.* ii. 884; *Decne in DC. Prodr.* viii. 658; *Schlechter in Journ. Bot.* 1898, 481. *S. maculata, Poir. Encycl. Suppl.* v. 234. *S. mixta, J. Donn, Hort. Cantab.* ed. 4, 53, *not of Masson. Orbea maculosa, Haworth, Syn. Pl. Succ.* 37; *G. Don, Gen. Syst.* iv. 119.

Origin unknown : *cultivated specimens* !

Described from living plants. There is no evidence that *S. maculosa* has ever been found in a wild state; J. Donn states that it was introduced in 1799, but gives no description; Jacquin merely states that he received it from England under the name of *S. maculosa.* In ed. 4 of Donn's *Hort. Cantab., S. maculosa* disappears from the list and *S. mixta* takes its place, and according to the testimony of old specimens at Kew it was cultivated as *S. mixta*, but is not *S. mixta*, Masson. From all this, however, I strongly suspect that *S. maculosa* is a hybrid raised from seed probably produced by *S. mixta*, and that Donn having made that discovery, altered the name to *S. mixta.* In the same manner hybrid *Stapelias* are in cultivation at the present period under the name of their seed-parent, although quite different from it.]

[60. **S. tridentata** (Schultes, Syst. Veg. vi. 49, and G. Don, Gen. Syst. iv. 117, name only); according to the testimony of dried specimens at Kew and in the Berlin Herbarium, cultivated at Berlin between 1806 and 1812, this is exactly like *S. maculosa* in the size, form and coloration of the corolla, and the corona is the same, but differs from that in having no hairs upon the disk and annulus and the lobes are ciliate with clavate hairs ½ lin. long, the spots on the lobes are also smaller and more numerous.

ORIGIN UNKNOWN : probably a hybrid raised in Europe, *cultivated specimens* !]

61. **S. Barklyi** (N. E. Br. in Hook. Ic. Pl. t. 1909); plant 3–4 in. high, branching at the base; stems erect, ½–¾ in. square, with obtuse angles and stout spreading conical teeth 2–3 lin. long, minutely puberulous; flowers 1–2 together near the base of the stems; pedicels stout, 2–4 in. long, glabrous to the eye, but with a very minute and rather sparse pubescence; sepals 4½ lin. long; corolla 5–6 in. in diam. with the lobes extended, rotate, with a thick solid annulus 1 in. or more in diam. on the disk, glabrous on the back and on the slightly rugose inner surface of the lobes, with the annulus and disk around it loosely villous with long soft purple hairs and the lobes ciliate with similar hairs, purple-brown, marked with numerous transverse pale yellow lines on the disk and basal ⅔ of the lobes, with their apical third entirely dark purple-brown; lobes 1¾–2 in. long, about 1½ in. broad, ovate, acute or acuminate; annulus stout, solid, with 5 slight broad crenations formed by 5 shallow grooves radiating from the centre, "buff, with yellowish transverse wrinkles" (*Barkly*); outer corona-lobes ascending-spreading, 3½–4 lin. long, 1 lin. broad, linear-oblong, acuminate, channelled down the face, yellow, dotted or mottled with purple-brown; inner corona-lobes 2-horned, purple-brown; outer horn 1½–2 lin. long, free to the base, ascending, laterally flattened, subulate or deltoid-subulate, entire or toothed behind or at the apex; inner horn 3–3½ lin. long, subulate, recurving from about the middle. *Schlechter in Journ. Bot.* 1898, 479.

WESTERN REGION : Little Namaqualand ; near Ookiep, *Morris* (*Barkly*, 31)!

Described from a specimen in fluid and a coloured drawing.

62. **S. irrorata** (Masson, Stap. 12, t. 9); stems erect, decumbent at the base, 1½–3½ in. long, ⅓–½ in. thick excluding the teeth, obtusely 4-angled, glabrous, green; teeth conical, acute, spreading, 2–3 lin. long; flowers solitary (always?) at the base of the stems; pedicels about 1 in. long, glabrous; sepals ovate, acute, glabrous; corolla about 2¾ in. in diam., pentagonal and acutely pointed in bud, glabrous and smooth outside, rugose on the glabrous inner face; disk depressed or saucer-shaped, purplish according to the figure; annulus none; lobes spreading, ¾–1 in. long, 7–8 lin. broad,

ovate, acuminate, not ciliate, pale yellow spotted and dotted with blood-red (purple-brown?); corona? *Willd. Sp. Pl.* i. 1291; *Pers. Syn. Pl.* i. 280; *Poir. Encycl.* vii. 388; *Ait. Hort. Kew. ed.* 2, ii. 94; *Schultes, Syst. Veg.* vi. 29; *Spreng. Syst. Veg.* i. 841; *Dietr. Syn. Pl.* ii. 887; *Decne in DC. Prodr.* viii. 655. *Podanthes irrorata, Haw. Syn. Pl. Succ.* 33; *G. Don, Gen. Syst.* iv. 118.

SOUTH AFRICA: without locality, *Masson.*

This species only appears to differ from *S. verrucosa* by the corolla being destitute of an annulus. It has not been refound by any other collector, and no specimen of it appears to have been preserved.

63. **S. verrucosa** (Masson, Stap. 11, t. 8); stems erect, decumbent at the base, 1½–3 in. long, 3–5 lin. thick (excluding the teeth), obtusely 4-angled, with spreading conical acute teeth 1½–3 lin. long, glabrous, green; flowers 1–3 together near the base of the young stems, developing successively; pedicels ½–1¼ in. long, spreading, glabrous; sepals 2½–5 lin. long, lanceolate or ovate-lanceolate, acuminate, glabrous; corolla in bud conically pointed, with a depressed-ovate pentagonal basal part, when expanded 1¾–2½ in. in diam., lobed to more than half-way down, with the united part saucer or shallowly basin-shaped, 2–3 lin. deep outside, having a distinct slightly raised pentagonal solid annulus surrounding the corona, about ½ lin. high, convex at the top, with 5 distinct channels radiating from the centre to its angles; lobes recurved-spreading, 8–10 lin. long, 6–8 lin. broad, deltoid-ovate, very acute or acuminate, not ciliate; all parts glabrous or with some short erect hairs at the very base under and concealed by the corona, smooth outside, very rugose with small papilla-like tubercles or very short irregular transverse ridges on the inner surface, pale yellow, covered with small dark blood-red spots, smaller and more crowded on the annulus, with the area around the corona entirely dark purple or purple-brown; outer corona-lobes horizontally or slightly deflexed-spreading, about ¾ lin. long and broad, subquadrate, acutely bifid, with (always?) a minute tooth at the base of the broadly rounded notch, dark chocolate, with yellowish margins? inner corona-lobes about 1 lin. long from the shoulder, subhorizontally incumbent upon the backs of the anthers and slightly longer than them, but not or scarcely erect at the tips, ovate-lanceolate, acuminate, slightly gibbous on the back at the base, yellow, with the gibbosity and probably the margins purple-brown. *Willd. Sp. Pl.* i. 1291, *and Enum. Pl. Hort. Berol.* 284; *Pers. Syn. Pl.* i. 280; *Poir. Encycl.* vii. 387, *and in Dict. Sc. Nat.* 1. 392; *Ait. Hort. Kew. ed.* 2, ii. 94; *Schultes, Syst. Veg.* vi. 29; *Link, Enum. Pl. Hort. Berol.* i. 256; *Spreng. Syst. Veg.* i. 840, 841; *Dietr. Syn. Pl.* ii. 887; *Decne in DC. Prodr.* viii. 655. *Podanthes verrucosa, Haw. Syn. Pl. Succ.* 33; *G. Don, Gen. Syst.* iv. 118.

VAR. β, **pulchra** (N. E. Br.); stems 3–4½ lin. thick excluding the teeth, sometimes mottled with dull purple; spots on the corolla varying from purple to dark purple-brown; annulus $\frac{1}{2}$–$\frac{3}{4}$ lin. high, purple at the pentagonal bottom of the cup, without distinct channels radiating from the centre to the angles; outer corona-lobes obtusely or acutely bifid, without a minute tooth in the rounded notch, dark chocolate, with or without yellowish margins, shining; inner corona-lobes $\frac{1}{2}$–1 lin. long from the shoulder, ovate and obtuse to lanceolate and acuminate, shorter to slightly longer than the anthers, sometimes slightly ascending but not meeting at the tips, yellow, with the margins and a stripe from the gibbosity down the back chocolate; follicles $4\frac{1}{2}$–5 in. long, $\frac{1}{2}$ in. thick, fusiform, acuminate, smooth, glabrous, striped with purple-brown; seeds about 3 lin. long and $2\frac{1}{4}$ lin. broad, ovate, flat, smooth, light brown, with a broad somewhat inflated ochreous-brown shining margin; otherwise as in the type. *S. pulchra, Schultes, Syst. Veg.* vi. 28; *Dietr. Syn. Pl.* ii. 886; *Decne in DC. Prodr.* viii. 656. *S. verrucosa, Jacq. Stap. t.* 18; *Loud. Encycl. Pl.* 200, *fig.* 3310; *Wien. Ill. Gart. Zeit.* 1894, 234, *fig.* 31, *and (a copy of the same) Rev. Hort. Belg.* 1895, xxi. 77, *fig.* 13 (*bad figures*); *N. E. Br. in Hook. Ic. Pl. under t.* 1923; *Schlechter in Engl. Jahrb.* xviii. *Beibl.* 45, 14, *and Journ. Bot.* 1898, 484 (*excl. syn. S. irrorata, Masson*), *not of Masson. S. irrorata, Lodd. Bot. Cab. t.* 127, *not of Masson. Podanthes pulchra, Haw. Syn. Pl. Succ.* 32; *G. Don, Gen. Syst.* iv. 117.

VAR. γ, **robusta** (N. E. Br.); stems very stout, 5–8 lin. thick excluding the teeth, green, sometimes mottled with dull purplish; flowers as in var. β, with a purple-brown area at the bottom of the cup formed by the annulus, otherwise as in the type. *S. verrucosa, Curtis, Bot. Mag. t.* 786. *Podanthes pulchra, var.* β, *Haw. Syn. Pl. Succ.* 33. *P. pulchra, var. major, Sweet, Hort. Brit. ed.* 1, 278. *P. pulchra, var. verrucosa, G. Don, Gen. Syst.* iv. 118.

VAR. δ, **punctifera** (N. E. Br.); stems 4–5 lin. thick excluding the 2–4 lin. long teeth, mottled and striped with dull purplish; corolla very pale or creamy yellow, marked with very small spots or minute pale purple dots; outer corona light purple; otherwise as in var. β.

VAR. ε, **pallescens** (N. E. Br.); corolla with a distinct cup or basin-like tube about 4 lin. deep, measured outside; lobes much acuminate, very pale yellow, with small purple-brown or brownish-red dots; otherwise as in var β.

VAR. ζ, **roriflua** (N. E. Br.); corolla with a distinct cup or basin-like tube 4–5 lin. deep outside, light yellow or pale greenish-yellow, spotted with larger and darker (sometimes "almost black," *Pillans*) spots than in var. ε; annulus often (always?) adnate to the wall of the tube and somewhat resembling a shelf-like thickening of it around the corona; otherwise as in var. β. *S. roriflua, Jacq. Stap. t.* 19; *Willd. Enum. Pl. Hort. Berol.* 285; *Poir. Encycl. Suppl.* v. 234; *Schultes, Syst. Veg.* vi. 30; *Link, Enum. Pl. Hort. Berol.* i. 257; *Spreng. Syst. Veg.* i. 840 (*excl. syn.*); *Kerner, Hort. Semp. t.* 322; *Decne in DC. Prodr.* viii. 656. *S. roriflora, Dietr. Syn. Pl.* ii. 886, *excl. syn. S. rugosa, Wendl. Coll. Pl.* ii. 41, *t.* 52 (*not of Jacq.*). *S. wendlandiana, Schultes, Syst. Veg.* vi. 39; *Decne in DC. Prodr.* viii. 659. *Podanthes roriflua, Sweet, Hort. Brit. ed.* 1, 278; *G. Don, Gen. Syst.* iv. 118. *Piaranthus rorifluus, Decne in DC. Prodr.* viii. 664. *Orbea wendlandiana, Schultes, Syst. Veg.* vi. 834; *Sweet, Hort. Brit. ed.* 1, 277; *G. Don, Gen. Syst.* iv. 120.

VAR. η, **conspicua** (N. E. Br.); stems 4–5 lin. thick, excluding the 2–3 lin.-long teeth, green; corolla about $1\frac{3}{4}$ in. in diam., with the united part saucer- or shallowly basin-shaped 2–$2\frac{1}{2}$ lin. deep outside, light yellow, marked with rounded dark purple-brown spots about 1 lin. in diam., much larger than in any other variety; otherwise as in var. β.

SOUTH AFRICA: without locality, *Masson*! Var. β: *Pillans*, 645! *cultivated specimen*! Var. δ: *cultivated specimens*!

COAST REGION: Var. β: Uniondale Div.; rocky hill near Haarlem, *Burchell*, 5022! Bathurst Div.; Port Alfred, *Mrs. Hutton*! and in *Herb. Pillans*, 654! 656! Albany Div.; Hell Poort and other places near Grahamstown, *MacOwan*! *Pillans*, 15! 30! 190! *Cooper*, 1534! Komgha Div.; near Komgha, *Flanagan*, 1696 (ex *Schlechter*)! Cathcart Div.; half a mile south of Waku Station,

Pillans, 65 ! Kaffraria, *Bowker*, 5 ! Var. δ: Alexandria Div. ; "Bellevue," *Brockle-bank in Herb. Pillans*, 655 !

CENTRAL REGION : Var. β : Somerset Div. ; near Somerset East, *Bowker*! *MacOwan*, 2177 ! Graaf Reinet Div. ; near Graaf Reinet, *Bolus*, 716 ! Var. γ : Somerset Div. ; Glen Avon Estate, 2 miles east of Somerset East, *Pillans*, 604 ! Var. ε : same locality as var. γ, *Pillans*, 56 ! Var. ζ: same locality as var. γ, *Pillans*, 152 ! 189 ! Var. η : same locality as var. γ, *Pillans*, 192 !

KALAHARI REGION : Var. β : Griqualand West, *Arnot (Barkly, 20)*! *cultivated specimens* from the Diamond Fields, *MacOwan*, 2256 !

Masson's type specimen of *S. verrucosa* is preserved in fluid at the British Museum and differs from all other specimens I have seen in the 5 very marked channels radiating to the angles of the annulus, they are represented in Masson's figure by 5 black rays ; no other collector appears to have found it. The commonest form of the species seems to be var. β which Jacquin has figured as *S. verrucosa*, which varies considerably in the depth of the yellow ground-colour and the darkness of the spotting.

64. S. pulchella (Masson, Stap. 22, t. 36) ; stems like those of *S. variegata*, 2–4 in. high, erect or decumbent at the base, $\frac{1}{3}$–$\frac{1}{2}$ in. thick, obtusely 4-angled, with stout conical acute spreading teeth, glabrous, green ; flowers usually 3 or more together, successively developed near the base of the stems ; pedicels 1–1$\frac{1}{4}$ in. long, glabrous ; sepals 1$\frac{1}{2}$–2$\frac{1}{2}$ lin. long, ovate to lanceolate, acute or acuminate, glabrous ; corolla 1$\frac{1}{4}$–2$\frac{1}{4}$ in. in diam., smooth on the back, minutely tuberculate-rugulose and glabrous on the inner face, not ciliate, sulphur-yellow, covered with numerous purple-brown dots and specks, which are smaller and more crowded upon the annulus, and the lobes sometimes narrowly edged with purple-brown ; disk shallowly saucer-shaped, with a solid more or less pentagonal annulus 4–7 lin. in diam., raised $\frac{1}{3}$–$\frac{2}{3}$ lin. above the surface, and a pentagonal pale purplish area margined with brighter purple-red or occasionally entirely sulphur-yellow at the bottom of its cup ; lobes 5–10 lin. long, 5–8 lin. broad, ovate or deltoid-ovate, very acute or acuminate, very spreading or recurved ; outer corona-lobes 1$\frac{1}{4}$–1$\frac{1}{2}$ lin. long, $\frac{1}{2}$ lin. broad, linear or oblong-linear, obtuse, truncate, emarginate or minutely 3-toothed at the apex, dark purple-brown with a pale yellowish spot or somewhat ∧ shaped mark below the apex and sometimes along the margins ; inner corona-lobes unequally 2-horned, pale yellow spotted with dark purple-brown, or purple-brown with yellow markings ; outer horns ascending, $\frac{1}{4}$–$\frac{1}{2}$ lin. long, conical or conical-subulate, obtuse ; inner horns $\frac{3}{4}$–1$\frac{1}{4}$ lin. long, connivent-erect or with recurving tips, filiform, varying from distinctly clavate to not at all thickened at the apex. *Willd. Sp. Pl. i. 1290 ; Pers. Syn. Pl. i. 280 ; Poir. Encycl. vii. 387, and in Dict. Sc. Nat. 1. 394 ; Ait. Hort. Kew. ed. 2, ii. 93 ; Schultes, Syst. Veg. vi. 30 ; Spreng. Syst. Veg. i. 838 ; Dietr. Syn. Pl. ii. 885 ; Decne in DC. Prodr. viii. 655 ; Loud. Encycl. Pl. 200, fig. 3312 ; N. E. Br. in Gard. Chron. 1882, xviii. 199 ; Schlechter in Journ. Bot. 1898, 482, excl. syn. ; Berger in Monatsschr. Kakt. xiv. 127. Podanthes pulchella, Haw. Syn. Pl. Succ. 33 ; G. Don, Gen. Syst. iv. 118.*

SOUTH AFRICA : without locality, *cultivated specimens* !

Coast Region : Uitenhage Div. ; near the bank of the Zwartkops River in the vicinity of Zwartkops Station, *Rabjohn* ! Port Elizabeth Div. ; sand-dunes along the coast south of Port Elizabeth, *Herb. Pillans*, 84 ! Alexandria Div. ; sandflats, *Brocklebank in Herb. Pillans*, 136 !

This species by its 2-horned inner corona-lobes and small solid annulus completely connects the groups to which *S. verrucosa* and *S. variegata* respectively belong.

[65. **S. cupularis** (N. E. Br. in Gard. Chron. 1897, xxii. 45); stems like those of *S. variegata*, 3–8 in. long, 4–7 lin. thick ; flowers 1–3 together ; corolla in bud flattened, with a short obtuse point, pentagonal, with the sinal angles projecting and slightly recurved ; when expanded about 2 in. in diam., slightly rugose on the glabrous inner face, pale lemon-yellow (rather lighter on the annulus), thickly covered with dark purple-brown spots, which are often confluent into irregular lines so that the yellow ground forms labyrinthine lines between them ; disk saucer-shaped ; annulus about $\frac{2}{3}$ in. in diam., $\frac{1}{4}$ in. deep, nearly circular, cup-shaped, with an erect acute margin ; lobes recurved, 8–9 lin. long and about the same in breadth, ovate, acute, ciliolate with minute clavate hairs ; outer corona-lobes $2\frac{1}{2}$ lin. long, deeply bifid with parallel slightly diverging teeth and a minute tooth at the base of the notch, pale greenish-yellow, dotted at the apex and down the middle with dark purple-brown and with a spot of it at the base ; horns of the inner corona-lobes subequal, both clavate at the apex, pale yellow, dotted with purple-brown, outer horn slightly spreading.

Origin : a hybrid raised in Europe, *cultivated specimen* !

Described from living plants. This is doubtless a hybrid between *S. mutabilis*, Jacq., and some form of *S. variegata*, Linn.]

[66. **S. angulata** (Tod. Hort. Bot. Panorm. i. 54, t. 13, fig. 3) ; stems about $2\frac{1}{2}$ in. high, like those of *S. variegata*, L. ; corolla about $2\frac{1}{2}$ in. in diam., purple-brown, marked to the tips of the lobes with numerous labyrinthine transverse yellowish lines ; lobes ovate, spreading or recurved, minutely ciliate with clavate hairs ; annulus pentagonal or subcircular, with an erect or scarcely recurved margin, light yellowish, dotted with purple-brown ; outer corona-lobes more or less narrowed and 2–3-toothed at the apex ; inner corona-lobes 2-horned, pale yellowish, minutely dotted with purple-brown ; horns filiform, clavate, subequal or the outer shorter.

Origin : a hybrid raised in Europe, *cultivated specimen* !

Described from a living plant. *S. mutabilis*, Jacq., is in all probability one of its parents. The yellowish markings on the corolla are narrower in the specimen seen than in Todaro's figure.]

67. **S. namaquensis** (N. E. Br. in Gard. Chron. 1882, xviii. 648, including var. *minor*); stems procumbent or decumbent, $1\frac{1}{2}$–$3\frac{1}{4}$ in.

long, $\frac{3}{4}$ in. thick excluding the teeth, obtusely 4-angled, with very
stout conical acute spreading teeth $2\frac{1}{2}$–$3\frac{1}{2}$ lin. long, glabrous, green,
prettily marked with irregular purple stripes where exposed to the
sun; flowers 1–4 together, successively developed near the base or
middle of the young stems; pedicels $1\frac{1}{4}$–$1\frac{1}{2}$ in. long, 2 lin. thick,
glabrous, purplish, striped with darker purple-red; sepals about
$\frac{1}{4}$ in. long, ovate, acuminate, glabrous; corolla in bud depressed-
pentagonal with a very acuminate point, when expanded 3–4 in. in
diam., flat, with recurving lobes and a large very prominent solid-
looking annulus, having the margin recurved so as to be nearly
circular in transverse section, and the bottom of its cup densely
covered with short erect stiff purple-brown hairs; back smooth and
glabrous; inner surface very rugose with crowded transverse
papillate ridges on the lobes and minutely tuberculate-rugose on the
annulus, pale greenish-yellow, everywhere covered with dark purple-
brown thick transverse lines or rounded or transverse or confluent
spots or labyrinthine markings, those on the annulus as large as or
smaller than those on the lobes; lobes 1–$1\frac{1}{4}$ in. long, $\frac{7}{8}$–1 in. broad
at the base, broadly ovate, shortly or long-acuminate, not ciliate;
outer corona-lobes ascending, $3\frac{1}{2}$–4 lin. long, linear or linear-lanceolate, entire, acute, yellow, dotted with purple-brown; inner corona-
lobes connivent-erect, with recurved tips, simple, 2–$2\frac{1}{2}$ lin. long,
filiform, clavate at the apex, slightly gibbous at the base, but
without an outer horn; follicles subparallel, 5–6 in. long, $\frac{3}{4}$ in.
thick, fusiform, acuminate, glabrous; seeds not seen. *N. E. Br. in
Hook. Ic. Pl. t.* 1908; *Schlechter in Journ. Bot.* 1898, 482.

VAR. *β*, ciliolata (N. E. Br. in Gard. Chron. 1882, xviii. 648); corolla shortly
or minutely ciliate with simple or subclavate white or white and dark purple
hairs mixed, spotting variable; outer corona-lobes entire, acute; otherwise as in
the type. *N. E. Br. in Hook. Ic. Pl. t.* 1908, *fig. B. S. ciliolata, Rüst in
Monatsschr. Kakt.* vi. 43, *not of Todaro.*

VAR. *γ*, bidens (N. E. Br.); corolla-lobes minutely or shortly ciliate with
simple white hairs, sometimes almost absent; outer corona-lobes rather deeply
bifid, with parallel teeth, with or without a minute tooth in the notch; otherwise
as in the type.

VAR. *δ*, tridentata (N. E. Br. in Gard. Chron. 1882, xviii. 648); corolla
shortly or minutely ciliate as in var. *β*; spotting variable; outer corona-lobes
minutely and regularly or irregularly 3-toothed, or entire and truncate or very
obtuse at the apex; otherwise as in the type. *N. E. Br. in Hook. Ic. Pl.
t.* 1908, *fig. C. S. tridentata, Rüst in Monatsschr. Kakt.* vi. 43, *not of Schultes.*

WESTERN REGION: Little Namaqualand (all varieties); Komkam, *Rich in
Herb. Pillans,* 16! and without precise locality, *Barkly,* 6! 64! 64 bis! Var. *β*:
Komkam, *Rich in Herb. Pillans,* 167! and without precise locality, *Barkly,* 38!
Var. *γ*: without precise locality, *Bolus!* Var. *δ*; Ookiep, *Templeman in Herb.
Pillans,* 7! and without precise locality (*Morris,* 7) *Barkly! Scully!*

This species according to Sir Henry Barkly is common in Namaqualand, and
from the specimens I have seen of it would appear to be as variable in colour,
ciliation and outer corona-lobes as *S. variegata,* no two gatherings of it which I
have examined being quite identical. Var. *δ* is described from living plants sent
to Kew by Sir Henry Barkly in 1874; the flowers remain expanded for
3–4 days.

68. S. lepida (Jacq. Stap. t. 43); plant very like *S. variegata*, glabrous in all parts, compactly branched; stems 2–3 in. long, 4–5 lin. thick, glabrous; flowers 1–2 together near the base of the stems, successively developed; pedicels $\frac{2}{3}$–1 in. long; sepals 2–2$\frac{1}{2}$ lin. long, ovate, acute or acuminate; corolla rather small, about 1$\frac{1}{2}$ in. in diam., with a distinct annulus on the disk; inner face transversely rugose on the lobes, granulate-tuberculate on the annulus, glabrous and not ciliate, sulphur-yellow, covered with rather small irregularly scattered dark purple-brown spots, without intermingling lines, those on the paler annulus smaller than those on the lobes, which are about $\frac{1}{2}$ in. long, broadly ovate, acute; annulus with a recurved-spreading acute margin; outer corona-lobes 1$\frac{1}{2}$ lin. long, $\frac{3}{4}$–1 lin. broad, oblong, emarginate or bifid, sometimes with a minute tooth at the base of the notch, very pale, greenish or greenish-yellow, with or without a central suffused and dotted stripe extending to about $\frac{2}{3}$ of the way up or some dots around the teeth; inner corona-lobes 2-horned, pale yellow or greenish, with or without purple-brown dots; inner horn 1$\frac{1}{2}$–1$\frac{2}{3}$ lin. long, erect, recurving at the slightly clavate apex; outer horn ascending-spreading, half as long as the inner or shorter, subulate. *Willd. Enum. Pl. Hort. Berol.* i. 280; *R. Br. in Mem. Wern. Soc.* i. 25; *Hornem. Hort. Hafn.* i. 248; *Poir. Encycl. Suppl.* v. 231; *Schultes, Syst. Veg.* vi. 30; *Link, Enum. Pl. Hort. Berol.* i. 256; *Spreng. Syst. Veg.* i. 838; *Dietr. Syn. Pl.* ii. 885; *Decne in DC. Prodr.* viii. 661; *Schlechter in Journ. Bot.* 1898, 481. *S. limosa, Salm-Dyck, Hort. Dyck.* 266. *Podanthes lepida, Haw. Syn. Pl. Succ.* 34. *Orbea lepida, Haw. Suppl. Pl. Succ.* 13; *G. Don, Gen. Syst.* iv. 121.

There are excellent specimens of this species in Haworth's Herbarium at Oxford. The stems as represented by Jacquin (as in so many other cases) are abnormally very much elongated; when grown under proper conditions, as evidenced by Haworth's specimens, they are short, erect and compact. *S. lepida* is closely allied to *S. variegata*, but its flowers are very much smaller and have a different appearance. It does not appear to have been refound since Scholl collected it.

69. S. variegata (Linn. Sp. Pl. i. 217); plant glabrous in all parts, freely branching at the base; stems erect from a decumbent base, 2–6 in. long, 4–5 lin. (in some varieties and hybrids up to $\frac{1}{2}$ in.) square, very obtusely 4-angled, with conical acute spreading teeth, 1$\frac{1}{2}$–2 lin. long, often having a minute tooth (*stipule*) on each side of the withering point, green, often mottled all over or tinted with purple at the tips; flowers 1–5 together at the base of the young stems, developing successively; pedicels 1–2$\frac{1}{4}$ in. long, 1–1$\frac{1}{2}$ lin. thick; sepals 2$\frac{1}{2}$–4 lin. long, ovate or ovate-lanceolate, acute or acuminate; corolla 2–3 in. in diam.; disk with a pentagonal annulus $\frac{3}{4}$–$\frac{7}{8}$ in. in diam., having a recurved acute-edged rim; lobes $\frac{2}{3}$–1 in. long, $\frac{1}{2}$–$\frac{7}{8}$ in. broad, ovate, acute or shortly acuminate, flat, radiately spreading or slightly recurved, 5–7-nerved ; back smooth, green, suffused on the lobes and nerved with purplish ; inner surface rugose with crowded irregular transverse ridges on the lobes and with small crowded granule-like tubercles on the annulus, otherwise glabrous and without a trace of minute ciliation on the lobes, pale greenish-yellow with dark purple-brown spots arranged in 6–7 longitudinal rows or sometimes irregularly scattered, besides a series along the margins, intermingled with slender lines of the same colour, but not producing a dark effect on the basal part of the lobes, paler yellow with much smaller spots or dots of purple-brown on the annulus ; outer corona-lobes ascending-spreading, 2–2¾ lin. long, linear-oblong, with parallel sides, minutely and obtusely 3-toothed at the apex, with the longer teeth not more than ⅛–⅙ of the length of the lobe, pale yellow, dotted or dusted with purple-brown and with a square spot of the same colour at the base ; inner corona-lobes 2-horned, pale yellow, dotted with purple-brown ; horns filiform, inner about 1½ lin. long, connivent-erect with recurved clavate and minutely tuberculate tips, outer suberect or ascending, straight, slightly clavate, rather shorter than or about as long as the inner. *Linn. Syst. Nat. ed.* 12, ii. 194. *S. normalis, Jacq. Stap. t.* 42 ; *Schultes, Syst. Veg.* vi. 39 ; *Link, Enum. Pl. Hort. Berol.* i. 256 ; *Spreng. Syst. Veg.* i. 838 ; *Dietr. Syn. Pl.* ii. 885 ; *Decne in DC. Prodr.* viii. 660. *S. woodfordiana, Schultes, l.c.* 41 ; *Link, l.c.* 257 ; *Decne, l.c.* 661. *Orbea normalis, Schultes, l.c.* 834 ; *G. Don, Gen. Syst.* iv. 120. *O. woodfordiana, Haw. Syn. Pl. Succ.* 42 ; *G. Don, l.c.* 121.

(The following references may belong here or partly to the varieties noted below :—*S. variegata, Miller, Gard. Dict. ed.* 8 ; *Burm. Fl. Cap. Prodr.* 7 ; *Rottb. Bot. Udstrakte Nytte,* 62, *fig.* 10 ; *Poir. Tabl.* ii. 325, *t.* 178, *fig.* 1 ; *Encycl.* vii. 388, *and Dict. Sc. Nat.* 1. 391 ; *Ait. Hort. Kew. ed.* 1, i. 309 *and ed.* 2, ii. 95 ; *Willd. Sp. Pl.* 1292, *and Enum. Pl. Hort. Berol.* 283 ; *Moench, Suppl. Meth. Pl.* 314 ; *Pers. Syn. Pl.* i. 280 ; *Herb. Gén. Amat.* ii. *t.* 125 ; *Drapiez, Herb. Amat.* i. *t.* 41 ; *Schultes, Syst. Veg.* vi. 37 ; *Link, Enum. Pl. Hort. Berol.* i. 256 ; *Spreng. Syst. Veg.* i. 838 ; *Brongn. in Ann. Sc. Nat.* xxiv. 268, 279, *t.* 14 ; *Schlechter in Journ. Bot.* 1898, 483, *partly. Orbea variegata, Haw. Syn. Pl. Succ.* 40.— *Fritillaria crassa promontorii bonæ spei &c., Stapel in Theophrast. Hist. Pl.* (1644), 335, *with fig. Apocynum humile aizoides &c., Hermann, Hort. Acad. Lugd. Bat. Cat.* 52 *and* 53, *with fig.* ; *Ray, Hist. Pl.* (*ed.* 1688) ii. 1903 ; *Pluk. Almagest. Bot.* 37, *with fig. Asclepias aizoides aphylla &c., Moris. Hist. Pl.* iii. 611, § 15, *t.* 3, *fig.* 4. *Crassa Rivini, Rupp. Fl. Jen. ed.* 1718, 26–27. *Stapelia denticulis ramorum extrorsum prominentibus, Linn. Hort. Cliff.* 77. *Stapelia denticulis ramorum patentibus, Royen, Fl. Leyd. Prodr.* 409. The plant figured as the "Small Cape Fritillary," "*Apocynum fritillaricum minus,*" in *Petiver, Opera,* i. (*Gazoph. dec.* ix.) 10, *Cat. no.* 450, *t.* 90, *fig.* 4, may be intended for some form of this species.)

Coast Region : Cape Div. ; Table Mountain or Lion Mountain, *Prior* ! and *cultivated specimens* ! Variations (enumerated below) according to Mr. Pillans, occur on the shore at Fish Hoek, *Pillans,* 141 ! on rocks south of Simons Town, and Houts Bay, northern slopes of Table Mountain, *Pillans,* 108 ! and 112, 114, 128 (ex *Pillans*), the Lion's Head, Lion's Back, western shores of Robben Island. Piquetberg Div. ; near Piquetberg, *Pillans,* 199 ! Worcester Div. ; near Worcester. Paarl Div. ; Paarl Mountain. Caledon Div. ; near Hermanus. Riversdale Div. ; near Riversdale. Mossel Bay Div. ; Mossel Bay.

Stapelia variegata is the oldest known member of the tribe *Stapelieæ* and would appear to have been introduced into cultivation in Holland about 1640, since the first mention of the plant I can find is in Stapel's edition of Theophrastus *Historia Plantarum,* 1644.

The type specimen of *S. variegata* in Linnæus' Herbarium consists of a well-preserved flower of the particular form of this variable species described above, the inner corona-lobes have been removed from the specimen, but the outer (which afford one of the chief characteristics of this form) are well preserved. It is the plant figured by Jacquin as *S. normalis,* and the same as the type specimen of *Orbea woodfordiana,* Haw., in Haworth's Herbarium at Oxford. On Linnæus' specimen the spots on the corolla-lobes are arranged in longitudinal rows exactly as figured by Jacquin. On Haworth's they are rather indistinct, but seem more scattered, this variation, however, I have seen on different flowers produced by the same plant.

In its flowers *S. variegata* is extremely variable, the stems of the various forms being very similar and often indistinguishable. The flowers vary in the shape of the mature (but not young) bud, coloration, ciliation of the lobes, flatness of the disk, shape of the annulus, form and toothing of the outer corona-lobes, and length and direction of the outer horn of the inner corona-lobes. Upon the various combinations of these characters, numerous forms have been described and cultivated as distinct species. But some of these supposed specific characters vary in different flowers produced by the same individual, sometimes in the same, at others in different years. As distinct varieties are often crossed by insect agency, innumerable variations have thus arisen in their native habitat and in European gardens. Until recently I had supposed that most of these different forms were local variations or races, and that only one form grew in the vicinity of Cape Town, but I am informed by Mr. Pillans that several forms often grow intermingled within quite a limited area—for example, he states that he "gathered 17 different variations of *S. variegata* on Robben Island in an area of 80 by 100 yards. All these had distinct coronal and colour distinctions. Above Cape Town [on Table Mountain] there are quite as many, and on the Lion's Rump rather fewer, I think."

For those forms which have been fully described and figured as distinct species I here give a synoptic key and enumerate their distinctive characters, but many variations of them merge into one another. Besides these, there are a large number of forms, distinct from any described, growing wild, which have not received names, but I refrain from describing these (except in 2 cases), because they are merely hybrid forms. In the *Monatsschrift für Kakteenkunde,* vi. 35–43, Dr. Rüst has published a synoptic key to 58 species and 27 varieties of the group to which *S. variegata* belongs. The characters given in the key are unfortunately quite inadequate for identification and there are no descriptions, but Dr. Rüst has very kindly lent to me his admirable drawings of all the forms he has enumerated. With the exception of *S. Barklyi* and *S. namaquensis* and its two varieties, all are forms of the different varieties of *S. variegata* described below or of *S. mutabilis* or hybrids between these and other forms, which have been raised in Europe, most of them in the nursery of Messrs. Dammann & Co. at Naples. Some are arranged among the varieties here recognised and the remainder in alphabetical order at the end, as I have not seen specimens. All the forms enumerated in the key have the margin of the annulus very spreading or recurved except in 72, *S. albicans,* in which it is only slightly recurved.

Corolla entirely blackish-purple or very dark purple-
　　brown with a few irregular transverse, winding or
　　longitudinal yellowish lines and spots producing a
　　somewhat marbled appearance:
　　Corolla 1¾–2 in. in diam. ; inner corona-lobes with the
　　　　outer horn not more than half as long as the inner　(p)　*marmorata.*

Corolla 2½–3¼ in. in diam. ; inner corona-lobes with
the outer horn nearly or quite as long as the
inner :
Corolla-lobes ovate ; ground colour blackish-purple... (q) *atropurpurea.*

Corolla-lobes deltoid-acuminate ; ground colour dark
purple-brown　　...　　...　　...　　...　　... (r) *atrata.*

Corolla-lobes pale greenish-yellow with crowded transverse thin purple-brown lines (not spots) on the basal
½–⅔ and dots or small spots on the apical part ;
annulus paler with small spots, its rim flattish (see
also var. i, *pallida*)　　...　　...　　...　　...　　... (g) *horizontalis.*

Corolla 2–3¾ in. in diam., pale yellow or greenish-yellow,
brighter on the annulus, marked all over with dots or
spots (sometimes confluent or connected by transverse thin lines) or long labyrinthine or short thick
lines of purple-brown :
Inner corona-lobes with the outer horn half or less
than half as long as the inner horn (see also note
under var. *bufonia* respecting *S. orbicularis*,
Andr.) ; corolla darkly marked with very confluent
spots ...　　...　　...　　...　　...　　...　　... (l) *brevicornis.*

Inner corona-lobes with the outer horn from more than
half as long to as long as the inner horn :
Corolla-lobes minutely but distinctly ciliate with
rather thick or clavate fixed (not vibratile)
hairs ; spots on the annulus as small as or
smaller than those on the lobes :
Teeth of the outer corona-lobes widely diverging :
Corolla about 2 in. in diam. ; lower ⅔ of lobes
labyrinthinally marked with thick brownish-
crimson lines broken up into spots towards
the tips　　...　　...　　...　　...　　... (70) *divergens.*

Corolla 2½–3 in. in diam. ; lobes labyrinthinally
marked with thick dark purple-brown lines
and confluent spots ; annulus pale yellow
with light purple-brown dots　　...　　... (71) *scutellata.*

Corolla 2½–3 in. in diam., spotted all over,
with or without a few slender transverse
lines interspersed on the base of the lobes　(e) *clypeata.*

Teeth of the outer corona-lobes not or but slightly
diverging :
Rim of the annulus light yellow, without
spots ; corolla with 5 purple-brown rays on
the disk extending to and beyond the
sinuses　　...　　...　　...　　...　　... (s) *Prometheus.*

Rim of the annulus more or less spotted ; corolla
without coloured rays on the disk :
Corolla 2–2½ in. in diam. ; lobes marked
with short thick transverse lines or
elongated spots and small rounded spots :

Margin of annulus very slightly recurved,
not forming a flat-topped rim ... (72) *albicans.*

Margin of annulus horizontally spreading,
forming a flat-topped rim ... ... (73) *hanburyana.*

Corolla $2\frac{1}{2}$–$3\frac{3}{4}$ in. in diam., light coloured ;
lobes with numerous rounded spots,
those on the basal half connected by a
network of slender transverse purple-
brown lines ... ... (f) *mixta.*

Corolla about 2 in. in diam., dark coloured
from the irregular spots being more or
less confluent ... ... (m) *conspurcata.*

Corolla-lobes not ciliate or the cilia reduced to
exceedingly minute and often papilla-like hairs,
rarely more than twice as long as thick :
Spots very numerous on each lobe, small or dot-
like, those on the annulus more minute ... (i) *pallida.*

Spots on the lobes not dot-like :
Spots not numerous on each lobe, large,
irregular or confluent, rich crimson-brown,
brighter (nearly blood-colour) and not
smaller on the annulus :
Corolla about 2 in. in diam. ; outer corona-
lobes deeply bifid, without a tooth in the
notch ... ... (o) *picta.*

Corolla $2\frac{1}{2}$–3 in. in diam. ; outer corona-lobes
3-toothed at the apex ... ... (n) *læta.*

Spots many on each lobe, of moderate size,
distinct or but slightly confluent except on
the basal part, irregularly scattered or in
longitudinal rows, those on the annulus
small or dot-like, all dark purple-brown,
never crimson-brown or blood-colour :
Base of the corolla-lobes more darkly coloured
than elsewhere from the spots being
confluent, or very distinctly connected
by or confluent with numerous transverse
dark purple-brown lines :
Annulus with a flattish rim ; corolla $2\frac{1}{2}$–$3\frac{1}{2}$
in. in diam. :
Annulus with 5 notches and 5 very
broad crenations ; outer corona-lobes
equally 3-toothed at the apex ... (h) *rugosa.*

Annulus pentagonal or 10-crenate ; outer
corona-lobes bifid with or without a
small tooth in the notch ... ... (f) *mixta.*

Annulus with a convex rim, not normally
notched nor crenate ; corolla $2\frac{1}{2}$–3 in.
in diam. :

Outer corona-lobes about 3 times as long
    as they are broad at the middle,
    tapering to the tips or with parallel
    sides, minutely notched, very shortly
    bifid or very minutely 3-toothed at
    the apex ... ... (k) *retusa.*

Outer corona-lobes 2–2½ times as long as
    they are broad at the middle,
    usually rather deeply bifid with
    parallel teeth and with or without a
    minute tooth in the notch :
Corolla in mature (not young) bud flat-
    topped, without a point, when ex-
    panded flat on the back of the disk (a) *trisulca.*

Corolla in mature bud conical, acute,
    when expanded not quite flat on
    the back of the disk ... ... (d) *bufonia.*

Base of the corolla-lobes not or scarcely more
    darkly coloured than elsewhere ; spots
    separate or not very conspicuously con-
    nected by slender purple-brown lines :
Outer corona-lobes not much longer than
    broad, shortly bifid ... ... (b) *marginata.*

Outer corona-lobes 2–3 times as long as
    broad :
Expanded corolla quite flat on the back
    of the disk ; outer corona-lobes bifid,
    with parallel obtuse teeth ... ... (c) *planiflora.*

Expanded corolla more or less convex on
    the back of the disk :
Outer corona-lobes bifid from ⅓ to
    more than half-way down, with
    diverging or subparallel teeth and
    with or without a minute tooth in
    the notch ... ... (e) *clypeata.*

Outer corona-lobes minutely 3-toothed
    at the apex, with the middle
    tooth shortest, all obtuse and not
    more than ⅛ to ⅙ of the length of
    the lobe ; ground colour of corolla
    lemon- or faintly greenish-yellow (69) **variegata.**

Outer corona-lobes very slightly
    notched or very minutely and
    equally 3-toothed at the apex ;
    ground colour of corolla light
    yellow or rather bright sulphur-
    yellow ... ... (j) *Curtisii.*

VAR. a, **trisulca** (N. E. Br.) ; corolla in mature (not when younger) bud flat-
or round-topped, not pointed ; disk flat at the back ; spots on the lobes irregular
or more or less in longitudinal rows, more crowded at the base, with rather thick
lines between them, producing a somewhat dark effect ; outer corona-lobes rather
deeply bifid, with a minute tooth in the notch ; otherwise as in the type.
S. trisulca, J. Donn, Hort. Cantab. ed. 3, 43 ; *Jacq. Stap. t.* 33. *S. variegata,
Blanc, Hints on Cacti, ed.* 1891, 96, *with fig. S. normalis, Lindl. Bot. Reg. t.* 755,
not of *Jacq. Orbea trisulca, Haw. Syn. Pl. Succ.* 39.

COAST REGION : Cape Div. ; near Cape Town, *cultivated specimens* !

Var. b, **marginata** (N. E. Br.); corolla-lobes narrowly margined and covered with small irregularly scattered spots of dark purple-brown, without intermingling lines or a suffusion of purple-brown at their base around the subpentagonal annulus, which is of a paler yellow with rather small dots of purple-brown; outer corona-lobes not much longer than broad, oblong, shortly bifid, with a broad notch and subparallel or very slightly diverging obtuse teeth, pale yellowish dotted only on the teeth with purple-brown and with 3 spots of the same colour at the base, "the middle one oblong, yellow in the centre, the lateral shorter, clavate" (*Willdenow*); horns of the inner corona diverging, with the outer subulate, obtuse; otherwise as in the type. *S. marginata, Willd. Enum. Pl. Hort. Berol. Suppl.* 13; *Schultes, Syst. Veg.* vi. 39; *Link, Enum. Pl. Hort. Berol.* i. 257; *Decne in DC. Prodr.* viii. 659; *Schlechter in Journ. Bot.* 1898, 481. *S. planiflora, var. marginata, Willd. Enum. Pl. Hort. Berol.* 284; *Poir. Encycl. Suppl.* v. 232. *Orbea marginata, Schultes, Syst. Veg.* vi. 834; *G. Don, Gen. Syst.* iv. 120. *O. planiflora, var. marginata, G. Don, l.c.* 120.

I have seen a specimen of this from Herb. Link in the Berlin Herbarium, but have not seen it alive.

[*S. marginata*, Rüst in *Monatsschr. Kakt.* vi. 37, not of Willd., is a hybrid, with a corolla about 3 in. in diam.; lobes minutely but distinctly ciliate with (apparently) simple hairs, dark purple-brown along the margins and on the apical half, with the centre of the basal half yellow, spotted with purple-brown, and the disk around the annulus purple-brown; annulus circular, light yellow, densely spotted with purple-brown in the cup, almost without spots on the rim; outer corona-lobes oblong-linear, bifid to ⅓ of the way down, pale yellow, dusted on the teeth and down the centre with purple-brown and having a spot at the base.]

Var. c, **planiflora** (N. E. Br.); corolla acute in bud; disk flat on the back when expanded; spots on the lobes irregularly scattered, without or with inconspicuous slender purple-brown lines intermingled with them; outer corona-lobes bifid, teeth not spreading; inner corona-lobes with subequal horns, the outer ascending-spreading, slightly or not at all clavate at the apex; otherwise as in the type. *S. planiflora, Jacq. Stap. t.* 40; *Willd. Enum. Pl. Hort. Berol.* 284; *Poir. Encycl. Suppl.* v. 232; *Schultes, Syst. Veg.* vi. 38; *Lodd. Bot. Cab. t.* 191; *Link, Enum. Pl. Hort. Berol.* i. 256; *Decne in DC. Prodr.* viii. 659; *N. E. Br. in Journ. Linn. Soc.* xvii. 169. *S. variegata, Jacq. Miscell.* i. 27, t. 4, *and Amer. Gewäch.* i. t. 94; *DC. Plant. Grass. t.* 149; *Tratt. Thesaur. t.* 18, *not of Linn. S. mutabilis, Hulle in Rev. Hort. Belg.* 1889, 193, *with fig., not of Jacq. Orbea planiflora, Haw. Suppl. Pl. Succ.* 12; *G. Don, Gen. Syst.* iv. 120.

Coast Region: Cape Div.; Table Mountain, *cultivated specimens*!

In the *Monatsschr. Kakt.* vi. 61, Weingart records a case in which this plant produced during one year stems variegated with white, but the following year only normally green stems. He failed to propagate the variegated form.

Var. d, **bufonia** (N. E. Br. in Hook. Ic. Pl. under t. 1907, p. 2); corolla acutely pointed in bud; disk under the annulus shallowly basin-like; spots irregularly scattered or more rarely in longitudinal rows, with the basal part of the lobes very darkly coloured from being rather densely covered with slender purple-brown lines between and connecting the spots, sometimes so crowded and confluent that that part appears to be dark purple-brown with a very irregular network of slender greenish-yellow lines; annulus 5-angled or circular; outer corona-lobes bifid to ¼ or ⅓ of the way down with the sides and teeth parallel, without a tooth in the notch; otherwise as in the type. *S. bufonia (by error buffonia), J. Donn, Hort. Cantab. ed.* 3, 43; *Jacq. Stap. t.* 35, 36; *Willd. Enum. Pl. Hort. Berol.* 283; *Poir. Encycl. Suppl.* v. 232; *Schultes, Syst. Veg.* vi. 40; *Spreng. Syst. Veg.* i. 838 (*excl. syns.*); *Dietr. Syn. Pl.* ii. 885; *Decne in DC. Prodr.* viii. 660. *S. Bufonis, Lodd. Bot. Cab. t.* 332. *S. buffoniana, G. Don, Gen. Syst.* iv. 117. *S. beffoniana, Schultes, l.c.* 49. *S. bisulca. J. Donn, Hort. Cantab. ed.* 3, 43; *Schultes, Syst. Veg.* vi. 37; *Link, Enum. Pl. Hort. Berol.* i. 256; *Decne* *in DC. Prodr.* viii. 659 ; *Schlechter in Journ. Bot.* 1898, 479. *S. orbiculata, J. Donn, l.c.* 43. *S. orbicularis, Lodd. Bot. Cab. t.* 811 ; *Fl. des Serres,* xii. 187, *t.* 1281 ; *Le Jardin,* 1892, 175. *S. ophiuncula, Haw. Syn. Pl. Succ.* 27 ; *Schultes, Syst. Veg.* vi. 27 ; *G. Don, Gen. Syst.* iv. 117 ; *Decne in DC. Prodr.* viii. 663 (*name only*). *S. ophioncula, Schlechter in Journ. Bot.* 1898, 482 (*name only*). *S. bidentata, Salm-Dyck, Hort. Dyck.* 266. *S. monstrosa, Steud. Nom. ed.* 2, ii. 631 (*name only*). *S. ciliolulata, Tod. ex Rüst in Monatsschr. Kakt.* vi. 38. *Orbea bisulca, Haw.,* and *O. bufonia, Haw. Syn. Pl. Succ.* 39, 40 ; *G. Don, Gen. Syst.* iv. 119, 120.

COAST REGION: Cape Div.; Table Mountain, &c., *Barkly,* 45 ! 60 ! 61 ! *MacOwan,* 2250 ! *Pillans* ! *cultivated specimens* ! Oaklands, *Jameson* !

Jacquin's figure of *S. bufonia* on t. 35 is very badly coloured and does not agree with his description, which states that the inner face of the corolla is " entirely dirty yellow, with scattered blackish spots and transverse striæ of the same colour." Under cultivation I have found this typical dark-flowered form, in which the spots and lines are confluent, sometimes produces on the same individual much lighter coloured flowers resembling those figured by Jacquin as a variety of *S. bufonia* on t. 36 ; this lighter coloured form is also much commoner than the darker variation. *S. monstrosa* and *S. ophiuncula* are undescribed, but plants that were in cultivation under those names many years ago belong to this variety ; *S. monstrosa* merely has fasciated stems, and *S. ophiuncula* is a condition with long stems, which become short when exposed to full sunlight. *S. bisulca* is a slight form with the outer corona-lobes not very deeply notched, but this character varies, as in one flower of Haworth's type of *Orbea bisulca* they are much more deeply notched.

S. bifolia (Schultes, Syst. Veg. vi. 49, and G. Don, Gen. Syst. iv. 117), of which no description is given, is probably an error for *S. bufonia.*

S. orbicularis (Andr. Bot. Rep. vii. t. 439 ; Poir. Encycl. Suppl. v. 233 ; Schultes, Syst. Veg. vi. 40 ; Decne in DC. Prodr. viii. 660 ; Schlechter in Journ. Bot. 1898, 482. *Orbea orbicularis,* Haw. Syn. Pl. Succ. 40 ; G. Don, Gen. Syst. iv. 120). This illustration which has puzzled every student of Stapelias, I believe to be a bad figure of *S. bufonia,* badly coloured in the same way as Jacquin's, by representing the ground colour as brown with transverse yellow lines instead of a yellow ground with purple-brown spots, and the inner corona-lobes with very short outer horns. I have never seen a flower like it and believe the corona-lobes to have been either damaged or imperfectly developed, and as Andrews states that he had only seen the plant in Loddiges' collection and Loddiges' own figure of *S. orbicularis* (an excellent one of the common form of var. *bufonia*) distinctly represents the outer horns as long as in typical *bufonia,* I think there is little doubt that Andrews' figure is intended for that plant. There are forms or hybrids in cultivation which have light coloured flowers as in var. *planiflora,* but have the shallowly basin-shaped disk of var. *bufonia.*

S. hispida, Horn ex Rüst in Monatsschr. Kakt. vi. 37, is a slight form of var. *bufonia,* with the spots rather fewer and those on the annulus somewhat larger than usual, the annulus is also irregularly margined with purple-brown, leaving an irregular few-spotted light yellow ring on the rim.

S. natalensis, Rüst, l.c. 37, is another form of var. *bufonia* or hybrid with darkly coloured flowers, from the spots being rather crowded and those on the annulus larger than usual.

S. atrata, var. *tigrina,* Dammann ex Rüst, l.c. 38, has no resemblance in colour to *S. atrata,* Tod., but is a variation or hybrid of var. *bufonia,* with the spots showing some tendency to arrangement in longitudinal rows ; annulus very light yellow with small round purple spots ; outer horn of the inner corona-lobes erect.

[*S. bisulca* (Rüst in Monatsschr. Kakt. vi. 39, not of Donn); this is a hybrid, quite distinct from *S. bisulca*, Donn; corolla 2½ in. in diam.; lobes ovate, shortly acuminate, not ciliate, pale greenish-yellow, very thickly covered with irregular purple-brown spots, giving the appearance of being reticulated with the yellow colour, entirely purple-brown on the disk; annulus circular, dark purple-brown with a few yellow specks; outer corona-lobes oblong-linear, 3-toothed at the apex, with the middle tooth smallest, pale yellow, dotted on the teeth and down the centre with purple-brown, without a basal spot; inner corona-lobes with the outer horn erectly ascending.

VAR. *proboscidea* (Rüst, l.c. 39); corolla 3 in. in diam.; lobes acuminate into rather long tail-like tips, with fewer and more rounded spots than in the above form and arranged in longitudinal rows, the disk around the annulus and base of the lobes entirely (or suffused with) purple-brown; annulus subpentagonal, pale yellowish, thickly covered with rounded spots of a lighter purple-brown than those on the lobes; corona as in the preceding form.]

VAR. e, **clypeata** (N. E. Br.); corolla-lobes with or without a microscopic ciliation, pale greenish-yellow or sulphur-yellow, with the spots irregularly scattered or in longitudinal rows, some of the lower occasionally confluent, but without or with very few slender purple-brown lines among them and the basal part usually not darker than elsewhere; outer corona-lobes variably bifid from ⅓–⅔ of the way down, with the teeth varying from stout to slender and from parallel to widely diverging, with or more often without a minute tooth in the notch; inner corona-lobes with the horns subequal or the outer shorter than the inner, both clavate or the outer scarcely thickened at the apex. *S. clypeata, J. Donn, Hort. Cantab. ed. 3, 43, and Jacq. Stap. t.* 34; *Schultes, Syst. Veg. vi.* 40; *Spreng. Syst. Veg. i.* 838, *excl. syn.*; *Dietr. Syn. Pl. ii.* 885; *Decne in DC. Prodr. viii.* 660. *S. Bufonis, Sims, Bot. Mag. t.* 1676; *Lodd. Bot. Cab. t.* 332? *S. quinquenervis, Schultes, Syst. Veg. vi.* 37; *Decne in DC. Prodr. viii.* 658. *S. variegata, Jacq. Stap. t.* 39; *Thunb. Prodr.* 46; *Fl. Cap. ed.* 2, ii. 170, *and ed. Schultes,* 241; *N. E. Br. in Journ. Linn. Soc. xvii.* 169, *and in Hook. Ic. under t.* 1907, *not of Linn. Orbea clypeata, Haw. Suppl. Pl. Succ.* 13; *G. Don, Gen. Syst. iv.* 120. *O. quinquenervia, Haw. Syn. Pl. Succ.* 38. *O. quinquenervis, Loud. Hort. Brit.* 97; *G. Don, Gen. Syst. iv.* 119. *O. bufonia, Haw. Suppl. Pl. Succ.* 13 (*not of Syn. Pl. Succ.*).

COAST REGION : Cape Div.; Lion Mountain and Table Mountain, *Barkly,* 3! *Pillans,* 100! and *cultivated specimens*!

The type specimen of *Orbea quinquenervia*, Haw., in Haworth's Herbarium at Oxford is identical with *S. clypeata*, J. Donn, and of Jacq. But between this and the plant figured by Jacquin as *S. variegata* I can find no valid distinction, the only character is that in typical *S. clypeata* the teeth of the outer corona are more divergent and sometimes, but not always, more slender than in *S. variegata*, Jacq., but I have seen flowers with every possible gradation between these two forms and also some in which the teeth are perfectly parallel on some of the lobes and very widely diverging on others of the same flower! The small tooth in the notch is also occasionally either present or absent in different flowers on the same plant or even in the same flower.

VAR. f, **mixta** (N. E. Br.); corolla 2½–3¾ in. in diam., pale greenish-yellow, paler on the annulus, uniformly covered on the lobes with rather small purple-brown spots more or less connected on the basal half or beyond by slender purple-brown lines between the rugosities or subconfluent, and on the annulus with smaller spots or dots; disk flat or nearly so, annulus varying from nearly circular to distinctly pentagonal or somewhat 10-crenate, with a rather broad flattish-convex rim; lobes with or without a ciliation of minute thick white hairs; outer corona-lobes linear-oblong, with parallel sides, variably 2–3-toothed at the apex, yellow, dotted with purple-brown on the apical half, some of the dots extending down the margins and others down the centre to a larger spot at the base; inner corona-lobes with horns of equal length, the outer ascending; otherwise as in the type. *S. mixta, Masson, Stap.* 23, *t.* 38 ; *Willd. Sp. Pl.* i. 1292 ; *Pers. Syn. Pl.* i. 280 ; *Poir. Encycl.* vii. 388 ; *R. Br. in Mem. Wern. Soc.* i. 25 ; *Schultes, Syst. Veg.* vi. 36 ; *Decne in DC. Prodr.* viii. 658. *Orbea mixta, Haw. Syn. Pl. Succ.* 38 ; *Schultes, Syst. Veg.* vi. 834 ; *G. Don, Gen. Syst.* iv. 119.

SOUTH AFRICA : without locality, *Masson.*
COAST REGION : Robertson Div. ; near Robertson, *Pillans,* 124 !

Masson's figure of this plant is very badly coloured, as he represents the ground-colour of the corolla-lobes as purple-brown marked with irregular yellow lines and describes them as " purple, with transverse yellow rugosities ; " this is really the case, but at the same time the effect produced is that the ground colour is yellow, spotted and lined with purple-brown.

[*S. mixta* and var. *pentagona,* Rüst in Monatsschr. Kakt. vi. 39, seem to be hybrid forms of var. *bufonia,* with a large annulus, but smaller flowers than in typical var. *mixta* ; the lobes are coloured as in the lighter forms of var. *bufonia* and the annulus is of a pale sulphur colour thinly dotted with purple brown on the rim and densely in the cup, very distinctly pentagonal in var. *pentagona* ; outer corona-lobes entire to bifid, with short parallel or diverging teeth, pale yellow, dotted with purple-brown on the apical part and down the centre, with (or in var. *pentagona* without) a basal spot ; outer horn of the inner corona-lobes ascending.]

VAR. g, **horizontalis** (N. E. Br.) ; corolla dull greenish-yellow on the lobes, with numerous closely placed transverse lines (but no spots) on the basal half, numerous dots or very small spots and often a central line on the apical part and a series of small contiguous spots around the margin, all dark purple-brown ; annulus pentagonal with a rather broad flattish-convex rim, finely granulate-rugose, paler than the lobes, with numerous small round purple-brown spots and some slender lines between the rugosities ; disk flat or very slightly saucer-shaped under the annulus ; lobes not ciliate ; outer corona-lobes linear oblong, bifid to ⅓ of their length, with parallel or more rarely diverging teeth ; outer horn of the inner corona-lobes straight, horizontally spreading, as long as or slightly longer than the inner horn, both clavate at the apex ; otherwise as in the type. *S. horizontalis, N. E. Br. in Hook. Ic. Pl. t.* 1907.

SOUTH AFRICA : cultivated in Cape Town Botanic Garden, origin unknown, *Barkly,* 4 ! and *cultivated specimens* !

When alive this is readily distinguished from all other forms of *S. variegata* known to me by the very distinct coloration, flat-looking annulus and the rather finely rugose surface. Its native locality is unknown, but the Bruintjes Hoogte plant distributed by MacOwan and Bolus as *S. horizontalis* may be correctly named ; I have not seen living specimens of it. This variety was never common in English gardens and seems to have disappeared from cultivation here and in South Africa, I have not seen it for over 20 years.

VAR. h, **rugosa** (N. E. Br.) ; corolla greenish-yellow with numerous small scattered spots and transverse irregular lines of dark purple-brown on the lobes and paler yellow dotted with dark purple-brown on the annulus ; disk flat on the back ; annulus with a rather broad flattish rim, pentagonal, with 5 slight grooves radiating from the centre to a slight notch at each angle ; outer corona-lobes linear-oblong, subequally and acutely 3-toothed at the apex ; otherwise as in the type. *S. rugosa, J. Donn, Hort. Cantab. ed.* 3, 43 ; *Jacq. Stap. t.* 41 ; *Willd. Enum. Pl. Hort. Berol.* 284 ; *Poir. Encycl. Suppl.* v. 232 ; *Schultes, Syst. Veg.* vi. 33 ; *Link, Enum. Pl. Hort. Berol.* i. 256 ; *Spreng. Syst. Veg.* i. 838 ; *Dietr. Syn. Pl.* ii. 885 ; *Decne in DC. Prodr.* viii. 656. *Orbea rugosa, Sweet, Hort. Brit. ed.* 1, 277 ; *Loud. Hort. Brit.* 96. *Tridentea rugosa, Schultes, Syst. Veg.* vi. 850 ; *G. Don, Gen. Syst.* iv. 118.

ORIGIN : stated to be a native of South Africa, cultivated in England in 1804 according to Donn. I have not seen it.

[*S. rugosa* and var. *coronata* (Rüst in Monatsschr. Kakt. vi. 40, not of Jacquin) are hybrid forms unlike *S. rugosa*, Jacq., with corollas 2½ in. in diam. and ovate acuminate lobes, not ciliate ; in *rugosa* yellow, with irregular purple-brown spots and very slender lines between them ; annulus circular, paler yellow than the lobes, but very thickly covered with rather large confluent purple-brown spots ; outer corona-lobes oblong-linear, 3-toothed, with the middle tooth minute, pale yellow, with a purple central stripe and darker dots on the apical part ; inner corona-lobes with the outer horn spreading.

VAR. *coronata* ; lobes pale sulphur-yellow, irregularly variegated with dark purple-brown spots and markings, with slender lines between them ; annulus circular, with the margin and inner part crimson-brown leaving a very irregular ring of light yellow between ; corona as above, but the outer corona-lobes narrowed at the teeth.]

VAR. i, **pallida** (N. E. Br. in Hook. Ic. Pl. under t. 1907, p. 2) ; corolla acuminate in bud, when expanded with the disk nearly or quite flat on the back ; lobes very pale greenish-yellow or bright lemon-yellow, covered with very numerous dark purple-brown dots or small spots (scarcely half as large as those on the type), usually without (but occasionally with) some slender not very conspicuous lines between the rugosities at the base, or rarely with clear borders and the central area covered with labyrinthine transverse purple-brown lines, both forms occasionally occur upon the same branch ! annulus nearly round or very distinctly pentagonal ; outer corona-lobes bifid, with or without a minute tooth in the notch and the teeth parallel or somewhat diverging, all varieties sometimes produced on the same plant ; outer horns of the inner corona-lobes equalling or rather shorter than the inner, ascending, usually but not always slightly thickened at the apex ; otherwise as in the type.

COAST REGION : Cape Div. ; northern slopes of Table Mountain, *Pillans*, 116 ! cultivated in Port Elizabeth Botanic Garden and believed to come from an eastern province, *Barkly*, 2 ! and *cultivated specimens* ! *Pillans*, 74 !

VAR. j, **Curtisii** (N. E. Br. in Hook. Ic. Pl. under t. 1907, p. 3) ; corolla pale greenish-yellow or lemon-yellow, with rather numerous and usually rather small purple-brown spots irregularly scattered or in longitudinal rows ; annulus circular, rather paler with smaller spots ; outer corona-lobes entire and very obtuse or sub-truncate, emarginate, or very minutely and equally 3-toothed at the apex, sometimes in the same flower! yellowish-green, dotted with purple-brown ; outer horn of the inner corona-lobes about as long as or shorter than the inner, ascending, slightly thickened at the apex. *Stapelia Curtisii, Schultes, Syst. Veg.* vi. 38 ; *Decne in DC. Prodr.* viii. 659. *S. variegata, Sims, Bot. Mag. t.* 26, not of *Linn. S. inodora, Decne, l.c.* 661. *Orbea Curtisii, Haw. Syn. Pl. Succ.* 40 ; *G. Don, Gen. Syst.* iv. 120. *O. inodora, Haw. Suppl. Pl. Succ.* 12 ; *G. Don, l.c.* 121 ; *Schlechter in Journ. Bot.* 1898, 481.

SOUTH AFRICA : without locality, but probably near Cape Town, *cultivated specimens* !

There are forms which I cannot otherwise distinguish from this having the outer corona-lobes more or less bifid at the apex. I have not seen a specimen of *Orbea inodora*, Haw., but he only distinguishes it from var. *Curtisii* by its "smaller stems and retuse-emarginate" outer corona-lobes, both very variable characters. Sir Henry Barkly's specimen (57) which I formerly quoted for this, has the outer corona-lobes much more deeply notched than in the type.

[*S. Curtisii*, Rüst in Monatsschr. Kakt. vi. 38, is not the above plant, but a hybrid, raised in Europe, probably between some form of *S. variegata*, var. *atropurpurea* and var. *bufonia*, with a corolla about 2 in. in diam., very dark purple-brown or violet-brown on the lobes, with a few irregular transverse and marginal yellowish markings and the circular annulus light yellow with rounded purple-brown spots ; outer corona-lobes bifid to ⅓ of the way down and tapering at the apex, pale yellow, dotted on the teeth and with a central stripe of purple-brown ; outer horn of the inner corona-lobes suberect.

VAR. k, **retusa** (N. E. Br.) ; corolla flat on the back, coloured much as in var. *bufonia* but less darkly at the base of the lobes and the spots on the pentagonal annulus usually fewer, larger and more irregular in form ; outer corona-lobes rather long, linear or tapering from base to apex, with a very small notch or very shortly bifid or sometimes very minutely 3-toothed at the tips, pale yellowish-green, marked at the tips and down the centre with dots or very minute specks of purple-brown and a rather paler square spot at the base ; both horns of the inner corona-lobes clavate, the outer rather shorter than the inner, ascending-spreading ; otherwise as in the type. *S. retusa, Schultes, Syst. Veg.* vi. 41 ; *Decne in DC. Prodr.* viii. 660 ; *Schlechter in Journ. Bot.* 1898, 483. *Orbea retusa, Haw. Syn. Pl. Succ.* 41 ; *G. Don, Gen. Syst.* iv. 120.

SOUTH AFRICA : without locality, *cultivated specimens* !

[*S. retusa* (Rüst in Monatsschr. Kakt. vi. 37, not of Schultes) is a hybrid form with a corolla 2½ in. in diam., pale yellow, rather sparsely covered with dark brownish-crimson spots more or less arranged in irregular longitudinal rows on the lobes and confluent on the disk in an irregular dark ring around the annulus ; lobes somewhat elongated ovate, acuminate, minutely ciliate with pale yellowish or white hairs ; annulus small, circular, rather brighter than the lobes with scattered round spots as large as those on the lobes ; outer corona-lobes extending to or beyond the margin of the annulus, deeply bifid, with diverging teeth, pale yellow, with a purple-brown stripe down the basal part and minute dots on the apical half ; outer horn of inner corona-lobes somewhat spreading.]

VAR. l, **brevicornis** (N. E. Br.) ; corolla 2¼–2½ in. in diam. ; disk saucer-shaped ; lobes not ciliate, thickly covered on a greenish-yellow ground with dark purple-brown irregular or transverse spots, which become more or less confluent into irregular transverse or thick lines on the basal part, with somewhat labyrinthine yellow lines between them ; annulus varying from nearly circular to pentagonal, dark purple-brown with a few irregular yellowish vein-like markings or 1 or 2 irregular yellowish concentric lines on its rim, from the rather large spots being very crowded or subconfluent into an irregular ring, and in the cup part from the much smaller spots or dots being densely crowded or subconfluent ; outer corona-lobes ⅛ in. long, oblong, bifid to ⅓ or ½ of the way down, with parallel teeth, with a few rather large subconfluent blackish- or very dark purple-brown dots on the apical half, a transverse yellow band at the middle, and a transverse purple-brown spot or band at the base ; inner corona-lobes with the outer horn very short, about ¼–⅓ as long as the inner, very spreading, subulate, subacute or obtuse, both horns irregularly banded or marked with blackish- or very dark purple-brown and yellow ; otherwise as in the type.

SOUTH AFRICA : locality unknown, *Pillans*, 47 !

Described from a living plant.

VAR. m, **conspurcata** (N. E. Br.) ; corolla rather small, about 2 in. in diam., flattish on the back under the pentagonal or suborbicular annulus ; lobes minutely but distinctly ciliate nearly to their tips with subclavate white and purple-brown hairs, rather darkly coloured from the irregular spots being more or less confluent and often covering the greater part of the lobe ; outer corona-lobes varying from entire and acute to deeply bifid with parallel teeth or minutely and equally 3-toothed at the apex, even in different flowers on the same plant ; inner corona-horns subequal or the outer shorter and ascending-spreading, acute or slightly clavate ; otherwise as in the type. *S. conspurcata, Willd. Enum. Pl. Hort. Berol.* 284 ; *Jacq. Stap. t.* 32 ; *Poir. Encycl. Suppl.* v. 231 ; *Schultes, Syst. Veg.* vi. 39 ; *Spreng. Syst. Veg.* i. 838 ; *Link, Enum. Pl. Hort. Berol.* i. 256 ; *Decne in DC. Prodr.* viii. 660 ; *Schlechter in Journ. Bot.* 1898, 479. *S. ciliolata, Tod. Ind. Sem. Hort. Panorm.* 1861, 10, *name only. S. obliqua, Willd. Enum. Pl. Hort. Berol. Suppl.* 13 ; *Schultes, Syst. Veg.* vi. 35 ; *Link, Enum. Pl. Hort. Berol.* i. 257 ; *Decne in DC. Prodr.* viii. 657 ; *Schlechter in Journ. Bot.* 1898, 482. *Orbea conspurcata, Schultes, Syst. Veg.* vi. 834 ; *G. Don, Gen. Syst.* iv. 120. *Tromotriche obliqua, Sweet, Hort. Brit. ed.* 1, 278 ; *G. Don, Gen. Syst.* iv. 119.

COAST REGION: Cape Div.; Table Mountain, *Pillans*, 71! and *cultivated specimens*!

Described from living plants. Some specimens of this variety cultivated as *S. ciliolata* are much less ciliate than others.

VAR. n, **læta** (N. E. Br.); corolla 2½–3 in. in diam., light bright lemon-yellow, marked with large irregular rich crimson-brown spots on the lobes and with large spots of a brighter crimson on the annulus, those at the base of the lobes more or less confluent; outer corona-lobes usually distinctly narrowed towards the apex, but sometimes with parallel sides, 3-toothed at the apex, with the middle tooth varying from minute to as long as the outer teeth, but always more slender, pale greenish-yellow, dotted to the base of the teeth with purple-brown and with a rectangular spot at the base and sometimes a transverse spot just below the teeth of the same colour; inner corona-lobes with the outer horn rather shorter than the inner, ascending. *S. picta, N. E. Br. in Hook. Ic. Pl. under t.* 1907, *p.* 4, *not of Donn*.

SOUTH AFRICA: locality unknown, cultivated in Cape Town Botanic Garden, *Barkly*, 23! *Pillans*, 24! *MacOwan and Bolus, Herb. Norm.* 920! and *cultivated specimens*!

Described from living plants.

VAR. o, **picta** (N. E. Br.); stems rather slender, 3–4 lin. thick; corolla rather small, about 2 in. in diam., marked with large irregular blotches or confluent spots of very dark crimson-brown upon a pale yellow ground, which forms irregular transverse and longitudinal lines between them; annulus rather lighter in colour than the lobes, with brighter spots; outer corona-lobes oblong-linear, divided nearly or quite to half-way down into 2 parallel linear teeth, otherwise as in the type. *S. picta, J. Donn, Hort. Cantab. ed.* 3, 43; *Sims, Bot. Mag. t.* 1169; *Ait. Hort. Kew. ed.* 2, ii. 94; *Poir. Encycl. Suppl.* v. 233, 234; *Allg. Teutsch. Gart. Mag.* vii. 413, *t.* 42 (*ex Schultes*); *Link, Enum. Pl. Hort. Berol.* i. 256. *S. anguinea, Jacq. Stap. t.* 37; *Lodd. Bot. Cab. t.* 828; *Link, l.c.* 257. *S. anguinea and S. picta, Schultes, Syst. Veg.* vi. 41; *Decne in DC. Prodr.* viii. 660, 659 *and* 661. *Orbea anguinea and O. ,picta, Haw. Syn. Pl. Succ.* 41, 42; *G. Don, Gen. Syst.* iv. 120. *O. anguina, Loud. Hort. Brit.* 96.

SOUTH AFRICA: without locality, cultivated specimens dated 1813, 1829 and 1832 in *Herb. Kew*! and *Herb. Haworth*!

VAR. p, **marmorata** (N. E. Br.); corolla rather small, 1¾–2 in. in diam.; lobes broadly ovate, subacuminate, not ciliate, blackish-purple, with a few irregular transverse and longitudinal pale yellow markings, annulus scarcely paler than the lobes, with a few irregular yellow markings and dots; outer corona-lobes with parallel sides, bifid to about ⅓ of the way down, with parallel teeth, sulphur-yellow with a stripe down the centre and some dots on the teeth dark purple-brown; inner corona-lobes pale yellow, dotted with purple-brown, with the outer horn very short, not more than half as long as the inner horn, subulate, not clavate, erect. *S. marmorata, Jacq. Stap. t.* 38; *Hornem. Hort. Bot. Hafn.* ii. 959; *Schultes, Syst. Veg.* vi. 40; *Link, Enum. Pl. Hort. Berol.* i. 257; *Spreng. Syst. Veg.* i. 838; *Dietr. Syn. Pl.* ii. 885; *Decne in DC. Prodr.* viii. 660. *Orbea marmorata, Schultes, Syst. Veg.* vi. 834; *G. Don, Gen. Syst.* iv. 120.

SOUTH AFRICA: without locality, cultivated specimen in *Herb. Haworth*!

S. lunata, Dammann, and vars. *umbrosa* and *minuta* (Rüst in Monatsschr. Kakt. vi. 43) are either hybrid forms or slight variations of *S. variegata*, var. *marmorata*, differing only as follows:—*S. lunata*, corolla-lobes minutely ciliate, very dark purple-brown or blackish-purple with a pale yellow crescent-shaped marking and 2–3 specks near the apex of each lobe and a pair of yellowish marks at each sinus; annulus rather lighter than the lobes, speckled with yellow. Var. *umbrosa*, corolla-lobes not ciliate, somewhat reticulated with whitish-yellow at the tips only; annulus with larger whitish markings; outer corona-lobes with dots at the apical part only, no central stripe. Var. *minuta*, corolla-lobes minutely ciliate, with a pale yellowish marking something like a flattened ring with 2 teeth on each; annulus with a few small yellowish marks; outer corona-lobes with the teeth slightly diverging.

S. marmorata, Rüst, l.c. 38, is totally different from *S. marmorata*, Jacq., and merely a hybrid form of *S. variegata* with a corolla 2¾ in. in diam., minutely ciliate on the lobes, which are margined and marked with a few large spots and blotches of dark purple-brown; annulus paler with a few smaller purple-brown spots.

VAR. q, **atropurpurea** (N. E. Br.); corolla 2½–3¼ in. in diam., intense blackish-purple with a few irregular yellowish markings, which are sometimes chiefly on the apical at others on the basal half of the lobes, or sometimes form an irregular or broken circle on each lobe; lobes broadly ovate, shortly acuminate; annulus pentagonal or subcircular, with or without 5 pairs of slightly raised ridges radiating from the centre to the angles, usually with a few spots or irregular marks of yellowish, but sometimes entirely blackish-purple, outer corona-lobes 2½–3 lin. long, oblong or linear-oblong, variable, shallowly to rather deeply bifid, with or without a minute tooth at the base of the notch (sometimes in the same flower !), or subequally 3-toothed at the apex, with the teeth parallel, pale yellow, dotted or marked with dark purple-brown at the apex, with the dots sometimes extending down the middle, or with a pale purplish spot at the base; inner corona-lobes with the outer horn as long as the inner, slightly thickened at the apex, ascending-spreading; otherwise as in the type. *S. atropurpurea, Salm-Dyck, Hort. Dyck.* 372; *Decne in DC. Prodr.* viii. 659; *Schlechter in Journ. Bot.* 1898, 479. *S. marmorata, Hulle in Rev. Hort. Belg.* 1889, 193, *with fig.*

SOUTH AFRICA : without locality ; *cultivated specimens* !
COAST REGION: Cape Div. ; northern shores of Robben Island, *Pillans*, 82 ! 132 ! and *cultivated specimens* !

I have not seen an authentic specimen of *S. atropurpurea*, Salm-Dyck, but the described plant fairly well accords with the original description, and I received a living plant of it over 30 years ago from a correspondent who informed me that he had it years before from the Continent as *S. atropurpurea*. It is very nearly related to var. *marmorata* but larger, and varies very considerably in the markings on the corolla. This particular form is, I understand from Mr. Pillans, confined to Robben Island.

S. Scylla, Dammann, Cat. 1894–1895, 120 and 109, fig. 77 ; Wien. Ill. Gart. Zeit. 1894, xix. 234, fig. 30, is a variety or hybrid form with about 3 irregular longitudinal yellowish markings on each lobe.

[*S. atropurpurea* (Rüst, not of Salm-Dyck) and vars. *rosea* and *Pan* (Rüst in Monatsschr. Kakt. vi. 40); these are hybrids, showing no connection with *S. atropurpurea*, Salm-Dyck ; the first is a hybrid form very similar to *S. variegata*, var. *trisulca*, with a corolla 3¼ in. in diam. ; lobes similarly coloured, but rather darkly spotted and the ground colour of the annulus is overspread or suffused with purple-brown, with round spots of a darker purple-brown.

VAR. *rosea* has a corolla 2¾ in. in diam. with ovate subacute lobes, not ciliate, very pale yellowish, with irregular blood-red spots and numerous fine lines between them, the spots on the disk around the annulus more or less confluent ; annulus pentagonal, paler than the lobes, with numerous round blood-red spots ; outer corona-lobes, linear-oblong, 3-toothed at the apex, with the middle tooth smallest, pale yellow, with purple-brown dots on the apical part and down the centre and a spot at the base ; inner corona-lobes with the outer horn suberect.

VAR. *Pan* has a corolla 3¼ in. in diam. ; lobes elongate-ovate, acute or acuminate, not ciliate, light yellow, densely and evenly spotted with dark purple-brown ; annulus pentagonal, suffused with brownish-red, with rather large round purple-brown spots and some yellow specks ; corona as in var. *rosea*.]

[VAR. r, **atrata** (N. E. Br.); corolla 3–3¼ in. in diam., very flat on the back, very minutely ciliate on the deltoid-acuminate (not ovate) lobes; inner surface (including the annulus) dull purplish-brown, somewhat indistinctly marked with rather large spots of a darker purple-brown, interspersed with a few yellowish spots or markings, chiefly on the terminal half of the lobes and annulus; annulus pentagonal, with 5 pairs of slightly raised ridges radiating from the centre to the angles; outer corona-lobes bifid, with the teeth parallel or slightly diverging, yellowish, densely dotted with purple-brown on the apical part and with a cuneate purple-brown spot at the middle; inner corona-lobes with subequal horns, the outer ascending, not clavate; otherwise as in the type. *S. atrata, Tod. Hort. Panorm.* i. 50, *t.* 13, *fig.* 1.

ORIGIN : raised in Europe, *cultivated specimens* !

This may be only a colour variation of var. *atropurpurea* or possibly a hybrid derived from it.]

[*S. atrata*, var. *rufescens*, Dammann ex Rüst in Monatsschr. Kakt. vi. 39, is a hybrid bearing no resemblance to *S. atrata*, Tod.; corolla 2¼ in. in diam.; lobes ovate, acute, not ciliate, yellow, narrowly margined and spotted with purple-brown, with the basal spots much larger than the rest and those on the disk around the annulus confluent, those on the upper part are arranged in rows with very slender longitudinal purple-brown veins between them; annulus circular, thickly covered with purple-brown blotches, leaving narrow irregular yellowish spaces between; outer corona-lobes oblong-linear, shortly bifid, pale yellow, with a pale purple-brown spot at the base and dots on the apical part; inner corona-lobes with the outer horn ascending.]

[VAR. s, **Prometheus**, Dammann ex Rüst in Monatsschr. Kakt. vi. 37, is a hybrid form with the corolla-lobes minutely ciliate, and the dark purple-brown area of confluent spots on the disk around the annulus extending in 5 rays to the sinuses, otherwise spotted much as in var. *bufonia*; annulus pentagonal, light yellow, without spots on the rim, but densely dusted with purple-brown in the cup; corona as in var. *bufonia*.]

[**70. S. divergens** (N. E. Br. in Gard. Chron. 1905, xxxvii. 49); habit, stems, pedicels and sepals as in *S. variegata*, and the flower similar in form; corolla in bud shortly conical from a broad flattened pentagonal base, acute, when expanded about 2 in. in diam., slightly rugulose on the inner face, light yellow, labyrinthinally marked on the disk and lower ⅔ of the lobes with thick irregularly ramified brownish-crimson lines, which become broken up into spots on the apical part of the lobes; annulus bright canary-yellow, with rather crowded brownish-crimson round spots in the rugulose basin-like part, but none on the smoother horizontally spreading rim; disk nearly flat; lobes about 9 lin. long and 7 lin. broad, deltoid-ovate, very acute, minutely ciliate with simple thick hairs to their tips; outer corona-lobes divided for ⅔ of the way down into 2 straight slender widely diverging teeth, 2–2½ lin. apart from tip to tip, pale yellow, dusted on the teeth and marked at the notch and base with a spot of dark crimson-brown; inner corona-lobes pale yellowish, dusted with crimson-brown, with the inner horn 2 lin. long, clavate, and the outer 1⅓ lin. long, erectly spreading, scarcely clavate.

ORIGIN : a hybrid raised in Europe, *cultivated specimens* !

Described from a living plant.]

[71. **S. scutellata** (Tod. Hort. Panorm. i. 52, t. 13, fig. 2) ; corolla $2\frac{1}{2}$–3 in. in diam., slightly basin-like under the annulus and somewhat convex on the back ; lobes ciliolate nearly to their tips with minute thick or subclavate hairs, light yellow, marked to their tips with dark purple-brown transverse spots and thick irregularly confluent wavy lines ; annulus with a recurved-spreading margin, of a paler yellow, with small purplish-brown dots and spots ; outer corona-lobes bifid, with diverging or parallel teeth, sometimes with a minute tooth in the notch ; horns of the inner corona subequal ; otherwise as in *S. variegata*, Linn.

ORIGIN : raised from seed in Europe, doubtless a hybrid, *cultivated specimen* !]

[72. **S. albicans** (Sprenger in Dammann, Cat. 1894, 58 and 53, fig. 51) ; stems like those of *S. variegata*, L., whitish or partly whitish and partly green, often becoming entirely green ; corolla. 2–$2\frac{1}{2}$ in. in diam., dark purple-brown at the tips of the lobes and along the margins, the remainder of the rugose inner surface yellow, with elongated transverse spots or thick broken lines of dark purple-brown ; lobes ovate, acute or shortly acuminate, ciliate with minute clavate hairs about $\frac{1}{4}$ lin. long ; annulus with its margin very slightly recurved ; outer corona-lobes linear-oblong, varying from acute to 3-toothed at the apex ; inner corona-lobes 2-horned, horns filiform, the inner clavate and rather longer than the spreading simple outer horn. *Wien. Ill. Gart. Zeit.* 1894, 234, *fig.* 29.

ORIGIN : A hybrid raised by Messrs. Dammann & Co. at Naples ; *cultivated specimens* !

Described from a living plant received from Mr. Sprenger, and raised from seeds of one of Todaro's species ; as the flower is much like that of *S. angulata*, Tod., in general character, that may have been its seed-parent.]

[*S. alba*, mentioned by Rother and others in *Monatsschr. Kakt.* v. 13, 15, and 30, without a description, is perhaps only another name for *S. albicans*. I have not seen it.]

[73. **S. hanburyana** (Berger and Rüst in Monatsschr. Kakt. ix. 6, fig. 7) ; stems like those of *S. variegata*, about 5–6 lin. thick ; pedicels about $1\frac{1}{4}$ in. long ; sepals $2\frac{1}{2}$–3 lin. long, $1\frac{1}{2}$ lin. broad, ovate, acute or acuminate ; corolla $2\frac{1}{4}$–$2\frac{1}{2}$ in. in diam., not very rugose on the inner surface ; lobes very spreading or recurving, about 10 lin. long and 8 lin. broad at the base, ovate, somewhat acuminate, ciliolate with minute subclavate hairs, pale greenish-yellow, evenly marked all over with rather small transverse purple-brown spots or short lines and narrowly edged with the same colour, the tips rather darker than the rest from the furrows between the rugosities being traversed with fine purple-brown lines ; annulus very flat-topped with a horizontally spreading margin, whitish-yellow, much paler than the lobes, marked above and beneath with small rounded purple-brown spots which are often confluent in short chains ; outer corona-lobes oblong or broadened upwards, even in the same flower, bifid or sometimes subtruncate at the apex, pale yellow, thickly dotted on the apical half and with an irregular purple-brown blotch at the base; inner corona-lobes with both horns slightly clavate at the tips and the outer erect, as long as the inner, pale yellow without dots on the basal part, elsewhere dotted with purple-brown. *N. E. Br. in Gard. Chron.* 1908, xliv. 169 *and* 167, *fig.* 65.

ORIGIN : a hybrid from some variety of *S. variegata ; cultivated specimens* !

Described from a living plant from Sir Thomas Hanbury's garden at La Mortola.]

Hybrids raised in Europe, not included in the key.

[74. **S. amœna** (Rüst in Monatsschr. Kakt. vi. 40) ; corolla 3 in. in diam. ; lobes broadly ovate, shortly acuminate, not ciliate, sulphur-yellow, with rather few irregularly scattered dark purple-brown spots and very slender lines between them ; annulus very large, $1\frac{1}{4}$ in. in diam., pentagonal, paler than the lobes, with small scattered round spots and dots of dark blood-red or purple-brown ; outer corona-lobes linear-oblong, slightly narrowed at the apex, bifid with a minute tooth in the notch, pale yellow with a dark purple or purple-brown spot at the base and dots on the apical part ; inner corona-lobes with the outer horn ascending.]

[75. **S. angulosa** (Tod. ex Rüst in Monatsschr. Kakt. vi. 38) ; stems rather stout ; corolla about 2 in. in diam., light yellow (paler on the circular annulus) densely covered with purple-brown dots or very small spots, those on the annulus smaller ; lobes not ciliate ; outer corona-lobes linear-oblong, bifid to $\frac{1}{3}$ of the way down, pale yellow, dusted all over with purple-brown, without a basal spot ; outer horn of inner corona-lobes suberect.

VAR. **Charybdis** (Dammann ex Rüst in Monatsschr. Kakt. vi. 40) ; corolla $2\frac{1}{4}$ in. in diam. ; lobes ovate, acute, minutely ciliate, light yellow, thickly covered with irregular or elongated transverse purple-brown spots ; annulus circular, paler than the lobes, with rather crowded short purple-brown lines ; outer corona-lobes oblong, bifid with a minute tooth in the notch, greenish on the apical half, dotted with purple-brown, pale yellowish with a central purple stripe below ; inner corona-lobes more greenish tinted than usual, with the outer horn ascending.

VAR. **Kreusa** (Dammann ex Rüst in Monatsschr. Kakt. vi. 39) ; corolla 2 in. in diam. ; lobes ovate, acute, not ciliate, dark purple-brown with some scattered yellow specks and dots ; annulus circular, pale yellowish, thickly dotted with blood-red or purple-brown ; outer corona-lobes bifid, with converging teeth and a minute tooth in the notch, pale yellow, with an elongated purple-brown spot at the base and dots on the apical half ; inner corona-lobes with the outer horn much shorter than the inner, suberect.

VAR. **Nemea** (Dammann ex Rüst in Monatsschr. Kakt. vi. 38) ; corolla 2 in. in diam. ; lobes ovate, very short, subacuminate, ciliate with short clavate hairs, dark purple-brown, with the basal $\frac{2}{3}$ speckled with yellow ; annulus apparently solid, with an erect obtuse (not recurving) margin, crenately 5-lobed, pale yellowish, with purple-brown dots and minute irregular lines ; outer corona-lobes oblong-linear, bifid, pale yellow with a purple-brown spot at the base and dots down the centre and on the teeth ; inner corona-lobes with the outer horn about half as long as the inner horn.

Var. **Thi** (Dammann ex Rüst, l.c. 37), only differs in having larger and much fewer spots on the lobes, and a spot at the base of the outer corona-lobes.

S. radiata (Rüst, l.c. 38, not of Sims), is another form similar to *S. angulosa*, var. *Thi*, with large spots on the corolla-lobes, and the outer corona-lobes narrowed at the apex and less deeply bifid, without a basal spot.]

[**76. S. Burtinii** (Rüst in Monatsschr. Kakt. vi. 40) ; stems rather stout ; corolla $2\frac{1}{2}$ in. in diam. ; lobes ovate, acuminate, not ciliate, pale yellow with very numerous crowded small ‚rounded purple-brown spots ; annulus somewhat 5-lobed, with the lobes or angles alternating with the corolla-lobes, pale sulphur-yellow or whitish-yellow, with a few small scattered purple-brown spots ; outer corona-lobes oblong-linear, bifid at the apex, pale yellow, dotted at the apex and down the centre with purple-brown, no basal spot ; inner corona-lobes with the outer horns nearly as long as the inner, suberect.]

[**77. S. chlorotica** (Rüst in Monatsschr. Kakt. vi. 43) ; stems rather stout ; corolla $1\frac{3}{4}$ in. in diam. ; lobes deltoid-ovate, acute, not ciliate, sulphur-yellow, with small light purple spots and dots, those on the disk and base of the lobes larger and more crowded than the rest ; annulus circular, paler than the lobes, with irregular reddish or pale purplish lines and dots mostly radiating from the centre to the margin ; outer corona-lobes oblong, bifid, with slightly diverging teeth, pale yellowish, with 2 small streaks of minute purple dots down the centre, without a basal spot ; inner corona-lobes pale yellow, unspotted, with the outer horn much shorter and stouter than the inner.]

[**78. S. decora** (Rüst in Monatsschr. Kakt. vi. 40, not of Masson) ; this is a hybrid, totally different from *S. decora*, Masson ; corolla $2\frac{1}{2}$ in. in diam. ; lobes deltoid-ovate, acute, not ciliate, light yellow, with the disk around the annulus and 5 broad bands radiating to the sinuses of a reddish-brown, the whole spotted with dark purple-brown, with the spots on the centre of the basal part of the lobes irregularly transversely elongated and larger than the others, many of which are dot-like ; annulus rather small, $\frac{2}{3}$ in. in diam., apparently without a recurved margin, 10-crenate, pale yellow, dotted with purple-brown ; outer corona-lobes linear-oblong, bifid at the apex, pale yellow, dotted on the teeth and down the centre with purple-brown, no basal spot ; inner corona-lobes with the outer horn somewhat spreading.]

[**79. S. discolor** (Rüst in Monatsschr. Kakt. vi. 37, not of Todaro) ; corolla 2–$2\frac{1}{2}$ in. in diam. ; lobes entirely dark purple-brown, without spots, ciliate with clavate hairs $\frac{1}{2}$ lin. long; annulus circular, pinkish, with small round purple-brown spots or dots; outer corona-lobes oblong-linear, narrowed at the apex on the outer margins of the teeth, bifid to $\frac{1}{3}$ of the way down, pale yellow, dusted with purple-brown on the teeth and down the centre, with or without a spot at the base; outer horn of the inner corona-lobes spreading.

VAR. **Bellona** (Dammann ex Rüst, l.c. 38); this only differs from var. *Electra* by the ground colour of the corolla-lobes being of a light purplish tint, the outer corona-lobes tapering at the apex from the base of the teeth and dotted all over the teeth as well as down the centre, the inner corona-lobes more thickly dotted and their outer horn nearly as long as the inner.

VAR. **Electra** (Dammann ex Rüst in Monatsschr. Kakt. vi. 38); corolla about 2 in. in diam.; lobes ovate, acute, pale yellow, thickly covered with purple-brown (or dark crimson-brown?) irregular spots, those on the disk at the base of the lobes more or less confluent; annulus circular, paler than the lobes, with smaller rounded spots; outer corona-lobes bifid to about half-way down, light yellow, dotted with purple-brown on the inner edges of the parallel teeth and half-way down the centre, without a basal spot; inner corona-lobes rather thinly and minutely dotted with purple-brown, with the outer horn rather short, suberect.

VAR. **Medusa** (Dammann ex Rüst, l.c. 39) is like var. *Bellona*, but the corolla-lobes are not ciliate and the ground colour is of a darker purplish tint; annulus pinkish-white or sulphur-white with small round purple spots; corona as in var. *Bellona*, but the outer lobes do not taper at the apex.]

VAR. **Muley Hassan** (Dammann ex Rüst, l.c. 37) only differs in having the annulus more thickly spotted.

VAR. **tricolor** (Rüst, l.c. 39); stems more cylindric than in the other forms, very slightly or scarcely grooved between the rounded angles; corolla as in the above varieties, but the lobes narrowly margined with dark purple-brown and their basal half shaded with dull purple; annulus light rosy-purplish with rather few small dark purple-brown round spots or dots; outer corona with very diverging teeth, which are dotted along their inner margins and the dots continued down the centre; both horns of the inner corona-lobes thickly dotted and the outer nearly as long as the inner.]

[80. **S. erecta** (Rüst in Monatsschr. Kakt. vi. 40); corolla 2 in. in diam.; lobes ovate, acute, not ciliate, light yellow, rather evenly spotted with purple-brown, with the spots on the disk around the annulus confluent in a narrow ring; annulus circular, pale yellow, densely dotted with purple-brown; outer corona-lobes oblong-linear, 3-toothed at the apex, with the middle-tooth very small, pale yellow, dotted on the apical half and down the centre with purple-brown, no basal spot; inner corona-lobes with the outer horn erectly ascending.]

[81. **S. luxurians** (Dammann ex Rüst in Monatsschr. Kakt. vi. 39); corolla $3–3\frac{1}{4}$ in. in diam.; lobes elongated ovate-deltoid, acuminate, not ciliate, entirely dark purple-brown, with traces of yellowish on the rugosities on the basal part; annulus pentagonal, entirely dark purple-brown; outer corona-lobes linear-oblong, 3-toothed at the apex, with the middle tooth smallest, pale yellow,

with a purple-brown spot at or rather above the middle and dots
on the apical part, no spot at the base; outer corona-lobes with
subequal horns, the outer ascending.

This is similar to *S. atrata*, Tod.]

[82. **S. micrantha** (Rüst in Monatsschr. Kakt. vi. 38); corolla about
2⅓ in. in diam.; lobes somewhat attenuate-deltoid-ovate, acuminate,
pale greenish-yellow, irregularly margined and marked with purple-
brown spots more or less confluent in a few longitudinal or transverse groups; annulus circular, light yellow, densely covered with
purple-brown confluent spots in the cup and very few on the rim, so
that it appears to have an interrupted yellow ring around it; outer
corona-lobes oblong-linear, minutely notched at the apex, pale
yellow, dotted at the apex and a few dots at the base; inner
corona-lobes with both horns dotted to the base and the outer horn
suberect.

This scarcely differs from *S. mirabilis*, Dammann, except in the colouring of the
annulus and corona.]

[83. **S. mirabilis** (Dammann ex Rüst in Monatsschr. Kakt. vi. 38);
corolla 2–2⅓ in. in diam.; lobes somewhat attenuate-deltoid or
deltoid-ovate, acuminate, pale yellow, with purple-brown margins
and rather few large irregular spots, those on the disk around the
annulus confluent, the others interspersed with slender purple-
brown lines; annulus circular, paler, thickly covered with small
light purple-brown spots; outer corona-lobes oblong, shallowly
emarginate or notched at the apex, pale yellow, without markings
except some minute dots on the very small teeth; inner corona also
only dotted at the tips of the horns, of which the outer are
suberect.]

[84. **S. multiflora** (Rüst in Monatsschr. Kakt. vi. 39, not of
DC.); this is a hybrid form of *S. variegata* and totally different
from *S. multiflora*, DC.; corolla 2¼ in. in diam.; lobes deltoid-ovate,
acuminate, minutely ciliate, pale yellow, edged and variegated with
irregular dark purple-brown markings and a blotch near the tips of
a lighter purple-brown, entirely dark purple-brown on the disk
around the annulus, which is circular, of a paler yellow than the
lobes, with purple-brown spots all round the margin and in the
cup, leaving a circle of yellow almost unspotted between; outer
corona-lobes linear-oblong, shortly bifid, with a minute tooth in the
notch, pale yellow, dotted with purple-brown down the centre and
on the teeth; inner corona-lobes with the outer horn ascending-
spreading.

The flowers are represented as being subsolitary and apparently not more
abundantly produced than in other forms and hybrids of *S. variegata*.]

[85. **S. munbyana** (Rüst in Monatsschr. Kakt. vi. 40); a hybrid
very similar to the lighter coloured forms of *S. variegata*, var.

bufonia, but the spots at the base of the lobes not or scarcely confluent and the outer corona-lobes have a minute tooth in the notch
at their apex and no basal spot.]

[86. **S. muricata** (Rüst in Monatsschr. Kakt. vi. 37) ; corolla about
$2\frac{1}{4}$ in. in diam. ; lobes ciliate with short clavate hairs, pale yellow,
thickly covered with small purple-brown spots, intermingled with
slender transverse lines of the same colour, scarcely darker on the
disk around the annulus, which is circular, densely dotted with
purple-brown in the cup, pale yellow without markings on the
border ; outer corona-lobes extending nearly or quite to the margin
of the annulus, bifid to half-way down, with slightly diverging
teeth, pale yellow, with a short basal stripe and dots on the teeth
of purple-brown ; outer horn of inner corona-lobes suberect.]

[87. **S. panifolia** (Rüst in Monatsschr. Kakt. vi. 38) ; corolla
$2\frac{1}{2}$ in. in diam. ; lobes pale yellow, with rather few irregular purple-
brown spots and numerous very slender purple-brown lines between
the rugosities, not ciliate; annulus rather large, pentagonal, paler
than the lobes, with small scattered round purple-brown spots;
corona as in *S. variegata*, var. *bufonia*.]

[88. **S. parvipunctata** (Rüst in Monatsschr. Kakt. vi. 39) ; corolla
$3\frac{1}{4}$–$3\frac{1}{2}$ in. in diam. ; lobes ovate, acuminate, not ciliate, pale
greenish-yellow, thickly covered with small purple-brown spots and
dots, with a suffused purple-brown area on the disk around the
annulus extending in 5 rays to the sinuses; annulus pentagonal,
paler than the lobes, with small round purple or purple-brown
spots; outer corona-lobes oblong-linear, 3-toothed at the apex, with
the middle tooth minute, pale yellow with a purple-brown spot at
the base and dots on the apical part ; inner corona-lobes with the
outer horn nearly as long as the inner, erectly ascending.]

[89. **S. purpurea, var. nigrescens** (Rüst in Monatsschr. Kakt.
vi. 43) ; corolla $2\frac{1}{4}$ in. in diam. ; lobes ovate, long-acuminate,
minutely ciliate, blackish-purple with a short pale yellow irregular
longitudinal streak at their tips; annulus circular, blackish purple,
without markings ; outer corona-lobes bifid, with slightly diverging
teeth, pale yellowish with blackish-purple teeth and some purple
specks near the base ; inner corona-lobes with blackish-purple horns
and the basal part pale yellow, outer horn half as long as the inner,
subulate.]

[90. **S. putida** (Berger in Monatsschr. Kakt. xv. 159) ; stems
$1\frac{1}{2}$–2 in. long, 5–6 lin. thick ; corolla-lobes yellowish, with large
purple-brown spots and a red line along the margins, not ciliate ;
annulus with a much recurved margin, purplish-brown, obscurely

spotted with darker ; outer corona-lobes sublinear, bifid or 3-toothed
at the apex, with straight teeth, the middle one minute ; otherwise
as in *S. variegata.*]

[91. **S. rectiflora** (Rüst in Monatsschr. Kakt. vi. 37) ; corolla
about 2 in. in diam., with a shallow cup-like disk and spreading
lobes, sulphur-yellow, with rounded and not very numerous purple-
brown spots having a tendency to be confluent in longitudinal
rows, and slender purple-brown lines between the rugosities ;
annulus circular, paler, dotted with purple-brown ; outer corona-
lobes linear-oblong, bifid to about $\frac{1}{3}$ of the way down, with parallel
teeth, pale yellow, dotted with purple-brown at the apex only, no
basal spot ; outer horn of the inner corona-lobes spreading.]

Var. *viridula,* Rüst, l.c. 37, only differs in having the corolla more densely
spotted and the spots on the disk at the base of the lobes confluent in a continuous
ring around the annulus.]

[92. **S. salmiana** (Rüst in Monatsschr. Kakt. vi. 39) ; corolla
3–3¼ in. in diam. ; lobes ovate, shortly acuminate, shortly ciliate,
light yellow, with numerous irregular purple-brown spots, those at
the base and on the disk around the annulus confluent ; annulus
circular, of a paler yellow than the lobes, densely covered with
moderately large purple-brown spots ; outer corona-lobes linear-
oblong, 3-toothed at the apex, with the middle tooth minute, pale
yellow, dotted on the teeth with purple-brown, with or without
2 small spots near the base ; inner corona-lobes only dotted on the
apical half of their horns ; outer horn spreading.

This is very similar to *S. variegata,* var. *trisulca,* differing chiefly by its ciliate
corolla-lobes.]

[93. **S. sanguinea** (Rüst in Monatsschr. Kakt. vi. 38, not of
Pasq.) ; corolla 2½ in. in diam. ; lobes minutely or very shortly
ciliate, blackish-brown or violet-brown, with a few yellowish dots or
small spots on their basal part and a few pale yellow lines meander-
ing in various directions on the apical half and usually separating
one large and two small spots from the rest of the brown colour ;
annulus circular, yellowish with small round purple-brown spots ;
outer corona-lobes bifid, pale yellow, with a central stripe and
some dots on the teeth of purple-brown ; inner corona-lobes with
the outer horn suberect.]

[94. **S. Sisyphus** (Dammann ex Rüst in Monatsschr. Kakt.
vi. 38) is very similar to *S. variegata,* var. *atrata,* and is probably a
hybrid derived from it or of similar origin ; corolla-lobes ovate,
acuminate, very dark purple-brown or violet-brown with a crescent-
like marking and a few dots of yellow at their tips, not ciliate ;
annulus circular, of a much lighter purple-brown with yellow dots

and very small slender irregular lines on the inner part of the rim ;
outer corona-lobes shortly bifid, light yellow, with a central stripe
and some dots on the teeth dark purple-brown ; inner corona-lobes
with dark purple-brown horns ; inner horn dusted with yellow at
the apex.]

[95. **S. umbilicata** (Rüst in Monatsschr. Kakt. vi. 40) ; corolla
2¼ in. in diam. ; lobes ovate, acute, not ciliate, sulphur-yellow, with
violet-brown rounded spots, those on the disk around the annulus
confluent ; annulus subpentagonal, pale yellow, with small round
blood-red spots ; outer corona-lobes linear-oblong, narrowed at the
bifid apex, pale yellow, with a purple-brown spot at the base and
dots on the teeth and down the centre ; inner corona-lobes with the
outer horn suberect.]

[96. **S. Uspenskyi** (Rüst in Monatsschr. Kakt. vi. 38) is either a
variation or hybrid form of var. *bufonia*, with the spots at the
base of the lobes less confluent and transversely much elongated ;
lobes short, broad, and more abruptly acuminate than usual ;
annulus rather large, pentagonal ; outer corona-lobes without a spot
at the base and the outer horn of the inner corona-lobes spreading.]

Imperfectly known species.

97. **S. canescens** (Haw. Syn. Pl. Succ. 26 ; Schultes, Syst. Veg.
vi. 26 ; G. Don, Gen. Syst. iv. 117 ; Decne in DC. Prodr. viii. 663 ;
Schlechter in Journ. Bot. 1898, 479) ; name only, without
description.

98. **S. cordata** (Haw. Syn. Pl. Succ. 26 ; Schultes, Syst. Veg.
vi. 26 ; G. Don, Gen. Syst. iv. 116 ; Decne in DC. Prodr. viii. 663 ;
Schlechter in Journ. Bot. 1898, 479) ; name only, without
description.

99. **S. emarginata** (Breit. Hort. Breit. 504) ; name only, without
description.

100. **S. gemmifera** (Salm-Dyck, Hort. Dyck. 266) ; name only,
without description, but placed in the same group as *S. variegata*,
Linn., and allies.

101. **S. fasciculata** (Thunb. Prodr. 46) ; stems about 1 in. long,
fasciculate, decumbent, teretely subhexagonal, with spreading acute
tubercles, glabrous, reddish at the apex ; pedicel as long as the
stem ; calyx (perianth of *Thunberg*) very deeply 5-partite ; segments
1 lin. long, ovate, acute, glabrous, persistent ; follicles 2, about
4 in. long, pedunculate (i.e. tapering below into a sort of stalk),

erect, acute, thickest at the middle, brownish-purple. *Thunb. Fl. Cap. ed. 2, ii. 170, and ed. Schultes, 241 ; Poir. Encycl. vii. 385 ; N. E. Br. in Journ. Linn. Soc. xvii. 168 ; Schlechter in Journ. Bot. 1898, 480. Piaranthus? fasciculata, Schultes, Syst. Veg. vi. 10 ; G. Don, Gen. Syst. iv. 113. P. ? fasciculatus, Decne in DC. Prodr. viii. 650.*

Central Region : Calvinia Div. ; Hantam Hills, near Calvinia, *Thunberg* !

There is no specimen in Thunberg's Herbarium named *S. fasciculata,* but one sheet, without a name, contains 2 pieces of stem of a *Stapelia* which so exactly agree with his description of *S. fasciculata* that I think there can be no doubt they form the type of his description of that plant. One of the stems bears 2 follicles 4 in. long on an erect pedicel 1⅜ in. long. As there are no flowers it is impossible to determine the genus to which it belongs ; however, I think it is almost certainly not a species of *Stapelia,* but probably either a *Huernia, Duvalia* or *Piaranthus.*

102. **S. flavicomata** (Haw. Suppl. Pl. Succ. 8) ; stems numerous, 4-angled, slender, with equal angles, pubescent ; leaf-bearing teeth very tumid and contiguous ; flowers not seen. *G. Don, Gen. Syst. iv. 116 ; Decne in DC. Prodr. viii. 663 ; Schlechter in Journ. Bot. 1898, 480.*

South Africa : in cultivation before 1810 ex *Haworth* !

Stated to be near *S. glandulifera,* Haw. (i.e. *S. glanduliflora,* Masson), but with more numerous shorter and thicker stems, with larger and more contiguous teeth.

103. **S. papillosa** (DC. Pl. Grass. ex Desf. Tabl. ed. 2, 92 ; Schultes, Syst. Veg. vi. 49 ; G. Don, Gen. Syst. iv. 117 ; Schlechter in Journ. Bot. 1898, 482) ; name only, without description. The name has not been found in published copies of De Candolle, Pl. Grasses.

104. **S. plicata** (Salm-Dyck, Hort. Dyck. 267) ; name only, but placed in the same group as *S. variegata* and allies.

105. **S. trifolia** (Breit. Hort. Breit. 505) ; name only, without description.

106. **S. verticillata** (Schultes, Syst. Veg. vi. 49, and G. Don, Gen. Syst. iv. 117) ; name only, without description.

107. **S. virgata** (Schultes, Syst. Veg. vi. 49, and G. Don, Gen. Syst. iv. 117) ; name only, without description.

LII. PIARANTHUS, R. Br.

Calyx 5-partite. *Corolla* usually deeply 5-lobed and rotate when fully expanded, rarely with a campanulate or cup-shaped tube, velvety or pubescent (rarely glabrous) on the inner surface. *Corona* arising from the staminal column, simple ; lobes 5, opposite the anthers, incumbent upon them with or without erect tips or rarely erect, dorsally produced or expanded into a truncate minutely tuberculate or denticulate crest. *Staminal column* arising from the base of the corolla, short ; anthers free, oblong, without appendages at their apex, incumbent upon the dilated top of the style. *Pollen-masses* solitary in each anther-cell, subhorizontal, pellucid along the inner margin near the apex, attached in pairs by very short caudicles to minute excrescences on the sides of the narrow pollen-carriers. *Follicles* and *seeds* not seen.

Very dwarf succulent leafless herbs, with watery juice ; stems decumbent or ascending, flowering at or near the apex or middle ; flowers in pairs or fascicles, erect, small or of moderate size.

DISTRIB. : Species 11, endemic.

This genus was founded in 1811 by Robert Brown upon *Stapelia pulla* and *S. punctata*, Masson, and characterised as having no outer corona ("*staminal corona simple, 5-leaved, with the leaflets toothed on the back*"). But this character must have been made from *S. punctata*, as *S. pulla* has a most distinct outer corona. In 1812, Haworth, recognising that *S. pulla* and *S. punctata* represented two distinct genera, unfortunately referred *S. pulla* to *Piaranthus*, and founded his genus *Obesia* upon *S. punctata*, *S. decora* and *S. geminata*, Masson, but ascribed to *Obesia* and *Piaranthus* identical characters. Matters have been further confused by the erroneous description of *Obesia* given by Decaisne in *DC. Prodr.* viii. 661, and by Bentham and Hooker having placed *Piaranthus* as a synonym of the totally different *Podanthes*, Haw., whilst finally Dr. Schlechter in *Journ. Bot.* 1898, 478–479 has united *Piaranthus* with *Caralluma*. The genus however is quite distinct from all others and very easily recognised from living plants by the corona and peculiar habit. Some of the species are rather closely allied, and although fairly easy to recognise when alive, I find are difficult to tabulate, whilst dried specimens are exceedingly difficult to discriminate unless exceptionally well dried. In some cases the flowers of the same species vary very considerably in colour and sometimes in ciliation, and distinct varieties might easily be mistaken for different species, but Mr. Pillans informs me that the different variations grow together and are connected by a series of intermediate forms.

Fully expanded corolla* with a distinct campanulate or
    cup-shaped tube :
  Corolla-lobes more than twice as long as broad,
    "papillose" (puberulous ?) on the inner surface,
    whitish, dotted with blood-red ... ... (1) **punctatus.**
  Corolla-lobes scarcely longer than broad, glabrous,
    blackish-purple ... ... (11) **grivanus.**
Fully expanded corolla rotate or subrotate without a
    distinct tube, velvety-puberulous or pubescent on
    the inner surface.
  Corona-lobes much longer than the anthers, with the
    tips connivent-erect, recurving, or erect and
    doubly curved :

---

* Caution is necessary with regard to this character, as I have occasionally seen upon cultivated plants of species belonging to the tubeless series, flowers that never fully expanded and had a very short cup-like tube. Also if fully expanded tubeless flowers are preserved in formalin, they invariably close up in a campanulate manner and the united part forms a short cup-like tube.

Corolla-lobes about 4½–5 lin. long :
  Corolla very pale yellowish or creamy, dotted with
    purple or crimson ... ... (2) **cornutus.**
  Corolla bright clear greenish-yellow, with small
    rounded dark purple-brown spots ... ... (8) **pulcher.**
Corolla-lobes 2–2½ lin. long, straw-yellow, without
  markings ... ... (10) **parvulus.**
Corona-lobes shorter to longer than the anthers and
  closely incumbent upon them, crossing or very
  slightly upcurved at the tips :
Corolla-lobes 5–9 lin. long :
  Corolla pale yellow or dull greenish-yellow without
    markings ... ... (4) **Pillansii.**
  Corolla yellow, ochreous or greenish-yellow, dotted,
    spotted or transversely marked all over with
    blood-red or purple :
    Crest of corona-lobes entire or subdenticulate at
      the dorsal margin, not tuberculate or ridged
      on the top ... ... (3) **geminatus.**
    Crest of corona-lobes truncate or toothed at the
      dorsal margin, tuberculate or with ridges on
      the top ... ... (4) **Pillansii,** var. β.
  Corolla yellow or greenish-yellow, dotted with
    dark purple-brown; crest of corona-lobes
    toothed at the rather acute-edged dorsal
    margin, with slight ridges on the top ... (5) **decorus.**
  Corolla dark crimson ? or purple with greenish-
    yellow transverse lines and markings on the
    basal part ... ... (4) **Pillansii,** var. γ.
Corolla-lobes 3½–4½ lin. long :
  Corolla pale pinkish-purple, finely marked with
    pale yellow transverse lines; corona yellow,
    without markings ... ... (6) **disparilis.**
  Corolla yellow and marked all over with purplish-
    crimson spots or transverse lines, or pale or
    dark purplish-crimson with yellow or creamy-
    white transverse lines on the basal part;
    corona orange-yellow with purple-brown
    markings ... ... (7) **fœtidus** and vars.
Corolla-lobes 2½–3½ lin. long, whitish spotted with
  dark purple-brown ... ... (9) **comptus.**

See also *Stapelia fasciculata,* Thunb., which may belong to this genus.

1. **P. punctatus** (R. Br. in Mem. Wern. Soc. i. 23); stems decumbent or ascending, 1–2 in. long, ½–¾ in. thick, oblong or somewhat clavate, very obtusely 4-angled, with 4–5 small tubercle-like teeth along each angle, glabrous, dull green; flowers 2–4 together above the middle of the stems, erect; pedicels ½–1 in. long, glabrous, purplish; sepals lanceolate, acute, glabrous; corolla with a distinct shortly campanulate tube 1½–2 lin. long; lobes about 4 lin. long, 1½ lin. broad, lanceolate, acute, spreading, papillate (puberulous?) on the inner face, whitish (faintly tinged with pink according to the figure), dotted with blood-red. *Sweet, Hort. Brit. ed.* 1, 278;

G. Don, Gen. Syst. iv. 113 ; *Decne in DC. Prodr.* viii. 650 ; *N. E. Br. in Journ. Linn. Soc.* xvii. 163, and *Gard. Chron.* 1879, xii. 9. *P. punctata, Schultes, Syst. Veg.* vi. 9. *Stapelia punctata, Masson, Stap.* 18, *t.* 24 ; *Willd. Sp. Pl.* i. 1289, *Pers. Syn. Pl.* i. 279 ; *Poir. Encycl.* vii. 386 ; *Ait. Hort. Kew. ed.* 2, ii. 92 ; *Link, Enum. Pl. Hort. Berol.* i. 258 ; *Spreng. Syst. Veg.* i. 841 ; *Dietr. Syn. Pl.* ii. 887. *Obesia ? punctata, Haw. Syn. Pl. Succ.* 43. *Caralluma punctata, Schlechter in Journ. Bot.* 1898, 478.

WESTERN REGION : Little Namaqualand, *Masson.*

Only known to me from Masson's figure and description.

2. **P. cornutus** (N. E. Br.) ; stems procumbent or ascending, $\frac{1}{2}$–$1\frac{1}{2}$ in. long, $\frac{1}{2}$–$\frac{2}{3}$ in. thick, globose or oblong, very obtusely 4-angled, with 3–5 tubercle-like teeth along the angles, greyish- or glaucous-green ; flowers in pairs above the middle or at the tips of the stems, erect ; pedicels $\frac{1}{4}$–$\frac{3}{4}$ in. long, glabrous ; sepals $\frac{1}{8}$ in. long, lanceolate, acute, glabrous ; corolla very deeply lobed, without a distinct tube ; lobes about 5 lin. long, $1\frac{1}{2}$ lin. broad at the base, lanceolate, acuminate, glabrous on the back, velvety-puberulous on the inner face, very pale yellowish or whitish, dotted with purple (or crimson ?) ; corona-lobes about 1 lin. long, narrowly linear-lanceolate or lanceolate-subulate, incumbent at the base upon the backs of the anthers and produced beyond them into connivent-erect acute tips, dorsally produced at the base into a subquadrate crest, truncate and tuberculate-denticulate at the top, yellow, without markings. *P. decorus ? N. E. Br. in Hook. Ic. Pl., under t.* 1924.

VAR. β, **grandis** (N. E. Br.) ; corolla-lobes 6–7 lin. long, colour unknown ; otherwise as in the type.

CENTRAL REGION : Var. β : Victoria West Div. ; near Victoria West, *Barkly,* 25 bis !
WESTERN REGION : Little Namaqualand, *Barkly,* 25 !

This is one of the plants which in *Hook. Ic. Pl.* under *t.* 1924, I referred doubtfully to *P. decorus,* but upon comparison with that species I now find that the corona is quite different.

3. **P. geminatus** (N. E. Br. in Journ. Linn. Soc. xvii. 163) ; stems procumbent or decumbent, 1–$1\frac{3}{4}$ in. long, $\frac{1}{2}$–$\frac{3}{4}$ in. thick, very obtusely or obscurely 4-angled with about 4 minute teeth $\frac{1}{2}$ lin. long along each angle, glabrous, light green ; flowers in pairs or solitary (rarely 3 or 4 together) at the tips of the stems ; pedicels $\frac{1}{4}$–$\frac{1}{2}$ in. long, glabrous ; sepals $1\frac{1}{2}$–2 lin. long, ovate or ovate-lanceolate, acuminate, glabrous ; corolla 1–$1\frac{1}{4}$ in. in diam., rotate, without a tube ; lobes widely spreading, 5–7 lin. long, about 2 lin. broad at the base, lanceolate or gradually tapering to the acute apex, with revolute margins, glabrous on the back, velvety-puberulous all over the inner face, not ciliate, yellow, dotted with blood-red ; corona-

lobes about 1'lin. long, linear-lanceolate or narrowly-oblong, acute or slightly toothed at the apex, closely incumbent upon the backs of the anthers and shortly exceeding them but not connivent-erect, dorsally produced at the base into a spreading crest entire or minutely subdenticulate at the dorsal margin, flat or nearly so and not tuberculate on the top, the entire corona appearing (viewed from above) like a star with 5 short rounded lobes connected at the base, yellow. *N. E. Br. in Gard. Chron.* 1879, xii. 9 ; *K. Schum. in Engl. und Prantl, Pflanzenfam.* iv. ii. 277. *Stapelia geminata, Masson, Stap.* 18, *t.* 25 ; *Willd. Sp. Pl.* i. 1290, *and Enum. Pl. Hort. Berol.* 282 ; *Pers. Syn. Pl.* i. 280 ; *Poir. Encycl.* vii. 386 ; *R. Br. in Mem. Wern. Soc.* i. 25 ; *Jacq. Stap. t.* 16 ; *Bot. Mag. t.* 1326 ; *Lodd. Bot. Cab. t.* 300 ; *Ait. Hort. Kew. ed.* 2, ii. 92 ; *Schultes, Syst. Veg.* vi. 42 ; *Link, Enum. Pl. Hort. Berol.* i. 257 ; *Spreng. Syst. Veg.* i. 840 ; *Dietr. Syn. Pl.* ii. 886 ; *Decne in DC. Prodr.* viii. 661. *Obesia geminata, Haw. Syn. Pl. Succ.* 42 ; *Sweet, Hort. Brit. ed.* 1, 278 ; *G. Don, Gen. Syst.* iv. 121. *Podanthes geminata, Nicholson, Dict. Gard.* iii. 172. *Caralluma geminata, Schlechter in Journ. Bot.* 1898, 478.

SOUTH AFRICA: without locality, *Masson,* cultivated specimens in *Herb. Kew.* ! and *Herb. Haworth* !

Described from a specimen cultivated in 1813, probably introduced by Masson.

4. **P. Pillansii** (N. E. Br.) ; tufted ; stems 1–1½ in. long, 5–8 lin. thick, decumbent, oblong or somewhat clavate-oblong, very obtusely 4- (rarely 5-) angled or subcylindric, glabrous, dull light green, mottled or tinted with dull purple, angles not or scarcely tuberculate, with very minute apiculus-like teeth, having a more minute tooth on each side at the base ; flowers usually in pairs at the middle or upper part of the young stems, erect ; pedicels ¼–⅔ in. long, glabrous ; sepals ⅛ in. long, lanceolate, acute, glabrous ; corolla rotate, lobed nearly to the base, 1¼–1⅓ in. in diam. ; lobes very spreading, ½–⅔ in. long, about 2–2½ lin. broad at the base, narrowly lanceolate, acute, convex from the margins being recurved, glabrous on the back, very shortly velvety-puberulous on the inner face, not ciliate, dull greenish-yellow or pale yellow without spots ; corona-lobes rather more than 1 lin. long, closely incumbent upon the backs of the anthers and exceeding them, with their tips crossing each other but not becoming erect, lanceolate, acute, entire or somewhat 3-toothed at the apex, produced at the base into a truncate or very obtuse dorsal crest, minutely tuberculate or with ridges on the top, and the dorsal margin neither acute-edged nor toothed, yellow, without markings.

VAR. β, **inconstans** (N. E. Br.) ; corolla 1⅓–1½ in. in diam. ; lobes 7–9 lin. long, 2 lin. broad at the base, narrowly lanceolate or lanceolate-attenuate, densely dotted or transversely marked with light purple on an ochreous ground, sometimes so minutely as to look light pinkish-brown ; tips of the corona-lobes not crossing and the dorsal margin of the crest sometimes toothed ; otherwise as in the type.

VAR. γ, **fuscatus** (N. E. Br.); corolla $1\frac{1}{4}$–$1\frac{1}{3}$ in. in diam.; lobes about $\frac{1}{2}$ in. long, $2\frac{1}{2}$ lin. broad, lanceolate, acuminate, dark purple (or dark crimson?) with very numerous slender irregular transverse and labyrinthine greenish-yellow lines and markings; tips of the corona-lobes not crossing each other; otherwise as in the type.

COAST REGION: Oudtshoorn Div.; hills near Oudtshoorn, *Pillans*, 691! Willowmore Div.; near Willowmore, *Marloth*, 4376! Var. β: Oudtshoorn Div.; hills near Oudtshoorn, *Pillans*! Var. γ: Oudtshoorn Div.; nine miles along the road from Oudtshoorn to George, *Pillans*, 686!

According to Dr. Marloth "the flowers have a strong scent of Valerian." Partly described from flowers in fluid.

5. P. decorus (N. E. Br. in Journ. Linn. Soc. xvii. 163); stems decumbent, 1–$1\frac{1}{2}$ in. long, or under cultivation sometimes very much longer, $\frac{1}{2}$–$\frac{3}{4}$ in. thick, obscurely or very obtusely 4-angled, oblong, with 3–5 small tubercle-like teeth along each angle, glabrous, greyish-green, or under cultivation green, slightly glaucous; flowers in pairs above the middle of the stems, erect; pedicels $\frac{1}{2}$–$\frac{3}{4}$ in. long; sepals about $1\frac{1}{2}$ lin. long, ovate-lanceolate, acute or acuminate, glabrous; corolla 1–$1\frac{1}{4}$ in. in diam., rotate, without a distinct tube, lobed to $\frac{3}{4}$ of the way down; lobes 5–6 lin. long, 2 lin. broad at the base, lanceolate, acute, with revolute margins at the apical part, glabrous on the back, velvety-puberulous on the inner face, not ciliate, yellowish, dotted with dark purple-brown; corona-lobes about 1 lin. long, closely incumbent upon the backs of the anthers and shorter than or slightly exceeding them, lanceolate or ovate, varying (even in the same flower) from acute to irregularly toothed at the apex, dorsally produced at the base into a short transversely rectangular crest, truncate or irregularly toothed at the rather acute-edged dorsal margin, with slight ridges from the edge to the inflexed part of the lobe, yellow, not spotted. *P. serrulatus, N. E. Br. l.c., and P. decorus and P. serrulatus, N. E. Br. in Gard. Chron. 1879, xii. 9. P. decorus, K. Schum. in Engl. und Prantl, Pflanzenfam. iv. ii. 277. Stapelia decora, Masson, Stap. 19, t. 26; Willd. Sp. Pl. i. 1290; Pers. Syn. Pl. i. 280; Poir. Encycl. vii. 387; Ait. Hort. Kew. ed. 2, ii. 93; Schultes, Syst. Veg. vi. 42; Spreng. Syst. Veg. i. 840; Dietr. Syn. Pl. ii. 886; Decne in DC. Prodr. viii. 661; Loud. Encycl. Pl. 200, fig. 3332. S. serrulata, Jacq. Stap. t. 17; Willd. Enum. Pl. Hort. Berol. 286; Poir. Encycl. Suppl. v. 234; Schultes, Syst. Veg. vi. 47; Spreng. Syst. Veg. i. 840; Link, Enum. Pl. Hort. Berol. i. 257; Dietr. Syn. Pl. ii. 886; Decne in DC. Prodr. viii. 658. Obesia decora, Haw. Syn. Pl. Succ. 43; Sweet, Hort. Brit. ed. 1, 278; G. Don, Gen. Syst. iv. 121. O. serrulata, Sweet, Hort. Brit. ed. 1, 278. Orbea decora, Steud. Nom. ed. 2, ii. 222. Caralluma decora and C. serrulata, Schlechter in Journ. Bot. 1898, 478. Caruncularia? serrulata, G. Don, Gen. Syst. iv. 122. Caruncularia serrata, Decne in DC. Prodr. viii. 658 (wrongly quoted as Haw. by Decne).*

SOUTH AFRICA: without locality, *Masson, cultivated specimen*!

Described from a specimen cultivated in 1813 (probably introduced by Masson), in which I find that whilst most of the corona-lobes are entire at the apex, some of them are toothed, as represented in Jacquin's figure of *Stapelia serrulata*, and I have no doubt that that plant and *S. decora*, Masson, are one species. As Jacquin received from England many of Masson's plants this may have been one of them.

6. P. disparilis (N. E. Br.) ; stems not seen, " almost cylindrical, same colour as the flowers " (*Pillans*), probably similar to those of *P. comptus* or *P. geminatus* ; pedicels ¼–½ in. long, glabrous ; sepals 1 lin. long, lanceolate, acuminate, recurved at the tips, glabrous ; corolla rotate, without a distinct tube, 9–10 lin. in expanse ; lobes ⅓ in. long, 1¾ lin. broad at the base, lanceolate, acuminate, convex from the margins being slightly recurved, glabrous on the back, velvety-puberulous on the inner face, not ciliate, " pale pinkish-purple, finely marked with transverse pale yellow lines " (*Pillans*) ; corona-lobes ¾ lin. long, closely incumbent upon the backs of the anthers and shortly exceeding them, with the tips very slightly turned up, yellow, without markings, lanceolate or linear-oblong, entire or more or less distinctly 3-toothed at the apex, with the middle tooth much the longest and acute, dorsally produced at the base into a subquadrate crest, truncate and distinctly toothed at the top.

CENTRAL REGION: Laingsburg Div. ; under bushes between Ladismith and Laingsburg, *Pillans*, 57 ! 617 !

The flowers of this species are similar to those of *P. Pillansii*, but are very much smaller and the crest of the corona-lobes is distinctly toothed, not merely minutely tuberculate. They emit an " odour like bad vinegar," according to Mr. Pillans.

7. P. fœtidus (N. E. Br.) ; stems tufted, ⅓–1½ (or under cultivation up to 2½) in. long, ⅓–⅔ in. thick, oblong, ovoid or globose obtusely and sometimes indistinctly 4–5-angled, tuberculate-toothed with the tubercles tipped with a small acute apiculus having a minute tooth on each side at its base, glabrous, green to grey-green, clouded or mottled with dull purple where exposed to the sun ; flowers 1–6 (but often 2) together near the top or at the middle of the stems, sometimes opening in pairs ; pedicels ¼–¾ in. long, erect, glabrous ; sepals 1½–2 lin. long, lanceolate or ovate-lanceolate, acuminate, with recurved tips, glabrous ; corolla ovoid to subglobose in bud, 7–11 lin. in diam. when fully expanded, rotate, without a distinct tube, glabrous and green or tinted with purplish on the back, evenly pubescent and neatly marked all over the inner face with short transverse purplish-crimson lines or lines and spots on a yellow ground, or sometimes the tips of the lobes are almost entirely purplish-crimson ; lobes 3½–4½ lin. long, 1¾–2⅔ lin. broad, ovate or ovate-lanceolate, acute, convex above, concave beneath, from the recurving margins, usually without but sometimes with a sparse ciliation of clavate vibratile hairs ; corona-lobes 1–1¼ lin. long, closely incumbent upon the backs of the anthers, and equalling or shortly exceeding them, linear-lanceolate to oblanceolate, acute, entire or minutely denticulate at the apex, with raised lines and minute tubercles on the back, dorsally dilated at the base into a subquadrate crest, truncate and denticulate-tuberculate on the top, deep orange-yellow, with the margins, the raised lines on the back and the tubercles of the crest all dark purple-brown, odour very strong, carrion-like, penetrating and perceptible at a short distance from the plant.

VAR. β, **multipunctatus** (N. E. Br.); corolla-lobes marked all over with small rounded dark purplish-crimson dots, otherwise as in the type.

VAR. γ, **pallidus** (N. E. Br.); corolla-lobes pale purple with creamy-white transverse lines and spots on the basal two-thirds, otherwise as in the type.

VAR. δ, **purpureus** (N. E. Br.); corolla-lobes bright deep purplish-crimson, with faint dull yellow transverse markings on the basal half, otherwise as in the type.

VAR. ε, **diversus** (N. E. Br.); corolla-lobes dark purplish-crimson, with creamy-yellow transverse lines and spots on the basal two-thirds; pubescence on the lower two-thirds of the lobes consisting of short greyish hairs and on the apical part of much longer dark purple simple and slightly clavate hairs mixed; otherwise as in the type.

COAST REGION : Bedford Div.; nine miles east-south-east of Bedford, *Pillans*, 165 ! Var. β: from the same locality growing with the type, *Pillans*, 185 !

CENTRAL REGION : Graaff Reinet Div. ; " Wheatlands," *Pillans*, 697 ! 699 ! near Graaff Reinet, *cultivated specimen* ! Var. β : Graaff Reinet Div. ; " Wheatlands," *Pillans*, 695 ! Vars. γ to ε : Graaff Reinet Div. ; western aspects across the Sundays River, east of Graaff Reinet ; var. γ, *Pillans*, 111 ! var. δ, *Pillans*, 690 ! " Wheatlands," *Pillans*, 698 ! var. ε, *Pillans*, 142 !

KALAHARI REGION : Griqualand West ; Griquatown, *Thompson* in *Herb. Pillans*, 133 !

This species is very variable in coloration, the above being some of the chief variations.

The type is described from a living plant which flowered with me in 1878, sent to Kew by Dr. Bolus.

8. **P. pulcher** (N. E. Br.); stems tufted, ½–1 in. (or more?) long, ⅓–½ in. thick, subglobose or oblong, obtusely 4-angled, with small tubercle-like teeth, glabrous, greyish-green or somewhat glaucous ; flowers 1–4 (often 2) together, near the middle or tips of the stems ; pedicels erect, 3–6 lin. long, stout, thickened upwards, glabrous ; sepals 2 lin. long, linear-lanceolate, acuminate, glabrous ; corolla subrotate, without a distinct tube, ¾ in. in diam., glabrous and dull greenish-brown on the back, inner face velvety-pubescent with white and purple hairs, bright clear greenish-yellow, marked all over with small rounded dark purple-brown spots ; lobes not spreading horizontally, 4½–5 lin. long, 2 lin. broad at the base, thence tapering in a nearly straight line to the apex, with revolute margins ; corona-lobes 1 lin. or rather more in length, linear-subulate, incumbent on the backs of and much exceeding the anthers, with short subdenticulate acute connivent-erect tips, dorsally produced at the base into a quadrate crest, flat and papillate on the top, clear deep yellow, without markings.

SOUTH AFRICA : without locality, but possibly from Colesberg Div., *Shaw* !

Described from a living plant sent to me by Dr. Shaw in 1876.

9. **P. comptus** (N. E. Br. in Hook. Ic. Pl. t. 1924, B); stems tufted, erect or decumbent, ¾–1 (or under cultivation up to 3) in. long, 4–7 lin. thick, obtusely 4-angled, with the teeth along the angles tubercle-like, apiculate, sometimes rather obscure, dull green or greyish-green, subglaucous; flowers 1–4 (often 2) together, between the angles, near or at the apex or sometimes near the middle of the stem, erect; pedicels 2–6 lin. long, glabrous; sepals ¾–1½ lin. long, ovate or lanceolate, acuminate, glabrous; corolla rotate, 7–9 lin. in diam., glabrous and dull greenish or greenish-brown outside, shortly pubescent all over the inner face, not ciliate, whitish, marked all over with small round dark purple-brown spots; disk flattish; lobes 2½–3½ lin. long, 2 lin. broad at the base, ovate-lanceolate, acute or acuminate, slightly convex from the margins being slightly recurved; corona-lobes closely incumbent on the backs of the anthers and not prolonged beyond them, ¾–1 lin. long, linear or linear-lanceolate, acute, obtuse or subdenticulate at the apex, dorsally expanded at the base into a truncate minutely tuberculate or more or less bifid crest, with or without a small warted tubercle between the teeth at the top, entirely yellow or dotted with purple-brown; odour somewhat sour, only perceptible close to the flower.— *K. Schum. in Engl. und Prantl, Pflanzenfam. iv. ii. 277. Caralluma compta, Schlechter in Journ. Bot. 1898, 479.*

Var. *β*, **ciliatus** (N. E. Br.); corolla-lobes ciliate with vibratile clavate hairs, otherwise as in the type.

Central Region: Beaufort West Div.; near the Gamka River, *Burke!* *Zeyher*, 1145! Prince Albert Div.; near Prince Albert, *Marloth*, 4586! Karoo at Grootfontein, *Dickson*! *Barkly*, 58! 71! 73! *Pillans*, 91! and *cultivated specimens*! Var. *β*: Prince Albert Div.; Grootfontein, *Pillans*, 186! Laingsburg Div.; near Laingsburg, *Chalwin*!
Western Region: Little Namaqualand, *Herb. Pillans*, 90!

Described from living plants.

10. **P. parvulus** (N. E. Br.); stems decumbent or ascending, ¾–1¾ in. long, 4–7 lin. thick, oblong or ovoid-oblong, obtusely 4-angled, with 3–5 tubercle-like apiculate teeth ¾–1½ lin. long along each angle, greyish-green or perhaps slightly glaucous (sometimes dull light green under cultivation), tinged or mottled with purple; flowers gradually developing up to 12 in number from the same point, but seldom more than 2 or 3 open together in the same fascicle, often 2–3 fascicles to a stem, at the middle or top; pedicels erect, ¼–½ in. long, rather slender, glabrous; sepals ¾ lin. long, lanceolate, acute, glabrous; corolla rotate, without a distinct tube, 5–6 lin. in diam.; lobes very spreading, 2–2½ lin. long, 1–1¼ lin. broad at the base, deltoid-lanceolate, acute, glabrous on the back, velvety-puberulous on the inner face, not ciliate; straw-yellow, without spots; corona-lobes ¾–1 lin. long, much overtopping the staminal column, in some plants connivent with recurved-hooked tips, in others erect with the upper half doubly curved, first backwards and downwards and then

upwards or outwards and then inwards, acute or irregularly 2–3-toothed at the apex, with a thick convex or truncate dorsal crest at their base, minutely tuberculate all over its top.

CENTRAL REGION: Laingsburg Div.; near Matjesfontein, *Pillans,* 130! three miles south-west of Laingsburg, *Pillans,* 672! Prince Albert Div.; near Grootfontein, *Pillans.*

11. **P. grivanus** (N. E. Br. in Hook. Ic. Pl. t. 1924, A); branches 1½–2 in. high, forking in all directions; branchlets scarcely angular but rather composed of aggregated tubercles, with a white spine at the end of the teeth" (*Barkly*); according to Miss Barkly's drawing the stems are 4-angled, with 3–4 conical finely pointed teeth 2–3 lin. long along each angle; flowers solitary (always ?) in the grooves between the angles below the middle of the stems; pedicels not seen, "very short" (*Barkly*); sepals 2½ to nearly 3 lin. long, 1 lin. broad at the base, ovate, acuminate, with recurved tips, glabrous; corolla " somewhat campanulate, a full inch from point to point, interior deep purple, almost black, with rough striæ across; exterior nearly as green as the calyx, streaked with dark perpendicular stripes" (*Barkly*); in the only dried flower seen it is about ¾ in. in diam., with a short and apparently broadly campanulate or cup-shaped tube, and deltoid-acute lobes 3–3½ lin. long and 2¾–3 lin. broad at the base, glabrous on both sides and not ciliate; corona-lobes about 2 lin. long, 1–1¼ lin. broad, ovate (not quite as figured), acuminate, closely incumbent upon the backs of the anthers and exceeding them, dorsally produced at the base into a short transverse crest, which is apparently flattened on the top and slightly crenate on the subtruncate hind margin, " reddish-brown" (*Barkly*). *K. Schum. in Engl. und Prantl, Pflanzenfam.* iv. ii. 277. *Caralluma grivana, Schlechter in Journ. Bot.* 1898, 479.

KALAHARI REGION: Griqualand West; Griva, *Arnot,* 6 (*Barkly,* 11)!

I have only seen one flower of this rather remarkable species, the rest of the description being compiled from the figure and description sent to Kew by Sir Henry Barkly in 1876, the plant not having been refound since then. Possibly when better known it may have to be removed from this genus.

LIII. DUVALIA, Haw.

Calyx 5-partite. *Corolla* rotate, deeply 5-lobed, with the disk raised into a cushion-like ring (annulus) around and supporting the outer corona and its margin reflexed; lobes narrowly linear-lanceolate to ovate, folded longitudinally backwards (replicate) into vertical plates or the basal half more or less expanded. *Corona* double, arising near the top of the staminal column, stipitate; outer corona flat, entire, more or less pentagonal rarely 10-angled, resting on the rim of or on the sides of the cup formed by the annulus; inner corona-lobes turgid, ovoid, more or less pointed at each end, subhorizontal with the dorsal point usually somewhat raised and the inner closely incumbent on the backs of the anthers and sometimes longer than them, but not produced into erect points. *Stamens* arising at the base of the corolla, united into a tube around the ovary and adnate to the dilated part of the style; anthers without a terminal appendage, inflexed on the table-like top of the style. *Pollen-masses* solitary in each anther-cell, horizontal or ascending, pellucid on one margin, attached by very short broad caudicles to lateral wing-like expansions of the pollen-carrier. *Follicles* erect, smooth. *Seeds* with a tuft of hairs at one end.

Succulent leafless herbs of very dwarf habit; stems decumbent or erect, in many of the species occasionally subterranean, with the tips rising to the surface, 4–6-angled, with spreading teeth, each tooth tipped with a minute deltoid-ovate or subulate acute rudimentary leaf, having a minute denticle (*stipule*) on each side at its base ; flowers in small clusters or cymes near the base or middle of the young stems, sometimes solitary.

DISTRIB. Species about 16, mostly South African, 1 in Tropical Africa and 1 in Arabia.

Corolla-lobes (at least at the base) and annulus pubescent
  on the inner face :
  Corolla-lobes 2½–5 lin. long, with a soft villous
    pubescence 1–1½ lin. long ...   .   (4) **elegans**
                                                                  and vars.

  Corolla-lobes 3–5½ lin. long, with a pubescence ⅛–½ lin.
    long ...   .   (6) **pubescens**
                                                                  and var.

Corolla-lobes glabrous on the inner face :
  Corolla-lobes ciliate with long clavate vibratile hairs,
    which easily become detached and are often
    absent from dried specimens :
    Annulus covered with soft spreading purple hairs
      1–2½ lin. long ; corolla 1¼–2 in. in diam.   ...   (3) **Corderoyi.**
    Annulus varying from almost glabrous to distinctly
      puberulous on the rim with hairs ¼–½ lin. long ;
      corolla ⅞–1¼ in. in diam. :
      Stems 6-angled ; corolla-lobes expanded and about
        4–5 lin. broad at the base, very smooth and
        shining ...   .   (1) **polita.**
      Stems 4–5-angled ; corolla-lobes replicate from
        base to apex into vertical plates   ...   ...   (8) **reclinata.**
  Corolla-lobes ciliate with clavate fixed or but slightly
    vibratile and non-clavate hairs mixed   ...   ...   (9) **hirtella** vars.

  Corolla-lobes ciliate (at least at the base) with non-clavate
    fixed hairs only, which are sometimes minute and
    scarcely visible except under a lens :
    Corolla-lobes expanded and 4–5 lin. broad at the
      base ; outer corona light yellow...   ...   ...   (2) **Pillansii.**
    Corolla-lobes openly replicate and 1½–2½ lin. across
      at the base, replicate into vertical plates at the
      apex ; outer corona reddish-brown   ...   ...   (5) **modesta.**
    Corolla-lobes closely replicate to the base into
      vertical plates, with the part at the sinuses
      reflexed vertically from the annulus or under it
      so as to nearly or quite touch the pedicel :

Corolla cream-colour, with pale purple tips to the
2–2½ lin.-long lobes ; cilia minute ... ... (13) **parviflora.**

Corolla-lobes apparently olive-brown or purple-
brown, 2½–3¾ lin. long ; annulus whitish,
spotted all over with purple-brown ; cilia
minute ... ... (12) **maculata.**

Corolla entirely dark chocolate or with a greenish
or ochreous ring or markings on the rim of
the annulus just around the outer corona :
Outer and inner corona white ; annulus 1½–1¾
lin. in diam. ; lobes ¾–1 lin. deep, with
scarcely perceptible cilia ... ... (14) **angustiloba.**

Outer corona green ; annulus 2½ lin. in diam. ;
lobes 1 lin. deep, with distinct cilia ... (7) **cæspitosa.**

Outer corona brownish-red or with yellowish
angles :
Corolla-lobes 4–5½ lin. long ; cilia usually
distinct ... ... (9) **hirtella.**

Corolla-lobes 3–4 lin. long ; cilia always
minute ... ... (11) **compacta.**

Corolla-lobes entirely without cilia (see also above, in
some of which they may easily be overlooked),
closely replicate to the base into vertical plates
1½–2 lin. deep ... ... (10) **radiata.**

1. **D. polita** (N. E. Br. in Gard. Chron. 1876, vi. 130) ; stems
rather elongated, lax, decumbent, 1½–2½ in. long, about ½ in. thick,
obtusely 6-angled, with subulate-pointed teeth 2½–4 lin. long ;
flowers 3–4 together, successively developed ; pedicels about 1 in.
long, rather slender, purplish ; sepals about ¼ in. long, lanceolate-
subulate, dingy green ; corolla 1–1¼ in. in diam., dull green on the
back, rich purplish-chocolate on the inner face, very smooth and
shining on the basal ⅔ of the lobes, not shining elsewhere ; lobes
erectly spreading, 4–5 lin. long and nearly as broad, ovate, acuminate, with the sides but slightly folded back, glabrous on both
sides, ciliate on the basal half with vibratile clavate purple hairs
⅔ lin. long ; annulus about ⅓ in. in diam. and 1 lin. high, pentagonal,
minutely puberulous ; outer corona 2½ lin. in diam., pentagonal,
chocolate-red ; inner corona-lobes dull orange-red. *Bot. Mag. t.* 6245 ;
K. Schum. in Engl. und Prantl, Pflanzenfam. iv. ii. 277 ; *Schlechter
in Journ. Bot.* 1898, 477. *D. dentata, N. E. Br. in Kew Bulletin,*
1895, 265, *and in Dyer, Fl. Trop. Afr.* iv. i. 502.

SOUTH AFRICA ? without locality or name of collector, *cultivated specimens* !

Also in Tropical Africa.

Described from a living plant. This now appears to be lost to cultivation. See
also 16, *D. transvaalensis*, Schlechter. When I described *D. dentata* I had unfortunately entirely overlooked *D. polita.*

2. **D. Pillansii** (N. E. Br.) ; stems ¾–2 in. long, ⅓–½ in. thick,
ovoid to oblong, obtusely 4–6-angled ; angles with conical acute
tubercle-like teeth 1½–2 lin. long ; flowers 2–3 together, successively
developed near the middle of the young stems ; pedicels usually

about $\frac{1}{4}$ (rarely up to $\frac{1}{2}$) in. long ; sepals $1-1\frac{1}{2}$ lin. long, ovate or ovate-lanceolate, acuminate ; corolla in bud deeply and acutely 5-lobed at the flat and rather thin basal part, abruptly acuminate into a very acute 5-grooved cone, when expanded $1-1\frac{1}{6}$ in. in diam., dark chocolate passing into light purple on the top of the annulus and fading into a narrow white ring around the corona ; lobes ascending-spreading, with recurving tips, 4–5 lin. long and as much in breadth at the base, broadly deltoid-ovate with the sides but slightly folded back at the base, rather abruptly acuminate and replicate at the apical part, glabrous on both sides, ciliate on the basal half with fine simple purple hairs $1-1\frac{1}{2}$ lin. long ; annulus $1-1\frac{1}{3}$ lin. high and $3\frac{1}{2}-4$ lin. in diam. at the top, with the sides much sloping outwards, obscurely pentagonal, glabrous ; outer corona about $2\frac{1}{4}$ lin. in diam., pentagonal, light canary-yellow as are also the inner corona-lobes.

Var. β, **albanica** (N. E. Br.) ; corolla-lobes suberect ; annulus slightly raised above the margin of the outer corona.

Coast Region: Var. β : Albany Div. ; vicinity of Grahamstown, *Pillans*, 19 ! and *cultivated specimen* !

Central Region : Aberdeen Div. ; near Aberdeen Road, *Pillans*, 42 !

Described from a living plant, and flowers preserved in fluid.

3. **D. Corderoyi** (N. E. Br. in Bot. Mag. under t. 6245) ; stems $\frac{1}{2}-1\frac{1}{4}$ in. long, about $\frac{3}{4}$ in. thick, globose or oblong, with 6 very obtuse or somewhat obscure tuberculate-dentate angles, dull green, tinged with purple where exposed to the sun ; flowers 2–4 together at about the middle or near the base of the young stems ; pedicels $\frac{1}{2}-1$ in. long, 1 lin. thick ; sepals $2-2\frac{1}{2}$ lin. long, lanceolate, acute, glabrous ; corolla $1\frac{1}{4}-2$ in. in diam. ; lobes very spreading, $\frac{1}{2}-\frac{2}{3}$ in. long, sometimes closely replicate for $\frac{2}{3}$ of their length with the basal part expanded, sometimes closely replicate only at the tips and openly at the other part, glabrous, ciliate along the basal $\frac{2}{3}$ with long vibratile clavate purple hairs, light or dull olive-green, tinged towards the tips with darker, or dull reddish-brown, some-times with a faint green tint ; annulus 5–6 lin. in diam., obscurely 5-angled, densely covered with long soft purple hairs ; outer corona $3-3\frac{1}{2}$ lin. in diam., pentagonal, dull brick-red, paler at the angles ; inner corona-lobes ovoid, dorsally obtuse, with the inner apex narrowed into a short linear point, buff-coloured. *N. E. Br. in Hook. Ic. Pl. under t.* 1925 ; *K. Schum. in Engl. und Prantl, Pflanzenfam.* iv. ii. 277 ; *Schlechter in Journ. Bot.* 1898, 477. *Stapelia Corderoyi, Hook. f. in Bot. Mag. t.* 6082.

South Africa : without locality, *cultivated specimens* !

Central Region : Middelburg Div. ; near Conway, *Pillans*, 191 ! Aliwal North Div. ; near the Orange River, *Burke* !

Described from living plants. This is one of the finest species of the genus, it seems to vary considerably in the colour of the corolla-lobes, and I have seen both colours described above upon the same plant ; in the *Botanical Magazine* the colour is far too bright and the bud is a total misrepresentation.

4. D. elegans (Haw. Syn. Pl. Succ. 44) ; stems decumbent and ascending, sometimes descending under the ground, $\frac{3}{4}$–$1\frac{3}{4}$ in. long, 4–6 lin. thick, oblong, with 4–5 obtuse tuberculate-denticulate angles, glabrous, dull green or purplish-tinted ; flowers 2–3 together near the base of the stems, developing successively ; pedicels 5–10 lin. long, glabrous ; sepals 1–$1\frac{1}{2}$ lin. long, lanceolate, acute, glabrous ; corolla $\frac{1}{2}$–$\frac{3}{4}$ in. in diam., rather flat, dark purple-brown, shining, pilose all over the inner face with rather long soft purple hairs 1–$1\frac{1}{2}$ lin. long, glabrous on the back ; lobes $2\frac{1}{2}$–3 lin. long, $1\frac{3}{4}$–$2\frac{1}{4}$ lin. broad at the base, ovate or deltoid-ovate, acute or shortly acuminate, flattish at the basal part, slightly replicate towards the apex ; annulus not very evident, scarcely raised above the level of the fold of the lobes ; outer corona $2\frac{1}{4}$–$2\frac{1}{2}$ lin. in diam., almost covering the annulus, nearly circular, obscurely pentagonal or rarely shortly 5-lobed, nearly flat, dark red-brown ; inner corona-lobes pale brownish-yellow. *G. Don, Gen. Syst.* iv. 121 ; *N. E. Br. in Hook. Ic. Pl. under t.* 1925 ; *Schlechter in Journ. Bot.* 1898, 476. *D. jacquiniana, Sweet, Hort. Brit. ed.* 1, 276 ; *G. Don, l.c.* 121 ; *Schlechter in Journ. Bot.* 1898, 476. *Stapelia elegans, Masson, Stap.* 19, *t.* 27 ; *Willd. Sp. Pl.* i. 1282 ; *Pers. Syn. Pl.* i. 278 ; *Poir. Encycl. Meth.* vii. 381, *and in Dict. Sc. Nat.* 1. 391 ; *Bot. Mag. t.* 1184 ; *Ait. Hort. Kew. ed.* 2, ii. 88 ; *Schultes, Syst. Veg.* vi. 44 ; *Lodd. Bot. Cab. t.* 1651 ; *Decne in DC. Prodr.* viii. 662. *S. radiata, Jacq. Stap. t.* 12 ; *Willd. Enum. Pl. Hort. Berol.* 285 ; *Poir. Encycl. Suppl.* v. 234 : *Dietr. Syn. Pl.* ii. 884 ; *Hornem. Hort. Bot. Hafn.* ii. *Suppl.* 30 ; *Spreng. Syst. Veg.* i. 837, *not of Sims or Link. S. jacquiniana, Schultes, Syst. Veg.* vi. 45 ; *Decne in DC. Prodr.* viii. 662. *S. Jacquini, Loud. Encycl. Pl.* 202.

VAR. β, **seminuda** (N. E. Br.) ; stems 2–4 in. long ; corolla having a frosted appearance when viewed with a lens (*Pillans*) ; lobes without hairs on the apical half, those on the margins greyish, flat, wavy, the others purple ; otherwise as in the type.

VAR. γ, **namaquana** (N. E. Br.) ; corolla-lobes $3\frac{1}{2}$–5 lin. long, replicate nearly to the base ; annulus very distinctly raised above the level of the fold of the lobes.

SOUTH AFRICA : Karoo, cultivated specimens probably introduced by Masson ! and *collector unknown* !

COAST REGION : Robertson Div. ; near Ashton, *Pillans,* 75 ! Var. β : Riversdale Div. ; near Riversdale, *Pillans,* 682 !

WESTERN REGION : Var. γ : Little Namaqualand, *Barkly,* 34 !

Described from living plants. The type form of this plant (introduced by *Masson*) is identical with the figure in the *Botanical Magazine* and Jacquin's figure of *Stapelia radiata.* Jacquin received this plant from England under the erroneous name of *S. radiata* and evidently had not seen the figure of the true *D. radiata* (*Stapelia radiata,* Bot. Mag. t. 619) since he quotes no reference to that species and also figures the latter as *S. replicata.*

5. D. modesta (N. E. Br.) ; stems $\frac{1}{2}$–1 in. (or more ?) long, $\frac{1}{3}$–$\frac{1}{2}$ in. thick, ovoid or oblong, with 4–5 obtuse tuberculate-toothed angles, glabrous ; flowers 2–3 together at the middle of the young stems, developing successively; pedicels $\frac{1}{3}$–$\frac{1}{2}$ in. long, glabrous; sepals 1–1$\frac{1}{2}$ lin. long, lanceolate, acute, glabrous; corolla $\frac{1}{2}$–$\frac{3}{4}$ in. in diam., dark chocolate; lobes 2$\frac{1}{2}$–3 lin. long, with the margins closely replicate at the apical half and reflexed-spreading at the basal part, ovate when flattened out, acuminate, ciliate for half of their length with soft fine simple purple hairs $\frac{1}{2}$–1 lin. long, otherwise quite glabrous; annulus about 2$\frac{1}{2}$ lin. in diam., $\frac{1}{2}$ lin. high, obtusely pentagonal, glabrous; outer corona 1$\frac{1}{2}$–2 lin. in diam., obtusely pentagonal, apparently dark reddish-brown; inner corona-lobes apparently reddish.

Central Region : Aberdeen Div. ; near Aberdeen Road, *Pillans,* 35 ! Somerset Div. ; near Pearston, *Pillans,* 35 !

6. D. pubescens (N. E. Br.); stems decumbent, $\frac{1}{2}$–2 in. long, $\frac{1}{3}$–$\frac{2}{3}$ in. thick, obtusely 4–5-angled, dull dark green; angles with horizontal stout conical teeth 1–2 lin. long ; flowers 2–4, successively developed at the middle or upper part of the young stems ; pedicels $\frac{1}{3}$–$\frac{1}{2}$ in. long, about 1 lin. thick ; sepals 1–1$\frac{1}{2}$ lin. long, ovate, very acute or acuminate; corolla $\frac{2}{3}$ to nearly 1 in. in diam., pubescent on the whole of the inner face, but usually more densely on the annulus than on the lobes, with short soft simple hairs $\frac{1}{8}$–$\frac{1}{2}$ lin. long, dark chocolate ; lobes very spreading, with recurved tips, 3–4 lin. long, replicate to the base into vertical plates 1–1$\frac{1}{4}$ lin. deep at the base, not ciliate ; annulus 3$\frac{1}{2}$–4 lin. in diam., 1–1$\frac{1}{3}$ lin. high, very obtusely or obscurely pentagonal, with vertical sides, and the sinuses between the lobes reflexed to the pedicel; outer corona 2$\frac{1}{4}$–2$\frac{1}{2}$ lin. in diam., pentagonal, dull reddish-brown.

Var. *β,* **major** (N. E. Br.) ; flowers from near the tips of the stems ; pedicels $\frac{1}{2}$–$\frac{3}{4}$ in. long ; sepals 1$\frac{3}{4}$ lin. long, lanceolate-attenuate ; corolla about 1$\frac{1}{4}$ in. in diam. ; lobes 5–5$\frac{1}{2}$ lin. long, replicate into vertical plates 1 lin. deep, thinly pubescent on the basal half, glabrous on the apical half ; otherwise as in the type.

Western Region: Little Namaqualand, *Ayres in Herb. Pillans,* 94 ! also *cultivated specimen* ! Var. *β* : Little Namaqualand, *Scully* !

The type described from a living plant cultivated at Cambridge in 1881.

7. D. cæspitosa (Haw. Syn. Pl. Succ. 45) ; stems $\frac{2}{3}$–1$\frac{1}{2}$ in. long, 5–7 lin. thick, ovoid or oblong, obtusely 4–5-angled, green ; angles tuberculate-dentate ; flowers 1–2 together from the middle or near the base of the young stems ; pedicels 4–5 lin. long, glabrous ; sepals about 1 lin. long, lanceolate, acute, glabrous ; corolla $\frac{2}{3}$–$\frac{3}{4}$ in. in diam. ; lobes horizontally spreading, closely replicate to the base into vertical plates, about 3–3$\frac{1}{2}$ lin. long and 1 lin. in depth, glabrous, ciliate on the basal half with very fine and rather long simple purple hairs, dark chocolate ; annulus 2$\frac{1}{2}$ lin. in diam. with a microscopic puberulence, entirely dark chocolate, or according to Jacquin, with a green ring around the outer corona ; outer corona 1$\frac{2}{3}$–1$\frac{3}{4}$ lin. in diam., pentagonal, green ; inner corona-lobes yellowish. *G. Don, Gen. Syst.* iv. 121 ; *Schlechter in Journ. Bot.* 1898, 476. *Stapelia cæspitosa, Masson, Stap.* 20, *t.* 29 ; *Willd. Sp. Pl.* i. 1282,

and *Enum. Pl. Hort. Berol.* 286 ; *Jacq. Stap. t.* 11 ; *Pers. Syn. Pl.* i.
278 ; *Poir. Encycl. Meth.* vii. 381, *and in Dict. Sc. Nat.* 1. 391 ; *Ait.
Hort. Kew. ed.* 2, ii. 88 ; *Schultes, Syst. Veg.* vi. 44 ; *Link, Enum. Pl.
Hort. Berol.* i. 257 ; *Spreng. Syst. Veg.* i. 837 ; *Dietr. Syn. Pl.* 884 ;
Decne in DC. Prodr. viii. 662.

South Africa : Karoo region, *Masson,* cultivated specimens dated 1806, 1812,
1813 (probably from plants introduced by *Masson*), in *Herb. Haworth* ! and *Herb.
Kew.* !

The flowers as figured by Jacquin appear to be too large. This species has not
been refound since Masson introduced it, and appears not to exist in cultivation.

8. **D. reclinata** (Haw. Syn. Pl. Succ. 44); stems $\frac{3}{4}$–3 in. (up to
4 in., *Masson,* and sometimes up to 1 ft., *Jacquin*) long, 5–6 lin.
thick, with 4–5 obtuse tuberculate-toothed angles, dull green ;
flowers 1–3 together near the base or middle of the young stems ;
pedicels $\frac{1}{2}$–1 in. long ; sepals $1\frac{1}{2}$–$2\frac{1}{4}$ lin. long, lanceolate-attenuate ;
corolla $\frac{7}{8}$–$1\frac{1}{4}$ in. in diam., dark chocolate or with the rim of the
annulus around the corona greenish, rather shining ; lobes 4–$5\frac{1}{2}$ lin.
long, very spreading, replicate to the base into vertical plates
$1\frac{1}{2}$–2 lin. deep, fringed on the basal half with vibratile clavate
purple hairs 1–$1\frac{1}{4}$ lin. long, otherwise glabrous ; annulus $3\frac{1}{2}$–4 lin.
in diam. and about $1\frac{1}{2}$ lin. high, obtusely pentagonal, glabrous, puberulous or very shortly pubescent on the rim ; outer corona obtusely
pentagonal, orange-brown or dull brownish-red ; inner corona-lobes
dull orange or brownish-orange. *G. Don, Gen. Syst.* iv. 121 ;
N. E. Br. in Hook. Ic. Pl. under t. 1925 ; *Schlechter in Journ. Bot.*
1898, 476. *D. propinqua, Berger in Monatsschr. Kakt.* 1904, 24.
Stapelia reclinata, Masson, Stap 19, *t.* 28 ; *Willd. Sp. Pl.* i. 1282,
and Enum. Pl. Hort. Berol. 285 ; *Pers. Syn. Pl.* i. 278 ; *Poir.
Encycl. Meth.* vii. 380 ; *Ait. Hort. Kew. ed.* 2, ii. 88 ; *Jacq. Stap.
t.* 14 ; *Schultes, Syst. Veg.* vi. 43 ; *Link, Enum. Pl. Hort. Berol.*
i. 257 ; *Spreng. Syst.* i. 837 ; *Dietr. Syn. Pl.* ii. 884 ; *Decne in DC.
Prodr.* viii. 662 ; *Loud. Encycl. Pl.* 200, *fig.* 3333. *S. radiata, Link,
Enum. Pl. Hort. Berol.* i. 257, *not of Sims nor Jacq.*

Var. β, **bifida** (N. E. Br.) ; annulus very shortly pubescent with purple hairs
on the rim ; inner corona-lobes often bifid at the inner apex ; otherwise as in the
type.

Var. γ, **angulata** (N. E. Br.) ; corolla-lobes about $\frac{1}{3}$ in. long, replicate into
vertical plates $1\frac{1}{4}$ lin. deep ; annulus 1 lin. high, obtusely 5-angled, of a lighter
purple-brown than the lobes, with 5 small straw-coloured patches around the outer
corona (*Pillans*), minutely pubescent all over the rim ; outer corona with
10 distinct angles and 10 flat sides between them, "straw-colour tinted with
purple ; inner corona rich cream colour" (*Pillans*) ; ciliation and all other
characters as in the type.

South Africa: without locality, *cultivated specimens* !
Coast Region : Var. γ : Ladismith Div. ; without precise locality, *Pillans,*
615 !
Central Region : Somerset Div. ; Karoo, *Barkly,* 51 ! 67 ! around Somerset
East, *MacOwan,* 2232 ! *Barkly,* 53 ! Graaff Reinet Div. ; hills near Graaff Reinet,
2600 ft., *Bolus,* 54 ! Aberdeen Div. ; near Aberdeen Road, *Pillans,* 37 ! Var. β :
Somerset Div. ; Glen Avon Estate near Somerset East, *Pillans,* 27 !

The stems as figured by Jacquin are very abnormal and entirely due to some condition of cultivation, being excessively "drawn up;" usually when grown under glass they are about ¾–1½ in. long, but some years I have had them grow to a length of 3 in. or perhaps more; if grown in the open air they do not attain such a length. I am indebted to Mr. Berger for a portion of his type of *D. propinqua*, which I find is identical with this species.

9. **D. hirtella** (Sweet, Hort. Brit. ed. 1, 276); stems subglobose to oblong, usually ¾–2 (sometimes up to 3) in. long, ½–¾ in. thick, obtusely 4–5- (rarely 6-) angled, tuberculate-toothed, glabrous, dull green or tinged or mottled with purple; flowers 1–5 together at the middle or upper part of the young stems, progressively developed; pedicels ½–1 in. long, glabrous; sepals 1⅓–2 lin. long, lanceolate, acuminate, glabrous; corolla ¾–1⅙ in. in diam., entirely dark chocolate or almost blackish; lobes 4–5½ lin. long, very spreading, lanceolate, acute, closely replicate quite to the base into vertical plates 1⅓–2 lin. deep at the base, glabrous, ciliate on the basal ⅓–½ of their length with simple hairs ½–¾ lin. long; annulus 3½–4 lin. in diam. and about ¾ lin. high, pentagonal, pubescent all over the top of the rim with purple hairs ¼–½ lin. long; outer corona about 2½ lin. in diam., obtusely pentagonal or nearly circular, dull brownish-red; inner corona-lobes dull yellow; follicles 3¼–6 in. long, slightly diverging, narrowly fusiform, acute, smooth, streaked with purple-brown. *G. Don, Gen. Syst.* iv. 122; *N. E. Br. in Hook. Ic. Pl. under t.* 1925; *K. Schum. in Engl. und Prantl, Pflanzenfam.* iv. ii. 277; *Schlechter in Journ. Bot.* 1898, 476. *Stapelia hirtella, Jacq. Stap. t.* 10; *Willd. Enum. Pl. Hort. Berol.* 285; *Haw. Syn. Pl. Succ.* 26, *and Suppl.* 9; *Poir. Encycl. Suppl.* v. 232; *Schultes, Syst. Veg.* vi. 26; *Spreng. Syst. Veg.* i. 837, *excl. syn.*; *Link, Enum. Pl. Hort. Berol.* i. 257; *Dietr. Syn. Pl.* ii. 884, *excl. syn.*; *Decne in DC. Prodr.* viii. 662, 663. *S. cæspitosa, DC. Pl. Grass. t.* 148, *not of Masson. S. cæspitosa, var. hirtella, Loud. Encycl. Pl.* 202. *S. cymosa* ("*Hofg.*" *in Berlin Herb.!*) *Hort. ex Schultes, Syst. Veg.* vi. 49, *and G. Don, Gen. Syst.* iv. 117. *S. reclinata, Sims, Bot. Mag. t.* 1397, *not of Masson.*

VAR. β, **obscura** (N. E. Br.); corolla-lobes ciliate with simple or simple and very slightly clavate fixed or subvibratile hairs intermingled, varying in different specimens from ¼–1 lin. long; annulus 3½–4½ lin. in diam., ¾–1⅓ lin. high, varying from glabrous to puberulous on the rim with minute hairs much shorter than those on the type, often marked around the margin of the outer corona with a speckling or ring of dull ochre; otherwise as in the type.

VAR. γ, **minor** (N. E. Br.); corolla-lobes 3½ lin. long, ciliate with simple and fixed or very slightly vibratile clavate hairs intermixed.

SOUTH AFRICA : without locality, *cultivated specimens*! Var. β, *cultivated specimens*!

COAST REGION : Ladismith Div.; without precise locality, *Pillans*, 665! also var. β, *Pillans*, 638!

CENTRAL REGION : Laingsburg Div.; north-western slopes of the Klein Zwart Berg, at Witte Poort, *Pillans*, 646! Prince Albert Div.; near Prince Albert, *Marloth*, 4585! Var. β : Laingsburg Div.; near Matjesfontein, *Pillans*, 13! 81! 641! *MacOwan* (cultivated specimen)! along the main road from Laingsburg to Ladismith near Witte Poort, *Pillans*, 673! 674! near Nuy, *Pillans*, 665! Var. γ : Worcester Div.; without precise locality, *Pillans*, 628!

Described from living plants, and specimens preserved in fluid. This is one of the commonest species in cultivation and varies in minor details very considerably. Specifically *D. hirtella* is closely related to *D. radiata,* and is usually easily discriminated by the very distinct cilia on the corolla-lobes, but is clearly connected with that species by some forms of var. *obscura* in which the cilia are quite minute and scarcely noticeable except under a lens, possibly these really represent *D. radiata,* see note under that species.

10. **D. radiata** (Haw. Syn. Pl. Succ. 45); stems 1–2 in. long, $\frac{2}{3}-\frac{3}{4}$ in. thick, globose or oblong, obtusely 4–5-angled; angles with stout conical acute teeth; flowers usually 2 together, developing in succession near the base of the stems; pedicels $\frac{2}{3}$–1 in. long; sepals lanceolate, acute or acuminate; corolla 10–14 lin. in diam., dark chocolate; lobes 4–5 lin. long, very spreading, closely replicate to the base into vertical plates $1\frac{1}{2}$–2 lin. deep, glabrous and not ciliate; annulus 4–5 lin. in diam., apparently about 1 lin. high, nearly circular, glabrous; outer corona about 3 lin. in diam., obscurely pentagonal or nearly circular, reddish-brown; inner corona-lobes dull yellowish or with a reddish tinge. *G. Don, Gen. Syst.* iv. 122; *K. Schum. in Engl. und Prantl, Pflanzenfam.* iv. ii. 277; *Schlechter in Journ. Bot.* 1898, 476, *excl. all syns. not here quoted. D. replicata, Sweet, Hort. Brit. ed.* 1, 276; *G. Don, Gen. Syst.* iv. 122; *Schlechter in Journ. Bot.* 1898, 477. *Stapelia radiata, Sims, Bot. Mag. t.* 619; *Ait. Hort. Kew. ed.* 2, ii. 93; *Schultes, Syst. Veg.* vi. 45; *Lodd. Bot. Cab. t.* 831; *Decne in DC. Prodr.* viii. 663, *not of Jacquin, Willdenow, Link, Hornemann or Sprengel. S. replicata, Jacq. Stap. t.* 15; *Poir. Encycl. Meth. Suppl.* v. 232; *Willd. Enum. Pl. Hort. Berol.* 286; *Schultes, Syst. Veg.* vi. 45; *Link, Enum. Pl. Hort. Berol.* i. 258; *Spreng. Syst. Veg.* i. 837, *excl. syns.; Dietr. Syn. Pl.* ii. 884; *Decne in DC. Prodr.* viii. 662.

South Africa: without locality; introduced into cultivation in 1799 ex *Loddiges.*

I am quite unable to determine if this is really a distinct species or identical with a minutely ciliated form of *D. hirtella,* var. *obscura* noted under that plant. According to all authors, *D. radiata* is quite destitute of cilia on the corolla-lobes, and Jacquin (who figures it under the name of *Stapelia replicata*) represents the annulus also as glabrous. The only example I have seen that might possibly be this species is a specimen in Kew Herbarium named "*Duvalia radiata*" and dated 1813, but this differs in having a few minute cilia on the lobes and the annulus is puberulous on the rim. The cilia might easily be overlooked without the aid of a lens, so if this specimen really represents the plant figured by Sims (on which the species is founded), then the minutely ciliated form of *D. hirtella,* var. *obscura* also belongs to it. The stems of *D. replicata* as represented by Jacquin are obviously abnormally elongated, owing to some condition of cultivation. Concerning *Stapelia radiata,* Jacq., see a note under *D. elegans,* Masson.

11. **D. compacta** (Haw. Syn. Pl. Succ. 46, and Suppl. 14); stems $\frac{1}{2}$–1 (rarely 2) in. long, $\frac{1}{2}$ in. thick, globose or oblong, 4–5-angled, the angles formed of series of rounded apiculate tubercles, dull green; flowers 1–5, successively developed at about the middle of the young stems; pedicels $\frac{1}{2}-\frac{3}{4}$ in. long; sepals 1–$1\frac{1}{4}$ lin. long,

lanceolate-ovate, acute; corolla $\frac{2}{3}$–$\frac{3}{4}$ in. in diam., entirely dark chocolate; lobes very spreading, with slightly recurving tips, $\frac{1}{4}$–$\frac{1}{3}$ in. long, lanceolate, acute, replicate to their base into vertical plates about 1 lin. deep, glabrous, ciliate at the base only with a few very minute simple hairs, only visible under a lens; annulus small, $2\frac{1}{4}$–$2\frac{1}{2}$ lin. in diam. with a narrow rim around the corona, glabrous or very minutely puberulous; outer corona $1\frac{1}{2}$–$1\frac{3}{4}$ lin. in diam., obscurely pentagonal, less flat than in many species, brownish-red; inner corona-lobes dull orange-red. *G. Don, Gen. Syst.* iv. 122; *Schlechter in Journ. Bot.* 1898, 476. *D. mastodes, Sweet, Hort. Brit.* ed. 1, 276; *G. Don, l.c.* 122; *Schlechter, l.c.* 477. *Stapelia compacta, Schultes, Syst. Veg.* vi. 46; *Decne in DC. Prodr.* viii. 662. *S. mastodes, Jacq. Stap. t.* 13; *Spreng. Syst. Veg.* i. 837; *Dietr. Syn. Pl.* ii. 884; *Decne in DC. Prodr.* viii. 663. *S. mustodes, Link, Enum. Pl. Hort. Berol.* i. 258, *under S. punctata. S. mastodis, St. Lag. in Ann. Soc. Bot. Lyon,* vii. 135.

South Africa: without locality, cultivated specimens in *Herb. Haworth*! and *Herb. Kew*!

Described from living plants. According to Haworth this species was introduced into cultivation before 1800 and may still exist in English gardens, but I have not seen it during the past ten years. It differs from all other species I have seen by the broad and short tubercles on the stems being more closely placed and more abruptly apiculate. Its flowers closely resemble the minutely ciliated forms of *D. hirtella,* Sweet, but are much smaller.

12. **D. maculata** (N. E. Br.); stems tufted, decumbent, $\frac{1}{2}$–$1\frac{1}{4}$ in. long, $\frac{1}{3}$–$\frac{1}{2}$ in. thick, oblong, 4–5-angled, glabrous, dull green; angles rounded, with very spreading or slightly recurving very acute teeth 2–$2\frac{1}{2}$ lin. long, having a minute tooth on each side at the middle; flowers 4–8 or more together, successively developed at the middle or towards the base of the young shoots; pedicels $\frac{1}{2}$–$\frac{2}{3}$ in. long, rather slender, glabrous; sepals 1–$1\frac{1}{4}$ lin. long, deltoid-subulate, acute, glabrous; corolla 7–10 lin. in diam., with the sinuses reflexed to the pedicel; lobes $2\frac{1}{2}$–$3\frac{3}{4}$ lin. long, replicate to the base into vertical plates $1\frac{1}{4}$–$1\frac{1}{3}$ lin. deep at the base, very spreading, very acute, apparently olive-brown or purple-brown, perhaps indistinctly mottled(?), glabrous, with an exceedingly minute ciliation on the basal part; annulus $2\frac{1}{2}$–$3\frac{1}{4}$ lin. in diam., obscurely pentagonal, with its sides sloping underneath it (not vertical), and the rim rising considerably above the margin of the outer corona so as to form a sort of basin containing the corona, minutely puberulous, whitish, very distinctly spotted all over with purple-brown; outer corona $1\frac{1}{2}$ lin. in diam., obtusely pentagonal, apparently yellow; inner corona-lobes $\frac{1}{2}$ lin. long, $\frac{1}{3}$ lin. broad, ovoid, dorsally obtuse, apparently yellowish.

Central Region: Aberdeen Div.; near Aberdeen Road, *Pillans,* 31!

This is the only species I have seen in which the outer corona is sunk much below the level of the rim of the annulus. Described from specimens in fluid and others dried.

13. **D. parviflora** (N. E. Br.); stems about 1 in. long and $\frac{1}{2}$ in. thick, oblong, very obtusely and somewhat obscurely 5(–6 ?)-angled, light green, mottled with dull purple, faintly glaucous, angles rather obscurely indented above each minute tooth, or obscurely and distantly tessellate-tuberculate; flowers 4–5 together, successively developed near the apex of the stems; pedicels $\frac{1}{6}$–$\frac{1}{3}$ in. long; sepals $\frac{3}{4}$–1 lin. long, ovate-lanceolate, acuminate; corolla about 5–7 lin. in diam., "cream-colour, with the apical half of the lobes pale purple" (*Pillans*); lobes 2–2$\frac{1}{2}$ lin. long, very spreading, closely replicate to the base into vertical plates $\frac{3}{4}$ lin. deep, glabrous, ciliate with a few very minute simple cream-coloured hairs at the base only; annulus 1$\frac{1}{2}$–2 lin. in diam., raised about $\frac{1}{4}$ lin. above the level of the lobes, obtusely pentagonal, glabrous, the top quite covered (without leaving a margin) by the equally large outer corona; "outer and inner coronas straw-coloured, anthers purple" (*Pillans*).

CENTRAL REGION: Laingsburg Div. ; at Witte Poort, on the main road between Laingsburg and Ladismith, *Pillans*, 621 !

14. **D. angustiloba** (N. E. Br. in Gard. Chron. 1883, xx. 230); densely tufted; stems $\frac{1}{2}$–1 in. long, $\frac{1}{2}$–$\frac{3}{4}$ in. thick, subglobose or oblong, obtusely 4–5-angled, dull green; angles tuberculate-dentate; flowers 5–20 or more together in gradually developed cymes or fascicles, often on stout peduncles, at about the middle of the younger stems; pedicels 1–1$\frac{1}{4}$ in. long, rather slender, glabrous; sepals 1$\frac{3}{4}$ lin. long, reflexed, lanceolate-subulate; corolla very acutely conical and 5-grooved in bud, when expanded $\frac{3}{4}$–1 in. in diam., glabrous, with an exceedingly minute ciliation at the base of the lobes, only visible under a lens, dark chocolate; lobes 4–5 lin. long, horizontally spreading, very narrowly linear-lanceolate, closely replicate from base to apex into vertical acute plates about 1 lin. in depth at the base, with an impressed line along the fold; annulus .1$\frac{1}{2}$–1$\frac{3}{4}$ lin. in diam., not much raised above the level of the lobes, with a very narrow rim; outer corona reduced to a mere margin and together with the inner corona resembling a miniature crown, pure white at first, becoming dirty white in a few days, the incumbent tips of the inner corona-lobes very acute. *N. E. Br. in Hook. Ic. Pl. t.* 1925 ; *K. Schum. in Engl. und Prantl, Pflanzenfam. iv. ii.* 277 ; *Schlechter in Journ. Bot.* 1898, 477 ; *Wien. Illustr. Gartenz.* 1896, 215, *fig.* 33.

CENTRAL REGION ? From the Karoo, near or on the way to the Diamond Fields, *Dickson* (*Barkly*, 33) ! and *cultivated specimens* !

Described from living plants.

Imperfectly known species.

15. **D. concolor** (Schlechter in Journ. Bot. 1898, 477); stems exactly tetragonous, with large teeth, glabrous; corolla 5-fid, blackish-purple; annulus broad, solid, pubescent; lobes lanceolate,

replicate to the base into vertical plates, very spreading, glabrous. *Stapelia concolor, Salm-Dyck, Hort. Dyck.* 372 ; *Decne in DC. Prodr.* viii. 662.

The origin of this species is unknown, and so far as I am aware no specimen of it exists, as I have been informed that no specimens of the Stapelias described by Salm-Dyck were preserved nor drawings made of them.   Salm-Dyck states of it : "This very singular species occurs in gardens under the name of *Stapelia barbata,* by the form of its branches and the unopened flowers it would seem to belong to *Huernia,* but is altogether related to the section *Duvalia.*   It is near *Stapelia reclinata,* but differs by its thicker exactly tetragonous branches and flowers twice as large."   It may be a native of Namaqualand.

16. D. transvaalensis (Schlechter in Engl. Jahrb. xx. Beibl. 51, 54) ; stems erect, up to $2\frac{1}{3}$ in. long, about 5 lin. thick, 6-angled, glabrous, with spreading teeth at the angles ; flowers few together, from below the middle of the stems ; pedicels up to $\frac{3}{4}$ in. long, glabrous ; sepals $\frac{1}{4}$ in. long, linear-lanceolate, very acute, glabrous ; corolla-lobes erect, $\frac{1}{2}$ in. long, ovate-triangular, acuminate, glabrous on both sides ; outer corona-lobes connate into an entire fleshy ring ; inner corona lobes fleshy, rhomboid in outline, attenuate-beaked towards the apex, obtuse. *Schlechter in Journ. Bot.* 1898, 477 ; *N. E. Br. in Dyer, Fl. Trop. Afr.* iv. i. 503.

KALAHARI REGION : Transvaal ; sandy places near Klipdam, 4600 ft., *Schlechter,* 4498.

Also in Tropical Africa.   In the *Journal of Botany* above quoted, Dr. Schlechter refers *D. dentata,* N. E. Br., as a synonym of this, but without having compared the specimens.   His description, however, makes no mention of the very striking vibratile cilia nor the minutely puberulous annulus of that species.   Should they upon comparison prove to be identical, *D. transvaalensis* then becomes a synonym of *D. polita,* N. E. Br.

17. D. glomerata (Haw. Syn. Pl. Succ. 46, and Suppl. 14) ; stems prostrate, glomerate ; flowers about 4 together ; corolla rather large, blackish-red, paler at the base of the annulus ; lobes replicate, except at the apex, naked (probably glabrous or without cilia is intended) ; annulus very obtuse, subrepand, pubescent when viewed with a lens. *G. Don, Gen. Syst.* iv. 122 ; *Schlechter in Journ. Bot.* 1898, 476.   *Stapelia glomerata, Schultes, Syst. Veg.* vi. 46 ; *Decne in DC. Prodr.* viii. 664.

SOUTH AFRICA : without locality, introduced into cultivation before 1808, ex *Haworth.*

18. D. lævigata (Haw. Syn. Pl. Succ. 46, and Suppl. 14) ; corolla-lobes horizontal ; annulus very large. *G. Don, Gen. Syst.* iv. 122 ; *Schlechter in Journ. Bot.* 1898, 476.   *Stapelia lævigata, Schultes, Syst. Veg.* vi. 46 ; *Decne in DC. Prodr.* viii. 664.

SOUTH AFRICA : without locality, introduced into cultivation before 1808, ex *Haworth.*

Stated to be allied to *D. tuberculata,* Haw., and *D. radiata,* Haw., to which it is very similar, but has more horizontally spreading petals and a larger annulus.

19. D. tuberculata (Haw. Syn. Pl. Succ. 46, and Suppl. 13);
stems 1–2 in. long, oblong, 4-angled, with large recurving teeth;
corolla-lobes replicate into vertical plates, blackish-rufescent, ciliate-
pubescent; annulus somewhat hairy; outer corona pale reddish;
inner corona-lobes rosy-white. *G. Don, Gen. Syst.* iv. 121. *Stapelia
tuberculata, Schultes, Syst. Veg.* vi. 46; *Decne in DC. Prodr.*
viii. 663.

SOUTH AFRICA: without locality, introduced into cultivation in 1774, ex
Haworth.

Stated to be somewhat smaller than *D. radiata,* Haw. Possibly not distinct
from *D. hirtella,* Sweet.

ORDER LXXXIX. **LOGANIACEÆ.**

(By D. PRAIN and H. A. CUMMINS.)

Flowers hermaphrodite, regular or slightly oblique. *Calyx* inferior,
4-lobed; tube usually short. *Corolla* funnel-shaped or salver-shaped,
rarely campanulate or rotate; lobes usually 4, occasionally 5, rarely
(*Anthocleista*) 8–16, imbricate, contorted or valvate. *Stamens* as many
as and alternate with the corolla-lobes; filaments filiform or rarely
dilated, free or rarely (*Anthocleista*) connate below, often short;
anthers usually dorsifixed, cells 2, parallel, dehiscing introrsely,
rarely (*Nuxia*) divaricate and confluent at apex. *Disk* 0 or rarely
(*Anthocleista*) fleshy. *Ovary* superior, 2-celled; style simple; stigma
terminal, conspicuous or small, entire or obscurely 2-lobed; ovules
∞, anatropous or amphitropous. *Fruit* a capsule septicidally
2-valved, valves again often partially loculicidally dehiscent; or
indehiscent, baccate. *Seeds* obovoid, oblong or globose, sometimes
by pressure angular or flattened; albumen fleshy or cartilaginous,
usually abundant; embryo straight, generally shorter than the
albumen; cotyledons semiterete or rarely (*Strychnos*) foliaceous.

Shrubs or trees; leaves opposite, less often whorled, rarely fasciculate, usually
entire or subentire; occasionally distinctly toothed or lobed; stipules usually
reduced to an interpetiolar line, occasionally distinct; flowers cymose or rarely
(*Gomphostigma*) racemose.

DISTRIB. Genera about 30, only 6 in South Africa; species about 400, mainly
tropical. Endemic genera 1, endemic species 15.

    * *Fruit a septicidally 2-valved capsule; corolla-lobes 4, imbricate; filaments free;
anthers dorsifixed; disk 0; cotyledons semiterete.*

    I. **Gomphostigma.**—Small shrubs. *Flowers* racemose; racemes terminal.
Calyx deeply lobed. *Corolla-tube* entire. *Filaments* long; anthers
exserted; cells parallel, distinct.

II. **Nuxia.**—Shrubs or small trees. *Flowers* cymose ; cymes in terminal corymbose panicles. *Calyx* shortly lobed. *Corolla-tube* circumscissile, the base persisting. *Filaments* long ; anthers exserted ; cells divaricate, confluent.

III. **Chilianthus.**—Shrubs or small trees. *Flowers* cymose, usually very small ; cymes usually dense, in terminal panicles. *Calyx* deeply lobed. *Corolla-tube* entire. *Filaments* usually long ; anthers exserted ; cells parallel, distinct.

IV. **Buddleia.**—Shrubs or small trees. *Flowers* cymose ; cymes usually dense in axillary heads or corymbs, or in terminal panicles. *Calyx* shortly lobed. *Corolla-tube* entire. *Filaments* short or very short ; anthers included ; cells parallel, distinct.

** *Fruit indehiscent, baccate ; calyx deeply lobed ; corolla-tube entire ; anthers exserted ; cells parallel, distinct ; flowers cymose.*

V. **Anthocleista.**—Trees (or shrubs, sometimes climbing). *Cymes* in terminal panicles ; flowers large. *Corolla-lobes* 8–16, contorted. *Filaments* connate below ; anthers linear, erect. *Disk* fleshy. *Ovary* 2-celled or from accessory dissepiments, 4-celled. *Pericarp* leathery. *Cotyledons* semiterete.

VI. **Strychnos.**—Shrubs, often climbing, or trees. *Cymes* axillary, less often terminal. *Corolla-lobes* 5 or 4, valvate. *Filaments* free ; anthers ovate, dorsifixed. *Disk* 0. *Ovary* 2-celled or, from absorption of dissepiment, 1-celled. *Pericarp* crustaceous. *Cotyledons* leafy.

I. GOMPHOSTIGMA, Turcz.

Calyx deeply 4-fid ; tube campanulate ; lobes ovate or oblong, imbricate. *Corolla* deeply 4-fid, subrotate-campanulate ; tube entire ; lobes wide-elliptic, imbricate, obtuse, as long as tube. *Stamens* 4, adnate to corolla-throat ; filaments filiform ; anthers oblong, exserted ; cells distinct, parallel. *Ovary* 2-celled, oblong, glabrous ; style filiform ; stigma terminal, dilated, faintly 2-lobed ; ovules many, several-seriate ; placentas linear. *Capsule* oblong, longer than calyx ; dehiscence septicidal ; valves ultimately splitting at apex, separating from and liberating the placentas. *Seeds* obovoid or angular ; testa loose ; albumen somewhat scanty ; embryo much shorter than albumen, straight, with semiterete cotyledons.

Stellately lepidote or tomentose, divaricate or erect branched shrubs or undershrubs ; leaves 1-nerved, sessile, opposite decussate, or fascicled, linear, elliptic or oval, small or very small ; stipules represented by a distinct interpetiolar ridge ; flowers opposite, pedicelled, in terminal simple or branched racemes, with sometimes the lowest flowers axillary.

DISTRIB. Species 2 ; one endemic, the other extending to Mozambique and Lower Guinea.

Spreading, divaricately branched ; leaves elliptic or
    ovate, 2–3 times as long as broad ... ... (1) **incanum.**

Erect, virgately branched ; leaves linear, less often sub-
    spathulate, 6–12 times as long as broad ... ... (2) **scoparioides.**

1. **G. incanum** (Oliv. in Hook. Ic. Pl. t. 1472); a spreading
divaricately branching undershrub; leaves opposite and in crowded
axillary fascicles, elliptic or ovate, subobtuse, densely stellately
lepidote on both surfaces, margin entire, 1–2 lin. long, $\frac{1}{2}$–$\frac{3}{4}$ lin.
wide; racemes 1–3 in. long, rhachis glabrescent; bracts wide-ovate,
connate round rhachis; pedicels 1–2 lin. long, the lowest bracteolate, stellate-lepidote below and glabrescent above the bracteoles,
upper pedicels glabrescent; calyx 1 lin. long; lobes ovate; corolla
4 lin. long; filaments 1$\frac{1}{2}$ lin. long.

CENTRAL REGION : Colesberg Div. ; near the Orange River, *Knobel* !

2. **G. scoparioides** (Turcz. in Bull. Soc. Nat. Mosc. xvi. 53); an
erect virgately branching shrub; leaves opposite, very rarely
fascicled, linear, acute, rarely subspathulate, subobtuse, glabrous
or sparingly grey-lepidote, entire or rarely remotely obscurely
toothed, 1–1$\frac{1}{2}$ in. long, 1$\frac{1}{4}$–1$\frac{1}{2}$ lin. wide, gradually diminishing
upwards; racemes 2–8 in. long, rhachis glabrous; bracts leafy
below, gradually smaller upwards, free; pedicels 1$\frac{1}{2}$–2$\frac{1}{2}$ lin. long, bracteolate, glabrous; calyx 1–1$\frac{1}{4}$ lin. long; lobes oblong; corolla white,
3–3$\frac{1}{2}$ lin. long; filaments 1$\frac{1}{2}$ lin. long; capsule 3–4 lin. long. *Benth.
in DC. Prodr.* x. 434; *Baker in Dyer, Fl. Trop. Afr.* iv. i. 511.
Buddleia virgata, Linn. f. Suppl. 123; *Thunb. Fl. Cap. ed. Schult.*
148.

COAST REGION : Uitenhage Div. ; by the Sundays River and Coega River, *Drège*,
8264a ! Queenstown Div. ; beside watercourses at Queenstown, 3500 ft., *Galpin*,
1667 ! by the river near Shiloh, *Baur*, 913 ! Zwartkei River, *Baur*, 51 ! British
Caffraria, *Mrs. Hutton, Cooper*, 344 !

CENTRAL REGION : Prince Albert Div. ; by the Gamka River, *Drège*, 8264b.
Somerset Div. ; near Somerset East, *Bowker*, 152 ! 178 ! by the Little Fish River,
MacOwan, 1148 ! Graaff Reinet Div. ; by the Sundays River, 2500 ft., *Bolus*,
62 ! Murraysburg Div. ; banks of streams near Murraysburg, 4000 ft., *Tyson*, 30 !
Richmond Div., *Burke* ! Albert Div., *Cooper*, 1370 ! Hopetown Div. ; by the
Orange River, *Burchell*, 2657 ! Prieska Div. ; banks of the Orange River, *Burchell*,
1634 ! by the Brak River, *Burchell*, 2122 !

WESTERN REGION : Little Namaqualand ; Orange River, near Verleptpram,
Drège, 8264c.

KALAHARI REGION : Orange River Colony ; *Cooper*, 973 ! 2869 ! Basutoland ;
Cooper, 2870 ! Transvaal ; near Lydenberg, *Wilms*, 1079 ! near Middelburg,
Wilms, 1079a ! Magalies River, *Burtt Davy*, 166 ! Daspoort, near Pretoria,
Miss Leendertz, 587 !

EASTERN REGION : Transkei ; Kreilis country, *Bowker* ! Griqualand East ; banks
of the Umzimkulu River, near Clydesdale, 2500 ft., *Tyson*, 2713 ! and in *MacOwan
Herb. Austr.-Afr.* 1224 ! Natal ; Mohlamba Range, 5000–6000 ft., *Sutherland* !
Biggers Berg, 4000 ft., *Sutherland* ! banks of the Umnweni River, 4000 ft., *Wood*,
3591 ! Drakensberg, near Newcastle, *Wilms*, 2174 ! and without precise locality,
Gerrard, 193 !

II. NUXIA, Lam.

Calyx shortly 4-fid; tube campanulate; lobes subtriangular, subvalvate. *Corolla* 4-fid; tube cylindric, included, circumscissile above
the persistent base; throat partially occluded by a ring of hairs;

lobes imbricate, spreading, obtuse or acute. *Stamens* 4, adnate to corolla-throat ; filaments filiform ; anthers ovate or oblong, exserted ; cells confluent at apex, divaricate. *Ovary* 2-celled, ovoid, usually silky ; style filiform ; stigma terminal, minute ; ovules many, several-seriate ; placentas linear or ovate. *Capsule* ovoid, obovoid or oblong, hardly exserted from calyx, dehiscence septicidal ; valves splitting at apex, separating from and liberating the placentas. *Seeds* oblong ; albumen fleshy ; embryo straight with semiterete cotyledons.

Glabrous or pilose erect shrubs or small trees ; leaves penninerved, shortly or distinctly petioled, opposite, decussate or 3-nately whorled, subcoriaceous or chartaceous ; stipules 0, interpetiolar ridge obsolete ; flowers cymose, cymes in dense or lax terminal thyrsoid panicles.

DISTRIB. Species about 30, mostly Tropical African, some in Madagascar, a few in South Africa, 4 of these being endemic.

Calyx chartaceous ; tube glabrous within ; corolla in bud
    obtuse at apex ; cymes in wide lax panicles 4–8 in.
    across ; leaves 3-nate, long-petioled ...   .   (1) **floribunda.**

Calyx coriaceous ; tube adpressed white silky within ;
    cymes in narrow usually dense panicles under 4 in.
    across ; leaves short-petioled :
Corolla in bud rounded and obtuse at apex ; lobes
    glabrous outside ; leaves opposite :
Pedicels considerably longer than calyx   ...   ..   (2) **gracilis.**

Pedicels never longer, mostly shorter than calyx ... (3) **dentata.**
Corolla in bud conical and acute or, if rounded, apicu-
    late at apex ; leaves 3-nate :
Flowers all distinctly pedicelled ; corolla-lobes silky
    outside ; leaves usually more or less pubescent :
Corolla under 2 lin. long, in bud rounded but
    apiculate at tip ; cymes dense   ...   ...   (4) **breviflora.**

Corolla over 2 lin. long, in bud conical and acute
    at tip ; cymes lax   ...   .   (5) **pubescens.**

Flowers all sessile or subsessile ; corona over 2 lin.
    long, in bud conical and acute at tip ; cymes
    dense :
Leaves more or less pubescent ; corolla-lobes
    almost always silky outside ...   .   (6) **congesta,** var.
        tomentosa.

Leaves glabrous ; corolla-lobes glabrous or very
    sparingly hairy outside   ...   .   (6) **congesta,** var.
        emarginata.

1. N. floribunda (Benth. in Hook. Comp. Bot. Mag. ii. 59) ; an erect tree 15–50 ft. high ; bark pale-grey ; leaves usually in 3-nate whorls, firmly papery, glabrous, oblong-elliptic, acute or acuminate, less often obtuse, base cuneate and entire, margin elsewhere remotely usually obscurely toothed, 2–6 in. long, $\frac{3}{4}$–$2\frac{1}{2}$ in. wide ; petiole $\frac{1}{2}$–$1\frac{1}{2}$ in. long ; cymes in ample panicles repeatedly dichotomously branched, 4–8 in. across ; bracts minute ; flowers sessile or the central of a cyme short-pedicelled ; calyx narrow-campanulate, 1 lin. long, thinly papery ; tube outside glabrous, within sparsely

beset with short silky hairs or quite glabrous ; corolla white, 1½ lin.
long, glabrous outside, in bud obtuse ; lobes narrow ligulate, obtuse,
longer than tube ; ovary glabrous. *DC. Prodr.* x. 435 ; *Pappe,
Silva Cap. ed.* 2, 30 ; *Fourcade, Rep. Natal For.* 1889, 115 ; *Sim,
For. Fl. Cap.* 274, *t.* 157 ; *Wood and Evans, Natal Plants,* i. 48,
t. 59.　*Chilianthus triphyllus, E. Meyer, Zwei Pfl. Docum.* 172.

SOUTH AFRICA : without locality, *Mund and Maire*!
COAST REGION : George Div. ; near George, *Prior*! Knysna Div. ; near the
Knysna River ford, *Burchell,* 5537! in the forest at Knysna, *Burchell,* 5448!
Bolus, 2145! Kaatjes Kraal, near Yzer Nek, *Burchell,* 5240! between Keurbooms
River and Bitou River, *Burchell,* 5295! Humansdorp Div. ; between Kromme
River and Gamtoos River, *Drège*! Storms River Forest, 500 ft., *Galpin,* 4333!
Uitenhage Div. ; between Van Standens River and Galgebosch, *Burchell,* 4763!
near Uitenhage, near streams, *Burchell,* 4241! banks of the Zwartkops River and
ravines of Van Standens Berg, *Zeyher*! Uitenhage, *Ecklon*! Port Elizabeth Div. ;
Krakakama, *Zeyher*!
EASTERN REGION : Natal; hills near Illovo River, *Gerrard,* 1893! Inanda,
Wood, 980! Western Zululand, *Baker in Herb. Evans,* 528.

2. N. gracilis (Engl. Jahrb. x. 243) ; a shrub 8–10 ft. high ; bark
brownish-grey ; leaves usually opposite, decussate, firmly papery,
glandular, oblong-lanceolate, subacute, base narrow-cuneate from the
middle and entire, margin elsewhere finely serrately toothed, 1–2 in.
long, 1½–3½ lin. wide ; petiole 1 lin. long ; cymes many, umbelliform in corymbose panicles ¾–2 in. across ; bracts narrow-spathulate or linear, 1½–2 lin. long, glandular-puberulous ; pedicels all
subequal, 2–2¼ lin. long, slender, glandular-puberulous ; calyx
campanulate, 1½ lin. long, coriaceous ; tube outside glandular-
puberulous, within densely adpressed white-silky ; corolla white,
2 lin. long, glabrous outside, in bud obtuse ; lobes ovate, obtuse,
rather shorter than tube ; ovary glabrous (or puberulous) ; capsule
1½ lin. long, sparsely finely pubescent with short hairs.

KALAHARI REGION : Griqualand West; Lower Campbell, *Burchell,* 1817!
Upper Campbell, *Burchell,* 1827! in stony places at 4000 ft., *Marloth*!

In Burchell's two gatherings the ovary is glabrous; his plants agree as
regards leaves, inflorescence, bracts, pedicels and calyx with that collected by
Marloth, which is represented at Kew by a specimen in fruit, where the capsules
are finely pubescent.

3. N. dentata (R. Br. in Salt, Abyss. Append. 63) ; a shrub
7–10 ft. high ; bark grey ; leaves usually opposite, decussate, papery,
glabrous, lanceolate or narrow-oblong, obtuse or subacute, base
narrow-cuneate from the middle and entire, margin elsewhere remotely
obscurely toothed, 1–3 in. long, usually 2–5 lin. wide, occasionally
wider ; petiole 1–4 lin. long ; cymes numerous in thyrsoid panicles
½–1 in. across ; bracts narrow-lanceolate, 1 lin. long ; terminal
pedicels 2 lin., lateral 1 lin. long ; calyx narrow-campanulate, 2 lin.
long, subcoriaceous ; tube outside glabrous, within densely adpressed
white-silky ; corolla white, 2½ lin. long, glabrous outside, in bud
obtuse ; lobes ovate-oblong, obtuse, rather shorter than tube ; ovary

silky. *DC. Prodr.* x. 435 ; *A. Rich. Tent. Fl. Abyss.* ii. 124 ; *Engl. Hochgebirgsfl. Trop. Afr.* 335 ; *Hiern in Cat. Afr. Pl. Welw.* i. 700 ; *Baker in Dyer, Fl. Trop. Afr.* iv. i. 513 ; *S. Moore in Journ. Bot.* 1903, 403 (*var. transvaalensis*). *N. oppositifolia, Benth. in DC. Prodr.* x. 435. *N. Schlechteri, Gilg in Engl. Jahrb.* xxxii. 140. *Lachnopylis oppositifolia, Hochst. in Flora,* 1843, 77 ; *A.DC. in DC. Prodr.* ix. 23.

KALAHARI REGION : Transvaal ; Komati Poort, 330 ft., *Schlechter,* 11738 ! Johannesburg, *Rand,* 1132 ! by the Mbetane River, near Leydsdorp, and Masetana River, near Shilovane, *Junod,* 1430 !

EASTERN REGION : Natal ; near the Nonoti River, *Wood,* 3578 ! and without precise locality, *Gerrard,* 716 !

Also in Tropical Africa.

4. N. breviflora (S. Moore in Journ. Bot. 1903, 403) ; a shrub ; bark grey, lacerate ; leaves usually in 3-nate whorls, coriaceous, densely stellate-pubescent on both surfaces, obovate or obovate-oblong, obtuse or retuse, base rounded entire, margin elsewhere entire or remotely serrate, 6–9 lin. long, 3–6 lin. wide ; petiole short ; cymes in rather dense corymbs $1\frac{1}{4}$ in. across ; bracts narrow-linear, $1\frac{1}{2}$ lin. long ; flowers pedicelled ; pedicels up to 1 lin. long ; calyx campanulate, 1 lin. long ; tube outside tomentose, within densely adpressed white silky ; corolla $1\frac{1}{2}$ lin. long, in bud rounded but shortly apiculate at the tip ; lobes hardly longer than tube, oblong-ovate, obtuse, silky outside, ovary silky.

KALAHARI REGION : Transvaal ; Johannesburg, northern escarpment of Witwaters Rand series, *Rand,* 712 !

5. N. pubescens (Sond. in Linnæa, xxiii. 84) ; a shrub or small tree, 5–20 ft. high ; bark grey, lacerate ; leaves usually in 3-nate whorls, subcoriaceous, glandular and sometimes densely, more often sparingly, pubescent on both surfaces, occasionally nearly, rarely quite glabrous, elliptic-oblong or ovate, acute or obtuse, base cuneate entire, margins elsewhere entire or remotely serrate, 1–$1\frac{1}{2}$ (rarely 2) in. long, $\frac{1}{2}$–$\frac{3}{4}$ (rarely 1) in. wide ; petiole 1–$1\frac{1}{2}$ lin. long ; cymes in rather lax corymbose panicles 2–4 in. across ; bracts spathulate or oblanceolate, 1 lin. long ; flowers all pedicelled ; pedicels slender, 1–$1\frac{1}{2}$ lin. long ; calyx narrow-campanulate, 2 lin. long, coriaceous ; tube outside glandular-pubescent or puberulous, within densely adpressed white silky ; corolla white, $2\frac{1}{2}$ lin. long, in bud conical acute ; lobes hardly as long as tube, oblong, acute, apex cucullate usually silky outside, inner face caruncled ; ovary silky in the upper half ; capsule 2 lin. long, densely white-silky. *N. congesta, var. brevifolia, Sond. in Linnæa,* xxiii. 83.

KALAHARI REGION : Transvaal ; Magaliesberg, *Burke,* 259 ! *Zeyher,* 1326 ! hills near Pretoria, 4100 ft., *McLea in Herb. Bolus,* 3102 ; *Kirk,* 30 ! Kudus Poort, *Rehmann,* 4665 ! *Miss Leendertz,* 626 ! summit of Saddleback Mountain, near Barberton, *Galpin,* 946 !

Almost as variable as regards tomentum as *N. congesta,* the species next described, to which it is very closely related but from which it is readily distinguished by its pedicelled flowers.

6. N. congesta (R. Br. in Salt, Abyss. Append. 63) ; a shrub or small tree, 5–20 ft. high ; bark grey, lacerate ; leaves usually in 3-nate whorls, casually opposite, subcoriaceous, glabrous on both surfaces, oblanceolate-oblong, subacute or obtuse rarely retuse, base cuneate entire, margin elsewhere entire or serrate, 2–3 (sometimes 4) in. long, 1–2 (rarely only ¾) in. wide ; petiole 1–2½ lin. long ; cymes in condensed thyrsoid panicles 1–3 in. across ; bracts narrowly spathulate or oblanceolate or linear, usually puberulous ; flowers sessile or the central of a cyme very shortly pedicelled ; calyx rather narrowly campanulate, 2 lin. long, coriaceous ; tube outside glandular, puberulous or pubescent, within densely adpressed white silky ; corolla white, 2¾ lin. long, in bud conical acute ; lobes hardly as long as the tube, oblong, acute, striate, outside uniformly silky, apex cucullate, inner face with a central thickened line ending above in a small caruncle ; ovary silky in the upper half ; capsule 2 lin. long, densely white silky. *Fresen. in Flora*, 1838, 606 ; *DC. Prodr.* x. 435 ; *A. Rich. Tent. Fl. Abyss.* ii. 123 ; *Engl. Hochgebirgsfl. Trop. Afr.* 335 ; *Baker in Dyer, Fl. Trop. Afr.* iv. i. 512. *Lachnopylis ternifolia, Hochst. in Flora*, 1843, 77 ; *A.DC. in DC. Prodr.* ix. 23.

VAR. β, **tomentosa** (Cummins) ; leaves oblanceolate-oblong or ovate-oblong or ovate, subacute, rarely obtuse, densely to sparingly pubescent beneath, mostly with stellate hairs, above sparingly stellate-pubescent or glabrous except on the main-nerves ; corolla-lobes outside uniformly silky or very rarely glabrous. *N. tomentosa, Sond. in Linnæa*, xxiii. 84.

VAR. γ, **emarginata** (Prain) ; leaves obovate-oblong or oblong, obtuse or retuse or emarginate, glabrous on both surfaces ; corolla-lobes outside quite glabrous or occasionally with a few white silky hairs. *N. emarginata, Sond. in Linnæa*, xxiii. 83. *N. congesta, Fourcade, Rep. Natal For.* 1889, 115 ; *Sim, For. Fl. Cap.* 275, *t.* 113, *fig.* I ; *not of R. Br.*

COAST REGION : Var. γ : Bathurst Div. ; near the Kowie River, below 500 ft., *Ecklon and Zeyher* ; near Port Alfred, *Burchell*, 4026 ! woods between the mouths of the Great Fish and Riet River, *MacOwan*, 297 !

CENTRAL REGION : Var. γ : Somerset Div.; *Bowker* !

KALAHARI REGION : Var. β : Transvaal ; Magaliesberg, *Zeyher* ! Woodbush, *Eastwood*, 5 ! mountains under Saddleback, near Barberton, 4000 ft., *Thorncroft in Herb. Wood*, 4161 ! Drakensberg, near Macamac, *McLea in Herb. Bolus*, 3014 !

EASTERN REGION : Var. β : Natal ; Esmont, 2300 ft., *Wood* ! Var. γ : Natal ; Ingoma, *Gerrard*, 1510 ! Inanda, *Wood*, 576 ! Bevaan Falls, *Wood*, 3190 !

There are no examples from South Africa which admit of being unquestionably referred to *Nuxia congesta*, R. Br., a species that in its typical form extends from the Shire Highlands to Abyssinia. The two varieties which occur in South Africa are readily distinguishable from each other and from the type, with which var. *tomentosa* agrees as to flowers and var. *emarginata* agrees as to leaves.

III. CHILIANTHUS, Burch.

Calyx deeply 4-fid ; tube very short ; lobes ovate or triangular, imbricate. *Corolla* 4-fid ; tube short, narrow-campanulate, entire ; lobes imbricate, spreading, oblong or ovate, obtuse. *Stamens* 4, adnate to corolla-throat ; filaments filiform ; anthers ovate, exserted ;

cells distinct, parallel. *Ovary* 2-celled, ovate, tomentose; style usually flexuous; stigma terminal, wide capitate or dilated; ovules many, few-seriate; placentas thickened. *Capsule* ovoid or oblong, exserted; dehiscence septicidal; valves splitting at apex, separating from and liberating the placentas. *Seeds* few, oblong, compressed and sometimes winged; albumen thin, fleshy; embryo straight.

Stellately tomentose or scurfy lepidote shrubs or small trees; leaves penninerved, petioled, opposite, decussate, subcoriaceous, chartaceous or herbaceous, entire or toothed; stipules represented by an interpetiolar ridge; flowers cymose, sweet scented, small or very small, usually very many; cymes in dense or somewhat lax terminal panicles.

DISTRIB. Species 4, all endemic.

Leaves lanceolate or oblong-lanceolate, narrow cuneate to
    base :
  Leaves lepidote-scurfy beneath; margins entire or
    remotely shallow-toothed; twigs more or less
    4-angled; filaments long; anthers far exserted ... (1) **arboreus.**

  Leaves densely rusty-pubescent beneath; margins
    finely closely crenate; twigs cylindric; filaments
    short; anthers little exserted   ...   . (2) **corrugatus.**

Leaves ovate, truncate or cordate, rarely wide cuneate at
    base, rusty or grey pubescent beneath; twigs
    cylindric; filaments long; anthers considerably
    exserted :
  Leaf-margins sinuately lobed, lobes again crenate   ... (3) **lobulatus.**

  Leaf-margins coarsely and irregularly simply toothed (4) **dysophyllus.**

1. **C. arboreus** (A.DC. in DC. Prodr. x. 435); an erect evergreen shrub, 6–10 ft. high; bark brownish; twigs 4-angled or occasionally crisply 4-winged; leaves coriaceous, when young sparingly scurfy but soon glabrous, rugulose above, beneath densely persistently rufous- or at times grey-scurfy, lanceolate or oblong-lanceolate, acuminate or acute, base cuneate from below middle, margin entire or remotely very shallowly toothed, 3–4 in. long, $2\frac{1}{2}$–6 lin. wide; petiole scurfy, $2\frac{1}{2}$–3 lin. long; cymes numerous in usually lax panicles 3–8 in. across; bracts $\frac{1}{2}$ lin. long, lanceolate, scurfy; pedicels $\frac{1}{4}$ lin. or less, scurfy; calyx campanulate, $\frac{1}{2}$–$\frac{3}{4}$ lin. long, scurfy; lobes triangular, as long as tube; corolla yellowish-white, $\frac{3}{4}$–$1\frac{1}{4}$ lin. long, outside puberulous; lobes longer than tube; filaments $\frac{3}{4}$–1 lin. long; anthers exserted beyond tips of corolla-lobes; capsule $\frac{3}{4}$ lin. long, scurfy. *Pappe, Silva Cap. ed.* 2, 30; *Fourcade, Rep. Natal For.* 1889, 115. *Chilianthus oleaceus, Burch. Trav.* i. 94; *Sim, For. Fl. Cap.* 276, *t.* 113, *fig.* II. *Scoparia arborea, Linn. f. Suppl.* 125; *Willd. Sp. Pl.* i. 653; *Thunb. Fl. Cap. ed. Schult.* 147. *Callicarpa paniculata, Lam. Encycl.* i. 563. *Buddlea salicifolia, Jacq. Hort. Schoenbr.* i. 12, *t.* 29, *not of Vahl. Buddleja saligna, Willd. Enum. Hort. Berol.* i. 159. *Nuxia saligna, Benth. in Hook. Comp. Bot. Mag.* ii. 59.

SOUTH AFRICA : without locality, *Forster* !

3 x 2

COAST REGION: Caledon Div. ; Boontjes Kraal, near Caledon, *Burchell,* 934 ! Swellendam Div. ; near Swellendam, *Pappe* ! Riversdale Div. ; by the Zoetemelks River, *Burchell,* 6805 ! between Gauritz River and Great Valsch River, *Burchell,* 6525 ! Mossel Bay Div. ; dry hills on the eastern side of Gauritz River, *Burchell,* 6434 ! George Div. ; Karoo between Gauritz River and Lange Kloof, *Zeyher* ! Knysna Div. ; near Knysna, *Burchell,* 5499 ! Uitenhage Div. ; Van Stadens River, *Drège,* 7889c ! banks of the Zwartkops River, *Zeyher,* 622 ! near Uitenhage, *Prior* ! Albany Div. ; Fish River, *Burke* ! between the source of the Kasuga River and Assegai Bush, *Burchell,* 4170 ! near Grahamstown, 2000 ft. *Atherstone* ! *Prior* ! Fort Beaufort Div. ; mountains by the Koonap River, *Baur,* 1103 ! British Kaffraria, *Cooper,* 315 !

CENTRAL REGION : Calvinia Div. ; between Grasberg River and Watervals River, *Drège,* 7889b ! Somerset Div. ; on Bosch Berg, *Burchell,* 3150 ! 3230 !

WESTERN REGION : Vanrhynsdorp Div. ; Ebenezer, *Drège,* 7889a !

KALAHARI REGION : Orange River Colony ; Olifantsfontein, *Rehmann,* 3507 ! Bechuanaland ; near the sources of Kuruman River, *Burchell,* 2494 ! Transvaal ; Heidelberg, *Miss Leendertz,* 1046 ! Houtbosch, *Rehmann,* 6009 !

EASTERN REGION : Griqualand East ; near Clydesdale, *Tyson,* 2055 ! and in *MacOwan and Bolus, Herb. Austr.-Afr.,* 1288 ! Natal ; Phoenix, *Wood,* 1108 ! Attercliffe, *Sanderson,* 360 !

2. C. corrugatus (A.DC. in DC. Prodr. x. 436) ; an erect shrub ; bark brown, fissured ; twigs cylindric, when young rusty-tomentose ; leaves firmly coriaceous, above glabrous shining closely and finely rugose, beneath densely rusty-tomentose, lanceolate, acute, base narrow cuneate, margin subrevolute, fine-crenate throughout, 2–3 in. long, $3\frac{1}{2}$–4 lin. wide ; petiole $\frac{1}{2}$ in. long, densely rusty-tomentose ; cymes numerous, few-flowered, globose, in pyramidal panicles $1\frac{1}{2}$ in. across ; bracts $\frac{1}{2}$ lin. long, lanceolate, densely tomentose ; pedicels $\frac{1}{4}$ lin. long, densely rusty ; calyx campanulate, 1 lin. long, densely rusty-tomentose ; lobes half as long as tube ; corolla white, 2 lin. long, outside pubescent, sparingly hirsute within ; lobes considerably shorter than tube ; filaments $\frac{1}{4}$ lin. long ; anthers very little exserted ; ovary and style stellate-tomentose. *Sim, For. Fl. Cap.* 276. *Nuxia corrugata, Benth. in Hook. Comp. Bot. Mag.* ii. 60.

CENTRAL REGION : Aliwal North Div. ; Witteberg Range, 5000–6000 ft., near streams, *Drège,* 3618 !

KALAHARI REGION : Orange River Colony ; near Harrismith, *Sankey,* 243 !

EASTERN REGION : Natal ; Greenwich Farm at Riet Vlei, *Fry in Herb. Galpin,* 2734 !

3. C. lobulatus (A.DC. in DC. Prodr. x. 436) ; a spreading shrub ; bark brown ; twigs cylindric, rusty-furfuraceous ; leaves subcoriaceous, above sparingly furfuraceous, rugulose, beneath densely rusty-furfuraceous, ovate or ovate-oblong, obtuse, base truncate or cordate, margin sinuately lobed, lobes 7–10 on each side again crenately toothed and sometimes undulate, $\frac{3}{4}$–3 in. long, $\frac{1}{3}$–$1\frac{1}{2}$ in. wide ; petiole 2–6 lin. long, rusty-furfuraceous ; cymes numerous, globose, in pyramidal panicles $1\frac{1}{2}$–3 in. long, $\frac{1}{2}$–$1\frac{1}{2}$ in. across ; bracts $\frac{1}{2}$ lin. long, lanceolate, rusty-furfuraceous ; pedicels obsolete ; calyx campanulate, $\frac{3}{4}$ lin. long, rusty-furfuraceous ; lobes half as long as tube ; corolla white, 1–$1\frac{1}{2}$ lin. long, outside

puberulous ; lobes oblong, obtuse, hardly as long as tube ; filaments
$\frac{1}{2}$ lin. long ; anthers exserted as far as tips of corolla-lobes ; capsule
$\frac{3}{4}$ lin. long, puberulous. *Sim, For. Fl. Cap.* 276. *Nuxia lobulata,
Benth. in Hook. Comp. Bot. Mag.* ii. 60.

SOUTH AFRICA : without locality, *Thom,* 391 ! 495 !

COAST REGION : Queenstown Div. ; on a height by the Klipplaats River, between
Shiloh and Table Mountain, *Drège,* 664b ! Mountain-top, Bowkers Park, Queenstown, 4750 ft., *Galpin,* 2558 !

CENTRAL REGION : Somerset Div. ; by the Little Fish River, *Burchell,* 3265 ! on
Bruintjes Hoogte, 4000 ft., *MacOwan,* 772 ! Craddock Div. ; Tarka River,
Zeyher ! near Cradock, *Cooper,* 502 ! Graaff Reinet Div. ; descent of the Voor
Sneeuw Berg, *Burchell,* 2847 ! near Graaff Reinet, 4200 ft., *Bolus,* 89 ! Beaufort
West Div. ; Nieuwveld Mountains, near Beaufort West, 3000-5000 ft., *Drège,*
664a ! Beaufort West, *Burke,* 524 ! Middelburg Div. ; between Wolve Kop and
Rhenoster Berg, *Burchell,* 2791 !

4. **C. dysophyllus** (A.DC. in DC. Prodr. x. 436) ; a diffuse or
climbing shrub, 5–12 ft. high ; bark brownish-grey ; twigs cylindric,
grey- or (less often) rusty-furfuraceous, sometimes glabrescent ;
leaves herbaceous, subrugulose above, grey or rusty woolly-pubescent
on both sides but more closely and persistently beneath, ovate or
ovate-lanceolate, acute, base truncate or very shortly and sometimes
unequally cuneate, very rarely subcordate and entire, margin elsewhere coarsely often irregularly toothed, 2–5 in. long, 1–2$\frac{1}{2}$ in. wide ;
petiole pubescent or glabrous, sometimes winged by the lamina
above, $\frac{1}{2}$–1$\frac{1}{4}$ in. long ; cymes numerous, open, in lax panicles 3–6 in.
across ; bracts $\frac{1}{2}$ lin. long, lanceolate, rusty- or grey-pubescent ;
pedicels $\frac{1}{4}$ lin. long, pubescent ; calyx campanulate, $\frac{3}{4}$ lin. long,
rusty- or grey-tomentose ; lobes longer than tube ; corolla yellowish-
white, 1$\frac{1}{2}$–1$\frac{3}{4}$ lin. long, outside puberulous ; lobes considerably
shorter than tube ; filament 1 lin. long ; anthers exserted beyond
tips of corolla-lobes ; capsule $\frac{3}{4}$ lin. long, puberulous. *Sim, For. Fl.
Cap.* 276. *Nuxia dysophylla, Benth. in Hook. Comp. Bot. Mag.*
ii. 60 ; *Sond. in Linnæa,* xxiii. 85 (*var. rufescens*).

COAST REGION : Uitenhage Div. and Neutral Territory, *Ecklon* ! Albany Div. ;
Blue Krantz, *Burchell,* 3647 ! near Grahamstown, 2000-2200 ft., *Atherston* !
MacOwan, 292 ! *Bowie,* 4 ! and *Glass in MacOwan, Herb. Austr.-Afr.,* 1638 ! Fort
Beaufort Div. ; Winterberg Range, *Zeyher* ! British Kaffraria ; *Cooper,* 34 !

KALAHARI REGION : Transvaal ; Rimers Creek, near Barberton, 3200-4000 ft.,
Galpin, 971 !

EASTERN REGION : Natal ; near Durban, *Gucimius* ! Umgeni River, *Gerrard,*
1023 ! Inanda, *Wood,* 705 ! Umhlote Valley, *Wood,* 608 ! sandy flat near Umbogo-
twine River, *Wood,* 3200 !

IV. BUDDLEIA, Linn.

Calyx snortly 4-fid ; tube campanulate ; lobes ovate or subtriangular, imbricate. *Corolla* 4-fid ; tube (in the South African
species) long, cylindric, salver-shaped ; lobes imbricate, ovate.
Stamens 4, adnate within corolla-tube ; filaments short or very short ;
anthers ovate or narrow-oblong, cordate or 2-lobed at base, included ;

cells distinct, parallel. *Ovary* 2-celled, ovate, pubescent; style usually flexuous; stigma (in the South African species) narrow-clavate laterally decurrent; ovules many, several-seriate; placentas thickened. *Capsule* ovoid or oblong or conical, exserted; dehiscence septicidal; valves entire or splitting at apex, separating from and liberating the placentas. *Seeds* many, oblong, compressed fusiform or discoid; testa lax and reticulate or close and sometimes winged; albumen fleshy; embryo straight.

Stellately pubescent or tomentose shrubs or small trees; leaves penninerved, petioled, opposite, decussate, subcoriaceous or herbaceous, entire or crenulate or toothed; stipules leafy or represented by an interpetiolar ridge; flowers cymose, often sweet-scented; cymes usually dense, in corymbose thyrsoid or strict terminal panicles.

DISTRIB. Species nearly 100, many in tropical and subtropical Asia and America; a few in tropical and South Africa, two of the South African species being endemic.

Stipules usually conspicuous, leafy, revolute-convex; calyx-teeth subtriangular, acute, half as long as tube; filaments longer than anthers :
  Leaves lanceolate, five times as long as broad, cordate or subhastate at base, finely crenulate, thinly pubescent and very finely rugose above; twigs somewhat 4-angled; corolla-throat hirsute with spreading hairs above the level of the stamens ... (1) **salvifolia.**

  Leaves oblong-lanceolate, ovate-lanceolate or ovate, two to three times as long as broad, cuneate at base, serrate or entire, glabrous above; twigs cylindric; corolla-throat not hirsute :
    Leaves thinly tomentose beneath, nerves and veins little impressed above; flowers yellowish-white (2) **auriculata.**

    Leaves densely pubescent beneath, nerves and veins deeply impressed above; flowers orange or salmon ... ... (2) **auriculata,** var. **euryifolia.**

Stipules represented by an interpetiolar ridge; calyx-teeth ovate, subobtuse, one-fourth the length of the tube; leaves smooth above, margin lobed or undulate or entire, neither serrate nor crenate : twigs cylindric; corolla-throat not hirsute; filaments shorter than anthers ... ... (3) **pulchella.**

1. **B. salvifolia** (Lam. Encycl. i. 513); an erect shrub, 5–10 ft. high, with spreading branches : twigs faintly 4-angled, rusty-tomentose : leaves firmly herbaceous, dark green, above thinly pubescent and finely rugose, beneath densely rusty- or at times hoary-tomentose, veins and nerves prominent, lanceolate, acuminate, base cordate or subhastate, margin very finely crenulate, $1\frac{1}{2}$–6 in. long, $\frac{1}{2}$–$1\frac{1}{4}$ in. wide; petiole $1\frac{1}{2}$–3 lin. long, densely tomentose : stipules usually leafy, revolute-convex, rounded, largest $2\frac{1}{2}$ lin. long, sometimes small, rarely reduced to an interpetiolar ridge; cymes numerous, few-flowered, capitate, in ovate-pyramidal panicles 6 in. long, 3–4 in. across; cyme-peduncles stout, densely tomentose, 3–4 lin. long; bracts 1–2 lin. long, linear or lanceolate; pedicels

obsolete ; calyx campanulate, 1 lin. long, densely pubescent ; teeth subtriangular, acute, half as long as tube ; corolla yellowish-white or buff with throat deep orange ; limb sometimes slightly tinged with lilac, 3–4 lin. long ; tube outside densely tomentose, within hirsute with simple hairs from the throat downwards ; free filaments rather longer than anthers ; anthers narrow-oblong ; capsule 1½ lin. long, densely tomentose. *Jacq. Hort. Schoenbr. i. 12, t. 28 ; Baker in Dyer, Fl. Trop. Afr. iv. i. 516. B. salviæfolia, Pappe, Silva Cap. ed. 2, 31 ; DC. Prodr. x. 444 ; Sim, For. Fl. Cap. 277, t. 114. Buddlea salviæfolia, Fourcade, Rep. Natal For. 1889, 115. Lantana salvifolia, Linn. Syst. ed. x. 1116.*

COAST REGION : Clanwilliam Div. ; Wupperthal, *Drège,* 699c ! Bull Hoek, 500 ft., *Schlechter* 8373 ! Caledon Div. ; Zwartberg, *Zeyher,* 3526 ! near Caledon, *Prior* ! George Div. ; about the source of the Keurbooms River in Long Kloof, *Burchell,* 5077 ! near George, 1000 ft., *Prior* ! *Schlechter* ! 2418 ! between Gauritz River and Long Kloof, *Ecklon* ! *Zeyher,* 1328 ! Knysna Div. ; Kaatjes Kraal, near Yzer Nek, *Burchell,* 5202 ! between Keurbooms River and Bitou River, *Burchell,* 5284 ! Uitenhage Div. ; between Galgebosch and Melk River, *Burchell,* 4766 ! banks of the Zwartkops River, *Zeyher,* 461 ! *Prior* ! Bedford Div. ; near Bedford, *Hutton* ! Queenstown Div. ; near Queenstown, *Cooper,* 193 ! Shiloh, *Baur,* 244 ! *Galpin,* 1623 ! Stockenstrom Div. ; Kat Berg, *Hutton* !

CENTRAL REGION : Somerset Div. ; on Bosch Berg, *Burchell,* 3226 ! *MacOwan,* 927 ! *Mrs. Barber,* 8 ! Graaff Reinet Div. ; Oude Berg, *Drège* ! Beaufort West Div. ; Nieuwveld Mountains, near Beaufort West, 3000–5000 ft., *Drège,* 699a ! Albert Div. : without precise locality; *Cooper,* 697 !

KALAHARI REGION : Basutoland, *Cooper,* 695 ! Transvaal ; Magaliesberg Range, *Burke,* 376 ! creeks near Barberton, *Galpin,* 481 ! Macamac Gold Fields, *McLea in Herb. Bolus,* 467 ! Lydenberg Dist., near Lydenberg and Spitzkop, *Wilms,* 1027 ! near Pretoria, *Miss Leendertz,* 187 !

EASTERN REGION : Tembuland ; Bazeia, *Baur,* 244 ! Pondoland ; between St. Johns River and Umtsikaba River, *Drège* ! Natal ; summit of Table Mountain, *Krauss,* 447 ! banks of the River Umtwalumi, *Gerrard,* 1850 ! and without precise locality, *Gerrard,* 1211 !

Also in Tropical Africa.

2. B. auriculata (Benth. in Hook. Comp. Bot. Mag. ii. 60) ; an erect shrub, 5–8 ft. high ; twigs cylindric ; leaves firmly herbaceous, glabrous, dark green, shining above with slightly impressed nerves and veins, beneath closely covered with a thin tawny or whitish tomentum with main-nerves prominent, oblong-lanceolate, less often ovate-lanceolate, acuminate, base cuneate and entire, margin elsewhere entire or finely serrate, 2–3 in. long, ¾–1 in. wide : petiole 2½–4 lin. long, thinly closely tomentose ; stipules leafy, revolute-convex, rounded, 1–2½ lin. long : cymes numerous, many-flowered, rather lax, opposite, forming short thyrsoid panicles leafy at least below, 2½–3 in. long, 1–1½ in. across ; cyme-peduncles slender, 2½–3 lin. long ; bracts 1 lin. long, lanceolate : pedicels short ; calyx campanulate, ½ lin. long, closely tomentose ; teeth subtriangular, acute, half as long as tube : corolla white with throat orange, 3–4 lin. long ; tube outside puberulous with stellate hairs, within sparingly pubescent with simple hairs below the anthers and especially along adnate portion of filaments ; free filaments rather longer than the ovate anthers. *DC. Prodr. x. 445.*

Var. β, **euryifolia** (Prain in Kew Bulletin, 1908, 162, name only); leaves subcoriaceous, above with deeply impressed nerves and veins, beneath closely rusty stellate-tomentose, with secondary venation as well as main-nerves prominent, ovate-lanceolate or ovate, acuminate, base rather wide cuneate or rounded entire, margin elsewhere distinctly acutely serrate, 2–2¾ in. long, 1–1¼ in. wide; petiole rusty stellate-pubescent; corolla orange-yellow with pale lilac limb or salmon-coloured, 2¼–2½ lin. long.

Coast Region: Fort Beaufort Div. ; Winterberg Range, *Ecklon*! Stockenstrom Div. ; Chumie Berg, *Ecklon*! Katberg, 4000 ft., *Hutton*! *Baur*, 874! *Shaw*!

Central Region: Somerset Div.; Bosch Berg, *Burchell*, 3157! 3179!

Kalahari Region: Var. β: Transvaal; Rimers Creek, near Barberton, 2900–3000 ft., *Galpin*, 970! *Thorncroft*, 146 (*in Herb. Wood*, 4825)! near Lydenberg, *Wilms*, 1030!

Eastern Region: Griqualand East; Pot River Berg, in Maclear District, 5500 ft., *Galpin*, 6771! Var. β: Natal; Tugela River, *Gerrard*, 1967! Griqualand East; in Umzimhlava woods, near Kokstad, 4000 ft., *Tyson in MacOwan and Bolus, Herb. Austr.-Afr.* 1287!

Although var. β differs markedly in general appearance, and further differs from *B. auriculata* as regards the colour noted for its corolla, the material available hardly justifies its recognition as a distinct species. The flowers are somewhat smaller but do not differ in any structural detail.

3. **B. pulchella** (N. E. Br. in Kew Bulletin, 1894, 389): an erect shrub, 5 ft. high ; twigs cylindric, hoary-pubescent soon glabrous : leaves herbaceous, rather pale green above thinly stellate-pubescent soon glabrous with slightly impressed main-nerves, beneath rather closely softly stellate-pubescent, mid-rib and main-nerves rather prominent, ovate, acute, base cuneate, margin entire or undulate or remotely and irregularly 1–2-jugately lobed, 1¼–3½ in. long, ½–1½ in. wide; petiole ⅓–½ in. long ; blade decurrent on the upper half or throughout, thinly stellate-pubescent ; stipules represented by a faint interpetiolar line ; cymes numerous, few-flowered, rather lax, opposite, forming short thyrsoid panicles leafy at least below, 2–3 in. long, 1–1½ in. across ; cyme-peduncles slender, 2½–3 lin. long ; bracts 1–1½ lin. long, linear ; pedicels short : calyx campanulate, 1½ lin. long, densely hoary-pubescent : teeth ovate, subobtuse, one-fourth as long as tube ; corolla reddish with throat orange and limb white, 4 lin. long ; tube outside stellate-puberulous, within glabrous above the stamens, pubescent with simple hairs below the anthers and especially along adnate portion of filaments free filaments shorter than and hidden by the ovate anthers. *Wood and Evans, Natal Plants*, i. 49, t. 60. *B. Woodii, Gilg in Engl. Jahrb.* xxiii. 201.

Eastern Region: Natal; near Kettle Fountain, *Cooper*, 1159! Inanda, *Wood*, 574! 1602! (*Herb. Natal*, 4086)! near York, 3300–4400 ft., *Wood*, 4869! and *cultivated specimens*!

B. lindleyana (Fortune in Lindl. Bot. Reg. Misc. 1844, 25), a Chinese species readily distinguished from the three above described by its cymes aggregated in a long raceme-like terminal thyrse with longer, curved corolla-tubes, is cultivated in South Africa in gardens and occurs sometimes as an escape ; e.g. Transvaal ; near Lydenberg, *Wilms*, 1028!

V. ANTHOCLEISTA, Afzel.

Calyx deeply 4-fid ; tube almost obsolete ; lobes very thickly coriaceous, orbicular, imbricate. *Corolla* 8–16-fid, firmly fleshy ; tube narrow campanulate ; lobes overlapping to right, twisted to left. *Stamens* 8–16, adnate to corolla-throat ; filaments flattened, below connate into a membranous ring ; anthers linear, exserted, erect, basifixed, sagittate ; cells distinct, parallel. *Ovary* 2-celled, or by development of placental dissepiments 4-celled, resting on a fleshy disk ; style short or long, slender ; stigma terminal, dilated, oblong-capitate or shortly cylindric ; ovules many ; placentas linear, involute once, or below twice, 2-fid. *Fruit* a globose or oblong 4-celled berry ; pericarp thickly coriaceous, hardening when dry. *Seeds* numerous, small, invested with pulp ; testa thin ; albumen cartilaginous ; embryo straight, shorter than albumen ; cotyledons semiterete.

Glabrous trees or shrubs, sometimes climbing, armed with axillary spines or unarmed ; leaves mostly near ends of shoots, large or very large, opposite, decussate, penninerved, petioled or sessile, entire ; leaf-sheaths auriculate or not, united by an interpetiolar ridge or connate laterally into a short sheathing cup ; flowers cymose ; cymes in lax terminal panicles, with thick peduncles and short stout pedicels ; bracts scale-like.

DISTRIB. Species about 20, mostly in Tropical Africa, a few in the Mascarenes ; one of the Tropical African species extending to South Africa.

1. **A. zambesiaca** (Baker in Kew Bulletin, 1895, 99) ; a tall unarmed little branched tree, 70 ft. high ; wood white, soft, hardening when kept ; bark grey ; leaves rather crowded towards ends of branches, firmly herbaceous, rather pale green, glabrous on both surfaces, obovate-oblong, obtuse, gradually narrowed from upper third to base, margin throughout minutely crisply crenulate, in young saplings 4½ ft. long, 15 in. wide, on branches of mature trees 9–12 in. long, 4–5 in. wide ; petiole very short, 3–4 lin. long, or obsolete ; leaf-sheath short but distinct, 6 lin. long, deep-channelled above, connate with opposite leaf-sheath into an ochreate cup 3–4 lin. deep ; cymes 3–5 times 3-furcate ; main-peduncles 7–8 in. long, secondary 3–3¼ in., tertiary 1¼–2 in. long, ultimate branches 6 lin. long ; bracts orbicular, squamous, coriaceous, 3–4 lin. long ; pedicels stout, 2–3 lin. long ; calyx 5 lin. long ; tube subobsolete ; lobes orbicular, thickly coriaceous ; corolla yellowish-white, firmly fleshy, 2¼ in. long ; tube narrowly infundibuliform-campanulate, 1⅓–1½ in. long, mouth 4 lin. wide ; lobes 12–13, narrow-oblong, obtuse, at length reflexed, 9 lin. long, 3–4 lin. wide ; stamens 12–13 ; filaments short ; anthers linear, 3 lin. long, sagittate, connective dorsal slightly produced, apiculus obtuse ; ovary glabrous, obovate, 4 lin. long ; style exserted, 1⅓ in. long ; stigma enlarged cylindric, obtuse, faintly 2-lobed at apex and faintly 6-grooved laterally, 1½ lin. long ; fruit ovoid subacute, 1½ in. long, 9 lin. in diam.,

4-celled ; seeds brown, shining ; testa finely reticulated. *Baker in Dyer, Fl. Trop. Afr.* iv. i. 540. *A. pulcherrima, Gilg in Engl. Jahrb.* xxx. 374, *t.* 17 ; *Baker in Dyer, Fl. Trop. Afr.* iv. i. 540.

Eastern Region : Swaziland ; Horo Forest, 1800 ft., *Leyson in Herb. Galpin,* 1358 !

Also in Tropical Africa.

VI. STRYCHNOS, Linn.

Calyx deeply 5- or 4-fid ; tube sometimes obsolete : lobes ovate or triangular, rarely narrow-linear, imbricate. *Corolla* 5- or 4-fid ; tube hypocrateriform or rotate or subcampanulate ; lobes valvate, spreading or reflexed, rarely suberect. *Stamens* 5 or 4, usually adnate to corolla-throat, rarely low down in corolla-tube : filaments filiform, usually short ; anthers ovate, exserted or rarely included : cells distinct, parallel. *Ovary* 2-celled, rarely by absorption of dissepiment 1-celled ; style straight ; stigma terminal, capitate : ovules usually several, rarely many, very rarely few, 2- or more-seriate. *Fruit* a globose berry, usually 2-celled, sometimes by absorption of partition, less often by abortion of a cell 1-celled ; pericarp crustaceous. *Seeds* usually several, rarely by abortion only 2 or 1, generally variously compressed, if very few or solitary globose or oval, embedded in pulp : albumen usually **cartilaginous** : embryo straight, shorter than the albumen : cotyledons flat, leafy, 3–5-nerved.

Trees or shrubs, erect (in all South African species) or climbing by hooked tendrils, usually glabrous, armed or unarmed ; leaves opposite, decussate, **margin** entire, 3-5- (rarely 7-) nerved, petioled ; stipules represented by an **interpetiolar** ridge or 0 ; flowers cymose ; cymes simple or in condensed or lax thyrsoid or corymbose panicles, axillary or terminal ; bracts small or very small.

Distrib. Species under 100 ; wide-spread in all tropical countries : a few species also subtropical ; four of the latter endemic in South Africa.

Cymes axillary or from old wood ; twigs never spiny ;
    calyx shorter than corolla-tube ; teeth ovate or
    triangular ; corolla-lobes spreading ; stamens inserted
    on corolla-throat ; anthers glabrous, exserted ; ovary
    2-celled :
    Corolla-lobes pubescent all over inner face, throat with
      a ring of hairs : calyx-tube as long as the triangular
      acute lobes ; filaments longer than the anthers ;
      flowers 5-merous ; leaves glabrous... ... ... (1) **Atherstonei.**
    Corolla-lobes glabrous within : calyx-tube very short
      or obsolete ; lobes ovate or orbicular, obtuse,
      with ciliate margins ; filaments not longer than
      anthers :
      Leaves quite glabrous on both surfaces :
      Leaves with pungent spinous tips, rigid and thickly
        coriaceous ; bark corky ; cymes condensed ;
        flowers 5-merous ; corolla-throat with a ring
        of hairs ... ... (2) **pungens.**
      Leaves unarmed, thinly coriaceous or chartaceous ;
        bark not corky ; cymes usually rather lax :
        Corolla-throat glabrous ; flowers 5-merous ... (3) **Henningsii.**

Corolla-throat with a ring of hairs :
    Ovary glabrous ; leaves mucronulate, not ex-
        ceeding 1½ in. in length ; filaments nearly
        as long as anthers ; flowers mostly
        5-merous, some 4-merous ... ... (4) **pauciflora.**
    Ovary pilose with silky white hairs in the
        upper half ; leaves not mucronulate, never
        less than 1¼ in. in length, usually 2–3
        (sometimes 4) in. long ; filaments very
        short ; flowers all 4-merous ... ... (5) **Gerrardi.**
Leaves softly pubescent on both surfaces, mem-
    branous ; corolla-throat with a ring of hairs ;
    flowers 4-merous ... ... (6) **dysophylla.**
Cymes terminal ; twigs often converted into spines ;
    calyx longer than corolla-tube ; teeth long lanceolate
    or linear or subulate ; corolla-lobes suberect ; stamens
    inserted near base of corolla-tube ; anthers united
    by an interlacing network of hairs, included and
    hidden by ring of hairs in corolla-throat ; ovary
    1-celled :
    Twigs densely pubescent ; leaves softly puberulous on
        both surfaces ; bracts and calyx-teeth pubescent ;
        corolla puberulous outside ... ... (7) **schumanniana.**
    Twigs glabrous ; leaves glabrous or with only tufts of
        hair in the angles between the nerves beneath ;
        bracts, calyx-teeth and corolla outside glabrous ... (8) **spinosa.**

1. **S. Atherstonei** (Harv. Thes. Cap. ii. 41, t. 164) ; a tree 20–25 ft. high, 1–1½ ft. in diam. ; twigs knotted ; leaves coriaceous, glabrous, dull green, ovate or obovate, obtuse or retuse, base narrow-cuneate, 3-nerved from very near base, 1–1½ in. long, 4–7 lin. wide ; petiole glabrous, 1–1½ lin. long ; cymes 3–5-flowered, 6–8 lin. long, as much across, axillary or from older wood ; peduncles 3 lin. long ; bracts ½ lin. long ; pedicels 1½ lin. long, glabrous ; calyx wide-campanulate, ¾ lin. long ; teeth 5, triangular, hardly longer than tube ; corolla greenish, 3 lin. long ; tube shorter than lobes ; lobes 5, at length spreading, outside glabrous, within pubescent with long hairs except at the callous tip ; stamens 5, adnate to corolla-throat, exserted ; filaments subulate, slender, longer than anthers ; ovary 2-celled ; ovules several in each cell ; berry 1-seeded, globose, 6–7 lin. in diam. ; seeds peltate, subcompressed. *Fourcade, Rep. Natal For.* 1889, 115 ; *Sim, For. Fl. Cap.* 273, *t.* 111, *fig.* I. *S. Baculum, Harv. Thes. Cap.* ii. 41. *S. decussata, Gilg in Engl. Jahrb.* xxviii. 121. *Atherstonea decussata, Pappe, Silva Cap. ed.* 2, 29.

COAST REGION : Port Elizabeth Div. ; Kakakama, *Zeyher,* 3368 ! Alexandria Div. ; Olifants Hoek, *Pappe* ! *Nightingale in MacOwan and Bolus, Herb. Norm. Austr.-Afr.* 1000 ! Bathurst Div. ; by the Kowie River, *Atherstone* !

EASTERN REGION : Natal ; near Durban, *Gerrard and McKen,* 847 ! Botanic Gardens, *Wood,* 1926 ! and without precise locality, *Gerrard,* 262 !

2. **S. pungens** (Solered. in Engl. und Prantl, Pflanzenfam. iv. ii. 40) ; a shrub or small tree ; bark corky, fissured ; leaves thickly coriaceous, glabrous, dull and rather pale green, ovate or oblong, acute

with a rigid spinous tip, base cuneate or rounded, 5-nerved, the outer
nerves slender submarginal arising from base, the pair next midrib
also from base or closely applied to midrib from $\frac{1}{2}$–1 lin. above base,
1$\frac{1}{2}$–2 in. long, $\frac{1}{2}$–1$\frac{1}{4}$ in. wide ; petiole glabrous, stout, $\frac{1}{4}$ lin. long ;
cymes 5–9-flowered, 5–7 lin. long, as much across, almost all from
below leaves ; peduncles obsolete ; bracts $\frac{1}{4}$ lin. long ; pedicels stout,
$\frac{1}{2}$ lin. long, pubescent ; calyx ovate, $\frac{1}{2}$ lin. long ; teeth 5, ovate,
obtuse, with ciliate margins ; tube obsolete ; corolla white, 3 lin.
long ; tube rather longer than lobes ; lobes 5, at length spreading,
outside glabrous, within with a ring of hairs at throat ; stamens 5,
adnate to corolla-throat ; filaments very short ; anthers exserted ;
ovary 2-celled ; style sparingly hairy at base : ovules several in each
cell ; berry globose, resting on accrescent calyx-lobes, when ripe
" as large as a cannon-ball " (*Buchner*) ; " 5 in. in diam."
(*Leendertz*). *Solered. in Engl. Jahrb.* xvii. 554. *S. occidentalis,*
Solered. in Engl. und Prantl, Pflanzenfam. iv. ii. 40.

Kalahari Region : Transvaal ; Magaliesberg Range, *Burke*, 56 ! *Zeyher*, 1185 !
McLea in Herb. Bolus, 5710 ! near Aapies River, 5300 ft., *Schlechter*, 3621 !
Aapies Poort, near Pretoria, *Rehmann*, 4161 ! hills near Pretoria, *Miss Leendertz*,
382 ! Jeppestown Ridge, near Johannesburg, 6000 ft., *Gilfillan in Herb. Galpin*,
6153 ! near Nylstroom, *Burtt Davy*, 2101 !

Also in Tropical Africa.

3. **S. Henningsii** (Gilg in Engl. Jahrb. xvii. 569) ; a tree 30–40
ft. high ; bark pale below, green above ; twigs knotted ; leaves
.thinly coriaceous, glabrous, dark green shining above, paler and
dull beneath, wide ovate, gradually narrowed in upper third to an
obtuse or subacute tip, base wide-cuneate or rounded, usually
gradually increasing in size upwards from base to apex of twig but
occasionally all remaining small throughout, 3- or 5-nerved from or
from very near the base, 1–3$\frac{1}{2}$ in. long, $\frac{3}{4}$–2$\frac{1}{2}$ in. wide : petiole
glabrous, 2–3 lin. long ; cymes simple, 3–5-flowered or once (some-
times twice) decussately branched, each branch 3–5-flowered,
axillary ; main peduncles 3–6 lin. long ; branches when present 2–3
lin. long ; bracts $\frac{1}{2}$ lin. long ; pedicels subobsolete ; calyx wide
campanulate, $\frac{3}{4}$ lin. long ; teeth 5, suborbicular, with ciliate margins ;
tube very short ; corolla white, at length becoming pale orange,
thick, 3 lin. long ; tube hardly as long as lobes : lobes 5, at length
spreading, quite glabrous without and within : stamens 5, adnate to
corolla-throat ; filaments very short, thick ; anthers exserted ;
ovary 2-celled, and style glabrous ; ovules solitary in each cell ;
berry globose, orange-red, when ripe 6 lin. in diam., almost always
1-seeded : seed globose or ovoid, 3 lin. in diam. *Wood, Natal*
Plants, iii. 24, *t.* 247 ; *Sim, For. Fl. Cap.* 273, *t.* 112. *S. Umbanda,*
Fourcade, Rep. Natal For. 1889, 116. *S. utilis, Sim ex Lister in*
Rep. Cape For. 1897, 98 ; *Sim, For. Fl. Cap.* 273.

Coast Region : King Williamstown Div. ; Perie Forest, *Scott Elliot*, 979 ! East
London Div. ; Quelegha Forest, *Hutchins in Herb. MacOwan*, 2884 ! Komgha
Div. ; woods near Komgha, *Flanagan*, 1102 !

EASTERN REGION: Pondoland, *Bachmann*, 1745; *Sim*. Natal; Tugela River, *Gerrard*, 1917! Umcomaas, *McKen*, 10! Berea, near Durban, *Wood*, 6672! 7978, and without precise locality, *Mrs. K. Saunders*!

4. **S. pauciflora** (Gilg in Engl. Jahrb. xxviii. 121); an erect shrub; twigs 4-angled; bark white; leaves chartaceous, glabrous, dark green shining above, paler and dull beneath, oval, subacute or acute, slightly softly mucronulate, base wide-cuneate or rounded, 3-nerved from or from very near the base, $\frac{3}{4}$–1$\frac{1}{2}$ in. long, 4–9 lin. wide; petiole glabrous, $\frac{1}{2}$–1 lin. long; cymes simple or branched, 3–15-flowered, axillary, 1$\frac{1}{2}$–4 lin. long, 1$\frac{1}{2}$–3 lin. across; peduncles 1 lin. long, puberulous; bracts $\frac{1}{2}$ lin. long; pedicels $\frac{1}{2}$ lin. long, slender, puberulous; calyx ovate, $\frac{1}{3}$ lin. long; teeth 5 or 4, ovate, with ciliate margins; tube subobsolete; corolla white, 1$\frac{1}{4}$ lin. long; tube hardly as long as lobes; lobes usually 5, sometimes 4, at length spreading, outside glabrous, within with a ring of hairs at throat; stamens usually 5, sometimes 4, adnate to corolla-throat; filaments about as long as anthers, filiform; anthers distinctly exserted; ovary 2-celled and style glabrous; ovules several in each cell; fruit not seen.

EASTERN REGION: Lourenço Marques, 100 ft., *Schlechter*, 11682!

5. **S. Gerrardi** (N. E. Br. in Kew Bulletin, 1896, 162); a tree 30–40 ft. high; bark grey; leaves thinly coriaceous or almost chartaceous, glabrous, dark green shining above, paler and dull beneath, elliptic or oblong or subobovate, obtuse or rarely subacute, base cuneate, margin faintly crisped, distinctly 3-nerved, lateral nerves closely applied to midrib from 2–4 lin. above the base, 1$\frac{1}{4}$–4 in. long, $\frac{1}{2}$–1$\frac{1}{4}$ in. wide; petiole glabrous, slender, 1–3 lin. long; cymes 3–5-flowered, axillary or from wood below leaves, 4–6 lin. long, as much across; peduncles 1–2 lin. long; bracts $\frac{1}{2}$ lin. long; pedicels $\frac{1}{2}$–1$\frac{1}{2}$ lin. long, glabrous; calyx campanulate, 1 lin. long; teeth 4, oblong,. with ciliate margins; tube subobsolete; corolla white or yellow, 3 lin. long; tube as long as lobes; lobes 4, at length spreading, outside glabrous, within with a ring of hairs at throat; stamens 4, adnate to corolla-throat; filaments subobsolete; anthers slightly exserted; ovary 2-celled, in the upper half pilose like the base of the style with long white hairs; ovules several in each cell; berry globose, when ripe yellow, 3 (less often 4) in. in diam.; seeds few, irregularly ovoid, subcompressed, 8 lin. long, 6 lin. wide, 4 lin. thick. *Wood and Evans, Natal Plants,* i. 16, *t.* 16. *S. Mackenii, Harv. MSS. in Mus. Kew.; Gerrard MSS. ex Fourcade in Rep. Natal For. 1889, 116. S. McKenii, Gerrard ex Wood and Evans, Natal Plants,* i. 16.

EASTERN REGION: Natal; Nonoti River, *Gerrard*, 1421! Umcomaas, *Gerrard and McKen*, 1422 (as to the leaves in Herb. Kew.)! near Durban, *McKen*, 726! in gardens, *Wood*, 1777! Berea, *Wood*, 5624! and without precise locality, *Cooper*! *Gerrard*, 774!

A note regarding the fruit is associated with the specimen of Gerrard and McKen, 1422, in the Herbarium at Kew. This speaks of the fruit as "small" and may indicate that either *S. Atherstoni* or *S. Henningsii* may have been confused with *S. Gerrardi* to which the leaves accompanying the note belong. There is no example of the fruit referred to in the note in the Kew Herbarium. According to Mr. Medley Wood the secondary branches of this *Strychnos* frequently grow vertically upwards, thus giving the tree a characteristic and readily recognisable appearance.

6. S. dysophylla (Benth. in Journ. Linn. Soc. i. 103); a shrub or low tree; bark dark brown; twigs pubescent; leaves membranous, dark green pubescent above, paler densely velvety beneath, ovate or obovate or suborbicular, rounded at apex, base cuneate or rounded, distinctly 3-nerved, lateral nerves closely applied to midrib from 2–4 lin. above base, often in addition a fairly distinct pair of nerves nearer the margin leave midrib close to base, $1\frac{1}{4}$–2 in. long, $\frac{3}{4}$–$1\frac{1}{2}$ in. wide; petiole velvety, $\frac{3}{4}$–1 lin. long; cymes 3–5-flowered, almost all on wood below leaves, 3–4 lin. long, as much across; peduncles short, velvety; bracts puberulous, $\frac{1}{2}$ lin. long; pedicels very short, velvety; calyx campanulate, 1 lin. long; teeth 4, oblong, with ciliate margins and sparsely pubescent outside; corolla white, $2\frac{1}{2}$ lin. long; tube rather longer than lobes; lobes 4, at length spreading, outside glabrous, within with a ring of hairs at throat; stamens 4, adnate to corolla-throat, filaments very short; anthers slightly exserted; ovary 2-celled; ovules several in each cell; berry globose, black, "sweet, well-tasted" (*Baines*). *S. randiæformis, Baill. Bull. Soc. Linn. Par. i. 246.*

Kalahari Region: Transvaal; on the red sand flats to the west of Blueberg and Hangklip Mountains, south of the Limpopo River, *Baines*!

Eastern Region: Delagoa Bay, *Forbes*! Katembe, *Schlechter*, 4615! Natal; Tugela, *Gerrard*, 1660!

7. S. schumanniana (Gilg in Warb. Kunene-Samb. Exped. 330); a shrub or small tree 8–25 ft. high; bark corky; twigs densely pubescent, often ending in rigid bare spines 3–4 lin. long; leaves herbaceous, pale green, dull and finely pubescent especially on the nerves on both surfaces, wide ovate, usually rounded, sometimes wide acuminate, almost always with a short soft mucro, base rounded or abruptly cuneate, usually 5-nerved occasionally 7-nerved from near base, sometimes 5–7-plinerved with lateral nerves closely applied to midrib for varying and irregular distances, at times as much as 10–12 lin. above base, $1\frac{1}{4}$–3 in. long, $\frac{3}{4}$–2 in. wide; petiole stout, densely pubescent, 1–3 lin. long; cymes terminal on young twigs, simple, 3–5-flowered, or branched, each branch 3–5-flowered; main peduncles 3–5 lin. long, densely pubescent; bracts subulate, 1 lin. long, pubescent; calyx campanulate, 2 lin. long; tube very short; teeth 5, linear, pubescent; corolla white, 3 lin. long; tube campanulate, nearly twice as long as lobes; lobes 5, hardly spreading, puberulous outside, within with a ring of long hairs at throat; stamens 5, inserted at base of corolla-tube, included and hidden by the hairs on corolla-throat; filaments filiform; anthers laterally connected by long interwoven hairs ; ovary 1-celled ; berry globose, 3 in. in diam. ; seeds many, flat, embedded in a sour or bitter pulp.

KALAHARI REGION : Transvaal ; Magaliesberg Range, at Wonderboom Poort, 5000 ft., *Schlechter,* 3630 ! Avoca, near Barberton, 1800 ft., *Galpin,* 895 !

8. **S. spinosa** (Lam. Ill. ii. 38) ; a shrub 8–10 ft. high ; twigs glabrous, often ending in rigid bare spines 6–8 lin. long ; leaves herbaceous, rather pale green and dull on both surfaces, glabrous above, beneath often with a patch of silky hairs in the angles between main-nerves and midrib, otherwise glabrous, obovate or suborbicular or ovate, usually rounded, sometimes acute, sometimes retuse at apex, almost always with a short soft mucro, base gradually or abruptly cuneate or rounded, usually 5-nerved, outer pair of nerves from near base, next pair usually closely applied to midrib from 2–4 lin. above base, $1\frac{1}{2}$–2 in. long, $\frac{3}{4}$–$1\frac{1}{4}$ in. wide ; petiole glabrous, 1–3 lin. long ; cymes terminal on young twigs, simple, 3–5-flowered, or once (less often twice) branched, each branch 3–5-flowered ; main-peduncles 2–3 lin. long, puberulous, branches $\frac{1}{2}$–1 lin. long ; bracts narrow lanceolate, $1\frac{1}{2}$ lin. long ; pedicels very short ; calyx campanulate, 2 lin. long ; tube very short ; teeth 5, lanceolate-subulate, glabrous or puberulous ; corolla greenish-white, 2 lin. long ; tube campanulate, as long as lobes ; lobes 5, hardly spreading, glabrous outside, within with a ring of long hairs at throat ; stamens 5, inserted at base of corolla-tube, included and hidden by the hairs on corolla-throat ; filaments filiform ; anthers laterally connected by long interwoven hairs ; ovary 1-celled ; ovules many on a central placenta ; berry globose, 3–4 in. in diam., at first green, at length yellow ; seeds many, flat, embedded in a sweet pulp. *Hiern in Cat. Afr. Pl. Welw.* i. 702 ; *Baker in Dyer, Fl. Trop. Afr.* iv. i. 536 ; *Sim, For. Fl. Cap.* 274, *t.* 111, *fig.* 2. *S. Vuntac, Boj. Hort. Maurit.* 205. *S. Lokua, A. Rich. Tent. Fl. Abyss.* ii. 53. *S. laxa, Solered. in Engl. Jahrb.* xvii. 554. *Caniram Vontac, Thouars in Dict. Sc. Nat.* vi. 428. *Brehmia spinosa, Harv. in Hook. Lond. Journ. Bot.* i. (1842) 26 ; *DC. Prodr.* ix. 18 ; *Baker, Fl. Maurit.* 235.

COAST REGION : Albany Div., *Bowker* !
KALAHARI REGION : Transvaal ; Crocodile Poort, near Barberton, *Galpin,* 1075 !
EASTERN REGION : Pondoland ; between Umtentu River and Umzimkulu River, *Drège* ! Natal ; near Durban, *Plant* ! *Wilms,* 2006 ! Inanda, *Wood,* 1084 ! and without precise locality, *Gerrard and McKen,* 591 ! *Gerrard,* 67 ! Delagoa Bay ; 3 miles north-west of Lourenço Marques, *Bolus,* 9704 !

Also in Tropical Africa.

This species is very nearly related to the preceding. The two belong to a group of forms recognised by the late Dr. Harvey as constituting a distinct genus. Whatever rank may be assigned to this group it will be seen from their characters that they differ more essentially from the remaining species of *Strychnos* than any species of *Chilianthus* does from the species included in the genus *Buddleia.* It is, however, more convenient, in a work like the present, to leave all four groups *Buddleia, Chilianthus, Strychnos* and *Brehmia* in the positions assigned to them in Bentham and Hooker's *Genera Plantarum.*

ORDER XC. **GENTIANEÆ.**

(By A. W. HILL and D. PRAIN.)

Flowers usually regular and hermaphrodite. *Calyx* inferior 4–6-lobed or -partite; lobes usually imbricate. *Corolla* gamopetalous: tube campanulate, funnel-shaped or cylindric, sometimes with a constricted limb; lobes 4–6, contorted, imbricate or induplicate-valvate. *Stamens* 4–6, inserted in the corolla-tube or on the corolla-throat; filaments filiform or dilated at the base; anthers dorsifixed or basifixed, sometimes spirally twisted, dehiscence longitudinal or at times by apical pores or slits. *Disk* obsolete or annular or 5 hypogynous glands. *Ovary* superior, 2-carpelled, cells 2 or 1; ovules on each placenta usually numerous; style simple; stigma entire or 2-lobed. *Fruit* usually a capsule, dehiscence usually septicidally 2-valved, sometimes partial and 4-valved; occasionally fruit subindehiscent. *Seeds* usually sessile, albuminous, sometimes winged; testa crustaceous or membranous; embryo small.

Annual or perennial herbs, rarely shrubs, nearly always glabrous; leaves usually opposite (in *Menyanthex* alternate); flowers in terminal cymes or fascicles; cymes sometimes paniculate, or axillary and fasciculate or solitary.

Species about 600, most plentiful in temperate regions throughout the globe; also on tropical mountains.

Tribe 1. EXACEÆ.—*Leaves* opposite. *Corolla-lobes* contorted. *Ovary* 2-celled.

 I. **Sebæa.**—*Disk-glands* between calyx and corolla 0. *Style* usually with a 2-glandular swelling near the base.

 II. **Exochænium.**—*Disk-glands* present between calyx and corolla. *Style* without a 2-glandular swelling.

Tribe 2. CHIRONIEÆ.—*Leaves* opposite. *Corolla-lobes* contorted, without nectaries. *Ovary* 1-celled; placentas intruded.

 III. **Orphium.**—More or less pubescent. *Calyx-lobes* obtuse, without keels. Crenulate *disk* between calyx and corolla. *Flowers* 5-merous in terminal cymes.

 IV. **Chironia.**—Glabrous. *Calyx-lobes* usually acute, keeled. No *disk* between calyx and corolla. *Flowers* 5-merous in terminal cymes; cymes sometimes reduced to a solitary flower, sometimes paniculate.

 V. **Enicostema.**—Glabrous. *Flowers* 5-merous, in axillary fascicles.

 VI. **Faroa.**—Glabrous. *Flowers* 4-merous, in terminal and axillary fascicles.

Tribe 3. SWERTIEÆ.—*Leaves* opposite. *Corolla-lobes* contorted, with basal nectaries on their inner face. *Ovary* 1-celled; placentas parietal.

 VII. **Swertia.**—The only South African genus.

Tribe 4. MENYANTHEÆ.—*Leaves* alternate or all radical. *Corolla-lobes* induplicate-valvate. *Ovary* 1-celled: placentas parietal.

 VIII. **Villarsia.**—*Fruit* partially 4-valved at apex, capsular.

 IX. **Limnanthemum.**—Fruit indehiscent or irregularly rupturing.

I. SEBÆA, Soland.

Calyx 4–5-lobed; tube usually short or very short; lobes ovate, acute or acuminate, keeled or winged. *Corolla* 4–5-lobed; tube cylindric, long or short; lobes oblong, spreading, contorted in æstivation. *Stamens* 4–5, inserted at or shortly below corolla sinuses; filaments exserted or included, sometimes very short; anthers with or without an apical gland, sometimes with two at the base. *Ovary* 2-celled; placentas axile; ovules many; style filiform, usually with a pair of papillate or glandular patches (apparently stigmatic) like a tubercular swelling near the base; stigma capitate, clavate or bilobed, sometimes almost confluent with the swellings. *Capsule* globose or ovoid, membranous or thinly coriaceous, septicidal, valves 2. *Seeds* many, minute and simple or larger and ridged or provided with frills.

Annual, biennial or perennial herbs; stems erect or more rarely procumbent, simple or branched with decurrent wings; leaves opposite, sessile, herbaceous, fleshy or coriaceous, sometimes forming a radical rosette; flowers small or of moderate size, yellow, less commonly white, arranged in terminal and axillary dichotomous cymes rarely solitary.

DISTRIB. Species about 100, chiefly in South Africa, a few in Tropical Africa, Madagascar, India, Australia and New Zealand.

Following the course adopted by Schinz in *Mitteil. Geogr. Ges. Lübeck,* xvii. 1903, the genera *Lagenias,* E. Meyer, and *Belmontia,* E. Meyer, have been merged into *Sebæa,* Soland., since the differences between the 3 original genera appear to be of too slight a nature to be of generic value. In Benth. and Hook. f. *Gen. Plant.* vol. ii., the genus *Lagenias* is included with *Sebæa,* but *Belmontia* is maintained and *Exochænium,* Griseb., is merged into it. In Engl. und Prantl, *Pflanzenfamilien,* iv. ii., Gilg keeps up *Sebæa, Lagenias* and *Belmontia* as distinct genera but unites Grisebach's *Exochænium* with the latter. In the present work *Exochænium* is maintained as a distinct genus (cf. Schinz in *Bull. Herb. Boiss.* 2ᵐᵉ sér. vi. 714). The sections of this genus have been defined and some critical notes on the species published in the *Kew Bulletin,* 1908, 317–336.

Section 1. TETRANDRIA. Calyx-segments, corolla-lobes and anthers, 4.

Leaves numerous:
　Leaves crowded towards the base to form a kind of
　　elongated basal rosette　...　...　...　...　(1) **capitata.**

　Leaves scattered all along the stems:
　　Stems stout; leaves ovate or cordate, coriaceous...　(2) **sclerosepala.**

　　Stems slender; leaves lanceolate or ovate, herba-
　　　ceous, subpetiolate　...　...　...　...　(3) **laxa.**

Leaves in 3–6 pairs:
　Leaves more or less fleshy, glaucous:
　　Calyx-segments broad, elliptic-obovate apiculate,
　　　without keel or wing　...　...　...　...　(4) **albens.**

　Calyx-segments with apical keel-wing:
　　Flowers collected into small capitula; bracts
　　　deltoid or ovate-deltoid:

　　　Flowers very small; anthers ¼ lin. long; style
　　　　¾–1 lin. long　...　...　...　...　(5) **minutiflora.**

Flowers small; anthers 1–1¼ lin. long; style
2½–3½ lin. long ... ... (6) **ambigua.**
Flowers in much-branched distinct cymes ... (7) **gibbosa.**
Calyx-segments with median keel-wing; bracts
more or less rhomboid-ovate ... ... (8) **glauca.**
Leaves herbaceous or slightly fleshy, bright green :
Calyx-segments keeled, or with a wing widest at
the middle :
Calyx-segments with a keel or very slight keel-
wing; style with a median swelling ... (9) **aurea.**
Calyx-segments with conspicuous wing ; style
without swelling ... ... (10) **Schlechteri.**
Calyx-segments with wings ½–¾ lin. broad, widest
towards the base :
Corolla-lobes equal in length to the tube;
anthers about ½ lin. long :
Stigma clavate, bilabiate ... ... (11) **ochroleuca.**
Stigma capitate ... ... (12) **Gilgii.**
Corolla-lobes longer than the tube ; anthers
¾–1 lin. long ... ... (13) **pallida.**

Section 2. PENTANDRIA. Calyx-segments, corolla-lobes and anthers 5.

Calyx-segments united to form a cylindrical tube for ¾
or more of their length ; stamens inserted in
corolla-tube ... ... (14) **compacta.**
Calyx-segments more or less free, or fused for not
more than ½ their length ; stamens inserted either
in sinuses or in corolla-tube :

Sub-section 1. ANNUÆ. Erect annuals with 2–6
(rarely more) pairs of leaves ; style usually with
biglandular swelling near the base ; stigma above
the anthers.
*Plants ½–6 (rarely 8) in. high ; stems simple, rarely
branching from the base :

Group 1. *Lageniades.* Leaves linear- or ovate-
lanceolate ; anthers inserted in the corolla-
tube ; style without biglandular swelling :
Filaments ¾–1½ lin. long ; stigma capitate ... (15) **pusilla.**
Filaments very short ; stigma shortly bilobed... (16) **rara.**

Group 2. *Filiformes.* Leaves minute, more or
less filiform ; stamens inserted in the corolla-
sinuses ; style usually with biglandular
swelling.
Calyx-segments united below to form a shallow
cup, wings broad ... ... (17) **mirabilis.**
Calyx-segments united below to form a shallow
cup, wings narrow :
Inflorescence branched ; flowers small ; leaves
ovate-lanceolate ... ... (18) **Junodii.**
Flowers usually borne singly, conspicuous ;
leaves linear ... ... (19) **filiformis.**
Calyx-segments united below into a conspicuous
cup, keeled ... ... (20) **exigua.**

Group 3. *Ovatæ.* Leaves ovate or cordate, usually conspicuous; stamens inserted in the corolla-sinuses or tube; style usually with biglandular swelling. (See also 39, *S. pentandra.*)

Stamens inserted in the corolla sinuses:
  Calyx-segments keeled... ... ... (21) **caladenia.**
  Calyx-segments winged, wings widest at the middle; anthers with only apical glands:
    Anthers $\frac{1}{4}$ lin. long; style $\frac{3}{4}$ lin. long ... (22) **acutiloba.**
    Anthers $\frac{3}{4}$–1 lin. long; style 2–3 lin. long (23) **grisebachiana.**
  Calyx-segments winged, wings widest at the base:
    Anthers without glands ... ... (24) **schizostigma.**
    Anthers with apical glands:
      Anthers $1\frac{1}{4}$–$1\frac{1}{2}$ lin. long; stigma and swelling separated ... ... (25) **saccata.**
      Anthers $\frac{1}{3}$–$\frac{1}{2}$ lin. long; stigma and swelling more or less confluent ... ... (26) **Zeyherii.**
    Anthers with apical and basal glands:
      Stigma elongate, ligulate... ... ... (27) **sulphurea.**
      Stigma shortly clavate, bilabiate ... (28) **scabra.**
Stamens inserted in the corolla-tube:
  Plants slender; leaves small; calyx-segments keeled:
    Anthers $\frac{1}{2}$ lin. long; style 2–3 lin. long; corolla-lobes erosely toothed ... ... (29) **erosa.**
    Anthers $\frac{1}{3}$ lin. long; style $1\frac{1}{2}$ lin. long; corolla-lobes entire ... ... (30) **pygmæa.**
  Plants robust; leaves conspicuous; calyx-segments widely winged at the base:
    Flowers large; stigma and swelling separated ... ... (31) **exacoides.**
    Flowers small; stigma and swelling confluent:
      Calyx-segment wings fringed with bristles; corolla-tube $2\frac{1}{2}$–$3\frac{3}{4}$ lin. long; lobes $1\frac{1}{2}$–2 lin. long ... ... (32) **micrantha.**
      Calyx-segment wings minutely pilose: corolla-tube $3\frac{1}{2}$–$4\frac{1}{2}$ lin. long; lobes $2\frac{1}{2}$–$3\frac{1}{2}$ lin. long ... ... (33) **intermedia.**

**Plants 4–12 in. or as much as 30 in. high, with usually a more or less conspicuous false rosette of leaves; stems usually much-branched from the base:

Group 4. *Rosulatæ.* Stamens inserted in the corolla-sinuses or tube.

Stamens inserted in the corolla-sinuses:
  Calyx-segments keeled:
    Anthers $\frac{1}{2}$ lin. long, with 3 glands:
      Style without swelling; stigma clavate (34) **Burchellii.**
      Style with swelling; stigma capitate ... (35) **Conrathii.**
    Anthers $1\frac{1}{2}$ lin. long, with large apical glands (36) **conspicua.**
  Calyx-segments winged, wings widest at the middle:
    Anther-glands inconspicuous; stigma capitate ... ... (37) **elongata.**

Anther-glands 3 ; stigma clavate :
    Distinct leaf-rosette ; leaves orbicular ... (38) **rotundifolia.**
    Radical leaves deciduous or absent, cauline
        leaves elliptic-ovate ... ... (39) **pentandra.**
Calyx-segments winged, wings widest towards
    base :
    Anthers without glands ; stigma $\frac{3}{4}$–1 lin.
        long, clavate-bilabiate ... ... (40) **macrostigma.**
    Anthers with 3 glands ; stigma small,
        clavate ... ... (41) **ramosissima.**
Stamens inserted in the corolla-tube ... ... (42) **primulina.**

Sub-section 2. PERENNES. Erect perennials or biennials, with more than 6 pairs of leaves ; stems branching from the base or from leaf-axils ; stamens inserted either in the corolla-sinuses or tube ; style usually with biglandular swelling near the base ; stigma above the anthers.

Group 1. *Erectæ.* Stems simple or branched from the base and terminated by definite, more or less compact inflorescences. (See also 65, *S. Rudolfii.*)

Plants 2–3 in. high ; leaves in 6–8 pairs ;
    inflorescences 1–5-flowered... ... ... (43) **vitellina.**
Plants 5 in.–2 ft. or more high ; leaves
    numerous ; inflorescences more than 5-
    flowered :
    Calyx-segments keeled :
    Corolla-lobes longer than the tube... ... (44) **Dregei.**
    Corolla-lobes shorter than the tube :
        Leaves ovate-lanceolate ; internodes long ;
            inflorescence compact, capitate ... (45) **Schoenlandii.**
        Leaves broadly orbicular-ovate ; inter-
        nodes short ; inflorescences more or
        less open ... ... (46) **sedoides.**
        Leaves broad, triangular-ovate ; inter-
        nodes short ; inflorescences of dense
        terminal capitula ... ... (47) **confertiflora.**
    Calyx-segments winged :
    Stamens inserted in the corolla-sinuses :
        Anthers $\frac{1}{2}$–$\frac{1}{3}$ lin. long ; style 2 lin. long (48) **acuminata.**
        Anthers $2\frac{3}{4}$–$3\frac{3}{4}$ lin. long ; style 4 lin.
            long ... ... (49) **macrantha.**
        Anthers 1–2 lin. long ; style 4–5 lin.
            long :
            Calyx-segments with a distinct wing ;
                stigma capitate ... ... (50) **erecta.**
            Calyx-segments with a narrow wing :
                Leaves narrowly ovate or elliptical ;
                style with swelling ; stigma
                shortly clavate... ... ... (51) **longicaulis.**
                Leaves orbicular-ovate ; style with-
                out swelling (or if present
                minute) ; stigma capitate ... (52) **grandiflora.**

Group 2. *Fastigiatæ.* Stems giving off numerous lateral branches ; inflorescences loose and somewhat paniculate.

†Stamens inserted in the corolla-sinuses :

Calyx-segments keeled or with a very narrow
keel-wing :
Plants much-branched from the base,
forming dense bushes, with numerous
axillary inflorescences :
Corolla-lobes longer than the tube :
Anthers with 3 conspicuous glands ... (53) **multiflora.**

Anthers without glands ... ... (54) **schinziana.**
Corolla-lobes equal in length to the
tube ; anther-glands minute ... (55) **macrophylla.**

Plants tall, branching above to form loose
terminal inflorescences :
Anthers with 3 minute glands ; stigma
shortly clavate ; corolla-lobes longer
than the tube ... ... (56) **crassulæfolia.**

Anthers with apical glands ; stigma
capitate :
Anthers $\frac{3}{4}$ lin. long ; filaments 1–1$\frac{1}{2}$ lin.
long ; corolla-lobes shorter than
the tube ... ... (57) **Brehmeri.**

Anthers 1–1$\frac{1}{4}$ lin. long ; filaments $\frac{1}{4}$–1
lin. long ; corolla-lobes equal in
length to the tube ... ... (58) **leiostyla.**

Calyx-segments with a well-developed wing :
Wings of the calyx-segments developed only
from the middle to the base of each
segment ; stigma bilobed, thick and
fleshy ... ... (59) **Macowanii.**

Wings of the calyx-segments developed
along the whole length of each segment ; stigma capitate or shortly
clavate :
Calyx-segments 4$\frac{1}{2}$–5 lin. long, overlapping widely at the base ... ... (60) **imbricata.**

Calyx-segments 3–4$\frac{1}{2}$ lin. long, not widely
overlapping at the base :
Anthers $\frac{3}{4}$ lin. long ; stigma capitate,
inconspicuous ... ... (61) **polyantha.**

Anthers 1–1$\frac{1}{2}$ lin. long ; stigma capitate or shortly clavate, distinct :
Corolla-lobes longer than the tube ... (62) **Rehmannii.**

Corolla-lobes equal in length to the
tube :
Leaves orbicular-ovate to subreniform, obtuse or apiculate ;
anthers 1–1$\frac{1}{2}$ lin. long ... (63) **hymenosepala.**

Leaves broadly-ovate, acute ;
anthers 1$\frac{1}{2}$–2 lin. long ... (64) **fastigiata.**
††Stamens inserted in corolla-tube ... ... (65) **Rudolfii.**

Sub-section 3. REPENTES. Creeping or rosette-form-
ing perennials with numerous spathulate or ovate
leaves ; flowers more or less sessile at the ends of
branches or borne on definite erect inflorescences.

Group 1. *Longistylæ.* Stamens inserted in corolla-
sinuses or tube ; style with biglandular swelling ;
stigma above anthers.

Cymes almost sessile, 1–5-flowered :
    Stamens in the corolla-sinuses ; anthers with
        a yellow apical gland ... ... (66) **repens.**
    Stamens inserted slightly below the corolla-
        sinuses ; anthers with a large black
        apical gland ... ... (67) **Marlothii.**
Cymes borne on erect stems ; stamens in the
    corolla-sinuses or just below them :
    Anthers with yellow apical glands and 2 basal
        glands ; filaments very short ... ... (68) **procumbens.**
    Anthers with large black apical glands ? minute
        basal glands ; filaments ½ lin. long ... (69) **thodeana.**

Group 2. *Brevistylæ.* Stamens inserted in the
    corolla-tube ; style without biglandular swell-
    ing ; stigma below the base of the anthers.
    Plants with a rosette of long spathulate leaves ;
        inflorescence erect ; anthers with a large
        black apical gland ... ... (70) **spathulata.**
    Creeping or rosette-forming plants ; flowers
        usually solitary, terminal, almost sessile ;
        anthers with 3 glands, yellow ... ... (71) **Thomasii.**

1. **S. capitata** (Cham. & Schlecht. in Linnæa, i. 193) ; annual,
erect, 4–8 in. high, simple or branching at the base ; leaves
herbaceous, ovate or elliptic, acute, basal leaves narrowed to a
short petiole, 5–7 lin. long, 2½–4½ lin. broad ; cymes compact,
fastigiate, 5–10-flowered ; branches 7–12 lin. long ; pedicels very
short, about ½ lin. long ; calyx-segments about 4 lin. long, boat-
shaped, ovate, acute, with a thickened wing ; corolla-tube 2–2½ lin.
long ; lobes ovate, subacute, 3 lin. long, about 2 lin. broad ; anthers
½ lin. long, glands absent ; filaments about 1 lin. long ; style 2 lin.
long, without glandular swelling ; stigma clavate (? biligulate).
*Griseb. Gen. et Spec. Gent. 166, and in DC. Prodr. ix. 53 ; Schinz in
Vierteljahrsschr. Naturf. Gesellsch. Zürich, xxxvii. 313 ; Gilg in Engl.
Jahrb. xxvi. 87 ; Schinz in Mitteil. Geogr. Ges. Lübeck, xvii.*
(1903) 15.

COAST REGION : Cape Div. ; summit of Table Mountain, *Burchell,* 561 !
Wolley Dod, 2122 ! 3500 ft., *Miss Kensit in Herb. Bolus,* 9356 ! *Mund and
Maire.*

2. **S. sclerosepala** (Gilg ex Schinz in Mitteil. Geogr. Ges. Lübeck,
xvii. 1903, 23) ; annual, erect, about 10 in. high, simple ; leaves
all cauline, crowded below, distant above the middle, increasing in
size from the base, ovate or cordate, acute, 4–5 lin. long, 3½–4½ lin.
broad, slightly coriaceous ; cymes compact, few-flowered ; branches
erect, about 7 lin. long ; pedicels very short, ½ lin. long ; calyx-
segments 4½ lin. long, 2 lin. wide, elliptic, acute, with wing ¼ lin.
wide, broadest at the middle ; corolla-tube 4 lin. long ; lobes 4 lin.
long, 2 lin. wide, ovate ; anthers nearly 1 lin. long ; filaments 1 lin.
long ; glands apparently absent ; style 3 lin. long, without glandular
swelling ; stigma clavate.

COAST REGION : George Div. ; on the Post Berg (Cradock Berg), near George, *Burchell,* 5897 ! Montagu Pass, *Rehmann,* 266 !

Sebæa sclerosepala shows striking similarity to *S. capitata* in certain particulars, and it seems possible that further collections may require the union of these two species. The few examples of these species in our herbaria show *S. sclerosepala* to be a taller plant than *S. capitata* with more scattered leaves and fewer flowered inflorescences, the flowers also are larger. In both species, however, the anthers are destitute of glands and the style has no glandular swelling.

3. S. laxa (N. E. Br. in Kew Bulletin, 1901, 128) ; annual, erect, lax, 4–10 in. high, branched ; leaves numerous, lanceolate or ovate, acute or acuminate, subpetiolate, 1–4 lin. long, $\frac{1}{2}$–$1\frac{3}{4}$ lin. broad ; cymes lax, 1–15-flowered ; branches about 5 lin. long ; pedicels 1–3 lin. long ; calyx-segments $1\frac{1}{2}$ lin. long, lanceolate, acute, with very narrow keel-wing ; corolla yellow ; tube $1\frac{1}{2}$–$1\frac{3}{4}$ lin. long ; lobes 2–$2\frac{1}{2}$ lin. long, 1 lin. broad, ovate-lanceolate, acute or subacute ; filaments $\frac{3}{4}$ lin. long ; anthers $\frac{1}{2}$ lin. long, with apical gland ; style $1\frac{1}{4}$ lin. long, with swelling, clavate. *Schinz in Mitteil. Geogr. Ges. Lübeck,* xvii. (1903) 21.

COAST REGION : Swellendam Div. ; Zuurbraak Mountain, 3000 ft., *Galpin,* 4337 ! Riversdale Div. ; on the Kampsche Berg, *Burchell,* 7089 !

4. S. albens (R. Br. Prodr. 452) ; annual, erect, unbranched, stout ; leaves in 3–5 pairs, ovate, obtuse or subacute, somewhat fleshy, 2–7 lin. long, $1\frac{1}{2}$–5 lin. broad ; cymes much-branched, many-flowered, forming a flat corymbose inflorescence ; pedicels very short ; calyx-segments elliptic-obovate, obtuse, apiculate, broadly rounded on the back, neither keeled nor winged ; corolla pale yellow or white ; tube $1\frac{1}{2}$ lin. long ; lobes $1\frac{1}{2}$–$3\frac{1}{2}$ lin. long, 1–2 lin. broad, oblong to elliptic, obtuse ; anthers $\frac{1}{2}$–$1\frac{1}{4}$ lin. long, with a large gland at the apex ; style 1–$2\frac{2}{3}$ lin. long, with a swelling rather below the middle ; stigma bilabiate. *Griseb. Gen. et Spec. Gent.* 171 ; *and in DC. Prodr.* ix. 53 ; *Cham. in Linnæa,* vi. 345 ; *E. Meyer, Comm.* 185 ; *Schinz in Vierteljahrsschr. Naturf. Gesellsch. Zürich,* xxxvii. 313 ; *Gilg in Engl. Jahrb.* xxvi. 88 ; *Schinz in Mitteil. Geogr. Ges. Lübeck,* xvii. (1903) 22. *Exacum albens, Linn. f. Suppl.* 123 (*excl. syn. E. pedunculatum, Linn.*). *Gentiana albens, Thunb. Prodr.* 48.

COAST REGION : Piquetberg Div. ; Piquetberg Range, near Groen Valley, below 1000 ft., *Drège* ! Malmesbury Div. ; near Groene Kloof (Mamre), 300 ft., *Bolus,* 4308 ! Cape Div. ; Salt River, *Burchell,* 682 ! Green Point, *MacGillivray,* 594 ! Cape Flats, 0–100 ft., *MacOwan, Herb. Austr.-Afr.* 1925 ! Fish Hoek Valley, *Wolley Dod,* 3439 ! 3439a ! by a creek beyond Paarden Island, *Wolley Dod,* 3311 ; Sand-downs, Capetown, *Prior* ! and without precise locality, *Harvey,* 612 ! *Thom,* 652 ! *Pappe* ! *Wallich* ! *Forster* !

5. S. minutiflora (Schinz in Bull. Herb. Boiss. iii. 413) ; annual, erect, 6–8 in. high, simple ; leaves ovate or cordate, subacute, 4–6 lin. long, 4–7 lin. wide, more or less fleshy, smaller near the base ; cymes corymbose, richly branched ; main branches $1\frac{1}{2}$

in. long; pedicels and ultimate branches very short, with flowers collected into small capitate cymules; larger bracts deltoid-ovate, smaller ovate; calyx-segments 2 lin. long, obovate or more or less cuneate, toothed, concave, with a short broad thick wing at the apex only; corolla white; tube 1–2 lin. long; lobes $\frac{3}{4}$–1 lin. long, $\frac{1}{2}$–$\frac{3}{4}$ lin. wide, ovate, obtuse; filaments very short; anthers $\frac{1}{4}$ lin. long, with an apical gland; style 1 lin. long, glandular thickening not noticeable; stigma capitate, minutely bifid. *Gilg in Engl. Jahrb.* xxvi. 88; *Schinz in Mitteil. Geogr. Ges. Lübeck,* xvii. (1903) 16.

COAST REGION : Cape Div. ; shore at Slang Kop, *Wolley Dod,* 3253 ! Riversdale Div. ; near Riversdale, 300 ft., *Schlechter,* 1701 ! Port Elizabeth Div. ; along the Baakens River, near Port Elizabeth, *Burchell,* 4340 ! Port Elizabeth, *E. S. C. A. Herbarium,* 71 !

6. **S. ambigua** (Cham. in Linnæa, vi. 346 and viii. 52); annual, erect, normally 3–6 in. rarely 12–16 in. high, simple or branching from the base; leaves in 4–5 pairs, fleshy, broadly ovate or cordate, subacute, about 6 lin. long, 6 lin. wide; cymes densely branched; flowers collected into cymose capitula, inflorescence corymbose; bracts deltoid-ovate or ovate, more or less acute; calyx-segments 2 lin. long, obovate or cuneate, rounded, erose, with a short broad thickened wing at the apex only; corolla-tube 2 lin. long; lobes 2 lin. long, 1–1$\frac{1}{2}$ lin. wide, ovate, rounded; filaments about 1 lin. long; anthers about 1 lin. long, with a large apical gland; style 2–3 lin. long, with a median glandular swelling not always distinct ; stigma shortly 2-lobed or subentire and clavate. *Griseb. Gen. et Spec. Gent.* 171 *and in DC. Prodr.* ix. 52 ; *Schinz in Vierteljahrsschr. Naturf. Gesellsch. Zürich,* xxxvii. 315 ; *Gilg in Engl. Jahrb.* xxvi. 88 ; *Schinz in Mitteil. Geogr. Ges. Lübeck,* xvii. (1903) 17 ; *A. W. Hill in Kew Bulletin,* 1908, 320. *S. aurea, R. Br., var. congesta, Eckl. et Zeyh. ex Drège in Linnæa,* xx. 195. *S. crassulæfolia, Zeyh., S. albens, Zeyh., and S. pallida, Zeyh. non E. Meyer, ex Schinz in Mitteil. Geogr. Ges. Lübeck,* viii. (1903) 17. *S. ambigua, var. gracilis, Cham. in Linnæa,* vi. 346 ; *Schinz in Mitteil. Geogr. Ges. Lübeck,* xvii. (1903) 18 *partly. S. ambigua, var. crassa, Cham. in Linnæa,* vi. 346 ; *Schinz in Mitteil. Geogr. Ges. Lübeck,* xvii. (1903) 18.

COAST REGION : Cape Div. ; in various localities near the sea shore on the Cape Peninsula, *Ecklon,* 77 ! *Burke* ! *Prior* ! *Harvey,* 613 ! *Milne,* 64 ! *MacGillivray,* 593 ! 593b ! *Grey* ! *Schlechter,* 7308 ! *Wolley Dod,* 2013 ! 3062 ! 3273 ! *Wilms,* 3475, partly !

7. **S. gibbosa** (Wolley Dod in Journ. Bot. 1901, 401) ; annual, 6–8 in. high, branching from the base ; leaves in 6–8 pairs, somewhat fleshy, broadly ovate, obtuse or subacute, 3–6 lin. long, 3–6 lin. broad, subcordate at the base ; cymes densely branched, ultimate cymes distinct ; branches 7–9 lin. long ; pedicels $\frac{1}{2}$–1$\frac{1}{2}$ lin. long ; bracts lanceolate, acute ; calyx-segments 1 lin. long, obovate-oblong, concave, obtuse or apiculate, with an apical rather broadly

gibbous keel-wing; corolla-tube 1 lin. long; lobes 1–1½ lin. long,
about ¾ lin. broad, ovate-oblong, obtuse; filaments ¾ lin. long;
anthers ½–¾ lin. long, with an apical gland; style about 2 lin. long,
with inconspicuous glandular swelling; stigma subclavate, entire.
A. W. Hill in Kew Bulletin, 1908, 321. *S. ambigua, var. gracilis,
Cham. in Linnæa*, vi. 346; *Schinz in Mitteil. Geogr. Ges. Lübeck*,
xvii. (1903) 18 *partly.*

COAST REGION : Cape Div. ; Muizenberg Vley, by the railway, *Wolley Dod*, 2332 !

This species resembles *S. ambigua*, Cham., in the character of its leaves and
calyx-segments, but differs markedly in its inflorescences and bracts, in which it
approaches more nearly to *S. aurea*, R. Br.

8. **S. glauca** (A. W. Hill in Kew Bulletin, 1908, 321); annual,
4–5 in. high, simple or branching from base; leaves in 4 pairs,
fleshy, broadly ovate, obtuse, 4–5 lin. long, 3–3½ lin. broad; cymes
densely branched forming a corymb; ultimate cymes crowded;
branches ¾–1¼ in. long; flowers almost sessile; bracts rhomboid-
ovate or narrowly ovate; calyx-segments obovate, concave, truncate,
erose, apiculate, 2–2⅓ lin. long, with a keel-wing broadest at the
middle, ¼–⅓ lin. wide; corolla-tube 2¼ lin. long; lobes 2¼ lin. long,
1½ lin. broad, ovate, subapiculate; filaments ⅓ lin. long; anthers
about ¾ lin. long, with a large apical gland; style about 2 lin. long,
with a median swelling; stigma bilabiate.

COAST REGION : Cape Div. ; roadside near Little Lions head, *Wolley Dod*,
3273a !

Similar to *S. ambigua*, Cham., in vegetative habit, but differs in the cymes not
being collected into dense heads, in the bracts and especially in the calyx-lobes
with their wings broadest at the middle, in which respect it approaches *S. pallida*,
E. Meyer. It is not improbable that this plant may be a hybrid between
S. ambigua and *S. pallida*.

9. **S. aurea** (R. Br. Prodr. 452); annual, 2–9 in. high, simple,
but sometimes branching from the base; leaves in 3–5 pairs,
herbaceous, broadly ovate or cordate, acute or subacute, basal pair
smallest, 1½–8 lin. long, 1–6 lin. broad; cymes very variable, but of
a densely branched corymbose type; branches very variable;
pedicels 2–3 lin. long; calyx-segments lanceolate or oblong-
lanceolate, concave, mucronate, 1–1½ lin. long, with a keel or very
slight keel-wing, broadest at the middle; corolla yellow to white;
tube 1–2 lin. long; lobes 1–2½ lin. long, ⅔–1¼ lin. wide, ovate-oblong,
subacute; filaments ¼–¾ lin. long; anthers ⅖–1 lin. long, with
a large apical gland; style 1–2½ lin. long, with a median
swelling; stigma entire and clavate or subcapitate or bifid with
broad lobes. *Griseb. Gen. et Spec. Gent. 167 and in DC. Prodr.
ix. 52 ; Cham. in Linnæa, vi. 346 ; E. Meyer, Comm. 185 ; Schinz in
Vierteljahrsschr. Naturf. Ges. Zürich, xxxvii. 315 ; Knoblauch in Bot.
Centralbl. lx. 324 ; Gilg in Engl. Jahrb. xxvi. 88 ; Schinz in Mitteil.
Geogr. Ges. Lübeck, xvii. (1903) 18 ; A. W. Hill in Kew Bulletin,
1908, 321. S. aurea, var. genuina, Schinz in Mitteil. Geogr. Ges.*

Lübeck, xvii. (1903) 19 ; *forma wurmbeana, E. Meyer, Comm.* 185
(*probably* =*var. sulphurea, Griseb. Gen. et Sp. Gent.* 167) ; *Schinz in
Mitteil. Geogr. Ges. Lübeck,* xvii. (1903) 19. *S. aurea, var. pallens,
Berg. in Griseb. Gen. et Spec. Gent.* 167 ; *Schinz in Mitteil. Geogr.
Ges. Lübeck,* xvii. (1903) 19 ; *forma cymosa, Schinz in . Mitteil.
Geogr. Ges. Lübeck,* xvii. (1903) 20 ; *forma gracilis, Schinz in Mitteil.
Geogr. Ges. Lübeck,* xvii. (1903) 20. *C. cymosa, Jaroscz, Pl. Nov.
Cap.* 10. *S. minima, Jaroscz, Pl. Nov. Cap.* 11. *Exacum sessile, Linn.
Sp. Pl. ed.* 1, 112, *partly,* ed. 2, 163, *partly* ; *not of Griseb. Gen. et Sp.
Gent.* 113. *Exacum aureum, Linn. f. Suppl.* 123 ; *Lam. Illustr.* i. 321,
t. 80, *fig.* 2. *Gentiana aurea, Thunb. Fl. Cap. ed.* 2, ii. 171.—
Centaurium minus aureum, etc., Pluk. Almagest. ii. 94, *t.* 275, *fig.* 3.
Centaurium angustifolium, etc., Burm. Rar. Afr. Pl. 206, *t.* 74, *fig.* 4.

VAR. β, **alata** (A. W. Hill in Kew Bulletin, 1908, 322), a well-marked varietal
form with usually tall stems and long internodes ; the calyx-segments have
pronounced wings broadest at the middle $\frac{1}{5}$–$\frac{2}{5}$ lin. wide.

SOUTH AFRICA : without locality, *Forster* ! *Wallich* ! *Thom* ! *Pappe* ! Var. β,
Drège ! *Pappe* !

COAST REGION : Clanwilliam Div. ; Wupperthal, *Drège* ! Tulbagh Div. ; near
Tulbagh, *Pappe* ! Worcester Div. ; Hex River Valley, 1700 ft., *Tyson,* 807 !
Cape Div. ; flats and hills around Cape Town, *Burchell,* 158 ! *Harvey,* 616 !
Ecklon, 732 ! 733 ! *Bolus,* 2876 ! *MacOwan and Bolus, Herb. Norm. Austr.-Afr.*
365 ; *Drège* ! *Wolley Dod,* 3330 ! 3111 ! 3204 ! *Wilms,* 3436 ! Slang Kop River,
Wolley Dod, 3254 ! Fish Hoek Valley, *Wolley Dod,* 3438 ! Simon's Bay,
MacGillivray, 592 ! *Wright* ! Riversdale Div. ; Heidelberg, 500 ft., *Galpin,* 4335 !
near Zoetemelks River, *Burchell,* 6731 ! George Div. ; near George, *Burchell,*
6008 ! 6060 ! Var. β : Clanwilliam Div. ; Blue Berg, *Drège* ! Cape Div. ; Sand-
flats between Tigerberg and Blueberg, *Drège* ! Lion Mountain, *Drège* ! Cape
Flats, *Zeyher,* 3420 ! Stellenbosch Div. ; near Stellenbosch, *Marloth,* 3441 !

A very variable plant in general appearance and also in the details of the flower ;
sometimes flowers with entire and 2-lobed stigmas occur on the same individual.
The forms which have been included under *S. aurea* proper can be easily distinguished from the variety var. *alata* by the absence of a distinct wing from the
calyx-segments.

10. S. Schlechteri (Schinz in Engl. Jahrb. xxiv. 454) ; annual,
erect, unbranched, 3–4 in. high, slender ; leaves in 4–5 pairs, ovate-
cordate or broadly ovate, subacute, $1\frac{1}{2}$–3 lin. long, $\frac{3}{4}$–$2\frac{1}{4}$ lin. broad ;
cymes small, simple, 3–7-flowered, compact ; branches $2\frac{1}{2}$ lin. long ;
pedicels very short ; calyx-segments about 2 lin. long, lanceolate,
acute, concave, with keel-wing $\frac{1}{2}$ lin. broad, broadest about the
middle ; corolla yellow ; tube $1\frac{2}{3}$–2 lin. long ; lobes $1\frac{1}{2}$–$1\frac{2}{3}$ lin. long,
$\frac{3}{4}$ lin. wide, ovate, subacute ; filaments $\frac{3}{4}$ lin. long ; anthers about
$\frac{1}{2}$ lin. long, with very small apical gland ; style $1\frac{1}{2}$ lin. long, swelling
not obvious ; stigma shortly clavate, bilabiate. *Schinz in Bull. Herb.
Boiss.* vi. 527 ; *Gilg in Engl. Jahrb.* xxvi. 88 ; *Schinz in Mitteil. Geogr.
Ges. Lübeck,* xvii. (1903) 21.

COAST REGION : Paarl Div. ; French Hoek, 3000 ft., *Schlechter,* 9307 !

11. S. ochroleuca (Wolley Dod in Journ. Bot. 1901, 400) ;
annual, erect, $1\frac{1}{2}$–$3\frac{1}{2}$ in. high, usually simple below, branching above

into a fairly dense more or less corymbose inflorescence ; leaves in
3–6 pairs, broadly ovate, acute, 3–5 lin. long, 2–4 lin. broad ; cymes
compact corymbose ; primary branches $\frac{1}{4}$ to $1\frac{1}{2}$ in. long ; pedicels
from almost sessile to 4 lin. long ; calyx-segments 2 lin. long, ovate,
acute, boat-shaped, mucronate, with a keel-wing $\frac{3}{4}$ lin. broad,
broadest below the middle ; corolla-tube about 2 lin. long ; lobes
2 lin. long, about 1 lin. broad, elliptic, obtuse, apiculate ; filaments
very short ; anthers $\frac{1}{2}$ lin. long, with an apical gland ; style $1\frac{1}{2}$ lin.
long, with a median swelling ; stigma clavate, more or less deeply
bilabiate. *Schinz in Mitteil. Geogr. Ges. Lübeck,* xvii. (1903) 20.

COAST REGION : Cape Div. ; Fish Hoek Valley, *Wolley Dod,* 3436 ! behind
Houtsbay hotel, *Wolley Dod,* 3270 ! west slope of Slang Kop, *Wolley Dod,* 3146 !
3252 ! near Smitswinkel bay, *Wolley Dod,* 3058 !

12. S. Gilgii (Schinz in Mitteil. Geogr. Ges. Lübeck, xvii. 1903,
27) ; annual, erect, unbranched, slender, 2–8 in. high ; leaves in
3–5 pairs, obovate, obtuse or subacute, 1–3 lin. long, $\frac{1}{2}$–$1\frac{1}{4}$ lin. broad,
slightly coriaceous ; cymes few-flowered, compact ; branches 3–5 lin.
long ; calyx-segments $1\frac{1}{2}$–$1\frac{3}{4}$ lin. long, obovate-oblong, slightly erose,
mucronate, with narrow keel-wing, most definite above middle ;
corolla-tube $1\frac{3}{4}$–2 lin. long ; lobes $1\frac{3}{4}$ lin. long, ovate, obtuse or
subacute ; filaments $\frac{2}{3}$–$\frac{3}{4}$ lin. long ; anthers $\frac{1}{3}$–$\frac{1}{2}$ lin. long, with an
apical gland ; style 1 lin. long, with swelling ; stigma capitate
A. W. Hill in Kew Bulletin, 1908, 322.

COAST REGION : Cape Div. ; near a stream on a flat on Muizenberg Mountain,
alt. 1300 ft., *Schlechter,* 150 !

13. S. pallida (E. Meyer, Comm. 185) ; annual, erect, 3–5 in.
high, simple or branching from the base ; leaves in about 4 pairs,
herbaceous, cordate or ovate, more or less acute, basal leaves
smaller, 2–4 lin. long, 2–4 lin. broad ; cymes usually consisting
of 2 main branches, more or less densely branched ; corymbose
branches 6–12 lin. long ; pedicels 2–3 lin. long ; calyx-segments
oblong-lanceolate, concave, acute or mucronate, 2–$2\frac{1}{2}$ lin. long, with
a keel-wing $\frac{1}{2}$–$\frac{3}{4}$ lin. wide, widest below the middle ; corolla white ;
tube $1\frac{3}{4}$–2 lin. long ; lobes $2\frac{1}{2}$–3 lin. long, $1\frac{1}{2}$–$1\frac{3}{4}$ lin. wide, ovate-
elliptic, subacute ; filaments about $\frac{1}{2}$ lin. long ; anthers $\frac{3}{4}$–1 lin.
long, with an apical gland ; style 2–$2\frac{3}{4}$ lin. long, with a median
swelling ; stigma bilabiate. *A. W. Hill in Kew Bulletin,* 1908, 323.
S. aurea, var. pallida, Schinz in Mitteil. Geogr. Ges. Lübeck, xvii.
(1903) 20.

COAST REGION : Tulbagh Div. ; Tulbagh Kloof, 300 ft., *Bolus* ! Malmesbury
Div. ; near Moorreesburg, *Bolus,* 9992 ! Cape Div. ; Fishoek Valley, *Wolley Dod,*
3437 ! Cape Flats, *Wolley Dod,* 425 ! Lion Mountain, *Drège* ! *Prior* ! *Bolus,* 7212 !
Flats between Cape Town and Tyger Valley, *Drège* ! Simons Bay, *Wright,* 98 !
Swellendam Div. ; mountains along the lower part of the Zondereinde River,
Zeyher, 1187, partly !

This species, which has been merged into *S. aurea* by Schinz, is no doubt very
closely allied to it, but is easily distinguished by the wide basal wings of the calyx-
segments and by the larger and more prominent corollas.

14. **S. compacta** (A. W. Hill in Kew Bulletin, 1908, 323);
perennial (?), cæspitose, densely virgately branched to form a many-
flowered inflorescence, 2–4 in. high ; leaves few, linear-lanceolate or
more rarely ovate-lanceolate, acute or acuminate, $1\frac{1}{2}$–$9\frac{1}{2}$ lin. long,
$\frac{1}{2}$–2 (rarely $2\frac{1}{2}$) lin. broad, somewhat fleshy ; bracts linear-lanceolate,
acute, about 6 lin. long ; calyx $6\frac{1}{2}$–$7\frac{1}{2}$ lin. long ; tube cylindrical or
slightly inflated ; segments 2 lin. long, lanceolate, acute, keeled ;
corolla-tube 5–6 lin. long ; lobes 3–$4\frac{1}{2}$ lin. long, $1\frac{1}{2}$–$2\frac{3}{4}$ lin. broad,
obovate-oblong, apiculate or subacute ; filaments very short, inserted
about $\frac{1}{4}$ lin. below sinuses ; anthers 1–$1\frac{1}{4}$ lin. long, with small apical
and minute basal glands sometimes absent ; style $3\frac{1}{2}$ lin. long, with
large swelling $\frac{1}{2}$–1 lin. long, 1 lin. above the base ; stigma shortly
bilabiate ; capsule ovoid.

CENTRAL REGION : Graaff Reinet Div. ; Dutoits (Farm ?) under Compass Berg
in cultivated ground, 5500–6000 ft., *Bolus,* 1853 ! Middelburg Div. ; near Middelburg, *Denoon,* 37 (in *Herb. Guthrie,* 1042) ! *Shaw* ! Colesberg Div. ; near Colesberg,
Shaw ! Colesberg Kopje, *Mrs. Barber,* 10 !

KALAHARI REGION : Griqualand West ; Dutoits Pan, near Kimberley, *Mrs.
Barber,* 21 ! Orange River Colony ; between Bloemfontein and Petrusburg, *Miss
Kensit in Herb. Bolus,* 12992 ; Bechuanaland ; by the Mashowa River, near
Takun, *Burchell,* 2252/4 !

A very distinct and striking species, apparently growing in the form of small
cushions covered with masses of flowers. The roots of this plant are very short
and thick and contain a mycorhizal fungus.

15. **S. pusilla** (Eckl. in Linnæa, vi. 346); annual, erect, simple,
sometimes branched, $\frac{1}{2}$–$1\frac{1}{2}$ in. high ; leaves few, lower very small,
upper as well as the bracts 3–4 lin. long, $\frac{1}{2}$–1 lin. broad, narrowly
ovate-lanceolate or elliptic, acute, somewhat fleshy ; inflorescences
3–5-flowered, with relatively conspicuous bracts ; branches 1–$1\frac{1}{2}$
lin. long ; pedicels $\frac{1}{2}$–$1\frac{1}{2}$ lin. long ; calyx-segments 2–3 lin. long,
elliptic or narrowly ovate-lanceolate, acute, keeled ; corolla-tube
$2\frac{1}{4}$–3 lin. long, cylindrical ; lobes $\frac{3}{4}$–1 lin. long, $\frac{1}{2}$ lin. broad, orbicular
or elliptic-obovate ; filaments $\frac{3}{4}$–$1\frac{1}{2}$ lin. long, inserted 1–$1\frac{1}{4}$ lin. above
the base of the tube ; anthers $\frac{1}{4}$–$\frac{1}{3}$ lin. long, with three small glands ;
style 1–2 lin. long ; stigma capitate. *Cham. in Linnæa,* viii.
53 ; *Griseb. Gen. et Spec. Gent.* 169 ; *Schinz in Bull. Herb. Boiss.*
2me *sér.* vi. 731. *Lagenias pusillus, E. Meyer, Comm.* 186 ; *Griseb.
in DC. Prodr.* ix. 54 ; *Schinz in Vierteljahrsschr. Naturf. Gesellsch.
Zürich,* xxxvii. 308.

COAST REGION : Clanwilliam Div. ; Cederberg Range, *Drège.* Paarl Div. ;
Paarl Mountain, 1000 ft., *Drège* ! Cape Div. ; Table Mountain, *Ecklon,* 731 ! Cape
Flats, *Harvey* ! Simons Bay, *Wright,* 95 ! and without precise locality, *Harvey,*
614 ! Caledon Div. ; Houw Hoek, 900 ft., *Schlechter,* 9376 partly !
WESTERN REGION : Vanrhynsdorp Div. ; Gift Berg, *Drège* !

16. **S. rara** (Wolley Dod in Journ. Bot. 1901, 401); annual,
erect, unbranched, $1\frac{1}{2}$–4 in. high ; leaves few, linear or linear-
lanceolate, subacute, 3–5 lin. long, $\frac{1}{2}$–1 lin. broad ; inflorescences
branched, 3–12-flowered ; branches about $2\frac{1}{2}$ lin. long ; pedicels

$\frac{1}{2}$ lin. long; calyx-segments 3–4 lin. long, narrowly lanceolate, acuminate, concave, with keel-wing $\frac{3}{4}$ lin. broad, broadest below the middle; corolla-tube $3\frac{1}{4}$ lin. long; lobes $1\frac{1}{4}$–$1\frac{1}{2}$ lin. long, $\frac{1}{2}$ lin. broad, elliptic, subacute; filaments very short, inserted $\frac{1}{3}$ lin. below the sinuses; anthers $\frac{1}{3}$–$\frac{1}{2}$ lin. long, glands minute or absent; style $1\frac{1}{2}$–$1\frac{3}{4}$ lin. long; stigma shortly bilobed. *Schinz in Bull. Herb. Boiss.* 2me *sér.* vi. 733.

SOUTH AFRICA: without locality, *Wallich.*
COAST REGION : Cape Div. ; Cape Flats between Uitvlugt and Duinefontein Road, *Wolley Dod,* 3413 ! Caledon Div. ; Houw Hoek, 900 ft., *Schlechter,* 9376 partly ! Caffraria, *Fraser* (in Edinburgh Herbarium) !

17. **S. mirabilis** (Gilg in Engl. Jahrb. xxvi. 92) ; annual, erect, unbranched, 3–8 in. high ; stem-wings glandular especially near the base of the stem ; leaves in 3–5 pairs, linear or linear-lanceolate, acute, 1–3 lin. long, about $\frac{1}{2}$ lin. broad, with a glandular patch on the back, upper leaves the larger ; cymes lax, spreading, 1–9-flowered ; branches up to $1\frac{3}{4}$ in. long ; pedicels 4–5 lin. long ; calyx-segments united below to form a shallow cup, 3–4 lin. long, ovate or elliptic, acuminate, with keel-wing about $\frac{1}{2}$ lin. broad, broadest at the base ; corolla-tube $2\frac{1}{2}$–$3\frac{1}{2}$ lin. long ; lobes 4–5 lin. long, $1\frac{3}{4}$–$2\frac{1}{4}$ lin. broad, elliptic-lanceolate, acute ; filaments $\frac{3}{4}$–$1\frac{1}{4}$ lin. long ; anthers 1–$1\frac{1}{4}$ lin. long, with small apical gland ; style $3\frac{1}{2}$–4 lin. long, swelling when present about 2 lin. above the base ; stigma clavate, variable in size up to $\frac{3}{4}$ lin. long, bilabiate, papillose on all sides. *Schinz in Mitteil. Geogr. Ges. Lübeck,* xvii. (1903) 35 ; *A. W. Hill in Kew Bulletin,* 1908, 324. *S. pratensis, Gilg in Engl. Jahrb.* xxx. 377, 378, *figs.* A–F*; Schinz in Mitteil. Geogr. Ges. Lübeck,* xvii. (1903) 38 *; Baker & N. E. Br. in Dyer, Flor. Trop. Afr.* iv. i. 550.

KALAHARI REGION : Orange River Colony, *Cooper,* 2756 ! Transvaal ; Spitz Kop, near Lydenburg, *Wilms,* 971 !
EASTERN REGION: Transkei ; near Tsomo, *Mrs. Barber,* 845 ! near Kentani, 1000 ft., *Miss Pegler,* 1187 ! Tembuland, Bazeia Mountain, 4000 ft., *Baur,* 621 ! Griqualand East ; "Woodlands" Farm, in Maclear District, 5600 ft., *Galpin,* 6772 ! Natal ; Mid-Illovo, 1000-2000 ft., *Wood,* 1884 !

S. mirabilis shows close relationship to *S. exigua* and *S. filiformis,* Schinz, in the glandular wings of the stem, the bilabiate stigma and in its general habit, it is distinguished chiefly by the winged calyx-segments, which are only slightly united at the base.

18. **S. Junodii** (Schinz in Bull. Herb. Boiss. iv. 442) ; annual, erect, slender, unbranched, 3–6 in. high ; leaves in 4–5 pairs, inconspicuous, ovate-lanceolate, acute, $\frac{1}{2}$–$1\frac{1}{2}$ lin. long ; cymes few-flowered, variable, lax ; branches when developed $1\frac{1}{4}$–$1\frac{1}{2}$ in. long ; pedicels 1–4 lin. long ; calyx-segments elliptic-lanceolate, with narrow wing, acuminate, $2\frac{1}{4}$ lin. long, slightly united at the base ; corolla-tube 2 lin. long ; lobes $1\frac{1}{2}$–2 lin. long, narrowly ovate, acute ; filaments $\frac{1}{2}$–1 lin. long ; anthers $\frac{1}{2}$ lin. long, with very small apical

gland ; style 2 lin. long, with slight swelling ; stigma shortly clavate. *Gilg in Engl. Jahrb.* xxvi. 93 ; *Schinz in Mitteil. Geogr. Ges. Lübeck,* xvii. (1903), 25.

KALAHARI REGION : Transvaal; Houtbosch Berg, in shady places, 6300 ft., *Schlechter,* 4767 !
EASTERN REGION : Natal ; Howick, in shady places, 3200 ft., *Schlechter,* 6783 !

19. **S. filiformis** (Schinz in Bull. Herb. Boiss. iii. 411) ; annual, erect ; stems slender, often hairy towards the base, $2\frac{1}{2}$–7 in. high, usually simple and terminated by a single flower, rarely 2-flowered ; leaves linear or linear-lanceolate, acute, about 1 lin. long, separated by long internodes ; calyx-segments 2–$3\frac{1}{2}$ lin. long, elliptic or ovate-lanceolate, acuminate, slightly fused together at the base and with narrow keel-wing ; corolla-tube 3–4 lin. long ; lobes 4 lin. long, 2–$2\frac{1}{4}$ lin. broad, ovate-lanceolate, acute ; filaments $\frac{1}{4}$–$\frac{1}{2}$ lin. long ; anthers $\frac{3}{4}$–1 lin. long, with slender elongated apical glands ; style $3\frac{1}{2}$–4 lin. long, with a large swelling up to 1 lin. long, 1 lin. above the base ; stigma bilobed, lobes $\frac{1}{2}$–$1\frac{1}{2}$ lin. long. *Gilg in Engl. Jahrb.* xxvi. 93 ; *Schinz in Mitteil. Geogr. Ges. Lübeck,* xvii. (1903), 26.

COAST REGION : Komgha Div. ; hills near Komgha, 2000 ft., *Flanagan,* 37 !
KALAHARI REGION : Transvaal; near Lydenburg, *Wilms,* 972 ! Tzaneen Estate, Zoutspan Berg, 4500 ft., *Evans,* 4014 !
EASTERN REGION : Transkei ; Cats Pass in Kentani Div., 600 ft., *Miss Pegler,* 1188 ! Griqualand East ; Leitsa footpath on the Drakensberg Range, Maclear Div., 7600 ft., *Galpin,* 6773 ! Natal ; Polela, 5000 ft., *Wood* (in Natal Herb.), 1884 !

20. **S. exigua** (Schinz in Mitteil. Geogr. Ges. Lübeck, xvii. 1903, 26) ; annual, erect, simple, rarely branching, 3–8 in. high ; wings of the stem often with short glandular hairs ; leaves linear-lanceolate, acute, 1–$4\frac{1}{2}$ lin. long ; cymes 1–3- (less commonly 5-) flowered ; branches $\frac{1}{2}$–2 in. long, spreading ; pedicels $\frac{1}{4}$–$\frac{3}{4}$ in. long ; calyx-segments $3\frac{1}{2}$–$4\frac{1}{2}$ lin. long, elliptic-lanceolate, acute or acuminate, keeled, united below to form a cup $1\frac{1}{2}$–2 lin. long ; corolla-tube $2\frac{1}{2}$–$3\frac{1}{2}$ lin. long ; lobes $3\frac{1}{2}$–$4\frac{1}{2}$ lin. long, about 2 lin. broad, broadly ovate, unguiculate, obtuse or subacute ; filaments variable, 1–$1\frac{1}{2}$ lin. long ; anthers 1–$1\frac{1}{2}$ lin. long, with minute apical glands sometimes almost invisible : style $2\frac{1}{2}$–$3\frac{1}{2}$ lin. long with swelling near the base ; stigma shortly bilabiate, with thick lobes. *S. linearifolia, Schinz in Vierteljahrsschr. Naturf. Gesellsch. Zürich,* xxxvii. 321 ; *Gilg in Engl. Jahrb.* xxvi. 93. *Chironia exigua, Oliver in Hook. Ic. Pl. t.* 1229.

CENTRAL REGION : Colesberg Div. ; near Naauwpoort, *Wenyon,* 30 ! Aliwal North Div. ; Elands Hoek, near Aliwal North, 4800 ft., *Bolus,* 12994 ! Barkly East ; near Rhodes, 6150 ft., *Galpin,* 2333 !
KALAHARI REGION : Griqualand East ; Diamond Fields, *Mrs. Barber,* 11 ! 22 ! Orange River Colony ; near Seven Fountains, *Burke,* 442 ! and probably from the same locality, *Zeyher,* 1191 ! Bloemfontein, *Rehmann,* 3799 ! Parys, *Grey College Herb.* 609 ! and without precise locality, *Mrs. Barber* ! Transvaal ; between Elands River and Klippan, *Rehmann,* 5062 ! near Johannesburg, *Galpin,* 1395 ! 1396 ! near Pietersburg, 4700 ft., *Schlechter,* 4361 !

This species may be distinguished by the campanulate calyx, the segments being fused for about $\frac{1}{3}$ of their length; *S. filiformis* also shows this character to a less marked extent; it can be distinguished from *S. exigua* by its more slender habit, and minute leaves, narrow keel-wings, the relatively large glands of the anthers and short anthers, filaments and the elongate bilabiate stigma. Wilms, 972 from the Northern Transvaal, which is one of the types of Schinz's species, shows certain affinities with the more slender forms of *S. exigua*. Both these species are closely allied to *Sebæa mirabilis*, but this latter species is distinguished by the presence of a conspicuous wing to the calyx-segments. From some of the specimens collected by Burke, however, it is not always easy to separate *S. exigua* from *S. mirabilis* by this character. In both, the stigmatic characters are similar, but *S. mirabilis* is usually the stronger plant. *S. filiformis* is also closely related to *S. Welwitschii* from Lower Guinea, and like that species and probably also *S. exigua* and *S. mirabilis* may be parasitic or possess mycorhiza.

21. S. caladenia (Gilg in Engl. Jahrb. xxvi. 89); annual, erect, slender, 2–3 in. high; leaves ovate-oblong, subacute, herbaceous, $2\frac{1}{2}$–$4\frac{1}{2}$ lin. long, $2\frac{1}{2}$–3 lin. broad; cymes lax, 5–7-flowered; branches 4 lin. long; pedicels $\frac{1}{2}$ lin. long; calyx-segments lanceolate; acuminate, with keel $2\frac{1}{4}$ lin. long; corolla-tube 3–$3\frac{1}{4}$ lin. long; lobes $1\frac{1}{2}$–$1\frac{3}{4}$ lin. long, $\frac{3}{4}$–1 lin. broad, ovate, obtuse; filaments $\frac{1}{2}$ lin. long; anthers 1–$1\frac{1}{2}$ lin. long, with apical and two basal glands; style $2\frac{1}{2}$ lin. long, with swelling; stigma bilabiate. *Schinz in Mitteil. Geogr. Ges. Lübeck*, xvii. (1903) 40.

Central Region: Calvinia Div.; Hantam Mountains, *Meyer*.

22. S. acutiloba (Schinz in Bull. Herb. Boiss. iii. 412); annual, erect, weak, unbranched, 2–4 in. high, with long internodes; leaves in 2–4 pairs, orbicular-ovate, acute, 1–2 lin. long, 1–2 lin. broad; cymes 3–5-flowered; branches slender, 2–3 lin. long; pedicels about $\frac{1}{2}$ lin. long; calyx-segments narrowly elliptic, acuminate, $1\frac{1}{2}$–$1\frac{3}{4}$ lin. long, with a distinct keel-wing widest about the middle; corolla-tube about $1\frac{1}{2}$ lin. long; lobes $\frac{3}{4}$–1 lin. long, $\frac{1}{2}$ lin. broad, ovate, acute; filaments $\frac{1}{4}$ lin. long; anthers $\frac{1}{4}$ lin. long, with very small apical gland; style $\frac{3}{4}$ lin. long; stigma and swelling confluent. *Gilg in Engl. Jahrb.* xxvi. 100; *Schinz in Mitteil. Geogr. Ges. Lübeck*, xvii. (1903), 39. *S. Tysonii, Schinz in Bull. Herb. Boiss.* 2me sér. viii. 703.

Coast Region: Knysna Div.; Zuur Vlakte, *Tyson*!
Eastern Region: Natal; near Clairmont, 60 ft., *Schlechter*, 3045! *Wood*, 4950!

S. Tysonii agrees exactly with *S. acutiloba*, but it is possible that there may be some mistake as to the locality written by MacOwan on the sheet of the former.

23. S. grisebachiana (Schinz in Vierteljahrsschr. Naturf. Gesellsch. Zürich, xxxvii. 322); annual, erect, simple or branched, 2–3 in. high; leaves in 4–6 pairs, the lowest separated by short internodes and giving an appearance of a leaf-rosette, $1\frac{1}{2}$–$2\frac{1}{2}$ lin. long and broad, ovate-orbicular, subacute; cymes few-flowered, compact; branches 1–5 lin. long; pedicels 1–2 lin. long; calyx-segments 2–$2\frac{1}{2}$ lin. long, lanceolate, acuminate, with narrow keel-wing, broadest about the middle; corolla-tube 2–3 lin. long; lobes

2½–3 lin. long, oblong, acute or obtuse; filaments ⅓ lin. long; anthers ¾–1 lin. long, with small apical gland; style 2–3 lin. long, with swelling below the middle; stigma clavate. *Gilg in Engl. Jahrb.* xxvi. 95; *Schinz in Mitteil. Geogr. Ges. Lübeck,* xvii. (1903) 36.

COAST REGION: George Div.; between Touw River and Kaymans River, *Burchell,* 5778! Montagu Pass, *Rehmann,* 264! Div. and locality?, *Krebs,* 233.

24. **S. schizostigma** (Gilg in Engl. Jahrb. xxvi. 93); annual, erect, usually unbranched, 2–5 in. high; leaves ovate-orbicular or subcordate, acute or shortly apiculate, 2–3½ lin. long, 2–3½ lin. broad; cymes 3–6-flowered; branches about 4 lin. long; pedicels about 2 lin. long; calyx-segments ovate-lanceolate, acute or acuminate, 2–2½ lin. long, with keel-wing broadest below middle, ½ lin. broad; corolla-tube 2–3 lin. long; lobes 2½–3 lin. long, ovate-lanceolate, subacute, 1–1½ lin. broad; filaments ¾ lin. long; anthers ¾–1 lin. long, without glands; style about 3 lin. long, with swelling; stigma large, ligulate, 1 lin. long. *Schinz in Mitteil. Geogr. Ges. Lübeck,* xvii. (1903) 36.

COAST REGION: Swellendam Div.; by the Kenko (Doorn) River, *Zeyher,* 1188! Riversdale Div.; Garcias Pass, 1400 ft., *Bolus,* 11347!

25. **S. saccata** (Schinz in Mitteil. Geogr. Ges. Lübeck, xvii. 1903, 25); annual, erect, unbranched, 2–6 in. high; leaves in 4–5 pairs, triangular-ovate, acute, 1–4 lin. long, ½–2½ lin. broad; cymes 1–5-flowered, compact; branches 2½–3½ lin. long; pedicels 2½–4 (rarely 6) lin. long; calyx-segments 2½–3 lin. long, ovate-lanceolate, acute, concave, with a keel-wing ½ lin. broad, broadest near the base; corolla-tube 3 lin. long; lobes 4 lin. long, 2 lin. broad, ovate, subacute; filaments 1 lin. long; anthers 1¼–1½ lin. long, with apical glands; style 2¾–3¼ lin. long, with swelling; stigma broadly ligulate, ½ lin. long.

KALAHARI REGION: Transvaal; Houtbosch Berg, 6800 ft., *Schlechter,* 4702! Secocoenis land in Middelburg district, *Gray,* 3767.

26. **S. Zeyherii** (Schinz in Vierteljahrsschr. Naturf. Gesellsch. Zürich, xxxvii. 325); annual, erect, simple or branched, 1½–8 in. high; leaves broadly or orbicular ovate, subacute, 2½–4 lin. long, 2½–3 lin. broad, sometimes crowded near the base; inflorescences compact, few- to many-flowered; branches ¼–1 in. long; pedicels ½–2½ lin. long; calyx-segments 2–2½ lin. long, narrowly elliptic, acuminate, with a broad thick wing, ½–¾ lin. broad, broadest near the base; corolla-tube 2¼–2¾ lin. long; lobes 1¾–2¼ lin. long, ¾ lin. broad, ovate, acute; filaments about ½ lin. long; anthers ⅓–½ lin. long, with minute apical glands; style 1¼–1¾ lin. long, with a large elongated swelling just below the stigma and more or less confluent with it; stigma clavate. *Gilg in Engl. Jahrb.* xxvi. 99; *Schinz in Mitteil. Geogr. Ges. Lübeck,* xvii. (1903), 38.

COAST REGION: Cape Div. ; Slang Kop, *Wolley Dod,* 3022! behind Wynber Butts, *Wolley Dod,* 3323! near Simonstown, *Wolley Dod,* 2844! near Smits-winkel Bay, *Wolley Dod,* 2930! Kenilworth, near Capetown, 70 ft., *Bolus,* 7924! Paarl Div.; near French Hoek, *Bolus*! Stellenbosch Div. ; Gordons Bay, *Bolus,* 9993! Caledon Div. ; near Hermanus, *Bolus*! Swellendam Div. ; near the Kenko (Doorn) River, *Zeyher,* 1188! Riversdale Div. ; between Garcias Pass and Muis Kraal, 1700 ft., *Bolus,* 11346!

This species resembles *S. schizostigma,* but can be distinguished by the character of the style and by the smaller anthers ; it has been confused with *S. micrantha* (*S. cordata,* E. Meyer, var. *micrantha,* Cham. & Schlecht), with which plant there is very close external resemblance, moreover they both grow on the Cape Peninsula. *S. micrantha* differs, however, in the possession of short stiff hairs on the edges of the large calyx-wings as well as in the position of the anthers in the corolla-throat.

27. S. sulphurea (Cham. & Schlecht. in Linnæa, i. 192); annual, erect, simple or branched, 1–6 in. high ; leaves broadly ovate or ovate, acute, 2–4 lin. long, 1–3 lin. broad, somewhat fleshy ; inflorescences 1- to many-flowered ; branches variable in length, often long and axillary ; pedicels 4–6 lin. long ; calyx-segments elliptic, acute, 3 lin. long, with keel-wing broadest at the base, about $\frac{1}{2}$ lin. broad ; corolla-tube 2–3 lin. long ; lobes 4–4$\frac{1}{2}$ lin. long, 1$\frac{1}{2}$–2 lin. broad, ovate, acute ; filaments $\frac{1}{4}$ lin. long ; anthers 1 lin. long, with large apical and two small basal glands ; style 3 lin. long, with swelling ; stigma ligulate, 1$\frac{1}{4}$ lin. long. *Cham. in Linnæa,* vi. 346 ; *Griseb. Gen. et Spec.* 168, *and in DC. Prodr.* ix. 53 ; *Schinz in Vierteljahrsschr. Naturf. Gesellsch. Zürich,* xxxvii. 319 ; *Gilg in Engl. Jahrb.* xxvi. 89.

COAST REGION: Cape Div. ; summit of Table Mountain, 3500 ft., *Ecklon,* 730! *Miss Kensit in Herb. Bolus,* 10749! tops of the Twelve Apostles, *Wolley Dod,* 3357! Steenberg Plateau to the summit of Constantia Berg, *Wolley Dod,* 3580! Caledon Div. ; Steenbrass River, in sandy places, 1500 ft., *Schlechter,* 5404! Walker Bay, near Hermanus, below 100 ft., *Bolus,* 9836!

28. S. scabra (Schinz in Mitteil. Geogr. Ges. Lübeck, xvii. 1903, 37) ; annual, erect, 3–6 in. high, simple or branched ; leaves broadly ovate or cordate, acute or subacute, 6–2$\frac{1}{2}$ lin. long, 6–2$\frac{1}{2}$ lin. broad ; inflorescences few- to many-flowered ; branches 3–6 lin. long ; pedicels 2–5 lin. long ; calyx-segments 3–5 lin. long, elliptic, acuminate, keel-wings very pronounced, widest at the base, $\frac{1}{2}$–1$\frac{1}{4}$ lin. broad ; corolla-tube 2$\frac{1}{2}$–3$\frac{1}{2}$ lin. long ; lobes 3–6 lin. long, 1$\frac{1}{2}$–3 lin. broad, elliptic-obovate ; filaments $\frac{1}{4}$–$\frac{1}{2}$ lin. long ; anthers 1–1$\frac{1}{2}$ lin. long, with large apical and two small basal glands ; style 2$\frac{1}{2}$–3$\frac{1}{2}$ lin. long, with swelling near the base ; stigma clavate, bilabiate. *S. pentandra, var. belmontioides, Schinz in Viertel-jahrsschr. Naturf. Gesellsch. Zürich,* xxxvii. (1891) 320.

COAST REGION : Cape Div. ; Cape Flats, near Doorn Hoogte, *Zeyher,* 1189a! Riversdale Div. ; near Riversdale, 450 ft.! *Schlechter,* 1711! *Bolus,* 11348! 11349!

29. S. erosa (Schinz in Bull. Herb. Boiss. 2me sér. vi. 728); annual, erect, simple or branching from the base ; stems slender,

3–6 in. high; leaves $1\frac{1}{2}$–$1\frac{3}{4}$ lin. long, $\frac{3}{4}$–$1\frac{1}{4}$ lin. broad, ovate below, ovate-lanceolate above, acute; cymes lax, few-flowered; branches $\frac{3}{4}$ in. long or less; pedicels 3–7 lin. long; calyx-segments $2\frac{3}{4}$ lin. long, connate below into a tube $\frac{1}{2}$–$\frac{3}{4}$ lin. long, ovate-lanceolate, acuminate, slightly keeled; corolla-tube $2\frac{1}{2}$ lin. long; lobes 2–$2\frac{1}{2}$ lin. long, 1–$1\frac{1}{2}$ lin. broad, obovate-oblong, cuspidate and erosely toothed; filaments very short; anthers $\frac{1}{2}$ lin. long, situated in the corolla-throat about $\frac{1}{2}$ lin. below the sinuses, with a conical apical gland $\frac{1}{4}$–$\frac{1}{2}$ lin. long; style 2–3 lin. long, with swelling near the base; stigma capitate-clavate, shortly bilobed.

KALAHARI REGION: Transvaal; in damp places near Brug Spruit, 4700 ft., *Schlechter*, 4119 (wrongly distributed as 2119)!

30. **S. pygmæa** (Schinz in Bull. Herb. Boiss. 2me sér. vi. 740); annual, erect, unbranched, slender, $1\frac{1}{2}$–$2\frac{1}{2}$ in. high; leaves ovate, acute or subacute, 4–6 pairs, $\frac{3}{4}$–1 lin. long, about $\frac{3}{4}$ lin. broad, herbaceous; cymes 2–5-flowered, lax; branches 4–7 lin. long; pedicels about 4 lin. long; calyx-segments narrowly elliptic, acuminate, keeled, 2–$2\frac{1}{2}$ lin. long, united below to form a short tube; corolla-tube 2–$2\frac{1}{2}$ lin. long; lobes $1\frac{3}{4}$–2 lin. long, $\frac{3}{4}$–1 lin. broad, obovate-oblong, cuspidate; filaments very short, inserted in tube $\frac{1}{2}$ lin. below the sinuses; anthers about $\frac{1}{3}$ lin. long, with relatively large apical gland; style $1\frac{1}{2}$ lin. long, with a medium swelling; stigma shortly clavate-ligulate. *A. W. Hill in Kew Bulletin*, 1908, 325.

KALAHARI REGION: Orange River Colony; top of Moolmans Kopje, at Zaaihoek in Harrismith district, 6500 ft., *Thode*. Transvaal; Houtbosch Berg, near Mamavulu, in damp places, 6800 ft., *Schlechter*, 4708!

This species is very closely related to *S. erosa*, Schinz, and is very probably only a smaller Alpine form, it differs principally in the smaller size of the anthers and style and in the absence of definite teeth to the corolla-segments.

31. **S. exacoides** (Schinz in Bull. Herb. Boiss. 2me sér. vi. 728); annual, erect, somewhat slender, 2–9 in. high, simple, more rarely branched; leaves in 3–4 pairs, broadly ovate, acute, cordate at the base, $2\frac{1}{2}$–7 lin. long, 2–7 lin. broad; internodes long; inflorescences few- to many-flowered; branches $\frac{1}{2}$–1 lin. long; pedicels $\frac{1}{4}$–$\frac{1}{2}$ lin. long; calyx-segments 6–7 lin. long, ovate, lanceolate, acuminate, with keel-wing $1\frac{1}{2}$–2 lin. broad, broadest at the base, strongly reticulate and rough, beset with minute hairs on the edge; corolla-tube 6–7 lin. long; lobes 4–6 lin. long, 2–3 lin. broad, broadly elliptic or obovate-unguiculate, apiculate or sometimes dentate; filaments very short, inserted in the tube $\frac{3}{4}$–1 lin. below the sinuses; anthers $\frac{3}{4}$–1 lin. long, with apical and two basal glands; style $5\frac{1}{2}$–6 lin. long, with swelling 1–$1\frac{1}{3}$ lin. long, 1–$1\frac{1}{3}$ lin. above the ovary; stigma about 1 lin. long, bilabiate always about or above the level of the anthers. *S. cordata, R. Br. Prodr.* 452; *Griseb. Gen. et Spec. Gent.* 164; var. *macrantha, Cham. & Schlecht. in Linnæa,*

i. 191 ; *Cham. in Linnæa,* vi. 345.　*Gentiana exacoides, Linn. Spec. Pl. ed.* 2, 332 ; *Thunb. Prodr.* 47, *and Fl. Cap. ed.* 2, ii. 172. *Exacum cordatum, Linn. f. Suppl.* 124.　*Belmontia cordata, E. Meyer, Comm.* 183 ; *Griseb. in DC. Prodr.* ix. 54 ; *Schinz in Vierteljahrsschr. Naturf. Gesellsch. Zürich,* xxxvi. 329 ; *Knoblauch in Bot. Centralbl.* lx. 326.—*Centaurium perfoliatum,* etc., *Burm. Rar. Afr. Pl. t.* 74, *fig.* 5.　*Centaurium capense, minus,* etc., *Seba, Mus.* i. *t.* 22, *fig.* 7.

SOUTH AFRICA : without locality, *Ecklon & Zeyher,* 653, *Thom,* 707 ! 734 ! *Harvey,* 617 ! *Pappe* ! *Forbes* !

COAST REGION : Malmesbury Div.; near Moorreesburg, 500 ft., *Bolus* ! near Groene Kloof (Mamre), 300 ft., *Bolus,* 4429 ! Paarl Div.; between Paarl and Lady Grey Bridge, *Drège* ! Cape Div.; various localities in the vicinity of Cape Town, *Spielhaus, Bowie* ! *Ecklon,* 729 ! *Zeyher,*.1189 ! *Hooker,* 437 ! *Cooper,* 2753 ! *Bolus,* 2875 ! 4429 ! *MacOwan,* 365 ! *Rehmann,* 759 ! 1708 ! *Wilms,* 3472 ! *Schlechter,* 1377 ! *Wolley Dod,* 169 ! 170 ! 3269 ! Simons Bay, *Wright* ! Stellenbosch Div.; Stellenbosch, *Sanderson,* 968 ! Caledon Div.; by the River Zondereinde, *Pappe* ! Swellendam Div.; near Swellendam, *Mund, Prior* ! Riversdale Div.; Garcias Pass, 1200 ft., *Galpin,* 4339 !

In the Dublin herbarium there are 2 specimens collected by Pappe in moist places on the Rivier Zonder Einde, which have dentate edges to the corolla-lobes, but otherwise agree with the typical form, in which a slight dentation of the petals is sometimes visible.

32. S. micrantha (Schinz in Bull. Herb. Boiss. 2me sér. vi. 739, partly) ; annual, erect, unbranched, 1–5 in. high ; leaves in 3–4 pairs, usually collected towards the base of the stem, broadly ovate, rounded or cordate at the base, obtuse or subacute, apiculate, slightly fleshy, 2½–4 lin. long, 2–4 lin. broad ; inflorescences compact, dense, conspicuous from the large calyx-wings ; branches 1–2 (more rarely 4–5) lin. long ; pedicels about 1 lin. long ; bracts ovate to linear-lanceolate, acute, fringed with short bristle-like hairs ; calyx-segments lanceolate, acuminate, 3–3½ lin. long, with a wing 1¼–1½ lin. broad, broadest at the base, strongly veined and fringed on the edge with short bristle-like hairs ; corolla-tube 2½–2¾ lin. long ; lobes 1½–2 lin. long, ½–¾ lin. broad, narrowly obovate or elliptic, acute, sometimes slightly erose ; filaments very short, inserted about ¼ lin. below the sinuses ; anthers about ¼ lin. long, with apical gland ; style 1¾–2 lin. long ; stigma clavate-bilabiate, confluent with the swelling. *S. cordata,* var. *micrantha, Cham. & Schlecht. in Linnæa,* i. 192 ; *Griseb. Gen. et Spec. Gent.* 165.　*Belmontia cordata,* var. *micrantha, Griseb. in DC. Prodr.* ix. 54 *partly. B. micrantha, Gilg in Engl. Jahrb.* xxvi. 102 *partly.—Centaurium perfoliatum, æthiopicum, Pluk. Almagest.* ii. 94, *t.* 275, *fig.* 4.

SOUTH AFRICA : without locality ; *Plukenet* (*in Herb. Sloane*) ! *Harvey,* 615 ! *Ecklon* !

COAST REGION : Cape Div.; above Oatlands House, *Wolley Dod,* 2842 ! Old road to Constantia Nek and behind Houts Bay Hotel, *Wolley Dod,* 3171 ! by Camps Bay Hotel, *Wolley Dod,* 1723 ; path near Smitswinkel Bay, *Wolley Dod,* 2931 ! Devils Peak, *Wilms,* 3473 ! Lion Mountain, 250 ft., *Schlechter,* 1376 ! Sea Point, 100 ft., *Bolus* ! Simons Bay, *Wright* ! Stellenbosch Div.; Gordons Bay, *Bolus,* 9994 ! Caledon Div.; Zwart Berg, 1200 ft., *Galpin,* 4338 !

33. S. intermedia (Schinz in Bull. Herb. Boiss. 2me sér. vi. 733);
annual, erect, 2½–4 in. high, usually simple ; internodes ½–¾ in. long,
leaves in 3 or 4 pairs, 3–5 lin. long, 2–6 lin. broad, broadly ovate or
cordate, subacute or acute, somewhat fleshy ; inflorescences regular,
compact, 3- to many-flowered ; branches about 5 lin. long ; pedicels
½–1 lin. long ; calyx-segments 3½–4 lin. long, ovate-lanceolate,
acuminate, with a wing ½–¾ lin. broad, broadest below the middle,
minutely pilose on the edge ; corolla-tube 3½–4½ lin. long ; lobes
2½–3 lin. long, ½–¾ lin. broad, elliptic or narrowly ovate-oblong,
unguiculate, acute ; filaments very short, inserted about ¼ lin. below
the sinuses ; anthers about ⅓ lin. long, with apical gland ; style 2–3
lin. long ; stigma 1 lin. long, confluent with the swelling. *S. cordata, var. intermedia, Cham. & Schlecht. in Linnæa, i. 191 ; Griseb.
Gen. et Spec. Gent. 165. Belmontia cordata, var. intermedia, Griseb.
in DC. Prodr. ix. 54 ; var. micrantha, E. Meyer, Comm. 183 ; Griseb.
in DC. Prodr. ix. 54 partly ; not of Cham. & Schlecht. Belmontia
intermedia, Knoblauch in Bot. Centralbl.* lx. 1894, 325 ; *Gilg in Engl.
Bot. Jahrb.* xxvi. 101.

Coast Region : Clanwilliam Div. ; hills near Brakfontein, 150 ft., *Schlechter*,
5287. Malmesbury Div. ; near Mooreesberg, 500 ft., *Bolus*, 9991 ! Paarl Div. ; by
the Berg River near Paarl, *Drège.* Cape Div. ; Lion Mountain, *Drège, Schlechter*
992, 1378 ! Signal Hill, *Wilms*, 3473 partly ! near Cape Town, *Harvey* ! Raapenberg, Mowbray, 40 ft., *Guthrie*, 176 ! Paarden Island, *Wolley Dod*, 3255 !

34. S. Burchellii (Gilg in Engl. Jahrb. xxvi. 89) ; annual or
biennial, erect, simple or branched from the base, 4–8 in. high,
with a radical leaf-rosette ; basal leaves obovate or elliptic-obovate,
subacute or obtuse, somewhat coriaceous; 5–8 lin. long, 2½–4 lin.
broad ; stem leaves ovate or triangular ovate, subacute, sometimes
cordate at the base ; cymes regularly branched, 9- to many-flowered,
branched, 5–8 lin. long, pedicels about ½ lin. long ; calyx-segments
obovate, rounded or slightly apiculate, thickened on the back,
2 lin. long, 1 lin. broad ; corolla-tube 2 lin. long ; lobes 1 lin. long,
½–1 lin. broad, obovate, obtuse ; anthers about ⅓ lin. long, almost
sessile, with three glands ; style ½–¾ lin. long, swelling not visible ;
stigma clavate. *Schinz in Mitteil. Geogr. Ges. Lübeck,* xvii.
(1903) 31.

Kalahari Region : Griqualand West ; at Griqua Town, *Burchell,* 1869 partly !
Transvaal ; between Waterval River and Zuikerbosh Rand, 4600 ft., *Schlechter*,
3497 !

Burchell's 1869 is the type of *S. Burchellii*, Gilg, but two plants have been
collected under this number, one of which belongs to this species, the other being
S. pentandra, E. Meyer.

35. S. Conrathii (Schinz in Mitteil. Geogr. Ges. Lübeck, xvii.
1903, 31) ; annual, erect, 6–11 in. high, simple or branched from
the base ; lowest leaf pairs very close making an apparent leaf-
rosette ; rosette leaves broadly obovate-oblong, often ¾–1½ in. long,
5–8 lin. broad, herbaceous ; stem leaves ovate, subacute 3–12 lin.

long; inflorescence a regular dichasium, many-flowered, with erect compact branches 6–9 lin. long; pedicels very short; calyx-segments elliptic, apiculate or acute, keeled, about 2 lin. long; corolla-tube 2 lin. long; lobes $1\frac{1}{4}$–$1\frac{1}{2}$ lin. long, $\frac{3}{4}$ lin. broad, rounded; filaments very short; anthers $\frac{1}{2}$ lin. long, with apical and two basal glands; style $\frac{3}{4}$–1 lin. long, with swelling; stigma capitate.

KALAHARI REGION: Transvaal; in damp places at Modderfontein, *Conrath,* 743!

36. **S. conspicua** (A. W. Hill in Kew Bulletin, 1908, 325); annual (?) with numerous erect or somewhat spreading flowering branches, 3–4 in. high; leaves obovate-oblong or spathulate, narrowing to form a petiole, 1–$1\frac{1}{4}$ in. long, 5–7 lin. broad, somewhat fleshy, crowded at the base to form a false leaf-rosette; inflorescences much-branched, usually many-flowered; branches 4–6 lin. long; pedicels about $\frac{1}{2}$ lin. long; calyx-segments 3–$3\frac{1}{2}$ lin. long, elliptic, acute, keeled; corolla-tube about 3 lin. long; lobes 5–6 lin. long, $1\frac{1}{2}$–$1\frac{3}{4}$ lin. broad, narrowly ovate, elongate, subacute or slightly apiculate; filaments $\frac{1}{2}$ lin. long; anthers $1\frac{1}{2}$ lin. long, with large apical glands; style 3–$3\frac{1}{2}$ lin. long, with swelling near base; stigma capitate.

KALAHARI REGION: Orange River Colony; in a marsh near Harrismith, 7000 ft., *Sankey,* 173!

37. **S. elongata** (E. Meyer, Comm. 184); annual, erect with a single stem 8–30 in. high, arising from a false rosette of closely packed basal leaves or more rarely a group of erect stems arising from the axils of the basal leaves; leaves at the base ovate-elliptic, subacute, or ovate, acute, $\frac{1}{2}$–$1\frac{3}{4}$ in. long, 2–10 lin. broad, the erect stems bear from 4–6 widely separated leaf pairs, triangular or elliptic-ovate, acute, cordate at the base; cymes terminal, much-branched and crowded forming a corymbose capitulum; branches 2–3 lin. long; pedicels $\frac{1}{2}$–1 lin. long; calyx-segments $2\frac{1}{2}$–$3\frac{1}{2}$ lin. long, narrowly obovate, acuminate with a narrow keel-wing broadest about the middle; corolla-tube $1\frac{1}{2}$–3 lin. long; lobes 2–4 lin. long, 1–$1\frac{1}{2}$ lin. broad, elliptic-ovate or obovate, subacute; filaments $\frac{1}{2}$–1 lin. long; anthers $\frac{1}{2}$–$1\frac{1}{2}$ lin. long, glands inconspicuous; style 1–4 lin. long, with swelling near the base; stigma capitate; capsule ovoid with the upper half thickened. *Gilg in Engl. Jahrb.* xxvi. 96; *Schinz in Mitteil. Geogr. Ges. Lübeck,* xvii. (1903) 41; *A. W. Hill in Kew Bulletin,* 1908, 325. *S. cuspidata, Schinz in Mitteil. Geogr. Ges. Lübeck,* xvii. (1903) 28.

COAST REGION: Riversdale Div.; Kampsche Berg, *Burchell,* 7085! mountains near Riversdale, *Schlechter,* 1840! Mossel Bay or Oudtshoorn Div.; Robinson Pass, *Bolus,* 12993! Knysna Div.; on a mountain near Roodemuur, between Plettenbergs Bay and Lange Kloof, 2000–2500 ft., *Drège,* 2827! Uniondale Div.; mountains near Avontuur, *Bolus,* 2402!

A distinct species with its false rosette of closely placed radical leaves and tall erect flowering stems. A considerable range of variation is shown in the size of the corolla and other parts of the flower, but there is a regular gradation of forms between the two extremes. The smaller-flowered forms were referred by Schinz to *S. cuspidata.*

38. **S. rotundifolia** (A. W. Hill in Kew Bulletin, 1908, 326); annual (?), with a radical leaf-rosette and erect flowering stem 6 in. high; leaves orbicular or orbicular-ovate, obtuse, $1\frac{1}{4}$ in. long, $1\frac{1}{8}$ in. broad, herbaceous, and 2 pairs of small leaves on the erect stem; inflorescence few-flowered, forming a small cymose umbel; branches 4–8 lin. long; pedicels 1 lin. long; calyx-segments 3 lin. long, elliptic-lanceolate, acute, with keel-wing about $\frac{1}{4}$ lin. wide, widest about the middle; corolla-tube about 4 lin. long; lobes 6 lin. long, $2\frac{1}{2}$ lin. broad, ovate-unguiculate, obtuse; filaments $\frac{3}{4}$ lin. long; anthers $1\frac{1}{3}$ lin. long, with apical and 2 basal glands; style 4–$4\frac{1}{2}$ lin. long, with swelling near the base; stigma clavate.

EASTERN REGION: Natal; Drakensberg, *Buchanan,* 31!

39. **S. pentandra** (E. Meyer, Comm. 184); annual, erect, simple or branching from the base or lower leaf-axils, 4–11 in. high; leaves in about 4 pairs, lower elliptic-ovate, obtuse, 7 lin.–$1\frac{1}{2}$ in. long, $\frac{1}{2}$–$\frac{3}{4}$ in. broad, upper leaves broadly triangular-ovate, acute; cymes compact, many-flowered; branches erect, $\frac{1}{2}$–1 in. long; pedicels 1 lin. long; calyx-segments about 3 lin. long, elliptic, acute, with keel-wing up to $\frac{1}{4}$ lin. broad, broadest at the middle; corolla-tube $3\frac{1}{2}$–4 lin. long; lobes $2\frac{1}{2}$–3 lin. long, $1\frac{1}{4}$–$1\frac{1}{2}$ lin. broad, ovate, obtuse; filaments about $\frac{1}{2}$ lin. long; anthers $1\frac{1}{4}$–$1\frac{1}{2}$ lin long, with large apical and two smaller basal glands; style $2\frac{1}{2}$–$3\frac{1}{2}$ lin. long, with swelling below the middle; stigma shortly clavate; ovary ovoid; seeds with narrow frills. *Schinz in Vierteljahrsschr. Naturf. Gesellsch. Zürich, xxxvii. 320; Gilg in Engl. Jahrb. xxvi. 90; Schinz in Mitteil. Geogr. Ges. Lübeck, xvii. 40; A. W. Hill in Kew Bulletin, 1908, 326. S. gariepina, Gilg in Engl. Jahrb. xxvi. 90; Schinz in Mitteil. Geogr. Ges. Lübeck, xvii. (1903) 41.*

CENTRAL REGION : Beaufort West Div.; near Beaufort West, 2700 ft., *Guthrie,* 3492! Richmond and Victoria West Div.; Nieuwveld and Uitvlugt, 3500–4000 ft., *Drège!* Graaff Reinet Div.; banks of watercourse near Graaff Reinet, 2500 ft., *Bolus,* 171! Somerset Div.; near Somerset East, *Bowker!* Colesberg Div.; Vanderwaltzfontein, near Colesberg, *Burke!* near Wonderheuvel, 4000 ft., *Drège.*

WESTERN REGION : Little Namaqualand; by the Orange River near Verleptpram, *Drège!* Steenberg, *Wyley!*

KALAHARI REGION : Griqualand West; at Griqua Town, *Burchell,* 1869 partly! near Kimberley, *Flanagan,* 1424!

EASTERN REGION : Transkei; between Gekau (Geua) River and Bashee River, *Drège,* 4920!

It has not been found possible to maintain both *S. pentandra* and *S. gariepina.* The specimens of *S. pentandra, a, b* and *c* of Drège, have been examined and are found to agree closely together. *S. pentandra* (*d*) of Drège has been placed by Gilg under *S. ramosissima.* The Drège specimen from Garip (Gariep River), which is not referred to in *E. Meyer's Comm.,* agrees closely with the type specimen of *S. pentandra.* In *E. Meyer, Comm.* i. 184, four localities, *a, b, c* and *d,* are given for *S. pentandra.* In *Meyer's Zwei Pfl. Documente,* 219, another locality, *c,* is referred to for *S. pentandra.*

The localities according to the latter publication should be as follows and not as in *Meyer's Comm.* :

(*a*) Winterveld, bei Limoenfontein und Groot Tafelberg, und Nieuweveld, zwischen Brakrivier und Uitvlugt.
(*b*) Wonderheuvel.
(*c*) Garip, Sandhügel am rechten Ufer des Flusses bei Verleptpram.
(*d*) Zuurebergen.
(*e*) Zwischen Gekau und Basche.

The specimen in the Berlin Herbarium labelled *Sebæa pentandra,* E. M. *b* Drège, has been given the wrong locality ; the *b* has been read *e* and the locality (inter Gekau et Basche) put on the label, it should be Wonderheuvel.

The type specimen of *S. gariepina,* Gilg in Meyer's Herb. at Lübeck, exactly equals the Kew specimen *S. pentandra,* E. Meyer, *a.* The Lübeck specimen is from Garip.

The Lübeck specimen from "Gekau und Basche, no. 4920" agrees with the Berlin specimen bearing the distribution label "*S. pentandra,* E. M. *b* Drège," which however is further labelled "Gekau et Basche," and these agree with the Kew specimen, Bowker, Somerset, which also equals the above.

40. S. macrostigma (Gilg in Engl. Jahrb. xxvi. 93); annual, erect, simple or branched from the base, 2–6 in. high ; leaves crowded at the base of the stem forming a kind of leaf-rosette, broadly ovate or orbicular-ovate, acute or shortly apiculate, herbaceous, $3\frac{1}{2}$–6 lin. long, 3–6 lin. broad ; cymes variable, 3- to many-flowered, lax and spreading or in dwarf forms compact ; branches in vigorous specimens 5–8 lin. long ; pedicels 2–7 lin. long ; calyx-segments $2\frac{1}{2}$–3 lin. long, narrowly lanceolate, acuminate, with keel-wing broadest near the base ; corolla-tube $2\frac{1}{2}$–3 lin. long ; lobes $2\frac{1}{2}$–3 lin. long, 1–$1\frac{1}{2}$ lin. broad, elliptic, obtuse ; filaments about $\frac{1}{2}$ lin. long ; anthers 1 lin. long, without glands ; style $2\frac{3}{4}$–3 lin. long, with swellings near the base ; stigma $\frac{3}{4}$–1 lin. long, clavate-bilabiate. *Schinz in Mitteil. Geogr. Ges. Lübeck,* xvii. (1903) 36. *S. humilis, N. E. Br. in Kew Bulletin,* 1901, 127.

COAST REGION : Riversdale Div. ; hills near Riversdale, 800 ft., *Bolus,* 11345 ! Albany Div. ; mountains around Grahamstown, 2000 ft., *Glass,* 1635 partly ! *Williamson* ! Beaumont, Fish River Rand, *Hutton,* 544 ! Queenstown Div. ; mountains near Queenstown, 4000 ft., *Galpin,* 1549 !

41. S. ramosissima (Gilg in Engl. Jahrb. xxvi. 91); annual, erect, usually with numerous branches from the base forming a broad many-flowered inflorescence ; leaves broadly ovate or cordate, acute, 4–7 lin. long, $3\frac{1}{2}$–5 lin. broad, somewhat coriaceous ; cymes much-branched, lax, spreading ; branches $\frac{3}{4}$–$1\frac{1}{4}$ in. long ; pedicels 5–7 lin. long ; calyx-segments $2\frac{1}{2}$ lin. long, narrowly ovate-lanceolate, acuminate, with narrow keel-wing, broadest rather below the middle ; corolla-tube $2\frac{1}{2}$–3 lin. long ; lobes 3–4 lin. long, $1\frac{1}{2}$ lin. broad, elliptical, subacute, sometimes toothed at the apex ; filaments about $\frac{1}{2}$ lin. long ; anthers about $1\frac{1}{4}$ lin. long, with well-marked apical and 2 basal glands ; style $2\frac{1}{2}$–3 lin. long, with swelling near the base ; stigma small, clavate. *Schinz in Mitteil. Geogr. Ges. Lübeck,* xvii. (1903) 35.

SOUTH AFRICA: without locality, *Harvey*, 618!

COAST REGION : Uitenhage Div. ; Zuurberg Range, *Drège*. Alexandria Div. ; Quaggas Flats, *Gill*! Bathurst Div. ; between Kaffir Drift and Port Alfred, *Burchell*, 3784! Albany Div. ; mountains around Grahamstown, 2000 ft., *Bolton*! *Glass in MacOwan, Herb. Austr.-Afr.* 1635 (at Kew, but not in the Berlin or Zürich Herbaria)! Komgha Div. ; hills near Komgha, 2000 ft., *Flanagan*, 1180 !

EASTERN REGION : Transkei ; around Kentani and near the coast,100–1000 ft., *Miss Pegler*, 490! Kreilis country, *Bowker*, 14 !

42. S. primulina (A. W. Hill in Kew Bulletin, 1908, 327); annual, with a radical leaf-rosette and an erect flowering stem $1\frac{1}{2}$–5 in. high ; leaves ovate, subacute, 4–8 lin. long, 2–$3\frac{1}{2}$ lin. broad, fleshy, becoming narrowed to form a petiole, the erect stems usually single with one or two pairs of ovate or narrowly ovate-lanceolate leaves ; cymes 2- to several-flowered ; branches 4–10 lin. long ; pedicels $\frac{1}{4}$–$\frac{1}{2}$ lin. long ; calyx-segments $3\frac{1}{2}$–4 lin. long, ovate-lanceolate, acute, united below to form a short cup, with a narrow keel-wing broadest about the middle ; corolla-tube 5 lin. long ; lobes $3\frac{1}{2}$–4 lin. long, $1\frac{1}{2}$–2 lin. broad, spathulate, obtuse ; filaments $\frac{1}{4}$ lin. long, inserted in the corolla-tube about $\frac{1}{3}$ lin. below the sinuses ; anthers 1–$1\frac{1}{2}$ lin. long with an apical and two basal glands ; style 4–$4\frac{1}{2}$ lin. long, with swelling 1 lin. above the base ; stigma clavate, more or less bilabiate.

KALAHARI REGION : Bechuanaland ; by the Moshowa River, near Takun, *Burchell*, 2252/5 ! between Kuruman and the Vaal River, *Cruickshank in Herb. Bolus*, 2540 !

A single small specimen in the Dublin Herbarium labelled ''Port Natal, Miss Owen'' appears to belong to this species, but the material is insufficient for exact determination.

43. S. vitellina (Schinz in Mitteil. Geogr. Ges. Lübeck, xvii. 1903, 38) ; perennial, erect, branched from the base, 2–3 in. high ; leaves cordate, acute, $2\frac{1}{2}$–$3\frac{1}{2}$ lin. long, 2–3 lin. broad, somewhat coriaceous, margins reflexed ; cymes few-flowered ; branches $2\frac{1}{2}$–4 lin. long ; pedicels about $\frac{1}{2}$ lin. long ; calyx-segments elliptic, acute, $2\frac{1}{2}$–3 lin. long, $\frac{3}{4}$ lin. broad, with a slight keel ; corolla-tube 3–4 lin. long ; lobes $2\frac{1}{2}$–4 lin. long, 2 lin. broad, obovate-elliptic, obtuse ; filaments $\frac{1}{2}$–$\frac{3}{4}$ lin. long ; anthers 1–$1\frac{1}{4}$ lin. long, with minute apical glands ; style $3\frac{1}{2}$–4 lin. long, with swelling ; stigma capitate.

EASTERN REGION : Natal ; plains near Catos Ridge, 3200 ft., *Schlechter*, 3259 ! near Durban, *Sutherland* ! and without precise locality, *Gerrard*, 91 !

44. S. Dregei (Schinz in Mitteil. Geogr. Ges. Lübeck, xvii. 1903, 51) ; perennial or biennial, erect, with stiff parallel branches, 10–14 in. high ; lower leaves broadly triangular-ovate, somewhat cordate at the base, subacute, up to 5 lin. long, 6 lin. broad, reflexed, some-what coriaceous with margins inrolled, upper leaves becoming narrowly triangular, acute ; cymes arranged in more or less compact narrowly oval heads ; branches about 5 lin. long ; pedicels $1\frac{1}{2}$–$2\frac{1}{2}$

lin. long ; calyx-segments narrowly ovate-lanceolate, acute or acuminate, $2\frac{1}{2}$–3 lin. long, sharply keeled ; corolla-tube $2\frac{1}{2}$–3 lin. long ; lobes 3–4 lin. long, about 1 lin. broad, narrowly obovate-oblong, obtuse or slightly cuspidate ; filaments 1–$1\frac{1}{2}$ lin. long ; anthers $1\frac{1}{4}$ lin. long, glands minute apparently 3 ; style $3\frac{1}{2}$ lin. long, with a small swelling ; stigma capitate, slightly clavate. *S. stricta, Gilg in Engl. Jahrb. xxvi. 90. S. crassulæfolia, Cham. & Schlecht. var. stricta, E. Meyer, Comm. 184.*

COAST REGION: Riversdale Div. ; Garcias Pass, *Bolus,* 11350 ! George Div. ; near George, 600 ft., *Schlechter,* 2408 ! Knysna Div. ; in marshy places, *Bowie* !

CENTRAL REGION: Graaff Reinet Div. ; near Graaff Reinet, *Wyley*! Aliwal North Div. ; Witteberg Range, 6000 ft., *Drège*!

45. S. Schœnlandii (Schinz in Bull. Herb. Boiss. 2me sér. vi. 741); perennial or biennial, erect, slender, 10–20 in. high, usually un-branched ; internodes relatively long $\frac{3}{4}$–$1\frac{1}{4}$ in. ; lower leaves broadly ovate, obtuse, 4–5 lin. long, 4–5 lin. broad, upper leaves ovate-lanceolate or elliptic, subacute, 3–4 lin. long, $1\frac{1}{2}$–$2\frac{1}{2}$ lin. broad, glaucous, somewhat fleshy ; inflorescence small, capitate, many-flowered, compact ; branches about 6 lin. long ; pedicels 1–2 lin. long ; calyx-segment 2–$3\frac{1}{2}$ lin. long, elliptic, acute, keeled ; corolla-tube $2\frac{1}{2}$–3 lin. long ; lobes 2–$2\frac{1}{2}$ lin. long, 1–$1\frac{1}{4}$ lin. broad, obovate-oblong or elliptic, subacute ; filaments very short ; anthers $\frac{1}{2}$–$\frac{3}{4}$ lin. long, situated either in the sinuses of the corolla or in the throat $\frac{1}{4}$–$\frac{1}{3}$ lin. below the sinuses ; apical glands present ; style 2–$2\frac{1}{2}$ lin. long, with swelling below the middle ; stigma capitate. *A. W. Hill in Kew Bulletin, 1908, 327. S. sedoides, Schinz in Mitteil. Geogr. Ges. Lübeck, xvii. (1903) 43 ; Schinz in Bull. Herb. Boiss. 2me sér. vi. 813, both partly.*

KALAHARI REGION : Orange River Colony, *Cooper,* 2751 ! Transvaal ; Witwaters Rand, *Hutton,* 880 !

EASTERN REGION : Natal ; Mount Moreland, 500 ft., *Wood,* 1386! and without precise locality, *Cooper,* 2750 partly! Zululand ; Ngoya, *Wylie in Herb. Wood,* 7370 ! 8497 !

This species is of interest as the distinction between *Sebæa* and *Belmontia* with regard to the anthers does not hold good. According to Schinz's type-specimen the anthers are in the throat about $\frac{1}{4}$ of a line below the sinuses of the corolla-lobes, but in Cooper, 2751, they occur in the sinuses as in *Sebæa* proper. *S. Schœnlandii* bears a certain external resemblance to *S. sedoides,* under which species two specimens have been placed, but it may be distinguished by the longer internodes. Cooper's 2750 at Edinburgh is *S. Rudolfii,* Schinz.

46. S. sedoides (Gilg in Engl. Jahrb. xxvi. 98) ; annual or biennial? erect, usually unbranched, 12–18 in. high ; internodes $\frac{1}{4}$–$\frac{1}{2}$ in. long ; leaves crowded, broadly cordate-orbicular to subreniform, rounded at the apex or shortly apiculate, 3–6 lin. long, 3–8 lin. broad, somewhat coriaceous ; cymes many-flowered, dense and compact, joining a terminal capitate or obconical inflorescence ; calyx-segments $2\frac{1}{4}$–3 lin. long, elliptic, acute, keeled or with a narrow keel-wing ; corolla-tube $2\frac{3}{4}$–$3\frac{1}{4}$ lin. long ; lobes 2–$2\frac{1}{2}$ lin.

long, $\frac{3}{4}$–1 lin. broad, obovate-elliptic, obtuse; filaments $\frac{1}{4}$ lin. long; anthers $\frac{2}{3}$–$\frac{7}{8}$ lin. long, with apical gland; style $2\frac{1}{2}$–$3\frac{1}{2}$ lin. long, with swelling below the middle; stigma capitate. *Schinz in Mitteil. Geogr. Ges. Lübeck*, xvii. (1903) 43.

EASTERN REGION: Tembuland; near Bazeia, 2000 ft., *Baur*, 32! near Elliot, *Bolus*, 10225! 10226! Griqualand East; mountains near Clydesdale, 2500 ft., *Tyson*, 2726! and in *MacOwan and Bolus, Herb. Norm. Austr.-Afr.* 1289! Natal; Tugela, *Gerrard*, 1983! Coastland, *Sutherland*! hills near Richmond Road, *Schlechter*, 6738! Riet Spruit, 3000–4000 ft., *Wylie in Herb. Wood*, 10227! Zululand; near Eshowe, *Maxwell in Herb. Bolus*, 9331!

47. S. confertiflora (Schinz in Mitteil. Geogr. Ges. Lübeck, xvii. 1903, 51); annual or biennial? simple or branched, 6–10 in. high; internodes $\frac{1}{2}$–$\frac{3}{4}$ in. long; leaves crowded, broadly triangular-ovate to cordate, subacute, 4–7 lin. long, 4–6 lin. broad, coriaceous; cymes many-flowered, forming a dense rounded capitulum; calyx-segments about $2\frac{1}{3}$ lin. long, elliptic, acute or apiculate, with keel; corolla-tube $2\frac{1}{2}$–$3\frac{1}{2}$ lin. long; lobes 2–$2\frac{1}{4}$ lin. long, 1 lin. broad, ovate-elliptic, subacute; anthers $\frac{7}{8}$–$1\frac{1}{4}$ lin. long; lobes sagittate at the base, apical gland, small; style 3 lin. long, with swelling near the middle; stigma capitate-clavate.

KALAHARI REGION: Transvaal; Spitz Kop, near Lydenburg, *Wilms*, 964! near Elandspruit Bergen, 6400 ft., *Schlechter*, 4000! Vlakfontein; near Amers Poort, *Burtt Davy*, 4049!

EASTERN REGION: Natal; Illovo, 2000 ft., *Wood*, 30! summit of Amajuba Hill, 8000 ft., *Burtt Davy*, 7747! *Mundy*, 3758. Swazieland, 4500 ft., *Challis*, 10662.

This species appears to be closely related to *S. sedoides*, but differs particularly in the larger anthers (*Wood*, 30) which are sagittate at the base, the leaves also are triangular- rather than orbicular-ovate, and the inflorescence is more definitely capitate than in *S. sedoides*.

48. S. acuminata (A. W. Hill in Kew Bulletin, 1908, 328); perennial, erect, simple, 20–26 in. high; leaves numerous, somewhat distant on the lower portion of the stem, crowded above, narrowly ovate-lanceolate below, linear-lanceolate above, acute, margins reflexed, somewhat coriaceous, $4\frac{1}{2}$–$6\frac{1}{2}$ lin. long, 1–2 lin. broad; cymes compact, many-flowered; branches 5–7 lin. long, slender; pedicels about $\frac{1}{2}$ lin. long; bracts linear; calyx-segments $3\frac{1}{2}$ lin. long, narrowly elliptic-lanceolate, acuminate, concave, with narrow keel-wing; corolla-tube 2–$2\frac{1}{4}$ lin. long; lobes $2\frac{1}{2}$ lin. long, $\frac{3}{4}$–1 lin. broad, elliptic-obovate, obtuse; filaments very short; anthers $\frac{1}{2}$–$\frac{1}{3}$ lin. long, with apical gland; style 2 lin. long, with swelling below middle; stigma capitate-clavate.

EASTERN REGION: Natal; near Boston, 3000–4000 ft., *Wood*! (in Herb. British Museum.)

49. S. macrantha (Gilg in Engl. Jahrb. xxvi. 94); perennial or biennial; stems stout, erect, unbranched, 20–25 in. high; leaves

broadly ovate or cordate, subacute or acute, coriaceous, about
7–8 lin. long, 5–9 lin. broad ; inflorescence regularly branched,
forming a loose corymbose panicle, 4–6 in. long, $2\frac{1}{2}$–4 in. in diam. ;
branches firm and about 1 in. long ; pedicels about 1 lin. long or
almost absent ; calyx-segments elliptic, acuminate, $3\frac{1}{2}$–5 lin. long,
with wing about $\frac{1}{4}$ lin. wide, widest about the middle ; corolla-tube
4–5 lin. long ; lobes $4\frac{1}{2}$–6 lin. long, $1\frac{1}{2}$–2 lin. broad, elliptic, subacute ; filaments $\frac{1}{4}$–$\frac{3}{4}$ lin. long ; anthers $2\frac{3}{4}$–$3\frac{3}{4}$ lin. long, with small
apical glands ; style 4 lin. long, with swelling ; stigma clavate.

KALAHARI REGION: Transvaal ; Spitz Kop, *Wilms*, 970.
EASTERN REGION: Natal ; Inanda, *Wood*, 866! hills at Sevenfontein,
3000–4000 ft., *Wylie in Herb. Wood*, 5214 !

50. S. erecta (A. W. Hill in Kew Bulletin, 1908, 328) ; perennial
or biennial ; stems erect, unbranched, about 18 in. high, springing
apparently from an underground creeping branched stem ; leaves
numerous, 4–5 lin. long, 3–$4\frac{1}{2}$ lin. broad, broadly ovate or cordate,
subacute or apiculate, somewhat coriaceous ; inflorescences terminal,
few-flowered ; branches 6–9 lin. long ; pedicels 1–2 lin. long ; calyx-
segments 5 lin. long, 2 lin. broad, ovate-oblong, apiculate, with keel-
wing $\frac{1}{4}$ lin. wide, widest about the middle ; corolla-tube 4 lin. long ;
lobes $5\frac{1}{2}$–6 lin. long, $2\frac{1}{2}$ lin. broad, obovate-oblong, obtuse ; filaments
1 lin. long ; anthers $1\frac{3}{4}$ lin. long, with small apical glands ; style
5 lin. long, with swelling near the base ; stigma capitate.

KALAHARI REGION : Transvaal ; Liliefontein in Carolina district, *Nicholson in
Transvaal Herbarium*, 4307 !

51. S. longicaulis (Schinz in Bull. Herb. Boiss. ii. 219) ;
perennial or biennial, erect, simple below, simple or branched above,
2 ft. or more high ; leaves ovate or orbicular-ovate, becoming elliptic-
ovate to narrowly ovate-lanceolate above, 4–8 lin. long, 3–7 lin. broad,
slightly coriaceous, subacute or acute ; cymes terminal, few-flowered,
compact ; branches 4–9 lin. long ; pedicels 3–6 lin. long ; calyx-
segments membranous, 4–5 lin. long, ovate-oblong, acute or acuminate
or apiculate, with a slight keel wing ; corolla-tube 3–4 lin. long ;
lobes 5–6 lin. long, 2–3 lin. broad, ovate-unguiculate, obtuse ;
filaments 1–$1\frac{1}{2}$ lin. long ; anthers $1\frac{1}{2}$ lin. long, with three very small
glands ; style $4\frac{1}{2}$ lin. long, with swelling near the base ; stigma
shortly clavate. *Gilg in Engl. Jahrb.* xxvi. 94 ; *Schinz in Mitteil.
Geogr. Ges. Lübeck,* xvii. (1903) 42 ; *A. W. Hill in Kew Bulletin,*
1908, 329. *S. Woodii, Gilg in Engl. Jahrb.* xxvi. 94 ; *Schinz in
Mitteil. Geogr. Ges. Lübeck,* xvii. (1903) 42. *S. crassulæfolia, var.
lanceolata, Schinz in Vierteljahrsschr. Naturf. Gesellsch. Zürich,*
xxxvii. (1891) 323. *S. macrosepala, Gilg in Engl. Jahrb.* xxvi. 91 ;
Schinz in Mitteil. Geogr. Ges. Lübeck, xvii. (1903) 48.

COAST REGION : King Williamstown Div. ; Perie Bush, *Scott-Elliot* !
EASTERN REGION : Griqualand East ; Mount Malowe, 5500 ft., *Tyson*, 3096 !
Natal ; in a swamp near Karkloof, 3000–4000 ft., *Wood*, 4447 ! Byrne, 3000 ft.,

Wood, 1844! near Weston, *Rehmann*, 7348! near Ladysmith, *Gerrard*, 835! Benvie, 4000–5000 ft., *Wylie in Herb. Wood*, 7875!

52. S. grandiflora (Schinz in Mitteil. Geogr. Ges. Lübeck, xvii. 1903, 45); perennial or biennial; stems erect, about 15 in. high, unbranched; internodes usually $\frac{3}{4}$–$1\frac{1}{4}$ in. long; leaves cordate to broadly orbicular-ovate, apiculate, 5–3 lin. long, 4 lin. broad, somewhat coriaceous, margins inrolled, dark green above, pale green below; cymes terminal, few-flowered, compact; branches about 4–6 lin. long; pedicels 1–2 lin. long; calyx-segments membranous, elliptic-ovate, apiculate or acuminate, with a narrow keel-wing, 4–$4\frac{1}{2}$ lin. long, $1\frac{1}{2}$ lin. broad; corolla-tube $3\frac{1}{4}$–$3\frac{1}{2}$ lin. long; lobes $3\frac{1}{2}$–$4\frac{1}{2}$ lin. long, 2–3 lin. broad, obovate-unguiculate, obtuse; filaments 1–$1\frac{1}{2}$ lin. long; anthers 1–$1\frac{1}{2}$ lin. long, with small apical gland; style $3\frac{1}{2}$–5 lin. long, swelling small, not always developed; stigma capitate.

KALAHARI REGION: Transvaal; Houtbosch Berg, in damp places, 6500 ft., *Schlechter*, 4768! Liliefontein in Carolina district, *Nicholson in Transvaal Herbarium*, 4308!

This species appears to be closely allied to *S. longicaulis* and *S. erecta*, both in floral and vegetative structure, and it is possible that a series of intermediate forms linking these species together may be found. In this species the leaves are ovate below becoming orbicular above, whilst in *S. longicaulis* they tend to be ovate below becoming lanceolate above—the stigma in *S. grandiflora* also is capitate.

53. S. multiflora (Schinz in Mitteil. Geogr. Ges. Lübeck, xvii. 1903, 44); perennial or biennial, erect, much-branched, about 12 in. high; leaves numerous, 3–$4\frac{1}{2}$ lin. long, 4–6 lin. broad, semicircular, apiculate, coriaceous, margins slightly reflexed; inflorescence much-branched, lax and spreading, many-flowered; branches 4–6 lin. long; pedicels about 4 lin. long; calyx-segments broadly keeled or keel becoming a narrow wing, 3–$3\frac{1}{2}$ lin. long, $1\frac{1}{2}$–$2\frac{1}{2}$ lin. broad, ovate or elliptic, acute; corolla-tube 3–$3\frac{1}{2}$ lin. long; lobes 4–$4\frac{1}{2}$ lin. long, 3 lin. broad, obovate-spathulate, rounded or subacute; filaments $\frac{1}{2}$ lin. long; anthers $1\frac{1}{2}$ lin. long, with a large apical and two large basal glands; style $3\frac{1}{2}$ lin. long, with swelling 1 lin. above base; stigma capitate-clavate.

CENTRAL REGION: Graaff Reinet Div.; Oude Berg, near Graaff Reinet, 4500 ft., *Bolus*, 171!

54. S. schinziana (Gilg in Engl. Jahrb. xxvi. 95); perennial, erect, stout, much-branched, stiff and bushy, about 8 in. high; leaves numerous, closely crowded together, broadly cordate or ovate, acute, coriaceous, margin reflexed, 3–$4\frac{1}{2}$ lin. long, 3–$4\frac{1}{2}$ lin. broad; stems much-branched above ending in few-flowered cymes; branches and pedicels about $1\frac{1}{2}$ lin. long; calyx-segments not always equal in size, 3 large and 2 small, 3–4 lin. long, ovate-lanceolate, acute, with narrow keel-wing; corolla-tube $3\frac{1}{2}$ lin. long; lobes 4–$4\frac{1}{2}$ lin. long, $2\frac{1}{2}$–$2\frac{3}{4}$ lin. broad, obovate, obtuse or subacute; filaments $\frac{3}{4}$ lin.

long; anthers 1¼ lin. long, glands absent; style 3–4 lin. long, with swelling 1–1½ lin. above the base; stigma capitate.

KALAHARI REGION : Transvaal; Devils Knuckles, *Wilms,* 965 ! *Burtt Davy,* 492 !
EASTERN REGION : Natal; Bushmans River, on the Drakensberg Range, *Evans,* 54 !

55. **S. macrophylla** (Gilg in Engl. Jahrb. xxvi. 96); perennial or biennial, erect, simple or branched, 10–1·1 in. high; leaves numerous, orbicular-ovate, acute, sometimes rounded, 5–7 lin. long, 6–9 lin. broad, somewhat coriaceous and varnished above; stem much-branched above to form a leafy dense-flowered inflorescence; actual cymes relatively few-flowered; branches about 4 lin. long; pedicels about 2 lin. long; calyx-segments elliptic-lanceolate, acute or acuminate, concave, with a narrow keel-wing, 3½–4½ lin. long; corolla-tube 3½–4½ lin. long; lobes 3½–4½ lin. long, 2–2½ lin. broad, spathulate, obtuse or subacute; filaments ⅓–¾ lin. long; anthers 1¼–1¾ lin. long, with apical and sometimes two basal glands? style 3½–4½ lin. long; stigma capitate, bilobed, sometimes appearing as very shortly clavate. *Schinz in Mitteil. Geogr. Ges. Lübeck,* xvii. (1903) 45; *A. W. Hill in Kew Bulletin,* 1908, 329. *S. hymenosepala, Schinz in Mitteil. Geogr. Ges. Lübeck,* xvii. (1903) 49 *partly. S. wittebergensis, Schinz in Mitteil. Geogr. Ges. Lübeck,* xvii. (1903) 45.

COAST REGION : King Williamstown Div. ; Buffalo, near Perie, 4000 ft., *Tyson,* 1047 ! Kaffraria, *Mrs. Barber* !
KALAHARI REGION : Orange River Colony ; Kadzi Berg in the Witteberg Range, *Rehmann,* 3999! Basutoland, *Cooper,* 713 ! Transvaal ; Drakensberg Range, near Pilgrims Rest, 5000 ft., *McLea in Herb. Bolus,* 171 ! 3099 !
EASTERN REGION : Griqualand East ; summit of Mount Currie, 6500 ft., *Tyson in MacOwan and Bolus Herb. Norm. Austr.-Afr.* 1289b !

56. **S. crassulæfolia** (Cham. & Schlecht. in Linnæa, i. 193); perennial or biennial, erect, slender, simple or branched, about 12 in. high; leaves numerous, the lower broadly orbicular-ovate or reniform, obtuse or apiculate, 2½–3 lin. long, 4–6 lin. broad, somewhat coriaceous, the upper leaves becoming narrowly triangular-ovate ; inflorescences lax, composed of numerous cymes at the ends of lateral branches ; cymes many-flowered, dense ; branches 4–6 lin. long; pedicels 1–2 lin. long; calyx-segments narrowly ovate-lanceolate, acute 2¾ lin. long, with sharp keel or very narrow keel-wing; corolla-tube 2½–3 lin. long; lobes 3–4 lin. long, 1–1½ lin. broad, obovate, oblong, obtuse; filaments ½–¾ lin. long; anthers 1–1¼ lin. long, glands minute, apparently 3 ; style 3–4 lin. long, with swelling; stigma shortly clavate. *Schinz in Vierteljahrsschr. Naturf. Gesellsch. Zürich,* xxxvii. 323 *partly; Griseb. Gen. et Spec. Gent.* 168 ; *Schinz in Mitteil. Geogr. Ges. Lübeck,* xvii. (1903) 50 *partly; A. W. Hill in Kew Bulletin,* 1908, 330. *S. crassulifolia, Griseb. in DC Prodr.* ix. 53 ; *Gilg in Engl. Jahrb.* xxvi. 97 *partly. S. hymenosepala, Schinz in Mitteil. Geogr. Ges. Lübeck,* xvii. (1903) 49 *partly.*

SOUTH AFRICA: without precise locality, *Ecklon*, 660.
COAST REGION: Knysna Div.; Plettenbergs Bay Poort, *Mund and Maire*.
Uitenhage Div.; near Uitenhage? 1000–3000 ft., *Zeyher*! *Ecklon and Zeyher*.
British Kaffraria, *Cooper*, 406!

57. S. Brehmeri (Schinz in Mitteil. Geogr. Ges. Lübeck, xvii.
1903, 50); perennial or biennial, erect, stout, branched, 15–20 in.
high; leaves numerous, broadly ovate to reniform, acute, 3–4 lin.
long, 4–6 lin. broad, coriaceous, varnished above, margins reflexed;
inflorescences composed of numerous lax branches bearing few- to
many-flowered cymes; branches irregular; pedicels 3–4 lin. long;
bracts numerous, narrowly ovate, acute; calyx-segments ovate-
oblong, acute or acuminate, 3–4 lin. long, with narrow keel-wing
broadest about the middle; corolla-tube $3\frac{1}{2}$–$4\frac{1}{2}$ lin. long; lobes 3–$3\frac{1}{2}$
lin. long, $1\frac{1}{2}$–2 lin. broad, obovate-oblong to obovate-orbicular,
unguiculate; filaments 1–$1\frac{1}{2}$ lin. long; anthers $\frac{3}{4}$ lin. long, with
apical gland; style 4–$4\frac{1}{2}$ lin. long, with inconspicuous swelling;
stigma capitate.

SOUTH AFRICA: without locality, *Bowie*!
COAST REGION: George Div.; near Touw River, *Burchell*, 5734! Knysna Div.;
near Knysna, *Burchell*, 5485! 5494!

This species is very similar to *S. crassulæfolia*, Cham. & Schlecht., but differs
in the generally larger floral structures and capitate stigma and also in the
numerous narrowly ovate acute bracts.

58. S. leiostyla (Gilg in Engl. Jahrb. xxvi. 97); perennial or
biennial, erect, simple or branched, stout, 12–15 in. high; leaves
cordate or ovate-orbicular, subacute, apex sometimes shortly apiculate, 4–9 lin. long, 4–7 lin. broad, somewhat coriaceous; cymes
many-flowered, compact, either arranged to form large capitate
inflorescences or borne in clusters at the ends of lateral shoots;
branches variable in length; pedicels 1–2 lin. long; calyx-segments
$2\frac{1}{2}$–$3\frac{1}{2}$ lin. long, elliptic, acute or acuminate, with a very narrow
keel-wing; corolla-tube $2\frac{1}{2}$–$3\frac{1}{2}$ lin. long; lobes $2\frac{1}{2}$–3 lin. long, $1\frac{1}{4}$–$1\frac{1}{2}$
lin. broad, obovate-oblong, subacute; filaments $\frac{1}{4}$–1 lin. long; anthers
1–$1\frac{1}{4}$ lin. long, with apical gland; style $2\frac{1}{2}$–$3\frac{1}{2}$ lin. long, with
swelling (sometimes absent); stigma capitate. *Schinz in Mitteil.
Geogr. Ges. Lübeck*, xvii. (1903) 32; *Baker and N. E. Br. in Dyer,
Fl. Trop. Afr.* iv. i. 548; *A. W. Hill in Kew Bulletin*, 1908, 330.
S. transvaalensis, Schinz in Mitteil. Geogr. Ges. Lübeck, xvii. (1903)
49. *S. sedoides, Schinz in Mitteil. Geogr. Ges. Lübeck*, xvii. (1903)
43 *partly*.

KALAHARI REGION: Orange River Colony; near Harrismith, *Sankey*, 175!
Bethlehem, *Richardson*! and without precise locality, *Cooper*, 2752! 2759!
Transvaal; between Trigards Fontein and Standerton, *Rehmann*, 6755! near
Pretoria, 4600 ft., *Schlechter*, 4157! Houtbosch Berg, 6400 ft., *Schlechter*, 4720!
Riet Spruit, *Krook*, 2016! near Vereeniging, *Burtt Davy*, 7767! Ermelo Experi-
mental Farm, 5400 ft., *Burtt Davy*, 7710!
EASTERN REGION: Natal; ranges 30–60 miles from the sea, 2000–3500 ft.,
Sutherland! Summit of Amajuba Hill, *Burtt Davy*, 7747b! and without precise
locality, *Gerrard*, 1983!

There appears to be no character to enable the two species *S. leiostyla* and *S. transvaalensis* both to be maintained, *S. leiostyla* has therefore a somewhat extended range from Nyassaland to Natal, but Buchanan, 270, which is the type of Gilg' species, appears to agree in all essentials with the other specimens quoted. The plant collected by Schlechter in the Houtbosch Berg, 4720, which forms the type of Schinz's *S. transvaalensis*, differs from the other examples in its more bushy habit and numerous lateral inflorescences, but in floral character there are no essential points of difference from the other specimens included under this species.

59. **S. Macowanii** (Gilg ex Schinz in Mitteil. Geogr. Ges. Lübeck, xvii. 1903, 47); annual or biennial, erect, simple below, branching above, about 5 in. high; leaves few, broadly ovate, acute, 5 lin. long, $4\frac{1}{2}$ lin. broad, herbaceous; cymes lax, few-flowered; calyx-segments $5\frac{1}{2}$ lin. long, $1\frac{1}{2}$ lin. broad, ovate-lanceolate, acute, with a keel-wing commencing about the middle and extending to the base of each segment; corolla-tube $4\frac{1}{2}$–5 lin. long; lobes 5 lin. long, $2\frac{1}{2}$ lin. broad, obovate-elliptic; filaments $\frac{1}{4}$ lin. long; anthers about $2\frac{1}{2}$ lin. long, with apical glands; style $3\frac{1}{2}$–$3\frac{3}{4}$ lin. long, with swelling below the middle; stigma bilobed, $\frac{2}{3}$–$\frac{3}{4}$ lin. long, thick and fleshy.

Coast Region: King Williamstown Div.; low hills around Kiug Williamstown, 1500 ft., *Tyson*, 2203!

A small single specimen with four flowers, conspicuous by the large calyx-segments, with the wing prominent only on the lower portion.

60. **S. imbricata** (A. W. Hill in Kew Bulletin, 1908, 331); perennial or biennial, simple or branched, erect, 7–10 in. high, stout; internodes short, 3–6 lin. long; leaves numerous, orbicular-ovate or cordate, apiculate, 4–$4\frac{1}{2}$ lin. long, $5\frac{1}{2}$–$4\frac{1}{2}$ lin. broad, somewhat coriaceous; inflorescences many-flowered, much-branched, forming a dense cymose corymb; branches 3–4 lin. long; pedicels about 1 lin. long; calyx-segments $4\frac{1}{2}$–5 lin. long, about $1\frac{1}{2}$–2 lin. broad, ovate-lanceolate, acuminate, overlapping widely at the base, with wing $\frac{1}{2}$ lin. broad, broadest at the base; corolla-tube 3–$3\frac{1}{4}$ lin. long; lobes $3\frac{1}{4}$–4 lin. long, $1\frac{1}{2}$ lin. broad, elliptic-obovate, apiculate; filaments $\frac{1}{4}$–$\frac{1}{3}$ lin. long; anthers $1\frac{1}{4}$ lin. long, with small apical gland; style $3\frac{1}{2}$–4 lin. long, with swelling near the base; stigma capitate-clavate.

Eastern Region: Natal; summit of Amajuba Hill, 8000 ft., *Burtt Davy*, 7747c!

This species is distinct, owing particularly to the dense corymbose inflorescence and the long calyx-segments, which overlap widely at the base; it shows some resemblance to *S. Rehmannii* and *S. polyantha* in floral structure.

61. **S. polyantha** (Gilg in Engl. Jahrb. xxvi. 95); annual or biennial, erect, richly branched above, 10–12 in. high; leaves ovate-orbicular or cordate, subacute, somewhat coriaceous, 3–5 lin. long, 4–5 lin. broad; inflorescences elongated, composed of 1–5-flowered axillary cymes; branches 1–2 lin. long; pedicels about 1 lin. long;

calyx-segments narrowly elliptic, acuminate, 3–3½ lin. long, with keel-wing ⅓ lin. broad, broadest about the middle ; corolla-tube 2¼–3 lin. long ; lobes 2½–2¾ lin. long, 1 lin. broad, obovate-lanceolate, acute ; filaments ½ lin. long ; anthers ¾ lin. long, with minute apical glands ; style 2½–3½ lin. long, with swelling ; stigma capitate, inconspicuous.

KALAHARI REGION : Transvaal ; Spitz Kop, near Lydenburg, *Wilms*, 963 ! mountains at MacMac, *Mudd* !

62. S. Rehmannii (Schinz in Vierteljahrsschr. Naturf. Gesellsch. Zürich, xxxvii. 1891, 322) ; annual or biennial, erect ; stems simple, about 8 in. high ; leaves somewhat widely separated ; internodes 7 lin.–1 in. long ; leaves broadly ovate, obtuse or subacute, 2½–4½ lin. long, 3–4½ lin. broad, herbaceous ; cymes relatively few-flowered, with erect branches ; pedicels about 1 lin. long ; calyx-segments 3 lin. long, with broad wing, broadest rather below the middle, ovate-lanceolate, acute ; corolla-tube 2¾–3¼ lin. long ; lobes 3½–4 (rarely 5) lin. long, oblong, obtuse ; filaments ½ lin. long ; anthers 1–1½ lin. long, with small apical gland ; style about 3 lin. long, with swelling ; stigma capitate. *Gilg in Engl. Jahrb.* xxvi. 95 ; *Schinz in Mitteil. Geogr. Ges. Lübeck,* xvii. (1903), 48.

KALAHARI REGION : Transvaal ; Houtbosch (Woodbush), *Rehmann*, 5925 !

In general appearance this species is very close to *S. hymenosepala*, Gilg, but differs in the longer internodes, more ovate herbaceous leaves ; only one anthergland appears to be present, whilst in *S. hymenosepala* from the Cape there appear to be two minute basal glands.

63. S. hymenosepala (Gilg in Engl. Jahrb. xxvi. 89) ; annual or biennial, erect, 6–11 in. high, usually branching from the base, with erect simple branches ; leaves numerous and crowded in the lower parts of the stem, broadly orbicular-ovate to subreniform, 3–4½ lin. long, 4–6 lin. broad, obtuse or apiculate, somewhat coriaceous ; inflorescence composed of compact many-flowered cymes borne on numerous axillary branches ; calyx-segments 2½–4 lin. long, ovate-lanceolate, acute, with membranous keel-wing ⅓–½ lin. wide, broadest below the middle ; corolla-tube 2½–3½ lin. long ; lobes about 3 lin. long, 1–2 lin. broad, spathulate, slightly cucullate ; filaments ½–¾ lin. long ; anthers 1–1½ lin. long, with apical and two minute basal glands ; style 3–3½ lin. long, with swelling ; stigma capitate or very shortly clavate. *Schinz in Mitteil. Geogr. Ges. Lübeck,* xvii. (1903) 49 *partly* ; *A. W. Hill in Kew Bulletin,* 1908, 331. *S. semialata, Gilg in Engl. Jahrb.* xxvi. 97 ; *Schinz in Mitteil. Geogr. Ges. Lübeck,* xvii. (1903) 46.

VAR. β, **grandiflora** (A. W. Hill in Kew Bulletin, 1908, 332), differs from the typical form in the more reniform-rounded leaves and long internodes in the upper parts of the stems, in the wider calyx-lobes and large orbicular-ovate corolla-lobes 4½ lin. long, 3 lin. broad. Cymes borne on long branches, few-flowered, somewhat pendulous.

COAST REGION : Alexandria Div. ; Zuurberg Range, *Bolus,* 9122 ! at Doorn Nek and Bontjes River, 2000–3000 ft., *Drège* ! Albany Div. ; mountains near Grahamstown, 2200 ft., *Galpin,* 375 ! *Atherstone,* 477 ! *Misses Daly and Sole,* 468 ! Stockenstrom Div. ; Katberg, *Shaw* ! Komgha Div. ; between Sandplaat and Komgha, 2200–3200 ft., *Drège,* 4921 !

EASTERN REGION : Transkei ; near Kentani, 1000 ft., *Miss Pegler,* 906 ! Kreilis Country, *Bowker* ! Pondoland ; near Umtamvuna, *Bachmann,* 1040 ; and without precise locality, *Krebs,* 232 ! Griqualand East ; marshy places near Klein Pot River, in Maclear Div., 4500 ft., *Galpin,* 6774 ! Var. β : Griqualand East ; Doodmans Krans (at the junction of the Witteberg and Drakensberg Ranges in Mount Fletcher Div.), 8850 ft., *Galpin,* 6776 !

From a careful comparison of the available material, there appears to be no good ground for maintaining both *S. semialata,* Gilg, and *S. hymenosepala,* Gilg. The principal difference in the description appears to be that in *S. hymenosepala* there are 3 glands to the anthers, whilst only an apical gland is found in *S. semialata.* This distinction, however, does not hold good, for the presence or absence of basal glands seems somewhat variable. *S. hymenosepala,* Gilg, is therefore adopted for the combined species ; it is very closely allied to *S. Rehmannii,* Schinz, from the Transvaal and perhaps should be united with that species. The internodes are shorter, however, in *S. hymenosepala* and the leaves more reniform. Both these latter species are very like *S. Rudolfii,* Schinz, in general appearance, but it is separated from them by the position and character of the anthers. Schinz in his monograph of the genus (*Mitteil. Geogr. Ges. Lübeck,* xvii. 1903, 49) includes some other numbers under *S. hymenosepala,* Gilg, but they do not appear to belong to this species and have been excluded. Shaw's specimen from the Kat Berg most nearly approaches var. *grandiflora.*

64. S. fastigiata (A. W. Hill in Kew Bulletin, 1908, 332) ; perennial or biennial, erect, stout, branched, 10–14 in. high ; branches given off all the length of the stem ; leaves numerous, somewhat crowded, broadly ovate, acute, 4–5 lin. long, 4–7 lin. broad, somewhat coriaceous and slightly varnished above ; inflorescences lax, composed of few-flowered cymes at the ends of lateral branches, branches leafy, inflorescence branches $3\frac{1}{2}$–$7\frac{1}{2}$ lin. long ; pedicels 5 lin. long ; bracts broadly ovate, acute or acuminate ; calyx-segments $3\frac{1}{2}$–$4\frac{1}{2}$ lin. long, ovate-lanceolate, acuminate, with keel-wing broadest at the middle, $\frac{1}{4}$–$\frac{1}{3}$ lin. wide ; corolla-tube about $3\frac{1}{2}$ lin. long ; lobes $3\frac{1}{2}$–4 lin. long, $1\frac{3}{4}$–$2\frac{1}{2}$ lin. wide, obovate ; filaments $\frac{1}{2}$–$\frac{3}{4}$ lin. long ; anthers $1\frac{1}{4}$–2 lin. long, with apical gland (? minute glands also) ; style $3\frac{1}{2}$–$4\frac{1}{2}$ lin. long, with swelling ; stigma capitate. *S. crassulæfolia,* Schinz *in Mitteil. Geogr. Ges. Lübeck,* xvii. (1903) 50 *partly, not of Cham. & Schlecht. S. hymenosepala, Gilg in Schinz, Mitteil. Geogr. Ges. Lübeck,* xvii. (1903) 49 *partly.*

COAST REGION : George Div. ; near George, *Prior* ! Uitenhage Div. ; Vanstadens Berg, *Burchell,* 4749 ! Albany Div. ; mountains near Grahamstown, 2000 ft., *Zeyher,* 205 ! *MacOwan,* 16 ! *Cooper,* 25 ! *Williamson* ! Signal Hill, near Grahamstown, *Schönland,* 16, Howisons Poort, *Schönland.* Bathurst Div. ; Port Alfred, *Miss Sole,* 468.

65. S. Rudolfii (Schinz in Bull. Herb. Boiss. 2me sér. vi. 741) ; annual, erect ; stems usually branching, 5–9 in. high ; leaves numerous, broadly ovate, obtuse, $2\frac{1}{2}$ lin. long, 2 lin. broad, somewhat coriaceous ; cymes rather densely corymbose, their branches 2 lin.

long or less ; pedicels $\frac{1}{2}$ lin. long ; calyx-segments 3–3$\frac{1}{2}$ lin. long, ovate - lanceolate, acute, with wide membranous keel-wing, $\frac{1}{3}$ lin. wide, broadest near the base ; corolla-tube 3–3$\frac{1}{2}$ lin. long ; lobes 2$\frac{1}{2}$–3 lin. long, obovate-oblong, obtuse or subacute, 1–1$\frac{1}{4}$ lin. broad ; anthers $\frac{3}{4}$–$\frac{7}{8}$ lin. long, with apical and very minute basal glands ; filaments very short, inserted in the corolla-throat about $\frac{1}{4}$–$\frac{1}{3}$ lin. below the sinuses ; style 2$\frac{1}{2}$–3 lin. long, with swelling near the base ; stigma capitate. *S. natalensis, Schinz in Bull. Herb. Boiss.* iv. 442 ; *Gilg in Engl. Jahrb.* xxvi. 95.

EASTERN REGION : Pondoland ; Fakus Territory, *Sutherland* ! Natal ; among rocks near Mount West, 5400 ft., *Schlechter,* 6819 ! Ranges 30–60 miles from the sea, 2000–3000 ft., *Sutherland* ! Noodsberg, *Wood,* 922 ! and without precise locality, *Sutherland* ! *Cooper,* 2750 (in Edinburgh Herbarium) !

66. S. repens (Schinz in Bull. Herb. Boiss. ii. 219) ; perennial ; stems slender, branched, creeping, cæspitose ; leaves broadly ovate or subreniform, obtuse or subacute, fleshy, about 2 lin. long, 1$\frac{1}{2}$–3 lin. broad, subsessile or petiolate ; cymes 1–3-flowered, almost sessile at the apex of leafy shoots ; calyx-segments 2$\frac{1}{2}$–3 lin. long, ovate, acute, keeled ; corolla-tube 2$\frac{1}{2}$–3 lin. long ; lobes 3–4 lin. long, 1$\frac{1}{2}$–2 lin. broad, obovate-oblong, obtuse ; filaments $\frac{1}{4}$–$\frac{1}{2}$ lin. long ; anthers 1–1$\frac{1}{4}$ lin. long, with large apical gland (orange when dry) ; style 3–3$\frac{1}{2}$ lin. long, with swelling near the base ; stigma capitate or shortly clavate, sometimes bilabiate. *Gilg in Engl. Jahrb.* xxvi. 98 ; *Schinz in Mitteil. Geogr. Ges. Lübeck,* xvii. (1903), 34. *S. Evansii, N. E. Br. in Kew Bulletin,* 1895, 27.

SOUTH AFRICA : without locality, *Schlechter,* 900a !

KALAHARI REGION : Transvaal ; in shade on the summit of a kloof at Ermelo, *Burtt Davy,* 1892 !

EASTERN REGION : Griqualand East ; Ingeli Mountain, 6000 ft., *Tyson,* 1378 ! Natal ; damp places on flat rocks by the Bushmans River, 6000–7000 ft., *Evans,* 56 ! and without precise locality, *Cooper,* 2761 !

67. S. Marlothii (Gilg in Engl. Jahrb. xxxviii. 83) ; perennial, procumbent with short stout creeping rhizomes ; leaves crowded at the ends of creeping stems, subrosulate or separated by short inter-nodes, obovate-orbicular, obtuse, 2$\frac{1}{2}$–3 lin. long, 2$\frac{1}{2}$ lin. broad, narrowed to form petioles 3–6 lin. long, fleshy ; cymes 3–5-flowered, sessile, buried amongst the leaves ; pedicels about 1 lin. long ; calyx-segments 2$\frac{1}{2}$–3 lin. long, elliptic, obtuse, apiculate, with a dorsal keel ; corolla-tube 3$\frac{1}{2}$–4 lin. long ; lobes 4–4$\frac{1}{2}$ lin. long, 1$\frac{1}{2}$ lin. broad, obovate-oblong, subacute ; filaments inserted slightly below the sinuses of the corolla ; anthers $\frac{3}{4}$ lin. long, with large conical apical gland, black when dry, $\frac{3}{4}$ lin. long ; style 3 lin. long, with swelling near the base, very slight or sometimes absent ; stigma capitate-clavate, bilobed. *Schinz in Bull. Herb. Bois.* 2me *sér.* vi. 737.

KALAHARI REGION : Orange River Colony ; summit of Mont aux Sources, 9500–10000 ft., *Mann,* 2886 ; *Bolus,* 12950 ! 10664 partly !

68. **S. procumbens** (A. W. Hill in Kew Bulletin, 1908, 333); perennial; stems branched, creeping, 2–4 in. long, terminating in erect flowering stems; leaves spathulate or orbicular-spathulate, 4–7 lin. long, 3–4 lin. broad, somewhat fleshy, narrowed below to form a short petiole; erect flowering stem 2–3 in. high, with 3–4 pairs of orbicular-ovate leaves; cymes capitate, dense, many-flowered; branches about 2 lin. long; pedicels very short; calyx-segments united below for a short distance, 3 lin. long, elliptic or obovate, apiculate, membranous with a thickened keel or narrow wing; corolla-tube 3 lin. long; lobes 3–4 lin. long, $1\frac{3}{4}$–2 lin. broad, spathulate, subacute; filaments very short, situated in the corolla sinuses or just below them; anthers $\frac{3}{4}$–1 lin. long, with apical gland (yellow) $\frac{1}{2}$ lin. long and 2 basal glands; style $2\frac{1}{2}$ lin. long, with swelling towards the base; stigma capitate.

KALAHARI REGION: Orange River Colony; summit of Mont aux Sources, 9500 ft., *Flanagan,* 2079 !

69. **S. thodeana** (Gilg in Engl. Jahrb. xxvi. 96); biennial or perennial, with a creeping rhizome, bearing rosettes of radical leaves and erect inflorescences or with creeping stems bearing somewhat scattered leaves which may or may not terminate in an inflorescence; leaves elongate or orbicular-spathulate, narrowed below to form almost a petiole in some cases, $\frac{1}{2}$–$1\frac{1}{2}$ in. long, $\frac{1}{3}$–$\frac{1}{2}$ in. broad, some-what coriaceous or fleshy; inflorescences 2–4 in. high, with usually 2 pairs of obovate subacute cauline leaves; cymes compact, several-flowered; branches 2–4 lin. long; pedicels very short; calyx-segments 3–4 lin. long, fused below for a short distance, elliptic, acute, with thickened keel; corolla-tube $3\frac{1}{2}$–4 lin. long; lobes 4–5 lin. long, $1\frac{1}{2}$–$1\frac{3}{4}$ lin. broad, narrowly elliptic or obovate-elliptic, unguiculate, subacute; filaments about $\frac{1}{2}$ lin. long, inserted just below the corolla sinuses; anthers $\frac{3}{4}$–$1\frac{1}{4}$ lin. long, with a large black conical apical gland $\frac{3}{4}$–1 lin. long and traces of basal glands; style 2–3 lin. long, with swelling near the base; stigma capitate or very shortly clavate. *Schinz in Mitteil. Geogr. Ges. Lübeck,* xvii. (1903) 47.

CENTRAL REGION: Barkly East Div. ; Ben McDhui, 9400 ft., *Galpin,* 6777 !
KALAHARI REGION: Orange River Colony; summit of Mont aux Sources, 10000 ft.. *Mann,* 2878 ; *Flanagan in Herb. Bolus,* 8216 !
EASTERN REGION : Natal; hill above Mjassute River Valley, near Emengweni, 6000–7000 ft., *Thode,* 67 ; Giants Castle, 8000 ft., *Guthrie in Herb. Bolus,* 4882 !

This species presents two very different facies according as it is found with a definite rosette of elongated leaves or with long creeping stem with the leaves separated by internodes 3–5 lin. long. In the latter case the plant resembles *S. repens* or *S. procumbens* very closely in the vegetative condition, but can be easily distinguished by the floral characters. In its rosette form it is very similar in many respects to *S. spathulata,* which possibly also has short creeping stems, the floral characters agree closely, especially the large glands to the anthers and the peculiar texture of the corolla. *S. thodeana* is distinguished from *S. spathulata* chiefly by the position of the anthers in the corolla-tube and by the character of the style.

70. **S. spathulata** (Steud. ex Griseb. in DC. Prodr. ix. 55); perennial or biennial; rhizome short, creeping; flowering stem erect, 5–12 in. high, springing from a large rosette of radical leaves; leaves elongate spathulate, obtuse, 1–5½ in. long, 5–9 lin. broad, somewhat fleshy, the erect flowering stem bears 2 or 3 pairs of obovate subacute leaves; inflorescence a short cymose raceme bearing 2–3 pairs of lateral cymes; branches 4–6 lin. long; cymes compact, 7–15-flowered; branches 2–3 lin. long; pedicels ½–1 lin. long; calyx-segments 4–4½ lin. long, slightly united below, elliptic-lanceolate, acute or acuminate, with a slight keel; corolla-tube 4½–6 lin. long; lobes about 3 lin. long, elliptic-obovate, unguiculate, 1¼–1½ lin. broad; filaments very short situated in the corolla-tube, about ¾–1 lin. below the sinuses; anthers ¾–1¼ lin. long, with dark chocolate-brown conical apical glands about 1 lin. long, anthers usually free in the bud and syngenesious in the open flower; style 1–1½ lin. long; stigma about ¾ lin. long, clavate-bilabiate, swelling just below the stigma and more or less confluent with apex of the stigma below the base of the anthers. *Schinz in Bull. Herb. Boiss.* 2me *sér.* vi. 732; *A. W. Hill in Kew Bulletin*, 1908, 334. *S. Flanaganii, Schinz in Bull. Herb. Boiss.* 2me *sér.* vi. 737. *Belmontia spathulata, E. Meyer, Comm.* 183; *Griseb. in DC. Prodr.* ix. 55. *B. Flanaganii, Schinz in Bull. Herb. Boiss.* iii. 413.

CENTRAL REGION: Aliwal North Div.; Witteberg Range, 7500 ft., *Drège*!
KALAHARI REGION: Basutoland; damp shady banks of Buffalo River, above the waterfall, 8100 ft., *Galpin*, 6778! top of Mont aux Sources, 9500 ft., and Ruellenberg, *Flanagan*, 2080!
EASTERN REGION: Natal; Giants Castle, 6000 ft., *Guthrie in Herb. Bolus*, 4881!

S. spathulata and *S. Flanaganii* appear to be the same thing, the original and only specimen of the former is in poor condition and has lost the large apical glands at the top of the anthers, as has also the Galpin specimen, the scar, however, is left, and as the plants agree in every other respect I have no hesitation in reducing *S. Flanaganii*, Schinz, to *S. spathulata*, Steud. *S. spathulata* shows close resemblance in many points to *S. thodeana*, the stigma and swelling, however, in *S. spathulata* are always below the base of the anthers and more or less confluent, whilst in *S. thodeana* the stigma is capitate and small and situated above the anthers with a small swelling near the base of the style below the anthers.

71. **S. Thomasii** (Schinz in Bull. Herb. Boiss. 2me sér. vi. 743); perennial or biennial; stems branched, procumbent, 5–9 in. long, sometimes rooting at the nodes or ascending ½–1½ in. high; leaves numerous, either separated by internodes 3–6 lin. long or closely compacted to form a radical rosette after the manner of *Gentiana verna*, ovate or orbicular-ovate, 4–8 lin. long, 2–6 lin. broad, rounded or subacute, slightly apiculate, coriaceous, varnished and dark green above, paler below; flowers Gentian-like, usually solitary at the apices of the stems; calyx-segments elliptic-lanceolate, acuminate, 5½–6 lin. long, thickened and keeled on the back, united below to form a short tube; corolla-tube cylindrical, 6–11 lin. long;

lobes 5–6½ lin. long, 3–4 lin. broad, ovate, acute, unguiculate;
filaments ½ lin. long, inserted 1½–3 lin. below the sinuses; anthers
1¼–1½ lin. long, with stipitate apical gland ⅓–½ lin. long and two
smaller basal glands; style about 2 lin. long; stigma 1–1¾ lin. long,
the apex below the base of the anthers; capsule ovoid; seeds small.
A. W. Hill in Kew Bulletin, 1908, 335. *Parasia Thomasii, S. Moore
in Journ. Bot.* 1901, 260, *and* 1907, 154.

COAST REGION : Fort Beaufort Div. ; top of the Winterberg Range, *Fraser* !
KALAHARI REGION : Orange River Colony; Mont aux Sources, 6500–10000 ft.,
Flanagan in Herb. Bolus, 8215 ! *Bolus*, 10664 partly ! and without precise locality,
Pateshall Thomas !
EASTERN REGION : Natal ; Tabamhlope, 6000 ft., *Wylie in Herb. Wood*, 10639 !
Giants Castle Pass, *Wylie in Herb. Wood*, 10639 ! Giants Castle, *Guthrie in Herb.
Bolus*, 4882 partly !

A very variable plant with either long procumbent stems or forming small
plants terminated by a single flower closely simulating *Gentiana verna*, both as
regards the form of the plant and the external appearance of the flower. In the
structure of the flower this species is closely allied to *S. spathulata*, Steud., both
in the anthers with their large apical glands and in the style, with its apex
situated below the base of the anthers, in which its swelling appears to be
confluent with the stigma or to almost entirely replace it. In this latter character,
as also in the glandular hairs between calyx and corolla, this species shows an
alliance with *Exochænium*.

II. EXOCHÆNIUM, Griseb.

Calyx 4–5-lobed ; tube very short ; lobes ovate-lanceolate,
acuminate, winged ; disc-scale ring between calyx and corolla.
Corolla 4–5-lobed ; tube suburceolate around ovary, funnel-shaped
or cylindric above constriction ; lobes ovate or oblong, spreading.
Stamens 4 or 5, inserted below corolla sinuses ; filaments included,
usually stout (except in the long-styled flowers) ; anthers with
apical glands, usually elongated, sometimes with two at the base (in
long-styled flowers anthers are syngenesious and open extrorsely).
Ovary 2-celled ; placentas axile ; ovules many ; style filiform, without glandular swellings ; stigma elongate clavate, covered with long
papillæ, usually situated below the base of the anthers or more
rarely on a level with them, in two or (?) more species the flowers
are heterostyled and long, and sometimes short- and mid-styled
flowers are found. *Capsule* globose or ovoid, coriaceous, septicidal ;
valves 2. *Seeds* many, small, simple.

Annual herbs apparently saprophytic or with mycorhiza ; stems erect, simple or
branched, with decurrent wings ; leaves opposite, sessile, herbaceous ; flowers
small or large, yellow, less commonly white, solitary or more rarely in few-flowered
terminal or axillary cymes.

DISTRIB. Species about 11, all Tropical African, one extending into South
Africa.

Some critical notes on this genus have been published in the *Kew Bulletin*,
1908, 336–341, with a plate.

1. E. grande (Griseb. in DC. Prodr. ix. 55); annual; stem erect, simple or branching above the base, very variable in size, 3–14 in. high; leaves ovate- or linear-lanceolate, acute or acuminate, the lower usually minute and scale-like, separated by more or less long internodes, upper becoming larger, 1–1½ in. long, with shorter internodes; flowers yellow, usually solitary; calyx-segments 5, rarely 4, 8–12 lin. long, ovate-lanceolate, acuminate, with a broad wing ½–1 lin. broad, broadest at the base; a ring of disc-scales occurs between the calyx and corolla; corolla-tube 8–11 lin. long, suburceolate around ovary, gradually funnel-shaped upwards from the constriction; lobes 4 or 5, broadly ovate, acute or apiculate, 6–8 lin. long, 3–5 lin. broad; flowers heterostyled; short-styled flowers: anthers 4 or 5, 1–1¼ lin. long, with large apical conical stipitate glands ⅔–1 lin. long, and two smaller basal glands; filaments stout, 2½–3 lin. long, inserted 3–3½ lin. from the base of the tube; style 1¾–2 lin. long; stigma clavate, 1¼–1½ lin. long, the top reaching about as far as the insertion of the filaments; long-styled flowers: anthers practically sessile, inserted about 3 lin. above the base of the tube, ⅔–1 lin. long, adhering together laterally to form a ring round the style, and thus opening extrorsely, with 3 glands, the apical ½–¾ lin. long, cylindrical or lanceolate, definitely stipitate; style 4½–5 lin. long; stigma 2½ lin. long, papillate, the stigmatic surface being free of the anthers; ovary globose-ovoid, deeply grooved, 2 lin. long, the upper ½ or ⅔ having the wall of the capsule strongly thickened, the lower part thinner; capsule opening probably as a pyxidium; seeds minute. Plants bearing both long- and short-styled flowers have been collected in the same localities. *Welw. in Trans. Linn. Soc.* xxvii. 49; *Schinz in Bull. Herb. Boiss.* 2me *sér.* vi. 745; *A. W. Hill in Kew Bulletin,* 1908, 337; *var. major, Schinz in Bull. Herb. Boiss.* 2me *sér.* vi. 802. *Belmontia grandis, E. Meyer, Comm.* 183; *Schinz in Vierteljahrsschr. Naturf. Gesellsch. Zürich,* xxxvii. 330; *Baker and N. E. Br. in Dyer, Fl. Trop. Afr.* iv. i. 1904, 553. *Sebæa grandis, Steud. Nom. ed.* 2, 1841, 550. *Parasia grandis, Hiern in Cat. Afr. Pl. Welw.* i. 707. *Parasia grandis, Hiern, var. major, S. Moore in Journ. Bot.* 1902, 384.

VAR. **homostylum** (A. W. Hill in Kew Bulletin, 1908, 338); agrees with the typical form as to the general vegetative characters, the plant being usually tall and more slender, with smaller leaves, in some cases the upper leaves being linear-lanceolate, 5 lin. long; flowers usually white, sometimes yellow; calyx-segments 4 or 5, 5–8 lin. long; corolla-tube 5–7 lin. long; lobes 4 or 5, 2½–4 lin. long, 1–2 lin. broad, ovate-lanceolate, acute; anthers 4 or 5, ½–⅔ lin. long, with smaller apical and basal glands; filaments stout, 1 lin. long, inserted 3–3½ lin. above the base of the corolla-tube; style 2½ lin. long; stigma about 1½ lin. long, on a level with the anthers; ovary as in the typical form. *Belmontia natalensis, Schinz in Bull. Herb. Boiss.* ii. 220. *Sebæa natalensis, Schinz in Bull. Herb. Boiss.* 2me *sér.* vi. 732.

KALAHARI REGION: Orange River Colony; Thaba 'Nchu, *Burke,* 205! *Miss Trollops!* Free State Flats, *Mrs. Barber!* between Rhenoster River and Vaal River, *Mrs. Barber,* 626! near Harrismith, 5500 ft., *Sankey,* 176! and without precise locality, *Zeyher,* 1192! *Cooper,* 2758! Basutoland; by the Caledon River,

Mrs. Barber, 20 ! *Hall.* Transvaal ; various localities, *Rehmann*, 5061, 6534, *Nelson*, 390 ! *Wilms*, 961 ! *Galpin*, 1291 ! *Schlechter*, 4137 ! *Rand*, 1227, *Burtt Davy*, 25 ! 1290 ! 7724 ! *Nicholson in Transvaal Herbarium*, 4309 ! *McLea in Herb. Bolus*, 5711 !

EASTERN REGION : Transkei ; Tsomo, *Mrs. Barber*, 856 ! Tembuland ; Bazeia, 2000 ft., *Baur*, 155 ! Pondoland ; between Umtentu River and Umzimkulu River, below 500 ft., *Drège* ! between St. Johns River and Umtsikaba River, *Drège.* Natal ; coast-land, *Sutherland* ! between Attercliff and Bothas Hill, 600–1500 ft., *Sanderson*, 98 ! Inanda, *Wood*, 10 ! Noods Berg, 2000 ft., *Wood*, 109 ! grassfields, &c., 2000 ft., *Wood*, 247 ; near Tugela, *Gerrard*, 782 ! and without precise locality, *Gerrard*, 253 ! *Sanderson*, 292 partly ! *Cooper*, 2754 ! Zululand ; Isandhlwana, *Patteshall Thomas.* Var. β : Natal ; near Durban, *Drège*, 4919 partly ! near Umzinto, *McKen*, 15 ! Clairmont, 20 ft., *Wood*, 6100 ; Noods Berg, *Wood*, 922 ! near Pinetown, 800 ft., *Wood*, 657 ; Eisdumbini, 1800 ft., *Wood*, 133 ! Inanda, *Wood*, 541 ! and without precise locality, *Sanderson*, 292 partly ! Zululand ; Ngoya, 1000–2000 ft., *Wood*, 9322 !

Also in Tropical Africa.

III. ORPHIUM, E. Meyer.

Calyx 5-lobed ; tube campanulate ; lobes oblong, unkeeled, separated from corolla by an annular crenulate disk. *Corolla* 5-lobed ; limb subrotate ; tube short ; lobes contorted. *Stamens* 5, declinate, inserted in corolla-throat ; filaments short, flattened at the base ; anthers linear-oblong, erect, at length twisted. *Ovary* 1-celled, ovoid-oblong ; placentas intruded ; ovules numerous ; style filiform ; stigma capitate, dilated, simple. *Capsule* narrow-ovoid, septicidal ; valves 2 with inflexed placentiferous margins. *Seeds* numerous, foveolate.

An erect virgately branching shrub, pubescent or rarely glabrescent ; leaves numerous, sessile, thick, linear or oblong ; flowers pink, showy, in terminal leafy cymes which are sometimes reduced to a single flower.

DISTRIB.   A single endemic species.

For critical notes on this genus see *Kew Bulletin*, 1908, 341.

1. **O. frutescens** (E. Meyer, Comm. 181, as to citation) ; shrubby ; stems ½–3 ft. long, much-branched ; branches virgate, leafy ; leaves opposite, decussate, flat, oblong-lanceolate or lanceolate or sublinear, obtuse or subacute, usually villous-pubescent on both sides, 1–1½ or rarely 2 in. long, 2–5 lin. wide ; flowers usually in 3-flowered, sometimes 5-flowered cymes, or solitary ; peduncles pubescent, ⅓–½ in. long ; calyx pubescent outside, 6 lin. long, divided half-way ; lobes obtuse or rarely subacute ; corolla rose-pink ; tube as long as calyx, narrow fusiform ; lobes obovate-obtuse, 10 lin. long, 5–6 lin. wide ; anthers twisted ; ovary 3 lin. long, acute ; capsule 4 lin. long, 2½ lin. wide. *Griseb. in DC. Prodr.* ix. 43. *O. frutescens, var. decussata, E. Meyer, Comm.* 181. *Chironia frutescens, Linn. Sp. Pl. ed.* 1, 190 ; *ed.* 2, 273 ; *Berg. Descr. Pl. Cap.* 45 ; *Lam. Ill. t.* 108, *fig.* 1 ; *Curt. Bot. Mag. t.* 37 ; *Thunb.*

Prodr. 35, *Flor. Cap.* ii. 110, *and in Trans. Linn. Soc.* vii. 250; *Pers. Syn. Pl.* i. 282; *Burch. Trav.* i. 37; *Griseb. Gen. et Sp. Gent.* 96. *C. decussata, Vent. Jard. Cels, t.* 31; *Pers. Syn. Pl.* i. 282; *Donn, Hort. Cantab. ed.* 6, 57; *Curt. Bot. Mag. t.* 707; *Reichb. Ic. Bot. Exot.* 16, *t.* 244. *C. grandiflora, Salisb. Prodr.* 137. *C. latifolia, Donn, Hort. Cantab. ed.* 2, 25. *C. frutescens, var. hirsuta, Cham. & Schlecht. in Linnæa,* i. 190, *mainly. C. orthostylis, Reichb. Ic. Bot. Exot.* 16, *t.* 245. *C. fruticosa, O. Kuntze, Rev. Gen. Pl.* ii. 432. *C. dianthiflora, Hort. ex Garden,* 1893, 213. *Roslinia frutescens, G. Don, Gen. Syst. Gard.* iv. 203.

VAR. β, **angustifolia** (Griseb. in DC. Prodr. ix. 43); leaves linear, subterete, often glabrous, sometimes sparsely villous-pubescent, 1–1½ in. long, 1½ lin. wide; peduncles and calyx glabrous or pubescent. *O. frutescens, E. Meyer, Comm.* 181, *as to specimens. O. frutescens, var. glabra, E. Meyer, Comm.* 181. *Chironia caryophylloides, Linn. Cent.* ii. 12; *Amœn. Acad.* iv. 308. *C. angustifolia, Curt. Bot. Mag. t.* 818. *C. frutescens, var. hirsuta, Cham. & Schlecht. in Linnæa,* i. 190 *partly. C. frutescens, var. glabra, Cham. & Schlecht. in Linnæa,* i. 190. *C. frutescens, var. angustifolia, Griseb. Gen. et Sp. Gent.* 96. *Roslinia angustifolia, G. Don, Gen. Syst.* iv. 203.

SOUTH AFRICA: without locality, *Sparrman*! *Oldenland*! *Osbeck*! *Spielhaus*! *Breutel*! *Thom,* 636! 687! *and cultivated specimens*!

COAST REGION: Cape Div.; Flats, sand-dunes and other places near Cape Town, *Petiver*! *Sonnerat*! *Thunberg*! *Burchell,* 8378! *Wallich*! *Lehmann*! *Hooker,* 523! *Harvey,* 611! *Miss Cole*! *Castelnau,* 235! 236! 418! *Milne,* 171! *Emerson*! *Pappe*! *Prior*! *Bunbury*! *MacOwan and Bolus, Herb. Norm.* 960! *Wolley Dod,* 352! Oaklands, *Jameson*! Camps Bay, near Little Lions Head, *Prior*! Stellenbosch Div.; Hottentots Holland Mountains, *Thunberg*! *Zeyher*! Caledon Div.; *Thom,* 1006! Swellendam Div.; between Grootvaders Bosch and Zuurbraak, *Burchell,* 7256! Riversdale Div.; near Riversdale, *Schlechter,* 2008! Uniondale Div.; Long Kloof, *Castelnau*! Var. β: Clanwilliam Div.; between Berg Valley and Zwartbast Kraal, *Drège*! between Lange Valley and Brandenberg, 600–1000 ft., *Drège*! Ruy Stream, near Lange Valley, *Zeyher,* 3426! Cape Div.; near Cape Town, *Sieber*! *Burchell,* 414! *Drège*! *Ecklon,* 79! *Wallich*! *Lehmann*! *Castelnau*! *Harvey,* 606! 608! Table Mountain, *Ecklon,* 647! near Rondebosch, *Burchell,* 729! Stellenbosch Div.; around Somerset West, *Ecklon,* 646! Riversdale Div.; near Kaffirkuils River, 500 ft., *Drège*!

The variety here recognised appears very distinct, but there are intermediates connecting it with typical *O. frutescens.* Within each variety there is considerable variation; in *O. frutescens* proper this is mainly restricted to differences in the breadth of the leaves, the character and amount of pubescence remaining uniform; in var. *angustifolia,* it is mainly in the degree of pubescence that variation takes place, the leaves remaining uniform in shape and size. It so happens that the three gatherings by Drège, cited by Meyer when establishing the genus *Orphium,* all belong to var. *angustifolia,* Griseb.

IV. CHIRONIA, Linn.

Calyx deeply 5-lobed; tube short campanulate, sometimes nearly obsolete; lobes lanceolate, rarely ovate, usually keeled. *Corolla* 5-lobed; tube short or very short; lobes spreading, contorted. *Stamens* 5, usually inserted on the corolla-throat, sometimes within the corolla-tube; filaments short; anthers linear-oblong, straight or spiral, exserted. *Ovary* 1-celled ; placentas 2-fid, little intruded ; ovules numerous ; style subulate ; stigma simple or 2-lobed. *Capsule* ovoid or subglobose, septicidal, 2-valved ; rarely (*C. baccifera*) fruit almost indehiscent, baccate. *Seeds* numerous, globose, foveolate.

Annual or perennial glabrous herbs, often branched ; leaves 1–3-nerved, usually numerous, sessile ; flowers usually pink, rarely purple, cymose ; cymes sometimes reduced to a single flower, sometimes paniculate.

DISTRIB. Species 33, in South Africa, Tropical Africa and Madagascar.

Critical notes on this genus have been published in the *Kew Bulletin*, 1908, 341–376.

Section 1. RŒSLINIA. Calyx 5-partite ; corolla-tube cylindric, narrowed under the limb ; anthers straight ; ovary obtuse ; fruit globose, baccate.

A single species ...   .   (1) **baccifera.**

Section 2. LINOCHIRON. Calyx 5-fid, rarely 5-partite ; corolla-tube narrow-campanulate ; anthers straight or faintly spiral ; ovary obtuse ; capsule obtuse.

Stems diffuse ; branches patent ; calyx-lobes broad, spreading :
  Calyx 5-partite, longer than corolla-tube ; corolla-lobes obovate or suborbicular   ...   .   (2) **arenaria.**

  Calyx 5-fid, not longer than corolla-tube ; corolla-lobes elliptic-lanceolate ...   .   (3) **Schinzii.**

Stems erect ; branches erect or ascending ; calyx 5-fid :
  Calyx longer than corolla-tube :
    Leaves narrow-linear or subulate :
      Calyx-lobes wide-triangular, obtuse or acute, erect, overlapping and subauriculate at the base   ...   .   (4) **emarginata.**

      Calyx-lobes narrow, acute or subobtuse, often slightly spreading, not subauriculate at the base   ...   .   (5) **gracilis.**

    Leaves narrow-spathulate ; calyx-lobes rounded at the apex, erect, not subauriculate   ...   ...   (6) **Zeyheri.**

  Calyx not longer than corolla-tube ; lobes triangular, acute, erect, not subauriculate ; leaves linear ...   (7) **linoides.**

Section 3. HETEROCHIRON. Calyx 5-partite ; corolla-tube cylindric ; anthers faintly spiral ; ovary subacute ; pedicels and calyx-lobes externally faintly puberulous.

A single species ...   .   (8) **Bansei.**

Section 4. PSEUDOSABBATIA. Calyx 5-fid ; corolla-tube cylindric ; limb slightly widened ; anthers from faintly to distinctly spiral ; ovary acute ; capsule pointed.

Stem-leaves in 4–8 pairs ; radical leaves few, usually soon vanishing ; panicles lax, their branches spreading ; anthers faintly spiral :
  Stem-leaves broad, ovate-lanceolate ; calyx-lobes rarely longer than the tube ; corolla-lobes ovate, obtuse, less often ovate-lanceolate, acute   ...   (9) **rosacea.**

  Stem-leaves narrow, lanceolate ; calyx-lobes longer than the tube ; corolla-lobes ovate-lanceolate, acute   ...   .   (10) **transvaalensis.**

Stem-leaves in 1–2 (rarely 3) pairs; radical leaves
many subrosulate, persisting; anthers distinctly
spiral :
  Panicles lax, their branches spreading; corolla-lobes
  usually entire; leaves obovate-spathulate or
  lanceolate ... ... (11) **palustris.**

  Panicles dense, their branches virgate; corolla-
  lobes usually erose; leaves lanceolate ... ... (12) **Krebsii.**

Section 5. PLOCANDRA. Calyx 5-sect; corolla-tube very short; limb
large campanulate; anthers distinctly spirally twisted; ovary acute; capsule
pointed.
  Primary peduncles short, never over ½ in., usually
  much shorter... ... ... (13) **humilis.**

  Primary peduncles long, never under ½ in., usually
  longer... ... ... (14) **purpurascens.**

Section 6. HIPPOCHIRON. Calyx 5-sect; corolla-tube cylindric, narrowed
under the limb; anthers straight or rarely distinctly spirally twisted; ovary
acute; capsule pointed.
Leaf-bases rounded or cordate :
  Stems erect, strongly angular; branches ascending;
  calyx longer than corolla-tube; anthers distinctly
  spiral ... ... (15) **Pegleræ.**

  Stems decumbent, terete; branches spreading; calyx
  not longer than corolla-tube; anthers straight (16) **peduncularis.**

Leaf-bases cuneate :
  Stems erect; leaf-margins not scabridulous :
    Stems simple or sparingly virgately branched
    above; leaves and flowers large ... ... (17) **jasminoides.**

    Stems much diffusely branched; leaves and flowers
    small :
      Leaves ⅓ in. long or less, not more than twice
      as long as broad ... ... (18) **serpyllifolia.**

      Leaves ¾ in. long or longer, at least four times
      as long as broad :
        Corolla-lobes acute ... ... (19) **laxa.**

        Corolla-lobes obtuse ... ... (20) **floribunda.**

  Stems below decumbent and rooting at the nodes;
  leaf-margins minutely scabridulous; corolla-
  lobes acute... ... ... (21) **maritima.**

Section 7. IXOCHIRON. Calyx 5-fid, viscid; corolla-tube cylindric, uniform;
anthers straight; ovary acute; capsule viscid, pointed.
Stems erect; leaves cuneate at the base :
  Calyx not longer than the corolla-tube, its lobes
  narrow, acute or acuminate :
    Leaves glandular-scabridulous, especially near the
    margins ... ... (22) **scabrida.**

    Leaves not glandular-scabridulous ... ... (23) **tabularis.**

  Calyx longer than the corolla-tube, its lobes broad,
  foliaceous ... ... (24) **tetragona.**

Stems decumbent or scandent, very long; leaves
cordate at the base; calyx not longer than corolla-
tube, its lobes narrow, acute ... ... (25) **melampyrifolia.**

1. C. baccifera (Linn. Sp. Pl. ed. 1, 190); stem distinctly angled, leafy, much-branched, $\frac{1}{2}$–3 ft. long; branches divaricate; leaves linear, acute, $\frac{1}{2}$–$\frac{3}{4}$ in. or (*elongata* form) 1 in. long, $\frac{1}{2}$–$\frac{3}{4}$ lin. wide; flowers solitary or 2–3, terminal, with peduncles $\frac{1}{4}$–$\frac{3}{4}$ or (*elongata* form) 1–1$\frac{1}{4}$ in. long; calyx 2$\frac{1}{2}$ lin. long, divided three-fourths; lobes ovate, subacute or acute; corolla-tube cylindric, rather longer than calyx; limb narrow; lobes obovate-oblong, obtuse, 2$\frac{1}{2}$–3 lin. long, 2 lin. wide; anthers straight; ovary subglobose, 2 lin. long; fruit baccate. *Linn. Sp. Pl. ed.* 2, 273; *Lam. Encycl. Meth.* i. 736; *Gærtn. f. Fruct.* ii. 156, *t.* 114, *fig.* 3; *Thunb. Prodr.* 35, *in Trans. Linn. Soc.* vii. 253 *and Flor. Cap. ed.* 2, ii. 107; *Willd. Sp. Pl.* i. 1070; *Pers. Syn.* i. 282; *Curt. Bot. Mag. t.* 233; *Ait. Hort. Kew. ed.* 2, ii. 7; *R. Br. Prodr.* 451; *Roem. & Schult. Syst. Veg.* iv. 203; *Burch. Trav.* i. 15, 31, 59; *Cham. in Linnæa,* vi. 345; *E. Meyer, Comm.* 180; *Griseb. Gen. et Sp. Gent.* 105 *and in DC. Prodr.* ix. 41; *Gilg in Ann. Nat. Hist. Hofmus. Wien,* xv. 65 *and in Engl. und Prantl, Pflanzenfam.* iv. 2, 78; *Schoch in Bot. Centralbl. Beih.* xiv. 188, *t.* 16, *fig.* 1. *C. baccifera, var. elongata, E. Meyer, Comm.* 180. *C. parviflora, Salisb. Prodr.* 136. *C. baccata, Hoffmg. Verz. Pflanz. Nachtr.* 211. *Rœslinia tetragona, Moench, Meth. Suppl.* 212. *Roslinia baccifera, G. Don, Gen. Syst.* iv. 203.

VAR. β, **grandiflora** (Griseb. Gen. et Sp. Gen. 105); leaves narrow-linear, $\frac{1}{3}$–$\frac{1}{2}$ in. long, $\frac{1}{4}$–$\frac{1}{3}$ lin. wide; calyx-lobes obtuse or subacute; corolla-lobes elliptic-oblong, subacute or obtuse, 5 lin. long, 2$\frac{1}{2}$ lin. wide. *Griseb. in DC. Prodr.* ix. 41.

VAR. γ, **Burchellii** (Prain in Kew Bulletin, 1908, 292); leaves narrow-oblong, obtuse or subacute, reflexed, $\frac{1}{3}$–$\frac{1}{2}$ in. long, 1$\frac{1}{4}$ lin. wide; calyx-lobes obtuse. *C. baccifera, var. dilatata, Schoch in Bot. Centralbl. Beih.* xiv. 189 *partly, hardly of E. Meyer.*

VAR. δ, **dilatata** (E. Meyer, Comm. 180); leaves subspathulate-oblong, obtuse, $\frac{3}{4}$–1 in. long, 2–3 lin. wide; calyx-lobes obtuse. *Griseb. in DC. Prodr.* ix. 41; *Schoch in Bot. Centralbl. Beih.* xiv. 189, *mainly.*

SOUTH AFRICA: without locality; *Oldenland! Oldenburg! Burmann! Commerson! Banks and Solander! Forster! Masson! Spielhaus! Hunnemann (Herb. Willdenow,* 4506)! *Hesse! Brogniart! Niven,* 12! *Bowie,* 2! *Sieber,* 187! *Ludwig! Harvey! Thom,* 285! *Pappe! Rehmann,* 1970! *Elliott! Kay,* 3779! and *cultivated specimens*! Var. β: *Thom*! Var. γ: *Osbeck! Harvey,* 607! and *cultivated specimens*!

COAST REGION: Clanwilliam Div.; Belle Fontein, Lange Valley, *Wallich*! Malmesbury Div.; near Hopefield, *Bachmann,* 989! 1808! Tulbagh Div.; New Kloof, 1000–2000 ft., *Drège*! Tulbagh Waterfall, *Ludwig*! Mitchells Pass, between Tulbagh and Ceres, *Wyley*! Worcester Div.; Hex River Valley, 1600 ft., *Tyson,* 759! near Worcester, *Rehmann,* 2492! Paarl Div.; Paarl, *Prior*! Cape Div.; various localities, *Sparrman! Thunberg! Osbeck! Sonnerat! Wahlberg! Spielhaus! Grondahl! Gaudichaud! Bergius! Lalande! Burchell,* 241! 774! *Bojer! Ecklon,* 31! 79! 650! 652! *Scholl! Verreaux,* 637! *Lehmann! Jameson! Jelinek,* 18! *Prior! Milne,* 182! *Bunbury! Rehmann,* 760! 1974! *Wright! Mrs. Elliott,* 54! 93! *Penther,* 2012! *Wilms,* 471! 3471! *Kuntze! Bolus,* 3378! *Wolley Dod,* 681! *Schlechter,* 16! *Brown! Barnard Fuller*! Stellenbosch Div.; Hottentots Holland, *Verreaux*! Riversdale Div.; without precise locality, *Rust,* 435! Oudtshoorn Div.; Cango, 2100 ft., *Bolus,* 12158! Mossel Bay Div.; Mossel Bay, *Fraser*! Great Brak River, *Schlechter,* 5755! Knysna Div.; Vlugt Valley, *Bolus,* 2395! Keurboom River, *Penther,* 2015!

Knysna, *Miss Newdegate*! and without precise locality, *Castelnau*, 582! Uitenhage Div.; Loeri River, *Penther*, 2025! Port Elizabeth Div.; near Port Elizabeth, *E.S.C.A. Herb.* 217! Bathurst Div.; near Port Alfred, 50 ft., *Galpin*, 286! Albany Div.; Grahamstown Flats, *Scott-Elliot*, 854! Var. β: Uitenhage Div.; near the Zwartkops River, *Drège*! *Zeyher*, 252! *Ecklon*, 651! Port Elizabeth Div.; Algoa Bay, *Prior*! near Port Elizabeth, *Mrs. Holland*, 47! *Miss West*, 31! Var. γ: Clanwilliam Div.; Alexanders Kloof, *Wallich*! Tulbagh Div.; New Kloof, *Drège*! Worcester Div.; Bains Kloof, *Wawra*, 56! Paarl Div.; Paarl Mountain, *Drège*, 1895! Cape Div.; near Cape Town, *Osbeck*! *Thunberg*! *Banks*! *Wallich*! *Krauss*! *Rehmann*, 1975! *Ecklon*! Simons Bay, *MacGillivray*, 620! Simons Town, *Wolley Dod*, 680! Riversdale Div.; Gouritz River, *Penther*, 2013! and without precise locality, *Rust*, 261! George Div.; Montague Pass, 1000 ft., *Marloth*, 2799! Uitenhage Div.; near Uitenhage, *Zeyher*! Bathurst Div.; mouth of the Great Fish River, *Burchell*, 3739! Port Alfred, *Haagner*, 74! Albany Div.; Assegai Bush, *Baur*, 1028! Fish River Heights, *Hutton*! Var. δ: East London Div.; East London, *Kuntze*! Komgha Div.; sea-shore near mouth of Kei River, *Flanagan*, 1146!

CENTRAL REGION: Aberdeen Div.; Camdeboo Berg, 4000–5000 ft. (*elongata* form), *Drège*!

EASTERN REGION: Var. γ: Transkei; near Gekau (Geua) River, below 1000 ft., *Drège*! Var. δ: Kaffraria; near coast, *Drège*, 4925! Transkei; Kentani, *Miss Pegler*, 373! Pondoland; without precise locality, *Bachmann*, 450! Natal; near Durban, *Gerrard*, 529! 698! *Grant*! *Sanderson*, 2010! *Wood*, 54! 857! 6371!

An example of typical *C. baccifera* in Herb. Delessert collected by Lemue is noted as being from near Takoon in Bechuanaland. No other specimen of *C. baccifera* appears to have been met with in the Kalahari Region; this locality is therefore for the present doubtful. The same remark applies to a specimen of *C. baccifera*, var. *dilatata* in Herb. Kew collected by Cooper (2763) and said to be from the Orange River Colony; this is a still more doubtful locality.

A rather variable species. Bowie terms it the "Tooth-ache Berry."

2. C. arenaria (E. Meyer, Comm. 180); stems faintly angled, leafy, ½–¾ ft. long; branches diffuse; leaves linear to lanceolate, subacute or acute, ½–¾ in. long, 1–1½ lin. wide; flowers solitary, terminal, their peduncles ⅓–1¼ in. long; calyx 4 lin. long, divided three-fourths; lobes wide-ovate, subacute or acute, slightly spreading; corolla-tube narrow-campanulate, shorter than calyx; lobes obovate or suborbicular, usually minutely apiculate, 4 lin. long, 3½–4 lin. wide; anthers straight; ovary wide-ovate, obtuse, 3 lin. long. *Schoch in Bot. Centralbl. Beih.* xiv. 213, *t.* 15, *fig.* 10.

VAR. β, **mediocris** (Prain in Kew Bulletin, 1908, 349); leaves narrow-oblong, obtuse or subacute, 2–3 lin. wide; flowers rather larger; corolla-lobes obtuse or retuse, never apiculate, 5 lin. long, 4½–5 lin. wide. *C. arenaria, Griseb. in DC. Prodr.* ix. 40, *hardly of E. Meyer. C. mediocris, Schoch in Bull. Herb. Boiss.* 2me *sér.* ii. 1011, *and in Bot. Centralbl. Beih.* xiv. 207.

COAST REGION: Clanwilliam Div.; Berg Valley, *Niven*, 11! between Pikeniers Kloof and Marcus Kraal, 1000–2000 ft. (mixed with var. β), *Drège*, 3058! Alexanders Kloof, *Wallich*! Lange Valley, in sandy places, *Zeyher*, 3428!

A very distinct species. The variety *mediocris*, considered by Schoch a species, was collected along with the type. Drège did not distinguish the two forms in the field, and Meyer distributed both as *C. arenaria*; Meyer's description applies, however, only to the narrow-leaved form; Grisebach's description, on the other hand, applies only to var. *mediocris*.

3. C. Schinzii (Schoch in Bull. Herb. Boiss. 2me sér. ii. 1012); stems angled, leafy, $\frac{1}{2}$–$\frac{3}{4}$ ft. long ; branches diffuse ; leaves wide-linear or narrow-elliptic, obtuse or subacute, $\frac{2}{3}$–$\frac{3}{4}$ in. long, 1–1$\frac{1}{2}$ lin. wide ; flowers solitary, terminal ; peduncles $\frac{1}{4}$ in. long ; calyx 2$\frac{1}{2}$ lin. long, divided half-way ; lobes ovate-lanceolate, acute, slightly spreading ; corolla-tube narrow-campanulate, as long as or slightly longer than the calyx ; lobes elliptic-lanceolate, subacute or obtuse, 4 lin. long, 1$\frac{1}{2}$ lin. wide ; anthers straight ; ovary ovoid, obtuse, 3 lin. long. *Schoch in Bot. Centralbl. Beih.* xiv. 199.

COAST REGION : Malmesbury Div. ; near Hopefield, *Bachmann,* 990 !

Only known from a single gathering ; most nearly allied to *C. arenaria,* of which it has somewhat the habit and appearance, but from which it differs in having shorter calyx-segments, a longer corolla-tube with narrower corolla-lobes and a narrower ovary.

4. C. emarginata (Jaroscz, Pl. Nov. Cap. 11); stem slightly angled, usually branching upwards, leafy, $\frac{1}{2}$–1 ft. long ; branches rather straight ; leaves narrow-linear or subulate, $\frac{1}{2}$–$\frac{3}{4}$ in. long, $\frac{1}{2}$–$\frac{3}{4}$ lin. wide ; flowers 2–3 or solitary, terminal ; peduncles $\frac{1}{3}$–$\frac{2}{3}$ in. long ; calyx 4–5 lin. long, divided half-way ; lobes wide-triangular, obtuse or subacute or acute, the outer subauriculately overlapping at the base ; corolla-tube narrow-campanulate, shorter than the calyx ; lobes oblong or ovate, retuse, 5$\frac{1}{2}$–8 lin. long, 3–4 lin. wide ; anthers straight ; ovary ovoid, obtuse, 3 lin. long. *C. linoides, Berg. Descr. Pl. Cap.* 43 (*excl. syn.*) ; *Burch. Trav.* i. 56 (*Rondebosch specimens*) ; *Griseb. Gen. et Sp. Gent.* 104 (*as to C. uniflora, Eckl. only*) ; *not of Linn. C. linoides, var., Lam. Encycl. Meth.* i. 736 *partly. C. linoides, var. Zeyheri, Griseb. in DC. Prodr.* ix. 41. *C. linoides, var. brevisepala, Schoch in Bot. Centralbl. Beih.* xiv. 203. *C. lychnoides, Cham. & Schlecht. in Linnæa,* i. 190 (*as to C. emarginata, Jaroscz, only*) ; *not of Berg. C. vulgaris, var. lychnoides, Cham. in Linnæa,* vi. 343. *C. uniflora, Eckl. ex Griseb. Gen. et Sp. Gent.* 105 ; *not of Lam. C. baccifera, Zeyh. ex Griseb. in DC. Prodr.* ix. 41 ; *not of Linn.*

SOUTH AFRICA : without locality, *Grubb in Herb. P. J. Bergius* ! *and cultivated specimen* !
COAST REGION : Piquetberg Div. ; Elands Berg, *Wallich* ! Malmesbury Div. ; near Hopefield, *Bachmann,* 991 ! 992 ! Tulbagh Div. ; Mitchells Pass between Tulbagh and Ceres, *Wyley* ! Cape Div. ; Cape Flats, *Oldenland* ! *Burmann* ! *Thunberg* ! *Nelson* ! *Banks and Solander* ! *Mrs. Elliott,* 55 ! *Scholl* ! *Ecklon,* 75 ! 348 ! 645 mainly ! *Pappe* ! *Zeyher,* 1197 mainly ! *Wallich,* 162 ! *Harvey,* 214 ! 609 ! *Reeves* ! *Lehmann* ! *Schulze,* 39 ! *Miss Cole* ! *Rehmann* ! *Wolley Dod,* 351 ! 681 ! *Schlechter,* 27 ! near Rondebosch, *Burchell,* 157 ! 720 ! *Wallich,* 405 ! at the base of Muizen Berg, 100 ft., *Bolus,* 3369 ! Simons Bay, *Prior* ! Caledon Div. ; near Genadendal, *Boental* ! Swellendam and George Div. ; without precise localities, *Bowie,* 1 !

Closely allied to *C. gracilis,* but readily distinguished by its subauriculately overlapping calyx-lobes. As in the somewhat parallel case of *C. tetragona* and *C. tabularis,* the question whether the differences exhibited by the calyces of these respective forms entitle them to specific rank is one that can only be answered by field-workers in South Africa.

5. **C. gracilis** (Salisb. ex Prain in Kew Bulletin, 1908, 293 ; not of Michx) ; stems slightly angled, usually branching upwards, leafy, $\frac{1}{3}-\frac{2}{3}$ ft. ; branches rather straight ; leaves narrow-linear or subulate, $\frac{1}{2}-1\frac{1}{4}$ in. long, $\frac{1}{3}-\frac{3}{4}$ lin. wide ; flowers 2–3 or solitary, terminal, with peduncles $\frac{1}{2}-1$ in. long ; calyx 4–5 lin. long, divided two-thirds to three-fourths ; lobes lanceolate or narrow-oblong, obtuse or acute, not subauriculate at base, often at length subpatent ; corolla-tube narrow-campanulate, shorter than calyx ; lobes oblong or ovate, obtuse or subacute, 4–6 lin. long, 3–3½ lin. wide ; anthers straight ; ovary ovoid, obtuse, 3 lin. long. *C. linoides, Thunb. Prodr. Pl. Cap.* 35, *in Trans. Linn. Soc.* vii. 252, *and Flor. Cap. ed.* 2, ii. 108 ; *Willd. Sp. Pl.* i. 1070 ; *Pers. Syn. Pl.* i. 282 (*excl. syn. Breyn.*) ; *Burch. Trav.* i. 19 (*Cape Flats loc. only*) ; *E. Meyer, Comm.* 179 (*Elands Kloof loc. only*) ; *Griseb. Gen. et Sp. Gent.* 104 *and in DC. Prodr.* ix. 41 (*excl. syn. C. uniflora, Eckl.*) ; *Schoch in Bot. Centralbl. Beih.* xiv. 202, *t.* 16, *fig.* 4 (*excl. syn. C. herbacea and C. uniflora and many cit.*) ; *not of Linn. C. linoides, var., Lam. Encycl. Meth.* i. 736 *partly. C. linoides, var. subulata, E. Meyer, Comm.* 180 ; *Schoch in Bot. Centralbl. Beih.* xiv. 203 (*excl. syn. var. Zeyheri*). *C. lychnoides, Cham. & Schlecht. in Linnæa,* i. 190 (*excl. syn. C. emarginata*) ; *not of Berg. C. vulgaris, var. intermedia, Cham. in Linnæa,* vi. 343 ; *not C. intermedia, Merat.*

VAR. β, **macrocalyx** (Prain in Kew Bulletin, 1908, 294) ; much-branched from the base ; leaves linear, acuminate, 1–1½ in. long, 1 lin. wide ; flowers usually solitary ; peduncles 1–1½ in. long ; calyx 6 lin. long, divided halfway ; lobes wide-triangular ; corolla-lobes oblong, obtuse or retuse, 9 lin. long, 4–5 lin. wide. *C. lychnoides, Thunb. Prodr.* 35, *in Trans. Linn. Soc.* vii. 252, *and Fl. Cap. ed.* 2, ii. 108 (*as to sheet δ in herb. propr. only*) ; *not of Berg.*

SOUTH AFRICA : without locality, *Sparrman* ! *Osbeck* ! *Burmann* ! *Krauss*, 458 ! *Lehmann* ! *Thom* ! *Reynaud* ! *Belanger* ! Var. β : *cultivated specimen* !
COAST REGION : Van Rhynsdorp Div. ; Wind Hoek, *Niven* ! Tulbagh Div. ; New Kloof, *Zeyher* ! Tulbagh Kloof, *Ecklon* ! Elands Kloof, 1000–1500 ft., *Drège*, 1894 ! Worcester Div. ; near Dutoits Kloof, 1500–2000 ft., *Drège* ! near Vogel Valley, *Ecklon* ! mountains near Worcester, *Rehmann*, 2491 ! Cape Div. ; Cape Flats, *Banks and Solander* ! *Roxburgh* ! *Halfer* ! *Wallich* ! *Bunbury* ! *Bergius* ! *Burchell*, 76 ! *Ecklon*, 642 ! *Zeyher*, 237 ! 1197 partly ! Tiger Berg, *Niven* ! Table Mountain, *Ecklon*, 40 ! Devils Mountain, *Bunbury* ! Caledon Div. ; Zwart Berg, *Zeyher*, 3424 ! near Caledon, *Pappe* ! hills near Grabouw, 1000 ft., *Bolus*, 4180 ! *Miss Kensit in Herb. Bolus*, 10482 ! between Brand Vlei and Villiersdorp, *Bolus*, 12997 ! Steenbrass River, 1150 ft., *Marloth*, 2848 ! Swellendam Div. ; without precise locality, *Thunberg* ! *Ludwig* ! Riversdale Div. ; near Garcias Pass, *Burchell*, 7147 ! Var. β : Cape Div. ; Cape Flats, *Sonnerat* ! *Wallich* ! *Hooker*, 607 ! *Ecklon*, 644 ! *Lalande* ! Constantia, *Mrs. Jameson* ! Simons Bay, *MacGillivray*, 619 ! *Milne*, 214 ! *Harvey* ! *Wright* ! Stellenbosch Div. ; hills near Hottentots Holland, *Thunberg* ! *Krauss* ! Knysna Div. ; salt marshes and moist plains, *Bowie*, 5 ! Plettenbergs Bay, *Bowie* !

In addition to the specimens recorded there is in Herb. Delessert a specimen of typical *C. gracilis* collected by Verreaux and marked '' Eutnage,'' and in Herb. Vienna another marked ''Port Natal, Poeppig.'' In the absence of further material it is safer, for the present, to leave Uitenhage and Natal as doubtful localities. The form here treated as var. *macrocalyx* is outwardly very distinct and is readily recognisable. As regards calyx it is intermediate between

C. gracilis and *C. emarginata*, but the lobes do not overlap at the base as in
C. emarginata. The characters of var. *macrocalyx* suggest that it may be a hybrid
between *C. emarginata* and *C. gracilis*, rather than an intermediate form.
Within *C. gracilis* proper two distinct forms may be recognised. Of these
one has shorter broader calyx-lobes not overlapping at the base ; this form was
included in *C. linoides* by Meyer and constitutes typical *C. linoides*, Schoch, not
of Linn. ; the other has long, narrow corolla-lobes and is *C. linoides*, Thunb., not
of Linn. and not of Berg., also *C. linoides*, var. *subulata*, E. Meyer and Schoch.
Both forms are included in *C. vulgaris*, var. *intermedia*, Cham., and in *C. linoides*,
Griseb., not of Linn., whose limitation is here adopted. In Niven's Tiger Berg
plant and in Burchell 7147 the calyx-lobes are long as in the second form, but
broad as in the first.

6. **C. Zeyheri** (Prain in Kew Bulletin, 1908, 295) ; stems slightly
angled, branched upwards, leafy, 1–1½ ft. long ; leaves narrowly
elliptic-spathulate, obtuse or subacute, 1½–2 in. long, 3 lin. wide ;
flowers solitary or in 3-flowered terminal cymes ; peduncles $\frac{1}{3}$–$\frac{3}{4}$
in. long ; calyx 3 lin. long, divided half-way ; lobes ovate, obtuse,
mucronulate ; corolla-tube narrow-campanulate, shorter than the
calyx ; lobes elliptic, obtuse, 5 lin. long, 3 lin. wide ; anthers
straight ; ovary ovoid, obtuse, 3 lin. long.

VAR. β, **angustifolia** (Prain l.c.) ; leaves narrowly lanceolate-spathulate, 1½ lin.
wide ; calyx-lobes ovate-lanceolate, obtuse. *C. linoides, E. Meyer, Comm.* 179
(*Wupperthal loc. only*); *not of Linn.*

COAST REGION : Clanwilliam Div. ; Companies Drift, Berg River Valley,
Zeyher, 1198 ! Var. β : Clanwilliam Div. ; Wupperthal, 1800 ft., *Drège* ! by the
Olifants River and near Brakfontein, *Ecklon* ! marshy soil and along river-banks
near Clanwilliam, *Leipoldt*, 360 !

A very distinct species. Ecklon's and Leipoldt's specimens are almost intermediate between the extremely broad-leaved form collected by Zeyher on which
the species is based, and the very narrow-leaved form issued by E. Meyer as
C. linoides. This species, unlike the other species of the section, dries black.

7. **C. linoides** (Linn. Sp. Pl. ed. 1, 189) ; stems slightly angled,
much-branched upwards, leafy, 2–3 ft. long ; branches spreading,
3–10 in. long ; leaves linear or narrow-lanceolate, acuminate, 1–1½
in. long, 1–1½ lin. wide ; flowers usually 2–5 towards ends of
branches on alternate peduncles, or solitary ; peduncles ½–1¼ in.
long ; calyx 3 lin. long, divided half-way ; lobes ovate-lanceolate,
acute, with paler submembranous margins ; corolla-tube narrow-
campanulate, as long as calyx ; lobes elliptic, obtuse, 5 lin. long,
3 lin. wide ; anthers straight or slightly spiral ; ovary ovoid, obtuse,
2½ lin. long. *Linn. Sp. Pl. ed.* 2, 272 ; *Lam. Encycl. Meth.* i. 736 ;
Moench, Meth. Suppl. 163 (*excl. syn. Berg.*); *Willd. Sp. Pl.* i. 1070
(*excl. syn. Berg. and syn. Thunb.*); *Sims, Bot. Mag. t.* 511 ; *Ait.
Hort. Kew. ed.* 2, ii. 7 ; *R. Br. Prodr.* i. 451 ; *Burch. Trav.* i. 15 ;
Drapiez, Herb. Amat. Fl. vi. 429, *with plate* ; *Cham. & Schlecht. in
Linnæa*, i. 191 ; *E. Meyer, Comm.* 179 (*excl. loc. Elands Kloof and
loc. Wupperthal*). *C. lychnoides, Berg. Descr. Pl. Cap.* 45 ; *Thunb.
Prodr.* 35, *in Trans. Linn. Soc.* vii. 252 *mainly* ; *Burch. Trav.* i. 19, 56.

C. vulgaris, var. linoides, Cham. in Linnæa, vi. 343.　*C. linoides,
var. longifolia, Griseb. Gen. et Sp. Gent.* 104 *and in DC. Prodr.* ix. 41.
C. ixifera, Hort. ex Garden, 1893, 213 *and* 1899, 265 ; *Kew Handlist
Tend. Dicot.* 167.　*C. Ecklonii, Schoch in Bull. Herb. Boiss.* 2me sér.
ii. 1013, *and in Bot. Centralbl. Beih.* xiv. 212.

SOUTH AFRICA : without locality ; *Grubb in Herb. P. J. Bergius*! *Wendland*
(*Herb. Willdenow,* 4505)! *and cultivated specimens* !

COAST REGION : Clanwilliam Div. ; Lamberts or Alexanders Kloof, *Wallich*!
near Ezelsbank, Cedarberg Range, *Drège*! Malmesbury Div. ; Zwartland, *Wallich*!
Nashkraal Hoek, near Hopefield, *Bachmann,* 993 ! Worcester Div. ; Bains Kloof,
Wawra, 17 ! Hex River Valley, 1500 ft., *Tyson,* 753 ! Cape Div. ; Flats and hills
near and around Capetown, *Sparrman*! *Oldenland*! *Osbeck*! *Burmann*!
Brogniart! *Sonnerat*! *Banks and Solander*! *Masson*! *Wallich*! *Burchell,* 46 ! 730 !
Gaudichaud! *Mund and Maire*! *Niven*! *Harvey,* 404 ! 665 ! *Lehmann*! *Bun-
bury*! *Pappe*! *Cooper,* 2808 ! *Boos*! *Scholl*! *Rehmann,* 1971 ! *Bolus,* 2877 !
Ecklon, 31 ! 64 ! *Bergius*! *Drège*! *Zeyher*! *Wolley Dod,* 803 ! *Schlechter,* 179 !
279 ! Stellenbosch Div. ; summit of Hottentots Holland mountains, *Thunberg*!
Riversdale Div. ; near Kaffirkuils River, *Drège,* 2242 ! George Div. ; with-
out precise locality, *Bowie,* 15 ! *Verreaux*! Knysna Div. ; near Knysna,
Pappe!

In Herb. Delessert there is a specimen of *C. linoides,* collected by Verreaux,
and noted by himself as being from " Eutnage." No other collector has met
with the species in the Uitenhage Div. ; the locality must therefore be considered
doubtful. In Herb. Dublin there is another specimen of *C. linoides,* collected
by Wyley (82), and noted by himself as being from " Namaqualand ; " this
locality, for the same reason, must also be considered doubtful.

A distinct, easily recognised and unusually uniform species ; the statements
to the contrary have arisen from the attempts to treat *C. gracilis, C. emarginata*
and *C. Zeyheri* as forms or varieties of *C. linoides.*

8. C. Bansei (Prain in Kew Bulletin, 1908, 295) ; stems slightly
angled, branched upwards, leafy, 1 ft. long (only part of stem seen) ;
branches mostly alternate, rather straight, 3–6 in. long ; leaves
lanceolate, acute, $1\frac{1}{4}$–$1\frac{3}{4}$ in. long, $2\frac{1}{2}$–3 lin. wide ; flowers 2–5 towards
ends of branches on alternate peduncles, or solitary terminal, with
peduncles $\frac{3}{4}$–$1\frac{1}{2}$ in. long, slightly puberulous towards apex ; calyx 6
lin. long, divided four-fifths ; lobes oblong, acute or subacute,
unkeeled ; the short tube faintly puberulous ; corolla-tube cylindric
as long as the calyx ; limb narrowed ; lobes elliptic, obtuse, 8–9 lin.
long, 6–7 lin. wide ; anthers faintly spiral ; ovary ovoid, subacute,
4 lin. long.

SOUTH AFRICA : without locality ; *cultivated specimen* !

No spontaneous example of this remarkable form has been met with in any
collection. The only specimen seen was collected by Banse in the Berlin Botanic
Garden, where it was grown as *C. barclayana* and believed to be from the Cape.
Its characters are to some extent a combination of those met with in *C. linoides*
whereof, though larger in all its parts, it has the general facies, and in *Orphium
frutescens* of which it has the corolla and with which it agrees in having unkeeled
calyx-lobes. The pubescence which, however, is very slight, is like that met with
in *Orphium,* but there is no trace of the intracalycine disk characteristic of that
genus.

9. C. rosacea (Gilg in Engl. Jahrb. xxvi. 104); stems terete, generally branching upwards, leafy throughout, $1\frac{1}{2}$–3 ft. long; radical leaves vanishing, cauline in 6–8 remote pairs, ovate-lanceolate, acute, base cuneate, 2–4 in. long, 9–12 lin. wide; flowers in lax terminal panicles with branches $1\frac{1}{2}$–2 in. and peduncles 1–$2\frac{1}{2}$ in. long, spreading; calyx $2\frac{1}{2}$–3 lin. long, divided half-way; lobes ovate, acuminate; corolla-tube cylindric rather longer than calyx; lobes ovate, obtuse to ovate-lanceolate, acute, entire, 9–10 lin. long, 3–4 (rarely 5) lin. wide; anthers slightly spiral; ovary oblong, acute, $3\frac{1}{2}$ lin. long. *Schoch in Bot. Centralbl. Beih.* xiv. 229 (*as to Natal plant only*). *C. maxima, Schoch in Bull. Herb. Boiss.* 2me sér. ii. 1014 *and in Bot. Centralbl. Beih.* xiv. 220. *C. peduncularis, Benth. et Hook. f. Gen. Pl.* ii. 805, *not of Lindl.*

EASTERN REGION: Griqualand East; Clydesdale, *Schlechter*, 6618 *partly*! Pondoland; swampy stream-banks near Sangmeister, *Bachmann*, 1038! Natal, in swampy places; near Durban, *Sanderson*, 39! near the coast, *Wood*, 983! Nonoti, 500 ft., *Wood*! near the Tugela River, 500 ft., *Wood*, 3950! near Phœnix, 250 ft., *Schlechter*, 3151! 3154! Zululand; Inyezaan, below 1600 ft., *Wylie in Herb. Wood*, 596! 5948! 5928! Umlalaas, below 1000 ft., *Wylie in Herb. Wood*, 8453! and without precise locality, *Mrs. McKenzie*! *Gerrard*, 1475!

Gerrard's 1475 in Herb. Kew serves as a connecting link between this species, of which it has the broad leaves, and *C. transvaalensis*, of which it has the long calyx-lobes. *C. maxima*, Schoch, is based on a specimen which has, on the other hand, the calyx-lobes rather shorter than usual, but does not otherwise differ from normal *C. rosacea*; it is to this last form that Gerrard 1475 in Herb. Dublin belongs.

10. C. transvaalensis (Gilg in Engl. Jahrb. xxvi. 106); stems terete, simple or sparingly branched, leafy throughout, 1–$2\frac{1}{2}$ ft. long; leaves few, radical, subrosulate, obovate-spathulate, obtuse or subacute, gradually narrowed to the base, $\frac{1}{2}$–$1\frac{1}{2}$ in. long, 2–4 lin. wide, soon vanishing; cauline in 4–8 remote pairs, lanceolate, acute, 1–3 in. long, 1–3 lin. wide; flowers in lax terminal panicles with branches 1–2 in. and peduncles $\frac{1}{2}$–$2\frac{1}{2}$ in. long, spreading; calyx 3–$3\frac{1}{2}$ lin. long, divided two-thirds; lobes ovate-lanceolate, acuminate; corolla-tube cylindric, as long as the calyx; lobes ovate-lanceolate, acute, entire, 6–8 lin. long, $2\frac{1}{2}$–$3\frac{1}{2}$ lin. wide; anthers slightly spiral; ovary oblong, acute, $3\frac{1}{2}$ lin. long. *Baker & N.E. Br. in Dyer, Fl. Trop. Afr.* iv. i. 555, 627; *Schoch in Bot. Centralbl. Beih.* xiv. 227; *Thonn. Blütenpfl. Afr. t.* 129. *C. palustris, Gilg in Warb. Kunenc-Samb. Exped.* 334, *not of Burch.*

KALAHARI REGION: Orange River Colony; Sanddrift Spruit, *Burke*, 298 in Herb. Brit. Museum! Transvaal; Magaliesberg Range, *Burke*! *Zeyher*, 1196! Crocodile River, *Burke*, 383! *Zeyher*! Johannesburg and vicinity, *Ommanney*, 72! *Gilfillan in Herb. Galpin*, 6232! *D'Éstourgies*! Pilgrims Rest, *Greenstock*! near Pretoria, 4000 ft., *McLea in Herb. Bolus*, 3101! *Burtt Davy*, 813! *Fehr*! Wonderboom Poort, *Rehmann*, 4526! *Miss Leendertz*, 450! Bronkhorst Spruit, near Middelburg, *Wilms*, 974a! near Lydenburg, *Wilms*, 974! river side at Highland Creek, Barberton, 2700 ft., *Galpin*, 779! Modderfontein, *Conrath*! Daspoort, *Miss Leendertz*! near Pietersburg, 4000 ft., *Bolus*, 10864! Mazol, *Eyles*, 527!

EASTERN REGION: Swaziland ; Vlei near Oshoek, *Burtt Davy*, 2845 !

Also in Tropical Africa.

Closely allied to *C. rosacea*, Gilg, and perhaps only a form of that species, but readily distinguished by its narrower leaves and longer calyx-lobes.

11. C. palustris (Burch. Trav. ii. 226); stems terete, scapiform, 1–2 ft. long ; leaves mostly radical, subrosulate, obovate-spathulate, obtuse or subacute, gradually narrowed to the base, 1½–2½ in. long, 6–9 lin. wide, persisting ; cauline in 1–2 remote pairs, oblanceolate, acute or obtuse, ½–2 in. long, 2–4 lin. wide ; flowers in lax terminal panicles with branches 2–3 in. and peduncles ½–2 in. long, spreading ; calyx 3–4 lin. long, divided half-way ; lobes linear-lanceolate, acuminate ; corolla-tube cylindric, as long as the calyx ; lobes elliptic, obtuse or subacute, usually entire, 6–8 lin. long, 3–4 lin. wide; anthers distinctly spiral ; ovary oblong, acute, 3½ lin. long. *Engl. Jahrb.* x. 243 ; *Gilg in Engl. Jahrb.* xxvi. 106 *and in Ann. Nat. Hist. Hofmus. Wien*, xv. 66. *C. palustris, var. radicata, Schoch in Bot. Centralbl. Beih.* xiv. 234. *Plocandra albens, var. radicata, E. Meyer, Comm.* 182. *P. palustris, Griseb. in DC. Prodr.* ix. 43 (*excl. syn. Chironia Krebsii*).

VAR. **foliata** (Prain) : radical leaves oblanceolate, 2–5 in. long, 4–6 lin. wide ; cauline lanceolate, acute, 1¼–2½ in. long, 2–3 lin. wide ; panicle branches 1–3 in. long ; peduncles ¼–¾ in. long. *C. palustris, Hook. f. in Bot. Mag. t.* 7101 ; *Schoch in Bot. Centralbl. Beih.* xiv. 233 ; *hardly of Burch. Plocandra albens, E. Meyer, Comm.* 182. *P. palustris, var. foliata, Griseb. in DC. Prodr.* ix. 43.

SOUTH AFRICA : without locality, *Zeyher*, 1194 ! *Holub*, 251 ! 4252 ! 5532 ! Var. β, *Mrs. Barber*, 19 !

COAST REGION : Queenstown Div. ; summit of Shepstone Berg, 5800 ft., *Galpin*, 1920 ! Kaffraria, *Fraser* ! Var. β : Cape Div. ; neighbourhood of Cape-town, *Castelnau*, 487 ! Bathurst Div. : banks of and near the Kowie River, *Zeyher* ! Albany Div. ; without precise locality, *Miss Bowker* ! King Williamstown Div. ; between Yellowwood (Kachu) River and Zandplaats, 1500–2000 ft., *Drège*, 4922 ! East London Div. ; near East London, grassy slopes by sea-shore, *Watson* ! *Galpin*, 7345 ! Komgha Div. ; near the mouth of the Kei River, 200 ft.. *Flanagan*, 1207 !

CENTRAL REGION : Wodehouse Div. ; near Mooi Plaats, *Drège* !

KALAHARI REGION : Griqualand West ; Griqua Town, *Burchell*, 1925 ! Orange River Colony ; Sanddrift Spruit, *Burke*, 298 ! near Harrismith, *Wood*, 4763 ! Bethlehem, *Richardson* ! and without precise locality, *Cooper*, 987 ! Bechuanaland ; near the sources of the Kuruman River, *Burchell*, 2510 ! *Marloth*, 1050 ! Trans-vaal ; between Trigardsfontein and Standerton, *Rehmann*, 6748 ! near Wilge River, 4600 ft., *Schlechter*, 4127 ! and without precise locality, *Sanderson* !

EASTERN REGION : Transkei ; marshes near Ibeka, 2500 ft., *Schlechter*, 6255 ! Griqualand East ; marshes near Kokstad, 4300–5000 ft., *Tyson*, 1676 ! *and in MacOwan and Bolus, Herb. Austr.-Afr.* 1291 ! Kumba, *Krook*, 2029 ! Natal ; Camperdown, 2000 ft., *Wood*, 463 ! *Haygarth in Herb. Wood*, 1958 ! near Glencoe, 4000–5000 ft., *Wood*, 5124 ! near Ladysmith, *Gerrard*, 670 ! and without precise locality, *Gerrard*, 730 ! *MacOwan* ! Var. β : Tembuland ; moist places near Bazeia, 2000 ft., *Baur*, 451 ! Transkei : swampy ground near Kentani, *Miss Pegler*, 305 !

In var. β are found combined the foliage of *C. Krebsii* and the floral characters of typical *C. palustris*.

12. C. Krebsii (Griseb. Gen. et Sp. Gent. 98); stems terete, scapiform, 1½–3 ft. long; leaves mostly radical, subrosulate, oblanceolate, obtuse or subacute, gradually narrowed to the base, 4–10 in. long, 4–7 lin. wide, persisting; cauline in 1–3 remote pairs, lanceolate, acute, 2–4 in. long, 2½–4 lin. wide; flowers in dense terminal panicles with branches ⅓–1½ in. and peduncles ¼–½ in. long, fastigiate; calyx 4 lin. long, divided half-way; lobes linear-lanceolate, acuminate; corolla-tube cylindric, as long as the calyx; lobes lanceolate, acute, usually somewhat erose, 5 lin. long, 1½–1¾ lin. wide; anthers distinctly spiral; ovary oblong, acute, 3½ lin. long. *Gilg in Engl. Jahrb.* xxvi. 106, *and in Ann. Nat. Hist. Hofmus. Wien,* xv. 66; *Baker & N. E. Br. in Dyer, Fl. Trop. Afr.* iv. i. 554. *C. densiflora, Scott Elliot in Journ. Bot.* 1891, 69; *Engl. Pflanzenw. Ost-Afr. C.* 314; *Gilg in Engl. und Prantl, Pflanzenfam.* iv. 2, 77; *Schoch in Bot. Centralbl. Beih.* xiv. 224, *t.* 16, *fig.* 6. *C. palustris, Knobl. in Bot. Centralbl.* lx. 329 *partly, not of Burch.*

SOUTH AFRICA : without locality, *Krebs* !

COAST REGION : Uitenhage Div. ; Great Winterhoek mountains, *Zeyher* ! and without precise locality, *Pappe* ! Bedford Div. ; near Bedford, *Mrs. Hutton* ! *Bennie* ! Stockenstrom Div. ; marshy ground on Lushington mountain, *Scully,* 165 ! Katberg. *Hutton* ! British Kaffraria, *Cooper,* 390 !

CENTRAL REGION : Somerset Div. ; marshy ground, Glenavon, *Scott Elliot,* 666 (626) ! Bosch Berg, 4700 ft., *MacOwan,* 1233 ! 1656 ! Queenstown Div. ; Winterberg Range, *Mrs. Barber,* 529 !

KALAHARI REGION, in marshy ground, 4000–6000 ft. : Orange River Colony ; near Harrismith, *Sankey,* 174 ! Transvaal ; Umlomati Valley, near Barberton, *Galpin,* 1131 ! near Brug Spruit, *Schlechter,* 3752 ! Athol, *M'Donald* ! Spitz Kop Gold Mine, *Wilms,* 975 !

EASTERN REGION : Griqualand East ; Newmarket, *Krook,* 2076 ! Natal, in marshy ground, 1800–5000 ft. ; Noodsberg, *Wood,* 121 ! 372 ! near Gillets, *Wood* ! near Howick, *Wood,* 5107 ! Newcastle, *Rehmann,* 7022 ! near Charlestown, *Wood,* 4694 ! near Emangweni, 4000 ft., *Thode,* 68 ! Dargle Farm, *Mrs. Fannin,* 69 ! and without precise locality, *Sanderson,* 446 partly !

Also in Tropical Africa.

13. C. humilis (Gilg in Engl. Jahrb. xxvi. 105); stems slightly angled, simple, leafy, 4–8 in. long; leaves rather small, radical sub-rosulate, lanceolate, acute, ½ in. long, 1 lin. wide; cauline in 3–6 remote pairs, linear, acute, ¾ in. long, ½ lin. wide; flowers 3–9 in terminal cymes with spreading branches ½–1 in. long, and peduncles very short ; calyx 5 lin. long, divided four-fifths ; lobes lanceolate, acuminate ; corolla-tube narrow-campanulate, green, shorter than the calyx ; lobes dark-purple, lanceolate or ovate-lanceolate, acuminate or acute, 6 lin. long, 2½–4 lin. wide ; anthers distinctly spiral ; ovary oblong, acute, 3 lin. long. *Schoch in Bot. Centralbl. Beih.* xiv. 230.

VAR. β, **Wilmsii** (Prain in Kew Bulletin, 1908, 350); stems more robust, often branched upwards, 1–1½ ft. long; radical leaves narrow-obovate, ¾–1 in. long, 2–3 lin. wide ; cauline 1–1½ in. long ; flowers in terminal panicles, with branches ¾–1½ in. long. *C. Wilmsii, Gilg in Engl. Jahrb.* xxvi. 105; *Schoch in Bot.*

Centralbl. Beih. xiv. 231.   *C. purpurascens, Rolfe in Oates, Matabeleland, ed. 2,*
404 ; *Schoch in Bot. Centralbl. Beih.* xiv. 219 *partly* ; *not of Benth. et Hook. f.*
C. humilis, Baker & N. E. Br. in Dyer, Fl. Trop. Afr. iv. i. 555, *hardly of Gilg.*

KALAHARI REGION : Bechuanaland ; neighbourhood of Takoon, *Lemue* ! Orange
River Colony ; Leeuw Spruit or Vredefort, *Barrett-Hamilton* ! Transvaal ;
Magaliesberg Range and Aapies River, *Burke,* 124 ! *Zeyher,* 1193 ! Crocodile
River, *Zeyher,* 1195 ! near Mooifontein, 5500 ft., *Schlechter,* 3552 ! Johannesburg,
Mrs. de Jongh in Herb. Galpin, 1475 ! Var. β : Transvaal ; near Pretoria,
4000 ft., *McLea in Herb. Bolus,* 3100 ! *Rehmann,* 4679 ! *Fehr* ! between Pretoria
and Johannesburg, *Scott Elliot,* 1356 ! near Johannesburg, *Ommanney,* 29 ! *Rand* !
Witwaters Rand, *Mrs. Hutton,* 926 ! Pilgrims Rest, *Greenstock* ! Heidelberg,
Miss Leendertz, 1028 ! near Modderfontein, *Conrath,* 737 ! near Wilge River,
4600 ft., *Schlechter,* 4122 ! Rustenburg, 4500 ft., *Miss Nation* ! sources of
Limpopo River, *Nelson,* 262 ! Nylstroom, *Nelson,* 283 ! between Porter and
Trigardsfontein, *Rehmann,* 6613 ! near Lydenburg, *Atherstone* ! Bronkhorst
Spruit, *Wilms,* 973 ! Nelspruit, *Rogers,* 290 ! Gold-fields, *Baines* ! Ermelo,
Burtt Davy, 2192 ! Houtbosch (Woodbush), *Rehmann,* 5828 ! 5829 ! 5928 !
Bolus, 11114 !

Also in Tropical Africa.

The variety here recognised is considered a distinct species by Gilg and Schoch.
It differs so little from true *C. humilis,* Gilg, that Baker and Brown are perhaps
justified in declining to treat it even as a variety.

14. C. purpurascens (Benth. et Hook. f. Gen. Pl. ii. 805);
stems distinctly angled, simple or branching upwards, leafy, 1½–2
ft. long ; leaves below obovate or obovate-lanceolate, 1½–2 in. long,
less often (form *Bachmannii*) lanceolate, 2–3 in. long, 3–6 lin. wide,
higher up lanceolate, 1½ in. or (form *Bachmannii*) linear-lanceolate
or linear, 2–3½ in. long, 1½–2 lin. wide ; flowers 3–5 in terminal
cymes often forming lax panicles with branches ascending, 1½–2½ in.
and peduncles ½–1 in. long ; calyx 6–7 lin. long, divided almost to
the base ; lobes linear-subulate ; corolla pink or purple ; tube
narrow-campanulate, shorter than the calyx ; lobes lanceolate,
acute, 6–8 lin. long, 1½–2 lin. wide ; filaments glandular at the
base ; anthers distinctly spiral ; ovary narrow-oblong, acute, 3½ lin.
long. *Gilg in Engl. Jahrb.* xxvi. 103 ; *Knobl. in Bot. Centralbl.* lx.
329 ; *Wood, Natal Pl.* iii. *part* 4, 19, *t.* 288 : *Schoch in Bot. Centralbl.*
Beih. xiv. 219.   *C. Bachmannii, Gilg in Engl. Jahrb.* xxvi. 103 ;
Schoch in Bot. Centralbl. Beih. xiv. 218.   *C. angolensis, Schoch in*
Bot. Centralbl. Beih. xiv. 228 (*as to Natal plant only*) ; *not of Gilg.*
Plocandra purpurascens, E. Meyer, Comm. 182 ; *Krauss, Beitr. Fl.*
Cap. und Natal. 122 ; *Griseb. in DC. Prodr.* ix. 44.

VAR. β, **Tysonii** (Prain in Kew Bulletin, 1908, 350) ; leaves larger, below oblong,
3–3½ in. long, 9–12 lin. wide, above lanceolate, 1½ in. long, 2–2½ lin. wide ; flowers
on secondary cyme-branches sometimes 4-merous. *C. Tysonii, Gilg in Engl. Jahrb.*
xxvi. 104 ; *Schoch in Bot. Centralbl. Beih.* xiv. 217.

VAR. γ, **impedita** (Prain, l.c.) ; flowers larger ; corolla-lobes ovate, acute, 10 lin.
long, 5 lin. wide.

EASTERN REGION : Pondoland ; near Bates (long-leaved form = *C. Bachmannii*),
Bachmann, 1037 ! Griqualand East ; near Clydesdale, 2500–4000 ft., *Schlechter,*
6618 ! and (mixed with *Bachmannii* form), *Tyson,* 2113 in Herb. Zurich ! Zuur-

berg, between Kokstad and Clydesdale, 5000 ft., *Tyson*, 1161 ! Natal ; between Umzimkulu River and Umkomaas River, *Drège*, 492? ! Umlaas, *Wood*, 876 ! near Durban, *Krauss*, 192 ! *Wood*, 141 ! *Gerrard*, 669 ! *Plant* (*Bachmannii* form), 46 ! 48 ! Clairmont, *Wood*, 1154 ! Coastland, *Sutherland* ! *Sanderson* (*Bachmannii* form), 170 ! and without precise locality, *Gerrard*, 463 ! *Sanderson*, 446 partly ! Zululand ; without precise locality, *Mrs. McKenzie* ! *Gerrard* (*Bachmannii* form), 550 ! Var. β. : Griqualand East ; near Clydesdale, 2500–4000 ft., *Tyson*, 2113 in Herb. Kew ! and in *MacOwan and Bolus, Herb. Austr.-Afr.* 1290 ! Var. γ : Natal ; Ixopo, *Krook*, 2028 !

We are unable to separate *C. Bachmannii* from *C. purpurascens* ; the two occur in the same localities and are connected by intermediates. We retain the variety *Tysonii* mainly because Dr. Gilg finds that Bolus, 1290 at Berlin, collected by Tyson, on which *C. Tysonii* is based, has all save the primary flowers of its cymes 4-merous. This is not the case with any specimen of this gathering in other herbaria ; some of the specimens, indeed, are true *C. purpurascens*, others are the *Bachmannii* form.

15. **C. Pegleræ** (Prain in Kew Bulletin, 1908, 297) ; stems distinctly angled, much-branched, erect, 1 ft. long ; branches ascending ; leaves wide-ovate, base rounded, subacute or obtuse, $\frac{1}{2}$–$\frac{3}{4}$ in. long, 4–5 lin. wide ; flowers terminal and in the upper axils ; peduncles $1\frac{1}{2}$–2 in.' long ; calyx 9–10 lin. long, divided almost to the base ; lobes linear-subulate ; corolla-tube narrow-cylindric, hardly as long as the calyx ; limb narrowed ; lobes ovate-lanceolate, acuminate, 4 lin. long, $1\frac{1}{2}$ lin. wide ; anthers distinctly spiral ; ovary narrow-oblong, acute, 4 lin. long.

EASTERN REGION : Transkei ; Kentani, in valleys at 1000 ft., *Miss Pegler*, 428 !

A very distinct species, most closely related to *C. peduncularis*, from which it is readily distinguished by the angled stems, smaller and thinner leaves, and much smaller flowers as well as by the spirally twisted anthers. In this last character the species agrees with those of the section *Plocandra*, where, however, though the calyx is as in *Hippochiron*, the corolla-tube is different, being shorter than the calyx and widened under the limb.

16. **C. peduncularis** (Lindl. in Bot. Reg. xxi. t. 1803) ; stems terete, leafy, much-branched, decumbent, 3–4 ft. long ; branches diffuse ; leaves ovate-lanceolate, base rounded or cordate, acuminate or acute, $1\frac{1}{4}$–$2\frac{1}{2}$ in. long, 6–9 lin. wide ; flowers 1–3 terminal and in the upper axils ; peduncles $1\frac{1}{2}$–3 in. long ; calyx 7 lin. long, divided almost to the base ; lobes linear or lanceolate, acuminate ; corolla-tube narrow-cylindric, rather larger than the calyx ; limb narrowed ; lobes ovate, acute or acuminate, 9 lin. long, 5–6 lin. wide ; anthers straight ; ovary narrow-oblong, acute, 6 lin. long. *Griseb. Gen. et Sp. Gent.* 100, *and in DC. Prodr.* ix. 39 ; *Gard. Chron.* 1888, iv. 324, *fig.* 42 ; *Scott and Breb. Ann. Bot.* v. 277 ; *Garden*, 1893, 212, *t.* 925 ; *Hook. f. in Bot. Mag. t.* 7047 ; *Knobl. in Bot. Centralbl.* lx. 329. *C. trinervis, Hort. ex Loud. Encycl. Pl. Suppl.* ii. 1306 ; *Paxt. Mag. Bot.* iii. 149, *with plate.* *C. trinervia, Hort. ex Ann. Fl. et Pomone*, 1834, 158 *with plate* ; *not of Linn.* *C. barclayana, Hort. Berol. ex Griseb. Gen. et Sp. Gent.* 100.

C. latifolia, E. Meyer, Comm. 178 ; *Schoch in Bot. Centralbl. Beih.*
xiv. 221 ; *not of Donn.   Eupodia purpurea, Raf. Fl. Tellur.* iii. 29.

SOUTH AFRICA : without locality, *cultivated specimens* !
COAST REGION : George and Knysna Divs. ; Outeniqualand and in Knysna
Forest, *Bowie,* 9 ! Uitenhage Div. ; near Uitenhage, *MacOwan,* 1058 ! Port
Elizabeth Div. ; near Port Elizabeth, *Burchell,* 4302 ! *Verreaux* ! *Drège.*
2238 ! *Prior* ! *Zeyher,* 3427 ! *Fraser* ! *Bolus* ! *Miss West,* 30 ! Algoa Bay, *Cooper,*
1476 ! *Prior* ! marshes at Krakakamma, 500 ft., *MacOwan,* 1058 ! Alexandria Div. ; Oliphants Hoek Forest, *Ecklon,* 640 ! Bathurst Div. ; between
Kasuga River and Port Alfred, *Burchell,* 3929 ! at Oliphantshoek on the Bushmans River, *Ecklon* ! near Port Alfred, *South* ! Albany Div. ; without precise
locality, *Miss Bowker* ! East London Div. ; near East London, *Watson* !

17. C. jasminoides (Linn. Pl. Afr. Rar. 9) ; stems faintly angular,
simple or little branched, remotely leafy, especially below, 2–3 ft.
long ; lower leaves ovate-lanceolate or lanceolate, subacute or acute,
rarely ovate, subobtuse, $1\frac{1}{4}$–2 in. long, 4–5 lin. wide, upper
lanceolate, acute or acuminate, $1\frac{1}{2}$–2 in. long, 3–4 lin. wide, glaucous
when young on both sides, persistently so beneath : flowers
occasionally solitary, usually few, sometimes 9–15 in subpaniculate
terminal cymes with ascending branches $1\frac{1}{2}$–$2\frac{1}{2}$ in., and peduncles
1–$1\frac{1}{2}$ in. long ; calyx 8 lin. long, divided almost to the base ; lobes
linear-subulate, rarely linear ; corolla-tube narrow-cylindric, as long
as the calyx ; limb narrowed ; lobes ovate-lanceolate, acuminate,
10–11 lin. long, 3–4 lin. wide ; anthers straight : ovary narrow-
oblong, acute, 5 lin. long. *Linn. Amœn. Acad.* vi. 84 ; *Sp. Pl. ed.*
2, 272 ; *Lam. Encycl. Meth.* i. 736 *and Ill. Gen. t.* 108, *fig.* 2 ; *Pers.
Syn. Pl.* i. 282 ; *R. Br. Prodr.* i. 451.   *C. nudicaulis, Willd. Sp. Pl.*
i. 1066, *hardly of Linn. f.   C. nudicaulis, var. elongata, Eckl. ex
Cham. in Linnæa,* vi. 344 ; *Griseb. Gen. et Sp. Gent.* 99, *and in
DC. Prodr.* ix. 39.   *C. lychnoides, E. Meyer, Comm.* 177 ; *Schoch in
Bot. Centralbl. Beih.* xiv. 209, *t.* 15, *fig.* 9 (*excl. syn. var.* **viminea**) ;
not of Berg., hardly of Linn.

VAR. β, **viminea** (Prain in Kew Bulletin, 1908, 351) ; stems remotely leafy ;
leaves all linear, paler green, $1\frac{1}{2}$–2 in. long, $1\frac{1}{2}$–2 lin. wide ; panicle-branches,
when present, 1–$1\frac{1}{2}$ in. and peduncles $\frac{1}{3}$–$\frac{1}{2}$ in. long.   *C. lychnoides, Linn. Mantiss.*
207 ; *Willd. Sp. Pl.* i. 1066 (*as to descr. and excl syn. Berg. and syn. Thunb.*) :
Schoch in Bot. Centralbl. Beih. xiv. 209 (*as to syn. var. viminea only*) ; *not of
Berg.   C. nudicaulis, var. viminea, Griseb. Gen. et Sp. Gent.* 99 *and in DC. Prodr.*
ix. 39 ; *Gilg ex Zahlbr. in Ann. Nat. Hist. Hofmus. Wien,* xv. 65.

VAR. γ, **multiflora** (Prain in Kew Bulletin, 1908, 351) ; stems rather closely
leafy throughout ; leaves all ovate-lanceolate, acute or acuminate, $1\frac{1}{2}$–$2\frac{1}{2}$ in. long,
4–6 lin. wide ; panicle-branches $1\frac{1}{2}$–$2\frac{1}{2}$ in. and peduncles 1–$1\frac{1}{2}$ in. long.   *C. cymosa,
Burm. f. Prodr. Cap.* 5 (*possibly*).   *C. nudicaulis, var. multiflora, Eckl. ex Schoch
in Bot. Centralbl. Beih.* xiv. 210.

VAR. δ, **tabularis** (Prain in Kew Bulletin, 1908, 351) ; stems rather closely
leafy below, naked upwards ; leaves ovate to ovate-lanceolate, subobtuse to acute,
1–2 in. long, 4–6 lin. wide.   *C. nudicaulis, Linn. f. Suppl.* 151 ; *Lam. Encycl.
Meth.* i. 737 ; *Thunb. Prodr.* i. 35, *in Trans. Linn. Soc.* vii. 249, *t.* 12, *fig.* 3 ;
Pers. Syn. Pl. i. 282 ; *R. Br. Prodr.* i. 451 ; *Roem. & Schult. Syst. Veg.* iv. 202 ;
E. Meyer, Comm. 177 ; *Griseb. Gen. et Sp. Gent.* 99 *and in DC. Prodr.* ix. 39 ;

Schoch in Bot. Centralbl. Beih. xiv. 200, *t.* 16, *fig.* 3. *C. nudicaulis, var. tabularis, Cham. in Linnæa,* vi. 344 (*not C. tabularis, Page*). *C. jasminoides, Burch. Trav.* i. 46.

SOUTH AFRICA: without locality, *Oldenland*! *Burmann*! *Forster*! *Harvey*! *Verreaux*! *Jacquin* (*Herb. Willdenow,* 4501)!
COAST REGION: Tulbagh Div.; Mosterts Berg, 2200 ft., *Bolus,* 5024! Worcester Div.; Dutoits Kloof, 2500–3500 ft., *Drège,* 1896! Paarl Div.; French Hoek, 2500 ft., *Schlechter,* 9262! Cape Div.; Cape Flats, *Swartz*! *Bergius*! *Sonnerat*! *Roxburgh*! *Ecklon,* 175! *Brown*! *Spielhaus*! *Wallich*! *Barnard*! *Fuller*! Sweet Valley Flats, *Wallich*! south of Constantia Berg, *Wolley Dod,* 2097! Simons Bay, *Prior*! near Simons Town, *Milne,* 150! Smitswinkel Stream, *Wolley Dod,* 2683! Swellendam Div.; Voormans Bosch, near Swellendam, *Zeyher,* 3429 partly! George Div.; mountains and borders of the forest, *Bowie,* 8! Var. *β*: Caledon Div.; Klein River Berg, *Ecklon,* 638! Zwart Berg, *Ecklon,* 528! Zoetemelks Valley, *Wallich*! Swellendam Div.; Buffeljagts River, *Penther,* 2021! River-dale Div.; Garcias Pass, 2500 ft., *Galpin,* 4341! Var. *γ*: Stellenbosch Div.; Hottentots Holland mountains, *Prior*! Swellendam Div.; Tradouw Berg, *Bowie,* 7! Voormans Bosch, near Swellendam, *Zeyher,* 3429 partly! *Ecklon,* 639! Riversdale Div.; upper part of Kampsche Berg, *Burchell,* 7061! 7128! Var. *δ*: Cape Div.; Table Mountain, 2600–3500 ft., *Oldenburg*! *Burmann*! *Sonnerat*! *Osbeck*! *Thunberg*! *Burchell,* 547! *Spielhaus*! *Bergius*! *Drège,* 7823! *Harvey*! *Bolus,* 4539! *Rehmann,* 757! *Schlechter,* 295! *Boos*! *Scholl*! *Penther,* 2017! *Knoop,* 60 partly! Wynberg, *Mund and Maire*! *Ecklon,* 174.

Ecklon enumerates *C. nudicaulis* among the plants found in the Uitenhage district (*South Afr. Quart. Journ.* i. 370). *C. nudicaulis,* Eckl., includes all the varieties defined above, and therefore is equivalent to *C. jasminoides,* Linn., as a whole. There are, however, no specimens of any variety of *C. jasminoides* from Uitenhage in the collections examined; the locality is therefore somewhat doubtful.

An extremely variable species as regards foliage, but remarkably uniform as regards flowers and fruit. The varieties here distinguished pass gradually one into the others.

18. C. serpyllifolia (Lehm. Ind. Sem. Hort. Hamb. 1828, 16); stems slightly angled, leafy, much-branched, $\frac{1}{2}$–1 ft. long; branches virgate; leaves ovate, obtuse or subacute, $\frac{1}{3}$ in. long, $2\frac{1}{2}$ lin. wide; flowers 2–3, or solitary, terminal; peduncles $\frac{1}{3}$–$\frac{3}{4}$ in. long; calyx $2\frac{1}{2}$ lin. long, divided three-fourths; lobes linear-subulate; corolla-tube narrow-cylindric, longer than the calyx; limb narrowed; lobes ovate-lanceolate, subacuminate, 4 lin. long, $1\frac{1}{2}$ lin. wide; anthers straight; ovary narrow-oblong, acute, 3 lin. long. *Lehm. in Linnæa,* v. 373; *Griseb. Gen. et Sp. Gent.* 106 *and in DC. Prodr.* ix. 41; *Gilg in Engl. und Prantl, Pflanzenfam.* iv. 2, 78; *Schoch in Bot. Centralbl. Beih.* xiv. 191 *partly. C. parvifolia, E. Meyer, Comm.* 178 *partly. C. ovata, Spreng. ex Griseb. in DC. Prodr.* ix. 41.

VAR. *β*, **laxa** (Griseb. Gen. et Sp. Gent. 106); leaves narrow-ovate, obtuse or subacute, $\frac{1}{3}$ in. long, $1\frac{1}{2}$ lin. wide, rather more distant; corolla-lobes lanceolate, acuminate. *Griseb. in DC. Prodr.* ix. 41; *Schoch in Bot. Centralbl. Beih.* xiv. 192 *partly. C. parvifolia, E. Meyer, Comm.* 178 *partly.*

VAR. *γ*, **microphylla** (Griseb. Gen. et Sp. Gent. 106); leaves ovate, acute or subacute, $\frac{1}{6}$–$\frac{1}{5}$ in. long, under 1 lin. wide; corolla-lobes lanceolate, acuminate. *Griseb. in DC. Prodr.* ix. 41; *Schoch in Bot. Centralbl. Beih.* xiv. 192. *C. serpyllifolia, Eckl. in South Afr. Quart. Journ.* i. 371, *hardly of Lehm. C. parvifolia, E. Meyer, Comm.* 178 *partly.*

South Africa: without locality, *cultivated specimens*!
Coast Region: Uitenhage Div.; near the Zwartkops River, below 100 ft., *Drège*, 2237! *Zeyher*, 165! 204! and without precise locality, *Verreaux*! Bathurst Div.; Port Alfred, *Atherstone*, 240! *South*! Var. β: Uitenhage Div.; near the Zwartkops River, 100 ft., *Zeyher*, 3431! *Ecklon*, 648! and without precise locality, *Verreaux*! Var. γ: Uitenhage Div.; Van Stadens Berg, 1000–1500 ft., *Drège*! between Krakakamma and Van Stadens Berg, *Zeyher*! *Ecklon*, 48! 649!

The varieties distinguished by Grisebach hardly deserve recognition on morphological grounds; var. β has all the appearance of a form of the typical plant growing among grass; var. γ that of a form growing in poor soil. The fruit has been stated by Schoch (*l.c.* p. 186) to be a berry; it is, however, a 2-valved capsule, like that of *C. laxa*, Gilg.

19. C. laxa (Gilg in Engl. Jahrb. xxvi. 105); stems slightly angled, leafy, much-branched, 1–1½ ft. long; branches ascending or spreading; leaves lanceolate, acuminate, ¾–1 in. long, 1–1½ lin. wide; flowers 2–3, or solitary, terminal; peduncles ½–1¼ in. long; calyx 3 lin. long, divided three-fourths; lobes linear-subulate; corolla-tube narrow-cylindric, longer than the calyx; limb narrowed; lobes ovate-lanceolate, subacuminate, 5–6 lin. long, 2–2½ lin. wide; anthers straight; ovary narrow-oblong, acute, 3 lin. long. *C. melampyrifolia, E. Meyer, Comm.* 177; *Schoch in Bot. Centralbl. Beih.* xiv. 208; *not of Lam. C. maritima, Griseb. in DC. Prodr.* ix. 39 (*as to syn. E. Meyer only*); *not of Eckl. C. Schlechteri, Schoch in Bull. Herb. Boiss.* 2me *sér.* ii. 1010 *and in Bot. Centralbl. Beih.* xiv. 214.

Coast Region: Komgha Div.; near Komgha, *Flanagan*, 12!
Eastern Region: Transkei; near the Bashee River, 2800 ft., *Schlechter*, 6282! near Butterworth, 2000 ft., *Bolus*, 10228! Tembuland; banks of the Umtata River, 400 ft., *Drège*, 4924! Griqualand East; Shawbury, 2000 ft., *Baur*, 229!

Nearly allied to *C. serpyllifolia*, Lehm., and only distinguished by its longer leaves and larger size. The two are related to each other much as *C. rosacea*, Gilg, and *C. transvaalensis*, Gilg, are related, an l it may be found, as the result of further field study, that they can only be distinguished as varieties. There is no character by which to separate *C. Schlechteri*, Schoch, based on Schlechter 6282, from *C. laxa* proper, based on Baur 229.

20. C. floribunda (Paxt. Mag. Bot. xi. 237); stems faintly angled, leafy, diffusely branched, 1–2 ft. long; leaves narrow-spathulate or lanceolate, acute, ¾–1 in. long, 2–3 lin. wide; flowers solitary, terminal; peduncles ½–1½ in. long; calyx 3 lin. long, divided three-fourths; lobes narrow-elliptic, acute; corolla-tube narrow-cylindric, as long as the calyx; limb narrowed; lobes obovate, obtuse, 5 lin. long, 4 lin. wide; anthers straight; ovary ovoid, acute, 3½ lin. long. *Paxt. Mag. Bot.* xii. 123, *with plate*; *Belg. Hortic.* 1860, 65, *with plate*; *Rev. Hortic. Belg.* 1907, 3, *with plate. C. Fischeri, Hort. Rolliss. ex Paxt. in Mag. Bot.* xi. 237 *and* xii. 123; *Schoch in Bot. Centralbl. Beih.* xiv. 215. *C. maritima, var.? frutescens, Griseb. in DC. Prodr.* ix. 39.

South Africa: without locality, *cultivated specimens*!

This distinct and very uniform species has been long known in European gardens, and is believed to have come from the Cape. There are however no specimens collected in South Africa in any of the herbaria examined. It is, as Grisebach has indicated, most nearly allied to *C. maritima*, but differs in calyx, in corolla-lobes and in having glands at the base of the filaments. The specimens seen by us show that it has been grown under various names in different collections : at Chelsea as *C. floribunda* ; Goettingen as *C. maritima* ; Rouen as *C. frutescens* ; St. Petersburg as *C. lychnoides* ; Berlin, Edinburgh and Lübeck as *C. Fischeri* ; Kew as *C. linoides* and *C. ixifera* ; Wurzburg as *C. baccifera.*

21. C. maritima (Eckl. in South Afr. Quart. Journ. i. 370, not of Willd.) ; stems distinctly angled, rooting at the nodes below, $\frac{1}{2}$–$1\frac{1}{2}$ ft. long ; branches suberect, remotely leafy, again branched ; leaves narrow-spathulate or lanceolate, subacute or acute, 1–$1\frac{1}{4}$ in. long, $1\frac{1}{2}$–$2\frac{1}{2}$ lin. wide ; flowers solitary, terminal ; peduncles 1–$1\frac{1}{2}$ in. long, occasionally 1–2 added in upper axils ; calyx 5 lin. long, divided three-fourths to four-fifths ; lobes linear, long-acuminate ; corolla-tube narrow-cylindric, as long as the calyx ; limb slightly narrowed ; lobes ovate-lanceolate, acute, 8 lin. long, 4–5 lin. wide ; anthers straight ; ovary narrow-ovoid, acute, $4\frac{1}{2}$ lin. long. *Griseb. Gen. et Sp. Gent.* 100 *and in DC. Prodr.* ix. 39 (*excl. syn. C. melampyrifolia*) ; *Schoch in Bot. Centralbl. Beih.* xiv. 216, *t.* 16, *fig.* 5. *C. lychnoides, Thunb. in Trans. Linn. Soc.* vii. 252 *partly ; not of Berg.*

SOUTH AFRICA : without locality, *Thunberg* !
COAST REGION : Cape Div. ; Fish Hoek, *Wolley Dod*, 650 ! Vyges Kraal River, *Wolley Dod*, 2391 ! sandy shore near Muizenberg, *Bunbury* ! salt meadows between Retreat and Muizenberg, 25 ft., *Schlechter*, 654 ! and without precise locality, *Harvey*, 610 ! *Verreaux* ! *Knoop*, 60 *partly* ! Port Elizabeth Div. ; Cape Recife, *Burchell*, 4384 ! *Ecklon*, 36 ! 641 ! along the coast, *E.S.C.A. Herb.* 291 ! *Hutton* ! sand hills, 50 ft., *Zeyher*, 793 ! 1199 !

A distinct and uniform species, most resembling *C. scabrida*, var. *ligulifolia*, but readily distinguished by its habit, its foliage and its quite different calyx.

22. C. scabrida (Griseb. Gen. et Sp. Gent. 103) ; stem angled, suberect, leafy, $\frac{1}{2}$–$\frac{3}{4}$ ft. long ; branches virgate ; leaves oblong or elliptic, obtuse or acute, $\frac{1}{2}$–$\frac{3}{4}$ in. long, 3–4 lin. wide, scabridulous especially on the margins ; flowers solitary, terminal ; peduncles $\frac{1}{2}$–$\frac{3}{4}$ in. long ; calyx 5 lin. long, divided half-way ; lobes lanceolate, acute, keeled ; tube keeled ; corolla-tube cylindric, longer than the calyx ; lobes elliptic-oblong, obtuse, 8 lin. long, 4–5 lin. wide ; anthers straight ; ovary ovoid, acute, 4 lin. long. *Griseb. in DC. Prodr.* ix. 40 ; *Schoch in Bot. Centralbl. Beih.* xiv. 205.

VAR. **ligulifolia** (Prain in Kew Bulletin, 1908, 297) ; leaves lanceolate-oblong to lanceolate, obtuse or acute, $\frac{1}{2}$–$\frac{3}{4}$ in. long, $1\frac{1}{2}$–3 lin. wide ; flowers 2–3 or solitary, terminal ; peduncles 1–2 in. long : calyx 6–7 lin long ; lobes lanceolate, acuminate ; corolla-tube as long as calyx ; stems sometimes long, prostrate, oftener short, suberect. *C. jasminoides, Cham. in Linnæa*, vi. 344 (*Cape Division specimens only*) ; *Knobl. in Bot. Centralbl.* lx. 328 (*as to C. viscosa, Zeyh. only*) ; *not of Linn. C. jasminoides, var.* β, *Banks ex Edw. Bot. Reg.* iii. *sub t.* 197. *C. jasminoides, var. lychnoides, Griseb. Gen. et Sp. Gent.* 102 (*as to Ecklon's plant only*) *and in DC. Prodr.* ix. 40 ; *not C. lychnoides, Berg. or Linn. C. viscosa,*

Zeyh. ex Griseb. in DC. Prodr. ix. 40. *C. tetragona, Schoch in Bot. Centralbl. Beih.* xiv. 196 (*as to C. viscosa, Zeyh. only*) ; *not of Linn. f. C. tetragona, var. linearis, Griseb. in DC. Prodr.* ix. 40 (*as to C. viscosa, Zeyh. only*) ; *not of E. Meyer.*

SOUTH AFRICA : without locality ; Var. β : *Burmann! Sonnerat! Buettner! Banks! Wallich! Nelson! Brown! Sieber! Thom,* 769 ! *Drège!*

COAST REGION : Swellendam Div. ; Rhinoceros Fontein near Sebastians Bay, *Gornot!* Var. β : Cape Div. ; Cape Flats, *Krauss! Rehmann,* 1972 ! near Cape-town, *Hesse!* Blue Berg, *Zeyher,* 1200 ! under Tiger Berg, near Riet Valley, *Ecklon,* 176 ! 262 ! near Durban Road, 100 ft., *MacOwan,* 96 ! and in *MacOwan and Bolus in Herb. Norm.* 961 ! Uitenhage Div. ; near Uitenhage, *Prior!*

The varieties recognisable differ very much, as do the two varieties of the nearly allied inland species *C. tetragona,* Linn. f., and *C. tabularis,* Page. It may transpire that this is only a littoral or sublittoral form of a wide-spread and protean species including all three.

23. C. tabularis (Page, Prodr. 121) ; stems distinctly angled, erect, leafy, $\frac{1}{2}$–$\frac{3}{4}$ ft. long ; branches virgate : leaves ovate or elliptic, acute, $\frac{1}{2}$–$\frac{3}{4}$ in. long, 3–4 lin. wide : flowers solitary or 2–3, terminal ; peduncles $\frac{1}{2}$–$1\frac{1}{2}$ in. long ; calyx 5–7 lin. long, divided half-way ; lobes lanceolate, acute, keeled ; tube strongly keeled or with narrow coriaceous wings ; corolla-tube cylindric, about as long as the calyx ; lobes elliptic-oblong, obtuse, 8 lin. long, 4–5 lin. wide ; anthers straight ; ovary ovoid, acute, 4 lin. long. *Steud. Nomencl. Bot. ed.* 2, i. 352. *C. jasminoides, Edw. Bot. Reg. t.* 197 ; *E. Meyer, Comm.* 179 ; *not of Linn. C. tetragona, Schoch in Bot. Centralbl. Beih.* xiv. 196 (*as to syn. Edw. Bot. Reg. only*) ; *not of Linn. f. C. tetragona, var. brevifolia, Griseb. in DC. Prodr.* ix. 40 (*as to syn. Edw. Bot. Reg. only*) ; *not of Gen. et Sp. Gent. C. tetragona, var. ovata, Schoch in Bot. Centralbl. Beih.* xiv. 197 (*as to* 9614 *Schlechter only*) ; *not of E. Meyer. Evalthe jasminoides, Raf. Fl. Tellur.* iii. 77.

VAR. β, **confusa** (Prain in Kew Bulletin, 1908, 298) ; stems 1–1½ ft. long ; leaves narrowly ovate-lanceolate to linear, 1–1¼ in. long, 1–2 lin. wide ; calyx-lobes lanceolate, acuminate. *C jasminoides, Cham. in Linnæa,* vi. 344 (*Caledon specimens only*) ; *Griseb. Gen. et Sp. Gent.* 101 *and in DC. Prodr.* ix. 40 (*as to description only*) ; *Gilg in Engl. und Prantl, Pflanzenfam.* iv. ii. 78 ; *Knobl. in Bot. Centralbl.* lx. 328 ; *Schoch in Bot. Centralbl. Beih.* xiv. 194 ; *not of Linn. C. tetragona, var. linearis, Schoch in Bot. Centralbl. Beih.* xiv. 197 (*as to* 4182 *Bolus only*) ; *not of E. Meyer.*

SOUTH AFRICA : without locality, *cultivated specimens* ! Var. β : *Drège! Hesse! Zeyher! Lehmann! Ludwig! Thom,* 808 ! and *cultivated specimens*!

COAST REGION : Stellenbosch Div. ; Lowrys Pass, 1000–2000 ft., *Drège!* Bredasdorp Div. ; hills near Elim, *Bolus,* 8578 ! Koude River, *Schlechter,* 9614 ! Swellendam Div. ; salt marshes, *Bowie!* Var. β : Caledon Div. ; mountains near Greitjes Gat, 1600 ft., *Bolus,* 4182 ! Houw Hoek, 1800 ft., *Ecklon! Schlechter,* 9393 ! Zwart Berg, *Pappe!* between Caledon and Pot River, *Ecklon!* Bavians Kloof, near Genadendal, *Ecklon! Gibbs in Herb. Bolus,* 10036 ! near Genadendal, *Prior! Pappe!*

Nearly allied to and, as Knoblauch suggests, perhaps not specifically separable from *C. tetragona,* Linn. f., from which however it is readily distinguished by its calyx with narrow, acute or acuminate, instead of wide, foliaceous lobes. The two vary similarly and equally as regards their leaves.

24. C. tetragona (Linn. f. Suppl. 151); stems distinctly angled, erect, leafy, $\frac{1}{2}$-1$\frac{1}{2}$ ft. long; branches virgate; leaves ovate or subelliptic, acute, $\frac{1}{2}$-$\frac{3}{4}$ in. long, 3-4 lin. wide; flowers solitary or 2-3, terminal; peduncles $\frac{1}{2}$-1$\frac{1}{2}$ in. long; calyx 7-9 lin. long, divided half-way; lobes wide triangular, subacute or acute, foliaceous, winged; tube with foliaceous wings; corolla-tube cylindric, shorter than the calyx; lobes elliptic-oblong, acute or obtuse, 8 lin. long, 5 lin. wide; anthers straight; ovary ovoid, acute, 4 lin. long. *Thunb. Prodr.* 35, *in Trans. Linn. Soc.* vii. 249, *t.* 12, *fig.* 2, *and Fl. Cap. ed.* 2, ii. 111 : *Willd. Sp. Pl.* i. 1071 ; *R. Br. Prodr.* i. 451 ; *Griseb. Gen. et Sp. Gent.* 102 *and in DC. Prodr.* ix. 40 ; *Knobl. Bot. Centralbl.* lx. 328 ; *Schoch in Bot. Centralbl. Beih.* xiv. 197. *C. tetragona, var. ovata, E. Meyer, Comm.* 179. *C. jasminoides, Thunb. Prodr.* 35, *in Trans. Linn. Soc.* vii. 251, *and Fl. Cap. ed.* 2, ii. 109 (*in small part, sheet* γ *partly*); *not of Linn.*

VAR. β, **linearis** (E. Meyer, Comm. 179); leaves narrow-elliptic or lanceolate or linear, $\frac{1}{2}$-1$\frac{1}{4}$ in. long, 1$\frac{1}{2}$-2$\frac{1}{4}$ lin. wide. *Griseb. in DC. Prodr.* ix. 40 (*excl. syn. C. viscosa, Zeyh.*); *Schoch in Bot. Centralbl. Beih.* xiv. 197 (*as to Drège* 7822 *only*). *C. tetragona, Schoch in Bot. Centralbl. Beih.* xiv. 196 (*as to syn. C. uniflora, Lam., and mainly as to specimens*); *hardly of Linn. f. C. tetragona, var. brevifolia, Griseb. Gen. et Sp. Gent.* 102 (*wholly*) *and in DC. Prodr.* ix. 40 (*excl. syn. C. jasminoides, Edw.*). *C. uniflora, Lam. Encycl.* i. 737, *and Ill. Gen. t.* 108, *fig.* 3 ; *Pers. Syn. Pl.* i 282. *C. jasminoides, Thunb. Prodr.* 35, *in Trans. Linn. Soc.* vii. 251, *and Fl. Cap. ed.* 2, ii. 109 (*as to specimens, sheet* β *and in part sheet* γ, *not as to descr.*); *Griseb. Gen. et Sp. Gent.* 101, *and in DC. Prodr.* ix. 40 (*as to Thunberg's Swellendam specimens, not as to descr.*) ; *not of Linn.*

SOUTH AFRICA : without locality ; *Burmann* ! *Masson* ! *Zeyher,* 162 ! Var. β : *Sparrman* ! *Masson* ! *Wallich* ! *Lehmann* !

COAST REGION : Cape Div. ; Table Mountain, *Fraser,* 52 ! Swellendam Div. ; near Puspus River, *Thunberg* ! Knysna Div. ; hills and glens near Plettenbergs Bay, *Bowie,* 10 ! 11 ! Uitenhage Div. ; between Van Stadens Berg and Klaasniemans Fontein, *Drège* ! between Van Stadens River and Bethelsdorp River, *Drège,* 2239 ! plains near Van Stadens River, *Ecklon* ! plateau near Van Stadens Berg, *Scott Elliot,* 262 ! near Witteklip, *MacOwan,* 1059 ! near Uitenhage, *Zeyher,* 321 ! *Penther,* 2031 ! Brak Fontein, *Zeyher,* 3425 ! Port Elizabeth Div. ; between Uitenhage and Algoa Bay, *Burchell,* 4284 ! Krakakamma, *Burchell,* 4536 ! near Port Elizabeth, *Drège* ! *Zeyher,* 3425 ! *Prior* ! *Miss West,* 29 ! Alexandria Div. ; Oliphants Hoek forest, *Ecklon,* 635 ! Albany Div. ; Sidbury, *Burke* ! Geilhoutboom, 800-1200 ft., *Drège* ! near Grahamstown, *Burke,* 83 ! Var. β : Stellenbosch Div. ; Lowrys Pass (Hottentots Holland Kloof), *Drège* ! Caledon Div. ; *Thom,* 961 ! Swellendam Div. ; without precise locality, *Thunberg* ! Knysna Div. ; Keurboom River, *Penther,* 2024 ! near Knysna, *Penther,* 2011 ! 2014 ! Uniondale Div. ; Long Kloof, *Drège,* 7822 ! Humansdorp Div. ; near Humansdorp, 150 ft., *Bolus,* 1566 ! *Schlechter,* 6029 ! Uitenhage Div. ; downs near Cape Recife, *Zeyher,* 750 ! *Prior* ! hills near Brakfontein, *Zeyher,* 321 ! plains near Winter Hoek Mountains and between Kromme River and Uitenhage, *Ecklon,* 636 ! near Enon, *Baur,* 1027 ! Port Elizabeth Div. ; sandhills near Port Elizabeth, *Prior* ! *Cooper,* 1458 ! 2764 ! *Miss Holland,* 49 ! Cape Recife, *Zeyher,* 3425! *E.C.S.A. Herb.* 224 ! Bathurst Div. ; near Kaffir Drift, *Burchell,* 3763 ! between Barville Park and the sea, *Burchell,* 4072 ! Port Alfred, *Hutton,* 598 ! *Atherstone,* 238 ! *Haagner in Herb. Conrath,* 738 ! Albany Div. ; by the stream at Grahamstown, *Burchell,* 3561 ! *Misses Daly and Sole,* 537 ! *Bolton* ! without precise locality, *Miss Bowker* !

As to the limitation of var. *linearis,* E. Meyer, we have followed Meyer. It

includes, however, two easily recognisable forms :—(1) with linear leaves up to
1 in. long ; (2) with narrow-elliptic leaves, $\frac{1}{2}$ in. long. The former is var. *linearis* as
limited by Schoch and is the plant distributed but not described as *C. jasminoides*,
Thunb., not of Linn. The latter is the plant described and figured by Lamarck
as *C. uniflora*, and is the plant originally intended by Grisebach as his var.
brevifolia. It is to this form that Schoch has restricted, in intention, the name
C. tetragona ; the original *C. tetragona*, Linn. f., is however, the form which
Meyer and Schoch have termed var. *ovata.* These various forms are connected by
intermediates, and *C. tetragona* as a whole is perhaps too closely related to
C. tabularis ; some of Schoch's specimens, quoted under *C. tetragona*, belong to
C. tabularis.

25. C. melampyrifolia (Lam. Ill. Gen. i. 479) ; stem distinctly
angled, decumbent or scandent, leafy, 5–6 ft. long ; branches
divaricate ; leaves lanceolate or ovate, base cordate, subamplexicaul,
acuminate at the revolute tip, $\frac{1}{2}$–1 in. long, 2–6 lin. wide ; flowers
solitary, terminal ; peduncles $\frac{3}{4}$–2 in. long ; calyx 4 lin. long,
divided half-way ; lobes ovate-lanceolate, acuminate, keeled ; tube
keeled ; corolla-tube cylindric, rather longer than the calyx ; lobes
ovate-oblong or oblong, subacute or acute, 9–12 lin. long, 4–6 lin.
wide ; anthers straight ; ovary ovate-oblong, acute, 5 lin. long.
Pers. Syn. Pl. i. 282 ; *Poir. Encycl. Suppl.* ii. 233 ; *Roem. and
Schult. Syst. Veg.* iv. 201. *C. lychnoides, Lam. Encycl.* i. 736 ; *not
of Berg. C. jasminoides, Thunb. Prodr.* i. 35, *in Trans. Linn. Soc.*
vii. 251, *and Fl. Cap. ed. 2*, ii. 109 (*as to descr. wholly, and as to sheet a
in Herb. propr.*) ; *Willd. Sp. Pl.* i. 1066 ; *Griseb. Gen. et Sp.
Gent.* 101 *and in DC. Prodr.* ix. 40 (*as to Krebs' and Willdenow's
plant cited, but excl. descr.*) ; *not of Linn. C. perfoliata, Eckl. in
South Afr. Quart. Journ.* i. 370 ; *Griseb. Gen. et Sp. Gent.* 104
and in DC. Prodr. ix. 40 ; *Gilg in Ann. Nat. Hist. Hofmus.
Wien*, xv. 65 *and in Engl. und Prantl. Pflanzenfam.* iv. 2, 78 ;
Schoch in Bot. Centralbl. Beih. xiv. 205 ; *not of Salisb. C. speciosa,
E. Meyer, Comm.* 178. *C. glutinosa, Paxt. Mag. Bot.* xv. 245, *with
plate.*

South Africa : without locality ; *Oldenburg ! Burmann ! Commerson ! Sonnerat !
Masson ! Mund and Maire*, 261 ! *Scholl ! Krebs*, 231 ! *Fleuron (Herb. Willdenow.*
4500) !
Coast Region : Caledon Div ; near Caledon, *Prior* ! Swellendam Div. ;
Tradouw Berg, *Bowie*, 4 ! and without precise locality, *Thunberg* ! Riversdale
Div. ; without precise locality, *Rust*, 314 ! Mossel Bay Div. ; Robinson Pass,
2500 ft., *Bolus* ! George Div. ; between Touw River and Kaymans River,
Burchell, 5772 ! ne r George, *Prior* ! *Bolus*, 12996 ! hills near Silver River,
Schlechter, 5869 ! *Penther*, 2026 ! Montagu Pass, *Rehmann*, 267 ! *Penther*, 2027 !
between George and Knysna, *Pappe* ! and without precise locality, *Bowie*, 3 !
Knysna Div. ; mountains near Plettenbergs Bay, *Bowie*, 4 ! and near the Great
Forest, *Castelnau*, 583 ! *Bolus*, 1921 ! Uitenhage Div. ; Van Stadens Berg,
Burchell, 4710 ! *Zeyher*, 193 ! *Pehlon*, 488 ! 634 ! Zuur Berg Range, 1500–
2000 ft.. *Drège* ! Port Elizabeth Div. ; Algoa Bay, *Forbes*, 80 ! 91 ! *Prior* !
near Port Elizabeth, *E.S.C.A. Herb.* 446 ! Albany Div. ; near Grahamstown,
Bolton ! *MacOwan*, 71 ! *Schönland*, 130 ! *Zeyher*, 193 ! 3430 ! *Scott Elliot*, 786 !
Galpin, 173 ! *Hutton* ! *Williamson* !

A very distinct species which will probably still be maintained even if it be
found necessary to unite *C. tetragona*, *C. tabularis* and *C. scabrida*. In the
herbarium two forms are readily recognisable ; one, which is the plant described

by Thunberg and cited by Grisebach (*Krebs*, 231) as *C. jasminoides*, and is also
the plant described by Ecklon as *C. perfoliata*, dries to a pale brown or straw-
colour; the other, which is *C. melampyrifolia*, Lam., also *C. speciosa*, E. Meyer,
and *C. glutinosa*, Paxt., dries blackish. There is, however, no morphological
character by which the two can be distinguished; both vary equally as regards
width of leaf.

V. **ENICOSTEMA**, Blume.

Calyx 5-toothed; tube campanulate; teeth lanceolate. *Corolla*
5-lobed; tube elongate, lower-half cylindric, upper narrowly funnel-
shaped; lobes small, lanceolate. *Stamens* 5, inserted at the middle
of the corolla-tube; filaments filiform, with a small double-hooded
scale at the base; anthers erect, straight, acute. *Ovary* 1-celled;
placentas little intruded; ovules numerous; style short, subulate,
stigma capitate. *Capsule* oblong, septicidal; valves 2. *Seeds*
numerous, globose, foveolate.

Erect perennial herbs; leaves numerous, opposite; flowers numerous, small, in
axillary clusters.

DISTRIB. Species 4, one each in Tropical America, Tropical Africa and
Madagascar with the following which extends to Tropical Africa and Asia.

1. **E. littorale** (Blume, Bijdr. 848); stem branching from the
base, glabrous, 4–18 in. high; leaves glabrous, in many pairs, sessile,
linear or lanceolate, usually acute, occasionally in the lower half of
the stem oblong, sometimes obtuse, or rarely obovate retuse, $\frac{3}{4}$–3 in.
long, $1\frac{1}{2}$–4 lin. wide; flowers in sessile axillary clusters throughout
the stem; calyx-tube $\frac{1}{2}$ lin. long; lobes $\frac{2}{3}$–1 lin. long, $\frac{1}{3}$ lin. wide,
ovate to linear-lanceolate, acuminate or acute, slightly spreading at
the tips, imbricate or valvate at the base, with narrowish submembranous margins; corolla-tube $2\frac{1}{2}$ lin. long; lobes 1–$1\frac{1}{4}$ lin. long;
style slightly thickened upwards. *Griseb. in DC. Prodr.* ix. 66;
Hook. f. Fl. Brit. Ind. iv. 101; *Klotzsch in Peters, Reise Mossamb.
Bot.* 271; *Baker & N. E. Br. in Dyer, Fl. Trop. Afr.* iv. i. 563.
E. verticillare, Baill. Hist. Pl. x. 131 *partly. E. verticillatum,
Engl. Pfl. Ost-Afr. C.* 313 *partly; Knobl. in Bot. Centralbl.* lx.
333. *Gentiana verticillata, Linn. f. Suppl.* 174, *not of Linn.
G. verticillaris, Retz. Obs.* ii. 15. *Slevogtia orientalis, Griseb. in DC.
Prodr.* ix. 65. *Hippion hyssopifolium, Spreng. Syst.* i. 589; *Griseb.
Gen. et Sp. Gent.* 134. *H. verticillatum, O. Kuntze, Rev. Gen. Pl.*
428; *Hiern in Cat. Afr. Pl. Welw.* i. 711 *partly, perhaps not of
F. W. Schmidt.—Centaurion angustifolium, floribus ex alis sessilibus,
Burm. Rar. Afr. Pl.* 206, *t.* 74, *fig.* 3.

KALAHARI REGION: Transvaal; Boshveld, Klippan, *Rehmann,* 5280! in grassy
plains near Pietersburg, 4000 ft., *Bolus,* 10865! Naboomfontein, 4300 ft.,
Schlechter, 4306! near Barberton, 2900 ft., *Thorncroft,* 400 (*Herb. Wood,* 4503)!
Thorncroft, 3026!

EASTERN REGION: Delagoa Bay; Matolla, 30 ft., *Schlechter,* 11697!

Also in Tropical Africa and South-eastern Asia.

VI. **FAROA**, Welw.

Calyx 4-lobed ; tube short, campanulate ; lobes acute or obtuse. *Corolla* 4-lobed ; tube not exceeding the calyx ; limb with 4 scales ; lobes spreading. *Stamens* 4, inserted at the mouth of the corolla-tube ; filaments filiform ; anthers oblong, minute. *Ovary* 1-celled ; placentas slightly intruded ; ovules numerous ; style subulate ; stigma 2-lobed or simple. *Capsule* subglobose, septicidal ; valves 2. *Seeds* numerous, subglobose, foveolate.

Annual, rarely perennial, branching herbs ; leaves usually numerous, sessile ; flowers in dense terminal or axillary clusters, minute.

DISTRIB. Species about 10, all Tropical African, the one here described also extending into our area.

1. F. salutaris (Welw. in Trans. Linn. Soc. xxvii. 45, t. 17) ; biennial or perennial ; stem 2–3 in. high, branching at the base ; branches slender, erect ; leaves glabrous, radical tufted, oblanceolate, obtuse, base cuneate, 6–9 lin. long, $1\frac{1}{2}$–$2\frac{1}{4}$ lin. wide, cauline opposite, 2–3-jugate, lanceolate, 2–6 lin. long, 1–$1\frac{1}{2}$ lin. wide ; flowers densely clustered ; clusters 8–10 lin. across, usually all terminal, at times a second axillary cluster, closely subtended by pairs of leaves ; pedicels glabrous, 1–2 lin. long ; calyx 1–$1\frac{1}{4}$ lin. long ; lobes erect, ovate-acute, submembranous, midrib thickened, green, slightly gibbous and slightly prolonged at the tip ; corolla-tube $1\frac{1}{4}$ lin. long ; lobes 1 lin. long, lanceolate, acute ; filaments longer than the corolla-lobes ; style $1\frac{3}{4}$–2 lin. long, slender ; stigma minutely 2-lobed. *Engl. Hochgebirgsfl. Trop. Afr.* 336 ; *Hiern in Cat. Afr. Pl. Welw.* i. 710 ; *Knobl. in Bot. Centralbl.* lx. 330 ; *Gilg in Baum, Kunene-Samb. Exped.* 333 ; *Baker & N. E. Br. in Dyer, Fl. Trop. Afr.* iv. i. 569.

KALAHARI REGION : Basutoland ; Machacha Mountain, *Bryce* !

Also in Tropical Africa.

VII. **SWERTIA**, Linn.

Calyx 4–5-lobed ; tube short or obsolete ; lobes acute or obtuse. *Corolla* 4–5-lobed ; limb subrotate ; tube very short and broad ; lobes acute or obtuse, with solitary or paired basal glandular nectaries. *Stamens* 4–5, inserted in the corolla-sinuses, shorter than the lobes ; filaments subulate or flattened ; anthers oblong or ovate, versatile. *Ovary* 1-celled ; placentas slightly intruded ; ovules numerous ; style short or obsolete ; stigma 2-lobed. *Capsule* oblong or lanceolate ; valves 2. *Seeds* numerous, minute, compressed, often winged.

Annual or perennial herbs ; leaves numerous, cauline opposite, sessile or very shortly petioled, radical narrowed to a petiole ; flowers in terminal corymbose or paniculate cymes ; corolla blue, yellow or white.

DISTRIB. Species about 90, rather more than one-third Tropical African, a few in Madagascar and in Europe, the greater number Asiatic.

1. S. stellarioides (Ficalho, Pl. Uteis, 225) ; annual ; stem $\frac{1}{2}$–1 ft. high, simple or branching above ; branches ascending, corymbose ; leaves glabrous, radical lanceolate, obtuse, narrowed to a short petiole, $\frac{1}{3}$–$\frac{3}{4}$ in. long, $1\frac{1}{2}$–2 lin. wide, cauline linear-lanceolate to linear, obtuse or subacute, $\frac{1}{3}$–1 in. long, $\frac{2}{3}$–2 lin. wide ; flowers 5-merous, in simple or panicled subumbellately 3–8-flowered cymes ; pedicels $\frac{1}{4}$–1 in. long ; outer two calyx-lobes larger than the others, all lanceolate or oblong-lanceolate, acute, $1\frac{1}{3}$–$3\frac{1}{3}$ lin. long, $\frac{1}{2}$–$\frac{2}{3}$ lin. wide ; corolla-lobes oblong-obovate, obtuse, $3\frac{1}{2}$–5 lin. long, $1\frac{1}{2}$–$1\frac{2}{3}$ lin. wide, white streaked with purple within, basal nectaries paired, oblong, ciliate ; stamens 2 lin. long ; ovary oblong, hardly narrowed to the very broad obtuse sessile 2-lobed stigma which is broader than long ; valves of capsule obtuse. *Hiern in Cat. Afr. Pl. Welw.* i. 711 ; *Baker & N. E. Br. in Dyer, Fl. Trop. Afr.* iv. i. 581. *S. Welwitschii, Engl. Hochgebirgsfl. Trop. Afr.* 339 ; *Gilg in Engl. Jahrb.* xxvi. 110. *Adenopogon stellarioides, Welw. Syn. Explic.* 27.

KALAHARI REGION : Orange River Colony ; marshes near Harrismith, 7000 ft., *Sankey*, 68 ! Transvaal, in marshy places ; Donkershoek, *Rehmann*, 6529 ! Houtbosch Berg, 6500 ft., *Schlechter*, 4699 ! near Lydenberg, *Wilms*, 967 ! Lomatie Valley, 4000 ft., *Galpin*, 1228 !

EASTERN REGION : Natal ; damp places near Van Reenen, 7500 ft., *Schlechter*, 6991 !

Also in Tropical Africa.

VIII. VILLARSIA, Vent.

Calyx deeply 5-lobed ; tube short, campanulate. *Corolla* 5-lobed ; tube short ; lobes spreading, induplicate-valvate. *Stamens* 5, adnate to the corolla-tube ; filaments filiform ; anthers linear, sagittate, versatile. *Ovary* surrounded by 5 minute hypogynous glands, 1-celled ; placentas parietal, thickened ; ovules usually numerous ; style short or long ; stigma 2-lobed ; lobes usually rather large. *Capsule* subglobose, opening by 4 apical valves. *Seeds* few or numerous ; testa crustaceous, smooth and shining, or muriculate, or hispidulous.

Herbs, usually palustrine ; stems simple with leaves reduced and scale-like or obsolete, or branching with a few cauline leaves ; radical leaves long-stalked, entire or sinuately toothed ; flowers white or yellow, cymose ; cymes sometimes paniculate.

DISTRIB. Species about 10 ; all save the one here described Australian.

1. V. ovata (Vent. Choix, 9, t. 9) ; tufted, usually partly submerged ; rootstock long, usually stout, creeping ; leaves coriaceous, mostly radical, 3–12 in. long ; blade ovate, base rounded or truncate very rarely cuneate or subcordate, obtuse, entire, 1–3 in. long, $\frac{1}{3}$–$2\frac{1}{2}$ in. wide ; petiole stout, sheathing at the base ; stem scapiform, central, usually branched, 3 in.–$1\frac{1}{2}$ ft. long ; cauline leaves short-

petioled or sessile, base cuneate, much smaller ; flowers in racemiform or corymbose cymes at the ends of the branches ; peduncles
$\frac{1}{2}$–$\frac{3}{4}$ in. long; calyx 4 lin. long, divided almost to the base ; lobes
elliptic, acute ; corolla yellow ; tube campanulate, rather shorter
than the calyx ; lobes oblong, retuse, apiculate at the base of the
sinus, with crenate-fimbriate margins, hirsute at the base within,
6 lin. long, 3 lin. wide ; hypogynous glands yellow ; capsule subglobose ; seeds 2, subglobose, obtusely carinate, faintly muricate.
Griseb. Gen. et Sp. Gent. 337 *and in DC. Prodr.* ix. 136 ; *E. Meyer,
Comm.* 186. *Menyanthes ovata, Linn. f. Suppl.* 133. *M. capensis,
Thunb. Prodr.* 34. *Renealmia capensis, Houtt. Handleid.* viii. 335,
t. 47, *fig.* 1.

SOUTH AFRICA : without locality, *Grey* !
COAST REGION : Tulbagh Div. ; marshes near New Kloof, 2000–3000 ft., *Drège* !
Worcester Div. ; Bains Kloof, 2700 ft., *Schlechter,* 9174 ! Paarl Div. ; Klein
Drakenstein Mountains, 500 ft., *Drège* ! Cape Div. ; Table Mountain, *Forster* !
Burchell, 562 ! *Ecklon,* 825 ! *Sieber* ! *Zeyher* ! *Harvey,* 213 ! 238 ! *Milne,* 174 !
MacGillivray, 608 ! *Prior* ! near Wynberg Reservoir, *Wolley Dod,* 167 ! marsh
south of Vlagge Berg, *Wolley Dod,* 166 ! Muizenberg, *Prior* ! Stellenbosch Div. ;
Hottentots Holland, *Prior* ! Caledon Div. ; New Kloof, Houw Hoek Mountains,
Burchell, 8110 ! Swellendam Div. ; by the Zondereinde River, near Appels Kraal,
Zeyher, 3435 ! Uniondale Div. ; Long Kloof, marshes at Sevenfontein River,
Burchell, 4887 ! and by Wagenbooms Stream, *Bolus,* 2404 ! Uitenhage Div. ;
Zuurberg Range, 2500 ft., *Drège* !

IX. **LIMNANTHEMUM**, S. M. Gmel.

Calyx deeply 5–6-lobed ; tube very short, campanulate. *Corolla*
5–6-lobed ; tube short ; lobes spreading, induplicate-valvate.
Stamens 5–6, inserted at or below corolla-throat ; filaments short ;
anthers versatile. *Ovary* surrounded by 5–6 minute hypogynous
glands, 1-celled ; placentas parietal ; ovules usually numerous :
style short or long : stigma 2-lobed. *Capsule* globose, ovoid or
oblong, indehiscent or irregularly rupturing. *Seeds* few or
numerous ; testa crustaceous, smooth or tuberculate, sometimes
subcarinate.

Herbs, aquatic ; stems erect or stolon-like, with alternate or subopposite leaves ;
leaves orbicular, elliptic or ovate, usually deeply notched at the base, margin
entire, crenate or dentate ; flowers white or yellow, solitary or in pairs or fascicles
at the nodes.

DISTRIB. Species about 20 in all tropical and temperate regions.

1. **L. thunbergianum** (Griseb. Gen. et Sp. Gent. 345) : tufted,
usually submerged : stems (false petioles) very variable, 2 in.–2 ft.
long, $\frac{3}{4}$–2 lin. thick ; leaves floating, orbicular with deep acute
basal sinus, coriaceous, 1–5 in. wide ; flowers 5-merous, in fascicles
of 10–25, close to or from $\frac{1}{4}$–1 in. below the blade ; pedicels stout,
$\frac{3}{4}$–2 in. long : calyx 4 lin. long, divided almost to the base ; lobes
oblong or lanceolate, obtuse or acute ; corolla white or yellow ;

tube campanulate; rather shorter than the calyx, with 5 tufts of hairs above the middle within; lobes oblong-lanceolate, acute, with ciliate margins and long hairs on the inner face, 7 lin. long, 4 lin. wide; hypogynous glands subquadrate, minutely ciliate; capsule ellipsoid, about as long as the calyx; seeds 6–18, compressed, $\frac{3}{4}$ lin. long, $\frac{1}{2}$ lin. thick, subcarinate, smooth, slightly shining, grey with darker mottling. *Griseb. in DC. Prodr.* ix. 139; *Gilg in Baum, Kunene-Samb. Exped.* 335; *Wood & Evans, Natal Pl.* i. 29, t. 34; *Baker & N. E. Br. in Dyer, Fl. Trop. Afr.* iv. i. 584. *L. forbesianum, Griseb. Gen. et Sp. Gent.* 345 *and in DC. Prodr.* ix. 139 *partly. L. ecklonianum, Griseb. Gen. et Sp. Gent.* 346 *and in DC. Prodr.* ix. 140. *Villarsia indica, E. Meyer, Comm.* 186, *not of Vent. Menyanthes indica, Thunb. Prodr.* 34, *not of Linn.*

COAST REGION: Cape Div.; in pools on Cape Flats, *Ecklon*! *Bunbury*! *Verreaux*! Humansdorp Div.: Kromme River, *Burchell*, 4872! Uitenhage Div.: Zwartkops River, *Ecklon*! *Zeyher*, 505! *Prior*! Witte River, near Enon, *Drège*! *Baur*, 1043! pools at Galgebosch, *MacOwan*, 1937! near Uitenhage, *Zeyher*! Zuur Berg Range, *Mrs. Barber*, 431! Albany Div.; in a river south of Signal Hill, Grahamstown, 1200 ft., *Galpin*, 2917! Bothas Hill, *Baur*, 1043! Kaffraria, *Baur*!

KALAHARI REGION: Transvaal; Belfast, *Burtt Davy*, 1271!

EASTERN REGION: Transkei; Kreilis Country, *Bowker*! Natal; near Durban, *M‘Ken*, 757! *Wood*, 83! Inanda, *Wood*, 66! Blackwater, Claremont, *Sanderson*, 524!

Also in Tropical Africa.

ADDENDA AND CORRIGENDA.

ERICACEÆ.

THE following key to the genera is to be substituted for that given upon pages 3 and 4. The bracts mentioned refer only to those which are upon the pedicel or come away with the calyx, not those persistent upon the flowering axis.

    * *Corolla and stamens hypogynous ; ovary superior.*

  † Ovary 2-4-celled (in *Coccosperma* sometimes 1-celled and in *Erica* rarely 8-celled) ; fruit sometimes 1-celled by abortion.

  ‡ Ovules 2 or more in each cell. Corolla 4-lobed or 3-4-toothed.

  § Flowers small or large, distinctly pedicellate. Ovary 3-4- (rarely 8-) celled. Fruit 3-4- (rarely 8-) valved. Seeds very small or minute, with a thin testa.

    a. *Stamens normally 8, rarely 6-7.*

I. **Erica.**—*Bracts* 3, very rarely 1 or 0. *Calyx* equally 4-partite, rarely 4-lobed. *Corolla* of various size and shape. *Stamens* free or the anthers cohering. *Stigma* various.

II. **Philippia.**—*Bracts* 0. *Calyx* unequally 4-lobed or 4-partite. *Corolla* very small. *Stamens* with free or connate filaments and anthers. *Stigma* large, peltate or saucer-shaped.

    b. *Stamens normally 4, occasionally 5-6.*

III. **Ericinella.**—*Bracts* 0. *Calyx* unequally 3-4-partite or 3-4-lobed. *Corolla* very small. *Stamens* free or with connate anthers. *Stigma* peltate or broadly obconic.

IV. **Blæria.**—*Bracts* 3, rarely 2. *Calyx* equally 4-lobed or 4-partite. *Corolla* small, tubular or campanulate. *Stamens* free. *Stigma* simple or peltate.

  §§ Flowers very small, subsessile or minutely pedicellate. Fruit indehiscent, containing 2-4 large seeds, flat on one side, very convex on the other with a thick hard testa.

XVIA. **Coccosperma.**—*Bracts* 0. *Calyx* unequally 4-lobed. *Corolla* campanulate or cup-shaped, with incurved lobes. *Stamens* 4-8 ; filaments and anthers connate. *Ovary* 1-2-celled, with 2 collateral ovules in each cell, often clinging together and appearing as one ; stigma large, peltate or funnel-shaped.

  ‡‡ Ovule solitary in each cell. (See under *Coccosperma*.)

  § Stamens 5-8. Corolla 4-lobed or 4-toothed.

  ‖ Bracts 0. Anthers connate, notched at the apex.

XVII. **Salaxis.**—*Calyx* unequally 4-lobed. *Corolla* very small, campanulate or cup-shaped. *Stamens* 8, rarely 6 ; filaments at first variably connate. *Stigma* large, funnel-shaped.

4 c 2

|||| Bracts 3.   Anthers free, bipartite.

a. *Calyx equally 4-partite, with campanulately erect segments.   Anthers dorsifixed close to the base.*

VII. **Eremia.**—*Leaves* spreading, not woolly.   *Calyx* ciliate or hairy, but not woolly.   *Stamens* 8, included, equalling or slightly exceeding the corolla. *Stigma* simple or minutely 4-lobed.

VIII. **Hexastemon.**—*Leaves* densely imbricate and together with the calyx white-woolly.   *Stamens* 6, much exserted.

b. *Calyx equally 4-lobed, nearly flat and almost square in outline.   Anthers basifixed.*

VIIΛ. **Platycalyx.**—*Corolla* subglobose or globose-ovoid, contracted at the mouth.   *Stamens* 6 or occasionally 5-7.   *Stigma* simple.

§§ Stamens 4, very rarely 3 ; filaments and anthers free.
|| Corolla 4- (rarely 3-) lobed.
¶ Anthers included or rarely slightly exceeding the corolla, about as broad as long (see also *Thoracosperma puberulum*) ; cells perfectly free, dorsally attached to the dilated or crutch-like apex of the filaments or very shortly tapering into a slender attachment at the base, separated or contiguous.

IX. **Grisebachia.**—*Bracts* 3.   *Calyx* equally or unequally 4-lobed $\frac{1}{2}$–$\frac{3}{4}$ of the way down or 4-partite, very conspicuously ciliate, often with long hairs.   *Corolla-tube* constricted at or above the middle or tubular or funnel-shaped.   *Stigma* minute, simple or capitate.

¶¶ Anthers more or less exserted, longer than broad, firmly basifixed ; cells not perfectly free ; filaments not dilated at the apex.

a. *Calyx-tube 0 to $\frac{1}{3}$ as long as the lobes.*

V. **Coilostigma.**—*Bracts* 0.   *Calyx* unequally 4-partite ; tube 0.   *Corolla* ovoid or cylindric.   *Anthers* bipartite.   *Stigma* peltate or crater-like.

VI. **Thoracosperma.**—*Bracts* 1 or 3.   *Calyx* rather inconspicuous or minute, equally and deeply 4-lobed or 4-partite, not clothed with long hairs.   *Corolla* ovoid or subcampanulate.   *Anthers* bipartite.   *Stigma* small, simple or subpeltate.

X. **Acrostemon.**—*Bracts* 3.   *Calyx* conspicuous, equally 4-lobed nearly or quite to the base, clothed with long or rarely gland-tipped hairs.   *Corolla* tubular or ovoid-tubular.   *Anther-cells* connate at the base.   *Stigma* simple or capitate.

b. *Calyx-tube usually much longer than or at least equalling the 4 equal teeth, 4-angled or 8-ribbed, thin, coriaceous or fleshy.*

XI. **Simocheilus.**—*Bracts* 0, 1 or 3.   *Corolla* tubular, tubular-campanulate or funnel-shaped.   *Stigma* minute, simple, thickened or capitate.

|||| Corolla equally 2-lobed, tubular or funnel-shaped.   Anthers exserted, basifixed, bipartite.

XIV. **Aniserica.**—*Bracts* 0.   *Calyx* campanulate or tubular-campanulate, equally 4-toothed or 4-lobed.

XV. **Sympieza.**—*Bracts* 0 or rarely 2.   *Calyx* of the lower or of all the flowers much flattened dorsally, 2-edged, 2-lobed ; of the upper flowers often sharply 3-4-angled, 3-4-lobed.

†† Ovary with 4 compressed angles, 1-celled with 4 ovules pendulous from the top of a central placenta free from the wall of the ovary.

X_A. **Thamnus.**—*Bracts* 3. *Calyx* campanulate, equally 4-toothed. *Corolla* obovoid or suburceolate. *Stamens* 4, free, exserted. *Capsule* 4-valved.

††† Ovary 1-celled. Ovule solitary, pendulous. Corolla 4-lobed or 4-toothed. (See under *Coccosperma*.)
‡ Stamens 8, included.

XIII_A. **Eremiopsis.**—*Bracts* 3. *Calyx* deeply and equally 4-lobed. *Corolla* campanulate. *Stamens* free ; anther-cells dorsifixed, separated by the crutch-like apex of the filaments. *Style* bent down upon the side of the ovary ; stigma simple.

XVI. **Lepterica.**—*Bracts* 0. *Calyx* equally 4-lobed. *Corolla* minute, obconic. *Staminal filaments* connate at the base ; anthers connate. *Style* stout, soon enlarging and blending with the young fruit so that the large peltate stigma appears sessile.

‡‡ Stamens 3-4, rarely 5.
§ Corolla mostly 1-2¼ (occasionally ¾) lin. long, 3-4-lobed. Anthers always exserted when mature, free. Stigma small, simple, thickened or capitate. (See also *Acrostemon eriocephalus*.)

XII. **Syndesmanthus.**—*Bracts* 0, 1 or 3. *Calyx* variable in shape, equally 3-4-toothed ; tube with or without 3-4 angles, thin or coriaceous, scarcely or not at all enlarging in fruit.

XIII. **Anomalanthus.**—*Bracts* 3. *Calyx* at first very small, campanulate, equally 4-toothed, fleshy, becoming much enlarged, very thick and ovoid cylindric or subglobose in fruit, rarely 4-angled.

§§ Corolla minute, mostly ¼-⅔ (rarely ¾) lin. long. Anthers included or partly exserted, free or connate. Stigma large, peltate or crater-like.

XVIII. **Scyphogyne.**—*Bracts* 0. *Calyx* unequally or occasionally equally 3-4-lobed, often angular, not fleshy.

*** Corolla and stamens arising from the middle or lower part of the ovary, which is half-inferior to it, but free from the calyx.*

XIX. **Lagenocarpus.**—*Bracts* 0. *Calyx* obconic or campanulate, unequally 4-toothed. *Corolla* minute, campanulate. *Stamens* 7-8, connate, included. *Ovary* 1-celled, with 4-5 ovules pendulous from near the apex of a central free placenta.

ERICA.   In the key to the genus delete the following :—
page 12, line 35 : —(in 44 ζ unknown)
page 13, lines 37-40 :—Corolla about 19 lin. long, nearly
glabrous, red   ...   . (45) **curviflora,** var. ζ.
page 14, lines 18-19 :—Leaves glabrous ; corolla about 19
lin. long ...   . (45) **curviflora,** var. ζ.

45. Erica curviflora (Linn.).   From the geography on p. 72, line 11, delete :—Var. ζ, *Mund & Maire* !

94. Erica tubercularis (Salisb.).   Add to the synonymy :—*Erica inclyta, Soland. ex Salisb. in Trans. Linn. Soc.* vi. 330.

117. Erica heliophila (Guthrie and Bolus). This should be
E heleophila (Guthrie and Bolus).

196. Erica Mundii (Guthrie and Bolus). Delete the following : -
page 151, line 9 from the bottom :—or even all the leaves, etc.
page 152, line 6 :—mountains near Voormans Bosch, *Zeyher*, 3258 !
page 152, lines 8–12 :—The different aspect of the broad- or narrow-leaved forms is
      sometimes puzzling ; but the characters of the flowers are
      fairly constant. . . . The narrow-leaved forms are
      probably the result of a drier season.

209. Erica polifolia (Salisb.). Add to the synonymy :—*Blæria
caduca*, *Thunb.? Diss. Blæria*, 10.

218. Erica nudiflora (Linn.). Add to the synonymy :—*Erica
pusilla*, *Thunb. Prodr.* 70, *and Fl. Cap. ed. Schultes*, 347. *Blæria
nudiflora*, *Thunb. Diss. Blæria*, 6.

230a. Erica (§ Pyronium) recta (Bolus in Trans. S. Afr. Philos.
Soc. xvi. 397) ; branches and branchlets very straight, erect, and
approximate, the former pale cinereous, smooth and nearly glabrous,
7–8 in. or more long, the latter always axillary and verticillate in
threes, internodes from 3 lin. to $1\frac{1}{3}$ in. long, slightly spreading and
then straightly ascending, simple or again sparingly and similarly
branched, closely leafy to the summit, puberulous, $\frac{1}{2}$–3 in. long ;
leaves 3-nate, erect, closely imbricate, the younger twice longer, the
older $\frac{1}{4}$ or $\frac{1}{3}$ longer, than the internodes, very shortly petiolate, oblong
or broadly linear, subobtuse, glabrous, sulcate below, margins naked,
very slightly inflexed, 1–$1\frac{3}{4}$ lin. long ; flowers terminal, normally
3-nate, sometimes 2, or solitary at the ends of the branchlets, few,
spreading or deflexed, rosy, discoloured (yellowish) on the limb of
the corolla ; pedicels stoutish, white-tomentose, rosy, 1–$1\frac{1}{2}$ lin. long ;
bracts all closely approximate, slightly spreading, lanceolate,
scarious below, tipped with a foliaceous or coloured callus, about
1 lin. long ; sepals like the bracts but broader and somewhat
unequal, lanceolate to ovate, the callous tip more distinctly
channelled and mostly rosy, $1\frac{1}{3}$ lin. long, from $\frac{1}{2}$–$\frac{2}{3}$ the length of the
corolla ; corolla ovate-urceolate, throat slightly contracted, mouth
rather glabrous, $2\frac{1}{2}$–$2\frac{3}{4}$ lin. long ; segments spreading, semiorbicular,
minutely ciliolate, about $\frac{1}{4}$ the length of the tube ; anthers semi-
exserted (probably at length fully exserted) lateral, longitudinally
narrow-semiovate, smooth, $\frac{3}{4}$ lin. long, aristate at the base ; pore
$\frac{1}{2}$ the length of the cell ; awns free, straightly deflexed, about $\frac{1}{3}$ the
length of the cell ; style exserted, stoutish ; stigma capitellate ; ovary
globose, glabrous, lobed ; ovules in each cell numerous, very minute.

Coast Region : Ladismith Div. on the Little Zwartberg Range near Vaartwel
and the Gamka River, 3000 ft., *Marloth*, 3993 !

Structurally near to *E. unilateralis*, but with larger flowers, free anther-awns
and a glabrous ovary. The remarkably straight ternate branches and very regular
erect leaves would, if they should prove constant, distinctly characterise this
species ; but the material at our disposal is scanty.

241. Erica læta (Bartl.). Add as a synonym :—*E. rubens, Lodd. Bot. Cab. t.* 557 ?

256. Erica subdivaricata. (Berg.). Add to the synonymy :—*E. paniculata, Wendl. ex Steud. Nomencl. ed.* 2, i. 577.

286. Erica odorata (Andr.). Add to the synonymy :—*E. Beaumontia, Andr. Heathery, t.* 253, *and Col. Heaths, t.* 222. *E. beaumontiana, Rolliss. ex Lodd. Bot. Cab. t.* 1686.

329. Erica microcodon (Guthrie & Bolus). For (*Ecklon & Zeyher* ?) in Cape Govt. Herb. ! read *Zeyher,* 3258 !

340. Erica incurva (Wendl.). Add to the synonymy :—*Blæria bruniæfolia, G. Don, Gen. Syst.* iii. 805.

343. Erica turmalis (Salisb.). Add synonym :—*Blæria turmalis, G. Don, Gen. Syst.* iii. 805.

358. Erica physantha, var. β, aristulata (Bolus) ; anthers shortly aristulate.

Coast Region: Prince Albert Div. ; rocky places in the Zwartberg Pass, 5000 ft., *Bolus,* 11596 !

Interesting as being the solitary exception to muticous anthers in this section. In every other respect the specimens agree with Burchell's type.

388. Erica chlamydiflora (Salisb.). In the synonymy after *E. viscaria, Roxb. ex Salisb. l.c.* add *not of Linn.*

1. Philippia leeana (Klotzsch). Add to the synonymy :—*Salaxis leeana, Dietr. Syn. Pl.* ii. 1260.

3. Philippia Chamissonis (Klotzsch). Add to the synonymy :— *Salaxis Chamissonis, Dietr. Syn. Pl.* ii. 1260.

1. Ericinella multiflora (Klotzsch). Add to the synonymy :— *Salaxis multiflora, Dietr. Syn. Pl.* ii. 1261.

2. Blæria fuscescens (Klotzsch). Add to the synonymy :—*Blæria trigona, Wendl. ex Steud. Nom. Bot. ed.* 2, i. 208.

7. Blæria purpurea (Linn. f.). Add to the synonymy :—*Blæria glabella, Wendl. Collect.* ii. 47, *t.* 55.

COILOSTIGMA, Klotzsch. The last two lines of the key on p. 327 should read :—

Branchlets subflexuose ; corolla less than 1 lin. long ...   (3) **zeyherianum.**
Branchlets straight ; corolla more than 1 lin. long   ...   (4) **dregeanum.**

4. Coilostigma dregeanum (Klotzsch in Linnæa, xii. 235). The following description should replace that given on p. 328 :—Branchlets slender, erect, very little spreading from the erect straight branches, densely puberulous, greyish; leaves 3-nate, $1\frac{3}{4}-2\frac{1}{2}$ lin. long with the petiole, erect, imbricate or shorter than the internodes, linear, acute, glabrous, minutely and irregularly subdenticulate; flowers in axillary and terminal clusters of 3–7, mostly crowded at the ends of the branchlets; pedicels up to $\frac{1}{4}$ lin. long; bracts none; calyx unequally 4-partite, the shorter sepals about $\frac{1}{3}$ and the longer (leaf-like) about $\frac{1}{2}$ as long as the corolla, linear, obtuse, glabrous, minutely ciliate; corolla nearly $1\frac{1}{4}$ lin. long, oblong-ovoid, contracted at the mouth, glabrous; teeth broader than long, rounded, incurved; stamens 4; anthers partly or wholly exserted, $\frac{1}{2}$ lin. long, oblong with parallel sides and a broad apical notch, truncate at the base, spurless; ovary compressed-orbicular, 2-celled, glabrous; stigma peltate, 4-angled, with a central projection. *Benth. in DC. Prodr.* vii. 708.

SOUTH AFRICA : without locality, *Drège,* 7753 !

Described from the type in the Berlin Herbarium. This may be only a form of *C. zeyherianum,* but more material is required to decide. It seems to differ in the straighter less spreading stems and branchlets and longer corolla. The main stem of the type is rather stouter than that of the specimens of *C. zeyherianum* I have seen, but the branchlets are quite as slender as in that, but not flexuose.

1. Thoracosperma paniculatum (Klotzsch). Add to the synonymy :—*Simocheilus quadriflorus, Benth. ex Steud. Nom. Bot. ed.* 2, ii. 589 (*an error for S. quadrifidus, Benth.*).

EREMIA, D. Don. To the generic characters add :—Anthers free, dorsifixed close to the base; cells free, contiguous, opening by lateral elongated pores.

PLATYCALYX, N. E. Br. To the generic characters add :— Anthers basifixed, bipartite.

GRISEBACHIA. Insert in the key on p. 338 after (14) plumosa :—

Calyx-lobes ciliate with long gland-tipped hairs;
  ovary glabrous ... ... (14a) **Pentheri.**

7. Grisebachia dregeana, var. vestita (Zahlbr. in Ann. Nat. Hofmus. Wien, xx. 43); leaves persistently whitish-pubescent and with a gland-tipped hair at the apex; calyx-lobes oblong-lanceolate, a little smaller and narrower than in the type; corolla $1\frac{1}{3}$ lin. long.

COAST REGION : Clanwilliam Div. ; Olifants River Valley, *Penther,* 2917.

14a. Grisebachia Pentheri (Zahlbr. in Ann. Nat. Hofmus. Wien, xx. 42, t. II. fig. iii. 14–22); a shrublet about 10 in. high; branchlets at first hairy, becoming glabrous, with a brownish-grey bark; leaves 3-nate, oblong or oblong-ovate, obtuse or subacute, slightly tuberculate and beset with rigid hairs; flower-heads 4–5 lin. in diam., 8–16-flowered, terminal; pedicels about $\frac{1}{2}$ lin. long, glabrous; bracts 3, unequal, subscarious with green thickened tips, the middle one broadly ovate, all ciliate with gland-tipped hairs, otherwise glabrous ; calyx 1 lin. long, obconic-campanulate, 4-angled, scarious, lobed to less than half-way down, glabrous, ciliate on the broadly triangular acute lobes with long gland-tipped hairs ; corolla up to 2 lin. long, constricted near the top of the hairy (puberulous?) tube : lobes $\frac{1}{3}$ as long as the tube, broadly triangular, glabrous ; stamens 4, not exceeding the corolla-lobes ; filaments shortly hairy : anthers with diverging scabrous cells ; ovary glabrous : style much exserted : stigma capitate.

COAST REGION : Clanwilliam Div. ; Elandsfontein, *Penther*, 2925.

The leaves are described as rather more than $1\frac{1}{4}$ in. (3·5 cm.) long and $\frac{1}{2}$ lin. broad, but in the figure of natural size are represented as about $\frac{3}{4}$ lin. long. The corolla-tube is also represented as glabrous.

OLEACEÆ.

For the generic key on p. 479, substitute the following :—

I. **Jasminum.**—Shrubs. Corolla salver-shaped ; lobes 5–12, contorted. Stamens included. Fruit a twin berry.

II. **Schrebera.**—Shrubs. Corolla salver-shaped. Stamens exserted. Fruit a woody capsule splitting lengthwise.

III. **Menodora.**—Small under-shrubs. Corolla funnel-shaped. Fruit of 2 globose membranous capsules splitting across.

IV. **Olea.**—Trees or shrubs. Corolla deeply 4-lobed ; lobes induplicate-valvate. Fruit a drupe.

6a. Olea macrocarpa (C. H. Wright) ; branches terete, slender, greyish, lenticillate ; leaves lanceolate tapering to both ends, 2 in. long, $\frac{1}{2}$ in. wide just below the middle, dark shining green above, paler and finely and minutely dotted beneath, glabrous, thickened and revolute at the entire margins ; petiole 4 lin. long, channelled above, verrucose beneath ; panicles terminal on the branches, many-flowered, $1\frac{1}{2}$ in. in diam. ; calyx 1 lin. in diam., cupular, shortly and bluntly 4-dentate, puberulous on the edge ; corolla nearly 3 times as long as the calyx ; lobes oblong, obtuse, cucullate ; fruit oblong, 9 lin. long, 4 lin. in diam.

KALAHARI REGION : Transvaal ; Zoutpansberg, "*D.F.O.*," 4329 ! forest near Pilgrims Rest, *Grenfell*, 869 !

9. Olea listeriana (Sim ex Lister in Rep. Conserv. For. Cape for 1897, 98, without description).

The "Umgogunya" of the Kaffrarian Coast Forests.

ASCLEPIADEÆ.

In the generic key, p. 524, line 36. for

††Corona of 5 broad bifid lobes, etc.,

read :—

††Corona of 5 dorsally flattened bifid or entire lobes, with a long linear-filiform or linear-lanceolate appendage on the inner face,

and strike out the other two lines as printed.

Periploca multiflora (Burm. Prodr. Fl. Cap. 7) ; name only, without description.

Woodia singularis. The key-characters of this species on p. 561 should read :—

> Corona-lobes subquadrate with square shoulders or
> rhomboid and angular above the middle, with a
> thick incurved obtuse middle point   ...   ...   **(3) singularis.**

Woodia singularis (N. E. Br.). P. 563, line 10 from the bottom, insert after the word " shoulders " :—or rhomboid-oblong and angular above the middle. Line 9 from the bottom, for " and about 1 lin. long," read :—about 1–1¼ lin. long.

XYSMALOBIUM. For the key-characters of (11) **orbiculare**, on p. 565, substitute :—

> Leaves usually oblong or elliptic, rarely lanceolate or
>   ovate, glabrous ; stem stout, simple :
>   Umbels numerous and racemosely arranged along.
>   the stem ...   .   **(11) orbiculare.**
>   Umbels 2, rising to the same level ...   .   **(11a) Woodii.**

3. Xysmalobium carinatum (N. E. Br.).   Add to the synonymy : *Gomphocarpus carinatus, Schlechter in Journ. Bot.* 1904, 258, *in note.*

9. Xysmalobium undulatum (R. Br.).   The locality "Alexandria Div. ; Zuurberg Range, *Burke !* " should read :—

Central Region : Steynsburg Div. ; Zuurberg Range, *Burke !*

11a. Xysmalobium Woodii (N. E. Br.) ; stem solitary, about 6–8 in. high, 2 lin. thick, simple, puberulous (at least along 2 broad lines) on the upper internodes, glabrous below ; leaves in 3 pairs on the specimen seen, with the middle pair 4½ in. long, 2 in. broad, the upper and lower pairs smaller, oblong or the lower pair ovate-oblong, obtuse or acute, truncate or rounded at the base, glabrous on both sides ; petioles 1–2 lin. long ; umbels 2, rising to the same level, 1½–1⅔ in. in diam., 16–20-flowered : peduncle of the lower umbel 1¾ in. and of the upper ⅔ in. long, puberulous, erect ; bracts 1–1½ lin. long, subulate-filiform ; pedicels about 5 lin. long, minutely puberulous ; sepals 2½–3 lin. long, linear-lanceolate, acute, minutely puberulous ; corolla lobed nearly to the base, subcampanulate or the lobes campanulately spreading and 4–5 lin. long, 2¼–2½ lin. broad, oblong or elliptic-oblong, obtuse, concave, glabrous on both sides and not ciliate, apparently white with purple tips and the base on the back also purplish ; corona-lobes arising ⅓ lin. up the staminal column, erect, contiguous, 1¼ lin. long and nearly as broad, subquadrate, very obtuse, cordate or with a notch at the subtruncate base, thick and fleshy, apparently rounded on the back, flattish, with a rather deep longitudinal groove on each side of the mid-line down the inner face; staminal column $1\frac{3}{4}$ lin. long: anther-appendages erect or erectly-spreading under the large dilated 5-crenate flattened style-apex, which partly overhangs the corona-lobes.

EASTERN REGION: Natal; near Van Reenen, 5000–6000 ft., *Wood*, 10830!

This very distinct species is very unlike any other in the genus.

3. Periglossum kassnerianum (Schlechter). Delete from the geography :—*Miss Pegler*, 1022! as this number rightly belongs to **Parapodium simile.**

SCHIZOGLOSSUM. In the key-characters of (80) **virgatum** on p. 595 for "corolla-lobes $\frac{1}{4}$ lin. long," read :—Corolla-lobes $\frac{3}{4}$ lin. long.

Schizoglossum. In the key on p. 596, the paragraph next after (56) **biflorum** should read :—

Corona-lobes ovate or oblong-ovate, tapering or abruptly contracted into a point shorter to longer than the appendage.

Schizoglossum. The key-character of (64) **glabrescens** on p. 597 should read :—Stem with 10–30 leafy nodes, instead of 10–16 as printed.

Schizoglossum. The key-characters on p. 599 relating to (81) **loreum** should be altered to :—

Corolla-lobes puberulous on the inner face :
    Corolla greenish ; stem often branched ... ,... (64) **glabrescens.**
    Corolla blackish-purple ; stem simple ... ... (81) **loreum.**

73. Schizoglossum parvulum (Schlechter). Line 2 from the bottom, after "Eastern Region" insert :—Var. β.

ASCLEPIAS. For the key-characters on p. 665, **** Corona-lobes without a keel, flap, horn, &c., as printed, substitute :—

****Corona-lobes without a keel, flap, horn or other process
    (but sometimes puberulous) within the cavity :
    Umbels sessile, 2 lateral, 1 terminal ; leaves linear... (21a) **reenensis.**
    Umbels all distinctly pedunculate :
    Leaves remarkably crisped-undulate, etc. ... (42) **crispa.**
    Leaves not undulate, etc.

21a. Asclepias reenensis (N. E. Br.) ; tuber $\frac{3}{4}$–1 in. long, 4–5 lin. thick ; stems solitary, erect, about 10 in. high and $\frac{1}{3}$ lin. thick at the base in the only specimen seen, puberulous all round ; leaves below the flowering part in 6 pairs, erect, subsessile, 7–15 lin. long, $\frac{1}{2}$–$\frac{3}{4}$ lin. broad, linear, acute, with revolute margins, puberulous on both sides ; umbels 2 lateral and 1 terminal, sessile, 2–5-flowered ;

pedicels 3–5 lin. long, slender, shortly villous-pubescent; sepals about 1½ lin. long, narrowly lanceolate-attenuate, very acute, pubescent; corolla lobed nearly to the base, "brown" (*Wood*): lobes spreading, 2½ lin. long, 1¼ lin. broad when flattened out, narrowly elliptic, minutely notched at the acutely pointed apex, narrowed at the base, replicate or with reflexed sides, pubescent on the back, glabrous on the inner face; corona-lobes arising ⅓ lin. up the staminal column rather more than 1 lin. high and ¾ lin. broad across the side, compressed-cucullate, with the dorsal (lower) margin convexly curved and incurved at its apex and the sides produced into erect deltoid obtuse teeth rising to the level of or slightly exceeding the staminal column, without a tooth or process in the cavity, glabrous, "yellow and white" (*Wood*); staminal column about 1½ lin. long.

EASTERN REGION: Natal; near Van Reenen, 5000–6000 ft., *Wood*, 8635!

The affinity of this species is with *A. cognata*, N. E. Br., *A. flava*, N. E. Br., and *A. schizoglossoides*, Schlechter, but the corona-lobes are different in form and have no process within the cavity. In habit and general appearance, apart from the size and structure of the flowers, it greatly resembles and might be easily mistaken for a *Schizoglossum* allied to *S. Buchanani*, N. E. Br., or *S. pilosum*, Schlechter.

CEROPEGIA. In the key-characters on p. 807 after (17) **radicans**, for "Corolla-tube 5–9 lin. long," read:—Corolla-tube 6–10 lin. long.

Ceropegia. For the key-characters of (22) **Caffrorum** as printed on p. 808, read:—

> Stem twining or more rarely straight and pendulous or prostrate, seldom bearing tubers; leaves often longer than broad:
> Stem ½–1⅛ lin. thick; corolla-lobes entirely blackish-purple, ciliate, otherwise glabrous ... ... (22) **Caffrorum**.
> Stem ⅓–¾ lin. thick; corolla-lobes green, with an oblique blackish-purple band near the middle, pilose with long deflexed purple hairs on the inner surface ... ... (22a) **barbertonensis**.

22a. Ceropegia barbertonensis (N. E. Br.); rootstock a tuber; stem slender, ½–¾ lin. thick, twining, glabrous, sometimes producing tubers at the nodes; leaves fleshy, but thin, rather rigid when alive; petiole 3–6 lin. long, marked with a slender blackish-purple ring at their base; blade 7–14 lin. long, 3–14 lin. broad, ovate to lanceolate, with a cordate, subcordate, rounded or cuneate base, acute, wavy or sometimes flat at the margins, glabrous, deep green or variegated with pale greenish along the veins above, pale green, with a darker midrib beneath, not shining; veins not conspicuous on either side; peduncles lateral at the nodes, 2–3½ lin. long, 1–2-flowered, glabrous; pedicels 2½–3 lin. long, glabrous; sepals 1 lin. long, subulate, glabrous; corolla-tube 9–10 lin. long, slightly curved, ellipsoid-inflated and $2\frac{1}{2}$ lin. in diam. at the base, cylindric and 1 lin. in diam. above, enlarging to $2\frac{1}{2}$–3 lin. in diam. at the funnel-shaped mouth, outside glabrous, nearly white at the base, very pale greenish above, inside thinly covered with deflexed white hairs except in the inflated base, where it is glabrous and marked with numerous minute greyish-green tubercles; lobes connivent-erect, connate at the tips, 3–$3\frac{1}{2}$ lin. long, somewhat spathulate-linear from a deltoid base, being slightly broadened at the tips, as is best seen when in bud, closely replicate, keeled down the inner face and there covered with long deflexed purple hairs, glabrous on the back, light green at the base, rather darker green at the upper part, with an oblique transverse blackish-purple band just below the middle and a few dots below it ; outer corona white, with 5 rounded sinuses or slight pockets alternating with 5 deltoid obtuse teeth $\frac{1}{3}$ lin. long adnate to the back of the base of the inner corona-lobes and rising to about the level of the staminal column ; inner corona-lobes white, $1\frac{1}{2}$ lin. long, flat, tapering below into a stalk-like base and from about the middle into fine subulate points, slightly divergent-erect, straight, not recurved at the tips, free or sometimes all connate at the middle and forming a sort of narrowly funnel-shaped basket above the staminal column.

KALAHARI REGION : Transvaal ; Woodbush, *Swierstra*, 3990 ! from near Barberton, a living cultivated plant sent to Kew by *Mr. W. E. Gumbleton* !

35. Brachystelma Galpinii (N. E. Br.). Specimens have now been received which confirm the doubt expressed on p. 862 that *B. pallidum* is only a slight form of *B. Galpinii*, and should be considered as a synonym of that species, scarcely worth retaining even as a variety.

15. Sebæa pusilla (Eckl.), var. **major** (A. W. Hill); much branched above the base, 3–$4\frac{1}{2}$ in. high ; branches slender, 1–3-flowered ; leaves 3–4 lin. long, 1–$1\frac{1}{2}$ lin. broad, elliptic-ovate or linear, acute ; calyx-segments 2–$2\frac{1}{2}$ lin. long, lanceolate ; corolla-tube 4–$4\frac{1}{2}$ lin. long ; lobes $1\frac{1}{2}$ lin. long, $\frac{3}{4}$–1 lin. broad, obovate, subacute ; filaments $1\frac{1}{4}$–$1\frac{1}{2}$ lin. long, inserted 2 lin. above the base of the tube, stouter and more sharply bent over at the apices than in type ; anthers $\frac{3}{4}$ lin. long ; style 2 lin. long.

COAST REGION: Clanwilliam Div. ; Banks of the Oliphants River near Clanwilliam, in muddy soil, *Leipoldt*, 654 ! in Herb. Albany Museum, Grahamstown.

INDEX.

[SYNONYMS ARE PRINTED IN *italics*.]

CORRIGENDA.

Page
5, line 22, for *Ericaceæ* read *Ericcæ*.
12, ,, 35, *for* 44 *read* 45.
19, ,, 5, for **heliophila** read **heleophila**.
50, ,, 18, for *Kuntze* read *Kunze*.
108, ,, 16, from the bottom, for *heliophila* read *heleophila*.
110, ,, 10, ,, ,, ,, ,, **heliophila** read **heleophila**.
288, ,, 17, ,, ,, ,, ,, *galliiflora* read *galiiflora*.
523, ,, 12, ,, ,, ,, ,, *Stamens* read *Stems*.
557, ,, 20, for *Massoni* read *Massonii*.
865, ,, 6, from the bottom, after **ANISOTOMA** *add* Fenzl.
984, ,, 4, ,, ,, ,, *for* Jacq. *read* J. Donn

Lightning Source UK Ltd.
Milton Keynes UK
UKOW03f2111130714

235008UK00001BB/28/P